Frontiers in Neutron Capture Therapy

Volume 1

Frontiers in Neutron Capture Therapy

Volume 1

Edited by

M. Frederick Hawthorne

Kenneth Shelly

University of California, Los Angeles
Los Angeles, California

and

Richard J. Wiersema

Neutron Therapies LLC
San Diego, California

Springer Science+Business Media, LLC

Library of Congress Cataloging-in-Publication Data

Frontiers in neutron capture therapy/edited by M. Frederick Hawthorne, Kenneth Shelly, and Richard J. Wiersema.
p. ; cm.
"Proceedings of the Eighth International Symposium on Neutron Capture Therapy for Cancer, held September 13–18, 1998, in Los Angeles, California"—T.p. verso.
Includes bibliographical references and index.
ISBN 978-1-4613-5478-9 ISBN 978-1-4615-1285-1 (eBook)
DOI 10.1007/978-1-4615-1285-1
1. Boron-neutron capture therapy—Congresses. I. Hawthorne, M. Frederick. II. Shelly, Kenneth. III. Wiersema, Richard J. IV. International Symposium on Neutron Capture Therapy (8th: 1998: Los Angeles, Calif.)
[DNLM: 1. Neutron Capture Therapy—Congresses. 2. Neoplasms—radiotherapy—Congresses. WN 250.5.R3 F935 2001]
RC271.N48 .F76 2001
616.99'40642—dc21

2001020257

Proceedings of the Eighth International Symposium on Neutron Capture Therapy for Cancer, held September 13–18, 1998, in Los Angeles, California

ISBN 978-1-4613-5478-9

Originally published by Kluwer Academic / Plenum Publishers, New York in 2001
Softcover reprint of the hardcover 1st edition 2001

http://www.wkap.nl/

10 9 8 7 6 5 4 3 2 1

A C.I.P. record for this book is available from the Library of Congress

In Memoriam

Professor Börje Larsson, 1931–1998
(Photo courtesy of Jörgen Carlsson)

Börje Larsson was selected to be the Hatanaka Lecturer at the Eighth International Symposium on Neutron Capture Therapy for Cancer in recognition of his contributions to the field of NCT and radiotherapy in general. Unfortunately, he was unable to attend due to illness, and shortly after the Symposium he passed away at age 67.

Larsson was born on June 3rd 1931 in Stockholm, Sweden. His education included a BSc in biology at Stockholm University in 1955 and a MSc in electrical engineering at the Royal Institute of Technology in Stockholm, 1958. He received his docotrate from the University of Uppsala in 1962. Larsson was assistant professor (1966–1979) and professor of physical biology (1979–1991) in Uppsala. From then until 1998 he was professor of medical radiation biology at Zürich University, Switzerland. Larsson's interdisciplinary approach to science became apparent early in his career. While still a student, he was invited to work on biomedical applications for the synchocyclotron in Uppsala at the internationally known Gustaf Werner Institute (now a part of the

Department of Radiation Sciences and the The Svedberg Laboratory). Larsson's careful experimental preparation paved the way for the world's first proton irradiation of a cancer patient in 1957. He continued to be a driving force in the work on irradiation of tumor patients, developing new dosimetric methods and analyzing the biological effects of protons, and he participated in new and planned accelerator centers for cancer treatment in USA, Europe and Japan.

Larsson became one of the most prominent pioneers in the development of radiation surgery. His contributions included studies on both the physical and biological aspects of radiation surgery, with clinical work on the treatments of various brain diseases including Parkinson's disease and arteriovenous blood vessel malformations. Larsson's technical leadership led to the development of radiation surgical equipment (the "gamma knife").

Larsson took an early interest in boron neutron capture therapy. He actively pursued the development of BNCT in Switzerland and Sweden, and particularly in recent years contributed to the prospect that the research reactor in Studsvik, Sweden, may be reconstructed to enable clinical treatments. Special landmarks of international recognition are his developmental work on neutron radiography and for being the initiator of the European concerted action for BNCT which in later years, through the action of Prof. Detlef Gabel and others, led to the clinical BNCT research at the Petten research reactor in the Netherlands.

Besides being interdesciplinary, Larsson's approach to science was international in scope. He had numerous international collaborators, and was an adviser to many large international accelerator and reactor centers which were developed or reconstructed for biomedical purposes. He also served as guest researcher at many universities. His contributions were awarded with several national and international prizes over the years, most notably The Guldkuggen for technical development of the gamma knife, as well as the prestigious Fairchild prize, presented in Boston for developmental work in radiation surgery. Larsson also organized many international scientific meetings. He served as the President of the International Society of Neutron Capture Therapy just prior to M. F. Hawthorne, and hosted the Seventh International Symposium on Neutron Capture Therapy for Cancer in Zürich.

Börje Larsson is survived by his wife and two children, and will be missed by his relatives, friends, and colleagues. With his death, the research community has lost a valuable collaborator and a unique inspiration whose scientific contributions will leave their mark on many disciplines for years to come. Larsson was a creative researcher who through his generosity and friendliness enabled many to share his ideas and gain advantages from his achievements. For all this, we wish to express our appreciation and sincere gratitude.

Prof. M. Frederick Hawthorne
Department of Chemistry and Biochemistry
University of California Los Angeles, USA

Prof. Jörgen Carlsson
Department of Oncology, Radiobiology, and Clinical Immunology
Uppsala University, Sweden

PREFACE

The Eighth International Symposium on Neutron Capture Therapy for Cancer convened September 13–18, 1998 at the Sheraton Grande Torrey Pines Hotel in La Jolla, California. This meeting site was selected in order to enhance nascent interactions of the NCT community with the internationally recognized biotechnology companies located in La Jolla. In addition, this venue provided warm and sunny weather beside the Pacific Ocean. The Eighth International Symposium accommodated approximately 400 delegates and accompanying persons. A total of 275 lectures and posters were presented. The Seventh (Zurich) and Eighth Symposioa were quite similar in size and general content although more clinical data were available at the time of the 1988 Symposium. Clinical trials are presently in progress, planned or in temporary suspension in the United States, Japan, The Netherlands, Finland, Sweden, Eastern Europe and South America. Neutron capture therapy is clearly advancing and the Ninth Symposium, to be held in Osaka in 2000, should provide further evidence of this progress. However, at this time investigators in NCT research can be justly pleased with the quality and wide scope of the papers published in these Proceedings.

The organizers of the La Jolla symposium hope the participants were rewarded both scientifically and socially. However, the illness of the Hatanaka Lecturer, Professor Börje Larsson, cast a shadow over the event. The organizers thank Dr. Nigel Crompton for his delivery of the Hatanaka Lecture on behalf of Professor Larsson.

We conclude by thanking all authors, participants and accompanying persons for their contributions and we look forward to the Ninth Symposium.

M. Frederick Hawthorne
Richard J. Wiersema
Kenneth Shelly

ACKNOWLEDGMENTS

The editors wish to express their gratitude to those whose assistance was vitally important in the planning and presentation of the Symposium:

Renate Alberts
Ann Bates
Margaret Bird
Jeanne Douglas
Debra A. Feakes
Patrick R. Gavin
Robin Green
Diana Hawthorne
Colin Higgs
Reiko Matsuoka
David W. Nigg
Darren Ricciardi
Sherry Vatter
Chris Wiersema
Jean Wiersema

Symposium Sponsors

The Symposium was made possible by the generous financial support from the following organizations:

United States Department of Energy
National Cancer Institute, National Institutes of Health
University of California Los Angeles

Additional support was provided by the following commercial sponsors, for which we are most grateful:

Neutron Therapies (San Diego, CA)
General Atomics (San Diego, CA)
Callery Chemical (Pittsburgh, PA)
U.S. Borax (Valencia, CA)
Strem Chemical (Newburyport, MA)
Ion Beam Applications (Louvain-la-Nueve, Belgium)

Hiroshi Hatanaka Lecture

Monochromatic x-rays, slow neutrons and radionuclides for targeted radiotherapy—present developments and prospects, *Börje Larsson (presented by Nigel E.A. Crompton)*

Ralph G. Fairchild Awards for Young Scientists

Judit Benczik, *University of Helsinki*
Cynthia Chuang, *Massachusetts Institute of Technology*
Andrea Wittig, *University of Essen*

CONTENTS

DOSIMETRY

FAST NEUTRON STUDIES

VOLUME 1: MEDICINE AND PHYSICS

INTRODUCTORY OVERVIEWS

1

MONOCHROMATIC X-RAYS, SLOW NEUTRONS, AND RADIONUCLIDES FOR TARGETED RADIOTHERAPY

Present Developments and Prospects

Börje Larsson[1,2]* and Nigel E. A. Crompton[3]

[1]Institute for Medical Radiobiology of the Zurich University and the Paul Scherrer Institute
CH-5232 Villigen PSI
Switzerland
[2]The Unit for Biomedical Radiation Sciences
Uppsala University
PO Box 531
S-751 21 Uppsala
Sweden
[3]Radiation Medicine
Paul Scherrer Institute
CH-5232 Villigen PSI
Switzerland

INTRODUCTION

A general breakthrough in the prevention or therapy of neoplastic diseases is not likely to occur, before the potential of the "modern molecular biology" can be fully exploited. Whatever the prospects for such a major scenario change, the desired development will certainly require decades of basic and clinical research at many frontiers. Without a doubt, neoplastic disease will remain a major human and socioeconomic problem in the foreseeable future: one in three children born now are likely to suffer from life-threatening tumours during their lifetime.

An increasing fraction of these future cancer patients will be treated with advanced radiation techniques, with or without concomitant surgery, using present-day systems for clinical management. Awaiting the wonders that the "new biology" will bring, there are

*Deceased, 12th October, 1998.

Frontiers in Neutron Capture Therapy, edited by Hawthorne *et al.*
Kluwer Academic / Plenum Publishers, New York, 2001.

two promising lines of development that may soon lead to better treatment results: the first, *continued improvement of "macrotumour therapy"* is concerned with the reduction of the total mass of malignant cells through local treatment of known primary or secondary macroscopic tumours, by surgery, radiosurgery or stereotactic radiotherapy: the other, the *introduction of new principles for "microtumour therapy"* aims at efficient, systemic treatment of radiologically invisible, infiltrating or disseminated microscopic growth with selectively cell-seeking and cell-killing agents. Combination of these strategies ("stereotactic conformal irradiation plus targeted therapy") could lead to radical improvement of the treatment results.

This research program gains originality through our ambition to exploit activation of stable nuclides or atoms as a means of producing short-range ionizing radiation selective in the target cell populations studied. Our aim is to understand and exploit receptor-mediated internalization of therapeutic compounds. The present contribution is divided into three projects. In part A, activation is brought about by slow neutrons, in part B by monochromatic photons. In part C, the scientific and clinical validity is studied by use of relevant *in vitro* models based on target cells isolated from specimens obtained by biopsy or surgery. The techniques developed for this purpose permit study of receptor-mediated cellular uptake of biomolecular vehicles (growth factors, hormones, antibody derivatives, *etc.*) loaded with selected nuclidic or atomic species, as well as multiparameter bioassays of radiation response induced by neutron or photon activation.

(A) REGIONAL THERAPY FOR MICROTUMOURS WITH keV NEUTRONS

Neutron capture therapy is a classic, but yet rarely used modality conceived early in the history of radiotherapy research. This *bimodal therapy* is based on the use of non-radioactive cell-seeking molecular vehicles loaded with the stable boron isotope, ^{10}B, that can be activated without risk at a time when it has been taken up by the tumour cells and cleared from the blood and intercellular space. This principle is gaining revived attention as a consequence of important developments in physics and chemistry and a deeper understanding of the biological aspects of cellular uptake of various vehicle molecules. We are contributing to this international endeavour by exploiting cold and thermal neutron beam facilities for topographic analysis of boron distributions and radiation response in experimental models and clinical specimens as outlined below (part C).

External, conformal radiotherapy is near its perfection. A case in point is functional radiosurgery with protons which is establishing itself in the more advanced radiotherapy clinics. It has moved a long way since the early preclinical model studies in the 50-ies where 3–7 mm size lesions in the internal capsule of a goat were treated. With optimized patient set-up clinical use of mm-sized proton beam for functional radiosurgery and benign neoplasms (including arteriovenous malformations (AVMs)) is also possible. Early AVM treatments in Stockholm with the "gamma knife" efficiently eradicated small AVMs and such treatments have now reached the state of efficient clinical routine as was summarized by Prof. Steiner in his Charlottesville statement.

From the early studies it was apparent that endothelial radiation damage was the cause of the vascular obliteration. Today this is readily confirmed in histological studies using high-resolution phase contrast microscopy. However, larger AVMs are much more difficult to treat because clinically effective doses to extended healthy tissue areas cause complications. It is the reason d'etre for targeted therapy for AVMs directed against endothelial structures, a theme which attracted Hiroshi Hatanaka (for this reason, and for many others, personal and scientific, we are very happy to honor his memory by giving this memorial lecture). Prof. Hatanaka had the ability to pursue a straightforward clinical course without loosing sight of the basic research aspects.

We also contemplated NCT for non-cerebral AVMs, such as large AVMs in the extremities which can otherwise be irradiated with conventional photon methods. One possibility, which was actively pursued, was to load the blood with BSH in order to selectively kill the endothelial cells. Although far from realizing the dream of NCT for vascular pathologies the clinical results were not too bad, although there was no proof of permanent cure. Awaiting objective evaluation, the results have not been published so far. Another more recent development in NCT has been boron neutron capture synovectomy, BNCS, the treatment of rheumatoid arthritis with localised injections of boron into the diseased joints. Promising results in animal models have been observed (Yanch *et al.*).

In this manner we have identified an important principal role of NCT that has been governing our research strategy in the 90-ies: to introduce boron as a model for any one targeting principle, such as the less innocent and more difficult-to-accomplish alpha particle therapy based on radioactive nuclides with astatine-211 as the most interesting option. The alphas and their tracks from both BNCT and astatine-211 can be visualised in a cellulose nitrate detector and are in principle the same.

(B) STEREOTACTIC RADIOSURGERY WITH MONOCHROMATIC X-RAYS

Photon activation therapy (PAT) based on monochromatic X-rays is a new concept that allows most precise localization of radiation dose to a radiosurgical target. The radiation effect can be augmented in target cells by tumour-seeking biomolecules loaded with iodine or other heavier atoms, a procedure that leads to the production of highly cell-killing Auger-electron cascades *in situ* by K-edge activation, provided the Auger emitter has found its way to the chromosomal DNA of the target cells. The potentialities of this therapy modality are obvious in stereotactic radiosurgery aiming at the selective destruction of small volumes of grey or white matter, in cases of "functional brain disease" (such as movement disorders or intractable pain), or elimination of small pathological structures in cases of small, circumscribed pathological targets (such as benign tumours or vascular malformations).

We are interested in activation therapy based on halogens and gadolinium in parallel with the NCT based on boron and gadolinium. There are obvious bridges between these fields, with additional links such as radionuclide diagnosis and therapy. The boron hole phenomenon also requires mention. This is caused by the atomic sized hole left due to the disintegration of the boron following the neutron capture event. It presents a fascinating field of radiobiological inquiry. Its region of action has the dimensions of high LET particle tracks. The targeting compounds used for PAT should locate the nuclides very close to the DNA. The Auger electron spectrum of photon-activated iodine-127 indicates that most of the electrons released have flight paths less than 100 nm. (Flight paths of electrons of 50 keV energy are about 200 nm and electrons of 10 keV are about 100 nm.)

The authors enjoying a first class flight (upgraded!) on a business trip to Miami.

We have initiated a program for PAT where what seems relevant from the point of view of other NCT proponents. Photon-activation radiosurgery appears to be an attractive, realistic principle based on free-electron Compton scattering. A JAG laser, as is commonly employed for eye surgery; or a "bedroom" sized free electron laser, using its own synchrotron light, could be employed. Our Monte Carlo calculations (J. Stepanek *et al.*) show that an energy-efficient production of a narrow forward-directed, monoenergetic photon beam suitable for K-edge activation of stable iodine can be obtained. These conditions compare favourably to those of NCT, based on boron or gadolinium, especially when small target areas are concerned as in the present stage of development. The photon beam would be produced based on a machine concept, borrowed from a technical development at Brookhaven (Dr. Blomberg). A similar machine is being promoted by the firm R_xD, inc. who have a contribution at this meeting (K. Frankel *et al.*).

Clearly, it is possible to build small electron rings for the production of suitable beams for radiosurgery in the hospital environment, and the potential for the practical use of such devices for different clinical applications should be carefully analyzed. In diagnostic radiology seemingly attractive applications have been identified, but it is still an open question whether monochromatic photons will be useful for treatment of deep-lying target volumes larger than a few cm^3. An accelerator device used to produce and circulate electrons of a few-hundred MeV could be used, not only for the production of narrow photon beams through free electron Compton collisions but also its bending magnets could at the same time be useful for the production of visible and ultraviolet light. So, a multitude of possibilities in spectroscopy and/or microscopy could be considered, for example in the chemical or cytological analysis of body fluids and cells from individual patients (for rapid multielement determinations, quantitative analytical microscopy, fluorescence-assisted cell sorting, *etc.*).

Auger electron radiography and radiobiology could be effectuated elegantly, at high sensitivity and resolution, especially when glancing incidence conditions are realized (such as with the cell culture system that now is being implemented as described below). Glancing incidence eliminates photoelectron production in the substrate, this is because the photons do not penetrate deeply enough to produce significant numbers of photoelectrons. It is being realized by choosing silicon as a substrate for cell cultivation and photon radiography.

Thanks to the physical situation thus described, there is an intimate relation between NCT and PAT, especially when experiments and therapy will be based on receptor-mediated cellular uptake, and DNA-targeting. This is amply illustrated by our trick to facilitate suppression of protons from neutron-capturing nitrogen (in NCT) and photoelectron production from macroelements (in PAT). For microscopic analysis of subcellular compound localization we incinerate the tissues or cells under study to remove most of the macroelements (except innocent oxygen). The incineration effectively burns off hydrogen, carbon, nitrogen and sulphur to leave an ash ghost of oxides and salts. The technique preserves structures even at the cellular and molecular level. Cold incineration at the tissue level e.g. of intestinal villus, leaves samples comparable to histological sections. At the subcellular level, DNA structures such as the polytene chromosomes of Drosophila, due to phosphorus are visible.

To assist the communities with interest in the two forms of activation therapy, NCT and PAT, respectively, we have launched a program for use of the new cold neutron facility at the Paul Scherrer Institute in Switzerland (the spallation neutron source SINQ). Our starting point was near 20-year old experience from the use of various slow neutron beam facilities in neutron capture radiobiology and radiography. A well-known example is the topographic analysis of BSH in animal tumour models. An interesting observation made during those studies was the 1000-fold enrichment of boron in plant material compared to animal tissues. Such techniques have proven themselves valuable in distribution studies at the tissue level. Our present ambition is to offer analogous means for studies of cellular uptake and subcellular distribution of boron-carrying ligands in experimental and clinical tumour specimens.

The importance of reducing the background from neutron capture in natural nitrogen-14 is obvious. Although single nitrogen protons are trivial of themselves, they cause problems because of the large numbers produced. A further reduction of parasitic signals is possible by using practically clean fluences of cold neutrons as they become available at the end of neutron beam guides. A computerized target station is being installed at the SINQ facility. The idea is to make this resource available "by correspondence" to the NCT community. Our contribution, "G. Kuehne *et al.*", at this meeting describes details of this project. Our team welcomes your inquiries! The suppression of the proton background, by use of a polycarbonate matrix, makes the alpha and lithium tracks stand out more clearly, permitting higher-resolution 3-D analysis of the particle trajectories in subcellular boron distribution studies.

(C) A SYSTEM FOR THE ASSAY OF CELLULAR MICROPHARMAKOKINETICS AND TREATMENT RESPONSE IN TISSUE SPECIMENS FROM INDIVIDUAL PATIENTS

A cell cultivation technology is being developed which permits convenient, high-sensitivity analysis of boron distributions, for example in cell populations prepared from patient biopsy or surgery specimens. The details of this approach were presented in a lecture given by Dr. Crompton earlier at this meeting (Crompton *et al.*). An important prerequisite was the cultivation of experimental and clinical cell specimens on silicon wafers surfaced with agarose and spotted with palladium: the Palladium island technique. The cultivated cells stick selectively to the 300 μm diameter palladium islands. When such tumour cell colonies are left to grow further, they produce 3-D microtumours which can be used for a vast spectrum of investigations including micropharmacokinetic studies of cell targeting processes. A single adenocarcinoma cell deposited on such a palladium island will grow to produce a microclone. Mixtures of tumour and stroma cells from biopsies, when cloned on separate palladium islands, produce well-defined subpopulations of monoclonal tumour, fibroblasts, or other cell types.

A special cell cultivation disc (the CCCD) has been constructed to exploit the palladium island technique for studies of tumour cell heterogeneity in terms of radiobiological and micropharmacological parameters. Our system for the study of radiation response in cohorts of single cells isolated from surgical biopsy specimens is based on the "Compact Cell Culture Disc" (CCCD), a planar sandwich structure designed to permit cultivation and microscopic study of a multitude of single cells and their progeny. This device contains a large number of microchambers in the form of shallow "micopits", permitting cultivation, microcloning, instrumental or visual observation, and eventual storage of hundreds to thousands separately confined single cells or monoclonal cellular colonies. The internal micropit environment is determined by the permeability characteristics of a covering foil. Each micropit thus forms a closed system that functions as a biological microlaboratory inaccessable to virus or bacteria.

The aim is to permit accurate and reproducible, computer-aided assays of microscopic and physical-chemical parameters in large sets of cellular specimens isolated from patient specimens, under ideal *in vitro* conditions. The CCCD makes it possible to ascertain optimized and uniform supply of respiratory gasses, nutrients, antibiotics, chemical reagents, *etc.* Macromolecular or biological medium components could be injected before the micropits are closed or by injection through the covering foil. After cultivation and observation, the cells can be fixed or frozen, without having to be removed, from their CCCD confinement. The single cell cultivation "pits" are produced by chemical micromachining. The thin quartz windows, produced by oxidation of the silicon surface, provide a substrate for cell growth which is also suitable for advanced microscopy (e.g. confocal laser scanning microscopy and confocal laser scanning macroscopy) and for microradiography.

The CCCD and the associated system for Computer-Aided Seeding, Microcloning, Analysis and Telemetric Evaluation ("CASE-MATE") offers particularly advantageous conditions for assay of receptor-mediated cellular uptake of biomolecular vehicles (growth factors, hormones, antibody derivatives *etc.*) loaded with selected nuclidic or atomic species, as well as the multiparameter bioassays of the radiation response induced by neutron or photon activation. The development of an airtight, CD-format housing for these CCCDs permits long-term cultivation, transport, frozen storage and exposure to neutrons at "distant" sources.

In the context of NCT, we hope in this way to contribute to a better understanding of the internalization process and subsequent conditions for DNA targeting. In particular the "dog-sledge-load" targetting principle. An example of this principle is the use of EGF as a "dog" to guide the "sledge" to the tumour, the use of dextran as a "sledge" to carry the "load", and the use of usually cytotoxic agents such as boron compounds as the "load". A long-term strategy is to combine the study of boron compound micropharmacokinetics with quantitative scintigraphy (PET) for purposes of tumour therapy planning and evaluation. Iodine-124 could be singled out as a most promising radionuclide for such purposes, in NCT as well as in PAT. In both cases, reliable analytical methods could be developed for the study of dispersed cell specimens from clinical specimens, under practically and ethically favourable conditions (this may be of interest for various studies of cell targeting, not only in cases of receptor-mediated endocytosis.

Many clinically relevant parameters for tumour therapy can be assessed using the method helping to realize a "Holy Grail" of modern medicine: individualization of patient therapy. The broad spectrum of assays available gives rise to the concept of a miniature "predictive assay mall". Cell morphology *in vitro* provides excellent conditions for micro-pharmakokinetic distribution analyses (with or without background staining). Non-toxic autofluorescence techniques provide a means to analyse the cells over time in the living state, particularly important for growth kinetic studies and the like. The hope is that studies of the kind referred to will facilitate the clinical validation process for various patient categories. Our team at PSI works in intimate cooperation with Uppsala; the Carlsson group, and the group responsible for the medical X-ray beam line at ESRF in Grenoble. In this context we must report that *Per Spanne*, for many years our eminent research partner in Grenoble lost his life in the recent Swiss Air Flight SR111 catastrophe near Halifax off the coast of Nova Scotia. We would like to take this opportunity to express our thanks for what Per Spanne has done for our projects and other medical-technical applications based on synchrotron radiation in the birth-phase of this new specialty. His memory remains with us as a fine scientist and a wonderful human being.

(D) IMPORTANCE OF THE PROGRAM

Once implemented clinically, the new methods would widen the indications for radiosurgery, in patients with primary malignant brain tumours or multiple brain metastases. They would also lead to the use of lower stereotactic radiosurgery dosage with less risk for complications caused by overdose to healthy tissues, an important goal to be strived for, in cases of larger benign tumours, arterio-venous malformations or radiosurgery for functional disorders based on selective attack on healthy oligodendroglia or vascular endothelium in the target volume. In a longer perspective, the accumulating experience in the use of receptor-mediated cellular uptake of biomolecules would be of value for other applications of targeted tumour therapy as well, for example in patients with neuroendocrine tumours in the abdomen or tumours with amplified expression of hormone or growth factor receptors.

The Hatanaka Lecture was an honor bestowed on Prof. Börje Larsson for his prestigious contributions and research in the field of Neutron Capture Therapy. Boerje did much to promote the field at national and international levels and was the president of the organizing committee of the extremely successful 7th NCT Symposium in Zurich in 1996. Due to health complications associated with his fatal illness he was unable to attend the meeting in La Jolla and requested his friend and associate PD Dr. Nigel E. A. Crompton to hold the lecture in his stead. Boerje is sadly and sorely missed. He passed on this greeting to the attendees of the 8th International Symposium.

Professor Börje Larsson wishes you every success as you strive towards improving cancer treatments with activation therapy techniques such as BNCT, BNCS and PAT. He thanks the Organizers for the honour to be able to hold the Hatanaka Lecture and he thanks the participants of the 8th International Symposium on Neutron Capture Therapy for Cancer for their interest and their recognition of these unique therapy modalities.

2

CURRENT TRENDS AND PROGRESS IN THE PHYSICS OF BNCT

David W. Nigg

Idaho National Engineering and Environmental Laboratory
P.O. Box 1625, Idaho Falls
ID 83415-3855
USA

1. INTRODUCTION

The first Phase I human trials of epithermal-neutron BNCT for high-grade primary brain tumors (Glioblastoma Multiforme) were initiated at Brookhaven National Laboratory (BNL) and at the Massachusetts Institute of Technology (MIT) in 1994. Phase I studies of epithermal-neutron BNCT for melanoma were also initiated at MIT in 1994. All of these trials feature the use of boronated phenylalanine (BPA) as the boronated pharmaceutical. More recently, Phase I clinical trials of epithermal-neutron BNCT for glioblastoma, using BSH, were initiated in a collaborative European program centered at the Petten research reactor facility in the Netherlands. Initiation of additional epithermal-neutron trials at a few other facilities worldwide is anticipated in the near future. Applications of fast-neutron therapy with BNCT augmentation are also under investigation at the University of Washington and at a few other fast-neutron facilities worldwide.

2. NEUTRON SOURCES FOR BNCT

In addition to the original three reactor-based epithermal neutron sources at the institutions noted above, there are several other research reactor facilities worldwide that have recently been successfully backfitted with filtered epithermal-neutron beam extraction ports of various types. There are also plans for construction of epithermal-neutron beams at a number of other research reactor installations at various times in the future. A few accelerator-based neutron sources for BNCT research have also been constructed and several others are planned.

Frontiers in Neutron Capture Therapy, edited by Hawthorne *et al.*
Kluwer Academic / Plenum Publishers, New York, 2001.

2.1. Reactor-Based Neutron Sources for BNCT

The 3 MW Brookhaven Medical Research Reactor, BMRR, was modified for epithermal-neutron BNCT approximately 10 years ago. An aluminum oxide (Al_2O_3) moderator-filter is used to produce the present epithermal-neutron beam. As of September 1998 over 40 patients had been treated at the BMRR epithermal-neutron facility. Plans are in place to upgrade this facility by installation of a fission converter region and a redesigned filter section composed of aluminum and aluminum oxide. This new beam is expected to have significantly improved quality and intensity.

Current clinical trials at the MIT Research Reactor (MITR-II) are conducted using the medical irradiation room below the core of the reactor. An epithermal-neutron beam is delivered to the treatment position through a vertical filter-moderator unit. A high-intensity, low background, horizontal epithermal-neutron beam extraction facility is currently under construction at MITR-II. This upgrade uses a fission converter, located outside the reactor vessel and driven by thermal neutrons from the reactor, as the neutron source. After filtration by aluminum, polytetrafluoroethylene (Teflon^{TM}), cadmium, and lead, the new beam will be collimated to the patient irradiation position in a new medical irradiation room. The new beam will have a much-improved level of neutronic performance. It should offer therapeutic ratios (ratio of tumor to normal tissue dose) for brain tumor applications that are approximately twice as great as are produced by the current vertical beam, all other factors being equal. Also, the intensity will be such that irradiation time can be as short as a few minutes and it will be possible to perform dose rate studies simply by operating the new beam below its maximum rated intensity.

The Petten epithermal-neutron beam was constructed in 1990 using the HB11 beam port of the 45 MW High Flux Reactor (HFR) operated by the Commission of the European Communities, Joint Research Center, Petten, in the Netherlands. Because of limited access to the core, the beam exit port is approximately 5 meters from the core centerline. The neutron spectrum of the beam is tailored via a series of materials whose primary action is by neutron filtration rather than moderation. As a result of this filtering process and the large distance from the core the HFR beam has a remarkable degree of monodirectionality, which serves to enhance the therapeutic ratio that can be obtained, especially at large depths in the treatment volume. Collaborative European human clinical trials of BNCT for glioblastoma began at the Petten facility in 1997.

The FiR1, operated by the Technical Research Centre of Finland (VTT) is a 250 kW TRIGA-II^{TM} pool reactor. To obtain a high quality epithermal-neutron beam from such a low-power neutron source, a patented high-efficiency filter/moderator material with a composition of 69% AlF_3, 30% Al, and 1% LiF was developed by VTT. This material was used to construct a 75-cm thick filter to produce an epithermal-neutron beam in 1996. Measurements made at this time validated the design calculations, which had predicted a beam having a particularly high intensity per MW of reactor power, with minimal fast-neutron and gamma components. In 1997 the filter length was reduced to 63 centimeters to increase the intensity and to increase the average neutron energy somewhat. A bismuth collimator was also added at that time to provide enhanced directionality of the beam. The FiR1 facility offers the first conclusive demonstration of the utility of low-power, passively safe TRIGA^{TM} research reactors as useful epithermal-neutron sources for BNCT applications. Clinical trials of BNCT for glioblastoma using BPA as the boron carrier are expected to begin at this facility in the near future.

The BNCT research group associated with the Nuclear Research Institute (NRI), in the Czech Republic has been active for many years. This group has constructed an epithermal-neutron beam facility at the LVR-15 research reactor. The filter-moderator region is composed of alternating layers of aluminum and aluminum fluoride with a total thickness of 61 centimeters (35 cm Al, 26 cm AlF_3). This facility, like the Petten facility, features a long collimator, which will give the beam excellent directionality. Clinical trials of BNCT for glioblastoma using BSH as the boron carrier are anticipated.

A new medical irradiation facility at JRR-4, operated by the Tokai Research Establishment of the Japanese Atomic Energy Research Institute (JAERI) has been constructed. This new facility offers both thermal and epithermal beam configurations. The neutron beam line consists of three parts. The first part is inside the core tank and is composed of aluminum blocks. The second part is a heavy water tank, about 110 centimeters thick. On the upstream side is an aluminum beam filter followed by four compartments that can be separately filled with heavy water to provide various degrees of neutron moderation and corresponding spectral tuning of the final neutron beam. The third part of the beamline is a bismuth gamma shield followed by a collimator. This facility will support a wide range of preclinical and clinical BNCT studies.

The Heavy Water Neutron Irradiation Facility of the Kyoto University Reactor (KUR) was remodeled in 1996, primarily for neutron capture therapy research. An Al/D_2O filtering and moderating region first produces an epithermal spectrum. This can be used directly or it can be further moderated by D_2O in the spectral shifter region to produce a hyperthermal or a fully thermal neutron beam. As of March 1998 a total of nine brain tumor patients had been treated with BNCT using the new facility. Three were treated with a thermal beam and six were treated using a mixed thermal-epithermal beam.

Veterinary radiation oncology researchers at the Washington State University (WSU) School of Veterinary Medicine have made major contributions to the understanding of normal tissue tolerance in BNCT. Studies conducted by the WSU group have historically used the epithermal-neutron beams available at Brookhaven and at the Petten facility. More recently, attention has been focused upon the development of a local epithermal-neutron beam facility for BNCT research and boronated pharmaceutical screening in large animal models at WSU. A collaborative effort of WSU and the Idaho National Engineering and Environmental Laboratory (INEEL) has resulted in the development of a design for an epithermal beam extraction port to be installed in the thermal column region of the TRIGA™ research reactor at WSU. Construction of this beam, which is similar in design to the FiR1 beam, was started in 1998 with initial testing and operation for animal applications anticipated in late 1999.

The McClellan Nuclear Radiation Center (MNRC) is located near Sacramento, California, on the grounds of McClellan Air Force Base. The MNRC operates a 2 MW TRIGA™ reactor that began operation in 1990, primarily to provide a highly collimated thermal neutron beams for radiographic examination of aircraft components. Modification of this facility to provide an additional dual spectrum thermal and epithermal neutron beam port for clinical BNCT research and for other medical and non-medical applications was initiated in 1997. This project is being done in collaboration with the University of California at Davis (UCD), with technical support from the General Atomics Company and from the INEEL. Since the Washington State University epithermal-neutron beam described previously is intended for animal research only, it is anticipated that the McClellan/UCD facility will constitute the next new full-scale reactor-based clinical BNCT research installation in the U.S.

In an interesting recent development, the National Atomic Energy Commission of Argentina has organized a multi-institutional research project involving the construction of an epithermal-neutron beam facility at the RA-6 reactor in Bariloche. This project involves several interrelated activities focused upon the development of BNCT in Argentina. The RA-6 beam is similar in some aspects to the planned Brookhaven beam upgrade. It employs an optimized combination of aluminum oxide and aluminum for neutron filtering. This beam was recently brought into operation, and construction of additional beam facilities in Argentina is planned.

A number of other reactor-based epithermal-neutron beam concepts have been proposed over the years and are now in various stages of realization. Of particular interest are the beam facility at the TAPIRO reactor in Italy, which is expected to become available in the near future, and a planned facility in Sweden that will feature two beams, one where the spectrum can be varied from thermal to epithermal as needed for research and a second, purely epithermal, beam for clinical applications. Other facilities are planned by researchers in Taiwan, France, and South Korea.

2.2. Accelerator Neutron Sources for BNCT

Two types of accelerator neutron sources are of interest for BNCT research and clinical trials. The first group of sources is composed of existing clinical fast-neutron facilities, which can be modified for exploration of NCT-augmented fast-neutron therapy at minimal additional cost. The second group of accelerator neutron sources for BNCT is composed of various developmental facilities designed to produce an epithermal neutron beam for BNCT as the primary therapy. These latter types of systems are believed in many quarters to be more deployable to a routine hospital environment than would be the case for new dedicated reactor facilities.

The Neutron Therapy Center at the University of Washington (UW) Medical Center offers an illustrative example of the use of an existing accelerator-based neutron facility for clinical studies of BNCT-augmented fast-neutron therapy. It was the site of a recent collaborative INEEL/UW effort to explore methods for optimizing the performance of the UW system for NCT-enhanced fast-neutron therapy applications. The goal of this project was to develop a new target for the UW proton cyclotron that produces a neutron beam having essentially the same fast-neutron physical depth-dose curve as the current system, but with a significantly-increased low-energy spectral component. This would be expected to yield an increased fraction of BNCT enhancement relative to the total therapeutic dose. The effort was successful and *in-vivo* clinical testing of BNCT-enhanced fast-neutron therapy for canine lung tumors using the new UW neutron beam was recently initiated.

Early work by the BNCT group at The Ohio State University (OSU) has provided much of the basis for later development of epithermal-neutron sources for BNCT using low-energy heavy particle accelerators. There are many interrelated design factors to consider with such systems. The group at OSU has continued their efforts over the years and has published a number of variations on their original design. Recent work by this group has included low-power prototype testing as well as various efforts to consider how a deployable hospital-based system might be designed.

The MIT Laboratory for Accelerator Beam Applications (LABA) is built around a 4 MeV tandem electrostatic accelerator that can produce either a proton or deuteron beam. It has a maximum rated power level of 10 kW and can therefore produce particle beams having currents in the range of up to about 4 mA, depending on the desired particle energy. The charged particle beam from the LABA tandem accelerator passes through a shielding wall and thence through an adjustable bending magnet that directs the beam to any of several experiment stations. Each of these stations can have its own target and neutron beam tailoring assembly. The LABA facility can be used for innovative BNCT-related research in the areas of accelerator target design and moderator-filter design. In addition, this facility can provide sufficient neutron flux levels for meaningful BNCT radiobiological research. One such initiative in this latter area involves the exploration of the possible application of BCNT techniques to the treatment of rheumatoid arthritis.

Researchers at the University of Birmingham in the United Kingdom have developed a slightly different variation for the design of a low-energy proton-beam-based accelerator epithermal neutron source for BNCT research. This concept features a neutron beam extraction arrangement that is at right angles to the axis of the proton beam impinging on the target. In the previous concepts, the incoming charged particle beam axis is coincident with the outgoing neutron beam axis. The use of the non-coaxial approach stems from certain practical considerations and it provides some neutronic advantages. The Dynamitron accelerator that drives the Birmingham facility presently operates reliably at a proton current of approximately 1 mA and it will be upgraded to produce 2 mA. Ultimately, it is planned to install a new ion source to produce a current of 5 mA but it is anticipated that clinical trials will be initiated at Birmingham using an operating current of 2 mA.

Researchers at the E.O. Lawrence Berkeley National Laboratory (LBL) and the University of California have initiated construction of a high-power accelerator-based epithermal-neutron source. In this system an electrostatic quadrupole accelerator is used to produce a proton beam that is directed onto a large-area water-cooled lithium target. Moderation and filtering of the resulting neutron source is performed in an assembly composed of aluminum and aluminum fluoride. A distinguishing feature of the neutron spectrum that is calculated for this device is that it is peaked near the high-energy end of the desirable epithermal-neutron range. This is expected to yield an improvement in the therapeutic ratio compared to what is produced by an epithermal neutron beam having a flat (in lethargy space) spectrum over the epithermal energy range. The neutron flux intensity that is anticipated to be achievable with the proposed LBL device is comparable to the intensity of several current reactor-based sources.

One final proton-accelerator-based concept for an epithermal-neutron source useful for BNCT applications is also currently under investigation by researchers at Idaho State University and MIT as well as in Russia. This idea involves using an incident proton beam that has an energy that is just above the threshold for neutron production in the target. In the case of a lithium target the threshold energy is 1.88 MeV and in the near-threshold concept the incident proton beam would have an energy in the range of 1.92 MeV. Under these conditions the angular distribution of the neutron source emitted from the target is highly forward-peaked due to so-called "kinetic collimation", an inherent feature of the kinematics of the neutron production interaction near the threshold. Under these conditions the spectrum of the emitted neutrons is quite soft and may require little or no moderation and filtering to produce an acceptable therapeutic ratio in clinical applications. Difficulties with this concept include the fact that the neutron yield is much smaller than is the case with the accelerator concepts that involve higher energy protons. This is counteracted, however, by the reduced filtering requirements. There are other questions involving stability and target design for this concept, but if it can be realized in a practical and deployable device, it could offer a low-cost system for hospital deployment.

Finally, the INEEL, in collaboration with Idaho State University, has also been investigating the feasibility of an alternate concept for an accelerator-based source of epithermal neutrons for BNCT that features a two-stage photoneutron production process driven by an electron accelerator. On the basis of neutronic performance, the proposed photoneutron device could also offer a promising alternate approach to the production of epithermal neutrons for BNCT.

3. COMPUTATIONAL DOSIMETRY AND TREATMENT PLANNING

Although the radiation dose in epithermal-neutron BNCT is to a large extent chemically targeted rather than geometrically targeted, it is still necessary to carefully optimize the geometric configuration of the incident neutron beam or beams using pre-treatment computations. In BNCT-enhanced fast neutron therapy there is an even greater necessity for careful treatment planning due to the requirement that both the geometrically-targeted fast neutron component of the therapeutic dose as well as the BNCT component be optimized to the extent practicable.

Treatment planning systems suitable for BNCT are typically built around a computational module for calculating the solution of the 3-D Boltzmann transport equation for neutral particles using the Monte Carlo stochastic simulation approach. Several methods for constructing patient geometry for the calculations from medical images have been successfully demonstrated.

The first of the geometric reconstruction methods used in BNCT employs what may be described as a "voxel reconstruction" technique. In this approach each plane of medical image data is partitioned into squares, each of which is assigned material properties based on average CT density or some other appropriate criterion. The square regions generated in each plane are mathematically stacked to construct a large 3-D array of parallelepipeds that is then input to a Monte Carlo radiation transport code. The various tissue compartments (including tumor regions of interest) are thereby represented as combinations of adjacent parallelepipeds. This is the approach taken in the MacNCTPLAN BNCT treatment planning system used in the clinical trials at MIT.

A useful extension to the voxel construction method described here has been reported by researchers in Australia. In this approach, a variable voxel size has been introduced. The voxels used for the geometric reconstruction can have dimensions on the order of a few millimeters in regions having complex structures or where there are other reasons for requiring a more detailed geometric description. Outside of these regions the voxels can be larger and fewer in number to conserve computation time.

The second method for direct construction of geometry from medical image data is based on the calculation of free-form surfaces fitted to the various tissue compartments. A surface reconstruction capability based on the use of non-uniform rational B-splines (NURBS) is used in the BNCT Radiation Treatment Planning Environment (BNCT_rtpe) software system developed by the INEEL, in collaboration with Montana State University. This system is used in the current clinical trials at Brookhaven and Petten. With this method, one first specifies a set of so-called "control points" outlining the anatomical regions of interest (e.g., skin, skull, brain, target volume, etc.) on each computer-displayed medical image plane. These control points are then fitted with closed NURBS curves and the region outlines for all image slices are combined to produce detailed equations describing the 3-D surfaces that enclose each volume of interest. The surface equations generated in the B-spline region reconstructions, in conjunction with appropriate region material descriptions, completely describe the problem and are subsequently used in a Monte Carlo radiation transport calculation performed by a specialized module developed specifically for use with BNCT_rtpe.

Although the treatment planning systems currently in clinical use for BNCT research have served the purpose well, they still suffer from somewhat long computation times (2–3 CPU-hours per field), relative to standard photon therapy planning software. This is largely due to the need for complete, rather than approximate 3D solutions to the relevant transport equation. As a result, greater computational speeds for BNCT treatment planning must generally be achieved through the application of improved numerical techniques rather than through the simplification of the governing transport equation. Recent efforts at the INEEL, in collaboration with Montana State University, have been directed toward this goal. This has resulted in a new paradigm for this type of calculation and the subsequent creation of a new treatment planning system for BNCT, denoted as SERA (Simulation Environment for Radiotherapy Applications). SERA is currently in initial clinical testing in connection with the trials at Brookhaven, and it is expected to ultimately replace the present BNCT_rtpe system.

The SERA treatment planning system employs a pixel-by pixel uniform volume element ("univel") reconstruction of the patient geometry. Fast line rasterization methods, implemented largely with integer arithmetic, are then used to allow rapid particle tracking through the univel geometry. The univel approach retains the geometric fidelity of the surface-based NURBS geometry reconstruction method but the execution time for the transport computations is reduced by a factor of between five and ten. SERA also features some much-improved image handling and display tools as well as several new three-dimensional geometry and dose distribution visualization tools as shown in Figure 1.

4. DISCUSSION

Significant progress has been made over the past several years in the supporting technologies that are essential for the ultimate success of neutron capture therapy. Three clinically useful epithermal-neutron beams with acceptable neutronic quality are now in routine use for human trials and recent developments promise a new generation of clinical epithermal-neutron sources that will have near-ideal neutronic properties. Treatment planning systems that are well suited for NCT applications are likewise now in routine use. Hospital-deployable accelerator-based epithermal-neutron sources, which were only in the conceptual stages not too long ago, are now on the verge of successful demonstration.

There are also several important initiatives to extend the use of NCT techniques beyond the traditional applications to brain tumors and melanoma. Interest in the use of existing clinical fast-neutron facilities for the exploration of NCT-enhanced fast-neutron therapy for several difficult malignancies is beginning to accelerate worldwide. Potential applications of NCT techniques to non-malignant conditions such as certain forms of rheumatoid arthritis are also under investigation.

However, despite impressive accomplishments with the supporting technologies many fundamental challenges remain for BNCT. Most importantly, now that initial epithermal-neutron BNCT trials are underway and additional trials are anticipated in the near future, it is essential that all interested groups communicate effectively and work together in a complimentary manner. If BNCT can indeed improve the prognosis for one or more afflictions, then this must be clearly demonstrated as expeditiously and as

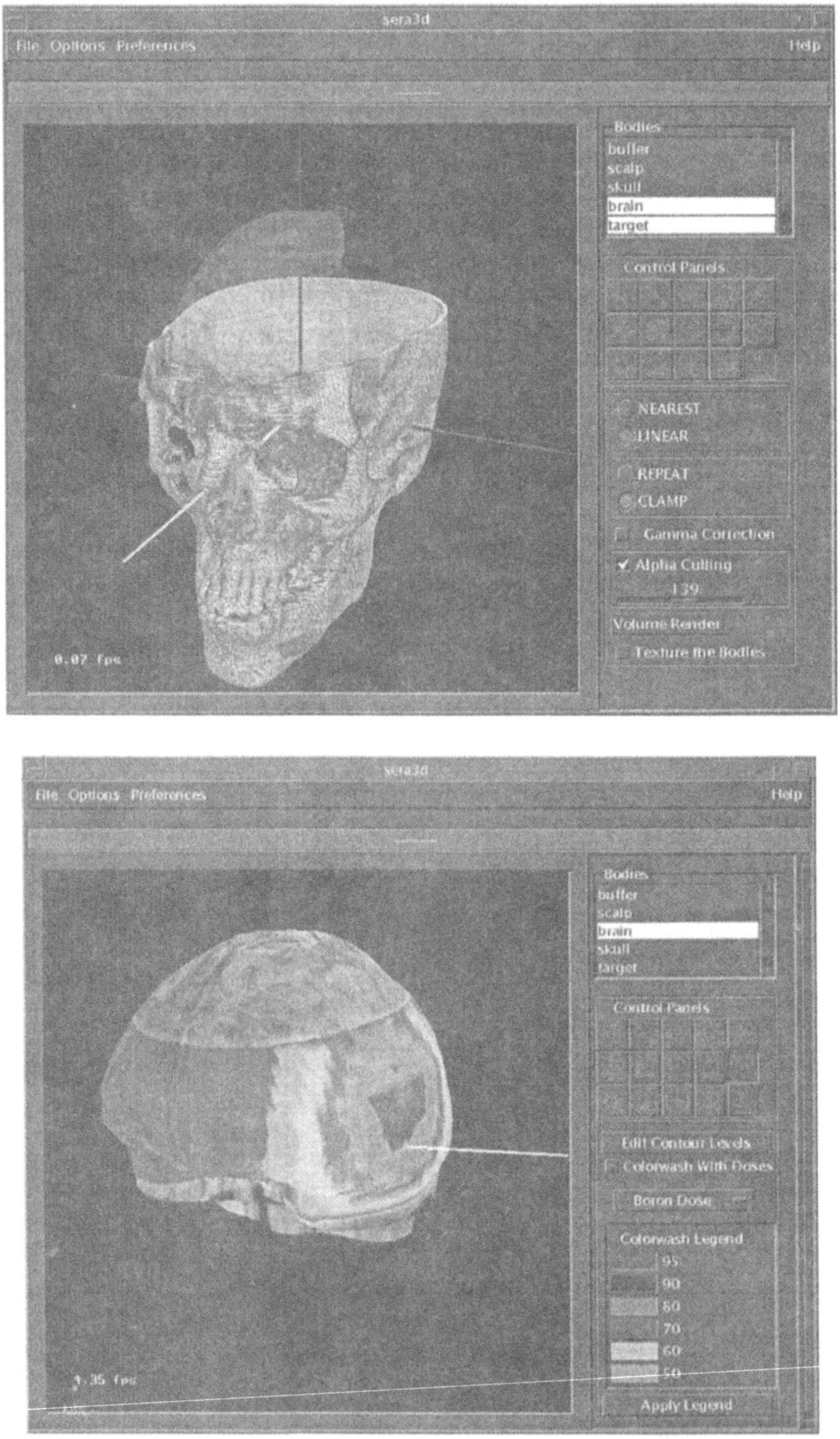

Figure 1. Three-dimensional anatomical reconstruction and BNCT dose visualization in the SERA system.

unequivocally as possible, in an independently reproducible manner, and according to accepted statistical standards. No single group is likely to do this alone. If the clinical development of epithermal-neutron NCT follows the pattern experienced with fast-neutron therapy, there will be no overnight miracles, but there will be steady incremental progress as information gained in multiple clinical trials is shared and used to improve the methods and tools. Also of absolutely essential importance will be the development and experimental application of continuously improving boron carriers. Deployable accelerator-based epithermal-neutron sources will also play an important role in the future. These are still in the developmental stage, however, and much of the necessary biological research can and should be accomplished now with the current generation of readily available reactor-based sources.

ACKNOWLEDGMENTS

The author wishes to express his sincere gratitude to the many institutions and individuals who supplied materials used in this plenary lecture.

The preparation of this article was sponsored by the United States Department of Energy, Office of Energy Research, under DOE Idaho Operations Office Contract Number DE-AC07-94ID13223.

3

CHEMISTRY AND PHARMACOLOGY OF AGENTS FOR BNCT

Debra A. Feakes

Department of Chemistry
Southwest Texas State University
San Marcos, Texas 78666

1. INTRODUCTION

Although boron neutron capture therapy (BNCT) is currently in clinical trials in several countries around the world, the continued development of BNCT for the treatment of cancer patients is dependent on the production of new compounds which exhibit biological characteristics exceeding that of the currently utilized disodium mercaptoundecahydro-*closo*-dodecaborate (BSH) or *para*-boronophenylalanine (*p*-BPA). Specifically, the compounds being developed should have an enhanced tumor uptake and retention, as well as exhibit a higher degree of specificity, particularly in terms of the tumor to blood boron ratio and tumor to normal brain boron ratio. Additionally, the degree of enhancement must be sufficiently significant to invest the resources necessary to produce the quantities required for clinical trials. Thorough reviews of the development of boron-containing compounds for application in BNCT have been published in recent years.[1,2] The focus of this review is the research which has been accomplished since the Seventh International Symposium on Neutron Capture Therapy for Cancer, held in Zurich, Switzerland in September, 1996.

Although a tremendous amount of research has been concentrated in the development of boron-containing macromolecules, such as monoclonal antibodies, recent research has emphasized the development of small boron-containing molecules. The new compounds can be categorized as: 1) derivatives of *p*-BPA and BSH, 2) boron-containing derivatives of known biomolecules, 3) boron-containing compounds designed for delivery by tumor-targeting delivery vehicles, and 4) porphyrin derivatives. Additionally, investigations into the application of the boron neutron capture reaction to other medical modalities, such as synovectomy for the treatment of rheumatoid arthritis[3] and dose enhancement for fast neutron therapy,[4] has received increased interest in the past several years.

Frontiers in Neutron Capture Therapy, edited by Hawthorne *et al.*
Kluwer Academic / Plenum Publishers, New York, 2001.

2. DERIVATIVES OF *p*-BPA AND BSH

2.1. Derivatives of *p*-BPA

The utility of *p*-BPA is limited by the low solubility of the compound. This difficulty has been alleviated, in part, by the utilization of the fructose complex in biological investigations. In an attempt to increase the solubility and decrease the uptake by the normal parenchyma, without affecting tumor uptake, a series of *p*-BPA derivatives which contain one to four hydroxyl groups connected in a cascading fashion (Figure 1) has been prepared.[5]

The biological activity of the new compounds was evaluated.[6] *In vitro* studies utilized the T98G human glioma cell line; *in vivo* studies utilized Fisher 344 rats intracranially implanted with 9L gliosarcoma. Survival studies, in which *in vitro* T98G cells were irradiated with thermal neutrons, indicated that the monohydroxylated and dihydroxylated derivatives of *p*-BPA were more lethal than the *p*-BPA control. The tetrahydroxylated derivative was less lethal. A subsequent evaluation of the dihydroxylated derivative of *p*-BPA in the Greene's melanotic melanoma cell line demonstrated analogous survival results. *In vivo* evaluation of the hydroxylated derivatives of *p*-BPA demonstrated that uptake of the compounds by the normal parenchyma decreased according to the number of hydroxyl groups.

2.2. Derivatives of $[B_{12}H_{12}]^{2-}$

Although BSH has been investigated as a potential agent for BNCT since the early 1960's,[7] few investigations have been completed regarding the derivative chemistry of this important molecule. Furthermore, little is known about the role of the sulfhydryl group in the tumor specificity of BSH. Recently, a series of derivatives of the $[B_{12}H_{12}]^{2-}$ ion which can be used as BNCT agents or as precursors to other compounds applicable to BNCT has been developed.[8] The first series of compounds (Figure 2), a group of alkylated and arylated anions, are prepared from the palladium catalyzed Grignard reaction of the monoiodinated $[B_{12}H_{11}I]^{2-}$ and provides a useful synthetic scheme for the preparation of boron-carbon bonds. The second series of derivatives (Figure 2), a group of hydroxylated derivatives, are prepared from the action of sulfuric acid on the $[B_{12}H_{12}]^{2-}$ anion. The hydroxyl substituent provides a moiety suitable for the attachment of a wide

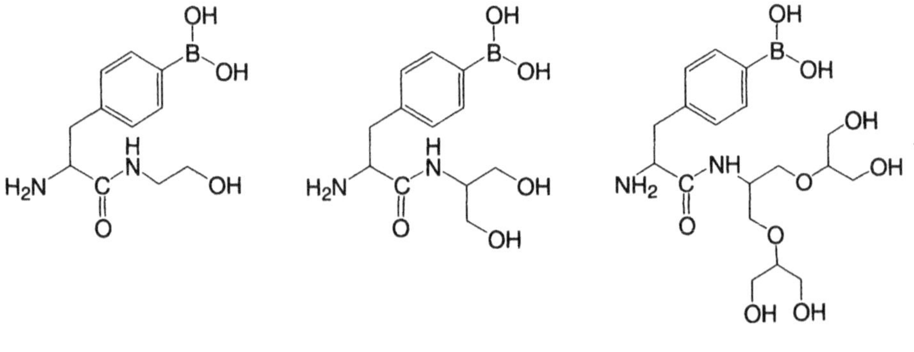

Figure 1. Hydroxylforms of *p*-BPA.

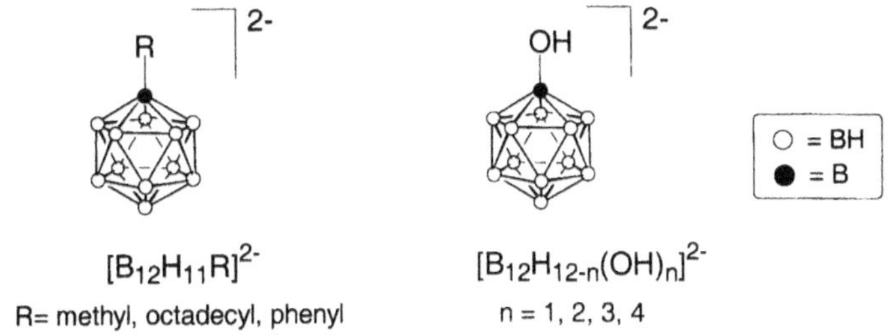

Figure 2. Derivatives of BSH.

variety of compounds useful for BNCT and may provide insight into the mechanism of uptake and retention of BSH.

3. BORON-CONTAINING DERIVATIVES OF KNOWN BIOMOLECULES

Many of the compounds currently under investigation are derived from molecules known to exhibit significant biological activity. One group of compounds being investigated by a number of researchers is a series of polyamines, predominantly putrescine, spermidine, and spermine (Figure 3). These polyamines are naturally occurring and exhibit a variety of promising biological properties: 1. they are know known to be essential for cell growth and differentiation, 2. elevated concentrations have been found in rapidly proliferating tumor cells, 3. cationic polyamines bind strongly to DNA through electrostatic interactions, and 4. derivatization of the polyamines has produced compounds which are more reactive than the parent.[9]

The polyamines have been derivatized with an *ortho*-carborane moiety at the terminal nitrogen position[9] (Figure 4) as well as at an internal nitrogen position (Figure 5).[9]

The derivatized polyamines have been evaluated in *in vitro* assays using F98 rat glioma cells.[10] Although DNA binding assays indicated that the *ortho*-carborane moiety hindered DNA binding, the internally alkylated spermine derivative exhibited greater binding than the unsubstituted spermidine compound. Carboranyl polyamines were taken up by the tumor cells in concentrations comparable to BSH and *p*-BPA, but at significantly lower media concentrations. The high level of toxicity exhibited by the *ortho*-carborane derivatives precluded further investigations. In particular, the

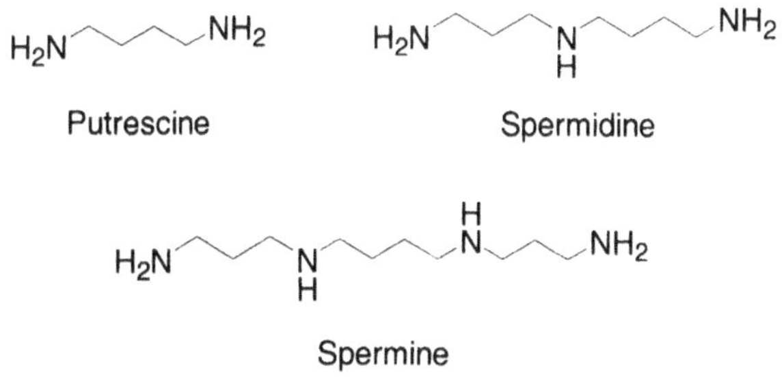

Figure 3. Naturally occurring polyamine compounds.

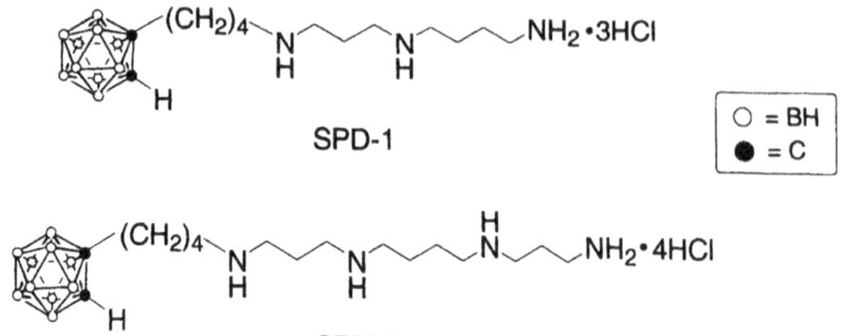

Figure 4. Terminally substituted *ortho*-carborane derivatives of the polyamine compounds.

terminally substituted polyamines were more toxic than the internally alkylated polyamines.

In an attempt to produce less toxic compounds, a second group of boron-containing polyamines were prepared (Figure 6).[11] In the first compound, the *ortho*-carborane substituent in the internally alkylated spermidine derivative was substituted with an ethylamine moiety and in the second compound, the *ortho*-carborane substituent has been replaced with a dihydroxyborylbenzyl group.

The preparation of *para*-carborane derivatives of the polyamine compounds has been reported (Figure 7).[12] An acridine residue has been incorporated into two of the derivatives because of the known potential of acridine to intercalate into DNA. The spermidine residue was incorporated into two of the derivatives to increase the water solubility of the *para*-carborane products and due to the known potential of spermidine to interact with DNA.

1-Aminocyclobutanecarboxylic acid (Figure 8) is a non-toxic compound known to be preferentially retained in intracerebral tumors.[13,14] Initially, two boron-containing derivatives of this compound were prepared. The first is a boronic acid derivative and the second is a *meta*-carborane derivative (Figure 9). A sugar derivative of the *meta*-

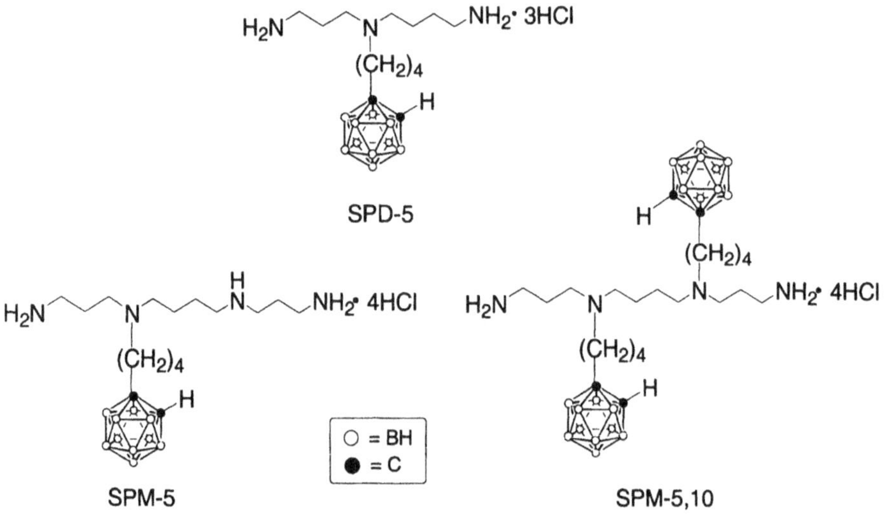

Figure 5. Internally substituted *ortho*-carborane derivatives of the polyamine compounds.

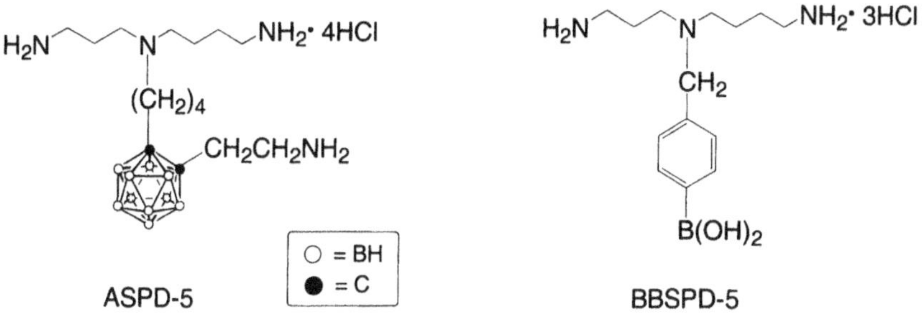

Figure 6. Boron-containing derivatives of the polyamine compounds.

carborane compound (Figure 9) was prepared because the *meta*-carborane derivative did not display the desired solubility.[13,14]

Preliminary *in vivo* investigations were performed using the boronic acid derivative and the sugar derivative in nude mice bearing implanted human glioblastoma.[15] The boronic acid derivative exhibited significant tumor retention over time at low injected doses. Both compounds exhibited significant tumor to normal brain boron concentrations. The *nido*-analogues of the *meta*-carborane compounds have also been prepared.

4. BORON-CONTAINING COMPOUNDS DESIGNED FOR DELIVERY BY TUMOR-TARGETTING DELIVERY VEHICLES

Unilamellar liposomes have been investigated as tumor-specific delivery vehicles for application in BNCT.[16,17,18] This modality has significant advantages since the boron-containing compounds are not required to exhibit any inherent tumor specificity. Additionally, a wide variety of boron-containing compounds can be incorporated within the liposomes. Hydrophilic boron-containing compounds can be encapsulated within the aqueous core of the liposomes while lipophilic boron-containing compounds can be embedded in the bilayer of the liposomes. Investigations to date have led to the

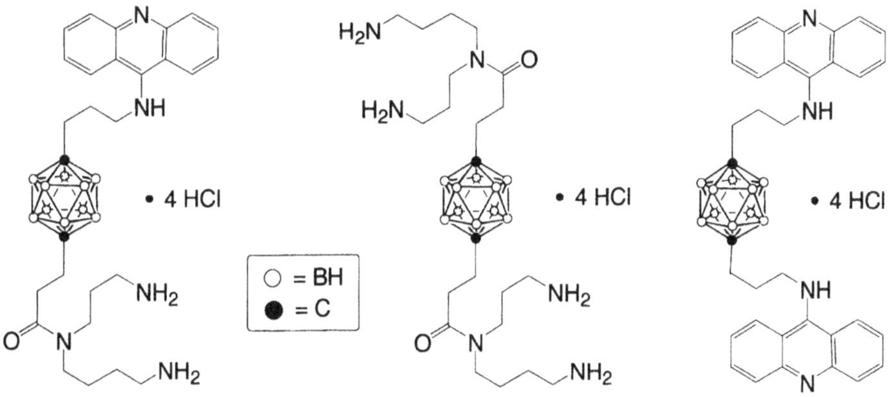

Figure 7. *Para*-carborane derivatives of the polyamine compounds.

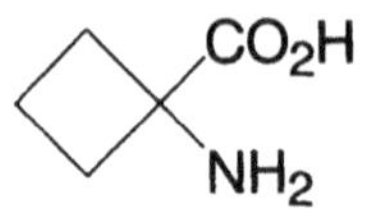

Figure 8. Structure of 1-aminocyclobutanecarboxylic acid.

development of a proposed mechanism for tumor cell retention based on the susceptibility of the boron-containing compound to nucleophilic attack or *in vivo* oxidation.[16] In an effort to investigate this hypothesis, as well as prepare a series of compounds for possible application to BNCT, a number of boron-containing compounds based on the parent $[B_{20}H_{18}]^{4-}$ anion have been prepared.

The *in vivo* biodistribution of the ammonio derivative of $[B_{20}H_{18}]^{4-}$, encapsulated in liposomes and investigated in mice bearing EMT6 mammary adenocarcinoma, is one of the most promising to date.[18] Cyclic voltammetry experiments indicated that this compound oxidized readily compared to other substituted and unsubstituted species. In an effort to examine this potential mechanism of tumor cell retention, the chemical oxidation of $[B_{20}H_{17}NH_3]^{3-}$ was investigated (Figure 10).[19]

Oxidation of the ammonio species using benzoquinone in acetonitrile yielded a unique μ-bridged species which incorporated the acetonitrile solvent. The produce can be reversibly protonated to produce the dinegative anion. Oxidation of the μ-bridged species yielded the oxidized anion with a uninegative charge. The electron-deficient, three-center, two-electron bonds in the oxidized anion make it susceptible to nucleophilic attack, a property proposed by Hawthorne and coworkers to enhance tumor retention.[18]

Similarly, the oxidation of the hydroxy derivative, $[B_{20}H_{17}OH]^{4-}$ has also been investigated (Figure 11).[20] Aqueous ferric ion oxidation yields the normal isomer of the dianion which contains the reactive electron-deficient, three-center, two-electron bonds. Oxidation of the reduced, substituted tetraanion using benzoquinone in acetonitrile yields a μ-oxo-bridged species. The μ-oxo-bridged species can be reversibly protonated.

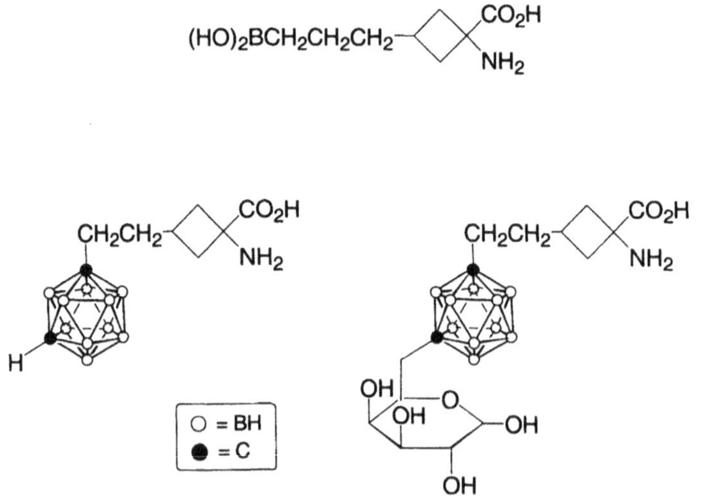

Figure 9. Boron-containing derivatives of 1-aminocyclobutanecarboxylic acid.

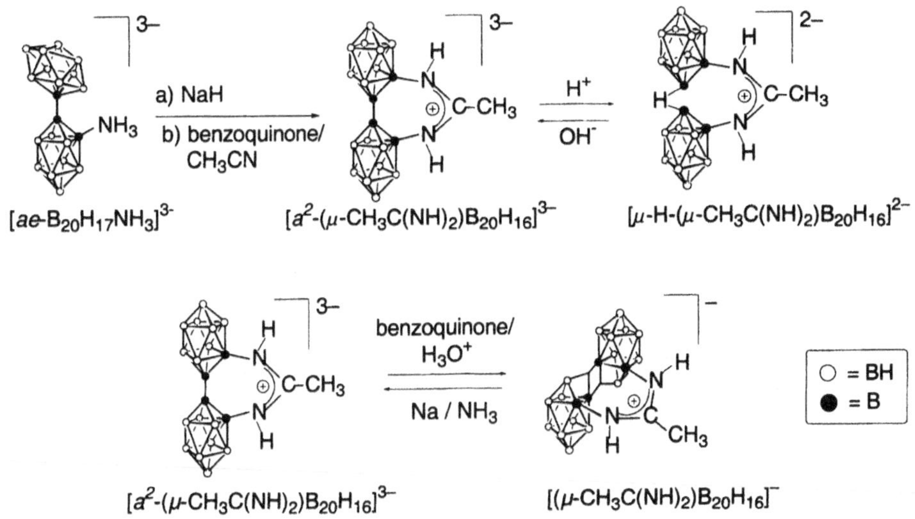

Figure 10. Oxidation sequence for the $[B_{20}H_{17}NH_3]^{3-}$ anion.

The preparation of the thiol derivative, $[B_{20}H_{17}SH]^{4-}$ (Figure 12) has been reported.[21] The thiol compound has the potential to interact directly with intracellular protein moieties at the thiol substituent and also has the potential to oxidize to more reactive species, such as $[B_{20}H_{17}SH]^{2-}$ or $[B_{20}H_{17}S\text{-}SB_{20}H_{17}]^{8-}$.

The biodistribution of the liposomally encapsulated compound, investigated in mice bearing EMT6 mammary adenocarcinoma, exhibits tumor accretion over the entire 48 hour time period, resulting in a final tumor boron concentration of 21 μg boron/g tumor and a final tumor to blood boron ratio of approximately 5.[22] The cellular toxicity and intracellular distribution of the compound are currently under investigation.

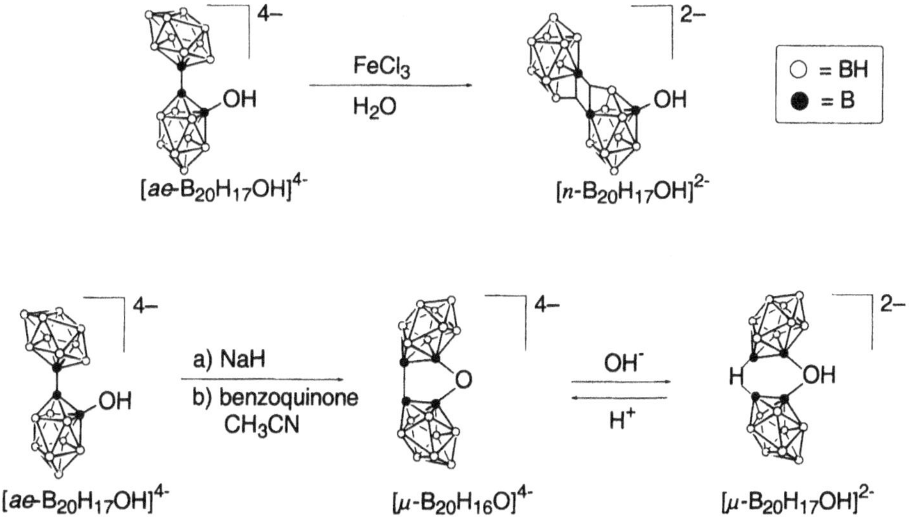

Figure 11. Oxidation sequence for the $[B_{20}H_{17}OH]^{4-}$ anion.

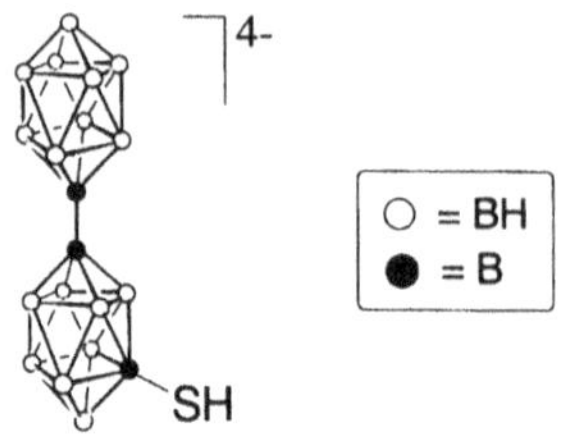

Figure 12. Structure of the reduced sulfur derivative, $[B_{20}H_{17}SH]^{4-}$.

The discovery of a novel polyhedral borane dianion, $[B_{22}H_{22}]^{2-}$ (Figure 13), during the preparation of decaborane from sodium borohydride led to the development of an optimized synthetic procedure.[23] The crystal structure of the dianion can be described as a fused polyhedron consisting of an open decaborane cage and a closed dodecaborane cage. The polyhedral borane cages are connected at two boron vertices. The cluster is stable and provides a wealth of potential derivative chemistry, applicable to BNCT, based on the known chemistry of decaborane, the $[B_{12}H_{12}]^{2-}$ cage, and carborane.

5. PORPHYRIN DERIVATIVES

Boron-containing porphyrin derivatives have been the subject of considerable interest, particularly with the promising results exhibited with BOPP by Kahl and coworkers.[24] A concerted effort has been made to synthesize and evaluate a series of boron-containing porphyrin compounds which vary in the degree of lipophilicity.[25] Two of the porphyrin starting materials (Figure 14) are obtained from *methylpheophorbid a* which is extracted from *Spirullina pacifica*. Derivatization of purpurin-18 is accomplished at the isocyclic anhydride and the propionic acid side chain. *Pyropheophorbid a* is derivatized at the vinyl substituent and the propionic methyl ester group.

One of the porphyrin derivatives (Figure 15) was derivatized by appending a BSH substituent at the propionic acid moiety. The lipophilicity of the porphyrin is adjusted by appending long side chains at the vinyl substituent.

The incorporation of both boron and gadolinium into a single compound (Figure 16) has been achieved and biological investigations are currently underway.[26] The basis for the synthesis is the ability to use the gadolinium center for a magnetic resonance imaging contrast enhancement agent and the boron center for boron neutron capture

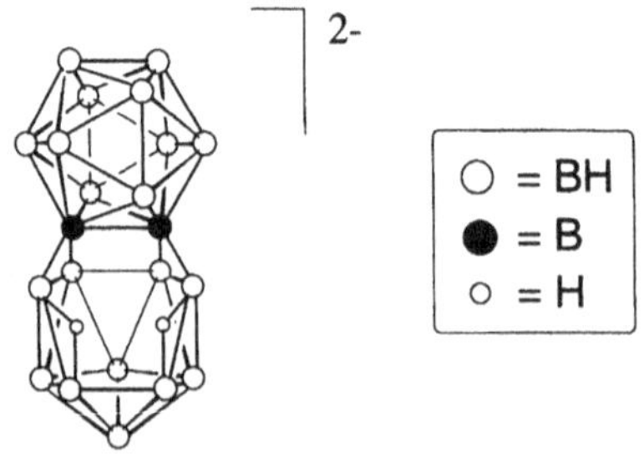

Figure 13. Structure of the $[B_{22}H_{22}]^{2-}$ dianion.

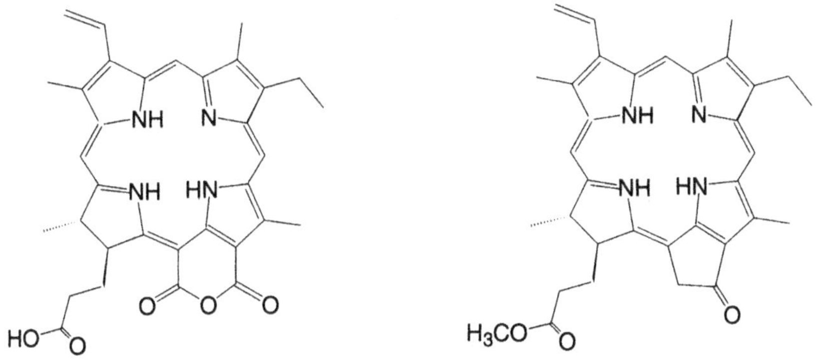

Figure 14. Porphyrin starting materials.

therapy. Although the gadolinium is unlikely to enhance the cytotoxicity of the neutron capture reaction unless the gadolinium is in close proximity to the nuclear DNA, the localization and concentration of boron in the tumor cells should be elucidated indirectly through the imaging process.

6. ADDITIONAL APPLICATIONS OF THE BORON NEUTRON CAPTURE REACTION

Recently, an increasing interest of the application of the boron neutron capture reaction as a dose enhancement for fast neutron radiotherapy has occurred.[4,27] A number of advantages of the use of BNCT for the augmentation of fast neutron radiotherapy have been delineated: 1) fast neutron therapy utilizes currently available medical facilities, 2) fast neutron therapy has the potential to be used for a wide variety of tumor systems, 3) fast neutron therapy is not necessarily limited by the requirement of BNCT to have boron-10 nuclei in every tumor cell, 4) fast neutron therapy can be dose enhanced using relatively simple boron-containing compounds, and 5) fast neutron therapy normal

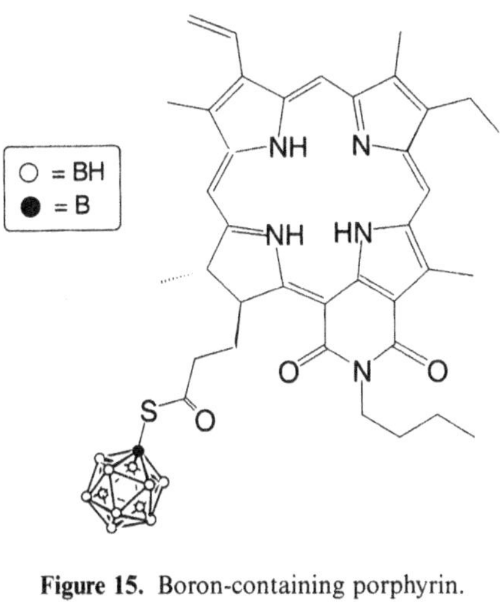

Figure 15. Boron-containing porphyrin.

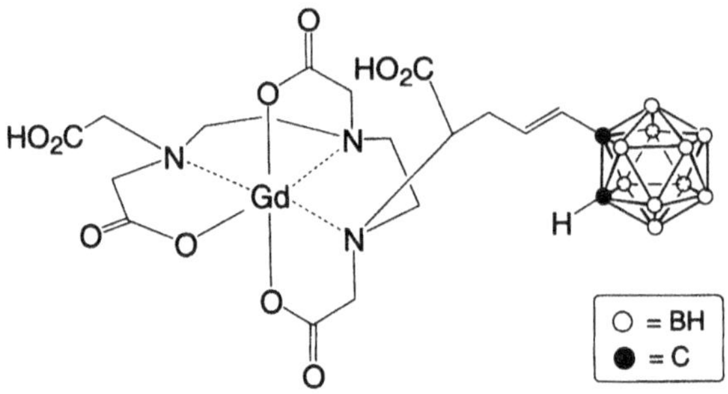

Figure 16. Gadolinium complex of a boron-containing compound.

tissue tolerances are already well-established.[4] In simple terms, the concept of augmentation is derived from the typical dose response curve where tumor response is plotted as a function of dose. This curve is characterized by a steep gradient at which time, the irradiation occurs. A small increase in the dose along the steep portion of the curve would, therefore, significantly enhance tumor response. *In vivo* investigations of fast neutron radiotherapy, enhanced by the boron neutron capture reaction, have been performed in dogs with spontaneous non-small cell lung cancer. Boron-10 enriched $Na_2[B_{10}H_{10}]$ was utilized in the trial with a goal to determine the feasibility and toxicity of the combined treatment protocols.[28]

7. CONCLUSIONS

Although impossible to review the vast amount of research which has been completed in the area of compound development for application in BNCT since the Seventh International Symposium on Neutron Capture Therapy for Cancer in the limited amount of space, several general trends can be delineated. In the past few years, there has been an increased emphasis on the development of low molecular weight compounds. Rather than simply making numerous compounds and incurring the expense of evaluating each new compound in biological protocols, there appears to be an increased effort to evaluate the *in vivo* biodistributions of the new compounds and use that information to design compounds with a high probability for success. Similarly, there are increased efforts to delineate the subcellular pathways and mechanisms of uptake and retention for the compounds currently being developed. Perhaps the one limitation which continues to hinder the progress of compound development for application in BNCT is the inability to compare new compounds to each other, as well as the currently utilized BSH and *p*-BPA, due to the varying biological protocols being utilized by the different research groups.

ACKNOWLEDGMENTS

The author would like to thank the members of the organizing committee for the opportunity to present this overview at the Eighth International Symposium on Neutron Capture Therapy and would like to acknowledge research support by awards from

Research Corporation (CC3913), the Robert A. Welch Foundation (AI1292), and Southwest Texas State University.

REFERENCES

1. M.F. Hawthorne, "The Role of Chemistry in the Development of Boron Neutron Capture Therapy of Cancer," *Angew. Chem. Int. Ed. Engl.* 32:950–984, 1993.
2. A.H. Soloway, W. Tjarks, B.A. Barnum, F. Rong, R.F. Barth, I.M. Codogni, and J.G. Wilson, " *Chem. Rev.* 98:1515–1562, 1998.
3. E. Binello, S. Shortkroff, A.G. Jones, C. Viveiros, G. Young, A. Davison, C.B. Sledge, and J.C. Yanch, "*In vitro* Analysis of ^{10}B Uptake for Boron Neutron Capture Synovectomy," *in*: Advances in Neutron Capture Therapy: Vol. II, Chemistry and Biology," B. Larsson, J. Crawford, and R. Weinrich, ed., Elsevier, Netherlands, pp. 609–613, 1997.
4. G.E. Laramore, "The Use of Neutrons in Cancer Therapy: A Historical Perspective Through the Modern Era," *Seminars in Oncology*, 24(6):672–685, 1997.
5. H. Nemoto, J. Cai, N. Asao, S. Iwamoto, and Y. Yamamoto, "Synthesis and Biological Properties of Water-Soluble p-Boronophenylalanine Derivatives. Relationship between Water Solubility, Cytotoxicity, and Cellular Uptake," *J. Med. Chem.* 38, 1673–1678, 1996.
6. M. Takagaki, K. Ono, Y. Oda, H. Kikuchi, H. Nemoto, S. Iwamoto, J. Cai, and Y. Yamamoto, "Hydroxylforms of *p*-Boronophenylalanine as Potential Boron Carriers on Boron Neutron Capture Therapy for Malignant Brain Tumors," *Cancer Research* 56:2017–2020, 1996.
7. A.H. Soloway, H. Hatanaka, and M.A. David, *J. Med. Chem.* 10:714, 1967.
8. T. Peymann, C.B. Knobler, and M.F. Hawthorne, unpublished results.
9. J. Cai, and A.H. Soloway, "Synthesis of Carboranyl Polyamines for DNA Targeting," *Tetrahedron Letters*, 37(52):9283–9286, 1996.
10. J. Cai, A.H. Soloway, R.F. Barth, D.M. Adams, J.R. Hariharan, I.M. Wyzlic, and K. Radcliffe, "Boron-Containing Polyamines as DNA Targeting Agents for Neutron Capture Therapy of Brain Tumors: Synthesis and Biological Evaluation," *J. Med. Chem.* 40:3887–3896, 1997.
11. A.H. Soloway, unpublished results.
12. H. Ghaneolhosseini, W. Tjarks, and S. Sjoberg, "Synthesis of Novel Boronated Acridines and Spermidines as Possible Agents for BNCT," *Tetrahedron*, 54(15):3877–3884, 1998.
13. R.R. Srivastava, R.R. Singhaus, and G.W. Kabalka, "1-Amino-3-[2-(1,7-dicarba-*closo*-dodecaboran(12)-1-yl)ethyl]cyclobutanecarboxylic Acid: A Potential BNCT Agent," *J. Org. Chem.* 62:4476–4478, 1997.
14. R.R. Srivastava, and G.W. Kabalka, "Syntheses of 1-Amino-3-[2-(7-(2-hydroxyethyl)-1,7-dicarba-*closo*-dodecaboran(12)-1-yl)ethyl]cyclobutanecarboxylic Acid and Its *nido*-Analogue: Potential BNCT Agents," *J. Org. Chem.* 62:8730–8734, 1997.
15. R.R. Srivastava, and G.W. Kabalka, unpublished results.
16. K. Shelly, D.A. Feakes, M.F. Hawthorne, P.G. Schmidt, T.A. Krisch, and W.F. Bauer, "Model Studies Directed Toward the Boron Neutron-Capture Therapy of Cancer: Boron Delivery to Murine Tumors with Liposomes," *Proc. Natl. Acad. Sci. USA*, 89:9039–9043, 1992.
17. D.A. Feakes, K. Shelly, C.B. Knobler, and M.F. Hawthorne, "$Na_3[B_{20}H_{17}NH_3]$: Synthesis and Liposomal Delivery to Murine Tumors," *Proc. Natl. Acad. Sci. USA*, 91:3029–3033, 1994.
18. D.A. Feakes, K. Shelly, and M.F. Hawthorne, "Selective Boron Delivery to Murine Tumors By Lipophilic Species Incorporated in the Membranes of Unilamellar Liposomes," *Proc. Natl. Acad. Sci. USA*, 92:1367–1370, 1995.
19. F. Li, K. Shelly, C.B. Knobler, and M.F. Hawthorne, "Formation of a Novel Amidinium-Bridged Polyhedral Borane Ion by Incorporation of an Acetonitrile Solvent Molecule," *Angew. Chem. Int. Ed.* 37(13–14):1865–1867, 1998.
20. F. Li, K. Shelly, C.B. Knobler, and M.F. Hawthorne, "A New Isomer of the $[B_{20}H_{18}]^{2-}$ Ion: Synthesis, Structure, and Reactivity of *cis*-$[B_{20}H_{18}]^{2-}$ and *cis*-$[B_{20}H_{17}NH_3]^{-}$," *Angew. Chem. Int. Ed.* 37(13–14):1868–1871, 1998.
21. D.A. Feakes, F.R. Harris, D.K. Hathaway, and V.S. Morton, "Preparation of Hydrophilic Boron-Containing Compounds for Incorporation into Liposomes," *in*: Advances in Neutron Capture Therapy: Vol. II, Chemistry and Biology," B. Larsson, J. Crawford, and R. Weinrich, ed., Elsevier, Netherlands, pp. 95–100, 1997.

22. D.A. Feakes, R.C. Waller, D.K. Hathaway, and V.S. Morton, unpublished results.
23. N.S. Hosmane, A. Franken, G. Zhang, R.R. Srivastava, R.Y. Smith, and B.F. Spielvogel, "Synthesis and Crystal Structure of a Novel Fused Polyhedral Borane Dianion, $[B_{22}H_{22}]^{2-}$: Potential Precursor for Use in Boron Neutron Capture Therapy (BNCT) of Cancer," *Main Group Metal Chemistry*, 21(6):319–324, 1998.
24. R.G. Fairchild, S.B. Kahl, B.H. Laster, J. Kalef-Ezra, and E.A. Popenoe, *Cancer Res.* 50:4860, 1990.
25. S. Ruf, J. Osterloh, and D. Gabel, unpublished results.
26. Y. Yamamoto, "New Reactions of Carboranes and Synthesis of a Carborane-Gadolinium Complex for BNCT," *Special Publication of the Royal Society of Chemistry* (Advances in Boron Chemistry), 201:281–288, 1997.
27. G.E. Laramore, P. Wootton, J.C. Livesey, D.S. Wilbur, R. Risler, M. Phillips, J. Jacky, T.A. Buchholz, T.W. Griffin, and S. Brossard, "Boron Neutron Capture Therapy: A Mechanism for Achieving A Concomitant Tumor Boost in Fast Neutron Radiotherapy," *Int. J. Radiation Oncology Biol. Phys.* 28(5):1135–1142, 1994.
28. K.J. Stelzer, P.R. Gavin, R. Risler, H. Kippenes, M.F. Hawthorne, D.W. Nigg, and G.E. Laramore, unpublished results.

CLINICAL EXPERIENCES

4

THE HARVARD-MIT BNCT PROGRAM

Overview of the Clinical Trials and Translational Research

Paul M. Busse[1], Robert G. Zamenhof[2], Otto K. Harling[3], Irving Kaplan[1], Jody Kaplan[1], Cynthia F. Chuang[3], John T. Goorley[3], W. S. Kiger III[3], Kent J. Riley[3], Lujia Tang[2], Guido R. Solares[2], and Matthew R. Palmer[2]

[1]Joint Center for Radiation Therapy
Harvard Medical School
330 Brookline Ave.
Boston, Massachusetts 02215
[2]Department of Radiology
Beth Israel Deaconess Medical Center
One Deaconess Road
Boston, Massachusetts 02215
[3]Nuclear Reactor Laboratory
Massachusetts Institute of Technology
138 Albany St.
Cambridge, Massachusetts 02139

1. INTRODUCTION

The clinical use of BNCT originated in Boston largely through the efforts of Dr. William Sweet at the Massachusetts General Hospital and his collaborators at the Brookhaven National Laboratory (BNL) and the Massachusetts Institute of Technology (MIT). The early clinical trials were for patients with primary brain tumors but were unsuccessful largely due to the poor biodistribution of ^{10}B and thermal neutron beams that had limited penetration into tissue. Clinical and basic researchers recognized there would have to be a dramatic improvement in the amount of ^{10}B delivered to a tumor as well as an improvement in the biodistribution, i.e., an increase in the tumor to normal tissue ratio, before clinical studies could be resumed. In addition, nuclear engineers and physicists realized epithermal beams needed to be developed that would obviate many of the physical limitations inherent in the thermal beams that had previously been used. The culmination of this multifaceted preclinical work in biology, physics, ^{10}B quantification, treatment planning, and epithermal beam design was in September1994 when the first

Frontiers in Neutron Capture Therapy, edited by Hawthorne *et al.*
Kluwer Academic / Plenum Publishers, New York, 2001.

human subject was irradiated with epithermal neutron BNCT at the MIT Nuclear Reactor Laboratory. This subject was irradiated to the plantar surface of the foot following the administration of oral p-boronophenylalanine (BPA); two weeks later another patient with glioblastoma received a single field of epithermal neutron irradiation at BNL, with intravenously administered BPA-fructose as a ^{10}B carrier. Since then a number of clinical trials that are primarily phase I in nature have been conducted or are currently in progress at both institutions with epithermal neutron beams.[1] Over 60 subjects have been irradiated, the majority to the cranium for glioblastoma multiforme and in addition, 4 subjects have been irradiated to 5 sites at MIT for melanoma of the extremity.

The largest experience with BNCT for primary CNS tumors is from Japan where a number of universities, hospitals and reactors have collaborated since 1968. The clinical effort, initially led by Hatanaka[2] has been assumed by Nakagawa and colleagues and the clinical results are being meticulously analyzed.[3] These data will represent what can be expected from a different approach, i.e., BSH as a ^{10}B carrier, thermal neutrons, and an open craniotomy to improve the depth distribution. A long clinical and laboratory experience also exists in Japan in the treatment of melanoma with BNCT.[4,5] In Petten, the first 10 subject cohort of an EORTC phase I trial has been completed.[6] BSH is the ^{10}B carrier and the epithermal neutron irradiation is carried out in 4 fractions on consecutive days.

This paper describes the Harvard-MIT BNCT Program with emphasis on the design, conduct and early results of the two phase I clinical trials, the peripheral melanoma trial and the combined GBM/metastatic melanoma CNS trial. Data continually accrue and will ultimately lead to an early estimation of a maximally tolerated dose (MTD) to be followed by phase II and eventually phase III trials. Translational research has been and remains an integral part of the program. The focus of this work is on a more thorough understanding of the spatial and temporal pharmacokinetics of ^{10}B in normal tissue and tumor in order to optimize the clinical capability of the high-intensity fission converter beam (FCB) under construction at MIT. Additional preclinical work is directed toward other tumor types and anatomic locations in order to assess suitability to BNCT.

2. THE CLINICAL PROGRAM

The two trials in progress at Harvard-MIT are similar in purpose and are designed to determine a separate normal tissue endpoint and secondarily tumor response. Both protocols have undergone critical review and have been approved by the respective investigational review committees at the Beth Israel Deaconess Medical Center and MIT. Separate but similarly worded informed consents are obtained for each institution. Subjects must be 18 or older, meet a minimal performance criterion of 70 on the Karnofsky scale, and must not have received a prior course of radiation to the site of BNCT irradiation. The projected life expectance should be at least 3 months and subjects with a diagnosis of phenylketonuria are excluded. Following initial evaluation, subjects undergo a CT and an MRI with and without contrast for treatment planning purposes. These studies also serve as baseline studies for follow-up. Each subject has an individualized treatment plan developed with MacNCTPLAN.[1] This software program facilitates the determination of the best beam entry positions and generates isodose contours for normal tissue as well as tumor. Subjects are simulated clinically in the appropriate position for each beam entry

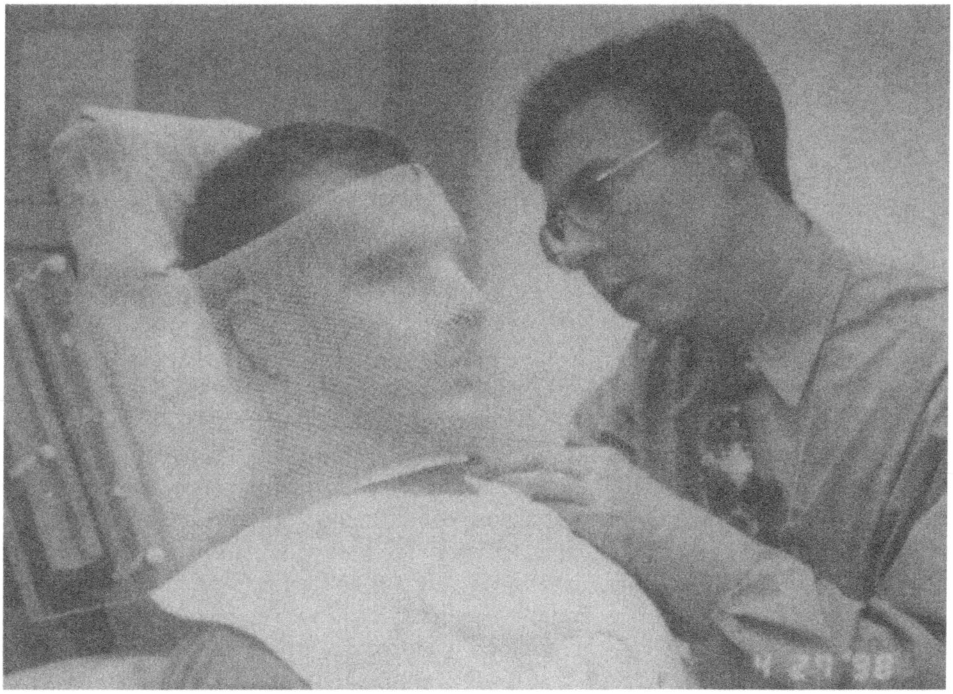

Figure 1. Clinical set-up of a subject at Harvard-MIT. The number of beams, entrance points and relative weighting are determined through treatment planning with MACNCTPlan. The subject is then placed in the appropriate position, either seated, prone or supine. An individualized mask is constructed for immobilization and to facilitate set-up.

position and immobilized with commercially available systems (Figure 1). All irradiations are performed at the MITR-II with the M-67 epithermal beam, the characteristics of which have been published.[7,8] ^{10}B is delivered as l-boronophenylalanine-fructose (BPA-f) through an intravenous infusion that ends 45 minutes before the start of irradiation. A detailed pharmacokinetic curve is generated for each subject and the final dosimetry is based on each subject's individual blood ^{10}B concentration curve. The radiation dose is expressed as RBE-cGy, which is the physical dose for each component in the beam weighted by its relative biological effectiveness (RBE).[3] The RBE for ^{10}B is expressed as the compound-RBE (CRBE) that also takes into account the distribution of the boron carrier molecule in tissue at the microscopic level. These values are derived from animal experiments and the measured distribution of ^{10}B in tumor and normal tissue (Table 1). The vital signs, EKG tracing and blood oxygenation are continuously followed from the initiation of infusion of BPA-f to when subjects are transported back to the hospital, where they continue to be monitored overnight. With rare exception, discharge is the following day. Follow-up is scheduled monthly with MRI scans whenever possible.

2.1. ^{10}B Pharmacokinetics

Because of the importance of ^{10}B in every aspect of BNCT a great deal of attention has been given to its proper quantification and an understanding of the pharmacokinetics. Normal tissue ^{10}B concentrations were considered to be the same as blood

Table 1. RBE values for radiation components in glioblastoma, brain, and melanoma

Tissue	^{10}B (1)	Fast neutron	Thermal neutron	Gamma
GBM (2)	3.8	3.2	3.2	0.5 (3)
Brain	1.3	3.2	3.2	0.5 (3)
Melanoma	4.0	3.2	3.2	0.5 (3)

1. RBE value for ^{10}B is adjusted for microdistribution in cells and referred to as C-RBE.
2. GBM: Glioblastoma Multiforme.
3. RBE is adjusted to account for prolonged irradiation duration and two fractions.

concentrations for treatment planning purposes. This was verified for skin by quantification of ^{10}B in normal skin by neutron-induced autoradiographic techniques.[9,10] There have been a number of changes in the way BPA has been administered since the first subject was irradiated in 1994. The original protocol for melanoma of the extremity called for BPA to be administered orally at a dose of 400 mg/kg. Subjects were not permitted to eat prior to BPA administration in order to facilitate BPA absorption. Despite the large amounts of BPA administered orally, the ^{10}B concentrations in the blood were relatively low, 3–5 ppm as were tumor concentrations, 10–15 ppm. Nevertheless, tumor samples had 3–4 times more ^{10}B than blood or normal skin and enough for a therapeutic gain of approximately 1.4.[5] In 1996, oral administration of BPA was abandoned and subjects in both protocols were administered BPA in the fructose complex (BPA-f) intravenously. Because of the fluid volume and hypertonicity, BPA-f has been administered through a central line placed the night before BNCT. The initial concentration was 250 mg/kg infused over 60 minutes. During an irradiation, the blood ^{10}B concentration was continuously monitored through a remote-sampling device that drew off a minimal volume from a distance of up to 25 feet and allowed for an uninterrupted analysis. Blood concentrations of ^{10}B were measured either through prompt gamma neutron activation analysis[11] (PGNAA) or inductively coupled plasma-atomic emission spectroscopy (ICP-AES). This produced a very characteristic pharmacokinetic curve in the blood where a peak concentration of approximately 30-ppm was achieved at the end of infusion (Figure 2). This was followed by a rapid decline thought to be redistribution and then by a gradual

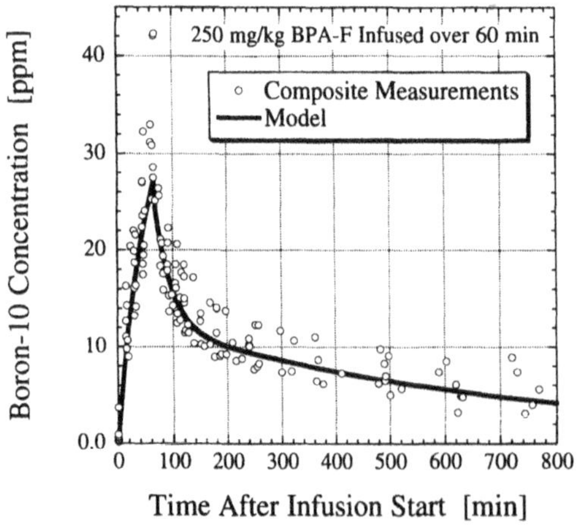

Figure 2. Pharmacokinetic curve of ^{10}B in blood. Aggregate data from subjects infused with 250 mg/kg BPA-f over 60 minutes. ^{10}B measured by either PGNAA or ICP-AES.

loss of ^{10}B attributable to renal excretion. All subjects were qualitatively similar, although there was some variability in the terminal slope. Eight subjects received 250 mg/kg over 60 minutes, after which the administered dose was increased in a step-wise fashion first to 300 mg/kg infused over 90 minutes and finally to 350 mg/kg infused over 90 minutes. This has led to an increase in the average ^{10}B concentration during irradiation and, at least theoretically, an increase in the ^{10}B in the tumor and concomitantly a higher tumor dose. There have been no untoward effects attributable to the BPA-f infusion; in particular, the vital signs have been stable and there has not been a clinically significant EKG abnormality. The upper limit of BPA-f that can be infused has yet to be determined. At present it seems this could be determined more by the volume that is infused than by the actual amount of BPA-f.

2.2. Melanoma of the Extremity

The first of the two clinical trials is for subjects with advanced stage melanoma of the extremities and is designed to determine the MTD of skin and underlying connective tissue to BNCT. BPA has been given through two different routes in this study with very different pharmacokinetics. Initially, the route of administration was orally at 400 mg/kg. The concentration of ^{10}B in the blood, skin, and in the tumor was low and variable; nevertheless, a favorable ratio was obtained between the concentration of ^{10}B in the melanoma nodules and the normal skin and blood. Four fractions were given on successive days to reach the target dose and 5 subjects were irradiated in this fashion to nodules in the lower extremities. One subject was given BPA by a different route of administration that is now considered standard, intravenously at 250 mg/kg infused over 1 hour (Table 1). In all cases, ^{10}B concentrations were measured at frequent intervals, and the blood ^{10}B concentration and the neutron flux determined the ^{10}B dose delivered to the melanoma nodule. The starting point in this trial was a dose of 1000 RBE-cGy prescribed to the point of maximum dose (D_{max}) just below the skin surface and escalated in step-wise increments of 250 RBE-cGy. The RBE values used for the dose calculation are shown in Table 1. Table 2 shows the subject characteristics; the median age was 65 (61–81), and all lesions were on the lower extremity. No subject had identifiable metastatic disease to liver, lung, or CNS at the time of irradiation. Most subjects had multiple palpable nodules. The clinical evaluation of normal tissue and tumor is shown in Table 3. Follow-up was variable, as some subjects underwent further therapy with conventional radiation or expired and were censored at the time of these events. There have been no occurrences, either acute or chronic, that would indicate proximity to an MTD. As a secondary but important endpoint, there have been two sites within the same subject that had a complete clinical and pathological response, meaning that the tumor could not be detected either by physical examination or following biopsies of the sites of original disease. Figures 3 and 4 show this subject before and over 1 year after BNCT. Punch biopsies were taken from both previous tumor sites and evaluated by standard hematoxylin and eosin staining as well as an immuno-stain against S-100 antigen. No tumor was present. The subject remains disease-free over 4 years and 2.5 years after BNCT to the first and second sites, respectively. Since melanoma has the undesirable phenotype of relative radioresistance to conventionally fractionated radiation therapy, this response represents an important proof of principal for BNCT and is at least suggestive of the therapeutic and selective nature of this form of radiation. These results also confirm the work of Mishima *et al.* who have reported good responses to BNCT in melanoma.[12]

Table 2. Patient characteristics, melanoma trial

Initials	Age	Primary site	Metastatic disease	Site irradiated	Tumor size
V.A.	65	R. Foot	None	Sole R. Foot	3 cm
G.H.	61	L Leg	None	L Leg	Mult. 2 cm
J.Y.	80	L. Leg	None	Lower L. Leg	Mult. 3 cm
P.D.	64	L. Toe	None	R. Med. Thigh	2–4 mm
J.Y.	81	L. Leg	None	Lower L Leg	Mult 3.5 cm

2.3. Glioblastoma and Intracranial Melanoma

The second trial at Harvard-MIT is also phase I but is designed to evaluate the tolerance of the central nervous system (CNS) to BNCT. Patient eligibility is restricted to a diagnosis of either glioblastoma multiforme or metastatic melanoma, a Karnofsky performance status of 70 or greater and no prior history of cranial irradiation. The dose is prescribed to the normal tissue D_{max}, a 1-cm volume that is centered in the dermal or subdermal tissue. The RBE values used in the dose calculation are shown in Table 1. Escalation was carried out in a step-wise fashion. The initial dose was 880 RBE-cGy which was increased by 10% every 3 subjects. At the time of this presentation, 18 subjects have been entered into the trial and 16 have been irradiated; two subjects were excluded due to decline in performance status. The median age was 56 (24–75). Five dose cohorts were completed: 880, 970, 1060, 1170, and 1280 RBE-cGy. Beam arrangements varied over the course of the trial. Initially, subjects were irradiated with a simple parallel-opposed beam arrangement; however, it became apparent that as with conventional therapy, considerable improvements to the tumor dose could be realized through the use of multiple non-coplanar treatment fields and this could be achieved within the constraints of dose prescription as originally defined in the protocol. An example of a 3-field irradiation for a relatively superficial frontal lobe tumor is seen in Figure 5. One to three fields have been used depending upon the location and size of the tumor. One of the characteristics of the M-67 beam is the relatively low flux, and consequently the irradiation times for each field could extend to 2–3 hours. Mid-way through the dose escalation steps, it was decided to perform the irradiations over two consecutive days with a repeat administration of BPA-f to facilitate patient comfort. This, in addition, may be advantageous as it may promote improved ^{10}B distribution among tumor cells in various stages of the cell cycle.

Table 3. Normal tissue reaction and clinical response

Initials	Follow-up	Skin acute	Skin chronic	Tumor response
V.A.	11 mo.	None	Slight hypopigmentation	Stable
G.H.	17 mo.	None	None	Stable
J.Y.	48 mo.	Mild (1)	None	Complete (2)
P.D.	10 mo.	None	None	Unable to evaluate (3)
J.Y.	28 mo.	None	None	Complete (2)

(1) Tumor nodule became red and desquamated.
(2) No tumor by physical examination and biopsy.
(3) Tumor nodules removed with biopsy.

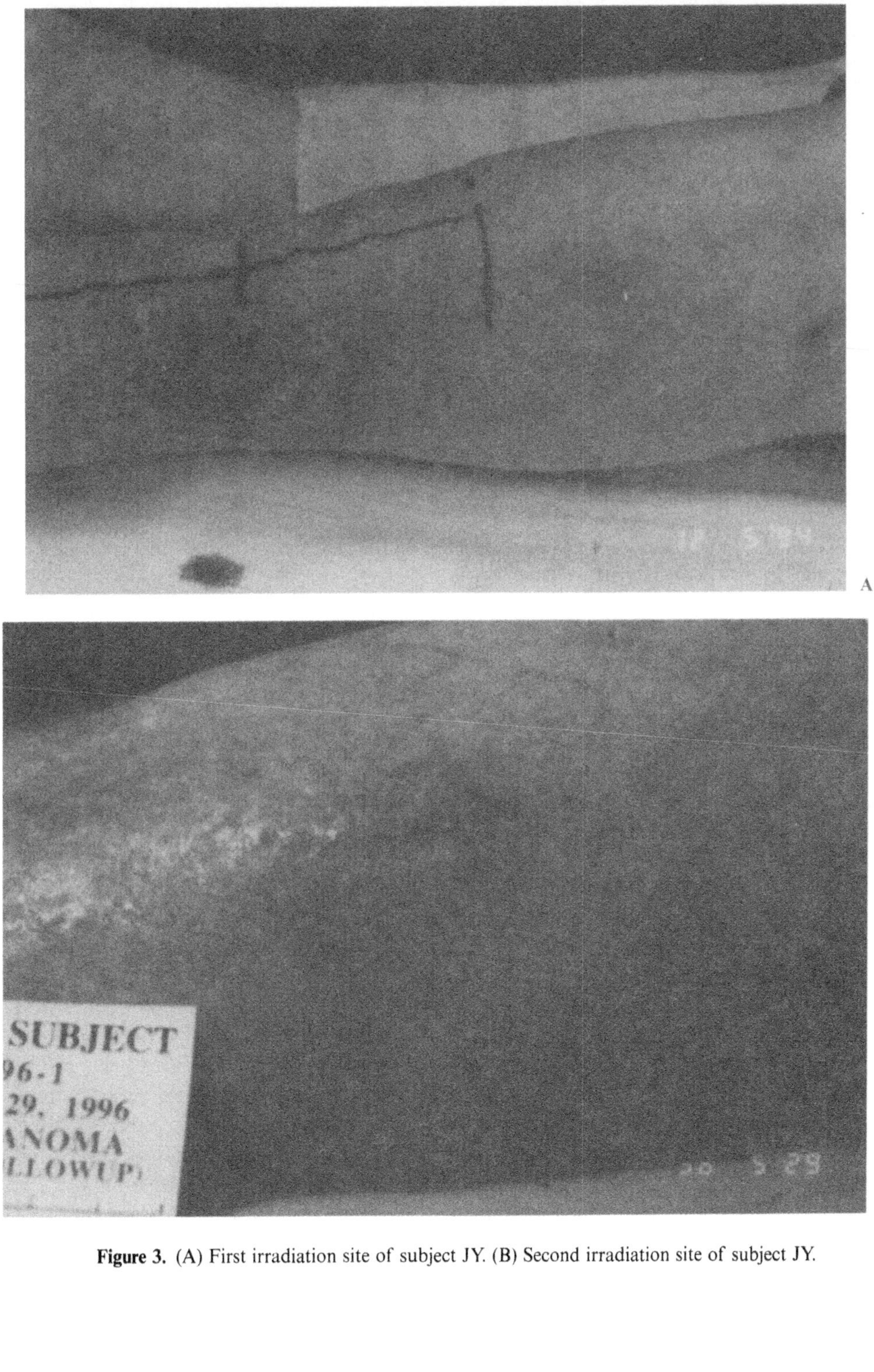

A

B

Figure 3. (A) First irradiation site of subject JY. (B) Second irradiation site of subject JY.

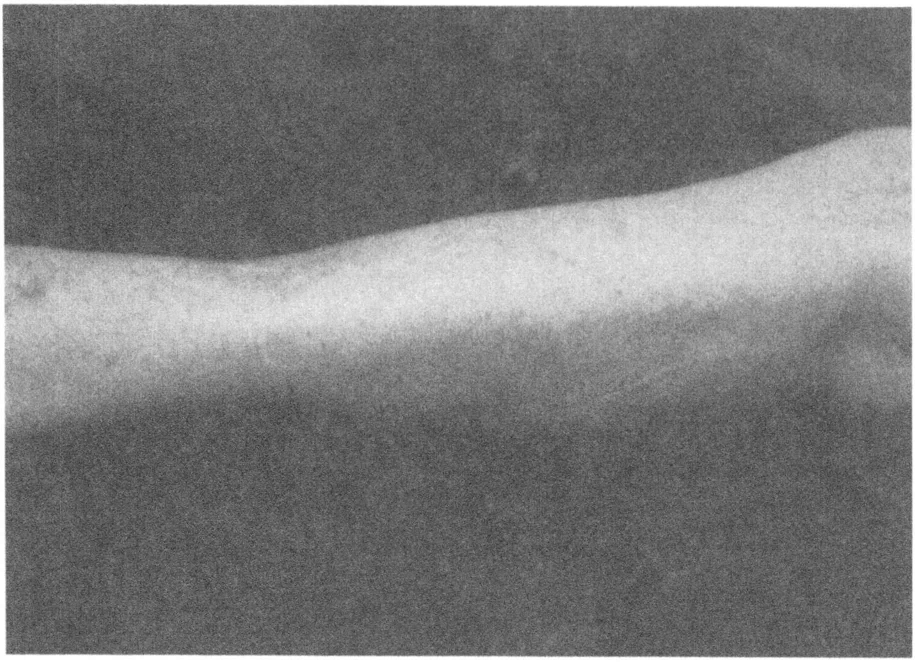

Figure 4. Subject JY at follow-up 2.5 years after irradiation of first site and 1 year after irradiation of second site. The entire lower extremity is free of disease 4 years and 2.5 years after BNCT irradiation of these two sites.

Because of the nature of the trial, there were no limitations on the location or extent of tumor. Consequently, the range of disease extends from a relatively superficial residuum of minimal disease to large volume gross disease extending to or across the midline. All lobes of the brain have been irradiated. As mentioned, treatment plans range from single en face fields to parallel opposed and unilateral 3-field arrangements. As a result, average normal brain and tumor doses do not reflect the systematic increase in prescription to the D_{max} (Figure 6A and B), but do gradually increase over the course of the escalation. The relationship between the volume-average whole brain dose and number of irradiation fields is shown in Figure 7 where it can be seen that a direct relationship exists between the complexity of the irradiation and average brain dose and a less-apparent relationship exists between the average brain dose and the D_{max}.

With respect to the primary intent of the trial, identification of MTD, there have been no signs or symptoms of radiation injury that would indicate an MTD had been reached. A number of anticipated and unanticipated acute effects have been noted, including temporary worsening of neurological symptoms, severe headache, nausea, parotiditis, and irritation of the oropharynx, which are enumerated in Table 4. All of these effects, with the exception of alopecia, have subsided with only minor medical intervention. One subject with a large amount of edema required intravenous mannitol in order to resolve symptoms of increased intracranial pressure. It should be noted that with rare exception, subjects were discharged from the hospital the following day and quickly returned to their preirradiation activities. Most but not all subjects were taking steroids (Dexamethasone) during BNCT; similarly, the use of antiseizure medication was variable. Intravenous or oral Ondansetron controlled nausea and vomiting. One subject who

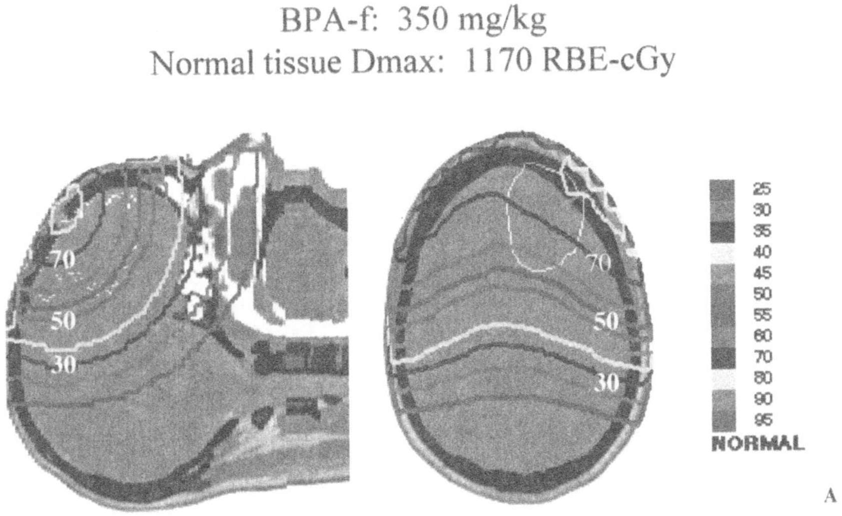

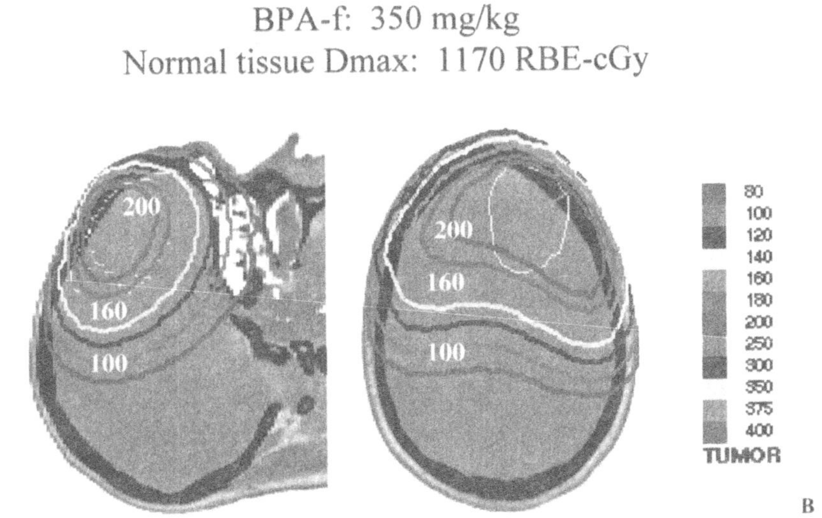

Figure 5. Example of three field irradiation for frontal lobe Glioblastoma. (A) Normal brain isodose contours representing percentage of D_{max} (1170 RBE-cGy). (B) Tumor isodose contours as a percentage of D_{max}. The dose received depends upon the type of tissue within the isodose volumes.

had a large unresected parietal tumor with peritumoral edema experienced a treatment-related death 5 days after BNCT (only 1 fraction out of a proposed 2-fraction irradiation had been delivered) secondary to a new ipsilateral thalamic infarct. This was not judged to be a generalizable dose-limiting event, although care has been taken to exclude subjects with large tumors and marginal performance status and to actively manage edema with a low threshold for mannitol diuresis. In these subjects Dexamethasone is prescribed at 10mg IV every 6 hours beginning 1 day prior to BNCT.

There was no dermal, bone, or normal brain toxicity as determined clinically, nor was there evidence of osteonecrosis, white matter changes or brain necrosis on MRI

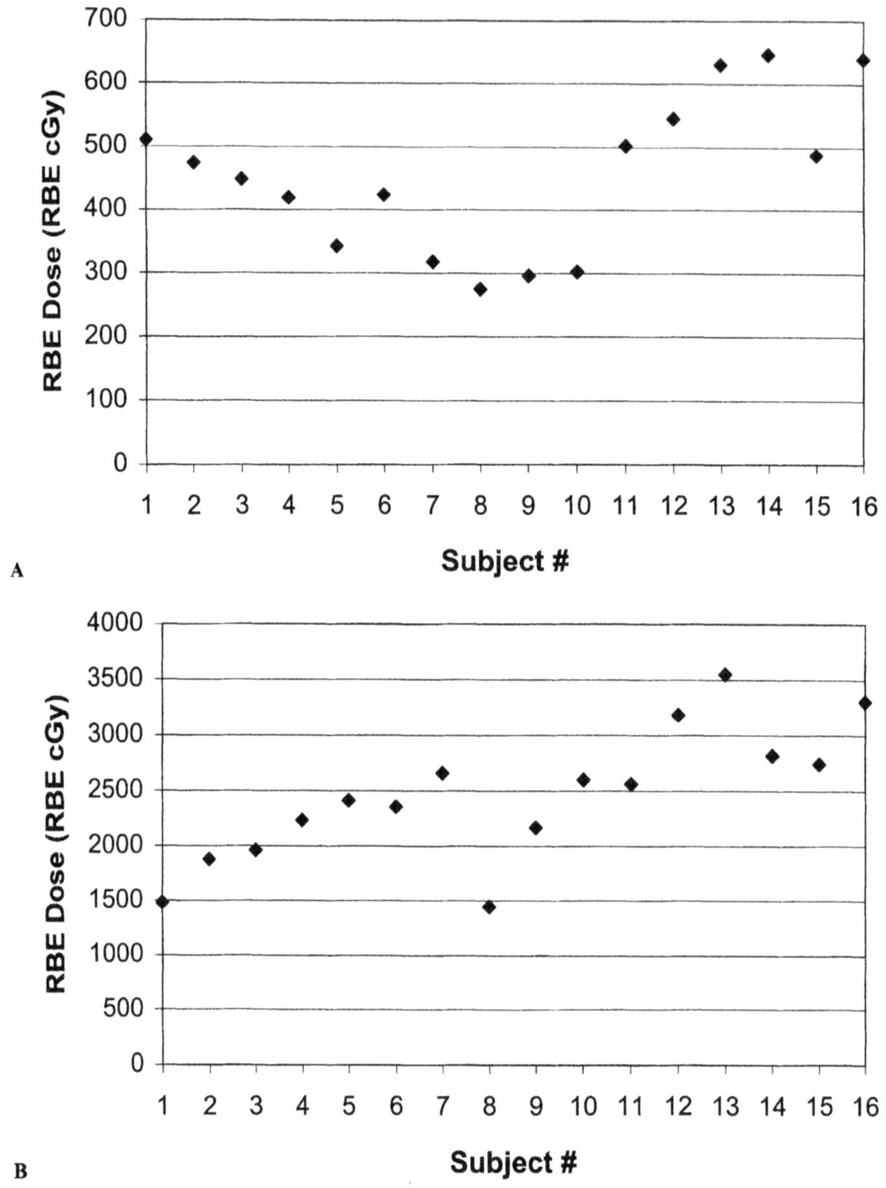

Figure 6. (A) Volume average normal brain dose by subject. (B) Volume average tumor volume dose by subject.

within the limited duration of follow-up. One subject with a large tumor experienced a progressive neurological decline over 3 months that was accompanied by progressive ipsilateral edema. Fourteen subjects either maintained or improved their performance status following BNCT. There are a number of ways to correlate clinical endpoints of toxicity with radiation dose to the CNS. Data on normal tissue D_{max} and average brain dose have already been presented. While easily quantified, they either lack clinical relevance (D_{max}) or are crude measurements (average brain dose) that incompletely express dose and most likely are not the normal tissue volumes that will ultimately determine the MTD for

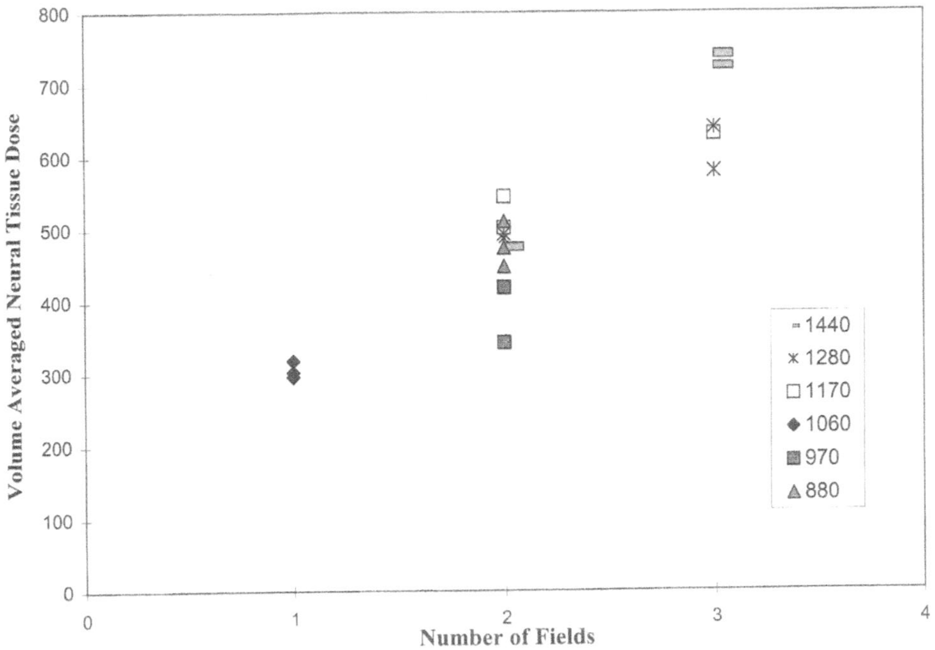

Figure 7. Relationship between the volume average brain dose and the number of irradiation fields.

multiple field-ipsilateral treatment plans. A better and certainly more complete representation of dose, the dose volume histogram (DVH) for brain and tumor volume has been generated for all subjects. An example of brain and tumor DVH for the 3 subjects who were irradiated to 1280 RBE-cGy is shown in Figure 8. The cumulative normal brain DVH data for all subjects in 5 dose cohorts are shown in Figure 9. These data are expressed on an absolute scale where brain volume is in ml and dose is in RBE-cGy. Since

Table 4. Immediate and Acute reactions following cranial BNCT

	No. of subjects
Immediate effects (1–2 days)	
Headache (mild-moderate)	7
Headache (severe)	1
Parotiditis	6
Nausea and Vomiting	6
Neurological change (transient)	3
Periorbital edema	1
Acute effects (within 31 days)	
Alopecia	14
Facial skin irritation	2
Fever of unknown origin	2
Oropharyngeal mucositis	1
Renal calculus	1
Xerostomia	1

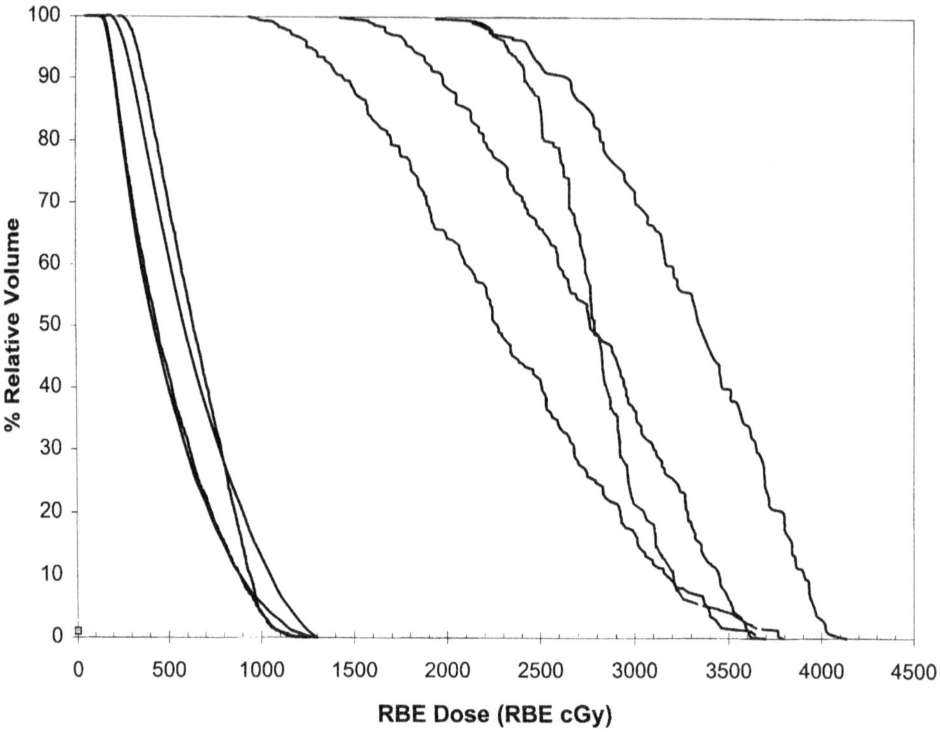

Figure 8. Dose volume histogram (DVH) for normal brain and tumor volume of the 3 subject cohort irradiated to 1280 RBE-cGy. The separation between the normal brain DVH and tumor volume DVH represents therapeutic gain from BNCT.

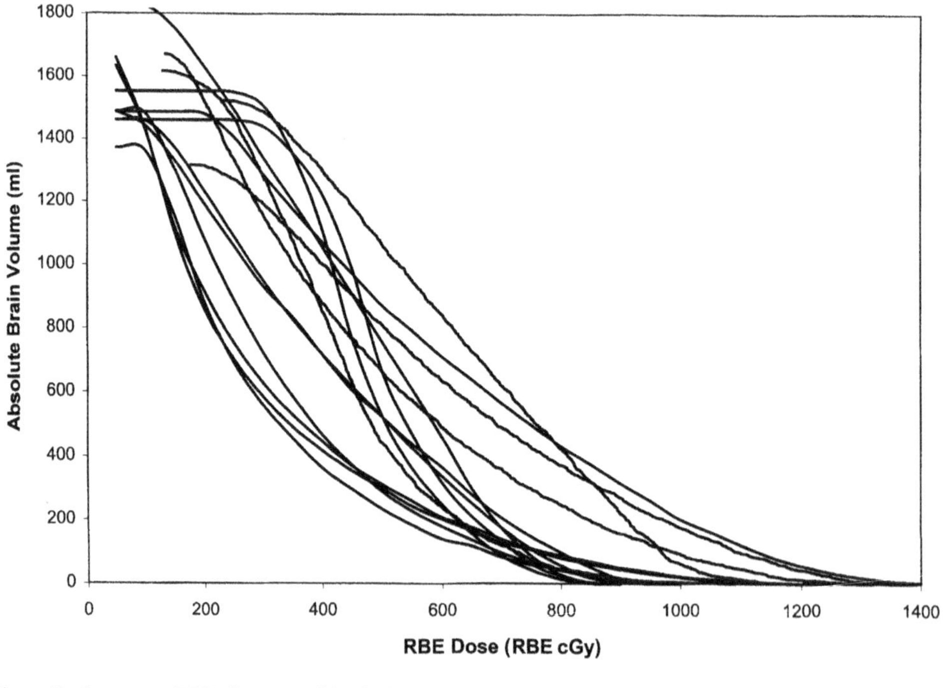

Figure 9. Aggregate DVH for normal brain in all subjects. Data are expressed on an expanded scale and in absolute units of volume and dose.

no late effects or limiting acute neurological effects have been noted, the dose-volume relationships displayed in Figure 9 are considered to be well tolerated.

Although tumor dose was not specifically defined by protocol, an effort was made to deliver as high a dose as possible while staying within the boundaries of normal tissue dose parameters. Due to the heterogeneity of tumor volumes and location, a wide range of doses to the tumor volume was seen. Nevertheless, there was a progressive increase in volume average tumor dose over the course of the trial (Figure 6B); however, in most cases, the minimum dose to the tumor volume dose not reach or exceed the theoretical minimum required to control glioblastoma as suggested by Laramore.[13] As expected, the tumor response has been mixed. Because of the difficulty in interpreting MRI images following surgery and BNCT, the time to progression is an unreliable endpoint except in the most unambiguous circumstance. In general, there has been a sustained interval of disease stability followed by progression in almost all subjects with GBM. One subject had regression of an area of contrast enhancement 2 months following BNCT (Figure 10) in which tumor and vascular necrosis was seen histologically at the site of previous rim enhancement. This subject died from intercurrent disease 3 months after BNCT. All failures were within or immediately adjacent to the original tumor. The Kaplan-Meier survival curve for all subjects with GBM is seen in Figure 11; the median survival is 12 months from the time of biopsy or resection and 11 months following BNCT.

The most clear-cut example of therapeutic efficacy of BNCT for intracranial disease is the irradiation of an unresected metastatic melanoma in the occipital lobe. Two posterior oblique fields were used and the average tumor dose was 2400 RBE-cGy. Serial contrast-enhanced MRI scans were obtained monthly at the Beth Israel Deaconess Medical Center, and over the course of 4 months, this tumor underwent a dramatic regression to the point of a complete radiographic response (Figure 12) that persisted until the subject died of widespread metastatic disease 1 year following BNCT. This subject's functional capacity was 100% until the final few weeks of life.

3. FUTURE CLINICAL WORK

The clinical utility of BNCT, as with any other form of radiation therapy, will depend upon the ability to deliver a tumoricidal dose to areas of gross and microscopic disease within an acceptable limit of injury to normal tissue. In the case of BNCT for primary brain cancer, the normal tissue of greatest interest is the central nervous system and related structures such as the eyes and optic tract. In order to achieve this, a sizable therapeutic gain must be realized and it will be derived from the ^{10}B reaction and its differential uptake in tumor versus normal tissue. Dose-volume histograms such as the one shown in Figure 8 demonstrate a sizable therapeutic gain, which is seen as the degree of horizontal separation between the normal brain and tumor volume DVH. Despite this separation, the dose to the tumor volume falls short of the 30 Gy theoretical minimum for control of glioblastoma as suggested by Laramore.[13] Strategies to increase tumor dose include simply increasing the amount of BPA administered, novel methods of BPA delivery,[14] the development of new drugs,[15] or further refinement of medical neutron beams.[16,17] This latter approach is currently underway at MIT-NRL[18] and uses a fission neutron source with a novel filter moderator assembly that will produce a horizontal beam of variable diameter with an epithermal flux of approximately 1.3×10^{10} neutrons-cm^{-2}-sec^{-1}. Most importantly, this new epithermal fission converter beam (FCB) will have negligible background components such as fast neutrons and gamma rays. By reducing

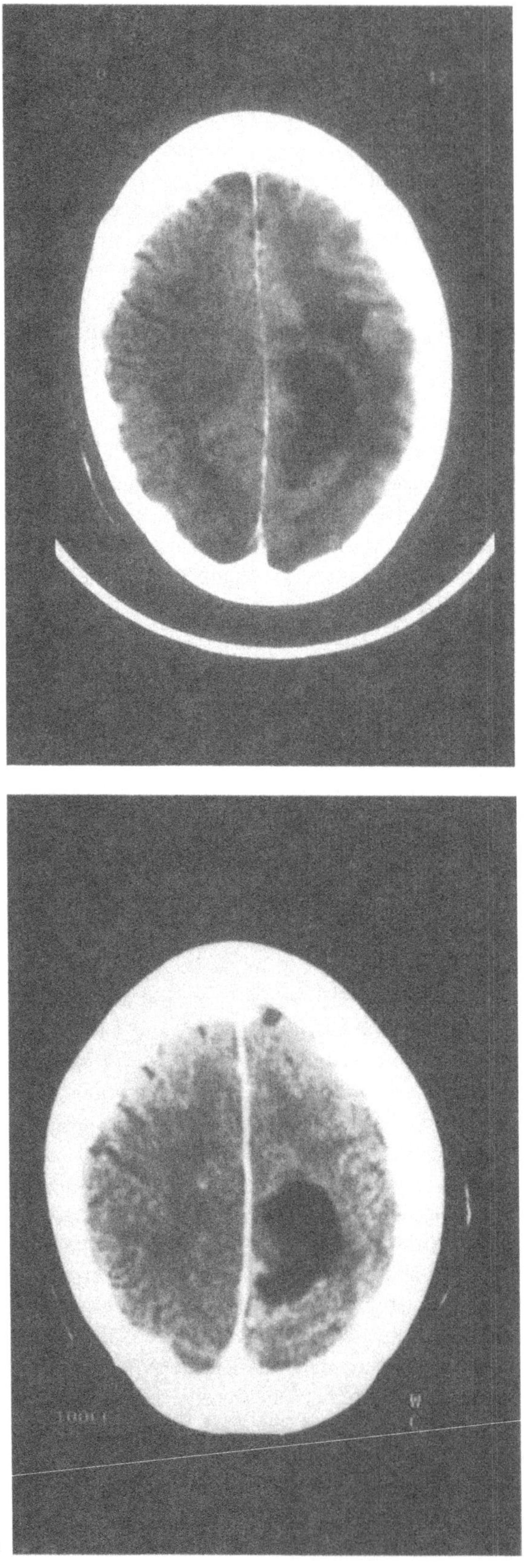

A

B

Figure 10. (A) Contrast enhanced CT image of subject with Glioblastoma. Tumor exhibits characteristic pattern of rim enhancement and central necrosis and was only biopsied prior to BNCT. (B) Contrast enhanced CT image of same subject 2 months after BNCT. There is resolution of the enhancing rim but residual-enhancing tumor medially.

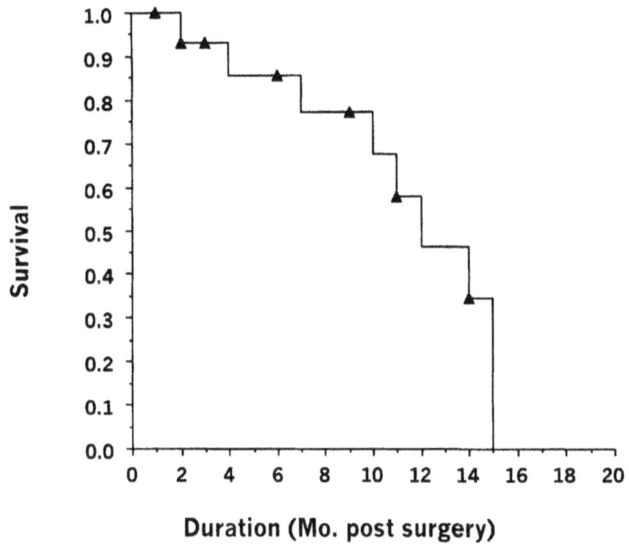

Figure 11. Kaplan Meier actuarial survival for BNCT subjects with Glioblastoma from the date of original surgery or biopsy. The median survival is 12 months.

these dose components to normal tissue and tumor, a greater proportion of dose will come from the ^{10}B reaction and result in a further increase in the therapeutic gain. The projected impact of these improvements are demonstrated in Figure 13, where the tumor DVH has a large shift to higher doses while the normal brain dose is essentially the same as with the current epithermal beam. The increased dose rate will lead to a potential reduction in the irradiation time by a factor of approximately 50, theoretically allowing a full course of BNCT to be completed in minutes. Because of the significant nature of this improvement both in beam quality and in intensity, definitive phase II trials will be initiated only after the FCB is ready for clinical use. Phase I work will continue until then, refining and solidifying the database for normal tissue effects from BNCT.

At the moment, only two tumor histologies are approved for clinical trials, glioblastoma and melanoma, and much is riding on the ability of BNCT to alter the natural history of these two notoriously resilient diseases. It is quite possible that other primary CNS tumors as well as epithelial and mesenchymal tumors elsewhere in the body are even better suited for BNCT. To that end, a program that will determine the biodistribution of ^{10}B in patients with tumors of various histologies and sites will run parallel to the phase I and II clinical trials, looking for favorable ^{10}B distributions and the possibility of expanding BNCT into other clinical areas. Exploratory experiments are currently being carried out examining the ^{10}B uptake in clinically relevant tissue culture cell lines exposed to BPA-f *in vitro*. A favorable ^{10}B tumor-to-blood ratio is not the only measure for potential clinical suitability; the most important ratio is that between tumor and immediately adjacent normal tissue. For example, the ^{10}B uptake in squamous cell carcinoma of the upper aerodigestive track may be high enough to consider possible BNCT, but if the normal mucosa of the oral cavity, oropharynx, and larynx also take up BPA then the therapeutic gain could be lost or at least abrogated.

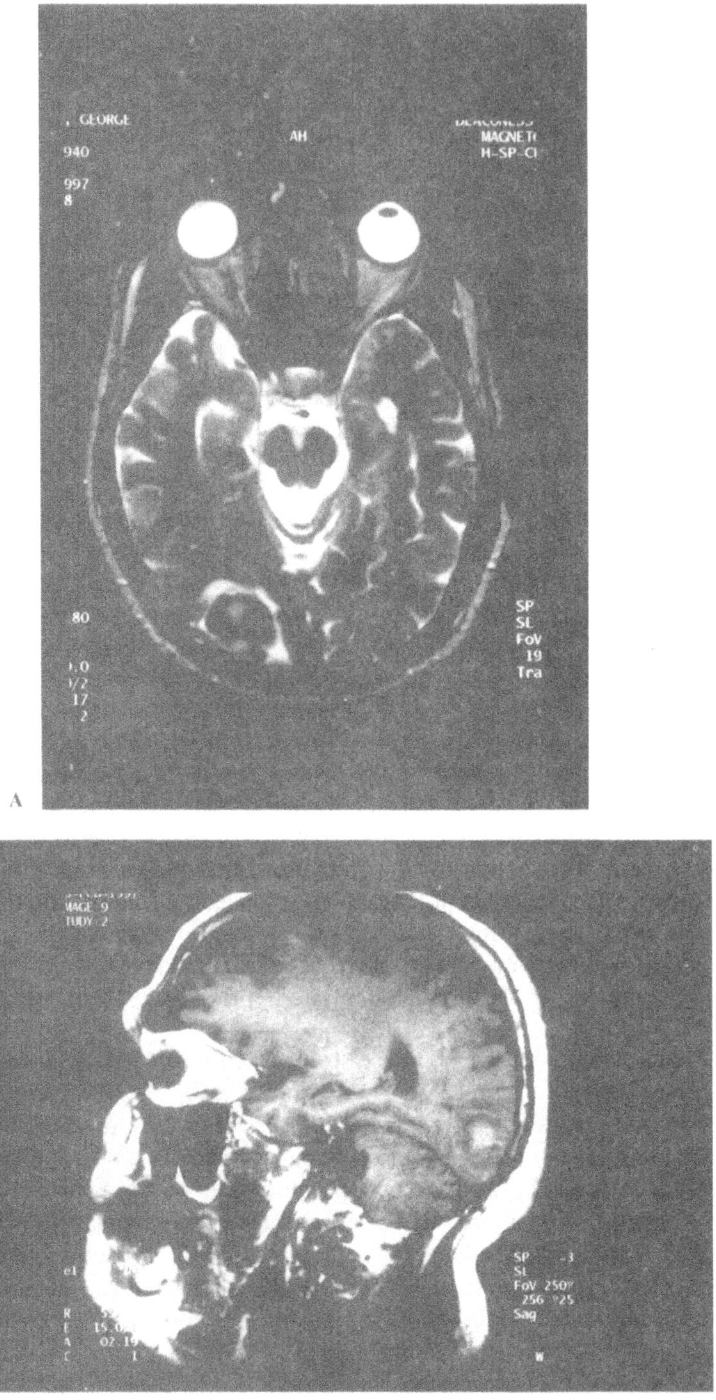

Figure 12. Complete tumor regression in subject irradiated with unresected occipital lobe melanoma. All images are contrast enhanced MRI. (A) Axial image before BNCT. (B) Sagittal image before BNCT. (C) Axial image 4 months after BNCT. (D) Sagittal image 4 months after BNCT.

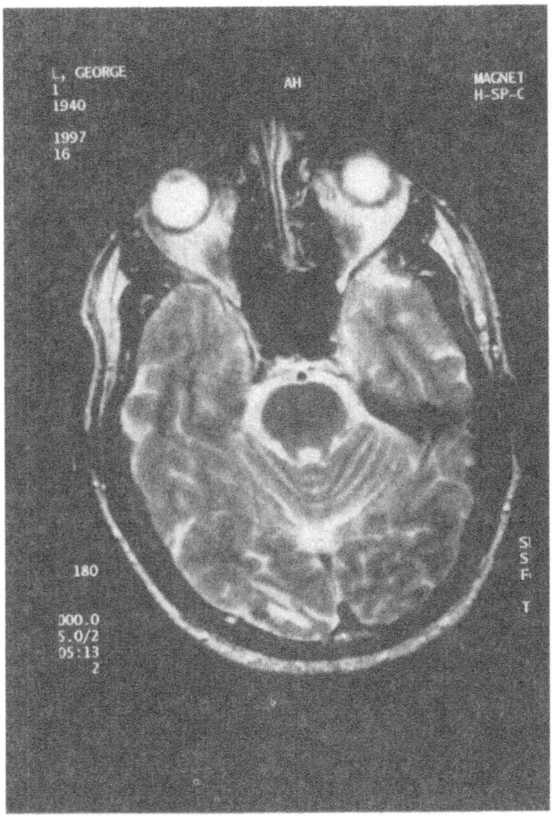

C

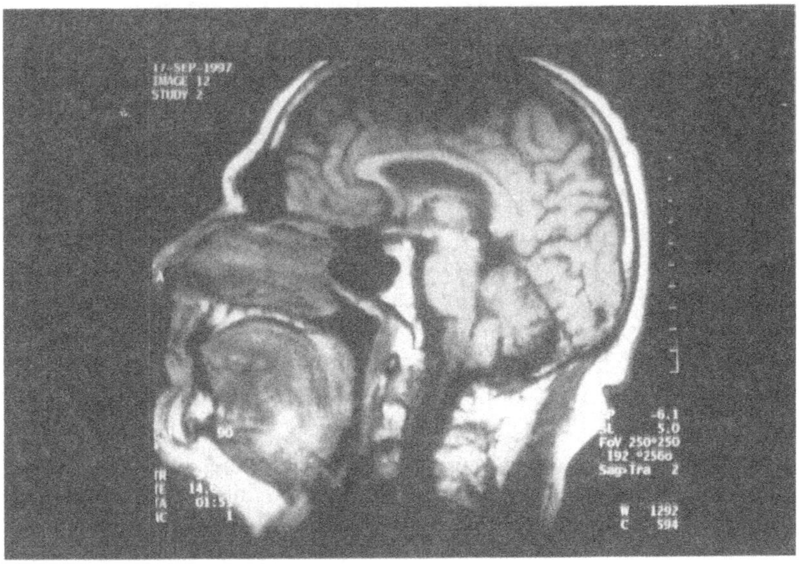

D

Figure 12. *Continued*

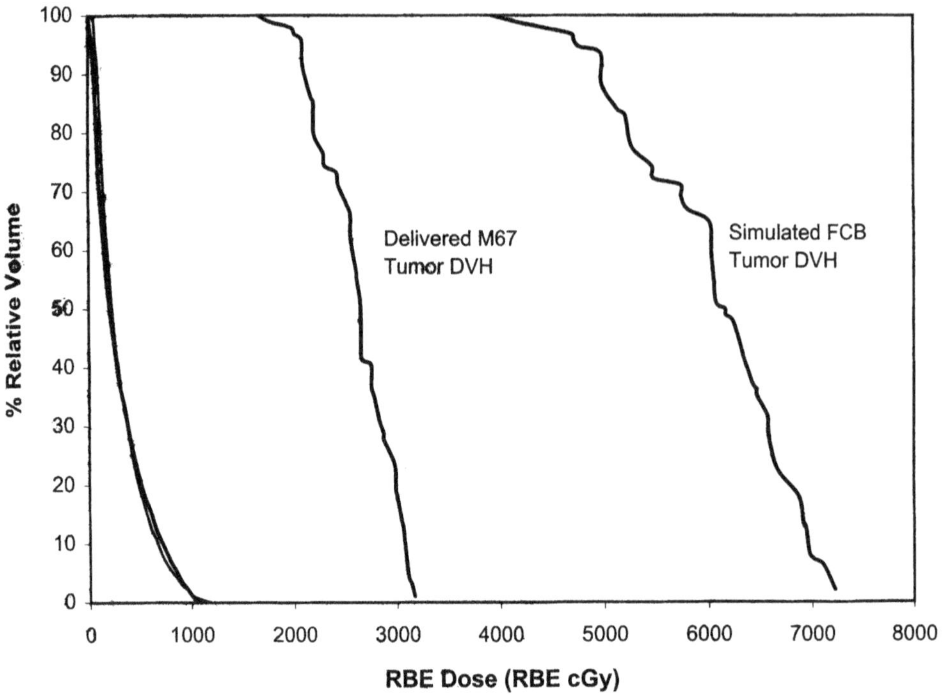

Figure 13. Comparison of normal brain and tumor volume DVH with M-67 and FCB epithermal neutron beams.

4. TRANSLATIONAL RESEARCH

As neutron sources become more refined through a reduction in unwanted dose components and the capability to deliver a full course of BNCT within a short treatment time, other questions become increasingly important such as the timing of the irradiation. What are the pharmacokinetics of BPA in tumor and normal tissue within the first several hours after infusion, and what is the optimal time to deliver the neutron irradiation in order to derive the maximum dose to tumor and minimum dose to normal tissue? The potential clinical ramifications of a neutron source like the FCB are considerable, but this opportunity for therapeutic improvement could go unrealized if our understanding of the behavior of BPA in tumor as well as in normal tissue does not keep pace with the advancements made in nuclear engineering. In order to increase our understanding of the biology, a program of translational research is ongoing at Harvard-MIT that has as its centerpiece the goal of optimal use of the FCB neutron source through an understanding the temporal and spatial pharmacokinetics of ^{10}B following delivery of intravenous BPA-f. This includes work with animal tumor models as well as the determination of the pharmacokinetics of BPA in human subjects utilizing a variety of analytical and computational methods. Continued refinements in software and hardware are also a priority in order to improve our treatment planning capabilities. This has a dual goal of improving treatment plans in order to deliver the highest dose to tumor bearing regions while minimizing dose to uninvolved normal tissue and developing treatment plans for BNCT to other disease sites. Data are also being generated on the microscopic distribution and uptake of ^{10}B in a variety of tumor cell lines *in vitro* using

HRQAR, a high resolution, neutron-induced autoradiographic method developed in our laboratory.[4]

4.1. *In Vivo* and *In Vitro* Experiments

The two currently available treatment-planning codes treat the brain as a homogeneous tissue with a uniform ^{10}B concentration. Since the brain is an extraordinarily complex structure, it is quite conceivable there may be anatomic differences in ^{10}B pharmacokinetics based on differences in blood flow or physiology that could have clinical implications for dosimetry. In order to test this hypothesis, a series of small animal experiments were designed and completed.[19] Previous work in this laboratory[20] with C57BL/6 mice and the murine glial tumor GL261 led to the selection of this system for the initial set of preclinical studies. The first set of experiments sought to determine the temporal and spatial distribution of ^{10}B in normal brain following a bolus intravenous injection of 250 mg/kg BPA-f. Pooled data are shown in Figure 14. The ^{10}B concentration in neural tissue varies with time after injection and is initially inversely related to the concentration of ^{10}B in the blood. In the first 60–70 minutes, brain tissue has much less ^{10}B than blood; however, this crosses over between 60 and 120 minutes, after which the brain concentration is 20–40% higher. Over a protracted irradiation, particularly if it extends before and after the crossover point, differences in the ratios will be offsetting,

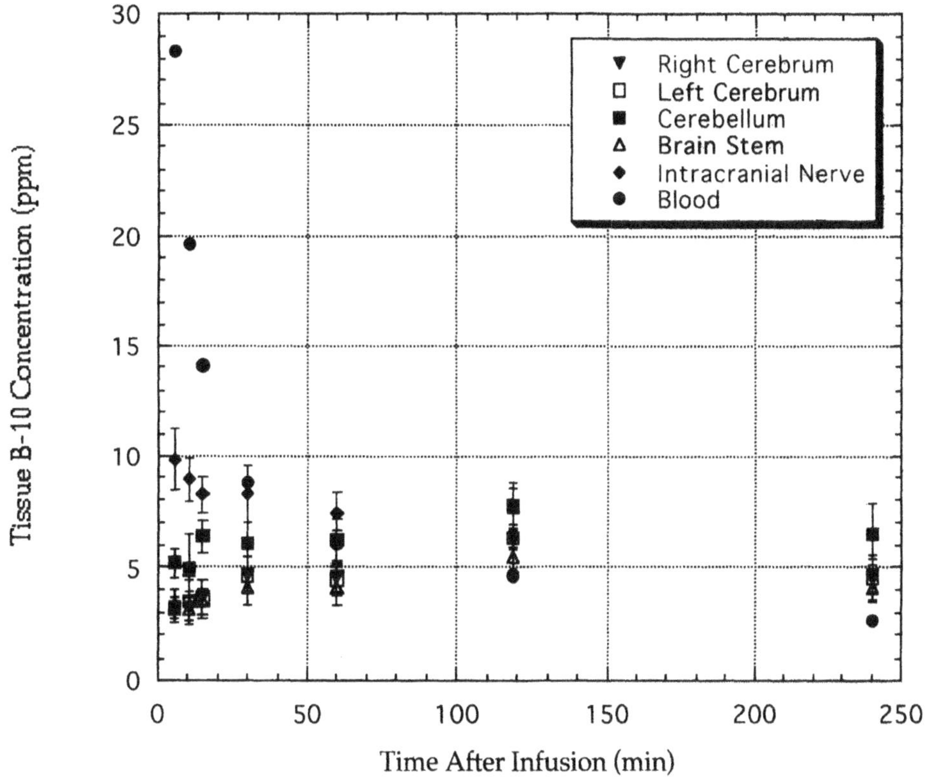

Figure 14. ^{10}B concentration as a function of time in blood and selected murine brain structures following a bolus injection of 350 mg/kg BPA-f.

minimizing the dosimetric impact; however, for short irradiations such as are proposed for the FCB, a difference in calculated and actual brain dose could be large, depending upon the timing of the irradiation, and will need to be further explored. Another important observation from this experiment is the subtle difference in the ^{10}B concentration within several sites within the CNS. Samples from the same animal were taken from cerebrum, cerebellum, brain stem and an intracranial nerve. The cerebellum has consistently higher ^{10}B concentration and there is a suggestion that this may be true for an intracranial nerve as well. This latter finding needs to be corroborated in a large animal model with better spatial resolution since the implications could be great if cranial nerves that are integral to the senses would receive a dose higher than average normal brain.

A similar set of experiments was conducted looking at the temporal changes in ^{10}B concentration in blood, normal brain (cerebrum) and tumor. 4×10^4 GL261 cells were inoculated into the cranium through stereotactic coordinates and incubated for approximately 2–3 weeks, after which a visible tumor was uniformly present. 250 mg/kg BPA-f was given by bolus injection and ^{10}B concentration was measured at various time points up to 240 minutes after injection. It can be seen in Figure 15 that GL261 quickly takes up BPA, and except for the earliest time point has a high but variable tumor to blood ratio. At all times, the concentration of ^{10}B in GL261 is considerably higher than

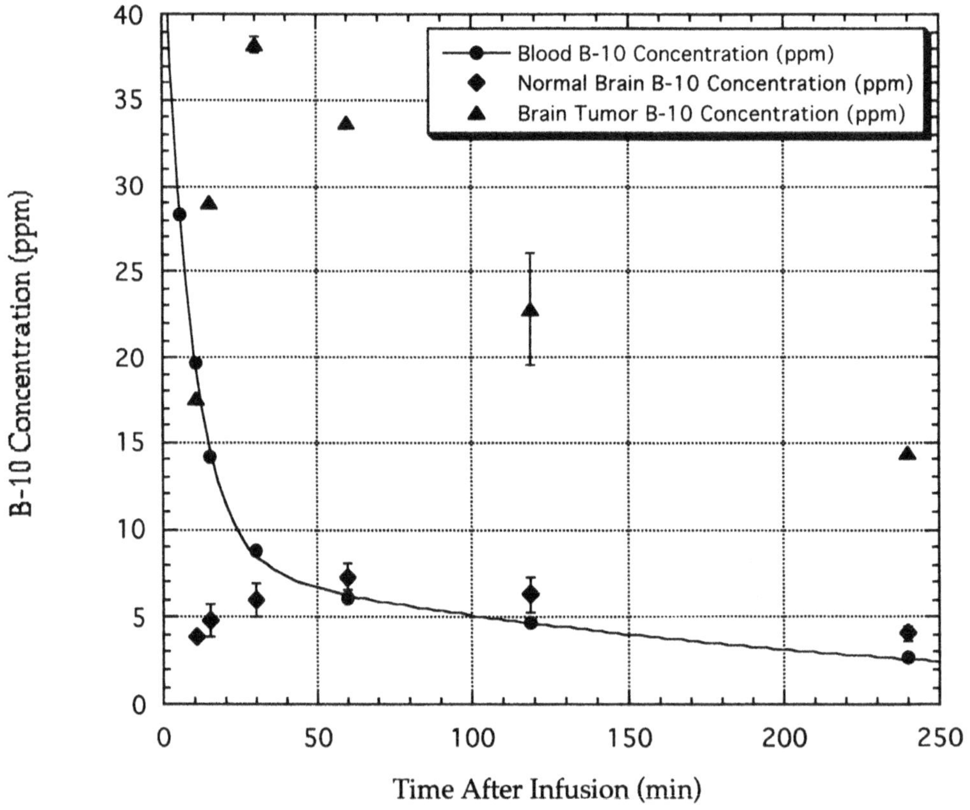

Figure 15. ^{10}B concentration as a function of time in blood, normal brain and intracerebral GL261 murine glial tumor following a bolus injection of 350 mg/kg.

in normal brain and the tumor to brain ratio also changes with time. Qualitatively these data are consistent with other experimental and clinically derived static ^{10}B measurements in glioblastoma and human normal brain,[21] but add the important element of a full time course. In order to gain additional insight into these pharmacokinetics, further experiments looking at B-16 melanoma and a human glioblastoma xenograph are in progress.

The two tumor types currently under clinical investigation, glioblastoma and melanoma, have similar properties in that they both take up enough BPA to make BNCT possible. Equally important are the surrounding normal tissues of brain and skin, which take up considerably less ^{10}B. Is this unique to these two histologies and anatomic sites or do other tumor histologies and types of normal tissue behave in the same way? Preliminary data presented by Chuang[22] suggests that the relationship may be more universal. These experiments used a well-characterized *in vivo* system of hepatic metastases from an intravenously injected suspension of human colon carcinoma cells grown *in vitro*. The experimental design was essentially the same as has been previously described; i.e., a bolus injection of BPA followed by the sampling of tumor nodules and normal liver for ^{10}B concentration. The data indicate that even in a setting where the normal tissue is highly vascularized and tumors grow as nodules, there is still a higher concentration of ^{10}B in adenocarcinoma than in liver. This finding has led to determine the BPA uptake characteristics of other cell types in order to generate a hypothesis for BNCT to other disease sites and histologies. There is some indication that other cell lines take up BPA but that the degree of uptake is variable.[23] These data need to be expanded and independently confirmed. Studies are underway looking at a range of human cell lines *in vitro*. A unique aspect of this work is the quantification of ^{10}B in cell cultures through track etch technology; a technique that could lend itself to the future screening of various tumor histologies to ascertain their suitability for BNCT.

4.2. Pharmacokinetic Modeling

The pharmacokinetics of BPA-f in blood have been described (Figure 2) and fully modeled in our laboratory with a two compartment system representing blood (plasma) and a general tissue compartment.[24] First order rate constants (k_1 and k_2) describe influx and efflux between the two compartments and a third rate constant (k_3) describes the loss of ^{10}B from the blood, presumably through renal excretion. A prediction from this model that would provide an explanation of the variability in the terminal portion of the curve is the relationship of k_3 to renal function. This is shown in Figure 16, where the k_3 is correlated with the calculated creatinine clearance. This model has been verified through its ability to accurately predict blood ^{10}B concentration in a variety of known conditions such as the infusion rate and concentration used at BNL. In order to derive a treatment plan that incorporates the best timing between BPA-f infusion and neutron irradiation, additional information is necessary, preferably at a similar level of accuracy. The *in vivo* animal experiments provide useful pharmacokinetic information on the broader issues of radiation timing and the relative ^{10}B concentrations of different regions of the brain, but even the best animal model is only a surrogate for human pharmacokinetic data. Information on the ^{10}B uptake of glioblastoma and normal brain are being generated by two groups using PET scanning with F-18 labeled BPA-f.[25,26] This non-invasive technology generates information on tumor location and static ratios of uptake as well as of kinetics. Based on PET images, Imahori[26] has developed a 3-compartment model that describes the kinetics of F-18 labeled BPA-f in brain and tumor. These kinetic parame-

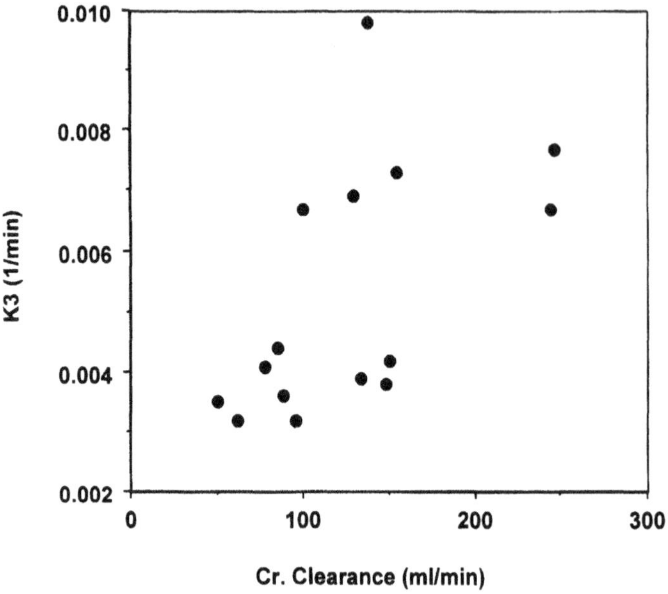

Figure 16. Relationship between calculated creatinine clearance and k_3. Creatinine clearance rates in ml/min and the rate constant in ml^{-1} for ^{10}B disappearance from blood (k_3) were calculated for each subject. The relationship supports a clinical observation that subjects with reduced renal function had higher average blood ^{10}B concentrations.

ters along with the parameters for blood derived by Kiger,[24] have been incorporated into a composite rendering of the complex relationships between blood, brain and tumor following an infusion of BPA-f (Figure 17). For an irradiation of short duration, it is not immediately clear when would be the best time for this to take place. In principle, the tumor ^{10}B concentration should be as high as possible but the best therapeutic ratio between tumor and brain may be after the brain peak. The tumor to blood ratio cannot be discounted because normal brain radiation effects could be due to vascular damage and the likelihood of this will be reduced with lower blood ^{10}B concentrations.

5. CONCLUSION

It is clear the clinical programs in BNCT including the Harvard-MIT BNCT Program have gathered considerable momentum since the last meeting in Zurich. As data accrue and higher doses are reached some of the anticipated effects on normal tissue and tumor should become evident and it will be incumbent upon all to share clinical data and make the most of these limited observations. In order to interpret any normal tissue effects or compare clinical information, a uniform standard of measurements and reporting of dose must be adopted. The emphasis on CNS tumors is important but it is equally important to move past the limited treatment paradigms that currently dominate this field and explore whether other tumor histologies and sites of disease may be even better suited to BNCT. Continued work on radiobiology, pharmacokinetics, and new compounds must continue and the payoff will be in steering BNCT toward its promise as a legitimate and efficacious therapeutic modality.

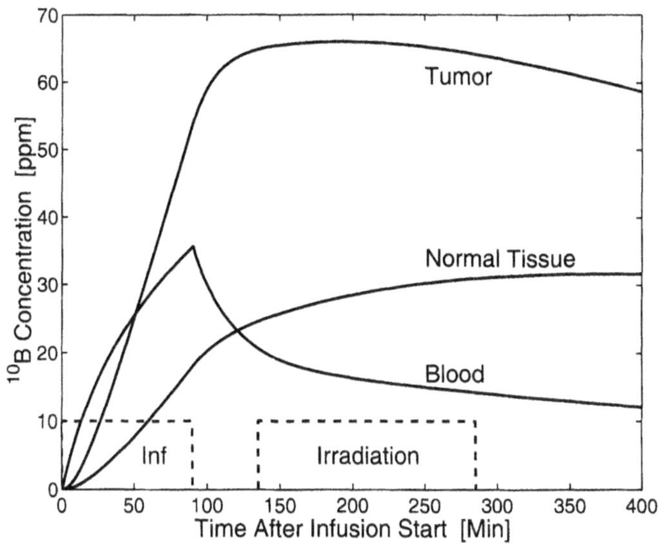

Figure 17. Predicted human blood, normal brain and Glioblastoma ^{10}B concentrations following an intravenous infusion of BPA-f. Pharmacokinetics are a determined using rate equations derived from blood data (Kiger) and PET imaging (Imahori).

ACKNOWLEDGMENTS

The authors would like to thank Dr. John Bernard and the MIT Nuclear Reactor Laboratory for reactor time and support provided by the reactor-sharing program. This work was supported by a grant from the United States Department of Energy: DE-FG02-96ER62193. Human subject transportation was provided by American Medical Response and the Fallon Ambulance Service.

REFERENCES

1. A.Z. Diaz, A.D. Chanana, J. Capala, M. Chadha, J.A. Coderre, E.H. Elowitz, J. Iwai, D.D. Joel, H.B. Liu, R. Ma, M. Shady, D.N. Slatkin, G.W. Tyson, and L. Wielopolski, Safety and efficacy of BNCT for glioblastoma multiforme: results from the initial dose escalation studies. *Eighth International Conference on Neutron Capture Therapy for Cancer.* LaJolla, Plenum Press, 1998.
2. H. Hatanaka, and N. Nakagawa, Clinical results of long-surviving brain tumor patients who underwent boron neutron capture therapy, *Int. J. Radiat. Oncol. Biol. Phys.*, 28:1061–1066, 1994.
3. Y. Nakagawa, and T. Kageji, Boron neutron capture therapy-recent aspects and new protocols in Japan. *Eighth International Conference on Neutron Capture Therapy for Cancer.* LaJolla, Plenum Press, 1998.
4. Y. Mishima, M. Ichihashi, S. Hatta, C. Honda, K. Yamamura, T. Nakagawa, H. Obara, J. Shirakawa, J. Hiratsuka, K. Taniyama, C. Tanaka, K. Kanda, T. Kobayashi, T. Sato, M.R. Ishida, Y. Ujeno, M. Takahashi, M. Abe, T. Nozaki, O. Aizawa, T. Matsumoto, T. Sato, H. Karashima, K. Yoshino, and H. Fukuda, First human clinical trial of melanoma neutron capture. Diagnosis and therapy, *Strahlenther. Onkol.*, 165:251–254, 1989.
5. Y. Mishima, Current clinical paradigms in melanoma BNCT. *Eighth International Conference on Neutron Capture Therapy for Cancer.* LaJolla, Plenum Press, 1998.
6. W. Sauerwein, R. Moss, K. Hideghety, F. Stecher-Rasmussen, M. deVries, H.J. Reulen, C. Gvtz, P. Paquis, F. Grochulla, K. Haselsberger, J. Wolbers, J. Rassow, J.P. Pingol, P. Watkins, C. Vroegindeweij, K. Ravensberg, S. Garbe, O.D. Wiestler, B. Turowski, F. Zanella, D. Touw, A. Siefert, R. Huiskamp, H.

Frankhauser, and D. Gabel, Status report on the european clinical trial of BNCT at Petten. *Eighth International Conference on Neutron Capture Therapy for Cancer*. LaJolla, Plenum Press, 1998.
7. O.K. Harling, K.A. Roberts, D.J. Moulin, and R.D. Rogus, Head phantoms for neutron capture therapy, *Med. Phys.*, 22:579–583, 1995.
8. R.D. Rogus, O.K. Harling, and J.C. Yanch, Mixed field dosimetry of epithermal neutron beams for boron neutron capture therapy at the MITR-II research reactor., *Med. Phys.*, 21:1611–1625, 1994.
9. G.R. Solares, and R.G. Zamenhof, A novel approach to the microdosimetry of neutron capture therapy. Part 1. High-resolution quantitative autoradiography applied to microdosimetry in neutron capture therapy., *Radiat. Res.*, 144:50–58, 1995.
10. C.S, Yam, Microdosimetric studies for neutron capture therapy and techniques for capture element selection. *Nuclear Engineering*. Cambridge, MA, Massachusetts Institute of Technology, 1995.
11. K.J. Riley, and O.K. Harling, An improved prompt gamma neutron activation analysis facility using a focused diffracted neutron beam., *Nucl. Instr. and Method. B*, 143:414–421, 1998.
12. Y. Mishima, C. Honda, M. Ichihashi, H. Obara, J. Hiratsuka, H. Fukuda, H. Karashima, T. Kobayashi, K. Kanda, and K. Yoshino, Treatment of Malignant Melanoma by single thermal neutron capture therapy with Melanoma-seeking 10B-compound, *Lancet*, 11:388–389, 1989.
13. G.E. Laramore, F.J. Wheeler, D.E. Wessol, K.J. Stelzer, and T.W. Griffin, A tumor-control curve for malignant gliomas derived from fast neutron radiotherapy data: implications for treatment delivery and compound selection. *in*: "Advances in Neutron Capture Therapy," Larsson B, Crawford J, and Weinreich R, eds., Elsevier, Amsterdam, 580–587, 1997.
14. R.F. Barth, W. Yang, J.H. Rotaru, M.L. Moeschberger, D.D. Joel, M.M. Newrocky, J.H. Goodman, and A.H. Soloway, Boron neutron capture therapy of brain tumors: enhanced survival following intracarotid injection of either sodium borocaptate or boronophenylalanine with or without blood-brain barrier disruption, *Cancer Res.*, 57:1129–1136, 1997.
15. A.H. Soloway, W. Tjarks, B.A. Barnum, F.-G. Rong, R.F. Barth, I.M. Codogni, and J.G. Wilson, The chemistry of neutron capture therapy, *Chemical Reviews*, 98:1515–1562, 1998.
16. S. Sakamoto, W.S. Kiger III, and O.K. Harling, Optimization of the neutronic design of the MIT fission converter beam. *in*: "Advances in Neutron Capture Therapy," Larsson B, Crawford J, and Weinreich R, eds., Elsevier, Amsterdam, 408–412, 1997.
17. W.S. Kiger III, and O.K. Harling, Design studies for the MIT fission converter beam. *in*: "Advances in Neutron Capture Therapy," Larsson B, Crawford J, and Weinreich R, eds., Elsevier, Amsterdam, 403–407, 1997.
18. W.S. Kiger III, S. Sakamoto, and O.K. Harling, Neutronic design of a fission converter-based epithermal neutron beam for neutron capture therapy, *Nuclear Science and Engineering*, 131:1–22, 1999.
19. C. Chuang, M. Palmer, R. Zamenhof, and P. Busse, Time-dependent biodistribution of BPA-f in normal and tumor brain tissues in mice. *Eighth International Conference on Neutron Capture Therapy for Cancer*. LaJolla, CA, Plenum Press, 1998.
20. S.C. Saris, G.R. Solares, D.E. Wazer, G. Cano, S.E. Kerley, M.A. Joyce, L.S. Adelman, O.K. Harling, H. Madoc-Jones, and R.G. Zamenhof, Boron Neutron Capture Therapy for murine malignant gliomas, *Cancer Res.*, 52:4627–4677, 1992.
21. E.H. Elowitz, R.M. Bergland, J.A. Coderre, D.D. Joel, M. Chadha, and A.D. Chanana, Biodistribution of p-boronophenylalanine (BPA) in patients with glioblastoma multiforme for use in boron neutron capture therapy, *Neurosurg*, 42:463–469, 1998.
22. C. Chuang, P. Busse, M. Palmer, P. Thomas, and R. Zamenhof, Selective tumor uptake of boronophenylalanine-fructose (BPA-f) in an animal model of hepatic colorectal metastasis. *Eighth International Conference on Neutron Capture Therapy for Cancer*. LaJolla, CA, Plenum Press, 1998.
23. J.A. Coderre, J. Capala, M. Makar, and A.Z. Diaz, Application of BNCT to other types of tumors. *in*: "Advances in Neutron Capture Therapy," Larsson B, Crawford J, and Weinreich R, eds., Elsevier, Amsterdam, 649–653, 1997.
24. W.S. Kiger III, M.R. Palmer, P.M. Busse, and R.G. Zamenhof, A pharmacokinetic model for the concentration of Boron-10 in blood following BPA-f administration in humans. *Eighth International Conference on Neutron Capture Therapy for Cancer*. LaJolla, CA, Plenum Press, 1998.
25. G.W. Kabalka, G.T. Smith, J.P. Dyke, W.S. Reid, C.P.D., Longford, T.G. Roberts, N.K. Reddy, E. Buonocore, and K.F. Hubner, Evaluation of fluorine-18-BPA-fructose for boron neutron capture treatment planning, *J. Nucl. Med.*, 38: ••, 1997.
26. Y. Imahori, S. Ueda, Y. Ohmori, T. Kusuki, K. Ono, R. Fujii, and T. Ido, Fluorine-18-labeled fluoroboronophenylalanine PET in patients with glioma, *J. Nucl. Med.*, 39:325–333, 1998.

5

BORON NEUTRON CAPTURE THERAPY FOR GLIOBLASTOMA MULTIFORME

Results from the Initial Phase I/II Dose Escalation Studies

Aidnag Z. Diaz[1,4], Arjun D. Chanana[1], Jacek Capala[1], Manjeet Chadha[2], Jeffrey A. Coderre[1], Eric H. Elowitz[3], Junichi Iwai[1], Darrel D. Joel[1], Hungyan B. Liu[5], Ruimei Ma[1], Noreen Pendzick[1], Nancy S. Peress[6], Magdy S. Shady[1], Daniel N. Slatkin[1], George W. Tyson[7], and Lucian Wielopolski[1]

[1]Medical Department
Brookhaven National Laboratory
Upton, New York
[2]Departments of Radiation Oncology and
[3]Neurosurgery, Beth Israel Medical Center
New York, New York
[4]Department of Radiation Oncology
Montefiore Medical Center
Bronx, New York
[5]McClellan Nuclear Radiation Center
McClellan AFB, California
[6]Departments of Pathology and
[7]Neurological Surgery
State University of New York
Stony Brook, New York

1. INTRODUCTION

Glioblastoma multiforme (GBM) accounts for about 80% of all adult malignant gliomas. Despite aggressive surgery, chemotherapy and radiation therapy, GBM continues to be one of the most intractable brain tumors. The median and 5-year survivals are typically 9 to 10 months and <5%, respectively.[1] For primary malignant brain tumors, the most powerful prognostic factors are histology, age and functional status.[2] Although GBM rarely metastasizes to other sites and organs, the tumor cells infiltrate deeply into

Frontiers in Neutron Capture Therapy, edited by Hawthorne *et al.*
Kluwer Academic / Plenum Publishers, New York, 2001.

the surrounding brain and may even reach the contralateral hemisphere. External beam photon radiation therapy cannot ablate the infiltrating GBM cells because radiation doses high enough to kill those cells would cause unacceptable necrosis of normal brain tissues. Boron neutron capture therapy (BNCT) has the potential to address this inadequacy through a more selective irradiation of GBM cells.

On September 13, 1994, improvements in the BMRR, in the delivery of BPA, and in the understanding of the radiobiology of BPA-mediated BNCT led to the initiation of closed-skull BNCT for human GBM at Brookhaven National Laboratory (BNL) using BPA-F and epithermal neutrons under US-FDA IND #43,317. The hypothesis to be tested was that BNCT as proposed for the Phase I/II clinical trial, would be safe and may be more effective than the conventional post-debulking radiation therapy in the treatment of selected patients with GBM. The basis for this hypothesis was that certain brain tumor patients were reported to have benefited from analogous forms of BNCT[3] and that results in animals[4] provided experimental evidence for effective control of gliosarcoma by BPA-mediated BNCT.

2. PATIENTS AND METHODS

BNCT was given to 41 patients with primary GBM. Because histologic tumor necrosis is a major prognostic variable in malignant supratentorial, astrocytic glioma, we employed the criteria of Nelson *et al.*[5] for the diagnosis of GBM. The following entry criteria were used: supratentorial, unilateral, unifocal and debulked GBM; deepest part of the contrast-enhancing tumor margin on the post-debulking craniotomy brain scan no more than ≈6.0 cm from the scalp surface; WBC $\geq$2500/mm^3; platelets $\geq$75,000/mm^3; creatinine $\leq$1.6 mg%; no evidence of congestive heart failure; Karnofsky performance status score $\geq$70; no history of phenylketonuria; no prior radiation therapy to the brain; no prior immunotherapy and/or systemic or intrathecal chemotherapy; and age $\geq$18 years. Informed consent was obtained from all patients.

2.1. Patient and Tumor Characteristics

The characteristics of 40 evaluable patients and their tumors are listed in Table 1. One non-evaluable patient developed symptoms and signs of craniospinal metastases by day 8 following BNCT and died of metastatic disease a few weeks following the completion of BNCT and was lost to follow-up.

2.2. Radiobiological Factors

To express the total BNCT dose in a common unit, and to compare BNCT doses with the effects of conventional photon radiation, appropriate multiplicative factors (RBE or CBE) times the physical absorbed radiation dose from each of the high-LET components of the BNCT dose are assumed to be additive. The total effective BNCT dose is then defined as the arithmetic sum of RBE- or CBE-corrected components using the unit Gy-Eq (gray-equivalent). In Table 2 the measured values for the biological effectiveness of the high-LET BNCT dose components are listed.

2.3. Boron Concentrations in Tissues

Boron analyses were performed by two methods: prompt-gamma-ray spectroscopy[6] and direct current plasma-atomic emission spectroscopy.[7] A method for estimating boron concentration in tumor cells based on measured blood boron concentrations has been described.[8] Using this method it is estimated that the average ^{10}B concentrations in GBM tumor cells is ~4 times greater than the concurrent blood ^{10}B concentration. In the calculation of the radiation dose to tumor cells, we assumed tumor-to-blood ^{10}B concentration ratio of 3.5 to 1.

2.4. Simulation of Patient Treatment Position, Neutron Beam Collimators, and BNCT Procedure

The treatment position simulation was usually carried out on the day before the scheduled BNCT[9,10] in a replica of the BMRR epithermal beam treatment room. This replica was built mainly to: 1) permit treatment position simulation in a room free of background radiation, 2) provide direct visualization of the beam entry point on the scalp, and 3) accustom the patient to the BNCT procedure. The desired position of the beam entry point on the scalp was achieved with most patients either in supine or lateral recumbence, except for 2 patients who were treated while seated. The apertures of the neutron beam collimator used during patient irradiation, was either 8 cm in diameter (14 patients) or 12 cm in diameter (26 patients). The collimators are mounted at the irradiation port to improve neutron penetration in the head, thereby increasing the thermal neutron flux at depth.[11]

On the day of BNCT, BPA-F was infused i.v. over 2 hours to deliver 250 mg (n = 29) or 290 mg (n = 11) BPA/kg body weight. Blood samples for ^{10}B determinations were obtained every 20 minutes during BPA-F infusion, just before the start of irradiation, in the middle, and at the end of BNCT irradiation. The reactor power during irradiation was 2 MW (14 patients) or 3 MW (26 patients). Twenty-five patients received single-field, 7 patients received bilateral, double-field and 8 patients received primarily ipsilateral, double-field irradiations. The duration of irradiation was determined by the time required to deliver the prescribed peak brain dose, which, in turn, depended upon the reactor power and the average blood ^{10}B concentration during irradiation. The average blood ^{10}B concentration during BNCT was calculated by linear extrapolation of the ^{10}B concentrations measured in the blood during the irradiation. The BNCT radiation doses for the contrast-enhancing tumor volume, target volume (tumor + 2 cm margin), whole brain, cerebral hemispheres, cerebellum and brain stem were calculated using the treatment planning software. The doses to other sites such as the basal ganglia, thalamus, retina, optic chiasm, oronasopharyngeal mucosae, parotid glands, external and middle ears, lacrimal glands, conjunctiva and scalp were estimated from the isodose contours superimposed on corresponding MRI brain scan images.

2.5. Post-BNCT Management and Follow-up

Following the completion of the BNCT irradiation the patient was observed at BNL for 1 to 2 hours, then transferred to the neurosurgical unit of a Collaborating Hospital for 48 hours of observation. After that, the patient was discharged to the care of the follow-up physician. The post-BNCT information was requested from

Table 1. Patient and glioblastoma multiforme characteristics[a]

Patient #	Age (yr)	Sex	KPS score	RPA class	GBM location	GBM diam. (cm)[b]	GBM vol. (cm^3)[c]	Minimum BNCT Dose (Gy-Eq) Tumor volume[d]	Target volume[e]	Survival from diagnosis (months)[f]	Dead
09-17-63R	50	F	90	5	left temporoparietal	3.50	18	27.84	16.18	14.70	Yes
09-18-59R	67	F	80	5	right parietal	5.30	29	31.90	10.43	14.70	Yes
09-18-90R	61	M	90	4	right parietal	3.90	18	26.80	13.54	16.71	Yes
09-19-03R	46	F	70	4	right parietooccipital	7.40	70	20.43	7.44	8.19	Yes
09-19-05R	68	M	90	5	right frontoparietal	3.90	19	24.27	8.50	9.51	Yes
09-19-14R	56	M	70	5	right frontoparietal	4.90	23	18.85	8.54	17.60	Yes
09-19-16R	75	F	70	5	right frontal	6.10	56	19.10	9.38	12.27	Yes
09-19-18R	69	F	70	5	left temporal	4.90	18	22.97	11.44	10.63	Yes
09-19-25R	46	F	90	3	right parietal	4.10	21	26.48	12.87	24.08	Yes
09-19-71R	46	F	100	3	left temporal	5.90	15	34.97	17.24	13.55	Yes
09-19-78R	57	M	80	5	left occipital	4.30	20	21.54	9.42	20.82	Yes
09-20-03R	50	F	80	5	left frontal	6.20	22	21.73	10.48	20.79	Yes
09-20-09R	64	F	80	5	right parietal	5.60	21	25.19	13.04	11.15	Yes
09-21-15R	48	M	80	4	right occipitoparietal	5.50	37	19.48	8.50	27.17	No
09-20-50R	57	F	70	5	right temporoparietal	7.20	70	26.59	18.00	10.53	Yes
09-20-53R	61	M	80	5	right parietooccipital	5.80	24	40.00	21.00	7.96	Yes
09-20-62R	53	M	90	5	left temporal	7.40	37	18.00	17.00	12.60	Yes
09-20-76R	43	M	70	4	left temporal	8.40	31	21.00	17.00	12.76	Yes
09-20-80R	48	F	70	4	left frontoparietal	8.70	83	28.00	17.00	6.41	Yes
09-20-98R	41	M	80	4	left frontal	3.00	4.9	35.00	19.00	13.09	Yes

09-21-04R	57	M	80	5	left occipital	5.00	29.6	32.00	18.00	21.81	Yes
09-21-16R	49	M	70	4	left parietal	5.40	18	43.00	22.00	6.05	Yes
09-21-34R	63	F	90	5	left parietooccipital	4.00	13	40.00	22.00	18.65	No
09-21-38R	57	M	90	5	right occipital	6.00	18	42.00	20.00	13.78	Yes
09-21-52R	47	F	80	4	left temporal	7.90	55	20.00	19.00	16.91	Yes
09-21-55R	39	M	70	4	right frontal	5.80	36	20.00	15.00	17.63	No
09-21-62R	59	M	90	5	right frontal	4.20	16	41.00	23.00	8.06	Yes
09-21-72R	40	M	70	4	right occipitoparietal	8.00	74	33.00	17.00	12.83	Yes
09-21-78R	53	M	80	5	right frontal	5.40	25	34.00	18.00	10.33	Yes
09-22-27R	54	M	90	5	left frontoparietal	2.50	8	44.00	24.00	12.20	No
09-22-37R	41	M	80	4	left temporal	2.00	2	50.00	35.00	12.34	No
09-22-40R	71	F	90	5	right parietal	4.70	18	51.00	29.00	10.00	Yes
09-22-99R	56	M	90	5	left temporal	4.80	11	36.00	25.00	11.68	Yes
09-23-32R	68	M	90	5	left temporal	5.90	18	41.00	25.00	8.59	Yes
09-23-46R	68	M	70	5	left parietal	3.00	9	54.60	29.00	9.67	Yes
09-23-57R	41	F	70	4	left temporoparietal	7.50	41	38.58	24.05	5.99	No
09-23-80R	79	M	70	5	left occipital	6.50	40	48.07	31.25	3.88	No
09-24-60R	40	M	70	4	right temporoparietal	7.20	59	29.71	16.53	0.99	No
09-24-76R	58	M	70	5	right temporal	9.20	61	27.10	21.80	0.69	No
09-25-37R	75	F	100	5	right temporoparietal	6.70	22	35.19	24.01	1.12	No

[a]KPS, Kanofsky performance status; RPA Recursive partitioning analysis (19); GBM, Glioblastoma multiforme; BNCT, Boron neutron capture therapy; Gy-Eq, gray-equivalent.
[b]Maximum post-operative tumor diameter.
[c]Post-operative contrast enhancing tumor volume.
[d]Minimum radiation dose to contrast-enhancing tumor volume.
[e]Minimum radiation dose to contrast-enhancing tumor volume plus the volume of a 2 cm-thick shell enclosing the tumor volume.
[f]survival from histopathological diagnosis.

Table 2. Measured values for the biological effectiveness of BPA-based BNCT radiation dose components[a]

Dose component	Biological effectiveness factor (RBE or CBE)	Study reference
$^{10}B(n,\alpha)^{7}Li$ reaction in		
CNS (spinal cord)	1.3	Morris *et al.*, 1994 Reference 15
tumor	3.8	Coderre *et al.*, 1993 Reference 16
scalp	2.5	Fukuda *et al.*, 1994 Reference 17
$^{14}N(n,p)^{14}C$ and $^{1}H(n,n')p$ reactions	3.2	Coderre *et al.*, 1993 Reference 16; Morris *et al.*, 1994 Reference 15
Gamma photons from the epithermal neutron beam and from the $^{1}H(n,\gamma)^{2}H$ reaction	1.0	

[a]BPA, boronophenylalanine; BNCT, boron neutron capture therapy; RBE relative biologic effectiveness; CBE, compound adjusted relative biologic effectiveness.

the follow-up physician at 1, 3, 8 and 12 weeks, every 3 months during the first 2 years of post-BNCT irradiation and every 4 months thereafter. The frequency of follow-up examinations, including brain scanning, was increased upon any suspicion of tumor progression.

2.6. Response Criteria

Evaluation of BNCT-mediated toxicity was the primary objective of this study and was documented from the post-BNCT reports provided by follow-up physicians. The patient information included history and physical examination, total and differential leukocyte counts, routine clinical tests of blood and urine, MRI brain scans, Mini Mental State scores, Karnofsky Performance Status scores, and acute and late BNCT mediated neurotoxicity scores. Grade 3 or grade 4 toxicity, if any, was to be scored as severe toxicity. Death directly related to BNCT was to be defined as grade 5 toxicity. The toxicity criteria and grading systems were based on the Cooperative Group Common Toxicity Criteria and Radiation Therapy Oncology Group (RTOG) and the European Organization for Research and Treatment of Cancer (EORTC) acute and late radiation morbidity criteria.[12] Local disease control and survival were secondary objectives of this study. Time to tumor progression and overall survival were defined as times from the date of histopathological diagnosis to the dates of GBM progression and death, respectively. Disease progression was defined as the post-BNCT increase in the MRI contrast-enhancing tumor volume, the appearance of new MRI contrast-enhancing lesion/s in the brain and/or progressive worsening of neurologic symptoms and signs attributable to tumor growth. Metabolic scans (mainly PET; ^{18}F-fluorodeoxyglucose) were done in 14 patients when GBM progression was suspected. Histopathological diagnosis of the tumor progression was obtained in those patients who were subjected to debulking of the progressive tumor. Tumor progression was scored as local or peripheral. Progression was scored as local when MRI and/or functional brain scans showed progression either in the contrast-enhancing tumor volume and/or in contiguity with the contrast-enhancing tumor volume. Other progression patterns were scored as peripheral.

3. RESULTS

The age of the 40 evaluable patients ranged from 39 yr. to 79 yr. (median = 55 yr.). There were 24 males and 16 females. The KPS score ranged from 70–100 (median = 80). There were 2, 13, and 25 patients in radiation therapy oncology group (19) recursive partitioning analysis (RPA) class III, IV and V, respectively. The maximum post-operative tumor diameter ranged from 2 to 9.2 cm (median = 5.5 cm). The post-operative tumor volume ranged from 2 to 83 cm^3 (median = 22 cm^3). The deepest part of the contrast-enhancing tumor margin on the post-operative brain scan ranged from 3.8 to 8.6 cm (median = 5.3 cm) from the scalp surface. In 13 patients the tumor involved corpus callosum, thalamus, basal ganglia and/or ependyma. In 25 and 15 patients the tumor debulking was reported to be gross total or subtotal, respectively. The interval from diagnosis to BNCT ranged from 0.8 to 5.7 months (median 1.6 months). No patient has been lost to follow-up. Twenty-eight and 12 patients had 1 and 2 pre-BNCT debulking craniotomies, respectively. The duration of post-BNCT follow-up ranged from 1.5 to 31.4 months (median = 9.1 months). At the cut-off time of analysis (September 1, tumor progression had been documented in 34 of 40 patients and 30 of these 34 patients had died.

3.1. Pre-BNCT BPA-F Infusions and Boron Concentration

The blood boron concentration reached a peak at the end of BPA-F infusion and was directly proportional to the dose of BPA. At the completion of the BPA-F infusion, blood boron levels decreased steadily with biphasic exponential clearance kinetics. The average blood boron concentration (mean ± sd) during BNCT irradiation was 12.2 ± 1.6 ppm (n = 29) or 14.1 ± 2.1 ppm (n = 11), following 250 mg and 290 mg BPA/kg body weight infusions, respectively.

3.2. BNCT Radiation Doses and Relative Risks

The duration of BNCT irradiations ranged from 38 to 73 min. The reactor power was either 2 or 3 MW. The normal brain peak dose rate ranged from 19.4 to 40.3 cGy-Eq/min. The peak and average brain doses ranged from 8.9 to 14.8 and 1.9 to 6.6 Gy-Eq, respectively. The minimal photon equivalent doses to tumor (tumor volume dose) and to any tumor cells infiltrating the target volume (target volume dose) ranged from 18 to 54.6 and 7.4 to 35 Gy-Eq, respectively. The calculated relative risks derived from surface dosimeters, for brain and whole-body ranged from 0.29 to 1.15 sieverts and 0.64 to 1.9 sieverts, respectively. The total whole body radiation doses from short-lived radioactive isotopes induced by BNCT were calculated to be <0.01 Gy.

3.3. Clinical Outcome: General, Quality of Life, and Brain Scans

The BPA-F infusion and the BNCT procedure were remarkably well tolerated. No patient complained of nausea, vomiting or headache during or soon after BNCT. All patients, except for two, as described in the following section, were discharged at the completion of the 48 hour observation period. All patients were maintained on prophylactic antiseizure medications. Attempts to wean the patients off steroids following BNCT led to the following results: 13 patients, 0 mg/day; 7 patients, 1–4 mg/day; 13 patients, 5–8 mg/day; 6 patients, no dose reduction; and 1 patient, increased dose. In most patients, the dose of dexamethasone had to be increased at the time of tumor progression. From completion of BNCT irradiation to the time of tumor progression (n = 34) or to the time of follow-up to date (n = 6) the KPS scores improved in 20 patients, did not change in 17 patients, and declined in 3 patients. The Mini Mental State scores improved in 4 patients, did not change in 32 patients and declined in 4 patients. Overall, with the possible exception of 3 patients, patients either improved or maintained their pre-BNCT functional status until tumor progression. MRI brain scans obtained within 48 hours of the BNCT irradiation did not show any significant interval change from the pre-BNCT brain scans in 38 of 40 patients, in 2 patients the scans showed increased interval edema with mass effect. Details about these 2 patients are provided in the following section. In brain scans covering the interval to tumor progression (n = 34) or to the time of the follow-up to date (n = 6) no significant change from pre-BNCT scans was seen in 31 patients and modest to moderate decrease in the size and/or the intensity of the enhancing lesion was seen in 9 patients.

3.4. Clinical outcome: Complications

Grade 1 erythema and alopecia were reported in all patients. Other scalp complications were reported in 7 patients from 2.9 to 24 months after BNCT and included pigmentation, depigmentation, patchy atrophy, dry desquamation and induration. The scalp complications required no specific therapy. The BNCT radiation doses to the scalp in these 7 patients ranged from 11.8 to 17.7 Gy-Eq (5.9 to 8.3 Gy). Complications involving the oral cavity were reported in 9 patients and included thrush, mildly dry mouth, slightly thickened saliva and altered perception of taste. The interval from BNCT to the date when the complication was first reported ranged from 0.2 to 2.0 months. The BNCT radiation doses in these 9 patients ranged from 1.2 to 9.8 Gy-Eq (0.8 to 6.3 Gy). Other reported complications were associated only with BNCT irradiations of temporal and frontal lobe tumors and included parotitis, tenderness of temporomandibular joint, mild otitis externa, mild otitis media, serous otitis media and conjunctivitis (Table 3). These complications were transient and responded promptly to appropriate medical therapy, except for the patient with serous otitis media, who required middle ear drainage.

Neurological events attributable to increased vasogenic edema and mass effect after BNCT appeared in 3 patients. One experienced sudden onset of slurred speech, nausea, vomiting, dizziness, drowsiness, expressive aphasia and mild hemiparesis about 24 hours following BNCT. An emergency CT brain scan revealed midline shift of a few mm. The patient responded promptly to intravenous dexamethasone. In the second, a preexisting mild expressive aphasia deteriorated to complete expressive aphasia within about 12 hours of BNCT. The MRI brain scan showed increased vasogenic edema and mass effect on the left lateral ventricle. The patient responded within 48 hours to increased oral dexamethasone. The third patient developed mild nausea, vomiting and somnolence about 12 hours following BNCT. The MRI brain scan did not show any significant change from the pre-BNCT brain scan. There was a favorable, prompt response to increased dexamethasone.

3.5. Post-BNCT Seizures

Post-BNCT seizures were reported in 8 patients; occurring from 24 to 120 hours after BNCT. In five patients, the seizures were similar to the pre-BNCT seizures and in 4 of these, blood levels of the antiseizure medications were found to be below

Table 3. Complications associated with boron neutron capture therapy of temporal and frontal lobe tumors[a]

Patient #	Tumor location	Beam	Complications	Interval from BNCT (months)[b]	BNCT doses[c] Physical (Gy)	Photon equivalent (Gy-Eq)
9-20-50	right temporo-parietal	bilateral	right serous otitis media,	0.39	9.5	17.2
			mild right otitis externa	2.70	7.0	14.2
9-20-62	left temporal	bilateral	mild left otitis externa	0.72	7.1	14.0
9-20-76	left temporal	bilateral	bilateral parotid &TMJ tenderness	0.86	6.3	8.8
9-21-55	right frontal	bilateral	bitemporal swellings,	0.03	6.3	8.8
			right conjunctivitis	0.07	5.2	7.2
9-22-99	left temporal	unilateral	mild left otitis externa	0.92	8.9	18.5
9-23-32	left temporal	unilateral	left conjunctivitis,	0.07	4.9	6.9
			otitis externa,	0.07	10.2	21.0
			otitis media	0.07	6.7	12.2
			parotitis	0.07	4.6	6.4
9-23-57	left temporo-parietal	unilateral	left parotitis	0.10	5.7	8.0

[a]BNCT, boron neutron capture therapy; Gy, gray; Gy-Eq, gray-equivalent; TMJ, temporomandibular joint.
[b]Interval from BNCT to the date when complication was first reported.
[c]Maximum estimated doses.

therapeutic range. Increase of antiseizure medication promptly controlled the seizures in 7 patients. In one patient the seizure activity was reported to have been mild and transient and required no treatment.

3.6. Tolerance of Normal Brain to BNCT Radiation and Radiation Necrosis

There is no evidence to date to suggest that double field BNCT posed any greater hazards than single-field BNCT. Brains from 9 patients who died from progressive GBM were available for gross and histopathological examinations. The interval from BNCT to death ranged from 4.4 to 14.4 months. The estimated maximum normal brain doses ranged from 12 to 14 Gy-Eq (8 to 10 Gy). The brain of one patient exhibited focal vascular fibrinoid necrosis of a non-tumorous superficial venule just lateral to the projected central axis of the neutron beam. This was the only extra-tumoral finding in the 9 necropsies that was attributable to radiation. Other postmortem findings included residual/recurrent tumor with necrosis, edema of the ipsilateral hemisphere and vascular changes of tumor blood vessels.

3.7. Tumor Progression and Survival

Tumor progression has been observed in 34 of 40 patients to date. The estimated median time to tumor progression from histopathological diagnosis in all patients is 30.3 weeks. Tumor progression was confirmed in 21 patients by histopathology (14 from debulked progressive tumors and 7 from tissues obtained at autopsy). In 13 patients, the diagnosis of tumor progression was based on worsening of neurological signs and symptoms directly attributable to the tumor progression and brain scans (MRI and PET = 6; MRI and SPECT = 1; MRI = 4). In 14 patients, anatomic and metabolic brain scans (^{18}FDG-PET, n = 12; ^{201}Thallium-SPECT, n = 2) were carried out at the time of suspected tumor progression and in 9 of these patients, histopathological reports of the debulked, progressive tumor were also available. The metabolic brain scans were no more reliable than the MRI brain scans in distinguishing between tumor progression and radiation necrosis.

Tumor progression was local in all cases (n = 34) but local progression was associated with peripheral progression in 19 cases (ipsilateral hemisphere = 13, contralateral hemisphere = 3, subtentorial = 3). We used post-operative MRI brain scans for determining eligibility and for treatment planning. Retrospective reviews of the available preoperative MRI brain scans indicate, however, that the zone of preoperative peritumoral vasogenic edema was a preferred zone for local tumor progression. Recommendations concerning the management of the tumor progression were made by the follow-up physicians. Thirteen patients received hospice-type comfort care only and 5 patients underwent tumor debulking with no additional treatment. In the remaining 16 patients, the following post-progression treatments were employed: debulking (DB) followed by fractionated photon radiotherapy (FRT) = 3; DB, FRT and chemotherapy (ChT) = 4; DB, stereotactic radiosurgery and ChT = 2; FRT and ChT = 1, DB and ChT = 2; and ChT only = 4.

Thirty patients are dead and 10 patients are alive as of the time of analysis: September 1, 1998. The estimated median times of survival from histopathological diagnosis for all patients (n = 40), RPA class IV patients (n = 13) and RPA class V patients (n = 25) are 12.4 months, 12.8 months and 12.3 months, respectively. There were 2 RPA class III patients and they survived for 13.6 and 24.1 months, respectively. The median survival of 13 patients over 60 yr. (12 RPA class V, 1 RPA class IV) and 9 patients over 65 yr. (9 RPA class V) were 10, and 9.7 months, respectively. There was no significant correlation between age, KPS, interval from presenting signs and symptoms to BNCT, number of pre-BNCT debulking craniotomies, tumor dimensions or BNCT radiation doses and time to tumor progression or survival time.

4. CONCLUSIONS

The BNCT procedure as employed to date has been relatively safe in all patients. We have not observed any grade 3 or grade 4 acute or early delayed radiation-induced normal brain toxicity. The palliation afforded by a single session of BNCT compares favorably with palliation provided by more conventional, protracted and relatively toxic treatments such as fractionated radiation therapy and adjuvant chemotherapy. We conclude that BNCT radiation dose-escalation to the normal brain tolerance threshold is indicated and that implementation of a fractionated BNCT trial for the evaluation of potential benefits of BPA retargeting would be desirable. The experience gained in the post-BNCT management of patients with subtotally debulked GBM supports the initiation of a BNCT trial for non-debulked GBM. Such a trial would be particularly relevant to patients who are either not good candidates for GBM debulking (as those with GBM in eloquent brain regions) or to elderly patients who derive little benefit from debulking of GBM.[13] Seven patients received fractionated photon radiation therapy and 2 patients received radiosurgical treatment for post-BNCT tumor progression without clinical evidence of acute or early delayed toxicity or histopathological evidence of radiation induced damage to normal brain (fractionated radiotherapy n = 2 radiosurgery n = 1). These observations and new information obtained from preclinical studies[14] open up the possibility of a BNCT clinical trial for post-BNCT or post-photon therapy recurrent GBM.

ACKNOWLEGMENTS

This work was supported by U.S./D.O.E., contract # DE-ACO2-98 CH 10886. Accordingly, the U.S. Government retains a non-exclusive, royalty-free license to publish or reproduce the published form of this contribution, or allow other to do so, for U.S. Government purposes.

REFERENCES

1. F.G. Davis, S. Freels, J. Grutsch, S. Barlas, and S. Brem, Survival rates in patients with primary malignant brain tumors stratified by patient age and tumor histological type: an analysis based on Surveillance, Epidemiology, and End Results (SEER) data, 1973–1991, *J. Neurosurg.*, 88:1–10, 1998.
2. W.J. Curran, CB Scott, J. Horton, J.S. Nelson, A.S. Weinstein, A.J. Fischbach, C.H. Chang, M. Rotman, S.O. Asbell, R.E. Krisch, and D.F. Nelson, Recursive partitioning analysis of prognostic factors in three radiation oncology group malignant glioma trials, *J. Nat. Cancer Inst.*, 85:704–710, 1993.
3. H. Hatanaka, and Y. Nakagawa, Clinical results of long surviving brain tumor patients who underwent boron neutron capture therapy, *Int. J. Radiat. Oncol. Biol. Phys.*, 28:1061–1066, 1994.
4. J.A. Coderre, T.M. Button, P.L. Micca, C. Fisher, M.M. Nawrocky, and H.B. Liu, Neutron capture therapy of the 9L rat gliosarcoma using the p-boronophenylalanine-fructose complex, *Int. J. Radiat. Oncol. Biol. Phys.*, 30:643–652, 1994.

5. J.S. Nelson, Y. Tsukada, D. Schoenfeld, K. Fulling, J. Lamarche, and N. Peress, Necrosis as a prognostic criterion in malignant supratentorial, astrocytic gliomas, *Cancer*, 52:550–554, 1983.
6. R.G. Fairchild, D. Gabel, B.H. Laster, D. Greenberg, W. Kiszenick, and P. Micca, Microanalytical techniques for boron analysis using the $^{10}B(n,\alpha)^{7}Li$ reaction, *Med. Phys.*, 13:50–56, 1986.
7. W. Bauer, P. Micca, and B. White, A rapid method for the direct analysis of boron in whole blood by atomic emission spectroscopy, *in*: "Advances in Neutron Capture Therapy", A.H. Soloway, R.F. Barth, and D.E. Carpenter, eds., Plenum Press, New York, pp. 403–407, 1993.
8. J.A. Coderre, A.D. Chanana, D.D. Joel, E.H. Elowitz, P.L. Micca, M.M. Nawrocky, M. Chadha, J.O. Gebbers, M. Shady, N.S. Peress, and D.N. Slatkin, Biodistribution of *p*-boronophenylalanine in patients with glioblastoma multiforme: Boron concentration correlates with tumor cellularity, *Radiat. Res.*, 149:163–170, 1998.
9. M. Chadha, J. Capala, J.A. Coderre, E.H. Elowitz, J. Iwai, D.D. Joel, H.B. Liu, L. Wielopolski, and A.D. Chanana, Boron neutron capture therapy (BNCT) for glioblastoma multiforme (GBM) using the epithermal neutron beam at the Brookhaven National Laboratory, *Int. J. Radiat. Oncol. Biol. Phys.*, 40:829–834, 1998.
10. L. Wielopolski, J. Capala, M. Chadha, N. Pendzick, and A.D. Chanana, Considerations for patient positioning in static beams for BNCT, *in*: "Advances in Neutron Capture Therapy, Volume I, Medicine and Physics," B. Larsson, J. Crawford, and R. Weinreich, eds., Elsevier, Amsterdam, pp. 357–360, 1997.
11. H.B. Liu, D.D. Greenberg, J. Capala, J.A. Coderre, and F.J. Wheeler, An improved neutron collimator for brain tumor irradiation in clinical boron neutron capture therapy, *Med. Phys.*, 23:2051–2060, 1996.
12. J.D. Cox, J. Stetz, and T.F. Pajak, Toxicity criteria of the radiation therapy oncology group (RTOG) and the European organization for research and treatment of cancer (EORTC), *Int. J. Radiat. Oncol. Biol. Phys.*, 31:1341–1346, 1995.
13. P.J. Kelly, C. Daumas-Duport, B.W. Sheithauer, and D.B. Kispert, Stereotactic histologic correlations of computed tomography-and magnetic resonance imaging—defined abnormalities in patients with glial neoplasms, *Mayo Clin. Proc.*, 62:450–459, 1987.
14. G.M. Morris, J.A. Coderre, J.W. Hopewell, P.L. Micca, and L. Wielopolski, Boron neutron capture therapy: Re-irradiation response of the rat spinal cord, *Radiother. Oncol.*, (in press).
15. G.M. Morris, J.A. Coderre, J.W. Hopewell, P.L. Micca, M.M. Nawrocky, H.B. Liu, and T. Bywaters, Response of the central nervous system to boron neutron capture irradiation: Evaluation using rat spinal cord model, *Radiother. Oncol.*, 32:249–255, 1994.
16. J.A. Coderre, M.S. Makar, P.L. Micca, M.M. Nawrocky, H.B. Liu, D.D. Joel, D.N. Slatkin, and H.I. Amols, Derivations of relative biological effectiveness for the high-LET radiation produced during boron neutron capture irradiations of the 9L rat gliosarcoma *in vitro* and *in vivo*, *Int. J. Radiat. Oncol. Biol. Phys.*, 27:1121–1129, 1993.
17. H. Fukuda, J. Hiratsuka, C. Honda, T. Kobayashi, K. Yoshino, H. Karashima, J. Takahashi, Y. Abe, K. Kanda, M. Ichihashi, and Y. Mishima, Boron neutron capture therapy of malignant melanoma using ^{10}B-paraboronophenylalanine with special reference to evaluation of radiation dose and damage to the skin, *Radiat. Res.*, 138:435–42, 1994

6

BORON NEUTRON CAPTURE THERAPY

Recent Aspect and New Protocol in Japan

Yoshinobu Nakagawa

Department of Neurosurgery
National Kagawa Children's Hospital
Kagawa 765, Japan

Boron neutron capture therapy in Japan was started by H. Hatanaka at HTR in 1968. Successively, four reactors (HTR, MuITR, KUR, and JRR-2) were authorized for medical use (Figure 1). He modified the method developed by W. Sweet. Furthermore, he decided to introduce BSH, which was discovered and tested by his coworker A. H. Solway in Boston, into the new trial as a boron compound. He placed BNCT as radio-surgery and combined surgical procedures and thermal neutron irradiations at the reactor. Hatanaka reported the outcome of the clinical trial in 1986 and showed significant improvement of the results, with long-term recurrence-free survivals in selected patients with malignant gliomas[1]. Many researchers not only in Japan but also in foreign countries were stimulated by his result and began to plan new basic and clinical trials. Following the standstill of MuITR in 1989, KUR and JRR-2 became new research facilities for BNCT in Japan. It was a fortunate timing for Dr. Oda and his group to start clinical trials at KUR. They modified the methods of Hatanaka and started BNCT using BSH in 1990. On the other hand Mishima started a study of a boron neutron capture therapy for melanoma in 1972 and developed BPA as a boron compound. BPA was introduced into treatment for brain tumor afterwards by Ueda in 1994. A new group led by Matsumura started clinical trials at JRR-2 in 1996 with BSH. We missed Dr. Hatanaka because of intracerebral hemorrhage and Dr. Nakagawa succeeded his group and has developed clinical trial using both KUR and JRR-2,[2,3]. Since 1968 to present, more than 200 patients with malignant intracranial tumors, especially malignant gliomas such as glioblastoma, anaplastic astrocytoma etc. were treated by BNCT in Japan (Table 1).

RECENT STANDARD TECHNIQUE OF BNCT

In order to improve the neutron penetration in the brain tissue, we debulk the brain tumor and make a cavity at preliminary operation one to two weeks before BNCT. Partial

Frontiers in Neutron Capture Therapy, edited by Hawthorne *et al.*
Kluwer Academic / Plenum Publishers, New York, 2001.

prompt gamma ray spectrometry during the irradiation. In order to measure the exact neutron flux at each point of interest, gold wires inserted in the tumor tissue are pulled out at 15–30 minutes after the full power operation of the reactor. It is possible to assess the exact neutron fluence at each point of interest. The plan for the remainder of the irradiation is then based on this up-to-the-minute data regarding boron concentration and neutron flux.

The timing of the neutron irradiation was designed according to the clinical analysis in our series. T. Kageji *et al.* reported detailed pharmacokinetics and boron uptake of BSH in the recently report 4. The mean boron concentration in the tumor was 25.8 ppm and the tumor to blood ratio (T/B) was 1.69 in our series. Neutron irradiation has been tried between 10 to 20 hours after a single infusion of BSH in our trial. The study showed a significant statistical correlation between boron uptake and time interval from the start of BSH infusion. Within the first 10 hours after BSH infusion, malignant glioma tissue contained high level of boron (30–60 ppm in tissue), however, at that time, since boron concentration in blood was also about two times higher, and the T/B ratio was below one. Following 15–20 hours after BSH infusion, the boron concentration in the tumor tissue showed above 20 ppm in 56% of malignant glioma patients and the T/B ratio was above one in 69%, above two in 38% of them. Twenty or more hours after infusion, boron was eliminated from the tumor and the concentration in the tumor tissue was below 10 ppm in two-thirds of the patients. Initial tumor concentration stayed below blood concentration within 10 hours after injection of BSH, and after 15–18 hours the T/B ratio was above one since boron clearance from tumor was slower than from blood. These data indicated that the neutron irradiation should be done around 15–20 hours after the BSH infusion. A positive tumor-to-blood ratio and a uniform tumor concentration around 10–40 ppm are needed for successful BNCT. On the other hand, subcellular boron-10 (BSH) localization in human glioblastoma was studied using laser microprobe mass analysis, and immunohistochemical study. We reported that boron-10 is found intracellularly in the tumor tissue whereas in healthy brain, all boron present is only in the blood vessels. Absorption of the boron in the tumor cells was revealed between 2 to 48 hours after BSH injection. Especially the staining at 6, 8, 12, 16 hours after the injection of BSH was significant. The other investigation using CR-39 and atomic force microscopy demonstrated BSH as multiple ultra small etch pits in the cytoplasm and nucleus of the tumor cells. The result shows possibility of microdosimetry.

CLINICAL RESULTS

We summarized clinical results of the patients treated by BNCT. The median survival time of glioblastoma, anaplastic astrocytoma and low grade astrocytoma reached 640 days, 1,811 days and 1,669 days, respectively 5. The result was not sufficient, however, the following factors became clear as an analysis of the outcome of the patients treated by BNCT. The important factors related to the result were radiation dose at the target point and maximum vascular radiation dose near the surface of the brain. Radiation planning was made according to the MRI or CT findings. The target point means the deepest lesion as an enhanced area of the tumor on MRI or CT. Radiation dose has been demonstrated by equivalent dose calculated using RBE. However, we believed the value of RBE of boron n-alpha reaction evaluated in the animal experiment was not suitable for the clinical trials in the human case. Most of the basic research was studied using small animals such as mice or rat. The experimental tumor model was made by intracerebral implantation of the tumor cells such as C-6 or 9-L. They are not human tumor cells and have original character and biology. Then we have not used the RBE in the recent series. At the standpoint of our concept of BNCT, each tumor cell should be destroyed by heavy particles, such as alpha particle or lithium nuclei. Of course we measured all components of the radiation, which include gamma ray. However, in the recent study, we demonstrated the radiation dose using physical dose of boron n-alpha reactions.

The tumor volume radiation dose of the last protocol was 10 Gy but increased up to 15 Gy in the new protocol. Surgical procedures and making a cavity played an important role not only to irradiate with sufficient neutron fluence but also to avoid radiation side effects. Radiation side effects or radiation necrosis was observed in 10.9% of our series. The factors related to radiation necrosis were maximum thermal neutron fluence on the brain surface and vascular dose. Therefore, we decided the maximum thermal neutron fluence on the surface of the brain below 2.0×10^3 n/cm^2. Vascular dose in the brain tissue near the surface of the brain must be controlled below 15 Gy. To improve the neutron penetration, we decided the introduction of mixed beams, which contain thermal neutron and epithermal neutron beams. KUR was reconstructed in 1997 and developed to be available mixed beams. Following the shutdown of the JRR-2, JRR-4 was renewed for medical use in 1998. Both reactors have capacity to yield thermal neutron beam, epithermal neutron beam and mixed beams. We compared the neutron fluences at the target point and on the surface of the cavity between the case treated by thermal neutron and case by mixed beam (Figure 3). Thanks to the cavity, neutron penetration was improved ca. 30% even if irradiated by thermal neutron. However, the case treated by mixed beam, the neutron fluence became 3 times of that studied by thermal neutron beam. Therefore, we decided to make new protocol.

Proposed protocol for malignant brain tumor in Japan 1998.

- Patients with glioma grade 3–4
- Less than 70 years of age
- No serious systemic disease
- Good general condition (KPS >70)

Procedures

- Neutron irradiation will be performed at KUR with mixed Thermal Neutron + BSH or at JRR-4 in Japan Atomic Energy Res. Inst. with Thermal Neutron (mode 2) + BSH
- 1–4 weeks before: Debulking surgery for histological diagnosis and mass reduction-Drip infusion of BSH (60 mg/Kg BW) through an intravenous route at 15 hours before irradiation
- One to two hours before: Craniotomy and tumor tissue sampling for analysis of boron concentration in tumor
- Neutron irradiation using mixed beam of thermal neutron and epithermal neutron beam
- Radiation time is decided according to the data of boron concentration in blood or tumor tissue and neutron flux at the target point and surface of the brain
- Follow up MRI or CT scan was studied every 3 month

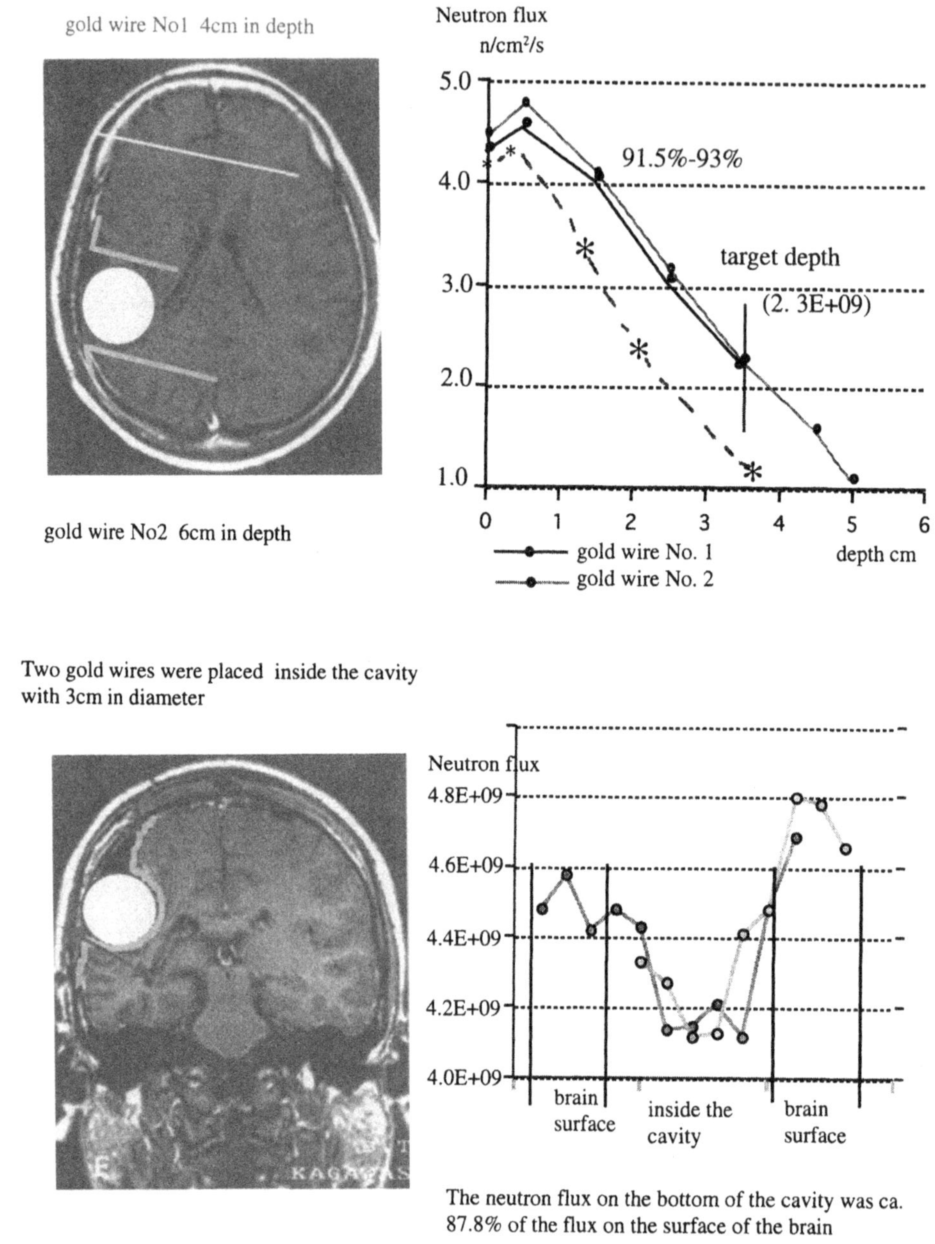

Figure 3. Neutron flux measured on the surface of the brain and cavity.

Radiation dose

- Minimum Tumor Volume Dose: 18 Gy*—
- Target Volume Dose: 15 Gy*
- Maximum Vascular Dose <10 Gy

---

* Physical dose of boron n-alpha reaction.

To improve the clinical results of BNCT, we need more detailed radiation dose in the tumor tissue and normal brain. We are developing computational dosimetry and treatment planning system for BNCT at JAERI. The detail of the study will be described elsewhere (Kumada *et al.* in this proceeding).

ACKNOWLEDGMENTS

It was hard lonely work, being the solitary group in the world continuing BNCT for more than 26 years. Hiroshi Hatanaka had been a great leader of the group. We missed him due to hypertensive intracerebral hemorrhage in 1994. It is however our great pleasure that a new clinical trial was re-started in USA and will be started in Europe.

This work has been continuously supported since 1969 up to 1992 by the Grants-In-Aid from Ministry of Education, Culture and Science, and from 1992 through 1998 by Grants-In-Aid from the Ministry of Health and Welfare of Japan.

Boron-10 compounds have been supplied by the Shionogi Research Laboratories since 1967. This research was carried out as a research program of the Musashi Institute of Technology, Research Reactor Institute of Kyoto University and Japan Atomic Energy Research Institute (JRR-2).

REFERENCES

1. H. Hatanaka: Boron-neutron capture therapy for tumors. In: Karim, ABM. & Laws, Jr. ER. (eds) Glioma, Springer-Verlag pp. 233–249 (1991).
2. Y. Nakagawa, and H. Hatanaka: Boron Neutron Capture Therapy—Clinical Brain Tumor Studies. Neuro Oncology (33), pp. 105–115 (1997).
3. H. Hatanaka, and Y. Nakagawa: Int. J. of Radiation Oncology Biology, Physics pp. 1061–1066 (1994).
4. T. Kageji, and Y. Nakagawa: Pharmacokinetics and boron uptake of BSH ($Na_2B_{12}H_{11}SH$) in patients with intracranial tumors. J. Neuro Oncology (33), pp. 117–130 (1997).
5. Y. Nakagawa: Recent study of Boron neutron capture therapy in patients with malignant brain tumor. Cancer Neutron Capture Therapy ed. by Yutaka Mishima, Plenum pp. 717–724 (1996).
6. Y. Nakagawa: What were important factors in patients treated by BNCT in Japan? Advances in Neutron Capture Therapy Volume 1, Medicine and Physics ed. by B. Larsson, Elsvier pp. 65–70 (1997).

STATUS REPORT ON THE EUROPEAN CLINICAL TRIAL OF BNCT AT PETTEN (EORTC PROTOCOL 11961)

W. Sauerwein[1], R. Moss[2], K. Hideghéty[1], F. Stecher-Rasmussen[3], M. De Vries[4], H.-J. Reulen[5], C. Götz[5], P. Paquis[6], F. Grochulla[7], K. Haselsberge[8], J. Wolbers[9], J. Rassow[1], J. P. Pignol[10], P. Watkins[2], C. Vroegindeweij[2], K. Ravensberg[3], S. Garbe[1], O. D. Wiestler[11], B. Turowski[12], F. Zanella[12], D. Touw[9], A. Siefert[5], R. Huiskamp[3], H. Fankhauser[13], and D. Gabel[14]

[1]Universität GHS Essen (D)
[2]European Commission Joint Research Centre Petten (NL)
[3]Netherlands Energy Research Foundation ECN Petten (NL)
[4]EORTC New Drug Development Office Amsterdam (NL)
[5]Klinikum Großhadern, Munich (D)
[6]Hôpital Pasteur CHU Nice (F)
[7]Zentralkrankenhaus Bremen (D)
[8]Karl-Franzens-Universität Graz (AU)
[9]Vrije Universiteit, Amsterdam (NL)
[10]CHR Mulhouse (F)
[11]Universität Bonn (D)
[12]Universität Frankfurt (D)
[13]CHU Vaudois Lausanne (CH)
[14]Universität Bremen

INTRODUCTION

In October 1997, after a preparatory phase of 8 years, the first patient in Europe was treated by BNCT at the High Flux Reactor (HFR) of the European Commission in Petten, The Netherlands.[1] The preparation for the first clinical BNCT trial in Europe included comprehensive scientific experimental and human investigations with the boron compound borocaptate sodium (BSH).[2,3] A multitude of open legal and administrative questions had to be overcome. Furthermore a complex structure had to be organized and the procedures had to be defined how to perform a multi-national, multi-institutional

Frontiers in Neutron Capture Therapy, edited by Hawthorne *et al.*
Kluwer Academic / Plenum Publishers, New York, 2001.

study using a nuclear reactor distant from hospital for patient treatment performed by an international team on a multidisciplinary base.[4] The international standards of clinical research as well as the accepted rules and definitions of conventional radiotherapy were applied, involving independent experts in order to assure the quality of the study and the quality of the performance of patient treatment. The goal was to execute the project to a very high standard as would be expected in the leading radiotherapy departments in Europe.[5]

TRIAL STRATEGY AND TRIAL DESIGN

A new drug without direct anti-cancer effect and a complex, new radiation modality are being investigated. The trial[1] is a phase I toxicity detection study, intended to establish a safe and reliable dose level for using BNCT. Phase I trials usually focus on determination of the qualitative (identifying the organ systems involved) and quantitative toxicity profile of a new *drug* agent in humans. In the case of BNCT, a binary treatment modality *both* components of the treatment, the *drug* and the *irradiation*, are acting alone and together. Therefore, *both* need to be tested together in a phase I scenario.[6] Phase II trials, in contrast, focus more on the anti-cancer activity of the drug (or treatment).

This trial investigates the toxic effects of BNCT on healthy tissue involving patients, suffering from glioblastoma multiforme, an incurable brain malignancy. The boron compound borocaptate sodium (BSH) is applied. For the first time anywhere in BNCT, a fractionated radiation scheme (the radiation and the drug are applied to the patient in several sessions) is used in this trial.

The maximum tolerated radiation dose and the dose limiting toxicity deriving from the irradiation itself or from the drug will be determined in defined conditions. Regarding the dose limiting toxicity, the local radiation toxicity due to BNCT of the irradiated region of the patient's head and the systemic toxicity due to BSH alone are to be detected. The radiation toxicity is recorded and reported as early radiation toxicity if it occurs within 90 days after the end of BNCT, and as late radiation toxicity if it occurs after 90 days after the end of BNCT. Patient survival is recorded as well as quality of life. These observations will influence the decision on whether further patients can be treated at a given dose level, and thus may lead to phase II trials.

Cohorts of 10 patients with glioblastoma undergo the following step-wise procedures with BNCT: Surgical removal of the tumor accompanied by a BSH uptake study; radiotherapy with BNCT instead of conventional radiotherapy for the tumor; evaluation of the toxicity. Patients are selected for whom no benefit from conventional treatment can be achieved, but who can be observed long enough to detect any adverse effects on brain tissue due to BNCT, or any unexpected high benefit in survival.

Eligibility Criteria

In this phase I trial, only patients suffering from glioblastoma multiforme with a good performance status (Karnofski ≥70) are eligible. They must be in between 50 and 70 years old. The tumor must be macroscopically removed by surgery without too much damage to the patient. The median survival for patients older the 50 years after optimal conventional treatment is about eight months, as indicated by multiple, large international studies. Such patients typically die from a recurrent tumor localized in the brain.

Surgery with BSH Uptake Study

Approximately 18 hours prior to operation, 100 mg/kg body weight BSH (not enriched) is infused.

Standard craniotomy for gross total resection is performed, in order to:

a. confirm the diagnosis of glioblastoma by the reference pathology center
b. debulk the tumor mass, offering the patient the benefit of a proven treatment
c. measure boron concentration in tumor, healthy tissues and in blood.

Radiation Therapy

The irradiation of the patient takes place at the 45-MW High-Flux Reactor (HFR) of the Joint Research Center of the European Commission, in Petten, The Netherlands. A medical wing and radiation room have been built there for that purpose.[7] The treatment will be performed in four fractions on four consecutive days.

Reasons for fractionation are:

a. The delivered total dose can be adjusted more closely to the prescribed dose (unwanted over- and underdosage in one fraction, due to unanticipated variation of ^{10}B concentration during the treatment can be corrected in the subsequent fractions).
b. There is a retargeting of BSH to tumor following repeated BSH administration.
c. Results of the dog experiments suggest a sparing of healthy tissue with fractionation.

A computer code that has been developed at INEEL[8] based on Monte Carlo calculations is used for treatment planning.

The starting dose was set at 80% of the dose which in a tissue tolerance study in dogs with a single fraction gave rise to a 50% incidence of neurological symptoms

Boron Compound Administration

For the first fraction BSH (>95% enrichment in ^{10}B) is administered at a concentration of 100 mg/kg body weight. The amounts to be administered at the second, third, and fourth fraction, is such that an average concentration of 30 ppm ^{10}B in blood is obtained for all fractions. The time between BSH infusion and irradiation is in a range of 12–18 hours.

Radiation Dose Escalation

The dose is to be increased by 10% over that of the previous cohort, at 6 months after the treatment of the last patient of that cohort, if no treatment related serious adverse events occur. If during this period at least two of the already irradiated patients show serious adverse event clearly arising from the treatment tested, the trial is closed.

PATIENT RECRUITMENT AND PERFORMANCE OF THE TREATMENT

The selection of patients takes place at six hospitals throughout Europe: in Graz, Austria; Amsterdam, The Netherlands; Bremen and Munich, Germany; Nice, France;

and Lausanne, Switzerland.[8] Patient registration began on July 25, 1997. Patients are entered into the study about four to five weeks before irradiation. A patient with a presumed glioblastoma enters surgery at one of the referral centers. The surgeons remove the entire tumor that is visible. After surgery, MRI images the result. The images are evaluated by the Radiodiagnostic Reference Center at the University of Frankfurt. If less the 70 percent of the initial tumor were removed, the patient is not eligible for BNCT. Having satisfied strict criteria, including the histological confirmation of glioblastoma multiforme, the patient, about four to five weeks following surgery, is sent to the Academic Hospital of the Free University in Amsterdam. On the first day of the planned week of treatment (Monday) the patient arrives at the Petten reactor for an orientation discussion and so that the radiotherapist and the treatment planners can precisely position the patient on the therapy table in the radiation room. This position has been calculated prior to patient's arrival using computerized tomography (CT) scans and special fixation devices ("masks") made after surgery by the radiotherapists of the referral hospitals. The positioning derives from the treatment plan, which is calculated at Petten. The treatment plan derives from the CT scans made from the top of the head to the chin. From these scans, at Petten a 3-D computer model is produced of the patient's head and brain with the tumor site. With this model, dose distribution can be calculated for different neutron beam directions. The radiotherapist then chooses the appropriate beam direction.

That same evening, back in Amsterdam, the patient receives the first BSH infusion. The morning after, on Tuesday, the patient is brought to Petten, and when the boron concentration in the blood is at the required level (30 ppm), radiation begins. The actions of the preceding 24 hours are repeated on each of the next three days. The patient therefore receives four doses of radiation for about 20 minutes each. After the radiation treatment, the patient returns to the referral center, where an elaborate follow-up schedule begins.[9]

ACTUAL STATUS OF THE TRIAL AND INTENDED CHANGES OF THE PROTOCOL

In August 1998 the last patient of the first cohort was treated. The observation period will end in March 1999. The first five patients were treated with only one beam, applying the maximal dose to the operated area ("tumor bed"). This setup was chosen to give maximal safety to the patients, during the period, in which the handling of the procedures could not be performed on a routine base. The next five patients were treated by two irradiation fields.[10] The maximal dose was applied to the operated area as well as to the "healthy" brain. This arrangement was developed to satisfy the following needs:

- The trial intends to establish a relationship between the observed effects and the delivered dose. If only the operated area would be irradiated, effects may be due to the operation alone or to the combination of surgery and radiotherapy. The maximal tolerated dose established under such circumstances would be useless for further phase II trials.
- Glioblastoma multiforme is a disease not localized exclusively in the area of the gross tumor. Tumor cells are spread throughout the brain. A curative approach in glioblastoma treatment needs to define the whole brain as target volume. The maximal tolerated dose investigated by this study needs therefore to be established for larger volumes.
- At the same time the potential benefit for the patients increases.

After the experiences made with the first ten patients, some changes of the protocol are intended: The actual upper age limit (70 years) will be suppressed, the performance status is a better criteria to select the appropriated patients. More important change will be the abolition of the pharmacokinetic study during surgery. The study was done to acquire individual data in order to predict the BSH delivery for the irradiation. It turned out, that the data obtained by these investigations were not necessary to perform the radiotherapy. The application of BSH on a one to one day base could be done in a way to reach in all patients a mean blood boron concentration of 30 ppm. By suppressing the BSH application prior to surgery, it will be possible to include patients after surgery (avoiding the difficult informed consent prior to a definitive diagnosis) and at the same time the hypothetical toxicity due to BSH can be reduced.

ACKNOWLEDGMENTS

The project is financially supported by the European Commission in the form of a Shared Cost Action from the Biomedical and Health Research Program (BIOMED II) of the EC Directorate-General XII, contract no. BMH4-CT96-0325, the participation of Medical Physicists became possible due to financial support of the Swiss Foundation "Neurochirugie 2001".

REFERENCES

1. European Organization for Research and Treatment of Cancer (EORTC) BNCT-Study Group. Postoperative Treatment of Glioblastoma with BNCT at the Petten Irradiation Facility. EORTC-Protocol 11961, EORTC, Brussels, 1997.
2. D. Gabel, D. Preusse, D. Haritz, F. Grochulla, K. Haselsberger, H. Fankhauser, C. Ceberg, H.-D. Peters, and U. Klotz, "Pharmacokinetics of $Na_2B_{12}H_{11}SH$ (BSH) in patients with malignant brain tumors as prerequisite for a phase I clinical trial of boron neutron capture," Acta Neurochirurgica (Wien), 139:606–612, 1997.
3. K. Philipp, R. Huiskamp, J. Casado, and R.L. Moss, "Normal tissue tolerance and fractionation effects in dogs of boron neutron capture irradiation with borocaptate sodium (BSH)." In: Y. Mishima (Ed.), Cancer neutron capture therapy, Plenum Press, New York and London, pp. 809–814, 1996.
4. W. Sauerwein. The clinical project at HFR Petten—a status report. In: B. Larsson, J. Crawford, and R. Weinreich (Eds.) Advances in Neutron Capture Therapy. Volume I, Medicine and Physics. Elsevier Science B.V. Amsterdam 77–82, 1997.
5. W. Sauerwein, R.L. Moss, J. Rassow, F. Stecher-Rasmussen, K. Hideghéty, S. Garbe, J.G. Wolbers, D. Gabel, P. Watkins, C. Vroegindeweji, and H. Sack. Quality Management at the European BNCT Center in Petten. In: M.F. Hawthorne, K. Shelly, R.W. Wiersema, (Eds.), "Frontiers in Neutron Capture Therapy," Kluwer Academic/Plenum Publishers, New York, 2001, pp. 105–111.
6. W. Sauerwein, K. Hideghéty, M. de Vries, J. Wanders, D. Gabel, H. Fankhauser, and R. Huiskamp. Conducting phase I clinical trial in binary treatment modality methodical questions for the evaluation of boron neutron capture therapy, Annals of Oncology Vol., 9, Suppl. 2. 1998.
7. R.L. Moss, F. Stecher-Rasmussen, K. Ravensberg, G. Constantine, and P. Watkins. Design, Construction, and Installation of an Epithermal Neutron Beam for BNCT at the High Flux Reactor at Petten. In: B.J. Allen, D.E. Moore, and B.V. Harrington (Eds.) Progress in Neutron Capture Therapy for Cancer, Plenum Press, New York, pp. 63–66, 1992.
8. D. Gabel, and W. Sauerwein. Approaching Clinical Trials of Boron Neutron Capture Therapy in Europe. In: H.D. Kogelnik, (Ed.), Progress in Radio-Oncology. Monduzzi Editore, Bologna, pp. 315–319, 1995.
9. W. Sauerwein, K. Hideghéty, D. Gabel, and R. Moss. European clinical trials of boron neutron capture therapy for glioblastoma. Nuclear News 41:54–56.

10. K. Hideghéty, W. Sauerwein, M. de Vries, F. Grochulla, C. Goetz, K. Haselsberger, P. Pacquis, J. Heimans, J. Wolbers, R. Moss, R. Huiskamp, F. Stecher-Rasmussen, J. Rassow, S. Garbe, H. Fankhauser, and D. Gabel. Report on the first patient group of the European phase I trial (EORTC Protocol 11961) at the High Flux Reactor Petten. In: M.F. Hawthorne, K. Shelly, R.W. Wiersema, (Eds.), "Frontiers in Neutron Capture Therapy," Kluwer Academic/Plenum Publishers, New York, 2001, pp. 105–111.

8

CASE STUDIES OF PATIENTS WITH GLIOBLASTOMA MULTIFORME TREATED WITH BORON NEUTRON CAPTURE THERAPY AT THE BROOKHAVEN MEDICAL RESEARCH REACTOR

E. H. Elowitz[1], J. Capala[2], J. A. Coderre[2], D. S. Cohen[1], A. Z. Diaz[2], D. D. Joel[2], and A. D. Chanana[2]

[1]Department of Neurosurgery
Beth Israel Medical Center
New York, New York
[2]Medical Department
Brookhaven National Laboratory
Upton, New York

1. INTRODUCTION

Over the past decade, research at Brookhaven National Laboratory (BNL) and other places in the world has led to several significant advances in boron neutron capture therapy (BNCT) technology. These include the development of epithermal neutron beams[1] as well as more tumor-specific boron compounds.[2] In 1994, a Phase I/II clinical trial of BNCT using intravenous p-boronophenylalanine-fructose (BPA-F) and epithermal neutrons was initiated to determine both the safety and possible efficacy of BNCT for glioblastoma multiforme.[3] As with any new therapy, a practical clinical experience with BNCT has been gained over time. Specific management issues seen in this trial have involved the use of steroid medication, early neurological deficits, seizure management, and BNCT-related necrosis.

2. MATERIALS AND METHODS

The entry criteria into the BNCT study were: GBM patients with tumors that were unilateral and within 6.0 cm from the scalp surface; age 18 or older; Karnofsky

Frontiers in Neutron Capture Therapy, edited by Hawthorne *et al.*
Kluwer Academic / Plenum Publishers, New York, 2001.

performance status ≥70; no prior irradiation or chemotherapy; and a prior craniotomy and debulking of tumor. BNCT was performed at BNL using the Brookhaven Medical Research Reactor at a minimum of three weeks following surgery. Appropriate informed consent was obtained. Treatment planning was performed with the software developed at the Idaho National Engineering Laboratory; contrast-enhanced MRI scan was used to determine the tumor volume (area of contrast enhancement) and target volume (tumor volume plus a 2-cm surrounding margin). Following BNCT, the patients were observed in hospital for a minimum of 48 hours and then followed with serial MRI or CT scans and physical examinations.

3. RESULTS AND DISCUSSION

Forty patients (28 males; 12 females) with GBM have been treated on the BNCT protocol. The median age is 56 (39–79 years) with a median Karnofsky performance status of 80 (70–100). The overall median survival has been approximately 13 months. Time to disease progression based on MRI scan was 31 weeks. The clinical course of the patients, both immediately following BNCT and in the months after, was closely followed.

3.1. Seizures

All patients with brain tumors are at an increased risk for seizures.[4] The same is true for any type of intracranial surgery. It is the general practice that most patients with glioblastoma multiforme be on anticonvulsant therapy, especially following craniotomy and resection of tumor. Even prior to BNCT, the majority of patients had been on anticonvulsant therapy. Following BNCT, eight patients had seizures ranging from one to five days post-treatment. Five of these eight patients had prior seizure histories. All patients responded to increased levels of anticonvulsant medication with no long-term adverse seizure-related complications. As a result of this experience, all patients must have therapeutic anticonvulsant levels prior to BNCT at BNL.

3.2. Early Neurologic Deterioration

Case 1

This 68-year-old man presented with an expressive aphasia and personality changes. CT scan revealed a 2.5 cm enhancing mass in his left temporal lobe with significant surrounding edema (Figure 1). The patient underwent a craniotomy and subtotal resection of tumor. The patient elected for BNCT. Just prior to BNCT, MRI scan revealed residual enhancing tumor with surrounding edema. The BNCT radiation doses were: tumor dose peak = 65.0 Gy-Eq and minimum = 41.0 Gy-Eq; target volume minimum = 25 Gy-Eq; brain dose peak = 12.6 Gy-Eq. The dose of BPA-F administered was 290 mg/kg body weight. While under observation in the hospital, 18 hours following BNCT, the patient was noted to have worsened expressive aphasia. An MRI scan revealed increased surrounding edema in the left temporoparietal region. The patient's steroids were then increased to dexamethasone 10 mg IV q4h with subsequent improvement in his aphasia. He was discharged from the hospital 72 hours following BNCT. An MRI scan three weeks post-BNCT revealed resolution of edema on steroids and the patient continued to have a clinical improvement. Two months post-BNCT, as the steroids were being weaned, MRI indicated the edema returning. Four months post-BNCT, MRI scan revealed significant

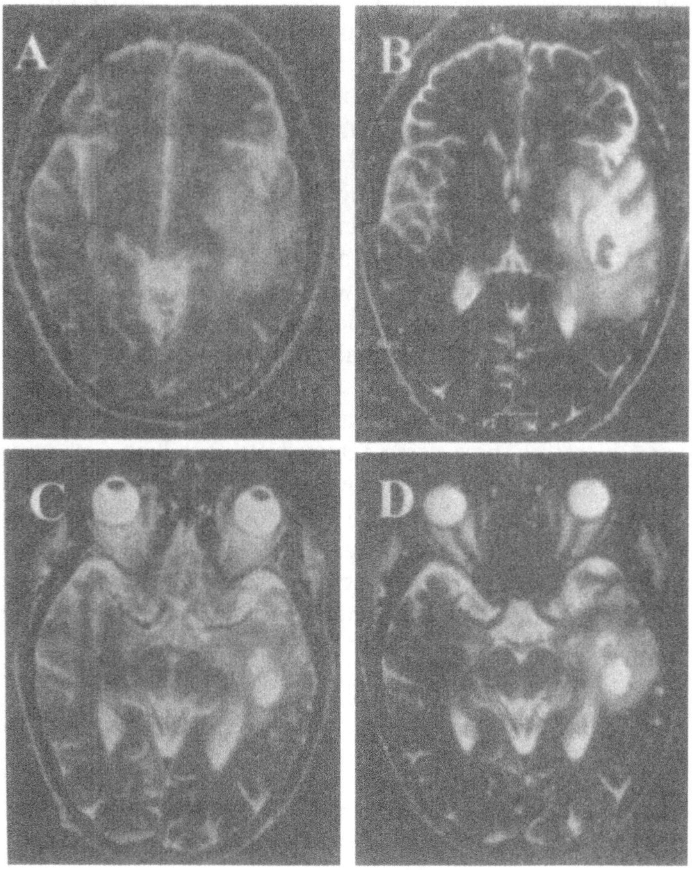

Figure 1. T2-weighted MRI scans of Case 1. A: Following surgical resection and prior to BNCT; B: Next day following BNCT with worsened peritumoral edema; C: 3 weeks post-BNCT with improvement in edema; D: 4 months post-BNCT with increased tumor size and edema.

increase in tumor volume as well as more edema. The patient was treated with external beam radiation therapy but expired seven months post-BNCT. An autopsy was performed, which revealed no evidence of radiation damage to the brain.

Case 2

This 48-year-old woman presented with a seizure, right arm paresis, and expressive aphasia. MRI scan revealed a large left parietal enhancing mass. A craniotomy and subtotal tumor resection was performed. The patient then selected BNCT for treatment. The BNCT radiation doses were: tumor dose peak = 54.0 Gy-Eq and minimum = 21.0 Gy-Eq; target volume minimum = 17 Gy-Eq; brain dose peak = 13.0 Gy-Eq. The dose of BPA-F administered was 250 mg/kg body weight. Approximately ten hours following BNCT she became lethargic with worsened aphasia and right hemiparesis. New MRI scan revealed moderate increase in edema and more mass effect on the lateral ventricle. The patient had a clinical improvement with high-dose dexamethasone, and was discharged from the hospital 72 hours following treatment. An MRI scan performed three months post-BNCT indicated increased size of tumor with distant brain involvement. No further treatment was elected, and the patient expired 5.5 months post-BNCT.

A total of three patients were noted to have mild early deterioration in neurologic function. One patient, in addition to the two case examples, had worsened symptoms but no change in the scan. Both Cases 1 and 2 had dominant hemispheric tumors with significant residual GBM. The MRI scans prior to BNCT demonstrated edema in eloquent regions; within 24 hours of BNCT both patients had worsened symptoms with increased edema seen on the imaging study. All three patients with mild early neurologic deterioration responded to increased doses of steroids.

BNCT is known to deposit a large amount of energy, in the form of alpha radiation, within tumor cells. It is no surprise, therefore, that patients with significant tumor burden at the time of BNCT may develop worsened edema and mass effect. If eloquent brain regions are involved, this edema is more likely to manifest itself symptomatically. Our experience to date, argues for maximal cytoreductive surgery prior to BNCT as a way of minimizing post-BNCT edema and mass effect.

Steroid administration has a significant role in decreasing peritumoral edema prior to and following surgery.[5] These cases demonstrate that high-dose steroids can help control edema post-BNCT. It has been the practice of BNL to pre-treat patients with steroids, generally dexamethasone, prior to BNCT, with an increased dose immediately before treatment and for 48 hours following. When adjusting the steroid dose for BNCT, it has been our observation that several factors should be taken into consideration: (1) The extent of residual tumor; (2) the extent of peritumoral edema; (3) dominant hemispheric tumors.

3.3. Necrosis Post-BNCT

Case 3

This 48-year-old man presented with confusion, dizziness, and a left hemianopsia. MRI scan revealed a large right occipital enhancing mass with significant edema, mass effect and midline shift (Figure 2). The patient underwent a craniotomy with resection of tumor. The patient was later treated with BNCT. The BNCT radiation doses were: tumor dose peak = 60.7 Gy-Eq and minimum = 19.5 Gy-Eq; target volume minimum = 8.5 Gy-Eq; brain dose peak = 13.1 Gy-Eq. The dose of BPA-F administered was 250 mg/kg body weight. The patient remained well following BNCT. Follow-up MRI scans at both six months and 12 months revealed decreased size of the tumor cavity, enhancement, and edema. MRI scans remained stable until 29 months post-BNCT, when a small area of enhancement with edema was demonstrated. Stereotactic biopsy revealed evidence of necrosis only with no tumor cells identified despite multiple biopsy samples. The patient remains clinically well and will be closely followed.

Case 4

This 63-year-old woman presented with a mild right hemiparesis. MRI scan revealed a large left parietal enhancing mass with surrounding edema. The patient underwent a craniotomy and resection of tumor and was later treated with BNCT. The BNCT radiation doses were: tumor dose peak = 68.0 Gy-Eq and minimum = 40.0 Gy-Eq; target volume minimum = 22 Gy-Eq; brain dose peak = 12.5 Gy-Eq. The dose of BPA-F administered was 250 mg/kg body weight. The patient remained clinically well with stable MRI scans until ten months post-BNCT, when a new nodule of enhancement was identified. The patient underwent a craniotomy and resection of this area of enhancement. Pathology revealed only necrosis with no tumor identified. Despite a stable clinical exam, MRI scan performed 14 months post-BNCT revealed a new area of enhancement more lateral

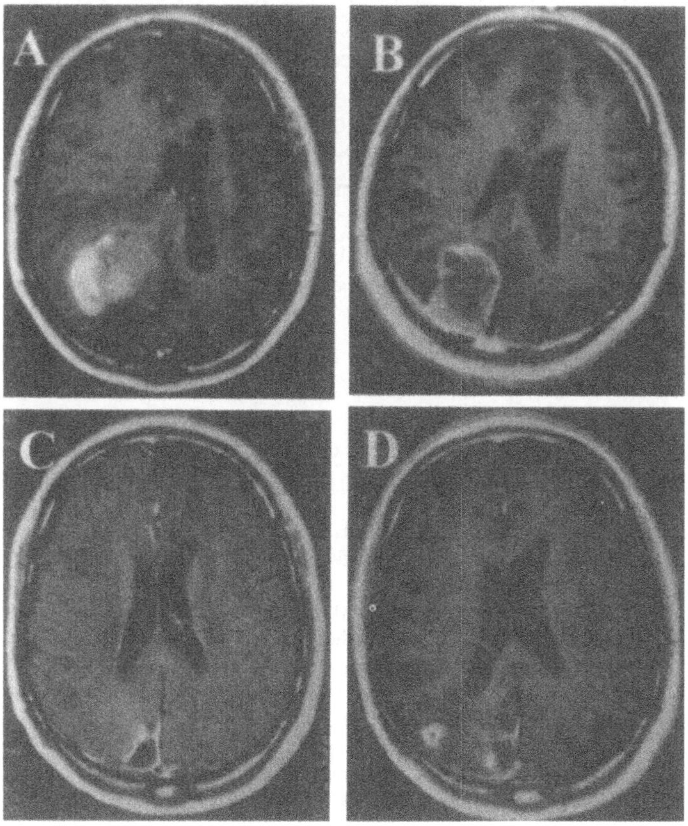

Figure 2. Contrast-enhanced T1-weighted MRI scans of Case 3. A: At presentation; B: Following surgical resecion and prior to BNCT; C: Stable at 1 year post-BNCT; D: 29 months post-BNCT with new area of enhancement

to the prior area of necrosis. A Thallium-SPECT scan was suggestive of tumor. The patient underwent a craniotomy and resection of this area. The pathology again revealed only necrosis with no evidence of tumor cells. At 20 months post-BNCT, the patient developed new right-sided sensory disturbance and MRI scan revealed increased enhancement with surrounding edema. An MRI spectroscopy was suggestive of necrosis. The patient was started on dexamethasone 16 mg/day with clinical improvement in her sensory complaints. Six weeks following the start of steroid treatment, an MRI scan revealed significant improvement in the enhancement as well as the edema. At present, the patient is being weaned from steroids and is alive 22 months post-BNCT with an excellent performance status.

Conventional external beam radiation is known to produce necrosis in approximately 5% of GBM patients.[6] Although the etiology of radiation necrosis is still uncertain, many authors have suggested that a vascular endothelial damage may mediate this necrotic response.[7] In this study, we observed post-BNCT necrosis in 2 of the 40 patients (5%). Post-BNCT necrosis has also been observed in the Japanese clinical BNCT trials.[8] In the current study, necrosis was seen in both patients within the area of the target volume, where the doses to normal brain (13.1 and 12.5 Gy-Eq) were at the calculated limit of tolerance to a single fraction of radiation. The patient in Case 3 is considered a long-term survivor and had no change in his MRI scan until the necrosis appeared 29 months post-BNCT. On the other hand, Case 4 had the necrosis manifest at ten months post-BNCT. Case 4 also had necrosis seen at 14 months and then presumably at 20 months post-BNCT. The last episode of increased contrast has been treated with oral steroids with an excellent clinical and radiographic response. The etiology of post-BNCT necrosis is unclear at present, but, as with conventional radiotherapy, may be related to an endothelial response. One patient responded well to moderate dose steroid therapy. It remains to be seen whether this will cause a long-term control of the necrosis. The prognosis for both of these patients with post-BNCT necrosis is also uncertain. They both will be followed very closely with clinical examination and radiographic studies.

REFERENCES

1. H.B. Liu, R.M. Brugger, D.D. Greenberg, D.C. Rorer, J.P. Hu, and H.M. Hauptman, Enhancement of the epithermal neutron bean used for boron neutron capture therapy, *Int J Radiat Oncol Biol Phys*, 28:1149–1156, 1994.
2. E.H. Elowitz, R.M. Bergland, J.A. Coderre, D.D. Joel, M. Chadha, and A.D. Chanana, Biodistribution of p-boronophenylalanine in patients with glioblastoma multiforme for use in boron neutron capture therapy, *Neurosurgery*, 42:463–469, 1998.
3. J.A. Coderre, E.H. Elowitz, M. Chadha, R.M. Bergland, J. Capala, D.D. Joel, H.B. Liu, D.N. Slatkin, and A.D. Chanana, Boron neutron capture therapy for glioblastoma multiforme using p-boronophenylalanine and epithermal neutrons: Trial design and early clinical results, *J Neurooncol*, 33:141–152, 1997.
4. M.S. Berger, and E. Keles, Epilepsy associated with brain tumors, *in*: "Brain Tumors," A.H. Kaye and E.R. Laws, Jr., ed., Churchill Livingstone, Edinburgh, pp. 239–246, 1995.
5. M. Bilsky, and J.B. Posner, Intensive and postoperative care of intracranial tumors, *in*: "Neurological and Neurosurgical Intensive Care," Raven Press, New York, pp. 309–329, 1993.
6. J.E. Marks, R.J. Baglan, S.C. Prassad, and W.F. Blank, Cerebral radio necrosis: incidence and risk in relation to dose, time, fractionation and volume, *Int J Radiat Oncol Biol Phys*, 7:243–252, 1981.
7. P.E. Valk, and W.P. Dillon, Radiation injury to the brain, *Am J Neuroradiol*, 12:45–62, 1991.
8. Y. Nakagawa, and H. Hatanaka, Boron neutron capture therapy: clinical brain tumor studies, *J Neurooncol*, 33:105–115, 1997.

9

PATTERNS OF TUMOR PROGRESSION FOLLOWING BNCT OF GLIOBLASTOMA MULTIFORME

Aidnag Z. Diaz[1,2], Arjun D. Chanana[1], Jacek Capala[1], and Ruimei Ma[1]

[1]Medical Department
Brookhaven National Laboratory
Upton, New York
[2]Department of Radiation Oncology
Montefiore Medical Center
Bronx, New York

INTRODUCTION

The recurrence patterns after conventional adjuvant treatment of glioblastoma multiforme has been analyzed in the past. The most consistent pattern seen in these studies is the predominance of local recurrence.[1,2,3] Hochberg and Pruitt were the first investigators to discover that the gross and microscopic tumors were within 2 cm of the contrast-enhancing tumor margin on pre-terminal CT scans in 83% of the patients.[3] This became the basis for target volume definition in radiotherapy. In the first several BNCT protocols, this target definition was adopted. However, there has been considerable consternation over the significance of peritumoral edema when planning radiation fields. Biopsy and autopsy studies from the Mayo Clinic and Duke University have shown infiltrating cells in edematous regions.[1,2] If so, peritumoral hypodense regions may be at risk for tumor recurrence and should be included in the treatment volume. However, this observation is still in dispute because the malignant nature of the infiltrating cells has not been proven. In the studies by Wallner *et al.* from Memorial Sloan Kettering Cancer Center, no tendency for tumor recurrence in edematous areas was shown.[4] The present paper reports on the recurrence patterns seen on our BNCT trial, in particular, with relation to regions of peritumoral edema.

PATIENTS AND METHODS

BNCT was given to 41 patients with primary GBM. Because histologic tumor necrosis is a major prognostic variable in malignant supratentorial, astrocytic glioma, we

Frontiers in Neutron Capture Therapy, edited by Hawthorne *et al.*
Kluwer Academic / Plenum Publishers, New York, 2001.

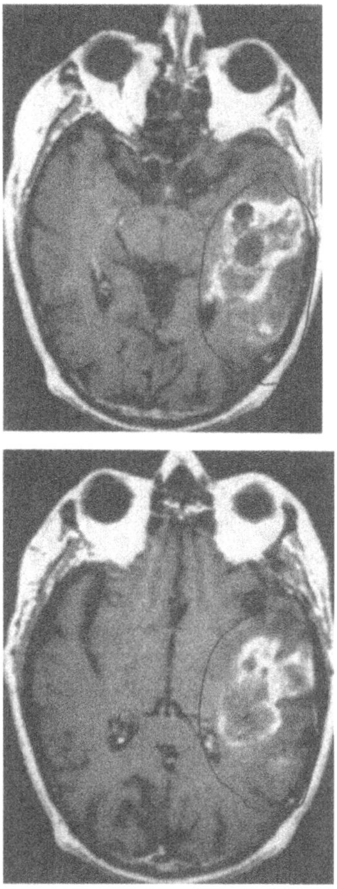

Figure 1. Predicted area at risk for recurrence, based on area of pre-operative peritumoral edema delineated by a white line, superimposed on actual recurrence.

employed the criteria of Nelson *et al.*[5] for the diagnosis of GBM. The following entry criteria were used: supratentorial, unilateral, unifocal and debulked GBM; deepest part of the contrast-enhancing tumor margin on the post-debulking craniotomy brain scan no more than ≈6.0 cm from the scalp surface; WBC ≥2,500/mm^3; platelets ≥75,000/mm^3; creatinine ≤1.6 mg%; no evidence of congestive heart failure; Karnofsky performance status score ≥70; no history of phenylketonuria; no prior radiation therapy to the brain; no prior immunotherapy and/or systemic or intrathecal chemotherapy; and age ≥18 years. Informed consent was obtained from all patients.

The characteristics of 40 evaluable patients and their tumors are listed in Table 1. One non-evaluable patient developed symptoms and signs of craniospinal metastases by day 8 following BNCT and died of metastatic disease a few weeks following the completion of BNCT and was lost to follow-up.

Time to tumor progression and overall survival were defined as times from the date of histopathological diagnosis to the dates of GBM progression and death, respectively. Disease progression was defined as post-BNCT increase in the MRI contrast-enhancing tumor volume, the appearance of new MRI contrast-enhancing lesion/s in the brain and/or progressive worsening of neurologic symptoms and signs attributable to tumor growth. Metabolic scans (mainly PET; ^{18}F-fluorodeoxyglucose) were done in 14 patients when GBM progression was suspected. Histopathological diagnosis of the tumor

Table 1. Patient and glioblastoma multiforme characteristics[a]

Patient #	Age (yr)	Sex	KPS score	RPA class	GBM location	GBM Diam (cm)[b]	GBM Vol. (cm^3)[c]	Minimum BNCT dose (Gy-Eq) Tumor volume[d]	Target volume[e]	Survival from diagnosis (months)[f]	Dead
09-17-63R	50	F	90	5	left temporoparietal	3.50	18	27.84	16.18	14.70	Yes
09-18-59R	67	F	80	5	right parietal	5.30	29	31.90	10.43	14.70	Yes
09-18-90R	61	M	90	4	right parietal	3.90	18	26.80	13.54	16.71	Yes
09-19-03R	46	F	70	4	right paietooccipital	7.40	70	20.43	7.44	8.19	Yes
09-19-05R	68	M	90	5	right frontoparietal	3.90	19	24.27	8.50	9.51	Yes
09-19-14R	56	M	70	5	right frontoparietal	4.90	23	18.85	8.54	17.60	Yes
09-19-16R	75	F	70	5	right frontal	6.10	56	19.10	9.38	12.27	Yes
09-19-18R	69	F	70	5	left temporal	4.90	18	22.97	11.44	10.63	Yes
09-19-25R	46	F	90	3	right parietal	4.10	21	26.48	12.87	24.08	Yes
09-19-71R	46	F	100	3	left temporal	5.90	15	34.97	17.24	13.55	Yes
09-19-78R	57	M	80	5	left occipital	4.30	20	21.54	9.42	20.82	Yes
09-20-03R	50	F	80	5	left frontal	6.20	22	21.73	10.48	20.79	Yes
09-20-09R	64	F	80	5	right parietal	5.60	21	25.19	13.04	11.15	Yes
09-21-15R	48	M	80	4	right occipitoparietal	5.50	37	19.48	8.50	27.17	No
09-20-50R	57	F	70	5	right temporoparietal	7.20	70	26.59	18.00	10.53	Yes
09-20-53R	61	M	80	5	right parietooccipital	5.80	24	40.00	21.00	7.96	Yes
09-20-62R	53	M	90	5	left temporal	7.40	37	18.00	17.00	12.60	Yes
09-20-76R	43	M	70	4	left temporal	8.40	31	21.00	17.00	12.76	Yes
09-20-80R	48	F	70	4	left frontoparietal	8.70	83	28.00	17.00	6.41	Yes
09-20-98R	41	M	80	4	left frontal	3.00	4.9	35.00	19.00	13.09	Yes
09-21-04R	57	M	80	5	left occipital	5.00	29.6	32.00	18.00	21.81	Yes
09-21-16R	49	M	70	4	left parietal	5.40	18	43.00	22.00	6.05	Yes
09-21-34R	63	F	90	5	left parietooccipital	4.00	13	40.00	22.00	18.65	No
09-21-38R	57	M	90	5	right occipital	6.00	18	42.00	20.00	13.78	Yes
09-21-52R	47	F	80	4	left temporal	7.90	55	20.00	19.00	16.91	Yes
09-21-55R	39	M	70	4	right frontal	5.80	36	20.00	15.00	17.63	No
09-21-62R	59	M	90	5	right frontal	4.20	16	41.00	23.00	8.06	Yes
09-21-72R	40	M	70	4	right occipitoparietal	8.00	74	33.00	17.00	12.83	Yes
09-21-78R	53	M	80	5	right frontal	5.40	25	34.00	18.00	10.33	Yes
09-22-27R	54	M	90	5	left frontoparietal	2.50	8	44.00	24.00	12.20	No
09-22-37R	41	M	80	4	left temporal	2.00	2	50.00	35.00	12.34	No
09-22-40R	71	F	90	5	right parietal	4.70	18	51.00	29.00	10.00	Yes
09-22-99R	56	M	90	5	left temporal	4.80	11	36.00	25.00	11.68	Yes
09-23-32R	68	M	90	5	left temporal	5.90	18	41.00	25.00	8.59	Yes
09-23-46R	68	M	70	5	left parietal	3.00	9	54.60	29.00	9.67	Yes
09-23-57R	41	F	70	4	left temporoparietal	7.50	41	38.58	24.05	5.99	No
09-23-80R	79	M	70	5	left occipital	6.50	40	48.07	31.25	3.88	No
09-24-60R	40	M	70	4	right temporoparietal	7.20	59	29.71	16.53	0.99	No
09-24-76R	58	M	70	5	right temporal	9.20	61	27.10	21.80	0.69	No
09-25-37R	75	F	100	5	right temporoparietal	6.70	22	35.19	24.01	1.12	No

[a]KPS, Kanofsky performance status; RPA, Recursive partitioning analysis (19); GBM, Glioblastoma multiforme; BNCT, Boron neutron capture therapy; Gy-Eq, gray-equivalent.
[b]Maximum post-operative tumor diameter.
[c]Post-operative contrast enhancing tumor volume.
[d]Minimum radiation dose to contrast-enhancing tumor volume.
[e]Minimum radiation dose to contrast-enhancing tumor volume plus the volume of a 2 cm-thick shell enclosing the tumor volume.
[f]Survival from histopathological diagnosis.

Table 2. Pattern of recurrence in 28 patients in which preoperative imaging was available

Recurrence within postoperative tumor	8
Recurrence within preoperative tumor and edema	15
Distal recurrence (outside the preoperative tumor and edema	5

progression was obtained in those patients who were subjected to debulking of the progressive tumor. Tumor progression was scored as local or peripheral. Progression was scored as local when MRI and/or functional brain scans showed progression either in the contrast-enhancing tumor volume and/or in contiguity with the contrast-enhancing tumor volume. Other progression patterns were scored as peripheral.

RESULTS

Tumor progression has been observed in 34 of 40 patients to date. The estimated median time to tumor progression from histopathological diagnosis in all patients is 30.3 weeks. Tumor progression was confirmed in 21 patients by histopathology (14 from debulked progressive tumors and 7 from tissues obtained at autopsy). In 13 patients, the diagnosis of tumor progression was based on worsening of neurological signs and symptoms directly attributable to the tumor progression and brain scans (MRI and PET = 6; MRI and SPECT = 1; MRI = 4). In 14 patients, anatomic and metabolic brain scans (^{18}FDG-PET, n = 12; ^{201}Thallium-SPECT, n = 2) were carried out at the time of suspected tumor progression and in 9 of these patients, histopathological reports of the debulked, progressive tumor were also available. There was considerable radiation-induced tumor necrosis in all evaluated patients. Tumor progression was local in all cases (n = 34) but local progression was associated with peripheral progression in 19 cases (ipsilateral hemisphere = 13, contralateral hemisphere = 3, subtentorial = 3). We used post-operative MRI brain scans for determining eligibility and for treatment planning. Retrospective review of 20 available pre-operative MRI brain scans indicates, however, that the zone of preoperative peritumor vasogenic edema was a preferred zone for local tumor progression as seen in table 2. Figure 1 shows an example of how closely the volume of recurrence resembles the volume encompassed by the preoperative peritumoral edema.

CONCLUSIONS

Glioblastoma patients treated with boron neutron capture therapy fail locally. The patients' pattern of recurrence after a single BNCT treatment is comparable to patterns of recurrence after standard therapy. The preoperative volume of tumor plus edema provides better coverage of the region at high risk for recurrence. Based on these observations and data from Kelly *et al.*,[2] in the new dose escalation protocol we define the target volume as the larger of pre-BNCT and preoperative tumor volume plus preoperative edema plus a 2-cm-thick shell that encloses it. This is in concordance with the definition of initial target volume used in the Radiation Therapy Oncology Group protocols 9305 and 9411. This change will enable us to optimize the treatment to the area most at risk for recurrence. It will also make the volumes used to report our data consistent with the volumes used in the most recent radiooncologic literature.

REFERENCES

1. P.C. Burger, E.R. Heinz, T. Shibata, *et al.*, Topographic anatomy and CT correlations in the untreated glioblastoma multiforme, *J Neurosurg* 68:698–704, 1988.
2. P.J. Kelly, C. Daumas-Duport, D.B. Kispert, *et al.*, Imaging-based stereotactic serial biopsies in untreated intracranial glial neoplasms, *J Neurosurg* 66:865–874, 1987.
3. F.H. Hochberg, and A. Pruitt, Assumptions in the radiotherapy of glioblastoma. *Neurology* 30:907–911, 1980.
4. K.E. Wallner, J.H. Galicich, G. Krol, *et al.*, Patterns of failure following treatment for glioblastoma multiforme and anaplastic astrocytoma, *Int J Radiat Oncol Biol Phys* 16:1405–1409, 1989.
5. J.S. Nelson, Y. Tsukada, D. Schoenfeld, K. Fulling, J. Lamarche, and N. Peress, Necrosis as a prognostic criterion in malignant supratentorial, astrocytic gliomas, *Cancer* 52:550–554, 1983.

10

INTERCOMPARISON OF THREE EPITHERMAL NEUTRON-BASED CLINICAL TRIALS OF BNCT

K. Hideghéty[1], W. Sauerwein[1], P. Busse[2], O. Harling[3], R. Zamenhof[4], A. Chanana[5], and J. Coderre[5]

[1]Strahlenklinik, Universitätsklinikum Essen
Germany
[2]Joint Center for Radiation Therapy
Harvard Medical School
USA
[3]Massachusette Institute of Technology Nuclear Reactor Laboratory
USA
[4]Department of Radiology
Beth Israel Deaconess Medical Center
USA
[5]Medical Research Center
Brookhaven National Laboratory, USA

A prerequisite for any clinical trial is a good idea, which is worth testing. The research work on Boron Neutron Capture Therapy, a special radiation therapeutic modality which theoretically allows the selective irradiation of tumor cells with high linear energy transfer radiation has reached the first step of clinical testing. To test a new therapy is a long term project, involving different types of clinical trials during its development.

At present, three clinical trials are in progress on cranial BNCT at the three available epithermal neutron sources; at Brookhaven and at MIT in the USA, and at Petten, in Europe. The common goal is to correlate delivered doses to observed biological effects. BNCT is a highly complex binary system; therefore an understanding of the definition of the delivered doses, and of the clinical tools employed to measure effect in the three separately developed trials is essential for the correct interpretation of the overall results.

Brookhaven National Laboratory (BNL) Brookhaven: protocol #4 **phase I/II** Study to test feasibility and safety of single fraction BNCT using BPA[2]

Harvard University-Massachusette Institute of Technology Nuclear Reactor Laboratory (H-MIT) Boston: protocol #46,175/#5 **phase I** Study to test feasibility and safety of BNCT using BPA, normal tissue toxicity, local tumor response in GBM and CNS melanoma[1]

Frontiers in Neutron Capture Therapy, edited by Hawthorne *et al.*
Kluwer Academic / Plenum Publishers, New York, 2001.

Table 1. Study design

Site	Total patient number	Pat./ cohort	BNCT	Changing parameters	Fractionation	Boron carrier
BNL	28	14	postop.	1 *vs* 2 fields depending on the tumor depth,	single	**BPA** escalated from 250 up to 495 mg/kg
H-MIT	not defined up to MTD	3	with measurable tumor	radiation dose escalated 10%/ cohort	single; 2–3 (field/fr.)	**BPA** escalated from 250 up to 350 mg/kg
HFR	40	10	postop.	radiation dose escalated 10%/ cohort	four	**BSH** defined constant blood ^{10}B concentration over the 4 fract.

High Flux Reactor (HFR) Petten: protocol 11961 **phase I** Study to test toxicity of BNCT using BSH, to determine the Maximal Tolerated Dose (MTD) and Dose Limiting Toxicity (DLT)[3-4]

All clinical groups aim to establish the safety of BNCT for brain tumors by defining the toxicity spectrum of the treatment, but there are basic differences between them in terms of the study design, scientific approach, primary and secondary objectives, tools of evaluation and definitions. The present work is restricted to the comparison of the written study designs; the dose prescription, definitions for treatment planning, dose reporting, source quality, dosimetry, treatment conditions, e.t.c. are topics for a separate publication.

Study design of the three ongoing clinical trials is summarized in the Table 1.

Drug pharmacokinetic and tissue distribution prior to BNCT is optional at BNL and H-MIT and part of the study at HFR.

The common primary endpoint is to detect radiation toxicity. The definition of the unacceptable morbidity differs considerably in the 3 trials. In Brookhaven the severe, RTOG grade 3–4 toxicity is defined as dose limiting toxicity, while in Boston significant MRI changes, visual field deterioration, severe acute or chronic morbidity are considered as stopping events. At HFR, treatment related life threatening, incapacitating serious adverse events are defined as unacceptable systemic or radiation morbidity.

As secondary endpoints, trends in Karnofsky performance status and in local tumor control are investigated at BNL, the local (short term) response on MRI at H-MIT, and local response and survival are recorded at HFR.

The **patient selection** is a crucial point for the outcome of the trial. Usually a phase I study population involves patients with diverse tumor types. In the three ongoing trials a special tumor type is selected: histologically confirmed glioblastoma multiforme and, in addition at H-MIT, radiologically documented CNS metastasis of melanoma malignum. The patients should have good performance status (Karnofsky index ≥70) in every ongoing study. Another important prognostic factor is that the age is limited at HFR ≥50 years, which defines a less favorable group than at BNL, where the patients above 40 years can be entered; the lower age limit at H-MIT is 18 years. Any comparison with historical control groups according to Curran classification may be misleading taking into consideration the different inclusion criteria at the three sites, which define special subgroups of glioblastoma patients, presumably with different prognosis than the larger, less homogeneous historical control. At BNL the deepest part of the supratentorial, unilateral, unifocal, partially or completely resected tumor must be <6 cm from the surface at BNCT. Furthermore patients are excluded if the precondition to achieve the prescribed minimal target dose without exceeding the peak normal brain dose is not fulfilled. At H-MIT unresectable or measurable residual tumor mass is required for the secondary endpoint, namely for the tumor response. For these patients life expectancy must be at least 3 months. In Europe patients are included with single, supratentorial, lobar lesion, outside major functional areas if the reference Neuroradiology Center confirms that >70% of tumor volume could be removed. In addition to a good and fast recovery from surgery and the ability to travel to The Netherlands is required. From the point of view of concomitant diseases and pretreatment the common intention is to exclude any therapy or illness, which could have any relevance on the BNCT morbidity. However there are considerable differences in the protocols, how it is defined, allowing latitude for the decision of the physician for formulating strictly described exclusion criteria.

At present the treatment (e.g. dose delivery) is not comparable at all, not only because of the different neutron sources, different planning systems and input data, different boron carrier and different conditions of boron compound administration, but because the treatment concepts and study philosophies differ.

The planning **target volume** defined at BNL based on the postoperative, planning CT as the actual contrast enhanced volume +2 cm shell around, can be remarkably different from the one delineated by the radiation oncologist at H-MIT using the contrast enhanced volume and using a 2 cm shell around the Gross Tumor Volume (GTV) on the preoperative MRI at HFR. However the relevance of the target definition is also remarkably different in the 3 trials. **The prescribed dose** is the minimal target *biological dose*, without exceeding the peak normal brain biological dose in Brookhaven and the normal tissue maximum biological dose in Boston (expressed in different units Gy-Eq; RBE-Gy). The radiation times vary both at BNL and H-MIT to achieve the prescribed biological doses at the measured blood boron concentrations. For treatment planning 3.5/1 tumor/blood and 1/1 healthy tissue concentrations/blood boron concentrations are assumed.

In Europe the leading *physical dose* component, the boron neutron capture dose, is defined at the "point" of the thermal neutron fluence maximum for the calculation of the radiation time. The assumed tissue boron concentration is equal to blood concentration = 30 ppm average over the 4 fractions. The amount and timing of the BSH infusion is changed every day in order to achieve the average 30 ppm blood boron concentration over the four consecutive days of the treatment.

The **radiation** performed at BNL is single-fraction with a collimator size of 12 cm diameter, at a Skin Wall Distance (SWD) = 0 from 1 field if the tumor margin >4–5 cm, or double fields. The patient head is fixed to the wall with tapes. At H-MIT the treatment is given with 1–3 fields using a collimator size of 15 cm, the SWD = 0. For practical reasons (long irradiation time of 2.5–3-hours) the BNCT is performed on two consecutive days (1–2 field/day) using 2 thermoplastic masks for fixation.

At HFR the BNCT is performed in four fractions on 4 consecutive days, collimator size is 12 cm, SWD = 30 cm, and multiple fields can be used. A thermoplastic mask is manufactured for coordinate system correlation and fixation of the patient's head.

At BNL mandatory anti-seizure **medication** is prescribed, so that the blood level should be in the therapeutic range and usually 4 × 4 mg/day Dexamethasone is given to the patients. Furthermore during the radiation the patients receive slight sedation by Valium or Lorazepam. At H-MIT no medication is prescribed, the patients are sedated by Valium or Lorazepam during the radiation. At HFR, 3 × 4 mg/day Dexamethasone must be given during BNCT.

Follow up is done by the referring physician at BNL and in the European trial and by the clinical investigator at Beth Israel-Deaconess Medical Center in Boston. The results are evaluated by the investigators in the USA-trials, whilst in Europe data collection, monitoring and data management are performed by NDDO and evaluated by independent experts and by the Study Center.

Apart from the common physical, neurological examination, blood, urine laboratory tests and MRI, at BNL mini mental score, KPS, steroid dependency and QL, at H-MIT full visual field, in Europe complete ophthalmologic, ENT, endocrine function and QL are followed. The grading of the adverse events is based on different scoring scales in the three ongoing studies using modified RTOG/EORTC early and late radiation morbidity scales, items from NCIC-CTC scales and modified SOMA categories. The image evaluation is done by the investigators in USA as the estimated percentage changes in the contrast enhanced tumor volume on post-BNCT scans and by the Reference Neuroradiology Center in Europe. In Europe the local response judged according to Zanella and MacDonald classification and by measuring the orthogonal dimensions of the tumor.

For the **management of tumor recurrence**, at BNL except BNCT fractionated photon radiotherapy, surgery, brachytherapy, photon or proton radiosurgery with or without chemotherapy, immunotherapy are allowed, at H-MIT any therapy is allowed.

In the European trial any form of radiotherapy is prohibited for the treatment of recurrance.

At the three reactor sites for clinical BNCT the first step has been taken to prove the safe clinical application of this highly complex binary radiotherapy modality for cranial tumors. As a secondary outcome, these early clinical trials might demonstrate that BNCT has some activity against brain tumors, but these trials can not conclude that it would be of therapeutic benefit to patients. This is the goal of the phase III trials. Further intensive research work is necessary to define the indication areas for BNCT in medicine. Investigation of alternative ways of clinical application of BNCT promises more knowledge and more experience in order to plan further research activities and to select better conditions for patient treatment. The basic condition for that is to coordinate the clinical research work in order to obtain precise, comparable data to be able to present common interpretations and common results on the feasibility of BNCT.

ACKNOWLEDGMENT

The European BNCT Study is financially supported by the European Commission in the form of a Shared Cost Action from the Biomedical and Health Research Programme (BIOMED II) of DG XII, contract No BMH4-CT96-0325 and by the Swiss foundation "Neurochirurgie 2001".

The US Department of Energy provided support for the work at H-MIT and BNL.

REFERENCES

1. P.M. Busse, R.G. Zamenhof, O.K. Harling, G. R. Solares, R. Tishler, M. Stevenson, T. Shafman, N. Coleman, and E. Fischer: Procedures for Phase-1 Study of Boron Neutron Capture Therapy for Glioblastoma Multiforme & Intracranial Melanoma at the New England Deaconess Hospital & The Massachusetts Institute of Technology.
2. A.D. Chanana: Boron Neuron-Capture Therapy of Glioblastoma Multiforme at the Brookhaven Medical Research Reactor—A Phase I/II Study (FDA IND #43,317) Protocol #4.
3. H. Fankhauser, D. Gabel, and W. Sauerwein: European Organization for Research and Treamtment of Cancer—BNCT Study Group Protocol 11961—Postoperative Treatment of Glioblastoma with BNCT at the Petten Irradiation Facility—Phase I Clinical Trial.
4. K. Hideghéty, W. Sauerwein, K. Haselsberger, F. Grochulla, H. Fankhauser, R. Moss, R. Huiskamp, D. Gabel, and M. de Vries (1999) Postoperative Treatment of Glioblastoma with Boron Neutron Capture Therapy at the Petten Irradiation Facility (EORTC Protocol 11961). Strahlenther Onkol 175:Suppl I.

11

REPORT ON THE FIRST PATIENT GROUP OF THE EUROPEAN PHASE I TRIAL (EORTC PROTOCOL 11961) AT THE HIGH FLUX REACTOR PETTEN

K. Hideghéty[1], W. Sauerwein[1], M. de Vries[3], F. Grochulla[4], C. Goetz[5], K. Haselsberger[6], P. Paquis[7], J. Heimans[8], J. Wolbers[9], R. Moss[10], R. Huiskamp[11], F. Stecher-Rasmussen[11], J. Rassow[2], S. Garbe[1], H. Fankhauser[12], and D. Gabel[13]

[1]Strahlenklinik, Universitätsklinikum Essen, Germany
[2]Institut für Medizinische Strahlenphysik
Universitätsklinikum Essen, Germany
[3]New Drug Development Office
Amsterdam, The Netherlands
[4]Klinik für Neurochirurgie
Zentralkrankenhaus Bremen, Germany
[5]Neurochirurgische Klinik
Klinikum Großhadern, Munich, Germany
[6]Karl-Franzens-Universität
Uni-Klinik für Neurochirurgie, Graz, Austrie
[7]Service de Neurochirurgie
Hôpital Pasteur CHU Nice, France
[8]Department of Neurology
Vrije Universiteit, Amsertam
[9]Department of Neurosurgery
Vrije Universiteit, Amsterdam
[10]EC Joint Research Centre, Petten
[11]Netherlands Energy Research Foundation
Petten, The Netherlands
[12]Neurochirurgien FMH
Lausanne, Swtzerland
[13]Fachbereich Chemie
Universität Bremen, Germany

Frontiers in Neutron Capture Therapy, edited by Hawthorne *et al.*
Kluwer Academic / Plenum Publishers, New York, 2001.

The first European clinical trial on BNCT (EORTC protocol 11961) started in 1997.[1] This is a phase I study, aimed to define the Maximum Tolerated Dose and Dose Limiting Toxicity in gliobastoma patients using BSH as boron carrier and fractionated radiation at the Epithermal Neutron Facility at Petten. Beside the evaluation of the systemic toxicity of BSH, and the early radiation toxicity in the irradiated volume, special emphasis has been placed on the late radiation morbidity. In addition survival will be observed. Up to this date 14 patients have been entered into the study, 12 males and 2 females at six Neurosurgery Centers in five European countries.

Average age of the patients at study registration was 62 years (51–74). The performance status at inclusion was very good, the median Karnofsky index was 90 (70–100). After the consent of the eligible patients to the study procedures they were registered for the first part of the trial: the boron uptake and pharmacokinetic study. Approximately 13–14 hours prior to surgery BSH were administered by 1 mg/kg/min speed iv.[2–5] Blood samples during 48 hours and tumor, skin, brain, muscle cerebro-spinal fluid and dura samples were taken during the surgery. The boron content was measured by ICP-AES spectroscopy.

Treatment

The results on the BSH pharmacokinetic and tissue uptake study of the first patient group performed during the surgery are analyzed elsewhere in this book or will be published separated. After recovery from surgery, if the Reference Neuropathology Center confirmed the diagnosis of glioblastoma multiforme, and the Reference Neuroradiology Center confirmed the extent of tumor removal (70% of the initial volume must be resected) and the blood boron concentration was higher or equal to 25 ppm, the eligible patients were prepared for BNCT. From the 14 patients 3 had partial, 4 subtotal, 7 complete tumor resection. The three patients with a remaining tumor volume larger than 30% of the initial tumor size had to be excluded. One more patient could not undergo BNCT because of an intercurrent infection and prolonged recovery after the surgery.

The first patient cohort (10 patients) has been treated by BNCT at the Petten Irradiation Facility. The preparation for the BNCT planning was started on a strictly defined, uniform way at the Referral Center. In the Radiotherapy Department a mask with defined landmarks on it was manufactured. Planning CT under fixed conditions was performed with the mask on the patient and the image data have been transferred electronically to Petten. The hard copies have been submitted to Essen. The treating radiotherapist in Essen indicated the target volume (contrast enhancing area on the preoperative MRI +2 cm shell around) and organs at risk on the hard copies, then they were sent to Petten. The planning were performed by the INEEL code. During the planning procedure the medical and the planning teams had intensive communication and much discussions.

In this trial the boron neutron capture absorbed dose D_B is prescribed in a physically defined point. This Dose Group Identification Point DGIP is located where the thermal fluence is a maximum for a given treatment plan. The calculation of the dose for treatment planning assumed a homogeneous boron concentration of 30 ppm in every tissue. For the other dose components a limiting maximum dose at the DGIP has been defined.[8–10]

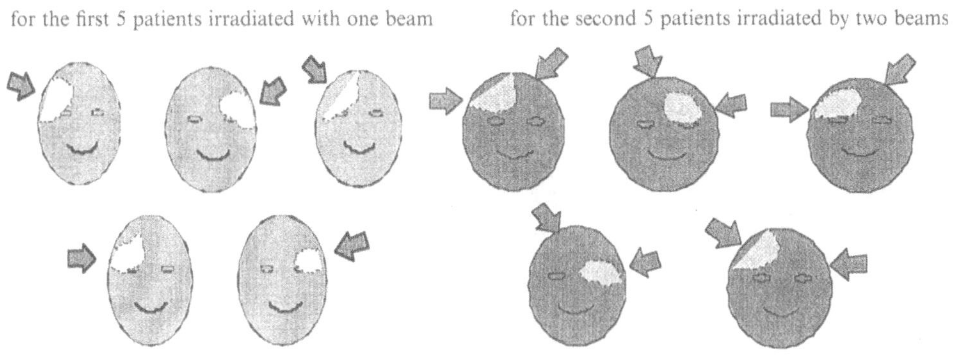

Figure 1. Scheme of the tumor localization and beam arrangement

The starting dose was set at 80% of the dose where neurological effects occurred in preclinical large animal experiments following a single fraction according to the interpretation of the healthy tissue tolerance study prior to design the clinical trial. The only variable parameter, the orientation of the patient's head to the beam (12 cm of diameter), at 30 cm Wall Skin Distance (WSD), was selected on the basis of the planning target volume localization. A single field was used for the treatment of the first 5 patients. The last 5 patients were irradiated with two beams (12 cm of beam diameter, 30 cm WSD), so that the configuration resulted two separated thermal neutron fluence peaks, one in the PTV in the operated area and one outside of the PTV, in "healthy" brain. Consequently a larger volume was irradiated in the second five patients but the boron neutron capture absorbed dose, which is defined for a cohort of patients, was the same 8,6 Gy calculated for the whole group of the 10 patients.

The initial tumor localization was in 3 cases temporal, in 3 patients frontal, in 2 patients occipital and in further 2 patients temporo-occipital. The average target volume was 135,7 cm^3 (range 29–213 cm^3).

Figure 1 shows the scheme of the tumor localization and beam arrangement. The first 5 patients were irradiated with a single beam, the second 5 patients with two tangential beams.

The patients traveled by public transport to Amsterdam, where they were hospitalized at the Neurosurgery Ward of the Vrije University during the week of BNCT. BNCT was performed in 4 fractions at the High Flux Reactor Petten.[6–7]

On the day before the day of radiation BSH had been administered i.v. at the speed of 1mg/kg/min. Two parameters were varied, the amount (11–100 mg/kg) and the time point of administration (10–25 hours prior to the radiation) in order to achieve an average of 30 ppm ^{10}B concentration in blood over the four fractions. The amount, start of the infusion and duration were prescribed each day after obtaining the actual pharmacokinetic data (from the regularly taken blood samples). In the first cohort of 10 patients an average 30,27 ppm (range 27,3–32,3 ppm) blood boron concentration could be achieved over the four fraction of BNCT. On the basis of the real boron concentration in the blood during the radiation the absorbed doses from the different physical dose components and the weighted dose were calculated and reported in defined points and volumes. In this study describing the absorbed dose of each dose component must always do reporting of the applied dose. In addition to a biologically weighted dose (D_{WU}) is reported, which

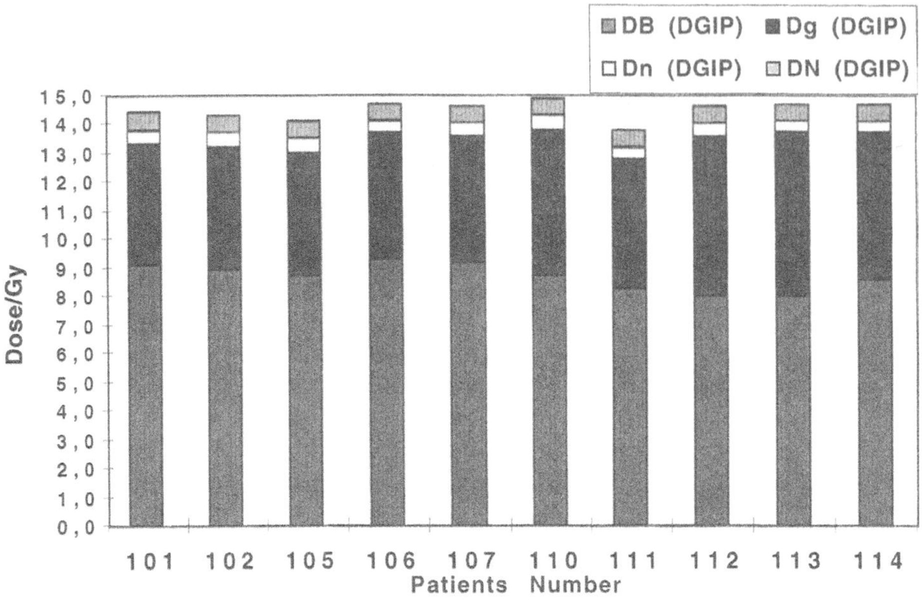

Figure 2. Absorbed doses in DGIP.

is the sum of the different dose components multiplied with their respective biological weighting factors: $D_{WU} = W_{n+N} \cdot D_{n+N} + W_g \cdot D_g + CF_{organ} \cdot D_B$

($W_{n+N} = 3.9$; $W_gC = 1$; CF = 0.37 for the tumor and CF = 0.81 for other tissues)

The doses are reported in

- defined points
  - *DGIP*: thermal fluence maximum for a given treatment plan
  - *Dose prescription point*: one point at the center of the former tumor area (In order to obtain a dose which can be topographical identified in the target region, the point at the geometrical middle of the PTV is defined for this study.
  - at the indicated points relevant to the organs at risk: *optic chiasm*, *eyes*, *ears*, *pituitary gland*, *salivary glands*, and the *skin* at beam entrance and exit.
- defined volumes
  - *planning target volume* and the *whole brain volume* including the target.

Doses—volume histograms are calculated in the defined volumes, furthermore the minimum, maximum, average, median and modal (D_{min}, D_{max}, $D_{average}$, D_{median}, D_{modal}) of each dose component and the biological weighted dose is reported for evaluation of doses adverse event relationship.

The absorbed doses of the BNCT of the first patient cohort, calculated on the basis of the actual delivered monitor units and measured boron concentration in the blood assuming equal homogeneous boron distribution in the human tissues, are shown in the Figures 2–5 in some of above defined points and volumes as examples.

In this clinical study the systemic toxicity due to BSH alone and the local radiation toxicity due to BNCT of the irradiated region of the patient's head are detected. The radiation toxicity is recorded and reported as early radiation toxicity in 90 days after the end of BNCT, and as late radiation toxicity after it.[11]

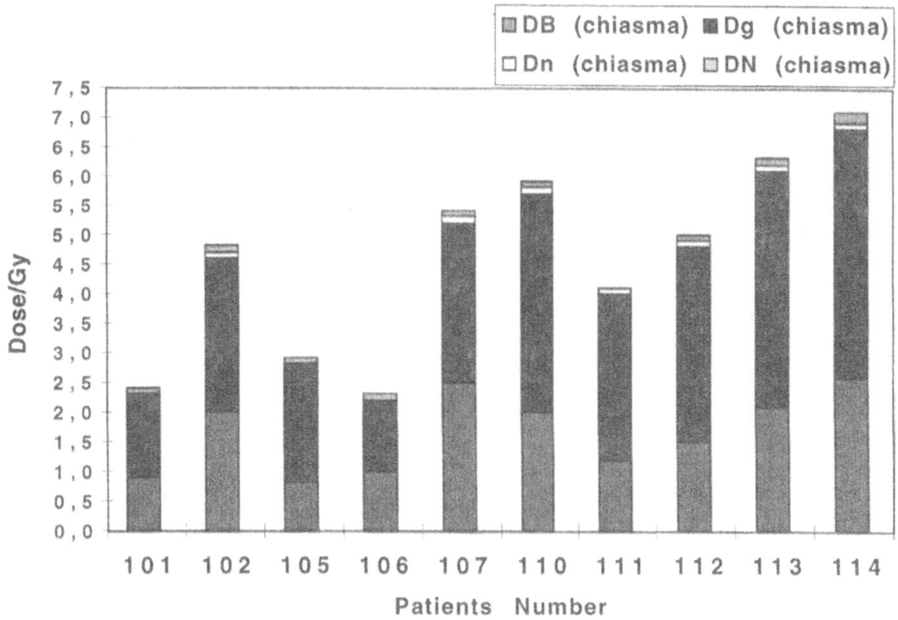

Figure 3. Absorbed doses in chiasma.

Four different toxicity scales are used for grading the events. The NCIC-CTC Grading of Toxicity is used for grading the drug related systemic toxicity as well as for detecting the early toxicity of irradiated organs. The RTOG/EORTC Early Toxicity Criteria for Eye, Ear, Skin, Mucous membrane, salivary gland, CNS are to be followed for the irradiation site, from the first day of the treatment through day 90. The RTOG/EORTC Late Toxicity Criteria for Eye, Ear, Irradiation Site, Skin,

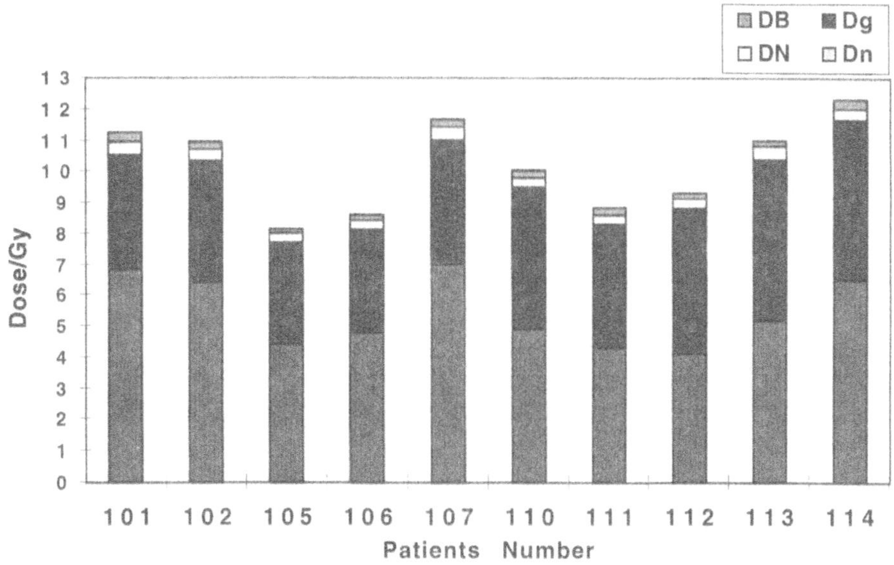

Figure 4. Median absorbed dose in the target volume.

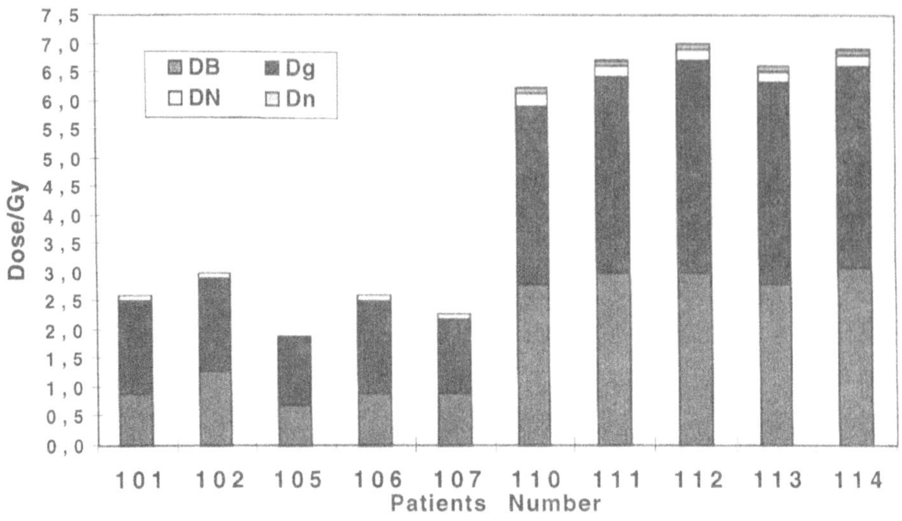

Figure 5. Absorbed dose in the whole brain median values.

Subcutaneous tissue, Mucous membrane, Salivary gland, Brain, Neurologic are supplemented by relevant items from the SOMA LENT (Late Effect on Normal Tissue) categories. The baseline examination for detection of toxicity due to BNCT had been carried out prior the patient's travel to the Netherlands to be treated by BNCT.

Each day of BNCT on the afternoon the systemic toxicity including the laboratory values (blood chemistry, hematology) changes and the early radiation toxicity were checked.

Further toxicity checks were performed in the Referring Centers. The result of the examination are reported on study case report forms and forwarded to NDDO, which provides the monitoring and data management for the study.

As a preliminary impression we can state that the treatment can be performed in a multinational approach under the conditions created at Petten. Compared to other treatment modalities in oncology, the acute toxicity seems to be acceptable: One event of serious toxicity had been observed in one patient, which is possibly related to BSH, who developed Grade IV agranulocytosis, which was resolved within 36 hours. Slight erythema, focal alopecia, taste change, headache and behavior changes were observed in the following 3 months after the end of BNCT. Up to date the late toxicity could not be evaluated, because the monitoring visits will be performed 6 months after the end of BNCT. The follow up is going on, the results on toxicity detection will be published based on monitored data.

If during a period of six months after the treatment of the tenth patient none or one patient shows treatment-related serious adverse events, the radiation dose will be increased by 10%, and this new dose level for the next cohort of ten patients will be opened in 1999.

ACKNOWLEDGMENT

The European BNCT Study is financially supported by the European Commission as Demonstration Project in the form of a Shared Cost Action from the Biomedical and Health Research Program (BIOMED II) of DGXII, contract no BMH4-CT96-0325 and by the Swiss foundation "Neurochirurgie 2001."

REFERENCES

1. Fankhauser, H., Gabel, D., and Sauerwein, W. European Organization for Research and Treamtment of Cancer—BNCT Study Group Protocol 11961—Postoperative Treatment of Glioblastoma with BNCT at the Petten Irradiation Facility—Phase I Clinical Trial.
2. Fankhauser, H., Stragliotto, G., and Zbinden, P. Borocaptate sodium (BSH) pharmacokinetics in glioma patients. Gabel, D.: Moss, R.L. Boron Neutron Capture Therapy. Toward Clinical Trials of Glioma Treatment. New York: Plenum Press; 1992:155–162.
3. Gabel, D. Radiobiological considerations concerning the development of compounds for boron neutron capture therapy. Strahlentherapie und Onkologie. 1993; 169:65–70.
4. Haritz, D., Gabel, D., and Huiskamp, R. Clinical Phase-I Study of $Na_2B_{12}H_{11}SH$ (BSH) in Patients with Malignant Glioma as Precondition for Boron Neutron Capture Therapy (BNCT). International Journal of Radiation Oncology Biology Physics. 1994; 28:1175–1181.
5. Haselsberger, K., Radner, H., and Pendl, G. Boron neutron capture therapy: Boron biodistribution and pharmacokinetics of $Na_2B_{12}H_{11}SH$ in patients with glioblastoma. Cancer Research. 1994; 54:6318–6320.
6. Hideghéty, K., Sauerwein, W., Moss, R., Stecher-Rassmussen, F., Ravensberg, K., de Vries, M., Watkins, P., Vroegindeweij, C., Garbe, S., Siefert, A., Grochulla, F., Goetz, C., Haselsberger, K., Pacquis, P., Heimans, J., Wolbers, J., Wiestler, O.D., Turowski, B., Zanella, F., and Rassow, J. Study preparation and patient treatment experience in the first European clinical trial on BNCT. Proceedings of the First E.U.-U.S. Workshop on Human BNCT Trials, Lund, Sweden 1998.
7. Moss, R.L., Casado, J., Ravensberg, K., Stecher-Rasmussen, F., and Watkins P. The completed BNCT facility at HFR Petten. Advances in Neutron Capture Therapy Vol.1 1997 Elsevier Science pp. 331–341.
8. Sauerwein, W. The clinical project at HFR Petten—A status report. Advances in Neutron Capture Therapy Vol.1 1997 Elsevier Science pp. 77–82.
9. Sauerwein, W., Hideghéty, K., Rassow, J., Götz, C., Gabel, D., Moss, R., Stecher-Rasmussen, F., and de Vries, M. Bor-Neutronen-Einfangtherapie. Eine neue Behandlungsmodalität auf dem europäischen Prüfstand, Strahlentherapie und Onkologie Band 174 1, 1998.
10. Sauerwein, W., Hideghéty, K., de Vries, M., Reulen, H.J., Grochulla, F., Goetz, C., Haselsberger, K., Pacquis, P., Wolbers, J., Wiestler, O.D., Gabel, D., Zanella, F., and Moss, R. Boron Neutron Capture Therapy (BNCT) for the Treatment of Glioblastoma (EORTC protocol 11961), Radiotherapy and Oncology Vol. 48 suppl. 1, 1998.
11. Sauerwein, W., Hideghéty, K., de Vries, M., Wanders, J., Gabel, D., Fankhauser, H., and Huiskamp, R. Conducting phase I clinical trial in binary treatment modality methodical questions for the evaluation of boron neutron capture therapy, Annals of Oncology Vol. 9, Suppl. 2, 1998.

12

QUALITY MANAGEMENT AT THE EUROPEAN BNCT CENTER IN PETTEN

W. Sauerwein[1], R. L. Moss[2], J. Rassow[1], F. Stecher-Rasmussen[3], K. Hideghéty[1], S. Garbe[1], J. G. Wolbers[4], D. Gabel[5], P. Watkins[2], C. Vroegindeweij[2], and H. Sack[1]

[1]Department of Radiotherapy
University of Essen, Germany
[2]HFR Unit, Joint Research Centre
European Commission, Petten
The Netherlands
[3]Business Unit Nuclear Energy
ECN Petten
The Netherlands
[4]Department of Neurosurgery
VU Amsterdam
The Netherlands
[5]Department of Chemistry
University of Bremen, Germany

INTRODUCTION

Boron Neutron Capture Therapy is based on the ability of the isotope ^{10}B to capture thermal neutrons and to disintegrate instantaneously producing the high LET particles, He and Li nuclei, with a high kinetic energy of about 2.5 MeV and a very short range in tissue of about 10 μm [$^{10}B(n,\alpha)^{7}Li$]. If such reactions can be produced selectively in tumour cells, an effective new modality for cancer treatment would be available. In the 1950's and early 1960's, BNCT trials in the USA were effectively a failure.[1] Nevertheless, in the late 60's, Hatanaka in Japan demonstrated that BNCT does benefit patients.[2] However, the reported results were difficult to interpret, because the treatment was not carried out in a controlled manner. In 1994, new trials started in the USA (at Brookhaven National Laboratory BNL[3] and Massachusetts Institute of Technology MIT[4]) involving glioblastoma (at BNL and MIT) and melanoma (at MIT). Meanwhile in Europe in the late 1980's, effective research into introducing BNCT began following the injection of financial support from the Biomedicine and Health Research programme (BIOMED I) of the European Commission. This has lead recently to the start of clinical trials of BNCT for glioblastoma multiforme, with the aim to establish a safe radiation dose for the treatment of brain tumours.[5] It is the first time that a clinical application has had to be realised on a real multi-national scale, whereby a unique facility available for BNCT is localised in one country (The Netherlands) and is operated by an international team of experts under the leadership of a German radiotherapist, treating patients coming from different European countries. It was therefore necessary that special structures had to be created with the support of administrations from different countries, who had to find and adapt solutions within existing laws that had never foreseen such a situation. Furthermore, due to the fact that a new drug, a new radiation beam and a new facility would be used, special efforts had to be made on quality assurance, in order to demonstrate that the set-up at the facility and the personnel involved complied with similar practices in conventional radiotherapy. In this article, a short description of the organisational structure and a more detailed description of the quality assurance aspects for the project are given.

ORGANISATIONAL STRUCTURE AND ADMINISTRATIVE OBSTACLES

The trial has been formulated such that patients are entered into the study from 6 different hospitals from 5 different countries (Austria, France, Germany, Switzerland and The Netherlands). The treatment is performed by the University of Essen (Germany) at the High Flux Reactor (HFR) Petten, which is owned by the European Commission and located in The Netherlands. During the period of treatment, patients are hospitalised at the Academic Hospital of the Free University (AZVU) in Amsterdam. The monitoring of the trial is performed by the New Drug Development Office (NDDO) of the European Organisation for Research and Treatment of Cancer (EORTC). The study is financed as a Shared Cost Action by the European Commission, within the BIOMED II Programme.[6] The treatment in Petten is carried out in co-operation with the Joint Research Centre, JRC, owners of the reactor, and the Netherlands Energy Research Foundation (ECN), under the overall clinical responsibility of the Department of Radiotherapy of the University of Essen which also provides the Medical Physicist. The co-operation of these 3 institutions, their different tasks and responsibilities has been agreed by contract.

To obtain the approval for such a complex multi-national project was extremely difficult and time consuming. This was primarily due to the problem of identifying the appropriate authorities in the different countries. Even the ministries themselves who deal with health policy, could not answer our questions clearly. No European approach is available due to the fact that medical applications fall under national law and that there is no harmonization on the European level.

Some of the issues which had to be solved are listed briefly here:

- licensing of the reactor as a facility for patient treatment,
- licensing of a facility which is not part of a hospital to irradiate patients,
- reconciling the different points of view of different ethics committees in different countries,
- gaining approval of the study protocol by different review boards at different levels in a multitude of institutions,

- gaining local approval on safety aspects, both nuclear and traditional, at the reactor site,
- handling a non-registered drug to be used in different countries following the study protocol,
- licensing of foreign physicians (EU and non-EU) to treat patients in The Netherlands, being themselves staff members of a non-Dutch institution,
- enabling a non-Dutch Medical Physicist to be responsible and liable for Medical Physics at the HFR Petten,
- identifying the different actions performed by persons coming from different institutions in different countries in order to establish and delineate the responsibility, and hence liability, towards the patient; furthermore to describe the tasks of all participants, and to create and approve the appropriate agreements and contracts to define such structures,
- applying the appropriate rules for radio-protection of the patients and the staff, respecting both German and Dutch regulations,
- establishing the EORTC BNCT Study Group,
- regulating the execution of the study protocol as well as the operation of the facility by appropriate Standard Operating Procedures respecting the rules of Good Clinical Practice,
- obtaining insurance for patients following different national procedures,
- building up the local infrastructure for patient care, including all foreseen emergencies,
- concluding contracts, subcontracts, associated contracts, collaboration agreements, etc. with all involved parties, following the rules established by the European Commission for Shared Cost Actions.

In the Netherlands alone, the following governmental bodies (with Dutch abbreviations in brackets) had to be involved:

- Ministry of Health, Welfare and Sport (VWS)
- Ministry of Economic Affairs (EZ)
- Ministry of Social Affairs (SZW)
- Ministry of Environment (VROM)
- Ministry of Foreign Affairs (BZ)
- Health Inspectorate for the province of North Holland
- Mayor's Office of the Community of Zijpe.

In the other countries, as well as on the European level, there were similar interactions which had to be done separately without any possibility of co-ordination.

GENERAL ASPECTS OF QUALITY ASSURANCE AND SAFETY ASSESSMENTS

BNCT at the HFR Petten will be performed respecting the European, National Dutch and whenever it is possible, the National German rules of safety and quality assurance for nuclear research reactors, for radio-protection, for radiotherapy and for clinical trials. In particular, quality assurance of safety provisions and functional performance characteristics conform to the most recent concepts and regulations of IEC publications and/or DIN standards for medical electron accelerators and treatment planning systems or—as far as possible—transferred analogously.

All relevant procedures concerning the performance of BNCT in Petten and the execution of the clinical trial are described by Standard Operation Procedures which are in possession of each participant of the Shared Cost Action.

The Irradiation Facility at Petten

The HFR is a 45 MW, materials testing reactor,[7] with the prime objective to perform experiments on nuclear fuels and materials destined for the European civil nuclear power programmes. In recent years, the reactor has increased its area of application into nuclear medicine, in particular radioisotope production, as well as BNCT.

At one of the eleven horizontal beam tubes located around the reactor, a specially designed irradiation room has been built dedicated to BNCT.[8] The facility has been constructed to reflect as close as is reasonably possible in such an environment, the standards and practices associated with conventional radiotherapy centres, such as cameras, intercoms, microphone, electro-optical laser positioning devices, therapy table and instrumentation to monitor the patient's pulse and oxygen pressure of the blood.

The electrical installation has been carried out according to the Dutch safety regulations of NEN 3134(21) on medical installations of Class S2.

Additional instrumentation, critical to the operational safety of the facility, includes monitoring and controlling the critical structural components and radiation components, such as the beam shutters, the radiation beam, the radiation levels in and around the facility. The room has been designed with the safety of the patient and personnel uppermost.

The performance standard of the required work and operational equipment, especially with respect to its accuracy, reliability and quality have been implemented to conform with accepted standards in quality assurance and control.

The design of the whole facility has been reviewed critically by the local Reactor Safety and Experimental Assessment Committees, which have the mandate to judge a facility on both its nuclear and conventional safety aspects, including reactor safety and radio-protection of the personnel. The working environment has been reviewed and positively assessed by the appropriate national body of the Ministry of Social Affairs (SZW). The complete facility has undergone a local Quality Audit, as part of the JRC's mission to become a licensed QA site according to ISO 9001 during 1998. Finally the facility had a site-visit by an independent physician with personal expertise in clinical applications of BNCT.

Dosimetry

Dosimetry guidelines, as followed in conventional treatment centres, apply to photon or fast neutron facilities. For BNCT facilities, where an epithermal neutron beam is used, the beam includes fast neutrons, as well as gammas, which come from both the beam itself and from activation of the in-beam material. In human tissue containing boron-10 compounds, the beam can be broken down into four main dose components, all with different relative biological effectiveness: the boron neutron capture absorbed dose, the nitrogen neutron capture absorbed dose, the fast neutron absorbed dose and the gamma ray absorbed dose.

To qualify the beam in Petten, regular free beam measurements are performed on a monthly basis using a multi-foil packet consisting of 12 activation foils. During patient treatment, on the first day of the treatment week, in-phantom activation foils and twinned ionisation chamber measurements are performed, which are used to calibrate the on-line monitors, including the in-beam monitors and the in-phantom pn-detector. On the succeeding days of treatment, more phantom measurements are performed using the pn-detector, twinned ionisation chambers and the in-beam monitors, which are all normalised to the first day's measurements.

Beam Monitoring and Beam Shutter Control

The safety concept is the same as that established for conventional radiotherapy purposes, but adapted for the special situation at the reactor. All safety systems are doubled by an independent second device acting in case of failure of the first. In the situation at HFR Petten, some of the important devices have even more than one back-up system. As an important example, the beam shutter control is described here in more detail.

The beam monitoring system includes four beam monitors: two ^{235}U fission chambers (neutron counters) and two GM-tubes (gamma ray counters), which are located in the fixed beam collimator, downstream from the main beam shutter and before the sliding gamma shutter. The 2 fission chambers together with the corresponding electronics systems, which includes two computer systems, form 2 independent monitoring systems.

The automatic opening and closing of the beam shutters is controlled by the fission chambers, according to the pre-set number of monitor counts which correspond to the required boron dose delivered at the DGIP (dose group identification point)[5] a patient. Both fission chambers are pre-set to close the shutters, which is automatically triggered. Fission chamber nr.1 acts as the principal counter, with nr.2 as the back-up. As an additional back-up for beam shutter closure, a timer is available.

In addition, if necessary, the beam shutters can be closed by means of a push-button on the beam shutter operations panel. If this fails, due to electrical failure, the beam shutters can also be closed by means of manual mechanical devices.

Radio-Protection

To conform with the Dutch regulations on radio-protection, an authorised Radio-Protection Committee for BNCT has been formed. The committee has the prime task to review and advise, on a half-yearly basis, the radio-protection methods used for BNCT. If need be, this advice is transmitted to any external authority. The Committee consists of members from each discipline in the BNCT group, and is chaired by an independent expert in radio-protection.

Due to the fact that German staff from Essen University Hospital need to work at Petten, German regulations on radio-protection, especially application of §20 StrSchV (Strahlenschutzverordnung) regulating the activities in foreign institutions, had to be followed. A special (German) licence for Essen University was the basis for a contract between Essen University and JRC in which supervision of the staff, personal dosimetry, rules of behaviour, etc. are defined.

BNCT Treatment Planning

Quality control of the treatment planning system is defined in one of the Standard Operating Procedures. The program is located on 2 separate workstations. Updated versions are first implemented on one workstation and fully tested prior to installation on the second, back-up workstation. Two standard test cases, i.e. a standard patient and standard phantom, are calculated to check for possible non-conformance. Once per month, the versions on both workstations, whether updated or not, are run for the two standard test cases. For each patient, an overview of the program and the treatment planning results are completed and included in the patient's treatment planning file.

Patient Positioning System

A positioning system has been developed which allows the preparation of the treatment at the patient referring hospital and the planning in Petten to be controlled and to be easily reproducible when positioning the patient in the irradiation room.

Prompt Gamma Ray Analysis System

The mean blood-boron concentration during the treatment of the patient is determined by means of prompt gamma ray analysis.[9] A quality control of the facility is performed on the first day of each treatment week, when the resolution of the detector is checked by a ^{60}Co source, and the function and accuracy of the measuring device is checked through means of calibration samples. The measured blood samples are later validated by ICP-AES, which is available at Petten.[10] Furthermore, one of the calibration samples is measured by ICP-AES.

Training of the Staff

Prior to treatment of the first patient a training programme has been carried out, whereby all procedures and actions necessary to perform BNCT have been simulated in "dummy runs". Special attention was paid to emergency situations including both technical and medical failures, as well as reactor scrams. The training for emergency situations will be repeated on a regular basis and at least once at the end of the 6 months observation period prior to the start of the treatment of a new cohort of patients.

THE CLINICAL TRIAL

The first clinical trial in Europe to apply BNCT was designed as a phase I study to establish a safe and reliable dose level for brain irradiation at HFR Petten using the boron compound BSH.[5] The trial is conducted under the principles of the Declaration of Helsinki. The study is performed according to the current recommendations of Good Clinical Practice for Trials on Medicinal Products in the European Community issued by the Commission of the European Communities in its final version, 1991; furthermore the guidelines for Good Clinical Practice[11] in its updated version (actually ICH topic E 6) will be applied whenever possible.

Study Protocol and Monitoring

The protocol has been reviewed by the EORTC Protocol Review Committee and was accepted on 5th August 1996, being a protocol of the EORTC BNCT Study Group. The protocol was reviewed and accepted not only by the Ethics Committees of each participating hospital but also reviewed by a special ethics committee established to give advice to the Dutch Government (KEMO).

The independent monitoring of the trial is performed by the EORTC New Drug Development Office NDDO. Evaluation of the patient's Quality of Life is performed by the EORTC Quality of Life Study Group. A Supervisory Board composed of 4 independent specialists (2 radiotherapists, 1 neurologist, 1 radiobiologist) has been established by the EORTC BNCT Study Group. This board acts as the review body for dose escalation steps, quality assurance of the study, including radiotherapy performed at Petten and gives advice on matters of clinical, ethical and technical bearing. This board will also give a report to the Dutch Ministry of Health after a treatment period of 2 years.

A Reference Pathologist will review all histopathological samples. A Reference Radiologist will review all radiodiagnostic images of the brain (CT and MRI) starting with the diagnosis and during the whole follow-up period.

Study Medication

For the study, the compound BSH (^{10}B-enriched >95%) is used. Supplying companies must produce the compound according to a master file, and have a written procedure for preparation and quality control of the final product and its intermediates. These procedures were supplied to all the national agencies with which BSH is registered as investigational drug for this trial. The chemical compound is imported to The Netherlands by the Pharmacy of the hospital of the Free University Amsterdam (AZVU), where quality checks are carried out prior to shipping the drug to the investigators. Purity of BSH is monitored for each batch as well as absence of pyrogenicity. The content for BSH per vial is controlled in 1 vial/batch, the degree of ^{10}B-enrichment is verified. Each year after initial delivery of a batch, another vial of BSH is analysed for absence of degradation products prior to further use of the batch.

Documentation

Data is handled following GCP guidelines. All documents directly related to the trial (i.e. patient identification codes, protocol, documentation, case report forms, approvals) and the documents directly related to the irradiation will be retained at the Study Centre in Essen for 30 years.

WELFARE OF THE PATIENTS AND MEDICAL SAFETY ASPECTS

The patients are informed by the treating neurosurgeons about all aspects of BNCT treatment and the clinical trial. In order to facilitate such a task, an information brochure with photographs showing the Petten facility is provided to the patients.

Prior to travelling long distances from their home hospitals, the patients will be examined by their own neurosurgeon. They will bring with them their own medicaments for 10 days, in order to avoid lack or exchange of medication which may not always be available in the same preparation in the foreign country. During their stay in the Netherlands, they will be under continuous medical supervision. They are hospitalised at the AZVU in Amsterdam in a neuro-surgical special care unit with a dedicated nurse to continously care for the patient and, amongst other things, is able to understand the patients mother tongue. The nurse will accompany the patient to Petten and back to the hospital. A medical doctor is continuously present during the stay of the patient at the HFR. A medical environment at the reactor has been created by fully equipping for outpatient treatment and medical emergencies. A landing place for a helicopter is available in the direct neighbourhood of the Medical Wing. In case of emergencies, a helicopter is available at the AZVU hospital to bring the patient back to the hospital.

The Medical Wing at Petten is decorated with photographs and posters of the cities and regions of the recruiting centres, thus giving the patient some familiar points in a foreign and strange environment. Relatives of the patients are encouraged to accompany them. Accommodation for relatives is organised by VU hospital with accommodation possibilities at the hospital itself.

CONCLUSION

The first demonstrations of BNCT were thwart with danger and were damaging to the patient. Likewise, the clinical trial at Petten takes place in a nuclear research reactor, which apart from being conducted in a non-hospital environment, is *per se* known to be dangerous. It is therefore of the utmost importance that special attention is given to safety, beyond normal rules, and to the training of staff.

In itself, the trial is also an unusual Phase I study, introducing a new drug with a new therapy modality, with really an unknown dose-effect relationship. Therefore, this trial must follow optimal procedures, which underscore the quality and qualified manner of performance.

It has also been shown that it is possible in Europe to carry out a multi-national cooperation in medical applications, thus opening the way to use unique, expensive facilities in Europe.

ACKNOWLEDGMENT

The clinical trial is financially supported by the European Commission in the form of a Shared Cost Action from the Biomedical and Health Research Programme (BIOMED II) of DGXII, contract no BMH4-CT96-0325.

REFERENCES

1. Slatkin, D.N., "A history of boron neutron capture therapy of brain tumours", Brain, 114, 1991.
2. Hatanaka, H. *et al.*, "Clinical experience of boron-neutron capture therapy for gliomas—a comparison with conventional chemo-immuno-radiotherapy", in Boron-Neutron Capture Therapy for Tumours, Ed. H. Hatanaka, Niigata, Japan: Nishimura, 1986.
3. Coderre, J.A., "BNCT for malignant glioma using BPA", Proc. 6th Int. Symp. on Neutron Capture Therapy, Kobe, Japan, November 1994; Ed. Y.Mishima, Plenum Press, New York, 1995.
4. Madoc-Jones, H., Zamenhof, R., Solares, R., and Harling, O., "Clinical trial begins at MIT/NEMC", idem.
5. Sauerwein, W., Gabel, D., and Fankhauser, H., "Glioma BNCT. Postoperative treatment of glioblastoma with BNCT at the Petten irradiation facility. 1. Phase I clinical trial", EORTC Protocol 11691, March 1997.
6. Gabel, D., "Approach to Boron Neutron Capture Therapy in Europe: Goals of a European Collaboration on Boron Neutron Capture Therapy", Proc. 2nd. Eur. Particle Accelerator Conf., Nice, 1990.

7. Ahlf, J., Zurita, A., eds., "High Flux Reactor (HFR) Petten, Characteristics of the Installation and the Irradiation Facilities", EUR 15151 EN, 1993.
8. Moss, R.L., Casado, J., Ravensberg, K., Stecher-Rasmussen F., and Watkins, P., "The Completed BNCT Facility at the HFR Petten", 7th Int. Symp. on Neutron Capture Therapy, Zurich, September 1996; Ed. B. Larsson, J. Crawford, Plenum Press, New York, 1997.
9. Raaijmakers, C.P.J., Konijnenberg, M.W., Dewit, L., Haritz, D., Huiskamp, R., Philipp, K., Siefert, A., Stecher-Rasmussen, F., and Mijnheer, B.J., "Monitoring of blood-10 B concentration for boron neutron capture therapy using prompt gamma-ray analysis", Acta Oncologica 34, 517–523 (1995).
10. Gregoire, V., Begg, A.C., Huiskamp, H., Verrijk, R., and Bartelink, H., "Selectivity of boron carriers for boron neutron capture therapy: pharmacological studies with borocaptate sodium, L-boronophenylalanine and boric acid in murine tumors.", Radiother.Oncol. 27:46–54, 1993.
11. Bohaychuk, W., and Ball, G., "Good Clinical Research Practices, An indexed reference to international guidelines and regulations, with practical interpretation", GCRP Publications, January 1994.

13

PSYCHOLOGICAL ASPECTS OF THE PATIENT TREATMENT BY BNCT AT PETTEN

K. Hideghéty[1], W. Sauerwein[1], R. Moss[2], M. A. Späth-Borst[3], J. Cassado[2], K.-H. Cloes[2], S. Blume[3], J. Heimans[4], and J. Wolbers[5]

[1]Strahlenklinik, Universitätsklinikum Essen
Germany
[2]EC Joint Research Centre Petten
The Netherlands
[3]Wetenschaps an Technologiedynamica
Universiteit van Amsterdam
The Netherlands
[4]Department of Neurology Vrije Universiteit
Amsterdam, The Netherlands
[5]Department of Neurosurgery Vrije Universiteit
Amsterdam, The Netherlands

The conduction of a phase I trial especially with incurable cancer involves a psychological burden for both sides: the patients and the staff as well. In addition the trial in progress is a multi-disciplinary, multi-national, multi-institutional study, outside of the traditional medical structure. The effectiveness of the team functioning is variable ranging from productive interdisciplinary collaboration to fragmented service co-ordination loaded with high emotional costs, tension, conflicts and frustrations. On the other hand due to the particular circumstances at Petten[5–6] both the BNCT-team members and the patients can gain paramount psychological benefit from the new emerging interdisciplinary field that use techniques from previous separate scientific disciplines. It considers key organisational, professional and interpersonal challenges that need to be analysed in order to address the possible ways of improvement if the present level of teamwork is to be enhanced or strengthen the present available advantages.[1]

TRANS-EUROPEAN PATIENT TREATMENT

A trans-national, cross-cultural and multi-institutional structure had to be organised respecting the study design, historical legacies, the geographical, political and legal conditions in Europe.

Frontiers in Neutron Capture Therapy, edited by Hawthorne *et al.*
Kluwer Academic / Plenum Publishers, New York, 2001.

The Europe-Wide Hospital

A trans-national hospital has been created with Neurosurgery Departments, where the patients are entered into the trial and operated, including Local Radiotherapy Units, which take care on patient preparation for BNCT planning, at Amsterdam, Bremen, Graz, Munich, Nice; a Neuropathology Department in Bonn; a Neuroradiology Department in Frankfurt; a Treating Radiotherapy Department in Essen; an Outpatient Radiotherapy Unit in Petten, a Pharmacy and an Inpatient Department in Amsterdam in order to create the availability for all European citizens of the unique European treatment facility located at Petten, in The Netherlands. To make functioning this "European Clinic" a trans-European, trans-cultural communication system had to be developed.[5–6,9–10]

PSYCHOLOGICAL IMPACT OF THE TREATMENT STRUCTURE AT A NUCLEAR REACTOR SITE IN A NON-MEDICAL ENVIRONMENT

At Petten security measures exist, such as guards and security fencing, which could be frightening for a patient. A picture booklet has been prepared and shown to the patients explaining the procedures and special circumstances, in order to decrease the anxiety from an unknown environment with strict regulation.

The medical infrastructure and appropriate conditions for an outpatient clinic were established at the nuclear reactor site by separating the treatment area from the other part of the reactor. A simple, familiarly decorated building, the so called "Medical Wing" next to the reactor, creates a unique ambience. Special attention had been placed on psychological aspects by choosing the colours, furniture, supplementary equipment, decoration. There are pictures from the hometowns of the patients hanging around the walls.

Nevertheless, while in The Netherlands, the patients are subjected to contrasting circumstances. They are hospitalised in an intensive care unit of the Neurosurgery Department at the Academic Hospital of Free University in Amsterdam, which is equipped for acute emergency care. They are then transported through beautiful, peaceful countryside each day to Petten.

For safety reasons, as dictated by the local regulation and consequently the practical procedures a lot of staff members get direct personal contact with the patients and their relatives or accompanying persons. The reactor people are familiar with having visitors. In their eyes the patient is a guest, a visitor. This behaviour has a large impact on the patients, who lose the "patient" status very quickly and have a metamorphosis to become an honoured visitor of a nuclear facility.

THE EMOTIONAL COSTS FOR THE STAFF IN THE PRESENT CIRCUMSTANCES

The staff members at Petten, whilst being experts of different fields, coming from various cultural backgrounds, and belonging to different institutes, have themselves some cross-cultural communication barriers, which under the present structure has little possibility for relief.

The basic conditions for outpatient treatment had been created involving different institutions (two of them without medical background). In addition to the participants are coming from different countries, from different culture, they are belonging to different institutions with different hierarchical structure, with different type of work organisation, having different social background. The definition of interests, the goals, and the motivation of the participating institutions and individuals differ basically from each other. Furthermore these definitions are changing globally by the time due to entering of new parties or specialists and by obtaining supplementary information and results.

The multi-institutional basis for clinical research proved to be a fragile, temporarily solution, which is overloaded with misunderstandings and frustrations. The maintenance of it is very time consuming, the execution of the tasks is uncontrollable if the participants belong to different structures with different aims and interests.

For medical work at long term, it will be absolutely mandatory to establish a defined structure accepted by all participants allowing constructive collaboration, increasing the quality of the work, and respecting the safety of the patients.

It was demonstrated that a multinational, multi-cultural medical network could be organised in Europe, providing the availability of a special, unique treatment modality, which requires qualified experts with special education and skills, special materials, high technology or expensive apparative conditions with appropriate infrastructure. If the interdisciplinary co-operation had a stable, well defined, appropriate structure the participants could gain remarkable benefit through a comprehensive interpersonal connection network, learning the different approaches of the different disciplines and through the experiences of the cross-cultural communication.[4,7,11] The staff members at Petten had large enthusiasm to have the possibility to take part in a medical project to do something direct for patients. Generally they have deep human approach with high empathy towards the patients, but on the other hand they have overestimated emotional expectations regarding the outcome of the BNCT. On a cognitive level it is accepted that the patients have no high probability of gaining individual benefit from a phase I study. However on an emotional basis there is an urge to go towards "treatment optimisation" for an individual patient, without taking into account that such optimisation can only be done on the basis of facts and knowledge, which have to be collected by the ongoing and joined trials. The high personal impact and assistance in the presence of patients and family members can be exhausted after a certain time, leading to disappointment and even to a "burned-out syndrome". The majority of the staff members are rather reluctant against the necessary medical hierarchy. Some of them have a denial of the fatal malignancy; first of all due to the protocol design, only patients with very good performance status can be treated, therefore the bad prognosis is not evident and may seem to be incredible. There is an unavoidable necessity to prevent the negative turn of the positive emotional approach of the staff members in time!

AM I A PATIENT OR A RESEARCH SUBJECT?

This question is rarely clearly formulated, yet it lies behind the patient's dilemmas and hesitation in consenting to enter a trial. After a thorough, compassionate presentation of their situation, patients with glioblastoma will usually consent to be included in the trial of a new therapy modality: for most patients, uncertainty and hope linked to the new method is preferable to the definitively grim prognosis that may be offered by palliative conventional radiotherapy alone. Patients included in phase I trials rarely realise that the main objective of the trial is not to test the efficacy of the modality, but to study its side effects, pharmacokinetics and late toxicity, applied under defined conditions. Consent to participate in a phase I trial and to all the associated additional investigations, frequent visits to the clinic, and eventual side effect often stems from false hopes that the patient has a chance for a positive response—when this chance is really less than 5%. However altruistic some patients may be with respect to medical science and the need for clear and scientifically valid answers to the dilemmas of modern medicine, the patient is always still the patient, with his or her unique individuality and interests. The patient then may be expected to consent to a trial only if he is convinced that this offers him better chances, or at least no worse a chance than choosing one of the standard treatment options.[2–3,8] In the case of BNCT it seems to be that the researchers themselves have the same unrealistic expectations. It solves the conflict, how to tell the truth, because the investigator himself is convinced about the high probability of achievement individual benefit, prolonged survival for the patients. Consequently there are no discrepancies between the oral- and metacommunication on the BNCT during the patient information. Yet it gives a release for the treating physician to offer something instead of presenting an entirely hopeless situation for a relative well being patient. On the other hand this unrealistic approach can lead to complete disappointment, to the loss of enthusiasm, and after all to a refusal to continue the participation in clinical research work on BNCT. In any case this attitude and the consequent behaviour of the investigators facilitate the decision of the patients to take part in a phase I study.

Patients are coming from different countries to a non-familiar medical surroundings, speak foreign language, have different human behaviour after facing their own hopeless situation, and have undergone radical brain surgery—a lot of psychological reactions, depression, denial, can be imagined. However, according to the study selection criteria they should have an excellent performance status after the recovery from the surgery at arrival to The Netherlands. They have a feeling of a kind of euphoria, because of the disappearance of their preoperative neurological, disabling symptoms, and to get rid of the medical environment. On the other hand the fears are continuously present of the unknown treatment, of the recurrence, of the near and far future. The first ten patients, who were treated at HFR did not show the signs of depression, and all of the (except one) had a very supportive familial-social background. Some special common feature of the patients' behaviour could be observed during week of BNCT:

- very good co-operation
- request for detailed information
- strong support from the family, touchable love between the couple, good communication within the families
- high capacity to be integrated into the BNCT team, as active participant of a research work
- pretension not to be anxious, nervous at arrival prior to the first visit of the facility
- real relief from fears after the first fraction of the treatment
- behavioural changes during the four fractions: change of their role to be a patient to be a "visitor" during the treatment week.
- expression a lot of positive emotion towards the staff at Petten

The problems which can be raised by the different language, different behaviour, different customs, different circumstances could be solved in a European trans-national patient treatment. Even under the present conditions the patients could gain paramount emotional support from the staff members. They have psychologically a clear advantage to get some hope in a hopeless situation, to get the possibility to participate actively in the procedures of the treatment, to involve the family members as well, to get back the dignity of a healthy human being at Petten.

ACKNOWLEDGMENT

The European BNCT Study is financially supported by the European Commission in the form of a Shared Cost Action from the Biomedical and Health Research Programme (BIOMED II) of DGXII, contract no BMH4-CT96–0325 and by the Swiss foundation "Neurochirurgie 2001"

REFERENCES

1. O. Barr. "Interdisciplinary Teamwork: Consideration of the Challenges," British Journal of Nursing, 1997, pp. 1005–1010.
2. E.J. Emanuel, and W. Bradford Patterson. "Ethics of Radomized Clincal Trials", Journal of Clinical Oncology, 1996, pp. 365–371.
3. B.A. Faulkner. "Care of Oncology Patients: caring for the carers", Completement-Ther Nurs-Midwifery", 1996, pp. 59–61.
4. B. Feinberg, and I. Feinberg. "Overall Survival of the Medical Oncologist: a New Outcome Measurement in Cancer Medicine", Cancer, 1998, pp. 2047–2056.
5. K. Hideghéty, W. Sauerwein, R. Moss, F. Stecher-Rassmussen, K. Ravensberg, M. de Vries, P. Watkins, C. Vroegindeweij, S. Garbe, A. Siefert, F. Grochulla, C. Goetz, K. Haselsberger, P. Pacquis, J. Heimans, J. Wolbers, O.D. Wiestler, B. Turowski, F. Zanella, and J. Rassow. Study preparation and patient treatment experience in the first European clinical trial on BNCT. Proceedings of the First E.U.-U.S. Workshop on Human BNCT Trials, Lund, Sweden, 1998.
6. K. Hideghéty, W. Sauerwein, K. Haselsberger, F. Grochulla, H. Fankhauser, R. Moss, R. Huiskamp, D. Gabel, and M. de Vries. "Postoperative Treatment of Glioblastoma with Boron Neutron Capture Therapy at the Petten Irradiation Facility (EORTC Protocol 11961)", Strahlenther Onkol 175, 1999, Suppl I, pp. 111–114.
7. N.K. Kwak. "A Human Resource Planning Model for Hospital/Medical Technologists: an Analytic Hierarchy Process Approach", J-Med-Syst., 1997, pp. 173–187.
8. T.F. Massoud, G.J. Hademenos, W.L. Young, E. Ago, J. Pile-Spellman, and F. Vinuela. "Principles and Philosophy of Modeling in Biomedical Research", FASEB-J, 1998, pp. 275–285.
9. W. Sauerwein, K. Hideghéty, J. Rassow, C. Götz, D. Gabel, R. Moss, F. Stecher-Rasmussen, and M. de Vries. "Bor-Neutronen-Einfangtherapie. Eine neue Behandlungs-modalität auf dem europäischen Prüfstand", Strahlentherapie und Onkologie, 1998, Vol. 174, Sondernr. 1, PV301.
10. W. Sauerwein, K. Hideghéty, M. de Vries, J. Wanders, D. Gabel, H. Fankhauser, and R. Huiskamp. Conducting phase I clinical trial in binary treatment modality methodical questions for the evaluation of boron neutron capture therapy, Annals of Oncology, 1998, Vol. 9, Suppl. 2.
11. J.W. Swanson. "Building a Successful Team through Collaboration", Nurs-Manage, 1 1997, pp. 173–187.

14

AT THE THRESHOLD OF CLINICAL TRIALS

The Status of the Finnish BNCT Project

Markus Färkkilä[1], Carita Aschan[4,10], Iiro Auterinen[7], Judit Benczik[9], Pekka Hiismäki[7,8], Juha Jääskeläinen[2], Eija Järviluoma[6], Heikki Joensuu[3], Merja Kallio[1,5], Leena Kankaanranta[3], Mika Kortesniemi[5,7], Antti Kosunen[14], Petri Kotiluoto[7], Martti Kulvik[1,5,11], Juha Laakso[5], Seppo Pakkala[14], Merja Rasilainen[6], Seppo Salmenhaara[7], Sauli Savolainen[4], Tiina Seppälä[5,7,10], Tom Serén[7], Marjatta Snellman[9], Marika Suominen[12], Mikko Tenhunen[3], Matti Toivonen[15], Leena Tähtinen[12], and Jyrki Vähätalo[13]

Departments of [1]Neurology
[2]Neurosurgery, [3]Oncology, [4]Radiology,
[5]Clinical Research Institute and [6]Hospital Pharmacy
Helsinki University Central Hospital
FIN-00029 HYKS, Finland
[7]VTT Chemical Technology
Technical Research Centre of Finland (VTT)
Espoo, Finland
[8]Radtek Inc. Helsinki, Finland
[9]Faculty of Veterinary Medicine
[10]Department of Physics
[11]Institute of Biotechnology
[12]Laboratory of Organic Chemistry
[13]Laboratory of Radiochemistry
University of Helsinki
Helsinki, Finland
[14]NC-therapy Inc., Helsinki, Finland
[15]Radiation and Nuclear Safety Authority (STUK)
Helsinki, Finland

The aim of the Finnish BNCT-project is to start BNC-treatments of malignant brain tumors. The first clinical trial is planned to start in early 1999 at the treatment facility of the 250 kW FiR 1 TRIGA research reactor. Excellent patient treatment facilities

Frontiers in Neutron Capture Therapy, edited by Hawthorne *et al.*
Kluwer Academic / Plenum Publishers, New York, 2001.

have been built at the reactor which is located only 5 km from the Helsinki University Central Hospital making the treatment facility very easy to reach.

The Finnish BNCT-project started in the early 1990's.[1] The epithermal neutron beam at FiR 1 has been created using the new FLUENTAL™ neutron moderator material. The beam has one of the world's top performances with high intensity and purity: the treatment times are shorter and the fast neutron and gamma contamination are the lowest compared to other epithermal BNCT facilities.[1] The neutron beam has been characterized using foil activation analyses to determine the neutron fluence rate and energy spectrum.[3] A dosimetry chain has been produced using dose and neutron fluence measurements in phantoms as well as a calibrated source description for the dose planning. These together with the calibrated neutron beam monitoring system enables exact dose delivery to the patient.

The Finnish BNCT-project will use 4-dihydroxyborylphenylalanine (BPA) as an anionic fructose complex (BPA–F) as boron carrier. Every BPA-F infusion solution will be prepared for individual patient basis and chemical and pharmaceutical testing for each batch will be performed. Boron biodistribution studies precede clinical BNCT trials. Boron concentrations in blood samples and tissue specimens are measured by ICP-AES and ICP-MS.[4] We have developed analytical and preparative techniques for BPA and other aromatic amino acids using HPLC with UV, radioactivity and mass detectors. Synthetic pathways to produce BPA and its more lipophilic congeners are under evaluation.

Boron measurement systems utilizing inductively coupled plasma-atomic emission spectrometry (ICP-AES), inductively coupled plasma-mass spectrometry (ICP-MS), proton induced gamma emission (PIGE) and secondary ion mass spectrometry (SIMS) have been investigated and developed for pharmacokinetic preclinical and clinical studies. For quick, accurate boron measurements both ICP-MS and especially ICP-AES are appropriate to be used for on site pharmacokinetic boron monitoring during BNCT treatments.[4] SIMS is being tested as a research tool for localizing boron measurements which would be prerequisite for good understanding of BNCT for different boron carriers and biological systems.

The healthy tissue tolerance study with dogs as a part of biological characterization of the beam was completed by April 1998: the follow-up continues.[3] The treatment planning system is based on the computer software developed by the Idaho National Engineering and Environmental Laboratory (INEEL). An individual treatment plan was calculated for each dog.[2] Due to head size differences, doses would have been significantly different for the dogs in the same dose group, if a single dose plan had been utilized. Beam-only dogs were irradiated with single field to weighted doses 13.1 Gy (W), 14.1 Gy (W) and 16 Gy (W). Four dogs were irradiated with single field after BPA-F infusion for an estimated dose of 14.1 Gy (W). In the first MRI scans in July 1998 one beam only dog that received 16 Gy (W) had visible changes in the brain. A reference study with photon irradiated dogs has reached the end of follow-up.

The licensing process for the medical use of the beam and for the patient treatments has been initiated. The first patient trial for glioblastoma is similar to the treatments given in Brookhaven National Laboratory, New York, with BPA-F as boron carrier. The starting dose of BPA-F is planned to be 370 mg/kg of BPA given as an intravenous 3 hour infusion. The starting average dose to normal brain will be 8 Gy (W) for the first patients. After sufficient follow-up for the first patients the dose will be raised if deemed desirable.

The project is lead by a multidisciplinary board. Dr. Färkkilä has acted as the chairman of the board; he is also the responsible leader of the project. The group has the advantage of involvement of clinicians who have been active in other brain tumor trials.

NC-Therapy Inc. is the applicant of the radiotherapy license and will take the responsibility for the treatment of patients after a first phase of clinical trials. Radtek Inc. was established in order to finance part of the investments at the reactor site. VTT continues as the licensed reactor operator.

The project has been and is financed, in addition to the individual institutions participating, with science grants from the University of Helsinki, the Finnish Academy of Sciences and development grants from the Technology Development Centre, Finland.

REFERENCES

1. I. Auterinen, P. Hiismäki, P. Kotiluoto *et al.* Metamorphosis of a 35 years old TRIGA reactor into a modern BNCT facility. (these proceedings).
2. S. Savolainen, I. Auterinen, T. Serén *et al.* Dosimetric chain for the dogs irradiated in the epithermal neutron beam at the Finnish BNCT facility. (these proceedings).
3. J. Benczik, T. Seppälä, J. Hopewell *et al.* Large animal model for healthy tissue tolerance study in BNCT (these proceedings).
4. M. Kulvik, J. Laakso, J. Vähätalo, and R. Zilliacus. On-line boron-10 determinations from blood samples by ICP-MS. (these proceedings).

15

AN AUTOPSY REPORT OF A GLIOBLASTOMA CASE AFTER BORON NEUTRON CAPTURE THERAPY

Comparison of MRI and Autopsy Findings

Kei Nakai[1], Akira Matsumura[1], Yasushi Shibata[1], Tetsuya Yamamoto[1], Shingo Takano[1], Tadao Nose[1], Yukinori Inadome[2], and Masayuki Noguchi[2]

[1]Department of Neurosurgery
[2]Department of Pathology
Institute of Clinical Medicine
University of Tsukuba
Tsukuba City, Ibaraki
Japan

1. INTRODUCTION

Clinical trials of BNCT for glioblastoma have been underway around the World. In Japan, the new reactor JRR-4 has been now reconstructed especially for therapeutic irradiation. From the clinical point of view, pathological studies are very important in understanding post BNCT clinical and MRI findings. However, pathological findings on target lesions, which received BNCT have scarcely been reported.[1] We report an interesting and rare autopsy case of giant cell glioblastoma and compare MRI findings with autopsy findings, and discuss histopathlogical changes after BNCT.

2. CASE PRESENTATION

2.1. Case: 45-Year-Old Male

His complaint was severe headache. MRI revealed a right temporal cystic tumor. There was a massive tumor at the anterior part of well-enhanced cystic wall. At the first operation, the tumor was macroscopically totally removed.

Frontiers in Neutron Capture Therapy, edited by Hawthorne *et al.*
Kluwer Academic / Plenum Publishers, New York, 2001.

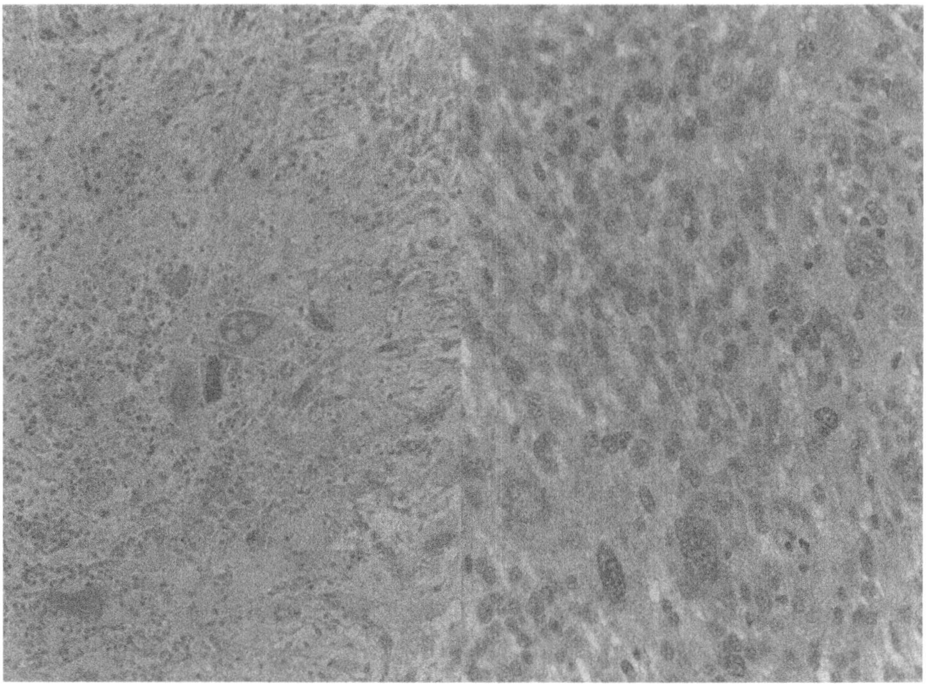

Figure 1. Photomicrograph of a specimen at the first operation showing giant cell (left ×200) and mitoses (right ×400). Hematoxylin and eosin stain.

2.2. Histpathological Findings

Histpathological diagnosis of the specimen, which was obtained at the operation, was giant cell glioblastoma (Figure 1).

Histpathological specimen revealed a giant form of the cells, which had multiple nuclei and diffusely eosinophillic cytoplasm. On the other hand, another part of the same specimen showed small cells, which revealed mitoses.

2.3. Clinical Course

One month after the tumor removal, the patient underwent BNCT using BSH with open craniotomy at Japan Atomic Energy Research Institute (JRR-2).

Neutron fluence was directly measured during neutron irradiation with fine gold wires, which were placed on the brain surface and along the tumor cavity horizontally and vertically, which was debulked at the first operation, and inserted into the brain from the bottom of the cavity. According to the Dr. Nakagawa's criteria,[2] neutron irradiation time was decided by directly measured neutron flux. The surface vascular dose from $^{10}B(n, \alpha)^7Li$ reaction was limited to 10 Gy physical dose. After irradiation, total neutron fluence on the surface of the brain was 1.6 to $2.1 \times 10^{13}\,n/cm^2$ and $5.8 \times 10^{12}\,n/cm^2$ at 4 cm depth, and 6.4×10^{12} or more, at the edge of tumor cavity. Plasma boron concentration at the irradiation was 20 ppm. Tumor boron concentration was not obtained, because there was macroscopically no residual tumor.

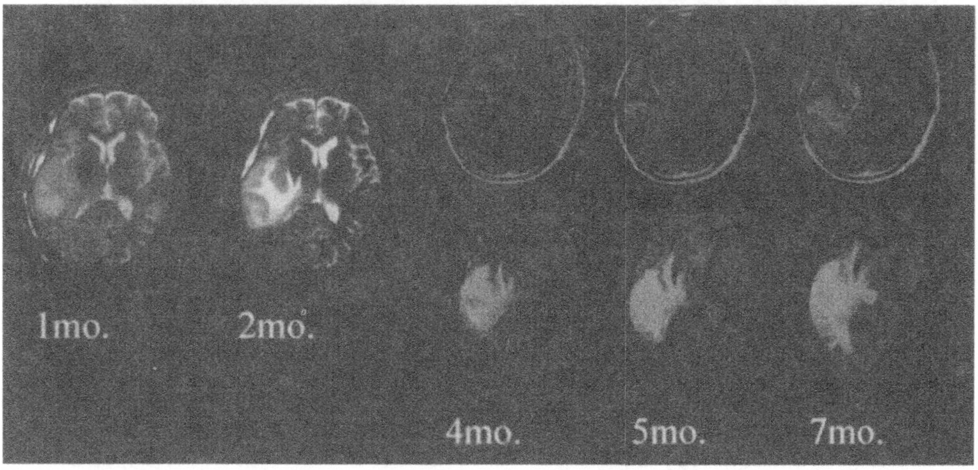

Figure 2. Post BNCT serial T2 weighted image (1, 2 mo., lower image of 4, 5, 7 mo.) and T1 weighted enhanced image (upper image of 4, 5, 7 mo.).

2.4. MRI Findings

One month after BNCT, T1 weighted image showed slightly low intensity areas with temporal lobe swelling. Tumor was not detected, but small enhancement area existed in the temporal base.

Two months after BNCT, high intensity changes appeared on T2 weighted images, which remained until 3 months (Figure 2). Clinically, the patient showed no neurological deterioration and his performance state remained well. Three to 7 month after BNCT, T2 intensity changes extended to frontal and occipital and brain swelling was increasing. Four months after BNCT, an enhancement area appeared and gradually enlarged (Figure 2).

Eleven months after BNCT, patient's consciousness gradually deteriorated. MRI showed dissemination of the tumor and diffuse brain swelling (Figure 3). Seeding tumors appeared at right frontal, the brain stem, medial side of the right temporal lobe, in the posterior horn of the left lateral ventricle. The patient expired because of brain herniation, 1 year following the first operation.

2.5. Pathological Findings

2.5.1. Macroscopic Findings. Brain surface of the right temporal lobe was shown to be yellowish necrosis, and there was local recurrence at medial side of right temporal lobe (Figure 4).

Dissemination of the tumor was also found in the right frontal, in the posterior horn of the left lateral ventricle, and around the brain stem, which were seen in MRI.

2.5.2. Microscopic Findings. The tumor contained two cell types (Figure 5). One was small round cell, many of which showed mitoses. Another one was a spindle shaped cell. These cells were uniform and showed low pleomorphism. The pathological diagnosis of autopsy was glioblastoma small cell variant which was different from the initial histopathological diagnosis.

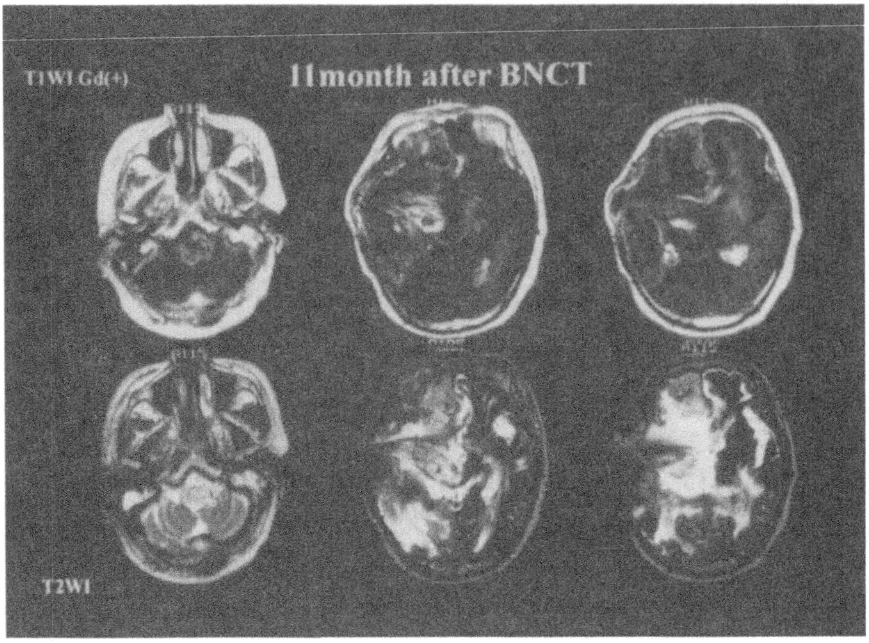

Figure 3. 11 month following BNCT. T1 weighted enhanced (upper) and T2 weighted (lower) showing dissemination of the tumor and diffuse brain swelling.

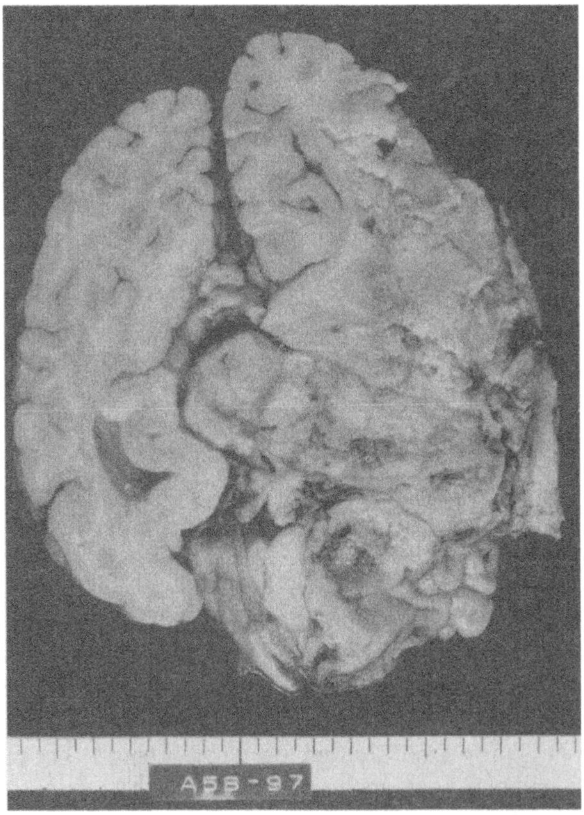

Figure 4. A macroscopic photograph of the brain. Right side of brain is located to right.

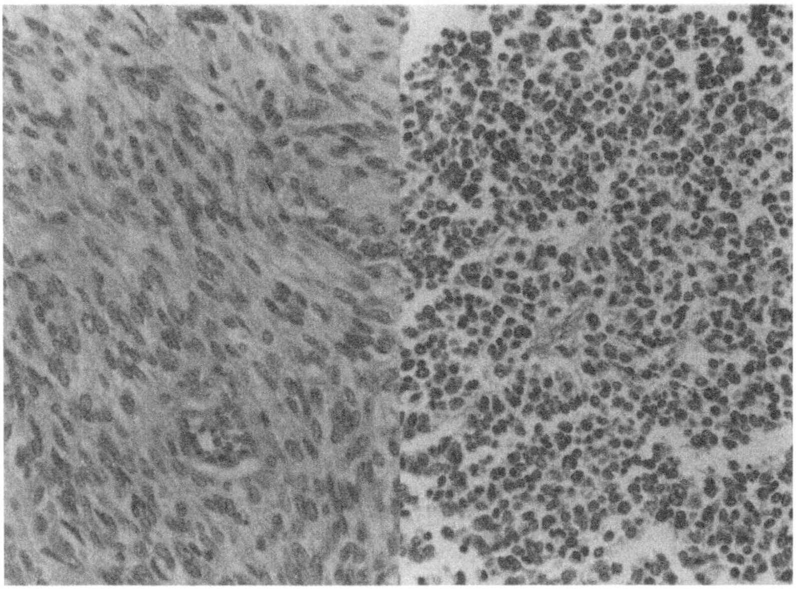

Figure 5. Photomicrograph of a specimen at the autopsy showing round cells (left) and spindle cells (right). ×400, hematoxylin and eosin stain.

3. DISCUSSION

3.1. Tumor Pathology Changes after BNCT

This case is characterized by histological change between the first operation and autopsy.

The original tumor was giant cell glioblastoma. Giant cells disappeared and histopathological diagnosis changed to glioblastoma small cell variant at post mortem examination. This suggests that pathological change was induced after BNCT.

Concerning the cause of pathological changes, we assume the difference in radio sensitivities between the giant cells and the smaller cells may play a role.

Kawano *et al.*[3] reported on the immunoreactivity of giant cell glioblastoma with a p53, a tumor suppresser gene, with a labeling index from 23 to 68% in giant cells and up to 19% in small cells. Hence it is thought that small cells, which express p53 wild type, should be more radiosensitive than giant cells, which express p53 mutant type. Another report by Peraud *et al.*[4] described that there is no apparent difference in p53 accumulation. In our case, p53 immunohistochemical staining failed to show any specific features, probably due to specimen fixation problems. Giant cells disappeared after BNCT, which could not be explained by different p53 expression among two cell types.

In general, giant cells frequently have a diameter of up to 100 micrometer, possessing a large amount of cytoplasm and multiple nuclei. The different concentration of boron compounds or heterogeneous microdistribution in a cell also may play a role in dose distribution. This is supported by microdosimetry model that showed dose distribution changes by the position of the nucleus within a cell.[5] To prove this hypothesis, subcellular boron distribution studies or alpha tracking method should be performed in such giant cell glioblastoma.

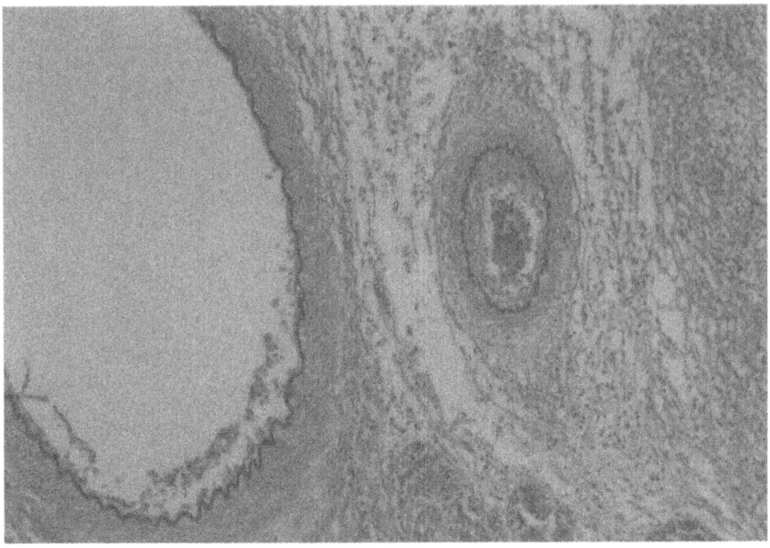

Figure 6. Photomicrograph showing vascular endothelial proliferation at smaller vessel. ×100.

3.2. Vascular Pathological Findings

At the margin of the radiation necrosis area, there were reactive astrocytes, perivascular fibrin exudation around the vessels and endothelial proliferation in the vessels. One previous report mentioned that all sizes of vessels were affected.[1] In contrast in our case, this endothelial proliferation were especially seen in small or middle-sized vessels. On the other hand large vessels were less involved and a single layer of endothelium was spared (Figure 6). The different behavior of endothelial changes regarding vessel size remains unclear. Microdistribution of alpha particles may differ in different lumen size. Theoretical calculations[6] and animal experimental models[7] supported this histpathological difference. Computed dosimetric simulation in vessels of different size may clarify the pathogenesis of this phenomenon.

3.3. MRI Findings

From MRI findings alone, it is often difficult to differentiate tumor recurrence and radiation necrosis. In our case, we also could not clarify the period when the tumor recurred. Retrospectively, changes in enhancement pattern after a certain period should be regarded as recurrent tumor. The enlargement of high intensity on T2 weighted image was one of the indicative signs of tumor recurrence. Post operative faint enhancement on T1 weighted image was mainly margin of the necrosis and the dense enhancement, which appeared later, was local tumor recurrence. Thereby changing enhancement patterns suggests tumor recurrence. Another diagnostic modalities such as thallium SPECT, PET or MRS may be performed in differentiating the necrosis and recurrence.

4. CONCLUSION

Difference in response to BNCT between the giant cells and the smaller cells occurred.

There is some possibility that differences in boron microdistribution between these two cell types may have affected the histopathological changes in our case.

In BNCT, endothelial proliferation was predominantly seen in the small vessels, but the cause for this difference is unknown.

If we could find out the mechanisms, we may prevent vascular damage, and improve the therapeutic result of BNCT.

MRI study was difficult to differentiate the radiation necrosis and recurrence. It is noteworthy if the enhancement pattern changed to estimate the patient's prognosis.

REFERENCES

1. A.K. Asbury, R.G. Ojemann, S.L. Nielsen, and W.H. Sweet, Neuropathologic study of fourteen cases if malignant brain tumor treated by boron-10 slow neutron capture radiation, J. Neuropathol. Exp. Neurol., 31(2), pp. 278–303, 1972.
2. K. Kitamura, Y. Nakagawa, T. Minobe, T. Masuda, and T. Kageji, Radiation necrosis and related factors after boron neutron capture therapy, in: Advances in Neutron Capture Therapy Vol. II Chemistry and Biology, Larsson, Crawford and Weinreich ed., Elsevier, Amsterdam, pp. 679–682, 1997.
3. H. Kawano, T. Kubota, K. Sato, T. Goya, S. Arikawa, and S. Wakisaka, Immunohistochemical study of giant cell in glioblastoma, Clin Neuropathol., 14: pp. 118–123, 1995.
4. A. Peraud, K. Watanabe, K.H. Plate, Y. Yonekawa, P. Kleihues, and H. Ohgaki, p53 mutations versus EGF receptor Expression in Giant Cell Glioblastomas, J. Neuropath. Exp. Neurol., 56: pp. 1236–1241, 1997.
5. C. Vroegindeweij, F.J. Wheeler, F. Stecher-Rasmussen, R. Huiskamp, and D. Gabel, A Microdosimetry model for BNCT analysis, in: "Cancer Neutron Capture Therapy", Mishima Y ed., Plenum Press, New York, pp. 485–493, 1996.
6. R.A. Rydin, O.L. Deutsch, and B.B. Murray, The effect of geometry on capillary wall dose for boron neutron capture therapy, Phys. Med. Biol., 21: pp. 134–138, 1976.
7. G.M. Morris, J.A. Coderre, A. Bywaters, E. Whitehouse, and J.W. Hopewell, Boron neutron capture irradiation of the rat spinal cord: Histopathological Evidence of a vascular-mediated pathogenesis, Radiation Research, 146:313–320, 1996.

16

LONG TERM FOLLOW UP IN PATIENTS WHO UNDERWENT BORON NEUTRON CAPTURE THERAPY

Yoshinobu Nakagawa[1], Pooh Kyonghon[1], Katsuji Kitamura[2], Teruyoshi Kageji[2], and Takashi Minobe[3]

[1]Department of Neurosurgery, National Kagawa Children's Hospital Zentsuji 765, Japan
[2]Department of Neurosurgery, Tokushima University Tokushima 770, Japan
[3]Japan Foundation for Emergency Medicine Tokyo, Japan

INTRODUCTION

BNCT is considered as an intra cellular internal radiation therapy. Alpha particles (4He) and recoiling lithium-7 (7Li) nuclei from the reaction between boron-10 and thermal neutrons have a high linear energy transfer (LET) and an associated high relative biological effectiveness (RBE). The great potential advantage of BNCT is its cell level selective destruction of tumor tissue without significant damage to healthy brain tissue. It is well known that for a successful brain tumor treatment, it is essential to secure a sufficient radiation dose (enough alpha particles). This depends on an adequate accumulation of boron in the tumor cells and a satisfactory neutron fluence at the target area.[1] Following the discontinuance of clinical trials in USA, renewal of BNCT was started by H. Hatanaka using a new boron compound, BSH ($Na_2B_{12}H_{11}SH$) at HTR (Hitachi training reactor) in Japan, 1968.[2] The present report is a review of long-term follow up study in the patients with malignant brain tumors treated by boron neutron capture therapy.

METHODS AND MATERIAL

Since 1968, we have treated 173 patients and performed boron-neutron capture therapy (BNCT) using 5 reactors (HTR; Hitachi training reactor, JRR-3; Reactor of Japan Atomic Energy Research Institute, MTIR; Reactor of Musashi Institute of Technology, KUR; Research Reactor Institute of Kyoto University, JRR-2; Reactor of Japan

Frontiers in Neutron Capture Therapy, edited by Hawthorne *et al.*
Kluwer Academic / Plenum Publishers, New York, 2001.

1 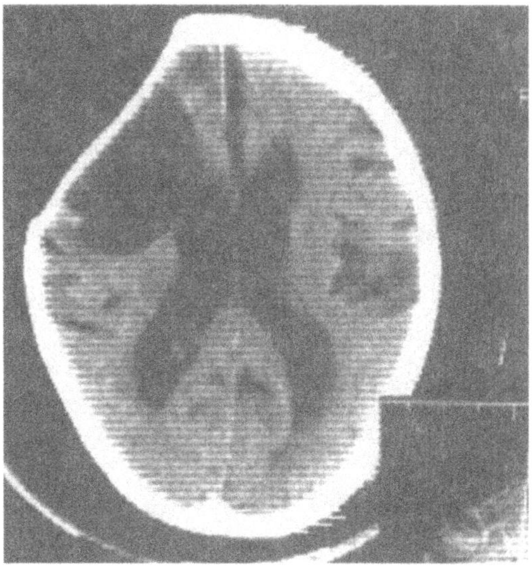

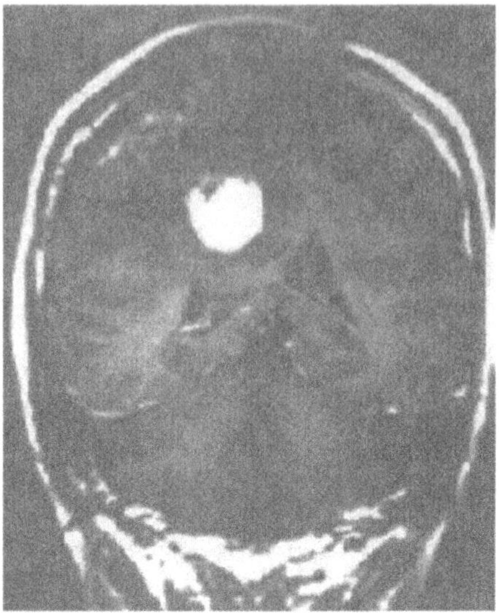 2

Figure 1. CT scan studied 16 years after BNCT. There was no abnormal findings which suspect recurrence of the tumor.
Figure 2. The recurrent tumor was located in the outside of 10 Gy range.

Atomic Energy Research Institute) in Japan. There were 81 patients with glioblastoma, 44 patients with anaplastic astrocytoma and 16 patients with low grade astrocytoma (grade 1 or 2). There were 32 patients with other types of tumor. To investigate the prognostic factors, which correlate to the result of BNCT, we divided the patients into two groups and analyzed the data. One group (group 1): the patients who lived more than 3 years. The other group (group 2): the patients who lived less than 3 years. We analyzed histology of the tumors, age of the patients, radiation time, boron concentration in blood, neutron fluences on the surface of the brain at target point target depth, tumor volume dose in each group.

RESULTS

The overall response rate demonstrated by CT or MRI in the patients with glioma was 60%. Ten patients (12%) of glioblastoma, 22 patients (32%) of anaplastic astrocytoma and 9 patients (56%) of low grade astrocytoma lived more than 3 years. Six patients (5 glioblastoma and one anaplastic astrocytoma) died within 90 days after BNCT. Seven patients (three glioblastomas and four anaplastic astrocytomas) lived more than 10 years. One patient out of seven died in recent two years. The cause of his death was insenescence. There were no abnormal findings demonstrated by follow up CT (Figure 1). The other patients died of recurrence of the tumor 7 years after BNCT (Figure 2). The relapse of the tumor occurred outside the radiation range. Out of 141 patients with glioma treated by BNCT, 33 lived or have lived longer than 3 years after BNCT. There are one patient with brain stem glioma and five patients with meningioma, 3 AVM and 2 other kinds of tumor (Table 1). As prognostic factors, grading of the tumor, ages of the patients and target depth were proved as important factors. The most important factor was tumor volume radiation dose demonstrated by boron n-alpha reaction. The tumor volume was calculated on CT or MRI findings. Twenty-eight patients were treated before the

Table 1. Histological diagnosis and patients lived more than 3 years treated by BNCT (1968–1998)

Glioblastoma	10/81*
Anaplastic astrocytoma	14/44
Lowgrade astrocytoma	9/16
Brain stem glioma	1/8
Meningioma	5/6
PNET	0/6
Metastatic brain tumor	0/5
AVM	3/4
Others	2/4
	Total 44/173 cases

44 patients lived more than 3 years/no. of patients.
*Another 4 patients are still alive.

induction of CT, therefore the patients were excluded in this study. The tumor volume radiation dose in the patients with grade 2 glioma were 11.5 Gy (group 1) vs. 6.7 Gy (group 2), 15.6 Gy (group 1) vs. 11.8 (group 2) in grade 3 glioma and 18.2 Gy (group 1) vs. 9.8 Gy (group 2) in glioblastoma patients (Table 2).

DISCUSSION

The outcomes of patients with astrocytoma grade 2 and grade 3 were better than glioblastoma.[3,4] Glioblastoma is a most poorly differentiated glioma and considered as a carcinoma of the brain. It usually grows in the white matter of the cerebrum and rapidly invades into the normal brain tissue with multiple directions before the time of diagnosis. It is believed that once the enhanced mass completely disappeared on MRI or CT after treatment, however, malignant glioma recurs from the infiltrated tumor cells a few years later. Most of the patients with such an invasive glioma, not only glioblastoma but also anaplastic astrocytoma and low grade astrocytoma are beyond the point of curative surgical removal of the tumor because of the risk of damage to the surrounding normal brain tissue. Therefore the target volume radiation dose is one of the important factors to get the better results. According to the data of the patients who lived more than 10 years after BNCT, physical dose at the target point (= tumor volume radiation dose) was more than 15 Gy in patients with anaplastic astrocytoma and 18 Gy in glioblastoma patients. As for such findings, glioblastoma lets us suppose that infiltrate range into the healthy tissue is wider than anaplastic astrocytoma. If effective dosage of thermal neutron beam can be radiated more widely, better result of BNCT should be obtained.

Table 2. Comparison of tumor volume radiation dose demonstrated by physical dose of boron n-alpha reaction

	Good outcome (lived more than 3 years)	Poor outcome
Grade 2	11.5 ± 1.2 (n = 6)	6.7 ± 3.0 (n = 5)
Grade 3	15.6 ± 9.3 (n = 13)	11.8 ± 7.8 (n = 21)
Grade 4	18.2 ± 9.4 (n = 7)	9.8 ± 7.1 Gy (n = 61)

It is well known that there are two independent parameters for yielding alpha particles. One parameter is the thermal neutron fluence rate in the tumor and in the surrounding tissue. The other parameter is the boron concentration in the tumor and other healthy tissue in its vicinity. To improve the neutron penetration, we decided on the introduction of mixed beams, which contain thermal neutron and epithermal neutron beams. KUR was reconstructed in 1997 and developed to be available with mixed beams. Following the shutdown of the JRR-2, JRR-4 was renewed for medical use in 1998. Both reactors have capacity to yield thermal neutron beam, epithermal neutron beam and mixed beams. The neutron flux became 2–3 times with target point by using surgical procedure and mixed beam in the recent cases. Detailed pharmacokinetics and boron uptakes of BSH were recently reported in the other paper.[5,6] Subcellular boron-10 (BSH) localization in human glioblastoma was studied using laser microprobe mass analysis by Haselsberger *et al.* They reported that boron-10 was found only in the nuclei of neoplastic cells but not in other cell compartment. Localization of BSH was also demonstrated using immunohistochemical method and atomic force microscope. Immunohistochemical study showed marked staining of BSH in the cytoplasm and nucleus of the tumor cells 4 hours after injection of BSH. Positive immuno reaction was scarcely seen in the normal brain tissue around the tumor. Examination using atomic force microscope demonstrated multiple ultra small etch pits dimensionally in the cytoplasm and nucleus.[7] These findings indicated that BSH is selectively absorbed in the tumor cells especially in cytoplasm and nucleus. Next step of our trial is developing microdosimetry and making more detail radiation planning clinical cases.

ACKNOWLEDGMENTS

This work has been continuously supported since 1969 up to 1992 by the Grants-In-Aid from Ministry of Education, Culture and Science, and from 1992 through 1998 by Grants-In-Aid from the Ministry of Health and Welfare of Japan.

Boron-10 compounds have been supplied by the Shionogi Research Laboratories since 1967. This research was carried out as a research program of the Musashi Institute of Technology, Research Reactor Institute of Kyoto University and Japan Atomic Energy Research Institute (JRR-2).

REFERENCES

1. Y. Nakagawa. Boron Neutron Capture Therapy—Clinical Brain Tumor Studies-. Neuro Oncology (33), pp. 105–115 (1997).
2. H. Hatanaka, and Y. Nakagawa. Int. J. of Radiation Oncology Biology, Physics, 28, pp. 1061–1066 (1994).
3. Y. Nakagawa. Advances in Neutron Capture Therapy Volume 1, Medicine and Physics ed. by B. Larsson, Elsvier, pp. 65–70 (1997).
4. T. Kageji, and Y. Nakagawa. J. Neuro Oncology, (33) 117–130 (1997).
5. T. Kageji, B. Otersen, D. Gabel, R. Huiskamp, Y. Nakagawa, and K. Matsumoto. Biochemical et Biophysica Acta 1391 377–383 (1998).
6. Y. Nakagawa, K. Pooh, M. Sone, T. Kageji, M. Nakaichi, H. Takahashi, K. Amemiya, and D. Gabel, Determination of BSH ($Na_2B_{12}H_{11}SH$) absorbed in tumor tissue, *in* "Frontiers in Neutron Capture Therapy," M.F. Hawthorne, K. Shelly, R.W. Wiersema, eds., Kluwer Academic/Plenum Publishers, New York, 2001, pp. 933–937.
7. K. Amemiya, H. Takahashi, M. Nakagawa, N. Yasuda, M. Yamamoto, Y. Nakagawa, T. Kageji, M. Nakaichi, and K. Ogura. Nucl. Instr. and Meth. B 159 75–80 (1999).

17

A CASE OF ASTROCYTOMA

Twelve Year History after BNCT

Shuji Kamano

Neurosurgical Services, Inagi Municipal Hospital
Omaru, Inagi City
Tokyo 206-0801, Japan

1. INTRODUCTION

Sweet *et al.* started clinical trials of BNCT in 1951.[1,2] However, the trials were discontinued after 37 cases were treated in the U.S.A. due to discouraging results. After that, Hatanaka in Japan started the renewed BNCT.[3,4,5] He treated 120 patients between 1968 and 1992. I had joined the Hatanaka group from 1973 to 1987. Meanwhile, one of my patients had received BNCT in 1987 and is alive for 12 years. I would like to review this case. During these years, in 1994, he received his second operation for clinical exacerbation and enlargement of the tumor mass. At that time, viable tumor cells (see End Note) are rarely seen. Currently, no evidence of tumor development has been noted. His recovery has been gradual. At this time CT findings are stable, and no recurrence has been demonstrated.

2. REPORT OF THE CASE

On May 26th, 1987 a 39-year-old man suddenly felt dizzy and had spastic sensations in the occipital region while on a train. He lost consciousness and had left hand convulsions. He recovered his consciousness when the emergency corps had arrived. He was found lying on a bench at a station platform. Shortly thereafter, he was carried to a local hospital. On the next day, he was referred to our hospital.

On admission he showed no evident neurological deficits. The CT revealed a left frontal low-density area without obvious enhancement. The MRI showed a distinct low-signal area in the left frontal lobe. On July 19th 1987, he was operated on using a left frontal craniotomy, suspecting a low-grade astrocytoma. A pinkish elastic well-demarcated tumor was extirpated. The pathological diagnosis was Astrocytoma Grade

Frontiers in Neutron Capture Therapy, edited by Hawthorne *et al.*
Kluwer Academic / Plenum Publishers, New York, 2001.

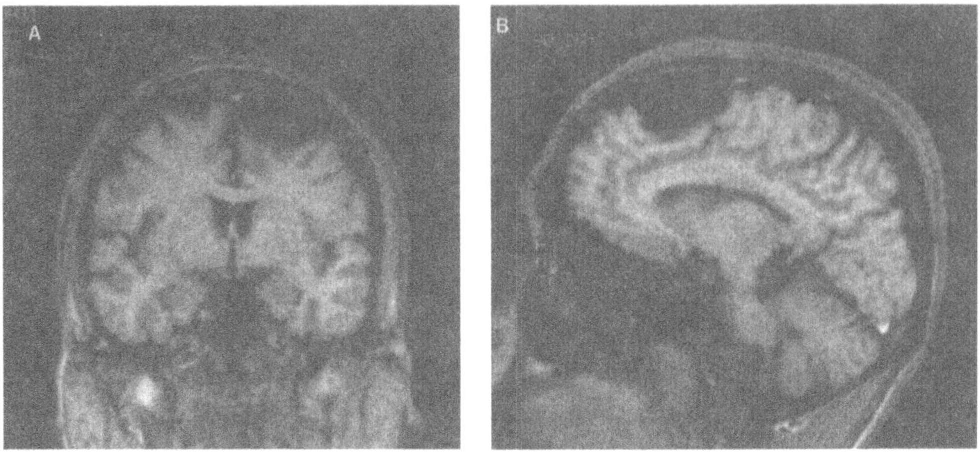

Figure 1. MRI upon the onset prior to the first operation on July 19, 1987.

II with the increased cellularity of astrocytes abundant in various sized round nuclei but without any evident pleomorphic features. Postoperatively, he showed mild grade nominal aphasia and an elevation of GOT and GPT.

Shortly after his recovery, he was transferred to Teikyo University Hospital on July 7th for the purpose of receiving Boron Neutron Capture Therapy under the supervision of Professor Hatanaka. On July 10th he received BNCT using the previous craniotomy site. At that time the thermal neutron fluence on the apex of the left frontal lobe was estimated to be 1.1×10^{13} nvt.; therefore surface fluence was calculated to be 1.58×10^{13} nvt. The detailed techniques of BNCT performed at the reactor in Musashi Institute of Technology during those periods can be found elsewhere.[3,4,5] Upon discharge on July 27th, he was prescribed anticonvulsants to be taken daily. In January of 1988 he returned to full time work, six months after the first operation.

He was said to have had central retinitis of the left eye on March 20th 1991, which was thought to be the side effect of radiation. As the CT showed, on September 16th 1993, the edematous areas surrounding the calcification had increased in size. Administration of prednisolone 20 mg daily was started successively and was effective to some extent. Several times thereafter, generalized convulsions occurred and varieties of anticonvulsants were increasingly added.

As his neurological condition deteriorated progressively, he was readmitted on November 17th 1994 in a wheel chair. Upon admission, he showed anisocoria, right hemiparesis and dysphasia, but he could walk. After admission he was restless while in bed and having nightmares. The CT on admission showed a mass effect with midline shift. The MRI revealed several irregular ring enhancements surrounding the low-signal area. He received the second operation on November 30th 1994. When the dura mater was opened, the craniotomized area was occupied with an amorphous hard mass, which was removed en bloc subtotally. The residual hard rim was also carefully removed. In accordance with the CT findings, the cut surface of the tumor mass was also amorphous and had some calcified parts. The microscopic observation of the surface area showed increased cellularity of glial cells without neuron cells and these features are reported to be consistent with Astrocytoma Grade II with no prominent atypical tumor cells. Although, these glial cells showed markedly altered appearance, i.e., there were enlarged

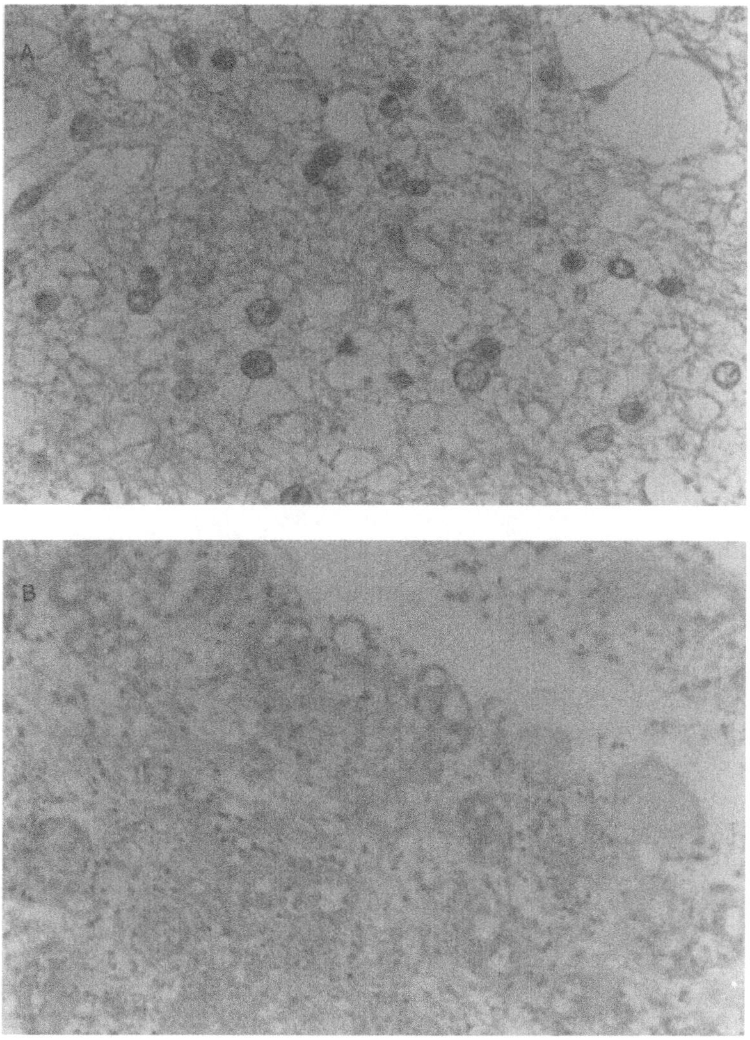

Figure 2a and b. Histology of the first operation (×400 & ×100, HE).

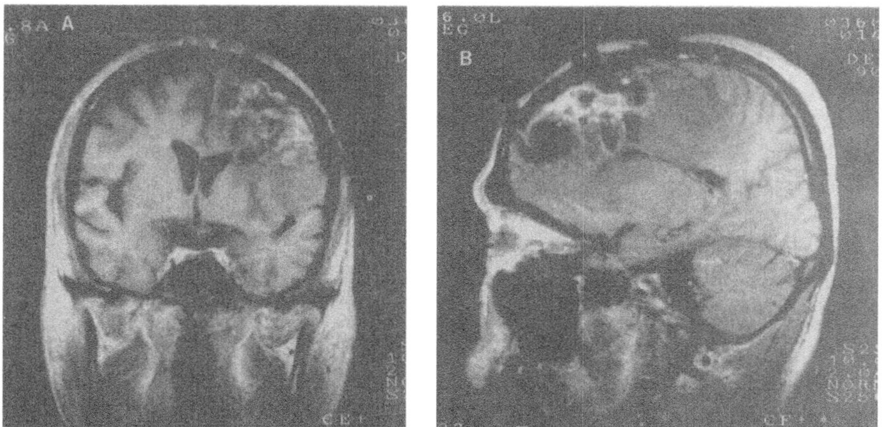

Figure 3. MRI on Nov. 22, 1994 prior to the second operation.

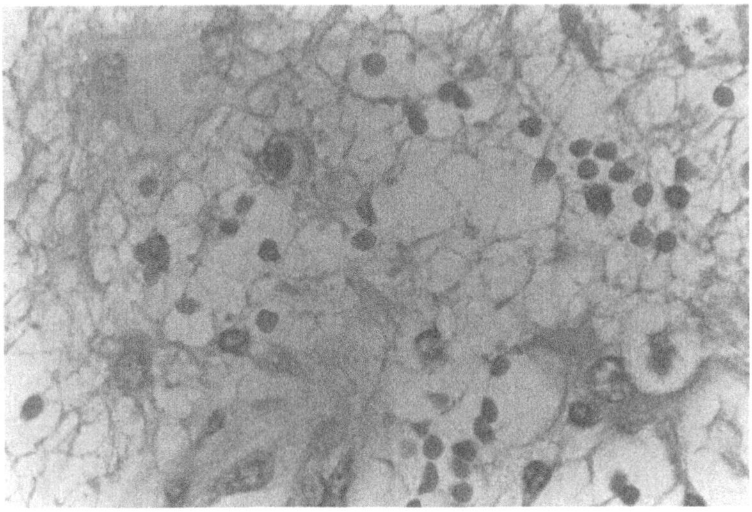

Figure 4. Histology of the second operation (×400, HE).

gigantic nuclei, condensed nuclei with disappearance of cytoplasms, remnant substrates with loss of nuclei or cytoplasms, etc. The cores of the mass contained abundant hyalin substances without cellular elements and among which small calcified substances, weakly stained by PAS, were found.

Postoperatively his speech disturbance and hemiparesis were worsened. By December 12th his recovery was sufficient enough to allow ambulation. Shortly thereafter he became gradually depressed and was found to have a large cyst with an enhanced round rim at the tumor site. A cyst tapping was done and the xanthochromic viscous fluid was obtained. The laboratory results showed pleocytosis (4,834/mm^3) and an abundance of protein (4.3 g/l). No bacilli were detected by the culture of the specimen. Therefore, the cyst punctures were repeated four times until Jan. 12th 1995. These procedures were effective in that the size of the cyst diminished markedly until the end of February. As he was in a relatively apathetic state despite the CT findings, doses of the anticonvulsants were carefully decreased even though concentrations of the anticonvulsants were within optimal treatment dose limits. As time progressed he became more active and vigorously worked out in physical therapy. Since discharge on Sept. 5th 1995, he has gradually shown improvement in motor activity and speech.

Currently, he has returned to full time employment in physical labor. The CT images now show that the tumor cites are replaced with the enlarged anterior horn of the left lateral ventricle. His CT findings have remained stable for the past 5 years.

3. DISCUSSION

This case showed a rather well defined tumor of low grade. The development of a space-occupying lesion was noticed five years after treatment. The impression of the reoperation was in agreement with the median survival (4.5–9.1 years)[6,7,8,9,10,11] of Astrocytoma Grade II after surgery and radiotherapy. At that time, the effectiveness of BNCT in this case appeared not different from conventional radiation as far as the interval for the recurrence was concerned. From the histological viewpoint, the pathological findings

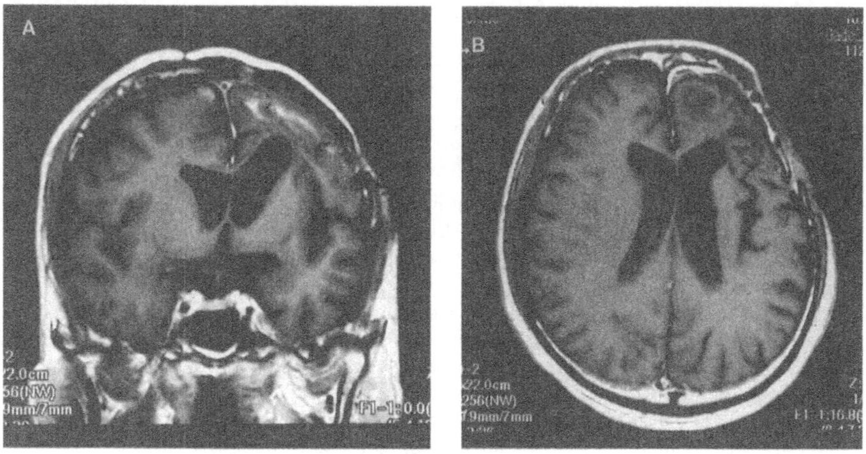

Figure 5. Gd enhanced MRI on July 23, 1998.

taken at the second operation were mostly in accordance with Grade III of Oboshi's criteria for evaluation of therapeutic effects of radiation,[12] which is employed at the Japan National Cancer Center Hospital. The Grade III implies an expression of the effectiveness of radiation therapy, that is, the tissue shows markedly altered appearance, in which presumably non-viable tumor cells are present singly or in small clusters and viable cells (see End Note) are hardly seen.

He has shown no sign of recurrence in the last 5 years. His CT findings have remained stable for 5 years and the tumor cites have been replaced with the enlarged anterior horn of the left lateral ventricle.

The efficacy of the radiation therapy for the treatment of low-grade astrocytoma is still in dispute. Moreover, an uptake study of the Boron compound to low-grade Astrocytoma reported an insufficient uptake of the boron compound for BNCT.[13] Fankhauser[14] proposed exclusion of low-grade astrocytoma for future trials. In spite of that, the case presented here shows some histological effects of radiation necrosis. Even if a low-grade Astrocytoma is sometimes said to be a benign Astrocytoma, less than two thirds of the patients with this tumor are able to survive more than five years.[6,7,8,9,10,11] Therefore, heavy particle radiation, including BNCT, should be considered as one of the therapeutic options for treatment of low-grade astrocytomas.

END NOTE

The "viable cells" mean the cells whose cellular configuration is fairly well preserved and include those displaying swelling of nucleus and cytoplasm, vacuolation, multiple nuclei, etc., since there is no proof that these changes are irreversible or lethal to the cells (cited from Shimosato *et al.*[9]).

ACKNOWLEDGMENT

I would like to acknowledge to the late Professor Hiroshi Hatanaka for the years spent under his tutorage at Teikyo University. I thank Masao Matsutani, Professor of

Saitama Medical School for helpful comments in pathohistology and Chad D. Nelson for preparing this paper.

REFERENCES

1. W.H. Sweet, and M. Javid, The possible use of neutron capturing isotopes such as boron-10 in the treatment of neoplasms. I. Intracranial tumor. J Neurosurg 9:200–209, 1952.
2. W.H. Sweet, Early history of development of boron neutron capture therapy of tumors. J Neurooncol 33:19–26, 1997.
3. H. Hatanaka, S. Kamano, K. Amano, S. Hojo, K. Sano, S. Egawa, and H. Yasukochi, Clinical experience of boron-neutron capture therapy for gliomas—A comparison with conventional chemo-immuno-radiotherapy. In: "Boron-neutron capture therapy for tumors." H. Hatanaka ed. Nishimura, Niigata, pp. 349–379, 1986.
4. H. Hatanaka, and Y. Nakagawa, Clinical results of long surviving brain tumor patients who underwent boron neutron capture therapy. Int J Radiat Oncol Biol Phys 28:1061–1066, 1994.
5. Y. Nakagawa, and H. Hatanaka, Boron neutron capture therapy. Clinical brain tumor studies. J Neurooncol 33:105–115, 1997.
6. H.J. Eyre, J.J. Crowlen, J.J. Townsend, J.R. Eltringham, R.A. Morantz, S.F. Schulman, J.M. Quagliana, and M. Al-Sarraf, A randomized trial of radiotherapy versus radiotherapy plus CCNU for incompletely resected low-grade gliomas: A southwest oncology group study. J Neurosurg 78:909–914, 1983.
7. P. Janny, H. Cure, M. Mohr, N. Heldt, F. Kwiatkowski, J.J. Lemaire, R. Plague, and R. Rozan, Low grade supratentorial astrocytomas. Management and prognostic factors. Cancer 50:1937–1945, 1994.
8. B.M. McCormick, D.C. Miller, G.N. Budzilovich, G.J. Voorhees, and J. Ransoff, Treatment and survival of low-grade astrocytoma in adult 1977–1988. Neurosurgery 31:636–642, 1992.
9. J.M. Piepmeier, Observation on the current treatment of low-grade astrocytic tumors of the cerebral hemispheres. J Neurosurg 67:177–181, 1987.
10. R. Soffieti, A. Chio, M.T. Giordana, E. Vasario, and D. Schiffer, Prognostic factors in well-differentiated cerebral astrocytomas in the adults. Neurosurgery 24:686–692, 1989.
11. A.C.Whitton, Low grade glioma of the cerebral hemisphere in adults: A retrospective analysis of 88 cases. Int J Radiat Oncol Biol Phys 18:783–786, 1990.
12. Y. Shimosato, S. Oboshi, and K. Baba, Histolgical evaluation of effects of Radiotherapy and chemotherapy for carcinomas. Jap J Clin Oncol 1:19–35, 1971.
13. G. Stragliotto, and H. Fankhauser, Biodistribution of boron sulfhydril for boron neutron capture therapy in patients with intracranial tumors. Neurosurgery 36:285–293, 1995.
14. H. Fankhauser, P.R. Gavin, and G. Stragliotto, Proposal of a treatment protocol for boron neutron capture therapy of supratentorial gliomas. Strahlenther Onkol 169:71–75, 1993.

TREATMENT PLANNING AND CLINICAL CONSIDERATIONS

18

TREATMENT PLANNING FOR THE FIRST GROUP OF PATIENTS IN THE EUROPEAN GLIOMA TRIAL AT THE HFR PETTEN

C. Vroegindeweij[1], P. Watkins[1], K. Hideghéty[2], W. Sauerwein[2], and R. L. Moss[1]

[1]Joint Research Centre
European Commission
Petten, The Netherlands
[2]Universitätsklinikum, Essen
Germany

1. INTRODUCTION

As part of the overall treatment planning system, the "bnct-rtpe/rtt_MC" treatment planning system developed by INEEL has been implemented, validated and used to perform the planning of BNCT irradiations of human patients. Together with this system, several tools have been developed at JRC to meet the requirements defined by the responsible radiotherapist. The definitions used in the treatment planning are described in the phase I protocol[1] and the Standard Operation Procedures.[2] This paper gives a short summary of the methodology used in the Petten treatment planning as well as an overview of the delivered doses to the patients.

2. METHODS

Preparation of data for the treatment planning—The MRI scans before and after tumour removal, as well as the planning CT scans are made in the referring hospital. The hardcopies are sent to the coordinating medical centre in Essen by express post and the images of the CT scans are sent by Internet to JRC. After defining the target region and the organs-at-risk on the CT scans, the Essen clinicians send the hardcopies to JRC. The data sent by Internet has to be converted into the correct format for importing into the treatment planning program.[3]

Prescribed parameters—In the ongoing phase I trial at Petten, the planning procedure is very complex and requires an intensive effort to obtain the individual topography that can be used as input for the program and which satisfy the requirements of the

Frontiers in Neutron Capture Therapy, edited by Hawthorne *et al.*
Kluwer Academic / Plenum Publishers, New York, 2001.

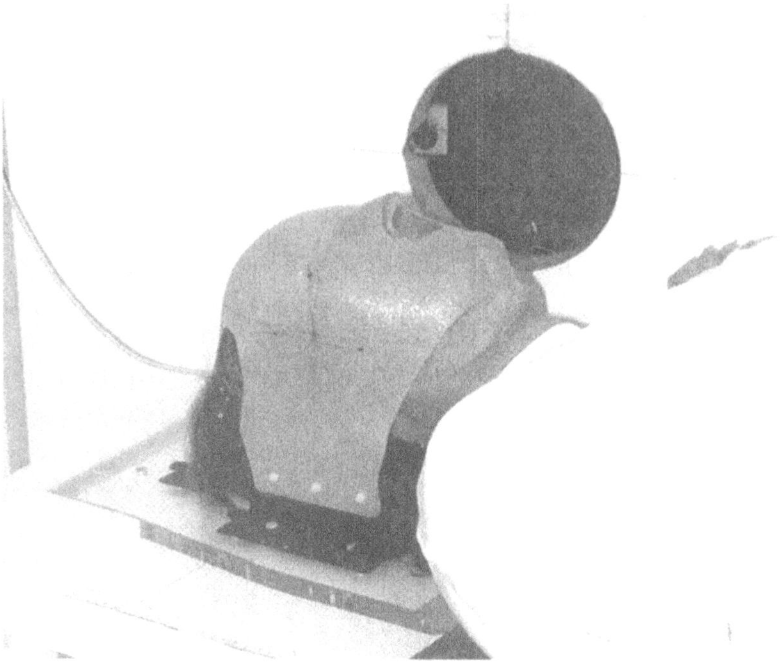

Figure 1. Patient position during treatment.

radiotherapists. After defining the volumes of interest, the planning is as similar as possible for each patient in order to obtain a clear outcome of the phase I healthy tissue tolerance study. All ten patients in the first dose group have been treated with a 12-cm diameter beam at a distance of 30 cm from the beam exit. All patients have been positioned in lateral recumbency. The first five patients have been treated with a single beam directed (if possible) towards the middle of the target. The second five patients have been treated with two beams generating two thermal maxima. A typical patient position is showed in Figure 1. An example of the beam arrangement as displayed by the computer for a single and a two beam irradiation are given in Figures 2 and 3.

3D-reconstruction—The 3D reconstruction is made in the program "bnct_rtpe" as developed by INEEL.[3] In the 3D reconstruction of the head, 3 volumes are defined: the

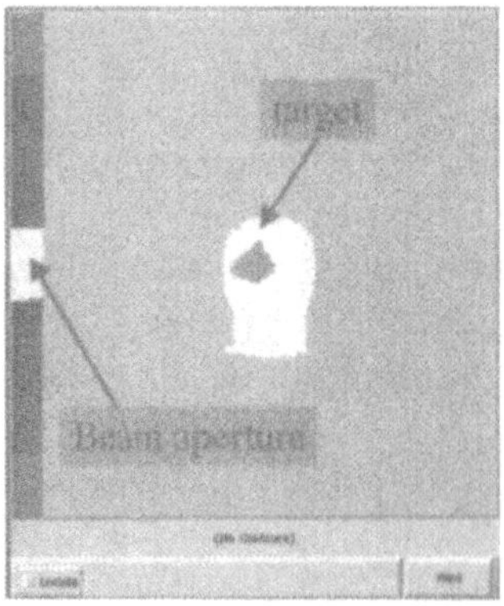

Figure 2. Frontal cross section of the beam arrangement for a patient using a single beam towards middle of the target.

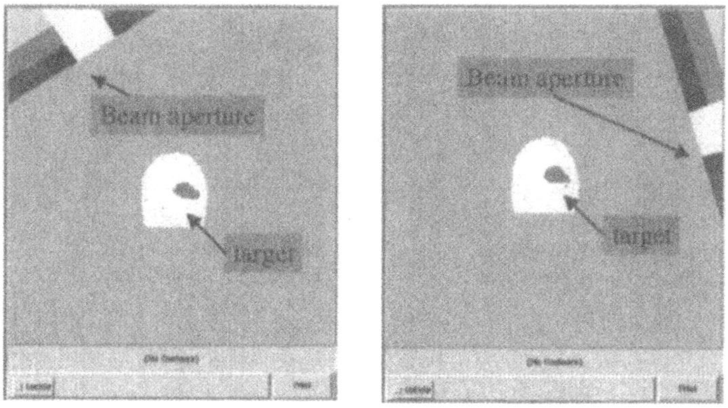

Figure 3. Frontal cross-section of the beam arrangement for a patient using two beams.

target (which is the pre-operative tumour with a safety margin of 2 cm), the brain and the remaining part of the head. The material composition in the healthy brain and in the target volume are chosen identically, as after surgery the volume of tumour in the target is very small. The material compositions used are according to ICRU-46.[4]

Transport calculations—The particle transport calculations are performed with the computer code "rtt_MC".[5] In the transport calculations a homogeneous boron concentration of 10 ppm is assumed in the whole brain, while no boron is defined in the other parts of the head. The neutron source is defined at the interface of the filter and the collimator, which is roughly 50 cm from the beam exit. The beam collimator and the treatment room are modelled (see Figure 4). Besides the dose-volume histograms and the dose depth curves, the dose at several organs-at-risk such as eyes, ears and glands as well as beam edge doses at the skin were separately calculated to estimate the possible radiation risks.

Post-processing and visualisation—A program called "dose" has been developed by JRC (using partly the INEEL tools "histo" and "simplot") to decrease the manual workload in the post-processing of the data generated by "rtt_MC". The program creates a

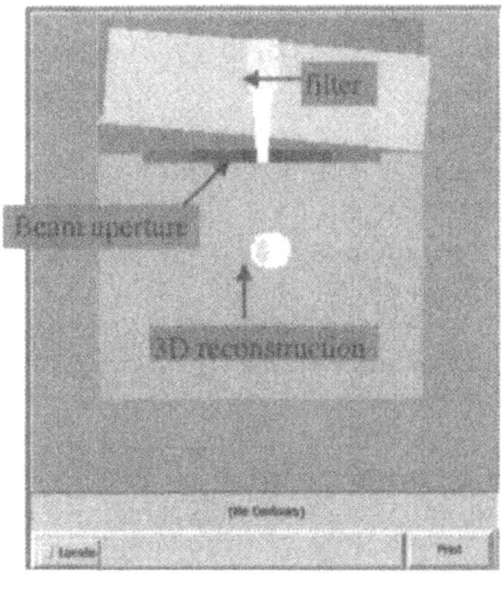

Figure 4. Horizontal cross-section of the treatment room and patient model as used in the Monte Carlo program.

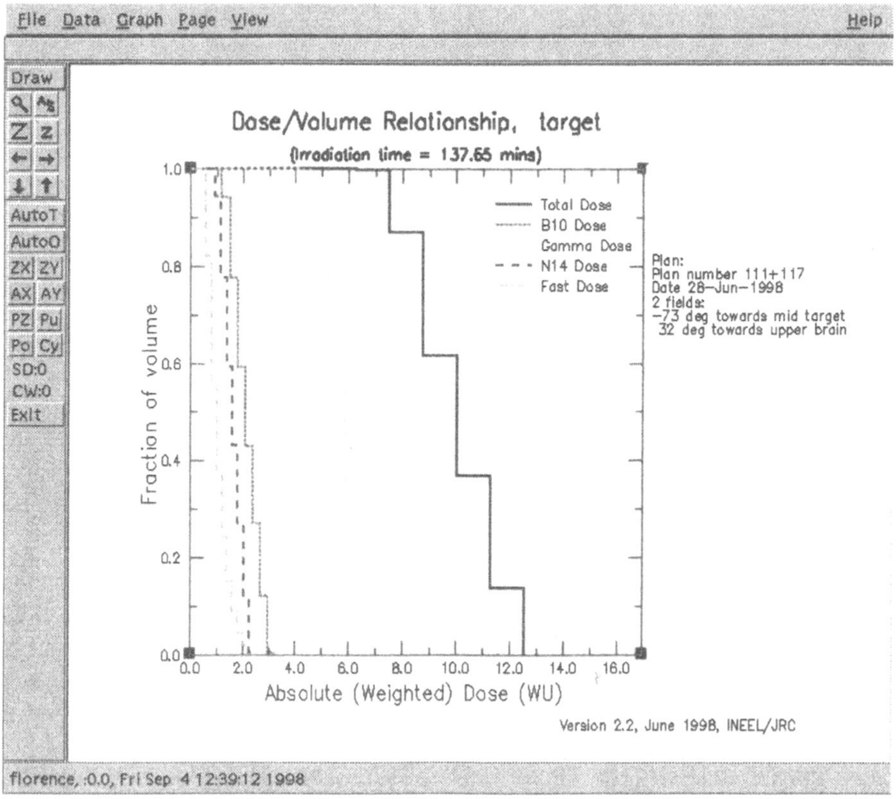

Figure 5. Example of a dose volume histogram for a patient treated with 2 beams.

file with the calculated relative doses at the points of interest and the maximum, minimum, average, mean and median dose in different regions. It also calculates the irradiation time needed to obtain the required dose, and shows automatically the dose-volume histograms (see Figure 5) and dose depths curves on the computer screen. In the output of "dose", information about the patient and the planning are automatically placed.

Determine treatment parameters—A spreadsheet has been developed to generate the absolute planned doses at the organs-at-risk (Table 1), coordinates needed for the positioning,[7] records of the delivered dose per fraction and total over all fractions for the measured boron concentration and actual number of monitor units, and the report forms for our control board (Table 2). The input for this spreadsheet is created by the program "dose" and use the measured boron concentration and the monitor units actually delivered.

Reports and documentation—Three reports are produced during planning and treatment. Firstly, information needed to decide which beam configuration will be used during treatment has to be produced. Several configurations are analysed and as a minimum the dose-volume histograms in the target and "normal" brain, the dose depth curves along the beam centre lines, the planned dose at the organs-at-risk and the iso-dose curves for the boron, total and weighted dose are calculated. These calculations are made with low statistics to save computer time. After a certain beam configuration is chosen, a so-called

Table 1. Example of the planning summary

Patient name					PRE - IRRADIATION PLANNING		Spreadsheet version TP6.5
Date of birth	.	final calculation			Weighting factor used for DWU		BNCT Rtpe version 2.2f
Date of calculation	.				W_α	1	rtt MC version 106 (5 Aug '97)
Plan number	5_final				W_N	3.9	source definition Oct '97
Planning Directory	.				W_n	3.9	florence.jrc.nl
Beams	B11 A000L73X003Y074Z057	coll = 12 cm	WSD = 30 cm	weight = 0.50	CF	0.37/0.81	Planner: Corine Vroegindeweij
	B17 A000R32X-11Y-37Z132	coll = 12 cm	WSD = 30 cm	weight = 0.50	D_B @ DGIP	0.062 Gy/min	Power: 45.00 MW
					planned irr. time	137.65 min	Boron conc: 30.00 ppm

signature Planner			signature MP				signature PCI					
		mat. (ICRU-46)	DB (Gy)	Dα (Gy)	DN (Gy)	Dn (Gy)	DT (Gy)	CF	DWU (WU)	coordinates [cm] x	y	z
Maximum dose	(protocol)		8.6	5.8	0.6	0.9						
MAD (DGIP)		brain	8.6	5.7	0.6	0.4	15.3	0.37	12.6			
Prescription point		brain	5.8	5.5	0.4	0.2	11.8	0.37	9.8			
Skin doses:												
Beam entrance	B11	tissue	3.3	3.8	0.4	0.9	8.4	0.81	11.5	0.3	7.2	5.7
Beam edge	B11	tissue	1.4	2.5	0.2	0.4	4.4	0.81	5.8	0.3	7.3	-0.6
Beam edge	B11	tissue	2.5	2.8	0.3	0.5	6.0	0.81	7.7	0.3	4.7	11.2
Beam exit	B11	tissue	0.1	1.3	0.0	0.0	1.4	0.81	1.5	0.3	-7.9	0.9
Beam entrance	B17	tissue	3.3	3.0	0.4	0.7	7.3	0.81	9.7	-1.1	-3.5	13.0
Beam edge	B17	tissue	1.6	2.0	0.2	0.4	4.1	0.81	5.4	-1.1	-7.6	8.2
Beam edge	B17	tissue	1.9	2.1	0.2	0.3	4.5	0.81	5.8	-1.1	3.6	12.8
Beam exit	B17	tissue	0.3	1.4	0.0	0.0	1.8	0.81	1.9	-1.1	6.5	-3.2
Left eye		tissue	2.0	1.7	0.2	0.1	4.1	0.81	4.7			
Right eye		tissue	0.4	1.0	0.0	0.0	1.5	0.81	1.6			
Left ear		tissue	3.0	3.7	0.3	0.1	7.1	0.81	7.8			
Right ear		tissue	0.6	2.1	0.1	0.0	2.8	0.81	3.0			
Optic chiasm		brain	2.3	4.0	0.1	0.1	6.6	0.81	6.8			
Pituitary gland		brain	2.1	3.8	0.1	0.1	6.1	0.81	6.4			
Left salivary gland		brain	1.9	2.1	0.1	0.1	4.1	0.81	4.4			
Right salivary gland		brain	0.1	1.1	0.0	0.0	1.2	0.81	1.2			
Hot spots			not existing									

Table 2. Example of the forms sent to the monitoring board

Glioma Treatment at Petten with BNCT Phase I Trial

Case Report Form Report on BNCT

Patient ID

Dates of BNCT

Fraction	1	2	3	4
Date (dd/mmm/yy)	.	.	.	.

Collimator 12 cm

Boron concentration 27.9 ppm (averaged over all four fractions)

Target

	x-coord	y-coord	z-coord	D_B	D_g	D_N	D_n	D_T	D_{WU}
Dose Group Identification Point	0.5	4.5	6.5	8.0	5.7	0.6	0.4	14.7	12.4
Prescription Point	0.3	2.6	4.3	5.4	5.5	0.4	0.2	11.4	9.6

Organs at risk

		D_B	D_g	D_N	D_n	D_T	D_{WU}
Skin doses:							
Beam entrance	B11	3.0	3.8	0.4	0.9	8.2	11.3
Beam edge	B11	1.3	2.5	0.2	0.4	4.3	5.7
Beam edge	B11	2.3	2.8	0.3	0.5	5.8	7.6
Beam exit	B11	0.1	1.3	0.0	0.0	1.4	1.5
Beam entrance	B17	3.0	3.0	0.4	0.7	7.1	9.5
Beam edge	B17	1.5	2.0	0.2	0.4	4.0	5.3
Beam edge	B17	1.7	2.1	0.2	0.3	4.4	5.7
Beam exit	B17	0.3	1.4	0.0	0.0	1.8	1.9
Left eye		1.9	1.7	0.2	0.1	3.9	4.6
Right eye		0.4	1.0	0.0	0.0	1.5	1.6
Left ear		2.7	3.7	0.3	0.1	6.9	7.7
Right ear		0.5	2.1	0.1	0.0	2.8	3.0
Optic chiasm		2.1	4.0	0.1	0.1	6.4	6.7
Pituitary gland		2.0	3.8	0.1	0.1	6.0	6.2
Left salivary gland		1.7	2.1	0.1	0.1	4.0	4.3
Right salivary gland		0.1	1.1	0.0	0.0	1.2	1.2
Hot spots		not existing	-	-	-	-	-

Target Volume 84 cm3

		DB	Dg	DN	Dn	DT	DWU
Dmax		8.0	5.9	0.6	0.5	14.7	12.5
Dmin		2.5	3.8	0.2	0.1	6.8	6.0
Dmedian		5.2	5.2	0.4	0.2	11.0	9.3
Daverage		5.3	5.1	0.4	0.2	11.0	9.4
Dmodal		4.4	5.4	0.3	0.1	9.5	8.1

Brain without Target Volume 1397 cm3

		DB	Dg	DN	Dn	DT	DWU
Dmax		8.3	5.9	0.6	0.8	15.0	13.0
Dmin		0.1	0.5	0.0	0.0	0.7	0.7
Dmedian		2.8	3.5	0.2	0.1	6.6	5.8
Daverage		3.2	3.4	0.2	0.2	6.9	6.0
Dmodal		1.2	4.3	0.1	0.0	3.6	4.4

signature PCI

"final treatment planning report" is produced. This report is based on a calculation with better statistics and contains the beam position in a 2-dimensional reconstruction (frontal cross section, as in Figures 2 and 3), pre-treatment planning (doses at organs-at-risk), positioning information, dose-volume histograms for normal brain, target and whole brain, dose-depth curves for each beam axis, iso-dose curves for all dose-components on the geometrical reconstruction of the head at the slice closest to the thermal neutron peak, and iso-dose curves for the boron, total and weighted dose every centimetre axially.

3. CONCLUSIONS

The treatment planning of the first 10 patients has been successfully performed using the INEEL treatment planning software and new developed tools at JRC. The patients treated with two beams under the given condition have a larger brain volume receiving a high dose than the single beam irradiations. The average dose is roughly doubled for two beams than for a single beam, while the maximum and minimum doses in the normal brain remain the same.

ACKNOWLEDGMENT

This work is supported by the European Commission through the BIOMED II programme.

REFERENCES

1. Protocol 11961: European Organisation for Research and Treatment of Cancer, BNCT Study Group, Postoperative Treatment of Glioblastoma with BNCT at the Petten Irradiation Facility, Phase I Clinical Trial, 5 August 1996.
2. R. Moss, F. Stecher-Rasmussen, W. Sauerwein, D. Gabel, P. Watkins, and C. Vroegindeweij, "Standard Operating Procedure followed by the Petten BNC-Team during Patient Treatment", these proceedings.
3. P. Watkins, and C. Vroegindeweij, "Medical Image Transfer and Conversion for BNCT Treatment Planning at Petten", these proceedings.
4. D.E. Wessol, R.S. Babcock, F.J. Wheeler, G.J. Harkin, L.L. Voss, and M.W. Frandsen, "BNCT_Rtpe: BNCT Radiation Treatment Planning Environment User's Manual Version 2.2", http://id.inel.gov/cart/rtpe-manual/, February 21, 1997.
5. ICRU-report 46: "Photon, Electron, Proton and Neutron Interaction Data for Body Tissues", International Commission on Radiation Units and Measurements, Bethesda, Maryland, 1992.
6. F.J. Wheeler, "Radiation Transport in Tissue by Monte Carlo", EGG-BNCT-11178, January, 1994.
7. P. Watkins, C. Vroegindeweij, and K. Hideghéty, "Patient Positioning at the HFR Petten", these proceeding.

19

QUALITY ASSURANCE FOR CLINICAL DOSIMETRY OF THE EUROPEAN TRIAL ON BNCT IN PETTEN

J. Rassow[1], F. Stecher-Rasmussen[2], K. Appelman[2], S. Garbe[1], J. Morrissey[3], R. Moss[3], K. Ravensberg[2], F. Roserot de Melin[3], W. Sauerwein[1], C. Vroegindeweij[3], and P. Watkins[3]

[1]University Hospitals Essen, Germany
[2]Netherlands Energy Research Foundation ECN
Petten, The Netherlands
[3]Joint Research Centre
Institute for Advanced Materials
Petten, The Netherlands

SUMMARY

The Council Directive on health protection 97/43/EURATOM requires for radiotherapy quality assurance programs for performance and safety characteristics including acceptance and repeated tests. For BNCT at the nuclear reactor in Petten such a programme is developed on the basis of IEC publications for medical electron accelerators. A comparison is given for requirements for accelerators and BNCT units indicating items which are not transferable, equal, additional or different.

1. INTRODUCTION

The Council Directive on health protection 97/43/EURATOM[1] of the European Union is based on the Recommendations of ICRP Report 60[2] and requires explicitly appropriate quality assurance programs for performance and safety of radiotherapy units including testing of performance characteristics on a regular basis. Especially for medical electron accelerators such programmes are adopted internationally as IEC Publications.[3,4,5] For the European BNCT trial at the nuclear reactor in Petten such a quality assurance programme was developed based as near as possible on the accelerator procedures. The fundamental differences of dosimetry for conventional photon/electron therapy and fast neutron and BNCT therapy must be taken into account. It can be shown

Frontiers in Neutron Capture Therapy, edited by Hawthorne *et al.*
Kluwer Academic / Plenum Publishers, New York, 2001.

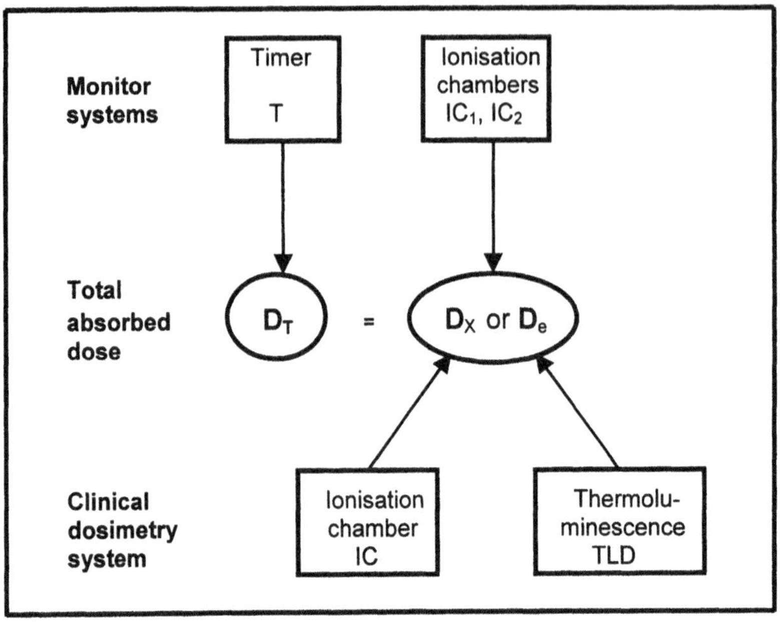

Figure 1. Scheme of dosimetry for medical electron accelerator (D_X = absorbed dose for X-rays, D_e = absorbed dose for electrons, ⟶: strong dependence).

that several items of accelerators are not applicable for BNCT and others are additional, while a long data series is transferable.

2. FUNDAMENTAL DIFFERENCES OF CLINICAL DOSIMETRY FOR MEDICAL ACCELERATORS; FAST NEUTRON THERAPY AND BNCT FACILITIES

2.1. Medical Electron Accelerator (Figure 1)

The typical dosimetrical aspects are:

- The primary dose component (D_X or D_e) has the same LET as the secondary dose component (D_e or D_X). A differentiation between both components is, therefore, irrelevant
- The sum of primary and secondary dose components are directly measurable by monitor and clinical dosimetry systems

2.2. Cyclotron Based Fast Neutron Therapy Facility (Figure 2)

- Two dose components (D_n and D_g) of different LET are relevant
- Essentially only a combination of both dose components is measured by monitors, but in clinical dosimetry systems both relevant dose components are to be measured separately
- The TLD two-peak method indicates separately both dose components of different LET based on the different LET repines of the two main glow peaks as shown by Rassow[6]

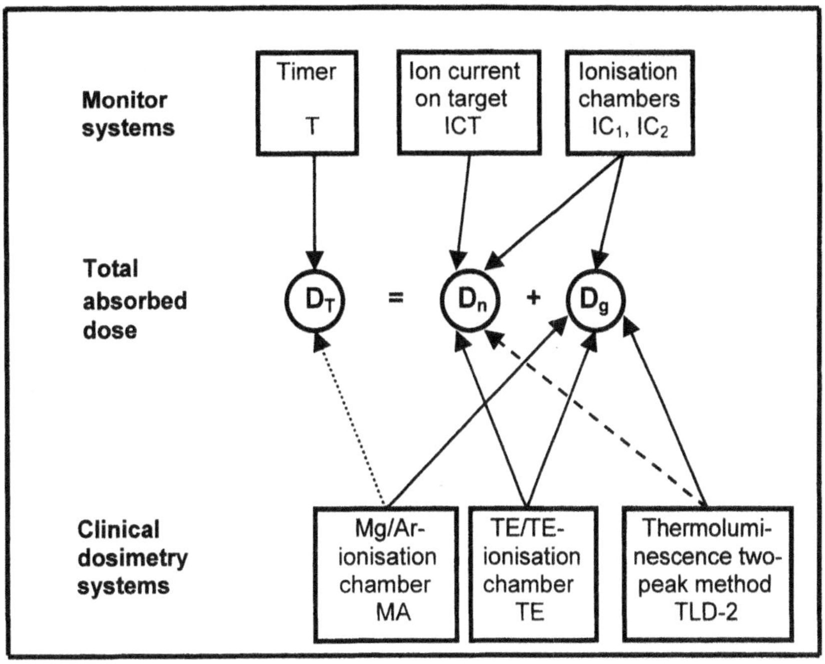

Figure 2. Scheme of dosimetry for medical cyclotron for fast neutron therapy (D_n = neutron absorbed dose, D_g = gamma ray absorbed dose, ⟶: strong dependence, - - →: weak dependence, - - →: very slight dependence).

2.3. BNCT at Nuclear Reactor Facility (Figure 3)

- Four dose components (D_B, D_g, D_n, D_N) of different LET are relevant
- Two dose components (D_B, D_N) are only calculable via measured thermal neutron fluence, but not directly measurable with respect to the particle range (a few μm)
- The main (reference) dose component $D_B(c_B)$ is directly depending on boron-10 concentration c_B and calculable only based on the assumption, that the measured c_B in blood is equal to the c_B in target cells (the real inter- and intracellular distribution of BSH-molecules is on principle unknown)

3. ORDER OF MAGNITUDE OF DOSE COMPONENTS AND ACHIEVABLE BORON-10 CONCENTRATIONS C_B IN PATIENTS

3.1. Order of Magnitude of Dose Components (D_B, D_g, D_n, D_N) at Representative Points in Patients (Table 1)

- All four dose components are calculated for many points of interest in the planning target volume, the treated volume and at organs of risk
- The dose components are given as orientation only for the dose group identification point (DGIP: 2.4 cm depth, point of maximum thermal neutron fluence) and the target prescription point (TPP: centre of target volume)

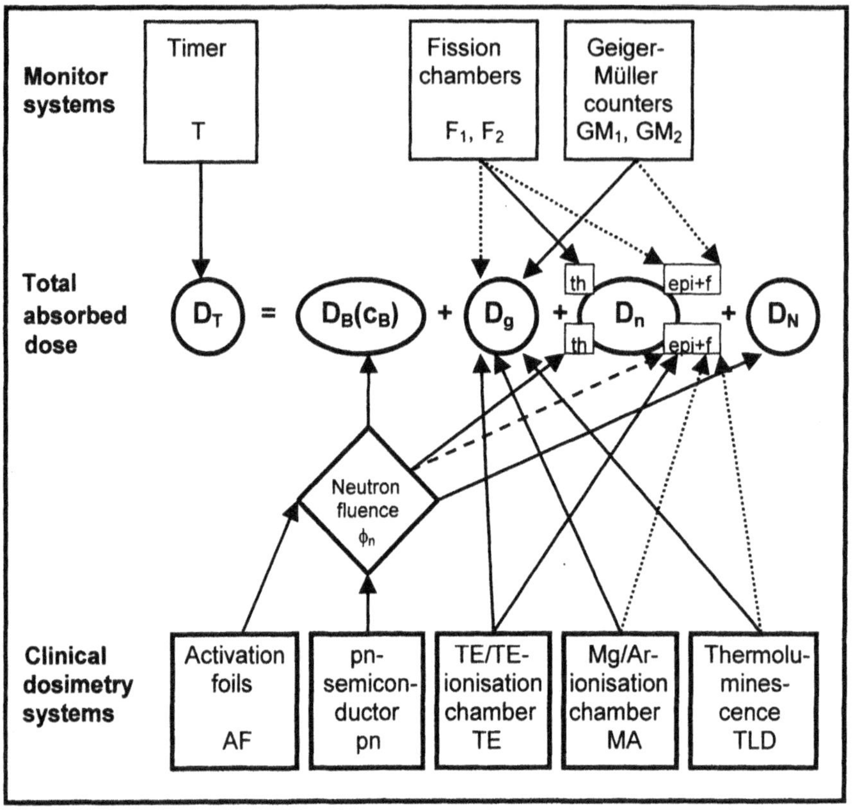

Figure 3. Scheme of dosimetry for BNCT at a nuclear reactor. (D_B = boron-10 neutron capture absorbed dose, c_B = boron-10 concentration, D_g = gamma ray absorbed dose, D_n = neutron absorbed dose, D_N = nitrogen (nuclear reaction) absorbed dose, th = thermal neutron sub-component, epi + f = epithermal + fast neutron sub-components, ⟶: strong dependence, - - →: weak dependence, - - →: very slight dependence).

Table 1. Calculated absorbed dose components for BNCT treatment of a brain tumour by single-beam technique (patient example A) and double-beam technique (patient example B) for real boron-10 concentrations c_B with collimator ϕ = 12 cm at dose group identification point (DGIP) and at target prescription point (TPP). Reference absorbed dose is D_B (c_B = 30 ppm) = 8.6 Gy at DGIP

Patient / specification point	D_B / Gy	D_g / Gy	D_n / Gy	D_N / Gy	D_T / Gy	D_{WU} / WU
A / DGIP	8.70	4.33	0.50	0.58	14.11	11.76
/ TPP	4.81	3.72	0.13	0.32	8.98	7.25
B / DGIP	8.32	4.54	0.41	0.58	13.84	11.46
/ TPP	5.12	4.33	0.18	0.36	9.98	8.30

Dose components		Weighting factors
D_B : boron-10 absorbed dose	D_T : total absorbed dose	w_B = 0.37 (brain)
D_g : gamma ray absorbed dose	D_{WU}: biologically weighted dose	= 0.81 (other organs)
D_n : (fast) neutron absorbed dose		w_g = 1.00
D_N : nitrogen absorbed dose	WU : weighted units	w_n = w_N = 3.9

Table 2. Average Boron Concentration c_B for Irradiation Durations t per Beam (Planned $c_B = 30\,ppm$)

Patient	A	B	C	D	E	F	G	H	I
c_B (ppm)	32.2	30.6	30.4	32.3	32.3	30.4	29.0	27.3	27.9
nr. of beams	1	1	1	1	1	2	2	2	2
t (min)	72.5	73.9	74.6	73.1	72.0	68.3	64.9	70.0	68.9

- The total absorbed doses D_T are given in comparison to the biologically weighted doses D_{WU} together with the weighting factors (w_B, w_g, w_n, w_N) according to EORTC Protocol 11961[7]

3.2. Representative Depth Dose Distribution of the Four Dose Components

- The depth d_{max} of absorbed dose maximum and the half value depth d_{50} are different for all four dose components
- d_{max} and d_{50} are increasing according to the sequence D_n, D_N, D_B, D_g
- The depths d_{max} and d_{50} for the total absorbed dose D_T are about 2.4 cm and 6.8 cm respectively

3.3. Actual Boron-10 Concentrations c_B Achieved for the First Nine BNCT Patients (Table 2)

- The maximum deviation Δc_B of boron-10 concentration c_B, averaged for all four fractions for each patient, is 7% (standard deviation for the determination of c_B is about 3%)
- The percentage deviations of c_B and the boron-10 neutron capture absorbed dose $D_B(c_B)$ are identical

3.4. Beam Sequence for Two-beam Treatments (Table 3)

- The uniformity of boron-10 concentration c_B for both beams of beams A and B for two-beam treatments is best for the sequence AB | BA | BA | AB (thus: for final calculation of D_B the actual average c_B can be taken for both beams)

Table 3. Mean boron-10 concentration c_B (in the midst of a single-beam irradiation) in blood for 2-beam BNCT treatment (patient example C) (planned $c_B = 30\,ppm$)

Fraction nr.	1	2	3	4	all
c_B / ppm for sequence AB|AB|AB|AB					
beam A	33.28	26.09	33.35	30.78	30.88
beam B	32.85	25.75	32.00	29.53	30.03
c_B / ppm for sequence AB|BA|BA|AB					
beam A	33.28	25.75	32.00	30.78	30.45
beam B	32,85	26.09	33.75	29.53	30.46

Table 4. Functional performance characteristics according to IEC 976[3]. Exceptions for BNCT facilities compared to medical electron accelerators

Omission of all dependencies, which are not relevant for actual BNCT treatments, i.e. dependencies on:
• radiation type • radiation (nominal) energy • absorbed dose rate • dose monitor stability (throughout the day and after high dose irradiation) • flattening and wedge filter • (continuously variable) radiation field • angular position of gantry • angular position of beam limiting device • moving beam technique • isocentre • indication of distances from focus and/or isocentre • light field indication • zero positions of rotational scales • isocentric and eccentric rotation of patient table • linear scales of patient table

4. FUNCTIONAL PERFORMANCE CHARACTERISTICS FOR BNCT FACILITIES COMPARED TO MEDICAL ELECTRON ACCELERATORS

- Several performance characteristics show for medical electron accelerators dependencies on irradiation and operational parameters, which are not relevant for BNCT facilities (Table 4)
- Several dosimetric performance characteristics, which are directly measurable for the primary dose component (X-rays or electrons) of medical electron accelerators are not transferable reasonably to BNCT facilities with four dose components (which are not all measurable). These performance characteristics can only be evaluated on the basis of calculation results (Table 5)
- Requirements of movements of the patient table are neither important nor is the test procedure of the isocentric patient table directly transferable to the actual one used for BNCT (Table 5)

5. SAFETY REQUIREMENTS FOR BNCT FACILITIES COMPARED TO MEDICAL ELECTRON ACCELERATORS

- Several safety requirements for medical electron accelerators are not relevant for BNCT facilities (Table 6)
- Most of the safety requirements for equipment functions are identically applicable for all teletherapy facilities including BNCT (Table 7)
- All safety requirements against stray radiation are equally applicable for all teletherapy facilities including BNCT (Table 8)

Table 5. Functional performance characteristics according to IEC 976[3] as relevant for BNCT treatment (recommended tolerance values taken from IEC 977[4] or, if energy dependent, to be declared by the manufacturer)

Performance characteristic	Medical electron accelerator	BNCT at nuclear reactor
• Dose monitoring system (two totally independent detectors for the sole or at least the most two relevant dose components) • reproducibility • proportionality • stability of calibration throughout the week	• two transmission ionisation chambers (X-radiation, electrons) • 0.5% *(6-monthly)* • 2.0% *(6-monthly)* • 2.0% *(weekly)*	• two fission chambers [f_1, f_2] (neutrons) • two Geiger-Müller counters [GM1, GM2] (gamma-radiation) • 0.5% *(weekly)* • 0.5% *(6-monthly)* • 2% *(weekly, short: daily)*
• Depth absorbed dose characteristics • depth absorbed dose maximum • penetrative quality (depth of 80% dose) • relative surface dose	• maximum deviation from declared value • (energy dependent) cm • 3 % or 3 mm *(short: weekly)* • (energy dependent) % *(all: 6-monthly)*	• maximum deviation from declared value • 2.4 cm (thermal neutrons) • (calculated for all four dose components) mm • (calculated for all four dose components) % *(all: for each patient)*
• Uniformity of radiation field • square radiation fields (ratio maximum to minimum dose in the flattened area in 10 cm depth of phantom) • symmetry within square radiation fields • maximum ratio of absorbed dose anywhere to central axis in plane of depth dose maximum • penumbra of radiation fields (distance of 80% to 20% dose points)	• (for X-rays) • 106 % (≤ 30 cm x 30 cm) • 110 % (> 30 cm x 30 cm) • 103 % *(all: 6-monthly, short: weekly)* • 107 % (≤ 30 cm x 30 cm) • 109 % *(type test)* (> 30 cm x 30 cm) • (energy dependent) mm *(6-monthly)*	• (for all dose components) • (calculated for all four dose components) % • (calculated for all four dose components) % • (calculated for all four dose components) % • (calculated for all four dose components) mm *(all: for each patient)*
• Indication of central axis for entry X-ray beam entry electron radiation exit X-ray beam	• 2 mm • 4 mm • 3 mm *(all: monthly)*	• 2 mm *(for each patient)*
• Movements of patient table • maximum difference of table height for different load	• 5 mm *(monthly, if important)*	• 2 mm *(for each patient)*

Table 6. Safety requirements according to IEC 60601-2-1 Ed.2[5]. Exceptions for BNCT facilities compared to medical electron accelerators

Omission of all requirements, which are not relevant for actual BNCT treatments, i.e.:
• absorbed dose rate (excess to > 2 $\dot{D}max$ resp. > 10 $\dot{D}max$) • monitoring of distribution of absorbed dose D • selection and display of radiation type • selection and display of energy • selection and display of stationary and moving beam therapy • selection and display of targets or other movable radiation beam production devices • selection and display of field flattening, beam scattering and wedge filters • selection and monitoring of other beam distribution systems (e.g. scanning) • other moving parts • gantry • radiation head • patient support system • beam applicators and beam modifying devices • attachment of accessories • retractable radiation beam shield • pneumatic and hydraulic power • adventitious ionizing radiation (thermionicvalves)

Table 7. Safety requirements for equipment functions according to IEC 60601-2-1 Ed.2[5] as relevant for BNCT treatment. (Interlock causes prevention of irradiation, if a device is not correctly preselected, displayed or adjusted. Automatic termination is caused, if safety tolerance limit is exceeded)

Requirement	Medical accelerator	BNCT at nuclear reactor
• Monitoring absorbed dose (capable for termination)	• double monitoring system: ± 5% deviation	• double monitoring system: ± 5% deviation
• Controlling timer	• 120% t_{plan} or $\Delta t = 0.1$ min (the greater)	• 102% t_{plan} ($\Delta t \approx 0.4$ min)
• Control of equipment use	• control by key (or pass word) • treatment room door switch	• the same (as for medical electron accelerators)
• Starting conditions	• key use • preselection of all operating parameters: ready state • all interlocks closed • start by operator only	• the same
• Interruption	• possible for later continuation without re-selection of operating parameters	• the same
• Termination	• stop of irradiation (and movements) with return to preparatory state (re-selection necessary) • automatic stop if operating parameters are adjusted	• the same
• Abnormal termination	• specific display of cause • warning • no further use possible without key	• the same
• Programmable electronic sub-systems (PESS)	• specific requirements	• the same

Table 8. Safety requirements against stray radiation according to IEC 60601-2-1 Ed.2[5] as relevant for BNCT treatment

Requirement	Medical electron accelerator	BNCT at nuclear reactor
• Protection against stray radiation	• Specific requirements (for X-radiation and electron irradiation)	• Automatic actions for gross malfunction of Ar-filter • Ratio of indicated values from GM counter to fission chamber
• Leakage radiation through beam limiting devices and in patient plane (area M: ϕ = 2 m)	• 2% D_{max} (at central axis in phantom at normal treatment distance [NTD] for 10 cm x 10 cm radiation field) • average 0.5% D_{max} in plane M (with shielded residual aperture)	• 8% D_{max} (free beam measurement with TE/TE ionisation chamber for collimator Φ = 12 cm) • 0.7% D_{max} (without shielded residual aperture, thus with backscatter)
• Leakage radiation outside area M in patient plane	• maximum 0.2% D_{max} • average 0.1% D_{max} • maximum 0.05% D_{max} (neutrons) • average 0.02% D_{max} (neutrons)	• < 0.001 $D_{T,max}$ (for closed primary and secondary shutter) ($\hat{=}$ < 2 μSv/h anywhere in the treatment room)
• Leakage radiation outside patient plane	• 0.5% D_{max} in 1 m distance from electron path way and reference axis • 0.05% D_{max} (neutrons)	

• Emission of ionizing radiation after termination due to induced radioactivity	• dose equivalent integrated between 10 s and 10 min after termination
	  • 10 μSv in 5 cm from surface of equipment
	  • 1 μSv in 1 m from surface of equipment
		• dose equivalent integrated between 10 s and 10 min after termination
		  • 10 μSv in 5 cm from surface of phantom
		  • 0.9 μSv in 1 m from surface of phantom

	• dose rate equivalent averaged between 10 s and 3 min after termination
	  • 200 μSv/h in 5 cm from surface of equipment
	  • 20 μSv/h in 1 m from surface of equipment
		• dose rate equivalent averaged between 10 s and 3 min after termination
		  • 75 μSv/h in 5 cm from surface of phantom
		  • 7.5 μSv/h in 1 m from surface of phantom

6. TEMPORAL STABILITY OF SOME FUNCTIONAL PERFORMANCE CHARACTERISTICS

6.1. Spectral Neutron Energy Fluence ("Energy")

- The stability of the spectral neutron energy fluence is demonstrable by the constancy of the ratio of reaction rates of different activation foils for the energy range of intermediate and fast neutrons
- No long-term trend can be seen over the last six years
- Small shifts of the absolute values are probably caused by small distinctions of the reactor cycle, not by the Ar-filter filling

Table 9. Determination of spectral neutron fluence in free beam—results of measurements with a set of 12 different activation foils

Date → Reaction ↓	Neutron energy range	920403 (reference)	930611	940503	950301	971022	980622
^{197}Au (n,γ)	intermediate	1	0.95	0.87	1.04	1.06	1.03
^{115}In (n,γ) Cd		1	0.99	0.87	1.02	1.07	0.97
^{45}Sc (n,γ) Cd		1	0.97	0.89	1.04	1.04	1.01
^{197}Au (n,γ) Cd		1	0.96	0.86	1.06	1.07	1.03
^{186}W (n,γ) Cd		1	0.95	0.87	1.03	1.04	1.00
^{238}U (n,γ) Cd		1	0.94	0.85	1.05	1.06	1.00
^{139}La (n,γ) Cd		1	0.95	0.88	1.04	1.06	1.02
^{55}Mn (n,γ) Cd		1	0.94	0.89	1.04	1.05	1.01
^{63}Cu (n,γ) Cd		1	0.95	0.86	1.05	1.06	1.02
Average			0.96	0.87	1.04	1.06	1.01
Stand. deviat.			1.40 %	1.43 %	1.11 %	0.91 %	1.87 %
^{115}In (n,n') Cd	fast	1	0.99	0.92	1.03	1.09	1.14
^{58}Ni (n,p) Cd		1	0.92	0.89	0.99	1.03	1.02
^{27}Al (n,α) Cd		1	0.93	0.86	0.94	0.98	0.97
Average			0.94	0.89	0.99	1.03	1.04
Stand. deviat.			4.14 %	3.73 %	4.53 %	5.41 %	8.74 %

6.2. Temporary Stability of Calibration of the Radiation Monitors

- The stability of the calibration of radiation detectors used as monitors or additionally is indicated by the "calibration factors" (ratio of the indication of the detector versus the indication of activation foils [thermal neutron fluence] as reference system)
- The absolute indication shows the systematic decrease of about 6% of the thermal neutron fluence within the reactor cycle of 24.7 days
- The stability of all detector types is about equal within 1–2%

7. CONCLUSIONS

- Clinical dosimetry is more complex for BNCT treatments compared to those by medical electron accelerators (four dose components of different LET, two of them only calculable compared to one directly measurable dose).
- Quality assurance for BNCT depends on less irradiation and operational parameters, because many performance and safety characteristics of medical electron accelerators are not relevant for BNCT facilities.
- The reproducibility of applied absorbed doses is equal to those applied by medical electron accelerators despite principle uncertainties of absolute absorbed doses at the cellular level for BNCT
- Safety requirements for equipment functions and against stray radiation are equally fulfilled for all types of teletherapy equipment including BNCT facilities.
- *In-vivo* measurements with ^{55}Mn activation foils for all 4 fractions for each of the 10 treated patients demonstrate a maximum standard deviation of 4% (caused mainly by contact problems of the foils at the patient surface and positioning shifts within the radiation field by small patient movements).

ACKNOWLEDGMENT

The clinical trial is financially supported by the European Commission in the form of a Shared Cost Action from the Biomedical and Health Research Programme (BIOMED II) of DGXII, contract no BMH4-CT96-0325 and by the Swiss Foundation "Neurochirurgie 2001".

REFERENCES

1. Council Directive 97/42/EURATOM of 30th June 1997 on Health Protection of Individuals Against the Dangers of Ionizing Radiation in Relation to Medical Exposures, Replacing the Directive 84/466/EURATOM, Amtsblatt der Europäischen Gemeinschaften Nr. L180/22 DE 9th July 1997.
2. ICRP Report 60, Recommendations of the International Commission on Radiological Protection, Publication 60, Pergamon Press, Oxford and New York, 1991.
3. IEC 976, Medical Electrical Equipment: Medical Electron Accelerators—Functional Performance Characteristics. International Standard IEC 976:1989-10. Central Office, Geneva.
4. IEC 977, Medical Electrical Equipment: Medical Electron Accelerators in the Range 1 MeV to 50 MeV—Guidelines for Functional Performance Characteristics. Technical Report IEC 977:1989-10. Central Office, Geneva.
5. IEC 60 601-2-1 Ed.2, Medical Electrical Equipment—Part 2-1: Particular Requirements for the Safety of Electron Accelerators in the Range 1 MeV to 50 MeV. Final Draft International Standard (FDIS) 60 601-2-1: 1998-02-27.
6. J. Rassow, TLD-300 detectors for separate measurement of total and gamma absorbed dose distributions of single, multiple, and moving beam neutron treatments—a new method of clinical dosimetry for fast neutron therapy. Technical Report Series IAEA, Vienna, 1984, pp. 141–173.
7. EORTC-Protocol 11961: Postoperative Treatment of Glioblastoma with BNCT at the Petten Irradiation Facility, Phase I Clinical Trial. European Organization for Research and Treatment of Cancer, BNCT Study Group, accepted by EORTC-PRC 1996-08-05.

20

MEDICAL IMAGE TRANSFER AND CONVERSION FOR BNCT TREATMENT PLANNING AT PETTEN

P. Watkins and C. Vroegindeweij

Joint Research Centre
European Commission,
Petten, The Netherlands

1. INTRODUCTION

The Phase I clinical trials now underway as part of the European Concerted Action on BNCT require that patients from medical centres in various countries within Europe come to Petten for irradiation in the BNCT facility of the High Flux Reactor (HFR).

Treatment planning calculations for the irradiations are performed at Petten using the *bnct_rtpe/rtt_MC*[1] system developed by INEEL. These calculations assist in the determination of the irradiation time and the orientation of the patient. The treatment planning system requires as input a 3D geometrical description of the head, brain and target of the individual patients. These data are generated from CT images of the patient's head, which are displayed on a computer and then used to define the appropriate regions.

To accomplish this, image data must be transferred to Petten and converted to an image format compatible with the treatment planning system, namely QSH.[2] This task is complicated by the different data formats used in the various medical centres, different imaging equipment, different levels of informatics expertise and varying national requirements for data security. In addition there is a need to transfer generated graphical image data (e.g. dose contour plots, target and brain definitions) back to the coordinating medical centre in Essen for comments and approval.

2. IMAGE TRANSFER

To use the *bnct_rtpe/rtt_MC* system the image data must be available in QSH format. This poses a number of technical problems that had to be overcome.

Six different centres within Europe provide medical image data (CT). These have various imaging equipment with different technical capabilities and generate data in many formats. To try and ensure consistency of data from the various centres the type of data

Frontiers in Neutron Capture Therapy, edited by Hawthorne *et al.*
Kluwer Academic / Plenum Publishers, New York, 2001.

required is defined in a Standard Operating Procedure (SOP) to include; number & spacing of images, field of view (FOV), location of markers, image format, etc.

Image data is transferred to Petten at least two weeks prior to patient irradiation. Image data is usually passed to Petten via the Internet using FTP to a specially created server. This FTP server has been designed with password protection, allows access from designated domains only and is "write only" to avoid the different medical centres from reading data from elsewhere. Data on magnetic media (DAT) has also been received on occasions.

The ability to use data encryption (PGP) for the transfer process and storage at Petten has also been foreseen if requested. Some national legislation, but not all, requires this.

3. DATA FORMATS

Many different formats are used for medical image data. Some "standards" with a published definition, for example, ACR-NEMA, DICOM, QSH etc. but others are proprietary. The image format used within *bnct_rtpe/rtt_MC* is QSH, which is not widely used by medical centres. Most of the CT data passed to Petten has been a dialect of ACR-NEMA (similar to DICOM).

Invariably medical image data consists of two components. First there is the "header" data which consists of general information regarding the patient, imaging parameters, imaging sequence and individual slices. Secondly there is the bitmap image data. These are described below in more detail for the ACR-NEMA and QSH formats and shown in Figure 1. Other data formats used within the European BNCT collaboration have included various proprietary "block" systems (again header plus pixel data) and raw 8 or 16 bit pixel maps.

3.1. ACR-NEMA Format

ACR-NEMA format image data usually consists of a single file per CT slice. Each file contains header data followed by an image bitmap. The header contains both ASCII and binary data stored as numbered "data items" within numbered "data groups". The bitmap image data is usually 512 × 512 pixels with each pixel represented by 16 bits of data. However many variations on this basic structure are possible and have been observed at Petten.

The official ACR-NEMA format definition only defines the header information for the even-numbered data groups. The odd-numbered data groups are available for use by the imaging equipment manufacturers and are not documented.

3.2. QSH Format

The QSH format consists of two files for any number of slices. The "qhd" header file is written in pure ASCII text. The second, "qim", file contains the bitmap images for all of the slices. For *bnct_rtpe* the image data used is 256 × 256 pixels with 16 bits per pixel.

4. THE "CONVERTOR" CODE

Although commercial software is available for converting between image formats, few can handle medical data. No package was found to process all the medical formats

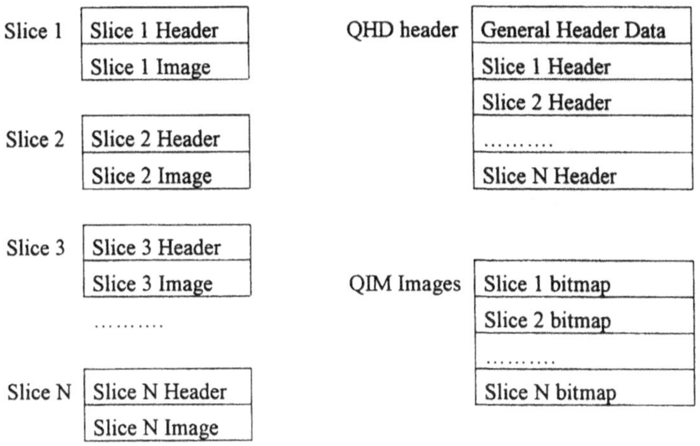

(a) ACR-NEMA Format (b) QSH Format

Figure 1. Diagrammatic view of ACR-NEMA and QSH image data formats.

passed to Petten as well as QSH. Furthermore there was a need to perform specialist manipulations of the slice data that was not a normal option for such software. Consequently our own software, "*convertor*", was developed to perform the image processing.

The task of the conversion software is to process the sequence of supplied CT images and generate the two appropriate QSH files. To achieve this, the header and pixel data of the CT images have to be separated, decoded and then re-written as new files in the QSH format.

The conversion software must be capable of processing CT data in many formats with little user intervention. This target has been achieved. However the software must also handle many practical problems. Typically:

- The data may require "byte swapping" depending on the computer systems used in the CT acquisition. This is corrected automatically for ACR-NEMA data.
- One medical centre provides nominal ACR-NEMA data but with header and pixel data in separate files. The software must combine the two files automatically.
- Often data is in ACR-NEMA format but the even-numbered data groups of the header contain little information. This has been moved to the undocumented odd-numbered data groups and has to be extracted piecemeal.
- Pixel data is usually stored in 16 bits, but only 12 bits are significant. In some cases a non-trivial compaction algorithm is used to reduce the size of the pixel data. This is permitted with the standard ACR-NEMA format definition. Unpacking the 12 bit data into the 16-bit form required by QSH is complex.
- CT sequences are often generated in several steps so that the file numbering is unreliable. Therefore the correct slice sequence must be determined from the header data. The slice spacing may vary when the acquisition process is interrupted and has to be corrected for. The CT sequence supplied may contain extra images in an unwanted orientation. These have to be rejected. Slices may be duplicated in some manner, possibly with the same pixel data but with different imaging parameters in the header.
- Some crucial data (e.g. Field-of-View—FOV) may be missing and has to be added by hand.

Numerous checks of the CT image data are performed, in particular to ensure that the FOV and slice spacing are constant, but also to ensure that the data meets the particular needs of the treatment planning system.

To deal with these problems a two-pass process is used in "*convertor*". The first pass validates, sorts and selects the required slices and the second pass generates the actual QSH data files. The first pass of the conversion software also generates a thumbnail overview of the slice sequence that allows a rapid examination of the slices selected.

5. PIXEL WINDOWING

Problems were found in selecting a pixel window to display the CT image data. Typically raw pixel data values vary between 0 and 4,096 for a given slice. However in the brain the pixel data may range over a very few values making it difficult to observe any detail. Commercial systems have sophisticated and automatic methods of selecting the appropriate pixel window to optimise the display but these are not currently available in *bnct_rtpe*.

The problem is made worse by the highly contrasting fiducial markers on the patient mask used for CT imaging. These markers are needed to accurately match the image data with the physical geometry. The pixel values for these markers are very high and the CT systems automatically scale the pixel data for each slice to compensate. Consequently slices without fiducial markers may have pixel values that range from 0 to 2,048 only. Thus the pixel window needed to view data in soft tissue changes for each slice.

One solution to this problem was to adjust the raw pixel data in the conversion code. This was rejected since the adjustment could not be implemented simply for all data formats. Moreover modification of the image data could have medical and legal implications.

6. GRAPHICAL DATA TRANSFER TO MEDICAL CENTRES

Once the CT image data is available in Petten and the data converted to QSH format the treatment planning calculations are performed. The definitions of the brain and skin regions are undertaken at Petten. However the target region in the brain and the "organs at risk" have to be defined by medical staff at Essen. At present this is done on hard copies of the CT image data which are sent directly from the medical centres to Essen. These hard copies are then forwarded to Petten with the appropriate regions marked and then entered into the *bnct_rtpe* model by hand.

There is no possibility at present to simplify this awkward approach. Ideally it should be possible for the medical staff to define the target region and organs at risk remotely, by accessing the treatment planning software. This is not yet practical. Alternatively the image data and region definition files need to be transferred to Essen and software tools developed to allow the medical staff to make simple checks and additions to the model.

The treatment planning calculations are performed at Petten but the decision as to the actual irradiation fields to use is made by the medical staff in Essen. To assist the medical staff considerable amounts of data have to be transferred to Essen including dose distribution contour plots, dose volume histograms and spreadsheet data. Transferring this data is a laborious and time-consuming task at present. In many cases bitmap window dumps have to be produced by hand, stored as local files and then transferred using FTP.

The execution of similar BNCT projects between a number of widely distributed centres would be greatly helped by developments in the possibilities of data transfer and remote operation. Some of these activities could become important future developments for *bnct_rtpe*.

7. CONCLUSIONS AND REMARKS

Image data from a number of medical centres has been transferred successfully to Petten.

Conversion of the data to the QSH format has been achieved for a number of input image data formats. A computer code, "*convertor*", has been developed for this purpose. The "*convertor*" code is now being extended and a GUI interface added.

Use of the Internet for image transfer is the preferred option and has been used successfully. The security implications have been addressed and encryption of data is available if required.

Transfer of graphical data to the medical centres remains very rudimentary. The development of more integrated and interactive systems is a major requirement for the future.

REFERENCES

1. D.E. Wessol, R.S. Babcock, F.J. Wheeler, G.J. Harkin, L.L. Voss, and M.W. Frandsen, "BNCT_*Rtpe*: BNCT Radiation Treatment Planning Environment User's Manual Version 2.2", *http://id.inel.gov/cart/bnct_rtpe-manual/*, February 21, 1997.
2. G.Q. Maguire Jr., and M.E. Noz, QSH: a Minimal but Highly Portable Image Display and Handling Toolkit, Computer Methods & Programs in Biomedicine, Vol. 27, pp. 229–240, 1988.

21

PATIENT POSITIONING AT THE HFR PETTEN

P. Watkins[1], C. Vroegindeweij[1], S. Garbe[2], and K. Hideghéty[2]

[1]Joint Research Centre
European Commission
Petten, The Netherlands
[2]Universitätsklinikum
Essen, Germany

1. INTRODUCTION

The epithermal beam at the Petten BNCT facility has now been used for the treatment of 10 human patients. One consideration for the patient treatments was to develop a method to position the patients accurately and quickly in the beam according to predefined parameters.

The Petten beam is rather different from most other operational BNCT facilities in that it is virtually mono-directional. Consequently the spatial attenuation down the beam line is small and the therapy position has been chosen so that the patient's head is 30 cm from the final aperture. All patient irradiations have been performed with the patient lying on their back on a purpose built therapy table.

The positioning procedure that has been developed has tried to minimise the inconvenience to the patient yet provide an accurate and reproducible means of locating the correct irradiation position.

2. SEQUENCE OF EVENTS FOR A BNCT IRRADIATION

The sequence of events for patient treatment in the present BNCT phase I trial at Petten is described in detail elsewhere.[1] In summary the sequence consists of:

(a) Patient enrollment by a referring medical centre after medical investigation, MRI scans, etc.
(b) Patient operation at the medical centre to debulk tumour.
(c) Post-operative CT imaging of patient at the medical centre.
(d) Transfer of CT data to Petten followed by conversion to appropriate format.[2]
(e) Treatment planning calculations at Petten,[3] using BNCT_Rtpe,[4] to evaluate dose distributions.

Frontiers in Neutron Capture Therapy, edited by Hawthorne *et al.*
Kluwer Academic / Plenum Publishers, New York, 2001.

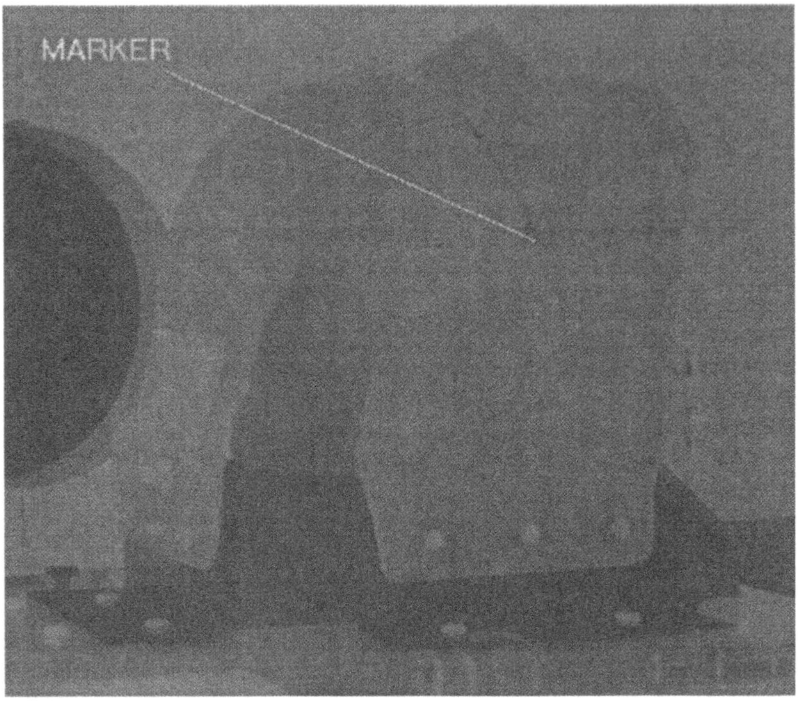

Figure 1. Mask used for patient irradiation at Petten.

(f) Selection of irradiation fields, irradiation times and patient position. These are defined by the radiotherapist, with support from the treatment planning calculations.
(g) Transport of patient to Amsterdam (Academic Hospital Free University).
(h) Treatment of the patient at Petten over a 5 day period. Patient is transported daily from Amsterdam to Petten, treated and then returned to Amsterdam. Day 0 is used for preparatory work. The irradiations are performed in 4 fractions on Days 1 to 4 with 1 or 2 fields per fraction.

3. PATIENT MASK AND TREATMENT PLANNING

For the CT imaging at the medical centre the patient is fixed in a mask, see Figure 1. Attached to either side of the mask are two fiducial markers—wires in the form of a cross. These markers are clearly visible on the CT images as well as being present physically on the mask. They provide the link between the physical geometry (mask plus patient) and the calculational model. The same mask is used to fix the patient's head stationary during the irradiation.

The treatment planning calculations use these same CT images to define the patient geometry. Thus the co-ordinate system used for the treatment planning (TP) has axes parallel to the CT image slices with an origin at the slice centre. This co-ordinate system will, in general, be translated and rotated slightly with respect to the physical co-ordinate system of the patient plus mask. However the co-ordinates of the two fiducial markers are easily identified in the TP system. Furthermore the treatment planning calculations also generate the co-ordinates of the field centre-line entrance and exit positions in the CT system. These co-ordinates have to be converted into physically meaningful points on the mask itself.

In practice the mask contains cut outs and will not be in close contact everywhere with the patient's head. Consequently the physical centre-line field entrance and/or exit positions may not fall on the mask itself or could be displaced from the patient's head by an undetermined distance. To avoid these difficulties a different procedure has been developed.

4. THE POSITIONING FRAME

To avoid working directly with the complicated curved geometry of the mask a simple set of (X,Y,Z) co-ordinates is introduced in the form of a rigid cuboidal metal frame, see Figure 2. This frame consists of graduated sides with two moveable perspex sliders and two moveable pointers. Both sliders and pointers hang from the upper horizontal parts of the frame and are free to move in the X and Y directions.

The patient's mask is fixed on a table at the centre of the frame. Relative to the axes of the frame the mask plus table can be translated, rotated and also locked in position. The two moveable pointers are used to position the fiducial markers accurately so that the co-ordinate system of the frame can be aligned with that of the TP system.

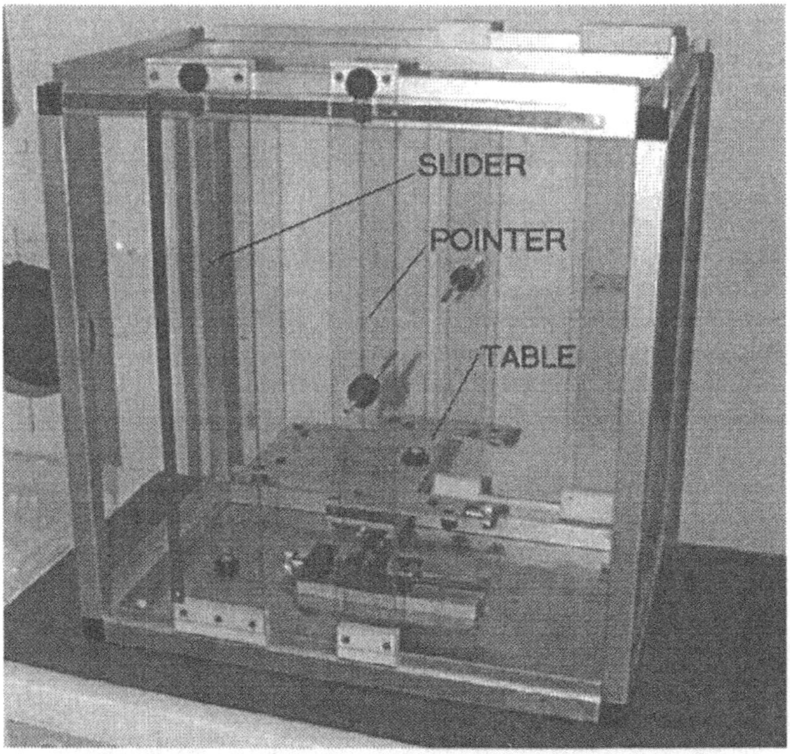

Figure 2. Patient positioning frame.

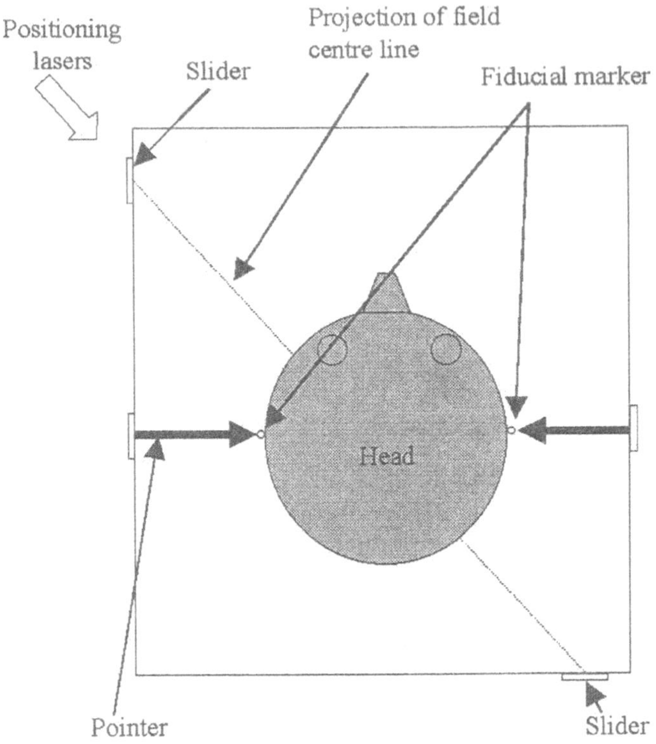

Figure 3. Positioning Frame details.

The line joining the irradiation field centre-line entrance and exit positions derived from the treatment planning system in the TP co-ordinate system may then be extrapolated to intersect the sides of the frame, see Figure 3. These extrapolated field centre-line positions are recorded on the moveable perspex sliders that are then locked in position. Next, the whole assembly of mask, frame and sliders is taken to the therapy room. Here the lasers used to define the physical beam centre lines are aligned with the entrance and exit points on the perspex sliders. Once this is accomplished the projected vertical and horizontal laser lines are physically drawn onto the mask thereby accurately defining the irradiation field.

In practice the use of only two markers during the positioning process leaves one degree of freedom undetermined. Effectively the mask could be rotated through an arbitrary angle about the axis of the beam line. However this motion is constrained by the patient geometry and even further restricted by the requirement to perform the irradiation with the patient in a prone position. Consequently the error introduced is minimal although for more complex patient positions it is planned to add a third marker to completely remove this problem.

All of the positioning information, the extrapolated entrance and exit points on the frame and the locations of the fiducial markers, are calculated automatically in the Treatment Planning Spreadsheet (TPS). This spreadsheet takes data from a number of sources, including the treatment planning calculations, to derive various data used in the patient irradiations.

5. POSITIONING DURING A PATIENT TREATMENT

For each patient treatment the centre lines of the selected irradiation fields are marked onto the mask on Day 0 using the positioning frame and the procedure described above. The patient's presence is not required for this process. For the actual patient irradiations, Days 1 to 4, the mask is positioned in the beam without the patient present, prior to the irradiation. This is achieved by moving the therapy table so that the field centre-lines on the mask are aligned with the centre-line lasers of the therapy facility.

The position is checked and corrected once the patient has been placed on the therapy table and the mask fixed around their head. The distance of the mask to the final aperture is measured and adjusted to 30 cm. Just prior to each irradiation the radiotherapist performs an independent check of the position using the centre-line entrance and exit positions relative to known clinical features.

6. CONCLUSIONS AND REMARKS

At Petten a positioning system has been devised to accurately locate human patients in the BNCT beam. This has been used for actual patient irradiations and found to be effective and efficient. The system makes use of a simple cuboidal positioning frame that avoids the need to work directly with the complex geometry of the patient or patient mask. Using the positioning frame and the beam centre-line lasers of the therapy room the various irradiation fields can be accurately marked on the patient mask.

Future developments of the positioning system are now being planned. The positioning rig will be rebuilt to improve user friendliness, better cope with incident fields at all angles and further simplify the positioning process. The data evaluated in the treatment planning spreadsheet (TPS) will be revised and simplified.

The construction of a "simulator room" is being planned. The intention is to create a room with a set of lasers that mimic the beam centre line lasers of the therapy room. Drawing the field centre lines on the mask in the positioning frame would then take place here. This room would be outside of the reactor, thereby reducing the use of the therapy room itself. Additional features, such as projection of the field onto the mask for visualisation purposes will also be available.

REFERENCES

1. Sauerwein, W., Gabel, D., and Fankhauser, H., "Glioma BNCT. Postoperative treatment of glioblastoma with BNCT at the Petten irradiation facility. 1. Phase I clinical trial", EORTC Protocol 11691, March 1997.
2. Watkins, P., and Vroegindeweij, C., "Medical Image Transfer and Conversion for BNCT Treatment Planning at Petten".
3. Vroegindeweij, C., Watkins, P., Hideghéty, K., Sauerwein, W., and Moss, R.L., "Treatment planning for the first group of patients in the European Glioma trial at the HFR Petten", these proceedings.
4. Wessol, D.E., Babcock, R.S., Wheeler, F.J., Harkin, G.J., Voss, L.L., and Frandsen, M.W., "BNCT_*Rtpe*: BNCT Radiation Treatment Planning Environment User's Manual Version 2.2", *http://id.inel.gov/cart/bnct_rtpe-manual/*, February 21, 1997.

22

TREATMENT PLANNING BASED ON DYNAMIC POSITRON EMISSION TOMOGRAPHY FOR BORON NEUTRON CAPTURE THERAPY FOR GLIOBLASTOMA MULTIFORME

T. L. Nichols, G. W. Kabalka, G. T. Smith, J. P. Dyke, D. Longford, and K. F. Hubner

The University of Tennessee Medical Center
Knoxville, Tennessee 37920

BNCT is a form of brachyradiotherapy based on the nuclear fission that results from the interaction of a boron-10 atom with a thermal neutron.[1,2,3] Lithium ions and alpha particles are produced in this annihilation reaction and each deposits a large amount of energy within 10 microns of the original boron atom. The advantage of BNCT is that the individual components are non-toxic to the cell: boron is non-toxic in the chemical forms utilized in BNCT and the neutron beam is of sufficiently low energy (0.5 eV–30 keV) such that radiation necrosis is minimized in normal tissue. The clinical success of BNCT depends on accurate treatment planning with respect to dosimetry. However, treatment planning protocols are themselves dependent on accurate knowledge of the quantity of boron present in both the targeted lesion as well as in the non-targeted tissue and the characteristics of the neutron beam. These data were not readily available to researchers at the time of the earlier BNCT clinical trials.[4]

In recent years, significant advances have been made in the development of tumor selective boron-containing agents[5] as well as in the modification of the nuclear reactors so that they can supply neutrons of appropriate energy.[6] In fact, clinical trials are now underway for the treatment of both glioblastoma multiforme (GBM) and metastatic malignant melanoma.[7,8,9] The recent development of PET[10,11] and MRI[12] techniques for *in vivo* imaging of boronated agents in humans can be used to provide more accurate biodistribution data for input into treatment planning protocols now being developed at Brookhaven National laboratory (BNL), the Idaho National Engineering and Environmental Laboratory (INEEL), and the Massachusetts Institute of Technology in collaboration with Harvard University (MIT/H). This is important because a significant fraction of the people in the U.S. who will die this year from glioblastoma multiforme and melanoma could be candidates for BNCT treatment.

We developed a method to prepare the fructose complex of fluorine-18 labeled *para*-boronophenylalanine (^{18}F-BPA-F),[10] an analogue of the agent currently being used

Frontiers in Neutron Capture Therapy, edited by Hawthorne *et al.*
Kluwer Academic / Plenum Publishers, New York, 2001.

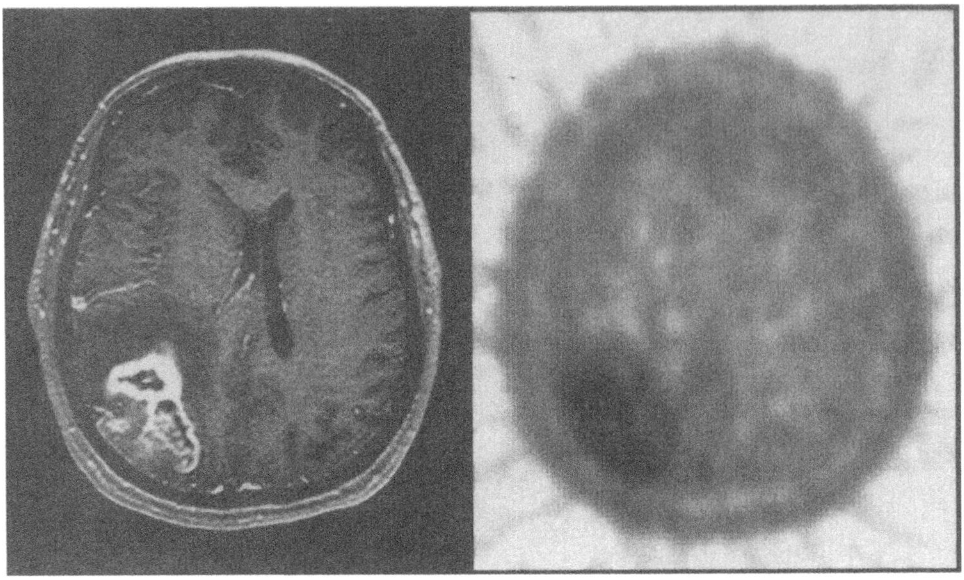

Figure 1. Pre-operative MRI (left) and PET (right) of GBM patient.

in BNCT clinical trials at both BNL and MIT/H. Fluorine-18 is a short-lived, positron emitting radionuclide commonly used in clinical PET. To evaluate the potential of PET in BNCT treatment planning, ^{18}F-BPA-F was administered to GBM and to malignant metastic melanoma patients who were then examined utilizing PET. The *in vivo* biodistribution data that we obtained are being utilized in the pretreatment planning protocols being developed at both BNL (in collaboration with the INEEL)[13] and at MIT/H[14] and could play a major role in increasing the efficacy of BNCT in the clinical environment.

As an example, in one patient, the tumor-to-normal brain concentration ratio of ^{18}F-BPA-F was 3.4:1 at 52 minutes post injection. As can be seen, the preoperative PET scan revealed an area of intense uptake which corresponds to the location of the lesion in the patient's MRI, Figure 1. Thus PET confirms that the therapeutic BNCT agent will be taken up by the target tissue.

In addition to demonstrating that the BNCT agent localized in the tumor, the data provided information not previously available for treatment planning. For example, the Monte Carlo programs utilized at INEEL, BNL and MIT/H are based on the assumption that the boron distribution in the tumor is homogeneous. Boron uptake in surrounding tissue is also assumed to be homogeneous although at a lower level. The PET data clearly demonstrate that the uptake of ^{18}F-BPA-F is far from homogeneous. A plot which illustrates the inhomogeneity of the ^{18}F-BPA-F distribution both in the tumor and surrounding tissue in the GBM patient referred to in Figure 1 is presented in Figure 2.

The effect of this PET data on the calculated dosimetry for that patient is presented in Figure 3. Clearly, the use of the *in vivo* PET data markedly affects the dosimetry calculations. The isodose curves are more tightly gathered about the tumor. Behind the tumor (with respect to the parieto-occipital neutron beam) shadowing is apparent which results in a decreased dose in a critical area. The shadowing effect occurs because of increased uptake of the neutron beam in the tumor as compared to regions around the

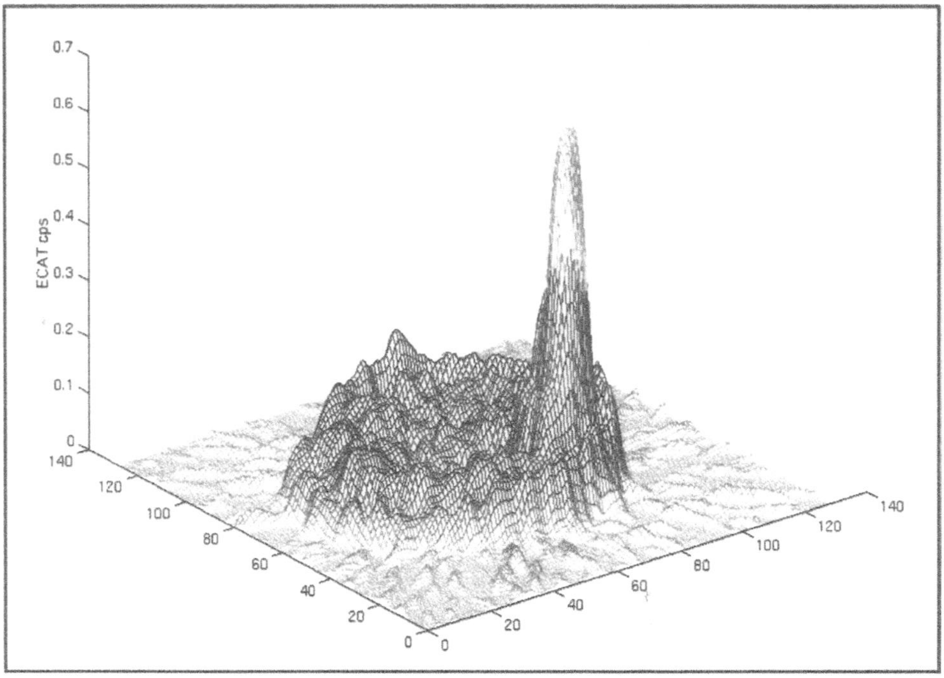

Figure 2. A three-dimensional plot of the ^{18}F-BPA-F activity in a glioblastoma patient in one of the axial planes.

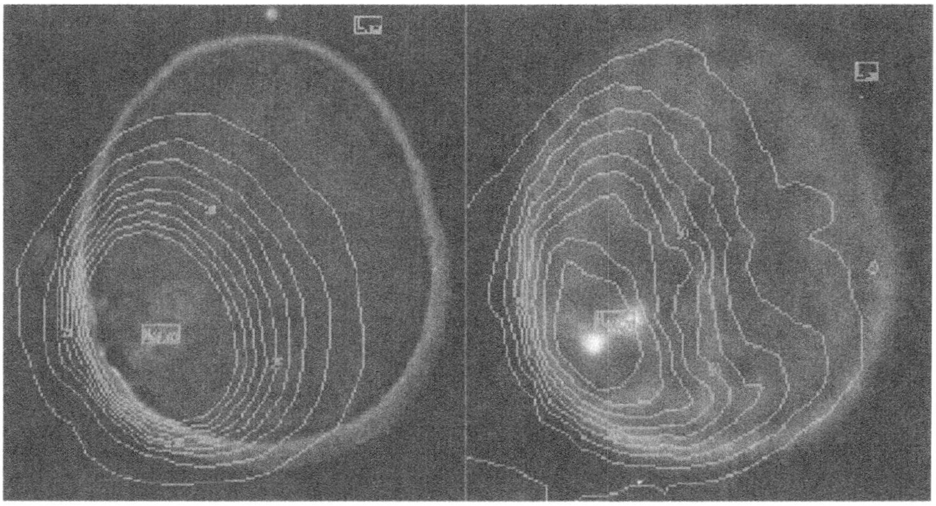

Figure 3. Left: Total BNCT dose contours generated from MRI data. Right: Total BNCT dose contours generated from PET data.

tumor. Thus, there are fewer neutrons behind the main tumor mass to interact with other boron containing tissue. Significantly, the clinical course of this patient most closely mirrored the PET-based dosimetry calculations with recurrence of the GBM in the shadowed region.

ACKNOWLEDGMENT

We wish to thank the U.S. Department of Energy, Lockheed Martin, Idaho Technologies Company (URC) and The Robert H. Cole Foundation for support.

REFERENCES

1. R. Gahbauer, N. Gupta, T. Blue, J. Goodman, R. Barth, J. Grecula, A.H. Soloway, W. Sauerwein, and A. Wambersie, Boron Neutron Capture Therapy: Principals and Potential, *Recent Results Cancer Res.*, 150:183, 1998.
2. M.R. Hawthorne, New Horizons for Therapy Based on the Boron Neutron Capture Therapy Reaction, *J. Microencapsul.* 4:174, 1998.
3. G.W. Kabalka, Recent Developments in Boron Neutron Capture Therapy, *Ex. Opin. Ther. Patents*, 8:545, 1998.
4. W.H. Sweet, Early History of Development of Boron Neutron Capture Therapy of Tumors, *J. Neuroncol.* 33:19, 1997.
5. A.H. Soloway, W. Tjarks, B.A. Barnum, F.-G. Ronj, R.F. Barth, I.M. Codogni, and J.G. Wilson, The Chemistry of Neutron Capture Therapy, *Chem. Rev.*, 98:1515, 1998.
6. M. Chadha, J. Capala, J.A. Coderre, E.H. Elowitz, J. Iwai, D.D. Joel, H.B. Lium, L. Wielopolski, and A.D. Chanana, Boron Neutron-Capture Therap. (BNCT) for Glioblastoma Multiforme (GBM) Using the Epithermal Neutron Beam at the Brookhaven National Laboratory, *Int. J. Radiat. Oncol. Biol. Phys.*, 40:829, 1998.
7. J.A. Coderre, E.H. Elowitz, M. Chadha, R. Bergtland, J. Capala, D.D. Joel, H.B. Liu, D.N. Slatkin, and A.D. Chanana, Boron Neutron Capture Therapy for Glioblastoma Multiforme Using *p*-Boronophenylalanine and Epithermal Neutrons: Trial Design and Early Clinical Results. *J. Neurooncol.* 33:141, 1997.
8. P.M. Busse, I. Kaplan, R.G. Zamenhof, O.K. Harling, J. Kaplan, C. Chuang, J. Goorley, W.W. Kiger III, K. Riley, L. Tang, G.R. Solares, and M.R. Palmer, BNCT for Glioblastoma Multiforme and Intracranial Metastatic Melanoma: Clinical Results of Harvard-Massachusetts Institute of Technology (MIT) Phase I Trial, *in* "Frontiers in Neutron Capture Therapy" M.F. Hawthorne, K. Shelly, and R.J. Wiersema, Plenum Press, New York, in press.
9. Y. Nakagawa, and H. Hatanaka, Boron Neutron Capture Therapy. Clinical Brain Tumor Studies, *J. Neurooncol*, 33:105, 1997.
10. G.W. Kabalka, G.T. Smith, W.S. Reid, C.P.D. Longford, T.G. Roberts, N.K. Reddy, E. Buonocore, and K.F. Hubner, Evaluation of Fluorine-18-BPA-Fructose for Boron Neutron Capture Treatment Planning, *J. Nucl. Med.*, 38:1762, 1997.
11. Y. Imahori, S. Ueda, Y. Ohmori, T. Kusuki, K. Ono, R. Fujii, and I. Tatsuo, Fluorine-18-Labeled Fluoroboronophenylalanine PET in Patients with Glioma, *J. Nucl. Med.* 39:325, 1998.
12. G.W. Kabalka, C. Tang, and P. Bendel, The Role of Boron MRI in Boron Neutron Capture Therapy, *J. Neuro-Oncol.*, 33:153, 1997.
13. D.W. Nigg, F.J. Wheeler, D.E. Wessol, J. Capala, and M. Chadha, Computational Dosimetry and Treatment Planning for Boron Neutron Capture Therapy, *J. Neurooncol.* 33:93, 1997.
14. R.G. Zamenhof, Microdosimetry for Boron Neutron Capture Therapy: A Review, *J. Neurooncol*, 33:81, 1997.

23

ACCELERATION OF MCNP-BASED TREATMENT PLANNING AND THE APPLICATION FOR *IN VIVO* BORON DETERMINATION BY THE GAMMA-RAY TELESCOPE

W. F. A. R. Verbakel, J. A. A. Verwey, F. Stecher-Rasmussen, and A. Hogenbirk

Netherlands Energy Research Foundation (ECN)
P.O. Box 1, 1755 ZG Petten
The Netherlands

1. INTRODUCTION

Monte Carlo calculation is a frequently used method for treatment planning for BNCT. In MacNCTPlan from Harvard/MIT,[1] the calculational part is performed with Monte Carlo N-Particle (MCNP). As the lengthy calculation times make the standard MCNP inconvenient for routine application, acceleration of the computational part is preferable. The acceleration of the treatment planning program makes it also useful to perform at the same time a simulation of prompt gamma-ray measurements with the gamma-ray telescope.[2,3] This has been proved by adding the description of a germanium detector in the MCNP model that is made by MacNCTPlan.

2. PROBLEM

Dose calculations with MCNP for a treatment planning program has some advantages above the use of programs which are based on for example the pencil-beam approach (interpolation of measurements in a phantom, irradiated with a pencil-beam):

- it is a simulation of physical particle transport, taking into account the real shapes and structures of a patient's body;

Frontiers in Neutron Capture Therapy, edited by Hawthorne *et al.*
Kluwer Academic / Plenum Publishers, New York, 2001.

- MCNP is a validated code for neutron, gamma-ray and electron transport;
- it is also appropriate to model a gamma-ray detector or activation foils.

Disadvantages are:

- cell to cell transport is very time-consuming, which makes high resolution computations unfeasible in practice;
- instead of the real dose, to limit the calculation time to some extend, only the kerma is calculated by energy dependent flux times energy dependent kerma-factor, which is a dose approximation assuming electron equilibrium (the range of the electrons is not included).

Until now, the present MacNCTPlan allows only a geometry mesh with a basic cell of 1 cm^3. After the calculation, the results are interpolated between these cells. In future, this will be insufficient. A higher resolution will be required, leading to a larger number of smaller cells. In the case of irradiation of a patient's head, the deviation between gamma-ray dose and kerma will not be so important. Conversely, if surface tumours are treated, the error made by kerma calculation will become significant, and real energy deposition has to be calculated. This is the dose due to ionisation losses of electrons, which is only possible by additional simulation of electron transport. The addition of both the smaller geometry mesh and electron transport, however, would lead to extremely longer calculation time in standard MCNP.

3. SOLUTION

A way to reduce the calculation time is the introduction of a particle transport accelerator that accelerates the computational part of particle transport through a CT-based lattice without influencing the physics. In standard MCNP, the tracking of particles from cell to cell is very time consuming. At every new cell, it is checked whether the material changes. In fact, many neighbouring cells consist of the same material, meaning that this check is unnecessary if these cells are grouped together. As a solution, MCNP has been expanded with a module to decrease the calculation time by making use of the MRR-method,[4] in which adjacent cells of the same material description are added together to a maximum rectangular volume. This way the number of cells and thus the number of cell boundary crossings of a particle can be reduced considerably.

Besides the acceleration of particle transport through the lattice, MCNP has been modified to allow determination of the real dose due to photons by also tracking the electrons that are created at photon interactions. Finally, the dose is determined by calculation of the energy deposition of the electrons. In the described case of an epithermal neutron beam on a head composed of 11,025 cells of 1 cm^3, the calculation time has been decreased with a factor 1.5 compared to the standard MCNP.

The third component of the modification is an increase of the tallying resolution without any additional computational effort. In our present MCNP-version it is possible to use a user-defined tallying grid which does not need to correspond with the geometry grid. In this case, the tallying grid is 2.5 mm × 2.5 mm × 1 cm, which is a factor 16 finer than in the standard version.

Figures 1 and 2 show the difference in result between an interpolation of a 1 cm^3 grid at the left, and a tallying grid of 0.0625 cm^3 at the right. The figure of the

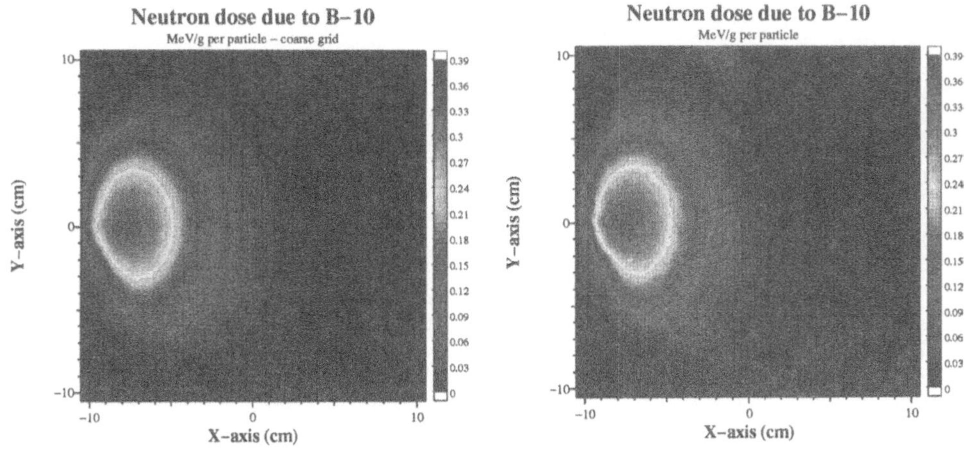

Figure 1. Neutron dose calculated with the coarse (left) and the fine (right) tallying grid.

gamma-rays has not yet been performed with electron transport due to the fact that this was only developed recently. The finer tallying grid shows much more structure, which may become important for sensitive organs.

4. SIMULATION OF THE GAMMA-RAY TELESCOPE

By means of measurements of the gamma-ray telescope,[3] it is possible to determine the average boron concentration averaged over a specific volume of the patient's head. A gamma-ray detector, which is located above a small hole in the roof of the therapy room, is able to measure the gamma-rays which are emitted at neutron capture in the patient's head. Besides the determination of the boron concentration, these measurements can also serve as an on-line dose monitoring system during the irradiation. The

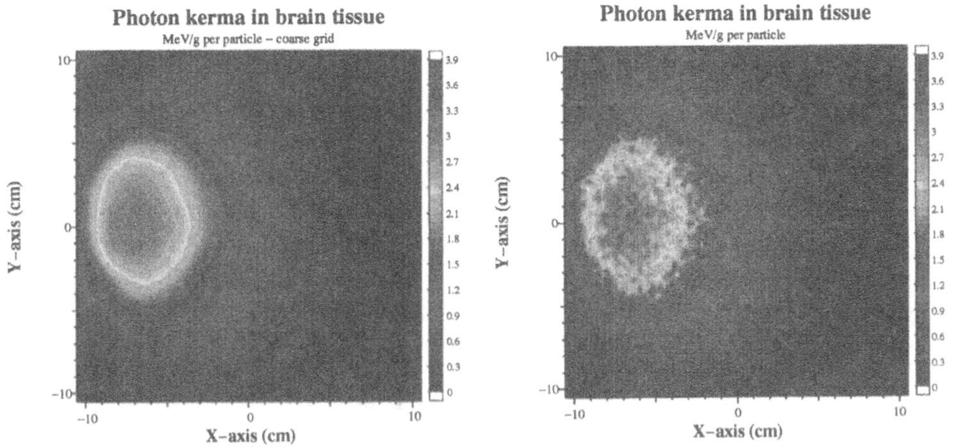

Figure 2. Gamma-ray kerma calculated with the coarse (left) and the fine (right) tallying grid.

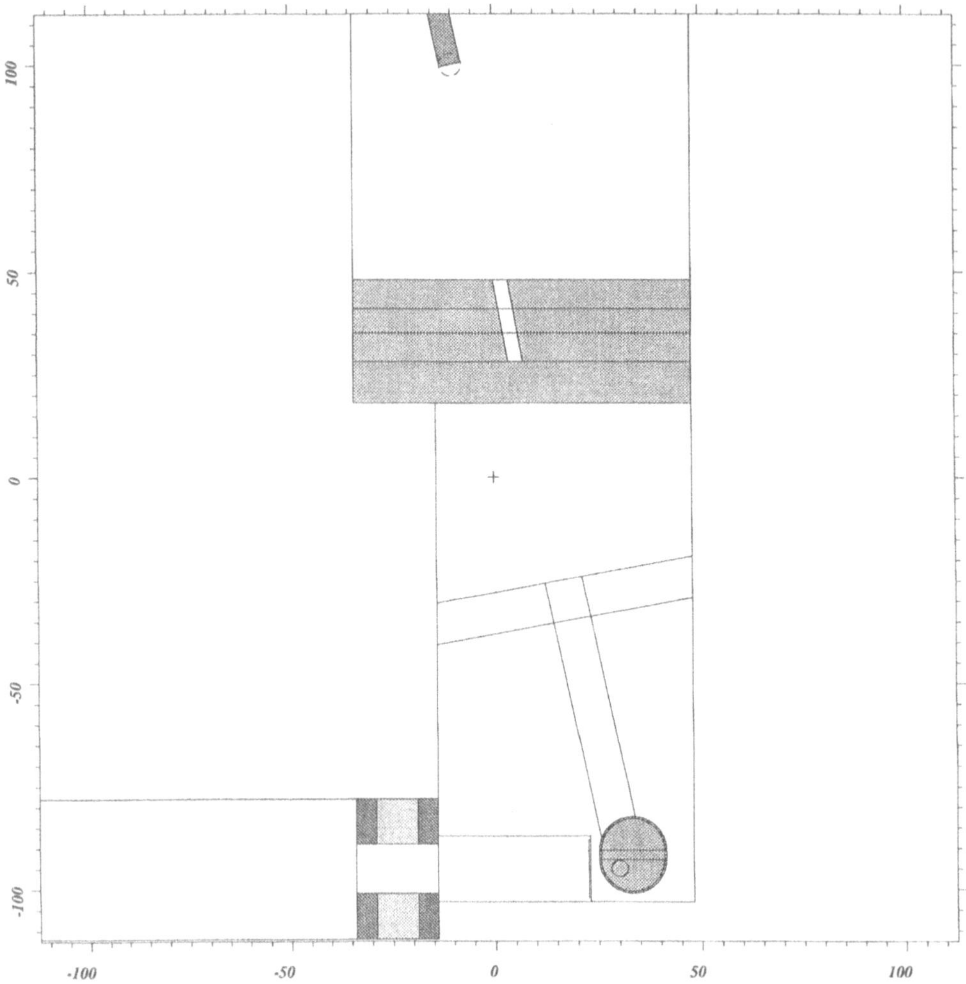

Figure 3. MCNP-model of a phantom head in a neutron beam with a DXTRAN-sphere and detector above a small hole in the ceiling of the therapy room. The beam diameter is 12 cm and the DXTRAN-sphere has a diameter of 5 cm.

position dependent emission of gamma-rays are a measure of the position dependent reaction rate in the head. In order to gain insight in the contributions to the measurements of several parts of the head, simulations of the gamma-ray telescope with the patient's head should be performed. MCNP allows to model the detector and possible collimators in the treatment planning model of every specific patient. By use of special variance reduction techniques, the gamma-ray spectrum as measured by a germanium detector is calculated. It required the use of a F8 pulse height tally of the HPGe-detector in combination with the DXTRAN-technique by which the gamma-rays are tracked to the detector. Addition of this feature appears to require hardly any (5%) additional computation time. A detector which views a diameter of 15 cm of the patient's head with 13 ppm ^{10}B requires 20 minutes computing time to obtain a boron line with an uncertainty of 3%. In these 20 minutes, 100,000 histories were tracked. Figure 3 shows the MCNP model of a phantom head with the gamma-ray telescope above the roof of the therapy room.

In the future, the combination of a calculated B/H gamma-ray count rate and the on-line gamma-ray measurements performed with a HPGe-detector enable to reconstruct the actual boron concentration averaged over healthy tissue volume that is viewed.

5. CONCLUSION

The ECN-developed tracking and tallying module of MCNP shows the following 3 improvements in one calculation:

- 16 times more tallying points
- photon dose calculation instead of kerma
- factor 1.5 reduction of calculation time.

Standard MCNP (on a DEC Alpha) required 203 minutes computing time for the tracking of 1 million particles. With the modified MCNP, the computing time was 140 minutes. In spite of this reduction, a computing time of 140 minutes is still very long for clinical applications of BNCT. Future computer hardware developments will help to reduce the computing time much more. Furthermore, a reduction of the size of the cells (to the order of CT-pixel-size), will increase the number of geometry cells with more than a factor 50. In standard MCNP this would also increase the computing time with a large factor. Conversely, the new developed computing algorithm as used for this study will be more efficient if the cell size is smaller and less different materials are required. It is expected that this reduction in amount of materials may even lead to a smaller number of MRR volumes than with a smaller number of cells with more different materials, leading to even shorter computing times.

ACKNOWLEDGMENT

The calculations have been performed with MCNP-model based on a CT-scan of a patients head. This model has been supplied by the Harvard/MIT-group of Dr. R.G. Zamenhof. We are very thankful for W.S. Kiger to provide us both the MCNP-model of a patient and the MacNCTPlan.

REFERENCES

1. Zamenhof, E. Redmond, M.S. Solaris *et al.*, Monte Carlo-based treatment planning for boron neutron capture therapy using custom designed models automatically generated from CT data, *Int. J. Radiation Oncology Biol. Phys.*, 35:383–397, 1996.
2. Verbakel, F. Stecher-Rasmussen, A gamma-ray telescope for on-line reconstruction of low boron concentrations in a human head during BNCT, *in:* "Advances in Neutron Capture Therapy, Vol 1", B. Larsson, J. Crawford, R. Weinreich, ed., Elsevier Science, Amsterdam, 1997, pp. 122–127.
3. Verbakel, F. Stecher-Rasmussen, In vivo determination of boron in patients during BNCT in Petten by means of the gamma-ray telescope, *in:* "Proceedings of the Eight International Symposium on Neutron Capture Therapy", 1998, (these proceedings).
4. K. Ogawa, S. Takahashi, Y. Satori, Description of an object in Monte Carlo simulations, *IEEE Trans. Nucl. Sc.* 44:1521–1526, 1997.

24

INTEGRATED MEDICAL IMAGE REGISTRATION, PATIENT POSITIONING, AND PATIENT MONITORING FOR CRANIAL BNCT

M. R. Palmer[1], W. S. Kiger[2], C. Zuo[1], L. P. Panych[3], C. R. G. Guttmann[3], R. G. Zamenhof[1], and P. M. Busse[4]

[1]Harvard Medical School
Beth Israel Deaconess Medical Center
Department of Radiology
[2]Massachusetts Institute of Technology
Nuclear Reactor Laboratory
[3]Harvard Medical School
Brigham and Women's Hospital
Department of Radiology
[4]Harvard Medical School
Joint Center for Radiation Therapy

INTRODUCTION

Ideal treatment planning for irradiation of brain tumors using BNCT involves a combination of temporal and spatial targeting. Spatial targeting, or target volume selection, is accomplished by CT and MR imaging during treatment planning. CT offers the highest quality geometric accuracy and serves as the geometric model for Monte Carlo simulations in our treatment planning system.[1] Contrast enhanced MR offers the best description of the tumor and surrounding areas of edema. In addition, in the future, PET scanning may be employed to aid in identifying the target temporally.

Stereotactic devices are commonly employed in imaging for neurological procedures. There is however, always a trade-off to be made in terms of comfort, degree of invasiveness, and the accuracy of the device. Few stereotactic devices are adaptable to all three imaging modalities as well as therapy.

The output of a treatment plan consists of an expression, for each field, of the beam axis entry point and direction in CT-based coordinates. Without an external reference, positioning the patient for therapy is a challenge and prone to appreciable errors. The ideal stereotactic frame for BNCT would be minimally invasive, serve to register MR,

Frontiers in Neutron Capture Therapy, edited by Hawthorne *et al.*
Kluwer Academic / Plenum Publishers, New York, 2001.

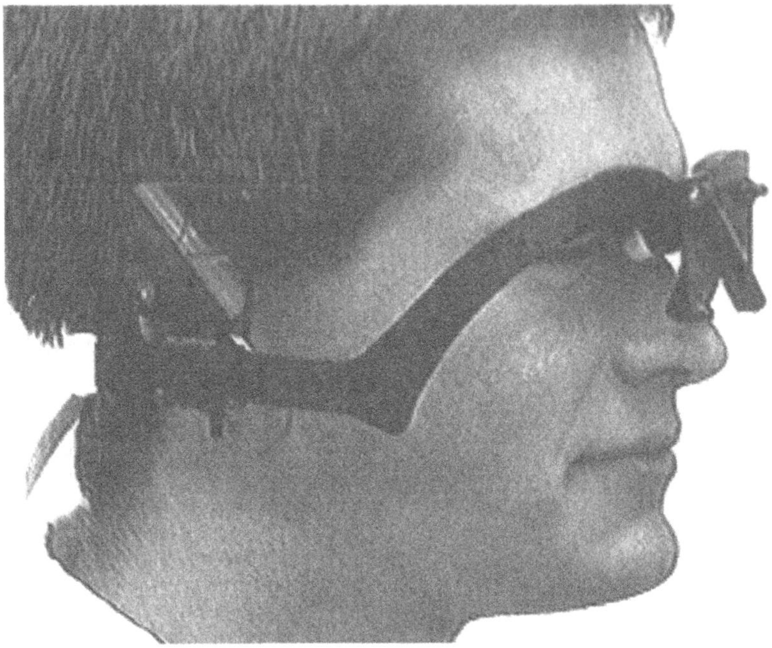

Figure 1. Anatomark stereotactic head frame. Three blocks are located on the frame—one above each ear piece and one in front of the eyes—to support the cylindrical fiducial markers.

CT and PET images and provide a reference system for patient positioning and monitoring. In the current study, we explore the potential of a new non-invasive stereotactic device for application in BNCT.

MATERIALS AND METHODS

The Anatomark stereotactic frame[2] is a non-invasive, comfortable registration device that can be used in MR, CT and PET imaging. As shown in Figure 1, the frame rests on the nasion and inside the ears and is attached to the head using straps much like a pair of goggles. Three blocks are located on the frame—one above each ear and one in front of the eyes. Each block supports two cylinders arranged as displaced crosses and filled with contrast medium so as to be detectable in each imaging modality. In addition the blocks contain external reference marks that can be used in conjunction with 3D digitization devices.

Axial sections through the marking system exhibit displaced circular marks on medical images. Figure 2 illustrates the resulting CT and MR images taken from the same subject.

An experiment was designed to determine the accuracy of locating fiducial marks on the scalp of a volunteer. Fourteen vitamin E capsules were stuck to a rubber cap or the subject's skin as shown in Figure 3. The Anatomark frame was also placed on the subject and the position of each frame reference mark and the tip of each capsule was recorded by a three-dimensional digitization device (Miniscribe 3D, Immersion Corporation). Next the subject was placed in a GE MR scanner and a sequence of

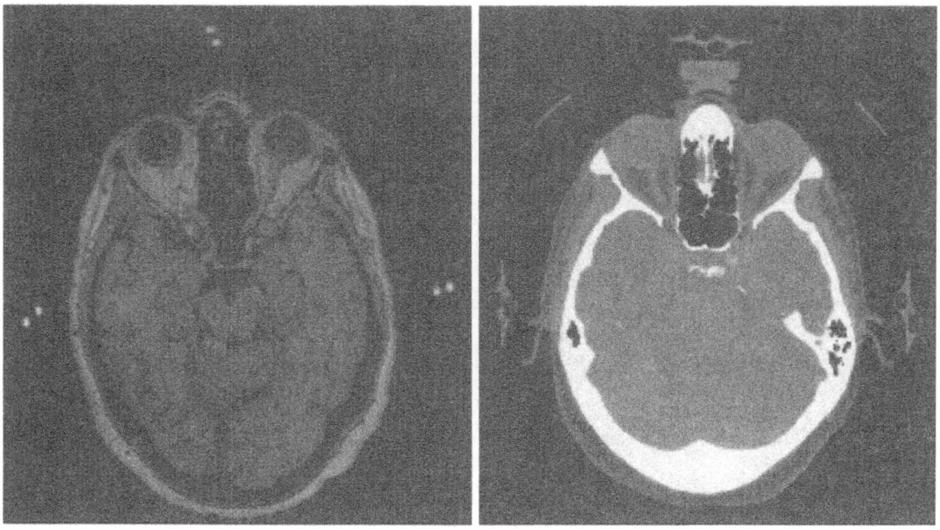

Figure 2. MR and CT Images from the same subject. Fiducial marks show up on MR (left) as paired bright circular regions and on CT (right) as paired low-density circular regions.

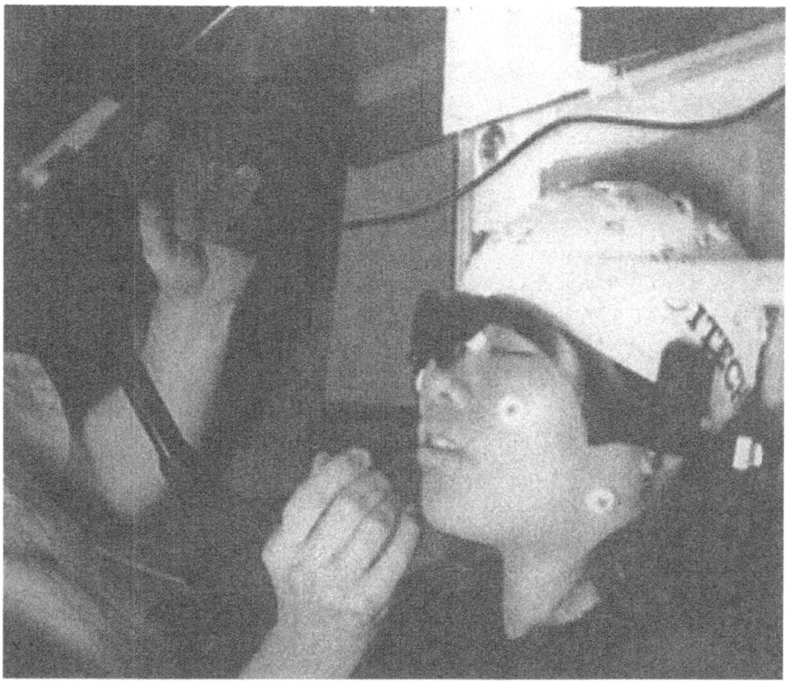

Figure 3. Experimental arrangement: fiducial markers and mechanical digitazation system used to measure the position of each mark.

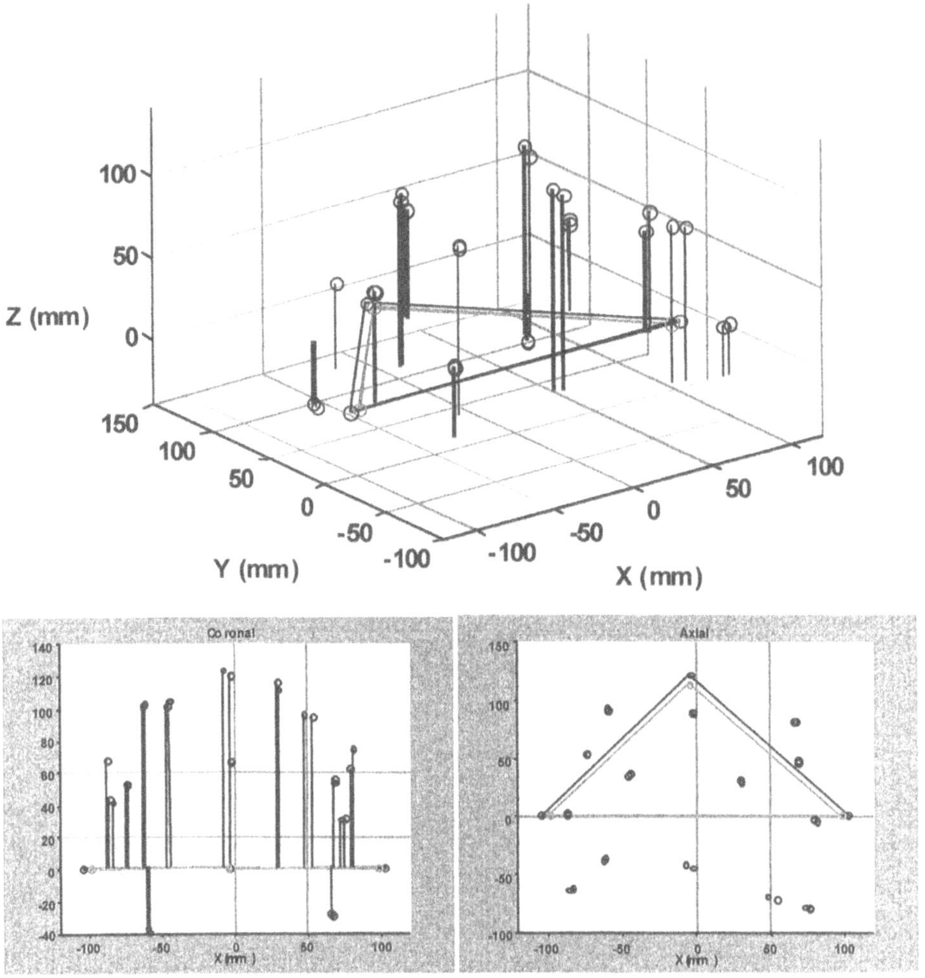

Figure 4. Comparison of fiducial marker position identification using external measuring device and MR imaging. The top graph shows the full three-dimensional set of points located by each method. The bottom two graphs show projections onto coronal (left) and axial (right) planes.

124 images was acquired at a slice separation of 1.5 mm and an in-plane pixel separation of 1.0 mm. A standard image analysis program was used to locate the tip of each capsule and estimate the position of frame reference marks within the MR data set. The frame reference marks in each measurement set were used to form the basis coordinate system as in.[3]

RESULTS

The results of fiducial mark measurement are shown on the graphs of Figure 4. Also shown are locations of frame reference marks as measured by the digitizer and by imaging and these are joined by lines to form triangles in space. The fact that the image-measured triangle is inscribed by the externally-measured triangle is due to the fact that image-based reference position is defined as mid-way between the two cylinders which is 5 mm inside the external reference mark.

Mean square deviation between these two types of measurements is 4 mm. This discrepancy is biased by a couple of measurements near the back of the head and is probably a result of marker movement between digitization and MR imaging. Excluding measurements from those two points reduces the RMS discrepancy to less than 1.5 mm.

CONCLUSION

The routine application of this device to BNCT would involve the use of the stereotactic frame in MR, CT (and possibly PET) treatment planning imaging followed by image post-processing for co-registration. Extension of these methods to patient positioning would require the use of an external 3D digitization device (mechanical arm such as we've used here, or another system such as an optical or ultrasonic measurement system). This is the focus of ongoing work in our laboratory.

ACKNOWLEDGMENTS

This work was supported by a grant from the Department of Energy, DE-FG02-97ER62193. The authors acknowledge the help of Cynthia Chuang and Tim Goorley in acquiring data for the study.

REFERENCES

1. R.G. Zamenhof, E. Redmond II, G.R. Solares, D. Katz, S. Kiger, and O.K. Harling. "Monte Carlo based treatment planning for boron neutron capture therapy using custom designed models automatically generated from CT data." Int. J. Rad. Oncol. Biol. Phys. 32:383–397, (1996).
2. Anatomark/Interneuron Corporation, 99 Hayden Avenue, Lexington, MA 02173.
3. K. Oshio, L.P. Panych, C.R.G. Guttmann. A simple Noninvasive Stereotactic Device for Routine MR Head Examinations, J Comput. Assist. Tomography, 20(4):588–591, (1996).

TREATMENT PLAN EVALUATION IN BORON NEUTRON CAPTURE THERAPY

N. Gupta[1], R. A. Gahbauer[1], and T. E. Blue[2]

[1]Division of Radiation Oncology
Arthur G. James Cancer Hospital and Research Institute
The Ohio State University
Columbus, Ohio 43210
[2]Nuclear Engineering Program
Department of Mechanical Engineering
The Ohio State University
Columbus, Ohio 43210

1. INTRODUCTION

The improvement of technology has seen the advent of modem radiotherapy techniques, such as stereotactic radiosurgery, static and dynamic conformational radiotherapy, and intensity modulated radiotherapy. To use such techniques effectively and safely, a new generation of fully three-dimensional treatment planning programs have been developed. Such systems use multiple patient images (CT, MRI, etc.) and allow fall 3-D calculation of doses, using non-coplanar beams and improved calculation models. Such detailed and accurate calculation and treatment planning techniques have made it possible to escalate doses to normal tissues to higher values and thereby treat the tumor to higher radiation doses with greater confidence.

The use of 3-D treatment planning systems generates extensive dose distribution data for each patient, and evaluating such data and comparing competing treatment plans is no longer as trivial as for plans developed by 2-D planning systems. Hence, different ways of representing the data have been developed to aid visualization and evaluation of such treatment plans. Efforts are also in place to relate treatment plans to biological effects.

In BNCT, clinical dose calculations are usually performed using a much coarser grid (larger voxel sizes) than the above systems, because of the calculation time restrictions imposed by the Monte Carlo calculations used.[1,2] Absorbed doses to normal brain from the epithermal beam can be calculated in 3D space with good accuracy; but, due

Frontiers in Neutron Capture Therapy, edited by Hawthorne *et al.*
Kluwer Academic / Plenum Publishers, New York, 2001.

to uncertainties in boron distribution and the derived RBE's, errors in calculated boron doses and RBE-weighted doses may be quite large, making it very difficult to even try to relate treatment plans to biological effects. The goal of this paper is to identify parameters that are meaningful in plan evaluation in BNCT and parameters that are not.

2. PARAMETERS USED IN TREATMENT PLAN EVALUATION FOR EXTERNAL BEAM TREATMENT PLANS

Various parameters have been suggested by different groups involved in 3D treatment planning for photon and charged particle beams as a means of evaluating treatment plans. Some of these parameters are used today to subjectively evaluate plans, and other parameters (especially the ones that try to predict biological effects) are still under investigation. It has been proposed to combine these parameters using some criteria to numerically score treatment plans as a basis of computer optimization of 3D treatment plans.

Some of the suggested parameters include the following.

Dose Volume Histograms (DVH):[3,4] Currently used widely as a plan comparison tool, DVHs may be plotted in cumulative or differential histograms and give an idea of existence of hot or cold spots (without information of their exact location) for the Planning Target Volume and desired structures at risk.

Tumor and Normal Tissue dose statistics: D_{ave} (average dose), V_{eff} (effective volume), V_D (volume receiving dose greater than D Gy), D_{min} (minimum dose), D_{max} (maximurn dose), etc. Normal Tissue Complication Probabilities (NTCP)[5,6] correlates dose distribution to biological end point of Normal Tissue effects. Different empirical and mechanistic models for NTCP have been proposed and are being investigated.

Tumor Control Probabilities (TCP):[7,8] TCP correlates tumor doses to biological end point of Tumor Control.

Multi-institution clinical trials involving the use of conformal external beam radiotherapy with escalating treatment doses are currently accruing as much data possible. This includes all planning data in an electronic form to try to evaluate treatment outcome and complication data with the above parameters, or any other parameter that may be developed in the future. These protocols also follow the internationally recommended standards of prescribing and reporting doses.

3. THE SCENARIO IN BNCT

The task of evaluating different treatment plans in BNCT is very ambiguous at this point. Some of the problems associated with this task are attributable to limitations in calculating tumor doses with any level of certainty (especially the crucial boron dose) and are due in part to an incomplete knowledge of dose-effect relationship for the mixed high and low LET radiation fields in normal brain.

The canine healthy tissue irradiation data from Brookhaven and Petten show evidence of volume effects and differences in structure sensitivity.[9–12] This is borne out by the fact that observed normal tissue effects in most dogs are at locations much deeper than the location of the peak dose in most cases. Hence, in BNCT, the problems of evaluating plans and relating them to biological effects are further complicated. To generate tolerance data and dose-effect relationships to aid plan evaluation in BNCT, the treatment planning data from patients currently being treated under dose escalation protocols need to be generated in an institution-independent fashion.

Unlike external photon beams, which are practically mono-directional, epithermal neutron beams generally have a great angular spread. In addition, epithermal neutron beams thermalize within the first few collisions in tissue and diffuse laterally due to scatter. Hence, field shaping using an epithermal neutron beam to conform to the tumor is not possible, and BNCT relies on boron compound localization in tumor to achieve dose conformation to the tumor. Since neutron fluxes fall off gently with depth and off-axis, normal tissue dose distributions in BNCT will not have sharp dose gradients within structures, as observed in high energy x-ray beams in which doses may fall off in the beam penumbra by 60% within a few millimeters.

4. CALCULATION OF BNCT DOSE STATISTICS

Clinical BNCT treatment planning is currently performed using Monte Carlo based treatment planning systems. In order to compute plans in a reasonable time frame, some limitations are imposed in the calculation. Firstly, the voxel sizes chosen for evaluation of doses must be large enough for statistical convergence of the Monte Carlo calculations in a reasonable time. Secondly, the number of detectors, or the "calculation matrix" must be spaced far enough apart to have a manageable number of detectors for quicker calculations. Therefore, calculations with grid spacing of 2 to 3 mm (as used commonly in 3D photon beam planning) are just not possible.

As a result of such limitations, the CT or MRI image voxels are collapsed to obtain a coarser voxel structure of larger voxels by averaging material compositions for use in BNCT treatment planning. However, doing this severely limits one's ability to calculate routinely and with reasonable accuracy structure doses and dose statistics for smaller substructures within the brain, due to the relatively few calculation points within a substructure and the fact that voxels may partially cover multiple adjacent structures. The above limitations imposed by using large voxel sizes in BNCT dose calculations are illustrated in Figure 1. Figure 1a shows the effect of average tissue types (bone, brain, voids, etc.) by collapsing adjacent voxels of a CT scan to obtain 1 cm^3 voxels, which results in the loss of anatomical details. Figure 1b shows the limitations of trying to use large voxels to calculate dose statistics in brain substructures. The Zubal head phantom with head and brain substructures shown in different colors is shown, with a grid corresponding to approximately 1 cm × 1 cm pixel sizes superimposed on it to show how few voxels would lie entirely within any substructure of interest.

5. DOCUMENTATION AND EVALUATION OF BNCT TREATMENT PLANS

The above limitations of BNCT treatment planning will make attempts at reporting structure doses very inaccurate, although such reports (if accurately calculated and reported) might be very helpful in correlating dose effect relationships in the future for patients currently being treated using this treatment modality. To calculate structure doses with reasonable accuracy and to preserve the structure volumes, it is essential that smaller voxel sizes be used to reconstruct the patient geometry for treatment planning calculations. This does not mean that doses in all voxels must be tallied. In fact, since the dose

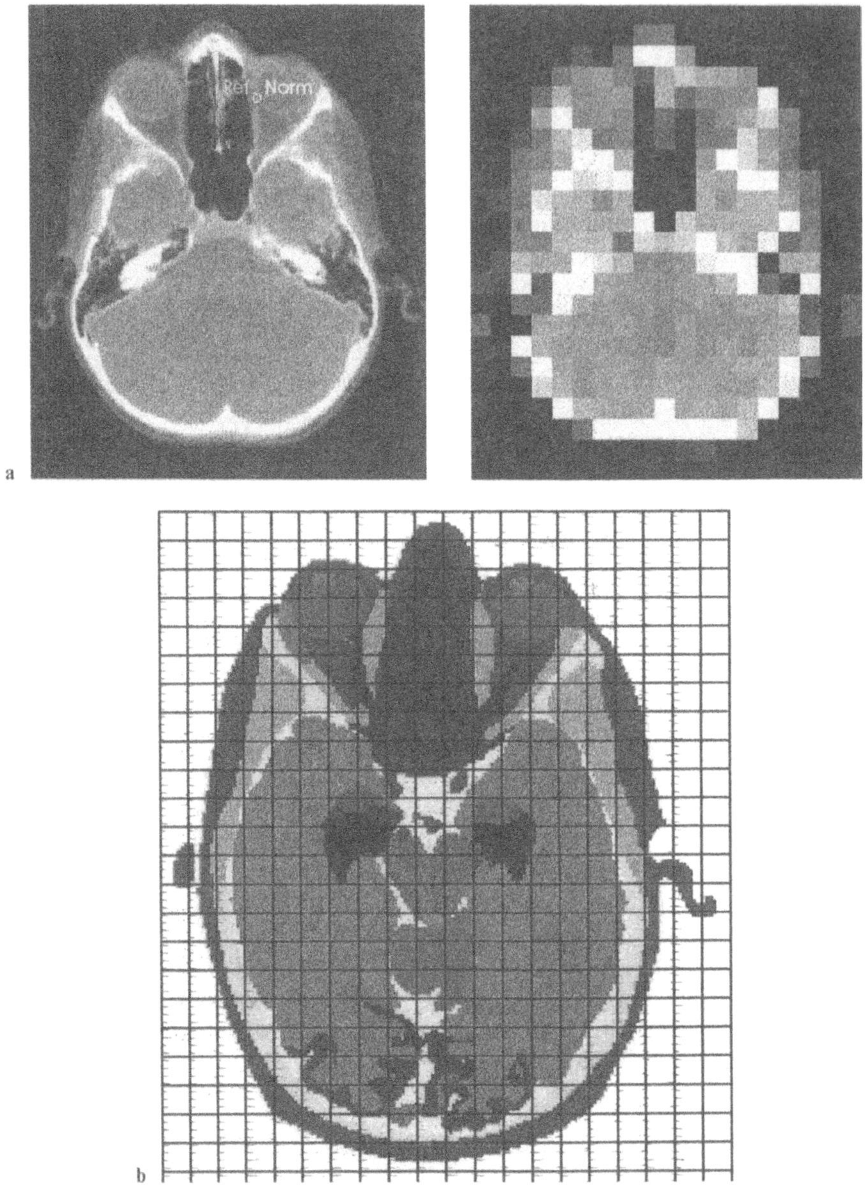

Figure 1. a) The effect of average tissue types (bone, brain, voids, etc.) by collapsing adjacent voxels of a CT scan to obtain 1 cm^3 voxels resulting in the loss of anatomical details is illustrated; b) The limitations of trying to calculate dose statistics in brain substructures using large voxels is illustrated by this figure. The Zubal head phantom with head and brain substructures shown in different shades of gray is show, with a grid corresponding to approximately 1 cm × 1 cm pixel sizes is superimposed on it, to show how few voxels would lie within any substructure of interest.

distributions for epithermal beams fall off gently in all directions, one may be able to calculate doses to smaller structures as a whole by clubbing all the voxels a structure encompasses into one tally (since doses within the structure would be uniform). For larger structures, one can assign groups of voxels as tallies and get reasonable accuracy in short time frame, while also getting dose statistics within the structure.

The parameters that seem reasonable to calculate and report, assuming uniform boron concentrations in normal brain, are:

1. Global maximum, minimum and average doses in the brain (physical doses for each dose component and the total physical and RRE-weighted doses);
2. Maximum and minimum physical dose to the tumor from each dose component;
3. Average physical and RBE-weighted doses to an agreed upon list of substructures of the brain and other structures of interest around the brain, to correlate structure tolerance to delivered doses, as more data is accrued with radiosurgery and other brain tumor treatment modalities;
4. Dose-volume histograms for physical and RBE-weighted doses for whole brain and/or each hemisphere and for larger structures within the brain.

The importance of reporting physical doses and the RBE's used to calculate the RBE-weighted doses cannot be overemphasized; so that, in the future, as better estimates of RBE's for different components are available, one can re-evaluate the plans easily in a consistent fashion.

Any plan evaluation parameter involving tumor dose estimates is not appropriate for BNCT at this time, since the level of uncertainty involved in such a parameter is very high. Some of the problems associated with calculating tumor doses are the following:

1. The target is not very well defined and usually encompasses gross tumor and some margins to cover microscopic disease, which includes normal brain which theoretically has much lower boron concentrations;
2. There is no reasonable way today of measuring the macroscopic distribution of ^{10}B within the target, and assumptions of uniform ^{10}B distribution within the heterogeneous tumor are just not correct. The variation of ^{10}B uptake within the tumor is a major contributor to dose inhomogeneity within the tumor;
3. The micro-distribution of ^{10}B, and thereby the micro-dosimetry for ^{10}B, for solid brain tumors is unknown.

6. CONCLUSIONS

As BNCT proceeds into dose escalation trials and the accrual of useful data becomes crucial, it is very important to use an agreed upon format to calculate and report dose distributions for patients treated in different institutions. The RTOG Conformal Therapy protocols (which have adopted the ICRU recommendations) provide a good example of the data that should be calculated and reported to be able to determine dose-effect relationships in the future. However, not all data required by the RTOG dose escalation protocols can be calculated for BNCT in a reasonable time frame for every patient. It is recommended that dose statistics for the whole brain and average structure doses for a selected group of structures be calculated and reported. Both physical (absorbed) and RBE-weighted doses (using the current best estimates of RBE's) should be reported, so that the data may be reprocessed as better estimates of RBE are available later. With the current near dearth of dose-effect data, the above parameters may serve as parameters for evaluating plans for BNCT.

REFERENCES

1. R. Zamenhof, E. Redmond, G. Solares, D. Katz, K. Riley, S. Kiger, and O. Harling, Monte Carlo-based treatment planning for boron neutron capture therapy using custom designed models automatically generated from CT data, *Int J Radiat Oncol Biol Phys.*, 35(2):383–397, (1996).
2. D.W. Nigg, P.D. Randolph, and F.J. Wheeler, Demonstration of three-dimensional deterministic radiation transport theory dose distribution analysis for boron neutron capture therapy, *Med Phys*, 18(1):43–53, (1991).
3. R.E. Drzymala, R. Mohan, L. Brewster, J. Chu, M. Goitein, W. Harms, and M. Urie, Dose-volume histograms, *Int J Radiat Oncol Biol Phys*, 21(1):71–78, (1991).
4. A. Niemierko, and M. Goitein, Dose-volume distributions: a new approach to dose volume histograms in three-dimensional treatment planning, *Med Phys*, 21(1):3–11, (1994).
5. B. Emami, J. Lyman, A. Brown, L. Coia, M. Goitein, J.E. Munzenrider, B. Shank, L.J. Solin, and M. Wesson, Tolerance of normal tissue to therapeutic irradiation. [Review], *Int J Radiat Oncol Biol Phys*, 21(1):109–122, (1991).
6. J.T. Lyman, Complication probability as assessed from dose-volume histograms, *Rad Res Suppl*, 8:S13–S19, (1985).
7. M. Goitein, The probability of controlling an inhomogeneously irradiated tumor, *in*: "Evaluation of Treatment Planning for Particle Beam Radiotherapy," Radiotherapy Development Branch and Radiation Research Program: (eds.), National Cancer Institute, Bethesda, (1987).
8. A. Niemierko, and M. Goitein, Implementation of a model for estimating tumor control probability for an inhomogeneously irradiated tumor, *Radiotherapy & Oncology*, 29(2):140–147, (1993).
9. P.R. Gavin, S.L. Kraft, L.R. Wendling, and D.L. Miller, Canine spontaneous brain tumors—a large animal model for BNCT, *Strahlentherapie und Onkologie*, 165(2–3):225–228, (1989).
10. P.R. Gavin, S.L. Kraft, C.E. DaHaan C.D. Swartz, and M.L. Griebenow, Large animal normal tissue tolerance with boron neutron capture therapy, *Int J Radiat Oncol Biol Phys*, 5:1099–1106, (1994).
11. P.R. Gavin, R. Huiskamp, F.J. Wheeler, S.L. Kraft, and C.E. DeHaan, Large animal normal tissue tolerance using an epithermal neutron beam and borocaptate sodium, *Strahlentherapie und Onkologie*, 169:48–56, (1992).
12. P.R. Gavin, S.L. Kraft, R. Huiskamp, and J.A. Coderre, A review: CNS effects and normal tissue tolerance in dogs. [Review] [25 refs], *Journal of Neuro-Oncology*, 33(1–2):71–80, (1997).

26

A COMPARISON OF TWO BNCT TREATMENT PLANNING PROGRAMS

MacNCTPlan and BNCT_RTPE

T. Goorley[1], F. Wheeler[2], J. Capala[3], W. S. Kiger III[1], M. Palmer[4], and R. Zamenhof [4]

[1]Nuclear Reactor Laboratory
Massachusetts Institute of Technology
[2]Idaho National Engineering and Environmental Laboratory
[3]Brookhaven National Laboratory
[4]Department of Radiology
Beth Israel Deaconess Medical Center Medical Department

1. INTRODUCTION

There are two current ongoing phase-I BNCT clinical trials in the United States for GBM,[1,2] one in Europe,[3] and several protocols in Japan.[4] The two U.S. sites, Harvard/MIT and Brookhaven National Laboratory (BNL), employ two different treatment planning systems, MacNCTPlan[5,6] and BNCT_RTPE,[7,8] developed respectively at Harvard/MIT, and Idaho National Engineering and Environmental Laboratory (INEEL). Both of these programs generate patient specific models from CT or MRI images which are used in neutron and photon transport calculations of dose rates within the patient's head (or any other body part). These programs also allow visualization of dose rate distributions by superimposing the isodose-rate contours over the patient's CT or MRI images, and construction of the dose volume histograms (DVHs) allowing the medical physicist and radiation oncologist to compare combinations of radiation beam orientations and weights. These two programs differ in the method of model composition and synthesis, user interface, neutron and photon cross sections, and Monte Carlo algorithms used in radiation transport calculations. Each of these differences is a potential source of discrepancies in calculated dose rates. The comparison of these two systems, since they are also used by other international BNCT researchers, is a step toward standardization of BNCT treatment planning.

Frontiers in Neutron Capture Therapy, edited by Hawthorne *et al.*
Kluwer Academic / Plenum Publishers, New York, 2001.

In MacNCTPlan, the radiation oncologist typically defines a region of interest (ROI) based on a contrast enhanced MRI image. This ROI and the thresholded pixel intensities of a registered CT data set are used to construct a voxelized head model. Segmented images are volume averaged into 1 cm^3 voxels in 20% increments of the four materials: air, bone, soft tissue, and tumor. MCNP,[9]* the transport engine for MacNCTPlan, is then used to calculate the dose rates to each of the 1 cm^3 voxels in the 21 by 21 by 25 cm model. BNCT_RTPE uses a model bounded by a nonuniform rational B-splines (NURBS), where a set of user defined points overlaying patient MRI images create a closed set of NURBS surfaces. Concentric surfaces bound regions of space containing scalp, bone, brain and the ROI. Rtt_MC is the Monte Carlo algorithm used for dose calculations in BNCT_RTPE. Both MacNCTPlan and BNCT_RPTE allow 3-D manipulation of the images to visualize irradiation beam entry and exit locations and calculated isodose contours.

2. MATERIALS AND METHODS

In order to quantify the dose discrepancies attributed to each of the above differences, a series of increasingly complex comparisons was used. Initially, tissue, bone and scalp compositions and kinetic energy release per mass (KERMA) factors were directly compared. The effects from different radiation transport algorithms, KERMA factors, and reaction cross sections were probed by comparing dose vs. depth profiles along the axis of a water-filled cylinder and a water-filled ellipsoid based on the Snyder head phantom. Since the representation, or model, of the ellipsoid used for particle transport is distinct in each of the two treatment planning systems, model construction was compared. The calculated depth vs. dose profiles, based on a three dimensional smooth surface ellipsoid like that used in BNCT_RTPE, was compared to the curve based on a voxelized representation of the same ellipsoid, constructed by MacNCTPlan. Both calculations were performed with MCNP using identical beam orientations, materials, KERMAs and transport physics. For the clinical test case, identical patient CT data and boron concentrations were used in MacNCTPlan and BNCT_RTPE as they would at their respective institutions; different elemental compositions, KERMAs, model construction and particle transport algorithms, were used to calculate isodose contours and dose volume histograms (DVHs) for normal tissue and tumor.

2.1. KERMA Comparisons

The neutron or photon KERMA factors are used to calculate dose. KERMAs are based on the cross section of a reaction of interest, the isotopic composition, and the energy released from that reaction. The cross sections used by MCNP and rtt_MC for transport are based on the Evaluated Nuclear Data File version 6 (ENDF/B6). Figure 1 compares the neutron brain tissue KERMAs used by MCNP and rtt_MC assuming the same tissue compositions. Neutron KERMA factors for individual reactions are given for the rtt_mc data, while the Harvard/MIT KERMAs represent the sum of all relevant reactions in brain tissue.

The data shown in Figure 1, even when plotted on a log-lin plot, indicates very close agreement between the two data sets. Figure 1 illustrates the different conceptual

---

* MCNP is a trademark of the Regents of the University of California, Los Alamos National Laboratory.

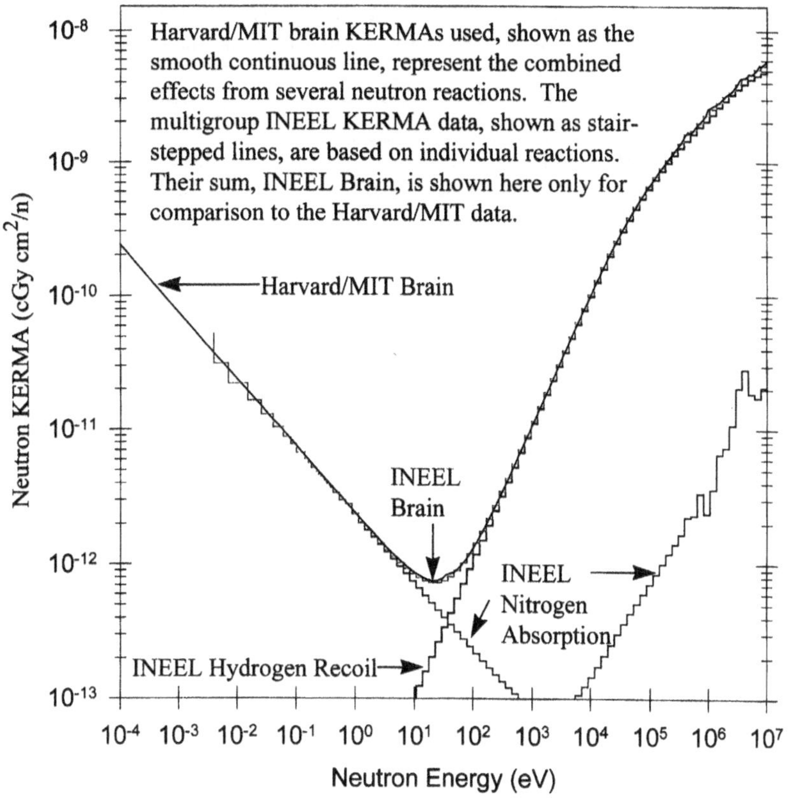

Figure 1. Comparison of Harvard/MIT and INEEL neutron KERMAs for brain tissue.

approaches of the two systems. In MacNCTPlan, all neutron reactions other than boron absorption, less than 0.5 eV constitute thermal neutron dose and are tallied as such. All reactions above 0.5 eV contribute to fast neutron dose. rtt_MC identifies KERMA deposited by the nitrogen neutron absorption and hydrogen recoil reactions individually. Since the sum of the INEEL hydrogen recoil and nitrogen absorption KERMAs are very close to the Harvard/MIT KERMAs, and both systems use the same RBE values, this conceptual difference results in no differences in dose calculations. Neither the gamma nor boron KERMAs comparisons show significant discrepancies. Figure 1 also illustrates an importance difference in the transport calculations performed by MCNP and rtt_MC. MCNP uses continuous energy cross section data, which uses log-log interpolation between two of several hundred cross section data points, while rtt_MC uses multigroup cross sections with ninety-four energy bins.

2.2. Elemental Composition

Although the elemental compositions were assumed to be the same for the KERMA comparison, each of the two institutions currently uses a different elemental composition for bone, tissue and skin. Table 1 shows the weight percent fractions of the primary elements in skin, bone and brain.

Harvard/MIT models the elemental composition of all soft tissue as that of brain, while BNL assumes a nitrogen concentration 1.5 times higher in scalp. BNL nitrogen

Table 1. Elemental tissue compositions used by BNL, Harvard/MIT and the European BNCT collaboration for BNCT treatment planning calculation

	Skin			Bone			Brain			Air
Element	BNL	MIT	Petten	BNL	MIT	Petten	BNL	MIT	Petten	MIT
ρ (g/cm^3)	1.07	1.047	1.09	1.61	1.5	1.61	1.047	1.047	1.04	0.0013
H	10.39	10.59	10.00	4.99	5.00	5.00	10.56	10.59	10.70	—
C	23.74	13.96	20.40	21.14	14.00	21.20	13.95	13.96	14.50	—
N	2.69	1.84	4.20	3.99	4.00	4.00	1.84	1.84	2.20	77.78
O	62.98	72.69	64.50	43.38	45.00	43.50	72.59	72.69	71.20	22.22
Na	—	—	0.20	0.10	—	0.10	0.14	—	0.20	—
P	—	0.39	0.10	8.08	11.00	8.10	0.39	0.39	0.40	—
Cl	0.21	0.14	0.30	0.28	—	—	0.14	0.14	0.30	—
K	—	0.39	0.10	—	—	—	0.39	0.39	0.30	—
Ca	—	—	—	17.55	21.00	17.60	—	—	—	—
Mg	—	—	—	0.20	—	0.20	—	—	—	—
S	—	—	0.20	0.30	—	0.30	—	—	0.20	—

doses in scalp are therefore 1.5 times higher. The European BNCT Collaboration uses a nitrogen content 1.2 times higher in the brain than both Harvard/MIT and BNL. MacNCTPlan models the air surrounding the patient's head, while BNCT_rtpe does not, but this and the differences in elemental compositions have little effect on the transport calculations. The dose differences, however, will vary directly with differences in nitrogen concentrations.

3. DOSE CALCULATION COMPARISONS

The effects of the differences in KERMAs and transport calculations described above were investigated using the simplistic case of a water cylinder coaxial with a cylindrical, monodirectional, monoenergetic 10 keV neutron beam modeled with both systems. Peak volume averaged doses calculated along the central axis agreed well, shown in Table 2, as did off-peak values. The second test to identify dose discrepancies resulting from different model methods used an ellipsoidal water phantom based on the Snyder head phantom. For both the cylinder and ellipsoid smooth surface models that used a monoenegetic neutron beam, the peak doses agree well, indicating slight effect from different KERMAs and transport algorithms. The ellipsoid model was also tested with a "generic" epithermal neutron beam consisting of three constant lethargy energy regions of 1% thermal neutrons, 97% epithermal neutrons (from 0.5 eV to 10 keV), and 2% fast neutrons up to 2 MeV, and a ^{60}Co photon beam. When the neutron spectrum was applied, only the fast neutron dose was significantly discrepant, by 17.5% near the entrance point; the dose profiles calculated with the ^{60}Co photon beam was exceedingly close. For both tests with the voxelized model, all peak doses are different by about 12%. Percent differences indicate the difference of the Harvard/MIT values from the INEEL values, with a negative difference indicating the Harvard/MIT value is higher.

3.1. Isodose Contour and Dose Volume Histogram Comparisons

The final comparison, based on a Harvard/MIT patient CT data set, was used to investigate the dose discrepancy caused by the cumulative effects of the KERMA, model

Table 2. Percent differences in peak doses from MCNP and rtt_MC dose comparison studies

Case	Incident beam	Modeling	Boron	Fast	Nitrogen	Gamma
Cylinder	10 KeV Neutron	Surface	1.2	−0.50	4.8	−2.5
Ellipsoid	10 KeV Neutron	Surface	2.2	4.2	6.1	4.2
Ellipsoid	10 KeV Neutron	Voxelization	8.3	12	3.7	14.5
Ellipsoid	Epithermal N Spectrum	Surface	4.8	−17.5	8.1	7.5
Ellipsoid	Epithermal N Spectrum	Voxelization	9.6	12.2	−12.9	12.8
Ellipsoid	^{60}Co Gamas	Surface	—	—	—	0.11
Ellipsoid	^{60}Co Gamas	Voxelization	—	—	—	0.16

construction, tissue composition, and transport algorithm differences between MacNCTPlan and BNCT_RTPE. The tumor was explicitly defined in the original CT data; the same tumor definition was used in the two treatment planing programs. Using the same generic epithermal beam, same boron concentrations for tissue and tumor, 13 and 45.5 ppm respectively, and the same RBE values for both planning systems, the peak tissue dose calculated was 7.34 RBE cGy for BNCT_RTPE and 7.01 RBE cGy with MacNCTPlan, a 4% difference. The maximum tumor doses were 39.8 RBE cGy and 38.2 RBE cGy respectively. Figure 2 shows the isodose contours superimposed on the CT image intersecting the beam axis. MacNCTPlan tissue isodose contours are shown in black, while BNCT_rtpe tissue isodose contours are shown in white. Isodose contour labels are the percentage of their respective global maximum. The smallest circular outline is the tumor. The figure to the right is an enlargement of the high dose region.

The isodose contours agree well, with differences of only a few mm. The tissue isodose contours generated by BNCT_RTPE are only superimposed over the brain, while MacNCTPlan tissue isodose contours are superimposed over the brain, bone and scalp. The dose volume histograms are also similar in shape, indicating that the dose distributions calculated are similar.

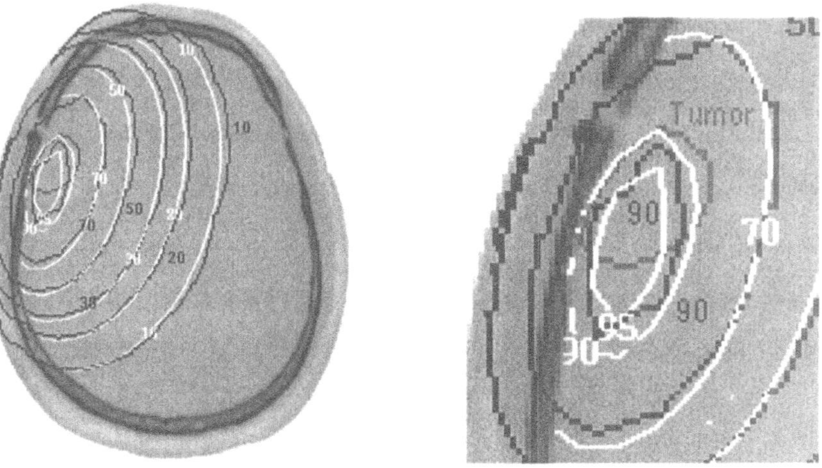

Figure 2. MacNCTPlan and BNCT_RTPE isodose contours for Harvard/MIT subject 97–4.

4. CONCLUSIONS

This paper presents a comparison of the two BNCT treatment planning programs currently used in U.S. clinical trials, MacNCTPlan and BNCT_RTPE. The effects of model synthesis, transport physics, materials compositions, and KERMAs were compared. While the designed test cases indicated greater than 10% dose discrepancies for the voxelized model, the more realistic clinical test only yielded a four percent difference in peak dose and a slight difference in the dose distribution and dose volume histogram. Since this final comparison includes the modeling differences, either the modeling differences were erroneously found, or two effects essentially cancel. Further investigation of the voxelization discrepancy is planned. Since the peak doses and dose distributions calculated in the clinical case are comparable, then patient doses from the Harvard/MIT and BNL protocols are directly comparable.

ACKNOWLEDGMENTS

The Department of Energy supported this work though DOE grant #DE-FGO2-97ER62193 and #DE-F02930017-3L. We would also like to thank Dr. Chuck Wemple from INEEL for his assistance with some of the MCNP calculations, Dr. Paul Busse and Dr. David Nigg for their impetus for this research, and Dr. Wolfgang Sauerwein for the materials data used by the European BNCT Collaboration.

REFERENCES

1. P. Busse, I. Kaplan, R. Zamenhof, O. Harling, J. Kaplan, C. Chuang, J. Goorley, W.S. Kiger III, K. Riley, L. Tang, G. Solares, and M. Palmer, " BNCT for Globlastoma Multiforme and Intracranial Metastatic Melomana: Clinical Results of the Harvard/MIT Phase I Trial. (these proceedings)
2. A.Z. Diaz, A.D. Chanana, J. Capala, M. Chadha, J.A. Coderre, E.H. Elowitz, J. Iwai, D.D. Joel, H.B. Liu, R. Ma, N. Pendzick, N.S. Peress, M.S. Shady, D.N. Slatkin, G.W. Tyson, and L. Wielopolski, Boron neutron capture therapy for glioblastoma multiforme: results from the initial phase I/II dose escalation studies. (these proceedings)
3. K. Hideghety, W. Sauerwein, M. de Vries, F. Grochulla, C. Goetz, K. Haselsberger, P. Pacquis, J. Heimans, J. Wolbers, R. Moss, R. Huiskamp, F. Stecher-Rasmussen, J. Rassow, S. Garbe, H. Fankhauser, and D. Gabel," Report on the First Patient Group of the European Phase I Trial (EORTC Protocol 11961) at the High Flux Reactor PETTEN (these proceedings).
4. Y. Nakagawa, P. Kyonghon, K. Kitamura, T. Kageji, and T. Minobe, "Long Term Follow Up In Patients Who Underwent Boron-Neutron Capture Therapy" (these proceedings).
5. R.G. Zamenhof, E. Redmond II, G.R. Solares, D. Katz, W.S. Kiger III, and O.K. Harling, "Monte Carlo based treatment planning for boron neutron capture therapy using custom designed models automatically generated from CT data," Int. J. Rad. Oncol. Biol. Phys. 35:383, 1996.
6. W.S. Kiger III, R.G. Zamenhof, G.R. Solares, E.L. Redmond II, and C. Yam. "MacNCTPlan: An Improved Macintosh-Based Treatment Planning Program," Trans. Am. Nuc. Soc. 75:38, 1996.
7. D.W. Nigg, F.J. Wheeler, D.E. Wessol, J. Capala, and M. Chadha, "Computational dosimetry and treatment planning for boron neutron capture therapy". J. Neuro-Oncol. 33:93–104, 1997.
8. D.E. Wessol, R.S. Babcock, F.J. Wheeler, G.J. Harkin, L.L. Voss, and M.W. Frandson, "BNCT••rtpe: BNCT Radiation Treatment Planning Environment Users Manual Version 2.2", url http://id.inel.gov/cart/rtpe-manual/
9. J.F. Briesmester, Ed. "MCNP-a general Monte Carlo N—particle transport code," v. 4B, Los Alamos National Laboratory report LA-12625-M Version 4B, March 1997.

27

THREE DOSE CALCULATION CODES APPLIED TO NEUTRON TRANSPORT IN BNCT

Juha S. Lampinen[1,2*], Tiina Seppälä[1,3], Iiro Auterinen[4], and Sauli Savolainen[2]

[1]Department of Physics
University of Helsinki
P.O.Box 9
FIN-00014 University of Helsinki, Finland
[2]Department of Radiology
[3]Clinical Research Institute
Helsinki University Central Hospital, Finland
[4]VTT Chemical Technology
Espoo, Finland

1. INTRODUCTION

The three dimensional BNCT_Rtpe/rtt_MC[1] is used for treatment planning in the proposed patient irradiations at the Finnish boron neutron capture therapy (BNCT) facility FiR 1. Two other radiation transport calculation codes used mainly in the characterization of the treatment beam, are DORT (Two-Dimensional Discrete Ordinates Transport Code)[2] and MCNP (Monte Carlo N-Particle Transport Code System).[3]

Both BNCT_Rtpe/rtt_MC and MCNP are based on the Monte Carlo method, while DORT solves the Boltzmann transport equation throughout a deterministic algorithm. Both Monte Carlo applications use similar radiation transport methods, the differences are in the geometry and tally specifications; BNCT_Rtpe/rtt_MC can use medical images (MR/CT) in constructing the geometry, in MCNP the tally output options are more flexible. There are, however, special front end applications for MCNP, which allow the user to construct the geometry from medical images, but these are not included in the standard MCNP distribution. Both systems, MCNP and BNCT_Rtpe/rtt_MC, can use the cross section data from the ENDF/B (IV/V) library.

In DORT neutrons and photons are transported to certain discrete directions through each cell of space mesh in the R-Z geometry from the reactor core to the phantom. The solution of the transport equation is iterated until fluxes will converge to

*Correspondance: Juha Lampinen, Department of Physics, P.O.Box 9, FIN-00014 University of Helsinki, Finland, E-mail: juha.lampinen@helsinki.fi, Tel: +358-9-191 8370, Fax: +358-9-191 8378.

Frontiers in Neutron Capture Therapy, edited by Hawthorne *et al.*
Kluwer Academic / Plenum Publishers, New York, 2001.

the defined accuracy. In this model the BUGLE-80 cross section library of 47 neutron and 20 photon energy groups are used.

In this work we have compared thermal neutron flux distributions in depth and tranverse direction, calculated with these three codes.

2. MATERIALS AND METHODS

The comparisons were done with calculations of the thermal neutron flux, E < 0.414 eV. The source specifications of the Monte Carlo runs, i.e. BNCT_Rtpe/rtt_MC and MCNP, were taken from DORT calculations.[4] Thermal neutron flux distributions in depth as well as in tranverse direction (profile) at 25 and 60 mm depths were calculated using all the three above mentioned systems.

The geometry of the problem, derived directly from MCNP input file with SABRINA code, is presented in Figure 1.

Tally geometries used in DORT and MCNP calculations were discs of 1 mm thickness and 5 mm of radius. In BNCT_Rtpe/rtt_MC, a cubic tally volume of 1 cm^3 is used as a default.

The neutron fluxes calculated with DORT and MCNP, i.e. flux in relation to number of particles emerging from the source plane, were further transformed to correspond real world values by multiplying the fluxes by the number of neutrons emerging from the source plane in FiR 1 reactor. in case of BNCT_Rtpe/rtt_MC, this transformation is done before calculations. In MCNP simulations, variance was reduced by using geo-metry splitting and russian roulette.[3] No variance reduction methods were used in BNCT_Rtpe/rtt_MC calculations.

3. RESULTS

Thermal neutron flux distributions in depth, calculated with each software, are presented in Figure 2. Differences in depth data between the calculation codes are shown in

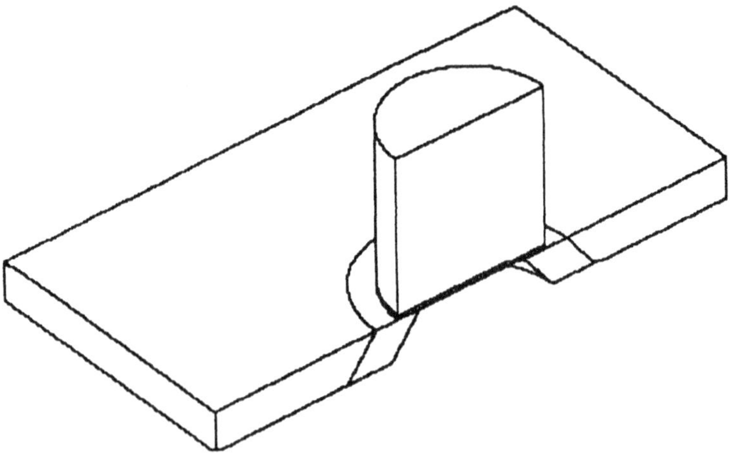

Figure 1. The geometry used in the neutron flux calculations. The illustration is cut in half. The cylindrical water phantom with a lucite face (5 mm) is shown in the middle. The beam collimator is made of two materials.

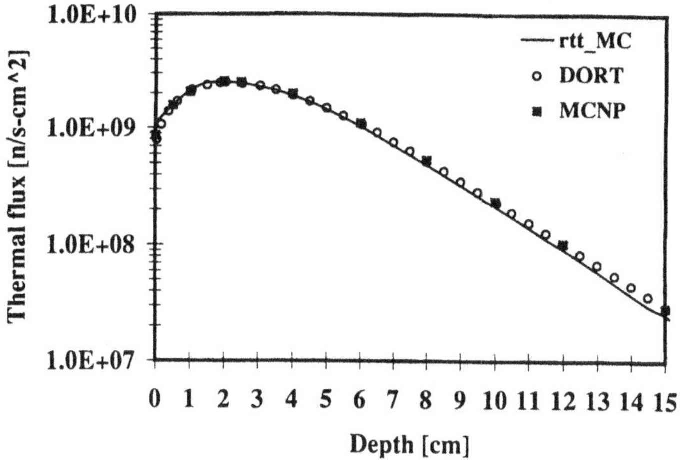

Figure 2. Thermal neutron flux distributions in depth in a cylindrical water phantom. For clarity, data for BNCT_Rtpe/rtt_MC is presented as a line connecting the discrete data points.

Figure 3. Profiles at 25 and 60 mm depths are given in Figure 4. The statistical uncertanties associated with results calculated with MCNP are less than 1% for all data points.

4. DISCUSSION

In this work, thermal neutron flux distributions, depth curves and profiles, calculated with three different code systems, DORT, MCNP and BNCT_Rtpe/rtt_MC were compared.

In general, the results are in good agreement. For example, the thermal fluxes calculated with BNCT_Rtpe/rtt_MC, MCNP and DORT, at 25 mm depth in the central axis are within 2%, but at the phantom surface the differences may be as high as 40%. At deeper depths, the fluxes given by BNCT_Rtpe/rtt_MC disagree with DORT and MCNP. For example, from the depth of 80 to 150 mm, the mean disagreement is about 10% (ratio of BNCT_Rtpe/rtt_MC and DORT), while mean MCNP/DORT ratio is 2%. The trend is visible in Figure 2. Explanation for this may be bad statistics in case of BNCT_Rtpe/rtt_MC at deeper calculation points.

In case of profiles, BNCT_Rtpe/rtt_MC, MCNP and DORT agree within 3%, even when the distance from the central axis is 80 mm.

The tally volume used in the BNCT_Rtpe/rtt_MC calculations is an 1 cm^3 cube. In areas, where the flux gradient is steep, the flux averaged over the cube is different from the flux scored in a smaller volume. In practise, at FiR 1 beam, the effect of this averaging used in BNCT_Rtpe/rtt_MC, leads to overestimation of the thermal flux by 40% compared to DORT calculations (30% compared to MCNP) when the tally is situated in the phantom surface (at 0 mm).

If treatment plans calculated with BNCT_Rtpe/rtt_MC are verified with activation or TLD measurements, there will be different results in areas of steep flux gradients, e.g. near air, tissue and bone-interfaces. Because of this, it is difficult to validate BNCT_Rtpe/rtt_MC calculated treatment plans with measurements. Therefore, the validation of calculations requires smaller tally volumes, especially near the interfaces.

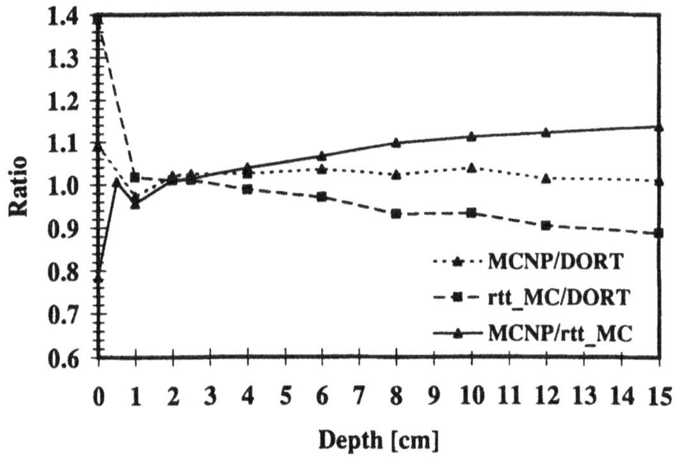

Figure 3. Ratios of depth distribution results of each calculation code pair.

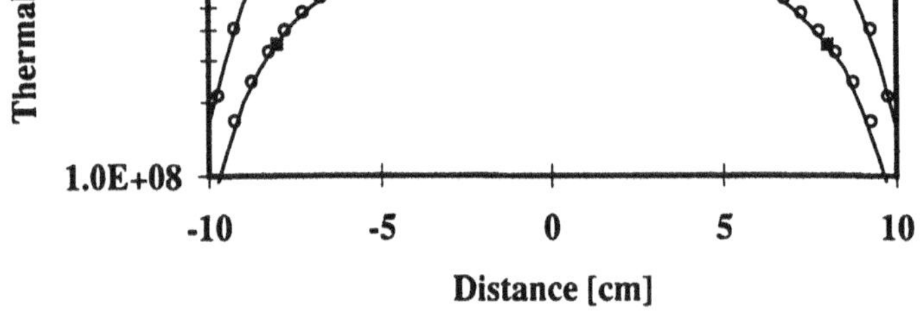

Figure 4. Tranverse distributions of the thermal neutron flux in a cylindrical water phantom at 25 (upper) and 60 mm depth. For clarity, data for BNCT_Rtpe/rtt_MC is presented as a line connecting the discrete data points.

REFERENCES

1. D.E. Wessol, *et al.*, http://id.inel.gov/cart/rtpe-manual.
2. W.A. Rhoades, and R.L. Childs, The DORT Two-Dimensional Discrete Ordinates Transport Code, *Nucl. Sci & Eng*, 99:88–89, 1988.
3. J.F. Briesmeister ed., "MCNP-A General Monte Carlo N-Particle Transport Code", Los Alamos National Laboratory Report, LA-12625-M 1997.
4. T. Seppälä, T. Serén, and I. Auterinen, Source characterisation for the rtt_MC treatment planning program at FiR1, *these proceedings (submitted)*.

28

SOURCE CHARACTERISATION FOR THE rtt_MC TREATMENT PLANNING PROGRAM AT FiR 1

Tiina Seppälä[1,2,3], Tom Serén[3], and Iiro Auterinen[3]

[1]Clinical Research Institute
Helsinki University Central Hospital
[2]Department of Physics
University of Helsinki
[3]VTT Chemical Technology
Technical Research Centre of Finland

1. INTRODUCTION

The epithermal neutron-photon treatment beam model is one of the basic elements in assuring the reliability of the doses in the treatment planning of the boron neutron capture therapy (BNCT). The beam model includes the nuclear reactor to generate the fission neutrons, the moderator to moderate the fission neutrons to epithermal neutrons with a small fast neutron and photon contamination, the collimator to collimate the epithermal neutrons and some surrounding materials, like part of a concrete shielding. The source model includes the source description at a plane at a known location near the exit plane of the collimator.

The Finnish BNCT facility (FiR 1) beam[1] has been modelled with the DORT (Two-Dimensional Discrete Ordinates Transport) code.[2] DORT was chosen based on the good experience, that INEEL has had in the modelling of the BMRR epithermal beam.[3] There have been three different geometrical configurations at FiR 1. The beam models have the same model of the core, but variable moderator thickness and either a delimiter or a collimator. The beam models have been verified by several measurement campaigns: 75 cm moderator thickness with the simple delimiter FiR(P75)[4,5] and with the collimator FiR(K75),[6,7] and 63 cm moderator thickness with the collimator FiR(K63).[6–10]

The neutron-photon source description from the DORT calculations is defined for rtt_MC, which is the Monte Carlo calculation module in the treatment planning software BNCT_Rtpe[11] developed at the Idaho National Engineering and Environmental

Correspondence: Tiina Seppälä, M.Sc., VTT Chemical Technology, P.O. Box 1404, FIN-02044 VTT, Finland.

Frontiers in Neutron Capture Therapy, edited by Hawthorne *et al.*
Kluwer Academic / Plenum Publishers, New York, 2001.

Laboratory (INEEL). This software is proposed for the patient treatment planning in the Finnish BNCT project.

The aim of this study was to define a source model for rtt_MC from the accurately defined DORT beam model, so that the rtt_MC treatment planning model has a good agreement with the activation measurements in the phantoms.

2. MATERIAL AND METHODS

2.1. DORT Beam Model

The DORT beam model FiR(K63) that will be used for patient treatment is used in this study. The 67-group coupled P_3 neutron-photon BUGLE-80 cross section library was used in the calculations. A forward-biased quadrature set (D_{166}) was used for accurate angular calculation at the beam-line from the core to the collimator output. The model has cylindrical geometry. The longitudinal central axis of the model passes from the reactor core via the centre of the FLUENTAL™ moderator to the centre of the collimator. The composition of the core was calculated from the burn-up of the fuel rods and the biased core loading (fresh fuel in the direction of the BNCT moderator) was taken into account. The model was calculated in two parts. The first part of the model, which is seldom changed, included the core, graphite reflector and water, and was a radiation source for the second part that included the moderator, collimator, phantoms and concrete. The spectra in the DORT model were verified with multifoil measurements at the bismuth face, at the 14 cm "free" beam exit and at the 11 and 14 cm source plane (50 mm inwards from the exit of the beam aperture) with a water phantom in place.[6] The neutron and gamma distributions in the phantoms were verified with Au/Mn activation foil, twin ionisation chamber and TLD measurements.[8]

2.2. Source Model for rtt_MC

For the source description the directional intensities of the neutrons and photons in the 67-group neutron and photon energy structure from the DORT model were averaged over a radial plane perpendicular to beam direction. The place for the source plane was chosen between the bismuth collimator and the phantom, where the epithermal neutron fluence was most homogeneous in the radial direction (Figure 1).

Two source planes for rtt_MC were defined from the DORT model 50 and 75 mm upstream from the exit of the beam aperture. The multifoil measurements were performed at the −50 mm source plane for 11 and 14 cm apertures and the calculated spectra were adjusted with the LSL-M2 code.[6] The adjustments were rather small (Figure 2). The adjusted source description at the depth −50 mm and two unadjusted source descriptions at the depth of −50 and −75 mm were defined for source models of the treatment planning program.

2.3. Verification of the Source Model

The source model was verified primarily with Au and Mn activation measurements in cylindrical phantoms that consist of three brain tissue substitute materials (Liquid B,[12,13] water and PMMA). Diluted Au-Al foils (1%, Ø 12 mm × 0,2 mm) obtained from ECN Petten were irradiated. The activity of the foils was measured with Ge detector

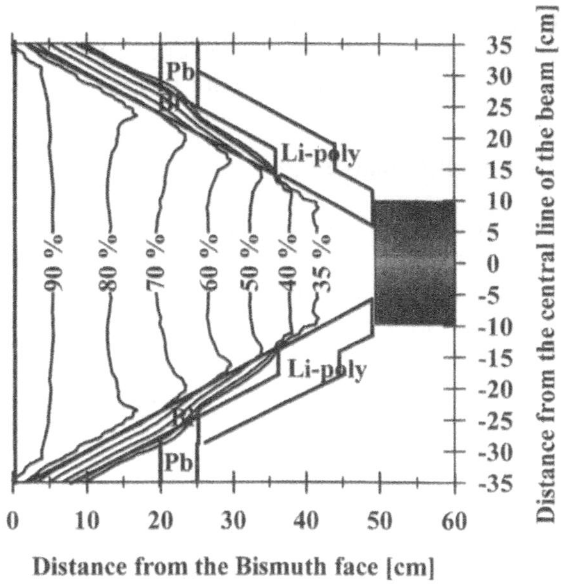

Figure 1. The calculated epithermal neutron fluence distribution inside the collimator with a cylindrical water phantom in front of the aperture. Bi is the conical bismuth collimator, Pb the lead layer and Li-poly the enriched lithiated polyethene.

system and the SAMPO program was used to determine the saturation activities and reaction rates.[5]

A new cross section file, Act.sigma, was included in rtt_MC with this option. The program uses the option to calculate the $^{197}Au(n,\gamma)$ and $^{55}Mn(n,\gamma)$ reaction rates in the model. The cross sections in 640 groups with uncertainties for $^{197}Au(n,\gamma)$ and $^{55}Mn(n,\gamma)$ reactions were taken from IRDF-90 library. They were condensed into a 94 group

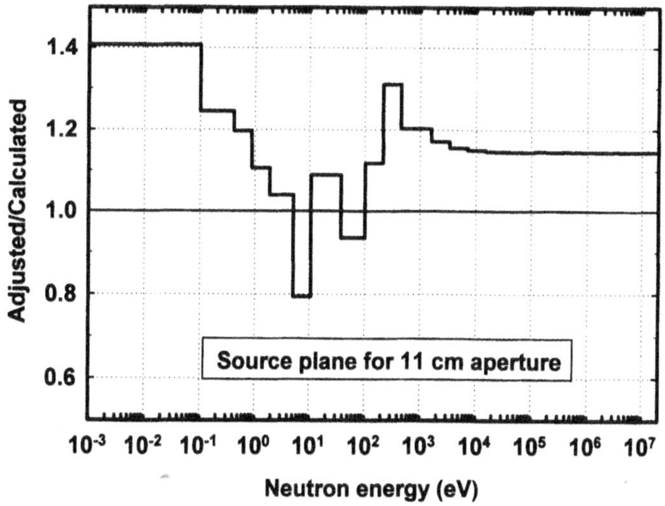

Figure 2. The ratio of adjusted and calculated neutron spectra at the −50 mm source plane for 11 cm beam. The uncertainties of the adjusted spectrum at the source plane were 23% for thermal neutrons, 8% for epithermal neutrons and 33% for fast neutrons.

structure using the program FLXPRO from the LSL-M2 package.[5] A standard 640-group weighting spectrum was used with a Watt spectrum for fast neutrons, a 1/E spectrum for epithermal neutrons and a Maxwellian spectrum for thermal neutrons.

2.4. Dose to Monitor Unit Connection

The intensity scaling and the link from the dose to the beam monitor units was obtained for rtt_MC calculations from the ^{197}Au reaction rate measurements in the phantom at the reference conditions. The calculated dose rate $\dot{D}^{calc}_{\mathrm{Ref}}$ at the reference monitor unit rate in a patient or a phantom is

$$\dot{D}^{calc}_{\mathrm{Ref}} = \frac{r^{meas}_{Au-197}(\dot{MU}_{\mathrm{Ref}})}{r^{calc}_{Au-197}(250\,kW)} * \dot{D}^{calc}(250\,kW) \tag{1}$$

where $r^{meas}_{Au\text{-}197}(\dot{MU}_{\mathrm{Ref}})$ is the measured Au-197 reaction rate at the reference monitor unit rate and $r^{calc}_{Au\text{-}197}(250\,kW)$ the calculated Au-197 reaction rate at the 250 kW reactor nominal power at the reference point and geometry (at the depth of 2.0 cm in the central axis of the cylindrical PMMA phantom). $\dot{D}^{calc}(250\,kW)$ is the calculated dose rate at the 250 kW reactor nominal power in a patient or a phantom. The number of monitor units MU to be administrated in a phantom or a patient irradiations is

$$MU = \frac{D}{\dot{D}^{calc}_{\mathrm{Ref}}} * \dot{MU}_{\mathrm{Ref}} \tag{2}$$

where D is an aimed dose in a patient or a phantom and $\dot{MU}_{\mathrm{Ref}}$ is the reference monitor unit rate.

3. RESULTS AND DISCUSSION

The DORT FiR(K63) beam model had very good agreement with measurements.[8] At the thermal neutron maximum the calculated ^{197}Au(n,γ) reaction rates compared to the measured ones were 4–6% higher in PMMA and water phantoms and 10–11% higher in Liquid B with 11 or 14 cm beams. According to these results the DORT beam model was used for source descriptions in the source models of rtt_MC. The source descriptions and models for rtt_MC were made for the all beam apertures separately to assure to a good accuracy of dose calculations. The percentage ratio of the calculated and measured ^{197}Au reaction rates in the water phantom with the three source models for 11 cm aperture are shown in Figure 3. The unadjusted source descriptions at the depth of −50 mm had the best agreement with measurements. The intensities of the source descriptions were scaled at the reference point and geometry in PMMA with the ratio of measured and calculated thermal neutron fluence, that is a basic result of rtt_MC. PMMA was chosen to reference material, because the small geometric uncertainty. The scaling factor from the ratio of measured and calculated thermal neutrons was 1.4% smaller than our final scaling factor defined with the ratio of ^{197}Au reaction rates (Eq. 1). The ratio of the thermal neutron fluences was used at first because at that moment rtt_MC didn't yet calculate reaction rates. This was the beam model used in the dose planning in our healthy

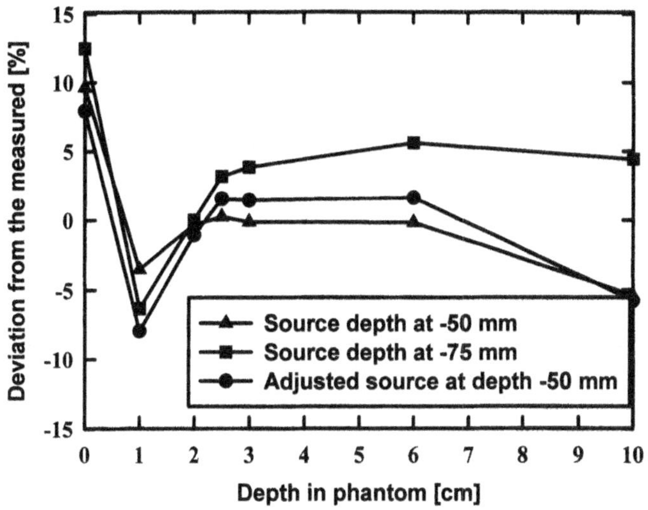

Figure 3. The percentage deviation of the calculated Au-197 reation rates (rtt_MC) from the measured ones (foils) with three source descriptions and modelled for 11 cm aperture in longitudinal axis of the cylindrical water phantom.

tissue tolerance study.[13] Rtt_MC overestimates the reaction rates 10–25% at the surface of the phantom. This was expected according to the comparison of the calculation codes.[14] The overestimation at the surface was only 10% (Figure 3), because the DORT model underestimated the thermal neutrons (Figure 2). Inside the phantom the difference from the measurements was ±5% (1SD) (Figure 3). Also using this beam model calculated total neutron doses (nitrogen + fast dose) and gamma doses compared to ionisation chamber measurements[8] were inside the measurement uncertainties.

The ratio of measured and calculated Au-197 reaction rates, which primarily are proportional to the thermal and epithermal neutron fluence rates, were chosen for scaling of the calculated values, because the measurements of reaction rates are independent compared to calculations and at the moment ^{197}Au reaction rates can be measured with the lowest uncertainty. An important reason for the choice of this scaling method was also that most of the dose components are induced from the thermalised neutrons i.e. boron, nitrogen and gamma dose and the fast neutron and the gamma dose from the beam are small at FiR 1. The ratio of measured and calculated Mn-55 reaction rates was 3% higher than the Au-197 ratio at the reference point and geometry. The uncertainty of measured reaction rates in phantom were 4%.

4. CONCLUSION

When the beam is modelled accurately enough, as with the two dimensional DORT model described here and verified in the phantoms, the calculated results can be used for the source description to the rtt_MC treatment planning software with only minor intensity scaling and with minimal spectral adjustment. The source model defined from this beam model that was only intensity scaled was used in the dose planning in our healthy tissue tolerance study in 1998.

ACKNOWLEDGMENTS

Authors wish to gratefully thank David Nigg and Floyd Wheeler at INEEL of their assistance in the DORT and rtt_MC modelling. The work was supported by the Technology Development Centre of Finland, Academy of Finland, VTT Chemical Technology and University of Helsinki.

REFERENCES

1. I. Auterinen, P. Hiismäki, P. Kotiluoto *et al.*, Metamorphosis of a 35 years old Triga reactor into a modern BNCT facility (in these proceedings).
2. W.A. Rhoades, and R.L. Childs, The DORT Two-Dimensional Discrete Ordinates Transport Code. *Nucl Sci & Engr* 99:88–89, 1988.
3. F.J. Wheeler, D.K. Parsons, B.L. Rushton *et al.*, Epithermal Neutron Beam Design for Neutron Capture Therapy at the Power Purst Facility and Brookhaven Medical Research Reactor, *Nuclear Technology* 92:106–117, 1990.
4. Kaita, T. Serén, and I. Auterinen, First characterisation of the Finnish epithermal neutron beam using activation detectors, Advances in Neutron Capture Therapy, Volume I, pp. 531–4, 1997.
5. D.W. Nigg, Y.D. Harker, J.K. Hartwell *et al.*, Collaborative Spectral Characterization of the Finnish Epithermal-Neutron Beam Facility for BNCT, in: INEEL BNCT Research Program Annual Report 1996, INEL/EXT-97-00319, April 1997.
6. T. Serén, I. Auterinen, and T. Seppälä, Spectrum measurements and calculations in the epithermal neutron beam at the FiR 1 BNCT facility, in: 15th European TRIGA Conference, VTT Symposium 197, Technical Research Centre of Finland, Espoo 1999.
7. A. Kosunen, M. Kortesniemi, H. Ylä-Mella *et al.*, Twin ionization chambers for dose determination in phantom in an epithermal neutron beam *Radiation Protection Dosimetry* 81(3) (1999) 187–194.
8. M. Kortesniemi, A. Kosunen, T. Serén *et al.*, Phantom dose distributions in the beam of Finnish BNCT facility (in these proceedings).
9. D.W. Nigg, Y.D. Harker, J.K. Hartwell *et al.*, Collaborative Neutronic Performance Characterization of the FiR 1 Clinical Epithermal-Neutron Beam Facility for BNCT, INEEL BNCT Research Program Annual Report 1998 (VENHUIZEN, J.R. Ed.) INEL/EXT-99-00293, Idaho Falls (1999) 13–38.
10. T. Serén, M. Kortesniemi, C. Aschan *et al.*, A Tale of Two Beams—Comparison of the Radiation Fields at the BMRR and FiR 1 Epithermal BNCT Facilities, Medical & Biological engineering & computing, 37, suppl. 1 (1999) 396–397.
11. D.E. Wessol, R.S. Babcock, F.J. Wheeler *et al.*, BNCT_Rtpe: BNCT Radiation Treatment Planning Environment User's Manual, Version 2.2, http://id.inel.gov/cart/rtpe-manual, February 21, 1997.
12. T. Seppälä, J. Vähätalo, I. Auterinen *et al.*, Brain tissue substitutes for phantom measurements in NCT, in: "Advances in Neutron Capture Therapy Vol 1", Medicine and Physics, Elsevier Science B.V., 188–191, pp. 1997.
13. T. Seppälä, J. Vähätalo, I. Auterinen *et al.*, Modelling of brain tissue substitutes for phantom materials in neutron capture therapy (NCT) dosimetry, Radiat. Phys. Chem., 55 (1999) 239–246.
14. J. Benczik, T. Seppälä, J. Hopewell *et al.*, Large animal model for healthy tissue tolerance study in BNCT (in these proceedings).
15. J. Lampinen, T. Seppälä, I. Auterinen *et al.*, Three dose calculation codes applied to neutron transport in BNCT (in these proceedings).

29

"CARONTE", A TREATMENT PLANNING SYSTEM BASED ON REAL MACROSCOPIC BORON DISTRIBUTION AND MCNP-4A CODE

Results and Discussion

N. Cerullo[1] and G. Daquino[2]

[1]DITEC—Genova University
Via All'Opera Pia, 15/A I-16145
Genova, Italy
[2]DCMN—Pisa University
Via Diotisalvi, 2 I-56100
Pisa, Italy

1. INTRODUCTION

The evaluation of the effectiveness of BNCT depends on the ability of knowing doses and fluences in tumor and healthy tissues associated with a BNCT irradiation on a patient. Up to now, the solution of this problem is difficult, particularly in the case of brain tumors.

Therefore, the Treatment Planning System (TPS) calculation procedure is very important and it must be able to simulate brain irradiation, starting from the real macroscopic distribution of boron concentrations, both in healthy and in tumor tissues. Current TPSs, using anatomic structures and tumor images, obtained by non-invasive methods, as CT (Computed Tomography) and MRI (Magnetic Resonance Imaging) and others, plot a model of the organ, inserting only two different boron concentrations, one in the tumor and the other in the surrounding healthy tissues. Therefore a more accurate boron distribution reconstruction is required. This consideration has induced us to develop a new TPS, able to follow the previous statement. In order to perform this task, it has been created an original system, named "CARONTE", to be applied, up to now, to the brain tumors, that reads **the real and accurate macro-scopic boron distribution information in tumor and healthy tissues**, furnished by PET data,[1] and inserts them into the brain model, to be used by the Monte Carlo code calculations.

This method was tested, at first, on a phantom model ad hoc prepared, and then applied to a real glioblastoma case, using transaxial PET slices kindly sent by Kyoto Center.[1]

Frontiers in Neutron Capture Therapy, edited by Hawthorne *et al.*
Kluwer Academic / Plenum Publishers, New York, 2001.

2. THE "CARONTE" TREATMENT PLANNING SYSTEM

The "CARONTE" TPS consists of a set of programs able to perform different tasks; the main ones are listed below:

- creation of a brain model,[2] which takes into account the variable skull thickness;
- filling this volume through the slices containing boron distribution data (gram/cm^3), obtained from different transaxial PET scans;
- creation of the input file, related to this brain model, for a MCNP-4A simulation;
- reading of the output file of the MCNP-4A. The results can be plotted both in a bidimensional format (isodose and isofluence surfaces on PET planes) and in a graphic of the dose and fluence values along an user-defined axis (on PET planes).

We have used, as the "brain" and "bone" materials present in the model, those indicated by Zamenhof.[3]

The physical doses considered by default in CARONTE are due to the following main reactions:

1) $^{10}B(n,\alpha)^{7}Li$ (therapeutic dose); 2) $^{14}N(n,p)^{14}C$;
3) $^{1}H(n,n')^{1}H$ (in the case of epithermal beam); 4) $^{1}H(n,\gamma)^{2}H$.

Besides, in order to obtain the biological doses one has to consider the RBE values, from time to time, through a CARONTE optional subroutine. In this way it is possible to use in the calculations the most recent results related to determination of the RBE values.

3. CARONTE APPLICATIONS AND DISCUSSION

Initially CARONTE system has been applied to a test model[4] having two region (tumor and healthy tissue), containing different values of boron concentration (see Figure 1a). Different simulations have been performed through this test model, changing, in particular, two fundamental parameters: the Source-to-Skin Distance (SSD) and the neutron spectrum.

For example, utilizing an epithermal neutron spectrum (a rough approximation of Petten HFR spectrum with a collimator of 12 cm), an SSD equal to 30 cm and a BPA infused dose of 2,500 mg, the isodose surfaces, reported in Figure 1b, have been obtained. The arrows indicate the coming direction of neutron flux. The dose values, associated to the colors of the figure, are specified. It must be noted that the number of digits in the figures comes directly from the MCNP output and it doesn't indicate necessarily the calculation uncertainties.

The isodose surfaces are radial decreasing, like those obtained by various previous simulations performed with other software system (Bnct_rtpe, etc.).[5] This situations is far from the real one, because of the non-uniform boron distribution, registered by the PET scans on a glioblastoma multiforme case; the central transaxial slice of the model, constructed with these data,[1] is reported in Figure 2. Here the red zones indicate the highest

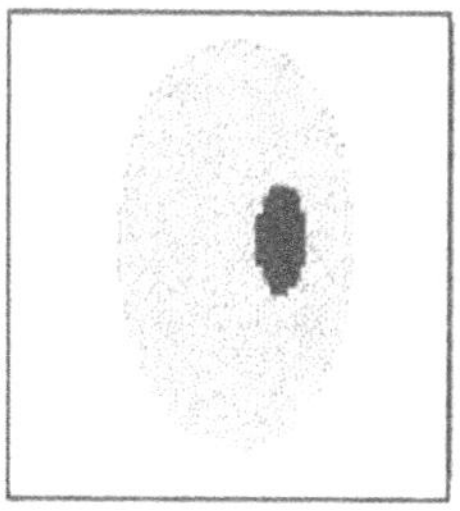

a)

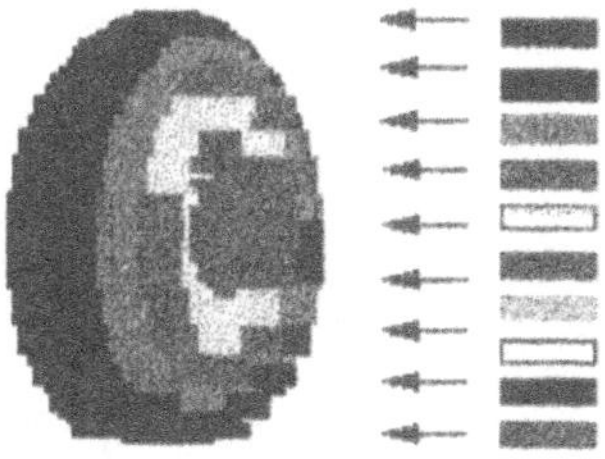

b)

Figure 1. a) Central slice inserted in the phantom model;[4] b) MCNP-4A simulation results with an epithermal neutron spectrum coming from the right (arrows direction).[2]

boron concentration and the color scale[6] used is the same utilized to report the results (as in Figure 1). Applying CARONTE on this brain model, different simulations, changing the same parameters used in the test trials (SSD and the neutron spectrum), have been performed. The irradiation flux comes from the left in about all the simulations, as the main tumor region is positioned in the left hemisphere. In particular it has been recognized, by the simulation results, that SSD parameter seems to have a minor contribution to the boron dose in the case of a collimated beam (as HFR). Figure 3 (*) represents the boron dose along the central axis of irradiation, related to five different neutron spectra.

It is noted that the best spectra are the monoenergetic ones, but also the great difficulties to produce them are known. However the epithermal neutron flux seems to be very efficient under this point of view. The boron dose values registered are related to a BPA infused dose equal to 2,500 mg. In Figure 4 the bidimensional distribution of boron dose for an epithermal neutron flux coming from the left side and the SSD equal to 30 cm is represented. It can be noted the better resolution of the results, if compared to the uniform boron distribution case, and the fact that the right region, even if characterized by an high boron concentration, is scarcely hit by the epithermal flux coming from the left.

In fact the dose values are due both to boron concentration and to neutron flux level. Therefore, in this case, a POP (Parallel Opposed) irradiation with epithermal flux

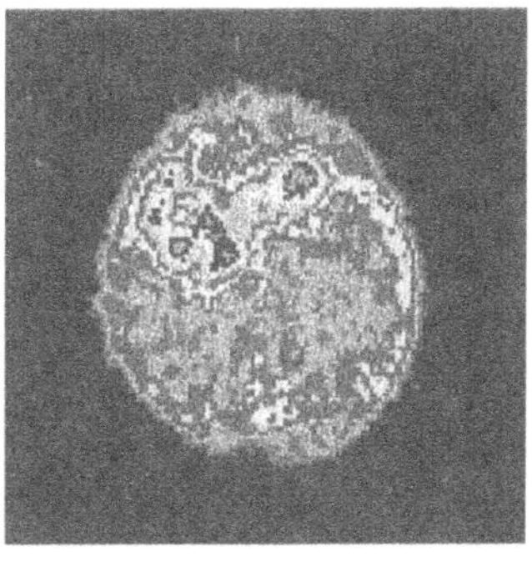

Figure 2. Central PET slice for the Kyoto glioblastoma multiforme case.[1] Spatial distribution: 8.6 mm.

(*) To understand the correct behavior of the dose values versus distance, in Figures 3 and 5, it is necessary to start the x-axis from the right side.

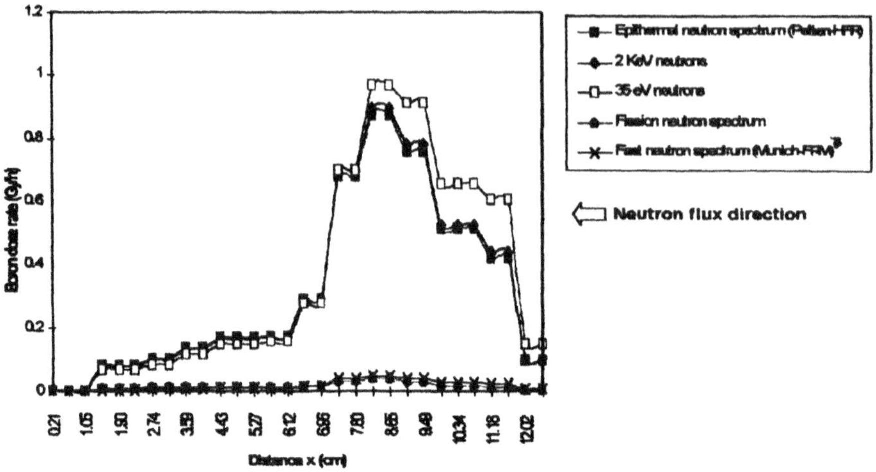

Figure 3. Boron dose, along central axis, obtained with five different neutron spectra. The ^{10}B spatial distribution is a function of the non-uniformity, but the basic distribution is the same reported in Figure 2.

has been performed. The behaviors of physical boron dose in the case of right and left epithemal irradiation (under the same BPA dose infused) are compared in Figure 5 (*). The asymmetry of the dose distribution in the POP irradiation case is due to the fact that also the tumor geometry isn't symmetric. Therefore, it is possible to establish the optimal directions of irradiation, performing different CARONTE simulations. Besides other irradiation configurations have been tested in relation to this cancer case.[6]

The computational time can be estimated in about 8 hours, using a 133 MHz PC-Pentium. Anyway, a research effort is in progress, in order to reduce this parameter.

4. EXTENDING CARONTE APPLICATION TO MRI AND PGRA

Recently some researchers[8] have tried to perform an MRI imaging not based on the ^{10}B (or ^{11}B), but on the group B-H (present in the BSH compound) magnetic properties. Therefore, it has been tried to extend CARONTE application to MRI, utiliz-

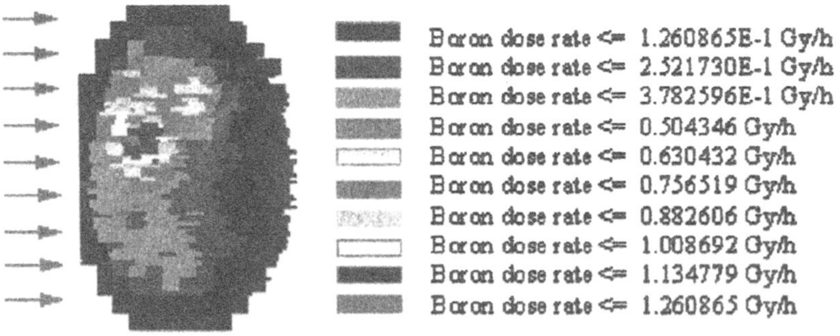

Figure 4. Isodose surfaces (arrows indicate the direction of the incoming neutron flux). In the right side the color scale and associated dose values are reported.

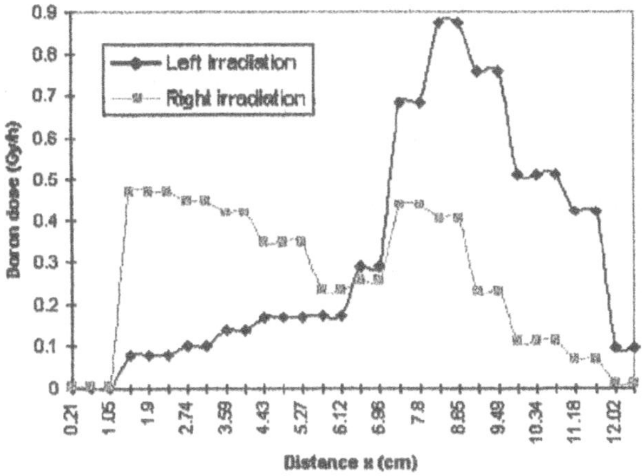

Figure 5. Boron dose in the case of left and right epithermal irradiation.

ing four MRI transaxial brain scans to detect the boron dose distribution into the brain model. The work is still in progress. This development process of CARONTE can be extended to PGRA (Prompt Gamma Ray Analysis) transaxial brain scans and we hope to upgrade the system (for 2.1 version) to this method of acquisition of boron distribution and concentration.

5. CONCLUSIONS

In this paper it has been briefly presented the CARONTE treatment planning system, its main properties and applications. The most relevant results have also been presented. It has to be pointed out that the code is still in a rough graphic state, which has to be improved in order to make its use more easy and fast. However, another development objective is the CARONTE application in the case of boron distribution data furnished by brain transaxial MRI or PGRA scans, but this activity needs further studies and deepening.

ACKNOWLEDGMENTS

Thanks are due to:

- the Kyoto University and, in special way, to Dr. Satoshi Ueda for the kind collaboration. Without his contribution this work would not have been possible;
- Dr. R. Guzzardi of Clinical Physiology Institute of CNR (Pisa) and Mr. O. Sorace, of his working group, for the very useful information and aid on PET use;
- SORIT s.r.l. (Pisa) for precious contribution to MCNP acquisition and use;
- Ms F. Mariani of DCMN of Pisa University for fruitful suggestions and advice about PET graphics.

REFERENCES

1. S. Ueda *et al.*, Positron Emission Tomography—Boron Neutron Capture Therapy System and Clinical Application to Patients with Malignant Glioma, *in* "Advances in Neutron Capture Therapy", B. Larsson ed., Elsevier Science Publ., Zurich, 1996, pp. II-280–285.
2. N. Cerullo, and G. Daquino, Tumour and Healthy Tissues Dose Calculation using Monte Carlo Code and PET Information on Boron Distribution in Tissues, *in* "Boron Neutron Capture Therapy, Oncological, Radiological, Chemical and Radiation Physics Aspects in the BNCT Technique. Present and Future in Italy", G. Tornielli ed., INFN Acts, Padua (Italy), 1998, pp. 55–62.
3. R. Zamenhof *et al.*, Monte Carlo-based Treatment Planning for Boron Neutron Capture Therapy using Custom designed Models automatically generated from CT Data, *Int. Journ. of Rad. Onc. Biol. Phys.*, 35:383–397, 1996.
4. N. Cerullo, and N.G. Daquino, Use of the Monte Carlo Code in Neutrons Behaviour Simulation in BNCT (Boron Neutorn Capture Therapy), *in* "Technologies for the New Century", ANS, Nashville (TN), 1998, pp. II-236–243.
5. D.W. Nigg, and D. Eng, Methods for Radiation Dose Distribution Analysis and Treatment Planning in Boron Neutron Capture Therapy, *Int. J. On Rad. Oncol. Biol. and Phys.*, 28:1121–1134, 1994.
6. G. Daquino, Degree Thesis in Nuclear Engineering, *Pisa University*, a.y. 1996–1997.
7. G. Pfister, G. Hehn, and F. El-Husseini, Optimization of fast neutron spectra available for neutron capture therapy, *Strahlentherapie und Oncologie*, 165:107–109, 1989.
8. B. Maraviglia, ^{10}B Compounds NMR Imaging for BNCT, *in* "Boron Neutron Capture Therapy, Oncological, Radiological, Chemical and Radiation Physics Aspects in the BNCT Technique. Present and Future in Italy", G. Tornielli ed., INFN Acts, Padua (Italy), 1998, pp. 45–47.

30

CASE-MATE: COMPUTER-AIDED CELL SEEDING, MICROCLONING, ANALYSIS, AND TELEMETRIC EVALUATION APPLIED TO NEUTRON CAPTURE THERAPY

N. E. A. Crompton[1], J. Gobrecht[2], C. Higgs[3], G. Kuehne[4], E. Lehmann[4], H. Reist[5], J. Stepanek[6], T. Suda[6], and B. Larsson[6]

[1]Radiation Medicine, Paul Scherrer Institute (PSI)
CH-5232 Villigen-PSI, Switzerland
[2]Micro- and Nanostructure Laboratory, PSI
[3]Synchrotron Light Source, PSI
[4]Spallation Source Division, PSI
[5]Division of Accelerator Physics, PSI
[6]Institute for Medical Radiobiology, PSI

INTRODUCTION

The disintegration of boron nuclei into alpha particles and lithium nuclei makes NCT a high LET therapy modality. There are various indications for the use of high-LET in radiation, especially when single doses are administered. In many situations this advantage is a mixed blessing: high-LET radiations help to reduce radiation-resistance effects caused by the cell cycle or differences in oxygenation levels, but at the cost of a linear dose-response curve. However, it is the very nonlinear nature of dose-response curves which is the reason for the success of fractionated radiation therapy. The nonlinear component of radiation response results from physiological responses mounted by the cell to deal with radiation damage. In an accompanying paper (Crompton *et al.*) we demonstrate that apoptosis, programmed cell death, is clearly induced by NCT. Therefore, cells are capable of mounting a physiological response to NCT-induced damage. It is vital to quantify the magnitude of this contribution in order to predict the efficiency of NCT and to understand better therapeutic mechanisms of action, and we hope to accomplish this using the CASE-MATE system.

Computer-**a**ided cell **se**eding, **m**icrocloning, visual or instrumental **a**nalysis, **t**elemetric **e**valuation (**CASE-MATE**) and eventual storage (*in vitro* or after fixation) of hundreds to thousands of separately confined biological specimens, in the form of single living cells or monoclonal cellular aggregates, will be possible using a novel microplate technology based on a compact disc **CD** for cell culture. The Compact Cell Culture Disc, **CCCD**, is a new invention, that permits microscopic and biochemical studies of individual "cultivars" (single cells, small multicellular organisms or tumour explants) in small

Frontiers in Neutron Capture Therapy, edited by Hawthorne *et al.*
Kluwer Academic / Plenum Publishers, New York, 2001.

or large numbers under optimized conditions. Once implanted with biological material to be studied, it constitutes a microlaboratory that allows control of the chemical and physical milieu of the individual cultivars. When sealed, it could be handled in non-sterile environments, for transportation, microscopic observation and for temporary or permanent storage.

The design is based on a new principle: cultivation microchambers in the form of sandwiches allowing manual or automatic computer-assisted isolation, cultivation, characterization and archiving of single proliferating cells and their progeny. A multitude of applications have been conceived, e.g. in molecular and cellular genetics, microbiology, tumour biology, toxicology, pharmacology and radiobiology. The CCCD could be adapted to automatized routine applications, in biotechnology, clinical medicine or environmental toxicology. It could probably find a place in biological teaching as well. In particular, it permits increased capacity, precision, and reproducibility in procedures aiming at "cell cloning", i.e. isolation, cultivation, characterization and archiving of single proliferating cells or their isogenetic progeny.

Our primary aim is to produce a microstructured "compact cell cloning disc" (CCCD) that would permit coordinated cultivation and observation of hundreds to thousands of single cells and their progeny in the form of "single microclones" consisting of up to ca 1000 genetically identical cells. In the most important clinical applications foreseen, cells derived from patient biopsies are to be seeded, manually or by means of a cell sorting device, in individual "pits", 1,6 mm or 0,8 mm wide, that permit anchorage-dependent cells to attach themselves on optically transparent layers of palladium, 0,3 mm in diameter, deposited on a thin agarose layer. Microscopic observation is today made with visible or UV light by reflection or fluorescence microscopy, but an additional important ambition is to provide the substrate with thin quartz or plastic windows to permit ideal optical conditions, also for measurements in the transmission mode at all wavelengths suitable for 2D- or 3D-microscopy.

Procedures have to be developed that would permit fabrication of chosen sandwich components and assembling of CCCDs under clean-room conditions typical for microtechnology laboratories. The basic steps would for example include surface finishing of the wafer, oxidation or nitridation, lithography, wet- or dry etching, metalization, sterilization etc. Some of these processing steps are fairly well established, others (e.g. etching with defined undercut inside the window) will need further development in cooperation with specialists. Introduction of more cost effective alternatives such as polyamide structures should also be strived for.

THE CCCD—A 10 CM SILICON-BASED COMPACT CELL CULTURE DISC

The CCCD is designed to permit the implantation of single cells individually, each one in its own separate micropit. The individual micropits consist of shallow (typically 50–300 micrometer deep), cylindrical depressions defined by holes (800 or 1,600 micrometer in diameter) in an "upper wafer" that faces geometrically conform optically transparent quartz windows in the "lower wafer". If desired, the micropit could be provided with a coated spot offering cellular anchorage, for example a "palladium island",[1] centrally deposited on the agarose-covered quartz window, to permit optimal conditions for microscopic observation of attached single cells or cell clones.

Chemical Micromachining

The silicon discs were produced using standard procedures established in microtechnology. The substrate material used was n-type 100mm diameter single crystal wafers, (100)– orientated, 250 µm thick and polished on both sides. A terminal oxide >1 µm was grown and a second layer of silicon nitride (~300 nm) added by low-pressure chemical vapour deposition (LPCVD). Openings on one side of the wafer were defined by photolithographic techniques. After a dry etch of the nitride/oxide layer on one side, the micropits were formed by etching through the wafer until the other oxide/nitride layer, which forms the transparent bottom of the pit was reached. Etching was done by two methods:

a) in KOH following standard procedures.[2] Pits produced by similar methods have been described before.[3]
b) by dry etching with a high-rate, anisotropic etching process developed by the STS Corporation. The advantage of this method is that a higher density of pits can be realised per unit area because of the almost vertical sidewalls produced by this process.

Control of the CCCD Internal Environment

The internal microchamber environment is determined by the permeability characteristics of the cover foil. Macromolecular or biological medium components could be injected before the micropits are closed or by injection through the cover foil. In the specific project referred to below as an application example, the cells will be cultivated in a medium derived from the patient's own serum, to simulate the environmental conditions of the tumour *in vivo*. For that purpose, only a few ml of the patient's blood is required.

Embedment of the CCCD in a CD-ROM Environment

The 10 cm CCCD could be embedded in a 12 cm CD structure. A ca. 5 mm broad strip in the CD periphery has been reserved for mechano-optical files giving specimen data in the form of a virtual bar code, experimental protocols, instructions for disc positioning, microscopic settings etc. (to facilitate computer-aided seeding, data assignment and telemetric evaluation). The CD structure can also provide a common reservoir of low-molecular weight medium components for the microchambers of the embedded CCCD.

COMPUTER-AIDED SEEDING, MICROCLONING, ANALYSIS, AND TELEMETRIC EVALUATION (CASE-MATE)

An important advance the CASE-MATE system brings, not only to NCT but radio-oncology in general, is the ability to measure biological parameters and simultaneously take tumour heterogeneity into account. Major challenges to targeted NCT are to bring sufficient amounts of the sensitizer, e.g. boron, selectively to the cancer cells and to reduce background therapy dose as much as possible. Pretreatment evaluation of boron distribution, tumour-genetic characteristics, and also normal-tissue radiosensitivity should employ parameters that can all be addressed using the CASE-MATE system. Furthermore, evaluation of tumour clonogens and cell-death fraction using microcloning and cytotoxicity assays are possible, and the role of genetic instability in clinical tumor eradication and tumour promotion as a secondary consequence of NCT can be studied. To date the design and preliminary testing of the prototype CASE-MATE system has been performed. Successful primary cultivation and microscopic analyses of single cells and cell clones isolated from clinical biopsy specimens have also been performed.

As an application example now under development, we employ confocal microscopy-based measurements permitting evaluation of the prerequisites for targeted cancer therapies in individual patients, through studies of pharmacokinetics and biological response in cell populations isolated from biopsy or surgical specimens. This new means for improved tumour therapy evaluation and treatment planning will serve a clinical community interconnected through an IT network similar to the networks for macroscopic treatment planning that are now implemented to serve the Swiss and Italian hadrontherapy communities.

DISCUSSION AND CONCLUSIONS

In the fields of molecular and cellular genetics, microbiology, oncology, radiobiology, pharmacology, *etc.* there is a need to culture individuals cells, cell clones and even multicellular microorganisms in order to study their growth, morphology and molecular-genetic, biochemical or biophysical characteristics.[4,5] The system described permits the isolation, continued culture, and investigation of single cells, clones, and other biological microspecimens in large numbers. It was especially designed for computer-aided investigations of growth, and the structural and functional characteristics displayed during the development of monoclonal cultures.

During the preparations for these investigations, different anchorage-dependent cell lines were cultured on palladium islands. A fundamental idea behind the CCCD design is the placement of such cell-attracting "coated spots" within a honeycomb of micropits which can be closed and sealed with a common cover. To the selected cells in any given micropit, specific nutrients and agents can be given in order to investigate the effects on cell population growth. A pair of superimposed, circular 0.3 mm thick silicon wafer forms the basis of the CCCD. Its diameter can be varied from 5 to 20 cm depending on the number of micropits required. This structure can best be compared with a standard compact disc filled with a honeycomb of micropits.

The individual micropits of the CCCD permit the study of individual cells and cell clones, in both small and large numbers, i.e. from single cells up to a few thousand cells, under optically optimized conditions. The covered micropit (the "microchamber") with its defined medium and growth conditions forms a sterile microlaboratory providing conditions for control and examination of the individual cells and their growth under favourable optical conditions. Control of both the chemical and physical milieu of the cells is possible. Autofluorescence or the fluorescence from chemical markers provide important measurement parameters that can preferably be analysed by confocal laser scanning microscopy adapted to the optical features of the CCCD microchambers. Both the whole CCCD and the individual growth chambers can be maintained under sterile or non-sterile conditions; for short term, long term, and permanent experimentation, transport and storage.

The CASE-MATE system that is now under development will employ confocal microscopy-based measurements permitting evaluation of the prerequisites for NCT in individual patients, in terms of boron cytopharmacokinetics and radiobiological response of cell populations isolated from biopsy or surgical specimens. This new means for improved tumour therapy evaluation and treatment planning will serve a clinical community interconnected through an IT network similar to the networks for macroscopic treatment planning that are now implemented to serve the Swiss and Italian hadrontherapy communities.

ACKNOWLEDGMENTS

We are indebted to R. Widmer of IBM Research, Zurich (CH), for etching some wafers with the STS process and to Microcloning CCCD AB, Jönköping (S) for financial support.

REFERENCES

1. W. Burkard, and B. Larsson, Towards individualiztion and optimization of radiation therapy: biological behaviour of tumor cells derived from human tumour biopsies. In U. Amaldi and B. Larsson (Eds.), 'Hadrontherapy in oncology', Amsterdam, Elsevier, 1994.
2. D.L. Kendall, C.B. Fleddermann, and K.J. Malloy, In: The Mechanical Properties of Semiconductors and Semimetals, Vol 37. Ed. K.T. Faber, and K.J. Malloy, Academic Press, 1992.
3. G. Mayer, and J.M. Koehler, Micromechanical Compartments for Biotechnical Applications: Fabrication and Investigation of Liquid Evaporation. Proc. Eurosensors X, Leuven, Belgium, 1996.
4. H. Andersson, T. Baechi, M. Hoechl, and C. Richter, Autofluorescence of living cells. J. Microscopy, 1998, 191, 1–7.
5. J. MARELIUS, M. Sc. Thesis, Uppsala University, Uppsala, Sweden, 1995.

31

STANDARD OPERATING PROCEDURES FOLLOWED BY THE PETTEN BNC-TEAM DURING PATIENT TREATMENT

R. L. Moss[1], F. Stecher-Rasmussen[2], W. Sauerwein[3], D. Gabel[4], P. Watkins[1], and C. Vroegindeweij[1]

[1]HFR Unit, Joint Research Centre
European Commission, Petten
[2]Netherlands Energy Research Foundation
ECN, Petten, The Netherlands
[3]Universitätsklinikum
Essen, Germany,
[4]Dept. of Chemistry
Universität Bremen, Germany

1. INTRODUCTION

In July 1997, the first European clinical trial on BNCT (EORTC 11961) was officially opened. The first patient was subsequently treated soon after in October. The treatment takes place at the High Flux Reactor (HFR) at Petten in The Netherlands.[1] The approach to treat a patient at a nuclear research reactor requires a complex organisational structure, which is further exacerbated by the involvement of technical staff unaccustomed to dealing with patients, as well as working alongside clinicians.

The clinical aspects of BNCT, such as patient recruitment, surgery, boron uptake, patient transfer to the Netherlands (Amsterdam), and treatment at Petten, are reported by the clinicians elsewhere at this Symposium.[2,3]

The purpose of this paper is to present the procedures adopted and carried out by the Technical Staff at Petten. This includes prior training exercises under the guidance of the medical staff. Furthermore, the technical staff not only serve as a supporting role to the clinicians but demonstrate that Standard Operating Procedures (SOP), following the guidelines of Good Clinical Practice,[4,5] are also applicable to the technical steps required to perform BNCT treatment at Petten.

2. STANDARD OPERATING PROCEDURES

Taking into account the complexity of BNCT as a completely new interdisciplinary treatment modality, with the concomitant complicated technical and medical

Frontiers in Neutron Capture Therapy, edited by Hawthorne *et al.*
Kluwer Academic / Plenum Publishers, New York, 2001.

Table 1. Chapters of the SOP manual

Chapter	Item
2	General
106	Documentation
107	Language
108	Code for information release to press, radio and television
109	Addresses of investigators etc.
110	PTB clock time
2	Ethics
3	Preparation for a study
4	Treatment preparation and surgery
5	Treatment at the HFR and follow-Up
6	Monitoring
7	Management of study medication
8	Toxicity and follow-up
9	Closure of a clinical study

infrastructure, as well as the complex organisational regulatory issues, the details of all special operations and procedures concerning the conduct of this trial have been written in Standard Operating Procedures following the guidelines of Good Clinical Practice (GCP). An SOP is a formal document describing the procedures to be followed in accomplishing a particular task within the trial. The SOP is written so that the procedures are practicable and can be executed by the staff. The SOP should also be reviewed and amended when required. In this latter respect, the SOPs are updated at least once per year and must be approved by the study management. The investigator of every institution notifies their personnel about the update of an SOP, distributes copies to the appropriate persons, and sees to it that only the up-to-date version of an SOP is used. Outdated versions of SOPs are filed at the Study Centre in Essen and at the University of Bremen.

The current version of the dossier of SOPs used in the trial at Petten, contains descriptions of some 55 procedures. The SOPs are grouped into the Chapters, as listed in Table 1. The sections of Chapter 1 are given to indicate that even the administrative rules require their own SOPs.

The procedures to perform the treatment at Petten are provided in Chapter 5 of the SOP manual, see Table 2. Each SOP has a responsible author and responsible (1–3) persons who check and countersign the procedure, following the guidelines of GCP.

3. PATIENT TREATMENT IN STEP-BY-STEP SOP'S

Following recruitment of a patient at one of the six recruitment centres, the technical route is described in a chronological order of procedures, as shown in the flow scheme in Figure 1. The relevant SOP number for each step is also indicated, whilst its title is given in Table 2.

4. REPORTING OF THE DAY'S ACTIVITIES- SOP 514

As an example of a technical SOP, the day's technical activities are reported in SOP 514. The treatment, with in particular the irradiation parameters, such as, on-line monitors, reference phantom measurements, patient dosimetry, beam characteristics (argon

Table 2. Technical SOP's

Chapter 1 General	
101	Documentation
102	Language
103	Code for information release to press, radio and television
Chapter 3 Treatment Preparation and Surgery	
308	Task descriptions and responsibilities of the technical staff
Chapter 4 Treatment Preparation and Surgery	
402	Image transfer and conversion
404	Treatment planning by bnct_rtpe/rtt_MC
405	Quality control treatment planning system
412	Preparation for travel of patient to Amsterdam
Chapter 5 Treatment at the HFR and Follow-Up (see below)	
Chapter 7 Management of Study Medication	
704	Boron Determination with ICP-AES
Chapter 5	
501	Free beam characterization
502	Neutron beam facility maintenance
503	Communication between facility operator and medical staff
504	On-line beam monitoring and beam shutter control
505	BNCT-Wing at the HFR Petten—provisions and procedures
506	Neutron beam facility operation
507	Boron analysis by prompt gamma ray spectroscopy
508	Positioning of patient
509	In-vivo dosimetry
510	Calculation of beam monitor units and irradiation time
512	Timing of BSH infusion in relation to projected treatment time and amount to be administered
513	Health physics check of patient and hospital staff
514	Irradiation summary
515	Reactor hall evacuation
516	Transport of patients
517	Security checks for patients and accompanying personnel at the JRC site and the reactor
519	Patient preparation for BNCT at Petten
520	Beam monitor calibration
521	Preparation for planning of BNCT
522	Actions to be taken in the case of patient emergency
523	Emergency case

system, etc.), personnel present and radioprotection measurements, are all summarised daily in the Irradiation Summary report.

For an irradiation "experiment" at the HFR, as well as for post-treatment monitoring, it is required that the radiation performance is recorded in an easily readable format. The daily Irradiation Summary report requires some 35 signatures.

5. CONCLUDING REMARKS

BNCT is a radiotherapy modality that requires a considerable effort and cooperation from non-medical personnel. To conform with medical practice, as closely as possible, all technical steps, as well as the medical steps, have been described by SOPs following the guidelines of GCP.

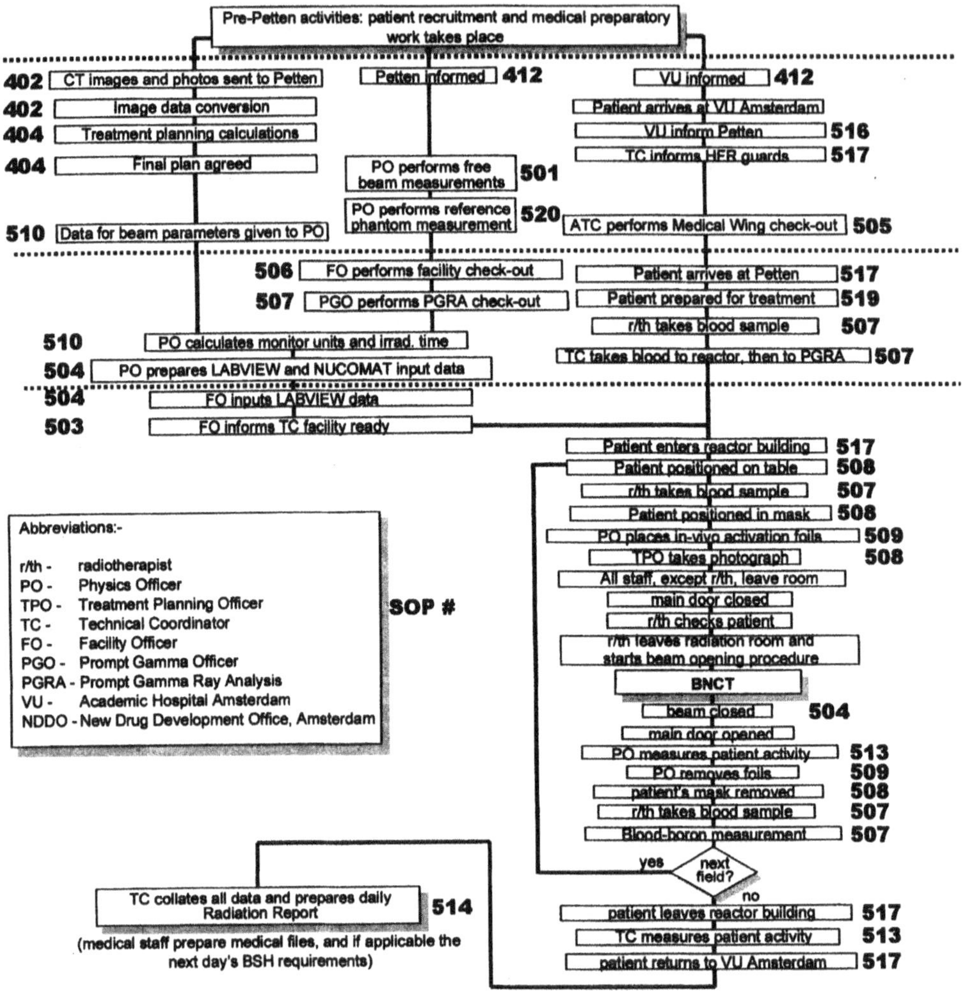

Figure 1. Step-by-step procedures, with corresponding SOP number, for BNCT treatment at Petten.

Despite the aim to establish and conform with medical practice, it was apparent during the first patient treatments, that many SOPs were inadequate and some required considerable updating. On completion and subsequent analyses of the first cohort of patients (Autumn 1998), a major review of all SOPs will begin.

ACKNOWLEDGMENTS

The clinical trial of BNCT for Glioblastoma at the HFR Petten is financially supported by the European Commission's Biomedicine and Health Research Programme (BIOMED II).

REFERENCES

1. R.L. Moss, J. Casado, K. Ravensberg, F. Stecher-Rasmussen, and P. Watkins, "The Completed BNCT Facility at the HFR Petten", *in* Proc. 7th Int. Symp. on Neutron Capture Therapy, Zurich, September 1996; Ed. B. Larsson, J. Crawford, Plenum Press, New York, 1997.
2. W. Sauerwein, R. Moss, K. Hideghéty, F. Stecher-Rasmussen, M. de Vries, H.-J. Reulen, C. Götz, P. Pacquis, F. Grochulla, K. Haselsberger, J. Wolbers, J. Rassow, J.P. Pignol, P. Watkins, C. Vroegindeweij, K. Ravensberg, S. Garbe, O.D. Wiestler, B. Turowski, F. Zanella, D. Touw, A. Siefert, R. Huiskamp, H. Fankhauser, and D. Gabel, "Status Report on the European Clinical Trial of BNCT at Petten", Report on the first group of patients in the European phase I trial (EORTC Protocol 11961) at the High Flux Reactor in Petten", these proceedings.
3. K. Hideghéty, W. Sauerwein, M. de Vries, F. Grochulla, C. Goetz, K. Haselsberger, P. Pacquis, J. Wolbers, J. Heimans, R. Huiskamp, R. Moss, D. Gabel, and H. Fankhauser, "Report on the first group of patients in the European phase I trial (EORTC Protocol 11961) at the High Flux eactor in Petten", these proceedings.
4. Good Clinical Practice for Trials on Medicinal Products in the European Community, for Proprietary Medicinal Products [CPMP] EEC 111/3976/88-EN, July 1990.
5. CMP Working Party on Efficiency of Medical Products, EEC Note for Guidance: Good clinical practice for trials on medical products in the European Community. Pharmacology & Toxicology 1990, 67, 361–372.

32

PREDICTING BORON-10 CONCENTRATIONS IN NORMAL BRAIN AND GBM FROM BLOOD MEASUREMENTS

M. R. Palmer[1], C. Chuang[2], W. S. Kiger[2], P. M. Busse[3], and R. G. Zamenhof[1]

[1]Department of Radiology
Beth Israel Deaconess Medical Center
Harvard Medical School
[2]Nuclear Reactor Laboratory
Massachusetts Institute of Technology
[3]Joint Center for Radiation Therapy
Harvard Medical School

INTRODUCTION

In boron neutron capture therapy (BNCT), radiation dose delivered to the target and surrounding tissues is a complex function of neutron energies, fluxes, tissue compositions, boron concentrations and relative biological effectiveness (RBE) factors. Of these contributions, the last two are likely the cause of the greatest error in dose estimates. Current practice is to assume that ^{10}B concentration in normal and tumor tissue following intravenous administration of ^{10}BPA-fructose is related statically to ^{10}B concentration in the blood. Factors of 1 : 1 for normal and 3.5 : 1 for tumor tissue to blood ratios are the typical assumption. Similarly, RBE factors for the ^{10}B capture reaction are assumed to be static.

Both of these assumptions are oversimplifications. ^{10}B in tissue, like any other substance administered via the blood, should be expected to exhibit wash-in and washout behavior characteristic of the infusion schedule and the cellular transport mechanisms. The possibility of dynamic RBE factors is somewhat related to the issue of ^{10}B transport because of the extremely short range of secondary particles. It is likely that the probability of normal or tumor cell death due to a ^{10}B capture reaction varies with the exact location of the ^{10}B atom within the cell or extracellular space. To some degree, we should expect the microdistribution of ^{10}B to evolve as it is transported and possibly metabolized within the various tissues.

The transport of ^{10}BPA has been studied by at least two groups using positron emission tomography (PET) with the fluorinated analog ^{18}F-BPA[1, 2]. The pharmacokinetics are fairly well described by a three compartment model. In the current work, we employ this model and the actual temporal profiles of blood concentration measured

Frontiers in Neutron Capture Therapy, edited by Hawthorne *et al.*
Kluwer Academic / Plenum Publishers, New York, 2001.

from patients undergoing therapy in the Harvard/MIT BNCT phase I clinical trials to predict brain tissue boron concentrations.

METHODS

Imahori *et al.* have proposed a three compartment model to describe the uptake of ^{18}F-BPA in the human brain. This model is shown in Figure 1. A single plasma compartment, C_p, drives the system consisting of two tissue compartments labeled A_1 and A_2 in the figure. Imahori has studied a number of normal and GBM patients and found rate constants K_{1-4} as: [0.011, 0.025, 0.033, 0.009] for normal tissue and [0.041, 0.033, 0.019, 0.011] for GBM.

In the Phase-I clinical trials at Harvard-MIT, GBM patients currently receive ^{10}BPA-fructose at a dose of 350 mg/Kg body weight infused at a constant rate over a period of 90 minutes. Administration of neutron therapy begins 45 minutes after the end of the infusion. Typical treatments are two and a half hours in duration. Whole blood is sampled at 15 minute intervals during infusion and treatment.

The time course of ^{10}B in the blood following infusion of ^{10}BPA-fructose has been studied extensively in our laboratory and is described elsewhere in these proceedings. The blood curve exhibits a characteristic biphasic exponential washout with striking consistency across subjects.

This model has been further investigated in our laboratory by application to experimental data obtained from mice injected with ^{10}BPA-fructose. Those results are reported elsewhere in these proceedings.

RESULTS

Bi-exponential curves fit to data from three of our subjects were convolved with the model equations to produce tissue ^{10}B predictions. The three blood curves were chosen as the extrema, highest and lowest average ^{10}B concentration, and a typical average ^{10}B concentration. The resulting tissue curves are shown in the graph of Figure 2. In

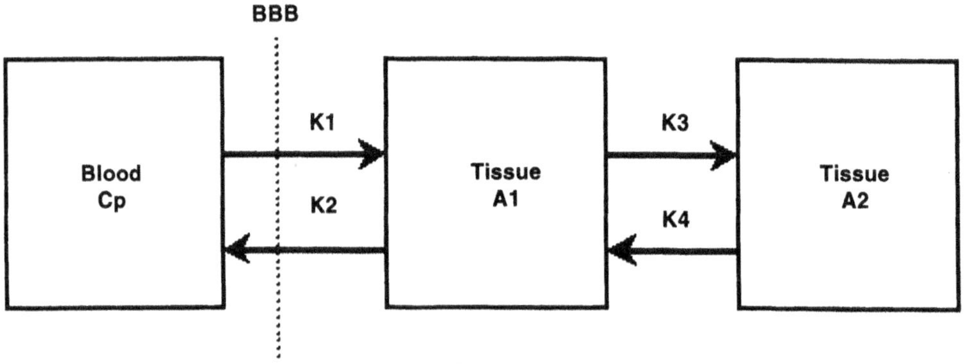

Figure 1. Three-compartment model proposed for human PET studies with ^{18}F-BPA. The model consists of a single blood compartment with plasma concentration C_p driving two tissue compartments, A_1 and A_2.

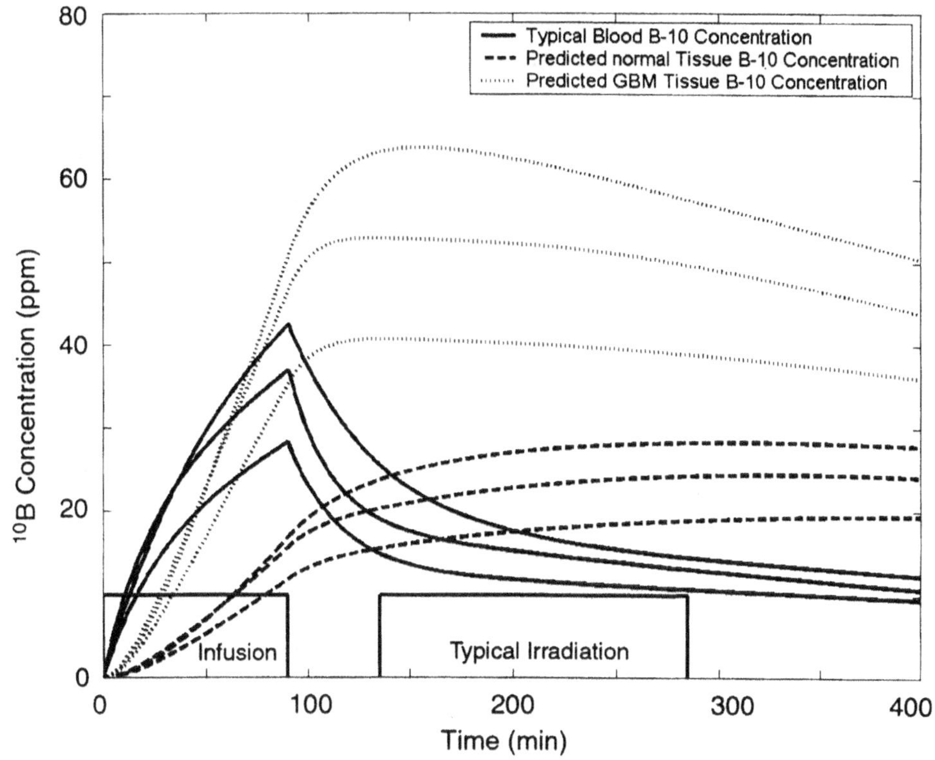

Figure 2. Model predictions for normal and tumor tissue response to typical blood input functions. The three blood curves (solid lines) which peak at the end of infusion are bi-exponential fits to data collected at 15-minute intervals from three subjects in the Phase I clinical BNCT trial at Harvard-MIT. Total normal (dashed lines) and tumor (dotted lines) tissue concentrations corresponding to these input functions are shown.

addition, corresponding normal to blood, tumor to blood, and tumor to normal concentration ratios are shown Figure 3.

Blood ^{10}B concentration reaches a peak at the end of infusion then tapers with the characteristic biphasic exponential washout. Total normal tissue ^{10}B concentration reaches a broad peak, several hours after the end of infusion while tumor ^{10}B concentration peaks about 30 minutes after the end of infusion. Increased blood ^{10}B concentration results in increased tissue ^{10}B concentration as might be expected. From the graphs of Figure 3, the ratio of tumor to normal ^{10}B concentration decreases with time from the very start of infusion. Of course the actual concentrations at small times are low.

^{10}B concentrations for blood and individual tissue compartments are shown in the graphs of Figure 4. These are normalized to the concentration of each component occurring midway through treatment. This normalization step is performed because distribution volumes for individual tissue compartments are unknown. The blood compartment of course reaches a relative peak early—at the end of infusion—the first tissue compartment soon after, and finally the second tissue compartment. With much reduced treatment times, such as we expect from new facilities currently under construction [3], the timing of irradiation could become very important. If the two tissue compartments in the model correspond to physical compartments on the order of the cell, compound RBE factors might well be different for irradiation just after the end of infusion, versus irradiation two to three hours later.

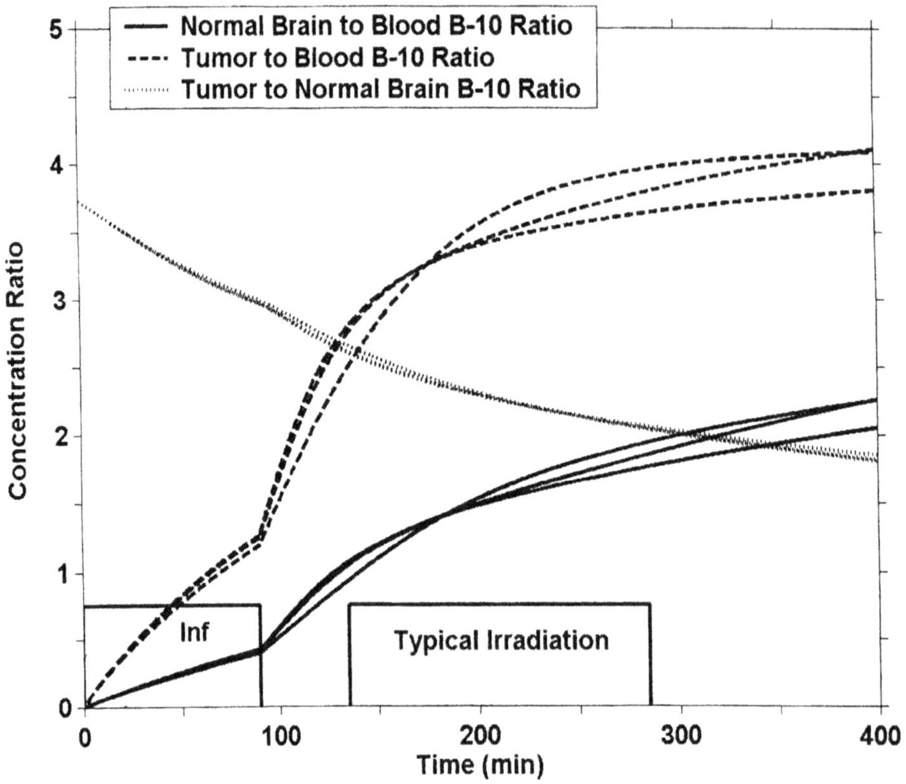

Figure 3. Tissue responses expressed as tumor to blood (dashed lines), normal to blood (solid lines), and tumor to normal (dotted lines) ^{10}B concentration ratios.

CONCLUSION

In BNCT with ^{10}BPA-fructose, methods to determine ^{10}B concentrations and estimate compound RBE factors have been relatively crude. In this work we have explored the application of a three-compartment model to describe the transport of ^{10}B from blood to various neural tissues. The model has been developed and validated for human PET studies with ^{18}F-BPA.

Model predictions show a total ^{10}B concentration in normal brain tissue that increases with time for at least three hours post 90-minute infusion and ^{10}B concentration in tumor tissue that peaks about 30 minutes post infusion then gradually decreases. Tumor to blood ^{10}B concentration ratios during a typical irradiation at Harvard-MIT are predicted as 3 to 4, while normal tissue to blood ^{10}B concentration ratios are between 1 and 2. The ratio of tumor to normal ^{10}B concentration is a monotonically decreasing function of time.

Total tissue ^{10}B accumulation is the sum of two compartments in the model—compartments which may be related to different physical spaces. If that is indeed the case, and if the compound RBE factor changes with the microdistribution of ^{10}B then these results suggest that the choice of when to deliver a short, high-intensity irradiation may be rather important. ^{10}B concentration in the second tissue compartment peaks in concentration several hours after the end of a 90-minute infusion.

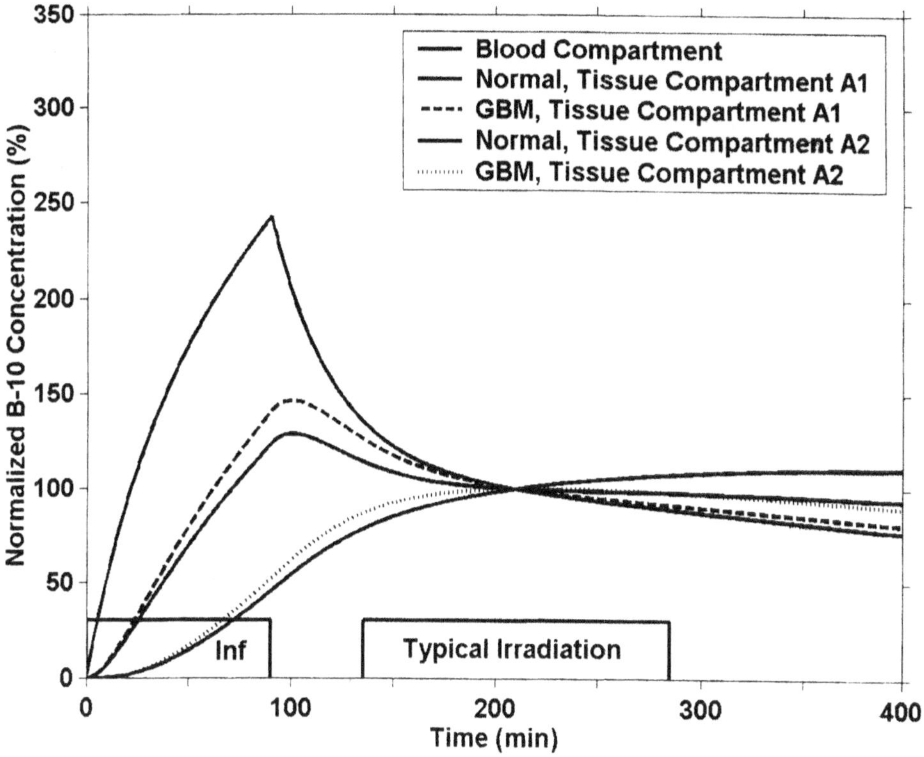

Figure 4. Relative Contribution as a function of time is illustrated by these curves normalized at a point midway through a typical irradiation.

ACKNOWLEDGMENTS

This work was supported by a grant from the Department of Energy, DE-FG02-97ER62193. The authors acknowledge the help of Jody Kaplan and Lujia Tang in acquiring data for this study.

REFERENCES

1. Y. Imahori, S. Ueda, Y. Ohmori, T. Kusuki, K. Ono, R. Fujii, and T. Ido. "Fluorine-18-labeled fluoroboronophenylalanine PET in patients with glioma." J Nucl Med. 1998 Feb; 39(2):325–333.
2. G.W. Kabalka, G.T. Smith, J.P. Dyke, W.S. Reid, C.P.D. Longford, T.G. Roberts, N.K. Reddy, E. Buonocore, and K.F. Hubner. "Evaluation of fluorine-18-BPA-fructose for boron neutron capture treatment planning." J Nucl Med. 1997 Nov; 38(11):1762–1767.
3. W.S. Kiger III, S. Sakamoto, and O.K. Harling, "Neutronic design of a fission converter-based epithermal neutron beam for neutron capture therapy," Nucl. Sci. and Eng., 131:1–22, 1999.

A PHARMACOKINETIC MODEL FOR THE CONCENTRATION OF BORON-10 IN BLOOD FOLLOWING BORONOPHENYLALANINE-FRUCTOSE ADMINISTRATION IN HUMANS

W. S. Kiger III[1], M. R. Palmer[2], R. G. Zamenhof[2], and P. M. Busse[3]

[1]Nuclear Reactor Laboratory
Massachusetts Institute of Technology
[2]Department of Radiology
Beth Israel Deaconess Medical Center
[3]Joint Center for Radiation Therapy
Harvard Medical School

1. INTRODUCTION

In the BNCT clinical trials currently underway, estimates of normal tissue ^{10}B dose are based on ^{10}B concentration in blood as a surrogate for normal tissue.[1,2] Furthermore, in the trials in the USA which use the boronophenylalanine-fructose complex (BPA-F) as the boron delivery agent, a temporally constant tumor to blood uptake ratio of approximately 3.5–4 to 1 is assumed.[3–5] These static tissue to blood uptake ratios, although measured by Coderre *et al.* 0.5 to 1.5 hours after the end of infusion,[6,7] do not represent the dynamic ^{10}B loading and washout behavior expected in these tissues and observed in animal models[8] and in PET studies of humans.[9,10] We believe that these assumptions do not accurately reflect the real situation and dosimetry. However, these assumptions are unavoidable at the present time due to the difficulty of measuring ^{10}B concentrations *in vivo* and the fact that measurements of the ^{10}B C-RBEs are referenced to concentrations of ^{10}B in blood rather than in tissue.[11,12]

The capability to model the concentration of ^{10}B in blood following infusion of the boron delivery agent (BPA-F) is one of the first steps toward modeling the dynamic behavior of ^{10}B and providing more accurate dosimetry. It allows two means of studying the dosimetry. First, by using the conventional assumptions for uptake noted above, the infusion schedule may be manipulated to produce optimum dosimetry conforming to the blood-based dosimetry protocols currently in place. Second, by coupling a model for ^{10}B in blood (or specifically plasma) with a pharmacokinetic model for normal tissue and tumor, such as that developed by Imahori *et al.*,[10] the dynamic behavior of ^{10}B and the resulting dosimetry in the relevant tissues may be analyzed for arbitrary infusion schedules.[13] Both methods allow the selection of optimum infusion schedules and optimum irradiation times (i.e., temporal targeting) without the need for expensive and time-

Frontiers in Neutron Capture Therapy, edited by Hawthorne *et al.*
Kluwer Academic / Plenum Publishers, New York, 2001.

consuming pharmacokinetic studies. When high-intensity epithermal neutron beams such as the fission converter beam under construction at the MIT Nuclear Reactor Laboratory become available,[14,15] temporal targeting will become a much more important issue than it is at the present time.

In this paper, we propose an open 2 compartment model for ^{10}B in blood following BPA-F infusion in humans. The model is derived using data from pharmacokinetic studies of patients enrolled in the Harvard–MIT Phase 1 clinical trials of BNCT for intracranial tumors (GBM and melanoma) and subcutaneous malignant melanoma of the extremities. The model is validated by predicting pharmacokinetic profiles resulting from infusion schedules different from those used to derive the model, including patient data from studies at BNL.

2. METHODS AND MATERIALS

In the H-MIT phase 1 clinical trials of BNCT, 12 patients have received IV BPA-f administrations of 250 mg/kg body mass infused over approximately 60 minutes through a central line into the jugular vein. After FDA and IRB approval to administer up to 350 mg/kg was granted, 2 patients were administered 300 mg/kg infused over approximately 90 minutes (2 infusions each), and 10 patients received 350 mg/kg infused over 90 minutes (2 infusions each).

The first 3 patients who received IV BPA-f received test doses of BPA-f prior to irradiation to allow full measurement of ^{10}B pharmacokinetic profiles. The need for test doses was obviated in subsequent patients by using remote blood sampling techniques during the irradiations. A 10 m long sterile sampling line penetrates the thick shielding walls of the medical irradiation room to access the patient's central line, allowing blood sampling without interruption of the irradiation.

During each BNCT irradiation, a pharmacokinetic study is performed. Patient blood is sampled as described above every 15 minutes during the infusion and for at least 2 hours postinfusion. Thereafter, blood samples are taken every 30 minutes until the end of irradiation, at which time the sampling frequency is reduced to once every 3 hours until the patient is discharged from the hospital or returns to the Nuclear Reactor Laboratory for additional irradiations.

Boron in blood is quantified during and after BNCT irradiations using Prompt Gamma Neutron Activation Analysis (PGNAA)[16] and/or Inductively Coupled Plasma Atomic Emission Spectroscopy (ICP-AES).[17,18] Uncertainties in the measurements are typically 5%. The ^{10}B blood profile collected during the irradiation is used in conjunction with beam calibration data, treatment planning data, and the beam monitors' histories to calculate the dose delivered to normal tissue and thus determine the duration of irradiation. Pharmacokinetic studies were performed on all 24 patients who have received BPA-F resulting in a high-quality data set of reasonable size.

3. ANALYSIS

We have employed an open two-compartment model, shown in Figure 1, to predict pharmacokinetic profiles resulting from BPA-F infusion. The first compartment, the central compartment, represents the blood and very well-perfused tissues with a ^{10}B

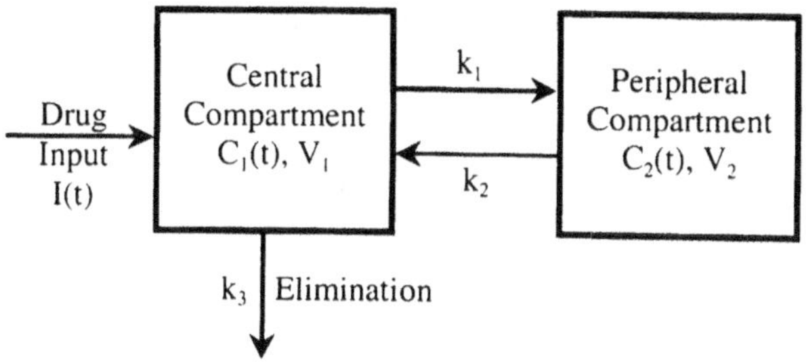

Figure 1. Open two-compartment model for predicting ^{10}B concentration in blood following BPA-F infusion. The drug is infused into the central compartment, which represents the blood and very well-perfused tissues. Tracer exchange between the central compartment and the peripheral compartment, representing less well-perfused tissues, and renal elimination from the central compartment are governed by first order rate constants k_1, k_2, and k_3 as shown.

concentration $C_1(t)$ and a volume of distribution V_1. The second, peripheral, compartment represents regions of the body less well-perfused, e.g., the brain. BPA-F is infused into the central compartment at a rate defined by the infusion (input) function $I(t)$ which has units of mg/kg min. Exchange of the tracer between the two compartments and renal elimination from the central compartment are governed by rate constants k_1, k_2, and k_3 as indicated. Transport is assumed to be first order because results from Coderre *et al.* indicate the linear behavior of this system[6] and the relatively low concentrations currently infused do not appear to saturate the transport mechanisms responsible for the metabolism of BPA-f. Good mixing is assumed in each compartment to make the problem mathematically tractable. However, we do not mean to imply a homogeneous distribution in each compartment. Implicit in the model are assumptions that the system is linear and time-invariant, allowing superposition of responses to different inputs at different times.

Writing a mass balance on the tracer for the two compartments leads to Eqn. 1. The factor f which multiplies $I(t)$ is a conversion factor between mg BPA-F and mg ^{10}B.

$$\frac{d}{dt}\begin{bmatrix} C_1(t) \\ C_2(t) \end{bmatrix} = \begin{bmatrix} -(k_1+k_3) & k_2\dfrac{V_2}{V_1} \\ k_1\dfrac{V_1}{V_2} & -k_2 \end{bmatrix} \begin{bmatrix} C_1(t) \\ C_2(t) \end{bmatrix} + \begin{bmatrix} \dfrac{f}{V_1} I(t) \\ 0 \end{bmatrix} \tag{1}$$

As expected, solution of (1) shows the system response as the convolution of the impulse response with the input:

$$\begin{bmatrix} C_1(t) \\ C_2(t) \end{bmatrix} = \frac{f}{\lambda_1-\lambda_2} \begin{bmatrix} \dfrac{1}{V_1}[(\lambda_1-k_2)e^{-\lambda_1 t}+(k_2-\lambda_2)e^{-\lambda_2 t}] \\ \dfrac{k_1}{V_2}(e^{-\lambda_2 t}-e^{-\lambda_1 t}) \end{bmatrix} \otimes I(t) \tag{2}$$

where the eigenvalues, λ_1 and λ_2, are given by

$$\lambda_1, \lambda_2 = \frac{1}{2}\left[(k_1+k_2+k_3) \pm \sqrt{(k_1+k_2+k_3)^2 - 4k_2k_3}\,\right] \tag{3}$$

For the most usual case of a top-hat (square pulse) infusion, of length T_0 and constant rate I_0, the input may be represented by

$$I(t)=\begin{Bmatrix} I_0, & 0<t\leq T_0 \\ 0, & t>T_0 \end{Bmatrix}=I_0(h(t)-h(t-T_0)) \tag{4}$$

where $h(t)$ is the Heaviside step function. Convolution of this input function with the impulse response yields the pharmacokinetic profiles for the first compartment:

$$C_1(t)=\begin{Bmatrix} \dfrac{I_0 f}{(\lambda_1-\lambda_2)V_1}\left[\dfrac{\lambda_1-k_2}{\lambda_1}(1-e^{-\lambda_1 t})+\dfrac{k_2-\lambda_2}{\lambda_2}(1-e^{-\lambda_2 t})\right], & 0<t\leq T_0 \\ \dfrac{I_0 f}{(\lambda_1-\lambda_2)V_1}\left[\dfrac{\lambda_1-k_2}{\lambda_1}(e^{\lambda_1 T_0}-1)e^{-\lambda_1 t}+\dfrac{k_2-\lambda_2}{\lambda_2}(e^{\lambda_2 T_0}-1)e^{-\lambda_2 t}\right], & t>T_0 \end{Bmatrix} \tag{5}$$

The three rate constants and volume of distribution for the central compartment were determined from each patient's pharmacokinetic profile by weighted least-squares fitting of the model prediction to ^{10}B concentration measurements in blood using MATLAB (The Mathworks, Natick, MA). The volume of distribution for the peripheral compartment (V_2), and hence $C_2(t)$ itself, are undetermined because there are no observations of $C_2(t)$.

4. RESULTS AND DISCUSSION

The pharmacokinetic profiles show a rapid rise in blood boron concentration during BPA-F infusion followed by biphasic exponential clearance with relatively little variability between patients. For 1 hour infusions of 250 mg/kg, peak ^{10}B concentrations at the end of infusion are typically ~27 μg/g. For infusions of 350 mg/kg over 90 min, presently the standard infusion schedule used in the H-MIT clinical trials, ^{10}B concentrations at the end of infusion are typically ~32 μg/g. The mean half-lives of the fast and slow phases of the clearance profile, believed to result from redistribution of the drug in the body and renal elimination, are 0.34 ± 0.13 and 9.0 ± 2.5 hours, respectively.

Figure 2a shows measured pharmacokinetic profiles and fitted model results for an H-MIT patients infused with 250 mg/kg BPA-F over approximately 1 hour. As indicated by this graph, generally the agreement between the fitted model and measurements is very good. Furthermore, for cohorts of similar patients (i.e., of similar health status) inter-patient variability in pharmacokinetic profiles is low, although nonnegligible. Table 1 lists the pharmacokinetic model parameters measured for 21 patients enrolled in the H-MIT clinical trials. Moreover, we have noticed that patients of better physical condition seem to have more rapid ^{10}B clearance rates. We are now seeking correlation between model parameters and clinical laboratory results, e.g., creatine clearance which measures glomerular filtration rate.

Data for 3 patients was excluded from the analysis of Table 1. The infusion for patient 96–4 was interrupted by an infusion pump malfunction. Analysis of the biodistribution curves for patients 96–2 and 97-X give rate constants and volumes of distributions distinctly different (> 2 standard deviations) from the others; these data were therefore excluded as outliers. These differences may arise from the fact that, unlike the other patients, these two received test infusions while undergoing surgical procedures.

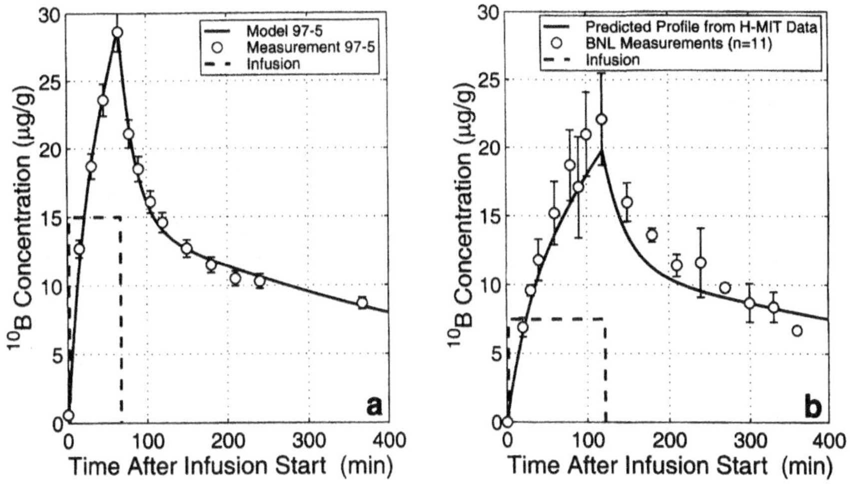

Figure 2. a. Measured pharmacokinetic profiles with fitted model results for an H-MIT patient infused with 250 mg/kg BPA-F over approximately 1 hour. b. Comparison of the average measured pharmacokinetic profile for a cohort of 11 patients infused with 250 mg/kg BPA-F over 2 hours at BNL[19] and the corresponding profile predicted by the model using rate constants derived from H-MIT patient data with different infusion schedules.

Table 1. Infusion parameters and measured pharmacokinetic model parameters. Data for patient 96-4 was not analyzed due to disruption of the infusion by a pump malfunction

Patient	BPA-F Dose mg/kg	Infusion Length min	k_1 min^{-1}	k_2 min^{-1}	k_3 min^{-1}	V_1 kg/kg
96-1 JY	310	60	0.019	0.009	0.0033	0.272
96-2 SJ*	250	60	0.125	0.024	0.0142	0.102
96-3 GM	250	60	0.009	0.004	0.0033	0.233
97-1 NM	250	60	0.025	0.011	0.0047	0.178
97-2 TT	250	62	0.017	0.008	0.0032	0.311
97-3 GC	250	65	0.035	0.012	0.0063	0.214
97-4 GR	250	65	0.018	0.010	0.0069	0.237
97-5 IR	250	65	0.032	0.013	0.0067	0.184
97-X HS*	250	60	0.048	0.011	0.0103	0.117
97-6 CM	250	67	0.019	0.008	0.0042	0.280
97-7 JT	250	60	0.032	0.012	0.0043	0.183
97-8 GB	300	90 & 87	0.023	0.011	0.0037	0.273
98-1 WB	300	102 & 102	0.029	0.012	0.0046	0.204
98-2 MD	350	82 & 92	0.025	0.013	0.0074	0.232
98-3 GW	350	92 & 90	0.023	0.013	0.0045	0.232
98-4 FL	350	90 & 90	0.021	0.007	0.0068	0.226
98-5 MM	350	90 & 90	0.018	0.009	0.0056	0.244
98-6 WV	350	119 & 119	0.028	0.009	0.0063	0.162
98-7 CR	350	90 & 89	0.014	0.005	0.0038	0.241
98-8 MP	350	93 & 88	0.028	0.009	0.0079	0.194
99-1 CM	350	91 &100	0.018	0.011	0.0053	0.247
99-2 GJ	350	90 & 90	0.021	0.008	0.0036	0.297
99-3 AK	350	88 & 90	0.023	0.014	0.0075	0.295
Mean			0.023	0.010	0.0052	0.235
Std. Dev.			0.006	0.003	0.0016	0.042
Coef. of Var.			28%	27%	30%	18%
Minimum	250	60	0.009	0.004	0.0032	0.162
Maximum	350	119	0.035	0.014	0.0079	0.311

*Parameters for patients 96-2 and 97-X were excluded from the statistical analysis at the bottom of Table 1 for reasons discussed above.

However, other explanations are possible, the primary being that the model rate constants may be coupled.

To demonstrate the validity of this model, a pharmacokinetic profile predicted with this model was compared with a measured pharmacokinetic profile produced by an infusion schedule different from those with which the impulse response function was derived. The mean rate constants and volume of distribution reported in Table 1 for the H-MIT patients were used in the model (Eq. 5) to predict a biodistribution curve for a 2 hour infusion of BPA-F at a rate of 125 mg/kg·hr. Comparison of the predicted pharmacokinetic profile with the average measured pharmacokinetic profile for 11 patients infused under these conditions (125 mg BPA-f/kg·hr for 2 hr) at BNL,[19] shown in Figure 2b, demonstrates good agreement, indicating the validity of the model for predicting the average pharmacokinetic response of a cohort of patients. Also, comparing average rate constants and volumes of distribution for cohorts of H-MIT patients divided according to infusion schedule shows good agreement and provides further evidence of the validity of this model; the mean rate constants for the different infusion schedules differ by less than 1 standard deviation.

A model of this type, with appropriately modified rate constants, would be useful in analysis of pharmacokinetic data from animal models.

5. CONCLUSIONS

An open 2-compartment model has been developed for predicting ^{10}B concentrations in blood following BPA-F infusion in humans. The model is derived using data from pharmacokinetic studies of patients enrolled in the Harvard–MIT Phase 1 clinical trials of BNCT. The validity of the model for predicting average pharmacokinetic response for a cohort of patients has been demonstrated.

ACKNOWLEDGMENTS

This research was supported by the U.S. Department of Energy under contract DE-FG02-97ER62193. Dr. Jeffrey Coderre provided the pharmacokinetic profile of the BNL patients used for model validation. The authors would like to thank Kent J. Riley, Jody Kaplan, Cynthia F. Chuang, and Lujia Tang for technical assistance with blood boron analyses.

REFERENCES

1. K. Ono, S. Masunaga, Y. Kinashi, M. Takagaki, T. Kobayashi, Y. Imahori, S. Ueda, and Y. Oda, "The Dose Planning of BNCT for Brain Tumors," in Cancer Neutron Capture Therapy, edited by Mishima, Plenum Press, New York, 1996, pp. 563–569.
2. C. Vroegindeweij, P. Watkins, K. Hideghety, W. Sauerwein, and R.L. Moss, "Treatment Planning for the First Group of Patients in the European Glioma Trial at the HFR Petten," presented at the Eighth International Conference on Neutron Capture Therapy for Cancer, La Jolla, CA, 1998 (in press).
3. P.M. Busse, R.G. Zamenhof, O.K. Harling, I. Kaplan, J. Kaplan, C.F. Chuang, J.T. Goorley, W.S. Kiger III, K.J. Riley, L. Tang, G.R. Solares, and M.R. Palmer, "The Harvard-MIT BNCT program: overview of the clinical trials and translational research," presented at the Eighth International Conference on Neutron Capture Therapy for Cancer, La Jolla, CA, 1998 (in press).

4. P.M. Busse, R.G. Zamenhof, H. Madoc-Jones, G. Solares, W.S. Kiger II, K.J. Riley, C.F. Chuang, G. Rogers, and O.K. Harling, "Clinical follow-up of patients with melanoma of the extremity treated in a phase I boron neutron capture therapy protocol," in Advances in Neutron Capture Therapy, edited by B. Larsson, J. Crawford, and R. Weinreich, Elsevier, Amsterdam, 1997, pp. 60–64.
5. M. Chadha, J. Capala, J.A. Coderre, E.H. Elowitz, J. Iwai, D.D. Joel, H.B. Liu, L. Wielopolski, and A.D. Chanana, Boron neutron-capture therapy (BNCT) for glioblastoma multiforme (GBM) using the epithermal neutron beam at the Brookhaven National Laboratory, "International Journal of Radiation Oncology, Biology, Physics," 40:829–834, 1998.
6. J.A. Coderre, A.D. Chanana, D.D. Joel, E.H. Elowitz, P.L. Micca, M.M. Nawrocky, M. Chadha, J.O. Gebbers, M. Shady, N.S. Peress, and D.N. Slatkin, Biodistribution of boronophenylalanine in patients with glioblastoma multiforme: boron concentration correlates with tumor cellularity, "Radiation Research," 149:163–170, 1998.
7. E.H. Elowitz, R.M. Bergland, J.A. Coderre, D.D. Joel, M. Chadha, and A.D. Chanana, Biodistribution of p-boronophenylalanine in patients with glioblastoma multiforme for use in boron neutron capture therapy, "Neurosurgery," 42:463–468; discussion 468–469, 1998.
8. C. Chuang, M. Palmer, R. Zamenhof, and P. Busse, "Time-dependent biodistribution of BPA-f in normal and tumor brain tissues in mice," presented at the Eighth International Conference on Neutron Capture Therapy for Cancer, La Jolla, CA, 1998 (in press).
9. G.W. Kabalka, G.T. Smith, J.P. Dyke, W.S. Reid, C.P.D. Longford, T.G. Roberts, N.K. Reddy, E. Buonocore, and K.F. Hubner, Evaluation of fluorine-18-BPA-fructose for boron neutron capture treatment planning, "Journal of Nuclear Medicine," 38:1762–1767, 1997.
10. Y. Imahori, S. Ueda, Y. Ohmori, T. Kusuki, K. Ono, R. Fujii, and T. Ido, Fluorine-18-labeled fluoroboronophenylalanine PET in patients with glioma, "J Nucl Med," 39:325–333, 1998.
11. J.A. Coderre, M.S. Makar, P.L. Micca, M.M. Nawrocky, H.B. Liu, D.D. Joel, D.N. Slatkin, and H.I. Amols, Derivations of relative biological effectiveness for the high-let radiations produced during boron neutron capture irradiations of the 9L rat gliosarcoma *in vitro* and *in vivo*, "Int J Radiat Oncol Biol Phys," 27:1121–1129, 1993.
12. G.M. Morris, J.A. Coderre, J.W. Hopewell, P.L. Micca, M.M. Nawrocky, H.B. Liu, and A. Bywaters, Response of the central nervous system to boron neutron capture irradiation: evaluation using rat spinal cord model, "Radiother Oncol," 32:249–255, 1994.
13. M.R. Palmer, C.F. Chuang, W.S. Kiger III, P.M. Busse, and R.G. Zamenhof, "Predicting Boron-10 Concentrations in Normal Brain and Glioblastoma from Blood Measurements," presented at the Eighth International Conference on Neutron Capture Therapy for Cancer, La Jolla, CA, 1998 (in press).
14. W.S. Kiger III, S. Sakamoto, and O.K. Harling, Neutronic design of a fission converter-based epithermal neutron beam for neutron capture therapy, "Nuclear Science and Engineering," 131:1–22, 1999.
15. S. Sakamoto, W.S. Kiger III, and O.K. Harling, Sensitivity Studies of Beam Directionality, Beam Size and Neutron Spectrum for a Fission Converter-Based Epithermal Neutron Beam for Boron Neutron Capture Therapy, "Medical Physics," 26:1979–1988, 1999.
16. K.J. Riley and O.K. Harling, An improved prompt gamma neutron activation analysis facility using a focused diffracted neutron beam, "Nuclear Instruments and Methods in Physics Research B," 143:414–421, 1998.
17. S.R. Tamat, D.E. Moore and B.J. Allen, Determination of boron in biological tissues by inductively coupled plasma atomic emission spectrometry, "Anal Chem," 59:2161–2164, 1987.
18. K.J. Riley, "Improved Boron 10 Quantification via PGNAA and ICP-AES," Nuclear Engineer's Thesis, Massachusetts Institute of Technology, 1997.
19. J.A. Coderre, E.H. Elowitz, M. Chadha, R.M. Bergland, J. Capala, D.D. Joel, H.B. Liu, D.N. Slatkin, and A.D. Chanana, Boron Neutron Capture Therapy for Glioblastoma Multiforme using p-Boronophenylalanine and Epithermal Neutrons, "Journal of Neuro-Oncology," 33:141–152, 1997.

34

QUALITY CONTROL OF $NA_2B_{12}H_{11}SH$, A DRUG FOR BORON NEUTRON CAPTURE THERAPY IN EORTC TRIAL 11961

Detlef Gabel[1], Daan Touw[2], Finn Stecher-Rasmussen[3], René Huiskamp[3], and Wolfgang Sauerwein[4]

[1]Department of Chemistry
University of Bremen, Germany
[2]Academic Hospital Free University Amsterdam
The Netherlands
[3]Netherlands Energy Research Foundation ECN
Petten, The Netherlands
[4]University of Essen
Germany

1. INTRODUCTION

The boron drug $Na_2B_{12}H_{11}SH$ (BSH) is presently being used in a clinical Phase I trial for glioma treatment with boron neutron capture therapy (BNCT). This study is carried out in The Netherlands at the High Flux Reactor of the European Commission. Patients are recruited from six neurosurgical centers in five European countries.

The drug BSH has been described for the first time in 1967. Therefore it has no patent rights. As a consequence, industry is unwilling to invest a great amount of money and effort in the production, quality control, and certification of the drug. Therefore quality control and certification is carried out by the pharmacy of the Academic Hospital of the Free University Amsterdam. Due to the limited stability in the administration solution, the drug has to be made up freshly every day, in the hospital where it is to be administered.

The chemical itself can be prepared through a multi-step procedure. A Drug Master File for the preparation has been registered by some of the potential suppliers with the Food and Drug Administration of the United States.

2. QUALITY REQUIREMENTS

BSH as drug for EORTC trial 11961 must conform to a number of quality criteria before it can be used in the trial. The tests are performed independently from the certificate of analysis from the supplier. They are performed by the pharmacy of the VU

Frontiers in Neutron Capture Therapy, edited by Hawthorne *et al.*
Kluwer Academic / Plenum Publishers, New York, 2001.

hospital as the EORTC-approved drug dispensing unit, ECN Petten and commercial laboratories.

Requirements for BSH in the study EORTC 11961 have been set to the following: The contents of each vial has to be 1.0 ± 0.1 g BSH (water-free weight). The drug must be 98% pure, as judged by high pressure liquid chromatography monitored at 220 nm. The isotopic enrichment has to be 95% boron-10 or higher. The sample must contain no more than 0.025 IE pyrogens per mg BSH.

3. TESTS FOR REQUIREMENTS

3.1. Identity

The identity of the material is checked by recording an infrared spectrum of the material in a KBr pellet. The spectrum is compared to spectra on file, obtained from authentized material. It should be borne in mind that the precise position of the resonances depends on the actual degree of enrichment.

3.2. Purity by HPLC

Purity is checked by high pressure liquid chromatography (HPLC), on an RP 18 column. The solvent system is methanol/water 70 : 35, 10 mM tetrabutylammonium HSO4 pH 6.8. The eluent is monitored at 220 nm. The material is acceptable when the area of the non-solvent peaks not representing BSH is less than 2% of that of BSH.

This method cannot detect the presence of salts, or of other material which is not UV-active.

3.3. Content of Active Substance

The content of a representative vial is checked by dissolving the total amount contained in the vial and analyzing the solution for boron, using ICP-AES. If other boron-containing substances are present, they would give rise to erroneous results. The amount required is that corresponding to 1.0 ± 0.1 g BSH (with the actual amount of enrichment).

Elemental analysis of the material for boron, hydrogen and sodium would eliminate the problem associated with the possible presence of other boron-containing compounds. It must be borne in mind, however, that elemental analysis of boron compounds is notoriously difficult.

3.4. Degree of Isotopic Enrichment

The degree of isotopic enrichment is measured by comparing the results of the ICP-AES analysis in section 3.3 with prompt gamma ray analysis for boron-10. Experience has shown that this comparison is, however, not always very reliable.

An alternative would be mass spectrometry. Here, the theoretical pattern of the isotope distribution of 12 boron atoms, being either boron-10 or boron-11, can be calculated and compared to the measured spectrum. The most suitable ionization method is fast atom bombardment FAB.

Table 1. Elemental analysis of different BSH batches from two different suppliers

Boron-10 enriched						Boron natural	
Supplier		1	2	2	2		2
Sample		1	1	2	3		4
Weight percent of elements							
	Theory					Theory	
Carbon	0.0	1.0	0.2	0.2	0.2	0.0	0.6
Hydrogen	5.8	6.1	4.4	6.3	6.3	5.5	5.5
Sulfur	15.3	14.8	*10.4*	*11.1*	*10.7*	14.6	14.1
Boron	57.1	57.5	*43.3*	*51.0*	58.7	59.0	57.8
Sodium	21.9	20.2	21.6	21.2	20.4	20.9	20.6
Accepted for trial		Yes	n.t.	*no*	*no*		n.t.

n.t.: not tested completely. Values given in italics are off by more than 2 percentage points.

3.5. Absence of Pyrogens

The absence of pyrogens is tested with a dynamic Limulus endotoxin test. The test should be performed such that no inhibition due to BSH is observed. BSH inhibits the Limulus test.

Figure 2 shows that the Limulus endotoxin assay is inhibited by BSH, to a degree which appears to be a function of the BSH concentration, but not of the endotoxin concentration. Therefore, a test at higher BSH concentrations is possible.

The need for testing for pyrogens could be eliminated if the material is heated to 180°C for 30 minutes. BSH does not decompose under such conditions.

3.6. Sterility

Just prior to infusion, the infusion solution is passed through a disposable sterile filter. Therefore, no requirements for sterility of the substance are necessary. If a heating step would be included for removal of pyrogens, the material would be sterile without further treatment.

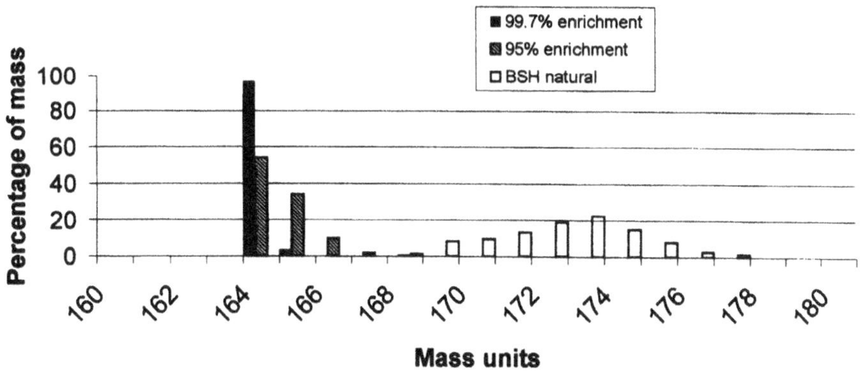

Figure 1. Isotope distribution of $Na_2B_{12}H_{11}SH$, measured by Fast Atom Bombardment Mass Spectrometry in negative mode. Data are given as the percentage of the individual mass in relation to the total mass over all isotope compositions.

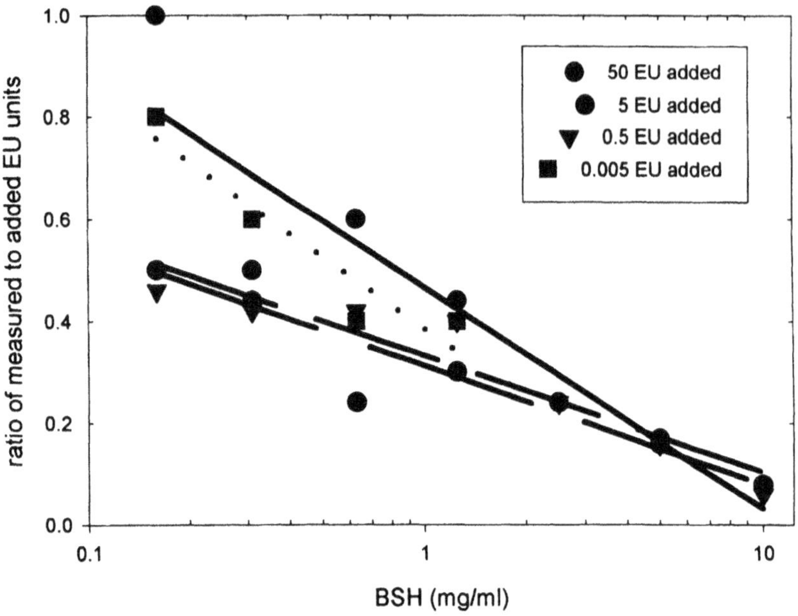

Figure 2. Limulus endotoxin text of BSH. The measured amount of endotoxin is given as percentage of the known amount of added endotoxin, in the presence of the amount of BSH indicated.

4. CONCLUSION

The quality control of BSH and other boron-containing drugs poses problems different from, and additional to, the requirements for other drugs. The reasons for this lie in the necessity to analyze both the total amount of boron and the amount of boron-10 present in the drug.

ACKNOWLEDGMENTS

The work is supported by the European Commission through the Biomed II program.

THE PHARMACOKINETICS OF BSH FOLLOWING SINGLE AND MULTIPLE ADMINISTRATIONS

Detlef Gabel[1], Els Winter-van Kampen[2], Finn Stecher-Rasmussen[2], Wolfgang Sauerwein[3], Katalin Hidéghety[3], Frank Grochulla[4], Claudia Goetz[5], Klaus Haselsberger[6], Philippe Paquis[7], John Wolbers[8], and Daan Touw[8]

[1]Department of Chemistry
University of Bremen
D-28359 Bremen, Germany
[2]ECN Petten
The Netherlands
[3]Department of Radiotherapy
University of Essen, Germany
[4]Neurosurgical Clinic
Zentralkrankenhaus St. Juergenstrasse
Bremen, Germany
[5]Department of Neurosurgery
Klinikum Grosshadern
Munich, Germany
[6]Department of Neurosurgery
Karl-Franzens-Universitaet Graz
Austria
[7]Service de Neurochirurgie
Hôpital Pasteur, Nice, France
[8]Department of Neurosurgery
Free University Hospital
Amsterdam, The Netherlands

1. INTRODUCTION

In EORTC trial 11961 "Glioma Treatment with BNCT", the treatment with BNCT is given in four fractions, on four consecutive days. Prior to each fraction, BSH ($Na_2B_{12}H_{11}SH$) is infused in amounts such that 30 ppm boron is reached as average over the four fractions. Before the BNCT treatment, BSH is administered at the time point of

Frontiers in Neutron Capture Therapy, edited by Hawthorne *et al.*
Kluwer Academic / Plenum Publishers, New York, 2001.

surgery, in order to gain data on the pharmacokinetics of BSH in the individual patient. On this basis, and assuming that each dosage of BSH is purely additive and that the pharmacokinetics of BSH at surgery is a good prediction of the pharmacokinetics at treatment, an individual prescription scheme can be established beforehand.

2. METHODS

2.1. BSH Administration at Surgery

At the time of surgery, 100 mg BSH per kg body weight was administered intravenously during a period of 100 minutes. Blood samples were taken at the following time points: 3, 6, 9, 11, 12, 13, 14, 18, 24, 48 hours, as well as at the time point of tumor removal (around 12 to 14 hours after the BSH administration. The blood samples were shipped frozen to ECN Petten, and analyzed for boron by ICP-AES. The degree of enrichment used in the study varied.

2.2. BSH Administration During BNCT

BNCT was administered in four individual fractions, on four consecutive days. For BNCT at the HFR Petten, BSH was administered around 15 hours prior to the planned irradiation. The boron concentration in blood during treatment (averaged over all four fractions) was to be 30 ppm. The first dose of BSH consisted of 100 mg/kg, the subsequent doses were adjusted downward in order not to exceed the average boron concentration of 30 ppm for the four fractions. In addition, the time point of irradiation could be chosen such that the boron concentration had dropped further, if required. Blood was sampled and analyzed for boron by prompt gamma ray analysis.

2.3. Pharmacokinetics Analysis

The boron concentrations of the samples obtained during surgery were used to derive the parameters of the following equation

$$c = c_1 \cdot e^{-k_1 \cdot t} + e^{-k_2 \cdot t} \tag{1}$$

with c being the concentration at time t, and c_1, c_2, k_1, and k_2 constants[1]. The fitting of the measured data to equation 1 was carried out using the SigmaPlot® non-linear curve fitting procedure. The concentration values were weighted with c^{-2}.

The constants were taken to generate curves of projected boron concentrations during treatment, assuming that each dose behaved independent of each other, and that the constants c_1 and c_2 were proportional to the amount administered.

3. RESULTS

The blood samples taken during surgery allowed to derive values for the parameters of equation 1 in all but one of the fourteen patients who had received BSH at surgery. For one patient, blood samples at times longer than 48 hours would have been required in order to obtain a reasonable fit.

For the other patients, the amount of boron in the blood found during treatment was either in reasonable agreement with the predicted values, or was higher. In none of the cases, the amount of boron was found to be lower than the predicted values. For all patients, the boron concentrations directly after the end of each administration were higher than predicted by equation 1, probably due to the fact that the pharmacokinetics are described best by a three-compartment model. The data collected during surgery do not allow the analysis for more elaborate models.

Examples of the pharmacokinetics are found in Figures 1 and 2.

4. CONCLUSIONS

Experience has shown that the pharmacokinetics of BSH at surgery does not predict very well the boron concentrations actually achieved during treatment. In most cases, boron concentrations were higher than predicted. This can be due to the special circumstances of the surgery (with fasting and loss of liquid), due to the possible existence of deeper compartments which are not taken into account in a two-compartment model, and to a possible non-additivity of the boron dosages.

On the basis of optimal treatment for an individual patient, pharmacokinetics data for BSH need not to be collected at surgery. Rather, an individual case-to-case and day-to-day decision is fully adequate.

ACKNOWLEDGMENTS

The work is supported by the European Commission through the Biomed II program.

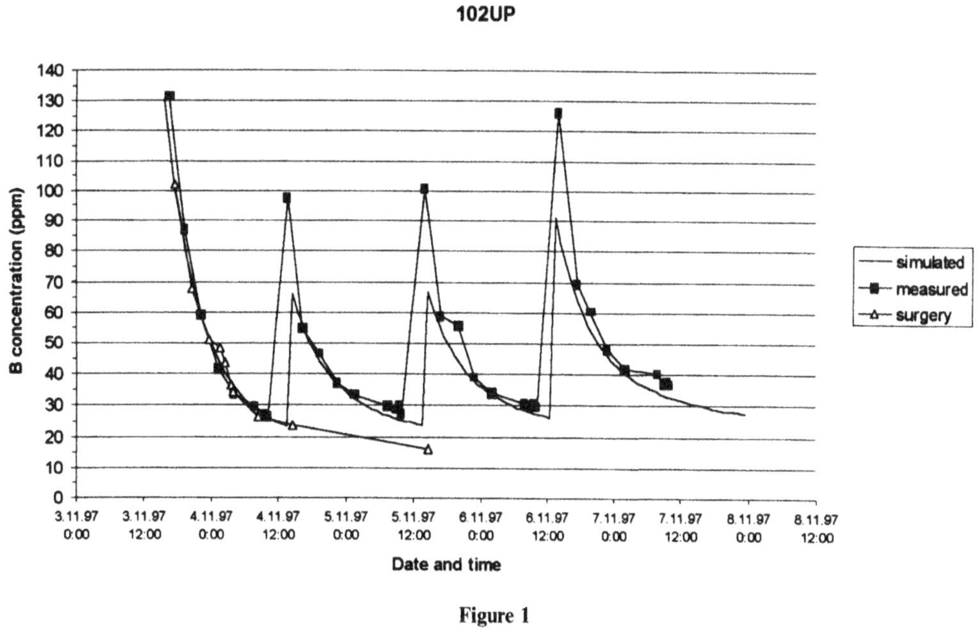

Figure 1

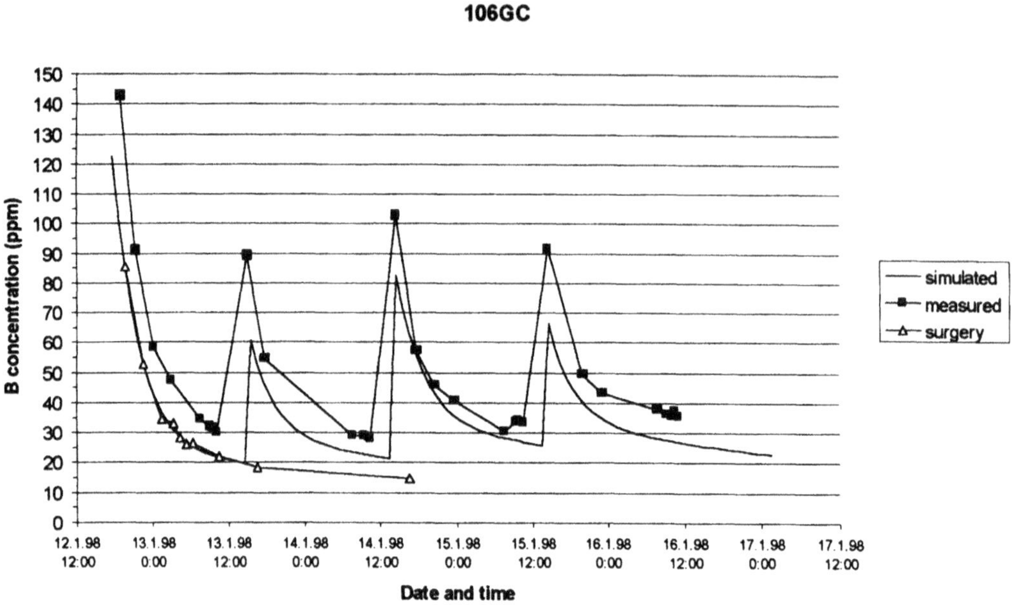

Figure 2

REFERENCE

1. Gabel D, Preusse D, Haritz D, Grochulla F, Haselsberger K, Fankhauser H, Ceberg C, Peters H-D, Klotz U. Pharmacokinetics of $Na_2B_{12}H_{11}SH$ (BSH) in patients with malignant brain tumours as prerequisite for a Phase I clinical trial of boron neutron capture. *Acta Neurochir.* 139:606–612, 1997.

REACTOR NEUTRON SOURCES AND FACILITIES

METAMORPHOSIS OF A 35 YEAR-OLD TRIGA REACTOR INTO A MODERN BNCT FACILITY

Iiro Auterinen[1], Pekka Hiismäki[1], Petri Kotiluoto[1], Rolf J. Rosenberg[1], Seppo Salmenhaara[1], Tiina Seppälä[1,2,3], Tom Serén[1], Vesa Tanner[1], Carita Aschan[2,5], Mika Kortesniemi[2,3], Antti Kosunen[4], Juha Lampinen[2], Sauli Savolainen[2,5], Matti Toivonen[4], and Petteri Välimäki[2]

[1]VTT Chemical Technology
P.O. Box 1404, FIN-02044 VTT, Finland
[2]Department of Physics
University of Helsinki
[3]Clinical Research Institute
Helsinki University Central Hospital
[4]STUK—Radiation and Nuclear Safety Authority
[5]Department of Radiology
Helsinki University Central Hospital

1. INTRODUCTION

Using FiR 1, a 250 kW TRIGA reactor as a neutron source for BNCT was screened as a viable option in 1990, as the future of the reactor was questioned. By the initiative of the medical radio isotope group at VTT the Finnish BNCT project started incorporating early on also the medical and medical physics sides.[1,2] At first a thermal neutron source only was considered conceivable, but quite soon it was realized that using aluminum-aluminum fluoride moderator a high quality epithermal neutron source was quite feasible.[3] After gaining support both from the medical community as well as from private and state financing sources a decision was made in 1994 to accomplish the necessary changes in the reactor and the reactor hall.

Now the Finnish BNCT facility (see Figure 1) at FiR 1 in Otaniemi, Espoo is close to completion for patient trials.

Frontiers in Neutron Capture Therapy, edited by Hawthorne *et al.*
Kluwer Academic / Plenum Publishers, New York, 2001.

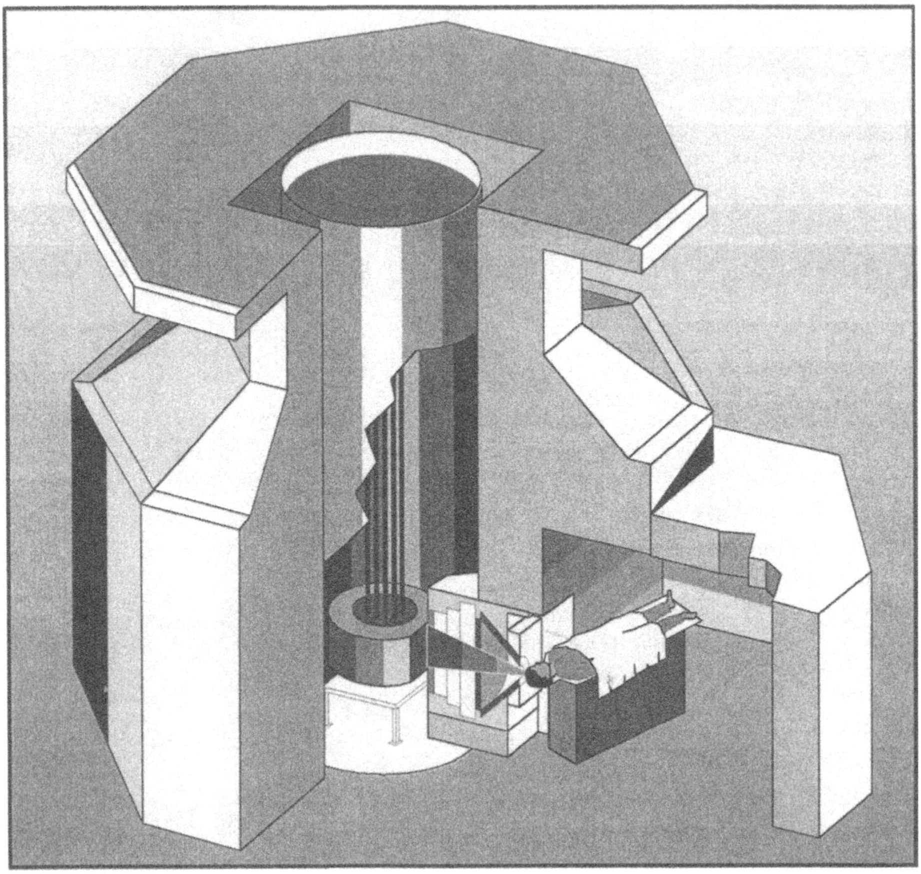

Figure 1. Schematic drawing of the BNCT facility at FiR 1.

2. DEVELOPMENT OF A NEW EPITHERMAL NEUTRON MODERATOR

Theoretical work using one and three dimensional discrete ordinates codes ANISN and TORT with the BUGLE-80 cross section data and analysis of availability of possible technical materials led to the conclusion that an optimized mixture of aluminum and fluorine would be one of the best practical choices. Combination of aluminum with fluorine instead of oxygen gives clearly better moderation performance for BNCT.[3,4]

The new neutron moderator material FLUENTAL™ is a composition of AlF_3 (69%m) and aluminum (30%m) and LiF (1%m).[5] A manufacturing process based on the hot isostatic pressing (HIP) technique was developed for this material producing solid blocks with practically 100% density with 3 g/cm^3. After identification of the theoretical potential of FLUENTAL™ it took till summer 1994 before the first small batch of the material was produced with proper process parameters in laboratory scale at VTT. In 1995 the process was scaled up to large 1,000 kg batches and in March 1996 enough material had been produced for the epithermal beam at FiR 1.

The graphite from the thermal column of FiR 1 had been removed and a support structure for a moderator container had been installed in summer 1995. Now in April 1996 a 75 cm thick FLUENTAL moderator with a bismuth gamma shield was installed. In this basic configuration the 70 by 70 cm wide bismuth face acted as source plane for the free field. The epithermal neutron field was characterized by the Finnish BNCT-dosimetry team and by visiting teams from NRI (Rez, Czech Republic) and INEEL (Idaho, USA).[6,7] These measurements confirmed that the epithermal neutron field had good spectral characteristics and sufficient intensity for clinical application.

3. THE EPITHERMAL BEAM AT FIR 1

3.1. Beam Construction

To be able to direct the epithermal neutrons for irradiation of brain tumors, a beam collimator was designed. One of the main goals was to create a favorable thermal field distribution within the target volume in the tissue, and especially deep penetration was emphasized. The positioning of the patient in the beam had to be flexible and parts of the patient outside the target volume had to be shielded from the direct irradiation of the beam.

For the design of the collimator the beam facility was modeled in cylindrical (R-Z) geometry using the two-dimensional DORT code. Field distributions in a cylindrical and an ellipsoidal target phantom produced by different collimator configurations were compared. For construction materials bismuth, lead and lithium (both natural and enriched in Li-6) polyethylene were considered. A conical form was chosen for the collimator. A bismuth inner cone with optimized thickness reflects neutrons back to the beam. Outside it lithium polyethylene is limiting the field by attenuating neutrons of all energies. Around the exit aperture Li-6-enriched polyethylene was used to avoid creation of thermal neutrons into the target. As a compromise between the flexible patient positioning and favorable field distribution in the phantom a cone with a 60 degree full angle was found to be optimal. The diameter of the beam exit aperture can be selected by the number of aperture collars in place from 20 cm down to 8 cm in steps of 3 cm.

The collimator was added to the basic configuration with the 75 cm thick moderator in September 1996 and its performance was characterized with free field and phantom measurements. Then after the renovation of the reactor hall the moderator was shortened to 63 cm by removing one of the three adjacent moderator blocks and replacing it with a thinner one. At this stage the lead shield which attenuates the gamma streaming through the concrete around the beam system was rebuild. Also lithium plastic shielding around the beam was extended and improved. The current configuration is shown in Figure 2.

3.2. Beam Monitoring System

An on-line neutron beam and reactor monitoring system[8] has been constructed as an integral part of the facility and is used for control of exact neutron delivery for the treatment. The epithermal neutron beam is monitored at the bismuth gamma shield after the moderator. The detectors are three pulse mode U235-fission chambers for epithermal neutron fluence rate and one current mode ionization chamber for gamma dose rate.

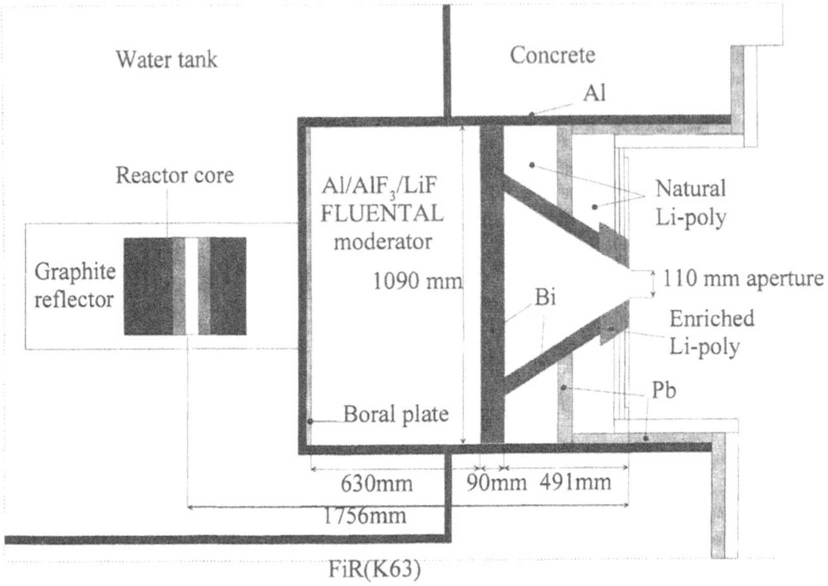

Figure 2. The epithermal neutron irradiation facility at FiR 1 in its current configuration with a 63 cm thick FLUENTAL™ moderator.

The detector signals are monitored on line with a virtual instrumentation (LabView, NI, USA) based PC-program, which records and displays the actual count rates and total counts of the detectors in the beam (Figure 3). Also reactor in-core power instrumentation and control rod positions are monitored via another LabView application. The main purpose of the monitoring system is to provide a dosimetric link to the dose in patient during the treatment, as the fission chamber count rates have been calibrated to the induced thermal neutron fluence rate and to the absorbed dose rate at reference conditions in a tissue substitute phantom.[9]

3.3. Beam Performance

A calculational model for the epithermal neutron beam has been created using the two-dimensional transport code DORT.[9] The model has been verified by several measurement campaigns.[9,10,11] The model accurately produces neutron and gamma fields both in free beams and in phantoms. Based on the DORT model source descriptions for the BNCT_Rtpe/rtt_MC treatment planning code and the MCNP code have been created.[9,12]

For validation and adjustment of the calculated neutron spectra activation detectors with different response characteristics, most of them from ECN Petten, were used. The detector foils were very thin or diluted in Al to avoid self-shielding effects. To measure the fast-neutron flux Ni, ^{54}Fe and Nb were irradiated at the Bi surface over a longer period. The measured reaction rates and calculated neutron spectra with uncertainties were used as input to the least squares adjustment code LSL-M2.[13] The cross section library was mainly based on IRDF-90 (IAEA). Generally the adjustments were moderate,[9] which confirms the good quality of the transport calculations. The linearity assumption in LSL-M2 was also not violated. As an example the adjusted and calculated neutron

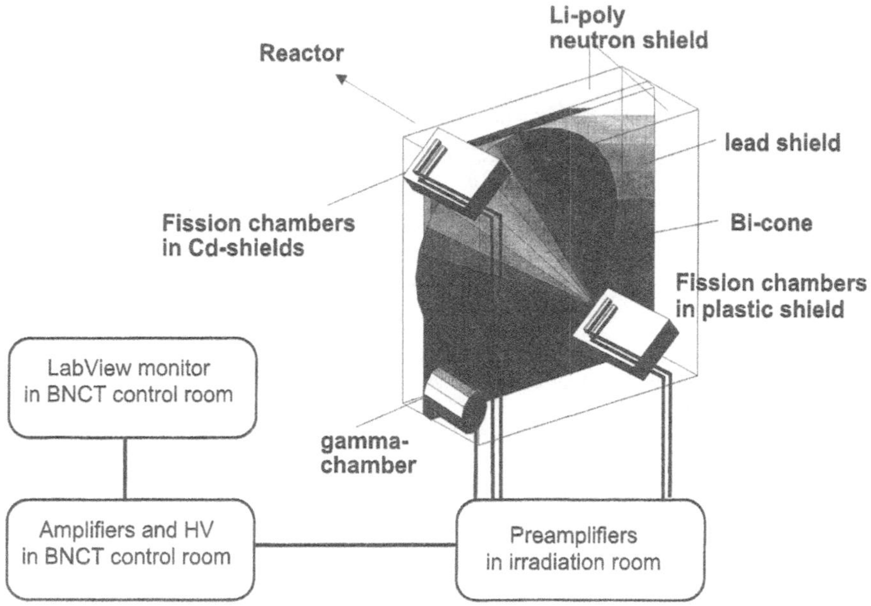

Figure 3. The beam monitor system of the epithermal neutron irradiation facility.

spectra of the FiR(K63) beam at the 14 cm diameter aperture plain is given in Figure 4. The uncertainties are of the order 20%, 5% and 30% for the thermal, epithermal and fast flux, respectively. The independent measurements performed by INEEL team at the current beam, FiR(K63), support also these results.

The main beam parameters of the two collimated beam configuration together with a comparison to three other existing BNCT facilities is given in Table 1.

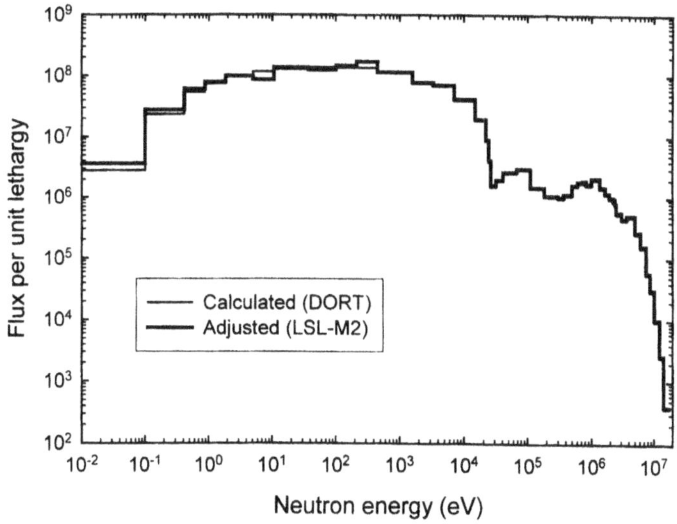

Figure 4. Adjusted and calculated neutron spectra of the FiR(K63) beam at the 14 cm diameter aperture plain.

Table 1. Intercomparison of BNCT facilities

Reactor/location	Reactor power [MW]	Epithermal flux [$cm^{-2} \times s^{-1}$]	Neutron current/ flux	Fast neutron dose/ epithermal fluence [$Gy/10^{13} cm^{-2}$]	Photon dose/ epithermal fluence [$Gy/10^{13} cm^{-2}$]
FiR 1/Finland (K75)[1]	0.25	0.5×10^9	0.79	1	0.5
FiR 1/Finland (K63)[2]	0.25	1.1×10^9	0.77	2	0.5
BMRR/USA[3]	3	0.8×10^9	0.80	5	2.0
MITR/USA[4]	5	0.2×10^9	0.55	13	13
HFR/Netherlands[5]	45	0.3×10^9	0.95	10	8.6

[1]Values refer to the center of the 14 cm diam. exit aperture plane for 75 cm thick FLUENTAL-moderator (measured value) [October 1996].
[2]Values refer to the center of the 14 cm diam. exit aperture plane for 63 cm thick FLUENTAL-moderator (measured value) [September 1998].
[3]Values refer to the center of the 12 cm diam. exit aperture plane. [Med. Phys. **23** (12), Dec. 1996].
[4]Estimated from data, given in Nucl. Sci. Eng. **110**, 330–348, (1992).
[5]HB11/BNCT Epithermal Neuron Beam. Parameters at 1 cm from the exit of the final aperture. [beampara, 17/3/94].

Human treatment simulations using the dose planning program BNCT_Rtpe /rtt_MC show that treatment times will be similar and the tumor dose somewhat higher compared to the current BNL facility when using BPA as boron carrier. The brain tolerance of 12.6 Gy (W) in a 1 cm^3 volume as applied in the BNL BNCT protocol #4 will be achieved in 36 minutes in one 14 cm diameter field and in 64 minutes when two opposite fields are used (Figure 5). Here boron concentration ratio brain to blood is assumed to be 1 with 15 ppm boron-10 in blood. For the total dose the weighted dose concept is used.[14]

4. THE RADIATION THERAPY CLINIC

4.1. BNCT-Facilities

The reactor building underwent a major renovation in winter 1996–97. The ground floor of the reactor hall was provided with a new entrance, easily accessible by any patient vehicle, a radiotherapy control room and rooms for patient preparation and laboratories.

A heavy concrete shielded therapy room was built outside the former thermal column of the reactor. Criteria were easy and clean assembly at reactor, possibility of removal part of the ceiling for maintenance work at the beam—for example when changing moderator thickness or bismuth collimator. Also optimized shielding for both gamma and neutrons was looked for. Therefore heavy concrete was selected as the basic material and by casting the concrete into steel tubes good structural strength was combined with high density, 4.35 g/cm^3. The achieved high density ensured that with the 80 cm wall and roof thickness the dose rate levels outside are well within the design objectives and actually lower than outside the original biological shield.[15]

The inner walls, ceiling and floor were lined with 30 mm thick lithium plastic elements. The elements are of in house design with variable lithium content depending on

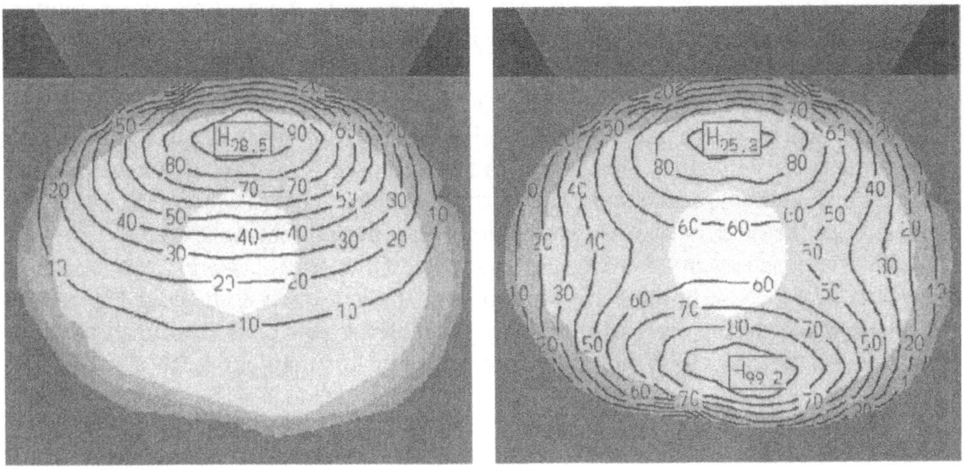

Figure 5. Weighted total isodoses for tumor ($c_{B\text{-}10}$ = 52.5 ppm) one (left) and two (right) 14 cm epithermal neutron fields when the healthy tissue dose is limited to 12.6 Gy (W) ($c_{B\text{-}10}$ = 15 ppm); ($w_{B,tumor}$ = 3.8, $w_{B,healthy\ tissue}$ = 1.3, w_{γ} = 1.0, w_N = 3.2, $w_{fast\ n}$ = 3.2, c_N = 1.84%, c_H = 10.57%). With one field irradiation time is 36 min and 100% in tumor = 71.7 Gy (W). With two fields irradiation time is 32 min + 32 min and 100% in tumor = 69.0 Gy (W).

if thermal neutron capture or fast neutron attenuation is desired. Lithium was selected as the neutron absorbing isotope throughout the beam collimator and irradiation room to enable background free boron prompt gamma monitoring in the future.

Several cubic meters of the concrete biological shield had to be cut away in order to allow positioning of the patient close enough to the reactor core in proper bodily orientation. Limiting factor in the removal was the stability of the original biological shield. Structural strength analysis performed by Olli Majamäki (IVO Power Engineering Ltd) showed that the thermal column could be widened so far that a suspending arch is still formed over the irradiation position. Therefore the cutting was limited to halfway of the side faces.

4.2. The Reactor

The top of the reactor tank was separated from the reactor hall in order to confine contamination in case of a leakage from irradiation samples or fuel elements. The ventilation of the building, emergency power supply system, heat exchangers and the secondary cooling circuit of the reactor including cooling towers were completely redesigned and rebuilt.

The reactor core loading and control rod operations were optimized for maximum epithermal beam output.

4.3. BNCT Facility Operation

A BNCT irradiation is started by manual reactor start-up and terminated by manual scram. In a BNCT irradiation the reactor is operated in full 250 kW thermal power. During irradiation the reactor power is maintained at the specified level by the reactor automation, but the initial starting procedure has to be done manually by the reactor operator. Beam monitor units are calculated from the fission counter readings. When the specified amount of beam monitor units is reached a manual reactor scram is initiated.

A rather rapid operation is achieved without beam shutter. The timing in patient irradiations will be comparable to that at BMRR. Startup after closing the irradiation room door takes 3 minutes. The patient can be removed 5 minutes after reactor scram without any recordable additional doses to the dosimeters carried by the personnel. Measurements of the dose rates in the irradiation room during reactor operation show that removal of the patient in an emergency case even if the reactor would keep running at full power would cause to the person bringing the patient out a dose about 1 mSv. Measurements made during reactor scram show that if the reactor is scrammed when the person enters the irradiation room the dose would be less than one tenth of that.

5. DISCUSSION

The first epithermal beam configuration at FiR 1 (K75) produced a very high purity beam with a fast neutron contamination one fourth of the BMRR (BNL, USA) facility. But with a similar current to flux ratio the epithermal flux was 40% lower than at BMRR. So it was decided to increase the intensity to shorten the irradiation times. Now with the FiR(K63) beam the irradiation time is very close to that at BNL so equal timing can be achieved if desired. Due to the rapid kinetics of BPA-F this was considered to be important. Also improved patient comfort due to shorter irradiations was appreciated. The fast neutron dose was designed to remain still under the BNL value. In the future the irradiation times can be made even shorter by increasing the reactor power, e.g. to double.

The epithermal beam is well characterized and a dosimetric system has been created for dose planning and control in patient irradiations.[14] As a final test of the beam and the facility prior to patient irradiations a healthy tissue tolerance study has been performed.[16] Patient fixation system is under development.

The expenditure of designing and accomplishing the construction work described was about 4.4 Million euros, half of that was caused by the renovation of the reactor hall and the other half has been spent to the research, development and construction of the epithermal beam. The epithermal irradiation facility was constructed by VTT under contract with the Radtek Inc., Espoo. Radtek is a company formed to combine private capital and state technology development funding (TEKES-Technology Development Centre Finland, Sitra) for this purpose.

REFERENCES

1. P. Hiismäki, I. Auterinen, and M. Färkkilä, The Finnish BNCT Program, An Overview, *In*: M. Kallio, and I. Auterinen (Eds.) Proceedings of the CLINCT BNCT Workshop. Helsinki: Helsinki University of Technology series TKK-F-A718, 2–4, 1994.
2. S. Savolainen, I. Auterinen, M. Kallio *et al.*, The Finnish boron neutron capture therapy program—an overview on scientific projects, Excerpta Medica. International Congress Series 1132, Vol. 1, 342–347, 1997.
3. I. Auterinen, and P. Hiismäki, Epithermal BNCT neutron beam design for a Triga II reactor, *In:* H. Soloway, and R.F. Barth (Eds.) Advances in Neutron Capture Therapy. Proceedings of the 5th International Symposium on Neutron Capture Therapy. Columbus, OH, September 14–17, 1992. New York: Plenum Press 81–84, 1993,.

4. I. Auterinen, P. Hiismäki, T. Serén, and T. Seppälä, Creating an Epithermal Neutron Field for BNCT Using a Triga Reactor and an Aluminum Fluoride Composite as Neutron Moderator, *In*: Proceedings of the 1998 ANS Radiation Protection and Shielding Division Topical Conference, Technologies for the New Century, April 19–23, 1998, Nashville, Tennessee, USA. ANS 1998, Vol. II, pp. 77–83.
5. P. Hiismäki, and I. Auterinen, Patent FI-92890 (1995).
6. K. Kaita, T. Serén, and I. Auterinen, First characterisation of the Finnish epithermal neutron beam using activation detectors, Excerpta Medica. International Congress Series 1132, Volume I, pp. 531–4, 1997.
7. D.W. Nigg, Y.D. Harker, J.K. Hartwell *et al.*, Collaborative Spectral Characterization of the Finnish Epithermal-Neutron Beam Facility for BNCT, *in*: INEEL BNCT Research Program Annual Report 1996, INEL/EXT-97-00319, April 1997.
8. V. Tanner, I. Auterinen, J. Helin, and S. Savolainen, On line neutron beam monitoring of the Finnish BNCT Facility, Nucl. Instrum. (in press).
9. T. Seppälä, T. Serén, and I. Auterinen, Source characterization for the rtt_MC treatment planning program at FiR 1, (these proceedings).
10. M. Kortesniemi, A. Kosunen, T. Serén *et al.*, Phantom dose distributions in the beam of Finnish BNCT facility, (these proceedings).
11. A. Kosunen, M. Kortesniemi, H. Ylä-Mella *et al.*, Twin ionization chambers for dose determination in phantom in an epithermal neutron beam, Radiat. Prot. Dosim. (in press).
12. J. Lampinen, T. Seppälä, I. Auterinen, and S. Savolainen, Three dose calculation codes applied to neutron transport in BNCT, (these proceedings).
13. F.W. Stallmann, LSL-M2: A Computer Program for Least-Squares Logarithmic Adjustment of Neutron Spectra, NUREG/CR-4349, ORNL/TM-9933, Oak Ridge March 1986.
14. S. Savolainen *et al.*, Dosimetry chain for the dogs irradiated in the epithermal neutron beam at the Finnish BNCT facility, (these proceedings).
15. P. Kotiluoto, P. Hiismäki, I. Auterinen, T. Seppälä, and C. Aschan, Shielding design and calculations for the Finnish BNCT facility, (these proceedings).
16. J. Benczik *et al.*, Large animal model for healthy tissue tolerance study in BNCT, (these proceedings).

37

STATUS OF THE MEDICAL IRRADIATION FACILITY AT JRR-4

T. Kishi, Y. Torii, H. Kumada, K. Yokoo, N. Ohhashi, and F. Sakurai

Department of Research Reactor
Tokai Research Establishment
Japan Atomic Energy Research Institute
Tokai-mura, Naka-gun, Ibaraki 319-11
Japan

INTRODUCTION

A medical irradiation facility for Boron Neutron Capture Therapy (BNCT) was installed at JRR-4 in June 1998.

JRR-4 was constructed for the purpose of shielding design study for nuclear ship. It was a swimming pool type research reactor using light water for its moderator and coolant. Its maximum thermal power is 3.5 MW and first criticality on January 28, 1965. The operation of JRR-4 using HEU fuel was terminated at the beginning of 1996, and was resumed using LEU fuel with the medical irradiation facility in July 1998.

The new facility at JRR-4 was planned to be capable of providing both thermal and epithermal neutron beams with enhanced neutron fluxes. The conventional heavy water tank and thermal column facility was remodeled into the new medical irradiation facility. A medical treatment room and patient-monitoring area were also prepared adjacent to the irradiation facility.

NEW MEDICAL IRRADIATION FACILITY

General Arrangements

The general arrangements of the facility are shown in Figure 1. The irradiation room and the adjacent medical treatment room are located on the basement. Patients are received at the entrance of the administration building and are transferred to the basement of reactor building using elevator.

Frontiers in Neutron Capture Therapy, edited by Hawthorne *et al.*
Kluwer Academic / Plenum Publishers, New York, 2001.

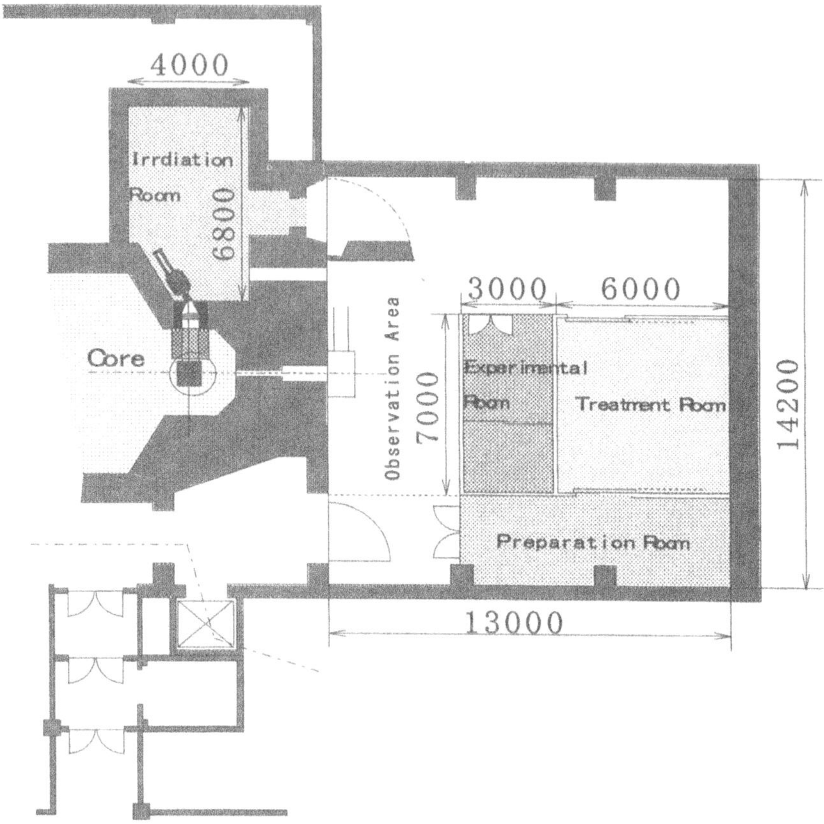

Figure 1. General arrangement of the medical irradiation facility.

Details of the Facility

The cross section of the medical irradiation facility is shown in Figure 2. The heavy water tank plays a role of neutron beam moderator with aluminum and heavy water. The heavy water layers are separated and optimum heavy water thickness can be chosen for both thermal and epithermal beams according to the need of each treatment. The tank is arranged with a cadmium shutter.

The beam experimental elements consist of: a) bismuth filter for reducing gamma-ray dose, b) lead and graphite for beam reflectors, c) LiF and B4C for decreasing activation of structural materials, d) LiF collimator at beam port, and e) borated polyethylene and lead blocks for shielding around the beam port. The beam port is not equipped with a beam shutter.

The irradiation angle of patients is possible to adjust within 90 degree to left side, and 60 degree to right side.

The medical treatment room and patient-monitoring area were also prepared adjacent to the irradiation facility. Patient-monitoring area was installed TV monitor, neutron and gamma rays monitor in the irradiation room.

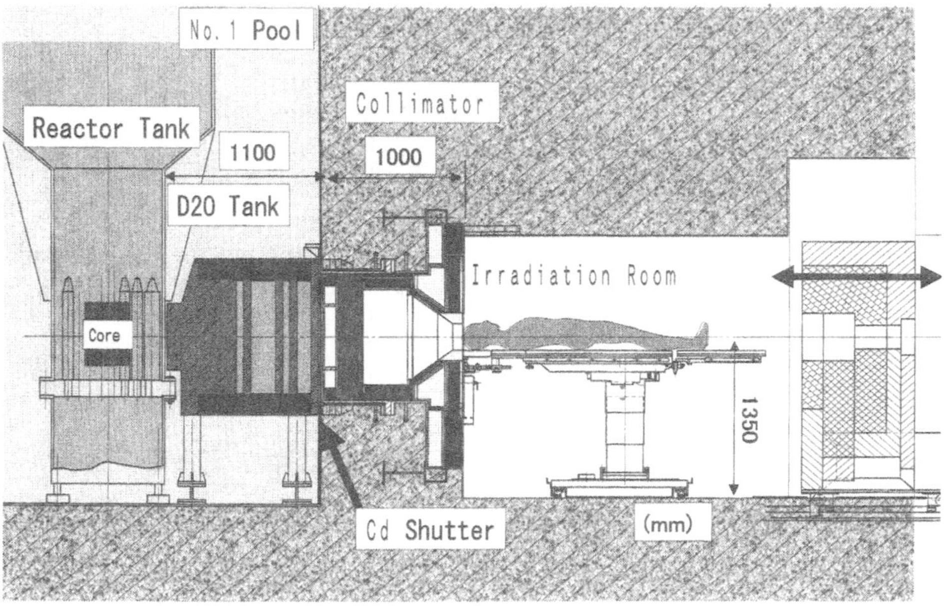

Figure 2. Cross-section of the medical irradiation facility.

Operation of the Facility

The rated output of the reactor is 3.5 MW, and the operation mode is daily operation (6 h/day). During the reactor operation for medical irradiation another utilization of the reactor is not foreseen. The facility can provide a wide variety of neutron beams. The change over of neutron spectrum is realized by changing the thickness of heavy water in the heavy water tank, by inserting/removing a thin cadmium plate in the beam line, and by changing the thickness of graphite reflector in the beam hole.

BEAM PERFORMANCE

The basic policy of the beam design is to provide a variety of neutron beams by easy change of configuration mentioned above. While thermal neutron beams are needed to continue the conventional BNCT in Japan, epithermal neutron beams are requested by users for the treatment of deep-seated tumors.

The objectives of the beam design were set as follows for free beam model:

(1) Thermal neutron flux at beam port (Thermal Beam Mode): $>2 \times 10^9$ n/cm^2/sec.
(2) Epithermal neutron flux at beam port (Epithermal Beam Mode): $>1.5 \times 10^9$ n/cm^2/sec.
(3) Gamma dose contamination: $<3 \times 10^{-13}$ Gy/n·cm^2.
(4) Fast neutron contamination: $<8 \times 10^{-13}$ Gy/n·cm^2.
(5) Size of beam port: 220 mm × 220 mm.
(6) Estimated Irradiation time: <2 h.

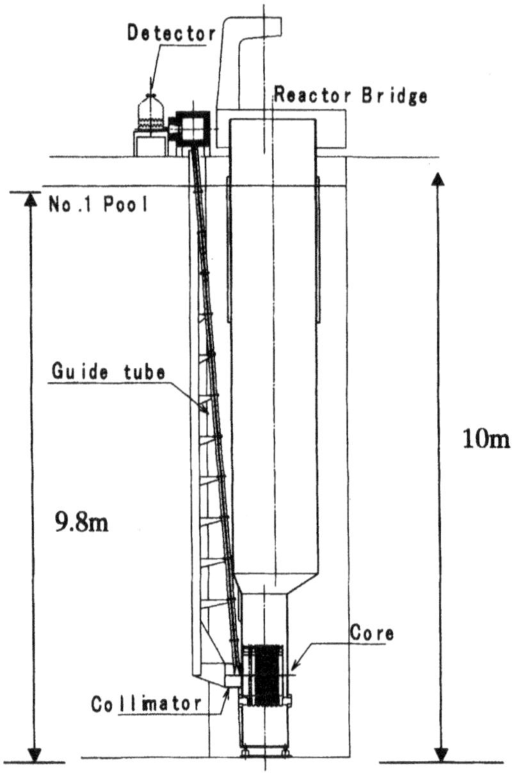

Figure 3. The prompt gamma analysis device.

Aluminum thickness of 75 cm was chosen to reduce fast neutron contamination in epithermal neutron beams, while remaining total heavy water thickness of 33 cm was enough to provide an almost equivalent thermal neutron beam as JRR-2 beam.

The thickness of the heavy water layer can be arbitrary chosen from 0 cm to 28 cm by 4 cm step. The maximum thickness is 33 cm.

BORON CONCENTRATION MEASUREMENT

The measurement of the boron concentration is important for deciding the irradiation time at BNCT. The measurement of the boron concentration using prompt gamma-ray analysis (PGA) device and/or inductively coupled plasma atomic emission spectroscopy (ICP-AES). The prompt gamma analysis device can measure the boron concentration in a short time. We was installed PGA in JRR-4. The prompt gamma analysis device is shown in Figure 3.

CONCLUSIONS

The full power operation of the JRR-4 was resumed with LEU fuel in October 1998. We are planning some experiments to measure the neutron fluxes and physical doses for irradiation filed of the medical irradiation facility in the end of 1998. The measurement of neutron flux and the gamma-ray dose is performed with a free beam and the PMMA phantom and the beam performances compared with the calculation value are confirmed. We are also planning experiments to measure PGA device.

We will be installed ICP-AES to measure the boron concentration.

JRR-4 will be started medical irradiation to patient in the beginning of 1999.

38

CLINICAL PROTOCOL USING NEW NEUTRON IRRADIATION FACILITY (JRR4) AT JAERI

Akira Matsumura[1], Yasushi Shibata[1], Tetsuya Yamamoto[1], Shingo Takano[1], Yoshinori Hayakawa[2], Kei Nakai[1], Yoshinobu Nakagawa[3], Fumio Sakurai[4], Yoshiya Torii[4], Kenji Yokoo[4], Hiroaki Kumada[4], Toshiaki Kishi[4], Kazuyoshi Yamamoto[4], Yasuyuki Akine[2], and Tadao Nose[1]

[1]Department of Neurosurgery and [2]Proton Medical Center
University of Tsukuba, Ibaraki, Japan
[3]Department of Neurosurgery
National Kagawa Children's Hospital
Kagawa, Japan
[4]Japan Atomic Energy Research Institute at Tokai
Ibaraki, Japan

1. INTRODUCTION

After the shutdown of the former JRR2, which had been used for boron neutron capture therapy (BNCT) in clinical cases, in summer 1998 a new neutron irradiation facility (JRR4) was installed as a medical reactor for BNCT at the Japan Atomic Energy Institute (JAERI), Tokai, Ibaraki. This JRR4 facility is a conversion of the former high-fuel uranium-loaded nuclear reactor to low-fuel uranium-loaded reactor. This facility is designed mainly for BNCT. There are medical facilities and equipment including an operating room and fully installed anesthetic apparatus to allow safe and effective intraoperative BNCT. An outline of this facility is presented in this chapter.

2. FACILITY AND TRIAL DESIGN

2.1. Facility Design

There are three types of beam available at this facility, which allow for several therapeutic modifications. Thermal mode (II), which is a neutron spectrum similar to the

Mailing address and reprint requests: Akira Matsumura, M.D., Department of Neurosurgery, Institute of Clinical Medicine, University of Tsukuba, 1-1-1 Tennodai, Tsukuba, Ibaraki 305-8575, Japan, Tel: +81-298-53-3220, 3168, Fax: +81-298-53-3214, E-mail: matsumur@md.tsukuba.ac.jp.

Frontiers in Neutron Capture Therapy, edited by Hawthorne *et al.*
Kluwer Academic / Plenum Publishers, New York, 2001.

former JRR2 thermal port, can be used for intraoperative BNCT. The proposed thermal neutron flux is 0.85×10^9 n/cm^2/sec with a cadmium ratio of 43. Thermal mode (I) is a mixed beam of thermal (2.3×10^9 n/cm^2/sec) and epithermal neutrons (0.88×10^9 n/cm^2/sec) associated with a higher amount of epithermal beam contamination than with thermal mode (II). This mode has a cadmium ratio of 5.2 and a gamma ratio of 2.5×10^{-13} Gy cm^2/n. The epithermal mode is expected to provide epithermal neutron flux of 1.7×10^9 n/cm^2/sec. The physical background is described in detail elsewhere[1] (also in Kishi *et al.* in these proceedings).

2.2. Computer Dosimetry System

A new neutron computer dosimetry system, now under development at JAERI, will provide three-dimensional dosimetry distribution and dose-volume histogram. This system is designed to cope with open craniotomy BNCT in which skin, skull and air void could be taken into account for calculation of neutron distribution. In the future, patient positioning may be calculated with the system linked to the operating table. The details of this system are described elsewhere (Kumada *et al.* in these proceedings).

2.3. Medical Facility

The entire facility is constructed to facilitate BNCT, including open craniotomy and general anesthesia (Figure 1). The operating room is clean and ventilated, and sterile water is provided for hand-washing before an operation, which is the same standard as maintained in the usual operating room in the hospital. General anesthesia is provided in the operating room as well as in the irradiation room. In the irradiation room, anesthetic tubes are introduced through a small tunnel from the sidewall of the irradiation room. The patient is watched via video monitors, ECG, blood pressure, and pulse oxymeter with either a wireless or a cable monitoring system. The upper part of the operating table (the patient's bed) could be detached from the lower part of the table, which facilitates transporting the patient from the operating room to the irradiation room. A second lower part is installed in the irradiation room, and attaches to the mobile upper part. The operating table itself can be tilted in different directions, which makes positioning the patient easy. The head is fixed with a three-point pin, carbon headframe, and modification of the irradiation field is finally adjusted with variable-fixation joints.

2.4. Clinical Trial Design

Previously, we have proposed inclusion criteria[2] for patients, including:

Patients with glioma grade 3–4
Under 70 years of age
Good general condition (KPS > 70)
No prior chemotherapy or radiation
Tumor not deeper than 4 cm from the surface
No serious systemic disease

With the new facility, however, the inclusion criterion of tumor depth may be changed to include cases with deeper lesions, since the mixed and epithermal beams may deliver sufficient neutron dosages to deeper regions.

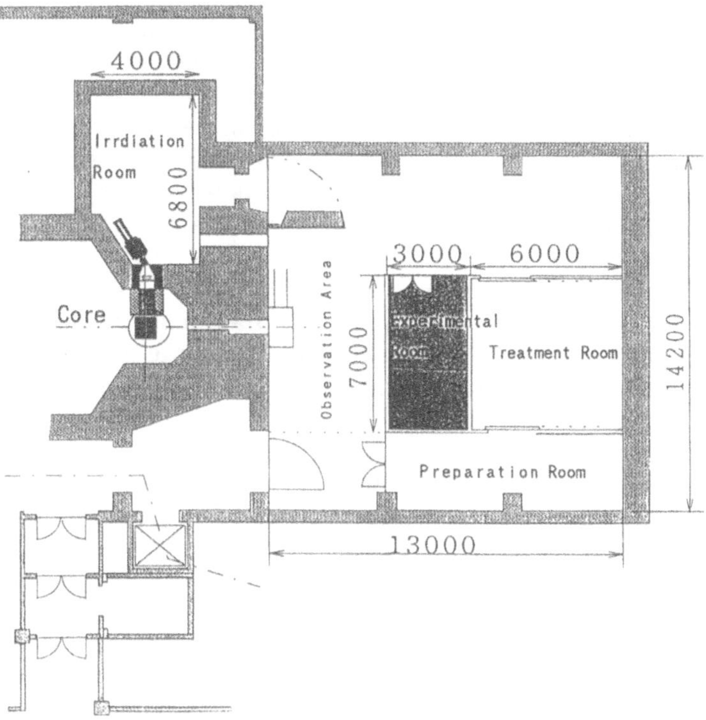

Figure 1. Outline of JRR4 facility design.

Our new clinical protocol will begin with intraoperative BNCT using the thermal mode (I) (mixed beam) in order to improve the neutron flux at deeper regions and reduce the relative flux on the surface of the brain. The neutron flux measurement using the gold wire will provide precise information about the neutron flux, and these data will be compared and calibrated with the neutron computation dosimetry system.

The ultimate goal of this facility will be non-operative BNCT using epithermal beam mode, which will be introduced after basic *in vitro* and *in vivo* animal study. Before doing non-operative BNCT, intraoperative epithermal BNCT may be considered based on the physical data on beam profiles.

3. DISCUSSION

The former JRR2 reactor at JAERI provided only thermal neutrons for BNCT, in which an open craniotomy was needed to get enough of a neutron dose into the deeper regions, and to prevent radiation injury to the skin. The thermal neutron has poor penetration into brain tissue, so that the clinical outcomes in deep-seated gliomas were worse than the outcomes for superficially located tumors. Even in the superficial tumors, a complete cure could be obtained in only a few cases.[3]

The new JRR4 irradiation facility could provide three modes of neutron, namely thermal, mixed thermal, and epithermal neutron modes. After obtaining the physical data in these three modes, a biological assessment should be performed to see the effect on the living cells and tissue. In the first step of restarting the clinical trial, we planned to start with intraoperative BNCT for three reasons. The first reason is that we will start with either the thermal or the mixed-beam mode since we have no basic data of epithermal neutrons. The second reason is that we wish to optimize the neutron computer dosimetry system by comparing the actual data using the gold wire measurement, which will be obtained during intraoperative BNCT. After we have proven the reliability of the computer dosimetry system, we could then step into non-operative BNCT. The final reason is to enhance the neutron flux in the deeper region by performing intraoperative BNCT. We have been doing intraoperative BNCT for years, so that there are no technical difficulties arising from intraoperative BNCT. Unless we switch to the fractionated BNCT, BNCT is usually performed as a single irradiation. To get more effective dose distribution, intraoperative BNCT may have an advantage over non-operative BNCT. The advantages and disadvantages of intra- and non-operative BNCT relative to various factors are summarized on Table 1.

With intraoperative BNCT, the skin is pulled away from the irradiated area, and the air void may improve neutron penetration without increasing the surface dose. Additionally, we have the chance to measure the actual boron concentration in the tumor tissue. The difficulty of dosimetry in BNCT in the earlier period was that there was no prompt-gamma system for immediate boron concentration analysis, which made dose planning for treatment difficult. Using prompt-gamma and an operative specimen, we may have more concrete data sets to analyze the irradiation effect, particularly in cases with BSH (borocaptate sodium). Unlike BPA, BSH is known to have large scatterings of tumor boron concentrations even in the same histological diagnosis.[4] There will be an overestimation of the irradiated dose in some cases if we uniformly use a tumor-to-blood ratio of 1. The overestimation of the calculated dose may lead the therapeutic evaluation to discouraging results, since in such cases the irradiation dose is estimated as higher than the actual dose given to the tumor. Thus, we think it is imperative to measure the boron concentration at the time of irradiation.

In conclusion, there are many ways to improve neutron capture therapy. Intraoperative BNCT still has a role in this context. The new JRR4 has a fully equipped operating room and anesthetic capabilities so that intraoperative BNCT can be performed safely and effectively. The clinical trial is directed to use the advantages provided by such medical equipment to improve therapeutic efficacy and prove basic data, such as data obtained from the computer dosimetry system, for future non-operative BNCT.

Table 1. Comparison of advantages and disadvantages of intraoperative and non-operative BNCT

	Intraoperative BNCT	Non-operative BNCT
Localization	direct	indirect
Dosimetry	direct (1dimensionD) indirect (2 to 3D)*	indirect (2 to 3D)*
Boron measurement	direct & indirect	indirect only
Fractionation	impossible	possible
Open craniotomy	necessary	not necessary
Neutron beam available	thermal, mixed, epithermal	epithermal

*With treatment planning system.

ACKNOWLEDGMENTS

Part of this research was supported by the Inter-University Program for the Joint Use of JAERI facilities. This study was also supported in part by the University of Tsukuba Research Projects and by a Research Grant-in-Aid for Scientific Research from the Japanese Ministry of Education, Science and Culture (07557091).

REFERENCES

1. K. Yokoo, T. Yamada, F. Sakurai, T. Nakajima, N. Ohhashi, and H. Izumo, A new medical irradiation facility at JRR-4, *in* "Advances in Neutron Capture Therapy", B. Larsson, J. Crawford, and R. Weinrich ed. Vol. I, Elsevier, Amsterdam, pp. 326–330, 1997.
2. A. Matsumura, Y. Shibata, T. Yamamoto, T. Yamada, H. Fujimori, K. Nakai, Y. Nakagawa, Y. Hayakawa, M. Isshiki, and T. Nose, The University of Tsukuba BNCT research group: first clinical experiences at JAERI, *in* "Advances in Neutron Capture Therapy", B. Larsson, J. Crawford, and R. Weinrich ed. Vol. I, Elsevier, Amsterdam, pp. 46–50, 1997.
3. H. Hatanaka, and Y. Nakagawa, Clinical results of long-surviving brain tumor patients who underwent boron neutron capture therapy. *Int J Radiat Oncol Biol Phys* 28:1061–1066, 1994.
4. H. Fankhauser, G. Stragliotto, and P. Zbinden, Borocaptate sodium (BSH) pharamacokinetics in glioma patients, *in* "Boron Neutron Capture Therapy. Towards Clinical Trials of Glioma Treatment", D. Gabel, and R. Moss ed., Plenum, New York and London, pp. 155–162, 1992.

39

UPGRADE OF THE EPITHERMAL NEUTRON BEAM USING ^{235}U FISSION PLATES AT BROOKHAVEN MEDICAL RESEARCH REACTOR (BMRR)

D. Rorer[1], F. Patti[1], R. Ma[2], A. D. Chanana[2], and H. B. Liu[3]

[1]Reactor Division,
[2]Medical Department
Brookhaven National Laboratory
Upton, NY 11973-5000
[3]McClellan Nuclear Radiation Center
McClellan AFB, CA 99562-2504

1. INTRODUCTION

Effective treatment of very deep intracranial tumors with BNCT requires a high intensity epithermal neutron beam and/or greater relative accumulation of boron in the tumor to deliver sufficient dose at depth. Since it is unlikely that a significantly improved boron delivery agent will become available for clinical use in the near future, the best approach for achieving an effective treatment of such tumors appears to be to improve the neutron beam. An upgraded epithermal beam facility, using ^{235}U fission plates, has been designed for the BMRR and construction is now beginning.

2. CONVERTER DESIGN

The design of the new beam incorporates a subcritical sheet-like assembly of fissionable material (~20% enriched ^{235}U) located external to the reactor, to use those neutrons which already have been thermalized (and which therefore are a liability in terms of epithermal beam purity) to produce fission spectrum neutrons.[1] This "rejuvenated" beam is then input to the epithermal filter/moderator, resulting in significant gains in epithermal neutron flux while minimizing beam contamination from fast and thermal neutrons and the capture gamma rays resulting from those neutrons.

Frontiers in Neutron Capture Therapy, edited by Hawthorne *et al.*
Kluwer Academic / Plenum Publishers, New York, 2001.

The converter and the filter/moderator are located in a semicylindrical shutter which moves up and down in the neutron beam. When the shutter is down, the converter plates are positioned out of the beam, below the level of the treatment room floor. In this position, high density concrete and steel block neutrons and gamma radiation from the reactor, and the converter plates are shielded by the concrete floor of the treatment room. With the shutter raised, the converter elements and the filter/moderator intercept and convert the neutrons from the reactor into an epithermal beam. The arrangement of the shutter, converter plates, and filter/moderator with respect to the core of the reactor is depicted in Figure 1. An isometric view of the shutter, converter, and filter/moderator is presented in Figure 2.

Fission converter plates were obtained as surplus items from the decommissioned JANUS reactor. These plates consisted of 2.5 mm metal foils of uranium which had been cross-rolled to the desired thickness and then clad in a 0.25 mm stainless steel jacket which was evacuated and back-filled with helium at approximately 0.3 atm pressure. Because of the uncertainty of the integrity of the original cladding, it was decided to encapsulate

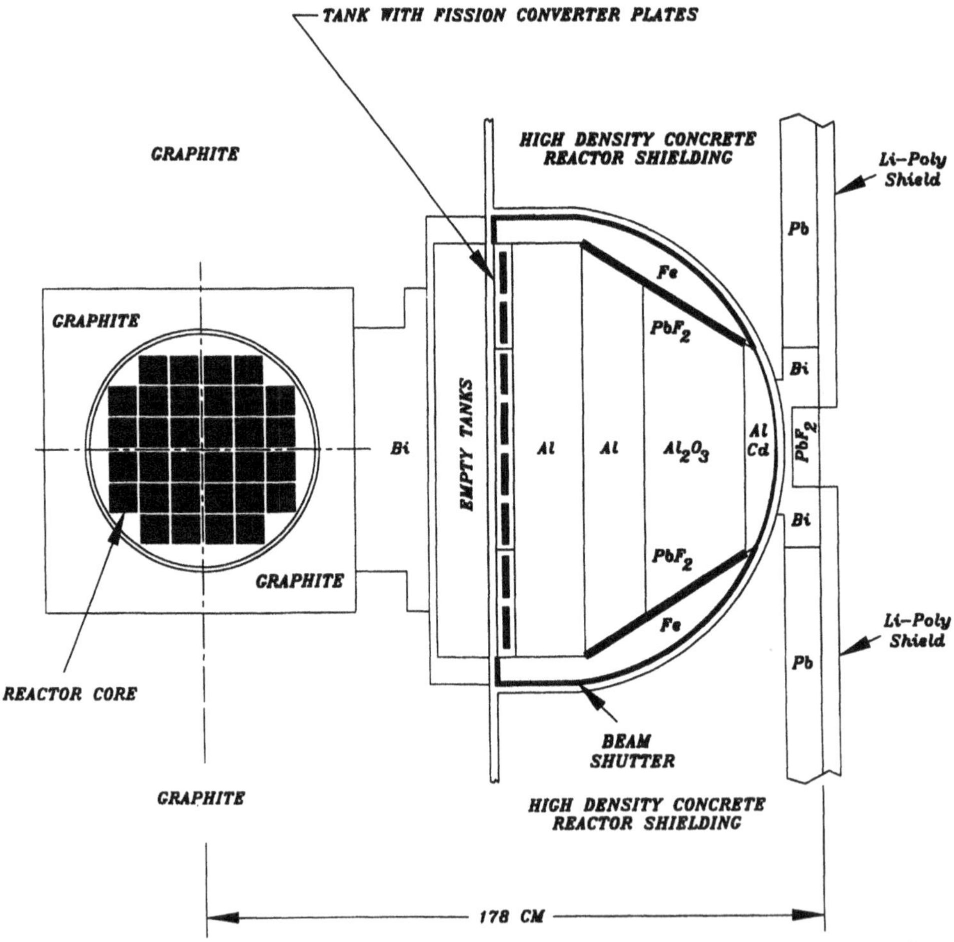

Figure 1. Plan view of the BMRR core, epithermal beam shutter, fission converter assembly, and filter/moderator.

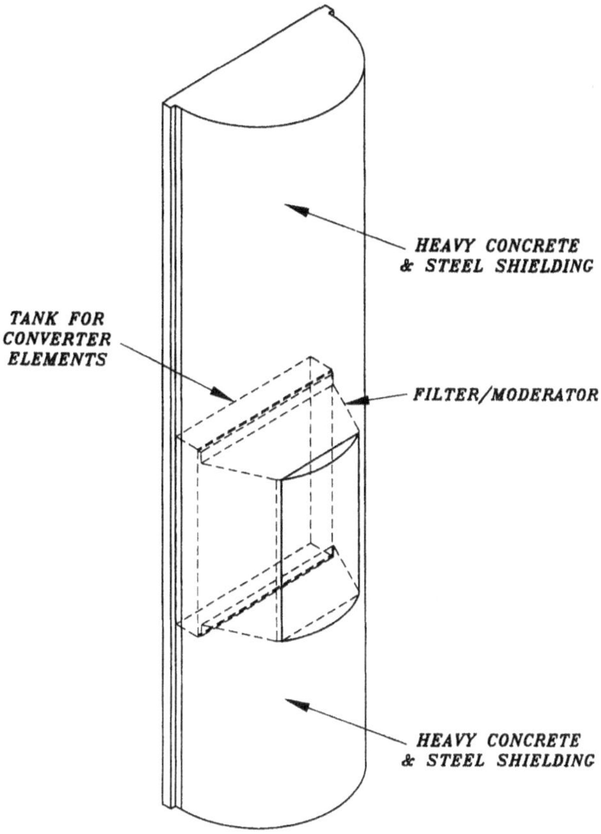

Figure 2. Isometric view of the shutter, converter tank, and filter/moderator.

these plates in helium-filled secondary containment enclosures. The enclosures, fabricated from 5,000 series aluminum alloy, each contain two plates to form a fission "element". The element enclosures are designed with sufficient clearance around the plates to allow for plate growth under irradiation, but the clearance is kept small enough to assure adequate heat transfer from the fission plates to the outside walls. The enclosures are sealed with full-penetration electron-beam welds and are back-filled with helium to ~0.5 atm pressure.

Four converter elements are located in a thin, flat vertical tank containing circulating light water, which removes the heat generated by fissioning of the uranium in the converter plates. The four elements are placed side by side to form a planar source of fission neutrons 99 cm high by 96 cm wide. The 12.7 mm thick walls of the tank are designed to withstand the hydrostatic head that will occur when the tank is filled to overflowing. The distance between the walls is maintained at a clear distance of 38.1 mm by three rows of spacer rails in the tank. The spacers also serve to guide and hold the converter elements in place, maintaining a small gap between elements. This tank provides a thin "sheet" of flowing water, 4.8 mm on the front and the back sides of the converter elements, to provide adequate heat transfer at reasonable flow rates (3.5 l/s), while limiting the thermalization of the epithermal neutrons by light water moderation.

A filter/moderator is located downstream of the converter assembly, and uses an arrangement of aluminum plates and 11 layers of 15 × 15 × 2.5 cm aluminum oxide tiles, similar to the filter/moderator currently in use at the BMRR, to slow the fast neutrons to epithermal energies and absorb neutrons of other, unwanted energies. These plates and tiles are contained in a welded aluminum container shaped in the form of a truncated pyramid with a rectangular base. This entire assembly is positioned on its side, with the base facing the converter plates, and is bolted in place immediately downstream of the converter tank. This filter/moderator assembly can be replaced in the future if it is desired to use a different material,[2] such as AlF_3.

3. FISSION CONVERTER COOLING SYSTEM

A small cooling system has been designed to remove 80 KW of heat generated by the fission plates when they are exposed to the thermal flux of 7.5×10^{11} n/cm^2-s of the BMRR at 3 MW maximum power. The cooling system consists of a U-tube heat exchanger, a centrifugal pump, and a partial flow cleanup system. Light water is forced to flow upward in the tank, past each of the four converter elements, into the upper portion of the tank, which functions as a discharge plenum and surge volume. There, water from each of the four element channels mixes together and overflows through two outlet pipes leading to the centrifugal pump. This pump then circulates the water through the heat exchanger, where it is cooled by the same secondary coolant source that cools the reactor itself. From there it returns to the fission converter tank through four inlet lines, one located under each converter element. Because the tank travels up and down with the shutter, all lines connecting the tank to the rest of the cooling system are flexible stainless steel bellows surrounded by braided stainless steel wire. Primary water purity is maintained by a cartridge-type filter and mixed bed demineralizer arranged in a bypass loop.

Instrumentation is provided to monitor primary water flow rate, water temperature, and water level in the Fission Converter Cooling System. In addition to providing visual indication of these parameters, the instrumentation also provides logging functions and alarms, and initiates automatic actions to protect the Fission Converter.

Flow measurements are used to close the shutter automatically if flow falls below a preset limit. The shutter close signal is derived from two-out-of-three logic using the outputs of one flow sensor in each of the two discharge pipes, and the sum of four flow sensors located in each of the four inlets to the tank. If the shutter does not close within a preset time period after receiving the automatic close signal, a scram signal is sent to the reactor.

Thermocouples at the inlet and outlet of the heat exchanger provide signals to a one-out-of-two logic circuit which will also signal the shutter to close if the temperature exceeds a preset value. Again, if the shutter fails to close within a preset time, a scram of the reactor is initiated. Finally, two level sensors in the converter tank will trip the primary pump in the Fission Converter Cooling System if the water level in the tank drops below a preset value.

4. TRANSFER OF CONVERTER ELEMENTS

Provisions must be made for safe removal of activated fission converter elements after they have been irradiated. Converter elements must be taken out of the converter

Table 1. Comparison of beam parameters before and after the Fission Converter upgrade

Epithermal neutron beam facility	Epithermal neutron flux n/cm²-s	Fast neutron dose per epithermal neutron cGy-cm²/n	Gamma dose per epithermal neutron cGy-cm²/n
Existing	$2.7 \pm 0.14 \times 10^9$	$4.3 \pm 0.4 \times 10^{-11}$	$1.0 \pm 0.1 \times 10^{-11}$
Upgraded	$19.0 \pm 1.0 \times 10^9$	$2.3 \pm 0.2 \times 10^{-11}$	$0.6 \pm 0.06 \times 10^{-11}$

tank if the fission converter assembly is to be removed from the reactor for any reason, such as shutter maintenance, replacement of fission plates, or exchange of shutters between the epithermal and thermal beam facilities. Access to the elements is provided by removing the upper section of the shutter and the top cover of the converter tank. A hand-operated winch and a remotely actuated coupling device are used to pull an individual 33 kg converter element out of the converter tank into a lead shielded transfer cask. Locating pins are used to assure precise alignment of the cask above the selected converter element. The transfer cask is used for transporting the individual converter plates from the reactor to a larger, heavier storage cask which can accomodate six irradiated converter elements in a critically safe configuration.

5. EXPECTED PERFORMANCE AND CONCLUSIONS

The beam design was modeled using the MCNP code,[3] verified by benchmark measurements of the existing thermal neutron flux near the planned region of the shutter. The comparison between computed and measured flux values for the existing beam (which has only a filter/moderator, but no fission converter plates) shows agreement within 10%.

Table 1 shows a comparison of beam parameters at the location of the patient for existing and upgraded designs. The epithermal neutron flux is expected to be increased by a factor of 7, while contamination from fast neutrons and gamma rays should be reduced by at least 40%.

It is concluded that the use of a fission converter placed outside the reactor, close to the patient, is a highly effective means for significantly increasing the epithermal flux available for BNCT, simultaneously reducing the relative contribution to overall dose from the unwanted background of fast neutrons and gamma rays, while eliminating the need for higher reactor operating power and increased reactor fuel consumption.

ACKNOWLEDGMENT

The authors would like to acknowledge the important contributions to the design by Eugene Hu, Tom Joos, Ray Karol, Robert Priest, and Paul Tichler of the BNL Reactor Division. Discussions with Prof. Otto Harling of MIT were also very helpful. This work was supported by the U.S. Department of Energy, Contract # DE-AC02-98CH10886.

REFERENCES

1. H. Reif, R. Van Heusden, and G. Perlini, Generating epithermal neutron beams for BNCT at TRIGA reactors, *in*: "Advances in Neutron Capture Therapy," Plenum Press, New York, pp. 85–88, 1993.
2. I. Auterinen, and P. Hiismaki, Epithermal neutron beam design for a TRIGA II reactor, *in*: "Advances in Neutron Capture Therapy," Plenum Press, New York, pp. 81–84, 1993.
3. Breismeister, J.F., "MCNP4B2- A Monte Carlo N-particle Transport Code System", CCC-660, Los Alamos National Laboratory, Los Alamos, New Mexico, 1997.

TRIGA® FUEL BASED CONVERTER ASSEMBLY DESIGN FOR A DUAL-MODE NEUTRON BEAM SYSTEM AT THE MCCLELLAN NUCLEAR RADIATION CENTER

H. B. Liu[1,2], J. Razvi[3], R. Rucker[3], R. Cerbone[3], M. Merrill[3], W. Whittemore[3], D. Newell[2], S. Autry[1], W. Richards[2], and J. Boggan[1]

[1]Department of Neurological Surgery
University of California-Davis
Sacramento, CA 95817
[2]McClellan Nuclear Radiation Center
McClellan AFB, CA 95652
[3]Sorrento Electronics (General Atomics)
San Diego, CA 92186

INTRODUCTION

The McClellan Nuclear Radiation Center (MNRC) is located 8 miles from downtown Sacramento, and 10 miles from the University of California Davis Medical Center. The MNRC's reactor is a TRIGA reactor design with light water as the primary coolant and moderator and graphite as the reflector. It began operation in January 1990 and was upgraded (June, 1997) to operate powers up to 2 MW, making it the youngest and the highest powered TRIGA reactor in the United States.

The reactor was originally designed to provide highly collimated thermal neutron radiographic beams for non-destructive inspection of aircraft parts, which is facilitated by four beam tubes having a tangential orientation to the reactor core. Because of the 20 to 30 degrees upward vertical orientation of the beam tubes, it is possible to provide a horizontal beam with full exposure to the reactor core. During the upgrade in 1997, a portion of the reactor bulk shield was cut and now filled with removable shielding blocks, allowing direct and wide access to the reactor core. The space occupied by light water

® TRIGA is a registered trademark of GA.

Frontiers in Neutron Capture Therapy, edited by Hawthorne *et al.*
Kluwer Academic / Plenum Publishers, New York, 2001.

between beam tubes within the reactor vessel allows a converter assembly and filtration/moderation materials to be inserted, while in the cut cavity, designs to accommodate additional filtration/moderation materials were made. The ultimate objective is to provide a dual-mode neutron beam system capable of delivering both epithermal and thermal neutron beams for use in neutron capture therapy (NCT) research and clinical applications.

METHOD

Monte Carlo code MCNP, version 4A, with continuous cross section data ENDF/B-V, was used for the design computation process. Benchmark experiments were carefully accomplished in the reactor core and outside the graphite reflector where a uranium converter assembly was designed. Thermal neutron flux comparisons between experimental measurements (using diluted bare and Cd/^{6}Li-polymer covered Au-Al foils and wires, Rh-SPNDs, and Hf-SPNDs) and MCNP computations were made for the existing configuration within the reactor vessel. Based on consistent comparisons to assure the accuracy of simulation, a uranium converter assembly design using existing TRIGA fuel elements was carefully examined. A design goal of reaching a K_{eff} factor equal or close to 0.90 was set to maximize the source of fission neutrons for subsequent filtration and moderation to provide both epithermal and thermal neutron beams for NCT irradiations in the same facility and also maintain the operational safety.

RESULTS

The design process is summarized in the following. (See Figures 1 and 2 for specified regions of interest regarding the designed facility).

A. Verification of Existing Core Simulation

The calculations started with criticality analyses (K-code computations) of the existing core and the surrounding structures within the reactor vessel. Both heterogeneous and homogeneous core configurations were simulated and calculations indicate an insignificant difference of the thermal neutron flux distributions in the reactor core center and outside the graphite reflector. Therefore, the homogeneous core configuration was used for all subsequent evaluations to simplify and speed up the calculations. By adjusting the amount of control rod material B_4C, the base case represents a case which the K_{eff} factor equals 1.0014 ± 0.0003 for a run of 2,000 cycles with 3,000 particles per cycle. The calculated thermal neutron flux in the reactor core center was 4.8×10^{13} n/cm^2.sec at 2 MW operating power, 7% higher than the measured value 4.5×10^{13} n/cm^2.sec. Outside the graphite reflector where the uranium converter assembly is planned, extensive measurements with SPNDs (self-powered neutron detectors) were made to assess the thermal neutron flux distributions covering a fan area 50 degree spread and 8″ off the core center line. The average thermal neutron flux was measured to be 3.9×10^{11} n/cm^2.sec at 2 MW operating power (at a distance 3.25″ away from the graphite reflector), which is the same as the calculated value. The peak flux was measured as 4.9×10^{11} n/cm^2.sec compared to 4.6×10^{11} n/cm^2.sec by calculation.

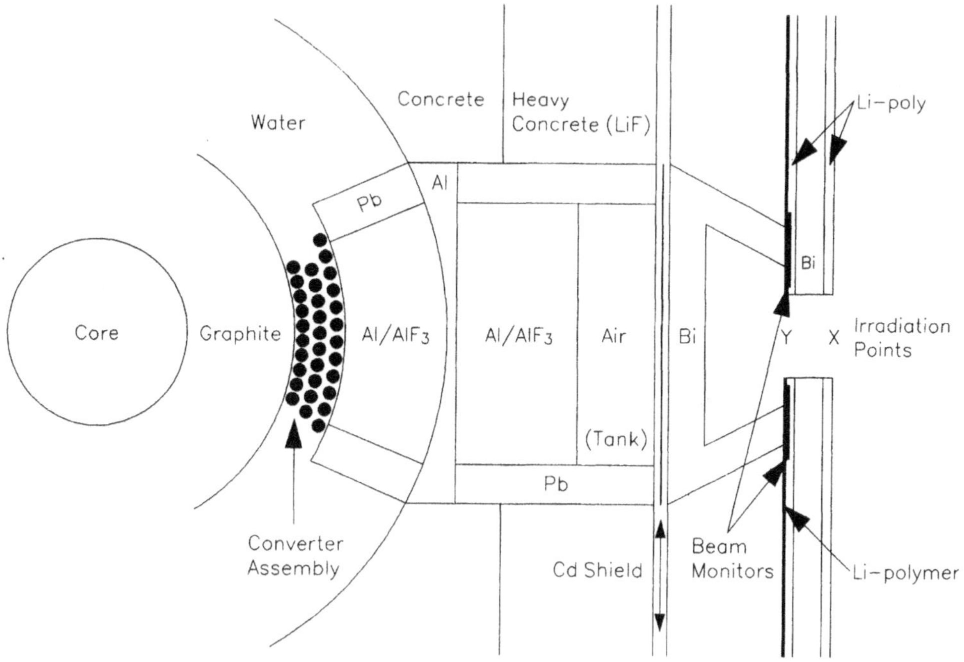

Figure 1. Horizontal cross sectional view of epithermal neutron beam mode.

B. Evaluation of Core Reactivity Impact from Assembly Support and In-Tank Insert

The second K-code evaluation was to include an aluminum assembly support and a stationary Al/AlF_3 in-tank insert inside the reactor vessel. The Al/AlF_3 (30/70 wt%) was selected to be the filtration/moderation material for extracting epithermal neutrons. During the design process, three rows of TRIGA fuel elements were determined to assure the K_{eff} factor will approach the design goal 0.90. In this evaluation, the space for the TRIGA fuel elements is occupied by light water to simulate the converter fuel elements are removed. The K_{eff} factor was found to be 1.0016 ± 0.0003, merely 0.02% increase (within the statistical errors of calculations). The average thermal neutron fluxes over the 15″ length (active length of TRIGA fuel element) in light water at the fuel element locations were calculated as 17×10^{11}, 5.3×10^{11}, and 1.4×10^{11} n/cm^2.sec at 2 MW operating power for the first, second, and third row, respectively.

C. Evaluation of Core Reactivity Coupling from Converter Assembly

The third K-code evaluation was made with the three rows of TRIGA fuel elements in place. The fuel elements are so-called the TRIGA 30/20 type; 30 wt% of uranium in U-ZrH mixture, uranium enriched to <20% U-235, and added with 0.9 wt% Er. In this configuration, the K_{eff} factor became 1.0025 ± 0.0003, or 0.11% higher than that of the base case. The average thermal neutron fluxes became 8.3×10^{11}, 6.8×10^{11}, and 5.5×10^{11} n/cm^2.sec at the first, second, and third rows, respectively. The dramatic flux changes

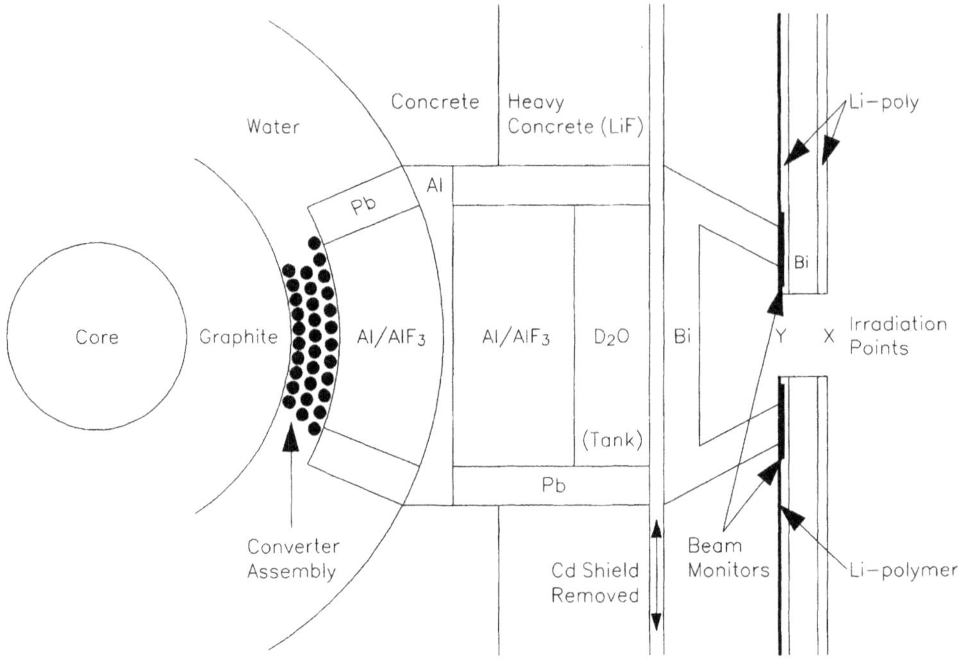

Figure 2. Horizontal cross sectional view of thermal neutron beam mode.

(–51%, +28%, +293%) compared to the second K-code evaluation are results due to a combination of neutron flux suppression (uranium), neutron conversion (thermal to fission neutrons), neutron multiplication (fission to thermal neutrons), and also the replacement of water with U-ZrH. The individual heating of each fuel element ranges from 2.1 to 4.8 kW per element, and the total heating is about 110 kW of 32 fuel elements. The average heating is 3.4 kW per element.

D. Evaluation of Multiplication Factor of Subcritical Assembly

Together with the third K-code evaluation, to better understand the designed subcritical converter assembly, the fourth K-code evaluation was run for the subcritical assembly alone, i.e. the core was defueled with no uranium. After a series of testing, the K_{eff} factor equals 0.89.

E. Dual-Mode Neutron Beam System Design (Neutron Contribution from Subcritical Converter Assembly)

Once the converter assembly design is completed, sufficient information is available for subsequent evaluations of the neutron beams. Based on information from the previous K-code evaluations, a fixed source, contrast to the K-code calculations, was made to try out designs for both epithermal and thermal neutron beams; maximizing the intensity and quality for both neutron beams. The final configurations of the design are illustrated in Figures 1 and 2, horizontal cross sectional views for epithermal and thermal neutron beam extractions, respectively.

Table 1. Key data for the designed neutron beam system

Converter assembly	1. 32 TRIGA 30/20 fuel elements with 0.9 wt% Er.
	2. ≈0.5 cm water gap between fuel elements.
	3. $K_{eff} = 0.89$.
	4. Total heating = ≈110 kW (2.1 to 4.8 kW per fuel element).
Downstream beam center line	5. 6.4 cm-thick Al. (structural support)
	6. 59.6 cm-thick Al/AlF_3 (30/70 wt%) composite.
	7. 18 cm-thick tank to be filled with air or D_2O.
	8. 0.1 cm-thick Cd.
	9. 12 cm-thick Bi.
	10. 50 cm-long beam path to X point; ≈220 cm away from core center or ≈150 cm away from assembly center.

F. Dual-Mode Neutron Beam System Design (Neutron Contribution from Reactor Core)

The extensive moderation of fission neutrons originated from the reactor core significantly limits their contribution to both neutron beams. Evaluations were made to assess this neutron contribution and, as expected, the contribution is insignificant and only accounts for a few percentages to the extracted neutron beams.

Key data for the designed neutron beam system are listed in Table 1.

DISCUSSION AND CONCLUSION

As see in Table 2, two excellent neutron beams would be available in the same facility for NCT research and clinical applications. Compared to the existing and well documented BMRR (Brookhaven Medical Research Reactor) beam (Ref. 1), the designed epithermal neutron beam would be high in intensity and low in contamination. Based on the current clinical trials at BNL (Brookhaven National Laboratory), a full-dose NCT irradiation could be shortened to <15 min using the designed epithermal neutron beam at the MNRC if a higher radiation dose rate to patients were proven to be safe. The quality of the beam would be excellent, not only decreasing the unnecessary whole body

Table 2. Comparisons of beam parameters between the neutron beams designed for the MNRC and the neutron beams operated at the BMRR

Epithermal neutron beam at X point	MNRC design	BMRR (Ref. 1)
$(J/\Phi)_{epi}$	0.74	0.80
ϕ_{epi} (n/cm^2.sec)	5.0×10^9	0.84×10^9
$K_{fast\ neutrons}/\phi_{epi}$ [cGy/(n/cm^2)]	2.0×10^{-11}	4.8×10^{-11}
$K_{gamma\ rays}/\phi_{epi}$ [cGy/(n/cm^2)]	0.6×10^{-11}	2.0×10^{-11}
Thermal neutron beam at Y point	MNRC Design	BMRR (Ref. 2)
$(J/\Phi)_{th}$	0.63	0.56
ϕ_{th} (n/cm^2.sec)	5.6×10^9	51×10^9
$K_{fast\ neutrons}/\phi_{th}$ [cGy/(n/cm^2)]	0.2×10^{-11}	2.8×10^{-11}
$K_{gamma\ rays}/\phi_{th}$ [cGy/(n/cm^2)]	0.7×10^{-11}	0.8×10^{-11}

radiation exposure to patients, but increasing the treatment dose to tumor under the same normal brain tolerance limit implemented at BNL.

As for the designed thermal neutron beam at the MNRC, though much lower in intensity when compared to the existing BMRR beam (Ref. 2), the designed intensity is sufficient to deliver a total thermal neutron fluence $1.0 \times 10^{13}\,n/cm^2$ for an implanted tumor 0.5 cm deep in a rat brain within 40 minutes. This will assure the rat therapeutic assay could be performed within a reasonable time for evaluating various chemical compounds attached with ^{10}B and/or ^{157}Gd, or other nuclides. The quality of the beam would be far superior to the existing thermal neutron beam at the BMRR, making it possible for multiple rat irradiations simultaneously (ex. four rats at a time).

ACKNOWLEDGMENT

This research was supported by the U.S. Department of Energy with University of California, Davis..

REFERENCES

1. H.B. Liu, D.D. Greenberg, J. Capala, J.A. Coderre, F.J. Wheeler, "An improved neutron collimator for brain tumor irradiations in clinical boron neutron capture therapy," Med. Phys. **23(12)**:2051–2060, 1996.
2. H.B. Liu, "Design of a small-animal thermal-neutron irradiation facility at the Brookhaven Medical Research Reactor," Nucl. Tech. **115**:311–319, 1996.

BNCT FILTER DESIGN STUDIES AT THE CEA-SACLAY ISIS RESEARCH REACTOR

F. Colomb, H. Carcreff, and C. Morin

CEA-SACLAY, DRN/DRE/SIREN
91191 Gif-sur Yvette Cedex
France

INTRODUCTION

In the frame of BNCT research, studies were performed at the French CEA center of Saclay, on epithermal neutron beams at ISIS research reactor. ISIS is a pool type reactor. Its nominal power reaches 700kW. Thermal and fast neutron fluxes are about 2×10^{12} n.cm^{-2}.s^{-1} and 2×10^{11} n.cm^{-2}.s^{-1} on the first external row around the core.

The aim of the CEA study was to do a theoretical study on materials chosen to get an epithermal neutron beam. A beam tube was specially designed to test different materials usable as shields. The device geometry was designed to limit thermal and fast neutron crossing. The first step was to validate a code scheme on the device without shields. On the second step, this validated code scheme was used to simulate shields inside the tube.

An experimental method was developed to determine epithermal neutron spectra. Such a method allows an approximation of the spectrum shape without using calculation codes. It can be used to analyze the response of different filter configurations. We used this method to validate transport calculation schemes by comparison between computed and measured spectra. This method is based on measurements of reaction rates provided by activation foils.

1. PRESENTATION OF THE DEVICE

The tube is constituted of two cylinders of aluminum covered inside with boroflex (boron polyurethane). The two cylinders are 50cm long. The first tube is 50cm in diameter and the second tube is smaller with a 20cm diameter. These tubes are closed by

Frontiers in Neutron Capture Therapy, edited by Hawthorne *et al.*
Kluwer Academic / Plenum Publishers, New York, 2001.

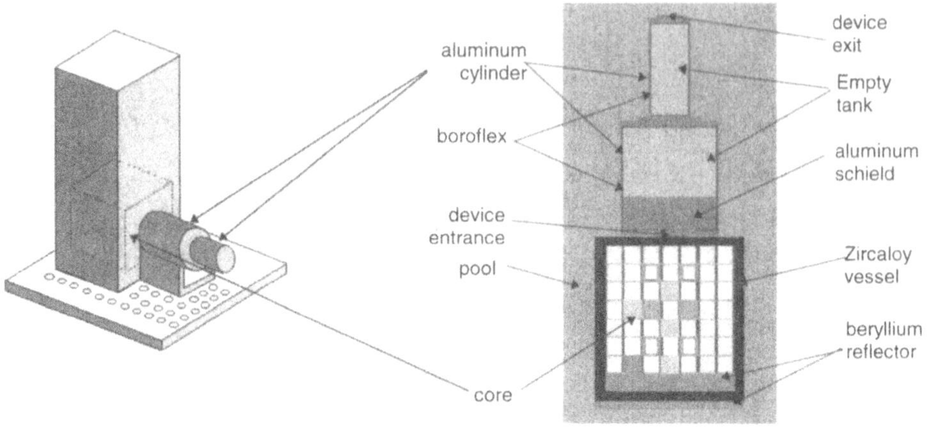

Figure 1. Reactor and device geometry.

shields of aluminum. Each shield is 2 cm thick. A 16 cm thick aluminum shield was fixed inside the first tube. This tube can be embedded in the external grid of the reactor. The second device was designed to put different shields inside and test them. The additionnal tube is smaller and farther from the core than the first tube and so is, less activated. It can be attached to the back of the first tube. To easily change shields inside, this second one is easily removable. So the idea was to keep the first device in water and emerge only the second one to change shields.

2. VALIDATION OF THE DEVICE SIMULATION

Our code scheme was validated at different steps of the device construction:

- on the first empty tube[1]
- on the first tube with aluminum shields
- on the device with the both tube.

To validate our code scheme, calculated spectra at the extremities of the device were compared to experimental ones. Spectra were calculated with the Monte Carlo code TRIPOLI, developed by the CEA.[3]

Experimental spectra were deduced from measurements of reaction rates provided by activation foils.[2] Flux levels were estimated in each energy range corresponding to each probe response using a matrix method. This method was validated in a location of the ISIS reactor where the neutron distribution is well known.

Figure 2 shows activation foils and energy groups used to get experimental spectra.

Bared gold was used for the thermal range, a set of resonance foils covered with cadmium for epithermal range, and threshold foils for the fast range.

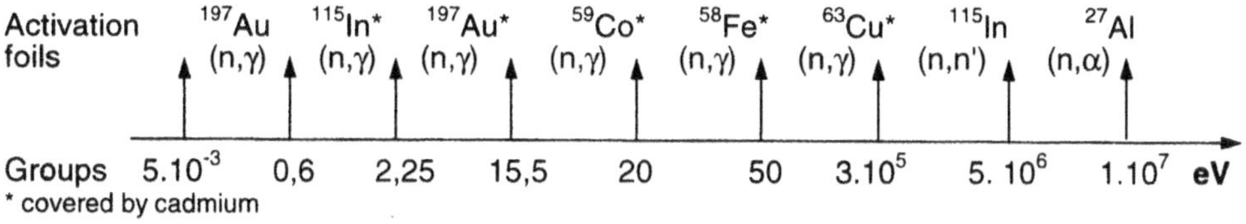

Figure 2. Activation foils and energy groups used to obtain experimental spectra.

Figure 3 presents the comparison between calculated and experimental spectra.

A good agreement between experimental and calculated spectra at the extremity of the device was achieved. Our code scheme reliabaly simulated the response of the device without the shield in the second tube. So, we used this validated code scheme to simulate, in a second step, various materials with different thickesses.

3. TEST OF DIFFERENT MATERIALS USABLE AS SHIELDS

A lot of materials were simulated in the second cylinder to design a neutron beam with an energy between in 1 eV and 10 keV. It appeared that titanium and iron cut fast neutrons but reduced epithermal neutrons as well. Iron presents a cross section window at 27 keV, but this energy is too high to be used for BNCT beam. Scandium presents a cross section window at 2 keV, but it has to be combined with other materials which cut neutrons with a energy higher than 10 keV. Sulfur and aluminum are the best materials to get an epithermal neutron beam. They cut fast neutrons and favour epithermal neutron crossing. The use of aluminum fluoride is very interesting too, it combining the properties of fluorine and aluminum. Aluminum presents a cross section window at 24 keV. Because of its resonance at this energy, fluoride cuts the neutrons with an energy in the vinicity of 24 keV.

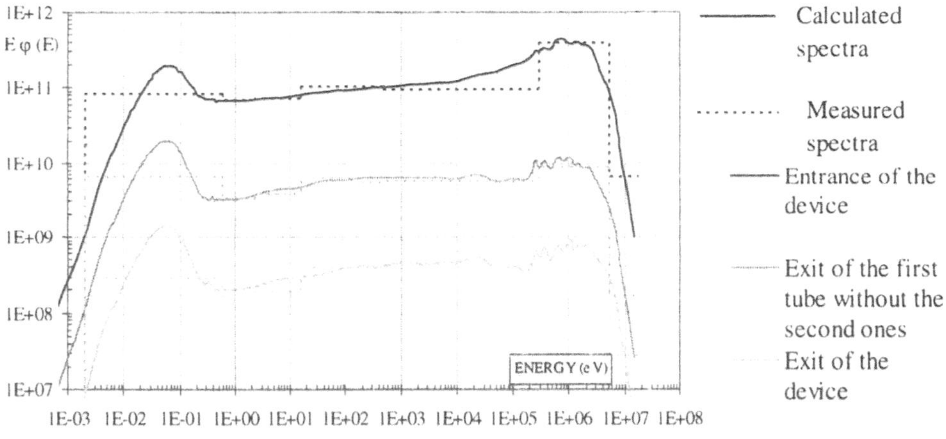

Figure 3. Comparison between calculated and experimental spectra.

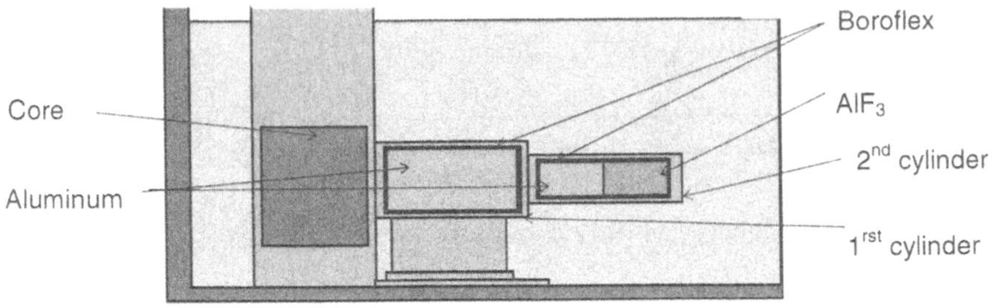

Figure 4. Final device.

The best spectra was obtained with a combination of aluminum and aluminum fluoride shields. Figure 4 presents this final device.

In this final configuration, the first tube is filled with aluminum. In the second tube, an aluminum fluoride shield (22 cm thick) follows an aluminum shield (20 cm thick).

Figure 5 shows the comparison of the entrance spectrum and the exit spectrum of this filter. These two spectra are normalized on the same total flux.

CONCLUSION

The aim of this study was to test different materials usable as shields to get an epithermal neutron beam. A code scheme was validated on a device where shields were tested afterwards. The best configuration was obtained with a combination of aluminum and aluminum fluoride.

In this study, we looked at the device response towards neutrons. In a near future we intend to optimize the filter response towards gamma radiation. The simulation will be carried out with a neutron-gamma Monte Carlo code.

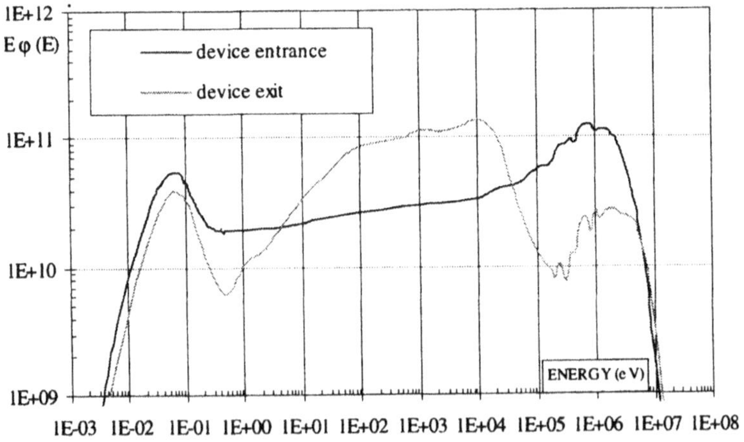

Figure 5. Comparison of spectra at the extremities of the filter.

REFERENCES

1. F. Colomb, G. Ermont "Study of an epithermal neutron beam for neutron capture therapy: Experimental validation of the Monte Carlo simulation", J. Chim. Phys. (1998) 95, 698–702.
2. F. Colomb, G. Ermont "Epithermal neutron spectra determination based on activation foils measurements", J. Chim. Phys. (1998) 95, 703–707.
3. T. Vergnaud "TRIPOLI-3: Code de Monte Carlo polycinetique", Rapport CEA-DMT 97/293.
4. J.C. Yanch, X-L. Zhou, R.E. Shefer, R.E. Klinkowstein, G.L. Brownell "Design of an accelerator-based epithermal neutron beam for boron neutron capture therapy" Plenum press, New York, 1992.

42

PREPARATION OF A BEAM TUBE OF THE PORTUGUESE RESEARCH REACTOR TO BE USED IN BNC AND OTHER ACTIVITIES

A. Ramalho[1], I. F. Gonçalves[1], I. C. Gonçalves[1], J. Salgado[1], M. Castro[2], and A. F. P. Fernandes[1]

[1]Instituto Tecnológico e Nuclear
Est. Nacional 10/2685 Sacavém, Portugal
[2]Centro de Estudos de Farmacologia Experimental e Clínica
Av. Egas Moniz/1600 Lisboa, Portugal

1. INTRODUCTION

Work is in progress aiming at the preparation of a beam tube to be used in various applications among which the ones in the area of BNC. In this context it is worth referring the use of a phantom to study the distribution of dose at the irradiation positions, the irradiation of biological samples, and boron determinations in biological samples using the PGNAA technique.

The MCNP code has been used to study several combinations of neutron filters to enhance the epithermal component of the neutron beam.

The shielding was also calculated with the MCNP code and is designed so that access, to the location where the beam emerges, can be gained under radiation controlled operation, with the reactor running at full power.

2. OBJECTIVES

The RPI is a 1 MW swimming pool type reactor similar to various others designed by AMF Atomics using, at present, 93% enriched uranium.

The reactor is provided with seven beam tubes, six radials (4 of 15.2 cm and 2 of 20.3 cm diameter) and one through tube (diameter of 10.2 cm) and with a thermal column with two access ports (one vertical and the other horizontal).

The reactor is Be reflected in three sides with the fourth one adjacent to the thermal column which is shielded from the reactor core by a 12.7 cm Pb plate and reaches the external surface of the biological shielding.

Frontiers in Neutron Capture Therapy, edited by Hawthorne *et al.*
Kluwer Academic / Plenum Publishers, New York, 2001.

As thermal neutrons have a low penetration in organic matter, epithermal neutrons are necessary to reach deeply seated malignancies to be treated by the BNCT technique.[1]

To pursue the work in this area at RPI, an epithermal beam is needed. With this purpose a beam tube is being prepared which will allow for the selection from two different beam openings and to use neutron beams with or without thermal component.

3. DESCRIPTION OF THE TUBE

The general layout of the beam tube arrangement is shown in the Figure 1, which represents schematically the components located inside the tube itself and the external shield, as well as its relative position to the reactor core.

Starting from the core side one finds: the neutron filter (Al); a 3 cm thick lead shield; an inner collimator ($\Phi = 5$ cm and length = 23 cm); a Cd shutter; the external collimator which includes the beam shutter and beam diameter selector ($\Phi = 2$ cm and $\Phi = 5$ cm, respectively).

The external shield is a hollowed parallelepiped shown in Figure 2, with an inner part of a hydrogenous material followed by a 1 mm Cd and a concrete enclosure.

The experimental cavity is (40 × 40 × 125)cm and can be accessed from the top. The design is such that the access to the experimental cavity is possible for short periods with the reactor running at full power, once the beam shutter is closed.

The system is provided with inhibitors and warnings to assure that it is operated safely.

Changing the filters is not complex and in order to increase the neutron flux it may be desirable to remove them completely; in this case the cavity cannot be accessed with the reactor running at full power.

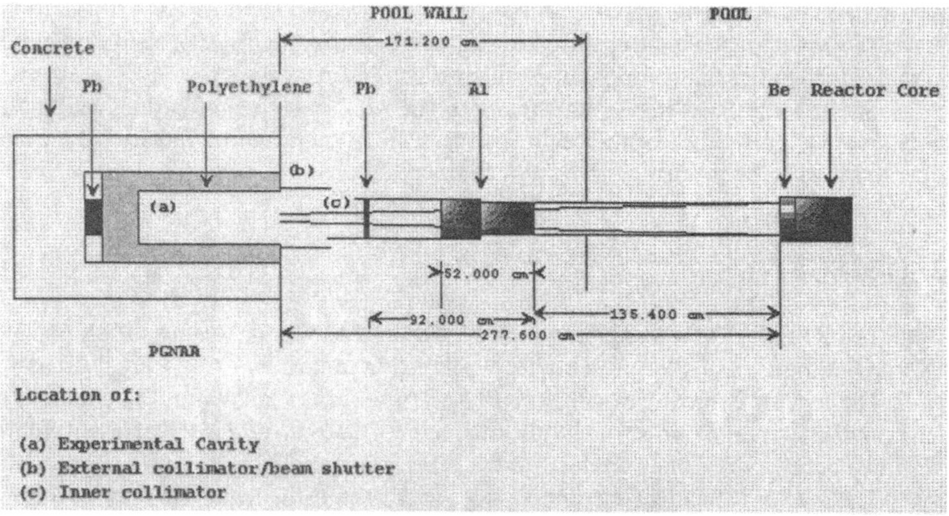

Figure 1. General view of the tube with filter and external protection.

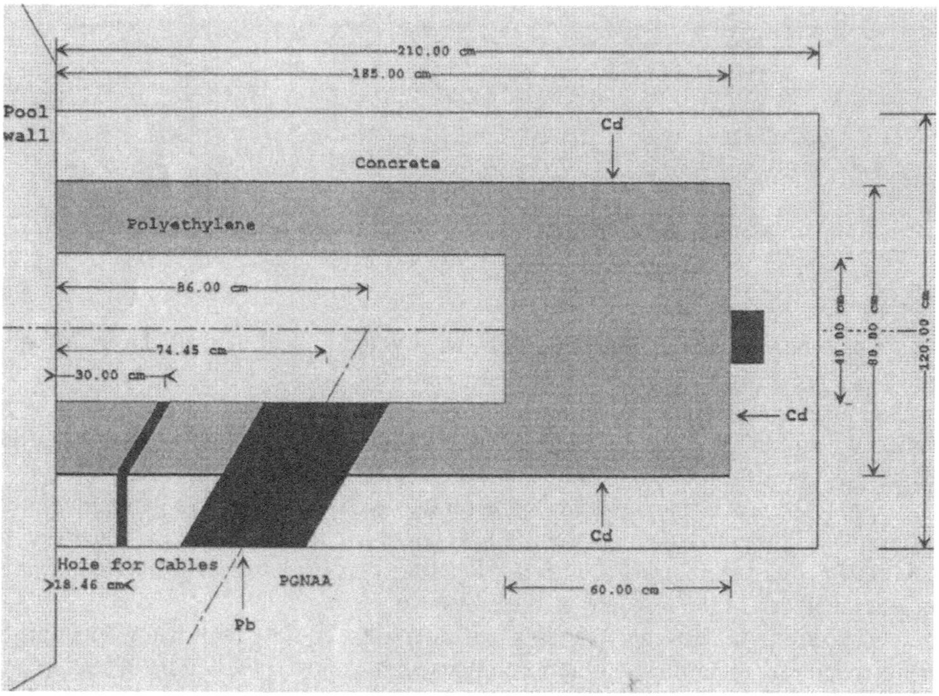

Figure 2. External protection of the beam tube.

4. CALCULATION METHODS

The studies of the filter arrangements to create an epithermal beam with low fast neutron contamination and the shielding arrangements, have been confirmed using the MCNP code[2]. The code was run in an Alfa Station, and the nuclear cross sections used are those from ENDF-VI. For each situation the typical running time to achieve a statistical error less than 5% is 48 hours.

The code capability for point detector calculation is used. This feature is a deterministic estimate of the flux at a certain point that takes into account the probability of scattering in the right direction, the attenuation by the materials and the effect of the distance. The contribution to the point detector is made at every source or collision event.

The input source to the program was derived from reactor core calculations using the CITATION code.[3] It consists of a plan source with the size of the beam entrance area and with uniform intensity. The results were obtained for the energy groups that have been considered by others,[4] i.e.: thermal $E_n < 0.625\,eV$; epithermal $0.625\,eV < E_n < 10\,keV$; fast $E_n > 10\,keV$. In order to speed up the calculations we did not considered the thermal neutrons of the source.

5. RESULTS

Several arrangements of filter materials were tested to find out the best combination in what concerns the epithermal flux intensity and the epithermal/fast flux ratio.

Table 1. Neutron fluxes, expressed in $n.cm^{-2}.s^{-1}$, for several filter arrangements

Case	Filter arrangement	ϕ_{th}	ϕ_{epi}	ϕ_{fast}	ϕ_{epi}/ϕ_{fast}
1	52 cm Al	1.68 E + 07	5.84 E + 06	3.37 E + 06	1.73 E + 00
2	30 cm Al + 22 cm Al_2O_3	1.08 E + 06	4.37 E + 05	7.67 E + 05	5.7 E − 01
3	30 cm Al + 22 cm AlF_3	2.02 E + 05	1.14 E + 05	2.73 E + 05	4.18 E − 01
4	52 cm Al_2O_3	3.56 E + 04	1.41 E + 04	5.71 E + 05	2.47 E − 02
5	52 cm AlF_3	1.14 E + 05	6.62 E + 04	1.36 E + 05	4.87 E − 01

Combinations of Al, Al_2O_3 and AlF_3 were tested for different thickness, of which the results presented in Table 1 are typical.

The calculations performed so far indicate that a 52 cm Al filter would give the highest ϕ_{epi} and the best ϕ_{epi}/ϕ_{fast} ratio. Further work is under way, particularly regarding the use of fluoride materials which are being selected by other teams.[5]

It is to be noted that, as indicated before, in order to reduce the gamma intensity, a 3 cm thick Pb filter is also included. Under these circumstances the expected thermal ($E < 0.625\,eV$) and epithermal ($0.625\,eV < E < 10\,keV$) neutron fluxes at the beam exit are the ones included in the first line of the table.

Concerning the external shielding it is interesting to note that, being adequate for the intended use, when the maximum total dose rate at any place will be of the order of 1 mrem/h, it will need, as the calculations indicate, some reinforcement if the filters are removed. In this case the points of maximum dose rate, which will be due to the fast neutrons, will be located in the plane perpendicular to the beam, passing through the centre of the experiment.

It should be noted that the spatial distribution of the dose rate is somewhat asymmetric, with a rate of decrease slower in the front side of the beam, than in the backside, i.e. the reactor side. The experiment was simulated with spheres of lead and polyethylene with various diameters.

In the context of BNC, this beam will be used in experiences with biological materials and to determine boron concentrations in biological samples, by Prompt Gamma Neutron Activation Analysis.

6. FUTURE WORK

In the context of BNC, the beam will be used in dosimetry work in phantoms, in testing of beam monitoring equipment and in PGNAA.

In a totally different area, it is planned to continue work on the determination of H in iron, based in neutron dispersion and active neutron detection method.

REFERENCES

1. Moss, R.L., Aizawa, O., Beynon, D., Brugger, R., Constantine, G., Harling, O., Liu, H.B., Watkins, P. "The requirements and development of neutron beams for neutron capture therapy of brain cancer" Journal of Neuro-Oncology 33, 24–40, 1997.
2. Briesmeister, J.F., ed., "MCNP—A General Monte Carlo n-Particle Transport Code", Version 4A, LA-12625-M, 1993.

3. Fowler, T.B., Vondy, D.R., Cunninggham, G.W., "Nuclear Reactor Core Analysis Code: CITATION", ORNL-TM-2496, Rev. 2, 1971.
4. Moss, R.L., "Status of the Boron Neutron Capture Therapy Project at the HFR, JRC Petten", Proceedings of the CLINCT BNCT Workshop, Helsinki, 1993, TKK-A718, 1994.
5. Savolainen, S., Auterinen, I., Kallio, M., *et al.*, "The Finnish boron neutron capture therapy program—an overview on scientific projects", Advances in Neutron Capture Therapy vol. I Medicine and Physics, 342–347, Borje Larsson, John Crawford and Regin Wienreich, ed. 1997.

NCT FACILITY DEVELOPMENT AND BEAM CHARACTERISATION AT THE RA-6 REACTOR

H. R. Blaumann, O. Calzetta Larrieu, J. M. Longhino, and A. F. Albornoz

Centro Atómico Bariloche
Comisión Nacional de Energía Atómica
Argentina

1. INTRODUCTION

The RA-6 reactor located at Bariloche Atomic Center, is a pool type one with 500 kW of nominal power and U 90% enriched fuel owned and operated by C.N.E.A.

The epithermal beam facility was designed [1] and built replacing the old thermal column (internal and external). Figure 1 shows a plant view of the complete facility including the material composition of the neutron moderator, the port and the external shield.

2. MONTE CARLO SIMULATION

The main features of the calculation process were:

- Coupled neutron-gamma calculation with MCNP4B [2] and cross sections based on ENDFB6 data library.
- Point detector tallies at the beam center and at several positions near the external shielding (for neutrons and photons).
- A detailed neutron and photon source in the core was obtained through a KCODE calculation.
- Neutron spectrum in 29 energy groups was calculated at the beam center (2 thermal / 15 epithermal / 12 fast).
- Photon doses were calculated by using the ICRP-21 flux-to-dose rate convertion factors. The calculated photon doses rate at the beam center agree within 20% with the measured ones.

Frontiers in Neutron Capture Therapy, edited by Hawthorne *et al.*
Kluwer Academic / Plenum Publishers, New York, 2001.

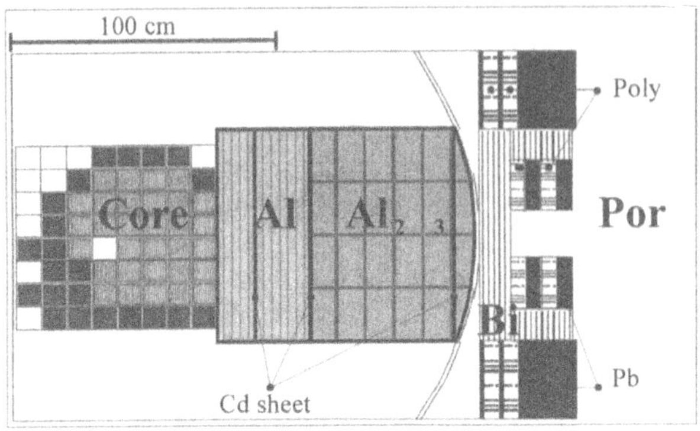

Figure 1. Plant view of the epithermal facility.

3. FREE BEAM MEASUREMENTS AND SPECTRUM ADJUSTMENT

3.1. Neutron and Gamma Characterization

Multiple activation detectors (Diluted, 0.1 mm, bare and Cd covered foils of Mn, Au, Cu, In and pure, 0.127 mm, Cd covered foils of Sc, Ag, In), with different energy response, were irradiated at the beam center, and the induced activities were measured by gamma spectroscopy for neutron energy characterization.

The gamma dose rate was measured at the beam center, with TLD's 700 and paired ionization chambers.

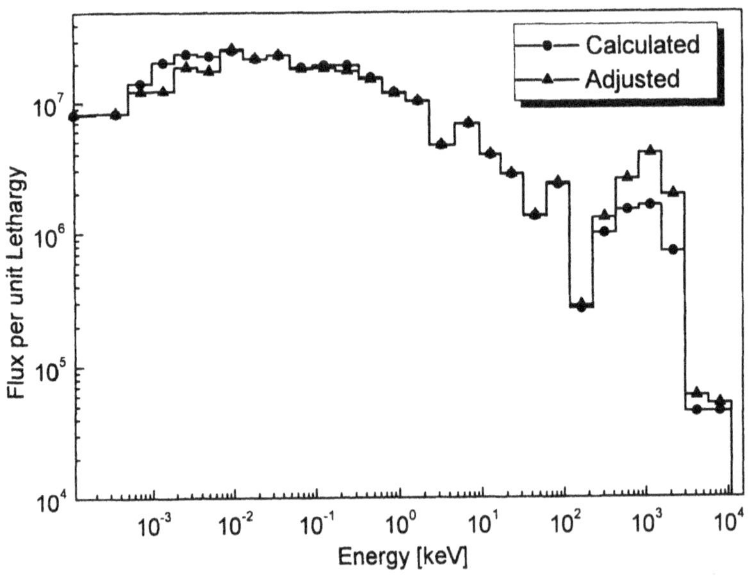

Figure 2. Calculated and adjusted neutron flux.

Table 1. Free beam parameters

Epithermal flux (0.5 eV–10 keV)	(0.32 ± 9%) 10^9 n/cm^2 sec
Fast neutron dose (>10 keV)/n_{epi}	(11.3 ± 16%) 10^{-13} Gy/n cm^2
photon dose / n_{epi}	(7.5 ± 13%) 10^{-13} Gy/n cm^2
thermal flux (<0.5 eV)/n_{epi}	0.07 ± 22%

3.2. Spectrum Adjustment

Figure 2 shows the calculated and adjusted neutron flux at the beam centre.

Measured activities and calculated spectrum were adjusted with the NMF-90 package [3]. Some of the results and remarks were:

- The IRDF-90/NMG-G was upgraded, including self-shielded data for pure epithermal reactions.
- Groupwise input uncertainties were evaluated conservatively, regarding their statistical errors, and the group to group correlations were assumed exponential on a lethargy exponential scale.
- Reaction rate uncertainties were lower than 10% and actual correlations were considered and evaluated.
- A reasonable value (1.3) for χ^2/N resulted from the adjustment.
- Large (~2) adjustment factors resulted in few groups of the fast energy region.
- Sensitivity analysis was perfomed but integrated flux values remained between their estimated errors.

3.3. Free Beam Parameters

The free beam parameters evaluated from the adjusted spectrum and group calculated kerma factors are showed in the Table 1.

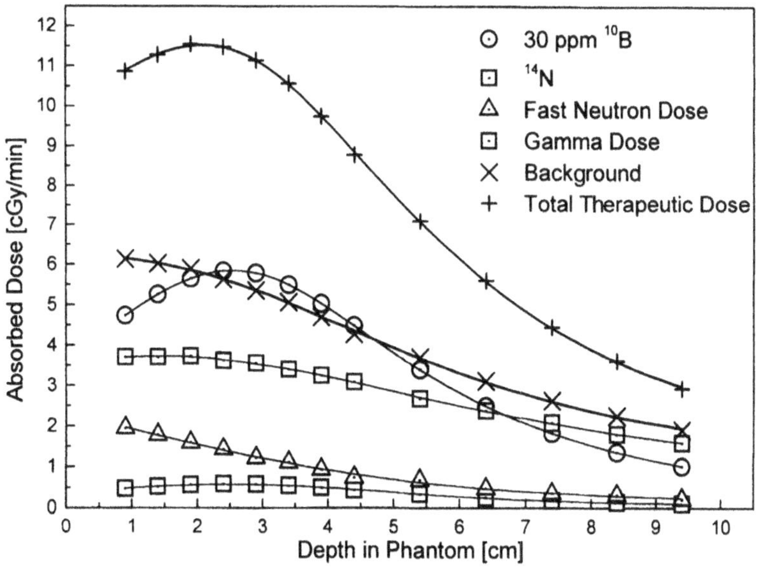

Figure 3. Absorbed dose in phantom at central line.

Table 2. Uncertainties for photon and fast neutron dose in phantom (3 cm depth)

Source	Uncertainty in photon dose (%)	Uncertainty in fast neutron dose (%)
Electrometer	1	1.5
Calibration of graphite chamber	1	3.5
Calibration of TE chamber	0.1	4.5
Relative sensitivity of graphite chamber	2.5	6
Relative sensitivity of TE chamber	0.1	5
Thermal response of graphite chamber	1.5	5
Thermal response of TE chamber	0.2	22
Positioning of chambers	1.3	2.6
Reactor power	1	1
Displacement correction factor	0.5	0.5
Thermal flux	0.5	4
Total uncertainty	**4**	**24**

4. IN PHANTOM MEASUREMENTS

To make these measurements we used a 17.3 cm diameter and 20.5 cm long cylindrical water phantom.

The gamma dose and the fast neutron dose rates inside the phantom were evaluated using the paired ionization chambers method. The thermal neutron flux was measured using bare and Cd covered gold foils. The N^{14} and B^{10} dose rates were calculated through the measured thermal neutron flux.

Figure 3 shows the absorbed dose in the center axis of the phantom.

The estimated uncertainties for gamma and fast neutron dose at 3 cm depth in phantom are showed in Table 2.

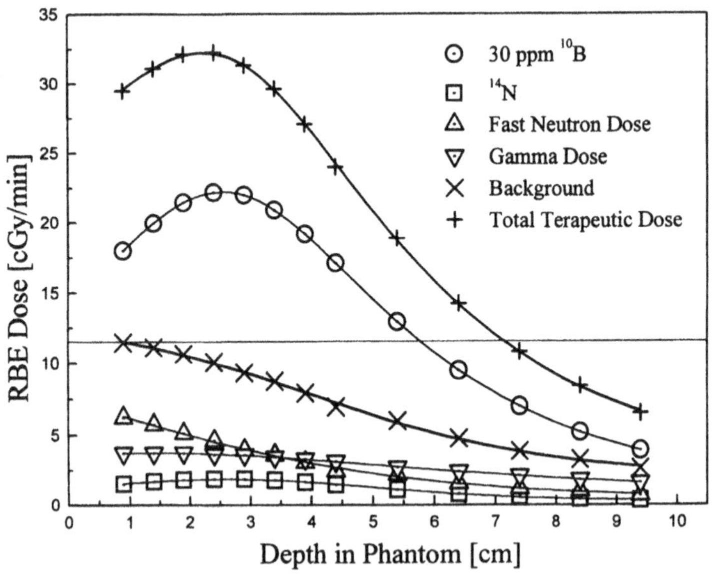

Figure 4. RBE dose in phantom at center line.

Table 3. Expected free beam parameters

Epithermal flux (0.5 eV–10 keV)	$1.1 * 10^9$ n/cm^2 sec
Fast neutron dose (>10 keV)/n_{epi}	$8.0 * 10^{-13}$ Gy/n cm^2
photon dose / n_{epi}	$1.5 * 10^{-13}$ Gy/n cm^2
thermal flux (<0.5 eV)/n_{epi}	0.03

Considering as RBE factors for gamma dose, fast neutron dose, N^{14} dose and B^{10} dose 1, 3.2, 3.2 and 3.8 [4] respectively, and without considering B^{10} in the normal tissue, the usually considered beam quality factors, i.e., Advantage Depth (AD), Advantage Ratio (AR) and Advantage Depth Dose Rate (ADDR), are:

AD (max) = 7.2 cm; AR = 3.1 and ADDR = 11.4 cGy/min.

Figure 4 shows the RBE dose in the center axis of the phantom.

5. FUTURE UPGRADE

During November 1998 we will replace the cylindrical port for a conical one. Due to this change, and if the thickness of the Bi shield is increased from 10 to 16 cm, the free beam parameters given in Table 3 are expected to be reached.

REFERENCES

1. Bustos D., Calzetta Larrieu O., and Blaumann H. Epithermal beam in the RA-6 reactor. In: Larsson B., Crawford J., Weinreich R. (eds) Advances in Neutron Capture Therapy. Volume 1, Medicine and Physics. Amsterdam: Elsevier Science, 1997; 420–423.
2. Briesmeister J.F. (ed). MCNP-A General Monte Carlo N-Particle Transport Code, Version 4B LA-12625-M, UC705 and UC700, March 1997.
3. Kocherov N.P. Neutron Metrology File NMF-90. IAEA-NDS-171, January 1996.
4. Capala J. *et al.* Radiation doses to brain under BNCT protocols at Brookhaven National Laboratory In: Larsson B., Crawford J., Weinreich R. (eds) Advances in Neutron Capture Therapy. Volume 1, Medicine and Physics. Amsterdam: Elsevier Science, 1997; 51–55.

A NCT FACILITY DESIGN AT HANARO

Byung Jin Jun* and Byung Chul Lee

HANARO Center, Korea Atomic Energy Research Institute
P. O. Box 105, Yusong, Taejon, 305-600
Korea

1. INTRODUCTION

A BNCT facility is under design using the IR beam port of HANARO, 30 MW open-tank-in pool type research reactor. Since all beam tubes in HANARO are optimized for low energy neutron experiments—tangential tubes with small nose area at the position of peak thermal neutron flux in the heavy water reflector, the filtered epithermal neutron flux is high enough neither for NCT research nor for patient treatment. If proper filters are used, however, reasonably high thermal neutron flux with low fast neutron and gamma contamination could be obtained, which is proper for NCT research and superficial cancer treatment.

The IR beam tube is chosen considering the space and convenience for patient positioning and transportation as well as interference with other experiments. Single crystals of silicon and bismuth at liquid nitrogen temperature located at the exit of beam port, filter fast neutrons and gamma-rays. A water shutter located inside of the beam tube permits the continued reactor operation regardless of NCT activity. Intensive calculations have been carried out changing dimensions and materials to search an optimum condition.[1,2] Now the design is almost fixed. The expected free neutron flux below 10 keV at the center of irradiation position is 2.6×10^9 n/cm^2-s, that above 10 keV is 1/410 of thermal flux, and gamma dose is 114 cGy/hr.

Meanwhile a detailed design for a prompt gamma neutron activation analysis (PGNAA) system is under progress for the on-site evaluation of boron concentration in the biological samples. Since the guided cold neutrons are not available at least for a few years, it is decided to use thermal neutrons. A method diffracting some part of white beam before it is filtered for neutron spectrometry is chosen so as to minimize

* E-mail: bjjun@nanum.kaeri.re.kr.

Frontiers in Neutron Capture Therapy, edited by Hawthorne *et al.*
Kluwer Academic / Plenum Publishers, New York, 2001.

background.[3] Now the basic design is almost fixed and more than 1×10^8 n/cm^2-s of thermal neutron flux at the sample position with sufficiently low background is expected.

An alternative method obtaining almost pure thermal neutron beam for NCT is studied. The appropriately focussed diffracted beam may have much better beam quality and quantity for thermal neutron NCT than filtered beam. Its detailed study is remained for future work.

2. PROPOSED DESIGN AND BEAM CHARACTERISTICS AT THE IRRADIATION POSITION

Figure 1 shows the proposed layout of NCT irradiation room in the HANARO reactor hall. The IR beam port is the only possible place for NCT at HANARO. While all seven beam tubes have similar beam quality optimized for low energy neutron experiments as tangential tubes with small nose area at the position of peak thermal neutron flux in the heavy water reflector, the IR beam port is closest to the exit of reactor hall, has no competition with other purposes at this moment, and has space to install an irradiation room.

Use of epithermal neutrons for NCT was given up from the beginning, because the unfiltered high energy neutron flux itself is not high enough, and installation of a fission converter is extremely difficult unless the reflector tank is replaced by a new one. It was decided to search a method obtaining highest thermal neutron flux with minimum

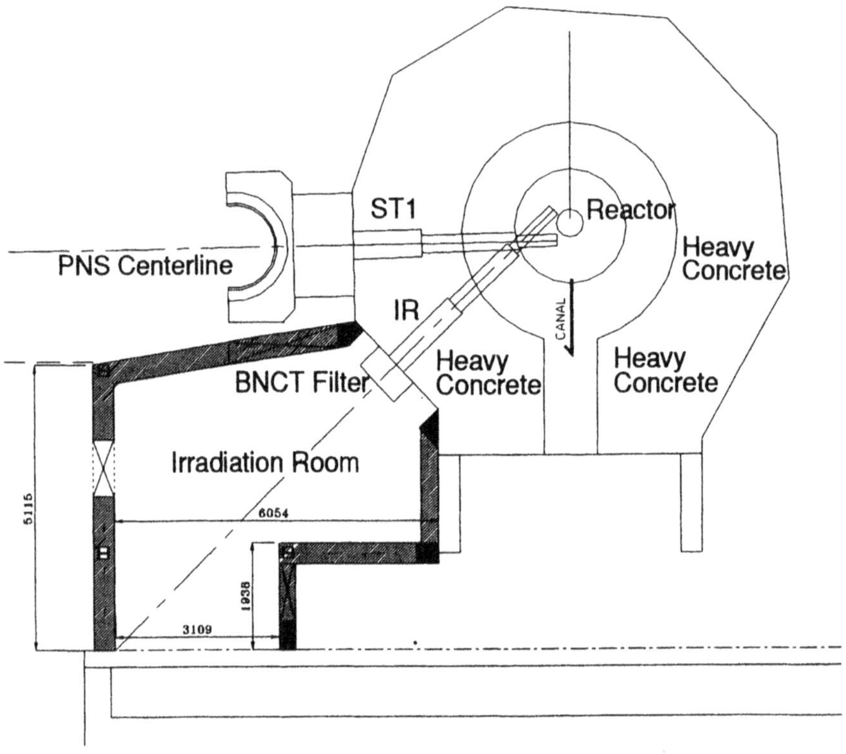

Figure 1. Proposed layout of NCT irradiation room in HANARO.

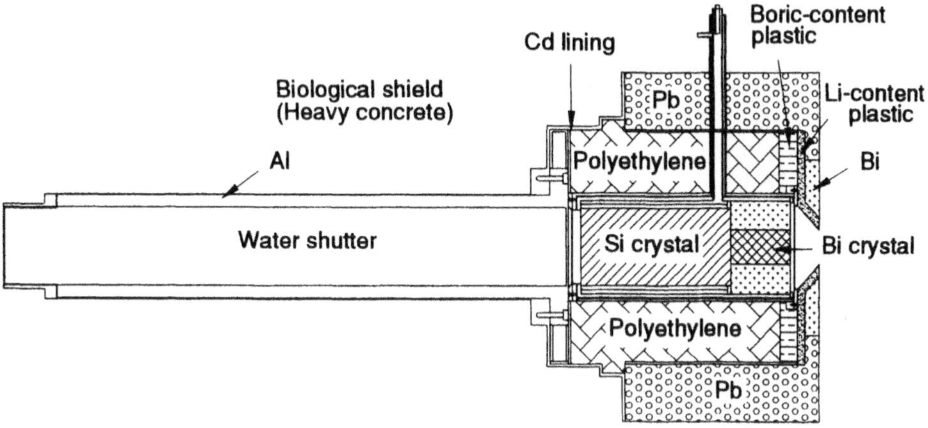

Figure 2. Proposed NCT facility at the IR beam tube in HANARO.

contamination of fast neutrons and gammas, for the basic research on NCT and superficial cancer treatment.

The design targets at the irradiation position of HANARO NCT facility are:

- thermal neutron flux (E < 10 keV), $\Phi_{th} > 3 \times 10^9$ n/cm^2-s
- dose from fast neutrons and gammas $<2 \times 10^{-10}$ RBE-cGy/unit Φ_{th}.

The proposed NCT beam extraction system is shown in Figure 2. A water shutter, 170 cm in length, is inside of the beam tube. When it is filled with water, all neutrons disappear in the water, and remaining gammas are shielded by the filter located at the exit of the beam port. The neutron beam is easily controlled by just filling in or draining water, by which the reactor is continuously operated regardless of NCT activity.

A silicon single crystal of 40 cm in length and 20 cm in diameter is chosen for fast neutron filtering, because silicon has slightly higher fast neutron cross-section than thermal, the thermal cross-section becomes much lower when it is crystalized, and large single crystals are commercially available from the semiconductor industry. This silicon also attenuates gammas coming from the reactor but it generates secondary gammas as well. Gammas are finally filtered by bismuth of 15 cm in thickness and 24 cm in diameter. Only the central part of 10 cm in diameter is composed of single crystal blocks and remaing part is normal bismuth. Since single crystals of silicon and bismuth have very low thermal neutron cross-section, forward peaked beam at the irradiation position can be obtained. The surrounding normal bismuth scatters neutrons and enhances neutron flux in the region of irradiation.

Both of silicon and bismuth filters are cooled by liquid nitrogen to enhance thermal neutron penetration.[4] The vacuum gap at the liquid nitrogen jacket is minimized as 2 mm in order to minimize radiation leakage. Mockup experiment shows that a 30 liter dewar can maintain liquid nitrogen for more than a day after the filter is sufficiently cooled down.

Since the majority of fast neutrons are scattered at the front part of silicon filter, it is surrounded by 10 cm thick polyethylene, thin cadmium plate, and then thick lead. Materials emitting less secondary gammas such as boron or lithium containing plastics and bismuth are selected for the shielding part near the outside surface where irradiation objects will be located. Material compositions of boron and lithium containing plastics are given in Table 1.

Table 1. Material compositions for boric- and Li-content plastics

Nuclide	Boric-content plastic	Li-content plastic
B4C	66%	—
B-10	10.28%	—
B-11	41.37%	—
C	14.35%	—
C	25%	20.5%
H	3%	2.2%
O	6%	5.8%
Li-6	—	17.1%
F	—	54.3%
Density	1.68 g/cc	1.89 g/cc

The neutron beam quality is analyzed by MCNP4B.[5] Calculated flux and dose at the irradiation position for free beam condition are given in Table 2. Though the thermal neutron flux is slightly lower than the aforementioned design target it is acceptable. As the forward current within polar angle of 5° is about 69%, it is an effective beam to irradiate tumorous area. The fast neutrons and gammas are sufficiently low compared to the thermal neutron flux level. When a water phantom is placed, the peak flux which is slightly higher than surface flux occurs at about 0.5 cm depth. This peak thermal neutron flux is about 8×10^9 n/cm^2-s. With this beam characteristic, it is anticipated that the irradiation time for superficial cancer is about 1 hour.

3. PGNAA SYSTEM DESIGN

The PGNAA system is an effective tool not only for the boron concentration measurement but also for multielement analysis. It is well known that the guided cold neutron beam is the best for this purpose from the view point of low background and high sensitivity. However, the installation of a cold neutron source and neutron guide tubes in HANARO is not certain within a few years. Therefore a method minimizing background with reasonably high thermal neutron flux and least interface with already planned or ongoing experiments, is searched.

The ST1 beam port which is located just beside NCT irradiation room is chosen for PGNAA system installation. It is intended to change the white beam direction by 90° upward by pyrolytic graphite (PG) monochromators before the beam passes through a filter for a polarized neutron spectrometer (PNS). It diffracts only upper and lower parts of the beam which is not used by the backside spectrometer. The sample position for

Table 2. Beam characteristics at the irradiation position for free beam condition in the proposed HANARO NCT facility

Beam characteristics	Design value
Thermal neutron flux, Φ_{th}	2.6×10^9 n/cm^2-s
Fast neutron flux, Φ_f	6.3×10^6 n/cm^2-s
Thermal-to-fast ratio, Φ_{th}/Φ_f	410
Thermal current-to-flux ratio	0.69
Fast neutron dose	5 cGy/hr
Gamma dose	114 cGy/hr

PGNAA is about 1.5 m above the beam port and measuring system is sitting on the shield of PNS. If all (0,0,2n) modes of Bragg's diffraction by appropriate mosaic spread PG are utilized and the focussing technique is utilized, the beam flux at the sample position is estimated to be more than 1×10^8 n/cm^2-s. Since the beam for the PGNAA is completely away from the main beam, sufficiently low background is expected.

4. POSSIBILITY OF BEAM DIFFRACTION FOR NCT

The technique of beam diffraction could be used to obtain almost pure thermal neutron beam for NCT. It is calculated that the reflected flux by PG has maximum value when the Maxwellian flux is diffracted by 10° (20° change of beam direction). It is about 10% of incident Maxwellian flux when a 0.8° mosaic spread and 0.4 cm thick PG is used. The reflectivity is proportional to the square root of mosaic spread and thickness, but the beam divergence after reflection is proportional to mosaic spread. Therefore optimization is needed for the best mosaic spread. The incident beam diameter of IR beam port is 25 cm. Therefore, if the whole beam is focussed to the surface of 10 cm in diameter after 10° reflection, the focussing effect is about 4 times. This simple calculation means that the neutron flux at the irradiation position is about 40% of incident beam (about 6.4×10^9 n/cm^2-s) when simply 0.8° mosaic spread and 0.4 cm thick PGs are used. It is more than two times of currently designed value, the beam is almost 100% forward beam almost free from fast neutrons and gammas, and it is not necessary to cool by liquid nitrogen. The only problem but the most important one is that a lot of very expensive PGs are needed. Since this alternative method needs more study, the work based on the aforementioned current design will be continued.

ACKNOWLEDGMENTS

This work was performed with the support of Ministry of Science and Technology of Korean government.

REFERENCES

1. Byung Chul Lee, Byung Jin Jun, and Ji Bok Lee, "Preliminary Design for BNCT in the HANARO," *Proc. 1998 ANS Topical Conf. on Radiation Protection and Shielding*, April 19–23, 1998, Nashville, USA, American Nuclear Society, pp. II-395–400.
2. Byung Jin Jun, and Byung Chul Lee, "Design of a BNCT Facility at HANARO," *Proc. 6th Mtg. of the Int. Group on Research Reactors*, April 29-May 1, 1998, Taejon, KOREA, Korea Atomic Energy Research Institute, pp. 245–250.
3. Hee Dong Choi, Soo Hyun Byun, and Byung Jin Jun, "Design of a Prompt Gamma Neutron Activation Analysis System at HANARO," *Proc. 6th Mtg. of the Int. Group on Research Reactors*, April 29-May 1, 1998, Taejon, KOREA, Korea Atomic Energy Research Institute, pp. 449–453.
4. A.K. Freund, Cross-sections of Materials Used as Neutron Monochromators and Filters, *Nucl. Inst. Methods*, Vol. 213, pp. 49–501, 1983.
5. J.F. Briesmeister, Editor, "MCNP-A General Monte Carlo N-Particle Transport Code, Version 4B," Los Alamos National Laboratory, LA-12625-M, 1997.

BEAM COLLIMATOR DESIGN FOR THE EPITHERMAL NEUTRON BEAM AT THE TSING HUA OPEN-POOL REACTOR (THOR)

Hong-Ming Liu

Nuclear Science and Technology Development Center
National Tsing Hua University
Hsinchu 30043
Taiwan

INTRODUCTION

In Taiwan, a program of epithermal neutron beam installation has been progressed since 1995. The thermal column of the Tsing Hua Open-pool Reactor (THOR) is selected as the site to be modified as an epithermal neutron beam. THOR is a 1 MW, light water moderated and cooled reactor of the TRIGA conversion type. By replacing the removable portion of the graphite blocks in the thermal column with a properly designed filters, the modification of the epithermal neutron beam was completed in 1997. Figure 1 shows the horizontal cross-section of the THOR epithermal neutron beam facility. The purpose of the lead and poly-B used at the end of the thermal column was to lower the background dose of the irradiation room. The filters were composed of Al and AlF_3 for extracting the epithermal neutron while the Bi was used for reducing the gamma ray component. The distance between the Bi face to the irradiation port is 16 cm. This space was left for beam collimator design with a cross-section of $40 \times 40\,cm^2$. Before progressing to the animal experiments and clinical trials by using this epithermal neutron beam, a beam collimator has to be installed to provide a suitable beam size, and to lower the background dose of the irradiation room from different components of radiation. The purpose of this research is aimed at providing a convenient beam collimator and to lower the background dose without reducing the epithermal neutron flux at the irradiation port.

METHODS AND MATERIALS

To optimize the beam collimator, the MCNP4A with ENDF/B-V cross section were used in this study. All the calculations were carried out on an IBM RISC 6,000/370 work

Frontiers in Neutron Capture Therapy, edited by Hawthorne *et al.*
Kluwer Academic / Plenum Publishers, New York, 2001.

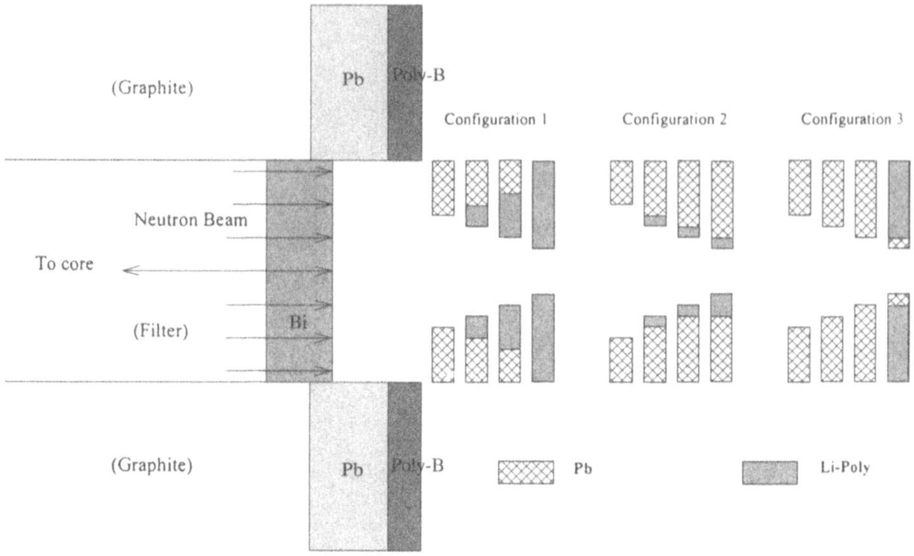

Figure 1. Horizontal section of the THOR epithermal neutron beam facility.

station. These calculations were run in a fixed source mode for a full-core geometry and more than 4,000,000 neutrons sampled from the Watt fission spectrum in the fuel elements were tracked in each run to provide good statistical error (below 5%) for the tallies in the cells located at the irradiation port.

Figure 2 shows the proposed collimator for the THOR epithermal neutron beam. The proposed collimator was composed by several concentric circles which made of either

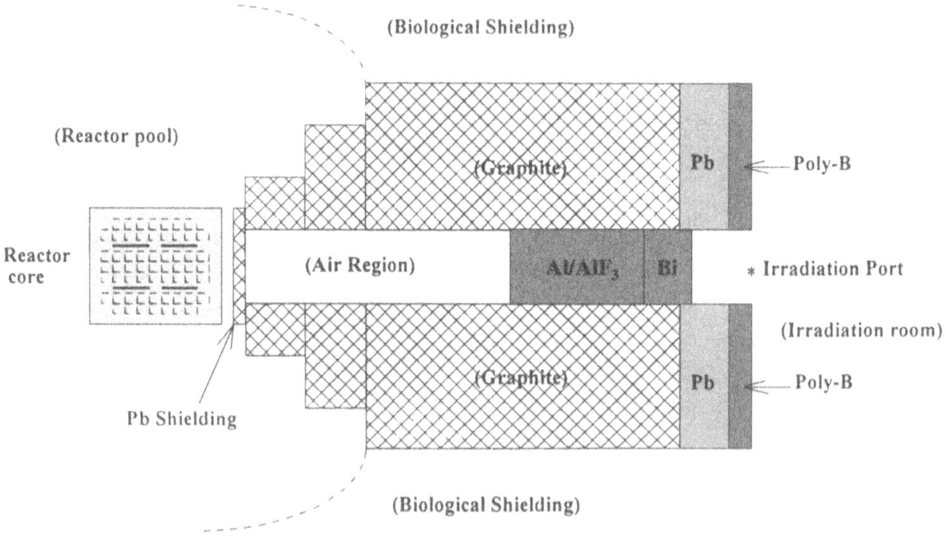

Figure 2. Proposed collimator for the THOR epithermal neutron beam with three typical configurations.

Table 1. Computed comparisons of beam collimator performances

	Irradiation port (in Air)			Background dose*	
	Ψ_{epi} (n/cm^2-sec)	$D_{n,fast}$ (cGy-cm^2/epi.n)	D_γ (cGy-cm^2/epi.n)	$D_{n,total}$ (cGy)	D_γ (cGy)
Existing beam	3.19×10^8	5.45×10^{-11}	11.63×10^{-11}	179	480
Configuration 1	1.78×10^8	5.52×10^{-11}	4.72×10^{-11}	64	53
Configuration 2	2.53×10^8	5.93×10^{-11}	4.82×10^{-11}	95	92
Configuration 3	3.83×10^8	5.33×10^{-11}	2.35×10^{-11}	65	84

*The background dose is specified for that there would be the same (3.6×10^{12} n/cm^2) epithermal neutron fluence at the irradiation port.

by ^{6}LiF dispersed in polyethylene (Li-Poly) or by lead. With a high content of H to moderate the fast neutrons and a high content of ^{6}Li to absorb thermal neutrons with minimal gamma ray production, Li-Poly is considered to be a suitable material for reduction the fast neutron dose. Also, the lead is a typical material for gamma ray shielding.

By using these two materials (Li-poly and lead) for the beam collimator designs, the proposed collimator was separated into 4 pieces that the thickness of each piece is 4 cm with a cross section of 40×40 cm^2. Each piece of the collimator will be drilled to open a central cylindrical space that can be fit into various concentric rings of designed materials. Such construction can be changed very easy for different experimental purpose. Three typical configurations for our beam collimator were also shown in Figure 2.

RESULTS AND CONCLUSIONS

To compare the performances of the epithermal neutron beam with and without the collimator, the average epithermal neutron flux within 8 cm in diameter at the center of the irradiation port and the dose induced by the fast neutron and gamma ray per epithermal neutron will be listed. Also, the total dose outside the beam hole (12 cm in diameter) will be specified as the background dose of the irradiation room.

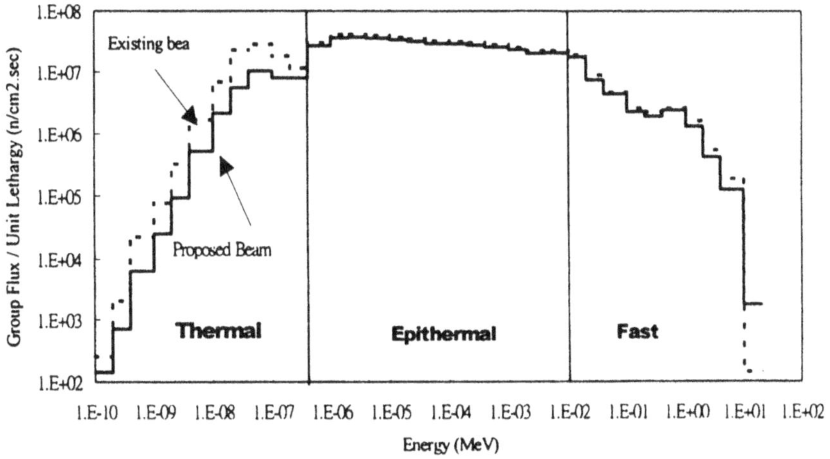

Figure 3. Compare the spectrum of the THOR epithermal neutron beam.

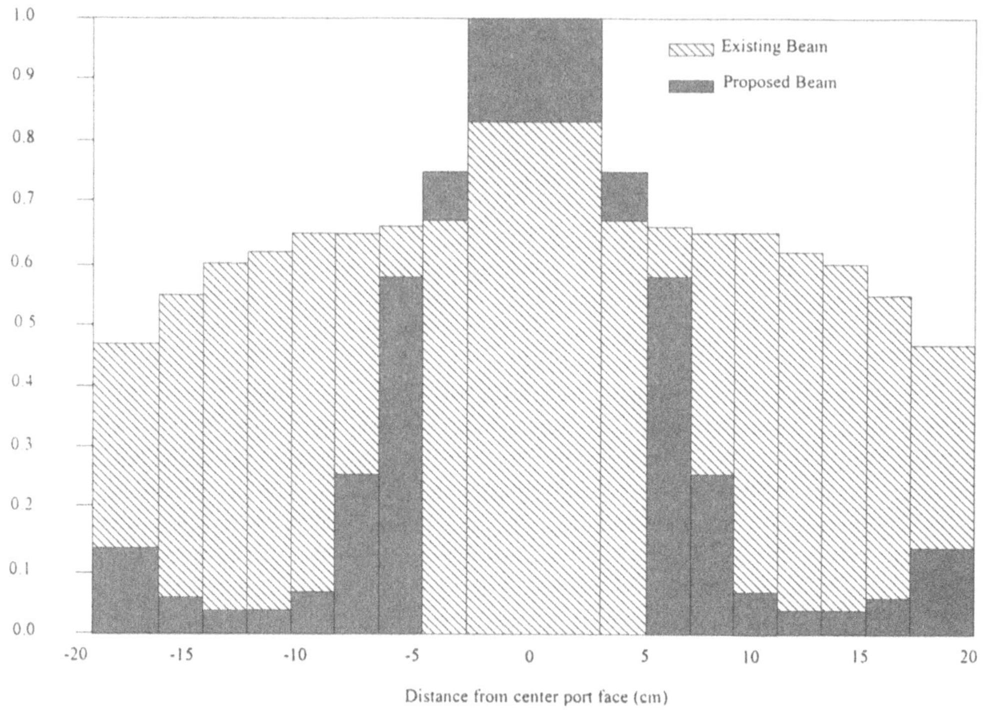

Figure 4. Epithermal neutron intensity profile across the irradiation port face

The MCNP simulation results are shown in Table 1. It is evident from this table that only configuration 3 can enhanced the epithermal neutron flux by 20%. The fast neutron dose per epithermal neutron at the irradiation port almost keeps constant but the gamma ray dose can be effectively reduced by adding the beam collimator. The background dose shown in Table 1 is specified for that there would be the same (3.6×10^{12} n/cm^2) epithermal neutron fluence at the irradiation port. It is very clear from the results that the total background dose with collimator is considerably lower than the total background dose without collimator.

Figure 3 shows the neutron flux spectrum at the irradiation port for both the existing beam and the proposed beam (Configuration 3). The epithermal neutron flux for both beams is considerably greater than the fast neutron flux, leading to a low fast neutron dose. The thermal neutron flux of the proposed beam is small than the existing beam due to the use of Li-Poly. Figure 4 shows the epithermal neutron intensity profile across the irradiation port face. The existing beam is considerably flat than the proposed beam. That is the reason why the background dose of the proposed beam is much lower than the existing beam (Table 1).

The results of this research teach us that if the beam collimator has large opening inside (face to the core) and small opening outside (face to the irradiation port) for lead and then add a piece of Li-Poly at the end of the collimator (configuration 3), the epithermal neutron flux can be enhanced by 20% while the background dose defined here can be effectively reduced more than 70%. At the same time, the change of this collimator is very easy for different experimental purpose.

ACKNOWLEDGMENTS

The work is supported by National Science Council of Taiwan, Republic of China, under the contract no. NSC87-2212-E-007-065.

REFERENCES

1. MCNP: Monte Carlo N-Particle Transport Code System, Version 4A, Oak Ridge National Laboratory, RSIC CCC-200, 1993.
2. H.M. Liu, "Installation of an epithermal neutron beam at the Tsing Hua Open-pool Reactor (THOR)," Proceeding of the 5th Asian Symposium on Research Reactor, Korea, 1996.
3. H.B. Liu *et al.*, "Improved apparatus for neutron capture therapy of rat brain tumors," Int. J. Radiation Oncology Biol. Phys., Vol. 28, No. 5, pp. 1167–1173, 1994.
4. F.J. Wheeler *et al.*, "Physics design for the Brookhaven Medical Research Reactor epithermal neutron source," in Neutron Beam design, Development, and Performance for Neutron Capture Therapy, edited by O.K. Harling *et al.*, pp. 83–95, Plenum Press, NY, 1990.

46

THE USAGE OF THE KUR ADVANCED CLINICAL IRRADIATION SYSTEM FOR NCT UNDER THE KUR CONTINOUS OPERATION

Tooru Kobayashi, Yoshinori Sakurai, Keiji Kanda, and Koji Ono

Research Reactor Institute
Kyoto University, Kumatori-cho
Sennan-gun, Osaka 590-0494, Japan

1. INTRODUCTION

The remodeling construction of the KUR Heavy Water Neutron Irradiation Facility was completed at the end of March 1996, mainly for neutron capture therapy (NCT).[1] The remodeled facility is available for the several "irradiation modes" whose neutron energy spectra are from almost pure thermal neutrons to epi-thermal neutrons, using the heavy water spectrum shifter and the thermal neutron filters of cadmium and boral. Clinical and experimental irradiations under the KUR continuous operation are possible using the radiation shield system and the equipment for the support of NCT clinical irradiations, such as a Remote Patient Carrier System with clinical collimators. We call the remodeled Heavy Water Facility and its accessory equipment, generally, "the KUR Advanced Clinical Irradiation System".

After the remodeling, 12 medical irradiations, 4 for the thermal neutron irradiation and 8 for the mix irradiation of thermal and epi-thermal neutrons, were already performed as of October 1998. All these clinical irradiations carried out with the start and stop of the KUR operations. The clinical irradiation under the KUR continuous operation will start in 1998. Also, in the near future, epi-thermal neutron irradiation will be started. We will be able to have the more opportunity of the clinical irradiation, and select more adequate date, time and irradiation condition.

2. THE KUR ADVANCED CLINICAL IRRADIATION SYSTEM

Figure 1 shows the outline of the KUR Advanced Clinical Irradiation System. The new irradiation room is consisted by heavy concrete blocks and the "entrance-shielding

Frontiers in Neutron Capture Therapy, edited by Hawthorne *et al.*
Kluwer Academic / Plenum Publishers, New York, 2001.

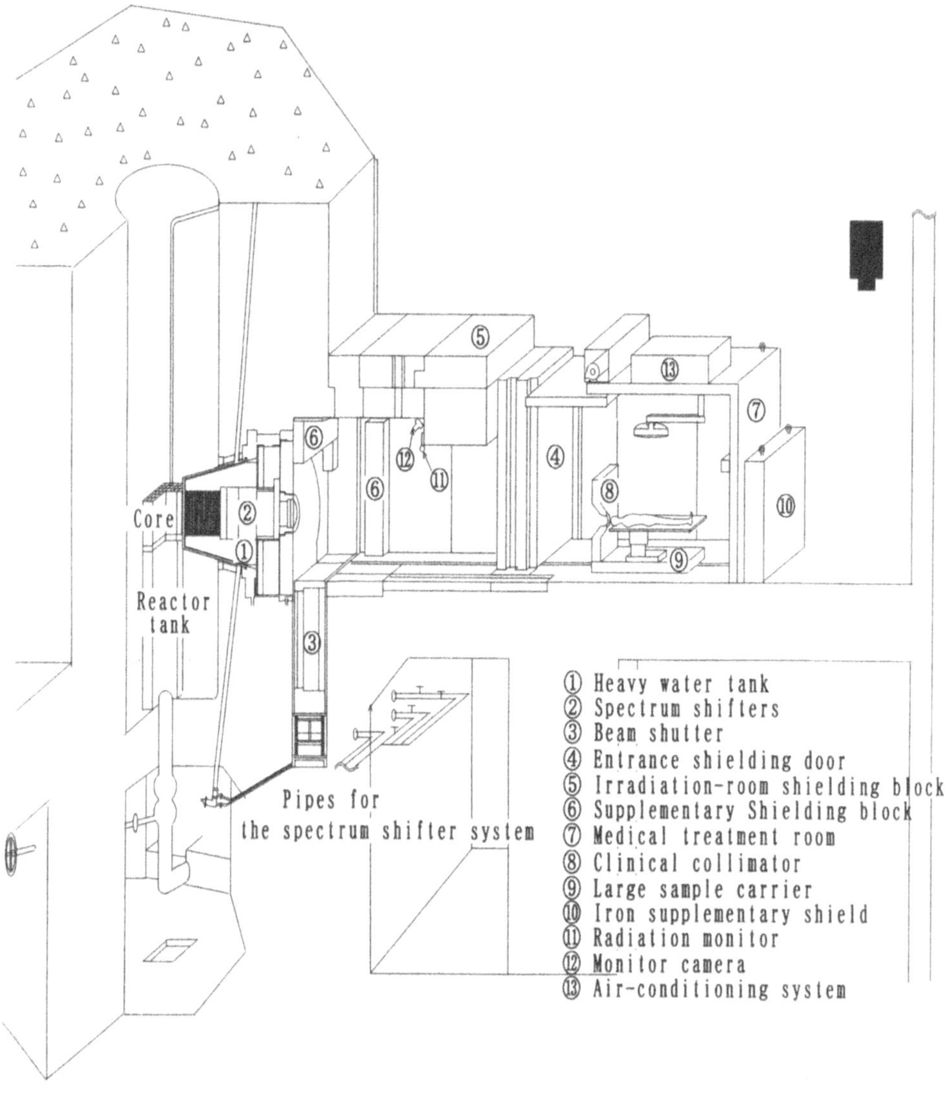

Figure 1. The KUR Advanced Clinical Irradiation System.

door." Its inner volume is $2.4^{w} \times 2.4^{h} \times 3.6^{l} m^{3}$. To make the dose equivalent rate an acceptable level at the working area in the irradiation room even at the 5 MW KUR operations, the radiation shielding system was installed. The system consists of (i) the heavy water shutter and the spectrum shifter, (ii) the thermal neutron filters, (iii) the beam shutter and the entrance shielding door, and (iv) the supplementary radiation shield added in the irradiation room. In the viewpoint of the safety, the "open" operation of the entrance-shielding door is interlocked for the conditions of the beam shutter "close", and the heavy water shutter and spectrum shifter "full".

The "Medical Treatment Room" was constructed adjacent to the irradiation room. In the Medical Treatment Room, a sterilizing air-conditioning system, and two non-shadow lights, *etc.*, are equipped. The physicians can operate on patients just before and

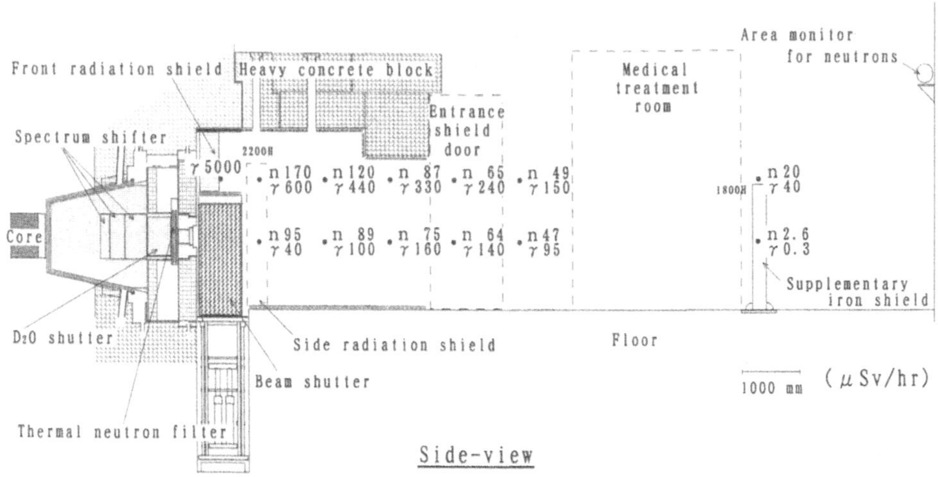

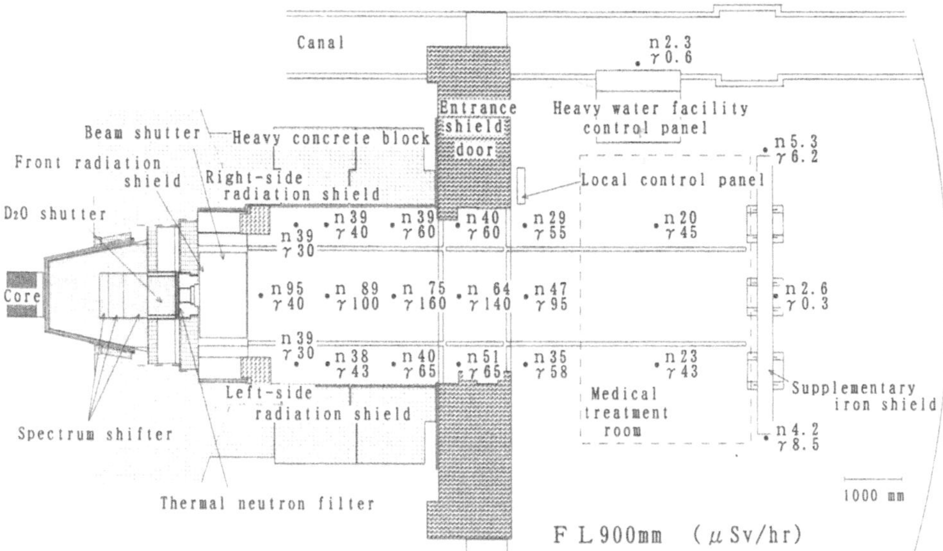

Figure 2. Dose exposure distributions inside and around the irradiation room working the Radiation Shield System under 5 MW KUR operation.

after clinical irradiations. The "Large Sample Carrier" was newly made as a remote-patient carrier system. A clinical bed and a clinical collimator can be loaded on the carrier that runs on the rail. Thus, the physicians can easily set up for the clinical irradiations and carry a patient between the irradiation room and the clinical treatment room. Three kinds of clinical collimator are available for the several irradiation conditions such as thermal neutron irradiation, epi-thermal neutron irradiation and 15-degree angled irradiation.

Table 1. Operating times for the several equipment

	Capacity (l)	Time for supply	Time for drain	Note
Heavy water tank	1,500	14 min.	23 min.	
Spectrum shifter 1	38.5	48 sec.	32 sec.	Drain at the same time: 160 sec.
Spectrum shifter 2	77.6	70 sec.	42 sec.	Supply at the same time: 180 sec.
Spectrum shifter 3	115.4	101 sec.	75 sec.	
Heavy water shutter	80.6	73 sec.	57 sec.	
	Moving range (cm)	Weight (t)	Time for drive (sec.)	
Boral filter	62	0.01	90	
Cadmium filter	62	0.01	90	
Beam shutter	165	7	60	
Entrance door	220	60	106	
Large sample carrier	90	3	60	

3. NCT UNDER THE KUR CONTINUOUS OPERATION

The procedure for a NCT clinical irradiation during a continuous operation is as follows: (i) the beam shutter is closed, (ii) the entrance shielding door is opened, (iii) the large sample carrier with patient moves into the irradiation room, (iv) the entrance shielding door is closed, (v) the beam shutter is opened, and (vi) the large sample carrier moves to the irradiation position. After the clinical irradiation, the reverse process is used.

Figure 2 shows the dose exposure distributions inside and around the irradiation room under the 5 MW KUR continuous operations for the entrance-shielding door "full-open". At the standard work area in the irradiation room, the dose equivalent rate was 75 cSv/hr for neutrons and 160 cSv/hr for gamma rays. So, the admittance time per a week should be limited within 4 hours.

It is realistic that the timing of the irradiation start and finish under the KUR continuous operation is decided for the minimum influence by the irradiation start and finish processes. Table 1 shows the operating times for the several equipment. It is important that the moving time of the large sample carrier is about 1 minute for the 90-cm distance between the waiting and irradiation positions. From these processes and the time characteristics, the standard timings of the irradiation start and finish are decided to be the arrival time of the sample at the irradiation position, and the departure time from the irradiation position, respectively. Due to the equipment operating times, total time from "putting a patient into the irradiation room" to "taking out from the room" is about 15 minutes as well as the irradiation time. The intensities of neutrons and gamma rays at the waiting position in air is almost 1/9–1/25 and 1/5–1/10, respectively, comparing with the irradiation position. These intensities change for the irradiation mode and geometry.

4. INFLUENCES ON THE REACTOR STABILITY OF THE MODE CHANGE

The influence of the irradiation mode change was experimentally confirmed on the KUR control and stability, such as the linear power "Lin-N", the logarithm power

"Log-N", the period monitor, the safety power channels and the thermal power. For the open and close of the thermal neutron filters, no influence was observed. For the supply and drain of heavy water in the spectrum shifters, the following results were obtained; (1) two safety power channels change about 0.02 MW (0.5%), (2) thermal powers estimated from both the primary and secondary coolant systems change about 0.04 MW (1%), (3) these changes are minus for the supply and plus for the drain, and (4) these changes are observed within a few minutes after the mode change. It was found that the control stability and safety of the KUR are maintained by the mode changes.

ACKNOWLEDGMENTS

The authors thank the Reactor Operation and Management Group for their cooperation in the measurements for the reactor control characteristics and Mr. Takaaki Yoshimoto for his helpful assist in the dose measurement.

REFERENCES

1. T. Kobayashi, K. Kanda, Y. Fujita, Y. Sakurai, and K. Ono, The Upgraded of the Heavy Water Facility of the Kyoto University Reactor for Neutron Capture Therapy, in: "Advances in Neutron Capture Therapy, Vol. I, Medicine and Physics", B. Larsson, J. Crawford, and R. Weinreich, ed., Elsevier, Amsterdam, 1997, pp. 321–325.

47

FEASIBILITY OF THE UTILIZATION OF BNCT IN THERMALIZING COLUMN OF TRIGA REACTOR

Marko Maučec*

"Jožef Stefan" Institute
Reactor Physics Division
Ljubljana, Slovenia

1. INTRODUCTION

At least twenty new patients fall ill for brain glioblastoma multiforme (GBM) in Slovenia every year. Therefore, the development of irradiation facility for clinical use is almost indispensable alternative to the epithermal neutron beam for "in-vitro" experiments on tumor cells or laboratory animals, already installed in the radial channel of the "Jožef Stefan" Institute (JSI) TRIGA reactor.[1] A unique reactor configuration has an important advantage: so called "dry cell" (see Figure 1), primarily devoted for a pool-type storage of a TRIGA reactor spent fuel elements could be appropriate for the patient irradiation room. This contribution presents the Monte Carlo suitability study of the thermalizing column (TC)/dry cell (DC) for epithermal neutron beam installation. The general purpose Monte Carlo code MCNP4B[2] was used for the modeling of the TRIGA reactor, together with the continuous cross-section libraries ENDF/B-VI and $S(\alpha,\beta)$ scattering data from ENDF/B-IV.

2. SUITABILITY STUDY OF FISSION CONVERTER AS A NEUTRON FLUX AMPLIFIER

With the small-power TRIGA reactors the recommended therapeutic limit level of epithermal neutron flux Φ_{epi} is not easy to achieve without substantial refurbishment of the thermal column. But utilizing the fission converter, the satisfactory high epithermal

* E-mail: *marko.maucec@ijs.si.*

Frontiers in Neutron Capture Therapy, edited by Hawthorne *et al.*
Kluwer Academic / Plenum Publishers, New York, 2001.

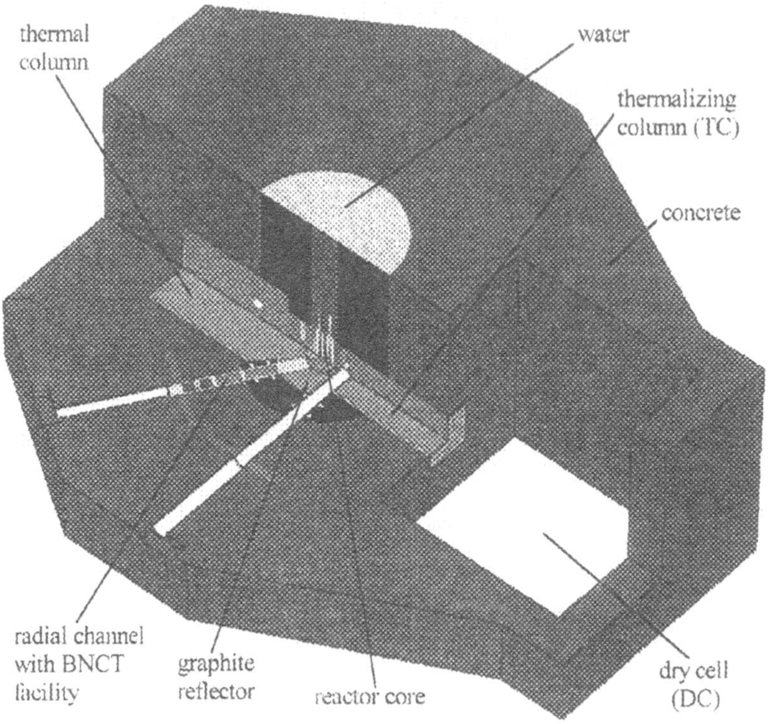

Figure 1. The 3D Monte Carlo model of the TRIGA Mark II research reactor.

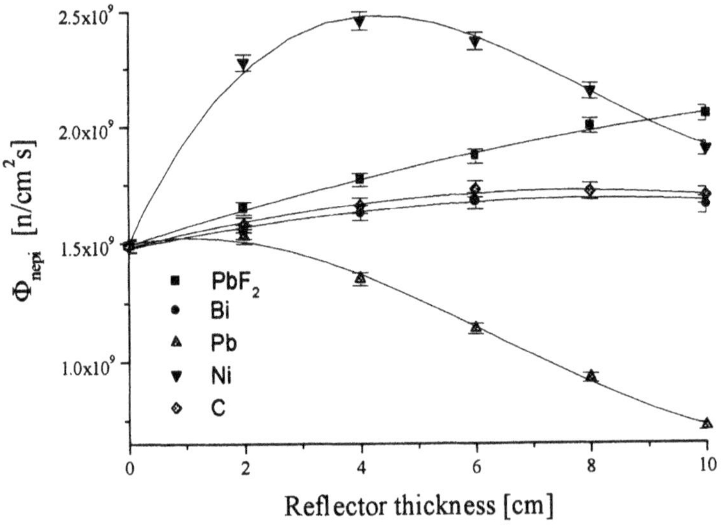

Figure 2. Epithermal neutron flux for different thickness of reflector material.

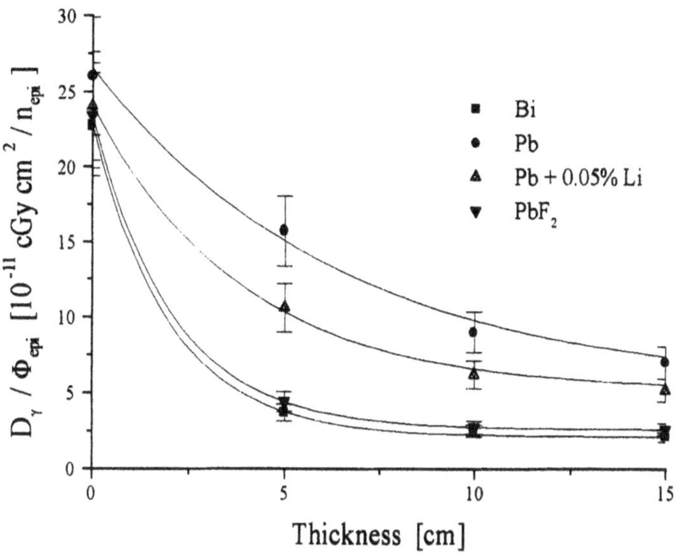

Figure 3. Specific gamma dose for different thickness of gamma shield.

neutron flux can be obtained.[3] In my study the fission plate with 1,468 g of total U mass (20% enriched) was encountered. The fission heat release in fissile material was estimated to be 2.3 kW with the temperature gradient of approx. 13.5 °C/s. The fission plate enhances epithermal neutron flux Φ_{epi} for a factor of 2.4 and specific fast neutrons dose for approx. 18%. On the other hand the specific gamma dose (D_γ/Φ_{epi}) in irradiation point is decreased for a factor of 2.3, since the major contribution of gamma rays comes from the reactor.

3. PARAMETRIC STUDIES OF COMPONENTS FOR BNCT FACILITY

3.1. Reflector Materials

It was estimated,[3] that the epithermal neutron flux can even be doubled, if the appropriate materials (Ni in that study) surrounding the filters are applied. In addition to Ni, Pb, Bi, C and PbF_2 were compared in case of my study (Figure 2) with thickness altered in following steps: 0, 2, 4, 6, 8 and 10 cm.

Except for the PbF_2, the optimum Φ_{epi} curves with the optimum reflector thickness can be observed, since increasing the reflector thickness reduces the active moderating cross-section area. Monotonic increase of Φ_{epi} with increasing thickness can easily be justified, since PbF_2 was estimated by Liu[4] as an alternative to Al_2O_3 and AlF_3 as independent epithermal neutron filter.

3.2. Gamma Shielding

The following materials were considered: Bi, Pb, PbF_2 and Pb + 0.05 wt% ^{6}Li, introduced by Liu.[5] Attenuation of Φ_{epi} and efficiency of gamma rays absorption (D_γ/Φ_{epi}) was calculated for following different thickness: 0, 5, 10 and 15 cm (Figure 3).

Table 1. Moderation properties of selected epithermal neutron filter materials (slab thickness 50 cm)

Material/density (g/cm^3)	Φ_{term} ($10^7 n/cm^2s$)	Φ_{epi} ($10^9 n/cm^2s$)	D_{nfast}/Φ_{epi} ($\times 10^{-13} Gy cm^2$)
Al_2O_3 (2.3)	0.9	2.1	17.0
Al_2O_3 (3.9)	1.0	1.4	4.7
AlF_3 (2.8)	0.9	2.2	3.5
FLUENTAL® (3.0)	0.0	0.9	1.8
PbF_2 (8.2)	0.8	2.2	2.5
Teflon®—CF_2 (2.2)	1.3	1.6	5.2
Therapeut. lim. values	—	1.0	5.0

Average relative errors: Φ_{term}—±7%, Φ_{epi}—±4%, D_{nfast}—±7% and D_{nfast}/Φ_{epi}—±11%.

Mainly from technical reasons and substantially lower cost, Bi was selected as the material for gamma shield. The thickness was estimated to be 10 cm, as this is the value where specific gamma dose (D_γ/Φ_{epi}) decreases below the recommended therapeutic limit of $3 \cdot 10^{-13}$ Gy cm^2/n$_{epi}$ (see Figure 3).

3.3. Epithermal Neutron Filters

Following materials were considered in calculations: Al_2O_3 ($\rho = 2.3$ and $3.9 g/cm^3$), AlF_3 ($\rho = 2.8 g/cm^3$), FLUENTAL™, PbF_2 and Teflon™(CF_2). The moderation efficiency (enhancement of Φ_{epi} with decrease of D_{nfast}) was probed with the MC transport of neutrons through the 50-cm thick slab of material without reflector (Table 1).

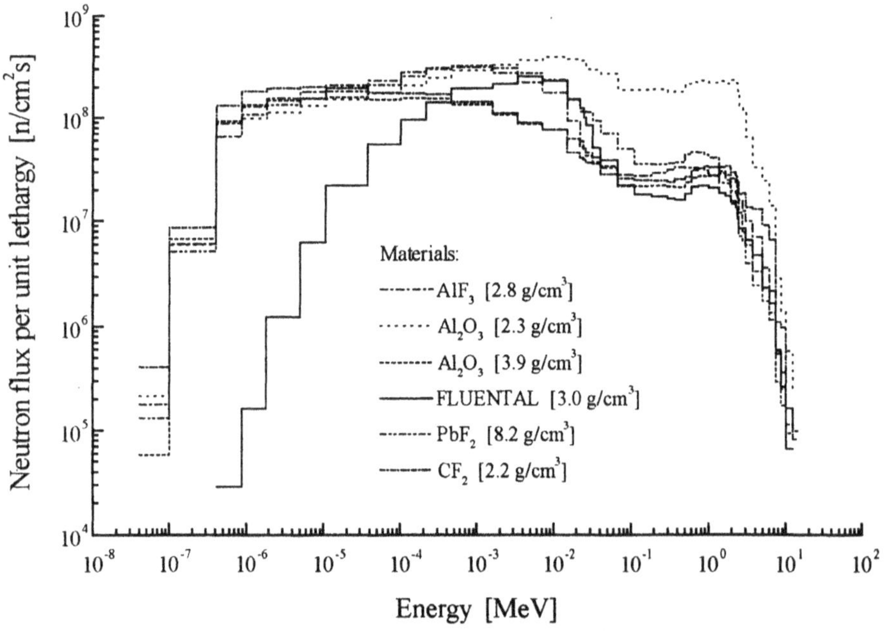

Figure 4. Neutron flux/unit lethargy for different materials of epithermal neutron filters.

PbF_2 and pure AlF_3 outperform other applied materials, when the level of epithermal neutron flux is considered. On the other hand FLUENTAL™ provides superior spectrum shifting to the epithermal energies range (0.4 eV < E < 10 keV) and 30% lower specific fast neutron dose with excellent thermal neutron cutoff (Figure 4).

4. PROPOSED CONFIGURATION OF BNCT IRRADIATION FACILITY

The proposed configuration (see Figure 5) of epithermal neutron irradiation facility for BNCT treatments in TC/DC of TRIGA reactor has the following characteristics:

- epithermal filter—50 cm of FLUENTAL™,
- thermal neutron absorber—0.05 cm of Cd,
- reflector—10 cm of PbF_2,
- collimator—PbF_2 cone,
- gamma shield—10 cm of Bi,
- neutron shield—6 cm of Li_2CO_3-poly,
- neutron source—250 kW TRIGA reactor and 2.3 kW fission plate.

With the proposed configuration epithermal neutron flux of $3.7 \cdot 10^8\,n/cm^2s$ (±3%) can be achieved. The purity of neutron beam, D_{nfast}/Φ_{epi} is $2.6 \cdot 10^{-13}\,Gy\,cm^2$ (±10%) and D_γ/Φ_{epi} is $1.7 \cdot 10^{-13}\,Gy\,cm^2$ (±6%), is comparable to the most of proposed or existing

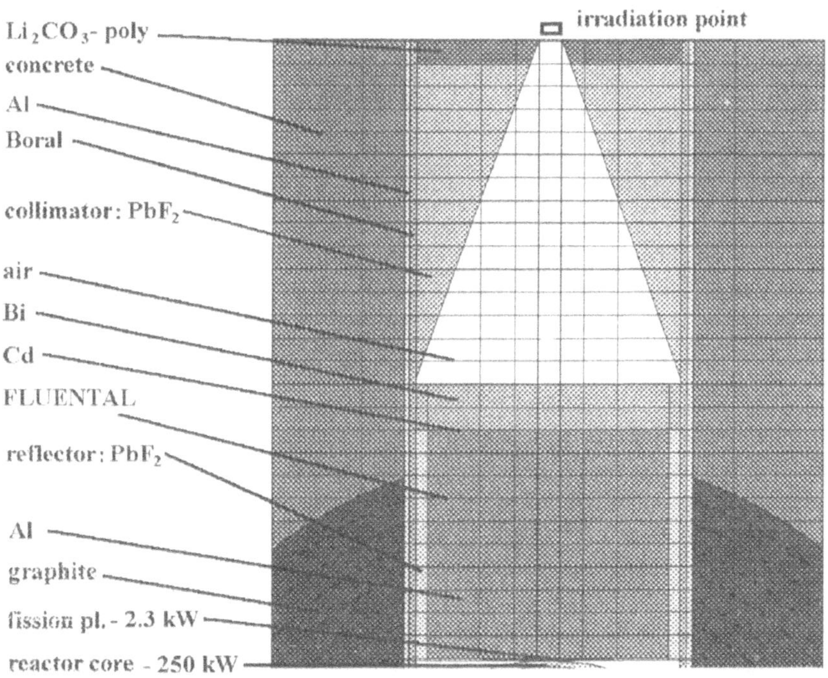

Figure 5. MCNP model of the proposed BNCT facility.

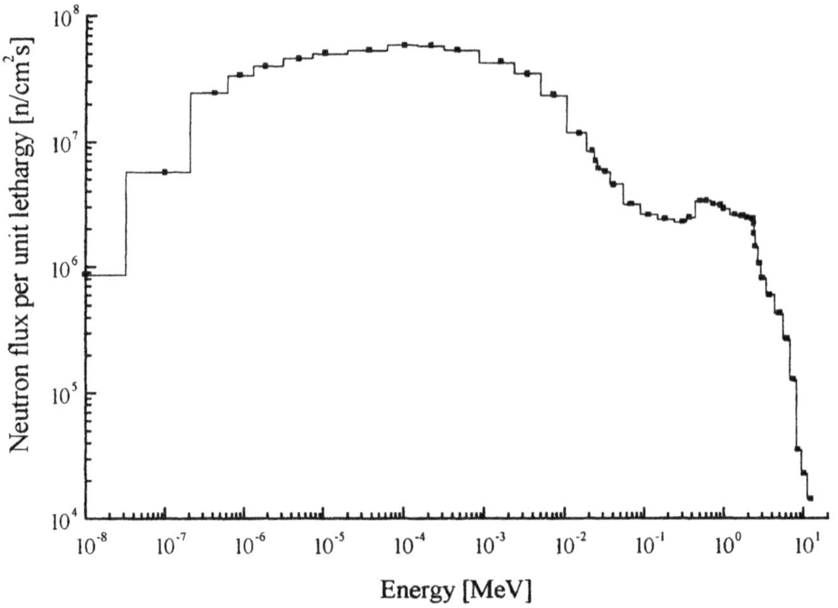

Figure 6. Neutron flux/unit lethargy for proposed configuration.

BNCT facilities. A very good spectrum shifting towards the epithermal energies is observed (see Figure 6): thermal flux Φ_{term} ($E < 0.4\,eV$) presents 1.5%, epithermal Φ_{epi} ($0.4\,eV < E < 10\,keV$) even 95% and fast flux Φ_{fast} ($E > 10\,keV$) 3.5% of integral neutron flux ($0\,eV < E < 20\,MeV$). The collimation of neutron beam is satisfactory: J_{epi}/Φ_{epi} is approx. 0.6.

5. CONCLUSIONS

The results of Monte Carlo calculations prove the thermalizing column of the TRIGA reactor as a quite appropriate for the installation of efficient irradiation facility for a clinical BNCT treatments. The refurbishment of the thermalizing column for the purposes of BNCT clinical treatments requires some significant improvements on the reactor itself and accompanying equipment:

- the upgrade of TRIGA reactor from current 250 kW to 750 kW would almost be mandatory. In this case the recommended therapy limit of Φ_{epi} ($10^9\,n/cm^2s$) can be exceeded for 10%, keeping the treatment times around 50 min. If the reactor retains the existing power level, the treatment times should be substantially longer (up to 2.5 h).
- the application of the fission converter for the purposes of BNCT irradiation in TC/DC, would require the installation of air based cooling system, with capacity of $500\,m^3/h$. Even in the case of reactor power raise up to 750 kW, the temperature remains below uranium safety limit (640 °C).

REFERENCES

1. M. Maučec, and B. Glumac, Development of irradiation facility for BNCT at the JSI TRIGA Research Reactor, in "Advances in Neutron Capture Therapy", Vol. I, Medicine & Physics, ELSEVIER, pp. 370–376, 1997.
2. J. Briesmeister, Ed., MCNP—A General Monte Carlo N-Particle Transport Code, Ver. 4B, LA-12625-M, March 1997.
3. H. Rief et al., Generating Epithermal Neutron Beams for BNCT at TRIGA Reactors, draft report prepared for *Pavia Workshop on New Clinical Applications and Neutron Sources for BNCT*, June 13–14, 1991.
4. H. Liu, PbF_2 compared to Al_2O_3 and AlF_3 to produce an epithermal neutron beam for radiotherapy, *Med. Phys.*, 23(2), 279–280, 1996.
5. H. Liu et al., Enhancement of the Epithermal Neutron Beam used for BNCT, *Int. J. Radiation Oncology. Biol. Phys.*, 28(5), 1149–1156, 1994.

THE CHARACTERISTICS OF THE UPDATED HEAVY WATER FACILITY OF THE KYOTO UNIVERSITY REACTOR (II) (NEUTRON ENERGY SPECTRA FOR SEVERAL IRRADIATION MODES)

Yoshinori Sakurai, Tooru Kobayashi, and Katsuhei Kobayashi

Research Reactor Institute
Kyoto University, Kumatori-cho
Sennan-gun, Osaka 590-0494
Japan

1. INTRODUCTION

The Heavy Water Neutron Irradiation Facility of the Kyoto University Reactor (KUR-HWNIF) was remodeled at March 1996 mainly for the advance on neutron capture therapy (NCT).[1] The remodeled facility is available for the several neutron energy spectra from almost pure thermal neutrons to epi-thermal neutrons, using the heavy water spectrum shifter, and the thermal neutron filters of cadmium and boral. After the remodeling, eighteen medical irradiations, five for the thermal neutron irradiation and thirteen for the mix irradiation of thermal and epi-thermal neutrons, were already performed as of Feburuary 2000.

The neutron energy spectra at the standard irradiation position (2 cm distance from the bismuth surface) for the three following irradiation modes; "OO-0000-F" (standard mixed neutron irradiation mode), "CO-0000-F" (standard epi-thermal neutron irradiation mode) and "OB-0000-F" were determined. The notation rule for the irradiation mode is described in detail in Ref. 1. For these irradiation modes, all heavy water spectrum shifter and heavy water shutter tanks are empty, so the neutron intensity is the largest among all of the irradiation modes.

2. THE UPDATED FACILITY

Figure 1 shows the updated facility. The epithermal neutron control region of aluminum and D_2O (80%/20%) is added to the inside of the heavy water tank to adjust the

Frontiers in Neutron Capture Therapy, edited by Hawthorne *et al.*
Kluwer Academic / Plenum Publishers, New York, 2001.

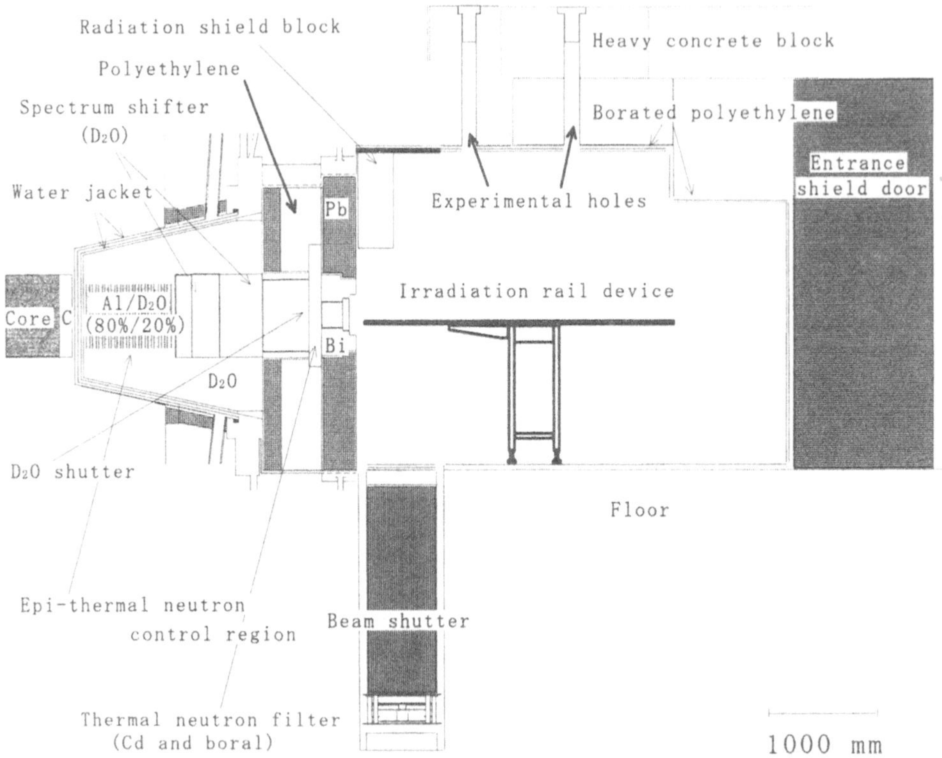

Figure 1. The KUR Heavy water neutron irradiation facility (side view).

neutron energy spectrum and to enhance the epi-thermal neutron component. The neutron energy spectrum shifter of D_2O is installed in the outside of the Al-D_2O region. The neutron energy spectrum shifter consists of three parts whose thickness are 10, 20 and 30 cm, in that order from the core-side. The thickness of the D_2O layer can be controlled from 0 cm to 60 cm by every 10 cm. A heavy water shutter (30 cm-thick) is also placed at the outside of the spectrum shifter. Outside of the heavy water shutter, two kinds of thermal neutron filters are added to control the thermal neutron component of the neutron beam. One is a cadmium filter of 1 mm thickness, and the other is a boral filter of 6.4 mm thickness. The apertures of the filters can be controlled from 0 mm to 620 mm.

3. DETERMINATION METHOD FOR NEUTRON ENERGY SPECTRUM

From the viewpoints of the improvement on the evaluation accuracy and following the dose estimation for the biomedical uses, the neutron spectrum determination was performed for the two separate energy regions; thermal neutron region, and epithermal and fast neutron region. For the thermal neutron region, the flux was mea-

Table 1. Nuclear reactions utilized in the adjusting by NEUPAC

Resonance reaction			
Nuclear reaction	Main resonance peak energy (eV)	Nuclear reaction	Main resonance peak energy (eV)
$^{115}In(n,\gamma)^{116m}In$	1.46	$^{197}Au(n,\gamma)^{198}Au$	4.91
Threshold reaction			
Nuclear reaction	Threshold energy (MeV)	Nuclear reaction	Threshold energy (MeV)
$^{115}In(n,n')^{115m}In$	1.2	$^{27}Al(n,p)^{27}Mg$	4.4
$^{47}Ti(n,p)^{47}Sc$	2.2	$^{24}Mg(n,p)^{24}Na$	6.8
$^{58}Ni(n,p)^{58}Co$	2.8	$^{27}Al(n,\alpha)^{24}Na$	7.2
$^{54}Fe(n,p)^{54}Mn$	3.1	$^{48}Ti(n,p)^{48}Sc$	7.6

sured with the activation method using bare gold foil and cadmium-covered foil in the ordinary manner. The thermal neutron spectrum was assumed to be the Maxwellian distribution of 40 degrees centigrade corresponding with the absorbed dose estimation in a living body.

For the epithermal and fast neutron region, the spectrum was determined by multi-activation foil method with adjusting code "NEUPAC".[2] Table 1 shows the nuclear reactions utilized in the adjusting. The initial input spectrum of 144 groups for NEUPAC was made by the interpolation of the calculated spectrum of 11 groups by "DOT 3.5" with "ENDF/B-IV" and "JENDL-3" libraries.[3] The thermal neutron spectrum was assumed to be a Maxwellian distribution of 40 degree centigrade. The spectrum shape of the fast neutron region was approximated by the Watt-like spectrum. For the epithermal neutron region, a linear approximation was adopted.

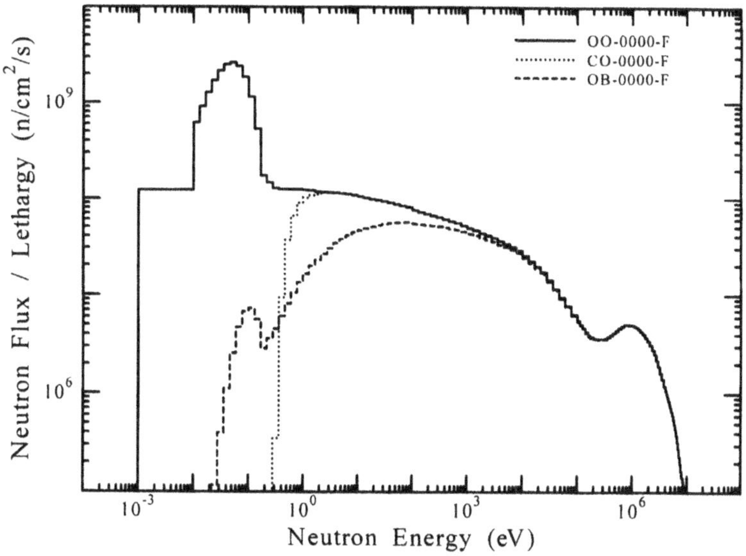

Figure 2. Determined neutron energy spectra for the OO-0000-F, CO-0000-F and OB-0000-F modes.

Table 2. Three-group neutron flux and Kerma absorbed dose rate for tissue

OO-0000-F mode				
Neutron groups	Neutron energy range (eV) Upper	Lower	Neutron flux ($n/cm^2/s$)	Kerma absorbed dose rate (cGy/hr)
Fast	—	1.0E + 4	4.7E + 07	1.6E + 2
Epi-thermal	1.0E + 4	6.0E − 1	8.0E + 08	2.3E + 1
Thermal	6.0E − 1	—	5.0E + 09	2.4E + 2
CO-0000-F mode				
Neutron groups	Neutron energy range (eV) Upper	Lower	Neutron flux ($n/cm^2/s$)	Kerma absorbed dose rate (cGy/hr)
Fast	—	1.0E + 4	4.7E + 07	1.6E + 2
Epi-thermal	1.0E + 4	6.0E − 1	7.3E + 08	2.2E + 1
Thermal	6.0E − 1	—	3.0E + 07	2.4E − 1
OB-0000-F mode				
Neutron groups	Neutron energy range (eV) Upper	Lower	Neutron flux ($n/cm^2/s$)	Kerma absorbed dose rate (cGy/hr)
Fast	—	1.0E + 4	4.6E + 07	1.6E + 2
Epi-thermal	1.0E + 4	6.0E − 1	3.9E + 08	1.9E + 1
Thermal	6.0E − 1	—	1.6E + 07	5.4E − 01

Combining the above two spectra at the "cadmium cut-off " energy (0.6 eV in this report), the final neutron energy spectrum was obtained. The uncertainty of this determination method was thought to be within about 20%.

4. RESULTS AND DISCUSSIONS

4.1. Determined Neutron Energy Spectrum and Absorbed Dose

Figure 2 shows the determined neutron energy spectra for the OO-0000-F, CO-0000-F and OB-0000-F modes. The difference between the CO-0000-F and the OB-0000-F, depends on the energy dependencies for the neutron penetration of the cadmium filter and the boral filter. Table 2 shows the Kerma absorbed dose for tissue at the standard irradiation position.[4] The composition of tissue was assumed to be H: 11.1%, C: 12.6%, N: 2.0%, O: 74.3%.

4.2. Depth Dose Distribution in a Living Body under NCT

For the confirmation of the determined neutron energy spectra, the depth dose distributions in a human head under NCT were calculated by the simulation calculations using DOT 3.5. Figure 3 shows the depth distribution of neutron fluxes along the central axis in a head phantom of 18 cm diameter and 20 cm length for the OO-0000-F mode. The experimental results using activation foils and wires are also shown in this figure. The calculated and experimental results are normalized by thermal neutron flux at the 2 cm depth from the phantom surface.

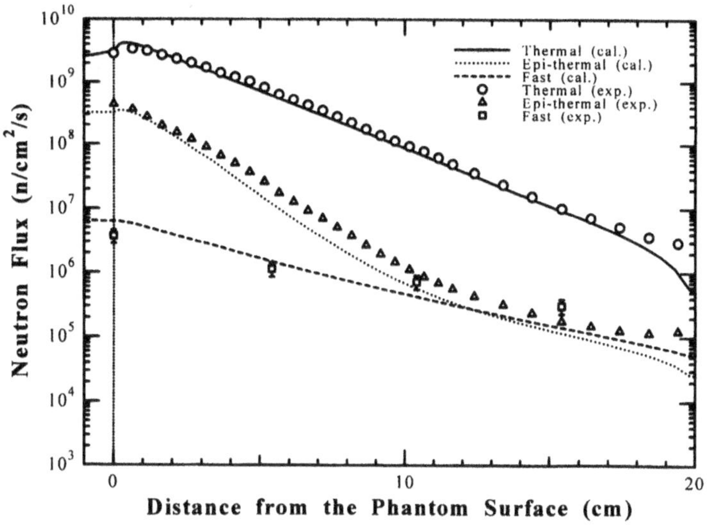

Figure 3. Depth neutron flux distributions in the water phantom for the OO-0000-F mode.

The calculation and experimental distributions are in good agreement for the thermal neutron flux. But the slope of the experimental ones are more gentle than the calculation ones for the epi-thermal and fast neutron fluxes. From these results, it is guessed that the actual fast neutron beam is more forward-directional than the calculated one.

ACKNOWLEDGMENTS

The authors thank Prof. Koji Ono of Kyoto University Research Reactor Institute (KURRI) for invaluable advice from the side of radiation biology.

REFERENCES

1. T. Kobayashi, K. Kanda, Y. Fujita, Y. Sakurai, and K. Ono, The upgraded of the Heavy Water Facility of the Kyoto University Reactor for neutron capture therapy, in "Advances in Neutron Capture Therapy, Vol. I, Medicine and Physics," B. Larsson, J. Crawford, and R. Weinreich, ed., Elsevier, Amsterdam, pp. 321–325, 1997.
2. T. Taniguchi, N. Ueda, M. Nakazawa, and A. Sekiguchi, "Neutron Unfolding Package Code 'NEUPAC-83'", NEUT Research Report 83-10, Department of Nuclear Engineering & Nuclear Engineering Research Laboratory, The Faculty of Engineering University of Tokyo, 1983.
3. W.A. Rhoades, and F.R. Mynatt, " DOT-III, Two Dimensional Discrete Ordinates Transport Code", ORNL-TM-4280, 1973.
4. R.S. Caswell, J.J. Coyne, and M.L. Randolph, Kerma factors for neutron energies below 30 MeV", Radiat. Res. 83:217–254, 1980.

THE TEST-TYPE HYPER-THERMAL NEUTRON GENERATOR FOR NEUTRON CAPTURE THERAPY

Yoshinori Sakurai and Tooru Kobayashi

Research Reactor Institute
Kyoto University, Kumatori-cho
Sennan-gun, Osaka 590-0494
Japan

1. INTRODUCTION

The total neutron cross-section of water, which is a main component of human tissue, monotonously decreases from about 100 barn to about 40 barn as the neutron energy increases from thermal neutron region to nearly 0.5 eV, and it is almost constant to be about 40 barn from nearly 0.5 eV to nearly 100 keV. Considering this energy characteristics of the water cross-section, we proposed the application of the thermal neutrons shifted to higher temperature, namely higher energy, which we named "hyper-thermal neutrons",[1] to neutron capture therapy (NCT).

For the incidence of hyper-thermal neutrons with a Maxwellian distribution of a few thousand K, (i) the dose depth distribution of thermal neutrons is expected to be intermediate between thermal neutron incidence and epi-thermal neutron incidence, (ii) the neutron controllability is expected to be as good as that of the thermal neutron incidence, and moreover (iii) the contamination of fast neutrons is negligible level. So to speak, hyper-thermal neutron irradiation takes advantages of both thermal neutron irradiation and epi-thermal neutron irradiation.[2]

2. A TEST-TYPE HYPER-THERMAL NEUTRON GENERATOR

On the basis of the parametric survey results for high temperature scatterers and those sizes,[1] the design study for a hyper-thermal neutron generator was performed and a "test-type" generator was made for trial. The main design concepts were as follows; (1) the total size of the test-type generator, especially the thickness of the "path" for neutron

Frontiers in Neutron Capture Therapy, edited by Hawthorne *et al.*
Kluwer Academic / Plenum Publishers, New York, 2001.

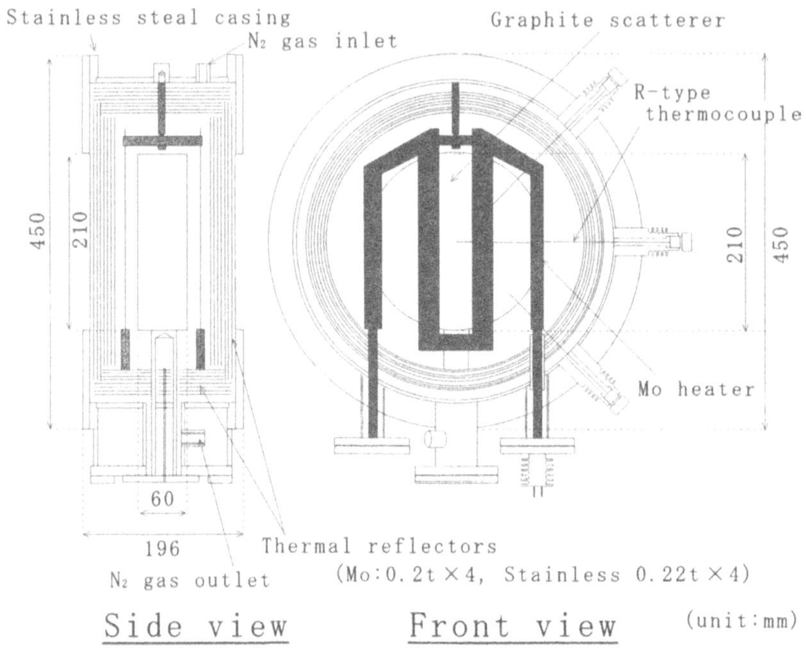

Figure 1. High temperature scatterer of the test-type hyper-thermal neutron generator.

beam, is as small as possible, using the structural material of low neutron-absorption, in order to obtain high neutron intensity. (2) the maximum attainable temperature is as high as possible, maintaining safety and controllability of the heating system, and (3) the high temperature scatterer is selected from the viewpoint of safety-first.

Figure 1 shows the enlargement of the high temperature scatterer part of the test-type hyper-thermal neutron generator, and the outline of the test-type generator is summarized in Table 1. A graphite plate of 6 cm in thickness and 21 cm in diameter was selected as the scatterer from the standing-points of its stability at high temperature, handling simplicity and easy maintenanceability. Stainless steel was selected as main

Table 1. Outline of the test-type hyper-thermal neutron generator

Available temperature	20 °C (293 K, room temperature)—almost 1,200 °C
Irradiation field size	210 mm diameter
Scatterer	Graphite of 60 mm thickness and 210 mm diameter Possible to replace with the other scatterer
Casing	Stainless steel (Outer size: 196 mm thickness, 450 mm diameter)
Control and heating method	Program controller (300 °C/hr under the heating operation) Electric heating using Mo heaters (12 kW)
Heat remove	Thermal reflection using Mo reflectors of 0.2 mm thickness × 4 and stainless steal reflectors of 0.22 mm thickness × 4 for one side Air-cooling using two fans
Heat measurement	Three R-type thermo-couples (Pt-Pt' 1.3%Rh) 0 cm, 5 cm and 9 cm from the central axis in the graphite scatterer.
Cover gas	N_2: 2.5 l/min under the heating operation 10 l/min under the cooling operation

material for the structural frame and the casing. The outer size of the stainless casing is 19.6 cm in thickness and 45 cm in diameter. The central part of 21 cm diameter, which corresponds to the irradiation field area (same as the scatterer area) is thin to be 0.2 mm. The inside of the stainless casing is filled with nitrogen gas under the employment, in order to prevent the oxidization of the graphite scatterer. The nitrogen gas also plays a role as cooling gas. The maximum attainable temperature is 1,200 °C (almost 1,500 K).

From the results of the heating-up tests, it was confirmed that there was no problem for thermal stability, controllability, safety and electrical power supply etc. at high temperature employment. It was found that the differences among the temperatures at the three points (0 cm, 5 cm and 9 cm from the central axis) in the high temperature scatterer were within 3%, and that the temperature at the scatterer center was stably attainable to be 1,200 °C.

3. IRRADIATION CHARACTERISTICS OF THE TEST-TYPE GENERATOR

3.1. Neutron Energy Spectrum

As an irradiation characteristics estimations for the test-type hyper-thermal neutron generator, the neutron energy spectrum dependent on the scatterer temperature was measured by time-of-flight (TOF) method employing the neutron generator of the electron linear accelerator (LINAC) at Kyoto University Research Reactor Institute (KURRI). At the KURRI-LINAC, neutrons are generated due to the (γ,n) reactions of the Bremsstrahlung X-ray at a tantalum metal target (water-cooled), and thermalized due to the light water.

Figure 2 shows the measured neutron energy specra for the graphite scatterer temperatures of 25 °C (room-temperature), 600 °C and 1,200 °C. The spectra were normal-

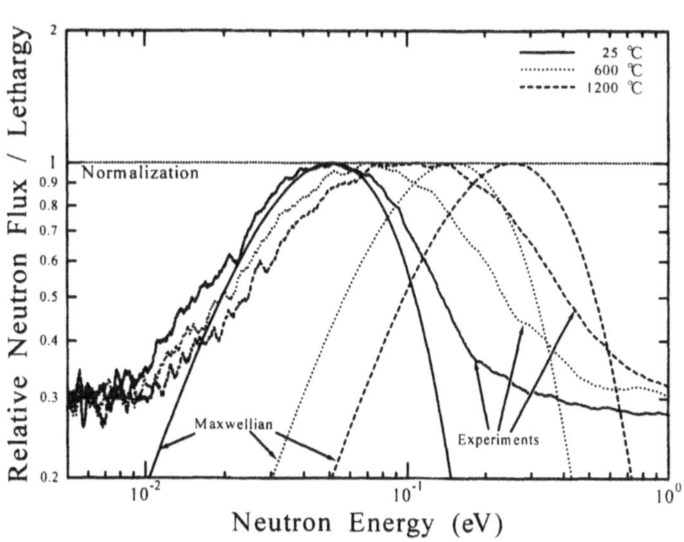

Figure 2. The relative comparison of the measured neutron energy spectra for the scatterer temperatures.

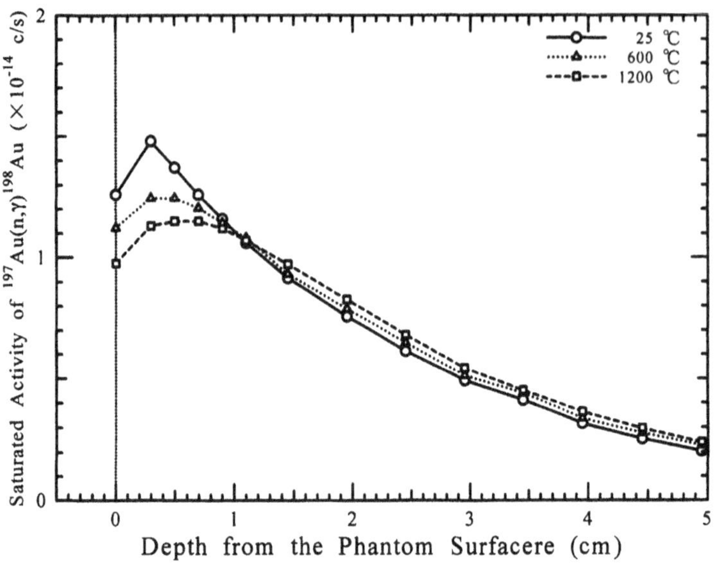

Figure 3. The depth distributions of saturated activity of $^{197}Au(n,\gamma)^{198}Au$ reaction along the central axis in a water phantom for the scatterer temperatures.

ized to be 1 at the spectrum peaks. It is shown that the neutron energy spectrum shifts to higher energy as the scatterer temperature is larger. For reference, the Maxwellian distributions of the respective graphite temperatures are shown in Figure 2. The shapes and peak energy of the measured spectra were broader and lower comparing to the Maxwellian distributions. The main reasons were thought to be as follows; (i) the high temperature scatterer of the test-type generator was surrounded with stainless steel and molybdenum as the casing, the thermal reflectors and the heaters, and (ii) the size of the test-type generator is 6 cm in thickness and 21 cm in diameter, a little thinner and smaller than that for the parametric survey calculations, so the up-scatter performance is a little smaller.

3.2. Thermal Neutron Dose Distribution

On the assumption of the application of hyper-thermal neutron irradiation to actual NCT, a phantom experiment was performed at the KUR Heavy Water Neutron Irradiation Facility (HWNIF)[3] in order to estimate thermal neutron dose distributions in a water phantom (18 cm in diameter, 20 cm in length). Figure 3 shows the saturated activities of $^{197}Au(n,\gamma)^{198}Au$ reaction along the central axis of the water phantom for the graphite scatterer temperatures of 25 °C, 600 °C and 1,200 °C.

Neutron absorbed dose distributions for brain tissue mainly depend on the reaction distributions of the 1/v materials such as $^{1}H(n,\gamma)^{2}D$, $^{14}N(n,p)^{14}C$, $^{10}B(n,\alpha)^{7}Li$ *etc.* in the thermal neutron region. The cross section of the $^{197}Au(n,\gamma)^{198}Au$ reaction also accords to the 1/v energy characteristics below about 1 eV. So, Figure 3 is thought to relatively correspond to the neutron absorbed dose distributions in the water phantom.

The peak position is deeper, the peak height is lower, and the distribution shape is more gentle as the scatterer temperature is higher. At the deeper than 2 cm, the distributions are almost the same for the scatterer temperatures. The controllability of hyper-thermal neutron incidence for the dose distribution in a human body can be confirmed. Assumed that the measured values are normalized at the peak position, the dose depth distribution for the scatterer temperature of 1,200 °C is about 1 cm improved comparing to the 25 °C scatterer temperature.

4. SUMMARY

A test-type hyper-thermal neutron generator was designed and made for trial, on the basis of the fundamental study results mainly by simulation calculations. From the estimation results of the generator characteristics by the experiments at the KURRI-LINAC and the KUR-HWNIF, the prospect about the feasibility of the "hyper-thermal neutron irradiation field for NCT" was opened.

ACKNOWLEDGMENTS

The research was mainly supported by Grant-in-Aid for scientific research (B:08458125) funded from the Ministry of Education, Science, Sports and Culture in Japan.

REFERENCES

1. Y. Sakurai, T. Kobayashi, and K. Kanda, Hyper-thermal neutron irradiation field for neutron capture therapy, Nucl. Instr. Meth. B 94:433–440, 1994.
2. Y. Sakurai, T. Kobayashi, and K. Kanda, A fundamental study on hyper-thermal neutrons for neutron capture therapy, Phys. Med. Biol. 39:2217–2227, 1994.
3. T. Kobayashi, K. Kanda, Y. Fujita, Y. Sakurai, and K. Ono, The upgraded of the Heavy Water Facility of the Kyoto University Reactor for neutron capture therapy, in "Advances in Neutron Capture Therapy, Vol.I, Medicine and Physics", B.Larsson, J.Crawford, and R.Weinreich, ed., Elsevier, Amsterdam, 1997, pp. 321–325.

WHAT ARE THE BEST ENERGIES FOR NEUTRON PRODUCTION FOR NCT?

J. F. Crawford* and D. Suwannakachorn†

P.S.I. and I.M.R., CH-5232 Villigen
Switzerland

1. INTRODUCTION

The $^{7}Li\ (p,n)\ ^{7}Be$ reaction is a promising source of neutrons for NCT. Most workers have taken the proton beam energy to be 2.5 MeV, to take advantage of the higher neutron yield resulting from the resonance in the reaction cross-section at 2.25 MeV. However it is well-known in the NCT community that the problem of moderating the neutrons to the keV energies required for therapy depends strongly on the initial energy of the neutrons. Specifically, it is very different for fission neutrons, whose mean energy is about 2.6 MeV, than it is for the lower energy neutrons from ^{7}Li. This suggests that the question of the most suitable proton energy should be addressed. The topic has been the subject of informal discussion within the community, work along these lines has been discussed and carried out by others [1,2,3,4] and it seemed worthwhile to investigate the question more carefully.

2. SIMULATIONS

Simulations were done using the well-known program MCNP[5] as follows:

- For convenience the neutron moderator was taken to be *BeO*. Other minimally-absorbing moderators can be expected to behave similarly.
- The $^{7}Li\ (p,n)\ ^{7}Be$ reaction was simulated[6] using the kinematics for three convenient proton energies:

* E-mail: John.Crawford@psi.ch.
† E-mail: Dusadee@psi.ch.

Frontiers in Neutron Capture Therapy, edited by Hawthorne *et al.*
Kluwer Academic / Plenum Publishers, New York, 2001.

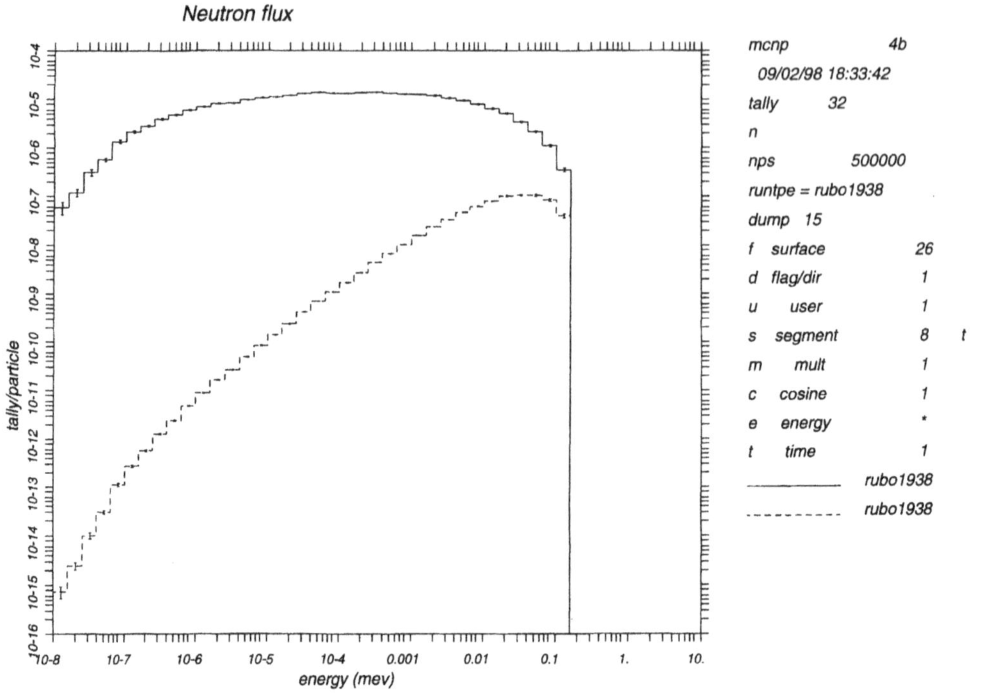

Figure 1. Neutron flux and flux multiplied by energy for $E_p = 1.93$ MeV.

- 1.93 MeV, just above the maximum proton energy that produces an ambiguous kinematic solution.
- 2.25 MeV, at the centre of the resonance maximum in the cross-section.
- 2.50 MeV, for comparison with previous work.

The moderator was in the form of a cylinder of *BeO* of diameter 40 cm, loaded with 2% natural *Li* by number to absorb thermal neutrons. The proton beam was taken to be coaxial with the cylinder, with a point neutron source on the axis. Neutrons were tallied at the downstream face of the moderator. The essence of the study was to vary the thickness of the moderator between the ^7Li source and the exit face of the moderator, studying the properties of the emerging neutron beam.

The criteria used to assess the neutron beam were based on those developed by the BNCT Concerted Action of the EU in the run-up to the patient treatments now under way at Petten, NL. The criterion for the beam energy was that the dose-weighted mean energy should not exceed 8 keV. Because of the behaviour of the dose response in the relevant range of neutron energies, this corresponds to a physical mean energy of about 1/3 of this, i.e. 2.67 keV.

The beam intensity and mean energy were tallied. Because of the kinematics of the ^7Li reaction, the question of contamination neutrons above 0.8 MeV can be ignored under the conditions of the simulations. For simplicity, γ contamination was also ignored, on the assumption that it can be reduced by thin bismuth shielding with little effect on the neutron beam.

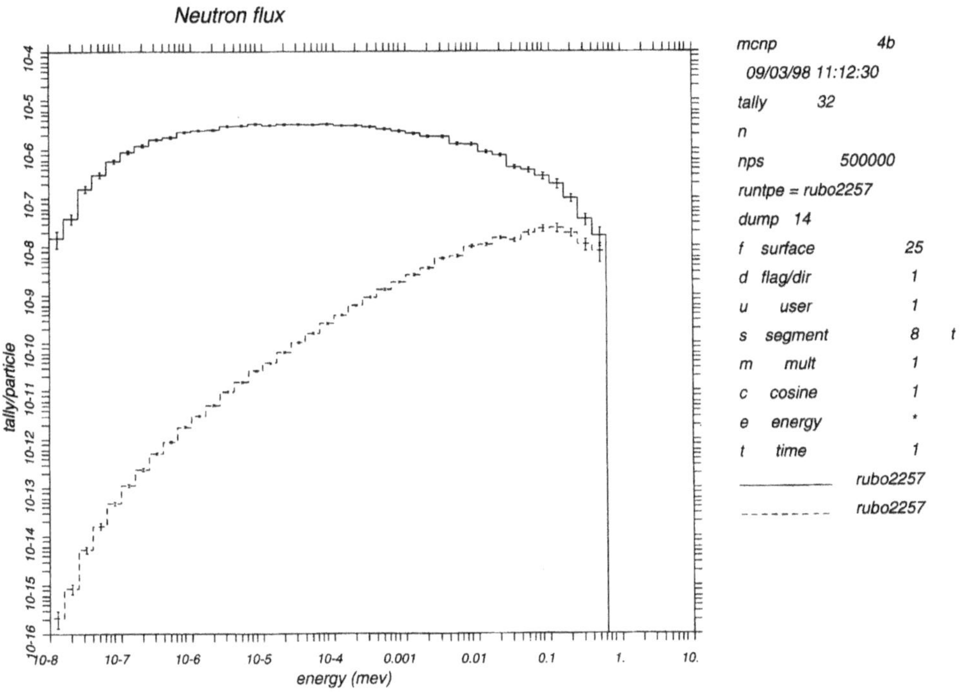

Figure 2. Neutron flux and flux multiplied by energy for $E_p = 2.25\,\mathrm{MeV}$.

3. RESULTS

Spectra for $E_p = 1.93$, 2.25 and 2.50 MeV are shown in Figures 1–3, which show flux and flux multiplied by energy; the ratio of the areas under these histograms is then the mean energy. The effect of E_p on the maximum neutron energy can be clearly seen.

Results are summarised in Table 1. It can be seen that:

1. The *BeO* thickness required to reach <3 keV varies with E_p.
2. The neutron flux per incident neutron is considerably higher at low E_p.
3. Maximum flux × yield peaks near $E_p = 2.25\,\mathrm{MeV}$.

Fluxes are averaged over the 40 cm moderator diameter, and as quoted appear to be too low for therapy. On the other hand, the neutron flux is higher near the centre of the moderator. The conclusion is that a beam energy of about 2.20–2.25 MeV is more effective than the 2.5 MeV assumed so far by most of the BNCT community.

Table 1. Beam properties as a function of E_p. Statistical errors were typically 1% on the fluxes, several % on the mean energies

Proton beam energy	1.93	2.25	2.50	MeV
Neutron yield	0.078	0.84	1.59	per 10^4 beam protons
Moderator thickness	11.75	19.25	27.25	cm *BeO*
Neutron flux	2.65	0.63	0.13	cm^{-2} per 10^4 neutrons produced
Neutron mean energy	2.7	2.8	2.5	keV
Flux × yield	2.07	5.33	2.00	cm^{-2} per 10^9 beam protons

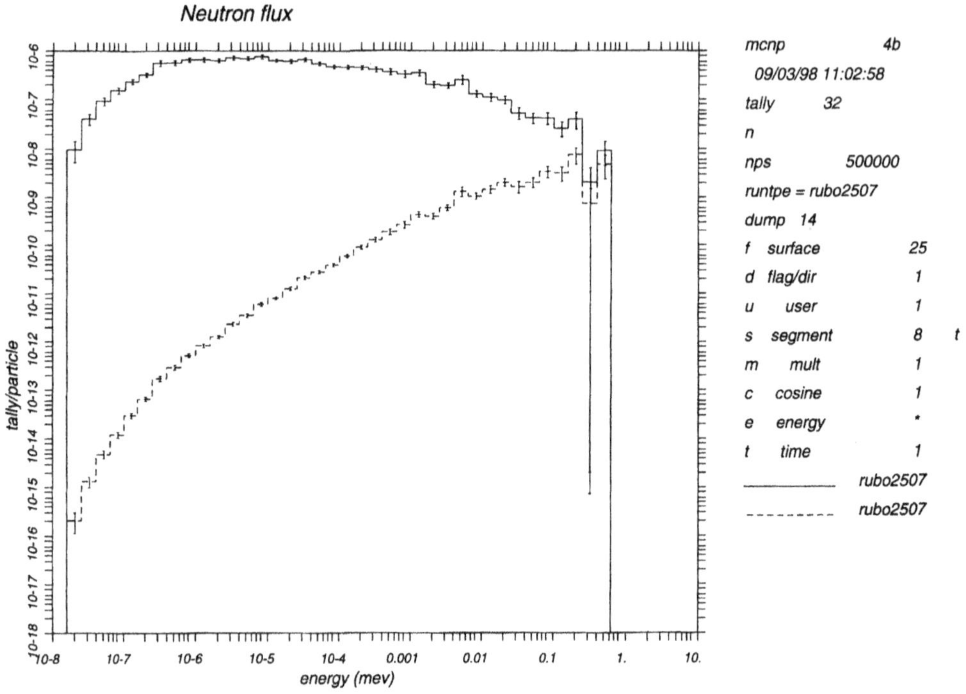

Figure 3. Neutron flux and flux multiplied by energy for $E_p = 2.50\,\text{MeV}$.

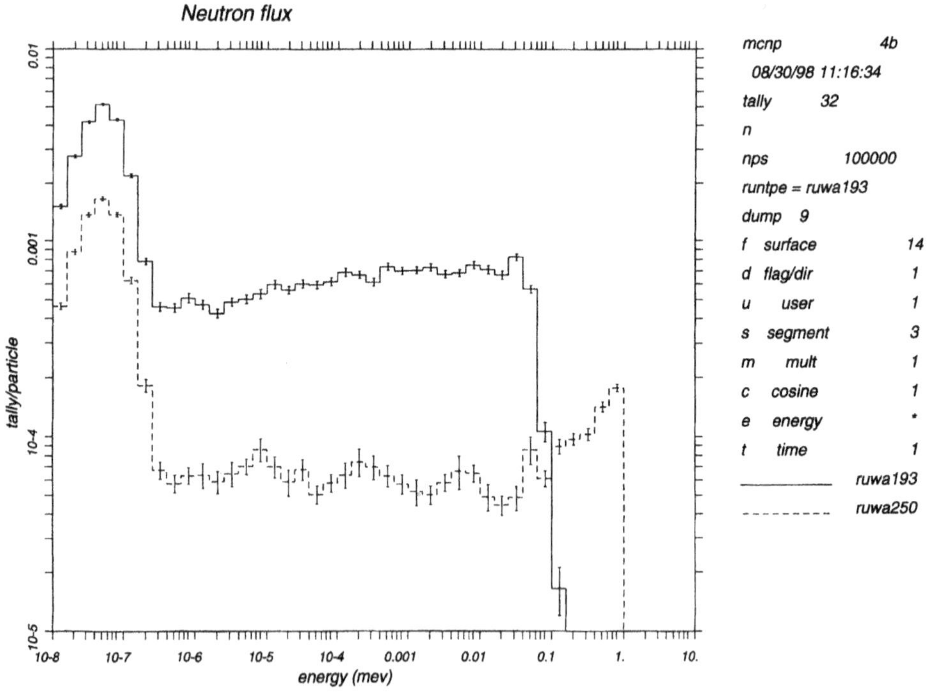

Figure 4. Neutron flux near a ^7Li source for $E_p = 1.93$ and $2.50\,\text{MeV}$.

Another application is the use of interstitial accelerator sources.[7] An example is shown in Figure 4. The neutron fluence three to four cm to one side of a ^{7}Li source is plotted for two different proton beam energies: 1.93 and 2.50 MeV. As can be seen, the thermal neutron fluence is higher and the contamination above about 0.1 Mev is zero at the lower proton energy.

In practice, the monoenergetic reactions assumed in this study are out of the question because an infinitely thin ^{7}Li target as well as a precisely monoenergetic proton beam would be needed. The best that could be achieved would be a beam that entered a target of finite thickness at above the optimum energy and exited below it, in such a way that the mean effective energy was optimised.

REFERENCES

1. R.J. Kudchadker *et al.*, "Advances in Neutron Capture Therapy", B. Larsson, J. Crawford, and R. Weinreich (Editors), Volume I, pp. 516–521.
2. V.N. Kononov *et al.*, ibid., pp. 528–532.
3. W.T. Chu *et al.*, ibid., pp. 533–537.
4. D.L. Bleuel, R.J. Donahue, B.A. Ludewigt, and J. Vujic, Medical Physics 25(9):1725–1734 (1998).
5. "MCNP—A General Monte Carlo N-Particle Transport Code, Version 4B", J.F. Briesmeister (Ed), Los Alamos National Laboratory Report LA-12625-M, (1997).
6. J.B. Marion, and J.L. Fowler (Eds.), Fast Neutron Physics, Interscience Publishers, New York, 1960, pp. 133–176.
7. H. Song *et al.*, "Advances in Neutron Capture Therapy", B. Larsson, J. Crawford, and R. Weinreich (Editors), Volume I, pp. 543–548, and references therein.

THE USE OF POLYTETRAFLUOROETHYLENE (TEFLON ®) CONTAINING FILTER/MODERATORS FOR EPITHERMAL NEUTRON BEAM PRODUCTION

Otto K. Harling, Gordon Kohse, Kent J. Riley, and Shuichi Sakamoto

MIT Nuclear Reactor Laboratory
138 Albany Street
Cambridge, MA 02139

ABSTRACT

We have investigated the suitability of fluorinated hydrocarbon polytetrafluoroethylene (PTFE) for application in the filter/moderator of an epithermal beam. We show that the neutronics performance, cost, and fabricability of Al and PTFE compares favorably with other good filter/moderator materials such as Al/Al_2O_3, Al/C, and Fluental ®. In this study, we have carried out irradiation tests to evaluate the performance of PTFE in the mixed gamma and fast neutron environment expected for the MIT Fission Converter epithermal neutron beam application. The irradiation stability was found to be more than adequate for the planned application.

2. INTRODUCTION

The nuclear properties of fluorine make it a desirable component of filter/moderators used in the production of epithermal beams for neutron capture therapy. However, materials used for beam filtration and moderation must also be physically and chemically stable in the beam environment and possess adequate mechanical properties. We have investigated the suitability of polytetrafluoroethylene, (PTFE, or Teflon ®) for this application.

Frontiers in Neutron Capture Therapy, edited by Hawthorne *et al.*
Kluwer Academic / Plenum Publishers, New York, 2001.

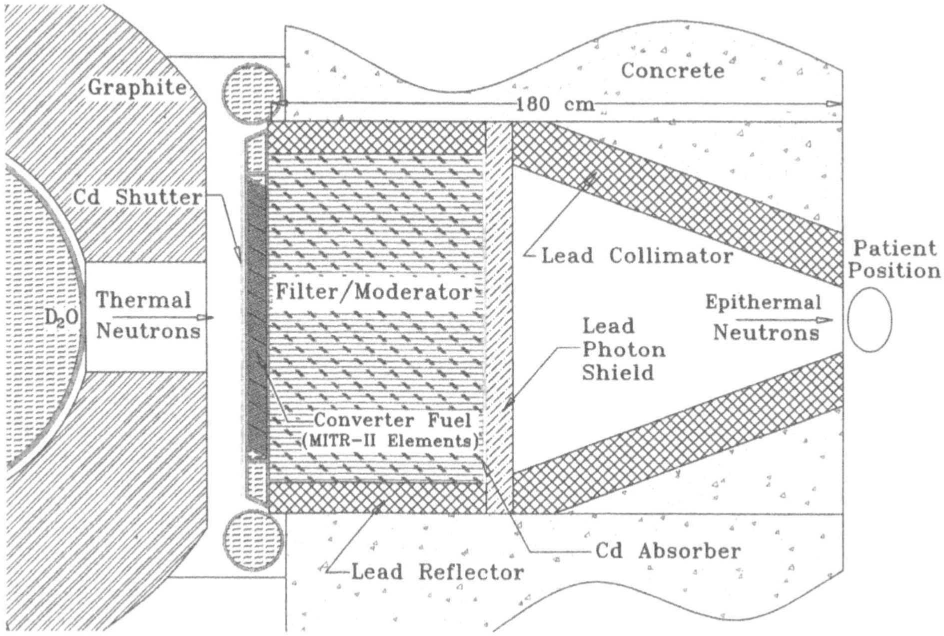

Figure 1. Plan view of the MIT fission converter based epithermal neutron beam facility. The designs reported here use eleven partially burned MITR-II fuel elements (312 g ^{235}U per element), D_2O coolant, and assume a reactor power of 5 MW. Other important beam components are indicated in the figure above.

3. NEUTRONIC PERFORMANCE

The neutronic performance of PTFE was evaluated by Monte Carlo radiation transport calculations (MCNP 4a) for use in the design of the MIT fission converter based beam (FCB).[1,2,3] Figure 1 shows the plan view of the MIT FCB which was used as the model to evaluate the neutronic performance of various filter/moderator materials and arrangements. Table 1 shows representative results of these calculations taken from References 2 and 3. The results in Table 1 show that high epithermal fluxes can be obtained with negligible specific fast neutron and gamma ray contaminations for a variety

Table 1. Summary of the calculated performance for the MIT fission converter based epithermal beam (see Figure 1) for various filter/moderator arrangements*

Filter/Moderator	ϕth n/cm^{2}s	ϕepi n/cm^{2}s	ϕf n/cm^{2}s	J_{epi}/ϕ_{epi}	D_{fn}/ϕ_{epi} cGycm^2/n	D_γ/ϕ_{epi} cGycm^2/n
83 cm Al – 17 cm Al_2O_3	9.28E + 08	1.32E + 10	4.28E + 08	0.65	1.69E – 11	6.39E – 12
	2.1%	1.3%	3.2%	1.8%	2.2%	2.8%
66 cm FLUENTAL	1.45E + 09	1.31E + 10	1.88E + 08	0.66	1.17E – 11	1.08E – 11
w/o LiF	0.6%	0.5%	1.3%	0.6%	0.7%	0.8%
80 cm Al – 17 cm $(CF_2)n$	1.07E + 09	1.38E + 10	4.04E + 08	0.64	1.63E – 11	7.80E – 12
	1.9%	1.3%	2.8%	1.8%	2.1%	2.4%
96 cm Al – 12 cm C	1.06E + 09	1.20E + 10	3.83E + 08	0.63	1.73E – 11	8.67E – 12
	1.2%	0.9%	1.9%	1.2%	1.5%	1.3%

*The statistical uncertainty of each quantity (one standard deviation) is listed below the value as a percentage.

Table 2. Cost Estimates for the filter/moderator materials in the MIT FCB. Various designs, all providing good neutronic performance

Filter/moderator arrangement	Cost of filter/moderator ($US)
96 cm Al/12 cm Graphite	$25,400†
80 cm Al/13 cm Teflon ®	$28,500
83 cm Al/17 cm Al_2O_3	$107,000
66 cm Fluental ®	$271,000*

†—Assumes no cost for reactor grade graphite, which is usually available as surplus material.
*—A license for human irradiations may be required.
Prices are calculated for a F/M with a height and width of 105 cm and length as needed to achieve comparable performance (see Table 1). Prices (in $US) include cost for cutting material to proper size

of filter/moderator materials. The Teflon® containing filter/moderator produces results which are comparable to other good filter/moderator materials including a proprietary patented composite of AlF_3 (69 m-%), Al(30 m-%) and LiF [Fluental®] which has been shown to have excellent properties for this application. Other researchers have also shown that PTFE is a useful component in epithermal beam filter/moderators.[4] While for the MIT FCB application, a fixed minimum distance (1.8 m) is available between the neutron source and the patient position, in other designs it may be desirable, e.g., from intensity considerations, to shorten the source to patient position distance. If shorter source to patient position distances are desired, the length of the filter/moderator plays an increasingly important role. Column 1 of Table 1 gives the thicknesses of the filter moderator. Fluental® is by far the shortest at 66 cm, Al/$(CF_2)n$ and Al/Al_2O_3 are comparable at 97 and 100 cm and with Al/C the longest moderator at 108 cm.

4. COST ESTIMATES FOR FILTER/MODERATORS

It is interesting to compare the estimated costs of various filter moderator materials in the MIT FCB application. Table 2 provides this information. There are significant cost advantages for Al/C or Al/Teflon ®, these are a factor of four less expensive than Al/Al_2O_3 and an order of magnitude less expensive than the proprietary Fluental®. Furthermore, licensing costs for Fluental ® could add to the cost of using Fluental®.

Flexibility and ease of fabrication in constructing a filter/moderator favors Al/C and Al/Teflon® since these materials are easily machined into the desired shapes and modifications can easily be achieved to optimize filter moderator performance. Fluental®, a refractory material, can be expected to be the most costly to modify by machining.

5. IRRADIATION PERFORMANCE OF TEFLON®

The irradiation stability of Al and graphite is more than adequate for this application. Aluminum oxide has been used successfully by the Brookhaven group in their current epithermal neutron beam facility. Fluental® has been tested for its irradiation

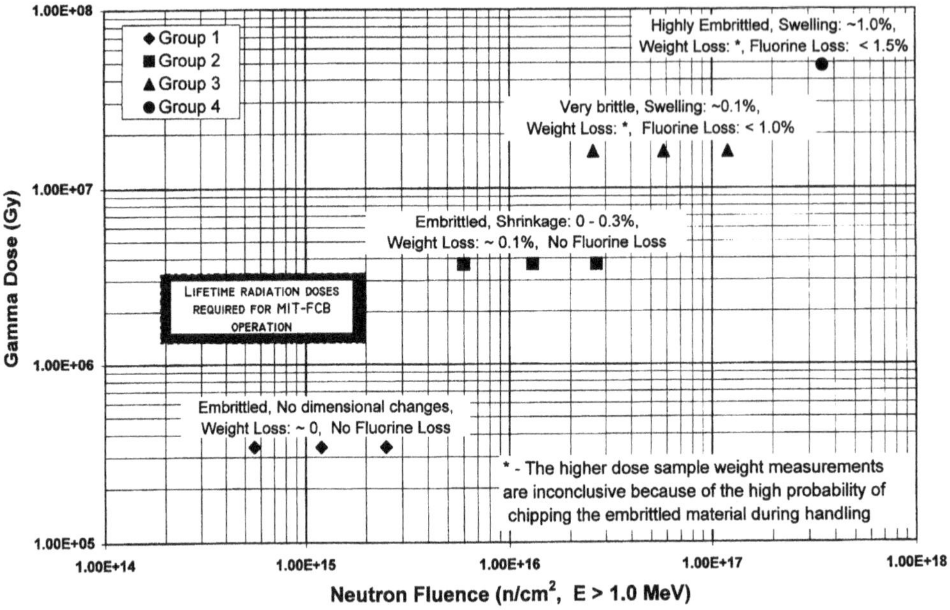

Figure 2. Irradiation performance of the Polytetrafluoroethlyene (PTFE) samples irradiated at the MITR-II. The box on the right represents the range of gamma dose and neutron fluence the PTFE will be exposed to over its lifetime in the MIT FCB application. Data points represent the gamma dose and neutron fluence that samples received and the nearby text describes the changes that were observed at each dose level.

resistance and there should be no *a priori* reason to expect that it will not perform satisfactorily in the epithermal beam application for NCT. Teflon® (PTFE) on the other hand, is known to be sensitive to irradiation with significant change in mechanical properties at gamma doses as low as E + 04 Gy. Some irradiation performance data[5] has been developed for mixed neutron and gamma radiation fields which indicates that PTFE may be acceptable for the filter/moderator application in NCT beams.

We have carried out a series of irradiations in the MIT Research Reactor to evaluate the suitability of PTFE for use in the MIT FCB. Samples of PTFE were irradiated in mixed fields of fast neutrons and gamma rays to several dose levels. The estimated fluence and dose for ten years of operation, at an operating capacity factor of 3%, in the MIT-FCB are 2 E + 15 n/cm^2 [E > 1 MeV] and 1.5 – 3.5 E + 06 Gy from gamma rays. Bars of virgin grade Teflon®, about 0.6 cm square by 10 cm long were irradiated in a pneumatic tube irradiation facility near the MITR-II's core tank. The samples received a mixed dose from fast neutrons and gamma rays. Dose rates were much higher, four orders of magnitude, during this test than those expected in the filter moderator application in order to provide data in a reasonable time. However, the ratio of fast neutron flux to gamma dose rate covered the range expected for the MIT/FCB.

The essential criteria for adequate performance of Teflon in the fission converter beam are basic physical integrity and chemical stability. In addition, although swelling can be accommodated in the design, it must be reasonably well quantified to ensure reliable performance. To assess these characteristics, the samples were weighed and the through-sheet thickness was measured before and after irradiation. Visual observation and subjective physical tests (bending, cleaving) were also performed. Neutron activation analysis for fluorine assay was also carried out.

The results of the irradiation tests are summarized in Figure 2 for several groups of samples which have been exposed to increasing levels of gamma dose and fast neutron fluence. The expected exposures for the MIT FCB are shown in the rectangle on the left hand side of the figure. Samples were highly embrittled even at the lower doses. Nevertheless, their physical integrity remained adequate for the application. Weight loss measurements and neutron activation measurements of fluorine loss showed no significant losses even at doses much higher than expected in the MITR application. Swelling or shrinkage was negligible at doses expected in the application and reached ~1% at doses one to two orders of magnitude higher than expected for the MIT FCB.

ACKNOWLEDGMENT

This work was supported by the US DOE, Contract No. DE FG 02-97ER 62489. Shuichi Sakamoto also gratefully acknowledges the Science and Technology Agency of Japan for supporting the work he carried out while at MIT.

REFERENCES

1. W.S. Kiger, III, S. Sakamoto, and O.K. Harling, "Neutronic Design of a Fission-Converter Based Epithermal neutron Beam for Neutron Capture Therapy," Nuclear Science and Engineering,: 131, 1–22, (1999).
2. W.S. Kiger, III, "Neutronic Design of a Fission Converter-Based Epithermal Beam for Neutron Capture Therapy," Masters of Science and Nuclear Engineer Thesis, Massachusetts Institute of Technology, Cambridge, MA, May 1966.
3. S. Sakamoto, "Sensitivity Studies of the Neutronic Design of a Fission Converter-Based Epithermal Beam for Boron Neutron Capture therapy," Masters of Science and Nuclear Engineers Thesis, Massachusetts Institute of Technology, Cambridge, Massachusetts, June 1997.
4. D.W. Nigg, H.E. Mitchell, V.D. Harker, and J.F. Harmon, Computational and experimental studies of an electron accelerator based epithermal photoneutron source facility for boron neutron capture therapy, in "INEL BNCT Research Program Annual Report 1995", Idaho National Engineering Laboratory, pp. 5–15 and pp. 27–45.
5. Private communication with David W. Nigg, November 24, 1997.

52

THREE-DIMENSIONAL MEASUREMENT OF THERMAL NEUTRON ABSORBED DOSE FOR BNCT EXPERIMENTAL PROGRAM, PERFORMED IN A FAST RESEARCH REACTOR

G. Gambarini[1,3], S. Agosteo[2,3], P. Marchesi[1], E. Nava[4], A. Pecci[1], and G. Rosi[5]

[1]Università degli Studi di Milano
Dipartimento di Fisica, Milano, Italy
[2]Dipartimento di Ingegneria Nucleare
Milano, Italy
[3]INFN Sezione di Milano, Italy
[4]Università degli Studi di Bologna
Dipartimento di Fisica, Bologna, Italy
[5]ENEA/INN-TEC
S. Maria di Galeria, Roma, Italy

1. INTRODUCTION

For treatment planning it is very important to perform preliminary measurements of the spatial distribution of absorbed dose with good resolution and reliability. To this purpose computer simulations are commonly utilised, supported by experimental measurements performed through small dosimeters placed at various locations in a tissue-equivalent phantom.

The aim of this work is to set up tissue-equivalent phantoms acting by themselves as continuous dosimeters in which the spatial distribution of absorbed dose can be measured by means of a suitable imaging technique. The contributions of the different components of the radiation field can be discriminated by means of differential measurements. Here the proposed method is described and the first results are presented, obtained by irradiating phantom-dosimeters with thermalized neutrons in a fast research reactor.

Frontiers in Neutron Capture Therapy, edited by Hawthorne *et al.*
Kluwer Academic / Plenum Publishers, New York, 2001.

Table 1. Composition of the dosimeter gels utilised in the experiment

Compound	Chemical formula	Concentration	
		FriXy-gel	FriXy-gel + N (2.2%)
Ferrous ammonium	$Fe(NH_4)_2(SO_4)_2 \cdot 6H_2O$	0.5 mM	0.5 mM
Xylenol orange	$C_{31}H_{27}N_2Na_5O_{13}S$	0.11 mM	0.11 mM
Sulphuric acid	H_2SO_4	25 mM	25 mM
Urea	CH_4N_2O	—	0.84 M
Agarose seaplaque	$(C_{12}H_{14}O_5(OH)_4)_n$	1%	1%

2. DOSIMETRY TECHNIQUE

The dosimetric phantom is obtained by incorporating a chemical dosimeter into a tissue-equivalent material in form of gel. The chemical dosimeter is a ferrous sulphate solution, which is the main component of the standard Fricke dosimeter, and the gelling agent is the polysaccharide Agarose SeaPlaque. In ferrous sulphate solutions, ionising radiation induces a conversion of ferrous ions Fe^{2+} to ferric ions Fe^{3+}, and the conversion yield is proportional to the absorbed dose.[1,2] The metal ion indicator Xylenol Orange is added to the gel composition in order to perform optical analysis. In presence of such compound, ferrous oxidation results in the formation of complexes, which present an absorption band of visible light centred at 585 nm. This light absorption has shown to be linearly correlated to the absorbed dose.[3,4] Some other chemical compounds are added, too, with the purpose of achieving the tissue-equivalence for the radiation of interest, as explained in section 4. The composition of the dosimeter gels utilised in this experiment is reported in Table 1.

This gel dosimeter, containing the Fricke solution and Xylenol Orange, will be named FriXy-gel.

The proposed technique for the 3-D measurement of absorbed dose is based on dose imaging by means of a CCD chamber.[4] To this purpose, the phantom to be analysed is composed of a set of piled-up gel layers. Each layer consists of a stratum of gel within two transparent polyethylene or mylar films, held by a frame. The gel layers are 3 mm thick. This was done with the aim of achieving both reliability of image detection and simplicity of sample preparation. After exposure of the whole phantom to ionising radiation, each layer is promptly imaged with the CCD camera. From the 2-D images, the 3-D image of the absorbed dose is reconstructed by means of suitable software. The imaging apparatus consists of a light diffuser, on which gel layers are located, and a CCD camera. Here an interference filter is placed between the 50-mm camera lens and the CCD detector, to match the wavelength (585 nm) of the absorption peak. Each acquired transmittance image includes two stripes of transmittance standards, with different optical densities associated to different dose values, for converting the grey levels into values of relative dose. In order to get precise values, the conversion of grey levels into dose is accomplished for each gel preparation. This calibration is performed by analysing samples having the same shape, exposed to known γ-doses.

3. THE TAPIRO REACTOR

The TAPIRO reactor, that is located in the ENEA Casaccia Center near Rome, is a highly enriched uranium-235 fast neutron facility, commonly utilised for fast reactor

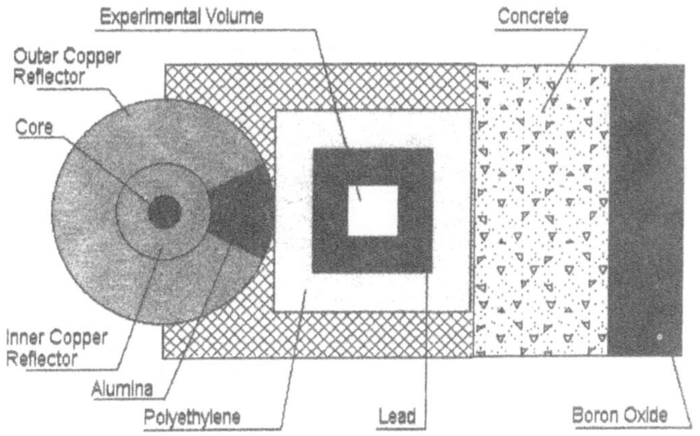

Figure 1. Moderating and shielding structure set up in the thermal column of the TAPIRO reactor.

shielding experiments, blanket studies, biological effects of fast neutrons and so on. The nominal power is 5 kW (thermal) and the maximum neutron flux is $4 \cdot 10^{12}$ n/cm^2 s.

In this experiment, we have utilised the thermal column. Here the graphite blocks were removed and a polyethylene structure was set up to thermalize neutrons. This structure, whose section is shown in Figure 1, has a cubic shape. It is made with polyethylene blocks, with 15 cm thickness. A 10-cm thick lead γ-shield was located inside the polyethylene, in order to have a very low γ background in the irradiation volume, which is a cubic cavity with sides of about 18 cm. The γ background of this reactor is very low, but the shielding is necessary in order to screen the γ-rays coming from the nuclear reactions produced by neutrons in the polyethylene. The thermal neutron flux in this thermal irradiation volume, at the maximum reactor power, is $7 \cdot 10^7$ n/cm^2 s.

4. EXPERIMENTAL RESULTS

For a good treatment planning in BNCT, besides the dose due to the presence of ^{10}B, it is convenient to determine the dose delivered by thermal neutrons in surrounding tissue without boron, and the reactor background dose, when it is not negligible.

Thermal and epithermal neutrons are not directly ionising particles, and the delivered dose is due to the secondary particles ejected in the nuclear reactions. So for tissue-equivalence the isotopic composition of the tissue-substitute has to be the same as that of the tissue to be simulated, at least for the isotopes giving the main contributions to the absorbed dose, that is H and N, for the reactions ^{1}H(n,γ)^{2}H (σ = 0.33 b) and ^{14}N(n,p)^{14}C (σ = 1.81 b). For small volumes, the main contribution is that from nitrogen but for large volumes, the hydrogen contribution becomes predominant. Owing to the different biological efficiency of the produced radiations, a measurement of the total dose is meaningless and a discrimination of contributions is necessary. Our approach is that of performing differential measurements with three different gel compositions. From a gel having the isotopic composition of the desired tissue, the total dose in healthy tissue is imaged. A gel with the same composition, depleted of nitrogen, gives the absorbed dose due to the γ-rays generated in the nuclear reaction of thermal neutrons with H.

Table 2. Elemental composition of the two gels and of adult brain

	H	N	C + O	Others
Brain (adult)	10.7	2.2	85.7	1.4
FriXy-gel	11.1	~0	88.8	0.1
FriXy-gel + N (2.2%)	10.9	2.2	86.8	0.1

In both gels also γ-background and recoil protons could give contributions but in our irradiation volume their relative amount, evaluated with Monte Carlo method, are negligible. A gel containing ^{10}B in the proper amount and shape gives the therapy dose. Previous calibrations of each gel are necessary for all kinds of involved radiation, for the comparative analysis of the results obtained.

In previous works,[5–8] imaging measurements of a Fricke-infused gel-phantom containing 40 μg/g of ^{10}B were performed in a medical NMR scanner. Isodose curves have been obtained with a bad resolution. Moreover, in phantoms not containing ^{10}B, the resolution of the technique was not sufficient for allowing isodose determinations. The technique described above, based on optical measurements, has appeared to be more promising. Therefore, to verify the feasibility of the technique, we have enquired the possibility of measuring doses in gels without boron. An exposure to thermal neutrons has been performed in the described shielded volume, in the thermal column of the TAPIRO nuclear reactor. The thermal neutron fluence at the entrance of the phantom, measured with activation technique, was $1.17 \cdot 10^{12}$ n/cm^2, comparable to that utilised for the therapy. The exposed phantom was made up of a polyethylene cylinder (16 cm diameter, 14 cm height) with a coaxial cylindrical hole (6 cm diameter). In the hole, a set of FriXy-gel layers (3 mm thick) were arranged, alternating gels with the two compositions reported in Table 1.

As it is shown in Table 2, where the elemental composition of the two gels and of adult brain (ICRU 46)[9] is reported, the first gel has a good tissue-equivalence for brain tissue and the second one is nitrogen depleted.

In Figure 2 the dose values obtained experimentally in the central axis of the phantom are plotted versus depth. In order to facilitate the visualisation of the small but radiobiologically significant difference in dose deposition, two profiles are sketched on the experimental values. The results show good consistence with those evaluated in a similar geometry by J. J. Fitzgerald.[10] For further validation, TLD measurements in a polyethylene phantom were performed and the results have confirmed that this technique is promising.

CONCLUSIONS

FriXy-gel dosimeters of a suitable composition, imaged with a CCD camera, have shown to give the possibility of 3-D determination of the absorbed dose in BNCT with the feasibility of discriminating between different dose contributions. In fact, from differential measurements in phantoms having the compositions here described, the therapeutic dose from ^{10}B, the dose in the surrounding tissue from protons ejected in the nitrogen reaction and the dose from hydrogen capture γ-rays and from recoil protons can be separated. As said before, in the irradiation volume of this experiment

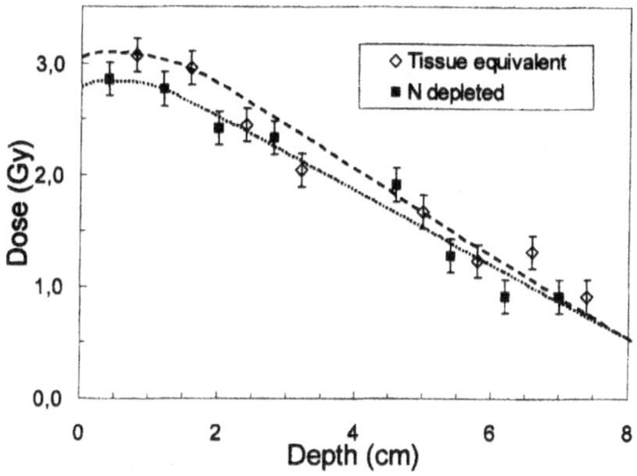

Figure 2. Depth-dose profiles in a FriXy-gel phantom exposed to thermal neutrons: ◇dose measured in the gel with tissue-equivalence for thermal neutrons; ■ dose measured in the gel depleted of nitrogen.

the contribution of fast neutrons to the absorbed dose is negligible. On the contrary, in the coming epithermal column the contribution of recoil protons will be significant; owing to its different RBE, a proper separation is therefore necessary. The promising results of this experiment have encouraged undertaking the study of proper gel composition aimed at separating this contribution too. This research has recently started.

ACKNOWLEDGMENTS

This work was partially supported by the Istituto Nazionale di Fisica Nucleare (Italy).

REFERENCES

1. R.J. Schulz, A.F. De Guzman, D.B. Nguyen, and J.C. Gore, Dose-response curves for Fricke-infused gels as obtained by nuclear magnetic resonance, *Phys. Med. Biol.*, 35:1611–1622, 1990.
2. G. Gambarini, S. Arrigoni, M.C. Cantone, N. Molho, L. Facchielli, and A.E. Sichirollo, Dose-response curve slope improvement and result reproducibility of ferrous-sulphate-doped gels analysed by NMR imaging, *Phys. Med. Biol.*, 39:703–717, 1994.
3. A. Appleby and A. Leghrouz, Imaging of radiation dose by visible color development in ferrous-agarose-xylenol orange gels, *Med. Phys.*, 18:309–312, 1991.
4. G. Gambarini, G. Gomarasca, R. Marchesini, A. Pecci, L. Pirola, and S. Tomatis, Three-dimensional determination of absorbed dose by spectrophotometric analysis of Ferrous-Sulphate Agarose gel, *Nucl. Instrum. and Meth.*, A 422:643–648, 1999.
5. G. Gambarini, S. Arrigoni, M. Bonardi, M.C. Cantone, D. de Bartolo, S. Desiati, L. Facchielli and A.E. Sichirollo, A system for 3-D absorbed dose measurements with tissue-equivalence for thermal neutrons, *Nucl. Instr. and Meth.*, A 353:406–410, 1994.
6. G. Gambarini, C. Birattari D. Monti, M.L. Fumagalli, A.Vai, P. Salvadori, L. Facchielli and A.E. Sichirollo, Fricke-infused Agarose gel phantoms for NMR dosimetry in Boron Neutron Capture Therapy and Proton Therapy, *Radiat. Prot. Dosim.*, 70:571–575, 1997.
7. G. Gambarini, D. Monti, M.L. Fumagalli, C. Birattari and P. Salvadori, Phantom dosimeters examined by NMR analysis: a promising technique for 3-D determination of absorbed dose, *Appl. Radiat. Isot.*, 48:1477–1484, 1997.

8. G. Gambarini, Three-dimensional determination of absorbed dose by NMR analysis of a tissue-equivalent phantom-dosimeter, *in* "Advances in Neutron Capture Therapy, Volume I, Medicine and Physics", B. Larsson, J. Crawford, R. Weinreich, eds., Elsevier Science, Amsterdam, 1997, pp. 208–211.
9. ICRU Report 46, Photon, electron, proton and neutron interaction data for body tissues, 1992.
10. J.J. Fitzgerald, Mathematical theory of radiation dosimetry, Gordon and Breach, New York, 1967, p. 479.

VALIDATION OF THE SPECIFICATION OF THE EPITHERMAL NEUTRON BEAM FOR BORON NEUTRON CAPTURE AT THE HFR PETTEN

C. Vroegindeweij, P. Watkins, and R. L. Moss

Joint Research Centre, European Commission
P.O. Box 2, 1755 ZG Petten
The Netherlands

1. INTRODUCTION

The beam of the BNCT facility at the HFR in Petten has a mixed neutron-gamma field, which in order to be applied for BNCT applications, must be thoroughly characterised through comparison between measurements and calculations. In particular, the beam source definition needs to be properly validated. Prior to October 1997, the epithermal neutron beam in Petten was extensively characterised by both measurements and calculations.

This validation study led to an adjustment in the divergence of the neutron beam reported previously as well as a decrease of the high energy neutron intensity. This paper gives an overview of the validation and its outcome.

2. MATERIALS AND METHODS

2.1. Dose and Fluence Rate Calculations

To achieve acceptable computer simulation times the calculations were divided into two. First, a detailed calculations of the whole reactor core and the beam filters was performed to generate the source term at the interface of the filter and the collimator. Second, calculations of dose and fluence rates in patients, phantoms or free beams use this source term. Because the incident neutron source is defined before the collimator, a source probability function is obtained which is independent of the collimator size. The

Frontiers in Neutron Capture Therapy, edited by Hawthorne *et al.*
Kluwer Academic / Plenum Publishers, New York, 2001.

first calculation (source term) is performed with MCNP,[1] the dose and fluence rates can be calculated by MCNP or the INEEL treatment planning software rtt_MC.[2]

2.2. Normalisation of Calculations

The normalisation factor in the dose and fluence calculations is based on a reference measurement since the complexity of the reactor makes it very difficult to calculate the absolute neutron density of the source term with a reasonable uncertainty. Also the daily fluctuation of beam intensity is taken into account via the reference measurement.

The reference measurement is characterised by:

Reference foil set:	AuAl and MnNi
Reference phantom:	Cubic PMMA phantom ($15 \times 15 \times 15\,cm^3$)
Reference depth:	2 cm depth at beam centre line.
Reference phantom position:	Beam exit—phantom distance = 30 cm
Reference reaction rate:	$^{55}Mn(n,\gamma)$

The daily change in beam intensity is taken into account by the ratio between the Mn reaction rates on the day of interest and the reference day (16th January 1995)

2.3. Preliminary Studies

A preliminary validation study by Peter Watkins[3] using MCNP showed a reasonable agreement between calculated and measured quantities. Floyd Wheeler and Yale Harker performed a second validation study using the treatment planning software, which resulted in small adjustments to the source specification. The differences between calculations and measurements were acceptable for the 12 and 15 cm apertures but unacceptable for the 8 cm aperture. Both validation studies were based on only a small portion of all the performed measurements. It was decided to perform a parameter study on the beam divergence and the spectrum and to repeat the validation study, using more data for a better agreement with all apertures.

2.4. Sensitivity Analysis

The influence of the beam divergence on the calculations was determined by performing all calculations with different beam divergence angles (the divergence angle is defined as the half angle of a cone). Five different angles were used; viz. 2, 4, 6, 8, and 10 degrees for all 3 collimators. Three slightly different spectra were considered: the standard spectrum, the standard spectrum with a reduced fast neutron intensity, and an adjusted spectrum based on free-beam measurements (the adjustment was based on a least square fit of the free beam measurements en the calculations).

2.5. Measurements for Sensitivity Analysis

The sensitivity analysis was based on foil activations in phantom and in the free beam. In the cubic PMMA reference phantom, two different sets were used: the standard foil packet of manganese-nickel and gold-aluminium (see Table 1), which was irradiated

Table 1. Dimensions of the foils of the standard foil packet

	Diameter [mm]	Thickness [mm]	Density [g/cm^3]
Au-Al (5 wt.% Au)	20	0.20	2.87
Mn-Ni (12 wt.% Ni)	20	0.10	7.46

at different depths in the standard phantom, and a rectangular foil package of manganese-nickel, gold-aluminium, nickel and aluminium (see Table 2) at 2 cm depth. The phantom was irradiated at a distance of 30 cm from the beam exit using three different collimator exit diameters (8, 12 and 15 cm). The foil packets were irradiated separately to prevent one foil packet influencing the other.

The ^{55}Mn(n,γ) and ^{197}Au(n,γ) reactions monitor the thermal and the lower part of the epithermal neutron energy range. The ^{58}Ni(n,p) reaction and the ^{27}Al(n,α) reaction are both threshold reactions: Ni atoms react with neutrons with an energy higher than 0.6 MeV, while the threshold for Al lies around 3 MeV.

In addition to measurements in phantom, a foil stack was irradiated in the free beam. Apart from the first Au foil, the foil stack was partly covered by 1 mm cadmium to absorb the thermal neutron component of the beam. The foils were placed at the exit of the beam in the centre of the beam opening using the 15 cm aperture. The composition of the foil stack is given in Table 3.

2.6. MCNP Model for Sensitivity Analysis

The foils were modelled in detail to take into account the distortion of the neutron fluence by the presence of the foils. The calculations were also performed for each posi-

Table 2. Dimensions of the rectangular foil packet used to determine fast neutron intensity at 2 cm depth

	Number of foils	Dimensions [mm]	Density [g/cm^3]
AuAl (5 wt.% Au)	1	20 × 19 × 0.2	2.87
MnNi (12 wt.% Ni)	1	20 × 10 × 0.1	7.46
Ni	8	20 × 19 × 0.25	8.90
Al	2	20 × 19 × 1.0	2.70

Table 3. Description of the foil packet used in free-beam measurements

	Cd	Composition		Dimensions [mm]	Volume [mm^3]	Mass [mg]	Density [g/cm^3]	MCNP [g/cm^3]	Sensitive energy range
^{197}Au(n,γ)	before	1% Au	99% Al	Ø12; d = 0.2	2.261E + 01	62.316	2.756	2.723	2–10 eV
^{115}In(n,γ)	yes	0.2% In	99.8% Al	Ø12; d = 0.2	2.261E + 01	65.638	2.903	2.703	1–10 eV
^{45}Sc(n,γ)	yes	99.99% Sc		Ø12; d = 0.1	1.130E + 01	42.884	3.794	3.500	0.1–600 eV
^{197}Au(n,γ)	yes	1% Au	99% Al	Ø12; d = 0.2	2.261E + 01	62.186	2.751	2.723	2–10 eV
^{186}W(n,γ)	yes	1% W	99% Al	Ø12; d = 0.2	2.261E + 01	61.284	2.711	2.723	10–20 eV
^{238}U(n,γ)	yes	22.83% ^{238}U	77% Al	Ø12; d = 0.1	1.130E + 01	37.878	3.351	3.364	5–40 eV
^{139}La(n,γ)	yes	5% La	95% Al	Ø12; d = 0.2	2.261E + 01	61.866	2.736	2.700	5 eV–1 keV
^{55}Mn(n,γ)	yes	1% Mn	99% Al	Ø12; d = 0.2	2.261E + 01	61.653	2.727	2.717	3 eV–1 keV
^{63}Cu(n,γ)	yes	10% Cu	90% al	Ø12; d = 0.2	2.261E + 01	64.351	2.846	2.902	5 eV–10 keV
^{115}In(n,n')	after	100% In		Ø12; d = 0.1	1.130E + 01	82.279	7.279	7.310	>0.32 MeV
^{58}Ni(n,p)	after	100% Ni		Ø20; d = 1.0	3.140E + 02	2,713.770	8.643	8.900	>0.453 MeV
^{27}Al(n,α)	after	100% Al		Ø20; d = 1.03	3.234E + 02	863.690	2.670	2.700	>3.62 MeV

tion separately. The calculations were performed with the MCNP computer code version 4A.

The calculations for the rectangular foil set (fast neutron component) were performed in two steps in order to obtain a low statistical uncertainty for the calculated fast neutron intensities. The first calculation determined the gold and manganese reaction rates, in the second calculation the nickel and aluminium reaction rates were obtained. This was possible, because the latter two reactions only occur for neutrons with an energy higher than a certain threshold. Statistical uncertainties in the calculated reaction rates were smaller than 5%.

2.7. Methodological differences between rtt_MC and MCNP

After comparing the measured and the MCNP-calculated reaction rates, the difference between rtt_MC and MCNP was investigated. The following list indicates where the differences exist in the input files for rtt_MC and MCNP:

1. the description of neutron source: rtt_MC uses a cosine distribution to sample the surface source (radial distribution), MCNP uses a flat distribution to sample the source surface
2. the models describing the apertures and the treatment room
3. the cross sections and KERMA libraries.

2.8. Quantitative Comparison between rtt_MC and MCNP

The reference configuration as described above is used to compare the following parameters:

Thermal fluence rate	(E < 0.414 eV)
Epithermal fluence rate	(0.414 eV < E < 9.12 keV)
Fast fluence rate	(E > 9.12 keV)

Positions of the calculated quantities were at 2, 3, 5, 10 cm depth along the beam centre line.

3. RESULTS

3.1. Standard Spectrum

For both foil sets in phantom, a 6, 8 or 10 degree beam divergence angle gave for the 8 cm aperture a deviation of more than 20% between calculations and measurements while for the 12 and 15 cm the differences were within 5% with the exception of the gold reaction rate at 2 cm depth. Smaller angles led to a much better agreement. But the gold reaction rates at 2 cm depth for all apertures and beam angles remained too high (up to 17). For the free beam foil packet, the 8 or 10 degrees angles gave an overestimation of the free-beam reaction rates by a factor of at least 2. The smaller beam angles gave ratios between calculations and measurements which were comparable to the ratios in the phantom. Overall the 2 degree beam angle gave the best agreement.

3.2. Effect of Reduced Fast Neutron Intensity

The effect of a reduction in the fast neutron intensity on the calculated manganese and gold reaction rates in phantom was negligible (Mn < 2% and Au < 5%). However the calculated nickel and aluminium reaction rates were much closer to the measured ones. The final reduction depends on the beam angle, larger angles needing a smaller reduction.

3.3. Effect of the Adjusted Spectrum

Changes in the spectrum influenced the ratio between the measured and calculated Au reaction rates in phantom considerably, while the influence on the manganese reaction rate was small. As with the other two spectra, the large angles gave an overestimation of the free-beam reaction rates by a factor of at least 2 and the smaller angles gave ratios between calculations and measurements which were comparable to the ratios in the phantom. Again the 2 degree angle gave the best agreement.

3.4. Adjustments on Source Term

Based on the outcome of this comparison, the beam divergence has been reduced to 2 degrees uniformly for all neutron energies. Also the fast neutron intensity has been reduced: above 0.3 MeV by a factor of 2 and above 7 MeV by a factor of 4. The same set of calculations was performed with the adjusted source term. Calculated reaction rates are now much closer to measured ones. Most of the differences between calculations and measurements are within statistical uncertainties. The previous large differences, such as the disagreement between calculations and measurements for the 8 cm aperture, the Au reaction rate at 2 cm depth and the fast neutron intensity have all been eliminated.

3.5. Results of Comparison MCNP and rtt_MC

The difference between the calculated quantities using MCNP or rtt_MC were relatively small. The thermal fluence rate was within the statistical uncertainties for the 12 and 15 cm aperture. For the 8 cm beam, the thermal fluence rate calculated with rtt_MC was 8% higher than that calculated with MCNP. For the epithermal fluence rates, rtt_MC and MCNP results were the same within statistical uncertainties. For the fast neutron fluence rate, rtt_MC was approx. 14% higher than MCNP. The differences were somewhat larger at 10 cm depth in the PMMA phantom.

4. DISCUSSIONS AND CONCLUSIONS

The new beam description gives a much better agreement with the measured reaction rates, both free beam and in phantom, than the old one. The average relative difference between calculations and measurements has been reduced by a factor of 2 and most of the differences are now within statistical uncertainties. Also the results for the 8 cm beam are now in reasonable agreement with the measurements, so that the 8 cm aperture can also now be used for patient irradiations. The results of this study gives further confidence that the new beam description is a better description of the neutron distribution free in air, in phantoms and in patients and that the agreement is perfectly acceptable for use in patient trials for all apertures. However, it should be pointed out that further improvement of the beam definition is still possible.

In further development of the source definitions the following items will be considered:

1. In the new (as it was in the old) beam definition the beam divergence is chosen independently from the neutron energy. In practice, there is a correlation between the neutron energy and the beam divergence. This will be further investigated.
2. The measured neutron dose with ionisation chambers is the only available indication of the correctness of the beam spectrum between 1 keV and 0.5 MeV, hence more effort should be put into a better understanding of these measurements. A large problem is that the effect of these relatively large detectors on the measured quantities is not known for the BNCT epithermal beam. The influence can be considerable for the neutron dose (up to 20% over the detector volume). Other measurement techniques, such as proton recoil, should be used to increase the knowledge of the epithermal energy range.
3. The fast neutron source description will be improved. Although the agreement of the In, Al and Ni reaction rates are improved, this agreement is not optimal.
4. The description of the incident gamma beam is a very rough approximation. Only a relatively small amount of effort has been put into the measurement of the gamma spectrum.

ACKNOWLEDGMENT

The authors would like to thank Adri Paardekooper, Karel Aaldijk and Wim Voorbraak from the Netherlands Energy Research Foundation for performing the foil measurements.

REFERENCES

1. Briesmeister, J.F., 'MCNP-4A, A General Monte Carlo Code for neutron and photon transport, version 4A', Los Alamos Report LA-12625, 1993.
2. Wheeler, F.J., 'Radiation Transport in Tissue by Monte Carlo, version X02', Idaho National Engineering Laboratory report no. EGG-BNCT-11178, Jan., 1994.
3. Watkins, P., *et al.*, 'Nuclear Characterisation of the HFR Petten BNCT Facility', In: Advances in Neutron Capture Therapy, Ed. A.H. Soloway *et al.*, Plenum Press, New York, 1993, 59–65.

54

DETERMINATION OF THE GEOMETRIC AND SPECTRAL CHARACTERISTICS OF BNCT BEAM (NEUTRON AND GAMMA-RAY)

M. Marek, L. Viererbl, J. Burian, and B. Jansky

Nuclear Research Institute Rez
CZ 25068 Rez, Czech Republic

INTRODUCTION

Knowledge of the basic parameters of the BNCT epithermal neutron beam is one of the essential conditions ensuring the beam optimal therapeutic use. It is necessary for us to have appropriate tools for determining the beam geometry, absolute values of the dosimetry quantities and spectral characteristics for both neutrons and gamma rays. As far as the Czech BNCT project[1] is concerned, the following set of the detectors is used:

- the beam geometry—the radial and angular distributions of the beam:
  Semiconductor detector with natural Li converter—thermal and epithermal (with Cd) neutrons,
  ^{238}U and ^{232}Th fission chambers—fast neutrons
  Semiconductor detector—gamma rays
- the dosimetry values of the beam:
  Activation foils—the fast, epithermal and thermal neutron fluence rates, neutron spectrum,
  Bonner spheres—the fast, epithermal and thermal neutron fluence rates, neutron spectrum,
  Al-P glass TLD—the gamma absorbed dose,
- Spectral beam values:
  Scintillation spectrometer—the fast neutron spectrum
  Hydrogen proportional spectrometer—the fast neutron spectrum
  Scintillation spectrometer—the gamma ray spectrum

Frontiers in Neutron Capture Therapy, edited by Hawthorne *et al.*
Kluwer Academic / Plenum Publishers, New York, 2001.

1. DETECTORS

1.1. Semiconductor Detector with Natural Li Converter

A small semiconductor detector with a Li converter (Si-Li) has been especially designed for the thermal and epithermal neutrons measurement both in the beam and in a phantom.[2] The detector has a cylindrical shape and is made from Al and polyethylene so that it can be regarded as a Tissue equivalent. Its dimensions of 14 mm in diameter and 17 mm length are limited by dimensions of Si wafer which is used (7.5 × 7.5 mm and 0.3 mm thickness). The thermal neutron efficiency of the detector (defined as rate of pulses [s^{-1}] recorded per unit of neutron fluence rate [$cm^{-2} s^{-1}$] incident on detector) depends on the amount of ^{6}Li in the converter. That is why detectors with a reasonable response in the neutron fluence rate up to $10^9 cm^{-2} s^{-1}$ can be designed. The detectors were calibrated in a thermal neutron field (a polyethylene cylinder driven by ^{252}Cf source with total emission of 10^9 n/s). Up to now we have been using the detectors with efficiencies from $2.452 \times 10^{-3} \pm 2.3\%$ (1 STD) and $1.927 \times 10^{-5} \pm 4.1\%$ (1 STD) cm^2.

The same type of the detector but without any converter is used for the relative gamma ray measurement. It is supposed to be calibrated using the ^{60}Co source.

1.2. ^{238}U, ^{232}Th Fission Chambers

The fission chambers for the fast neutron detection were produced in Russia.[3] They are cylindrical with outer dimensions of 7.5 × 71 mm and their inner surface is covered with a solid compound containing either ^{238}U or ^{232}Th of total weight of 6 mg. The length of the cover is 10 mm (3 mg/cm^2 density). The declared efficiency of the fast neutron detection is 2×10^{-6} and $6 \times 10^{-7} cm^2$ respectively. If calibrated using the ^{252}Cf source the responses were determined $1.8 \times 10^{-6} \pm 4.5\%$ (1 STD) and $5.6 \times 10^{-7} \pm 5.2\%$ (1 STD) cm^2 respectively. The chambers are used only for relative measurements of the beam profile for the time being.

1.3. Activation Foils

The set of the reactions covers the thermal (^{197}Au (n,γ) ^{198}Au) and epithermal (in Cd^{115}In (n,γ) ^{116}In, ^{45}Sc (n,γ) ^{46}Sc, ^{197}Au (n,γ) ^{198}Au, ^{238}U (n,γ) ^{239}U, ^{186}W (n,γ) ^{187}W, ^{139}La (n,γ) ^{140}La, ^{55}Mn (n,γ) ^{56}Mn, ^{63}Cu (n,γ) ^{64}Cu) neutron energy range up to 5 keV and the fast neutron energy range above ~0.5 MeV (^{115}In (a,n′) ^{115m}In, ^{58}Ni (n,p) ^{58}Co, ^{27}Al (n,α) ^{24}Na). All the foils have the diameter of 12 mm. The resonance foils are encapsulated in 1 mm thick cadmium. The uncertainties for reaction rates are between 1.5–3% (1 STD).

Data from the activation measurement are used as an input into an adjustment procedure[4] which provides a means for combining them with a calculated neutron spectrum resulting in determining an optimal estimation of the thermal, epithermal and fast neutron fluence rates and their uncertainties. The SAND II[5] and BASACF[6] codes with the IRDF-90[7] and DOSCROS84[8] cross-section data library are used for the adjustment. The BASACF code supposing the normal distribution of the activities, spectra and cross sections is based on the Bayesian approach to the problem solution. The program needs all statistical information including the covariance matrixes of the activities, spectrum and cross sections to be entered. In case the covariance is not procurable the data could be generated using the Gaussian method according to the recomendation of

ASTM.[4] The uncertainties for fluence rates determination are in the interval of 5–20% (1 STD).

1.4. Bonner Spheres Spectrometer

The Bonner spheres detector[3] consists of a cylindrical 4 × 8 mm Li(Eu) scintillation thermal neutron detector and a set of polyethylene spheres having 2″, 3″, 4″, 5″, 8″, 10″ and 12″ in the diameter. As the beam is usually smaller then 15 cm in the diameter only first four spheres and the bare detector are used for the spectral measurements. The advantage of the spectrometer is that the 90% response intervals of the spheres continuously cover the epithermal part of the neutron energy range.

Responses of the spheres were calculated taking into consideration the beam geometry using the MCNP[9] code and the ENDF/B-VI[10] cross section library. The detector was calibrated in the neutron field of a ^{252}Cf neutron source and in a leakage spectrum from a D_2O sphere of 20 cm in diameter with the ^{252}Cf neutron source installed in the center of the sphere. Estimated uncertainties in the detectors counting rates are ~5% (1 STD).

The spectrum adjustment procedure is the same as in the case of the activation foils. The neutron dosimetry data related to both the activation foils and the Bonner spheres are usually combined and unfolded together.

1.5. Al-P Glass TLD

A standard type of TLD is used to get absolute information about the gamma absorbed dose rate in the BNCT beam. The used Czech Al-P glass[3] TLDs are cylindrical of $\phi 8 \times 1$ mm. They are placed into cylindrical Pb boxes of 0.5 mm wall thickness and a central hole of 3.3 mm in the diameter. The response of the detector lies in the energy range 25 keV–7.5 MeV and the detector can measure the gamma absorbed dose up to 10 Gy. The thermal neutron correction factor derived in a thermal neutron field[11] is 5×10^{-13} Gy cm^2. The estimated uncertainties in the photon absorbed dose determination is 20% (1 STD).

1.6. Scintillation Spectrometer for Neutrons

The scintillation spectrometer is used with a cylindrical stilbene crystal for the detection of the fast neutrons[12] with energies above ~0.5 MeV. The size of the crystal can be change from $\phi 10 \times 10$ mm to $\phi 40 \times 40$ depending on the neutron fluence rate and the neutron to gamma ray ratio. Generally, the greater the neutron fluence rate and/or the smaller neutron to gamma ratio are, the smaller the size of the crystal is more appropriate.

A discrimination of the signal, caused by recoiled protons produced as a result of the elastic neutron scattering by hydrogen nuclei, from those caused by the Compton electrons produced in the crystal and in its cover due to the interactions of gamma rays, is based on a different shape of the scintillation pulses (the PSD method).[12] Because of the wide dynamic range of the scintillation pulses resulting from the non-linearity of the light production in the organic scintillator, it is necessary to measure the spectrum divided in two parts. One part covers energy interval (~0.5, 3) MeV and the second (1.5, 12) MeV. An energy calibration is performed by measuring Compton electrons from a ^{23}Na gamma source with two energy lines 0.511 and 1.28 MeV. The detector was calibrated with ^{252}Cf and PuBe sources.[12]

Reconstruction of the neutron spectrum from the proton recoil spectrum is based on the Johnson algorithm[13] that uses the direct differentiation technique. Correction on multiple neutron scattering on hydrogen, proton escape and neutron interactions with carbon nuclei are included into the unfolding process.[12]

The uncertainty of the resulted neutron spectrum depends on the uncertainty of the energy calibration, which is ~2% (1 STD), uncertainty of the light output versus proton energy, estimated as ~1.5% (1 STD), detector efficiency, which is ~2% (1 STD), systematic uncertainty of 2–3% (1 STD) and statistical error connected with the method of unfolding. The total uncertainty in the neutron spectrum is estimated to vary from ~7% (1 STD) at neutron energy of 1 MeV to 15% (1 STD) at 10 MeV.

1.7. Hydrogen Proportional Spectrometer

A spherical hydrogen gas filled proportional counter presented by Jansky[14] is used to measure fast neutron spectra lying in the energy interval (60, 1,500) keV. To cover such a wide energy range, three counters filled with hydrogen of different pressures (100, 300 and 1,000 kPa) are used. The diameter of all the detectors is 40 mm. A calibration of the spectrometer was performed on the basis of measurement of leakage neutron spectra from a iron spheres of 20, 30 and 50 cm in diameter and D_2O filled sphere of 30 cm in diameter having a ^{252}Cf neutron source installed into the center of the sphere. Energy calibration of the chamber was performed in an iron-filtered beam driven by a ^{252}Cf neutron source using both the minims in the iron cross sections at 138, 166, 217, 271, 311, 352, 463 and 634 keV and maxims 140, 190, 504 and 691 keV. The proton recoil spectrum obtained by measurement is unfolded to the neutron spectrum using a direct differentiation algorithm with a correction related to the double neutron interaction with hydrogen nuclei and to the proton escape from the detector volume. The uncertainty of the resulted neutron spectrum depends on the uncertainty of the energy calibration, which is 1–2% (1 STD), detector efficiency, which is 2–3% (1 STD), and statistical error.

1.8. Scintillation Spectrometer for Gamma Rays

The scintillation spectrometer with the cylindrical stilbene crystal of $\phi 40 \times 40$ mm can be also used to measure the gamma ray spectra in the energy range of (0.4, 12) MeV. The PSD electronic system enables to separate the gamma events in the detector from those resulting from the neutron reactions. The measured Compton electron spectrum is unfolded to the gamma spectrum using the direct differentiation algorithm. Because of a difference between the ideal and real response functions the resulting gamma ray spectrum is corrected with an empirical correction matrix according the method described by Trykov.[15] Estimated error in the photon fluence rate is 20% (1 STD).

2. RESULTS OF MEASUREMENTS

The main parameters of the LVR-15 beam are presented in the Table 1. The activation foils and Bonner sphere spectrometer were used to adjust the MCNP calculated beam spectrum. The epithermal (<0.5 eV, 10 keV>) rate Φ_e and fast neutron fluence

Table 1. The LVR-15 beam parameters (1 STD)

Neutron fluence rate	2.02×10^8	±9.6%	$cm^{-2}s^{-1}$
Φ_e	1.82×10^8	±8.3%	$cm^{-2}s^{-1}$
Φ_f	1.90×10^7	±12%	$cm^{-2}s^{-1}$
Fast neutron kerma rate			
Stilbene + Bonner Spectrometer	0.774	±12%	Gy/h
Gamma absorbed dose rate			
Radiameter[16]	0.620	±15%	Gy/h
TLD	0.600	±25%	Gy/h
Stilbene spectrometer	0.558	±14%	Gy/h

(>10 keV) rate Φ_f and fast neutron kerma rate were determined from the adjusted spectrum. The gamma absorbed dose rate was measured with TLD and radiameter Eberline[16] and calculated from the differential measurement.

The LVR-15 beam was 14 cm in diameter and its epithermal neutron radial profiles were measured with the Si detector in Cd cover at distances z = 1, 6, 11 and 15 cm and radii of 0, 3.5, 7, 10, 13 and 17.5 cm (Figure 1). Results show that the epithermal fluence rate decreases twice between z = 0 and 10 cm. The axial profile was measured also inside the collimator (Figure 1). Statistical error of each point is better than 1% (1 STD).

The same detector as above was used to describe the epithermal neutron distribution at the VTT Espoo Triga reactor. At that time none collimator was installed as the experiment was aimed of showing the sufficient intensity and purity of the source. So that the distribution was flat and decrease was nearly linearly dependent on the distance from the source surface (Figure 2). Statistical error of each point is better than 1% (1 STD).

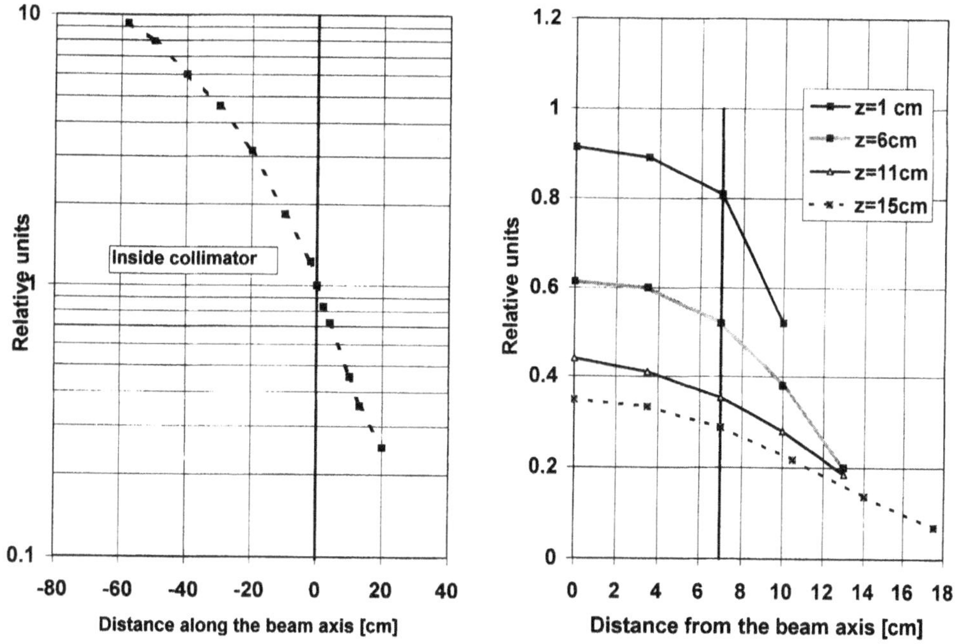

Figure 1. The axial and radial profile of the LVR-15 beam. The epithermal flux measured with Si-Li detector in Cd.

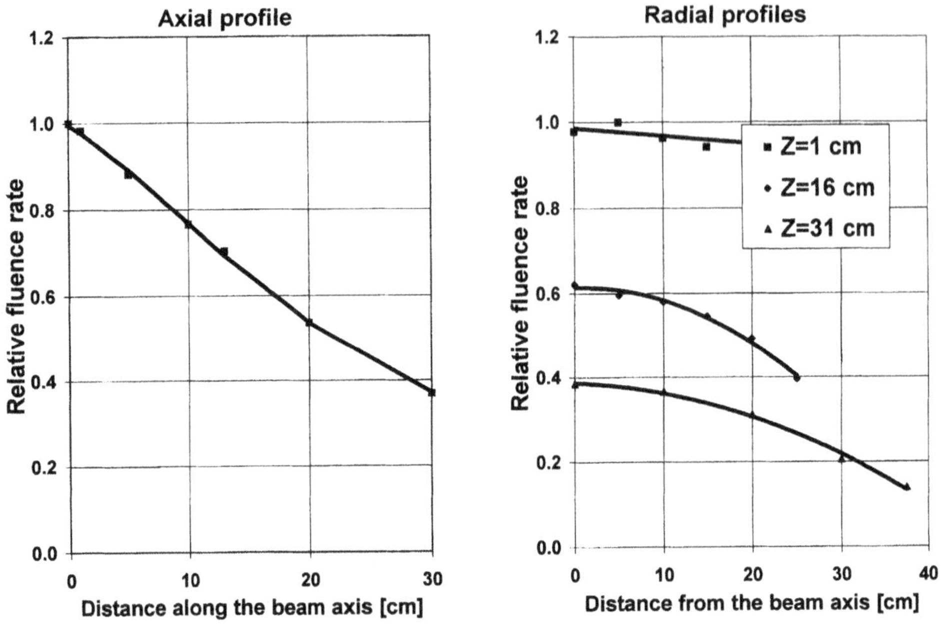

Figure 2. Axial and radial profiles at VTT Espoo beam.

At the LVR-15 reactor the fast neutron behavior was studied also. The results of the measurements with ^{238}U fission chamber is in the next Figure. Statistical error of each point is better than 1% (1 STD). As it was expected the fast portion of the beam is much better collimated then the epithermal one (Figure 3).

Spectral measurements were performed both at the VTT and NRI reactors. Neutron dosimetry data from activation foil measurement with the Bonner spectrometer

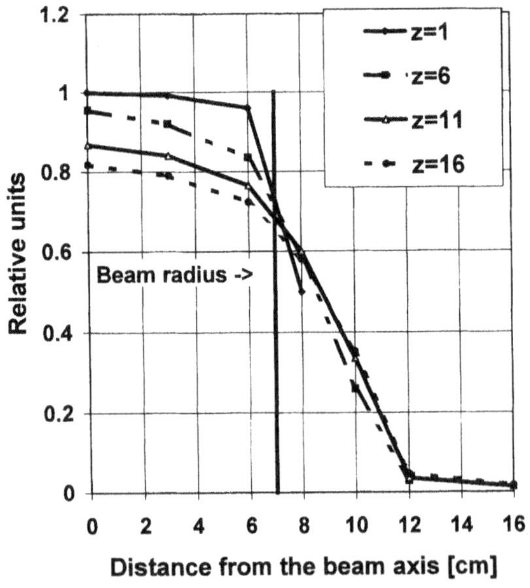

Figure 3. Fast neutron profile at the LVR-15.

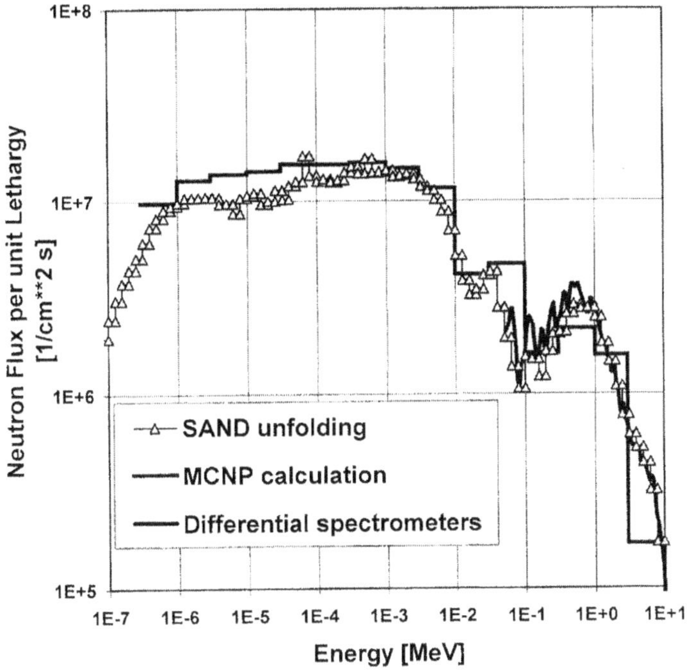

Figure 4. Neutron spectrum at LVR.

responses were combined for the LVR-15 reactor. They were adjusted to a MCNP calculated neutron spectrum. Standard deviation between measured and calculated detector responses of 9.6% show quite good agreement between the experimental and theoretical approaches (Figure 4). In the figure we can see also a comparison with differential method (stilbene and hydrogen spectrometers) results. The agreement seems to be good.

Only the Bonner spectrometer and the stilbene spectrometer were used to describe the neutron spectrum at the TRIGA reactor. A calculated spectrum[17] adjusted to the Bonner spectrometer data were compared with the calculated spectrum and with the spectrum resulted from the activation foil adjustment presented by Kaita[17] (see Figure 5). Standard deviation between measured and calculated detector responses was 9.8%.

The gamma ray spectrum was measured at the LVR-15 beam with the 40 × 40 mm stilbene crystal. The result can be seen in the Figure 6. A presence of hard gamma rays transported directly from the reactor core can be recognised at the spectrum (the energy regions 5–6 and 8 MeV).

3. CONCLUSIONS

The Si detector, because of its small dimensions, variable efficiency and on-line application, seems to be an appropriate detector for the epithermal and thermal neutrons and the gamma rays beam geometry determination.

Due to a broad energy gap between the resonance and threshold activation reactions the Bonner sphere spectrometer having response from eV's to MeV's is a quite good

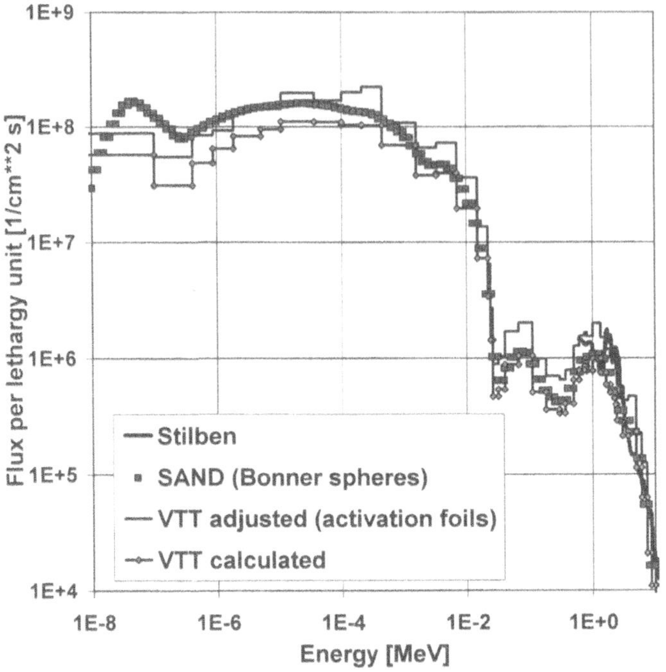

Figure 5. Neutron spectrum at VTT.

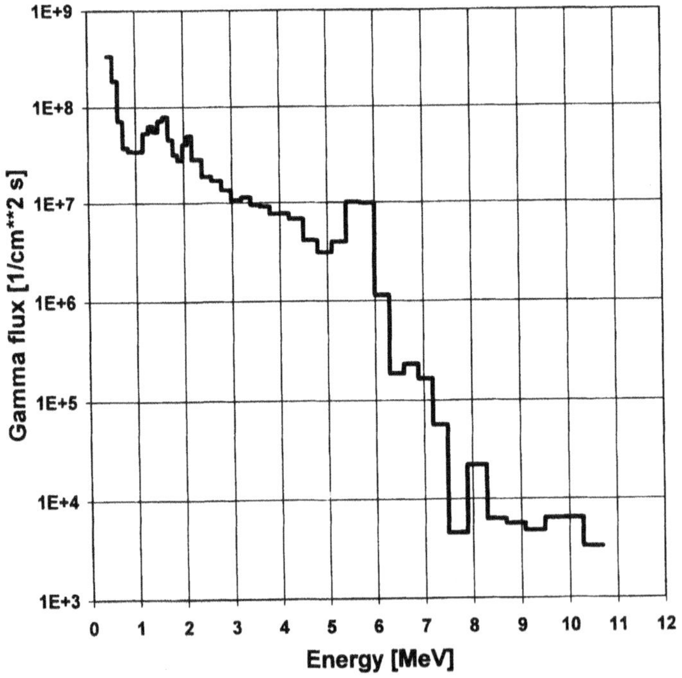

Figure 6. Gamma spectrum at LVR.

tool for adjusting calculated neutron spectrum of the beam. Its disadvantage—high efficiency—is supposed to be improved by replacing the contemporary thermal neutron detector with a semiconductor one with a low efficiency to thermal neutrons.

The differential spectrometers (stilbene and hydrogen proportional) can give direct information about the detailed structure and absolute value of the neutron spectrum above ~50 keV. The same is valid for the gamma ray stilbene spectrometer. The disadvantage of all of them is their high efficiency and therefore they can be used only at a lower power of the beam source.

ACKNOWLEDGMENT

This work was supported by contracts GACR 202/97/K038 and IGA-MZ (CR) No. 4834-2

REFERENCES

1. Burian J., Marek M., Rataj J., Prokes K., Petruzelka L., Mares V., and Stopka P.: The BNCT facility at NRI reactor in the Czech Republic, Proceedings of 7th Int. Symp. On NCT for Cancer, Zurich, Sept. 1996, Elsevier, Vol. I, p. 391.
2. Marek M.: Small Si-Li Detectors for Thermal Neutron On-line Measurements, *J. The Safety of Nuclear Energy*, 3 (41), 7/8, p. 165–168, 1995.
3. Marek M., and Burian J.: Free beam and phantom dosimetry methods for boron neutron capture therapy, *J. Nucleon*, 1 (1994), 25–27.
4. ASTM Designation E 944-89: Standard Guide for Application of Neutron Spectrum Adjustment Methods in Reactor Surveillance.
5. Tichy M.: The Program BASACF. PC Version. Description and User's Guide. Institute of Radiiation Dosimetry, Prague, March 1990.
6. Marek M.: PC Version of SAND II Code, NRI 9451 R,D, September 1991.
7. Kocherov N.P., and McLaughlin P.K.: The International Reactor Dosimetry FILE (IRDF-90), IAEA-NDS-141, Rev. 2, October 1993, IAEA, Vienna.
8. Zijp W.L., *et al.*: Cross-section Library DOSCROS84 (in a 640 group structure of the SAND-II type), ECN-160, October 1984, Petten.
9. "MCNP- A General Monte Carlo N-Particle Transport Code (Version 4A)", LA-12625, J.F. Briesmeister, Ed., Los Alamos National Laboratory (1993).
10. ENDF-201, ENDF/B-VI Summary Documentation, edited by Rose P.F., Brookhaven National Laboratory Report BNL-NCS-1741, fouth edition, October 1991.
11. Spurny F.: Methods of outside irradiation dosimetry and their application (in Czech). PhD thesis. Czechoslovak Academy of Sciences, Prague, 1984.
12. Marek M.: Fast neutron scintillation spectrometer with Stilbene crystal (in Czech). PhD thesis. Czechoslovak Academy of Sciences. Prague. 1990.
13. Johnson R.H.: Improvements in the differentiation method of spectrum unfolding, *J. Nuclear Instrumentation and Methods*, 179, 301–307, 1981.
14. Jansky B.: Fast neutron spectra measurements with proportional counters, Fast neutron spectra reference (in Czech), PhD thesis, Nuclear Research Institute, Rez, 1991.
15. Kuchtevic V.I., Trykov O.A., and Trykov L.A.: Scintillation spectrometer with single crystal (in Russian), Atomizdat, Moskow, 1971.
16. FH 41 Radiameter, Health Physics Instruments, Eberline Instruments GmBH, Erlangen, Germany.
17. Kaita K., Seren T., and Auterinen I.: First Characterisation of the Finnish Epithermal Neutron Beam Using Activation Detectors, Proceedings of 7th Int. Symp. On NCT for Cancer, Zurich, Sept. 1996, Elsevier, Vol. I.

BNCT BEAM QUALITY MAXIMIZATION

E. Greenspan and Y. Karni

Department of Nuclear Engineering
University of California
Berkeley, CA 94720

ABSTRACT

The Beam Shaping Assembly (BSA) optimization code SWAN is applied for establishing (1) the maximum attainable beam quality dependence on the BSA thickness; (2) the tradeoff between the maximum attainable beam quality and the dose-equivalent delivered, per source neutron, to the tumor; (3) the sensitivity of the above characteristics to the BSA optimization criterion used; and (4) the maximum beam quality attainable from four different neutron sources: ^{7}Li(2.5 MeV p, n), Be(19 MeV p, n), D-D and D-T. It is found that: (1) The thicker the BSA is, the higher can be the beam quality. However, (2) the specific beam intensity drops nearly exponentially with the BSA thickness. (3) Hard spectrum neutron sources can provide as good a beam quality as soft spectrum neutron sources. (4) For a given beam quality, the ^{7}Li(2.5 MeV p, n) source can deliver the highest dose-equivalent per source neutron while the Be(19 MeV p, n) source can deliver the highest dose-equivalent per beam ion as well as per beam power. (5) When accounting for constraints imposed by state-of-the-art technology, the beam quality which could be provided by both proton accelerator systems considered is, practically, the same for the same treatment time. The beam quality attainable from the state-of-the-art fusion neutron sources considered is significantly lower. (6) Technological improvements which are anticipated within the next few years are predicted to bring the beam quality attainable from the ^{7}Li(2.5 MeV p, n), Be(19 MeV p, n) and D-T sources to, essentially, the same value for the same treatment time. Hence (7) the choice of the best neutron source to use for BNCT should be determined by economic considerations. (8) The optimal BSA composition varies with the BSA thickness, brain conditions and optimization criterion. (9) The most desirable BSA optimization criterion appears to us to be maximizing the dose-equivalent delivered per source neutron to the tumor for a given beam quality. (10) The definition of the figure-of-merit to be used can have a significant effect on the outcome of the BSA optimization. The constraint on the maximum permissible skin dose has also a significant impact on the optimization outcome. (11) Al/AlF_3

Frontiers in Neutron Capture Therapy, edited by Hawthorne *et al.*
Kluwer Academic / Plenum Publishers, New York, 2001.

may, or may not be the most effective primary neutron slowing down material for the BSA of a $^7Li(2.5\,MeV\ p, n)$ neutron source; under certain conditions ^7LiF and MgF_2 may be preferable.

1. INTRODUCTION

A variety of facilities for medical and other applications of nuclear radiation require the design of a "Beam Shaping Assembly" (BSA). Its purpose is to shape the spectrum and/or geometry of the beam for maximizing its effectiveness. The purpose of this paper is to illustrate the SWAN methodology for optimal BSA designs. The illustration pertains to Boron Neutron Capture Therapy (BNCT).

It has recently been shown[1] that: (1) the success probability of Boron Neutron Capture Therapy (BNCT) strongly depends on the clinical quality of the neutron beam. (2) The beam quality of three of the most advanced BNCT facilities, all using reactor neutrons, is not good enough for treating deep-seated tumors by irradiation from one side. As the beam quality strongly depends on the design of the Beam Shaping Assembly (BSA), it is important to identify the optimal BSA design.

The BSA optimization can be very complex as it may involve dozens of independent variables. Conventional nuclear design tools and practices are not effective for such an optimization, as they have to be applied numerous times to check different combinations of the independent variables the user may wish to try. They do not give the user direct indication on promising constituents and on their preferred location in the BSA. Neither do they give the user an indication on how close to the optimum is the design he/she is considering. Recently we developed[2] an intelligent method for identifying the optimal BSA design; it is incorporated in the code SWAN.[2,3] The "intelligence" of SWAN is based upon information it derives using first-order perturbation theory. This extra information guides a systematic and efficient search for the optimal composition of a BSA of a given thickness.

The purpose of this paper is to apply SWAN for: (1) establishing the maximum attainable beam quality dependence on the BSA thickness; (2) establishing the tradeoff between the maximum attainable beam quality and the equivalent dose delivered, per source neutron, to the tumor (to be referred as the "specific beam intensity"); (3) establishing the sensitivity of the above characteristics to the BSA optimization criterion used; and (4) comparing the maximum beam quality attainable from different neutron sources. One school of thought in the BNCT community is of the opinion that the softer the neutron source spectrum, the higher will be the beam quality it can deliver. As pointed out recently,[4,5] we'll show here more convincingly that this opinion may not be justified.

2. METHODOLOGY

2.1. Definitions

Figure 1 gives a typical dose-equivalent profile across the brain with a tumor located between 8 and 8.5 cm deep. We define:

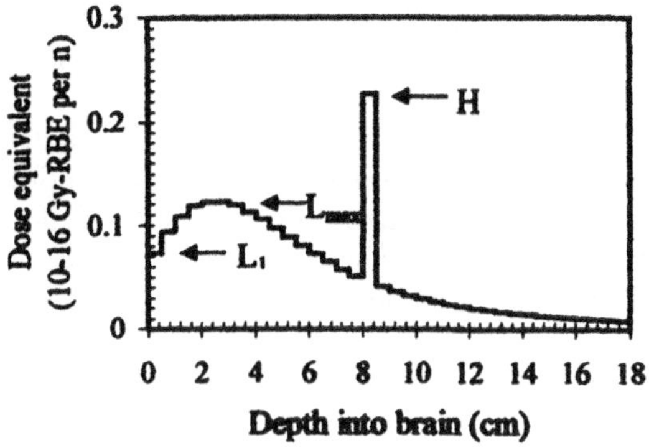

Figure 1. Typical dose-equivalent distribution across the brain.

H — the dose-equivalent delivered to the tumor per source neutron; we shall refer to it as the "specific beam intensity".

L_{max} — the maximum dose-equivalent delivered per source neutron to the normal tissue. Its location varies with the BSA thickness and composition, as well as with the neutron source spectrum.

L_l — the dose-equivalent delivered per source neutron to the skin. The L_{max}/L_l ratio varies with the BSA thickness and composition, neutron source type, etc.

2.2. Optimization Statement

Two constraints are usually imposed on the BSA design: (1) The maximum dose-equivalent imparted to the normal tissue and to the skin will not exceed a given value, to be denoted by, respectively, NTMD (Normal Tissue Maximum Dose-equivalent) and SMD (Skin Maximum Dose equivalent; it is usually set to be smaller than the NTMD). (2) The dose-equivalent delivered to the tumor will be large enough to kill all the tumor cells.

The BNCT facility design goal is to provide a beam that will be able to kill even the deepest seated tumor within a reasonable treatment time (say, up to one hour) without exceeding the constraints on the dose-equivalent to the normal tissue (NTMD & SMD). We shall refer to this "tumor control dose" as TCD. The values assigned to the TCD and NTMD dictate the optimization goal: H/L_{max} is to be $\geq$ TCD/NTMD. The treatment time constraint, T, imposes a limit on the acceptable H for a given neutron source, as H(Gy-RBE/n) S(n/s) 3600 (s/h) T(h) $\geq$ TCD. When the irradiation of a tumor that is located at the brain center is performed from two opposite sides, H/L_{max} nearly doubles, so that H(Gy-RBE/n) S(n/s) 3600 (s/h) T(h) $\geq \sim$0.5 TCD.

Correspondingly, SWAN is set to address either one of the following types of BSA optimization problems. For a given neutron source spectrum (possibly also geometry), BSA thickness, boron concentrations and dose to dose-equivalent conversion factors, find the space dependent concentration of a selected set of BSA constituents that will either maximize H/L_{max} or maximize H for a predefined value of H/L_{max}. The first optimization problem is useful for identifying upper bounds on attainable beam qualities and of the dependence of these bounds on different design variables, such as BSA thickness and neutron source spectrum. The second optimization problem is the more realistic for situations in which the neutron source strength is given, as it takes into account the treatment time constraint.

It is not clear to us that H/L_{max} is the best optimization criterion. L_{max} is applied to a small brain volume whereas clinical treatment designers consider[6] the integrated dose delivered to the tissue over the brain volume exposed to "high dose"—say, between 90% L_{max} and L_{max}. Hence, in order to get an indication on the sensitivity of the optimal BSA design and performance on the definition of the constraint used to measure the acceptable normal tissue dose, we are also considering H/L_1 as the BSA optimization goal. We anticipate that the optimal BSA will be somewhere in between that obtained when maximizing H/L_{max} to that obtained when maximizing H/L_1.

2.3. The SWAN Code

SWAN is composed of two major modules:[2,3] (1) The one-dimensional discrete ordinates code ANISN for calculating the neutron flux and one or more adjoint functions across the system, and (2) The code SWIF which uses the flux and adjoint functions to calculate material "replacement effectiveness functions" (REFs) and uses these REFs to guide an iterative search of the optimal system composition. The REF of material i in zone z with respect to performance parameter P expresses the effect, on P, of a replacement of a unit volume of reference material R in z by an equal volume of material i. The REF is calculated in SWAN using first-order perturbation theory. Only one SWAN run, involving the solution of one transport and one adjoint equation, is required for generating the REFs pertaining to H for any number of candidate constituents.[2,3] Generation of the set of REFs for H/L requires the solution of one additional adjoint equation.

The optimization is done in two steps: (1) Search for promising constituents; by comparing the REFs of candidate constituents. (2) Finding that distribution of the selected constituents that will maximize P. Step 2 is done iteratively until convergence; after each iteration SWAN recalculates the REFs and uses them for another composition variation. This process is repeated for each source type, BSA thickness, boron concentration in the brain (if it varies) and definition of the location for measuring L.

SWAN is presently restricted to the simulation of one-dimensional BSA. Nevertheless, the SWAN methodology is applicable to two-, and three-dimensional systems. As the current SWAN is one dimensional, the BNCT facility is modeled as concentric spherical shells around a spherical neutron source, 0.5 cm in radius surrounded by the BSA, air gap and "brain". The thickness of the BSA spherical shell is a parametric variable. The air gap shell is 0.5 cm thick. The spherical shell representing the brain is taken to be 18 cm thick. The tumor is presented as a spherical shell located between 8 to 8.5 cm into the brain. The source neutrons angular distribution is assumed to be isotropic.

The above is a gross approximation of the actual geometry; it does not benefit from the use of a good reflector and from the use of a collimator between the BSA and the brain. These two omissions are expected to underestimate H/L for a given BSA thickness. The presentation of the tumor as a spherical shell is also expected to cause an underestimation of H/L. So does the assumption of isotropic neutron sources. Consequently, the results to be presented should only be interpreted as indicating trends. Nevertheless,

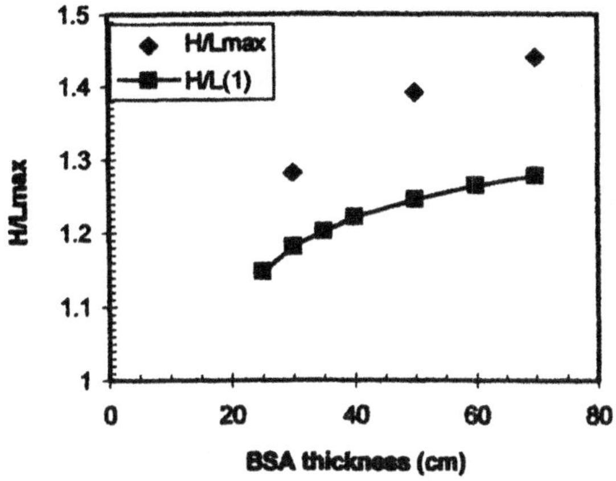

Figure 2. Effect of BSA thickness on maximum attainable beam quality for a ^{7}Li(2.5 MeVp,n) source and MgF_2, Pb and ^{6}LiF BSA.

a recent independent optimization of BSA for D-T neutron sources carried out with the 3-D code MCNP[6] provides a strong verification of the trends identified with the 1-D SWAN. MCNP calculations[7] also provided confirmation of the optimal BSA design suggested by SWAN for a Be(19 MeV p, n) source.

3. MAXIMUM ATTAINABLE BEAM QUALITY

3.1. Problem Definition and Assumptions

In this section we apply SWAN to establish the following: (1) The maximum attainable beam quality dependence on the BSA thickness. (2) The tradeoff between the specific beam intensity, H, and the maximum attainable beam quality, H/L. (3) The dependence of the above on the optimization criterion. (4) The dependence of the optimal BSA composition on the optimization criterion.

The neutron source considered for this study is based on the ^{7}Li(p,n) reaction induced by 2.5 MeV protons; it provides a relatively soft spectrum—the most energetic neutrons are emitted at approximately 0.8 MeV. The tumor is assumed to be 8 to 8.5 cm deep. The boron concentrations, RBEs and compound factors used in this study were taken from Ref. 8: For tumor—45.5 ppm ^{10}B and compound factor 3.8, for normal tissue—13 ppm and 1.3. A RBE of 3.2 was assumed for charged particle kinetic energy deposition, whereas 1.0 is used for the RBE of gamma energy deposition. These values differ from the ones used in earlier SWAN optimizations.[2,3]

3.2. Results

Figure 2 shows the effect of BSA thickness on the maximum attainable beam quality for the two optimization criteria, while Figure 3 gives the corresponding specific beam intensities. The constituents used for these BSA are MgF_2, ^{7}LiF, Pb and ^{6}LiF. MgF_2

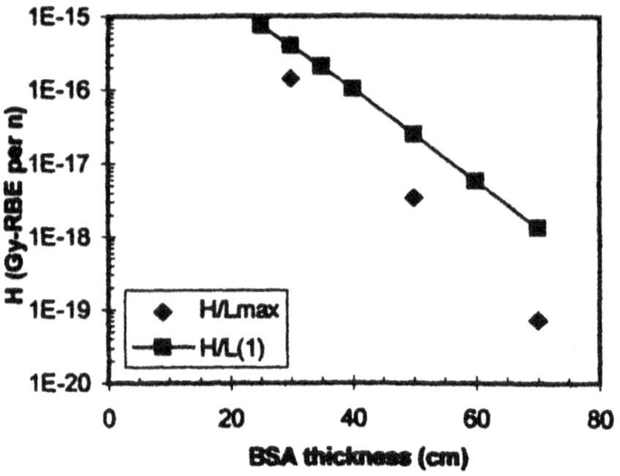

Figure 3. Effect of BSA thickness on the specific beam intensity for a ^{7}Li(2.5 MeVp,n) source and MgF_2, Pb and ^{6}LiF BSA.

was found[9] to be the primary BSA constituent offering the highest H/L for a given BSA thickness. It is slightly superior to ^{7}LiF and AlF_3. Figure 4 illustrates the dependence of the volume fraction of the BSA constituents on the optimal BSA thickness while Figure 5 shows the sensitivity of the optimal ^{6}LiF volume fraction on the optimization criterion. Figure 6 compares the neutron spectrum in air beyond optimal BSA of different thickness, while Figure 7 illustrates the dependence of the dose-depth distribution on the optimization criterion.

The following trends are found as the BSA is made thicker: (1) Maximum attainable H/L increases while the gradient of H/L vs. BSA thickness declines. (2) Specific beam intensity (H) sharply drops. (3) L_{max}/L_l increases (not shown in attached figures), i.e., L_{av}/L_{max} goes down, where L_{av} is the average dose imparted to the normal tissue. (4) Pb and ^{6}LiF volume fraction increases. (5) Neutron flux-in-air spectrum becomes narrower. For a given BSA thickness, using L_{max} for maximizing H/L leads to the following trends, vs. using L_l: (6) Higher H/L_{max}; (7) Lower H; (8) Smaller L_{max}/L_l. (9) Harder and

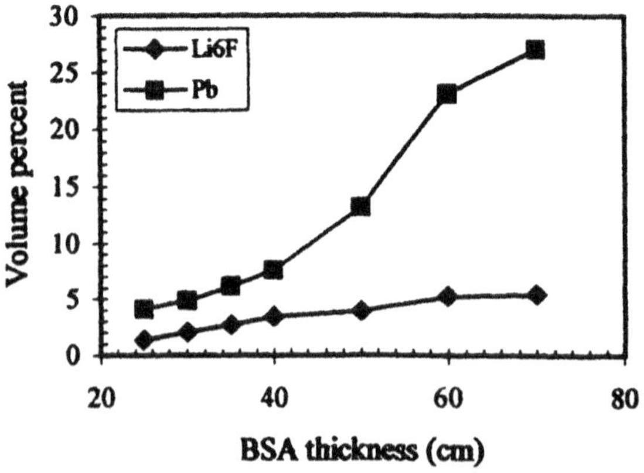

Figure 4. Lead and ^{6}LiF volume fraction dependence on the optimal BSA thickness. H/L_l maximization.

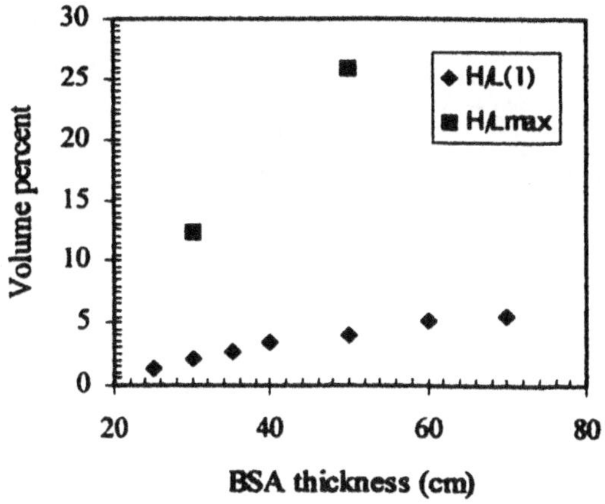

Figure 5. Sensitivity of the optimal ^{6}LiF volume fraction to the optimization criterion.

narrower flux-in-air spectrum (not shown in attached figures). (10) Higher ^{6}LiF and Pb (latter not shown) volume fraction.

4. COMPARISON OF DIFFERENT ACCELERATOR NEUTRON SOURCES

4.1. Problem Definition and Assumptions

In this section we are applying SWAN to perform a systematic reliable comparison of the potential of different neutron sources for BNCT. For a comparison to be reliable, one has to first find the real optimal BSA for each source, using a consistent set of

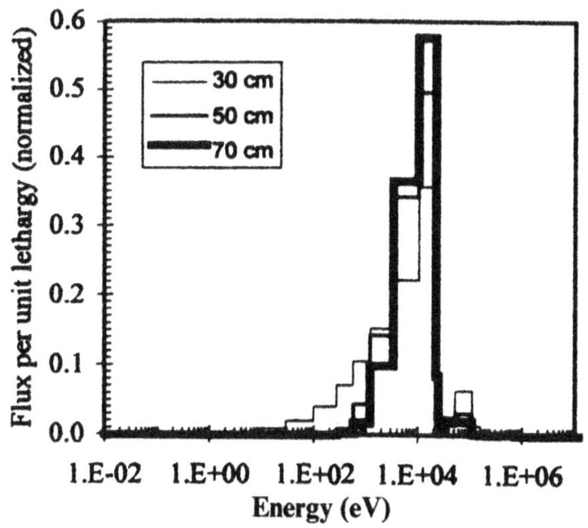

Figure 6. Optimal neutron spectrum in air dependence on optimal BSA thickness. H/L_{max} optimization.

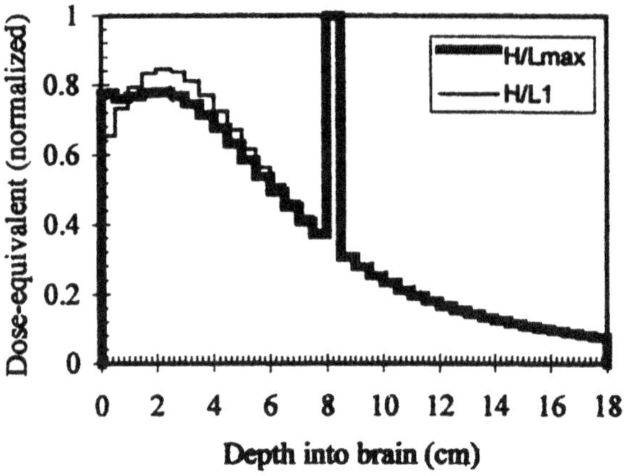

Figure 7. Dose-depth distribution dependence on the optimization criterion. 30 cm BSA.

assumptions. Such a comparison is very difficult to come by without a tool like SWAN. This is especially so when considering high neutron energy sources; the optimal BSA for such hard spectrum sources may require half a dozen or so constituents.[4,5] The specific question addressed is whether or not soft spectrum neutron sources are advantageous over hard spectrum neutron sources.

Four neutron sources of widely different spectra are considered; they are based on the following reactions: (1) ^{7}Li(p,n) induced by 2.5 MeV protons; it provides $E < 0.8$ MeV neutrons. (2) D-D; it provides 2.45 MeV neutrons. (3) Be(p,n) induced by 19 MeV protons; it provides ~1 MeV to ~15 MeV neutrons. (4) D-T; it provides 14 MeV neutrons.

The different sources are compared on equal beam quality basis. The beam quality is measured by H/L_{max}. For a given beam quality we compare the neutron sources in five different ways: (1) By their specific beam intensity per source neutron—the parameter H, now to be denoted as H_1. (2) By their specific beam intensity per source ion—H_2; the required accelerator beam current is inversely proportional to H_2. (3) By their specific beam intensity per source ion beam energy—H_3; the required accelerator beam power is inversely proportional to H_3. (4) By the maximum dose-equivalent rate that existing technology can deliver to the tumor. (5) By the maximum dose-equivalent rate that could be delivered to the tumor by the technology expected to be available within few years.

The beam-ion to neutron conversion factors assumed are 1.44×10^{-4}, 5.94×10^{-8}, 9.10×10^{-3}, and 3.13×10^{-5} for, respectively, neutron source type 1, 2, 3 and 4 defined above. The present day technology constraints assumed for the two proton accelerators is a maximum beam current of 20 mA and 3 mA. The former corresponds to the LLNL designed ESQ[10] whereas the latter refers to current designs of proton cyclotrons.[11,12] The corresponding near term technology constraints were assumed to be 50 mA and 10 mA. The former is set by the power handling ability of the ^{7}Li target—here assumed to be 125 kW. The latter is based on predicted cyclotron current.[11] For the D-T source the technology constraint was taken to be 2×10^{13} n/s; this is the highest yield achieved so far—at Sandia National Laboratory.[13] For near term technology we assumed, after Ref. 12, 5×10^{14} n/s. The maximum attainable D-D source intensity is assumed to be a factor of 50 lower.

As the BSA thickness and composition necessary for providing a given H/L_{max} strongly depend on the neutron source spectrum, the BSA thickness is used as a parametric variable. For each source and brain condition we apply SWAN to find the maximum beam quality attainable from BSA of a number of different selected thicknesses. The materials considered as candidate constituents include D_2O, ^6LiF, ^7LiF, Be, BeO, MgF_2, Al, AlF_3, Al/AlF_3, Al_2O_3, Fe, Pb and W.

As before, the tumor is assumed to be 8 to 8.5 cm deep. The boron concentrations, RBEs and compound factors used in this study were taken from Ref. 8: For tumor—45.5 ppm ^{10}B and compound factor 3.8; for normal tissue—13 ppm and 1.3. A RBE of 3.2 was assumed for charged particle kinetic energy deposition, whereas 1.0 is used for the RBE of gamma energy deposition.

4.2. Maximizing H_l for a given H/L_{max}

Before embarking upon the comparison of the four accelerator neutron sources, we explored the following question: will the application of SWAN to maximize H/L_{max} for a given BSA thickness result in maximum H_l for the specific neutron source under consideration? Consider two BSA consisting of different sets of constituents, to be denoted as Type A BSA and Type B BSA. Suppose that Type A offers the highest H/L_{max} for a given BSA thickness. Based on the findings of Sec. 3 we can expect Type B to offer the same H/L_{max} for a larger BSA thickness. The amplitude of H_l depends, for a given neutron source, on a couple of factors: (a) the distance between the source and the tumor; H_l is expected to vary like the inverse of the square of this distance; and (b) the attenuation of the dose-equivalent by the BSA constituents. In principle it may turn out that the overall attenuation of Type B BSA will be sufficiently smaller than that of Type A BSA so that the combined effect of attenuation and solid angle will be to the favor of Type B BSA; i.e., that $H_l(B) > H_l(A)$ for the same H/L_{max}.

We investigated such a possibility by comparing the performance of a BSA made of Al/AlF_3 + Pb + ^6LiF with that considered in Sec. 3. Both use the ^7Li(2.5 MeV p, n) neutron source. Figure 8 compares the performance of MgF_2 and Al/AlF_3 as the major constituents of the BSA. For BSA thickness smaller than 35 cm, MgF_2 offers higher beam quality (H/L_{max}). However, for the same beam quality, the specific beam intensity (H_l) offered by Al/AlF_3 is higher. It is concluded that (a) for the "brain conditions" assumed in this study, Al/AlF_3 is preferable to MgF_2 (as well as to ^7LiF), and (b) maximum H for a given H/L_{max} constraint is a more realistic optimization goal than just maximizing H/L_{max}. Nevertheless, the comparisons that follow are based on H/L_{max} maximization. Using Al/AlF_3 instead of the ^7LiF found to maximize H/L_{max} for the other three neutron sources did not result in a higher H_l.

4.3. H/L_{max} vs. H Tradeoff

The data presented in the two parts of Figure 8 can be worked out into H_l versus H/L_{max} dependence. Such a tradeoff is shown in Figure 9 for the four neutron sources considered. Figures 10 and 11 show a similar tradeoff but in terms of, respectively, H_2 and H_3.

It is observed that the ^7Li(2.5 MeV p, n) source offers the highest H_l as long as the beam quality $H/L_{max} < 1.4$. For higher H/L_{max} values the most energetic neutron sources take the lead. Notice that the slope of the H_l vs. H/L_{max} curves for the two softer neutron sources is different and steeper than that of the two hardest neutron sources. It should be realized that the H/L_{max} scale is correlated with the BSA thickness scale; a higher H/L_{max} implies a thicker BSA. Another interesting observation is that the Be(19 MeV p,

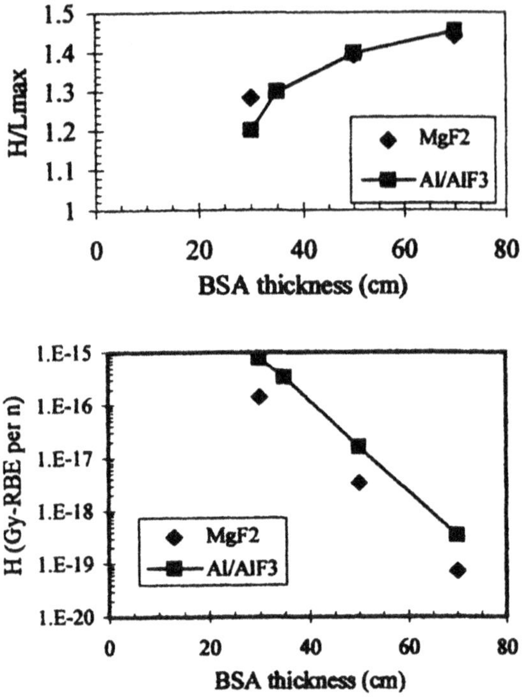

Figure 8. Comparison of optimal BSA using Al/AlF$_3$ Vs. MgF$_2$; ^{7}Li(2.5 MeV p, n) source.

n) and D-T neutron sources are almost identical, when measured in terms of H$_1$. The D-D neutron source is by far, the inferior of all.

The Be(19 MeV p, n) source offers the highest H$_2$ (Figure 10). That is, the 19 MeV p accelerator current required to deliver a given dose to the tumor in a given time is, by far, the smallest. The D-D neutron source requires much higher current.

When compared on per ion beam power requirement, the three non D-D neutron sources show comparable performance, with the Be(19 MeV p, n) neutron source having

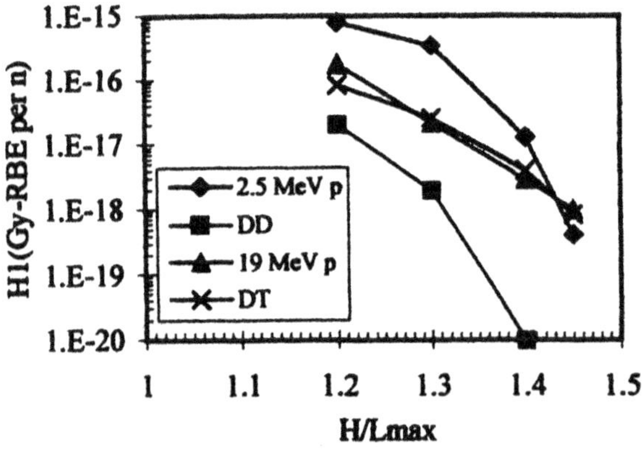

Figure 9. Specific beam intensity per source neutron vs. beam quality attainable by the 4 neutron sources.

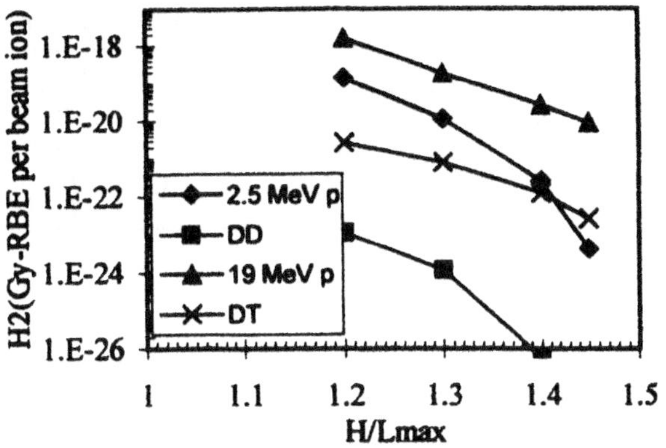

Figure 10. Specific beam intensity per beam ion vs. beam quality attainable by the 4 neutron sources.

a slight lead. The H_3 value of the ^7Li(2.5 MeV p, n) source falls behind that of the two most energetic neutron sources for high values of H/L_{max}; the higher the H/L_{max}, the less attractive becomes the ^7Li(2.5 MeV p, n) source.

It need be realized that the optimal BSA thickness and composition identified by SWAN strongly depend on the neutron source spectrum. For identical beam quality the BSA thickness for the harder spectrum sources is nearly twice that for the ^7Li(2.5 MeV p, n) source. Also, SWAN is asking for significant amounts of Pb and Fe in the source side of the BSA for the hard spectrum sources (in addition to a fluoride, usually ^7LiF, and some Pb and ^6LiF closer to the brain side).

4.4. Limits Imposed By Accelerator Technology

Let us introduce now one extra parameter—the maximum neutron yield rate, S_{max} (n/s), that can be delivered by the different accelerator—target systems based on present day technologies. Using the relationship H_{min}(Gy-RBE/n) S_{max}(n/s) 3600 (s/h) T(h) = TCD (See Sec. 2), and assuming TCD/T = 30 (Gy-RBE)/h, we get the requirement for the minimum $H_1(H_{min})$ corresponding to the present day technology limit on S_{max}. Table 1 compares the resulting H_{min} of the four types of accelerators. Combining this

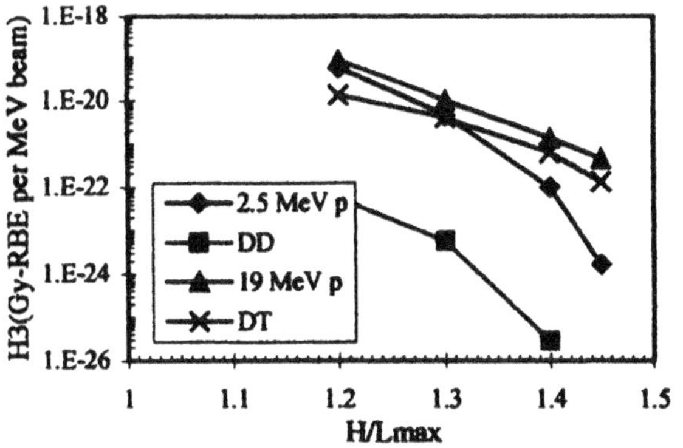

Figure 11. Specific beam intensity per beam energy vs. beam quality attainable by the 4 neutron sources.

Table 1. Minimum H_l that can be delivered by different accelerator neutron sources

Neutron source type	N/p	Max. current	Max. n/s	Min. H_l Gy-RBE/n
^{7}Li(p, n)	1.44E – 04	20 mA	1.8E + 13	4.63E – 16
Be(p, n)	9.10E – 03	3 mA	1.7E + 14	4.89E – 17
D-T	3.13E – 05		2.0E + 13	4.17E – 16
D-D	5.94E – 08		4.0E + 11	2.08E – 14

information with that of Figure 9 gives a correlation between the maximum H/L_{max} achievable by each of the accelerator types. This correlation is established in Figure 12 and summarized in Table 2. It is observed that the ^{7}Li(2.5 MeV p, n) and the Be(19 MeV p, n) neutron sources of present day technology can offer the same beam quality for the same treatment time. The D-T and, much more so, D-D neutron sources lag behind.

Next we compare the beam quality that might be attained from the different neutron sources in the near future (say, within 5 years). We assume that technological advancements will enable to increase the neutron source intensity. Table 3 gives the near future source intensities we assumed (based on Refs. 10, 11 and 14 for, respectively, ^{7}Li(p,n), Be(p,n) and D-T), along with the predicted beam quality. It is found that the predicted beam quality is practically the same for the three neutron source technologies examined. The beam power requirement for the D-T source is the highest of the three.

Finally we examined the consequences of increasing the neutron yield of the different sources by using a fast neutron amplifier, as described in Ref. 15. Table 4 summarizes our preliminary estimates for state-of-the-art accelerator technology. For the high energy neutron sources an amplification of nearly 50 was found possible. For the low energy source the amplification is close to half that much. It is concluded that the high energy neutron source Be(19 MeV p, n) has a potential of providing a higher beam quality, for the same treatment time, than the soft spectrum neutron source ^{7}Li(2.5 MeV p, n). Alternatively, for the same beam quality, the treatment time with the Be(19 MeV p, n) source need be shorter. The performance of the D-T source is somewhat inferior to the other two.

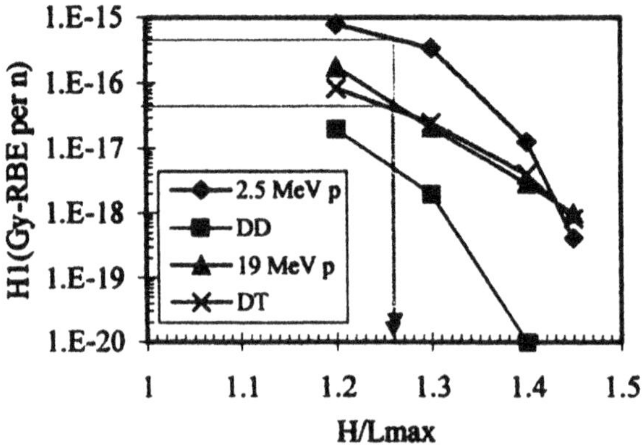

Figure 12. Determination of the best beam quality that can be delivered by different accelerator neutron sources.

Table 2. Maximum H_l/L_{max} that can be delivered by different neutron sources of present day technology

Neutron source type	Max. current or n/s	Max H/Lmax	BSA Thick. (cm)	Power on target (kW)
^{7}Li(p,n)	20 mA	1.26	33	50
Be(p,n)	3 mA	1.26	59	57
D-T	2E + 13	1.0	40	20
D-D	4E + 11	<<1.0		20

Table 3. Maximum H_l/L_{max} deliverable by accelerator neutron sources of near future technology

Neutron source type	Max. current or n/s	Max H/Lmax	BSA Thick. (cm)	Power on target (kW)
^{7}Li(p,n)	50 mA	1.32	38	125
Be(p,n)	10 mA	1.33	73	190
D-T	5E + 14	1.33	80	500

5. DISCUSSION

As the results presented in this paper pertain to simplified 1-D systems, they should not be taken as exact in absolute values but rather as indicating correct trends. These trends were verified with 3-D MCNP calculations.[6,7] The beam qualities obtained in this study may exceed the best beam qualities BNCT facilities have been designed to have. Consider, for example, Figure 2. Typical BSA thickness called for a ^{7}Li(2.5 MeV p, n) neutron source is on the order of 30 cm. Figure 2 shows that the beam quality can significantly increase (on the expense of reducing the specific beam intensity (Figure 3) by going to thicker BSA. The thicker BSA enables to make the in-air spectrum narrower, i.e., peaking near the optimal energy (Figure 6).

Analysis of the cost effectiveness of BNCT facilities based on different neutron sources was out of the scope of this work. We would like to mention, nevertheless, that the Be(19 MeV p, n) source and, if developed to generate ~5×10^{14} n/s, the D-T source appear more suitable than the ^{7}Li(2.5 MeV p, n) source for the production of radioisotopes of need for the hospital for two reasons: (a) They offer approximately an order of magnitude higher neutron fluxes (and of a wider neutron energy range); and (b) Their BSA provides a much better shield to the BNCT patient from the decay of the medical isotopes to be produced since it is thicker (at least twice) and has a significantly higher average atomic number. The isotope production may take place concurrently with the BNCT.

Table 4. Maximum H_l/L_{max} deliverable by accelerator neutron sources of present day technology driving a fast subcritical assembly

Neutron source type	Max. current or n/s	Max H/Lmax	BSA Thick. (cm)	Power on target (kW)
^{7}Li(p,n)	20 mA	1.39	53	50
Be(p,n)	3 mA	1.45	110	57
D-T	2E + 13	1.33	80	20
D-D	4E + 11	<1.0		20

6. CONCLUDING REMARKS

Following is a summary of the trends we found: (1) The thicker the BSA is, the higher can be the beam quality. However, (2) the specific beam intensity drops exponentially with the BSA thickness. Typically, $d(H/L_{max})/dH_l < \sim -0.01$ [Gy-RBE per n]$^{-1}$. (3) Hard spectrum neutron sources can provide as good a beam quality as soft spectrum neutron sources. (4) For a given beam quality, the ^7Li(2.5 MeV p, n) source can deliver the highest dose-equivalent per source neutron while the Be(19 MeV p, n) source can deliver the highest dose-equivalent per beam ion as well as per beam power. (5) When accounting for constraints imposed by state-of-the-art technology, the beam quality which could be provided by both proton accelerator systems considered is, practically, the same for the same treatment time. The beam quality attainable from the state-of-the-art fusion neutron sources considered is significantly lower. (6) Technological improvements which are anticipated within the next few years are predicted to bring the beam quality attainable from the ^7Li(2.5 MeV p, n), Be(19 MeV p, n) and D-T sources to, essentially, the same value for the same treatment time. Hence (7) the choice of the best neutron source to use for BNCT should be determined by economic considerations. (8) The optimal BSA composition varies with the BSA thickness, brain conditions and optimization criterion. (9) The most desirable BSA optimization criterion appears to us to be maximizing H for a given value of H/L (i.e., for a given beam quality). (10) The definition of L to be used for the H/L figure-of-merit can have a significant effect on the outcome of the BSA optimization. The constraint on the maximum permissible skin dose has also a significant impact on the optimization outcome. Hence, it is desirable that the clinical treatment community and BSA design community will collaborate on defining the most desirable optimization criterion and constraints. (11) Al/AlF_3 may, or may not be the most effective primary neutron slowing down material for the BSA of a ^7Li(2.5 MeV p, n) neutron source; under certain conditions ^7LiF and MgF_2 may be preferable.

REFERENCES

1. F.J. Wheeler *et al.*, "Reactor Beam Calculations to Determine Optimum Delivery of Epithermal Neutrons for Treatment of Brain Tumors," *Trans. Am. Nucl. Soc.*, 77:42, 1997.
2. Y. Karni, and E. Greenspan, "Using SWAN for Optimizing the Neutron Source Assembly Design for BNCT Facilities," *Proc. 1st Int. Workshop on Accelerator-Based Neutron Sources for BNCT*, Jackson, Wyoming, Sept. 11–14, 1994.
3. E. Greenspan, and Y. Karni, "The "Swan" Approach to Neutron Beam Optimization for BNCT and Other Applications" Proc. 1998 Int. Conf. on the Physics of Nuclear Science and Technology, LI., N.Y., October 5–8, 1998.
4. Y. Karni, E. Greenspan, J. Vujic, and B. Ludewigt, "Optimal Beam-Shaping Assemblies for BNCT Facilities Using 2.5 MeV p on ^7Li or 19 MeV p on Be," *Trans. American Nuclear Society*, 73:29, 1995.
5. E. Greenspan, Y. Karni, J. Vujic, and B. Ludewigt, "BNCT Beam Shaping Sensitivity to Neutron Source Spectrum," *Trans. Amer. Nuc. Soc.*, 77:44, 1997.
6. J.M. Verbenke, J. Vujic, and K.N. Leung, "Investigation of $^2H(d,n)^3H$ and $^3H(d,n)^4He$ fusion reactions as alternative neutron sources for BNCT," Presented at the *8th International Symp. On Neutron Capture Therapy for Cancer*, La Jolla, CA, Sept. 13–18, 1998.
7. D.L. Bleuel, LBNL, Personal Communication, 1997.
8. A.D. Chanana, "Boron Neutron Capture Therapy of Glioblastoma Multiforme at the Brookhaven Medical Research Reactor—a Phase I/II Study," FDA IND # 43,317, January 2, 1996.

9. Y. Karni, E. Greenspan, and J. Vujic, "Perturbation Theory Guided Design of Beam Shaping Assemblies,"*Trans. Amer. Nucl. Soc.*, 77:354, 1997.
10. B. Ludewigt *et al.*, "An Epithermal Neutron Source for BNCT Based on an ESQ-Accelerator," *Proc. Topical Mtg. on Accelerator Applications of Nuclear Technology*, pp. 489–494, Albuquerque, N.M., Nov. 16–20, 1997.
11. Dr. J.P. Modrillion, Nice, France, Personal communication, June 1998.
12. J.R. Stevenson, and P.L. Larabie, "Research and Commercial Isotope Production at TRIUMF," *Trans. Amer. Nucl. Soc.*, 77, 1997.
13. F.M. Bacon *et al.*, "High Heat Flux Target for Intense Neutron Source," *Review of Scientific Instruments*, 55(1), January 1984.
14. Dr. K-N Leung, LBNL, Personal Communication, August 1998. See also K.N. Leung *et al.*, "A Sealed-Accelerator-Tube Neutron Generator for Boron Neutron Capture Therapy Application," Presented at the *8th International Symp. On Neutron Capture Therapy for Cancer*, La Jolla, CA, Sept. 13–18, 1998.
15. Y. Karni, E. Greenspan, and J. Vujic, "Accelerator Driven Fission Neutron Source for BNCT," *Proc. ANS Embedded Topical Mtg. On Accelerator Applications '97*, Albuquerque, N.M. Nov. 1997.

A BIOLOGICAL COMPARISON OF NEUTRON BEAMS USED FOR BNCT RESEARCH

C. Mansfield[1], J. W. Hopewell[2], T. D. Beynon[1], and R. Huiskamp[3]

[1]School of Physics and Astronomy
University of Birmingham
Edgbaston, Birmingham, B15 2TT, UK
[2]Research Institute
Oxford University
Oxford, OX3 7LJ
UK
[3]Netherlands Energy Research Foundation
ECN, Petten, The Netherlands

ABSTRACT

It is important for the BNCT community to be able to compare and predict the biological effects of mixed radiation fields at various centres. A quantitative comparison of the beams would allow more valid discussions of radiobiological and clinical data between centres. If a biological system could be irradiated in various BNCT facilities and its radiation response analysed such a comparison will be possible. If the beams at each centre can thus be compared the preclinical and radiobiological studies performed at other centres will not have to be repeated at new centres, such as at Birmingham, and comparison of clinical data will be more meaningful.

This report describes the design of an *in vitro* experiment which fulfils these criteria. Cell survival will be determined in relation to the depth in a phantom following irradiation in a BNCT beam. To overcome the problems of the variable repair of sub-lethal damage from the gamma radiation component of any beam during the irradiation period all irradiations will be performed at 4 °C. The different components of the total dose rate at each position in the phantom will be estimated using paired ionisation chambers and activation foils. To ensure that a valid comparison is being made, the beam survival curves will be compared to x-ray survival curves produced at each centre in this study.

Frontiers in Neutron Capture Therapy, edited by Hawthorne *et al.*
Kluwer Academic / Plenum Publishers, New York, 2001.

METHODS

Irraditaions will be performed at:

The High Flux Reactor (HFR) at the Joint Research Council, Petten, the Netherlands;
The Brookhaven Medical Research Reactor (BMRR), Brookhaven National Laboratory, New York, USA;
The Dynamitron beam at Birmingham University, UK;
Possibly the FiR, Technical Research Centre, Finland.

These centres were chosen to compare with the Birmingham beam and each other as they have already carried out *in vitro*[1,2] and *in vivo*,[3,4] radiobiological studies.

For each beam cells are irradiated at depth in a phantom. Batches of cells from the same stock are kept frozen and a fresh batch will be used for each centre. In this way the inherent radiosensitivty of the cells should be kept uniform throughout the study.

At each depth the various dose components are measured and survival curves constructed. Each such survival curve are then normalised to an x-ray survival curve performed at that centre. This will minimise any differences between experimental conditions at each centres.

This article will discuss the tissue culture and dosimetry procedures employed, and the design of the phantom. A method of overcoming the "dose rate related effect" will also be examined.

The Dose Rate Related Effect

The importance of the dose components varies with depth in a phantom. As depth increases the flux of neutrons decreases faster than the flux of photons. At depths greater than a few centimetres the gamma dose becomes important. However, this is still very low—0.05 Gy/min at the HFR at Petten.[5] Since repair of sublethal damage is possible following photon irradiation some damage at these depths would normally be repaired during long irradiations. This effect will be greater for long irradiations, which produce a low surviving fraction than for short irradiations. This would make a comparison between exposures extremely complex.

To overcome this problem all irradiations are performed at 4 °C. At this temperature repair of sublethal radiation damage is inhibited and so cell survival will be independent of dose rate. For the cell line that will be used, V79, reproducible plating efficiencies can be maintained even after cells have been kept on ice for 24 hours. The x-ray irradiations are also performed at 4 °C.

Tissue Culture Procedures

A cell line was needed which was not only well established and researched, but also hardy and with a short cell cycle. The Chinese hamster line V79 satisfied these conditions, and was obtained from the European Collection of Cell Cultures. A cell bank was built up within a few passages of receiving the cell sample and a new vial is used for each set of experiments. This ensures that all cells will be irradiated after a similar number of passages and so will be genetically similar.

All cells are kept in exponential growth in standard conditions. An attempt was made to irradiate the cells adhered to the side of a T25 flask. In this way the depth at which the cells would be irradiated can be exactly determined. However, this was found to be an extremely time consuming and inaccurate method. The most detrimental step was determining the cell density for each flask. This was done using a hemocytometer after irradiation and before the cells were plated out. This meant the cells were out of the incubator for an unnecessarily long time and that each cell density was only determined to within ±20%.

An alternative method was tested and has proved to be far more accurate and reproducible. Approximately 10^6 cells are seeded into a T175 flask the afternoon before irradiation and incubated overnight. Immediately prior to irradiation the cells are trypsinised and the cell density determined. Aliquots of the cells are placed in 1.5 ml centrifuge tubes, sealed with sterile film and placed on ice before and after irradiation. Following irradiation the appropriate number of cells are seeded into tissue culture dishes. The cells are then incubated for 7 days, fixed, stained and the colonies of greater than 50 cells are counted.

This method was tested at Birmingham University using 250 kVp x-rays and was found to be reproducible and quick. This method will now be used for the neutron and x-ray irradiations at each of the BNCT centres.

DOSIMETRY

To compare the beams the physical dose rates at each depth must be determined. The thermal neutron component are estimated using activation foils placed inside a centrifuge tube filled with medium. The dose rates from the epithermal neutron, fast neutron and the gamma components are determined using paired ionisation chambers placed at the appropriate depth. At all centres use is made of the beam monitoring systems to ensure that the dose rate given can be related to that measured. Since it is not physically possible to place ionisation chambers at all depths in the phantom MCNP simulations are performed to complete the dosimetric evaluation.

PHANTOM DESIGN

The phantom does not have to be exactly comparable to a head, since the values of RBE calculated will be purely comparative. However, it is important that the phantom has similar neutron transport properties—similar atomic composition and dimensions as a head. Water is the obvious choice as a tissue equivalent liquid. Low Z materials, such as Perspex and polyethylene also have suitable properties.

The phantom is based on a design that has been previously used for this type of *in vitro* irradiation. This makes the study more comparable to these previous investigations. Cylindrical phantoms, of length 23.0 cm and diameter 16.6 cm, have previously been used by the Petten group[2] and the Brookhaven National Laboratory project.

A phantom was constructed by the University of Birmingham workshop. It comprises of a cylinder, sealed at one end and with a detachable lid at the other. It is water tight and the base is flat to ensure stability. Holes were drilled into the top of the phantom to accommodate ion chambers. A rod was placed along the horizontal axis of the

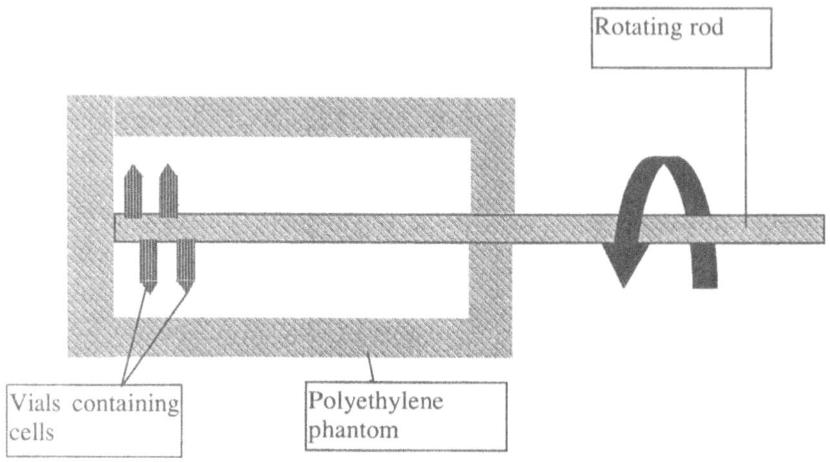

Figure 1. Birmingham phantom.

phantom with slots into which the centrifuge tubes could be placed. A motor is used to rotate the rod at 2 rpm to keep the cells from sinking to the bottom of the tube during long irradiations. This is illustrated in Figure 1.

FUTURE WORK

A series of neutron irradiations will be performed at the HFR, the BMRR, the FiR, and the Dynaimtron beam. Dosimetry measurements will be taken at each centre and each experiment will be simulated using MCNP. X-ray survival curves will be performed at each centre.

Values for RBE at each depth in the phantom will be calculated, enabling more valid comparisons of *in vivo* experiments and patient data between centres.

ACKNOWLEDGMENTS

I would like to thank all the centres involved with this project for all their help and support. I would also like to thank the CRC Institute for Cancer Studies, University of Birmingham, for the allowing me to use their tissue culture facilities.

REFERENCES

1. J.A. Coderre, M.S. Makar, P.L. Micca, M.M. Nawrocky, H.B. Liu, D.D. Joel, D.N. Slatkin, and H.I. Amols, Derivations of relative biological effectiveness for the high LET radiations produced during boron neutron capture irradiations of the 9L rat gliosarcoma *in vitro* and *in vivo*, *Int J of Radiat Oncol, Biol, Phys*, 27(5):1121–1129, 1993.
2. R. Huiskamp, A.C. Begg, V.G.A. Gregoire, R. Verrijk, D. Gabel, A. Siefert, and R.L. Moss, Radiobiology studies at Petten: Status on cell culture, mice and dog experiments, *in* Progress in Neutron Capture Therapy for Cancer, ed. B.J. Allen *et al.*, Plenum Press, New York, 1992, pp. 485–488.
3. P.R. Gavin, S.L. Kraft, R. Huiskamp, and J.A. Coderre, A review: CNS effects and normal tissue tolerance in dogs, *J of Neuro Oncol*, 33(1–2):59–70, 1997.

4. J. Benczzik, T. Seppälä, J. Hopewell, M. Snellman, I. Auterinen, R. Joensuu, U. Abo Ramadan, S. Savolainen, M. Färkkilä, H. Joensuu, M. Kallio, M. Kulvik, and M. Tenhunun, Large animal model for healthy tissue tolerance study in BNCT, *in* "Frontiers in Neutron Capture Therapy," M.F. Hawthorne, K. Shelly, R.W. Wiersema, eds., Kluwer Academic/Plenum Publishers, New York, 2001, pp. 1233–1238.
5. C.P.J. Raaijmakers, and M.W. Konijnenberg, Determination of dose components in phantoms irradiated with an epithermal neutron beam for boron neutron capture therapy, *Med Phys*, 22(3):321–329, 1995.

57

A COMPARISON OF NEUTRON FIELDS FOR BNCT

Jeffrey E. Woollard[1], Thomas E. Blue[1], and Jacek Capala[2]

[1]Nuclear Engineering Program
The Ohio State University
206 W. 18th Avenue
Columbus, OH 43210
[2]Medical Department
Brookhaven National Laboratory
Upton, NY 11973

1. INTRODUCTION

Previously, we have developed the in-phantom neutron field assessment parameters T and D_{Tumor} for the evaluation of epithermal neutron fields for use in BNCT.[1] These parameters are based on an energy-spectrum-dependent neutron normal-tissue RBE[2] and the treatment planning methodology of Gahbauer and his co-workers,[3] which includes the effects of dose fractionation.

In this paper, these neutron field assessment parameters are used to evaluate several designs of an Accelerator-Based Neutron Source (ABNS) for BNCT and the Brookhaven Medical Research Reactor (BMRR) epithermal neutron beam[4] (hereafter called the BMRR-ENB), in order to judge the suitability of an ABNS for BNCT. The BMRR-ENB was chosen as the basis for comparison, because it is presently being used in human clinical trials of BNCT[5] and because it is the standard to which other neutron beams are most often compared.

2. BACKGROUND

This section provides a brief review of our in-phantom neutron field assessment parameters. The development of these parameters is presented in Woollard *et al.*[1] One neutron field assessment parameter is the treatment time, T. It is the total time required for a BNCT treatment, including all treatment fractions. Since the total dose delivered to the tumor is limited by the tolerance of the surrounding normal tissue, the treatment

Frontiers in Neutron Capture Therapy, edited by Hawthorne *et al.*
Kluwer Academic / Plenum Publishers, New York, 2001.

time is defined as the time required to escalate the normal tissue RBE-dose to the tolerance of the normal brain. The treatment time is given by the following expression,

$$T = \frac{ETDL_m}{\left[\dot{D}_\gamma + \left(\frac{ETDL_m \cdot ETDH_l}{ETDL_l \cdot ETDH_m}\overline{RBE(E_n)_{norm}}\dot{D}_n + RBE_B CF \dot{d}_B [B]\right)\right]_{d_{max}}}. \quad (1)$$

The second neutron field assessment parameter is the high-LET absorbed-dose to the tumor, denoted as D_{Tumor}. D_{Tumor} is the product of the high-LET absorbed-dose rate at the tumor location and the treatment time. D_{Tumor} is given by the following expression,

$$D_{Tumor} = \frac{\left[\dot{D}_n + \dot{d}_B [B] R_{t/b}\right]_{Tumor\ Location}}{\left[\dot{D}_\gamma + \left(\frac{ETDL_m \cdot ETDH_l}{ETDL_l \cdot ETDH_m}\overline{RBE(E_n)_{norm}}\dot{D}_n + RBE_B CF \dot{d}_B [B]\right)\right]_{d_{max}}} ETDL_m \quad (2)$$

The symbols in the above expressions for T and D_{Tumor} are defined as follows:

[B]: ^{10}B concentration in blood in ppm.

$RBE_B CF$: The product of the RBE and the compound factor (CF) for the boron absorbed dose. The $RBE_B CF$ accounts for the unknowns (such as boron localization within normal tissues for a compound) which influence the biological effectiveness (relative to low-LET radiation) of the boron absorbed dose in normal tissue indexed to the boron absorbed dose to blood.[6]

d_{max}: The depth for which the limiting RBE-dose to normal tissue is reached.

$\dot{D}_n$: Neutron absorbed dose rate at the specified location in the brain.

$\dot{D}_\gamma$: Gamma-ray absorbed-dose rate at the specified location in the brain.

$\dot{d}_B$: Specific (*per ppm of ^{10}B*) boron absorbed-dose rate in the blood at the specified location in the brain, assuming that charged-particle equilibrium exists within the blood.

$ETDL_m$: Estimated Tolerance Dose. $ETDL_m$ is the Central Nervous System (CNS) normal-tissue RBE-dose at tolerance for a mixed field of high- and low-LET radiations for some fractionation scheme with m fractions.

$ETDH_m$: The CNS absorbed-dose at tolerance for high-LET radiations delivered in m fractions.

$R_{t/b}$: Ratio of ^{10}B concentration in tumor to ^{10}B concentration in blood (assuming macroscopically uniform ^{10}B concentrations within the tumor).

$\overline{\mathbf{RBE(E_n)_{norm}}}$: the kerma-averaged neutron RBE,[1] given by;

$$\overline{RBE(E_n)_{norm}} = \frac{\int RBE^b_{norm}(E_n) k_n(E_n) \phi(E_n) dE_n}{\int k_n(E_n) \phi(E_n) dE_n} \quad (3)$$

where $k_n(E_n)$ is the energy dependent neutron kerma factor for brain tissue (cGy cm^2); $\phi(E_n)$ is the energy dependent neutron flux at the irradiation point ($cm^{-2} s^{-1} keV^{-1}$)

and $RBE^{b}_{norm}(E_n)$ is the energy dependent normal-tissue neutron RBE for human adult brain tissue normalized to the neutron RBE measured in large animal normal tissue tolerance studies performed at the Brookhaven Medical Research Reactor (BMRR).

3. ANALYSIS

In addition to the BMRR-ENB, several ABNS designs were evaluated for this analysis. The designs included the OSU design of an ABNS for BNCT, which consists of a 25 cm D_2O moderator, a 30 cm thick Li_2CO_3 reflector and a 10 cm thick Li_2CO_3 delimiter (hereafter called the OSU-ABNS). This ABNS design is described in detail in Woollard *et al.*[1] The next design consisted of a 34 cm thick Al/AlF_3 moderator, a 30 cm thick Al_2O_3 reflector and a 13.3 cm thick ^{6}Li-polyethylene delimiter (hereafter called the AlF_3-Al_2O_3 ABNS). The final design consisted of a 22 cm thick ^{7}LiF moderator, a 30 cm thick Al_2O_3 reflector and a 13.3 cm thick ^{6}Li-polyethylene delimiter (hereafter called the ^{7}LiF-Al_2O_3 ABNS). The final two ABNS designs are described in detail in Bleuel *et al.*[7]

For these ABNS designs, neutrons are produced via the ^{7}Li(p,n) reaction by bombarding a Li-7 target with protons. For each ABNS design, a 10 mA beam of protons was assumed to be incident on the Li-7 target. The energy of the protons, however, varied for each design. Proton energies of 2.5 MeV, 2.4 MeV and 2.3 MeV were assumed for the OSU-ABNS, the AlF_3-AL_2O_3 ABNS and the ^{7}LiF-AL_2O_3 ABNS designs, respectively. For the BMRR-ENB, it was assumed that the BMRR was operating at a reactor power of 3 MW.

The neutron field assessment parameters were calculated in a $14 \times 14 \times 14\,cm^3$ Lucite cube phantom located in the irradiation port of each facility. Calculation of T and D_{Tumor} required values for the neutron absorbed dose rate, the gamma-ray absorbed-dose rate, the specific (*per ppm of ^{10}B*) boron absorbed-dose rate, and the kerma-averaged normal tissue neutron RBE. These values were calculated as a function of depth along the centerline of the phantom for the ABNS designs using the Monte-Carlo radiation transport code MCNP4B.[8] These calculations were performed on an IBM RS6000 workstation. ENDF/B-V continuous energy cross section data were used in the calculations. The thermal neutron scattering function $S(\alpha,\beta)$ for polyethylene was used for the Lucite cube phantom.[4] The values were calculated as a function of depth along the centerline of the phantom for the BMRR-ENB using Monte-Carlo based treatment-planning software.[9]

The evaluation of T and D_{Tumor} requires that some assumptions be made regarding the fractionation scheme and the Estimated Tolerance Dose (ETD) for brain for low-LET ($ETDL_m$) and high-LET ($ETDH_m$) radiation delivered in m fractions. For this analysis, we have assumed that the patient is treated with four fractions (m = 4), in five days. According to Gahbauer,[6] for treatment with four fractions in five days, $ETDL_4$ = 2,300 cGy and $ETDH_4$ = 1,000 cGy. Also, according to Gahbauer, for treatment with one fraction, $ETDL_1$ = 1,200 cGy and $ETDH_1$ = 600 cGy.

The evaluation of T and D_{Tumor} also requires that some assumptions be made about boron concentrations and localization. For this analysis we have assumed that the ^{10}B concentration in blood in ppm, [B], equals 30 ppm and that the ratio of ^{10}B concentration in tumor to ^{10}B concentration in blood, $R_{t/b}$, equals 1.3, values which are typical for the B-10 delivery agent BSH.[10] Also, according to a recent evaluation of the BMRR dog data, it was assumed that for BSH the product of the RBE and the compound factor

Table 1. Calculated treatment times for each neutron source evaluated

Neutron source	Treatment time (min)	Time per fraction[a] (min)
AlF_3-Al_2O_3 ABNS	149	37
^7LiF-Al_2O_3 ABNS	88	22
OSU-ABNS	214	54
BMRR-ENB (3MW)	91	23

[a]Assumes a four fraction treatment scheme.

(CF) for the boron absorbed dose (RBE_BCF) is 0.27 for the endpoint of late changes in the magnetic resonance images.[11]

4. RESULTS

For the calculated absorbed dose rate distributions and with the above assumptions regarding fractionation, tolerances, RBEs, compound factors and boron concentrations, the in-phantom neutron field assessment parameters were calculated for each neutron source. The calculated treatment times (T) are presented in Table 1. Also shown in Table 1 are the corresponding treatment times per fraction for a four fraction treatment scheme.

Curves of D_{Tumor} versus tumor depth along the phantom centerline are presented in Figure 1. These curves indicate that the quality of the neutron fields for the ABNS designs and the BMRR-ENB are comparable. The value of D_{Tumor} for the ABNS designs is essentially equal for tumor depths ranging from 0 to approximately 1.5 cm and the value of D_{Tumor} for the OSU-ABNS is largest for tumor depths ranging from approximately 1.5 to 14 cm.

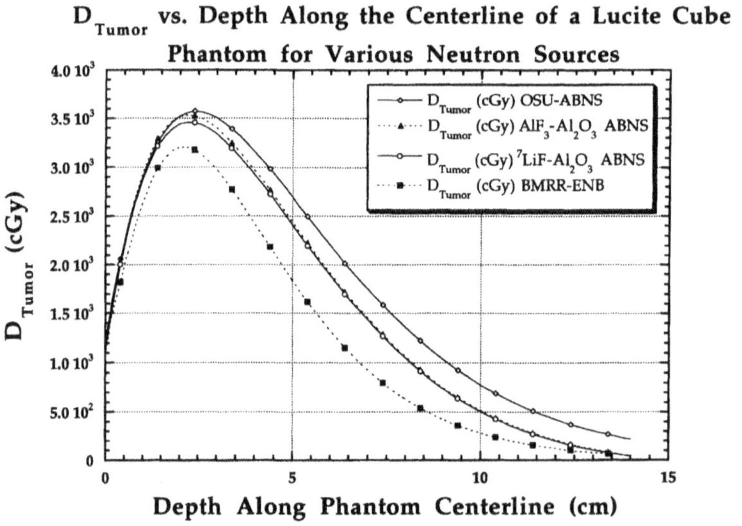

Figure 1. Calculated values of D_{Tumor} versus tumor depth for each neutron source evaluated.

5. CONCLUSION

When using the in-phantom neutron field assessment parameters T and D_{Tumor} to evaluate neutron fields for BNCT, it is desirable to maximize the value of D_{Tumor} while keeping the treatment time, T, reasonable. On the basis of the calculated values of T and D_{Tumor}, the neutron fields produced by the ABNS designs are comparable with the neutron field produced by the BMRR-ENB. The larger value of D_{Tumor} at depth for the OSU-ABNS should not be viewed as indicating that the OSU-ABNS beam has superior beam quality. Rather, it should be viewed as a consequence of the fact that the OSU-ABNS and the other neutron sources were compared using the parameters with which the OSU-ABNS was optimized.

REFERENCES

1. J.E. Woollard, T.E. Blue, N. Gupta, and R.A. Gahbauer, "Development and Application of Neutron Field Optimization Parameters for an Accelerator-Based Neutron Source for Boron Neutron Capture Therapy", Nucl. Tech., **155**:100–112, (1996).
2. T.E. Blue, J.E. Woollard, N. Gupta, and R.A. Gahbauer, "Beam Design and Evaluation for BNCT," Proc. of First Inter. Workshop on Accelerator-Based Neutron Sources for Boron Neutron Capture Therapy, Jackson, WY, Sept. 11–14, 1994, INEL Report Conference-940976, pp. 197–212, (1995).
3. N. Gupta, R.A. Gahbauer, T.E. Blue, and A. Wambersie, "Dose Prescription in Boron Neutron Capture Therapy," Int. J. Radiat. Oncol. Biol. Phys., **28**:1157–1166, (1994).
4. H.B. Liu, D.D. Greenberg, J. Capala, and F.J. Wheeler, "An Improved Neutron Collimator for Brain Tumor Irradiations in Clinical Boron Neutron Capture Therapy", Med. Phys., **23**:2051–2060, (1996).
5. J.A. Coderre, E.H. Elowitz, M. Chadha, R. Bergland, J. Capala, D.D. Joel, H.B. Liu, D.N. Slatkin, and A.D. Chanana, "Boron Neutron Capture Therapy for Glioblastoma Multiforme using p-boronophenylalanine and Epithermal Neutrons: Trial Design and Early Clinical Results", J. Neuro. Oncol., **33**:141–152, (1997).
6. R.A. Gahbauer, R.G. Fairchild, J.H. Goodman, and T.E. Blue, "RBE in Normal Tissue Studies", *Boron Neutron Capture Therapy: Toward Clinical Trials of Glioma Treatment*, D. Gabel and R. Moss (ed), Plenum Press, N.Y., 123–128, (1992).
7. D.L. Bleuel, S.V. Costes, R.J. Donahue, B.A. Ludewigt, and J. Vujic, "Neutronics Analysis of Three Beam-Filter Assemblies for an Accelerator-Based BNCT Facility", Proceedings of the Topical Meeting on Nuclear Applications of Accelerator Technology, Albuquerque, NM., Nov. 16–20, 1997, pp. 502–507, (1997).
8. J.F. Briesmeister (ed.), "MCNP–A General Monte Carlo N-particle Transport Code, Version 4A," LA-12625, Los Alamos National Laboratory (1993).
9. D.W. Nigg, F.J. Wheeler, D.E. Wessol, J. Capala, and M. Chadha, "Computational Dosimetry and Treatment Planning for Boron Neutron Capture Therapy", J. Neuro. Oncol., **33**:93–104, (1997).
10. H. Frankhauser, and P.R. Gavin, "Summing Up: Clinical Papers", in *Advances in Neutron Capture Therapy*, A.H. Soloway, R.F. Barth, and D.E. Carpenter (ed), Plenum Press, N.Y., 799–805, (1993).
11. P.R. Gavin, S.L. Kraft, R. Huiskamp, and J.A. Coderre, "A Review: CNS Effects and Normal Tissue Tolerance in Dogs", J. Neuro-Oncol., **33**:71–80, (1997).

THE BNCT PROJECT IN THE CZECH REPUBLIC BEFORE THE START OF CLINICAL TREATMENT

J. Burian[1], M. Marek[1], J. Rataj[1], K. Prokes[2], F. Novy[2], F. Tovarys[3], V. Dbaly[3], J. Honzatko[4], I. Tomandl[4], and O. Kriz[5]

[1]Nucl. Res. Inst. Rez, plc (NRI)
[2]Dept. of Oncology, Faculty Hospital of Charles University, Prague
[3]Dept. of Neurosurgery Hospital Na Homolce, Prague
[4]Nucl. Phys. Inst., Ac. Sci. CR, Rez
[5]Katchem Ltd. Prague

INTRODUCTION

The BNCT group of specialists from different scientific fields (physics, medicine, chemistry, radiobiology and pharmacology) in Czechia decided as a first priority to initiate the glioma clinical trial at the LVR reactor of NRI. The principal aim now is to prepare the Phase I trial, to establish maximum tolerated dose by healthy tissue when irradiating with BSH drug. The activity has been concentrated in this final stage also on verifying of epithermal beam characteristics and on improvement of the parameters of the beam and on the finishing of the construction the irradiation room itself. The treatment protocol was prepared and put forward to corresponding authorities for approval.

LVR-15 REACTOR

The development and construction of BNCT facility were realized at LVR reactor of NRI Rez. The LVR-15 is a light water reactor with enriched fuel (36%) and standard thermal power 10 MW. This, on commercial basis running reactor, is used for material testing experiments at water loops and rigs, for radio-pharmaceuticals production, irradiation of silicon crystals and for basic and applied research experiments at horizontal channels, and for BNCT as a source of epithermal neutrons.

EPITHERMAL BEAM

A beam of epithermal neutrons has been constructed at the LVR using empty space of thermal column as described.[1] A main drawback of our arrangements is a rather long

Frontiers in Neutron Capture Therapy, edited by Hawthorne *et al.*
Kluwer Academic / Plenum Publishers, New York, 2001.

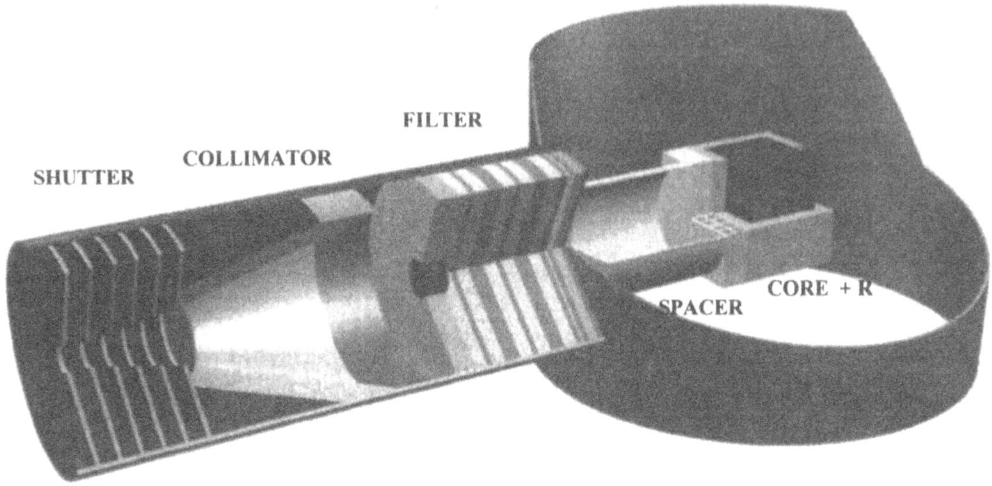

Figure 1. Configuration of epithermal beam at LVR-15 geometry.

distance between the core and irradiation point (about 4 m). Computer codes DORT (discrete ordinates) and MCNP (Monte Carlo) are used for the design of configuration of our system. Several assemblies have been experimentally tested with the aim to determine parameters both of the free beam and the beam inside the phantom. The techniques used for measurement are described in the separate paper of this symposium.[2] Typical configuration of epithermal neutron beam in the LVR-15 geometry is demonstrated in Figure 1.

Fast neutrons escaping the core and reflector are transported through the empty inner shutter (can be filled by water) to block of filters. Epithermal neutrons are collimated and transferred through the outer shutter to the irradiation point. The block of filters in Figure 1 consists of B_4C thin layer, lead-5cm, ten layers aluminum and aluminum fluoride, total thickness 61 cm (35 cm Al, 26 cm AlF_3), lead-11 cm. There is 1 cm Ti and B_4C thin layer just behind Al—C collimator. The characteristics of the beam measured in July 1998 are:

Free beam	$\Phi_{epi} = 1.82\ 10^8\,n/cm^2s$	$D_{f.n.}/\Phi_{epi} = 12\ 10^{-11}\,cGy\,cm^2/n$
	$\Phi_{fast} = 0.19\ 10^8\,n/cm^2s$	$D_\gamma/\Phi_{epi} = 9.3\ 10^{-11}\,cGy\,cm^2/n$
	$\Phi_{tot} = 2.02\ 10^8\,n/cm^2s$	
Phantom	$\Phi_{th} = 4.82\ 10^8\,n/cm^2s$	$D_{f.n.} = 0.625\,Gy/h$
at depth 2.5 cm	$D_\gamma = 1.87\,Gy/h$	

The parameters are lower than optimum and we are still trying to improve them. The one of promising possibilities is to remodel the edge of core where the air spacers could be exchanged by burnt fuel elements.

The irradiation room, size 4 × 4 × 2.2 m (see Figure 2), was built of concrete blocks. Thickness of the wall and ceiling is 40 cm, the inner surface is covered with boronated polyethylene. The irradiation room is equipped with TV camera, intercom, laser optical devices, patient treatment table. There are in the control room TV monitor for patient observing, PC with monitoring dosimetry system, beam operating console, communication device for patients.

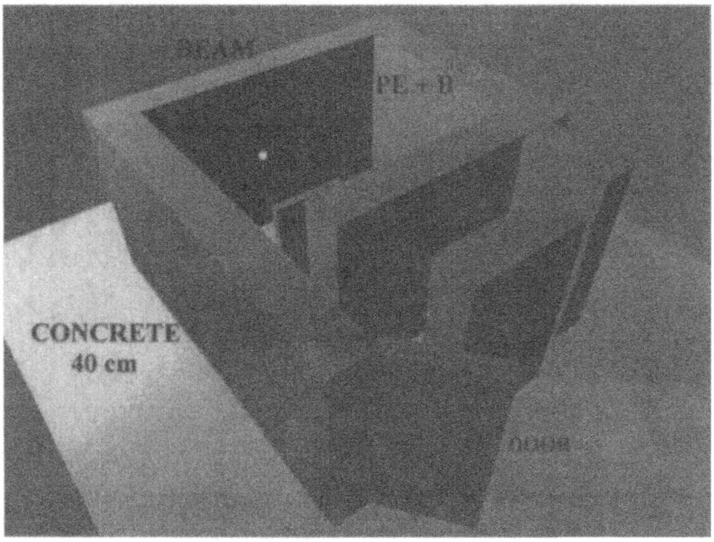

Figure 2. Irradiation room (without ceiling).

BORON CONCENTRATION MEASUREMENT

A prompt γ-ray analysis (PGA) facility for determination of ^{10}B in biological samples was built inside the reactor hall. The 6-meter mirror neutron guide provides the neutron flux 2.8 n/cm^{2}s in the target position at the reactor power of 8 MW. It is possible to reach relative efficiency 25% for detection of characteristic 478 keV γ line HPGe. The device is shielded from a γ-ray and neutron background in reactor hall by a combined shielding made of Pb and Li_2CO_3. The experimental set-up makes it possible to measure 1.0 or 0.5 ml liquid samples in teflon vials at the present time. The $^{10}B/H$ γ-ray signal ratio is used for determination of ^{10}B concentration.

The sensitivity of this facility is 4.9 counts within 478 keV Doppler broadened peak per 1 μg of ^{10}B. The facility was tested with water solutions of BSH and BPA as well as with blood solutions of BSH (see Figure 3). A good agreement was obtained between the values of ^{10}B concentration by the PGA method and the values from the ICP method.

TREATMENT PLANNING

Perfect treatment planning is one of the most important part of the BNCT treatment. It was decided to accept the internationally-recognized treatment planning software from INEEL and Harvard/MIT.

The ratio of boron concentration in blood to healthy tissue is obtained by analyzing of the tissue samples during neurosurgery operation if the BSH infusion was applied 12 hours earlier. The blood samples are taken and analyzed also during irradiation and according to the actual boron concentration the irradiation time determined in treatment planning is to be corrected.

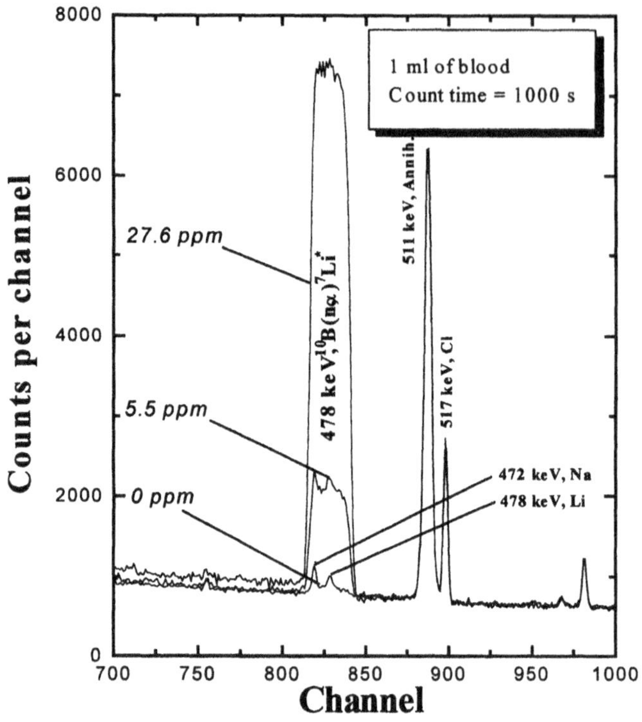

Figure 3. The spectra for boron concentration measurement in blood samples.

PROTOCOL

The Protocol specifying treatment of glioblastoma with BNCT at the LVR has been prepared in details.

Patient Selection

Main inclusion (preoperative) criteria—age 40–60; Karnofsky index ≥70; MRI (CT) diagnosed tumor, presumptive glioblastoma, unilateral, subcortical, subtotally resectable; informed consent. Inclusion (postoperative) criteria—histology of glioblastoma multiforme (G IV) resection of at least 70% of the tumor volume (MRI); Karnofsky index ≥70. Exclusion criteria—any other malignant tumor; prior anticancer treatment for the present tumor; previous brain surgery (other than tumor biopsy); partial tumor resection; adverse reaction to BSH; severe disease of e.g. heart, lung, liver, kidney.

Sampling of Tissues for Boron Analysis During Surgery

Preoperative period is used for study of BSH pharmacokinetics, BSH is given to the patient 12 hours before operation, blood concentration is measured at regular intervals. Samples of skin, dura mater, normal brain and tumor tissue will be taken at the time of operation and BSH concentration will be measured.

Administered Amount

The amount of BSH to be administered is directed by the fact that the better results in BNCT will be achieved with more boron present. At the LVR project BSH will be administered at a amount 100 mg/kg body weight.

Fractionation

Considerable part of irradiation is provided by high LET radiation during BNCT treatment and the damage of healthy brain tissue is not practically repaired. Sparing of the tissue with fractionation is more effective for low LET component of radiation, however considering the low dose rate of this component, we calculate that the repairing of healthy tissue is taking place during process of irradiation and therefore has been decided to accomplish the treatment in one fraction. For the sake of simplicity and simple evaluation of the side irradiation effect it was decided that the irradiation will be accomplished from one field predominantly in spite of the fact that long time irradiation is not comfortable for patient.

Starting Dose

We decided on the base of published results to use 7.1 Gy-Eq as a protocol starting dose for brain healthy tissue with system of increasing the dose by steps 8.8, 10.6, 11.8 and 13 Gy-Eq. Three patients will be irradiated in each dose level. The following weighting factors are used for defining the starting dose: fast neutrons and protons from Nitrogen reaction—3.3, boron reaction in white brain matter—0.37, boron reaction in tumor 0.81, gamma radiation—1.

Continuation at the Phase II Trial

The aim of Phase I trials is to establish the save dose for healthy brain tissue. There is a chance that patients may benefit from BNCT procedure even for the first steps of starting dose. The therapeutical effect is supposed not to be neglect for final steps. The Phase II investigating the effectiveness of this new treatment modality might go on in parallel to final dose steps, if no unacceptable adverse events are observed (neurological symptoms, MRI detectable lesions). The survival of patients will then be principal criterion in the Phase II.

BORON COMPOUND

The domestic supplier Katchem Ltd. produces BSH (as well as L-BPA). The quality of the product is in the agreement with Test of quality control asked in EU project. Katchem BSH was administered also for two patients treated in Petten.

CONCLUSIONS

The start of clinical trials is supposed to be in 1999, the group of 15 patients will be treated in this Phase I trial with the system of dose escalation. The approval of Protocol and other necessary documents by corresponding authorities is in the progress now.

ACKNOWLEDGMENTS

This work was supported by Contract GACR 202/97/K038 and IGA—MZ (CR) No 4834-2.

REFERENCES

1. J. Burian, M. Marek, J. Rataj, K. Prokes, L. Petruzelka, V. Mares, and P. Stopka: The BNCT facility at NRI reactor in the Czech Republic. Proceedings of 7th Int. Symp. on NCT for Cancer, Zurich, September 1996, Elsevier, Vol. I, p. 391.
2. M. Marek, L. Viererbl, and J. Burian: Determination of the geometric and spectral characteristics of BNCT beam (neutrons and gamma-rays), paper presented at 8th Int. Symp., La Jolla.

59

THE RUSSIAN PROJECT ON NEUTRON CAPTURE THERAPY FOR CANCER

V. F. Khokhlov[1], V. N. Kulakov[1], K. N. Zaitsev[2], Yu. S. Riaboukhine[3], S. P. Yarmonenko[3], O. N. Deineko[4], and V. N. Komienko[5]

[1]Russian State Scientific Center, Institute of Biophysics
46, ul. Zhiopisnaja, 123182 Moscow, Russia
[2]Moscow State Institute of Engineering Physics
31 shosse Kashirckoe, 115409 Moscow, Russia
[3]Cancer Research Centre, Russian Academy of Medical Sciences
24 shosse Kashirckoe, 115478 Moscow, Russia
[4]Association for Invalids of Radiation Accidents and Damages (INRAD)
46, ul. Zhivopisnaja, 123182, Moscow, Russia
[5]N. N. Burdenko Institute of Neurosurgery
5 ul. Fadeeva, 125047 Moscow, Russia

INTRODUCTION

Investigations in neutron capture therapy (NCT) for cancer were initiated in our country in the early 1960's. Large-scale radiobiological investigations were performed in Obninsk (Prof. Yu. S. Riaboukhine). Boron compounds were developed successfully in well known scientific institutes: the A. N. Nesmeyanov Institute of Organo-element Compounds and the N. S. Kurnakov Institute of General and Inorganic Chemistry of the Russian Academy of Sciences, the State Scientific Research Institute of Chemistry and Technology of Organo-element Compounds, the Institute of Biophysics of the USSR Ministry of Health, and others.

In 1993 the studies of the efficiency of the gadolinium-containing preparation Magnevist® (Schering AG, Berlin)[1] commenced with the support of the German Institute of Diagnostics

The aim of this project, using the considerable experience gained in previous years, is to develop technology for NCT of malignant tumors. This effort was initiated with the support of the Ministry of Science and Technologies and the Ministry of Health Protection of the Russian Federation and is financed by the Russian Foundation for Technological Development and by the Moscow Government.

The scientific program has three goals:

Frontiers in Neutron Capture Therapy, edited by Hawthorne *et al.*
Kluwer Academic / Plenum Publishers, New York, 2001.

1. To design an optimum system for irradiation of tumor-bearing laboratory animals and phantoms;
2. To study the secondary dose effects of boron- and gadolinium-containing agents in tumor;
3. To investigate the efficacy of NCT for melanoma and another tumors, using boron- and gadolinium-containing agents either alone and in combination.

REACTOR

It is supposed to use the research reactor, IRT MEPhI (Figures 1–2), in Moscow.

The first experiments on an irradiating has been carryed out on the canal K-2 on a background of a potent concomitant gamma radiation of a beam (1.2 cGy/s at low level of thermal neutrons flux ($\sim 10^8$ n/cm^2·s). The express biological protection, system of monitoring and system of accessory equipment was not arranged.

In the course of the realization of the first stage of the Project (years 1997–1998) the irradiation facility at the sliding channel K-4 was set in action (Figures 3, 4). Its position with respect to the center of the reactor active zone and a special design of the collimator substantially decreased the dose rate of the secondary γ-radiation in the beam of the K-4 channel (to 0.14 cGy/s), and allowed to increase the thermal neutron flux up to 109 n/cm^2·s. The main characteristics of the IRT MEPhI reactor canal K-4 are presented in the Table 1.

PHANTOM/ANIMAL DOSIMETRY

Gamma-radiation was monitored by detectors on the basis of manganese-activated alumophosphate glass with diameter of 8 mm and thickness of 1 mm^2. The thermal neutron fluxes were measured using detectors based on ^{6}Li-enriched lithium-containing glass.[2] The fast neutron fluxes were measured by track detectors on the basis of ^{237}Np.[3]

At the following stages of investigations the attempt will be realized to carry out conversion electrons doses measurements by thin TTLD-580 detectors[4] (thin-layer thermoluminescence detector on the basis of MgB_4O_7 with diameter of 15 mm and thickness of about 80 µm). For this purpose the special container simulating a tumor in the phantom is made in which fastening thin detectors is stipulated.

A system for calculation and measurement of irradiation doses in phantoms and in tumor-bearing animals has been designed and implemented.

Distributions comparable to the experimental doses were calculated using the MCNP-4a code.[5]

BORON/GADOLINIUM DELIVERY AGENTS

First, boron- and gadolinium-containing agents for NCT have been prepared, i.e., the original NCT drug—Dipentast[6] and substances containing gadolinium and boron, such as the complex of **Gd** with diethylenetriaminepentaacetic acid (H_5dtpa) modified by the amino-undecahydro-*closo*-dodecaborate anion (AUDB).[7] Also, a new method of preparing boronophenylalanine has been developed.

Figure 1. Reactor IRT MEPhI.

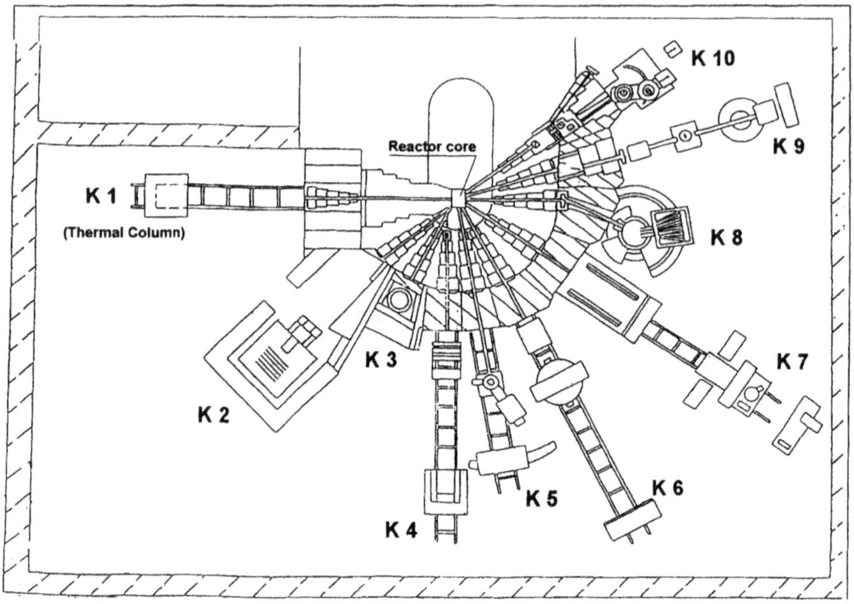

Figure 2. The scheme of experimental canals on a reactor IRT MEPhI.

Figure 3. K-4—channel at the IRT MEPhI reactor for radiobiological experimentation. The radiation protection block is moved out. The device for fixation of small laboratory animal is visible.

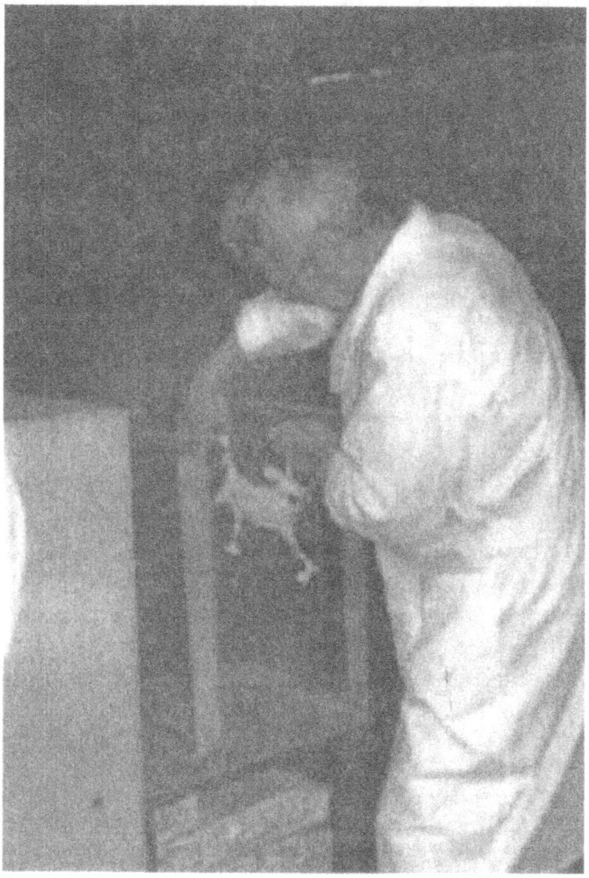

Figure 4. Fixing a rat in a radiobiological experiment.

The results of the biological studies allowed to select the following preparations: an original medical form of the di-sodium salt of gadopentat—**Gd**dtpaNa_2 (1) developed by our group; the Gd3+ complex with modified H_5dtpa: **Gd**dtpaR_1R_2 (2,3), where 2 – R_1 = H; R_2 = AUDB; 3 – R_1 = R_2 = AUDB.

Complexes 1 and 2 in the physiological conditions (pH 7.0–7.5) are negatively charged (–2). Complexes 2 and 3 each contain two neutron-absorbing atoms—**B** and Gd.

We also used in the studies L-^{10}**B**-boronphenylalanine (^{10}**B**-BPA), manifactured by BBI (USA), and produce according to the scheme developed in our group.

The medical form for water-soluble agents the water solution with pH of about 7.2; water insoluble agents were prepared as oil emulsions. The suspension on the base of

Table 1. Features of the beam K-4 canal for NCT

Reactor power, MW	Thermal neutron flux density, $n/cm^2 \cdot s$	Epithermal neutron flux density, $0.1\,eV < E < 10\,KeV$, $n/cm^2 \cdot s$	Fast neutron flux density, $E > 2{,}4\,MeV$ $n/cm^2 \cdot s$	Dimension of the beam port, mm	γ-Radiation dose rate cGy/s
2.5	$0.9 \cdot 10^9$	$0.23 \cdot 10^8$	$0.65 \cdot 10^7$	from 30 to 150	0.14

gadopentat was prepared consisting of 13–20 μm particles of the denaturasted serum albumine (SA) modified by gadopentat: SA-**G**ddtpaNa.

PRE-CLINICAL STUDIES

The pre-clinical tests of the agents mentioned above were carried out on the cell suspensions and rats. At the first stage the efficiency of the boron- and gadolinium-containing agents in NTC was estimated, and at the second stage—their safety.

The experimental sarcomas Jensen and C-45 (rats), and melanoma B-16 (mice) were used in experiments.

The tumor cultures were obtained from Cancer Research Center (Russian Academy of Medical Sciences, Moscow).

The agents efficiency in NTC was estimated *in vitro* and *in vivo*. In the *in vitro* experiments the cell suspensions of sarcoma C-45 or melanoma B-16, in which the agent was added in concentration of 13.750 ppm, were preliminarily neutron irradiated, implanted into the animal's thigh, and the tumor development was observed. In the *in vivo* experiments the agent solution was introduced into the tumor at the concentration of 13,750 ppm, and then the tumor was neutron irradiated.

Radiobiological and dosimetric investigations were carried out using this experimental beam. GdNCT in the case of Jensen and C-45 sarcomas was found to be highly effective. Therapeutic efficiency of Magnevist and Dipentast was in the range 73–85% in these experiments (Figure 5).

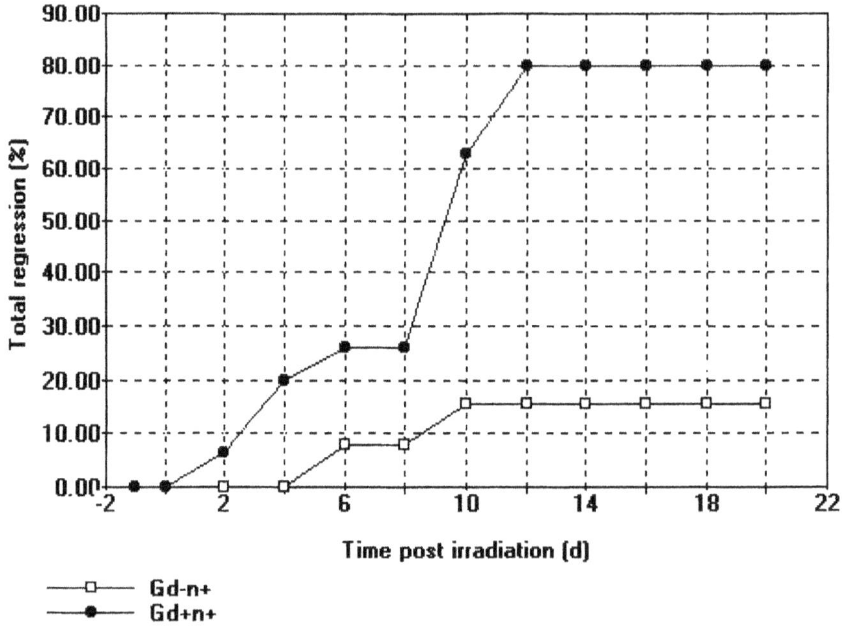

Figure 5. Relative number of nonpalpable tumors as a function of time (d) post-treatment Tumor regression for the groups neutron irradiated 20 min in the absence (Gd-n+, n = 13) and presence (Gd+n+, n = 16) of 13,750 ppm gadolinium from both experiments.

It was found that the therapeutic efficiency of the agents 1 and 3 is about 75–85%. The efficience of the agent 2 was somewhat lower (about 60%). Using of oil emulsion and suspension, and of neutral complexes instead of charged ones decreases the rate of removal of the substance from the tumor. However, application of emulsions and suspensions lowers the therapeutic efficiency of the agent due to its nonuniform distribution in the tumor.

Values of LD-50 for Dipentast and Magnevist were practically the same. Agents 1–3 are also characterized by low toxicity.

At present the pre-clinical tests of NTC commenced on dogs with spontaneous melanoma for the samples ^{10}**B**-BPA and Dipentast.

PLANS

In the next two years, we intend to reconstruct the K-1 canal of the reactor (Figure 2) for clinical studies, to continue the development of new boron- and gadolinium-containing NCT agents, and to implement pre-clinical trials to treat spontaneous melanoma using these new NCT agents.

The K-1 channel reconstruction will include: shifting of its axis (sliding channel), reduction of its length, and changing of the collimator design. After reconstruction the thermal neutron flux will amount up to $2 \cdot 10^9$ n/cm^2 and the gamma-ray dose rate < 0.07 cGy/s. These characteristics will provide acceptable time limits of local irradiating of the patient (approximately 90 minutes) for achievement of integral fluence of thermal neutrons $\sim 10^{13}$ n/cm^2. After reconstruction all radiobiological investigations and researches of efficacy of various drugs (on animal) will be carried out on the canal K-4, and medical on the canal K-1. (Russian sanitary rules overlapping experimental (on animals) and medical researches are prohibited).

The works on perfecting irradiation base will be prolonged both in a direction of enriching of the characteristics of a neutron beam, and on solution of a problem of equipment of express interfacing locations for preparation of the patient and his transporting in irradiation box on an special seat.

We intend to investigate the efficiency of application in NTC of the following compounds: Gd complexes with the modified H_5dtpa, including also the boron-containing compounds (negatively charged and neutral).

It is supposed also for perspective compounds to develop the medicinal forms, with approaching for NCT pharmacokinetics characteristics: a sufficient retention in a tumor tissue ($t_{1/2} \sim 60$ minutes) and full elimination from an organism during 1–2 day.

REFERENCES

1. V.F. Khokhlov, P.N. Yashkin at all. "Neutron Capture Therapy with Gadopentate Dimeglumine: Experiments on Tumor-Bearing Rats", "Academic Radiology", 1995, No 2, pp. 392–398.
2. I.A. Bochvar, T.I. Gimadova, I.B. Keyrim-Markus, *et al.* "The Method of IKS Dosimetry," Atomizdat, Moscow, 1977, pp. 41–49, 58.
3. S.N. Kraitor. "Dosimetry for Radiation Accidents," I.B. Keyrim-Markus (ed), Atomizdat, Moscow, 1979, pp. 167–246.
4. V.F. Khokhlov, A.I. Shaks *et al.* Dose distribution in ^{157}Gd and ^{10}B containing phantoms subsequent to irradiation with thermal neutrons. (these proceedings).

5. Briesmeister J.F. MCNP-4A General Monte-Carlo N-Particle Transport Code. Version 4A, Los Alamos National Laboratory report, LA-12625, 1995.
6. N.K. Sviridov, P.V. Sergeev, V.N. Kulakov *et al.*, "Investigation of the effect of the NMR contrast agents and their substances on the proton spin-lattice relaxation in the reticulosarcoma-45 of rats *ex vivo* and its visualization", Voprosi onkologii (in Russian), 1999, v.45, No 2, p. 167.
7. V.N. Kulakov, V.I. Bregadze, V.F. Khokhlov *et al.* Design of Boron- and Gadolinium-containing agents for NCT", (these proceedings).

ACCELERATORS AND OTHER NEUTRON SOURCES

60

DEVELOPMENT OF AN ACCELERATOR-BASED BNCT FACILITY AT THE BERKELEY LAB

B. A. Ludewigt, D. L. Bleuel, W. T. Chu, R. J. Donahue, J. Kwan, L. L. Reginato, and R. P. Wells

Ernest Orlando Lawrence Berkeley National Laboratory
Berkeley, CA 94720

1. INTRODUCTION

The clinical success of boron neutron capture therapy (BNCT) depends on both the selective concentration of the boron carrying drug in the tumor cells and the quality of the available neutron beam. For the treatment of glioblastoma multiforme (GBM) the ability of the neutron beam to deeply penetrate is of great importance.

At Berkeley Lab an accelerator-based neutron source is being developed which will deliver neutron beams optimized for their clinical application and suitable for deployment at hospitals. The goal of the BNCT program at Berkeley Lab is to construct an accelerator-based BNCT facility and to make it available for clinical trials to be conducted in collaboration with the UCSF Cancer Center. Important progress has been made over the past 2 years. Neutronics studies have been performed to determine accelerator and target requirements and to predict the quality and the clinical attributes of the neutron beam. The work on the two most critical components, the neutron production target and the accelerator power supply, has reached an advanced state.

2. ACCELERATOR-BASED BNCT FACILITY

The combination of unique accelerator expertise, past experience in high-LET radiotherapy, and the availability of a shielded treatment room provide an excellent opportunity for developing and constructing an accelerator-based BNCT facility at Berkeley Lab. After evaluating a number of nuclear reactions and accelerator options[1] we chose the $^{7}Li(p,n)^{7}Be$ reaction at proton energies between 2.3 and 2.5 MeV for the neutron production. The low primary neutron energies from this reaction help to produce epithermal neutron beams with near optimal energy spectra.

In the BNCT facility under construction, the proton beam will be generated by an electrostatic quadrupole accelerator capable of producing a proton beam current of 50 mA. The beam is then guided through the beam transport line into the treatment room

Frontiers in Neutron Capture Therapy, edited by Hawthorne *et al.*
Kluwer Academic / Plenum Publishers, New York, 2001.

where it impinges on the neutron production target. The target consists of a 90 μm thin layer of lithium deposited on a metal cooling backing. It is followed by the moderator. Both, moderator and target, are surrounded by reflector material for preventing neutrons not going forward from escaping and reflecting them back towards the patient. A delimiter placed between the moderator-reflector assembly and the patient defines the neutron beam diameter and shields the patients from unwanted radiation. Starting from clinical requirements for the treatment of deep-seated brain tumors we have optimized the epithermal neutron beam and determined the required proton current and other beam parameters.[2]

3. HIGH CURRENT PROTON ACCELERATOR

An accelerator which can produce a high proton beam current is essential for achieving high quality neutron beams and short treatment times. Over the last decade a special dc- accelerator, the electrostatic-quadrupole (ESQ) accelerator, has been developed for the heavy ion fusion program at Berkeley Lab. This dc-accelerator features a column of electrostatic quadrupoles instead of thick apertures and is ideally suited to produce very high-current ion beams in the desired energy range.[3] The focussing field in this structure is decoupled from the axially accelerating field making it possible to achieve strong focussing without exceeding the axial breakdown limit. Furthermore, the strong transverse field prevents electrons from down streaming and causing arc-downs. A cross section of the accelerator indicating the 13 module ESQ-column and the power supply structure is schematically shown in Figure 1.

In addition to the acceleration column itself, the power supply is crucial for accelerating the desired beam currents. Because it would be difficult to generate tens of mA using a conventional, capacitively-coupled power supply, a new type of power supply is being developed and built. It is based on the concept of an air-core coupled transformer.[4] As can be seen in Figure 1 a primary coil and 55 secondary coils supplying high voltages

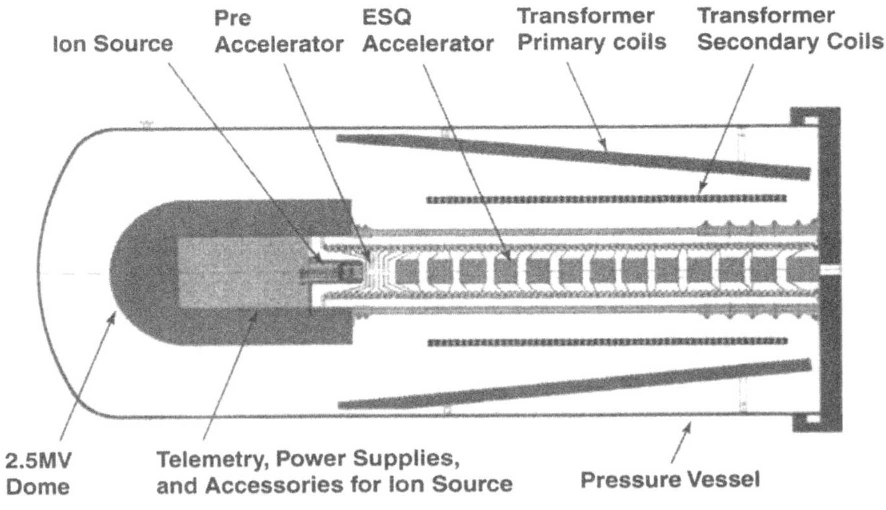

Figure 1. Cross section of 2.5 MeV, 50 mA ESQ-accelerator.

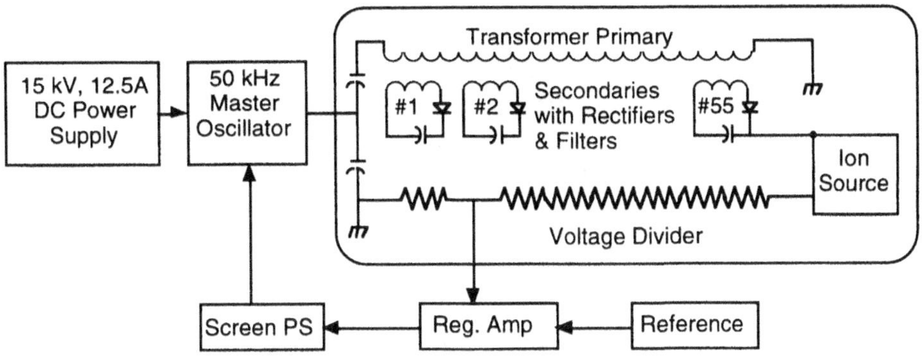

Figure 2. Air-core transformer power supply.

to the ESQ electrodes are placed inside the pressure vessel. The electrical structure of the power supply is schematically depicted in Figure 2. The master oscillator generates a 50 kV sinusoidal voltage on the primary coil. Through inductive coupling a high voltage is induced on each of the 55 secondaries. Rectifier and filter circuits produce the desired dc voltage. The output voltage can be set by a reference from 0 to 3 MV and can be adjusted to within 1%. The maximum beam output power is 125 kW or 50 mA at 2.5 MeV with an oscillator efficiency of 67%. The primary coil has been installed inside the pressure tank and two-thirds of the secondary coils along with rectifier and filter circuits have been fabricated.

A multicusp source, which can deliver positive hydrogen ion beams with a monatomic fraction of higher than 90%,[5] is mounted in the high voltage dome. Radio-frequency induction discharge provides reliable and long-life source operation. The pressure tank, which is about 6 m long and 2.4 m in diameter, and other parts were taken from a decommissioned accelerator. Future hospital-based ESQ-accelerators enclosed in a steel tank filled with SF_6 could be more compact, approximately 3.5 m long.

4. NEUTRON PRODUCTION TARGET

A metallic lithium target for the production of neutrons via the $^7Li(p,n)^7Be$ reaction at high beam currents requires efficient cooling. A 2.5 MeV, 50 mA proton beam deposits a heat-load of 125 kW onto the target while the melting point of lithium is only 179C. We have developed target designs which can handle such heat-loads and make it possible to utilize the accelerator's capability. A lithium layer, thick enough for the protons to loose enough of their energy to drop below the reaction threshold of 1.89 MeV, is deposited onto a metal cooling backing. Not making the layer thicker than necessary minimizes the heating of the lithium itself. The backing is convectively cooled through a large number of narrow cooling passages cut into the metal plate. Using finite element modeling the cooling panel geometry, such as size and spacing of the cooling channels, has been optimized.

A prototype target panel was made out of aluminum and its thermal performance was tested at the Plasma Materials Test Facility at Sandia National Laboratory, Albuquerque, NM. Using a scanned electron beam a well-defined heat-load was deposited on the surface of the panel. Temperature measurements confirmed the modeling results[6]

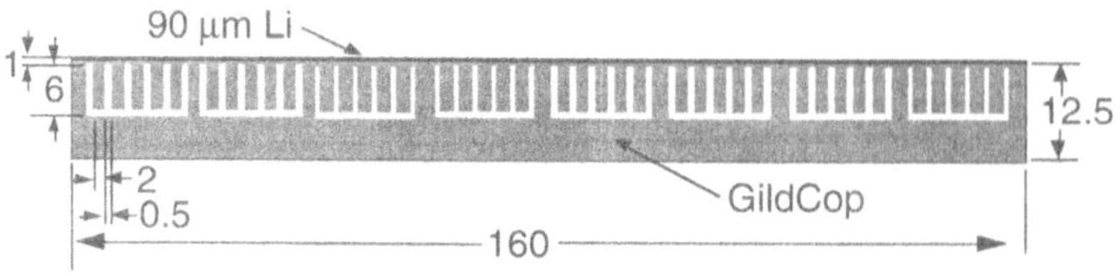

Figure 3. Target cross section (units in mm).

indicating that the target can handle up to 600 W/cm^2. Two aluminum panels mounted in V-shape geometry at 30% in respect to the beam axis satisfy the cooling requirement. However, copper, although less preferable in terms of its radiation transport properties, has greatly superior mechanical properties. Finite element modeling has shown that significantly improved cooling can be achieved without creating undo thermal stresses when using Glidcop™ (99% copper content), a material which combines excellent thermal properties with high strength. The cross section of the target panel is shown in Figure 3. Such a target can handle a heat-load of 1 kW/cm^2 while keeping the lithium temperature below 150C. If the beam is spread over a circular area of about 13 cm in diameter, this target can handle a heat load of 125 kW as deposited by a 50 mA, 2.5 MeV proton beam. Neutronics modeling has shown that target sizes up to 15 cm in diameter can be used without degrading the quality of the neutron beam and with only a small penalty in the total neutron flux. A further advantage of the target design is the small total amount of water in the target panel.

5. NEUTRON BEAM MODELING

The accelerator-based neutron source will produce a high neutron flux. In contrast to previous studies, which focussed on maximizing the epithermal neutron flux,[7–9] this source provides the opportunity to optimize the quality of the neutron beam by improved moderation and filtering. In-air parameters are not sufficient for comparing and evaluating neutron beams for BNCT and in-phantom distributions must be used. For our studies we chose the dose to the midline of the brain as the figure-of-merit. Maximizing the dose to that point leads to the most penetrating neutron beams which, we believe, are best suited for the treatment of deeper seated brain tumors. Through modeling the radiation transport through moderator, reflector and filter assembly and head phantom the best energy spectrum was found for an Al/AlF_3 moderator.[2] In contrast to other neutron beams designed for BNCT,[7–10] the neutron current spectrum measured at the exit of the moderator displays a pronounced peak at about 15 keV with a sharp falloff towards higher energies and a somewhat softer falloff towards lower energies.[2]

The different shape of the neutron energy spectrum translates into a remarkable increase in penetration as can be seen when comparing in-phantom distributions for the accelerator produced beam with the ones calculated for the beam at the Brookhaven Medical Research Reactor (BMRR). Boron concentrations, compound factors, and relative biological effectiveness (RBE) values from the BMRR dose calculation protocol[11] were used for all dose calculations. The values are: normal tissue ^{10}B concentration: 13 ppm; normal tissue compound factor: 1.3; tumor ^{10}B concentration: 45.5 ppm; tumor compound factor: 3.8; proton-recoil RBE: 3.2; nitrogen capture RBE: 3.2; and γ RBE:

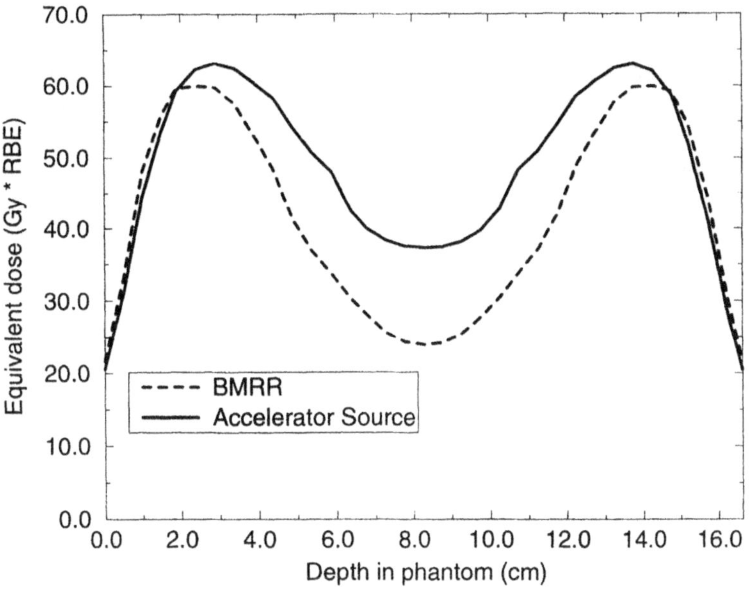

Figure 4. Depth distributions of RBE-weighted tumor doses along the beam axis for the accelerator-produced beam in comparison to the BMRR beam.

1.0. In addition, the same delimiter, 12 cm diameter opening and 13 cm thick, and the same distance from moderator surface to phantom as for the BMRR beam were assumed.[2] This approach made a direct comparison with the existing BMRR beam possible. Figure 4 shows tumor depth dose distributions along the beam axis calculated for two parallel-opposed beams. The RBE-weighted tumor doses for both beams represent a treatment in which a maximum RBE-weighted dose of 12.5 RBE-Gy (tolerance dose) was given to the normal brain tissue. No further normalization was applied to the curves in Figure 4. The dose calculations were based on CT data describing a specific head. The difference between the two beams is striking. Around the midline of the brain, in this case at a depth of 8 cm, the beam from the accelerator neutron source is able to deliver a roughly 50% higher dose than the BMRR beam.

Further improvement in the quality of the accelerator-generated neutron beam is possible. The energy spectrum can be further optimized by using a different, less moderating reflector material, e.g., lead instead of Al_2O_3. In addition, the directionality can be improved by changing the delimiter and increasing the distance between moderator and patient. However, both of these changes lower the neutron flux and, therefore, the dose rate. Our modeling[2] results indicate that for the present design a proton beam current of 20 mA produces a neutron flux of $1 \cdot 10^9\,n/(s\,cm^2)$ on the surface of the phantom. The treatment time, i.e., the time until the normal tissue dose limit is reached, is about 40 min.

6. SUMMARY AND CONCLUSIONS

Important progress has been made towards the construction of an accelerator-based BNCT facility. The power supply for the high-current accelerator is nearing completion and a lithium target has been designed and tested which can handle the 125 kW heat-load deposited by a 2.5 MeV, 50 mA proton beam. Our neutronics studies have demonstrated the importance of optimizing the neutron beam, in particular its energy spectrum, for maximizing the dose to deep seated tumors near the midline of the brain. For example, in cases in which the minimum target dose at BMRR is currently 21 to 25 RBE-Gy one may be able to deliver 32 to 37 RBE-Gy. Experience from a clinical GBM trial with fast neutrons suggests that a RBE-weighted dose of 30 RBE-Gy is needed to achieve a tumor control probability of 50%.[12] This indicates that the tumor control probability may jump from almost zero for doses from 21 to 27 RBE-Gy to 75% and higher for doses above 32 RBE-Gy which are obtainable with a more penetrating beam.

ACKNOWLEDGMENTS

This work is supported by the Director, Office of Energy Research of the U.S. Department of Energy under Contract No. DE-AC03-76SF00098.

REFERENCES

1. Ludewigt B.A. Clinical Requirements and Accelerator Concepts for BNCT. Proceedings of the Particle Accelerator Conference 1997, Vancouver, BC, Canada, May 1997.
2. Bleuel D.L., Donahue R.J., Ludewigt B.A., and Vujic J. Med. Phys. 1998; 25(9):1725–1734.
3. Kwan J.W., Andersson O.A., Reginato L.L., Vella M.C., and Yu S.S. A 2.5 MeV electrostatic quadrupole DC-accelerator for BNCT applications. Nucl. Instrum. Methods Phys. Res. 1995; B99:710–712.
4. Reginato L.L., Ayers J., Johnson R., Peters C., and Steveson R. Designing power supplies for 2.5 MeV, 100 mA d.c. for BNCT. Proceedings of the 14th International Conference on the Application of Accelerators in Research and Industry, Denton, Texas, November, 1996 (AIP Press, New York, 1997), Vol. II:1313–1315.
5. Leung K.N. Ion Sources for High Purity Ions. Proceedings of the 14th International Conference on the Application of Accelerators in Research and Industry, Denton, Texas, November, 1996 (AIP Press, New York, 1997), Vol. II:1203–1205.
6. Ludewigt B.A., Chu W.T., Donahue R.J., Kwan J.W., Phillips T.L., Reginato L.L., and Wells R.P. An Epithermal Neutron Source for BNCT based on an ESQ Accelerator. Proceedings of the Topical Meeting on Nuclear Applications of Accelerator Technology, Albuquerque, NM, Nov. 1997:489–494.
7. Wang C.K., Bleu T.E., and Gahbauer R. A neutronic study of an accelerator-based neutron irradiation facility for boron neutron capture therapy. Nucl. Technol. 84, 1989:93–107.
8. Yanch J.C., Zhou X.L., Shefer R.E., and Klinkowstein R.E. Accelerator-based epithermal neutron beam design for neutron capture therapy. Med. Phys. 1992; 19:709–721.
9. Allen D.A., and Beynon T.D. A design study for an accelerator-based epithermal neutron beam for BNCT. Phys. Med. Biol. 1995; 40:807–821.
10. Becker G.K., Harker Y.D., Miller L.G., Anderl R.A., and Wheeler F.J. Neutron spectrum measurements in the aluminum oxide filtered beam facility at the Brookhaven Medical Research Reactor. In: Neutron Beam Development and Performance for Neutron Capture Therapy. Eds. O. Harling, J.A. Bernard, R.G. Zamenhof. Plenum Press, New York, 1990.
11. Chanana, A.D. Boron neutron-capture therapy of glioblastoma multiforme at the Brookhaven Medical Research Reactor, A Phase I/II study, FDA IND #43317, Protocol #4. Med. Dep., Brookhaven Natl. Lab., Upton, NY 11973-5000, January, 1996.
12. Laramore G.E., Wheeler F.J., Wessol D.E., Stelzer K.J., and Griffin T.W. A BNCT Tumor Control Curve for Malignant Glioma from Fast Neutron Therapy Data: Implications for Beam Delivery and Compound Selection. In: Advances in Neutron Therapy, Elsevier 1997; Vol II:580–587.

61

THE RELATIONSHIP BETWEEN BNCT NEUTRON BEAM PARAMETERS AND TREATMENT PLANNING FIGURES-OF-MERIT

D. A. Allen and T. D. Beynon

School of Physics and Astronomy
University of Birmingham
B15 2TT, UK

1. INTRODUCTION

The task of designing an epithermal neutron beam for BNCT is not an easy one, especially since the design goals are often ill-defined. Most people agree that the epithermal neutron fluence rate should be "as high as possible" but in any case between 10^8 and $10^9\,cm^{-2}\,s^{-1}$. Similarly, there is general consensus that the fast neutron and γ-ray dose rates per unit epithermal neutron fluence should be "as low as possible" and figures such as "less than about $8 \times 10^{-13}\,Gy\,cm^2$ for neutrons and $2 \times 10^{-13}\,Gy\,cm^2$ for photons" have been used. These figures are not universally accepted, which is not surprising since there is no scientific basis for the adoption of these limits.

The ultimate purpose of a BNCT neutron beam is to deliver a therapeutic dose to tumour tissue within a clinically acceptable treatment time. This must be achieved whilst maximising the ratio of the biologically-weighted dose delivered to tumour and that delivered to healthy tissue. It is these goals, therefore, which must direct our design work. However, this is not straight forward, since the concepts of in-phantom dose depth distributions, treatment advantage ratio and advantage depth are not so easily applied to the design process as the single figure empty beam parameters. It is the purpose of this paper to explore the relationship between the in-phantom treatment planning figures-of-merit and the empty beam parameters.

2. EMPTY BEAM PARAMETERS AND TREATMENT PLANNING FIGURES-OF-MERIT

The following parameters have been used to characterise the neutron beam with no phantom present (empty beam).

Frontiers in Neutron Capture Therapy, edited by Hawthorne *et al.*
Kluwer Academic / Plenum Publishers, New York, 2001.

The epithermal neutron fluence rate

$$\Phi_u = \int_{0.4eV}^{10keV} \phi_n(E)dE \tag{1}$$

the neutron kerma per unit epithermal neutron fluence

$$G_n = \int_0^\infty K_n(E)\phi_n(E)dE \Big/ \Phi_u \tag{2}$$

and photon kerma per unit epithermal neutron fluence

$$G_p = \int_0^\infty K_p(E)\phi_p(E)dE \Big/ \Phi_u \tag{3}$$

Where K_n and K_p are the neutron and photon kerma coefficients respectively and $\phi_n(E)$ and $\phi_p(E)$ are the neutron and photon fluences.

The following figures-of-merit were used to characterise in-phantom performance of the neutron beam. They are taken from Wallace *et al.*[1] and from the current Brookhaven clinical trials:

Treatment Time—the time required to deliver a maximum photon-equivalent dose to healthy tissue of 12.6 Gy-Eq.

Peak Therapeutic Ratio—therapeutic ratio (TR) at any depth in the phantom is the dose delivered to a tumour at that depth divided by the maximum healthy tissue dose. The peak TR is the maximum in the TR curve.

Advantage Depth (AD)—the depth into phantom at which the therapeutic ratio falls to 1.0.

3. METHOD OF INVESTIGATION

The basic moderator assembly configuration is described elsewhere.[2] Three different types of moderator design have been investigated with various moderator depths. These are i) D_2O moderator (poisoned with 0.3% ^6Li) with graphite reflector, ii) AlF_3/Al/LiF moderator with graphite reflector and iii) AlF_3/Al/LiF moderator with nickel reflector. The AlF_3/Al/LiF moderator had the following composition-43.3at% Al, 55.9at% F and 0.9at% Li. For each design, as the depth is increased Φ_u and G_n will decrease. G_p will also change.

We have used the Monte Carlo code MCNP4b to simulate the transport of neutrons and photons through the assembly, using the source and tally method previously described.[3] In addition to empty beam simulations, we have also calculated the dose-depth distributions for each depth of moderator in a standard MIRD phantom,[4] which includes brain, skull, scalp and the rest of the body as part of the simulation. From these distributions the treatment planning figures-of-merit were calculated. In calculating the treatment times we have assumed a proton beam current of 5 mA.

The relative biological effectiveness (RBE), compound biological effectiveness (CBE) factors and ^{10}B concentrations which were assumed in calculating the photon-equivalent dose distributions were the same as those used in reference 2. A typical set of in-phantom dose-depth curves can be seen in Figure 1, which also illustrates the advantage depth.

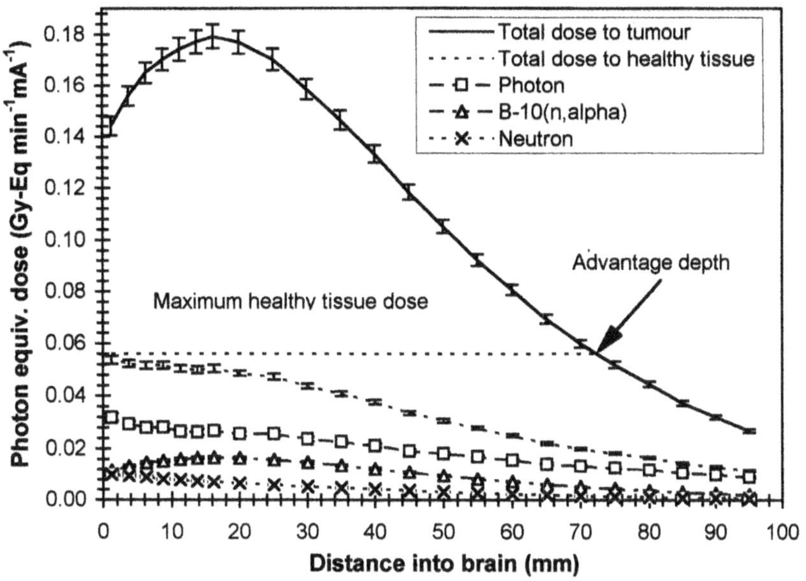

Figure 1. Typical biologically-weighted dose components in-phantom, showing the total healthy tissue and tumour doses and indicating where the advantage depth occurs.

4. RESULTS

4.1. Treatment Time and Empty Beam Epithermal Neutron Fluence Rate

Figure 2 shows how the treatment time, determined by the maximum permissible dose to healthy tissue, is determined by the empty beam epithermal neutron fluence rate Φ_u. In general, the higher the epithermal neutron fluence rate, the shorter the treatment time, but the expected inverse relationship does not quite hold true. This is because the treatment time is determined by the maximum dose to healthy tissue and the fast neutron dose near the phantom surface is very high for small moderators (high fluence rate designs), which causes the plots in this figure to curve upwards at high fluence rate.

4.2. Peak Therapeutic Ratio and Neutron Kerma Per Unit Epithermal Neutron Fluence

Figure 3 is a plot of the peak therapeutic ratio as a function of the neutron kerma per unit epithermal fluence in the empty beam. There is a strong negative correlation between peak TR and G_n for each of the moderator systems studied. The lower the neutron kerma in the beam, the better the achievable TR. For an increase in G_n of 10^{-12} Gy cm^2 the peak TR falls by approximately 0.6. However, for the same G_n we can achieve an improved TR by using AlF_3/Al/LiF as a moderator rather than by using D_2O. Therefore, simply calculating the empty beam parameter, G_n, is not sufficient to determine the best moderator system—we need in-phantom simulations.

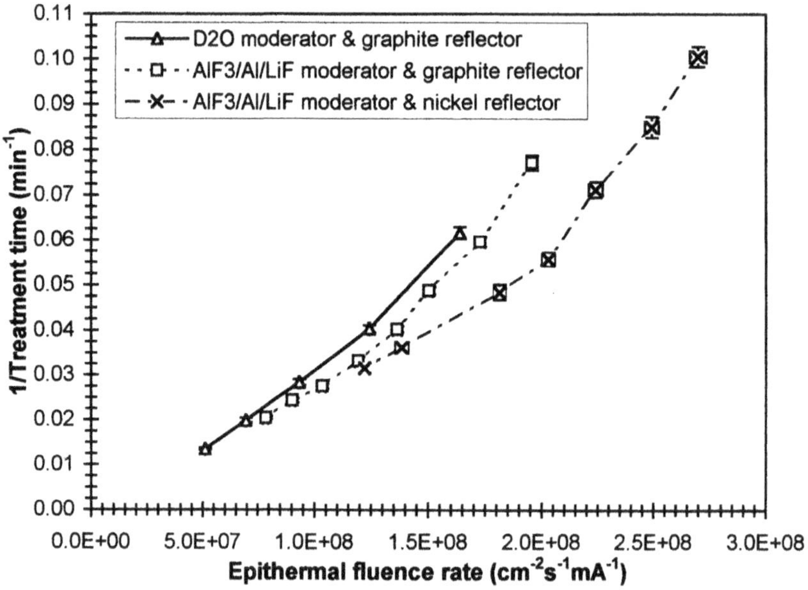

Figure 2. The relationship between treatment time and the empty beam epithermal neutron fluence rate. Each point on these plots represents a different thickness of moderator.

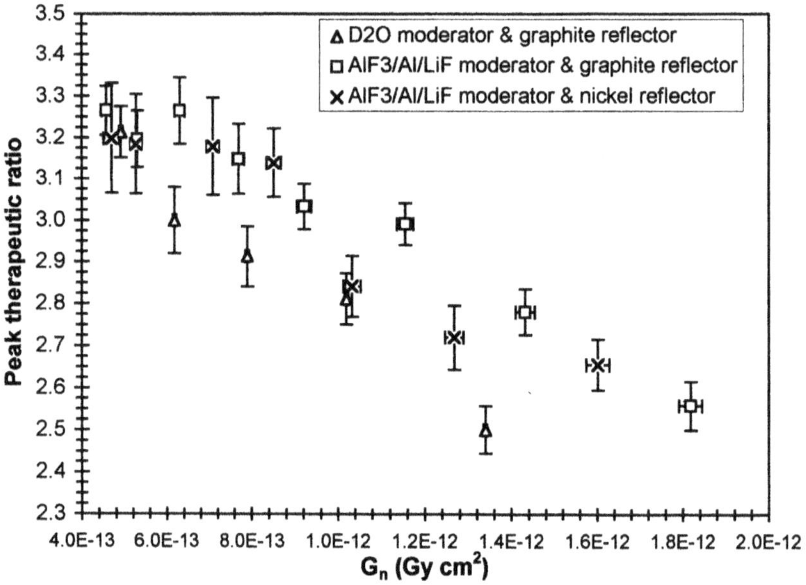

Figure 3. A plot of peak therapeutic ratio in-phantom against empty beam G_n. Each point on these plots represents a different thickness of moderator.

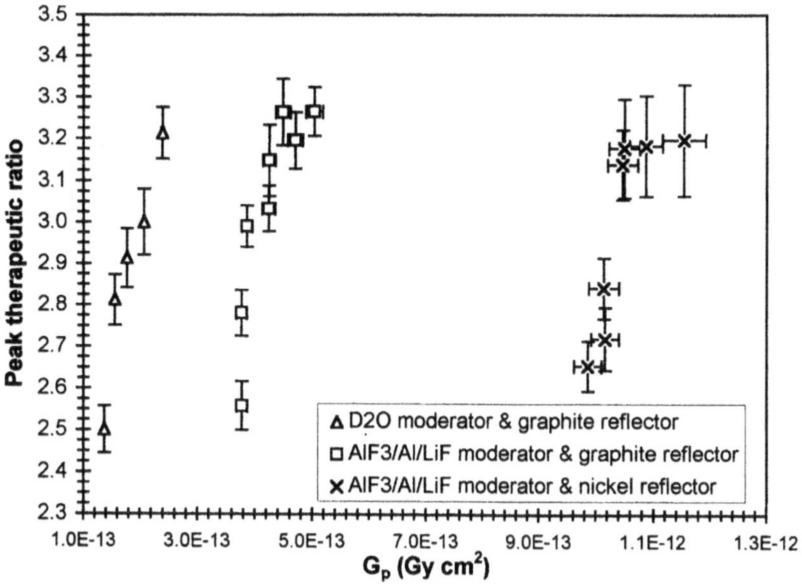

Figure 4. A plot of peak therapeutic ratio in-phantom against empty beam G_p. Each point on these plots represents a different thickness of moderator.

4.3. Peak Therapeutic Ratio and Photon Kerma Per Unit Epithermal Neutron Fluence

Figure 4 is a plot of peak TR against empty beam G_p. This plot indicates an apparent improvement in moderator performance as the photon contamination level increases. This is clearly nonsense. The anomaly arises because by changing the moderator design we are changing both G_p and G_n together. We need a method of assessment which can alter these parameters independently. MCNP allows us to do this. This plot also shows us how unimportant G_p is as a moderator performance assessment parameter at this level of photon contamination. The photon contamination level in the nickel reflected system is almost an order of magnitude worse than that for D_2O, yet the peak TR achieved is very similar.

4.4. How Important Is Photon Contamination in the Beam?

Since MCNP enables the user to selectively switch off photon production within any region of the simulation, it is possible to separate the in-phantom photon dose into two components. That which is due to γ-rays generated within the phantom (which are unavoidable) and that which is due to γ-rays coming from the moderator assembly. The latter can be minimised by judicious design.

Using combined moderator and phantom Monte Carlo simulations and a spreadsheet, we have artificially altered the level of photon contamination (characterised by the beam parameter G_p) by multiplying that part of the in-phantom dose distribution which is due to moderator-produced photons. In this way in-phantom dose distributions and the treatment planning figures-of-merit are simply recalculated for each level of photon contamination in the beam. This recalculation is, therefore, independent of any changes in the neutron spectrum and fluence which would accom-

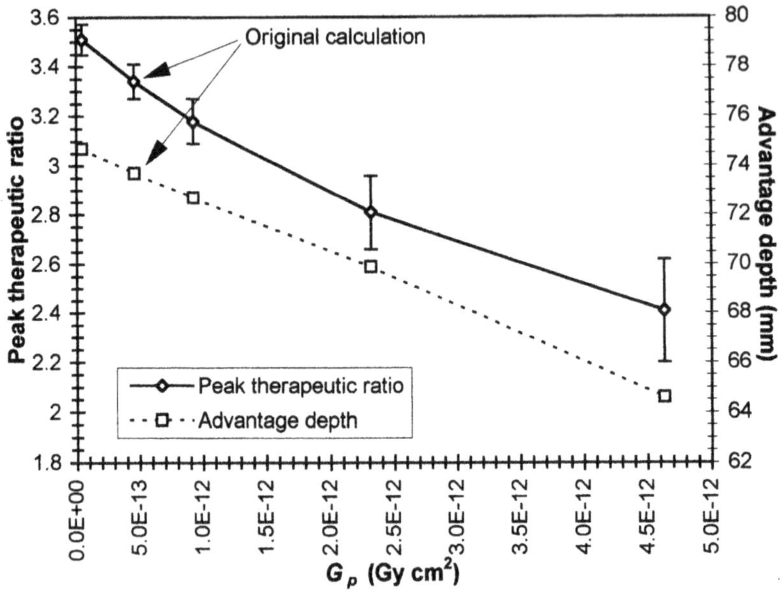

Figure 5. Plots of peak therapeutic ratio and advantage depth against the external photon contamination level (G_p).

pany any real moderator design changes. Figure 5 shows the results. For an increase in Gp of 10^{-12} Gy cm^2, the peak TR falls by approximately 0.24. We can say, therefore, that peak TR is approximately 2.5 times less sensitive to changes in G_p than to changes in G_n.

5. CONCLUSIONS

It is possible to relate the parameters which characterise an empty BNCT neutron beam to treatment planning figures-of-merit derived from in-phantom dose-depth curve calculations. The following relationships appear to be true.

There is an approximate inverse relationship between the expected treatment time and the epithermal neutron fluence rate in the empty beam.

There is a strong relationship between the maximum achievable therapeutic ratio and the neutron dose in the beam. As the neutron dose in the beam (G_n) increases the peak (and overall) TR decreases for all moderator types. This is reflected in a similar trend for advantage depth.

In contrast, the effect of the photon contamination in the beam is much less important (at this level of contamination). We can therefore say that a major design goal in designing a beam for BNCT should be to keep G_n as low as possible. However, this is subject to being able to deliver a sufficiently high epithermal neutron fluence rate to achieve an acceptable treatment time. We have assumed that γ-rays make no effective contribution to the treatment of tumours.

The level of photon contamination in the empty beam is of secondary importance and could be allowed to be about three times higher than the neutron dose before it begins to significantly compromise the treatment planning figures-of-merit.

In summary, it is worthwhile compromising with the level of photon contamination in the beam in order to improve the beam's neutron characteristics. That is why the Birmingham BNCT group is now considering a moderator based on AlF_3/Al/Li rather than D_2O.

REFERENCES

1. S.A. Wallace, J.N. Mathur, and B.J. Allen, Treatment planning figures of merit in thermal and epithermal boron neutron capture therapy of brain tumours, Phys. Med. Biol., 39:897–906, 1994.
2. D.A. Allen, T.D. Beynon, and S. Green, Design for an accelerator-based orthogonal epithermal neutron beam for boron neutron capture therapy, accepted for Med. Phys., 25, 1998.
3. D.A. Allen, and T.D. Beynon, A design study for an accelerator-based epithermal neutron beam for BNCT, Phys. Med. Biol., 40:807–821, 1995.
4. W.S. Snyder, M.R. Ford, G.G. Warner, and H.L. Fisher, Estimates of absorbed fractions for monoenergetic photon sources uniformly distributed in various organs of a heterogeneous phantom, J. Nucl. Med., MIRD Supplement no. 3, Pamphlet 5, 1969.

62

DEVELOPMENT OF A HIGH-POWER WATER-COOLED BERYLLIUM TARGET FOR USE IN ACCELERATOR-BASED BORON NEUTRON CAPTURE THERAPY

Brandon W. Blackburn[1], Jacquelyn C. Yanch[2], and Robert E. Klinkowstein[3]

[1]Department of Nuclear Engineering
[2]Department of Nuclear Engineering and Whitaker College of Health Sciences
Massachusetts Institute of Technology
Cambridge, MA 02139
[3]Newton Scientific Inc.
Cambridge, MA 02141

1. INTRODUCTION

The most likely alternative to reactor based therapy for both BNCT and BNCS(Boron Neutron Capture Synovectomy), will be accelerator-based neutron sources. Their size, cost, and licensing requirements make accelerators an attractive and feasible candidate for clinical use. Research at MIT's Laboratory for Accelerator Beam Applications (LABA) has focused on various aspects of both BNCT and BNCS. The accelerator used at LABA is a 4.1 MeV tandem accelerator capable of accelerating a beam of protons, deuterons, and alpha particles with an energy resolution of about 100 eV.

It is estimated that beam powers of up to 30–40 kW[1,2] could be needed for BNCT and 3–5 kW[3] will be needed for BNCS in order to make therapy times reasonable based on target therapeutic doses of 15 RBE-Gy in BNCT and 100 RBE-Gy in BNCS. Under these conditions a BNCT or BNCS target could be subjected to an average power density (q") as high as 1–2 kW/cm^2. Cooling of a target subjected to such high power densities poses problems that must be overcome if a viable target is to be designed. A simple solution to this problem would be to spread the beam over a large target area to reduce the power density. In both BNCT and BNCS, however, it is necessary for the target to be placed in a moderator/reflector assembly. A target diameter smaller than the moderator diameter is desirable in order to limit neutron leakage from the sides of the moderator.

Frontiers in Neutron Capture Therapy, edited by Hawthorne *et al.*
Kluwer Academic / Plenum Publishers, New York, 2001.

If the size of the target were not as restrictive, the heat removal problem could simply be solved by spreading the beam out over a large area and cooling the target with conventional convective flow. Due to the constraints of the beamline and their respective moderator/reflector assemblies, the targets in use at LABA are currently limited to 15.5 cm^2 for BNCT and 5.75 cm^2 for BNCS.

Lithium and beryllium have to date been the most widely investigated target materials for accelerator neutron production. For BNCT, lithium based targets provide a high flux of relatively soft neutrons compared to beryllium targets which thereby lessens the amount of neutrons lost during the moderation process. Despite this need for extra moderation, beryllium has become the preferred target material at LABA due to its superior thermal and mechanical properties. This paper describes the development and testing of beryllium targets for both BNCT and BNCS at LABA. These targets both utilize a submerged jet-impingement cooling configuration with light water as the working fluid.

2. TARGET DESIGN

Due to its favorable mechanical and thermal properties, beryllium is well suited as a high power target material for neutron production. It has a high melting point of 1,289 °C, while its thermal conductivity of 200 W/m-K at 300 K is comparable to that of aluminum. It does not react violently with water as does lithium, and it can be machined and fabricated into a working target much more easily than lithium. Submerged jet impingement was considered the most effective means by which to cool a high power target. This cooling configuration consists of injecting a high velocity jet of water through a static region of coolant before striking the center of the target normal to its surface.[4] Conventional convective cooling simply passes a stream of coolant over the target surface. One advantage of submerged jet impingement is that the maximum coolant rates are achieved at the center of the target where the beam strikes. A second advantage over simple convective cooling is that the dynamic pressure created by the fluid striking the target raises the saturation temperature of the coolant and helps delay the critical heat flux (CHF) crisis. Boiling can substantially improve the heat transfer of a convective system as long as CHF is not exceeded. At some point, the heat flux becomes so high that a vapor surface coats the back of the target and the heat transfer decreases abruptly. Failure of the target once CHF has been exceeded is usually prompt and catastrophic.

The BNCT target consists of a circular beryllium disk with an area of 15.5 cm^2 and is 0.254 cm thick. This target is housed in a stainless steel assembly which allows water to enter through the back before returning out the sides as shown in Figure 1.

The BNCS target assembly consists of a beryllium plug which has a target area of 5.75 cm^2 and is 1.905 cm thick. In order to fit into the moderator this plug is positioned in a series of three concentric cylinders which allow the water to enter along the beam axis. It is redirected and formed into a jet by a Teflon plug before striking the target and returning back along the beamline axis. Figure 2 illustrates the complete BNCS target assembly placed inside its moderator/reflector.

3. EXPERIMENTAL PROCEDURE

Both target assemblies have been tested under varying power densities and with different coolant flowrates. Quantifying the heat removal ability of the targets was

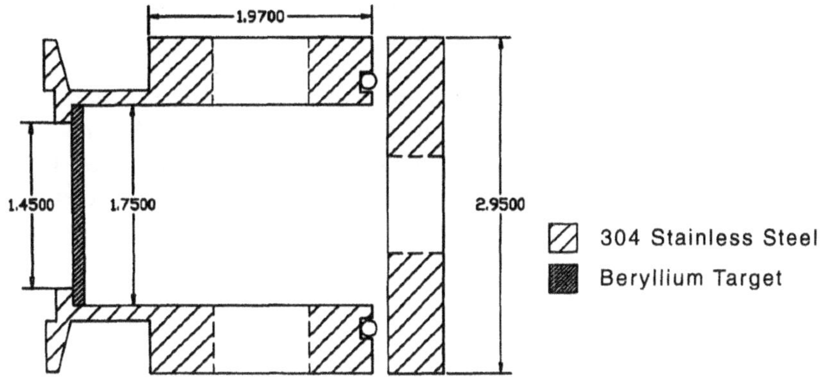

Figure 1. Illustration of the BNCT target housing. The coolant jet enters from the back, strikes the target and exits through the sides. All dimensions in inches.

accomplished by measuring the heat flux (q″), target surface temperature (Ts), coolant temperature (Tc), and jet velocity (V). Equations 1 and 2 could then be used to measure the heat transfer coefficient (h) as a function of q″ and Reynolds number (Re). The thermocouples used to measure the target surface temperature were kept to the smallest possible size so that they did not interfere with the coolant flow.

$$q'' = h(Ts\text{-}Tc) \tag{1}$$

$$Re = VD/\upsilon \tag{2}$$

where D is the jet diameter; and v is the kinematic viscosity of the coolant.

For the BNCS target, all tests were conducted on the LABA tandem accelerator, and most tests have been conducted during use as the neutron producing source for BNCS rabbit irradiation experiments.[5] During these experiments beam current was

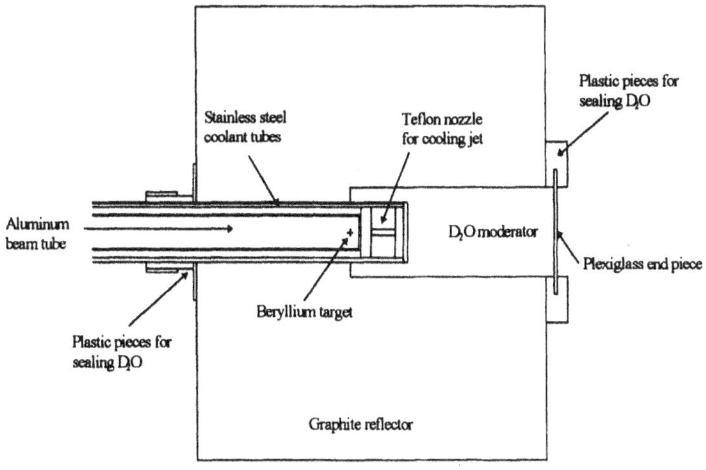

Figure 2. Illustration of the BNCS target assembly in graphite moderator/reflector. In order to estimate scale, the diameter of the Al beam tube is 1.375″.

150 μA with a nominal diameter of 1.5 cm. The BNCT target has been tested both on and off the accelerator in order to test heat removal under variable power density and coolant flow conditions. On the accelerator, the target was tested at currents up to 250 μA with a nominal beam diameter of 2 cm. In order to test the BNCT target without the associated neutron production, an identical assembly was used which allowed the beryllium target material to be interchanged with aluminum or stainless steel. The stainless steel was necessary so that resistive heating could be used to test heat flux levels in excess of 1 kW/cm^2 with a large electrical power supply.

4. RESULTS

The heat transfer rates reported for both targets represent h values calculated at the stagnation point at the center of the target. Radial variation in the heat transfer coefficient is small since the ratio of the target diameter to jet diameter is small. In both targets, there is a clear indication that the heat transfer improves for Re 2,300–4,000 which marks the onset of turbulence. It was found that the nozzle to target spacing, Z/D (nozzle to target distance/nozzle diameter), was relatively unimportant for large Reynolds numbers. Heat transfer rates peaked at a Z/D of about 1.

Both targets require that a pump be used to generate sufficient coolant flow rates. For the BNCS target, a $^1/_2$ hp centrifugal pump is sufficient to deliver up to 6 GPM through the system generating jet velocities of up to 12 m/s. Tests on the BNCS target were carried out with flowrates up to 3 GPM. At this flowrate, heat transfer coefficients of 4×10^4 W/m^{2}K were achieved. If the target were to be run with the maximum flowrate of 6 GPM, the target would be capable of handling 2–4 kW of total power.

Because of the high flowrates needed in the BNCT target, a 15 hp pump was necessary to generate the required jet velocities. Due to pump limitations, the highest velocity that could be achieved was 24.13 m/s at a flow rate of 87 GPM through a 1.70 cm nozzle. For the BNCT target, the peak local heat flux measured under these flow rate conditions was 5.07 kW/cm^2, while the heat transfer coefficient was calculated to be 2.58×10^5 W/m^{2}K. Averaged over the effective target surface, the heat flux was 1.5 kW/cm^2.

Heat transfer coefficients calculated using Eq.1–2 for both targets were found to scale according to $Re^{0.6\text{–}0.65}$ depending on heat flux levels and nozzle diameter over a range of Reynolds numbers from 5×10^4–5×10^5. With this relation, the Reynolds number can be used to estimate target temperatures for various flow rates once the heat flux is known. Extrapolating to an ultimate jet velocity of 35 m/s for a 1.70 cm nozzle, 1.9 kW/cm^2 could be removed with this cooling configuration. The fact that the conductivity of stainless steel is much less than beryllium would most likely improve these numbers resulting in even lower target temperatures than were measured. It is therefore reasonable to predict that a total power of 30 kW or more could be safely handled by this cooling configuration in a beryllium BNCT target.

5. CONCLUSIONS

Submerged jet impingement has been shown in two different targets to be an effective method of removing large amounts of heat from a small target area. This type of cooling configuration could be used to cool targets subjected to 1–2 kW/cm^2 without damage or failure. With the current pumping configuration the BNCS target is capable of handling 2–4 kW of total beam power. This could be increased simply by replacing the current pump with a higher power pump. With the assembly illustrated in Figure 2, a 2.6 MeV deuteron beam of 2 mA could deliver 100 RBE-Gy in less than 15 minutes assuming boron concentrations of 1,000 ppm in the synovium.[6] The BNCT target has been tested at heat loads of up to 1.5 kW/cm^2 with jet velocities of 24 m/s. The heat transfer coefficient was seen to scale with $Re^{0.6-0.65}$. Using this scaling factor, and extrapolating from 24 m/s to 35 m/s, it is predicted that this kind of target could withstand up to 30 kW of beam power. For 3.7 MeV protons on beryllium a 30 kW beam could deliver 13 RBE-cGy/min-mA to a tumor at a depth of 4 cm. At this rate 15 RBE-Gy could be delivered in 14.2 minutes with a boron loading of 30 ppm in the tumor.[7]

ACKNOWLEDGMENT

This work was conducted with funding from the US Deptarment of Energy under grant DE-FG02-89ER60874.

REFERENCES

1. Yanch, J.C., R.E. Shefer, R.E. Klinkowstein, W.B. Howard, H. Song, B. Blackburn, and E. Binello, "Research in Boron Neutron Capture Therapy at MIT LABA, " in Proceedings of the Fourteenth International Conference, Denton, Texas, 1996, *Applications of Accelerators in Research and Industry*, Part Two, pp. 1281–1284.
2. Yanch, J.C., R.E. Shefer, R.E. Klinkowstein, H. Song, B.Blackburn, and W .B. Howard, "Use of the $^9Be(d,n)$ Nuclear Reaction for BNCT: Simulation and Experimental Measurement," Proceedings of the 7th International Symposium on Neutron Capture Therapy for Cancer, Elsevier Science, 1997.
3. Yanch, J.C., R.E. Shefer, and E. Binello, "Design of Low-energy Neutron Beams for Boron Neutron Capture Synovectomy," in *International Conference: Neutrons in Research and Industry*, George Vourvopolous, Editor, 1997, pp. 31–40.
4. Pais, M., J. Leland, W. Chang, and L. Chow, "Single-Phase Heat Transfer Characteristics of Submerged Jet Impingement Cooling Using JP-5," in Proceedings of the Conference on Thermal Phenomena in Electronic Systems, 1994, pp. 178–183.
5. Yanch, J.C., R.E. Shefer, S. Shortkroff, E. Binello, D. Gierga, A.G. Jones, G. Yong, and A. Davison, "Progress in Boron Neutron Capture Synovectomy for the Treatment of Rheumatoid Arthritis", to be published in Proceedings of the Eighth International Symposium on Neutron Capture Therapy for Cancer, La Jolla, Sept. 1998.
6. Gierga, D.P, J.C. Yanch, and R.E. Shefer, "Development and construction of a neutron beam line for accelerator-based boron neutron capture synovectomy," *Medical Physics*. **27**, 203–213, Jan. 2000.
7. W. Howard, "Accelerator Based Boron Neutron Capture Therapy," MIT Ph.D. Thesis. 1997.

A SEALED-ACCELERATOR-TUBE NEUTRON GENERATOR FOR BORON NEUTRON CAPTURE THERAPY APPLICATION

K. N. Leung[1], Y. Lee[1], J. M. Verbeke[1,2], J. Vujic[2], M. D. Williams[1], L. K. Wu[1], and N. Zahir[1,2]

[1]Lawrence Berkeley National Laboratory
[2]Nuclear Engineering Department
University of California
Berkeley Berkeley, CA 94720

ABSTRACT

Radio-frequency (RF) driven ion sources are being developed in Lawrence Berkeley National Laboratory (LBNL) for sealed-accelerator-tube neutron generator applications. By using a 2.5-cm-diameter RF-driven multicusp source and a computer designed 100 keV accelerator column, peak extractable hydrogen current exceeding 1 A from a 3-mm-diameter aperture, together with H^+ yields over 94% have been achieved. These experimental findings together with recent moderator design will enable one to develop compact 14 MeV neutron generators based on the D-T fusion reaction. In this new neutron generator, the ion source, the accelerator and the target are all housed in a sealed metal container without pumping. With a 120 keV and 1 A deuteron beam, it is estimated that a treatment time of ~45 minutes is needed for boron neutron capture therapy.

1. INTRODUCTION

The RF-driven multicusp ion source developed at Lawrence Berkeley National Laboratory has found numerous applications ranging from neutral beam injection systems for fusion reactors to particle accelerators, proton therapy machines and ion implantation systems[1]. Such sources are simple to operate, have long lifetimes, high gas efficiencies and provide high density plasmas with high monatomic species yields. These characteristics make the RF-driven ion source a viable candidate for the next generation of compact, high-output, sealed-tube neutron generators, utilizing the fusion of deuterium and tritium, or deuterium and deuterium.

Frontiers in Neutron Capture Therapy, edited by Hawthorne *et al.*
Kluwer Academic / Plenum Publishers, New York, 2001.

Recently, LBNL has developed a compact, sealed-accelerator-tube neutron generator capable of producing a neutron flux in the range of 10^9 to 10^{10} D-T neutrons per second.[2–5] The ion source, a miniaturized variation of earlier RF-driven multicusp ion sources, is designed to fit within a 5-cm-diameter borehole. Typical operating parameters include repetition rates up to 100 pps, with pulse widths between 10 and 80 μs (limited only by the available RF power supply) and source pressure as low as 5 mTorr. In this configuration, peak extractable hydrogen current densities exceeding 1 A/cm^2 with H^+ ion yields over 94% have been achieved. From this output, a D-T neutron yield of 10^9 neutrons per second can be projected. Simple scaling of the ion source and extraction aperture size could bring the neutron output even higher.

These experimental findings together with recent moderator design[6] will enable one to develop compact 14 MeV neutron generators based on the D-T fusion reaction. In this new system, the ion source, the accelerator and the target are all housed in a sealed metal container without pumping. With a 120 keV and 1 A average D^+ beam current, it is estimated that a treatment time of ~45 minutes is needed for boron neutron capture therapy (10^{14} neutrons/sec). This article describes the design and characteristics of the new neutron generator.

2. NEUTRON TUBE DESIGN

The main components of the sealed D-T neutron tube are the ion source, the 120 kV accelerator column, the water-cooled target and the vacuum system. Figure 1 is a schematic diagram of the sealed D-T neutron generator. It is a scale up version of the compact neutron tube that LBNL has developed. The characteristics of this neutron generator are as follows.

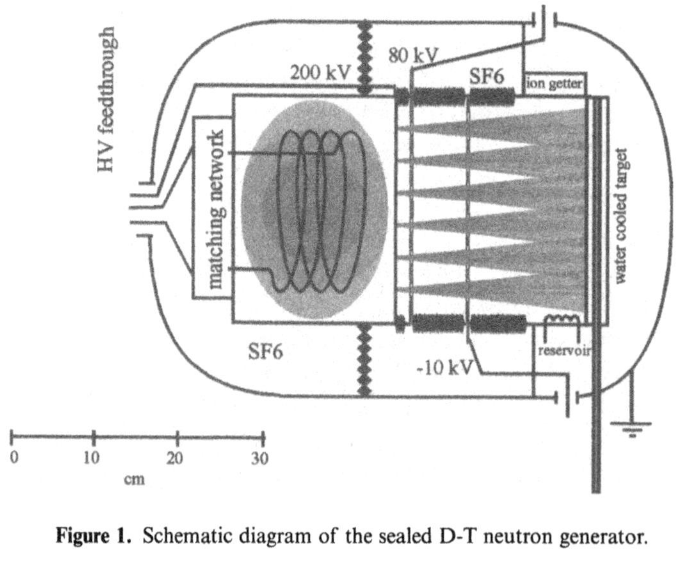

Figure 1. Schematic diagram of the sealed D-T neutron generator.

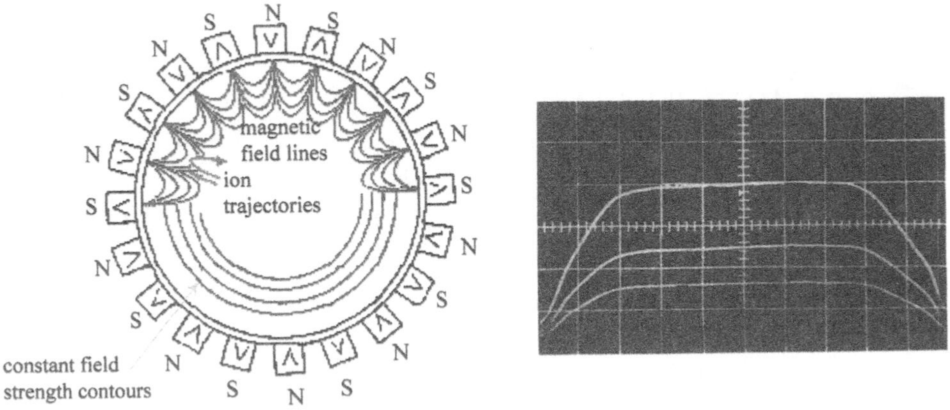

Figure 2. Cross-sectional view of the source chamber and radial plasma density profiles.

2.1. The Ion Source

In order to achieve a neutron yield of 10^{14} neutrons/sec, a large 20-cm-diameter multicusp source together with a multi-aperture extraction system to produce an ion beam current of 1 A, accelerated to 120 kV, and impinging on a well-cooled target is required. The multicusp generator is a 20-cm-diameter cylindrical stainless-steel chamber surrounded with columns of samarium-cobalt magnets. Figure 2 shows a cross-sectional view of the source chamber and the radial plasma density profiles. Apart from an area of about 5 cm from the chamber wall, the plasma density is very uniform in the central region. Thus, one can form multiple beamlets with good ion optics to enhance the extractable current.

If the extraction area is 100 cm^2 and assuming 50% transparency, then the current density at the extraction apertures will be ~20 mA/cm^2. This current density is much lower than other cw or long pulse ion beam system such as the neutral beam injectors now operated in the tokamak fusion reactors. In order to enhance the monatomic ion species D^+ or T^+, a permanent magnet filter will be installed in the source chamber.[3] The filter configuration can be optimized to produce high D^+ and T^+ concentration in the extracted beam which in turn can increase the neutron output.

The plasma is produced by RF induction discharge. In order to deliver RF power to the plasma, a coupler in the form of a multi-turn induction coil is used. The RF antenna coil is fabricated from copper tubing which is normally coated with a thin layer of porcelain for electrical insulation from the conducting plasma. The source is closed-off at one end with a back flange carrying the necessary feedthroughs for the antenna, and a capacitance manometer to monitor source operating pressure. The source will be operated at a background pressure of about 1–2 mTorr.

We plan to operate the source in a pulsed mode with a 10% duty factor at a current density of 200 mA/cm^2. This will produce an average beam current of 1 A on the target. The RF input power for the beam pulse is now higher and the atomic species fraction should improve. The RF power supply is a broad band power amplifier driven at 2 MHz by a signal generator. A pulse generator gates the amplifier so as to generate plasma pulses with durations of 1 ms and repetition rates of 100 Hz. An impedance matching network connects the RF power supply to the antenna coil. The purpose of this network is to

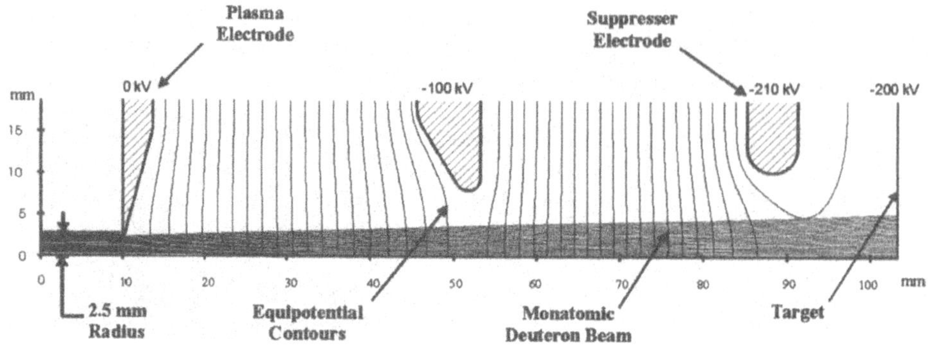

Figure 3. D^+ ion trajectories for a 200 kV accelerator column.

match the nominal 50 Ω output impedance of the RF amplifier to the typical low impedance (~1 Ω) of the plasma-antenna load.

2.2. The 120 kV Accelerator Column

The beam extraction system which closes off the other end of the ion source chamber, contains three electrodes with multiple 3-mm-diameter apertures. The shape, the separation and the voltage distribution on these electrodes are designed by using the I-GUN computation code. Figure 3 shows the D^+ ion trajectories for a 200 kV accelerator column. In this arrangement, each beamlet clears all the electrodes. The beamlets expand after they exit from the last electrode and therefore the power can be spread uniformly across the target. The presence of a suppressor electrode will prevent electrons from accelerating back to the source chamber. Thus source and electrode damages caused by high energy backstreaming electrons can be avoided.

2.3. The Water-Cooled Target

The target is a water-cooled copper plate covered by a layer of titanium which can absorb deuterium and tritium atoms efficiently. The power density on the target due to the impinging D^+ and T^+ ions is ~1.2 kW/cm^2 which can be easily handled by a modest water flow-rate. In order to maximize the D-T fusion reaction on the target, we plan to operate the multicusp source with a 50%–50% mixture of deuterium and tritium. Our past experience indicates that the target can be loaded by the beam in a very short period of time.

The entire tube will be operated with an ambient pressure of ~2 mTorr. Low gas pressure is essential for reducing both charge exchange and high-voltage break-down in the accelerator column. In order to control the source operating pressure, some getter pumps will be installed inside the tube. If higher gas pressures are needed, the getter pumps can be heated to release some deuterium and tritium gases. The sealed neutron tube is operated in a safe environment. Since there is no weak component, the lifetime of the tube should be very long. Higher neutron flux can be achieved either by increasing the extraction area or by operating the source with higher discharge power. Epithermal neutrons are needed for BNCT applications. By coupling the sealed neutron tube with a properly designed moderator,[6] one can generate the desired neutron energy spectrum.

ACKNOWLEDGMENTS

We would like to thank members of the Plasma and Ion Source Technology Group at LBNL for all the technical assistance. This work is supported by the U.S. Department of Energy under Contract No. DE-AC03–76SF00098.

REFERENCES

1. K.N. Leung, Rev. Sci. Instrum., *67*, 1302, (1996).
2. L.T. Perkins, P.R. Herz, K.N. Leung, and D. Pickard, Rev. Sci. Instrum., *65*, 1186, (1996).
3. L.T. Perkins, *et al.*, Rev. Sci. Instrum., *67*, 1057 (1996).
4. L.T. Perkins, C.M. Celata, K.N. Leung, D.S. Pickard, R. Vilaithong, and M.D. Williams, Proc. of the 1997 Particle Accelerator Conference, Vancouver, Canada.
5. L.T. Perkins *et al.*, Proceedings of the 14th ICAARI, Denton, TX, Nov 1996.
6. J.M. Verbeke, J. Vujic, and K.N. Leung, Investigation of $^2H(d,n)^3He$ and $^3H(d,n)^4He$ fusion reactions as alternative neutron sources for BNCT, *in* "Frontiers in Neutron Capture Therapy," M.F. Hawthorne, K. Shelly, R.W. Wiersema, eds., Kluwer Academic/Plenum Publishers, New York, 2001, pp. 483–488.

64

CONSTRUCTION OF AN ACCELERATOR-BASED NEUTRON BEAMLINE FOR TREATMENT OF ARTHRITIC JOINTS

David P. Gierga[1], Jacquelyn C. Yanch[1], Ruth E. Shefer[2], and Brandon W. Blackburn[1]

[1]Department of Nuclear Engineering
Massachusetts Institute of Technology
Cambridge, MA USA
[2]Newton Scientific, Inc.
Cambridge, MA USA

1. INTRODUCTION

Boron Neutron Capture Synovectomy (BNCS) is a proposed application of the $^{10}B(n,\alpha)Li$ nuclear reaction for the treatment of rheumatoid arthritis. In rheumatoid joints, the synovium becomes grossly inflamed, which leads to painful joint dysfunction. The goal of BNCS is the complete ablation of the synovium in order to relieve the pain and disability of the patient. BNCS is a two part modality in which a boron-labeled compound is injected directly into the arthritic joint prior to neutron irradiation. The therapeutic dose for a BNCS treatment is assumed to be 10,000 RBE-cGy, based on an empirical estimate of the effective clinical dose in beta-particle radiation synovectomy studies.[1] This dose is significantly higher than the therapeutic doses necessary for reproductive cell death, the goal of boron neutron capture therapy (BNCT) treatments of cancers such as melanoma or glioblastoma multiforme. In BNCS, enough dose must be delivered to the synovial cells in order to achieve functional cell death. Additionally, since the boron compound is injected directly into the joint, levels of boron uptake are orders of magnitude higher in BNCS, compared to BNCT, which normally uses a systemic administration of the boron agent. Boron uptake in the synovium is assumed to be 1,000 ppm for the purposes of the calculations presented in this paper. This high uptake has been verified in *in vivo* experiments.[2]

Frontiers in Neutron Capture Therapy, edited by Hawthorne *et al.*
Kluwer Academic / Plenum Publishers, New York, 2001.

2. METHODS

2.1. Monte Carlo

Monte Carlo simulations were used in designing beams for the treatment of arthritic human knee and finger joints, as well as rabbit knee joints. A concentric knee phantom was used, based on magnetic resonance images of arthritic joints; values for human finger and rabbit knee joints were scaled based on these dimensions. The outer diameters of the phantoms were 8.7 cm for the human knee, 3.0 cm for the rabbit knee, and 2.2 cm for the human finger. Calculations were done using the Monte Carlo code MCNP4B.[3] Boron was explicitly modeled in the synovium to account for flux depression and photon production. Although approximate, RBE values were used to scale the physical dose rates. For this reason, RBE factors similar to those used in the BNCT community were used: 4.0, 3.8, 1.0 for ^{10}B reaction products, neutrons, and photons, respectively. Tissue dose rates were determined by summing the thermal neutron, fast neutron, boron, and photon doses. Treatment times are calculated for a 10,000 RBE-cGy dose to the entire synovium. In the design process, only beams resulting in skin doses of less than 800 RBE-cGy were considered. This very conservative limit corresponds to the threshold for temporary and mild skin reddening.[4]

2.2. Neutron Producing Charged Particle Reactions

A variety of accelerator-based charged particle neutron sources were considered. The $^{7}Li(p,n)$ reaction at 2.5 MeV[5] is attractive as a prolific source of relatively low energy neutrons. Practical difficulties arise in constructing and using high powered lithium targets, as a result of lithium's low melting point and reactive chemical properties. Based on these engineering considerations, initial efforts have focused on beryllium targets. Beryllium based reactions investigated in this study include the $^{9}Be(p,n)$ reactions at 4.0 and 3.7 MeV proton energy,[6] and the $^{9}Be(d,n)$ reactions at 2.6 and 1.5 MeV proton energy.[7,8]

2.3. Rear and Side Reflector Optimization

Beam performance can be improved by placing a reflector behind and around the irradiated joint (see Figure 1). Such a reflector will enable the beam to be utilized in a more efficient manner, as some of the beam will be redirected toward the target region instead of simply exiting the phantom. Additional moderation of the neutron beam is also possible. Systematic comparisons of various reflector configurations were done using a 2.6 MeV $^{9}Be(d,n)$ neutron source and a human knee phantom.

2.4. Effect of Boron Uptake in Synovium

To assess the dependence of therapy time and skin dose on boron uptake, simulations were performed over a wide range of uptake levels using the 2.6 MeV $^{9}Be(d,n)$ neutron source and a human knee phantom. Boron uptake levels were varied from 1 to 20,000 ppm, with the boron explicitly modeled in all cases.

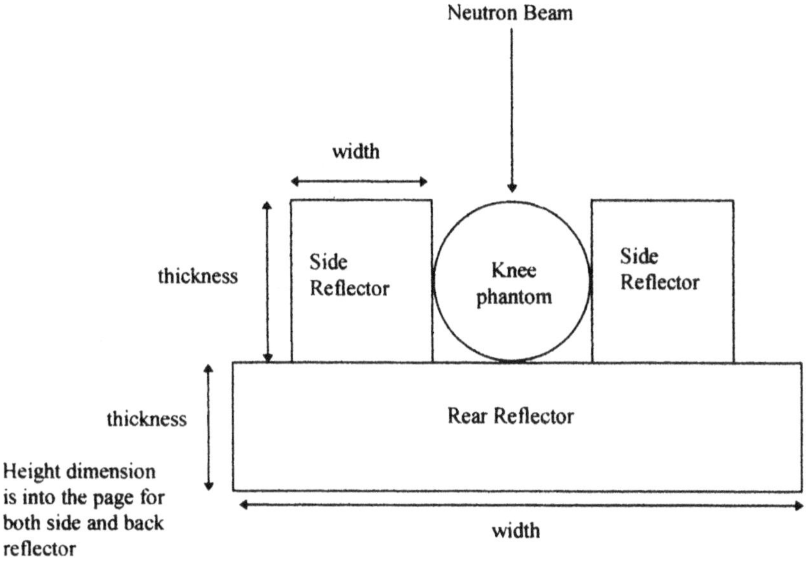

Figure 1. Schematic diagram of the knee phantom with side and rear reflectors.

3. RESULTS AND DISCUSSION

3.1. Beamline Assembly

Results of the design study led to the construction of a dedicated BNCS beamline at the Laboratory for Accelerator Beam Applications (LABA) at MIT. The beamline consists of a 9 cm diameter, 23 cm length D_2O moderator surrounded by an 18 cm thick graphite reflector. The accelerator target assembly can be placed at any position along the central axis of the reflector/moderator assembly, which allows greater flexibility in the choice of beam characteristics. The neutron target is a single piece of beryllium which is mounted on the end of a small aluminum beam tube. The beryllium target is electrically insulated from the aluminum tube, and target current and temperature are measured by a thermocouple attached to the center of the target, on the coolant side of the beryllium. The target is cooled using the submerged jet-impingement technique, described in Blackburn *et al.*[9] The aluminum target tube is surrounded by two concentric coolant tubes. Light water coolant enters the outer tube, and is forced through a teflon plug, which leads to a jet of water impinging on the target. Water then flows between the inner coolant tube and the aluminum target tube before exiting the assembly. A cross sectional view of the beamline assembly, taken from the Monte Carlo model, is shown in Figure 2. The entire beamline assembly is mounted on the high current tandem electrostatic accelerator in operation at LABA.

3.2. Rear and Side Reflector Studies

A variety of reflector configurations were considered, with therapy times and skin doses (therapy parameters) used as comparison criteria. First, comparisons were done for a reflector behind the irradiated joint. It was shown that, for a graphite reflector,

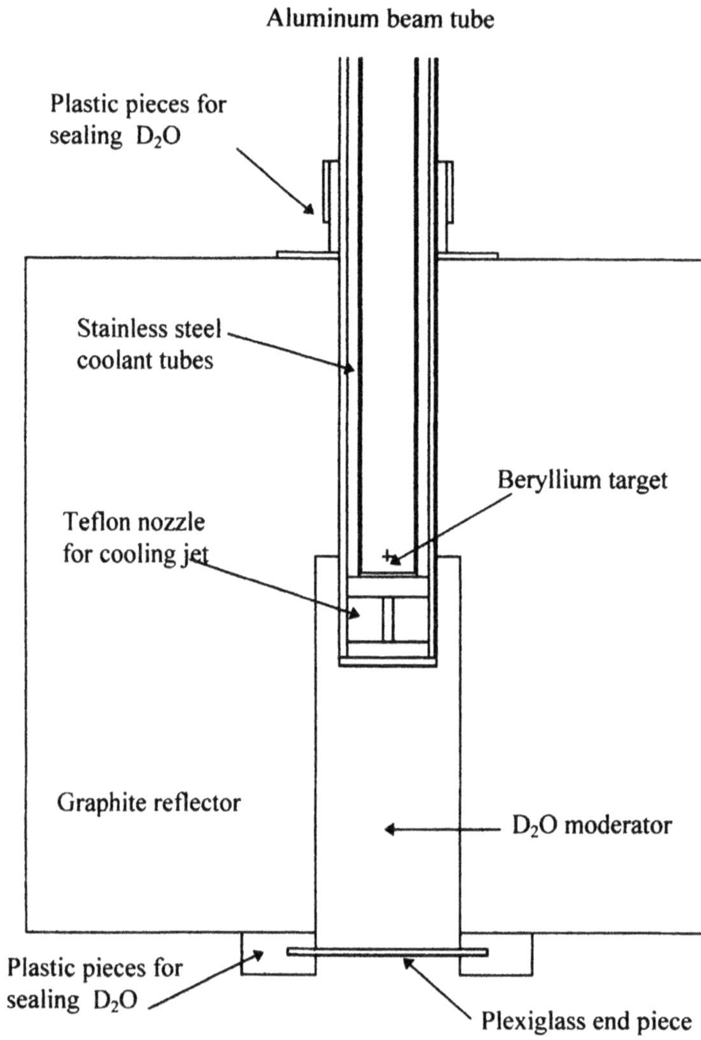

Figure 2. Schematic diagram of the BNCS beamline assembly.

increasing the reflector thickness was beneficial up to 10 cm. The width and height of the rear reflector were also varied in unison. Once the dimensions of the width and height reached about 45 cm, there was no further improvement in the therapy parameters. Overall, by introducing a reflector behind the joint, a factor of 2 improvement in the therapy parameters (i.e. a reduction in the therapy time, bone dose, and skin dose) is seen for single beam irradiations, and about 20–30% improvement for parallel opposed beam irradiations. A side reflector was also added to the model. The height of the side reflector was varied up to 30 cm. A side reflector of this height leads to another 20–30% improvement in therapy parameters. The total improvement in the final rear and side reflector configuration compared to a case with no such reflectors is a factor of 2–3 for single beams, and a factor of 1.5–2 for parallel opposed beam irradiations. Note that for rabbit knee simulations, the size of the back and side reflectors was modified to accommodate the smaller dimensions of the rabbit. The back reflector dimensions were $10 \times 20 \times 40\,cm^3$, while the side reflector dimensions were $3 \times 10 \times 4.3\,cm^3$.

3.3. Effect of Boron Uptake in Synovium

Figure 3 illustrates the effect of boron uptake in synovium on therapy time and skin dose. As the amount of boron in the target tissue increases, the therapy becomes more attractive. Therapy times, based on a 1 mA beam current, do not become reasonable until the uptake is several hundred ppm. Skin doses are enormous for low levels of boron uptake, and do not fall below the 800 RBE-cGy skin dose limit until the uptake is roughly 300–400 ppm, for parallel opposed irradiations. Note that for the parallel opposed case, the treatment time is the total treatment time for two sequential, parallel opposed irradiations.

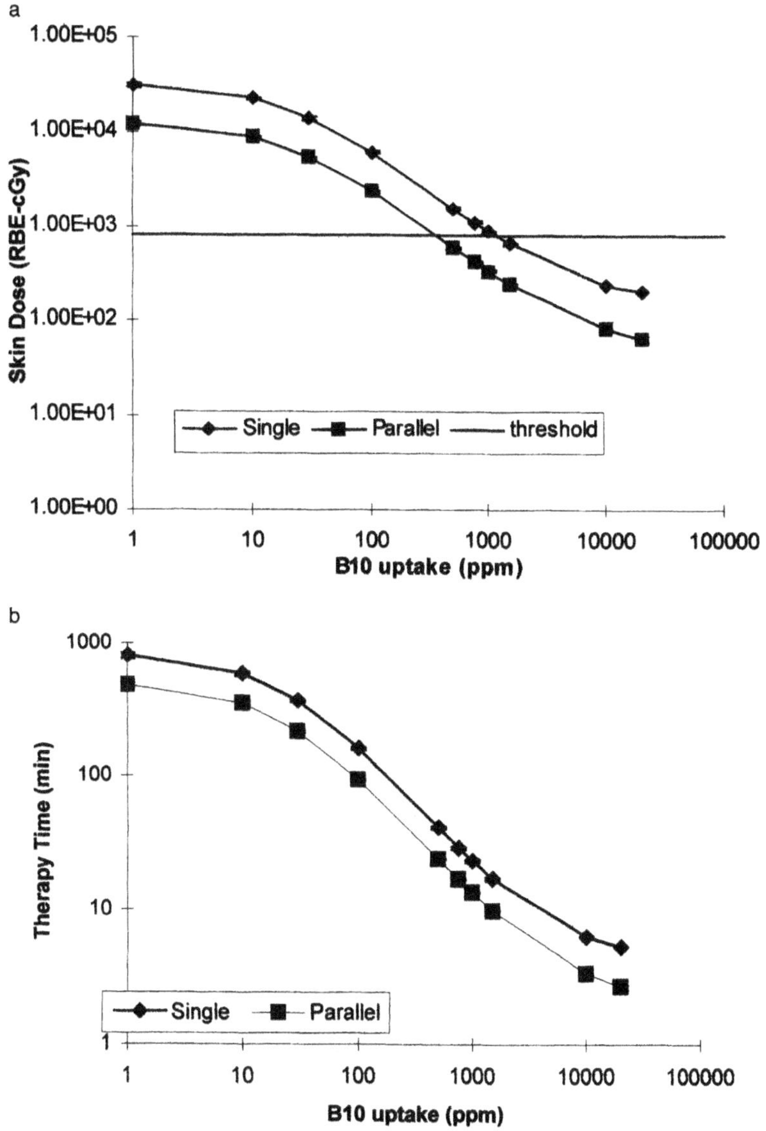

Figure 3. (a) Skin dose and (b) therapy time as a function of synovial boron-10 uptake.

Table 1. BNCS therapy parameters

Configuration 1,000 ppm ^{10}B uptake	Moderator length (cm)	Therapy time (min)	Skin dose (RBE-cGy)	Bone ratio
Human knee		Parallel beam		
2.6 MeV $^9Be(d,n)$	23	13	335	38
1.5 MeV $^9Be(d,n)$	20	48	586	22
4.0 MeV $^9Be(p,n)$	15	7.3	203	73
3.7 MeV $^9Be(p,n)$	15	10	190	79
2.5 MeV $^7Li(p,n)$	15	7	129	108
Human finger		Parallel beam		
2.6 MeV $^9Be(d,n)$	23	6.1	227	42
1.5 MeV $^9Be(d,n)$	20	24	406	24
4.0 MeV $^9Be(p,n)$	15	3.6	161	62
3.7 MeV $^9Be(p,n)$	15	4.8	136	69
2.5 MeV $^7Li(p,n)$	15	3.3	91	108
Rabbit knee		Single beam		
2.6 MeV $^9Be(d,n)$	23	10	418	30
1.5 MeV $^9Be(d,n)$	22	48	760	17
4.0 MeV $^9Be(p,n)$	15	5.9	334	48
3.7 MeV $^9Be(p,n)$	15	8.3	280	59
2.5 MeV $^7Li(p,n)$	15	5.6	189	89

3.4. Final Therapy Parameters

Therapy parameters have been calculated for the optimized side and rear reflector geometry. These results are summarized in Table 1. Treatment times are given in minutes, based on a 1 mA beam current. All of the configurations listed in Table 1 are viable for BNCS. The $^7Li(p,n)$ based beams give the best results, with the shortest therapy times, and the lowest skin and bone doses. Beams using the $^9Be(p,n)$ reaction, however, give results comparable to $^7Li(p,n)$. The $^9Be(d,n)$ reactions result in higher neutron energies, resulting in higher skin doses, and require more neutron moderation, resulting in longer therapy times. The 1.5 MeV $^9Be(d,n)$ reaction, however, can be accessed with a lower energy, and thus less costly, accelerator.

4. SUMMARY

A neutron beamline is now operational for BNCS experiments at MIT LABA. This beamline has been used in two series of rabbit irradiations to date.[10] Monte Carlo calculations have been carried out to determine therapy parameters for human knee, human finger, and rabbit knee treatments. The addition of reflector behind and around the joint has been shown to decrease therapy times and skin doses. Treatment times are about 40 minutes or less for all joints (for a 1 mA beam current), with the skin doses well below the prescribed limits. Experiments at LABA are ongoing to examine the viability of BNCS as a treatment for rheumatoid arthritis.

ACKNOWLEDGMENT

This work has been supported by a grant from the National Institutes of Health (Award Number 1R43AR44680-02).

REFERENCES

1. E. Deutsch, J.W. Brodack, K.R. Deutsch, Radiation Synovectomy Revisited, *Eur J Nucl Med* 20:1113–1127, 1993
2. E. Binello, S. Shortkroff, R.E, Shefer, A.G. Jones, A. Davison, T. Thornhill, and J.C. Yanch, Analysis of *in vivo* uptake for boron neutron capture synovectomy, *in* "Frontiers in Neutron Capture Therapy," M.F. Hawthorne, K. Shelly, R.W. Wiersema, eds., Kluwer Academic/Plenum Publishers, New York, 2001, pp. 1063–1066.
3. J.F. Briesmeister, ed., MCNP—A General Monte Carlo N-Particle Transport Code, Version 4B, Los Alamos National Laboratory Report LA-12625-M, 1997.
4. A.H.W. Nias, An Introduction to Radiology, John Wiley & Sons, New York, 1990.
5. H. Liskien, and H. Paulsen, Neutron Production Cross Sections and Energies for the reaction $^{7}Li(p,n)^{7}Be$ and $^{7}Li(p,n)^{7}Be^{*}$, *Atomic Data and Nuclear Data Tables*, 15:57–84, 1985.
6. W.B. Howard, J.C. Yanch, S.M. Grimes, T.N. Massey, S.I. Al-Quaraishi, D.K. Jacobs, and C.E. Brient, Measurement of the $^{9}Be(p,n)$ thick target spectrum for use in accelerator-based BNCT, *Med Phys* 23:1233–1235, 1996.
7. J.W. Meadows, The thick target Be(d,n) neutron spectra for neutron energies between 2.6 and 7.0 MeV, Argonne National Laboratory Nuclear Data and Measurement Series, ANL/NDM-124, 1991.
8. J. Guzek, Ph.D. thesis, University of Witwatersrand, South Africa, 1998.
9. B.W. Blackburn, J.C. Yanch, and R.E. Klinkowstein, Development of a high-power water cooled beryllium target for use in accelerator-based boron neutron capture therapy, *Med Phys*, 25:1967–1974, 1998.
10. J.C. Yanch, S. Shortkroff, R.E. Shefer, E. Binello, D. Gierga, A.G. Jones, G. Young, C. Vivieros, and B. Blackburn, Progress in the development of boron neutron capture synovectomy for the treatment of rheumatoid arthritis, *in* "Frontiers in Neutron Capture Therapy," M.F. Hawthorne, K. Shelly, R.W. Wiersema, eds., Kluwer Academic/Plenum Publishers, New York, 2001, pp. 1389–1397.

65

USE OF THE ^{9}Be(d,n) NUCLEAR REACTION FOR ACCELERATOR-BASED BORON NEUTRON CAPTURE THERAPY

S. M. White[1], R. E. Klinkowstein[2], J. C. Yanch[1], K. D. Held[3], and K. J. Riley[1]

[1]Massachusetts Institute of Technology
Department of Nuclear Engineering
Cambridge, MA
[2]Newton Scientific Inc.
Cambridge, MA
[3]Massachusetts General Hospital
Department of Radiation Oncology
Boston, MA

1. INTRODUCTION

Accelerator-based boron neutron capture therapy (AB-BNCT) is currently being studied as an alternative to reactor-based BNCT at MIT's Laboratory for Accelerator Beam Applications (LABA).[1] A tandem accelerator designed by Newton Scientific, Inc. has been installed at LABA,[2] and an associated switching magnet permits simultaneous setup of a number of experimental end stations. Previous work has indicated that the ^{9}Be(d,n) reaction may be promising for use in AB-BNCT.[3] These early studies investigated the use of deuterons at 2.6 MeV. This beam showed potential but would not be clinically usable due to the large fast neutron component. A softer neutron spectrum can be produced using lower deuteron energies. If the reaction could be utilized at low deuteron energies to produce a useful therapy beam for AB-BNCT, it would be accessible to small, inexpensive, low energy accelerators, as well as to higher energy accelerators such as the tandem electrostatic accelerator at LABA. Deuteron energies from 1.3 MeV to 1.8 MeV have now also been examined. The work presented here examines the trade-off between treatment time (neutron yield) and advantage depth (neutron energy spectrum) in an attempt to determine the most useful deuteron energy using the ^{9}Be(d,n) reaction with LABA's accelerator and experimental setup.

Frontiers in Neutron Capture Therapy, edited by Hawthorne *et al.*
Kluwer Academic / Plenum Publishers, New York, 2001.

1.1. Experimental Setup

The accelerator used in these experiments is a high current tandem accelerator built by Newton Scientific, Inc. that can accelerate protons or deuterons up to 4.1 MeV.[4] The moderator/reflector assembly used consists of a 24 cm diameter, 27 cm long D_2O moderator surrounded by an 18 cm thick lead reflector.[1] The accelerated deuteron beam impinges upon the beryllium target at the end of the beam tube, 1 cm within the D_2O moderator, and produces energetic neutrons. These neutrons are moderated by the heavy water through elastic scattering reactions with deuterium and oxygen. Neutrons leaving the moderator may then interact with the brain phantom. The ellipsoidal head phantom has major outer axes of 13.6, 19.6, and 16.6 cm to simulate a standard-sized human brain,[5] and is on-loan from the MIT Reactor BNCT Group.

2. MIXED FIELD DOSIMETRY

Mixed field dosimetry is used to experimentally evaluate dose profiles resulting from various neutron beams in a water-filled phantom. The dual ionization chamber method is used to separate the neutron and photon dose components, and the cadmium difference method provides a measurement of the thermal flux.[6] Doses are calculated by using fluence to dose conversion factors. Finally, dose rate vs. depth curves can be generated. This mixed field dosimetry method has been studied extensively by the BNCT group at the MIT Reactor,[6] and is used as the dosimetry method for the reactor-based BNCT human clinical trials currently being undertaken at MIT,[5] and in Petten, The Netherlands.[7] In the present work, measurements were taken at the surface and at depths of 1, 2, 3, 4, 6, 8, and 10 cm along the brain phantom centerline.

3. FIGURES OF MERIT

Three criteria were chosen to evaluate the experimental and simulated dosimetry results. These were the advantage depth, the tumor dose rate at a depth of four centimeters, and the ratio of the fast neutron dose rate to the total tumor dose rate at a depth of one centimeter. A large advantage depth, large tumor dose rate, and small ratio of fast to thermal dose rates are desirable. The tumor dose rate is selected at a depth of four centimeters to examine treatment of a deep-seeded tumor, whereas the ratio of the fast neutron dose rate to the total tumor dose rate is selected at a depth of one centimeter to examine the amount of damage to the skin by fast neutrons.

3.1. Experimental Results

As deuteron energy was increased, the tumor dose rate correspondingly improved due to the neutron yield increase. However, the advantage depth decreased, and the ratio of the fast neutron dose rate to the thermal neutron dose rate at a depth of 1 cm increased.[8] Bombarding the beryllium target with 1.5 MeV deuterons produces the optimum beam in terms of a high tumor dose rate, a significant advantage depth, and a low fast to thermal neutron dose rate ratio. This (Figure 1) beam delivers a total tumor dose rate of 8 RBE-cGy/min-mA at the advantage depth of 5.9 cm, assuming tumor boron concentration of 40 ppm and RBE values of 1.0 for photons, 3.2 for neutrons, and 3.8 for boron in tumor tissue.

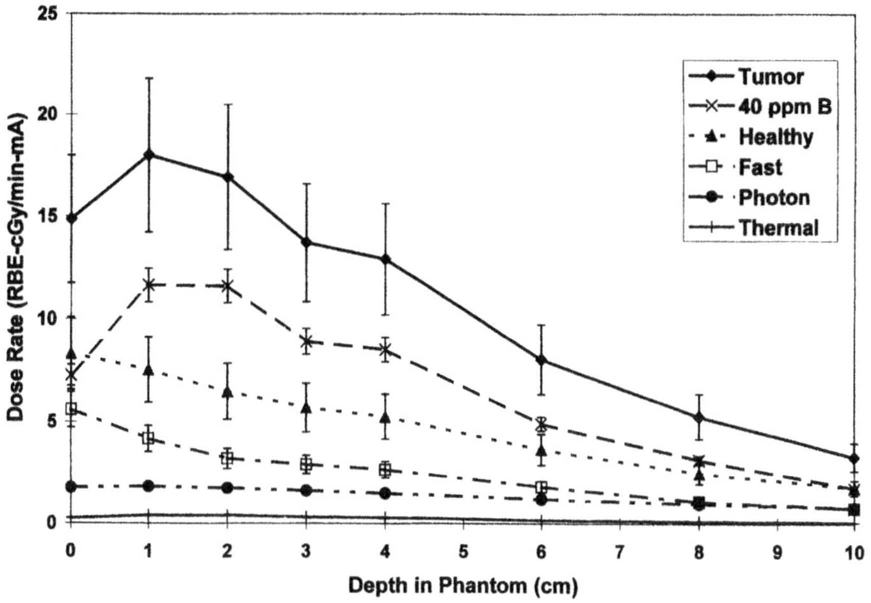

Figure 1. 1.5 MeV d-Be results.

3.2. Comparison with Simulation

Beam dosimetry has been modeled using the code MCNP, a general Monte Carlo N-Particle transport code.[9] The geometry of the moderator/reflector assembly used experimentally in this work was previously modeled at LABA,[1] and was used in this simulation. The ellipsoidal brain phantom was approximated by modeling a right-circular cylinder. This approximation is valid for simulating doses along the phantom's central axis.[1] Spectral data for the ^{9}Be(d,n) reaction at 1.4 MeV were obtained from Whittlestone.[10] Simulated and experimental results for the fast neutron dose rate is in very good agreement (see Figure 3) using a rough estimate of the neutron yield made by scaling the MCNP predictions of thermal neutron dose rate per source neutron to match the experimental thermal neutron dose rate measurements (see Figure 2).

4. BIOLOGICAL DOSIMETRY

Biological effects as a function of depth will be used to further characterize the suitability of the ^{9}Be(d,n)-based epithermal beam. An initial experiment using xrs-6 cells (DNA double strand break repair deficient Chinese hamster ovary cells) was performed. The xrs-6 cells were plated in T25 flasks, which were then placed at depths of 0, 2.5, 5.0, 7.5, and 10.0 cm within a water phantom and irradiated with the moderated 1.5 MeV d-Be mixed field beam. A curve of cell survival versus dose was constructed and is shown in Figure 4. The non-RBE weighted mixed field dosimetry results were used to determine the dose delivered at each depth. Xrs-6 cells were also irradiated using a 250 kVp x-ray machine, and these survival data are also shown in Figure 4.

As seen in Figure 4, cell survival after irradiation with the mixed field beam is similar to the survival using the x-ray beam, which is an unlikely event given the greater

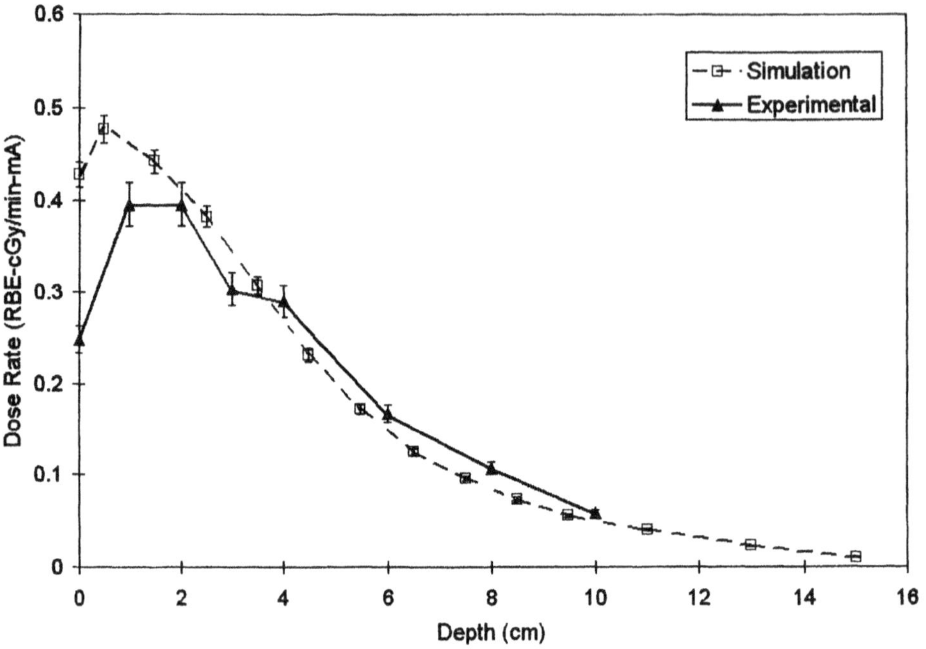

Figure 2. Comparison of 1.5 MeV d-Be experimental and simulated results—dose rate due to thermal neutrons.

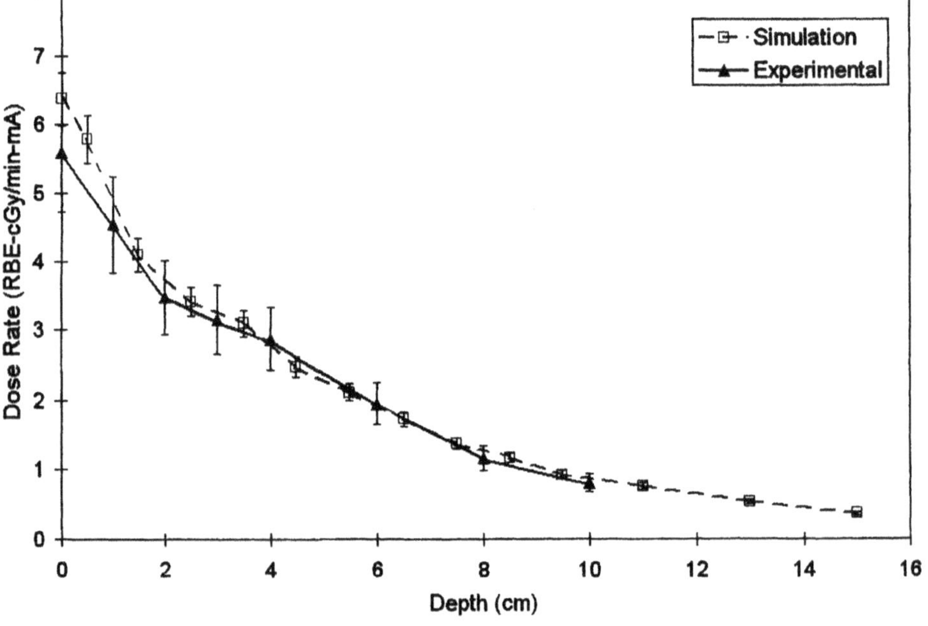

Figure 3. Comparison of 1.5 MeV d-Be experimental and simulated results—dose rate due to fast neutrons.

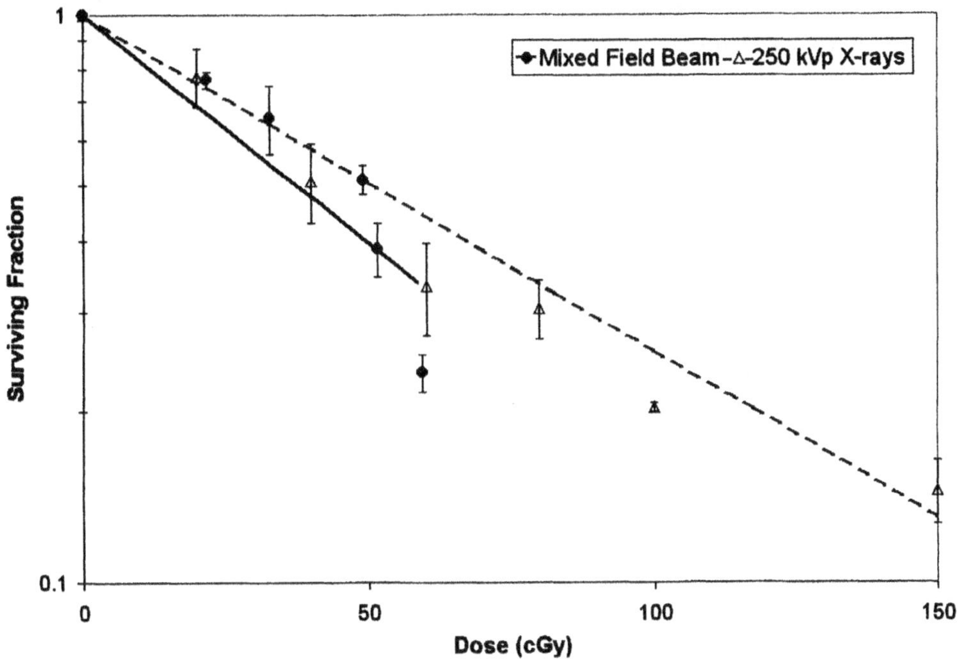

Figure 4. Preliminary xrs-6 survival vs. dose results.

biological effectiveness of neutrons relative to photons. The dose due to the mixed field beam may have been overestimated due to the difficulties in obtaining an accurate estimate of deuteron current on target. Alternatively, uncertainty in the estimates of the fast neutron dose rate obtained via the dual ionization chamber method may also be possible. Detailed investigation of both of these possibilities is underway and is described below.

5. CONCLUSIONS AND FUTURE WORK

All deuteron energies investigated produced a beam that, once moderated, appears viable for AB-BNCT. However, two issues still need to be addressed in these experiments. The first is to determine accurately the neutron to gamma response ratio (B/A ratio) of the tissue equivalent ionization chamber. The B/A ratio is unknown for neutron energies less than 50 keV.[6] The value used in these experiments potentially overestimates the fast neutron dose rate. Future work includes production of low-energy neutrons (<50 keV) using a variety of near-threshold reactions so that the B/A ratio can be determined experimentally.

The second issue to be addressed is the accuracy of the measurement of current on target. The target configuration is currently being modified so that losses of secondary electron current will be minimized.

Beam-dependent RBE values versus depth will be obtained from the biology experiments. The xrs-6 experiment will be repeated, and results will be compared with the same experiment using Chinese hamster ovary cells that are not repair deficient. A dose rate experiment will also be performed to determine the extent of repair that occurs during the mixed field irradiation.

ACKNOWLEDGMENT

This work has been supported by a grant from the U.S. Department of Energy (Award Number DE-FGO2-89ER60874).

REFERENCES

1. J.C. Yanch, X.-L. Zhou, R.E. Shefer, and R.E. Klinkowstein, Accelerator-based epithermal neutron beam design for neutron capture therapy, *Med. Phys.*, 19:709–721, 1992.
2. R.E. Shefer, R.E. Klinkowstein, J.C. Yanch, and W.B. Howard, Tandem electrostatic accelerators for BNCT, *in*: "Proceedings of the 1st International Workshop on Accelerator-Based Neutron Sources for Boron Neutron Capture Therapy," INEL CONF-940976, Jackson, WY, 1994, pp. 89–97.
3. J.C. Yanch, R.E. Shefer, R.E. Klinkowstein, H. Song, B. Blackburn, and W.B. Howard, Use of the ^{9}Be(d,n) nuclear reaction for BNCT: simulation and experimental measurement, *in*: "Advances in Neutron Capture Therapy, Volume I, Medicine and Physics," B. Larsson, J. Crawford, R. Weinreich, eds., Proceedings of the Seventh International Symposium on Neutron Capture Therapy for Cancer, Elsevier Science, Amsterdam, 1997, pp. 564–569.
4. R.E. Klinkowstein, R.E. Shefer, J.C. Yanch, W.B. Howard, H. Song, E. Binello, B.W. Blackburn, J.L. Daigle, S.M. Sears, C.H. Goldie, and R.J. Ledoux, Operation of a high current tandem electrostatic accelerator for boron neutron capture therapy, *in*: "Advances in Neutron Capture Therapy, Volume I, Medicine and Physics," B. Larsson, J. Crawford, R. Weinreich, eds., Proceedings of the Seventh International Symposium on Neutron Capture Therapy for Cancer, Elsevier Science, Amsterdam, 1997, pp. 522–527.
5. O.K. Harling, K.A. Roberts, D.J. Moulin, and R.D. Rogus, Head phantoms for neutron capture therapy, *Med. Phys.*, 22:579–583, 1995.
6. R.D. Rogus, Design and dosimetry of epithermal neutron beams for clinical trials of boron neutron capture therapy at the MITR-II reactor, *in*: Ph.D. Thesis, Nuclear Engineering, Massachusetts Institute of Technology, Cambridge, MA, 1994.
7. C.P.J. Raaijmakers, M.W. Konijnenberg, H.W. Verhagen, and B.J. Mijnheer, Determination of dose components in phantoms irradiated with an epithermal neutron beam for boron neutron capture therapy, *Med. Phys.*, 22:321–329, 1995.
8. S.M. White, Beam characterization for accelerator-based boron neutron capture therapy using the ^{9}Be(d,n) nuclear reaction, *in*: S.M. Thesis, Nuclear Engineering, Massachusetts Institute of Technology, Cambridge, MA, 1998.
9. J.F. Briesmeister, MCNP—A general monte carlo n-particle transport code, Version 4B, Los Alamos National Laboratory, Los Alamos, NM, 1997.
10. S. Whittlestone, Neutron distributions from the deuteron bombardment of a thick beryllium target, *J. Phys. D: Appl. Phys.*, 10:1715–1723, 1977.

AN EXPERIMENTAL PROGRAM FOR BNCT USING A THERMAL ACCELERATOR-BASED SOURCE AND THE EPITHERMAL FACILITY AT THE TAPIRO REACTOR

S. Agosteo[1], P. Colautti[2], K. W. Burn[3], S. Cevolani[3], F. d'Errico[4], E. Fantuzzi[5], G. Fortuna[2], G. Gambarini[6], A. Pisent[2], G. Rosi[7], C. R. Rossi[8], R. Tinti[3], and G. Tornielli[9]

[1]Dipartimento di Ingegneria Nucleare
via Ponzio 34/3, 20133 Milano, Italy
and INFN Sezione di Milano, Italy
[2]INFN, Laboratori Nazionali di Legnaro
via Romea 4, 35020 Legnaro, Italy
[3]ENEA-ERG-SIEC
via Martiri di Monte Sole 4
40129 Bologna, Italy
[4]Yale University School of Medicine
Department of Therapeutic Radiology
Division of Radiological Physics
333 Cedar Street, HRT-219, New Haven, CT 06510
[5]ENEA-AMB-PRO-IRP
via dei Colli, 16, 40136 Bologna
[6]Università degli Studi di Milano, Dipartimento di Fisica
via Celoria 16, Milano, Italy
and INFN Sezione di Milano
[7]ENEA/INN-TEC, via Anguillarese 301
00060 S. Maria di Galeria, Rome, Italy
[8]Dipartimento di Scienze Oncologiche e Chirurgiche
Università di Padova, Clinica Chirurgica II
via Giustiniani 2, Padova, Italy
[9]Dipartimento di Fisica, Università di Padova
35131 Padova, Italy

Frontiers in Neutron Capture Therapy, edited by Hawthorne *et al.*
Kluwer Academic / Plenum Publishers, New York, 2001.

1. INTRODUCTION

An accelerator-based source of thermal neutrons for treating cutaneous melanoma is foreseen in the framework of the SPES project,[1] proposing a facility for exotic nuclei production at the Laboratori Nazionali di Legnaro (LNL) in Italy. The goal of the project, scheduled for the next five years (1999–2003), is the construction of a 10 MeV proton/deuteron accelerator capable of delivering a current of 30 mA, consisting of a 5 MeV Radio-Frequency Quadrupole (RFQ) and of a Drift-Tube Linac (DTL). Part of the beam (1–3 mA) delivered by the RFQ to the linear accelerator will be used for neutron production for BNCT, via the $^{9}Be(p,n)^{9}B$ and/or the $^{9}Be(d,n)^{10}B$ reactions. The BNCT facility which will be built at the LNL is intended to be in the framework of the TERA program for the realization of hadrontherapy facilities in Italy.[2]

In the future, the linear accelerator will be upgraded to 100 MeV. Presently, an intense R&D program on high intensity accelerator techniques and targetry is in progress, financially supported in the framework of the Italian project for accelerator-driven systems. In particular, the RFQ prototype in aluminium is already available at LNL for radio-frequency measurements and the final version is planned to be ready in two years. A schematic layout of the facility proposed at LNL is shown in Figure 1, together with the existing accelerators (ALPI, XTU TANDEM) and the related experimental halls.

In the period necessary for the RFQ construction, an experimental program for the optimization of the neutron moderator, the source characterization, microdosimetry of the therapeutic beam and phantom dosimetry will be carried out. The BNCT experimental facility,[3,4] already available at the Van De Graaff accelerator of the LNL (max.

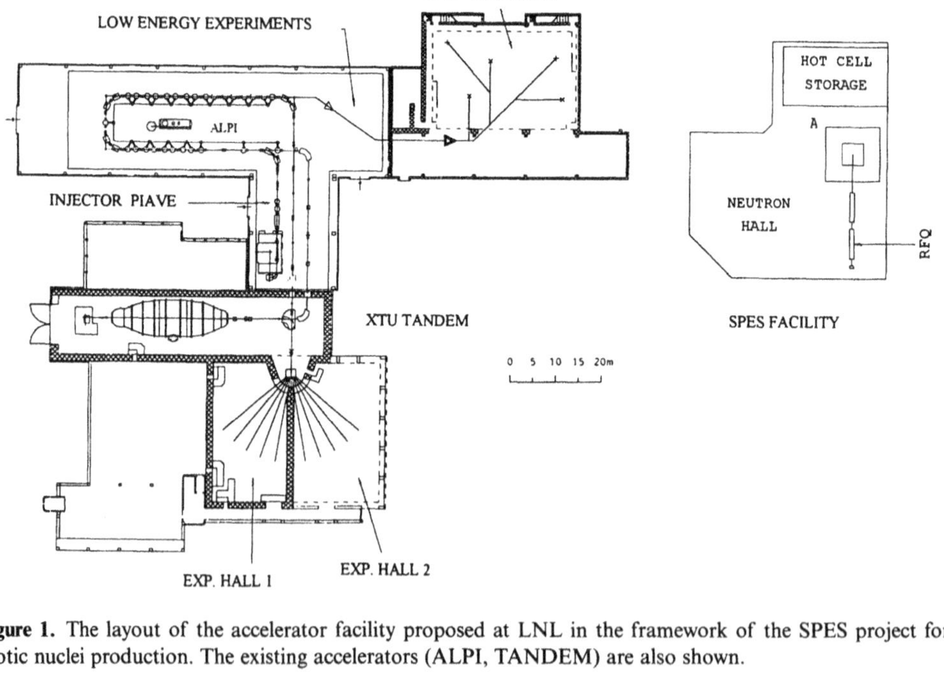

Figure 1. The layout of the accelerator facility proposed at LNL in the framework of the SPES project for exotic nuclei production. The existing accelerators (ALPI, TANDEM) are also shown.

energy 7 MeV) will be employed for this purpose. It should be pointed out that this accelerator-based facility, used for the characterization of the irradiation field for BNCT of the explanted liver and skin melanoma, was conceived for 7 MeV deuterons on beryllium. The target is contained in a heavy water moderator in turn enclosed in a graphite structure. The heavy water container is in aluminium. It was demonstrated[4] that in this case the thermal neutron fluence necessary for the treatment of melanoma (i.e. $10^9 cm^{-2} s^{-1}$) can be achieved with a deuteron current of about 0.6 mA. Only the characterization of the irradiation beam and microdosimetry can be performed at the Van De Graaff accelerator of the LNL, because it delivers a maximum current of 3 μA. Phantom dosimetry will be carried out with FriXy gels,[5] whose sensitivity requires the neutron fluence values necessary for BNCT and therefore cannot be irradiated using the Van De Graaff at present. The thermal facility is already available and the epithermal facility at the TAPIRO reactor near Rome (Italy) will be available in a few months. These will be used for setting up the FriXy gel dosimeters.

2. CUTANEOUS MELANOMA IN ITALY

Although no reliable data are available for the whole Italian country, the incidence of cutaneous melanoma in the Veneto region (the region to which Legnaro belongs) is 8–9% of all tumors. The current standard treatment is surgical excision, with a disease-free margin of 1 cm for T1 and T2 tumors, 2 cm for T3 and 3 cm for T4. Sentinel node biopsy is undertaken for each lesion thicker than 1 mm in order to evaluate the risk of metastases. If lymphonode metastasis is confirmed at histology, a formal therapeutic dissection is performed. Melanoma cells may not only spread to regional nodes, but may also cluster and grow along the lymphatic vessels, becoming palpable sub-cutaneous nodules in about 5% of the cases. These nodules are considered "satellites" if they are confined within 2 cm of the primary lesion and "in-transit metastases" if they extend farther. The management of these lesions is a challenging issue and presently the best results are achieved by lipophascectomy for head and neck, trunk and limb-root tumors, while limb lesions are amenable to hyperthermic antiblastic perfusion.[6] The 40% of the patients undergo local recurrence after these well-standardized treatments and do not respond to systemic chemotherapy. Therefore they can be considered for BNCT.

3. THE ACCELERATOR-BASED SOURCE AT LNL

The $^{9}Be(p,n)^{9}B$ reaction at 5 MeV will be considered, because it has the highest neutron yield at this proton energy.[7] Other proton energies will be investigated together with the use of deuteron beams, in research aimed at optimizing the moderator dimensions versus the projectile energy and current.

The design of the target cooling system is quite critical, because a power density of about $5 kW cm^{-2}$ (produced by 5 MeV protons at a current of 3 mA with a beam radius of 1 cm) has to be removed from a site not easy to access (the target is contained in a pipe inserted in the D_2O moderator). Moreover, the structures of this system should not affect the characteristics of the therapy beam. Preliminary studies, based on a rotating target and on a primary coolant loop of heavy water are in progress and an experimental prototype is foreseen.

The thermal neutron fluence required for the treatment is $10^{13}\,cm^{-2}$ and the maximum irradiation time is 1–2 hours, based on the melanoma BNCT trials performed in Japan.[8] The cumulative physical dose due to prompt gamma rays and fast neutrons should be kept below 2 Gy for the whole treatment, which compares with about 20 Gy (physical dose) delivered to the tumour by the $^{10}B(n,\alpha)^{7}Li$ reaction products. The maximum depth of the lesion which will be considered for treatment is 3 cm while the size of irradiation field will be $10 \times 10\,cm^2$.

4. THE TAPIRO REACTOR

A brief description of the TAPIRO reactor facility in Rome, Italy, will be given here, because it will be utilized for the research on the FriXy gel dosimeters and on microdosimetry. It should be pointed out that the utilization of this reactor for BNCT of brain tumors[9] is the subject of another project.

The TAPIRO reactor is a highly enriched ^{235}U fast neutron facility. Since 1971, it has been used for fast reactor shielding experiments, blanket studies, biological effects of fast neutrons, etc. The nominal power is 5 kW (thermal) and the maximum neutron fluence rate (at the core centre) is $4 \times 10^{12}\,cm^{-2}\,s^{-1}$. The reactor has a cylindrical core (diameter 12.58 cm, height 10.87 cm) made of an alloy of uranium (98.5% weight fraction) and molybdenum (1.5%), totally enveloped by a copper reflector. 22.1 kg of uranium are contained in the core. The reactor is shielded by borate concrete 170 cm thick.

The epithermal column for BNCT of brain tumors is under construction. A schematic view of the epithermal column is shown in Figure 2. A sector of the copper reflector was removed and substituted with alumina. The epithermal spectrum shifter is in contact with the alumina sector of the reflector and consists of alternate layers of

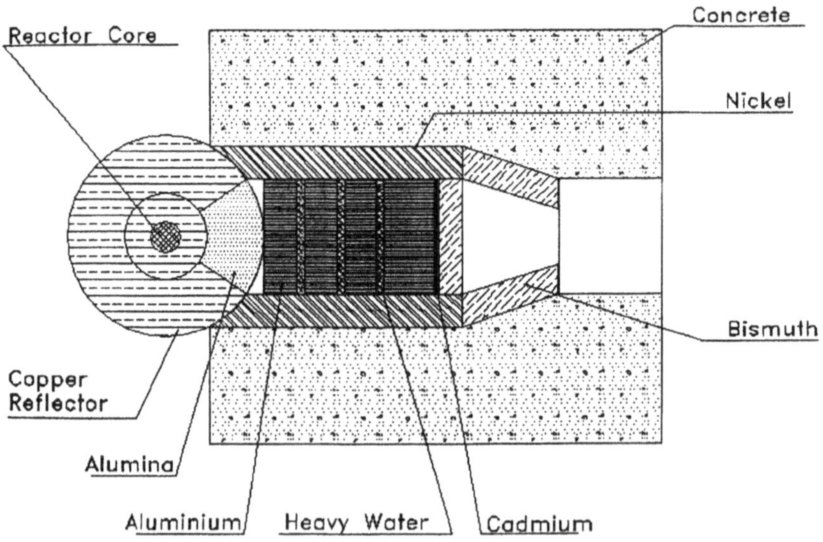

Figure 2. Schematic view of the epithermal facility for BNCT of brain tumor under construction at the TAPIRO reactor near Rome, Italy.

aluminium and heavy water, a cadmium foil and a bismuth gamma shield. The entire column is surrounded by a nickel reflector. The final collimation is provided by a bismuth cone shielded by concrete. An epithermal fluence rate of $6 \times 10^8 \, cm^{-2} s^{-1}$ can be achieved with this structure. A thermal column for the characterization of dosimeters for BNCT of skin melanoma is also available, providing a thermal fluence rate of about $7 \times 10^7 \, cm^{-2} s^{-1}$. A detailed description of this structure, made of reactor-grade graphite and polyethylene, with gamma ray filters in lead, is given in ref. 5.

5. THE EXPERIMENTAL PROGRAM

5.1. Characterization of the Accelerator-Based Source

Monte Carlo simulations will be performed to check if the moderator dimensions of the experimental facility available at LNL have to be modified. For this purpose, the double differential distribution of neutrons produced by a thick beryllium target with 5 MeV protons, given in ref. 10, will be utilized, together with the data measured by a group of the Istituto Nazionale di Fisica Nucleare in Bari.[11] The characterization of the therapy beam will be performed at the Van De Graaff accelerator of the LNL, exploiting the experience gained with the source based on the $^{9}Be(d,n)^{10}B$ reaction.[12] In particular, the thermal fluence uniformity in the irradiation field will be measured with activation foils and BF_3 and ^{3}He detectors. Neutron spectrometry will be performed with ^{4}He recoil detectors, a set of Bonner spheres and superheated bubble detectors, while the gamma ray dose will be measured with TLDs and ionization chambers.

5.2. Microdosimetry

BNCT radiation dosimetry is complex, since it has to take into account the dose due both to the external radiation field and to the reaction products of $^{11}B^*$ (the excited isotope of boron created by the absorption of a thermal neutron by the isotope ^{10}B) and of $^{15}N^*$ (the excited isotope of nitrogen created by thermal neutron absorption by ^{14}N). Living cells experience radiation events of very different LET, ranging from a few tenth of keV/μm (2.2 MeV gamma rays) to about 300 keV/μm (840 keV ^{7}Li ions). Moreover, the neutron energy distribution and consequently the radiation LET spectrum depend on depth in tissue. The LET spectrum can even change with time, since the accelerator-based BNCT beam features can not be assumed constant. Since radiation of different LET has different biological and therapeutic effectiveness, a simple dose monitoring is not enough to optimize the therapeutic treatment. It is necessary to monitor the relative variation of all the different LET components.

The accelerator-based BNCT therapeutic beam will be monitored with a Tissue-Equivalent Proportional Counter (TEPC). TEPC is the instrument of choice to measure microdosimetric spectra,[13] including fluctuations of absorbed energy in gas volumes of tissue-equivalent diameter of 1 μm or less, at a density of $1 \, g cm^{-3}$.

For experimental studies with the experimental facility of the LNL, a modified design of such a counter will be used.[14] The SHONKA A-150 wall and the filling gas mixture will be loaded with ^{10}B in order to simulate the micro-dose spectrum experienced by the cell. Using ^{10}B only in the wall, or only in the gas cavity, or in both will allow the measurement of the energy fluctuations experienced by the biological structure (simulated by the TEPC cavity) when the living cell has been loaded only in the membrane and

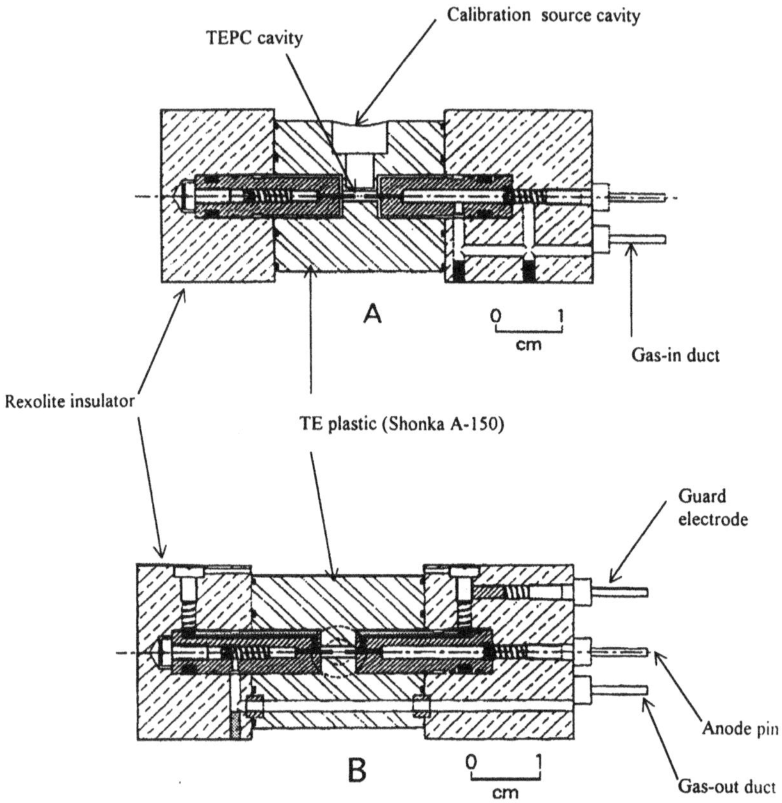

Figure 3. Sectional views of the microdosimeter with a sensitive volume of $1\,mm^3$ constructed at the LNL. Section A: side view. Section B: top view.

cytoplasm, or only in the nucleus or in both. Kota and Maugham[15] showed that it is possible, by changing only the simulated volume in a TEPC, to obtain information about the probability of lethal effect on a cell derived from different boron loadings, inside the nucleus or in the cytoplasm.

The future therapeutic facility will use intense beams to treat a patient. A typical therapeutic beam requires a neutron fluence rate of $10^9\,cm^{-2}\,s^{-1}$, which would give rise to about 10^7 events per second in a counter[16] of usual dimensions (a few cubic centimeters). With such intense beams, the monitoring can be performed only with mini-counters with a $1\,mm^3$ cavity, giving rise to reasonable counting rates of about $10^4\,s^{-1}$.

The horizontal and vertical sections of the LNL mini-TEPC with $1\,mm^3$ cavity are shown in Figure 3. The gas-in duct and the big T-shaped cavity in which a ^{244}Cm α-source can be inserted to calibrate the counter are shown in section A. Just beneath the source cavity, one can see the small cylindrical cavity which is the $1\,mm^3$ sensitive volume of the counter. In section B the electrode connections and the gas-out duct are shown.

This counter was designed for monitoring a high energy proton therapeutic beam. However, it can be constructed with a borated wall 1 mm thick, to monitor a BNCT therapeutic beam and to investigate the beam quality variation a few millimeters behind the skin surface.

5.3. Phantom Dosimetry

The three dimensional measurements of absorbed dose in phantom will be performed by the technique described in ref. 5, utilising tissue-equivalent phantoms acting as continuous dosimeters. The phantom material is FriXy gel, a suitable solution whose main component is the Fricke dosimeter, incorporated into an Agarose gel. The imaging technique is based on a visible light transmittance measurement, in a proper wavelength interval (560–600 nm), performed with a high sensitivity and high resolution CCD camera.

From the comparative analysis of the results obtained irradiating phantoms of different isotopic composition, various contributions to the absorbed dose will be determined. Besides the therapeutic dose due to the thermal neutron reaction with ^{10}B, the dose delivered in tissue without boron will also be measured, discriminating between two contributions: the dose from the reaction $^{14}N(n,p)^{14}C$ and the cumulative dose from $^{1}H(n,\gamma)^{2}H$ and from background.

To verify the feasibility of the described determinations, preliminary measurements have been performed on phantoms exposed in the thermal facility of the TAPIRO reactor. The results are promising, as is discussed in ref. 5 for the depth-dose profiles of the two contributions mentioned above, obtained in a tissue-equivalent gel not loaded with boron.

REFERENCES

1. L.B. Tecchio (on behalf of the SPES Study Group), The radioactive ion beams facility project for the Legnaro laboratories, *in*: "Proceedings of the Eight International Conference on Heavy Ion Accelerator Technology", Argonne, Illinois, 1998.
2. U. Amaldi, and M. Silari eds., "The TERA Project for Oncological Hadrontherapy", INFN-LNF, Frascati, Italy, 1994.
3. S. Agosteo, G. Bodei, P. Colautti, M.G. Corrado, F. d'Errico, S. Monti, M. Silari, and R. Tinti, An accelerator-based thermal neutron source for BNCT, *in*: "Advances in Neutron Capture Therapy, Vol. I—Medicine and Physics", B. Larsson, J. Crawford, and R. Weinreich, eds., Elsevier Science, Amsterdam, The Netherlands, 1997 pp. 483–489.
4. S. Agosteo, M.G. Corrado, P. Colautti, S. Monti, R. Tinti, and G. Tornielli, An accelerator-based neutron source for BNCT of skin melanoma, *in*: "Proceedings of the International Conference on Nuclear Data for Science and Technology", G. Reffo, A. Ventura, and C. Grandi, eds., Editrice Compositori, Bologna, Italy, 1997 pp. 1677–1681.
5. G. Gambarini, P. Marchesi, E. Nava, A. Pecci, and G. Rosi, Three-dimensional measurements of thermal neutron absorbed dose for BNCT experimental program, performed in a fast research reactor, this issue.
6. C.R. Rossi, M. Foletto, and A. Vecchiato, Management of cutaneous melanoma Mo: state of the art and trends, *Eur. J. Cancer*, 33:2302–2312, 1997.
7. M.R. Hawkesworth, Neutron radiography: equipment and methods, *Atomic Energy Review*, 15:169–220, 1977.
8. Y. Mishima, Melanoma and nonmelanoma neutron capture therapy using gene therapy: overview, *in*: "Advances in Neutron Capture Therapy, Vol. I—Medicine and Physics", B. Larsson, J. Crawford, and R. Weinreich, eds., Elsevier Science, Amsterdam, The Netherlands, 1997 pp. 10–25.
9. F. Casali, S. Agosteo, M. Alfinito, K.W. Burn, P. Chirco, E. Nava, G. Rosi, M. Rossi, and M. Zanarini, Design of a facility for boron neutron capture therapy, by MCNP, in a fast reactor, *in*: "Proceedings of the International Conference on Nuclear Data for Science and Technology", G. Reffo, A. Ventura, and C. Grandi, eds., Editrice Compositori, Bologna, Italy, 1997 pp. 1688–1692.
10. T.N. Massey, S.M. Grimes, C.E. Brient, S.I. Al-Quraishi, D.K. Jakobs, R.T. Wheeler, J.E. O'Donnel, J.E. Oldendick, W.B. Howard, J. Yanch, D.L. Smith, and B.J. Micklick, Measurements of stopping target spectra from the $^{9}Be(p,n)$ and the $^{9}Be(d,n)$ reactions, *in*: "Proceedings of the International Conference

on Nuclear Data for Science and Technology", G. Reffo, A. Ventura, and C. Grandi, eds., Editrice Compositori, Bologna, Italy, 1997 pp. 1655–1657.

11. E. Bisceglie, P. Colangelo, N. Colonna, G. Maggipinto, G. Raso, P. Santorelli, and V. Variale, Energy optimization of epithermal neutron beams for accelerator-based BNCT, this issue.
12. S. Agosteo, P. Colautti, M.G. Corrado, F. d'Errico, M. Matzke, S. Monti, M. Silari, and R. Tinti, Characterisation of an accelerator-based neutron source for BNCT of explanted livers, *Radiat. Prot. Dosim.*, 70:559–566, 1997.
13. International Commission on Radiation Units and Measurements. "Microdosimetry—ICRU Report 36," International Commission on Radiation Units and Measurements, Bethesda, Maryland, 1983, pp. 1–118.
14. V. Conte, L. De Nardo, P. Querini, P. Colautti, and G. Tornielli, Microdosimetry for BNCT, *in*: "Boron Neutron Capture Therapy, Proceedings of the Workshop held in Padua 24–25 February 1998", G. Tornielli, and P. Colautti, eds., INFN/TC-98/12 and TERA98/2, 1998, pp. 63–66.
15. C. Kota, and R.L. Maughan, Microdosimetric analysis of absorbed dose in boron neutron capture therapy, *Radiat. Prot. Dosim.*, 70:555–558, 1997.
16. C.S. Wuu, H.I. Amols, P. Kliauga, L.E. Reinstein, and S. Saraf, Microdosimetry for boron neutron capture therapy, *Radiat. Res.*, 130:355–359, 1992.

INVESTIGATION OF $^2H(d,n)^3He$ AND $^3H(d,n)^4He$ FUSION REACTIONS AS ALTERNATIVE NEUTRON SOURCES FOR BNCT

J. M. Verbeke[1,2], J. Vujic[1], and K. N. Leung[2]

[1]Nuclear Engineering Department
University of California, Berkeley
[2]Lawrence Berkeley National Laboratory
Berkeley, CA 94720

ABSTRACT

Beam-shaping assemblies have been designed to moderate high energy neutrons from the fusion reactions $^2H(d,n)^3He$ (DD) and $^3H(d,n)^4He$ (DT) for use in Boron Neutron Capture Therapy. The low yield of DD is an obstacle for the treatment of Glioblastoma Multiforme. Our analysis shows that the high-energy neutrons from DT can be moderated to epithermal energy range without reducing the neutron flux to a negligible level. With the optimal beam-shaping assembly design for DT, a 1 A deuteron beam with energy of 120 keV leads to a treatment time of 50 minutes. The dose near the center of the brain obtained with this configuration is about 50% higher than the dose from a typical spectrum produced by the Brookhaven Medical Research Reactor (BMRR), and is comparable to the dose obtained by other accelerator-produced neutron beams.

1. INTRODUCTION

The main goal in this study is to identify some alternative accelerator-based reactions that could result in a simple and inexpensive accelerator and target system while satisfying all of the requirements for boron neutron capture therapy (BNCT).

The need for epithermal neutron beams with energy distribution peaking around 10 keV[1,2] led different groups to investigate reactor- and accelerator-based neutron sources. Reactions such as $^7Li(p,n)$, $^9Be(p,n)$ and $^9Be(d,n)$ are currently being investigated as accelerator-based neutron sources for BNCT because the energy of neutrons is of the order of hundreds of keV.[2,3,4] The drawback of these reactions is the high energy of the

Frontiers in Neutron Capture Therapy, edited by Hawthorne *et al.*
Kluwer Academic / Plenum Publishers, New York, 2001.

protons and deuterons required, which varies usually between 2.3 and 4.0 MeV depending on the reaction, requiring a large accelerator and treatment facility.

To the best of authors knowledge, the reactions $^2H(d,n)^3He$ (DD) and $^3H(d,n)^4He$ (DT) have not been investigated so far for BNCT, because of the high energy of the neutrons produced, 2.43 and 14.1 MeV respectively. Provided these neutrons can be moderated to therapeutically useful regions without large losses, the high neutron yields for relatively low deuteron beam energy (100–200 keV) is a definite advantage over other accelerator-based neutron sources. A 120 keV deuteron beam yields approximately 3.7×10^{11} n/sec/A and 10^{14} n/sec/A for DD and DT respectively.[5] This paper investigates the feasibility of generating neutrons for BNCT using the DD and DT reactions. It focuses on two points: designing a moderator to decrease neutron energies to epithermal regions and determining the dose distribution in a head phantom. Monte-Carlo codes are used for both tasks.

2. NEUTRON BEAM-SHAPING ASSEMBLY (BSA)

In order to obtain a therapeutically useful epithermal neutron beam, the BSA needs to be optimized. The neutron transport through the BSA is simulated using the Monte-Carlo code MCNP.[6] After moderation, a second Monte-Carlo code BNCT_RTPE[7] is used to calculate the dose distribution in a head phantom.

The neutron source is characterized as follows. Monoenergetic neutrons are emitted isotropically and uniformly from a 5 cm-diameter spotsize on a 40 cm-diameter target. Source neutrons enter a cylindrical BSA with a monoenergetic energy distribution.

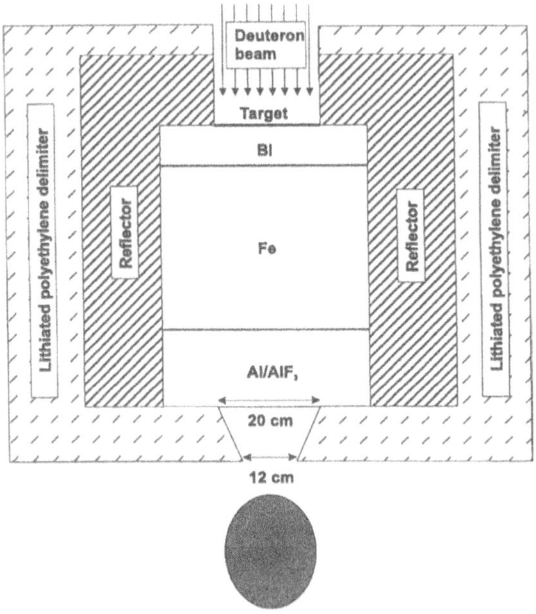

Figure 1. Beam-shaping assembly (BSA).

They travel through the cylinder composed of several layers of different materials until they reach the other side where the patient is located.

An accepted figure-of-merit used to measure the neutron beam quality is the dose equivalent to the tumor at the centerline of the brain, and it is based on biological criteria. The tumor dose at a depth of 8 cm is used as a figure-of-merit in our study. This dose is limited by Brookhaven National Laboratory's (BNL) clinical trial protocol[8] which suggests that the local equivalent dose to the healthy tissues must not exceed 12.5 Gy. The treatment time will thus be limited by this value.

For the dose computation with BNCT_RTPE,[7] concentrations of 13 μg/g of ^{10}B in healthy tissues and 45.5 μg/g of ^{10}B in tumor are used. The individual dose components (boron dose, gamma dose, nitro-dose and fast dose) are combined to yield estimates of total doses per neutron to healthy tissue and tumor, using the following equation:

$$D_{total} = CF * D_B + RBE_N * D_N + RBE_r * D_r + RBE_\gamma * D_\gamma$$

where the following assumptions are made: the normal tissue and tumor compound factors CF are 1.3 and 3.8, respectively; proton-recoil reactions RBE_r: 3.2; nitrogen capture reaction RBE_N: 3.2; RBE_γ: 1.0. The ^{10}B concentrations, compound factors and RBE factors were taken from values used in BNL's protocol.[8]

2.1. Optimal BSA and Dose Distribution for DD

The optimal combination of materials that we came up with for DD was 30 cm of ^7LiF and 18 cm of 40%Al/60%AlF_3. The BSA is 25 cm in diameter and surrounded by a thin (0.5 mm) layer of ^6LiF and a thick 30 cm Al_2O_3 reflector. A detailed description of the geometry can be found in Ref. 9.

The neutron spectrum after moderation and the equivalent doses to the healthy tissues (with its components) are shown in Figure 2. The equivalent tumor dose to the desired depth of 8 cm with this moderation is 18 Gy-Eq. The equivalent skin dose is 8.9 Gy-Eq. Accounting for the neutron yield of DD—which is 8.5×10^9 n/sec/mA for a 400 keV beam[10]—this moderation would lead to a treatment time of 840 hours for a 5 mA deuteron beam. Even with a 1 A deuteron beam, the treatment time would be more than 4 hours. This unacceptably large treatment time could potentially be reduced by using several beams, by increasing the beam intensity, or by improving the beam shaping assembly.

2.2. Optimal BSA and Dose Distribution for DT

A few simulations for different combinations of materials showed that all neutrons above 1 MeV contribute not only to the fast dose at the surface of the brain, but also to the fast dose throughout the head, including the point of maximum tissue dose which limits the treatment time. The BSA optimization emphasis was thus on decreasing the high energy neutron flux to a level as low as possible.

The best BSA design we came up with was composed of thick layers of bismuth, iron, 40%Al/60%AlF_3, and thin layers of ^6LiF and lead. The BSA is surrounded by a 20-cm thick lead reflector. The large cross-sections of bismuth and lead for (n,2n) reactions are used beneficially to compensate for the neutron losses by leakage and absorption during the moderation. Iron is used to reduce the fast neutron energies from 14 MeV to

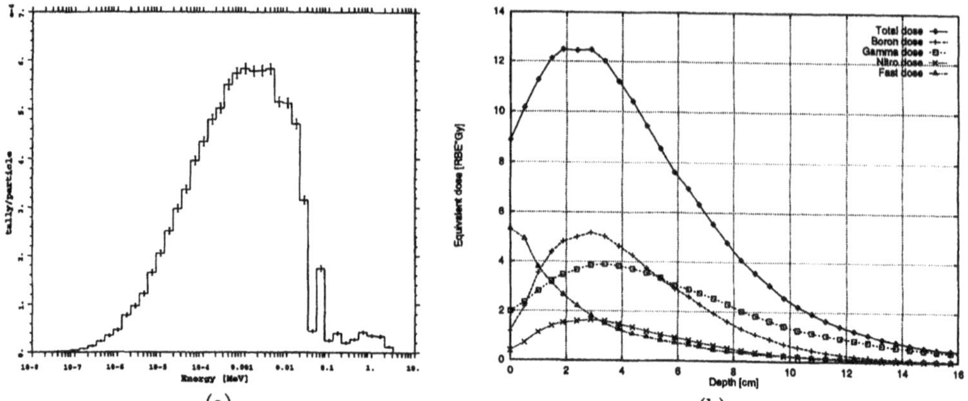

(a) (b)

Figure 2. (a) Neutron energy distribution after moderation and (b) total dose to the healthy tissues with the different dose components in the head phantom.

around 1 MeV. The mixture 40%Al/60%AlF_3 is interesting in the sense that the elastic scattering resonances of aluminum supplement exactly the ones of fluorine from 27 keV up to the high energy tail, except for a narrow energy range around 70 keV. This resonance structure will preferentially reduce the number of neutrons above 27 keV. Lithium fluoride and lead are finally used to reduce the thermal neutron flux and to decrease the photon flux respectively at the end of the moderator.

The neutron spectrum after moderation and the components of the doses to the healthy tissues are shown in Figure 3. With this moderation, the equivalent tumor dose in the head phantom at 8 cm depth is 21.9 Gy-Eq and the equivalent dose to the skin is 9.6 Gy-Eq. With a 120 keV deuteron beam, a 1 A deuteron beam intensity would lead to a treatment time of 50 minutes.

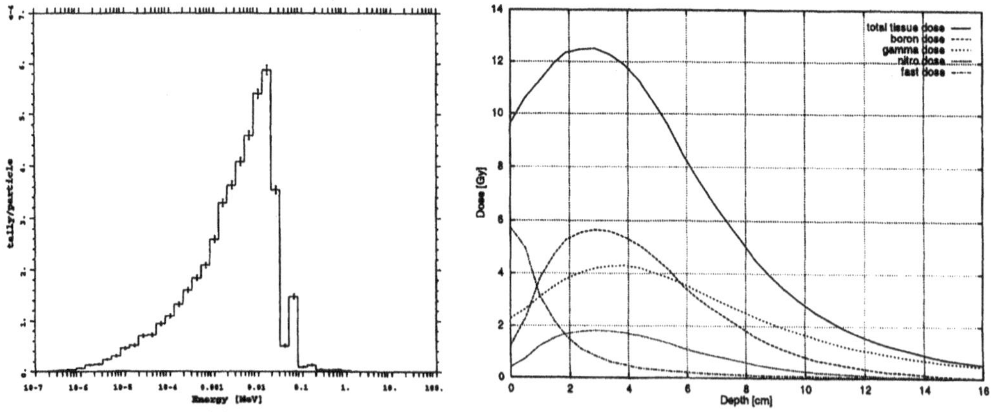

(a) (b)

Figure 3. (a) Neutron energy distribution after moderation and (b) total dose to the healthy tissues with the different dose components in the head phantom.

Table 1. Comparison of treatment characteristics for different neutron sources

Neutron source	$^{3}H(d,n)^{4}He$	$^{2}H(d,n)^{3}He$	$^{7}Li(p,n)^{7}Be$	$^{7}Li(p,n)^{7}Be$	BMRR
Moderator	Bi/Fe Al/AlF_3	^{7}LiF Al/AlF_3	^{7}LiF (Ref. 2)	Al/AlF_3 (Ref. 2)	(Al_2O_3) (Ref. 11)
Proton energy [Mev]	0.12	0.4	2.3	2.4	—
Proton current [mA]	1,000	5,000	20	20	(3 MW)
Treatment time [min]	50	50	40	54	39
Equ. Tumor dose (max) [Gy-Eq]	64.8	60.1	64.3	65.1	61.6
Equ. Tumor dose (5 cm) [Gy-Eq]	51.8	42.9	50.5	51.4	38.6
Equ. Tumor dose (8 cm) [Gy-Eq]	21.9	18.0	21.4	22.3	14.5
Advantage depth [cm]	9.5	9.0	9.5	9.5	8.4

2.3. Summary and Comparison with Other Neutron Sources

Table 1 summarizes the main characteristics of the tumor dose distribution for DD, DT and other neutron sources proposed or used for BNCT. The optimized spectrum from the DT neutron source produces about 50% higher dose near the center of the brain than the currently used reactor spectrum at BMRR.

3. CONCLUSIONS

Two fusion reactions (DD and DT) have been studied as possible neutron sources for BNCT. The low neutron yield of the DD reaction is an obstacle for its use in this particular therapy. Neutrons from the DT reaction could be moderated to around 10 keV without reducing the neutron flux to a negligible level. An equivalent tumor dose of 21.9 Gy-Eq could be achieved at a depth of 8 cm in the head phantom, which is about 50% higher than doses obtained by a neutron beam currently used at BMRR for clinical trials. For a treatment time of 50 minutes, the deuteron beam requirements are 120 keV and 1 A. Taking advantage of the small sizes of the ion sources and accelerators, several beams could easily be used in parallel and treatment times could be reduced to a fraction of an hour.

ACKNOWLEDGMENTS

This work is partly supported by the US Department of Energy under contract number DE-AC03-76SF00098.

REFERENCES

1. J.C. Yanch, X.-L. Zhou, and G.L. Brownell, A Monte Carlo Investigation of the Dosimetric Properties of Monoenergetic Neutron Beams for Boron Neutron Capture Therapy, *Radiation Research*, 126:1–20, 1991.
2. D.L. Bleuel, R.J. Donahue, B.A. Ludewigt, and J. Vujic, Designing Accelerator-Based Epithermal Neutron Beams for BNCT, *Medical Physics*, 25(9):1725–1734, 1998.
3. C.K.C. Wang, T.E. Blue, and R. Gahbauer, A Neutronics Study of an Accelerator-Based Neutron Irradiation Facility for BNCT, *Nuclear Technology*, 84:93–107, 1989.

4. Y.C. Yanch, X.L. Zhou, R.E. Shefer, and R.A. Klinkowstein, Accelerator-Based Epithermal Neutron Beam Design for Neutron Capture Therapy, *Med. Phys.*, 19:709–721, 1992.
5. L.A. Shope, Theoretical thick target yields for the D-D, D-T, and T-D nuclear reactions using the metal occluders Ti and Er and energies up to 300 keV, SC-TM-66-247, Sandia Laboratory, Albuquerque, July 1966.
6. "MCNP-A A General Monte Carlo N-Particle Transport Code, Version 4B", J.F. Briesmeister, Ed., LA-12625-M, Los Alamos National Laboratory, 1997.
7. D. Nigg, Methods for Radiation Dose Analysis and Treatment Planning in BNCT, *Int. J. of Rad. Onc. Bio. Phy.*, 28:1121–1134, 1994.
8. A.D. Chanana, Boron Neutron Capture Therapy of Glioblastoma Multiforme at the Brookhaven Medical Research Reactor, A Phase I/II Study, FDA IND #43317, Protocol #4, Medical Department, Brookhaveen National Laboratory, Upton, NY 11973, January 1996.
9. J.M. Verbeke, S.V. Costes, D. Bleuel, J. Vujic, and K.N. Leung, Designing an Epithermal Neutron Beam for BNCT for the Fusion Reactions $^{2}H(d,n)^{3}He$ and $^{3}H(d,n)^{4}He$, proceedings of the International Conference on the Physics of Nuclear Science and Technology, Long Island, NY, October 5–8, 1998, pp. 1600–1607.
10. High Voltage Engineering Corporation, Burlington, Massachusetts.
11. J.A. Coderre, R. Bergland, J. Capala, M. Chadha, A.J. Chanana, E. Elowitz, D.D. Joel, H.B. Liu, and D. Slatkin, Boron Neutron Capture Therapy for glioblastoma multiforme using p-boronophenylalanine and epithermal neutrons—Trial design and early clinical results, *J. Neuro-Oncol.*, 33:93–104, 1997.

SPECTRAL CHARACTERIZATION OF ACCELERATOR-BASED BEAMS FOR BNCT USING NEUTRON ACTIVATION FOILS

L. G. Murphy[1], J. C. Yanch[1], and D. W. Nigg[2]

[1]Department of Nuclear Engineering
Massachusetts Institute of Technology
Cambridge, Massachusetts
[2]Idaho Engineering and Environmental Laboratory
Idaho Falls, Idaho

1. INTRODUCTION

Spectral characterization measurements at the Massachusetts Institute of Technology Laboratory for Accelerator Beam (MIT LABA) facility were conducted using an activation foil spectrometry method designed by the Idaho National Engineering and Environmental Laboratory (INEEL). This method has been previously used to characterize the reactor-based BNCT beams at Brookhaven National Laboratory[1] and the Technical Research Center of Finland[2] and the proton-cyclotron-based fast neutron radiotherapy facility at the University of Washington School of Medicine.[3] MIT LABA is for the first time applying INEEL's activation foil spectrometry method to accelerator-based BNCT/BNCS beams. The INEEL method combines both experimental measurements and computer simulations to calculate the neutron spectrum of an epithermal beam over various energy regions. Several foil materials are selected with primary resonance peaks that cover the epithermal range of interest (0.5 eV to 10 keV). The foils are activated by the neutron beam, and their induced activities are measured using standard gamma spectroscopy methods and converted to reaction rates. The Monte Carlo N-Particle Transport Code[4] (MCNP) is used to take into account the neutron flux perturbation through each foil material, determining the effective cross section of each foil under each energy region. This allows the use of "thick" foils rather than traditional "thin" foils.[5] Finally, two different spectral unfolding techniques are used to determine the energy-dependent neutron flux from the measured reaction rates of the foils.[2,5,6]

Frontiers in Neutron Capture Therapy, edited by Hawthorne *et al.*
Kluwer Academic / Plenum Publishers, New York, 2001.

2. MIT LABA FACILITY DESCRIPTION

MIT LABA has a tandem electrostatic accelerator capable of accelerating protons, deuterons, and alphas over a wide range of energies into five beam ports.[7] One of these beamlines is dedicated as a BNCT beam.[8] The BNCT beam currently has a beryllium target capable of utilizing Be^9(d,n) or Be^9(p,n) reactions, a lead reflector, and heavy water moderator assembly.

3. MEASUREMENT METHODS AND MATERIALS

The following sections describe the INEEL activation foil spectrometry method[1,2] in detail as it is applied to LABA's accelerator-based BNCT beam. Neutrons emitted by the ^{9}Be(d,n) reaction at a deuteron energy of 1.5 MeV were moderated by a 27 cm × 24 cm diameter D_2O cylinder surrounded by 18 cm of lead. The dosimetric performance of this neutron beam in phantom has been previously reported.[9] While this beam is inferior in advantage depth to beams based on both the ^{7}Li(p,n) and the ^{9}Be(p,n) reactions, it appears to be clinically useable and has the advantage that it can be accessed by low energy accelerators.

3.1. Activation Foil Measurements

The foil materials selected to characterize the neutron beam are indium, gold, copper, manganese, and tungsten. The interaction characteristics for each foil material are listed in Table 1. In previous applications of the INEEL characterization method, cobalt foils have also been used.[1,2] After analysis of this material, cobalt was found unsuitable for use in characterizing LABA's accelerator-based beam because of its relatively long half-life of 5.3 years and relatively small cross section (850 barns) under the primary resonance peak.

All foils are 0.5″ in diameter and range in thickness from one to five mils (0.0254 mm to 0.127 mm). For both the epithermal and fast neutron measurements, the foils were irradiated in stacks of five surrounded by a cadmium cover to suppress the foils' response to thermal neutrons. Each foil stack contains five foils of the same material. For labeling purposes, the foils in each stack are numbered one through five, with Foil 1 being the closest to the incident neutron beam and Foil 5 being the furthest. Prior to the

Table 1. Activation foil interactions used in LABA's measurements

Foil material and interaction	Energy of primary response	Cross section (barns)	Induced activity	Gamma decay energy of interest	Half-life of gamma decay
^{115}In(n,γ)	1.46 eV	29,000	^{116}In^m	1,294 keV 1,097 keV 417 keV	54.2 min.
^{197}Au(n,γ)	5 eV	27,400	^{198}Au	412 keV	64.7 hr.
^{186}W(n,γ)	18 eV	15,400	^{187}W	686 keV	23.9 hr.
^{55}Mn(n,γ)	340 eV	45.7	^{56}Mn	847 keV	2.6 hr.
^{63}Cu(n,γ)	580 eV	412	^{64}Cu	511 keV (annihilation)	12.7 hr.
^{115}In(n,n′)	339 keV (threshold)	N/A	^{115}In^m	336 keV	4.5 hr.

experiments, the foil thickness needed for each material was determined using the MCNP simulation, so that the response of foil 3 is linearly independent of foil 1, allowing both measurements to be used in the same spectral unfolding matrix. These calculations relied on the use of "thick" foils so that the neutron flux would be significantly depleted under the resonance peak by the time it reaches Foil 3.

The thermal and epithermal neutron measurements were conducted using the resonance absorption reactions of the indium, gold, copper, manganese, and gold foils. In addition, a stack of uncovered gold foils was used to derive the thermal neutron flux. These six foil stacks were placed inside a "foil wheel" designed specifically by INEEL for simultaneous irradiation of foils. The foil wheel is made from Teflon and can hold seven foil stacks.

Fast neutron measurements used inelastic scatter (threshold) reactions in indium foils and resonance reactions in indium and copper foils. Cadmium-covered stacks of indium and copper foils were individually irradiated inside a hollow boron sphere (ID = 2.5 cm, OD = 5 cm). The boron sphere, provided by INEEL, is close to 100% ^{10}B and is designed to suppress the thermal and epithermal neutron flux, such that the foils interact primarily with fast neutrons. At LABA, the In(n,n′), In(n,γ), and Cu(n,γ) reactions were examined in this configuration. The foil wheel and boron sphere were supported by a Teflon holder designed specifically for use in front of the BNCT beam at LABA. Both assemblies were held in position by a standard tripod, which ensured reproducible positioning.

3.2. Reaction Rate Measurements

After irradiation, the induced activities of the foils were counted using standard HP(Ge) detectors and a multi-channel analyzer. Only Foils 1 and 3 in each packet in the foil wheel were individually counted, while all 5 foils in each packet irradiated in the boron sphere were counted together and taken as one datum point. The gamma peaks corresponding to the activation reactions of interest were analyzed to determine the volume-averaged reaction rate per atom for each foil, using Equation 1, based on the net counts (C – B), decay constant (λ), detector efficiency (ε), branching ratio (b_γ), irradiation time (t_i), count time (t_c), wait time (t_w), number density (N), and the foil volume (V).

$$R = \frac{(C-B)\lambda}{\varepsilon b_\gamma (1-e^{-\lambda t_i})(1-\lambda t_c)e^{-\lambda t_w} NV}\left(\frac{reactions}{atom \cdot \sec}\right) \qquad (1)$$

3.3. MCNP Model and Simulations

MCNP4B[4] was then used to accurately model the experimental layout in front of the BNCT beam. The source input was placed at the location of the beryllium target in the model and was taken from the experimentally measured 1.5 MeV Be(d.n) neutron spectrum from Guzek.[10] Two MCNP simulations were run for each experimental setup (foil wheel, indium foils in boron sphere, and copper foils in boron sphere). One run tallied the number of (n,γ) or (n,n′) reactions in each foil within each energy region. The other run tallied the unperturbed neutron flux in the location of each foil, with the foil packet and foil wheel or sphere assembly materials replaced with air. Dividing the tally results of both runs gives the effective microscopic cross-section for each foil within each energy region.

4. RESULTS

The spectral characterization experiments using the 1.5 MeV $^{9}Be(d,n)$ reaction were conducted between May and August 1998. These experiments included three foil wheel irradiation runs at 75–115 μA for 60–75 minutes. There were uncertainties in the current-on-target measurements for the beam, a common problem for accelerator-based beams. These uncertainties affect only the magnitude of the neutron flux and yield and not shape of the spectrum. As a result, all measured reaction rates were normalized to the measured accelerator beam power of one irradiation run (115.3 μA for 62.5 minutes) in order to obtain comparable data. This artificial normalization can be replaced with an absolute normalization when the current uncertainties are resolved.

The experimentally measured reaction rates and computer-generated effective cross sections were directly unfolded to find the flux in each energy region. Two separate direct unfolding techniques were used: full range and independent point.[2] Each method used the relationship in Equation 2 to solve for the flux in each energy region. The Student Version of MATLAB Version 4[11] was used for all matrix manipulations, based on the method described in References 3 and 6.

$$\begin{bmatrix} a_{11} & a_{12} & \cdots & a_{1NG} \\ a_{21} & a_{22} & \cdots & a_{2NG} \\ \vdots & \vdots & \cdots & \vdots \\ \vdots & \vdots & \cdots & \vdots \\ a_{NF1} & a_{NF2} & \cdots & a_{NF\,NG} \end{bmatrix} \begin{bmatrix} F_1 \\ F_2 \\ \vdots \\ F_{NG} \end{bmatrix} = \begin{bmatrix} R_1 \\ R_2 \\ \vdots \\ \vdots \\ R_{NF} \end{bmatrix} \tag{2}$$

where NF is the number of reactions, NG is the number of energy groups, a_{ij} is the MCNP-calculated effective cross section for reaction i in energy group j, R_i is the measured volume-averaged reaction rate per atom for reaction i, and Fj is the unfolded flux in energy group j. Figure 1a shows the directly unfolded unperturbed neutron spectrum found from the full range method compared to the MCNP-generated neutron spectrum. This method used the experimental data from all nine measured reactions (of the foil closest to the beam, Foil 1) to solve for the flux in six energy regions over the full energy range. Figure 1b shows the neutron spectrum found from the independent point method. This method used the measured reaction rates to separately solve for the flux only under the primary energies of each specific reaction. Using Equation 2, one system of equations was solved to obtain each point (line) using the experimental data from Foils 1 and 3 (front and center foils in the foil stack). The copper data from the foil wheel were not used because its first and third foil responses were not sufficiently independent. The in-air yield of the BNCT beam was found by integrating under the directly unfolded neutron spectrum in Figure 1a. The total neutron flux was calculated to be 1.07×10^{7} (±4.2%) n/cm^{2}sec at 115.3 μA (measured current).

5. DISCUSSION AND CONCLUSION

MIT LABA found that INEEL's activation foil spectrometry method could be effectively applied to an accelerator-based BNCT beam. In addition, the results presented in Figure 1 show very good agreement between the experimentally measured and MCNP

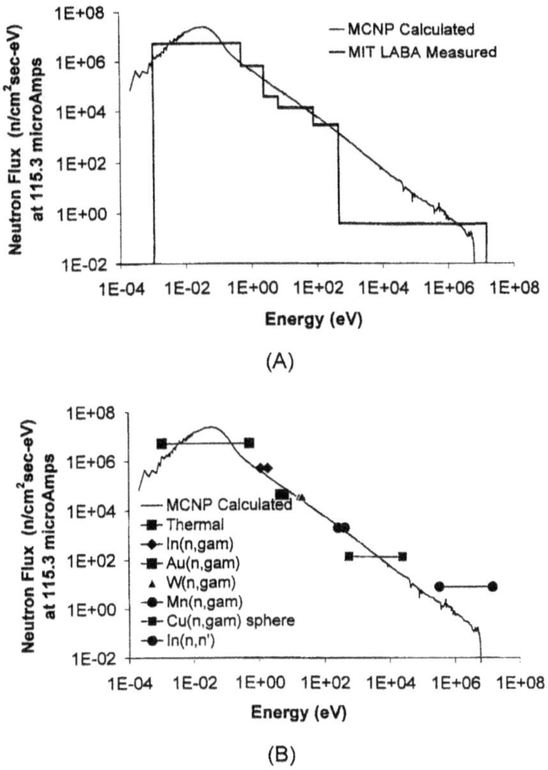

Figure 1. Directly unfolded neutron spectra for the MIT LABA BNCT beam using the 1.5 MeV Be(d,n) reaction at 115.3 microAmps (measured). The measured spectral data were unfolded by the full range (A) and independent point (B) methods. The magnitude of the neutron spectra is affected by the current measurement uncertainties. When the absolute-to-measured current ratio is determined, multiplying the spectra by this factor will give the absolute magnitude of the neutron flux.

calculated neutron spectra. The only discrepancy in the presented data is the fast neutron region of Figure 1b. This discrepancy and the accelerator beam current uncertainties are under investigation. Even though the absolute magnitudes of the neutron flux and yield can not yet be determined, they can be easily calculated once the accelerator current uncertainties are resolved. Spectral characterization of LABA's BNCS beam using this method is planned for the immediate future.

ACKNOWLEDGMENTS

This work was supported by the Idaho National Engineering and Environmental Laboratory University Research Consortium Program.

REFERENCES

1. Y.D. Harker, R.A. Anderl, G.K. Becker, and L.G. Miller, "Spectral Characterization of the Epithermal Neutron Beam at Brookhaven Medical Research Reactor," *Nucl. Sci. Eng.*, 110:355–368, 1992.

2. D.W. Nigg, Y.D. Harker, J.K. Hartwell, C.A. Temple, T. Seppala, T. Seren, K. Kaita, and I. Auterinen. "Collaborative Spectral characterization of the Finnish Epithermal-Neutron Beam Facility for BNCT," in: *INEEL BNCT Research Program Annual Report—1996*, INEEL/EXT-97-00319, 1997.
3. D.W. Nigg, C.A. Wemple, J.K. Hartwell, Y.D. Harker, J.R. Venhuizen, and R. Risler. "Validation of Neutron Production and Beam Line Modeling for the University of Washington Fast-Neutron Facility", in: *INEEL BNCT Research Program Annual Report—1997*, INEEL/EXT-98-00327, 1998.
4. MCNP—A General Monte Carlo N-Particle Transport Code, Version 4B," LA-12625-M, J.F. Briesmeister, Ed., Los Alamos National Lab, 1997.
5. G.F. Knoll, "Radiation Detection and Measurement," 2nd Ed., John Wiley & sons, New York, 1989, pp. 703–711.
6. D.W. Nigg, and Y.D. Harker, "Method Used for Direct Unfolding of Neutron Spectra from Activation Data," in: *INEEL BNCT Research Program Annual Report—1997*, INEEL/EXT-98-00327, 1998.
7. R.E. Shefer, R.E. Klinkowstein, J.C. Yanch, and W.B. Howard, "Tandem Electrostatic Accelerators for BNCT," *Proceedings of the First International Workshop for Accelerator-Based Sources of BNCT*, INEL CONF-940976, 86–97, 1994.
8. J.C. Yanch, X.-L. Zhou, R.E. Shefer, and R.E. Klinkowstein. "Accelerator-based Epithermal Neutron Beam Design for Neutron Capture therapy," *Med. Phys.* 19(3):709–721, 1992.
9. S.M. White, R.E. Klinkowstein, and J.C. Yanch, "Accelerator-Based BNCT using the Be(d,n) Nuclear Reaction," Transactions of the American Nuclear Society, TANSAO, Vol. 77:50–51, 1997.
10. J. Guzek, Ph.D. Thesis, University of the Witwatersrand, Johannesburg, South Africa, in preparation.
11. MathWorks, Inc. "MATLAB, High-Performance Numeric Computation and Visualization Software: Reference guide," MathWorks, Natick, Massachusetts, 1992.

PRELIMINARY STUDY ON ACCELERATOR-BASED NEUTRON FIELDS FOR NCT

Au-Activity Distributions in a Water Phantom with Two Types of Accelerators

Kenichi Tanaka[1], Yoshinori Sakurai[2], Tooru Kobayashi[2], Yoshinobu Nakagawa[3], Satoru Endo[4], and Masaharu Hoshi[4]

[1]Department of Nuclear Engineering
Kyoto University, Yoshida-Honmachi
Sakyo-ku, Kyoto 606-8501, Japan
[2]Kyoto University Research Reactor Institute
Kumatori-cho, Sennan-gun, Osaka 590-0494, Japan
[3]Department of Neurosurgery
National Kagawa Children's Hospital
Zentsuji-cho, Zentsuji 765-8501, Japan
[4]Research Institute for Radiation Biology and Medicine
Kasumi, Minami-ku, Hiroshima 734-8553, Japan

1. INTRODUCTION

Accelerator-based neutron irradiation systems for neutron capture therapy (NCT) have been studied mainly in the USA and EU since the beginning of the 1980's. Compared to reactor-based neutron irradiation systems, accelerator-based neutron irradiation systems have several advantages, such as the probability of the construction near hospitals, the simplicity for starting and stopping, the flexibility of the irradiation direction, etc. Considering the yield and the energy of produced neutrons, $^{7}Li(p,n)^{7}Be$ (Q-value: −1.644 MeV, threshold: 1.881 MeV) with around 2.5 MeV protons and $^{9}Be(p,n)^{9}B$ (Q-value: −1.852 MeV, threshold: 2.059 MeV) with around 4.0 MeV protons have mainly been investigated as the neutron production reactions.[1,2]

As the first step, we performed experiments with 2.5 MeV protons for a ^{7}Li target and 4.2 MeV protons for a ^{9}Be target. The purposes were as follows: (1) To investigate the influence of the kind of target system and moderator thickness on the obtained

Frontiers in Neutron Capture Therapy, edited by Hawthorne *et al.*
Kluwer Academic / Plenum Publishers, New York, 2001.

Table 1. Conditions of the experiments

Nuclear reaction	$^{7}Li(p,n)^{7}Be$	$^{9}Be(p,n)^{9}B$
Proton energy	2.5 MeV	4.2 MeV
Average current	420 μA	20 μA
Beam size and shape	20 mm × 20 mm, square	20 mm × 10 mm, ellipse
Target material	Li metal ($^{7}Li > 99\%$)	Be metal (^{9}Be 100%)
Target size	23 mm^{d} × 150 μm^{t}	60 mm^{d} × 3 mm^{t}
Type of accelerator	Shenkel	Cyclotron
Location	Research Institute for Radiation Biology and Medicine, Hiroshima, Japan	Sumitomo Heavy Industries Examination & Inspection, LTD, Ehime, Japan

neutron flux and its distribution in a water phantom, (2) To make a rough estimation of the proton current necessary for treatments, (3) To get data to check the simulation method, (4) To estimate thermal neutron flux using the change of the self shielding factor dependent on the diameter of Au wire instead of the Cd ratio which is not applicable in a patient's body due to the large thermal neutron absorption and the large gamma-ray production by Cd.

Recently the utilization of near-threshold reactions has been proposed for the ^{7}Li targe.[3] In this case, the produced neutrons are less energetic, which can make the moderators smaller or unnecessary. We intend to investigate the accelerator-based neutron irradiation system using near-threshold reactions with the simulation method checked in this study and other experiments which will be held.

2. EXPERIMENTAL METHOD

Table 1 shows the conditions of the experiments. The experimental geometries with a cylindrical moderator are shown in Figures 1a and b for the Li target and the Be target, respectively. Three kinds of moderators, whose thicknesses were 20, 25, 30 cm, were provided. The moderator casings were made of 2 mm-thick stainless steel, whose inner diameter, namely the moderator diameter, was 20 cm.

Neutron fluxes were measured with the activation analysis using Au wires (0.25 mm^{d} and 0.70 mm^{d}) and Au foils (10 mm^{d} × 0.10 mm^{t}) which were naked and/or

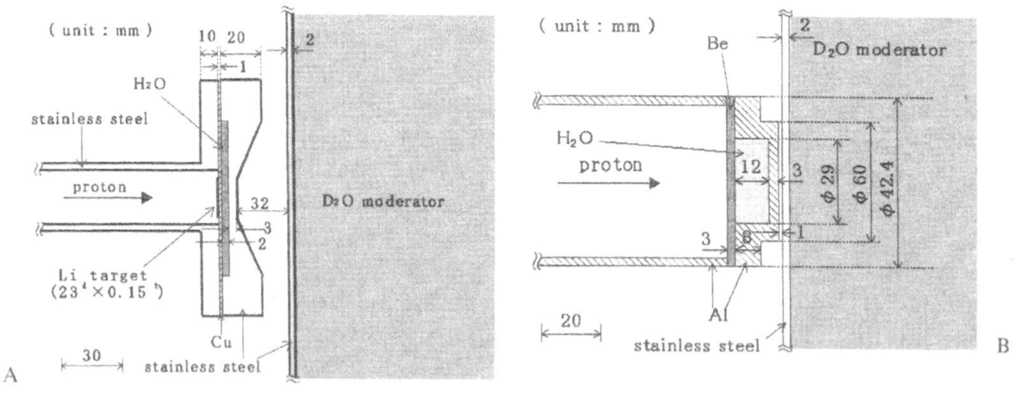

Figure 1. Geometries around the targets.

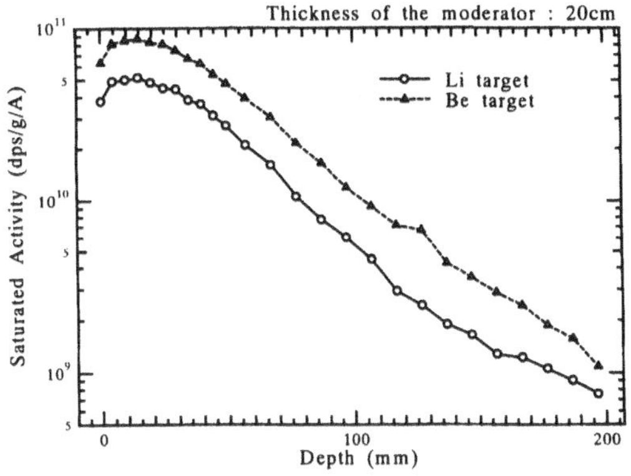

Figure 2. Saturated activity distributions of the 0.25 mm diameter Au on the centerline of the phantom.

Cd-covered. The Cd-covers for the Au foils were 0.70 mm in thickness and 17 mm in diameter. The Cd-covers for the wires were tube-shaped whose inner diameter was 1 mm and outer diameter was 2 mm. In the "free-in-air" experiments, the naked and Cd-covered Au foils were placed on the back surface of the moderator at the distance of 20 mm from the center. In the phantom experiments, the 0.25 mm diameter Au wire was set on the centerline of the phantom and the 0.70 mm diameter Au wire was set on the line 5 mm distant from the centerline. The cylindrical phantom was made of 2 mm-thick acrylic acid resin and filled with natural water. The outer diameter was 20 cm and the length was 20 cm.

3. RESULTS AND DISCUSSIONS

Figure 2 shows the saturated activity distributions of the 0.25 mm diameter Au wires on the centerline of the phantom. The saturated activity distribution for the Be target is larger than that for the Li target. This is consistent with that the Be target with 4.2 MeV protons produces more neutrons than the Li target with 2.5 MeV protons, which we can understand by the cross section of (p,n) reaction and the energy spectra of the

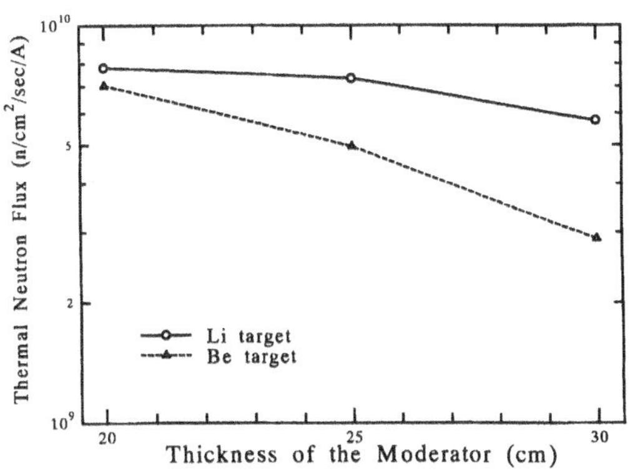

Figure 3. Thickness dependencies of the thermal neutron flux behind the moderators.

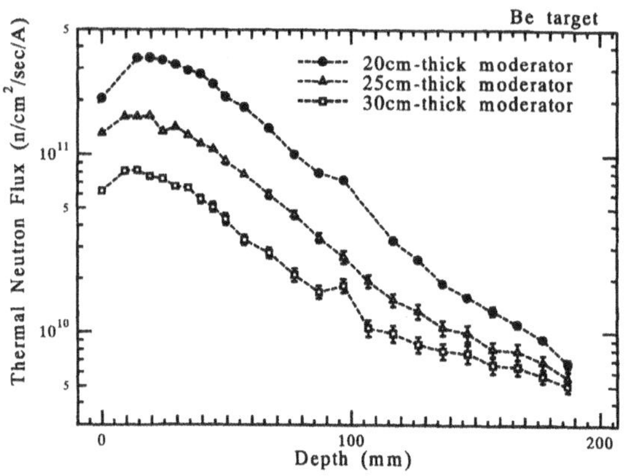

Figure 9. Thermal neutron flux distribution on the centerline of the phantom.

Figure 6 shows the calculational and experimental saturated activity distributions for the Li target. The gentle slopes are seen for the experiment and the calculation with the walls, while they are not seen for the calculations without the walls. From this, the gentle slopes are expected to be derived not from the incidental fast neutrons but from the neutrons scattered by the walls. Here, it is considered that the shield against the neutrons scattered by the walls are valid for reducing the whole body exposure. The distributions calculated with the walls are similar to those calculated without the walls up to the depth of 10 cm in the phantom. So the experimental geometry without the shield against the scattered neutrons are expected to be available for measuring the dose distributions up to the depth of 10 cm. A remarkable dependency of the saturated activity distribution on the moderator thickness is not seen. In the viewpoint of the neutron intensity, the 20 cm-thick moderator is the most hopeful of these three moderators.

Figure 7 shows the ratio of the saturated activity for the 0.70 mm diameter Au wire to that for the 0.25 mm diameter Au wire for the Li target. Most of the Au activation is derived from thermal neutrons and 4.9 eV resonance neutrons. But, dependently on the diameter of Au wire, the self shielding factor for the thermal neutrons changes less than that of the 4.9 eV resonance peak neutrons. Therefore the ratio of the Au wire saturated activity for the 0.70 mm diameter to that for the 0.25 mm diameter for the thermal neutrons is larger than that for resonance neutrons. This may be available for the estimation of thermal neutron flux without the Cd ratio. In Figure 7, the ratios for the three moderators are similar and the distributions are almost constant. It seems that the ratio of the thermal neutrons is too large for the differences of the ratio of the saturated activity to be recognized by this method in this experiment.

Figure 8 shows the distributions of the epi-thermal neutron flux per unit lethargy near 4.9 eV on the centerline of the phantom for the Be target calculated assuming a 1/E spectrum. Figure 9 shows the distributions of the thermal neutron flux on the centerline of the phantom for the Be target. Supposing that ^{10}B density is 30 ppm and treatment time is 30 minutes, the necessary thermal neutron flux to get physical dose of 15 Gy is about 3.5×10^9 n/cm^2/s. Roughly estimating from Figure 9, the necessary proton current is about 10 mA.

REFERENCES

1. C.K. Chris Wang, *et al.*, "Neutronic Study of an Accelerator-based Neutron Irradiation Facility for Boron Neutron Capture Therapy", Nucl. Tech, 84:93–107, 1989.
2. Gang Yue, *et al.*, Measurements of neutron energy spectra from a thick beryllium target with incident proton energies from 3 to 4 MeV, "Advances in Neutron Capture Therapy Volume I, Medicine and Physics", Elsevier, Amsterdam, 1996, pp. 505–510.
3. R.J. Kudchadker, *et al.*, Balancing neutron yield and moderator dimensions in acelerator neutron sources, "Advances in Neutron Capture Therapy Volume I, Medicine and Physics", Elsevier, Amsterdam, 1996, pp. 516–521.
4. D.A. Tattam, *et al.*, Developmental and preliminary neutron fluence measurements in the Birmingham BCT beam, "Advances in Neutron Capture Therapy Volume I, Medicine and Physics©, Elsevier, Amsterdam, 1996, pp. 472–476.

70

EXPERIMENTAL IN-PHANTOM DOSIMETRY OF THE ^{7}Li(p,n) REACTION NEAR THRESHOLD

David P. Gierga[1], Susan M. White[1], Robert E. Klinkowstein[2], Charles L. Lee[1], and Jacquelyn C. Yanch[1]

[1]Department of Nuclear Engineering
Massachusetts Institute of Technology
Cambridge, MA
[2]Newton Scientific, Inc.
Cambridge, MA

1. INTRODUCTION

The ^{7}Li(p,n) nuclear reaction, near threshold, has been proposed by a number of authors[1-4] as a possible neutron source for boron neutron capture therapy (BNCT) applications. The ^{7}Li(p,n)^{7}Be nuclear reaction has a threshold of 1.881 MeV. Although the neutron yield decreases as the proton energy decreases, there may be several advantages in using ^{7}Li(p,n) near threshold neutron sources. Firstly, at proton energies near threshold, the maximum neutron energies that result will be quite low, which can relax beam moderation requirements. Since less moderation is necessary, a smaller fraction of neutrons will be lost from attenuation through the moderator. This may offset the reduced neutron yield at proton energies near the reaction threshold. Secondly, near the reaction threshold, neutrons are emitted in a narrow, forward-directed cone, so efficient use can be made of the neutron yield. Previous work in this area has primarily been theoretical in nature, with the exception of the work of Lee *et al.*[4] The objective of the work described here was to experimentally evaluate the dosimetric feasibility of using unmoderated neutron beams based on the ^{7}Li(p,n) reaction at a number of energies near threshold. These experiments have been performed at the Laboratory for Accelerator Beam Applications (LABA) at the Massachusetts Institute of Technology. In order to examine several proton energies, it is important to be able to easily change the energy of the incident particle and to be able to access a wide range of energies. Since the neutron yield is strongly varying near the reaction threshold, it is important to ensure that the proton energy is well known and remains constant during the experiment. A supplemental objective of these experiments was therefore to examine the capabilities of the LABA accelerator in terms of energy stability.

Frontiers in Neutron Capture Therapy, edited by Hawthorne *et al.*
Kluwer Academic / Plenum Publishers, New York, 2001.

2. METHODS

2.1. Accelerator

The accelerator used in this series of experiments was the 4.1 MeV tandem electrostatic accelerator[5] at MIT LABA. The machine was designed by Newton Scientific, Inc. (NSI) and generates proton, deuteron, or alpha particle beams. The maximum power level for the accelerator is 10 kW. Beam currents of up to 4 mA can be generated from the multi-cusp, high current negative ion source, in combination with magnetic suppression of secondary particles in the accelerating tube. The magnetic suppression leads to very low bremsstrahlung emission rates, so a minimal amount of accelerator shielding is required. The machine is 4.3 m long and weighs approximately 1,000 kg. A switching magnet installed on the accelerator directs the beam to any of five beam ports.

2.2. Lithium Target

The neutron target was prepared by pressing a 1.5 mm thick pure lithium foil on to a stainless steel disk. The target was prepared in an argon environment to minimize oxide and nitride formations on the lithium surface, and the lithium surface was scraped off prior to mounting. The target assembly was then mounted on a gate valve, sealed in argon, placed on the accelerator beamline, and evacuated. An aluminum aperture was placed immediately upstream of the target, which served to limit the diameter of the impinging proton beam to 1.9 cm. The dosimetry experiments were run at very low currents (~6 μA), so an elaborate cooling system was not necessary and the target could be air-cooled.

2.3. Energy Calibration and Stability

In near threshold experiments, since the neutron yield is a strongly varying function of proton energy, it is of great importance to have very accurate knowledge of the proton energy. The proton energy was therefore calibrated by observing the neutron yield as a function of proton energy, and then extrapolating back to zero yield and setting this energy to the known reaction threshold energy. In order to examine energy stability, the accelerating potential (terminal voltage) was continuously monitored with a vibrating reed, fast response, generating volt meter (GVM). The terminal voltage was feedback stabilized during accelerator operation to compensate for voltage variations due to beam current loading and electronic drifts.

2.4. Dosimetry Techniques

The depth dose response in phantom was experimentally determined for three proton energies: 1.89, 1.90, and 1.92 MeV. A brain phantom on loan from the MIT Reactor Group was used in these experiments.[6,9] The phantom is a water-filled, ellipsoidal head phantom with a thin acrylic shell. The dimensions of the major ellipsoidal axes are 13.6 cm, 19.6 cm, and 16.6 cm for the x, y, and z axes, respectively. The dual ionization chamber technique in conjunction with gold foil activation analysis was used to calculate the various dose components as a function of depth in phantom. This method of mixed field dosimetry is discussed in detail by Rogus.[6] One ion chamber is sensitive to both neutrons and photons, while the other is very sensitive to one radiation component, but has a low sensitivity to the other component. Tissue equivalent (Far West IC-18) and graphite ion chambers (Far West IC-18G) were used in this experiment. The tissue equivalent ion chamber provides nearly equal sensitivity to neutrons and photons, while the graphite ion chamber is primarily sensitive to photons. The cadmium difference method is used to provide a measure of the thermal neutron flux. Cadmium covered and bare ^{197}Au foils are irradiated and then counted using a germanium detector. Cadmium has an extremely high capture cross section for thermal energies up to about 0.4eV. The cadmium covered foils, therefore, are activated only by neutrons of energy greater that 0.4eV, while the bare foils are activated by neutrons of all energies. Once the thermal neutron flux is determined by analyzing the cadmium covered and bare foils, the thermal flux is multiplied by appropriate kerma factors[7,8] to determine the thermal neutron and boron dose rates. Boron uptake in the tumor was assumed to be 40ppm, compared to boron uptake in healthy tissue of 11ppm. Physical dose rates are scaled by the following RBEs: 3.8 for the boron dose, 3.2 for the fast and thermal neutron doses, and 1.0 for the photon dose.[10]

3. RESULTS

3.1. Energy Calibration and Stability

Monitoring of the GVM signal indicated that the terminal voltage was regulated to better than 100 volts RMS during beam irradiations. This energy variation leads to about a 1.6% to 0.1% uncertainty in the neutron yield at proton energies of 1.89 and 1.92 MeV, respectively. Energy calibration data are shown in Figure 1.

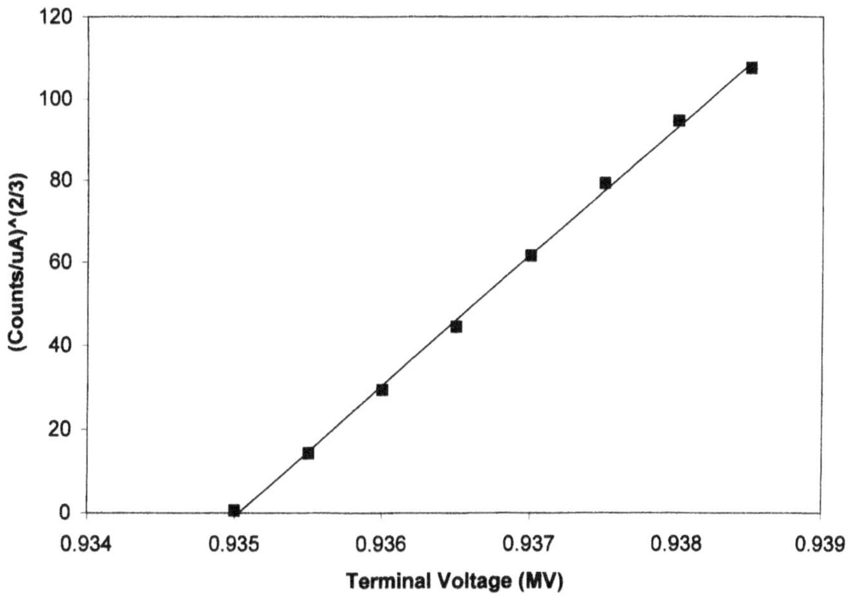

Figure 1. Accelerator energy calibration: determination of terminal voltage corresponding to $^{7}Li(p,n)$ reaction threshold.

A linear curve can be generated by plotting the neutron levels to the 2/3 power, since the neutron output scales as the proton energy to the (3/2) power. The energy for zero neutron output was found from the curve and set to the known reaction threshold energy.

3.2. Depth Dose Results

Depth dose curves are presented for proton energies of 1.89, 1.90, and 1.92 MeV. The tumor dose rate is the RBE-weighted sum of the photon, thermal neutron, fast neutron, and 40 ppm boron dose rate. The healthy tissue dose rate is the RBE-weighted sum of the photon, thermal neutron, fast neutron, and 11 ppm boron dose rate. The error bars reflect the uncertainty in the mixed-field dosimetry technique (15.5% for fast neutrons, 7.8% for photons, 5.9% for thermal neutrons),[11] and the variation in the terminal voltage setting (ranges from 0.1% to 1.6%, depending on proton energy).

4. DISCUSSION AND SUMMARY

The depth dose results lead to several conclusions. The photon doses, while roughly constant with proton energy, become relatively more important as the proton energy approaches the reaction threshold. The primary contribution to the photon dose is from the $^{7}Li(p,p')$ source gamma. No attempt was made in this project to minimize the photon production in the target by making it a sufficient thickness so that the protons only interact in the lithium until their energy falls below the reaction threshold; this will be necessary in any viable near threshold $^{7}Li(p,n)$ neutron source. Although no surface data points are shown on the curves, it is obvious that the advantage depth of these beams (proton energies 10–40 keV above threshold) is essentially zero. The fast neutron component is

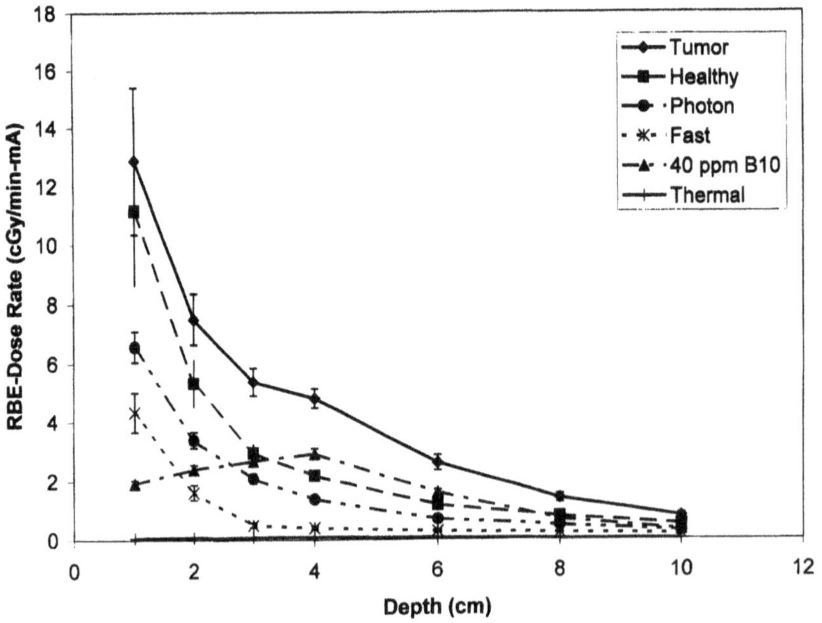

Figure 2. In phantom depth dose results for $^{7}Li(p,n)$, $E_p = 1.89$ MeV.

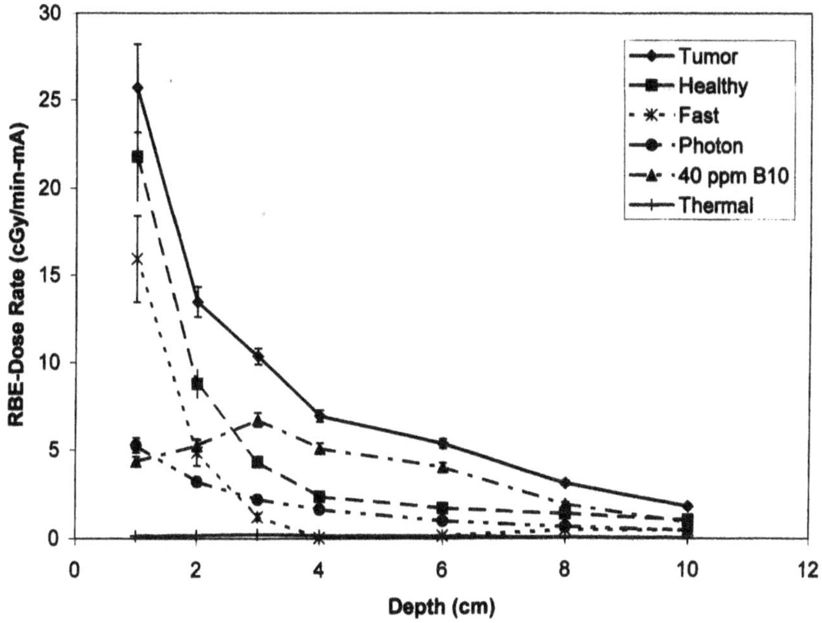

Figure 3. In phantom depth dose results for $^{7}Li(p,n)$, $E_p = 1.90$ MeV.

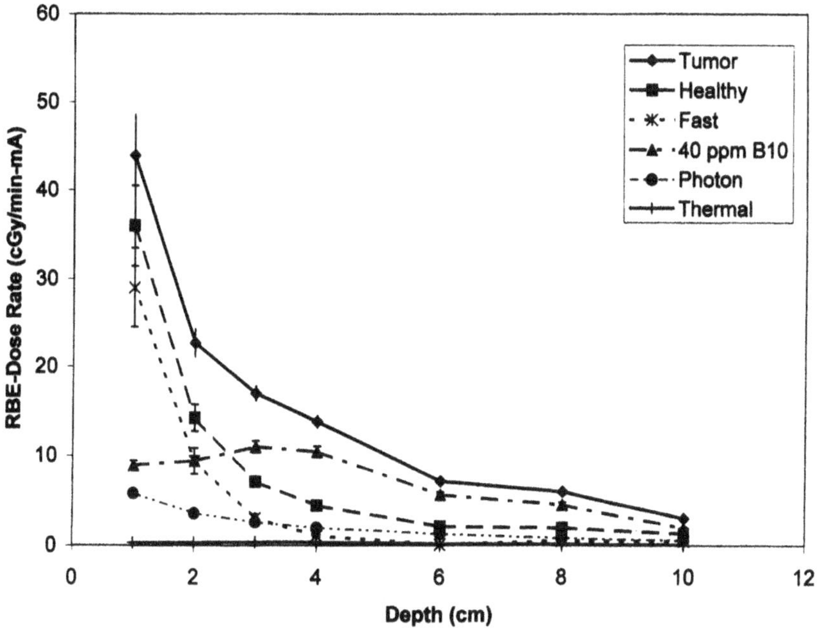

Figure 4. In phantom depth dose results for $^{7}Li(p,n)$, $E_p = 1.92$ MeV.

very high for shallow doses. This implies that unmoderated near threshold ^{7}Li(p,n) neutron sources are not a viable option for BNCT. These conclusions do not apply to moderated near threshold neutron sources.[4] Moderated beam characteristics can be estimated from the unmoderated depth dose curves by assuming the first 4 cm of the phantom, for example, actually are part of a moderator assembly, with the phantom surface starting at 4 cm, and the 5 cm depth point on the curve corresponding to a phantom depth of 1 cm after 4 cm of moderation. Using this methodology for the ^{7}Li(p,n) reaction at 1.92 MeV, the advantage depth is estimated to be roughly 5–6 cm, with an advantage depth total tumor dose rate of about 5 RBE-cGy/min/mA. These values have not been optimized, but demonstrate the potential for moderated near threshold ^{7}Li(p,n) neutron sources, and confirm the findings of Lee[3] that near threshold beams may be of practical importance for shallow tumors.

ACKNOWLEGMENT

This work has been supported by a grant from the U.S. Department of Energy (Award Number DE-FGO2-89ER60874).

REFERENCES

1. V.N. Kononov, V.I. Regushevsky, and N.A. Soloviev, Accelerator-based intense and directed neutron source for BNCT, IPPE-2577, Obninsk, 1996.
2. R.J. Kudchadker, J.F. Kunze, and J.F. Harmon, Accelerator Designs that move close to the target and eliminate much moderator, Proceedings of the 7th International Symposium on Neutron Capture Therapy for Cancer, Zurich, Switzerland, 1996.
3. C.L. Lee, The Design of an Intense Accelerator-Based Epithermal Neutron Beam Prototype for BNCT Using Near-Threshold Reactions, Ph.D. Thesis, Massachusetts Institute of Technology, August 1998.
4. C.L. Lee, X.-L. Zhou, R.J. Kudchadker, F. Harmon, and Y. Harker, Feasibility of a near-threshold BNCT neutron source based on phantom dosimetry, These Proceedings.
5. R.E. Klinkowstein, R.E. Shefer, J.C. Yanch, W.B. Howard, H. Song, E. Binello, B.W. Blackburn, J.L. Daigle, S.M. Sears, C.H. Goldie, and R.J. Ledoux, Operation of a high current tandem electrostatic acclerator for boron neutron capture therapy, Proceedings of the 7th International Symposium on Neutron Capture Therapy for Cancer, Zurich, Switzerland, 1996.
6. R.D. Rogus, O.K. Harling, and J.C. Yanch, Mixed field dosimetry of epithermal neutron beams for boron neutron capture therapy at the MITR-II research reactor, *Med Phys* 21:1611–1625, 1994.
7. R. Caswell, J. Coyne, and M. Randolph, Kerma factors of elements and compounds for neutron energies below 30 MeV, *Int J Appl Radiat Isot* 33:1227–1262, 1982.
8. R. Zamenhof, B. Murray, G. Brownell, G. Wellum, and E. Tolpin, BNCT for treatment of cerebral gliomas. 1: Theoretical evaluation of the efficacy of various neutron beams, *Med Phys* 2:47–60, 1975.
9. O.K. Harling, K.A. Roberts, D.J. Moulin, and R.D. Rogus, Head Phantoms for neutron capture therapy, *Med Phys* 22:579–583, 1995.
10. J.A. Coderre, and G.M. Morris, The radiation biology of boron neutron capture therapy, *Rad Res* 151:1–18, 1999.
11. S.M. White, Beam Characterization for Accelerator-Based Boron Neutron Capture Therapy using the ^{9}Be(d,n) Nuclear Reaction, M.S. thesis, Massachusetts Institute of Technology, 1998.

FEASIBILITY OF A NEAR-THRESHOLD BNCT NEUTRON SOURCE BASED ON PHANTOM DOSIMETRY

C. L. Lee[1,2], R. J. Kudchadker[1], F. Harmon[1], X.-L. Zhou[2], and Y. D. Harker[3]

[1]Department of Physics
Idaho State University
Pocatello, ID 83209
[2]Department of Nuclear Engineering
Massachusetts Institute of Technology
Cambridge, MA 02139
[3]Lockheed Martin Idaho Technologies Company
INEEL, Idaho Falls, ID 83415

Near-threshold boron neutron capture therapy (BNCT) is an accelerator-based concept that produces neutrons from thick lithium targets using proton beam energies only tens of keV above the ^{7}Li(p,n)^{7}Be reaction threshold. Proton energies in this range lead to lower neutron yields than a higher proton energy, such as 2.5 MeV, but lower energy neutrons are produced and hence less moderation is required. This allows thinner moderators that place the patient closer to the neutron source. A summary is presented here of calculations and experiments that have been performed that demonstrate the feasibility of near-threshold neutron sources for BNCT. A model for predicting near-threshold differential neutron yields from thick targets of lithium metal and lithium compounds was developed. Neutron yields from this model were used as neutron sources for Monte Carlo (MCNP) simulations of a head phantom. Calculated dose components were experimentally verified using an acrylic phantom. Initial dose calculations using treatment planning software, which indicate near-threshold neutron sources are competitive with existing reactor BNCT beams, are also presented.

1. INTRODUCTION

Neutron source designers for accelerator boron neutron capture therapy (BNCT) historically have worked to develop beams of the greatest possible penetrability,

Frontiers in Neutron Capture Therapy, edited by Hawthorne *et al.*
Kluwer Academic / Plenum Publishers, New York, 2001.

attempting to match the treatment capabilities of reactors performing BNCT. However, most accelerator BNCT designs require currents of 10 to 50 mA.[1,2] This limits these designs to a small number of large research centers that can afford such technology. A fundamentally different paradigm for accelerator BNCT neutron source development is necessary, if a primary design goal is accessibility by a large range of hospitals. Less penetrating neutron beams are acceptable in this case, provided this restriction is mediated by an increase in portability, a decrease in capital costs, and enhanced flexibility in patient care. Near-threshold BNCT offers this combination of benefits.

One of the unique aspects of BNCT is the requirement of low fast neutron contamination in the neutron beams used for treatments, while still producing a large epithermal neutron flux. This requirement leads to near-threshold BNCT, an accelerator concept based on the special properties of the $^7Li(p,n)^7Be$ reaction.[3,4] In particular, the threshold of this reaction is quite low (1.88 MeV), it produces relatively low energy neutrons (30 keV at threshold), and the thick target yield is very large (4.3×10^{10} neutrons/mC for 1.93 MeV protons and 8.9×10^{11} neutrons/mC for 2.5 MeV protons). Of interest in near-threshold BNCT is the fact that the (p,n) cross section increases from zero to more than 260 mbarns within only 50 keV of the reaction threshold. For this reason, it is possible to get large neutron yields with proton energies only tens of keV above threshold. Thick target yields produced with near threshold protons have the advantage of lower neutron energies, requiring less moderation to reach epithermal energies. In addition, neutrons produced near threshold are more forward directed than those produced at higher proton energies.

A systematic study of the feasibility of near-threshold BNCT has been performed under joint work between Idaho State University (ISU), the Massachusetts Institute of Technology (MIT), and Idaho National Engineering and Environmental Laboratory (INEEL). This work consisted of improved methods for calculating thick target neutron yields near threshold; Monte Carlo simulations of neutron and photon transport for a variety of moderator, reflector, and target backing materials and dimensions; target heat removal designs using fins; and experimental validation within each of these areas. A unified design concept was developed from these results. Incorporation of this design into BNCT treatment planning software indicates that the resulting neutron beam is capable of the treatment of a significant fraction of glioblastoma multiforme (GBM) brain tumors.

2. BEAM DESIGN

Modeling near-threshold BNCT target assemblies requires accurate energy- and angle-dependent thick target neutron yields. It is desirable that the method for calculating these near-threshold yields is also capable of calculating neutron yields for higher proton energies in a self-consistent manner, in order to facilitate comparisons between different proton energies. In addition, the low melting point and high reactivity of lithium may limit its utility as an accelerator target for medical applications. Therefore, modeling of certain lithium compounds is desirable. Previously, no existing computer codes satisfied the requirements spelled out above. A calculational method has been developed and implemented into a computer code that satisfies these criteria.[5,6] This code is used to create neutron source descriptions for use in the Monte Carlo N-particle (MCNP) computer code.[7]

Using MCNP, the effects of beam parameters including proton energy, moderator material and thickness, and thermal neutron and photon shielding on figures of merit such as advantage depth, advantage ratio, and advantage depth dose rate were evaluated.[8]

Each dose component was weighted by an appropriate relative biological effectiveness (RBE), and healthy tissue and tumor boron uptake levels were based on experimentally determined values for boronophenylalanine (BPA). MCNP simulations indicate that light water (H_2O) is the ideal moderator, enriched ^{6}LiF and Pb are preferred thermal neutron and photon shielding materials, and effective neutron beams can be achieved with proton energies from 1.91 MeV to 1.97 MeV. For proton energies below 1.91 MeV, neutron yields are too low to allow reasonable treatment times, while for proton energies above 1.97 MeV, too much moderation is required to sufficiently reduce the fast neutron component of the beam. These results were used to develop target designs capable of treating tumors located within 6 cm of the head surface in less than one hour with accelerator proton currents of 5 mA or less. A schematic of a representative near-threshold neutron beam using 1.95 MeV protons is given in Figure 1. The figure is to scale. The light water moderator is 3 cm thick with a radius of 12 cm, and the reflector material is aluminum oxide. Cooling is provided by light water flowing through a series of copper fins behind the target. The total mass of target, moderator, shielding, and target cooling is less than 6 kg, making it possible to mount this system on a standard hospital treatment gantry.

Neutron flux calculations have been compared with experimental results using a cylindrical, acrylic phantom.[9] The phantom was designed and constructed to permit almost unlimited freedom of detector placement inside the phantom, so that spatial resolution is only limited by detector size and experimental time constraints. Fast neutron and photon doses are measured using the paired ionization chamber method with small

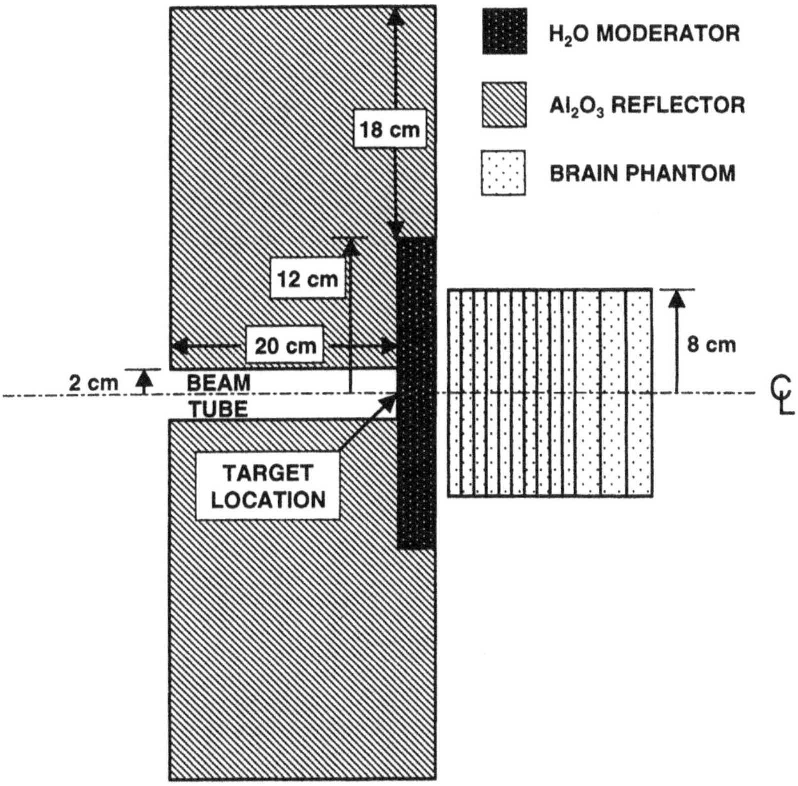

Figure 1. Schematic representation of a near-threshold neutron source using 1.95 MeV protons.

tissue equivalent and graphite chambers, and thermal neutron fluxes are measured using small (1.69 cm diameter) boron trifluoride (BF_3) proportional counters, as well as indium and gold foils.[10]

3. TREATMENT PLANNING

Initial simulations of GBM treatments using near-threshold beams were performed using the BNCT Radiation Treatment Planning Environment (BNCT_RTPE) and Radiation Transport in Tissue by Monte Carlo (rtt_MC) software.[11] Due to the radial non-uniformity of the neutron flux, both neutron and photon fluxes were divided into two radial zones. The first zone consisted of neutrons and photons emitted within a 4.25 cm radius of the beam centerline, the full width at half maximum of the epithermal neutron flux exiting the moderator unit. The second radial zone consisted of neutrons and photons produced between 4.25 cm and 12 cm of the beam centerline. The neutron beam is produced with the geometry given in Figure 1. The center of the tumor was located 3 cm beneath the right posterior surface of the patient head, with an approximate radius of 1 cm. A target volume was defined to have a thickness of 2 cm around the tumor. Single-beam, single-fraction irradiation was considered for these initial studies. For comparison, this tumor was also treated with a model of the Brookhaven National Laboratory (BNL) neutron beam used in human clinical trials.

Dose volume curves for healthy brain, target, and tumor volumes for both neutron beams are shown in Figures 2 and 3. The boron concentration was 13 ppm in healthy brain and 45.5 ppm in target and tumor volumes. Dose equivalents, determined by weighting each physical dose component with the appropriate RBE values for BPA, were adjusted to give identical reference doses (total dose equivalent to the maximum thermal neutron flux location) of 12.4 Gy-Eq. An integrated accelerator current of just over 60 mA-min was necessary to achieve this reference dose level. The healthy brain average

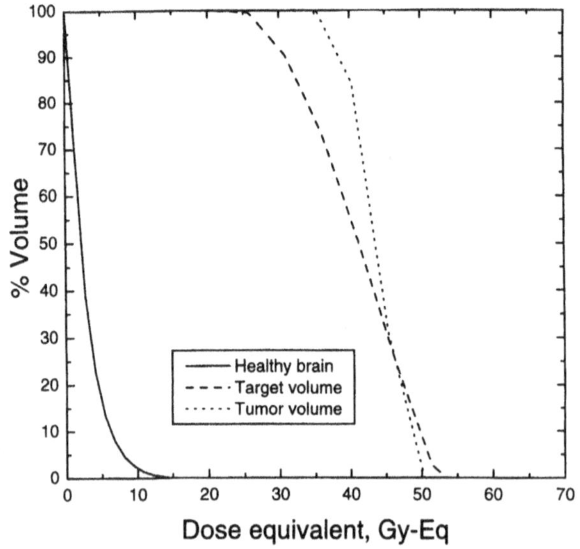

Figure 2. Dose volume curves for a near-threshold neutron beam produced with 1.95 MeV protons.

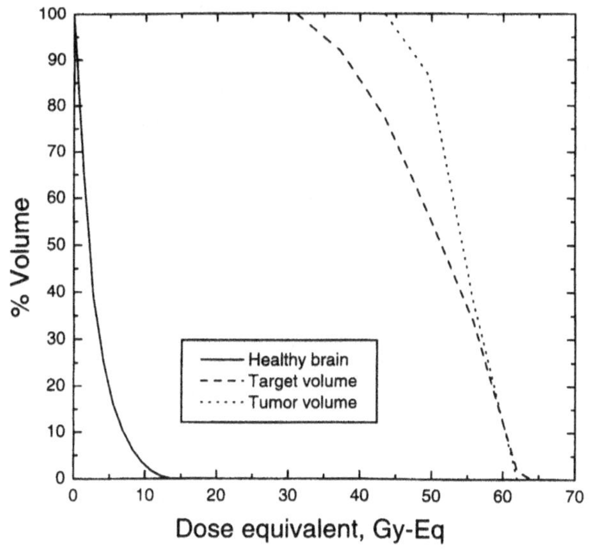

Figure 3. Dose volume curves for a model of the Brookhaven National Lab BNCT treatment beam.

dose equivalent was 2.7 Gy-Eq for the near-threshold beam, slightly lower than the corresponding BNL value of 2.9 Gy-Eq. While target and tumor dose equivalents for the near-threshold beam were slightly lower than for the BNL beam, they were sufficient for treatment of this patient. It should be emphasized that this was the initial evaluation of near-threshold neutron beams using treatment planning software, and improvements in target and tumor dose equivalents are expected.

4. SUMMARY AND FUTURE WORK

An overview of calculations and supporting experimental investigations of neutron source, moderator, and cooling designs for near-threshold BNCT has been presented. Initial evaluation of a near-threshold neutron beam based on these studies, using standard BNCT treatment planning software, has shown that these beams are capable of treating a wide variety of GBMs. Near-threshold beams, while having lower penetrability than some other accelerator-based BNCT concepts, have relatively modest accelerator current requirements, and the small size, mass, and cost of target, moderator, and reflector materials make the concept ideal for widespread hospital use. Future work will investigate near-threshold beam efficacy in greater detail using treatment planning software, including evaluation of multiple-beam irradiation. In addition, the utility of near-threshold beams for other NCT applications such as boron neutron capture synovectomy (BNCS) will be studied.

REFERENCES

1. M.C. Dobelbower, and T.E. Blue, A Monte Carlo analysis of the effect of increasing target diameter for an accelerator-based neutron source for BNCT, *in:* "Proceedings of the Topical Meeting on Nuclear

Applications of Accelerator Technology," American Nuclear Society, Albuquerque, NM USA, 1997, pp. 495–501.

2. B.A. Ludewigt, W.T. Chu, R.J. Donahue, J. Kwan, T.L. Phillips, L.L. Reginato, and R.P. Wells, An epithermal neutron source for BNCT based on an ESQ-accelerator, *in:* "Proceedings of the Topical Meeting on Nuclear Applications of Accelerator Technology," American Nuclear Society, Albuquerque, NM USA, 1997, pp. 489–494.
3. R.J. Kudchadker, J.F. Kunze, and J.F. Harmon, Accelerator designs that move close to the target and eliminate much moderator, *in:* "Advances in Neutron Capture Therapy," B. Larsson, J. Crawford, and R. Weinreich, eds., Elsevier Science B. V., Zurich, Switzerland, 1997, pp. 516–521.
4. X.-L. Zhou, and C. Lee, Analysis of epithermal neutron production by near-threshold (p,n) reactions, *Appl. Radiat. Isot.*, 48:1571–1575, 1997.
5. C.L. Lee, "The design of an intense accelerator-based epithermal neutron beam prototype for BNCT using near-threshold reactions," Ph.D. thesis, Massachusetts Institute of Technology, 1998.
6. C.L. Lee, and X.-L. Zhou, Thick target neutron yields for the ^{7}Li(p,n)^{7}Be reaction near threshold, *Nucl. Instr. Meth. B*, 152:1–11, 1999.
7. J.F. Briesmeister, ed., "MCNP-A general Monte Carlo N-particle transport code, Version 4A," LA-12625-M: Los Alamos National Laboratory, 1993.
8. C.L. Lee, X.-L. Zhou, R.J. Kudchadker, F. Harmon, and Y.D. Harker, A Monte Carlo dosimetry-based evaluation of the $Li^7(p,n)Be^7$ reaction near threshold for accelerator boron neutron capture therapy, *Med. Phys.*, 27:192–202, 2000.
9. C.L. Lee, X.-L. Zhou, J.F. Harmon, R.W. Bartholomay, Y.D. Harker, and R.J. Kudchadker, Thermal neutron flux mapping in a head phantom, *Nucl. Instr. Meth. A*, 422:106–110, 1999.
10. R.D. Rogus, O.K. Harling, and J.C. Yanch, Mixed field dosimetry of epithermal neutron beams for boron neutron capture therapy at the MITR-II research reactor, *Med. Phys.*, 21:1611–1625, 1994.
11. D.W. Nigg, F.J. Wheeler, D.E. Wessol, C.A. Wemple, R. Babcock, and J. Capala, Some recent developments in treatment planning software and methodology for BNCT, *in:* "Advances in Neutron Capture Therapy," B. Larsson, J. Crawford, and R. Weinreich, eds., Elsevier Science B. V., Zurich, Switzerland, 1997, pp. 91–94.

72

ANALYSIS OF ACCELERATOR BASED NEUTRON SPECTRA FOR BNCT USING PROTON RECOIL SPECTROSCOPY

L. Wielopolski[1], H. Ludewig[2], J. R. Powell[2], D. Raparia[3], J. G. Alessi[3], and D. I. Lowenstein[3]

[1]Department of Applied Science
[2]Department of Applied Technology
[3]Alternating Gradient Synchrotron
Brookhaven National Laboratory
Bldg. 490-D, Upton, NY 11973

INTRODUCTION

Boron Neutron Capture Therapy (BNCT) is a promising binary treatment modality for high-grade primary brain tumors (glioblastoma multiforme, GM) and other cancers.[1-3] BNCT employs a boron-10 containing compound that preferentially accumulates in the cancer cells in the brain. Upon neutron capture by ^{10}B energetic alpha particle and tritons released at the absorption site have the potential to kill the cancer cell. In order to gain penetration depth in the brain Fairchild[4,5] proposed, for this purpose, the use of energetic epithermal neutrons at about 10 keV. Phase I/II clinical trials of BNCT for GM are underway at the Brookhaven Medical Research Reactor[6] (BMRR) and at the MIT Reactor, using these nuclear reactors as the source for epithermal neutrons.

In light of the limitations of new reactor installations, e.g. cost, safety and licensing, and limited capability for modulating the reactor based neutron beam energy spectra, alternative neutron sources are being contemplated for wider implementation of this modality in a hospital environment. For example, accelerator based neutron sources offer the possibility of tailoring the neutron beams, in terms of improved depth-dose distributions, to the individual and offer, with relative ease, the capability of modifying the neutron beam energy and port size.

In previous work new concepts for compact accelerator/target configurations were published.[7,8] In this work, using a Van de Graaff accelerator we have explored different materials for filtering and reflecting neutron beams produced by irradiating a thick Li target with 1.8 to 2.5 MeV proton beams. However, since the yield and the maximum

Frontiers in Neutron Capture Therapy, edited by Hawthorne *et al.*
Kluwer Academic / Plenum Publishers, New York, 2001.

neutron energy emerging from the Li-7(p,n)Be-7 reaction increase with increase in the proton beam energy, there is a need for optimization of the proton energy versus filter and shielding requirements to obtain the desired epithermal neutron beam. The MCNP-4A computer code was used for the initial design studies that were verified with benchmark experiments using a proton recoil spectroscopy detection system. Comparison was also made between in phantom $^{10}BF_3$ readings made at the BMRR and those made at the RARAF facility using a Van de Graaff accelerator.

EXPERIMENTAL SET-UP AND ANALYSIS

The experiments were carried out on a Van de Graaff accelerator at the Radiological Research Accelerator Facility (RARAF) at Nevis Laboratory, Columbia University. Measurements were made using a cylindrical 3 atm gas proportional proton recoil (PR) counter and a $^{10}BF_3$ counter. The PR counter was filled with partial pressures of 95.23% H_2, 4.76% CH_4 and 0.01% ^3He for internal energy calibration purposes. Neutron spectroscopy was performed in air with the PR counter axis parallel to the central beam axis and in contact with the beam port. A small $^{10}BF_3$ counter was used in a phantom in the RARAF and the BMRR facilities for independent confirmation purposes.

The geometrical configuration of the target, filter and reflector assembly is shown in Figure 1. This configuration was used for the Monte Carlo calculations and the experiments. In an alternative configuration the Be reflector was replaced with two Teflon and Fe disks, and in the third configuration both Mo and Be reflectors were replaced with two larger Teflon and Fe disks. In addition there was about a 1 mm thick water layer for target cooling purposes. The model calculates the neutron energy spectrum at the beam port and is used to estimate the current required to deliver an epithermal neutron flux of about 10^{12} n/cm²/hr required to treat a patient to a maximum peak dose in the normal brain to about 13.6 Gy in one hour. The calculations were performed at several proton

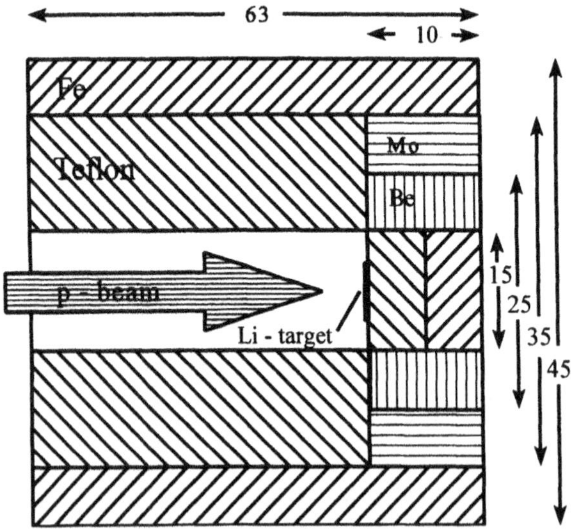

Figure 1. Schematic configuration of the filter assembly with a water cooled (not shown) Li target. All the dimensions are in cm.

energies, although the main effort was directed to proton energies of 2.5 MeV. The MCNP code was run for 5 million histories and the results were normalized per one source neutron. To derive the beam current, a thick Li target neutron yield of 1.4×10^{-4} neutrons per proton at 2.5 MeV was used.

The measured pulse height distributions (PHD) were analyzed using the PSNS code.[9] The code is based on the characteristics of the n-p scattering kernel in which neutron energy transfer to the proton is from zero to the maximum neutron energy. This enables one to relate the measured PHD $M(E)$ to the neutron spectrum $\phi(E)$ through the following equations,[10] Eq. 1

$$M(E) = N_H T \int_{E}^{E_{max}} dE_n \frac{\sigma(E_n)}{E_n} \phi(E_n)$$

where E is the neutron energy, N_H is the number of hydrogen atoms in the active volume of the counter, T is the counting time in seconds and $\sigma(E)$ is the scattering cross section. Thus the neutron spectrum is unfolded by differentiating the measured PHD, Eq. 2, instead of the alternative unfolding scheme that involves matrix inversion.

$$\phi(E_n) = -\frac{E}{N_H T \sigma(E)} \frac{dM(E)}{dE}$$

RESULTS

PHDs measured for bare, without moderator, thin and thick Li targets using the PR counter and proton energy of 2.3 MeV, are shown in Figure 2. The neutron spectra resulting from unfolding of the PHDs shown in Figure 2 are shown in Figure 3.

The unfolded neutron spectra for the three filter/reflector configurations mentioned before and a bare target without filtration are shown in Figure 4 for a 2.5 MeV proton beam. The agreement between the measured spectrum of the Teflon iron filter configuration and the calculated one can be seen in Figure 5. These two spectra were normalized to the same area in the energy interval between 35 and 200 keV. In this energy range the contributions to the measured PHD due to background gamma radiation, that contributes below 35 keV, and wall effects that potentially may occur due to high energy neutrons, at the high end of the spectrum, are minimal. The maximum neutron energy due to 2.5 MeV protons is about 787 keV.

The epithermal neutron fluxes measured from the unfolded neutron spectra in the energy interval between 5 keV (at present the lower energy limit of the PHD) and 35 keV were normalized per mC of proton charge. From these values estimates of the currents required to deliver a neutron fluence of 10^{12} n/cm^2 during one hour of treatment were obtained. These estimates were corrected, using ratios obtained from the MCNP calculations, to include epithermal neutrons from 1 eV up to 5 keV to be included in the current predictions. For the filter/reflector configurations considered here the mean correction factor is about 2.5. This correction factor should not be confused with the agreement between these predicted values, based on experiments, and the estimated currents based on the MCNP calculations as they are summarized in Table 1. The agreement is in the range of 20% to 70% and these results are similar to the lower values reported in the literature.[11]

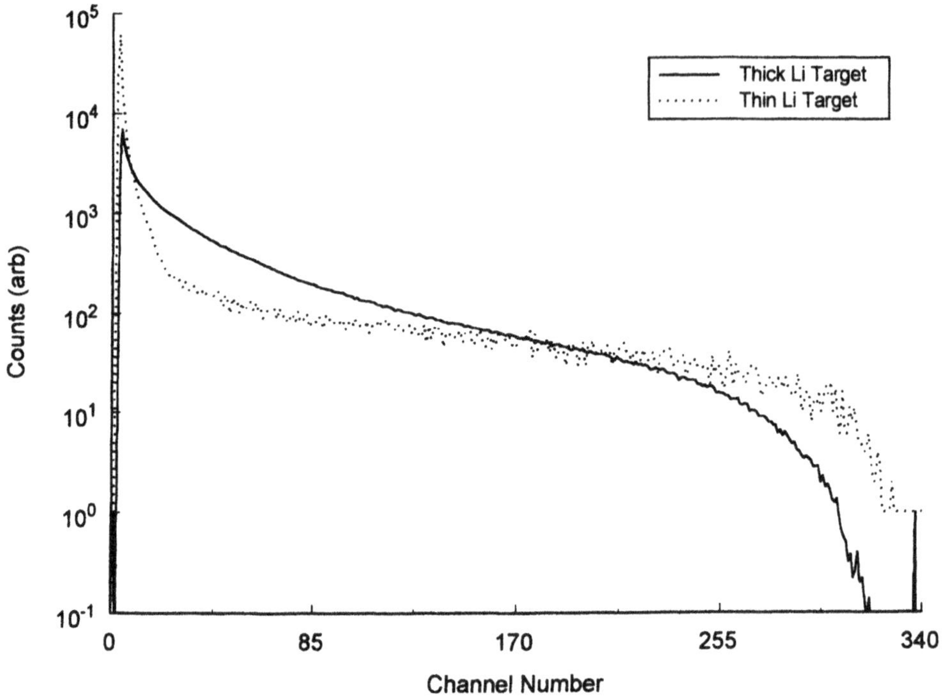

Figure 2. Pulse height distributions measured from bare thin and thick Li targets irradiated with a 2.3 MeV proton beam.

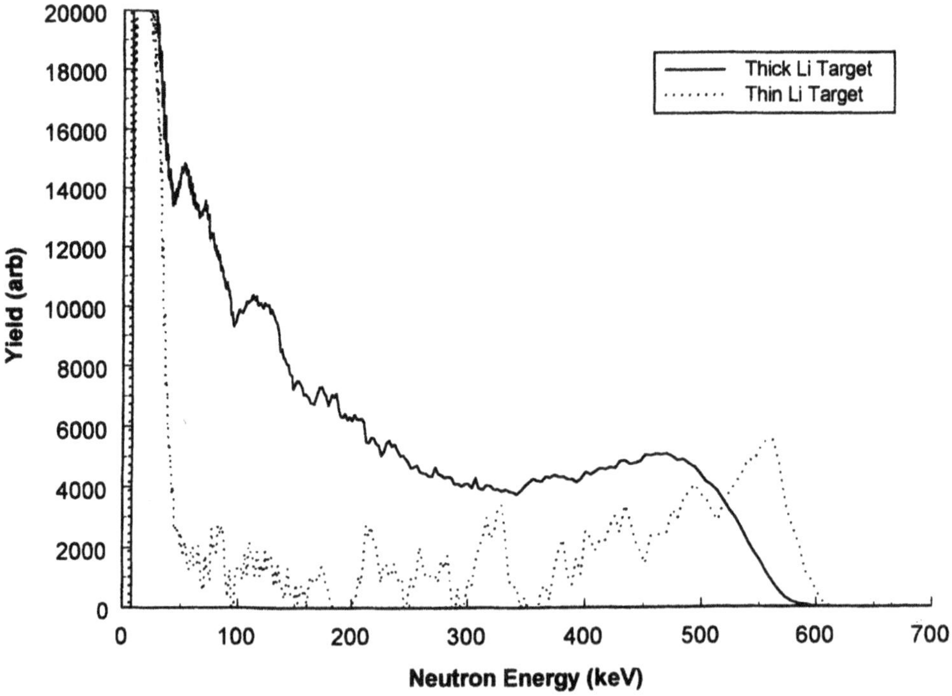

Figure 3. Unfolded neutron spectra obtained from PHDs shown in Figure 2.

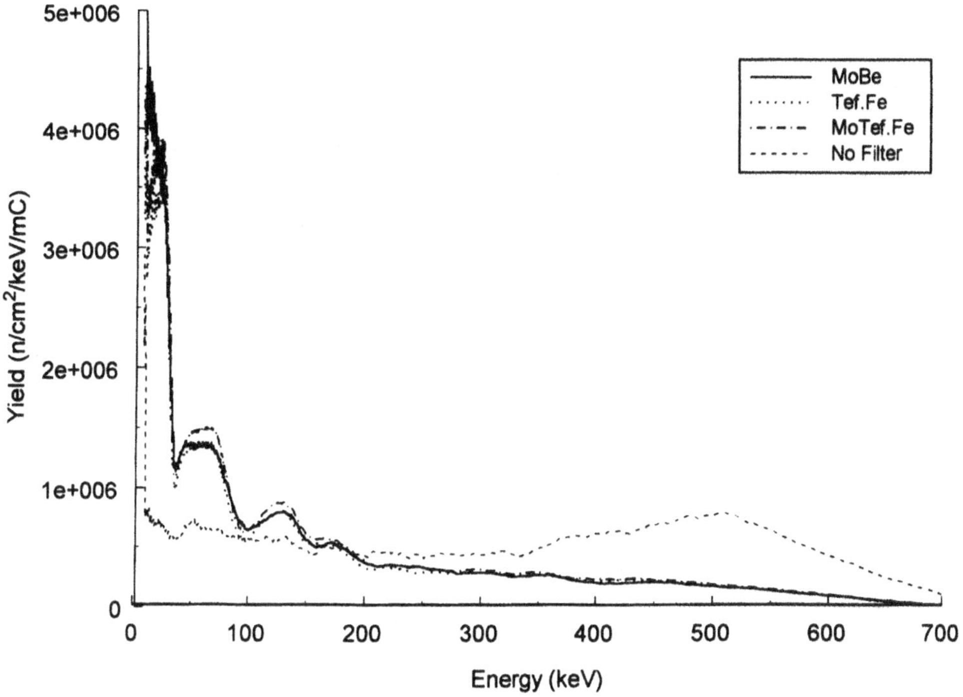

Figure 4. Unfolded neutron spectra for different filter/reflector configurations together with a neutron spectrum from a bare source. Proton energy is 2.5 MeV.

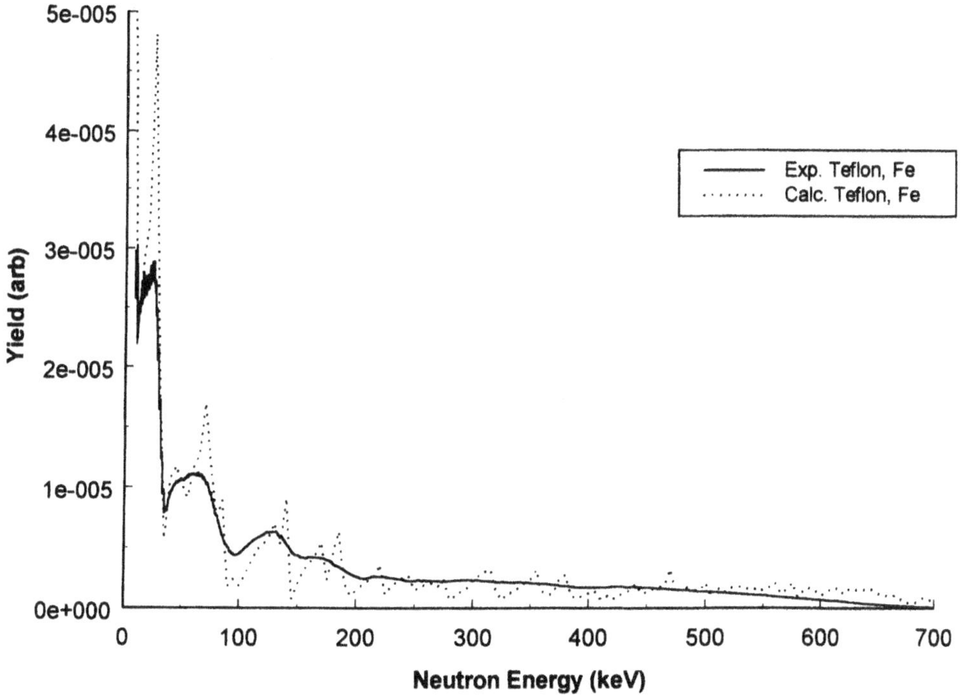

Figure 5. Comparison between measured and calculated spectra at a proton energy of 2.5 MeV.

Table 1. The estimated proton beam currents required to deliver a fluence of 10^{12} n/cm^2 during one hour based on the measurements and calculations of the epithermal fluxes between 1 eV and 35 keV

Filter	Current inferred from measurements (mA)	Current inferred from MCNP calculations (mA)
No filter	12.60	—
Teflon, Fe no reflector	8.96	—
Teflon, Fe	2.22	1.28
Mo, teflon, Fe	2.20	1.32
Mo, Be	1.62	1.33

In order to obtain a partial independent confirmation of the predicted currents, a small $^{10}BF_3$ counter was used for measurements in phantom at the depth of maximum thermal flux. These measurements were taken in the epithermal port of the BMRR, where clinical trials are being carried out at present, and at the RARAF accelerator facility. The phantom consisted of a $14 \times 14 \times 14\,cm^3$ polyethylene cube with holes to accommodate the $^{10}BF_3$ counter at various locations, and proper plugs for filling the unused holes. Based on the counts obtained from the $^{10}BF_3$ counter the estimated proton beam current required to deliver the same count rate as at BMRR at 3 MW, at a typical operating power during patient treatment, was about 1 mA for 2.5 MeV proton beam and about 6 mA for the 1.9 MeV proton beam. Earlier results using a small $^{10}BF_3$ counter in the BMRR facility have been presented elsewhere.[12] In addition, lateral scans in air and in phantom at the RARAF facility showed that the neutron beam resulting from 2.5 MeV protons is wider, as expected, and flatter than that from 1.9 MeV protons.

SUMMARY AND CONCLUSIONS

The objective of the present work was to validate MCNP calculated neutron spectra with those determined experimentally, and no attempts were made to calculate the doses that are essential for a full beam characterization. Confirmation of the model used in the Monte Carlo calculations is vital in developing and characterizing therapy facility designs and treatment planning systems. The neutron energy spectra are of particular importance in accelerator-based BNCT when investigating the use of lithium or other elements as possible target materials. The neutron spectra are needed: 1) as a source input file for transport calculations to develop suitable moderator and filter assemblies, 2) for evaluating accelerator beam current requirements and the target cooling schemes, 3) for evaluating facility shielding requirements, and 4) for proper in phantom dose calculations. The presented results show a good agreement between the measured and the calculated spectra. At present the low energy edge, that has been measured using the PR detector, is about 5 keV. However, neutron spectroscopy using PR detectors can be extended to about 1 keV. Based on these preliminary measurements, a proton beam of 2.5 mA, conservatively 5 mA, appears to be adequate for use in BNCT. This current is lower than that in excess of 20 mA that was claimed in the in the past. Such a beam will deliver an epithermal flux of 10^{12} n/cm^2 during a one hour irradiation, with a large beam port of about 10 cm in diameter.

ACKNOWLEDGMENTS

The authors would like to thank Dr. Gerhard Randerf-Pehrson and Stephen Marino from the Radiological Research Accelerator Facility at Columbia University for their assistance in carrying out the experiments, and David E. Alburger for his measurements at the BMRR. This project has been funded by the Laboratory Directed Research and Development funds under the auspices of the U.S. Department of Energy Contract No. DE-AC02-98CH10886).

REFERENCES

1. P.A. Zahl, and F.S. Cooper, Physical and biological considerations in the use of slow neutrons for cancer therapy. Radiology 37, 673–682, 1941.
2. W.H. Sweet, The uses of nuclear disintegration in the diagnosis and treatment of brain tumors. New England Journal of Medicine, 245, 875–878, 1951.
3. D.N. Slatkin, A history of boron neutron capture therapy of brain tumors. Brain, 114, 1609–1629, 1991.
4. R.G. Fairchild, Development and dosimetry of an epithermal neutron beam for possible use in neutron capture therapy. Phys. Med. Biol., 10, 491, 1965.
5. R.G. Fairchild, J. Kalef-Ezra, S.K. Saraf, S. Fiarman, E. Ramsey, L. Wielopolski, B.H. Laster, and F.J. Wheeler, Installation and testing of an optimized epithermal neutron beam at the Brookhaven Medical Research Reactor (BMRR), in: "Neutron Beam Design, Development and Performance for Neutron Capture Therapy," ed; O. Harling, and R. Zamenhof, Plenum Press, New York 1989, pp. 185–199.
6. J.A. Coderre, E.H. Elowitz, M. Chadha, R. Bergland, J. Capala, D.D. Joel, H.B. Liu, D.N. Slatkin, and A.D. Chanana, Boron neutron capture therapy for glioblastoma multiforme using *p*-boronophenylalanine and epithermal neutrons: Trial design and early clinical results. J. Neuro-Oncology 33, 141–152, 1997.
7. J.R. Powell, H. Ludewig, M. Todosow, and M. Reich, Proceedings of the ASME-JSME 4th International Conference on Nuclear Engineering -ICONE-4, pp. 493–502, 1996.
8. J.R. Powell, H. Ludewig, M. Todosow, and M. Reich, Applications of Accelerators in Research and Industry, pp. 1309–1312, AIP Press, New York, 1997.
9. E.F. Bennett, and T.J. Yule, *ANL-7763*, 1971.
10. N. Tsoulfanidis, "Measurements and Detection of Radiation," ch. 14, Hemisphere Publishing Corporation Washington, New York, London, 1983.
11. R.J. Kudchadker, J.F. Kunze, and J.F. Harmon, "Advances in Neutron Capture Therapy," pp. 516–521, Elsevier, Amsterdam-Lausanne-New York-Oxford, 1997.
12. D.E. Alburger, D. Raparia, and M.S. Zucker, Med. Phys., 25, 1735–1738, 1998.

ENERGY OPTIMIZATION OF EPITHERMAL NEUTRON BEAMS FOR ACCELERATOR-BASED BNCT

E. Bisceglie, P. Colangelo[1], N. Colonna[1], G. Maggipinto[2], V. Paticchio[1], G. Raso[1], P. Santorelli[1], and V. Variale[1]

[1]Istituto Nazionale di Fisica Nucleare
Sezione di Bari, Italy
[2]Dipartimento di Fisica
Università di Bari
via Amendola n.173
I-70126 Bari, Italy

ABSTRACT

The optimal neutron energy for the treatment of deep-seated tumors, in particular Glioblastoma Multiforme, is still the subject of debate within the scientific community, due to its important implications on the choice of the most appropriate neutron source for the therapy. We present an analysis of the therapeutic gain as a function of the neutron energy, performed with the aim of identifying the optimal features of the neutron beam. A method for the beam quality evaluation is also discussed, which allows for a fast and meaningful comparison between different epithermal neutron beams produced by nuclear reactors and by accelerator-based neutron sources.

INTRODUCTION

The possible success of Boron Neutron Capture Therapy, a two-component modality for cancer treatment, relies on the use of neutron beams of suitable intensity and spectral features.[1] An important aspect currently being investigated is the identification of the optimal neutron energy for the treatment of deep-seated tumors (such as Glioblastoma Multiforme). This problem is particularly important in view of the present perspectives of producing therapeutic beams using accelerator-based facilities. In this case, in fact, the optimal energy distribution could be achieved by choosing the

Frontiers in Neutron Capture Therapy, edited by Hawthorne *et al.*
Kluwer Academic / Plenum Publishers, New York, 2001.

appropriate neutron-producing reaction as well as the material and geometry of the beam shaping assembly (moderator, reflector, filter etc.).

With the aim of identifying the criteria for the optimization of the neutron energy for BNCT, we performed simulations of the dose deposition in the brain for monoenergetic neutron beams in a wide energy range, and analyzed significant figures of merit. Based on these results, we developed a method to evaluate the quality of a realistic epithermal neutron beam of known energy profile, thus allowing to perform fast and meaningful comparisons between different neutron-producing reactions and various moderator configurations.

We report here the results of this set of simulations and, in particular, we present an analysis of the therapeutic gain as a function of the energy of the incoming neutrons. We also describe a method for a fast evaluation of the beam quality, and present a comparison between different therapeutic neutron beams.

MONTE CARLO SIMULATIONS AND FIGURES OF MERIT

Because of the nature of neutron interactions, BNCT analyses heavily rely on the use of Monte Carlo simulations.[2] In particular, simulations can be used to study the optimal neutron energy for the treatment of deep-seated tumors. As an example, by analyzing the dose distribution produced by monoenergetic neutron beams, Allen *et al.*[3] have shown the advantages of using an epithermal neutron beam as compared with a thermal one. This conclusion was based on the comparison between the dose distribution produced by a monochromatic neutron beam of 35 eV with a thermal beam.

In this work, we have extended the analysis of ref.[3] to higher neutron energies, in an attempt to better characterize the optimal neutron energy for the therapy. While the results of Allen *et al.* are obtained with an MCNP simulation code, we have decided to follow a slightly different approach for the calculations of the dose delivered by neutrons in biological tissues. In the code we used, based on the simulation tool GEANT[4] developed at CERN, the neutron interactions are simulated by the MICAP routine[5] according to tabulated experimental cross-sections (ENDF-B/6.1). The code has been chosen for its versatility in the definition of materials and geometry. The most important feature of this type of simulations, however, consists in the possibility of directly calculating the energy released in the tissues by the charged particles produced by the neutron interactions, whereas in different codes the dose produced by the neutron beam is calculated employing appropriate Kerma factors.

Assuming specific concentrations of ^{10}B for both normal and tumor tissues, the boron capture reactions were considered, and the alpha particles produced by the subsequent decay were followed. The energy lost by charged particles produced by neutron and photon interactions was determined in a quasi-continuum fashion, according to calculated and tabulated stopping powers.

For a meaningful comparison, we performed simulations in the same conditions chosen by Allen *et al.*[3] In particular, we considered a standard model of the skull and brain, with a tumor of the size of 1 cm, positioned at a depth of 5 cm from the brain surface; ^{10}B concentrations of 10 and 43 ppm in healthy and tumor tissues, respectively, were assumed; RBE factors of 1.6 and 2.3 for protons, and alpha and ^{7}Li particles, respectively, were also considered. To quantitatively study the optimal neutron energy for the therapy, we performed a set of calculation for monochromatic neutron beams in a wide energy range. For some selected energies, our results were checked against the results of

MCNP calculations, performed in the same conditions; agreement within a few percent in the dose distribution was observed.

Figure 1 shows the results obtained for monochromatic neutron beams at energies of 0.025 eV, 100 eV, 1 keV and 10 keV. The total dose is represented by the solid symbols, while the histogram shows the dose from ^{10}B capture interactions only.

Among the different figures of merit (such as the advantage depth, the therapeutic depth, etc.), we chose to characterize the optimal energy for the treatment of deep-seated tumors by means of the therapeutic gain (TG), defined as the ratio between the average dose deposited in the tumor region and the maximum dose delivered to normal tissues. For an accurate determination of TG, a fitting procedure was employed on the Monte Carlo results. In particular, the dose distribution to healthy tissues was fitted with a 4th order polynomial, up to 12 cm depth and excluding the tumor region; the results of the fit are shown by the continuos curves in Figure 1. The values of the therapeutic gain for the selected energies are reported in the figure. The maximum dose to normal tissues is determined from the result of the fit, while the average dose to tumor is obtained by integrating over the tumor region.

The analysis of the therapeutic gain versus the energy of the incoming neutron beam, at fixed tumor geometry, is shown in Figure 2a. The results clearly show the possibility of choosing an optimal energy configuration. In fact, in correspondence of the assumed ^{10}B concentration and tumor depth, the therapeutic gain shows a sharp maximum between 1 and 10 keV. Neutrons of lower energies mainly thermalize before the tumor location, while high energy neutrons are responsible of a large dose to the skin and the brain surface, due to recoiled protons. As a consequence, the therapeutic gains rapidly drops for neutron energies above 25 keV. For this reason, the optimization of the neutron

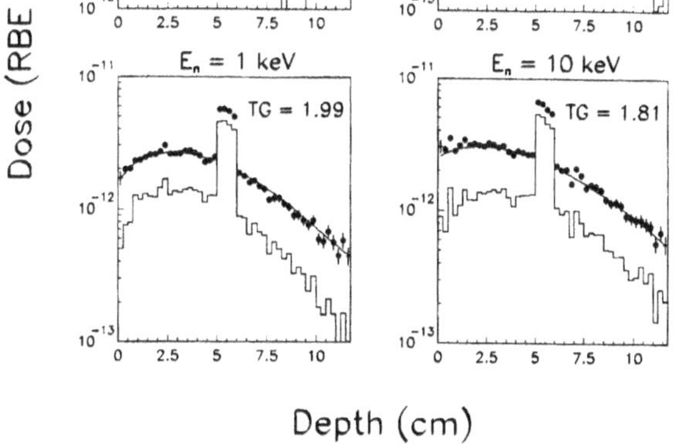

Figure 1. Dose distribution as a function of the depth in the brain for monoenergetic and unidirectional neutron beams of four different energies. The solid symbols represent the released total dose; the histograms represent the dose produced by particles following neutron capture by ^{10}B. The therapeutic gain TG is also reported for the various cases. The continuous curves represent a polynomial fit described in the text.

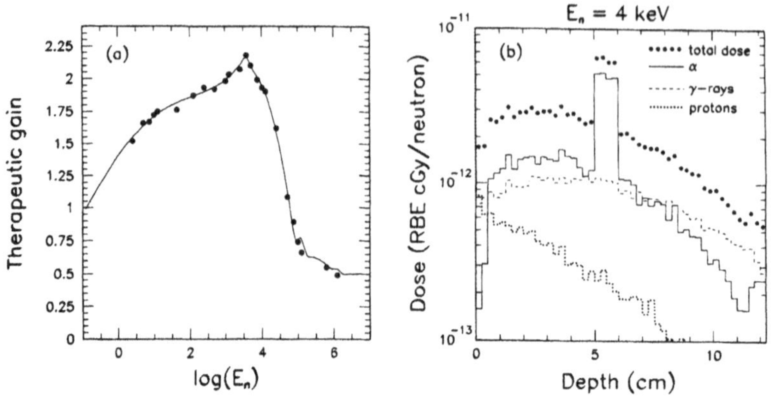

Figure 2. (a): Therapeutic gain as a function of the neutron energy for a monoenergetic and unidirectional neutron beam. The solid symbols represent the results of the simulations performed for several neutron energies between 25 meV and 1 MeV; the solid line represents a polynomial fit. (b): Total dose (solid symbols) and different components of the dose released by a neutron beam of the optimal energy of 4 keV. The different neutron-induced reactions contributing to the total dose are described in the text.

beam requires a minimization of the high energy component. The dose distribution for the "optimal" neutron energy of 4 keV is shown in Figure 2b, together with the various components of the dose: the one corresponding to the α following the neutron capture by ^{10}B, the dose produced by the 2.2 MeV γ-rays produced in the H radiative capture, and the dose produced by recoiled protons and by the 0.6 MeV protons emitted after ^{14}N capture.

A FAST METHOD FOR BEAM QUALITY EVALUATION

For a realistic neutron beam, characterized by a composite energy profile, the determination of the overall therapeutic gain can be obtained without performing a complete simulation of the dose deposition in the brain. The method we propose is based on the observation that the dose released by a neutron beam of composite energy spectrum can be decomposed in the contributions of the different energy components. An analytical expression of the dose distribution D(E,x), as a function of the neutron energy E and of the depth x, is obtained by fitting the dose distribution for monoenergetic beams in the variable x using a 4th order polynomial, whose coefficients are expressed as a function of the neutron energy. The results of this procedure are shown in Figure 1 for the selected neutron energies. In the whole energy region considered in the simulations (that is for neutrons of $0.025\,eV < E_n < 1\,MeV$), the analytical representation of D(E,x) accurately reproduces the behavior of the simulated dose distributions. The therapeutic gain is also polynomially fitted as a function of the neutron energy, and the result is shown by the curve in Figure 2a.

For a given normalized neutron spectrum dn(E)/dE, the overall dose distribution is obtained by integrating D(E,x) over the energy, up to 10 MeV, weighted by the neutron energy spectrum:

$$D(x) = \int D(E,x)\frac{dn(E)}{dE}dE \tag{1}$$

In correspondence to the tumor location, the dose is calculated according to the energy-dependent therapeutic gain TG(E) of Figure 2, using the formula:

$$D_T = \int TG(E) D_m(E) \frac{dn(E)}{dE} dE \tag{2}$$

where $D_m(E)$ is the maximum dose to the normal tissues for a given neutron energy E. The results of this procedure are shown in Figure 3 for two different epithermal neutron beams.

The solid histogram in Figure 3a shows the energy spectrum of a neutron beam that would be obtained from an accelerator-based BNCT facility by using the $^7Li(p,n)^7Be$ reaction, and a 22 cm thick ^7LIF moderator.[6] The overall dose distribution corresponding to this beam, as calculated by applying eqs. (1) and (2), is shown by the solid curve in Figure 3b. The possibility of predicting the dose distribution for any neutron beam, without performing lengthy simulations, represents a convenient result of the method here presented, although some uncertainties may be present, due to the absence of terms related to the neutron beam directionality and irradiation field size. The ratio of the dose to tumor and to normal tissue allows to immediately calculate the overall therapeutic gain. In the case of the accelerator-produced neutron beam, the therapeutic gain turns out to be equal to 1.73. The dashed histograms in Figure 3a and 3b refer to the neutron beam currently used for clinical trials at the Brookhaven Medical Research Reactor. The overall therapeutic gain for this beam is TG = 1.56, sensibly lower than the one that could be obtained by an accelerator-based facility. The reason for this difference is the lower penetrability of the BMRR epithermal neutron beam, as evident from the rapid decrease of the dose distribution as a function of the depth in the brain.

We applied our method to evaluate the quality of an epithermal neutron beam that could be obtained by using the reaction $^{13}C(d,n)^{14}N$ at E_d = 1.5 MeV, recently investigated as a possible neutron source for accelerator-based BNCT facility.[7] We also investigated, using this method, the optimal beam shaping assembly for the same

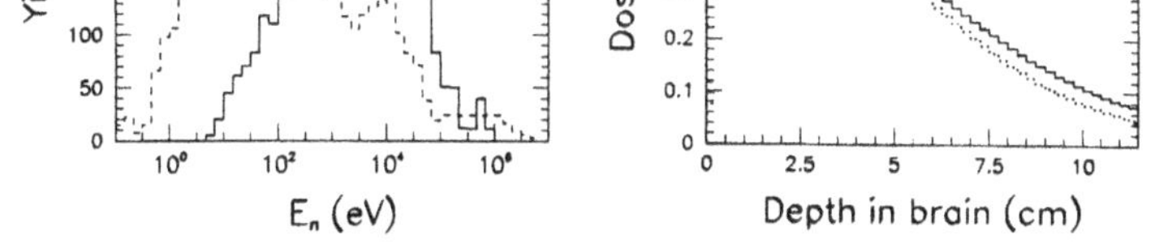

Figure 3. (a): Energy distribution for neutrons produced in the ^{7}Li(p,n) reactions at 2.5 MeV, after a 22 cm thick LiF moderator (solid line), compared with the energy distribution of the therapeutic neutron beam currently used in clinical trials at the Brookhaven Medical Research Reactor. (b): Overall dose distribution produced by the two considered neutron beams, as calculated using the method described in the text. The corresponding values of the therapeutic gain TG are also indicated.

reaction. The maximum therapeutic gain would be obtained, in this case, using a 28 cm thick Teflon moderator $(CF_2)_n$ and an Al_2O_3 reflector. The overall therapeutic gain obtained for such a beam would be TG = 1.68, a value lower than the one corresponding to the $^7Li(p,n)^7Be$ reaction, but higher than the one relative to the BMRR beam. More experimental data on the energy spectrum of neutrons emitted in the reaction are however necessary before concluding on the applicability of the $^{13}C(d,n)$ ^{14}N reaction to accelerator-based BNCT.

CONCLUSIONS

By analyzing the therapeutic gain in a wide range of neutron energies, we showed that the maximum therapeutic effect for a 5 cm deep tumor in the brain would be achieved by a monoenergetic neutron beam of energy between 1 and 10 keV, for currently assumed concentrations of ^{10}B in normal and tumor tissues. For non-monoenergetic beams, the overall therapeutic gain can be estimated by running a complete simulation of the dose deposition. Alternatively, we presented a method that allows to evaluate the beam quality of a given epithermal neutron beam of known energy spectrum. The method relies on a fitting procedure and on a weighted integral of the doses to normal and tumor tissues; it can be conveniently applied to optimize the material and geometry of the moderator for a given neutron source, and to perform a comparison between the epithermal beams obtained from different neutron-producing reactions.

Our analysis supports the conclusion that epithermal neutron beams produced by accelerators are characterized by a larger overall therapeutic gain with respect to the one obtainable using the neutron beams currently available from nuclear reactors. Among different neutron-producing reactions, the $^7Li(p,n)^7Be$ at 2.5 MeV leads to epithermal beams with the highest TG, although a good-quality neutron beam could also be obtained using the reaction $^{13}C(d,n)^{14}N$. A systematic study of the features of various neutron-producing reactions and of several moderator configurations is currently being performed.

REFERENCES

1. For a recent review see: R.F. Barth, A.H. Soloway, and R.M. Brugger, Boron Neutron Capture Therapy of Brain Tumors: Past History, Current Status and Future Potential, *Cancer Investigations*, 14(6): 534–550, 1996.
2. F.J. Wheeler, D.E. Wessol, and R. Babcok, Improvements in patient treatment planning systems, *in* Cancer Neutron Capture Therapy, edited by Y.Mishima, 1996, pp. 289–294.
3. S.A. Wallace, J.N. Mathur, and B.J. Allen, "Treatment planning figures of merit in thermal and epithermal boron neutron capture therapy of brain tumors", *Phys. Med. Biol.*, 39:897–906, 1994.
4. R. Brun *et al.*, GEANT3, Report CERN DD/EE/84.1, 1986.
5. J.O. Johnson, and T.A. Gabriel, Report TM-10340, ORNL, 1988.
6. D.L. Bleuel, R.J. Donahue, and B.A. Ludewigt, "On Optimizing the $^7Li(p,n)$ Proton Beam Energy and Moderator Material for BNCT", proceedings of the 14th International Conference on the Application of Accelerators in Research and Industry, Denton, Tx, Nov. 1996; LBNL-39057
7. N. Colonna *et al.*, "Measurements of low-energy (d,n) reactions for BNCT", *Med. Phys.* 26:793–798, 1999.

74

APPLICATIONS OF CALIFORNIUM-252 AND THE CALIFORNIUM USER FACILITY FOR NEUTRON CAPTURE THERAPY RESEARCH

R. C. Martin[1], C. W. Alexander[1], C. M. Simmons[1], and D. L. Marsh[2]

[1]Chemical Technology Division
Oak Ridge National Laboratory
Oak Ridge, TN 37831-6385
[2]Nuclear Engineering Department
The University of Tennessee
Knoxville, TN 37996-2300

1. INTRODUCTION

Neutrons for neutron capture therapy (NCT) are potentially available from three sources: nuclear reactors, accelerators, and ^{252}Cf radioisotope sources. A small number of reactors are currently or will soon become available for NCT. The costs of building and operating a reactor are too large to justify NCT-only usage. Accelerators which can provide effective NCT treatments are nearing realization, but the potential cost, except for boron neutron capture synovectomy, is sobering. The epithermal flux from available ^{252}Cf neutron sources is too low for conventional NCT, but ^{252}Cf can potentially provide a valuable contribution in several ways: (1) provide a neutron flux for testing NCT pharmaceuticals such as boron-containing compounds *in vitro*; (2) provide a neutron flux for skin and near-surface neutron therapy; (3) provide a neutron driver for a subcritical multiplying assembly (SMA) for epithermal beam generation; and (4) provide miniature sources for neutron brachytherapy (i.e., inserting the source inside the body into the tumor location), which can be used in conjunction with NCT pharmaceuticals. Although an appropriately designed SMA could provide a hospital-based neutron source for NCT at a limited number of sites, neutron brachytherapy represents the greatest medical potential of ^{252}Cf in bringing relatively low-cost neutron therapy to a large number of clinical locations.

Frontiers in Neutron Capture Therapy, edited by Hawthorne *et al.*
Kluwer Academic / Plenum Publishers, New York, 2001.

2. CALIFORNIUM-252 AND THE CALIFORNIUM USER FACILITY

The radioisotope ^{252}Cf spontaneously fissions with an average 2.645-year half-life. Californium-252 is routinely encapsulated as compact, portable, intense neutron sources, with 1 mg of ^{252}Cf emitting 2.314×10^9 neutrons/s. The photon dose from ^{252}Cf sources is typically an order of magnitude lower than the neutron dose. The primary industrial applications of ^{252}Cf are neutron radiography, neutron activation analysis for elemental analysis and process control, nuclear and waste assays, and reactor start-up sources. The primary international source of ^{252}Cf is the Radiochemical Engineering Development Center (REDC) at Oak Ridge National Laboratory (ORNL), Tennessee, USA. The REDC stores and processes the national inventory of ^{252}Cf for the U.S. Department of Energy (DOE), with annual production averaging 250 mg. The ^{252}Cf inventory at the REDC, up to several hundred milligrams, is typically the largest available at any one location in the world.

The Californium User Facility for Neutron Science (CUF), a part of the REDC, was established in 1996 to promote research with and industrial applications of ^{252}Cf neutron sources.[1] The CUF provides a unique opportunity for working with small intense neutron sources and designing experiments with very flexible irradiation configurations. The CUF is unique in having two uncontaminated hot cells for hands-on experimental setup, after which neutron sources can be transferred into cells for irradiation. Up to 50 mg ($>10^{11}$ neutrons/s) may be routinely used for irradiation experiments in an uncontaminated hot cell. Irradiations with up to several hundred milligrams may be conducted in the CUF water-filled storage pool. Notable experiments conducted at the CUF include land-mine detection experiments, irradiation performance testing of solid-state detectors for use at the CERN accelerator, dosimetry experiments for neutron brachytherapy, and irradiation of cancer cells to evaluate experimental boronated pharmaceuticals for boron NCT (BNCT).[2]

Table 1 provides estimates of thermal (<0.5 eV) and fast neutron fluxes available from several existing configurations of neutron moderators within the CUF, as determined using the Monte Carlo N-Particle (MCNP)-4B neutron transport code.[3] Thermal fluxes between 10^8 and $10^9\,cm^{-2}\,s^{-1}$ and fast fluxes $>10^9\,cm^{-2}\,s^{-1}$ are readily achievable for purposes of experimentation. Calculations for the large polyethylene block configuration

Table 1. Irradiation configurations and MCNP flux estimates

Irradiator design	Distance from source (cm)	Neutron flux[a] ($10^9\,cm^{-2}\,s^{-1}$)			Thermal/fast flux ratio
		Thermal[b]	Fast[b]	Total	
Polyethylene + H_2O 8 × 8 cm cylinder	3	0.17	1.2	1.4	0.14
Polyethylene block 26 × 26 × 21 cm	3	0.69	1.4	2.1	0.49
Polyethylene block 97 × 41 × 21 cm	6.8	0.59	0.27	0.86	2.2
Polyethylene block 97 × 41 × 21 cm	13.5	0.11	0.025	0.135	4.4
In-pool (H_2O)[c]	6.35	0.45	0.37	0.82	1.2

[a]All fluxes normalized to 50 mg of ^{252}Cf.
[b]Thermal neutrons defined as having energies ≤0.5 eV; fast neutrons, >0.5 eV.
[c]Use of available ^{252}Cf inventory can increase in-pool fluxes.

indicate that the epithermal flux component (0.5 eV to 10 keV) is comparable to the thermal (i.e., ~$6 \times 10^8 cm^{-2} s^{-1}$ at 6.8 cm and $1.5 \times 10^8 cm^{-2} s^{-1}$ at 13.5 cm of polyethylene for a 50-mg ^{252}Cf source). Within the ^{252}Cf storage pool, a storage lattice position surrounded by 100 mg of ^{252}Cf receives a thermal flux of ~$9 \times 10^8 cm^{-2} s^{-1}$ with thermal-to-fast ratio of 1.2.

3. NEUTRON BRACHYTHERAPY AND BNCT-BOOSTED BRACHYTHERAPY

For many years, ^{252}Cf brachytherapy has been used to treat cancer, primarily for skin, head and neck, and gynecological tumors and for muscular sarcomas.[4] Californium-252 brachytherapy can often provide more effective treatment of radioresistant tumors than conventional photon sources. Treatments in the United States have used lower-intensity ^{252}Cf sources, typically ≤30 µg of ^{252}Cf manually loaded into patients, while Eastern European countries have demonstrated remotely afterloaded high-dose-rate (HDR) brachytherapy with physically larger sources containing up to 1 mg of ^{252}Cf.[5] To date, ~3,400 patients have been treated using ^{252}Cf brachytherapy, ~1,500 of them in the United States. Of the ~1,200 patients in the United States whose clinical treatments can provide comparative data to conventional radiotherapy, the average 5-year survival rates for ^{252}Cf brachytherapy were ~75% compared with ~45% for conventional methods.[6]

In 1998 the REDC reestablished ^{252}Cf medical source fabrication in the United States and produced its first sources for use in manual loading. Efforts at the REDC are now focused on production of miniature HDR sources which can be coupled to a remote afterloader unit, analogous to photon HDR brachytherapy systems.[1] Remote afterloading will greatly reduce exposure to clinical personnel and permit use of ^{252}Cf sources with intensities an order of magnitude greater than those currently in use. As a result, treatment times can be reduced from hours to minutes. Source outer diameters ~1 mm will permit treatment of glioblastomas.

Boron augmentation of dose near a ^{252}Cf brachytherapy source is insignificant because of the large fast neutron dose. One study indicates neutron dose rates of 5.8 cGy $h^{-1} \mu g^{-1}$ of ^{252}Cf at a radial distance of 0.5 cm from a source and 1.9 cGy $h^{-1} \mu g^{-1}$ at 1 cm.[7] However, at greater distances the boron contribution becomes significant, increasing tumor dose by 7.7% at 5 cm and 23.1% at 10 cm, assuming 50 ppm ^{10}B in a water-equivalent phantom. The advantage of BNCT-boosted brachytherapy is the improved killing of cancerous cells away from the neutron source, such as those represented by metastases, while the fast neutrons provide a killing mechanism which is not dependent on thorough impregnation of boron into the cancerous cells.

4. BNCT PHARMACEUTICAL TESTING AT THE CUF

A recent BNCT experiment investigated the uptake of several experimental boronated compounds by living human lung cancer cells, followed by irradiation at the CUF with a thermal neutron flux of $2 \times 10^8 cm^{-2} s^{-1}$ in a mixed neutron field using 28 mg of ^{252}Cf.[1,2] The irradiation of the experimental compounds, two nucleosides and two amino acids developed by G. W. Kabalka at the University of Tennessee, Knoxville, clearly differentiated the cell-killing potential of the compounds and helped to identify those most promising for further study. The irradiation protocol developed for this experiment can be conveniently adapted to a standardized procedure for evaluating the relative cell-killing effectiveness of experimental NCT pharmaceuticals in a neutron flux.

5. CALIFORNIUM-DRIVEN SUBCRITICAL MULTIPLYING ASSEMBLIES

Kim *et al.*[8] concluded that 1 g of ^{252}Cf could provide a useable epithermal beam for NCT but that a more realistic (in terms of availability) 100-mg source inside a ^{235}U SMA could provide a feasible source of neutrons for BNCT. Californium-252—driven SMAs have been previously demonstrated. An SMA at DOE's Mound Facility which used approximately 2 kg of ^{235}U provided a 28X increase in the neutron source intensity over that of a bare ^{252}Cf source.[9] A thermal flux of $10^9\,cm^{-2}s^{-1}$ was achieved using only a 5-mg ^{252}Cf source. A major advantage of the SMA was its moderate operating costs, as operators were required only during irradiation setup and initiation. Although additional safety margin might be designed into a therapeutic unit, thus reducing the multiplication factor, the overall concept is simpler in design and maintenance compared with other neutron sources for BNCT.

Researchers at the REDC are currently investigating multiplier designs which could provide comparable neutron emissions with less ^{235}U, thereby addressing the safeguards concerns of using ^{235}U and potentially reducing the large ^{252}Cf masses required for a hospital-based unit. An item of current interest is the practicality of installing an SMA at the CUF to provide a prototype therapeutic unit, taking advantage of the existing infrastructure and the hundreds of milligrams of ^{252}Cf inventory to reduce SMA requirements, ^{235}U masses, and operational costs.

6. CALIFORNIUM-252 SOURCE AVAILABILITY

In addition to source usage at the CUF, the Californium Industrial/University Loan Program provides cost-effective ^{252}Cf neutron sources for research at the user's facility. Direct purchase of 100 mg of ^{252}Cf would be prohibitive at current costs and would require source replacement every several years due to isotope decay. However, loans of ^{252}Cf sources for noncommercial medical research could provide a very cost-effective means of therapy using an SMA. Sources can be loaned from DOE without charges for the isotope and often from existing inventory without source fabrication charges. Sources with intensities exceeding 10^{10} neutrons/s can be loaned for <$25,000 and smaller sources for less. University and medical loans of small sources ~10^7 neutrons/s incur only container transport costs. Loaned sources are returned to ORNL after use without disposal concerns. Sources with intensities up to 10^{11} neutrons/s require fabrication, but for a loan the total fabrication and delivery costs remain <$50,000. Agencies or subcontractors of the U.S. government as well as universities are eligible for source loans. The Californium Loan Program is typically limited to institutions within the United States, although with DOE approval qualified foreign institutions can receive loaned sources.

Commercial applications do not qualify for source loans. Current DOE isotope charges are $56 per microgram of ^{252}Cf and $56,000 per milligram, plus encapsulation, handling, and shipping fees. ORNL cannot sell sources in competition with those commercially available, but specialized and very intense sources (typically >10^{10} neutrons/s) can be supplied by ORNL. Brachytherapy and other ^{252}Cf medical sources for clinical trials can be loaned from DOE until a commercial product becomes available.

7. SUMMARY AND FUTURE WORK

The CUF is versatile for NCT-related experiments, and the potential applications of its world-class ^{252}Cf inventory remain to be explored. A ^{252}Cf-driven SMA for hospital-based therapy is achievable in principle based on previous SMA experience and on computer simulation for BNCT usage. The number of feasible SMA treatment centers would be very limited based on current rates of ^{252}Cf production (each center might require nearly 25% of the current annual ^{252}Cf production), but this limitation could be addressed by (1) increased annual ^{252}Cf production if mandated by DOE and (2) technological improvements in SMA design, which could decrease ^{252}Cf requirements by ~50% compared with current expectations for a hospital-based unit. The REDC is currently exploring an advanced SMA design to address these technological issues, while the CUF is conducting preliminary evaluations of the feasibility of incorporating a prototype SMA design into its hot cell infrastructure which, using the large REDC ^{252}Cf inventory, could be expected to provide a therapeutically useful and very cost-effective BNCT beam.

The future availability of ^{252}Cf HDR brachytherapy represents a useful complement to current clinical BNCT. Neutron brachytherapy can treat more types of tumors as well as glioblastomas, has been clinically tested in thousands of patients, and can be conveniently combined with the use of boronated pharmaceuticals for clinical testing. Although the use of boronated pharmaceuticals is not essential for effective brachytherapy treatments, such combined therapy can provide a significantly broadened patient profile available for BNCT treatment to improve clinical trial statistics of boronated pharmaceuticals prior to U.S. Food and Drug Administration approval. With the expected total cost for the ^{252}Cf HDR source and remote afterloader unit at <$500,000, ^{252}Cf HDR brachytherapy has the potential of making neutron therapy widely available to large numbers of patients.

ACKNOWLEDGMENT

This work was supported by the U.S. Department of Energy under contract DE-AC05-96OR22464.

REFERENCES

1. R.C. Martin, T.E. Byrne, and L.F. Miller, Biomedical neutron research at the Californium User Facility for Neutron Science, *J. Radioanal. Nucl. Chem.*, 236:5–10, 1998.
2. T.E. Byrne, L.F. Miller, G.W. Kabalka, and R.C. Martin, A new simplified system for the evaluation of BNCT pharmaceuticals, *in:* "Frontiers in Neutron Capture Therapy," M.F. Hawthorne, K. Shelly, R.W. Wiersema, eds., Kluwer Academic/Plenum Publishers, New York, 2001, pp. 955–957.
3. R.C. Martin, E.H. Smith, D.C. Glasgow, E.A. Jerde, D.L. Marsh, and L. Zhao, Neutron activation analysis at the Californium User Facility for Neutron Science, *in:* "Technologies for the New Century," American Nuclear Society, La Grange Park, Illinois, 1998, pp. II-442 to II-449.

4. Y. Maruyama, J.G. Wierzbicki, B.M. Vtyurin, and K. Kaneta, Californium-252 neutron brachytherapy, *in:* "Principles and Practice of Brachytherapy," S. Nag, ed., Futura Publishing, Armonk, New York, 1997, pp. 649–687.
5. J.G. Wierzbicki, ed., "Californium-252: Isotope for 21st Century Radiotherapy," Kluwer Academic Publishers, Dordrecht, The Netherlands, 1997.
6. M.J. Rivard, Harper Hospital, Wayne State University, personal communication, September 1998.
7. J.C. Yanch, and R.G. Zamenhof, Dosimetry of ^{252}Cf sources for neutron radiotherapy with and without augmentation by boron neutron capture therapy, *Radiat. Res.*, 131:249–256, 1992.
8. J.K. Kim, J.C. Yanch, and M.J. Wilson, Californium-based epithermal neutron beams for neutron capture therapy, *in:* "Advances in Neutron Capture Therapy," A.H. Soloway, R.F. Barth, and D.E. Carpenter, eds., Plenum Press, New York, 1993, pp. 131–134.
9. J.D. Hastings, Performance of the Mound Facility californium multiplier, *in:* "Neutron Radiography," J.P. Barton, and P. von der Hardt, eds., D. Reidel Publishing Co., Dordrecht, Holland, 1983, pp. 695–697.

NEUTRONIC DESIGN OF A ^{252}Cf-^{235}U EPITHERMAL IRRADIATION FACILITY APPLIED TO BNCT

A. P. Andrade and T. P. R. Campos

Nuclear Eng. Depart.
Federal Univ. of Minas Gerais
842, Contorno Avenue, Room 912
Belo Horizonte, MG, Brasil

1. INTRODUCTION

The neutron generators supporting to the clinical trials are usually the fission research reactors[1,2] and the high-current proton accelerators.[3,4] In both case, special systems of moderator, collimators, filters and shielding are optimally assembled in order to match the requirements of the BNCT treatment. Those requirements are the high intensity of the epithermal neutron component, the low level of contamination, and the radiation shielding criteria. However, it is not always possible to fulfill all engineering and physical requirements due to the adaptation of a beam port of a nuclear research reactor to accomplish a medical facility. It has often required large changes in the reactor core and in the configuration of the structural building, investing the high cost. This paper addresses about an intermediary alternative, cheap and simple, irradiation facility that is able to deliver a good quality epithermal neutron beam with low background of fast neutrons and gamma rays. Herein, a conceptual design is presented and neutronic simulations are carried out. The proposed facility is composed of a set of ^{252}Cf sources coupled with special systems of moderator, collimators, filters and shielding. A geometric and material-choice optimization is made in order to maximize the epithermal neutron flux and minimize the fast neutron and gamma ray doses in the irradiation port. Monte Carlo N-Particle transport code—MCNP[7] is used to estimate and optimize the main geometrical and material-choice parameters of the moderators, collimators, filters and shielding systems. The irradiation port is able to deliver an epithermal neutron flux at in the range of 4 eV up to 40 keV in a forward directional neutron current. In order to increase the intensity of the epithermal neutron flux in the beam port, a configuration loaded with

Frontiers in Neutron Capture Therapy, edited by Hawthorne *et al.*
Kluwer Academic / Plenum Publishers, New York, 2001.

^{252}Cf-^{235}U is proposed in a subcritical neutronic fashion. This configuration presents an additional fuel fissile mass—^{235}U coupled with the ^{252}Cf neutron sources. It can deliver epithermal neutron flux up to 10^8n/cm^{2}s with low background of fast neutrons and gamma rays comparable with the figure-of-merits of the beam ports of the *MITR-II*,[2] *BMRR*[1] and *Petten* facility, presented in literature. The present facility may give opportunity to other groups to continue their work in BNCT.[6,9]

2. THE PROPOSED ^{252}Cf AND ^{252}Cf-^{235}U IRRADIATION FACILITY

The optimal configuration of the proposed ^{252}Cf irradiator presents cylindrical geometry with a coaxial aluminum filter placed in the center canal for the extraction of the epithermal neutron beam. The irradiator is composed of 10 ^{252}Cf sources of 3 mg (1.6 Ci). A mixed field of neutrons and gamma rays is created from the fission of ^{252}Cf isotope and the disintegration of the fission products, approaching 6.9×10^{10} n/s and 3.96×10^{11} gamma-rays/s[7]. These sources have a radial distribution on the half height of the irradiator embed in cylinder heavy water moderator. The optimal radius is 15 cm for the source distribution and 5 cm for the central aluminum canal. External to the moderator tank of D_2O, a 15 cm thick bismuth shielding attenuates the undesirable gamma rays emitted from the own ^{252}Cf source and helps the neutron reflection on the walls. After the bismuth shielding, the irradiator is involved in a 35 cm thickness radiological shielding composed of light water with ^6Li isotope at the concentration of 1/3 for light water molecule. In order to protect the irradiator and its components from the external influence and to help shielding the irradiator, an external lead sheet of 5 cm thickness covers the irradiator. A thin sheet of lithium fluoride (enriched to 93% of ^6Li) is introduced between the D_2O moderator and the aluminum. The aluminum and LiF filters are placed in order to reduce the influence of the fast and thermal neutrons in the therapeutic beam. Collimators of bismuth and lithiated polyethylene (containing 45 wt% of lithium carbonate enriched to 93% in ^6Li) makes the neutron and gamma ray flux distributions centered in a cylinder irradiation port of 20 cm diameter. The ^{252}Cf-^{235}U configuration, embed on the D_2O moderator, presents two concentric cylindrical sheets of fissile fuel placed near the ^{252}Cf neutron sources. The fuel is composed of a homogeneous mixture of uranium-zirconium hydride alloy containing 8.4 wt% uranium enriched to 20% in ^{235}U. The hydrogen-to-zirconium atom ratio is 1.6, and the physical density of the fuel is 5.95 g/cm^3. The total ^{235}U mass contained in this configuration is 1.75 Kg.

3. MCNP SIMULATION

The optimal configuration for the ^{252}Cf irradiator was obtained from various simulations using MCNP4A for several three-dimension geometries of moderator, collimators, filters and shielding systems, applying materials like Pb, Bi, C, Al, LiF, ^6Li, ^{10}B, S, Fe, Si, H_2O, D_2O, lithiated polyethylene, high density concrete, air and the homogenized alloy of UZrH (20% enriched to ^{235}U). The neutron and gamma ray spectra emerging on the central aluminum canal for the various mediums and for its geometric dimensions were evaluated in all calculations. The nuclear data libraries ENDL-85 and ENDF/B-VI were used for the MCNP continuous-energy simulations. The ^{252}Cf O_2 neutrons sources emit neutrons with energy between 1.4 up to 8.5 MeV, according to the spectrum of the

^{252}Cf neutrons source produced by the Amersham Corporation.[7] Gamma rays with energy in the range between 0.5 and 6.5 MeV emitted due to the fission of the ^{252}Cf isotope and the decay of its fission products were considered in the transport of photons exclusively. Tallies of currents and fluxes of neutrons and gamma rays in the various surfaces and cells were classified in 7 photon groups in the range of 0 up to 8 MeV, and 3 neutron groups, a thermal group in the range of 0 up to 4 eV; an epithermal group in the range of 4 eV up to 40 keV; and a fast group in the range of 40 keV up to 15 MeV. The statistical error for the convergence of the tallies was required to be less than 10% and after refined to 5%. The criticality calculations of the proposed high flux ^{252}Cf-^{235}U irradiator were also computed in MCNP. The k_{eff} multiplication factor was estimated over 300 cycles, counting 1,500 particles per cycle.

4. RESULTS

The major neutronic characteristics of the epithermal neutron beams, found in the literature,[2,4,5,8] and estimated for the irradiation port of the proposed ^{252}Cf irradiators are tabulated in Table 1. Low levels of fast neutron and gamma ray doses are found in the both proposed configurations. The irradiator facility composed of only ^{252}Cf neutron sources presents a higher value of the figure-of-merit J/ϕ that represents the convergence of the beam (defined as the ratio of the neutron flux and current incident on the external surface of the irradiation port). However, it also presents low neutron intensity, requiring the use of additional fissile material in order to obtain a practicable beam for using in BNCT. The mixed configuration of ^{252}Cf-^{235}U presents an epithermal neutron flux up to 10^8 n/cm^2.s achieving an amplification factor of 9.8 for epithermal neutrons in relation to the previous configuration without fuel, maintaining the low rates of fast neutron and gamma ray doses, and the directionality factor of the therapeutic beam. The criticality factor, k_{eff}, for this mixed configuration was estimated to 0.95885 ± 0.00131, keeping the irradiator facility in subcritical condition. The reduction of fuel mass, or reduction of the numbers of Cf sources can adjust the k_{eff} to a more safe position approaching less than 0.90. The power of the central core with the k_{eff} at 0.95885 ± 0.00131 is 0.1 kWatts.

The ^{252}Cf source configuration presents a neutron spectrum composed by 12% of thermal neutrons, 76% of epithermal neutrons and 12% of fast neutrons. The ^{252}Cf-^{235}U configuration presents a neutron spectrum composed by 7% of thermal neutrons, 80% of epithermal neutrons and 13% of fast neutrons. In the both configurations present a good ratio of epithermal and fast neutrons higher than 6, in the irradiation port. The

Table 1. Major parameters of the epithermal neutrons beam ports applied to NCT in *MIT*, *BNL*, and *Petten*, and those of the proposed ^{252}Cf and ^{252}Cf-^{235}U irradiators[2,4,5,8]

Reactor	ϕ_{epi} [n/cm^2.s]	D_f/ϕ_{epi} [Gy.cm^2/n]	D_f [cGy/h]	D_γ/ϕ_{epi} [Gy.cm^2/n]	D_γ [cGy/h]	J/ϕ
BMRR	1.8×10^9	4.3×10^{-13}	2.8×10^2	1.3×10^{-13}	8.4×10^1	0.67
MITR-II	2×10^8	13×10^{-13}	9.4×10^1	14×10^{-13}	1.0×10^2	0.55
Petten	3.3×10^8	10.4×10^{-13}	1.2×10^2	8.4×10^{-13}	1.0×10^2	>0.8
^{252}Cf + 1.75 Kg ^{235}U	1.1×10^8	32×10^{-13}	1.3×10^2	20×10^{-13}	7.9×10^1	0.998
^{252}Cf Irradiator	1.1×10^7	31×10^{-13}	1.2×10^1	16×10^{-13}	6.49	0.98

undesirable dose of gamma rays, in the mixed ^{252}Cf-^{235}U configuration, is maintained in acceptable limits.

5. CONCLUSION

Both configurations of the proposed irradiator, ^{252}Cf and ^{252}Cf-^{235}U, present reasonable epithermal neutron spectra with low background of fast neutrons and gamma rays. It suggests a suitable use of the spontaneous fission ^{252}Cf neutron sources coupled with ^{235}U to the assembly of neutron irradiators. In spite of the low neutrons intensities obtained in the proposed ^{252}Cf irradiator configuration, this configuration already represents a good, safe and cheap neutron source to be applied in low-dose-rate neutron capture therapies. It may be applied in a fractionated treatment planned in various sections. The proposed configuration of a ^{252}Cf-^{235}U epithermal neutron facility has the drawback of the constant presence of the fissile fuel and its fission products. However the subcriticality and the expected simple procedures of maintenance and operation provide an attractive alternative as an epithermal neutron source. The ^{252}Cf-^{235}U and ^{252}Cf epithermal neutron facilities seem a good efficient neutron source able to be applied to neutron capture therapies in hospitals. Heat transfer analysis and final neutronic evaluations have been proposed for future work. Also, safe considerations for the fissile fuel coupled with ^{252}Cf source, toxicity, and analysis of costs for construction, operation and maintenance of both the proposed ^{252}Cf configurations are required in order to certify its viability on neutron capture therapies.

REFERENCES

1. H.B. Liu, Enhancement of the epithermal neutron beam used for boron neutron capture therapy. *International Journal of Oncology Biology and Physics*, 28(5):1149–1156, 1994.
2. E.L. Redmond II, J.C. Yanch, and O.K. Harling, Monte Carlo simulation of the Massachusetts Institute of Technology Research Reactor. *Nuclear Technology*, 106:1–14, 1994.
3. J.F. Evans, and T.E. Blue, Shielding design of a treatment room for an accelerator-based epithermal neutron irradiation facility for BNCT. *Health Physics*, 71(5):692–699, 1996.
4. J.E. Woolard, Development and application of neutron field optimization parameters for an accelerator-based neutron source for BNCT. *Nuclear Technology*, 115:100–113, 1996.
5. P. Watkins, MCNP Calculations for the design and characterization of the Petten BNCT epithermal neutron beam. *Progress In Neutron Capture Therapy For Cancer*, New-York, 1992. Plenum Press, pp. 71–73.
6. I.F., Chaves, and T.P.R. Campos, Dosage Evaluation on Boron Neutron Capture Therapy, *Brazilian Jour. Cancer*, 42(3):151–160, 1996.
7. AMERSHAM. Radiation sources for industrial gauging and analytical instrum. *Amersham Corporation*, 1997.
8. H.B. Liu, Design of neutron beams for neutron capture therapy using a 300-kW slab TRIGA reactor. *Nuclear Technology*, 109:314–325, 1995.
9. T.P.R. Campos, and I.F. Chaves, Radiotherapy coupling boron neutron capture therapy (BNCT) and Cf-252 brachytherapy. *Eur. J. Cancer*, 31A:197, Sup.5., 1997.

MEASUREMENT OF FAST NEUTRON, GAMMA, AND THERMAL NEUTRON DOSE AROUND A Cf-252 SOURCE

Somsak Wanwilairat[1,2], Thiraphat Vilaithong[1], Vicharn Lorvidhaya[2], Rainer Schmidt[3], and Wilfried Hoffmann[4]

[1]Fast Neutron Research Facility
Department of Physics, Faculty of Science
Chiang Mai University, Chiang Mai 50200, Thailand
[2]Radiation Oncology Section
Department of Radiology, Faculty of Medicine
Chiang Mai University, Chiang Mai 50000, Thailand
[3]Department for Radiation Oncology
University Hospital Eppendorf
University of Hamburg
Martinistr. 52, 20241 Hamburg, Germany
[4]Department of Physics
University of Wuppertal
42097 Wuppertal, Germany

INTRODUCTION

The greater biological effectiveness of neutrons compared to photons is increasing the interest in using neutron sources in brachytherapy. Because of the achievable neutron intensity and the half-life ^{252}Cf is very attractive for this type of clinical application.

At the University Hospital in Chiang Mai a high intense ^{252}Cf has been installed for clinical application in brachytherapy. Before the source can be used in medical trials, an elaborate experimental investigation on the dose distributions of neutrons and photons has to be carried out. As it is intended to use the neutron source for bulky tumors, i.e. advanced stage of cervical carcinoma, the use of ^{252}Cf in brachytherapy may be improved by adding a considerable amount of dose via the boron neutron capture reaction (BNCT) of thermal neutrons with ^{10}B located in the tumor. Therefore the flux of thermal neutrons has to be evaluated for the ^{252}Cf-source.

Frontiers in Neutron Capture Therapy, edited by Hawthorne *et al.*
Kluwer Academic / Plenum Publishers, New York, 2001.

Table 1. Characteristic parameter of ^{252}Cf -source

Active length [mm]	6.0
Exterior length [mm]	12.0
External diameter [mm]	5.5
Wall thickness [mm]	1.1
Actual weight of ^{252}Cf [μg]*	100.1

*calculated from the weight specified in 1990.

Different types of detectors are used to determine the total dose, photon dose and the thermal neutron flux in different positions in a water phantom where the ^{252}Cf-source is positioned close to the center.

METHODS AND MATERIALS

The characteristic parameters of the source uses for the dose measurements are described in Table 1. The measurements were performed in a water tank of the dimension (length * width * height): 53 cm * 56 cm * 40 cm with 1 cm thick lucite walls. The Cf-source was housed in a cylindrical lucite tube with an outer diameter of 9 mm. The closest distance between source and chamber was 16 mm, defining the distance between the center of the source and chamber, respectively.

Three different ionization chambers were used, each produced by Wellhoefer Dosimetry (Germany) in a comparable design. The inner volume of the thimble chambers was 0.32 cc. The TE/TE chamber was made out of A150 and was operated with TE gas (composition by partial pressure: 64.4% CH_4, 32.4% CO_2, 3.2% N_2). The second chamber (Mg/Ar) consisted out of magnesium, the third chamber (MgB/Ar) was a magnesium chamber of identical design additionally coated with 3 μm ^{10}B of 92% purity internally (see Table 2), both chamber are flushed with argon-gas of 99.995 purity.[1] All chambers are calibrated in a ^{60}Co-beam in terms of absorbed dose to water.

The chambers have a gas-flow provision and during the measurements a constant gas-flow of 17 cc/min. was used. The gas was flushed through a 15 m long copper pipe to adjust the temperature of the gas to room conditions.

As the Mg/Ar-chamber could be operated only close to the source where the induced charge was considerably greater as the resolution of the electrometer, an GM-counter (GM2, Far West Technology, Goleta, USA) was employed to determine the photon dose additionally. This counter could be operated only at large distances to the source where the dose rate and the dead time of the counter are related such that the dead time correction was always below 2%. Both photon sensitive detectors were used

Table 2. Dose values at different depths

Depth	3 cm	5 cm	10 cm
Total dose [cGy/h * μg]	0.336	0.109	0.026
Gamma dose [cGy/h * μg]	0.143	0.056	0.016
Neutron dose [cGy/h * μg]	0.192	0.053	0.008
Thermal dose [cGy/h * μg]	0.0015	0.0012	0.0004
50 ppm ^{10}B dose [cGy/h * μg]	0.0284	0.0217	0.0075
Rel. ^{10}B enhancement [%]	8	20	28

in an overlapping region between 5 to 10 cm distance to the source, but no adjustments to fit the dose distributions were necessary.

The calibration of the MgB/Ar-chamber for a homogeneous flux of thermal neutrons with 37 meV was performed at the standard laboratory PTB in Germany.[1]

For the conversion of neutron flux into dose in muscle tissue the kerma factor was taken from ICRU-Report 44.[2]

The data to convert the thermal neutron fluence into an additional dose by use of ^{10}B with a concentration of 50 ppm by use of kerma factors are adapted form a paper from Konijnenberg *et al.*[3] From these data the reading of the MgB/Ar-chamber can be converted into neutron flux, absorbed dose from thermal neutrons, and boron dose originating for thermal neutrons.

The positioning of the chambers was performed with an accuracy of 1 mm. This corresponds to a different in the dose of 5% close to the source. For the calibration of the TE/TE- and Mg/Ar-chamber an accuracy of 1% was assumed.

The directional response of the TE/TE and Mg/Ar chamber is almost homogeneous. For the MgB/Ar ionization chamber the directional response for thermal neutrons was determined and has to be taken into account, the directional response of the chamber was integrated in 4π according to the investigation given in.[1]

$$R(4\pi) = 0.835\,R(90^\circ)$$

where R(90°) is the response of the chamber to laterally incident thermal neutrons. If the thermal neutron field is not homogeneous the directional response of the chamber could introduce some error that was estimated to be less than 4%.

The energy response of the MgB/Ar-chamber was investigated in,[1] too. For the evaluation of the thermal neutron flux the energy spectrum would have to be known. But as long as the measured charge, generated in the chamber almost exclusively by reaction with the boron coating, will be used to calculate dose enhancement by boron, the spectrum does not have to be known. So only differences in the spectrum during calibration and measurement will affect the boron calculations. These differences are supposed to introduce an error of 5%. A summary of the estimated uncertainties is given in Table 3.

Simulation calculations were performed by use of the Monte Carlo Code MCNP, version 4a. The $^{10}B(n,\alpha)^7Li$ reaction rates on the detector surfaces were calculated. The neutron cross section data were taken from FSXLIB-J3R2 files of the JENDL-3.2 library.[5,6] Each MCNP simulation used 100,000 source neutrons and delivered a result with a statistical accuracy of 5% or better.

Table 3. Estimated uncertainties in dose calculations [%]

Item	Neutron dose	Photon dose	Thermal neutron
Position in phantom	5	5	5
Displacement correction	1.5	1.5	0.5
Directional response	1	1	4
Temperature, pressure, gas flow	0.5	0.5	0.5
Electrometer	1	1	0.5
Photon calibration	1	1	1
Neutron calibration	5	—	5
Total error	7.5	5.5	8.2

The MgB/Ar- and TE/TE-chamber were used to measure the thermal and total dose distribution along the source in 2, 5 and 10 cm distances (longitudinal scans), transversal to these scans at the same distances (transversal scans). From these scans the center of the source was determined (Figure 1). Depth scans were performed perpendicular to the longitudinal and transversal scans on the axis of the source. The distances between source and chamber were defined as the distance from the center of the source to the center of the chamber. The minimal distance that could be realized was 1.6 cm.

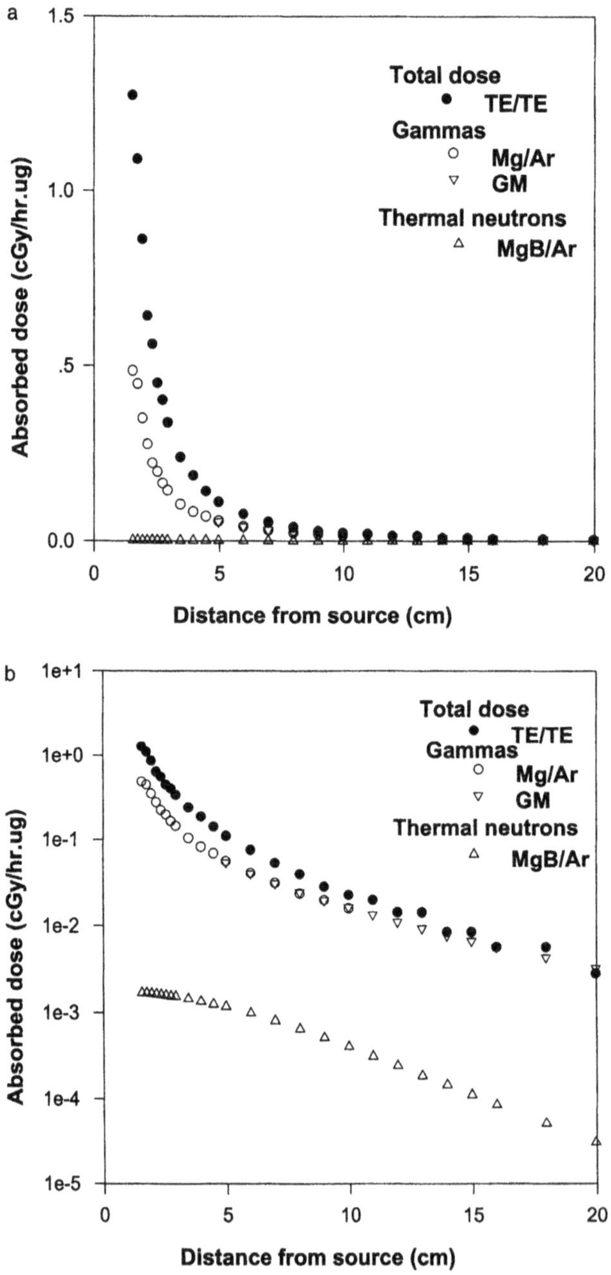

Figure 1. Measured total dose, gamma dose and thermal neutron dose in water as a function of distance from the source in, a) linear scale, b) logarithmic scale.

After determining the center of the source by longitudinal and transversal scans the Mg/Ar- and TE/TE-chamber were used to measure the total dose, the Mg/Ar-chamber and the GM-counter were applied for the gamma dose. A maximal distance to the source at which a sufficient current was produced by the Mg/Ar-chamber to give an accurate measurement was 10 cm. Contrarily the GM-counter could not be operated at close distances to the chamber because of the dose rate dependent dead time. The closest the GM counter was positioned to the source was 5 cm so that there was an overlap of measurements with the neutron insensitive devices between 5 and 10 cm distance. Though the photon calibration of the GM-counter and the Mg/Ar-chamber were obtained independently the evaluation of the photon component in the overlap region gave almost identical results with a mean difference of 1%.

RESULTS

From the MgB/Ar-chamber the thermal neutron dose was determined for different distances to the source. The results are given in Figure 1a on a linear and on Figure 1b on a logarithmic scale.

From the measurements with the ionization chambers and the GM-counter the relative contribution of neutrons and photons to the total dose at different distances to the source can be evaluated. From Figure 2 it can be seen that the behavior can be linearly interpolated in a region close to the surface. It can be stated that above 5 cm distance from the source the photon dose is larger than the neutron dose.

The scattering of the data in Figure 2 for large distances from the source is attributed to the low current from the TE/TE-chamber. As the measured data are influenced

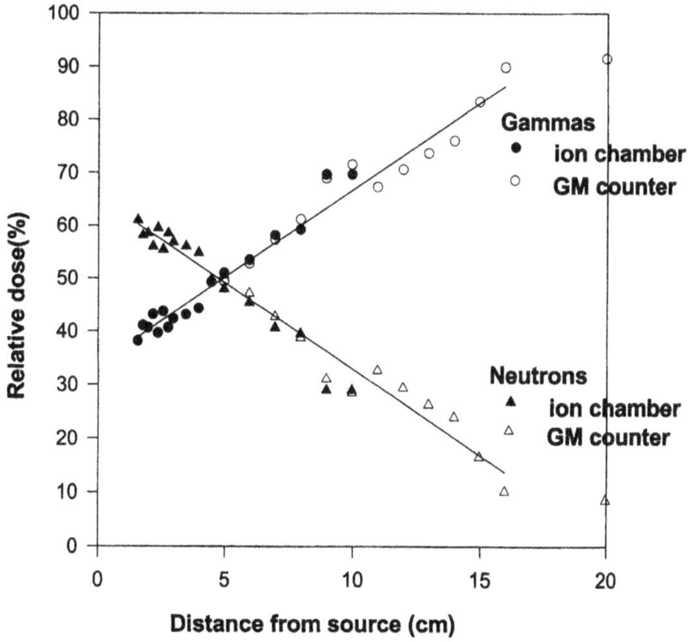

Figure 2. Relative dose contribution of fast neutron and gammas (a) and thermal neutrons (b) to the total dose as a function of distance from the source.

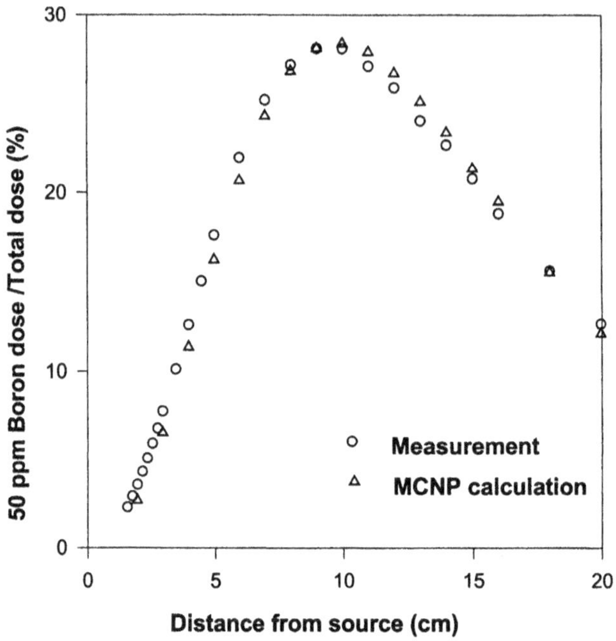

Figure 3. Comparison of the relative thermal neutron dose distribution as measured using MgB/Ar-chamber and calculated with MCNP-code, normalized to the maximum value of the MgB/Ar-chamber.

by the resolution of the electrometer, an analytical fit through the total dose data was performed. The fit of this function is shown in Figure 1 (solid line). At greater distances the accuracy of the total dose is influenced by the unfavorable dose rate in relation to the instrumental resolution of the electrometer.

For the consideration of the use of ^{10}B it will be important to evaluate the additional dose. The thermal neutron dose given in Figure 1 is converted into the boron dose for 50 ppm ^{10}B by use of the kerma factors. The relative enhancement is evaluated by relating the boron dose to the total dose at the corresponding position. The total dose is calculated according to fit function to avoid an overestimation of the inaccuracy in the total dose described above. In Figure 3 these calculated data are compared to the MCNP simulation. The calculated data are normalized to the measured value at the maximum. Therefore this comparison is presenting a consistency of the relative change of the boron enhancement with distance to the source.

In Table 2 the different dose components at 3, 5 and 10 cm depth are summarized. It can be concluded that the additional use of 50 ppm ^{10}B will raise the dose by up to 28%.

DISCUSSION

The use of paired ionization chambers to separate the neutron and photon dose in the field of a ^{252}Cf-source or an epithermal reactor beam is described by several authors.[7,8] Here a GM-counter has been applied to extend the measurements of the photon dose to larger distances from the source and a boron coated ionization chamber with magnesium wall material was use to detect the thermal neutron component in the mixed beam. The measured distributions of the neutron and gamma dose show beside from the $1/d^2$ dependence an increase of the relative contribution of gammas to the total dose due to the 2.2 MeV gamma rays produced when the thermalized neutrons are captured by hydrogen nuclei. The relative fast neutron contribution decreases. The measurements show that at 5 cm distance from the source the ratio of gamma and neutron dose is one. This is in good agreement with earlier calculations by Yanch and Zamenhof.[9] Above 16 cm distance 90% of the total dose is caused by gammas only. For treatment planning purposes the relative neutron decrease and the relative gamma increase can be described by a linear function for distances between 2 and 15 cm.

The thermal dose measurements and the corresponding determination of the 50 ppm ^{10}B dose show that the dose due to the presence of Boron-10 increases almost linearly up to a distance of 9.5 cm and then decreases. The maximum dose increases due to 50 ppm ^{10}B is 28% of the total dose. The shape of this depth dose distribution is in very good agreement with the MCNP Monte Carlo calculation.

If the results are compared to the calculations of Yanch and Zamenhoff[9] it can be noted that their 50 ppm ^{10}B dose for 3 cm depth is 0.031 cGy/(h $*$ μg), which is in very good agreement with the values of 0.028 cGy/(h $*$ μg) of this study considering that their calculations were done for a different type of source. Their maximum increase in tumor dose for 50 ppm ^{10}B of 27.7% is in very good agreement with the measurements presented here, the peak position, however, was measured to be closer to the source namely at 9.5 cm depth in water. Above 10 cm distance the ^{10}B dose decreases rapidly. The results of this measurement now allow starting treatment planning in brachytherapy using BNCT.

ACKNOWLEDGMENT

This work was supported by the Thailand Research Fund (TRF) and the German Academic Exchange Service (DAAD).

REFERENCES

1. L. Luedemann, T. Matzen, M. Matzke, R. Schmidt, and W. Scobel: *Determination of the thermal neutron flux in a fast neutron beam by use of a boron-coated ionization chamber*. Med. Phys. 22, 1743–1747, 1995.
2. ICRU Report 44: *Tissue Substitutes in Radiation Dosimetry and Measurements. International Commission on Radiation Units and Measurements*, Bethesda, USA, 1988.
3. M.W. Konijnenberg, L.G.H. Dewit, B.J. Mijnheer, C.P.J. Raajmakers, and P.R.D. Watkins: Dose Homogeneity in boron Neutron Capture Therapy Using an Epithermal Neutron Beam. Radiat. Res. 142, 327–339, 1995.
4. J.F. Briesmeister: MCNP—A general Monto Carlo N-Particle transport code. Los Alamos National Laboratory report LA-12625-M, 1993.
5. K. Kosako, F. Maekawa, Y. Oyama, Y. Uno, and H. Maekawa: FSXLIB-J3R2: A Continuous Energy Cross Section Library for MCNP Based on JENDL-3.2. JAERI-Data/Code 94-20, 1994.
6. T. Nakagawa, K. Shibata, S. Chiba, T. Fukahori, Y. Nakajima, Y. Kikuchi, T. Kawano, Y. Kanda, T. Osawa, H. Matsunobu, M. Kawai, A. Zukeiran, T. Watanabe, S. Igarashi, K. Kosako, and T. Asami: *Japanese Evaluated Nuclear Data Library Version 3 Revision-2:JENDL-3.2., J.* Nucl. Sci. Technol., 32, pp. 1259–1271. 1995.
7. L.L. Anderson: Cf-252 physics and dosimetry. Nuclear Science Applications 2, 273–281, 1986.
8. R.D. Rogus, O.K. Harling, and J.C. Yanch: *Mixed field dosimetry of epithermal neutron beams for boron capture therapy at the MITR-II research reactor*. Med. Phys. 21, 1611–1625, 1994.
9. J.C. Yanch, and R.G. Zamenhof: *Dosimetry of ^{252}Cf Sources for Neutron Radiotherapy with and without Augmentation by Boron Neutron Capture Therapy*. Radiat. Res. 131, 249–256, 1992.

CALCULATED BORON NEUTRON CAPTURE DOSE ENHANCEMENT IN ICRU 44 BRAIN WITH ^{252}Cf

Mark J. Rivard[1], Laurie E. Gaspar[2], and Jacek G. Wierzbicki[3]

[1]Department of Radiation Oncology
Tufts University School of Medicine
New England Medical Center #246
750 Washington Street
Boston, Massachusetts 02111
[2]Department of of Radiation Oncology
University of Colorado Health Sciences Center
4200 East 9th Avenue
Denver, Colorado 80262
[3]Cancer Treatment Center
Saint Mary's Medical Center
830 South Jefferson Road
Saginaw, Michigan 48602

1. INTRODUCTION

As ^{252}Cf is a fast neutron emitter of relatively low average energy, there is potential to augment ^{252}Cf brachytherapy with boron neutron capture (BNC) dose enhancement for treatment of malignant disease. ^{252}Cf neutrons moderate within the human tissue and may be captured through the $^{10}B(n,\alpha + g)^{7}Li$ nuclear reaction, Q = 2.79 MeV. Following this nuclear capture, a 477.6 keV photon is emitted 93.7% of the time by relaxation of the excited ^{7}Li nucleus; however, the large majority of locally absorbed high-LET dose is deposited by the alpha particle and lithium ion. Consequently, it is possible that, with ^{10}B-loaded drugs which have affinity towards malignant tumor cells, ^{252}Cf brachytherapy may benefit from BNC dose enhancement. Calculations of the BNC dose enhancement are made for a variety of ^{10}B loadings and sizes of phantom material

Frontiers in Neutron Capture Therapy, edited by Hawthorne *et al.*
Kluwer Academic / Plenum Publishers, New York, 2001.

composed of, as an illustration, of brain tissue, as well for tumors of diameter ranging from 2 to 6 cm. Additionally, the moderated ^{252}Cf neutron energy spectrum is examined to demonstrate significant perturbation of the thermal neutron flux by increasing ^{10}B loadings.

2. MATERIALS AND METHODS

In this calculative study, MCNP[1] was used in a distributed computing environment[2–4] for brisk computation of the neutron energy spectrum and calculation of boron neutron capture reactions (BNCR). To permit accurate thermal neutron transport, the MCNP solid-state S(α,β) neutron scattering library (lwtr.01t) was used to model low-energy neutron transport on hydrogen.[4] Throughout, the calculative geometry was spherical with a point ^{252}Cf source positioned centrally. The entire volume was divided into radial bins using spheres of varying radii to permit spatial resolution. Phantom diameters ranged from 10 to 30 cm, and the ^{252}Cf neutron energy spectrum was modeled with an isotropic Maxwellian distribution with T = 1.42 MeV.[5]

Neutron transport was calculated with ICRU 44 brain material and a mass density of 1.04 g/cm^3. This material was loaded with uniform concentrations of ^{10}B ranging from 1 to 500 μg ^{10}B per gram of brain tissue. Unlike the approach of Zamenhof *et al.*[6] where neutron fluxes were convolved with kerma to dose conversion factors, the BNCR dose was calculated using a dose deposition formalism as presented in Equation 1 where only the high-LET dose was determined.

$$D_{BNCR}(r, S, L) = 133.488 \frac{(2.34\ NCR)}{\rho} \quad (1)$$

where:

$D_{BNCR}(r,S,L)$ =	high-LET BNCR dose rate	[cGy/h-μg]
r =	radius from phantom center	[cm]
S =	^{252}Cf source strength	[μg]
L =	μg ^{10}B loading per gram brain tissue	[dimensionless]
133.488 =	conversion factor, MeV/g-neutron to cGy/h-μg	[dimensionless]
2.34 =	high-LET energy deposited by α and ^6Li nuclei	[MeV/NCR]
NCR =	calculated neutron capture reaction rate	[NCR/neutron-cm^3]
ρ =	mass density of ICRU 44 brain	[g/cm^3]

An additional studies was performed which examined BNCR dose as a function of radius in which 10 μg ^{10}B per gram of brain tissue (10 ppm) was uniformly distributed throughout a 15 cm diameter head phantom. A centrally positioned tumor was modeled with a ^{10}B loading of 30 ppm with tumor diameter ranging from 2 to 6 cm. These loadings were used to model a clinical tumor environment and ^{10}B tumor loadings with a tumor : healthy brain ^{10}B specificity ratio of 3 : 1. For all studies, the neutron energy spectrum at each radii was divided into energy bins from 1 meV to 15.85 MeV with 10 energy bins per decade of energy. This scheme permitted regular bin spacing when examining the moderated neutron energy spectrum in a log-log manner.

3. RESULTS

3.1. Calculated BNCR Dose Rate for a Variety of ^{10}B Loadings and Phantom Sizes

For a brain phantom with a diameter of 15 cm, the impact of ^{10}B loading and radial position on the calculated BNCR dose rate was determined. Figure 1 presents the calculated BNCR dose rate [cGy/(h-ppm ^{10}B-mg ^{252}Cf)] for radii of 0.5, 1, 2, 3, 4, and 5 cm where straight lines are provided to guide the eye between calculated values. When displayed in this manner, it is evident that for increasing ^{10}B loadings, the BNCR dose rate did not increase linearly. This effect was most apparent for the larger radii where the proportion of thermal neutrons was highest, yet the impact of thermal neutron burnup and ^{10}B self-shielding was greatest. For clinically relevant ^{10}B loadings of 10 and 30 ppm, the degree of non-linearity was −5% and −14%, respectively for all distances, 0.25 to 7.0 cm, examined. The impact of phantom size upon the calculated BNCR dose rate was examined at radii of 0.5, 1, 2, and 5 cm in Figure 2 with a ^{10}B concentration of 30 ppm. At a radius of 5 cm, the BNCR dose rate increased 47%, from 8.5 to 12.5 cGy/(h-mg ^{252}Cf), when examining phanta with diameters of 15 and 30 cm, respectively. At a radius of 1 cm, there was a 74% increase in BNCR dose rate for diameters of 10 and 30 cm. Evidently, the presence of scattering material increased the BNCR dose rate due to inscatter of thermal neutrons which would have otherwise exited the phantom volume.

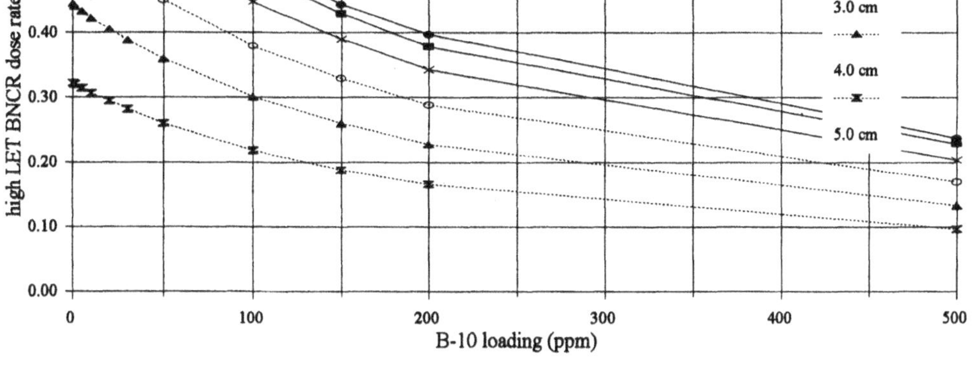

Figure 1. Impact of ^{10}B loading on calculated BNCR dose rate (15 cm diameter brain phantom).

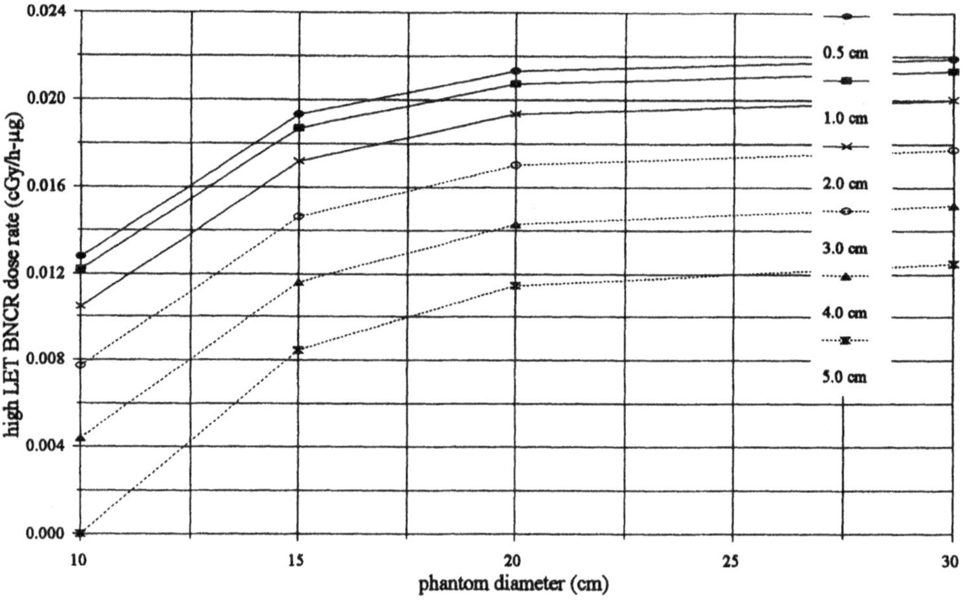

Figure 2. Impact of phantom size on calculated BNCR dose rate in brain loaded with 30 ppm ^{10}B.

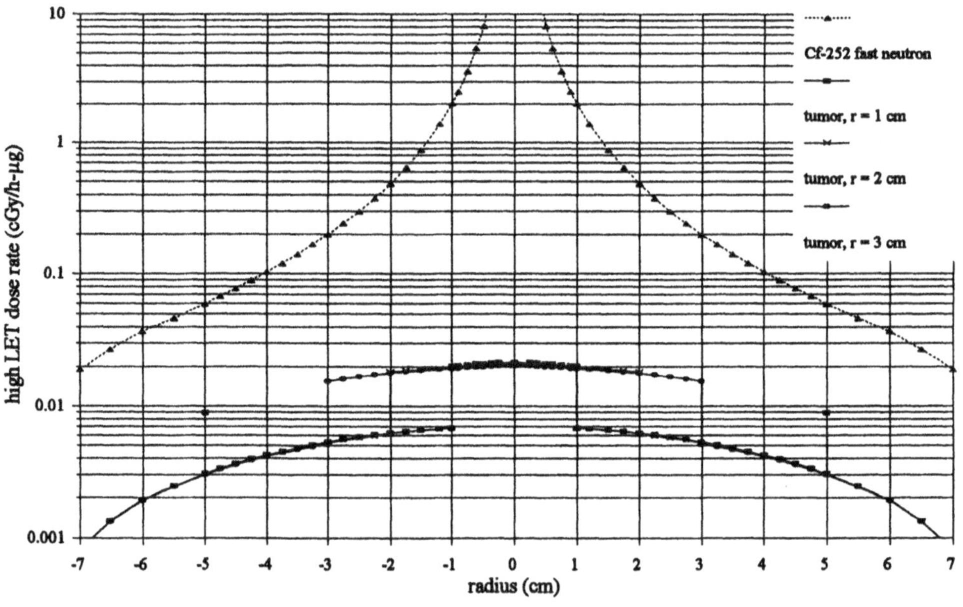

Figure 3. High LET dose distribution for a BNCR enhanced ^{252}Cf brachytherapy tumor model.

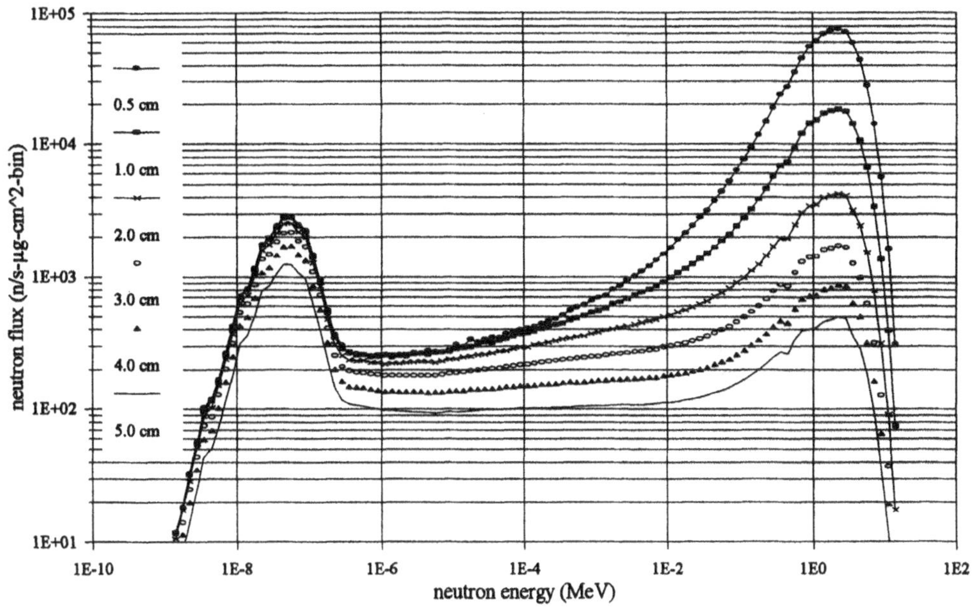

Figure 4. Moderated neutron energy spectrum at 0.5, 1.0, 2.0, 3.0, 4.0, and 5.0 cm (30 ppm ^{10}B).

3.2. Calculated BNCR Dose Rate for a Clinical Tumor Model

High-LET dose distributions for BNCR enhanced ^{252}Cf brachytherapy are presented for tumors with diameters ranging from 2 to 6 cm in Figure 3. For comparison, the ^{252}Cf fast neutron dose rate is also shown. For a small quantity of microscopic disease positioned at a radius of 5 cm with a ^{10}B loading of 30 ppm, there was a 9.5% high-LET BNCR dose rate enhancement at this point. This 9.4% increase at 5 cm was obtained by adding the initial ^{252}Cf fast neutron dose rate, 0.0590 cGy/(h-mg ^{252}Cf), to the difference in high-LET BNCR dose rates between normal and tumor tissues, 0.0030 and 0.0089 cGy/(h-mg ^{252}Cf), for total high-LET dose rates of 0.0620 and 0.0679 cGy/(h-mg ^{252}Cf), respectively.

3.3. Calculated ^{252}Cf Neutron Energy Spectra

For a 15 cm diameter brain phantom with a 30 ppm ^{10}B loading, Figure 4 illustrates the impact on the neutron energy spectrum for varying radii. The gradual decrease of neutrons with thermal energies, compared to the near r^{+2} fall-off of the fast component, is apparent. The neutron energy spectrum at 5 cm is displayed in Figure 5 for the same phantom; however, the ^{10}B loading varies from 1 to 500 ppm. Finally, in Figure 6 are the neutron energy spectra at a radius of 3 cm with a 30 ppm ^{10}B loading for phanta with diameters from 10 to 20 cm.

4. DISCUSSION

4.1. Clinical Impact of BNCR Dose Enhancement on ^{252}Cf Brachytherapy

To assay the clinical utility of BNCR dose enhancement at 5 cm from the ^{252}Cf source, one must multiply the relative biological effectiveness (RBE) values for each type

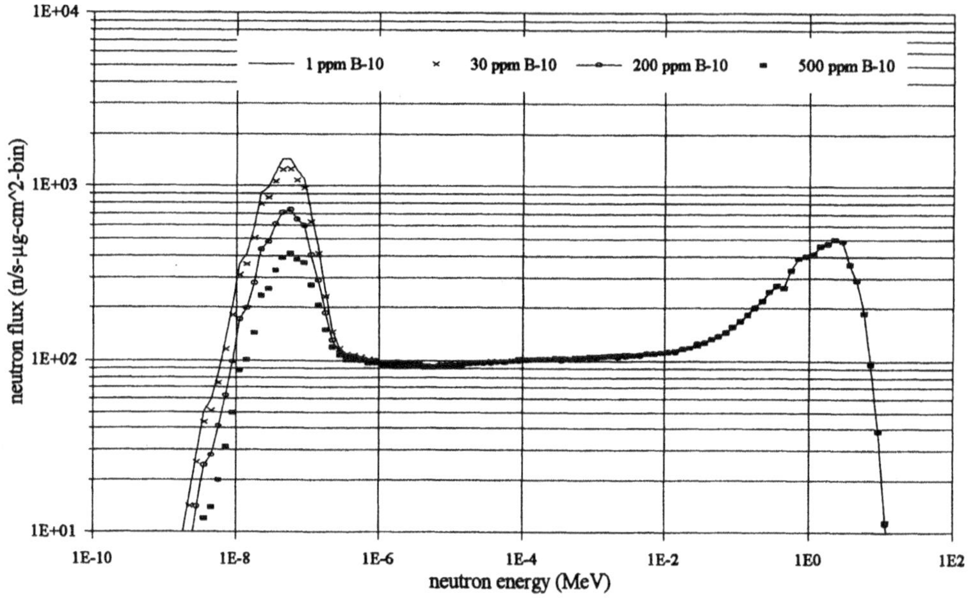

Figure 5. Moderated neutron energy spectrum for ^{10}B loadings of 1 to 500 ppm at 5 cm.

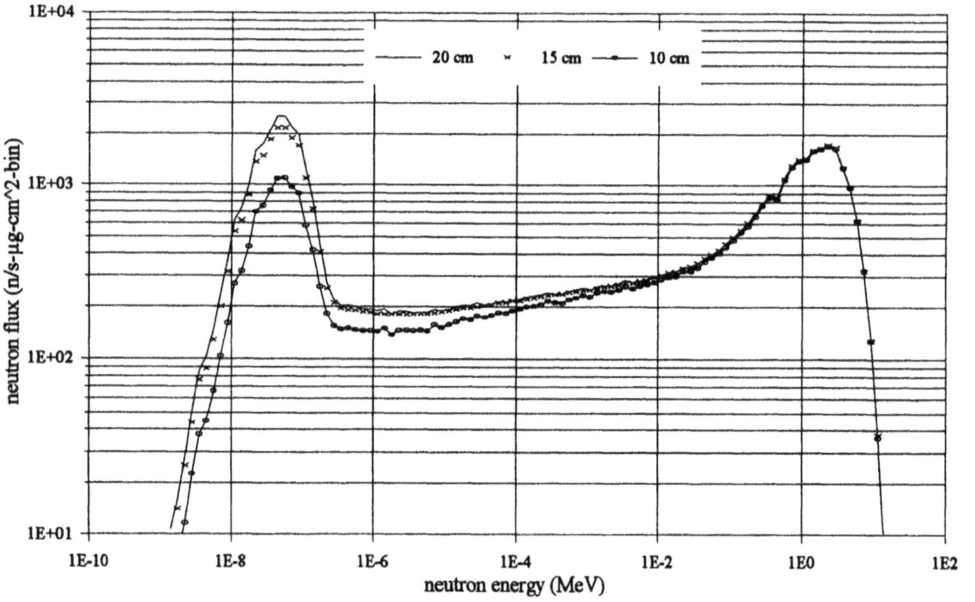

Figure 6. Moderated neutron energy spectrum for a variety of phantom sizes at 3 cm.

of high-LET radiation with the absorbed dose to obtain the weighted dose or dose-equivalent [Gy]. Using a ^{252}Cf fast neutron RBE of 3.5 for the both healthy brain and microscopic tumor cells,[7–9] and a BNCR RBE of 1.3 and 3.8 for healthy brain and microscopic tumor cells, respectively,[10,11] the total healthy and microscopic tumor high-LET dose-eq rates are derived as 0.210 and 0.240 cGy-eq/h-μg, respectively. If one assumes the low-LET ^{252}Cf photon and $^{1}H(n,\gamma)^{2}H$ dose distributions are approximately similar to that of the fast neutron dose rate, and that their intensities relative to the ^{252}Cf fast neutron dose rate are about 40% and 1%, respectively,[12] then the ^{252}Cf low-LET photon dose to healthy brain and microscopic tumor cells at 5 cm is 0.0242 cGy/(h-μg ^{252}Cf). The final low-LET dose component is from the 478 keV photons. If the $^{10}B(n,\gamma)$ dose distribution is approximately similar to that of the $^{10}B(n,\alpha)^{7}Li$ dose rate, and a relative intensity of 13% is taken by estimating the mean chord length and energy deposition by 0.48 MeV photons, then the $^{10}B(n,\gamma)$ dose rate to healthy brain and microscopic tumor cells at 5 cm is 0.0004 cGy/h-μg ^{252}Cf. The sum of the low-LET dose rates to both healthy brain and microscopic tumor cells is 0.025 cGy/h-μg. Summing the low-LET photon doses with the high-LET doses yields a total BNCR dose-eq of 0.235 and 0.265 cGy-eq/h-μg for the healthy brain and microscopic tumor cells, respectively. Thus, the dose-eq enhancement is 12.7% at 5 cm in a 15 cm brain phantom with a ^{10}B loading of 30 ppm and a drug specificity in tumor:healthy tissue of 3:1. Using this methodology, dose-eq enhancements of 0.6% and 6.5% were obtained at distances of 1 and 3 cm, respectively. The clinical significance of the ^{10}B enhancement for ^{252}Cf brachytherapy was minimal considering the ^{10}B loadings and RBE factors used in this analysis. If dugs are developed which could deliver an order of magnitude greater ^{10}B loading with improved tissue specificity, then ^{10}B enhanced ^{252}Cf brachytherapy may be clinically feasible.

An additional Monte Carlo run was performed with a 50 cm diameter D_2O reflector surrounding a 15 cm diameter head (30 ppm brain) phantom. At 6 cm radius, there was a 15.4% enhancement in the ^{252}Cf fast neutron dose rate and 77.3% enhancement in the BNCR dose rate. If one excludes all BNCR dose enhancement from the healthy tissue dose, as would be the case for a perfect tumor seeking drug, then the total high-LET dose enhancement at 5 cm for healthy tissue and microscopic disease were 8.6% and 13.0%, respectively, with a therapeutic ratio of 1.04.

5. CONCLUSION

Calculations of the ^{10}B enhanced ^{252}Cf brachytherapy were performed using MCNP. Though there was significant moderation of the ^{252}Cf fast neutrons, as evidenced by the neutron energy spectra, the magnitude of the BNCR dose enhancements were small compared to the ^{252}Cf fast neutron dose. As expected, ^{10}B loadings up to 500 ppm did not significantly alter the ^{252}Cf fast neutron energy spectrum. Attempts at improving the thermal neutron flux through addition of a D_2O moderator caused minimal enhancement. Though the neutrons emitted by ^{252}Cf are of relatively low-energy compared with other fast neutron sources, the fast neutron dose dominated over the BNCR dose enhancement for ^{10}B loadings less than 50 ppm. Consequently, ^{252}Cf does not appear to be a useful source for BNCR enhanced radiotherapy using the aforementioned ^{10}B loadings (less than 50 ppm) and RBE factors used in the derivation of this conclusion.

REFERENCES

1. J.F. Briesmeister, Monte Carlo N-Particle transport code system, *in*: "MCNP4B User's Manual," LANL, Los Alamos, 1997.
2. A. Geist, A. Beguelin, J. Dongarra, W. Jiang, R. Manchek, and V. Sunderam. "PVM: Parallel Virtual Machine—A User's Guide and Tutorial for Networked Parallel Computing," The MIT Press, Cambridge, MA, 1994.
3. F. Van den Heuvel, M.J. Rivard, J.G. Wierzbicki, and D.P. Ragan, Implementation of distributed computing for Monte Carlo simulations using PVM in a low tech environment, *in*: "Medical Physics 1997," Obninsk, 1997, pp. 90–91.
4. M.J. Rivard, J.G. Wierzbicki, and F. Van den Heuvel, Calculations of the ^{252}Cf neutron spectrum in water for various positions and loadings of ^{10}B and ^{157}Gd, *in*: "American Nuclear Society Radiation Protection and Shielding Division: Technologies for the New Century," D.T. Ingersoll ed., ANS Inc., La Grange Park, IL, 1998, pp. 211–218.
5. L.L. Anderson, Status of dosimetry for ^{252}Cf medical neutron sources, *Phys. Med. Biol.*, 18:779–799, 1973.
6. R.G. Zamenhof, B.W. Murray, G.L. Brownell, G.R. Wellum, and E.I. Tolpin, Boron neutron capture therapy for the treatment of cerebral gliomas: I. theoretical evaluation of the efficacy of various neutron beams, *Med. Phys.*, 2:47–60, 1975.
7. L.A. Marjina, V.N. Kiseleva, M.I. Nechushkin, V.N. Chekhonadsky, and G.P. Elisyutin, The results of treatment of patients with endometrial cancer and carcinoma cervix using ^{252}Cf HDR, *in*: "Californium-252: Isotope for 21st Century Radiotherapy," J.G. Wierzbicki ed., Kluwer Academic Publishers, Netherlands, 1997, pp. 115–130.
8. O.N. Denisenko, V.N. Ivanov, V.A. Kozlov, V.O. Sidorchenkov, A.A. Omarov, I.M. Chernichenko, and V.N. Chekhonadsky, Dosimetry and treatment planning in high dose rate ^{252}Cf brachytherapy, ibid, 221–231.
9. A.G. Konoplyannikov, Biological effects of gamma-neutron radiation ^{252}Cf or fission neutrons from the BR-10 reactor on tumor and normal cells and tissues, ibid, 257–261.
10. G.M. Morris, J.A. Coderre, J.W. Hopewell, P.L. Micca, N.M. Nawrocky, H.B. Liu, and A. Bywaters, Response of the central nervous system to boron neutron capture irradiation: Evaluation using rat spinal cord model, *Radiother. Oncol.*, 32:249–255, 1994.
11. J.A. Coderre, D.D. Joel, P.L. Micca, N.M. Nawrocky, and D.N. Slatkin, Control of intracerebral gliosarcomas in rats by boron neutron capture therapy with p-boronophenylalanine, *Radiat. Res.*, 129:290–296, 1992.
12. J.G. Wierzbicki, M.J. Rivard, and W.A. Roberts, Physics and dosimetry of clinical ^{252}Cf sources, *in*: "Californium-252: Isotope for 21st Century Radiotherapy," J.G. Wierzbicki ed., Kluwer Academic Publishers, Netherlands, 1997, pp. 25–53.

DOSIMETRY

SEMICONDUCTOR DOSIMETRY IN BNCT

Present and Future

Anatoly B. Rosenfeld

Radiation Physics Group
University of Wollongong
Northfields Ave.
Wollongong, NSW, 2522, Australia

1. INTRODUCTION

The problems of neutron and gamma dosimetry in epithermal Boron Neutron Capture Therapy (BNCT) is quite complicated due to the wide epithermal neutron spectra 0.5–10 keV, fast neutron contribution up to 14 MeV and photons from the reactor and induced in the phantom due to H(n,γ)H reaction. Furthermore, contributions from high LET products arising from B-10(n,α)Li-7 and N-14(n,p)C-14 demand microscopic dose measurements.

Phantom dosimetry is important in the prediction of depth dose distributions in mixed radiation fields. Separation of neutron and gamma dose components is normally performed using paired ionization chamber techniques. The simulation of boron dose requires the thermal neutron flux depth distribution. Activation analysis of gold foils is widely used for measuring this thermal neutron flux. For the measurement of high LET components the microdosimetry approach using gas proportional counters has been used in several BNCT facilities.

This paper is a short review of the development and possible application of semiconductor dosimeters in phantoms for macro and microdosimetry and thermal neutron flux measurements based on work performed at the University of Wollongong. Such detectors offer the prospect of improved spatial resolution and real-time measurements in contrast to the previously mentioned traditional measurements.

2. INTEGRAL SEMICONDUCTOR DOSIMETRY IN BNCT

Two different types of sensors, p-i-n diodes, and Metal Oxide Semiconductor Field Effect Transistor (MOSFET) detectors have been used in phantom measurements.

Frontiers in Neutron Capture Therapy, edited by Hawthorne *et al.*
Kluwer Academic / Plenum Publishers, New York, 2001.

2.1. P-I-N Neutron Dosimeter

The principle of operation of p-i-n neutron dosimeters is based on radiation damage in silicon under neutron bombardment, which results in a change in the forward voltage drop of the diode. It has been shown that such dosimeters are tissue equivalent for fast neutron dosimetry (such as Cf-252 fission spectra)[1] and have a wide dose range.[2] The small size of p-i-n diodes of about $1\,mm^3$ minimizes radiation field perturbation in the phantom. This dosimeter is passive and can be read many times without deterioration of accumulated dose information.

However, testing of p-i-n dosimeters in a Lucite phantom at Petten's epithermal BNCT beam indicated that such dosimeters cannot be used for tissue dose in neutron fields with intermediate and low energy components such as epithermal BNCT[3] because the ratio of silicon to tissue damage KERMA is not constant at low neutron energies. For BNCT applications, p-i-n diodes may still be used for verification of Monte-Carlo simulation of the neutron spectra in different points of the phantom[4] based on a calculation of neutron KERMA in silicon and comparison with experiment.

2.2. MOSFET Dosimeters

The principle of operation of a MOSFET dosimeter is based on ionization of the gate silicon oxide followed by buildup of positive charge on the Si-SiO_2 interface. The shift of the voltage drop between source and drain under constant read-out current is proportional to accumulated dose. MOSFET dosimeters have the advantage of an extremely small dosimetric volume (thickness of SiO_2 about $1\,\mu m$) and consequently a high spatial resolution. Our group has verified that MOSFET probes may be used for gamma dosimetry in BNCT following testing in a Lucite phantom at the Petten and BNL BNCT facilities. For accurate gamma dose measurements the MOSFET detector should be used in active mode (positive gate voltage of 5 V during irradiation) which increases the sensitivity to about 5 mV/cGy and maintains good linearity with dose.

The experiments on MOSFET dosimeters in epithermal neutron beams demonstrate the importance of careful selection of packaging materials to minimize activation. Thermal neutron contribution to the MOSFET response may be reduced by using a LiF cap but such a cap can reduce the γ field by decreasing the photon contribution from H(n,γ)H reactions near to the detector. These problems may be solved by placing the MOSFET chip on a TE plastic substrate and avoiding the use of gold in the chip interconnect construction.[5] The complete device size is about $0.5 \times 0.5\,mm^2$ and the device may be used for real-time online measurements.

Another application of MOSFET dosimeters is in boron dose measurements in a phantom using a paired method. Two MOSFET detectors were produced one of them covered by a B-10 converter. The difference in response will correspond to the dose due to the boron fission products (alpha particle and lithium ion). In contrast to thermal neutron flux measurements that use foil, this method for boron dose contribution uses the boron capture reaction directly and thus is not influenced by differences in energy dependence of the cross-section of other neutron capture reactions.

The detector thickness of the paired MOSFET dosimeter is less then 2 mm. The device can be used under full reactor power with simultaneous measurement of gamma dose and the relative boron response. The measurements of the relative boron response of a MOSFET paired detector in a $15 \times 15 \times 15\,cm^3$ Lucite phantom at the BNL BNCT facility compared well with Monte-Carlo calculations[7] with good agreement in the depth of the maximum boron response found. Note, that the paired MOSFET detectors provide a real boron-10 response whilst Monte-Carlo calculations based on cadmium cut-off the neutron energy at 0.5 eV so that one should not always expect to obtain a good coincidence between the measurement and such theoretical calculations.

The tissue equivalence of a SiO_2 dosimetric volume for high-energy photons has been demonstrated in medical LINACs. However, calibration data obtained using photon beams cannot be used for the TE MOSFET response to high LET products (α, Li-7) due to the higher recombination present in high LET plasma tracks.

In summary, both p-i-n and MOSFET dosimeters have high spatial resolution in comparison to ionization chambers and satisfy the Bragg-Gray cavity theory.

3. COUNT MODE SEMICONDUCTOR DOSIMETRY IN BNCT

Count mode operation typically involves coupling thermal neutron converter materials to diode type detectors. Applications in the literature[8,9] have used B-10(n,α)Li-7 and Li-6(n,α)H-3 converters coupled to surface barrier detectors to measure the free air thermal fluence. The disadvantage in this early work is the overlapping of Li-7 (E~0.84 MeV) and piled-up pulses from the γ photons under high flux conditions. Some improvement in the discrimination of these events may be achieved by using the Li-6(n,α)H-3 reaction where the total energy of the released particles is about 4.8 MeV.

Two types of detectors that improve upon this early work have recently been investigated by our group and tested at BNL. They are based on $3 \times 3\,mm^2$ ion-implanted silicon detectors covered by B-10 and U-235 thermal neutron converters. The advantage of the U-235 converter lies in the high energy of the fission products (up to 90 MeV) which is well above the gamma component. Furthermore, this detector allows us to measure the absolute thermal neutron flux without preliminary calibration since we can determine the number of U-235 atoms by measuring the natural alpha activity of the U-235. Figure 1 shows the U-235 fission fragment spectra obtained at a depth of 11 cm in a Lucite phantom at BNL under a reactor power of 20 kW. The figure also shows the

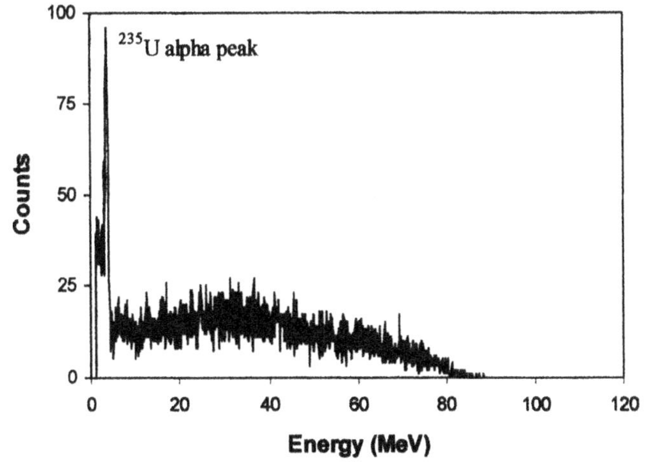

Figure 1. P-i-n diode measurement of U-235 fission fragment spectra obtained at a depth of 11 cm in a Lucite phantom at BNL under a reactor power of 20 kW. Natural alpha particle peak is seen at 4–5 MeV.

natural alpha peak at around 4–5 MeV. Depth distributions for thermal neutron fluence obtained with this detector were in good agreement with absolute data calculated by Monte-Carlo simulations.

Although this detector has many advantages, it has less radiation hardness then the B-10 and Li-6 detectors due to the higher damage of the fragments. For operation under full reactor power, the size of the detector should be reduced along with the mass of the U-235 converter.

4. SILICON-BASED MICRODOSIMETRY IN BNCT

Experimental microdosimetry is a useful tool for measuring the radiobiological properties of a beam. The traditional microdosimetric detector is the proportional gas counter with a typical volume of 1 cc although recent developments have produced a sensitive volume of 0.0027 cc.[10] The smaller volume improves spatial resolution and pile-up problems encountered due to the large size. A novel approach to microdosimetry was developed using high-resolution quantitative auto-radiography and 2D- Monte-Carlo simulation.[11] This method takes into account a realistic cellular structure and boron spatial distribution.

Silicon microdosimetry offers the possibility of providing a tool for simultaneously measuring the absolute thermal neutron flux and microdosimetric spectra at a high spatial resolution and at full reactor power. The approach involves modeling biological cells by a two dimensional array of silicon diode cells.[12] The manufacture of silicon diodes with cellular (10 μm) or subcellular dimensions is easily performed using standard integrated circuit technology with feature sizes of 0.3 μm now available. Such an approach also offers the possibility of varying the localization of boron (using ion-implantation techniques) within the silicon cell to best approximate the distribution under real cellular conditions.

The issues of tissue equivalence and sensitive volume definition have been extensively studied by our group. The silicon detector was found to be tissue equivalent for high LET secondary products in BNCT provided appropriate geometrical scaling of the detector volume is performed.[13] The first prototype silicon microdosimeter consists of an array of 4800 diodes each with junction sizes of $10 \times 10 \times 2\,\mu m^3$ and total cell size of $30 \times 30 \times 2\,\mu m^3$. The device, constructed using Silicon-On-Insulator (SOI) technology, is shown in Figure 2. The sensitive volume has been characterized using alpha broadbeam spectroscopy, ion microbeam imaging and 3D semiconductor device simulation software.[14]

The microdosimeter may be used in neutron capture applications due to the use of boron as a dopant in the p+ regions of the diode that surrounds the central n+-p diode junction. By analogy with a biological cell the p+ region is representative of the cytoplasm surrounding the n+ junction or nucleus. Thus we can measure a microdosimetric spectra along with the absolute thermal neutron flux since we know precisely the number of B-10 atoms in the device (from ion-implantation and device construction details).

Microdosimetric spectra have been obtained at three neutron therapy facilities: the fast neutron therapy facility at the superconducting cyclotron, Gershenson Radiation Oncology Center, Detroit, the epithermal neutron irradiation facility at the 3 MW Brookhaven Medical Research Reactor (BNL) and the boron neutron capture facility at Kyoto University Reactor (KUR).

Figure 2. An SEM photo of the silicon microdosimeter showing the metal tracks connecting the diode cells in parallel. The image is about 24 μm across with the $10 \times 10\,\mu m^2$ diode junction underneath the metal track contact on the right.

The amount of boron is sufficiently high to sensitize the detector to thermal neutrons. The spectra shown in Figure 3 compares the microdosimetric spectra of the KUR and BNL beams at similar depths.[15] The BNL reactor power was 300 kW whilst the KUR measurement was performed at a full reactor power of 6 MW. The normalized microdosimetric spectra clearly indicate a peak at 200 keV/μm corresponding to the alpha and lithium boron fission products. The KUR measurement indicates a higher contribution due to recoil protons (<100 keV/μm) from fast neutron reactions. This measurement

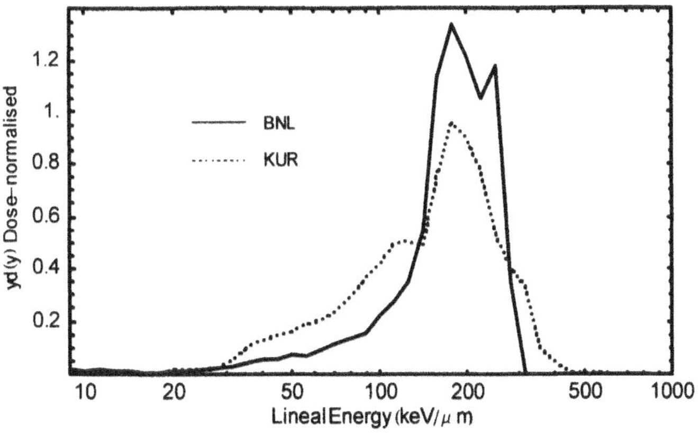

Figure 3. Comparison of BNCT microdosimetric spectra from KUR (1 cm, water phantom, thermal mode beam, reactor power = 6 MW) and BNL (1.3 cm depth in Lucite phantom, reactor power = 300 kw). Equal areas under different regions of the function *yd(y)* correspond to equal doses.

demonstrates the devices utility for comparing beam characteristics and identifying microdosimetric components.

The silicon microdosimeter allows one to measure on-line and simultaneously thermal neutron flux, fast neutron dose (recoil protons) and microdosimetric spectra. The device avoids field perturbation, operates at full reactor power, and is capable of high resolution mapping. Further developments in silicon microdosimeter will use new microelectronic technology to generate small fully isolated cells with a well-defined sensitive volume. Integration of the readout electronics, particularly the pre-amplifier will significantly improve low energy sensitivity and noise performance.

5. CONCLUSION

The family of semiconductor dosimeters based on MOSFET and p-i-n diode structures are very useful in BNCT for in phantom measurements of boron response, gamma dose and thermal flux measurements. They are small, feature low voltage operation and are simple in operation.

A new approach to microdosimetry has been investigated using a SOI based diode array. This device is capable of simultaneous measurement of microdosimetric spectra and absolute thermal neutron fluence. Future work using the latest microelectronic processing technology may enable the development of nano-scale structures for nanodosimetry applications.

REFERENCES

1. H.J. Delafield, and A.H. Reading, Silicon diode measurements for monoenergetic neutrons and critical assemblies (H.P.R.R. and VIPER), United Kingdom Atomic Energy Authority Report, Harwell, April, 1981.
2. A.B. Rosenfeld, *et al.*, P-I-N diodes with a wide measurement range of fast neutron doses, Radiation Protection Dosimetry, 33:175–178, 1990.
3. M.G. Carolan, S.A. Wallace, B.J. Allen, A.B. Rosenfeld, H.A. Nariaty, F. Stecher-Rosmussen, J.N. Mathur, C.P.J. Raaijmakers, and M.W. Konijnenberg, Calculations of epithermal beam dose distribution in phantoms, Proceedings of the sixth international symposium on NCT for cancer, Kobe, Japan, 303–309, 1994.
4. M.G. Carolan, A.B. Rosenfeld, J.N. Mathur, and B.J. Allen, Characterization and use of MOSFET γ dosimeter and silicon PIN diodes neutron dosimeter for epithermal neutron beam dosimetry, Proceedings of the seventh international symposium on NCT for cancer, 192–197, 1997.
5. A.B. Rosenfeld, M.G. Carolan, G.I. Kaplan, B.J. Allen, and V.I. Khivrich, MOSFET dosimeters: The role of encapsulation on dosimetric characteristics in mixed gamma-neutron and megavoltage X-ray fields., IEEE Trans. Nucl. Sci., 42:1870–1877, 1996.
6. A.B. Rosenfeld, G.I. Kaplan, M.G. Carolan, B.J. Allen, R.L. Maughan, M. Yudelev, C. Kota, and J. Coderre, Simultaneous macro-micro dosimetry with MOSFETs, IEEE Trans. Nucl. Sci., 43:2693–2700, 1996.
7. A.B. Rosenfeld, G.I. Kaplan, B.J. Allen, J. Coderre, T. Kobayashi, R.L Maughan, C. Kota, M. Yudelev, V.I. Khivrich, and P.G. Litovchenko, Semiconductor detectors for thermal neutron flux measurements in phantoms used for epithermal BNCT and fast neutron therapy, *in* "Frontiers in Neutron Capture Therapy," M.F. Hawthorne, K. Shelly, R.W. Wiersema, eds., Kluwer Academic/Plenum Publishers, New York, 2001, pp. 1175–1180.
8. Y. Hayakawa, S. Harasawa, A. Nakamoto, K. Amano, H. Hatanaka, and J. Egawa, Simultaneous monitoring system of thermal neutron flux for boron neutron capture therapy, Radiat. Res., 75:243–251, 1978.

9. S. Harasawa, A. Nakamoto, Y. Hayakawa, J. Egawa, O. Aizawa, T. Nozaki, T. Minobe, and H. Hatanaka, Improved monitoring system for of neutron flux during boron neutron capture therapy, Radiat. Res., 88:187–193, 1981.
10. J. Burmeister, C. Kota, R.L. Maughan, K. Riley, O. Harling, L. Wielpolski, R. Ma, and J. Coderre, Microdosimetry studies at the BMRR and the MITR-II using miniature tissue-equivalent proportional counters, *in* "Frontiers in Neutron Capture Therapy," M.F. Hawthorne, K. Shelly, R.W. Wiersema, eds., Kluwer Academic/Plenum Publishers, New York, 2001, pp. 581–586.
11. G.R. Solares, and R.G. Zamenhof, A novel approach to the microdosimetry of neutron capture therapy. Part I. High-resolution quantitative autoradiography applied to microdosimetry in neutron capture therapy, Radiat. Res., 144:50–58, 1995.
12. A.B. Rosenfeld, G.I. Kaplan, M.G. Carolan, B.J. Allen, R. Maughan, M. Yudelev, C. Kota, and J. Coderre, Simultaneous macro-micro dosimetry with MOSFETs, IEEE Trans. Nucl. Sci., 43:2693–2700, 1996.
13. P.D. Bradley, and A.B. Rosenfeld, Tissue Equivalence Correction for Silicon Microdosimetry Detectors in Boron Neutron Capture Therapy, Med. Phys., In press.
14. P.D. Bradley, A.B. Rosenfeld, K.K. Lee, D. Jamieson, and S. Satoh, Charge collection and radiation hardness of a SOI microdosimeter for space and medical applications, IEEE Trans. Nucl. Sci., 45: In press, 1998.
15. P.D. Bradley, A.B. Rosenfeld, B.J. Allen, T. Kobayashi, Y. Sakurai, C. Kota, R.L. Maughan, M. Yudelev, J. Coderre, and J. Capala, Application of silicon diode arrays for microdosimetry in BNCT and FNT, *in* "Frontiers in Neutron Capture Therapy," M.F. Hawthorne, K. Shelly, R.W. Wiersema, eds., Kluwer Academic/Plenum Publishers, New York, 2001, pp. 615–621.

REPORTING A BNCT IRRADIATION

Application of the ICRU Recommendations to the Specific Situation in BNCT

R. Gahbauer[1], N. Gupta[1], T. Blue[1], D. Carpenter[1], W. Sauerwein[2], and A. Wambersie[3]

[1]The Ohio State University
Columbus, Ohio
[2]Universitatsklinikum Essen
Essen, Germany
[3]Clinique Universitaires, St. Luc
Brussels, Belgium

1. INTRODUCTION

If BNCT were performed in a single, unique facility and if its science were unique, concepts and terminology specific to this new modality could be developed. However, BNCT is a form of radiation therapy. To benefit from the vast accumulated experience with all types of radiation therapy and to ensure a common understanding and allow a fruitful exchange of information between different centers, clear, well defined, unambiguous, and universally accepted concepts and terminology are needed. The International Commission on Radiation Units (ICRU) has provided the framework for this communication exchange by defining concepts and terminology and publishing recommendations for reporting external beam therapy (ICRU Reports 29 and 50) and brachytherapy (ICRU Reports 38 and 58). The aim of this presentation is to discuss the application of the ICRU recommendations to the specific problems encountered in BNCT.

2. DEFINITION AND PRESCRIPTION OF VOLUMES

Before the selection of any treatment modality (surgical or radiation), a Gross Tumor Volume (GTV) must be defined by physical examination or imaging. Around the GTV, there is a volume containing microscopic extensions of disease with a probability estimated from oncological experience; this is the Clinical Target Volume (CTV). These volumes must

Frontiers in Neutron Capture Therapy, edited by Hawthorne *et al.*
Kluwer Academic / Plenum Publishers, New York, 2001.

be adequately encompassed by the treatment. They are pure oncological concepts, and their delineation is part of the basic treatment prescription. Just as in surgery, normal tissues can limit the extent and/or amount of treatment that can be safely delivered. Therefore, these normal tissues must also be delineated before treatment is begun.

To adequately deliver treatment to the target volume(s) and protect the normal tissues at risk, margins must be planned for the physical limitations of a given modality or for daily setup uncertainties. These margins are contained within the Planning Target Volume (PTV). During delivery of radiation to the PTV, additional volumes receive some amount of radiation and therefore are defined by ICRU for reporting purposes (Treated Volume, Irradiated Volume).[1]

3. DOSE PRESCRIPTION AND REPORTING

The treatment prescription is the sole responsibility of the radiation oncologist. The ICRU recommendations emphasize minimum requirements for the reporting of dose. The dose must be reported at the ICRU reference point that must be clinically relevant, clearly and unambiguously defined, accurately measurable, and not near a steep dose gradient. The best estimate of the minimum and maximum doses in the Planning Target Volume should also be recorded. The reporting of any additional information is of course strongly encouraged, e.g., Dose Volume Histograms, etc.

A point in the geometrical center of the PTV often adequately meets the requirement for the ICRU reference point in conventional radiation therapy. Consensus on a meaningful reference point in BNCT must still be developed.

One of the main difficulties in reporting BNCT is that the absorbed dose at a given point is the result of the contribution of several (at least four) radiation qualities: i.e., gamma rays, neutrons, high-LET, and irradiation resulting from the disintegration of boron atoms.

Also boron dose is a function of boron concentration and thermal neutron fluence. Boron concentration can be accurately determined in blood before and during radiation and thus can be estimated in normal tissues with fair accuracy. In tumor, the uncertainty of macroscopic boron concentrations is large, and the relevant boron distribution at the microscopic level (in all tumor cells) is unknown. Therefore, the only possible approach is to assume a uniform boron concentration (derived form the measured blood boron concentration) to estimate the absorbed dose level.

The criteria for the ICRU reference point may therefore best be met at a normal tissue point in or outside the PTV. A consensus on the definition of the ICRU reference point in BNCT is needed. At present, tumor doses can only be estimated from known patterns of drug distribution and measurement of blood concentrations. It is important that physical measurements and assumptions (e.g., tumor/blood ratios, actual blood concentration) be clearly and separately reported in any dose statement. In addition, an international consensus on the dosimetric protocols used is urgently needed (e.g., tissue composition, etc.).

4. BIOLOGICALLY EQUIVALENT DOSE, COMPOUND FACTORS, RBE

Each of the radiation qualities encountered in BNCT has its own "RBE" for a given biological endpoint and irradiation condition. The quantity, "absorbed dose," at a given point is used to specify the amount of the therapeutic irradiation at that point and is the relevant quantity commonly used for the different techniques, e.g., external beam therapy with photons, electrons, neutrons, protons, etc. The absorbed dose at a point is defined as the "mean energy imparted" per unit of mass in a volume surrounding that point. The averaging process implies that a large number of ionizing particles are crossing the volume of interest. There is a strict relationship between absorbed dose and cell lethality (i.e., the linear-quadratic relationship). In BNCT, the situation is more complex because four different radiation components contribute to the absorbed dose at a given point:

- gamma rays from the reactor and from the hydrogen capture;
- the nitrogen capture dose;
- fast neutron dose;
- the so-called boron dose, which is the absorbed dose due to the alpha particles, resulting from the capture of thermal neutrons by boron atoms followed by the disintegration of the boron atoms.

The concept of absorbed dose can be applied to the first three of these, as in conventional therapeutic radiation modalities; but doing this has limitations in the case of the boron dose, since the boron incorporation, and thus the number of alpha particles, is not uniform at the cellular level. It can and probably does vary between zero and many from one cancer cell to another. In such conditions, the "averaging process" implied in the definition of the concept of absorbed dose is no longer meaningful.[2]

Strictly speaking then, in BNCT an RBE (dose ratio) cannot be defined; and, therefore, biological weighting factors, e.g., Compound Factor or weighted unit (WU),[3] have been introduced to describe globally the variance in biological effectiveness due this uneven and not readily measurable microdistribution.[4-8]

In clinical situations, it is only possible to average the boron concentration over a large population of cells and thus derive a kind of average dose. If tumor cells do not contain boron or contain less boron than normal tissue, any increase in neutron fluence will yield no therapeutic benefit and will disproportionately increase the toxicity in normal tissues. Therefore, determining compound factors or modifying factors to establish a therapeutic gain calculation is irrelevant, since the paucity of boron in a given tumor cell population would limit therapeutic gain with respect to normal tissues, even if the average boron concentration indicates a significant gain.

Consensus on the use and reporting of biological weighting factors must be developed. At this point, all physically measurable parameters must be separately reported. For example, the same "equivalent dose" at different blood boron concentrations is associated with a very different dose distribution and depth spectrum. Table 1 illustrates some of the differences among facilities.

5. CONCLUSIONS

A. In BNCT, doses to normal tissues can be more accurately determined than doses to the malignant cells because of the large differences in boron incorporation in malignant cells.

B. Given our present methods, tumor doses should be reported as best estimates with a clear statement of the underlying assumptions: e.g., tumor to blood ratios, homogeneous distribution, etc.

Table 1. Examples of Current Usage

	Facility using BPA	Facility using BSH
Prescribing of dose	Peak BNCT dose (Tumor): Peak dose (normal brain): limit . . . Minimum Target Volume dose: —Dose escalation by equivalent dose at variable boron concentrations —No ceiling on dose components	Prescription point: central portion of PTV Dose Group Identification Point (DGIP): where the thermal fluence is maximum Maximum Accepted Dose (MAD): dose to DGIP (protocol specified dose group) Boron Blood concentration: part of prescription (as similar aspossible) Absorbed dose of each component has ceiling
Reporting of dose	Normal tissue points: —Component doses reported —Equivalent doses used for protocol specification and dose escalation Tumor Dose, "BNCT dose" calculation: —Measure average blood/boron concentration during radiation —Assume Tumor/Blood ratio of . . .	Normal tissue points: —Biological equivalent dose: Weighted Dose Unit = WU = ("triple U") [measurable parameters, dose, and boron are reported, then weighting factors are applied] $D_{wu} = W_{n+N} \cdot D_{n+N} + W_g \cdot Dg + CF_{organ} \cdot D_{boron}$
Biological weighting factors for:	Boron: Tumor endpoint CNS, spinal cord Rat skin moist desquamation Rat skin dermal necrosis Human skin moist desquamation Protons, recoil and capture Photons	Boron: Normal brain Other organs Protons, recoil and capture Photons

C. It is essential that all centers involved in BNCT adopt the same definitions of terms and concepts, as well as the same dosimetric protocols (for relevant exchange of information).

D. In all cases, all physical dose components, measured boron-10, and compound concentration(s) should be reported.

E. In addition, for the exchange of information, comparison of results, and simply to improve credibility, it is important that the BNCT community adopts, whenever possible, the general terms and concepts currently used by the radiation therapy community. In that respect, the ICRU's definitions of volumes and its recommendations for reporting treatment (the specification points) should be followed or adapted (and supplemented when necessary) to apply specifically to BNCT.

REFERENCES

1. Prescribing, recording and reporting photon beam therapy, "ICRU Report 50: International Commission on Radiation Units and Measurements," ICRU, Bethesda, MD., 1993.
2. R. Gahbauer, N. Gupta, T. Blue, J. Goodman, R. Barth, J. Grecula, A.H. Soloway, W. Sauerwein, and A. Wambersie, Boron neutron capture therapy: Principles and potential, in: "Recent Results in Cancer Research," vol 150., A. Wambersie and H. Engenhart, eds., Springer-Verlag, Berlin, 1998, pp. 183–209.

3. T.E. Blue, N. Gupta, and J.E. Wollard, A calculation of the energy dependence of the RBE of neutrons," Phys. Med. Biol., 38:1693–1712, 1993.
4. W. Sauerwein, J. Rassow, and B. Mijnheer, Considerations about specification and reporting of dose in BNCT, in: "Advances in Neutron Capture Therapy," B. Larsson, J. Crawford, and R. Weinreich, eds., Elsevier Science B.V., Amsterdam, 1997, pp. 531–534.
5. R.A. Gahbauer, R.G. Fairchild, J.H. Goodman, and T.E. Blue, Can relative biological effectiveness be used for treatment planning in boron neutron capture therapy? in "Tumor Response Monitoring and Treatment Planning," A. Breit, ed., Springer, Berlin, 1992.
6. N. Gupta, R.A. Gahbauer, T.E. Blue, and A. Wambersie, Dose prescription in boron neutron capture therapy," Int. J. Radiat. Oncol. Biol. Phys., 28(5):1157–1166, 1994.
7. F.J. Wheeler, Microdosimetric calculations for boron-drug compound factors for boron neutron capture therapy, Trans. Am. Nucl. Soc., 65:147, 1992.
8. T. Kobayashi, and K. Kanda, Analytical dose calculation in cell nucleus from spherical ^{10}B containing region and a concept of RBE for neutron capture therapy, in: "Advances in Neutron Capture Therapy," A.H. Soloway, R.F. Barth, and D.E. Carpenter, eds., Plenum Press, New York, 1993, pp. 217–220.

80

RADIATION RISKS FROM *p*-BORONOPHENYLALANINE-MEDIATED BNCT FOR GLIOBLASTOMA MULTIFORME

R. Ma[1], J. Capala[1], J. Coderre[1], A. Z. Diaz[1], D. Greenberg[1], H. B. Liu[2,3], D. N. Slatkin[1], and A. D. Chanana[1]

[1]Medical Department
Brookhaven National Laboratory
Upton, NY 11973
[2]Department of Neurological Surgery
University of California-Davis
Sacramento, CA 95817
[3]McClellan Nuclear Radiation Center
McClellan AFB, CA 95652

1. INTRODUCTION

Boron neutron capture therapy (BNCT) is a binary treatment modality that requires selective delivery of a ^{10}B-labeled compound to a tumor and slow neutron irradiation of the tumor-bearing tissues. The capture of thermalized neutrons by ^{10}B releases high linear-energy-transfer (LET) alpha (α) and ^{7}Li particles via the $^{10}B(n,\alpha)^{7}Li$ reaction. These particles can kill or render non-clonogenic ^{10}B-rich tumor cells while sparing boron-poor tissues. A phase I/II clinical trial of *p*-boronophenylalanine (≥95% ^{10}B enriched)-mediated BNCT for GBM is being conducted using the Brookhaven Medical Research Reactor's clinical epithermal neutron irradiation facility.

During BNCT treatment, a patient is exposed to a mixed field radiation, primarily neutrons and gamma-rays. Whole body radiation risks need to be assessed. Because there are no official quantitative guidelines at present for long-term whole body risk from exposure to diagnostic or therapeutic mixed-field radiation, 1990 recommendations of the International Commission on Radiological Protection (ICRP 60)[1] for occupational exposure were used, although they were developed to assess similar risks in a different population.

Frontiers in Neutron Capture Therapy, edited by Hawthorne *et al.*
Kluwer Academic / Plenum Publishers, New York, 2001.

2. METHODS

The probability of both deterministic and stochastic effects of radiation in exposed individuals depends on the type and energy of the radiation, the tissue or organ irradiated as well as physical absorbed doses. For the calculations pertaining to long-term whole-body radiation risk from BNCT, the radiation weighting factors, also called quality factors (QFs), are assumed to be 1 for the gamma component of the radiation and 20 for the non-gamma (i.e. fast neutron dose and doses from $^{10}B(n,\alpha)^{7}Li$ and $^{14}N(n,p)^{14}C$ reactions) components of the radiation field. The contribution of an exposed organ or tissue to risk is calculated by multiplying the equivalent dose with its tissue-weighting factor. The tissue weighting factors conform to the current ICRP standard for occupational exposure to radiation.[1] The sum of the tissue radiation risk weighting factors is 1.05 for female and 0.95 for male. The total risk designated as "effective dose" with the unit "sievert, Sv" is normalized for a female by the multiplicative factor 0.952 to derive a gender-neutral index of radiation risk.[2] The effective dose is then calculated as the sum of the weighted equivalent doses in a specified list of radiation sensitive organs and tissues.[1]

In order to provide post-BNCT verification of the dose, ^{7}Li-enriched LiF, thermoluminescent dosimeters (TLD) and gold wires are used to measure gamma absorbed dose and thermal neutron fluences, respectively. Three TLDs and three Au wires are taped on the patient's skin surface at each of the following four locations: midline over the forehead, the cricoid cartilage, the xiphoid process and the umbilicus. The whole body radiation risk is estimated in three steps. First, the thermal neutron fluences and the gamma doses at skin surface are calculated based on the dosimeter readings. Results are used to calculate the absorbed doses (Gy) at those locations. Absorbed doses from the $^{10}B(n,\alpha)^{7}Li$ reactions are based on the blood ^{10}B concentration. Second, average absorbed doses at several body locations that are used as reference zones, computed by Monte Carlo simulations are normalized to absorbed doses derived from the dosimeters. The reference zones include brain, thyroid (deep), xiphoid (superficial), xiphoid (deep, axial), umbilical (superficial), umbilical (deep, axial), and pubic zone (superficial), as shown in Table 1. For the brain, the BNCT treatment planning software[3] is used while for extracranial sites, Monte Carlo neutron and photon transport code (MCNP)[4] is used. The factors used to calculate the doses at deep locations from the dosimetric data of body surface are derived from MCNP simulations. Based on the absorbed doses, the total equivalent doses in the reference zones are calculated. Third, the equivalent dose of an exposed

Table 1. Absorbed and equivalent doses of the reference zones

Reference zone (adult)	Absorbed dose(Gy) G(γ)	Absorbed dose(Gy) D(non-γ)	Equivalent dose (GyxQF) D(non-γ) × 20 + G(γ) × 1
Brain (average)	G(brain)	D(brain)	A
Thyroid (average; deep)	G(cricoid) × 1.2	D(cricoid) × 1.9	B
Xiphoid zone (superficial)	G(xiphoid)	D(xiphoid)	C
Xiphoid (deep, axial)	G(xiphoid) × 1.2	D(xiphoid) × 1.9	D
Umbilical (superficial)	G(umbilical)	D(umbilical)	E
Umbilical (deep, axial)	G(umbilical) × 1.2	D(umbilical) × 1.9	F
Pubic zone (superficial)	G(umbilical)^2/D(xiphoid)	D(umbilical)^2/D(xiphoid)	G

Table 2. Whole-body radiation risk estimate

Organ/Tissue	Equivalent dose Gy × QF	Tissue weighting factor W_T	Relative contribution to risk (Sv)[a] Gy × QF × W_T/1.05 (0.95)
Brain	A	0.025	0.61 (0.29–1.0)
Breast (female)	(B + C + D)/3	0.10 (female) 0.0 (male)	0.031 (0.0–0.19)
Lung	(C + D)/2	0.12	0.059 (0.022–0.137)
Red marrow	(C + D)/2	0.12	0.059 (0.022–0.137)
Thyroid	B	0.05	0.058 (0.027–0.176)
Stomach	F	0.12	0.041 (0.18–0.103)
Colon	F	0.12	0.041 (0.18–0.103)
Gonads	G	0.20	0.026 (0.01–0.067)
Esophagus	(C + D)/2	0.05	0.025 (0.009–0.057)
Liver	F	0.05	0.017 (0.007–0.043)
Urinary bladder	(F + G)/2	0.05	0.012 (0.005–0.030)
Bone surface	C	0.01	0.0038 (0.001–0.009)
Skin	C	0.01	0.0038 (0.001–0.009)
Other, non-brain	(C + D)/2	0.025	0.012 (0.001–0.029)
Total	ΣGy × QF	1.05 (female) 0.95 (male)	1.0 (0.64–1.90)

[a]Mean ± sd (range).

organ or tissue is estimated based on the equivalent doses of the reference zones, as shown in Table 2, column 2. The contribution of an exposed organ or tissue to risk is then calculated by multiplying the equivalent dose with its tissue-weighting factor. Then the total relative risk designated as "effective dose" is calculated as the sum of the weighted equivalent doses in all the ICRP-designated critical tissues and organs.

During BNCT, a very small portion of normal stable nuclides in the body, mainly ^{23}Na, ^{37}Cl, ^{31}P, and ^{127}I, are activated to produce radioactive nuclides. The absorbed doses from the decay of these radioactive nuclides total less than 0.01 Gy, which is negligible by comparison with absorbed doses from other sources.

3. RESULTS AND DISCUSSION

The thermal neutron fluences and gamma doses obtained from the dosimeters on the forehead, over the cricoid and xiphoid cartilages and at the umbilicus are listed in Table 3 for 38 BNCT patients. The effective doses as well as the relative contribution to risk of radiosensitive organs and tissues are summarized in Table 2, column 4. For 38 patients treated by BNCT, the effective doses averaged 1.0 Sv (range 0.6 to 1.9 Sv), of which contributions from brain irradiation averaged 0.6 Sv (range 0.3 to 1.0 Sv) i.e., 60% of the average effective dose.

Table 3. Thermal neutron fluences and gamma doses recorded by surface dosimeters

Dosimeter location[c]	Thermal neutron fluence (n/cm^2 × 10^{10})[a]	Gamma dose (cGy)[a]
Forehead	10.3 ± 18.3 (0.5–90.6)	60.8 ± 22.3 (24.0–126.7)
Cricoid cartilage	1.1 ± 0.7 (0.4–4.6)	38.3 ± 13.8 (21.8–75.0)
Xiphoid cartilage	0.6 ± 0.3 (0.2–1.7)	19.6 ± 6.1 (7.4–37.6)
Umbilicus	0.4 ± 0.2 (0.1–1.0)	9.6 ± 3.8 (3.8–25.3)

[a]Mean ± sd (range).

Based on ICRP 60, the average effective dose of ~1 Sv corresponds to ~5% increase in the long-term probability of cancer induction. The combined risks for the induction of non-fatal cancer and heritable effects are estimated to be less than 2%. These risks are generally considered acceptable in adults with usually inexorably fatal illness of rather short duration, such as cerebral glioblastoma multiforme.

ACKNOWLEDGMENTS

This work was supported by the DOE contract # DE-AC02-98CH10886.

REFERENCES

1. ICRP Publication 60, 1990 Recommendations of the International Commission on Radiological Protection, Pergamon Press, New York, 1991.
2. D.N. Slatkin, J.A. Kalef-Ezra, K.E. Balbi, L. Wieloposki, and J.F. Rosen, L-line x-ray fluorescence of tibial Lead: correction and adjustment of radiation risk to ICRP 60. *Radiation Protection Dosimetry*, 42:319–322, 1992.
3. D.W. Nigg, F.J. Wheeler, D.E. Wessol, J. Capala, and M. Chadha, Computational dosimetry and treatment planning for boron neutron capture therapy. *J. Neuro-Oncol.*, 33:93–104, 1997.
4. J.F Briesmeister, MCNP—A general Monte Carlo code for neutron and photon transport, *Report number LA-7396-M, Revision 2*, Los Alamos National Laboratory, Los Alamos, NM, 1986.

81

NOMINAL EFFECTIVE RADIATION DOSES DELIVERED TO THE TUMOR AND DIFFERENT PARTS OF THE NORMAL BRAIN DURING BNCT FOR GLIOBLASTOMA MULTIFORME

J. Capala[1], J. A. Coderre[1], A. Z. Diaz[1,2], R. Ma[1], and A. D. Chanana[1]

[1]Brookhaven National Laboratory
Upton, NY
[2]Montefiore Medical Center
Bronx, NY

1. INTRODUCTION

The mixed radiation field produced during BNCT comprises radiations with different linear energy transfer (LET) characteristics and different efficacies in biological systems (relative biological effectiveness: RBE). The four major components of the BNCT radiation dose are: boron neutron capture (BNC) dose, gamma dose (originating primarily from neutron capture in hydrogen), fast neutron dose, and the nitrogen neutron capture (NNC) dose. Because of different attenuation of gamma rays and neutrons with various energies, the relative contributions of these components to the total dose vary with depth in tissues. In addition, the short range of BNC reaction products makes the ^{10}B microdistribution particularly important and a compound biological effectiveness (CBE) factor must be used to express the BNC dose in photon gray-equivalent (Gy-Eq) units.[1] Evaluation of the radiation doses delivered during BNCT combines data on spatial distribution of gamma and neutron fluxes, the ^{10}B concentration and RBE/CBE factors. We report the radiation doses delivered to the tumors (post-operative contrast enhancing volumes), target volumes (tumor + a 2-cm margin), and different parts of the brains of 41 glioblastoma multiforme (GBM) patients treated with BNCT at the Brookhaven Medical Research Reactor (BMRR).[2]

Frontiers in Neutron Capture Therapy, edited by Hawthorne *et al.*
Kluwer Academic / Plenum Publishers, New York, 2001.

2. MATERIALS AND METHODS

Irradiations were carried out at the epithermal port of the BMRR using collimators with an aperuture of either 8 cm or 12 cm diameter.[3] Treatment planning procedures developed for clinical BNCT trials at BMRR have been described.[4] Single- or double-field (bilateral or quasiunilateral) irradiation geometry was applied.[5] The BNCT treatment planning system developed at the Idaho Engineering and Environmental Laboratory[6] was used for dose calculations. Homogenous boron concentrations in the brain and tumor cells, respectively, equal and 3.5 fold higher than that in the blood was assumed for dose calculations. The nominal effective doses are expressed in Gy-Eq units using RBE factors of 1 and 3.2 for gamma and non-BNC high-LET components (fast neutrons + NNC), respectively, and CBE factors of 1.3 and 3.8 for BNC reaction products from *p*-boronophenylalanine (BPA) in the normal brain and tumor, respectively.[7]

3. RESULTS

3.1. Doses to the Brain

Radiation doses delivered to the normal brain using various irradiation geometries are listed in Table 1. Peak normal brain dose (dose to 1 cm^3 of the normal brain exposed to the highest dose) ranged from 8.9 to 14.8 Gy-Eq. The predominant component of the peak dose was the BNC dose, contributing from 34% to 51% of the total dose, followed by gamma dose (25%–37%), fast neutron dose (11%–21%), and NNC dose (11%–16%). The volume-averaged brain doses were the direct consequence of the prescribed peak dose and the irradiation geometry, and ranged from 1.9 to 6.6 Gy-Eq, of which 38%–51% was gamma dose, 25%–40% BNC dose, 12%–15% fast neutron dose, and 9%–11% NNC dose.

Table 2 summarizes the average doses delivered to cerebral hemispheres, cerebellum and brain stem. In all cases, the highest normal brain doses were delivered to ipsilateral

Table 1. Radiation doses delivered to the normal brain using various irradiation geometries. The numbers represent mean value and the range (minimum—maximum) of the doses expressed in photon equivalent (Gy-Eq) units

	Single-field		Double-field, 12-cm collimator	
	8-cm collimator	12-cm collimator	Bilateral	Quasi-unilateral
Number of patients	15	11	7	8
Peak doses				
Total	**11.1 (8.9–13.8)**	**13.3 (12.3–14.8)**	**12.6 (12.0–14.2)**	**12.5 (12.1–12.9)**
BNC	4.5 (3.8–5.5)	5.4 (4.6–7.5)	4.7 (4.1–5.6)	5.6 (5.0–6.0)
Gamma	3.1 (2.4–4.1)	4.1 (3.6–4.7)	4.2 (3.8–4.6)	3.7 (3.5–4.2)
NNC	1.5 (1.1–1.9)	1.8 (1.5–2.0)	1.7 (1.5–1.8)	1.6 (1.4–1.9)
Fast neutron	2.0 (1.5–2.5)	1.9 (1.5–2.5)	2.0 (1.5–2.5)	1.6 (1.4–1.9)
Average doses				
Total	**2.3 (1.9–2.9)**	**3.6 (3.2–4.1)**	**5.0 (4.1–6.6)**	**5.2 (4.7–6.0)**
BNC	0.7 (0.6–1.0)	1.1 (1.0–1.5)	1.5 (1.1–2.0)	1.8 (1.4–2.1)
Gamma	1.0 (0.9–1.2)	1.6 (1.4–1.8)	2.3 (1.8–3.0)	2.2 (1.9–2.5)
NNC	0.2 (0.2–0.3)	0.4 (0.3–0.5)	0.5 (0.4–0.7)	0.5 (0.5–0.6)
Fast neutron	0.3 (0.3–0.4)	0.5 (0.4–0.6)	0.7 (0.6–0.9)	0.7 (0.6–0.8)

Table 2. Average radiation doses delivered to different parts of the brain using various irradiation geometries. The numbers represent mean value and the range (minimum—maximum) of the doses expressed in photon equivalent (Gy-Eq) units

	Single-field		Double-field, 12-cm collimator	
	8-cm collimator	12-cm collimator	Bilateral	Quasi-unilateral
Number of patients	15	11	7	8
Ipsilateral cerebral hemisphere				
Total	**5.4 (3.9–6.7)**	**5.5 (4.0–6.3)**	**6.0 (5.3–6.8)**	**7.5 (6.4–8.4)**
BNC	1.9 (1.2–2.6)	1.9 (1.2–2.6)	1.9 (1.6–2.2)	2.9 (2.2–3.4)
Gamma	2.2 (1.7–2.6)	2.2 (1.7–2.5)	2.6 (2.2–2.9)	2.8 (2.4–3.1)
NNC	0.6 (0.4–0.8)	0.6 (0.5–0.8)	0.7 (0.6–0.8)	0.8 (0.7–1.0)
Fast neutron	0.8 (0.5–1.0)	0.8 (0.6–0.9)	0.8 (0.7–0.9)	1.0 (0.8–1.2)
Contralateral cerebral hemisphere				
Total	**2.0 (1.2–3.1)**	**2.1 (1.4–3.5)**	**4.4 (3.3–7.1)**	**3.2 (2.8–4.1)**
BNC	0.5 (0.2–0.9)	0.6 (0.2–1.0)	1.2 (0.8–2.2)	0.9 (0.6–1.3)
Gamma	1.1 (0.8–1.4)	1.1 (0.8–1.7)	2.1 (1.6–3.1)	1.6 (1.4–2.0)
NNC	0.2 (0.1–0.3)	0.4 (0.1–0.9)	0.5 (0.3–0.8)	0.3 (0.2–0.4)
Fast neutron	0.2 (0.1–0.4)	0.3 (0.2–0.5)	0.6 (0.5–1.0)	0.4 (0.3–0.5)
Cerebellum				
Total	**2.1 (0.9–4.2)**	**2.6 (1.0–5.2)**	**4.3 (1.4–6.1)**	**5.2 (3.2–8.9)**
BNC	0.5 (0.1–1.3)	0.7 (0.1–1.7)	1.2 (0.3–2.1)	1.7 (1.2–2.3)
Gamma	1.1 (0.6–2.0)	1.3 (0.7–2.3)	2.1 (0.9–3.0)	2.2 (1.4–3.0)
NNC	0.2 (0.0–0.4)	0.3 (0.0–0.6)	0.6 (0.3–0.9)	0.6 (0.3–1.7)
Fast neutron	0.2 (0.1–0.5)	0.3 (0.1–0.7)	0.5 (0.2–0.8)	0.7 (0.3–1.9)
Brain stem				
Total	**2.1 (1.2–3.8)**	**2.4 (1.5–3.6)**	**4.7 (2.6–6.9)**	**3.9 (3.1–4.9)**
BNC	0.5 (0.2–1.1)	0.6 (0.2–1.0)	1.2 (0.6–1.9)	1.1 (0.8–1.3)
Gamma	1.2 (0.8–2.1)	1.4 (0.9–1.9)	2.5 (1.5–3.6)	2.0 (1.6–2.6)
NNC	0.2 (0.1–0.4)	0.2 (0.1–1.4)	0.4 (0.2–0.7)	0.3 (0.2–0.4)
Fast neutron	0.2 (0.1–0.4)	0.3 (0.2–0.4)	0.5 (0.3–0.7)	0.4 (0.3–0.5)

cerebral hemispheres. They varied with the irradiation geometry and ranged from 5.4 ± 0.9 Gy-Eq for single-field irradiation to 7.5 ± 0.7 Gy-Eq for double-field, quasi-unilateral irradiations. The relative contributions of BNC, gamma, NNC and fast neutrons were 35 ± 4%, 40 ± 3%, 11 ± 0.6%, and 14 ± 0.8%, respectively. The highest volume-averaged dose delivered to the whole cerebral hemisphere was 8.4 Gy-Eq. Doses to the contralateral cerebral hemispheres ranged from 1.2 to 7.1 Gy-Eq, and the gamma dose component dominated, contributing from 43% to 67% of the total dose, while 17% to 35% resulted from BNC. Average doses to the cerebellum were in the range from 0.9 Gy-Eq (single-field irradiation) to 8.9 Gy-Eq (double-field, quasi-unilateral irradiation). Again, the gamma dose component dominated (34%–71%) followed by BNC (13%–37%), fast neutron dose (10%–21%) and NNC (6%-10%). Average doses to the brain stem ranged from 1.2 to 6.9 Gy-Eq, of which 57 ± 5%, 25 ± 5%, 10 ± 0.6% and 8 ± 1.3% was gamma, BNC, fast neutrons and NNC dose, respectively.

3.2. Doses to the Tumor Cells within the Tumor and Target Volumes

Average and minimum radiation doses delivered to tumor cells in the tumor and target volumes are listed in Table 3. The average and minimum tumor doses were in the ranges from 36 to 75 Gy-Eq and from 18 to 55 Gy-Eq, respectively. Most of the doses

Table 3. Radiation doses delivered to tumor and target volumes using various irradiation geometries. The numbers represent mean value and the range (minimum—maximum) of the doses expressed in photon equivalent (Gy-Eq) units

	Single-field		Double-field, 12-cm collimator	
	8-cm collimator	12-cm collimator	Bilateral	Quasi-unilateral
Number of patients	15	11	7	8
Average tumor doses				
Total	**48 (36–59)**	**47 (37–75)**	**51 (42–65)**	**50 (38–63)**
BNC	42 (29–52)	41 (32–68)	45 (37–59)	44 (32–56)
Gamma	3.9 (3.1–4.6)	3.1 (2.6–3.9)	3.5 (2.5–4.2)	3.3 (2.6–4.0)
NNC	1.4 (1.1–1.8)	1.3 (1.1–1.6)	1.4 (1.1–1.8)	1.3 (1.1–1.5)
Fast neutron	1.4 (1.0–1.7)	1.3 (1.0–1.6)	1.4 (1.1–2.2)	1.4 (1.1–1.6)
Minimum tumor doses				
Total	**31 (18–43)**	**30 (19–55)**	**37 (27–51)**	**30 (19–48)**
BNC	26 (14–36)	26 (16–49)	32 (22–45)	26 (16–42)
Gamma	3.1 (2.2–3.8)	2.5 (1.8–3.5)	2.9 (1.9–3.5)	2.5 (2.0–3.3)
NNC	0.9 (0.5–1.3)	0.8 (0.5–1.3)	1.0 (0.7–1.6)	0.8 (0.5–1.1)
Fast neutron	0.7 (0.4–1.0)	0.7 (0.4–0.9)	0.8 (0.6–1.6)	0.7 (0.5–1.0)
Average target doses				
Total	**32 (24–38)**	**46 (40–60)**	**38 (30–44)**	**48 (40–55)**
BNC	28 (21–34)	41 (34–55)	31 (24–37)	42 (34–59)
Gamma	2.5 (2.0–3.0)	3.6 (3.1–4.1)	3.7 (3.5–4.1)	3.5 (3.2–3.9)
NNC	0.9 (0.7–1.1)	1.3 (1.2–1.6)	1.2 (1.0–1.3)	1.2 (1.1–1.4)
Fast neutron	1.0 (0.8–1.1)	1.3 (1.1–1.7)	1.2 (1.0–1.4)	1.2 (1.1–1.4)
Minimum target doses				
Total	**11 (7–17)**	**22 (18–35)**	**18 (15–23)**	**24 (16–31)**
BNC	9 (6–15)	19 (14–30)	15 (11–19)	20 (13–27)
Gamma	1.4 (1.0–1.8)	2.5 (2.1–3.0)	2.6 (2.2–3.0)	2.1 (0.2–2.7)
NNC	0.3 (0.2–0.5)	0.6 (0.5–1.1)	0.6 (0.5–0.6)	0.6 (0.4–0.7)
Fast neutron	0.3 (0.2–0.4)	0.5 (0.4–0.9)	0.5 (0.5–0.6)	0.5 (0.4–0.6)

(81%–91% of the average dose and 76%–90% of the minimum dose) originated from BNC reaction. The minimum doses delivered to tumor cells within the target volume were in the range from 7 to 35 Gy-Eq, of which 74%–94% was BNC dose.

4. DISCUSSION

These results represent the radiation dose escalation implemented in clinical BNCT Phase I/II trials at Brookhaven National Laboratory (BNL). Since the first protocols, which used an 8-cm collimator and single field irradiations, the average brain dose has been gradually increased more than 100% by: 1) increasing the size of collimator to 12-cm; 2) increasing the prescribed peak normal brain dose to 12.6 Gy-Eq; and 3) employing two-field irradiations. Accordingly, doses to various part of the brain have been increased, reaching levels as high as 8.9 Gy-Eq, 8.4 Gy-Eq, and 6.9 Gy-Eq for cerebellum, cerebral hemisphere, and brain stem, respectively. It is noteworthy that only 35% of the ipsilateral dose and approximately 25% of the doses to other selected brain structures (Table 2) are BNC doses. The remaining fraction of the dose depends solely on the purity of the epithermal beam and time of irradiation and would be delivered to normal tissues even in absence of ^{10}B. Therefore, to improve the therapeutic ratio, not only should the neutron beam quality be improved and the tumor to normal tissue boron concentration ratio be increased, but also the amount of boron in tumor cells must be high enough to limit the time of neutron irradiation needed to deliver a lethal radiation dose to these cells. In spite of the significant increase during the dose escalation studies, the doses delivered to the normal brains of our patients remained considerably lower than doses proved to be safe in preclinical canine brain tolerance studies and conventional radiotherapy in humans. These observations support further dose escalation. In order to increase the dose to tumor cells in the target volume, in the next phase of the dose escalation protocol at BNL, patients will receive peak and average brain doses of up to 15 Gy-Eq and 11 Gy-Eq, respectively, provided that the average doses to critical structures are within the limits of tolerance.

More than 74% of the minimum target dose originated from BNC reaction. Therefore, the assumptions regarding CBE factors and boron concentration in tumor cells dispersed in the normal brain of the target volume are crucial for estimation of nominal values of doses delivered these cells. In this report, uniform concentration of boron in each tumor cell within the target volume was assumed. This assumption, however, remains to be fully validated.[8] Additional studies to address the accuracy of the current CBE factors utilized in this report are also under way in our laboratory.

REFERENCES

1. J.A. Coderre, M.S. Makar, P.L. Micca, M.M. Nawrocky, H.B. Liu, D.D. Joel, D.N. Slatkin, and H.I. Amols, Derivations of relative biological effectiveness for the high-LET radiations produced during boron neutron capture irradiations of the 9L rat gliosarcoma *in vitro* and *in vivo*. *Int. J. Radiat. Oncol. Biol. Phys.* 27:1121–1129, 1993.
2. A.Z. Diaz, A.D. Chanana, J. Capala, M. Chadha, J.A. Coderre, E.H. Elowitz, J. Iwai, D.D. Joel, H.B. Liu, R. Ma, N. Pendzick, N.S. Peress, M.S. Shady, D.N. Slatkin, G.W. Tyson, and L. Wielopolski, Boron neutron capture therapy for glioblastoma multiforme: results from the initial phase I/II dose escalation studies, *in* "Frontiers in Neutron Capture Therapy," M.F. Hawthorne, K. Shelly, R.W. Wiersema, eds., Kluwer Academic/Plenum Publishers, New York, 2001, pp. 61–72.
3. H.B. Liu, D.D. Greenberg, J. Capala, and F.J. Wheeler, An improved neutron collimator for brain tumor irradiations in clinical boron neutron capture therapy. *Med. Phys.* 23:2051–2060, 1996.
4. J. Capala, M. Chadha, J.A. Coderre, A.Z. Diaz, H.B. Liu, R. Ma, D.E. Wessol, F.J. Wheeler, L. Wielopolski, and A.D. Chanana, Boron Neutron therapy of glioblastoma multiforme: Treatment planning implementation in dose escalation trials. *Int. J. Radiat. Oncol. Biol. Phys.*, 1998 (submitted).
5. J. Capala, A.Z. Diaz, R. Ma, and A.D. Chanana, Optimization of double-field boron neutron capture therapy for unilateral, supratentorial glioblastoma multiforme, Program and Abstracts of the Eighth International Symposium on NCT for Cancer, La Jolla, CA, 1998, p. 134.
6. D.W. Nigg, F.J. Wheeler, D.E. Wessol, J. Capala, and M. Chadha, Computational dosimetry and treatment planning for boron neutron capture therapy. *J. Neuro-Oncol.* 33:93–104, 1997.
7. J.A. Coderre and G.M. Morris, Review article: The radiation biology of boron neutron capture therapy. *Rad. Res.*, 151, 1–18 (1998).
8. D.R. Smith, S. Chandra, J.A. Coderre, D.D. Joel, D.N. Slatkin, A.D. Chanana, E.H. Elowitz, M.M. Nawrocki, P.L. Micca, and G.M. Morrison, Ion microscopy imaging of boron from *p*-boronophenylalanine (BPA) in surgically acquired samples of human tumor tissue, *in* "Frontiers in Neutron Capture Therapy," M.F. Hawthorne, K. Shelly, R.W. Wiersema, eds., Kluwer Academic/Plenum Publishers, New York, 2001, pp. 899–903.

82

MICRODOSIMETRY STUDIES AT THE BMRR AND THE MITR-II USING MINIATURE TISSUE-EQUIVALENT PROPORTIONAL COUNTERS

J. Burmeister[1], C. Kota[1], R. L. Maughan[1], K. Riley[2], O. Harling[2],
L. Wielopolski[3], R. Ma[3], and J. Coderre[3]

[1]Karmanos Cancer Institute
Gershenson Radiation Oncology Center
Harper Hospital and Wayne State University
Detroit, Michigan
[2]Massachusetts Institute of Technology
Cambridge, Massachusetts
[3]Medical Department
Brookhaven National Laboratory
Upton, New York

1. INTRODUCTION

Conventional dosimetry methods for BNCT measure the macroscopic absorbed dose, but give no information about its microscopic nature or radiation quality. A low pressure tissue-equivalent proportional counter (TEPC) is capable of providing this information by supplying the complete single event charged particle spectrum depositing the dose in a simulated microscopic site. Miniature TEPCs have been designed to overcome the limitations in spatial resolution and dose rate of conventional Rossi counters. They are small enough to provide excellent spatial resolution and to be used in the high flux beams used for BNCT with a total deadtime of only a few percent and negligible pulse pile-up. A pair of TEPCs, made of A-150 tissue-equivalent plastic and A-150 loaded with 200 μg/g ^{10}B, with sensitive volumes of 0.01 cm^3 have been constructed and characterized. Another TEPC has been constructed to investigate the effect of differences between A-150 plastic and brain tissue-equivalent plastic on the secondary charged particle spectrum. Collaborative arrangements have been made with Brookhaven National Laboratory and Massachusetts Institute of Technology to make microdosimetric measurements at the BMRR and the MITR-II. The results of these measurements are presented.

Frontiers in Neutron Capture Therapy, edited by Hawthorne *et al.*
Kluwer Academic / Plenum Publishers, New York, 2001.

2. TEPC MICRODOSIMETRY

TEPC dosimetry requires only two measurements to obtain the gamma, neutron, and boron neutron capture dose components at a given point. The TEPCs are filled with propane based tissue-equivalent counting gas at a low pressure to simulate a microscopic site in tissue. A single TEPC measurement yields the complete energy spectrum of secondary charged particles depositing the absorbed dose in this simulated microscopic site. A second measurement made with an identical TEPC loaded with ^{10}B yields the spectrum of charged particles depositing the BNC dose. This dosimetry system requires fewer measurements and has smaller estimated uncertainties for photon and neutron dose measurement than conventional dosimetry techniques,[1,2] and can be used as an independent verification of the accuracy of BNCT dosimetry methods currently in use.

3. DETECTOR DESIGN AND CHARACTERIZATION

The proportional counters constructed for this study have a 2.5 mm diameter right cylindrical collecting volume. This size was chosen based on the count rate capabilities of the TEPCs and data collection system and the beam intensities in which they were proposed to be operated. It allows an increase in particle flux by a factor of 20 over a standard ½″ Rossi counter while maintaining the same deadtime. The tissue-equivalent (TE) detector has a wall made of A-150 tissue-equivalent conducting plastic. The boron loaded (TE + B) detector has a wall made of A-150 loaded with 200 μg/g ^{10}B. The 200 μg/g loading was chosen to provide good statistics for the BNC dose without significantly attenuating the thermal fluence in the wall. A-150 plastic was designed to match the hydrogen and nitrogen content of muscle tissue, and therefore is not well matched to brain tissue. This difference in hydrogen and nitrogen content will affect the thermal fluence within the detecting volume. More importantly, the 580 keV protons from the $^{14}N(n,p)^{14}C$ reaction contribute a significant fraction of the neutron dose in the epithermal beam. For this reason, a plastic named A-181 was formulated to match the hydrogen and nitrogen composition of ICRU #44 brain tissue. This plastic was used in the construction of the brain tissue-equivalent (BTE) detector. The ratio of kerma factors for brain tissue to A-181 is very close to unity for nearly the entire range of neutron energies present in the epithermal beam.

Calibration of the detectors is based on features of the lineal energy spectrum. All detectors have been thoroughly characterized in a ^{252}Cf radiation field. Measured gamma, neutron and BNC dose components agree well with other measurements and calculations.[3–5] A detailed explanation of the calibration and characterization of the detectors is presented elsewhere.[6]

4. RESULTS

Measurements were made at the MITR-II Research Reactor using the MIT ellipsoidal water-filled head phantom. A microdosimetric spectrum separated into its gamma and neutron components is shown in Figure 1. The relative response of the TE + B detector may be seen in Figure 2. The shape of the secondary electron spectrum from the gamma dose was assessed at a depth of 10 cm where there exists a minimal contribution from the neutron dose. This gamma spectrum shape was then used to extract the gamma component from the spectra at other depths. The remainder of the spectrum from the

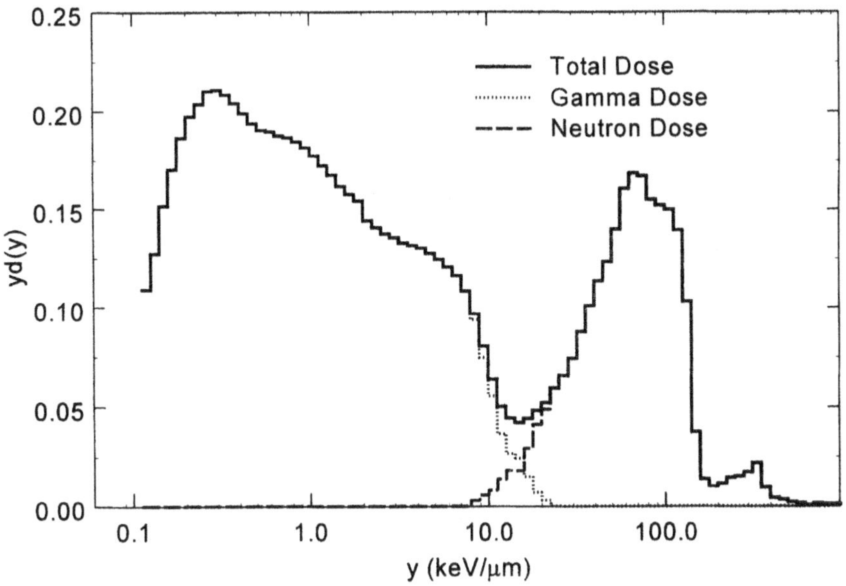

Figure 1. Lineal energy spectrum in the MITR-II beam at a depth of 2 cm along the central axis in the MIT ellipsoidal head phantom. (Plot represents dose per unit logarithmic interval of lineal energy density, y, against logarithm of y.)

TE detector was labeled as the neutron dose. The BNC dose was measured by subtracting the TE detector response from the TE + B detector response. Central axis results for gamma and neutron doses are presented in Figure 3. These are shown in comparison with measurements made using conventional techniques.[1] Measurements were also made in the BMRR Research Reactor using the BNL acrylic cube phantom. Spectral shapes of the gamma and neutron dose components were very similar to those obtained in the

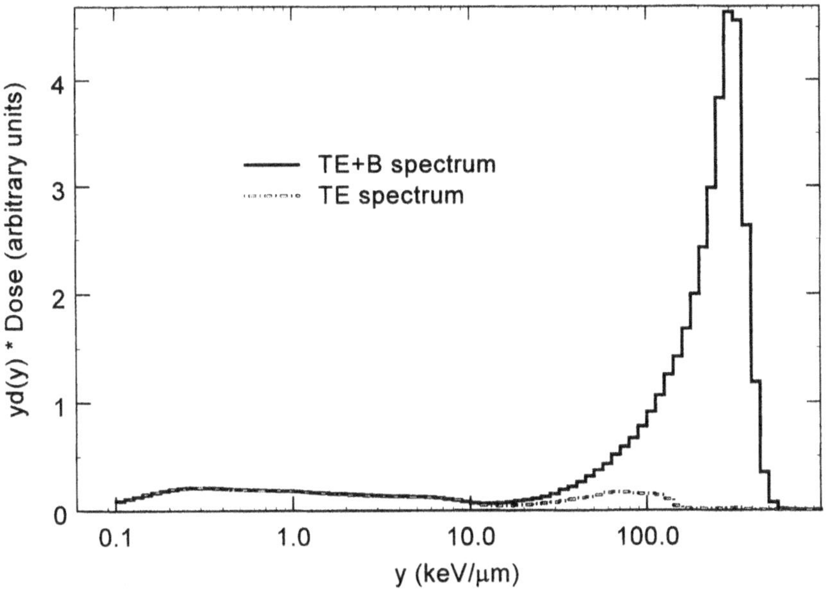

Figure 2. Comparison of lineal energy spectra measured with the TE and TE + B detectors in the MITR-II beam at a depth of 2 cm.

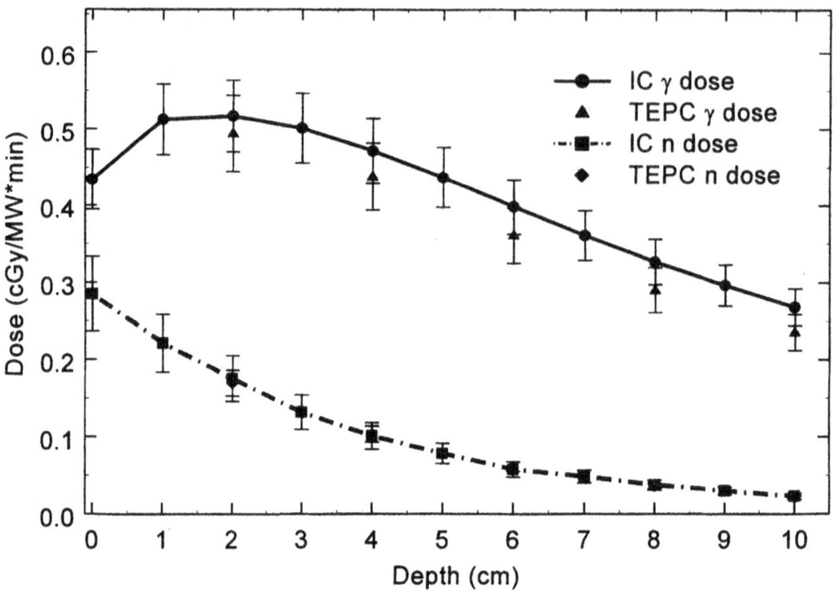

Figure 3. Comparison of gamma and neutron absorbed doses to brain tissue in the MITR-II beam measured using TEPCs and ionization chambers (IC). (The ionization chamber measurement technique is described by Rogus, *et al.*[1])

MITR-II epithermal beam. However, there were significant differences in their relative contributions as a function of depth. Central axis absorbed dose rates for the gamma and neutron components are shown in Figure 4 in comparison with calculations made for the same arrangement using the Monte Carlo treatment planning system.[7] The TEPC-measured BNC doses at both facilities are shown in Table 1 in comparison to the foil activation and Monte Carlo results of MIT and BNL, respectively.

5. DISCUSSION

BNC doses measured with the TEPCs in ^{252}Cf agree well with other dosimetry methods, however, there appears to be a systematic difference in the BNC results at MIT and BNL shown in Table 1. The thermal neutron fluence in the epithermal beam is significantly higher than that in the ^{252}Cf field used for calibration. However, if there were a dose rate effect on the BNC dose, one would expect to see the same effect on the gamma dose measured by the TE + B detector. This is not observed. The discrepancy in BNC doses at both facilities is currently under investigation.

Results from the BTE detector showed a significantly lower neutron absorbed dose than that measured with the TE detector. This difference in neutron dose results from the difference in nitrogen content between the two plastics. Nitrogen capture doses have been calculated using the BNL Monte Carlo treatment planning system for the experimental setup used at the BMRR. Scaling the nitrogen capture dose by the nitrogen contents of A-150 and A-181 allows a calculation of the change in neutron dose. This calculated difference shows good agreement with that observed between measurements with the TE and BTE detectors.

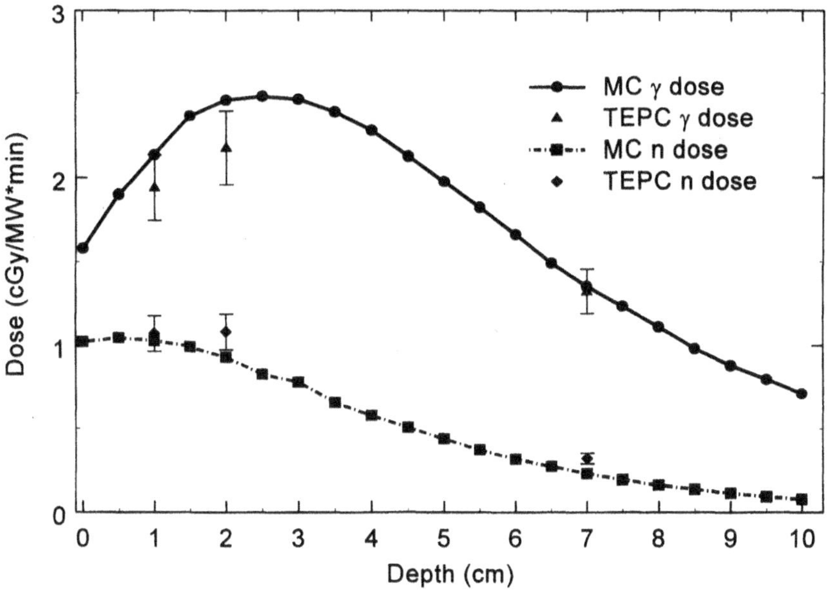

Figure 4. Comparison of TEPC measurements with Monte Carlo (MC) calculations for gamma and neutron absorbed doses to brain tissue in the BMRR beam. (The computational dosimetry technique is described by Nigg, *et al.*[7])

6. CONCLUSIONS AND FUTURE DIRECTIONS

Miniature TEPCs were designed and constructed for microdosimetry of BNCT. These detectors were used to make measurements at the MIT and BNL BNCT facilities. The gamma and neutron components of the absorbed dose measured with the TEPCs showed good agreement with results from conventional techniques in the MITR-II beam. There appear to be small, systematic differences between the TEPC measurements and Monte Carlo calculations for gamma and neutron doses at the BMRR. The total gamma plus neutron dose measured with the TEPCs agrees well with calculations, indicating a possible dose component separation problem. This technique allows a direct measurement of the BNC dose using the boron nucleus. However, the BNC dose measured with

Table 1. BNC doses for 30 ppm ^{10}B in comparison to foil activation measurements at MIT and Monte Carlo calculations at BNL[a]

	MIT BNC doses			BNL BNC doses		
Depth (cm)	TEPC	Foil activation	TEPC FA	TEPC	Monte carlo	TEPC MC
1	—	—	—	4.74 ± 0.46	7.13	0.66
2	0.556 ± 0.054	0.607 ± 0.045	0.92	6.01 ± 0.58	9.33	0.64
4	0.423 ± 0.041	0.507 ± 0.038	0.83	—	—	—
6	0.263 ± 0.026	0.325 ± 0.024	0.81	—	—	—
7	—	—	—	2.02 ± 0.19	2.63	0.77
8	0.142 ± 0.014	0.164 ± 0.012	0.87	—	—	—
10	0.071 ± 0.007	0.084 ± 0.006	0.85	—	—	—

[a]All doses are to brain tissue in units of cGy/MW*min and measurements were made as described in the text.

the TEPCs showed a systematic difference on the order of 15% and 30% in comparison to results found by MIT and BNL, respectively. Reasons for this discrepancy are still under investigation. Future measurements are planned at the BNCT facilities at Petten and in Finland, as well as return trips to MIT and BNL. These measurements will allow an intercomparison of the dosimetric techniques and microdosimetric spectra of each facility. Furthermore, results may be correlated with existing radiobiological data in an attempt to create a model which would allow the extraction of cell survival from the measured microdosimetric spectra. Further discussion on the utility of microdosimetric spectra in the analysis of absorbed dose in BNCT is provided by Kota.[8]

ACKNOWLEDGMENTS

This work has been supported in part by the U.S. Department of Energy, Grant # DE-FG02-96ER62217.

REFERENCES

1. R.D. Rogus, O.K. Harling, and J.C. Yanch, Mixed field dosimetry of epithermal neutron beams for boron neutron capture therapy at the MITR-II research reactor, Med Phys, 21(10):1611–1625, 1994.
2. C.P.J. Raaijmakers, M.W. Konijnenberg, H.W. Verhagen, and B.J. Mijnheer, Determination of dose components in phantoms irradiated with an epithermal neutron beam for boron neutron capture therapy, Med Phys, 22(3):321–329, 1995.
3. J. Yanch, and R. Zamenhof, Dosimetry of ^{252}Cf sources for neutron radiotherapy with and without augmentation by boron neutron capture therapy, Rad Res, 131:249–256, 1992.
4. J.G. Wierzbicki, Y. Maruyama, and A.T. Porter, Measurement of augmentation of ^{252}Cf implant by ^{10}B and ^{157}Gd neutron capture, Med Phys, 21(6):787–790, 1994.
5. C. Kota, "Microdosimetric Considerations in the use of the Boron Neutron Capture Reaction in Radiation Therapy," Ph.D. Thesis, Wayne State University, Detroit, USA, 1996.
6. J. Burmeister, C. Kota, and R.L. Maughan, Paired miniature TEPCs for dosimetry in high flux epithermal neutron capture therapy beams, Nuc Inst and Meth in Phys Res, A422:606–610, 1999.
7. D.W. Nigg, F.J. Wheeler, D.E. Wessol, J. Capala, and M. Chadha, Computational dosimetry and treatment planning for boron neutron capture therapy, J. of Neuro-Oncology, 33:93–104, 1997.
8. C. Kota, J. Burmeister, and R.L. Maughan, Utility of measured microdosimetric single event spectra in interpreting quality of radiation fields used in BNCT and BNCEFNT, *ibid.*

RADIATION DOSIMETRY FOR NCT FACILITIES AT THE BROOKHAVEN MEDICAL RESEARCH REACTOR

N. E. Holden, J.-P. Hu, D. D. Greenberg, and R. N. Reciniello

Brookhaven National Laboratory
Upton, New York, USA

1. INTRODUCTION

Brookhaven Medical Research Reactor (BMRR) is a 3 mega-watt (MW) heterogeneous, tank-type, light water cooled and moderated, graphite reflected reactor, which was designed for medical and biological studies and became operational in 1959. Over time, the BMRR was modified to provide thermal and epithermal neutron beams suitable for research studies. NCT studies have been performed at both the epithermal neutron irradiation facility (ENIF) on the east side of the BMRR reactor core and the thermal neutron irradiation facility (TNIF) on the west side of the core.[1] Neutron and gamma-ray dosimetry performed from 1994 to the present in both facilities are described and the results are presented and discussed.

2. MODIFICATION

In recent years, the epithermal neutron flux has been enhanced by a series of changes to the BMRR. The shutter between the reactor core and the ENIF has been modified with the addition of aluminum and aluminum oxide to down scatter the fast neutrons into the electron-volt (eV) or low kilo-electron-volt (keV) energy range. The number of ^{235}U fuel elements has increased from 28 to 32 and four of the newer elements have been placed on the east side of the core facing the ENIF. This has resulted in a 50% increase in the epithermal neutron flux.

A modification project[2] replacing the epithermal shutter with a fission converter plate shutter is underway, which should increase the epithermal neutron flux by a factor of seven and the epithermal/fast neutron flux ratio by a factor of two.

Frontiers in Neutron Capture Therapy, edited by Hawthorne *et al.*
Kluwer Academic / Plenum Publishers, New York, 2001.

3. NEUTRON AND GAMMA-RAY DOSIMETRY

Clinical Boron Neutron Capture Therapy (BNCT) dose escalation research protocols were initiated at BMRR in September 1994. A total of 41 glioblastoma multiforme patients have received BNCT. With the NCT treatment, a measurement program began to determine neutron fluxes and the gamma-ray dose rate at various locations in the ENIF and in the TNIF. Bare and cadmium covered gold foils, various threshold detector foils have been used to determine neutron fluxes. Thermoluminescent dosimeters (TLDs) and test TLD badges with ^{6}LiF and ^{7}LiF chips have been utilized to provide experimental data on the neutron dose rates and gamma-ray dose rates. TLD test badges use LiF:Mg, Ti as the thermoluminescent material in the form of solid chips. Three of the chips are TLD-700 material (enriched in ^{7}Li to 99.93%) and one chip is TLD-600 material (enriched in ^{6}Li to 95.6%). The badges also have filters of plastic, copper and thin aluminized mylar film of various thicknesses to separate and measure mixed fields of neutrons, beta particles and gamma-rays.

MCNP code[3] Monte Carlo calculations have been run at various locations throughout the ENIF and the TNIF, to obtain energy dependent spatial distributions of gamma-rays and neutrons from thermal energies (0.01–0.4 eV) up to fast energies (0.1–10.0 MeV). MCNP has been used to mock up the geometry of the core, reflector, the epithermal port and the thermal port construction and the collimators.

Radiation workers are monitored via the TLD badge (read once a month) and a self-reading dosimeter during each experiment. An early concern was whether workers would be subject to a significant dose rate from working with patients who were being irradiated. These workers prepare the room and the patient, operate the shutter, locate the patient in the proper position in the room and move the patient in and out of the room during the treatment. The reactor power level is $\leq$10 KW, whenever workers are in the treatment room. Gamma-ray doses for the representative key personnel involved[4] in the care of the first 12 patients receiving BNCT are: Responsible Physician, 27. μGy; Radiation Oncologist, 25. μGy; Nurse, 32. μGy, Medical Physicist, 23. μGy; and Health Physicist, 14. μGy.

4. EPITHERMAL ROOM DOSIMETRY RESULTS

In both shielded rooms, the neutron beam is located 3 feet above the floor and in the center of the reactor wall. Radiation shielding in both beam ports end with shields of bismuth. Removable collimators are mounted on the bismuth face to collimate the neutron beams. For ENIF, experimental comparisons have been made with the patient in place, with a tissue equivalent head phantom in place to mock up the patient's head position and with the room empty to study the impact of the patient on the background flux and dose rate values. Beam center of the ENIF has a six inch thick collimator of polyethylene and Li_2CO_3 (93% enriched in ^{6}Li, 45% by weight). The collimator has a concave cavity 20 cm in diameter on the reactor side tapering to 12 cm on the room side. At the opening in the collimator, the 3 MW power level neutron fluxes are: thermal neutron flux is $\approx 3 \times 10^7$ n/cm^2/s, the epithermal flux is $\approx 6 \times 10^8$ n/cm^2/s and the fast flux is $\approx 2 \times 10^7$ n/cm^2/s. There is reasonable agreement between the MCNP calculations and the various bare and cadmium covered gold and threshold foil measurements used to determine the neutron flux values.

A series of experiments on neutron and gamma-ray dose rates were performed in the ENIF between October 1997 and January 1998. The results measured using the TLD

Table 1. Epithermal room neutron and gamma-ray dose rates

	Empty room		Phantom in place		Patient in place	
Location in room	N Sv/h	γ cGy/h	N Sv/h	γ cGy/h	N Sv/h	γ cGy/h
Door	0.21	1.5	0.13	0.8	0.14	1.3
North window	0.32	3.2	0.19	2.5	0.18	3.7
South window	0.20	1.7	0.11	0.9	—	—
3′ Opp. wall	0.45	5.5	0.13	3.2	0.13	4.7
8′ Opp. wall	0.41	4.1	0.15	3.2	0.15	4.0
2′, 3 o'clock	0.35	4.8	0.29	4.8	0.52	6.8
2′, 6 o'clock	0.45	4.5	0.28	3.0	0.65	8.5
2′, 9 o'clock	0.32	6.4	0.24	5.1	0.27	9.4
2′,12 o'clock	0.26	15.	0.23	23.	0.30	24.
4′, 3 o'clock	0.29	2.5	—	—	0.14	7.8
Collimator	2.66	138.				

personnel monitoring badges with ^{6}LiF and ^{7}LiF chips are given in Table I. During an earlier October 1995 measurement using an Eberline ASP1 "rem-ball" neutron sensitive instrument, the 3 MW neutron dose rate (30 cm from the collimator) in the ENIF was determined to be 1.0 Sv/h. An NRC ADM multi-purpose gamma sensitive instrument determined the 3 MW gamma dose rate at the same 30 cm from the collimator to be 1.0 Gy/h. These values can be compared to the 2.66 Sv/h and 1.38 Gy/h values in Table 1. The 3 MW values from Table 1 were measured at the collimator face rather than 30 cm away and should be larger due to the geometric factor, which may have a 1/r or 1/r^2 dependence. This is indeed the case.

From Table 1, it can be noted that the neutron and gamma-ray dose rates fall off rapidly as you move out from the center of the beam. For the empty room, there is a reduction in the neutron dose rate at the reactor face by a factor of 5 to 10 in all directions (3 o'clock, 6 o'clock, 9 o'clock, and 12 o'clock) at a distance of two feet from the beam center, compared to the collimator center results. There is a similar reduction of 10 to 30 in the gamma-ray dose rate. For the area above the collimator, there appears to be much less gamma-ray shielding, resulting in a much larger gamma-ray dose rate under all conditions.

By comparing the data with the patient in place to the data for the empty room, there is obviously neutron scattering back from the patient to the face of the reactor. However, the scattering of the gamma radiation back to the reactor face is larger. Although there is a significant reduction in the neutron dose rate at the 3 foot beam center position on the wall opposite the patient, there appears to be much less change in the gamma-ray dose rate. This would indicate that there is much more absorption of neutrons than gamma radiation by the patient, which is, of course, the preferred situation. It might be noted that the head phantom absorbs more gamma radiation than does the patient.

In Table 2, cadmium ratios for gold foils are listed for various locations, (Cd Ratio = bare foil activity/cadmium covered foil activity). ^{113}Cd has a very large absorption cross section for thermal neutrons, 2.04×10^{-20} cm^2 (20,400 barns),[5] due to a large capture resonance at a neutron energy[6] of 0.178 ev. Since cadmium absorbs most thermal neutrons, the cadmium ratio will correspond to a ratio of thermal and epithermal to epithermal neutron reactions. Cadmium ratios are larger at the opposite wall indicating a higher percentage of thermal neutrons relative to epithermal neutrons reach those points, when the

Table 2. Comparison of cadmium ratios in the epithermal room

Location in room	Empty room cd-ratio	Phantom in place cd-ratio	Patient in place cd-ratio
Door	2.1	1.9	2.0
North window	1.5	1.7	1.6
South window	1.5	1.5	1.6
3′ Opp. wall	1.2	1.3	1.4
8′ Opp. wall	1.3	1.2	1.5
2′, 3 o'clock	1.3	1.3	1.4
2′, 6 o'clock	1.1	1.2	1.3
2′, 9 o'clock	1.0	1.4	1.3
2′,12 o'clock	1.5	1.5	1.5
4′, 3 o'clock	1.5	1.5	1.8
Collimator	1.0		

patient is in place compared to an empty room. The initial beam has an order of magnitude larger epithermal to thermal neutron flux. This effect would be due to the loss of epithermal neutrons from slowing down in the patient and increasing the thermal component.

The calculated values were only available for the reactor shield wall at the 3 o'clock, 6 o'clock, 9 o'clock, and 12 o'clock locations and at the collimator because the flux was too low at other locations in the room to obtain statistically meaningful results. The calculated values with uncertainties in the range from 7% to 12% were 1.04, which agreed with the measurement except for the 3 o'clock and 12 o'clock positions on the reactor shield wall.

5. THERMAL ROOM DOSIMETRY RESULTS

There is a collimator mounted on the bismuth face in the TNIF. It is made of a mixture of polyethylene and Li_2CO_3 (≈93% enriched in ^{6}Li and 45% by weight). The collimator is 10.2 cm thick with a conical cavity, 12 cm diameter on the reactor side tapering to 2 cm on the side facing the shielded room. Neutron and gamma-ray dose rate data in the thermal room are listed in Table 3. The collimated beam exits the reactor inside a

Table 3. Thermal room neutron and gamma-ray dose rates

Location in room	Neutron dose rate mSv/h	Gamma-ray dose rate cGy/h
Door Internal	0.54	0.096
Door External	0.06	—
Window Internal	64.7	1.8
Window External	0.06	0.06
Collimator Face	300,000	2,000
Collimator 30 cm	60,000	400
Backwall Internal	52.	9.6
Backwall External	2.2	0.36
Enclosure Internal	2,210	144
Enclosure External	115	53.S

polyethylene shielded enclosure. The results give an indication of the effectiveness of the shielded viewing window, the entrance door, the polyethylene enclosure and the back wall of the treatment room for reducing both neutron and gamma-ray dose rates. At the collimator face, 3 MW power level neutron flux values are: thermal flux $\approx 1.6 \times 10^{10}$ n/cm^2/s, the epithermal flux is $\approx 1. \times 10^9$ n/cm^2/s and the fast flux is $\approx 4 \times 10^8$ n/cm^2/s. There is general agreement between the calculated and measured values.

6. CONCLUSIONS

Modifications have been made to the BMRR to significantly increase the available epithermal neutron flux to a patient in clinical trials of BNCT. The above data indicate that the flux and dose rate are concentrated in the center of the beam and that the patient absorbs neutrons rather than gamma radiation. As previously noted even with the increasing flux values, gamma-ray dose received by the attending personnel has remained minimal. Further work on measuring and calculating neutron and gamma-ray flux and dose rates are planned in the future. Neutron flux values in the center of the thermal port and epithermal port beams have been characterized with an agreement between the measurements and the calculations.

ACKNOWLEDGMENT

This research was carried out under the auspices of the US Department of Energy Contract No. DE-AC02-98CH10886.

REFERENCES

1. H.B. Liu, J. Capala, D.D. Joel, R.M. Brugger, and D.C. Rorer, Advances in neutron capture therapy, Vol. I, "Medicine and Physics", B. Larsson, J. Crawford and R. Weinreich, ed., Elsevier Science B.V., 1997, pp. 311–315.
2. D. Rorer, F. Patti, R. Ma, A.D. Chanana, and H.B. Liu, Upgrade of the epithermal neutron beam using ^{235}U fission plates at brookhaven medical research reactor (BMRR), in "Frontiers in Neutron Capture Therapy", M.F. Hawthorne, K. Shelly, and R.J. Wiersema, ed., Plenum Publishing Corporation, New York, 1999.
3. MCNP-4B2 "Monte Carlo N-Particle Transport Code System (CCC-660)", contributed by Los Alamos National Laboratory, January 29, 1977, distributed by the Oak Ridge Radiation Safety Information Computational Center.
4. N.E. Holden, D.C. Rorer, F.J. Patti, H.B. Liu, R.N. Reciniello, and A.D. Chanana, Technical aspects of boron neutron capture therapy at the BNL medical research reactor, in "Nuclear Data for Science and Technology", G. Reffo, A. Ventura, and C. Grandi, ed., Italian Physical Society, Bologna, Italy, 1997, pp. 1674–1676.
5. N.E. Holden, Neutron scattering and absorption properties, in "Handbook of Chemistry and Physics", 78th Edition, 1997–1998, D.R. Lide, ed., CRC Press, Inc. Boca Raton, 1997, pp. 11.147–11.163.
6. S.F. Mughabghab, M. Divadeenam, and N.E. Holden, Neutron Cross Sections, Vol. 1 "Neutron Resonance Parameters and Thermal Cross Sections", Part A, Z-1-60. Academic Press Inc., New Yrok, 1981.

IN VIVO DETERMINATION OF BORON IN PATIENTS DURING BNCT IN PETTEN BY MEANS OF THE GAMMA-RAY TELESCOPE

W. F. A. R. Verbakel and F. Stecher-Rasmussen

Netherlands Energy Research Foundation (ECN)
P.O.Box 1, NL-1755 ZG Petten
The Netherlands

1. INTRODUCTION

This study shows the feasibility of measuring an average boron concentration in the head of the patient during the BNCT irradiation by capture γ-ray spectroscopy. These measurements can be performed with the γ-ray telescope: a HPGe-detector outside the therapy room, which measures the γ-rays emitted by the head of the patient through a small hole in the ceiling. Until now, in the Petten BNCT Phase I study, 10 patients with a glioblastoma multiforme have been irradiated in dose group 1. They were injected with the boron compound BSH about 12 hours before treatment. The tumour has been debulked a few weeks before irradiation. Before and after the irradiation, the boron concentration in blood samples is determined by prompt γ-ray spectroscopy at the neighbouring beam tube HB7 from blood samples. For proper dosimetry it is also essential to know the distribution of ^{10}B in the head during the therapy.

2. INTEGRAL MEASUREMENTS

In the proceedings of 1996 it was already proved that a macroscopic average boron concentration in a phantom can be determined by measuring the 478 keV prompt γ-rays which are emitted in 93% of the boron neutron capture reactions.[1] Since then, clinical trials have started and measurement data from the 10 patients has been collected. Figure 1 shows the present situation where an HPGe-detector located above the ceiling of the

Frontiers in Neutron Capture Therapy, edited by Hawthorne *et al.*
Kluwer Academic / Plenum Publishers, New York, 2001.

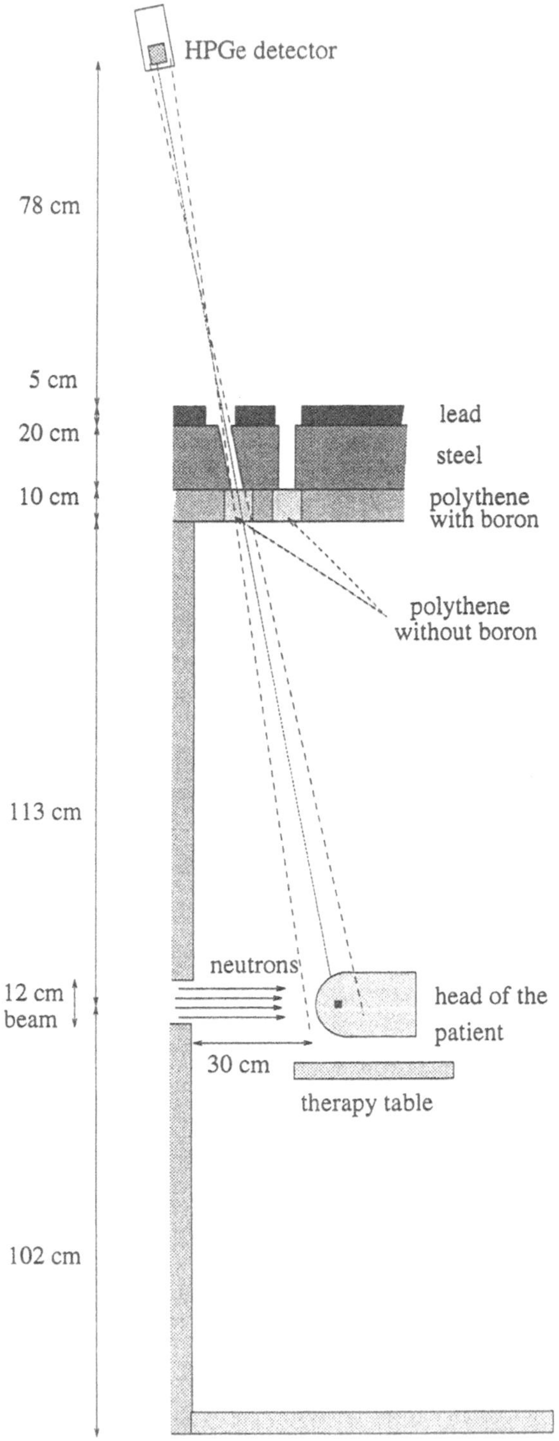

Figure 1. Vertical cross section of the therapy room with the integral detector system.

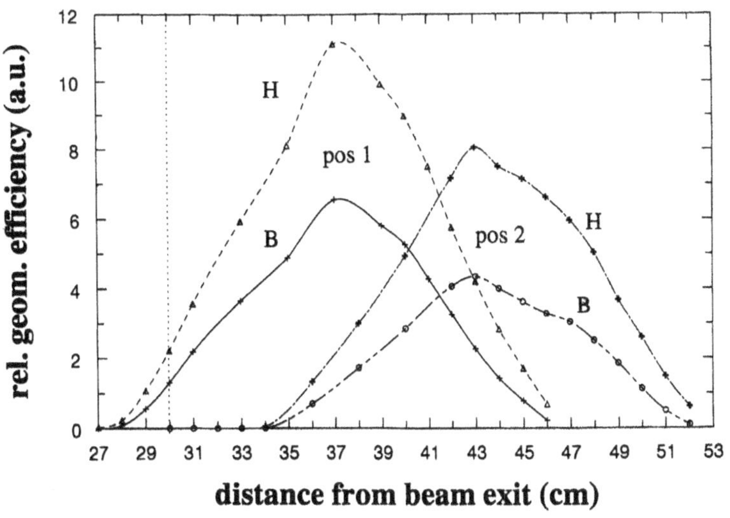

Figure 2. Geometric efficiency of the two detector positions, calculated with MCNP.

therapy room can measure the γ-rays originating from a large part of the head of the patient. As only an average boron concentration of a large volume can be reconstructed, these are called integral measurements. In spite of the presence of a large amount of boron polythene inside the therapy room, no boron-line is detected from this shielding. The therapy table provides a very weak boron line for which a correction is made in the measurements. Figure 2 shows the geometric efficiency of the detector in normal position (pos1) and in shifted position (pos2). Since the previous (debulked) tumour region approximately ranges from 32 to 37 cm from the beam exit, pos2 will measure hardly any tumour region if the neutron beam is directed to the tumour.

3. BORON DETERMINATION

A single measurement takes 1 minute live time, which corresponds up to 3 minutes real time due to the high dead time of 50 to 70%. Figure 3 shows the clear appearance of the B-, H- and Cl-lines in the spectrum. Since the boron count rate (*B*) depends on the thermal neutron distribution, the hydrogen count rate (*H*) will be used as a reference for the determination of the boron concentration. The average boron concentration in the head, $[B_{head}]$ is determined from the *B*/*H*-ratio by

$$B/H = \varepsilon \cdot T_{poly} \cdot T_{head} \cdot 0.93 \cdot 0.010 \cdot [B_{head}]$$

where a correction is made for the difference in detector efficiency for B and H γ-rays (ε), the difference in transmission through the polythene shielding (T_{poly}) and the difference in transmission through the head itself. Calculations with data from the treatment planning program, show that this latter factor varies between 0.55 and 0.78 for different patient positions. The factor 0.010 is the ratio of production of B and H γ-rays if there is 1 ppm boron in the head.

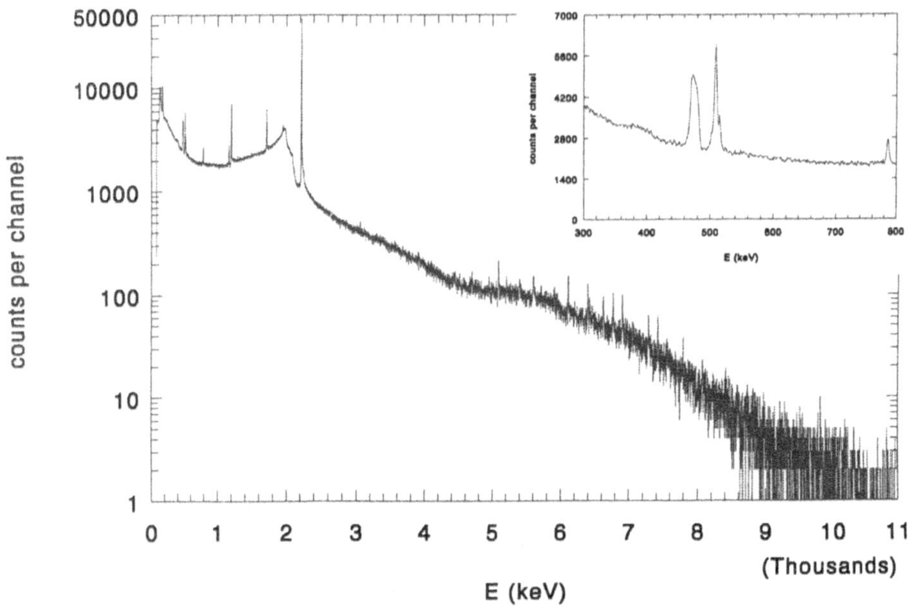

Figure 3. An integral measurement on a patient during BNCT, measured for 60 s live time.

4. RESULTS FROM ONE PATIENT

In spite of the fact that the boron concentration in blood decreases almost 10% during one irradiation, there appears no significant decrease of the *B*/*H*-ratio for all 10 patients. For the results in Table 1, the *B*/*H*-ratio's have been averaged for each day. The results show that the average boron concentration in the head, as measured with the γ-ray telescope, is proportional to the concentration in the blood. In order to investigate which components of the head contribute to the boron concentration, the irradiation of this patient's head has been simulated with MCNP4A. The calculated *B*/*H*-ratio's are obtained for 6 ppm ^{10}B in healthy tissue, 41 ppm in a 0.5 cm thick shell as skin and

Table 1. Results of boron measurements for one specific patient[a]

		MCNP					Average
B_{bl} *(ppm)*		41		14	30.5	27.6	28.3
Beam 1							
Pos1, *B*/*H*		0.154	0.141	0.062	0.116	0.105	
Pos2, *B*/*H*		0.112	0.119	0.045	0.085	0.080	
B_{head}/B_{bl} (%)	P1	22.4		25.5	22.5	22.5	23.3
(±10–15%)	P2	16.6		17.6	15.1	17.2	16.6
Beam 2							
Pos1, *B*/*H*		0.115	0.134	0.047	0.088	0.083	
Pos2, *B*/*H*		0.091	0.119	0.038	0.075	0.068	
B_{head}/B_{bl} (%)	P1	15.5		17.6	15.6	16.3	16.3
(±10–15%)	P2	12.1		13.2	13.3	13.2	13.0

[a]The correction factor T_{head} has been calculated for both beams (depth of beam, which is defined as the attenuation distance from centre of beam 1, is 12.5 cm, respectively 9 cm for beam 2).

Table 2. Reconstructed average boron concentration in the measured volume divided by the blood concentration[a]

	Patient	10		9		8		7		6	
B_{head}/B_{bl} (%)	Beam	1	2	1	2	1	2	1	2	1	2
Pos1		24	19	38	15	23	16	21	11	27	25
Pos2		18	16	30	15	17	13	16	10	23	29
Pos1/Pos2		1.34	1.20	1.24	1.04	1.40	1.25	1.29	1.08	1.17	0.86

[a]The relative standard deviation is between 10 and 20% for all reconstructions.

41 ppm in a "previous tumour region" of 2 cm radius. If there would be an equal boron concentration in the whole head, the contribution of the modelled skin and tumour would be 4% and 4.4% for Beam1 Pos1, 2.1% and 1.1% for Beam1 Pos2, 4.8% and 0.3% for Beam2 Pos1, and 1.2% and 1.8% for Beam2 Pos2. As a result, it can be concluded that the contribution of the tumour can almost be neglected except for Beam1 Pos1, and the contribution of the skin is certainly important for detector Pos1 for both beams.

5. RESULTS FOR 5 PATIENTS

Table 2 shows the results of the reconstruction for 5 patients. Some of the very high values can be caused by a large contribution of the skin if the depth of the beam is small. It can be concluded that B_{head}/B_{bl} ranges between 0.10 and 0.25 for all measurements that are not viewing the previous tumour region. If this region is in the field of view, the values are on average 15% higher. The higher boron concentration with Pos2 compared to Pos1 for patient 6 Beam 2 are caused by the fact that for this patient, the previous tumour region was in the centre of the field of view of Pos2, which was not the case for the other patients.

6. COMPARISON WITH LITERATURE

The boron compound BSH is supposed not to have an uptake by healthy tissue.[2] This would infer that the average boron concentration in healthy tissue divided by the boron concentration in blood is equal to the blood fraction. This blood fraction, averaged over white and grey matter, is approximately 4%. Including the blood in the large vessels, the blood fraction of a head will be at maximum 6–7%. Conversely, the measurements provide values ranging from 10% to 25%. For a more extended comparison, other experimental values for BSH are presented below.

For rabbits, B_{head}/B_{bl} has been measured with PGRA on samples taken at several times after infusion.[3] The results are 0.06 (10 minutes), 0.16 (1 hour), 0.20 (3 h), 0.27 (6 h) and 0.24 (24 h), all with an uncertainty between 14 and 20%. For rats the results are 0.17 and 0.13 (±40–60%), both measured 2.5 hours after infusion.[4] For patients the results[5] are 0.09 (3 h), 0.14 (6 h) and 0.22 (12 h) ±50%. Samples taken at surgery from the Petten patients provided B_{head}/B_{bl}-ratio's of 0.43, 0.16, 0.39, 0.21, 0.15 and 0.19. However, it is not impossible that some samples contained partly tumour or necrotic tissue. It is noteworthy that all above mentioned values are in the same range as the "healthy tissue" measurements of the γ-ray telescope.

7. CONCLUSION

For the first group of patients, the dose distributions in the head are calculated with a treatment planning program, assuming an even boron concentration in the head equal to the concentration in blood. This is an overestimate since it is expected that there is no boron uptake by the healthy tissue. Accurate dose determination requires information about the boron concentration in the healthy tissue and in the tumour. Measurements with the γ-ray telescope prove that on-line boron determination in patients during BNCT is possible. As the integral measurements are averaged over a large volume, the exact interpretation of the results requires Monte Carlo simulations and extended phantom measurements to understand the contributions of different head compartments. Conversely, the results show directly that there is more boron in the head than expected from the blood fraction of the head if there would be no boron uptake by healthy tissue for BSH. The results are consistent with measurements of surgery samples in spite of the large uncertainty of this data. Furthermore, it is striking that the B/H-ratio shows no significant decrease during the irradiation. These results raise the question: where is the boron in the tissue? There may also be a larger boron concentration in the plexus choroidius or in the liquor. Is there an uptake in the interstitial space between the cells? To answer these questions, the γ-ray telescope needs further development like position dependent measurements. The introduction of a moveable collimator allows detection of boron, only originating from a vertical cylinder of 3 cm diameter. This reduces the contribution of the skin and allows selective healthy tissue measurements.

ACKNOWLEDGMENT

The authors want to acknowledge the BIOMED II program of the European Union (BMH4-CT96-0325).

REFERENCES

1. W. Verbakel, and F. Stecher-Rasmussen, A gamma-ray telescope for on-line reconstruction of low boron concentrations in a human head during BNCT, *in*: "Advances in Neutron Capture Therapy, Vol 1", B. Larsson, J. Crawford, R. Weinreich, ed., Elsevier Science, Amsterdam, 1997, pp. 122–127.
2. D. Gabel, BSH as a boron carrier for glioma BNCT: overview and update, *in*: "Advances in Neutron Capture Therapy, Vol 2", B. Larsson, J. Crawford, R. Weinreich, ed., Elsevier Science, Amsterdam, 1997, pp. 149–152.
3. K. Mukai, and Y. Nakagawa, Prompt gamma-ray spectroscopy for in-vivo measurement of boron-10 concentration in rabbit brain tissue, *Neurol. Med. Chir.*, 35:855–860, 1995.
4. W. Yang, Boron Neutron Capture Therapy of brain tumours: enhanced survival following intracarotid injection of sodium borocaptate with or without blood-brain barrier disruption, *Int. J. Radiation Oncology Biol. Phys.*, 37:663–672, 1997.
5. V. Horn, Clinical phase I biodistribution study of borocaptate sodium, *in*: "Advances in Neutron Capture Therapy, Vol 2", B. Larsson, J. Crawford, R. Weinreich, ed., Elsevier Science, Amsterdam, 1997, pp. 205–209.

MCNP ENHANCEMENTS, PARALLEL COMPUTING, AND ERROR ANALYSIS FOR BNCT[a]

T. Goorley[1], G. McKinney[2], K. Adams[2], and G. Estes[3]

[1]Nuclear Reactor Laboratory
Massachusetts Institute of Technology
[2]X-CI, Los Alamos National Laboratory
[3]X-TM, Los Alamos National Laboratory

1. INTRODUCTION

The Boron Neutron Capture Therapy (BNCT) treatment planning procedure used by the Harvard/MIT BNCT Program relies on MCNP[a] to calculate dose rates throughout a patient specific model for NCT irradiations. Since MCNP transport calculations are a time consuming portion of the treatment planning process, their acceleration greatly improves treatment planning efficiency. Source code augmentations and implementation of parallel computing on Windows NT computers have greatly decreased the time needed for these dosimetry calculations. A statistical uncertainty analysis verified that the appropriate number of source particles, which varies with the irradiation beam orientation, was being used for treatment planning.

2. MCNP ENHANCEMENTS

MCNP is a well benchmarked, general purpose radiation transport program with extensive capabilities.[1] Some of these capabilities are useful for BNCT dose calculations, including modeling the spatial, energy and angular distributions of the irradiation beam, and calculating the flux depression caused by explicitly increased boron concentrations in the modeled tumor.[2] MCNP can also calculate the dose rates from the thermal and fast neutron reactions in tissue, primarily from nitrogen absorption and hydrogen recoil

[a] MCNP is a trademark of the Regents of the University of California, Los Alamos National Laboratory.

Frontiers in Neutron Capture Therapy, edited by Hawthorne *et al.*
Kluwer Academic / Plenum Publishers, New York, 2001.

Table 1. Single CPU MCNP wall-clock run times in minutes for 10^4 or 10^6 source particles

	NPS = 10,000		NPS = 1,000,000	
Voxel geometry	Neutron	Gamma	Neutron	Gamma
Lattice	839	86	83,000*	8,400*
Lattice with tracking enhancement	704	86	69,000*	8,400*
Lattice with tally enhancement	4.4	2.8	86.9	20.7
Lattice with both enhancements	4.4	3.2	85.6	20.4

respectively, neutron absorption by boron, and gamma ray interactions, to every voxel of a lattice model based on the patient specific CT data.[3,4] Some of the other features included in MCNP are not needed for these BNCT calculations, and their removal significantly reduces run time. The enhancements assume that the lattice is a rectangular parallelepiped, and simplifies lattice tracking calculations. The lattice tallying calculations are also accelerated by removing energy bin tallies, although the DE and DF cards, which are used in BNCT dose rate calculations to convert energy dependent neutron or photon fluxes to KERMA rates, are kept intact. The tracking improvements account for a factor of 1.2 runtime reduction for neutrons, while the tallying improvements account for a factor of 190. The executable created from the combination of these two modifications is called MCNPBNCT. While MCNPBNCT will not pass the MCNP test suite, which is used to ensure that MCNP was properly installed, MCNPBNCT was verified by running the same input deck as the unmodified MCNP and comparing the results, which were within statistical uncertainty for the converged result from 10 million particles.

Table 1 shows wall-clock runtimes, in minutes, associated with one beam orientation for Harvard/MIT subject 97-4. The asterisks denote estimated times, while the other values are actual times with errors of a few seconds. The wall-clock runtime is slightly greater than the sum of the MCNP startup time, cp0, and the final MCNP runtime, ctm. Wall clock run times are important to clinical treatment planning situations, where the time between dose rate data analysis is a limiting factor.

As can be seen from Table 1, the dominant time savings comes from modifications of the tally routine in MCNP for this calculation. Only for runs with more than 3 million particles do the tracking modifications contribute to runtime reduction when combined with the tally modification. For shorter runs, the wall-clock runtimes for MCNPBNCT, as shown on the last line in Table 1, are dominated by the setup times (cp0): 2.73 minutes for neutron source particle runs and 2.36 minutes for gamma source particle runs. The estimated wall clock run times were made by adding the startup time to the linearly scaled particle tracking time. The patch files needed to create MCNPBNCT are available from the first author.

3. PARALLEL COMPUTING WITH MCNP

MCNP has the capability to simultaneously run over a heterogeneous computing network, using a program called Parallel Virtual Machine (PVM), developed at Oak Ridge National Laboratory.[5,6,7] Both software programs were recently modified to allow IBM PCs running the operating systems Windows NT to be linked together, along with the other supported systems (UNIX Sun Solaris, SGI Irix, PC Linux, etc.), allowing the MCNP transport calculation runtimes to be reduced by a factor equivalent to the number of computers being used.[8]

Table 2. Parallel MCNPBNCT wall-clock runtimes

Number of CPUs used	Number of spawned tasks	NPS = 10,000	NPS = 1,000,000
1	Not run in parallel	7.6	106
2	2	15	99
3	2	8	64
3	3	12	59
4	3	8	44

The primary obstacle in the development of a Windows NT parallel version of MCNP was the mixed language programming, where real numbers, integers and character strings had to be passed correctly between MCNP, written in FORTRAN, and PVM, written in C. This required compilers able to compile each program's source code, and a common linker able to link together each compiler's object files. The recent release of Digital Visual FORTRAN 5.0, compatible with Microsoft Visual C/C++ 4.0, fulfilled these requirements. Although several difficulties were encountered, attributed to the beta status of version of PVM 3.4 beta 6 for Windows, modifications were developed that allow MCNP4B to effectively pass the MCNP parallel test suite.

Table 2 shows the wall clock runtimes, in minutes, for MCNPBNCT running in parallel under Windows NT on up to four 200 MHz Pentium Pro computers. The third and fourth CPU used were on a dual CPU motherboard in one computer.

Various options in PVM and MCNP can be used to reduced the total wall-clock runtime of the dose rate calculation. Although the MCNP master process, which performs startup calculations and directs spawned tasks, typically does not run any particles, it does use a portion of the computational power, constantly checking to see whether the subtasks have finished. The master task can be run on its own CPU, when the number of spawned tasks is less than the number of linked CPUs available. As shown in lines three and four of Table 2, it is more efficient to have the master on its own CPU only for short runs. Since tasks spawned by MCNP are assigned sequentially through the PVM configuration listing, it is possible to make a PVM configuration of dual CPU mother board computers followed by single processor computers, and have MCNP assign two processes only to the dual CPU computers. In MCNP4B, the tasks command line option allows the optional use of a negative number of spawned processes, which disables PVM load balancing,[9] providing some runtime reduction for a homogeneous computing network. The MCNP PRDMP card controls the communication frequency between spawned and master processes.[1]

4. VOXEL DOSE RATE STATISTICAL UNCERTAINTY ANALYSIS

In addition to calculating the dose rates to each voxel, MCNP also calculates the statistical uncertainty associated with each voxel for each dose component. It is important that high dose regions, particularly around the region of interest, such as the tumor and peak dose rate location, have statistical uncertainties low enough to ensure statistically significant comparisons between different beam orientations. Typically the statistical uncertainty associated with these high dose region is 5% for scoping runs, and 1% for final planning runs. Figure 1 shows the dose rate and associated statistical uncertainty

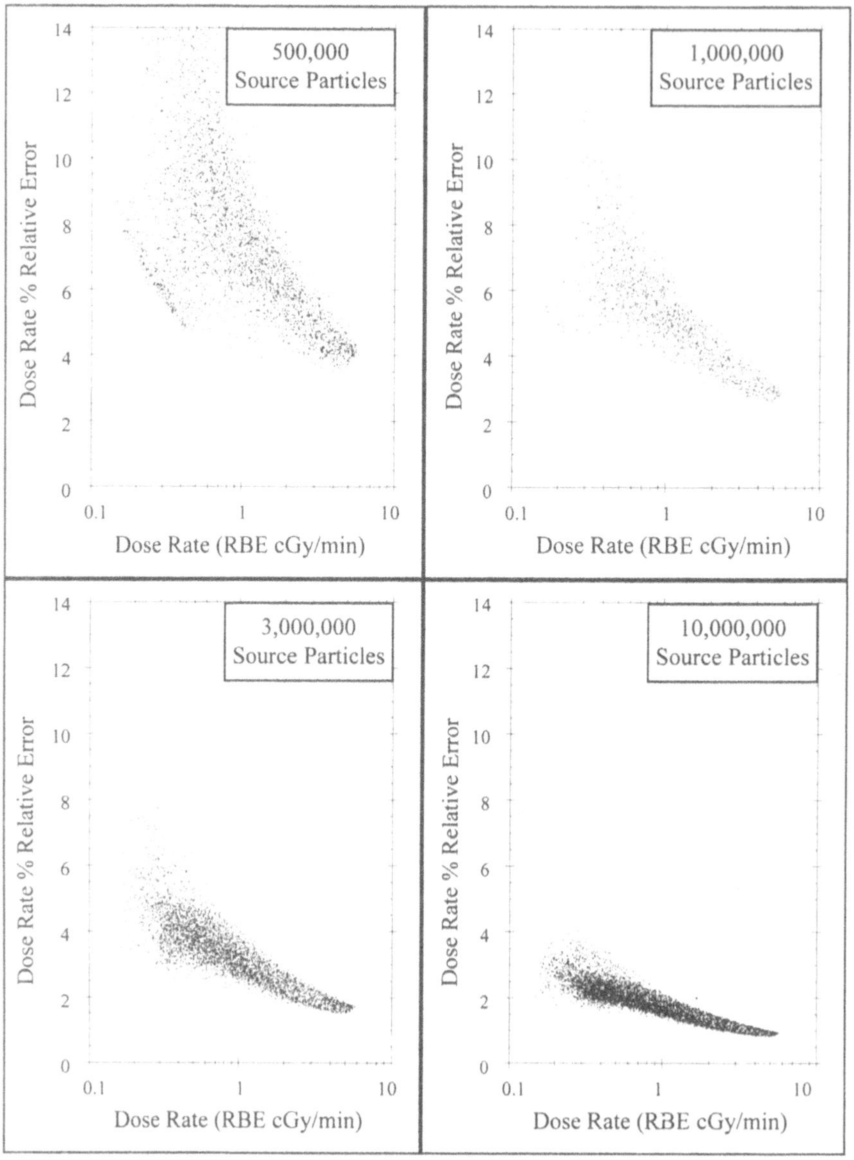

Figure 1. Dose rate statistical uncertainty for various number of source particles.

for various number of source particles for Harvard/MIT BNCT Subject 97-3. Each dot represents a single voxel.

As expected, higher dose rate regions have more particles passing through them, and have lower statistical uncertainties. When the region of interest is near the beam entrance, it will be in a high dose rate region, and will require fewer number of source particles (~1 Million) to achieve a 5% uncertainty. When the entrance location is located far from the region of interest, as in the case of an contralateral beam, between 1 and 3 million particles must be run to achieve the same amount of statistical uncertainty. An example of a statistically significant contribution from a contralateral beam to a primary beam's high dose rate region is: the dose rate to the region of interest from a contralateral beam is 0.5 RBE cGy/min +/– 0.05 RBE cGy/min, while the primary beam is in 5 RBE cGy/min +/– 0.25 RBE cGy/min. If the primary beam calculation had been run with fewer source particles, its peak dose rate statistical uncertainty would be the same as the second beam's dose rate contribution to that point.

5. CONCLUSIONS

This paper presents three approaches to reduce the total wall clock runtimes of MCNP dose rate calculations. MCNP modifications which removed unnecessary calculations, particularly unnecessary energy bin tallying, significantly reduced wall-clock runtimes. Modifying the existing PVM capability within MCNP to work with Windows NT allowed additional office computers to be used during treatment planning calculations, further reducing the calculation time. A statistical uncertainty analysis was used to determine whether excessive number of source particles were being run, but the appropriate number was already being used. The total wall clock runtime of a typical one million source particle beam dose rate calculation running on four 200 MHz Pentium Pro CPUs was reduced to forty-four minutes.

ACKNOWLEDGMENTS

This research was supported by grants from the U.S. Department of Energy to the Radiation Transport Group at Los Alamos National Laboratory, with subsequent subcontracts to Harvard/MIT. The authors would like to thank the Harvard/MIT BNCT team, including Stead Kiger, Dr. Matthew Palmer and Dr. Robert Zamenhof for their support. The MCNP enhancements for BNCT were, in part, first proposed by LANL in a paper presented at the Fifth International Symposium on Neutron Capture Therapy.[10]

REFERENCES

1. J.F. Briesmeister, Ed., "MCNP—A General Monte Carlo N-Particle Transport Code," v. 4B, Los Alamos National Laboratory report LA-12625-M Version 4B, March 1997.
2. R.G. Zamenhof, E.L. Redmond II, G.R. Solares, D. Katz, W.S. Kiger III, and O.K. Harling. "Monte Carlo-Based Treatment Planning for Boron Neutron Capture Therapy Using Custom Designed Models Automatically Generated from CT Data." *Int. J. Radiation Oncology Biol. and Phys.*, 35[2]:383–397, 1996.
3. T. Goorley, G.W. McKinney, K. Adams, and G.P. Estes. "MCNP Speed Advances for Boron Neutron Capture Therapy." Trans. ANS Rad. Prot. Shield. Div. Topical Conf. Apr. 19–23 1998, 2, 95.
4. W.S. Kiger III, R.G. Zamenhof, G.R. Solares, E.L. Redmond II, and C. Yam. "MacNCTPlan: An Improved Macintosh-Based Treatment Planning Program," *Trans. Am. Nuc. Soc.*, 75:38, 1996.
5. Al Geist, A. Benguelin, J. Dongarra, R. Manchek, W. Jiang, and V. Sunderman, "PVM: Parallel Virtual Machine—A User's Guide and Tutorial for Networked Parallel Computing," The MIT Press, Cambridge, 1994.
6. J. Dongarra, A. Geist, R. Manchek, and V. Sunderam, "Integrated PVM Framework Supports Heterogeneous Network Computing", *Comp. Phys.*, 7:166–175, 1993.

7. G.W. McKinney, "Parallel Processing Monte Carlo Radiation Transport Codes," Pro. 8th ICRS, Arlington, TX, Apr 24–28 1994.
8. G.W. McKinney *et al.*, "Multiprocessing MCNP on an IBM RS/6000 Cluster," *Trans. Am. Nucl. Soc.*, 68:212, 1993.
9. G.W. McKinney, "A Practical Guide to Using MCNP with PVM," *Trans. Am. Nucl. Soc.*, 71:397, 1994.
10. Guy P. Estes and William M. Taylor, "Potential MCNP Enhancements for NCT", Advances in Neutron Capture Therapy, Albert H. Soloway, Roth F. Barth and David E. Carpenters (Eds.), Plenum Press, 1993.

CALCULATION OF ABSORBED DOSE DISTRIBUTIONS USING REMOVAL-DIFFUSION THEORY FOR BNCT TREATMENT PLANNING

B. Albertson[1], J. Niemkiewicz[2], T. E. Blue[1], and N. Gupta[1]

[1]The Ohio State University
Columbus, Ohio
[2]Grant/Riverside Methodist Hospitals
Columbus, Ohio

1. INTRODUCTION

Successful implementation of Boron Neutron Capture Therapy (BNCT), requires a knowledge of the radiation dose distributions in the tissue being treated. In order to plan a BNCT treatment, a treatment planning system is necessary, which is accurate, as well as, fast enough to calculate dose distributions in minutes. Most BNCT treatment planning has focused on Monte Carlo[1,2] or Discrete Ordinates[3] techniques. Removal-diffusion theory may provide an alternative calculation technique that may be both accurate enough and fast enough for BNCT treatment planning. This work investigates the use of removal-diffusion theory for BNCT to calculate neutron flux distributions and, subsequently, neutron absorbed dose distributions, within a patient's head.

2. METHODS

2.1. Theory

In removal-diffusion theory, the neutron flux is thought of as consisting of two components, a collided flux and an uncollided flux. To calculate the collided neutron flux, removal diffusion theory mates a standard diffusion theory calculation with a source term of first collided neutrons, which are removed from an uncollided neutron field by their first collision.[4] The collided neutron flux is then summed with the uncollided neutron flux

Frontiers in Neutron Capture Therapy, edited by Hawthorne *et al.*
Kluwer Academic / Plenum Publishers, New York, 2001.

to find the total neutron flux. The one-dimensional multi-group removal-diffusion equation is shown as Eq. 1 for the neutron flux (ϕ_g) in energy group g.

$$-D_g \frac{d^2\phi_g}{dx^2} + \Sigma_{R,g}\phi_g = \sum_{g'=1}^{g-1} \Sigma_{s,g'\to g}\phi_{g'} + \sum_{g'=1}^{g} P_{s,g'\to g} I_{g'} e^{-(P_{t,g'}x)} \qquad (1)$$

where,

D_g = The diffusion coefficient for collided neutrons in energy group g (cm).

$\Sigma_{R,g}$ = The macroscopic removal cross section for collided neutrons in energy group g (cm^{-1}).

$\Sigma_{s,g'\to g}$ = The macroscopic cross section for scattering of collided neutrons from energy group g′ to energy group g (cm^{-1}).

$P_{s,g'\to g}$ = The macroscopic cross section for uncollided neutrons scattering from group g′ to become collided neutrons in energy group g (cm^{-1}).

$I_{g'}$ = The incident uncollided neutron flux in energy group g′ ($cm^{-2}sec^{-1}$).

$P_{t,g'}$ = The macroscopic total cross section for uncollided neutrons in energy group g′ (cm^{-1}).

For this paper, the four-group energy structure shown in Table 1 was used for both the uncollided and collided neutrons. A monodirectional source of uncollided neutrons was assumed to be perpendicularly incident on a two-dimensional water phantom. All calculations assumed an incident neutron flux of 1 ($cm^{-2}sec^{-1}$) with a 1/E energy spectrum in groups 1 through 3. There were no source neutrons born in the thermal group. The phantom was infinite in one transverse dimension. The other transverse dimension (17 cm) and the axial dimension (15 cm) were chosen to be similar to the dimensions of a human head.

2.2. Removal Parameters

The calculation of the first collided source is accomplished in removal-diffusion theory by transporting external source neutrons by ray-tracing techniques to their site of first collision. The removal parameters, $P_{t,g'}$ and $P_{s,g'\to g}$, which are used in the ray-tracing calculation were determined by fitting the results of MCNP[5] Monte Carlo calculations to appropriate exponential forms for the simple case of a monodirectional beam (traveling in the + x-dir) perpendicularly incident on a semi-infinite (finite in x, infinite in y and z) slab. The parameters, which were determined in this manner, are appropriate for the calculation of the first collided source distribution for more general source angular distributions.

2.3. Diffusion Parameters

The diffusion calculations were performed using the diffusion code DIF3D.[6] The parameters which must be input to this code (diffusion parameters) were determined by

Table 1. Neutron energy group structure

Energy group	Lower energy (eV)	Upper energy (eV)
1	1.0×10^3	3.0×10^3
2	10	1.0×10^3
3	0.4	10
4	0	0.4

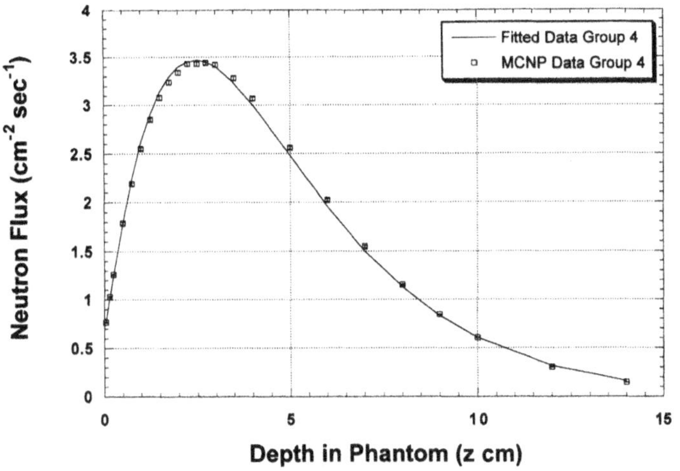

Figure 1. Curve fit of flux data form solution to the 4-group neutron removal-diffusion equation to Monte Carlo neutron flux data from MCNP for energy group 4.

fitting the analytical solution to the four-group one-dimensional neutron removal-diffusion equation to the neutron flux data which was obtained for the same geometry by Monte Carlo calculations. Specifically, the analytical solution was obtained for the one-dimensional, semi-infinite slab geometry, which is described above in Section 2.2. Since in this geometry the slab dimension (x) is parallel to the incident neutron beam axis, henceforth in this paper the corresponding solution is called the "axial" solution and the corresponding diffusion parameters are called the "axial" diffusion parameters. Specifically, axial transport cross-sections, extrapolation distances, and scattering and absorption cross-sections were obtained by fitting for each of the 4 energy groups. The curve fit for the one-dimensional axial geometry for group 4 is shown in Figure 1. The curve fits for the other three groups fit equally well.

The axial diffusion parameters were implemented, along with the first collided source, in DIF3D. The implementation of these parameters, especially the first collided source, in a DIF3D calculation is what comprises a removal-diffusion theory calculation. From this point on, a DIF3D calculation, with the appropriate first collided source and diffusion parameters, will be referred to as a removal-diffusion theory calculation. A transverse extrapolation distance was also required for the removal-diffusion theory calculation. The transverse extrapolation distance was calculated using MCNP and was determined from the value of the neutron flux and its slope at the phantom boundary along a transverse plane perpendicular to the beam axis at a depth of 2.25 cm in the two-dimensional water phantom described in Section 2.1.

2.4. Absorbed Dose Rate Calculation

With all of the necessary removal and diffusion parameters determined, as specified in Section 2.2 and 2.3, a removal-diffusion theory calculation was performed for the two-dimensional geometry which is described in Section 2.1. Kerma factors were applied to the removal-diffusion theory neutron flux data to calculate the sum of the neutron and boron absorbed dose rate distributions ($\dot{D}_g$) for each of the four energy groups. These absorbed dose rates were then summed over the four energy groups to give the total absorbed dose rate ($\dot{D}$). Equations 2 and 3 show the mathematical formalism for the calculation of $\dot{D}$. A Monte Carlo calculation was performed to calculate the same quantities in the same two-dimensional geometry for comparison purposes.

$$\dot{D}_g = k_g^n \phi_g + k_g^{B-10} \phi_g \tag{2}$$

$$\dot{D} = \sum_{g=1}^{4} \dot{D}_g \tag{3}$$

where,

$\dot{D}_g$ = The absorbed dose rate in group g (cGy sec^{-1}).
ϕ_g = The neutron flux in group g (cm^{-2} sec^{-1}).
k_g^n = The group averaged neutron kerma factor (cGy cm^2).
k_g^{B-10} = The group averaged boron—10 kerma factor (cGy cm^2).
$\dot{D}$ = The tot.al absorbed dose rate summed over all groups (cGy).

3. RESULTS

The axial and transverse total absorbed dose rate profiles are shown in Figures 2 and 3, respectively, as calculated with MCNP and as calculated using removal-diffusion theory. The transverse profile is for a depth of 2.25 cm. Both the axial and transverse profiles assume a boron-10 concentration of 10 ppm. On these figures, the MCNP calculations include bars to indicate +/– 5% from the mean of the MCNP data. These bars are not calculated error values. Rather these bars are intended as a guide to the reader in determining if the DIF3D calculations are sufficiently accurate. The total absorbed dose rate at each detector location in the phantom, as calculated using MCNP, had a relative error less than 3.5%. This relative error required approximately 6,300 minutes of computer time on an IBM RS/6000 workstation, which resulted in 11 million neutron histories. The DIF3D calculation performed on the same computer required only 5 minutes.

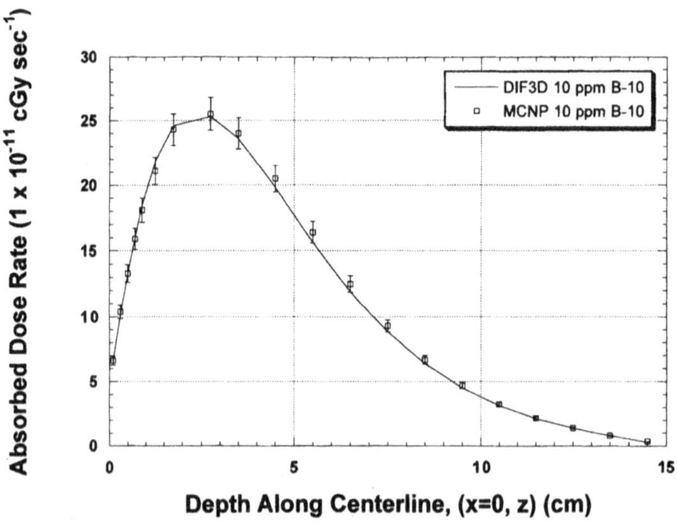

Figure 2. Axial total absorbed dose rate profile in a two-dimensional water phantom along the phantom centerline. Plot shows comparison between absorbed dose rates calculated using MCNP and DIF3D.

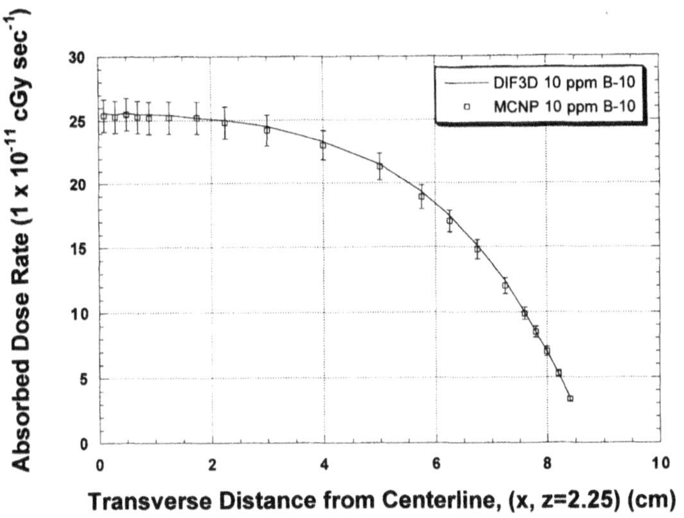

Figure 3. Transverse total absorbed dose rate profile in a two-dimensional water phantom at a depth in phantom of 2.25 cm. Plot shows comparison between absorbed dose rates calculated using MCNP and DIF3D.

4. CONCLUSIONS

Removal-diffusion theory as implemented using DIF3D may provide the speed in computation necessary for treatment planning in BNCT, while performing calculations on a very fine spatial scale. As shown in the data presented, the removal-diffusion theory calculations agree to within +/– 5% of Monte Carlo calculations for the phantom and locations in the phantom that were investigated. Therefore, one may conclude that removal-diffusion theory may be sufficiently accurate for BNCT treatment planning, given the proper choice of removal and diffusion parameters.

REFERENCES

1. F.J. Wheeler, M.L. Griebenow, D.E. Wessol, and D.W. Nigg, "Analytical Modeling for Neutron Capture Therapy," Strahlentherapie und Oncolgie, **165**, 186 (1989).
2. R.G. Zamenhoff, S. Clement, K. Lin, C. Lui, D. Ziegelmiller, and O.K. Harling, "Monte Carlo Treatment Planning and High-Resolution Alpha Track Autoradiography for Neutron Capture Therapy, "Strahlentherapie und Oncolgie, **165**, 188 (1989).
3. D.W. Nigg, P.D. Randolph, and F.J. Wheeler, "Demonstration of Three-Dimensional Deterministic Radiation Transport Theory Dose Distribution Analysis for Boron Neutron Capture Therapy," Medical Physics, **18(1)**, 43 (1991).
4. J. Niemkiewicz, and T.E. Blue, "The Use of Removal-Diffusion Theory to Calculate Neutron Flux Distributions for Dose Determination in Boron Neutron Capture Therapy," Proc. 7th International Symposium on Neutron Capture Therapy for Cancer, B. Larson, J. Crawford, and R. Weinreich, Eds., Zurich, Switzerland, Sept. 4–7 1996.
5. J. Briesmeister, Ed., "MCNP—A General Monte Carlo N-Particle Transport Code, Version 4B," LA-12625-M (March 1997).
6. K.L. Derstine, "DIF3D: A Code to Solve One-, Two-, and Three-Dimensional Finite Difference Diffusion Theory Problems," Argonne National Laboratory, ANL-82-64, Argonne (1984).

THE DEVELOPMENT OF A COMPUTATIONAL DOSIMETRY SYSTEM FOR BNCT AT JAERI

H. Kumada[1], Y. Torii[1], K. Saito[1], Y. Yamaguchi[1], A. Matsumura[2], Y. Nakagawa[3], and F. Sakurai[1]

[1]Japan Atomic Energy Research Institute
Ibaraki, Japan
[2]Institute of Clinical Medicine
University of Tsukuba
Ibaraki, Japan
[3]National Kagawa Children's Hospital
Kagawa 765-8501, Japan

INTRODUCTION

A new medical irradiation facility has been installed in JRR-4. It was designed to permit selection of neutron energies from thermal neutron to epithermal range by changing the thickness of the heavy water layers. This facility then allows us to continue the thermal neutron irradiation protocol now performed in JRR-2, and to initiate the research and development of an epithermal neutron beam more suitable for deep brain tumor. Dosimetric measurements in the deep brain are not possible; therefore software simulation is being developed to estimate the dosimetry in these structures. The current status of the development of this new Computational Dosimetry System (JCDS) at JAERI will be described.

DESCRIPTION

The JCDS can be summarized by 3 steps (Figure 1). Preprocessing creates a numerical model for the Monte Carlo code calculations. This is followed by the process of dose calculation, which calculates neutron and gamma ray dose distributions. Post processing generates the displays of the calculation results.

Frontiers in Neutron Capture Therapy, edited by Hawthorne *et al.*
Kluwer Academic / Plenum Publishers, New York, 2001.

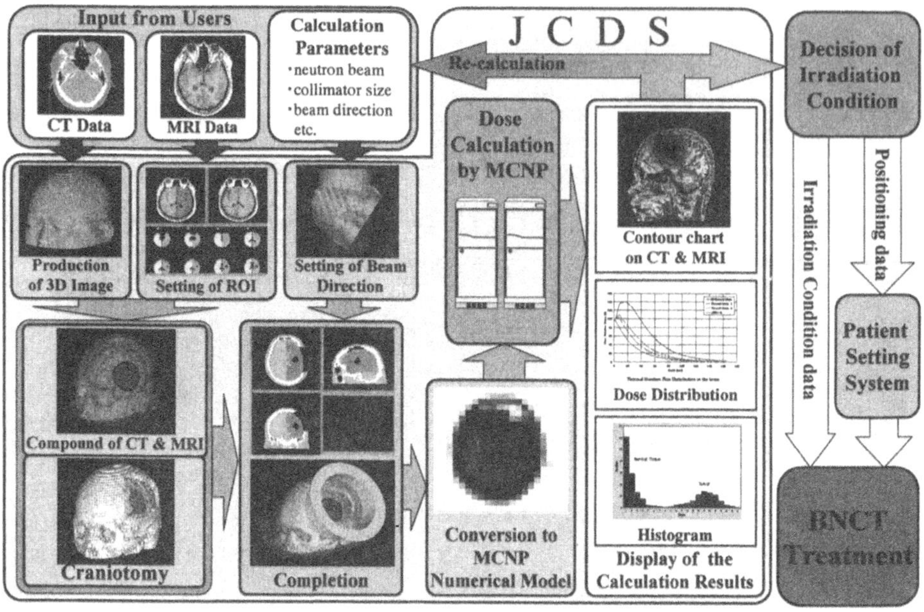

Figure 1. Flow of computational dosimetry system for BNCT at JRR-4.

Preprocessing

Preprocessing creates detailed numerical models for Monte Carlo code calculation. The numerical models are produced from calculation condition and a 3-D head image of the patient. First, users choose the calculation parameters necessary for dose calculation, e.g., neutron beam mode, collimator size, beam direction, boron concentration, etc.

The 3-D head image transferred from CT and MRI images by using the DICOM protocol, which is the standard format for medical image data transport.

The tissue components such as soft tissue, bone and air can be automatically selected from the CT data set. On the other hand, MRI allows differentiating among normal tissue and for example tumor (Figure 2). It is therefore used to specify the tumor volume and the region of interest (ROI). A margin of 2 cm is added automatically to define the target volume. This 3-D model is editable to simulate the surgical processes, such as skin flap opening and bone removal.

The Process of Dose Calculation

Utilizing the numerical models from preprocessing, the neutron and gamma ray distributions are calculated by MCNP Monte Carlo code.

Post Processing

The dose distributions so obtained will be displayed on the CT and MRI image as a contour (Figure 3). Total dose to normal tissue, tumor and target volume and any ROI are displayed.

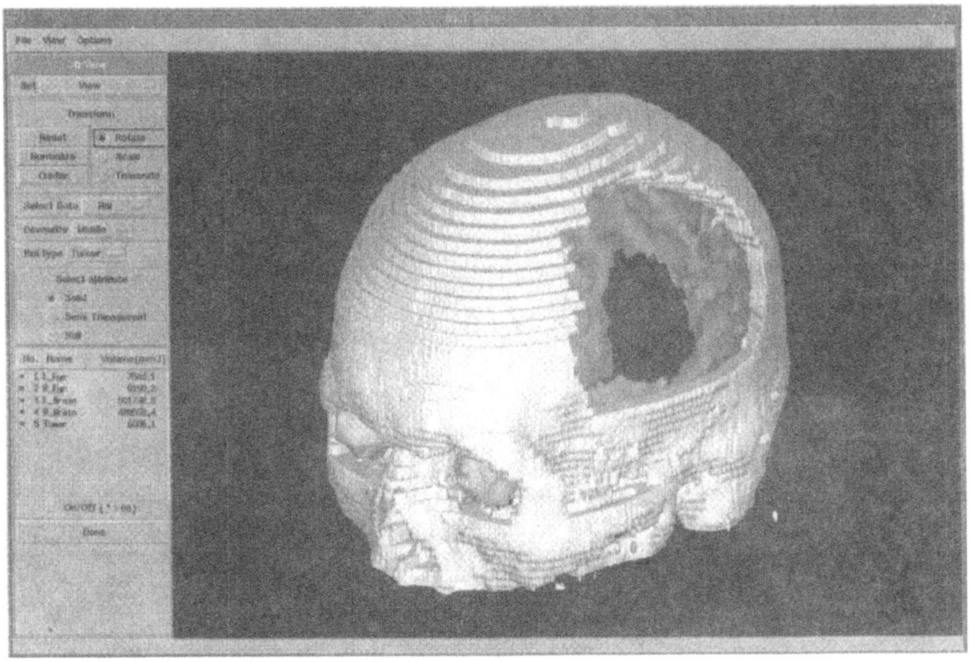

Figure 2. Processing of 3 Dimension Model Conduction.

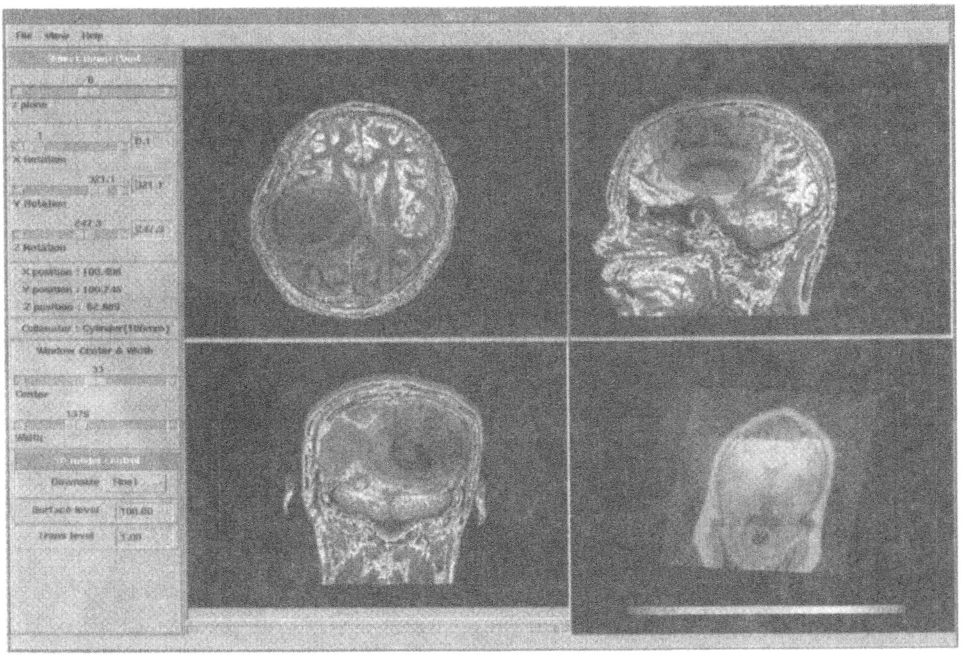

Figure 3. Display of the Dose Calculation Results.

SYSTEM CHARACTERISTICS

The advantage of JCDS is as follows. This system will be designed to work with the UNIX and Windows operating System. For the purpose of efficient utilization, the system will operate on a network. Virtual CT or MRI images are available simultaneously. The 3-D head image can be manipulated to simulate surgical procedures such as the opening of a flap or the removal of the bone. Therefore it can be used for BNCT with craniotomy. The irradiation position data of a patient obtained can be transferred to a Patient Setting System, which permits the accurate positioning and immobilization of the patient in the irradiation room.

CONCLUSION

A Computational Dosimetry System (JCDS) has been developed at JAERI. After completion of a beta-version of this system in this fiscal year, it will be verified by measurements data in phantom and by direct measurements in those patients treated by a thermal neutron beam with craniotomy.

88

APPLICATION OF SILICON DIODE ARRAYS FOR MICRODOSIMETRY IN BNCT AND FNT

Peter D. Bradley[1]*, Anatoly B. Rosenfeld[1], Barry J. Allen[2], Tooru Kobayashi[3], Yoshinori Sakurai[3], Chandrasekhar Kota[4], Richard L. Maughan[4], Mark Yudelev[4], Jeffrey Coderre[5], and Jacek Capala[5]

[1]Radiation Physics Group
University of Wollongong
Northfields Ave., Wollongong, NSW, 2522, Australia
[2]St. George Hospital, Cancer Care Centre
Gray St, Kogarah, NSW, 2217, Australia
[3]Reactor Research Institute
Kyoto University
Kumatori-cho, Sennan-gun, Osaka, 590-04, Japan
[4]Gershenson Radiation Oncology Center
Harper Hospital
Wayne State University, Detroit, Michigan
[5]Medical Department
Brookhaven National Laboratory
Upton, New York, 11973

1. INTRODUCTION

Dosimetry in boron neutron capture therapy (BNCT) and fast neutron therapy (FNT) is complicated by the number of varying radiation components incident at a cellular scale. The separation of varying LET components in radiation fields, using microdosimetric methods, is important due to the varying biological effectiveness of these components. The traditional tool of experimental microdosimetry for medical applications is the proportional gas counter.[1] Wuu *et al.*[2] first applied a Rossi type proportional counter to BNCT dosimetry. This has been followed by other studies by Kota[3] and Maughan *et al.*[4] Although generally considered the best currently available detector, the

* Email correspondence: peter_bradley@uow.edu.au.

Frontiers in Neutron Capture Therapy, edited by Hawthorne *et al.*
Kluwer Academic / Plenum Publishers, New York, 2001.

proportional gas counter has several shortcomings. These include wall effects, a relatively large physical size, phase effects due to measurement in a gaseous phase, and an inability to simulate an array of cells.[1,5]

This paper demonstrates the first application of silicon p-n junction SOI (Silicon-On-Insulator) arrays for microdosimetry in radiation oncology. The detector removes all of the previously mentioned problems associated with proportional gas counters. Furthermore, silicon diode array microdosimetry exploits the small size, rapid development, and current manufacturing capabilities of integrated circuit technology. The device is capable of simultaneously measuring thermal neutron flux and microdosimetric spectra. We provide a brief description of the device and its characterization followed by examples of applications in FNT and BNCT.

2. MATERIALS AND METHODS

2.1. Device Description

The device consists of a matrix of 4,800 (40 × 120) diodes with each diode having junction dimensions of 10 × 10 μm (comparable to biological cell dimensions). The spacing between diode junctions is 10 μm and all diodes are connected in parallel to improve collection statistics. The test structures were fabricated on bonded SOI wafers with thickness 2 μm. The N+ and P+ silicon layers with depths 0.2 and 0.5 μm were constructed by arsenic and boron implantation's at 30 keV and 5×10^{15} ions/cm^2. The impurity concentration of the P silicon was 1.5×10^{15} cm^{-3}.

The detector is connected to a spectroscopic system in which charge collection induced by radiation interactions is recorded by a multi-channel analyzer.

2.2. Device Characterization and Performance

To date three main problems have impeded the development of silicon based microdosimetry. Specifically the sensitive volume definition is obscured by charge diffusion, a method of converting microdosimetric spectra in silicon to an equivalent tissue measurement is required and the device has poor low energy sensitivity. We have characterized the performance of our prototype device in relation to these issues.

2.2.1. Sensitive Volume. A crucial problem in silicon based microdosimetry is the construction of a well defined sensitive volume. The sensitive volume has been characterized using alpha broadbeam spectroscopy, ion microbeam imaging and 3D semiconductor device simulation software. A brief summary of results is presented here with details provided in an earlier paper.[6]

The SOI technology provides a well-defined sensitive depth of 2 μm since the charge collection is limited by the insulating layer. However, analysis of the microbeam measurements indicates that lateral diffusion significantly affects the charge collection behavior of the device as shown in Figure 1. This has two main effects on the dimensions of the sensitive volume:

1. The sensitive volume width and length are increased beyond the junction dimensions

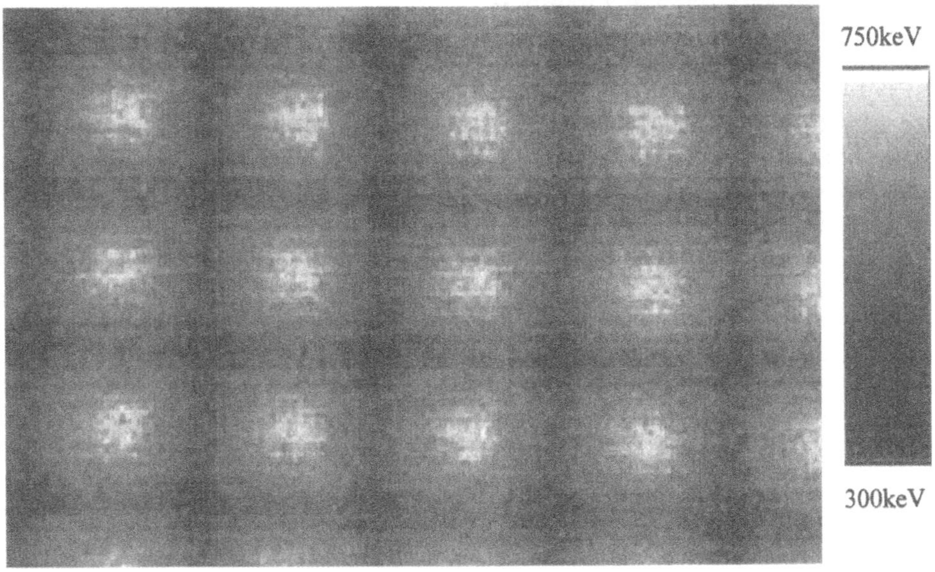

Figure 1. Microbeam image (2 MeV alpha) of 10 × 10 μm 2 μm SOI array. Black corresponds to lowest charge collection amplitude and white is highest amplitude.

2. Outside of the depletion area, charge collection occurs via diffusion and is therefore less efficient then within the junction. Therefore, the charge collection depth, on average, is less then the SOI thickness.

Sensitive volume dimensions are required in order to calculate the mean chord length needed for microdosimetric spectra. In our case, the diode sensitive volume may be approximated by a rectangular parallelepipeds of dimensions 1.7 × 30 × 30 μm. A future device should use lateral isolation barriers or deep etching to isolate individual cells and eliminate lateral diffusion.

2.2.2. Tissue Equivalence. The tissue equivalence of silicon for microdosimetric measurements was analysed by considering the ranges of particles relevant to boron neutron capture therapy. It was shown in an earlier paper[7] that the microdosimetric spectra obtained in silicon for a given volume is equivalent to that obtained in a tissue volume with all linear dimensions scaled by 1/0.63. This relationship was verified using Monte Carlo techniques.

2.2.3. Noise Performance. Experiments performed at the Kyoto University Reactor indicate that the device currently has a low energy noise threshold of around 25 keV/μm. The device was shielded using aluminium foil and connected by a 1 m coaxial cable to a Canberra 2003T charge sensitive preamplifier. It is expected that a noise level as low as 1–5 keV/μm could be achieved by integrating the preamplifier on the same chip as the detector. Such a device would ensure sensitivity to protons with LET in the range 20–100 keV/μm. The design of an integrated pre-amp and detector is currently underway.

2.3. Irradiation Setup and Phantoms

Microdosimetric spectra have been obtained at three neutron therapy facilities: the fast neutron therapy facility at the superconducting cyclotron, Gershenson Radiation Oncology Center (GROC), Detroit, the epithermal neutron irradiation facility at the 3 MW Brookhaven Medical Research Reactor (BNL)[8] and the boron neutron capture facility at Kyoto University Reactor (KUR).

A lucite phantom was used at BNL and GROC whilst a water phantom was used at KUR. The Lucite cubic phantom consisted of blocks 2.5 cm thick arranged to give a cube of side 15 cm. One block was designed with a recess for insertion of the microdosimeter probe. For the head water phantom used at KUR, a 3D remote controlled actuator system was used to accurately position the microdosimeter (+/–2 mm) and minimize personnel radiation exposure during measurements. Measurements shown in this paper were performed along the central axis of the beam. A 1-meter BNC compatible coaxial cable runs through the probe to connect the microdosimeter to a preamplifier.

3. CLINICAL RESULTS

3.1. BNCT Microdosimetry with Simultaneous Thermal Neutron Flux Depth Distribution

The device p+ region contains boron introduced via ion-implantation. The p+ region defines the perimeter of the diode cell and surrounds the central n+ diode junction as shown in Figure 2. By analogy with a biological cell the p+ region is representative of the cytoplasm surrounding the n+ junction or nucleus. Energy deposited closer to the nucleus will have a greater cell killing effectiveness and similarly, in the detector, energy deposited in the junction has a higher charge collection efficiency. Note that the device is still a crude approximation to a cell since cell morphology is irregular and our diode cell size is somewhat larger than a typical biological cell.

The amount of boron is sufficiently high to sensitize the detector to thermal neutrons. The spectra shown in Figure 3 compares the microdosimetric spectra of the KUR and BNL beams at similar depths. The BNL reactor power was 300 kW whilst the KUR

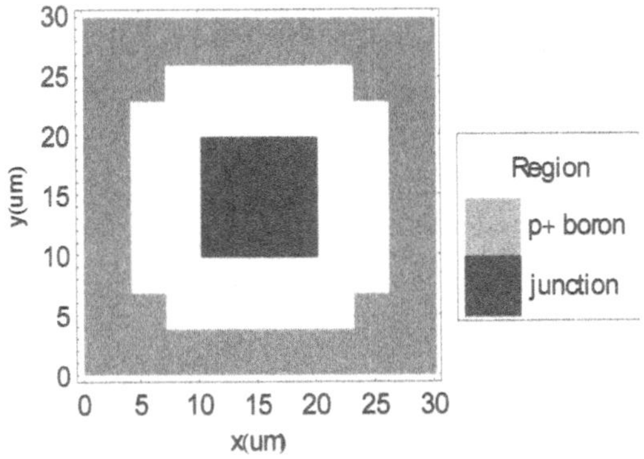

Figure 2. Plan view of a single diode cell showing the p+ boron ion-implantation surrounding the central diode junction.

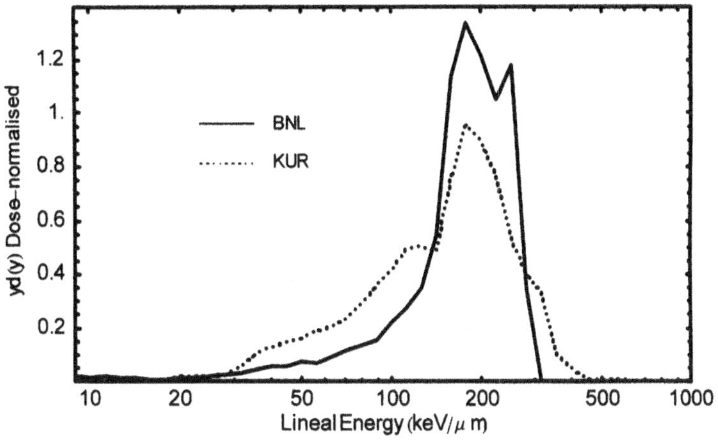

Figure 3. Comparison of BNCT microdosimetric spectra from KUR (1 cm, water phantom, thermal mode beam, reactor power = 6 MW) and BNL (1.3 cm depth in Lucite phantom, reactor power = 300 kw). Equal areas under different regions of the function *yd(y)* correspond to equal doses.

measurement was performed at a full reactor power of 6 MW. The normalized microdosimetric spectra clearly indicate a peak at 200 keV/μm corresponding to the alpha and lithium boron fission products. The KUR measurement indicates a higher contribution due to recoil protons (<100 keV/μm) from fast neutron reactions. This measurement demonstrates the devices utility for comparing beam characteristics and identifying microdosimetric components.

Since we know the ion-implantation and mask details, we can calculate the number of boron-10 atoms within the device (Na = 2.22×10^{13} atoms). Given the boron cross-section to thermal neutrons (σ = 3,837 barns) and measuring the boron-10 count rate (B in cnts/sec) we can then calculate the thermal neutron flux (ϕ in n/cm^2/s) using the following equation.

$$\phi = B/(Na\ \sigma) = 1.173 \times 10^7\ B \tag{1}$$

The calculated neutron depth distribution at KUR, compared to theoretical calculations, is shown in Figure 4. The excellent comparison indicates that the device is capable of simultaneous microdosimetric and thermal neutron depth distribution measurements at full reactor power. Such a measurement is not possible with a proportional gas counter or any other device.

3.2. FNT Microdosimetry with Measurement of Thermal Neutron Component

In thermal neutron enhanced fast neutron therapy it is necessary to measure the thermal neutron component of the beam. In this case we cover the device with a 1% B-10 Lucite coating to increase the devices sensitivity to thermal neutrons. Figure 5 shows the microdosimetric spectra with and without the coating along with the difference. The difference spectra may be used to estimate thermal neutron fluxes. The "no coat" spectra shows significant high LET components. These arise from inelastic and elastic nuclear reactions within the silicon that produce heavy ion recoil products such as Si, Mg and

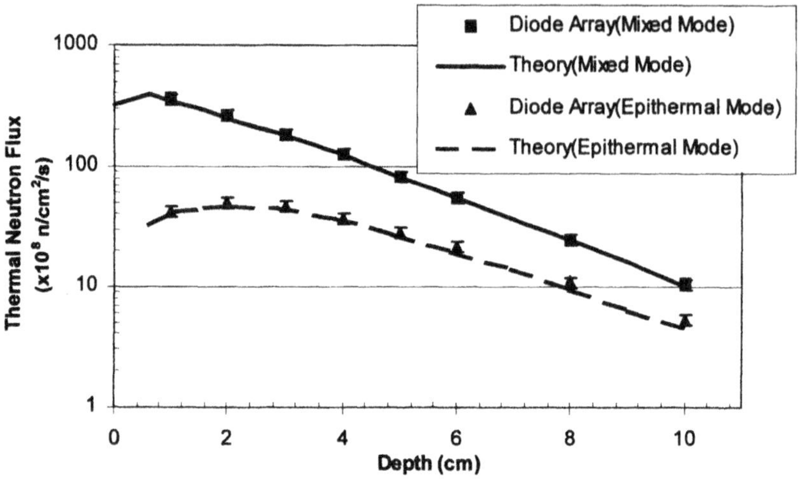

Figure 4. Thermal neutron flux depth distribution at KUR measured using the microdosimeter and compared to Monte-Carlo theoretical calculations. Mixed mode and thermal mode refer to differing beam filtering configurations.

Al ions. The sensitivity of the device used in this experiment was too high to measure the proton peak at around 20 kev/µm. This problem may be solved by noise reduction techniques such as locating a low noise preamplifier immediately adjacent to the detector. Work is currently underway to analyze the microdosimetric spectra obtained by the silicon microdosimeter in a fast neutron beam.

4. CONCLUSION

The silicon diode array is intended as an alternative to the proportional gas counter for experimental microdosimetry. The diode array is a low cost system without the need

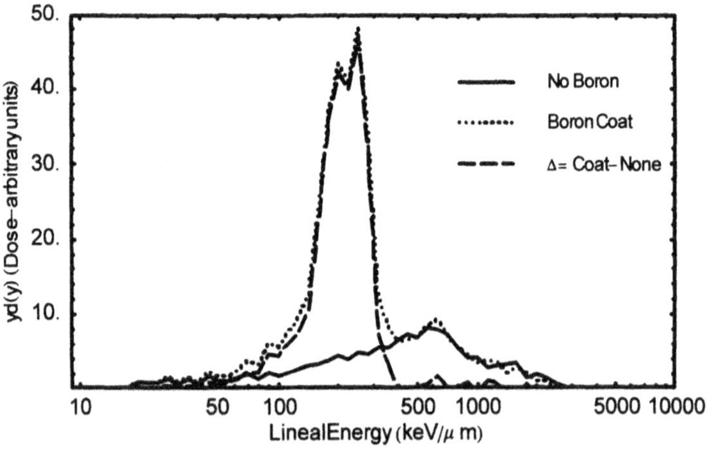

Figure 5. Microdosimetric spectra obtained at a depth of 6.4 cm in the Lucite phantom exposed to a 20 MeV fast neutron beam, Gershenson Radiation Oncology Center, Detroit.

for gas flow technology and high voltage supplies. The diode array offers such advantages as a measurement in the solid phase and thus removal of wall effects, simulation of a two dimensional array of cells for improvement in collection statistics, small physical size and operation at clinical reactor power settings. Furthermore, the possibility exists that future devices may be constructed with sizes and morphology more nearly representative of biological cells, specific boron implantation in various cell/diode regions and simultaneous macrodosimetry and thermal neutron flux measurements on the same device. The construction of such devices may exploit the rapid development and current manufacturing capabilities of silicon based integrated circuit technology.

In response to the failure of traditional microdosimetry to adequately represent realistic tissue architecture and boron distribution, Solares and Zamenhof[9] have developed high-resolution quantitative autoradiography. This technique is difficult and time consuming, although it is currently the most realistic microdosimetric technique available. However, the improvements in traditional microdosimetry possible using silicon diode arrays may help to bridge the gap between proportional counter microdosimetry and the more realistic autoradiography method.

ACKNOWLEDGMENTS

The authors would like to thank the Australian Nuclear Science and Technology Organisation, Gernot Heiser at University of New South Wales for assistance with semiconductor modeling, the University of Melbourne for microbeam measurements and Fujitsu Research Laboratories for providing the diode array samples. We are greatly indebted to the staff at KUR for the BNCT measurements, the staff of BNL medical research reactor for their support and the staff at Gershenson Radiation Oncology Center, Harper Hospital for their generous time and effort. This work was funded by Australian National Health and Medical Research Council Grant No 960804 and an Australian Department of Industry, Science and Technology grant.

REFERENCES

1. International Commission on Radiation Units and Measurements, "Microdosimetry (ICRU report; 36)," ICRU, Maryland, 1993.
2. C.S. Wuu, H.I. Amols, P. Kliauga, L.E. Reinstein, and S. Saraf, Microdosimetry for Boron Neutron Capture Therapy, *Radiat. Res.*, 130:355–359, 1992.
3. C. Kota, "Microdosimetric considerations in the use of the boron neutron capture reaction in radiation therapy", PhD, Wayne State University, 1996.
4. R.L. Maughan, M. Yudelev, and C. Kota, A microdosimetric study of dose enhancement in a fast neutron beam due to boron neutron capture therapy, *Phys. Med. Biol.*, 38:1957–1961, 1993.
5. H.H. Rossi, and M. Zaider, "Microdosimetry and Its Applications," Springer, London, 1996.
6. P.D. Bradley, A.B. Rosenfeld, K.K. Lee, D. Jamieson, and S. Satoh, Charge collection and radiation hardness of a SOI microdosimeter for space and medical applications, *IEEE Trans. Nucl. Sci.*, 45: In press, 1998.
7. P.D. Bradley, and A.B. Rosenfeld, Tissue Equivalence Correction for Silicon Microdosimetry Detectors in Boron Neutron Capture Therapy, *Med. Phys.*, In press-1998.
8. P.D. Bradley, A.B. Rosenfeld, B.J. Allen, J. Corderre, and J. Capela, Performance of silicon microdosimetry detectors in boron neutron capture therapy, *Radiat. Res.*, In press 1998, submitted Jan 1998.
9. G.R. Solares, and R.G. Zamenhof, A novel approach to the microdosimetry of neutron capture therapy. Part I. High-resolution quantitative autoradiography applied to microdosimetry in neutron capture therapy., *Radiat. Res.*, 144:50–58, 1995.

SHIELDING DESIGN AND CALCULATIONS FOR THE FINNISH BNCT FACILITY

Petri Kotiluoto[1], Pekka Hiismäki[1], Iiro Auterinen[1], Tiina Seppälä[1,2,4], and Carita Aschan[2,3]

[1]VTT Chemical Technology
[2]Department of Physics
University of Helsinki
[3]Department of Radiology
[4]Clinical Research Institute
Helsinki University Central Hospital

1. INTRODUCTION

In the radiation shielding design of the Finnish Boron Neutron Capture Therapy (BNCT) facility,[1] protecting the facility personnel from radiation exposure and to limit the whole body dose of the patient have been the primary safety concerns. Before the construction work of the irradiation room with heavy concrete walls and roof was started, the technical designs were verified by calculations. A shielding design limit of dose rate under 20μSv/h was set for laboratory area outside the irradiation room. For the shielding design purposes different materials and material thicknesses were tested by modelling of the materials and geometry and calculating the corresponding radiation levels. Shielding design calculations were made by two-dimensional radiation transport code DORT.[2]

In this paper the calculation model and results are presented. After the construction work of the Finnish epithermal BNCT facility, radiation protection measurements have been made to ensure safe radiation levels. These measurements are also reported in this paper.

2. DORT MODEL OF THE FINNISH IRRADIATION ROOM

For the two-dimensional radiation transport code the complicated three-dimensional geometry of the irradiation room had to be simplified to a cylindrical

Frontiers in Neutron Capture Therapy, edited by Hawthorne *et al.*
Kluwer Academic / Plenum Publishers, New York, 2001.

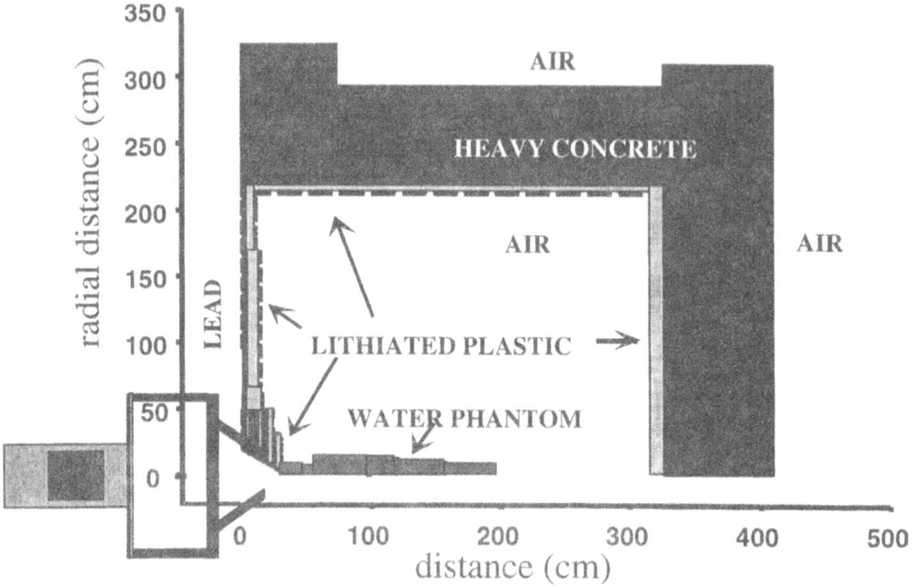

Figure 1. Two-dimensional cylindrical symmetric DORT geometry used in shielding design calculations with different shielding materials: lead, lithiated plastic, bismuth beam collimator and heavy concrete walls. In this model aperture size is 14cm and a water phantom is included. For clarity also reactor core, moderator and bismuth cone are drawn.

symmetric form (Figure 1). The cylindrical DORT beam model was extended to contain shielding around the beam exit and the heavy concrete walls of the irradiation room, with inside lining by neutron absorbing elements. Also a cylindrical symmetric water phantom (modification of the BOMAB[3] phantom) was constructed to present the effect of the patient to the dose rates. This water phantom was based on the volumes of an adult male human (from ICRP23[4] and Nutrition Canada[5]). Construction of the phantom model lead to remarkable simplifications for the real human geometry.

Two different operating situations were modelled: one with the maximum aperture size 200mm with no damping materials in the beam (for instance patient, treatment table or phantom) and the second with aperture size 140mm and with the water phantom representing the patient. BUGLE-80 cross section library was used with 47 neutron energy groups and 20 gamma groups. Direction quadrature set of only 48 angles was utilized in order to save computer time, which resulted in noticeable ray efects, but still produced suitable results for shielding design purposes.

3. NEW SHIELDING MATERIALS

The walls and the roof of the irradiation room were made from square profile steel tube elements filled with heavy concrete (Figure 2). The steel alone is responsible for carrying all the loads, which made it possible to optimize the density of the concrete and allowed omitting its strength. The total average density of 4.35g/cm^3 of the roof and walls was enough to limit the wall thickness to 80cm (Figure 3), implemented from two layers of elements.

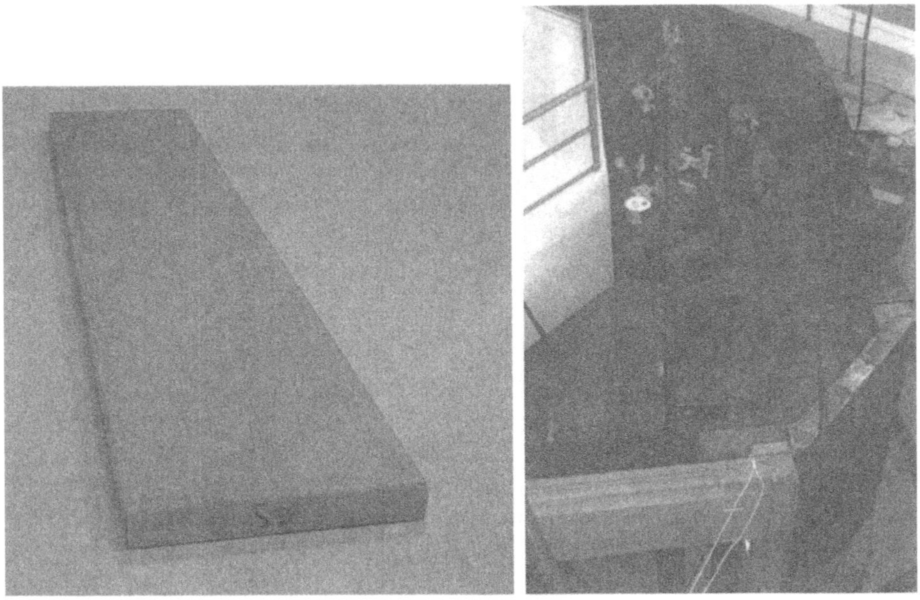

Figure 2. Left: special composite of lithium carbonate, polyethylene powder and paraffin wax encapsulated in 1 mm thick PET shell, with element size of 1,250 × 225 × 32 mm. Right: construction work of the irradiation room with two layers of steel tube elements, filled with density optimized heavy concrete.

For the lining of the walls and the ceiling of the therapy room a new shielding material was developed. Design goals were easy adjustment of the lithium/hydrogen ratio as well as good castability. The goals were met by a three component composite made from lithium carbonate and polyethylene powders and paraffine wax. Above 60 degrees centigrade the mixture formed a plaster, which was cast in one millimeter thick polyester envelopes to make 1,250 × 225 × 30 mm elements (Figure 2). In direct beam hit area 30% lithium carbonate concentration by volume was used in order to both moderate fast

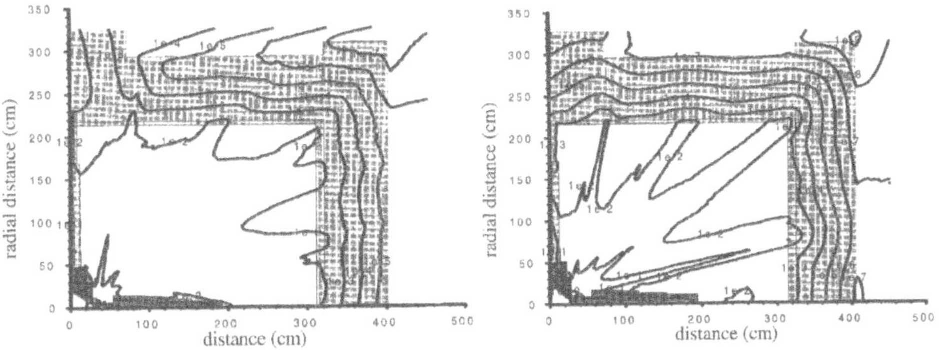

Figure 3. Left: gamma isodoses (Sv/h) with water phantom and aperture size 14 cm. Right: fast neutron isodoses (Sv/h) with water phantom and aperture size 14 cm. Both calculations show 80 cm thickness of the wall sufficient for shielding.

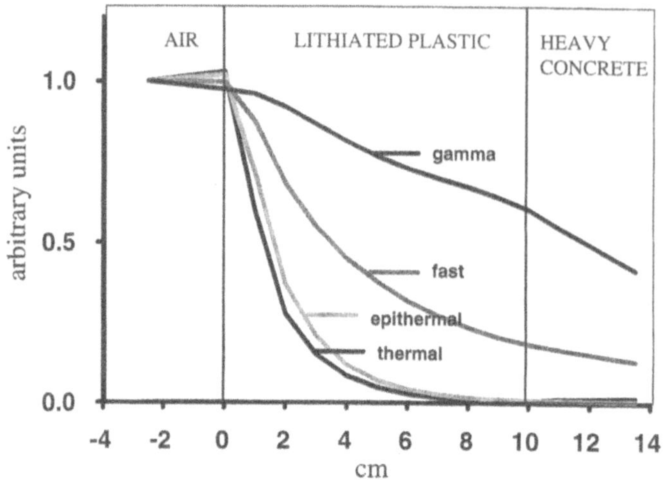

Figure 4. Calculated fluence rates of different radiation components (in arbitrary units) in the lithium carbonate (30%) polymer composite, in the direct beam hit area.

neutrons effectively and still capture the thermal ones (Figure 4), but elsewhere 50–60% lithium carbonate concentration by volume was used for maximum absorption of thermal neutrons.

4. ESTIMATED WHOLE BODY DOSE OF THE PATIENT

A cylindrical symmetric water phantom modification of the BOMAB phantom was constructed, based on the volumes of an adult male human. Both gamma and neutron dose rates were calculated to ICRU46[6] adult male brain tissue. Quality factors 1 and 20 were used for gamma and neutron radiation, respectively. Radiation risk tissue weighting factors were taken from ICRP60.[7] With this extremely simplified human phantom model a crude estimate of the whole body dose of the patient was obtained, being 0.6 Sv for the estimated treatment time of 1 h.

5. MEASUREMENTS FOR RADIATION PROTECTION

Measurements have been made to verify the calculations and the sufficiency of the implemented shielding structures and materials. Inside the irradiation room, TLD measurements have been made for gamma radiation with MCP-7s (^{7}LiF:Mg, Cu, P) detectors.[8] Outside the irradiation room both neutron and gamma dose rates have been measured with standard radiation survey meters. Comparison between the measurement results and the calculations show, that the constructed cylindrical symmetric DORT model has produced valid results both around the beam exit and outside the irradiation room walls (Figure 5). Gamma dose rate outside the heavy concrete walls has been less than 1 μSv/h, the calculated value being 1 μSv/h.

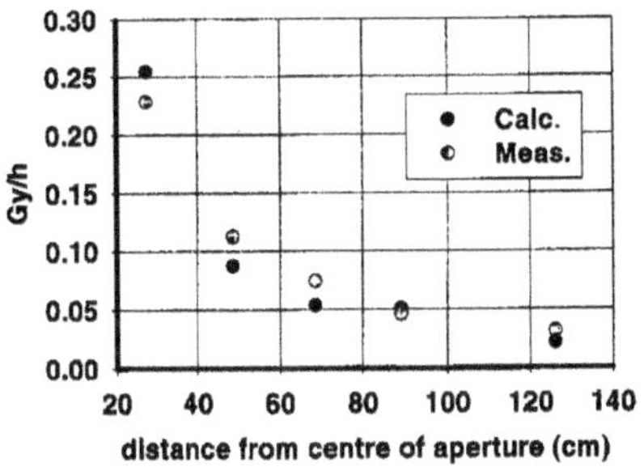

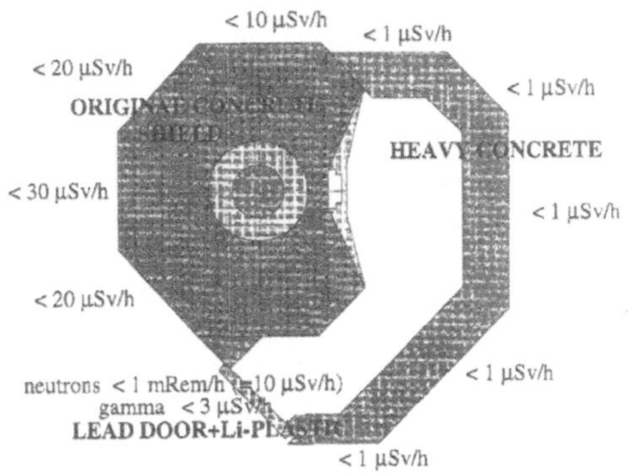

Figure 5. Left: comparison between the TLD gamma measurements and calculations in the beam aperture surroundings with water phantom and aperture size 14 cm. Right: measured dose rates outside the irradiation room for gamma and neutron radiation with no damping materials in the beam (for instance patient, treatment table or phantom).

6. DISCUSSION

DORT modelling has worked quite well for shielding design at the Finnish BNCT facility. However, the simplified DORT model did not allow to evaluate the design for the lead door.

If one desires to calculate the whole body dose of the patient more detailed, MCNP[9] modelling would be needed. DORT modelling does not allow positioning of the patient in different angles because of the cylindrical symmetry and also leads to remarkable simplifications for the human geometry. The whole body dose estimate given in this paper should therefore be considered with caution.

7. CONCLUSIONS

New shielding materials have been developed and succesfully used for the Finnish BNCT facility. An 80 cm thick density optimized heavy concrete wall allowed enough design margin. Safety criterion of dose rate under 20 μSv/h has been achieved outside the irradiation room and verified with standard radiation survey meters.

ACKNOWLEDGMENTS

The financial support from the Technology Development Centre, the Finnish Academy, University of Helsinki, and Helsinki University Central Hospital are acknowledged.

REFERENCES

1. S. Savolainen, I. Auterinen, M. Kallio *et al.*, Advances in Neutron Capture Therapy Vol 1, *Elsevier Science*, p. 342, 1997.

2. W.A. Rhoades, and R.L. Childs, The DORT Two-Dimensional Discrete Ordinates Transport Code, *Nucl. Sci & Engr.* 99:88–89, 1988.
3. G.H. Kramer, L. Burns, and L. Noel, The BRMD BOMAB phantom family, *Health Physics*, Vol 61, 1991.
4. International Commission on Radiological Protection, Report of the Task Group on Reference Man, ICRP Publication No. 23, 1975.
5. Nutrition Canada, Anthropometry report: height, weight and body dimensions, *Health and Welfare Canada*, 1980.
6. International Commission on Radiation Units and Measurements, Photon, Electron, Proton and Neutron Interaction Data for Body Tissues, ICRU Report 46, 1992.
7. International Commission on Radiological Protection, 1990 Recommendations of the International Commission on Radiological Protection, ICRP Publication No. 60, 1991.
8. C. Aschan, M. Toivonen, S. Savolainen, T. Seppälä, and I. Auterinen, Epithermal neutron beam dosimetry with TL dosemeters for boron neutron capture therapy, *Radiat. Prot. Dosim.* (in press).
9. J.F. Briesmeister, ed., MCNP-A General Monte Carlo N-Particle Transport Code, Los Alamos National Laboratory Report, LA-12625-M 1997.

NCT TREATMENT ROOM DOSE DETERMINATIONS USING THE MONTE CARLO ADJOINT SHIELDING (MASH) METHOD

R. A. Rydin, T. R. Hubbard, and R. U. Mulder

Department of Mechanical, Aerospace and Nuclear Engineering
University of Virginia
Charlottesville, VA 22901

1. INTRODUCTION

For the first time, a methodology is available that can accurately calculate dose rates in parts of the patient's body and in the treatment room during NCT treatment of a brain tumor.[1] Existing methods concentrate only on the head. A modified version of the MASH code package,[2] which has been extensively validated against air-over-ground irradiation experiments performed on armored vehicles at Aberdeen Proving Ground, MD,[3] was developed for this application. The MASH methodology for NCT treatment planning and facility design and licensing[4] consists of an RZ geometry forward discrete-ordinates (S_n) flux calculation through a complicated but regular and symmetric beamport filter into a crude room model (DORT), coupled over a suitable surface to a Monte Carlo adjoint calculation (MORSE) of a geometrically complicated human body model seated in a treatment room. By placing the adjoint dose source sequentially in various positions in the body and/or treatment room, accurate dose determinations can be made at virtually any position in the room. Thus, it is possible to evaluate the relative doses to organs other than the tumor for variations in patient orientation and size, local shielding, and room characteristics.

2. VALIDATION

The BMRR epithermal neutron irradiation facility (ENIF) at Brookhaven National Laboratory (BNL) was modeled in RZ geometry (see Figure 1) with DORT based upon drawings and descriptions of the facility.[5] This is a fairly severe test for a discrete coupled

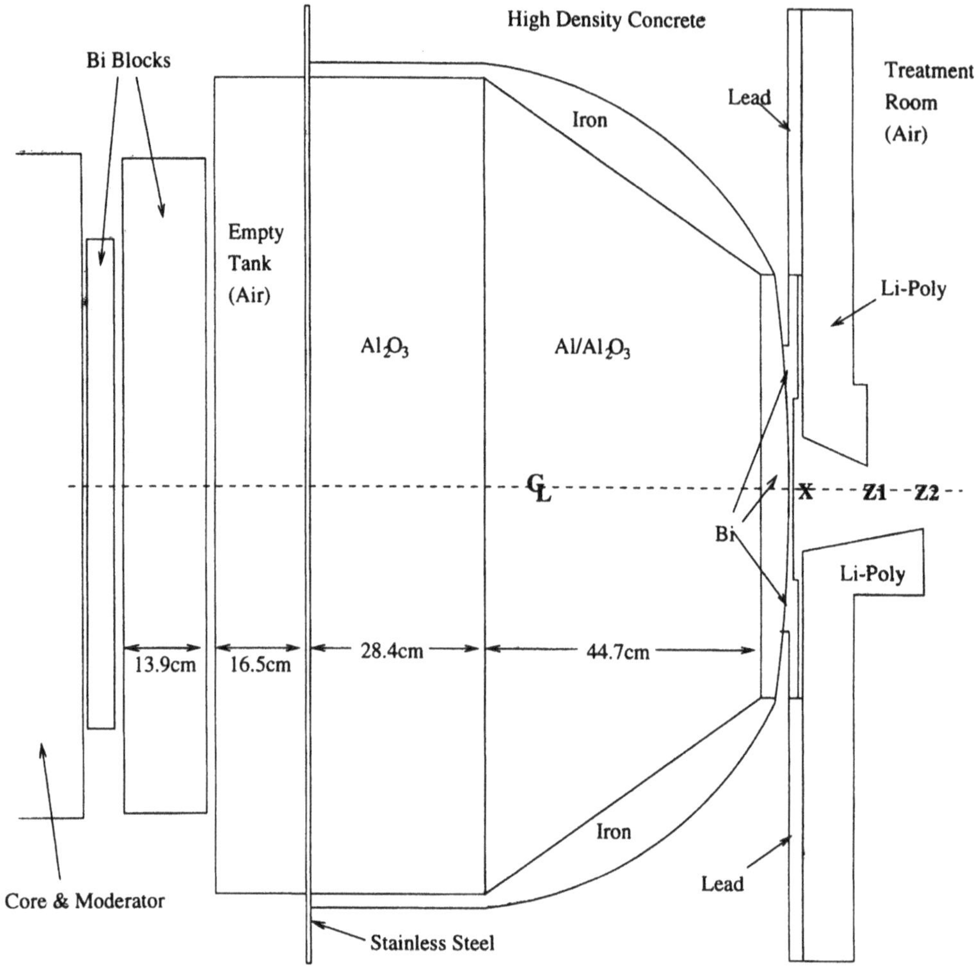

Figure 1. Computational model of the BMRR ENIF with beam shutter open. The upper half depicts the existing collimator (Z1) and the lower half depicts the new collimator (Z2). The in-air measurement point is located at X.

neutron-gamma radiation transport calculation, and was meant primarily to serve as a qualitative validation of the modeling procedure and cross section libraries. A standard reactor fission spectrum and the BUGLE93 library were used. Comparisons were made to BMRR flux and dose measurements[6] at three locations in the beamport collimator, as shown in Table 1, where R_j is a dose ratio figure of merit. When normalized at position X, the results are in fairly good agreement (of the order of 30%) at position Z1 with the short collimator, and at position Z2 with the long collimator.

Subsequent MASH calculations were made of the thermal neutron flux vs depth in a phantom head model placed in the ENIF. A comparison of the MASH results and the BNL results from a standard forward Monte Carlo MCNP model are given in Figure 2. They agree quite well and serve to validate the subsequent MASH calculations. Note that full forward Monte Carlo calculations of the entire patient plus treatment room are difficult to do accurately.

Table 1. Comparison of BMRR measurements (E) with DORT calculations (C) normalized at X ϕ_{epi}

	Position X		Position Z1		Position Z2	
Quantity	E	C	E	C	E	C
ϕ_{epi} (n/cm^{2}s × 10^9)	2.7 ± 0.16	2.7	1.4 ± 0.084	1.1	0.84 ± 0.05	0.72
D_{fastn} (cGy/s)	0.117 ± 0.018	0.118	0.064 ± 0.010	0.047	0.039 ± 0.006	0.031
D_γ (cGy/s)	0.027 ± 0.001	0.027	0.022 ± 0.001	0.016	0.017 ± 0.001	0.013
R_j merit ratio	0.56	0.55	0.72	0.62	0.80	0.63

3. TREATMENT ROOM CALCULATIONS

The DORT beam-port model included a cylindrical head phantom as a beam scatterer and a concrete treatment room wall to provide backscatter. The coupling surface was a large cylindrical box entirely surrounding the patient, but nested inside the walls of the treatment room and cutting off the head phantom. One of the many MASH models of a human patient is shown in Figure 3. Calculations were made for various positions of the body relative to the orientation of the beam. There appears to be little benefit to the trunk and extremities by rotating the body out of the beam line! This was a rather surprising finding, but the head acts like a scattering source, and neutrons and gamma rays simply stream outward to the rest of the body which is fixed in distance from

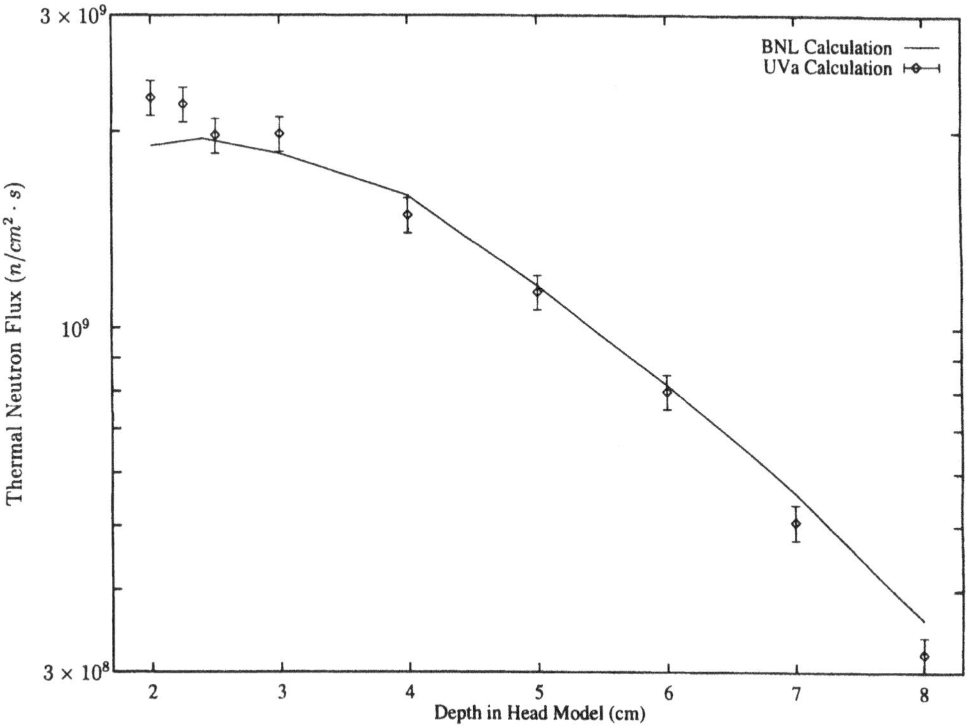

Figure 2. Calculated thermal neutron flux as a function of depth in a head model irradiated by the BMMR beam. The BNL calculations were performed with MCNP and verified experimentally. The University of Virginia (UVa) calculations were performed with MASH, one point at a time.

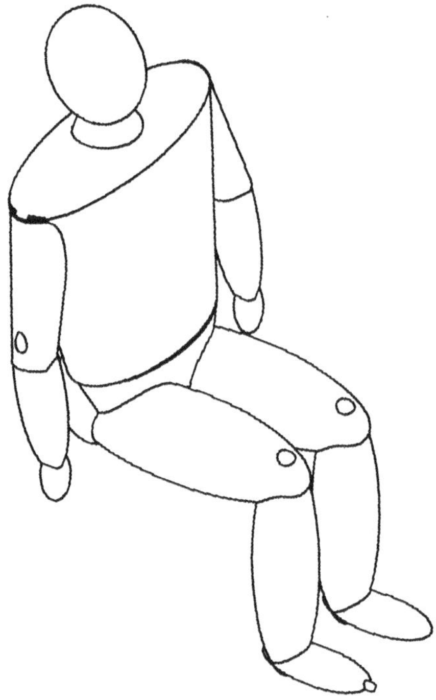

Figure 3. Combinatorial geometry model of a human adult male in the sitting orientation.

the head. At most there is some self-shielding in the body itself. Small patients receive larger relative doses to other organs than large patients, because the distances are smaller.

Of great importance is the judicious application of local shielding in the form of lithiated (0.0086 at/b-cm of Li-6) polyethylene. For these comparison calculations, a crude rectangular box modeled of commercially available (Reactor Experiments, Inc.) lithiated plastic sheets was placed around most of the body, which was in a supine position. As already noted, the primary goal was to shield the rest of the body and the room from radiation leaving the head! As shown in Table 2, a significant reduction of the dose to the heart and gonads can be achieved in this manner. The fractional standard deviations were typically 10%. Only the head and feet were sticking out of the shield box.

It was also found that some care is needed to avoid ray effects in the DORT part of the calculation if the far reaches of the treatment room are of importance. With this proviso, doses anywhere in the room can now be predicted with the patient present.

Table 2. Body position dose rates for various thicknesses of box-like lithiated plastic shields

Location	Dose rate (cGy/s) no shield	Dose rate (cGy/s) 1 cm shield	Dose rate (cGy/s) 2 cm shield
Head axial centerline, 10 cm deep	0.076	0.065	0.071
Head axial centerline, 13 cm deep	0.046	0.054	0.044
Head axial centerline, 16 cm deep	0.030	0.029	0.026
Center of heart	0.018	0.0042	0.0040
Center of gonad	0.024	0.0096	0.0074
Center of hand	0.034	0.023	0.018
Center of foot (outside of shield)	0.016	0.020	0.025

4. REQUIEM FOR THE UVAR

Calculations done at the University of Virginia[7] had shown that the UVAR had the potential to be a world-class NCT treatment facility. Unfortunately, the UVAR ceased operation in July 1998, and is now being decommissioned.

REFERENCES

1. T.R. Hubbard, R.A. Rydin, and T.D. Burns, "A coupled deterministic/stochastic method for dose computations in a filtered BNCT beam". *In Proceedings of the Seventh International Symposium on Neutron Capture Therapy for Cancer*, Zurich, Switzerland, September 1996. International Society for Neutron Capture Therapy.
2. J.O. Johnson, Ed., M.B. Emmett, and W.A. Rhoades *et al.*, "A user's manual for MASH 1.0—a monte carlo adjoint shielding code system,"—ORNL/TM—11778. Technical report, Oak Ridge National Laboratory, March 1992.
3. R.A. Rydin , C.R. Heimbach, and H. Caton, "The Effects of Holes and Gaps on the Shielding Effectiveness of a Lined Steel Box", *In Proceedings of the ANS Topical Meeting on Advances in Nuclear Engineering Computation and Radiation Shielding*, Santa Fe, NM, April 1989.
4. T.R. Hubbard, "A coupled deterministic/stochastic method for computing neutron capture therapy dose rates," PhD Dissertation in Nuclear Engineering, University of Virginia, August 1997.
5. R.G. Fairchild, and J. Kalef-Ezra *et al.*, "Installation and testing of an optimized epithermal neutron beam at the Brookhaven Medical Research Reactor (BMRR)." O.K. Harling, editor, *Neutron Beam Design, Development, and Performance for Neutron Capture Therapy, 1990.*
6. H.B. Liu, D.D. Greenberg, and J. Capala, "An improved neutron collimator for brain tumor irradiations in clinical boron neutron capture therapy." *Medical Physics*, **23**, December (1996).
7. T.D. Burns, T.R. Hubbard, R.A. Rydin, and A.B. Reynolds, "Monte Carlo and deterministic analysis of a University of Virginia BNCT Facility," In *Proceedings of the International Meeting for Neutron Applications*, Crete, 1996.

RADIOPROTECTION FOR PATIENTS AND STAFF DURING BNCT AT THE HFR PETTEN

R. L. Moss[1], F. Stecher-Rasmussen[2], K. Appelman[2], J. Lutgerink[2], A. Keverling-Buisman[2], J. Rassow[3], and W. Sauerwein[3]

[1]HFR Unit, Joint Research Centre
European Commission
Petten, The Netherlands
[2]Netherlands Energy Research Foundation, ECN
Petten, The Netherlands
[3]Universitätsklinikum
Essen, Germany

1. INTRODUCTION

The European clinical trial (EORTC 11961) of BNCT for glioblastoma patients started at Petten in October 1997. The treatment of a patient and the potential exposure of personnel to ionising radiation require by the national Nuclear Energy Law, that the JRC (as licence holder of the HFR) must ensure that radiological protection and monitoring of all personnel, including external staff, is provided and that the correct radioprotection measures are taken and followed.

Due to the structure of the European trial, where the treatment takes place at a facility in the Netherlands under the responsibility of clinicians from Germany, it had to be demonstrated that measures taken satisfy both German and Dutch radioprotection laws. To respect both laws, a BNCT radioprotection committee was formed under the chairmanship of an independent radioprotection expert, with members representing all disciplines in the trial. A contractual agreement had to be signed between the German institute (University of Essen) and JRC Petten to guarantee that procedures to be followed complied with German radioprotection regulations (Strahlenschutzordnung §20).

During treatment, personnel must wear standard personal dosimeters, as per normal for radiation workers. The radiotherapist wear in addition, finger dosimeters and personal dosimeters from their own institute. A special nuance of BNCT is the fact

Frontiers in Neutron Capture Therapy, edited by Hawthorne *et al.*
Kluwer Academic / Plenum Publishers, New York, 2001.

that both the patient and the supporting treatment tools, such as mask and therapy table, become radioactive during treatment. As such a special Standard Operating Procedure,[1] approved by the BNCT Radioprotection Committee, was written whereby measurements of the patient and surrounds at regular intervals after treatment is checked and an appropriate form completed and reported to the BNCT Radioprotection Committee.

2. RADIOPROTECTION COMMITTEE

To conform with the Dutch regulations on radio-protection, a Radio-Protection Committee for BNCT has been formed. The committee has the prime task to review and advise, on a half-yearly basis, the radio-protection methods used for BNCT. If need be, this advice is transmitted to any external authority. The Committee consists of members from each discipline in the BNCT group, and is chaired by an independent expert in radio-protection.

Due to the fact that German staff from Essen University Hospital need to work at Petten, German regulations on radio-protection, especially application of the radio-protection decree: §20 StrSchV (Strahlenschutzverordnung), which regulates the activities of German staff in foreign institutions, had to be contractually agreed. The decree defines regulations on supervision of the staff, personal dosimetry, rules of behaviour, etc.

Radio-protection includes the issuing of personal dosimeters (type: universal dosimeter) to all staff, finger or ring dosimeters to the radiotherapists, and pen dosimeters to participants classified as visitors, eg. nurse(s) and relatives of the patient. Furthermore it is necessary to measure and record all material in and out of the reactor and perform activation measurements on all material used in patient treatment. The patient is an exceptional case, as they are exempted from radioprotection regulations. As such, it is not required that a personal dosimeter is issued to the patient. However, following treatment, the patient is monitored for radioactivity. So far, the reported radiation doses received by the staff are well below the allowable limits.

3. PATIENT ACTIVATION MEASUREMENTS

3.1. Standard Measurements

BNCT of the patient results in neutron activation of a number of naturally occurring elements in the irradiated volume, i.e. the head. As a result, activation measurements of the patient's head are taken on 3 separate occasions using a standard portable dose ratemeter (Type: PDR1). The first measurement is some 1–2 minutes directly following treatment, the second when the patient leaves the reactor building (5–10 minutes post-BNCT) and lastly, just prior to the patient leaving Petten to return to the hospital in Amsterdam (30–45 minutes post-BNCT). Measurements are taken both at contact and at 30 cm distance from the head. The results have been compiled for all 10 patients. As examples, the results of measurements on two patients are given in Table 1. Patient A received one field of radiation and Patient B, two fields. The contact doses, hence maximum levels, are schematically shown for all patients in Figure 1.

Table 1. Example of radiation measurements taken on 2 of the 10 of patients

Patient	Fraction	Date	Start of irradiation	End of irradiation	Activation (μSv/h) [contact (at 30 cm)]					
					After	Time	Wing	Time	Leaving	Time
A	1	13-jan-98	9:45	10:02	45 (20)	10:05	30 (5)	10:10	10 (2)	10:38
A	2	14-jan-98	10:05	10:24	45 (20)	10:27	25 (6)	10:30	9 (<3)	10:55
A	3	15-jan-98	9:50	10:08	50 (30)	10:10	17 (4)	10:15	9 (2)	11:12
A	4	16-jan-98	10:36	10:54	45 (20)	10:55	25 (6)	11:05	9 (<5)	11:55
B	1	23-jun-98	9:40	10:29	40 (20)	10:32	9 (2)	10:40	7 (<2)	10:50
B	2	24-jun-98	9:46	10:32	45 (20)	10:34	9 (2)	10:37	5 (<2)	11:10
B	3	25-jun-98	9:28	10:15	25 (7)	10:17	9 (2)	10:20	4 (<2)	11:20
B	4	26-jun-98	9:25	10:12	40 (8)	10:14	10 (2)	10:16	7 (2)	11:00

Peak levels, i.e., at contact and directly after radiation, are of the order of 40–50 μSv/h, falling to less than 10 μSv/h some 30–50 minutes after treatment. Whether a patient received 2 fields (and hence twice the duration of radiation) makes little difference. The remaining activity is predominately due to ^{24}Na only (half-life = 15 h). Activity due to other elements have much shorter half-lives, hence do not contribute or are not additive to the final levels.

3.2. Activation Measurements using a Portable Gamma Analyser

To ascertain and confirm which elements contributed the most to activation of the patient's head, gamma-ray spectrometry measurements have been performed on 2 of the 10 patients. The equipment or counting chain consists of an EG&G HPGe detector (relative efficiency and FWHM at 1.33 MeV of Co-60: 12.7% and 1.71 keV, type: 26N-1602C), an EG&G 459 high voltage power supply, an EG&G 572 amplifier, an ECN

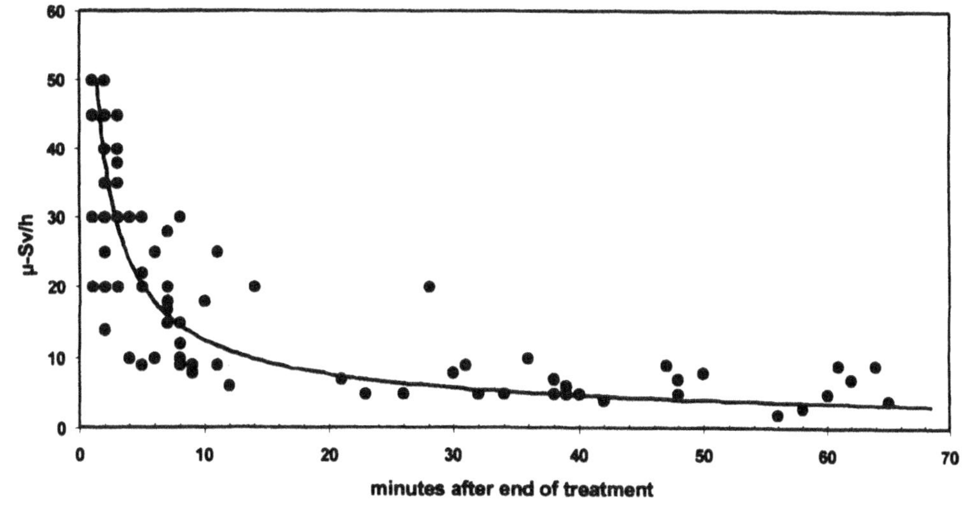

Figure 1. Patient activation measurements for all patients, as a function of time after radiation.

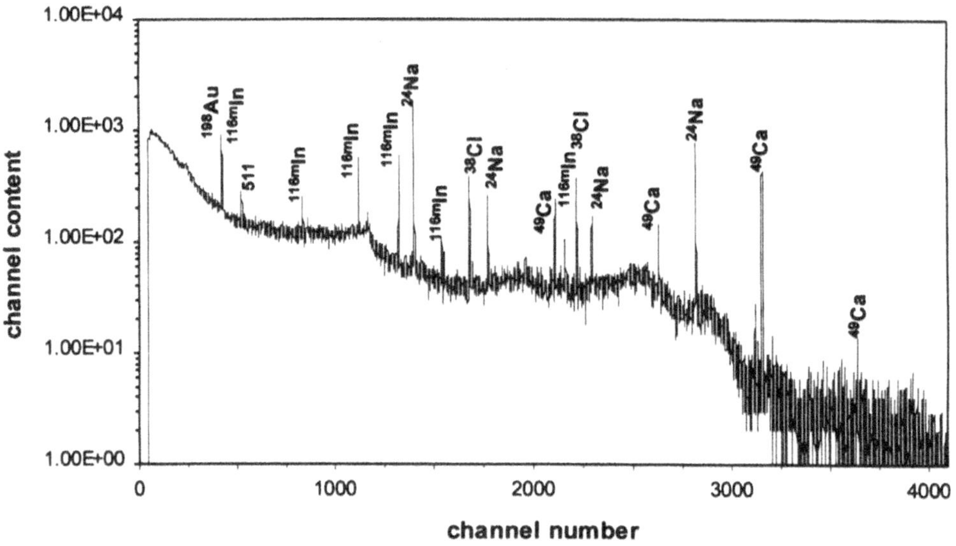

Figure 2. Gamma spectrum taken of one of the 10 patients soon after BNCT.

portable power supply and a Canberra Accuspec interface, mounted in a Toshiba 3200 SX laptop computer.

Shortly after the treatment was finished, the patient was placed on a chair in the BNCT-Wing, where the portable spectrometer had been set-up. Prior to the arrival of the patient, background measurements were taken. The resulting spectra, with and without the patient, were analysed.

The predominate isotopes were, as expected, identified as ^{38}Cl, ^{49}Ca and ^{24}Na. However, in one patient, the isotopes ^{198}Au and ^{116m}In, were also present. The gamma spectrum for this second patient is shown in Figure 2. It was concluded that the former isotope was due to (this) patient's gold filling and the latter isotope, assumably, due to one of the drugs taken by the patient. It should be noted that only those radioisotopes which are gamma emitters have been measured.

4. REPORTING OF MEASUREMENTS

As well as the patient, neutron activation occurs in the supporting treatment tools, such as the patient's positioning mask and the therapy table. To comply with the radio-protection regulations, a special Standard Operating Procedure (SOP) was written in which all measurements could be reported. The SOP, approved by the BNCT Radio-protection Committee, records the measurements on the patient and surroundings at regular intervals after treatment. All measurements are recorded on a form, and forwarded to the Technical Coordinator, and included in the daily Irradiation Summary report (SOP 514).

ACKNOWLEDGMENTS

The clinical trial on BNCT for glioblastoma at the HFR Petten is financially supported by the European Commission's Biomedicine and Health Research Programme (BIOMED II).

REFERENCE

1. R.L. Moss, F. Stecher-Rasmussen, W. Sauerwein, D. Gabel, P. Watkins, and C. Vroegindeweij, "Standard Operating Procedures followed by the Petten BNC-Team during patient treatment", these proceedings.

REAL-TIME ABSORBED DOSE ESTIMATION SYSTEM FOR BNCT BY PG-SPECT

Tooru Kobayashi[1], Yoshinori Sakurai[1], and Masayori Ishikawa[2]

[1]Research Reactor Institute, Kyoto University
Kumatori-cho, Sennan-gun, Osaka 590-0494, Japan
[2]Graduate School of Energy Science, Kyoto University
Sakyo-ku, Kyoto 606-8501, Japan

1. INTRODUCTION

Boron neutron capture therapy (BNCT) has been regularly performed since 1990 at KURRI,[1,2] and 72 patients were treated as of October 1998. The absorbed dose in BNCT at our institute is determined according to the following three steps. (1) Thermal neutron fluxes are monitored at some points in the irradiated part with the activation method using gold wires, during the first 15 minutes of the clinical irradiation. (2) The ^{10}B concentration in the samples of the patient blood and tissue are measured by prompt gamma-ray analysis (PGA),[3] and the concentrations in the tumor part and normal tissue are estimated combining with the data obtained from the other BNCT experiments. (3) the absorbed dose at an interested part is calculated from the measured or estimated thermal neutron flux and ^{10}B concentration. On this dose estimation method, it takes at least about 30 minutes to decide the irradiation time.

In order to decide the irradiation conditions more accurately and more effectively, the direct estimation of the absorbed dose distribution is desirable for BNCT using on-line and 3-D imaging system under a clinical irradiation. Single photon emission computed tomography (SPECT) technique using prompt gamma rays is thought to be an "ideal" dose estimation system to realize a real-time dose-distribution estimation during a BNCT clinical irradiation. It is named PG-SPECT (Prompt Gamma ray—SPECT), which is a combination technique of PGA and SPECT.

2. PG-SPECT SYSTEM FOR BNCT

In an actual BNCT clinical irradiation, the gamma rays from $^{10}B(n, \alpha\ \gamma)^7Li$ and $^1H(n, \gamma)^2D$ reactions, are thought to occupy a main part of the generated gamma rays in

Frontiers in Neutron Capture Therapy, edited by Hawthorne *et al.*
Kluwer Academic / Plenum Publishers, New York, 2001.

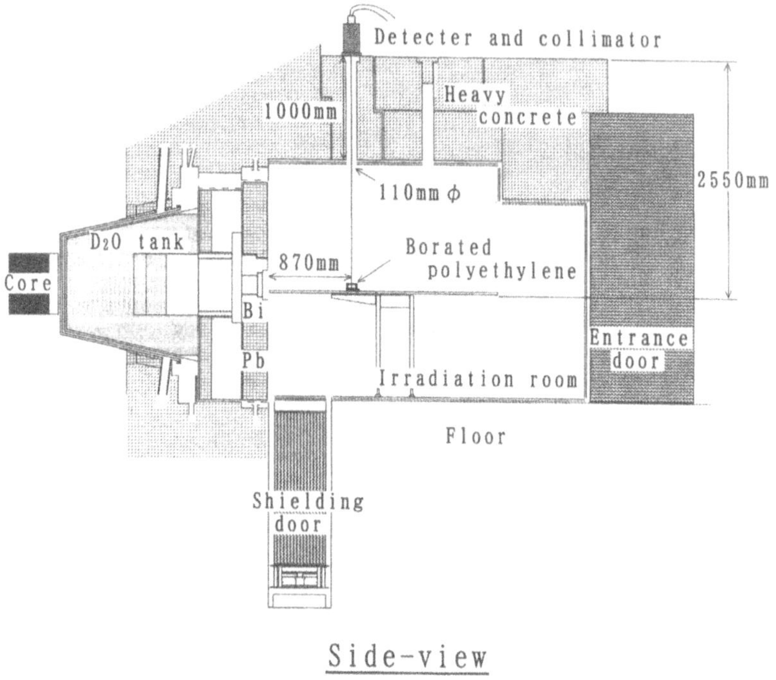

Figure 1. Geometry of the fundamental experiment for the PG-SPECT system at the KUR-HWNIF.

the human body.[3,4] The yield distribution of the $^{10}B(n, \alpha\ \gamma)^{7}Li$ gamma rays corresponds to the ^{10}B enhancement distribution in human body. Both the $^{1}H(n, \gamma)^{2}D$ and $^{14}N(n, p)^{14}C$ cross sections almost accord to 1/v energy characteristics. On the assumption that ^{1}H and ^{14}N atoms homogeneously exist in human body, the yield distribution of the $^{1}H(n, \gamma)^{2}D$ gamma rays corresponds to that of the $^{14}N(n, p)^{14}C$ reactions, and almost thermal neutron flux distribution.

The PG-SPECT system for BNCT needs to measure the gamma rays from human body separately, under high-level background of gamma rays and neutrons during a BNCT clinical irradiation. In this system, many detectors with high energy resolution, such as semiconductor detectors, are needed for 3-D and real-time dose estimation. The smaller view field of collimator is better for high-resolution SPECT image. The higher detection efficiency is better for short-time effective measurement. These conditions are contrary. For the realization of this system, the following subjects are important; (i)

Table 1. The present measurement conditions by the PG-SPECT system for a BNCT clinical irradiation

Sampling time	30 min.
Box-cell size	1 cm × 1 cm × 1 cm
Minimum thermal neutron flux	5×10^{8} n/cm^2/s
Minimum ^{10}B concentration	10 ppm
Minimum gamma-ray yield from $^{10}B(n, \alpha\ \gamma)^{7}Li$	1.1×10^{6} n/s/cm^3

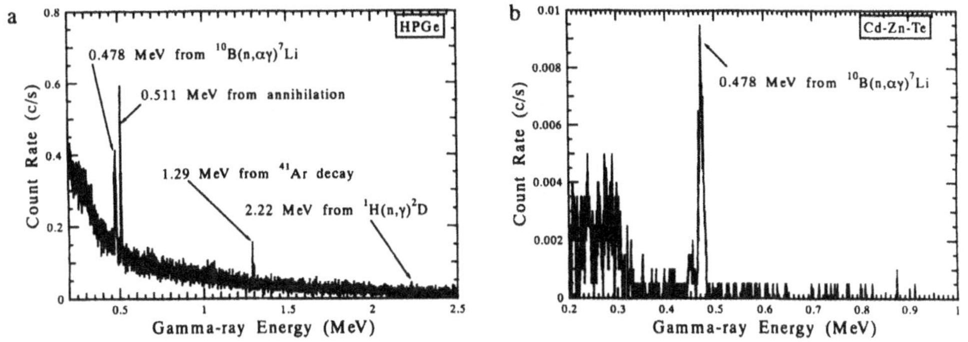

Figure 2. Measured gamma-ray spectra from the B_4C polyethylene plate at the KUR-HWNIF by a) the HPGe and b) CdZnTe semiconductor detectors.

suitable detector type, size and number, (ii) effective collimation, and (iii) effective radiation shield mainly against the background gamma-rays. These subjects are strongly connected each other. For the present, measurement conditions for the PG-SPECT system during a BNCT clinical irradiation are assumed to be as shown in Table 1.

3. DETECTOR FOR PG-SPECT SYSTEM

As many detectors will be handled in some assemblies with collimators and radiation shields, the expected performance of the detector for the PG-SPECT system is as follows: (i) sufficient energy-resolution, (ii) sufficient detection-efficiency for 0.478 keV $^{10}B(n, \alpha\ \gamma)^7Li$ gamma ray and 2.22 MeV $^1H(n, \gamma)^2D$ gamma ray, (iii) small size and (iv) maintenance simplicity. For these requirements, we have an idea of applying CdTe and HPGe semiconductor detectors.

It is easy to get high-efficiency HPGe semiconductor detectors on commercial base. However, cooling with liquid nitrogen *etc.* is indispensable under employment, because the influence of the thermal-induced leakage current is large due to the small energy gap of 0.7 eV.

As the energy gap of CdTe crystal is larger to be 1.47 eV, it can be employed as a semiconductor detector even at room temperature. Moreover, the energy transfer per unit thickness due to photo-electric effect in CdTe is about 5.5 times larger than that in Ge, for the both energies of 0.478 and 2.22 MeV. In the viewpoint of the detector downsizing and the maintenance simplicity, CdTe semiconductor detector is adequate for the PG-SPECT system.

Table 2. Characteristics of the semiconductor detector systems employed in the experiment

Detector	HPGe (GEM-10185-S)	CdZnTe (XR-100T-CZT)
Semiconductor size	41.8 mm diameter × 48.6 mm thickness	3 mm × 3 mm × 2 mm thickness
Temperature	−196 °C	−23 °C
Cooling method	Liquid N_2	Peltier cooler

4. AN EXPERIMENT FOR FUNDAMENTAL STUDY

An experiment was performed at the KUR Heavy Water Neutron Irradiation Facility (HWNIF), for confirming the fundamental characteristics of HPGe and CdTe semiconductor detectors for the PG-SPECT system. Figure 1 shows the experimental geometry. The detector was put outside of the irradiation room, and the prompt gamma rays of $^{10}B(n, \alpha\ \gamma)^7Li$ and $^1H(n, \gamma)^2D$ reactions, which occurred in the irradiation room, were measured. A water phantom and a B_4C polyethylene plate were used as the prompt gamma-ray emitters. The distance between the gamma-ray emitter and the detector was about 2.5m. The prompt gamma rays were collimated with tungsten collimation system whose aperture is 3mm in diameter and 80mm in length. Table 2 shows the characteristics of the detector systems employed in this experiment. The CdZnTe ($Cd_{0.9}Zn_{0.1}Te$) semiconductor is improved in the hole mobility by doping of Zn to CdTe.

Figures 2a and b show the measured gamma-ray energy spectra for the B_4C polyethylene plate by the HPGe and CdZnTe detectors, respectively. As the gamma rays with over 1 MeV did not appear in the CdZnTe detector, the range over 1 MeV is omitted in Figure 2b. For the HPGe detector, two prompt gamma-ray peaks at 0.478 MeV from the $^{10}B(n, \alpha\ \gamma)^7Li$ reactions and 2.22 MeV from the $^1H(n, \gamma)^2D$ reactions, an annihilation gamma-ray peak at 0.511 MeV, and a decay gamma-ray peak from ^{41}Ar decay in air at 1.29 MeV, were measured as shown in Figure 2a. For the CdZnTe detector, only the $^{10}B(n, \alpha\ \gamma)^7Li$ gamma-ray peak was measured as shown in Figure 2b. One of the reasons is that the CdZnTe detector size is too small to measure the higher-energy gamma rays such as the ^{41}Ar decay and the $^1H(n, \gamma)^2D$ prompt gamma rays.

Here, notice that the annihilation gamma rays are not shown in Figure 2b but shown in Figure 2a. The reason is thought to be as follows. The annihilation gamma rays were generated isotropically around the detector, especially in the collimator. So, the ratio of the annihilation gamma rays to the $^{10}B(n, \alpha\ \gamma)^7Li$ gamma rays is thought to be small for the small-size CdZnTe detectors employed in this experiment. From this fact, we can expect to design an exclusive detector for the $^{10}B(n, \alpha\ \gamma)^7Li$ gamma rays with small influence of the unnecessary gamma rays coming from the surroundings.

5. CONCLUSION

The applicability of HPGe and CdTe semiconductor detectors for the PG-SPECT system were evaluated. It was confirmed that the small-size CdZnTe detector with the Peltier cooling is adequate for this system in the viewpoints of downsizing, maintenance simplicity, and low sensitivity for the unnecessary gamma rays. The present research subjects are followings:

(1) exclusive CdZnTe detector for $^{10}B(n, \alpha\ \gamma)^7Li$ prompt gamma rays,
(2) effective collimation and radiation shield system for the exclusive CdZnTe detector, and
(3) effective detector for $^1H(n, \gamma)^2D$ prompt gamma rays.

ACKNOWLEDGMENTS

This research was partially supported by Grant-in-Aid for scientific research (A:07558186) funded from the Ministry of Education, Science, Sports and Culture in Japan.

REFERENCES

1. T. Kobayashi, K. Kanda, Y. Ujeno, and M.R. Ishida, Biomedical Irradiation System for Boron Neutron Capture Therapy at the Kyoto University Reactor, in: "Neutron Beam Design, Development, and Performance for Neutron Capture Therapy", O.K. Harling, *et al.*, ed., Plenum Press, New York, 1990, pp. 321–339.
2. T. Kobayashi, K. Kanda, Y. Fujita, Y. Sakurai, and K. Ono, The Upgraded of the Heavy Water Facility of the Kyoto University Reactor for Neutron Capture Therapy, in: "Advances in Neutron Capture Therapy, Vol. I, Medicine and Physics", B. Larsson, J. Crawford, and R. Weinreich, ed., Elsevier, Amsterdam, 1997, pp. 321–325.
3. T. Kobayashi, and K. Kanda, Microanalysis System of ppm Order ^{10}B Concentrations in Tissue for Neutron Capture Therapy by Prompt Gamma ray Spectrometry, Nucl. Instr. and Meth. 204:525–531, 1983.
4. W.F.A.R. Verbakel, and F.M. Stecher-Rasmussen, A Gamma Ray Telescope for On-Line Measurements of Low Boron Concentrations in a Head Phantom for BNCT, Nucl. Instr. and Meth. A 394:163–172, 1997.

UTILITY OF MEASURED MICRODOSIMETRIC SINGLE EVENT SPECTRA IN INTERPRETING QUALITY OF RADIATION FIELDS USED IN BNCT

Chandrasekhar Kota, Jay Burmeister, and Richard Maughan

Gershenson Radiation Oncology Center
Karmanos Cancer Institute
Harper Hospital and Wayne State University, Detroit, MI, 48201

1. INTRODUCTION

The absorbed dose in BNCT on a cellular level is due to (1) secondary electrons produced predominantly by gamma rays, (2) protons produced by neutrons and the $^{14}N(n_{th},p)^{14}C$ reaction, and (3) alpha particles and Li ions produced by the BNC reaction. The relative contribution of electrons, protons and BNC reaction products changes with position within a beam in a patient/phantom and between different beams, leading to different biological effectiveness for the same absorbed dose. Additionally, (1) the proton energy spectrum is expected to change due to the rapidly degrading neutron energy spectrum with depth and changes in the relative contributions of recoil protons and protons from the $^{14}N(n_{th},p)^{14}C$ reaction, and (2) the BNC reaction products' spectrum is expected to depend on the cellular location of the ^{10}B atoms.

It is non-trivial to measure dose from different components separately, and the methods of such measurements have a bearing on how the dose is interpreted for biological effectiveness. Traditionally, most experimental measurements strive to separately determine the gamma ray and neutron absorbed dose by using paired detectors with different sensitivities to these two radiations. The BNC dose is calculated from the thermal fluence measured using foil activation techniques. The total effective dose is then calculated using the following equation:

$$D_{mix} = D_{gamma} * RBE_{gamma} + D_{neutron} * RBE_{neutron} + D_{bnc} * RBE_{bnc} \quad (1)$$

Since no epithermal beam exists without the concomitant gamma rays, determination of RBE_{gamma} and $RBE_{neutron}$ is only approximate. Furthermore, there is no easy way

Frontiers in Neutron Capture Therapy, edited by Hawthorne *et al.*
Kluwer Academic / Plenum Publishers, New York, 2001.

to determine changes in the RBE of the neutron dose component within and between different BNCT beams. The concept of compound factors has been introduced to account for the effect of the cellular location of the ^{10}B carrying drug on the RBE.

In this paper we present a biophysical approach to measuring the total dose and its effectiveness in BNCT. This approach is based on using the proportional counter microdosimetric technique to measure the secondary charged particle spectrum of the total dose, and to use it with a biophysical or phenomenological model of radiation damage to estimate the biological effectiveness.[1]

2. METHODS AND MATERIALS

Proportional counter microdosimetry[2] with Rossi counters has traditionally been used to estimate the biological effectiveness of fast neutron beams. We have adapted and refined the technique to provide accurate absorbed dose measurements and microdosimetric spectra of the gamma (electron), neutron (proton) and BNC reaction products, in high intensity BNCT beams. Several small volume (0.01 cc) proportional counters with tissue (or brain) equivalent plastic walls (with and without ^{10}B) have been constructed and used to make measurements in BNCT beams. The details of counter construction and measurements are reported in an accompanying paper in these proceedings.[3]

The basic quantity of measured by these counters is the energy lost by individual charged particles (events) in simulated microscopic volumes, typically ranging from 0.25 μm to 4 μm. This energy loss is described in terms of lineal energy "y", defined as the energy deposited in the simulated volume divided by the mean chord length of the volume. Summation of the energy loss by all events gives the absorbed dose. Figure 2 in reference 3 shows the secondary charged particle spectra (single event spectra, or lineal energy spectra) displayed in the yd(y) vs log(y) format which gives the fractional contribution to the dose under different lineal energy intervals. This figure shows the spectra measured by paired (TE/TE + ^{10}B) counters. In the TE counter spectrum, the contribution to the absorbed dose from the secondary electrons, in the lineal energy interval from ~0.1 keV/μm to ~15 keV/μm is seen to be well separated from the contribution from protons in the lineal energy interval from ~15 keV/μm to ~150 keV/μm. In the TE + ^{10}B counter spectrum, the additional contribution from the BNC reaction products in the lineal energy range of ~20 keV/μm to ~500 keV/μm is clearly visible.

The data presented in this figure can be used in several different biophysical/phenomenological models to estimate the biological effectiveness of the absorbed dose in BNCT. Two different approaches are elucidated below: *1). Microscopic mechanistic model of radiation damage*: In this approach, the amount of energy deposited in the sensitive volume of the cell (most likely the nucleus), is calculated from the single event spectra at different dose levels. This quantity, which is the multi-event specfic energy is then convoluted with a microscopic effectiveness function to obtain the radiation damage or biological effectiveness. The microscopic effectiveness function represents the probability of cell inactivation being caused by a given amount of energy distribution in the cell sensitive volume. Differences in measured single event spectra might produce different multi-event specific energy distributions, which in turn would result in variations in the calculated biological effectiveness. *2). Macroscopic phenomenological biological weighting function:* In this approach, correlated radiobiological data and microdosimetric single event spectra data in several (and sufficiently different) neutron beams are used to deconvolve a biological weighting function representing the variation in biological effect (BE) with lineal energy. The BE of a new neutron beam can then be determined

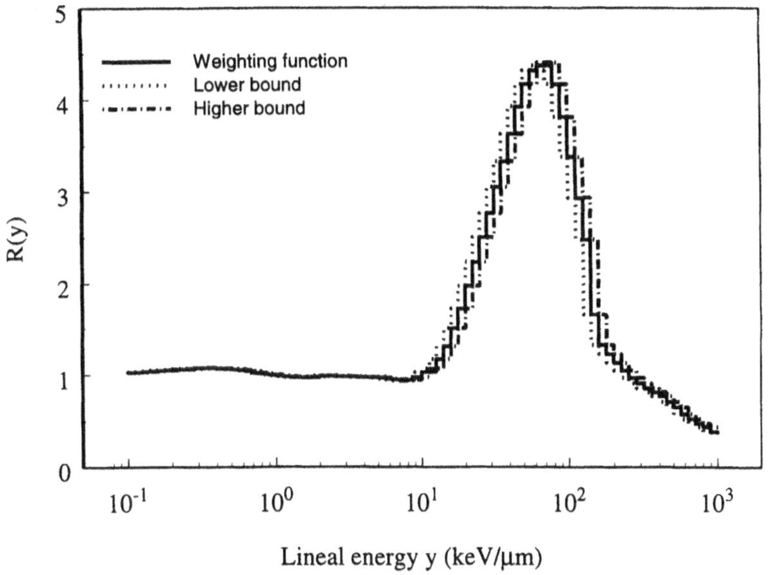

Figure 1. (a). Biological weighting function for early effects in mice crypt cell regeneration at a dose level of 8 Gy. (b) lower bound, and (c) upper bound to this function representing variations due to site size and other systematic differences.

by convolving its single event spectrum with the weighting function. This BE would be valid only for the specific end point and biological system represented by the weighting function.

In the present work, we have restricted our analysis to using a biological weighting function to determine changes in the RBE of the neutron component within and between different BNCT beams. Microdosimetric measurements were made in BNCT beams at BNL, MIT, and in a "preliminary design" beam at the University of Birmingham. The single event spectra in these beams are similar to those measured in fast neutron therapy, with differences in the relative contributions from different lineal energy intervals. Based on this similarity, we have used a macroscopic biological weighting function, calculated by Loncol *et al.*,[4] for early effects in mice crypt cell regeneration at a dose level of 8 Gy. Since these microdosimetric data were measured for a 2 μm site size, while the present measurements are with either a 0.5 μm site or a 1 μm site, the weighting function was accordingly modified to represent these variations in site sizes, and to further accommodate any systematic differences in the different sets of measurements. Figure 1 shows the original weighting function, and the upper and lower bounds used to account for the differences discussed above. The measured single event spectra in different beams were convolved with the three curves to obtain the range of corresponding possible RBE values.

3. RESULTS AND DISCUSSIONS

Figure 2 shows the single event spectra of neutron dose component along the central axis of the MIT beam. Very small variations in the spectra at different depths are observed. Similar spectra in beams at BNL and Birmingham demonstrated small variations as well. This data can be qualitatively interpreted as demonstrating insignificant variation in the RBE of the neutron dose component within these different beams. This

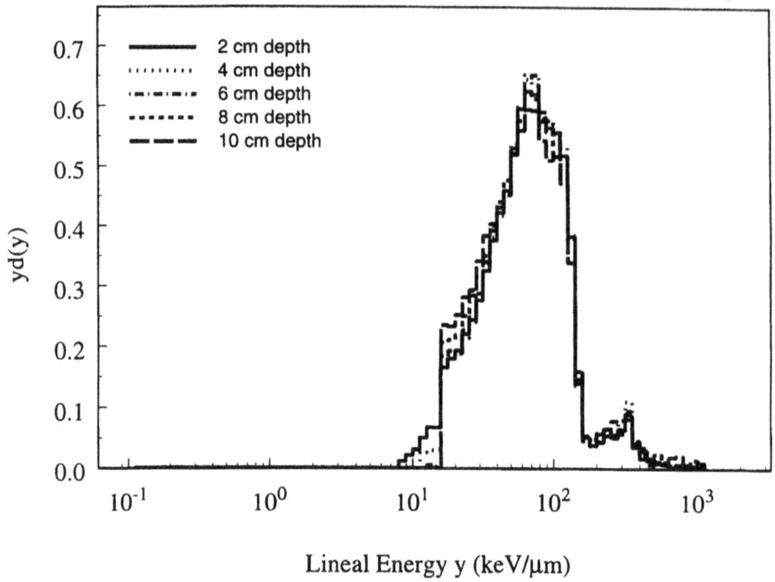

Figure 2. Single event spectra of the neutron dose components as a function of depth in the MIT beam.

result is contrary to the expectation of significant changes in neutron RBE due to a rapidly degrading neutron energy spectrum in the epithermal beams. Possible explanations are: (i) the dominant change in the neutron spectrum is in a energy interval which contributes insignificantly to the kerma, (ii) the simulated site sizes used for the measurements represent Bragg-Gray cavities so that the dose is measured correctly, but are too large for all the secondary charged particles to be considered "crossers", (iii) protons from the $^{14}N(n_{th},p)^{14}C$ reaction, whose lineal energy spectrum is invariant with position, contribute significantly to the total proton spectrum.

The neutron single event spectra in the 3 different BNCT beams are shown in Figure 3. Qualitatively, a small difference in the spectra from BNL and MIT is observed. The significant difference in the spectrum in the Birmingham beam is most likely due to difference in the site size used in this measurement. A quantitative measure of the RBE is obtained by convolving these spectra with the biological weighting functions in Figure 1. In beams at MIT, BNL and Birmingham, neutron RBE is calculated to be ~3.48, 3.4, and 3.8 respectively, at all points in phantom with variations within uncertainties of the calculation. In the Birmingham beam, the higher RBE is due to smaller site size used for making measurements. Using the upper bound of the R(y) vs y function (which corresponds to the smaller site size) changes this value to ~3.54, which is consistent with those calculated for the MIT and BNL beams. It was observed that the actual RBE values calculated using the different curves were slightly different, but their variations within and between the different beams remained unaffected.

4. CONCLUSIONS

Measured single event spectra in 3 different BNCT beams indicate small RBE differences between the neutron dose components in these beams and no variations within

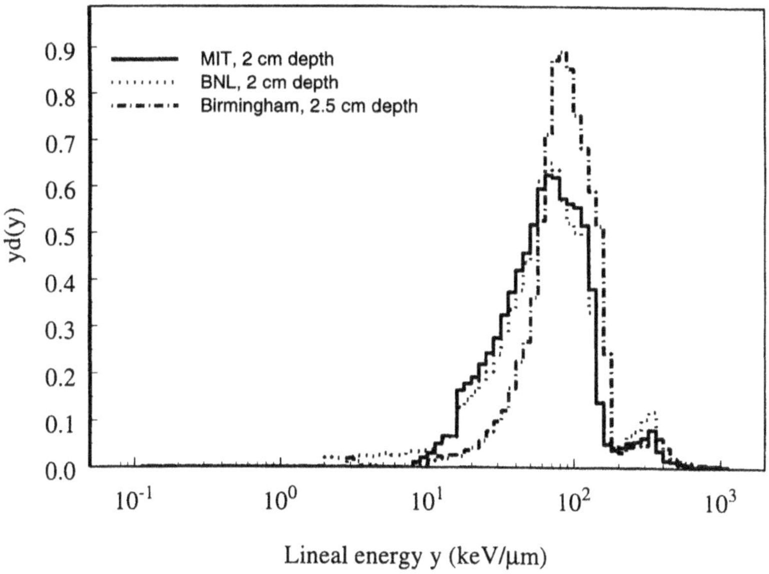

Figure 3. Single event spectra of the neutron dose component at 2 cm depth in the beams at MIT, BNL and Birmingham.

the beams. The current method is useful to estimate RBE variations and differences, rather than absolute values.

ACKNOWLEDGMENTS

Work supported in part by a grant from the U.S. Department of Energy Grant # DE-FG02-96ER62217. We are grateful to the staff at MIT, BNL and Birmingham for their help in acquiring the microdosimetric data.

REFERENCES

1. Microdosimetric considerations in the use of the boron neutron capture reaction in radiation therapy. Ph.D. Thesis. Wayne State University, 1996.
2. Rossi, H.H., and Rosenzweig, W., "A device for the measurement of dose as a function of specific ionization," Radiology, 64, 404–411, 1955.
3. J. Burmeister, C. Kota, R.L. Maughan, K. Riley, O. Harling, L. Wielpolski, R. Ma, and J. Coderre, Microdosimetry studies at the BMRR and the MITR-II using miniature tissue-equivalent proportional counters, *in* "Frontiers in Neutron Capture Therapy," M.F. Hawthorne, K. Shelly, R.W. Wiersema, eds., Kluwer Academic/Plenum Publishers, New York, 2001, pp. 581–586.
4. Loncol T., Cosgrove V., Denis J.M., Gueulette J., Mazal A., Menzel H.G., Pihet P., and Sabattier R. "Radiobiological effectiveness of radiation beams with broad LET spectra: microdosimetric analysis using biological weighting functions", *Proceedings of the eleventh symposium on microdosimetry*, Radiation Protection Dosimetry Vol 52 No. 1–4, 1994.

94

CELL CLUSTER MODELS FOR MICRODOSIMETRY

Petteri Välimäki[1,4], Juha Lampinen[1,3], Antti Kuronen[2], and Sauli Savolainen[1,3]

[1]Department of Physics
University of Helsinki
[2]Lab. of Comp. Eng.
Helsinki University of Technology
[3]Department of Radiology
Helsinki University Central Hospital
[4]Medical Engineering Center
Helsinki University Central Hospital

1. INTRODUCTION

The most important dose component in boron neutron capture therapy (BNCT) is the one resulting from the $^{10}B(n,\alpha)^{7}Li$ reaction. This dose is more tumor than normal tissue selective, due to the accumulation of the boron carrier compound in the tumor. The total weighted dose in BNCT is a sum of physical dose components multiplied by the weighting factors of each component. The weighting factors are related to the relative biological effectiveness (RBE) of each radiation quality (LET characteristics) and, in case of boron, also to the pharmacokinetics (microdistribution) of the boron compound. The weighting factor of the radiation dose is inaccurate as it is a function of the boron compound and the tissue characteristics. Due to the short range of the α-induced dose, the boron distribution should be known in cellular—and favorably—in subcellular level to define the accurate weighting factors in different tissues for the boron dose.[1]

Microdosimetry has been defined as the science that deals with the spatial, temporal, and energy-spectral distributions of energy imparted in cellular and subcellular biological structures and the relationship of such distributions to biological effects.[2] Macrodosimetry is traditionally defined dealing with the scale of whole tissue. Individual cells, chromatin fiber and DNA are the domains of interest in microdosimetry. So far in BNCT microdosimetry the correlation of the proposed models with experimental

Frontiers in Neutron Capture Therapy, edited by Hawthorne *et al.*
Kluwer Academic / Plenum Publishers, New York, 2001.

radiobiological data is lacking.[3] The microdosimetry protocols of the BNCT are still under construction. One aspect of the problem is the cellular level energy deposition in tumor and healthy cells.

Cell cluster models are needed for microdosimetry calculations. One of the proposed cell cluster models for BNCT-microdosimetry is the so-called close-packed cubic geometry model in which the spherical cells are located in a symmetrical lattice structure.[4] Another cubic geometry model approach with spherical cells is to use body-centered cubic arrays presented by Kassis *et al.*[5] They used two kinds of geometries: the tube geometry relevant in *in vitro* studies and simulated tissue geometry of possible interest for a highly cellular organ. Cluster construction using optimization methods has been used by Charlton and Utteridge.[6] They filled a cube at random with spherical tumor cells and let the cells migrate towards the origin. Charlton *et al.*[7] have constructed boxes with stem cell at the center and spherical fat cells of a given radius placed in the box at random. Cells were allowed to overlap each other. Charlton[8] has also presented a model based on an exceptional ellipsoidal shapes of the cells and nuclei. Verrijk *et al.*[9] have used an array of $3 \times 3 \times 3$ cuboidal cells with dynamic geometrical variables.

The aim of the study was to develop a new cell cluster model to characterize both normal and cancerous tissues for microdosimetry calculations in radioimmunotherapy as well as in BNCT. In the model, single cells are located asymmetrically in the three-dimensions in the nearest possible position related to the origin.

2. MATERIALS AND METHODS

2.1. Cell Dimensions

Tumor cells are frequently characterized by a concentrically located spherical nucleus. Normal tissue cells in situ show typically long, amorphous shapes.[4] There is a large variation of the cell radii: average cell radius was 20% larger than the peak value of the radius distribution according to flow cytometry study of C-18 cells,[10] fat cells in the marrow of diameters up to 200 μm and human stem cells 8–16 μm[7], Lorimore *et al.*[11] suggest a minimum stem cell diameter of 7 μm. In the model calculations spherical or ellipsoidal cell shapes have been assumed for both the healthy and tumor cells with different radii.

2.2 Cluster Model

A new cell cluster model constructed by *CellPacker*, a computer program developed in study, is based on a systematic way of piling spherical cells in the system of spherical coordinates. The program aims to locate each cell as near to the center of the spherical tissue as possible by changing the spherical coordinates in turns. At every suitable point, *CellPacker* calculates the distances to the central points of the earlier located cells. If all the distances are greater than the sums of the radii and the intercellular distances of the respective cells, there is enough room for the cell to be located. The coordinates of the point found are then tabulated and utilized later in calculating the distances.

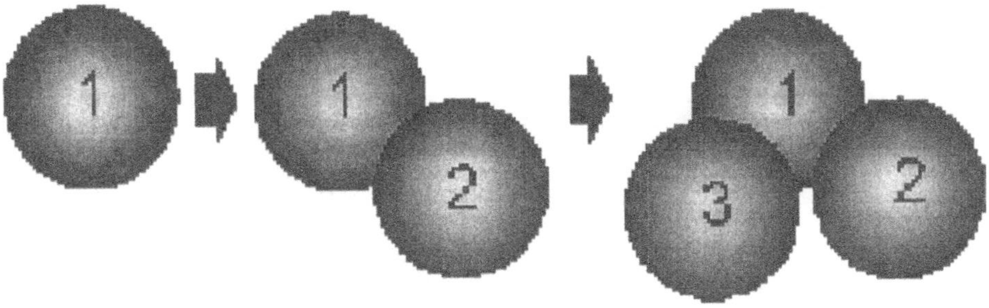

Figure 1. The origin set in the central point of the first cell the program goes through all the capable values of the φ-coordinate.

The program uses spherical coordinates while seeking a location for the central point of the cell. First it goes through all the capable values of the Ω-coordinate in one round with certain fixed values of the υ- and r-coordinates (Figure 1).

After that the value of the υ-coordinate is changed (Figure 2) and in this way the whole surface of the sphere with a certain radius is explored. Then the r-coordinate is increased and the process starts from the beginning.

In every examined point the program calculates the distances to the tabulated central points of the already laid cells. If the distances are equal to or greater than all the sums of the radii and the intercellular distances of the respective cells, there is enough space for the cell and the coordinates of its central point are tabulated.

In Figure 3 a cell cluster constructed by *CellPacker* is shown. The cross sections of the cell cluster model of *CellPacker* and the conventional close-packed cubic geometry model are shown in Figure 4. Compared to the closed-packed cubic model, the *CellPacker* model achieves higher density around the tumor tissue boundary.

3. DISCUSSION AND CONCLUSION

The shape and the size of single tumor and normal cells play an important role in the modeling. There is a large variation of the cell radii even among the cells of the same tissue. Cell size in general can easily be altered in cluster model computer programs, but the program should also be able to alter geometry parameters randomly during a

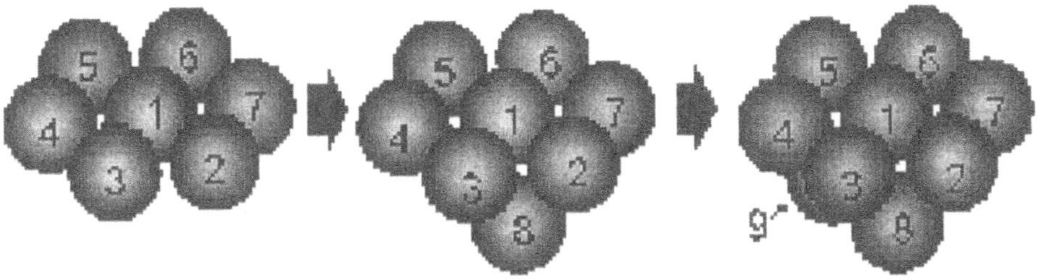

Figure 2. Examination of the capable φ-coordinates with increased θ-angle.

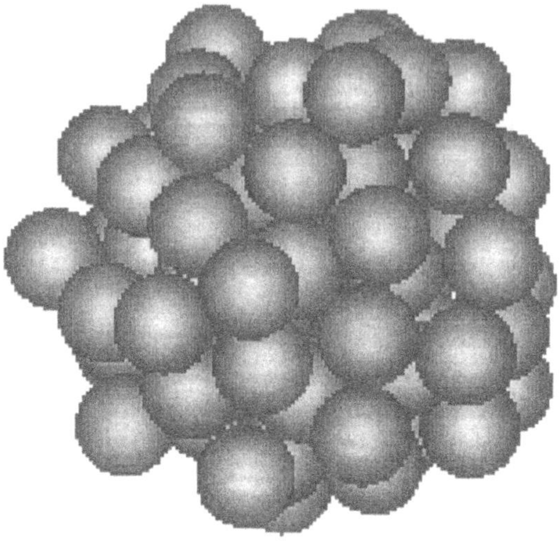

Figure 3. Cell cluster of 30 μm tumor cells and 75 μm normal cells constructed by *CellPacker*.

simulation as Verrijk *et al.* propose.[9] Verrijk *et al.* themselves have developed a model with various possibilities to get variation to the tissue dimensions, but the model deals only with small tissues of 27 cells. Charlton and Utteridge[6] have used cell radii chosen at random from a distribution of 5–7 μm in half micrometer intervals in the ratio 1:2:4:2:1.

The most often used way of modeling the cells and nuclei as concentric spheres is unsatisfactory in that the shape of the nuclei of tumor cells can be highly irregular.[6] Tumor cells though are frequently characterized by a concentrically located spherical nucleus. Normal tissue cells in situ show typically long, amorphous shapes.[4] Computational and modeling difficulties have so far hindered the use of the more complex forms.

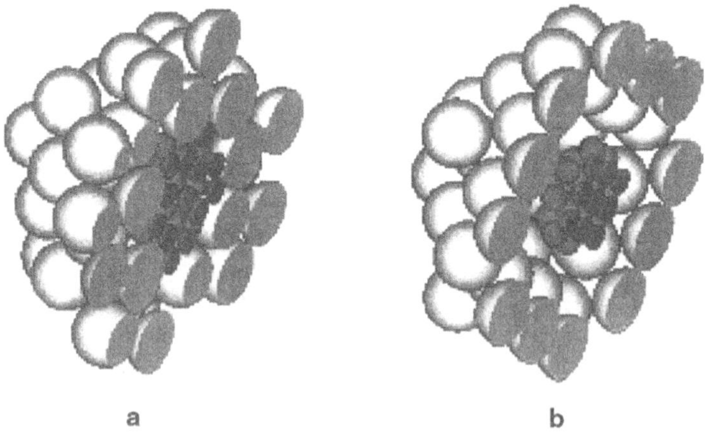

Figure 4. Cross sections of the cell cluster model of *CellPacker* (a) and the conventional close-packed cubic geometry model (b).

As an exception, Charlton has developed a model describing cells and nuclei as ellipsoids.[8] In the fat cell model by Charton *et al.*[7] the spherical structures of the tissue are partly broken by reciprocal overlapping spheres.

All the previously mentioned models have a way to adjust the intercellular space. Thus the simulation of the lowly cellular tissue with evenly distributed cells is not a problem. Instead, the modeling of the highly cellular tissue presupposes cells to be packed as closely to each other as possible. References to filling percentages of the real tissues seem though still be missing in the microdosimetric publications. Makrigiorgos *et al.*[12] state qualitatively that the filling percentage 74 of the close-packed cubic geometry is close to what cytological studies reveal for some human tissue. The basic microdosimetric conclusions of their study were not however strongly affected by the choice of the body centered cubic geometry filling percentage 68[5] instead of the close-packed one. CellPacker achieves the filling percentage of 64, but compared to the closed-packed cubic model, it achieves higher density around the tumor tissue boundary where big spheres are packed around smaller ones (Figure 3). This has also a locally significant microdosimetric effect.[13] The maximal filling percentages of the other methods seem to be missing in the microdosimetric publications.

In the future the analytical microdosimetric study will move on more to subcellular level. Charlton's and Utteridge's optimization model,[6] Verrijk *et al.*'s model[9] and Charlton's ellipsoidal model[8] have already simulated nuclei in the cell. The CellPacker model is presently lacking this feature, but is expandable also to that direction.

ACKNOWLEDGMENTS

The work was supported by the Academy of Finland and University of Helsinki.

REFERENCES

1. S. Savolainen, I. Auterinen, C. Aschan *et al.*, Dosimetry chain for the dogs irradiated in the epithermal neutron beam at the Finnish BNCT facility, *these proceedings (submitted)*.
2. F.H. Attix, "Introduction to Radiological Physics and Radiation Dosimetry", Wiley-Interscience Publ., New York, 1986, p. 501.
3. R.G. Zamenhof, Microdosimetry for boron neutron capture therapy: a review, *J. Neuro-Oncol.*, 33:81–92, 1997.
4. J. Rassow, F. Pöller, F. Steinberg and P. Meiner, Physical and tumor biological aspects and calculation model of dosage in boron neutron capture therapy (BNCT), Strahlentherapie und Onkologie, 169:7–17, 1993.
5. A.I. Kassis, S.J. Adelstein, C. Haydock, and K.S.R. Sastry, Thallium-201: An experimental and a theoretical radiobiological approach to dosimetry, *J. Nucl. Med.*, 24:1169–1175, 1983.
6. D. Charlton, and T. Utteridge, Monte Carlo modelling of packed cells used to evaluate dose in BCT, in "Proceedings of the Seventh International Symposium on BNCT for Cancer", B. Larsson, J. Crawford, and R. Weinreich, ed., Elsevier, 1997, pp. 243–247.
7. D.E. Charlton, T.D. Utteridge, and A.H. Beddoe, Microdosimetry of haemopoietic stem cells irradiated by alpha particles from the short-lived products of 222Rn decays in fat cells and haemopoietic tissue, *Int. J. Radiat. Biol.*, 69:585–592, 1996.
8. D.E. Charlton, Energy deposition in small ellipsoidal volumes by high-LET particles: application to thermal neutron dosimetry, *Int. J. Radiat. Biol.*, 59:827–842, 1991.
9. R. Verrijk, R. Huiskamp, A.C. Begg, F.J. Wheeler, and P.R.D. Watkins, A comprehensive PC-based computer model for microdosimetry of BNCT, *Int. Radiat Biol.*, 65:241–253, 1994.
10. T.G. Stinchcomb, and J.C. Roeske, Analysis of survival of C-18 cells after irradiation in suspension with chelated and ionic bismuth-212 using microdosimetry, *Radiation Research*, 140:48–54, 1994.

11. S.A. Lorimore, D.T. Goodhead, and E.G. Wright, Inactivation of haemopoietic stem cells by slow α-particles, *Int. J. Radiat. Biol.*, 63:655–660, 1993.
12. G.M. Makrigiorgos, S.J. Adelstein, and A.I. Kassis, Limitations of conventional internal dosimetry at the cellular level, *J. Nucl. Med.*, 30:1856–1864, 1989.
13. J. Lampinen, P. Välimäki, A. Kuronen, J. Stepanek, and S. Savolainen, Cluster models in cellular level dosimetry, *Acta Oncolociga* (in press).

MEASUREMENTS OF PHANTOM DOSE DISTRIBUTIONS AT THE FINNISH BNCT FACILITY

Mika Kortesniemi[1,2]*, Antti Kosunen[3], Carita Aschan[2,5], Tom Serén[4], Petri Kotiluoto[4], Matti Toivonen[3], Petteri Välimäki[2], Tiina Seppälä[1,2], Iiro Auterinen[4], and Sauli Savolainen[5]

[1]Clinical Research Institute
Helsinki University Central Hospital
P.O.Box 105, FIN-00029 HYKS, Finland
[2]Department of Physics
University of Helsinki
[3]STUK—Radiation and Nuclear Safety Authority
Helsinki, Finland
[4]VTT Chemical Technology
Technical Research Centre of Finland
[5]Department of Radiology
Helsinki University Central Hospital

1. INTRODUCTION

The Finnish BNCT facility with a new epithermal beam is close to complete.[1] The neutron spectrum of the beam were determined by calculations and activation measurements free in air.[2] Consequently, as the FiR 1 BNCT beam was renewed, also the new beam model FiR(K63) to be used by the actual dose planning system was created and verified.[3] The verification was done by comparison of the calculated and the measured dose distributions in phantoms using different dosimetric methods which together can provide appropriate reliability.

The basic question in the verification of the BNCT dosimetry is the experimental determination of the neutron fluence and the dose distributions which are generated by

* Email: mika.kortesniemi@huch.fi.

Frontiers in Neutron Capture Therapy, edited by Hawthorne *et al.*
Kluwer Academic / Plenum Publishers, New York, 2001.

the neutron beam in a human-like material and geometry. Interpretation of the detector responses is related to the reliability of the calculated neutron spectrum in the phantom. These questions served as a motivation to the experimental approach with prevailing and novel methods of the BNCT dosimetry.

The other objective for this work was to find out simple and efficient methods for quality control at FiR 1. As the BNCT facility will be in regular operation simple and efficient methods, providing reliable estimates of the beam quality and the dose characteristics are required. Thereafter profound considerations should be made, for example, on phantom materials and geometries. Also the automation of the quality control routines have to be developed.

2. METHODS

Several different phantoms were used in this study. The construction of the liquid phantom is partly shown in Figure 1. The liquid phantom system consists of a small changeable phantom extension and a large phantom base pool. The pool includes the 3D transport mechanism for the detectors. Using the transport mechanism the detectors were moved to the precise locations inside the phantoms. The movements of the detectors were controlled automatically by the computer program.

For the liquid phantoms the shapes covered cylinder, doghead and Snyder. The Snyder phantom, simulating the human head (which is in Figure 1 attached to the pool), is 20cm long (extension from the base pool) and it has the ellipsoidal shape and size defined by Snyder W. *et al.*[4] The cylinder phantom is 20cm long and 20cm in diameter. The phantom wall (PMMA) thickness was 5mm. The doghead phantom has a complex,

Figure 1. The Snyder extension phantom attached to the phantom base pool. The 3-D transport mechanism is carrying the TE(TE) ionization chamber inside the phantom liquid.

natural shape and it is 22cm long. The wall thickness for doghead and Snyder was 2mm. A solid (24cm long) PMMA cylinder was used separately from the liquid phantoms. In the PMMA cylinder all measurements were made at the central axis. Detectors were inserted in a hole in the PMMA phantom and special fit parts were used to attain the desired depth and to exclude air between the phantom material and the detector.

The simple PMMA phantom was tested for the use as a quality control phantom for checking the output of the FiR 1 beam. On the contrary, the complex doghead phantom was used to test the radiation transport calculations.

The same series of measurement points was used in all measurements. The depth dose rates were measured in the central axis of the phantoms, aligned parallel to the beam axis. The measurements were also performed horizontally and vertically in the transverse direction, across the phantoms. Those profile curves were determined at two depths: at 25mm and at 60mm.

The phantom materials were chosen according to practicality and resemblance to the brain tissue. Liquid B, PMMA and water were used separately where the liquid B was used as a brain tissue substitute material, consisting of water, alcohol, urea and inorganic salts.[5]

Four different detector types were used to determine the neutron and the gamma dose characteristics in the phantoms. A pair of ionization chambers were applied according to the twin ionization chamber method.[6] The Exradin Mg-chamber was used with extra pure Ar gas to measure the gamma dose rate and the Exradin TE (tissue equivalent) —chamber with TE gas to measure the combined gamma-neutron dose rate.[7,8] The composition of TE gas was: N_2 3.2%, CO_2 32.4% and CH_4 64.4% vol. The chamber volume was 0.5 cubiccm for both of the chambers. The neutron fluence was determined by Au and Mn activation foils.[2] Thermoluminescent detectors were used to measure the

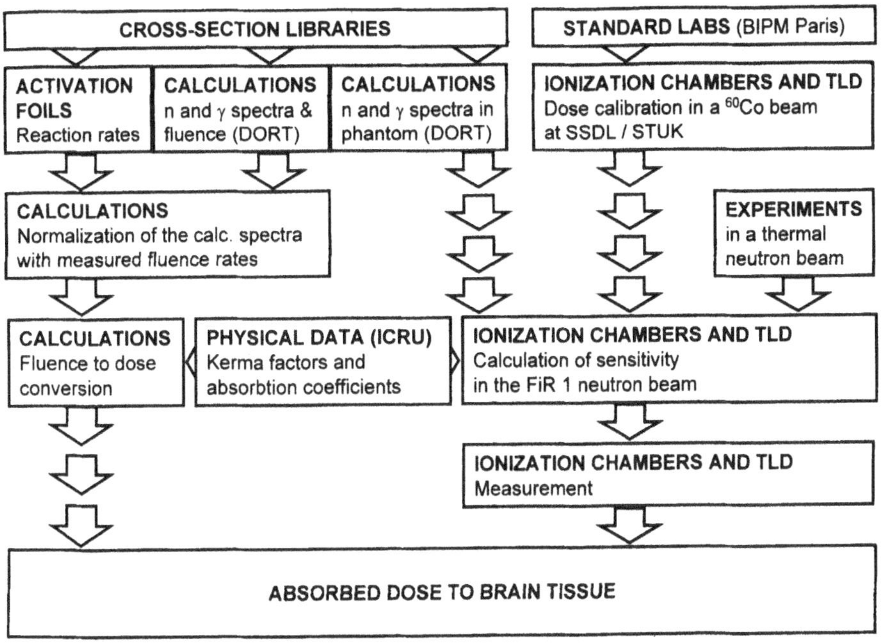

Figure 2. The relation and use of different dosimetric methods, extending from the international dose standard to the determined absorbed dose.

Table 1. The overall uncertainties (1 S.D.) of the measured gamma dose and neutron dose or fluence expressed in percents[8,9,10]

Detector	Neutron dose or fluence	Gamma dose
Au foil	4	not applicable
Ionisation chambers	21	6
TLD's	16	<20*
Si(Li) diode	2**	not used

*Uncertainty of thermal neutron sensitivity of the MCP-7s TL detectors[10] dominates the total uncertainty of TLD's which is at its highest (approx. 20%) at the depth of thermal neutron fluence maximum.
**Uncertainty of the relative Li reaction rate.

gamma dose and the relative neutron fluence. MTS-Ns (LiF:Mg,Ti) and MCP-7s (^{7}LiF:Mg,Cu,P) TL detectors (TLD Niewiadomski & Co., Krakow, Poland) were used for neutron and gamma detection, respectively.[9,10] Also, a small Si(Li) diode detector was used to measure the relative Li reaction rate.[11] The calculations were made with DORT code.[3] The relation of different dosimetric methods is presented in Figure 2.

3. RESULTS

The dose rate distributions resulting from the ionization chamber and TLD measurements in the central axis of the cylindrical liquid B phantom are shown in Figure 3 together with the calculated depth dose rate curves. The uncertainties of the dosimetric methods are presented in Table 1.

The phantom shape comparison showed minor differences between the used phantom shapes. The Snyder featured a deeper dose maximum compared with the other shapes. The phantom material comparison showing the gamma dose rate distributions in the cylindrical phantom is presented in the Figure 4. The profile curves demonstrated a symmetric dose rate distributions in the used depths in the phantoms.

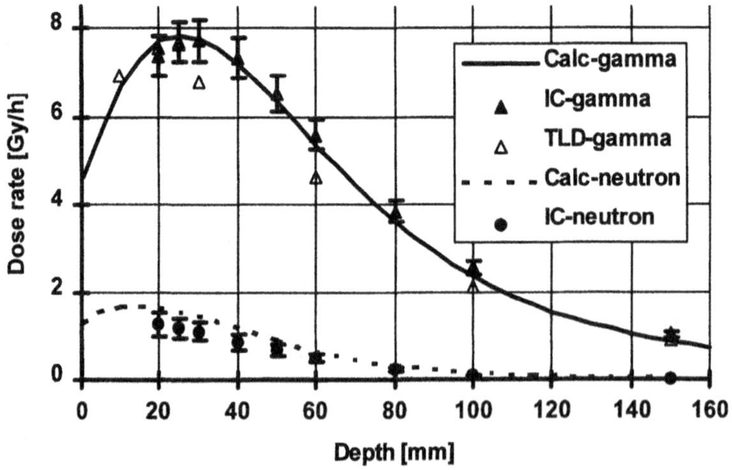

Figure 3. Calculated and measured (IC and TLD) dose rate distributions in the cylindrical liquid B phantom.[8] The nominal power of the reactor was 250 kW.

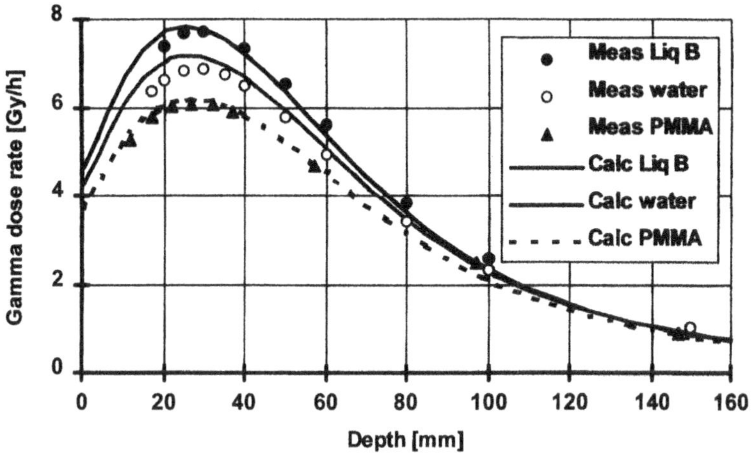

Figure 4. Calculated and measured (IC) gamma dose rate distributions in the cylindrical phantoms of different materials.[8] The nominal power of the reactor was 250 kW.

A clear difference in the maximum gamma doses can be observed in Figure 4. The differences are in obvious congruence with the calculated curves. The elemental composition provides an explanation to the distinctive shapes as water and liquid B contain 14% more hydrogen than PMMA and the liquid B contains chlorine where the other materials do not.[5] Compared to the other phantoms the deeper dose maximum in the Snyder phantom can be regarded as a consequence of its distinctive convex shape compared with the others.

4. CONCLUSIONS

The present beam model of the FiR 1 BNCT facility based on DORT calculations is supported by the experimental approach. The results of the gamma dose determinations between the different dosimetric methods were consistent within an acceptable level. Still, the large uncertainties related to the neutron dose measurements imply the need of improvements in the ionization chamber technique.

However, the ionization chambers are well suited for quality control of the beam stability in the practice of the measurements. Also, the solid PMMA cylinder is suitable as a part of the daily quality control with practical and accurately repeatable phantom set-up.

International quidance for dosimetry and intercomparisons between BNCT facilities are essential to achieve comparability of treatment results and to support the multidisciplinary approach in the BNCT community.

ACKNOWLEDGMENTS

The financial support from the Technology Development Centre, the Finnish Academy, University of Helsinki and Helsinki University Central Hospital is acknowledged.

REFERENCES

1. Auterinen I. *et al.* Metamorphosis of a 35 year old triga reactor into a modern BNCT facility, in: "Frontiers in Neutron Capture Therapy," M.F. Hawthorne, K. Shelly, R.W. Wiersema, eds., Kluwer Academic/Plenum Publishers, New York, 2001, pp. 267–275.
2. Serén T. *et al.* Spectrum measurements and calculations in the epithermal neutron beam at the FiR 1 BNCT facility. 15th European TRIGA Conference, Espoo, 1998.
3. Seppälä T. *et al.* Source characterisation for the rtt_MC treatment planning program at FiR 1, in: "Frontiers in Neutron Capture Therapy," M.F. Hawthorne, K. Shelly, R.W. Wiersema, eds., Kluwer Academic/Plenum Publishers, New York, 2001, pp. 219–224.
4. Snyder W. *et al.* Estimates of absorbed fractions for monoenergetic photon sources uniformely distributed in various organs of a heterogenous phantom, MIRD, *J. Nucl. Med. Suppl.*, No.3. Pamphlet 5, 1969.
5. Seppälä T. *et al.* Brain tissue substitutes for phantom measurements in NCT, in: "Advances in Neutron Capture Therapy Vol 1", Medicine and Physics, Elsevier Science B.V., 1997, pp. 188–191.
6. ICRU, Clinical neutron dosimetry—Part I: Determination of absorbed dose in a patient treated by external beams of fast neutrons, in: " ICRU Report 45", International Commission on Radiation Units and Measurements, Bethesda MD., 1989.
7. Kosunen A. *et al.* The twin ionisation technique in quality control of an mixed epithermal neutron and gamma beam for NCT, in: "Advances in Neutron Capture Therapy Vol 1", Medicine and Physics, Elsevier Science B.V., 1997, pp. 182–187.
8. Kosunen A. *et al.* Twin ionisation chambers for dose determinations in phantom in an epithermal neutron beam, *Radiat. Prot. Dosim.* (in press).
9. Aschan C. *et al.* Epithermal neutron beam dosimetry with TL dosemeters for boron neutron capture therapy, *Radiat. Prot. Dosim.* (in press).
10. Aschan C. *et al.* Experimental correction for thermal neutron sensitivity of gamma-ray TL dosemeters irradiated at BNCT beams, *Radiat. Prot. Dosim.* (submitted).
11. Pospisil S. *et al.* Si Diode as a Small Detector of Slow Neutrons, *Radiat. Prot. Dosim.*, 46:115–118 (1993).

RADIOPROTECTION DOSIMETRY IN THE EUROPEAN TRIAL ON BNCT STUDIED BY AN ANTHROPOMORPHIC PHANTOM

S. Garbe[1], J. Rassow[1], C. Vroegindeweij[2], R. Moss[2], K. Hideghéty[1], W. Sauerwein[1], C. Kessler[3], and F. Stecher-Rasmussen[3]

[1]Klinik und Poliklinik für Strahlentherapie
Universitätsklinikum Essen, Germany
[2]Joint Research Centre (JRC)
European Commission
Petten, The Netherlands
[3]Energie Onderzoek Centrum (ECN)
Petten, The Netherlands

1. INTRODUCTION

Different parts of the patient body are exposed indirectly during BNCT treatment of glioblastoma multiforme, depending on the geometrical alignment of the patient to the beam. Successful clinical application requires applicable solutions for the shielding of the organs at risk. At present during the irradiation, there is no way to measure the dose components *in vivo* in other parts of the body apart from the surface.

In order to investigate and to improve this aspect of medical dosimetry in the European clinical trial on BNCT, an anthropomorphic Alderson-RANDO® phantom system[1] has been introduced. With the use of this phantom system all dose components can be determined at any location in the body.

The neutron spectrum and the γ-ray spectrum are measured dependent on depth and location in an absolute and relative manner by the means of:

- Mg(Ar) ionisation chamber for the γ-ray absorbed dose (D_g)
- TE(TE) ionisation chamber for the fast neutron absorbed dose + γ-ray absorbed dose (D_n)
- Neutron activation monitors for the thermal neutron fluence (Φ_{th})
- Li-covered PN-diode for the thermal neutron fluence (Φ_{th})

Frontiers in Neutron Capture Therapy, edited by Hawthorne *et al.*
Kluwer Academic / Plenum Publishers, New York, 2001.

In order to simulate possible patient irradiation situations, different geometrical set-ups, shielding, material compositions and filtering have been studied.

The result is a three-dimensional dose distribution in the anthropomorphic phantom.

2. EXPERIMENTAL

The beam line set-up of the HB11 at the High Flux Reactor in Petten is described in reference 2. The experimental set-up parameter for this experiment is a collimated beam with a diameter of 120 mm, the distance between the opening and the phantom surface (SSD) is chosen during the treatment to 300 mm. The phantom is placed on the therapy table and the head is fixed to the head-holder using the red cushion described in.[2,3]

The investigated geometrical arrangements of the phantom placed on the therapy table with respect to the beam direction are four different angles of incidence 0°, 10°, 45° and 90° respectively. The Alderson-RANDO® phantom itself consists of an anthropomorphic shaped body separated in 51 slices with a thickness of 25 mm, more details could be found in reference 1.

The investigations are performed for all four angles of incidence at 20 positions along the central (body) axis (from the top of the scalp to thigh), at 33 positions on four cross profiles in central body plane varying the spacing between 15 mm and 25 mm. The same positions are chosen above and below the central body plane, while the displacement of these planes is 25 mm in both directions perpendicular to the central plane. The rotation axis of the phantom is set to center of the head.

The experimental dosimetry is performed at all positions using two different detector systems PN-diode and paired ionisation chambers consisting of TE/TE- and Mg/Ar-ionisation chamber. Activation foils analysis was made at three positions along the central body axis. Using the nomenclature as recommended by the ICRU[4] the specifications of the used paired ionisation chamber detector systems are:

> Mg/Ar-chamber: Exradin-chamber, Type M2, Lisle, IL, M182, calibrated in a ^{60}Co γ-ray source at Petten. The ionisation chamber (IC) material is A-150 plastic.
> TE/TE-chamber: Exradin chamber, Type T2, Lisle, IL, T138, calibrated in a ^{60}Co γ-ray source at Petten is used. The Ionisation chamber consists of A-150 plastic and the TE-gas consists of 3.2% N_2, 32.4% CO_2 and 64.4% CH_4.

The ionisation-chamber readings are corrected for air-pressure and temperature. The equation used for conversion of the ionisation-chamber reading into dose according to ICRU[4] is:

$$R_u = h_u * D_g + k_u * D_n + k_u' * \Phi_{th}$$

$$R_t = h_t * D_g + k_t * D_n + k_t' * \Phi_{th}$$

where D_g is the γ-ray absorbed dose, D_n the neutron absorbed dose and Φ_{th} the thermal neutron fluence, the subscript u refers to detectors with low sensitivity for neutron and t to detectors with equal sensitivity. In Table 1 the values used for k and h for calculation are demonstrated.

The values used in this article are normalised to the monitor reading of fission-chamber 1, described in.[2] In order to determine the thermal neutron fluence activation

Table 1. The values k and h used for determining the relative sensitivity of the ionisation chambers to neutrons and γ-rays

	h_u, h_t	K_u, k_t	k'_u, k'_t [10^{-12} Gycm^2]
Mg/Ar	0.99	0.00	0.213
TE/TE	1.22	0.87	0.434

foil analysis had been performed using AuAl (5 wt% Au) and MnNi (88 wt% Mn) foil of 9 mm × 20 mm × 0.1 mm. Where no activation analysis had been performed the measured PN-value is taken for Φ_{th}.

For the investigation of the shielding influence a non optimized borated polyethylene brick wall of 50 mm thickness has been placed between the head and shoulder area of the phantom. In order to see the difference in the profiles along and across the body axis in front and behind the shielding, measurements had been performed with all three detector systems.

3. RESULTS

As results some profiles of the neutron absorbed dose, the γ-ray absorbed dose and the thermal neutron fluence along the central body axis are shown. In Figure 1 the dose profiles are shown of D_n, D_g and Φ_{th} for the worst case (an unlikely treatment case) where the beam is parallel (0°) to the body axis (the beam is entering at the scalp of the body).

In Figure 2 the dose profiles for D_g, D_n and PN-diode are shown for a representative treatment case, using a beam which is directed 45° to central body axis (the beam is entering at the scalp of the body).

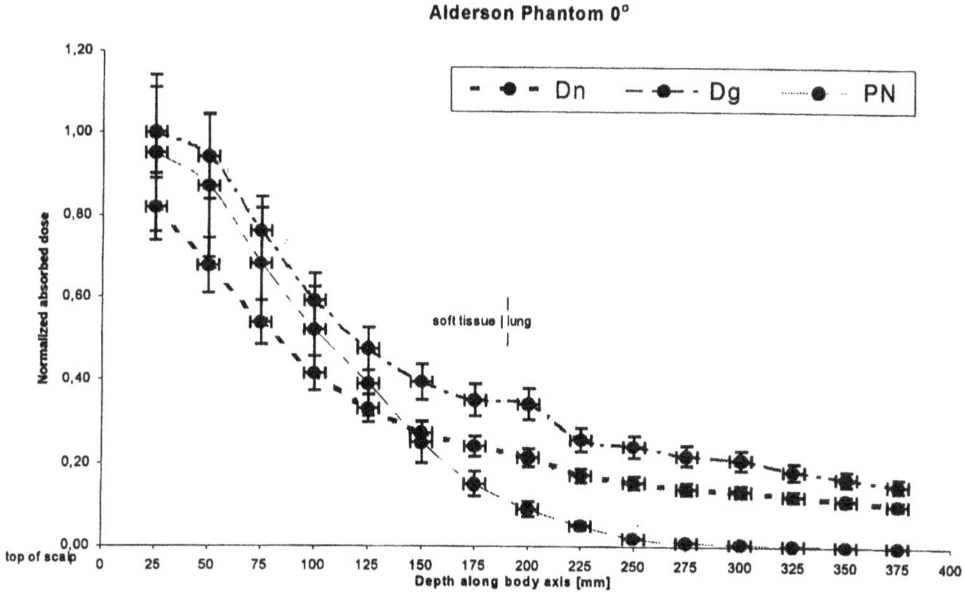

Figure 1. In this figure the dose profiles for the γ-ray absorbed dose (D_g), the neutron absorbed dose (D_n) and the thermal neutron fluence (Φ_{th}) in the 0° geometry (beam parallel to central body axis) are shown.

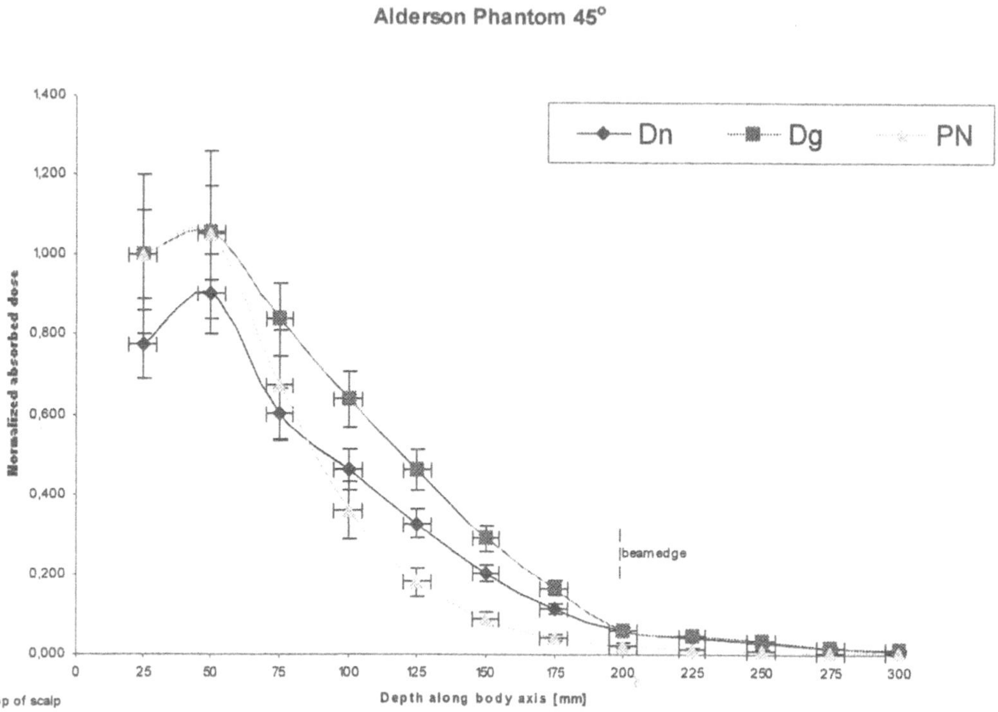

Figure 2. In this figure the dose profiles for the γ-ray absorbed dose (D_g), the neutron absorbed dose (D_n) and the thermal neutron fluence (Φ_{th}) in the 45° geometry (beam directed 45° to central body axis) is shown.

The influence of a borated polyethylene shielding in the 0° geometry for the neutron absorbed dose is shown in Figure 3 as a cross profile in the lung area.

4. DISCUSSION

The measured γ-ray absorbed dose in this mixed field beam is significantly influenced by the thermal neutron flux, therefore an uncertainty of $\Delta D_g \approx 10\%$ is found. This high uncertainty is mainly introduced by the rather rough derivation of Φ_{th} from the PN-signal in order to determine the γ-ray absorbed dose (D_g). A possible solution in order to reduce the uncertainty is to introduce a Li-cover to the IC.

The measured D_n (fast neutron absorbed dose) therefore is not influenced, it is directly derived from the Mg/Ar-IC and TE/TE-IC-signal ($\Delta D_n \approx 8\%$).

As shown in Figure 1 the neutron absorbed dose along the patient body axis especially in the lung region is reduced to about 10% in the worst case (at 0°) and less in (1%) the regular case. The γ-ray absorbed dose showed the same behaviour as the neutron absorbed dose.

In the cross profile (in central plane) a broadening of the beam shape occurs with depth (as expected), whereby the broadening of the neutron absorbed dose is smaller compared to the γ-ray absorbed dose.

It could be found that the cross profiles in the upper and lower plane showed a deviation in the total absorbed dose D_t ($D_t = D_g + D_n$) at the lung and other cavities.

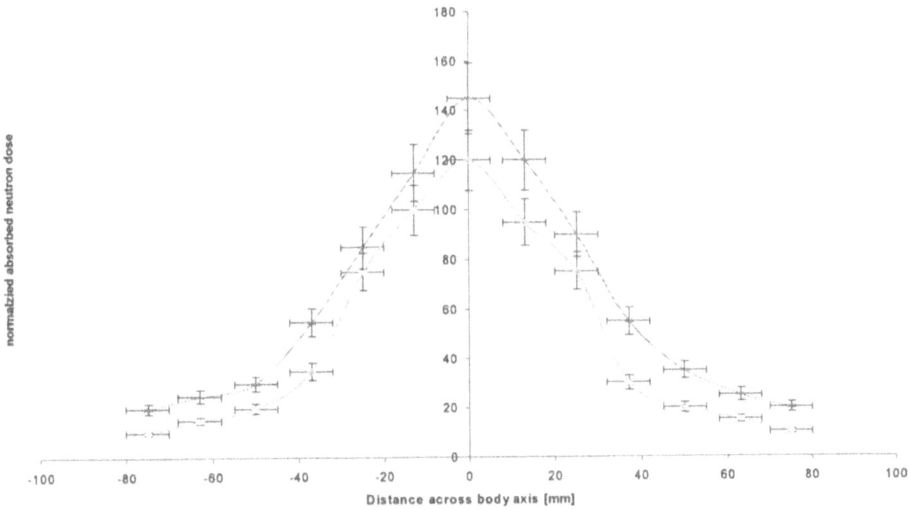

Figure 3. The influence of a borated polyethylene shielding placed between the head and the shoulder to the neutron absorbed dose D_n at the lung area measured by a cross profile.

The preliminary non optimized shielding with polyethylene and borated-polyethylene introduces a dose reduction of the total absorbed dose of 15% at the maximum. The shape of the beam is narrowed in all dose components.

5. CONCLUSIONS

The Alderson Rando® phantom offers an excellent possibility to study the spatial dose distribution and the exposure of the entire body during a glioblastoma treatment of a mixed beam at HB11 in Petten. It could be shown with respect to the radioprotection of the patient, that the shielding of the body and the organs at risk, have a benefit of 15% in the maximum total absorbed dose. But there are remaining problems, how a real dose determination could be performed and how the main dose component (Boron absorbed dose) in BNCT caused by the $^{10}B(n,\alpha)^{7}Li$-reaction could be measured.

In order to get an improved approach to a more real 3D-Investigation with higher accuracy and a better dose determination, further investigation using an advanced TLD technique[6] in the phantom will be evaluated. For the study of the influence of the shielding, improved shielding devices will be investigated as well as shielding calculation including backscatter effects, etc. This calculation will be based on MCNP-calculations[7] using on CT-scans of the phantom.

ACKNOWLEDGMENT

The clinical trial is financially supported by the European Commission in the form of a Shared Cost Action from the Biomedical and Health Research Programme

(BIOMED II) of DGXII, contract no BMH4-CT96-0325 and by the Swiss Foundation "Neurochirurgie 2001".

REFERENCES

1. P.C. Shrimpton *et al.*, The tissue-equivalence of the Alderson Rando anthropomorphic phantom for X-rays of diagnostics qualities, Phys. Med. Bio. 26, 1, 1981, 133–139.
2. R. Moss *et al.*, The requirement and development of neutron beams for neutron capture therapy of brain cancer, Journal Neuro-Onco. 33, 1997, 27–40.
3. P. Watkins *et al.*, Patient positioning at the HFR Petten, *in* "Frontiers in Neutron Capture Therapy," M.F. Hawthorne, K. Shelly, R.W. Wiersema, eds., Kluwer Academic/Plenum Publishers, New York, 2001, pp. 179–183.
4. ICRU report 45, Clinical neutron dosimetry part I, International Commission on Radiation Units and Measurements, Bethesda MD, 1989.
5. C.P.J. Raijmaakers, "Dosimetry and Treatment Planning for Boron Neutron Capture Therapy", Thesis, Amsterdam.
6. C. Kessler *et al.*, Application of thermoluminescence dosimeters to mixed neutron-gamma dosimetry for BNCT, *in* "Frontiers in Neutron Capture Therapy," M.F. Hawthorne, K. Shelly, R.W. Wiersema, eds., Kluwer Academic/Plenum Publishers, New York, 2001, pp. 1165–1173.
7. C. Vroegindeweij *et al.*, Validation of the description of the epithermal neutron beam for boron Neutron capture at the HFR Petten, *in* "Frontiers in Neutron Capture Therapy," M.F. Hawthorne, K. Shelly, R.W. Wiersema, eds., Kluwer Academic/Plenum Publishers, New York, 2001, pp. 375–380.

DEVELOPMENT OF A NEUTRON ENERGY-BIASED IN-AIR FIGURE-OF-MERIT FOR PREDICTING IN-PHANTOM BNCT NEUTRON BEAM CHARACTERISTICS*

D. L. Bleuel[1,2], R. J. Donahue[1], B. A. Ludewigt[1], and J. Vujic[2]

[1]Ernest Orlando Lawrence Berkeley National Laboratory
Berkeley, CA 94720
[2]Department of the Nuclear Engineering
University of California
Berkeley, CA 94720

1. INTRODUCTION

Maximizing the efficacy of a Boron Neutron Capture Therapy (BNCT) modality requires improvements in two areas: development of better boron-delivering drugs and better tailoring of the epithermal neutron beam. Previous attempts at the latter have not always been successful in predicting optimal in-phantom results, often assuming that all neutrons within a certain "useful" energy range are equally valuable for BNCT purposes. Some studies have recognized that higher energy epithermal neutrons (~10 keV) and more forward-directed neutrons provide more penetrating beams. This results in higher tumor doses at centerline phantom depths. The exact effect of neutron beam energy spectrum shaping for BNCT has not been thoroughly explored. One reason for this has been the usage of inappropriate and often misleading in-air figures-of-merit.

A few previous studies[1] have examined the effects of monoenergetic neutron beams, but have not fully applied these results to help direct the design of an optimal neutron beam spectrum. In this report we examine the in-phantom dose characteristics of nearly monoenergetic neutron beams (with narrow energy bins) in the range from thermal to 14

* This work was supported by the Director, Office of Energy Research, Nuclear Physics Division of the Office of High Energy and Nuclear Physics, of the U.S. Department of Energy under Contract No. DE-AC03-76DF00098.

Frontiers in Neutron Capture Therapy, edited by Hawthorne *et al.*
Kluwer Academic / Plenum Publishers, New York, 2001.

MeV, using the MCNP Monte Carlo code, and apply the results to predict the relative depth-dose characteristics of BNCT neutron sources. By scrutinizing these beams on a per-neutron basis, the benefits and disadvantages of each beam's spectrum can be determined and used to guide future neutron spectrum development. Additionally, by computationally unfolding the energy-dependent dose responses with an input spectrum, in-phantom figures-of-merit may be applied successfully to an in-air spectrum. This procedure produces, in essence, an in-air figure-of-merit that predicts in-phantom beam characteristics.

2. FIGURES-OF-MERIT

Many different figures of merit have been used to evaluate the characteristics of neutron beams for BNCT. No one parameter has yet emerged clearly as the most appropriate. These can be grouped into two catagories: in-air figures-of-merit and in-phantom figures-of-merit. The most popular in-air figures-of-merit in the BNCT community appear to be D_f/Φ_{epi} and D_γ/Φ_{epi}, which are the ratios of, respectively, the in-air fast dose and the in-air photon dose to the epithermal neutron flux. Although popular, these parameters do very little towards evaluating the clinical effectiveness of a particular beam and there are no clear parameters on what values are acceptable. It is generally assumed that the higher these values, the better the beam. However, these figures-of-merit only pit the worth of a loosely-defined range of neutrons, called "epithermal," against the effects of the unwanted fast and gamma doses at the surface of the skin. They do not necessarily relate to the actual treatment doses, which are usually limited by a maximum normal tissue dose at some depth in the brain, not at the skin surface.

Ideally, a figure-of-merit should reflect the in-phantom dose that can be delivered to a tumor volume, either real or hypothetical, within clinically-defined limits such as the maximum allowable tissue dose and/or a skin dose limit. A correct evaluation involves in-phantom analysis, which is often not used since it is computationally time consuming and the results of different studies may not be comparable unless each use exactly the same phantom geometry and boron-delivering drug parameters.

Table 1 shows how D_f/Φ_{epi} and D_γ/Φ_{epi} can be misleading predictors of in-phantom behavior. Two neutron sources are evaluated: the beam currently used in clinical trials at the Brookhaven Medical Research Reactor (BMRR),[2] and an accelerator-based source designed at LBNL. The LBNL design is an improvement on previous[3] designs, with modifications including a lead reflector, and a cylindrical moderator consisting of 28 cm of

Table 1. Figures-of-merit for two neutron beams[a]

	Figure-of-merit	BMRR	LBNL
In-Air	D_f/Φ_{epi} [Gy/(n/cm^2)]	4.8×10^{-13}	8.9×10^{-13}
	D_γ/Φ_{epi} [Gy/(n/cm^2)]	2.0×10^{-13}	3.2×10^{-13}
In-Phantom	RBE Equiv. tumor dose (8 cm) (Gy/Eq.)	14.5	23.4
	Advantage depth (cm)	8.4	9.9
	RBE Equiv. D_f at $D_{tis.}$(Max.) (Gy-Eq.)	2.05	0.632

[a]The LBNL beam has a higher fast dose in-air than the BMRR beam, but a lower in-phantom fast dose contribution to $D_{tis.}$(Max.), as well as superior in-phantom qualities. Equivalent doses are calculated assuming a maximum normal tissue tolerence of 12.5 Gy-eq. and using BMRR-defined clinical RBEs and compound factors.

Fluental (69% AlF_3, 30% Al, 1% LiF) followed by 6 cm of Teflon. The neutron source for this design is the ^{7}Li(p,n) reaction, utilizing 2.4 MeV protons. The in-phantom figures-of-merit demonstrate the superiority of the more penetrating LBNL beam, but the in-air figures-of-merit predict the opposite. Examination of the depth-dose response in Figure 1 shows why this is true. While the LBNL beam has a significantly higher in-air fast dose than the BMRR beam, the neutrons that contribute to this dose have much lower energies than the neutrons in the BMRR beam. Thus, these neutrons are absorbed more quickly and actually contribute a lower dose component to the treatment-limiting maximum tissue dose than in the BMRR beam. Also, the more penetrating LBNL beam provides a deeper point of maximum tissue dose, where the fast component is a smaller fraction of the total dose. Therefore, these in-air figures-of-merit are of limited use in determining the quality of a particular neutron beam.

Pursuit of a more useful in-air figure-of-merit remains an elusive, yet highly desirable goal. Since the determination of an in-phantom dose response often involves lengthy Monte Carlo simulations, much computer time can be saved if optimal beams could be predicted from the in-air neutron spectrum. This requires that an appropriate in-phantom figure-of-merit be identified to predict the clinical efficacy of in-air beams.

Our beam optimization goal has been to produce the most penetrating beam possible that delivers the highest tumor dose to a tumor at the center of the brain for fixed limits on maximum tissue dose and skin dose. The invasive, non-localized character of glioblastoma multiforme encourages the use of a whole-brain approach to treatment rather than limiting the treatment to variably defined tumor volumes. Because the center of the brain is the most difficult part to treat, virtually all cases benefit from

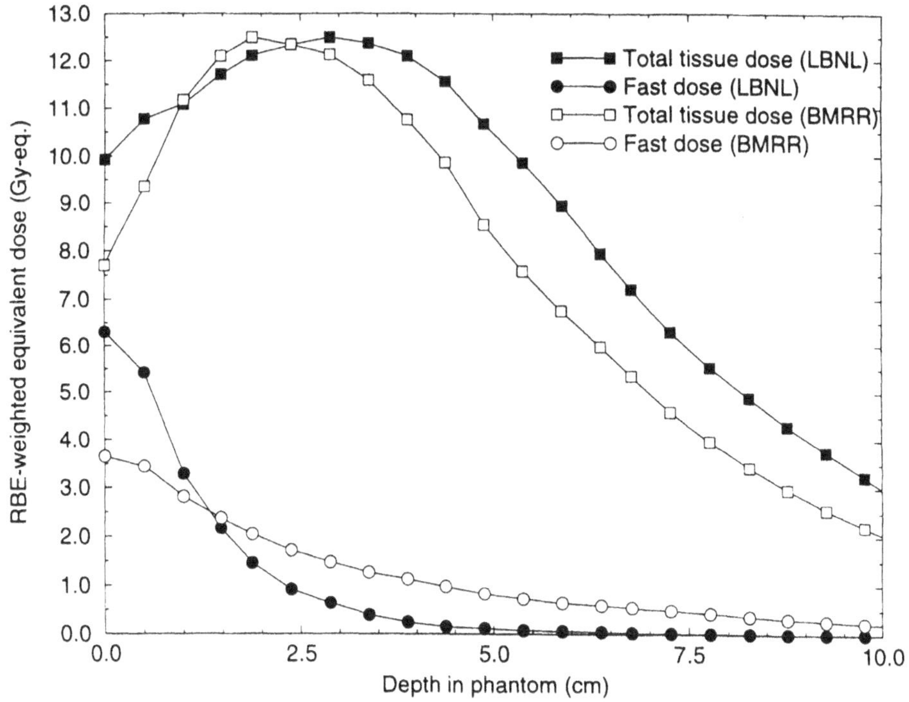

Figure 1. Total tissue dose response and fast dose component of BMRR and LBNL beams.

maximizing the dose at this point. However, it is understood that certain easier-to-treat cases of extremely shallow and localized tumors may not require a maximally penetrating beam.

With these aims in mind, the most appropriate in-phantom figure-of-merit was chosen to be TG_{bm}, the Therapeutic Gain at the brain midpoint, defined as the ratio of the RBE-weighted tumor dose at the brain midpoint to the maximum RBE-weighted tissue dose:

$$TG_{bm} = \frac{D_{tumor}(Midpoint)}{D_{tis.}(Max.)} \quad (1)$$

$$D_{tumor/tis.} = CF_{tumor/tis.} \cdot D_B + RBE_N \cdot D_N + RBE_H \cdot D_H + RBE_\gamma \cdot D_\gamma \quad (2)$$

where D_B, D_N, D_H, and D_γ refer respectively to the physical doses due to the $^{10}B(n,\alpha)$ boron, $^{14}N(n,p)$ nitrogen, $^{1}H(n,n')$ proton-recoil, and the γ dose via the $^{1}H(n,\gamma)$ capture reaction. The physical doses were multiplied by a corresponding compound factor (CF) or relative biological effectiveness (RBE). Boron concentrations and compound factors were used as established[4] for the boron compound p-boronphenylalanine (BPA) along with RBEs as established in the dose calculation protocol for clinical trials at the BMRR: normal tissue ^{10}B concentration: 13 ppm; normal tissue compound factor: 1.3; tumor ^{10}B concentration: 45.5 ppm; tumor compound factor: 3.8; RBE_N: 3.2; RBE_H: 3.2; and RBE_γ: 1.0. Because TG_{bm} is a ratio, it is independent of an actual defined limit on $D_{tis.}(Max.)$.

With TG_{bm} established as an appropriate in-phantom figure-of-merit for gauging beam quality, an in-air parameter can be sought to predict this quantity as a function of the neutron spectrum.

3. NEUTRON ENERGY DOSE RESPONSE

3.1. Modeling

The first step in predicting the dose response of any neutron energy spectrum was modeling a large number (~100) of nearly monoenergetic neutron energy distributions through a head phantom using the MCNP[5] Monte Carlo code. Each simulation consisted of a monodirectional neutron beam with a narrow, uniform energy distribution, incident onto the full surface of a head phantom. A modified Snyder[6] head phantom was used, consisting of skin, bone, and brain in the geometry with surfaces described by the following equations:

$$\left(\frac{X}{6.5}\right)^2 + \left(\frac{Y}{6}\right)^2 + \left(\frac{Z}{9}\right)^2 = 1 \quad (3)$$

$$\left(\frac{X+1}{8.3}\right)^2 + \left(\frac{Y}{6.8}\right)^2 + \left(\frac{Z}{9.8}\right)^2 = 1 \quad (4)$$

$$\left(\frac{X+1}{8.8}\right)^2 + \left(\frac{Y}{7.3}\right)^2 + \left(\frac{Z}{10.3}\right)^2 = 1 \quad (5)$$

The modification to the Snyder phantom was the addition of a half centimeter layer of skin as used by Liu *et al.*[7] Elemental compositions for brain, skin, and skull were taken from ICRU46.[8] The midpoint of this phantom is at 7.3 cm. Narrow energy bins were chosen instead of monoenergetic beams in order to assure that all energies between thermal and 14 MeV would be represented and that no resonances in material transport would be neglected or artificially dominate the response. The width of each bin was chosen to be equilogarithmic, with ten bins per energy decade.

For each simulation, the dose response for each reaction in Equation 2 was determined along the centerline of the beam through the brain. In addition, the dose responses for the oxygen and carbon neutron reactions were determined and added to Equation 1. At very high neutron energies (>~1 MeV), the dose due to reactions with oxygen can approach nearly 10% that of the otherwise dominant proton recoil dose, due to resonances in oxygen's neutron cross section. While carbon's dose contribution is negligible, its dose response was added for completeness. For both these doses, an RBE of 3.2 was assumed.

In this fashion, a large database was constructed from which dose as a function of energy, position, element, and RBE can be determined for a wide parallel neutron beam. Using this database, the dose response of any input monodirectional neutron spectrum can be instantly calculated without the need for a separate Monte Carlo simulation. These input neutron spectra can be either realistically generated spectra or hypothetical beams.

Some effort was made to develop a simple algebraic representation of the dose response as a function of the above parameters through curve fitting. However, it was not possible to develop a simple equation without introducing large errors compared to the accuracy of the original dose response database. For this reason, this option is not used and all in-air neutron spectra are directly folded into the energy-dependent dose response database. This ensures the highest degree of accuracy.

3.2. Moderator Optimization

This new method of evaluating an in-air neutron beam spectrum was initially tested by comparing the TG_{bm} prediction against that calculated with the INEEL Monte Carlo treatment planning code, BNCT_RTPE.[9] There are many differences in these two models, including elemental composition, phantom geometry, beam size, and angular distribution. Therefore, the values of TG_{bm} calculated with each method were expected to be quite different. However, each method should produce the same results in judging different neutron spectra against each other. Figure 2 shows that for two different moderators (Al/AlF_3 and D_2O), the same thickness of moderator is predicted as optimal, independent of the model chosen. It should be noted that because the database includes data only from monodirectional neutrons, which are inherently more penetrating, it cannot be used to produce accurate absolute doses and can only be used to gauge different neutron spectra of similar J/Φ against each other. In particular, if there are additional constraints on treatment other than a maximum tissue dose (e.g., surface dose or treatment time), great care must be taken to not misinterpret the results given by the database. For instance, previous LBNL optimization constraints have placed a limit of 10.0 Gy-eq on the surface dose, in addition to a maximum tissue dose constraint of 12.5 Gy-eq. In beams where BNCT_RTPE calculates a surface dose of 10.0 Gy-eq, the database response is only 8.5 Gy-eq. This seems to be a consistent underrepresentation for beams with a typical J/Φ of ~0.8. This should be considered when a skin dose constraint is applied. To be fully useful in predicting dose responses, however, the database should be expanded to allow for angular dependence.

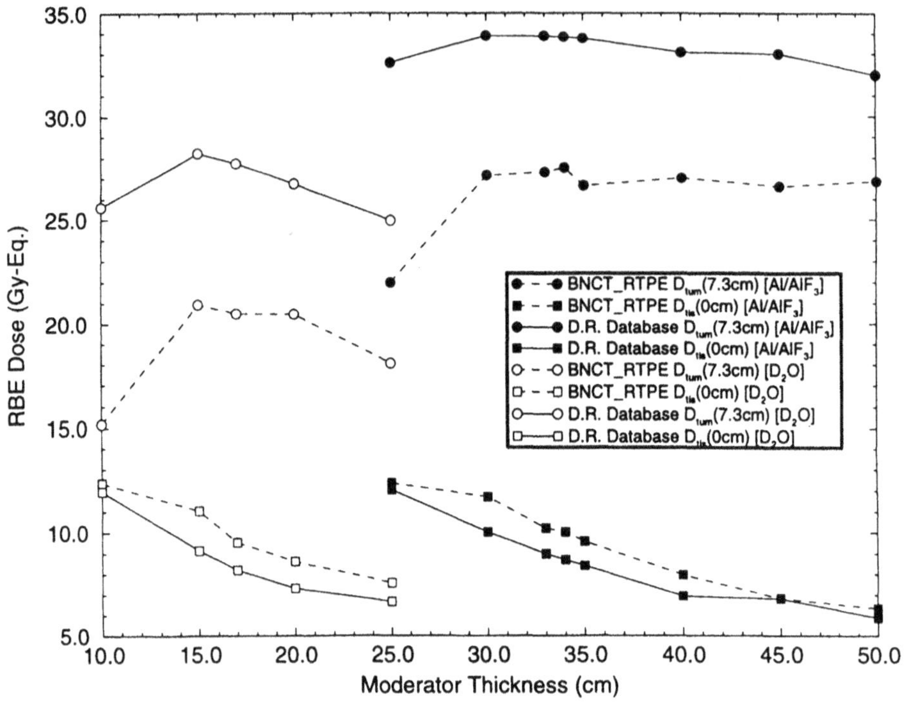

Figure 2. Comparison of using BNCT_RTPE or the Dose Response (D.R.) Database to optimize the moderator thickness of either D_2O or Al/AlF_3. The maximum normal tissue tolerance was limited to 12.5 Gy-Eq.

3.3. The Ideal Neutron Spectrum

Using a Monte Carlo code such as BNCT_RTPE or MCNP can take many hours or days of computer time, whereas evaluating an in-air neutron spectrum with the predetermined energy-dependant dose response database is instantaneous. Because the database allows instantaneous evaluation of a neutron energy spectrum, it becomes a powerful tool in determining the effect of changing existing neutron beams, or in quickly evaluating hypothetical neutron spectra.

This ability to instantly evaluate any input spectrum allows investigation into determining an "ideal" neutron spectrum for BNCT, which would produce the maximum value of TG_{bm}. Though such an "ideal" neutron spectrum may not be physically realizable, knowing its characteristics allows one to tailor real neutron beams to more closely match it. This can provide insight as to what gains may be made from further tailoring/filtering of existing beams and whether those gains are large enough to warrant corresponding losses in total neutron flux.

At first, it appears the relative worth of a given neutron energy can be determined by simply plotting TG_{bm} as a function of neutron energy, as in Figure 3. However, this is not necessarily a valid method of gauging the contribution of a given neutron energy to TG_{bm} in a real beam, which would contain a full spectrum of energies. TG_{bm} is dependent on the position of the point of maximum tissue dose, which can differ from beam to beam. In particular, for many monoenergetic beams, the point of maximum tissue dose may be at the skin surface. The contribution from this energy to the dose at the maximum

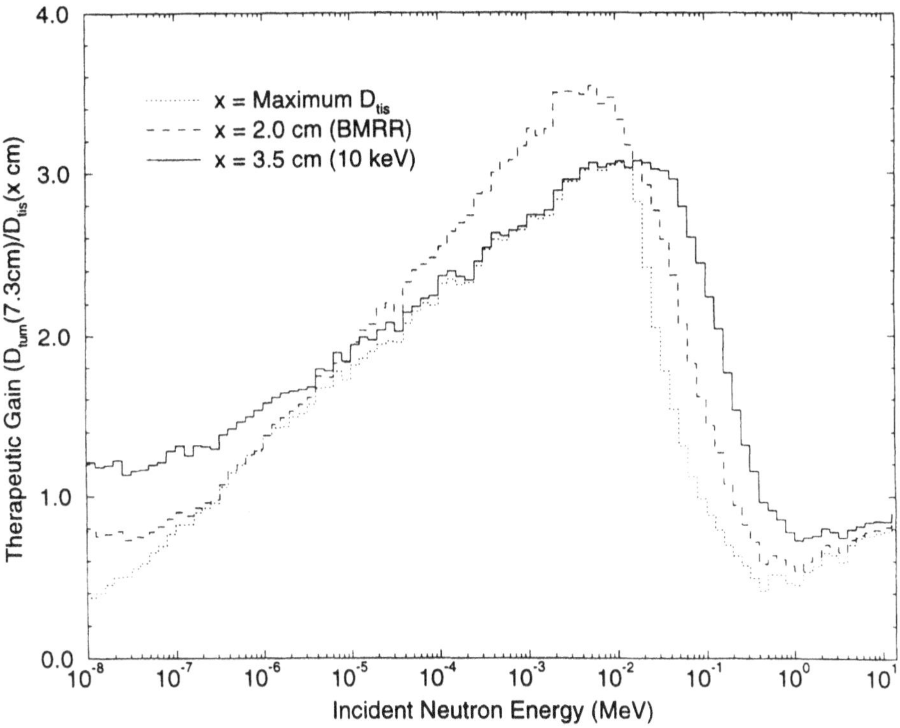

Figure 3. Therapeutic Gain at the brain midpoint as a function of neutron energy, representative of dose responses with peaks at 2.0 cm, 3.5 cm, or where produced by the nearly monoenergetic beam.

tissue dose in a real beam, which may be at up to 3.5 cm depth, will be vastly different. This becomes problematic in that neutrons of a particular energy will be of different worth to different beams. For instance, 40 keV neutrons might lower the TG_{bm} of a beam with a tissue dose maximum at 2.0 cm, but the same energy neutrons may raise TG_{bm} in a beam with a dose maximum at 3.5 MeV.

Therefore, TG_{bm}(2.0 cm) and TG_{bm}(3.5 cm) have also been plotted in Figure 3. These values are defined as in Equation 1, with the exception that D_{tis}(Max.) is assumed to be at 2.0 cm and 3.5 cm, respectively. These values were chosen as they represent two extremes in maximum dose position. The BMRR tissue dose peaks at 2.0 cm, while the maximum depth at which monodirectional beams peak are at 3.5 cm. If the LBNL spectrum were monodirectional, it would peak close to 3.5 cm. It's actual peak is at 2.9 cm, the TG_{bm} curve for which looks similar to that of a 3.5 cm peak. By plotting TG_{bm} this way, the dose contribution to different beams with different dose maxima for each neutron energy can be seen.

An obvious feature of Figure 3 is the maximum in TG_{bm}, which occurs at around 10 keV. In the most penetrating neutron beams (Max. = 3.5 cm), this maximum extends from about 4 keV to about 40 keV. However, it never rises any higher. It was thought that certain higher energy neutrons (~40 keV–100 keV) might actually provide higher penetrability to a neutron beam and therefore be advantageous in small quantities, so long as there are not too many to adversely contribute to the point of maximum tissue dose or change its position. However, the fact that the 3.5 cm line never rises above its maximum at around 10 keV suggests that TG_{bm} can never be increased above this value of about 3.

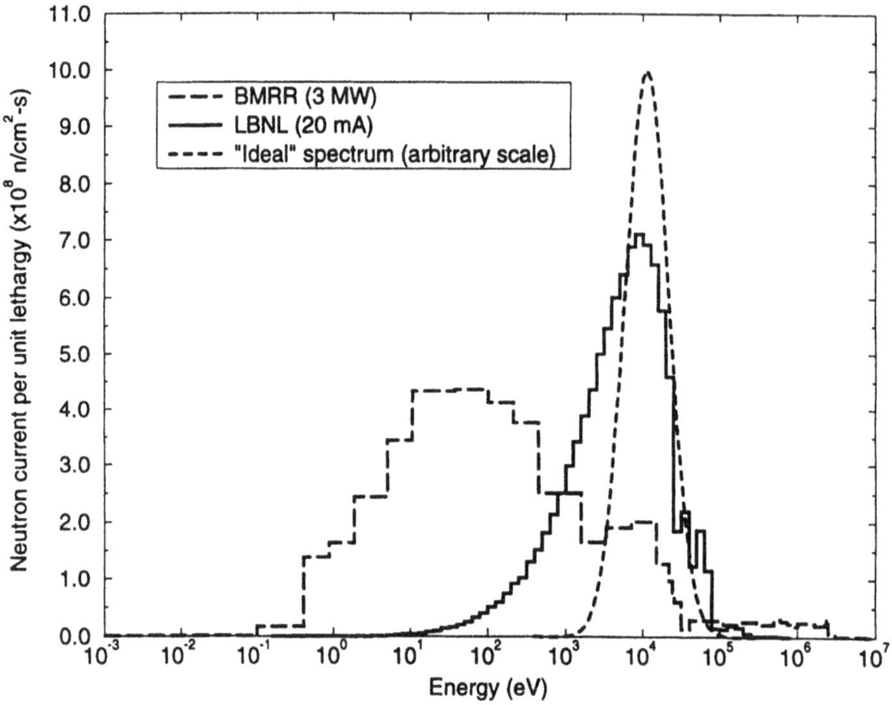

Figure 4. Neutron energy spectra for the BMRR beam, an LBNL accelerator-based design, and an ideal spectrum which produces the highest TG_{bm} possible.

A 10 keV neutron beam yields the highest TG_{bm} obtainable, though some width in energy can be tolerated with no loss in quality.

Also notable is the lower energy maximum in the 2.0 cm line, which rises above the maximum in the 3.5 cm line. Because these energies individually have TG_{bm} maxima at 3.5 cm, their contribution to the dose at 2.0 cm is lower. It is nearly impossible, however, to conceive of a neutron spectrum which is composed primarily of neutrons at this energy, yet which has a maximum at 2.0 cm and this artifact cannot be exploited to produce a higher TG_{bm} than 3.

The ideal neutron spectrum for BNCT is therefore shown, as in Figure 4, to be a spread of energies between ~4 keV and ~40 keV, so long as the number of neutrons above ~20 keV is kept low enough to not adversely affect the position of D_{tis}(Max.). Also shown are the BMRR beam and the LBNL accelerator-produced Fluental/Teflon-moderated beam. The LBNL beam is quite close to the ideal spectrum, demonstrating that there are few advantages to be gained by further alteration of the energy spectrum. The dose response database predicts that the BMRR, LBNL, and ideal spectra yield, respectively, a TG_{bm} of 2.0, 2.9 and 3.0. Again, these values assume that each spectrum is made purely monodirectional and only reflect the relative differences in energy spectrum.

4. CONCLUSIONS

This study has demonstrated the strong importance of the shape of the neutron energy spectrum for usage in BNCT, which we have taken advantage of in shaping a nearly ideal spectrum. Our accelerator-based spectrum produces a monodirectional TG_{bm} that is within 5% of the maximum that can be realized, leading us to conclude that future beam shaping should focus on improving directionality and beam intensity. The dose response database has proven a valuable tool in instantaneous evaluation of an in-air spectrum to predict in-phantom parameters. However, to be fully applicable, it must be expanded to account for the angular distribution of a neutron beam. It would also benefit from expansion to allow for variable beam sizes and phantom geometries, after which it can be distributed for public use in optimizing neutron beam designs.

REFERENCES

1. Yanch, J.C., Zhou, X-l., and Brownell, G.L., "A Monte Carlo Investigation of the Dosimetric Properties of Monoenergetic Neutron Beams for Neutron Capture Therapy," Radiation Research 126, 1–20 (1991).
2. Coderre, J.A., Bergland, R., Capala, J., Chadha, M., Chanana, A.J., Elowitz, E., Joel, D.D., Liu, H.B., and Slatkin, D., "Boron Neutron Capture Therapy for Glioblastoma Multiforme using p-boronophenylalanine and Epithermal Neutrons–Trials Design and Early Clinical Results," J. Neuro. Oncol. 33, 93–104 (1997).
3. Bleuel, D.L., Donahue, R.J., Ludewigt, B.A., and Vujic, J., "Designing Accelerator-based Epithermal Neutron Beams for Boron Neutron Capture Therapy," Medical Physics 25(9), 1725–1734 (1998).
4. Chanana, A.D., "Boron Neutron-Capture Therapy of Glioblastoma Multiforme at the Brookhaven Medical Research Reactor, A Phase I/II Study, FDA IND 43,317, Protocol 4," Med. Dep. Brookhaven Natl. Lab., Upton, NY 11973-5000, January, 1996.
5. Briesmeister, J.F., MCNP—A General Monte Carlo N-Particle Transport Code, Version 4B, Los Alamos National Lab, 1997.
6. Snyder, W.S., Fisher, Jr., H.L., Ford, M.R., and Warner, G.G., "Estimates for Absorbed Fractions for Monoenergetic Photon Sources Uniformly Distributed in Various Organs of a Heterogeneous Phantom (Appendix B)," J. Nucl. Medicine, MIRD Supplement No. 3, Pamphlet 5 (1969).
7. Liu, H.B., Greenberg, D.D., and Capala, J., "An Improved Neutron Collimator for Brain Tumor Irradiations in Clinical Boron Neutron Capture Therapy," Medical Physics 23(12), 2051–2061 (1996).
8. "Photon, Electron, Proton and Neutron Interaction Data for Body Tissues," ICRU Report 46, February 28 (1992).
9. Nigg, D.W., Wheeler, F.J., Wessol, D.E., Capala, J., and Chada, M., "Computational Dosimetry adn Treatment Planning for Boron Neutron Capture Therapy," J. Neuro-Oncol., 33, 93–104 (1997).

REFERENCE PHANTOM DESIGN FOR THE BIRMINGHAM BNCT PROJECT

C. Wojnecki[1]* and S. Green[2]

[1]School of Physics and Astronomy
University of Birmingham
Edgbaston, Birmingham B15 2TT, UK
[2]Department of Medical Physics
Queen Elizabeth Hospital
Edgbaston, Birmingham B15 2TH, UK

INTRODUCTION

In-phantom parameters have been shown to be important in the design of epithermal neutron beams for BNCT and rectangular phantoms have proved to be useful for this purpose. However it is imperative to use suitable head phantoms for accurate treatment planning. In order to benchmark treatment planning systems, several requirements have to be fulfilled:

The head phantom should model as closely as possible the human head with respect to size, shape and position.

The tissue substitutes used must be neutronically similar to the tissues that constitute the human head. Consequently, the mass density and radiation transport properties should be as close as possible as those of the skull and the brain. The most important parameter for neutron transport is the hydrogen atom density. Hydrogen attenuates and thermalises the incident epithermal neutron flux by elastic scattering and in addition, the thermal neutron capture reaction $^{1}H(n,\gamma)\,^{2}H$ produces a significant photon dose.

The design of the head phantom should permit the measurement of the dose components at many locations.

* E-mail address: cecile@aps7.ph.bham.ac.uk.

Frontiers in Neutron Capture Therapy, edited by Hawthorne *et al.*
Kluwer Academic / Plenum Publishers, New York, 2001.

1. DESIGN OF A HEAD PHANTOM

The design of our experimental head phantom is based on the head model proposed by Snyder *et al.*[1] In this model, the brain and the skull are modelled as homogeneous regions inside non-concentric ellipsoids. It approximates the size and shape of an average brain without including anatomical details.

In Snyder head model, the brain is the volume enclosed by the ellipsoid:

$$(x/6)^2 + (y/9)^2 + (z/6.5)^2 - 1 \leq 0$$

The skull is the volume between the above surface and the non-concentric ellipsoid:

$$(x/6.8)^2 + (y/9.8)^2 + [(z+1)/8.3]^2 \leq 1$$

All dimensions are in units of centimetres. The x-axis is defined as the lateral axis, the y-axis is the anterior-posterior axis and the z-axis is the superior-inferior axis.

As mentioned earlier, the main objective for designing a head phantom is to match hydrogen densities in order to simulate the neutron slowing down properties of the human skull and brain. Table 1 summarises the compositions and densities of the actual body tissues and of some selected tissue substitutes.[2]

In designing the experimental head phantom, we aim at reproducing the mathematical model as closely as possible. Therefore, an ellipsoidal head phantom was designed with outer dimensions equal to those given by the reference Snyder head model.

Initially, two different designs were drawn to compare the radiation transport with the Snyder head model to that obtained with the experimental phantom. The first design that is similar to the head phantom developed by the MIT/Harvard group[3] is an ellipsoidal acrylic shell (2 mm thick) attached to an acrylic base equipped with portholes to permit the insertion of detectors into the volume of water. Since lateral irradiations of the human head are planned for the treatment of Glioblastoma Multiforme, the ellipsoidal shell is truncated on a plane parallel to the mid-brain line at a distance of 3.5 cm beyond the centre line of the shell. A side view of this design is shown on Figure 1a. This is referred to as the "base design".

In the second design, the same ellipsoidal acrylic shell is attached to a water tank (30 × 30 × 50 cm^3) whose size would permit the remote-controlled movements of detectors in order to draw a continuous three-dimensional dose map. Figure 1b shows a side view of this geometry. In later discussion, this is referred to as the "tank design".

Table 1. Densities and elemental compositions of the considered materials

	Brain	Water	Skull	Acrylic plastic
Density (g·cm^{-3})	1.04	1.00	1.61	1.19
H density (g·cm^{-3})	0.1107	0.111	0.075	0.096
H (% by mass)	10.7	11.2	5	8
C (% by mass)	14.5	—	21.2	60
N (% by mass)	2.2	—	4	—
O (% by mass)	71.2	88.8	43.5	32

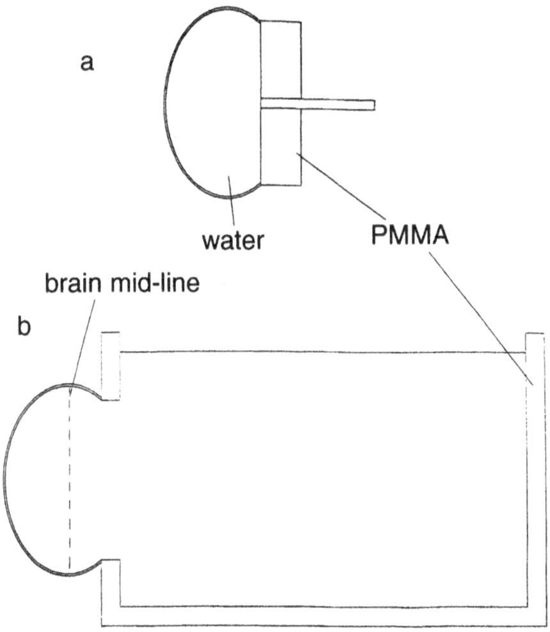

Figure 1a and b. Side views of the base design and the tank design.

The Monte Carlo code MCNP4a developed by Los Alamos National Laboratory was used to model the transport of neutrons and neutron-generated photons within the various geometries from the epithermal neutron source available at the University of Birmingham.[4]

2. RESULTS

2.1. Thermal Neutrons

The results of the MCNP calculations obtained for the thermal neutron fluence rates are illustrated in Figure 2.

The graph shows a significant difference between the thermal flux generated in the experimental phantoms and that produced in the Snyder model. This is mainly due to the thickness and hydrogen density of the materials used: along the lateral axis, the thickness of the skull in the mathematical model is 0.8 cm and our experimental phantoms have a shell of 0.2 cm thick. Thus, in our phantoms the 0.8 cm of bone with a hydrogen density of 0.075 $g \cdot cm^{-3}$, are replaced with 0.2 cm of acrylic plastic and 0.6 cm of water with hydrogen densities of 0.096 and 0.111 $g \cdot cm^{-3}$, respectively.

2.2. Gamma-ray Photons

The thermal neutron capture reaction $^{1}H(n,\gamma)^{2}H$ produces a significant photon fluence which gives non-localised dose. The photon doses calculated along the central beam axis at different depths are given in Figure 3.

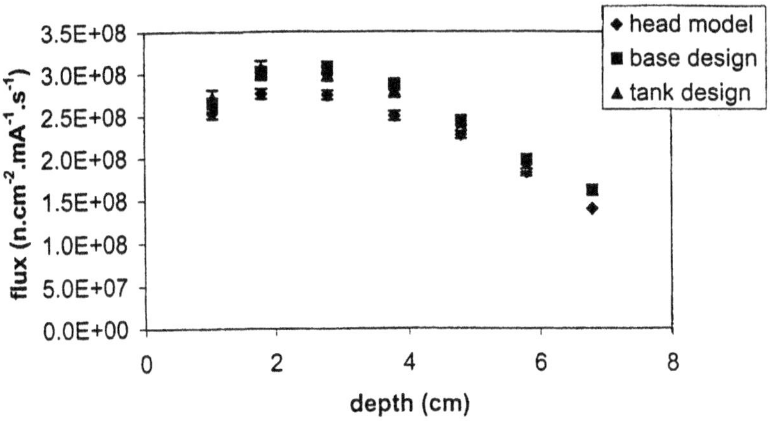

Figure 2. Comparison of the thermal neutron fluence rates along the beam axis in various geometries.

It can be seen that a large disparity between the results obtained in the two experimental phantoms exist. This significant difference can be explained by the presence of the large volume of water attached to the ellipsoidal head phantom in the tank design. A large amount of gamma rays are returned from the water tank and contribute to the flux inside the ellipsoidal shell.

3. CONSTRUCTION OF THE HEAD PHANTOM

The base design has been preferred and an optimised ellipsoidal head phantom has been designed under technical considerations.

The construction of the head phantom consists of several stages. A mould is shaped from wood to the outer dimensions of the Snyder's head model. This mould is then filled with plaster of Paris to make the model. A semi-ellipsoidal shell made of acrylic plastic is vacuum-formed using a technique similar to that used to form beam direction shells

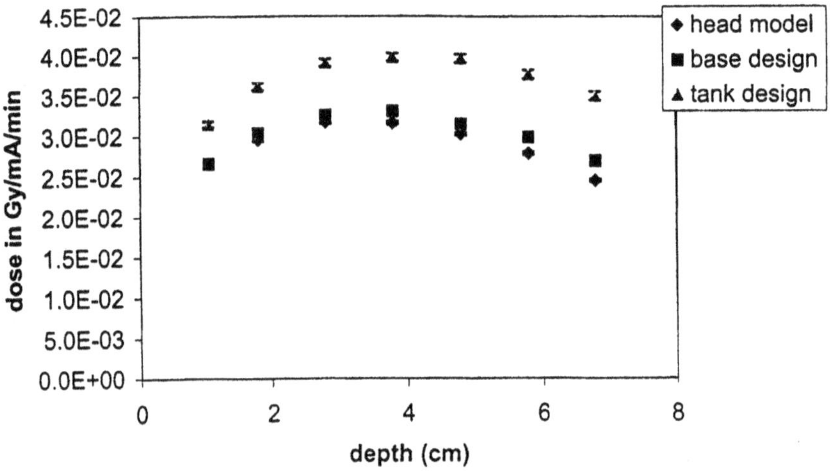

Figure 3. Comparison of photon doses along the beam axis in various geometries.

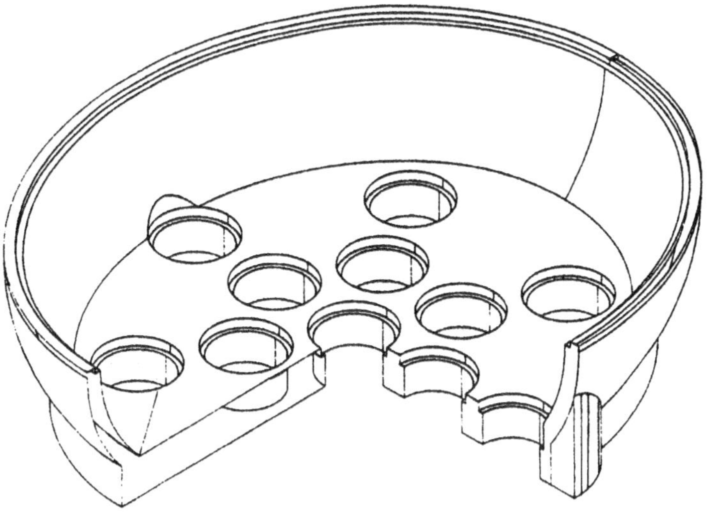

Figure 4. Technical sketch showing the machining from a single PMMA block.

in Radiotherapy Departments. In order to avoid water leaks between the shell and the acrylic base, the truncated semi-ellipsoidal part and the first few centimetres of the base are machined from a single block of acrylic plastic, as shown by Figure 4. To allow the use of detectors, thirteen holes are drilled out of the base and fitted with guide tubes made out acrylic plastic rods.[3] Hollow watertight tubes are machined to be inserted from the base of the phantom.

CONCLUSION

The thermal neutron fluxes produced in our experimental phantoms have shown similar slowing down properties between the two designs. However, the photon doses obtained in the phantoms imply that the base design models more closely the Snyder head model. The difference between the properties of the experimental phantom chosen and the human skull and the brain can be taken into account when the experimental isodose contours are used to calibrate the treatment planning software. Measurements made in this phantom will validate the calculations performed by three treatment planning systems: the MacNCTPLAN developed by the Harvard/MIT group, the AUTOVOX program developed by the Sydney group, and the SERA code developed by the INEEL group.

ACKNOWLEDGMENTS

Great thanks are due to Phil Cooper from the Radiotherapy Mould Room, Steve Brookes and his team from the School of Physics Mechanical workshop for their advice and effort towards the optimisation of the ellipsoidal head phantom.

REFERENCES

1. W.S. Snyder, H.L. Fisher, M.R. Ford, and G.G. Warner, "Estimates of absorbed fractions for mono-energetic photon sources uniformly distributed in various organs of a heterogeneous phantom." Journal of Nuclear Medicine, suppl. 3, pamphlet 5, 1969.
2. ICRU reports 44 and 46.
3. O.K. Harling, K.A. Roberts, D.J. Moulin, and R.D. Rogus, "Head phantoms for Neutron Capture Therapy." Medical Physics vol 22(5):579–583, 1995.
4. D.A. Allen, and T.D. Beynon, "A design study for an accelerator-based epithermal neutron beam for BNCT." Phys. Med. Biol. vol 40:807–821, 1995.

ABSORBED DOSE ESTIMATES WITH A HIGH RESOLUTION VOXEL-BASED HEAD PHANTOM

Jeffrey F. Evans[1], Thomas E. Blue[1], and Nilendu Gupta[2]

[1]Nuclear Engineering Program
Department of Mechanical Engineering and
[2]Division of Radiation Oncology
Arthur James Cancer Hospital & Research Institute
The Ohio State University
Columbus, OH 43210

1. INTRODUCTION

This work describes a high resolution voxel-based computer model of a human head and its use for calculating absorbed doses to the structures of the brain. Particularly, we are interested in knowing if such structure doses can be used for optimizing the neutron field of the accelerator-based neutron source[1] that is being designed at The Ohio State University (OSU). We are also investigating the uses of structure doses for evaluating BNCT treatment plans.

The motivation for this work arose from the results of the healthy dog irradiation studies conducted at Brookhaven National Laboratory (BNL). Gavin *et al.*[2] demonstrated evidence of volume effects and differences in sensitivity among the structures of the dog brain irradiated with the neutron beam from the Brookhaven Medical Research Reactor. This experimental evidence for differing structure sensitivities from the BNL studies corresponds well with that of other experimental models, where photon or proton beams were used to irradiate the brains of large animals.[3] Additionally, the wealth of clinical data from the past 30 years not only provides strong evidence of differing sensitivities to radiation among the structures of the human brain, but also indicates that these differences in sensitivity need to be taken into consideration when planning a radiation treatment.[4,5,6,7]

While definitive clinical tolerance data for brain structures are scarce amongst the literature,[8,9] there are increasing signs that medical researchers are rectifying this situation.[5,10,11] It is our thought that, as data from clinical trials are gathered, knowledge of the absorbed doses to brain structures may contribute to the optimization of neutron fields and treatment planning for BNCT. What follows is a description of the voxel-

Frontiers in Neutron Capture Therapy, edited by Hawthorne *et al.*
Kluwer Academic / Plenum Publishers, New York, 2001.

based Zubal head phantom[12] and the MCNP[13] model that we created to calculate the absorbed doses, to the structures of the brain, from a unilateral irradiation with the OSU accelerator-based neutron source. The initial results of this calculation are also presented.

2. MODELS

2.1. The Zubal Head Phantom

The Zubal head phantom is a voxel-based, i.e., digital, 3D model of a human head that can be displayed and manipulated on a computer. It was created by a research group, led by Dr. George Zubal, at the Imaging Science Research Laboratories, Yale University School of Medicine.[12] A set of 124 MR transverse images was made of the head of a healthy male volunteer. Next, members of the medical staff at Yale, using a high resolution bitpad and display, manually identified and outlined sixty-one anatomical structures and tissues on the images. Finally, the full set of images, along with the anatomical contour information, was processed into a 256 × 256 × 128 one-byte array (an 8 MB binary file).

The head phantom can be accessed as a one-dimensional array of integer numbers ranging from 0 to 125. Each number in the array corresponds to a voxel, i.e., volume element, within the 3D lattice of the phantom. The value of a number in the array identifies to which anatomical structure the voxel "belongs," and its position in the 1D array determines the location of the voxel within the lattice. The dimensions of a voxel are $1.1 \times 1.1 \times 1.4\,mm^3$. The region of the head that is represented by the Zubal phantom spans from ear to ear, from tip of the nose to the back of the head, and from the top of the head to the base of the skull at the second cervical vertebrae. Many structures of the brain are identified: prefrontal, frontal, parietal, occipital, and temporal lobes, motor cortex, white matter, corpus callosum, hippocampus, uncus, caudate nucleus, insula cortex, putamen, internal capsule, septum pellucidium, thalamus, amygdala, globus pallidus, cerebellum, pons, medulla oblongata, optic nerve, and pituitary gland.

2.2. Processing the Phantom for MCNP

The Zubal head phantom can be viewed and easily processed using NIH Image.[14] There was a significant amount of void space in the 8.4 million voxel lattice. This empty space was trimmed from the model, reducing the lattice to a set of 171 × 217 × 120 voxels (a 4.2 MB binary file). While MCNP is able to process and run a lattice problem of this size, the RAM requirements are exceptionally large (160 MB).

To ease the RAM requirements, the number of structures was reduced by combining the physiologically unimportant tissues into singly identified structures, e.g., the lateral ventricles, 3rd and 4th ventricles, cerebral fluid, and cerebrospinal fluid were identified separately in the 4.2 MB phantom; they were relabeled as cerebral fluid. As another example, there were regions of blood pool identified throughout the 4.2 MB phantom. These regions were relabeled to match their surrounding tissue. All in all, the original sixty-one structures were reduced to thirty-nine. Additionally, the lattice was folded into a set of 85 × 109 × 120 voxels (a 1.1 MB binary file), with voxel volumes of $2.2 \times 2.2 \times 1.4\,mm^3$. This was done with an in-house processing code. Once the folded lattice model was translated to MCNP format, only 48 MB of RAM were necessary to run a calculation.

2.3. A Unilateral Irradiation Model for MCNP

Elemental compositions for twelve tissue types were used for the MCNP model: adipose tissue, eye lens, muscle, skeleton-cranium, skeleton-spongiosa, skeleton-cartilage, and skin were taken from ICRU Report 46,[15] while brain-gray matter, brain-white matter, cerebrospinal fluid, eyes, and spinal cord were taken from Duck.[16] The sinuses were assumed to be filled with air. ENDF/B-VI neutron cross sections were used for all elements. The ENDF S(α,β) thermal neutron scattering treatment for H in light water was used for brain (gray and white matter), cartilage, cerebrospinal fluid, lens, eyes, muscle, skin, and spinal cord tissues, while the S(α,β) treatment for H in polyethylene was used for adipose tissue.

MCNP was run in "n p" mode, i.e., source neutrons with photon generation. The neutron source was modeled to approximate the physical properties of the neutrons emanating from the treatment port of the moderator assembly for the accelerator-based neutron source that is being designed at OSU:[1,17] a circular surface (d = 25 cm) with the neutrons emitted monodirectionally from, normal to, and uniformly over the surface, and with a 1/E neutron energy spectrum with zero photon contamination. The probability density function that describes the 1/E energy spectrum is defined as

$$f(E)=\begin{cases}\dfrac{1}{E\cdot\ln(E_{max}/E_{min})} & E_{min}\le E\le E_{max}\\ 0 & \text{otherwise}\end{cases} \qquad (1)$$

where E is neutron energy and E_{min} and E_{max} equal 4×10^{-7} MeV and 0.33 MeV, respectively.

The phantom was situated such that the right side of the head was located 5 cm from the source surface with the beam centerline passing through the geometrical center of the brain. The source neutrons were incident normally upon the right side of the head. To account for scattering effects, the Cristy/Eckerman adult male mathematical phantom[18] was "attached" to the Zubal head phantom so as to represent an entire human body lying supine during the irradiation. The space surrounding the phantom was filled with air.

The volume-averaged neutron kerma, photon absorbed dose, and boron kerma were calculated for all of the structures of the brain using MCNP's tally type 4 (fluence) multiplied by the appropriate response function. Neutron kerma factors and photon fluence-to-absorbed dose factors were taken from ICRU Report 46;[15] the boron reaction kerma factors were supplied by Woollard.[19] It was assumed that the calculated neutron and boron kermas were equal to the corresponding absorbed doses. Of important note is that absorbed doses were calculated separately for the left and right sections of structures that exist in both hemispheres of the brain, e.g., the absorbed dose to the left and right sections of the parietal lobes were calculated separately, while the absorbed dose to the corpus callosum was calculated for this structure as a whole.

3. RESULTS AND DISCUSSION

A viable executable of MCNP was compiled, tested, and verified on a Power Macintosh 7500/323 MHz 750 with 161 MHz backside cache. 20 million source histories required approximately 2 days of processing time. All estimated structure absorbed doses were shown to be statistically reliable: relative errors were <3% (<10% is deemed to be statistically reliable by LANL XTM[13]).

Table 1. RBE-dose to selected structures for a single fraction, right lateral irradiation

Structure	RBE-dose (RBE-cGy)[a]
Left frontal lobe	300
Right frontal lobe	838
Left motor cortex	167
Right motor cortex	1,180
Left parietal lobe	244
Right parietal lobe	874
Left occipital lobe	312
Right occipital lobe	712
Left temporal lobe	233
Right temporal lobe	1,110
Left insula cortex	255
Right insula cortex	1,080
Left hippocampus	318
Right hippocampus	884
Pons	544
Pituitary gland	577
Left optic nerve	391
Right optic nerve	677
Left lens	248
Right lens	986
Left lacrimal gland	120
Right lacrimal gland	1,280
Left parotid gland	160
Right parotid gland	1,100
Volume-weighted RBE-dose to the whole brain	586

[a]Normalized to a source of 7.1×10^{14} emitted neutrons.

Table 1 lists the estimated RBE-dose to selected structures, plus the volume-weighted RBE-dose to the whole brain. From the results reported by Gabel *et al.*,[20] the neutron RBE was assumed to be 3.9, the compound factor of the $^{10}B(n,\alpha)^{7}Li$ reaction for BSH in brain tissue was assumed to be 0.37, and the average ^{10}B concentration in blood was assumed to be 30 ppm. Additionally, the photon RBE was assumed to be 1. The results in Table 1 were normalized to a source of 7.1×10^{14} emitted neutrons. For perspective, 7.1×10^{14} emitted neutrons are equivalent to a neutron flux of $1 \times 10^{9}\,cm^{-2}s^{-1}$ delivered in 24 min.

Evident in Table 1 is the asymmetry in dose to the left and right sections of the structures. With a single fraction, right lateral irradiation, and assuming all other parameters are equal, the dose to the right section of any given structure may exceed the dose to its left counterpart by a factor of 1.3–6, and in the case of the lacrimal and parotid glands, by a factor as large as 10. The estimated RBE-doses to the right motor cortex, right temporal lobe, and right insula cortex are 11.8, 11.1, and 10.8 RBE-Gy, respectively, while the volume-weighted RBE-dose to the brain is 5.9 RBE-Gy. To put these values in perspective, Gavin *et al.*[21] assumed that healthy dogs, irradiated at BNL and Petten, and presenting MR changes without neurologic symptoms, received an average RBE-dose of at least 12.8 RBE-Gy, presumably delivered to only one hemisphere. From the same studies,[21] it was assumed that dogs, displaying clinical signs indicative of lethal brain necrosis, received at least 14.9 RBE-Gy to a critical volume of the brain. Returning to the results of this work and assuming equal sensitivities to those of the dogs in the review by Gavin *et al.*,[21] and neglecting volume effects, the implications are that, for the irradiation scenario described in this work, the normal tissue tolerance will first be reached in structures before the tolerance for whole brain is reached. Thus, there is reason to speculate that treatment time may be limited by the doses to certain structures of the brain.

These speculations, of course, only apply to a unilateral irradiation with the OSU accelerator-based neutron source. We are planning to calculate and analyze results for a bilateral irradiation, also. While we must emphasize that we are not recommending the use of structure doses for BNCT treatment planning, we are, however, actively investigating their use for treatment plan evaluation and for optimizing the design of the neutron field of our accelerator-based neutron source.

ACKNOWLEDGMENT

This research was performed under appointment to the Laboratory Graduate Research Participation Program, which is administered by the Oak Ridge Institute for Science and Education (ORISE) for the United States Department of Energy. Funding and research facilities were supplied by the Radiation Internal Dose Information Center, ORISE, Oak Ridge, TN and in small part by DOE grant DE-FG02-93ER61612.

REFERENCES

1. J.E. Woollard, T.E. Blue, N. Gupta, and R.A. Gahbauer, Development and application of neutron field optimization parameters for an accelerator-based neutron source for boron neutron capture therapy, *Nucl Technol*, 115:100–113, 1996.
2. P.R. Gavin, S.L. Kraft, C.E. DeHaan, C.D. Swartz, and M.L. Griebenow, Large animal normal tissue tolerance with boron neutron capture, *Int J Radiat Oncol Biol Phys*, 28(5):1099–1106, 1994.
3. J.R. Fike, and G.T. Gobbel, Central nervous system radiation injury in large animal models, *in*: "Radiation Injury to the Nervous System," P.H. Gutin, S.A. Leibel, and G.E. Sheline, eds., Raven Press, New York, 1991, pp. 113–135.
4. R.C. Miller, R.L. Foote, R.J. Coffey, D.A. Gorman, J.D. Earle, P.J. Schomberg, and R.W. Kline, The role of stereotactic radiosurgery in the treatment of malignant skull base tumors, *Int J Radiat Oncol Biol Phys*, 39(5):977–981, 1997.
5. M.M. Urie, B. Fullerton, H. Tatsuzaki, S. Birnbaum, H.D. Suit, K. Convery, S. Skates, and M. Goitein, A dose response analysis of injury to cranial nerves and/or nuclei following proton beam radiation therapy, *Int J Radiat Oncol Biol Phys*, 23(1):27–39, 1992.
6. J.D. Slater, M. Austin-Seymour, J. Munzenrider, S. Birnbaum, R. Carroll, A. Klibanski, P. Riskind, M. Urie, L. Verhey, and M. Goitein, Endocrine function following high dose proton therapy for tumors of the upper clivus, *Int J Radiat Oncol Biol Phys*, 15(3):607–611, 1988.
7. W.M. Saunders, G.T.Y. Chen, M. Austin-Seymour, J.R. Castro, J.M. Collier, G. Gauger, P. Gutin, T.L. Phillips, S. Pitluck, R.E. Walton, and S.R. Zink, Precision, high dose radiotherapy. II. helium ion treatment of tumors adjacent to critical central nervous system structures, *Int J Radiat Oncol Biol Phys*, 11(7):1339–1347, 1985.
8. T.E. Schultheiss, L.E. Kun, K.K. Ang, and L.C. Stephens, Radiation response of the central nervous system, *Int J Radiat Oncol Biol Phys*, 31(5):1093–1112, 1995.
9. S.A. Leibel, and G.E. Sheline, Tolerance of the brain and spinal cord to conventional irradiation, *in*: "Radiation Injury to the Nervous System," P.H. Gutin, S.A. Leibel, and G.E. Sheline, eds., Raven Press, New York, 1991, pp. 239–256.
10. R. Santoni, N. Liebsch, D.M. Finkelstein, E. Hug, P. Hanssens, M. Goitein, A.R. Smith, D. O'Farrell, J.T. Efird, B. Fullerton, and J.E. Munzenrider, Temporal lobe (TL) damage following surgery and high-dose photon and proton irradiation in 96 patients affected by chordomas and chondrosarcomas of the base of the skull, *Int J Radiat Oncol Biol Phys*, 41(1):59.68, 1998.

11. J. Debus, E.B. Hug, N.J. Liebsch, D. O'Farrell, D. Finkelstein, J. Efird, and J.E. Munzenrider, Brainstem tolerance to conformal radiotherapy of skull base tumors, *Int J Radiat Oncol Biol Phys*, 39(5):967–975, 1997.
12. I.G. Zubal, C.R. Harrell, E.O. Smith, and A.L. Smith, Two dedicated software, voxel-based, anthropomorphic (torso and head) phantoms, *in*: "Proceedings of the International Workshop held at the National Radiological Protection Board, Chilton, UK, on 6 and 7 July, 1995," P.J. Dimbylow, ed., NRPB, Chilton, 1996, pp. 105–111.
13. J.F. Briesmeister, ed., "MCNP—A General Monte Carlo N-Particle Transport Code," Los Alamos National Laboratory, LA-12625-M, Version 4B, March 1997.
14. W.S. Rasband, and D.S. Bright, NIH Image: a public domain image processing program for the Macintosh, *Microbeam Analysis Society Journal*, 4:137–149, 1995.
15. International Commission on Radiation Units and Measurements, "Photon, Electron, Proton and Neutron Interaction Data for Body Tissues," ICRU Report 46, ICRU, Bethesda, MD, 1992.
16. F.A. Duck, "Physical Properties of Tissue—A Comprehensive Reference Book," Academic Press, London, 1990.
17. C.K.C. Wang, T.E. Blue, and R. Gahbauer, A neutronic study of an accelerator-based neutron irradiation facility for boron neutron capture therapy, *Nucl Technol*, 84:93–107, 1989.
18. M. Cristy, and K.F. Eckerman, "Specific Absorbed Fractions of Energy at Various Ages from Internal Photon Sources. I. Methods," Oak Ridge National Laboratory, ORNL/TM-8381/V1, 1987.
19. J.E. Woollard, private communication, June 1998.
20. D. Gabel, K.H.I. Philipp, and R. Huiskamp, The compound factor of $NA_2B_{12}H_{11}SH$ and the relative biological effectiveness of recoil protons for neurological damage in boron neutron capture therapy, *in*: "Cancer Neutron Capture Therapy," Y. Mishima, ed., Plenum Press, New York, 1996, pp. 815–821.
21. P.R. Gavin, S.L. Kraft, R. Huiskamp, and J.A. Coderre, A review: CNS effects and normal tissue tolerance in dogs, *J Neurooncol*, 33:71–80, 1997.

FAST NEUTRON STUDIES

NEUTRON CAPTURE THERAPY (NCT) ENHANCEMENT OF FAST NEUTRON RADIOTHERAPY AT THE UNIVERSITY OF WASHINGTON

Past, Present, and Future Directions

George E. Laramore and Keith J. Stelzer

Department of Radiation Oncology
University of Washington Medical Center
Box 356043, Seattle, WA 98195-6043

1. INTRODUCTION

Fast neutrons in the energy range of several MeV to several tens of MeV are a form of high linear energy transfer (LET) radiation. Their depth-dose properties are comparable (or perhaps slightly inferior) to those of megavoltage photon irradiation and one depends upon their advantageous radiobiological properties to more effectively sterilize certain classes of tumors.[1] The first patient was treated with fast neutrons by Stone and coworkers[2] in 1938 using an early cyclotron at the Crocker Radiation Laboratory (now the Lawrence Berkeley Laboratory). Between then and 1943 a total of 240 patients were treated using low-energy beams by today's standards. Almost all of these patients had very advanced tumors and some had received prior radiotherapy using the orthovoltage beams of that era. There were few long term survivors and those patients that did survive had severe radiation sequellae. Interest in this form of cancer treatment waned until the 1950s when radiation biologists learned to grow mammalian cells in culture and achieved a better understanding of the radiobiological properties of fast neutrons. A subsequent review of Stone's initial work showed that the appropriate relative biological effectiveness (RBE) factors corresponding to Stone's treatment regimens were substantially larger than had been determined from early animal experiments which resulted in excessive doses being given to these early patients.

In the 1960s clinical trials for fast neutron radiotherapy were resumed by Catterall and coworkers[3] at the Hammersmith Hospital in London, England. It was early

Frontiers in Neutron Capture Therapy, edited by Hawthorne *et al.*
Kluwer Academic / Plenum Publishers, New York, 2001.

favorable reports from these studies, namely that many patients with advanced head and neck tumors responded amazingly well to this form of therapy, that led directly to a world-wide, clinical research effort in fast neutron radiotherapy which burgeoned in the 1970s and early 1980s. In the 1970s the great majority of patients treated with fast neutrons received their treatment using laboratory-based facilities that were originally constructed for physics research purposes. These facilities generally could produce only a fixed, horizontal beam and were restricted to fairly simple field blocking devices. Simulation and verification of treatment fields was difficult. In the late 1970s the National Cancer Institute of the United States recognized the deficiencies of the early neutron therapy centers, and it became obvious that units having the same degree of technological sophistication as modern hospital megavoltage therapy units were required to adequately evaluate fast neutron radiotherapy. Contracts were ultimately awarded to 4 centers for the design and construction of such facilities. A review of the clinical trials carried out at these centers and elsewhere throughout the world has been recently conducted by Laramore[4] and the current thinking as to those tumor types and attendant clinical situations for which fast neutron radiotherapy offers an advantage over conventional low LET radiation is summarized in Table 1. A listing of the locations of centers currently offering fast neutron radiotherapy is also given in Ref. (4).

Normal tissue side effects make it difficult to attempt to improve tumor control further by simply escalating the dose and most modern neutron treatment facilities already use isocentric gantries, variable beam shaping collimators, and 3-dimensional treatment delivery techniques. A neutron capture therapy (NCT) boost offers the potential of significantly increasing the biologically-effective dose delivered to the tumor on a cell-by-cell basis and hence not only improving local control in those situations where fast neutron radiotherapy has been proven to be advantageous, but also extending the indications for fast neutron radiotherapy to other more common tumors listed in Table 1.

Table 1. Summary of those tumor types and clinical situations where fast neutron radiotherapy offers a therapeutic advantage over conventional photon/electron radiotherapy. The more "plus" signs, the stronger the indication for fast neutrons[a]

Site/Histology	Relative Indication
Most Promising	
Salivary Gland Tumors (inoperable, gross residual disease)	++++
Prostate Cancer (locally-advanced)	++
Sarcomas of soft tissue, bone, cartilage (inoperable, gross residual disease)	+
Equivocal	
Non-small cell lung cancer (inoperable)	+/–
Massive cervical adenopathy from squamous cell head & neck cancer	+/–
No Advantage	
Head & Neck (squamous cell)	–
Malignant gliomas of brain	–
Esophageal cancer	–
Pancreatic cancer	–
Bladder cancer	–
Cervix cancer	–

[a]Reprinted from Laramore.[4]

Due to space limitations this manuscript will emphasize work done at the University of Washington. However, research in this area is also taking place at other centers throughout the world.[5,6,7,8,9]

2. DESCRIPTION OF PROCESS

Apart from resonance effects, the neutron capture cross section for ^{10}B and other elements scales approximately as $E^{-1/2}$, where E is the neutron energy. This makes the capture process less efficient at high energy; for example, the capture cross section of a 1 MeV neutron is approximately 6,000 times smaller than for a thermal neutron. However, the fast neutron beam is moderated as it penetrates tissue producing a small, thermalized component. Waterman *et al.*[10] were the first to recognize that this component had the potential of augmenting the therapeutic effectiveness of a fast neutron radiotherapy beam if ^{10}B could be selectively incorporated into tumor cells. Following their approach, the effective "slow" neutron component in the University of Washington fast neutron therapy beam was measured by activating ^{23}Na and then scaling relative to ^{10}B via the ratio of the respective cross sections.[11] Measured intensity profiles for a large field at a 5 cm depth in a water phantom are shown in Figure 1. The "slow" neutron dose has been scaled to correspond to 1 μg/g of ^{10}B in the absorbing medium.

While the "slow" neutron profile follows that of the "fast" component reasonably well for this large field size, it tends to fall off much more rapidly for smaller field sizes due to scattering out of the primary beam region.[11] Even for large field sizes, the energy deposited by the "slow" neutron component is only about 0.1% of the energy deposited by the total beam (per μg/g of ^{10}B). To achieve a 10% gain thus requires a selective tumor concentration of 100 μg/g of ^{10}B. However, even smaller ^{10}B concentrations can achieve clinical significance with currently available ^{10}B carrier agents due to two multiplicative factors. The first relates to a high RBE factor because of the very high LET of the emitted ^{4}He and ^{7}Li particles. The second multiplicative factor is due to the steepness of the dose response curve and the fact that fast neutron radiotherapy alone reaches the steep portion of the curve. This is illustrated in Figure 2 which shows that a small incremental increase, Δ, in tumor dose can result in a large change, Γ, in tumor control probability.

Using experimental data from Gabel *et al.*[12] for the V79 cell line and the Poisson model for tumor control put forth by Porter,[13,14] Laramore *et al.*[11] estimated that if the NCT enhancement effect could achieve as little as 1 additional log of tumor cell killing,

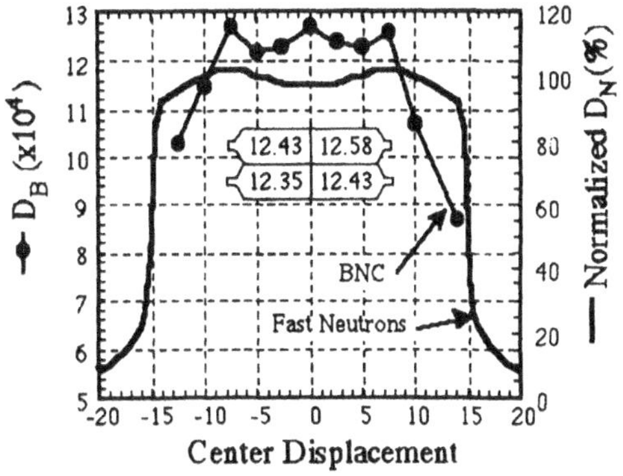

Figure 1. Beam intensity profiles for the "fast" and "slow" components for a 30 cm × 30 cm field for the standard University of Washington therapy neutron beam in a water phantom at 5 cm depth. The beam is generated using a 50.5 MeV p → Be reaction. Reprinted from Laramore *et al.*[11]

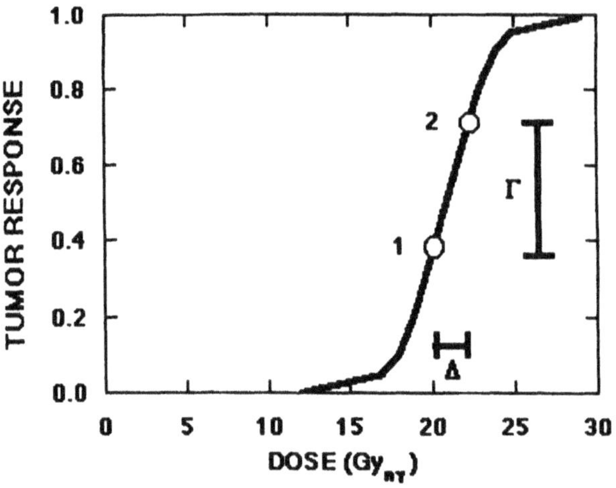

Figure 2. Schematic illustration of dose response curve for a tumor treated with fast neutron radiotherapy plus a NCT boost dose enhancement. Reprinted from Laramore *et al.*[11]

it had the potential of increasing an unenhanced tumor control probability of $0.2 \rightarrow 0.85$ and of $0.5 \rightarrow 0.93$. The model saturates (e.g., produces too high an estimate) for large increases in the tumor control probability but nevertheless, it appears that the effect should be large enough to be clinically significant.

3. EXPERIMENTAL STUDIES

To test the accuracy of the model, *in vitro* measurements of the enhanced cell killing were performed using the V79 cell line.[11] Both ^{10}B-enriched boric acid and cesium mercaptoundecahydro-*closo*-dodecaborate were used as the boron carriers. A detailed comparison between the experimental results and the predictions of the model showed agreement to within a factor of 2.

Radiobiological studies on NCT enhancement of fast neutron radiotherapy were performed using the 36B10 glioma model in Fisher rats.[15] The 36B10 cells were harvested from *in vitro* cultures and then 4×10^5 cells were injected subcutaneously into six sites on the dorsal flank of each rat. Tumors were allowed to grow 28–33 days to a median size of 1.4 g prior to neutron irradiation. ^{10}B-enriched L-*para* -boronophenylalanine (BPA) as the fructose complex was injected intraperitoneally. Injections took place 3 hours and 1.5 hours prior to irradiation with 1 g/kg of BPA being injected each time. Details of the pharmacokinetics and radiation techniques are described by Buchholz *et al.*[15] Surviving fractions at each radiation dose level were determined by an *in vitro* colony-formation assay. Cell survival data is shown in Figure 3 with each point representing the mean value for 6–7 different tumor nodules.

The mean ^{10}B concentration in excised test nodules was $68.4 \pm 19.8\,\mu g/g$. For fast neutron doses of 5.0, 7.5, and 10.0 $Gy_{n\gamma}$, the amount of NCT enhancement was statistically-significant at $p < 0.005$ (Student's t-test), while for a dose of 12.50 $Gy_{n\gamma}$, the difference was only borderline significant at $p = 0.08$. This may be because of the increased cell killing by the fast neutrons alone at the highest dose level. Curve fitting techniques using the linear-quadratic model showed changes in α and β from 0.23 to 0.41 $Gy_{n\gamma}^{-1}$ and 0.01 to 0.02 $Gy_{n\gamma}^{-2}$, respectively, with the NCT enhancement. The dose modification factor at the 0.01 survival level was 1.31 ± 0.09.

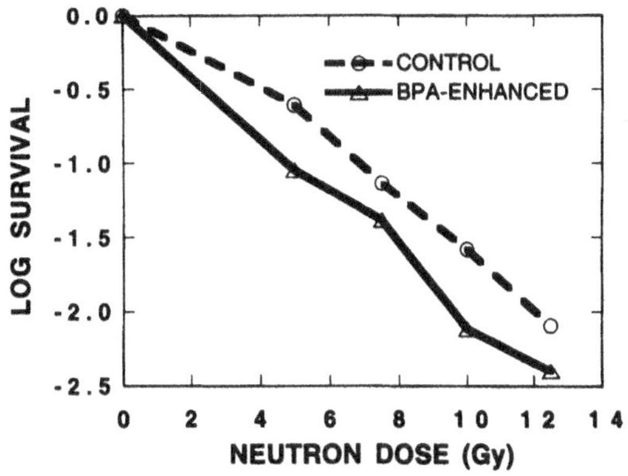

Figure 3. Enhanced cell killing with intraperitoneal BPA in the 36B10 Fisher rat glioma model. Replotted from Buchholz *et al.*[15]

Next the technique was applied to a human melanoma test system using orally-administered, ^{10}B-enriched BPA for a patient with multiple melanoma skin nodules.[16] Although a dose of 500 mg/kg of BPA was given, the average ^{10}B tumor concentration was only 11.379 µg/g. This was approximately 3 times that of the blood and subcutaneous normal tissue. A special radiation configuration was developed using a large 20 cm × 20 cm field that grazed the skin surface such that the fast neutron dose was restricted to an effective 3 cm × 3 cm field. The NCT-enhanced dose was approximately 0.2% of the total beam energy per µg/g ^{10}B concentration which is approximately twice that produced using a more traditional field configuration. A detailed description of the irradiation procedure and the experiment itself can be found in a paper by Laramore *et al.*[16] Two tumor nodules were treated to a fast neutron dose of 10 $Gy_{n\gamma}$ in a single fraction; one with NCT enhancement via BPA and one without this enhancement. This was felt to be near the maximum dose that could be given without causing skin necrosis. NCT enhancement gave a more rapid tumor response but both of the treated nodules eventually regrew before the patient succumbed to leptomeningeal tumor spread. Autopsy data showed no significant differences in regards to radiation changes in the subcutaneous tissues.

4. ONGOING STUDIES AT THE UNIVERSITY OF WASHINGTON

The choice of a human tumor system for future clinical studies depends upon three principle factors: (i) there must be some evidence that treatment with fast neutrons alone approaches the steep portion of the dose response curve as indicated in Figure 2, (ii) a ^{10}B carrier must be identified that will concentrate in tumor preferentially in comparison to the critical dose-limiting normal tissue for that particular situation, and (iii) the tumor must be common enough to generate commercial interest in order to produce the drug under good manufacturing practice (GMP) conditions. Non-small cell lung cancer appears to be a good candidate at this juncture. In 1998 the American Cancer Society estimates that there will be approximately 171,500 new cases of primary lung cancer in the United States and approximately 160,100 deaths.[17] We would expect about 60% of new primary lung cancers to be non-small cell. Early stage tumors will be surgically treated assuming that there are no medical contraindications and patients with known metastatic disease at the time of presentation are not candidates for curative therapy. We thus estimate that approximately 40,000 cases per year of non-small cell lung cancer in the United States will be potential candidates for a more effective radiotherapeutic treatment.

There have been two randomized, clinical trials for inoperable, non-small cell lung cancer. The first was a three armed study comparing conventional photon radiation vs. mixed (neutron/photon) radiation vs. neutron radiation alone.[18] The complication rate was higher on the neutron only arm (perhaps due in part to the relatively low energy neutron beams in use at that time), and there was no difference in overall survival. A second randomized trial was conducted using modern hospital-based facilities which compared a neutron only regimen with conventional photon irradiation.[19] For the entire cohort of patients there was no difference in overall survival. However, there was a statistically-significant survival advantage for the subset of patients having squamous cell histology. There was also a nonsignificant trend toward increased survival for patients of all histologies with favorable prognostic factors (no pleural effusion, not T4 or N3 in stage, weight loss less than 5% of normal body weight). Except for skin and subcutaneous changes, acute and late toxicity were comparable in the two treatment arms. There are also two retrospective analyses that report apparently improved survival with neutron radiation for patients having Pancoast tumors.[20,21] The overall results are suggestive of improved survival in the subgroup of patients with non-small cell lung cancers that are not prone to early distant metastases; it is only for such patients that improved local/regional control might be expected to translate into improved survival.

When one reviews in detail the pattern of local failure in the patients on the randomized trial, a common site of failure was adjacent to the spinal column where the high RBE for neural tissue damage by neutrons limited the tumor dose that could be safely given. An agent under investigation, disodium decahydrodecaborate ($Na_2B_{10}H_{10}$), which we will refer to as GB-10, is of interest in this setting. It is essentially non-toxic, hydrophillic, does not cross the blood-brain barrier (or the membrane that similarly protects spinal tissue), and is highly stable in an aqueous solution. This compound was first synthesized by Hawthorne and co-workers.[22] Early animal studies by Soloway *et al.*[23] showed an LD_{50} of 685 mgB/kg in Swiss albino mice and a very low toxicity profile in humans was noted by Sweet *et al.*[24] Unpublished data of Gavin in beagle dogs indicates a maximum tolerance dose (no effect) of 500 mgB/kg of body weight making it much less toxic than BSH which has a similar biodistribution profile.

Dose escalating studies using the University of Washington fast neutron radiotherapy beam and GB-10 are ongoing for canines with spontaneous non-small cell lung cancer.[25] Radiation doses are being prescribed to the midplane considering only the fast neutron component. Currently we are using 4 weekly fractions with a large field and a small field being treated each fraction. The large field includes the spinal cord while the small field excludes it. The spinal cord dose is being maintained at 9 $Gy_{n\gamma}$ and the dose delivered through the small field varied. (Note that we following the convention of ICRU 50 and specify the dose delivered to beam isocenter and utilize 3-dimensional dosimetric calculations to determine the doses given throughout the tumor and surrounding normal tissues.[26] The subscript "$n\gamma$" means that the γ-contaminent is included in the specified dose.) Two dogs have been treated to a total dose of 18 $Gy_{n\gamma}$ and in the future we anticipate escalating the dose to 20 $Gy_{n\gamma}$. The GB-10 is being administered via a continuous infusion (25 mgB/kg over 1 hour) prior to treatment with a maintenance infusion (12.5 mgB/kg) during treatment. With this technique it appears possible to have tumor boron doses in excess of 100 μg/g while the dog is being treated. For details of the experimental procedure and results to date, the reader is referred to the paper by Stelzer *et al.*[25]

5. FUTURE WORK AT THE UNIVERSITY OF WASHINGTON

Work is continuing in collaboration with the Idaho National Engineering and Environmental Laboratory[27] to optimize the "slow" neutron component of the therapy beam without losing its favorable attributes for fast neutron radiotherapy. A layered design using beryllium on tungsten appears attractive both in Monte Carlo simulations and experimental measurements. We have been using this target in canine experiments and it has performed well.

An investigational new drug (IND) application has been submitted for pharmacokinetic and biodistribution studies in patients who are undergoing surgical treatment for non-small cell lung cancer and glioblastoma multiforme of the brain. Patients will be studied with loading and maintenance infusions. We anticipate extending the trials to patients with soft tissue sarcomas and squamous cell tumors of the head and neck. If the biodistribution data so warrant, we will then embark upon a phase I therapy trial.

REFERENCES

1. E.J. Hall. "Radiobiology for the Radiologist (ed 4)", Lippincott, Philadelphia, 1992. See in particular Chapts. 9 and 14.
2. R.S. Stone. Neutron therapy and specific ionization. *Am. J. Roentgenol.* 59:771–785, 1948.
3. M. Catterall. The treatment of advanced cancer by fast neutrons from the Medical Research Council's cyclotron at Hammersmith Hospital, London. *Eur. J. Cancer Clin. Oncol.* 10:343–347, 1974.
4. G.E. Laramore. The use of neutrons in cancer therapy: A historical perspective through the modern era. *Sem. Oncol.* 24:672–685, 1997.
5. W. Siegler, W. Saurwein, and C. Streffer. Fast neutrons from the Essen cyclotron can be used successfully for neutron capture experiments *in vitro*. *Strhlenther. Onkol.* 165:210–212, 1989.
6. L. Luderman, T. Matzen, M. Matzke, R. Schmidt, and W. Scobel. Determination of the thermal neutron flux in a fast neutron beam by use of a boron-coated ionization chamber. *Med. Phys.* 22:1743–1747, 1995.
7. R.L. Maughan, C. Kota, and J.D. Foreman. Feasability of boron neutron capture enhancement of fast neturon therapy utilizing a superconducting cyclotron, *in* "Advances in Neutron Capture Therapy, Vol. I: Medicine and Physics". B. Larsson, J. Crawford, R. Weinreich, eds, Elsevier, Amsterdam, 1997, pp. 490–495.
8. K. Langen, A.J. Lennox, T.K. Kroc, and P.M. De Luca, *in* "Advances in Neutron Capture Therapy, Vol. I: Medicine and Physics". B. Larsson, J. Crawford, and R. Weinreich, eds, Elsevier, Amsterdam, 1997, pp. 501–504.
9. J.P. Pigot, A. Courdl, P. Paquis, N. Iborra-Brassart, M. Chatel, N. Bretea, W. Sauerwein, D. Gabel, and P. Chauvel. Neutron capture potentiation in a fast neutron beam leads to a potential therapeutic window for unresectable glioblastomas, *in* "Advances in Neutron Capture Therapy, Vol. II: Chemistry and Biology". B. Larsson, J. Crawford, and R. Weinreich, eds, Elsevier, Amsterdam, 1997, pp. 519–523.
10. F.M. Waterman, F.T. Kuchnir, L.S. Skaggs, D.K. Bewley, B.C. Page, and F.H. Attix. The use of ^{10}B to enhance the tumor dose in fast-neutron radiotherapy. *Phys. Med. Biol.* 23:592–602, 1978.
11. G E. Laramore, P. Wootton, J.C. Livesey, D.S. Wilbur, R. Risler, M. Phillips, J. Jacky, T.A. Buchholz, T.W. Griffin, and S. Brossard. Boron neutron capture therapy: A mechanism for achieving a concomitant tumor boost in fast neutron radiotherapy. *Int. J. Radiat. Oncol. Biol. Phys.* 28:1135–1142, 1994.
12. D. Gabel, S. Foster, and R.G. Fairchild. The Monte Carlo simulation of the biological effect of the ^{10}B (n, α) ^{7}Li reaction in cells and tissue and its implication for boron neutron capture therapy. *Radiat. Res.* 111:14–25, 1987.
13. E.H. Porter. The statistics of dose/cure relationships for irradiated tumors. Part I. *Br. J. Radiol.* 53:210–227, 1980.
14. E.H. Porter. The statistics of dose/cure relationships for irradiated tumors. Part II. *Br. J. Radiol.* 53:336–345, 1980.

15. T.A. Buchholz, J.S. Rasey, G.E. Laramore, J.C. Livesey, L. Chin, R. Risler, D.K. Hamlin, A.M. Spence, and T.W. Griffin. Concomitant boron neutron capture therapy during fast neutron irradiation of a rat glioma. *Radiology* 19:863–867, 1994.
16. G.E. Laramore, T.W. Griffin, R. Risler, P. Wootton, and D.S. Wilbur. Fast neutron radiotherapy and boron neutron capture therapy: Application to a human melanoma test system. *Bull. Cancer (Paris)* 83S: 191s–197s, 1996.
17. S.H. Landis, T. Murray, S. Bolden, and P.A. Wingo. Cancer statistics, 1998. *CA Cancer J. Clin.* 48:6–29, 1998.
18. G.E. Laramore, M. Bauer, T.W. Griffin, F.J. Thomas, F.R. Hendrickson, M.H. Maor, B.R. Griffin, J.P. Saxton, and L.W. Davis. Fast neutron and mixed beam radiotherapy for inoperable non-small cell carcinoma of the lung. *Am. J. Clin. Oncol.* 9:233–243, 1986.
19. W-J. Koh, J.M. Krall, L.J. Peters, M.H. Maor, G.E. Laramore, C.M. Burnison, L.W. Davis, S. Zink, and T.W. Griffin. Neutron vs photon radiation therapy for inoperable regional non-small cell lung cancer: Results of a multicenter randomized trial. *Int. J. Radiat. Oncol. Biol. Phys.* 27:499–505, 1993.
20. K. Sawada, S. Fukuma, Y. Seki, F. Tanaka, I. Ishida, K. Matsumura, M. Nakamo, K. Iida, S. Sata, Y. Umegaki, H. Tsunemoto, and S. Morita. Clinical experience in patients with Pancoast tumors treated by fast neutron radiotherapy. *Gan. No. Rinsho.* A7:111–114, 1983.
21. R. Komaki, C.F. Mountain, J.M. Holbert, A.S. Garden, R. Shallenberger, J.D. Cox, M.H. Maor, V.F. Guinee, and B. Samuels. Superior sulcus tumors: Treatment selection and results for 85 patients without metastasis (M_0) at presentations. *Int. J. Radiat. Oncol. Biol. Phys.* 19:31–36, 1990.
22. M.F. Hawthorne, and A.R. Pitochelli. The reactions of bis-acetonitrile decaboranes with amines. *J. Am. Chem. Soc.* 81:5519, 1959.
23. A.H. Soloway, R.L. Wright, and J.R. Messer. Evaluation of boron compounds for use in neutron capture therapy of brain tumors. I. Animal investigations. *J. Pharm. Exp. Therap.* 134:117–122, 1961.
24. W.H. Sweet, A.H. Soloway, and R.L. Wright. Evaluation of boron compounds for use in neutron capture therapy of brain tumors. II. Studies in man. *J. Pharm. Exp. Therap.* 137:263–266; 1962.
25. K.J. Stelzer, P.R. Gavin, R. Risler, H. Kippenes, M.F. Hawthorne, D.W. Nigg, and G.E. Laramore. Boron neutron capture-enhanced fast neutron therapy (BNC/FNT) for non-small cell lung cancer in canine patients, *in* "Frontiers in Neutron Capture Therapy," M.F. Hawthorne, K. Shelly, R.W. Wiersema, eds., Kluwer Academic/Plenum Publishers, New York, 2001, pp. 735–739.
26. ICRU 50. Prescribing, recording, and reporting photon beam therapy. ICRU Press, Bethesda, MD, 1993.
27. D.W. Nigg, C.A. Wemple, R. Risler, J.K. Hartwell, and Y.D. Harker. Modifications of the University of Washington neutron radiotherapy facility for optimization of neutron-capture-enhanced fast neutron therapy, *in* "Frontiers in Neutron Capture Therapy," M.F. Hawthorne, K. Shelly, R.W. Wiersema, eds., Kluwer Academic/Plenum Publishers, New York, 2001, pp. 729–734.

PROGRESS TOWARDS BORON NEUTRON CAPTURE ENHANCEMENT OF FAST NEUTRON THERAPY AT THE HARPER HOSPITAL NEUTRON THERAPY FACILITY

Richard L. Maughan, Chandrasekhar Kota, Jay Burmeister, Kenneth J. Levin, Paul J. Chuba, Laurie E. Gaspar, and Jeffrey D. Forman

Gershenson Radiation Oncology Center
Karmanos Cancer Institute
Harper Hospital and Wayne State University
3990 John R., Detroit, Michigan 48201

1. INTRODUCTION

The neutron beam at the Gershenson Radiation Oncology Center is produced in a unique gantry mounted superconducting cyclotron.[1] A 48.5 MeV deuteron beam is incident on an internal beryllium target and a beam current of up to 15 μA produces a fast neutron dose rate of 48 cGy per minute. The beam current is presently limited by the power dissipation characteristics of the target, but a relatively simple target modification will allow this limit to be increased by a factor of 3–3.5. Although the thermal neutron fluence is reasonable compared to that produced in a reactor, the physical dose enhancement due to the boron neutron capture reaction is only 2 to 4% for a 15 cm × 15 cm field at depths between 2.5 and 15 cm with a ^{10}B concentration of 50 μg g^{-1} (Figure 1). Therefore, in order to have a fast neutron beam which is suitable for boron neutron capture enhanced fast neutron therapy (BNCEFNT) it is necessary to increase the thermal neutron component of the beam relative to the fast neutron dose by filtration and/or moderation or some other strategy. Using a high Z target to increase the yield of low energy neutrons through spallation reactions may be one option, although it is desirable to retain the fast neutron characteristics of the beam. In particular a fast neutron component with a reasonable depth dose characteristic is required if the fast neutron component of the dose is going to be effectively exploited. It has been shown that beams with a suitable mix of dose components (i.e. 50–60% of the biological effect due to fast

Frontiers in Neutron Capture Therapy, edited by Hawthorne *et al.*
Kluwer Academic / Plenum Publishers, New York, 2001.

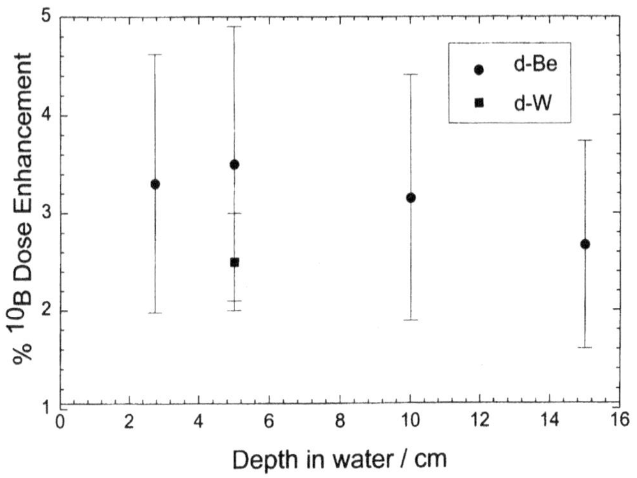

Figure 1. Results of dose enhancement measurements in a d(48.5)-Be (circles) and a d(48.5)-W (square) fast neutron beam for a 15cm × 15cm with 50μg g^{-1}. Measurements were made using the dual microdosimetric counter technique described in the text.

neutrons, 10% due to gamma rays and 25–35% due to boron neutron capture products) can be produced by using an iron or tungsten filter.[2,3] These filters reduce the average energy of the neutron beam such that the depth of the 50% isodose is reduced from 13.6 cm to 9–11 cm for a 10 cm × 10 cm field. This beam penetration is comparable to that of a ^{60}Co beam and is sufficient for the treatment of glioblastoma multiforme. With a filtered beam arrangement an increased beam current is necessary to produce a clinically useful BNCEFNT beam is achievable.

The present paper investigates the potential advantages of using a tungsten (high Z) target with a 48.5 MeV deuteron beam and also addresses the issue of how the target design can be modified to allow for the higher incident beam currents necessary to give realistic treatment times if a filtered beam is used for BNCEFNT.

2. MATERIALS AND METHODS

Measurements have been made of the boron dose enhancement obtained when a tungsten target is used in place of the beryllium target normally used for fast neutron therapy. These measurements were made using the dual tissue-equivalent proportional counter method in which two counters which are identical except for the composition of their outer walls. These walls are constructed from A150 tissue equivalent plastic (TEP) but in one counter the walls are loaded with 50 μg g^{-1} of ^{10}B. An enhanced α/^{6}Li peak is observed in the event size spectra measured with the ^{10}B loaded counter and by subtracting this from the spectra measured with the "normal" A150 TEP it is possible to calculate the α/^{6}Li dose. Fast neutron and γ ray dose can also be calculated. This method of dosimetry has been described in detail elsewhere.[4,5] Measurements were made at an SSD of 183 cm and 5 cm depth in a 30 × 30 × 30 cm^{3} water phantom with a 15 cm × 15 cm field size.

3. RESULTS

The result of the dual counter measurement made with the deuteron beam incident on a tungsten target is shown in Figure 1. For comparison data taken with the dual counters when the deuteron beam was incident on a beryllium target under similar conditions at depths of 2.75, 5, 10 and 15 cm in water are also plotted in the figure. The measured enhancement with a tungsten target is not significantly different from that observed with the beryllium target; within the uncertainties.

4. DISCUSSION

4.1. Tungsten Target Neutron Yield

The expectation had been that there would be an increase in the number of spallation reactions produced when the deuteron beam was incident on the high Z tungsten target relative to the low Z beryllium target, with a corresponding increase in the number of low energy and thermal neutrons. This effect is certainly seen when protons are incident on tungsten rather than beryllium and is being exploited by Nigg *et al.* who describe a layered target of tungsten and beryllium for use with the 50 MeV proton beam at the University of Washington in these proceedings.[6] However, at a given particle energy, the neutron yield from the deuteron on beryllium reaction on a thick target is very much greater than the yield seen from the proton on beryllium reaction on the same target.[7] In fact, there is an approximately 6 times greater yield; most of these extra neutrons originate from the deuteron stripping reaction. Hence, only about one sixth of the neutrons originate from within the target nucleus. It is presumably the spectrum of the neutrons produced within the target nuclei which is effected by the higher Z of the target. Thus, the effect of the target Z on the overall neutron spectrum produced by deuterons incident on a thick target is small because this spectrum is predominated by the deuteron stripping interactions. Hence, the deuteron on tungsten and deuteron on beryllium spectra are very similar in shape. This can be seen in the spectral measurements of Meulders *et al.*[8] at energies above about 3–4 MeV for a 50 MeV deuteron beam incident on beryllium and tantalum targets (Z[Ta] = 73, Z[W] = 74). The fact that the measured boron neutron capture enhancement is the same for both the tungsten and beryllium target suggests that the d-W(48.5) reaction does not yield relatively more low energy and thermal neutrons in comparison to the d-Be(48.5).

4.2. New Target Design

Although tungsten and beryllium both yield approximately the same boron neutron capture enhancement, the overall neutron yield from the beryllium target is approximately 5–6 times greater per μA of incident beam. This makes the neutron beam produced by the d-Be reaction more suitable for modification by filtration in order to produce a beam for BNCEFNT since the neutron flux per μA is considerably greater. With a deuteron produced fast neutron beam earlier studies of the feasibility of BNCEFNT showed that a sufficient RBE weighted dose rate could be obtained for therapy from a neutron beam produced by 50 μA of 48.5 MeV deuteron beam incident on a thick beryllium target and filtered by 25 cm of iron.[2] The target presently used for fast neutron therapy with the Harper Hospital superconducting cyclotron is capable of dissipating only ~750 W of

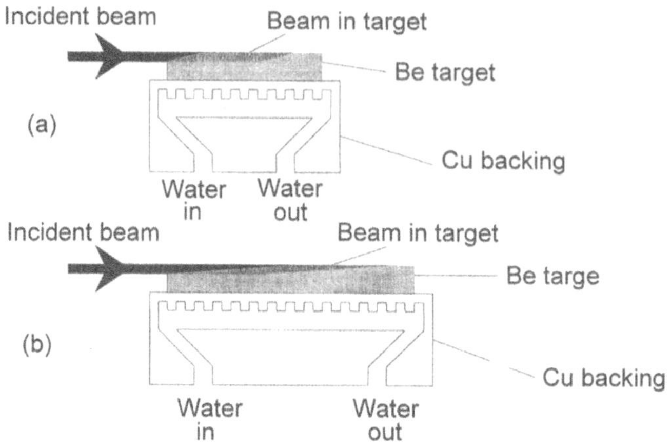

Figure 2. Schematic diagram showing how the existing beryllium target (a) can be modified to increase the "cooling area" area of the beam as shown in (b). This area increase is achieved by increasing the glancing angle on the leading edge of the target and increasing the target length.

power (i.e., the beam is limited to 15 μA by cooling requirements). The radius gain per turn in the cyclotron is only about 0.7 mm giving an effective beam width is 0.7 mm and its height is ~10 mm. An internal target with a glancing angle of 20.3° on the incident edge is used to spread the beam power over an area of about 1 cm^2. This geometry is shown schematically in Figure 2(a). The target is comprised of beryllium (15.9 × 20.1 × 3.2 mm) brazed onto a copper backing, the backside of which is channeled and cooled by chilled water. The deuteron beam travels parallel to the copper-beryllium braze so that cooling is achieved over a large area of the Cu-Be interface, which is determined by the glancing angle and the range of the deuterons in beryllium, as indicated in Figure 2. The 48.5 MeV deuteron beam penetrates about 8.1 mm into the beryllium.

The target is mounted on a vacuum interlock so that it can be removed from the cyclotron for inspection or replacement without breaking the vacuum in the acceleration chamber. The target assembly inserts through a 63.8 mm diameter hole in the cyclotron magnet yoke. It is possible to redesign the beryllium target head with a shallower glancing angle and to spread the incident beam over a wider area as shown schematically in Figure 2(b). With this arrangement the "cooling area" can be increased by a factor of 3 to 3.5 and it should, therefore, be possible to use incident beam currents of 50 μA, thus, making the production of a filtered d-Be(48.5) beam with sufficient intensity for BNCEFNT feasible. A target of this design is under construction.

5. CONCLUSIONS

There is no advantage in using a tungsten target for BNCEFNT in place of the beryllium target normally used as the neutron producing target with the 48.5 MeV deuteron beam since the BNC enhancement observed with the two targets is identical (Figure 1). The beryllium target power dissipation can be improved by a simple design change in which the glancing angle at which the beam strikes the target is reduced from ~20° to ~6°. This modification should allow the target to handle a 50 μA beam, which is sufficient for clinical BNCEFNT if a ^{10}B concentration of $50\,\mu g\,g^{-1}$ can be achieved in the tumor.

REFERENCES

1. R.L. Maughan, W.E. Powers, and H.G. Blosser, A superconducting cyclotron for neutron radiation therapy, *Med. Phys.*, 21:779–785, 1994.
2. R.L. Maughan, C. Kota, and J.D. Forman, A prototype BNCT/BNCEFNT facility utilizing the superconducting cyclotron at Harper Hospital, *in* "Advances in Neutron Capture Therapy, Volume I. Medicine and Physics," B. Larsson, J. Crawford, and R. Weinreich, eds., Elsevier, Amsterdam, 1998, pp. 490–495.
3. C. Kota, R.L. Maughan, J. Burmeister, and J.D. Forman, A modified fast neutron beam for boron neutron capture enhanced fast neutron therapy, *in* "Advances in Neutron Capture Therapy, Volume I. Medicine and Physics," B. Larsson, J. Crawford, and R. Weinreich, eds., Elsevier, Amsterdam, 1998, pp. 496–500.
4. R.L. Maughan, C. Kota, and M. Yudelev, A microdosimetric study of dose enhancement in a fast neutron beam due to boron neutron capture, *Phys. Med. Biol.*, 38:1957–1961, 1993.
5. C. Kota, and R.L. Maughan, A dosimetry system for boron neutron capture therapy based on the dual counter microdosimetric technique, *Bulletin du Cancer—Radiothérapie*, 83(Suppl. 1):173–175, 1996.
6. D.W. Nigg, C.A. Wimple, R. Risler, J.K. Hartwell, and Y.D. Harker, Modification of the University of Washington neutron radiotherapy facility for optimization of neutron-capture-enhanced fast neutron therapy, *ibid.*
7. International Commission on Radiation Units and Measurements (ICRU), Report 45, "Clinical Neutron Dosimetry Part I. Determination of Absorbed Dose in a Patient Treated by External Beams of Fast Neutrons", ICRU, Bethesda, 1989, pp. 9–13.
8. J.P. Meulders, P. Leleux, P.C. Marq, and C. Pirart, Fast neutron yields and spectra from targets of varying atomic number bombarded with deuterons from 16 to 50 MeV, *Phys. Med. Biol.*, 20:235–243, 1975.

CHARACTERIZATION OF A MODIFIED FAST NEUTRON BEAM FOR BORON NEUTRON CAPTURE ENHANCEMENT OF FAST NEUTRON THERAPY

J. Burmeister, C. Kota, M. Yudelev, and R. L. Maughan

Karmanos Cancer Institute
Gershenson Radiation Oncology Center
Harper Hospital and Wayne State University
Detroit, Michigan 48201

1. INTRODUCTION

The use of the boron neutron capture reaction to provide a localized tumor boost in fast neutron therapy was originally proposed by Waterman *et al.*[1] This treatment modality may be referred to as boron neutron capture enhanced fast neutron therapy (BNCEFNT).[2] Many fast neutron facilities are currently investigating the possibility of treating with BNCEFNT. Investigations are underway at the Harper Hospital cyclotron produced d(48.5)+Be fast neutron therapy facility in an attempt to produce a suitable modified beam for BNCEFNT.[3,4,5] Characterization of this modified beam is in progress using both proportional counter and ionization chamber techniques with results presented.

2. MODIFIED BEAM

Substantial work has been done to modify the cyclotron produced d(48.5)+Be fast neutron beam to yield a suitable therapeutic gain for BNCEFNT. These efforts include moderation of the beam with aluminum, lead, steel, and tungsten. A beam moderated with 25 cm of steel was chosen based on measurements to determine the tumor and normal tissue dose rates as a function of position and boron concentration. This beam arrangement has been presented by Kota *et al.*[5] In order to determine whether this beam will provide favorable characteristics for treatment with BNCEFNT, measurements are being made to obtain the beam data necessary to perform treatment planning for this technique.

Frontiers in Neutron Capture Therapy, edited by Hawthorne *et al.*
Kluwer Academic / Plenum Publishers, New York, 2001.

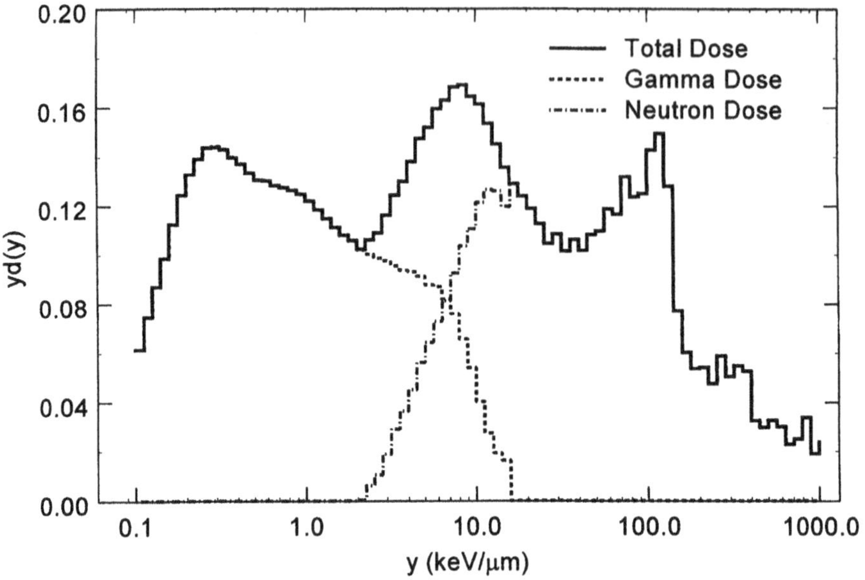

Figure 1. Lineal energy spectrum in the modified fast neutron beam proposed for BNCEFNT. This spectrum was measured at a depth of 2.5 cm along the central axis of a $15 \times 15\,cm^2$ field. (Plot represents dose per unit logarithmic interval of lineal energy density, y, against logarithm of y.)

3. TEPC MICRODOSIMETRY

Measurements have been made using paired miniature tissue equivalent proportional counters (TEPCs). The TEPC measurements are based on the dual counter method. One TEPC is constructed of A-150 tissue-equivalent plastic and the other of A-150 loaded with an appropriate amount of ^{10}B. Characteristics of these detectors are presented elsewhere.[6,7] The gamma and neutron absorbed doses may be extracted from lineal energy spectra measured with the A-150 TEPC by fitting a representative gamma spectrum to the mixed field spectrum. The BNC dose may be obtained by subtracting the A-150 spectrum from the boron-loaded A-150 spectrum. Point measurements have been made in the moderated beam proposed for BNCEFNT. Figure 1 shows the moderated beam lineal energy spectrum at a depth of 2.5 cm in water separated into its gamma and neutron components. Changes in the modified beam characteristics with depth are shown in Figure 2. Finally, Figure 3 illustrates the relative BNC dose enhancement for 200 μg/g ^{10}B.

4. PROPOSED IONIZATION CHAMBER MEASUREMENTS

TEPC microdosimetry provides the complete single event spectrum of charged particles depositing the absorbed dose. However, this information comes at the expense of the time required to individually process each charged particle event. This technique therefore becomes impractical when detailed information such as depth-dose curves and beam profiles is needed. This information is not easily obtained using foil activation methods, either. The amount of time required to obtain thermal flux data for the

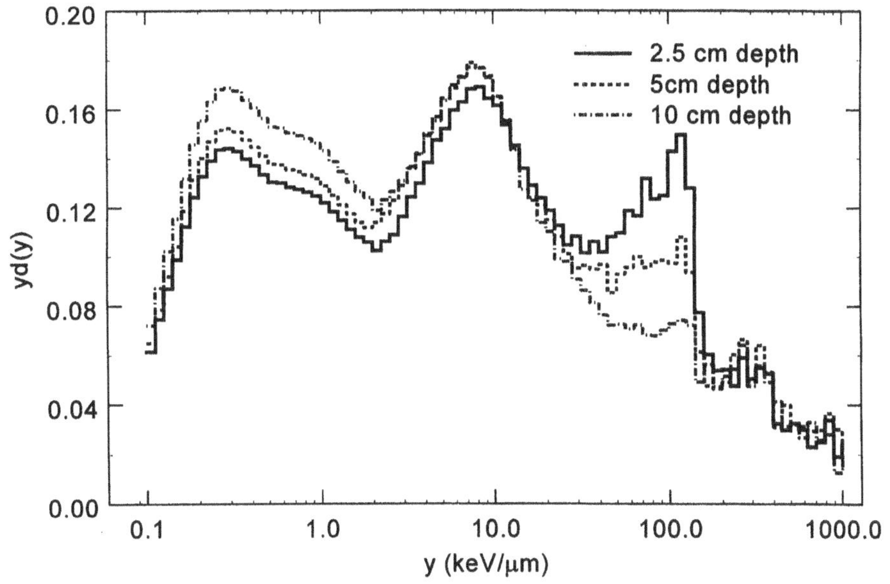

Figure 2. Changes in lineal energy spectra with depth along the central axis of the 15 × 15 cm² field.

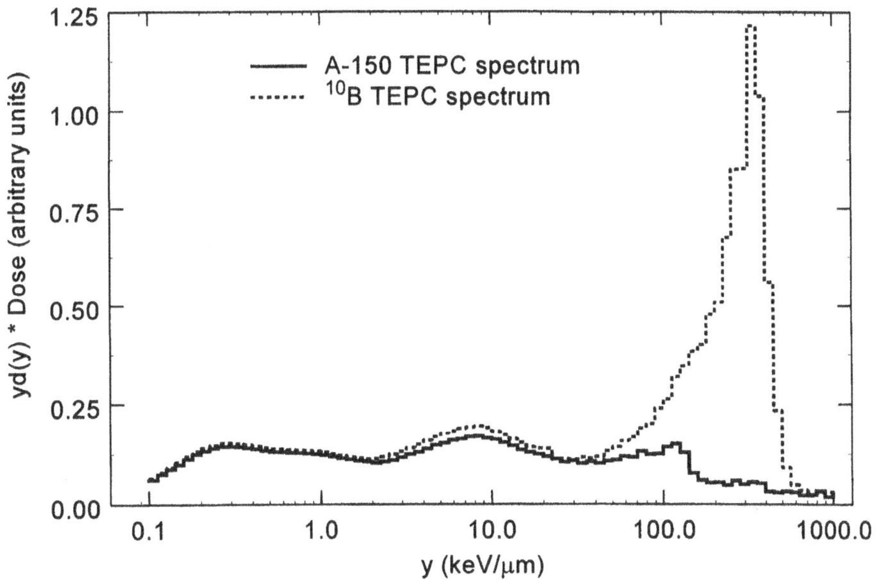

Figure 3. Comparison of lineal energy spectra at a depth of 2.5 cm measured with the A-150 and ^{10}B loaded TEPCs. (The plots are not normalized to unit dose to illustrate the relative sizes of the spectra.)

modified beam using foil activation or TEPC techniques becomes prohibitively large due to the amount of information necessary for treatment planning purposes, especially when more than one modified beam is to be assessed. For this reason, a pair of small volume magnesium ion chambers was built to obtain the thermal neutron flux. Based on the concept presented by Lüdemann *et al.*,[8] one of the magnesium chambers, denoted the Mg(B) chamber, was loaded with ^{10}B. This chamber has an inner wall lined with a 25 μm thick boron-loaded foil and its response is therefore dominated by ionizations produced by products of the boron neutron capture reaction. The contribution to the response of the Mg(B) chamber from gamma rays and fast neutrons is not negligible. For this reason, a matching Mg chamber was built. The relative boron neutron capture dose may be obtained by subtracting the Mg chamber response from the Mg(B) chamber response. Furthermore, when corresponding measurements are made with a tissue-equivalent (A-150) ion chamber, one may obtain the gamma and neutron components of the mixed field. The dual counter TEPC technique has proven to be an excellent tool for BNC dose calibration of the ionization chamber system. These chambers may be scanned in a water phantom to quickly obtain the beam data required for treatment planning for BNCEFNT in the modified fast neutron beam.

The characteristics of both chambers are very similar to commercially available magnesium chambers. They have been dose calibrated versus a Farmer chamber in a ^{60}Co radiation field. Their fast neutron sensitivity has been measured in the unmodified d(48.5)+Be fast neutron beam. They have also been calibrated for thermal neutron sensitivity in a thermal port at the Ford Nuclear Reactor at the University of Michigan. The characterization of this set of ionization chambers is complete, and a more thorough characterization of the modified beam using these chambers is planned.

5. RESULTS

Table 1 shows the RBE-weighted gamma, neutron, and BNC dose components at three depths along the central axis of the modified beam. Therapeutic gains of approximately 20% are achievable at depths required to treat targets within the brain. Based on work presented by Maughan *et al.*,[4] a therapeutic gain of approximately 25% would provide a substantial therapeutic window for treating tumors not controllable with conventional therapy. The RBE-weighted tumor dose rate at a depth of 2.5 cm is approximately 3.3 cGy/min with the existing target. A new target design proposed in an accompanying paper[9] would allow a dose rate of greater than 11 cGy/min, making the use of BNCEFNT clinically feasible.

Table 1. RBE-weighted absorbed doses in cGy/MU to brain tissue extracted from the TEPC measurements shown in Figures 1–3[a]

Depth (cm)	Gamma	Neutron	BNC (7.5 ppm)	BNC (30 ppm)	Therapeutic gain factor
2.5	0.0134	0.0515	0.0045	0.0180	1.19
5.0	0.0128	0.0387	0.0041	0.0163	1.22
10.0	0.0113	0.0291	0.0016	0.0064	1.11

[a]Boron-10 concentrations of 30 ppm and 7.5 ppm were assumed for tumor and normal tissue, respectively. RBE values of 3.8, 3.2, and 1.0 were used for the BNC, fast neutron, and gamma doses, respectively.

6. CONCLUSION AND FUTURE DIRECTIONS

TEPC dosimetry of a BNCEFNT beam offers the advantage of providing the complete single event charged particle spectrum in a simulated microscopic site. This yields information about the radiation quality as well as information about the microscopic nature of the absorbed dose. However, for a thorough characterization of the beam, a quicker method is needed. The three ion chamber method (A-150, Mg, and Mg(B) chambers) yields quick and accurate results for the gamma, neutron and BNC dose components. The set of ion chambers has been characterized and will be used to collect the data necessary for treatment planning for BNCEFNT. A description of the treatment planning method is presented by Kota.[10]

ACKNOWLEDGMENTS

This work has been supported in part by the U.S. Department of Energy, Grant # DE-FG02-96ER62217.

REFERENCES

1. F.M. Waterman, F.T. Kuchnir, L.S. Skaggs, D.K. Bewley, B.C. Page, and F.H. Attix, The use of ^{10}B to enhance the tumour dose in fast-neutron therapy, Phys. Med. Biol. 23:592–602, 1978.
2. C. Kota. "Microdosimetric Considerations in the use of the Boron Neutron Capture Reaction in Radiation Therapy," Ph.D. Thesis, Wayne State University, Detroit, USA, 1996.
3. R.L. Maughan, C. Kota, and M. Yudelev, A microdosimetric study of the dose enhancement in a fast neutron beam due to boron neutron capture, Phys. Med. Biol. 37:1957–1961, 1992.
4. R.L. Maughan, C. Kota, and J.D. Forman, Feasibility of boron neutron capture enhancement of fast neutron therapy utilizing a superconducting cyclotron, *in*: "Advances in Neutron Capture Therapy, Volume 1," B. Larsson, J. Crawford, and R. Weinreich, ed., Elsevier Science, Amsterdam, 1997, pp. 490–495.
5. C. Kota, R.L. Maughan, J. Burmeister, and J.D. Forman, A modified fast neutron therapy beam for boron neutron capture enhanced fast neutron therapy (BNCEFNT), *in*: "Advances in Neutron Capture Therapy, Volume 1," B. Larsson, J. Crawford, and R. Weinreich, ed., Elsevier Science, Amsterdam, 1997, pp. 496–500.
6. J. Burmeister, C. Kota, and R.L. Maughan, Paired miniature TEPCs for dosimetry in high flux epithermal neutron capture therapy beams, Nuc Inst and Meth in Phys Res, A422:606–610, 1999.
7. J. Burmeister, C. Kota, R.L. Maughan, K. Riley, O. Harling, L. Wielopolski, R. Ma, and J. Coderre, Microdosimetry studies at the BMRR and MITR-II using miniature tissue equivalent proportional counters, *ibid.*
8. L. Lüdeman, T. Matzen, M. Matzke, R. Schmidt, and W. Scobel, Determination of the thermal neutron flux in a fast neutron beam by use of a boron-coated ionization chamber, Med. Phys. 22:1743–1747, 1995.
9. R.L. Maughan, C. Kota, J. Burmeister, K.J. Levin, P.J. Chuba, L.E. Gaspar, and J.D. Forman, Progress towards boron neutron capture enhancement of fast neutron therapy at the harper hospital neutron therapy facility, *ibid.*
10. C. Kota, J. Burmeister, R.L. Maughan, K. Levin, P. Chuba, L. Gaspar, and J.D. Forman, Treatment planning for boron neutron capture enhanced fast neutron therapy, *ibid.*

MODIFICATIONS TO rtt_MC FOR FAST NEUTRON THERAPY TREATMENT PLANNING

C. A. Wemple, F. J. Wheeler, and D. W. Nigg

Idaho National Engineering and Environmental Laboratory
Idaho Falls, Idaho

1. INTRODUCTION

Fast neutron radiotherapy (FNT) has emerged as a favored modality for treatment of certain cancers, and FNT-enhanced neutron capture therapy (NCT) is a subject of increasing research interest. Unlike the current situation with clinical applications of epithermal neutron NCT, which rely primarily on nuclear reactors for the neutron source, FNT neutron sources are generated using particle accelerators. This yields a neutron source spectrum with a peak energy very near the particle beam energy, which is several orders of magnitude higher energy than used for epithermal NCT. As with all medical applications of radiation, it is important that the dosimetric parameters of such neutron sources be well characterized, and that patient-specific treatment planning computations be performed for each clinical application.

To accommodate the physics of these high-energy neutron sources, modification and expansion of the neutron transport code rtt_MC[1] was undertaken, in order to expand the utility of this well-known epithermal NCT treatment planning tool into the realm of FNT and NCT-augmented FNT. This modification comprised four phases: (1) cross section acquisition, processing, and formatting; (2) development of new neutron tracking routines; (3) development of proton (recoil hydrogen nucleus) dose computation routines, to accurately model the surface and entry region dose distribution; and (4) development of photon production and tracking routines, to model the dose distribution from photons produced by high-energy neutron reactions on biological nuclei.

2. CROSS SECTIONS

Generally, the medical applications of FNT are limited to neutron energies below 100 MeV, and typically are in the 20–66 MeV range. Conventional cross sections for reactor applications, which account for nearly all of the evaluated data at worldwide data

Frontiers in Neutron Capture Therapy, edited by Hawthorne *et al.*
Kluwer Academic / Plenum Publishers, New York, 2001.

repositories, are only considered for the energy range below 20 MeV, and are most accurate below 10 MeV. Thus, a new series of evaluations for biologically-important nuclides had to be commissioned. Evaluations conducted at Los Alamos National Laboratory (LANL),[2–4] based mostly on theoretical optical model calculations, for beryllium, carbon, nitrogen, oxygen, aluminum, silicon, calcium, iron, tungsten, lead, and ^{238}U, merged the conventional data below 20 MeV with the new evaluated cross sections, to produce self-consistent neutron interaction and photon production cross section data sets for neutron energies below 100 MeV. The NJOY[7] code suite was used by the XTM group at LANL to process data for nine nuclides[6] (Be, C, O, Al, Si, ^{40}Ca, Fe, W, and ^{238}U), with several corrections[6] to the evaluations made during the processing.

Some local post-processing of these libraries was necessary before application in rtt_MC. Since the data is identical to previously-existing cross section data below about 20 MeV, the data below the defined upper energy considered by the NCT version of rtt_MC (16.9 MeV) were deleted. This preserved the original calculation path for epithermal NCT applications. Gamma production data were translated from individual energy-angular distributions for each reaction to a unified neutron-gamma production matrix, with 50 equiprobable gamma energy bins for each neutron energy. This allows a much faster, simpler, and equally accurate description of the gamma field produced by neutron reactions during irradiation.

Additionally, isotopic gamma cross sections for energies up to 100 MeV were processed locally from ENDF/B-VI data using the NJOY code. An 86-group gamma cross section library in the RAFFLE format[8] was produced using auxiliary codes created explicitly for this purpose. Gamma kermas were mixed from these isotopic data to yield the kermas for ICRU brain tissue[9] with 1 ppm ^{10}B. A new coupled neutron-gamma library was created using these new gamma data, and the existing rtt_MC neutron data below 16.9 MeV.

3. PARTICLE TRACKING ROUTINES

The basic neutron transport methodology employed by the epithermal rtt_MC routines was preserved for the fast neutron routines. Neutron histories are grouped into batches, with each batch considered to be statistically independent. This is assured by re-initializing the random number generator at the start of each batch, and by limiting the size of each batch to prevent any random number sequence overlap. Only scattering, defined as any reaction with a neutron in the exiting channel, and absorption, defined as any non-scattering reaction, are considered for each neutron history. The standard rtt_MC spatial tracking routines are used unaltered for the ultra-fast neutron tracking, and only minor modifications were made to the tallies to accommodate these changes.

The main difference from the standard tracking method is in the energy tracking. The standard routines use a multi-group, mixed material cross section method, and the ultra-fast routines employ a continuous energy, isotopic cross section method. For each neutron history, an initial set of isotopic cross sections for each material is computed at the initial energy. These cross sections are then mixed to form the material cross section, which is used in the spatial tracking. When a collision occurs, the material cross sections are used to determine whether this is a scattering or absorption reaction. Absorption terminates the particle history. For a scattering reaction, the isotope is selected, and then the determination of the type of scattering reaction (elastic or inelastic) is made. Then, the scattering reaction is simulated, with the energy and direction of the exiting neutron(s) calculated according to the physics of the particular reaction. The process then reverts back to the cross section computation at the new energy, and continues until the particle is killed, or leaks out of the system. If the particle energy drops below 16.9 MeV, then the standard rtt_MC energy tracking system is followed for the remainder of the particle history.

Gamma production occurs at each neutron collision. The gammas produced by ultra-fast neutron collisions are banked to a file, for use during the normal gamma transport process. After the gammas produced by the epithermal neutron tracking routines are processed, a separate routine to process this file is called. This routine follows exactly the same transport process as the normal gamma transport routine, with the gamma source provided from the file.

This separation of methods for the ultra-fast and epithermal paths has several advantages. Foremost, it preserved the independent path for the epithermal calculations, which means that previous calculations are unaffected by the addition of the new calculation path. Thus, results calculated by prior versions of the code can be reproduced. In addition, it simplified the task of incorporating the new routines into the existing code. It also allowed the application of independent convergence acceleration techniques to the ultra-fast path, without affecting the epithermal calculations.

4. PROTON RECOIL DOSE CALCULATION

The calculation of the dose caused by recoiling hydrogen nuclei (protons) from elastic scattering of fast neutrons was added to accurately model the dose distribution, which does not necessarily follow the KERMA distribution, especially near the beam entry. The details of the method are given elsewhere.[10,11] The method calculates the exiting proton energy and direction from the physics of elastic scattering. This information is then fed to a simple proton transport routine, which uses a simple energy deposition model based on a continuous slowing down method using tabulated proton ranges in matter. These tabulations[12] contain an average path length traversed by a proton slowing down in matter, and a detour factor to account for non-straight line motion.

Dose is tallied on an arbitrary cartesian coordinate voxel mesh, corresponding with the mesh used for the other tallies in rtt_MC. Boundary crossings, both for the voxel mesh and the rtt_MC geometry, are accounted for in the proton tracking. Total energy deposited by a proton in a voxel is determined by the path length traversed in the voxel, and the proton path length from the tabulations. This path length is calculated at proton birth, and at any boundary crossings that entail a material change. The path length is dependent only on the material and the proton energy. It should be noted that this method is sufficient only for macrodosimetric applications. More rigorous charged-particle transport methods must be utilized for microdosimetry calculations.

5. BIASED ULTRA-FAST NEUTRON MODE

An unfortunate consequence of the improved accuracy gained from the proton recoil dose calculation is poor convergence of the proton dose tally results. This is caused by the short path length of protons in matter. In an effort to improve the proton recoil dose tally statistics, a biased ultra-fast neutron mode was added to rtt_MC. This new mode is analogous to the biased fast neutron mode already employed for the epithermal calculations. In this calculation path, only the source above the ultra-fast neutron cutoff energy (16.9 MeV) is considered, and tracked in a highly-optimized transport mode, in which only scattering on hydrogen is considered, and no gamma production is allowed. The weight of each particle is adjusted appropriately at each collision, and the proton recoil dose calculation is performed for the resulting recoil nuclei. Because of the increased efficiency of this mode, a much larger number of neutron histories can be followed.

6. BENCHMARKING

To thoroughly test the modifications to rtt_MC for FNT described in the previous section, it was necessary to define a series of benchmark problems to exercise the new code segments and physics modules. This was approached in two phases: comparisons of simple, theoretical models with standard code results; and comparisons with measured values in experimental phantom irradiations. The simple, idealized model calculations were compared with both the MCNP[5] and DORT[13] radiation transport codes, which are the international "gold standards" for transport code benchmarking. Experimental data were obtained from phantom irradiations at the University of Washington fast neutron therapy facility,[14,15] and dose/depth data compared with the rtt_MC calculations.

6.1. Computational Benchmarks

A simple model was constructed of a large, cubic water phantom, 30 cm on a side, with a plane source of neutrons 1 cm from the cube, and parallel to the free surface. Four different sources were examined: two isotropic sources with uniform energy distributions in the ranges 10–12.2 MeV, and 45–50 MeV; and two monodirectional (normal to the source plane) sources with the same energy distributions. For each source energy-angle configuration, two different source apertures (5 × 5 cm square and 20 × 20 cm square) were examined. In all cases, dose or kerma along the phantom centerline, and three-group neutron fluxes at the phantom centerline were tallied. The three neutron energy groups spanned the ranges 0–0.414 eV (thermal), 0.414–20 MeV, and 20–50 MeV. Results were calculated for all four cases with both rtt_MC and MCNP, and with DORT for the monodirectional plane sources only.

Typical results, in this case, for the monodirectional, high-energy, large aperture source, are shown for the dose/depth and kerma/depth behavior in Figure 1. The dose/depth and kerma/depth results, as well as the thermal flux results, show good agreement between the three codes, with particularly excellent agreement between rtt_MC and MCNP. The small wiggles in the rtt_MC results are caused by statistical variations in the proton recoil dose, which is very difficult to converge. Similar agreement was observed for the other models examined.

The main sources of the differences between rtt_MC and DORT are the cross section data, and the DORT physics model, which only computes kerma, and not dose. The DORT data comes from the HILO86[16] library, which is more suited to high-energy spallation facilities, and has a poor representation for thermal neutrons. For this reason, DORT results for the thermal flux are not presented.

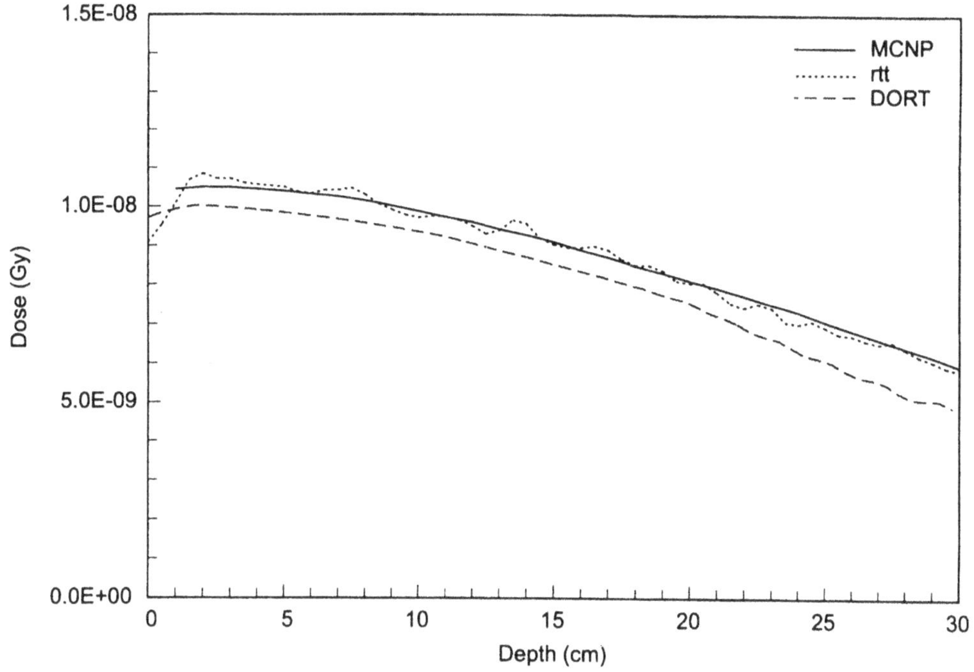

Figure 1. Dose-depth curves for the 45–50 MeV, monodirectional source in the large water phantom.

6.2. University of Washington Accelerator Measurements

6.2.1. Lucite Phantom Experiments. The standard INEEL lucite phantom was used for these measurements. This phantom is 12.7 cm in diameter, and 18.1 cm in height, and has various penetrations for placement of foils and wires. For these measurements, the center of the planar free surface was placed at the isocenter. The boron dose rate (cGy/s/ppm) along the phantom centerline was measured using borosilicate glass beads and copper/gold wires, placed at 4 cm intervals down the center of the phantom.

The boron kerma rate at depth in the phantom was computed using the MCNP code, in addition to the dose calculations made with rtt_MC. The results of these dose and kerma rate calculations, and the measured dose/depth profiles, are shown in Figure 2. The MCNP/rtt_MC comparison for dose is the most pertinent in this case, and excellent agreement was obtained. The disparity for the measured and calculated doses is primarily caused by room return in the measured values. The computer code models could not include the large concrete treatment room structure, so the room return is not accounted for in the calculated results.

6.2.2. Large Water Phantom Experiments. The standard UW water phantom was used for these measurements. This phantom is a 30 cm cube of light water in a thin-walled cubic plastic container. For the measurements, the center of the free surface of the water

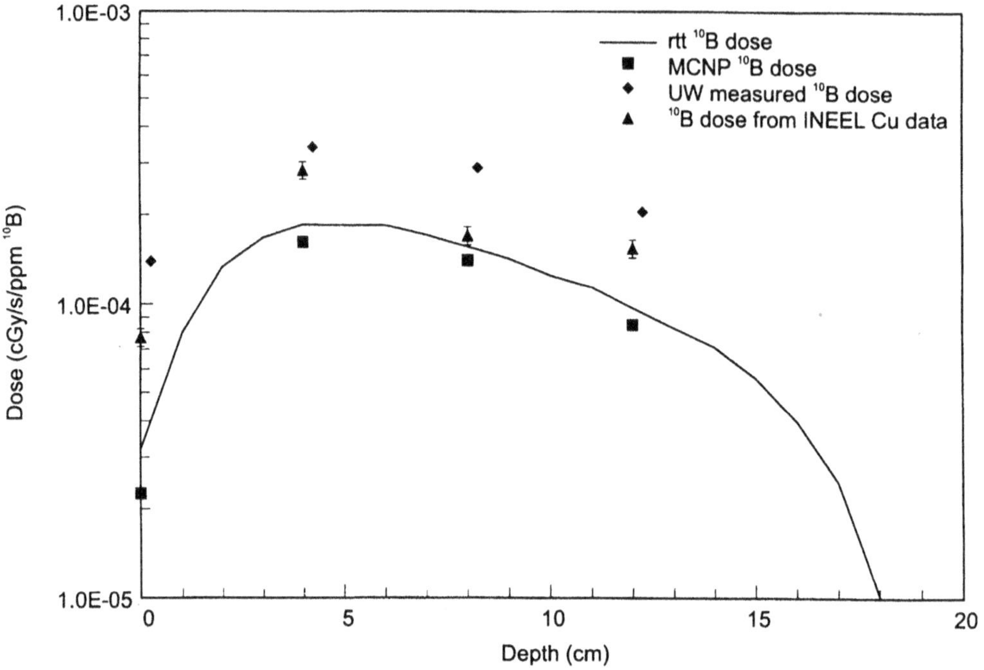

Figure 2. Boron dose-depth curves for the 7.5 × 7.5 cm University of Washington beam at 65.8 μA in the INEEL lucite phantom.

phantom was placed at the isocenter. Total dose rates were measured at depth in the phantom for three different field sizes—small (7.5 cm), medium (10 cm) and large (20.5 cm) squares. Total dose rates at depth were calculated using the rtt_MC code, and total kerma rates with the MCNP code.

Figure 3 shows the dose and kerma rate comparisons for the 10 cm, projected field size. In general, these results exhibit good agreement, with excellent consistency of the two calculated results. There is a trend towards underestimating the total dose in rtt_MC for larger field sizes, and, conversely, a trend to overestimate the total dose for smaller field sizes. This results from the approximations made in the modelling of the facility collimator. The actual collimator is a doubly-hinged, multi-leaf iron collimator, with a flattening filter included between the target and collimator exit. To project a square field, the leaves will have a small angle from the vertical. The model could not easily account for this double-hinging affect, and thus the collimator walls were modelled as a vertical rectangular parallelepiped. The flattening filter is actually a two-piece filter, with each piece having a flat surface, and a shaped surface resembling a highly-damped wave function. The shaped surfaces are negative images of each other, and would form a simple rectangular parallelepiped if brought together. In the collimator model, the filter was approximated as a simple rectangular parallelepiped. Scoping calculations to determine the affect of the angular distribution at the target on the calculated rtt_MC dose profile in phantom show that this can have a major influence for relatively small changes in the angular distribution. It follows from this that complete accuracy in modelling the structures between target and isocenter is necessary. Accounting for this lack of modelling fidelity, the results display good agreement.

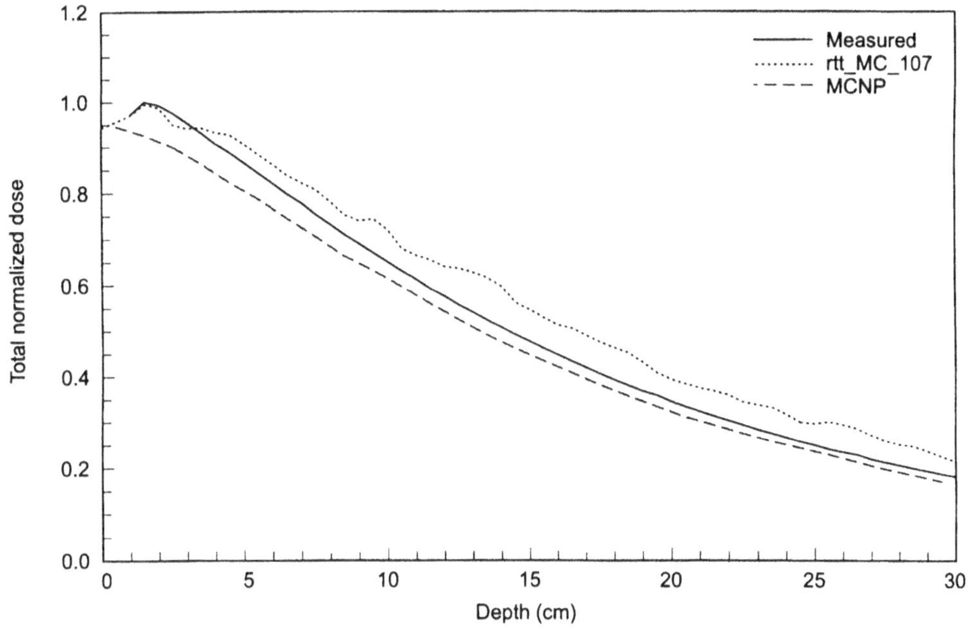

Figure 3. Dose-depth curves for the 10 × 10cm University of Washington beam in the large water phantom.

7. SUMMARY AND CONCLUSIONS

A significant expansion of the rtt_MC capabilities has been completed. The addition of the ultra-fast neutron and gamma transport capability allows treatment planning for FNT and mixed FNT-NCT treatments to be considered by the BNCT_rtpe code suite. In addition, several useful extensions and improvements were added to the epithermal NCT modules, without altering the existing calculation path. Testing of the modifications to the rtt_MC code for fast neutron therapy treatment planning was conducted, using comparisons to both measured dose rates, and calculations for idealized phantom models. These comparisons displayed excellent agreement, and demonstrate that these modifications can accurately make dose predictions for fast neutron therapy treatment planning.

REFERENCES

1. F.J. Wheeler, Radiation Transport in Tissue by Monte Carlo, EGG-BNCT-11178, Idaho National Engineering Laboratory, January 1994.
2. P.G. Young, *et al.*, Transport Data Libraries for Incident Proton and Neutron Energies to 100MeV, LA-11753-MS, Los Alamos National Laboratory (1990).
3. M.B. Chadwick, L.J. Cox, P.G. Young, and A.S. Meigooni, Calculation and Evaluation of Cross Sections and Kerma Factors for Neutrons Up To 100MeV on Carbon, UCRL-JC-121303, Lawrence Livermore National Laboratory (1995).
4. M.B. Chadwick, and P.G. Young, Calculation and Evaluation of Cross Sections and Kerma Factors for Neutrons up to 100MeV on 16O & 14N, UCRL-JC-122435, Lawrence Livermore National Laboratory (1995).

5. J.F. Briesmeister, ed., MCNP—A General Monte Carlo N-Particle Transport Code, Version 4A, LA-12625-M, Los Alamos National Laboratory (1993).
6. R.C. Little, Summary Documentation for the 100XS Neutron Cross Section Library (Release 1.0), XTM:95–259, Los Alamos National Laboratory, October 1995.
7. R.E. MacFarlane, and D.W. Muir, The NJOY Nuclear Data Processing System, Version 91, LA-12740-M, Los Alamos National Laboratory (1994).
8. R.A. Grimesey, D.A. Millsap, and J.R. Johnson, ENDRAFL; ENDF/B Multigroup Cross Section Libraries for the RAFFLE Monte Carlo Code, EGG-PHYS-6149, Idaho National Engineering Laboratory, December 1982.
9. Photon, Electron, Proton, and Neutron Interaction Data for Body Tissues, ICRU Report 46, International Commission on Radiation Units and Measurements, February 1992.
10. C.A. Wemple, and D.W. Nigg, Method for Calculation of Proton Recoil Dose in MCNP, *ANS Transactions*, 75, 144–145 (1996).
11. C.A. Wemple, Documentation of Proton Dose Calculation Capability, CAW-01-96, Idaho National Engineering Laboratory, January 1996.
12. Stopping Power and Ranges for Protons and Alpha Particles, ICRU Report 49, International Commission on Radiation Units and Measurements, May 1993.
13. W.A. Rhodes, and R.L. Childs, An Updated Version of the DOT-4 One- and Two-Dimensional Neutron/Photon Transport Code, ORNL-5851, Oak Ridge National Laboratory (1982).
14. D.W. Nigg, C.A. Wemple, J.K. Hartwell, Y.D. Harker, J.R. Venhuizen, and R. Risler, *V*alidation of Computational Methods for Treatment Planning of Fast-Neutron Therapy Using Activation Foil Techniques, ANS Radiation Protection and Shielding Division Topical Meeting, Nashville, TN, April 19–23, 1998.
15. R. Risler, University of Washington Medical Center, private communication.
16. R.G. Alsmiller, J.M. Barnes, and J.D. Drischler, Neutron-Photon Multigroup Cross Sections for Neutron Energies ≥400 MeV (Revision 1), ORNL/TM-9801, Oak Ridge National Laboratory, February 1986.

TREATMENT PLANNING FOR BORON NEUTRON CAPTURE ENHANCED FAST NEUTRON THERAPY

Chandrasekhar Kota, Jay Burmeister, Richard Maughan, Kenneth Levin, Paul Chuba, Laurie Gaspar, and Jeffrey Forman

Gershenson Radiation Oncology Center
Karmanos Cancer Institute
Harper Hospital and Wayne State University
Detroit, Michigan 48201

1. INTRODUCTION

Clinical data in fast neutron therapy suggests that the normal brain tolerance dose for fractionated whole brain irradiations with fast neutrons is about 14 Gy while the tumor control dose is about 21 Gy for glioblastoma multiforme.[1] If 6 Gy equivalent of fast neutron dose can be delivered to the tumor by the boron neutron capture (BNC) dose, a therapeutic window opens up for the treatment of these highly malignant tumors. In previous studies, we have shown that by passing the d(48.5) + Be fast neutron beam through 25 cms of steel, we create a beam that is capable of meeting the above requirement.[2,3] Our current need is to create patient treatment plans with this beam to verify its clinical utility. Existing Monte Carlo based treatment planning algorithms for BNCT are not readily applicable to BNCEFNT because of the lack of reliable cross-section data for the higher energy neutrons encountered in these beams. Consequently, we have investigated the following treatment planning strategy for BNCEFNT.

2. TREATMENT PLANNING ALGORITHM FOR BNCEFNT

In BNCEFNT, the absorbed dose is from neutrons, gamma rays and the BNC reaction. We need to calculate physical doses from each component separately, weight by RBE (if desired) and sum to obtain a cumulative effective dose distribution (which could be expressed relative to fast neutron dose) for plan evaluation. Additionally, for the BNC dose component, we need to calculate dose to normal brain and tumor, which have

Frontiers in Neutron Capture Therapy, edited by Hawthorne *et al.*
Kluwer Academic / Plenum Publishers, New York, 2001.

different ^{10}B concentrations. The existing treatment planning method (algorithm) for fast neutron therapy (FNT) is based on the "differential scatter air ratio" concept, commonly employed in external beam photon treatment planning. In this algorithm, the dose in the patient is decomposed into two components: (1) dose from primary radiation in the patient in each voxel, and (2) dose from scattered radiation reaching that voxel from elsewhere. Clinical implementation is based on relative dose measurements in a water phantom, and absolute calibration at a reference distance in air (tissue air ratio: TAR concept) or in phantom (tissue phantom ratio: TPR concept).

This algorithm currently works satisfactorily for FNT, and recent experiments have indicated that heterogeneity corrections in FNT can also be implemented using density-scaling methods. We have experimentally determined that the neutron (including the nitrogen thermal neutron capture dose) and gamma dose components in BNCEFNT beams follow depth dose curves and dose profiles similar to those in FNT and conventional photon therapy. Consequently, they can be modeled using the differential SAR concept. This work deals with the feasibility of adapting the existing algorithm for planning the BNC dose component. While the thermal fluence has depth dose characteristics similar to FNT beams, the profiles are ill defined due to the diffuse nature of the thermal neutron interactions. However, unlike in BNCT beams, the thermal neutrons are predominantly produced in the phantom as a result of fast neutron moderation. Consequently, it might be possible to separate the thermal fluence into primary and scatter components. Based on these observations, we feel that the existing algorithm could be adapted to plan for BNCEFNT.

The feasibility of this approach can be relatively easily evaluated by comparing the depth dose and profile data of the thermal fluence with that encountered in fast neutron and photon beams. Similarity between these data would suggest a high likelihood of success for our approach. The following input data are required for this algorithm: (1) Percent depth dose curves for a number of field sizes extending to maximum field size intended for use, (2) Beam dose profiles at different depths for the largest field size intended for use and profiles at other field sizes for evaluation, and (3) Absolute output measurements at reference depth for the above field sizes.

To gather the required data, we designed and constructed a Mg ion chamber with an inner surface containing ^{10}B to measure the effective thermal fluence for the BNC reaction. The details of the construction and calibration this chamber have been described elsewhere.[4] The ionization response of this chamber compared to a regular Mg ion chamber was found to be at least ~20 times greater. Since the chamber response is due to ^{10}B interactions, the measured data directly reflects the relative ^{10}B dose distributions, eliminating the need for any corrections. The chamber was used with a commercially available water phantom scanning system to measure the relative depth dose and profile data. Absolute ionization measurements at reference depth were made with a calibrated electrometer. These absolute measurements were further verified by point measurements of ^{10}B dose made with paired A-150/A-150 + B proportional counters.[5]

3. RESULTS AND DISCUSSIONS

Figure 1 shows the measured, physical percent depth dose (PDD) curves for a range of field sizes. The PDDs follow trends commonly observed in photon beams and fast neutron beams.

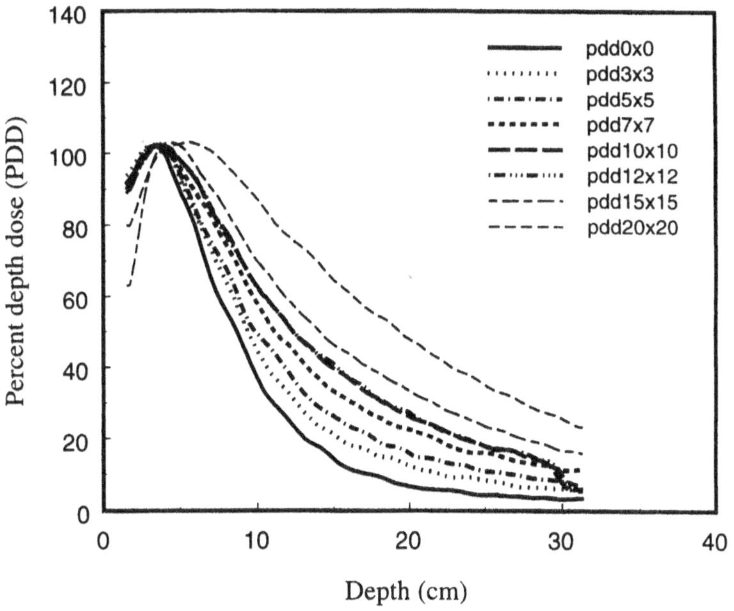

Figure 1. Percent depth dose (PDD) for field sizes ranging from 0 cm × 0 cm (closed collimator to 20 cm × 20 cm.

Figure 2 shows the beam profiles at 5 cm depth measured for a range of field sizes. These profiles are normalized to the absolute output at that depth for the individual field sizes. There is a significant difference between these profiles and those encountered in FNT beams. The thermal fluence in the phantom with the collimator completely closed is significant, and comparable to that measured in the open fields. These thermal neutrons are probably produced by leakage and energy-degraded neutrons transmitted through the collimator. Consequently, for all field sizes, the measured profiles demonstrate significant background outside the field, as defined by the collimator opening corrected for divergence. This phenomenon is exacerbated for small fields, whose absolute output is low because of decreased phantom scatter.

Figure 3 shows the profiles with the closed collimator profile (transmission, or background) subtracted, and each normalized to 100%. These profiles, although not as well defined as those in FNT or photon beams, display trends observed in these beams.

This suggests that the algorithm, with a few modifications, might be able to model the BNC dose component. For neutron or photon beam, the algorithm uses a single value of transmission factor and another single parameter describing the penumbra shape. In our case, we will probably need to vary the transmission as a function of field size to account for its change. Figure 4 compares the measured and calculated penumbra for a 20 cm × 20 cm field size, demonstrating that the model does work for a single field size.

The thermal neutrons are primarily produced in the phantom, so the concept of TAR cannot be used. However, tissue phantom ratios (TPR) could be used instead, with the assumption that collimator scatter of the primary neutron fluence only weakly affects the thermal fluence distribution in the phantom, which is predominantly due to phantom scatter. So, collimator scatter need not be evaluated separately as usually required for this

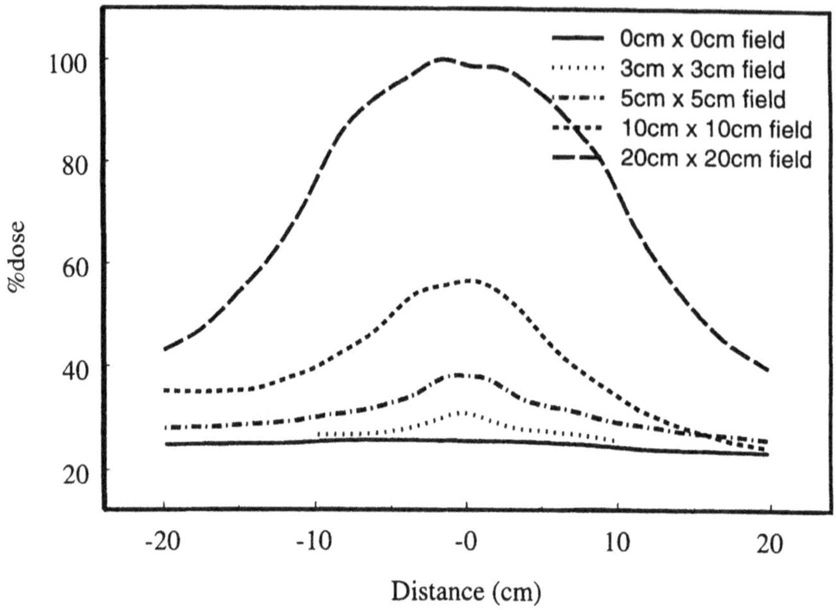

Figure 2. Beam profile data for thermal fluence measured at 5 cm depth in various field sizes and normalized to absolute output.

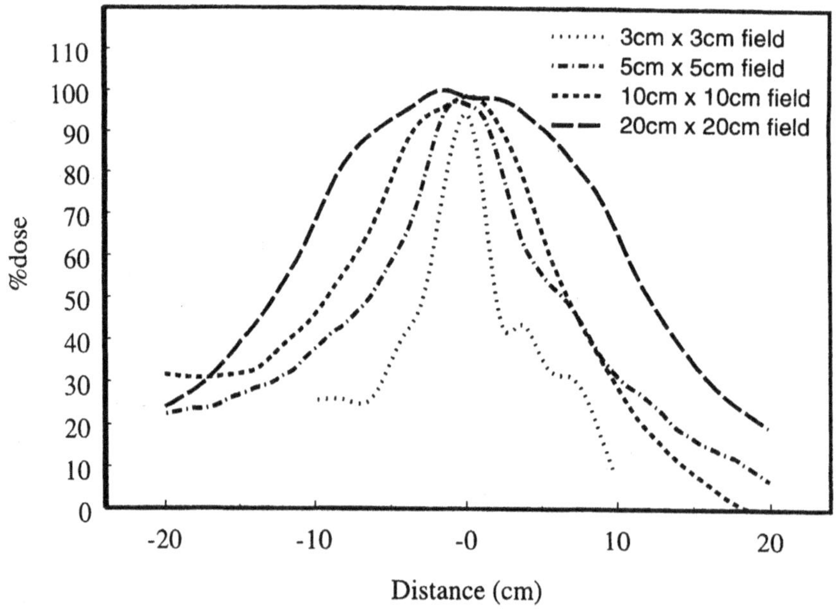

Figure 3. Background subtracted beam profiles for different field sizes, each normalized to its maximum value as 100%.

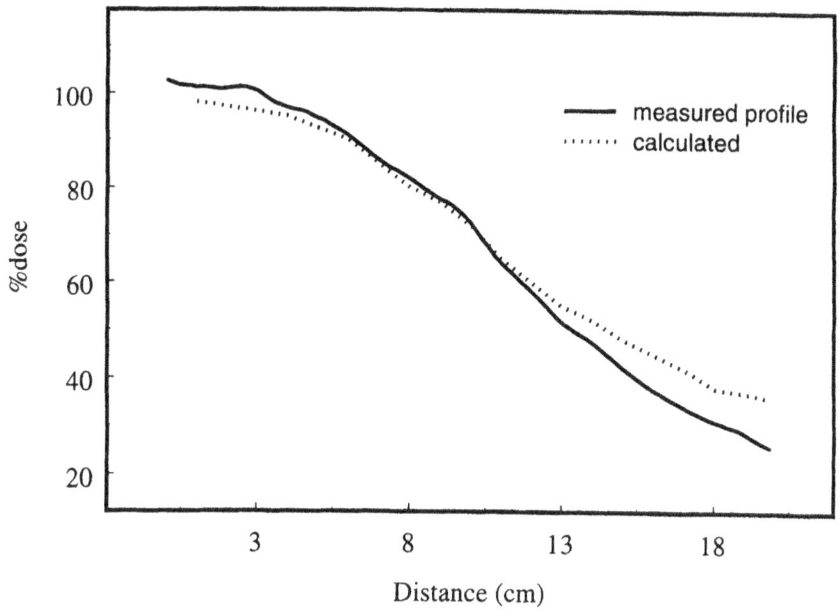

Figure 4. Comparison of measured and calculated beam profiles at 10 cm depth for 20 cm × 20 cm field, demonstrating the reasonably good fit of the algorithm.

algorithm, but instead combined with output factors. We have made dose calibration measurements at a reference depth of 5 cm, which is beyond the buildup region.

4. CONCLUSIONS

Existing treatment planning system for FNT, based on a differential scatter model seems promising for BNCEFNT use. While the measured PDDs follow the same trends with field size as photons or fast neutrons, measured profiles show ill-defined penumbras, especially at larger field sizes. Significant transmission through the collimator necessitates some modifications to the existing algorithm, which are deemed feasible.

ACKNOWLEDGMENTS

We would like to thank Tony He and Weinji Peng for their valuable assistance with the treatment planning algorithm.

REFERENCES

1. Kota C., "Neutron radiotherapy for brain tumors: A retrospective TCP and NTCP analysis", Int. J. Rad. Onc. Biol. Phys., p. 752, Vol 46, No. 3, 2000 (abstract).
2. Maughan R.L., Kota C., and Forman J.D. "Feasibility of boron neutron capture enhancement of fast neutron therapy utilizing a superconducting cyclotron", Advances in Neutron Capture Therapy, Vol I., Elsevier Science, 1997.

3. Kota C., Maughan R.L., Burmeister J., and Forman J.D. "A modified fast neutron therapy beam for boron neutron capture enhanced fast neutron therapy (BNCEFNT)", Advances in Neutron Capture Therapy, Vol I., Elsevier Science, 1997.
4. Burmeister J., Kota C., and Maughan R.L., "Paired Mg and Mg(B) ionization chambers for measurement of boron neutron capture dose in neutron beams", Medical Physics, p. 2482, 26(11), 1999.
5. Burmeister J., Kota C., and Maughan R.L., "Characterization of a modified fast neutron beam for boron neutron capture enhancement of fast neutron therapy", Submitted to *Proceedings of the Eight International Symposium on Neutron Capture Therapy for Cancer*.

105

MODIFICATION OF THE UNIVERSITY OF WASHINGTON NEUTRON RADIOTHERAPY FACILITY FOR OPTIMIZATION OF NEUTRON-CAPTURE-ENHANCED FAST NEUTRON THERAPY

D. W. Nigg[1], C. A. Wemple[1], R. Risler[2], J. K. Hartwell[1], Y. D. Harker[1], and G. E. Laramore[2]

[1]Idaho National Engineering and Environmental Laboratory
Idaho Falls Idaho
[2]University of Washington
Department of Radiation Oncology
Seattle Washington

1. INTRODUCTION

The Fast-Neutron Therapy (FNT) facility at the University of Washington (UW) is the focus for an ongoing collaborative investigation of BNCT-enhanced fast-neutron therapy for certain lung tumors and possibly for other malignancies. This paper describes the design and testing of a modified neutron production target assembly that has been developed to provide improved performance of the UW system for applications involving BNCT-enhanced FNT.

In the UW facility protons are accelerated in a cyclotron to an energy of 50.5 MeV. The resulting proton beam is directed by a series of magnets and focusing devices onto a 10.5 millimeter thick beryllium target located in the treatment head of an isocentric gantry system shown in Figure 1. Neutrons produced in the target are subsequently directed through flattening and wedge filters and then through a multileaf collimator to produce the desired shaped neutron field at the isocenter. The proton beam can also be directed into a fixed horizontal treatment head assembly in an adjacent room. This treatment head is shown in Figure 2. It is not gantry-mounted but it is otherwise basically identical to the head of the gantry system used for routine therapy. It can be used as an experimental facility for animal irradiations, testing of various proposed modifications of the treatment head hardware, etc.

Frontiers in Neutron Capture Therapy, edited by Hawthorne *et al.*
Kluwer Academic / Plenum Publishers, New York, 2001.

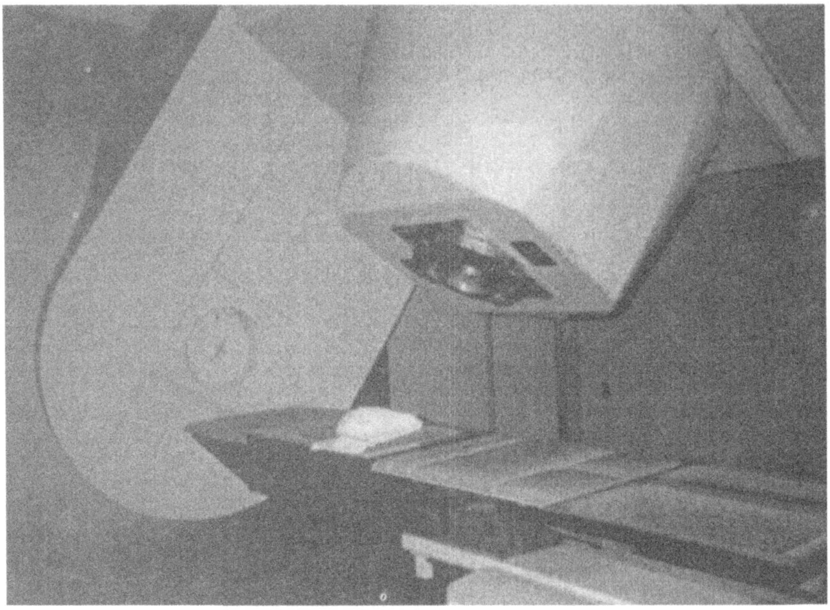

Figure 1. Fast-neutron therapy gantry at the University of Washington.

Figure 2. Fixed horizontal fast-neutron therapy head at the University of Washington.

2. METHODS

Radiation transport computations to model the treatment head of the UW system were performed using coupled Monte Carlo charged particle and neutron transport methods. Proton transport through, and neutron generation within, the neutron production target was modeled using the LAHET code system.[1] Transport of the neutrons from the target through the filter-collimator section of the treatment head and thence to the isocenter was modeled using the well-known MCNP Monte Carlo transport code.[2] Extensive calculations were performed to develop a design for an alternate neutron production target useful for fast neutron therapy with BNCT augmentation. The goal was to arrive at a target design capable of producing a neutron beam that yields essentially the same fast-neutron physical depth-dose curve as is the case with the current system, but which also has a significantly-increased low-energy spectral component, thereby yielding an increased fraction of BNCT enhancement relative to the total therapeutic dose. Although these objectives conflict to a large extent, two laminated-target designs that were expected to provide the desired performance characteristics were successfully identified. The first proposed design features a three-layer target composed of beryllium, carbon, and tungsten, while in the second design only beryllium and tungsten are used.

Measurements of the neutronic performance of a prototype neutron production target based on the second of the two proposed new designs (beryllium-tungsten) were recently completed. This modified target was installed in the horizontal treatment head for testing. It was composed of a 5-millimeter layer of beryllium, followed by a 2.5-millimeter layer of tungsten, with a water-cooled copper backing. Free-field neutron spectrum measurements were performed for the beam produced by the new target using extensions of standard activation-foil techniques, previously validated using the gantry beam.[3] Performance of the new beam in a water phantom was measured using a tissue-equivalent ion chamber to characterize the fast-neutron depth-dose curve and sodium activation in soda-lime glass beads to characterize the thermal-neutron flux (and thus the expected boron neutron capture dose) as a function of depth.

3. RESULTS AND DISCUSSION

The results of the various measurements were quite consistent with expectations based on the design calculations for the modified target. The measured free-field spectrum of the neutron field produced at the isocenter by the modified target, unfolded from ten different activation foil responses using a direct least-squares technique[4] is shown in Figure 3 along with the corresponding measured spectrum of the field produced by the standard target. It should be noted here that the incident proton currents were slightly different for the two targets, as indicated in the figure. The new spectrum features an enhanced low-energy tail relative to the spectrum of the beam produced by the standard target, but it has essentially the same neutron flux above 40 MeV, per unit proton current, and a reduced spectral component in the 10–40 MeV range. The neutron depth-dose curve in a large water phantom is essentially the same for both targets, as required. This is shown for a 10×10 centimeter incident neutron field in Figure 4. It is important to recognize that the measured data in Figure 4 show the depth-*dose* distribution while the calculated data are for the depth-*KERMA* distribution. Accordingly there will be differences

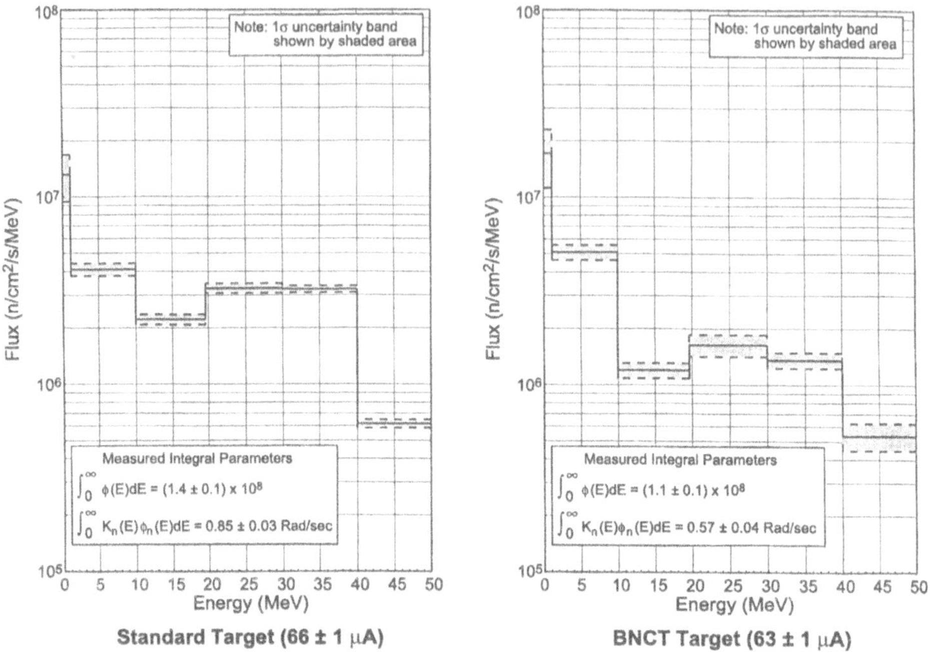

Figure 3. Measured free-field neutron source spectra for the University of Washington Fast-Neutron Therapy Facility.

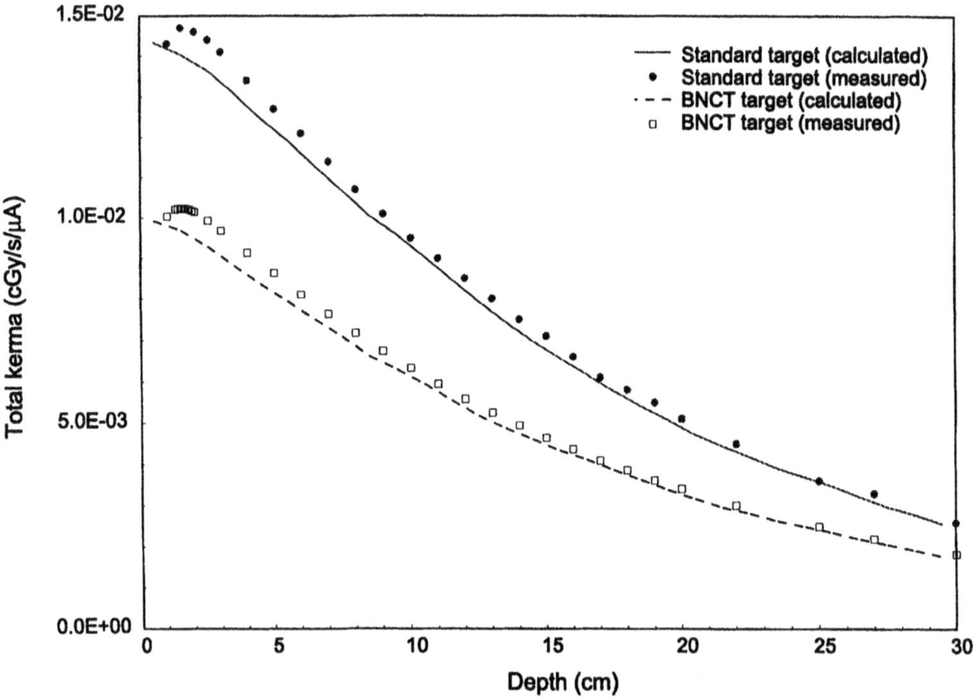

Figure 4. Comparison of total KERMA/depth profiles for old and new UW targets, 10 × 10 cm square beam.

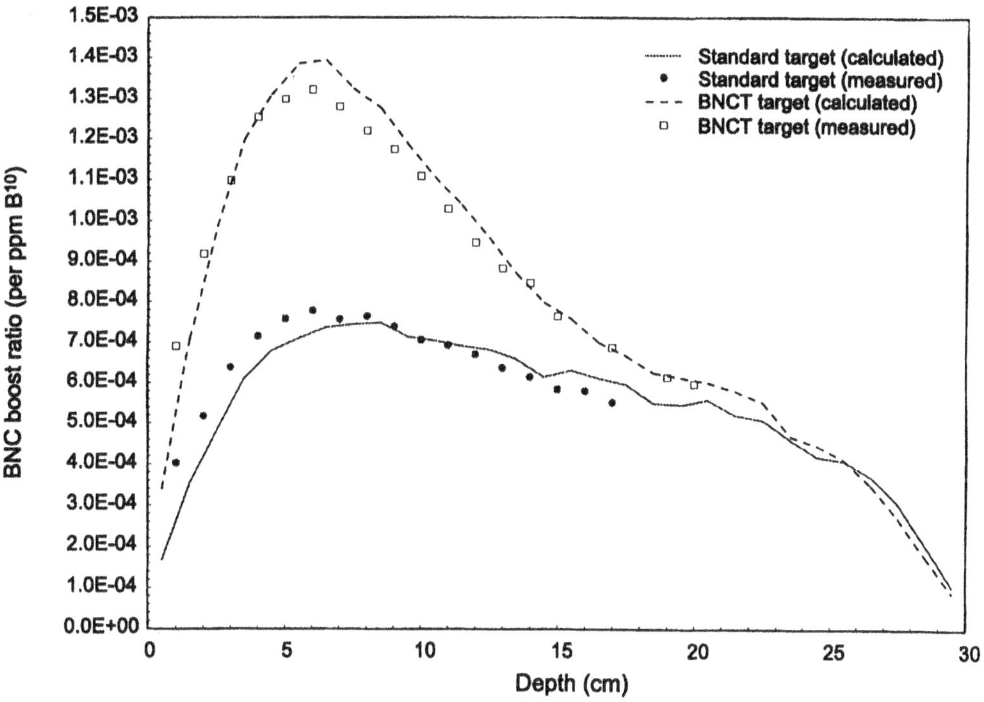

Figure 5. BNC boost ratio for old and new UW targets, 10 × 10cm square beam.

between the calculated and measured data near the surface due to the lack of charged-particle equilibrium for recoil protons in this region.

The levels of BNCT enhancement achievable with each of the two targets are shown in Figure 5. In this figure the "boost-ratio" is defined as the boron KERMA per ppm ^{10}B that is produced at a certain depth, divided by the total dose generated by the incident beam at the same depth with no boron. Thus, for example, with a boron concentration of 100 ppm at a depth of about 6 centimeters, there would be an increase in total dose over that produced by the beam alone of 13–14% with the new target and only about 7% with the standard target. It may be noted also that the boost ratio is field size dependent. Other measurements performed during this study showed, for example, that the BNCT enhancement could be as high as 22% at 100 ppm ^{10}B with the new target and a 20 × 20 cm field.

It should be noted at this point that while the beam produced by the modified UW neutron production target produces the same *physical* depth-dose distribution as is the case with the standard beam, it does not automatically follow that the *RBE-weighted* depth-dose distribution will also be the same. The modified beam has a somewhat softer overall spectrum in the free field and at shallow depths in tissue. The flux-weighted average energy of the new beam is therefore somewhat lower than that of the standard beam, especially within 5–10 centimeters of the surface. Although the high-energy components of the two beams are essentially the same, some neutrons have been shifted from the intermediate-energy range to the low-energy tail in the case of the new beam. Using data cited by Bewley[5] it may be estimated that this downward shift in average beam energy could be expected to produce an increase in the RBE of the UW beam by as much as 7–8% near the surface, but only about 3–5% at depths beyond about 10 centimeters. This

RBE change, while relatively small, may be of some significance when planning treatments using the new beam.

The modified neutron production target thus offers an improved BNCT-enhancement component with minimal sacrifice of the desirable fast-neutron depth-dose characteristics of the neutron beam produced by the standard UW neutron production target. *In vivo* clinical testing of BNCT-enhanced fast-neutron therapy for canine lung tumors using this neutron beam was recently initiated.[6]

ACKNOWLEDGMENTS

This work was sponsored by the United States Department of Energy, Office of Energy Research, under DOE Idaho Operations Office Contract Number DE-AC07-94ID13223.

REFERENCES

1. Prael, R.E., and Lichtenstein, H., The LAHET Code System, LA-UR-89-3014, Los Alamos National Laboratory, USA, 1989.
2. Breismeister, J.F., MCNP—A General Monte Carlo N-Particle Transport Code, Version 4A, LA-12625-M, Los Alamos National Laboratory, USA, 1993.
3. Nigg, D.W., Wemple, C.A., Hartwell, J.K., Harker, Y.D., and Venhuizen, J.R., Validation of Neutron Production and Beam Line Modeling for the University of Washington Fast-Neutron Radiotherapy Facility, In: INEEL BNCT Research Program Annual Report—1997, INEEL-EXT-98-00327, 1998. (*http://id.inel.gov/cart*)
4. Nigg, D.W., and Harker, Y.D., Method Used for Direct Unfolding of Neutron Spectra from Activation Data, In: INEEL BNCT Research Program Annual Report—1997, INEEL-EXT-98-00327, 1998. (*http://id.inel.gov/cart*)
5. Bewley, D.K., The Physics and Radiobiology of Fast-Neutron Beams, Adam Hilger, Bristol,England, 1989.
6. Steltzer, K.J., Gavin, P.R., Risler, R., Kippenes, H., Hawthorne, M.F., Nigg, D.W., and Laramore, G.E., Boron Neutron Capture-Enhanced Fast Neutron Therapy for Non-Small Cell Lung Cancer in Canine Patients, In: "Frontiers in Neutron Capture Therapy," M.F. Hawthorne, K. Shelly, R.W. Wiersema, eds., Kluwer Academic/Plenum Publishers, New York, 2001, pp. 735–739.

BORON NEUTRON CAPTURE-ENHANCED FAST NEUTRON THERAPY (BNC/FNT) FOR NON-SMALL CELL LUNG CANCER IN CANINE PATIENTS

K. J. Stelzer[1], P. R. Gavin[2], R. Risler[1], H. Kippenes[2], M. F. Hawthorne[3], D. W. Nigg[4], and G. E. Laramore[1]

[1]Department of Radiation Oncology
University of Washington Medical Center
Seattle, Washington
[2]Department of Veterinary Clinical Sciences
Washington State University
Pullman, Washington
[3]Department of Chemistry and Biochemistry
University of California
Los Angeles, California
[4]Idaho National Engineering and Environmental Laboratory
Idaho Falls, Idaho

1. INTRODUCTION

The clinical utility of boron neutron capture (BNC) with thermal/epithermal neutron beams has been limited to tumors located at shallow depths, such as in the brain and skin. Capture of the thermal neutron component within a predominantly fast neutron (FN) beam could selectively enhance cytotoxicity in deeply located tumors, such as lung cancer, thereby increasing the availability of this treatment to far greater numbers of cancer patients.[1–4] Even combining concurrent chemotherapy and photon radiation followed by surgery for treatment of locally advanced non-small cell lung cancer (NSCLC), local recurrence was detected in 40% of treatment failures.[5] With improved effectiveness of systemic therapies, local control of bulky disease at the primary site becomes increasingly important for improving median survival and increasing probability of cure. The purpose of this ongoing study was to determine the feasibility and toxicity of BNC/FNT in a trial for dogs with spontaneous NSCLC.

Frontiers in Neutron Capture Therapy, edited by Hawthorne *et al.*
Kluwer Academic / Plenum Publishers, New York, 2001.

2. METHODS

Eligibility criteria included a pathological diagnosis of NSCLC, gross or microscopic residual disease after biopsy or surgery, and consent of the owner. Treatment fields were designed by simulation based on postoperative computerized tomography (CT) scans, and fields were aligned to the skin surface using lasers mounted in the treatment room. Disodium decahydrodecaborate (99% boron-10 enriched) was infused beginning 1 hour prior to FN treatment. The dosage of boron-10 was 25 mg/kg administered by continuous infusion over 1 hour prior to treatment (0.42 mg boron-10/kg/min × 60 min) followed by continuous infusion at half that rate through completion of the treatment session. Study of dogs undergoing surgery for spontaneous lung tumors have found that the tumor concentration of boron-10 after disodium decahydrodecaborate infusion is approximated by the blood concentration (Gavin, unpublished data). The same is true for normal lung tissue based on wet weight, but the concentration in aerated lung would be expected to be lower. Concentration of boron-10 in skin with infusion of this agent in dogs undergoing surgery for brain and lung tumors has been similar to blood concentration. Fast neutron treatment was administered with a clinically-used cyclotron-generated 50.5 MeV proton beam on a beryllium target, producing a fast neutron beam with an average energy of 22 MeV, at 150 cm source to skin distance. Dogs were treated in the prone position using general anesthesia. Dog #3 was treated with a new target assembly to enhance the thermal neutron component within tissue (e.g., 0.115 Gy from BNC per 1 Gy FN dose at 9 cm depth through a 10 × 10 cm field at 100 ppm boron-10). Radiation doses were prescribed to the midplane at central axis, considering only the FN component, through opposed lateral fields to 18 Gy in 4 weekly fractions. Two pairs of square opposed lateral fields were treated at each session, with each pair receiving 2.25 Gy in a session. The first pair (10 × 10 to 16 × 16 cm) covered the primary tumor along with hilar and mediastinal lymph nodes, and included 10–16 cm of spinal cord to model the standard dosing in humans (Figure 1). The second (boost) pair (6 × 6 to 10 × 10 cm) excluded the spinal cord by shifting the fields anteriorly and/or by reducing the field size, and covered the region of residual tumor. Boron-10 was measured in blood samples taken at the start and finish of each treatment session.

3. RESULTS

To date, 3 dogs have been treated. Dog #2 (mixed breed) was euthanized after only 1 fraction due to diagnosis of pleural metastases. Dogs #1 (greyhound) and 3 (mixed breed) completed the entire course of BNC/FNT.

3.1. Dog #1

Histopathology from the original tumor demonstrated extensive lymphatic space invasion. In fact, the gross residual disease treated by BNC/FNT consisted of 2 nodules visible on a postoperative CT scan, one in each lung, encompassable in a single 6 × 6 cm boost field. The mean (+/− one standard deviation) of boron-10 blood concentrations for the 4 fractions was 177 +/− 35 ppm. A thoracic CT scan 3 days following completion of treatment showed complete response of the gross residual disease. There was no evidence of esophagitis as evidenced by weight gain for approximately 1 month after

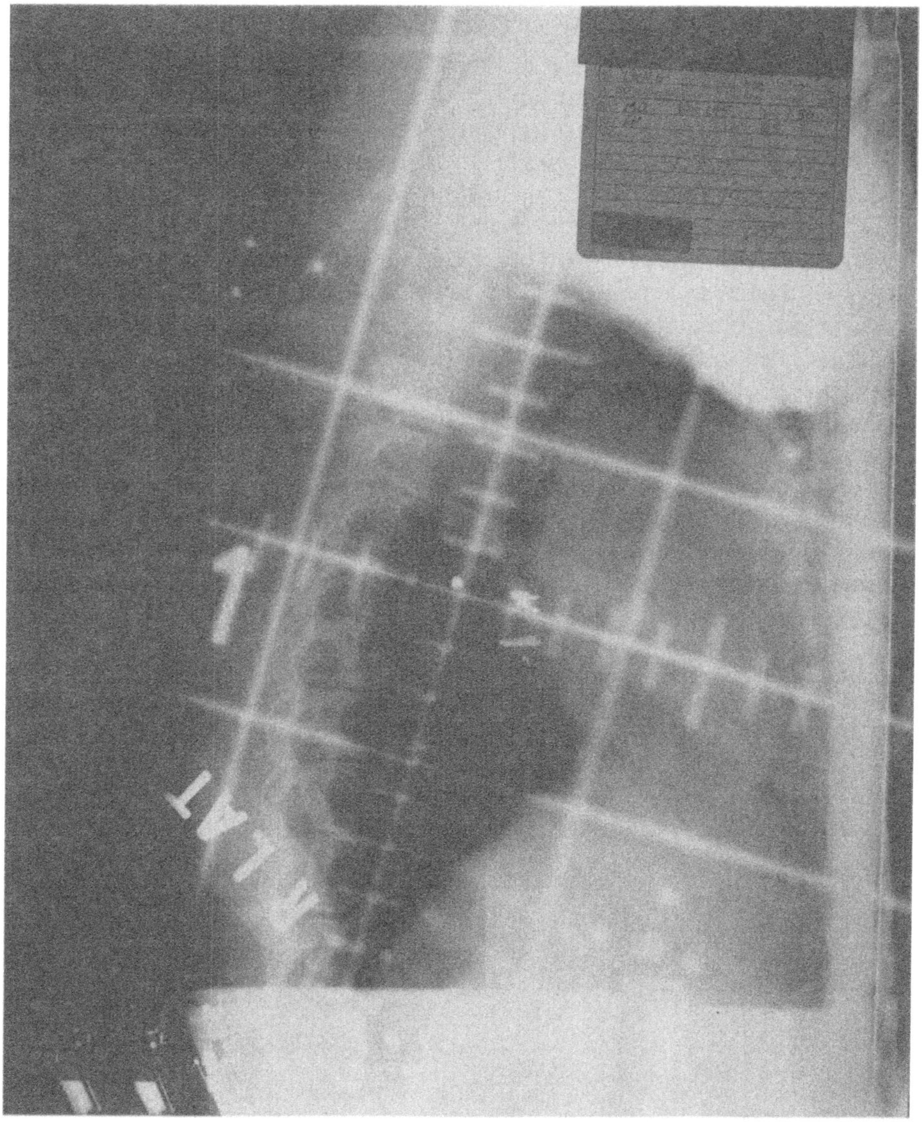

Figure 1. Simulation fields (10 × 10 cm) for Dog #3. The surgical clips represent the site of a microscopically-involved margin along the right mainstem bronchus. Note inclusion of the spinal cord within this field.

completing treatment. There was focal moist desquamation within a region traumatized by shaving to mark the radiation field. This dog was euthanized 2.5 months after completing treatment due to generalized decline. There were no symptoms suggestive of spinal cord injury or radiation pneumonitis. Necropsy revealed tumor with extensive lymphovascular spread and focal areas of alveolar septal thickening and transudate consistent with pneumonitis. Some specimens of lung, presumably outside the treatment volume, did not show pneumonitis. Spinal cord from within the radiation field was histologically normal.

3.2. Dog #3

There was a microscopically positive surgical margin at the right mainstem bronchus. Fluorodeoxyglucose positron emission tomography after surgery and prior to BNC/FNT revealed activity at that site consistent with residual tumor. The mean (+/– s.d.) of boron-10 blood concentrations for the 4 fractions was 109 +/– 3 ppm. Two weeks after completing treatment, dry desquamation was observed. Weight remained stable through treatment. One month after treatment, the dog developed a dry cough associated with a slight decrease in activity, but weight remained within 1 kg of baseline. Three months after treatment, a moist cough developed with low-grade fever, and these symptoms responded to antibiotics. A CT scan revealed changes in the lung parenchyma that corresponded to the radiation field and were consisted with radiation pneumonitis (Figure 2).

Pericardial fluid was also visible caudal to the radiation field. There was no evidence of recurrent tumor on that CT scan. Six months after treatment, the dog was euthanized because of rapid decline with cough, dyspnea, anorexia, and vomiting. There were no signs of spinal cord injury on exam. Necropsy was performed at a community veterinary facility and did not provide optimal sampling of tissues of interest. Lung specimens showed no evidence of tumor, but did have acute vascular congestion and

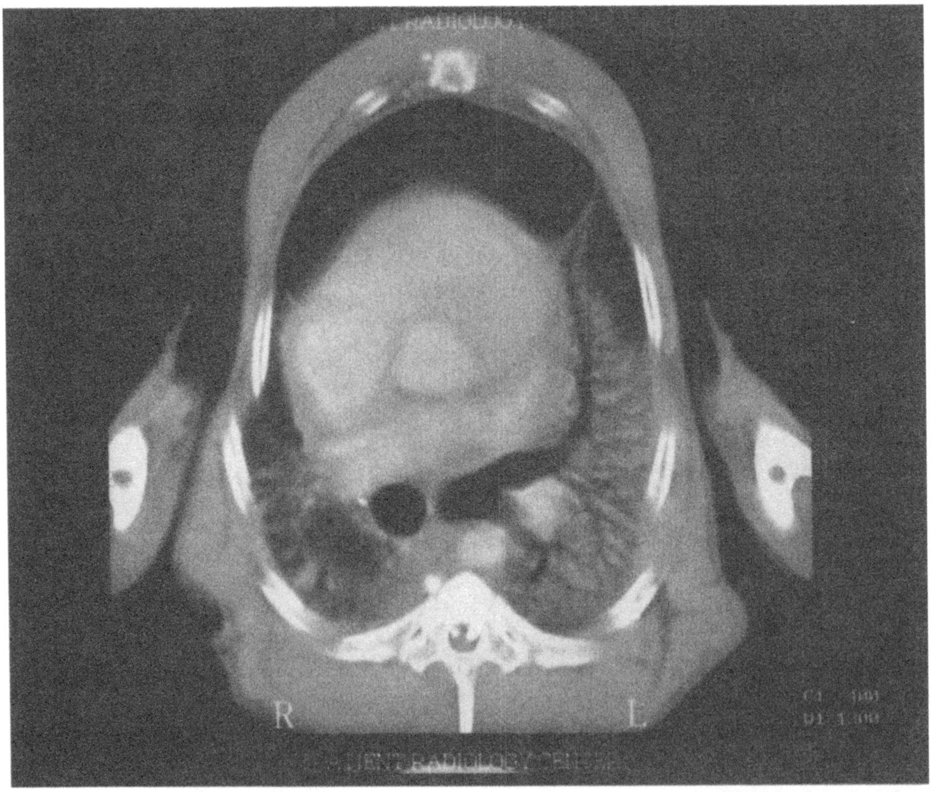

Figure 2. Computerized tomography scan 3 months after treatment. The pulmonary infiltrate and air bronchograms were within the radiation field. Note the surgical clip at the right mainstem bronchus for correlation with Figure 1.

hemorrhage reflecting terminal cardiovascular collapse. The absence of lung parenchyma fibrosis in light of the CT scan suggests that the lung tissue sampled at necropsy was outside the radiation field. There was a focal 2×0.6 cm hemorrhage with fibrosis in the atrial myocardium with macrophages containing hemosiderin, suggestive of chronic bleeding into the atrial wall. Immunohistochemistry staining of the heart tissue did not reveal tumor cells.

4. CONCLUSIONS

Fast neutron therapy augmented with boron neutron capture may be feasible in canine patients with NSCLC. In the dogs treated to date, the level of skin and pulmonary toxicity was no more severe than that which is commonly observed for human patients undergoing fast neutron radiation therapy at this facility for NSCLC. There was no evidence of significant acute esophagitis and no symptoms or histologic evidence of spinal cord injury. The etiology of the myocardial hemorrhage remains unknown. Approximately 60% of the heart volume received a fast neutron dose of 9 Gy and 30% of the heart volume received a fast neutron dose of 18 Gy. Of note, Dog #1 had a similar volume of heart treated, but showed no evidence of injury on histopathology. Although the primary focus of this study is on toxicity, the complete response of gross disease observed in Dog #1 immediately after completion of therapy is encouraging.

For Dog #3, the physical dose from the BNC component in tumor was estimated as 2.3 Gy (total physical dose of high linear energy transfer radiation approximately 20.3 Gy). Because of the steep nature of the dose-response curve for fast neutrons, a selective tumor augmentation with high linear energy transfer radiation of this magnitude has the potential for a substantial increase in the probability of tumor control.[3,4] We are planning future treatments of canine patients with escalation of the fast neutron dose to 20 Gy while maintaining spinal cord dose at 9 Gy.

REFERENCES

1. G.E. Laramore, P. Wooton, J.C. Livesey, D.S. Wilbur, R. Risler, M. Phillips, J. Jacky, T.A. Buchholz, T.W. Griffin, and S. Brossard. Boron neutron capture therapy: a mechanism for achieving a concomitant tumor boost in fast neutron radiotherapy, *Int. J. Radiat. Oncol. Biol. Phys.*, 5:1135–1142, 1994.
2. W.-J. Koh, J.M. Krall, L.J. Peters, M.H. Maor, G.E. Laramore, C.M. Burnison, L.W. Davis, S. Zink, and T.W. Griffin. Neutron vs photon radiation therapy for inoperable regional non-small cell lung cancer: results of a multicenter randomized trial, *Int. J. Radiat. Oncol. Biol. Phys.*, 27:499–505, 1993.
3. G.E. Laramore, and K.J. Stelzer. Neutron capture therapy (NCT) enhancement of fast neutron radiotherapy: past, present, and future directions, *in* "Frontiers in Neutron Capture Therapy," M.F. Hawthorne, K. Shelly, R.W. Wiersema, eds., Kluwer Academic/Plenum Publishers, New York, 2001, pp. 695–702.
4. K.J. Stelzer, K.L. Lindsley, P.S. Cho, G.E. Laramore, and T.W. Griffin. Fast neutron radiotherapy: the University of Washington experience and potential use of concomitant boost with boron neutron capture, *Radiation Protection Dosimetry*, 70:471–475, 1997.
5. K.S. Albain, V.W. Rusch, J.J. Crowley, T.W. Rice, A.T. Turrisi, J.K. Weick, V.A. Lonchyna, C.A. Presant, R.J. McKenna, D.R. Gandara, H. Fosmire, S.A. Taylor, K.J. Stelzer, K.R. Beasly, and R.B. Livingston. Concurrent cisplatin/etoposide plus chest radiotherapy followed by surgery for stages IIIA (N2) and IIIB non-small-cell lung cancer: mature results of Southwest Oncology Group phase II study 8805, *J. Clin. Oncol.*, 13:1180–1892, 1995.

BNCEFN OPTIMISATION WITH LEAD BLOCKS COLLIMATION AND GRAPHITE EMBEDDING

The Monte Carlo Simulation of Various Fast Neutron Beams Intercomparison

Jean-Philippe Pignol[1], Philippe Paquis[2], Roland Sabattier[3], Arlene Lennox[4], Thomas Kroc[4], Dan Jones[5], Jakobus Slabbert[5], Cheikh M. B. Diop[6], Gilbert Blondiau[3], and Noël Breteau[3]

[1]Département de radiothérapie
Institut Gustave-Roussy
94805 Villejuif cedex, France
[2]Service de Neurochirurgie
Hôpital Pasteur
06002 Nice cedex 1, France
[3]CERI-CNRS
3A rue de la Ferrollerie
45071 Orléans Cedex 2, France
[4]Fermi National Accelerator Laboratory
MS301, Batavia, Illinois 60510
[5]National Accelerator Centre
ZA Faure 7131 South Africa
[6]SERMA, Commissariat à l'Energie Atomique
91191 Gif sur Yvette cedex, France

1. INTRODUCTION

Several authors have demonstrated the feasibility and efficiency of Boron Neutron Capture Enhancement of Fast Neutron (BNCEFN) at their facility.[1-5] The two major difficulties of this technique are to incorporate a large amount of ^{10}B selectively into the tumor (>100 μg/g) and to optimize neutron thermalization in the tissues. Unfortunately, it has been also demonstrated that the number of thermal neutrons per fast neutron Gy ($n_{th}.Gy^{-1}$) is inversely proportional to fast neutron energy, and strongly proportional to the size of fast neutron irradiation field.[3-5] Thus, a dramatic decrease of thermal

Frontiers in Neutron Capture Therapy, edited by Hawthorne *et al.*
Kluwer Academic / Plenum Publishers, New York, 2001.

neutron flux is seen when the irradiation fields are reduced to conform the target volume.

For the purpose of treating glioblastoma with BNCEFN, the use of a lead collimator placed in close contact with the head, combined with an appropriate scattering medium, facilitates proper collimation of the fast neutron beam and causes an increase, rather than a decrease, in the thermal neutron flux.[5] Moreover, with this technique, the number of $n_{th}.Gy^{-1}$ appears to be independent to the fast neutron irradiation field size.[6]

To study the relative magnitude of these effects in different fast neutron beams, Monte Carlo simulations, using the combination of the FLUKA and MCNP codes,[7] were performed using the following facility geometries: p(66)+Be(40) National Accelerator Centre (South Africa), p(66)+Be(49) without hardening filter Fermi National Accelerator Laboratory (Fermilab) (USA), and p(34)+Be Orleans (France).

2. MATERIAL AND METHODS

The primary neutron yield from each facilities target was first calculated with the three dimensional charged particle FLUKA source code.[8] Input data included accurate descriptions of each target layer (beryllium, graphite, copper . . .) in cylindrical geometries. The neutron productions were given in neutrons per source proton. They were calculated in the forward direction, in 0° to 15° polar angles, following a 24 bins energy grid ranging from 2.8×10^{-4} MeV to the highest neutron energy available. Standard deviations less than 5% were achieved by simulating 2×10^6 proton interactions for each target configuration.

The neutron transport was simulated with the MCNP Monte Carlo code,[9] using ENDF B-VI cross sections.[10] The three dimensional configuration of each fast neutron facility was described in the input files. The first set of simulations used a parallelepiped water phantom placed at the appropriate SSD to test the influence of the irradiation field size on thermal neutron enhancement. Calculations were done for $10 \times 10\,cm^2$ and $20 \times 20\,cm^2$ fields. A second set of simulations was made to evaluate the relative contribution of the lead blocks collimator, using the same phantom irradiated with the widest primary collimated field available, and with a 20 cm thick lead ($\rho = 11.35\,g.cm^{-3}$) additional collimator delimiting a $10 \times 10\,cm^2$ field. A third set of simulations was made to quantitize the effect of the full BNCEFN optimization, i.e., using lead blocks collimation and a graphite bolusing media (Figure 1). In the first case, a spherical phantom was placed at the appropriate SSD, embedded in a graphite parallelepiped and placed behind lead blocks with the primary collimator were set to the maximum field size. In the second case, the phantom was placed in air while the primary collimators delimited a $5 \times 5\,cm^2$ field. A Thermal Optimization Parameter (TOP) was defined as the ratio of thermal neutron flux per fast neutron Gy at 5 cm depth in the phantom, for the different optimization configurations.

For each configuration a standard deviation close to 2% was achieved by simulating the transport of 2×10^7 source neutrons. In the phantoms, 20×20 tally grids of $1\,cm^3$ cubes were described on a plane parallel to the beam axis, in order to get a realistic image of the thermal neutron flux and of the fast neutron Kinetic Energy Released in Material (KERMA). The KERMA associated with the fast neutrons were calculated using heating numbers given by the LLDOS library,[7] and the thermal neutrons were defined as neutrons having maximum energy of 10^{-7} MeV.

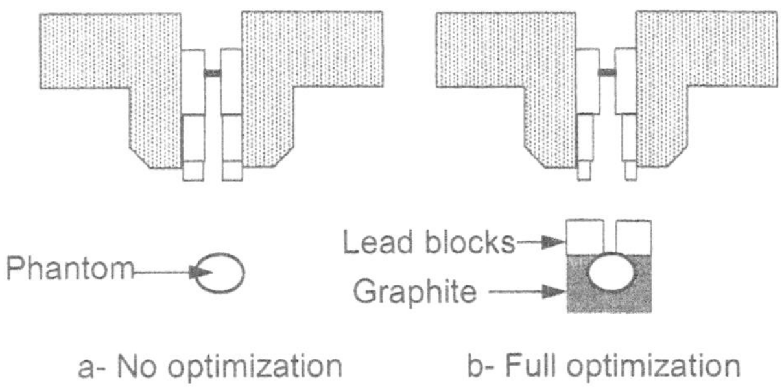

Figure 1. Schematic representation of the two configurations simulated for each fast neutron beam facility, in order to test the impact of BNCEFN optimization with additional lead blocks collimation and graphite bolusing embedding.

3. RESULTS

Table 1 shows the influence of the irradiation field size and the fast neutron mean energy on the number $n_{th}.Gy^{-1}$ calculated at 5 cm depth in the phantom for the three facilities. For each beam, the thermal neutron flux is reduced of a factor 2 when the irradiation field is reduced from $20 \times 20\,cm^2$ to $10 \times 10\,cm^2$. The thermal neutron flux also decrease when the fast neutron energy increase. Note that even though they have the same proton energies, Fermilab's mean fast neutron energy is lower than the NAC's, because of the larger amount of energy absorbed in the beryllium and the lack of hardening filter.

Figure 2 shows the contribution of the lead blocks collimator on the thermal neutron yield, in the phantom, and on the beam axis. The intercomparison between the NAC and the Orleans beams shows that the impact of lead collimation is higher for the higher energy beam. The Fermilab curves shows shapes closer to the Orleans than the NAC beam.

Table 2 shows the results of the lead blocks collimator and graphite embedding optimization evaluation. Happily, when a $5 \times 5\,cm^2$ primary collimated field is used to irradiate a cylindrical water phantom, the influence of the fast neutron energy disappears, i.e., very similar amount of thermal neutrons per fast neutron Gy are calculated. When a full optimization is used, nearly the same amount of n_{th}/Gy are found at 5 cm depth in the phantom.

Finally, in the full optimization configuration TOP close to 20 are calculated for each beam. That means that, assuming a ^{10}B concentration within the of 100 μg/g of ^{10}B, a physical dose enhancement of 22% is calculated for the full-optimization arrangement while this enhancement is only of 1.1% in the no-optimization set.

Table 1. Influence of fast neutron energy and field size on the thermal neutrons per fast neutron Gy measured at 5 cm within a large parralelepiped water phantom

Facility	10 × 10	20 × 20	Ratio
Orleans p(34)+Be	10.4 $10^9 n_{th}.Gy^{-1}$	21.1 $10^9 n_{th}.Gy^{-1}$	2.02
Fermi Lab p(66)+Be(49)	5.08 $10^9 n_{th}.Gy^{-1}$	11.2 $10^9 n_{th}.Gy^{-1}$	2.21
NAC p(66)+Be(40)	3.86 $10^9 n_{th}.Gy^{-1}$	8.37 $10^9 n_{th}.Gy^{-1}$	2.16

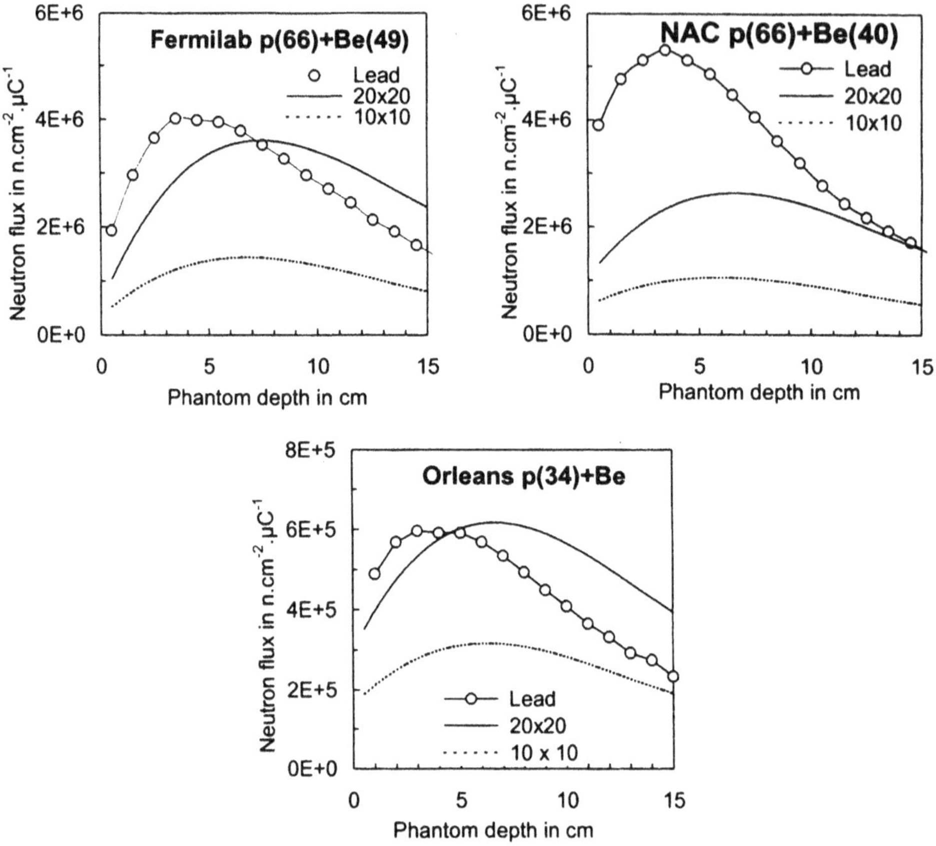

Figure 2. Thermal neutron flux calculated within the phantom for the three fast neutron facilities. Three configurations were studied: primary collimated 10 × 10 cm^2 or 20 × 20 cm^2 fields, and a 10 × 10 cm^2 field shaped with the additional collimator.

4. DISCUSSION

Although FLUKA is a source code dedicated to high energy physic, and that ENDF B-VI cross sections are only fully reliable up to 20 MeV (with a linear interpolation over 20 MeV), a previous comparison to experimental thermal flux measurement at the Nice facility has shown that the FLUKA-MCNP code combination yields reliable results for BNCEFN.[7]

As has been previously demonstrated by other authors,[3,4,5] the thermal neutron flux measured in a phantom depends strongly on the field size and the mean fast neutron energy. But this dependence is observed when measurements are made in a large scattering medium such as a parallelepiped phantom. It disappears when smaller irradiation fields and no scattering media are used. This finding suggests that the mean fast neutron energy influence is strongly associated with neutron scattering.

Our results also demonstrate that the use of heavy metal collimators causes a greater enhancement for higher energy beams like the NAC. For the Fermilab, despite the fast neutron are also generated with 66 MeV proton, the larger amount of energy

Table 2. Evaluation of the impact of the of BNCEFN optimization with additional lead blocks collimation and graphite reflector for various fast neutron beam facility[a]

Facility	No optimization	Full optimization	TOP
Orleans p(34)+Be	1.07 $10^9 n_{th}.Gy^{-1}$	21.6 $10^9 n_{th}.Gy^{-1}$	20.2
Fermilab p(66)+Be(49)	1.04 $10^9 n_{th}.Gy^{-1}$	18,8 $10^9 n_{th}.Gy^{-1}$	18.1
NAC p(66)+Be(40)	0.90 $10^9 n_{th}.Gy^{-1}$	20.5 $10^9 n_{th}.Gy^{-1}$	22.8

[a]The number of thermal neutrons per ast neutron Gy calculated at 5 cm depth within a cylindrical water phantom are reported. A Thermal Optimization Parameter (TOP) is calculated between the no-optimization and the full-optimization configuration.

absorbed in the beryllium and the lack of hardening filter, and finally the greater amount of low energy neutrons might explain that the thermal neutron yield curves in the phantom are closer to the Orleans ones. Looking at the final enhancement, very close TOP values are calculated for each beams when the full optimization is used. The fact that the lead collimation improvement on thermal flux is more important for the harder fast neutron beams, and that the same n_{th}/Gy are found for the three facilities for a cylindrical phantom irradiated in air, suggest that the impact of the graphite bolusing material is more important for the lower energy fast neutron beams.

The use of lead blocks collimation and graphite scattering medium appears to be a very powerful and a robust solution for BNCEFN, regardless of the initial fast neutron characteristics for the range of the neutron beam energies used in this study. The full optimization condition is basically a secondary lead collimator which also acts as a scatterer source of thermal neutron. When no graphite is used (Figure 2), the $n_{th}.Gy^{-1}$ ratio decreases with depth after a peak at 4 cm depth. When graphite embedding is used, this ratio remains roughly constant with depth.[6]

Compared to standard FN dosimetry, the treatment plans shows significant but acceptable modifications:[6] the 40%, 60% and 80% isodoses are shifted 1.5 cm closer to the phantom surface; a transmission of 10% is found behind 20 cm of lead. The photon

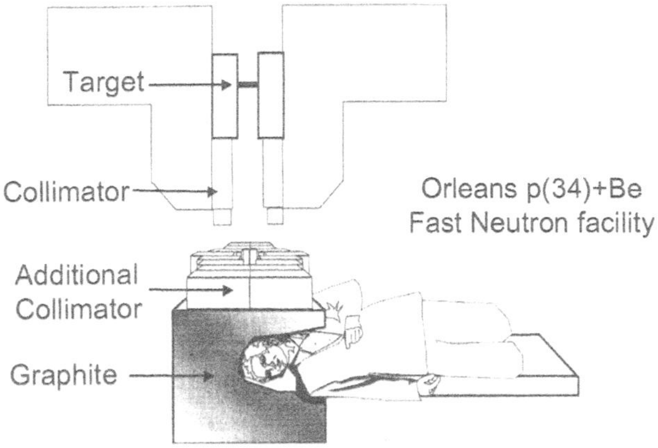

Figure 3. Patient set-up using the additional lead collimation device and the graphite embedding at the Orleans facility.

transport was not implemented in our calculations, to avoid excessive time cost of MC simulation. However, supplementary simulations shows that an 11% additional dose deposition is expected in all the phantom volume from the photon generated in the graphite and the lead additional collimator.

Figure 3 shows the design of the additional collimator and graphite embedding set-up planned at the Orleans facility. Patients would be positionned of the irradiation table, with the head accurately fixed by a mask. The irradiation field would be shaped by an customized additional lead collimator, placed at the right position using the beam light simulation. Finally, two graphite blocks, would be placed around the patient just before the FN irradiation to improve neutron scattering.

REFERENCES

1. W. Sauerwein, W. Ziegler, H. Szypniewski, and C. Streffer, Boron neutron capture therapy (BNCT) using fast neutrons: effects in two human tumor cell lines. *Strahlenther. Onkol.*, 166:26–29, 1990.
2. L. Lüdeman, T. Matzen, M. Matzke, R. Schmidt, and W. Scobel, Determination of the thermal neutron flux in a fast neutron beam by use of a boron-coated ionization chamber. *Med. Phys.*, 22:1743–1747, 1995.
3. G.E. Laramore, P. Wooton, J.V. Livesey, D.S. Wilbur, R. Risler, M. Phillips, J. Jacky, T.A. Buchholz, T.W. Griffin, and S. Brossard, Boron neutron capture therapy: A mechanism for achieving a concomitant tumor boost in fast neutron radiotherapy. *Int. J. Rad. Oncol. Biol. Phys.*, 28:1135–1142, 1994.
4. T.A. Buchholz, G.E. Laramore, K.J. Telzer, R. Risler, P. Wooton, and T.W. Griffin, Boron neutron capture enhanced fast neutron radiotherapy for malignant gliomas and other tumors. *J. Neuro Oncol.*, 33:171–178, 1997.
5. P. Paquis, J.P. Pignol, M. Lonjon, N. Brassart, A. Courdi, P. Chauvel, P. Grellier, and M. Chatel, Boron neutron capture enhancement (BNCE) of fast neutron irradiation for glioblastoma: increase of thermal neutron flux with heavy material collimation a theoretical evaluation. *J. Neuro Oncol.*, 41:21–30, 1999.
6. J.P. Pignol, P. Paquis, P. Cuendet, D. Gibon, C.M.B. Diop, and R. Sabattier, Beam collimation and patient contentious material Optimizations for ^{10}Boron Neutron Capture Enhancement of fast neutron (BNCEFN): definition of the optimum irradiation technique. *Int. J. Rad. Oncol. Biol. Phys.*, 43:1151–1159, 1999.
7. J.P. Pignol, P. Cuendet, N. Brassart, G. Fares, F. Colomb, C.M.B. Diop, R. Sabattier, A. Hachem, and G. Prevot, Combined use of FLUKA and MCNP-4A for the Monte Carlo simulation of the dosimetry of ^{10}Boron Neutron capture Enhancement of fast neutron. *Medical Physics*, 25:885–891, 1998.
8. A. Fasso, A. Ferrari, J. Ranft, and P.R. Sala, FLUKA: Performances and Applications in the Intermediate Energy Range, *in* <<Proceedings of the Specialists Meeting on Shielding Aspects of Accelerators Target and Irradiation Facilities>>. Arlington Texas April 28–29, 1994.
9. J.F. Briesmeister, <<MCNP—A general Monte Carlo code for Neutron and Photon transport version 4-A>>. Los Alamos National Laboratory LA-7396-M, 1986.
10. J.S. Hendricks, S.C. Frankle, and J.D. Court, <<ENDF/B-VI Data for MCNP>>. Los Alamos National Library LA-12891, 1994.

ON THE STUDY OF A LAMINATED BERYLLIUM/TUNGSTEN TARGET FOR BORON NEUTRON CAPTURE ENHANCED FAST-NEUTRON THERAPY

Richard Oberer, C.-K. Chris Wang, and Lisa Chiang

Nuclear Engineering and Health Physics Program
Georgia Institute of Technology
Atlanta, Georgia, 30332-0225

1. INTRODUCTION

Researchers at several fast neutron therapy (FNT) facilities have been working on the modification of their existing beams to implement **b**oron **n**eutron **c**apture **e**nhanced **f**ast **n**eutron **t**herapy (BNCEFNT). The goal is to achieve a tumor dose enhancement of a traditional FNT by 15 to 20% via $^{10}B(n,\alpha)$ reactions. Most of the existing FNT facilities produce neutrons by bombarding protons or deuterons onto a beryllium (Be) target. The neutrons produced from such a target are so energetic that only a very small percentage of them are thermalized to produce $^{10}B(n,\alpha)$ reactions. The effectiveness of the B-10 enhancement can be quantitatively studied via a parameter called *t*umor *d*ose *e*nhancement *f*actor (TDEF). It is defined as

$$TDEF = \frac{D_{B-10}}{D_n}\frac{RBE_{B-10}}{RBE_n} \tag{1}$$

where D_{B-10} is the absorbed dose due to $^{10}B(n,\alpha)$ reactions, D_n is the absorbed dose due to fast neutrons, RBE_{B-10} is the relative biological effectiveness of the absorbed dose due to $^{10}B(n,\alpha)$ reactions, and RBE_n is the relative biological effectiveness of the absorbed dose due to fast neutrons.

To increase TDEF, the general approach is to soften the neutron beam spectrum. A previously attempted method is to moderate the beam spectrum by placing a tungsten (W) block in the beam collimator. This method has a major disadvantage that the tungsten block significantly attenuates the original fast neutron beam intensity. Recently, Nigg

Frontiers in Neutron Capture Therapy, edited by Hawthorne *et al.*
Kluwer Academic / Plenum Publishers, New York, 2001.

et al. proposed the use of a laminated target to implement the BNCEFNT for the University of Washington Neutron Radiotherapy Facility.[1] The laminated target is comprised of a disk of beryllium followed by a disk of tungsten. This target produces both high-energy and low-energy neutrons. High-energy neutrons are produced as the incident protons (or deuterons) pass through the beryllium, and low-energy neutrons are produced as the protons (or deuterons) slow down and stop in tungsten. The advantage of the laminated target is that it produces abundant low-energy neutrons without a significant reduction of the intensity of high-energy neutrons.

The use of a softer neutron beam to increase TDEF, however, is not without a price. The trade-off turns out to be that as the incident beam spectrum becomes softer, the neutron dose drops more sharply with depth in tissue. That is, although a soft neutron beam does increase TDEF, it also causes more damage to normal tissue at the beam entrance. Therefore, for a complete evaluation of a BNCEFNT facility, one must consider this trade-off carefully. In this study, we performed Monte Carlo neutron transport calculations to obtain TDEF and neutron dose at various depths in a phantom for a laminated neutron production target. The study is intended to quantitatively assess the BNCEFNT of the neutron beam at the Fermi National Accelerator Laboratory, Neutron Therapy Facility (Fermi NTF).

2. DESCRIPTION OF THE TARGET

The laminated target considered in our computational study includes two geometric configurations and several different thickness combinations of beryllium and tungsten. In the first geometric configuration, the beryllium is immediately followed by tungsten. This configuration is herein referred as the "joined target." In the second configuration, the tungsten was moved 100 cm closer to the isocenter. The low-energy part of the neutron beam spectrum at the patient is thereby significantly increased. The second geometric configuration is herein referred as the "split target."

Various thickness combinations of beryllium and tungsten were considered. The targets were given names of the form BeXXW, where XX is the proton energy lost in the beryllium. For example, Be10W means that 10 MeV of proton energy of the 66 MeV proton beam is lost in the beryllium and the remaining energy of proton is lost in the tungsten. The other dimensions and placement of the laminated target are identical to the existing target at Fermi NTF for the joined target cases. In the case of Be49, no tungsten follows the beryllium. Target Be49 represents the existing Fermi NTF target previously described.

3. COMPUTATIONAL METHOD

The computations were performed in two steps. In the first step, a point neutron source was generated for input into the second step. The point neutron source was computed using either MCNPX[2] or LAHET.[3] The proton beam was modeled as a monoenergetic, monodirectional plane source impinging on the beryllium and tungsten target. The resulting neutron fluence was then tallied over the surface of a 50 cm sphere by energy and angle. In the second step, the angular and energy dependent neutron source was input into MCNP 4B,[4] which was used to transport this neutron point source through the collimator assembly into the phantom.

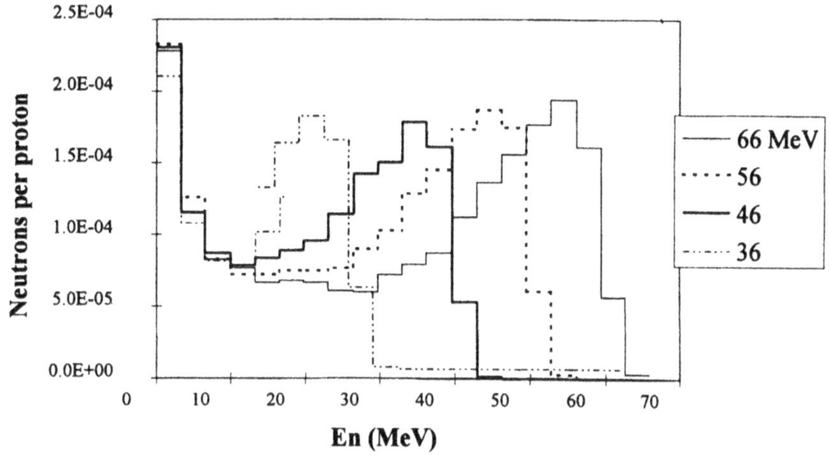

Figure 1. Neutron spectra from the various 5-MeV thick beryllium segments comprising the target.

Decoupling the computation into two or more steps has three disadvantages. First, it converts random errors in previous steps into systematic errors in subsequent steps. Second, information about the energy and angle of the neutron source is lost through binning of the data, and finally, it makes determination of the true uncertainty of the computation difficult.

4. RESULTS AND DISCUSSION

Figures 1 and 2 shows the neutron spectrum from each 5 MeV thick segment of the laminated target. The spectra shown are for the neutrons in a forward cone of 25.8 degrees ($\cos(\theta) > 0.9$). The spectra of the neutrons arising from the beryllium for several

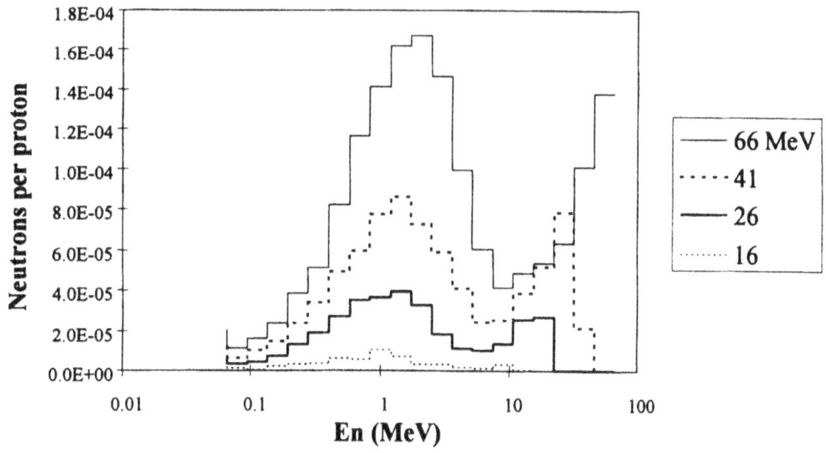

Figure 2. Neutron from the various 5-MeV thick tungsten segments comprising the target.

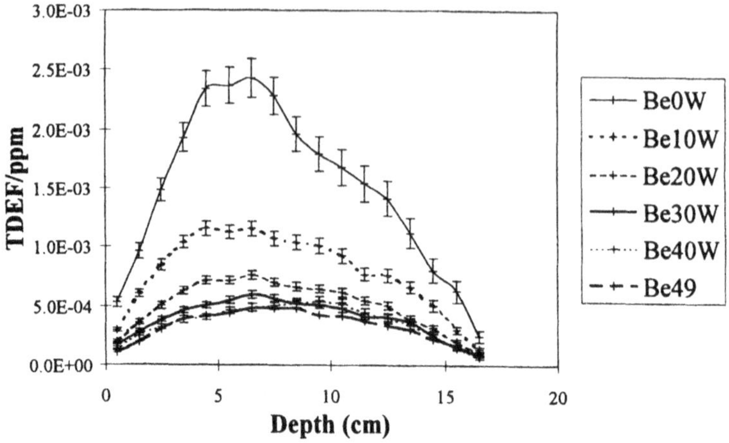

Figure 3. TDEF per ppm of B-10 as a function of depth in the phantom for the joined targets.

of the disks are shown in Figure 1. These neutrons tend to be very forwardly directed. The spectrum tends to peak near the energy of the impinging protons.

The neutrons emitted from the tungsten, on the other hand, tend to be nearly isotropic. The spectra are shown in Figure 2. There is a component centered around 1 MeV, which is independent of the proton energy.

The neutron spectrum for a particular target is compiled by adding up these individual components. For a physically realizable target all of the energy ranges must be included with no overlap. The strategy of the laminated target approach for an improved TDEF is to reduce the mid range neutron energies while augmenting the higher and lower ends of the spectrum.

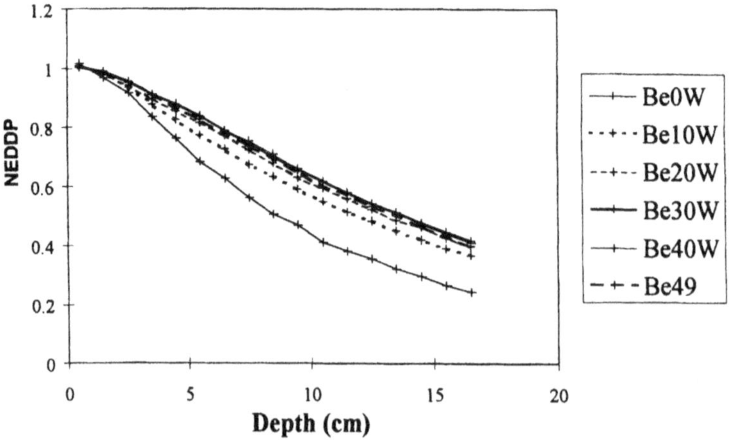

Figure 4. Normalized effective depth-dose profile in the phantom for $cRBE = 60$, joined target.

4.1. The Joined Target

TDEF was normalized to 1 ppm of B-10. In this study, TDEF per ppm of B-10 was computed for each of 16 (1 cm) cells in a phantom and is shown in Figure 3. As indicated in Eqn. 1, the dose due to photons generated from the neutrons was not included. Calculations showed that this photon dose was small compared to the neutron dose, D_n.

4.1.1. Normalized Effective Depth-Dose. To study the trade-off between TDEF and depth-dose distribution, we invented a new parameter called *n*ormalized *e*ffective *d*epth-*d*ose *p*rofile (NEDDP), which is the effective dose in cell *i* normalized to the effective fast neutron dose to the normal tissue at surface (i.e. cell 1). That is,

$$NEDDP = \frac{D_{i,n}RBE_n + cD_{i,B-10}RBE_{B-10}}{D_{1,n}RBE_n} \tag{2}$$

where c is the concentration of B-10 in the tissue in ppm and $D_{i,B\text{-}10}$ is the absorbed dose due to ^{10}B(n,α) reactions from 1 ppm of B-10 in cell *i*. Eq. 2 can be recast into

$$NEDDP = \frac{D_{i,n} + cRBE \cdot D_{i,B-10}}{D_{1,n}} \tag{3}$$

$$\text{where} \quad cRBE = \left(c\frac{RBE_{B-10}}{RBE_n} \right)$$

It is intended to represent the normalized tumor effective dose for tumors at any arbitrary depth in a patient. This normalization is justified by the fact that the dose to normal tissue at surface most likely will limit the dose that can be delivered to a tumor. Figure 4 shows the NEDDP for neutron beams produced from the various target

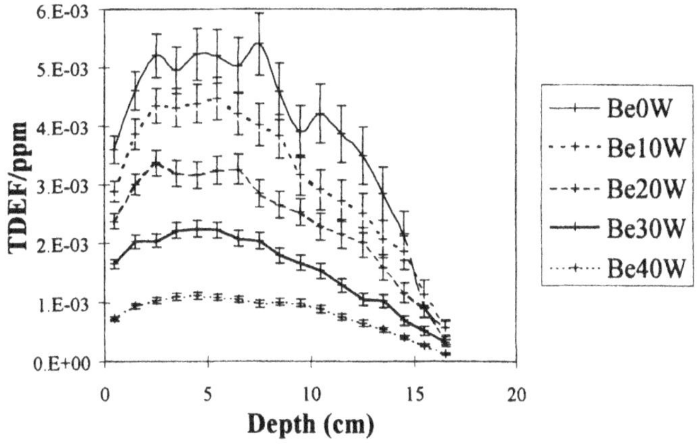

Figure 5. TDEF per ppm as a function of depth in the phantom for the split targets.

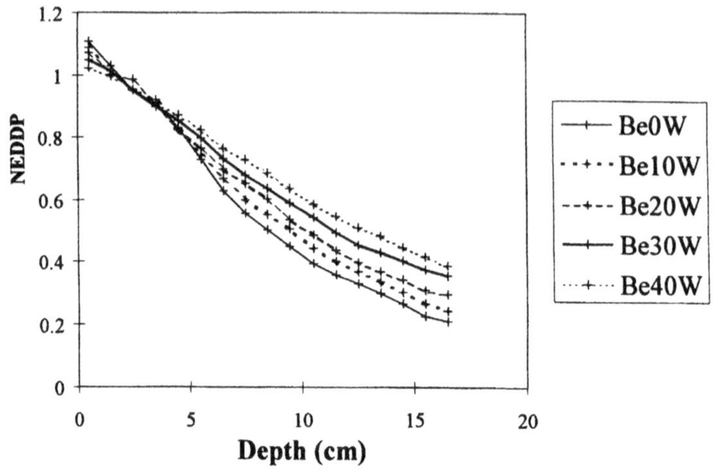

Figure 6. Normalized effective depth-dose profile in the phantom for $cRBE = 60$, split target.

thickness combinations for a nominal value of $cRBE = 60$. As shown, the NEDDP in a soft neutron beam drops more rapidly in a phantom than it does in a hard beam. The use of a soft neutron beam is therefore disadvantageous. In order for a soft neutron beam (i.e. the Be0W beam) to have an advantage over the hard beam (i.e. the Be49 beam), the normalized effective dose to a tumor in the soft beam must be greater than that in the hard beam. Our calculations indicate that this would require a *cRBE* value greater than 200. At this value the slight advantage extends only a few centimeters into the phantom. To extend the advantage to 10 cm, *cRBE* must be greater than 1,000 which is beyond the reach of all existing B-10 compounds.

4.2. Split Target

The results for the split target are shown in Figures 5 and 6. The NEDDP of the Be40W target is nearly the same as that of the joined target. However as the beryllium is decreased and tungsten added, the surface dose increases dramatically. The result is an even steeper NEDDP with the split target configuration.

The TDEF per ppm of B-10 is increased by a factor of two as shown in Figure 5, over that of the joined target shown in Figure 3. However the steeper NEDDP outweighs the benefit of TDEF from B-10. Figure 6 shows the NEDDPs in the phantom for $cRBE = 60$ for the split target.

5. CONCLUSION

Although the Be/W laminated target improves the TDEF by as much as a factor of five for the joined target or a factor of ten for the split target, the sacrifice in the depth-dose profile eliminates the advantage except for unrealistically high values of *cRBE*. In addition, the enhancement from the laminated target tends to peak at about 5 cm and vanishes at greater than 10 cm for deep-seated tumors.

REFERENCES

1. D.W. Nigg, C.A. Wemple, R. Risler, J.K. Hartwell, Y.D. Harker, and G.E. Laramore, Modification of the University of Washington Neutron Radiotherapy Facility for Optimization of Neutron-Capture-Enhanced Fast Neutron therapy, *in* "Frontiers in Neutron Capture Therapy," M.F. Hawthorne, K. Shelly, R.W. Wiersema, eds., Kluwer Academic/Plenum Publishers, New York, 2001, pp. 729–734.
2. R.E. Prael, and H. Lichtenstein, "User Guide to LCS: The LAHET Code System," Los Alamos National Laboratory report LA-UR-89-3014, 1989.
3. H.G. Hughes, R.E. Prael, and R.C. Little, "MCNPX—The LAHET/MCNP Code Merger," Los Alamos National Laboratory report XTM-RN(U)97-012-, 1997.
4. J.F. Briesmeister, "MCNP—A General Monte Carlo N-Particle Transport Code," Los Alamos National Laboratory report LA-12625-M, Version 4B 1997.

PROGRESS TOWARD FAST NEUTRON RELATIVE BIOLOGICAL EFFECTIVENESS DETERMINATION VIA PROTON BOMBARDMENT OF V79 CELLS

M. A. Sitek[1], J. C. Yanch[1], R. E. Klinkowstein[2], H. Song[1], and K. D. Held[3]

[1]Massachusetts Institute of Technology
Department of Nuclear Engineering
E25-330 45 Carleton Street
Cambridge, MA 02139
[2]Newton Scientific, Inc.
245 Bent St., Cambridge, MA 02141
[3]Massachusetts General Hospital
Department of Radiation Oncology
Boston, MA

1. INTRODUCTION

Conclusive data relating to the relative biological effectiveness (RBE) of the fast neutrons encountered in neutron capture therapies are limited. At the Massachusetts Institute of Technology's (MIT) Laboratory for Accelerator Beam Applications (LABA), experimental determination of fast neutron RBEs is underway using monoenergetic proton beams to simulate the recoil protons created following elastic collisions of neutrons in tissue. Proton beams of varying energies can be created by the high current tandem electrostatic accelerator at MIT LABA. V79 cells are grown on thin polymer films which also serve as the vacuum window for the accelerator beam port. RBE will be determined using three different cell survival levels as endpoints and 250 kVp x-rays as a reference radiation. The uniformity of the proton beam will be confirmed and the absorbed dose in the cells will be determined through the use of radiochromic film. The ultimate goal of the experiments is to determine fast neutron RBEs as a function of energy.

Significant progress toward RBE experiments has been made and will be the focus of this paper. In working towards RBE irradiation experiments, four main challenges

Frontiers in Neutron Capture Therapy, edited by Hawthorne *et al.*
Kluwer Academic / Plenum Publishers, New York, 2001.

have presented themselves: (1) the ability to obtain a low current, low energy proton beam uniformly over an irradiation area of 75 mm; (2) finding a material compatible with cell growth to also serve as a vacuum; (3) establishing a suitable irradiation procedure and geometry; and (4) how to determine the dose being delivered to the V79 cells via extremely short-range (submicron to micron) particles.

2. CHALLENGES

2.1. Low Current, Low Energy Uniform Proton Beam

The Newton Scientific, Inc. tandem accelerator at MIT LABA is utilized to accelerate protons to the desired energies and currents. The variable proton beam energy has a resolution of 0.1 keV. Low current proton beams down to 50 keV have been achieved reliably in vacuum. The accelerator beam current is measured with a programmable electrometer with a resolution of 0.1 fA. The minimum desired proton energy hitting the cells will ultimately depend on the thickness of the vacuum/cell window. The energy loss through this window will dictate the initial energy of the beam needed to achieve the desired energy to the target cells. With a 7.5 micron Kapton™ window, the cells will be irradiated by monoenergetic proton beams with energies in the range of 12 keV to 2.5 MeV. The total energy straggling of the protons through the graphite stripping foil of the accelerator and the thin vacuum window range from 90% for low energy protons down to 4% for those of higher energy. The stripping foil energy straggling only accounts for about 1.5% of the total straggling. The remainder of the straggling is from the vacuum window. Table 1 presents the energy straggling through a 7.5 micron Kapton™ window as a function of energy.

The ultra-low picoampere currents of the variable energy accelerator are generated by allowing protons to leak out of the accelerator's ion source. Further current reduction is achieved by defocusing the beam. To measure such low beam currents, electronic noise reduction techniques have been developed to be used in conjunction with an electrometer capable of measuring picoamperes of beam current.

To ensure beam uniformity, the beam is first centered and focused, then defocused and diffused. This step is important because it ensures that the center or most uniform part of the beam is striking the cells as opposed to the edge of the beam which is unstable and non-uniform. GafChromic® film, type MD-55 is used to verify beam uniformity. An image of the exposed film is converted into a digital form via a document scanner[1] and then analyzed by image processing software.

Table 1. Energy straggling through 7.5 micron Kapton™ window

Energy to window (keV)	Energy to Cells (keV)	Straggle (keV)	Percent straggle
604	10	9	90.00
614	25	14	56.00
625	47	16	34.04
635	70	17	24.29
645	94	17	18.09
655	117	17	14.53
675	163	17	10.43
695	205	16	7.80
795	390	18	4.62

2.2. Cell Growth Surface Compatible as a Vacuum Window

The link between the accelerator and the cell biology aspects of the experiments is the ultra-thin cell growth surface which serves as the vacuum window of the accelerator beam port. A window material strong enough to serve as a vacuum window for the accelerator and capable of allowing a sufficient number of cell colonies to grow is desired. The low energy range of the protons requires a thin window in order to minimize energy straggling. However, the thickness of the window needs to be balanced with the desired cell growth area and be able to hold vacuum (approximately 1.5E-6 torr). Several materials were investigated. Polyimides and silicon nitride were found to have the highest strength properties over a sufficient area.

2.2.1. Kapton™ Window. The polyimide, Kapton™, with a nominal thickness of 7.5 microns was obtained from DuPont's High Performance Film Division and tested as a vacuum window and cell growth surface. Stronger and thinner polyimides are available, but at considerable cost as compared to Kapton™. The 7.5 micron Kapton™ was chosen mainly to test the V79 cells' ability to attach and form colonies on a polyimide and to perfect the technique of mounting the polyimide to form a suitable vacuum window.

The Kapton™ film is mounted to 18-8 stainless steel fender washers with a diameter of 37 mm which is the approximate width of the accelerators beam port at the irradiation site. A commercially available glue, Spray Mount™, is used to bind the Kapton™ to the washer. The washer has an exposed area of 75 mm^2 (a 10 mm diameter hole), where the V79 cells are plated prior to irradiation. The Kapton™ is capable of withstanding vacuum over a larger area, but the deflection of the window while exposed to vacuum is substantial and less stable. Each Kapton™ window is pretreated with Cell-Tak™, a cellular adhesion protein matrix solution, to ensure cellular attachment throughout irradiation.

2.2.2. Cell Biology Aspects. V79 cells were chosen as the test cell line for the initial experiments because of their relatively short doubling time (10 to 12 hours) and their ability to survive in relatively harsh conditions. In addition, their radioresponsiveness is well understood and documented. The cells plating efficiency had to be tested with the materials that composed the Kapton™ washers in order to accurately develop cell survival curves. The three main components of the Kapton™ washer window were studied independently. The 18-8 stainless steel portion had no discernible effect on cell survival, as expected, since this type of steel is routinely used in biological applications. The Kapton™ alone also had no discernible effect on cell survival. The Spray Mount™ adhesive alone prohibited cell colony growth after four days. The combined stainless steel, Kapton™, and Spray Mount™ washer reduced the plating efficiency to 70% from 85% for a 60 mm tissue culture dish.

The irradiation area of 75 mm^2 required that the traditional 200 cell colonies after irradiation be modified to allow enough colonies to develop distinctly. The optimal final number of cell colonies was determined to be 100. The colonies are to be stained with methylene blue after 3.5 days which allows approximately 50 cells per colony to develop.

2.3. Irradiation Procedure and Geometry

Careful and precise irradiation of the cells is necessary to ensure the accuracy of the dose being delivered. Prior to irradiation, cells are plated on the Kapton™ window

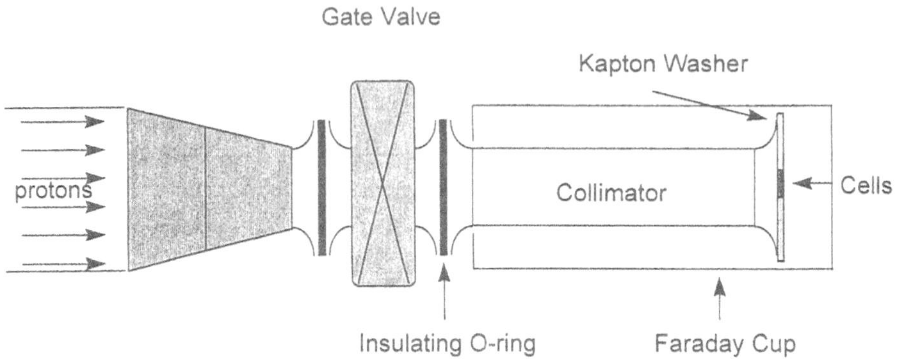

Figure 1. Irradiation geometry end station.

and allowed to attach for three hours. A cap filled with medium is then used to cover the cells to keep them moist during irradiation and handling. The cell assembly is then affixed to the end of the accelerator beam port by opening the manual gate valve (as seen in Figure 1). The air injected into the vacuum system is pumped out within five minutes. The manual gate valve, at that point, is open but prevents protons from striking the cells due to the presence of a blocking plate. The proton beam is then adjusted to the desired current and energy as measured on the blocking plate of the manual gate valve. With the manual gate valve open, the pneumatic gate valve (millisecond closing time) is used to control the irradiation of the cells to the desired proton fluence and hence dose.

2.4. Dose Verification

The absorbed dose to the V79 cells was calculated as a function of proton energy and beam current. The average cell thickness was taken to be 0.5 microns and the cell diameter to be 10 microns. The cells were assumed to be composed of only water. The ion slowing down code, TRIM (*Tr*ansport of *I*ons in *M*atter),[2] was used to determine the fraction of the proton energy that is deposited in the cell volume. Combining the energy deposited with the cell mass allowed for the determination of dose per particle at a given proton energy. The appropriate beam fluence was then calculated for the desired absorbed doses as a function of proton energy. Figure 2 presents a plot of the proton beam current as a function of energy for 300 rad. The plot can be linearly scaled for higher and lower doses. The minimum in the plot at 360 keV represents the energy at which the protons are no longer completely stopped in the cells.

Dose rates of approximately 0.12 ± 0.01 Gy/s are generated based on accelerator beam currents on the order of 6 ± 0.01 pA. GafChromic®, type MD-55, dosimetry media will be used to determine the dose delivered to the cell position as a function of particle fluence. GafChromic® film changes to darker shades of blue as a function of ionizing radiation dose. 250 kVp x-rays will be used to calibrate the film. The MD-55 film can then be proton irradiated to various doses by monitoring the estimated proton fluence on target. Thus, a relationship between absorbed dose and proton fluence on target can be developed.

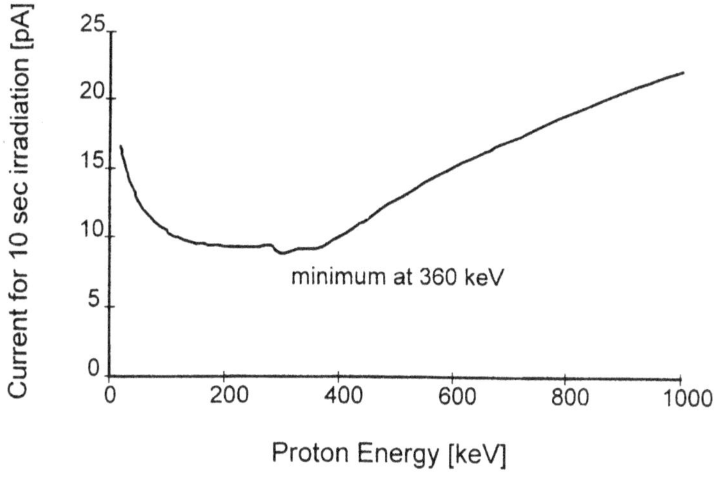

Figure 2. Current variation of proton energy for 300 rad.

3. FUTURE WORK

The only remaining obstacle to the construction of cell survival curves to determine RBEs is the calibration of the GafChromic® film. After analyzing the RBE data for the Kapton™ windows, thinner windows may be attempted to reduce the spread in the proton energies irradiating the cells. In addition, different cell lines may also be used for RBE determination, including human cell lines.

ACKNOWLEDGMENTS

This work is supported, in part, by participation in the United States Nuclear Regulatory Commission Graduate Fellowship Program.

REFERENCES

1. MA Stevens *et al.* "High-resolution dosimetry using radiochromic film and a document scanner." *Phy. in Med. & Bio.*, 41:2357–2365, 1996.
2. JF Ziegler *Transport of Ions in Matter*. v98.1, January 1998.

AUTHOR INDEX, Volume 1

KEYWORD INDEX, Volume 1

GPSR Compliance

The European Union's (EU) General Product Safety Regulation (GPSR) is a set of rules that requires consumer products to be safe and our obligations to ensure this.

If you have any concerns about our products, you can contact us on ProductSafety@springernature.com

In case Publisher is established outside the EU, the EU authorized representative is:

Springer Nature Customer Service Center GmbH
Europaplatz 3
69115 Heidelberg, Germany

Batch number: 10838292

Printed by Printforce, the Netherlands

Frontiers in Neutron Capture Therapy

Volume 2

Edited by

M. Frederick Hawthorne
Kenneth Shelly
University of California, Los Angeles
Los Angeles, California

and

Richard J. Wiersema
Neutron Therapies LLC
San Diego, California

Springer Science+Business Media, LLC

Library of Congress Cataloging-in-Publication Data

Frontiers in neutron capture therapy/edited by M. Frederick Hawthorne, Kenneth Shelly, and Richard J. Wiersema.
p. ; cm.
"Proceedings of the Eighth International Symposium on Neutron Capture Therapy for Cancer, held September 13–18, 1998, in Los Angeles, California"—T.p. verso.
Includes bibliographical references and index.
ISBN 978-1-4613-5478-9 ISBN 978-1-4615-1285-1 (eBook)
DOI 10.1007/978-1-4615-1285-1
1. Boron-neutron capture therapy—Congresses. I. Hawthorne, M. Frederick. II. Shelly, Kenneth. III. Wiersema, Richard J. IV. International Symposium on Neutron Capture Therapy (8th: 1998: Los Angeles, Calif.)
[DNLM: 1. Neutron Capture Therapy—Congresses. 2. Neoplasms—radiotherapy—Congresses. WN 250.5.R3 F935 2001]
RC271.N48 .F76 2001
616.99′40642—dc21

2001020257

Proceedings of the Eighth International Symposium on Neutron Capture Therapy for Cancer, held September 13–18, 1998, in Los Angeles, California

ISBN 978-1-4613-5478-9

Originally published by Kluwer Academic / Plenum Publishers, New York in 2001
Softcover reprint of the hardcover 1st edition 2001

http://www.wkap.nl/

10 9 8 7 6 5 4 3 2 1

A C.I.P. record for this book is available from the Library of Congress

CONTENTS

Volume 2: Chemistry and Radiobiology

CHEMISTRY AND SYNTHESIS

IMAGING AND ANALYSIS OF NUCLIDES

TISSUE TARGETING

PRECLINICAL DOSIMETRY

FUTURE DIRECTIONS AND EMERGING APPLICATIONS

VOLUME 2: CHEMISTRY AND RADIOBIOLOGY

CHEMISTRY AND SYNTHESIS

A CONCISE SYNTHESIS OF ENANTIOMERICALLY PURE L-(4-BORONOPHENYL)ALANINE FROM L-TYROSINE

Hiroyuki Nakamura, Masaru Fujiwara, and Yoshinori Yamamoto

Department of Chemistry
Graduate School of Science
Tohoku University
Sendai 980-8578, Japan

1. INTRODUCTION

4-Boronophenylalanine (BPA)[1] is a practical boron compound, which is clinically used not only for the treatment of malignant melanoma but also for that of brain tumor, on neutron capture therapy (NCT).[1-3] Since Mishima and co-workers[4] revealed that the *L*-form of BPA is more efficiently incorporated into melanoma cells than racemic one, the enantioselective synthesis of *L*-BPA has been required. Enriched *L*-BPA was prepared enzymatically through α-chymotrypsin hydrolysis[5] of the ethyl ester of racemic BPA synthesized by the traditional method.[1,6] In this case, 50% of another enantiomer (*D*-BPA) was recovered from the racemic material. Recently, two synthetic routes of *L*-BPA were reported. Asymmetric hydrogenation route gave *L*-BPA with enantiomeric excess of up to 88% (96% ee after recrystallization).[7] Another route using palladium-catalyzed coupling reaction of iodophenylboronate with the chiral organozinc derived from *L*-serine needed rather lengthy synthetic steps.[8] Herein we report a concise synthesis of enantiomerically pure *L*-BPA from *L*-tyrosine using palladium-catalyzed carbon-boron bond formation reaction (eq. 1).

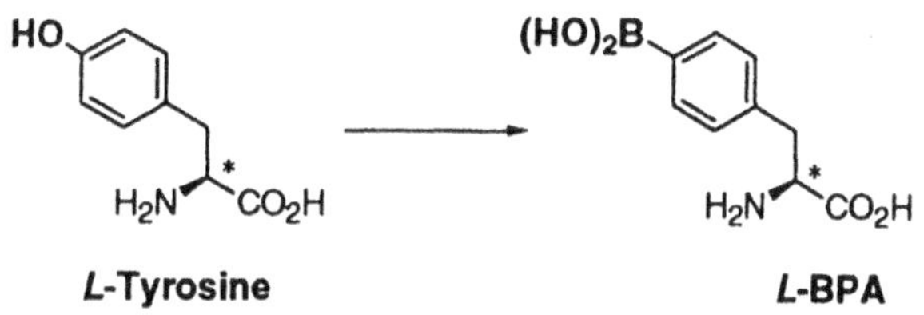

(1)

Frontiers in Neutron Capture Therapy, edited by Hawthorne *et al.*
Kluwer Academic / Plenum Publishers, New York, 2001.

2. RESULTS AND DISCUSSION

Palladium-catalyzed cross-coupling reaction of the pinacol ester of diboronic acid **1a** with haloarenes, which was developed by Miyaura, is a direct procedure for the synthesis of arylboronic esters from haloarenes (eq. 2).[9] Triflate **2** was readily prepared from *L*-tyrosine by the literature procedure.[10] The cross-coupling reaction of the triflate **2** with **1a** in dioxane proceeded very smoothly in a catalytic amount of $PdCl_2$(dppf) (1,1′-bis(diphenylphosphino)ferrocene), giving the corresponding arylboronic ester **3a** in quantitative yield. However, subsequent hydrolysis of pinacol ester **3a** with $NaIO_4$[11] gave an unseparable mixture of the boronic acid **4** and unidentified materials (Scheme 1). The removal of the Cbz (carbobenzoxy) protecting group by the treatment of **4** with either Pd-catalyst/H_2 or acidic reagents gave a mixture of the desired product (*L*-BPA) and unidentified materials, and separation of *L*-BPA from the mixture was difficult. The other hydrolysis method, based on transesterification with phenylboronic acid, was not effective for arylboronic ester **3a**.[11]

Thus, 1,3-diphenyl-1,3-propanediol **5** was chosen as an ester group of diboron since the boronic ester derived from this diol can be easily cleaved by hydrogenolysis. The reaction of the diol **5** and tetrakis(dimethylamino)diboron **6** proceeded very smoothly in THF at room temperature to give bis-(1,3-diphenyl-1,3-propanediolato)diboron **1b** in 72% yield (Scheme 2). The triflate **2** underwent the palladium-catalyzed cross-coupling reaction with the diboron **1b** in DMF at 100 °C, giving the corresponding arylboronic ester **3b** in 65% yield. The use of dioxane as a solvent gave **3b** in lower yield. The protective groups of **3b** were removed all at once under the hydrogenolytic condition with $Pd(OH)_2/C$, and *L*-BPA was obtained in 74% yield without racemization.

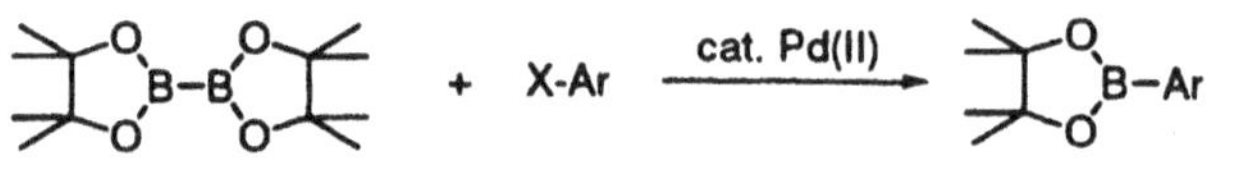

(2)

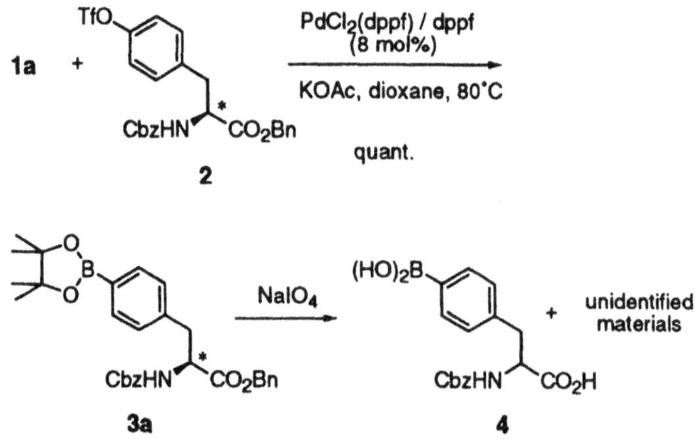

Scheme 1

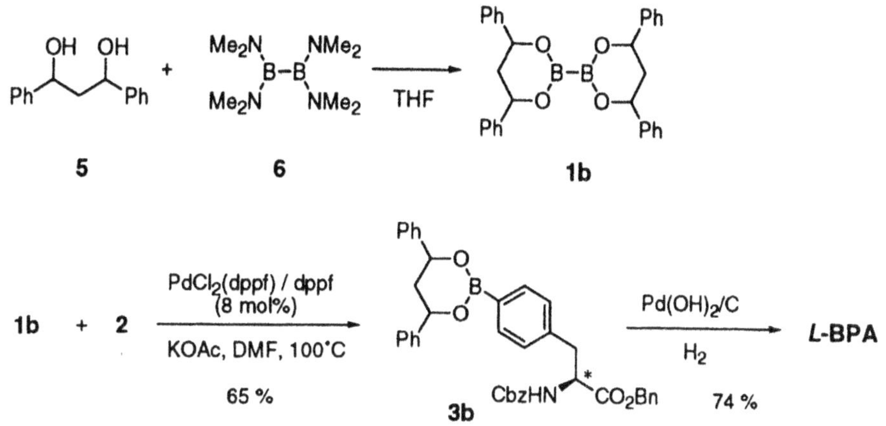

Scheme 2

3. CONCLUSION

The palladium-catalyzed cross-coupling reaction between C-OTf and B-B bonding is a key for the concise synthesis of *L*-BPA from *L*-tyrosine. We are now in a position to supply *L*-BPA, synthesized through short steps from naturally occurring *L*-tyrosine, for medical and biological research.

REFERENCES

1. H.R. Snyder, A.J. Reedy, and W.M. Lennarz, Synthesis of aromatic boronic acids. Aldehyde boronic acids and a boronic acid analog of tyrosine, J. *J. Am. Chem. Soc.*, *80*:835–838, 1958.
2. G.L. Locher, Biological Effects and the Therapeutic Possibilities of Neutrons, *Am. J. Roentgenol.*, *36*:1, 1936.
3. For recent reviews, see: (a) M.F. Hawthorne, The Role of Chemistry in the Development of Boron Neutron Capture Therapy, *Angew. Chem., Int. Ed. Engl.*, *32*:950–984, 1993. (b) Y. Yamamoto, Molecular Design, and Synthesis of B-10 Carriers for Neutron Capture Therapy, *Pure. Appl. Chem.*, *63*:423–426, 1991. (c) R.F. Barth, A.H. Soloway, and R.G. Fairchild, Boron Neutron Capture Therapy of Cancer, *Cancer Research*, *50*:1061–1070, 1990. (d) A.H. Soloway, W. Tjarks, B.A. Barnum, F.-G. Rong, R.F. Barth, I.M. Wyzlic, and J.G. Wilson, The Chemistry of Neutron Capture Therapy, *Chem. Rev.*, *98*:1515–1562, 1998.
4. Y. Mishima, M. Ichihashi, S. Hatta, C. Honda, K. Yamamura, and T. Nakagawa, New thermal neutron capture therapy for malignant melanoma: Melanogenesis-seeking ^{10}B molecule-melanoma cell interaction from *in vitro* to first clinical trial, *Pigment Cell Res.*, *2*:226–234, 1989.
5. D.C. Roberts, K. Suda, J. Samanen, and D.S. Kemp, Pluripotential amino acids I. (L)-p-Dihydroxyborylphenylalanine (L-Bpa) as a precursor of L-Phe and L-Tyr containing peptides; specific tritiation of L-Phe-containing peptides at a final step in synthesis, *Tetrahedron Lett.*, *21*:3435–3438, 1980.
6. M. Kirihata, T. Morimoto, and I. Ichimoto, A simple and improved synthesis of p-boronophenylalanine, a boron carrier for the boron-neutron capture therapy, *Biosci. Biotech. Biochem.*, *57*:1940–1941, 1993.
7. E.G. Samsel, Enantioselective synthesis of L-(–)-4-boronophenylalanine (L-BPA), *US Patent* 5157149, 1992.
8. C. Malan, and C. Morin, Synthesis of 4-borono-L-phenylalanine, *Synlett*, 167–168, 1996.
9. (a) T. Ishiyama, M. Murata, and N. Miyaura, Palladium(0)-catalyzed cross-coupling reaction of alkoxydiboron with haloarenes: a direct procedure for arylboronic esters, *J. Org. Chem.*, *60*:7508–7510, 1995. (b) M. Murata, S. Watanabe, and Y. Masuda, Novel palladium(0)-catalyzed coupling reaction of dialkoxyborane with aryl halides: convenient synthetic route to arylboronates, *J. Org. Chem.*,

62:6458–6459, 1997. (c) N. Miyaura, and A. Suzuki, Palladium-catalyzed cross-coupling reactions of organoboron compounds, *Chem. Rev.*, *95*:2457–2483, 1995.

10. W. Wang, N.U. Obeyesekere, and J.S. McMurray, Stereospecific synthesis of 4-carboxyphenylalanine and derivatives for use in Fmoc-based solid-phase peptide synthesis, *Tetrahedron Lett.*, *37*:6661–6664, 1996.
11. S.J. Coutts, J. Adams, D. Krolikowski, and R.J. Snow, Two efficient methods for the cleavage of pinanediol boronate esters yielding the free boronic acids, *Tetrahedron Lett.*, *35*:5109–5112, 1994.

THE COMPLEXATION OF *p*-BORONOPHENYLALANINE WITH ALCOHOLS OR CARBOXYLIC ACIDS STUDIED BY ZONE ELECTROPHORESIS

Yoshinori Kitaoka and Mitsue Kobayashi

Research Reactor Institute of Kyoto University
Noda, Kumatori-cho, Sennan-gun
Osaka 590-0494, Japan

1. INTRODUCTION

Recently many researchers have paid their attention to p-boronophenylalanine (p-BPA) as a boron carrier for boron neutron capture therapy. As p-BPA is sparingly soluble in water, the fructose-p-BPA complex has been used in the actual administration of p-BPA. The transportation or accumulation of p-boronophenylalanine in organic tissues or in cells largely depends on the interaction between the dihydroxyboryl moiety and functional groups of biological materials. The complexation of boric acid with alcohols has been studied by many workers from early days and been applied to chemical analysis.[1] During studies of the physicochemical nature of p-BPA, we first found the complexation of p-BPA with oxalic acid by the use of paper electrophoresis.[2] Since then, we have been studying the interaction between p-BPA or its related compounds and carboxylic acids using paper electrophoresis.[3,4] In this report, we would like to show electrophoretic behavior of p-BPA in alcohol or carboxylic acid supporting solutions. Based on these results, the charge state and chemical stability of the complexes of p-BPA in different pH solutions will be discussed.

2. APPARATUS AND PROCEDURES

The techniques and apparatus used were similar to those described in a previous paper.[2] p-BPA was purchased from Boron Biologicals (Raleigh, NC, USA). Alcohols, acids and other reagents were purchased from Wako Chemicals (Osaka, Japan). A spotting strip (Toyoroshi No.51A, 1 × 40 cm) was dipped in the supporting solution and the

Frontiers in Neutron Capture Therapy, edited by Hawthorne *et al.*
Kluwer Academic / Plenum Publishers, New York, 2001.

excess of supporting solution was removed by another filter paper. A 5 μl volume of standard solution (picric acid = PA) or sample solution was spotted at the three different positions (A, 5 cm to the cathodic side from the center of the strip; B, the center; C, 5 cm to the anodic side) on the strip. Six strips (three pairs) were set in paralleled in n-hexane in the migration chamber and the electrode cells. The cell solutions were 0.1 M NaCl solution. A constant electric voltage (1,000 v per 30 cm) was applied to two pairs for 30 min at constant temperature(ca. 20 °C). One pair was allowed to stand in the migration chamber without application of the electric voltage and was used to estimate the effect of capillary action. The p-BPA on the strip was detected by spraying with ninhydrin or curucumin solution.

3. RESULTS AND DISCUSSION

The movements due to capillary action under the electrophoresis conditions used (51 A filter paper, 30 min at ca. 20 °C) were +0.7 ± 0.2, 0.0 ± 0.2, –0.7 ± 0.2 (cm) at the A, B and C positions, respectively (positive movement towards the anodes, negative movement towards the cathode). The movements due to capillary action varied with the distances from the center of a strip. The movements due to electroosmotic flow varied with the pH-value of supporting solutions. The relation between the charge (Z) and pH-value of the supporting solution is theoretically expressed by a sigmoidal curve.[1] In the identical case, the mobility should be proportional to the ratio of the charge to the square root of molecular weight.

p-BPA is a derivative of an amino acid and is assumed to be cationic in acidic solutions, zwitterionic in neutral solution and anionic in alkaline solutions.

The migration distances observed in our method are usually a summation of capillary action, electro-osmotic flow and electrophoretic movement. As the effect of a capillary action on the migration is minimum at the center of a support, the migration distances of p-BPA or picric acid (PA) spotted at the center of a support are taken in this report.

In order to elucidate the bases of the interaction between the dihydroxyboryl moiety and hydroxy moiety, we picked ethylene glycol (EG) as a typical diol, and did inositol (Ino) as a typical hexol.

Observed migration distances of p-BPA in aqueous ethylene glycol (EG), inositol (Ino) and fructose (Fru) containing NaCl as electrolytes are shown in Figure 1. However, p-BPA did not migrate to the anodic side in any supporting solution mentioned above. This means that the negatively charged complex of p-BPA with EG, Ino or Fru was not formed.

Thus, we first dissolved p-BPA in a strongly alkaline solution and then reacted with equimolar amount of fructose. The resulting solution was investigated using a phosphate buffer solution (A, pH = 6.86) or phthalate buffer (B, pH = 4.01). Figure 2 shows the observed migration distances of the complexed p-BPA in these supporting solutions. A comet-like spot of p-BPA was first found at the anionic side in the phosphate buffer solution (A). This anionic movement seems to be due to the negatively charged complex of p-BPA with fructose. However, this complex was decomposed in acidic solution because the anionic movement was not observed in the phthalate buffer anymore.

Having found the complexation of p-BPA with citric acid,[3] we picked *trans*-acotinic acid and isocitric acid as tribasic carboxylic acids.

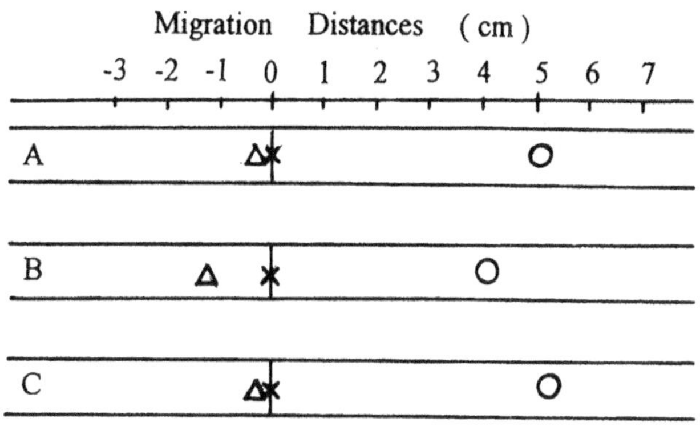

Figure 1. Observed migration distances of p-BPA in aqueous ethylene glycol, inositol and fructose containing NaCl. Electrophoresis conditions: sample; p-BPA (△); PA (■); 1000 V/30 cm, 30 min. ~20 °C, Toyoroshi No. 51A. Spotting position (x). Supporting solution: A, ethylene glycol (1) +0.1 M NaCl (1); B, inositol (1) +0.1 M NaCl (1); C, 0.1 M fructose + 0.1 M NaCl.

Observed migration distances of p-BPA in *trans*-acotinic acid, isocitric acid and citric acid are shown in Figure 3. The migration of p-BPA in citric acid was shown as a reference.[3] Cationic movements were observed in *trans*-acotinic acid while anionic migration was observed in isocitric acid. The anionic movement indicates that complexation of p-BPA with isocitric acid occurs.[6] Comparing the electrophoresis results of these tricarboxylic acids, we can discuss on the stereochemical structure necessary for the boron ester formation.

In order to know the stability of the complex of p-BPA with citric[3] or oxalic acid[2] in near neutral solution, we carried out the electrophoresis of the complexed p-BPA in phosphate buffer as shown in Figure 4.

The complex of p-BPA with citric acid showed anionic movement while that of p-BPA with oxalic acid did not show anionic movement. Thus, we assumed that the complex with oxalic acid decomposed.

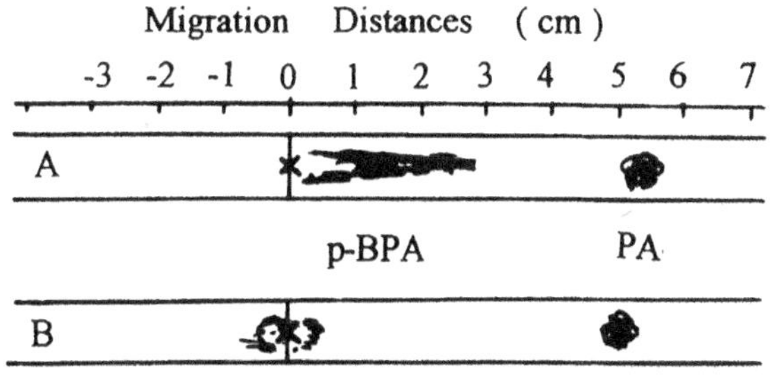

Figure 2. Observed migration distances of Fructose—p-BPA complex in phosphate and phthalate buffer solutions. Electrophoresis conditions: sample; p-BPA; PA; 1,000 V/30 cm, 30 min. ~20 °C, Toyoroshi No. 51A. Spotting position (x). A: 0.05 M p-BPA reacted with 0.05 M Fructose under alkaline soln., and migrated in Phosphate Buffer (pH = 6.86). B: 0.05 M p-BPA reacted with 0.05 M Fructose under alkaline soln., and migrated in Phthalate Buffer (pH = 4.01).

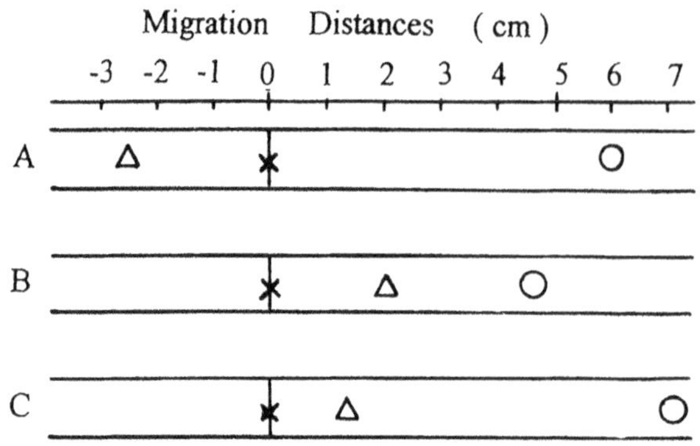

Figure 3. Observed migration distances of p-BPA in *trans*-Acotinic Acid, Isocitric Acid, and CitricAcid. Electrophoresis conditions: sample; p-BPA (△); PA (○); 1,000 V/30 cm, 30 min. ~20 °C, Toyoroshi No. 51A. Spotting position (x). Supporting solution: A, *trans*-Acotinic Acid; B, Isocitric Acid; C, Citric Acid.

4. SUMMARY

Electrophoresis is a unique chemical means for the study of the charge state of a compound in solution. The electrophoretic mobility depends on the ratio of a molecular weight to the charge. Thus, the mobility is closely related with the chemical reactions such as dissociation, association, complexation, and solvation in solution. The complexation found in this experiment was classified into two kinds, that is ether formation (B—O—R, Type A) and ester formation (B—O—CO—R, Type B), where B is a boron moiety and R is an alkyl moiety. Type A is the complex typically formed. The reaction of p-BPA with alcohols such as fructose results in the ether formation (Type A) which is stable under neutral and alkaline conditions. On the other hand, the reaction of p-BPA with oxalic, citric or isocitric gave another type of the complexation, that is ester formation (Type B). The literature described about the ester formation of dihydroxyboryl moiety is very few compared with that of the ether formation. The ester formation occurred in acidic conditions and under the large excess of acid. The existence of this complex was also confirmed with infrared spectroscopy.[5] The degree of this complexation largely

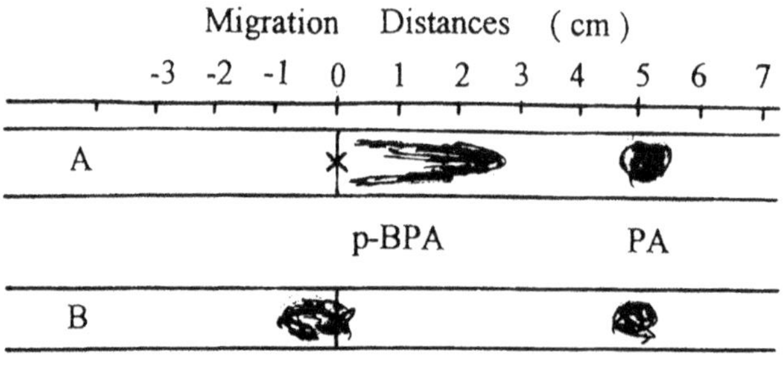

Figure 4. Observed migration distances of p-BPA complexed with Citric or Oxalic Acid in phosphate buffer. Electrophoresis conditions: sample; p-BPA; PA, 1,000 V/30 cm, 30 min., ~20 °C, Toyoroshi No. 51A. Spotting position (x). A: 0.05 M p-BPA reacted with 0.1 M Citric Acid, and migrated in phosphate buffer (pH = 6.86). B: 0.01 M p-BPA reacted with 0.1 M Oxalic Acid, and migrated in phosphate buffer (pH = 6.86).

depends on the electrostatic and stereochemical nature of the acid. Generally speaking, Type B complexes are more likely to occur in acidic sites within a body.

ACKNOWLEDGMENT

A part of this study was supported by Grant-in-Aid for Scientific Research from the ministry of Educations, Science, Sports and Culture, Japan.

REFERENCES

1. Y. Kiso, M. Kobayashi, and Y. Kitaoka, Chromatography in solution, "Analytical Chemistry of Phosphorus Compounds," M. Halmann, ed. John Wiley and Sons, New York, 1972, pp. 93–148.
2. Y. Kitaoka, M. Kobayashi, and Y. Ujeno, Studies of electrophoretic behaviour of p-dihydroxy-borylphenyl-alanine and related compounds by the three spot method, *J. Chromatogr.*, 585:36–372, 1991.
3. Y. Kitaoka, M. Kobayashi, and Y. Ujeno, Paper electrophoretic behavior of p-dihydroxyboryl-phenylalanine in aqueous solutions of di- and tri-carboxylic acids as supporting solutions, *Annu. Repts Res. Reactor Inst. Kyoto Univ.*, 25:95–99, 1992.
4. Y. Kitaoka, M. Kobayashi, K. Kawamoto, T. Morimoto, M. Kirihata, and I. Ichimoto, Study of the interaction of p-boronophenylalanine and its analogs with various carboxylic acids by using a zone electrophoresis, "Cancer Neutron Capture Therapy, Plenum Press, New York, 1996, pp. 127–130.
5. M. Kobayashi, Y. Kitaoka, Y. Tanaka, and K. Kawamoto, Electrophoretic behaviour and infrared spectra of dihydroxyboryl compounds in aqueous di- and tricarboxylic acids, *J. Chromatogr. A*, 678:351–358, 1994.

RECENT DEVELOPMENTS IN THE CHEMISTRY OF $[B_{12}H_{12}]^{2-}$

Toralf Peymann, Kenneth Shelly, and M. Frederick Hawthorne

Department of Chemistry and Biochemistry
University of California, Los Angeles
405 Hilgard Avenue
Los Angeles, California 90095-1569

1. INTRODUCTION

The dodecaborate(2–) cluster $[B_{12}H_{12}]^{2-}$, **1**, is less frequently found as the boron component in drugs designed for BNCT than the isoelectronic carboranes $C_2B_{10}H_{12}$. The carboranes contain C—H vertices that are easily modified by employing methods of organic chemistry.[1] Thus, it is facile to link these carborane clusters to tumor-targeting moieties. Furthermore, the B—H vertices of the carboranes differ in reactivity depending on their position relative to the electron-withdrawing carbon atoms. This difference allows for the selective modification of certain B—H vertices. The anion **1**, on the other hand, consists of 12 identical boron atoms. Therefore, it is difficult to control the degree of substitution and the regiospecifity of reactions performed on the icosahedral framework of **1**. For example, the diiodination of *o*–carborane gives exclusively the 9,12–isomer, whereas it leads to an isomeric mixture for **1**.[2] Relative to the carboranes, however, the dodecaborate(2–) cluster offers advantages such as its stability regarding basic degradation, water-solubility of its derivatives, and higher boron content.

To use the anion **1** as a building block in the syntheses of new BNCT drugs, we were interested in ways to derivatize **1**. Here we report the hydroxylation of **1** by aqueous sulfuric acid. Depending on reaction time, temperature, and acid concentration we were able to synthesize the anions $[B_{12}H_{12-n}(OH)_n]^{2-}$ (n = 1–4), **2**–**5**. The species obtained were isomerically pure.

Recently, we published a method to monoalkylate and -arylate **1** by the palladium-catalyzed reaction of $[B_{12}H_{11}I]^{2-}$ with Grignard reagents.[3] Thus, we obtained $[B_{12}H_{11}(n\text{–}C_{18}H_{37})]^{2-}$, **6**, a compound that resembles stearate because it contains a C_{18} alkyl chain and a charged head group, the cluster. We synthesized **6** because it may be incorporated in the unimellar bilayer as well as sequestered in the aqueous core of liposomes

Frontiers in Neutron Capture Therapy, edited by Hawthorne *et al.*
Kluwer Academic / Plenum Publishers, New York, 2001.

resulting in an effective vehicle for delivering a high boron dose to tumor cells. Here we describe the biodistribution of **6** in a BALB/c mice bearing EMT6 tumors via liposomal delivery.

2. RESULTS

We obtain the monohydroxyl species $[B_{12}H_{11}(OH)]^{2-}$, **2**, by heating Cs_2**1** in 38% sulfuric acid at 90 °C for 40 min. Its ^{11}B NMR spectrum displays the typical pattern of a monosubstituted B_{12} cluster with four signals at ppm 6.9 (s, B1), −14.4 (d, B2–B5), −16.6 (d, B6–B11), −22.7 (d, B12). Increasing the sulfuric acid concentration to 48% and heating Cs_2**1** at 90 °C for 90 min results in the dihydroxyl derivative $[B_{12}H_{10}(OH)_2]^{2-}$, **3**, which exhibits an ^{11}B NMR spectrum of five signals at ppm 3.4 (s; B1,7), −15.2 (d; B2,3), −17.4 (d; B4,6,8,11), −20.0 (d; B9,10), and −23.6 (d; B5,12). In the two-dimensional ^{11}B NMR spectrum of **3** (see Figure 1) the singlet attributable to the B—OH vertices at 3.4 ppm has three crosspeaks with the signals of the B—H vertices. This is only consistent with the structure of the 1,7–isomer.

Both the mono and the dihydroxy derivative **2** and **3** have been described before, but were obtained in a low-yield two-step synthesis.[4] The heating of Cs_2**1** at 110 °C in 48% sulfuric acid for 20 hours affords a trisubstituted derivative $[B_{12}H_9(OH)_3]^{2-}$, **4**. Its $\{^1H\}^{11}B$ NMR spectrum displays four signals at ppm 0.8, −18.0, −20.7, and −24.5. The

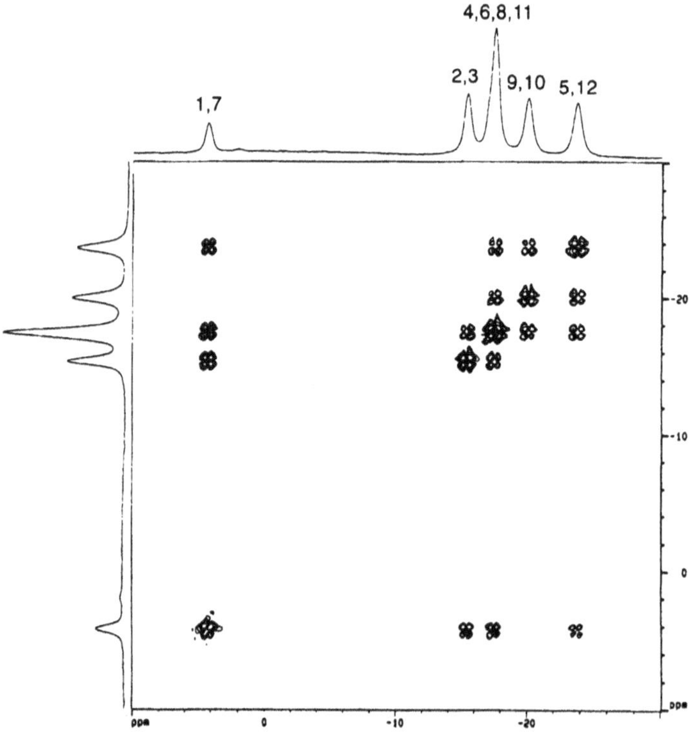

Figure 1. 2D ^{11}B NMR spectrum of 1,7–$[B_{12}H_{10}(OH)_2]^{2-}$, **3**.

structure of **4** is elucidated by its two-dimensional ^{11}B NMR spectrum. The singlet of the B—OH vertices has crosspeaks with each of the three signals of the B—H vertices, which corresponds to the 1,7,9 isomer. The stirring of Cs_2**1** in 57% sulfuric acid at 175 °C for five hours gives the tetrasubstituted $[B_{12}H_8(OH)_4]^{2-}$ anion, **5**. In the $\{^1H\}^{11}B$ NMR spectrum of **5** four signals at ppm −0.7, −2.3, −20.4, and −25.2 in a 2:2:4:4 ratio are present. The two downfield signals appear as singlets in its ^{11}B NMR spectrum and have no crosspeaks in the two-dimensional ^{11}B NMR spectrum, which is only consistent for the 1,2,8,10–isomer.

3. DISCUSSION

We developed a method to hydroxylate **1** by reacting it with aqueous sulfuric acid. The monosubstituted anion **2** was obtained in 78% yield. The dihydroxylation of **1** yielded only the 1,7–isomer **3**. The hydroxylation of **3** may produce four trisubstituted anions (1,2,4; 1,2,8; 1,7,9; 1,2,9). Assuming all remaining B—H vertices are equal, the formation of the 1,7,9 isomer **4** is only 20% probable; yet, species **4** is the only product formed. The third hydroxyl group favors the position *meta* to both of the previously substituted boron atoms. Putting a fourth hydroxyl substituent on the cluster may result in the formation of three isomers (1,2,7,10; 1,2,4,12; 1,2,8,10). In the first case, the fourth hydroxyl function is attached to the boron atom that is (*ortho*, *ortho*, *meta*) to the other three B—OH vertices. In the second isomer, the relevant boron atom is (*ortho*, *ortho*, *para*) to the *ipso*–boron atoms. In the third possible anion, the fourth B—OH vertex is (*ortho*, *meta*, *meta*) to the hydroxylated boron vertices. The 1,2,8,10–isomer **5** has the most *meta* contacts and is formed exclusively.

The hydroxyl function influences the substitution of the $[B_{12}H_{12}]^{2-}$ cluster in two ways. First, it deactivates the cluster towards further hydroxylation. To attach additional hydroxyl substituents at the cluster requires more drastic reaction conditions. Second, the hydroxyl group is a *meta* director. To date, only the multi-fluorination of the dodecaborate(2–) cluster has been explored. The authors also observed that *meta*-substitution was favored.[5] This *meta* effect may be common for electron-withdrawing substituents (e.g., chalkogens or halogens) on the icosahedron **1**. Further studies, however, will be necessary.

4. BIODISTRIBUTION OF $[B_{12}H_{11}(n\text{—}C_{18}H_{37})]^{2-}$, 6

The derivative **6** was obtained by the palladium-catalyzed alkylation of $[B_{12}H_{11}I]^{2-}$ with n—$C_{18}H_{37}MgBr$.[3] We studied the biodistribution of Na_2**6** in BALB/c mice bearing EMT6 tumors (n = 5–7 animals per timepoint) via liposomal delivery. The boron concentration was determined by ICP-AES. The results of this study are shown in Figure 2 (The variation is given as s.e.m. = standard deviation/$n^{1/2}$).

The highest boron concentration is found in the liver. This is not unusual for liposomal delivery of a boron-containing drug. Moreover, the boron concentration decreases rapidly with time. The tumor accumulation of boron reaches its maximum after 30 hours with about 35 ppm, which is considered the therapeutic dose for successful BNCT. At this timepoint, the tumor : blood ratio is about 6.

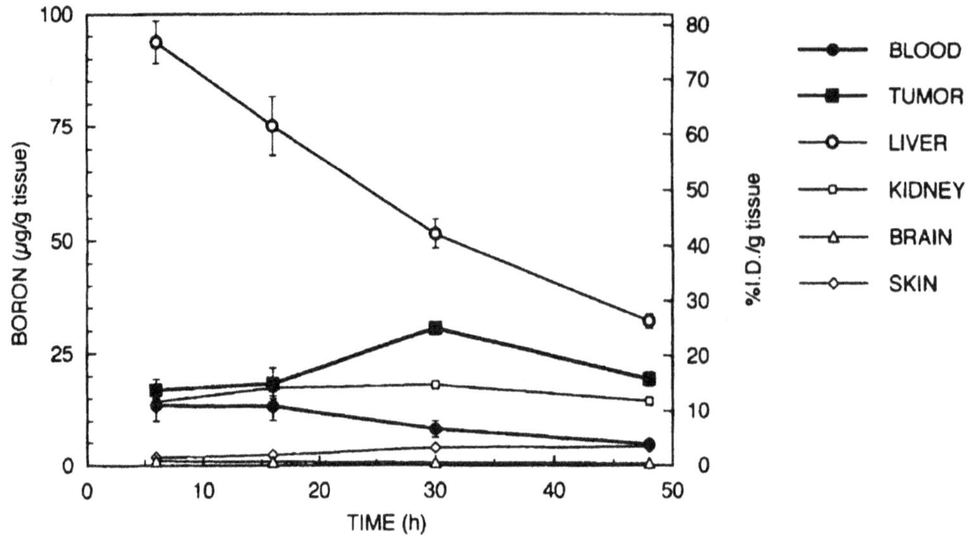

Figure 2. Biodistribution of $[B_{12}H_{11}(n\text{—}C_{18}H_{37})]^{2-}$, **6**.

ACKNOWLEDGMENTS

The authors wish to thank Dr. W. F. Bauer of the INEEL for the boron analyses of the tissue samples and Dr. P. R. Gavin and his group of Washington State University for performing the biodistribution experiments. This work was supported by the US Department of Energy (Grant No. DE-FG02-95ER61975).

REFERENCES

1. R.N. Grimes. *Carboranes*, Academic Press, New York, 1970.
2. W.H. Knoth, H.C. Miller, J.C. Sauer, J.H. Balthis, Y.T. Chia, and E.L. Muetterties, Chemistry of Boranes. IX. Halogenation of $B_{10}H_{10}^{-2}$ and $B_{12}H_{12}^{-2}$, *Inorg. Chem.*, 3:159–167, 1964.
3. T. Peymann, C.B. Knobler, and M.F. Hawthorne, Synthesis of Alkyl and Aryl Derivatives of *closo*-$B_{12}H_{12}^{2-}$ by the Palladium-Catalyzed Coupling of *closo*-$B_{12}H_{11}I^{2-}$ with Grignard Reagents, *Inorg. Chem.*, 37:1544–1548, 1998.
4. W.H. Knoth, J.C. Sauer, D.C. England, W.R. Hertler, and E.L. Muetterties, Chemistry of Boranes. XIX. Derivative Chemistry of $B_{10}H_{10}^{-2}$ and $B_{12}H_{12}^{-2}$, *J. Am. Chem. Soc.*, 86:3973–3983, 1964.
5. K.A. Solntsev, A.M. Mebel, N.A. Votinova, N.T. Kuznetsov, and O.P. Charkin, The Polyhedral $B_{12}H_{12}^{2-}$ Anion as a Three-Dimensional Aromatic System, *Koord. Khimiya*, 18:340–364, 1992.

SYNTHESIS OF SCHIFF BASES AND MONOALKYLAMINO DERIVATIVES ON THE BASE OF AMINO-*closo*-DODECABORATE(1-) ANION $[B_{12}H_{11}NH_3]^-$

Igor B. Sivaev[1,2], Aleksandr B. Bruskin[3], Vladimir V. Nesterov[1], Mikhail Yu. Antipin[1], Vladimir I. Bregadze[1], and Stefan Sjöberg[2]

[1]Nesmeyanov Institute of Organo-Element Compounds
Russian Academy of Sciences
Vavilov Str. 28, 117813, Moscow, Russia
[2]Department of Organic Chemistry
Institute of Chemistry
Uppsala University
P.O. Box 531, S-75121, Uppsala, Sweden
[3]Russian State Research Centre "Institute of Theoretical and Experimental Physics"
Bol'shaya Cheremushkinskaya Str. 25V, 117289, Moscow, Russia

INTRODUCTION

Dodecahydro-*closo*-dodecaborate(2-) anion $[B_{12}H_{12}]^{2-}$ as well as carboranes $C_2B_{10}H_{12}$ have been considered for a long time as promising boron moieties for boron neutron capture therapy.[1] The main advantages of the $[B_{12}H_{12}]^{2-}$ anion and its derivatives are their high water solubility as sodium salts and simple methods of the parent anion synthesis from ^{10}B-enriched raw material. The main problems of the $[B_{12}H_{12}]^{2-}$ synthetic chemistry compared to that of carborane are the absence of a distinguished reaction center due to its high, close to spherical, symmetry and high reactivity in respect to electrophiles that often give mixtures of products with various substitution degrees. To avoid this complication, the primary introduction of a reaction center (—OH, —SH, —NH_2, or —I) is necessary.

Study of the $[B_{12}H_{11}OH]^{2-}$ chemistry[2,3] showed that $[B_{12}H_{11}OH]^{2-}$ is a very weak nucleophile and its alkylation requires strong basic conditions that limit its synthetic use. Alkylation of the $[B_{12}H_{11}SH]^{2-}$ anion generally results in disubstituted sulfonium derivatives $[B_{12}H_{11}SR_2]^-$ that require the use of a special technique[4] to protect the sulfur atom.

Frontiers in Neutron Capture Therapy, edited by Hawthorne *et al.*
Kluwer Academic / Plenum Publishers, New York, 2001.

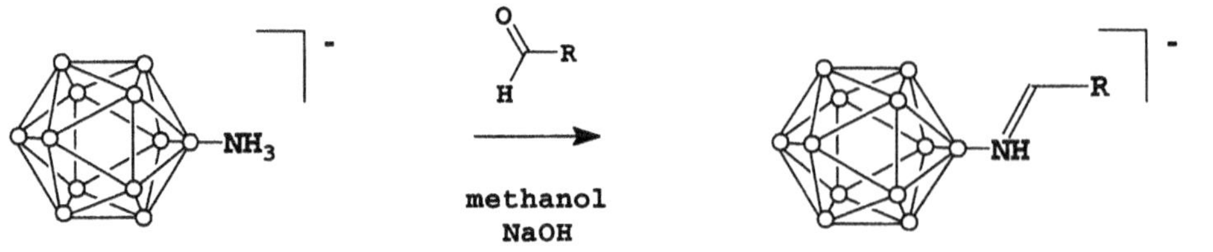

O = BH(B)

R = C_6H_5, 2-C_6H_4OMe, 4-C_6H_4OMe, 4-C_6H_4SMe
4-$C_6H_4NMe_2$, 4-C_6H_4NHCOMe, 4-C_6H_4CN,
4-C_6H_4Br, 4-C_6H_4Cl, 1-$C_{10}H_7$, 2-$C_{10}H_7$,
CH=CH-Me, CH=CH-Ph

Figure 1. Synthesis of Schiff bases.

The possibility of formation of B—C bond by the palladium-catalyzed coupling of $[B_{12}H_{11}I]^{2-}$ with Grignard reagents was demonstrated recently.[5]

Alkylation of the $[B_{12}H_{11}NH_3]^-$ anion gives, as a rule, mixture of alkylation products with different substitution degree $[B_{12}H_{11}NH_2R]^-$, $[B_{12}H_{11}NHR_2]^-$, and $[B_{12}H_{11}NR_3]^-$.[6,7]

In this paper we report a synthesis of Schiff bases derived from aminoundecahydro-*closo*-dodecaborate anion $[B_{12}H_{11}NH{=}CHR]^-$ and their reduction into monosubstituted amines $[B_{12}H_{11}NH_2CH_2R]^-$.

RESULTS AND DISCUSSION

Synthesis of Schiff Bases

The reaction of $(Bu_4N)[B_{12}H_{11}NH_3]$ with aldehydes in methanol in the presence of catalytic amounts of alkali gives in high yield *N*-protonated Schiff bases $[B_{12}H_{11}NH{=}CHR]^-$ (R = C_6H_5, 2—C_6H_4OMe, 4—C_6H_4OMe, 4—C_6H_4SMe, 4—$C_6H_4NMe_2$, 4—C_6H_4NHCOMe, 4—C_6H_4CN, 4—C_6H_4Br, 4—C_6H_4Cl, 1—$C_{10}H_7$, 2—$C_{10}H_7$, CH=CH—Me, CH=CH—Ph) (Figure 1). Generally after addition of sodium hydroxide solution to methanolic solution of $[B_{12}H_{11}NH_3]^-$ and aldehyde, the colour of the solution turned yellow and after stirring for 20–120 min at room temperature, the corresponding Schiff bases $(Bu_4N)[B_{12}H_{11}NH{=}CHR]^-$ precipitated. The Schiff bases obtained had a colour from yellowish to lemon-yellow.

The ^{1}H NMR spectra of the Schiff bases in dimethylsulfoxide-d_6 contain signals of the N—H protons of the protonated imine group in the range of 10.7–12.6 ppm and of the C—H hydrogens of the imine group at 8.0–9.5 ppm. The signals normally appear as doublets ($J \sim 19$ Hz), however sometimes they appear as broadened singlets. The signal of the protonated imine N—H hydrogen rapidly disappears after addition of D_2O (the doublet of the imine C—H hydrogen transforms into the singlet) and is solvent-dependent (for $(Bu_4N)[B_{12}H_{11}NH{=}CHC_6H_4{-}4{-}N(CH_3)_2]$, the N—H proton signal appears at 10.78 and 8.60 ppm in DMSO-d_6 and $CDCl_3$, respectively). The ^{13}C NMR spectra contain a signal of the imine carbon atom in the range of 164–170 ppm. The IR

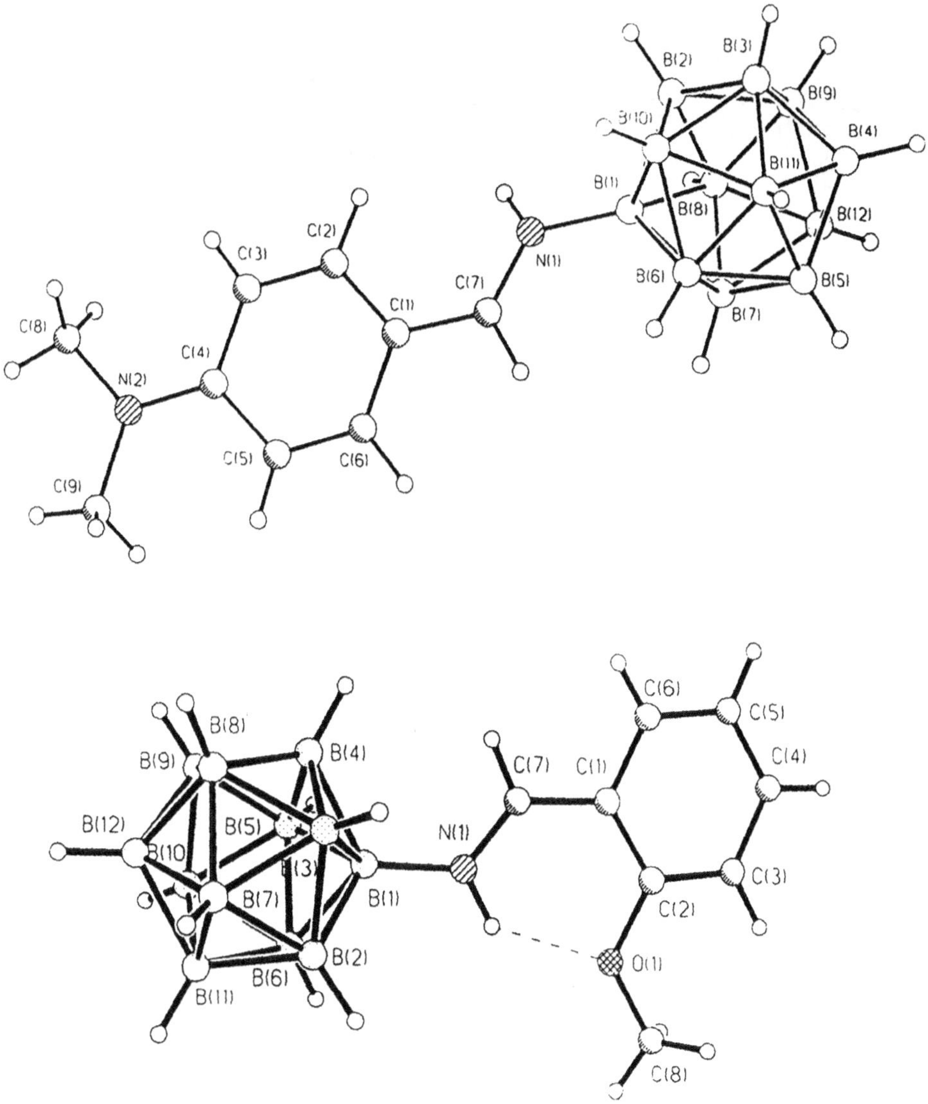

Figure 2. Structures of $[B_{12}H_{11}NH{=}CHC_6H_4{-}4{-}NMe_2]^-$ and $[B_{12}H_{11}NH{=}CHC_6H_4{-}2{-}OMe]^-$.

spectra of the Schiff bases contain a strong band of the C=N stretching at 1630–1650 cm^{-1} and bands of the N—H protons stretch vibrations which usually appear as two sharp bands approximately at 3250 and 3280 cm^{-1}.

Structure of $(Bu_4N)[B_{12}H_{11}NH{=}CHC_6H_4{-}4{-}NMe_2]$ (I) and $(Bu_4N)[B_{12}H_{11}NH{=}CHC_6H_4{-}2{-}OMe]$ (II)

Molecular crystal structures of $(Bu_4N)[B_{12}H_{11}NH{=}CHC_6H_4{-}4{-}NMe_2]{\cdot}CHCl_3$ and $(Bu_4N)[B_{12}H_{11}NH{=}CHC_6H_4{-}2{-}OMe]$ were determined using the method of single crystal X-ray analysis. Compound **I** crystallizes in the monoclinic space group $P2_1/c$ with Z = 4 and unit cell dimensions a = 11.465(6) Å, b = 21.314(7) Å, c = 16.625(6) Å,

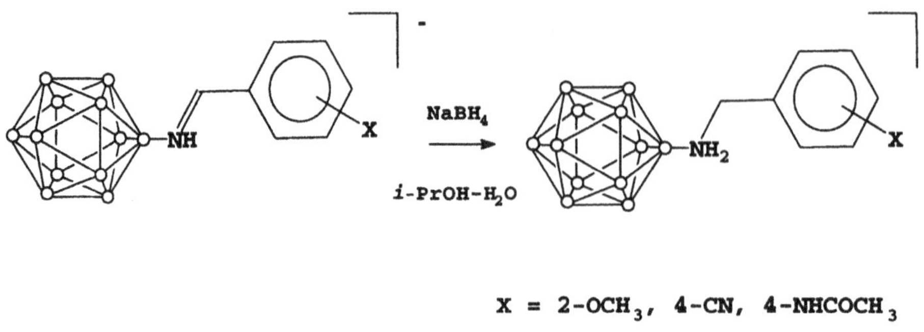

Figure 3. Reduction of Schiff bases.

ß = 101.38(3)°. Crystals **II** belong to the monoclinic space group P2(1)/c system, a = 10.282(2) Å, b = 20.272(5) Å, c = 17.052(3) Å, ß = 105.31(2)°, Z = 4. Both anions have *trans*-configuration (Figure 2). The lengths of the B—N bond are 1.517(7) and 1.525(4) Å, respectively for **I** and **II**, that is of the same order as in $(Ph_4As)[B_{12}H_{11}NH—4—C_5H_4N]*2CH_3CN$[8] (1.513(12) Å). The imine N═C bonds in **I** and **II** are 1.274(6) and 1.274(3) Å, respectively, which is typical for the C═N double bonds. The conformation of the substitutents in the anions studied is close to planar. Formation of strong intramolecular hydrogen bond between the imine hydrogen atom and the methoxy oxygen atom in *ortho*-position of benzene ring (H(N1) . . . O(1) 1.98(2) Å) takes place in the case of the structure **II**.

Reduction of the Schiff Bases

Reduction of the Schiff bases $(Bu_4N)[B_{12}H_{11}NH═CHAr]$ (Ar = 2—$C_6H_4OCH_3$, 4—C_6H_4CN, 4—$C_6H_4NHCOCH_3$) with sodium borohydride in aqueous isopropanol (1 : 3) at room temperature results in the corresponding amines. $(Bu_4N)[B_{12}H_{11}NH_2CH_2Ar]$ (Figure 3).

Functionalized Derivatives of the $[B_{12}H_{12}]^{2-}$ Anion

Using the mild method of synthesis of monoalkylamino derivatives of the *closo*-dodecaborate anion described above we prepared some functionalized boron compounds containing functional groups on the aromatic ring (Figure 4). Amine $(Bu_4N)[B_{12}H_{11}NH_2CH_2C_6H_4—4—NH_2]$ was prepared by deprotection of $(Bu_4N)[B_{12}H_{11}NH_2CH_2C_6H_4—4—NHCOCH_3]$ in refluxing 2 M ethanolic sodium hydroxide. Alkaline hydrolysis of $(Bu_4N)[B_{12}H_{11}NH_2CH_2C_6H_4—4—CN]$ in refluxing 2 M solution of sodium hydroxide in aqueous methanol (1 : 3) gave acid $(Bu_4N)[B_{12}H_{11}NH_2CH_2C_6H_4—4—COOH]$.

CONCLUSIONS

The method described for the attachment of the $[B_{12}H_{11}NH_3]^-$ anion to aldehydes followed by the reduction of the imines formed enables mild introduction of the $[B_{12}]$ cage into more complex organic molecules, including biologically active ones. Moreover,

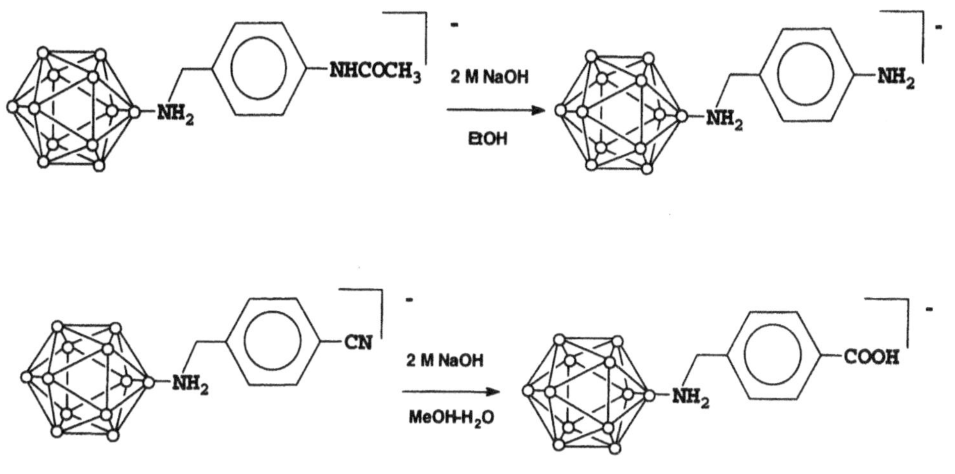

Figure 4. Functionalized derivatives of the $[B_{12}H_{12}]^{2-}$ anion.

considering the recently described methods of mild synthesis of *closo*-1,2—$C_2B_{10}H_{12}$ and *nido*-[7,8—$C_2B_9H_{12}]^-$ carborane derivatives by the reactions of aldehydes with 1,2—$C_2B_{10}H_{12}$[9] and $[B_{10}H_{13}OH]^{2-}$,[10] respectively, it is now possible to prepare a series of boron-containing compounds with different types of boron cages starting from the same aldehydes.

ACKNOWLEDGMENTS

The authors thank the Royal Swedish Academy of Sciences, the Swedish Cancer Society, and the Russian Foundation for Basic Researches for financial support.

REFERENCES

1. M.F. Hawthorne, *Angew. Chem. Int. Ed. Engl.*, 32:950–984, 1993.
2. T. Peymann, E. Lork, and D. Gabel, *Inorg. Chem.*, 35:1355–1360, 1996.
3. A.B. Bruskin, A.M. Orlova, I.B. Sivaev, A.A. Semioshkin, V.I. Bregadze, and S. Sjöberg, *In*: Advances in Neutron Capture Therapy, Vol. II, Chemistry and Biology; B. Larsson, J. Crawford, and R. Weinreich, eds.; Elsevier Science B.V., Amsterdam, pp. 46–50, 1997.
4. D. Gabel, D. Moller, S. Harfst, J. Röesler, and H. Ketz, *Inorg. Chem.*, 32:2276–2278, 1993.
5. T. Peymann, C.B. Knobler, and M.F. Hawthorne, *Inorg. Chem.*, 37:1544–1548, 1998.
6. T. Peymann, E. Lork, M. Schmidt, H. Nöth, and D. Gabel, *Chem. Ber.*, 130:795–799, 1997.
7. B. Grüner, B. Bonnetot, and H. Mongeot, *Collect. Czech. Chem. Soc.*, 62:1185–1204, 1997.
8. T. Koch, and W. Preetz, *Z. Naturforsch.*, 52B:939–942, 1997.
9. H. Nakamura, K. Aoyagi, and Y. Yamamoto, *J. Am. Chem. Soc.*, 120:1167–1171, 1998.
10. B. Brellochs, *Abstr. European Conf. Boron Chem.*, Barcelona, p. 63, 1997.

POTENTIAL BNCT AGENTS

Synthesis of Boron-Containing Pyrimidine Nucleoside and Approaches in the Synthesis of 4-borauracil

Jin-Cong Zhuo, Albert H. Soloway, Beverly A. Barnum, and Werner Tjarks

College of Pharmacy
The Ohio State University
Columbus, OH 43210

1. INTRODUCTION

Boron-containing nucleosides are attractive agents for BNCT[1] because they may be (1) taken up selectively into tumors due to the high mitotic rate of tumor cells *vs.* normal cells, (2) intracellularly converted to the corresponding nucleotides through phosphorylation by appropriate enzymes such as TK1 or dCK, and (3) possibly incorporated into tumor cell DNA thereby enhancing the cytotoxity of the neutron capture reaction.

The boron-containing delivery agent should be non-toxic, selectively target tumor cells and ideally localize within the nucleus. Several boron-containing nucleosides have been synthesized.[1] All these compounds have the boron moiety directly attached on a pyrimidine nucleus. We are very interested in the development of ring isosteres in which a carbonyl group on a pyrimidine such a uracil or thymine is replaced by a hydroxyboryl entity, i.e., 4-borauracil and 4-borathymine, respectively (Figure 1). We[2] and others[3a] have been successful in the synthesis of benzo-fused 4-borauracil, 2-R-1,2-dihydroxy-3-keto-2,4,1-benzodiazaborine [R = H, CH_3, C_6H_5] (Figure 2), and relative compounds.[3b] These results suggested that the boron analogues of the naturally-occuring bases may well be stable under physiological conditions. One question is whether the 4-borauracil and nucleoside of the benzo-fused 4-borauracil can be synthesized? In this paper, we report the successful synthesis of the protected nucleoside and our efforts toward the synthesis of 4-borauracil.

Frontiers in Neutron Capture Therapy, edited by Hawthorne *et al.*
Kluwer Academic / Plenum Publishers, New York, 2001.

R = H, Uracil
R = Me, Thymidine

R = H, 4-Borauracil
R = Me, 4-Borathymidine

Figure 1

2. RESULTS AND DISCUSSION

2.1. Synthesis of N-Methyl Benzoborauracil Nucleoside

The 2,4-(1*H*, 3*H*)-quinazolinediones, carbon analogs to these benzoborauracils, are known to form both ribofuranosyl and deoxyribofuranosyl nucleosides.[4] However, attempts to prepare the benzoborauracil nucleoside by standard procedures were not successful under various reaction conditions. Condensation of the TMS protected N-methyl benzoborauracil **1** and 1-O-Ac-2,3,5-tri-O-benzoyl-ribofuranose **2a** in MeCN with catalytic amounts of trimethylsilyl triflate or $SnCl_4$ did not afford the desired nucleoside **3** (Scheme 1). Nucleoside formation also did not occur in the reactions of tri-O-benzoyl-D-ribofuranosyl chloride with the TMS protected N-methyl benzoborauracil in MeCN with catalytic amounts of trimethylsilyl triflate, CuI, $ZnCl_2$ or $SnCl_4$.

It has been shown that ribose reacts with aniline and gives the corresponding N-phenyl sugar.[5] Similarly, condensation of D-ribose **4** with 2-aminophenylboronic acid gave the corresponding adduct **5** in good yield (Scheme 2). Treatment of **5** with methylisocyanate in MeCN did not give the corresponding nucleoside **6**. This may be due to the free hydroxyl groups in sugar moiety. The desired nucleoside, **9**, was obtained by the reaction of methylisocyanate with compound **8** in which the hydroxyl groups in the sugar were masked.

2.2. Approaches in the Synthesis of 4-Borauracil

Borauracil and borathymine are analogs of the naturally-occurring uracil and thymine, respectively. They may be potential boron delivery agents for BNCT. The

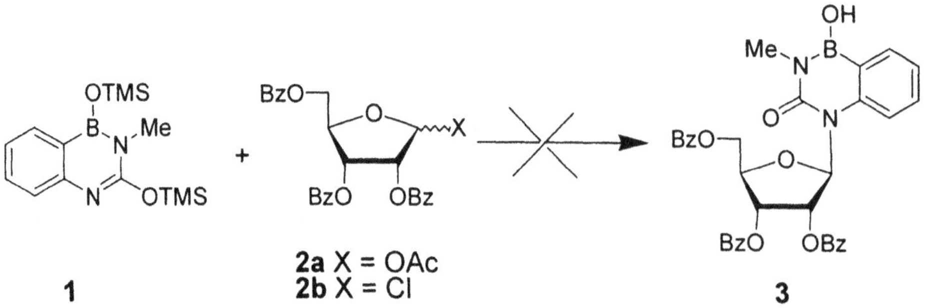

1 **2a** X = OAc **2b** X = Cl **3**

Scheme 1

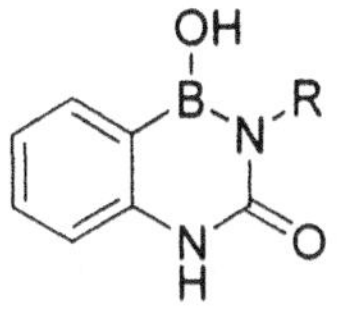

R=H, Me, Ph

Figure 2

successful synthesis of the nucleoside, **9**, demonstrates that it is worthy to explore the synthesis of borauracil and borathymine. Several possible ways for their synthesis are shown in Scheme 3.

2.2.1. Attempts to Synthesize Borauracil by Route A. The known compound, **10**, is the first example which contains N—C=C—B moiety and has been previously synthesized by the condensation of dipropyl(2-pyridylamino)borane and ethyl ethynyl ether.[6] Boron-containing ureas, which are known compounds,[7] can be considered as analogs of dipropyl(2-pyridylamino)borane. However, reaction of boron-urea **12a–b**, prepared *in situ* by reaction of ureas **11a–b** with tributylborane in THF, with ethyl ethynyl ether did not afford any desired products **13a–b** (Scheme 4).

2.2.2. Attempts to Synthesize Borauracil by Route B. As a model reaction, tributylaminoborane reacts smoothly with phenyl isocyanate in benzene and produces the expected product, **14**, in good yield. Similarly, treatment of compound **15** with excess

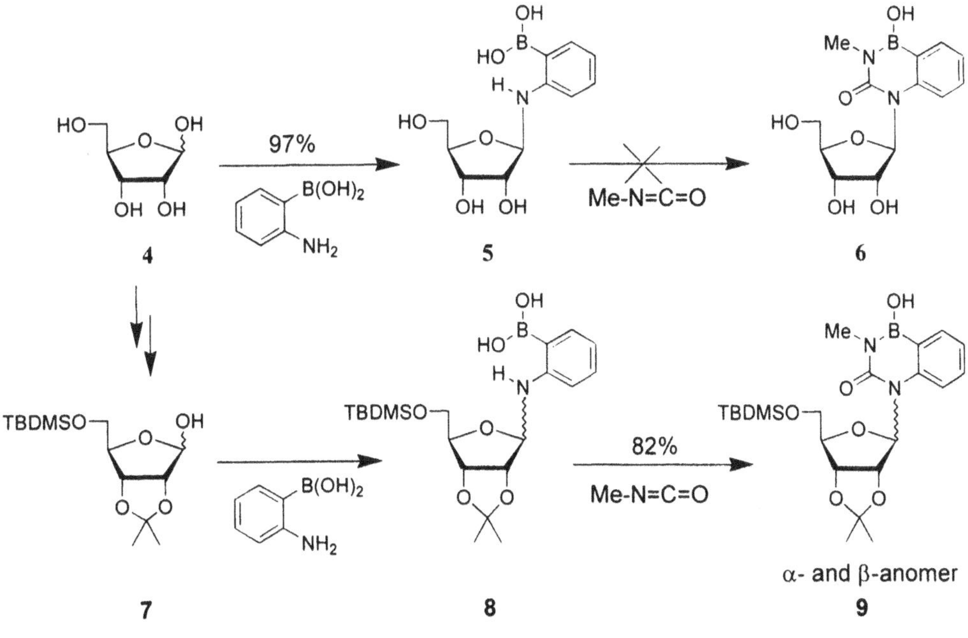

Scheme 2

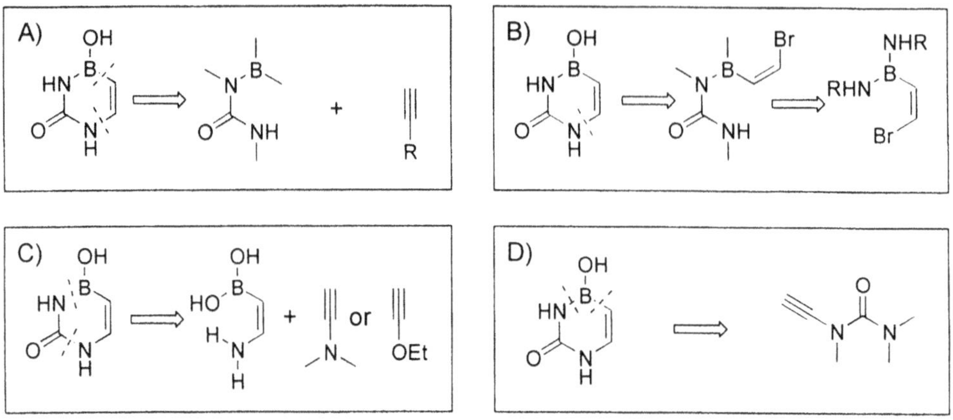

Scheme 3

butylamine in hexane at –78 °C gave the corresponding product **16**. However, reaction of **16** with phenyl isocyanate affords N-phenyl-N′-butylurea instead of the expected compound, **17** (Scheme 5).

2.2.3. Attempts to Synthesize Borauracil by Route C. Hydroboration of ethyl ethynyl ether in THF by catecholborane gave compound **18**. Attempts to produce compound **19** by treatment of **18** with various amines failed (Scheme 6).

Interestingly, hydroboration of N-methyl-N-phenylacetylene **20** in THF did not afford the boronated compound **21**. However, using ethyl ether as solvent we obtained the desired compound. Nevertheless, attempts to obtain **22** by hydrolysis of **21** resulted in the cleavage of the carbon-boron bond and none of the desired product. The easy loss of boron has been observed for some related compounds (Scheme 7).[8]

2.2.4. Attempts to Synthesize Borauracil by Route D. Since the above attempts have been unsuccessful our present efforts have focused on trials to synthesize the borauracil derivative by direct hydroboration of N-acetylenylurea. Reaction of compound **23** with

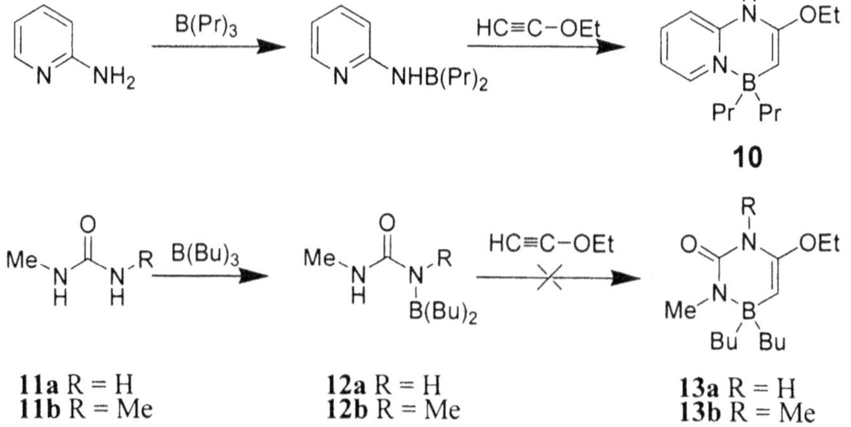

Scheme 4

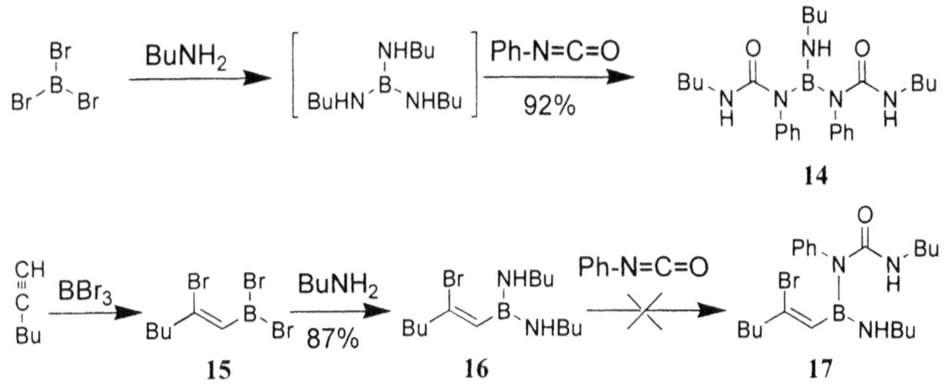

Scheme 5

catecholborane yielded **24** as the major product and **25** as the minor component (Scheme 8). The formation of **24** as the major product can be attributed to steric effects arising from the phenyl group. It is known that hydroboration is sensitive to steric interaction.[9] Therefore, it would be desirable to synthesize substituted N-acetylenylureas, **26**. They should be hydroboronated mainly at the terminal position and lead to **27**. Compounds **26** have not been reported in the literature and studies are underway to determine whether they are stable and can be synthesized.

3. CONCLUSIONS

1. This work demonstrates that a heterocyclic ring structure [—C═C—B(OH)—N—C(O)—N—] is generally stable.
2. The ribose nucleoside of benzoborauracil has been synthesized and shown to be chemically stable.
3. Compound **21** can be synthesized. This suggests that a B—CH═CH—N moiety which is an essential requirement for the synthesis of 4-borauracil and 4-borathymidine may be stable.
4. 4-Borauracil and its protected nucleoside may be synthesized by hydroboration of N-acetylenylureas **26**.

ACKNOWLEDGMENT

We acknowledge the support for our research from U.S. Department of Energy Grant (DE-FG02-90ER60972) and the Ohio State University Comprehensive Cancer Center Grant (P30 CA—16058).

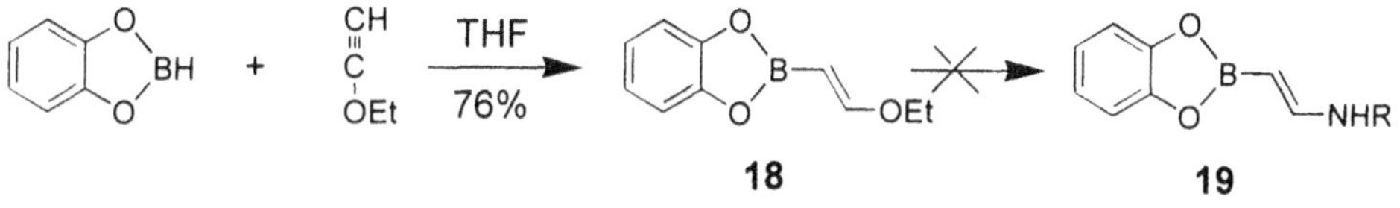

Scheme 6

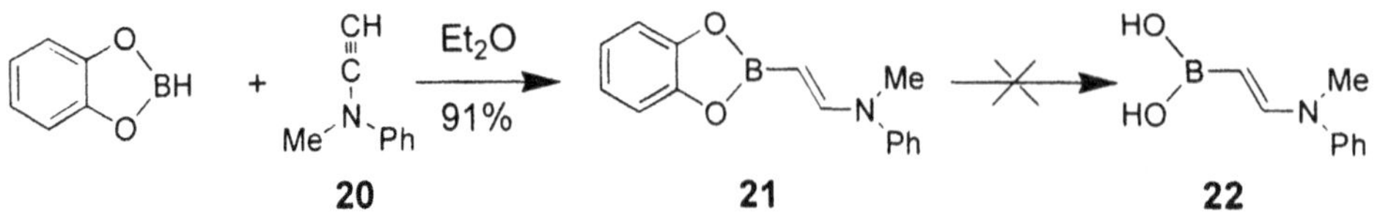

Scheme 7

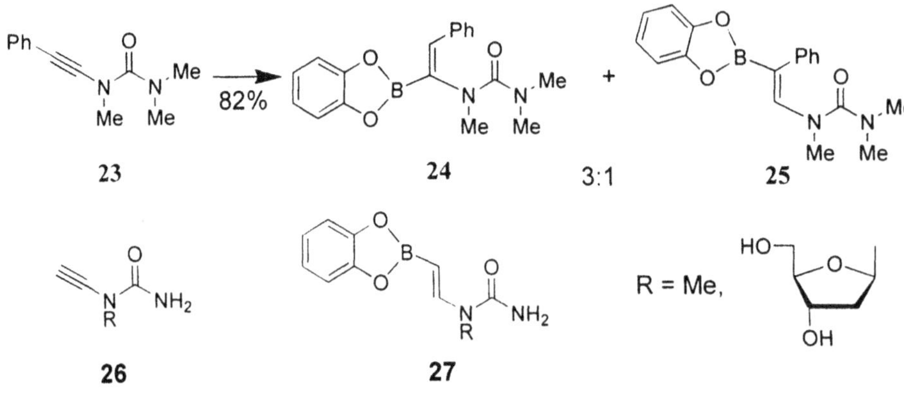

Scheme 8

REFERENCES

1. A.H. Soloway, W. Tjarks, B.A. Barnum, F.-G. Rong, R.F. Barth, I.M. Codogni, and J.G. Wilson, The Chemistry of Neutron Capture Therapy, *Chem. Rev.* 98:1515, 1998.
2. B.A. Barnum, J.C. Beeson, F.-G. Rong, A.H. Soloway, W. Tjarks, G.T. Jordan IV, and S.G. Shore, Synthesis of a New Boron-Heterocycle-A Uracil Analog: 2-R-1,2-dihydro-1-hydroxy-3-keto-2,4,1-benzodiazaborines, *Advances in Boron Chemistry*, Siebert, W. Ed., The Royal Society of Chemistry: Cambridge, UK, pp. 240–243, 1997.
3. (a) M.P. Hughes, and B.D. Smith, Enhanced Carboxylate Binding Using Urea and Amide-Based Receptors with Internal Lewis Acid Coordination: A Cooperative Polarization Effect, *J. Org. Chem.* 62:4492–4499, 1997. (b) M.P. Groziak, A.D. Ganguly, and P.D. Robinson, Boron Heterocycles Bearing a Peripheral Resemblance to Natural-Occuring Purines: Design, Synthesis, Structures, and Properties, *J. Am. Chem. Soc.* 116:7597–7605, 1994.
4. M.G. Stout, and R. Robins, The Synthesis of Some Quinazoline Nucleosides, *J. Org. Chem.* 33:1219–1225, 1968.
5. C. Chavis, C. De Gourcy, F. Dumont, and J.-L. Imbach, NMR Studies of D-Ribosylamines in Solution: Drivatives of primary Amines, *Carbohydrate Res.* 113:1–20, 1983.
6. V.A. Dorokhov, and B.M. Mikhailov, Organoboron Compounds CCLXXXV. Diakyl(2-pyridylamino)boranes, *Zh. Obshch. Khim.* 44:1281–1287, 1974.
7. Farbenfabriken Bayer A.-G., Brit.Patent, 919374, 1963.
8. D.S. Matteson, unpublished results, personal communication.
9. C.F. Lane, and G.W. Kabalka, Catecholborane. A New Hydroboration Reagent, *Tetrahedron*, 32:981–990, 1976.

BORYLATED FERROCENIUM COMPOUNDS

Potential New Drugs for BNCT?

L. Weissfloch[1,8], M. Nagl[2], M. Wagner[2], T. Probst[3], D. Preusse[4], P. Nilsson[5], R. Senekowitsch-Schmidtke[6], D. Gabel[4], J. Carlsson[5], K. Tempel[7], and M. Molls[1]

[1]Department of Radiation Oncology
Klinikum rechts der Isar
[2]Institute for Inorganic Chemistry
[3]Institute for Radiochemistry
Garching, TU Munich, Germany
[4]Department of Chemistry
University of Bremen, Germany
[5]Biomedical Radiation Sciences
Uppsala University, Sweden
[6]Department of Nuclear Medicine
Klinikum rechts der Isar
TU Munich, Germany
[7]Inst. of Pharmacology, Toxicology and Pharmacy
Faculty of Veterinary Medicine
LMU Munich, Germany
[8]Medical Department
Brookhaven National Laboratory
Upton, NY, USA

INTRODUCTION

In industrialized countries, cancer is the most frequent cause of death, apart from cardiovascular diseases. Surgery, radiotherapy and chemotherapy are the most common treatments, but the destruction of primary tumors by irradiation often fails due to radiosensitive tissues next to the neoplasm. Safe application of high radiation doses to malignant tumors is crucial in curative radiation treatment. The risk of normal tissue complications has to be low. To reduce the radiation dose, treatment schemes in which the tumor cells are selectively sensitized and subsequently destroyed by radiation are of

Frontiers in Neutron Capture Therapy, edited by Hawthorne *et al.*
Kluwer Academic / Plenum Publishers, New York, 2001.

special interest. In principle, boron neutron capture therapy (BNCT) fulfils these requirements. With regard to the biological problems, we focused our investigations on testing boron-containing compounds to accumulate in murine tumors, using a new class of ferrocenium derivatives as boron carriers for BNCT.

MATERIALS AND METHODS

Ferrocenium compounds have been acknowledged to accumulate in murine tumors. Beside this aspect that these compounds likely qualify for BNCT, this class of drugs has also a cell killing, cytostatic effect.[1,2] New derivatives were developed[3] and twelve of them plus Ferrocenium itself were tested in a mammary carcinoma (AT17) of the C3H-mouse, meanwhile (Figure 1). The details of tumor[4] and anaesthesia[5] have been published elsewhere. The mice received up to 300mg of the derivatives diluted in 0.5ml 0.9% NaCl intraperitoneally.[2] At certain points of time from 0.5 to 4 hours after injection tumor bearing mice were sacrificed. Tumor, liver, kidneys and brain, as well as samples of the thigh muscles and integral blood were taken. The plasma was elaborated. For all trials, samples were also taken from untreated animals as control groups. The boron contents of the samples were determined both by ICP-AES and ICP-MS. These analytical methods have been described in detail elsewhere.[6] Additionally, whole-body cryosections and alpha-particle sensitive film exposures were performed.

RESULTS AND DISCUSSION

Shortly after the i.p. injections of the Ferrocenium derivatives very different side effects were observed. All treated animals went through a short-term period of hyperactivity. Six Ferrocenium derivatives showed a very high neurological toxicity down to 20mg/kg b.w. These side effects did not occur during atropine protection. That means these compounds inhibit the acetyl-cholinesterase, but we stopped these testing. The other derivatives induced somnolence for several hours in the treated mice (Table 1).

After injection of 300mg/kg b.w. Ferrocenium 2b, the mice became also somnolent, but recovered totally to normal behaviour after 40 minutes. The determination of boron after injection of Ferrocenium 2b showed high boron content in spleen and liver, less in kidney, lung and muscle, and poor in integral blood (2.5 ± 1.5ppm) and tumor (0.9 ± 0.3ppm). But it is remarkable that boron content of 2.1 ± 0.9ppm could be found in the brain. That means that Ferrocenium 2b penetrates the blood-brain-barrier, which is a characteristic that may be advantageous in the treatment of astrocytomas and glioblastomas. To complete the overview on the biodistribution of this ferrocenium, cryosections and alpha-track etches are in evaluation.

CONCLUSION AND FUTURE DIRECTIONS

The AT17-tumor did not accumulate the Ferrocenium 2b in the tumor itself in amounts sufficient for BNCT. The toxicity of the compound must be reduced. But

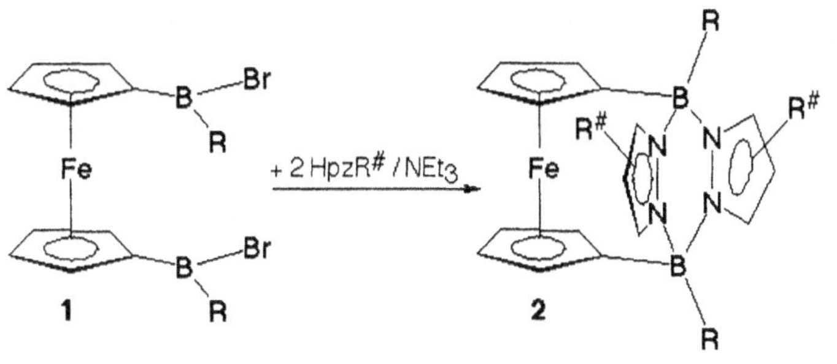

1	R
a	Br
b	Me
c	OEt
d	NC_4H_8

2	R	R#
a	Br	3,4,5-H
b	Me	3,4,5-H
c	Me	3,5-Me; 4-Br
d	Me	3,4,5-COOEt
e	OEt	3,4,5-H
f	$N_2C_3H_3$	3,4,5-H
g	Me	3,5-H;4-I
h	Me	3,4,5-COOMe

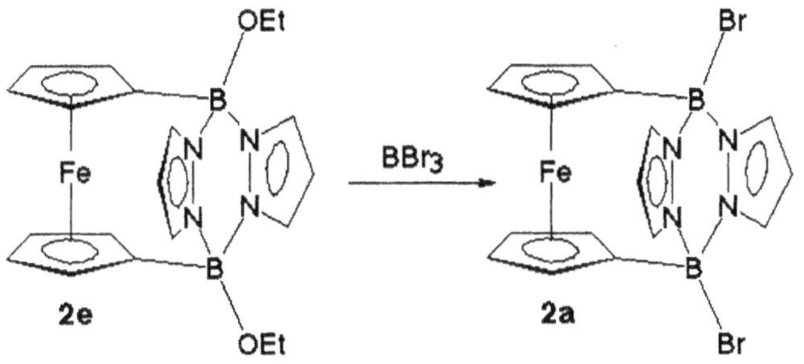

Figure 1

Ferrocenium 2b is interesting cause of its penetration of the blood-brain-barrier. Therefore the modification of Ferrocenium 2b, as well as further modifications of its administration will be a topic for research in the near future. To reduce the toxicity to a minimum, a substitution of a Methyl to an Isopropyl in the molecule should lead to a non-inhibition of the acetyl-cholinesterase.[7] The variation of the administration of this so modified compound with long-term infusion using osmotic pump could increase the drug's uptake. To proof the selectivity towards brain tumors the drug has to be tested in rodents, bearing xeno-transplanted glioblastomas or astrocytomas. Paramagnetic drugs such as ferrocenium compounds can be detected using a new screening method using

Table 1. Administration data and veterinarian diagnosis; all drug concentrations are given in [mg/kg] b.w.

Substance	Concentration	Diagnosis
2b	400	Short-time ataxia, somnolence up to 12 h
	300	Reduced activity up to 40 min
	200–50[a]	Reduced activity up to 40 min less pronounced
2c*	400	Reduced activity, dyspnoea, somnolent
	200	Reduced activity up to 4 h
	100	Inconspicuous behavior ("fit"), no side effects
2e	400–100[b]	Reduced activity up to 8 h
2f	400–25[c]	Immediately heaviest excitations, jumps, polypnoea (panting breathing), ataxia soon followed by paraplegia of hint limbs, finally vomiting-like symptoms with gas-bubbled bronchial secretion; euthanasia
2h[#]	400–25[d]	In the beginning short-timed screaming, rapidly becoming very somnolent, e.g. no reactions; euthanasia
4a	200	Immediate death post i.p.
	100	Death within less than 1 min, showing heaviest excitations, within a couple of minutes dead, see 100
	50	Hyperactivity, followed by dyspnoea (laboured breathing), later on
	25	increasing somnolence; euthanasia
4b	400–25[c]	See 2f
4c	400–25[c]	See 2f
5	200–25[e]	Immediate death post/during i.p.
6	400–25[d]	Immediate excitations, followed by dys- and hypopnoea inclusively increasing somnolence; euthanasia

*: results obtained by injection of a suspension; [#]: foam formation during solubilization; [a]: stepwise 200; 100; 50; [b]: stepwise: 400/200/100; [c]: stepwise: 400/200/100/50/25; [d]: stepwise: 400/300/200/100/50/25; [e]: stepwise 200; 50; 20.

NMR with high strength magnetic field.[8] This would allow a more rapid analysis of the drug biodistribution in living, anesthesized animals.

ACKNOWLEDGMENT

Financial support from the Bayerisches Staatsministerium fuer Landesentwicklung und Umweltfragen, Muenchen, and the Pinguin Foundation of the Henkel KG Company, Duesseldorf, is gratefully acknowledged.

REFERENCES

1. Köpf-Maier P., and Köpf H.: Non-Platinum-Group Metal Antitumor Agent: History, Current Status and Perspectives. *Chem. Rev.* 87:1137–52, 1987.
2. Köpf-Maier P., Köpf H., and Neuse E.W.: Ferroceniumsalze—die ersten tumorhemmenden Eisenverbindungen. *Angew. Chemie* 96(6):446–447, 1984.
3. Jäkle F., Priermeier T., and Wagner M.: Novel *ansa*-Ferrocenes with *o*-Phenylene-type Bridges by B-N Adduct Formation. *J. Chem. Soc. Chem. Commun.* 1765–1766, 1995.
4. Weissfloch L., Auberger T., Wagner F.M., Senekowitsch R., Tempel K., and Molls M.: Oxygen tensions in rodent tumors after irradiation with neutrons. *Adv. Exp. Med. Biol.* 411:255–260, 1997.
5. Flecknell P.: Anaesthesia of Common Laboratory Species—Small Rodents. *In: Laboratory Animal Anaesthesia.* Academic Press, London, pp. 160–171, 1997.

6. Probst T., Berryman N., Lemmen P., Weissfloch L., Auberger T., Gabel D., Carlsson J., and Larsson B.: A comparison of ICP-AES and ICP-MS with quantitative neutron capture radiography for the determination of boron in biological samples from cancer therapy. *J. Anal. At. Spectrom.* 12:1115–1122, 1997.
7. Personnel communications: Tempel K., and Petter A.: Inst. of Pharmacology, Toxicology and Pharmacy, Faculty of Veterinary Medicine, LMU Munich, Germany; Gabel D: Dept. of Chemistry, Univ. of Bremen, Germany, 1998.
8. Komoroski R.A.: *In vivo* nmr of drugs. *Analyt. Chem.* 60:1024–1033, 1994.

ENANTIOSELECTIVE SYNTHESES OF ^{10}B-ENRICHED L- AND D-CARBORANYLALANINE AND THEIR RADIOBIOLOGICAL EVALUATION IN V-79 CHINESE HAMSTER CELLS

Stephen B. Kahl[1], John J. Schaeck[1], Brenda Laster[2], and Lynn Warkentien[2]

[1]Department of Pharmaceutical Chemistry
University of California, San Francisco
[2]Medical Department
Brookhaven National Laboratory

INTRODUCTION

The single greatest impediment to widespread clinical use of BNCT is the development of boron compounds that deliver large amounts of boron to the tumor, are highly tumor selective while clearing from normal tissues and blood, and which have acceptably low systemic toxicity. Boronated amino acids represent one potentially useful approach to meeting these requirements. While BPA possesses some of these very desirable characteristics, its single boron atom and lack of significant aqueous solubility requires that very large doses of the fructose complex be administered to achieve minimal acceptable tumor boron and tumor:normal tissue ratios. Amino acids bearing the polyhedral carborane cage represent a potential source of compounds which, either in and of themselves or in a small tumor-seeking peptide, might provide alternatives to BPA. The carborane analog of phenylalanine, carboranylalanine or Car, has received special attention in this regard. First synthesized by Zakharkin *et al.*[1] and by Brattsev and coworkers[2] and later by Soloway's group as the racemic mixture,[3] Car has more recently been prepared stereoselectively.[4-6] Our group was the first to publish a truly stereoselective method of preparation by which both enantiomers could be made available.[7] Beginning with allyl-*o*-carborane and following an oxazolidinone approach pioneered by Evans *et al.*,[8] the 6 step synthesis produced either L- or D-Car in about 34% overall yield.

Though this synthesis is an elegant method, it suffers from the drawback that the carborane cage is introduced very early in the sequence in the form of allyl-*o*-carborane.

Frontiers in Neutron Capture Therapy, edited by Hawthorne *et al.*
Kluwer Academic / Plenum Publishers, New York, 2001.

Figure 1. Structures of L-Carboranylalanine (1a) and D-Carboranylalanine (1b).

This is a particular problem when preparing ^{10}B enriched material for radiobiological studies and eventually for BNCT. The material loss and waste of boron could be minimized if the carborane cage were introduced later in the synthesis and could be constructed from commercially available ^{10}B enriched precursors such as decaborane or *o*-carborane. Although allyl-*o*-carborane could be prepared from either of these starting materials, the overall yield of the desired amino acids would be decreased still further. Consequently, we sought other strategies to prepare ^{10}B-enriched carboranylalanine in an enantioselective manner. We report here two successful methods and preliminary radiobiological studies of their neutron sensitizing properties in V-79 Chinese hamster cells.

METHODS

Synthesis

L-Carboranylalanine **1a** was prepared by the method shown in Scheme 1 starting with the commercially available chiral auxiliary **2** in a modification of the procedure of Williams *et al.*[9] The enolate of oxazinone **2** was treated with propargyl bromide to form the masked propargyl glycine **3** in an unoptimized yield of ~70%. Standard reaction of ^{10}B-enriched decaborane with acetonitrile gave the enriched bis-ligand compound which

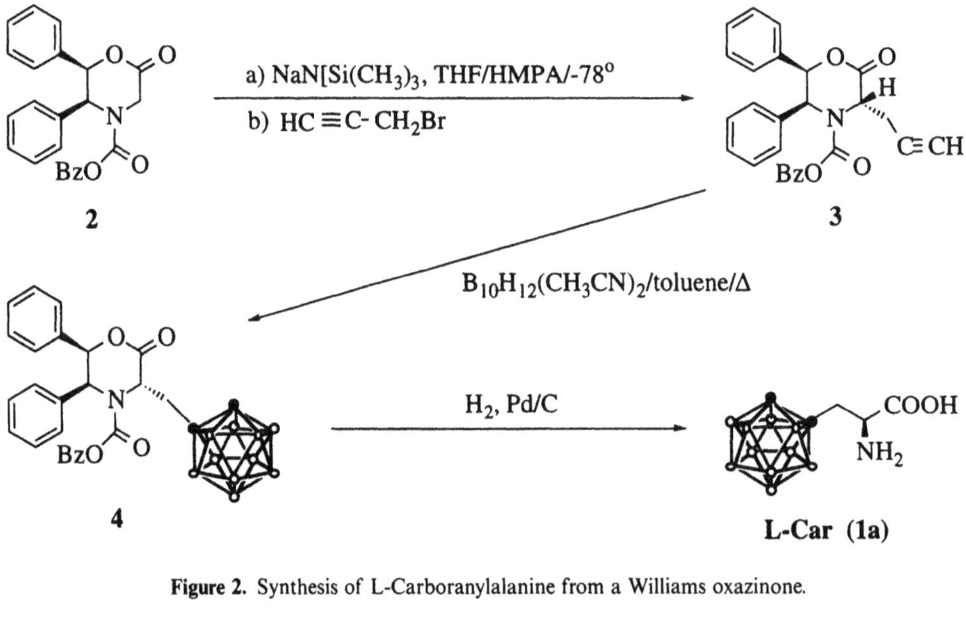

Figure 2. Synthesis of L-Carboranylalanine from a Williams oxazinone.

was then reacted with the chiral alkyne to produce carboranyl oxazinone **4** in ~60% yield. Exhaustive hydrogenation of **4** over Pd/C gave **1a** which was immediately isolated as the hydrochloride salt by treatment of an ethanolic solution of **1a** with hydrogen chloride gas. L-Car produced in this manner was identical in all respects to that provided by our earlier synthesis. The unoptimized yield of the reduction step is ~50% giving the enriched amino acid in an overall yield of ~30%. Efforts are currently underway to raise this yield as Williams reported typical yields in the range of 85–95%.[9] The opposite chiral auxiliary is also commercially available and can be used to produce the D-enantiomer, but another method was used for this purpose in the present study.

D-carboranylalanine **1b** was obtained by a modification of the oxazoldinone strategy of Evans (not shown). This approach is similar to that used in our earlier work, except that 4-pentynoic acid rather than 3-carboranyl propanoic acid was first attached to the oxazolidinone. The starting alkynoic acid was obtained by Jones oxidation of the corresponding alcohol. In our previous work, direct azidation of the carboranyl oxazolidinone was not possible and a two step method employing stereoselective bromination by NBS followed by an SN_2 displacement and inversion by tetramethylguanidium azide (TMGA) was used. Despite concerns as to the stability of the alkyne to NBS, this method was applied and found to be successful. The alkynyl bromide product was condensed with enriched bis-acetonitrile decaborane to give the expected carboranyl bromide in ~60% isolated yield. Azidation with inversion at the nascent α site provided the azide in ~80–90% yield which is similar to that found in our previous synthesis. Removal of the oxazolidinone auxiliary by transesterification of azido imide with titanium tetraisopropoxide and benzyl alcohol gave the azido benzyl ester which was reduced with H_2 to give the desired **1b**, again isolated as the hydrochloride salt. Combined yield of the last two steps amounted to ~65% with the overall yield of **1b** from the alkynyl bromide of 30–35%.

Cell Irradiations

V-79 Chinese hamster cells in logarithmic growth were propagated in Dulbecco's modified Eagle's media supplemented with 10% fetal bovine serum, 1% penicillin-streptomycin-fungisone, and 2.0 mM L-glutamine. Before irradiations, cells were pre-incubated with ^{10}B enriched L- or D-Car hydrochloride at various concentrations for either 4 or 16 hours. After incubation, cells were washed 3 times with boron-free PBS to remove free extracellular compounds and intracellular compound not bound to cell constituents. Cells were trypsinized, harvested in boron-free medium and suspended in growth medium at a density of 3.0×10^5 cells/ml. Irradiations were carried out at the thermal neutron port of the Brookhaven Medical Research Reactor. After irradiation for various times, cells were plated for undisturbed colony growth for 5 days. Colonies were then washed with PBS, fixed with ethanol, and stained with crystal violet prior to optoelectronic counting with an Artec colony counter.

RESULTS AND DISCUSSION

Synthesis

An enantioselective synthesis of Car in which the carborane cage was added or constructed at a late stage in the reaction sequence was the goal of this work. The sequence shown in Figure 2 meets that aim and provides L-Car in only 3 steps from readily available starting materials. Oxazinone chiral auxiliary **2** (and its enantiomer) can be purchased (Aldrich) or readily constructed in multi-gram quantities at low cost, and the *t*-Boc analogs are also available. The latter produce *t*-Boc protected amino acid products which can also be obtained by treatment of, for example, **1a** with Boc-dicarbonate. Conversion of **2** into masked propargyl glycine **3** was carried out in an unoptimized yield of ~65%. The yield of carboranyl oxazinone **4** from the chiral alkyne **3** was surprisingly high (~60%) considering the steric hindrance present in the alkyne. We have also since found that the yield of carboranyl oxazinone precursor to D-Car from the other enantiomeric auxiliary is ~68%. The final hydrogenation step which produces the desired **1a** proceeds with a yield of only about 50% compared to yields reported by Williams of typically >90%. Nevertheless, even the unoptimized overall yield of L-Car from **3** was ≤30% which is significantly better than what we would have obtained in our previous method had we started with $B_{10}H_{14}$. L-Car was immediately isolated as the hydrochloride salt since carboranylalanine has been reported to undergo spontaneous degradation to the *nido* form,[5] presumably through either intra- or intermolecular reaction with the amine group. In contrast to the very poor aqueous solubility of zwitterionic **1a**, the hydrochloride salt is very water soluble. Efforts are currently underway to optimize the hydrogenation step; a yield of only 80% in this step would provide L-Car (and presumably D-Car) in ~50% overall yield. We believe this new, shorter stereoselective route to the enantiomers of Car will eventually become the method of choice.

Cell Irradiations

Cell irradiation studies were carried out at Brookhaven in a manner similar to that published by us several years ago.[10] Incubations of V-79 Chinese hamster cells with both L- and D-Car at concentrations from 100–200 μg/ml medium demonstrated that neither enantiomer was cytotoxic at these concentrations; plating efficiencies of both treated and control cells were identical. This is in sharp contrast to an earlier report by Soloway and coworkers who found that racemic Car reduced [^{3}H]-TdR uptake in MRA 27 cells by 50% at a dose of ~60 μg/ml in the absence of β-cyclodextrin.[11] This disparity is not likely to have been caused by differences in cell lines nor by using racemic rather than enantiomerically pure Car, but perhaps may be due to the presence of small amounts of toxic impurities from their synthesis. Another possibility is that their *closo* amino acid mixture might have degraded to the *nido* compounds in solution, but this also unlikely as the *nido* amino acid was reported to be non-toxic at all concentrations up to 400 μg/ml. This disparity in findings also illustrates rather dramatically the danger of dismissing potential BNCT agents on the basis of a single study.

Irradiation of washed cells resuspended in boron-free medium followed by 5 days of colony growth and plotting of the resultant cell survival curves in all cases resulted in significant dose enhancement relative to controls. Radiobiological parameters derived from these curves are presented in Table 1. From the resulting cell survival curves and the data in Table 1, it is evident that both L- and D-Car are taken up and retained by the V-79 cells and are not removed by vigorous washing in boron-free medium. This is in clear contrast to L-BPA which is largely released from cells under similar conditions.[12,13] Surprisingly, increasing the L-Car pulse time from 4 to 16 hours resulted in a 28% decrease in the dose enhancement factor at the 10% survival level and a concomittant increase in the D_0 as can be seen in Table 1. Similar results were observed with the D enantiomer. In general, compounds which are retained intracellularly show an increase

Table 1. Effects of varying concentrations of ^{10}B-carboranylalanine on V-79 Chinese Hamster cell survival after 16 hour incubation

Sample (conc.)	α	β	D_0	Dose enhancement factor*
Control	0.13 ± 0.02	0.005 ± 0.001	4.3 ± 0.21	
D-Car (100 μg/ml)	0.23 ± 0.04	0.004 ± 0.006	3.6 ± 0.36	1.4 ± 0.63
D-Car (150 μg/ml)	0.40 ± 0.07	0.004 ± 0.011	2.4 ± 0.45	2.2 ± 0.15
D-Car (200 μg/ml)	0.48 ± 0.02	0.003 ± 0.003	2.0 ± 0.04	2.6 ± 0.07
L-Car (200 μg/ml)	0.37 ± 0.02	0.004 ± 0.001	2.3 ± 0.11	2.1 ± 0.06
L-Car (200 μg/ml)#	0.52 ± 0.03	0.004 ± 0.004	1.8 ± 0.05	2.9 ± 0.09

#Data from 4 hour incubation; *calculated at 10% survival.

in dose enhancement as the incubation time is increased. The source of this effect is unknown.

As can be seen from Table 1, increasing the exposure concentration for D-Car increases dose enhancement and results in a striking reduction in the D_0. The α-coefficient represents the linear portion of the cell survival curve while the β-coefficient reflects the shoulder. A high α/β ratio is consistent with the production of high LET damage to the cell. This ratio ranges from 58 to 160 for the Car data in Table 1. Another surprising aspect of our findings is the relatively small difference between L-Car and D-Car when incubated under similar conditions. For example, D_0 and dose enhancement factor for D-Car at 200 μg/ml and 16 hour incubation time show it to be a slightly better sensitizer than L-Car. The difference is small, but is greater than the experimental errors of the measurements. On the basis of its configuration, one would predict that L-Car would be more likely to be taken up and retained than D-Car as has been observed with the enantiomers of BPA.[14] Our observation suggests that rather than acting purely as an amino acid analog and undergoing carrier-mediated transport, Car may be taken up by a process by which its chiral configuration is not critical. Car is one of the most hydrophobic amino acids known and it is tempting to speculate that it becomes associated with the hydrophobic core of lipoproteins such as LDL and is taken up by receptor-mediated endocytosis of the LDL-Car complex. Further experiments will be needed to shed light on Car transport and localization. However, it is clear from our results that neither L- nor D-Car is toxic to cells in the absence of thermal neutrons, but that in their presence both enantiomers are very effective intracellularly localized neutron sensitizers. Our results open the door to further exploration of these non-natural amino acids by making them available in enantiomerically pure form by reasonable processes.

ACKNOWLEDGMENTS

We wish to thank the Department of Energy for its support of this work through grant DE-FG03-89ER60873.

REFERENCES

1. L.I. Zakharkin, A.V. Grebennikov, and A.I. L'vov *Izv. Akad. Nauk, SSSR Ser. Khim.* 106–112 (1970).
2. V.A. Brattsev, and V.I. Stanko *Zh. Obshch. Khim.* **39**, 1175–6 (1969).
3. I.M. Wyzlic, and A.H. Soloway *Tetrahedron Lett.* **33**, 7489–7490 (1992).

4. J.-L. Fauchere, O. Leukhart, A. Eberle, and R. Schwyzer *Helv. Chim. Acta* **62**, 1382–1395 (1979).
5. S. Sjöberg, J. Carlsson, P. Lindstrom, and J. Malmquist *Current Topics in the Chemistry of Boron*; G.W. Kabalka, Ed; Royal Society of Chemistry: Cambridge, U.K. pp. 172–176 (1994).
6. W. Karnbrock, H.-J. Musiol, and L. Moroder *Tetrahedron* **51**, 1187–1196 (1992).
7. P.A. Radel, and S.B. Kahl *J. Org. Chem.* **61**, 4582–4588 (1996).
8. D.A. Evans, D.C. Britton, J.A. Ellman, and R.L. Dorow *J. Am. Chem. Soc.* **112**, 4011–4030 (1990).
9. R.M. Williams, P.J. Sinclair, D. Zhai, and D. Chen *J. Am. Chem. Soc.* **110**, 1547–1557 (1988).
10. R.G. Fairchild, S.B. Kahl, B.H. Laster, J. Kalef-Ezra, and E.A. Popenoe *Cancer Res.* **50**, 4860–4865 (1990).
11. J.-H. Yong, R.F. Barth, J.H. Rotaru, I.M. Wyzlic, and A.H. Soloway *Anticancer Res.* **15**, 2039–2044 (1995).
12. B.D. Bennett, J. Mumford-Zisk, J.A. Coderre, and G.H. Morrison, *Radiat. Res.* **140**, 72–78 (1994).
13. V.P. Bond, B.H. Laster, and L. Wielopolski, *Radiat. Res.* **141**, 287–293 (1995).
14. J.A. Coderre, J.D. Glass, R.G. Fairchild, P.L. Micca, I. Fand, and D.D. Joel *Cancer Res.* **50**, 138–141 (1990).

A POTENTIAL NEW BNCT AGENT

Galactose Containing 1-Amino-3[2-(1,7-Dicarba-*closo*-Dodecaboran-1-YL)Ethyl]-Cyclobutanecarboxylic Acid

G. W. Kabalka, G. Li, and R. R. Srivastava

Department of Chemistry and Radiology
The University of Tennessee
Knoxville, Tennessee 37996-1600

Boron neutron capture therapy (BNCT) is a binary approach to cancer therapy.[1] In this protocol, boron-10 is delivered to the tumor tissue, and upon irradiation with thermal neutrons, cytotoxic α-particles and high energy lithium-7 ions are produced. The linear energy transfer (LET) of these heavy charged particles (which is on the order of one cell diameter) makes them lethal to the boron-containing cells. The success of this protocol is dependent upon the selective delivery of a sufficient quantity of boron-10 to the tumor tissues (10–30 μg boron-10/g of tumor with tumor to normal tissue ratio of 3:1). While various carrier molecules are under investigation as potential boron-delivery agents, the only drug currently in clinical trials in the United States is a boron-containing amino acid,[2] 4-dihydroxyborylphenylalanine (BPA).

Our investigation in the BNCT compound development area has focused on the unnatural, cyclic α-amino acid, 1-aminocyclobutanecarboxylic acid (ACBC), as a carrier molecule. This unnatural amino acid is known to be preferentially retained in the intracerebral tumors.[3] In addition, ACBC has been found to be nontoxic. We have identified the *meta*-carborane cage as a potential boron source because this dodecahedronboron cage is extremely stable under acidic and basic conditions.[4] Recently we reported the synthesis of a series of *meta*-carborane-containing ACBC derivatives, 1-amino-3[2-(1,7-dicarba-*closo*-dodecaboran(12)-1-yl)ethyl]cyclobutanecarboxylic acid[5] (**1**), and its less lipophilic analog 1-amino-3[2-(7-hydroxyethyl)-1,7-dicarba-*closo*-dodecaboran(12)-1-yl)ethyl]cyclo-butanecarboxylic acid[6] (**2**). However, these boron-containing molecules are highly lipophilic and therefore their uses are limited. In the present study, we wish to report the synthesis of a galactose containing, 1-amino-3[2-(6-deoxy-α/β-D-galactopyranos-6-yl)-1,7-dicarba-*closo*-dodecaboran(12)-1-yl)ethyl]cyclobutanecarboxylic acid (**3**), a less lipophilic analogue of **1**.

The title compound was synthesized as outlined in Scheme 1. The synthesis was initiated by the dialkylation of protected *meta*-carboranyl cyclobutanone **6**. Carboranyl-

Frontiers in Neutron Capture Therapy, edited by Hawthorne *et al.*
Kluwer Academic / Plenum Publishers, New York, 2001.

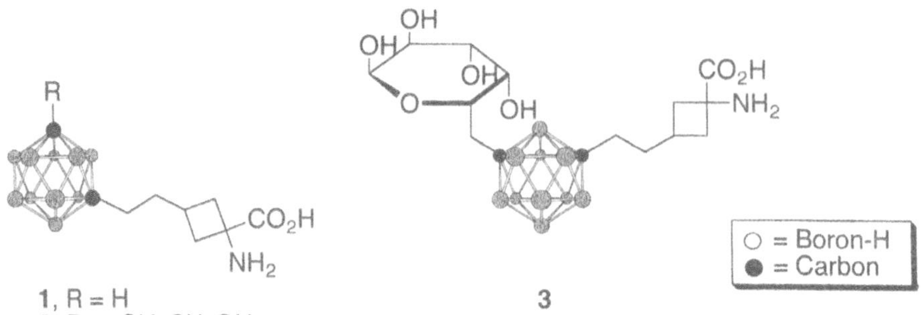

1, R = H
2, R = $-CH_2CH_2OH$
3

Figure 1. Boron-containing aminocyclobutanecarboxylic acids.

cyclobutanone **5** was synthesized from *meta*-carborane in three steps following the literature procedure.[5] The polyol was introduced by nucleophilic displacement of the triflate on the protected carbohydrate **7** using the anion of the *meta*-carborane derivative. Ketone **5** was protected as the hemithiol derivative by reaction with mercaptoethanol in the presence of a catalytic amount of p-toluenesulfonic acid[7] to obtain ketal **6** in 94% yield. Lithiation of **6** with butyllithium in tetrahydrofuran at room temperature, followed by reaction with the triflate of 1,2:3,4-di-O-isopropylidene-α-D-galactopyranose[8] **7**, provided **8** in 81% yield. Treatment of the isopropylidene protected galactopyranose-containing *meta*-carboranyl hemithioketal **8** with $HgCl_2$ and dilute NaOH in tetrahydrofuran[7] afforded the cyclobutanone derivative **9** in high yield (89%). The resultant carbohydrate containing *meta*-carboranyl cyclobutanone was then reacted with ammonium carbonate and potassium cyanide in a pressure tube.[9] Hydantoin **10** was formed in excellent yield (93%). In the next step, hydantoin **10** was heated in the presence of 2 *N* NaOH in a pressure tube to afford the amino acid, 1-amino-3[2-[7-(6-deoxy-1,2:3,4-di-O-isopropylidene-D-galactopyranos-6-yl)-1,7-dicarba-*closo*-dodecaboran(12)-1-yl]ethyl]cyclobutanecarboxylic acid (**11**), in almost quantitative yield (97%). In the final step of the synthesis, the isopropylidene protective groups were removed by the reaction with 4*N* HCl in tetrahydrofuran to obtain 1-amino-3[2-[7-(6-deoxy-α/β-D-galactopyranos-6-yl)-1,7-dicarba-*closo*-dodecaboran(12)-1-yl]ethyl] cyclobutanecarboxylic acid (**3**) as the hydrochloride salt in 95% yield.[10]

We report the successful preparation of a galactose containing, *meta*-carboranyl 1-aminocyclobutanecarboxylic acid derivative. Because of the presence of a highly hydrophilic galactose moiety, the aqueous solubility of the galactose-containing boronated ACBC **3** was considerably better than the parent *meta*-carboranyl-ACBC **1**. The compound is being investigated for potential use in BNCT.

ACKNOWLEDGMENT

The research was supported by the U. S. Department of Energy, the INEEL-URC Program, and the Robert H. Cole Foundation. We also wish to thank Jeff Coderre, Ph.D. and Peggy Mica (Brookhaven National Laboratory) as well as Carmen Lozzio, M. D. and Nancy Tuttle-Fuller, M. S. (University of Tennessee Medical Center, Knoxville) for their technical assistance.

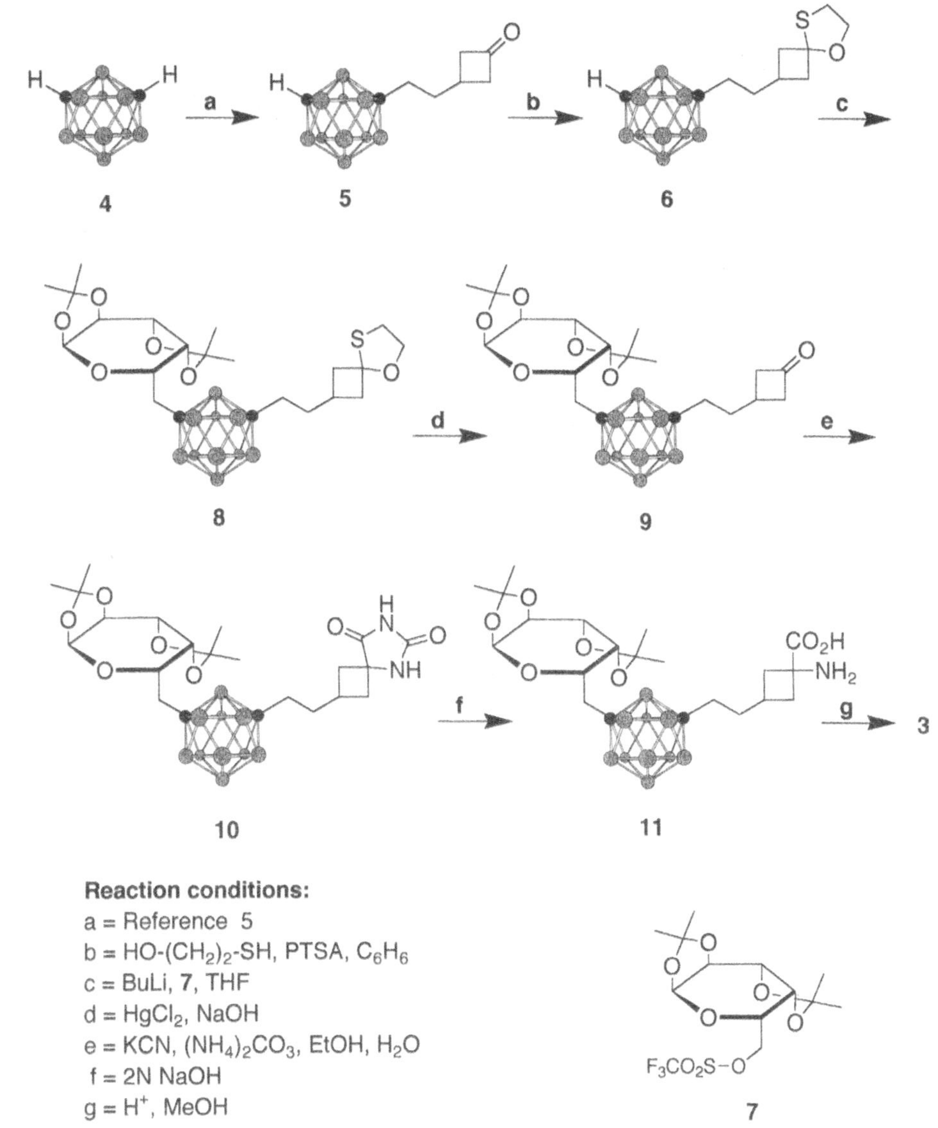

Scheme 1. Synthesis of galactose-containing *meta*-carboranyl-ACBC.

REFERENCES

1. A.H. Soloway, W.Tjarks, B.A. Barnum, F.-G. Ronj, R.F. Barth, I.M. Codogni, and J.G. Wilson, The Chemistry of Neutron Capture Therapy, *Chem. Rev.*, 98:1515, 1998.
2. J.A. Coderre, E.H. Elowitz, M. Chadha, R. Bergtland, J. Capala, D.D. Joel, H.B. Liu, D.N. Slatkin, and A.D. Chanana, Boron Neutron Capture Therapy for Glioblastoma Multiforme Using *p*-Boronophenylalanine and Epithermal Neutrons: Trial Design and Early Clinical Results, *J. Neurooncol.* 33:141, 1997.
3. K.F. Hubner, G.T. Smith, J.A. Thie, and E. Buonocore, Dynamic Positron Emission Tomography (PET) of Brain Tumors Using 1-Aminocyclobutane[C-11]carboxylic Acid (C-11 ACBC) and A 2-Component Model, *J. Nucl. Med.*, 33:922, 1992.
4. R.N. Grimes, "Carborane," Academic Press, New York and London, 1970.

5. R.R. Srivastava, R.R. Singhaus, and G.W. Kabalka, Synthesis of 1-Amino-3[2-(1,7-dicarba-closo-dodecaboran(12)-1-yl)ethyl]cyclobutanecarboxylic Acid: A Potential BNCT Agent, *J. Org. Chem.*, 62:4477, 1997.
6. R.R. Srivastava, and G.W. Kabalka, Synthesis of 1-Amino-3[2-(7-(2-hydroxyethyl)1,7-dicarba-*closo*-dodecaboran(12)-1-yl)ethyl]-cyclobutanecarboxylic Acid and Its *nido*-Analogue: Potential BNCT Agents, *J. Org. Chem.*, 62:8730, 1997.
7. D. Djerass, M. Shamma, and T.Y. Kar, Studies in Organic Sulfur Compounds. X. The Scope of the Raney Nickel Desulfurization of Cyclic Hemithioketals (1,3-Oxathiolanes and 1,3-Oxathianes), *J. Am. Chem. Soc.*, 80:4723, 1958.
8. R.W. Binkley, M.G. Ambrose, and D.G. Hehemann, Synthesis of Deoxyhalogeno Sugars. Displacement of the (Trifluoromethanesulfonyl)oxy (Triflyl) Group by Halide Ion, *J. Org. Chem.*, 45:4387, 1980.
9. H.R. Henze, and R.J. Speer, Identification of Carbonyl Compounds through Conversion into Hydantoins, *J. Org. Chem.*, 2:1316, 1942.
10. Analytical data of galactose-containing *meta*-carboranyl ACBC **3**: ^1H NMR (CD_3OD) δ 4.35 (d, J = 5.80 Hz, 1 H), 4.00–3.40 (m, 4 H), 2.66 (m, 2 H), 2.42 (m, 3 H), 2.20–1.75 (m, 4 H), 1.61 (m, 2 H). ^{13}C NMR (CD_3OD) δ 176.64, 98.18 (94.06), 77.35, 75.30, 74.96 (75.11), 74.84 (71.07), 73.62 (70.25), 72.97 (70.17), 55.91 (58.20), 40.09, 37.66 (38.47), 36.02(36.69), 35.33(35.80), 29.07 (28.85). Anal. Calcd. for $C_{15}H_{34}BClNO_7$: C, 37.22; H, 7.08; N, 2.89. Found: C, 36.97; H, 7.05; N, 2.67.

BENZOQUINONE OXIDATION OF THE $[ae\text{-}B_{20}H_{17}NH_3]^{3-}$ AND $[ae\text{-}B_{20}H_{17}OH]^{4-}$ IONS

Fangbiao Li, Kenneth Shelly, Carolyn B. Knobler, and M. Frederick Hawthorne

Department of Chemistry and Biochemistry
University of California at Los Angeles
405 Hilgard Avenue
Los Angeles, California 90095-1569

1. INTRODUCTION

The high boron content, water solubility, and rich derivative chemistry of polyhedral *closo*-borane anions make them attractive boron carriers for BNCT. The derivative chemistry of the polyhedral $[trans\text{-}B_{20}H_{18}]^{2-}$ anion[1] has been under investigation for the development of water-soluble boron-containing drugs for encapsulation within liposomes.[2] Several alkoxyl and ammonio derivatives have been produced by the reductive substitution of $[trans\text{-}B_{20}H_{18}]^{2-}$ with the appropriate nucleophile.[2–6] Subsequent ferric ion oxidation of the hydroxyl and alkoxyl substituted anions has been reported to produce several new species.[4,7] The oxidation of reduced eicosaborate species represents a highly valuable synthetic method for the synthesis of new twenty-boron derivatives.

In order to discover milder and/or nonaqueous oxidation conditions which could be developed for use with a variety of polyhedral borane species, we examined the reactivity of the organic oxidizing agent benzoquinone with ammonio- and hydroxyl-substituted eicosaborate derivatives. Benzoquinone and tetrachlorobenzoquinone in acetonitrile were previously found to be effective reagents for the oxidation of $[B_{10}H_{10}]^{2-}$ or $[e^2\text{-}B_{20}H_{18}]^{4-}$ to the $[trans\text{-}B_{20}H_{18}]^{2-}$ ion.[8]

2. MATERIALS AND METHODS

Synthetic reactions were performed under an atmosphere of dry nitrogen. Acetonitrile (MeCN) was distilled from CaH_2, and benzoquinine was sublimed prior to use. ^{11}B NMR spectra were obtained with a Bruker AM-500 instrument at 160 MHz, and were externally referenced to $BF_3{\bullet}Et_2O$ in C_6D_6 (peaks upfield of the reference are designated as negative).

Frontiers in Neutron Capture Therapy, edited by Hawthorne *et al.*
Kluwer Academic / Plenum Publishers, New York, 2001.

Dry $[Me_3NH]_3[ae\text{-}B_{20}H_{17}NH_3]$[2,6] (2.00 g, 4.6 mmol) was added to a stirred suspension of NaH (0.50 g, 21 mmol) in 100 mL of MeCN. Hydrogen gas was evolved, and the mixture was stirred for 1 h. The solution was filtered, and MeCN and Me_3N were removed from the filtrate in vacuo. The residue was dissolved in 100 mL dry MeCN. Benzoquinone (1.50 g, 14.1 mmol) was added, and the mixture was stirred for 3 h and filtered. Removal of the solvent in vacuo produced crude Na_3[**1**]. $^{11}B\{^1H\}$ NMR (ppm, H_2O): 6.7 (2B, BB_{ap}); −5.4 (2B, BH_{ap}); −13.7 (2B, BN); −22.1, −24.3, −27.3, −29.7 (14B total, BH_{eq}).

The crude Na_3[**1**] was dissolved in 50 mL of aqueous Me_4NCl (1 N). The addition of excess 1 N HCl produced a brown preciptate of crude $[Me_4N]_2$[**2**]. The precipitate was isolated by filtration, dissolved in 50 mL of MeCN/$CHCl_3$ (1 : 1) and passed through a silica gel column (230–400 mesh, 3 × 45 cm), eluting with the same solvent. The yellow eluent was collected and the solvent was removed in vacuuo. The residue was dissolved in 20 mL H_2O and 5 mL 1 N aqueous Me_4NCl was added to precipitate $[Me_4N]_2$[**2**]. The product was collected by filtration, purified by recrystallization from MeCN/isopropanol, and dried in vacuo to yield 0.25 g (12%) $[Me_4N]_2$[**2**] as a white powder. $^{11}B\{^1H\}$ NMR (ppm, H_2O): 9.3 (2B, BH_{ap}); −11.8 (2B, BHB_{ap}); −14.0 (2B, BN); −18.1, −22.9, and −25.2 (14B total, equatorial BH_{eq}).

Crude Na_3[**2**] (prepared from $[Me_3NH]_3[ae\text{-}B_{20}H_{17}NH_3]$ as described above) was dissolved in 50 mL water and the solution was acidified to pH 2 with 2 N HCl. The mixture was stirred for 5 min and filtered. Aqueous Me_4NCl (5 N) was added until precipitation was complete. The brown precipitate was collected by filtration, the resulting solid was dissolved in 40 mL of acetone/$CHCl_3$ (1 : 1) and passed through silica gel column (230–400 mesh, 3 × 45 cm), eluting with the same solvent. The yellow solution was collected and the solvent was removed in vacuo. The residue was dissolved in 20 mL H_2O and 5 mL aqueous 1 N Me_4NCl was added to precipitate crude $[Me_4N]_2$[**3**]. The product was purified by recrystallization from acetone/ether and dried in vacuo to yield 0.42 g (25%) $[Me_4N]$[**3**] as bright yellow crystals. $^{11}B\{^1H\}$ NMR (ppm, H_2O): 31.3 (2B, BH_{ap}); 15.6 (2B, B_{eq}); −0.5 (2B, BN); −4.4, −12.5, −16.7, −19.0 (12B total, BH_{eq}); −29.1 (2B, BH_{ap}).

Dry $[Me_3NH]_4[ae\text{-}B_{20}H_{17}OH]$[4,7] (2.00 g, 4.1 mmol) was added to a stirred suspension of NaH (0.50 g, 20.8 mmol) in dry MeCN. Hydrogen gas was evolved, and the mixture was stirred at room temperature for 1 h. The solution was filtered, and the solvent was removed in vacuo. The residue was dissolved in 100 mL dry MeCN. Benzoquinone (1.50 g, 14 mmol) was added, and the solution was stirred for 3 h. The mixture was filtered, and MeCN was removed in vacuo. The residue was dissolved in 50 mL water, and the solution was filtered again. The $\mathbf{5}^{4-}$ ion was identified from its ^{11}B NMR spectrum.[4]

The solution was acidified with 1 N HCl to bring the pH to 3. The mixture was stirred for 5 minutes and filtered. Aqueous 1 N Me_4NCl was added until precipitation was complete. The resulting white solid was collected by filtration, purified by recrystallization from MeCN/EtOH, and dried in vacuo to yield $[Me_4N]_2$[**6**] (1.17 g, 72%). The product was identified by its ^{11}B NMR and ESI-MS spectra.[4]

3. RESULTS AND DISCUSSION

3.1. Benzoquinone Oxidation of $[ae\text{-}B_{20}H_{17}NH_3]^{3-}$

Following the deprotonation of $[Me_3NH]_3[ae\text{-}B_{20}H_{17}NH_3]$ by NaH in anhydrous acetonitrile, benzoquinone oxidation produced the unexpected disubstituted derivative $[a^2\text{-}(\mu\text{-}MeC(NH)_2)B_{20}H_{16}]^{3-}$ ($\mathbf{1}^{3-}$, Figure 1).[9] The structure of this novel ion contains a

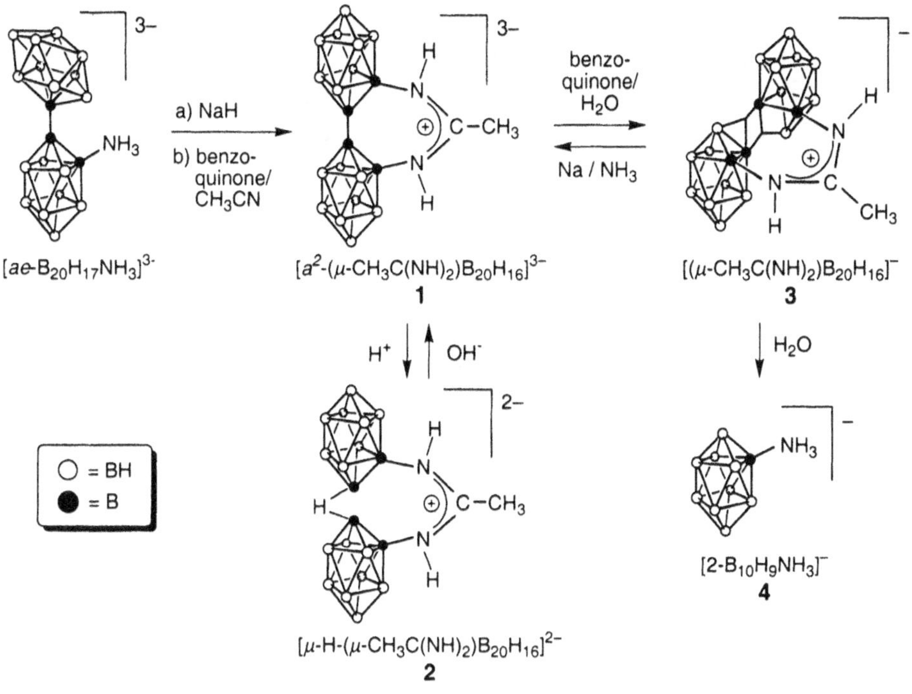

Figure 1. Benzoquinone oxidation of [*ae*-$B_{20}H_{17}NH_3$]$^{3-}$.

bridging amidinium group, which was formed due to the unexpected participation of the solvent during the oxidation reaction.

During the oxidation reaction, the [*ae*-$B_{20}H_{17}NH_3$]$^{3-}$ anion was first rearranged to the a^2-isomer and the ammonio group was deprotonated. Then the oxidant (benzoquinone) subsequently removed a hydride from the borane cage to produce a transient, positively charged boron vertex with an empty orbital. This reactive boron vertex then reacted with a molecule of the acetonitrile solvent. The nitrogen atom of the amino group attacked the activated nitrile carbon atom of the acetonitrile ligand and the amidinium group was formed. Acidification of an aqueous solution of $\mathbf{1}^{3-}$ produced the hydrogen and amidinium doubly-bridged [μ-H-(μ-MeC$(NH)_2$)$B_{20}H_{16}$]$^{2-}$ ion ($\mathbf{2}^{2-}$, Figure 1) which was isolated in 12% yield as its tetramethylammonium salt.

In acidic aqueous solution, the anion $\mathbf{1}^{3-}$ could be further oxidized by benzoquinone to produce the [(μ-MeC$(NH)_2$)$B_{20}H_{16}$]$^-$ ($\mathbf{3}^-$, Figure 2) ion which was isolated in 25% yield as the corresponding tetramethylammonium salt. The structure of [Me_4N][**3**] was determined by single crystal X-ray diffraction studies.[9] The sodium metal reduction of [Me_4N][**3**] in liquid ammonia regenerated the $\mathbf{1}^{3-}$ anion. The instability of $\mathbf{3}^-$ in aqueous solution renders it unsuitable for application as a BNCT agent. The hydrolysis of [Me_4N][**3**] in refluxing water produces the known [2-$B_{10}H_9NH_3$]$^-$ ion ($\mathbf{4}^-$, Figure 2), identified by ^{11}B NMR and electrospray ionization mass spectrometry (ESI-MS).

3.2. Benzoquinone Oxidation of [*ae*-$B_{20}H_{17}OH$]$^{4-}$

Using a similar method, the benzoquinone oxidation of [*ae*-$B_{20}H_{17}OH$]$^{4-}$ in dry CH_3CN produced the oxygen-bridged polyhedral borane anion [μ-$B_{20}H_{16}O$]$^{4-}$ ($\mathbf{5}^{4-}$, Figure 2).[9] Acidification of an aqueous solution of $\mathbf{5}^{4-}$ produced the anion [μ-$B_{20}H_{17}OH$]$^{2-}$ ($\mathbf{6}^{2-}$,

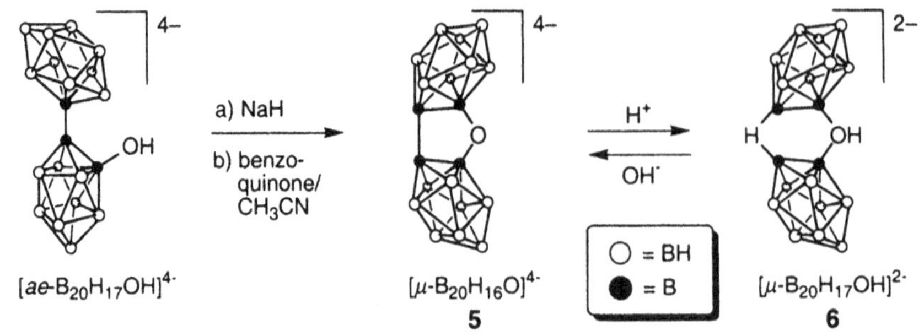

Figure 2. Benzoquinone oxidation of [ae-$B_{20}H_{17}OH]^{4-}$.

Figure 2), isolated as the tetramethylammonium salt. The mechanism of the formation of $\mathbf{5}^{4-}$ was believed to be similar to that of the formation of $\mathbf{1}^{3-}$. The deprotonation of the hydroxyl group increases the nucleophilic ability of the oxygen atom during the oxidation reaction. The product was characterized by its ^{11}B NMR and ESI-MS spectra. According to the ^{11}B NMR of the reaction mixture, only a trace amount of boric acid was produced during the reaction, which suggests benzoquinone is a milder oxidizing regent than ferric ion.

4. CONCLUSION

Benzoquinone is a valuable reagent for the oxidation of a variety of reduced $[B_{20}H_{18}]^{4-}$ derivatives. It is a mild oxidant that can be employed in both aqueous and organic solvents. The reaction of benzoquinone with [ae-$B_{20}H_{17}NH_3]^{3-}$ in acetonitrile yielded an unexpected product, [a^2-(μ-MeC(NH)$_2$)$B_{20}H_{16}]^{3-}$. Further oxidation of the obtained product in aqueous solvent resulted in the [(μ-MeC(NH)$_2$)$B_{20}H_{16}]^{-}$ ion. Benzoquinone oxidation of [ae-$B_{20}H_{17}OH]^{4-}$ in acetonitrile provided a new method for the direct synthesis of [μ-$B_{20}H_{17}OH]^{2-}$. The oxidation of other polyhedral borane anions using the organic oxidant benzoquinone in non-aqueous media may also lead to a variety of new species with potential utility in BNCT.

ACKNOWLEDGMENTS

This work was supported by the Office of Health and Environmental Research, U.S. Department of Energy (Grant No. DE-FG03-95ER61975).

REFERENCES

1. B.L. Chamberland, and E.L. Muetterties, Chemistry of boranes. XVIII. Oxidation of $[B_{10}H_{10}]^{2-}$ and its derivatives, *Inorg. Chem.*, 3:1450–1456, 1964.
2. D.A. Feakes, K. Shelly, C.B. Knobler, and M.F. Hawthorne, $Na_3[B_{20}H_{17}NH_3]$: Synthesis and liposomal delivery to murine tumors, *Proc. Natl. Acad. Sci. USA*, 92:3029–3033, 1994.
3. K. Shelly, D.A. Feakes, M.F. Hawthorne, P.G. Schmidt, T.A. Krisch, and W.F. Bauer, Model studies directed toward the boron neutron-capture therapy of cancer: Boron delivery to murine tumors with liposomes, *Proc. Natl. Acad. Sci. USA*, 89:9039–9043, 1992.

4. F. Li, K. Shelly, R.R. Kane, C.B. Knobler, and M.F. Hawthorne, Synthesis and structure of the polyhedral [μ-$B_{20}H_{17}OH]^{2-}$ borane anion containing both oxygen and hydrogen-bridged bonds, *J. Am. Chem. Soc.*, 118:6506–6507, 1996.
5. F. Li, K. Shelly, R.R. Kane, C.B. Knobler, and M.F. Hawthorne, The novel [n-$B_{20}H_{18}]^{2-}$ induced ring-opening of tetrahydrofuran by alkoxide anions, *Angew. Chem. Int. Ed.*, 35:2646–2649, 1996.
6. E.M. Georgiev, K. Shelly, D.A. Feakes, J. Kuniyoshi, S. Romano, and M.F. Hawthorne, Synthesis of amine derivatives of the polyhedral borane anion $[B_{20}H_{18}]^{4-}$, *Inorg. Chem.*, 35:5412–5416, 1996.
7. M.F. Hawthorne, R.L. Pilling, and P.M. Garrett, A study of the reaction of hydroxide ion with $[B_{20}H_{18}]^{2-}$, *J. Am. Chem. Soc.*, 87:4740–4746, 1965.
8. R.A. Watson-Clark, and M.F. Hawthorne, An exploratory evaluation of the reactions of organic oxidants with polyhedral borane anions, *Inorg. Chem.*, 36:5419–5420, 1997.
9. F. Li, K. Shelly, C.B. Knobler, and M.F. Hawthorne, Formation of a novel amidinium-bridged polyhedral borane ion by incorporation of an acetonitrile solvent molecule, *Angew. Chem. Int. Ed.*, 37:1865–1868, 1998.

BSMel: AN AROMATIC AMINO ACID ANALOG OF HIGH BORON CONTENT, EASILY PREPARED FROM B-10 ENRICHED AND STEREOCHEMICALLY PURE PRECURSORS

John D. Glass, Michael S. Makar, and Jeffrey A. Coderre

Medical Department
Brookhaven National Laboratory
Upton, NY, 11973

1. INTRODUCTION AND DISCUSSION

In the familiar format of BNCT for cancer, the cells of a malignant tumor are sensitized to thermal neutrons by systemic administration of a B-10 containing compound that is selectively taken up by malignant cells. The boron-loaded tumor cells are then killed by neutron irradiation within a defined target volume. Effective BNCT requires a minimum of about 30μg ^{10}B/g in the targeted malignant cells. To produce a useful therapeutic gain, there should be a differential of three or four fold between the boron concentration in malignant cells compared to that in the "normal" cells. This demanding pattern of boron delivery must be accomplished by a compound that has no serious toxicity to any vital tissue, either within the treatment volume or elsewhere in the body. Given these daunting requirements, it is remarkable that two compounds, BSH and L-Bpa, are currently in clinical trials, aspiring to the cure of glioblastoma multiforme. These gratifying results are hard won by many people over more than three decades. Hopefully, a careful choice of therapeutic targets, along with diligent attention to the literature of metabolite analog design, will allow new BNCT applications using new boron delivery molecules to reach clinical trial in a shorter time frame.

In a bone marrow purging format, systemic eradication of cancer is accomplished by high doses of radiation or chemotherapy which coincidentally destroy hemopoetic stem cells in the bone marrow. The responsibility of BNCT in this format is to selectively remove malignant cells from a sample of the patient's bone marrow, harvested before the high-dose radiation or chemotherapy, so that this cleansed marrow can be used to reconstitute the immune system. BNCT purging of malignant cells from bone marrow is fundamentally different from BNCT in which a boron-containing compound is delivered systemically to achieve selective killing of malignant cells in an irradiated area of the

Frontiers in Neutron Capture Therapy, edited by Hawthorne *et al.*
Kluwer Academic / Plenum Publishers, New York, 2001.

patient. In the purging mode, marrow cells are treated with the boron vector and with neutron irradiation outside the body. The amount of boron-containing compound required for therapy is less, and chemical toxicity problems are limited to the hemopoietic stem cells. Also, the geometry of the tissue irradiation zone can be manipulated to essentially eliminate problems of neutron beam attenuation. There is still a requirement for selective uptake of boron into malignant cells compared to "normal" cells, but the only normal cells that have to be spared are the hemopoietic stem cells.

In this study, a widely used metabolite analog (melphalan or 4-[bis(2-chloroethyl)amino]-L-phenylalanine) that kills tumor cells by alkylation (cross-linking) of double-stranded DNA was converted to a metabolite analog containing a high B-10 content for use as a boron delivery agent for neutron capture therapy. The new metabolite analog (BSMel) retains the capacity to be actively accumulated by some types of malignant cells, notably mouse leukemia and lymphoma cells. Studies are continuing in order to determine whether or not this compound can be used to purge leukemia and lymphoma cells from bone marrow while leaving hemopoeitic stem cells viable for use in autologous marrow transplants.

2. EXPERIMENTAL

2.1. Synthesis of ^{10}BSMel

^{10}BSH (880 mg) was dissolved in 30 mL 5% sodium bicarbonate. Melphalan (213 mg) was added and the solution was stirred vigorously in a stoppered tube. Within about 2 hr, most of the melphalan had dissolved and a fine white precipitate began to form. At the end of 48 hr the reaction mixture was placed in the refrigerator overnight to complete separation of a fine white solid. The precipitate was collected by centrifugation and dissolved in 30 mL water with heating to the boiling point. The hot solution was left to cool to room temperature and a small amount of insoluble material was removed by centrifugation. Acidification to pH 2.7 with HCl caused heavy precipitation of the product, which was completed by storage in the refrigerator overnight. The product was collected by centrifugation and washed with water resuspension and centrifugal sedimentation. The washed centrifugal pellet was frozen and dried by lyophilization to yield 120–160 mg of ^{10}BSMel. The compound contains a single dodecaborane cage rather than the two cages that would result from substitution of both arms of the nitrogen mustard group. From the known propensity of BSH to undergo dialkylations and from the symmetry of the proton NMR spectrum of the product, we tentatively conclude that BSMel is probably the bicyclic compound resulting from alkylation of a single BSH molecule by both arms of the nitrogen mustard group of a melphalan molecule.

BSMel Analytical data: Thin layer chromatography: The compound is a single homogeneous spot that is both ninhydrin positive and fluorescence quenching, in two solvent systems. System 1: Silica G; i-PrOH: H_2O: NH_4OH 70:15:15; $R_f = 0.61$ for BSMel; $R_f = 0.70$ for melphalan. System 2: Silica G; n-BuOH:acetic acid:water 4:1:1; $R_f = 0.52$ for BSMel; $R_f = 0.72$ for melphalan. Elemental analysis: Calc.: C 39.28; H 7.31; N 7.05; ^{10}B 28.98 (assumes 96% ^{10}B enrichment). Found: C 40.54; H 7.40; N 7.11; ^{10}B 28.1 (by prompt gamma). NMR: (D_2O) 7.24, 7.07 (AB quartet, 4H, Ar-H); 6.75 (br. s, NH); 4.0–4.2 (m, 2H, Ar-CH_2); 2.7–3.4 (m, 10H, CH_2); 0.7–2.1 (br. m, 11H, BH). Note: CH peak probably buried in DOH peak at ~4.8 ppm.

2.2. Measurement of ^{10}BSMel Uptake into Selected Tumor Cell Lines

Cultured tumor cells were equilibrated for 2hr with medium containing graded doses of boron-containing compound, then stripped of extracellular medium by centrifugation through an oil layer, as previously described.[1] B-10 was then measured in the cellular fraction by direct current plasma atomic emission spectroscopy.[1] Since boric acid has been shown to equilibrate freely between intracellular and extracellular volumes

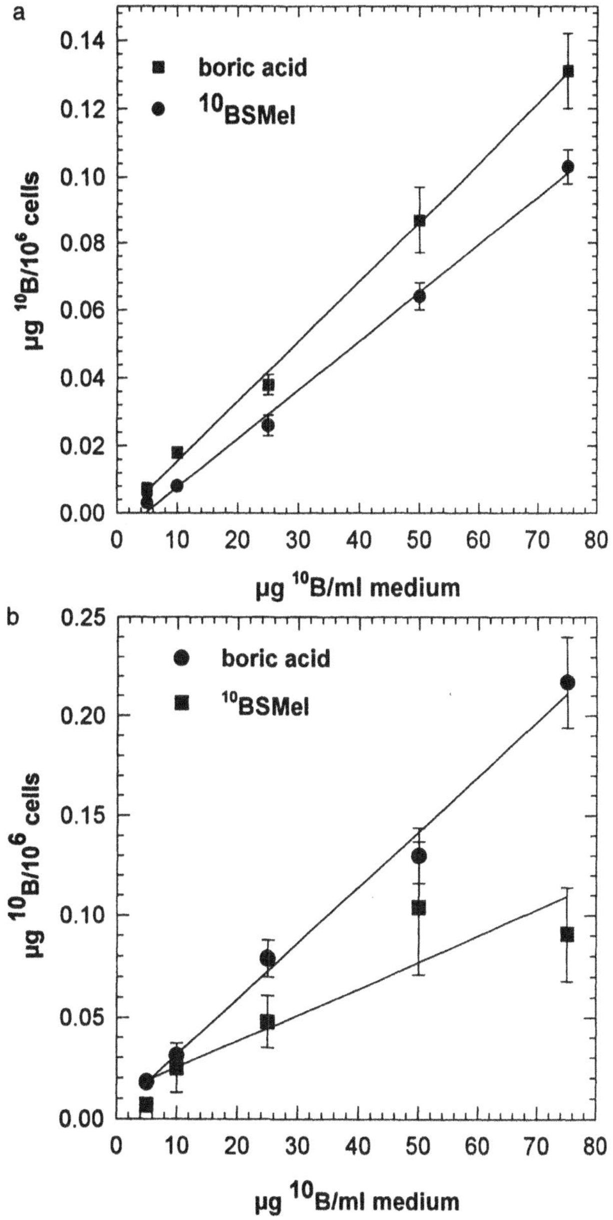

Figure 1. Intracellular accumulation of the boron compounds BSMel, BPA and boric acid in the (a) 9L rat gliosarcoma, (b) a pigmented murine melanoma, (c) a murine leukemia, and (d) a murine lymphoma. The slope of the boric acid line is used as a reference indicating uniform intracellular and extracellular boron concentration.

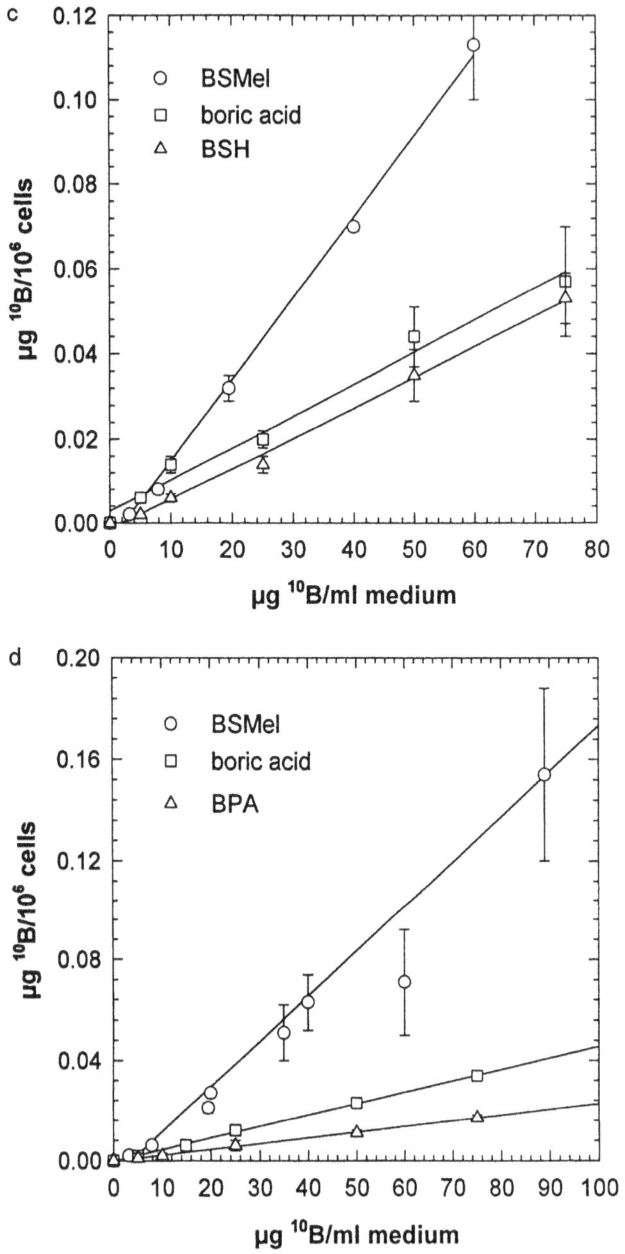

Figure 1. *Continued.*

under these experimental conditions, boric acid controls are used to establish a line representing no active accumulation and no exclusion of boron. A slope greater than that for boric acid represents active accumulation of boron above the concentration in the medium and a slope below that for boric acid represents exclusion of boron from the intracellular space.

^{10}BSMel is strongly taken up by L1210 mouse leukemia cells and by L5178 mouse lymphoma cells, but not by GS-9L rat gliosarcoma cells or by B16 mouse melanoma cells (Figure 1a–d). The uptake into the mouse leukemia and lymphoma cells is comparable to that reported elsewhere for Bpa uptake into rat GS-9L gliosarcoma cells,[1] which is adequate for BNCT in the rat brain tumor model.[2]

3. SUMMARY

The first stage in the development of a new BNCT modality, ex-vivo purging of bone marrow, has been accomplished. A boron containing analog of the anti-cancer agent melphalan has been synthesized. BSMel retains specificity for the targeted tumor cells: leukemia and lymphoma. Experiments to determine the effectiveness of BSMel-based BNCT against leukemia cells *in vitro*, and to measurement of the sensitivity of bone marrow stem cells to BSMel-based BNCT are underway.

REFERENCES

1. J. Capala, M.S. Makar, and J.A. Coderre, Accumulation of boron in malignant and normal cells incubated *in vitro* with boronophenylalanine, mercaptoborane or boric acid, *Radiat. Res.*, 146:554–560, 1996.
2. J.A. Coderre, T.M. Button, P.L. Micca, C.D. Fisher, M.M. Nawrocky, and H.B. Liu, Neutron capture therapy of the 9L rat gliosarcoma using the *p*-boronophenylalanine-fructose complex, *Int. J. Radiat. Oncol. Biol. Phys.*, 30:643–652, 1994.

CURRENT STATUS OF SELENIUM COUPLING TO HOECHST-CARBORANE

Deborah de F. dos Santos[1], Mario Argentini[2], Regin Weinreich[2], and Hans-Jürgen Hansen[1]

[1]Organisch-chemisches Institut der Universität Zürich
Winterthurerstrasse 190, CH-8057, Switzerland
[2]Paul Scherrer Institut, Labor für Radio-und Umweltchemie
CH-5232 Villigen-PSI, Switzerland

INTRODUCTION

This study explores the possibility of coupling an *o*-carborane moiety to a drug type displaying high DNA affinity, in order to introduce sufficient amounts of ^{10}B atoms into the tumor cell for efficiency in BNCT cancer treatment.

4-[5-(4-Methyl-1-piperazinyl)-2,5′-bi-1*H*-benzimidazol-2′-yl]phenol (**1**) also known as Hoechst 33258 constitutes the base of our chemical modifications. This DNA linker is a planar molecule that binds to the minor groove region of duplex DNA rich in A-T base pairs and widely used as a fluorescent DNA stain. The NH groups of the bisbenzimidazole system that form three-center hydrogen bonds with adenine and thymine are thought to maintain their bonding capacity when bulky *ortho*-carboranyl substituents replace the original phenolic group of Hoechst 33258.

RESULTS

Bisbenzimidazoles with methylpiperazinyl and carboranyl residues can only be prepared in convergent syntheses starting from the carboranyl extremity of the molecule as shown in *Scheme 1.*

The DNA-intercalators **7a**[5]**–c** labelled as positron-emitting boronated pharmaceuticals will allow studies at microscopic (NCR) and macroscopic (PET) levels. However, the introduction of a radionuclide for PET studies has to be carried out at the piperazinyl moiety which in Hoechst 33258 interacts with cytosine and guanine. Since no longer-lived (as required by uptake time in tumors) positron-emitting nuclides exist among the elements carbon, oxygen, hydrogen and nitrogen, our strategy is directed towards the replacement of the methylpiperazinyl group by a 1,4-selenazyl (selenomor-

Frontiers in Neutron Capture Therapy, edited by Hawthorne *et al.*
Kluwer Academic / Plenum Publishers, New York, 2001.

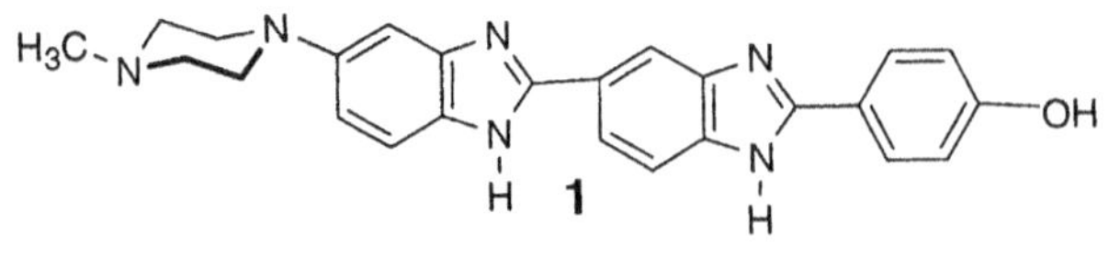

Figure 1. Hoechst 33258.

pholine) group with a 7.1 h ^{73}Se (65%, β^+) to be introduced in the last step of the synthesis.[6]

Encouraged by these results we synthesized the compounds **11a–c** by coupling N,N′-[4-(bis-2-hydroxyethyl)amino]-1,2-phenylenediamine (**10**) with **6a–c** (*Scheme 3*). **11a–c** were isolated and identified by ^{1}H-NMR spectroscopy.

The novel compounds **11a–c** as well as the previously synthesized **7a–c** may be considered as candidates for BNCT.

ANALYTICAL DATA

The ^{1}H NMR spectra were measured with a Varian Gemini 2000 and a Bruker ARX 300 both at 300 MHz. δ in ppm rel. to internal Me_4Si or referred to the residual ^{1}H present in deuterated solvents. *J* in Hz. Identification of compounds **7b** and **7c** as *closo-o*-carboranes were ascertained by mass spectra (MS) measured by ESI with a Fission VG TRIO 2000 at 40 eV and CI with a Finnigan MAT 90 (NH_3), with ions in *m/z* (rel.%). Spot-test with $PdCl_2$/HCl:[7] *nido-o*-carboranes stain immediately, *closo-o*-carboranes only after heating.

Compound **7a**.[5] **7b**. ^{1}H NMR (MeOH-d_4 + drops of TFA-*d*): 3.02 (s, 3H, H_3C—N—), 3.24 (t, $J = 12$, 2H, H_3C—N—CH_2), 3.38 (t, $J = 12$, 2H, H_3C—N—CH_2), 3.70 (d, $J = 13$, 2H, Ar—N—CH_2), 3.78 (s, 2H, $CH_2CB_{10}H_{10}CH$), 3.99 (d, $J = 13$, 2H, Ar—N—CH_2), 4.60 (s, 1H, $CB_{10}H_{10}CH$), 7.37 (d, $J_{4,6} = 2$, 1H, H-4), 7.47 (dd, $J_{4,6} = 2$, $J_{6,7} = 9$, 1 H, H-6), 7.64 (d, $J = 8.4$, 2H, ortho to $CH_2CB_{10}H_{10}CH$), 7.78 (d, $J_{6,7} = 9$, 1H, H-7), 8.12 (d, $J_{6',7'} = 9$, 1H, H-7′), 8.23 (d, $J = 8.4$, 2H, meta to $CH_2CB_{10}H_{10}CH$), 8.27 (dd, $J_{4',6'} = 1.5$, $J_{6',7'} = 8.7$, 1H, H-6′), 8.63 (s, $J_{4',6'} + J_{4',7'}$, 1H, H-4′). ESI MS: 567 (97), 566 (100), 565 (59), 564 (25).

7c. ^{1}H NMR (MeOH-d_4 + drops of TFA-*d*): 2.98 (s, 3H, H_3C—N—), 3.24 (t, $J = 12$, 2H, H_3C—N—CH_2), 3.33 (t, $J = 12$, 2H, H_3C—N—CH_2), 3.67 (d, $J = 11$, 2H, Ar—N—CH_2), 3.95 (d, $J = 12$, 2H, Ar—N—CH_2), 5.15 (s, 1H, $CB_{10}H_{10}CH$), 7.35 (d, $J_{4,6} = 2.1$, 1H, H-4), 7.41 (dd, $J_{4,6} = 2.1$, $J_{6,7} = 9$, 1H, H-6), 7.75 (d, $J_{6,7} = 9$, 1H, H-7), 7.94 (d, $J = 9$, 2H, Ar—H ortho to $CB_{10}H_{10}CH$), 8.14 (dd, $J_{4',7'} = 0.6$, $J_{6',7'} = 9$, 1H, H-7′), 8.23 (d, $J = 9$, 2H, Ar—H meta to $CB_{10}H_{10}CH$), 8.31 (dd, $J_{4',6'} = 1.8$, 1H, H-6′), 8.67 (dd, $J_{4',7'} = 0.6$, $J_{4',6'} = 1.8$, 1H, H-4′). ESI MS: 553 (100), 552 (97), 551 (57), 550 (25).

9. ^{1}H NMR ($CDCl_3$): 2.71 (AA′XX′, 4H, $N(CH_2)_2$), 3,78 (AA′XX′, 4H, $Se(CH_2)_2$), 6.80 (m, 3H, H-2 + H-4 + H-6), 7.24 (m, 2H, H-3 + H-5). CI-MS: 228 (100, $[M + H]^+$), 227 (16), 226 (49), 225 (21), 224 (19).

11a. ^{1}H NMR (MeOH-d_4): 3.55 (t, $J = 5.8$, 4H, N—$(CH_2)_2$), 3.77 (t, $J = 5.8$, 4H, $HO(CH_2)_2$), 4.55 (s, 2H, $OCH_2CB_{10}H_{10}CH$), 4.75 (s, 1H, $CB_{10}H_{10}CH$), 6.85 (dd, $J_{4,6} = 2$, $J_{6,7} = 8.8$, 1H, H-6), 6.88 (d, $J_{4,6} = 2$, 1H, H-4), 7.05 (AA′XX′, $J_A(J_X) = 8.9$, 2H, Ar—H ortho to $OCH_2CB_{10}H_{10}CH$), 7.42(d, $J_{6,7} = 8.8$, 1H, H-7), 7.63 (d, $J_{6',7'} = 8.5$, 1H, H-7′), 7.86 (dd, $J_{4',6'} = 1.5$, $J_{6',7'} = 8.5$, 1H, H-6′), 8.00 (AA′XX′, $J_A(J_X) = 8.9$, 2H, Ar—H meta to $OCH_2CB_{10}H_{10}CH$), 8.15 (d, $J_{4',6'} = 1.5$, 1H, H-4′).

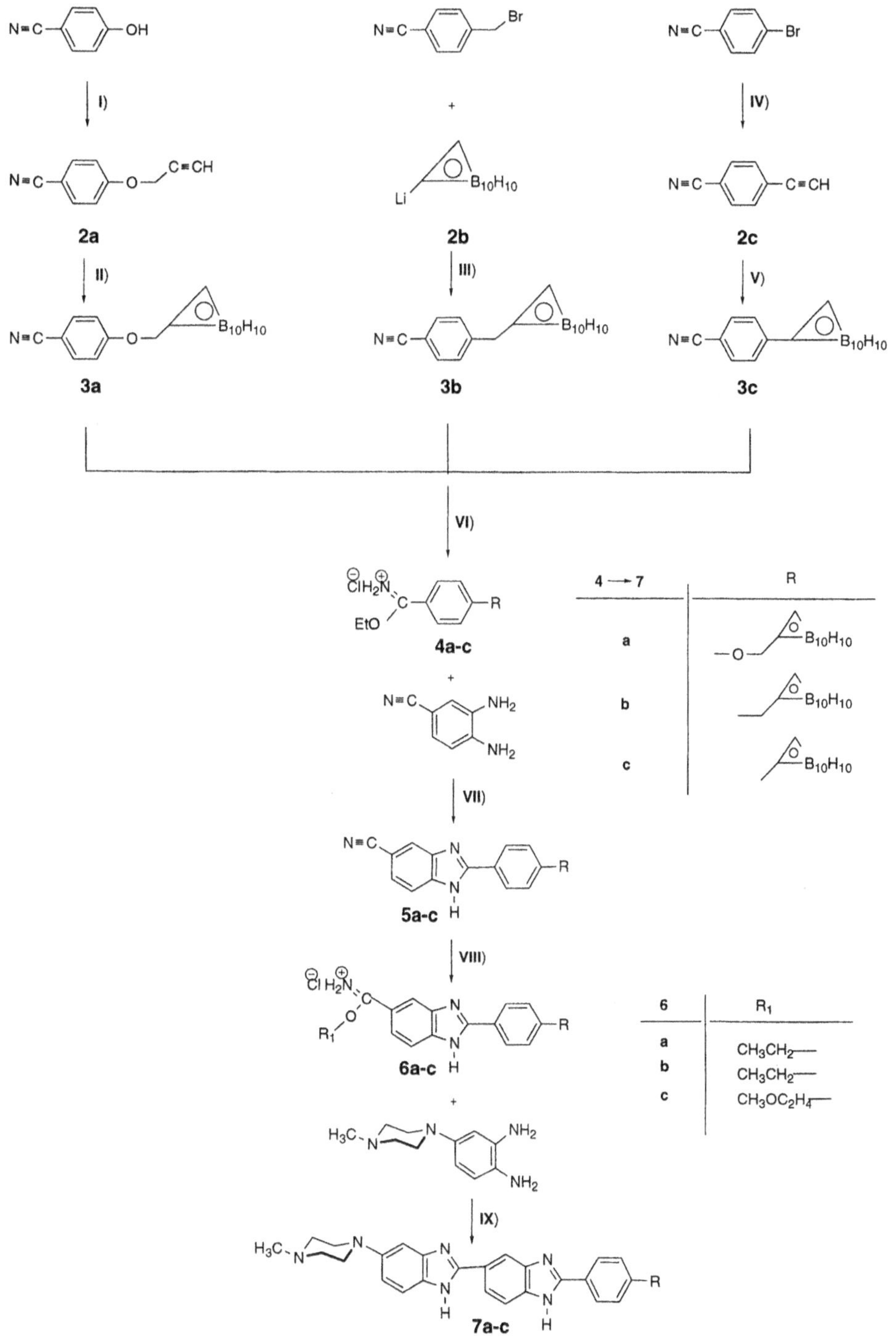

Scheme 1. *o*-Carboranyl derivatives of Hoechst 33258 synthesized by using convergent synthesis and identified by ^{1}H-NMR spectroscopy and MS. **I)** Propargyl bromide, K_2CO_3, acetone, reflux. **II)** 6,9-$(CH_3CN)_2B_{10}H_{12}$,[1] toluene, reflux.[2] **III)** *o*-Carborane, BuLi, THF, 30 min/0 °C, 12 h/25 °C.[3] **IV)** Ethynyltrimethylsilane, Et_3N, argon gas, CuI, $[(C_6H_5)_3P]_2PdCl_2$, 1 M aq. KOH, MeOH.[4a,b,c] **V)** 6,9-$(CH_3CN)_2B_{10}H_{12}$, toluene, reflux.[4b,c] **VI)** EtOH, dry HCl. **VII)** AcOH, argon gas, reflux. **VIII)** **5a**-EtOH-benzene/dry HCl, **5b**-EtOH, dry HCl, **5c**-2-methoxy-ethanol, dry HCl. **IX)** AcOH, argon gas, reflux.

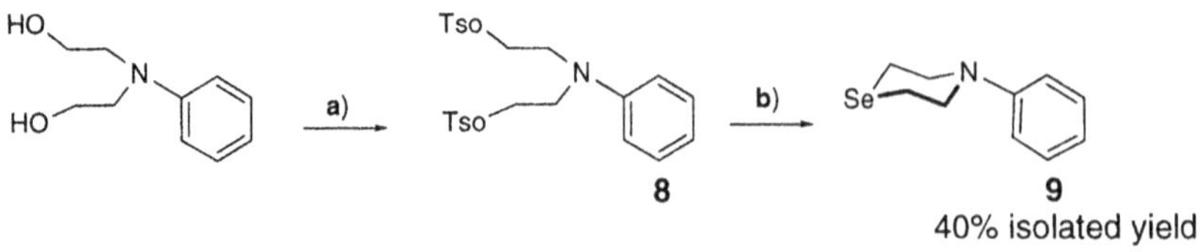

Scheme 2. Preparation of 4-Phenyl selenomorpholine (**9**) as a model compound following *Scheme 2*. **a)** CH_2Cl_2, Et_3N, TsCl (p-toluene sulphonyl chloride). **b)** Se°, $NaBH_4$, EtOH or Se°, $Li(C_2H_5)_3BH$, THF.

11b. ^1H NMR (MeOH-d_4): 3.58 (t, J = 5.8, 4H, N—$(CH_2)_2$), 3.63 (s, 2H, $CH_2CB_{10}H_{10}CH$), 3.77 (t, J = 5.8, 4H, $HO(CH_2)_2$), 4.48 (s, 1H, $CB_{10}H_{10}CH$), 6.88 (dd, $J_{4,6}$ = 2, $J_{6,7}$ = 8.8, 1H, H-6), 6.92 (d, $J_{4,6}$ = 2, 1H, H-4), 7.39 (AA′XX′, $J_A(J_X)$ = 8.3, 2H, Ar—H ortho to $CH_2CB_{10}H_{10}CH$), 7.45 (d, $J_{6,7}$ = 8.8, 1H, H-7), 7.71 (d, $J_{6',7'}$ = 8.5, 1H, H-7′), 7.95 (dd, $J_{4',6'}$ = 1.5, $J_{6',7'}$ = 8.5, 1H, H-6′), 8.09 (AA′XX′, $J_A(J_X)$ = 8.3, 2H, Ar—H meta to $CH_2CB_{10}H_{10}CH$), 8.25 (d, $J_{4',6'}$ = 1.5, 1H, H-4′).

11c. ^1H NMR (MeOH-d_4): 3.57 (t, J = 5.8, 4H, N—$(CH_2)_2$), 3.78 (t, J = 5.8, 4H, $HO(CH_2)_2$), 5.18 (s, 1H, $CB_{10}H_{10}CH$), 6.85 (dd, $J_{4,6}$ = 2, $J_{6,7}$ = 8.8, 1H, H-6), 6.91 (d, $J_{4,6}$ = 2, 1H, H-4), 7.45 (d, $J_{6,7}$ = 8.8, 1H, H-7), 7.70 (AA′XX′, $J_A(J_X)$ = 8.5, 2H, Ar—H ortho to $CB_{10}H_{10}CH$), 7.72 (d, $J_{6',7'}$ = 8.9, 1H, H-7′), 7.95 (d, $J_{6',7'}$ = 8.9, 1H, H-6′), 8.09 (AA′XX′, $J_A(J_X)$ = 8.5, 2H, Ar—H meta to $CB_{10}H_{10}CH$), 8.25 (s, 1H, H-4′).

FUTURE WORK

At this time trials to transform **11a–c** into the final compounds **13a–c** following the *Scheme 4* are under development.

ACKNOWLEDGMENTS

Thanks to Prof. Börje Larsson for the concept of this work and the enthusiastic support given.

We are indebted to the financial support provided by the Swiss Cancer Research (KFS 176-9-1995), to Paul Scherrer Institute and to the University of Zürich.

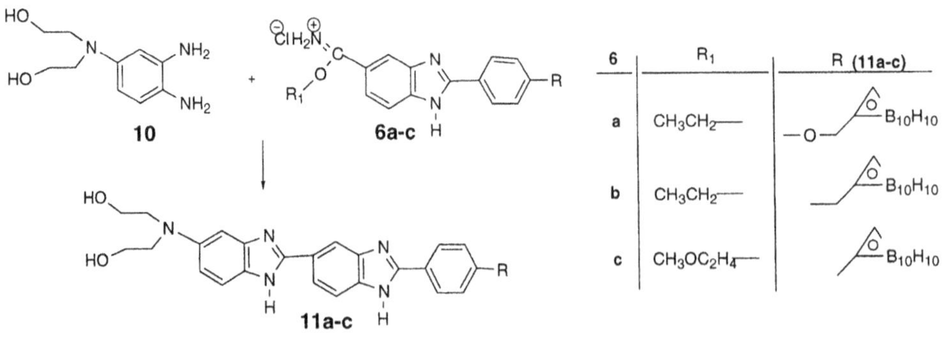

Scheme 3. Isopropanol, N_2 gas, reflux.

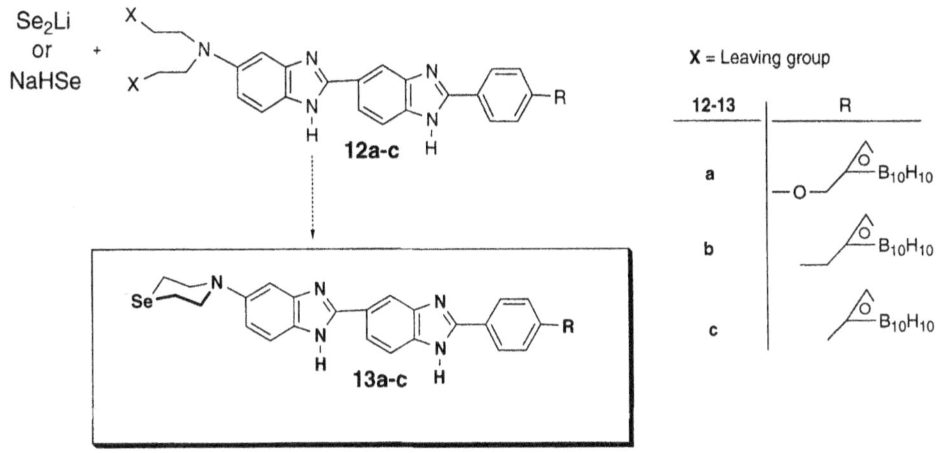

Scheme 4. Synthethic chemical approach for obtainting of Hoechst-Carborane labelled with selenium.

We thank Wella Cosmital SA, CH-1723, Marly for a generous gift of N,N′-[4-(bis-2-hydroxyethyl)amino]-2-nitroaniline.

REFERENCES

1. R.J. Schaeffer, A new type of substituted borane, *Am. Chem. Soc.*, 79:1006, 1957.
2. a) M.M. Fein, J. Bobinsky, N. Mayes, N. Schwartz, and M.S. Cohen, Carboranes. I. The preparation and chemistry of 1-Isopropenylcarborane and its derivatives (a new family of stable clovoboranes), *Inorg. Chem.*, 2:1111, 1963. b) T.L. Heying, J.W. Ager, S.L. Clark, D.J. Mangold, H.L. Goldstein, M. Hillman, R.J. Polak, and J.W. Szymanski, A new series of organoboranes. I. Carboranes from the reaction of decaborane with acetylenic compounds, *Inorg. Chem.*, 2:1089, 1963.
3. J. Cai, H. Nemoto, B. Singaram, and Y. Yamamoto, Facile reaction of o-Carboranyllithium reagents with functionalized alkyl halides, *Tetrahedron Lett.*, 19:3383–3386, 1996.
4. a) S. Takahashi, Y. Kuroyama, K. Sonogashino, and N. Hagihara, A convenient synthesis of ethynylarenes and diethynylarenes, *Synthesis*, August:627–630, 1980. b) M.D. Murphy, D.M.P. Mingos, and J.M. Forward, Synthesis of icosahedral carboranes for second harmonic generation. Part 1, *J. Mater. Chem.*, 3(1):67–76, 1993. c) D.P. Kelly, S.A. Bateman, R.F. Martin, M.E. Reum, M. Rose, and A.R.D. Whittaker, DNA binding compounds.V, Synthesis and characterization of boron-containing bibenzimidazoles related to the DNA minor groove binder, Hoechst 33258, *Aust. J. Chem.*, 47:247–252, 1994.
5. M. Argentini, D.F. Dos Santos, R. Weinreich, and H.-J. Hansen, Synthesis of an o-Carboranyl Derivative of 4-[5-(4-Methyl-1-piperazinyl)-2,5′-bi-1*H*-benzimidazol-2′-yl]phenol, *Inorg Chem.*, 37(23):6018–6022, 1998.
6. D.F. Dos Santos, "Contributions in labelling Hoechst compound 33258 with boron and selenium to be used in boron neutron capture therapy", Dissertation, Universität Zürich, 1996.
7. A.H. Soloway, W. Tjarks, B.A. Barnum, F.-G. Rong, R.F. Barth, I.M. Codogni, and J.G. Wilson, The Chemistry of Neutron Capture, Chem. Rev., 98:1515–1562, 1998.

NEW SYNTHESIS OF BORONATED CHLORINS WITH DIFFERENT LIPOPHILICITY

Silke Ruf, Jens Osterloh, and Detlef Gabel

University of Bremen
Department of Chemistry
Leobener Str. 28259 Bremen, Germany

INTRODUCTION

It is well known that porphyrins and chlorins accumulate selectively in tumor cells. Therefore binding BSH to porphyrins could be a successful approach to increase the tumor/blood- and the tumor/normal tissue-ratio of ^{10}B. The increase of these two ratios is an important supposition for successful BNCT treatment of tumor patients.

Previous investigations of Pandey *et al.*[1] show that cellular uptake of porphyrins is related to their lipophilicity. With regard to the cellular uptake these results suggest that an optimal lipophilicity exists for porphyrins. As a consequence the porphyrins used in this approach must contain two different reactive positions: one to bind BSH to the porphyrins and another one to optimize the lipophilicity.

As a further supposition it is required that the porphyrins can be obtained easily in great amounts and with high purity. Calculations demonstrate that a few gram of boronated porphyrins per BNCT treatment are necessary for one patient.[6] Of course high purities are required for any clinical application.

PORPHYRINS

For this approach the chlorins pyropheophorbide *a* (**4**), purpurin-18 (**3**) and purpurin-18-methyester (**5**) are selected as educts. All three chlorins are derivatives of chlorophyll *a* (**1**). Chlorophyll *a* (**1**) can be extracted from the alga *Spirulina pacifica.* This procedure represents a simple pathway to obtain the starting chlorins in great amounts without a long multi-step-synthesis. The pathway is described in Figure 1.

Starting with chlorophyll *a* it is possible to prepare the educts either in two steps (**3**), (**4**) or three steps (**5**). All steps are well known by literature,[2] do not require long reaction times and are not costly.

Frontiers in Neutron Capture Therapy, edited by Hawthorne *et al.*
Kluwer Academic / Plenum Publishers, New York, 2001.

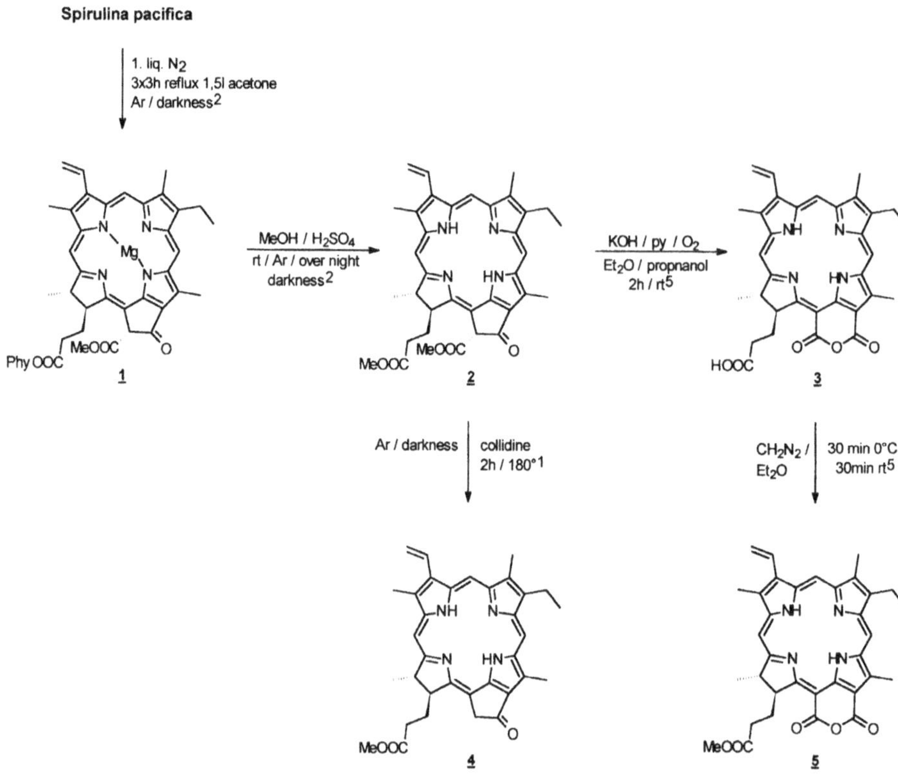

Figure 1. Synthesis of chlorins from *S. pacifica* extracts.

Pyropheophorbide *a* (**4**) contains a vinyl group which can be etherified by different alcohols after HBr is added to the double bond. With this step the lipophilicity can be varied. After cleaving the ester the propionic acid can be esterified with BSH. The propionic acid of purpurin-18 (**3**) can also be esterified with BSH. In this case variation of the lipophilicity should be possible by treating the isocyclic anhydride with different alkylamines. In purpurin-18-methylester (**5**) a carboxylic acid can be generated by opening the isocyclic anhydride with dialkylamines. The length of the alkyl chains determines the lipophilicity of the derivatives. Again the generated carboxylic acid could be esterified with BSH.

SYNTHESIS PATHWAYS

Synthesis of Boronated Alkylether Derivatives of Pyropheophorbide A

The overview about this pathway is given in Figure 2.

For these syntheses it was possible to develop an efficient method to activate the propionic acid. In previous investigations this step appeared to be difficult because the chlorins often underwent degradation or many products were obtained, which could not be separated.[3] We could show that oxalyl chloride is a suitable agent for activation of

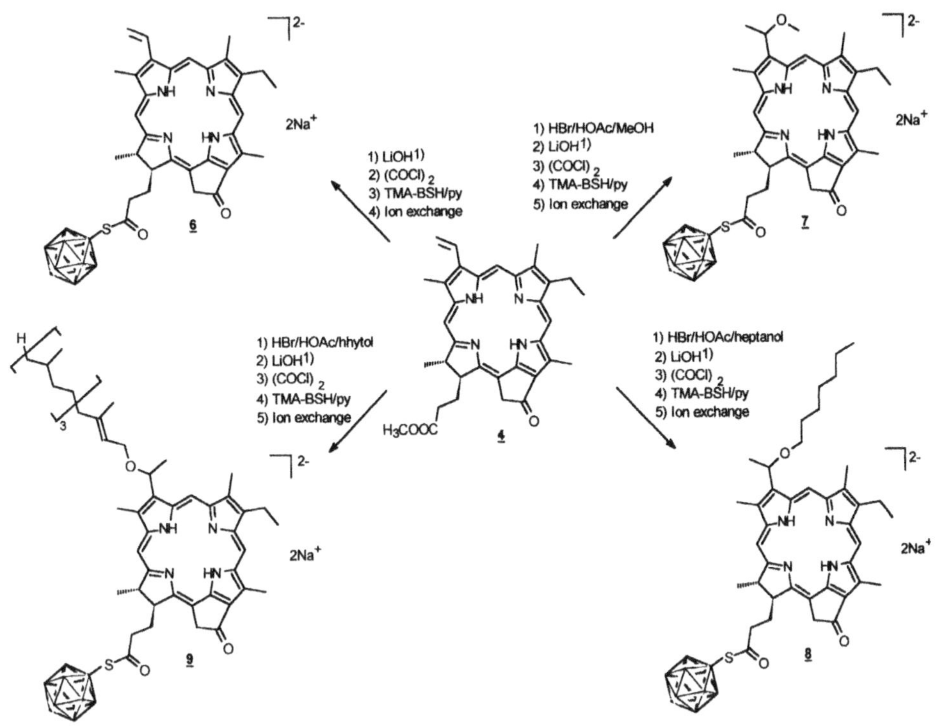

Figure 2. Synthesis of boronated chlorin derivatives.

carboxylic acids in chlorins under mild conditions. The chlorins had to be stirred in $(COCl)_2/CH_2Cl_2$ just for 20 minutes at room temperature under argon atmosphere. After the esterification with BSH the crude products were purified by column chromatography over both silica and RP-18. All boronated chlorin derivatives were characterized by ^{1}H-NMR, MS, IR and UV/Vis. Purities and different lipophilicities were demonstrated by HPL-chromatograms. Using the chromatography system described below the retention times of the derivatives increase with their lipophilicity. As it was assumed the lipophilicity increases with the length of the alkyl ether chain. The results are shown in Table 1. In addition the obtained yields are given related to the concentrations of pyropheophorbide *a* (**4**). With regard to the desired yield and purity it has to be established that the yields are insufficient but the purities are acceptable for further biochemical investigations.

With this synthesis pathway it is possible to obtain boronated chlorins (**6**)–(**9**) with variable lipophilicity. The subcellular localization of boronated chlorins (**6**), (**7**) and (**8**) was investigated by S. Steen-Ecke.[4]

Synthesis Pathways Starting with Derivatives of Purpurin-18 (3)

Starting with Purpurin-18 (3). Using the same procedure described above to activate the propionic acid side chain of (**3**) the esterification with BSH succeeds with a yield of 37%. Chromatography first over silica and then over RP-18 was necessary to purify 10. With this procedure, a purity of 98.1% (HPLC) was obtained.

Table 1. (Chromatography conditions: RP-18, gradient system eluent: 20% MeOH/80% H_2O 30 mM TEAF, 400 nm)

Chlorin	Retention time [min]	HPLC-purity [%]	Yield [%]
6	26′30	93.3	16
7	24′10	95.4	36
8	29′38	94.4	29
9	32′37/32′53 cis/trans	36.6/55.0	18 (isomeric mixture)

In the following step it was not possible to open the isocyclic anhydride with butylamine. MS data showed the loss of the BSH cluster. It is supposed that the chlorin is substituted by butylamine losing the boron cluster.

The treatment of purpurin-18 with BSH yielded no product that could be isolated, purified and characterized. MS data suggest the degradation of this chlorin.

Starting with Purpurin-18-Methylester (5). After opening the isocyclic anhydride with diethyl amine the generated chlorin underwent degradation during the activation with oxalyl chloride. This happened although the same mild conditions were used as described for the other successful esterifications.

CONCLUSION

For the activation of carboxylic acids in chlorins a suitable method was developed starting with pyropheophorbide *a* (**4**). With this method it is possible to synthesize boronated chlorins with a varying lipophilicity. Column chromatography of these products over silica and RP-18 succeeds with excellent purities. The educt can be obtained by simple derivatization of chlorophyll *a* (**1**), a natural chlorin which can be extracted from *Spirulina pacifica* in any amount.

ACKNOWLEDGMENT

The authors gratefully acknowledge the support for this investigation by the Deutsche Forschungsgemeinschaft and the Fonds der Chemischen Industrie.

REFERENCES

1. R.K. Pandey, A.B. Sumlin, S. Constantine, M. Aoudia, W.R. Potter, D.A. Bellnier, B.W. Henderson, M.A. Rodgers, K.M. Smith, and T.J. Dougherty, Alkyl Ether Analogs of Chlorophyll-a Derivatives: Part 1. Synthesis, Photophysical Properties and Photo-dynamic Efficacy, *Photochem. Photobiol.*, 64:194–204, 1996.
2. K.M. Smith, D.A. Goff, and D.J. Simpson, Meso substitution of Chlorophyll derivatives: Direct route for transformation of Bacteriopheophorbides d into Bacterio-pheophorbides c, *J. Am. Chem. Soc.*, 107:4946–4954, 1985.
3. A. Raitzig, Darstellung und Aufreinigung borierter Porphyrine und Phthalocyanine, Ph.D. thesis, University of Bremen 1997.

4. S. Steen-Ecke, F.-P. Montforts, and D. Gabel, Subcellular localization of BSH-containing and non-boron-containing chlorin derivatives in living cells using confocal laser scanning microscopy, *in* "Frontiers in Neutron Capture Therapy," M.F. Hawthorne, K. Shelly, R.W. Wiersema, eds., Kluwer Academic/Plenum Publishers, New York, 2001, pp. 883–886.
5. S.-J. Lee, N. Jagerovic, and K.M. Smith, Use of the chlorophyll derivative Purpurin-18 for synthesis of sensitizers for use in Photodynamic Therapy, *J. Chem. Soc. Perkin Trans. I*, 2369–2377, 1993.
6. J. Osterloh, Versuche zur Darstellung borhaltiger Purpurinderivate mit variabler Lipophilie, master thesis, University of Bremen 1998.

FLUORINATION OF *p*-BORONOPHENYLALANINE

Kazuo Yoshino[1]*, Yoshiro Inamoto[1], Tatsuo Ido[2], Hidetake Kakihana[3], Masamitsu Ichihashi[4], and Yutaka Mishima[5]

[1]Department of Chemistry, Faculty of Science
Shinshu University, Asahi, Matsumoto 390-8621, Japan
[2]Division of Radiopharmaceutical Chemistry
Cyclotron and Radioisotope Center
Tohoku University, Aoba, Sendai, 980, Japan
[3]Energy Research Center
Wakasa Bay, Tsuruga 914-0047, Japan
[4]Department of Dermatology
Kobe University School of Medicine
Kobe 650-0017, Japan
[5]Mishima Institute for Dermatological Research
Kobe 657-0063, Japan

INTRODUCTION

Using tumor-seeking property of *p*-Boronophenylalanine (BPA), BPA was fluorinated with ^{18}F, and ^{18}F-*p*-^{10}B-BPA has been used for positron emission tomography (PET).[1] But, chemical and biological properties of the ^{18}F-^{10}B-BPA are not completely understood. There are following merits for synthesis of ^{19}F-BPA. (1) Since some fluorinated compounds have better biological activity than not-fluorinated ones, there is a possibility that ^{19}F-BPA has better boron accumulation property than BPA. (2) Since ^{19}F-BPA is stable, the precise boron uptake experiments and the studies of the chemical properties can be carried out. (3) ^{19}F-NMR is applicable for the identification of the products of the fluorination reaction. In this paper, we report on the larger scale synthesis of ^{19}F-BPA for the chemical and the biological studies, and the identification of not only 2-^{19}F-BPA, but also 2,5-^{19}F-BPA.

Frontiers in Neutron Capture Therapy, edited by Hawthorne *et al.*
Kluwer Academic / Plenum Publishers, New York, 2001.

EXPERIMENTAL METHODS

BPA with natural abundance ratio of boron was used. The synthesis of ^{19}F-BPA with $^{19}F_2/CH_3COO^{19}F$ was carried out by the Ishiwata's method[2] without using ^{18}F. $CH_3COO^{19}F$ was obtained by passing F_2 gas (0.15% F_2/N_2 for 25mL/min. flow rate and 0.14% F_2/N_2 for 50mL/min. flow rate) through a CH_3COONa column (8mm i.d. × 150 mm or 300mm length). The reaction products were separated by HPLC using a Delta-Pak C_{18} cartridge (100Å6, 15µm, 25mm i.d. × 100mm length) with a pre-column cartridge (25mm i.d. × 10mm length). Both cartridges were inserted in an RCM 25 × 10 compression module (Waters). The mobile phase was 0.1% acetic acid solution, and the flow rate was 10mL/min. The elution profile was monitored using 215nm absorbance. JEOL LA 400 NB was used for NMR measurements. BF_3OEt_2 was used for the external reference of both ^{19}F and ^{11}B (δ = 0ppm).

RESULTS AND DISCUSSION

Ishiwata *et al.* have carried out the direct ^{18}F-fluorination of BPA using mixture gas of $^{18}F_2$ and $^{19}F_2$. The amount of BPA they used was 100µmol. We have increased the scale to 2,000µmol BPA. A typical HPLC elution profile for the obtained products is shown in Figure 1 together with that for the mixed solution of BPA, phenylalanine, *o*-, *m*-, and *p*-F-phenylalanines. Two new peaks appeared in the case of fluorination reaction product. They have also appeared[2] in the radiochromatogram of ^{18}F-fluorination product. Ishiwata *et al.* have identified the bigger peak (peak 1) as 2-F-BPA from ^{1}H-NMR and mass spectroscopic data. They have not identified the smaller peak (peak 2),

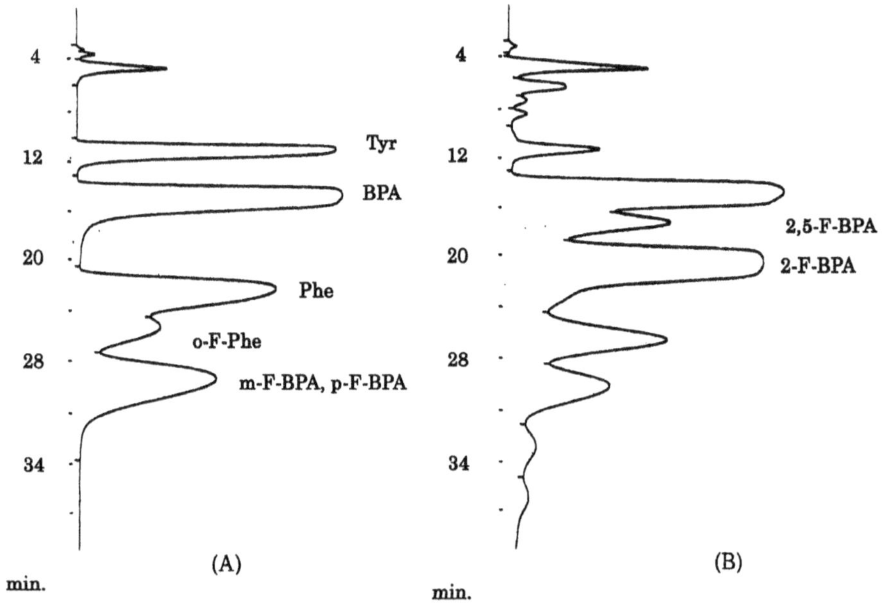

Figure 1. Chromatograms of the mixed solution of Tyr, BPA, Phe, *o*-F-Phe, *m*-F-Phe, and *p*-F-Phe (A) and the reacted solution of BPA with AcOF at 50mL/min. for 11h (B).

but assumed as 3-F-BPA. We have analyzed the two species by ^{1}H-, ^{19}F- and ^{11}B-NMR. The peak 1 was confirmed as 2-F-BPA (^{19}F-NMR, 33.8 ppm). In the case of the peak 2, we have assigned it to 2,5-F-BPA from the following reasons. a) The ^{1}H-NMR showed two benzene protons. b) Two ^{19}F-NMR signals (40.8 and 28.2 ppm) were observed, which indicates that two F atoms are not symmetrically substituted. c) Comparing the three ^{11}B-NMR chemical shift values of BPA, 2-F-BPA and the peak 2, the value of the peak 2 (27.7 ppm) is 1.8 ppm higher field shifted than that of BPA (29.5 ppm), and also 1.1 ppm higher field shifted than that of 2-F-BPA (28.9 ppm). Since the substituted F withdraws the electron on the benzene ring, the ring current decreases, which means the

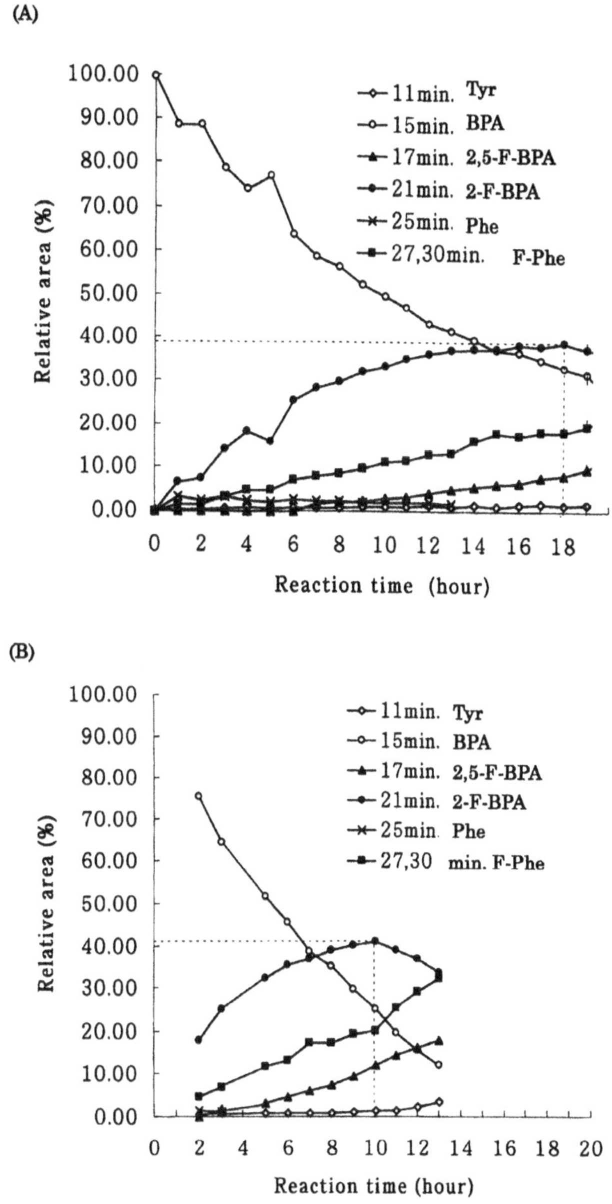

Figure 2. Time coarse of relative areas of BPA, 2-F-BPA, 2,5-F-BPA and other species. (A) Flow rate: 25 mL/min. (B) Flow rate: 50 mL/min. The relative areas were obtained by integrating each peak from a valley to next valley.

paramagnetic effect on ^{11}B decreases and the ^{11}B signal of F-substituted BPA appears at the higher field than that of BPA. Therefore, the most highly shifted ^{11}B signal of the peak 2 can be assigned to 2,5-F-BPA.

In the reaction system, we used 25 mL/min F_2 flow rate. Jewett *et al.* reported[3] that more rapid F_2 rate than 25 mL/min. decreases the product yield. This flow rate implies that the time required for 2,000 μmol BPA reaches to 24 hours. To reduce the reaction time, we have increased the flow rate to 50 mL/min. In this case, 30 cm CH_3COONa column was used. The result is shown in Figure 2 together with 25 mL/min. flow rate case. When the flow rate is 25 mL/min, the maximum yield was obtained at 18 hours. The same yield was obtained at 10 hours in the case of 50 mL/min. Therefore, the increase of the rate is quite effective. Comparing the yields of 2,5-F-BPA in both cases, the yield in the case of 50 mL/min. is higher than that in the case of 25 mL/min.

CONCLUSIONS

It has been demonstrated that the direct fluorination of BPA with $^{19}F_2/CH_3COO^{19}F$ yields not only 2-^{19}F-BPA but also 2,5-^{19}F-BPA. If 2,5-^{19}F-BPA has the accumulation property, we can use ^{18}F-fluorinated 2,5-F-BPA as a PET reagent in addition to 2-^{18}F-BPA.

REFERENCES

1. Y. Imahori, S. Ueda, Y. Ohmori, T. Kushiki, K. Ono, R. Fujii, and T. Ido, Fluorine-18-labeled fluoroboronophenylalanine positron emission tomographic studies in patients with gliomas for boron neutron capture therapy, *J. Nucl. Med.* 38:1762–1767, 1998.
2. K. Ishiwata, T. Ido, A.A. Mejia, M. Ichihashi, and Y. Mishima, Synthesis and radiation dosimetry of 4-borono-2-[18F]fluoro-D,L-phenylalanine: a target compound for PET and boron neutron capture therapy. *Appl. Radiat. Isot.*, 42:325–328, 1991.
3. D.M. Jewett, J.F. Potocki, and R.E. Ehrenkaufer, A gas-solid-phase microchemical method for the synthesis of acetyl hypofluorite, *J. Fluorine Chem.*, 24:477–484, 1984.

RADIOIODINATION TECHNIQUES FOR AROMATIC AMINO ACIDS

Possible Tracers for BPA

J. Vähätalo[1]*, M. Kulvik[2,3], S. Savolainen[4,5], and S.-L. Karonen[5]

[1]Laboratory of Radiochemistry
Department of Chemistry
University of Helsinki
[2]Institute of Biotechnology
Electron Microscopy
University of Helsinki
[3]Department of Neurology
[4]Department of Radiology
[5]Laboratory Department, Nuclear Medicine
Helsinki University Central Hospital

1. INTRODUCTION

Prevailing boron carriers for BNCT are disodium mercaptoundecahydro-*closo*-dodecaborate (borocaptate sodium, BSH) and 4-dihydroxyboryl phenylalanine (4-boronophenylalanine, BPA).[1,2] For clinical patient studies BPA is administred as an anionic fructose complex (BPA-F) infusions. As an aromatic amino acid without the fructose moiety BPA is a structural analogue for natural aromatic amino acid tyrosine. In BPA the hydroxyl group of tyrosine is substituted by the dihydroxyboryl group (or borono group), —$B(OH)_2$. The dihydroxyboryl group is known to be fragile in aromatic molecules and $B(OH)_2$–group can be substituted by electrophiles.[3,4] However, by direct electrophilic fluorination of BPA with [^{18}F]AcOF or [^{18}F]F_2 followed by HPLC separation, 4-borono-2-[^{18}F]fluorophenylalanine (^{18}FBPA) was prepared with radiochemical yields of 25–35% and with a radiochemical purity of over 99%.[5] Clinical patient

*Correspondance: Jyrki Vähätalo, Laboratory of Radiochemistry, Department of Chemistry, University of Helsinki, P. O. Box 55, FIN-00014 University of Helsinki, Finland. E-mail: jyrki.vahatalo@helsinki.fi; Tel. +358-9-191 40154; Fax +358-9-191 40121.

Frontiers in Neutron Capture Therapy, edited by Hawthorne *et al.*
Kluwer Academic / Plenum Publishers, New York, 2001.

studies have shown that positron emission tomography (PET) with fluorinated analogue of BPA is a valuable technique for prediction of the effectiveness of boron neutron capture therapy using BPA as a boron carrier.[6,7]

The success of ^{18}FBPA in clinical studies have raised the question whether iodinated BPA derivatives can be labelled directly by electrophilic methods. Weinreich *et al.* have reported the first study of radioiodinating of BPA with a positron emitter ^{124}I leading one labelled radioiodinated product. However, they do not have yet determined the exact structure of the product.[8] Iodogen^{TM} (1,3,4,6-tetrachloro-3α,6α-diphenylglycoluril) allows labelling with rapid contact time and with gentle conditions to the substance to be labelled.[9] Enzymatic radioiodination is a milder alternative technique than the use of chemical oxidants.[10] In order to improve tracers for boron neutron capture therapy, we tested possibilities to label 4-boronophenylalanine directly with radioiodine using gentle chemical oxidants. Because tyrosine is an structural analogue of natural tyrosine; we also tested a standard direct Iodogen^{TM}—radioiodination technique for natural tyrosine.

2. EXPERIMENTAL

BPA was purchased from Katchem Ltd. (Prague, Czech Republic), tyrosine from Merck (Darmstadt, Germany) and Na^{125}I from Amersham International plc (Buckinghamshire, England). Iodogen^{TM} was obtained from Sigma Chemical Co. (St. Louis, Missouri, USA). In a reaction vial coated with Iodogen^{TM} (0.347 μmol) was added 250 μl borate buffer, 0.5 μmol BPA or tyrosine in 0.5 M HCl, 37 MBq Na^{125}I, 20 nmol KI as iodine carrier. Lactosorbent was prepared by a method of Karonen.[11] In a reaction vial was added 20 μl lactosorbent suspension, 250 μl acetate buffer, 0.5 μmol BPA, 37 MBq Na^{125}I, 20 nmol KI as iodine carrier and 10 μl diluted perhydrol-solution. The contents were mixed and the reaction vial was incubated at room temperature from 1 to 30 minutes.

Analyses of the reaction mixtures were performed by radio-HPLC using a LKB Bromma HPLC pump 2150 and a LKB Bromma 2151 Variable Wavelength Monitor (LKB, Pharmacia Produkter, Sweden) with a detector wavelength of 254 nm. Sample injection was accomplished by a Rheodyne 7125 injector (L.P., Cotati, CA, USA). For continuous measurement of radioactivity, the outlet of the UV detector was connected to a NaI(Tl) well type scintillation detector. The effluent was monitored with associated electronics and software (Pico Technology). For analytical separations, a Spherisorb 10 ODS (250 × 4.6 mm) column (HPLC Technology, Inc, Cheshire, United Kingdom) was used with H_2O/EtOH/AcOH 96/1.5/2.5 (v/v/v) containing 2.5 g/l NH_4OAc, pH 4, as eluent, at flow rate of 2 ml/min.

We determined labelled products with commercially available non-radioactive reference compounds: 4-Iodo-D-phenylalanine (IPhe) was obtained from Aldrich Chemical Company Ltd. (Gillingham, Dorset, England), 3-iodo-L-tyrosine (ITyr) from Fluka Chemie AG (Buchs, Switzerland), D,L-phenylalanine (Phe), L-tyrosine (Tyr) from Merck (Darmstadt, Germany).

3. RESULTS

A direct electrophilic radioiodinating method using Iodogen^{TM} as an oxidant gave reproducible amounts of 4-[^{125}I]iodophenylalanine instead of radioiodinated BPA. A direct electrophilic radioiodination method using lactosorbent resulted in four radioiodinated aromatic amino acids. Two of the products were identified as 4-iodophenylalanine

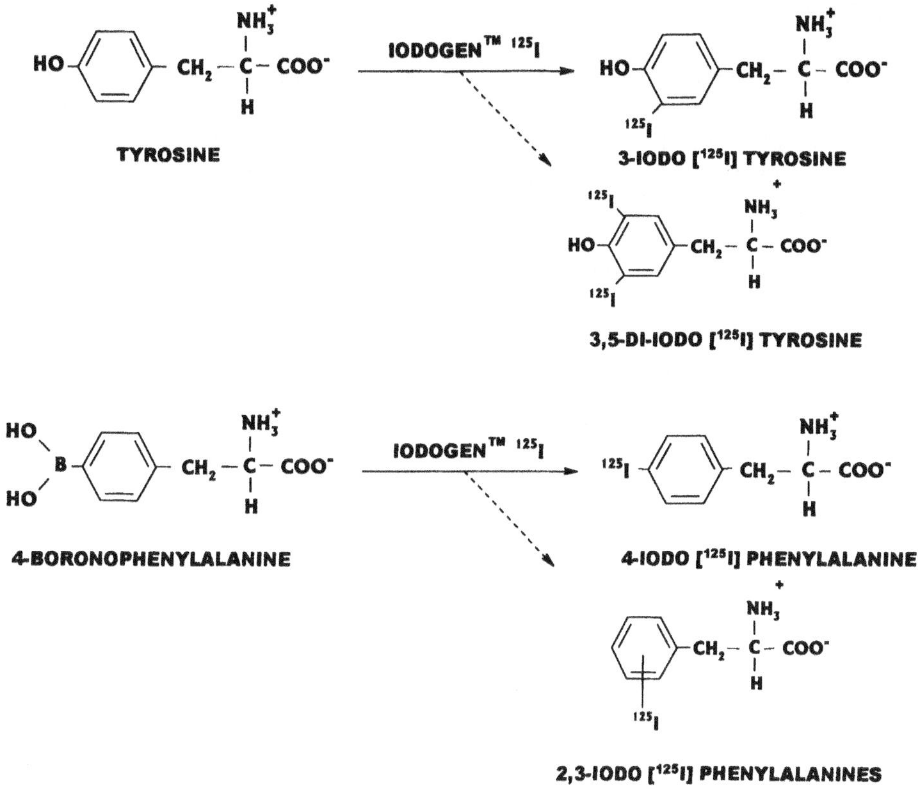

Figure 1. Radiolabelling technique for BPA and tyrosine using Iodogen™ as oxidant.

and 3-iodotyrosine. The others were presumed to be 2- and 3-iodophenylalanines. Deboronation of BPA leads also to the formation of two natural amino acids, phenylalanine and tyrosine. A direct electrophilic Iodogen™—technique for radioiodination of tyrosine gave an excellent radiochemical yield (>99%) 3-[^{125}I]iodotyrosine. A formation of a corresponding di-iodo compound was observed, but it was avoidable using suitable labelling conditions.

Table 1. Radiochemical yield for direct iodination using Iodogen™ (BPA and tyrosine) and lactosorbent (BPA) as oxidants

Method	Products	Yield
Iodogen™		
BPA	4-[^{125}I]iodophenylalanine	>90%
	2,3-[^{125}I]iodophenylalanines	5%
tyrosine	3-[^{125}I]iodotyrosine	99%
	3,5-di[^{125}I]iodotyrosine	1%
Lactosorbent		
BPA	4-[^{125}I]iodophenylalanine	30%
	2,3-[^{125}]iodophenylalanines	20%
	3-[^{125}I]iodotyrosine	5%

4. DISCUSSION

The dihydroxyboryl group, —$B(OH)_2$, in aromatic molecules is electron withdrawing and deactivates strongly direct electrophilic radiolabelling; it acts as a good leaving group: no radiolabelled BPA was observed using gentle oxidants and mild electrophilic radioiodination techniques investigated in this study. The hydroxyl group, —OH, is very different from the dihydroxyboryl group in direct electrophilic aromatic radiohalogenation: the OH-group is electron donating and activates the aromatic ring toward direct electrophilic labelling, resulting in efficient radioiodination techniques with excellent radiochemical yield. Based on the similar aromatic amino acid transport mechanisms to malignant and normal brain tissue, and the known metabolic fate of different aromatic amino acids, radiolabelled tyrosine is proposed to be used as a tracer for BPA.

ACKNOWLEDGMENTS

The financial support from the Technology Development Centre, University of Helsinki, the Finnish Academy and Helsinki University Central Hospital are acknowledged.

REFERENCES

1. C. Morin, The chemistry of boron analogues of biomolecules, *Tetrahedron*, 50:12521–12569, 1994.
2. A.H. Soloway, W. Tjarks, B.A. Barnum, F.-G. Rong, R.F. Barth, I.M. Codogni, and J.G. Wilson, The chemistry of neutron capture therapy, *Chem. Rev.*, 98:1515–1562, 1998.
3. G.W. Kabalka, A.R. Kunda, and K. Muralidhar, Synthesis of iodine-125 labelled aryl and vinyl iodides, *J. Labelled Compd. Radiopharm.*, 19:795–799, 1982.
4. J. Vähätalo, J. Tuominen, J. Kokkonen, O. Kriz, S.-L. Karonen, and M. Kallio, Trace Impurities Identified by High Performance Liquid Chromatography/Electrospray Mass Spectrometry in Two Different Synthetic Batches of 4-Boronophenylalanine, *Rapid Commun. Mass Spectrom.*, 12:1118–1122, 1998.
5. K. Ishiwata, T. Ido, A.A. Meija, M. Ichihashi, and Y. Mishima, Synthesis and radiation dosimetry of 4-borono-2-[^{18}F]fluoro-D,L-phenylalanine: a target compound for PET and boron neutron capture therapy, *Appl. Radiat. Isot.*, 42:325–328, 1991.
6. G.W. Kabalka, G.T. Smith, J.P. Dyke, W.S. Reid, C.P.D. Longford, T.G. Roberts, N.K. Reddy, E. Buonocore, and K.F. Hübner, Evaluation of Fluorine-18-BPA-Fructose for Boron Neutron Capture Treatment Planning, *J. Nucl. Med.*, 38:1762–1767, 1997.
7. Y. Imahori, S. Ueda, Y. Ohmori, T. Kusuki, K. Ono, R. Fujii, and T. Ido, Fluorine-18-Labeled Fluroboronophenylalanine PET in Patients with Glioma, *J. Nucl. Med.*, 39:325–333, 1998.
8. R. Weinreich, D.F. Dos Santos, L. Wyer, M. Argentini, and H.-J. Hansen, Labelling of BNCT drugs with longer lived positron emitters, *in*: "Advances In Neutron Capture Therapy, Vol. 2," B. Larsson, J. Crawford, and R. Weinreich, eds, Elsevier, Amsterdam, 1997, pp. 295–298.
9. M. Bourdoiseau, Iodine radiochemistry and radiopharmaceutical product labeling, *Nucl. Med. Biol.*, 13:83–88, 1986.
10. R.H. Seevers, and R.E. Counsell, Radioiodination techniques for small organic molecules. *Chem. Rev.*, 82:575–590, 1982.
11. S.-L. Karonen, Solid lactoperoxidase in the iodination of L-tyrosine and albumin, *Appl. Biochem. Biotechn.*, 6:119–128, 1981.

SYNTHETIC APPROACHES TO A NOVEL IODINATED ANALOGUE OF BPA

M. Suominen[1], J. Vähätalo[2], J. Yli-Kauhaluoma[3], and T. Hase[1]

[1]Laboratory of Organic Chemistry
Department of Chemistry
University of Helsinki
[2]Laboratory of Radiochemistry
Department of Chemistry
University of Helsinki
[3]VTT Chemical Technology

1. INTRODUCTION

Prevailing clinical boron carriers for BNCT are disodium mercaptoundecahydro-*closo*-dodecaborate (sodium borocaptate, BSH) and 4-dihydroxyboryl phenylalanine (4-boronophenylalanine, BPA). It has been shown that boron-10 uptake and concentrations in brain tumours *in vivo* can be estimated using positron emission tomography (PET) for detecting 4-borono-2-[^{18}F]fluorophenylalanine.[1] The success of the radiofluorinated analogue of BPA in basic research and clinical studies has raised the question whether electrophilic methods could be used for labelling BPA with radioiodine. Weinreich *et al.* reported the first study of radioiodinating of BPA with a positron emitter ^{124}I leading to one labelled product.[2] However, they had not yet determined the exact structure of the product.

We have evaluated corresponding direct radioiodination methods for aromatic amino acids including BPA. Radioiodinated products were separated and identified by reverse phase HPLC with UV and radioactivity detectors.[3] We determined the radioiodinated products with non-radioactive commercially available reference compounds. However, a stable analogue for radioiodinated BPA was not available and the target of this study was to prepare 4-borono-2-iodo-D,L-phenylalanine (IBPA).

2. EXPERIMENTAL

One approach in synthetising IBPA was as follows: In stage one, a Grignard reagent prepared from 4-bromotoluene was allowed to react with tri-*n*-butylborate to give

Frontiers in Neutron Capture Therapy, edited by Hawthorne *et al.*
Kluwer Academic / Plenum Publishers, New York, 2001.

4-tolylboronic acid.[4] This could also be purchased from Tokyo Casei. 4-Tolylboronic acid was nitrated with fuming HNO_3 in –40 °C to prevent dinitrotoluene formation. At this point, the product was crystallised from ice water and dissolved in a 1 M aqueous solution of NaOH.[5] If a more basic solution of NaOH was used, the boronic acid group was lost. The radical benzylic bromination was carried out in carbon tetrachloride with Br_2 during a five-day reaction with slight warming.[6] Due to a long reaction time, an equilibrium was reached and we always obtained equimolar amounts of 4-dibromomethyl-3-nitrophenylboronic acid **1** and 4-bromomethyl-3-nitrophenylboronic acid **2**. Alkylation with diethylacetamidomalonate in the presence of potassium *tert*-butoxide and subsequent catalytic (Pd-C) hydrogenation in absolute ethanol gave dimethyl acetamido(2-amino-4-boronobenzyl)malonate.[7,8]

3. DISCUSSION

Radical bromination of *ortho*-nitrated methylphenylboronic acid appeared to be very difficult. This is due to the electronic effects of the nitro group, which leads to the deactivation of the adjacent methyl group.

The products **1** and **2** were obtained in almost equimolar amounts after 4-day reaction using 2 to 3 equivalent of NBS. Since radical reactions are equilibrium reactions, long reaction times create problems in obtaining only the monobrominated product.

In addition, boronic acid group creates many problems because of its Lewis acid character. It is easily dissosiated from the ring in basic conditions. Furthermore, the boronic acid group adds to the deactivation of the ring thus contributing to the deactivation of the methyl group.

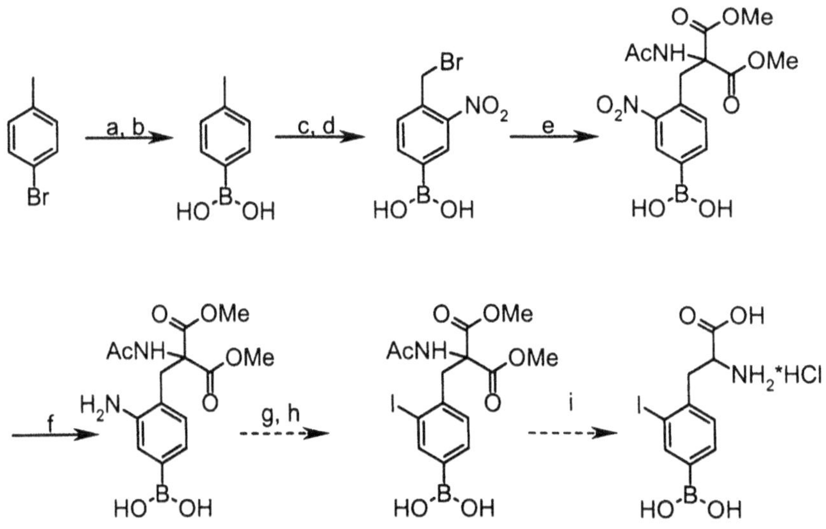

Figure 1. Preparation scheme of 4-borono-2-iodophenylalanine. a) Mg, Et_2O, Grignard reaction; b) $(n\text{-BuO})_3B$, Et_2O, –78 °C; c) fuming HNO_3, –40 °C; d) NBS or Br_2, peroxide or hν; e) dimethylacetamidomalonate, KOBu-t, DMF, 0 °C; f) PdC/H_2, EtOH; [g) $NaNO_2$; h) KI; i) HCl, synthesis is not yet completed].

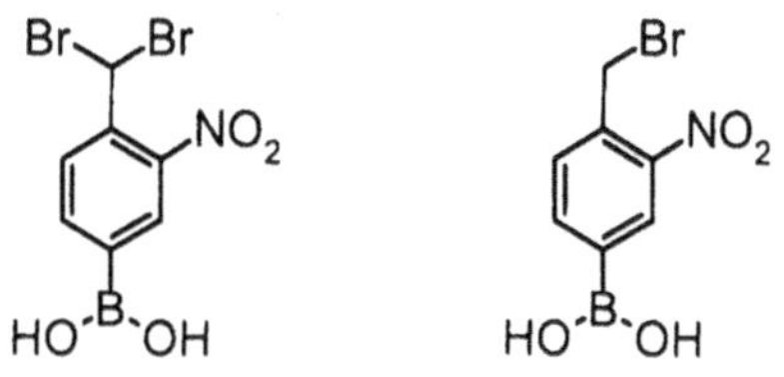

1 2

Figure 2. Structures of the two bromination products.

ACKNOWLEDGMENTS

The financial support from the Technology Development Centre, University of Helsinki, the Finnish Academy, VTT Chemical Technology and Helsinki University Central Hospital are acknowledged.

REFERENCES

1. Y. Imahori, S. Ueda, Y. Ohmori, T. Kusuki, K. Ono, R. Fujii, and T. Ido, Fluorine-18-labelled fluoroboronophenylalanine PET in patients with glioma, *J. Nucl. Med., Vol. 39*:325, 1998.
2. R. Weinreich, D.F. Dos Santos, L. Wyer, M. Argentini, and H.-J. Hansen, Labelling of BNCT drugs with longer lived positron emitters, *Advances In Neutron Capture Therapy, Vol. 2*:295, 1997.
3. J. Vähätalo, M. Kulvik, S. Savolainen, and S.-L. Karonen, Radioiodination techniques for aromatic amino acids; Possible tracers for BPA, *in "Program and Abstracts of the Eight International Symposium on Neutron Capture Therapy for* Cancer," E-66, p. 69, 1998.
4. F.R. Bean, and J.R. Johnson, Derivatives of Phenylboric Acid, Their Preparation and Action Upon Bacteria. II, Hydroxyphenylboronic Acids, *J. Am. Chem. Soc. 54*:4415, 1932.
5. K. Torsell, H. Meyer, and B. Zacharias, Zur Kenntnis der Arylborsäuren. II, *Arkiv för Kemi 10*:497, 1957.
6. K. Torsell, Zur Kenntnis der Arylborsäuren. III, *Arkiv för Kemi 10*:507, 1957.
7. H.R. Snyder, A.J. Reedy, and Wm. J. Lennartz, Synthesis of Aromatic Boronic Acids. Aldehydo Boronic Acids and a Boronic Acid Analog of Tyrosine, *J. Am. Chem. Soc. 80*:835, 1958.
8. I.D. Entwistle, A.E. Jackson, R.A.W. Johnstone, and R.P. Telford, Reduction of Nitro-compounds, *J. Chem. Soc. Perkin I* 443, 1977.

DESIGN OF BORON- AND GADOLINIUM-CONTAINING AGENTS FOR NCT

V. N. Kulakov[1], V. I. Bregadze[2], I. B. Sivaev[2], S. M. Nikitin[1], Yu. V. Gol'tyapin[1], and V. F. Khokhlov[1]

[1]Russian State Research Center "Institute of Biophysics"
Zhivopisnaya St. 46, 123182, Moscow, Russia
[2]A.N. Nesmeyanov Institute of Organoelement Compounds
Russian Academy of Sciences
Vavilov Str. 28, 117813, Moscow, Russia

Search of preparations for neutron capture therapy is carried out in the framework of the Russian project "Neutron Capture Therapy" in the following main directions: synthesis of gadolinium complexes; design of compounds containing boron and gadolinium atoms; investigation of the structure of the compounds prepared and creation of their optimum medicinal forms.

Complexes of gadolinium with diethylenetriaminepentaacetic acid (DTPA) are characterized by high stability and are widely used in medicinal diagnostics as magnetic resonance contrasting agents. We elaborated a new medicinal form of the gadolinium complex of composition $Na_2[Gd(DTPA)]$, which was named Dipentast.[1] Comparative characteristics of Dipentast and Magnevist® (Gadopentate dimegluminic salt) are presented in Table 1.

Structure investigations were carried out using the methods of NMR spectroscopy and time-differential perturbed angular $\gamma\gamma$ correlations (TDPAC) in combination with quantum-chemical calculations including optimization of the molecule geometry with determination of atom coordinates, bond lengths and angles. A similar approach was used to study the structures of complexes of various radionuclides.[2–6] Advantages of the TDPAC method are absence of restrictions for concentration of a sample and its state in wide temperature range; content of metal ions in a sample (~0.1 ml) can be approximately 10–12 g-ions. The radioactivity of a sample in this case is low, ~0.37 MBq (~10 μCi), which avoids radiation damage of the complex molecule. The structural studies performed revealed that the basic parameters of the gadolinium and indium complexes are similar, the geometry of coordination sphere of the gadolinium complex is distinguished from the geometry determined by X-ray structure analysis. So, the possibility of application of ^{111}In as radioactive marker in the pharmacokinetic studies of gadolinium complexes was assumed to be justified.

Frontiers in Neutron Capture Therapy, edited by Hawthorne *et al.*
Kluwer Academic / Plenum Publishers, New York, 2001.

Table 1. Basic characteristics of Dipentast and Magnevist®

	Magnevist®	Dipentast
Molecular weight, D	938	591
Concentration of substance, mmol/ml (mg/ml)	0.5 (469)	0.5 (296)
Gadolinium content, % wt.	17	26.5
pH of medicinal form	6.5–8.0	6.5–8.0
Viscosity, cP, at 20 °C	4.9	2.20–2.35
37 °C	2.9	(30 °C)
Density at 20 °C, g/cm^3	1.2	1.15–1.18
Stability constant, log K	22.5	22.5
LD_{50}, mmol/kg: for mouses intravenously	5–8	6.9
for rats intravenously	10	8.3
Removal from body	Complete	Complete
Average efficiency of spin-lattice relaxation ^1H in buffer solution (20 MHz, 30 °C, mmol/s)	7.8 ± 0.3*	7.6 ± 0.3*
Period of semi-removal, $T_{1/2}$, min. at injection in tumor	23 ± 5	48 ± 5
in muscle	18 ± 3	50 ± 8

The idea to combine the facilities of gadolinium and boron neutron capture therapy looks very promising and various efforts in this direction have been made. In order to prepare compounds containing both boron and gadolinium atoms we synthesized the boron-containing ligand by modification of the known DTPA ligand. The reaction of tetramethylammonium amino-undecahydro-*closo*-dodecaborate $(Me_4N)[B_{12}H_{11}NH_3]$ with DTPA dianhydride (molar ratio 1:3) in dimethylformamide in the presence of sodium hydride gave the monoamide derivative (Figure 1).

The design of medicinal forms of boron- and gadolinium-containing compounds had as a goal to decrease the rate of elimination of preparations from tumor. Medicinal forms were prepared by a neutralization of the charges of molecules and their transformation into weakly dissociating salts, and by preparation of prolonged medicinal forms, in particular colloidal solutions and suspensions. A neutralization of the charge slows down a rate of absorption and elimination of preparation in two times. A use of oil solutions leads to the same results. A sizes of a denatured albumin particles modificated by DTPA are 15–25μ. After intramuscular introduction an albuminous suspension is

Table 2. Calculated bond lengths and bond angles for the coordination sphere of the $[M(DTPA)]^{2-}$ (M = Gd, In) complexes

	d, Å			φ Deg	
Bond	M = Gd	M = In	Angle	M = Gd	M = In
O(27)—M	2.17	2.01	O(27)-M-N(15)	74.10	77.64
N(15)—M	2.32	2.20	O(23)-M-N(15)	80.73	82.85
O(23)—M	2.18	2.02	O(23)-M-N(12)	74.72	78.43
N(12)—M	2.32	2.16	N(12)-M-N(15)	79.11	82.80
O(2)—M	2.18	2.17	O(2)-M-N(12)	105.03	110.55
N(8)—M	2.32	2.21	O(2)-M-O(23)	68.82	69.31
			O(2)-M-N(8)	74.54	78.44
			N(12)-M-O(27)	148.16	157.23
			N(8)-M-N(15)	137.87	136.25

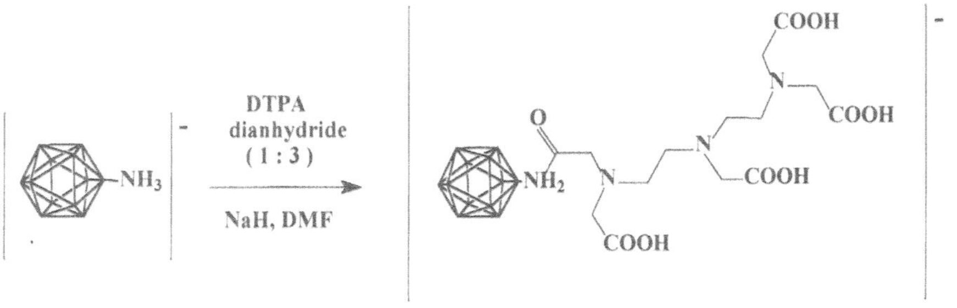

Figure 1. Scheme of synthesis of boron-containing derivative of DTPA ligand.

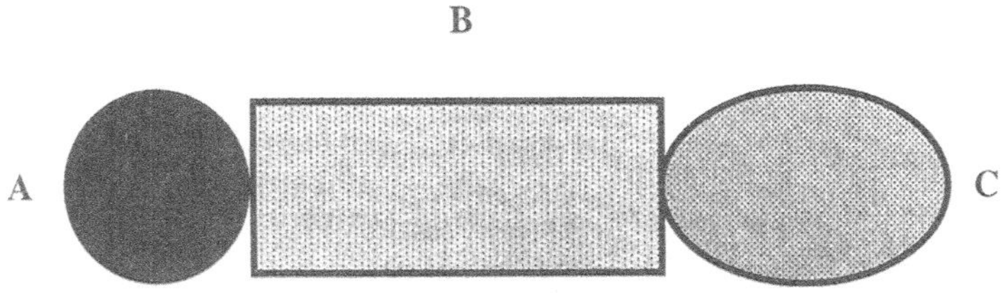

A - NCT-active part (^{157}Gd, ^{10}B)

B - property determining part;

C - tumor targeting part

Figure 2. General construction of NCT-agents.

removed very slowly. Oil solutions and suspension are allocated irregular in tumor. Aqueous medicinal forms providing delayed removal of NCT-agents from tumor were designed and tested experimentally on animals.

The results obtained allow a significant reduction in investigations directed toward the empirical search for NCT agents, and permit the design of new agents based on the scheme in Figure 2.

The instrumental and calculational methods worked out for structural investigations make it possible to correlate biological properties with the structure of NCT agents, and to use computer simulations for creating pharmaceuticals with optimal characteristics for NCT.

REFERENCES

1. V. Panov, N. Shimanovsky, V. Kulakov, *et al.*, Comparison of known and new MRI contrast media efficacy in the model system and in tissues of rats, *in:* Abstr. Int. Conf. «Experimental Pathological Processes, Magnetic Resonance Materials in Physics, Biology and Medicine», Suzdal, Russia, 1998, p. 64.
2. Z.Z. Axelrod, V.N. Kulakov, A.A. Sorokin *et al.*, Time-differential perturbed angular correlation of In-complexes used nuclear medicine, *J. Radioanalyt. Nucl. Chem. Letters*, 199:257–265, 1995.

3. Z.Z. Axelrod, V.N. Kulakov, A.A. Sorokin *et al.*, Stability of the In-ligands complexes to the electron capture decay after-effects, *in:* Abstr. X Int. Symp. on Hyperfine Interactions, Leuven, Belgie, 1995, p. 258.
4. V.N. Kulakov, S.M. Nikitin, N.A. Babushkina *et al.*, Spectroscopic studies of complexation in solutions: ^{111}In complexes with glycydimethylenephosphonic acid, *Russ. J. Coord. Chem.*, 23:829–835, 1997.
5. S.M. Nikitin and V.N. Kulakov Structure of technetium (5+) complexes with some tetradentate ligands, *Radiokhimiya* (in Russian), 39:401–406, 1997.
6. L.G. Shpinkova, V.N. Kulakov, A.A. Sorokin, *et al.*, Stability of ^{111}In-ligand complexes studied by TDPAC, *Z. Naturforsch.*, 53A:630–635, 1998.

IMAGING AND ANALYSIS OF NUCLIDES

ON-LINE BORON-10 DETERMINATION FROM BLOOD SAMPLES BY ICP-MS

M. Kulvik[1], J. Laakso[1], J. Vähätalo[2], and R. Zilliacus[3]

[1]Department of Neurology
Helsinki University Central Hospital
[2]Laboratory of Radiochemistry
University of Helsinki
[3]VTT Chemical Technology

INTRODUCTION

Accurate and fast boron analysis is essential in BNC-treatment of glioma patients. Standard boron analytical assay from biological samples are direct current plasma atomic emission spectrometry,[1] inductively coupled plasma–atomic emission spectrometry,[2] inductively coupled plasma–mass spectrometry[3] or prompt gamma-ray analysis.[4] Various standard and alternative techniques for boron analysis have been tested by the Finnish BNCT research group.[5]

Prompt-gamma neutron activation analysis is ideal in validating tissue boron concentrations simultaneously with the irradiation treatment, but in a BNCT station without a beam shutter other methods are required for pre-irradiation pharmacokinetic calculations.

ICP based methods have gained interest as sensitive and accurate means for boron determination in biological fluids. With ICP-MS in optimal conditions the detection limit of boron in serum sample solutions is 5 ppb. However, the use of these methods has been hampered by tedious and time-requiring sample pretreatment.

In the present study we report a fast sample pretreatment method for ICP boron analyses evaluate the results obtained with two independent methods, ICP-AES and ICP-MS, employing two different sample pretreatment techniques.

EXPERIMENTAL

During preclinical animal studies with intravenous BPA-F infusions thes system was set up, and the accuracy if the ICP-MS-boron-10 analysis procedure was tested.

Frontiers in Neutron Capture Therapy, edited by Hawthorne *et al.*
Kluwer Academic / Plenum Publishers, New York, 2001.

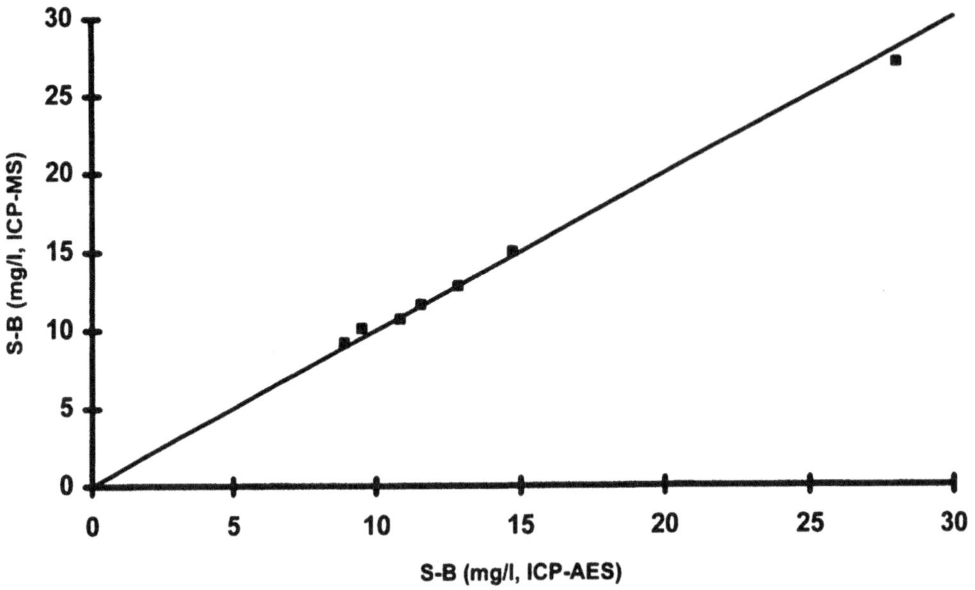

Figure 1. XY-plot of whole blood boron results obtained with ICP-MS and ICP-AES showing the individual data points and the line of identity.

As sample pretreatment, heparinized whole blood samples were taken after intravenous infusion of BPA-fructose. In order to hemolyze and deproteinize the samples 5 parts of 5% (w/w) trichloroacetic acid (TCA) was added to 1 part of the sample and mixed in. After centrfugation (5min. 4000 RMP) clear supernatant was taken for boron analyses with ICP-MS. For the ICP-AES determination aliquots of whole blood samples were digested with HNO_3 and H_2O_2.

A VG Plasma Quad 2+ instrument equipped with a Gilson autosampler and a V-groove nebulizer was employed. The same samples were also analyzed for boron by employing a Baird PS-4 inductively coupled plasma emission spectrometer.

RESULTS

The results of ICP-MS agrees with the results of ICP-AES (Figure 1). Regression analysis revealed that the correlation coefficient between the methods was 0.999. The mean absolute difference of ICP-MS results, in the concentration range studied, was +0.04mg/l with a relative variation in the range from −5.8% to 3.5% in respect of ICP-AES.

DISCUSSION

With TCA-precipitation the need for tedious sample pretreatments can be avoided. The time elapsed from drawing of blood sample to availability of analysis results was no more than 15 minutes. The linearity of analysis results suggest that ICP-based techniques are well-suited for both pharmacokinetic studies and, employing TCA-precipitation, for boron on-line monitoring in BNC-treatments.

REFERENCES

1. R.F. Barth, D.M. Adams, A.H. Soloway, E.B. Mechetner, F. Alam, and A.K. Annisuzzaman, *Anal. Chem.*, 63:890–893, 1991.
2. S.R. Tamat, and D.E. Moore, *Anal. Chem.*, 59:2161–2164, 1987.
3. A.M. Nyomora, R.N. Sah, P.H. Brown, and R.O. Miller, Fresenius *J. Anal. Chem.*, *357*:1185–1191, 1997.
4. C.P. Raaijmakers, M.W. Konijnenberg, L. Dewit, D. Hariz, R. Huiskamp, K. Philipp, A. Siefert, F. Stecher-Rasmussen, and B.J. Mijnheer, *Acta Oncol.*, 34:517–523, 1995.
5. S. Savolainen, J. Räisänen, J. Laakso, I. Ruokonen, R. Zilliacus, J. Likonen, U. Abo Ramadan, V. Eteläniemi, J. Callaway, and M. Kallio, "*Cancer Neutron Capture Therapy*," Yutaka Mishima, ed., Plenum Press: New York, 1996, pp. 189–195.

THE ROLE OF DYNAMIC POSITRON EMISSION TOMOGRAPHY FOR QUANTIFYING BORON-10 *IN VIVO* USING THE FRUCTOSE COMPLEX OF FLUORINE-18 LABELED BORONOPHENYLALANINE

G. W. Kabalka, G. T. Smith, T. L. Nichols, J. P. Dyke, D. Longford, and K. F. Hubner

Biomedical Imaging Center
The University of Tennessee Medical Center
Knoxville, Tennessee 37920

One of the major difficulties facing successful boron neutron capture therapy (BNCT) is the paucity of *in vivo* pharmacokinetic data.[1] Analytical techniques such as prompt gamma analysis, secondary ion mass spectrometry, and resonance ionization spectroscopy are simply not applicable to the *in vivo* detection of boron.[2] To address this problem, both positron emission tomography (PET)[3,4] and magnetic resonance imaging (MRI)[5] techniques for the *in vivo* detection of boronated compounds are being developed. Nuclear medicine diagnostic techniques such as PET have proven to be extremely valuable in evaluating a variety of physiological parameters including the pharmacokinetic behavior of therapeutic agents.[6,7,8] In fact, over nine thousand patients have been evaluated at The University of Tennessee utilizing PET since 1989.

We prepared the fluorine-18 labeled analogue (^{18}F-BPA-F) of the boronophenylalanine-fructose complex (BPA-F) which is currently being used in U.S. Phase I/II clinical trials and have evaluated six patients prior to their receiving BNCT at either Brookhaven National Laboratory (BNL) or at the Massachusetts Institute of Technology and Harvard (MIT/H). Based on our experience[3] and the Japanese studies,[4] PET provides information concerning the *in vivo* uptake and the overall pharmacokinetic behavior of the BPA-F in humans which is proving to be valuable in BNCT treatment planning (Figure 1).

The data was used to determine that the tumor-to-normal brain concentration ratio of ^{18}F-BPA-F was 3.4:1 at 52 min. post injection. This pre-operative PET scan revealed an area of intense uptake in the right posterior parietal region which corresponds to the location of the lesion as seen on CT and MRI.

Frontiers in Neutron Capture Therapy, edited by Hawthorne *et al.*
Kluwer Academic / Plenum Publishers, New York, 2001.

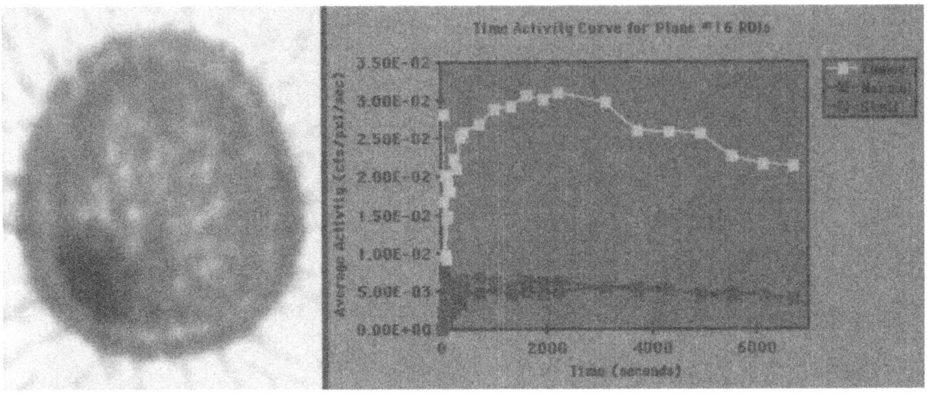

Figure 1. Pre-operative ^{18}F-BPA-F PET scan and time-activity uptake curves of a GBM patient.

The patient underwent two additional PET studies using ^{18}F-BPA-F. The first was carried out six weeks post craniotomy which showed a small residual focal uptake in the medial portion of the original tumor, corresponding to a site of known residual tumor as confirmed by the neurosurgeon. The PET data confirmed, for the first time, that residual tumor was present after resection and that the uptake of BPA-F in tumor remained selective. The results indicate that the biodistribution data obtained via PET might be more appropriate for planning BNCT treatment protocols than the biopsy data currently being utilized.

A PET study performed seven weeks after BNCT, Figure 2, proved even more valuable. The image revealed little or no uptake of ^{18}F-BPA-F at the original tumor site, however a ring of activity surrounded that site.

The fact that ^{18}F-BPA-F is no longer taken up at the original tumor site could be evidence of the success of BNCT. The uptake of ^{18}F-BPA-F around the tumor site may be due to necrotic activity. Further studies will be required to determine whether the uptake of ^{18}F-BPA-F is due to necrosis, residual tumor cells, or granulation tissue. Clearly PET can provide useful information in BNCT treatment planning and will be valuable in pharmacokinetic and tissue dosimetry studies. Further investigations are planned in

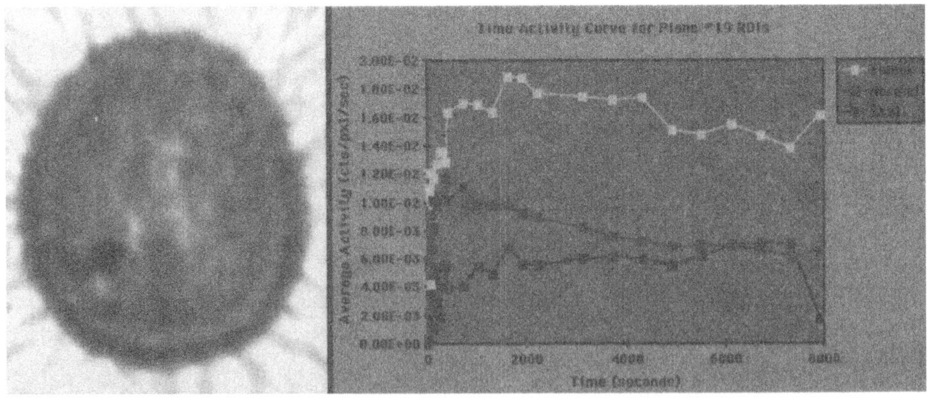

Figure 2. Post-operative ^{18}F-BPA-F PET scan of the same patient seven weeks after BNCT treatment. Uptake is in right posterior parietal region surrounding original tumor location.

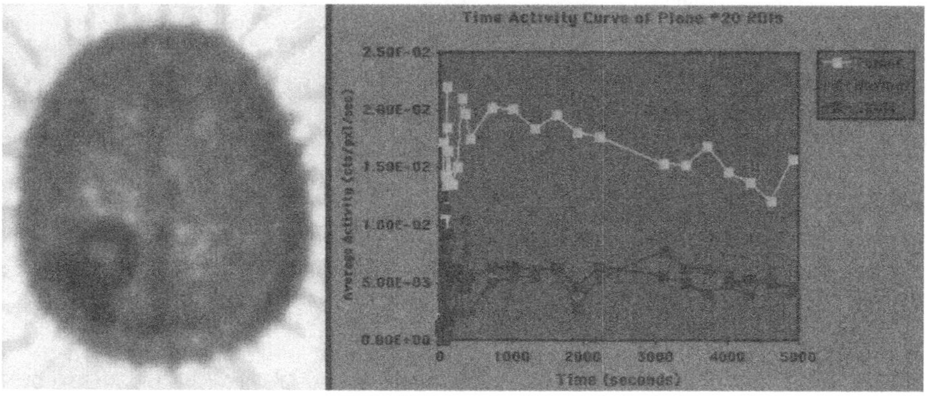

Figure 3. PET scans of post-BNCT patient using ^{18}F-BPA-F (left) and ^{11}C-ACBC (right).

collaboration with Brookhaven National Laboratory in which the *in vivo* distribution of both boron-10 and fluorine-18 will be determined by sequential PET imaging and biopsy.

The combination of ^{18}F-BPA-F and PET may prove to be valuable in identifying more effective BNCT agents. For example, we were recently able to carry out a serial PET investigation on a post-BNCT patient using both ^{18}F-BPA-F and carbon-11 labeled 1-aminocyclobutanecarboxylic acid, ^{11}C-ACBC. This agent is one member of a family of isotopically labeled, unnatural, alicyclic amino acids that were initially developed at Oak Ridge Associated Universities[9] and which are currently under investigation at The University of Tennessee Medical Center for use in imaging tumors.

In the current study we were able to compare ^{11}C-ACBC in a post BNCT patient who had also been imaged with ^{18}F-BPA-F a few days earlier. The results were striking in that the PET scans revealed that ^{11}C-ACBC was more tumor selective than ^{18}F-BPA-F, Figure 3. The selectivity of the ^{11}C-ACBC scan forms the foundation for our current synthetic studies focused on a new series of boronated, unnatural amino acids.

We are in the process of evaluating both ^{18}F-BPA-F and other amino acids for potential use in whole body PET. We feel that these agents may also localize in tumors other than glioblastoma multiforme and melanomas. This could be of interest for the application of BNCT to the treatment of cancers in other body sites if they meet the physical requirements of BNCT.[10,11,12]

ACKNOWLEDGMENT

We wish to thank the U.S. Department of Energy, Lockheed Martin Idaho Technologies Company (URC) and The Robert H. Cole Foundation for support.

REFERENCES

1. R. Gahbauer, N. Gupta, T. Blue, J. Goodman, R. Barth, J. Grecula, A.H. Soloway, W. Sauerwein, and A. Wambersie, Boron Neutron Captuere Therapy: Principals and Potential, *Recent Results Cancer Res.*, 150:183, 1998.
2. H.F. Arlinghaus, M.T. Spaar, R.C. Switzer, and G.W. Kabalka, Imaging of Boron in Tissue at the Cellular Level for Boron Neutron Capture Therapy, *Anal. Chem.* 69:3169, 1997.

3. G.W. Kabalka, G.T. Smith, W.S. Reid, C.P.D. Longford, T.G. Roberts, N.K. Reddy, E. Buonocore, and K.F. Hubner, Evaluation of Fluorine-18-BPA-Fructose for Boron Neutron Capture Treatment Planning, *J. Nucl. Med.*, 38:1762, 1997.
4. Y. Imahori, S. Ueda, Y. Ohmori, T. Kusuki, K. Ono, R. Fujii, and I. Tatsuo, Fluorine-18-Labeled Fluoroboronophenylalanine PET in Patients with Glioma, *J. Nucl. Med.* 39:325, 1998.
5. G.W. Kabalka, C. Tang, and P. Bendel, The Role of Boron MRI in Boron Neutron Capture Therapy, *J. Neuro-Oncol.*, 33:153, 1997.
6. M.E. Phelps, and J.R. Barrio, Effects of Large Neutral Amino Acid Concentrations on 6-[F-18]Fluoro-L-Dopa Kinetics, *J. Cereb. Blood Flow Metab.* 18:43, 1998.
7. N.D. Volkow, G.J. Wang, J.S. Fowler, R. Hitzemann, S.J. Gatley, S.S. Dewey, and N. Pappas, Enhanced Sensitivity to Benzodiazepines in Active Cocain-Abusing Subjects: A PET Study, *Am. J. Psychiatry*, 155:200, 1998.
8. E. Buonocore, K. Hubner, J. Collman, and G.W. Kabalka, Clinical Positron Emission Tomgraphy, Mosby Yearbook, Hanover, MD, 1991.
9. R.L. Hayes, L.C. Washburn, B.W. Wieland, T.T. Sun, R.R. Turtle, and T.A. Butler, Carboxyl-Labeled ^{11}C-1-Aminocyclopentanecarboxylic Acid, a Potential Agent for Cancer Detection, *J. Nucl. Med.* 17:748, 1976.
10. H. Yanagie, T. Tomita, H. Kobayashi, Y. Fujii, Y. Noaka, Y. Saegusa, K. Hasumi, M. Eriguchi, T. Kobayashi, and K. Ono, Inhibition of Human Pancreatic Cancer Growth in Nude Mice by Boron Neutron Capture Therapy, *Br. J. Cancer.*, 75:660, 1997.
11. J.P. Pignol, H. Oudart, P. Chauvel, W. Sauerwein, D. Gabel, and G. Prevot, Selective Delivery of ^{10}B to Soft Tissue Sarcoma Using ^{10}B-L-Borophenylalanine for Boron Neutron Capture Therapy, *Br. J. Radiol.*, 843:320, 1998.
12. R.A. Watson-Clark, M.L. Banquerigo, K. Shelly, M.F. Hawthorne, and E. Brahn, Model Studies Directed Toward the Application of Boron Neutron Capture Therapy to Rheumatoid Arthritis: Boron Delivery by Liposomes in Rat Collagen-Induced Arthritis, *Proc. Natl. Acad. Sci. U.S.A.*, 95:2531, 1998.

STRUCTURE DETERMINATION OF *p*-BORONOPHENYLALANINE-FRUCTOSE COMPLEX USING $^{1}J_{CC}$ COUPLING CONSTANTS

Miyuki Kitazawa and Kazuo Yoshino

Department of Chemistry, Faculty of Science
Shinshu University
Matsumoto 390-8621, Japan

INTRODUCTION

p-Boronophenylalanine (BPA) has been used for Boron Neutron Capture Therapy for malignant tumors. The solubility of BPA in water at blood pH value is very low. Yoshino *et al.*[1] have recommended BPA-fructose complex for injection into patients because of its high stability and solubility. Nowadays, the BPA-fructose complex is used in actual BNCT trails both for melanoma and brain tumor. To elucidate the high complex ability of fructose to BPA, more insights about the structural features of BPA-fructose complex are required. But, the detailed structure of BPA-fructose complex has not been obtained.

Recently, Norrild *et al.*[2] have succeeded in identifying the bonding sites of tolyl-boronic acid-fructose complexes by using $^{1}J_{CC}$ coupling constants. In this paper, we report a structural study on the BPA-fructose complexes with ^{13}C NMR using $^{13}C_6$-enriched fructose.

EXPERIMENTAL

Reagents

Uniformly $^{13}C_6$-labeled D-fructose (^{13}C; 99%, Cambridge Isotope Laboratory) was used without further purification. BPA was synthesized by the Snyder's method[3] with minor modification, and used after recystallization from water.

Frontiers in Neutron Capture Therapy, edited by Hawthorne *et al.*
Kluwer Academic / Plenum Publishers, New York, 2001.

Sample Preparation

Sample **A:** $^{13}C_6$-labeled D-fructose D_2O solution (0.22 M). Sample **B:** alkaline D_2O solution containing both BPA (0.37 M) and $^{13}C_6$-labeled D-fructose (0.15 M). The pH value of the H_2O solution containing same compositions of sample **B** was about 9. Sample **C:** neutral H_2O solution containing the same amounts of BPA and $^{13}C_6$-labeled fructose as sample **B**. The pH value of this solution was 7.4. All samples contained sodium trimethlylsilylpropionate as an internal reference.

NMR Measurements

^1H-decoupled ^{13}C NMR spectra was measured by JEOL LA 400NB at 22 °C. The resonance frequency was 100.40 MHz.

RESULTS AND DISCUSSION

^{13}C Chemical Shifts

Fructose exists in aqueous solution as complex equilibrium mixtures of five isomeric forms, α- and β-fructopyranose, α- and β-fructofuranoses and acyclic keto form. Among them β-fructopyranose, β- and α-fructofuranoses are major species (Figure 1). ^{13}C NMR spectrum of sample **A** (Figure 2 (a)) shows the signals of these three isomers. The spectrum of sample **B** is shown in Figure 2 (b). Comparing it with that of Sample **A**, it is clear that some additional peaks appeared. The previous ^{11}B NMR study showed[4] that more than 90% of BPA complexes with fructose under the same condition. Therefore, the additional peaks in Figure 2(b) can be attributed to the fructose moieties of BPA-fructose complexes. These peaks are composed of two types of signals, which means that two types of complex exist. Complex **1** with sharp peaks (from C2 to C6) presents mainly and Complex **2** with broad peaks (from C'2 to C'5) exists in minor amounts. It is well known that furanose rings show substantially higher chemical shift values of the ring carbon atoms compared with pyranose rings. The C3, C4, C5 carbons of fructose moiety of Complex **1** have rather high chemical shifts values (84.8, 82.1 and 87.6 ppm respectively), which indicates property of five-membered ring; i.e. fructofuranose. Complex **2** also have chemical shift values of the furanose form.

In the spectrum of Sample **C** (Figure 2 (c), pH 7.4) the signals of complexes became broader, in addition, more additional peaks appeared. Although we could not assign accurately the chemical shifts of them, the existence of the C5 peak of Complex **1** suggests that the complex formation of Complex **1** occurs even in the neutral pH solution.

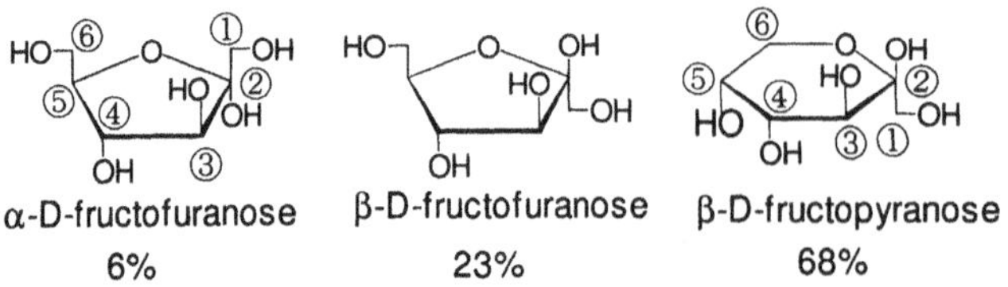

Figure 1. Structures and existence ratios (%) of the main isomers of fructose at equilibrium in D_2O at 25 °C[2].

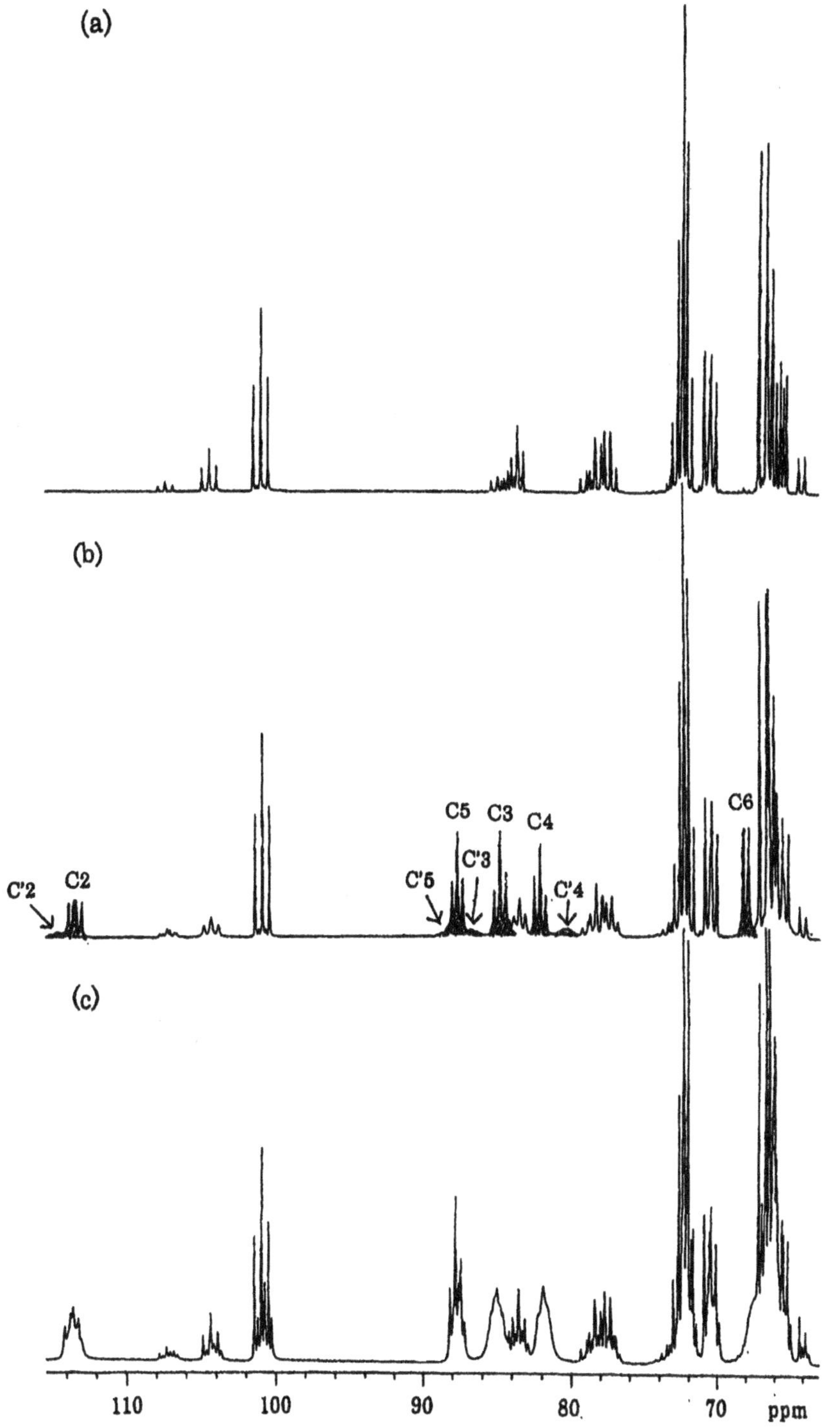

Figure 2. ^{13}C NMR spectra. (a) Sample **A:** $^{13}C_6$-enriched D-fructose D_2O solution (0.22 M), (b) Sample **B:** alkaline D_2O solution containing both BPA (0.37 M) and $^{13}C_6$-enriched D-fructose (0.15 M), (c) Sample **C:** neutral H_2O solution containing the same amounts of BPA and $^{13}C_6$-enriched fructose as sample **B**.

Table 1. $^1J_{CC}$ coupling constants (in Hz) for the fructofranoses and the fructose parts of BPA complex

	J_{C1C2}	J_{C2C3}	J_{C3C4}	J_{C4C5}	J_{C5C6}
β-D-fructofuranose	50.5	44.3	40.3[a]	38.9[a]	40.1
α-D-fructofuranose[a]	50.4	47.3	40	40	42.7
complex **1**	54.1	37.6	38.8	38	35.5

$^1J_{CC}$ Coupling Constants

The $^1J_{CC}$ values afford useful information on molecular structure and conformation. Serianni *et al.*[5] calculated effects of dihedral torsion angles on $^1J_{CC}$ coupling constants in ethylene glycol (H—O—C—C—O—H). They showed that $^1J_{CC}$ becomes minimum when C—C torsion approaches eclipsed configuration, whereas the maximal value is observed when the C—C torsion gets closer to anti configuration. Furthermore, C—O torsional angles showed the same trend, and afford greater effect on the size of $^1J_{CC}$ than the C—C torsional angles. Norrild *et al.* applied this method to determine the structures of glucose[6] and fructose[2] complexes with boronic acid. They found that $^1J_{CC}$ values become smaller for the binding sites in the complex and simultaneously larger at neighboring C—C fragments outside the 1,3-dioxolan rings. They explained the facts by decrease of C—C and C—O torsional angles of the —O—C—C—O— fragments within 1,3-dioxolan of the boronated esters. Table 1 shows the $^1J_{CC}$ values of the fructose moiety of Complex **1** together with the values of α- and β-D-fructofuranoses. The signals of Complex **2** could not give accurate $^1J_{CC}$ values because of its broadness and overlapping with other peaks. Table 1 shows that the $^1J_{C2C3}$ value of Complex **1** is smaller than the corresponding values of α and β-D-fructofuranoses, and also shows that the $^1J_{C1C2}$ value of Complex **1** is larger than the corresponding values of α- and β-D-fructofuranoses. From these two facts, it can be concluded that boric acid moiety of BPA complexes with the hydroxy groups of C2 and C3 of fructofuranose. In this case, C2—OH and C3—OH must locate on the same side of the ring, thus the type of fructose is β-D-fructofuranoses. Furthermore, Table 1 shows that the $^1J_{C5C6}$ of Complex **1** is smaller than the corresponding value of β-D-fructofuranose. This fact indicates that complex formation also involves C6 in such a way that O5—C5—C6—O6 dihedral angle decreases. From these results, it was found that Complex **1** is formed between BPA and β-D-fructofuranoses,

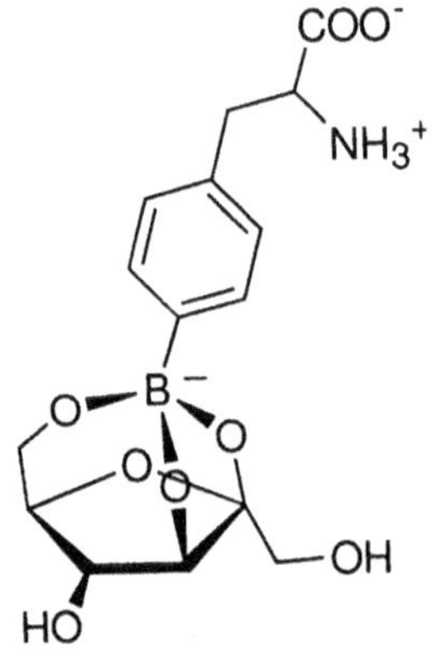

Figure 3. The structure of BPA-fructose complex present mainly in mild alkaline solution.

in which boron atom binds to the three oxygens at the sites of C2, C3 and C6 of furanose ring under mild alkaline condition as shown in Figure 3. This form is also supported by the NMR study by Bendel *et al.*[7]

CONCLUSION

The chemical shift study under neutral pH and mild alkaline conditions supported that BPA complexes with fructofuranose forms and the $^{1}J_{CC}$ coupling constant study have demonstrated that the BPA complexes with β-D-fructofuranose at it's sites of C2, C3, and C6 positions in mild alkaline solution. More investigations are required for the detailed structure in the neutral pH solution.

REFERENCES

1. Y. Mori, A. Suzuki, K. Yoshino, and H. Kakihana, Complex formation of *p*-boronophenylalanine with some monosaccharides, *Pigment Cell Research*, 2:273, 1989.
2. J.C. Norrild, and H. Eggert, Boronic acids as fructose sensors. Structure determination of the complexes involved using $^{1}J_{CC}$ coupling constants, *J. Chem. Soc., Perkin Trans.* 2:2583, 1996.
3. H.R. Snyder, A.J. Reedy, and W. Lennarz, Synthesis of aromatic boronic acids/aldehydo boronic acids and boronic acid analog of tyrosine, *J. Am. Chem. Soc.*, 80:835, 1958.
4. H. Kakihana, K. Yoshino, and Y. Mori, Relation between structures of *p*-boronophenylalanine related compounds and their usefulness for boron neutron capture therapy, *in*: "Advances in Neutron Capture Therapy," A.H. Soloway *et al.* eds., Plenum Press, New York, 1993, pp. 257–260.
5. I. Carmichael, D.M. Chipman, C.A. Podlasek, and A.S. Serianni, Torsional effects on the one-bond ^{13}C—^{13}C spin coupling constant in ethylene glycol: Insights into the behavior of $^{1}J_{CC}$ in carbohydrates, *J. Am. Chem. Soc.*, 115:10863–10870, 1993.
6. J.C. Norrild, and H. Eggert, Evidence for mono- and bisdentate boronate complexes of glucose in the furanose form. Application of $^{1}J_{C-C}$ coupling constants as a structural probe, *J. Am. Chem. Soc.*, 117:1479–1484, 1995.
7. P. Bendel, C. Anderson, and G.W. Kabalka, Structure of the BPA-fructose complex: A C-13 NMR investigation, *in*: "Frontiers in Neutron Capture Therapy," M.F. Hawthorne, K. Shelly, R.W. Wiersema, eds., Kluwer Academic/Plenum Publishers, New York, 2001, pp. 869–874.

SIMULTANEOUS DETERMINATION OF *p*-BORONOPHENYLALANINE AND BORIC ACID USING ^{11}B-NMR

Kazuo Yoshino, Yasuhito Tanaka[1], Hidetake Kakihana[2],
Masamitsu Ichihashi[3], and Yutaka Mishima[4]

[1]Department of Chemistry, Faculty of Science
Shinshu University
Asahi, Matsumoto 390-8621, Japan
[2]Energy Research Center
Wakasa Bay, Tsuruga 914-0047, Japan
[3]Department of Dermatology
Kobe University School of Medicine
Kobe 650-0017, Japan
[4]Mishima Institute for Dermatological Research
Kobe 657-0063, Japan

INTRODUCTION

We have found that *p*-Boronophenylalanine (BPA) can complex with melanin monomers and the BPA part of these complexes decompose into boric acid.[1] To evaluate these reactions, it is necessary to determine the concentrations of both BPA and boric acid in blood of melanoma-bearing patients. We have determined the concentrations of both species as follows.[2] The concentration of BPA was analyzed by the HPLC method, and the total boron concentration was obtained by the chemical method. Then, boric acid concentration was calculated by subtracting BPA concentration from the total boron concentration. These procedures are rather complicated and time-consuming. The ^{11}B-NMR chemical shift of BPA differs from that of boric acid and the signal area ratio of BPA : boric acid corresponds to the concentration ratio of these species. This fact means that there is a possibility to determine BPA and boric acid simultaneously by the NMR method. Recently, Matsunaga *et al.* have succeeded[3] in measuring the *in vivo* trace ^{11}B signal in plants. In this paper, we report on the basic study for ppm-order determination

Frontiers in Neutron Capture Therapy, edited by Hawthorne *et al.*
Kluwer Academic / Plenum Publishers, New York, 2001.

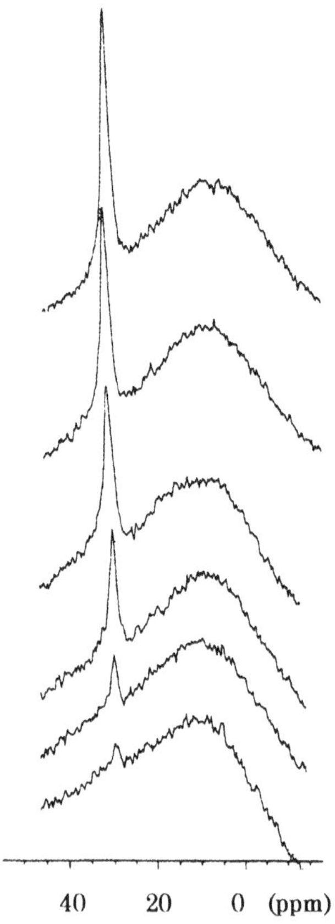

Figure 1. ^{11}B-NMR spectra of aqueous phenylboronic acid solutions. The boron concentrations are 10, 8, 6, 4, 2, 1 μg B/mL from top to bottom. The reference signal of BF_3OEt_2 at $\delta = 0.0$ is not shown.

of ^{11}B in aqueous solution of phenylboronic acid or BPA, and also an attempt to application of this method to the mixed solution of BPA and boric acid.

EXPERIMENTAL METHODS

JEOL LA400NB were used. The sample tube used was 5-mm quartz tube. ^{11}B-NMR measurements were carried out without a sample spin and field lock. BF_3OEt_2 ether solution with D_2O capillary was used for external reference. Every NMR measurements were carried out for 5 times, and the average value was used.

RESULTS AND DISCUSSION

Since the concentrations of BPA and boric acid in blood after intra-venous injection of BPA into tumor-bearing patient are very low (0–20 ppm B), it is necessary for NMR measurement to accumulate FT-NMR scanning. We have investigated the NMR measurement conditions to reduce the accumulation time. As a result, we have used the

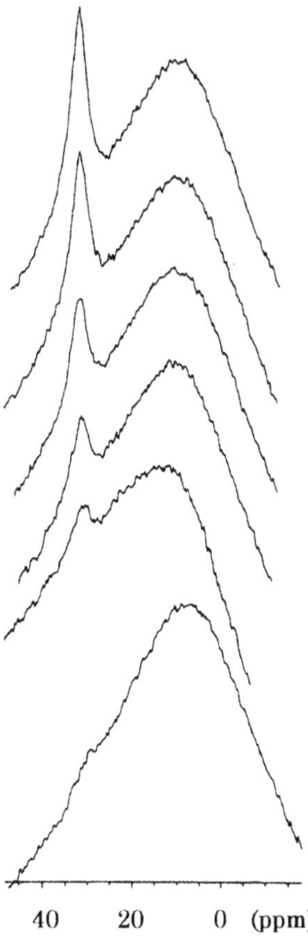

Figure 2. ^{11}B-NMR spectra of aqueous BPA solutions. The boron concentrations are 10, 8, 6, 4, 2, 1 μg B/mL from top to bottom. The reference signal of BF_3OEt_2 at $\delta = 0.0$ is not shown.

following parameters. Data points: 1,000, pw1: ca.90 degree pulse, PD:0.1 sec or 0.03 sec (minimum value), RGAIN:35(maximum value).

Figures 1 and 2 show the ^{11}B-NMR spectra of aqueous solutions of phenylboronic acid and BPA respectively. The concentration range was 1 to 10 μg B/mL. In the case of phenylboronic acid, the accumulation time for the 10,000 FT scans with 0.1 sec PD was 23 minute. In the case of BPA, since the signal is broader than that of phenylboronic acid, it requires more FT scans, which means longer accumulation time. Since the relaxation time for broad signals is very short, the pulse delay time can be reduced. Therefore, we have used 0.03 sec for PD, and increased FT scans to 100,000. Under this condition, the accumulation time was 1 hour and 50 minutes. The calibration curves were linear in both cases as shown in Figure 3, with the correlation factor of 0.9986 (phenylboronic acid) and 0.9998 (BPA) respectively.

Attempt of Simultaneous Determination of BPA and Boric Acid

The ^{11}B-NMR spectrum of the mixed solution of BPA and boric acid are shown in Figure 4. The signal of boric acid appeared at about 20 ppm which is 10 ppm higher than that of BPA. The signal areas of both signals were obtained by separating the signal area at 25 ppm. In the case of 5 μg B/mL boric acid, a calibration curve for BPA with 2–10

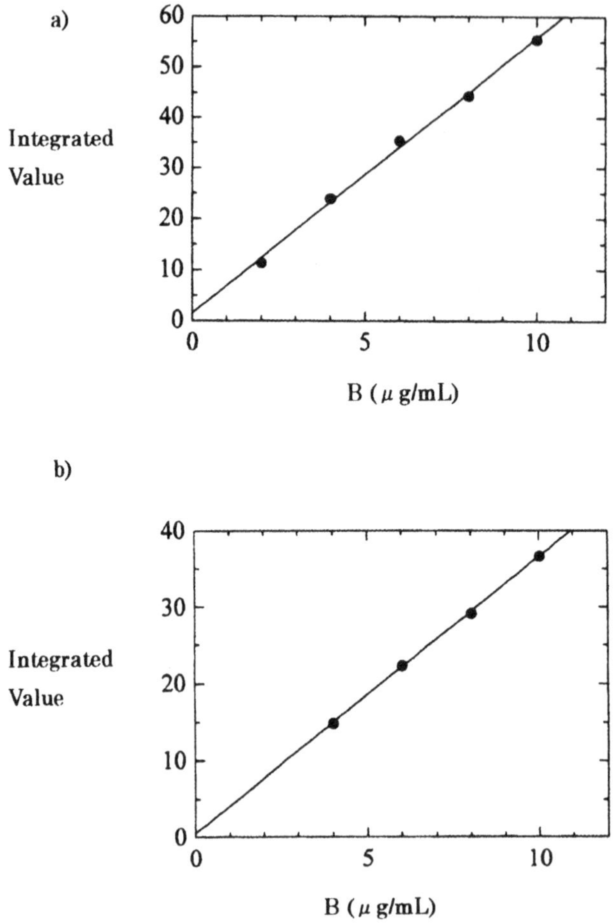

Figure 3. Calibration curves for phenylboronic acid (a) and BPA (b).

μgB/mL is shown in Figure 5. The extrapolating of the calibration line to zero boron became minus value. This is due to the overlapping of BPA signal and boric acid signal. If the signal is resolved into each signals completely, the line will pass the zero point. If NMR with more strong magnetic strength (ca. 500 MHz or 600 MHz) can be used, there seems to be no overlapping signals, and also FT-scan might be reduced. This implies that the analyzing time will be shorter.

CONCLUSIONS

The BPA concentrations in aqueous solution with 2–10 μgB/mL can be measured by ^{11}B-NMR within two hours. The method also has the potential for simultaneous determination of BPA and boric acid in blood of BNCT patients.

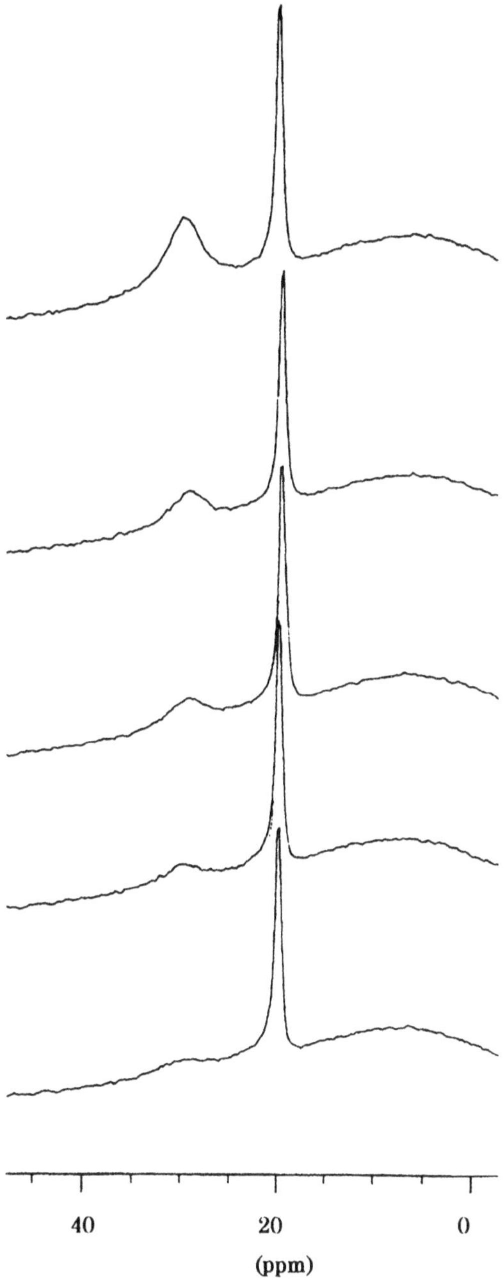

Figure 4. ^{11}B-NMR spectra of a mixed solution of BPA and boric acid. The boron concentration of boric acid is 5 μg/mL for all samples, and the boron concentration of BPA are 10, 5, 4, 3, 2 μg/mL from top to bottom.

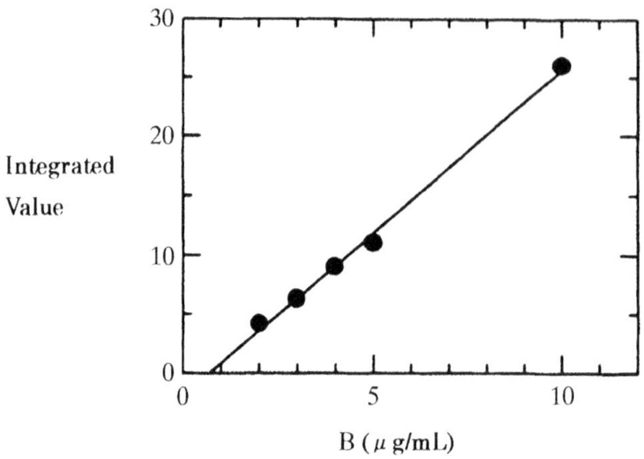

Figure 5. Calibration curve for aqueous BPA solution containing 5μg B/mL boric acid.

REFERENCES

1. K. Yoshino, Y. Mishima, M. Kimura, J. Hiratsuka, Y. Mori, S. Ito, and H. Kakihana, Capture of p-boronophenylalanine in malignant melanoma cells by complex formation with melanin monomers, DOPA, DHI and DHICA, *in* "Advances in Neutron Capture Therapy, Volume II, Chemistry and Biology", B. Larsson *et al.* eds., Elsevier Science B.V., Amsterdam, 1997, pp. 234–238.
2. K. Yoshino, R. Nakajima, A. Takaoka, Y. Mori, H. Kakihana, Y. Mishima, and M. Ichihashi, Determination of released boric acid *in vivo* after p-boronophenylalanine administration into melanoma-bearing subjects, *in* "Advances in Neutron Capture Therapy, Volume II, Chemistry and Biology", B. Larsson *et al.* eds., Elsevier Science B.V., Amsterdam, 1997, pp. 334–338.
3. T. Matsunaga, and T. Nagata, *In vivo* ^{11}B NMR observation of plant tissue. *Anal. Sci.* 11:889–892, 1995.

STRUCTURE OF THE BPA-FRUCTOSE COMPLEX

A C-13 NMR Investigation

P. Bendel[1], C. Anderson[2], and G. W. Kabalka[3]

[1]Department of Chemical Services, MR Center
Weizmann Institute of Science
Rehovot 76100, Israel
[2]Department of Chemistry
University of Tennessee
Knoxville, Tennessee, 37996

1. INTRODUCTION

The fructose complex of Boronophenylalanine (BPA) is commonly used for *in vivo* administration, due to its improved water solubility.[1] Until now, however, the detailed molecular structure of the BPA-fructose complex was not yet reported, although the structures of complexes between fructose and *p*-tolylboronic acid were recently published.[2] In this contribution, we present the results of ^{13}C NMR studies of aqueous solutions of BPA and fructose mixtures at varying concentration ratios. Our approach was to directly identify the fructose isomer interacting with BPA, using 2D exchange spectroscopy,[3] and propose the structure which most likely results from this interaction. Kitazawa and Yoshino[4] independently arrived at an identical structure by analyzing the $^{1}J_{CC}$ coupling constants in the NMR spectrum of the complex.

2. MATERIALS AND METHODS

^{10}B-enriched L-p-Boronophenylalanine (Boron Biologicals, Raleigh, USA), and D(-)Fructose (BDH, Poole, England) were used without further purification. Mixtures containing varying proportions of BPA and fructose were prepared by two approaches: For the first approach, aimed at achieving solutions with relatively high proportion of free BPA (out of the total BPA pool), BPA was dissolved directly in D_2O to saturation levels. The filtered solutions had pH values of about 7.1. Small amounts of a fructose stock solution (0.32 M) were added, and the pH of the solutions was titrated back to 7.0 with 0.2 N KOH. The final solutions contained about 3 mM total BPA, and comparable

Frontiers in Neutron Capture Therapy, edited by Hawthorne *et al.*
Kluwer Academic / Plenum Publishers, New York, 2001.

concentrations of total fructose. These solutions were used to verify the spectral assignments of free vs. complexed BPA, and to estimate the equilibrium ratios of the different mixture components under conditions of a relatively high free/complex BPA ratio. In the second approach, aimed at obtaining the concentrated solutions needed for ^{13}C NMR spectroscopy with sufficient signal-to-noise (S/N) ratio, BPA was added directly to the fructose stock solution, the pH was transiently raised to 9.5–10, and then lowered back to neutrality, after clear solutions were obtained. The reported pH values reflect directly the readings on the pH-meter, without correction for the measurement in D_2O solution.

^{13}C NMR spectra (1 D) were obtained from 10 mm samples, on a 9.4 T-magnet spectrometer (Bruker, Karlsruhe, Germany), using a standard "pulse-acquire" pulse sequence with continuous ^1H decoupling (at different power levels between and during accumulation of individual FID's). Pulses with 70° flip angle were applied every 3.8 sec., and between 800 and 20,000 scans (for the concentrated and dilute solutions, respectively) were signal-averaged. Chemical shifts are reported relative to Trimethylsilyl-propionic acid (TSP) as internal calibration standard at 0.0 ppm. Two-dimensional ^{13}C spectra were obtained using a 2 D phase-sensitive NOESY pulse sequence,[5] with bi-level proton decoupling, as described above for the 1 D spectroscopy. A mixing time of 300 ms, relaxation delay of 3.7 sec, 2 K increments of the evolution time, and 16 scans per increment were used.

The quantitative determination of absolute and relative solute concentrations was based on the integral measurement of assigned signals (using the standard manufacturer's deconvolution software). Absolute concentrations were based on the known weighed amounts and dilution factors for fructose, assuming that all the fructose, whether free or complexed, was fully dissolved and NMR detectable. The effect of partial saturation on the measured signal intensities was checked by acquiring (on one of the concentrated samples) spectra at three different repetition times, of 2.8 sec., 3.8 sec., and 5.3 sec, using 70° flip angle excitation pulses. The relative integrated peak areas were measured at each repetition time. It was found that the signal intensities for all the non-protonated carbon resonances (anomeric C2 sugar peaks, carbonyl and quaternary carbon BPA signals) increased with increasing repetition time, while the relative intensities of the protonated

Table 1. ^{13}C chemical shifts (ppm) for free BPA and the BPA-fructose complex, 30 °C, pH = 7.6

Carbon	BPA	BPA in complex	Fructose in complex
Carbonyl	176.7	177.1	
α	58.8	59.0	
β	39.3	39.1	
Aromatic ring C1	140.7	135.8	
Aromatic ring C2, C6	131.8	131.0	
Aromatic ring C3, C5	137.1	135.0	
Aromatic ring C4		147.6	
C1			66.3
C2			113.7
C3			85.2
C4			81.8
C5			87.8
C6			67.4

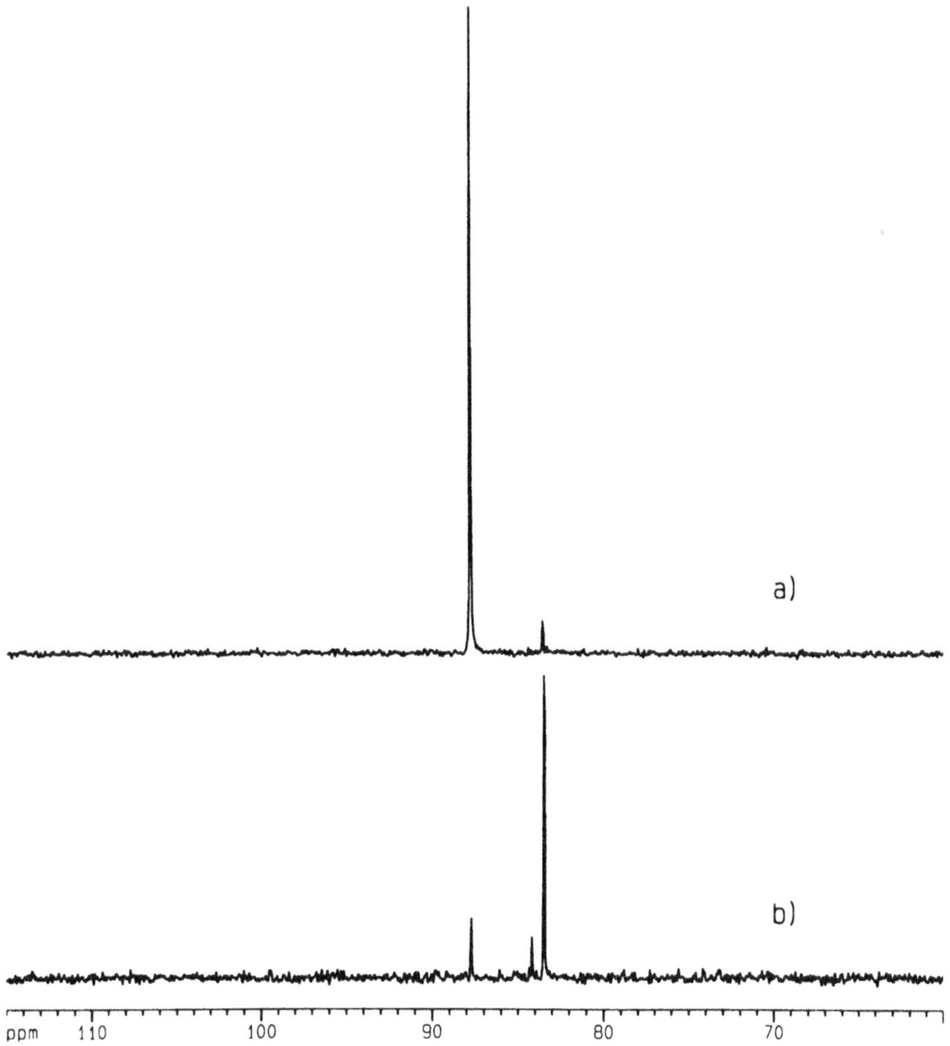

Figure 1. (a) Cross-section at the level of the C5 resonance of fructose in the complex (87.8 ppm), showing that the only connection is to the C5 resonance of β-furanose (83.5 ppm). (b) Cross-section at the level of the β-furanose C5 resonance, showing cross-peaks to the C5 resonances of fructose in the complex, and to the C5 resonance in α-furanose (84.2 ppm).

(CH— and CH_2— moieties) remained practically unchanged. We therefore concluded that, even if the signal intensities of the protonated carbons are still partially saturated for a 3.8 sec repetition time, the extent of this saturation is identical for each of the molecular species in the mixture. Therefore, the measured signal intensities should correctly reflect the solution concentrations. The relative solution concentrations were therefore based on the integrals of the assigned protonated carbon signals. For each molecule, we took the average of 2–4 peaks, depending on the number of signals that were sufficiently well resolved to permit reliable area measurement.

3. RESULTS AND DISCUSSION

3.1. Spectral Assignments

The ^{13}C chemical shifts of the free sugar isomers were identical to those reported by Norrild and Eggert.[2] The S/N in our spectra (at natural ^{13}C abundance) was not high enough to detect the minor isomers α-pyranose (≈2%), and the linear ketose (≈0.7%), but all peaks of the major isomers (β-pyranose, β-furanose, and α-furanose) were identified. The chemical shifts of the peaks belonging to the fructose moiety in the complex are listed in Table 1, along with the assignments for BPA, and BPA in the complex.

3.2. 2D Exchange Spectroscopy

Using the basic NOESY pulse sequence will generate cross-peaks at the coordinates of frequencies which are in some kind of dynamic exchange, if the exchange rate is slow compared to the difference in resonance frequencies (otherwise only a single, averaged peak would be observed), but fast (or comparable to) the spin-lattice relaxation rate $1/T_1$. In principle, cross-peaks will also appear between the frequencies of nuclei experiencing coupled dipolar relaxation (NOE), but this effect is completely negligible for ^{13}C at natural abundance. In all our 2D experiments, the only free fructose peaks showing connections with the corresponding sugar carbon peaks in the complex, were those belonging to the β-furanose isomer. This clearly indicates that BPA interacts predominantly with the β-furanose form of fructose for the formation of the BPA-fructose complex. In addition, cross-peaks between α- and β-furanose, and between free and complexed BPA, were detected. Figure 1 displays some representative 1D cross-sections through the 2D spectrum.

3.3. The BPA-fructose Complex Structure

A plausible structure which could be formed between BPA and β-fructofuranose is shown in Figure 2. It is similar to the predominant structure previously proposed for the complex between fructose and p-tolylboronic acid in aqueous solution.[2]

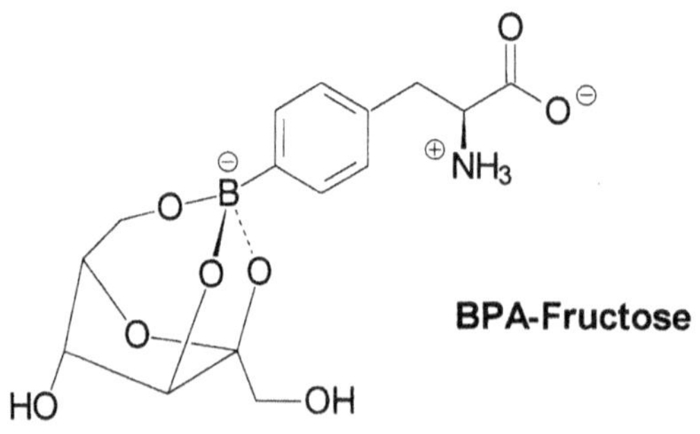

Figure 2. Structure of most abundant form of BPA-fructose complex.

3.4. Possible Dynamic Equilibrium between Different Complex Structures

Norrild and Eggert[2] found that 82% of the fructose- p-tolylboronic acid complex in aqueous solution (pD = 11, 1:1 monomer ratio) was in the tridentate form (as in Figure 2), while 18% (13% "exo-" + 5% "endo-isomers) were in a bidentate configuration, in which the C6 fructose carbon is disconnected from the boronic acid group. Our present results indicate that the fructose-BPA complex is probably similarly distributed between different structures, even though we only observe a single peak for each carbon resonance of the fructose in the complex. We base this conclusion on the following observations: a) the signals of the fructose carbons in the complex are all significantly broader than those of the free sugar carbons, and the line-widths of these broadened peaks decrease with increasing temperature; b) the line-widths (from 9 Hz for C5 to 66 Hz for C6) correlate with the differences in chemical shift between the tridentate and bidentate complexes, as reported by Norrild and Eggert[2] (from 0.69 ppm for C5 to 3.01 ppm for C6); c) the chemical shifts of the fructose-complex carbons in our spectra are all in-between the chemical shifts of the bi- and tridentate forms of the fructose- p-tolylboronic acid.[2] All these results strongly indicate that the signals observed in our spectra represent a dynamic equilibrium (in the fast-exchange limit) between at least two conformations of the complex. If we accept the chemical shifts of the different conformations from ref. 2, then the chemical shifts in our spectra can be used to calculate the proportion of the tridentate complex of BPA-fructose. The average result for the signals of C1, C2, C3, C4, and C6 is 0.81 ± 0.03. The result for C5 (0.71) slightly deviates from this average, but can be considered less reliable, due to the small chemical shift difference between the isomers.

3.5. Equilibrium Ratios of Constituents in BPA-Fructose Mixtures

We measured the relative proportions of free BPA, BPA in complex, the three main free fructose isomers, and fructose in complex in various BPA/fructose mixtures, in which the proportion of free BPA ranged from 4% to 61% (of the total BPA pool). In all these mixtures the distribution of the remaining free sugar was unchanged, and within experimental error of the values reported in the literature for aqueous fructose solutions.[6] In all the mixtures the ratio BPA : sugar in the complex was 1 : 1 (within experimental error). The apparent formation constant for the complex, from BPA and β-fructofuranose also appeared to be independent on the ratios at which BPA and fructose were mixed ($K_f = 6.8 \pm 0.5 \times 10^2 M^{-1}$, at pH = 7.0–7.6, 30°C).

ACKNOWLEDGMENTS

This research was supported by the U.S. Department of Energy, the Robert H. Cole Foundation, and the Israel Science Foundation founded by the Israel Academy of Sciences and Humanities. We also wish to thank M. Kitazawa and K. Yoshino from the Shinshu University, Matsumoto, Japan, for helpful discussions.

REFERENCES

1. Y. Mori, A. Suzuki, K. Yoshino, and H. Kakihana, Complex formation of p-boronophenylalanine with some monosaccharides, *Pigm. Cell Res.*, 2:273–277, 1989.

2. J. Chr. Norrild, and H. Eggert, Boronic acids as fructose sensors. Structure determination of the complexes involved using ${}^1J_{CC}$ coupling constants, *J. Chem. Soc., Perkin Trans.* 2:2583–2588, 1996.
3. J. Jeener, B.H. Meier, P. Bachmann, and R.R. Ernst, Investigation of exchange processes by two-dimensional NMR spectroscopy, *J. Chem. Phys.* 71:4546–4553, 1979.
4. M. Kitazawa, and K. Yoshino, Structure determination of *p*-boronophenylalanine fructose complex using ${}^1J_{CC}$ coupling constants, *in* "Frontiers in Neutron Capture Therapy," M.F. Hawthorne, K. Shelly, R.W. Wiersema, eds., Kluwer Academic/Plenum Publishers, New York, 2001, pp. 857–861.
5. D.J. States, R.A. Haberkorn, and D.J. Ruben, A two-dimensional nuclear overhauser experiment with pure absorption phase in four quadrants, *J. Magn. Reson.* 48:286, 1982.
6. W.J. Goux, Complex isomerization of ketoses: a ${}^{13}C$ NMR study of the base-catalyzed ring-opening and ring-closing rates of D-fructose isomers in aqueous solution, *J. Am. Chem. Soc.* 107:4320–4327, 1985.

PROTON MR SPECTROSCOPIC MEASUREMENT OF *p*-BORONOPHENYLALANINE FOR *IN VIVO* PHARMACOKINETIC STUDIES

A Preliminary Study

C. S. Zuo[1], P. V. Prasad[1], L. Tang[1], N. Adnani[2], and R. G. Zamenhof[1]

[1]Department of Radiology, Section of Radiological Physics
Beth Israel Deaconess Medical Center & Harvard Medical School
Boston, MA, U.S.A.
[2]Medical Physics Unit, Montreal General Hospital
Montreal, Canada

1. INTRODUCTION

The success of BNCT treatment is predicated on sufficient amounts of B-10 label being incorporated into tumor (or its afferent vasculature) concomitantly with lower amounts being incorporated into normal tissues and blood. With the proper fluence of neutrons delivered to the target tissues the radiation dose to tumor cells should be adequate to cause cytotoxicity while that to normal tissues should be within their tolerance range. To optimize the delivery of BNCT it is necessary to know the concentration of B-10 (or its carrier molecule) during and following its intravenous infusion.[1,2] Currently our protocol involves periodic central venous blood sampling followed by *ex-vivo* assay of the B-10 concentration by prompt-gamma neutron activation analysis.[3] This method, although highly accurate, does not provide corresponding B-10 concentration kinetics in normal brain and brain tumor. The use of NMR as a potential non-invasive technique for studying the pharmacokinetics of boronated delivery agents in BNCT has previously been proposed.[4-7] The studies cited have attempted to directly measure the concentrations of boron in biological tissues. Although naturally occuring boron is isotopically composed of approximately 20% B-10 and 80% B-11, the boron that is employed for actual BNCT therapy is typically 98–99% isotopically enriched in B-10. B-10 has a nuclear magnetic spin of 3 and a gyromagnetic ratio of 4.6 MHz/T. Its large quadrupole moment produces such large line broadening that B-10 has to be detected by non-conventional

Frontiers in Neutron Capture Therapy, edited by Hawthorne *et al.*
Kluwer Academic / Plenum Publishers, New York, 2001.

NMR techniques.[4] The proton-boron double resonance measurement requires dual RF channel operation, sophisticated acquisition scheme and parameter control as well as potentially high RF power deposition. These factors hinder the potential application of the double resonance technique to be used clinically. B-11, on the other hand, has a spin of 3/2 and a gyromagnetic ratio of 13.7 MHz/T. Although the line-broadening phenomenon is still problematic for the satellite transitions between $\pm 3/2 \leftrightarrow \pm 1/2$, the measurement of the boronated delivery agents BSH and BPA (containing naturally occurring boron) has been successfully demonstrated at the upper end of the typical range of physiological concentrations by observing the central transition $-1/2 \leftrightarrow +1/2$. If the measurement of boron as reported were to be made using the same delivery agents that are actually employed for BNCT therapy (i.e., 98–99% isotopically enriched in B-10) the method would have insufficient sensitivity to be useful. Rather than attempting to detect boron, we have explored the possibility of employing *proton* NMR to detect the protons associated with the B-10 carrier molecule BPA; in our case the neutral amino acid phenylalanine.

2. MATERIALS AND METHODS

2.1. Materials Used

(i) BPA-D_2O Phantom. B-10-enriched BPA (Boron Biologicals, Inc., Raleigh, NC, USA) was dissolved in deuterium oxide (D_2O) at concentrations of 0.3, 0.8, 2.3, 3.1, 4.7, & 9.3 mM, corresponding to 3–93 µg B-10/ml. When complete dissolution of the BPA proved difficult to achieve, Na_2CO_3 was added to adjust the pH to approximately 9 and thereby promote full dissolution.

(ii) Brain Metabolite Phantom. 12 mmole each of NAA, creatine, lactic acid, myoinositol, and 1.6 mmole of BPA were dissolved in 1,000 ml H_2O. The concentration of NAA was consistent with reported typical values while those of lactic acid and myoinositol are probably somewhat higher.[10]

(iii) Phenylalanine and Tyrosine Phantoms. Phenylalanine is anticipated to be present if complete boronation of the BPA fails to be achieved during manufacture, while tyrosine is anticipated to be present if the B-10 is removed from the aromatic ring of BPA (by metabolism or by neutron capture by B-10) leaving the aromatic ring intact. In order to examine these background components, 3 mg/ml each of phenylalanine & tyrosine (HPLC grade, Sigma Chemical Co., St. Louis, USA) were dissolved in 5% acetic acid.

(iv) Homogeneized Rat Brain/BPA Phantom. 350 mg/kg of BPA-fructose (the sugar with which BPA administered to live animals and humans is always complexed[9]) was administered by tail-vein injection to a Sprague-Dawley rat. One hour later the rat was killed and the brain removed and homogeneised. Independent measurements by PGNAA determined a BPA concentration in the homogeneized brain of 1.2 mM of BPA (12 µg/ml of B-10).

(v) Human Subject. Informed consent was obtained from a human subject with intracranial melanoma metastases who was undergoing a clinical BNCT protocol through the Harvard-MIT BNCT program. The MRI study was conducted 8 hours following the intravenous administration over 90 minutes of 350 mg/kg of BPA-fructose. At

that time the blood B-10 concentration (measured by PGNAA) was 10µg/ml (1.0mM BPA), and it is estimated that the corresponding tumor B-10 concentration was approximately 30–40µg/ml (3–4nM BPA).[8]

2.2. NMR Measurements of Phantoms

Proton NMR spectra were acquired using a 360MHz vertical bore spectrometer (Bruker Instruments, Billerica, MA, U.S.A.) and a 5mm probe. Data were collected from approximately 0.3ml aliquots of each sample contained in a 5mm NMR tube. Water suppression was performed using a 400ms long radio frequency (RF) pulse centered at water resonance frequency prior to a 3µs (approx. 15°) pulse for free induction decay (FID) collections. Other acquisition parameters were averages: 32; repetition time (TR): 4s; spectral size: 1024; spectral width: 3.6kHz. Data processing included apodization (6Hz Gaussian broadening), zero filling, Fourier transformation, and manual zero-order phasing.

2.3. NMR Measurement of Human Subject

The human NMR spectrum was acquired using a water pre-saturation, single-voxel stimulated echo sequence with TE of 30ms, voxel size $2 \times 1.7 \times 1.6 cm^3$, TR 1.5sec, and 128 signal averages for a total acquisition time of 3min 16sec on a 1.5 Tesla whole body MR scanner (Siemens medical system, Erlangen, Germany).

3. RESULTS

The chemical form of p-boronophenylalanine is shown in Figure 1. The four hydrogen atoms which have been identified as producing the BPA NMR signature reported here are labeled H_a and H_b.

Figure 2 shows the NMR spectrum of BPA in D_2O, which was used as the solvent for the aqueous phantoms in order to minimize the obscuration of any weak NMR signals by the presence of the very strong H_2O signal at 4.7ppm. The aromatic protons, H_a and H_b, produce resonances at 7.1 and 7.4ppm, as well as at 4.1 and 2.7 and at 2 and 3ppm on the other side of the water resonance. Additionally, the -CH- protons resonate at 3.4ppm and the benzylic $-CH_2-$ protons at 2.9 and 2.7ppm. No J-coupling splitting

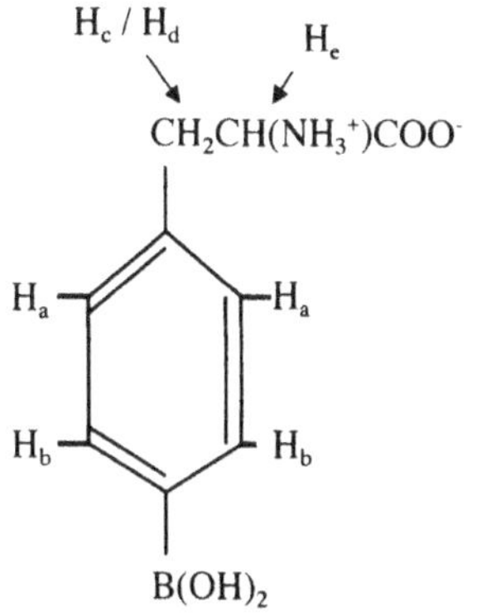

Figure 1. Molecular structure of p-boronophenylalanine (BPA). The molecule contains two types of magnetic-equivalent aromatic protons, H_a & H_b leading to two distinct resonances in the BPA NMR spectrum.

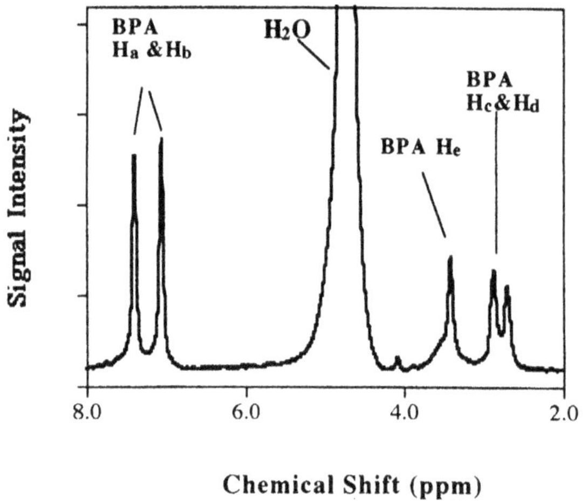

Figure 2. A proton NMR spectrum of 3.1 mM of BPA in D_2O solution. Residual of water resonance appears at 4.7 ppm, aromatic proton resonances are 7.1 & 7.4 ppm, —CH- at 3.4 ppm and —CH_2- at 2.9 & 2.7 ppm. A weak signal at 4.1 ppm may be related to NH_2.

was observed between the two aromatic protons. NMR spectra also were acquired of fructose, phenylalanine, and tyrosine in D_2O to identify any potential interferences with the BPA proton resonances; however, only the conclusions will be stated. The proton spectrum of BPA-fructose complex showed BPA proton resonance peaks at 7.1–7.4 ppm and at 2–3 ppm without interference from fructose, and at 2.7 and 4.1 ppm in superposition with fructose proton resonances. The phenylalanine and tyrosine NMR spectrum showed aromatic doublet peaks downfield from the water resonance, but the chemical shifts were slightly different from those of BPA: 6.7 and 7.0 ppm compared to 7.1 and 7.4 ppm for BPA; probably due to more electron shielding from —OH- compared to —$B(OH)_2$- in the BPA. Also the —CH- resonance shifted from 3.4 to 3.9 ppm and the —CH_2- resonances shifted from 2.9 and 2.7 ppm to 2.9 and 3.0 ppm respectively. This suggests that proton NMR can distinguish BPA from phenylalanine and tyrosine if they have comparable concentrations. The proton spectrum of the metabolite phantom showed the aromatic doublet due to BPA still to be clearly visible at 7.1–7.4 ppm. Although NMR resonance signals from the brain metabolites (most likely from the —NH- groups) were seen at 6.6 and 8.0 ppm there was no overlap with the aromatic

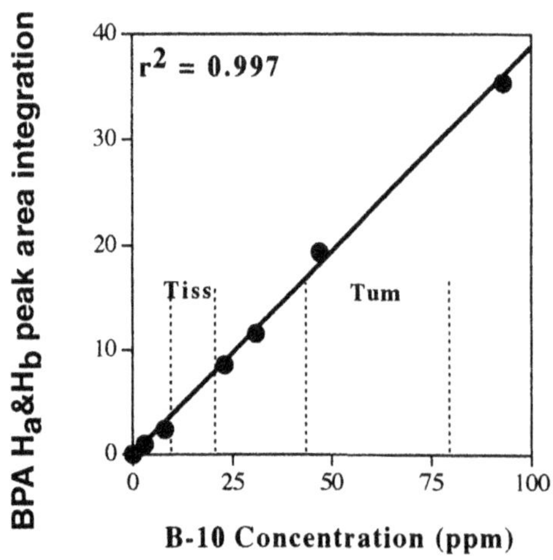

Figure 3. Integration of the peak areas of the BPA aromatic doublet H_a & H_b (7.1 & 7.4 ppm) *vs* B-10 concentration measured by PGNAA.

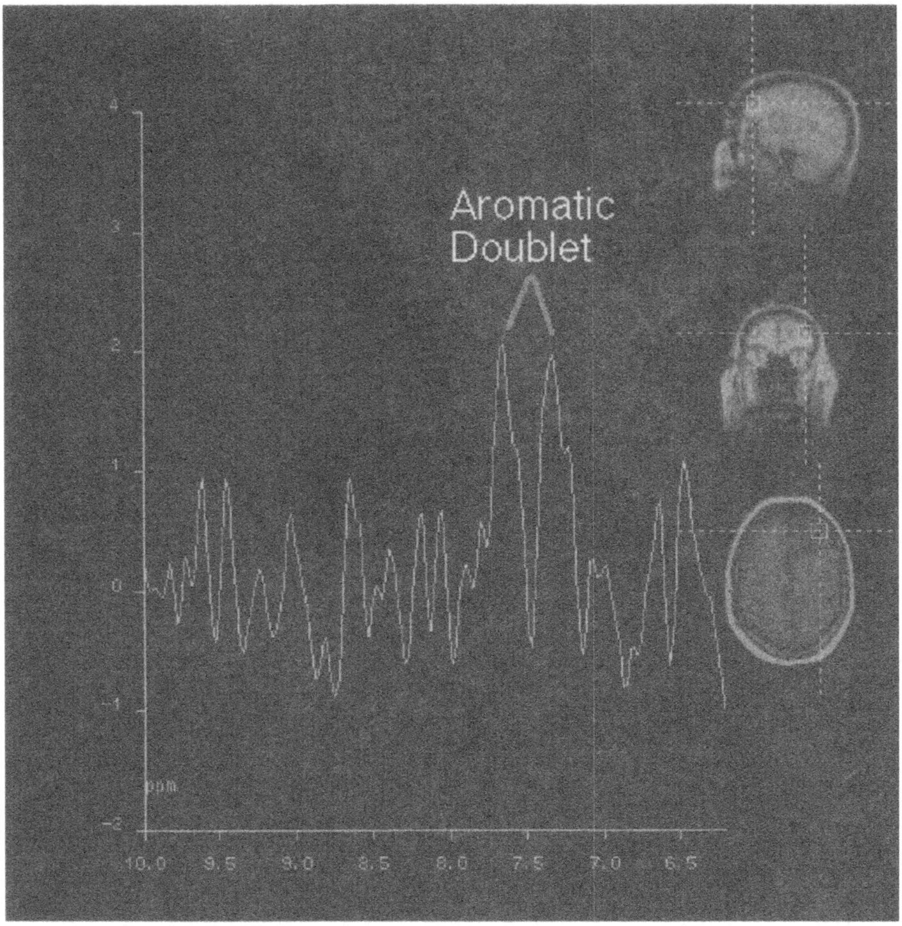

Figure 4. *In vivo* proton NMR spectrum of human metastatic melanoma in the brain 8 hours following IV administration of 350 mg/kg of BPA-fructose. The aromatic proton doublet is clearly seen at 7.1 & 7.4 ppm.

proton resonances from BPA. However, the other part of characteristic BPA spectral signature, between 4.1 and 2.7 ppm, was totally obscured by signals from the brain metabolites. A proton NMR spectrum acquired from a homogeneized rat brain containing BPA-fructose showed consistent features vis-a-vis interferences from brain metabolites, fructose, and other components of brain tissue. Nevertheless, the 7.1–7.4 ppm spectral range remained free of significant interferences and clearly showed the characteristic aromatic proton doublet. To summarize the results of the "interference" experiments, it would appear that the aromatic proton doublet at 7.1–7.4 ppm still remains largely clean in the presence of fructose, phenylalanine, tyrosine, and typical concentrations of brain metabolites. Figure 3 shows the results of the incremental BPA-fructose concentration experiment. The integration of the aromatic doublet demonstrated an excellent correlation ($r^2 = 0.997$) with the BPA concentrations independently measured by PGNAA in the range of 0.3 to 9.3 mM, corresponding to B-10 concentrations of 3 to 93 μg/ml. Typical BPA concentrations in blood and tissues under clinical conditions are 5–100 μg/ml B-10.

Figure 4 shows the proton NMR spectrum obtained from a human patient in a 1.5T clinical NMR scanner. The figure shows a 15 × 15 × 15mm MRS region-of-interest placed over the left frontal melanoma metastasis in the patient's brain in three orthogonal views.

The NMR spectrum clearly shows the aromatic doublet at 7.1–7.4ppm, albeit in the presence of a high degree of noise. Nevertheless, with only a 3-minute acquisition time and a tissue sampling volume of $3.4 cm^3$ the result is quite encouraging.

DISCUSSION AND CONCLUSIONS

For *in-vivo* NMR detection of exogenously administered boronated delivery agents it is desirable that the signals be sufficiently strong for precise detection and not be obscured by signals from other endogenous or exogenous compounds. Although we have identified several characteristic resonances in the proton spectrum of BPA in the range of 2.7–7.4ppm, the aromatic doublet at 7.1 & 7.4ppm has been specifically identified as being well suited for NMR quantitation of BPA *in-vivo* due to a lack of interference from endogenous and exogenous sources, as well as immunity from interference by the water suppression process. The other BPA resonance signals, between 2.0 and 4.1ppm, are essentially obscured by resonance signals from fructose and brain metabolites. The location of this doublet structure well away from the majority of the NMR resonance signals makes it amenable to selective excitation, which would be of particular significance if spatial mapping of BPA were to be implemented by the proposed method. The much higher gyromagnetic ratio of H-1 than either B-10 or B-11 greatly increases the BPA spectral signal intensity and hence its NMR detectability *in-vivo*. Moreover, as mentioned previously, B-10 and B-11 are not readily amenable to NMR measurement in the context of BNCT since in clinical BNCT >98% of boron is enriched in B-10 (with a very low gyromagnetic ratio) while B-11, despite its higher gyromagnetic ratio, is only present in physiologically minute concentrations as the result of its de-enrichment. A preliminary BPA NMR measurement conducted on a human subject using a 1.5T clinical scanner confirms that it is readily feasible to observe the aromatic proton doublet with an acquisition time of only 3 minutes and a tissue volume of $3.4 cm^3$ containing approximately 30–40μg/ml B-10 (3–4mM BPA). Our preliminary results therefore indicate that proton NMR could be a non-invasive and potentially useful technique for BPA quantification in BNCT.

ACKNOWLEDGMENTS

CZ was supported in part by the Whitaker Foundation (G-96-0047), PVP by a grant-in-aid from the American Heart Association and NIH (RO1 DK53221), and RGZ by a grant from the U.S. DOE (DE-FG02-96ER62193).

REFERENCES

1. D.E. Wazer, R.G. Zamenhof, and O.K. Harling *et al.*, "Boron Neutron Capture Therapy," in *Radiation Oncology: Technology and Biology*, edited by P.M. Mauch, and J.S. Loeffler (W.B. Saunders Company, Philadelphia, 1994).

2. R.F. Barth, A.H. Soloway, and R.G. Fairchild *et al.*, "Boron Neutron Capture Therapy for Cancer," Cancer **70**, 2995–3007 (1992).
3. Kent J. Riley, and Otto K. Harling, "Boron-10 Analyses Using Prompt Gamma Neutron Activation Analysis and ICP-AES," Trans Am Nuc Soc **75**, 31–2 (1996).
4. P. Bendel, J. Zilberstein, and Y. Salomon, "In vivo detection of a boron-neutron-capture agent in melanoma by proton observed 1H-10B double resonance.," Magn Reson Med **32**(2), 170–174 (1994).
5. P. Bendel, A. Frantz, and J. Zilberstein *et al.*, "Boron-11 NMR of borocaptate: relaxation and *in vivo* detection in melanoma-bearing mice.," Magn Reson Med **39**(3), 439–447 (1998).
6. K.M. Bradshaw, J.R. Hadley, and H. Buswell *et al.*, "B11 MRI studies of BSH pharmacokinetics in high-grade human brain tumors," in *Advances in neutron capture therapy Volume II, chemistry and biology, Proceedings of the seventh International symposium on neutron capture therapy for cancer*, edited by B. Larsson, J. Crawford, and R. Weinreich (Elsevier, Zurich, Switzerland, 1996), Vol. 2, pp. 261–266.
7. P.P. Tang, M.P. Schweizer, and K.M. Bradshaw *et al.*, "11B nuclear magnetic resonance studies of the interaction of borocaptate sodium with serum albumin," Biochem Pharmacol **49**(5), 625–632 (1995).
8. J.A. Coderre, A.D. Chanana, and D.D. Joel *et al.*, "Biodistribution of boronophenylalanine in patients with glioblastoma multiforme: boron concentration correlates with tumor cellularity.," Radiat Res **149**(2), 163–170 (1998).
9. Y. Mori, A. Suzuki, and K. Yoshino *et al.*, "Complex formation of p-boronophenylalanine with some monosaccharides.," Pigment Cell Res **2**(4), 273–277 (1989).
10. E.B. Cady, "Quantitative combined phosphorus and proton PRESS of the brains of newborn human infants," Magn Reson Med **33**(4), 557–563 (1995).

SUBCELLULAR LOCALIZATION OF BSH-CONTAINING AND NON-BORON-CONTAINING CHLORIN DERIVATIVES IN LIVING CELLS USING CONFOCAL LASER SCANNING MICROSCOPY

Stefanie Steen-Ecke, Franz-Peter Montforts, and Detlef Gabel

Department of Chemistry
University of Bremen, Germany

1. INTRODUCTION

Chlorins are of great interest for binary cancer treatment forms such as boron neutron capture therapy (BNCT) and photodynamic therapy (PDT) due to their physical and chemical properties. In the latter therapy these compounds serve as photosensitizers whereas in BNCT they are used as boron carriers. The highly selective accumulation of these substances in tumor tissue is important for both forms of therapy. The cellular uptake and intracellular localization of these molecules is dependent upon their structure of pendant side chains. To achieve better cellular uptake and tumor enrichment structurally different chlorins were synthesized, among them BSH-containing chlorins.

We investigated the cellular uptake and subcellular localization of these compounds *in vitro* by using confocal laser scanning microscopy (CLSM). We used two different cell lines to compare uptake and intracellular localization: C6 glioma cells and V79 hamster cells. The differences in cellular uptake and subcellular localization in relation to the structure of the chlorins will be discussed.

2. MATERIALS AND METHODS

The cells were subcultured twice a week in medium containing fetal calf serum (FCS). For subcellular localization 10^3 cells were seeded into a $4.2\,cm^2$ chambered coverglass containing 2 ml medium and left to incubate at 37 °C in a 5% CO_2-atmosphere for 12 h to ensure proper attachment to the coverglass. Chlorins 1, 2 and 3 were synthesized

Frontiers in Neutron Capture Therapy, edited by Hawthorne *et al.*
Kluwer Academic / Plenum Publishers, New York, 2001.

by S. Ruf.[1] Chlorins 4 and 5 were provided by F.-P. Montforts.[2] All chlorins were dissolved in DMSO/PBS-buffer and diluted in medium resulting in final concentrations of 30μM to 100μM. The chlorins were added to the cells in 2ml volumes. The incubation periods varied from 30min up to 24h. For comparison the following fluorescent markers were used to stain certain cellular compartments: acridine orange (AO) for acidic organelles, Neutral Red for lysosomes, Rhodamine-123 for mitochondria and 3,3′-dihexyloxacarbocyanine iodide ($DiOC_6$) for endoplasmic reticulum. After incubation cells were washed with PBS-buffer and immediatly viewed with a Zeiss Axioskop LSM 410 invert. The samples were excited with an argon-ion-laser with excitation wavelengths either at 488nm or 514nm.

3. RESULTS

The structures of the chlorins are shown in Figure 1. All chlorins were taken up by both cell lines within 30min. The fluorescence of the chlorins became stronger with increasing incubation time until 16h of incubation. Between 16h and 24h of incubation no differences were observed neither in distribution nor in the strength of fluorescence. No chlorin was found to be located inside a nucleus. Differences of distribution between C6 cells and V79 cells were observed.

Chlorins 1, 2, 3 and 4, all doubly negatively charged, showed similar distribution patterns in V79 cells. These chlorins show an inhomogenous distribution within the cytoplasm. A possible site of interaction could be the microtubuli of the cytoskeleton and the mitotic spindle (Figures 2 and 3). The interaction of porphyrins, with two negative charges, with microtubuli has already been observed by others.[3,4]

In C6 cells chlorins 1 and 2 displayed a distribution pattern which resembled that of the endoplasmic reticulum (Figure 4) whereas with chlorins 3 and 4 stronger fluorescence of the perinuclear region was observed (Figures 5 and 6). Chlorin 5 which carries no negative charge displayed a granularlike pattern throughout the cytoplasm in both cell lines (Figure 7).

ACKNOWLEDGMENTS

The authors gratefully acknowledge the support for this investigation by the Deutsche Forschungsgemeinschaft and the Fond der Chemischen Industrie.

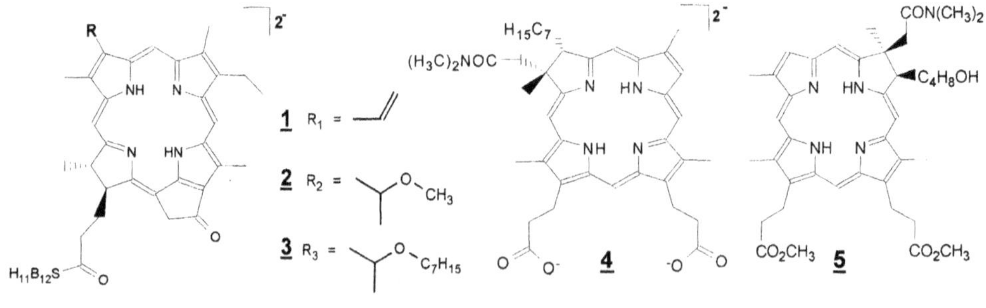

Figure 1. structures of the chlorins examined.

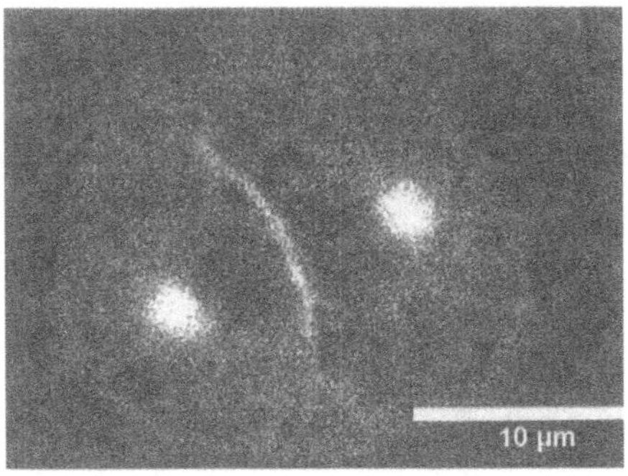

Figure 2. V79 cell, *2* after 30 min of incubation note the fluorescent spots near the nuclei and the fluorescent structure between them.

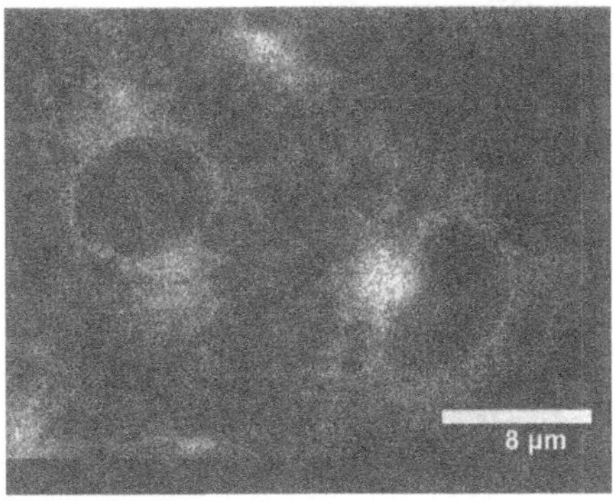

Figure 3. V79 cells, chlorin *3* after 8 h of incubation, note the fluorescent spots near the nuclei and the structures in the cytoplasm.

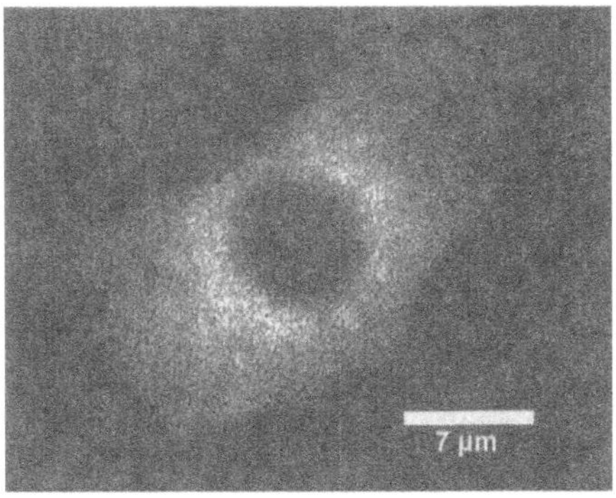

Figure 4. C6 cell, chlorin *1* after 2 h of incubation, the stainign pattern resembles that of the endoplasmic reticulum.

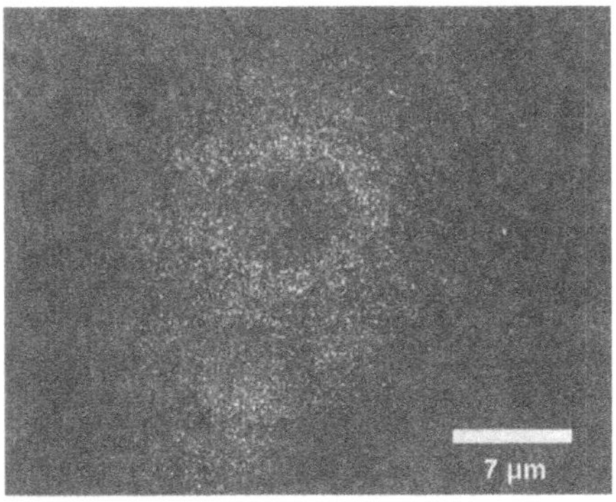

Figure 5. C6 cell chlorin *3* after 4 h of incubation shows also stronger fluorescence in the perinuclear region.

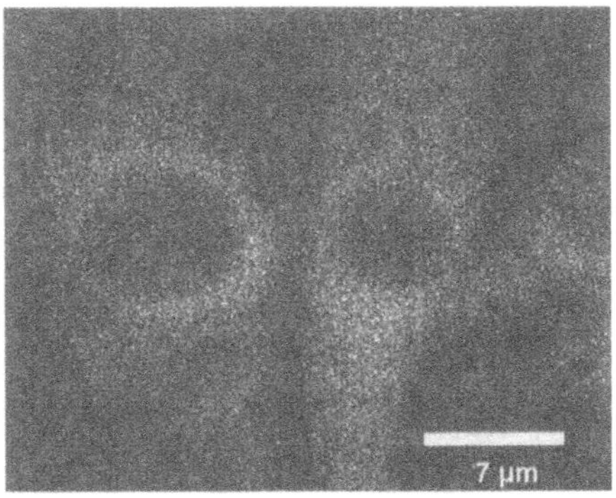

Figure 6. C6 cells, *4* after 16 h of incubation shows a strong fluorescence in the perinclear region.

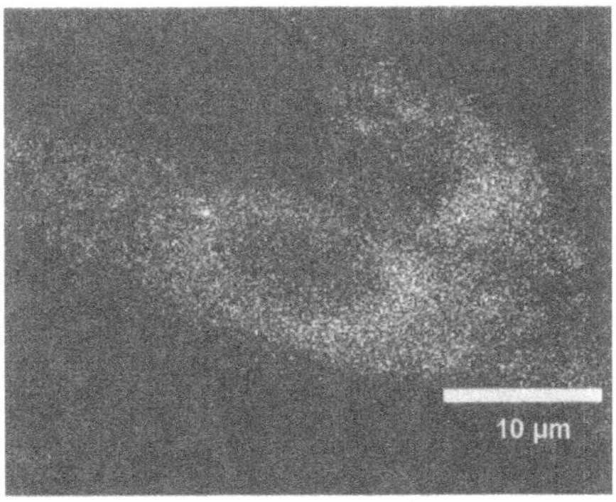

Figure 7. C6 cells, *5* after 4 h of incubation, chlorin *5* displayes in both cell lines a granular-like pattern throughout the cytoplasm.

REFERENCES

1. S. Ruf, J. Osterloh, and D. Gabel, New synthesis of boronated chlorins with different lipophilicity, *in* "Frontiers in Neutron Capture Therapy," M.F. Hawthorne, K. Shelly, R.W. Wiersema, eds., Kluwer Academic/Plenum Publishers, New York, 2001, pp. 825–829.
2. F.-P. Montforts, D. Kusch, F. Höper, S. Braun, B. Gerlach, H.-D. Brauer, G. Schermann, and J.G. Moser, Selective Synthesis and Characterization of Chlorins as Sensitizers for Photodynamic Therapy, *SPIE*, 2675:212–221, 1996.
3. K. Berg, The unpolymerized form of tubulin is the target for microtubule inhibition by photoactivated tetra(4-sulfonatophenyl)porphine, *Biochim. Biophys. Acta*, 1135:147–153, 1992.
4. J.W. Winkelmann, D. Arad, and S. Kimel, Stereochemical factors in the transport and binding of photosensitizers in biological systems and in photodynamic therapy, *J. Photochem. Photobiol. B: Biol.*, 18:181–189, 1993.

QUANTITATIVE ANALYSIS OF BORON IN CO-CULTURES OF HUMAN GLIOBLASTOMA T98G CELLS AND NORMAL CELL LINES WITHIN THE SAME FIELD OF VIEW OF ION MICROSCOPY IMAGING

Daniel R. Lorey II, Subhash Chandra, Duane R. Smith, and George H. Morrison

Ion Microscopy Laboratory
Department of Chemistry and Chemical Biology
Cornell University
Ithaca, New York, 14853-1301

1. INTRODUCTION

Compounds newly designed for use as therapeutic boron delivery agents in Boron Neutron Capture Therapy (BNCT) of cancer need to be examined not only for their subcellular localization characteristics in tumor cells, but also for their selectivity for tumor versus normal cells. This information on selectivity is especially critical for compounds designed to deliver boron specifically to tumor cells by targeting receptor sites on tumor cells, such as epidermal growth factor receptors. Strict cryogenic sample preparation, which preserves the native chemical and morphological makeup of the cell, permits a potentially powerful method for obtaining this information: tumor and normal cell lines can be mixed together and co-cultured *in vitro*, identified optically after cryogenic sample preparation based on their distinctive morphologies, and simultaneously and quantitatively analyzed for boron in the same ion microscopic field of view. Ion microscopy, based on secondary ion mass spectrometry (SIMS), provides visual images of isotopic distributions from H to U in a one-to-one relation to cell morphology with a spatial resolution of 0.5μm and sensitivities in the ppm to ppb range.[1,2] In the present study, we have employed ion microscopy to examine the feasibility of using mixed cultures by examining the subcellular distribution of boron from BPA-F and BSH in mixed cultures of the T98G human glioblastoma cell line, a tumor cell line derived from a human glioblastoma multiforme tumor, with two different normal cell lines—LLC-PK_1 porcine kidney epithelial cells and GM3448 human skin fibroblasts.

Frontiers in Neutron Capture Therapy, edited by Hawthorne *et al.*
Kluwer Academic / Plenum Publishers, New York, 2001.

2. MATERIALS AND METHODS

T98G human glioblastoma cells, LLC-PK_1 porcine kidney epithelial cells and GM3448 normal human skin fibroblasts were cultured separately. T98Gs and cells from either normal cell line were then well-mixed, seeded upon polished, high purity, semiconductor-grade silicon chips (about $1\,cm^2$) and maintained in the nutrient medium of the normal cell line. To study the uptake and intracellular distributions of boron in the mixed cultures, the co-cultured cells on chips were treated with 110µg/mL boron equivalent of BPA-fructose (BPA-F) or BSH for one hour. After the treatments, the cells were cryogenically prepared, freeze-fractured and freeze-dried.[3] The freeze-dried cells were then optically screened under reflected light Nomarski optics to find areas containing fractured cells in which the two cell types had mixed. These regions were then analyzed with a CAMECA IMS-3f ion microscope. Ion images of B, K, Na, Ca, and C were recorded with a CCD camera from the phosphor screen/microchannel plate assembly of the ion microscope and processed digitally for boron quantification of subcellular compartments using the carbon signal as an internal standard.[4]

3. RESULTS AND DISCUSSION

For co-cultures of tumor and normal cells to be successful, the two cell types must mix and grow together and fracture successfully during the cryogenic sample preparation required for quantitative ion microscopy imaging. Figure 1 shows two optical images of an area of a mixed culture of T98G and LLC-PK_1 cells that have been successfully fast-frozen, freeze-fractured, and then freeze-dried. The top photo shows a large area of LLC-PK_1 cells with a few adjacent T98G cells. The lower photo is a magnified view of this interface region (boxed in top photo), with two T98G cells indicated by arrows. The T98G cells are clearly distinguishable from the adjacent LLC-PK_1 cells due to their distinctive shape and often multi-nucleated morphology, whereas LLC-PK_1s have large, rounded single nuclei within a characteristically dome-shaped growth of cells together. After cryogenically prepared samples of co-cultured cells have been optically screened to locate and identify cells, those identical regions can then be analyzed with the ion microscope, thus providing a simultaneous elemental analysis of both normal and tumor cell lines. Similar morphological discrimination is seen for mixed cultures of T98G and human skin fibroblast GM3448 cells (not shown).

Figure 2 illustrates an example of an ion microscopic analysis of mixed cell cultures. An optical photo and ion images for Ca, B, K, Na, and C of the identical cells in a mixed culture of T98G and LLC-PK_1 cells that were treated with BPA-F are shown. Two T98G cells, at the right of the field of view, are clearly recognizable in the optical photo. Potassium, sodium and carbon are nearly homogeneous throughout the cell. For most of the cells, both T98G and LLC-PK_1, the ratio of intracellular K/Na concentrations is approximately 10, indicative of healthy, well-preserved cells. Thus, the mixing of cell lines during the experimental duration was not injurious to either cell type. The calcium distribution reveals low intensity nuclei against the higher intensities in the cytoplasm. The overall boron content of these healthy cells is approximately 130ppm with some dynamic distribution seen within the cells, which is similar to the amount of boron seen in T98G cells alone after the same treatment.[5] A mixed T98G/LLC-PK_1 analysis of subcellular BSH distribution (not shown), showed a non-preferential, nearly homogeneous boron concentration of approximately 110ppm, again similar to BSH-treated

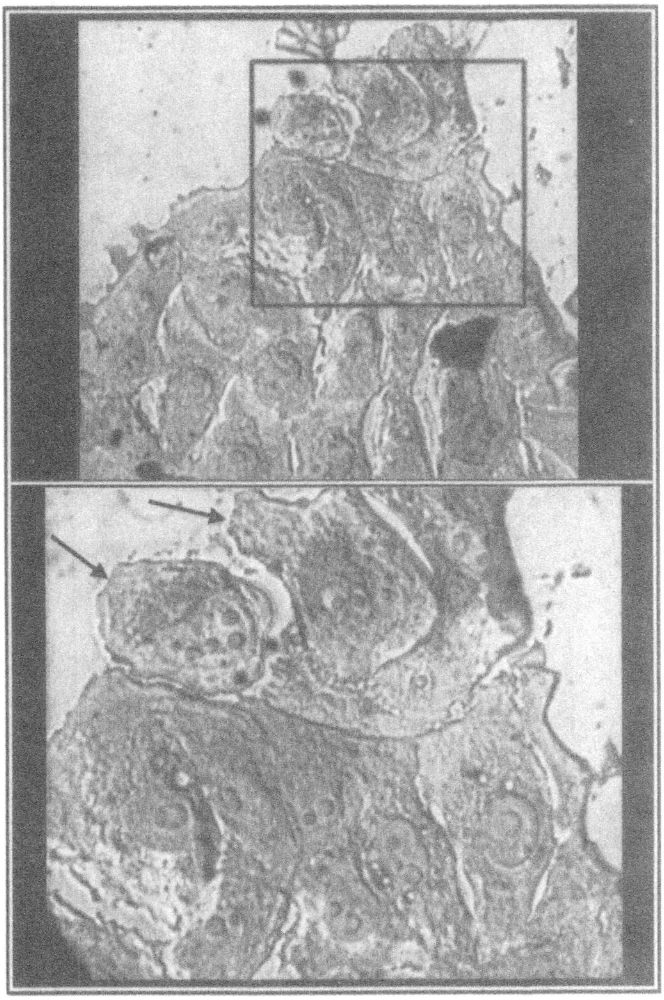

Figure 1. Reflected light Nomarski micrographs of cryogenically prepared co-cultured T98G and LLC-PK_1 cells. The boxed region at top is magnified in the lower image, and two T98G cells are indicated with arrows.

T98G cells alone.[5] We were also successful in co-culturing T98G cells with human skin fibroblasts, GM3448. The details of this study will be discussed elsewhere.

4. CONCLUSIONS

These two model systems have demonstrated that tumor and normal cell lines can be grown together in mixed cultures, successfully prepared using rigorous cryogenic sample preparation methods, and remain morphologically distinguishable. In conjunction with the powerful analytical tool of ion microscopy, this technique can provide much needed information about boron uptake patterns and preferences, subcellular boron localization, and compound toxicity effects in both tumor and normal cell lines simultaneously in the same ion microscopic field of view. The two compounds used

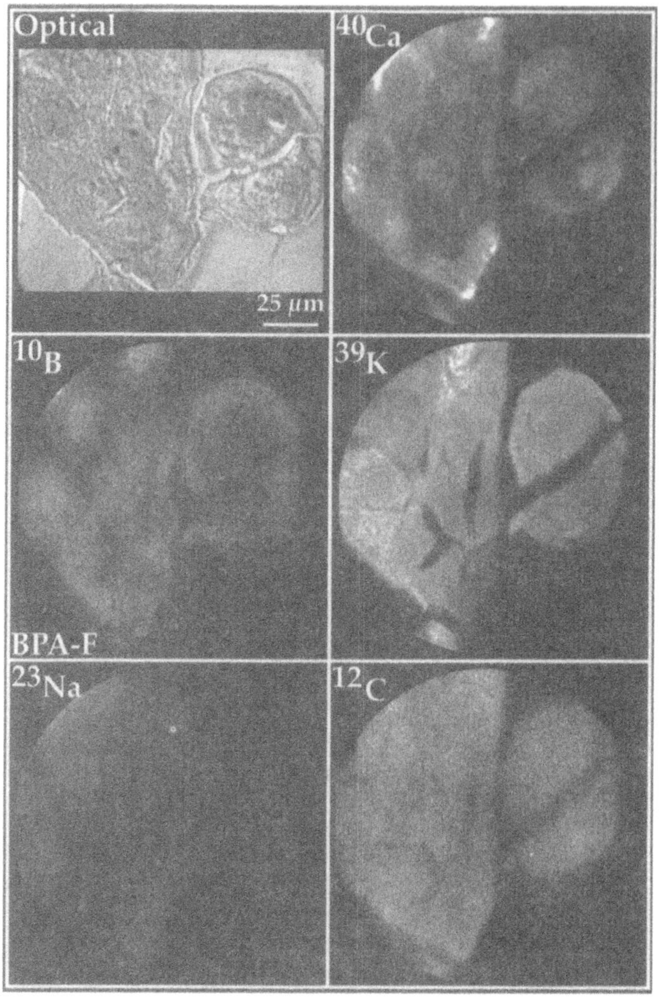

Figure 2. Optical photo and SIMS ion images of Ca, B, K, Na and C of co-cultured and cryogenically prepared T98G and LLC-PK_1 cells treated with 110μg/mL boron equivalent of BPA-F for 1 hour. Within a given ion image, brighter signals are indicative of higher concentration.

in this study demonstrate the feasibility of carrying out quantitative boron analyses in these mixed systems. The true potential for this exciting methodology will be realized in the evaluation of newly synthesized boronated compounds which are being developed to selectively deliver boron to tumor cells by targeting receptor sites on the cell surface. This methodology can be readily extended to co-cultures of other cell lines which may be more suitable for analyzing the behavior of particular BNCT drugs.

ACKNOWLEDGMENT

The authors gratefully acknowledge the U. S. Department of Energy (DOE) for supporting our BNCT ion microscopy program at Cornell University under contract #DE-FG02-91ER61138.

REFERENCES

1. G.H. Morrison, and G. Slodzian, Ion microscopy, *Anal. Chem.*, 47:932A–943A, 1975.
2. S. Chandra, and G.H. Morrison, Ion microscopy in biology and medicine, *Methods Enzymol.*, 158:157–179, 1988.
3. S. Chandra, G.H. Morrison, and C.C. Wolcott, Imaging of ion fluxes and elemental distribution in cultured cells by ion microscopy: A freeze-fracture methodology, *J. Microsc. (Oxford)*, 144:15–37, 1986.
4. W.A. Ausserer, Y.C. Ling, S. Chandra, and G.H. Morrison, Quantitative imaging of boron, sodium, potassium, calcium and magnesium in cultured cells with ion microscopy, *Anal. Chem.*, 61:2690–2695, 1989.
5. S. Chandra, D.R. Lorey, II, S.L. Lessig, D.R. Smith, and G.H. Morrison, Quantitative imaging of boron from BPA and BSH in glioblastoma cells with ion microscopy, *in*: "Advances in Neutron Capture Therapy: Vol. II, Chemistry and Biology," B. Larsson, J. Crawford, and R. Weinreich, eds., Elsevier, New York, 1997, pp. 315–320.

ION MICROSCOPY IMAGING OF BORON IN ANIMAL TISSUE SECTIONS FOR NEUTRON CAPTURE THERAPY

Duane R. Smith[1], Darrel D. Joel[2], Jeffrey A. Coderre[2], Gerard M. Morris[3], Subhash Chandra[1], Marta M. Nawrocky[2], Peggy L. Micca[2], Hemant Patel[3], and George H. Morrison[1]

[1]Department of Chemistry and Chemical Biology
Cornell University
Ithaca, New York 14853-1301
[2]Medical Department
Brookhaven National Laboratory
Upton, New York 11973
[3]Research Institute, The Churchill Hospital
University of Oxford
Oxford, United Kingdom

1. INTRODUCTION

Ion microscopy imaging[1] of animal tissue sections enables us to evaluate the *in situ* distribution of physiologically relevant elements, including the isotope ^{10}B, with cellular to subcellular resolution and with detection limits ranging from the low-ppm to ppb levels. An overview of three ion microscopy imaging studies is presented. The first two address the issue of the localization of ^{10}B from BPA-F in the mucosal and glandular tissues associated with the oral cavity. Of concern is the possibility that BNCT-related radiation damage in soft tissues of the head and neck may be dose-limiting. The third study explores the time dependence of BPA-F boron delivery by intravenous (iv) infusion. This study was undertaken in direct response to our previous ion microscopy imaging studies[1,2] which show a fixed, 2/1 partitioning of ^{10}B from BPA-F between tumor cells in the main tumor mass and tumor cells invading the normal brain.

2. ORAL MUCOSA AND PAROTID SALIVARY GLAND OF THE DOG

Figure 1 shows an optical (a) and sequential ion images of ^{39}K (b) and ^{10}B (c) from a frozen freeze-dried cryosection of the oral mucosa of a healthy dog. The dog was

Frontiers in Neutron Capture Therapy, edited by Hawthorne *et al.*
Kluwer Academic / Plenum Publishers, New York, 2001.

administered 500 mg BPA/kg body weight in an iv infusion over 2 hours. One hour post-infusion the dog was sacrificed. Visible within the optical image is a portion of the surface of the oral mucosa (black arrow), the epithelium (Ep) and the lamina propria (LP). The white arrows in the optical image point to the dark-stained basal cell layer of the Ep which repopulates the cells of the continuously eroding oral surface. The basal cell layer can be visualized as the line separating the Ep and the LP. The dotted line in the ^{39}K and ^{10}B ion images indicates the approximate location of the basal cell layer of the Ep. In the ^{39}K image, oval shaped cell nuclei with slightly higher intensities are discernible throughout the Ep. However, the ^{10}B image shows only selective nuclei (arrows) with higher boron accumulation characteristics and indicates a dynamic distribution of ^{10}B in cells lying between the basal cell layer and the outer surface (not visible) of the Ep. These nuclei have the highest wet weight concentration of ^{10}B (74.0 ± 16.0 μg/g (mean ± standard deviation)). However, they occupy less than 15% of the total volume of the Ep. For dosimetric calculations it is equally important to know the distribution and concentration of ^{10}B in the remainder of the oral mucosa. The gradient of boron intensity in the Ep (excluding the selective nuclei) and in the LP is minimal and the concentration of ^{10}B,

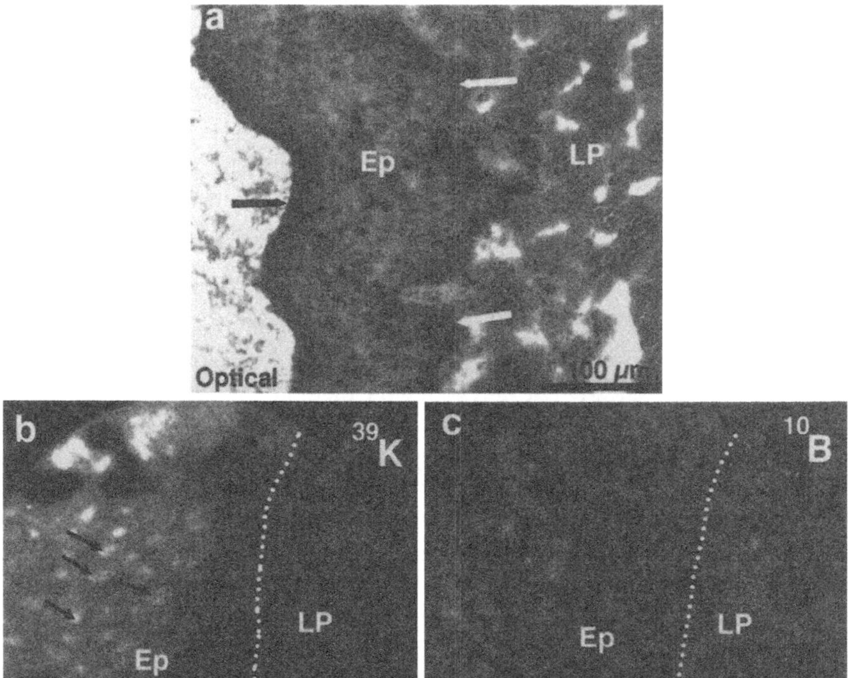

Figure 1. Optical (a) and sequential ion images of ^{39}K (b) and ^{10}B (c) from the oral mucosa of a healthy dog treated with BPA-F. Adjacent 4 μm-thick cryosections were used for optical and ion microscopy imaging. The cryosection for optical microscopy was stained with hematoxylin and eosin. The cryosection for ion microscopy was freeze-dried. The black arrows in (b) and (c) point to representative cell nuclei visible in the Ep.

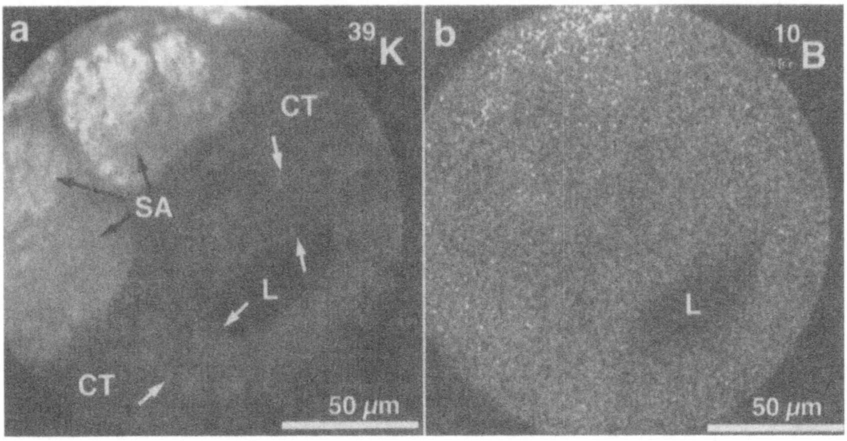

Figure 2. Sequential ion images of ^{39}K (a) and ^{10}B (b) from the parotid salivary gland of a dog treated with BPA-F.

averaged for both the Ep and LP, is 21.9 ± 5.9μg/g. The quantification of boron ion images is detailed elsewhere.[1,3]

Figure 2 shows sequential ion images of ^{39}K (a) and ^{10}B (b) from a cryosection of the parotid salivary gland from the same dog discussed above. The morphology of the parotid gland is evident from the gradient of the intensity in the potassium image. Visible is the lumen (L) of an intralobular duct surrounded by connective tissue (CT) and what are most likely cross-sections of serous acini (SA). The white arrows point to the inside and outside edges of the wall of the duct. The gradient of ^{39}K in the wall of the duct also reveals the rounded nuclei of columnar cells. In contrast to the potassium image, the gradient in the boron image is minimal and only the lumen of the duct is discernible. ^{10}B from BPA-F was found to be uniformly distributed in the punctate form throughout the parotid gland with an average concentration of 26.9 ± 8.6μg/g. Bulk analysis of this same tissue by atomic emission spectroscopy yielded a value consistent with the uniformity of the boron distribution.

3. BORON LOCALIZATION STUDIES IN THE VENTRAL TONGUE OF THE RAT

In a separate study[4,5] BPA-F was administered to male Fischer 344 rats and the ventral surface of the tongue was irradiated in the thermal neutron beam of the Brookhaven Medical Research Reactor. Blister or ulcer formation, chosen as the end-point of a dose-response, was evident 6–7 days post-irradiation. Ion microscopy imaging studies were initiated to determine the microlocalization and concentration of ^{10}B in the Ep, the LP and the muscles of the ventral tongue. Male Fischer 344 rats received 700mg BPA/kg body weight in a single intraperitoneal injection. In preparation for ion microscopy analysis, frozen specimens were oriented to take cryosections perpendicular to the length of the tongue approximately 1/2 to 1cm posterior to the tip. The pattern of BPA boron delivery to the Ep of the ventral tongue of the rat was similar to that found in the Ep of the oral mucosa of the dog (section 2). Between the basal cell layer and the ventral surface were cell nuclei with concentrations of ^{10}B as high as 40.0μg/g. The average concentration of ^{10}B in the basal cell layer was 21.6 ± 2.9μg/g. The concentration of ^{10}B in the lamina propria was found to be quite variable (8.0–75.0μg/g) yet independent of the ^{10}B content found in the contiguous Ep and muscle. The muscle tissue of the tongue had the lowest concentration of ^{10}B (15.6 ± 2.9μg/g).

4. ENHANCED BPA-F BORON DELIVERY TO TUMOR CELLS AFTER LONG TERM INTRAVENOUS INFUSION

Male Fischer 344 rats bearing the intracranial 9L gliosarcoma were administered BPA-F intravenously for 2 hours, 3 hours, 6 hours and 24 hours. (Note: The data for the 6-hour infusion is pending and is not shown here.) The rats were sacrificed one hour post-infusion. Ion microscopy imaging was used to evaluate the distribution of ^{10}B in the main tumor mass, the clusters of tumor cells invading the normal brain and the normal brain. Table 1 shows the results of the infusion study for 2 hours, 3 hours and 24 hours. The 2-hour infusion delivered the highest concentration of ^{10}B to the main tumor mass. The ratio of the concentration of ^{10}B in the main tumor mass to the concentration of ^{10}B in the clusters of tumor cells invading the normal brain is also the highest (2.7/1). The ratio decreases to 2.0/1 for the 3-hour infusion. However, further data is needed to determine if there is a statistical difference between the 2-hour and the 3-hour infusions. For the 24-hour infusion the concentration of ^{10}B changes dramatically and the partitioning of ^{10}B between the tumor cells in the main tumor mass and the clusters of tumor cells invading the normal brain becomes nearly 1/1. The concentration of ^{10}B in cells of the main tumor mass decreased 33% in comparison to the amount delivered in the 2- and 3-hour infusions. This may be predominantly related to the dose rate which decreased from 300 mg BPA/kg/hr for the 2-hour infusion to 125mg BPA/kg/hr for the 24-hour infusion. More significantly however, the amount of ^{10}B delivered to the clusters of infiltrating tumor cells increased 55%. The concentration of ^{10}B in the normal brain was fairly constant for the three infusions (20.2μg/g, 24.5μg/g and 23.8μg/g for 2 hours, 3 hours and 24 hours, respectively.).

5. CONCLUSION

In the dog and rat studies, ^{10}B from BPA-F was found to localize with high concentration in selective nuclei between the basal cell layer and the outer surface of the Ep. ^{10}B was uniformly distributed in the remainder of the oral mucosa of the dog. The

Table 1. Intravenous infusion studies

Dose rate (mg BPA/kg/hr)	Infusion time (hours)	$[^{10}B]_{MTM}$ (μg/g)	$[^{10}B]_{clusters}$ (μg/g)	$[^{10}B]_{MTM}/[^{10}B]_{clusters}$ a
300	2	83.4 ± 23.4	30.7 ± 11.9	2.7/1
250	3	74.8 ± 29.7	37.4 ± 7.2	2.0/1
125	24	53.0 ± 16.6	52.5 ± 14.5	1.0/1

Concentrations are given as wet weight values assuming 85% cell water content; concentration values are presented as (mean ± standard deviation); calculated from mean concentrations only. (MTM = main tumor mass.)

distribution of ^{10}B in the parotid gland of the dog was uniform throughout, with a concentration slightly higher than that found in the oral mucosa (26.9µg/g vs. 21.9µg/g, excluding the selective nuclei). The lamina propria on the ventral side of the rat tongue was found to have a variable concentration of ^{10}B with a pattern of distribution independent of that found in the Ep and the muscle tissue. Rats bearing intracranial 9L gliosarcomas received BPA-F by intravenous infusion for 2 hours, 3 hours, 6 hours (not shown) and 24 hours. The partitioning of ^{10}B between the main tumor mass and the clusters of tumor cells infiltrating the normal brain is nearly 1/1 after a 24-hour infusion. The dose rate and the length of the infusion are found to be significant factors in the delivery of ^{10}B. For the main tumor mass, the data suggests that the dose rate is predominant. However, the length of the iv infusion is the most significant factor for delivery of ^{10}B to clusters of tumor cells invading the normal brain. The concentration of ^{10}B in the normal brain was observed to be independent of the dose rate and the length of the iv infusion. These observations indicate possible benefits of longer iv infusions of BPA-F for BNCT. All three studies presented in this paper are ongoing.

ACKNOWLEDGMENT

This work was supported in part by the U.S. D.O.E. Contract #DE-FG02-91ER61138, U.S. D.O.E. Contract #DE-AC02-98CH10886 and The Cancer Research Campaign.

REFERENCES

1. D.R. Smith, S. Chandra, J.A. Coderre, and G.H. Morrison, Ion microscopy imaging of ^{10}B from *p*-boronophenylalanine in a brain tumor model for boron neutron capture therapy, *Cancer Res.*, 56:4302–4306, 1996.
2. D.R. Smith, S. Chandra, J.A. Coderre, R.F. Barth, W. Yang, L. Liu, J.L. Wilson, P.L. Micca, M.M. Nawrocky, J.H. Rotaru, and G.H. Morrison, Quantitative ion microscopy imaging of boron-10 in rat brain tumor models for BNCT, *in*: "Advances in Neutron Capture Therapy, Volume II, Chemistry and Biology", B. Larrson, J. Crawford, and R. Weinreich, eds., Elsevier, New York, 1997, pp. 308–314.
3. W.A. Ausserer, Y.-C., Ling, S. Chandra, and G.H. Morrison, Quantitative imaging of boron, calcium, magnesium, potassium, and sodium distributions in cultured cells with ion microscopy, *Anal. Chem.*, 61:2690–2695, 1989.
4. G.M. Morris, J.A. Coderre, and J.W. Hopewell, A rat model of oral mucosal response to boron neutron capture irradiation, *in these proceedings*: "Frontiers in Neutron Capture Therapy", M.F. Hawthorne, K. Shelly, and R.J. Wiersema, eds., Plenum, New York.
5. J.A. Coderre, G.M. Morris, R. Ma, P.L. Micca, D.T. Lombardo, and K.M. Youngs, Split dose BNCT irradiations of rat oral mucosa and rat brain tumor: Effects on tolerance and survival, *in*: "Frontiers in Neutron Capture Therapy," M.F. Hawthorne, K. Shelly, R.W. Wiersema, eds., Kluwer Academic/Plenum Publishers, New York, 2001, pp. 1279–1283.

ION MICROSCOPY IMAGING OF BORON FROM *p*-BORONOPHENYLALANINE IN SURGICALLY ACQUIRED SAMPLES OF HUMAN BRAIN TUMOR TISSUE

Duane R. Smith[1], Subhash Chandra[1], Jeffrey A. Coderre[2], Darrel D. Joel[2], Daniel N. Slatkin[2], Arjun D. Chanana[2], Eric H. Elowitz[3], Marta M. Nawrocky[2], Peggy L. Micca[2], and George H. Morrison[1]

[1]Department of Chemistry and Chemical Biology
Cornell University
Ithaca, New York 14853-1301
[2]Medical Department
Brookhaven National Laboratory
Upton, New York 11973
[3]Department of Neurosurgery
Beth Israel Medical Center
170 East End Avenue
New York, New York 10128

1. INTRODUCTION

Using ion microscopy[1] we are able to quantitatively image physiologically relevant isotopes, including ^{10}B, in cancerous and contiguous normal tissues with cellular to subcellular resolution and with detection limits in the low-ppm to ppb range. Over the past several years we have been using this technique to analyze tissue sections from animal brain tumor models,[1-3] as well as specimens prepared from tissue cultures,[4,5] to develop appropriate cryogenic sampling procedures to be applied to the *in situ* analyses of ^{10}B in human brain tumor specimens. An ion microscopy imaging study is presented which demonstrates the localization of ^{10}B from the fructose complex of *p*-boronophenylalanine (BPA-F) in tumor cells from a patient with glioblastoma multiforme. Patient A was infused intravenously for one hour to deliver 170 mg BPA/kg body weight. The craniotomy commenced one-half hour after the end of the infusion. Tissue specimens from patient A were frozen on dry ice, stored at –80 °C, and at a later date cryosectioned and freeze-dried for ion microscopy analysis. The boron concentration in

Frontiers in Neutron Capture Therapy, edited by Hawthorne *et al.*
Kluwer Academic / Plenum Publishers, New York, 2001.

tumor cells, calculated from the grayscale intensity of ion images, was found to have a dynamic distribution with an average approximately 3.5 times higher than the ^{10}B levels in the normal brain and blood. Ion images are shown to illustrate the complexities of cryogenic sample preparation for the analysis of ^{10}B in human brain tumor tissues at the microscopic level.

2. ION MICROSCOPY

Ion microscopy was performed using a CAMECA IMS-3f ion microscope, a direct imaging secondary ion mass spectrometer.[6] Frozen freeze-dried cryosections of human brain tumor tissue were analyzed using instrumental and image acquisition parameters similar to those described in reference 1. Images of potassium, sodium, boron, carbon, magnesium and calcium distributions were acquired sequentially to evaluate the ^{10}B delivery characteristics of BPA-F at the cellular level in human brain tumor tissues. Ion images were compared with optical images of adjacent, hemotoxylin/eosin stained cryosections to correlate chemical and histopathological morphology, respectively. The quantification of boron ion images is detailed elsewhere.[1,7]

3. UNDERSTANDING SAMPLE-PREPARATION-INDUCED ARTIFACTS

The nominal resolution of the CAMECA IMS-3f ion microscope is 0.5 µm and a typical image is acquired within a 150 µm field-of-view. The image magnification is approximately ×1,000. For large scale or "bulk" analytical techniques (i.e., atomic emission spectroscopy, etc.), microscopic artifacts, specifically those which result from the redistribution of physiologically relevant elemental gradients, contribute little to the total error of the analysis. However, bulk analytical techniques provide macroscopic results which can be assigned to cellular and subcellular compartments only by supposition. For microscopic techniques, artifacts which approach an instrument's resolution can produce significant analytical errors leading to erroneous results if ignored or improperly interpreted. In tissues and cultures, elemental gradients are easily altered under pathological conditions and/or through inappropriate sample preparation. Artifacts associated with sample preparation may result from techniques involving excessive handling, alteration of the chemical matrix (i.e., chemical fixation, plastic embedding, freeze-substitution, etc.) or those which employ inadequate cryopreservation. The latter, in conjunction with the pathological state, is the subject of this paper.

Figure 1(a) shows an optical image of a hematoxylin/eosin stained cryosection of surgically acquired tumor tissue from patient A. The portion of tissue shown is from the cortex adjacent to the main tumor mass. Both regions of tumor cell infiltration and relatively normal cortex were found in cryosections such as the one shown. The tissue shows characteristic signs of pathological and sample preparation induced damage. Dark-stained tumor cell nuclei in narrow bands of tissue surround large spongiform spaces. These spaces are due in part to edema, a pathological response to the vascular damage associated with the tumor growth. The arrow in the optical image points to an artery which is crushed, indicating that the tissue was damaged during sample preparation. The damage most likely occurred during freezing. Further evidence for this

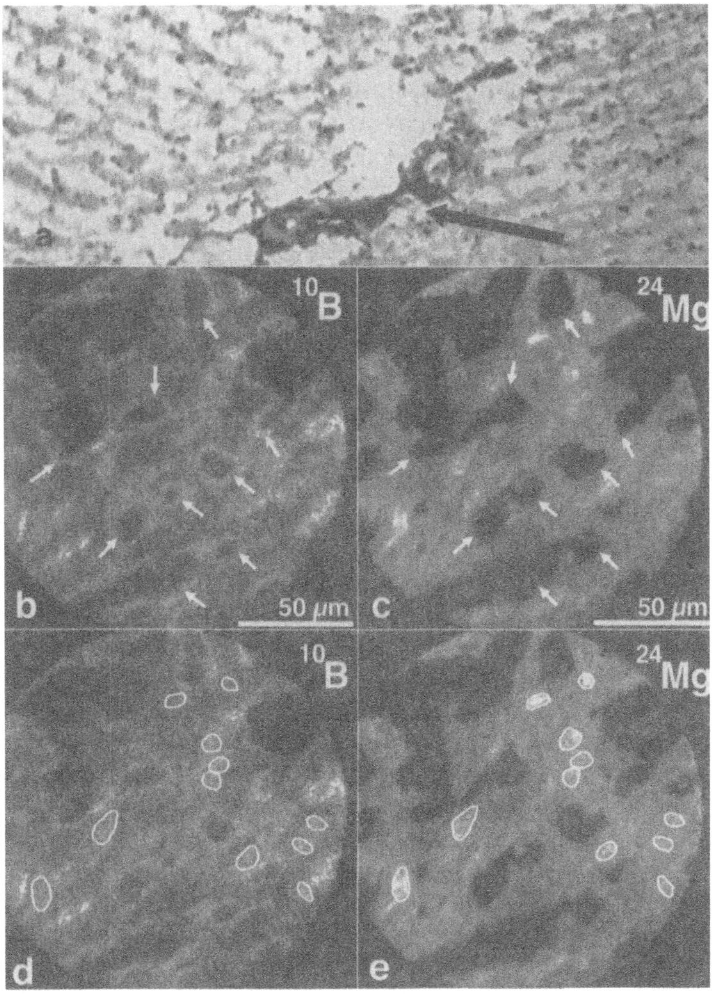

Figure 1. Optical (a) and sequential ion images of ^{10}B (b,d) and ^{24}Mg (c,e) from the edematous tumor bearing cortex of patient A. The tissue specimen was acquired during the biodistribution phase of Brookhaven National Laboratory's clinical trials of BNCT. Adjacent 4μm-thick cryosections were used for optical and ion microscopy imaging. The cryosection for optical microscopy was stained with hematoxylin and eosin. The cryosection for ion microscopy was freeze-dried. The arrow in (a) points to an artery which was crushed during freezing. (b,d) and (c,e) are the same images of ^{10}B and ^{24}Mg, respectively, but were illustrated differently for visual clarity. Arrows in (b) and (c) point to edges of voids. The regions circled in (d) and (e) are tumor cell nuclei.

comes from the "webby" or shrunken appearance of the tissue throughout the image. The tissue pieces acquired during surgery were put in plastic cryomolds, covered with Tissue-Tek® O.C.T. compound and placed on dry ice to freeze. The end result was slow freezing of the tissue. The effects of slow freezing are well known.[8] Freezing begins in the extracellular spaces. The slow ice crystal growth causes water to be drawn from the cells into the extracellular spaces. The intracellular spaces contract and deform as the extracellular spaces enlarge. The formation of large ice crystals in the extracellular spaces produces crush artifacts similar to those observed in and around the artery. Conversely, small "non-damaging" ice crystals form within intracellular spaces due to the reduced water content.

Figure 1 (b,c) are sequential ion images of ^{10}B and ^{24}Mg taken from the frozen freeze-dried cryosection adjacent to that shown in the optical image. Tumor cell nuclei are visible as regions of bright grayscale intensity in both ion images. A number of these have been circled for visual clarity and are shown in the respective panels (d,e). In both the ^{24}Mg and ^{10}B image, large, dark and irregular shaped voids are visible. The voids are consistent with the pathological response to vasogenic edema. Their expanded size and shape are also consistent with the effects of slow freezing and the formation of large ice crystals. Severe redistribution artifacts appear as halos (arrows in (b)) of bright grayscale intensity at the edges of the voids in the ^{10}B image. Similar gradients of intensity were observed for sequential images of ^{39}K and ^{23}Na (not shown). In comparison, the edges of the voids in the ^{24}Mg image are clean. The gradient of intensity of ^{10}B, ^{39}K and ^{23}Na around the edges of the voids are artifacts caused by segregation which most likely occurred when the tissue specimens were placed on dry ice to freeze. Potassium and sodium are highly diffusible in living systems. The extracellular spaces of the edematous brain tissue *in vivo* are filled with a fluid that is derived from a whole plasma component of blood.[9] The fluid has a high sodium and low potassium content, the opposite of that found in the intracellular matrix. The major source of sodium at the edges of the voids is most likely from the edema fluid. The major source of potassium is anticipated to be from the intracellular matrix. As water was drawn from the tissue, a significant quantity of potassium is free to diffuse across cell membranes. In contrast to what we have observed for potassium and sodium, magnesium shows little or no segregation at the edges of the voids (arrows in (c)). Magnesium is like potassium, being low in blood plasma in comparison to the total intracellular concentration. The majority of intracellular magnesium is known to be tightly bound[10] and, even if damage does occur during slow freezing, very little leakage of the analyte is expected. Boron is not a normal component of blood plasma, but in this case the patient received BPA-F prior to surgery and it is not unlikely that the ^{10}B concentration in the edema fluid is nearly that of the blood. From animal studies[1,2] it is anticipated that the ^{10}B content of the infiltrating human brain tumor cells should be higher than the blood or the plasma component of blood and the chemical potential from the gradient of concentration should direct boron from intracellular to extracellular spaces. However, earlier studies[4] in our laboratory provide evidence suggesting that the majority of the ^{10}B from BPA-F was not altered significantly even in damaged cells that showed reduced potassium and elevated sodium signals. Therefore we were able to assign concentration values to cell nuclei such as those circled in the ^{10}B image (d).

4. SUMMARY

This paper shows some of the microscopic artifacts which can arise when tissues are frozen slowly. The effects of the freezing process on the distribution and redistribution of physiologically normal ion gradients must be thoroughly understood since cryopreservation is a critical step in sample preparation for ion microscopy. Poor cryogenic preparations complicate the interpretation of ion gradients at the microscopic level and can lead to erroneous, possibly catastrophic, results if ignored or misinterpreted.

ACKNOWLEDGMENTS

Supported in part by the U.S. D.O.E. Contract #DE-FG02-91ER61138 and U.S. D.O.E. Contract #DE-AC02-98CH10886.

REFERENCES

1. D.R. Smith, S. Chandra, J.A. Coderre, and G.H. Morrison, Ion microscopy imaging of ^{10}B from *p*-boronophenylalanine in a brain tumor model for boron neutron capture therapy, *Cancer Res.*, 56:4302–4306, 1996.
2. D.R. Smith, S. Chandra, J.A. Coderre, R.F. Barth, W. Yang, L. Liu, J.L. Wilson, P.L. Micca, M.M. Nawrocky, J.H. Rotaru, and G.H. Morrison, Quantitative ion microscopy imaging of boron-10 in rat brain tumor models for BNCT, *in*: "Advances in Neutron Capture Therapy, Volume II, Chemistry and Biology", B. Larsson, J. Crawford, and R. Weinreich, eds., Elsevier, New York, 1997, pp. 308–314.
3. D.R. Smith, D.D. Joel, J.A. Coderre, G.M. Morris, S. Chandra, M.M. Nawrocky, P.L. Micca, H. Patel, and G.H. Morrison, Ion microscopy imaging of boron in animal tissue sections for neutron capture therapy, *in*: "Frontiers in Neutron Capture Therapy," M.F. Hawthorne, K. Shelly, R.W. Wiersema, eds., Kluwer Academic/Plenum Publishers, New York, 2001, pp. 893–897.
4. S. Chandra, D.R. Lorey II, S.L. Lessig, D.R. Smith, and G.H. Morrison, Quantitative imaging of boron from BPA and BSH in glioblastoma cells with ion microscopy, *in*: "Advances in Neutron Capture Therapy, Volume II, Chemistry and Biology", B. Larrson, J. Crawford, and R. Weinreich, eds., Elsevier, New York, 1997, pp. 315–320.
5. S. Chandra, D.R. Lorey, D.R. Smith, and G.H. Morrison, *In vitro* exposure of human T98G glioblastoma cells to mixed bnct drugs: Acquiring independent quantitative images of boron atoms delivered by ^{10}BPA-F and ^{11}BSH in the same cell by sims ion microscopy, *in*: "Frontiers in Neutron Capture Therapy," M.F. Hawthorne, K. Shelly, R.W. Wiersema, eds., Kluwer Academic/Plenum Publishers, New York, 2001, pp. 905–909.
6. S. Chandra, and G.H. Morrison, Ion microscopy in biology and medicine, *Meth. Enzymol.*, 158:157–179, 1988.
7. W.A. Ausserer, Y.-C. Ling, S. Chandra, and G.H. Morrison, Quantitative imaging of boron, calcium, magnesium, potassium, and sodium distributions in cultured cells with ion microscopy, *Anal. Chem.*, 61:2690–2695, 1989.
8. A.W. Robards, and U.B. Sleytr. "Low Temperature Methods in Biological Electron Microscopy," Elsevier, New York, 1985, p. 67.
9. G.A. Rosenberg. "Brain Fluids and Metabolism," Oxford University Press, New York, 1990, p. 179.
10. R.B. Martin, *in*: "Metal Ions in Biological Systems, Compendium on Magnesium and Its Role in Biology, Nutrition, and Physiology, Volume 26," H. Sigel, and A. Sigel, eds., Marcel Dekker, Inc., New York, 1990, pp. 1–13.

IN VITRO EXPOSURE OF HUMAN T98G GLIOBLASTOMA CELLS TO MIXED BNCT DRUGS

Acquiring Independent Quantitative Images of Boron Atoms Delivered by ^{10}BPA-F and ^{11}BSH in the Same Cell by SIMS Ion Microscopy

Subhash Chandra, Daniel R. Lorey II, Duane R. Smith, and George H. Morrison

Ion Microscopy Laboratory
Department of Chemistry and Chemical Biology
Cornell University
Ithaca, New York, 14853-1301

1. INTRODUCTION

The selective targeting with sufficient quantities of ^{10}B atoms to the tumor cells is critical for Boron Neutron Capture Therapy (BNCT) of cancer. At present there are two BNCT drugs, *p*-boronophenylalanine (BPA) and sodium borocaptate (BSH), approved for use in clinical trials. Since both BPA and BSH may deliver boron to tumor cells by different mechanisms, the concept of mixing the two compounds may have the merit of providing a unique approach for attaining a higher net accumulation of boron atoms in tumor cells than would otherwise be attained by an individual compound. Clinical applications of this mixed-drug approach would require that both drugs provide ^{10}B atoms. For laboratory research purposes, however, one can use a mixture such as ^{10}B labeled ^{10}BPA and ^{11}B labeled ^{11}BSH and an isotopic imaging technique of ion microscopy for studying intracellular accumulations of both ^{10}B and ^{11}B in the same cell. Ion microscopy, based on secondary ion mass spectrometry (SIMS),[1,2] is capable of quantitatively imaging the subcellular distribution of both ^{10}B and ^{11}B atoms delivered by two different BNCT agents in the same cell since the technique distinguishes isotopes based on their mass-to-charge (m/z) ratio. The exposure of cells to two BNCT compounds mixed together may result in the enhancement of boron uptake by tumor cells. In the present study, by using ^{10}B labeled *p*-boronophenylalanine-fructose (^{10}BPA-F) and ^{11}B labeled sodium

Frontiers in Neutron Capture Therapy, edited by Hawthorne *et al.*
Kluwer Academic / Plenum Publishers, New York, 2001.

borocaptate (^{11}BSH) as a mixed-drug model we were able to image the subcellular distribution of both ^{10}B and ^{11}B atoms independently in the same cell with ion microscopy. Such independent imaging of boron atoms from two mixed-drugs in the same cell introduces a novel approach for checking the boron delivery characteristic of two BNCT compounds combined together.

2. MATERIALS AND METHODS

Cells from T98G human glioblastoma cell line were grown *in vitro* on the polished surface of high purity N-type semiconductor grade silicon pieces. An electrically conducting cell growth substrate is a necessity for ion microscopy measurements since the sample is held at 4,500 volts in the sample chamber of the ion microscope. After 72 hr. of cell seeding the sister cells growing on different silicon pieces were subjected to three treatments. The first treatment included the exposure of cells to both ^{10}BPA-F and ^{11}BSH mixed together for 1 hr. in the concentration of 110 μg/ml boron equivalent of the individual drug. The second and third treatments included the exposure of T98G human glioblastoma cells to 110 μg/ml boron equivalent concentrations of the individual drugs, ^{10}BPA-F or ^{11}BSH, alone for 1 hr. After the designated treatments, the cells were cryogenically prepared for ion microscopy analysis that included flash freezing, sandwich-fracturing and freeze-drying.[3,4] The freeze-dried cells were photographed under reflected light Nomarski optics and gold coated to enhance electrical conductivity prior to ion microanalysis. A CAMECA IMS-3f ion microscope was used for subcellular ion microanalysis study. This instrument is capable of producing isotope images with a spatial resolution of 0.5 μm. Isotope images of ^{10}B, ^{11}B, ^{12}C, ^{39}K, ^{23}Na, and ^{40}Ca were recorded from the phosphor screen/microchannel plate assembly of the ion microscope via a charge-coupled device (CCD) camera and processed digitally for boron quantification of subcellular compartments using carbon as an internal standard.[5] The subcellular dry weight boron concentrations obtained from the analysis of freeze-dried cells were converted into wet weight concentrations by assuming 85% cell water content. Image quantification of the raw ^{10}B signals from ^{10}BPA-F in the mixed-drug treatment required an additional digital-image-processing step in order to subtract pixel-by-pixel the 20% natural abundance of ^{10}B contributed from the use of naturally abundant ^{11}BSH (the naturally abundant ^{11}BSH has 80% ^{11}B and 20% ^{10}B atoms). A similar correction has been described in details for ion microscopy imaging of magnesium isotopes.[6]

3. RESULTS AND DISCUSSION

Subcellular imaging of ^{10}B and ^{11}B in the same cell from the mixed ^{10}BPA-F and ^{11}BSH treatment is shown in Figure 1. The top left panel shows an optical photo of representative frozen freeze-dried human T98G glioblastoma cells. The cytoplasmic and nuclear compartments in individual cells are discernible. Ion microscopy images of ^{40}Ca, ^{10}B (delivered by ^{10}BPA-F) and ^{11}B (delivered by ^{11}BSH) from the same cells are shown in designated panels. For ion images, the level of brightness within an image indicates higher concentrations. The calcium distribution shown in the ^{40}Ca image unequivocally reveals the low intensity cell nucleus. The location of the nucleus is matched between the

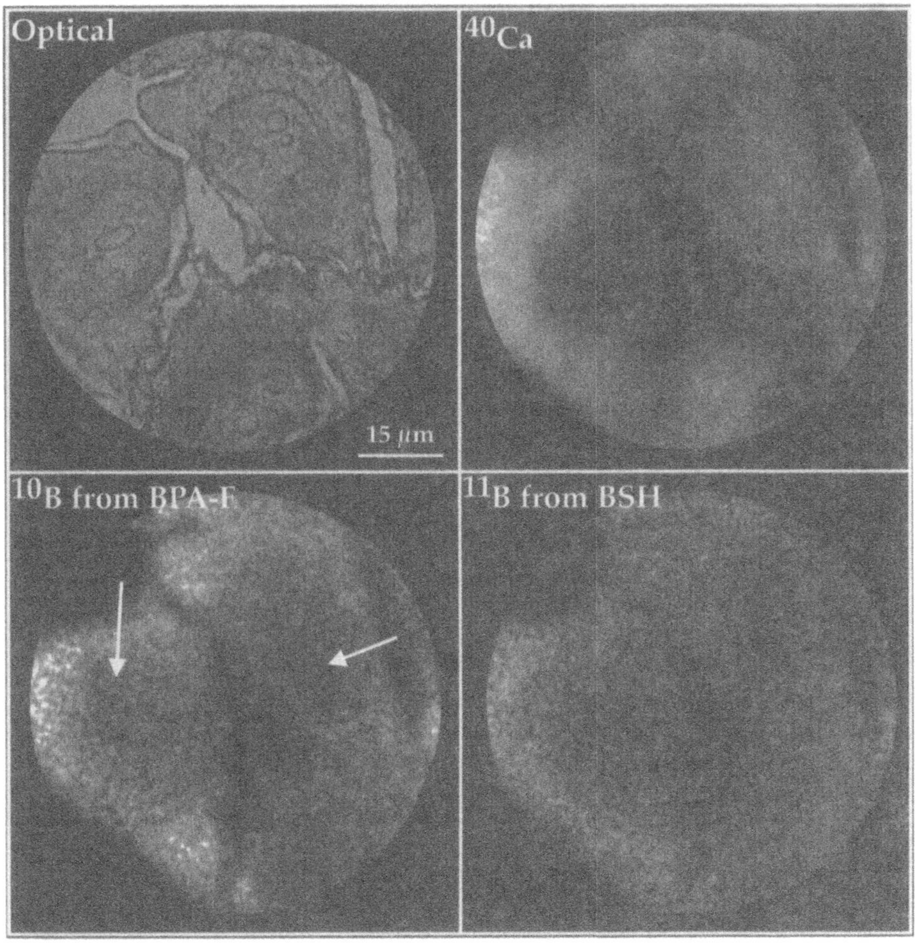

Figure 1. Ion microscopy subcellular localization of ^{10}B atoms from ^{10}BPA-F and ^{11}B atoms from ^{11}BSH in the same T98G human glioblastoma cell after 1 hr. exposure of these drugs mixed together. The designated panels in the image show an optical image of individual cells and their ^{40}Ca, ^{10}B and ^{11}B subcellular distributions revealed by ion microscopy.

optical photo and the calcium image. Since the cell cytoplasm contains calcium storing organelles like the endoplasmic reticulum and Golgi apparatus, the ^{40}Ca image distinctly reveals higher cytoplasmic intensities than nuclear regions.[7] The image of ^{10}B atoms delivered by ^{10}BPA-F shows minor differences between the nuclei and cytoplasm, with slightly lower intensities from a perinuclear cytoplasmic region indicated by arrows. This perinuclear region is most probably occupied by the Golgi apparatus.[7,8] Furthermore, in some regions of the cytoplasm there appears to be a punctate distribution of higher ^{10}B signals. The image of ^{11}B atoms delivered by ^{11}BSH reveals a minor heterogeneity throughout the cell without any noticeable gradient between nuclei and cytoplasmic compartments. In general, the ^{11}B distribution appears to be punctate throughout the cell. These cells also revealed high-^{39}K and low-^{23}Na signals suggestive of well preserved healthy cells (images not shown).

In the mixed drug treatment, the nuclear and cytoplasmic concentrations of ^{10}B from ^{10}BPA-F were 186 ± 50 and 178 ± 50 ppm wet weight (mean ± st. dev., n = 13 cells from four different sampling regions of ion microscopy imaging), respectively. These differences between the nuclear and cytoplasmic compartments are not statistically significant ($p < 0.05$). However, for these calculations we averaged the whole cytoplasm, including the perinuclear region containing much lower concentrations of ^{10}B (as shown in Figure 1). In the same cells, the ^{11}B concentrations from ^{11}BSH were 87 ± 29 ppm wet weight in the nuclei and 91 ± 24 ppm wet weight in the cytoplasm. The differences for ^{11}B between the nuclei and cytoplasm were not statistically significant ($p < 0.05$).

In treatments that included single drug exposures (either ^{10}BPA-F or ^{11}BSH), the ^{10}B and ^{11}B intracellular distributions and concentrations were comparable to observations discussed above for each drug in the mixed-drug treatment (images not shown here and the details will be discussed elsewhere due to page limitations of the proceedings). This implies that BPA-F and BSH mixed together have an additive effect on the net boron delivery to T98G cells under the conditions of the present study. Furthermore, the mixing of these two drugs have not affected their intracellular localization. These observations may have important implications in BNCT. A caution must be exercised in extrapolating these observations to animal models and human studies of *glioblastoma multiforme.* Unlike animal models, the cells in cultures lack the blood-brain-barrier and can be constantly exposed to the desired concentration of boron delivery agents. The strength of the present approach lies in its ability to assess boron distributions at a subcellular level and the boron delivery characteristic of two BNCT drugs combined together. Such observations made *in vitro* may provide the very first clue to the feasibility of mixing the two BNCT agents as a source for increasing the net boron content of tumor cells than would otherwise be attained by a single drug. The methodological approach reported here for imaging boron atoms from two different drugs can be easily extended to tissue studies of animal models of brain tumors.

ACKNOWLEDGMENTS

We gratefully acknowledge Dr. Jeffrey A. Coderre, Medical Department of Brookhaven National Laboratory, Upton, New York, for providing us ^{10}BPA-F for the study. This work was supported by the United States Department of Energy under a contract # DE-FG02-91ER61138.

REFERENCES

1. G.H. Morrison, and G. Slodzian, Ion microscopy, *Anal. Chem.*, 47:932A–943A, 1975.
2. S. Chandra, and G.H. Morrison, Ion microscopy in biology and medicine, *Methods Enzymol.*, 158:157–179, 1988.
3. S. Chandra, G.H. Morrison, and C.C. Wolcott, Imaging of ion fluxes and elemental distribution in cultured cells by ion microscopy: A freeze-fracture methodology, *J. Microsc. (Oxford)*, 144:15–37, 1986.
4. S. Chandra, and G.H. Morrison, Evaluation of fracture planes and cell morphology in complimentary fractures of cultured cells in the frozen-hydrated state by field-emission secondary electron microscopy: feasibility for ion localization and fluorescence imaging studies, *J. Microsc. (Oxford)*, 186:232–245, 1997.
5. W.A. Ausserer, Y.-C. Ling, S. Chandra, and G.H. Morrison, Quantitative imaging of boron, calcium, magnesium, potassium, and sodium distributions in cultured cells with ion microscopy, *Anal. Chem.*, 61:2690–2695, 1989.

6. S. Chandra, and G.H. Morrison, Evaluation of ^{26}Mg stable isotope as an *in vivo* tracer of magnesium transport for SIMS ion microscopy imaging studies, *J. Microsc. (Oxford)*, 188:182–190, 1997.
7. S. Chandra, E.P.W. Kable, G.H. Morrison, and W.W. Webb, Calcium sequestration in the Golgi apparatus of cultured mammalian cells revealed by laser scanning confocal microscopy and ion microscopy, *J. Cell Sci.*, 100:747–752, 1991.
8. S. Chandra, D.R. Lorey II, S.L. Lessig, D.R. Smith, and G.H. Morrison, Quantitative imaging of boron from BPA and BSH in glioblastoma cells with ion microscopy, *in*: "Advances in Neutron Capture Therapy: Volume II, Chemistry and Biology," B. Larsson, J. Crawford, and R. Weinreich, eds., Elsevier, New York, 1997, pp. 315–320.

A COMBINATION OF IMMUNOHISTOCHEMISTRY AND PHOTOELECTRON SPECTROMICROSCOPY FOR INVESTIGATIONS OF SUBCELLULAR BORON DISTRIBUTION

Michael Neumann[1], Gelsomina De Stasio[2,3], Benjamin Gilbert[2], José Redondo,[2] and Detlef Gabel[1]

[1]Department of Chemistry
University of Bremen, Germany
[2]Institut de Physique Appliqueé
Ecole Polytechnique Fédérale de Lausanne, Switzerland
[3]Instituto di Struttura della Materia del Consiglio Nazionale delle Ricerche, Italy

1. INTRODUCTION

We have combined immunohistochemistry and photoelectron spectromicroscopy with synchrotron radiation for investigations of the subcellular distribution and the chemical state of boron in human glioma tissue.

Microtome sections of glioma tissue from patients, infused with $Na_2B_{12}H_{11}SH$ (BSH) 22 to 24 h prior to surgery, were stained immunohistochemically with a nickel-diaminobenzidine-solution. The nickel in this method was used as a marker for the desired antigen. After incineration the stained tissue samples were analyzed with the MEPHISTO spectromicroscope. By parallel detection of boron (Figure 1) and nickel (Figure 2) it is possible to obtain high resolution maps of the distribution of BSH in the tissue sample. As an additional information the chemical state of the detected element can be revealed. Because of the small range of the secondary electrons this method is limited on the measurements of elements on the surface or ultra thin samples.

2. METHODS

Paraffin sections of tumor tissue were mounted on silicon wafers and stained immunohistochemically with diaminobenzidine-nickel using the peroxidase-

Frontiers in Neutron Capture Therapy, edited by Hawthorne *et al.*
Kluwer Academic / Plenum Publishers, New York, 2001.

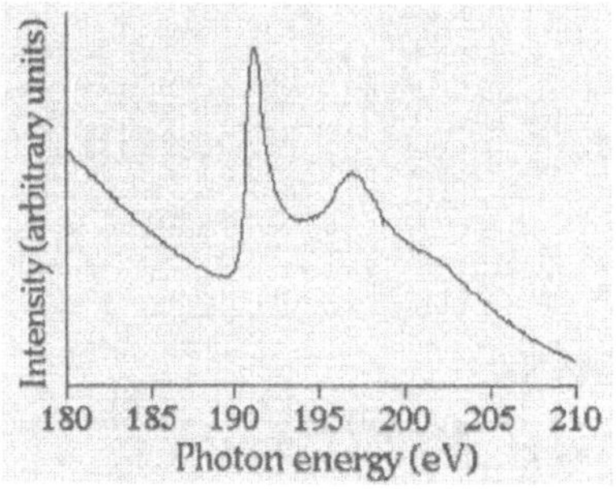

Figure 1. Total photoelectron yield spectrum of boron in BSH with the 1s-absorption maximum at 191 eV. The peak at 197 eV corresponds to phosphorus which is present in all natural tissues.

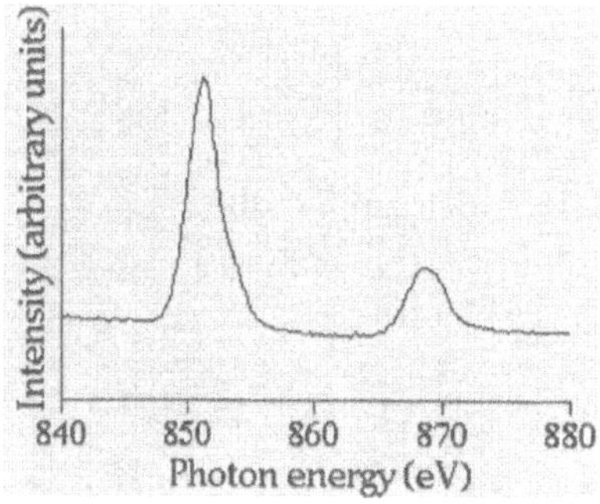

Figure 2. Total photoelectron yield spectrum of nickel with the characteristics 2p-peaks at 852 and 868 eV.

avidin-biotin-complex-method. At the same time, neighboring sections were mounted on glass and treated in the same way to provide reference samples for the optical microscope. The antibodies used for staining were either anti-BSH, anti-Ki-67 (a protein found in the nuclei of proliferating cells) or anti-von Willebrandt factor (located in blood vessel endothelia). After incineration the MEPHISTO spectromicroscope performs an elemental and chemical state microanalysis of the stained samples and produces images of the area under investigation. Using tunable soft x-rays, the MEPHISTO spectromicroscope

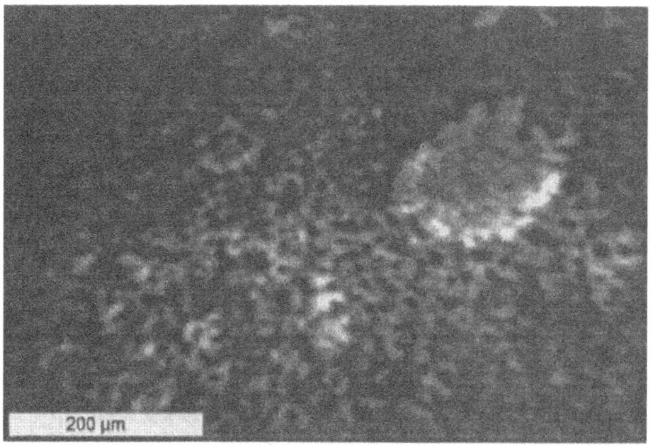

Figure 3. Tissue micrograph taken at 854 eV (the maximum of the nickel 2p 3/2 peak) of a tumor sample stained with nickel against von-Willebrandt factor.

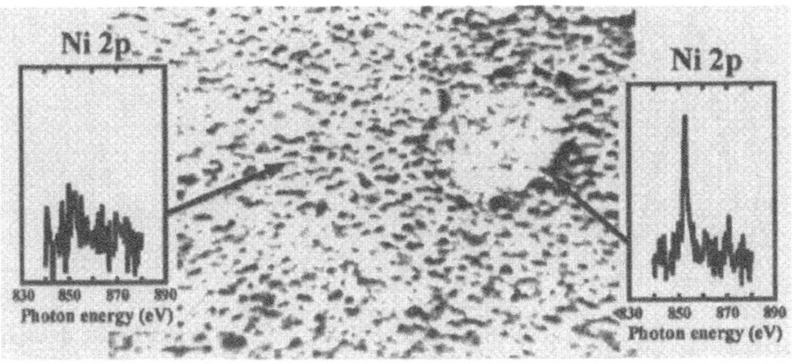

Figure 4. Nickel-distribution map corresponding to the tissue region of Figure 3.

acquires total photoelectron yield spectra. The field of view can be varied between 5 and 500 μm. The detection limit for boron has not been established yet but a value about 100 ppm is regular.[1] A comparison of the obtained images for boron and nickel leads to the subcellular distribution of BSH in the tissue samples.

3. RESULTS

After optimizing the staining procedure and the nickel concentration used in the staining, we were able to visualize the selected antigen with the photoelectron microscope (Figures 3–5). It was also possible to obtain information on the subcellular localization of BSH by simultaneous detection and comparison of boron and nickel (Figure 5).

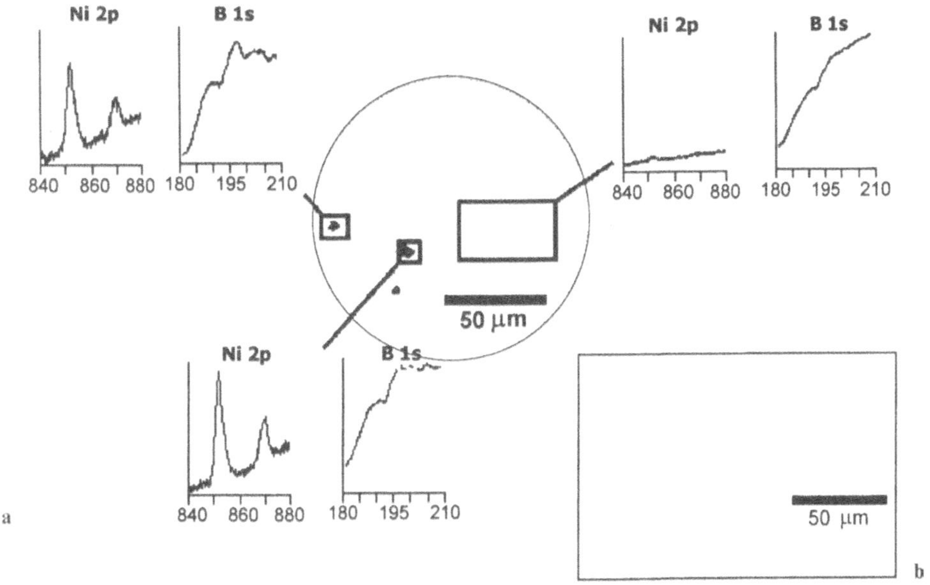

Figure 5. a) Nickel distribution map of a tissue sample stained against Ki-67 together with the absorption spectra of boron and nickel for the marked regions. The dark spots show the regions of highest nickel concentration. The spectra show that most boron can be found near the cell nuclei. b) Photograph of a neighboring tissue section which was treated in the same way. The distribution and size of the nuclei correspond to the nickel containing spots of a).

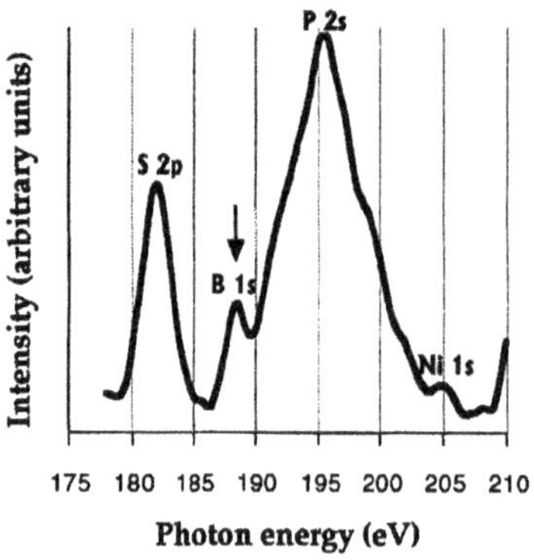

Figure 6. X-ray absorption spectra of a tissue sample stained with nickel against Ki-67. There is a peak at 188.3 eV (arrow) which corresponds to boron in a chemical state other than BSH or boron oxide.

In these investigation boron was detected often in the immediate environment of the cell nuclei (±2 μm). This was also observed by immunohistochemical studies of Otersen *et al.*[2]

Additionally, and unexpectedly, the same tumor tissue was also found to contain boron in a chemical state other than BSH (Figure 6). A reference absorption spectrum was acquired from BSH before and after ashing (Figure 7) which revealed that the origin of the chemical reaction of boron in the tissue is not the incineration process.

4. CONCLUSIONS

The combination of immunohistochemistry and photoelectron spectromicroscopy with synchrotron radiation is a useful method to investigate the subcellular distribution of BSH in glioma tissue.

This method demonstrate for the first time that the chemical state of BSH may be altered *in vivo* metabolism. The new boron compound formed has not been yet identified, but either a change of the boron cage or a binding to another molecule physiologically present in the tissue is plausible. Further studies are needed to clarify the nature of the boron compound which is responsible of the found signal at 188.3 eV.

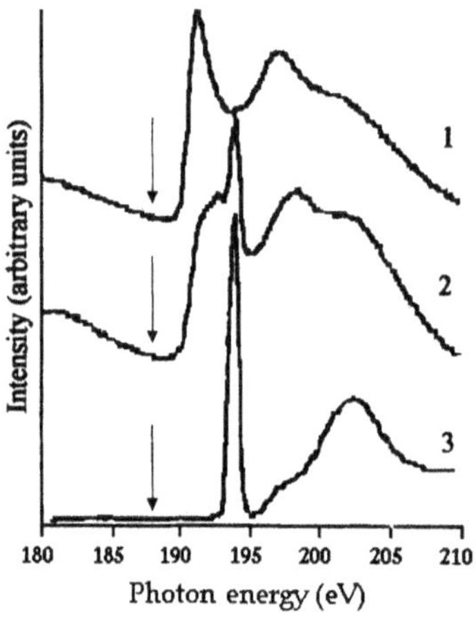

Figure 7. X-ray absorption spectra of a dried droplet of BSH solution, before (1) and after ashing (2). The sharp peak at 194 eV formed after ashing is due to the formation of boron oxide as confirmed by the spectrum of powdered B_2O_3 (3). The arrow marks the photon energy of the B 1s-peak in Figure 6.

ACKNOWLEDGMENTS

This work is supported by the Deutsche Forschungsgemeinschaft and the Fonds der Chemischen Industrie.

REFERENCES

1. De Stasio, G., Pochon, S., Lorusso, G.F., Tonner, B.P., Mercanti, D., Ciotti, M.T., Oddo, N., Galli, P., Perfetti, P., and Margaritondo, G.: Zinc uptake by brain cells: "surface" versus "bulk", *J. Phys. D: Applied Physics* 29:2209–2215, 1996.
2. Otersen, B., Haritz, D., Grochulla, F., Bergmann, M., Sierralta, W.D., and Gabel, D.: Binding and distribution of $Na_2B_{12}H_{11}SH$ on cellular and subcellular level in tumor tissue of glioma patients in boron neutron capture therapy; *J. Neuro-Oncol.* 33:131–139, 1997.

A METHOD FOR DETERMINING THE CONCENTRATION OF BORON IN BLOOD BY THE MEASUREMENT OF THERMAL NEUTRON FLUX DEPRESSION

S. L. Brooke[1,2], S. Green[1], and D. R. Weaver[2]

[1]Medical Physics Department
Queen Elizabeth Hospital
Edgbaston, Birmingham, B15 2TH, The United Kingdom
[2]School of Physics and Astronomy
University of Birmingham
Edgbaston, Birmingham, B15 2TT, The United Kingdom

1. INTRODUCTION

The investigation of a simple and inexpensive method for determining the concentration of ^{10}B in blood is described. This method involves the measurement of the flux depression of thermal neutrons as they pass through a boron-containing sample. The technique is of particular interest in Birmingham for two reasons: 1) Birmingham is an accelerator based facility where prompt-gamma ray measurements are less practicable than with reactor based facilities where multiple beams of neutrons may be extracted. 2) Various radioisotope neutron sources suitable for using with this technique are available at the University of Birmingham (1.1×10^{11} Bq Am/Be and a total of 1.85×10^{12} Bq Pu/Be). Although these sources are not of a high enough activity for prompt-gamma measurements[1] they may be adequate for a technique measuring the thermal neutron flux depression due to the ^{10}B in a sample.[2,3]

It is proposed to use this method as a verification of the technique to be used for routine assaying of boron in blood—inductively coupled plasma-atomic emission spectroscopy.

2. MATERIALS AND METHODS

The sample is placed between the neutron source and a thermal neutron detector in a moderator (for example as shown in Figure 1). Thermal neutrons are

Frontiers in Neutron Capture Therapy, edited by Hawthorne *et al.*
Kluwer Academic / Plenum Publishers, New York, 2001.

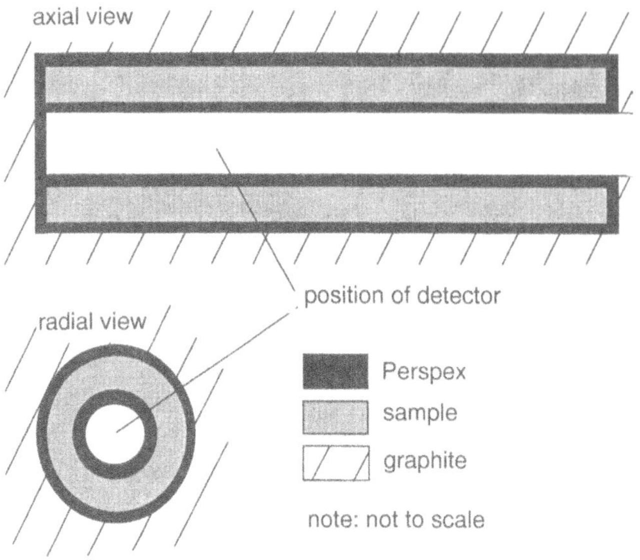

Figure 1. Example of a sample holder for thermal neutron flux depression measurements.

captured by ^{10}B atoms in the sample so reducing the flux reaching the detector. Measurements for a blank and ^{10}B containing sample are compared to calculate the flux depression due to the ^{10}B in the sample. Note that this technique can also be applied to the measurement of the concentration of any element with a high thermal neutron cross section e.g. ^{157}Gd.

The ability of Monte Carlo calculations (MCNP4b) to model the system adequately was verified by comparing experimental and calculated measurements. For measurements a half inch BF_3 gas proportional counter (Centronic 5EB40/13) was surrounded by an annulus of sample contained in a Perspex holder similar to that shown in Figure 1. The sample volume with this holder was 62.5 ml and extended the length of the detector. The samples were solutions of boron or gadolinium made up from the chemical compounds $Na_2B_4O_7.10H_2O$ and $Gd(NO_3)_3.6H_2O$ respectively. Solution concentrations of the isotopes of interest from 0 to 200 ppm were measured (isotopic concentrations calculated assuming the natural abundance of the relevant isotope). Measurements were performed in a graphite stack ($2.56 \times 1.58 \times 1.42$ m) with a 1.1×10^{11} Bq Am/Be source (Amersham type X.3). The measurement time was 5 minutes. This set-up was modeled using MCNP4b with samples of ^{10}B or $^{155}Gd/^{157}Gd$ in water. This comparison between experimental and calculated measurements is shown in Figure 2.

Monte Carlo calculations (MCNP4b) were then used to optimize the geometry of the sample and to calculate the sensitivity of thermal neutron flux depression measurements as a method of assaying the ^{10}B concentration in a solution. The half inch BF_3 gas proportional counter previously mentioned and a half inch ^3He gas proportional counter (Centronic 7.6He3/760/13) were modeled as the thermal neutron detectors. Although the maximum available source activity was taken to be that from a 1.85×10^{12} Bq Pu/Be source, the maximum source activity used in these calculations was limited by the maximum count rate that could be measured by the neutron detectors.

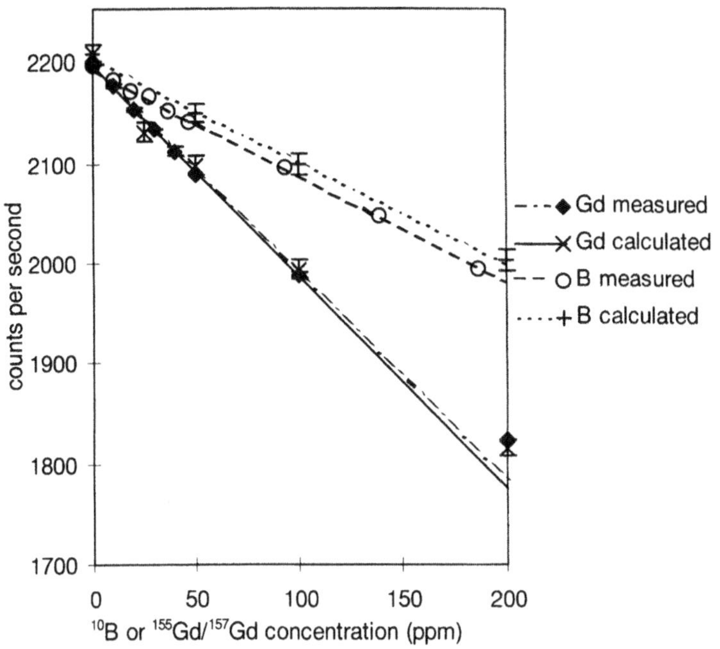

Figure 2. Comparison between the flux measured with a BF_3 gas proportional counter and calculated with MCNP4b. Error bars show one standard deviation on the measurements. Straight lines are fitted from 0 to 200 ppm for ^{10}B and from 0 to 100 ppm for $^{155}Gd/^{157}Gd$ (due to non-linearity at higher concentrations).

3. RESULTS

The optimum investigated sample geometry was found to be an annulus of sample surrounding the length of the active volume of the detector. The results described here were obtained in this geometry. The sensitivity of this technique is dependent on the volume of sample, neutron source activity, thermal neutron detector sensitivity and sampling time. It has been calculated that a 20 ml sample of 10 ppm ^{10}B solution can be measured with a half-inch diameter BF_3 gas proportional counter with a standard deviation of ±5% in under 30 minutes with a neutron source of activity 1.85×10^{12} Bq and in approximately 8 hours with a neutron source of activity 1.1×10^{11} Bq. The same sample can be measured with a half-inch diameter ^3He gas proportional counter with a standard deviation of ±5% in approximately 2 hours with a neutron source of activity 1.1×10^{11} Bq.

Figure 3 shows the concentration that can be measured with a half-inch BF_3 gas proportional counter with a standard deviation of ±10, ±5 and ±1% as a function of measurement time with a source activity of 1.85×10^{12} Bq. The sample geometry is a 20 ml volume annulus extending the sensitive volume of the detector.

4. DISCUSSION

This technique is inexpensive (as the neutron sources are already available at the University of Birmingham) and simple to perform. It is non-destructive to the sample allowing for retrospective re-analysis if required. In addition this technique can be applied

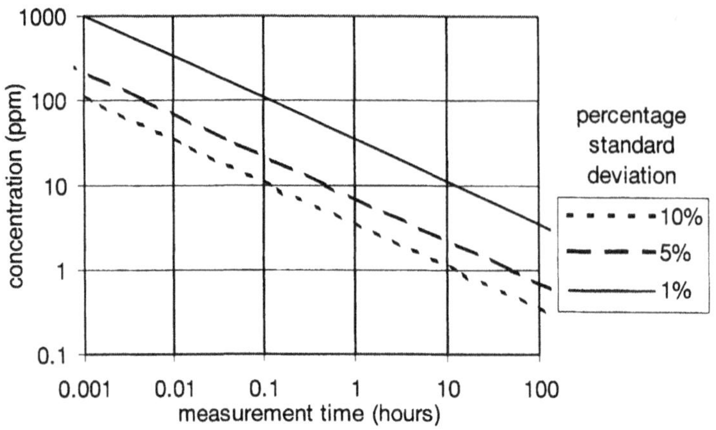

Figure 3. The minimum concentration of ^{10}B that can be measured with a BF_3 detector in a measurement time with a standard deviation of ±1, ±5 and ±10%.

to the concentration measurements for any element with a high thermal neutron capture cross section. Figure 2 shows that the technique is equally applicable to the determination of the concentration of Gd in solution and will have increased sensitivity to the measurement of ^{157}Gd in respect to ^{10}B due to the higher thermal neutron cross-section of this isotope. As a comparison to the results for ^{10}B it has been calculated that a 20 ml sample of 10 ppm ^{157}Gd solution can be measured with a half-inch diameter BF_3 gas proportional counter with a standard deviation of ±5% in under 4 minutes with a neutron source of activity of 1.85×10^{12} Bq.

Disadvantages of this technique include the use of high activity neutron sources which will raise radiological hazard problems that will need to be addressed. The required long sampling times and large sample volumes prohibit the use of this technique for the routine assaying of boron in blood for BNCT. It does however provide a viable simple and inexpensive method of validating the chosen routine technique.

A further sample geometry which will be investigated is a column of sample with the neutron source at one end and the thermal neutron detector at the other. This will increase the effective depth of sample between the source and detector but will require a small volume thermal neutron detector with high sensitivity; one possible detector that will be investigated is a ^{6}Li loaded glass scintillator.

It has been shown that the concentration of ^{10}B in a water sample can be measured by thermal neutron flux depression measurements. Although the required sampling times prohibit the use of this technique for routine assay of boron in blood at the expected concentrations it may be used as an independent check for the routine measurements. Further optimization of sample geometry and thermal neutron detector may improve the sensitivity of this technique.

ACKNOWLEDGMENTS

This project is funded through the New Blood Research Training Fellowship awarded by the NHS Executive (West Midlands).

REFERENCES

1. D.I. Ross. Various Studies Relating to Boron Neutron Capture Therapy, University of Birmingham, PhD thesis, 1992.
2. F. Bombardier. Thermal Neutron Dosimetry for Boron Neutron Capture Therapy, University of Birmingham, MSc thesis, 1996.
3. A. Meehan. Detection of Trace Amounts of Boron-10 for use in Boron Neutron Capture Therapy, University of Birmingham, MSc thesis, 1996.

NON-INVASIVE QUANTIFICATION OF GADOLINIUM UPTAKE IN TUMORS

M. A. Meyer, R. M. Brugger, J. C. Lattimer, K. Kutikkad, A. R. Schoone, and G. Ehrhardt

University of Missouri
Columbia, MO 65211

One of the most lingering challenges of neutron capture therapy (NCT) development is to be able to determine the uptake of the active capture element (B, Gd, Li, or U) without invasive biopsies and just before or during neutron irradiation. In current protocols, boronated phenylalanine (BPA) is injected intravenously into the patient during preliminary surgical debulking of the tumor. Tissue biopsies and blood samples are obtained at this time and uptake of the agent in the tumor and healthy tissue are measured. These are used several weeks later to indicate the uptake when a second injection of boronated phenylalanine (BPA) is made just prior to neutron irradiation. There is the assumption that a similar uptake occurs during the second injection as during the first injection.

Gadolinium (Gd) is a possible NCT element and its uptake can be imaged, but not readily quantified by magnetic resonance imaging (MRI). Due to complex non-linearities in signal intensity with progressively higher concentrations of gadolinium, MRI could not accurately determine Gd content within tumors after intravenous injection of Gd-DTPA or a Gd tagged tumor avid agent. For this reason, it was therefore questioned whether computerized axial tomography (CT) could be used to not only image gadolinium but also quantify its uptake within the brain.

METHODS

A GE CT 9800 and a GE High Speed Advantage CT with helical imaging capabilities were used for all imaging studies (General Electric, Madison, WI). Syringes filled with the intravenous contrast agent Gd-DTPA in serial dilutions were imaged at 3 and 7 mm thickness with CT and then regions of interest drawn within the tomographic cross sectional images, and sample measurements of radiodensity taken using Hounsfield units of measurements. Identical techniques were used to analyze clinical images from an iodine allergic patient who required contrast infusion to assess cerebral circulation

Frontiers in Neutron Capture Therapy, edited by Hawthorne *et al.*
Kluwer Academic / Plenum Publishers, New York, 2001.

while having a transient ischemic attack. Regions of interest were drawn within the middle cerebral arteries to measure the radio density after intravenous bolus injection of gadolinium contrast.

RESULTS AND DISCUSSION

A linear correlation between gadolinium content and CT radiodensity as measured in Hounsfield units was noted (Figure 1); a 1,500 ppm solution containing unenriched Gd gave a CT number of approximately 30 Hounsfield units.

Although iodinated contrast agents are at least twice as dense gadolinium based agents when studied by this way with CT, they have a safety profile that is very different from gadolinium based agents. On the whole, gadolinium based agents are tolerated far better than iodinated contrast, with a far smaller incidence for any type of allergic reactions. Knowing this quality along with the apparent CT radiodensity, gadolinium DTPA was used in an emergent clinical situation in a 63 year old female with a severe iodine allergy who developed brain stem ischemic symptoms one day after a cardiac bypass operation, and then rapidly became unconscious. Since MRI was contra-indicated in her recent post-operative state, gadolinium DTPA was given intravenously to assess the adequacy of the vertebral-basilar and middle cerebal arterial circulation using helical CT imaging techniques. One minute after the administration of 20 cc of contrast, which was well tolerated, the middle cerebal arteries were well visualized with an average enhancement of about 21 Hounsfield units (HU). The densities changed within the middle cerebral artery from 32 HU at baseline (Figure 2, left 7 mm thickness) to 53 HU at one minute after contrast (Figure 2, right, 3 mm thickness).

The usual injection of Magnevist (Gd-DTPA) is 0.1 mmol/kg, but toxicity testing indicates that large animals can tolerate greater than 5 mmol/kg.[1] From previous studies,[2] this would produce an uptake of Gd of 1,100 ppm in a brain tumor, which would show a CT number of 20 HU. Another study[3] has shown that 1,100 ppm of Gd in a tumor in a head phantom and irradiation with epithermal neutrons should deliver a tumorcidal dose. Development of this CT method of quantifying the uptake of Gd-DTPA is continuing with CT scans in animals developing spontaneous brain tumors, and in autopsy

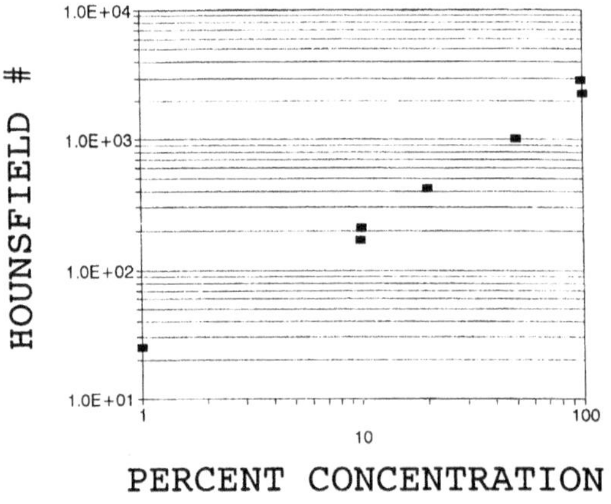

Figure 1. CT radiodensity as measured by Hounsfield units of syringe phantoms filled with progressively dilute solutions of gadolinium DTPA (Magnevist). Percent concentration reflects the degree to which the Magnevist has been diluted in water, from full strength at 100% down to a 1% dilution containing 1 cc Magnevist in a 100 cc total volume.

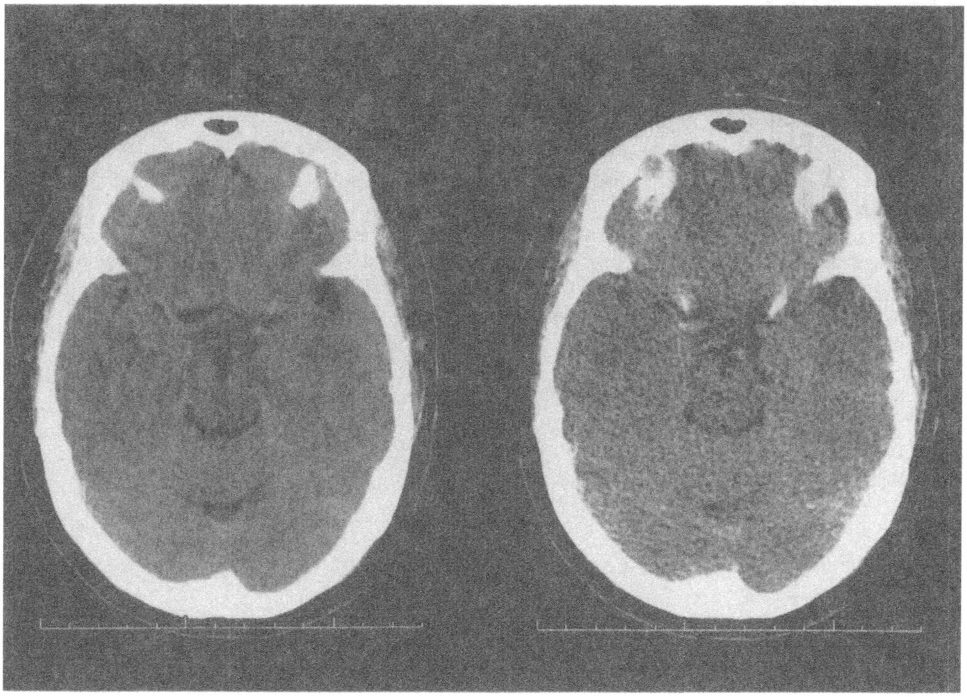

Figure 2. Axial CT sections of the brain from a 63 year old before (left) and one minute after the intravenous administration of gadolinium DTPA (right). The average rise in density within the proximal right middle cerebral artery was 21 Hounsfield units at one minute after contrast administration.

tissue scanned at 1 mm thick slices, and reconstructed three-dimensionally. It is expected that higher doses of gadolinium contrast agents could produce intra-tumoral concentrations of 100 ppm or more, as the standard dose given to the patient illustrated here displayed a transient anterial rise in concentration that exceeded 1000 ppm. Another approach to enhancing tumor concentrations would involve selective intra-arterial injections to the tumor by catheter studies.

CONCLUSIONS AND FUTURE DIRECTIONS

Our studies to date suggest that intravenous injection of Gd-DTPA at standard dosages used in MRI result in quantifiable radiodensities within major vessels when scanned by CT shortly after injection. Although this may have direct applicability to neutron capture therapy of arterio-venous malformations, which enhance strongly after Gd-DTPA administration, higher doses could be explored for the demonstration of adequate enhancement of brain tumors. The use of tumor avid compounds tagged with gadolinium such as Gd tagged porphyrins should also be explored by CT techniques in neutron irradiated nude mice bearing transplanted human gliomas. Preliminary studies in patients using gadolinium texcephyrin show high tumor uptake and may serve as a senistizer to standard external beam radiation therapy.[4]

REFERENCES

1. Magnevist data sheet, Berlex labs, Wayne, NJ.
2. Brugger, R., Gibbs, S., Shih, J.A., Pickup, S., Luo, X., Oro, J., and Laster, B. Estimated doses by Gd NCT in: Topics in Dosimetry and Treatment Planning for NCT; editors: Zamenhoff, K.; Solares, G.; Harling, O.; Advanced Medical Publishing, 1994.
3. Shih, J.A., and Brugger, R.M. Gadolinium as Neutron Capture Therapy Agents, Medical Physics 19(3) pages 733–744, 1992.
4. Rosenthal, D.I., Nurenberg, P., Becerra, C.R., Farenkel, E.P., Carbone, D.P., Lunn, B.L., Miller, R., Engel, J., Young, S., Miles, D., and Renschler, M.F. A Phase I Single-dose Trial of Gadolinium Texaphyrin (Gd-Tex), a Tumor Selective Radiation Sensitizer Detectable by Magnetic Resonance Imaging. Clinical Cancer Research. 5(4):739–745, 1999.

SUBCELLULAR BIODISTRIBUTION OF $Na_2B_{12}H_{11}SH$ (BSH) IN RAT GLIOMA MODEL

Teruyoshi Kageji[1], Shinji Nagahiro, Keizo Matsumoto[1], Birte Otersen[2], Detlef Gabel[2], Manekazu Nakaichi[3], and Yoshinobu Nakagawa[4]

[1]Department of Neurological Surgery
School of Medicine, The University of Tokushima
2-50, Kuramto-cho, Tokushima 770, Japan
[2]Department of Chemistry
University of Bremen
Box 330440, D-28334 Bremen, Germany
[3]Department of Veterinary Surgery
Faculty of Agriculture
Yamaguchi University
Yamaguchi, Japan
[4]Department of Neurologial Surgery
National Kagawa Children's Hospital
Zentsuji-cho, Zentsuji 765, Japan

1. INTRODUCTION

Mercaptoundecahydrododecaborate (BSH: $Na_2B_{12}H_{11}SH$) has been used clinically as a boron compound for boron neutron capture therapy (BNCT) in patients with malignant glioma in Japan since 1968 to present. BSH is characterized by the absence of toxic side effect and represents a promising boron carrier applied for the therapy of malignant glioma. In 1967, Sweet, Soloway and their colleagues found BSH to accumulate in a mouse glioma model. More recently, studies have shown that BSH is also accumulated into the tumor tissue in various rat glioma model. Hatanaka started the renewal of BNCT since 1968 in Japan, with BSH and thermal neutron. From 1968 to present, more than 150 patients with malignant intracranial tumors, especially malignant glioma were treated by BNCT using BSH in Japan. In Europe, BSH has been introduced as a boron carrier for clinical trial of BNCT from 1997.

Despite the clinical long use of BSH about 30 years, published data on the mechanism of boron uptake and the precise boron location in a cell structure are few and lack in standardization. This work is aimed to clarify some hypotheses about subcellular

Frontiers in Neutron Capture Therapy, edited by Hawthorne *et al.*
Kluwer Academic / Plenum Publishers, New York, 2001.

biodistribution of BSH in a rat glioma model by means of immunohistochemical approach. It will be investigated where BSH locates in a tumor cell. Furthermore, it will be investigated whether any of the following features could be responsible for subcellular localization of BSH such as time interval from BSH infusion to tumor tissue sampling.

2. MATERIALS AND METHODS

The Wister rats were used for this experiment. They were anesthetized and placed in stereotactic head holders. An intracerebral injection of 5.0×10^6 C6 glioma cells was introduced into the region of cerebral hemisphere. BSH was infused to the rats through the tail vein two weeks after implantation.

The host were 18 Wister rats. For this analysis, 50 mg ^{10}B/kg of BSH was given to the rats intravenously. Time duration was calculated from the start of BSH infusion.

Host rats were divided into six groups according to the time interval from BSH infusion to sacrifice; 1 hr, 4 hrs, 8 hrs, 16 hrs, 24 hrs, 48 hrs after the start of BSH infusion. The samples were immediately frozen in liquid nitrogen. The 7-μm thickness of the sections were cut by microtome for immunohistochemistry.

Antibodies against BSH were raised in goats. (the preparation of the antibodies was kindly performed by Prof. Dr. P. W. Jungblut, Max Planck Institute for Experimental Endocrinology, Hannover). These antibodies are specific against BSH and show no cross reactivity with other epitopes in healthy and tumor brain tissue.

Birte Otersen and Detlef Gabel have analyzed the microdistribution of BSH with anti-BSH antibody in human brain tissues. They reported that in healthy brain tissue, boron can only be detected in endothelial cell, in agreement with the non-measurable uptake of boron in healthy brain. In tumor tissue in malignant glioma, boron can be detected frequently in cell membrane, and often in cytoplasm. Equally, cell nuclei frequently accumulate boron.[6]

Immunohistochemical investigations were carried out by means of the ABC (avidin-biotin-peroxidase complex method) with different color development, or by using two different color-producing enzymes conjugated to the corresponding antibodies. In addition, serial sections are to be stained individually.

3. RESULTS

This immunohistochemical investigation with rat brain tumor model showed the following results: In an early period such as one hour after BSH infusion, strong staining was found around tumor vessels or in the interstitial space, and boron has already contacted with tumor cells. Some of tumor cells were stained with anti-BSH antibody in a whole cell, including cytoplasm and cell membrane. Around eight to 16 hours after infusion, staining in the interstitial space was likely to disappear, and we could find strong staining as dots in cell nucleus. And then, boron seemed to exist still in cell nucleus and cytoplasm after 48 hours after infusion (Figure 1).

4. DISCUSSION

Despite the long clinical use of BSH about 30 years, published data are few and lack on the mechanism of boron uptake and the precise boron location in cell structure.

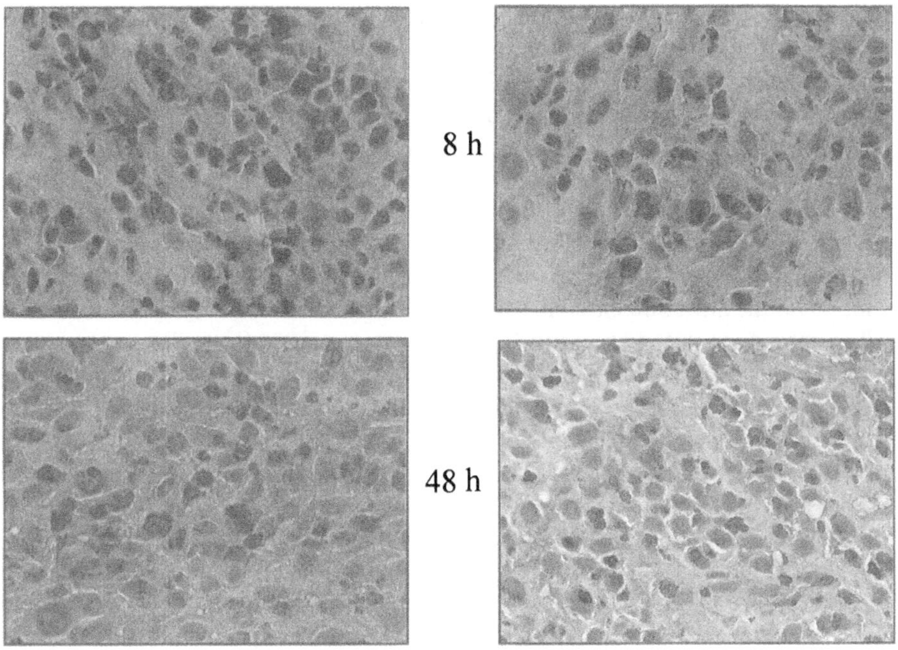

Figure 1. Immunohistochemical study with anti-BSH antibody in rat brain tumor model to analyze boron distribution in a tumor cell with time interval from infusion. (Rats were sacrifired 1, 8, 16, 48, hours after BSH infusion.)

It is reported that all boron present in tumor tissue is bound firmly to the tissue. To investigate the binding of BSH to human glioma tissue, QNCRs (quantitative neutron capture radiography) of fixed sections exposed to either formaldehyde vapor and formaline solution were compared to untreated sections. Boron was present in both preparations to the same extent.[6]

There were several reports about boron distribution in human brain tumor. By means of neutron-induced alpha-autoradiography, some investigators studied the intracellular localization of boron in human brain tumor. Direct observation of the tracks of alpha-particles in tumor tissue can confirm the actual transfer of energy. It was observed that the track density inside or outside the nuclear membrane was at least four times as high as for entire tumor tissue.[1] Subcellular fractionation showed that the boron compound was mostly partitioned in the soluble component of the cell, which was in the cytoplasm or the extracellular space (60%); a maximum of 40% were located in cell nuclei and mitochondria.[2] Laser microprobe mass analysis is a modern analytical method that combines laser ionization with time-of-flight mass spectroscopy. It has found that accidental boron contamination of samples were easily recognized on the basis of $^{10}B:^{11}B$ isotope ratio. In glioblastoma, boron-10 was found only in the nuclei of neoplastic cells but not in other cell components.[4] In *in vitro* studies with 9L gliosarcoma cells, most of the boron localized in the cytoplasm of the cell, but the nuclei contained only minimal concentration of boron.[5] Using the isotopic imaging technique ion microscopy, boron was more concentrated in the cytoplasm than in the nucleus but was not strongly localized within cells.[3,7,8] By means of specific antibodies against BSH, immunohistochemistry was performed. In tumor tissue, a strong intracellular staining is found, which is frequently observed on cell membranes. The whole cell body is uniformly stained, and it is found frequently that cell nuclei are stained intensively.[6]

The previous investigations about boron subcellular distribution of BSH revealed various results; some investigators report that boron locates only in cell nuclei, others reported boron located in cell membrane, cytoplasm and also cell nuclei. The author thinks that these various results could be due to the materials. Most of the previous investigators analyzed the boron distribution in tumor tissue with human samples. It is assumed that several factors have an influence on boron uptake in tumor tissue; total dose of infused BSH, route of infusion, tumoral histology and time interval from BSH infusion to tumor tissue sampling. We have a hypothesis about boron uptake mechanism in tumor cells. One is that time interval from the start of BSH infusion to tumor tissue sampling is the most influential in subcellular boron distribution; boron might travel from cell membrane to cytoplasm with time after interaction with a cell, consequently reaching cell nucleus. In case of human brain tissue, it is difficult to achieve similar conditions in samples since numbers of patients for malignant glioma are restricted. In case of rat brain tumor model, one can obtain enough tumor samples to analyze these hypotheses under the same conditions. It will be possible to study whether the hypothesis might be true or not by means of rat brain tumor model.

This immunohistochemical investigation with rat brain tumor model showed the following results: In an early period such as one hour after BSH infusion, boron was around tumor vessels or interstitial space, and has already contacted with tumor cells. Some of tumor cells were stained with anti-BSH antibody in a whole cell, including cytoplasm. Around eight to 16 hours after infusion, most of boron seemed to collect in cell nucleus and stained as dots. At that time, the staining of the interstitial space was very weak. Boron seemed to exist in a cell after 48 hours after infusion (Figure 1). As a result, we conclude the following theory about boron uptake. First, BSH passes through the disrupted blood-brain-barrier, and then interacts with cell membrane. Second, after binding to the membrane, boron goes into the cytoplasm, and then travels the nucleus. After infusion, it takes for a short time to reach cell nucleus, for example during one hour in this rat model. Boron was retained for a long time, more than 48 hours in the cell nucleus, and then washed out from a tumor cell.

Our own results presented here and the clinical boron measurements support the following hypothesis on boron uptake into malignant glioma cells: BSH can pass through the disrupted blood-brain-barrier (BBB) easily and can come in contact with tumor cells; there, BSH can bind on the extracellular surface of plasma membrane to choline residues of phosphatidylcholine and sphingomyelin, which are increased abnormally in tumor tissue. After binding to the plasma membrane, boron with choline residues may be internalized into the cell by endocytotic pathways and eventually travel to cell nuclei and cytoplasm. The passage through the disrupted BBB is essential for boron accumulation around tumor cells. The formation of stable, neutral ion pairs between the boron cluster and quaternary ammonium salts could be crucially responsible for binding, retention and internalization.

5. CONCLUSION

With immunohistochemical study in a rat glioma model, boron was recognized in a whole cell in an early period after venous infusion. Consequently, it moved from cell membrane and cytoplasm to nucleus with time, and retained there for a long time, at least 48 hours. This immunohistochemical investigation with anti-BSH antibody in rat glioma model supports the following hypothesis on boron uptake into malignant glioma cells: BSH can pass through the disrupted blood-brain-barrier (BBB) easily and can come in contact with tumor cells; there, BSH can bind on the extracellular surface of plasma membrane to choline residues of phosphatidylcholine and sphingomyelin, which are increased abnormally in tumor tissue. After binding to the plasma membrane, boron with choline residues may be internalized into the cell by endocytic pathways and eventually travel to cell nuclei and cytoplasm.

REFERENCES

1. Amano K.: Chapter VI. Boron-10-Mercaptoundecahydodecaborate distribution in human brain tumors as studied by neutron-induced alpha-autoradiography. In: Hatanaka H. (ed) Boron-neutron capture therapy for tumors. Nishimura, Niigata, 1986, pp. 107–115.
2. Ceberg C.P., Persson A., Brun A., Huiskamp R., Fyhr A.S., Persson B.R.R., and Salford L.G.: Performance of sulfhydryl boron hydride in patients with grade III and IV astrocytoma: a basis for boron neutron capture therapy. J Neurosurgery 83:79–65, 1995.
3. Chandra S., Lorey II D.R., Lessig S.L., Smith D.R., and Morrison G.H.: Quantitative imaging of boron from BPA and BSH in glioblastoma cells with ion microscopy. Advances in Neutron Capture Therapy. Volume II, Chemistry and Biology, pp. 315–320.
4. Haselberger K., Radner H., Gossler W., Schlagenhaufen C., and Pendl G.: Subcellular boron-10 localization in glioblastoma for boron neutron capture therapy with $Na_2B_{12}H_{11}SH$. J Neurosurg 81:741–744, 1994.
5. Nguyen T., Brownell G.L., Holden S.A., Kahl S., Miura M., and Teicher B.A.: Subcellular distribution of various boron compounds and implications for their efficacy in boron neutron capture therapy by Monte Carlo Simulations. Radat Res 133:33–40, 1993.
6. Otersen B., Haritz D., Grochulla F., Bergmann M., Sierralta W., and Gabel D.: Binding and distribution of $Na_2B_{12}H_{11}SH$ on cellular and subcellular level in tumor tissue of glioma patients in boron neutron capture therapy. J of Neuro-Oncology 33:131–139, 1997.
7. Smith D.R., Chandra S., Coderre J.A., Barth R.F., Yang W., Liu L., Wilson J.L., Micca P.L., Nawrocky M.M., Rotaru, J.H., and Morroson G.H.: Quantitative ion microscopy imaging of boron-10 in rat brain tumor models for BNCT. Advances in Neutron Capture Therapy. Volume II, Chemistry and Biology, pp. 308–314.
8. Zha X., Ausserer W.A., and Morrison G.H.: Quantitative imaging of a radiotherapeutic drugs, $Na_2B_{12}H_{11}SH$, at subcellular resolution in tissue cultures using ion microscopy. Cancer Res 52:5219–5222, 1992.

DETERMINATION OF BSH ($Na_2B_{12}H_{11}SH$) ABSORBED IN TUMOR TISSUE

Yoshinobu Nakagawa[1], Kyonghon Pooh[1], Michiko Sone[1], Teruyoshi Kageji[2], M. Nakaichi[3], Hiroshi Takahashi[4], K. Amemiya[4], and Detlef Gabel[5]

[1]Department of Neurosurgery
National Kagawa Children's Hospital
Kagawa 765, Japan
[2]Department of Neurosurgery
Tokushima University
Tokushima 770, Japan
[3]Department of Veterinary Surgery
Yamaguchi University
[4]Department of Quantum Engineering and System Science
School of Engineering
The University of Tokyo
[5]Department of Chemistry
University of Bremen

INTRODUCTION

Since 1968 we have tried BNCT using the combination of BSH ($Na_2B_{12}H_{11}SH$) and thermal neutron beam. However, it is still controversial that BSH is selectively absorbed in the tumor cells especially in the cell nucleus. In this study, in order to investigate the accumulation of absorbed BSH in the tumor cells, we intended to demonstrate them using alpha autoradiography, immunohistochemical study and atomic force microscope in experimental brain tumor model in rats.

MATERIALS AND METHODS

Experimental brain tumor model was made by stereotactic intracerebral transplantation of C-6 glioma cells (5.0×10^6) in Wister rats. After two–three weeks, BSH (50–60 mg B-10/kg) was injected into the peritoneal cavity. The rats were sacrificed at

Frontiers in Neutron Capture Therapy, edited by Hawthorne *et al.*
Kluwer Academic / Plenum Publishers, New York, 2001.

1 to 48 hours after the injection of BSH. The whole brains were removed and frozen at –80C. Boron concentration in blood, brain tumors and brain tissue was measured by prompt gamma ray spectrometry and/or ICP-AES.

Frozen brain was cut into 7–10 micrometer of thickness for alpha-autoradiography and immunohistochemical study. Antibodies against BSH could be raised in goats. These antibodies are specific against BSH and show no cross reactivity with other epitopes in healthy and tumor brain tissue. Immunohistochemical study was performed by ABC (avidin-biotin-peroxidase complex method) with different color development, or by using two different color-producing enzymes conjugated to the corresponding antibodies. In addition, serial sections were stained individually. One micrometer sections of the frozen brain with brain tumor tissue was made by cryostat for the examination by atomic force microscope. Each section was mounted on nitrocellulose film (CN-85), glass slides and CR39. Nitrocellulose film and CR39 were irradiated by thermal neutron beam to a fluence of 1.3–1.4 $\times$ 10^{13} neutrons/cm^2 at KUR. The films and CR39 were etched in NaOH, 70°C for 10 minutes and investigated microscopically by microscope and atomic force microscopy. The other frozen section was stained by Hematoxylin-Eosin staining and compared to the findings of atomic force microscopy.

RESULTS

Boron Concentration

The boron concentration in blood and brain tissue showed peak levels just after the injection of boron compound. Both values rapidly decreased up to 4 hours and then showed gradual decrease. The mean concentration of BSH in blood at four hours after the injection of BSH was 26.7 ppm.

Autoradiography

High density of alpha tracks was demonstrated in the area corresponding to the tumor tissue, whereas less pronounced density was observed in the normal brain tissue. Etched and enlarged pits were revealed over the cell bodies, and especially over the cell nucleus (Figure 1).

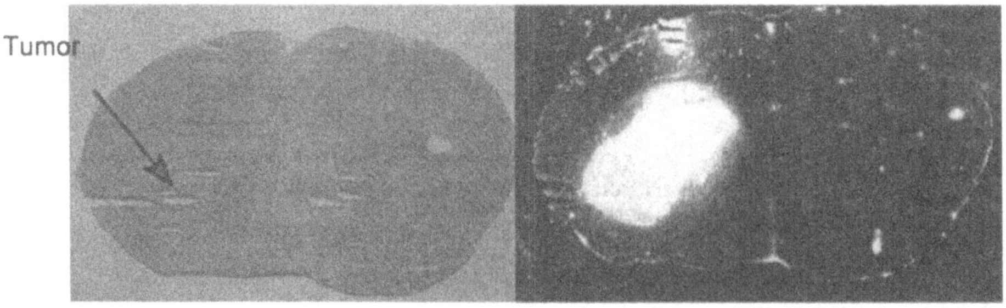

Figure 1. Experimental brain tumor model in rat. Hematoxylin Eosin staining in left. Autoradiography in right.

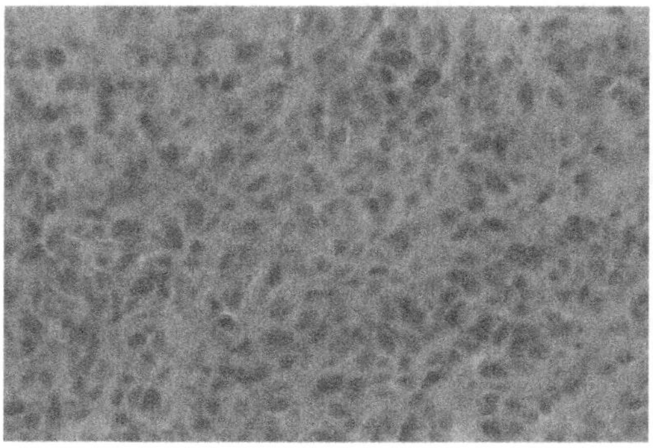

1hr after injection of BSH

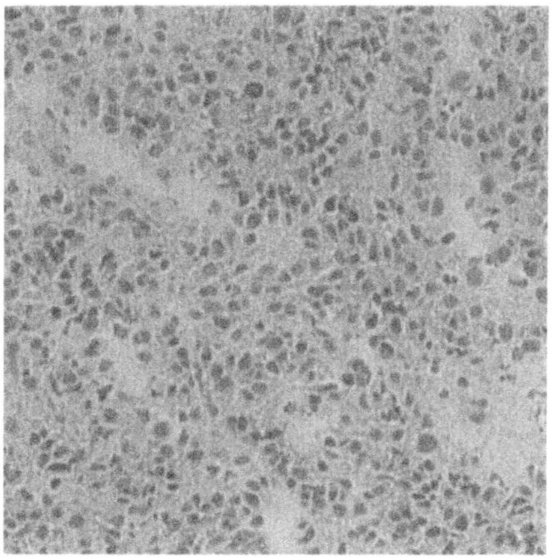

4hrs after injection of BSH

Figure 2. Immunohistochemical study revealed that SBH was markedly demonstrated in the cytoplasm and nucleus of the tumor cells. Positive immunoreaction was scarcely seen in the normal brain tissue around the tumor.

Immunohistochemical Study

One hour after BSH infusion, marked staining of BSH was demonstrated around tumor vessels or in the extracellular space. These findings indicated that boron has already contacted with tumor cells. Some of tumor cells were stained with anti-BSH antibody in a whole cell, including cytoplasm and cell membrane. Four hours after infusion, BSH was markedly demonstrated in the cytoplasm and nucleus of the tumor cells. Positive immuno reaction was scarcely seen in the normal brain tissue around the tumor (Figure 2).

Examination by Atomic Force Microscope

Multiple ultra small etch pits were three dimensionally demonstrated in the cytoplasm and nucleus. The typical diameter of the etch pits was below 1 μm. The etch pit depth was about several hundred nm.

These findings indicated that BSH is selectively absorbed in the tumor cells especially in cytoplasm and nucleus (Figure 3).

DISCUSSION

The immunohistochemical technique in this study was established in the Department of Chemistry, University of Bremen by Prof. Dr. Detlef Gabel and his collaborators. The antibodies against BSH could be raised in goats. The preparation of the antibodies was kindly performed by Prof. Dr. P. W. Jungblut, Max Planck Institute for Experimental Endocrinology, Harnover. D. Gabel *et al.* analyzed the microdistribution of BSH with the antibodies in human brain tissues. They reported that in healthy brain tissue, boron could only be detected in endothelial cell, in agreement with the

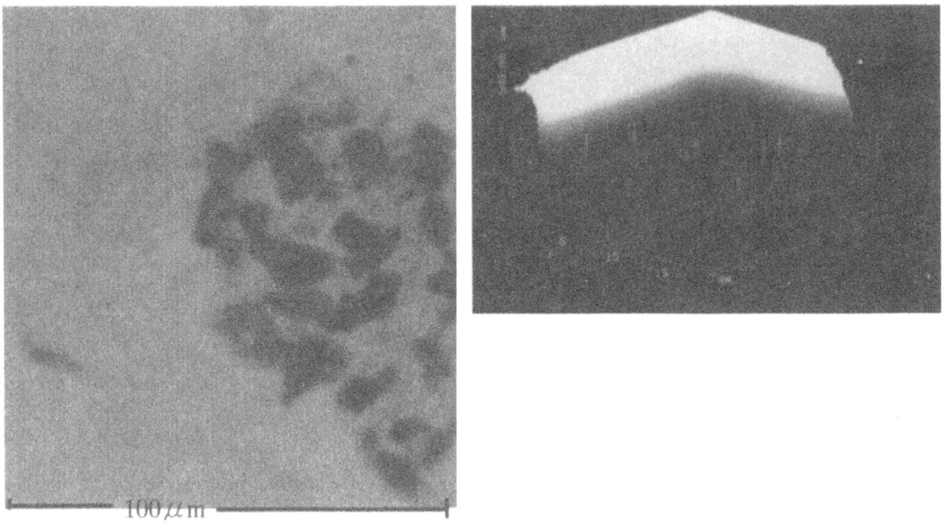

Figure 3. Multiple ultra small etch pits (blue dots) were demonstrated in the cytoplasm and nucleus (left). The radiographic image of a silver mesh obtained with AFM after neutron irradiation and etching (right).

non-measurable uptake of boron in healthy brain. In tumor tissue in malignant glioma, boron can be detected frequently in cell membrane, and often in cytoplasm. Equally, cell nuclei frequently accumulate boron. In this study we demonstrated the accumulation of BSH in the cytoplasm and nuclear membrane of the tumor cells at 4 hour after injection of BSH. The same findings were demonstrated by atomic force microscopy (AFM). AFM has been widely used for insulating material observation because of its high resolution and applicability to many different materials including transparent ones. AFM has also been applied to the readout of solid state nuclear track detectors (SSNTD) by several researchers and its excellent characteristics have been reported. In this technique, the SSNTD is etched for a very short time, about several minutes. The etched surface of the SSNTD is scanned by an AFM. Consequently very small etch pits of several hundred nm can be observed. The method can be used for the microdosimetry in BNCT.

The findings demonstrated in this study supported the result reported by Gabel *et al.* They supposed that BSH can pass through the disrupted blood-brain-barrier (BBB) mainly in the tumor tissue and/or damaged brain tissue around the tumor and can come in contact with tumor cells. BSH is absorbed selectively by tumor cells, however, to analyze micro-distribution in tumor cells, more detailed study using CR-39 and atomic force microscope is necessary.

REFERENCES

1. Kageji T., Kitamura K., Hatanaka K., and Nakagawa Y.: Advances in Neutron Capture Therapy. Volume II, Chemistry and Biology. Amsterdam, Elsevier Science B.V., 1997, pp. 197–201.
2. Kageji T., Nakagawa Y., Kitamura K., Matsumoto K., and Hatanaka H.: J of Neuro-oncology 33:117–130, 1997.
3. Kageji T., Otersen B., Gabel D., Huiskamp R., Nakagawa Y., and Matsumoto K.: Boron Neutron Capture Therapy. Biochimica et Biophysica Acta (in press).

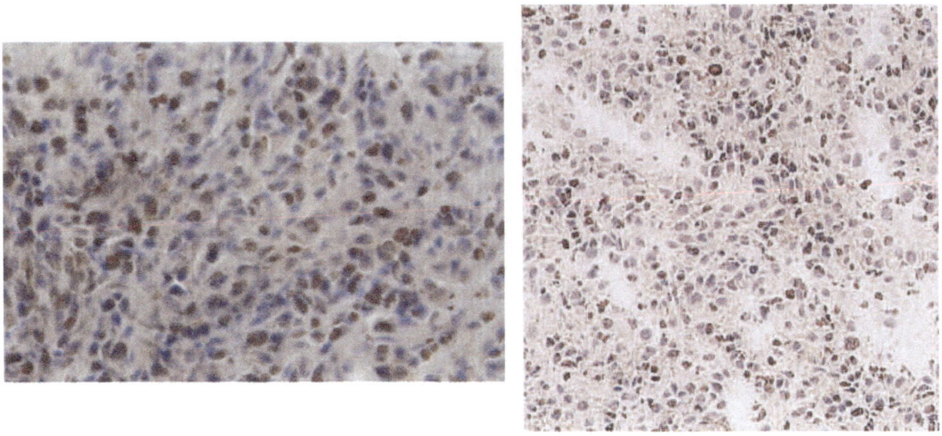

Figure 2. Immunohistochemical study revealed that BSH was markedly demonstrated in the cytoplasm and nucleus of the tumor cells. Positive immunoreaction was scarcely seen in the normal brain tissue around the tumor. (Left: 1 hr after injection of BSH; right: 4 hrs after injection of BSH.)

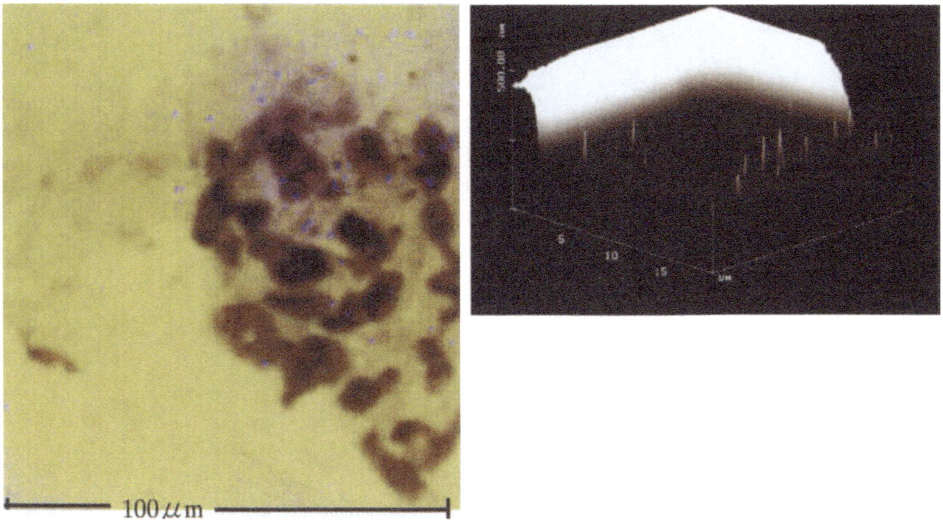

Figure 3. Multiple ultra small etch pits (blue dots) were demonstrated in the cytoplasm and nucleus (left). The radiographic image of a silver mesh obtained with AFM after neutron irradiation and etching (right).

4. Nakagawa Y., and Hatanaka H.: J of Neuro-oncology 33:105–115, 1997.
5. Nguyen T., Brownell G.L., Holden S.A., Kahl S., Miura M., and Teicher B.A.: Radat Res 133:33–40, 1993.
6. Otersen B., Haritz D., Grochulla F., Bergmann M., Sierralta W., and Gabel D.: J of Neuro-Oncology 33:131–139, 1997.
7. Amemiya K., Takahashi H., Nakagawa M., Yasuda N., Yamamoto M., Nakagawa Y., Kageji T., Nakaichi M., and Ogura K.: Nucl. Instr. and Meth. B 159 75–80 (1999).

DETECTION OF ^{10}B DISTRIBUTIONS IN HISTOLOGICAL SAMPLES BY NEUTRON CAPTURE AUTORADIOGRAPHY USING COLD NEUTRON BEAMS AND PHOTOLUMINESCENT IMAGING PLATES

J. Rant[1], M. Čemažar[2], G. Bayon[3], Y. Yanagie[4], J. Škrk[2], H. Kobayashi[5], D. Gabel[6], and G. Serša[2]

[1]J. Stefan Institute, Jamova 39
1000 Ljubljana, Slovenia
[2]Institute of Oncology, Zaloska c. 2
1105 Ljubljana, Slovenia
[3]CEA Saclay
F91191 Gif sur Yvette Cedex, France
[4]Institute of Medical Science
University of Tokyo
4-6-1 Shiroganedai, Tokyo 108, Japan
[5]Institute of Atomic Energy
Rikkyo University
2-5-1 Nagasaka, Kanagawa, 240-01, Japan
[6]Chemistry Department
University of Bremen
POB 330 440
D-28334 Bremen, Germany

1. INTRODUCTION

The Neutron Capture Autoradiography (NCAR) using various Solid State Nuclear Track Detectors (SSNTDs) is a well established method to detect and measure the distributions of ^{10}B in the ppm range on macroscopic and microscopic level in histological sections of tissues loaded with ^{10}B compounds used for BNCT.[1,2] The SSNTD techniques require relatively high neutron fluences, chemical etching and somewhat time consuming quantitative evaluation. Two exposures are normally needed—for B distribution visual-

Frontiers in Neutron Capture Therapy, edited by Hawthorne *et al.*
Kluwer Academic / Plenum Publishers, New York, 2001.

ization (10^{12}–10^{13} n/cm^2) and 2–3 orders of magnitude lower exposure for quantitative boron determination by various track counting techniques. Recently a new technique of NCAR using sensitive photoluminescent storage phosphors (Imaging Plates or IP) has been proposed to detect heavy charged particles from the $^{10}B(n,\alpha)^7Li$ reactions for the topography of ^{10}B distributions in histological sections.[3] The new method exploits excellent detection properties of IP systems such as very high detection sensitivity and quantum detection efficiency, broad linear response and dynamic range, very small image distortion, reusability of IP and possibilities of digital autoradiography with direct quantitative evaluation.[4,5] A severe problem for using IP in NCAR is their high sensitivity to all types of ionizing radiations. The background of direct and induced gamma-rays as well as of epithermal and fast neutrons has to be suppressed or filtered out of the neutron beam. To improve the signal/background ratio and to increase the detectability of ^{10}B we proposed the use of rather gamma ray free cold neutron beams for the IP-NCAR of ^{10}B distributions in histological samples.[6] In this work the first results of preliminary experiments in IP-NCAR with cold neutrons are presented.

2. METHODS AND MATERIALS

2.1. Neutron Capture Autoradiography

The procedure for NCAR with IP is practically the same as used in NCAR with SSNTDs. The freezed dry whole body histological sections of mice were put directly on the unprotected surface of the imaging plate, wrapped by light tight thin plastic cover and sealed after evacuating air by vacuum pump. By this way an intense tight contact between the IP and the sample was achieved. The use of thin light tight plastic foil as irradiation cassette is essential for IP NCAR as the IP have to be protected from the light (in particular UV) and all types of ionizing radiations. The use of thin low Z “casette” materials with a low neutron absorption cross section is important to minimize the production of neutron and gamma-ray induced secondary ionizing radiations in order to minimize the induced background of the IP. The wrapped IP/sample assemblies were air mailed to Saclay for irradiation and immediately returned back to IJS for reading. The time lag between the preparation of the sample/IP assembly and read out was typically 6–8 days and the time elapsed after the neutron irradiation was 4–6 days. During the time lag the signal faded for a factor ~2.5 and the cosmic/environmental radiation background increased for ~25 X!

2.2. Cold Neutron Source

The neutron irradiaions were performed on the neutron radiographic facility at the cold neutron guide tube G4 of the research reactor ORPHEE (Saclay, France). The cold neutron flux is approximately 9.10^8 n/cm^2 s and the average energy of the neutrons is 0.003 eV. The neutron beam with a square profile 25 mm by 150 mm is skewed for about 7% in both X and Y axis. The samples are irradiated by moving along the X-axis and the neutron fluences have to be corrected relevant to the irradiation position along the vertical axis. Neutron fluences varied between 1.0–5.0 10^9 n/cm^2. The gamma background is considered to be negligible.[7]

2.3. Imaging Plate Detection System

The imaging system used consists of FUJI Photo Film imaging plates type BAS IP-TR, FUJI BAS 1,500 laser scanner and an UV IP eraser. The digital images are handled, processed and analysed by a personal computer (Intel 486, 120 MHz, 28 M RAM, 1G hard disk) and printed by Epson Stylus Photo printer. The image handling, processing and analysing software is TINA 2.0 (Raytest, Germany) and Photoshop ADOBE 3.0. The IP TR type consists of a plastic support plate (12.7 cm × 12.7 cm) covered with 50 μm thick layer of crystals of photostimulable storage phosphor $BaFBr:Eu^{2+}$ contained in a plastic binder. The phosphor layer is intended for the registration of ^{3}H soft beta radiation and is hence unprotected and susceptible for surface damage and wetting. The thickness of the active layer of the phosphor is much thicker than the range of α and ^{7}Li particles in the phosphor (<10 μm). The BAS 1500 He-Ne laser (632.8 nm) scanner has a pixel size 100 μm × 100 μm, scanning speed 14 μs/pixel and 12 bit digitalization (4096 gray levels). Scanners with pixel size 25 μm × 25 μm and digitalization of 16 or 32 bits are commercially available. The inherent resolution of the system is worse than the pixel size, about 200 μm (with advanced scanners ~40 μm).

2.4. Standards and Histological Samples

The standard samples with known concentration of ^{10}B were prepared by pouring 10 μl drops of sodium borocaptate (BSH) solution in water on the filter paper (B impurity below detection limit 3 ppm) forming stains of ~1 cm^2. The B content of the BSH solutions was measured by plasma emission spectroscopy. The measured ^{10}B concentrations (in ppm) were: 841 ∓ 28, 248 ∓ 8.7, 84.6 ∓ 1.5, 25.2 ppm ∓ 0.32, 8 ∓ 0.3. Boron concentration on the filter paper were determined by digesting the paper samples in 65% HNO_3, diluting by pure water for factor 10 and measuring again by plasma emission spectroscopy. The measured ^{10}B concentrations were 483 ∓ 15.6, 176 ∓ 3.0, 73.5 ∓ 5.8, 19.1 ∓ 2.0 and 7.3 ∓ 4.2 ppm. The B detection limit was at 3 ppm.

In addition similar filter paper ^{10}B concentration indicators with rather unreliable and approximate concentration values of 24,000 ppm, 2,400 ppm (confirmed by PGAA), 240 ppm and 24 ppm were used.

Three different samples of whole body histological sections of mice were supplied by two of the co-authors. Sample 1a was a whole body section of the tumor bearing mouse sacrificed 3 h after the intravenous injection of ^{10}BSH solution (1.5246%, PGAA result) and sample 1b was a section of a mouse sacrificed 6 h after injection of ^{10}B conjugated liposome/PEG liposome (117 ppm). In the former sample high B concentration was found in the liver and to smaller extent also in the tumor and kidney. In the later case the B was localized only in the tumor at low concentration, as confirmed by subsequent NCAR with SSNTDs. The thickness of the samples was about 40 μm and were freezed dried mounted on the 3 M scotch tape. Details of the preparation of the samples can be found in the literature.[3,8] Sample 2 was a whole body section of the mouse with B16 melanoma, sacrificed 1 h after the intraperitoneal injection of 2 mg ^{10}BSH. The thickness of this sample was 50 μm.

3. RESULTS AND DISCUSSION

The surface distribution of B in each stain is not uniform and an avarage signal from the whole surface area of the stain has to be determined using tools of the TINA software and correlated with the B content. The average signal of photostimulated luminescence (PSL) above the average of the local background was found proportional to the B concentration of the relevant stain in the whole range of B concentrations. The stains with B content of about 70 ppm can be easily detected and the detection limit in this case is around 30 ppm, since the 19 ppm and 24 ppm stains can be observed only on the original image and the signal can hardly be discerned from the background noise. In IP NCAR the backround due to protons from neutron captures in nitrogen and heavy particls and gamma rays from neutron absorption in hydrogen, oxygen and Li and B impurities in tissue is rather high and signal/background ratio is much worse than with NCAR using SSNTDs. The SSNTDs are sensitive only for heavy charged particles above certain threshold energies and offering the possibility of particle track discrimination by properly selecting the etching conditions[1] and/or etchant[8,9] and in addition by automatic image analysis by measuring track dimesions, shape and their optical properties.[2,10] However, for complete visual and quantitative image evaluation in SSNTD NCAR two exposures are needed. Computerized track discrimination and counting requires fluences more than 2 orders of magnitude lower (~2 $10^9 n_{th}/cm^2$) than required for visual image observation ($>10^{12} n_{th}/cm^2$) in order to prevent track overlapping. With IP NCAR as digital autoradiographic method quantitative evaluation and visual observation can be performed simultaneously.

The IP NCAR images of two whole body sections of mices prepared to put high B concentration in the body organs clearly revealed the distribution of B in the organs and tissues as well as differences in B content of different organs. In NCAR autoradiography of sample 1a (Figure 1) the high B concentration is in liver and at much smaller concentration in tumor, bladder and intestines. Using an appropriate calibration curve for B concentration monitoring prepared in the same way as actual samples and using similar tissue standard it could be possible immediately to determine the B content in selected regions. The spatial resolution is fair (200 μm) but with modern scanners it can be improved down to ~40 μm, close to that of the whole body NCAR using SSNTD

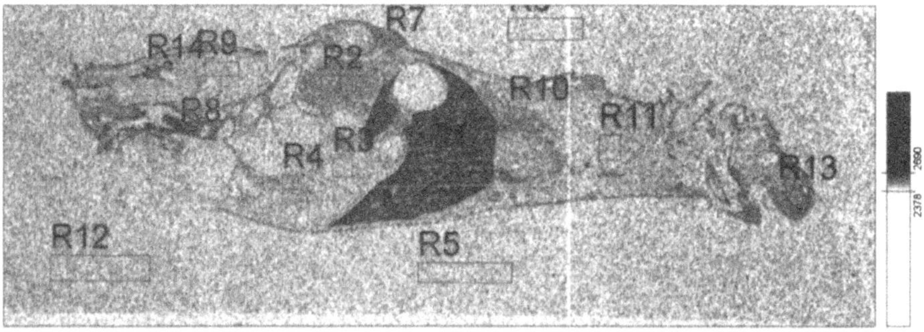

Figure 1. CNCAR of a mouse sample (Dr. Yanagie), sacrificed 3 hours after intravenous injection of B-10 BSH solution (1.5246%). Reactor ORPHEE (CEA, Saclay), exposure 5.10 + 9 n/cm2, FUJI type TR IP, scanned after 7 days waiting time! High B content is in liver (R1) and at much smaler concentration also in the tumor (R7).

(10–20 μm).[1,2] Reduction of the proton background is important to push the detectability of B by IP NCAR down to 10 ppm level. Further experiments should be repeated in optimal experimental conditions.

4. CONCLUSIONS

At present the digital NCAR using clean cold neutron beams and IP as detectors has a potential to detect ^{10}B in histological sections of tissue at the level of about few 10 ppm and with the spatial resolution of ~40 μm. Using calibration curves prepared by tissue standards with well known B content and in the same way as actual histological samples for BNCT studies the method enables immediate and simple quantitative analysis. The advantage of IP-NCAR vs. NCAR with SSNTDs should be the much lower neutron fluence (10^7–10^9 vs. 10^{10}–10^{13} n/cm^2 with SSNTDs), no intermediate chemical treatment (track etching) and direct and fast computational handling and evaluation of the digitized autoradiographic image. Further experiments in optimal experimental conditions are required to estimate accurately the detectability of B in tissues and to estimate the signal/background ratio. The use of IP with thin storage phosphor layer of about 10 μm should be prefered to decrease the γ-ray background.

REFERENCES

1. B. Larsson, D. Gabel, and H.G. Börner, Phys. Med. Biol., 29:361–70, 1984.
2. J.E. Woolard, T.E. Blue, J.F. Curran, M.C. Dobelbower, and H.R. Busby, Nucl. Sci. Eng., 110:96–103, 1992.
3. H. Yanagie, *et al.*, Neutron Radiography (5), Proc. 5. World conf. on neutron radiography, June 17–20, 1996, Berlin, DGZfP, Berlin, 1997, pp. 750–757.
4. J. Miyahara, K. Takahashi, Y. Amemiya, N. Kamiya, and Y. Satow, Nucl. Instr. Meth. in Phys. Res., A246:572–578, 1986.
5. M. Sonoda, M. Tahano, J. Miyahara, and H. Kato, Radiology, 148:148–53, 1983.
6. H. Yanagie, R. Ilic, J. Skvarč, and J. Rant, IJS internal report, Dec. 1997.
7. G. Bayon, CEA/CEN, Saclay, France, personal communication, august 1998.
8. G. Somogyi, and I. Hunyadi, Proc. 10th Int. Conf. "Solid State Nuclear Track Detectors", Lyon, France, 2–6 July 1979, (Eds., H. François *et al.*), Pergamon Press, 1979, pp. 443–452.
9. H. Yanagie, K. Ogura, T. Matsumoto, M. Eriguchi, and H. Kobayashi, Proc. 3. Int. Topical Mtg. on Neutron Radiography, March 16–19, 1998, Lucerne, to be published in Nucl. Inst. Meth.
10. J. Skvarč, R. Ilić, H. Yanagie, J. Rant, K. Ogura, and H. Kobayashi, Selective radiography with etched track detectors, manuscript (May 1998), to be published.

ALPHA-AUTORADIOGRAPHIC DETERMINATION OF ^{10}B CONCENTRATIONS IN CANCER BEARING MICE FOR BORON NEUTRON-CAPTURE THERAPY

Hironobu Yanagie[1], Koichi Ogura[2], Yasumasa Nonaka[1], Toshio Matumoto[3], Yoshinori Sakurai[4], Tooru Kobayashi[4], Kouji Ono[4], Masazumi Eriguchi[1], and Hisao Kobayashi[5]

[1]Department of Surgery, Institute of Medical Science
University of Tokyo, 4-6-1 Shiroganedai, Minato-ku, Tokyo 108-8639
[2]Physical Science Laboratories
College of Industrial Technology
Nihon University
1-2-1 Izumi-cho, Narashino, Chiba 275
[3]Metabolism and Pharmacokinetics Group
Omiya Research Laboratory
Nikken Chemicals Co. Ltd.
1-346 Kitabukuro-cho, Omiya, Saitama 33
[4]Kyoto University Research Reactor Institute
Kumatori, Osaka 590-0494
[5]Institute for Atomic Energy
Rikkyo University
2-5-1 Nagasaka, Yokosuka, Kanagawa, 240-01 Japan

Abstract

To confirm the effectiveness of the ^{10}B delivery system, it is required to achieve an accurate measurement of ^{10}B concentrations in biological samples with a sensitivity in the ppm range for boron neutron-capture therapy (BNCT). We applied CR-39 plastic track detectors to alpha-autoradiographic measurements of the ^{10}B biodistribution in sliced whole body samples of pancreatic cancer bearing mice. CR-39 can record the energetic heavy charged particles generating from the ^{10}B (n, α) ^{7}Li reaction, as well as neutron induced recoil protons which disturb alpha-radiographic imaging and alpha-track

Frontiers in Neutron Capture Therapy, edited by Hawthorne *et al.*
Kluwer Academic / Plenum Publishers, New York, 2001.

counting as background tracks. To desensitize unwanted proton tracks selectively we applied PEW (KOH + C_2H_5OH + H_2O) solution to the etching of CR-39 detector. The discrimination between alpha-tracks and recoiled proton tracks was made by track size selection method. This enabled to estimate quantitatively the distributions of ^{10}B concentrations within the tissue sections comparing with the standards.

INTRODUCTION

The cytotoxic effect of boron neutron capture therapy (BNCT) is caused by a nuclear reaction between ^{10}B and thermal neutrons. The resultant ^7Li particles and α particles are high LET (linear energy transfer) particles with relatively high biological efficiency. These particles (α and ^7Li) destroy cells within about 10-μm path length from the site of capture reaction. It is necessary to accumulate ^{10}B compounds selectively in tumor cells for selective cyototoxicity of tumor cells without affecting adjacent healthy cells.

Liposomes are useful drug carriers, and it is possible to carry a large amount of ^{10}B compound in a liposome, which can be delivered to a tumor cells. We have reported that ^{10}B atoms delivered by immunoliposomes are cytotoxic to human pancreatic carcinoma cells (AsPC-1) with thermal neutron irradiation *in vitro*.[1] The intratumoral injection of boronated immunoliposomes can increase the retention of ^{10}B atoms in tumor cells, causing tumor growth suppression *in vivo* under thermal neutron irradiation.[2]

It is required achieve an accurate measurement of ^{10}B distributions and concentrations in biological samples with a sensitivity in the ppm range for BNCT. We applied CR-39 (polyallyldiglycol carbonate) plastic track detectors to alpha-autoradiographic measurements of the ^{10}B biodistribution in sliced whole body samples of mice. To selectively desensitize undesirable proton tracks, we applied PEW (KOH + C_2H_5OH + H_2O) solution to the etching of CR-39 detector. The subsequent use of an alpha-track radiographic image analysis system enabled a discrimination between alpha-tracks and recoiled proton tracks was made by track size selection method. This enabled to estimate quantitatively the distributions of ^{10}B concentrations within the tissue sections comparing using suitable standards.

MATERIALS AND METHODS

Target Tumor Cell and Tumor Injection

A human pancreatic carcinoma AsPC-1 (1×10^7) cells were injected sub-cutaneously into the back of the male BALB/c nu/nu mice (Nihon SLC). At 10 to 15 days after injection, when the estimated tumor weight reached 100 to 300 mg, 0.2 ml of ^{10}B-liposome solution were injected intraveneouslly or intratumorally. The mice were sacrificed in 3 and 6 hours after the ^{10}B-liposome solution was injected. The procedures for the tumor implantation and the sacrifice of the animals were in accordance with approved guidelines of the Institutions Animal Ethics Committee.

Preparation of Liposomes Containing 10B-Compound

Freze dried pre-formed liposome (COATSOME EL-C-01 : Nichiyu liposome Co. Ltd.) is composed with L-α-dipalmitoyl phosphatidylcholine (26 μmoles), cholesterol (20 μmoles), and stealylamine (4 μmoles). One-half ml of 100 mM $Na_2{}^{10}B_{12}H_{11}SH$ (BSH) (Wako Chemical Co. Ltd.) solution was added to the COATSOME EL-C-01, and made the ^{10}B-liposome solution.

Preparation of Standard Samples and Sliced Mice Samples

The sacrificed mice were frozen at −60 °C. Subsequently, the frozen mice were cut sagittaly into 40 μm thick sections and put on the mending tape, freeze-dried at −20 °C in 2 weeks, and air dried in one more week.

Boron-containing standard samples were also prepared using drying filter paper sheets wetted by BSH solutions of five different ^{10}B concentrations of relative scale; 1, 10^{-1}, 10^{-2}, 10^{-3} and 10^{-4}, respectively, where the concentration of 1 corresponds to the concentration 1.58×10^4 ppm of ^{10}B compounds. The ^{10}B concentrations of ^{10}B-liposome solution and ^{10}BSH solutions were determined by the prompt γ-ray spectrometry at the Research Reactor Institute, Kyoto University.[3]

Neutron Irradiations

The whole-body sections are put in close contact with the CR 39 track etch detector plates (HARZLAS; Fukuvi Chemical Industry) using thin adhesive tape.

For visible observation of track image, thermal neutron flux of an order of 10^{12} n/cm^2 was required and the samples were attached onto the surface near the center hole of the thermal column output face (1.2 m × 1.2 m) in the TRIGA-II reactor of the Rikkyo University (RUR). The neutron flux has been measured to be 1×10^8 n/(cm^2 s) and Cd ratio was 6400.

For counting the tracks, the tangential beam port No. 2 with flux of 1.5×10^6 n/(cm^2 s) and with the Cd ratio of 2.0 was used. The total fluence was varied between 4.5×10^7 n/cm^2 and 2.7×10^{10} n/cm^2, depending on the exposure time.

Etching Procedure

For α-autoradiographic imaging including proton of tracks produced by ^{14}N (n, p) reaction, where ^{14}N is the biogenically abundant nuclide, the CR-39 detector plates were etched in a 7 M NaOH solution at 70 °C for 2 hours to reveal tracks.

For quantitative estimation of ^{10}B concentration, In order to selectively desensitize and eliminate undesirable proton tracks, we applied PEW solution (15 wt% KOH + 65 wt% C_2H_5OH + 20 wt% H_2O) at 50 °C for 8 minutes to the etching of the CR-39 detector plates.

RESULTS AND DISCUSSION

Track Etch Imaging and Desensitization of Proton Tracks

The samples and the detector plates of neutron capture autoradiographic images were exposed to 1.8×10^{12} n/cm^2 at the thermal column output face and were etched by the NaOH method and by the PEW method. In both cases, the CR-39 detector plates attached together with the set of standard samples containing 1.58×10^4 ppm of ^{10}B-compounds were exposed to thermal neutrons with a fluence of 4.5×10^8 n/cm^2. Two peaks of track size are appearing in case of the etching in NaOH solution as shown in Figure 1(a). The lower peak corresponds to proton tracks. The higher one is due to α- and Li tracks. In case of the etching using PEW solution, the greater part of proton tracks disappeared and only one peak due to alpha tracks was observed as shown in Figure 1(b). In addition to the desensitization effect, the PEW solution markedly increases the bulk etching rate and the quality of etched surface of the CR-39 plate. Consequently, the undesirable proton tracks for α-autoradiographic imaging were effectively desensitized as shown in Figure 1(b). Therefore, the discrimination between α-tracks and proton tracks have been easily achieved. This enabled to quantitatively estimate the ^{10}B concentrations within the tissue sections by comparison with the standard samples. Figure 2 shows the α-track densities under the calibration standard papers as functions of the neutron fluence and the assumed relative ^{10}B concentrations, where the relative ^{10}B concentration of 100 equals to 1.58×10^4 ppm of ^{10}B-compound. When samples were exposed to higher neutron fluences, the relation between track density and boron

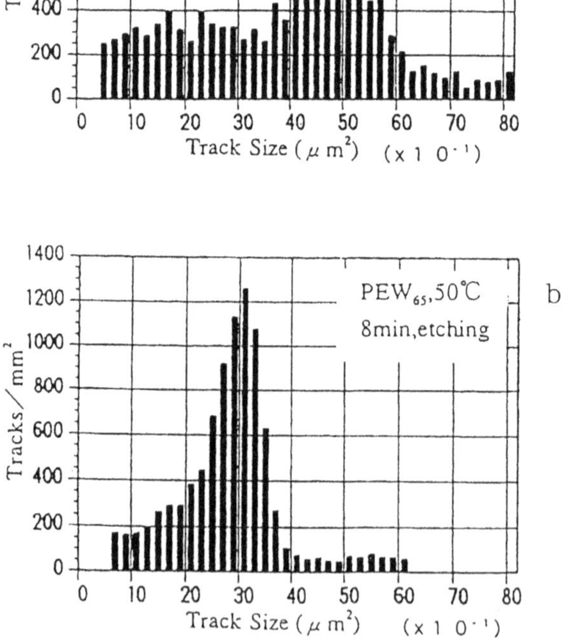

Figure 1. Track size distributions for CR-39 etched in (a) 7M-NaOH and (b) PEW solutions.

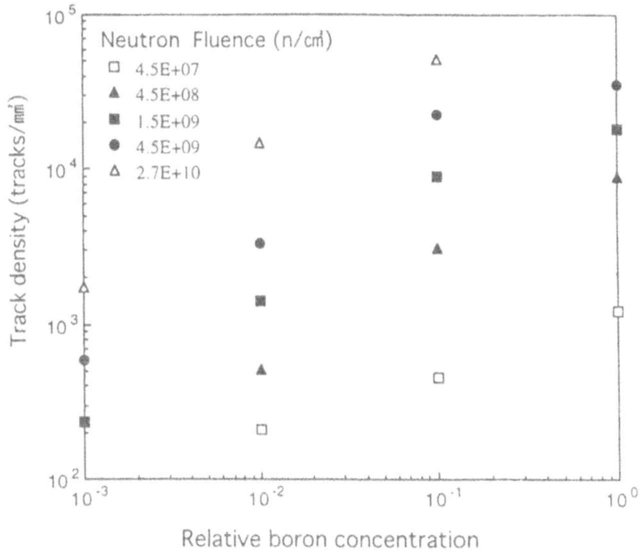

Figure 2. Track densities as function of the relative ^{10}B concentration and the neutron fluence. The concentration of 1 corresponds to the concentration 1.58×10^4 ppm of ^{10}B compounds.

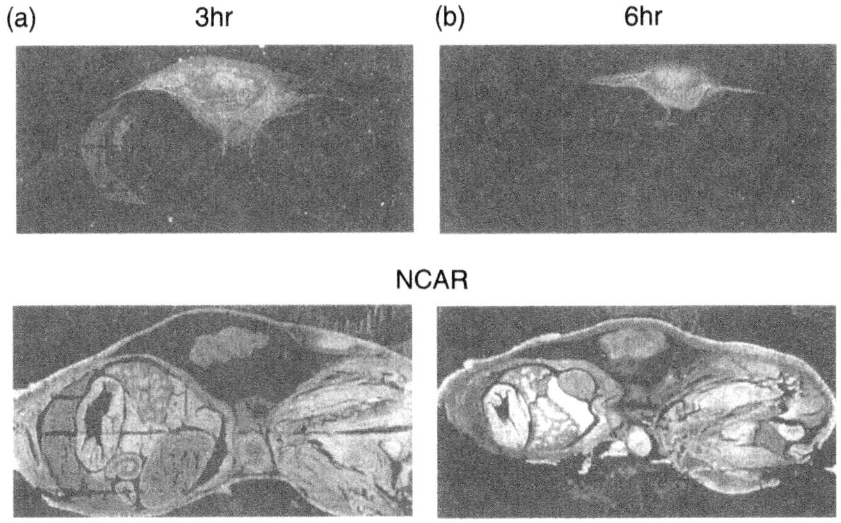

Figure 3. Quantitative whole-body boron distribution obtained by neutron capture autoradiography after intratumoral injection of ^{10}B-liposome solution. Neutron capture autoradiograph was prepared from the whole body section (40 μm thick) of a pancreatic cancer-bearing mouse. The mouse was sacrificed at (a) 3 hours and (b) 6 hours after the ^{10}B-liposome solution was injected. The neutron fluence needed to produce this image was 2.7×10^{10} n/cm^2. This CR-39 plates were etching in PEW solution at 50 °C for 8 minutes.

Table 1. Determination of ^{10}B concentration after intratumoral injection of ^{10}B entrapped cationic liposome solution

Post injection hours		Track density (No. / mm^2)	^{10}B concentration (ppm)
3	Tumor	$(3.10 \pm 0.24) \times 10^4$	695.2
	Liver	$(4.10 \pm 0.13) \times 10^3$	41.9
6	Tumor	$(1.90 \pm 0.59) \times 10^3$	268.6
	Liver	$(1.55 \pm 0.27) \times 10^2$	<10

concentration lost its linearity due to the increase in the number of overlapped tracks. According to Figure 2 and the quality of the images, it was found that the optimum neutron fluence for both the ^{10}B concentration measurements and the imaging in the interval between 15 ppm and 1,500 ppm ranges between 10^9 and 10^{10} n/cm^2.

Track Density Analysis

Figure 3 shows a whole body sections of neutron capture radiograph from a set of AsPC-1 pancreatic cancer bearing mice that have been intratumoral injections of 1.58×10^4 ppm of ^{10}B liposome solution. The sample of mice sacrified at 3 and 6 hours after injections were prepared and exposed to neutron beam of the fluence of 2.7×10^{10} n/cm^2. CR-39 plate after the etching in PEW solution at 50 °C for 8 minutes was scanned. Table 1 shows the track densities were measured in the tumor and in the liver. Observed track densities in the image of the tumor ranged from $(7.82 \pm 0.24) \times 10^3$ to $(5.48 \pm 0.23) \times 10^4$ tracks/mm^2 for the sample sacrificed at 3 hours after injection of ^{10}B-compound solution and from $(6.17 \pm 0.59) \times 10^3$ to $(3.15 \pm 0.68) \times 10^4$ tracks/mm^2 for the sample sacrificed at 6 hours after injection, respectively. Average track density in the image of liver sacrificed at 3 hours after injection was $(4.10 \pm 0.13) \times 10^3$ tracks/mm^2 and the track density of $(1.55 \pm 0.27) \times 10^2$ tracks/mm^2 was estimated in the portion of the liver sacrificed at 6 hours after injection.

From observed track densities and Figure 2, we can estimate the ^{10}B accumulations in the organs. ^{10}B accumulations of 2,050 ppm and 83.7 ppm were estimated for the strongly and weakly, concentrated part of the tumor at 3 hours after injection, respectively. This result promises that the CR-39 radiography using the track counting is possible to determine the micro- and fine structure, i.e. micro-autoradiography, of ^{10}B distribution in the tumor.[4]

CONCLUSION

Two types of techniques to describe ^{10}B distributions and area densities were successfully examined in this study. The NaOH method was applied to observe visually qualitative distribution of ^{10}B.

If we want to determine ^{10}B distribution more exactly, the NaOH method was no more suitable than the PEW method because of the interference of the proton tracks. Thus, ^{10}B accumulation in the tumor was clearly visible using the neutron capture autoradiography.

REFERENCES

1. H. Yanagie, T. Tomita, H. Kobayashi, Y. Fujii, T. Takahashi, K. Hasumi, H. Nariuchi, and M. Sekiguchi, Application of boronated anti-CEA immunoliposomes to tumour cell growth inhibition in *in vitro* boron neutron capture therapy model, Br. J. Cancer, 63, 522–526, 1991.
2. H. Yanagie, T. Tomita, H. Kobayashi, Y. Fujii, Y. Nonaka, Y. Saegusa, K. Hasumi, M. Eriguchi, T. Kobayashi, and K. Ono, Inhibition of human pancreatic cancer growth in nude mice by boron neutron capture therapy, Br. J. Cancer, 75, 660–665, 1997.
3. T. Kobayashi, and K. Kanda, Microanalysis system of ppm-order ^{10}B concentrations in tissue for boron neutron capture therapy by prompt gamma-ray spectroscopy, Nucl. Instr. Methods, 204, 525–531, 1983.
4. H. Yanagie, K. Ogura, T. Matsumoto, M. Eriguchi, and H. Kobayashi, Neutron capture autoradiographic determination of ^{10}B distributions and concentrations in biological samples for boron neutron capture therapy, Nucl. Instr. Methods in Physics Research A 424, 122–128, 1999.

PHARMACOLOGY AND COMPOUND EVALUATION

A NEW SIMPLIFIED SYSTEM FOR THE EVALUATION OF BNCT PHARMACEUTICALS

T. E. Byrne[1], L. F. Miller[2], G. W. Kabalka[3], and R. C. Martin[4]

[1]Roane State
Harriman, Tennessee
[2]Nuclear Engineering Department
[3]Chemistry Department
University of Tennessee, Knoxville, Tennessee
[4]Oak Ridge National Laboratory
Oak Ridge, Tennessee

A system for testing potential BNCT pharmaceuticals in cell cultures has been developed with the cooperation of Oak Ridge National Laboratory (ORNL), the University of Tennessee Chemistry Department and the University of Tennessee Nuclear Engineering Department. A BNCT test model has been established with the use of the human lung cancer cell line A 549. These cells were maintained in standard laboratory facilities and subjected to boronated chemicals. Following toxicity studies the human lung cancer cells were exposed to ^{252}Cf neutron sources provided by the Radiochemical Engineering Development Center (REDC) at ORNL. The isotope ^{252}Cf performs effectively for BNCT applications. The neutron spectrum is similar to that of a reactor fission source with an average energy of 2.1 MeV. A 50 mg source of ^{252}Cf moderated by water provides a source on the order of 1×10^9 thermal neutrons/cm^2/sec at a distance of 3 cm. The half-life of ^{252}Cf is 2.65 years, and thus may provide a simple and reliable source of neutrons for BNCT in locations without suitable nuclear reactors. The REDC of ORNL stores and processes the U.S. stockpile of ^{252}Cf.

The compounds examined were boron containing nucleosides and boron-containing amino acids (Figure 1). Chemical analysis for the measurement of total cellular boron was conducted at ORNL using inductively coupled plasma mass spectroscopy. Boron concentrations were measured against a calibration curve generated with a series of NIST standards. In nitric acid solutions containing boron, the detection limit was 0.4 ppb. In similar solutions containing cellular debris the boron detection limit was determined to be 1.0 ppb.

Calculations of cellular dosimetry in human lung cancer cells from sources of ^{252}Cf and intracellular ^{10}B concentrations have been made (Table 1). All experiments with thermal neutrons plus boronated pharmaceutical compounds displayed a greater cancer

Frontiers in Neutron Capture Therapy, edited by Hawthorne *et al.*
Kluwer Academic / Plenum Publishers, New York, 2001.

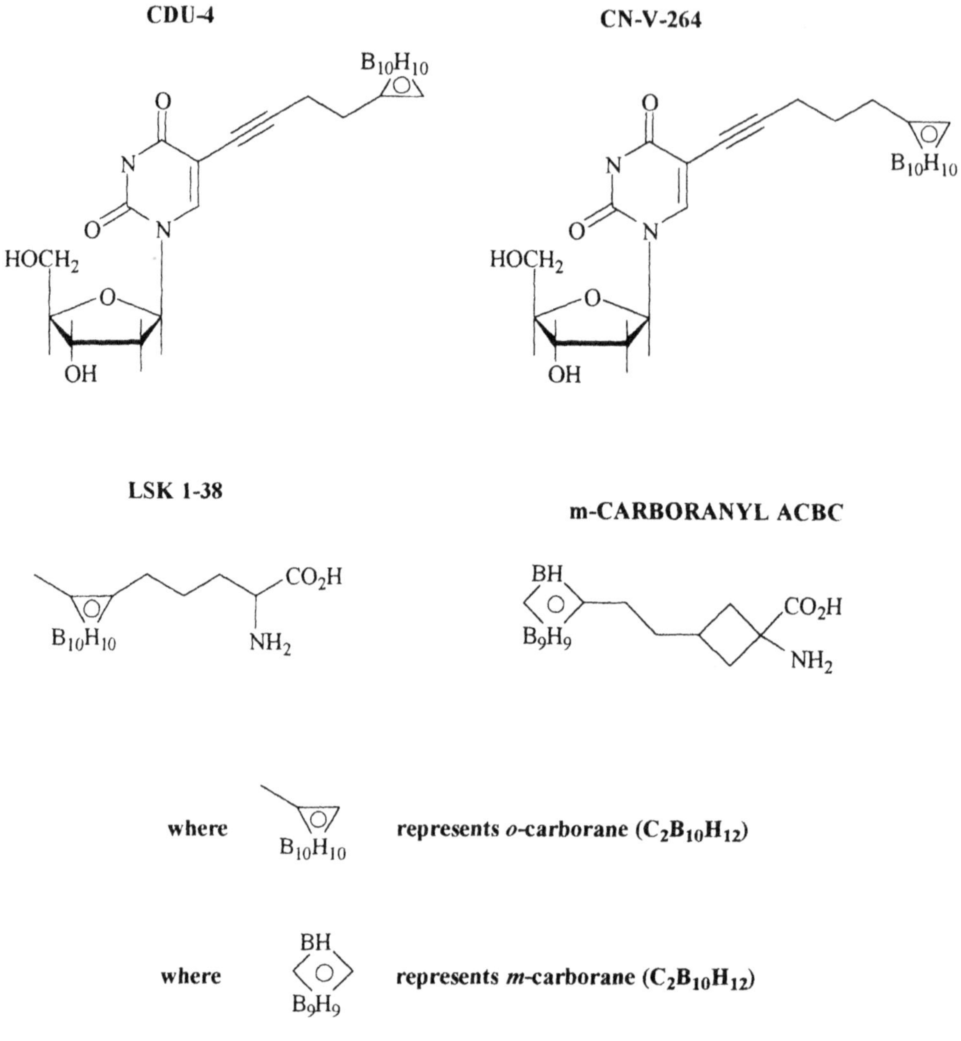

Figure 1. Boronated chemical structures.

Table 1. Uptake of boronated compounds by A549 lung cancer cells and calculated number of thermal neutron interactions per cell at four minute exposure

Compound	Boron in cell (ng B/cell)	Atoms of ^{10}B (atoms ^{10}B/cell)	Average number of interactions per cell
ACBC	24.6×10^{-3}	274×10^{9}	5.0
LSK 1-38	8.6×10^{-3}	96×10^{9}	1.7
CN-V-264	4.5×10^{-3}	50×10^{9}	0.9
CDU-4	0.9×10^{-3}	10×10^{9}	0.18

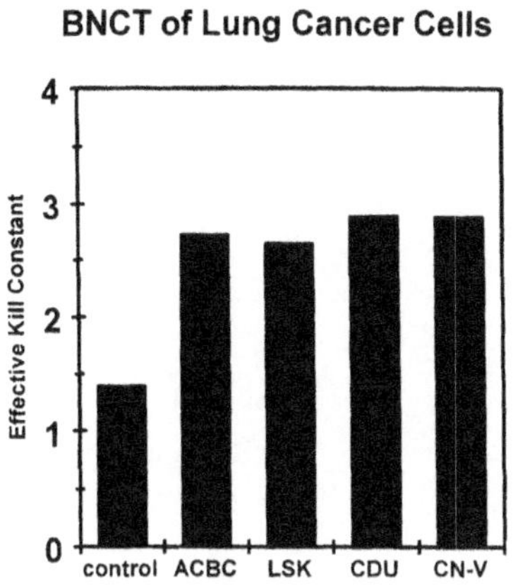

Figure 2. Bar graph of effective kill constants with various boron containing compounds versus controls.

cell killing effect than control experiments without boronated compounds (Figure 2). Thus, a cooperative program for the future evaluation of BNCT pharmaceutical agents is now in place.

In conclusion, the following objectives of this research have been achieved:

1. The cellular toxicity of the proposed BNCT compounds has been determined. The least toxic agent of the A549 lung cancer cell is ACBC followed by LSK 1-38, CDU-4 and the most toxic agent is CN-V-264.
2. The cellular uptake of the BNCT compounds has been determined. The greatest cellular uptake of B/cell occurred with ACBC at a concentration of 24.6×10^{-3} ng followed by LSK 1-38 with a concentration of 8.6×10^{-3} ng B/cell, CN-V-264 demonstrated an uptake concentration of 4.5×10^{-3} ng B/cell and CDU-4 displayed the lowest uptake concentration of 0.9×10^{-3} ng B/cell.
3. A 30 mg source of ^{252}Cf moderated by 2.5 cm of water and polyethylene produced a thermal neutron flux of approximately 2×10^8 neutrons/cm^2/sec. The REDC source allowed for the development of cellular survival curves in A549 lung cancer cells.
4. A system for testing potential BNCT pharmaceuticals in cell cultures has been developed with the cooperation of ORNL, the UTK Chemistry Department and the UTK Nuclear Engineering Department.

ACKNOWLEDGMENTS

We wish to thank the U. S. Department of Energy (DOE), Oak Ridge National Laboratory (ORNL) and Lockheed Martin Idaho Technology University Research Consortium (URC) for support of this research.

COMPARTMENTAL AND NON-COMPARTMENTAL METHODS IN STUDYING THE KINETICS OF BORON-10 AFTER BORONOPHENYLALANINE FRUCTOSE COMPLEX (BPA-F) INFUSION IN DOGS

Päivi Ryynänen[1,2], Sauli Savolainen[1,2], Judit Benczik[3,4], Martti Kulvik[5], Jyrki Vähätalo[6], and Marjatta Snellman[3]

[1]Department of Physics, University of Helsinki
[2]Department of Radiology, Helsinki University Central Hospital
[3]Faculty of Veterinary Sciences, University of Helsinki
[4]Clinical Research Institute, Helsinki University Central Hospital
[5]Department of Neurology, Helsinki University Central Hospital
[6]Laboratory of Radiochemistry, University of Helsinki

1. INTRODUCTION

One of the essential requirements for optimising BNC-treatment is to know the boron-10 concentration in tissues under irradiation. A model which describes the kinetics of boron-10 in the patient would be valuable. The primary aim of the present study was to create a model for the pharmacokinetics of boron-10 in dogs after 4-dihydroxy-borylphenylalanine fructose complex (BPA-F) infusion so that we could predict the boron-10 concentration in blood during the irradiation.[1] In addition we wanted to clarify if it is possible to design a simple compartmental model which could be used in defining the kinetic behaviour of boron-10 in the blood circulation.

2. METHODS

Nine Beagle dogs were divided into three different groups (A, B and C) according to the infusion time, surveillance time and the infused amount of BPA (Table 1). Blood samples were collected every 15 minutes during the first hour after the end of the infusion and then 2–5 samples were taken during the next two hours from the dogs in group

Frontiers in Neutron Capture Therapy, edited by Hawthorne *et al.*
Kluwer Academic / Plenum Publishers, New York, 2001.

Table 1. The division of the beagle dogs into groups according to the length of infusion, surveillance time and the infused amount of BPA

Group	Number of dogs	Surveillance time [min]	Amount of BPA [mg/kg]	Infusion time [min]
A	2	1,350–1,380	250	60
B	2	1,350–1,380	250	30
C	5	200–260	700	60

C. From the dogs in groups A and B the first few samples were taken every half an hour and continuing once an hour until 10 hours from the start of the infusion. Extra samples were taken 13 and 23 hours after the start of the infusion. From one dog in group B the samples were taken once an hour until 23 hours from the start of the infusion had passed. The boron-10 concentration in respective blood samples were measured using inductively coupled plasma-mass spectrometry, ICP-MS.[2] A bi-exponential model was built according to the test results from the dogs in groups A and B to estimate the duration of the irradiation for group C.

2.1. Data Analysis

2.1.1. Non-compartmental Analysis. Bi-exponential function was fitted to the observed data. The detected bi-exponential disappearance time concentration curve, $C_P(t)$, can be represented as:[3]

$$C_p(t) = A_1\exp(-g_1 t) + A_2\exp(-g_2 t), \quad (1)$$

where A_i and g_i are the constants and coefficients, respectively.

2.1.2. Compartmental Analysis. Two different three-compartmental models were investigated in this study. Both models contained plasma and depot pools. In the more simple model the third pool was urine, and there was only one-way flux between the plasma and urine pools (Figure 1). The other three-compartmental model consisted a kidney pool, which had a reverse flux from kidneys back to the plasma pool and a excretion route. Differential equations of created models were solved using iterative computer methods. The computer program used was SAAM (Simulation Analysis and Modelling), Version two.[4]

2.2. Statistical Analysis

In non-compartmental models the F-test was applied to measure the discrepancy between the fit and observed values. To compare the differences between the goodness of the fits of the different compartmental models the χ^2-test was adapted.

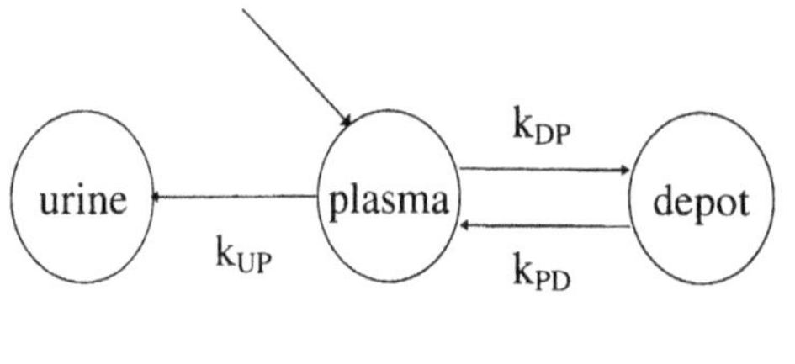

Figure 1. The three-compartmental open model used to fit plasma and urine boron-10 concentrations after BPA-fructose complex infusion. k = transfer coefficient; P = plasma; D = depot; U = urine.

3. RESULTS

3.1. Non-compartmental Models

The mean half lives of the fast component for groups A, B and C were 9.9 ± 14 minutes, 19 ± 10 minutes and 14 ± 4 minutes, respectively. The slow components for the same groups were 3.0 ± 3.9 hours, 9 ± 2 hours and 4.0 ± 1.9 hours, respectively. An example of the fitted two-exponential model is presented in Figure 2a. for one dog (group A). The obtained two-exponential model from the test results of groups A and B was adapted to the actual test results of group C. An example of this is represented in Figure 2b.

3.2. Compartmental Models

The data indicates that a three-compartmental model (plasma, tumour and urine or kidneys as compartments) describes the kinetics of boron-10 after a BPA-F infusion in dogs. Mean transfer coefficients for the dogs in group C, where the third compartment are the kidneys, and there is a flux from kidneys back to the plasma were from plasma to kidneys 0.05 ± 0.01, from depot to plasma 0.16 ± 0.11, from plasma to depot 0.10 ± 0.09 and from kidneys to plasma $(43 \pm 7)*10^{-4}$, all in units min^{-1}. The turnover time from the depot is 12 ± 8 min.

4. DISCUSSION

In this study the kinetics of boron-10 after BPA-F infusion were modelled with two separate modelling equipment; with exponential fit and with compartmental modelling. The result that the boron-10 disappearance curve is biexponential has also been reported by other studies.[5,6] According to both models boron-10 is rapidly excreted to urine.

From Figure 2b one can notice that the constructed model overestimates the contribution of the slow half live; consequently the curves of the models predict higher boron-10 concentrations than those measured. This is due to the different observation matrixes. However, the differences between the model based estimations of boron-10 concentration in the blood and the post-irradiation measured values were lower than 0.7 ppm.[7]

There is a need for a simple and fault tolerant model, which could be easily and quickly utilised in clinical practise. Simple three- and four-compartmental models have been used earlier in modelling other possible boron carrier compounds.[8,9] Models meant for daily clinical use are often greatly simplified, although the biological behaviour of tracers may be very complex.

The mean half lives of the slow components after the 30 minutes infusion are longer than after the 60 minutes infusion. It is to be noticed that the boron kinetics in one of the dogs in group A was different from that of the others. The surveillance time was the most significant factor in the modelling point.

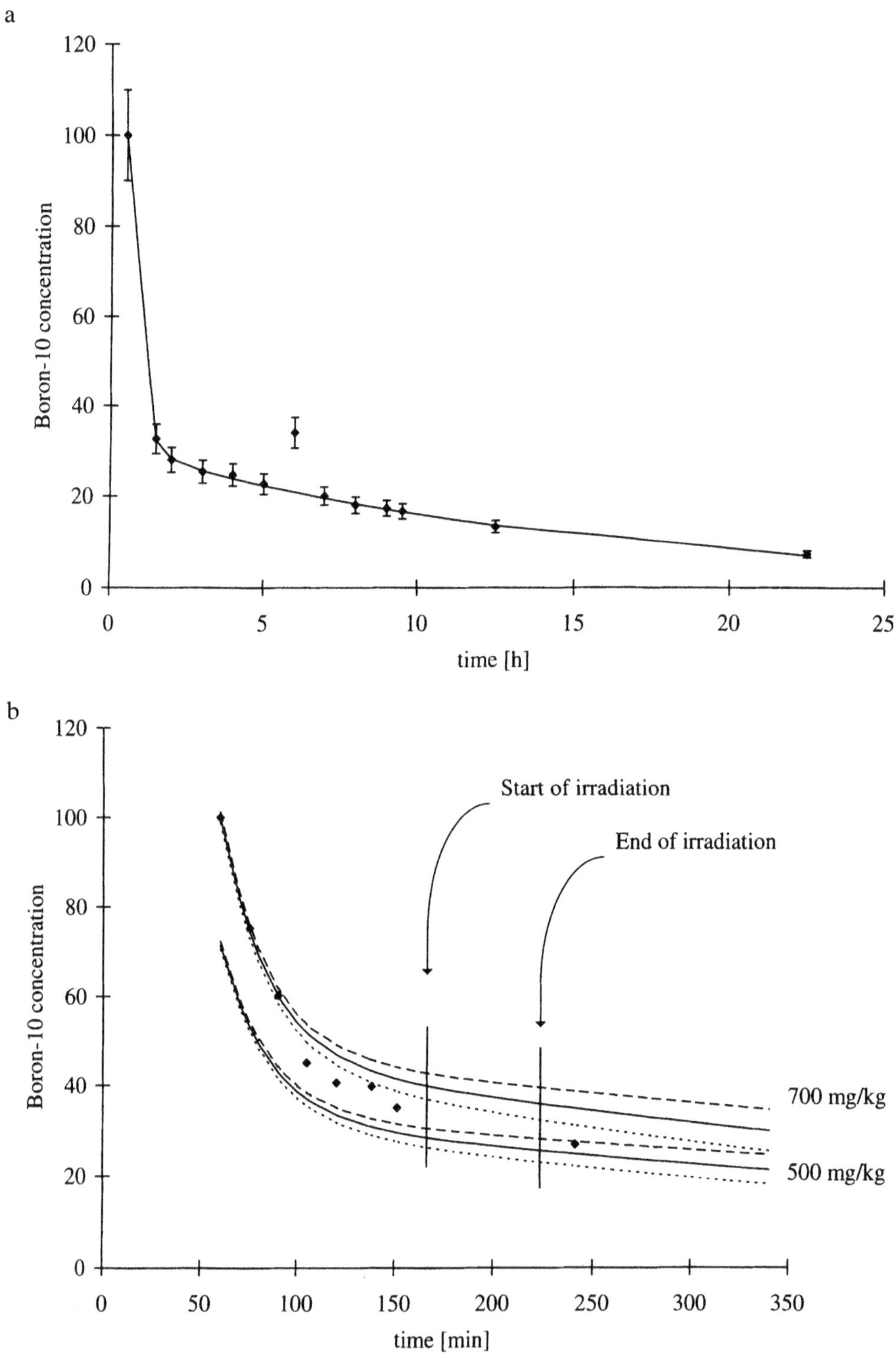

Figure 2. The plasma clearance of boron-10 after BPA-F infusion represented as a sum of the 2-exponential curve for one dog. Diamonds present the observed data values. The corresponding mean half-lives were 14.9 min for the fast and 10.5 h for the slow component (a). The obtained model from the test results of the group A and B (lines) and the actual test results from one dog in group C (diamonds). The corresponding mean half-lives for the slow component were 7.6 h (solid line), 10.5 h (long spot line) and 5.8 h (short spot line) (b).

5. CONCLUSIONS

Both non-compartmental and compartmental modelling are sufficient to describe the kinetics of BPA-F as estimated by measuring corresponding boron-10 levels. However, it is essential to have similar observation matrixes for the modelling and for the actual clinical treatments with patients. During the irradiation it is very difficult to directly measure the boron-10 concentration in the patient. It is of great importance to be able to estimate the actual boron-10 concentration in the tissue during irradiation, when considering the safety and effectiveness of the therapy.

ACKNOWLEDGMENTS

The work was supported by the Technology Development Centre of Finland, Academy of Finland, University of Helsinki, and the State Subsidy for University Hospitals, Helsinki University Central Hospital, research grant of the Departments of Radiology and Neurology. HUCH Pharmacy is acknowledged for technical support for preparing BPA-F infusion solutions.

REFERENCES

1. J. Benczik, T. Seppälä, and J. Hopewell *et al.*, Large animal model for healthy tissue tolerance study in BNCT, *in* "Frontiers in Neutron Capture Therapy," M.F. Hawthorne, K. Shelly, R.W. Wiersema, eds., Kluwer Academic/Plenum Publishers, New York, 2001, pp. 1233–1238.
2. M. Kulvik, J. Laakso, J. Vähätalo, and R. Zilliacus, On-line boron-10 determination from blood samples by ICP-MS, *in* "Frontiers in Neutron Capture Therapy," M.F. Hawthorne, K. Shelly, R.W. Wiersema, eds., Kluwer Academic/Plenum Publishers, New York, 2001, pp. 849–851.
3. N.A. Lassen, and W. Perl, "Tracer Kinetic Method in Medical Physiology," Raven Press, New York, 1979, pp. 1–189.
4. RFKA. "*SAAM II User Quide*," University of Washington, Seattle, 1994, pp. 1–326.
5. H. Fukuda, C. Honda, N. Wadabayashi, T. Kobayashi, K. Yoshino, J. Hiratsuka, T. Akaizawa, M. Ichihashi, and Y. Mishima, Pharmacokinetics of 10B-P-BPA in tumor, skin and blood: a study of BNCT for malignant melanoma, *in* Program and Abstracts of the Eighth International Symposium on NCT for Cancer, La Jolla, CA, 1998, p. 176.
6. W.S. Kiger, M.R. Palmer, R.G. Zamenhof, and P.M. Busse, A pharmacokinetic model for the concentration of B-10 in blood following boronophenylalanine-fructose administration in humans, *in* "Frontiers in Neutron Capture Therapy," M.F. Hawthorne, K. Shelly, R.W. Wiersema, eds., Kluwer Academic/Plenum Publishers, New York, 2001, pp. 249–255.
7. S. Savolainen, I. Auterinen, and C. Aschan *et al.*, Dosimetry chain for the dogs irradiated in the epithermal neutron beam at the Finnish BNCT facility, *in* "Frontiers in Neutron Capture Therapy," M.F. Hawthorne, K. Shelly, R.W. Wiersema, eds., Kluwer Academic/Plenum Publishers, New York, 2001, pp. 1245–1250.
8. C.P. Ceberg, A. Persson, A. Brun, R. Huiskamp, A.-S. Fyhr, S.B. Kahl, B.R.R. Persson, and L.G. Salford, Performance of sulfhydryl boron hydride in patients with grade III and IV astrocytoma: A basis for boron neutron capture therapy, *J Neurosurg*, 83:79–85, 1995.
9. C.P. Ceberg, A. Brun, S.B. Kahl, M.S. Koo, B.R.R. Persson, and L.G. Salford, A comparative study on the pharmacokinetics and biodistribution of boronated porphyrin (BOPP) and sulfhydrul boron hydride (BSH) in the RG2 rat glioma model, *J Neurosurg*, 83:86–92, 1995.

TIME-DEPENDENT BIODISTRIBUTION OF BPA-F IN NORMAL AND TUMOR BRAIN TISSUES IN MICE

Cynthia Chuang[1], Matthew Palmer[2], Robert Zamenhof[2], and Paul Busse[3]

[1]Nuclear Reactor Laboratory
Massachusetts Institute of Technology
Cambridge, MA 02139
[2]Department of Radiology
Beth Israel Deaconess Medical Center
[3]Joint Center for Radiation Therapy
Harvard Medical School
Boston, MA 02215

1. INTRODUCTION

With the completion of the fission converter epithermal neutron beam (FCB) at MIT, BNCT irradiations conducted by the Harvard-MIT BNCT program could be reduced from approximately 2 1/2 hours per beam to a few minutes.[1] Therefore, the timing of the irradiation becomes important. An optimal therapeutic gain is expected to be gained when patients are irradiated during a time window which has the best tumor/normal tissue ^{10}B concentration ratio, and the boron concentration is high. Since in the current clinical trial, blood ^{10}B concentration is used as a surrogate for normal brain tissue ^{10}B concentration, while the metabolism of BPA-f in tumor and normal brain tissues is not well understood, it is important to determine the biodistribution and pharmacokinetics of boron in these tissues. In addition, to be able to prescribe accurate radiation dose to radiation sensitive areas such as the optic chiasm, knowledge of the spatial biodistribution of BPA-f in these tissues is essential. A GL261 murine glioma tumor model is being used to study in detail the time-dependent biodistribution of BPA-f in tumor and several anatomic sites in the normal brain. Preliminary results indicate that there are slight differences in the temporal uptake of BPA-f in different regions of the normal brain, and that the pharmacokinetics of BPA-f in GL261 tumor and normal brain are vastly different.

Frontiers in Neutron Capture Therapy, edited by Hawthorne *et al.*
Kluwer Academic / Plenum Publishers, New York, 2001.

2. MATERIALS AND METHODS

2.1. The C57BL/6 Mouse Normal Brain Model

C57BL/6 mice were given a bolus injection of BPA-f (250 mg–350 mg BPA-f/kg body weight) through the tail vein using a 26G needle. Five mice each were sacrificed at 7 different time points (5, 10, 15, 30, 60, 120, and 240 min after injection). Cardiac puncture was performed to collect blood samples. The brain tissues were removed post mortem, and separated into right cerebrum, left cerebrum, cerebellum, brain stem and intracranial nerves.

2.2. The GL261 Mouse Glioma Tumor Model

Mouse glioma GL261 cells were maintained in Dulbecco's Modified Eagle Medium (GIBCO Laboratories Life Technologies Inc., Grand Island, NY) supplemented with 10% fetal bovine serum (FBS; SIGMA Chemical Company, St. Louis, MO), 1% L-Glutamine (GIBCO), 100 units/ml of penicillin G (GIBCO) and 100 mg/ml streptomycin[2] (GIBCO). For the initiation of intracranial tumor, the C57BL/6 mouse was first anesthetized (intraperitoneal Pentobarbital, 30 mg/kg body weight), and it's head was shaved. Using aseptic techniques, the scalp was incised, and the tumor cell suspension (2×10^4 cells in 0.02 ml of medium without serum) was injected into the mouse skull at a point 2 mm to the right of midline, and 2 mm above the bregma at a depth of 2.5 mm into the dura[3] using a 30-gauge needle. This technique resulted in a locally expanding tumor in all mice. Approximately 2–3 weeks after tumor implantation, the mice were given a BPA-f (350 mg/kg) bolus injection. They were sacrificed at 7 different time points, and blood, brain and tumor tissues were collected.

2.3. Boron Analysis

Boron analysis was performed by two methods: prompt gamma neutron activation analysis (PGNAA) and inductive coupled plasma-atomic emission spectroscopy (ICP-AES). The PGNAA method detects the secondary 478 keV photons produced from the $^{10}B(n,\alpha)^7Li$ reaction. Blood samples were pipetted into Teflon vials, and placed one by one manually into the 4DH3 beam port of the MIT-II Research Reactor. The measurement proceeded until less than 10% error in ^{10}B counts was obtained (most of the higher concentration samples had less than 5% error). System calibration was performed on each day of measurement using U.S. NIST ^{10}B-enriched boric acid. The blood sample ^{10}B concentrations were measured by the PGNAA method.

Inductive coupled plasma-atomic emission spectroscopy (ICP-AES) detects the characteristic de-excitation photons from all isotopes of boron and the number of detected photons is proportional to the amount of boron element present in the sample. Tissue samples were weighed; for every 50 mg of tissue, 0.15 ml of nitric acid: sulfuric acid (1 : 1) mixture was added, and digestion was allowed to proceed for an hour in a 60 degree C waterbath. 0.5 ml of 10% Triton X-100 was then added and the acid-digested solution was diluted to a final volume of 1 ml.[4] The solutions were filtered and its B concentration measured using ICP-AES. Standards were prepared in a similar manner using U.S. NIST ^{10}B-enriched boric acid. System calibration was performed on each day of measurement. The boron standards for the ICP-AES were also analyzed with the prompt gamma system to provide system cross-calibration.

2.4. Three-compartment Pharmacokinetic Modeling

Imahori and others[5] conducted cranial PET studies of human glioma subjects using ^{18}F- BPA. A 3-compartment model was used to explain the pharmacokinetics of the PET data, which consisted of a plasma compartment, an extracellular, and an intracellular compartment. From these PET data, the various rate constants were obtained for normal brain tissue and for Glioblastoma Multiforme. By using the model and rate constants the concentration of ^{10}B for normal brain and tumor tissues can be predicted from the blood data. The same normal tissue rate constants were applied to predict the normal brain tissue ^{10}B concentrations in mice based on the measured mouse blood curve, to see if mice have similar BPA-f metabolism to humans.

3. RESULTS

3.1. The C57BL/6 Mouse Normal Brain Model

Sixty-five C57BL/6 mice were used for the normal neural tissue spatial distribution study. Thirty received 250 mg BPA-f/kg body weight, and thirty-five received 350 mg/kg via a bolus tail vein injection. The 250 mg/kg cohort was sacrificed at 5, 10, 15, 30, 60, and 120 minutes after injection. The 350 mg/kg group was sacrificed at 5, 10, 15, 30, 60, 120, and 240 minutes after injection. The ^{10}B distributions in blood as a function of time for the two concentrations of BPA-f infused are shown in Figure 1. The mouse blood

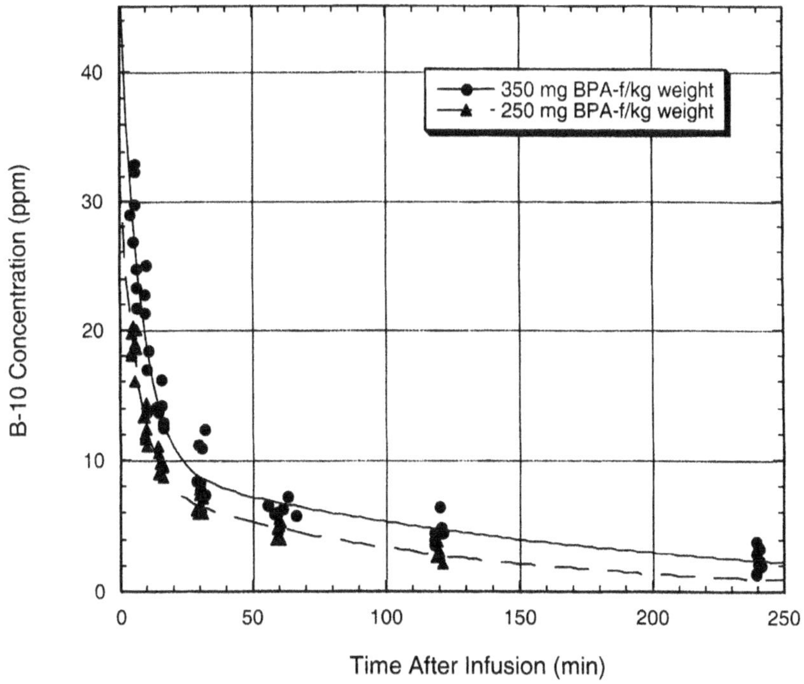

Figure 1. Boron concentration (ppm) in blood as a function of time following a bolus tail vein injection of either 250 mg BPA-f/kg body weight or 350 mg BPA-f/kg body weight.

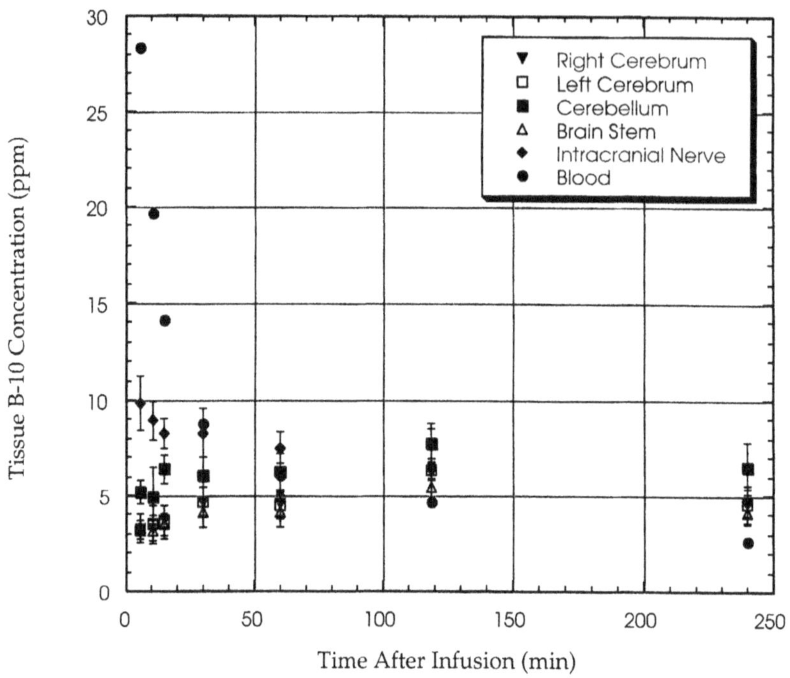

Figure 2. Boron concentration (ppm) in blood and various murine normal neural tissues as a function of time following a bolus tail vein injection of 350 mg BPA-f/kg body weight.

curve is remarkably similar to that of the human blood biodisitribution curve, but is scaled in time.

The neural tissues were removed at time of sacrifice, and separated into right cerebrum, left cerebrum, cerebellum, brain stem, and intracranial nerve. Figure 2 is the data from the 350 mg/kg group which shows that the intracranial nerve has a slightly higher ^{10}B concentration at all time points studied than the ^{10}B concentrations in cerebrum and brain stem. This finding will be verified in a larger animal model.

Figure 3 shows the ^{10}B distribution in blood and in normal neural tissues (averaged between right cerebrum, left cerebrum, and brain stem) as a function of time for the two concentration of BPA-f, with the 250 mg/kg data points scaled to 350 mg/kg. They demonstrate a linear relationship between the BPA-f dose administered and the ^{10}B concentrations obtained in blood and normal neural tissues.

3.2. The GL261 Mouse Glioma Tumor Model

For this set of experiments, mice bearing intracranial glioma tumors were injected with 350 mg BPA-f/kg body weight via a single bolus, and were sacrificed at the designated 7 time points. The blood, normal brain tissue and tumor were removed and analyzed for ^{10}B concentrations. Preliminary results show that the ^{10}B concentration achieved in glioma cells is quite high, up to 6 times that of normal brain tissue at 15 and 30 minutes after BPA-f injection. This time dependence of ^{10}B accumulation in glioma tumor and normal brain tissue is shown in Figure 4.

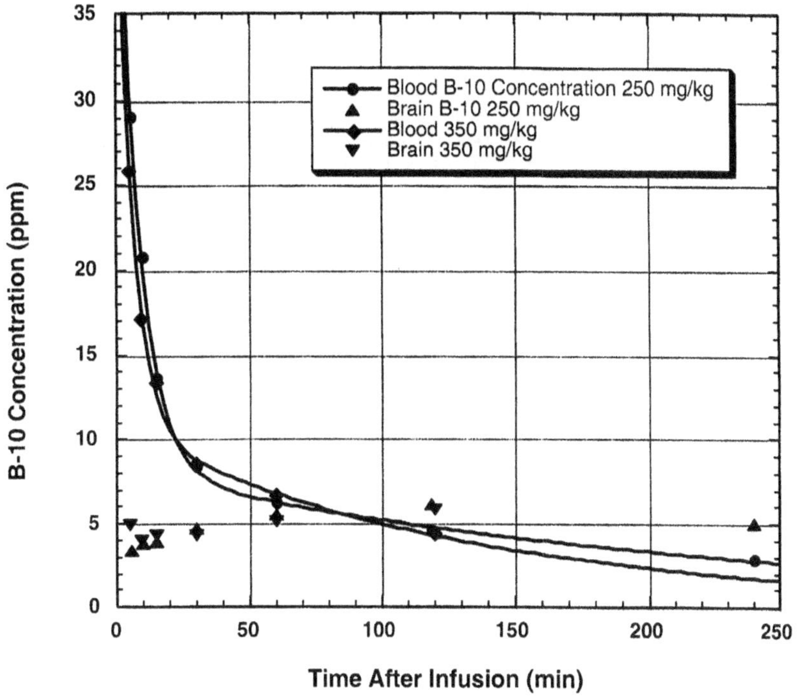

Figure 3. Blood and normal brain ^{10}B biodistribution (ppm) following a bolus tail vein injection of either 250 mg BPA-f/kg body weight or 350 mg BPA-f/kg body weight, with 250 mg/kg data scaled to 350 mg/kg.

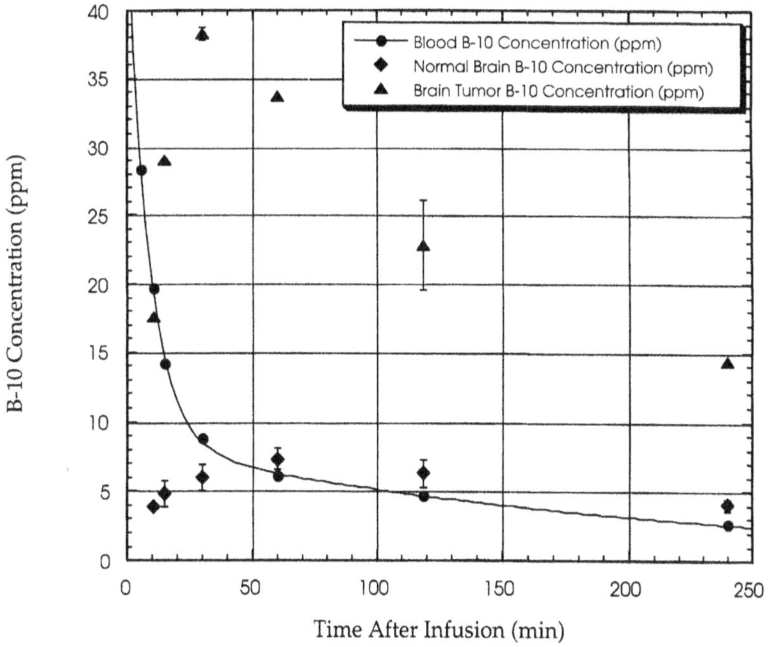

Figure 4. Boron concentration (ppm) in blood, brain and GL261 tumor as a function of time following a bolus tail vein injection of 350 mg BPA-f/kg body weight.

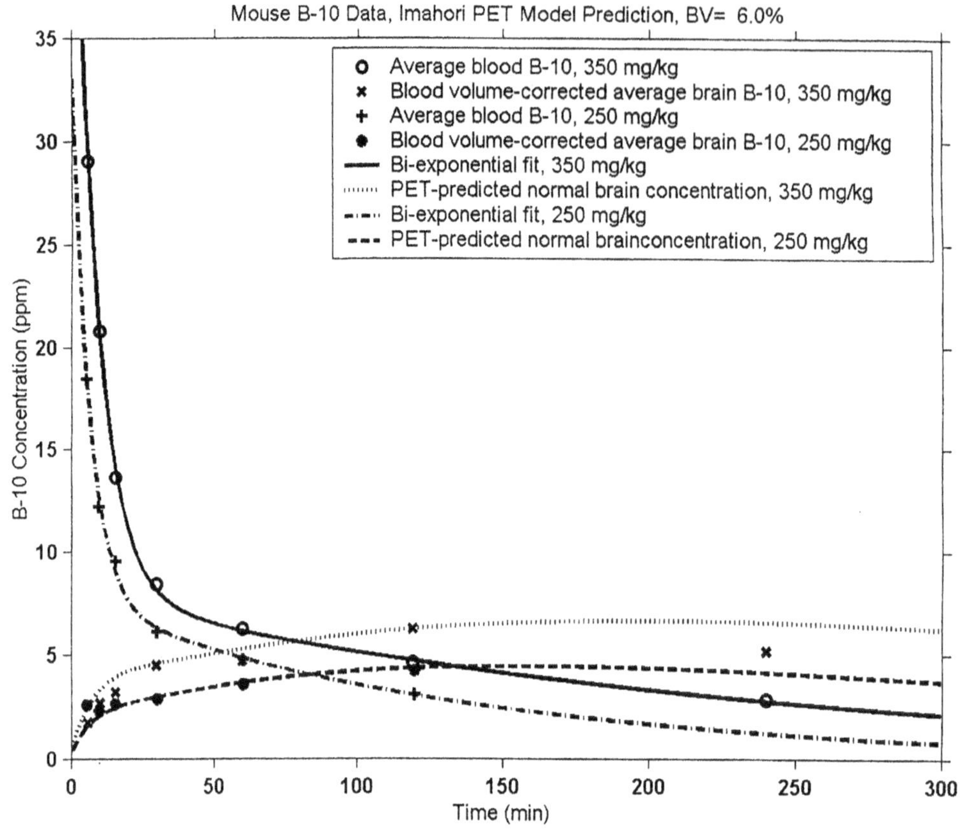

Figure 5. Predicted and experimental normal neural tissue ^{10}B concentrations.

3.3. Three-compartment Pharmacokinetic Modeling

The compartmental model proposed by Imahori tries to predict brain and GBM ^{10}B concentrations based on blood ^{10}B concentration. The rate constants obtained for normal human brain were used in conjunction with blood data from mice to try to predict normal mouse brain ^{10}B concentration. Figure 5 shows the predicted normal brain ^{10}B temporal distribution and the experimental data points. It is clear that the prediction matches well with the experimental data.

4. CONCLUSION

Several conclusions can be drawn from the normal neural tissue experiments. First, blood curves between mouse and human are remarkably similar in form, but scaled in time, i.e. they both exhibit bi-phasic exponential wash out. Secondly, blood and normal brain ^{10}B concentrations show a linear relationship with BPA-f dose. Third, preliminary results show that different parts of the brain have different levels of BPA-f uptake. This observation, if confirmed with further studies in larger animals with higher anatomical resolution, suggests that the brain should not be considered as a homogeneous organ for treatment planning.

For the mouse glioma model, preliminary results show that GL261 tumor has much higher ^{10}B concentration than normal brain for all time points studied, which is qualitatively consistent with human GBM biopsy data (4). The results confirmed that blood, normal brain and tumor all have quite different BPA-f pharmacokinetics. A detailed knowledge of the pharmacokinetic behaviors of these tissues could have important practical implications for treatment planning.

By using rate constants derived from human PET studies, the BPA-f uptake predicted by the three-compartment model can accurately model the pharmacokinetics of the normal brain ^{10}B concentrations in mice.

ACKNOWLEDGMENTS

We thank Peggy Micca from BNL and Kent Riley from MIT NRL for their initial technical instructions for boron analysis techniques. We also thank Dr. Lujia Tang from the BIDMC BNCT lab for her support in manufacturing BPA-f and maintaining the lab, David Northey from Dana Farber Cancer Institute for continuing help in handling animals, and Bill Dahlberg from Harvard School of Public Health for his expert advise on cell culture work. This research was supported by the Department of Energy under DOE grant # DE-FG02-97ER62193.

REFERENCES

1. W.S. Kiger, III, S. Sakamoto, and O.K. Harling, Neutronic Design of a Fission Converter-Based Epithermal Neutron Beam for Neutron Capture Therapy, *Nuclear Science and Engineering*, 131:1–22, 1999.
2. J. Yu, J. Burwick, G. Dranoff, and X. Breakefield, Gene Therapy for Metastatic Brain Tumors by Vaccination with Granulocyte-Macrophage Colony-Stimulating Factor-Transduced Tumor Cells, *Hum. Gene Ther.*, 8(9):1065–1072, 1997.
3. J. Yu, J. Burwick, G. Dranoff, and X. Breakefield, Gene Therapy for Metastatic Brain Tumors by Vaccination with Granulocyte-Macrophage Colony-Stimulating Factor-Transduced Tumor Cells, *Hum. Gene Ther.*, 8(9):1065–1072, 1997.
4. J. Coderre, A. Chanana, D. Joel, E. Elowitz, P. Micca, M. Nawrocky, M. Chadha, J. Gebbers, M. Shady, N. Peress, and D. Slatkin, Biodistribution of Boronophenylalanine in Patinets with Glioblastoma Multiforme: Boron Concentration Correlates with Tumor Cellularity, *Rad. Res.*, 149:163–170, 1998.
5. Y. Imahori, S. Ueda, Y. Ohmori, T. Kusuki, K. Ono, R. Fujii, and T. Ido, Fluorine-18-Labeled Fluoroboronophenylalanine PET in Patients with Glioma, *J. Nuclear Medicine*, 39:325–333, 1998.

BORON-10 DISTRIBUTION OF *p*-BORONOPHENYLALANINE IN THE EXPERIMENTAL BRAIN TUMOR IN THE RATS

Munekazu Nakaichi[1], Yuko Hori[1], Toshinari Hasegawa[1], Sanenori Nakama[1], Akira Takeuchi[2], and Yoshinobu Nakagawa[3]

[1]Department of Veterinary Surgery
Yamaguchi University, Yamaguchi
[2]College of Bioresource Sciences
Nihon University, Kanagawa
[3]Department of Neurosurgery
National Kagawa Children's Hospital
Kagawa, Japan

1. INTRODUCTION

Boron neutron capture therapy (BNCT) has been employed in the treatment of malignant brain tumor in human. In this type of radiation therapy, biodistribution of ^{10}B in tumor and peritumoral tissue is highly important because a range of α-particle produced by n-α reaction is quite short in biological tissue. Many boron compounds have been tested to achieve an optimal biodistribution for BNCT, and sodium borocaptate (BSH) has been believed to be the one giving such distribution.[1,2] BSH-based BNCT, therefore, has been employed in the treatment of glioblastoma. Another boron compound recently used in BNCT for brain tumor is para-boronophenylalanine (BPA). Although BPA is a standard boron compound for BNCT against melanoma, it is well known to accumulate in brain tumor tissue more extensively than BSH, indicating its therapeutic potential for glioblastoma.[3] Unlike BSH, however, BPA is considered to easily penetrate into normal brain tissue. A safety of BPA-based BNCT has not been proven and pre-clinical study to confirm its detailed biodistribution using experimental animals must be essential. The objective of this study is to measure the boron concentrations achieved by BPA in various organs in tumor-bearing rats, including brain tumor and other major structures around the head, and compare to those achieved by BSH.

Frontiers in Neutron Capture Therapy, edited by Hawthorne *et al.*
Kluwer Academic / Plenum Publishers, New York, 2001.

2. MATERIALS AND METHODS

Experimental brain tumor model was produced by intracranial transplantation of C6 glioma cells. Thirty-six three-week-old male Wistar rats were used in this study. A cell suspension of C6 with a concentration of 1.0×10^8/ml was prepared. Under general anesthesia, 5.0×10^6 of tumor cells (50 μl) were innoculated into the right hemisphere of the brain, the subcortical region to a depth of 5 mm from the dural surface. Two weeks after the tumor transplantation, the rats were randomly divided into two groups: one for BPA administration (n = 20) and the other for BSH administration (n = 16). Both boron compounds were administered to the tumor-bearing rats with a single dose of 50 mg of boron per kilogram body weight. BPA was administered as a fructose-complex form prepared using the method reported by Yoshino *et al.*[4] BPA was injected intraperitoneally, whereas BSH was administered intravenously from a tail vein. The rats were sacrificed at different time after the administration of the boron compounds (BPA: 1, 2, 4, 6, 8 hours, BSH: 1, 4, 8 hours) and the following samples were obtained to measure the boron concentration: blood, brain tumor, normal brain, scalp, temporal muscle and both eyes. Boron concentration were measured by inductively coupled plasma atomic emission spectrometry (ICP-AES).[5] The weighed samples were first chemically digested with a mixture of 70% of perchloric acid and 30% of hydrogen peroxide at 75 °C. The samples became colorless after one hour incubation with the chemical reagents and were filtered through a 0.45 millipore filter before being introducing into the ICP-AES instument.

3. RESULTS

The results of the biodistribution study of BSH in the various tissues of the tumor-bearing rats are shown in Figure 1. The peak levels were observed one hour after the intravenous injection of BSH in all the tissues examined. However, BSH was rapidly eliminated from the rats. The peak boron concentration in the tumor tissue was 28.79 ppm, but it decreased to 2.39 ppm in eight hours after the injection. Boron concentration in the blood was 95.76 ppm at its peak and it was higher than that in the tumor tissue throughout the experiment, giving the tumor/blood (T/B) ratio less than 1.0. In the normal brain, 5.17 ppm of the boron was detected one hour after the injection, however, no boron was detected after that.

Time sequential changes of the boron concentrations in various tissures after the BPA injection are shown in Figure 2. The boron concentrations achieved by BPA increased gradually after the intraperitoneal injection and reached a peak in four hours in all the tissues examined. The boron concentration in the brain tumor reached a peak of 42.10 ppm which was much higher than the peak level achieved by BSH, and at this time, T/B ratio exceeded 1.0. In contrast to BSH, however, considerable boron was detected in the normal brain tissue. It reached 21.57 ppm at its peak. Tumor/blood ratio and tumor/normal brain ratio obtained in this experiment are summarized in Table 1.

Furthermore, other major structures around the brain such as the scalp, temporal muscle and eyes also showed much higher boron concentration in comparison with BSH. The peak boron concentrations in the scalp, temporal muscle and eye were 40.29 ppm, 30.88 ppm and 23.46 ppm, respectively.

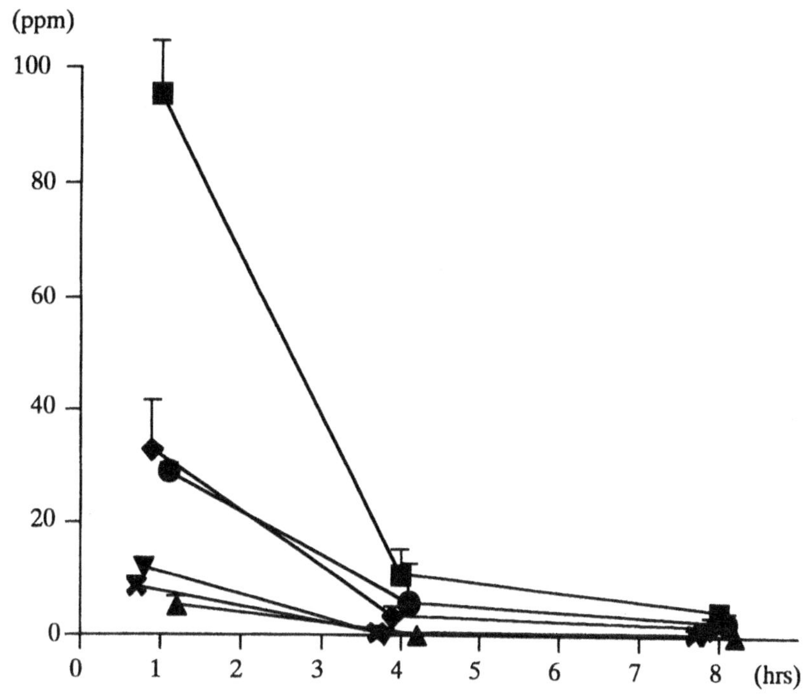

Figure 1. Time sequential changes of the boron concentration in the tumor-bearing rats after the single injection of BSH (50 mg boron/kg BW.) ■-blood, ●-tumor, ▲-normal brain, ◆-scalp, ▼-temporal muscle, ✖-eye.

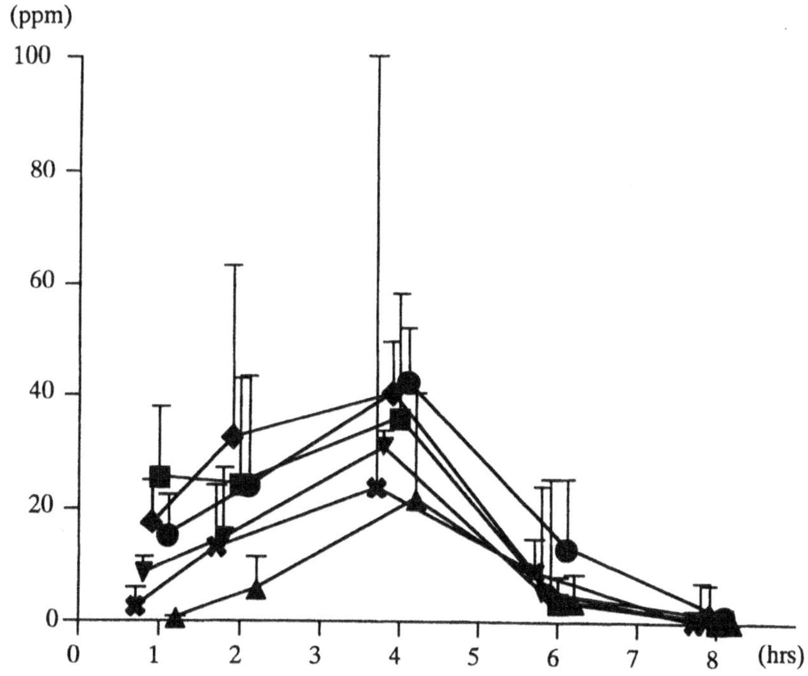

Figure 2. Time sequential changes of the boron concentration in the tumor-bearing rats after the single injection of BPA (50 mg boron/kg BW.) ■-blood, ●-tumor, ▲-normal brain, ◆-scalp, ▼-temporal muscle, ✖-eye.

Table 1. Tumor/blood (T:B) and tumor/normal brain (T:N) ratios for BSH and BPA. (∞: not evaluated because boron level in normal brain was zero)

Hours after adminstration		1	2	4	6	8
BSH	T:B	0.30	—	0.54	—	0.54
	T:N	5.57	—	∞	—	∞
BPA	T:B	0.54	0.99	1.18	3.95	3.29
	T:N	28.47	4.21	1.95	3.55	∞

4. DISCUSSION

The therapeutic efficacy of BNCT is mostly dependent on the boron concentration achieved in the tumor tissue. BPA is proven to accumulate into brain tumor tissue as well as melanoma tissue,[3] and a combination use of BPA and epithermal neutron beam is a recent trend in BNCT for glioblastoma. Also in this experiment, BPA was shown to accumulate in the C6 glioma tissue more extensively than BSH. Therefore, BPA is considered as a favorable boron carrier for brain tumor with regard to the tumor accumulation.

However, BPA-based BNCT has been carried out without craniotomy, taking advantage of increased penetration of epithermal neutron beam. In this type of BNCT, biodistribution of BPA around the head is considered much more important because some normal tissues such as scalp and temporal muscle would be unavoidably included in radiation field, whereas craniotomy has been routinely performed in BSH-based BNCT with the aims of delivering enough thermal neutrons to the deeper lesions in the brain and preventing the scalp and temporal muscle from radiation injuries. However, the reports on the BPA distribution in the surrounding tissues of the head were quite limited and detailed distribution of BPA still remained unclear.

In this experiment, BPA was administered to the rats intraperitoneally, whereas BSH intravenously. This is because of lower solubility of BPA even though the fructose-complex was utilized. The volume of the BPA solution for each tumor-bearing rat was too large to be given intravenously. For a biodistribution study of BPA given intravenously, large animal models should be developed.

Our experiment showed that BPA delivered much more boron to the surrounding normal brain tissues in comparison with BSH and peak levels of the boron in the scalp and temporal muscle were also much higher than those of BSH. These boron levels might be high enough to cause serious side effects if enough thermal neutrons were delivered.

BPA also accumulated in the eyes of the tumor-beraring rats. Since eye is well known to be highly sensitive to radiation, even a small amount of the boron in the eye has the possibility to cause serious side effcts. Therefore, caution for eye protection should be appropriately employed in the clinical use of BPA-based BNCT. As the whole eye tissue was used as the sample for boron measurement in this experiment, detailed distribution of BPA in the eye still remained unclear. Further experiments are needed to clarify the distribution of BPA in the eye for better understanding of the effects of BPA-based BNCT on eyes.

Since our biodistribution study demonstrated here was performed in small laboratory animals, there might be some differences when compared to the biodistribution of the compound in human. Especially, biodistribution and clearance of BSH in this tumor model were quite different from the previous results in human patients.[1,2] Therefore, our data shoud not directly apply to human patients. However, this experiment brought us the important findings on the BPA biodistribution as follows: (1) BPA can deliver the boron to the experimental brain tumor more extensively than BSH. (2) BPA, however, accumulates in the normal brain and other surrounding structures, such as the scalp, temporal muscle, and eyes, which have the possibility to be included in the radiation field. (3) Therefore, in the clinical trials with BPA-based BNCT for brain tumor, these features the of biodistribution of BPA should be cautiously considered.

REFERENCES

1. T. Kagegi, Y. Nakagawa, K. Kitamura, K. Matsumoto, and H. Hatanaka. Pharmacokinetics and boron uptake of BSH ($Na_2B_{12}H_{11}SH$) in patients with intracranial tumors, J. Neuro-Oncol., 33:117–130, 1997.
2. Y. Nakagawa, and H. Hatanaka. Boron neutron capture therapy-Clinical brain tumor studies, J. Neuro-Oncol., 33:105–115, 1997.
3. J.A. Coderre, T.M. Button, P.L. Micca, C.D. Fisher, M.M. Nawrocky, and H.B. Liu. Neutron capture therapy of the 9L gliosarcoma using the p-boronophenylalanine-fructose complex, Int. J. Radiati. Oncology Biol. Phys., 30:643–652, 1994.
4. K. Yoshino, A. Suzuki, Y. Mori, H. Kakihana, C. Honda, Y. Mishima, T. Kobayashi, and K. Kanda. Improvement of solubility of p-boronophenylalanine by complex formation with monosaccharides. Strahlentherapie 166:127–129, 1989.
5. S.W. Tamat, and D.E. Moore, Determination of boron in biological tissues by inductively coupled plasma atomic emission spectrometry, Anal. Chem., 59:2126–2154, 1987.

BPA UPTAKE AND EFFLUX THROUGH THE CELLULAR MEMBRANE

A. Wittig[1,2], W. A. Sauerwein[2], and J. A. Coderre[1]

[1]Medical Department
Brookhaven National Laboratory
Upton, New York, 11973
[2]Department of Radiation Oncology
University of Essen
45122-Essen, Germany

1. INTRODUCTION

The compound *p*-(dihydroxy)-boronophenylalanine (BPA) is currently being evaluated as a boron-carrier for BNCT in clinical trials.[1,2] Knowledge of the uptake and efflux mechanisms of BPA in and from tumor cells might allow a specific modification of those processes aiming at a higher intracellular boron concentration.

BPA is a boronated analogue of the neutral amino acid tyrosine and might therefore be taken up by the cells by an amino acid transport system. The three systems L-, A- and ASC are in particular responsible for the transport of neutral amino acids. In this study mainly the systems L and A have been investigated.

System L has a broad specificity for neutral amino acids with linear or branched side chains and aromatic structures such as tyrosine. The transport mechanism proposed for the system L is a sodium independent exchange mechanism, a substrate-coupled antiport. The driving force is a chemical gradient of amino acids preaccumulated in the cell. Trans-stimulation regulates the L-system, therefore amino acids preaccumulated in the cell activate this carrier. 2-Aminobicyclo-[2.2.1]-Heptane-2-Carboxylic Acid (BCH) is a specific substrate for the L-transport protein.[3]

System A accepts amino acids with unbranched, aliphatic side chains such as glutamine and proline. It is proposed that the transport mechanism for system A is a sodium symport and the driving force is the electrochemical potential of a sodium gradient. The activity of system A is regulated by trans-inhibition so that amino acids accumulated in cellular compartments lead to a decreased activity of this carrier. The synthetic model substrate for this system is 2-Aminoisobutyric Acid (MeAiB).[3]

Frontiers in Neutron Capture Therapy, edited by Hawthorne *et al.*
Kluwer Academic / Plenum Publishers, New York, 2001.

These characteristics were used to evaluate, if the amino acid carriers might be responsible for the BPA transport. To plan the experimental procedure it had to be taken into account that the activity of amino acid transport systems is dependent on factors such as the pH, the particular cell type, there are ionic and energy dependencies, exchange phenomena, metabolic inhibitors and other regulations. From cells in culture it is also known that these transport systems can be affected by the cell density, the availability of serum factors, composition and concentration of amino acids in the culture medium and in intracellular compartments and the cell proliferation. Establishing an *in vitro* system to investigate the cellular transport of BPA it was therefore required to control the environmental conditions for the cells very carefully and keep them constant during the experiment.

2. MATERIALS AND METHODS

The cells were depleted from intracellular amino acids by an incubation in Hank's balanced salt solution (Gibco BRL, Grand Island, NY, U.S.A., Cat. No. 11201) for one hour before the experiments started. This medium contains only salts and glucose, it is free of amino acids. For the experiments cells were also maintained in Hank's BSS. All chemicals were dissolved in this solution and the pH was adjusted to 7.4.

For the uptake experiments cells were incubated for 2–240 minutes in BPA (Centronic Croyden, England, Lot. No. P056 and Boron Biologicals, Raleigh, NC, U.S.A., Lot. No. 971203, ^{10}B enrichment 95 atom % by weight) in a concentration of 50 ppm ^{10}B (4.9 mmol/l BPA). To analyze the intracellular boron content direct-current plasma emission spectroscopy (DCP-AES)[4] was used after an oil filtration technique to separate cells from extracellular boron in the cell medium.[5]

For the efflux assay the cells were incubated for 2 hours in a medium containing BPA in a concentration of 50 ppm ^{10}B. The cells were washed two times to remove the BPA containing medium and then incubated in a boron free medium. After incubation times ranging from 2 to 120 minutes, the cells were harvested and separated from extracellular boron with the oil filtration technique. The boron content was also measured with DCP-AES.

3. RESULTS

The uptake of BPA in GS-9L cells at 37 °C was significantly higher than the uptake at 21 °C.

The BPA uptake was compared to the uptake of boric acid (Eagle Picher, Quapaw, OK, U.S.A., Lot. No. BL-EV-87-95-BA, ^{10}B enrichment: 95.91 atom % by weight). The uptake of BPA was more than twice the one of boric acid. In the malignant GS-9L cells, a rat gliosarcoma cell line, the BPA uptake was significantly higher than the uptake in V79 cells, which are fibroblastic cells.

The L-system for neutral amino acids was tested as potential carrier for BPA with the following experiment: cells were preincubated with BCH (Calbiochem®, La Jolla, CA, U.S.A., Lot. No. 800393) and after that exposed to BPA. In a second experiment cells were incubated simultaneously in BPA and BCH.

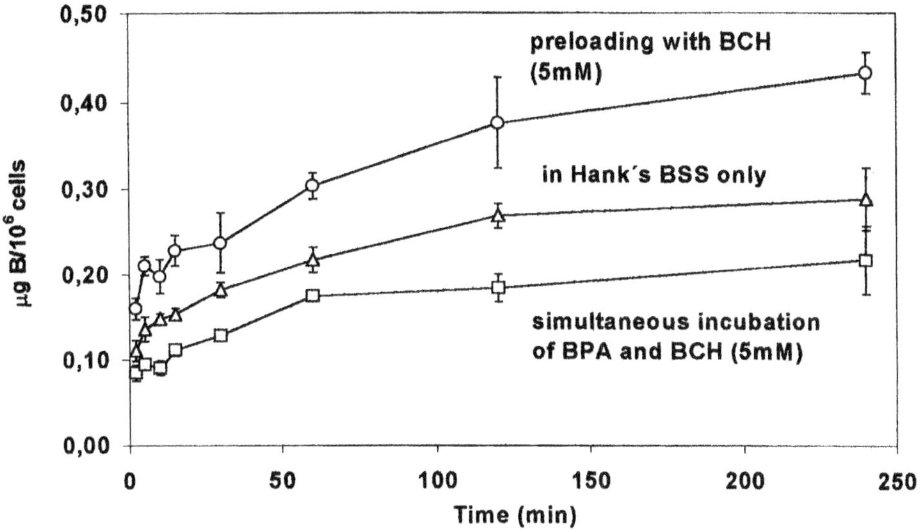

Figure 1. Time course of boron accumulation in GS-9L cells at 37 °C. B-compound: BPA (50 ppm ^{10}B ≈ 4.9 mmol/l BPA). Cells were preloaded for 1 hour with 5 mM BCH dissolved in Hank's BSS (○), or depleted of amino acids by an incubation in Hank's BSS for 1 hour before they were exposed to BPA (△), or incubated simultaneously with BCH and BPA dissolved together in Hank's BSS (❑). The symbols represent the mean ± SD of three individual experiments.

A preloading with BCH increased the uptake of BPA. The initial transport velocity was very high and the boron concentration in the cells after 240 minutes was nearly twice as high as the concentration without the preloading. The simultaneous incubation of BPA and BCH led to an inhibition of the BPA uptake.

To investigate if the A-system might be responsible for the BPA transport, cells were preloaded with MeAiB (Aldrich Chemical Company Inc., Milwakauee, WI, U.S.A., Lot. No. 14421JR) and then exposed to BPA. In a second experiment cells were incubated simultaneously in MeAiB and BPA. The preloading and the simultaneous incubation of BPA and MeAiB led to an increase of the intracellular boron concentration. In the preloading experiment the final boron concentration doubled in comparison to the BPA uptake in Hank's BSS only and corresponds therefore to the boron concentration in the cells after a preloading with BCH. A simultaneous incubation led only to a slight increase of the boron concentration in the cells.

The efflux of BPA from GS-9L cells was measured in αMEM and in Hank's solution. In αMEM BPA was washed out fast. After 120 minutes, hardly any boron remained in the cells. If this experiment was carried out in Hank's solution the efflux was much slower.

The BPA efflux was also temperature dependent. At 21 °C the efflux was slower in both media, but the difference was more significant for the experiment carried out in αMEM.

To investigate if an exchange mechanism might also be involved in the BPA efflux, GS-9L cells were exposed to BPA for two hours and then incubated in a Hank's solution which contained either BCH or MeAiB. The efflux was fast in both experiments, but the transport velocity was faster, if the experiment was carried out with BCH in the Hank's solution.

4. DISCUSSION

The results of the study confirm a certain selectivity of the BPA accumulation for malignant cells. The high BPA uptake compared to the uptake of boric acid, which is considered to be transported into the cells only by diffusion, indicates that an active process is involved in the BPA transport. The results of this study further support that BPA is transported in GS-9L cells by a temperature sensitive system with high affinity for L-system substrates. The existence of a sodium-independent L-amino acid transporter in astrocytes has been proven by other authors.[6]

BPA might be transported by the L-system because its transport characteristics correspond to the characteristics of the L-system. A preaccumulation of BCH in the cells might lead to a trans-stimulation of the transporter and an exchange of the intracellularly accumulated amino acid against extracellular BPA. For this reason the concentration of BPA in the cells is higher than without the preloading. The reduced boron content in the cells after a simultaneous incubation of BPA and the L-system substrates can be explained as a competitive inhibition of the BPA uptake.

It is unlikely that BPA is transported by the A-system. A preaccumulation of MeAiB and a simultaneous incubation of MeAiB and BPA led to an increased boron concentration in the cells. These results do not support the A-system to be responsible for the BPA transport.

The reason for this observation might be overlapping transport kinetics of system L and A substrates which is known, for example, for MeAiB and Asparagine.[7,8,9] Amino acid uptake by the L-system involves an exchange of intracellular for extracellular amino acids. It has been reported[10] that L-system substrates (Valine, BCH) and A-system substrates (Asparagine, MeAiB) involve a heteroexchange reaction in which the L-system substrate is taken into the cell by displacing an A-system substrate that exits the cell.[9,11]

Data suggest that the efflux of intracellularly accumulated BPA from the cells is a transport process involving an exchange mechanism. When efflux experiments were carried out in complete medium αMEM, containing a number of other amino acids, the rate of efflux was rapid. When efflux experiments were carried out in Hank's BSS, which contains no added amino acids, the rate of efflux was considerably reduced. When efflux experiments were carried out in Hank's BSS with added BCH or added MeAiB, the rate of efflux was again rapid. These results imply that the intracellular BPA is exchanged against amino acids present in αMEM or the MeAiB and BCH added to the Hank's BSS.

It remains to be proven that BPA is not metabolized in the cells and that ^{10}B which was measured in this study leaves the cell as BPA.

The results of this study support the hypothesis that BPA is transported by the L-system which can be further stimulated by amino acids preaccumulated in the cell by either system L or A.

ACKNOWLEDGMENTS

A. Wittig was supported by a grant of the Flora-Immerschitt-Stiftung, Deutscher Stifterverband # TS 181/06/97. J. A. Coderre is supported by the U.S. Department of Energy, Office of Biological and Environmental Research Contract # DE-ACO2-98CH10886. We gratefully acknowledge the excellent technical assistance of P. Micca, M. S. Makar and A. LoMonte.

REFERENCES

1. J.A. Coderre, E.H. Elowitz, M. Chadha, R. Bergland, J. Capala, D.D. Joel, H.B. Liu, D.N. Slatkin, and A.D. Chanana, Boron neutron capture therapy for glioblastoma multiforme using p-boronophenylalanine and epithermal neutrons: Trial design and early clinical results. *J. Neurooncol.*, 33:141–152, 1997.
2. P.M. Busse, R. Zamenhof, H. Madoc-Jones, G. Solares, S. Kiger, K. Riley, C. Chuang, G. Rogers, and O. Harling, Clinical follow-up of patients with melanoma of the extremity treated in a phase I boron neutron capture protocol. *in*: Advances in Neutron Capture Therapy, Volume I Medicine and Physics, B. Larsson, J. Crawford, R. Weinreich eds, Elsvier Sciences B.V., 1997, pp. 60–64.
3. M.S. Kilberg, Amino acid transport in isolated rat hepatocytes. *J. Membr. Biol.*, 69:1–12, 1982.
4. J.A. Coderre, A.D. Chanana, D.D. Joel, E.H. Elowitz, P.L. Micca, M.M. Nawrocky, M. Chadha, J.-O. Gebbers, M. Shady, N.S. Peress, and D.N. Slatkin, Biodistribution of boronophenylalanine in patients with glioblastoma multiforme: boron concentration correlates with tumor cellularity. *Rad. Res.*, 149:163–170, 1998.
5. J. Capala, M.S. Makar, and J.A. Coderre, Accumulation of boron in malignant and normal cells incubated *in vitro* with boronophenylalanine, mercaptoborane or boric acid. *Rad. Res.*, 146:554–560, 1996.
6. M.J. Tsai, and E.H.Y. Lee, Characterization of L-DOPA transport in cultured rat and mouse astrocytes. *J. Neurosci. Res.*, 43:490–495, 1996.
7. M.S. Kilberg, B.R. Stevens, and D.A. Novak, Recent advances in mammalian amino acid transport. *Annu. Rev. Nutr.*, 13:137–165, 1993.
8. M.A. Shotwell, M.S. Kilberg, and D.L. Oxender, Regulation of amino acid system L in CHO cells. *J. Biol. Chem.*, 257:2974–2980, 1982.
9. H.N. Christensen, C. de Cespedes, M.E. Handlogten, and G. Ronquist, Energization of amino acid transport studies for the Ehrlich ascites tumor cell. *Biochem. Biophys. Acta.*, 330:487–522, 1973.
10. R.K. Singh, C.A. Rinehart, J.P. Kim, S. Tolleston-Rinehart, L.F. Lawing, D.G. Kaufmann, and G.P. Siegal, Tumor cell invasion of basement membrane *in vitro* is regulated by amino acids. *Cancer Invest.*, 14(1):6–18, 1996.
11. H.N. Christensen, Role of amino acid transport and countertransport in nutrition and metabolism. *Physiol. Rev.*, 70:43–77, 1990.

A FURTHER MECHANISM STUDY OF BORON ACCUMULATION TO AND RELEASE FROM MELANOMA WITH *p*-BORONOPHENYLALANINE

Kazuo Yoshino[1], Mamiko Kamijo[1], Mototetsu Kimura[1], Yutaka Mishima[2], Hirofumi Kondo[2], Mitsutaka Kadota[2], Syosuke Ito[3], Kazumasa Wakamatsu[3], and Hidetake Kakihana[4]

[1]Department of Chemistry, Faculty of Science
Shinshu University, Asahi, Matsumoto 390-8621, Japan
[2]Mishima Institute for Dermatological Research
Kobe 657-0063, Japan
[3]School of Hygiene and Institute for Comprehensive Medical School
Fujita-Gakuen Health University
Toyoake, Aichi 470-1192, Japan
[4]Energy Research Center
Wakasa Bay
Tsuruga 914-0047, Japan

1. INTRODUCTION

p-Boronophenylalanine (BPA) has the boron delivery property to melanoma cells. We have previously proposed[1] the accumulation property of ^{10}B in melanoma cells is due to the complex formation of BPA with melanin monomer, DOPA, DHI, and DHICA which are highly exists in melanoma cells, and also the release mechanism is due to the decomposition of the complexes to boric acid. The mechanism is based on the findings obtained from the ^{11}B-NMR measurements for the mixed solution of BPA and these monomers. In the measurement of *in vivo* melanoma samples obtained from BPA-injected hamsters, two signals were observed. We have assigned one of two signals to free BPA and the other signal at 8 ppm to BPA-monomers. But, in the case of *in vitro* sample the chemical shift value of the complex is 10 ppm. This means that there is a possibility that the 8 ppm signal of *in vivo* sample is not the BPA-monomer complex. Therefore, we have considered that a further investigation is necessary for this BPA-melanin monomer system

Frontiers in Neutron Capture Therapy, edited by Hawthorne *et al.*
Kluwer Academic / Plenum Publishers, New York, 2001.

to find other species, which can complex with BPA. In this paper, we describe the kinetic ^{11}B-NMR observation results for the BPA-melanin monomer system.

2. EXPERIMENTAL METHODS

JEOL LA400 NB was used for ^{11}B-NMR measurements. BPA with natural abundance ratio of boron was used. BF_3OEt_2 was used for an external reference ($\delta = 0$). ^{11}B-NMR measurements of melanoma samples were carried out without a sample spin and field lock.

3. RESULTS AND DISCUSSION

The mixed pH 7 solution of 1 mM BPA and 0.5 mM Dopachrome was prepared, and ^{11}B-NMR measurement was carried out. As shown in Figure 1, immediately after the mixing, the complex signal at 10 ppm was observed, and the half of BPA decomposed to boric acid at 2 days after the mixing. As calculated from stoichiometry, this change was assigned to the complex formation of BPA with DHI and its decomposition to boric

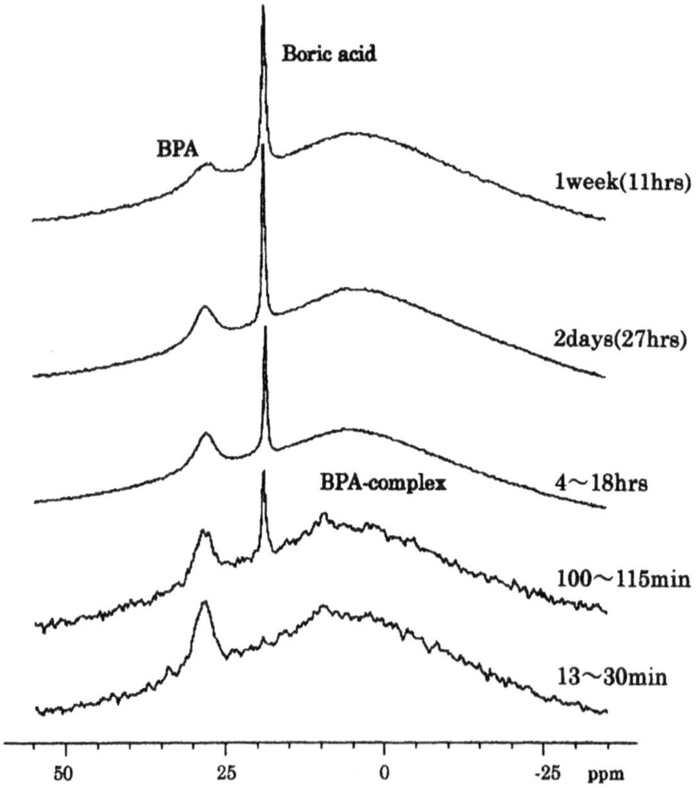

Figure 1. ^{11}B-NMR spectra of BPA and Dopachrome solutions (pH 7). BPA: 0.5 mM, Dopachrome: 1 mM. Two types of FT accumulation time are shown at the right side of each spectrum. Up to 14 hours: actual measurement time, After 14 hours: measurement time is shown in blanket. The reference signal of BF_3OEt_2 (= 0 ppm) is not shown.

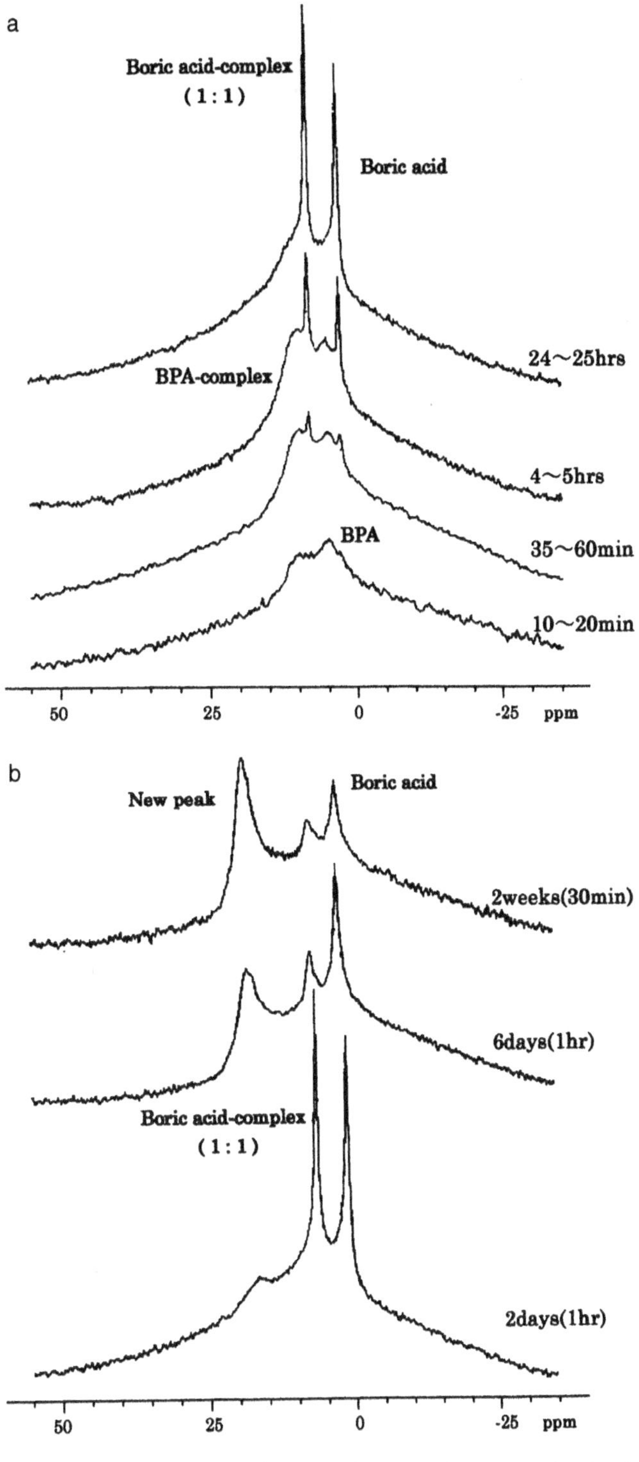

Figure 2. ^{11}B-NMR spectra of BPA and DHI-Dimer solution (MeOH : H_2O = 1 : 1, pH 11). BPA: 2 mM, DHI-dimer 1 mM. a) from mixing to 1 day, b) from 2 days to 2 week. FT accumulation time is shown as the same manner as Figure 1. The reference signal of BF_3OEt_2 (= 0 ppm) is not shown.

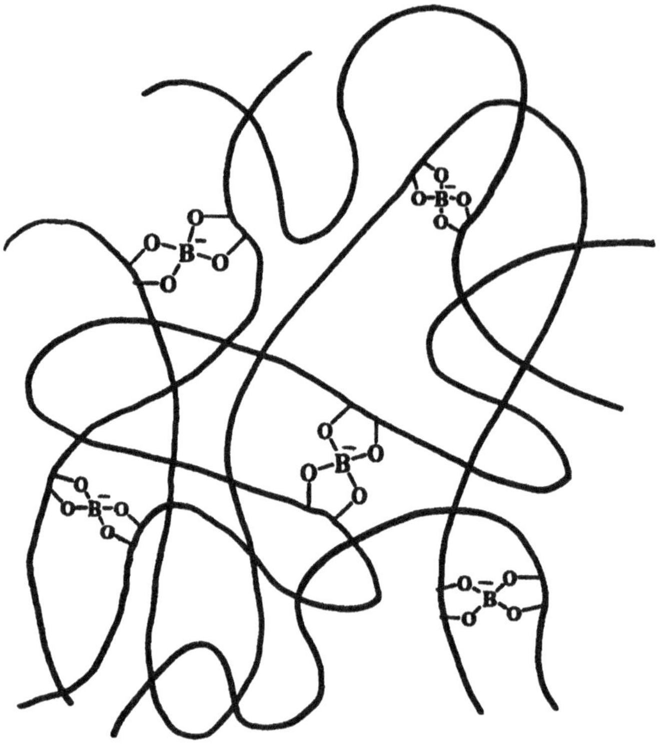

Figure 3. Proposed structure of melanin-incorporated boric acid.

acid. The spectrum gradually changed and finally all of BPA changed to boric acid. Since the mole ratio of BPA and Dopachrome was 2:1, the complete disappearance of BPA implies that there were other dihydroxy species was produced after the BPA-DHI complex is oxidized. The same mole ratio experiment was carried out at pH 8. In this case disappearance of BPA signal was faster than the case at pH 7. This is due to the stronger complex formation at pH 8. Recently, DHI or DHICA is found to be oxidized to form dimer or trimer before polymerization,[2] and these compounds have dihydroxy groups that can complex with BPA. Therefore, our ^{11}B-NMR finding implies that BPA complexes with dimer or trimer and again decomposes into boric acid by oxidation of the BPA-dimer complex or the BPA-trimer complex. To confirm this point, we have carried out ^{11}B-NMR experiment on BPA-dimer system. Since acetyl groups protected the dihydroxy groups of the DHI-dimer[3] used, it was insoluble in water. Therefore, we have dissolved it in methanol, and a potassium carbonate solution was mixed to remove the acetyl groups. The mole ratio of BPA and the dimer was 2:1, which implies that BPA:dihydroxy group ratio is 1:1. The ^{11}B-NMR spectrum change was shown in Figure 2. As expected, the signal of the BPA-dimer complex was observed as shown in Figure 2-a. Since the solution was alkaline, the signal of free BPA appeared at high field side of the complex signal. As continuing measurement, two sharp signals were observed. The signal at 8 ppm is that of the 1:1 dimer-boric acid complex, and the signal at 3 ppm is that of boric acid. At 24 hours after the mixing, the broad signals of both BPA-dimer complex and free BPA disappeared. The results have demonstrated that even DHI-dimer complexes with BPA and the complex is oxidized to yield boric acid. This fact explains the

^{11}B-NMR spectrum change of the 2:1 BPA-dopachrome system. Continuing the NMR measurement of the BPA-dimer solution, a spectacular change was observed as shown in Figure 2b. A new signal was appeared at 18 ppm. The signal slowly increased with time. On the other hand, the 1:1 dimer (or tetramer)-boric acid complex signal at 10 ppm decreased. This indicates that the boric acid complex species changed to a new species. Boric acid can complex with DOPA as 1:2 ratio, and the chemical shift of the 1:2 complex shifts to low field, (15.5 ppm).[4] Therefore, we assigned the new peak to the new type complex, in which both sides of boron are complexed with polymers. A possible structure is shown in Figure 3. Boron might be located between melanin polymer chains. Since the prepared solution contains methanol and alkali, we must carefully discuss with the obtained results. But, if melanin polymer capture boric acid produced from BPA, it is very important for the boron-capture mechanism in melanoma cells. Therefore, a further study using melanoma cells must be carried out.

4. CONCLUSION

^{11}B-NMR *in vitro* studies have demonstrated that not only melanin monomer but also melanin dimer capture BPA by complex formation between boronic acid group of BPA and diol parts of melanin monomer or dimer, and the BPA part of the complex decomposes to boric acid. In the methanol–alkali solution containing BPA and melanin dimer, a new ^{11}B-NMR peak was observed, which implies that the produced boric acid from decomposition of BPA may be incorporated in the melanin polymer.

REFERENCES

1. K. Yoshino, Y. Mishima, M. Kimura, J. Hiratsuka, Y. Mori, S. Ito, and H. Kakihana, Capture of *p*-boronophenylalanine in malignant melanoma cells by complex formation with melanin monomers, DOPA, DHI and DHICA, *in* "Advances in Neutron Capture Therapy, Volume II, Chemistry and Biology", B. Larsson *et al.* eds., Elsevier Science B.V., Amsterdam, 1997, pp. 234–238.
2. M. d'Ischia, A. Napolitano, and G. Prota, Oxidative polymerization of 5,6-dihydroxyindoles. Tracking the biosynthetic pathway to melanin pigments, *Gazzetta. Chimica. Italiana.*, 126:783–789, 1996.
3. M. d'Ischia, A. Napolitano, K. Tsiakas, and G. Prota, New intermediates in the oxidative polymerization of 5,6-dihydroxyindole to melanin promoted by the peroxidase/H_2O_2 system, *Tetrahedron*, 46:5789–5796, 1990.
4. K. Yoshino, M. Kotaka, M. Okamoto, and H. Kakihana, ^{11}B-NMR study of the complex formation of borate with catechol and L-dopa, *Bull. Chem. Soc. Jpn.*, 52:3005–3009, 1979.

ENHANCED UPTAKE OF BORONOPHENYLALANINE AND EFFICACY OF BNCT IN F98 GLIOMA BEARING RATS FOLLOWING ADMINISTRATION OF CEREPORT™

Weilian Yang[1], Rolf F. Barth[1], Joan H. Rotaru[1], Jeffrey A. Coderre[2], Marta M. Nawrocky[2], and Raymond T. Bartus[3]

[1]Department of Pathology
The Ohio State University
Columbus, OH 43210
[2]Medical Department
Brookhaven National Laboratory
Upton, NY 11973
[3]Alkermes, Inc.
Cambridge, MA 02139

1. INTRODUCTION

There is a broad consensus of opinion that the blood-brain and blood-tumor barriers significantly limit the movement of a wide variety of therapeutic agents from the vascular compartment into brain tumors.[1] Cereport™, previously known as RMP-7, is a synthetic nonapeptide and bradykinin analogue, that produces a transient, pharmacologically mediated opening the blood-brain barrier.[2] Cereport exerts its permeabilizing effects on capillary vascular endothelial cells within brain tumors by activating the bradykinin B2 receptor thereby, opening their tight intercellular junctions.[3] Data have been published on use of Cereport to enhance the delivery of methotrexate, carboplatin and a variety of cytokines (TNF-α, IL-2, IFN-γ),[4,5] and clinical trials to evaluate its efficacy in enhancing drug delivery in patients with brain tumors currently are in progress. We recently have reported that the intracarotid infusion of a hyperosmotic (1.373 mOsmol or 25%) solution of mannitol enhanced the tumor delivery of boronophenylalanine (BPA), and sodium borocaptate (BSH), and efficacy of BNCT in F98 glioma bearing rats.[6–8] The purpose of the present study was to determine if the tumor uptake of BPA

Frontiers in Neutron Capture Therapy, edited by Hawthorne *et al.*
Kluwer Academic / Plenum Publishers, New York, 2001.

and efficacy of BNCT could be enhanced by Cereport and to compare this with previous obtained data using osmotic BBB-D.

2. MATERIALS AND METHODS

2.1. Brain Tumor Model

The F98 glioma cell line was derived from an undifferentiated neoplasm, transplacentally induced by N-ethyl-N-nitrosourea, in an inbred CD Fischer rat and has been propagated *in vitro* and *in vivo* since 1971.[9] A stereotactic procedure, which previously has been used by us for BNCT studies of the F98 glioma,[10] was employed to implant the tumor intracerebrally into male Fischer rats weighing ~200–220 g.

2.2. Biodistribution Studies

F98 glioma bearing rats were anesthetized with 1% isoflurane following which a polyethylene catheter (PE-10) was inserted retrograde through the right external carotid artery to the bifurcation of the common carotid artery for i.c. administration of Cereport to the tumor bearing cerebral hemisphere, as described in detail elsewhere.[10] Cereport was infused i.c. at a dose of 0.1 µg/kg b.w./min over 15 minutes in a total volume of 1.5 ml using a constant flow pump. BPA was converted to a more soluble fructose complex and administered at a dose of 300–500 mg/kg b.w over 10 minutes at a rate of 120 µl/min or i.v. by bolus injection over 30 seconds via the penile vein. Excipient control animals received saline by i.c. infusion followed by i.c. or i.v. administration of BPA. At 0.5, 2.5, and 4 hours following infusion of Cereport and injection of BPA animals were euthanized with halothane and brains were removed and tumors and brain around tumor (BAT) were carefully dissected out from surrounding normal brain. Blood and other tissues (skin, muscle, liver and eye) also were obtained and each sample was weighed, and then frozen and stored at −20 °C until boron determinations were performed by means of direct current plasma-atomic emission spectroscopy (DCP-AES).[6]

2.3. Boron Neutron Therapy Capture Experiments

BNCT was initiated 14 days following stereotactic implantation of 1,000 F98 glioma cells. Rats were randomized into 6 experimental groups of 10–12 animals each, as follows: group 1, i.c. administration of BPA (500 mg/kg b.w.) and Cereport (0.5 µg/kg b.w.); group 2, i.c BPA and saline; group 3, i.v. BPA and i.c. Cereport; group 4, i.v. BPA and i.c. saline; group 5, i.c. Cereport and radiation; group 6, i.c. Cereport but otherwise was untreated. Since the optimum tumor to brain and blood ratios were observed at 2.5 hours following administration of BPA, this time point was selected as the interval between BPA administration and the initiation of BNCT. Irradiations were carried out at the Brookhaven Medical Research Reactor (BMRR). All irradiated rats were anesthetized with a 1.2/1 mixture of ketamine/xylazine and placed supine in a body shield-head stablizer, as described elsewhere.[7] The tumor implantation site was centered in the 1.15 cm diameter aperture of the neutron beam collimator and animals were irradiated for 8 Megawatt-minutes. A blood sample for boron determination was taken from each animal prior to irradiation. The therapeutic responses of the treated rats were evaluated by determining the prolongation of median (MeST) and mean (MST) survival times in days following implantation of tumor. The Wilcoxon-Gehan rank sum test and the Bonferroni method of multiple comparisons were applied to the survival data to test for significant differences between the treated groups and controls.

3. RESULTS AND DISCUSSION

3.1. Tissue Biodistribution Studies

Boron concentrations in tumor, brain and blood in F98 glioma bearing rats at 2.5 hours following i.v. or i.c. injection of BPA with or without i.c. infusion of Cereport are shown in Figure 1. The mean boron concentration ± standard deviation (SD) at 2.5 hours in tumor was 55.7 ± 9.6 μg/g after i.c. injection with Cereport *vs* 33.6 ± 3.9 μg/g without it, and 29.4 ± 9.9 μg/g for i.v. injection of BPA with Cereport *vs* 15.4 ± 3.5 without it ($p < 0.05$). The tumor boron concentrations in animals that received i.c. Cereport were increased 1.7X and 1.9X, respectively, compared to animals that received i.c. or i.v. BPA without Cereport. The best composite tumor to normal tissue ratios were observed at 2.5 hours after i.c administration of BPA and Cereport. The tumor: brain ratio (T:Br) was 6.3 and the tumor: blood ratio (T:Bl) was 7.7 compared to 4.4 and 5.1, respectively, for i.c. BPA without Cereport. Following injection of i.v. BPA with Cereport, the T:Br and T:Bl ratios were 3.9 and 4.5, respectively, compared to 2.6 and 2.3. A more detailed description of the effects of Cereport on the biodistribution of BPA in F98 glioma bearing rats is presented elsewhere.[12]

3.2. Survival Following BNCT

BNCT survival data are presented as Kaplan-Meier plots in Figure 2. Untreated control rats had a MST of 23 ± 3 days and irradiated controls had a MST of 28 ± 5 days. Animals that received i.c. BPA with Cereport and BNCT had a MST of 50 ± 16 days compared to 43 ± 6 days in animals that did not receive Cereport. Animals that received

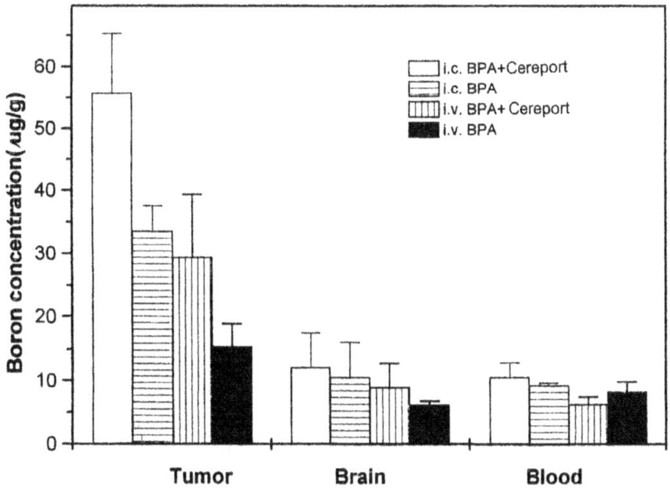

Figure 1. Boron concentrations at 2.5 hours following i.v. or i.c. injection of BPA or with or without i.c. Cereport. BST = brain surrounding tumor; Brain = ipsilateral (tumor-bearing) cerebral hemisphere.

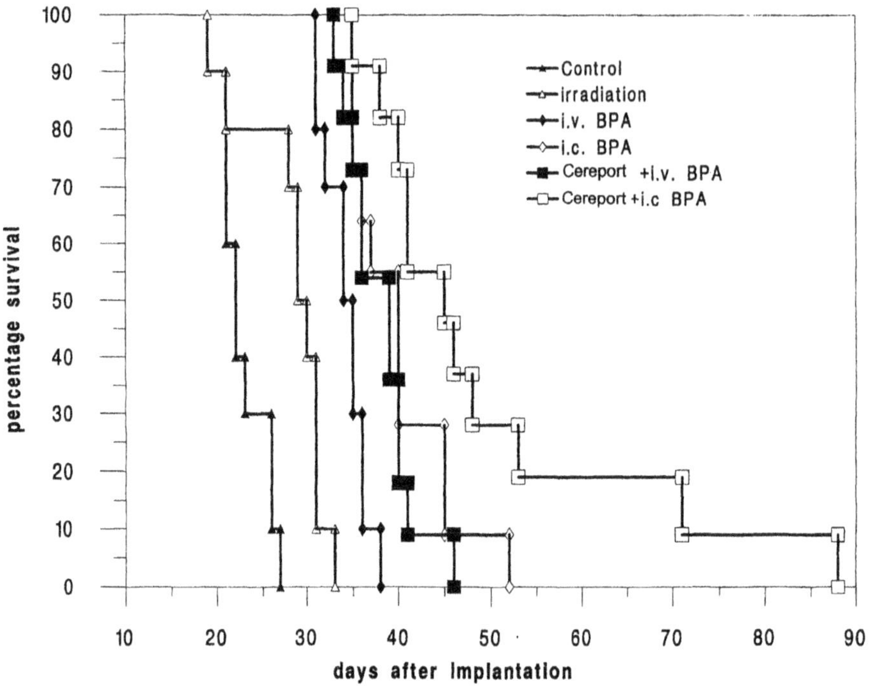

Figure 2. Kaplan-Meier survival plots for F98 glioma bearing rats following i.v. or i.c. injection of BPA and BNCT with or without Cereport.

i.v. BPA with i.c. Cereport had a MST 38 ± 4 days compared to 34 ± 3 days without Cereport. As determined by the Wilcoxon rank sum test, BNCT significantly increased the survival of F98 glioma-bearing rats when compared to untreated ($p < 0.001$) and irradiated controls ($p < 0.003$). The survival times associated with i.c. administration of Cereport followed by either i.c. or i.v. BPA compared to those observed were significant at the levels of $p = 0.04$ and $p = 0.05$, respectively.

4. DISCUSSION AND CONCLUSIONS

In the present study we have shown that the i.c. infusion of Cereport can significantly enhance tumor boron delivery and survival of F98 glioma bearing rats following BNCT compared to that of animals which did not receive Cereport. We previously have reported that tumor boron delivery and the efficacy of BNCT could be enhanced by hyperosmotic, mannitol induced BBB-D combined with i.c. injection of BPA. Although equivalent tumor brain values were obtained using this method and Cereport,[12] better survival data obtained with the former.[6,7] In contrast to mannitol-induced BBB-D, Cereport exerts it effects via a pharmacologic, receptor mediated mechanism that reportedly is more specific for vascular endothelial cells of tumor capillaries. This may be advantageous for the delivery of cytoreductive chemotherapeutic agents that also are neurotoxic to normal brain, but since BNCT is a binary system, normal brain boron levels only are important at the time of irradiation. Lower boron values were observed in brain surrounding tumor (BST) following administration of Cereport compared to those seen in animals that had mannitol induced BBB-D.[12] This may account for the better survival data obtained using the latter method. On the other hand, there were no deaths following administration of Cereport compared to a 10% mortality associated with mannitol induced BBB-D.[6] Further experimental studies are in progress in our laboratory to assess the use of Cereport to enhance boron delivery to brain tumors using both the F98 glioma model, and a previously employed model for melanoma metastatic to the brain.

ACKNOWLEDGMENTS

These studies have been supported by a grant from Alkermes, Inc., Cambridge, MA.

REFERENCES

1. N.H. Greig, Drug delivery to the brain by blood-brain barrier circumvention and drug modification, *In* Neuwelt EA, (ed), *Implications of the Blood-Brain Barrier and its Manipulation, Vol. I. Basic Science Aspects*, New York: Plenum Medical Book Co., 1989, pp. 311–367.
2. R.T. Bartus, P.J. Elliott, R.L. Dean, N.H. Hayward, T.L. Nagle, M.R. Huff, P.A. Snodgrass, and D.G, Blunt, Controlled modulation of BBB permeability using the bradykinin agonist, RMP-7. Exp. Neurol. 142:14–28, 1996.
3. K.L. Black, Biochemical opening of the blood-brain barrier. Adv. Drug Delivery Rev. 15:37–52, 1995.
4. S. Nakano, K. Matsukado, and K.L. Black, Enhanced delivery of cytokines and intercellular adhesion molecule 1(ICAM-1) expression in glioma by intracarotid infusion of bradykinin analog RMP-7. Neurol. Res. 19:501–508, 1997.
5. K.T. Matsukado, T. Inamura, S. Nakano, M. Fukui, R.T. Bartus, and K.L. Black, Enhanced tumor uptake of carboplatin and survival in glioma-bearing rats by intracarotid infusion of bradykinin analog, RMP-7. Neurosurgery 39:125–134, 1996.
6. R.F. Barth, W. Yang, J.H. Rotaru, M.L. Moeschberger, D.D. Joel, M.M. Nawrocky, J.H. Goodman, and A.H. Soloway, Boron neutron capture therapy of brain tumors: enhanced survival following intracarotid injection of either sodium borocaptate or boronophenylalanine with or without blood-brain barrier disruption. Cancer Res. 57:1129–1136, 1997.
7. W. Yang, R.F. Barth, J.H. Rotaru, D.D. Joel, M.M. Nawrocky, and J.H. Goodman, Enhanced survival of glioma bearing rats following boron neutron capture therapy with osmotic blood-brain barrier disruption and intracarotid injection of boronophenylalanine. J. Neurooncol. 32:59–70, 1997.
8. W. Yang, R.F. Barth, J. Rotaru, M. Moeschberger, D.D. Joel, M.N. Nawrocky, J.H. Goodman, and A.H. Soloway, Boron neutron capture of brain tumors: enhanced survival following intracarotid injection of sodium borocaptate with or without blood-brain barrier disruption. Int. J. Rad. Onc. Bio. Phys. 37:663–672, 1997.
9. R.F. Barth, Rat brain tumor models in experimental neuro-oncology: the 9L, C6, T9, F98, RG2 (D74), RT-2 and CNS-1 gliomas. J. Neurooncol. 36:91–102, 1998.
10. W. Yang, R.F. Barth, D.E. Carpenter, M.L. Moeschberger, and J.H. Goodman, Enhanced delivery of boronophenylalanine for neutron capture therapy by means of intracarotid injection and blood-brain barrier disruption. Neurosurgery 38:985–992, 1996.
11. R.F. Barth, D.M. Adams, A.H. Soloway, E. Mechetner, F. Alam, and A. Anisuzzaman, Determination of boron in tissues and cells using direct-current plasma atomic emission spectroscopy. Anal. Chem. 63:890–893, 1991.
12. R.F. Barth, W. Yang, M.L. Moeschberger, J.H. Goodman, and R.T. Bartus, Enhanced delivery of boronophenylalanine for neutron capture therapy of brain tumors using the bradykinin analog Cereport (receptor-mediated permeabilizer-7). Neurosurgery 44:350–359, 1999.

EFFECT OF DOSE AND INFUSION TIME ON THE DELIVERY OF *p*-BORONOPHENYLALANINE FOR NEUTRON CAPTURE THERAPY

Darrel D. Joel, Jeffrey A. Coderre, Peggy L. Micca, and Marta M. Nawrocky

Medical Department
Brookhaven National Laboratory
Upton, NY 11973

1. INTRODUCTION

In the clinical trial of BNCT of glioblastoma multiforme currently in progress at Brookhaven National Laboratory, *p*-boronophenylalanine (BPA) complexed with fructose (BPA-F) was administered to patients intravenously over a two-hour period at an initial dose of 250 mg (currently 290 mg) BPA/kg body weight.[1,2] Biodistribution studies indicated that this infusion regimen yields tumor-to-blood (T:B) and tumor-to-normal brain tissue boron concentration ratios of about 3.5 to 1.[3] Enhancement of either tumor boron uptake and/or the tumor-to-blood (T:B) boron concentration ratio would have the potential of significantly improving the therapeutic gain of BNCT. In this study, the rat 9L gliosarcoma model was used to examine the effect of the total dose of BPA and the infusion time on tumor boron concentration and T:B boron concentration ratios.

2. MATERIALS AND METHODS

2.1. The 9L Rat Gliosarcoma Model

Induction of subcutaneous tumors in male Fisher 344 rats weighing 200 to 250 g was done as previously described.[4] Biodistribution studies were done 11 days post-implantation, when the tumors were ~100 mg in mass. For initiation of intracerebral tumors, 1×10^4 cultured cells in 1 μl of medium were injected into the left frontal lobe of

Frontiers in Neutron Capture Therapy, edited by Hawthorne *et al.*
Kluwer Academic / Plenum Publishers, New York, 2001.

similar rats as previously described.[4] Biodistribution studies were done 14 days post-implantation when tumors ranged in weight from 40 to 50 mg. Intravenous infusion was achieved by inserting a silastic rubber cannula into the anterior facial vein with the tip of the cannula secured near that vein's entrance into the external jugular vein.[5] To directly perfuse intracerebral tumors via the internal carotid artery, the left common carotid artery was exposed at its bifurcation and a silastic rubber cannula was inserted into the external carotid artery until the tip of the cannula is situated at the bifurcation of the common carotid artery.[5] Continuous infusions were done following recovery from anesthesia using small animal infusion swivels (Harvard Apparatus, South Natick, MA) with spring tethers and calibrated syringe pumps. BPA-F was prepared as previously described.[6] Concentrations used in these studies ranged from 26.5 to 86.5 mg BPA/ml. Boron analysis was performed by either prompt gamma-ray spectroscopy[7] or direct current plasma-atomic emission spectroscopy (DCP-AES).[8,9]

3. RESULTS AND DISCUSSION

3.1. Blood and Tumor ^{10}B Concentrations after Infusion of BPA-F at a Dose-Rate of 125, 250, and 500 mg BPA/kg/hr for Various Time Periods

Twenty eight rats (4 separate experiments) bearing multiple subcutaneous 9L gliosarcomas were given a total dose of 250 mg BPA/kg body weight administered intravenously over a 2-hr period at a dose rate of 125 mg BPA/kg/hr. One hour after infusion, the blood boron concentration decreased from a mean of 12.8 ± 3.6 to 8.2 ± 2.9 μg ^{10}B/g (~36%); thereafter; blood boron levels remained relatively constant (see Table 1). In contrast the mean tumor boron concentration only decreased from 32.7 ± 5.7 to 30.6 ± 5.5 μg ^{10}B/g (~6.5%) in the first hour, but then decreased more rapidly. These data suggest that the optimal time for BNCT would be about 1 hr after infusion when the tumor boron levels are relatively high, resulting in the largest T:B ratios (3.6 to 1).

Two groups of four rats each were infused intravenously with BPA-F for 3 hours and 7 hours, respectively, at a dose rate of 125 mg BPA/kg/hr. Similar to the results obtained from the 2-hr infusion study, blood boron levels decreased by about 35% at 1 hour after the end of infusion while tumor boron concentrations decreased less rapidly yielding T:B ratios that were highest 1 to 3 hours after the end of infusion (see Table 1).

During the 7-hour infusion period, blood boron levels continued to increase during the first 5 hours and then remained constant for the next 2 hours. One hour after infusion, the blood boron concentration decreased from a mean of 16.2 ± 2.2 μg ^{10}B/g to a mean of 10.7 ± 2.3 μg ^{10}B/g (~34%). A further decrease was seen between 1 and 2 hours post-infusion, after which blood boron levels remained relatively constant. The T:B boron concentration ratios at 1, 2, and 3 hours post-infusion were 4.6, 5.0, and 5.1, respectively. These were the highest T:B boron concentration ratios observed with any intravenous infusion schedule tested.

Eight rats were infused intravenously with BPA-F for 2 hours at a dose-rate of 250 mg BPA/kg/hr. Tumor and blood ^{10}B concentrations were about 60% higher than those observed following a 2-hour infusion at a dose rate of 125 mg BPA/hr/kg (see Tables 1 and 2). Again, the percent decrease in blood boron concentration at 1 hour post-infusion was significantly greater than the decrease in tumor concentration, ~42% and

Table 1. Blood and tumor boron concentrations and tumor-to-blood ratio (T/B) after continuous intravenous infusion of BPA-F at a dose rate of 125 mg BPA/kg/hr for various time periods

Hours post-infusion	125 mg/kg/hr								
	2 hr-infusion (n = 28)			3 hr-infusion (n = 4)			7 hr-infusion (n = 4)		
	Blood	Tumor	T/B	Blood	Tumor	T/B	Blood	Tumor	T/B
0	12.8 ± 3.6	32.7 ± 5.7	2.6	16.7 ± 1.7	41.3 ± 2.2	2.5	16.2 ± 2.2	60.0 ± 8.2	3.7
1	8.2 ± 2.9	30.6 ± 5.5	3.7	10.5 ± 1.5	40.2 ± 3.0	3.8	10.7 ± 2.3	48.7 ± 2.2	4.6
2	8.0 ± 2.8	26.0 ± 5.1	3.3	9.0 ± 2.8	37.2 ± 2.1	4.1	9.0 ± 1.2	45.0 ± 13.1	5.0
3	7.4 ± 2.7	20.8 ± 4.6	2.8	6.9 ± 1.4	33.4 ± 4.4	4.8	7.7 ± 1.2	39.6 ± 3.4	5.1
4	6.2 ± 2.5	19.5 ± 4.4	3.1	6.0 ± 1.6	27.3 ± 5.1	4.6	8.8 ± 1.2	38.0 ± 8.6	4.3

~13% respectively, resulting in the highest T:B boron concentration ratios 1 and 2 hours after injection. It is of interest that the blood and tumor ^{10}B concentrations achieved with this dose rate (250 mg BPA/kg/hr) in rats were roughly equivalent to those obtained in humans infused with BPA-F for 2 hours at a dose rate of 125 mg BPA/kg/hr.[3]

One group of four rats bearing multiple subcutaneous tumors was infused i.v. for 3 hours and a second group of 4 rats (two rats bearing subcutaneous tumors and two rats bearing intracerebral tumors) was infused i.v. for 6 hours at a dose rate of 250 mg BPA/kg/hr. As observed with other BPA-F infusion regimens, blood boron concentrations decreased more rapidly during the first hour post-infusion than did the corresponding tumor boron concentrations (see Tables 1 and 2).

Four rats were infused i.v. for 2 hours at a dose-rate of 500 mg BPA/kg/hr. Blood and tumor ^{10}B concentrations were significantly higher than those obtained following a 2-hour infusion at a dose-rate of 250 mg BPA/kg/hr (see Table 2). However the T:B concentration ratio was similar in both groups.

3.2. Tumor and Blood Boron Concentration after Intracarotid Infusion of BPA-F

Ten rats bearing intracerebral tumors were infused for 1 hour with BPA-F at an infusion-rate of 125 mg BPA/kg/hr. Five rats were infused via the internal carotid (ic) artery and 5 rats were infused intravenously. Although blood boron concentrations were the same with both routes of infusion (7.6 ± 0.1 and 7.6 ± 0.3 µg/g) tumor boron concentrations were significantly higher following intracarotid infusion as compared to intravenous infusion (38.2 ± 4.5 vs 25.7 ± 4.7 µg/g) yielding T:B ratios of 5.0 and 3.4 respectively.

These studies indicate that tumor boron levels in excess of 60 µg/g can be readily obtained by either increasing the duration of infusion or by increasing the total dose of BPA given in a 2 hr period, however, the T:B boron concentration ratios remain relatively constant. Increasing the concentration of boron in the tumor, even in the absence of an increased T:B boron concentration ratio, provides some therapeutic gain with BNCT. Calculations suggest that a 50% increase in tumor boron concentration would, in principle, result in an increase of approximately 25% in the radiation dose to a tumor extending 6 cm deep in the skull while maintaining the same radiation dose to normal brain (therapeutic gain).[10] An increase in the therapeutic gain can also be achieved by

Table 2. Blood and tumor boron concentrations and tumor-to-blood ratio (T/B) after continuous intravenous infusion of BPA-F at dose rates of 250 and 500 mg BPA/kg/hr for various time periods

Hours post-infusion	250 mg/kg/hr									500 mg/kg/hr		
	2 hr-infusion (n = 8)			3 hr-infusion (n = 4)			6 hr-infusion (n = 4)			2 hr infusion (n = 8)		
	Blood	Tumor	T/B	Blood	Tumor	T/B	Blood	Tumor	T/B	Blood	Tumor	T/B
0	21.1 ± 4.6	52.7 ± 7.3	2.5	28.0 ± 3.5	64.3 ± 5.2	2.3	34.3 ± 2.3			36.1 ± 2.1	84.0 ± 3.2	2.3
1	12.3 ± 3.5	45.9 ± 6.8	3.7	18.5 ± 1.4	57.6 ± 4.8	3.1	21.8 ± 1.7	85.2 ± 17.6	3.9	19.0 ± 1.1	69.0 ± 5.3	3.6
2	10.5 ± 3.2	39.7 ± 6.3	3.8	13.9 ± 1.5	51.6 ± 5.4	3.7				17.3 ± 1.4	58.3 ± 4.9	3.4
3	9.6 ± 3.1	32.8 ± 5.7	3.4	14.9 ± 2.0	42.2 ± 5.7	2.8				14.0 ± 0.5	44.4 ± 4.1	3.2
4				11.5 ± 2.4	43.4 ± 5.5	3.8				12.7 ± 0.5	35.1 ± 6.5	2.8

increasing the T:B boron concentration ratio. When BPA-F was infused via the internal carotid artery (i.c.) a significant increase in the T:B boron concentration ratio was observed. A 1-hour i.c. infusion yielded an average tumor-to-blood ratio of 5.0 whereas samples from rats concurrently infused i.v. at the same dose yielded an average T:B ratio of 3.4. Calculations indicate this degree of increase in the T:B ratio would also result in an increased therapeutic gain exceeding 25%.[10]

It should be emphasized that tumor boron concentrations reported in these studies are average boron concentrations of gross tumor tissue. Such measurements do not provide any information on either individual cells within the tumor or on "nests" of cells invading brain tissue surrounding the main tumor mass. Smith *et al.*[11] reported that following a single intraperitoneal injection of BPA-F in rats, "nests" of tumor cells contained on the average less boron than cells in the main tumor mass. In contrast, "nests" of tumor cells from rats infused with BPA-F continuously for 24 hours had boron levels similar to cells in the main tumor mass (D. Smith, Cornell University, Ithaca NY; unpublished observations). It remains to be determined whether or not shorter infusion times, *i.e.*, 6 hrs, will result in adequate uptake of boron by infiltrating tumor cells.

ACKNOWLEDGMENTS

This research was supported in part by the U.S. Department of Energy, Office of Biological and Environmental Research under Contract DE-AC02-98CH10886.

REFERENCES

1. J.A. Coderre, E. Elowitz, M. Chadha, R. Bergland, J. Capala, D.D. Joel, H.B. Liu, D.N. Slatkin, and A.D. Chanana, Boron neutron capture therapy for glioblastoma multiforme using *p*-boronophenylalanine and epithermal neutrons: Trial design and early clinical results, *J. Neuro-Oncol.*, 33:141–152, 1997.
2. M. Chadha, J. Capala, J.A. Coderre, E.H. Elowitz, J. Iwai, D.D. Joel, H.B. Liu, L. Wielopolski, and A.D. Chanana, Boron neutron capture therapy (BNCT) for glioblastoma multiforme (GBM), using the epithermal neutron beam at the Brookhaven National Laboratory. *Int. J. Radiat. Oncol. Biol. Phys.*, 40:829–834, 1998.
3. J.A. Coderre, A.D. Chanana, D.D. Joel, E.H. Elowitz, P.L. Micca, M.M. Nawrocky, M. Chadha, J.-O. Gebbers, M. Shady, N.S. Peress, and D.N. Slatkin, Biodistribution of boronophenylalanine in patients with glioblastoma multiforme: Boron concentration correlates with tumor cellularity, *Radiat. Res.*, 149:163–170, 1998.
4. D.D. Joel, R.G. Fairchild, J.A. Laissue, S.K. Saraf, J.A. Kalef-Ezra, and D.N. Slatkin, Boron neutron capture therapy of intracerebral rat gliosarcomas, *Proc. Natl. Acad. Sci. USA*, 87:9808–9812, 1990.
5. D.D. Joel, D.N. Slatkin, P.L. Micca, M.M. Nawrocky, T. Dubois, and C. Velez, Uptake of boron into human gliomas of athymic mice and into syngeneic cerebral gliomas of rats after intracarotid indusion of sulfhydryl boranes, *in*: "Clinical aspects of neutron capture therapy," R. Fairchild, V. Bond, and A. Woodhead, eds., Plenum Press, New York, 1989, pp. 325–332.
6. J.A. Coderre, T.M. Button, P.L. Micca, C.D. Fisher, M.M. Nawrocky, and H.B. Liu, Neutron capture therapy of the 9L rat gliosarcoma using the *p*-boronophenylalanine-fructose complex, *Int. J. Radiat. Oncol. Biol. Phys.*, 30:643–652, 1994.
7. R.G. Fairchild, D. Gabel, B.H. Laster, D. Greenberg, W. Kiszenick, and P. Micca, Microanalytical techniques for boron analysis using the $^{10}B(n,\alpha)^{7}Li$ reaction, *Med. Phys.*, 13:50–56, 1986.
8. R.F. Barth, D.M. Adams, A.H. Soloway, E.B. Mechetner, F. Alam, and A.K.M. Anisuzzaman, Determination of boron in tissues and cells using direct-current plasma atomic emission spectroscopy, *Anal. Chem.*, 63:890–893, 1991.

9. W. Bauer, P. Micca, and B. White, A rapid method for the direct analysis of boron in whole blood by atomic emission spectroscopy, *in*: "Advances in Neutron Capture Therapy," A. Soloway, R. Barth, and D. Carpenter, eds., Plenum Press, New York, 1993, pp. 403–407.
10. J.A. Coderre, Boron neutron capture therapy, *in*: "Textbook of Radiation Oncology," S. Leibel, and T. Phillips, eds., W. B. Saunders Company, Philadelphia, 1998, pp. 1263–1277.
11. D.R. Smith, S. Chandra, J.A. Coderre, and G.H. Morrison, Ion microscopy imaging of ^{10}B from *p*-boronophenylalanine in a brain tumor model for boron neutron capture therapy, *Cancer Res.*, 56:4302–4306, 1996.

IN VITRO UPTAKE STUDY OF 5-CARBORANYL URIDINE AND ITS DERIVATIVES

Toshinari Hasegawa[1], Munekazu Nakaichi[1], Sanenori Nakama[1], Hiroyuki Nakamura[2], Yoshinori Yamamoto[2], Masao Takagaki[3], and Akira Takeuchi[4]

[1]Department of Veterinary Surgery
Yamaguchi University, Yamaguchi
[2]Department of Chemistry
Tohoku University, Sendai
[3]Research Reactor Institute
Kyoto University, Osaka
[4]College of Bioresource Sciences
Nihon University
Kanagawa, Japan

1. INTRODUCTION

Recently, much attention has been focused on synthesis of new tumor-seeking boron compounds to improve the therapeutic efficacy of BNCT, and boronated uridine is one being expected for favorable uptake by tumor cells by the pathways for the synthesis of DNA. Previous studies on boronated uridines[1] have shown that such compounds have high cytotoxicity in cultured cells, limiting their potential clinical use. In this experiment, 5-carboranyl uridine and its derivatives, characterized by high water solubility and low cytotoxicity, were investigated for their accumulation in cultured glial tumor cells. *In vivo* biodistributions of these boronated uridines in brain tumor-bearing rats were also investigated.

2. MATERIALS AND METHODS

2.1. Boron Compounds

Four kinds of boronated uridine, 5-carboranyl uridine (5B10U),[2] 5-hydroxy-methylcarboranyl uridine (5HB10U), 5-nido-carboranyl uridine (5B9TU),[2] and 5′-b-D-glycosyl-

Frontiers in Neutron Capture Therapy, edited by Hawthorne *et al.*
Kluwer Academic / Plenum Publishers, New York, 2001.

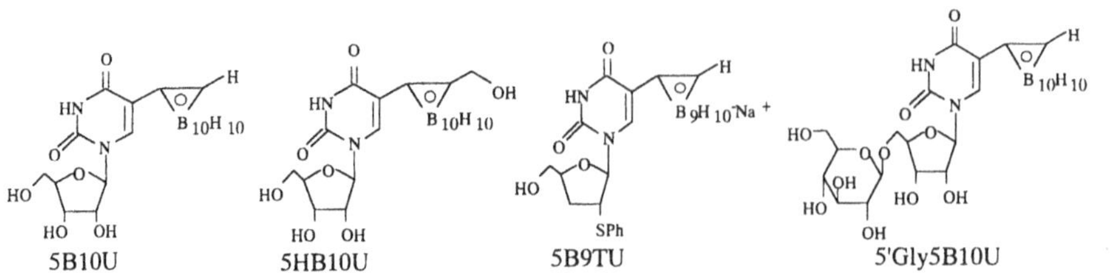

Figure 1. Chemical structures of 5-carboranyl uridine and its derivatives.

5-carboranyl uridine (5′Gly5B10U) were included in this study. Their chemical structures are shown in Figure 1. Sodium borocaptate (BSH) was also used as a control to the boronated uridines. All four boronated uridines contain natural boron, whereas BSH is boron 10-enriched.

2.2. Tumor Cells and Experimental Brain Tumor Model

C6 rat glioma cell line was used in this study. The cells were maintained in Eagle's MEM (Nissui Pure Industries, Osaka, Japan) supplemented with 10% fetal bovine serum (ICN biochemicals Japan Co., Ltd), gentamicin sulfate (50 mg/L), amphotericin B (250 µg/mL). For *in vivo* biodistribution studies, tumor-bearing rats were experimentally produced using C6 cells. Twenty-four three-week-old male Wistar rats were received intracranial transplantation of 5.0×10^6 of viable C6 cells under general anesthesia. The rats used for the biodistribution studies two weeks after the transplantation.

2.3. *In Vitro* Uptake Study

All the four boronated uridines and BSH were tested with respected to their accumulation in C6 cells. The cells in growth phase prepared in 150 mm^2 cell culture dishes were incubated in the culture medium containing various concentrations of the boron compounds. Boron concentrations in the cultured media were 10, 25, 40 ppm for 5HB10U, 5B9TU and BSH. Same concentrations were used for 5′Gly5B10U with an addition of two extra dose of 100 ppm and 200 ppm. For 5B10U, lower concentrations of 5, 10, 15 ppm were used because of the high cytotoxicity shown in the preliminary experiment. After incubation for the indicated hours (6, 12, 18, 24 hrs), the cells were washed 4 times and the number of the viable cells was counted by dye-exclusion method. The cells were centrifuged again and the supernatant was removed. Then the cells were chemically digested with a mixture of 70% $HClO_4$ and 30% H_2O_2 at 75 °C.[3] After one hour incubation with the chemical reagents, the samples became colorless and were diluted with DD water and filtered through a 0.45 µm filter before being introduced into the inductivity coupled plasma atomic emission spectrometry (ICP-AES) instruments.

2.4. Neutron Irradiation Study

The effects of *in vitro* thermal neutron irradiation after the exposure to 5HB10U and BSH were examined. Six hours after 25 ppm boron loading, the cells were washed four times with PBS and trypsinized to prepare a single cell suspension with a concentration of 2.0×10^4/mL. One mL of the suspension in columm-shape Teflon tubes was irradiated with thermal neutrons at the Reactor in Kyoto University (KUR, Osaka, Japan) at a neutron flux of 1.95×10^9 n/cm²/s. After the thermal neutron irradiation, a colony formation assay was performed to determine the surviving fraction of the irradiated cells. Three hundreds cells (150 μL of the irradiated suspension) were placed in three tissue culture dishes. Ten days later, the colonies were fixed with methanol and stained with 0.1% crystal violet for a quantitative visual inspection.

2.5. *In Vivo* Biodistribution Study

The biodistribution of 5B10U and 5HB10U were examined in tumor-bearing rats. Two weeks after the tumor transplantation, each boronated uridine was dissolved in 2% β-cyclodextrin and injected into the peritoneal cavity of the rats with a dose of 50 mg boron per kilogram body weight. Each three rats were sacrificed at 1, 4, 8 hours after the injection and weighed samples ranging from 200 to 400 mg were obtained from the following tissues: blood, brain tumor and normal brain. The boron concentrations of these samples were measured by ICP-AES as described above. Biodistribution of 5B9TU and 5′Gly5B10U was also examined at four hours after injection.

3. RESULTS AND DISCUSSION

3.1. *In Vitro* Uptake and Irradiation Study

The results of the *in vitro* uptake of carboranyl uridines are shown in Figure 2. Cellular uptake observed for 5HB10U and 5B9TU was much higher than that of BSH that has been the major compound in the clinical trials of BNCT. Twenty-five ppm boron loading for 24 hrs with 5HB10U and 5B9TU in the culture medium resulted in 81.65 ppm and 57.63 ppm values cellular uptake, respectively, whereas only 11.6 ppm was found for BSH. The boron uptake of these two compounds apparently showed a dose-dependent manner (Figure 2A). On the other hand, both 5B10U and 5′Gly5B10U showed a low cellular uptake. Because of the high cytotoxicity of 5B10U towards C6 cells, loading with

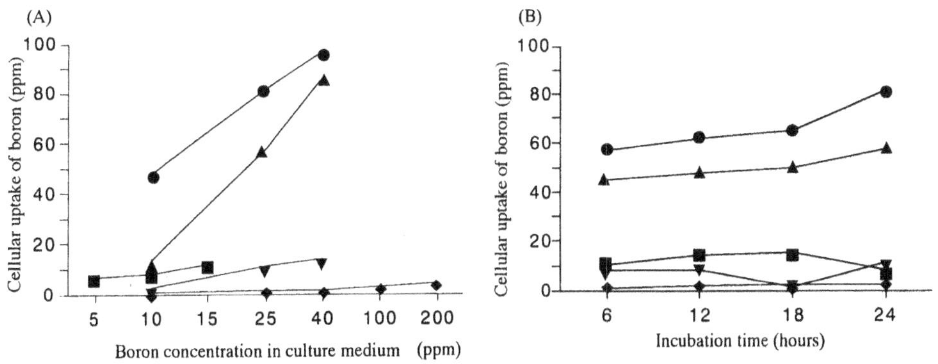

Figure 2. *In vitro* uptakes of carboranyl uridine and BSH by C6 glioma cells. (A) The cellular uptakes of 5B10U (■), 5HB10U (●), 5B9TU (▲), 5′Gly5B10U (◆), BSH (▼) are shown (incubation time; 24 hrs). (B) Time sequential changes of the cellular uptake of carboranyl uridines. (5B10U; 10 ppm boron loading, others; 25 ppm boron loading.)

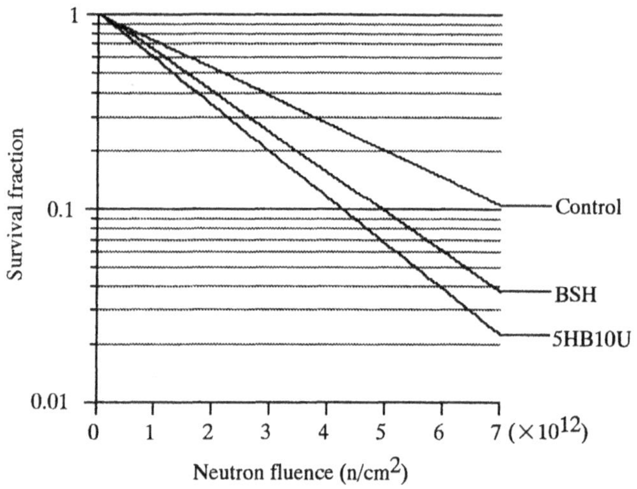

Figure 3. Surviving fraction of C6 cells after *in vitro* BNCT using 5HB10U and BSH. 5HB10U is assumed to be boron-10 enriched.

lower boron concentrations were performed. However, the cellular uptake of 5B10U was much lower than those found at the same dose for 5HB10U and 5B9TU. In contrast to 5B10U, 5′Gly5B10U had almost no cytotoxicity towards C6 cells. However, the cellular uptake was quite low despite loading the culture medium with up to 200 ppm boron.

Figure 2B shows the time-dependent uptake of the boronated uridines. Boron levels in the cells were relatively high after 6 hours incubation. These values were over 80% of those achieved after 24 hrs incubation. These findings mean that these boron compounds, especially 5HB10U and 5B9TU, were easily taken up by the cells in the early phase of the incubation.

In vitro thermal neutron irradiation after the exposure to 5HB10U showed some cell killing effect (Figure 3). Because BSH used in this study was boron-10 enriched, its cell killing effect was stronger than that of 5HB10U that contained natural boron. If we assume that 5HB10U is boron-10 enriched, the surviving fraction is estimated as shown in Figure 3. Boron concentrations in the samples after washing were not measured in this study because of the smaller volume of the samples. Based on these results, the two boronated uridines, 5HB10U and 5B9TU, were considered to have high potential to accumulate in C6 tumor cells and also considered worthy of being tested for their *in vivo* biodistribution.

3.2. *In Vivo* Biodistribution Study

Time-sequential changes of the boron concentration in various organs of the tumor-bearing rats were studied for 5B10U and 5HB10U (Figure 4). The peak level in most organs, including brain tumor, were achieved at 1 hour after the administration despite the intraperitoneal injection. This suggested that these compounds were rapidly absorbed from the peritoneal cavity. However, the boron concentrations in the blood were always higher than those in the tumor tissue. In other word, tumor/blood ratio (T/B ratio)

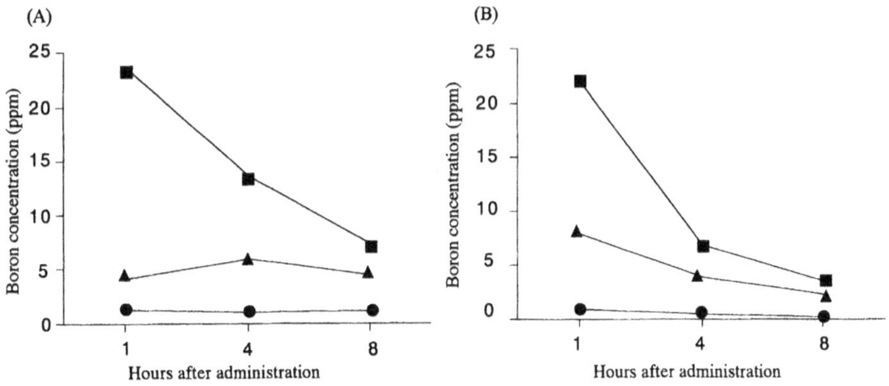

Figure 4. *In vivo* biodistribution of 5B10U(A) and 5HB10U (B) in tumor-bearing rats, (■-blood, ●-normal brain, ▲-tumor.)

of both boronated uridines did not exceed 1.0 throughout this experiment. Furthermore, the peak boron concentration in the tumor tissue delivered from 5B10U and 5HB10U were 5.97 and 7.93 ppm, respectively. These values were much lower than the boron concentration required for clinical BNCT. Although both boronated uridines hardly delivered boron to the normal brain tissue, which is considered advantageous for BNCT of glioblastoma, their biodistributions characterized by lower boron concentrations in the tumor tissues were far from the optimal for BNCT.

Since the total amounts of 5B9TU and 5′Gly5B10U available in this experiment were quite small, only limited biodistribution results for these two compounds could be obtained. Although we cannot reach to the definitive conclusion, the two boronated uridines might not accumulate in the tumor tissue *in vivo* as extensively as other boronated uridines. Boron concentrations achieved by 5B9TU and 5′Gly5HB10U in the tumors were 5.47 and 2.58 ppm, respectively, with the T/B ratio less than 1.0.

4. CONCLUSION

Although 5HB10U and 5B9TU showed favorable uptake by C6 glioma cells *in vitro*, the results of *in vivo* uptake studies of these compounds were disappointing. The synthesize of new tumor-seeking boron carrier with improved biodistribution pattern remains to be essential for an optimal protocol in BNCT.

ACKNOWLEDGMENT

This work has been carried out in part under the Visiting Researcher's Program of the Research Reactor Institute, Kyoto University.

REFERENCES

1. L. Liu, R.F. Barth, W. Tjarks, A.H. Soloway, and A.K.M. Amisuzzaman. *In vitro* and *in vivo* evaluation of carboranyl uridines as boron delivery agents for neutron capture therapy. Anticancer Res., 16:113–120, 1996.
2. Y. Yamamoto. Search for new tumor-seeking boron compounds in Japan, In "Advances in Neutron Capture Therapy. Volume II, Chemistry and Biology," B. Larsson, J. Crawford, and R. Weinreich, eds., Elsevier, Amsterdam, 1997, pp. 22–29.
3. S.R. Tamat, D.E. Moore, and B.J. Allen. Determination of boron in biological tissues by inductivity coupled plasma atomic emission spectrometry. Anal. Chem., 59:2161–2164, 1987.

EVALUATION OF CARBORANYL 2′-DEOXYURIDINE DERIVATIVES AS SUBSTRATES FOR HUMAN THYMIDINE KINASES 1 AND 2

Jianghai Wang[1], Anthony J. Lunato[2], Abul K. M. Anisuzzaman[2], Seiichiro Ikeda[3], Weihua Ji[2], Feng-Guang Rong[2], Staffan Eriksson[1], David H. Ives[3], Albert H. Soloway[2], and Werner Tjarks[2]

[1]Department of Veterinary Medical Chemistry
Swedish University of Agricultural Sciences
Uppsala, Sweden
[2]College of Pharmacy and
[3]Department of Biochemistry
The Ohio State University
500 W. 12th Ave
Columbus, OH 43210

1. INTRODUCTION

Cellular nucleoside kinases play a pivotal role in the use of nucleosides for cancer and antiviral therapy.[1] For BNCT, the cytosolic thymidine kinase (TK1) may be a particularly important target enzyme. TK1 activity is present in proliferating cells but is virtually absent from all quiescent cells.[2] Boron-containing thymidine derivatives, that are good substrates for TK1, may be entrapped in proliferating neoplastic cells after conversion to the monophosphate. Therefore, such agents may be excellent vehicles for the selective delivery of boron-10 to those compartments of a tumor consisting of viable cells. A substantial number of nucleosides modified with various boron moieties at different positions have been synthesized and evaluated biologically for use in BNCT.[3] Some thymidine derivatives were found to be phosphorylated *in vitro*[4] and in phosphoryl transfer assays with purified human thymidine kinases.[5] In these experiments, the observed rates of phosphorylation were generally low compared to natural nucleosides and did not distinguish between phosphorylation by TK1 and mitochondrial thymidine kinase (TK2) which appears to be equally active in proliferating and non-proliferating cells, thus providing no basis for selective uptake of boronated nucleosides in tumor cells.[6]

Frontiers in Neutron Capture Therapy, edited by Hawthorne *et al.*
Kluwer Academic / Plenum Publishers, New York, 2001.

This paper describes the evaluation of the new tethered carboranyl 2′-deoxyuridine derivatives **S-4**, **S-5**, **S-6**, **S-8**, **N-4**, **N-5**, **N-6**, and **N-7** in phosphoryl transfer assays with recombinant human TK1 and TK2, as well as with human TK1 purified from leukemic cells in comparison with several known boronated nucleosides.

2. METHODS

A detailed description of the syntheses of the new nucleosides as well as the expression and purification of human recombinant TK1 and TK2 will be published elsewhere. TK1 was also purified to near-homogeneity from chronic myelocytic leukemic cells concentrated by leukophoresis and provided by the Tissue Procurement Service of the Ohio State University Comprehensive Cancer Center by affinity chromatography as described previously.[7] The nucleosides were dissolved in DMSO prior to addition to the assay mixtures. In case of purified TK1, the final concentrations of the assay mixtures were: 2.5 mM nucleoside, 2.5% DMSO, 15 mC$_i$[γ-^{32}P]-ATP (10–15 Ci/mmol, DuPont Co.), 100 mM Tris-HCl, pH 8, 0.2 mM $MgCl_2$, 10 mM DTT, and 0.5 mg/ml BSA. Assay mixtures were incubated at 25 °C for 15 to 60 min. In case of recombinant TK1 and TK2, the final concentrations of the assay mixtures were: 0.1 mM nucleoside, 0.16–2.5% DMSO, 50 mM Tris-HCl, pH 7.6, 5 mM $MgCl_2$, 15 mM NaF, 125 mM KCl, 0.5 mg/ml BSA, 10 mM DTT, and 100 µM [γ-^{32}P]-ATP (Amersham). Assay mixtures were incubated at 37 °C for 10 to 30 min. In both cases, the phosphorylation products were analyzed by PEI-cellulose TLC using isobutyric acid : ammonium hydroxide : water (66 : 1 : 33) as the solvent system and quantified as described previously.[8]

3. RESULTS

The structures of all tested carboranyl nucleosides and the results of the phosphoryl transfer assays carried out with these agents are displayed in Figure 1 and Table 1, respectively. Saturating substrate concentrations were applied in the experiments. Thus, the obtained results express the maximal turnover number (V_{max}) from nucleoside to nucleotide for the saturated enzyme. These conditions may reflect appropriately the clinical situation where relatively large amounts of boronated nucleoside may have to be administered resulting in high physiological concentrations (and thereby enzyme saturation) necessary to achieve the therapeutic objective. Eight out of seventeen tested carboranyl 2′-deoxynucleosides were phosphorylated by TK2 at low rates. However, only the N-3 substituted nucleosides were phosporylated by both recombinant and purified TK1, the therapeutically-relevant thymidine kinase. The phosphorylation rates are comparable with those observed for e.g. Ara-C, AZT, and BrdUrd with salvage enzymes such as TK1 or deoxycytidine kinase (dCK).[6] In order to assure that the product spots observed on PEI-TLC plates are indeed phosphorylation products of the nucleosides, the 5′-monophosphate of **N-4** was synthesized and used as a reference in the TLC experiments. Synthetic 5′-monophosphate of **N-4** and enzymatically produced 5′-monophosphate of **N-4** possess identical R_f-values. In addition, the enzymatically produced 5′-monophosphate of **N-4** was converted to **N-4** by the action of alkaline phosphatase (calf intestine), as confirmed by TLC.

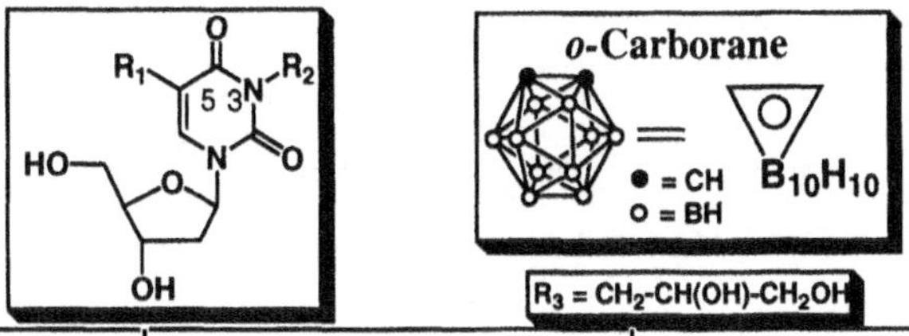

COMPOUND	R_1	R_2
S-4	$S-(CH_2)_4-$ $B_{10}H_{10}$	H
S-5	$S-(CH_2)_5-$ $B_{10}H_{10}$	H
S-6	$S-(CH_2)_6-$ $B_{10}H_{10}$	H
S-8	$S-(CH_2)_8-$ $B_{10}H_{10}$	H
CDU	$B_{10}H_{10}$	H
SUB-7	$(CH_2)_2-CO-O-(CH_2)_3-$ $B_{10}H_{10}$	H
UBN-7	$CH{=}CH_2-CO-NH-(CH_2)_3-$ $B_{10}H_{10}$	H
UB-6	$CH{=}CH_2-CO-O-(CH_2)_2-$ $B_{10}H_{10}$	H
UB-8	$CH{=}CH_2-CO-O-(CH_2)_4-$ $B_{10}H_{10}$	H
CDU-2OH	$-R_3$ $B_{10}H_{10}$	H
SUB-7-2OH	$(CH_2)_2-CO-O-(CH_2)_3-$ $-R_3$ $B_{10}H_{10}$	H
UB-7-2OH	$CH{=}CH_2-CO-O-(CH_2)_3-$ $-R_3$ $B_{10}H_{10}$	H
UB-8-2OH	$CH{=}CH_2-CO-O-(CH_2)_4-$ $-R_3$ $B_{10}H_{10}$	H
N-4	Me	$(CH_2)_4-$ $B_{10}H_{10}$
N-5	Me	$(CH_2)_5-$ $B_{10}H_{10}$
N-6	Me	$(CH_2)_6-$ $B_{10}H_{10}$
N-7	Me	$(CH_2)_7-$ $B_{10}H_{10}$

Figure 1.

4. CONCLUSION

TK1 tolerates bulky carboranylalkyl substituents at the N-3 position of thymidine, but it does not tolerate bulky carboranylalkyl(thio) groups at the C-5 position of 2′-deoxyuridine. The compounds with a shorter 5-S-tether between nucleoside and carborane cage appear to be the best substrates for TK2. However, the phosphorylation rates for TK2 are very low for all of the tested nucleosides. The data obtained in this study do not support the tether concept developed by us[5] and others[9] to reduce possible steric interference in the binding of carboranyl nucleosides with enzyme because of a too close proximity of the bulky carborane moiety to the nucleoside. To the best of our knowledge, substantial phosphorylation of thymidine derivatives with bulky groups at the N-3 position by TK1, as observed in this study, has not been reported as yet. The results

Table 1. Relative phosphorylation of 2-deoxyuridines by thymidine kinases 1 and 2

Compound	Recombinant TK1	Recombinant TK2	Purified TK1
Thd	1[a]	1[a]	nm
dUrd	0.55 ± 0.18	0.75 ± 0.05	1[b]
S-4	–	++	nm
S-5	–	+	–
S-6	–	+	nm
S-8	–	+	nm
CDU	–	+	nm
SUB-7	–	–	nm
UBN-7	–	–	nm
UB-6	–	–	nm
UB-8	–	–	nm
CDU-2OH	–	–	nm
SUB-7-2OH	–	–	nm
UB-7-2OH	–	+	nm
UB-8-2OH	–	–	nm
N-4	0.28 ± 0.03	+	0.20 ± 0.05
N-5	0.19 ± 0.10	–	0.35 ± 0.01
N-6	0.04 ± 0.02	–	0.32 ± 0.01
N-7	0.03 ± 0.03	+	0.34 ± 0.06

Mean ± SD values are based 3 experiments for purified TK1 and recombinant TK2 as well as 4 experiments for recombinant TK1.
a = For the recombinant enzymes, the obtained values for thymidine (Thd) were set to 1.
b = For purified TK1, the obtained value for 2′-deoxyuridine (dUrd) was set to 1.
(nm) = Not measured.
(–) = Not detectable.
(+) = A phosphorylation product could be detected but the amount was too small for an accurate quantification.
(++) = In case of **S-4**, the amount of the detected phosphorylation product appeared to be somewhat higher than for other compounds in the TK2 series but was still insufficient for an accurate quantification.

of our experiments will open new avenues in the future design of boronated 2′-deoxyuridines for BNCT and may also have impact in compound development for traditional chemotherapy of cancer and viral diseases.

ACKNOWLEDGMENT

We thank Dr. Raymond F. Schinazi for providing CDU and acknowledge CALLERY CHEMICAL COMPANY for providing decaborane. This work was supported by the U.S. Department of Energy grants DE-FG02-90ER60972 and DE-FG02-95ER62059 and the U.S. Public Health Service grants R01CA-53896 and P30CA16058.

REFERENCES

1. C.K. Chu, and D.C. Baker, "Nucleosides and Nucleotides as Antitumor Agents and Antiviral Agents," Plenum Press, New York, 1993.
2. L.J. Bello, *Expl. Cell Res.*, 89, 263–274, 1974.
3. A.H. Soloway, W. Tjarks, B.A. Barnum, F.-G. Rong, R.F. Barth, I.M. Codogni, and J.G. Wilson, *Chem. Rev.*, 98, 1515–1562, 1998.

4. R.F. Schinazi, N.M. Goudgaon, G. Fulcrand, Y. El Kattan, Z. Lesnikowski, G. Ullas, J. Moravek, and D.C. Liotta, *Int. J. Radiat. Oncol. Biol. Phys.*, 28, 1113–1120, 1994.
5. F.-G. Rong, A.H. Soloway, S. Ikeda, and D.H. Ives, *Nucleosides Nucleotides*, 16, 379–401, 1997. F.-G. Rong, A.H. Soloway, S. Ikeda, and D.H. Ives, *Nucleosides* Nucleotides, 14, 1873–1887, 1995.
6. E.S.J. Arnér, and S. Eriksson, *Pharm. Ther.*, 67, 155–186, 1995.
7. P. Gröbner, and P. Loidl, *J. Biol. Chem.*, 259, 8012–8014, 1984. Y.-C. Cheng, *Methods in Enzymol.*, 51, 346–354, 1991. D.H. Ives, J.P. Durham, and V.S. Tucker, *Anal. Biochem.*, 28, 192–205, 1969. D.H. Ives, *Anal. Biochem.*, 136, 416–420, 1984.
8. S. Eriksson, B. Kierdaszuk, B. Munch-Petersen, B. Öberg, and N.G. Johansson, *Biochem. Biophys. Res. Commun.*, 176, 586–591, 1991.
9. G.W. Kabalka, N. Kesavulu Reddy, and C. Narayana, *in*: "Cancer Neutron Capture Therapy," Y. Mishima, ed., Plenum Press, New York, 1996, pp. 157–161.

IN VIVO AND *IN VITRO* UPTAKE STUDY OF BORONATED PORPHYRINS FOR NEUTRON CAPTURE THERAPY

Akira Matsumura[1], Yasushi Shibata[1], Tetsuya Yamamoto[1], Fumiyo Yoshida[1], Kei Nakai[1], Isao Sakata[2], Susumu Nakajima[3], Masafumi Okumura[4], and Tadao Nose[1]

[1]Department of Neurosurgery
University of Tsukuba
Ibaraki, Japan
[2]Toyo Hakka Co Ltd.
Okayama, Japan
[3]Central Division of Surgery
Asahikawa Medical College
Hokkaido, Japan
[4]Schering Japan Research Institute
Osaka, Japan

1. INTRODUCTION

To improve boron neutron capture therapy (BNCT), a new tumor-seeking drug is required. Boronated porphyrins have been developed for boron carriers in boron neutron capture therapy (BNCT) with some positive results.[1–5] We have developed a boronated metalloporphyrin compound, STA-BX900, and previously reported its significance compared to borocaptate sodium (BSH).[3] To improve the delivery of such compounds to the tumor, two other types of boron porphyrin compounds, whose side chains were slightly modified, have been evaluated. The *in vivo* and *in vitro* uptake study of these drugs will be presented.

2. MATERIALS AND METHODS

2.1. Boronated Metalloporphyrins

Three types of boronated manganese metalloporphyrin, STA-BX900 (MW = 871.8), STA-BX909 (MW = 926.8), and STABX910 (MW = 944.8) have been developed.

Frontiers in Neutron Capture Therapy, edited by Hawthorne *et al.*
Kluwer Academic / Plenum Publishers, New York, 2001.

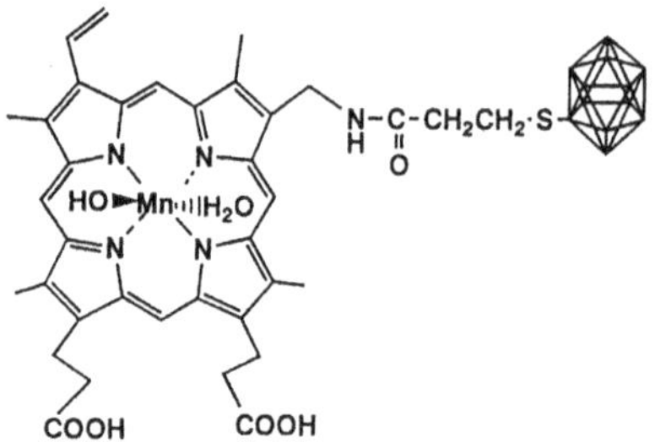

STA-BX900

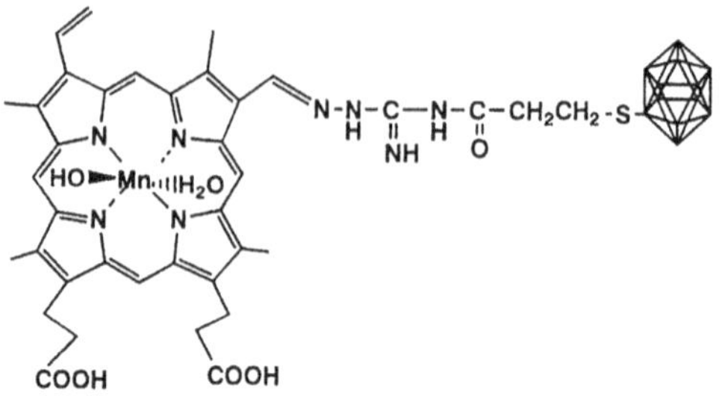

STA-BX909

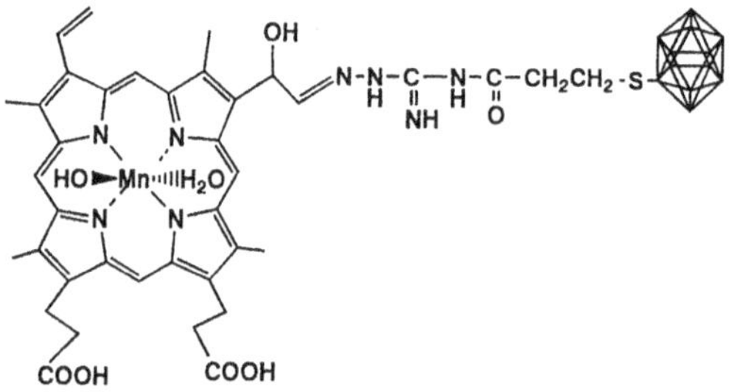

STA-BX910

Figure 1. Molecular formula of three types of boronated manganese metalloporphyrin, STA-BX900, STA-BX909 and STABX910.

The differences between these drugs are minor modifications of the side chains (Figure 1). These compounds were synthesized using ^{10}B enriched Cesium Borocaptate (Centronics Ltd.).

2.2. *In Vivo* Boron Concentration Analysis

For *in vivo* analysis, 9L gliosarcoma (1×10^4 cells/10 μl of 9L cells) was injected into each of four Fisher rat brains. Two weeks later, each rat received drugs via bolus intravenous injection, with a dose of 0.05 mmol/kg. Twenty-four hours after these injections, the rats were killed and tissues were collected and frozen. After the tissues were dissolved, boron concentrations were measured by inductively coupled plasma atomic emission spectroscopy (ICP-AES) analysis (Seiko SPS1200A, Japan).

2.3. *In Vitro* Boron Uptake Analysis

9L gliosarcoma cells were incubated in tissue culture flasks. BSH and STA-BX909 were added to the conditioned medium (^{10}B; 30 μg/ml). At 3, 6 and then at 24 hours (n = 6) later, the conditioned medium was washed twice with a medium containing no boron compounds. The cells were collected, counted, and further dissolved with 10% nitric acid in a microwave oven. The boron concentrations were measured by using ICP-AES.

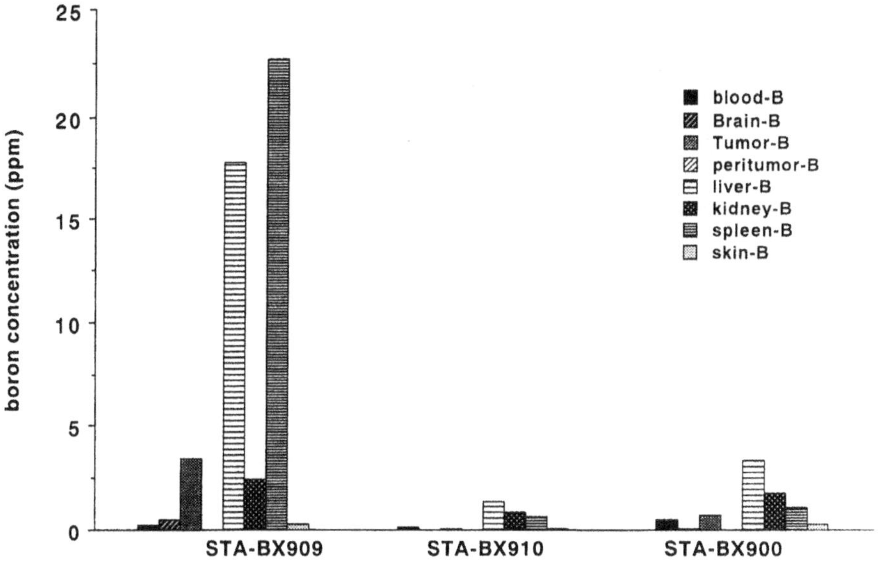

Figure 2. *In vivo* uptake study of STA-BX909, STA-BX910, and STA-BX900 on various tissues (24 hours after bolus injection of 0.05 mmol/kg compound).

3. RESULTS

3.1. *In Vivo* Uptake Study

The administration of three types of boronated metalloporphyrin revealed that boron concentration in the tumor was highest in STA-BX909 (3.48 ± 2.48 ppm; mean ± S.D.) followed by STA-BX900[3] (0.73 ± 0.65) and STA-BX910 (0.07 ± 0.1) (Figure 2). The tumor/blood ratio was highest in STA-BX909, followed by STA-BX900, and it was lowest in STA-BX910. The high tumor boron concentration correlated with the liver and spleen boron concentration. From these results, STA-BX909 was chosen for further *in vitro* study.

3.2. *In Vitro* Boron Concentration Analysis

Boron concentration remained low in BSH even after 24 hours. STA-BX909 showed increasing boron concentration as a function of time. Porphyrin accumulated in tumor cells and the increase in boron concentration indicated an active uptake (Figure 3). From these data, further radiobiological study was performed to prove the radiobiological effect in neutron capture therapy using STA-BX909[2] (in another chapter in these proceedings).

4. DISCUSSION

The boron compound used in the clinical trials in Japan and in European studies is borocaptate sodium (BSH).[6] BSH has the advantage of not crossing the blood-brain

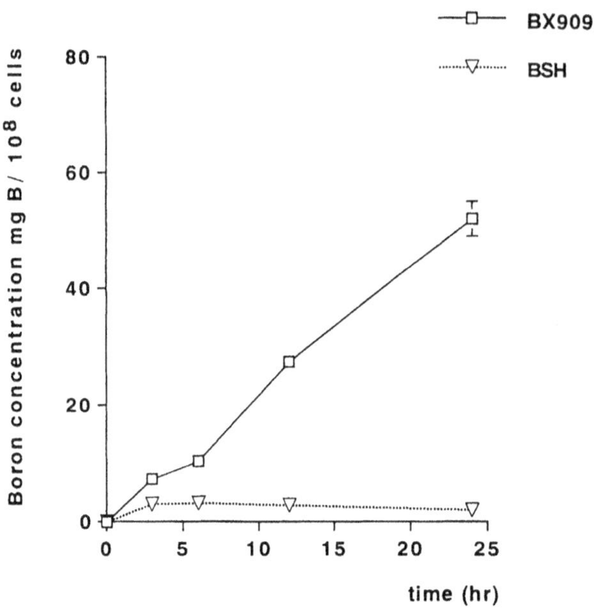

Figure 3. *In vitro* boron uptake in 9L gliosarcoma cell line with BSH and STA-BX909.

barrier, resulting in a specific boron accumulation in the tumor. It is also known that BSH accumulates throughout the cell cycle, which may cover all the tumor cells.[7] However, in BSH, the absolute boron concentration is not high enough, and there is no big difference between blood boron concentration and tumor boron concentration at the time of BNCT (approximately 12 hours after infusion of BSH). This lack of a major difference requires that irradiation doses be limited in order to prevent the radiation from damaging to normal tissue.[6]

Another compound, used in the U.S.[8] and in Japan,[9] is *p*-boronophenylalanine (BPA), which is a derivative of essential amino acid phenylalanine. This compound reaches a high tumor-to-blood ratio of up to 3:4. However, this compound crosses the blood-brain barrier, and boron uptake in the normal brain is similar to that in the blood concentration, and this similarity is also a limiting factor for effective treatment.[8] This compound has an another limitation, in which the quiescent cells are not much involved in BNCT.[7]

Our previous study revealed that metalloporphyrin ATX-S10 is taken up in all of the cell cycle in various glioma cell lines.[10] Although the metalloporphyrin used in this study was not boronated, this suggests that porphyrin has the advantage of covering all the cell cycle at the time of BNCT and also gaining the tumor-to-blood ratio.

Among STA-BX series compounds, STA-BX909 showed relatively high boron concentration in the tumor tissue. Porphyrin is known to be mediated into the cells through LDL receptors, and LDL receptors have been described as being up-regulated in the tumor tissue.[11] The high uptake in liver also suggests some relationship to the LDL metabolism of STA-BX909. STA-BX909 has an additional aminoguanidine in the side chain as compared to STA-BX900. This structure may have contributed to higher uptake of

STA-BX909 in the tumor tissue. On the other hand, STA-BX910 has an additional—OH compared to STA-BX909. This modification resulted in a decrease in tumor uptake of this compound. Although the exact mechanism should be further investigated, the hydrophilic or lipophilic properties may have influenced the tumor affinity markedly as shown in our study.

In our study with STA-BX909, we could reach 3.5 ppm boron in the tumor when 0.05 mmol/kg of STA-BX909 was given in a single shot intravenous infusion. Since there has been no toxicity study yet of our compounds, it should be examined whether an absolute therapeutic boron concentration could be obtained within a safe dose. BOPP[1] and tetracarboranyl phenylporphyrin[4] have four boron cages attached to the porphyrin cage; such modification will gain absolute boron concentration in the tumor while avoiding an increase in the dose administered. In future studies, such modifications should be considered in the STA-BX series.

ACKNOWLEDGMENTS

This study was supported in part by the University of Tsukuba Research Projects and by a Research Grant-in-Aid for Scientific Research from the Japanese Ministry of Education, Science and Culture (07557091).

REFERENCES

1. S.B. Kahl, M.S. Koo, B.H. Laster, and R.G. Fairchild, Boronated porphyrins in NCT: Results with a new potent tumor localizer. *Strahlenther Onkol*, 165:134–137, 1989.
2. A. Matsumura, Y. Shibata, T. Yamamoto, F. Yoshida, K. Nakai, M. Kiriya, N. Shimojo, Y. Hayakawa, T. Nose, S. Nakajima, I. Sakata, N. Miwa, and Y. Itoh, Radiobiological effects of boronated porphyrin (STA-BX909) as a compound for neutron capture therapy. An *in vitro* study, *in* "Recent Progress of Photodynamic Therapy. Proceedings of the 7th Annual Conference of JCIPA". T. Nishisaka ed., Ionics, Tokyo, 1997, pp. 125–130.
3. A. Matsumura, Y. Shibata, T. Yamamoto, H. Fujimori, K. Nakai, I. Sakata, N. Miwa, S. Nakajima, and T. Nose, Pharmacokinetic Study of Boronated Metalloporphyrin (STA-BX900) and BSH in the 9L Rat Glioma Model, *in* "Advances in Neutron Capture Therapy", B. Larsson, J. Crawford, and R. Weinrich ed. Vol. II, Elsevier, Amsterdam, 1997, pp. 158–162.
4. M. Miura, P.L. Micca, C.D. Fisher, C.R. Gordon, J.C. Heinrichs, and D.N. Slatkin, Evaluation of carborane-containing porphyrins as tumour targeting agents for boron neutron capture therapy. *Br J Rad*, 71:773–781, 1998.
5. T. Yamamoto, A. Matsumura, Y. Shibata, T. Yoshizawa, H. Fujimori, K. Nakagawa, K. Nakai, Y. Yoshii, T. Nose, I. Sakata, and S. Nakajima, MR imaging using boronated Mn-metalloporphyrin (STA-BX900) for BNCT, *in* "Advances in Neutron Capture Therapy", B. Larsson, J. Crawford, and R. Weinrich ed. Vol. II, Elsevier, Amsterdam, 1997, pp. 267–270.
6. H. Fankhauser, G. Stragliotto, P. Zbinden, Borocaptate sodium (BSH) pharamacokinetics in glioma patients, in: Boron Neutron Capture Therapy. Towards Clinical Trials of Glioma Treatment, D. Gabel, and R. Moss eds. Plenum Press, New York and London, 1992, pp. 155–162.
7. S. Masunaga, K. Ono, Y. Sakurai, M. Takagaki, T. Kobayashi, M. Suzuki, Y. Kinashi, and M. Akaboshi, Response of quiescent and total tumor cells in solid tumors to neutrons with various cadmium ratios. *Int J Radiation Oncology Biol Phys*, 41:1163–1170, 1998.
8. E.H. Elowitz, R.M. Bergland, J.A. Coderre, D.D. Joel, M. Chadha, and A.D. Chanana, Biodistribution of *p*-boronophenylalanine in patients with glioblastoma multiforme for use in boron neutron capture therapy. *Neurosurgery*, 42:463–8, 1998.
9. S. Ueda, Y. Imahori, Y. Ohmori, E. Yoshino, K. Ono, T. Kobayashi, M. Takagaki, Y. Oda, T. Ido, and Y. Mishima: Positron emission tomography and boron neutron capture therapy system to the patient

with malignant brain tumor- the first clinical trial using ^{10}B-BPA, *in* "Cancer Neutron Capture Therapy", Y. Mishima ed., Plenum Press, New York and London, 1996, pp. 823–827.

10. Y. Shibata, A. Matsumura, F. Yoshida, T. Yamamoto, K. Nakai, T. Nose, I. Sakata, and S. Nakajima, Cell cycle dependency of porphyrin uptake in glioma cell line: Flow cytometry analysis. *Cancer Lett*, 129:77–85, 1998.
11. S. Nakajima, T. Takemura, and I. Sakata, Tumor-localizing activity of porphyrin and its affinity to LDL and transferrin. *Cancer Lett*, 92:113–118, 1995.

TOXICOLOGY OF BOPP IN LARGE ANIMALS

Jay Tibbitts[1], John R. Fike[2], Kathleen R. Lamborn[2], Andrew W. Bollen[3], and Stephen B. Kahl[4]

[1]Department of Biopharmaceutical Sciences
[2]Brain Tumor Research Center-Department of Neurological Surgery
[3]Department of Pathology
[4]Department of Pharmaceutical Chemistry
University of California-San Francisco
San Francisco, CA 94143-044

1. INTRODUCTION

One of the most important characteristics of any therapeutic agent, including those for binary therapies such as BNCT and photodynamic therapy (PDT), is an acceptable level of toxicity. The boronated porphyrin BOPP (Figure 1) is a "third generation" candidate for BNCT treatment of cerebral gliomas and potentially for other cancers. In preclinical studies in mice, BOPP has been shown to be selectively taken up in the C6 implanted intracerebral glioma after both intravenous and intraperitoneal administration.[1] Doses as high as 200 mg/kg were readily tolerated with no apparent signs of morbidity and no mortality of animals. Thus the maximum tolerated dose (MTD) in mice is at least this high. Tumor : normal brain ratios as high as 400 : 1 and tumor : blood ratios of 11 : 1 were readily obtainable with concomitant tumor BOPP levels of >200 μg/g tumor equivalent to ~65 ppm tumor boron. Since BOPP is a boronated free base porphyrin, it also is a potential sensitizer for photodynamic therapy (PDT). In rats bearing the C6 glioma, BOPP (25 mg/kg iv) was found to be approximately one order of magnitude more effective than hematoporphyrin derivative (HpD) in producing the same degree of tumor necrosis depth while sparing normal brain.[2] In contrast, BOPP seems to be more toxic in rats than would be predicted on the basis of the mouse MTD. Recent studies in rats bearing the 9L gliosarcoma suggest that the maximum tolerated dose in rats is only ~70 mg/kg, a factor of approximately one-half what would be predicted on the basis of body surface area comparisons.[3] Significant tumor growth delay following treatment with BOPP followed by thermal neutrons has been observed in mice bearing the KHJJ melanoma.[4] Since larger animals are much more similar to humans than mice or rats, we undertook a full toxicological evaluation of BOPP in dogs.

Frontiers in Neutron Capture Therapy, edited by Hawthorne *et al.*
Kluwer Academic / Plenum Publishers, New York, 2001.

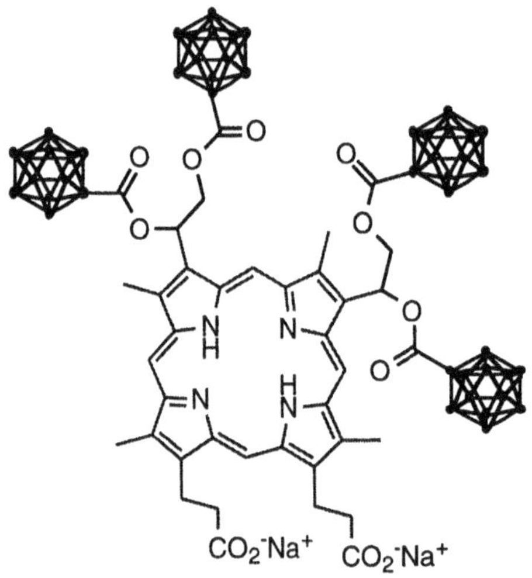

Figure 1. BOPP disodium salt. Carborane cages indicated by ⬢.

2. METHODS

Seventeen adult male beagle dogs, 1–2 years of age were fasted for 12 hours prior to the administration of BOPP. A dose of 35 mg/kg of the disodium salt dissolved in normal saline at pH 7.6 was given over 1–2 hours by i.v. in the right cephalic vein. This dose was chosen in hopes that it might approach the maximum tolerated dose or to at least elicit some adverse effects in order to predict what might be of concern in human studies. Groups of 4 dogs were studied for either 25 hours, 79 hours, 240 hours or 672 hours after infusion of BOPP. One subject was given control saline only and studied for 240 hours. Blood (by jugular venipuncture) and urine samples were obtained prior to and at multiple time points during the study. A full panel of clinical chemistry, hematology and urinalysis tests were performed on these samples. At the conclusion of the study, dogs were humanely euthanatized and a complete necropsy performed. Histopathologic evaluation was carried out on tissue sections prepared by standard methods. Boron content of plasma and tissue samples from 28 different sites were obtained by inductively coupled plasma atomic emission spectrometry (ICP-AES) at the Idaho National Engineering and Environmental Laboratory. Results of the boron pharmacokinetics analysis are currently in press.[5]

3. RESULTS

BOPP was very well-tolerated in all subjects with no mortality or obvious signs of serious morbidity. Clinically apparent adverse effects of BOPP administration were mild and transient, limited mostly to some early nausea and vomiting and mild photosensitivity on exposure to bright sunlight. Cephalic vein thrombosis was evident in ten of fourteen subjects examined subjects at the site of infusion, but these did not appear painful nor was there any evidence of distal edema or redness in the area of the thrombus. In dogs studied for 28 days, the thrombus was found to be resolved by 17–18 days. Cephalic vein lesions examined at autopsy showed evidence of organizing thrombi with vessel wall

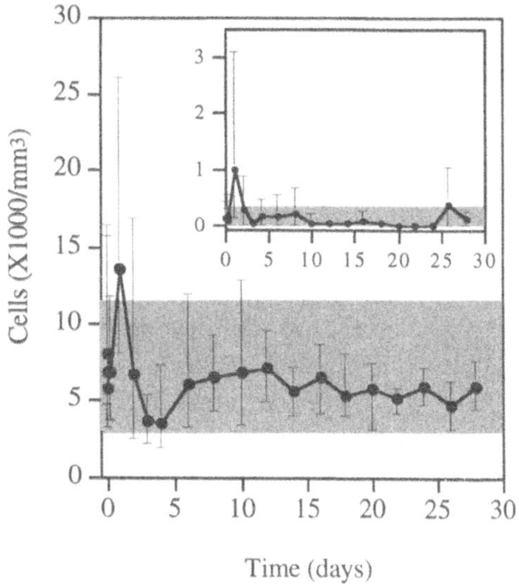

Figure 2. Mean segmented (main) and band (inset) neutrophil counts. Shaded areas indicate normal reference range.

inflammation. Histopathologic studies of lungs, kidneys and other tissues revealed no thrombi present, suggesting that this is a local phenomenon possibly related to the slow rate of infusion. There was a slight (~6–7%) decline in average weight, perhaps related to the nausea, and mild febrile response within the first 48 hours which resolved by 4–5 days. The latter was associated with an increase in circulating white blood cells, both segmented and band neutrophils, which also resolved within 48 hours as shown in Figure 2.

Figure 3 shows that average platelet levels decreased below the normal range on days 2 and 4 after BOPP infusion but remained in the normal range at all times beyond day 6. Three dogs exhibited significantly elevated ALT levels beginning at about day 6, and the same 3 subjects showed increased AST levels slightly earlier. Changes in liver enzyme levels, especially ALT, are sensitive indicators of damage to the liver, and Plasma

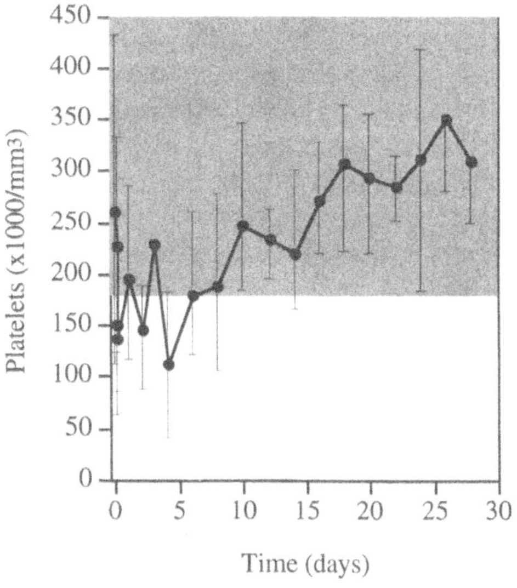

Figure 3. Mean platelet counts. Shaded areas show normal reference range for platelets.

bilirubin levels were slightly elevated in most subjects after BOPP administration, but *ex vivo* tests of blood chemistry revealed that BOPP caused a false increase in bilirubin levels. Other clinical chemistry parameters were essentially normal in all subjects.

4. DISCUSSION

In this study we have administered BOPP to normal beagle dogs at a dose of 35 mg/kg and found that the adverse effects elicited by BOPP are minimal. It should be explicitly noted here that this dose is equivalent to 200 mg/kg in a mice and 140 mg/kg in rats and represents a high dose for a porphyrin compound. It was chosen specifically with the expectation that it might be close to or at the maximum tolerated dose and that significant toxic effects might be elicited so as to serve as a guide to the kinds of toxicity which might be expected in human clinical trials. Clinically apparent side effects were mild and transient and included nausea, vomiting and skin photosensitivity in the nasal area when exposed to strong sunlight. None of these were severe enough to cause concern and all could potentially be easily controlled or eliminated in human patients. Weight loss was also mild and the animals appeared to have normal appetites. The mild early febrile response observed appeared to correlate with an early spike in circulating white blood cells, as shown in Figure 2. Such a response may have been a result of an early inflammatory response initiated by BOPP. The cephalic vein thrombosis at the site of infusion seen in several of the dogs appears to be a local effect which may be related to the slow rate and site of the infusion. Histopathologic evaluation of tissues such as the lungs, kidneys and other highly perfused organs at necropsy showed no evidence of thrombus formation in these tissues, suggesting that this effect is not systemic. In the 4 dogs followed for 28 days, the thrombus disappeared by about day 17, and there was little or no histopathologic evidence of cephalic vein damage in these dogs at necropsy.

The early thrombocytopenia shown in Figure 3 is particularly interesting since thrombocytopenia has also been a hallmark associated with administration of three *nido*-carboranyl porphyrins to rats and mice.[3,6] BOPP as administered contains four *closo* carborane cages but conceivably might undergo metabolic conversion to one or more *nido* cages considering the prolonged high liver boron levels observed in both rats[7] and dogs.[5] This magnitude of this effect, however, is not of major clinical concern. At no time did platelet counts become low enough to raise concern of bleeding risks and no dogs had any evidence of clinical clotting abnormalities.

While the adverse effects described above were observed in most or all of the subjects, some adverse effects were documented in only a few dogs. Chief among these were changes in plasma levels of the liver enzymes ALT (alanine aminotransferase) and AST (aspartate aminotransferase) which are indicators of liver damage. ALT is predominantly found in the liver and is released into the circulation very soon after liver insult, while AST is found in extrahepatic as well as hepatic tissues and is usually released a few days after injury. We found plasma levels of ALT elevated above normal levels in five dogs, but in four of the five the increase of ALT did not occur until about day 6. In 3 of these 4 the elevation above normal was significant while in the fourth it was minor. In the fifth dog, ALT increase showed a very different pattern, rising quite quickly after infusion and then declining to normal levels at about day 6. Only 3 of these 5 dogs showed elevations of AST above normal, one just barely before dropping into the normal range. These rises seemed to follow a biphasic pattern hitting an peak at about day followed by a drop and then a second, somewhat higher peak. The increases found in 2 of the dogs were found to correlate with histopathologic changes see in the livers of these dogs. These were described as mild periportal acute and chronic inflammation by Dr. Bollen, our veterinary pathologist. A third dog, however, showed an increase in liver enzymes that had returned to normal or near normal levels by 28 days. This animal had no histopathologic evidence of liver damage on autopsy. The time course of liver enzyme rises, particularly ALT, is particularly noteworthy. An acute, non-sustained injury would cause a rapid increase in ALT and AST, rather than the delayed onset observed in this study. The cause of this phenomenon may be related to the slow formation of a hepatotoxic metabolite, or the slow accumulation of BOPP or a metabolite in hepatocytes that resulted in toxicity. Alkaline phosphatase levels were not found to be increased in these subjects, indicating that there was little or no biliary system injury. It is also interesting to note that less than half the subjects studied for 10 days or more exhibited evidence of hepatic injury.

This study demonstrates conclusively that BOPP is not significantly toxic at this dose in dogs and provides valuable information regarding its potential for eliciting specific side effects in human trials. Obviously, the toxic effects of any agent are both dose and species dependent and the human dose of BOPP required for BNCT therapeutic effect is not yet known. This study provides evidence that doses in the range of those likely to be useful for BNCT treatments in humans may prove to be well-tolerated. Based on a body surface area comparison, the dose given to dogs in this study would be equivalent to ~200 mg/kg in mice and ~140 mg/kg in rats. Mice have been found to tolerate 200 mg/kg well, but 140 mg/kg in rats is approximately twice the lethal dose. The equivalent human dose would be ~15 mg/kg. BOPP was approved by the United States Food and Drug Administration in May for human trials as a PDT sensitizer for patients with glioblastoma. At the time of the presentation of this paper, 12 patients had been treated at the Royal Melbourne Hospital in a classical dose escalation study. No drug related toxicity has been observed in any of these patients. Of particular note is the lack of thrombus formation at any injection site, the absence of any incidence of thrombocytopenia and the lack of changes in serum levels of liver associated enzymes. These studies will undoubtedly continue to provide valuable toxicological and biodistribution information for BOPP as a BNCT agent in humans.

ACKNOWLEDGMENT

The authors would like to acknowledge the financial support of the National Institutes of Health (CA37961 to SBK and CA13525 to JRF), the American Foundation for Pharmaceutical Education (Graduate Fellowship to JT) and the Higgins Glioma Fund of the University of California, San Francisco.

REFERENCES

1. J.S. Hill, S.B. Kahl, A.H. Kaye, S.S. Stylli, M.-S. Koo, M.B. Gonzales, N.J. Vardaxis, and C.I. Johnson, *Proc. Natl. Sci. U.S.A.* **89**, 1785–1789 (1992).
2. J.S. Hill, S.B. Kahl, S.S. Stylli, Y. Nakamura, M.S. Koo, and A.H. Kaye, *Proc. Natl. Acad. Sci. U.S.A.* **92**, 12126–12130 (1995).
3. S.B. Kahl, D.D. Joel, M.M. Nawrocky, P.L. Micca, K.P. Tran, G.C. Finkel, and D.N. Slatkin, *Proc. Natl. Acad. Sci. U.S.A.* **87**, 7265–7269 (1990).

4. B.H. Laster, S.B. Kahl, L. Warkentien, and V.P. Bond. In *Advances in Neutron Capture Therapy*, pp. 535–539. Edited by A.H. Soloway, R.F. Barth, and D.E. Carpenter. New York, Plenum Press (1993).
5. J. Tibbitts, N.C. Sambol, J.R. Fike, and S.B. Kahl, *Pharm. Res.* (in press, 2000).
6. M. Miura, P.L. Micca, C.D. Fisher, C.R. Gordon, J.C. Heinrichs, and D.L. Slatkin, *Br. J. Radiol.* **71**, 773–781 (1998).
7. C.P. Ceberg, A. Brun, S.B. Kahl, M.S. Koo, B.R.R. Persson, and L.G. Salford, *J. Neurosurg.* **83**, 86–92 (1995).

BIODISTRIBUTION STUDY OF NOVEL DODECABORATE-PHTHALOCYANINES IN THE B-16 MOUSE MELANOMA MODEL

R. Spryshkova[1], E. Grigorieva[1], V. Riabkova[1], Zh. Spure[1], M. Naidenov[1], K. Zhizhin[2], E. Malinina[2], A. Chernyavsky[2], E. Luk'yanets[2], K. Solntsev[2], N. Kuznetsov[2], and G. Borisov[3]

[1]Cancer Research Centre of Russian Academy of Medical Sciences, Moscow
[2]Institute of General and Inorganic Chemistry of Russian Academy of Sciences, Moscow
[3]Russian Research Centre "Kurchatov Institute", Moscow, Russia

1. INTRODUCTION

Subsequent development of BNCT to considerable degree depends on the availability of the boronated compounds which could be more suitable for BNCT than BSH and BPA being clinically used. One of the directions of the compounds optimization might be an application of phthalocyanines as carriers for boron delivery to the tumour cells.[1]

This paper presents the results of biodistribution study of the novel phthalocyanines Co(II) and Al(III) coordination compounds containing $B_{12}H_{11}NH_2CH_2$-groups:

$Na_4[CoPc(CH_2NH_2B_{12}H_{11})_4]$ (1) $Na_4[(AlOH)Pc(CH_2NH_2B_{12}H_{11})_4]$ (3)
$Na_8[CoPc(CH_2NH_2B_{12}H_{11})_8]$ (2) $Na_8[(AlOH)Pc(CH_2NH_2B_{12}H_{11})_8]$ (4)

2. MATERIALS AND METHODS

2.1. Synthesis of Phthalocyanine-$B_{12}H_{11}NH_2CH_2$ Complexes

The synthesis of phthalocyanine Co(II) and Al(III) coordination compounds containing $B_{12}H_{11}NH_2CH_2$-groups were carried out according to the procedures described earlier.[2] The total synthetic scheme is shown in Figure 1.

Frontiers in Neutron Capture Therapy, edited by Hawthorne *et al.*
Kluwer Academic / Plenum Publishers, New York, 2001.

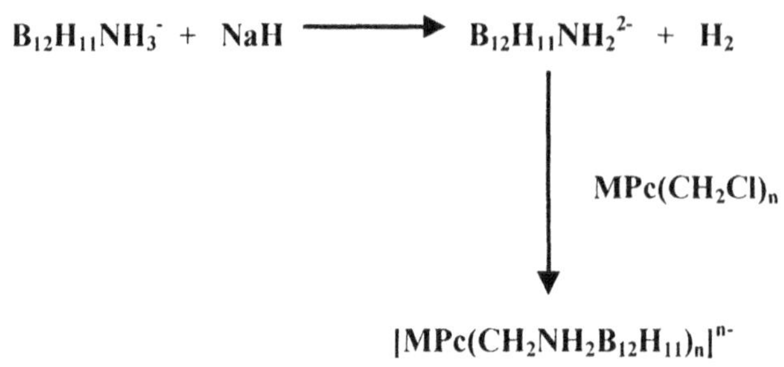

Figure 1. Synthesis of phthalocyanine-$B_{12}H_{11}NH_2CH_2$ complexes. Pc = $C_{32}H_{20-n}N_8$; n = 4 or 8; M = Co^{2+} or $[AlOH]^{2+}$.

The reactions were carried out in dimethylformamide solutions at 5 °C (reduced form). The final compound (Na^+ salts) yields were 90% and up of theory. The purity of the final products was examined by chemical analysis and spectroscopic methods and accounted for 99.5%. Sodium salts of the final anions are easily water-soluble.

2.2. Biological Studies

2.2.1. Tolerance. The physiological saline solutions (pss) of the compounds (cds) 1–4 were administered i.p. into C57Bl/6 male mice (20–23 g) in the following dose ranges (3 animals per one dose): 150–30, 150–19, 150.0–37.5, 150.0–12.5 µg of boron per gram of the animal body weight (µg B/g b.w.), respectively. The maximal dose which did not cause the animal's death within a week of observation was used for the biodistribution studies.

2.2.2. Biodistribution. The pss of the cds 1–4 in a volume 0.20–0.23 ml containing 30.0, 75.0, 75.0 and 52.7 µg B/g b.w., respectively, were injected i.p. into C57BL/6 male mice (20–23 g) 11 days after the hosts were inoculated s.c. with B-16 melanoma cells. Compound (cd) 1 was injected i.v. (sinus venous of the eye) in the dose 7.5 µg B/g b.w. as well. The mice were sacrificed by cervical dislocation at different times after injection. The tumour and adjacent tissues (skin, muscle and blood) were excised for boron determination. The tissues of the excretory organs—liver, lung, kidneys, as well as excrements, gall-bladder, duodenal content, and heart, spleen, brain and internal fat were also analyzed. The samples were prepared and analyzed for boron content as described previously.[3] The boron content in µg was calculated per gram of crude tissue.

3. RESULTS AND DISCUSSION

3.1. Tolerance of the Compounds

That the tolerance of cds 1–4 under the conditions of our experiments proved to be practically the same was unexpected. Their maximal tolerated doses at the i.p. injection were about 75 µg B/g b.w. These doses caused variable depression in experimental animals depending on the compound used. The visible symptoms of depression disappeared approximately in 24 h. Cd 3 had a minimal manifestation of toxicity while the toxical manifestation of cd 1 was the highest. After i.p. injection of cd 1 in the dose of

30 μg B/g b.w. there were no visible changes in animal's behavior whereas i.v. injection led to the animal's death within one day. The maximal tolerated dose was 7.5 μg B/g b.w. At the animal dissection after cds 1–4 injection it was found that the internal organs were dyed in blue-green colour of a varying intensity. The biodistribution data exhibits considerable boron uptake with fat which was more expressed for cd 1. The average boron contents (a.b.c.) within 144 h after cd 1 administration were 20.1–59.5 μg/g of fat or 67–198% of B injected per g b.w. The least ability to be absorbed by fat was seen with cd 3: the a.b.c. in fat was 9–40% as compared with B injected per g b.w. Cds 1–4 are firmly bound to fat as there was a small change of the boron content in fat up to 8 days of the observation. Besides, cds 2 and 4 at all doses indicated above precipitated in the abdominal cavity like small, intensely dyed, aggregates which were not observed 96 h after the cds injection. This property was more expressed for cd 2. After mixing of the 7 mM (7.5 mg B/ml) solution of this cd with the heparin containing murine blood in the ratio 1 : 2 the coloured sediment precipitated as well. Both the binding of the cds with fat and the preecipitation of the conglomerations leads to the decrease of the cds amount staying in blood and is, apparently, the reason of practically similar tolerance of the phthalocyanines studied after i.p. injection.

3.2. Biodistribution of the Compounds

An a.b.c. in tumour, blood, skin and muscle taken from not less than three animals at different times after injection of the cds 1–4 in percents of B injected per g b.w. are shown in Figures 2, 3.

3.2.1. Compounds 1 and 2. After i.p. injection of cd 1 in the dose of 30 μg B/g b.w. the boron content in the tumour was gradually increased (Figure 2A) and up to 24 h it achieved on the average 10.4 ± 13.4 μg/g of the tumour tissue. To 48–96 h the a.b.c. drops and accounts for 8–9 μg/g. Boron content in tumours 24 h after injection 3.2 (0.9–13.0),

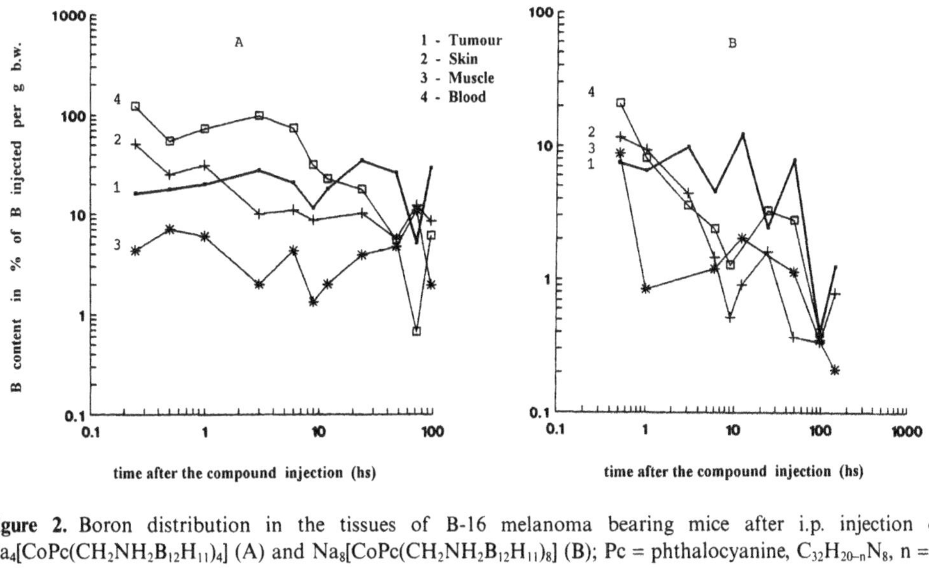

Figure 2. Boron distribution in the tissues of B-16 melanoma bearing mice after i.p. injection of $Na_4[CoPc(CH_2NH_2B_{12}H_{11})_4]$ (A) and $Na_8[CoPc(CH_2NH_2B_{12}H_{11})_8]$ (B); Pc = phthalocyanine, $C_{32}H_{20-n}N_8$, n = 4 or 8.

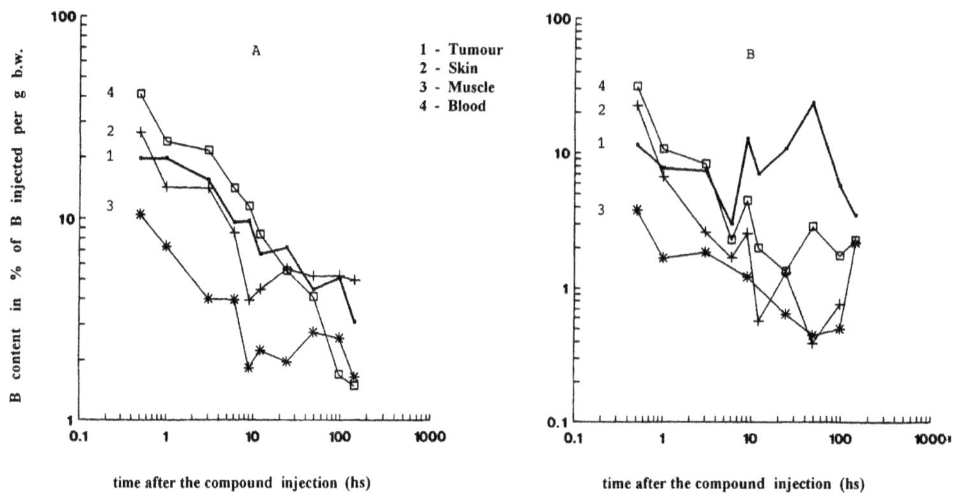

Figure 3. Boron distribution in the tissues of B-16 melanoma bearing mice after i.p. injection of $Na_4[(AlOH)Pc(CH_2NH_2B_{12}H_{11})_4]$ (A) and $Na_8[(AlOH)Pc(CH_2NH_2B_{12}H_{11})_8]$ (B); Pc = phthalocyanine, $C_{32}H_{20-n}N_8$, n = 4 or 8.

3.4 (2.0–30.0) and 8.7 (5.0–50.0) times larger than in blood, skin and muscles, respectively (Table 1).

After i.v. injection of cd 1 in the maximal tolerated dose (7.5 μg B/g b.w.) neither high boron content nor the optimal boron ratios between the tumours and the adjacent tissues were achieved. After i.p. injection of cd 2 (75 μg B/g b.w.) maximal a.b.c. in tumour equal to 9.2 ± 14.2 μg/g and 6.0 ± 12.8 μg/g was found at 12 and 48 h. Tumour-to-blood, skin and muscle average boron ratios were 5.4 (0.9–9.0), 13.5 (0.9–21.0) and 6.2 (2.0–10.0) for 12 h, and 2.8 (0.5–6.0), 21.2 (2.4–50.0) and 6.9 (3.0–10.0) for 48 h, respectively (Table 1).

For both cds a high dispersedness of the boron content in the tumours was observed: 24 and 96 h after the cd 1 injection it fluctuates from 4.0 to 36.0 μg/g and from 0.3 to 31.0 μg/g; 12 and 48 h after the cd 2 injection it ranges from 0.4 to 36.0 μg/g and from 0.2 to 34.0 μg/g, respectively. Only 10–15% of all tumours tested had heightened boron values while the rest of the tumours, irrespective of the time points, had a very small amount of boron. There was no dependence between boron content in tumours and their sizes and boron content in blood either.

Cds 1 and 2 were eliminated from the body by the all excretory organs but first of all by the liver. Within at least 6 days of the observation the boron content gradually decreased in the kidneys, blood, lung and other tissues while it not only decreased but even is increased in the liver. The spleen continues to retain boron within the observation period as well.

Thus, it appears from our experiments that cds 1 and 2 by their toxical and biodistribution data are not suitable for BNCT. The cds did not show a selective boron uptake by all tumours and they are neurotoxic and firmly dye tissues in blue-green colour.

3.2.2. Compounds 3 and 4. According to our findings cd 3 is not likely to be suitable for BNCT, at least not for melanoma. After i.p. injection (75 μg/g b.w.) the maximal a.b.c. in tumour was equal to 14.8± 11.0 μg/g was achieved in 1 h and then gradually decreased (Figure 3A). At this time, there is a high concentration of boron in the blood. Tumour-

Table 1. Tumour to adjacent tissues boron ratios in the B-16 melanoma bearing mice after i.p. injection of the phthalocyanine-$B_{12}H_{11}CH_2NH_2$ compounds

		Hours after the compound injection														
Compound	Tissue	0.25	0.5	1	3	6	9	12	24	48	72	96	120	144	168	192
	Tumour/blood	0.13	0.33	0.3	0.3	0.3	0.4	0.8	3.2	4.6	8.0	4.8	5.7	1.4	1.9	1.0
1	Tumour/skin	0.3	0.7	0.7	2.8	1.9	1.3	—	3.4	4.6	0.4	3.4	0.8	0.4	0.7	0.7
	Tumour/muscle	3.7	2.5	3.3	13.8	2.8	8.5	9.0	8.7	5.6	0.5	14.5	1.7	2.0	1.9	1.5
	Tumour/blood	—	0.4	0.8	2.8	1.9	0.8	5.4	0.8	2.8	—	1.0	—	2.8	—	—
2	Tumour/skin	—	0.6	0.7	2.2	3.1	2.0	13.5	1.5	21.2	—	1.2	—	1.6	—	—
	Tumour/muscle	—	0.9	7.8	>7.4	3.8	>0.8	6.2	>1.8	6.9	—	1.2	—	6.1	—	—
	Tumour/blood	—	0.5	0.8	0.7	0.7	0.9	0.8	1.3	1.1	—	2.9	—	2.1	—	—
3	Tumour/skin	—	0.7	1.4	1.1	1.1	2.4	1.5	1.3	0.9	—	1.0	—	0.6	—	—
	Tumour/muscle	—	1.9	2.7	3.9	2.4	5.2	2.9	3.6	3.8	—	2.0	—	1.9	—	—
	Tumour/blood	—	0.4	0.7	0.9	1.3	2.9	3.7	8.1	8.3	—	3.3	—	1.6	—	—
4	Tumour/skin	—	0.5	1.2	2.8	1.8	5.2	12.3	8.1	9.8	—	7.5	—	>1.9	—	—
	Tumour/muscle	—	3.1	4.6	3.9	>1.6	11.2	>3.7	19.0	16.2	—	10.0	—	1.6	—	—

1—$Na_4[CoPc(CH_2NH_2B_{12}H_{11})_4]$ 3—$Na_4[(AlOH)Pc(CH_2NH_2B_{12}H_{11})_4]$ Pc—phthalocyanine, $C_{32}H_{20-n}N_8$, n = 4 or 8
2—$Na_8[CoPc(CH_2NH_2B_{12}H_{11})_8]$ 4—$Na_8[(AlOH)Pc(CH_2NH_2B_{12}H_{11})_8]$

to-blood boron ratio equal to about 3 appears at 96 h when boron content in tumour makes up only 4 μg/g. The boron content in skin is close to that in tumour at all time points studied. Cd 3 is sufficiently eliminated from the body by the liver and kidneys. The boron retention in the liver and spleen was not observed.

In Figure 3B the data of the boron content in the murine tissues after cd 4 injection in the dose 52.7 μg/g b.w. are shown. The a.b.c. in tumour was higher than that in the blood, skin and muscle since 6 h after injection. The maximum a.b.c. in tumour equal to 12.5 ± 23.0 μg/g was found 48 h after the cd 4 administration. In the same time the boron ratios tumour-to-blood, skin and muscle were 8.3 (0.3–10.6), 9.8 (0.5–36.8) and 16.2 (1.0–47.2), respectively (Table 1). However, only 2 tumours from 8 tested had this time heightened amount of boron equal to 36.8 and 60.0 μg/g. The boron content in the rest of the tumours within the observation period ranged from 0.11 to 17.60 μg/g. Like cds 1–3, cd 4 is eliminated from the body first of all by the liver and kidneys. And if in the kidneys as well as in lung, spleen and blood the boron level gradually decreased the boron content in the liver tissue remained practically at the same level within a week of the observation. Cd 4 is not retained in the skin: since 3 h after cd injection the a.b.c. in it varies in the range of 0.2–1.4 μg/g.

This compound needs to be further studied as a potential candidate for BNCT (?).

4. CONCLUSION

Novel $B_{12}H_{11}CH_2NH_2$-containing Co(II) and Al(III) phthalocyanines have been developed. They have a good water solubility but are not possessed of an expressed selective tumour uptake in the melanoma model. The answer to the question whether cd 4 is suitable for BNCT or not requires further investigations especially in understanding the high boron uptake by some individual tumours.

ACKNOWLEDGMENTS

Financial support from INTAS funds is gratefully acknowledged.

REFERENCES

1. M.F. Hawthorne, The Role of Chemistry in the Development of Boron Neutron Capture Therapy of Cancer, Angew. Chem. Int. Ed. Engl., 32:950–984, 1993.
2. K.Yu. Zhizhin, E.A. Malinina, L.V. Goeva, A.S. Chernyavsky, S.V. Ivanov, E.A. Luk'yanets, K.A. Solntsev, and N.T. Kuznetsov, New boron-containing water-soluble phthalocyanines, Doclady Academii Nauk, 357(2):206–208, 1997.
3. R.A. Spryshkova, E. Yu. Grigirieva, M.G. Naidenov, J.H. Morris, V.A. Brattsev, G.I. Borisov, V.I. Riabkova, and A.S. Halansky, Biodistribution of $Na_2[B_{12}H_{11}SCN]$ in C57Bl/6 mice bearing melanoma B-16 and in Wistar rats bearing glioblastoma 1018, in "Advances in Neutron Capture Therapy," eds. B. Larsson, J. Crawford, and R. Weinreich, Excepta Medica, Amsterdam, 1997, pp. 253–257.

INVESTIGATIONS OF THE INTERACTION OF PHOSPHOLIPIDS WITH MERCAPTOUNDECAHYDRO-*CLOSO*-DODECABORATE (2-) (BSH) AND DODECAHYDRO-*CLOSO*-DODECABORATE (2-) ($B_{12}H_{12}^{2-}$)

Silke Lutz and Detlef Gabel

Department of Chemistry
University of Bremen
Bremen, Germany

1. INTRODUCTION

BSH has been used in boron neutron capture therapy for many years. Immunohistochemical investigations of Otersen *et al.*[1] have shown that BSH binds firmly to human glioma tissue and is present within cells. Otersen *et al.*[2] have given the following hypothesis: BSH interacts with lipids of the cell membrane and the uptake occurs by an endocytotic pathway. It is known that human brain tumors show an increased concentration of choline containing compounds in comparison to normal brain tissue.[3,4] Therefore the interaction of BSH with phosphatidylcholine, a component of the cell membrane, was investigated in this study. As additional model substance the cluster without the sulfhydryl group was also used.

2. SPECTROSCOPY IN ORGANIC SOLVENTS

Kageji *et al.*[5] found that, while BSH is insoluble in carbon tetrachloride (CCl_4), a BSH-phosphatidylcholine complex is soluble in CCl_4. In this study fourier transform infrared (FTIR) and nuclear magnetic resonance spectroscopy (NMR) was carried out with following samples: ethanol / water solutions of egg-yolk phosphatidylcholine (PC) and BSH or dodecaborate were incubated in a relation 1 : 1 for several hours and extracted into deuterochloroform ($CDCl_3$). The FTIR spectra [Figure 1] can be quantified. There are about 13 PC molecules per BSH and about 2 PC molecules per dodecaborate in the $CDCl_3$ phase. The proton-NMR spectrum of the dodecaborate sample shows that the

Frontiers in Neutron Capture Therapy, edited by Hawthorne *et al.*
Kluwer Academic / Plenum Publishers, New York, 2001.

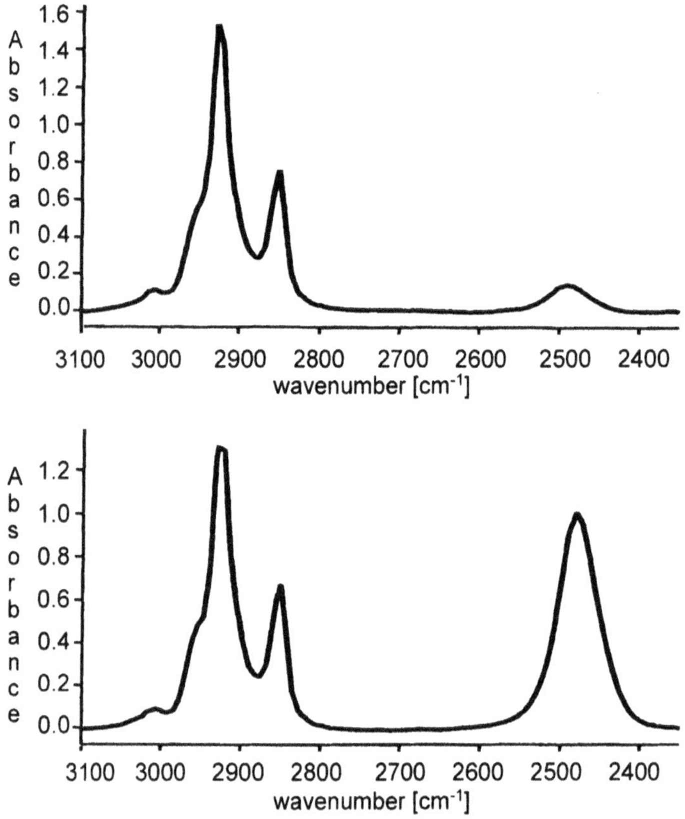

Figure 1. FTIR spectra of the phospholipid (PC) + BSH (top spectrum) and the phospholipid (PC) + $B_{12}H_{12}^{2-}$ (bottom spectrum) in $CDCl_3$. There are 13.5 PC molecules per BSH and 2.4 PC molecules per $B_{12}H_{12}^{2-}$ in the $CDCl_3$ phase.

signals of the phospholipid headgroups are clearly broadened. In general a decreased mobility of molecules leads to broadened NMR signals. In both boron containing samples some signals of the phospholipid headgroups are shifted. The singlet of the methyl protons of the choline ammonium group for example is shifted highfield [Figure 2]. The ^{11}B NMR spectrum of the dodecaborate sample shows a very broadened singlet. Normally dodecaborate gives a rather well defined singlet whereas substituted compounds show multiplets. These experiments lead to following conclusions:

- the phospholipid and the boron clusters interact
- a covalent binding between the phospholipid and the boron cluster can be excluded
- in the case of dodecaborate the signals of the NMR spectra are clearly broadened. The mobility of the lipid molecules and the boron cluster appears to be strongly decreased. A micellar structure is probable where a core of dodecaborate molecules is surrounded by PC.

3. NMR SPECTROSCOPY OF LIPOSOMES

Unilamellar liposomes are often used as model systems of biological membranes. Unilamellar vesicles of pure dipalmitoylphosphatidylcholine (DPPC) were prepared by

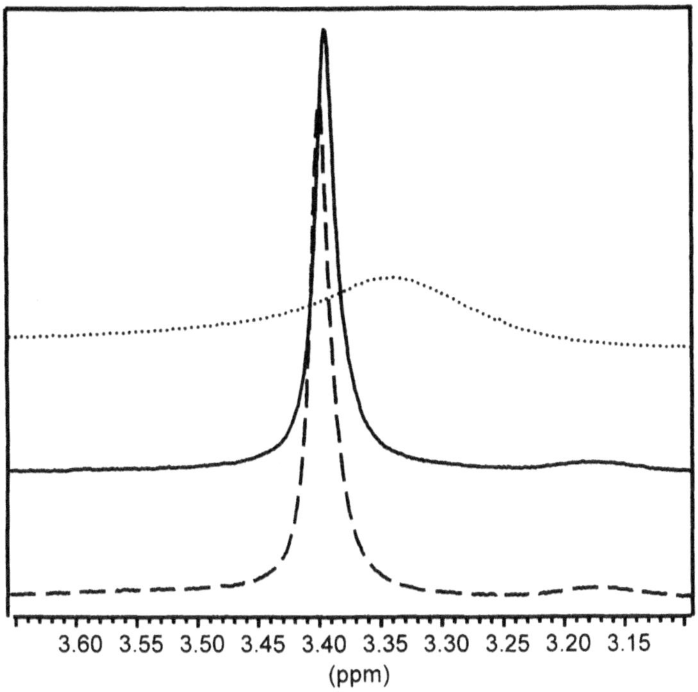

Figure 2. ^{1}H-NMR choline resonance in $CDCl_3$ of PC + $B_{12}H_{12}^{2-}$ (top spectrum . . . . . .), PC + BSH (middle spectrum ____) and PC without a boron compound (bottom spectrum ----).

sonicating a suspension of multilamellar vesicles with a titanium probe and subsequent ultracentrifugation. After the preparation in deuterium oxide BSH was added in different concentrations to the liposomes and the suspensions were incubated overnight. At higher BSH concentrations the turbidity and the viscosity of the suspensions increase. At a relation DPPC:BSH = 1:8 the initially fluid sample becomes a gel. Figure 3 shows the proton NMR signals of the choline methyl groups of the different liposome samples at 45 °C. The splitting of this signal is representative for small vesicles. It originates from the inner and outer headgroups of the bilayer and can be attributed to their difference in curvature. Increasing BSH concentrations lead to a shift of the outside choline methyl resonance to lower field. In addition the shoulder of the signals disappears and their intensity decreases. The change of the chemical shift shows an interaction of BSH at the headgroups of the outer side of the bilayer. At the same time the vesicles seem to become larger because the resonance of the inner choline methyl groups disappears. The decreasing intensities suggest that the vesicles aggregate and eventually become too large and immobile to yield NMR signals.

4. DIFFERENTIAL SCANNING CALORIMETRY OF LIPOSOMES

Measuring the phase transition of liposomes with differential scanning calorimetry renders a deeper insight into the interaction of vesicles with BSH. The results of the experiments in TRIS buffer are shown in Figure 4. The more BSH is added to the liposome suspension the lower is the phase transition temperature. This effect stops at a

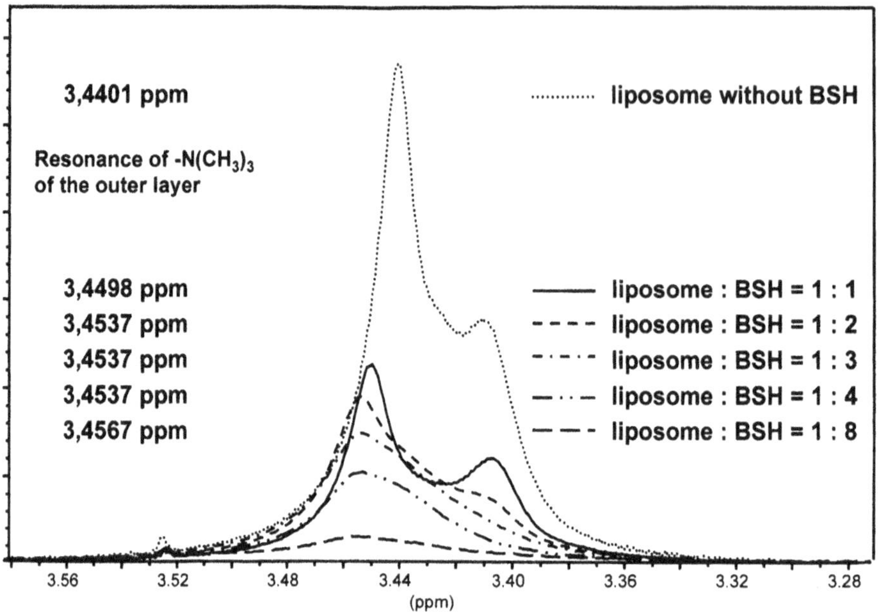

Figure 3. ^{1}H-NMR choline resonance of different liposome suspensions in D_2O.

certain point and the mixture with a relation DPPC : BSH = 1 : 8 has again a higher phase transition temperature. By integrating the peaks the enthalpy of transition can be obtained. The two samples with high BSH concentrations need much more energy for their transition compared to the other samples. The disappearence of the smaller peak at about 37 °C attributed to very small vesicles in the suspension is in correspondence to the results of the NMR experiment. A fusion to larger vesicles with a phase transition

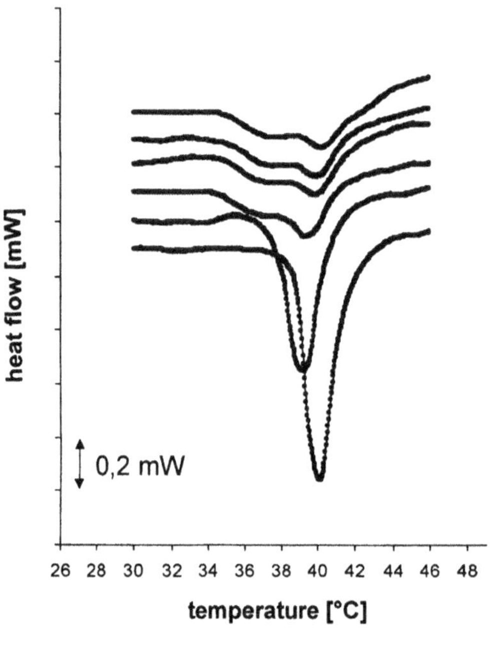

Figure 4. Differential scanning calorimetry measurements of different liposome suspensions in TRIS-buffer. Traces from top to bottom: liposome without BSH, liposome : BSH = 8 : 1 (molar ratio), liposome : BSH = 4 : 1, liposome : BSH = 1 : 1, liposome : BSH = 1 : 4, liposome : BSH = 1 : 8.

temperature of about 40 °C occurs in the case of high BSH concentrations. Aggregation could be responsible for higher transition enthalpies.

5. CONCLUSIONS

In conclusion interactions of BSH with phospholipids were detected by using several methods. Infrared and NMR-measurements demonstrate an interaction of egg-yolk phosphatidylcholine with BSH and dodecaborate in organic solvents. NMR and differential scanning calorimetry measurements show that BSH influences the physical and chemical properties of unilamellar liposomes. BSH changes the chemical shift, the intensity and the shape of the NMR signal of the choline methyl groups. In addition BSH influences the phase transition temperature and enthalpy of unilamellar liposomes. Fusion and subsequent aggregation of the liposomes is supposed. If such interactions occur on the surface of cells, biochemical changes of the membrane could influence endocytosis or allow diffusion of BSH through the membrane. Such mechanism could be responsible for the uptake of BSH in glioma cells which has been observed consistently.

ACKNOWLEDGMENTS

The authors gratefully acknowledge the support for this investigation by the Deutsche Forschungsgemeinschaft and the Fonds der Chemischen Industrie.

REFERENCES

1. B. Otersen, D. Haritz, F. Grochulla, M. Bergmann, W. Sierralta, and D. Gabel, Binding and distribution of $Na_2B_{12}H_{11}SH$ on cellular and subcellular level in tumor tissue of glioma patients in boron neutron capture therapy, *J. Neuro-Oncol.*, 33:131–139, 1997.
2. B. Otersen, W. Sierralta, D. Haritz, F. Grochulla, M. Bergmann, and D. Gabel, Cellular uptake mechanism of BSH in tumor tissue of glioma patients, in: *Advances in neutron capture therapy*, Proceedings of the seventh international symposium on neutron capture therapy for cancer, Ed. B. Larsson, J. Crawford, and R. Weinreich, Elsevier, Exerpta Medica,Volume 2:169, 1997.
3. D. Ott, J. Henning, and T. Ernst, Human brain tumors: assessment with *in vivo* proton MR spectroscopy, *Radiology*, 186:745–752, 1993.
4. E.D. Gotsis, K. Fountas, E. Kapsalaki, P. Toulas, G. Peristeris, and N. Papadakis, *In vivo* proton MR spectroscopy: the diagnostic possibilities of lipid resonances in brain tumors, *Anticanc. Res.*, 16:1565–1568, 1996.
5. T. Kageji, B. Otersen, D. Gabel, R. Huiskamp, Y. Nakagawa, and K. Matsumoto, Interaction of mercaptoundecahydrododecaborate (BSH) with phosphatidylcholine: relevance to boron neutron capture therapy, *Biochim. Biophys. Acta*, 1391:377–383, 1998.

MEMBRANE BINDING OF $Na_2B_{12}H_{11}SH$ (BSH) WITH PHOSPHATIDYLCHOLINE IN BORON NEUTRON CAPTURE THERAPY

Teruyoshi Kageji[1], Shinji Nagahiro[1], Keizo Matsumoto[1], Birte Otersen[2], Detlef Gabel[2], and Yoshinobu Nakagawa[3]

[1]Department of Neurological Surgery, School of Medicine
The University of Tokushima
2-50, Kuramto-cho, Tokushima 770, Japan
[2]Department of Chemistry
University of Bremen
Box 330440, D-28334 Bremen, Germany
[3]Department of Neurological Surgery
National Kagawa Children's Hospital
Zentsuji-cho, Zentsuji 765, Japan

1. INTRODUCTION

Several analytical methods have been used to determine boron levels in biological samples, such as inductively coupled plasma atomic emission spectrometry (ICP-AES), prompt-gamma spectroscopy and magnetic imaging of boron-11. Most of these require complicated procedures or long sample preparation times. Recently, Fourier transform infared (FTIR) spectrometry has proven to be a simple and rapid technique suitable for the quantitation of BSH in aqueous and non-aqueous solutions.[1–4] The analysis does not require complicated procedures, and the solution can be introduced directly into the spectrometer for measurement. For BSH, the B—H absorption band in aqueous solution occurs at 2,490–2,500 cm^{-1}. Lipid content can also be determined by FTIR using the C—H absorption in the region 2,850–2,970 cm^{-1}.[3,4]

With regard to the boron biodistribution in tumor cells, immunohistochemical investigation showed that BSH was firmly bound to brain tumor tissue, and that boron was found in the cytoplasm and in the nuclei of cells.[5] Despite the long use of BSH in clinical practice, little is known to date on the precise mechanism by which BSH accumulates in tumor cells. Cell membrane interaction has been suggested as one possible binding mechanism of BSH.[5] In this study, we used FTIR to determine boron levels in solution directly, and investigated the mechanism of BSH interaction with phosphatidylcholine.

Frontiers in Neutron Capture Therapy, edited by Hawthorne *et al.*
Kluwer Academic / Plenum Publishers, New York, 2001.

2. MATERIALS AND METHODS

2.1. Materials

Cesium mercaptoundecahydrododecaborate ($Cs_2B_{12}H_{11}SH$) was synthesized in the laboratory. L-α-phosphatidylcholine (from fresh frozen egg yolk) was obtained from Sigma Chemical Co. (Deisenhofen, Germany). Carbon tetrachloride (CCl_4), ethanol and phosphate-buffered saline (PBS) were of analytical grade from Riedel-de Haen (Seelze, Germany). All water used was distilled deionized water.

2.2. FTIR Spectrometry

All data were recorded with a Bio-Rad FTIR spectrometer Model FTS 20/80 with a Bio-Rad 3,200 Data Station, and a wide-range mercury cadmium telluride detector. For each determination, a resolution of $16\,cm^{-1}$ was used and 256 scans were collected. The spectra were obtained using a variable pathlength liquid sampling cell with calcium fluoride (CaF_2) windows and 0.025–1.0 mm spacers. In general, a pathlength of 1 mm was used for CCl_4 solutions and a pathlength of 50 μm was used for aqueous solutions. During the experiment, single-beam spectra of solvents were recorded as the background. Then the spectrum of the test sample was recorded against the background spectrum. Each spectrum was base-line corrected before quantitative evaluation.

Absorbance for BSH was measured at the maximum absorption in the wave number range 2,490–2,500 cm^{-1}, that for phosphatidylcholine at the maximum absorption in the wave number range 2,900–2,950 cm^{-1}. All concentrations are expressed as a concentration of original samples in CCl_4 or PBS.

2.3. Calibration Solutions

Stock solutions of BSH were prepared containing 22.8 mM in 1 M PBS, pH 7.4. Serial dilution of the stock solution in PBS were made to obtain the BSH concentrations ranging 0.05–1.14 mM. Stock solutions of phosphatidylcholine were prepared containing 13.5 mM in CCl_4, and dilution were made to provide solutions containing 0.14 – 0.68 mM phosphatidylcholine.

2.4. Analysis of Boron in Samples

2.4.1. Analysis of Boron in Stock Solutions for Calibration Curve. Each sample of stock solutions was injected into the cell directly, and examined by FTIR spectrometry in 50 μm pathlength for aqueous solutions and 1.0 mm pathlength for CCl_4 solutions.

2.4.2. Analysis of Boron in Mixed Solutions of Phosphatidylcholine with BSH. For analysis, we used an extraction method. Aliquots of the BSH stock solution (10.0 or 50.0 mM, solvent: PBS) were added to phosphatidylcholine (1.0, 2.0, 3.0, 4.0 mM, solvent: ethanol), to give molar ratios (BSH/PC) in incubation solutions between 0 and 6.0, and total volume 1.0 ml in each sample. Each sample was shaken by vortex for several minutes, and incubated for four hours at room temperature. Organic solvent was removed using a rotary evaporator, and the residue was dissolved in 1.0 ml of CCl_4. The samples were remained stationary for 30 minutes, and the supernatant of the solution was injected into the microcell and analyzed by FTIR spectrometry.

3. RESULTS

3.1. Calibration Curve

The IR spectrum of BSH in PBS displayed the B—H absorption band at 2,450–2,530 cm^{-1}. The relationship between the absorbance maximum at 2,490 cm^{-1} and the concentration of BSH was linear with the statistical regression equation:

$$[BSH(mM)] = 137.48 * Absorbance\ (pathlength = 50\,\mu m) - 0.048\ (R^2 = 0.985)$$

The IR spectrum of phosphatidylcholine (PC) in CCl_4 displayed the C—H absorption band at 2,850–2,970 cm^{-1}. The relationship between the absorbance maximum at 2,925 cm^{-1} and the concentration of phosphatidylcholine was linear with the statistical regression equation:

$$[PC(mM)] = 8.889 * Absorbance\ (pathlength = 1.0\,mm) + 0.044\ (R^2 = 0.998)$$

3.2. Interaction Between Phosphatidylcholine and BSH

Figure 1 shows the IR spectra of CCl_4 extracts of 4.0 mM BSH (spectrum 1), 4.0 mM phosphatidylcholine (spectrum 2), and a mixture of BSH (4.0 mM) and phosphatidylcholine (4.0 mM) (spectrum 3) worked up as described above. The IR spectrum of the CCl_4 extract of BSH displayed no absorbance maximum around 2,500 cm^{-1} and we concluded that only trace amounts of BSH could dissolve in CCl_4. The IR spectrum of phosphatidylcholine displayed absorbance maxima around 2,925 and 2,855 cm^{-1} due to the C—H band, and phosphatidylcholine could dissolve well in CCl_4 as expected. In the case of BSH-phosphatidylcholine extracts, we could find absorbance maxima due to both the C—H and the B—H band. The absorbance due to the B—H band increased with increasing BSH concentration to a certain level, and then leveled off at a molar ratio BSH/PC of 1:2 (Figure 2).

4. DISCUSSION

Immunohistochemical investigations revealed the following:[5] In healthy brain tissue, boron could only be detected in endothelial cells, in agreement with the non-measurable uptake of boron. In malignant brain tumor tissue, a strong intracellular staining was found, which was frequently observed on cell membranes. Often, the whole cell body was uniformly stained. Equally, cell nuclei were stained intensely. Furthermore, the binding of boron to tumor tissue is firm. The strong binding of BSH to human tumor tissue and the presence of boron and BSH within cells and cell nuclei is unexpected, since BSH is a small doubly negatively charged ion.

An explanation for this, unexpected behavior could lie in the formation of water-insoluble ion pairs of BSH with quarternary ammonium salts such as choline. We observed that BSH was practically insoluble in CCl_4, but phosphatidylcholine showed good solubility in nonpolar solvents such as CCl_4. With these different solubilities in CCl_4, we assumed that any BSH-phosphatidylcholine complex formed could be separated from free BSH by extraction. Our experiments show that, while BSH is not soluble in

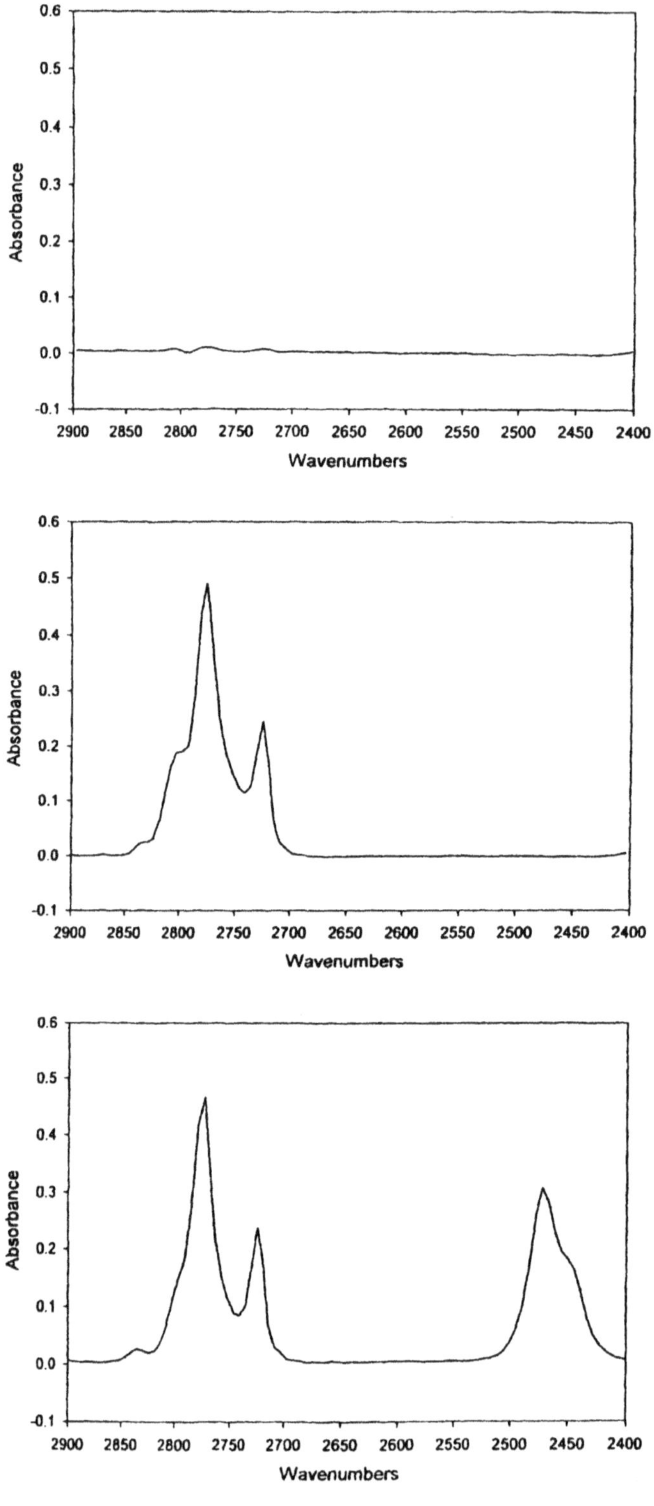

Figure 1. IR spectra of CCl_4 extracts of 4.0 mMol/l BSH (spectrum 1: upper), 4.0 mMol/l phosphatidylcholine (spectrum 2: middle) and a mixture of BSH (4.0 mMol/l) and phosphatidylcholine (4.0 mMol/l) (spectrum 3: lower).

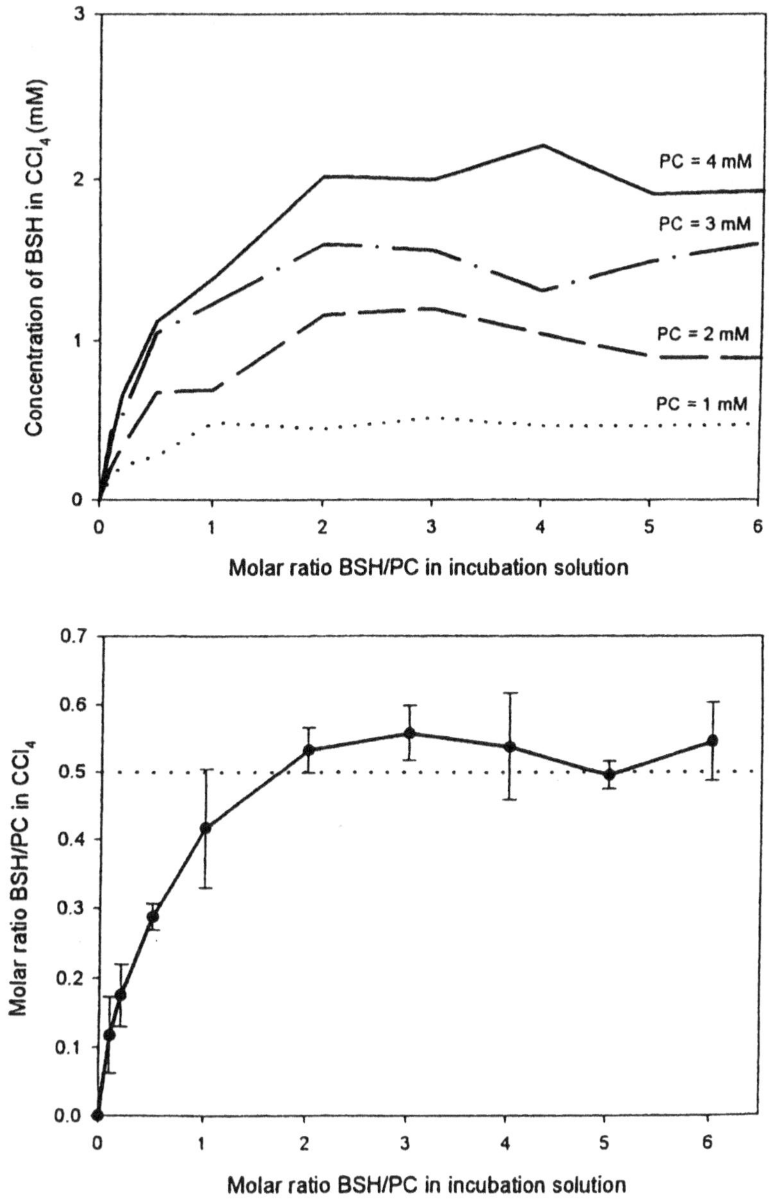

Figure 2. Interaction between BSH and phosphatidylcholine (PC). The relationship between the concentration of BSH in CCl_4 and varying molar ratios (BSH/PC) in the incubation solution (upper), and the relationship between the molar ratio (BSH/PC) in CCl_4 and varying molar ratios (BSH/PC) in the incubation solution (lower).

CCl_4, a BSH-phosphatidylcholine complex of one molecule BSH and two molecules phosphatidylcholine is soluble in CCl_4.

In experiments with varying molar ratios BSH/PC, the CCl_4 extract showed an increase of the molar ratio in the extract, up to a value of 1:2. The ratio did not increase further with increasing BSH excess.

Our own results presented here, support the following hypothesis on boron uptake into malignant glioma cells: BSH can pass through the disrupted blood-brain-barrier

(BBB) easily and can come in contact with tumor cells; there, BSH can bind on the extracellular surface of plasma membrane to choline residues of phosphatidylcholine and sphingomyelin, which are increased abnormally in tumor tissue. After binding to the plasma membrane, boron with choline residues may be internalized into the cell by endocytotic pathways and eventually travel to cell nuclei. The passage through the disrupted BBB is essential for boron accumulation around tumor cells. The formation of stable, neutral ion pairs between the boron cluster and quaternary ammonium salts could be crucially responsible for binding, retention and internalization.

REFERENCES

1. Lu D.R., and Munro C.: Quantitative analysis of mercaptoundecahydrododecaborate by Fourier transform infrared spectroscopy. Pharm Res 9(9), 1199–1202, 1992.
2. Sainl P., Lai J.C.K., and Lu D.R.: FT-IR measurement of mercaptoundecahydro-dodecaborate in human plasma. J Pharm Biomed Anal 12(9), 1091–1095, 1994.
3. Setiawan Y., Rise T., and Moore D.E.: Fourier transform infrared (FTIR) spectrometry for the assay of polyhedral boron compounds in plasma and pharmaceutical formulations. Pharm Res 11(5), 723–727, 1994.
4. Pidgeon C., Apostol G., and Markovich R.: Fourier transform infared assay of liposomal lipids. Anal Biochem 181, 28–32, 1989.
5. Otersen B., Haritz D., Grochulla F., Bergmann M., Sierralta W., and Gabel D.: Binding and distribution of $Na_2B_{12}H_{11}SH$ on cellular and subcellular level in tumor tissue of glioma patients in boron neutron capture therapy. J of Neuro-Oncology 33:131–139, 1997.

INVESTIGATION OF THE TOXICITY AND CELLULAR UPTAKE OF $Na_4[B_{20}H_{17}SH]$ IN EMT6 CELLS

R. Corey Waller[2], Debra A. Feakes[1], Jennifer Spinler[1], Greg Southard[2], and Gary M. Aron[2]

[1]Department of Chemistry
[2]Department of Biology
Southwest Texas State University
San Marcos, TX 78666

1. INTRODUCTION

The development of boron-containing compounds for application in the boron neutron capture therapy (BNCT) of cancer is not only dependent on the synthesis of appropriate compounds, but also the biological evaluation of tumor uptake and retention, toxicity, and pharmacokinetics. Unilamellar liposomes of a specific size and composition have been shown to deliver polyhedral borane compounds having little or no inherent tumor specificity to the tumor mass. Compounds with suitable reactivity are retained by the tumor in therapeutic amounts and other compounds, which lack suitable reactivity, are rapidly cleared from all tissues.[1,2] Based on the proposed mechanisms of retention, the thiol derivative, $Na_4[B_{20}H_{17}SH]$, was prepared for investigation.[3] This compound possesses two characteristics known to enhance tumor retention: the thiol substituent, analogous to the known $Na_2B_{12}H_{11}SH$ (BSH), and the potential to oxidize to a more reactive species, $Na_2[B_{20}H_{17}SH]$. Although the gross distribution of boron was elucidated from the murine biodistribution[4] of the encapsulated compound, the cellular toxicity and pharmacokinetics of the encapsulated polyhedral borane was unknown.

Evaluation of the cellular toxicity of compounds prepared for medical applications has traditionally been accomplished using Trypan Blue exclusion of non-viable cells, five to seven days after exposure to the compound.[5,6] In order to increase the efficiency of the toxicity evaluation, the MTT test[7] was utilized for the evaluation of the $Na_4[B_{20}H_{17}SH]$ toxicity. The MTT test is a colorimetric assay which utilizes the reduction of MTT (3-(4,5-dimethylthiazol-2-yl)-2,5-diphenyltetrazolium bromide) to formazan by mitochondrial dehydrogenases of metabolically active cells (Figure 1).[7] The MTT method takes only three days to complete and allows for multiple samples within the same experiment. As validation, the MTT results were correlated to results obtained using the standard Trypan Blue exclusion technique.

Frontiers in Neutron Capture Therapy, edited by Hawthorne *et al.*
Kluwer Academic / Plenum Publishers, New York, 2001.

The ability of a compound to be endocytosed *in vitro* indicates the potential effectiveness of the compound as a drug and potential problems with toxicity can be elucidated during the experiment. Cell cultures can also be used as a model to show the uptake and efflux potential of the compound.[5,6] The rate of uptake, as well as the retention time, can be utilized to design future compounds with a higher potential for success.

2. METHODS AND MATERIALS

2.1. Cell Cultures

Primary cultures of EMT6 mammary adenocarcinoma cells[8] were obtained from Jacki Granger at Washington State University. Cells were grown at 37 °C with an atmosphere of 5% CO_2 in Waymouth's basal medium (WBM, Sigma), supplemented with 10% fetal bovine serum (FBS, Sigma). Cells were passaged using CMF-1x buffer solution followed by Trypsin-EDTA (0.05% and 0.025%, respectively) every 60–72 hours. All reagents and cells were stored in boron-free vessels.

2.2. Preparation of Liposomal Suspensions

The liposomal suspensions were prepared by probe sonication of a 300-mg dried film composed of equimolar amounts of DSPC and cholesterol with 6.0 mL of the hydrating solution (160 mM in $Na_4B_{20}H_{17}SH$) at 65 °C for 20 min. Vesicles were separated from the remaining free polyhedral borane salt by elution through a column of Sephadex G-25 (medium) with isotonic HEPES buffered saline (HBS) buffer. The liposomal formulation was diluted with HBS to a lipid concentration of 23–24 mg/ml and sterilized by filtration through a 0.22-µm Millipore membrane. The integrity of the encapsulated borane salt was confirmed by ^{11}B NMR at 128 MHz. The volume-weighted mean vesicle diameters of the liposomes, obtained by dynamic light scattering, ranged from 30–60 nm.

2.3. Cytotoxicity of $Na_4[B_{20}H_{17}SH]$ Using Trypan Blue Exclusion Technique

EMT6 monolayers were grown in 60 mm culture dishes to a confluent monolayer. The media was decanted and replaced with serum-free medium containing liposomally encapsulated $Na_4[B_{20}H_{17}SH]$ ranging in concentration from 50–1,500 µM boron. Two control groups were incubated in serum-free medium containing liposomes encapsulating a hypertonic NaCl solution (400 mM). Each of the culture dishes was incubated with the encapsulated polyhedral borane for 90 minutes at 37 °C. The liposomal suspension was decanted from the culture dishes and the cells washed three times with Earl's balanced salt solution (EBSS). Cultures were trypsinized, pelleted, and resuspended at a concentration of 2×10^6 cells/mL in WBM with 10% FBS. Cell suspensions were plated at three different dilutions in 60 mm culture dishes and then allowed to incubate at 37 °C for 5 days, replacing media as needed. After five days, the cells from each of the dilutions were trypsinized and counted on a hemocytometer after treatment with Trypan Blue. Experiments were performed in triplicate.

2.4. Cytotoxicity of $Na_4[B_{20}H_{17}SH]$ Using MTT Test

EMT6 cells were grown in 75 mm culture flasks at 37 °C for 36 hours in WBM with 10% FBS. Confluent monolayers of cells were trypsinized, pelleted, and resuspended in fresh WBM with 10% FBS at a concentration of 1,000 cells/100 µL. Each well of a 96-well tissue culture plate was loaded with 100 µL of the cell suspension and incubated for 18 hours at 37 °C. After 18 hours, the media was decanted and the cells were rinsed twice with EBSS. Serum-free WBM containing liposomally encapsulated $Na_4[B_{20}H_{17}SH]$ at concentrations ranging from 100–1,500 µM boron was added to each well. A control well was incubated in serum-free WBM containing liposomes encapsulating a hypertonic NaCl solution (400 mM), but no boron-containing compound. The culture plates were incubated for 90 minutes at 37 °C. The medium was decanted and the cells rinsed with EBSS and fresh medium was applied. After 24 hours of incubation at 37 °C, the medium was decanted, 20 µL of MTT solution (400 µg/mL) in phosphate buffered saline (PBS) at pH 7.2 was added to each well and incubated for 1 hour at 37 °C. The formazan crystals were dissolved in 100 µL of acidic isopropanol with 4% Triton X-100. The absorbance of the blue-colored formazan was measured in a computer-controlled microplate reader (BioTek Instruments Number EL311). The absorbance at 690 nm was automatically subtracted from the absorbance at 540 nm. The optical density for each corresponding well was averaged and divided by the control and expressed as percent viability.

2.5. Cellular Uptake Investigation

EMT6 monolayers were grown in 75 mm culture flasks at 37 °C for 36 hours in WBM with 10% FBS. The cells were trypsinized, pelleted, and resuspended in serum-free WBM to a concentration of 10^6 cells/mL. Nine different 1.8 mL Eppendorf tubes were loaded with 1 mL of cell suspension. A 500 µM boron solution of liposomally encapsulated $Na_4[B_{20}H_{17}SH]$ was added to each tube, which were then incubated at various times (60, 50, 40, 30, 20, 10, 5, 3, and 1 minute). One minute before the end of the time period, cells were pelleted by centrifugation at 17,000 × g for 1 minute. The medium containing the boron-containing compound was quickly decanted and replaced with EBSS and vortexed. The resulting suspension was repelleted for 1 minute at 17,000 × g and washed twice with EBSS. Each cell pellet was sent to the Idaho National Engineering and Environmental Laboratory (INEEL) for total boron analysis using inductively coupled plasma atomic emission spectroscopy (ICP-AES). Experiments were performed twice.

3. RESULTS

3.1. Cytotoxicity Studies

The survival of EMT6 cells after exposure to 100–1,500 µM boron as liposomally encapsulated $Na_4[B_{20}H_{17}SH]$ was determined as a percent viability (Figure 1).

Both the MTT method and the Trypan Blue exclusion technique gave a CD_{50} (cytotoxic dose to 50% of the cells) close to 400 µM boron.

3.2. Cellular Uptake Studies

The uptake rate of a 500 µM solution of liposomally encapsulated $Na_4[B_{20}H_{17}SH]$ by EMT6 cells was 0.06 ppm/hour (Figure 2). The maximum amount of boron taken up by the cells was 3.36 µg B per 10^6 cells. Maximum boron uptake was achieved at the end of the experimental time frame of 60 minutes. There was an initial lag phase which lasted

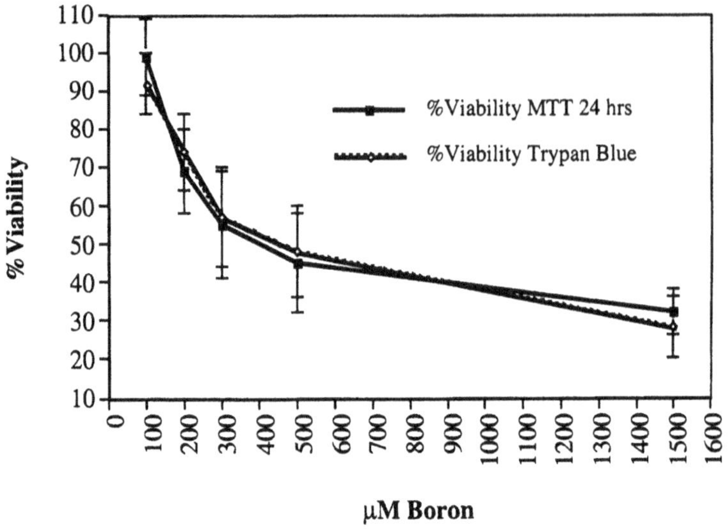

Figure 1. Graph of percent viability of EMT6 cells exposed to liposomally encapsulated $Na_4[B_{20}H_{17}SH]$ as a function of boron concentration.

approximately 3 minutes, followed by a rapid increase for the next 2 minutes. After this, a steady increase in uptake through the 60 minute time period was observed.

4. DISCUSSION

The thiol derivative of $[B_{20}H_{18}]^{4-}$, $[B_{20}H_{17}SH]^{4-}$, was prepared based on the likely reactivity of both the thiol substituent and the oxidative ability of this class of compounds. The biodistribution of the liposomally encapsulated $Na_4[B_{20}H_{17}SH]$ in mice bearing EMT6 mammary adenocarcinoma exhibited a high degree of tumor uptake

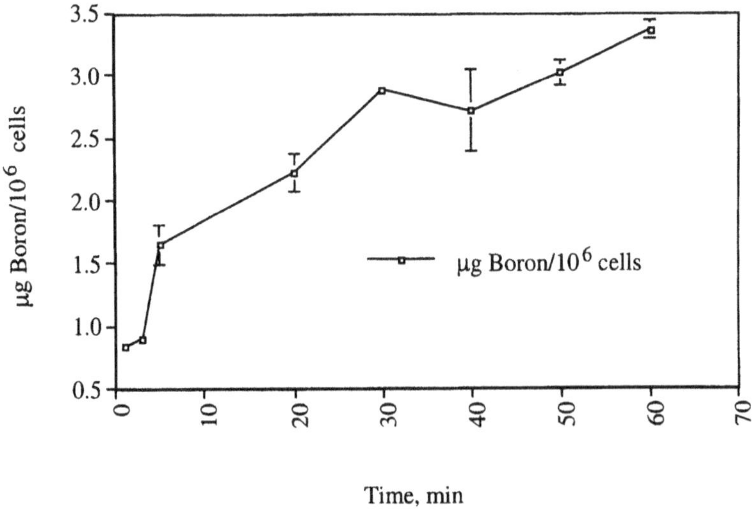

Figure 2. Concentration of boron as a function of time.

and retention.[4] Although the compound exhibited retention within the tumor system suitable for application in BNCT, the cellular toxicity and uptake characteristics were unknown.

The cytotoxic dose (CD_{50}) of the liposomally encapsulated $Na_4[B_{20}H_{17}SH]$ was determined using both the MTT and the Trypan Blue exclusion techniques. This dose was calculated to be approximately 400 µM boron. The two techniques exhibited a variance of less than 5%.

The cellular uptake of the lipsomally encapsulated $Na_4[B_{20}H_{17}SH]$ increased over the entire time period of 60 minutes. The maximum boron concentration at 60 minutes was determined to be 3.36 µg B per 10^6 cells. Longer time periods will be utilized in future investigations and a study of the efflux of the compound from the cells in boron-free media will be completed.

ACKNOWLEDGMENTS

This research was supported by awards from the Robert A. Welch Foundation (AI1292), the Dorothy L. Coker Research Fellowship (R. C. Waller), and Southwest Texas State University. The authors would like to gratefully acknowledge the generosity of Jacki Granger at Washington State University, who provided the EMT6 cell culture along with growth protocols, and Dr. William F. Bauer at the Idaho National Engineering and Environmental Laboratory, who provided total boron analysis.

REFERENCES

1. K. Shelly, D.A. Feakes, M.F. Hawthorne, P.G. Schmidt, T.A. Krisch, and W.F. Bauer, "Model Studies Directed Toward the Boron Neutron-Capture Therapy of Cancer: Boron Delivery to Murine Tumors with Liposomes," *Proc. Natl. Acad. Sci. USA*, 89:9039–9043, 1992.
2. D.A. Feakes, K. Shelly, C.B. Knobler, and M.F. Hawthorne, "$Na_3[B_{20}H_{17}NH_3]$: Synthesis and Liposomal Delivery to Murine Tumors, " *Proc. Natl. Acad. Sci. USA*, 91:3029–3033, 1994.
3. D.A. Feakes, F.R. Harris, D.K. Hathaway, and V.S. Morton, "Preparation of Hydrophilic Boron-Containing Compounds for Incorporation into Liposomes," *in*: Advances in Neutron Capture Therapy: Vol. II, Chemistry and Biology," B. Larsson, J. Crawford, and R. Weinrich, ed., Elsevier, Netherlands, 1997, pp. 95–100.
4. D.A. Feakes, R.C. Waller, D.K. Hathaway, and V.S. Morton, unpublished results.
5. T. Nguyen, G.L. Brownell, S.A. Holden, S. Kahl, M. Miura, and B.A. Teichert, "Subcellular Distribution of Various Boron Compounds and Implications for Their Efficacy in Boron Neutron Capture Therapy by Monte Carlo Simulations," *Radiation Research*, 133:33–40, 1993.
6. T. Nguyen, G.L. Brownell, S.A. Holden, and B.A. Teicher, "Intracellular Distribution of Various Boron Compounds for Use in Boron Neutron Capture Therapy," *Biochemical Pharmacology*, 45(1):147–155, 1993.
7. R. Pauwels, J. Balzarini, M. Baba, R. Snoeck, D. Schols, P. Herdewijn, J. Desmyter, and E. De Clercq, "Rapid and Automated Tetrazaolium-Based Colorimetric Assay for the Detection of Anti-HIV Compounds," *Journal of Virological Methods*, 20:309–321, 1988.
8. S. Rockwell, "*In Vivo-In Vitro* Tumor Systems: New Models for Studying the Response of Tumors to Therapy," *Laboratory Animal Science*, 27(5):831–851, 1977.

EVALUATION OF THE BINDING OF POLYHEDRAL BORANE COMPOUNDS TO INTRACELLULAR PROTEIN MOIETIES AND MONOMERIC AMINO ACIDS

R. Corey Waller[2], Jennifer Spinler[1], and Debra A. Feakes[1]

[1]Department of Chemistry
[2]Department of Biology
Southwest Texas State University
San Marcos, TX 78666

1. INTRODUCTION

The successful treatment of cancer by boron neutron capture therapy (BNCT) requires the selective concentration, and retention, of boron-containing compounds within malignant tumors. Previous research has demonstrated the effectiveness of unilamellar liposomes to deliver a wide variety of boron-containing compounds selectively to the tumor mass in small animal experiments.[1-3] Many water-soluble polyhedral boranes, such as $Na_2[B_{20}H_{18}]$[1] and $Na_4[B_{20}H_{17}SH]$,[4] have exhibited prolonged tumor retention, after liposomal delivery, in mice bearing EMT6 mammary adenocarcinoma. The retention exhibited by these compounds, however, is not common to all molecules in the polyhedral borane family and unreactive species, such as $Na_3[B_{20}H_{19}]$ and $Na_4[B_{20}H_{17}OH]$, are rapidly cleared from all tissues.[1] Although Hawthorne and coworkers have proposed that the difference in the tumor retention exhibited by this class of compounds is due to potential interactions with intracellular protein moieties (Figure 1),[1] the localization of the boron-containing compounds, as well as the potential protein interactions, on the subcellular level has not been determined.

2. METHODS AND MATERIALS

2.1. Reaction of $Na_2[B_{20}H_{18}]$ with Amino Acids and Choline

L-Tyrosine, L-cystine, choline, L-aspartic acid, and L-glutamic acid were purchased from Sigma Chemical Company. The general reaction procedure is as follows. An aqueous solution of $Na_2[B_{20}H_{18}]$ and the compound, in a 1 : 3 molar ratio, was combined

Frontiers in Neutron Capture Therapy, edited by Hawthorne *et al.*
Kluwer Academic / Plenum Publishers, New York, 2001.

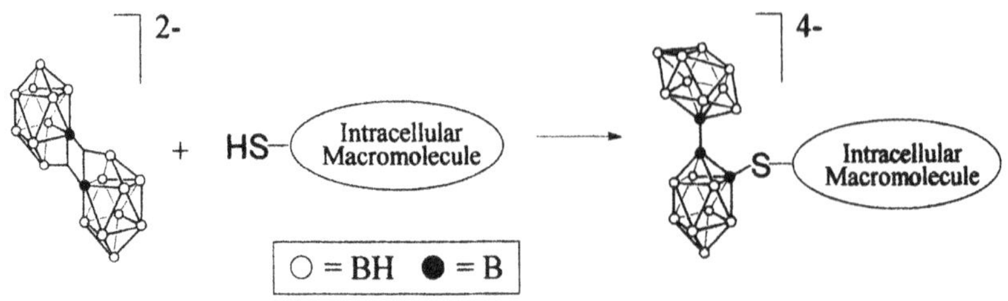

Figure 1. Proposed interaction of the normal isomer of $[B_{20}H_{18}]^{2-}$ with an intracellular protein moiety.

and the pH of the solution adjusted to 7.4–8.3, depending on the particular compound. The reaction vessel was placed in a 37 °C water bath for 72 hours and the reaction was monitored using ^{11}B NMR spectroscopy at 24 hour intervals. Products were isolated by precipitation as the potassium salt from ethanol, or the methyltriphenylphosphonium salt from water. The solids were filtered, dried *in vacuo*, and characterized by ^{11}B NMR, ^{1}H NMR, and IR spectroscopy.

2.2. Reaction of $Na_2[B_{20}H_{18}]$ and $Na_4[B_{20}H_{17}SH]$ with Protein Moieties and Evaluation by SDS-PAGE

Bovine serum albumin, β-galactosidase, cytochrome-C, G-actin, and poly-L-threonine were purchased from Sigma or Sigma-Aldrich Chemical Company. A solution of 100 μg of the specific protein residue in 2 mL of distilled water was placed in a reaction vessel and an amount of $Na_2[B_{20}H_{18}]$ or $Na_4[B_{20}H_{17}SH]$ equal to 100X the molar amount of the protein sample was added. The reaction mixtures were placed in a 37 °C water bath for 36 hours. Samples were removed from the water bath, concentrated to dryness, and brought up to a concentration of 1 μg/μL. An aliquot (20 μL) of the reaction sample was loaded onto a 5% stacking gel with a 10% separating gel and eluted for 1 hour at 250 mV. The resulting gel was stained with G-250 comassie blue and photographed.

3. RESULTS

3.1. Reaction of $Na_2[B_{20}H_{18}]$ with Amino Acids and Choline

Reaction of the electron-deficient, three-center, two-electron bonds of $Na_2[B_{20}H_{18}]$ are easily monitored by ^{11}B NMR spectroscopy. A mixture of $Na_2[B_{20}H_{18}]$ in distilled H_20 at 37 °C for 72 hours exhibited a normal $Na_2[B_{20}H_{18}]$ ^{11}B NMR spectrum, characterized by downfield peaks at approximately +29 ppm and +14 ppm as well as the cluster of five peaks spanning −8 ppm to −27 ppm. The reactions of $Na_2[B_{20}H_{18}]$ with tyrosine, cystine, and choline showed the loss of the two downfield signals, at +29 ppm and +14 ppm and the formation of signals in the upfield region of the spectrum, approximately +10 to −20 ppm, which is indicative of a reaction with $Na_2[B_{20}H_{18}]$. Reactions of $Na_2[B_{20}H_{18}]$ with glutamic acid and aspartic acid showed no peak shifts in the ^{11}B NMR spectrum.

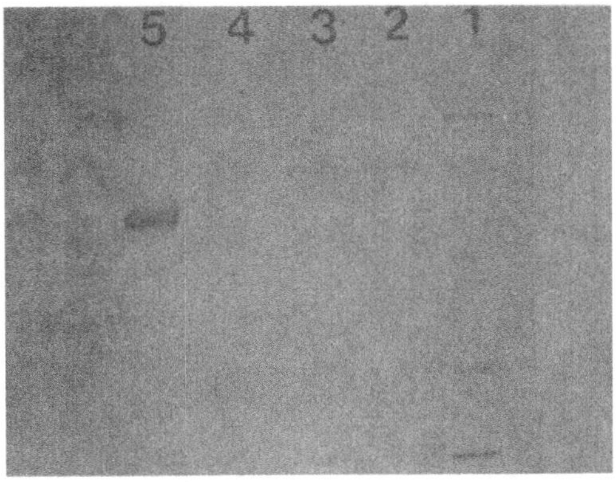

Figure 2. SDS-PAGE of the reaction between BSA and $Na_2[B_{20}H_{18}]$. Column 1: standard, Columns 2–4: BSA with $Na_2[B_{20}H_{18}]$ loaded at 30, 20, and 15 ug protein, respectively, Column 5: BSA control.

3.2. Reaction of $Na_2[B_{20}H_{18}]$ and $Na_4[B_{20}H_{17}SH]$ with Protein Moieties and Evaluation by SDS-PAGE

The reaction of $Na_2[B_{20}H_{18}]$ with BSA resulted in a significant increase in the molecular weight of the BSA, as depicted on the polyacrylamide gel (Figure 2). The ^{11}B NMR of the reaction mixture indicated the formation of substituted polyhedral borane species. Significant increases in the molecular weight of the protein as a result of the reaction between $Na_2[B_{20}H_{18}]$ and β-galactosidase (Figure 3), $Na_2[B_{20}H_{18}]$ and cytochrome-C (Figure 4), and $Na_4[B_{20}H_{17}SH]$ and cytochrome-C (Figure 4) were also observed. No change in molecular weight was observed in the reaction of $Na_2[B_{20}H_{18}]$ and G-actin.

The calculated molecular weight changes for BSA, β-galactosidase, cytochrome-C, as well as G-actin are listed in Table 1.

Figure 3. SDS-PAGE of the reaction between β-galactosidase and $Na_2[B_{20}H_{18}]$. Column 1: standard, Column 2: β-galactosidase control, Column 3: β-galactosidase and $Na_2[B_{20}H_{18}]$.

Figure 4. SDS-PAGE of the reactions between cytochrome-C/$Na_2[B_{20}H_{18}]$ and cytochrome-C/$Na_4[B_{20}H_{17}SH]$. Column 1: standard, Column 2: cytochrome-C and $Na_2[B_{20}H_{18}]$, Column 3: cytochrome-C control, Column 4: cytochrome-C and $Na_4[B_{20}H_{17}SH]$.

4. DISCUSSION

Determination of the possible intracellular interactions of polyhedral boranes is important if they are to be utilized for BNCT. Additionally, an enhanced understanding of these interactions enables the production of new compounds with a high potential for increased retention within the tumor. The reactivity of $Na_2[B_{20}H_{18}]$, a compound known to be retained by the tumor in mice bearing EMT6 mammary adenocarcinoma, with nucleophiles commonly found in cellular systems was investigated as an initial base for future studies. Compounds which contain strong nucleophilic substituents, such as choline, tyrosine, and cystine, reacted with $Na_2[B_{20}H_{18}]$. No reaction was observed with the control reaction. Glutamic acid and aspartic acid exhibited no reaction with $Na_2[B_{20}H_{18}]$, indicating that the nucleophilic sites of the carboxylic acids are not sufficiently strong to react with the electron-deficient, three-center, two-electron bonds present in $Na_2[B_{20}H_{18}]$.

Table 1. Molecular weight changes resulting from the reactions of the proteins with the polyhedral borane molecules

Protein	Calculated MW (d)	MW (d) after reaction with $Na_2[B_{20}H_{18}]$	MW (d) after reaction with $Na_4[B_{20}H_{17}SH]$	Change in MW (d)
BSA	50,000	56,000	x[4]	6,000
β-Galactosidase	77,000	81,000	x[4]	5,000
Cytochrome-C, a band[1]	◊[3]	◊[3]	13,000	−3,000
Cytochrome-C, b band[1]	16,000	16,000	16,000	No change
Cytochrome-C, c band[1]	◊[3]	27,000	27,000	11,000
Cytochrome-C, d band[1]	◊[3]	63,000	◊[3]	47,000
G-Actin, a band[2]	19,000	19,000	x[4]	No change
G-Actin, b band[2]	31,000	31,000	x[4]	No change

[1]See Figure 4 for band identification; [2]figure not included; [3]no band observed; [4]not included in experiment.

Reaction of $Na_2[B_{20}H_{18}]$ and $Na_4[B_{20}H_{17}SH]$ with a variety of proteins was successfully evaluated using SDS-PAGE techniques to monitor increases in protein molecular weight. The proteins used were selected based on their location within the tumor mass, as well as their relative abundance. Bovine serum albumin (BSA) was selected due to the presence of both hydrophobic and hydrophilic zones in its tertiary structure. Cytochrome-C is a highly reactive carrier molecule located in the mitochondria. G-Actin was evaluated due to the relatively high abundance in the cytosol as well as its hydro phobic properties and β-galactosidase was selected because of the relatively high abundance in lysozomes and the potential to be incorporated into the normal endocytic pathway.

Based on this initial investigation, the primary factor believed to enhance the binding of reactive polyhedral borane molecules to the protein is the hydrophilicity of the protein structure. Bovine serum albumin, cytochrome-C, and β-galactosidase have a substantial hydrophilic character and exhibited an increase in the molecular weight of the protein after reaction with the polyhedral borane molecules. The most lipophilic protein investigated, G-actin, exhibited no increase in molecular weight after reaction with the $Na_2[B_{20}H_{18}]$ molecule. Quantitation of the amount of boron incorporated by the protein structure will be evaluated by inductively coupled plasma-atomic emission spectroscopy (ICP-AES) in future experiments.

ACKNOWLEDGMENTS

This research was supported by awards from the Robert A. Welch Foundation (AI1292), the Dorothy L. Coker Research Fellowship (R. C. Waller), and Southwest Texas State University.

REFERENCES

1. K. Shelly, D.A. Feakes, M.F. Hawthorne, P.G. Schmidt, T.A. Krisch, and W.F. Bauer, "Model Studies Directed Toward the Boron Neutron-Capture Therapy of Cancer: Boron Delivery to Murine Tumors with Liposomes," *Proc. Natl. Acad. Sci. USA*, 89:9039–9043, 1992.
2. D.A. Feakes, K. Shelly, and M.F. Hawthorne, "Selective Boron Delivery to Murine Tumors by Lipophilic Species Incorporated in the Membranes of Unilamellar Liposomes," *Proc. Natl. Acad. Sci. USA*, 92:1367–1370, 1995.
3. D.A. Feakes, K. Shelly, C.B. Knobler, and M.F. Hawthorne, "$Na_3[B_{20}H_{17}NH_3]$: Synthesis and Liposomal Delivery to Murine Tumors," *Proc. Natl. Acad. Sci. USA*, 91:3029–3033, 1994.
4. D.A. Feakes, R.C. Waller, D.K. Hathaway, and V.S. Morton, unpublished results.

CREMOPHOR EL ALTERS UPTAKE, RETENTION, DISTRIBUTION, AND TOXICITY OF THE BORON CONTAINING ETHER LIPID *rac*-1-(9-*o*-carboranyl)nonyl-2-methyl-glycero-3-phosphocholine (B-Et-11-OMe)

Peter Lemmen[1], Lothar Weißfloch[2], Thomas Probst[3], Peter Luppa[4], Peggy L. Micca[5], and Michiko Miura[5]

[1]Institut für Organische Chemie und Biochemie, and
[3]Institut für Radiochemie
Technische Universität München
D-85747 Garching
[2]Klinik für Radiotherapie und Radiologische Onkologie, and
[4]Institut für Klinische Chemie und Pathobiochemie
Klinikum rechts der Isar
D-81675 München
[5]Medical Department
Brookhaven National Laboratory
Upton, NY 11973-5000

1. INTRODUCTION

We have proposed boron-containing ether lipids as boron carriers for BNCT; because these are expected to accumulate selectively in tumors.[1] These properties depend most probably on the absence of an ether lipid-cleaving enzyme, namely alkyl glycerol monooxidase (AGMO) in a variety of tumors.[2] Encouraged by these results we have synthesized the carborane substituted derivative **1** and analogs.[1]

Despite promising accumulation in cultured human glioma cells, U-343-MGa, and in murine melanoma cells, B16,[3] a solution of **1** in physiological saline was not sufficiently taken up by the murine experimental tumor AT17. Instead, the bulk of the compound was rapidly taken up by the liver and excreted via the intestines. We have attributed this fate to the particulated nature of aggregated species.[4] The delivery of boron-containing ether lipids as boron carriers for BNCT needs to be improved. Only then the anticipated relative accumulation by different metabolic fate can be effective.

Frontiers in Neutron Capture Therapy, edited by Hawthorne *et al.*
Kluwer Academic / Plenum Publishers, New York, 2001.

Cremophor EL is a non ionic surfactant prepared by reacting castor oil with ethylene oxide. This high molecular weight tenside (MW ca. 2,500 Da)[5] has been used to solubilize hydrophobic drugs including paclitaxel (taxol). It is known to reverse multi drug resistance in some cases.[6] Cremophor is tolerated by humans in doses of 20–40 ml/day for 20–30 consecutive days; in some cases however, anaphylactoid reactions have been observed.[6]

Cremophor has been used also to transport porphyrine derivatives in photodynamic therapy.[7] Boron containing porphyrins were transported into murine tumors with very favorable ratios tumor/blood and tumor/brain in the context of BNCT.[8] Solubilisation by Cremophor promoted incorporation of porphyrins into plasma low-density lipoprotein,[7] this phenomenon may be responsible for tumor localization. In this paper we studied the influence of Cremophor on the uptake and biodistribution of the boron-containing ether lipid **1**.

To this end we conducted three different experiments:

- a low dose distribution study at 250 mg B-Et-11-OMe/kg b.w.
- a dose escalation study to assess the maximum tolerated dose and biodistribution.
- the pharmacokinetics of **1**.

2. MATERIALS AND METHODS

rac-1-(9-*o*-Carboranyl)nonyl-2-methyl-glycero-3-phosphocholine (B-Et-11-OMe) (Figure 1) was prepared as described.[1] Cremophor EL and propylene glycol were obtained from Sigma.

Animals and tumor model: The serially transplanted murine mammary tumor EMT-6,[9] ca. 10^7 cells/0.1 ml, was transplanted into the back of female BALB/c mice. After 14 days tumors were grown to 150–450 mg.

Administration: B-Et11-OMe (5.4 mg) was stirred with warm Cremophor EL (0.5 ml) until complete dissolution. Propylene glycol (1 ml) and then 0.9% saline (15 ml) were added resulting in a clear solution containing 0.33 mg/ml B-Et-11-O-Me, 3% Cremophor and 6% propylene glycol.[10] In an analogous way solutions containing 2.7, 5.3, and 10.7 mg/ml B-Et-11-O-Me in 3% cremophor / 6% propylene glycol were prepared.

Boron analyses in tissue were performed by inductively coupled plasma atomic emission spectrometry (ICP-AES) and inductively coupled plasma mass spectrometry

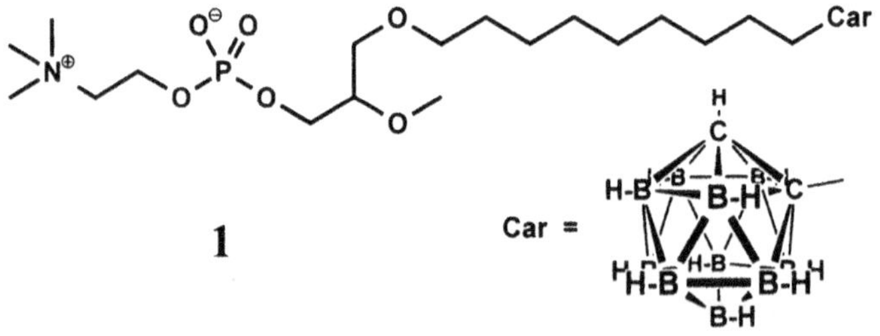

1

Figure 1. B-Et11-OMe.

(ICP-MS) as described[11] (Munich), or by direct current plasma-atomic emission spectroscopy (DCP-AES) as described[8] (Brookhaven).

Low dose distribution study (performed at BNL): B-Et-11-OMe (0.5 ml solution cont. 0.33 mg/ml/20 g b.w. each) was injected i.p. six times over a 2-day period leading to a total dose of 250 mg B-Et-11-OMe/kg b.w. Two groups of mice were euthanasized 15 hrs (6 mice, group A) and 39 hrs (6 mice, group B) after the last injection.

Under deep halothane anesthesia leading to euthanasia, right ventricular blood was collected in Microtainer tubes (Becton-Dickinson, Rutherford, NJ) containing EDTA for hematological analyses and boron measurements of whole blood and in Microtainer tubes containing lithium heparin for chemical analyses and enzyme essays in blood plasma. Tumor, brain and liver were sampled at necropsy for boron analyses.

Levels of GLU, BUN, BCR, ALT AST, ALP, and CPK were determined in blood plasma using Cobas Fara II microanalysis system (Roche, Nutley, NJ). Blood PLT were counted with a serono-Baker System 9000 (Serono-Baker, Allentown, PA) automated hematology microanalyzer.

Dose escalation study (performed in Munich): Group C was treated with 6 i.p. injections of B-Et-11-OMe in 3% cremophor / 6% propylene glycol (0.5 ml solution cont. 2.7 mg/ml/20 g b.w. each) leading to a total dose of 400 mg/kg b.w. (6 mice).

As the behavior of group **C** appeared completely normal, two further groups of 3 mice each were treated with 800 mg/kg b.w. B-Et-11-OMe (group D), and 1,600 mg/kg b.w. B-Et-11-OMe in 3% cremophor / 6% propylene glycol (group E). Group F: 3% cremophor / 6% propylene glycol (5 mice, control). Group G: 0.9% saline (5 mice, control) 39 hours after the last injection the mice were euthanasized by deep ether anesthesia and removal of right ventricular blood for chemical analysis and enzyme essays. Organs were inspected and collected for boron analysis.

Kinetic study: B-Et-11-OMe (0.5 ml solution/20 g b.w. cont. 5.3. mg/ml) was injected i.p., leading to a dose of 133 mg B-Et11-OMe/kg b.w. Groups of mice were euthanasized by deep ether anesthesia and cutting the aorta 2, 4, 8 (4 mice each), 16, and 32 hrs (5 mice each) later. Organs and blood were inspected and collected for boron analysis.

3. RESULTS AND DISCUSSION

The appearance and behavior of mice in groups A (250 mg/kg b.w.) through D (800 mg/kg b.w.), F and G remained normal before euthanasia. At necropsy, organs and tissue appeared normal. All parameters of renal and hepatic function were normal, they did not indicate significant alterations from controls.

Also there was no sign of toxicity of cremophor at the concentration used. This is in accord with literature. Cremophor has been used *in vivo* in different concentrations. At 9%, no statistical difference to the control group was seen.[8] When 20% cremophor and 6% propylene glycol were used, no gross increase in mortality was reported.[7] The behavior of group E (1,600 mg/kg b.w.) however indicated significant toxicity after the 4th injection. Two mice died after the 5th injection. The third was euthanasized due to morbidity.

Obviously, Cremophor significantly reduces the toxicity of B-Et11-OMe. In an earlier investigation, 50 mg/kg, administered as a single injection of a micellar solution of the pure compound, induced significant though transient toxicity.[4] In this study up to 133 mg/kg could be given six times successively, giving rise to a total dose of

Table 1. Boron content of selected tissue [mg/kg] 15 and 39 hrs. post final injection (Mean value ± standard deviation)

Group (n)	A (6)	B (6)	C (5)
Dose [mg/kg b.w.]	250	250	400
Time [hrs.]	15	39	39
Blood	0.30 ± 0.06	0.13 ± 0.04	1.40 ± 1.39
Liver	9.02 ± 4.28	4.50 ± 3.08	9.48 ± 0.49
Brain	4.87 ± 3.14	4.34 ± 2.58	2.00 ± 0.34
Tumor	6.05 ± 2.52	4.40 ± 1.61	1.76 ± 0.83

800 mg/kg b.w., (group D) without apparent negative effects. Further increase of the dose to 1,600 mg/kg (group E) however proved to be intolerable.

Boron contents of tissue are summarized in Tables 1 and 2.

The ether lipid as a pure substance was taken up and excreted again very rapidly by the liver. Maximum concentration was reached after 1.5 hours.[4] Using Cremophor as a vehicle both uptake and excretion took place later. After administration of 250 mg B-Et11-OMe maximum concentration was reached after 4 hours. After 39 hours 4.4 ppm B (8.8% of the injected dose) were still present. At higher concentration uptake and retention resemble those of pure substance. Tumor/blood-ratio increases with time also in these cases, however to a lower extend.

It should be noted, that under mediation of Cremophor also the brain takes up B-Et11-OMe. Applied as the pure substance, no significant amount was taken up into this tissue.

Most important, uptake and retention into tumor were changed: Following application of 250 mg/kg by means of Cremophor, 12% of injected dose were still present after 15 hours. 8.8% of injected dose were retained after 39 hours. This corresponds to tumor/blood-ratios of 20 and 34 after 15 and 39 hours respectively. However, after application of higher doses at constant Cremophor concentration uptake and retention time decrease, as observed for the pure substance. Tumor/blood boron concentration ratio increases with time but to a lesser extent. We interpret these results as follows: biological properties like toxicity, distribution and retention do not solely depend on the chemical nature of the compound but on the solvent vehicle as well. Lipophilic substances will form these transport vectors by aggregation. We expect the properties of these aggregated species to be dependent, *inter alia*, on the molar ratio of surfactant and lipophilic compound. At high ratio Cremophor/B-Et11-OMe the properties are dominated by

Table 2. Boron content of selected tissue [mg/kg] (median (range)) at different times after single injection of 133 mg/kg B-Et11-OMe

Time [h] (n)	2 (4)	4 (4)	8 (4)	16 (5)	32 (5)
Blood	30.5 (18.9–45)	18.9 (6.8–22.6)	9.15 (6.4–12.7)	0.56 (0.4–0.74)	0.27 (0.24–0.44)
Liver	73 (61–82)	153 (100–169)	90 (71–134)	5.8 (4.9–7.1)	5.1 (2.9–9.6)
Brain	0.35 (0.31–0.6)	0.54 (0.3–0.73)	1.12 (0.92–1.36)	1.05 (0.93–1.31)	0.35 (0.31–0.5)
Tumor	3.3 (2.6–15.1)	5.95 (3–9.1)	5.6 (4.2–8.1)	1.28 (0.99–1.83)	0.95 (0.38–1.35)
Skin	37.5 (29–67)	26.6 (7.4–66)	4.45 (4.1–6.1)	2.3 (1.3–3)	0.8 (0.6–2.5)
Muscle	5.5 (3.8–12)	2.72 (1.08–6.4)	5.05 (3.2–11)	18 (2.08–58)	0.49 (0.4–0.58)
Kidney	14.7 (11.9–19.2)	14.1 (8.8–16.1)	15.5 (13.4–18)	5.6 (3.8–6.9)	2.88 (2.62–3.51)
Spleen	11.1 (8.6–22.3)	8.4 (3.8–11.5)	5.05 (3.6–6.3)	3.17 (2.08–4.3)	1.27 (0.89–1.44)
Plasma	26.5 (16–49)	21.4 (8–22.2)	8.8 (5.3–14.1)	0.57 (0.38–1)	0.17 (0.15–0.17)

Cremophor; the compound is taken up and retained. At lower ratio (ca. 1) the properties resemble those of the pure compound.

A dose escalation study using constant ratio Cremophor/B-Et11-OMe is under way.

ACKNOWLEDGMENTS

Training in toxicity assessment by and helpful discussions with Dr. Daniel N. Slatkin are gratefully acknowledged. The work was supported by DFG (Le 553/3-1) and Fonds der chemischen Industrie (P.L.), Pinguinstiftung (T.P.), and Bayerisches StMLU (L.W.)

REFERENCES

1. Lemmen, P., and Werner, B., Synthesis of rac-1-(9-o-carboranyl)nonyl-2-methyl-glycero-3-phosphocholine (B-Et-11-OMe), a potential boron carrier for Boron Neutron Capture Therapy (BNCT), *Chem. Phys. Lipids*, 62:185–191, 1992.
2. Soodsma, F., Piantadosi, C., and Snyder, F., The biocleavage of alkyl glyceryl ethers in morris hepatomas and other transplantable neoplasms. *Cancer Res.*, 30:309–311, 1970.
3. Lindström, P., Olsson, P., Malmquist, J., Pettersson, J., Lemmen, P., Werner, B., Sjöberg, S., Olin, Å., and Carlsson, J., New carborane based compounds for boron neutron capture therapy. Binding and toxicity of ANC-1, DAC-1, and B-Et-11-OMe in cultured human glioma and mouse melanoma cells, *Anti-Cancer Drugs*, 5:43–52, 1994.
4. Lemmen, P., Weißfloch, L., Auberger, T., and Probst, T., Uptake and distribution of the boron-containing ether lipid B-Et-11-OMe in tumor bearing mice, *Anti-Cancer Drugs*, 6:744–748, 1995.
5. Sigma, Data Sheet C-5135, Cremophor EL.
6. Woodcock, M., Jefferson, S., Linsenmeyer, M.E., Crowther, P.J., Chojnowski, G.M., Williams, B., and Bertoncello, I., Reversal of the multidrug resistance phenotype with cremophor EL, a common vehicle for water-insoluble vitamins and drugs, *Cancer Res.*, 50:4199–4203, 1990.
7. Woodburn, K., Chang, C.K., Lee, S., Henderson, B., and Kessel, D., Biodistribution and PDT efficacy of a ketochlorin photosensitizer as a function of the delivery vehicle, *Photochem. Photobiol.*, 60:154–159, 1994.
8. Miura, M., Micca, P.L., Fisher, C.D., Heinrichs, J.C., Donaldson, J.A., Finkel, G.C., and Slatkin, D.N., Synthesis of a nickel tetracarboranylphenylporphyrin for boron neutron-capture therapy: Biodistribution and toxicity in tumor-bearing mice., *Int. J. Cancer*, 68:114–119, 1996.
9. Rockwell, S.C., Kallman, R.F., and Fajardo, L.F., Characteristics of a serially transplanted mouse mammary tumor and its tissue culture-adapted derivative, *J. Natl. Cancer Inst.*, 49:735–747, 1972.
10. Morgan, R., Garbo, G.M., Kreimer-Birnbaum, M., Keck, R.W., Chaudhuri, K., and Selman, S.H., Morphological study of the combined effect of purpurin derivatives and light on transplantable rat bladder tumors, *Cancer Res.*, 47:496–498, 1987.
11. Probst, T., Berryman, N.G., Lemmen, P., Weissfloch, L., Auberger, T., Gabel, D., Carlsson, J., and Larsson, B., Comparison of inductively coupled plasma atomic emission spectrometry and inductively coupled plasma mass spectrometry with quantitative neutron capture radiography for the determination of boron in biological samples from cancer therapy, *J. Anal. At. Spectrom.*, 12:1115–1122, 1997.

ANALYSIS OF *IN VIVO* UPTAKE FOR BORON NEUTRON CAPTURE SYNOVECTOMY

E. Binello[1], S. Shortkroff[2], R. E. Shefer[3], A. G. Jones[4], A. Davison[5], T. Thornhill[2], and J. C. Yanch[1]

[1]Department of Nuclear Engineering and Whitaker College of Health Sciences and Technology
Massachusetts Institute of Technology, Cambridge, MA
[2]Department of Orthopedic Surgery
Brigham & Women's Hospital, Boston, MA
[3]Newton Scientific Incorporated
Cambridge, MA
[4]Department of Radiology
Harvard Medical School, Boston, MA
[5]Department of Chemistry
Massachusetts Institute of Technology, Cambridge, MA

1. INTRODUCTION

Boron Neutron Capture Synovectomy (BNCS) has been proposed as a new application of the boron neutron capture reaction for the treatment of Rheumatoid Arthritis.[1] Rheumatoid Arthritis (RA) is an autoimmune disease of unknown origin, primarily characterized by an inflammation of the synovium, the membrane lining the joint space. Prolonged synovial inflammation results in the secretion of various enzymes which degrade cartilage and bone, eventually leading to the loss of joint function.

The primary treatment for RA is the administration of various drugs. However, a single joint may remain unresponsive to drug treatment, thereby requiring the physical removal of the synovium, termed synovectomy. Currently, the only option available in the United States is surgery, while in Europe and elsewhere, the injection of beta-emitters into the joint is also available.

Although BNCS is based on the same principle as BNCT for cancer, several differences exist such as the route of administration of the boron compound. In BNCS, the boron compound would be injected locally into the joint space, where it would be taken up by the synovium. The joint would then be irradiated with neutrons of a desired therapeutic energy range[2] inducing the boron neutron capture reaction. This approach

Frontiers in Neutron Capture Therapy, edited by Hawthorne *et al.*
Kluwer Academic / Plenum Publishers, New York, 2001.

has advantages over the existing options for synovectomy in that it is non-invasive and does not require the administration of radioactive substances.

Boron uptake by the synovium and by other parts of the joint are important parameters in the assessment of the potential of BNCS. This paper presents the investigation of boron uptake *in vivo* by the synovium, as well as by other parts of the joint, and compares these results with previously obtained results ex vivo.[2,3,4] Results of the *in vivo* study using $K_2B_{12}H_{12}$ as the boron compound were critical in determining whether to proceed to animal irradiations to test the therapeutic efficacy of BNCS.

2. METHODS

2.1. Antigen-Induced-Arthritis (AIA) Model

Healthy New Zealand white rabbits, weighing 3 to 4 kgs each, were purchased and allowed to acclimatize to their new surroundings for about 1 week before beginning the process of creating an arthritic condition local to their knees. The AIA rabbit model is widely used in the evaluation of various treatments for RA.[5] In this model, the rabbit undergoes a "first sensitization" where a total of 5 mg. of ovalbumin with Freund's complete adjuvant is injected subcutaneously. After three weeks the rabbit undergoes a "second sensitization" which is the same as the first, except that the adjuvant is Freund's incomplete. Then, three weeks later, following an intra-muscular injection of standard anesthetic cocktail, an intra-articular injection of 5 mg. of ovalbumin is delivered to the joint space. An arthus reactions occurs within 3 hours and proceeds to a full synovitis in 3 to 5 days. A strict pain management protocol is followed whereby buprenorphine (opiate) is administered via subcutaneous injections every 12 hours beginning from after the intra- articular injection of ovalbumin and continuing until the time of sacrifice.

2.2. Boron Uptake

On the third day after the arthritis induction, 0.25 mL of $K_2B_{12}H_{12}$, was injected intra-articularly. After varying times (0.25, 1, 4 and 24 hours), the rabbits were sacrificed and the knee joint dissected. A full biopsy was performed on the rabbits used for the 4 hour time point. Boron uptake was quantified via Prompt Gamma Neutron Activation Analysis (PGNAA) at the MIT Research Reactor.[6] The $^{10}B/^{1}H$ ratio (measuring the 478 keV gamma from boron capture and the 2.2 MeV gamma from hydrogen capture) was used to account for variations in the neutron flux. A calibration curve was constructed correlating $^{10}B/^{1}H$ ratios to boron concentrations and used to calculate uptake in the rabbit samples. Measurement errors were kept below 10%.

3. RESULTS AND DISCUSSION

Table 1 shows the boron uptake by the synovium obtained from an injection of 0.25 mL of 5,000 ppm of boron (total 1.25 mg.) in the form of $K_2B_{12}H_{12}$. There is a significant decrease in uptake with time. At the 15 minute time point, the average synovial concentration is about 10% of the initial injection concentration.

The highest value in the *in vivo* study is much lower than the values typically found in ex vivo studies where tissue samples, excised from human patients and incubated in

Table 1. *In vivo* boron uptake by the synovium

Time (hr)	Range (ppm)	Mean (ppm)
0.25	265–950	490 (n = 4)
0.50	35–190	130 (n = 3)
1.00	30–50	40 (n = 4)
4.00	<10	— (n = 3)
24.0	<1	— (n = 3)

boron-containing medium, retained about 20% to 50% of the original compound concentration.[2,3] If one considers that in a rabbit joint there is about 1 mL of fluid which is not aspirated before injection of 0.25 mL of boron compound, then the effective concentration of the compound is lower by approximately a factor of four, rendering the average uptake level observed at 15 minutes consistent with the ex vivo uptake levels at 1 hour. Another difference between the *in vivo* and the *ex vivo* studies is the uptake of boron as a function of incubation time. Ex vivo, a longer incubation results in higher uptake while *in vivo*, a longer incubation results in lower uptake. The *ex vivo* system does not take into account many factors found *in vivo* such as circulatory and lymphatic flow. In fact, as a first approximation to the *in vivo* situation, washout studies were performed[4] in which excised human tissue was incubated in boronated medium for 1 hour and then reincubated in boron-free medium for 30 minutes, 1 and 2 hours. No significant difference was found between the incubation times, which exhibited an average tissue retention of the initial boron concentration ranging from 18 to 25%. This implied that *in vivo* uptake levels would be substantially lower than *ex vivo*. At 15 minutes, the *in vivo* synovial uptake ranged from 5 to 20% of the initial incubation concentration with an average of about 10%, similar value to those found in the washout studies.

The other joint tissues, such as patella, tendon and ligament, had significantly lower uptake. The patella had a concentration less than 20 ppm at all times, while the tendon decreased from an average of 44 ppm at 15 minutes to less than 2 ppm at 24 hours. Finally, the boron concentration in the ligament decreased from an average concentration of 282 ppm at 15 minutes to less than 5 ppm at 24 hours. It should be noted that it is technically impossible to completely clear the synovium from the other joint tissues and that this creates a particular problem with low weight samples, such as the ligament, since even a small amount of boron can significantly affect the $^{10}B/^{1}H$ ratio. This may be the reason for the high measured value of ligament boron uptake, as compared to other joint tissues.

The biopsy results for the 4 hour rabbits (n = 3) revealed a boron concentration of less than 5 ppm for the lungs, kidneys, liver, spleen, bone marrow and blood. The urine had a boron concentration of 50 ppm, which, considering the large volume of the rabbit bladder (approximately 15 ml) and assuming homogeneity of the urine, could represent as much as 60% of the total amount of boron injected (total amount of boron in bladder 0.75 mg.).

4. CONCLUSIONS AND FUTURE WORK

Characterization of $K_2B_{12}H_{12}$ uptake by the synovium both *ex vivo* (human synovium) and *in vivo* (rabbit model) showed high boron concentrations. High uptake levels enable a very rapid delivery of dose (10,000 RBE-cGy is considered the therapeutic dose based on clinical trials of radiation synovectomy using β-emitters[7]). A moderated neutron beam, based on the 1.5 MeV Be(d,n) reactions and a particle current of 1 mA, results in a treatment time of 16 minutes.[8]

Neutron irradiations of AIA rabbits using $K_2B_{12}H_{12}$ have already begun at the Massachusetts Institute of Technology Laboratory for Accelerator Beam Applications (MIT LABA); early results show synovial necrosis.[9] *In vivo* uptake studies with $K_2B_{12}H_{12}$ are being continued at higher compound concentrations. Future work includes the analysis of uptake using macrophages and fibroblasts (the two main types of cells found in the synovium), as well as testing of newly synthesized boronated compounds as available.

ACKNOWLEDGMENT

This work was supported by a grant from the Idaho National Engineering and Environmental Laboratory (INEEL) University Research Consortium.

REFERENCES

1. L.S. Johnson, J.C. Yanch, S. Shortkroff, and C.B. Sledge, Temporal and spatial distribution of boron uptake in excised human synovium, *in*: "Proceedings The Sixth International Symposium on Neutron Capture Therapy for Cancer", Kobe (Japan), October/November 1994.
2. E. Binello, S. Shortkroff, A.G. Jones, C. Viveiros, A. Ly, C.B. Sledge, A. Davison, R.E. Shefer, and J.C. Yanch, Research in Boron Neutron Capture Synovectomy, *in*: "Proceedings of the 5th International Conference on Applications of Nuclear Techniques: Neutrons in Research and Industry", Society of Photo-Optical Instrumentation Engineers, Bellingham, 1997, pp. 68–71.
3. E. Binello, S. Shortkroff, C. Viveiros, G. Young, A.G. Jones, A. Davison, C.B. Sledge, and J.C. Yanch, *In vitro* analysis of boron-10 uptake for Boron Neutron Capture Synovectomy, *in*: "Advances in Neutron Capture Therapy", B. Larsson, J. Crawford and R. Weinrich, ed., Elsevier Science, Amsterdam, Volume 2: Chemistry and Biology, 1997, pp. 609–613.
4. E. Binello, S. Shortkroff, and J.C. Yanch, PGNAA of human arthritic synovium for Boron Neutron Capture Synovectomy, *in*: "Transactions of the 1997 Winter Meeting of the American Nuclear Society", Albuquerque, November 1997, TRANSAO 77 1-560 (1997), pp. 548–549.
5. L. Sokoloff, Animal models of Rheumatoid Arthritis, *Int Rev Exp Path* 26:107–145, 1984.
6. O.K. Harling, J.-M. Chabeuf, F. Lambert, and G. Yasuda, A prompt gamma neutron activation analysis facility using a diffracted beam, *Nucl Instrum Methods Phys Res* B83:557–562, 1993.
7. M.R. Zalutsky, P.P. Venkatesan, R.J. English, S. Shortkroff, C.B. Sledge, and S.J. Adelstein, Radiation Synovectomy with 165 Dy-FHMA: lymph node uptake and radiation dosimetry calculations, *Int J Nucl Med Biol* 12(6):457–465, 1986.
8. D.P. Gierga, J.C. Yanch, R.E. Shefer, and B.W. Blackburn, Construction of an accelerator-based neutron beamline for treatment of arthritic joints, *in*: "Frontiers in Neutron Capture Therapy," M.F. Hawthorne, K. Shelly, R.W. Wiersema, eds., Kluwer Academic/Plenum Publishers, New York, 2001, pp. 461–467.
9. J.C. Yanch, S. Shortkroff, R.E. Shefer, E. Binello, D. Gierga, A.G. Jones, G. Young, C. Vivieros, and B. Blackburn, Progress in the development of boron neutron capture synovectomy for the treatment of rheumatoid arthritis, *in*: "Frontiers in Neutron Capture Therapy," M.F. Hawthorne, K. Shelly, R.W. Wiersema, eds., Kluwer Academic/Plenum Publishers, New York, 2001, pp. 1389–1397.

TISSUE TARGETING

SLT-PARTICLES FOR TWO-STEP TARGETING IN BORON NEUTRON CAPTURE THERAPY

E. Bohl[1], J. Carlsson[1], K. Edwards[2], S. Sjöberg[3], and L. Gedda[1]

[1]Biomedical Radiation Sciences
Rudbeck Laboratory 75185 Uppsala
[2]Physical Chemistry and
[3]Organic Chemistry
Uppsala University
Uppsala, Sweden

1. INTRODUCTION

There are investigations proposing that liposomes, both stabilized and unstabilized, gather at sites of pathology, such as tumors, infections, and inflammations.[1-4] The site-specific uptake of liposomal substances ought to increase with a targeting agent on the liposome. One way to conduct the targeting is to use PEG (polyethylene glycol) as a spacer and conjugate the targeting agent to the terminus of the polymer.[5] This conjugated liposome can be named a SLT-particle (stabilized liposome with targeting ligand).

When using radionuclides to kill tumors the toxicity is highly increased if the radionuclides are delivered to the nucleus instead of evenly distributed throughout the cell.[6] Since the DNA is the target it is desirable to make sure the radionuclides enters the nucleus. To ensure nucleus specific delivery the idea of "two-step targeting" was born as an improvement of tumor targeting. The first targeting step is for the carrier, in this case the liposome, to find the tumor cell and be internalized. After internalization the carrier is degraded and the drugs are released in the cytoplasm. The second targeting step is then for the drugs to find their way to the nucleus. This can be done by using DNA-intercalating agents as drugs.

The results presented in this report are a preliminary study to determine the magnitude of the unspecific cellular uptake of liposome encapsulated boronated DNA intercalators. It is essential that the unspecific drug delivery is at a low level to ensure low damage to healthy tissue.

We have used the boronated DNA-intercalating agents WSA-1 (water-soluble boronated acridine 1) and WSP-1 (water soluble boronated phenanthridine 1) (Figure 1).[7] WSA-1 is composed of an intercalating acridine chromophore with a boron cage and

Frontiers in Neutron Capture Therapy, edited by Hawthorne *et al.*
Kluwer Academic / Plenum Publishers, New York, 2001.

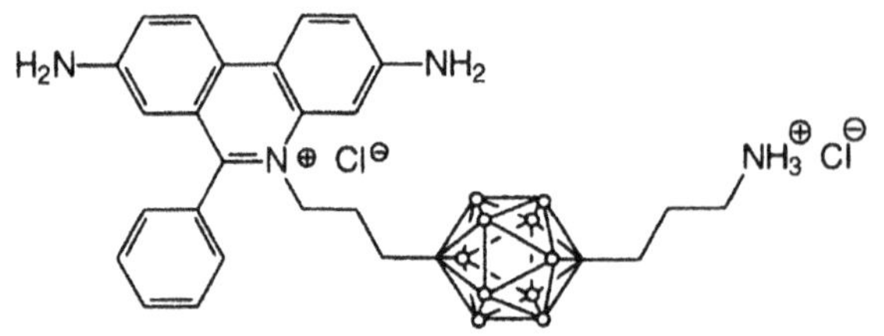

WSP-1 (M_r = 599.7)

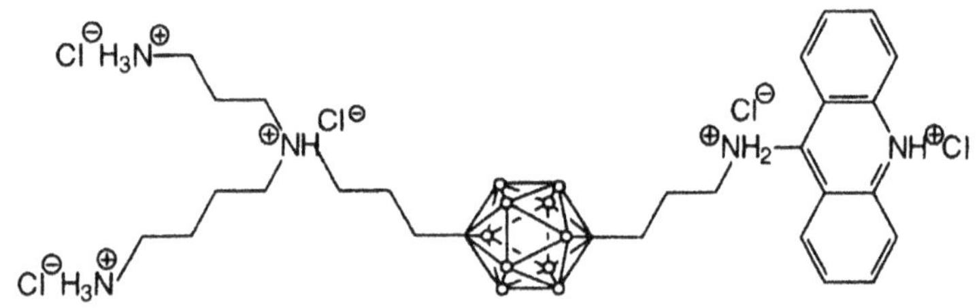

WSA-1 (M_r = 746.1)

Figure 1. Water soluble boronated phenanthridine (WSP-1) and water soluble boronated acridine (WSA-1) (o = BH).

a spermidine derivative. The spermidine part is added to enhance DNA binding properties of the compound and it also increases the hydrophilicity of the molecule.[7] The WSP-1 is a phenanthridinium derivative such as ethidium bromide, one of the most well known DNA-intercalating agents. These two compounds were chosen from a variety of synthesized boron cage bearing DNA-intercalators because they exhibit some appealing features.[8] With active loading of the liposomes the above compounds precipitate and form crystals due to the pH gradient during loading.

With WSA-1 and WSP-1 there have been results indicating that it is possible to load as much as 10^5 intercalating molecules into one liposome, that is 10^6 ^{10}B atoms. To acquire the desirable boron amount in the cells ($\approx 10^9$ atoms) approximately 1,000 liposomes have to be internalized into the cells. If the second targeting step works (i.e. the DNA intercalating agents manages to bind to DNA) even fewer liposomes per cell will be necessary.

2. MATERIALS AND METHODS

2.1. Liposomes

The liposome composition was DSPC:Chol:DSPE-PEG (55:40:5mole%). The liposomes were prepared by freeze thawing and extruding through a 100nm filter. The loading of the liposomes with WSA-1 and WSP-1 was conducted using pH gradients.

2.2. Monolayer Uptake

Uptake in monolayer U-343 glioma cells of WSA-1, WSA-1 in stabilized liposomes, WSP-1, and WSP-1 in stabilized liposomes was studied. The compounds were diluted in incubation medium to a concentration of 1 μg/ml ~0.2 ppm boron. For the first experiment the medium was removed after incubation, the cell dishes were washed with serum-free medium and trypsinized. The cell suspension was transferred to a centrifuge tube. The cell suspension was centrifuged. After thorough removal of the supernatant the cell pellet was weighed, dissolved in nitric acid and digested under heat and pressure.[9] Then the sample was diluted in Elga water and analyzed by ICP-MS (inductively coupled plasma mass spectroscopy).

In the second experiment the dishes were kept unwashed. The cells were mechanically removed from the plate and the cell suspension was moved to a centrifuge tube. A pellet was formed by centrifugation and the same procedure as above for boron content determination was performed.

2.3. Spheroid Uptake

The binding in DU-145 prostatic spheroids was studied using WSA-1, WSA-1 in stabilized liposomes, WSP-1 and WSP-1 in stabilized liposomes in a concentration of 5 μg/ml. The spheroids were incubated with 3 ml incubation medium in rotating siliconized small glass flasks. Spheroids were removed from the glass flasks, washed and fixed in cold 10% formalin for at least 1 h. The spheroids were dehydrated and embedded in paraffin and 5 μm sections were cut using a microtome. To achieve a quantitative measurement of the substance bound in the spheroids fluorescent images were analyzed regarding the mean fluorescent intensity at different depths.

2.4. Toxicity Test

The toxicity of the compounds was studied by conducting a clonogenic survival test. The compounds were diluted in culture medium to 0.5, 1, 5, 10 and 20 μg/ml and incubated with U-343 glioma monolayer cells for 24 hours. After incubation the culture medium was removed and the cells were centrifuged. The cell pellet was resuspended in 10 ml culture medium. After counting and dilution the cells were plated in triplicate into Petri dishes (10 cm) and further incubated for 14 days to allow colonies to form. The dishes were then washed, fixed and stained. The colonies were counted manually and the survival and uncertainty was calculated. The survival data were plotted into a chart and fitted into an exponential survival curve.

3. RESULTS AND DISCUSSION

3.1. Monolayer Uptake

The results from the ICP-MS analysis of the boron content in the cell samples are shown in Figures 2–3. Figure 2 shows that after two hours incubation the uptake of boronated substance in cell culture was considerably lower when WSA-1 was encapsulated in a liposome compared to the uptake of the free substance. The boron level in the samples incubated with stabilized liposomes containing WSA-1 was at the same level as the background. After 24 h incubation the boron analysis of the samples with free WSA-1 showed elevated uptake but the boron content in the stabilized liposomal sample was still on background level.

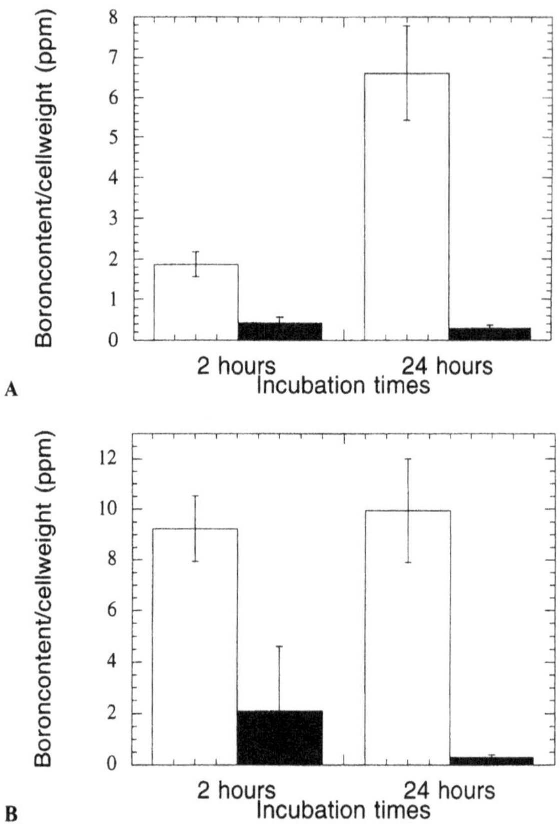

Figure 2. A. Boron concentration in washed cell samples incubated with free WSA-1 and WSA-1 in stabilized liposomes. B. Boron concentration in unwashed cell samples incubated with free WSA-1 and WSA-1 in stabilized liposomes (□ Free WSA-1; ■ WSA-1 in stabilized liposomes).

Figure 2b shows that the experiment with cells kept unwashed incubated with free substance exhibits a higher boron signal than the washed samples. In the samples with stabilized liposomes the boron level is still the same as the background.

For WSP-1 the same uptake pattern as for WSA-1 could be seen (Figure 3) at all times, markedly higher boron level in the samples incubated with free substance compared to the samples with liposomal substance. The boron level in the cell samples that were kept unwashed was slightly higher than for the washed samples (Figure 3b). In both experiments the uptake level of the liposomal substance was at background level.

3.2. Spheroidal Uptake

It is shown in Figure 4a that both free WSA-1 and stabilized liposomes filled with WSA-1 are able to penetrate the spheroid. The occurrence of liposomal WSA-1 even seems to increase in the necrotic region. Figure 4b show that liposomal WSP-1 penetrates the spheroid slightly better than free WSP-1 which mostly gather in the outer limits of the spheroid.

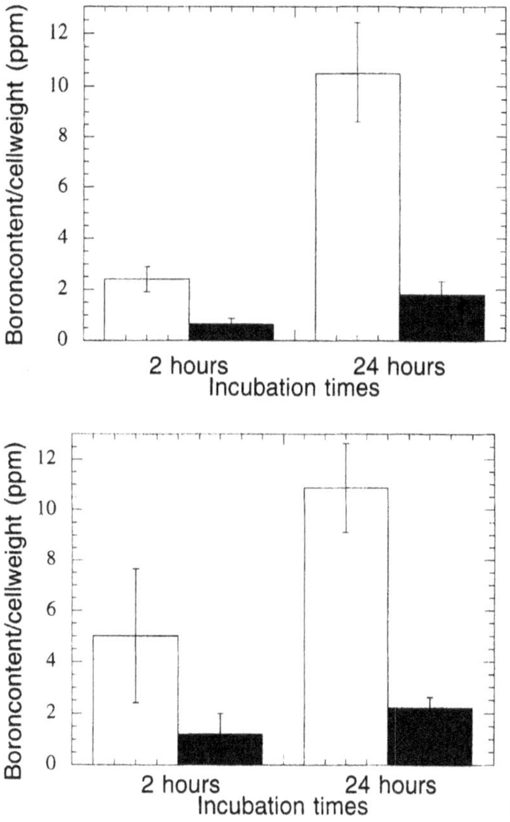

A

B

Figure 3. A. Boron concentration in washed cell samples incubated with free WSP-1 and liposomal WSP-1. B. Boron concentration in unwashed cell samples incubated with free WSP-1 and liposomal WSP-1 (□ Free WSA-1; ■ WSP-1 in stabilized liposomes).

The uptake level was generally low, the only distinct fluorescence exhibited was by free WSP-1. The fact that the liposomes were able to penetrate spheroids is good since we want the liposomes to penetrate tumors.

3.3. Toxicity Test

The clonogenic survival test showed (Figure 5) that after 24 h incubation the liposome encapsulated compounds did exhibit a less toxic behavior than the free compounds. The most toxic compound WSP-1 exhibited a large decrease in survival even at low concentrations. This is to be compared to WSP-1 in stabilized liposomes, which showed virtually no decrease in survival even at the highest concentration. Between WSA-1 and WSA-1 in liposomes there was a tendency for the free substance to be more toxic though the difference is not as large as for the WSP-1 samples.

The clonogenic survival toxicity results show that the toxicity decreases if the substance is encapsulated in a liposome. Both liposomal compounds have very low uptake in both monolayer cells and spheroids and are the least toxic compounds. These results correlate with other results, which are the higher uptake the more toxicity.[10] The result is

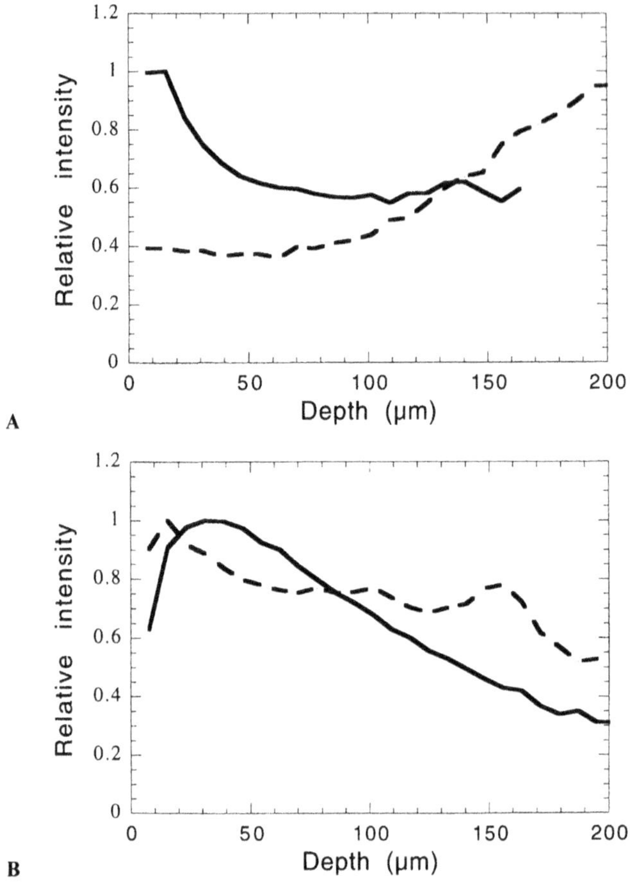

Figure 4. A. Penetration in spheroids of WSA-1 and WSA-1 in stabilized liposomes. B. Penetration in spheroids of WSP-1 and WSP-1 in stabilized liposomes (—— Free WSA-1; ------ WSA-1 in stabilized liposomes).

also very promising since one of the main goals with liposomal drug delivery is to reduce the toxicity of the drug in circulation.

4. CONCLUSION

Since the stabilized liposomes loaded with boronated DNA-intercalating compounds are to be used in two-step targeting the unspecific uptake should be low. Since we want the liposomes to enter the cells via receptor mediated endocytosis, stability and low toxicity are essential to keep healthy tissue undamaged. It is also very important that the liposomes are able to penetrate into spheroids since they are to be used in hunting tumors. The experiments conducted in this report have tried to show that the loaded stabilized liposomes exhibit these properties.

ACKNOWLEDGMENTS

This work was supported by the Swedish Cancer Society.

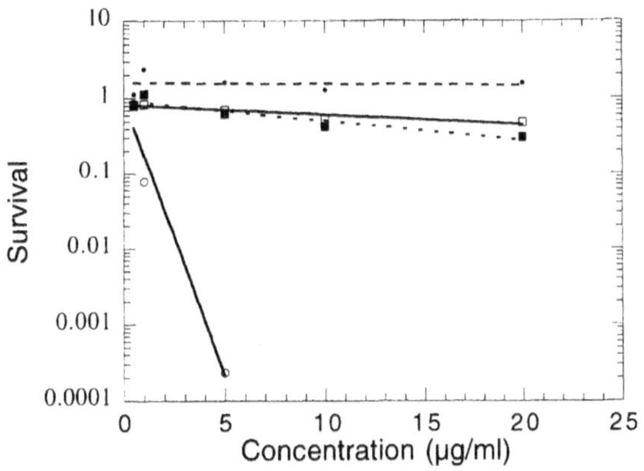

Figure 5. Survival as a function of concentration of the different compounds (—o— Free WSP-1; —□— WSP-1 in stabilized liposomes; --■-- Free WSA-1; ----WSA-1 in stabilized liposomes).

REFERENCES

1. K. Shelly, D.A. Feakes, M.F. Hawthorne, P.G. Schmidt, T.A. Krisch, and W.F. Bauer. Model studies directed toward the boron neutron-capture therapy of cancer: Boron delivery to murine tumors with liposomes. *Proc. Natl. Acad. Sci. USA*, 89:9039–9043, 1992.
2. D. Feakes, K. Shelly, and M.F. Selective boron delivery to murine tumors by lipophilic species incorporated in the membranes of unilamellar liposomes. *Proc. Natl. Acad. Sci. USA*. 92:1367–1370, 1995.
3. M.F. Hawthorne, and K. Shelly. Liposomes as drug delivery vehicles for boron agents. *Journal of Neuro-Oncology*. 33:53–58, 1997.
4. M. Woodle. Sterically stabilized liposome therapeutics. *Advanced Drug Delivery Reviews*, 16:249–265.
5. T. Allen, E. Brandeis, C. Hansen, G. Kao, and S. Zalipsky. A new strategy for attachment of antibodies to sterically stabilized liposomes resulting in efficient targeting to cancer cells. *Biochimia et Biophysica Acta*. 1237:99–108, 1995.
6. T. Hartman, and J. Carlsson. Radiation dose heterogeneity in receptor and antigen mediated boron neutron capture therapy. *Radiotherapy and Oncology*. 31:61–75, 1993.
7. H. Ghaneolhosseini. Boron Neutron Capture Therapy, Synthesis of Boronated Amines- and DNA Intercalating Agents for Potential Use in Cancer Therapy. Acta Universitatis Uppsaliensis, *Comprehensive summaries of Uppsala Dissertations from the Faculty of Science and Technology*, 356:62 pp. ISBN 91-554-4181-5, 1998.
8. L. Gedda. Boron Neutron Capture Therapy, Preclinical Studies of Compounds for Tumour Targeting. Acta Universitatis Uppsaliensis, *Comprehensive Summaries of Dissertations from the Faculty of Medicine*, 724:66 pp ISBN 91-554-4090-8. 1997.
9. P. Lindström, P. Olsson, J. Malmqvist, J. Pettersson, P. Lemmen, B. Werner, S. Sjöberg, Å. Olin, and J. Carlsson. New carborane-based compounds for boron neutron capture therapy: binding and toxicity of ANC-1, DAC-1 and B-Et-11-OMe in cultured human glioma and mouse melanoma cells. *Anti-Cancer Drugs*. 5:43–52, 1994.
10. H. Nemoto, J. Cai, and N. Asao. Synthesis and biological properties of water-soluble *p*-boronophenyl alanine derivatives. Relationship between water solubility, cytotoxicity, and cellular uptake. *J. Med. Chem.* 38:1673–1678, 1995.

APPLICATION OF ^{10}B ENTRAPPED LIPOSOME TO BORON NEUTRON CAPTURE THERAPY FOR PANCREATIC CANCER MODEL *IN VIVO*

Hironobu Yanagie[1], Koichi Ogura[2], Kazuo Maruyama[3], Yasumasa Nonaka[1], Toshio Matumoto[4], Yoshinori Sakurai[5], Tooru Kobayashi[5], Kouji Ono[5], Masazumi Eriguchi[1], and Hisao Kobayashi[6]

[1]Department of Surgery
Institute of Medical Science
University of Tokyo
4-6-1 Shiroganedai, Minato-ku, Tokyo 108-8639
[2]Physical Science Laboratories
College of Industrial Technology
Nihon University
1-2-1 Izumi-cho, Narashino, Chiba 275
[3]Department of Pharmaceutics
Teikyo University
1091 Sagamiko, Kanagawa 199-0195
[4]Metabolism and Pharmacokinetics Group
Omiya Research Laboratory
Nikken Chemicals Co. Ltd.
1-346 Kitabukuro-cho, Omiya, Saitama 33
[5]Kyoto University Research Reactor Institute
Kumatori, Osaka 590-0494
[6]Institute for Atomic Energy
Rikkyo University
2-5-1 Nagasaka, Yokosuka, Kanagawa, 240-01, Japan

1. INTRODUCTION

Liposomes have been investigated extensively as carriers for anticancer drugs in attempts to direct active agents to tumors or to protect sensitive tissues from toxicity.[1] It is

Frontiers in Neutron Capture Therapy, edited by Hawthorne *et al.*
Kluwer Academic / Plenum Publishers, New York, 2001.

necessary to accumulate ^{10}B compounds selectively in tumor cells for selective cyotototoxicity of tumor cells without affecting adjacent healthy cells in effective boron neutron capture therapy (BNCT). Liposomes are useful drug carriers, and it is possible to carry a large amount of ^{10}B compound in a liposome, which can be delivered to a tumor cells. We have reported that ^{10}B atoms delivered by immuno-liposomes are cytotoxic to human pancreatic carcinoma cells (AsPC-1) with thermal neutron irradiation *in vitro.*[2] The intratumoral injection of boronated immunoliposomes can increase the retention of ^{10}B atoms in tumor cells, causing tumor growth suppression *in vivo* under thermal neutron irradiation.[3]

Multilamellar vesicle immunoliposomes are easily phagocytized by the reticuloendotherial system (RES), so it is very difficult to accumulate the liposomal contents in the target cancers *in vivo*. Forssen et al. have reported that small and rigid unilamellar vesicles consisting of highly purified distearoyl-phosphatidylcholine and cholesterol can produce a 10-fold increased delivery of daunorubicin to solid tumors *in vivo*.[4]

In this study, we prepared the liposomes as the effective ^{10}B carrier. We performed a technique of neutron capture autoradiography using track etch detectors (SSNTDs) to qualitatively determine the ^{10}B biodistribution in whole body samples of mice. The tumor contained high level of boron until 12 hours after intravenous injection of ^{10}B-liposome solution by SSNTDs. Intravenous injection of liposome solution inhibited tumor cell growth with thermal neutron irradiation *in vivo.*

2. MATERIALS AND METHODS

2.1. Chemicals

Sodium salt of undecahydro-mercaptocloso-dodecaborate ($Na_2{}^{10}B_{12}H_{11}SH$: BSH) was obtained by Wako Chemical Co. Ltd. (Tokyo, Japan).

2.2. Target Tumor Cells and Mice

The human pancreatic carcinoma cell line AsPC-1, which produces carcinoembryonic antigen (CEA) was obtained from Dainihon Seiyaku Co. Ltd. (Osaka, Japan). AsPC-1 was maintained in RPMI 1640 medium (Hazleton Biologics, INC, Kansas, USA) supplemented with 10% fetal calf serum (Cell Culture Laboratories, Ohio, USA) and $100\,\mu g\,ml^{-1}$ kanamycin. All cultures were incubated in high moisture air with 5% CO_2 at 37 °C. The medium was changed routinely three times a week.

Male BALB/c nu/nu mice were obtained from Nihon SLC (Shizuoka, Japan) and used at 6 to 7 weeks of age. In each experiment, mice of similar age and weight were selected. Mice were housed in plastic cages and maintained in an air-conditioned room. The procedures for tumor implantation and sacrifice of the animals were in accordance with approved guidelines of the Institutions Animal Ethics Committee.

2.3. Preparation of Liposomes Containing ^{10}B-compound

Freeze dried pre-formed liposome (COATSOME EL-C-01) was obtained by Nichiyu liposome Co. Ltd. (Tokyo, Japan). The COATSOME EL-C-01 is composed with L-α-dipalmitoyl phosphatidylcholine (DPPC) (26 μmoles), cholesterol (CH) (20 μmoles), and stearylamine (SA) (4 μmoles). One-half ml of 100 mM ^{10}B-compound (BSH) solution was added to the COATSOME EL-C-01, and small-unilamellar vesicles sized to

125 ± 31 nm in mean diameter were prepared by rehydration with vortex dispersion. The liposome was centrifuged at 200,000 g for 20 min at 4 °C (Hitachi CS120, S100AT5 rotor) to remove free BSH, and then resuspended in phosphate buffered saline (pH 7.4). Liposome size was measured with a dynamic laser light scattering spectrophotometer (ELS-700, Otsuka Electronics, Tokyo). The liposome solution containing BSH is called ^{10}B-liposome solution hereafter.

2.4. Experimental Procedure

2.4.1. Preparation of Sliced Mice Samples. AsPC-1 cells (1×10^7) were injected subcutaneously into the back of male BALB/c nu/nu mice. At 10–14 days after injection, when the estimated tumor weight reached 100–300 mg, 0.2 ml of ^{10}B entrapped COATSOME EL-C-01 solution were injected intravenously. The mice were sacrificed 3, 6 and 12 hours after injection of the ^{10}B entrapped COATSOME EL-C-01 solution, and the mice were frozen at −60 °C. Subsequently, the frozen mice were cut saggitally into 40 µm thick sections putting on the mending tape, and freezed dried at −20 °C in 2 weeks, and air dried in one more week. Thus, the dried samples were prepared.

2.4.2. Preparation of Standard Samples. Boron-containing standard samples were also prepared using drying filter paper sheets wetted by BSH solutions of five different ^{10}B concentrations of relative scale; 1, 10^{-1}, 10^{-2}, 10^{-3} and 10^{-4}, respectively, where the concentration of 1 corresponds to the concentration 1.58×10^4 ppm of ^{10}B compounds. The ^{10}B concentrations of ^{10}B-liposome solution and ^{10}BSH solutions were determined by the prompt γ-ray spectrometry at the Research Reactor Institute, Kyoto University.[5]

2.5. Neutron Irradiations: Experiments for Neutron Capture Autoradiography

The whole-body sections are put in close contact with the CR 39 track etch detector plates (HARZLAS, Fukubi Chemical Industry) using thin adhesive tape. The set of mouse samples and standard samples were simultaneously exposed. For visible observation of track image, thermal neutron flux of an order of 10^{12} n/cm^2 was required and the samples were attached onto the surface near the center hole of the thermal column output face (1.2 m × 1.2 m) in the TRIGA-II reactor of the Rikkyo University (RUR). The neutron flux has been measured to be 1×10^8 n/(cm^2·s) and Cd ratio was 6,400.

2.6. Experiments for BNCT

Three hours after intravenous injection of ^{10}B-liposome solution (3,200 µg of ^{10}B), tumor bearing mice were irradiated for 37 min with 9 x 10^8 n/(cm^2·s) thermal neutrons (total fluence: 2×10^{12} n/cm^2) at the TRIGA-II atomic reactor of Rikkyo University. After irradiation, the effect of BNCT was calculated on the basis of tumor volume and morphological findings of the tumors at 3-day intervals.

2.7. Etching Procedure

For neutron capture alpha-autoradiograph (NCAR) including proton tracks produced by ^{14}N (n, p) reaction, where ^{14}N is the biologically abundant nuclide, the

CR-39 detector plates were etched in a 7 M NaOH solution at 70 °C for 2 hours to reveal tracks. If the fast neutron component is significantly present in the beam, α-autoradiographic image is also contaminated by the recoil protons due to the knock on hydrogen in the tissue and the detector plate itself. The track registration sensitivity of CR-39 detector depends on the restricted energy loss (REL) of incident particles. This means that the track sizes of α-particles are larger than that of protons. Therefore, we can discriminate α-tracks from the background proton tracks by means of a track size analysis using an image analysis system consisting of a CCD camera, a personal computer and an optical microscope.

3. RESULTS AND DISCUSSION

Clearance of liposomes from circulation *in vivo* depends on their rate of uptake by reticuloendothelial system (RES) cells in the liver and spleen. It is known that liposomes given small size and rigid composition is essentially stable in the circulation and show the prolonged circulation time and the relatively low RES uptake. More importantly, these rigid liposomes (100–200 nm in mean diameter) are able to traverse the endothelium of blood vessels in solid tumors and are extravasated into the interstitial spaces, resulting in significantly greater accumulation. Generally, the capillary permeability of the endothelial barrier in newly vascularized tumors is significantly greater than that of healthy tissue.[6,7] Under such conditions, liposomes of small size and longer half-life in the circulation will have more opportunity to traverse the tumor capillaries.

One of extremely important concerns in BNCT is the ^{10}B delivery system, for which liposomes are very attractive and interested. To escape this phagocytosis of RES, we have tried to prepare small unilamellar boronated liposomes which is composed with DPPC/CH/SA and sized to 125 ± 31 nm in mean diameter.

3.1. Track Etch Imaging

Figure 1 shows a whole-body sections of neutron capture radiograph from a set of AsPC-1 pancreatic cancer-bearing mice that have been intravenously given the injection of 3,200 μg of ^{10}B-liposome solution. The slices of sacrificed and frozen mice were prepared 3 and 12 hours after the injection. The NCAR image of the prepared slice and of the standard samples are shown in this figure, where the samples were exposed to 1.8×10^{12} n/cm^2 at the thermal column output face and the detector plates were etched in 7 M NaOH solution. It is readily apparent that the tumor contains high level of boron until 12 hours after injection. There are also areas within liver which contain high level of boron after 3 hour injection, however the concentration of boron is decreasing as the course of time. The visible image shows the ^{10}B rich regions at tumor site and at other organs, especially liver with different ^{10}B concentration depending on the time required after the injection of ^{10}B-liposome solution. The image can describe also organs in the whole body section by means of proton tracks with different contrast in comparison to the α track image.

3.2. Growth Inhibition of Tumor Treated with ^{10}B-liposome Solution

The effect of ^{10}B-liposome solution on AsPC-1 cell growth was examined with thermal neutron irradiation. The amount of ^{10}B in the liposome solution was 1.58×10^4 ppm of ^{10}B by the prompt γ-ray spectrometry. As shown in Figure 2, tumor

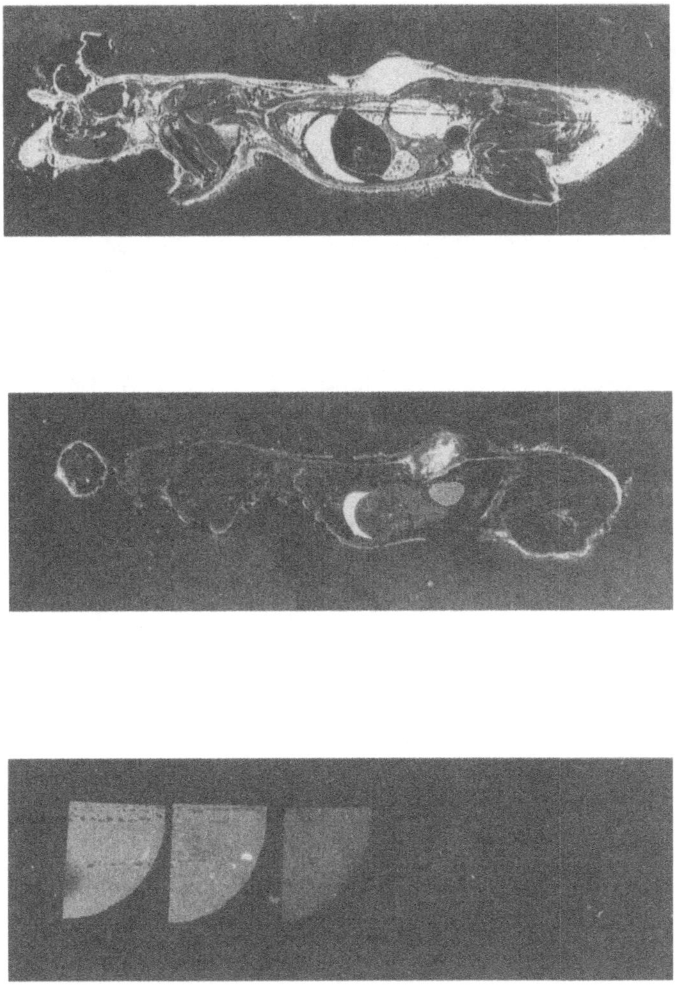

Figure 1. Neutron capture α-autoradiography of whole-body boron distribution after intravenous injection of ^{10}B-liposome solution in tumor bearing mouse. Neutron capture autoradiograph was prepared from the whole-body section (40 μm thick) of a pancreatic cancer-bearing mouse. The mouse was sacrificed at (a) 3 hours and (b) 12 hours after the ^{10}B-liposome solution was injected. The neutron fluence needed to produce this image was 1.8×10^{12} n/cm^2. The bright areas correspond to areas of high boron concentration. The standard samples filter paper sheets with different ^{10}B concentrations were simultaneously irradiated to the thermal neutrons.

growth after BNCT using intravenous injection of ^{10}B-liposome solution had been suppressed and the tumor had nor regrown and became necrotic. The control groups treated with thrmal neutron irradiation, or intravenous injection of ^{10}B-liposome solution showed no inhibition of growth, respectively. We found that the ^{10}B-liposome have the possibility of retention to the tumor cells and providing sufficient ^{10}B atoms for the BNCT. Transfection efficiency of liposome systems can be significantly improved when complexed with a ligand such as transferrin.[8] The presence of the ligand facilitates the entry of ^{10}B-compound into cells through receptor-mediated endocytosis. Maruyama *et al.* have demonstrated that polyethylene glycol binding liposomes of small size (about 100 nm mean diameter) and rigid lipid composition showed significantly greater accumulation in solid tumor.[9] We try to prepare boronated polyethylene glycol binding BSA

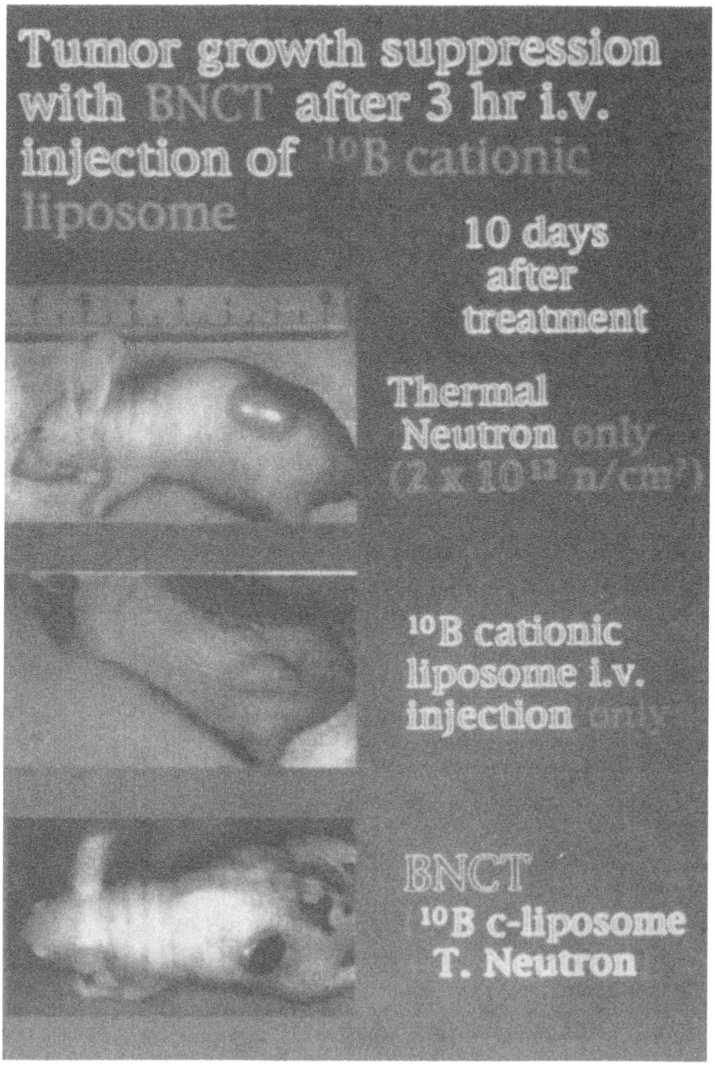

Figure 2. Three hours intravenous injection of ^{10}B-liposome solution, tumor bearing mice were irradiated with thermal neutron (total fluence: 2×10^{12} n/cm^2) at the TRIGA-II atomic reactor in Rikkyo University.

and immunoliposomes (stealth immunoliposomes), and to perform BNCT experiments *in vitro* and *in vivo*.[10] Experiments with these newly ^{10}B delivery systems are in progress for application of clinical BNCT trials.

4. CONCLUSION

Boron-10 accumulation in the tumor was clearly visible using the neutron capture autoradiography. The accurate measurement of ^{10}B distributions in biological samples with sensitivity in the ppm range is essential for evaluating the potential usefulness of various boron-containing compounds for BNCT. We also continue to study the microdosimetry of ^{10}B atoms using the micro-autoradiography of ^{10}B distribution in tumor cells. We will be able to apply these new technique for effective BNCT for cancer.

BNCT with intravenous injection of small and rigid liposome with a high content of ^{10}BSH has the ability to destroy malignant cells at the edge of the tumor mass. Experiments using systemic injection of intelligent ^{10}B delivery systems are in progress for selective targeting of ^{10}B atoms in tumors for the application of clinical BNCT.

REFERENCES

1. T. Tomita, M. Watanabe, T. Takahashi, K. Kumai, T. Tadakuma, and T. Yasuda, Temperature-sensitive release of adriamycin, an amphiphilic antitumor agent, from dipalmitoyl-phosphatidylcholine-cholesterol liposomes, Biochim. Biophys. Acta, 978, 185–190, 1989.
2. H. Yanagie, T. Tomita, H. Kobayashi, Y. Fujii, T. Takahashi, K. Hasumi, H. Nariuchi, and M. Sekiguchi, Application of boronated anti-CEA immunoliposomes to tumour cell growth inhibition in *in vitro* boron neutron capture therapy model, Br. J. Cancer, 63, 522–526, 1991.
3. H. Yanagie, T. Tomita, H. Kobayashi, Y. Fujii, Y. Nonaka, Y. Saegusa, K. Hasumi, M. Eriguchi, T. Kobayashi, and K. Ono, Inhibition of human pancreatic cancer growth in nude mice by boron neutron capture therapy, Br. J. Cancer, 75, 660–665, 1997.
4. E.A. Forssen, D.M. Coulter, and R.T. Proffitt, Selective *in vivo* localization of Daunorubicin small unilamellar vesicles in solid tumor, Cancer Res., 52, 3255–3261, 1992.
5. T. Kobayashi, and K. Kanda, Microanalysis system of ppm-order ^{10}B concentrations in tissue for boron neutron capture therapy by prompt gamma-ray spectroscopy, Nucl. Instr. Methods, 204, 525–531, 1983.
6. R.K. Jain, and L.E. Gerlowski, Extravascular transport in normal and tumor tissue. Crit. Rev. Oncol. Hematol., 5, 115–121, 1986.
7. H.F. Dvorak, J.A. Nagy, J.T. Dvorak, and A.M. Dvorak, Identification and characterization of the blood vessels of solid tumors that are leaky to circulating macromolecules. Am. J. Phathol., 133, 95–109, 1988.
8. D. Yu, A. Matin, W. Xia, F. Sorgi, L. Huang, and M.C. Hung, Liposome-mediated *in vivo* E1A gene transfer suppressed dissemination of ovarian cancer cells that overexpress HER-2/neu, Oncogene 11, 1383–1388, 1995.
9. K. Maruyama, T. Takizawa, T. Yuda, S.J. Kennel, L. Huang, and M. Iwatsuru. Targetability of novel immunoliposomes modified with amphipathic poly(ethylene glycol)s conjugated at their distal terminals to monoclonal antibodies. Biochim. Biophys. Acta 1234, 74–80, 1995.
10. H. Yanagie, T. Sato, H. Kobayashi, and M. Eriguchi, Application of boronated polyethyleneglycol binding-BSA to boron neutron capture therapy, Proceedings of the 5th International Conference on Applications of Nuclear Techniques, Washington, USA, SPIE, 2867, 101–104, 1997.

LDL RECEPTOR STATUS OF HUMAN GLIOBLASTOMA CELLS

E. A. Blakely[1], K. A. Bjornstad[1], L. Maletinska[1], D. E. Callahan[1], L. J. Knoff[1], D. F. Deen[2], and T. M. Forte[1]

[1]Life Sciences Division
Lawrence Berkeley National Laboratory
Berkeley, CA 94720
[2]Brain Tumor Research Center
University of California
San Francisco, CA 94143-0520

1. INTRODUCTION

Porphyrins are known to associate with plasma lipoproteins, particularly low density lipoproteins (LDLs), and are thus likely to be taken up into cells by the high affinity saturable LDL receptor mechanism.[1] We have previously reported that the boronated protoporphyrin, BOPP, is incorporated into the human glioblastoma cell line, SF 767 (derived from a glioblastoma multiforme biopsy specimen obtained from a grade 4 malignant glioma), and requires the presence of lipoproteins for maximum uptake into cells.[2] We concluded from these studies that, like other porphyrins, BOPP forms a complex with LDL before it is taken up into the SF 767 cells via the LDL receptor; however, the presence of saturable LDL receptors on SF 767 cells has not been previously demonstrated. We have therefore investigated whether the LDL receptor was present on exponentially growing SF 767 cells. We report here our initial results indicating relatively high levels of LDL-receptors per cell (over 200,000) from the SF-767 line.

2. MATERIALS AND METHODS

The glioblastoma cell line, SF-767, previously described by Basu *et al.*[3] was used. Classical LDL binding studies were carried out at 4 °C with ^{125}I-LDL in the presence and absence of a 50-fold excess of unlabeled LDL.[4] The latter provides information on the non-specific binding of LDL. To assure that LDL receptors were maximally upregulated, the cells were grown for 24 hr in medium containing lipoprotein deficient serum. The difference between total ^{125}I-LDL binding and non-specific binding provides information

Frontiers in Neutron Capture Therapy, edited by Hawthorne *et al.*
Kluwer Academic / Plenum Publishers, New York, 2001.

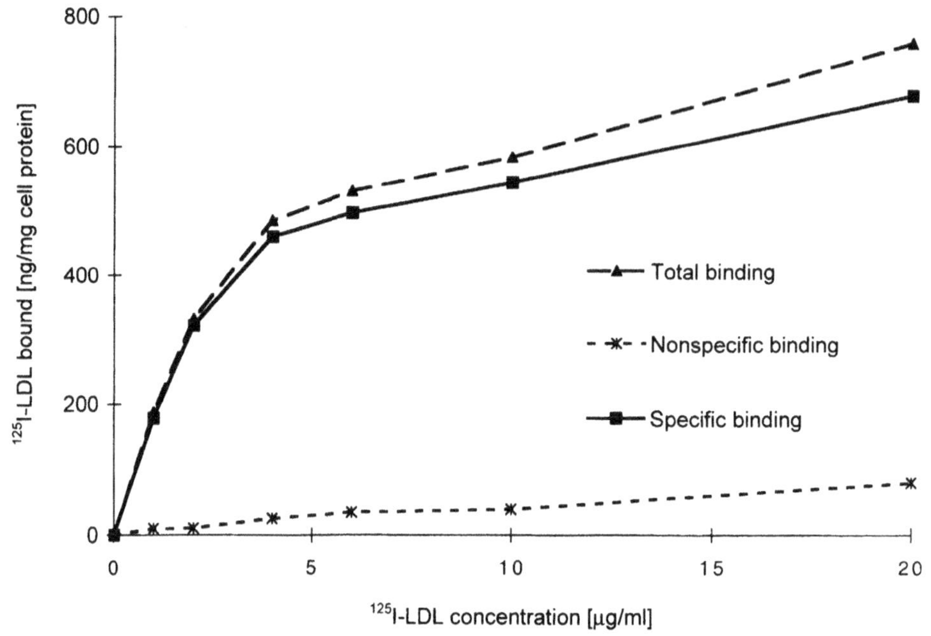

Figure 1. ^{125}I-LDL binding to SF 767 cells.

on the saturable, specific binding of LDL. LDLs were labeled by the Iodo-Bead method using the instructions provided by the manufacturer (Pierce, Rockford, IL).

3. RESULTS

Figure 1 is a representative example of four binding studies showing binding of LDL to SF 767 cells at 4 °C. The figure clearly demonstrates that there is a saturable and specific binding component where saturation occurs at approximately 760 ng ^{125}I-LDL/mg cell protein.

The Scatchard plot analysis shown in Figure 2 indicates high affinity binding of LDL to its receptor, where the K_d (binding affinity) is 4.9 nM. From these data we calculate that at 4 °C there are approximately 210,000 receptors per cell. To place the latter data in perspective with receptor number on other cell types, we have summarized the K_d and B_{max} (maximal LDL binding) data from several other cell types in Table 1. This information suggests that, compared to fibroblasts and the human hepatoma line, Hep G2, SF 767 cells have a greater B_{max}; however, the K_d is similar for fibroblasts and SF 767 cells and this affinity is greater than that found in Hep G2 cells. Overall, the data suggest that this glioblastoma cell line has large numbers of LDL receptors that provide a potential mechanism for targeting therapeutic agents to these cells.

4. DISCUSSION AND CONCLUSIONS

It is well known that white blood cells from patients with monocytic or myelomonocytic leukemia have markedly increased LDL degradation compared with cells from

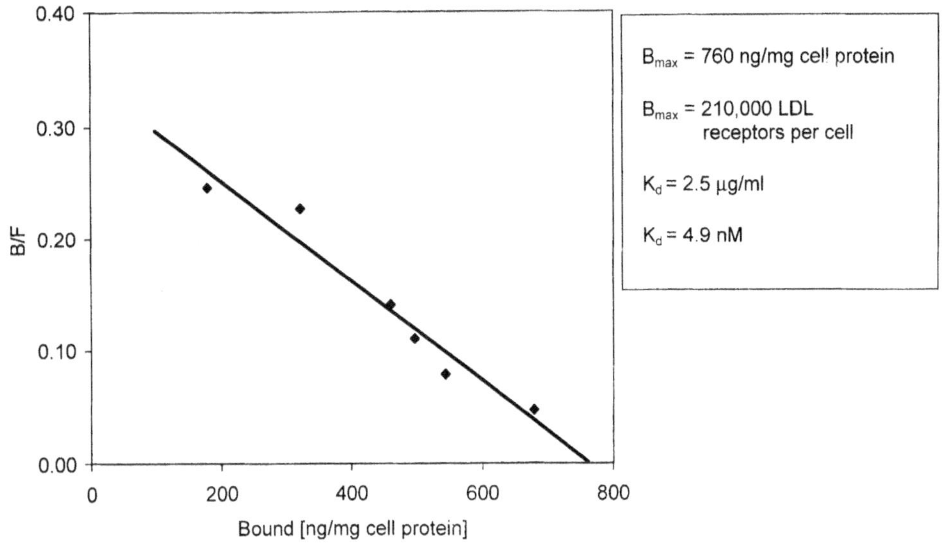

Figure 2. Scatchard plot analysis of ^{125}I-LDL binding to SF 767 cells.

control subjects.[8] The rapid receptor-mediated degradation of LDL in leukemia subjects is associated with a hypocholesteremic state in these patients. Interestingly, patients with lymphoblastic leukemia do not have a reduction in plasma cholesterol, nor do the white blood cells exhibit increased binding and uptake of LDL. These data suggest that some tumor cells have increased numbers of high affinity LDL receptors while others do not. High numbers of high affinity LDL receptors could be a mechanism for targeting anti-tumor drugs to specific cell types using LDL as the transport vehicle.

We have recently shown that SF 767 cells take up BOPP and LDL and that both components localize in the lysosome which is the compartment for the degradation of internalized LDL.[2] These findings are suggestive that the LDL receptor is the prime mechanism for delivery of BOPP to this glioblastoma cell. The intracellular localization of BOPP and LDL, however, provided no insights into LDL receptor affinity or receptor number. Without the latter information it is difficult to assess whether the use of LDL receptors for targeting anti-tumor pharmaceuticals (e.g., boronated and/or chemotherapy agents) is a potentially useful therapeutic strategy as suggested by Laster *et al.*[9] Our results indicate that the SF 767 cell line has relatively high numbers of high affinity LDL receptors. If elevated LDL receptors are a common feature of glioblastoma cells, increasing LDL receptor number by pharmacological means and

Table 1. LDL receptor data reported in the literature (binding at 4 °C)

Human cell type/tissue	B_{max} (ng/mg cell protein)	B_{max} (LDL receptors per cell)	K_d (nM)	Reference
Adult fibroblast/skin	45	7,500	4.5	Brown & Goldstein[5]
Neonate fibroblast/skin	120	100,000	2.8	Pitas et al.[6]
HepG2 line carcinoma/liver	520	n.d.	30.0	Hara, McCall & Forte[7]
Adult SF-767 line glioblastoma/brain	760	210,000	4.9	Present study

n.d.—Not determined.

devising boronated compounds that are recognized by the LDL receptor may be an efficient mechanism for targeting glioblastoma cells for BNCT. A comprehensive survey of LDL receptors on other glioblastoma cell lines is currently underway.

ACKNOWLEDGMENTS

This work was supported by the Director, Office of Energy Research, Office of Laboratory Policy and Infrastructure Management, of the U.S. Department of Energy under contract No. DE-AC03-76SF00098 and by NIH HL18574 (TMF).

REFERENCES

1. A.J. Barel, G. Jori, A. Perin, P. Romandini, A. Pagnan, and S. Biffanti, Role of high-, low-, and very low-density lipoproteins in the transport and tumor-delivery of hematoporphyrin *in vivo*, *Cancer Letters*, 32:145–150, 1986.
2. D.E. Callahan, T.M. Forte, S.M.J. Afzal, D.F. Deen, S.B. Kahl, K.A. Bjornstad, W.L. Bauer, and E.A. Blakely, Boronated protoporphyin (BOPP): Localization in lysosomes of the human glioma cell line SF-767 with uptake modulated by lipoprotein levels, *Int. J. Radiat. Oncol. Biol. Phys.*, 45:761–777, 1999.
3. H.S. Basu, M. Pellarin, B.G. Feuerstein, A. Skirahata, K. Samejima, D.F. Deen, and L.J. Marton, Interaction of polyamine analogue, 1,19-bis-(ethylamino)-5,10,15-triazanonadecane (BE-4-4-4-4), with DNA and effect on growth, survival, and polyamine levels in seven human brain tumor cell lines, *Cancer Res.*, 53:3948–3955, 1993.
4. T.L. Innerarity, R.E. Pitas, and R.W. Mahley, Lipoprotein-receptor interactions, *Methods Enzymol.*, 129:542–565, 1986.
5. M.S. Brown, and J.L. Goldstein, Familial hypercholesterolemia: Defective binding of lipoproteins to cultured fibroblasts associated with impaired regulation of 3-Hydroxy-3-Methylglutaryl coenzyme A reductase activity, *Proc. Natl. Acad. Sci. USA*, 71:788–792, 1974.
6. R.E. Pitas, T.L. Innerarity, K. Arnold, and R.W. Mahley, Rate and equilibrium constants for binding of apo-E HDL_c (a cholesterol-induced lipoprotein) and low density lipoproteins to human fibroblasts: Evidence for multiple receptor binding of apo-E HDL_c, *Proc. Natl. Acad. Sci. USA* 76:231–2315, 1979.
7. S. Hara, M.R. McCall, and T.M. Forte, Re-uptake of nascent low-density lipoproteins by HEpG2 cells, *Biochimica et Biophysica Acta*, 1168:199–204. 1993.
8. S. Vitols, G. Gahrton, A. Ost, and C. Peterson, Elevated low density lipoprotein receptor activity in leukemic cells with monocytic differentiation, *Blood*, 83:1186–1193, 1984.
9. B.H. Laster, S.B. Kahl, E.A. Popenoe, D.W. Pate, and R.G. Fairchild, Biological efficacy of boronated low-density lipoprotein for boron neutron capture therapy as measured in cell culture, *Cancer Res.*, 51:4588–4593, 1991.

THE ADVANTAGE OF PORPHYRIN FOR BNCT FROM THE POINT OF VIEW OF CELL CYCLE

Yasushi Shibata[1], Akira Matsumura[1], Fumiyo Yoshida[2], Tetsuya Yamamoto[1], Kei Nakai[1], Tadao Nose[1], Isao Sakata[3], and Susumu Nakajima[4]

[1]Department of Neurosurgery
[2]College of Medical Technology
University of Tsukuba
Tsukuba, Ibaraki, 305, Japan
[3]Toyo Hakka Co. Ltd.
Okayama, 719-03, Japan
[4]Department of Operative Surgery
Asahikawa Medical College
Asahikawa, Hokkaido, 078, Japan

1. INTRODUCTION

The potential for using boronated porphyrins for BNCT have been reported by some groups.[1-5] Our recent studies have also shown that porphyrin is one of the potential tumor targeting agents for BNCT.[6,7] The tumor specific uptake of porphyrin is well known, however the uptake of porphyrin regarding the cell cycle has not been reported. We examined tumor cell uptake of porphyrin in the relation with cell cycle using flow cytometry system in *in vitro* glioma cell lines.

2. MATERIALS AND METHOD

2.1. Cell Preparation

Four established brain tumor cell lines(A-172, U-251, 9L, ONS) were cultured in RPMI 1640 medium or Earle's MEM solution(Gibco, New York, USA) at 37°C in an atmosphere of 5% CO_2 in air. A-172 and U-251 are human glioma cell lines. 9L is rat gliosarcoma cell line. ONS is human medulloblastoma cell line. Cells were grown in petri dishes and detached with 0.25% trypsin solution. The cells were washed with phosphate buffered saline (PBS). Each cell suspension were centrifuged and pellets

Frontiers in Neutron Capture Therapy, edited by Hawthorne *et al.*
Kluwer Academic / Plenum Publishers, New York, 2001.

including about 1 × 10^6 cells were cultured in RPMI 1640 medium with the porphyrin ATX-S10-Na (13,17-bis(1-carboxypropionyl) carbamoylethyl-8-ethenyl-2-hydroxy-3-hydroxyiminoethylidene-2,7,12, 18-tetramethyl-porphyrin sodium, Toyo Hakka Co. Ltd., Okayama, Japan) at the concentration of 25 μg/ml of RPMI 1640 solution for 2 hours at 37 °C in an atmosphere of 5% CO_2 in air. From the previous studies, we have reached the results these concentration and incubation time is enough to obtain clear results. In order to be representative of the clinical situation, drug concentration and incubation time should be small enough. And each tumor cell lines have short doubling times. So short incubation time should be selected to evaluate each points of cell cycles. The molecular formula of ATX-S10-Na is $C_{42}H_{41}N_7O_{12}Na_4$, and its molecular weight is 927.79. Each cell suspension was centrifuged and the pellets were fixed with ice-cooled 70% ethanol and stored at 4 °C for several hours. Fixed cells were then stained with chromomycin A3 at the concentration of 20 g/ml for 1 hour at 4 °C. The cells were centrifuged and each pellet was suspended in PBS.

2.2. Flow Cytometric Analysis of the Tumor Cells

The 1 × 10^5 cells of each cell lines was analyzed with flow cytometry. The flow cytometric data acquisition was conducted using FACSort (Becton Dickinson). The activated wave length of chromomycin A3 is 400–460 nm (peak: 420 nm). The fluorescent wave length of chromomycin A3 is 490–610 nm (peak: 560 nm). The activated wave length of ATX-S10 is 400–417nm and the peak fluorescent wave length of ATX-S10 is 677 nm.[8] DNA content stained with chromomycin A3 was measured with FL2 range. The fluorescence of ATX-S10 was measured with FL3 range. Data were analyzed with Lysis 2 software. The cells in G0/G1 and G2/M phase were identified from the histogram of FL2.

3. RESULTS

The study revealed porphyrin uptake in all cells, regardless of the phase of the cell cycle. However those in G0/G1 phase showed moderate uptake of porphyrin and those in the G2/M phase showed higher uptake[9] (Figure 1, reproduced from the reference No. 9 with permission).

4. DISCUSSION

For an effective BNCT, boron uptake into tumor cells is essential. In the research of BNCT, boron uptake has been investigated in relation to whole tumor tissue. However the boron uptake and radiosensitivity of each cell line probably depends on the phase of the cell cycle. So the effect of BNCT should be examined for each cell line in different phase of the cell cycle. Borocaptate sodium(BSH) and Boronophenylalanine(BPA) are 2 major boron compounds used in BNCT. BSH is selectively uptaken by tumor tissue and does not cross the intact blood brain barrier.[10] The uptake of BSH by tumor cells is not related to the cell cycle. BPA, on the other hand, showed selective uptake by actively dividing cells.[11] The tumor control effect of BNCT using BPA was better than that using BSH.[12] This difference might be due to the more uptake of BPA by G2/M phase cells. However, BNCT using BPA does not include therapy to quiescent cell population.[11]

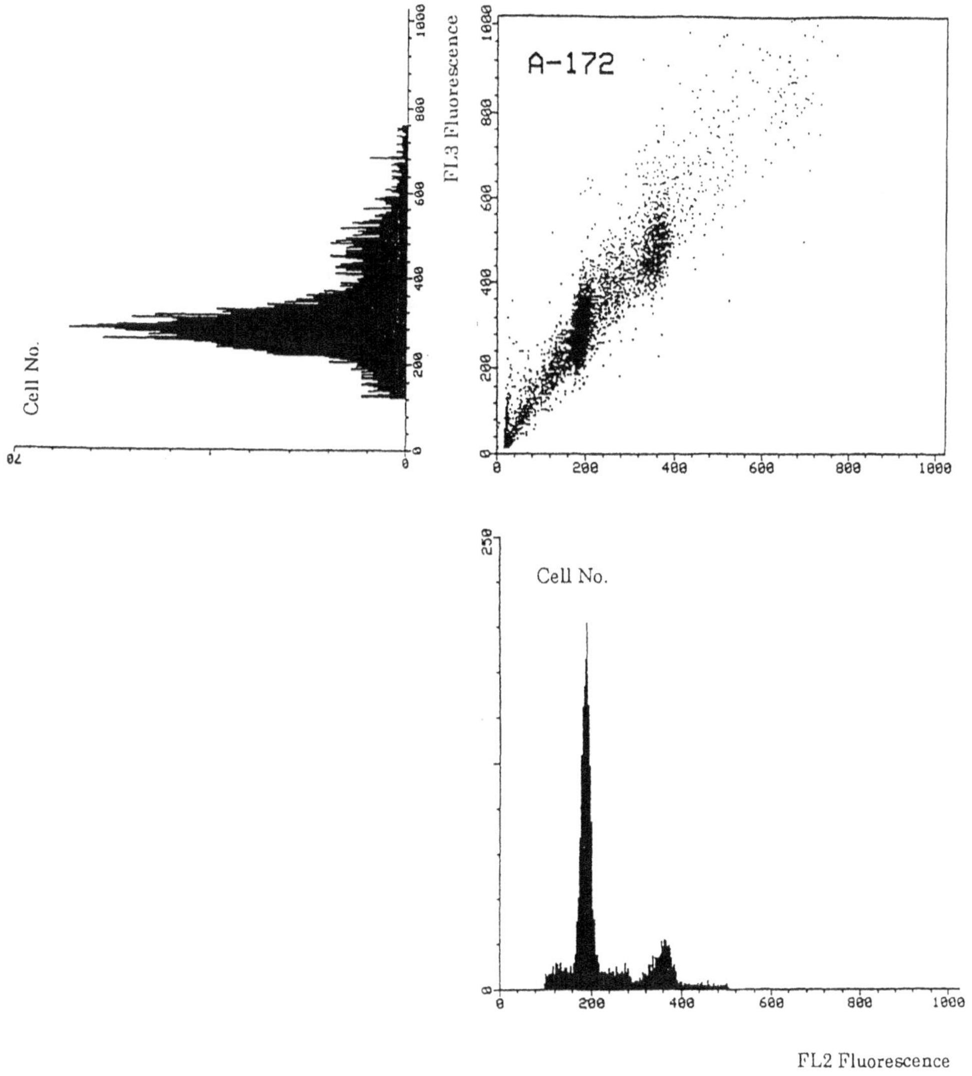

Figure 1. Showed results of A-172 cell in flow cytometry. Upper right figure is dot plot, and horizontal axis is FL2 fluorescence and vertical axis is FL3 fluorescence. Lower right figure is the histogram, and horizontal axis is FL2 fluorescence and vertical axis is cell number. Upper left figure is the histogram, and horizontal axis is cell number and vertical axis is FL2 fluorescence. These figures revealed 2 groups of cells in G0/G1 and G2/M phases and the cells in G2/M phase showed more uptake of the porphyrin.

Cancer therapy requires cytotoxic or cytocidal effects not only on proliferating G2/M cells but also on G0/G1 cells which may enter the active cell cycle. After selective death of cells in cells cycle, quiescent cells enter the proliferating phase, which is called recruitment.[12] BNCT is being performed as non-fractionated radiation therapy. Hence, the target of BNCT should include all cells at rest and in cell cycle. Our data demonstrated the cell cycle independent property of porphyrin uptake by tumor cells. The cells in G0/G1 phase showed moderate uptake of porphyrin and the cells in G2/M phase showed more uptake of porphyrin than the cells in G0/G1 phase. The uptake of porphyrin by each cell was almost proportional to the DNA content in each cell. These results support the therapeutic strategy of BNCT using porphyrin compounds. The targets of BNCT using porphyrin compounds are cells at rest and cells undergoing cell division. On BNCT using porphyrin compounds a more lethal effect is expected for cells in the G2/M phase.

REFERENCES

1. J.S. Hill, S.B. Kahl, A.H. Kaye, S.S. Stylli, M.S. Koo, M.F. Gonzales, N.J. Vardaxis, and C.I. Johnson, Selective tumor uptake of a boronated porphyrin in an animal model of cerebral glioma. Proc. Natl. Acad. Sci. USA, 89:1785–1789, 1992.
2. C.P. Ceberg, A. Brun, S.B. Kahl, M.S. Koo, B. R.R. Persson, and L.G. Salford, A comparative study on the pharmacokinetics and biodistribution of boronated pophyrin(BOPP) and sulfhydryl boron hydride(BSH) in the RG2 rat glioma model. J. Neurosurg 83:86–92, 1995.
3. J.S. Hill, S.B. Kahl, S.S. Stylli, Y. Nakamura, M.S. Koo, and A.H. Kaye, Selective tumor kill of cerebral glioma by photodynamic therapy using a boronated porphyrin photosensitizer, Proc. Natl. Acad. Sci. USA, 92:12126–12130, 1995.
4. M. Miura, P.L. Micca, C.D. Fisher, J.C. Heinrichs, J.A. Donaldson, G.C. Finkel, and D.N. Slatkin, Synthesis of a nickel tetracarboranylphenylporphyrin for boron neutron-capture therapy: biodistribution and toxicity in tumor-bearing mice. Int. J. Cancer, 68:114–119, 1996.
5. M. Miura, P.L. Micca, C.D. Fisher, C.R. Gordon, J.C. Heinrichs, and D.N. Slatkin, Evaluation of carborane-containing porphyrins as tumour targeting agents for boron neutron capture therapy, British J. Radiol, 71:773–781, 1998.
6. Y. Shibata, A. Matsumura, T. Yamamoto, K. Nakagawa, Y. Yoshii, T. Nose, I. Sakata, S. Nakajima, Y. Hayakawa, and K. Ono, Neutron capture therapy with a new boron-porphyrin compound in the rat 9L glioma model. J. Exp. Clin. Cancer Res, 17:285–289, 1998.
7. A. Matsumura, Y. Shibata, T. Yamamoto, F. Yoshida, T. Isobe, K. Nakai, Y. Hayakawa, M. Kiriya, N. Shimojo, K. Ono, I. Sakata, S. Nakajima, M. Okumura, and T. Nose, A new boronated porphyrin (STA-BX909) for neutron capture therapy: an *in vitro* survival assay and *in vivo* tissue uptake study. Cancer Letters, 141:203–209, 1999.
8. S. Nakajima, T. Takemura, and I. Sakata, Acoustic, fluorescent diagnosis of malignant lesions using by HAT-D01 and ATX-S10. SPIE, 2371:495–500, 1995.
9. Y. Shibata, A. Matsumura, F. Yoshida, T. Yamamoto, K. Nakai, T. Nose, I. Sakata, and S. Nakajima, Cell cycle dependency of porphyrin uptake in a glioma cell line. Cacer Letters, 129:77–85, 1998.
10. J.A. Coderre, J.D. Glass, R.G. Fairchild, U. Roy, S. Cohen, and I. Fand, Selective targeting of boronophenylalanine to melanoma in BALB/c mice for neutron capture therapy. Cancer Res., 47:6377–6383, 1987.
11. K. Ono, S. Masunaga, Y. Kinashi, M. Takagaki, M. Akaboshi, T. Kobayashi, and K. Akuta, Radiobiological evidence suggesting heterogeneous microdistribution of boron compounds in tumors: its relation to quiescent cell population and tumor cure in neutron capture therapy. Int. J. Radiation Oncology Biol. Phys., 34:1081–1086, 1996.
12. S. Masunaga, K. Ono, M. Mitsumori, and M. Abe, Alteration of radiosensitivity of quiescent cell populations in solid tumors irradiated with x-rays twice at various intervals. Jpn. J. Cancer Res., 84:1130–1135, 1983.

STRATEGIES TO IMPROVE THE EFFICACY OF NEUTRON CAPTURE THERAPY OF BRAIN TUMORS BY OPTIMIZING DELIVERY OF BORON COMPOUNDS

Rolf F. Barth, Weilian Yang, and Albert H. Soloway

The Ohio State University
Department of Pathology
165 Hamilton Hall
1645 Neil Avenue, Columbus, OH 43210
The Ohio State University
Department of Pharmacy
353 Parks Hall
500 W. 12th Avenue, Columbus, OH 43210

1. FACTORS AFFECTING BORON COMPOUND DELIVERY

As recently reviewed by us,[1,2] considerable effort has been directed towards the design and synthesis of low and high molecular weight boron delivery agents to target brain tumors for neutron capture therapy. Until recently,[3–7] however, little effort has been directed towards the development of strategies to maximize tumor boron uptake and concomitantly to minimize normal brain and blood levels. Greig[8] has reviewed the chemical, pharmacologic, and pharmacokinetic requirements for optimizing drug delivery to brain tumors. We[9] and Chen *et al.*[10] have addressed this subject from the perspective of boron compound delivery. Factors that may influence the delivery of boron compounds to brain tumors include: 1) the plasma concentration profile of the boron containing drug, which depends upon the amount and route of administration, 2) the permeability of the blood-brain-barrier (BBB) to the boron containing compound, and 3) blood flow within the tumor. One major determinant is the lipophilicity of the drug, and two of the chemotherapeutic agents that have been widely used for treating patients with brain tumors, bischloroethylnitrosourea (BCNU) and cyclohexylchloroethylnitrosourea (CCNU), have high octanol/water partition coefficients, and correspondingly high permeability values. In general, a high steady-state blood concentration will maximize brain uptake, while rapid clearance will reduce it, except in the case of intra-arterial drug administration. The intravenous route currently is being used clinically to deliver both

Frontiers in Neutron Capture Therapy, edited by Hawthorne *et al.*
Kluwer Academic / Plenum Publishers, New York, 2001.

sodium borocaptate (BSH)[11] and boronophenylalanine (BPA),[12] and although this is convenient, it may not be ideally suited for delivering low and high molecular weight agents to brain tumors.

1.1. Enhancing Drug Delivery to the Brain

Strategies to enhance the delivery of drugs to brain tumors broadly can be classified as invasive, pharmacologic or physiologic.[13] Invasive techniques that have been used clinically include direct intratumoral injection,[14,15] the implantation of sustained release polymers,[16,17] convection enhanced delivery,[18,19] and hyperosmotic (1.373 m Osmol) mannitol-induced disruption of the BBB.[20,21] Pharmacologic approaches include the use of small, lipid soluble molecules and liposomes with diameters <50 nm, that can traverse the BBB,[13] and the bradykinin agonist, Receptor Mediated Permeablizer-7 (RMP-7 or Cereport™), a synthetic nonapeptide that has been reported to selectively open the BBB within brain tumors.[6,23] Physiologic strategies, developed by Pardridge[13] and his co-workers, include the use of pseudo-nutrients such as insulin like growth factor-1 (IGF-1), cationic antibodies and chimeric peptides, all of which utilize specific membrane transport systems.

2. BREACHING THE BLOOD BRAIN BARRIER

The BBB normally prevents the passage of ionized, water soluble drugs that have molecular weights >180 Da.[8] Even if the main bulk of the brain tumor has a permeable vascular endothelium, the brain around tumor, which contains invasive tumor cells, has decreased permeability to various anti-tumor agents.[19,24] Endothelial cell clefts in microvessels of the tumor may vary widely from normal to abnormal, even within different regions of the same tumor, demonstrating the complexity of the tumor vasculature and the variability of the BBB within tumors.[25,26] Rapoport[20] and Neuwelt[21] have advocated the use of hyperosmotic mannitol-induced BBB-D to enhance the delivery of cytoreductive chemotherapeutic agents in patients with brain tumors. Neuwelt's most recent data suggest that this approach can significantly enhance the survival of patients with primary central nervous system lymphomas,[27] and children and young adults with nonglial primary brain tumors.[28] On the negative side, however, BBB-D combined with the intra-arterial administration of cytoreductive chemo-therapeutic agents has been associated with increased uptake by normal brain and a concomitant increase in neurotoxicity,[29] and it has been suggested that this technique is of limited clinical usefulness.[30] Pharmacodynamic studies have shown that agents with rapid blood clearance, moderate lipophilicity, and low neurotoxicity are more effectively delivered by the intra-arterial route.[31]

3. STUDIES WITH THE F98 RAT GLIOMA

3.1. Optimizing Delivery of BSH and BPA

As shown by our own recent experimental studies[3,4,6,7] enhancing the delivery of either BPA or BSH can have a significant effect both on increasing boron uptake of brain tumors, as well as the therapeutic efficacy of BNCT. Using the well characterized F98 rat glioma model,[32] we have shown that intracarotid (i.c.) injection of either BSH or BPA resulted in a 2 fold increase in tumor boron uptake compared to values obtained following intravenous (i.v.) injection, and this was further increased to 4-fold compared to i.v. injection if a hyperosmotic (1.373 mOsmol or 25%) solution of mannitol was infused via the internal carotid artery prior to injection of the compound.[3,4] Higher tumor boron levels were observed 1 hour after i.c. administration of either BPA or BSH, but these were associated with unacceptably high normal brain boron levels. On the other hand, by 2.5 hours, normal brain boron levels had fallen significantly, but tumor boron levels were still high. The time interval required to achieve an optimum tumor : brain ratio may provide a partial explanation for the failure of a number of investigators[24,26] to confirm that hyperosmotic mannitol-induced BBB-D could enhance the delivery of cytotoxic drugs in brain tumor bearing animals. As described in the following section, these increases in tumor boron uptake were associated with corresponding increases in mean (MST) and median (MeST) survival times of F98 glioma bearing rats following BNCT.

3.2. Boron Neutron Capture Therapy

One thousand F98 glioma cells were implanted stereotactically into the brains of syngeneic rats Fischer rats. Fourteen days later animals were given either BPA alone or in combination with BSH by i.v or i.c. injection with or without hyperosmotic mannitol induced BBB-D. BNCT studies were carried out by us at the Brookhaven Medical Research Reactor (BMRR). Survival times of animals that received either BPA or BSH i.c. with BBB-D were increased 188% and 117%, respectively, compared to irradiated controls.[3–5] The best survival data were obtained using both BSH and BPA in combination (Figure 1).[7] The MST of rats that had BBB-D and i.c. BPA and BSH was 140 days with a 25% cure rate (i.e. survival >1 year) compared to 73 days for i.c. injection, 41 days for animals that received the compounds i.v, and 24 days for untreated controls. It is noteworthy that this tumor heretofore has been completely refractory to all therapeutic modalities including chemo-, radio-, and immunotherapy.[32]

3.3. Studies with Cereport Induced BBB-D

We also have observed a similar enhancement in tumor uptake of BPA in F98 glioma bearing rats following i.c. administration of Cereport, (previously known as Receptor Mediated Permeabilizer-7 or RMP-7), which produces a transient, pharmacologically mediated opening of the BBB. Two series of biodistribution studies were performed in F98 glioma bearing rats that had received either i.v. or i.c. infusion of Cereport at doses ranging from 1.5–7.5 μg/kg body weight (b.w.) followed by i.c. (or i.v.) injection of BPA (300 mg/kg b.w.). Animals were euthanized at 0.5, 2.5, or 4 hours later. Averaging overall time points, i.c. infusion of Cereport significantly enhanced tumor boron uptake ($p = 0.0001$) *versus* the saline control. Tumor boron values were equivalent at 0.5 (36.0 μg/g) and 2.5 hours (38.5 μg/g) following i.c. administration of Cereport and BPA, and then decreased by 33% (25.7 μg/g) at 4 hours. These differences in tumor boron uptake were significant ($\alpha = 0.05$), when compared with i.c. BPA without Cereport at the corresponding times (22.6, 21.8 and 15.3 μg/g, respectively). Normal brain tissue from tumor and non-tumor bearing cerebral hemispheres and blood levels were not different from those measured in the saline controls. Again using the F98 glioma model, BNCT studies were carried out at the BMRR 14 days following intracerebral tumor cell implantation. Rats received BPA i.c. or i.v. dose of 500 mg/kg b.w. with or without i.c. or i.v. adminis-

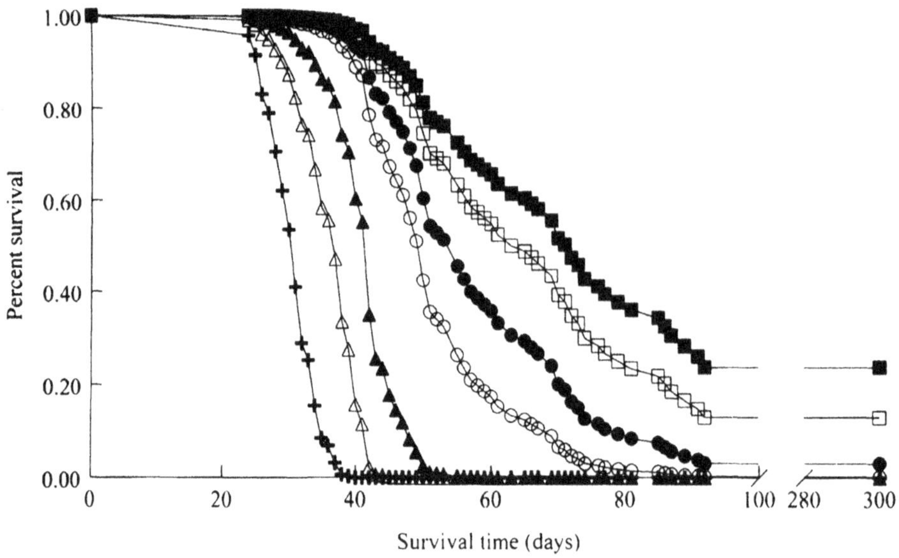

Figure 1. Enhancement of efficacy of BNCT by optimizing boron compound delivery. BNCT was performed at the BMRR 14 days after stereotactic implantation of F98 glioma cells into the brains of syngeneic Fischer rats. Animals were given either BPA alone or in combination with BSH by i.v. or i.c. injection with hyperosmotic mannitol-induced BBB-D. The survival times, in days after tumor implantation, are recorded for irradiated controls (+), BPA i.v. administration (△), BPA and BSH i.v. administration (▲), BPA i.c. administration (○), BPA and BSH i.c. administration (●), BPA i.c. administration with BBB-D (■), and BPA and BSH i.c. administration with BBB-D (■). Survival data have been plotted using Cox's proportional hazards model, which performs a simultaneous fit of the survival curves using all the data points with a partial likelihood approach. Therefore, the number of data points in each curve includes all of the death times, rather than only those animals in a specific group. A 25% cure rate was obtained using BPA and BSH i.c. administration with BBB-D, which constitutes the best survival data ever obtained with this previously uniformly lethal rat glioma model.[32]

tration of Cereport. Kaplan-Meier survival curves are shown in Figure 2. The MST of animals that received i.c. Cereport and BPA was 50 days compared to 43 days for i.c. and 34 days for i.v. BPA alone, 28 days for irradiated controls, and 24 days for untreated controls. These studies, described in detail elsewhere,[33] have established that i.c. Cereport not only can increase tumor uptake of BPA, but also enhance the efficacy of BNCT.

4. CLINICAL CORRELATES

4.1. Intra-Arterial Delivery

In Hatanaka's clinical studies, carried out between ~1968 and 1992, the majority of patients received BSH intra-arterially via either the vertebral or carotid arteries.[11] A recent analysis of these data, carried out by Kageji *et al.*[34] concluded that there were no significant increases in tumor boron content associated with intra-arterial compared to i.v. administration of BSH. Based on this, it was recommended that BSH be administered i.v. since this route is safer than the intra-arterial one. However, this conclusion was based on tumor brain values generally obtained at times greater than 12 hours following administration of BSH at which time tumor boron values were declining. Our findings, which

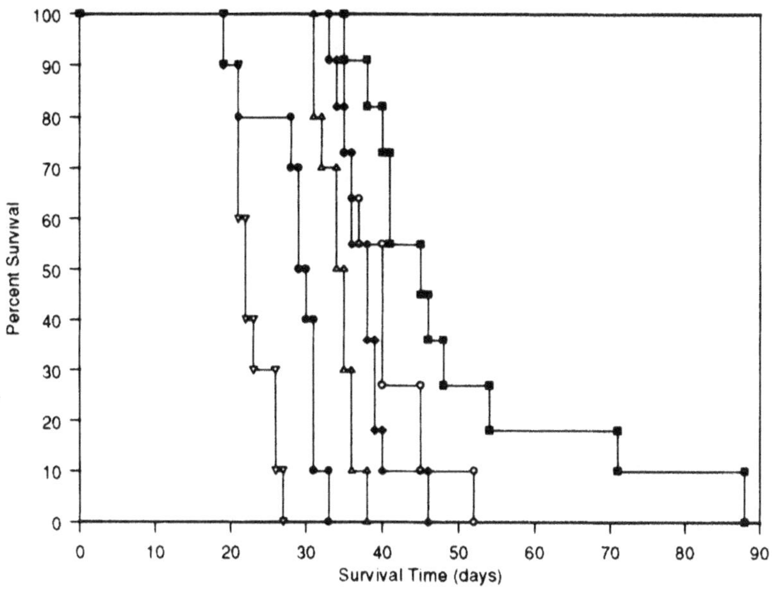

Figure 2. Kaplan-Meier survival curves of F98 glioma bearing rats following BNCT. Survival time in days following implantation is given for untreated controls (▽), irradiated controls (●), or animals that received i.v. BPA with Cereport (◆) or without Cereport (△), or i.c. BPA with (■) or without i.c. Cereport (○).

revealed a significant therapeutic gain associated with i.c. administration of either BSH or BPA with or without BBB-D, suggest that intra-arterial administration needs to be further evaluated, perhaps in larger animals such as dogs, and clinically in humans.

4.2. Blood-Brain Barrier Disruption

BBB-D has been used by Neuwelt *et al.* as a means to increase the uptake of low molecular weight cytoreductive chemotherapeutic agents that usually are ineffective against human brain tumors.[27,28] One of the problems associated with the use of BBB-D for this purpose, is the increased neurotoxicity that may result from exposure of critical regions within normal brain to highly toxic drugs.[30] Since both BSH and BPA, as well as new boron compounds that currently are being evaluated, are not neurotoxic, intra-arterial administration with or without BBB-D could further increase tumor boron uptake, especially by disseminated tumor cells within those parts of the brain that have an intact BBB. One of the significant advantages of BNCT is that the time interval between boron compound administration and irradiation can be adjusted in order to obtain maximum tumor uptake and the best composite tumor : brain and tumor : blood ratios.

5. OPTIMIZATION OF DELIVERY AND COMPOUND FACTOR

The terms compound factor[35] (CF) and, and more recently compound biological effectiveness (CBE)[36,37] have been introduced to emphasize the complexity of calculating an RBE value for the $^{10}B(n,\alpha)^{7}Li$ reaction. Our survival data for brain tumor bearing rats that received either BSH or BPA i.v. suggest that they had identical CFs or CBEs. Other previously published data,[37,38] including the RBE values that we have used in the present study, have assigned a much higher value to the RBE (CF or CBE) for BPA compared to BSH. Hiratsuka *et al.*,[39] using a tumor growth delay time assay, determined RBE's of 2.22 and 2.51 for the $B(n,\alpha)^7Li$ using BPA as the capture agent and Fukuda *et al.*[40] arrived at a value of 2.5, based on the skin response in humans. On the other hand, the survival data obtained following i.c. injection indicated that BPA was superior to BSH (MST 40 days *versus* 52 days, $p = 0.01$) and this was further enhanced by BBB-D (MST 83 days *versus* 52 days, $p = 0.04$), suggesting that the CF or CBE of BPA was greater than that of BSH. Previously reported data, including our own, illustrate the complexity associated with assigning an RBE, CF or CBE value to a boron compound, and this is discussed in detail in a recent review by Coderre and Morris.[41] In our studies[3] the increase of MeST and %ILS appeared to be directly proportional to the tumor boron concentrations, and independent of the compound infused and both parameters increased linearly as a function of boron concentration. Coderre *et al.*[38] reported that the apparent compound-dependent RBEs of BPA and BSSB (dimerized BSH), as determined by clonogenic cell survival following BNCT *in vivo*, were ~3.8 and ~1.2 respectively. Nevertheless, the long-term survival of rats treated with either BPA-based or BSSB-based BNCT was similar when the compounds were administered in amounts that yielded similar ^{10}B concentrations in the tumor. The authors suggested that this paradox may have been due to different compound-dependent mechanisms by which tumor cells were killed. This has been our rationale for using BSH and BPA in combination with one another for BNCT of the F98 glioma.[7] At this point in time, therefore, it would seem prudent to exercise a certain amount of caution before either pronouncing BPA superior or dismissing BSH as a boron delivery agent for BNCT. If our rat data can be extrapolated to the human, then i.v. BPA or BSH might be expected to produce similar clinical results, but i.c. BPA with or without BBB-D might be more effective than BSH. However, if the levels of BPA were higher than BSH in normal brain, this could have an adverse effect on its tolerance. Our present data[3-5] suggest that the CF or CBE are not only dependent upon the chemical composition and biochemical properties of the boron capture agent, but also on how it is delivered, which could significantly affect microdistribution. Since the radiation effects of the $^{10}B(n,\alpha)^7Li$ capture reaction occur at the level of individual cells, only microdosimetric techniques can provide a true estimate of the radiation that is delivered to the tumor. Since high grade gliomas may vary considerably in their rate of cell division and epidermal growth factor receptor (EGFR) expression, different strategies may be required for other low molecular weight boron containing compounds whose uptake is cell cycle dependent, or for high molecular weight agents such as monoclonal or bispecific antibodies or boronated proteins such as EGF. For example, we recently reported that direct intratumoral injection of boronated EGF[42] was both selective and effective in delivering potentially therapeutic amounts of boron to C6 EGFR positive intracerebral gliomas in rats compared to the extremely small quantities that were delivered to the tumor by the i.v. route. If, on the other hand, the C6 gliomas were wild-type and EGFR negative, only small amounts of the bioconjugate persisted in the tumors. In the future, optimization of boron compound delivery will be an integral part of the pre-clinical and clinical evaluation of new capture agents for BNCT.

6. CONCLUSIONS

What can realistically be expected clinically from BNCT at this time and in the near future? It remains to be determined whether clinical cures of glioblastomas can be achieved using either BSH or BPA as capture agents, but it is not unreasonable to hope that there might be a significant prolongation in life span and quality of life over that attainable with current therapeutic modalities for brain tumors. Furthermore, a change in the pattern of tumor reoccurrence from within a 2-cm zone adjacent to the surgical margins to more distant sites within the brain would further suggest therapeutic efficacy. Optimization of the route of boron compound delivery may have an important impact on this. Further studies are currently in progress in our laboratory to optimize the delivery of low and high molecular weight boron compounds to enhance the efficacy of BNCT.

REFERENCES

1. A.H. Soloway, W. Tjarks, B.A. Barnum, F-G. Rong, R.F. Barth, I.M. Codogni, and J.G. Wilson, The chemistry of boron neutron capture therapy. Chemical Rev 98:1515–1562, 1998.
2. R.F. Barth, A.H. Soloway, J.H. Goodman, R.A. Gahbauer, N. Gupta, T.E. Blue, W. Yang, and W. Tjarks, Boron neutron capture therapy of brain tumors: an emerging therapeutic modality. *Neurosurg* 44:433–451, 1999.
3. R.F. Barth, W. Yang, J.H. Rotaru, M.L. Moeschberger, D.D. Joel, M.M. Nawrocky, J.H. Goodman, and A.H. Soloway, Boron neutron capture therapy of brain tumors: enhanced survival following intracarotid injection of either sodium borocaptate or boronophenylalanine with or without blood-brain barrier disruption. *Cancer Res* 57:1129–1136, 1997.
4. W. Yang, R.F. Barth, J.H. Rotaru, M.L. Moeschberger, D.D. Joel, M.M. Nawrocky, J.H. Goodman, and A.H. Soloway, Boron neutron capture therapy of brain tumors: enhanced survival following intracarotid injection of sodium borocaptate with or without blood-brain barrier disruption. *Int J Radiat Oncol Biol Phys* 37:663–672, 1997.
5. W. Yang, R.F. Barth, J.H. Rotaru, D.D. Joel, M.M. Nawrocky, and J.H. Goodman, Enhanced survival of glioma bearing rats following boron neutron capture therapy with osmotic blood-brain barrier disruption and intracarotid injection of boronophenylalanine. *J Neuro-Oncol* 32:59–70, 1997.
6. R.F. Barth, W. Yang, M.L. Moeschberger, J.H. Goodman and R.T. Bartus, Enhanced delivery of boronophenylalanine for neutron capture therapy of brain tumors using the bradykinin analogue Cereport™ (RMP-7). *Neurosurgery* 44:351–360, 1999.
7. R.F. Barth, W. Yang, J.H. Rotaru, M.L. Moeschberger, C.P. Boesel, A.H. Soloway, D.D. Joel, M.M. Nawrocky, K. Ono, and J.H. Goodman, Boron neutron capture therapy of brain tumors: enhanced survival and cure following blood-brain barrier disruption and intracarotid injection of sodium borocaptate and boronophenylalanine. *Int J Radiat Oncol Biol Phys* 44:April, 2000.
8. N.H. Greig, Drug delivery to the brain by blood-brain barrier circumvention and drug modification, *in:* "Implications of the Blood-Brain Barrier and its Manipulation", Vol. 1: Basic Science Aspects, E.A. Neuwelt, ed., Plenum Press, 1989, pp. 311–367.
9. R.F. Barth, W. Yang, A.H. Soloway, and J.H. Goodman, Alternative approaches for the delivery of boron containing agents to brain tumors, *in:* "Advances in Neutron Capture Therapy", Vol. II, Chemistry and Biology, B. Larsson, J. Crawford, and R. Weinrich eds, Elsevier Science, B.V., 1997, pp. 216–224.
10. W. Chen, S.C. Mehta, and D.R. Lu, Selective boron drug delivery to brain tumors for boron neutron capture therapy. *Adv. Drug Del. Rev* 26:231–247, 1997.
11. Y. Nakagawa, and H. Hatanaka, Boron neutron capture therapy: clinical brain tumor studies. *J Neuro-Oncol* 33:105–115, 1997.
12. E.H. Elowitz, R.M. Bergland, J.A. Coderre, D.D. Joel, M. Chadha, and A.D. Chanana, Biodistribution of p-boronophenylalanine in patients with glioblastoma multiforme for use in boron neutron capture therapy. *Neurosurgery* 42:463–469, 1998.
13. W.M. Pardridge, Drug delivery to the brain. *J Cerebral Blood Flow Metabol* 17:713–731, 1997.
14. P. Riva, A. Arista, G. Franceschi, M. Frattarelli, A. Sturiale, N. Riva, M. Casi, and R. Rossitti, Local treatment of malignant gliomas by direct infusion of specific monoclonal antibodies labeled with ^{131}I: comparison of the results obtained in recurrent and newly diagnosed tumors. *Cancer Res* 55:5952–5956, 1995.

15. K.A. Walter, R.J. Tamargo, A. Olivi, P.C. Burger, and H. Brem. Intratumoral chemotherapy. *Neurosurgery* 37:1128–1145, 1995.
16. S. Valtonen, U. Timonen, P. Toivanen, H. Kalimo, L. Kivipelto, O. Heiskanen, G. Unsgaard, and T. Kuurne, Interstitial chemotherapy with carmustine-loaded polymers for high-grade gliomas: a randomized double-blind study. *Neurosurgery* 41:44–49, 1997.
17. R.J. Tamargo, R. Langer, and H. Brem, Interstitial drug delivery to the central nervous system using controlled release polymers: chemotherapy for brain tumors. *Methods in Neurosciences* 21:135–149, 1994.
18. D.M. Liberman, D.W. Laske, P.F. Morrison, K.S. Bankiewicz, and E.H. Oldfield, Convection-enhanced distribution of large molecules in gray matter during interstitial drug infusion. *J Neurosurgery* 82:1021–1029, 1995.
19. D.W. Laske, R.J. Youle, and E.H. Oldfield, Tumor regression with regional distribution of the targeted toxin TF-CRM107 in patients with malignant brain tumors. Nature Med 12:1362–1368, 1997.
20. S.I. Rapoport, and P.I. Robinson, A therapeutic role for osmotic opening of the blood-brain barrier, Re-evaluation of literature and of importance of source-sink relations between brain and tumor. *in:* "Pathophysiology of the Blood-Brain Barrier", B.B. Johnson, C. Owman, and H. Widner eds, Elsevier Science Publishers, B.V., 1990, pp. 167–181.
21. R.A. Kroll, and E.A. Neuwelt, Outwitting the blood-brain barrier for therapeutic purposes: osmotic opening and other means. *Neurosurgery* 42:1083–1100, 1998.
22. D.R. Groothuis, The blood-brain and blood-tumor barriers: a review of strategies for increasing drug delivery. *Neuro-Oncol* 2:45–59, 2000.
23. P.J. Elliott, H.J. Hayward, M.R. Huff, T.L. Nagle, T.L. Black, and R.T. Bartus, Unlocking the blood-brain barrier: a role for RMP-7 in brain tumor therapy. *Exp Neurol* 141:214–224, 1996.
24. R.G. Blasberg, and D.R. Groothuis. Chemotherapy of brain tumors: physiological and pharmacokinetic considerations, *in:* "Seminars in Oncology," Vol. 13, 1986, pp. 70–82.
25. R.K. Jain, Delivery of novel therapeutic agents in tumors: physiological barriers and strategies. *J Nat Cancer Inst* 81:570–576, 1989.
26. D.R. Groothuis, P.C. Warkne, P. Molnar, G.D. Lapin, and M.A. Mikhael. Effect of hyperosmotic blood-brain barrier disruption and transcapillary transport in canine brain tumors. *J Neurosurgery* 72:441–449, 1990.
27. S.A. Dahlborg, W.D. Henner, J.R. Crossen, M. Tableman, A. Petrillo, R. Braziel, and E.A. Neuwelt, Non-AIDS primary CNS lymphoma: first example of a durable response in a primary brain tumor using enhanced chemotherapy delivery without cognitive loss and without radiotherapy. *Cancer J Sci Amer* 2:166–174, 1996.
28. S.A. Dahlborg, A. Petrillo, J.R. Crossen, S. Roman-Goldstein, N.D. Doolittle, K.H. Fuller, and E.A. Neuwelt, The potential for complete and durable response in nonglial primary brain tumors in children and young adults with enhanced chemotherapy delivery. *Cancer J Sci Amer* 4:110–124, 1998.
29. W.R. Shapiro, R.M. Voorhies, E.M. Hiesiger, P.B. Sher, G.A. Basler, and L.E. Lipschutz, Pharmacokinetics of tumor cell exposure to [^{14}C] methotrexate after intracarotid administration without and with hyperosmotic opening of the blood-brain and blood-tumor barriers in rat brain tumors: a quantitative autoradiographic study. *Cancer Res* 48:694–701, 1998.
30. W.R. Shapiro, Is there a therapeutic role for osmotic breaching of the blood-brain barrier. *Ann Neurology* 24:680–681, 1988. (Letter)
31. C.D. Fenstermacher, and A.L. Cowles, Theoretic limitation of intracarotid infusion in brain tumor chemotherapy. *Cancer Treat Rep* 61:519–526, 1977.
32. R.F. Barth, Rat brain tumor models in experimental neuro-oncology: the 9L, C6, T9, F98, RG2 (D74), RT-2, and CNS-1 gliomas. *J Neuro-Oncol* 36:55–66, 1998.
33. W. Yang, R.F. Barth, R.T. Bartus, J.H. Rotaru, M.L. Moeschberger, A.K. Ferketich, M.M. Nawrocky, J.A. Coderre, and J.H. Goodman, Improved survival following neutron capture therapy of brain tumors using cereport (RMP-7) mediated blood brain barrier modulation to enhance delivery of boronophenylalanine. *Neurosurgery* (Submitted).
34. T. Kageji, Y. Matsumoto, Y. Nakagawa, and H. Hatanaka, Pharmacokinetics and boron uptake of BSH ($Na_2B_{12}H_{11}SH$) in patients with intracranial tumors. *J Neuro-Oncol* 32:117–130, 1997.
35. R.A. Gahbauer, R.G. Fairchild, J.H. Goodman, and T.E. Blue, RBE in normal tissue studies, *in:* "Neutron Capture Therapy", D. Gabel, and R. Moss eds, Plenum Press, New York, 1992, pp. 123–128.
36. G.M. Morris, J.A. Coderre, J.W. Hopewell, P.L. Micca, and C. Fisher, Boron neutron capture irradiation of the rat spinal cord: effects of variable doses of borocaptate sodium. *Radiotherapy Oncol* 39:253–259, 1966.

37. J.A. Coderre, M.S. Makar, P.L. Micca, M.M. Nawrocky, H.B. Liu, D.D. Joel, D.N. Slatkin, and H.I. Amol, Derivations of relative biological effectiveness for the high-LET radiations produced during boron neutron capture irradiations of the 9L gliosarcoma *in vitro* and *in vivo*. *Int J Radiat Oncol Biol Phys* 27:1121–1129, 1993.
38. J.A.Coderre, M.S. Makar, P.L. Micca, M.M. Nawrocky, H.B. Liu, D.D. Joel, D.N. Slatkin, and H.I. Amol, Derivations of relative biological effectiveness for the high LET radiations produced during boron neutron capture irradiations of the 9L gliosarcoma *in vitro* and *in vivo*. *Int J Radiat Oncol Biol Phys* 27:1121–1129, 1993.
39. J. Hiratsuka, M. Kono, and Y. Mishima, RBE's of thermal neutron capture therapy and $^{10}B(n,\alpha)^{7}Li$ reaction on melanoma-bearing hamsters. *Pigm Cell Res* 2:352–355, 1989.
40. H. Fukuda, Y. Mishima, J. Hiratsuka, C. Honda, N. Wadabayayashi, T. Kobayashi, K. Yoshino, H.K. Takahashi, Y. Abe, K. Kanda, and M. Ichihashi, BNCT of malignant melanoma: radiobiological analysis and data comparison with conventional radiotherapy, *in:* "Cancer Neutron Capture Therapy", Y. Mishima, ed., Plenum Press, New York, 1996, pp. 663–671.
41. J.A. Coderre, and G.M. Morris, The radiation biology of boron neutron capture therapy. *Radiat Res* 151:1–18, 1999.
42. W. Yang, R.F. Barth, D.A. Adams, and A.H. Soloway, Intratumoral delivery of boronated epidermal growth factor for neutron capture therapy of brain tumors. *Cancer Res* 57:4333–4339, 1997.

BORON-RICH OLIGOMERIC PHOSPHATE DIESTERS THAT TARGET THE CELL NUCLEUS *IN VITRO*

Mark B. Smuckler[1], Lufeng Guan[1], Akira Nakanishi[2], Harumi Kasamatsu[2], and M. Frederick Hawthorne[1]

[1]Department of Chemistry and Biochemistry
University of California
Los Angeles, California 90095
[2]Department of Molecular, Cell, and Developmental Biology
and the Molecular Biology Institute
University of California
Los Angeles, California 90095

1. INTRODUCTION

The development of tumor-targeting agents for BNCT which contain large numbers of ^{10}B atoms and are readily taken up by cells is of great interest.[1,2] Effective BNCT demands the uptake of ~10^8–10^9 randomly distributed ^{10}B atoms/tumor cell. Lesser amounts of ^{10}B are required for successful therapy when the boron is concentrated in the cell nucleus.[3] Recently, we have observed the selective accumulation of a number of homogeneous *nido*-carboranyl oligomeric phosphate diesters (*nido*-OPDs) in EMT6 tumors grown in BALB/c mice (unpublished results), but it has yet to be shown that these species accumulate in the cell nucleus. Here we report results which indicate that *nido*-OPDs are capable of delivering sufficient amounts of ^{10}B to the nucleus for effective BNCT.

Two structurally unique series of fluorescein-labeled OPDs (Figure 1), *nido*-$(CB)_5$ and both *closo*- and *nido*-$(G1)_n$ (n = 5,20), were synthesized[4,5] and studied for their ability to accumulate and be retained in the cell nucleus. The carborane cage in *nido*-$(CB)_5$ is part of the oligomeric backbone, while the cage in the G1-containing oligomers is tethered to the backbone. The *closo*-OPDs contain one negative charge per repeating unit, while the *nido* species contain two per unit. This imparts greater hydrophilicity to the *nido*-OPDs. Because of the OPDs' inability to enter cells under culture conditions, alternative methods for crossing the plasma membrane were employed. Two types of in vitro

Frontiers in Neutron Capture Therapy, edited by Hawthorne *et al.*
Kluwer Academic / Plenum Publishers, New York, 2001.

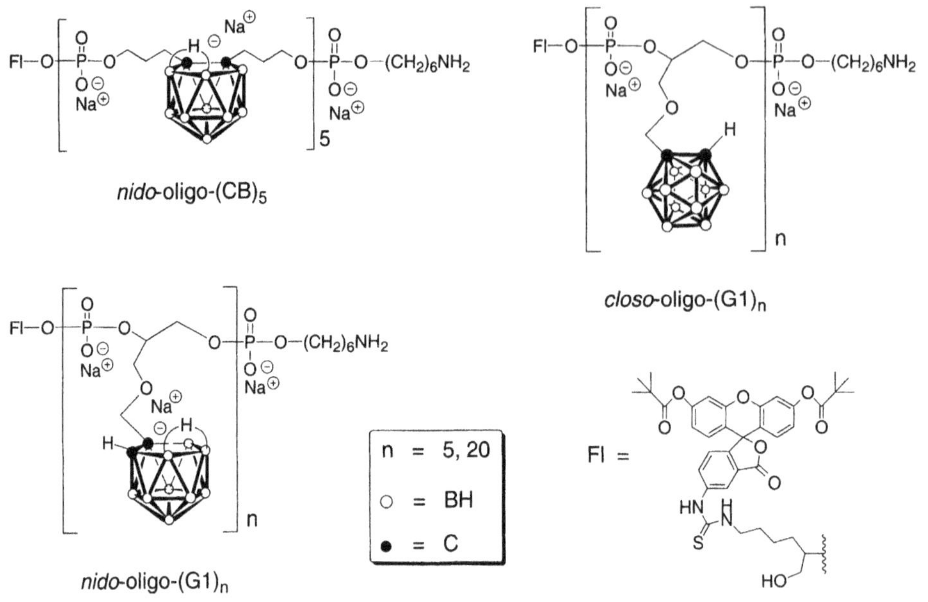

Figure 1. Structures of the fluorescein-labeled OPDs.

studies were conducted with these OPDs: (1) TC7 cells were microinjected[6] with OPDs to investigate their subcellular distribution in living cells, and (2) plasma membrane-permeabilized TC7 cells[7] were incubated with OPDs to asses cell-free nuclear entry of OPDs. TC7 cells, a subline of African green monkey kidney cells, were used in all experiments due to their apparent resilience to the process of microinjection and our desire to use a mammalian cell line. In these in vitro studies, $\sim 10^8$–10^9 boron atoms were administered per cell and the fluorescein subcellular localization was observed by epifluorescence microscopy.[8]

2. RESULTS AND DISCUSSION

Co-injection of fluorescein-labeled *nido*-$(CB)_5$ and rhodamine-labeled bovine serum albumin (BSA) into the cytoplasm of the TC7 cells resulted in nuclear localization of *nido*-$(CB)_5$ (Figure 2a) and non-specific subcellular distribution of BSA (Figure 2b). Because aqueous channels in the nuclear pore complex allow diffusion of small molecules (<20–40 kDa) in and out of the nucleus,[9] failure of the large BSA molecules (~66 kDa) to enter the cell nucleus suggests that small *nido*-$(CB)_5$ oligomers (2.7 kDa) enter the nucleus by diffusion.

When microinjected with either *nido*-$(CB)_5$ or *nido*-$(G1)_5$, nuclear uptake was observed within 10 min (Figure 2, c and d). Similar short-term nuclear localization occurred following microinjection of *nido*-$(G1)_{20}$ (Table 1), thus demonstrating the rapid nuclear accumulation of small (2.7–8.0 kDa) *nido*-OPDs in living cells. Moreover, after 24 h of incubation *nido*-$(CB)_5$ remained primarily in the nucleus (Figure 2e), while *nido*-$(G1)_5$ redistributed throughout the cytoplasm and nucleus (Figure 2f). This suggests that the location of the anionic carborane cage relative to the oligomer backbone affects the long-term (24 h) subcellular distribution of *nido*-OPDs.

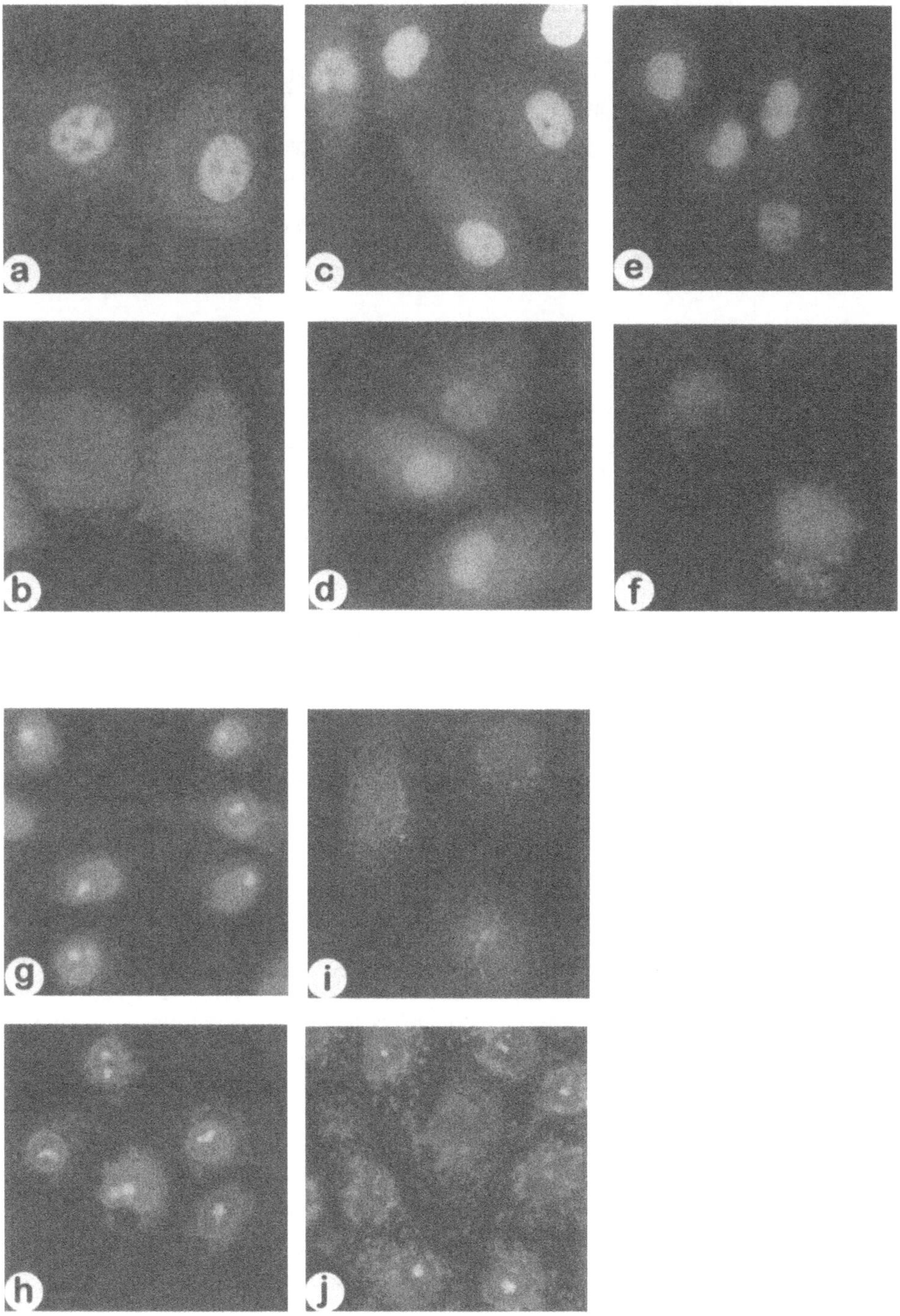

Figure 2. Subcellular localization of fluorescein-labeled OPDs *in vitro*. (Used with permission.) **a**, **b**, The same set of cells 2 h after co-injection of *nido*-$(CB)_5$ and rhodamine-labeled BSA. When viewed at their characteristic UV wavelengths *nido*-$(CB)_5$ appears in the nucleus (**a**), BSA appears in the cytoplasm (**b**). **c**, **d**, Rapid nuclear accumulation of *nido*-$(CB)_5$ (**c**) and *nido*-$(G1)_5$ (**d**) within 10 min after microinjection. **e**, **f**, Long-term nuclear retention of *nido*-$(CB)_5$ (**e**) and *nido*-$(G1)_5$ (**f**) in the cells 24 h after microinjection. **g**, **h**, Distribution of fluorescein in subcellular compartments in permeabilized cells incubated with *nido*-$(CB)_5$ (**g**) and *nido*-$(G1)_5$ (**h**). **i**, **j**, Subcellular distribution of *closo*-$(G1)_5$ in the cells 2 h after microinjection (**i**) and in permeabilized cells (**j**).

Table 1. Subcellular localization of fluorescein-labeled OPDs following microinjection

Fluorescein-labeled oligomers	Subcellular localization at ~2 h	Subcellular localization at ~24 h
Fl-*nido*-$(CB)_5$	Nucleus	Nucleus
Fl-*nido*-$(G1)_5$	Nucleus	Nucleus/Cytoplasm
Fl-*closo*-$(G1)_5$	Cytoplasm	Cytoplasm
Fl-*nido*-$(G1)_{20}$	Nucleus	ND*
Fl-*closo*-$(G1)_{20}$	Cytoplasm	ND*

*ND, not determined.

Cell-free studies using permeabilized cells further demonstrated the nuclear accumulation of *nido*-$(CB)_5$ and *nido*-$(G1)_5$. Selective permeabilization of the cell's plasma membrane with digitonin removes cytoplasmic components essential for active nuclear transport. Permeabilization of cells followed by incubation for 30 min with either *nido*-$(CB)_5$ or *nido*-$(G1)_5$ resulted in nuclear accumulation of *nido*-$(CB)_5$ and *nido*-$(G1)_5$, respectively (Figure 2, g and h). Unlike the more hydrophilic *nido*-OPDs, *closo*-$(G1)_5$ and *closo*-$(G1)_{20}$ accumulated non-specifically in vitro (Figure 2i, compare h and j; Table 1). Following microinjection (Figure 2i), and more obviously seen with permeabilized cells (Figure 2j), *closo*-$(G1)_5$ adhered to cytoplasmic components. These cell-free studies indicate diffusion across the nuclear membrane of the *nido*-OPDs, but not of *closo*-$(G1)_5$. The rapid accumulation of these small (2.7–8 kDa) *nido*-OPDs suggest that they have an affinity for nuclear components. Further experiments are necessary to elucidate the exact nature of the interaction of *nido*-OPDs with nuclear constituents.

3. CONCLUSIONS

Hydrophilic *nido*-OPDs are attractive as components of BNCT agents due to their selective nuclear uptake *in vitro*. Encapsulation of *nido*-OPDs in tumor cell-targeted liposomes should circumvent the compounds' inability to penetrate the plasma membrane and provide selective cell death. Apparently, the disposition of the anionic *nido*-carborane cages relative to the OPD chain influences the long-term (24 h) nuclear retention of microinjected *nido*-OPDs. This suggests a structural selectivity in the extent to which various *nido*-OPDs bind to macromolecules in the nucleus. Thus, a systematic examination of the nuclear accumulation of *nido*-OPD derivatives is necessary to determine optimal structural features for successful BNCT candidates. The structure-function relationships of the boron-rich *nido*-OPDs discussed here provide a sound starting point for the development of BNCT agents which cross the plasma membrane and target and retain in the cell nucleus.

ACKNOWLEDGMENTS

This work was supported by the National Cancer Institute (Grant No. 443835-FH-29617) and a pilot fund from UCLA Jonsson Comprehensive Cancer Center.

REFERENCES

1. A.H. Soloway, W. Tjarks, B.A. Barnum, F.-G. Rong, R.F. Barth, I.W. Codogni, and J.G. Wilson, The chemistry of neutron capture therapy, *Chem. Rev.*, 98:1515–1562, 1998.
2. M.F. Hawthorne, The role of chemistry in the development of boron neutron capture therapy of cancer, *Angew. Chem., Int. Ed. Engl.*, 32:950–984, 1993.
3. T. Hartman, and J. Carlsson, Radiation dose heterogeneity in receptor and antigen mediated boron neutron capture therapy, *Radiother. Oncol.*, 31:61–75, 1994.
4. R.R. Kane, K. Drechsel, and M.F. Hawthorne, Automated syntheses of carborane-derived homogeneous oligophosphates: reagents for use in the immunoprotein-mediated boron neutron capture therapy of cancer, *J. Am. Chem. Soc.*, 115:8853–8854, 1993.
5. K. Drechsel, C.S. Lee, E.W. Leung, R.R. Kane, and M.F. Hawthorne, Synthesis of new building blocks for boron-rich oligomers in boron neutron capture therapy (BNCT). I., *Tetrahedron Lett.*, 35:6217–6220, 1994.
6. J. Clever, and H. Kasamatsu, Simian virus 40 Vp2/3 small structural proteins harbor their own nuclear transport signal, *Virology*, 181:78–90, 1991.
7. S.A. Adam, R.A. Marr, and L. Gerace, Nuclear protein import in permeabilized mammalian cells requires soluble cytoplasmic factors, *J. Cell Biol.*, 111:807–816, 1990.
8. H. Kasamatsu, and A. Nehorayan, Intracellular localization of viral polypeptides during simian virus 40 infection, *J. Virol.*, 32:648–660, 1979.
9. D.J. Forbes, Structure and function of the nuclear pore complex, *Annu. Rev. Cell Biol.*, 8:495–527, 1992.

BIODISTRIBUTION OF TECHNETIUM-99m LABELED EPIDERMAL GROWTH FACTOR FOLLOWING INTRACAROTID INJECTION INTO RATS BEARING INTRACEREBRAL EGF RECEPTOR (+) GLIOMAS

Weilian Yang[1], Rolf F. Barth[1], Dianne M. Adams[1], Renee Leveille[2], and Jacek Capala[3]

[1]Department of Pathology
[2]Veterinary Clinical Sciences
The Ohio State University
Columbus, OH 43210 and
[3]Medical Department
Brookhaven National Laboratory
Upton, NY 11973

1. INTRODUCTION

The gene for epidermal growth factor receptor (EGFR) and its cell surface membrane expression is amplified in a majority of high grade gliomas,[1] suggesting that this receptor potentially may be a useful target for the selective delivery of ^{10}B to EGFR (+) brain tumors.[2] There has been increasing interest in the possible use of the EGFR as a molecular target for the delivery of boron compounds to EGFR (+) brain tumors for BNCT.[3] Two requirements must be fulfilled. The first, as recently described by us,[4] is the development of methodology to chemically construct boron-containing EGF bioconjugates that retain their *in vitro* and *in vivo* affinity for the EGFR and the second is the ability to selectively deliver sufficient amounts of boronated EGF (B-EGF) to the tumor.[5] Since exogenous EGF has a high propensity to localize in the liver and small amounts localize in the tumor following intravenous (i.v.) injection, optimization of delivery is critical to the successful use of B-EGF bioconjugates as targeting agents for brain tumors.[6]

We previously have reported on the biodistribution of ^{99m}Tc labeled EGF following intracerebral injection into $C6_{EGFR}$ glioma bearing rats and have shown selective tumor retention of the radioligand compared to that seen in EGFR negative (wildtype) tumors.[5]

Frontiers in Neutron Capture Therapy, edited by Hawthorne *et al.*
Kluwer Academic / Plenum Publishers, New York, 2001.

The purpose of the present study was to determine if intracarotid (i.c.) injection of the ^{99m}Tc-EGF bioconjugate, which would permit first pass uptake by the tumor, could enhance delivery compared to that seen following i.v. injection.

2. MATERIAL AND METHODS

2.1. Labeling of EGF with Technetium-99m

As previously described by us,[4] EGF first was modified with 2-iminothiolane (Traut's reagent) in order to introduce a thiol group into its aminoterminal end, following which excess Traut's reagent was removed using a Biospin 6 column (Bio-Rad, Hercules, CA). EGF-SH then was reacted with ^{99m}Tc sodium pertechnate[7] in a nitrogen purged reducing buffer containing 20 mM potassium monohydrophthalate, 10 mM potassium sodium tartate, and 1.5 mM stannous tartrate, pH 6.5. ^{99m}Tc binding was determined by thin layer chromatography.

2.2. Biodistribution of ^{99m}Tc labeled EGF

Ten thousand C6 wildtype or EGFR gene transfected glioma cells ($C6_{EGFR}$)[8] were implanted stereotactically into the caudate nucleus of the right cerebral hemisphere of male Fischer rats using a previously described procedure.[6] Four weeks following intracerebral implantation of tumor cells, C6 and $C6_{EGFR}$ glioma bearing rats were injected i.c. or i.v. with 10 μg (150–200 μCi) of ^{99m}Tc-EGF using the method described by us for intracerebral injection. The distribution of radioactivity in living animals was monitored by means of external scintigraphy using a gamma camera with a low energy collimator optimized for ^{99m}Tc, or by well counting following euthanization of animals at the termination of the study.

3. RESULTS

3.1. Biodistribution of ^{99m}Tc Labeled EGF

Tumor and normal tissue uptake of radioactivity following different routes of injection of ^{99m}Tc EGF in rats bearing C6 or $C6_{EGFR}$ gliomas are summarized in Table 1.

Four hours following i.c. injection of ^{99m}Tc-EGF, 0.34% of the injected dose per gram (% ID/g) tissue was localized in $C6_{EGFR}$ tumors, compared to 0.07% ID/g in animals bearing C6 wildtype gliomas. Hyperosmotic (1.373 mOsmol) mannitol induced disruption of the blood-brain barrier (BBB-D) followed by i.c. injection of ^{99m}Tc-EGF did not significantly increase tumor uptake compared to i.c. injection without BBB-D (0.45% *vs* 0.34% ID/g, $p > 0.1$). In contrast, normal brain uptake of radioactivity following BBB-D was much higher. Intracarotid administration significantly increased tumor uptake compared to i.v. injection (0.34 *vs* 0.14% ID/g, $p < 0.04$), but the uptake of ^{99m}Tc-EGF in the liver was ~5–7% and 12–20% ID/g in the kidneys following either i.c. or i.v. administration compare to 0.3–0.4% ID in tumors. These results are in agreement with previously reported data[5,6] that EGF had a high propensity to localize in the liver, spleen and kidneys. Unfortunately, i.c. injection alone of the EGF does not solve the problem of low tumor and high extraneous tissue uptake.

Table 1. Percentage of administered dose/gram of tumor and normal tissue at 4 hours following intravenous or intracarotid injection of ^{99m}Tc labeled EGF with or without BBB-D

Route	C6 Glioma	Tumor	Brain (R)	Brain (L)	Liver	Spleen	Kidney	Skin	Muscle	Blood
i.v.	EGFR	0.14 ± 0.02	0.05 ± 0.02	0.02 ± 0.01	6.28 ± 1.55	3.63 ± 1.62	20.62 ± 6.72	0.15 ± 0.04	0.06 ± 0.02	0.14 ± 0.06
i.c.	EGFR	0.34 ± 0.08	0.06 ± 0.02	0.03 ± 0.02	9.53 ± 1.91	6.72 ± 1.49	12.53 ± 4.93	0.16 ± 0.14	0.07 ± 0.05	0.18 ± 0.05
i.c. + BBB-D	EGFR	0.45 ± 0.22	0.12 ± 0.05	0.03 ± 0.01	5.83 ± 1.80	4.51 ± 2.78	18.35 ± 9.35	0.24 ± 0.22	0.07 ± 0.05	0.12 ± 0.03
i.c.	Wild	0.07 ± 0.02	0.04 ± 0.01	0.01 ± 0.002	4.07 ± 1.13	3.75 ± 0.93	15.98 ± 3.39	0.11 ± 0.04	0.03 ± 0.01	0.11 ± 0.06

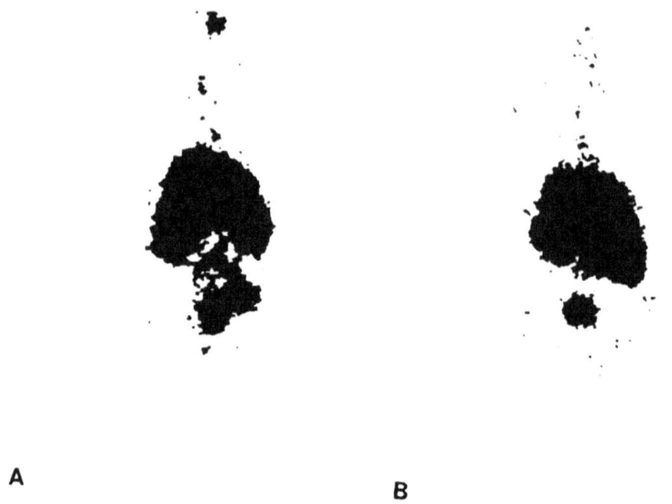

Figure 1. External gamma scintigraphy of rats bearing gliomas. Imaging was carried out 4 h following i.c. injection of ^{99m}Tc-EGF into $C6_{EGFR}$(A) and C6 wild-type(B) glioma-bearing rats. Tumors could be clearly visualized in $C6_{EGFR}$ but not C6 wildtype glioma bearing rats. High concentrations of radioactivity also were seen in the liver, spleen and kidneys.

3.2. Scintigraphic Studies

As shown by external gamma scintigraphy ^{99m}Tc-EGF could be clearly visualized in $C6_{EGFR}$ glioma bearing rats, whereas no such localization was evident in C6 wildtype tumor bearing animals (Figure 1).

The regions of interest (ROI) over the liver and kidneys revealed that >50% of the whole body radioactivity accumulated in these organs, and only 0.004% was seen in the tumor following systemic administration of ^{99m}Tc-EGF.

4. DISCUSSION

Our observations strongly suggest that it is highly unlikely that i.v. or i.c. injection alone will be useful for delivering high molecular weight agents such as B-EGF to brain tumors and that other approaches will be required. Although i.c. injection of ^{99m}Tc-EGF produced a 2–3X enhancement in tumor uptake compared to i.v. injection, this did not reduce the high uptake by liver, spleen and kidneys. Furthermore, hyperosmotic blood-brain barrier disruption did not significantly increase uptake compared to i.c. injection. We previously reported that direct intratumoral (i.t.) injection of B-EGF could deliver >15 ppm of boron to $C6_{EGFR}$ tumors and unless the B-EGF bioconjugate can be chemically modified so as to reduce its hepatic and splenic uptake, it is unlikely that systemic administration can be used to target brain tumors. Either direct i.t. injection or a variant of it, such as convection enhanced delivery will be required. Studies are in progress to assess the feasibility of this approach.

ACKNOWLEDGMENT

The studies described in this report have been supported by Department of Energy grants #DE-FG02-95ER62059 and DE-FG02-98ER62595. We thank Linda Carlos for secretarial assistance in the preparation of this manuscript.

REFERENCES

1. K. Schwechheimer, S. Huang, and W.K. Cavenee, EGFR gene amplification-rearrangement in human glioblastoma. Int. J Cancer, 62:145–148, 1995.
2. H. Modjtahedi, and C. Dean, The receptor for EGF and its ligands: expression, prognostic value and target for therapy in cancer (Review), Int. J. Oncol. 4:227–296, 1994.
3. J. Carlsson, L. Gedda, C. Gronvik, T. Hartman, A. Lindstrom, P. Lindstrom, P. Olsson, J. Ponten, S. Sjoberg, and B. Westermark, Strategy for boron neutron capture therapy against tumor cells with over-expression of the epidermal growth factor-receptor. Int. J. Radiation Oncol Biol Phys 30:105–115, 1994.
4. J. Capala, R.F. Barth, M. Bendayan, M. Lauzon, D. Adams, A. Soloway, R. Fenstermaker, and J. Carlsson, Boronated epidermal growth factor as a potential targeting agent for boron neutron capture therapy of brain tumors. Bioconjugate Chem. 7:7–15, 1996.
5. J. Capala, R.F. Barth, M.Q. Bailey, R.A. Fenstermaker, M.J. Marek, and B.A. Rhodes, Radiolabeling of Epidermal Growth Factor with ^{99m}Tc and *in vivo* localization following intracerebral injection into normal and glioma-bearing rats. Bioconjugate Chem. 8:289–295, 1997.
6. W. Yang, R.F. Barth, D.M. Adams, and A.H. Soloway, Intratumoral delivery of boronated epidermal growth factor for neutron capture therapy of brain tumors. Cancer Res 57:4333–4339, 1997.
7. B.A. Rhodes, Direct labeling of proteins with ^{99m}Tc, Nucl Med Bio. 18, 667–676.
8. R.A. Fenstermaker, J. Capala, A. Hujer, R.F. Barth, A. Hujer, H. Kung, and D. Kaetzel, Jr.: Implantable C6 glioma cell lines transfected with the epidermal growth factor receptor gene. Leukemia 9:Suppl 1:s106–s112, 1995.
9. L. Vinter-Jensen, J. Fokiaer, P.E. Jorgensen, J. Marqversen, M. Rehling, E.Z. Dajani, and E. Nexo, Tissue distribution of ^{131} I-labelled epidermal growth factor in the pig visualized by dynamic scintigraphy. J. Endorcrinol. 144:5–12, 1995.

BIOLOGICAL EVALUATION OF BORONATED POLYAMINES AS POTENTIAL AGENTS FOR BNCT

Jin-Cong Zhuo[1], Jianping Cai[1], Albert H. Soloway[1], Rolf F. Barth[2], Dianne M. Adams[2], Weihua Ji[1], Werner Tjarks[1], and Joan H. Rotaru[2]

[1]College of Pharmacy
[2]Department of Pathology
The Ohio State University
Columbus, Ohio 43210

1. INTRODUCTION

The use of BNCT as a cancer treatment modality has been well documented.[1,2] The boron-containing delivery agent should be non-toxic, selectively target tumor cells and ideally localize within the nucleus. Various boron-containing analogues of biologically active compounds, such as amino acids, peptides, nucleosides/nucleotides, porphyrins and DNA binder have been synthesized.[1] Polyamines, such as spermidine (**SPD**) and spermine (**SPM**) are essential for cell growth and cell differentiation in mammals. They interact electrostatically with DNA in a non-specific manner. Their concentration levels are increased in tumor cells, and within such cells, levels are especially high in the nucleus.[3,4] In addition, there seems to be a facilitated transport mechanism operative for polyamines into malignant cells.[5] For this reason, boron-containing polyamines have been considered as potential BNCT agents.[6]

We have previously designed and synthesized *N*-carboranyl tethered polyamine analogues of spermidine and spermine and have correlated their chemical structure with their *in vitro* toxicity, cellular uptake, and DNA binding properties.[6] The toxicity strongly depended upon the position of the carboranyl group on the polyamine scaffold. Polyamines substituted at the terminal amino groups (**SPD-1**, **SPM-1**) possessed a greater toxicity than those substituted at the central amino group (**SPD-5**, **SPM-5**). Nevertheless, even for the latter compounds, the observed *in vitro* toxicity was very high. To diminish the toxicity, we increased their hydrophilicity by modification of the *o*-carborane cages in the **SPD-5** and **SPM-5** structures by inserting 2-aminoethyl and 2,3-dihydroxylpropyl groups at the 2-position. These compounds, shown in Figure

Frontiers in Neutron Capture Therapy, edited by Hawthorne *et al.*
Kluwer Academic / Plenum Publishers, New York, 2001.

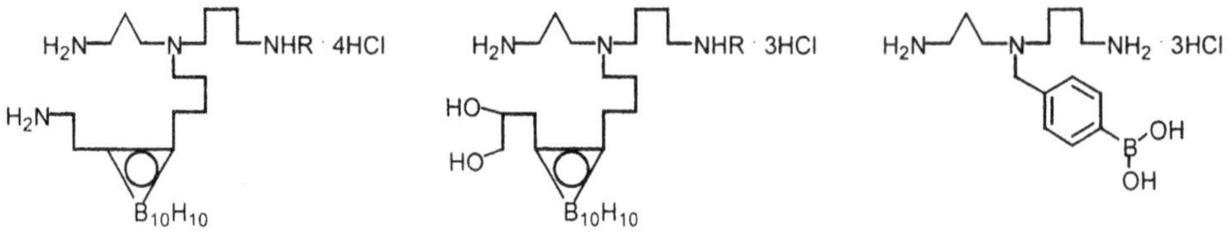

ASPD-5 R = H
ASPM-5 R = $\diagup\!\!\diagdown$NH$_2$ · HCl

DHSPD-5 R = H
DHSPM-5 R = $\diagup\!\!\diagdown$NH$_2$ · HCl

BBSPD-5

Figure 1

1, were designated as **ASPD-5**, **ASPM-5**, **DHSPD-5** and **DHSPM-5**. Another boronated polyamine with increased hydrophilicity was 4-(dihydroxyboryl)phenylmethyl spermidine (**BBSPD-5**) (Figure 1). The synthesis of these compounds will be presented in detail elsewhere.

2. RESULTS AND DISCUSSION

Two major questions regarding these boron-containing polyamines were whether these structures behaved biologically similar to SPD/SPM, and whether they possessed the requisite properties for BNCT agents. In order to determine this, the following studies have been undertaken: (1) their *in vitro* growth inhibitory effects using F98 rat glioma cells; (2) their ability to displace ethidium bromide from calf thymus DNA; (3) the *in vitro* cellular uptake using the F98 glioma cells; and (4) their *in vivo* biodistribution in mice with subcutaneous B16 melanoma and intracerebral GL261 glioma.

The *in vitro* toxicity of carboranyl polyamines **ASPD-5**, **ASPM-5**, **DHSPD-5**, and **DHSPM-5**, possessing a hydrophilic 2-aminoethyl or 2,3-dihydroxypropyl groups attached to the carborane moiety, were similar to those of **SPD** and **SPM** themselves. They were much less toxic than **SPD-5** and **SPM-5**. Especially, low toxicity was observed for the non-carboranyl analogue, **BBSPD-5**, with a 4-(dihydroxyboryl)phenylmethyl group at N-5 of the **SPD** backbone.

The binding of polyamine analogues to DNA was determined by an ethidium bromide displacement assay (Table 2). Such assays have been used previously in the evaluation of DNA binding capacities of polyamines.[7,8] All polyamine analogues, possessing 2-aminoethyl and 2,3-dihydroxylpropyl groups at the 2-position of the *o*-

Table 1. *In vitro* IC_{50} of polyamine analogues

Compound	IC_{50}	Compound	IC_{50}	Compound	IC_{50}
SPD	>100 μM[a,b]	**SPM**	>100 μM[a,b]	**DHSPD-5**	>100 μM[a]
SPD-5	22 μM[b]	**SPM-5**	22 μM[b]	**DHSPM-5**	>100 μM[a]
ASPD-5	>100 μM[a]	**ASPM-5**	>100 μM[a]	**BBSPD-5**	>5 mM[a]

The IC_{50} value is the concentration of the test compound required to produce a 50% reduction in F98 glioma cells: a) in sulforhodamine B (SRB) absorption (Skehan, P. *et al.*, *National Cancer Institute*, vol. 82(13), July 4, 1990); b) in the uptake of [^{3}H]TdR (Ref. 8).

Table 2. DNA binding assay

Compound	IC_{50} (μM)	Compound	IC_{50} (μM)	Compound	IC_{50} (μM)
SPD	105 ± 8.7	**SPM**	3.0 ± 1.7	**DHSPD-5**	66.7 ± 5.7
SPD-5	147	**SPM-5**	23	**DHSPM-5**	21.7 ± 3.8
ASPD-5	3.8 ± 0.6	**ASPM-5**	4.0 ± 0.9	**BBSPD-5**	115.0 ± 7.1

Polyamine displacement of ethidium bromide from calf thymus DNA. Ethidium bromide and DNA concentrations were 1.6 and 10μg, respectively. IC_{50} values are the concentration of polyamine required to decrease the fluorescence of the ethidium bromide-DNA complex by 50%.

carborane cages had stronger affinity for DNA than did the corresponding **SPD-5** and **SPM-5** analogues.[6] The enhancement of DNA affinity was observed for **ASPD-5**, **DHSPD-5** and **BBSPD-5**, as compared with natural **SPD**; whereas **ASPM-5** showed comparable DNA affinity relative to natural **SPM**, **DHSPM-5** had diminished DNA binding properties. The best DNA binder was **ASPD-5**. These results demonstrated that the polyamine analogues have DNA binding properties that are similar to those of the naturally-occurring compounds.

The *in vitro* cellular uptake of carboranyl polyamines by F98 glioma cells was determined by measuring cellular boron content by means of direct current plasma atomic emission spectroscopy (DCP-AES).[9] Uptake of the tested compounds was compared with **SPD-5** and **SPM-5**[6] (Table 3), and two clinically used agents **BSH** and **BPA**.[6] Cellular uptake of the new carboranyl polyamines was lower than that obtained for **SPD-5** and **SPM-5**, but it was still higher than that obtained for **BSH** and **BPA**.

The above research has shown that **ASPD-5** and **BBSPD-5** possessed low cellular toxicity, are good DNA binders, and have significant *in vitro* cellular uptake. Therefore, the *in vivo* biodistribution studies of the compounds were performed in two different tumor models, intracerebral GL261 glioma and subcutaneous B16 melanoma cells. These results are shown in Tables 4 and 5.

Compound **BBSPD-5** by our method of *in vivo* evaluation showed unmeasurable levels of boron in tumor and for this reason is not considered as being an appropriate BNCT agent.

Table 3. *In vitro* cellular uptake studies

Compound	C_{media} (μg of B/mL)	$C_{cellular}$ (μg of B/g of cells) 48 h[a]	C_{media} (μg of B/mL)	$C_{cellular}$ (μg of B/g of cells)	
				24 h[b]	48 h[a]
SPD-5	0.5	43	0.5	65	43
ASPD-5	0.5	19	5.4	51	116
DHSPD-5	0.5	9	5.4	51	104
SPM-5	0.5	64			
ASPM-5	0.5	15	5.3	82	229
DHSPM-5	0.5	20	5.3	60	83
BBSPD-5			5.1		114.5
BPA			50	81	
BSH			570	39	

F98 cells were incubated for (a) 48 h and (b) 24 h with the polyamine analogues at molar compound concentrations corresponding to the boron amounts indicated in the table. Cells were washed, counted, and digested. Boron concentration C_{media} and cellular were determined by DCP-AES (10^9 cells ≅ 1 g of cells).

Table 4. Biodistribution of boronated polyamines **ASPD-5** and **BBSPD-5** in mice with intracerebral glioma

	μg boron/gm tissue	
Tissue	ASPD-5[a]	BBSPD-5[b]
Tumor	7.0 ± 3.8	nm
Brain	1.6 ± 1.0	nm
Blood	3.4 ± 2.1	nm
Liver	51.3 ± 27.3	22.9 ± 10.9
Spleen	30.9 ± 13.2	nm
Kidney	112.0 ± 53.2	44.0 ± 12.6
Skin	6.3 ± 3.8	nm
Muscle	5.3 ± 1.9	nm
Eyes	nm	nm
T/Bl	1.7	
T/Br	4.4	

C57Bl mice were implanted intracerebrally with GL261 brain tumors. Animals were placed on a polyamine deficient diet and water containing 2% DFMO for 3 days. Alzet® pumps were implanted and compound infusion was carried out for 3 days while the polyamine deficient diet continued for this period. Animals were sacrificed at the end of this 3 day period, and tissues removed for boron analysis. 5 mice per group: compound given a) 30 mg compound/Kg per day; b) 10 mg boron/Kg per day.

Though the tumor boron levels arising from **ASPD-5** were insufficient of itself to be a useful BNCT agent, it remains to be determined whether it can be used as a component in a BNCT cocktail. Nevertheless, the boronated polyamines that have been synthesized to date do not appear to be very promising as shown by the very high concentrations observed in the nontarget tissues of liver, spleen and kidney compared with those in tumor.

Table 5. Biodistrubution of **ASPD-5** in mice with B16 melanoma

	μg boron/gm tissue	
Tissue	A	B
Tumor	16.2 ± 5.1	18.5 ± 2.8
Brain	2.0 ± 0.7	2.1 ± 1.1
Blood	5.3 ± 3.5	2.9 ± 1.1
Liver	76.8 ± 12.2	87.5 ± 12.3
Spleen	25.8 ± 14.5	37.7 ± 14.2
Kidney	157 ± 36.3	192.5 ± 27.3
Skin	4.4 ± 1.3	7.1 ± 1.8
Muscle	7.9 ± 3.1	9.2 ± 1.3
Eyes	8.1 ± 3.5	6.4 ± 0.5
T/Bl	3.1	6.3
T/Br	8.1	8.8

C57Bl mice were implanted subcutaneously with B16 melanoma. Animals were placed on a polyamine deficient diet and water containing 2% DFMO for 3 days. Alzet® pumps were implanted and compound infusion was carried out for 4 days while the polyamine deficient diet continued for this period. Animals were sacrificed at the end of this 4 day period, and tissues removed for boron analysis. 5 mice per group: (A) 30 mg/Kg compound; (B) 40 mg/Kg compound.

In conclusion, these *in vitro* biodistribution studies have ruled out **BBSPD-5** and probably **ASPD-5** as well, as potential BNCT agents.

ACKNOWLEDGMENT

We acknowledge the support for our research from U.S. Public Health Service Grants (R01 CA-71033 and P30 CA-16058) and the Comprehensive Cancer Center of The Ohio State University.

REFERENCES

1. A.H. Soloway, W. Tjarks, B.A. Barnum, F.-G. Rong, R.F. Barth, I.M. Codogni, and J.G. Wilson, The chemistry of neutron capture therapy. *Chem. Rev.* 98:1515–1562, 1998.
2. Y. Mishima (Ed.), "Cancer Neutron Capture Therapy", Plenum Press: New York, 1996.
3. E.S. Regate, S. Boggs, A. Grudziak, and M. Deutsch, Polyamines in brain tumor therapy. *J. Neuro-Oncol.*, 25:167–179, 1995.
4. C.W. Tabor, and H. Tabor, Polyamine. *Annu. Rev. Biochem*, 53:749–790, 1984.
5. N. Seiler, J.G. Delcros, and J.P. Moulinoux, Polyamine transport in mammalian cells. an update, *Int. J. Biochem. Cell Biol.*, 28:843–861, 1996.
6. J. Cai, A.H. Soloway, R.F. Barth, D.M. Adams, J.R. Hariharan, I.M. Wyzlic, and K. Radcliffe, Boron-containing polyamines as DNA targeting agents for neutron capture therapy of brain tumors: synthesis and biological evaluation. *J. Med. Chem.*, 40:3887–3896, 1997.
7. B.F. Cain, B.C. Baguley, and W.A. Denny, Potential antitumor agents. 28. Deoxyribonucleic acid poly-intercalating agents. *J. Med. Chem.*, 21:658–668, 1978.
8. H.S. Basu, H.C.A. Schwiegert, B.G. Feuerstein, and L.J. Marton, Effects of varaion in the structure of spermine on the association with DNA and the induction of DNA conformational changes. *Biochem. J.*, 269:329–334, 1990.
9. R.F. Barth, D.M. Adams, A.H. Soloway, *et al.* Determination of boron in tissues and cells using direct current plasma atomic emission spectroscopy. *Anal. Chem.*, 63:890–893, 1991.

PRECLINICAL DEVELOPMENT OF β-D-5-*o*-CARBORANYL-2′-DEOXYURIDINE (D-CDU) FOR THE TREATMENT OF MALIGNANT BRAIN TUMORS

Raymond F. Schinazi[1,2], Selwyn J. Hurwitz[1,2], Irina Liberman[1,2], Amy Juodawlkis[1,2], Junxing Shi[1,2], Dennis C. Liotta[1], Jeffrey Coderre[3], and Jeffrey Olson[1]

[1]Emory University and
[2]Veterans Affairs Medical Center
Atlanta, GA 30033
[3]Brookhaven National Laboratory
Upton, NY 11973-5000

1. INTRODUCTION

The major concerns in drug development for boron neutron capture therapy (BNCT) include the accumulation of sufficient boron in tumor tissue, while limiting exposure in neighboring non-tumor tissue. Since BNCT is a binary treatment where cellular toxicity is desired only following the irradiation of target tissue containing accumulated boron with epithermal neutrons, additional toxicity in the absence of neutrons may lead to undesirable side effects in non-targeted tissue. Therefore, compounds developed for BNCT should have low toxicity. The nucleoside derivative β-D-5-*o*-carboranyl-2′-deoxyuridine (D-CDU)[1] (Figure 1) was selected for development, since it was essentially non-toxic in cell culture and rodents, and contains ten boron atoms per molecule in a lipophilic *closo*-carboranyl moiety. This moiety enhances the lipophilicity of the nucleoside, resulting in an octanol:water partition coefficient (π) of 3,050 at pH 7.4,[2] which should result in increased permeability of the nucleoside through lipophilic membranes.

The cellular toxicity and pharmacology, pharmacokinetics, tissue distribution and *in vivo* toxicity were all considered in selecting D-CDU as a lead compound for the treatment of malignant brain tumors using BNCT. A pilot study was then initiated to test the efficacy of BNCT with D-CDU to treat malignant brain tumors by making use of rats bearing intracranial 9L glioma cell tumors.

Frontiers in Neutron Capture Therapy, edited by Hawthorne *et al.*
Kluwer Academic / Plenum Publishers, New York, 2001.

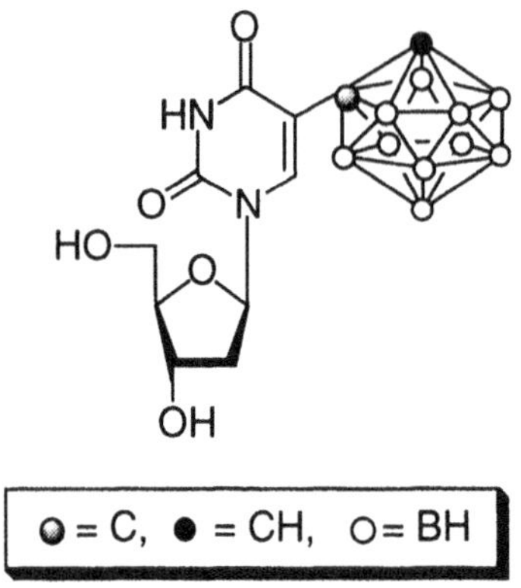

Figure 1. Structure β-D-5-*o*-carboranyl-2′-deoxyuridine (D-CDU).

2. RESULTS AND DISCUSSION

D-CDU had a low toxicity against CEM human lymphocyte (IC_{50} = 71 μM), U-251 human glioma (IC_{50} > 100 μM), and rat 9L glioma (IC_{50} = 83 μM) cells. Cellular accumulation of D-CDU (1 μM) was similar in human U-251 and rat 9L glioma cells with plateau levels of ~100 times the extracellular concentration based on measured cell volumes, and was reached within 10 min. About 5% of the D-CDU accumulated in the cells did not egress the cells readily when medium was replaced with drug free medium. This corresponds to a concentration of ≤5 times the original extracellular concentration.[3] The accumulation of D-CDU in CEM cells was insensitive to the nucleoside uptake inhibitors *S*-(4-nitrobenzyl)-6-thioinosine and dipyridamole, but was sensitive to the nucleoside-base inhibitor papaverine suggesting that D-CDU accumulates more like a base than a nucleoside.[3] Other nucleoside derivatives such as acyclovir have also been shown to enter cells through a similar nucleoside base accumulation process.[4]

D-CDU showed no noticeable toxicity when administered intraperitoneally (i.p.) at doses up to 150 mg/kg in male Fischer 344 rats (200–250 g). No abnormal behavior or loss in body weight was noted for 10 days. We have previously studied the two-compartment pharmacokinetics in rats injected i.p. with 25 mg/kg D-CDU. The α and β half-lives were 0.25 and 1.26 hr, respectively, and the plasma concentration at 1 hr post administration was 13 μM.[2] The serum concentrations at 1 hr of 2 additional rats given 5 mg/kg were 1.73 ± 0.3 μM, suggesting that the pharmacokinetics are linear in this dose range. Since the α half-life represents distribution of compound between tissues, equilibrium between D-CDU should be achieved within 2 hr (>4 α half-lives). A pilot study was initiated in which rats were irradiated with neutrons 2 hr after i.p. dosing.

Four groups of rats with intracranial 9L tumors were treated as follows: Group 1 (n = 6) was untreated; Group 2 (n = 6) received neutron irradiation only; Groups 3 (n = 6) and 4 (n = 7) received a single i.p. dose of 30 mg/kg and 150 mg/kg of D-CDU (20% ^{10}B enriched), respectively, 2 hr before neutron therapy. An additional set of animals (n = 3 per D-CDU group) were sacrificed at 2 hr to measure boron levels in the tumor and brain.

The collimator used for the brain tumor irradiations consisted of Li_2CO_3 dispersed in polyethylene, 93% enriched in ^6Li, 7.5% total Li by weight, 10.2 cm thick, with a conical aperture 12 cm in diameter at the reactor side tapering to 2 cm diameter at the rat

Table 1. Tissue levels at 2 hr after intraperitoneal dosing of D-CDU to rats and the median survival time after BNCT

D-CDU (i.p. dose)	μg B/g tumor[a] (mean ± SD)	Tumor/brain[a] ratio	Brain/blood[a] ratio	Median survival time (days)[b]
Untreated	0			20
Neutrons only (n)	0			32
30 mg/kg + n	2.3 ± 1.6	11.5 ± 3.6	0.10 ± 0.06	55
150 mg/kg + n	7.4 ± 1.3	6.8 ± 2.0	0.17 ± 0.02	35

[a]3 rats per group; [b]6 rats per group.

irradiation point.[5] The irradiation geometry and the platform to hold the rat have been reported. At 1 MW reactor power, the entrance thermal neutron flux was $7.17 \times 10^9 cm^{-2}s^{-1}$.[5] Assuming that the tumor extends from 3 mm to 7 mm beneath the skin surface, the geometrical mean of the thermal neutron fluence rate in the tumor volume was $5.1 \times 10^9 cm^{-2}s^{-1}$ and the minimum fluence rate was $4.4 \times 10^9 cm^{-2}s^{-1}$. At 3 MW reactor power, the mean dose rates to the tumor volume were: boron capture, 7.95 cGy/min/μg ^{10}B; fast neutrons, 27.0 cGy/min; total gamma, 13.8 cGy/min; nitrogen capture, 18.3 cGy/min. All neutron irradiations were for 24 MW-min (3 MW reactor power for 8 min).

Untreated rats had a median survival of 20 days and none survived longer than 27 days. Rats treated with neutrons alone survived significantly longer than untreated rats and had a median survival of 32 days ($p < 0.002$, logrank test, PHREG procedure Ver. 6. SAS Institute Inc, Cary, NC, 1996). The group receiving 30 mg/kg D-CDU survived significantly longer than the beam only group (median survival = 55 days, $p = 0.036$). The delay in mortality for the highest group (150 mg/kg D-CDU) was not significantly greater than the beam only group (median survival = 35 days). This result may be related to the lower selectivity of D-CDU at the highest dose, based on the tumor/brain and brain/blood ratios (Table 1).

This suggests that an optimal dose may exist for D-CDU between 30 and 150 mg/kg for 20% enriched D-CDU. It is anticipated that a 100% enriched product would deliver a 5-fold higher tumor level of B-10 for each molecule of D-CDU accumulated. If the decreased survival of rats receiving the higher dose results from a decreased specificity for brain tumor cells, then a 100% enriched product would be beneficial. D-CDU could then be given at a dose enabling sufficient B-10 to accumulate in brain tumors, while maintaining a high brain tumor/normal brain B-10 ratio, resulting in an increased therapeutic ratio for D-CDU. The concentration in tumors was only 7.4 μg B/g after a dose of 150 mg/kg.[6] This is below the threshold of 25–30 μg B/g that is required based on dosimetry studies performed using boric acid, which distributes relatively evenly in biological fluids and cellular compartments. However D-CDU is a more complex molecule than boric acid, since it is monophosphorylated in cells and may be reduced to the anionic *nido*-form. Therefore, it is likely that the cellular distribution of these D-CDU metabolites is probably not as uniform as boric acid and may require more complex dosimetry. Sub-cellular localization or complexation with cellular organelles and/or macromolecules may result in subcellular micro-environments where boron concentrations are higher than average. These studies are sufficiently promising to warrant the synthesis of 100% enriched D-CDU to further develop D-CDU as an agent for the neutron therapy of brain tumors. Additional pharmacodynamic studies are indicated to determine optimal dosing strategies for D-CDU for the treatment of malignant brain tumors.

ACKNOWLEDGMENTS

This work was supported in part by Department of Energy grants DE-FG02-96ER62156 and DE-AC02-98CH10888 and the Department of Veterans Affairs.

REFERENCES

1. R.F. Schinazi, N.M. Goudgaon, G. Fulcrand, Y. El Kattan, Z. Lesnikowski, G.V. Ullas, J. Moravek, and D.C. Liotta, Cellular pharmacology and biological activity of 5-carboranyl-2′-deoxyuridine, *Intl J Radiation Oncol Biol Phys.* 28(5):1113–1120, 1994.
2. V.R. Jarugula, R.F. Schinazi, G. Fulcrand-El Kattan, D.C. Liotta, and F.D. Boudinot, Pharmacokinetics of 5-carboranyl-2′-deoxyuridine in rats, *J Pharm Sci.* 83(12):1697–1699, 1994.
3. S.J. Hurwitz, L. Ma, A. Eleuteri, J. Wright, J. Moravek, and R.F. Schinazi, Cellular pharmacology of the D- and L-enantiomers of β-5-*o*-carboranyl-2′-deoxyuridine. *Nucleosides Nucleotides Nucleic Acids* 19:691–702, 2000.
4. W. Mahony, B. Domin, R. McConnell, and T.J. Zimmerman, Acyclovir transport into human erythrocytes. *J. Biol. Chem.* 263:9285–9291, 1998.
5. H.B. Liu, D.D. Joel, D.N. Slatkin, and J.A. Coderre, Improved apparatus for neutron capture therapy of rat brain tumors, *Int. J. Radiat. Oncol. Biol. Phys.* 28:1167–1173, 1994.
6. J.A. Coderre, and G.M. Morris, The radiation biology of boron neutron capture therapy. *Rad. Res.* 151:1–18, 1999.

SELECTIVE TUMOR UPTAKE OF BORONOPHENYLALANINE-FRUCTOSE (BPA-F) IN AN ANIMAL MODEL OF HEPATIC COLORECTAL METASTASES

Cynthia Chuang[1], Paul Busse[2], Matthew Palmer[3], Peter Thomas[4], and Robert Zamenhof[3]

[1]Nuclear Reactor Laboratory
Massachusetts Institute of Technology
Cambridge, MA 02139
[2]Joint Center for Radiation Therapy
[3]Department of Radiology
[4]Department of Cancer Biology
Beth Israel Deaconess Medical Center
Harvard Medical School
Boston, MA 02215

1. INTRODUCTION

In 1998 more than 150,000 Americans will be diagnosed with colorectal cancer, and over 60,000 will die from this tumor. At the time of diagnosis, 15–25% will have hepatic metastases, and another 20–30% will develop metastatic hepatic lesions subsequent to resection of the primary tumors. Treatment options for hepatic metastases include aggressive surgical resection, administration of regional chemotherapeutic agents, systemic chemotherapy, radiotherapy, or other palliative and experimental procedures. Despite well-defined pharmacological benefits of selected chemotherapeutics, there is no convincing survival advantage for many of the above mentioned techniques.[1,2,3,4] The median survival after diagnosis is typically 22–24 months, and has changed minimally over the past decade.

Colorectal metastases are hypovascular and can continue to grow under hypoxic conditions. Thus, the use of an actively transported boron-containing compound selectively localized in tumor and the high LET radiation produced by BNCT could be a useful therapeutic approach to this clinical problem. This approach has previously been considered by the BNCT group at the University of Pavia in Italy, where animal experiments were conducted using boric acid, sodium borocaptate (BSH), and ^{10}B-conjugated

Frontiers in Neutron Capture Therapy, edited by Hawthorne *et al.*
Kluwer Academic / Plenum Publishers, New York, 2001.

tetraphenyl porphyrin sulfonate (TPPS), and toxicity and biodistribution studies following intravenous injection of these compounds were performed.[5]

In the present study, a well-characterized animal model using the athymic mouse is being used to gain information on the biodistribution of boronophenylalanine-fructose (BPA-f) in hepatic colorectal metastases and normal liver tissue as a function of time. Results indicate that there is preferential uptake of BPA-f in hepatic colorectal metastatic cells relative to normal liver. This paper reports the results of a complete temporal biodistribution of BPA-f in blood, skin, hepatic colorectal metastasis and normal liver.

2. MATERIALS AND METHODS

2.1. The CX-1 Hepatic Colorectal Metastasis Model

Tumor cells are grown and harvested in collaboration with the Tumor Biology Lab of the Department of Cancer Biology at the Beth Israel Deaconess Medical Center. Human CX-1 colorectal carcinoma cells (10^7 cells/ml phosphate buffered saline in pH 7.4) is used. The CX-1 cells are maintained in RPMI 1640 medium (GIBCO Laboratories Life Technologies Inc., Grand Island, NY) with 10% fetal bovine serum (FBS; SIGMA Chemical Company, St. Louis, MO), 1% L-Glutamine (GIBCO), 100 units/ml of penicillin G (GIBCO) and 100 mg/ml streptomycin (GIBCO). After induction of anesthesia in the athymic nude mouse (intraperitoneal Pentobarbital, 30 mg/kg body weight), a small left subcostal incision was made and 0.1 ml of the cell suspension (a total of 10^6 cells) was injected into the spleen of the mice using a 26-gauge needle. Intrasplenic injection of tumor cells resulted in multiple hypovascular metastasis which were well established after 6 weeks.

Athymic nude mice bearing hepatic colorectal metastases were given a bolus injection of BPA-f (350 mg BPA-f/kg body weight) through the tail vein using a 26-gauge needle. This is the same dose as is currently being used in the BNCT clinical trials at our institutions. Five mice each were sacrificed at 7 different time points (5, 10, 15, 30, 60, 120, and 240 min after injection). Cardiac puncture was performed to collect blood samples. Samples of skin, tumor and normal liver tissues were removed after the death of the animal.

2.2. Boron Analysis

Boron analysis was performed by two methods: prompt gamma neutron activation analysis (PGNAA) and inductive coupled plasma-atomic emission spectroscopy (ICP-AES). The PGNAA method is a nuclear analytical technique that detects the 478 keV photons produced from the de-excitation of the lithium recoil nucleus resulting from the $^{10}B(n, \alpha)^7Li$ reaction. The number of counts integrated in the 478 keV peak is proportional to the number of boron atoms present in the sample. Samples were pipetted into Teflon vials, and placed one by one manually into the 4DH3 beam port of the MIT-II Research Reactor. The measurements proceeded until less than 10% error in ^{10}B counts was obtained (most of the higher concentration samples had less than 5% error). System calibration was performed on each day of measurement using U.S. National Institute of Standards and Technology ^{10}B-enriched boric acid. The blood sample ^{10}B concentrations were measured by PGNAA method.

Inductive coupled plasma-atomic emission spectroscopy (ICP-AES) is an atomic excitation technique. It detects the characteristic de-excitation photons from all isotopes of boron and the number of detected photons is proportional to the amount of boron element present in the sample. Tissue samples were weighed; for every 50 mg of tissue, 0.15 ml of nitric acid: sulfuric acid (1 : 1) mixture was added, and digestion was allowed to proceed for an hour in a 60 degree C waterbath. 0.5 ml of 10% Triton X-100 was then added and the acid-digested solution was diluted to a final volume of 1 ml.[6] The solutions were filtered and their B concentrations measured using ICP-AES. Standards were prepared in a similar manner using U.S. National Institute of Standards and Technology ^{10}B-enriched boric acid. System calibration was performed on each day of measurement. The boron standards for the ICP-AES were also analyzed with the PGNAA system to provide system cross-calibration.

3. RESULTS

Thirty five athymic mice bearing hepatic colorectal metastasis were used for the biodistribution studies. Most mice had multiple metastatic nodules while a few (5) had more than half of their livers consumed by tumor. After receiving a bolus injection of BPA-f (350 mg BPA-f/kg body weight) through the tail vein, they were sacrificed at the prescribed time points. Their bloods were drawn at sacrifice and measured using PGNAA. Samples of skin, normal liver, and tumor were collected after death and were analyzed by ICP-AES after acid-digestion. Figure 1 shows the time dependence of boron accumulation in blood, skin, normal liver, and tumor. It is clear that both the skin and the normal liver follow a similar time course as the blood, which is to be expected since they are both highly vascularized organs. Due to the hypovascular nature of the hepatic colorectal metastasis, the boron accumulation in the tumor lagged that of the normal liver, but quickly built up and surpassed that of the normal liver. By 15 min after the bolus injection the metastases had 1.5 times the boron concentration in blood, increasing to approximately 2.5 times by 60 min after injection, and decreasing again to about 1.5 times at 240 min. Similar trends were observed for tumor/normal liver, and tumor/skin. Tumor boron was approximately 1.3 times that in normal liver at 15 min, 1.8 times at 30 min, 2 times at 60 min and 120 min, and barely 1 at 240 min after the bolus injection.

4. CONCLUSION

The attainment of a tumor therapeutic ratio of greater than unity in liver, a highly vascularized organ, represents a challenge for BNCT. Nevertheless, in the present study encouraging results for BPA-f uptake in hepatic colorectal metastases were obtained. In the athymic mouse model, during 30–60 min after a bolus injection, the differential uptake of BPA-f in the hepatic metastases and normal liver tissue was approximately equal to 2 (T/NT). While this ratio is lower than that seen in experimental brain tumor models, it is encouraging given the biology of this setting. Additional studies will be necessary to derive the RBE's for the liver and the hepatic colorectal metastases. Once obtained it will be possible to generate detailed dosimetric calculations and to determine the suitability of BNCT for the treatment of hepatic colorectal metastasis.

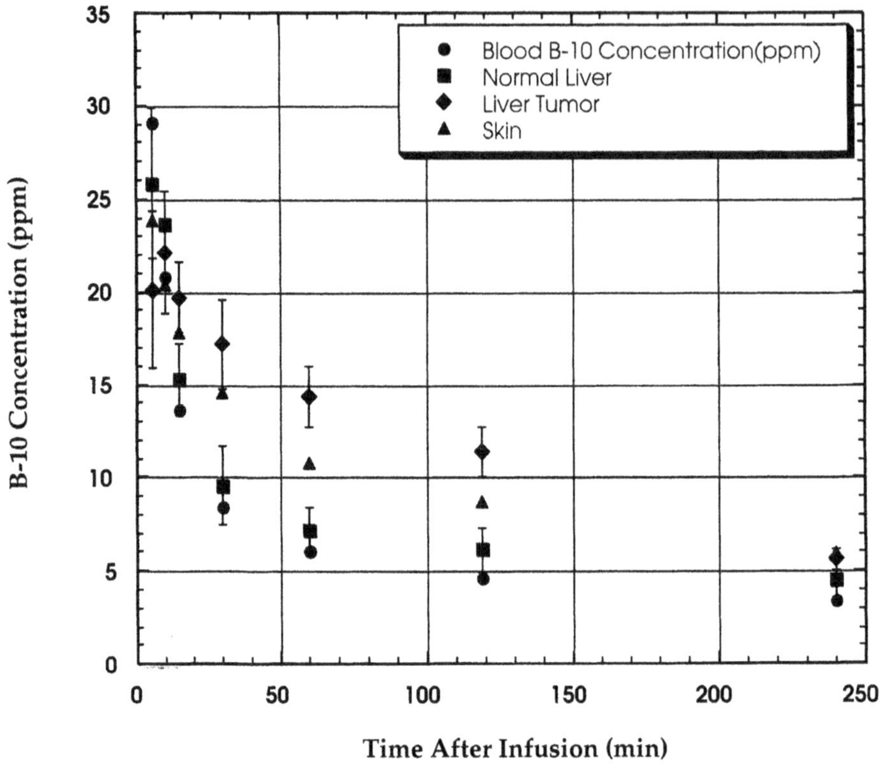

Figure 1. Boron concentrations in blood, skin, normal liver tissue and hepatic colorectal metastasis as a function of time following a bolus tail vein injection of BPA-f that delivered the equivalent of 350 mg BPA-f/kg body weight.

ACKNOWLEDGMENTS

We thank Peggy Micca from BNL and Kent Riley from MIT NRL for their initial technical instructions for boron analysis techniques. We also thank Dr. Osman Cay from BIDMC Dept. of Radiology for his preliminary exploration of this topic which leads to this work, and David Northey from Dana Farber Cancer Institute for continuing help in handling animals. This research was supported by the Department of Energy under DOE grant # DE-FG02-97ER62193.

REFERENCES

1. K. Hanazaki, N. Kawamura, H. Sodeyama, M. Wakabayashi, S. Yokoyama, Y. Sode, T. Miyazaki, and Y. Okazaki, Carcinoembryonic antigen and intra-arterial chemotherapy response of liver metastases, *Hepatogastroenterology*, 45(20):462–467, 1998.
2. M. O'Connell, D. Nagorney, A. Bernath, and G. Schroeder, Sequential intrahepatic fluorodeoxyuridine and systemic fluorouracil plus leucovorin for the treatment of metastatic colorectal cancer confined to the liver, *J. Clin. Oncol.*, 16(7):2528–2533, 1998.
3. B. Hill, T. Minard, and L. Brady, Mococlonal antibodies in the treatment of metastatic carcinoma to the liver: undated report of a pilot study including leukapheresis, *Am. J. Clin. Oncol.*, 21(3):291–293, 1998.

4. J. Seitz, P. Rougier, and M. Giovannini, Recent Advances in systemic chemotherapy of metastatic colorectal cancer, *Ann. Ital. Chir.*, 67(6):805–812, 1996.
5. T. Pinelli, S. Altieri, F. Fossati, and A. Zonta, Development of a method to use boron neutron capture therapy for diffused tumours of liver (Taormina Project), *in*: "Cancer Neutron Capture Therapy," Y. Mishima ed., Plenum Press, New York, 1996, pp. 783–794.
6. J. Coderre, A. Chanana, D. Joel, E. Elowitz, P. Micca, M. Nawrocky, M. Chadha, J. Gebbers, M. Shady, N. Peress, and D. Slatkin, Biodistribution of Boronophenylalanine in Patinets with Glioblastoma Multiforme: Boron Concentration Correlates with Tumor Cellularity, *Rad. Res.*, 149:163–170, 1998.

PRECLINICAL DOSIMETRY

BNCT MICRODOSIMETRY OF A RAT GLIOMA MODEL WITH BPA-F OR BSH

W. S. Kiger III[1], H. Patel[2], and R. G. Zamenhof[3]

[1]Nuclear Reactor Laboratory
Massachusetts Institute of Technology
[2]Research Institute
Churchill Hospital
University of Oxford
[3]Department of Radiology
Beth Israel Deaconess Medical Center
Harvard Medical School

1. INTRODUCTION

The BNCT microdosimetry technique developed in the Harvard-MIT BNCT Program involves both experimental and computational components. First, a neutron-induced alpha particle track etch technique developed in our laboratory termed High Resolution Quantitative Autoradiography (HRQAR) is used to measure the microscopic distribution of ^{10}B in a thin tissue section.[1] HRQAR allows simultaneous visualization of tissue histology and track etch pits, which sample the ^{10}B microdistribution. The polycarbonate track detector used in the technique is sensitive only to high LET particles (not protons) and has a linear response and sensitivity down to ~1 μg/g ^{10}B. By employing the actual boron microdistribution and tissue morphology obtained with the HRQAR technique in a 2D Monte Carlo simulation of charged particle transport and energy deposition in tissue, our approach to BNCT microdosimetry circumvents the simplifying assumptions usually required in microdosimetry, i.e., using elementary mathematical shapes and functions to represent the tissue architecture and boron distribution. This paper will discuss microdosimetric results obtained for 9L gliosarcoma in rats injected with boronophenylalanine-fructose (BPA-F) or $Na_2B_{12}H_{11}SH$ (BSH) using the HRQAR-based microdosimetry technique.

Frontiers in Neutron Capture Therapy, edited by Hawthorne *et al.*
Kluwer Academic / Plenum Publishers, New York, 2001.

2. METHODS AND MATERIALS

2.1. Sample Acquisition/Preparation

9L Gliosarcoma bearing Fischer F344 rats were injected intra-peritoneally with BPA-F or BSH. BPA-F was administered at a dose of 700 mg/kg body weight. Rats were sacrificed by decapitation at 180 min after BPA-F administration. BSH was dissolved in saline and administered at a dose of 200 mg/kg body weight. Rats were sacrificed by decapitation at 30 min after BSH administration. Tumor samples were extracted and frozen for sectioning and HRQAR analysis.

2.2. Histology and Boron Microdistribution

The tumor samples were sectioned at 4 μm thickness and the HRQAR procedure was performed as previously described with a few modifications.[1] HRQAR autoradiograms were imaged with a video camera mounted on a Leitz Orthoplan microscope. A 63 × objective, specially corrected for use without coverslips, provided especially sharp images of the plastic track detector surface containing the track etch pits and the hematoxylin and eosin (H & E) stained histology below. Still images were acquired using OpenLab imaging software (Improvision, Inc., Coventry, UK) running on a Power Macintosh computer with a framegrabber. Autoradiogram images were processed with MATLAB (The Mathworks, Natick, MA) and Adobe Photoshop to produce segmented images of the tracks (the ^{10}B microdistribution) and segmented images of the histology. Each of the nuclei in a segmented histological image was given a distinct color index to allow adjacent nuclei to be easily distinguished by the microdosimetry software.

2.3. Microdosimetry Analysis

Microdosimetry analysis is carried out with a 2D Monte Carlo simulation of the energy deposition of the ^{10}B reaction products in a thin tissue section. The simulation is implemented in MATLAB. Tissue microanatomy (the segmented tissue image) forms the geometric model for charged particle transport. The actual microdistribution (from the image of the tracks) or an assumed boron microdistribution used for comparison comprises the spatial distribution of the source for the charged particle transport simulation of the BNC reaction. Neutron capture event sites are sampled from the spatial distribution of the source. The trajectories of the ^{10}B reaction products are sampled isotropically and are antiparallel. The reaction products move in straight lines in the simulation, which is a good approximation for these heavy charged particles. Stopping power distributions in ICRU brain tissue[2] were calculated for the charged particles produced by each of the two pathways of the $^{10}B(n,\alpha)$ reaction using the charged particle transport code SRIM98.01.[3] The stopping power is assumed not to vary between different microanatomical structures in the microdosimetry simulations. To eliminate edge effects, i.e., transport off the edge of the image, source points were not allowed within a buffer along the 4 edges of the image. The width of this buffer was set equal to the maximum range of the most penetrating charged particle. Due to the high magnification, the buffer represents a significant fraction of the field and unfortunately reduces the sample size significantly.

Single event distributions for 4 microdosimetric parameters were tallied for charged particle hits to cell nuclei (i.e., nuclear traversals): track length, energy imparted, lineal energy,[4,5]* and dose-weighted LET[6] (DWLET). Traversals of different nuclei were tallied distinctly to avoid inappropriate summation. In addition to measuring the nuclear fraction, the fractional area occupied by nuclei in the image, which reflects nuclear volume fraction in the tissue, two additional parameters were tallied, hit frequency and hit fraction. Hit frequency is defined as the mean number of hits to distinct nuclei per neutron capture event. Hit fraction is the mean probability that a neutron capture event results in at least one hit.

In this microdosimetric technique, the principle of 2D surrogate modeling, derived from the mathematical field of stereology, is invoked to effectively obtain full 3D Monte Carlo calculations for the tissues of interest based on the actual microanatomy.[7,8] The principle of the 2D surrogate approach is that a single plane passing through the tissue (i.e., a histological section) which contains the cross sections of many randomly oriented cells is equivalent to sampling a single effective cell with a large number of randomly oriented planes. Thus, in our approach to BNCT microdosimetry, unlike the usual approach in which a single cell or a single cell in a matrix of its nearest neighbors is modeled, microdosimetric parameters are tallied over many cells to obtain microdosimetric data for a single effective cell, which, in an average sense, is representative of all cells in the tissue.

3. RESULTS AND DISCUSSION

Results of BNC microdosimetry simulations for 9L rat brain tumor infused with BPA or BSH are listed in Table 1. Frequency means of the 4 single event distributions are reported on a per hit basis (hit mean) and on a per neutron capture event basis (event mean). In addition to calculations using the actual ^{10}B microdistributions, simulations were performed with a uniform microdistribution for all samples and with extranuclear distributions for the BSH samples. The calculations using strictly extranuclear ^{10}B microdistributions were performed for the BSH samples because BSH is believed by some researchers to locate principally in the extracellular space. Because, in these small samples (72 μm by 55 μm fields), the nuclear fraction varies significantly, calculations using the uniform distribution provide a valuable baseline for comparison. Larger fields or larger numbers of fields would provide greater statistical rigor. Table 2 presents the actual to uniform microdosimetric parameter ratios for each sample. The ratios are used in order to normalize out the effect of different nuclear fraction in the different samples, enabling a comparison of the different microdistributions. In each simulation 400,000 to 800,000 ^{10}B neutron capture events were sampled so that the Monte Carlo statistical uncertainty in all reported quantities is very low, less than 0.3%.

Analysis of the microdosimetric parameter ratios showed that measured ^{10}B microdistributions for BPA and BSH in 9L gliosarcoma do not appear to produce microdosimetry significantly (±10%) different from a uniform microdistribution. An extranuclear ^{10}B microdistribution in 9L gliosarcoma, however, produces microdosimetric parameters 30 ± 15% lower than those for a uniform distribution, suggesting that the BSH microdistribution may not be strictly extranuclear as supposed. However, some researchers report a dynamic microdistribution of ^{10}B following BSH infusion in a rat

* Rather than use the usual definition of lineal energy, energy imparted divided by mean chord length of the sensitive site, we have employed a more physically relevant definition: energy imparted divided by track length.

Table 1. Microdosimetric parameters calculated for GS9L rat brain tumors with measured ^{10}B microdistributions for BPA or BSH or assumed uniform or extranuclear distributions[a]

Sample # 7.24.98	^{10}B microdistribution	Nuclear fraction	Hit freq.	Hit fraction		Track length µm	Energy imparted keV	Lineal energy keV/µm	DWLET $keV^2/µm$
#29D2Tum3	Actual BPA	0.351	1.576	0.864	Event mean	3.356	614.8	138.1	1.37E + 05
					Hit mean	3.681	674.3	151.5	1.50E + 05
	Uniform	0.351	1.525	0.854	Event mean	3.391	628.0	139.0	1.41E + 05
					Hit mean	3.761	696.7	154.3	1.56E + 05
#43D2Tum11	Actual BPA	0.424	1.764	0.912	Event mean	2.988	540.4	142.5	1.20E + 05
					Hit mean	3.167	572.7	151.0	1.27E + 05
	Uniform	0.424	1.815	0.907	Event mean	3.157	585.6	146.6	1.31E + 05
					Hit mean	3.348	621.0	155.5	1.39E + 05
#54D4Tum3	Actual BSH	0.274	0.941	0.660	Event mean	2.333	418.7	105.0	9.23E + 04
					Hit mean	3.281	588.8	147.7	1.30E + 05
	Uniform	0.274	0.912	0.643	Event mean	2.332	426.2	105.3	9.50E + 04
					Hit mean	3.349	612.1	151.2	1.36E + 05
	Extranuclear	0.274	0.698	0.541	Event mean	1.595	255.9	75.9	5.24E + 04
					Hit mean	2.773	444.8	131.9	9.10E + 04
#54D4Tum7	Actual BSH	0.349	1.455	0.860	Event mean	3.405	632.5	136.6	1.42E + 05
					Hit mean	3.792	704.4	152.1	1.58E + 05
	Uniform	0.349	1.446	0.864	Event mean	3.304	610.4	139.0	1.37E + 05
					Hit mean	3.663	676.5	154.1	1.51E + 05
	Extranuclear	0.349	1.154	0.773	Event mean	2.339	372.4	106.3	7.60E + 04
					Hit mean	2.882	459.0	131.0	9.37E + 04

[a] "Hit mean" parameters are averaged over only hits—traversals of distinct nuclei—excluding misses. "Event mean" parameters are averaged over all events, including misses. Monte Carlo statistical uncertainty in all quantities is less than 0.3%. DWLET denotes dose-weighted LET.

Table 2. Ratios microdosimetric parameters calculated for GS9L rat brain tumors with measured ^{10}B microdistributions for BPA or BSH or assumed extranuclear distributions relative to an assumed uniform distribution for the same sample[a]

Sample # 7.24.98	^{10}B Microdist.	Hit freq.	Hit fraction		Trach length μm	Energy imparted keV	Lineal energy keV/μm	DWLET $keV^2/\mu m$
#29D2Tum3	*BPA*	1.034	1.012	Event mean	0.990	0.979	0.993	0.973
	Uniform			Hit mean	0.979	0.968	0.982	0.962
#43D2Tum11	*BPA*	0.972	1.006	Event mean	0.947	0.923	0.972	0.912
	Uniform			Hit mean	0.946	0.922	0.971	0.911
#54D4Tum3	*BSH*	1.033	1.027	Event mean	1.001	0.982	0.998	0.971
	Uniform			Hit mean	0.980	0.962	0.977	0.951
	Extranuclear	0.766	0.841	Event mean	0.684	0.600	0.721	0.551
	Uniform			Hit mean	0.828	0.727	0.873	0.667
#54D4Tum7	*BSH*	1.007	0.995	Event mean	1.030	1.036	0.983	1.041
	Uniform			Hit mean	1.035	1.041	0.988	1.046
	Extranuclear	0.798	0.894	Event mean	0.708	0.610	0.765	0.557
	Uniform			Hit mean	0.787	0.678	0.850	0.619

[a]"Hit mean" parameters are averaged over only hits—traversals of distinct nuclei—excluding misses. "Event mean" parameters are averaged over all events, including misses. Monte Carlo statistical uncertainty in all quantities is less than 0.3%. DWLET denotes dose-weighted LET.

glioma model, with an initially uniform distribution and several (8–16) hours later a principally intranuclear distribution.[9] This finding is not inconsistent with the result of the present study. However, recent work by Chandra *et al.* indicates that a short (60 s) thaw of *in vitro* cell specimens produces significant intracellular migration of BSH and BPA.[10] The possibility that this migration also occurs in short thaws of tissue sections raises questions about our sample preparation technique. At one point in the HRQAR procedure, before irradiation, the tissue section is briefly thawed for a few seconds to allow the tissue section to adhere to the track detector and its substrate. The sample otherwise remains frozen at −20 °C or below before and during irradiation. The complete cryogenic rigor in sample preparation recommended by Chandra *et al.* may be difficult to attain.

In the present study, our HRQAR-based microdosimetry technique has been used to evaluate the microdistributions of two boron compounds. The microdosimetry analysis will be extended to include other heavy charged particles present in BNCT radiation fields, e.g. protons, and to allow calculation of absolute microdosimetric quantities.

4. CONCLUSIONS

Microdosimetry simulations suggest that the ^{10}B microdistributions in GS9L bearing rats injected with BPA or BSH do not significantly differ from a uniform distribution. However, lack of complete cryogenic rigor in sample preparation may have caused migration of ^{10}B.

ACKNOWLEDGMENTS

This research was supported by the U.S. Department of Energy under contract DE-FG02-97ER62193.

REFERENCES

1. G.R. Solares, and R.G Zamenhof, A Novel Approach to the Microdosimetry of Neutron Capture Therapy, Part I: High-Resolution Quantitative Autoradiography Applied to Microdosimetry in Neutron Capture Therapy, Radiat. Res., 144:50–58, 1995.
2. International Commission on Radiation Units and Measurements, "Photon, electron, proton, and neutron interaction data for body tissues," ICRU Report 46, Bethesda, Md., U.S.A, 1992.
3. J.F. Ziegler, SRIM98, "The Stopping and Range of Ions In Matter," IBM Research, Yorktown, NY (1998). http://www.research.ibm.com/ionbeams/SRIM/SRIMINTR.HTM
4. International Commission on Radiation Units and Measurements, "Microdosimetry," ICRU Report 36, Bethesda, Md., U.S.A, 1983.
5. H.H. Rossi, and M. Zaider, "Microdosimetry and Its Applications," Springer Verlag, New York, 1996.
6. G. Solares, W.S. Kiger III, and R. Zamenhof, "Microdosimetry Studies at the Harvard/MIT Phase-I Clinical Trial of Boron Neutron Capture Therapy," in Advances in Neutron Capture Therapy, edited by B. Larsson, J. Crawford, and R. Weinreich, Elsevier, Amsterdam (1997).
7. C.S. Yam, "Microdosimetric Studies for Neutron Capture Therapy and Techniques for Capture Element Selection," Ph.D. Thesis, Massachusetts Institute of Technology, 1995.
8. C.S. Yam, G.R. Solares, and R.G. Zamenhof, Verification of the Two-Dimensional Approach for NCT Microdosimetry, Trans. Am. Nucl. Soc., 71:142, 1994.
9. T. Kageji, S. Nagahiro, K. Matsumoto, B. Otersen, D. Gabel, M. Nakaichi, and Y. Nakagawa, Subcellular biodistribution of $Na_2B_{12}H_{11}SH$ (BSH) in a rat glioma model, *in* "Frontiers in Neutron Capture Therapy," M.F. Hawthorne, K. Shelly, R.W. Wiersema, eds., Kluwer Academic/Plenum Publishers, New York, 2001, pp. 927–931.
10. S. Chandra, D.R. Lorey, D.R. Smith, and G.H. Morrison, *In Vitro* exposure of human T98G glioblastoma cells to mixed BNCT drugs: acquiring independent quantitative images of boron atoms delivered by ^{10}BPA-F and ^{11}BSH in the same cell by SIMS ion microscopy, *in* "Frontiers in Neutron Capture Therapy," M.F. Hawthorne, K. Shelly, R.W. Wiersema, eds., Kluwer Academic/Plenum Publishers, New York, 2001, pp. 905–909.

DETERMINATION OF THE COMPOUND FACTORS OF BSH AND BPA FOR GLIAL PROGENITOR SURVIVAL IN A SMALL ANIMAL MODEL USING THERMAL AND EPITHERMAL NEUTRON IRRADIATION

H. Philippo[1], E. A. M. Winter-van Kampen[1], S. Spanjer[1], R. Huiskamp[1], and A. J. van der Kogel[2]

[1]Health Physics
Netherlands Energy Research Foundation ECN
Petten, The Netherlands
[2]Institute of Radiotherapy
University Hospital Nijmegen
Nijmegen, The Netherlands

1. INTRODUCTION

The response of normal central nervous system (CNS) after BNCT of malignant glioblastoma multiforme is primarily dependent on the microdistribution of the used boron carriers and the microdosimetry within the target structure. The two boron carriers presently used for BNCT are BPA and BSH have very different distribution patterns. With BSH, which does not cross the blood-brain-barrier (BBB) radiation damage to the endothelial cells and brain parenchyma is primarily coming from α-particles originating from $^{10}B(n,\alpha)^{7}Li$ events occurring in the lumen of a blood vessel. In contrast BPA homogeneously distributes from blood to the brain parenchyma resulting in similar concentrations in blood and brain parenchyma and thereby in a more homogeneous irradiation of endothelial cells and brain parenchyma by these α-particles. The overall biological effectiveness of the boron dose from the $^{10}B(n,\alpha)^{7}Li$ reaction with a certain boron carrier has been termed compound factor[1] which describes the product of the microdistribution of the compound and the RBE of $^{10}B(n,\alpha)^{7}Li$ reaction for the target tissue. With the presently clinically used beams large animals had to be used to study normal brain tolerance to obtain late tissue effects at an acceptable whole-body radiation dose. In the present investigation a small animal *in vivo/in vitro* model was used

Frontiers in Neutron Capture Therapy, edited by Hawthorne *et al.*
Kluwer Academic / Plenum Publishers, New York, 2001.

to determine the compound factors for BSH and BPA for glial progenitor survival in the normal brain parenchyma after epithermal neutron irradiation.

2. MATERIAL AND METHODS

Male adult Wistar rats, 280–320gr., age 10–14 weeks obtained from Harlan CPB. Neonatal rats for preparation of astrocytes type I-monolayers were obtained by purchasing pregnant female Wistar rats delivering their pups at our Laboratory. Rats were given either 50mg ^{10}B/kg b.w. as BSH or BPA as iv. bolus injection or saline and irradiated with a thermal neutron beam(LFR) under anesthesia or with the clinically used HB11 epithermal neutron beam(HFR) in a perspex phantom without anaesthesia. 18mm build-up material was used to thermalize epithermal neutrons at the location of the optic nerve. Twenty-four hr. after the irradiation, glial progenitor cells (O2-A progenitors) were isolated from the optic nerves of the thermal or epithermal neutron irradiated rats and clonogenic survival was determined *in vitro*. The clonogenic assay consists of three main procedures: a) preparation of monolayers of type-1 astrocytes from the cortex of neonatal rats, b) preparation of the glial progenitor cell suspension from the optic nerve and c) immuno fluorescent staining with A_2B_5 monoclonal antibodies and counting of surviving clones. Details of the procedure have been published previously.[2]

Dose components in the phantom using the HB11 beam were determined using MCNP calculations. Dosimetry for the thermal beam was performed with TE and Mg/Ar chambers.

The thermal fluence for both irradiation setups was determined using the gold wires activation method. Wires were placed at the position of the optic nerve.

In order to obtain the relative biological effectiveness (RBE) for the (epi)thermal beam, beam-only survival curves were compared with 300kV X-rays survival curves. A weighted regression analysis was carried out using linear and linear-quadratic fit using a statistical software package (Stata Release 3.1). Survival is represented by the equation $S = e^{(-\alpha D-\beta D\wedge 2)}$, in which S = survival, D = dose and α and β are constants. In cases where β is not significant from 0 or negative, a linear fit was applied.

The RBE of (epi) thermal beam, was calculated using the equation: $RBE_{beam} = dose_{X\text{-}rays}/dose_{beam}$.

The RBE of the proton component of the (epi) thermal beam, i.e., fast neutron recoil protons plus $^{14}N(n,p)^{14}C$ protons, was calculated using the equation: Proton $RBE_{beam} = (dose_{X\text{-}rays} - \text{Gamma } dose_{beam})/\text{Proton } dose_{beam}$.

To obtain the compound factors for BPA and BSH survival curves with compounds and (epi)thermal neutron irradiation were compared with 300kV X-rays survival curves. The compound factor was calculated using the equation: Compound factor = $(dose_{X\text{-}rays} - \text{total } dose_{beam})/dose_{boron}$.

3. RESULTS

Survival of glial progenitor cells after thermal or epithermal neutron irradiation in combination with either BSH, BPA or saline (beam only) are shown in Figures 1 and 2 respectively. Survival was best represented with a linear dose effect relationship. The X-rays survival curve obtained from previous experiments[3] is added.

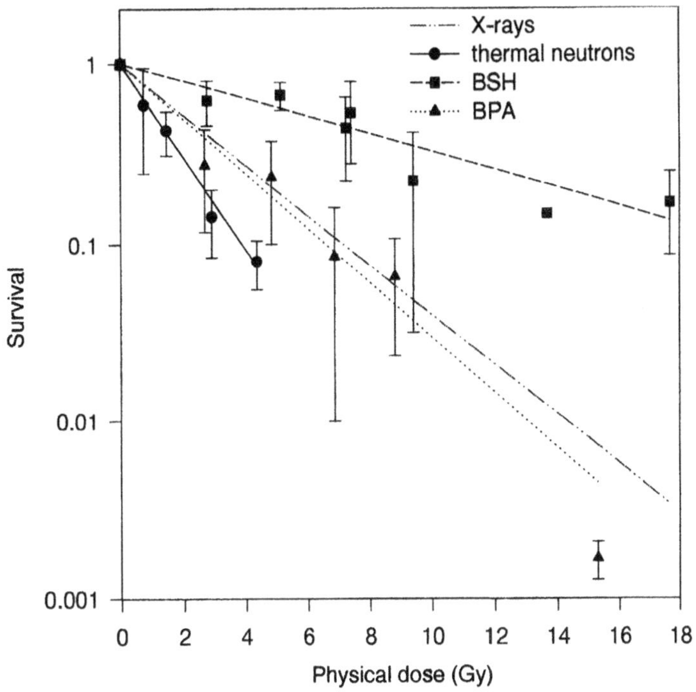

Figure 1. Thermal neutrons.

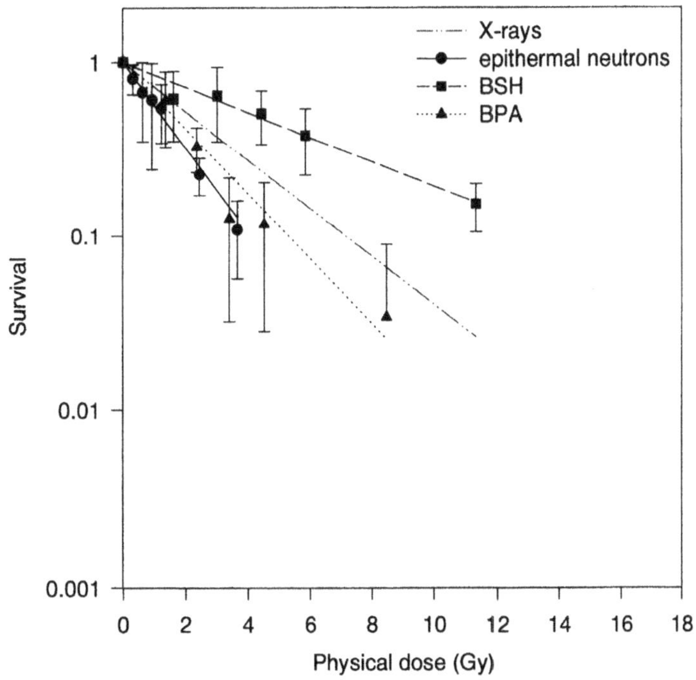

Figure 2. Epithermal neutrons.

Table 1. Beam and proton RBE's and compound factors for BPA and BSH for O-2A progenitor cell survival in optic nerve after thermal or epithermal neutron irradiation

	LFR Thermal neutrons	HB11 Epithermal neutrons
Beam RBE	1.84	1.75
Proton RBE	3.68	5.36
BPA compound factor	0.91	1.19
BSH compound factor	0.05	0.17

The results (Table 1) show that the RBE for glial stem cell survival after thermal neutron beam or epithermal neutron beam alone was equal to 1.84 and 1.75 respectively.

The RBE of the high LET part of these beams for glial stem cell survival was equal to 3.68 for the thermal beam and 5.36 for the epithermal beam.

The compound factor for BSH and BPA obtained with thermal beam irradiation was equal to 0.05 and 0.91 respectively. Using the epithermal beam the calculated compound factor for BSH was equal to 0.17 and for BPA equal to 1.19.

4. DISCUSSION

In the present investigation a small animal model was used in a thermal and epithermal beam to determine the RBE of the beams and the compound factors for BPA and BSH for glial progenitor survival in CNS. The results show that the RBE of the thermal and epithermal neutron beam are in the same order of magnitude, i.e., 1.8. However when the RBE of the proton dose component for both beams is calculated values of 3.6 and 5.4 are obtained for the thermal and epithermal beam respectively. The latter RBE value is considerably higher than the 1 MeV fission neutron RBE for glial progenitor survival, equal to 4.3, obtained with the same animal model.[3] In general fission neutron RBE's are the upper limit for high LET RBE's for a given tissue. This discrepancy can be attributed to either uncertainties in the proton dose determined for HB11 or a higher biological effectiveness of lower energy neutrons.

The compound factors determined for BPA, a compound which homogeneously is distributed in CNS, after thermal and epithermal neutron irradiation are comparable with those obtained for rodent and canine CNS necrosis.[4,5]

The compound factors determined for BSH, a compound which does not cross the blood-brain-barrier in CNS, after thermal and epithermal neutron irradiation are much lower than those obtained for rodent and canine CNS necrosis.[4,5] This clearly indicates that glial progenitor cells are not involved in the process of CNS necrosis and further implicates the CNS vasculature as primary target for radiation induced CNS necrosis.[6]

Although the presently used *in vivo/in vitro* O2-A clonogenic assay in small animals cannot be used to determine the compound factors for CNS necrosis after epithermal irradiation for compounds with a non homogeneous boron distribution in CNS, the model allows investigation of the influence of differences in microdistribution of new boron compound in CNS after epithermal neutron irradiation without having to use large animal models.

REFERENCES

1. R.A. Gahbauer, R.G. Fairchild, J.H.Goodman, and T.E. Blue, RBE in normal tissues, *in*: "Boron Neutron Capture Therapy; Toward Clinical Trials of Glioma Treatment", D. Gabel, and R. Moss, ed, Plenum Press, New York, 1992, pp. 123–128.
2. R.W.M. van der Maazen, I. Verhagen, and A.J. vander Kogel, An *in vitro* clonogenic assay to assess radiation damage in rat CNS glial progenitor cells, *Int. J. Radiat. Biol.*, 58:835–844, 1990.
3. H. Philippo, R. Huiskamp, E.A.M. Winter, B. Gharbaran, and A.J. van der Kogel, Age-dependent of radiosensitivity of glial progenitors for *in vivo* fission neutron and X-ray irradiation, *Radiat. Res.*, accepted for publication.
4. P.R. Gavin, S.L. Kraft, R. Huiskamp, and J.A. Coderre, A. Review: Cns effetcs and normal tissue tolerance in dogs, *J. Neuro-Oncol.*, 33:71–80, 1997.
5. G.M. Morris, J.A. Coderre, J.W. Hopewell, P.L. Micca, M.M. Nawrocky, H.B. Liu, and A. Bywaters, Response of central nervous system to boron neutron capture irradiation: evaluation using rat spinal cord model, *Radiother. Oncol.*, 32:249–255, 1994.
6. A.J. van der Kogel, B.J. Kleiboer, I. Verhagen, G.M. Morris, J.W. Hopewell, and J.A. Coderre, White matter necrosis of the spinal cord: studies on glial progenitor survival and selective vascular irradiation, *in*: "Radiation Research 1895–1995: Volume 2: Congress Lectures, Proceedings of the 10th International Congress of Radiation Research, Wurzburg, Germany", U. Hagen, D. Harder, H. Jung, and C. Streffer, ed, H. Sturtz, and A.G. Wurzburg, 1996, pp. 769–772.

ANALYTICAL EVALUATING METHOD OF $^{10}B(n,\alpha\gamma)^{7}Li$ PROMPT GAMMA-RAY SPECTRUM FOR MEASURING ABSORBED DOSE DISTRIBUTION IN BORON NEUTRON CAPTURE THERAPY

Masayori Ishikawa[1], Tooru Kobayashi[2], and Keiji Kanda[2]

[1]Graduate School of Energy Science
Kyoto University, Sakyo-ku
Kyoto 606-8501, Japan
[2]Research Reactor Institute
Kyoto University Kumatori-cho
Sennan-gun, Osaka 590-0494, Japan

1. INTRODUCTION

The purpose of this research is to establish a new absorbed dose estimation system. At present, the absorbed dose is estimated according to the following procedures in Kyoto University Research Reactor Institute(KURRI): 1) A neutron flux at a tumor is measured by activating gold wires. 2) ^{10}B concentration at the tumor is estimated from ^{10}B concentration of patient's blood by using the PGA(Prompt Gamma-ray Analysis) method.[1] 3) From them a boron reaction rate is calculated to estimate the absorbed dose. In this method, at least 15 minutes are necessary to irradiate gold wire and as total it takes about 30 minutes to estimate the absorbed dose. The absorbed dose estimated in this method is influenced by patient's condition and reactor condition. As another method of estimating the absorbed dose, it is considerable to use the prompt gamma-ray yielded from $^{10}B(n,\alpha\gamma)^{7}Li$ reactions. For example, this prompt gamma-ray is applied to SPECT (Single Photon Emission Computed Tomography). In other words, the absorbed dose distribution is made visible by counting prompt gamma-ray surround a patient. We name this absorbed dose estimation system as PG-SPECT(Prompt Gamma-ray Single Photon Emission Computed Tomography),[2] and which is under developing for clinical use.

We will use the Cd-Zn-Te semiconductor detector as detectors of PG-SPECT. The Cd-Zn-Te semiconductor detector has enough energy resolution to distinguish the gamma-ray as the signal from $^{10}B(n,\alpha\gamma)^{7}Li$ reaction and the scattered gamma-ray. But,

Frontiers in Neutron Capture Therapy, edited by Hawthorne *et al.*
Kluwer Academic / Plenum Publishers, New York, 2001.

there is a problem that the counting rate becomes small from following reasons. The first reason is that the counting rates decreases as much as the diameter of collimator is small and the length of collimator is long. A thick and long collimator which includes a shield against neutron and back ground gamma-ray is necessary for PG-SPECT because PG-SPECT will be arranged in the irradiation room. The second reason is that the detector of PG-SPECT should be made as small as possible to improve the accuracy of the reconstructed image because many detectors will be arranged in the limited space. As the comparatively high energy gamma-rays are measured with the small detector, detecting efficiency becomes small.

On the other hand, the irradiation time during Boron Neutron Capture Therapy (BNCT) is about 1–2 hours using the Heavy Water Neutron Irradiation Facility at KURRI. Generally we must estimate an absorbed dose rate at least within one hour to avoid excessive irradiation because the irradiation time is decided from the absorbed dose rate in radiotherapy. Moreover, since measurements from many directions by a limited number of detectors are necessary for the SPECT, measuring time in each direction becomes shorter. The situation of low counting efficiency and measuring in limited time means that a statistical error would be large. In this paper, we report a new analytical technique to reduce statistical error as to improve accuracy of PG-SPECT.

2. THEORY OF ANALYTICAL TECHNIQUE

Prompt gamma-rays yielded from $^{10}B(n,\alpha\gamma)^{7}Li$ reaction have an unique energy spectrum as shown in Figure 1 because of the Doppler Effect. We consider that this broadened spectrum has much information than the total count. If this information is applied to the error evaluation of the counting, a statistical error can be reduced. When the number of counts is increased, finally this spectrum shape should reach to the ideal spectrum function as shown in Figure 1 with a solid line. From this idea, true mean value can be estimated accurately by fitting observed energy spectrum with the ideal spectrum function.

In practical measurement, an energy spectrum is divided into several parts. Each part should have random counts related to the ideal spectrum function. Given that the number of divisions is α, the probabilities of each part are $p_1, p_2, \ldots, p_\alpha$ ($\Sigma p_i = 1$) and observed counts of each part are $n_1, n_2, \ldots, n_\alpha$ ($\Sigma n_i = N$), the estimated mean value $x(n_1, n_2, \ldots, n_\alpha)$ is evaluated by a least square fitting with the ideal spectrum function and is expressed as follows.

$$x(n_1, n_2, \ldots, n_a) = \frac{\sum_{i=1}^{a} p_i n_i}{\sum_{i=1}^{a} p_i^2} \tag{1}$$

Next, we examined the reliability of this evaluated value x. The probability $P(n_1, n_2, \ldots, n_\alpha)$ is expressed as follows.

$$P(n_1, n_2, \ldots, n_a) = \frac{N!}{n_1! n_2! \ldots n_a!} p_1^{n_1} p_2^{n_2} \ldots p_a^{n_a} \tag{2}$$

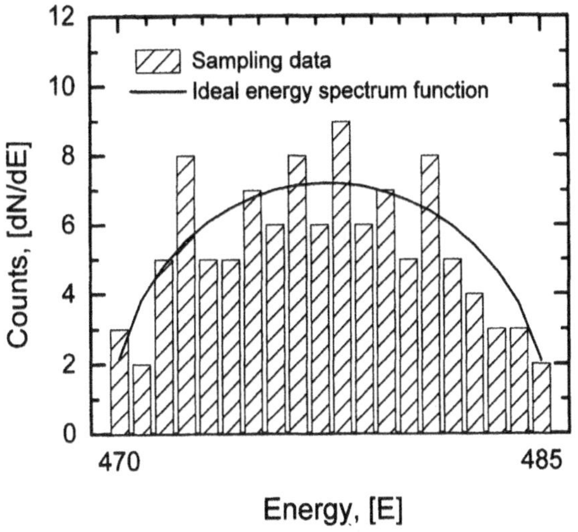

Figure 1. Unique energy spectrum.

From Eq. 1 and Eq. 2, variance V is expressed

$$V=\sum_{n_1=0}^{N}\sum_{n_2=0}^{N-n_1}\cdots\sum_{n_a=0}^{N-n_1-n_2-\dots-n_{a-1}}\frac{N!}{n_1!n_2!\dots n_a!}p_1^{n_1}p_2^{n_2}\dots p_a^{n_a}\cdot\left(\frac{\sum_{i=1}^{a}p_in_i}{\sum_{i=1}^{a}p_i^2}-N\right)^2 \tag{3}$$

Consequently, the variance V and the standard deviation σ can be expressed as follows.

$$V=N\left\{\frac{\sum_{i=1}^{a}p_i^3}{\left(\sum_{i=1}^{a}p_i^2\right)^2}-1\right\} \tag{4}$$

$$\sigma=\sqrt{N\left\{\frac{\sum_{i=1}^{a}p_i^3}{\left(\sum_{i=1}^{a}p_i^2\right)^2}-1\right\}}\cong 0.2\times\sqrt{N} \tag{5}$$

When α is three or more, σ becomes about $0.2\times\sqrt{N}$. As a conventional standard deviation[3] is equal to $\sqrt{N}$, this means that present analytical technique can reduce a standard deviation by about 80%. We name this new analytical technique as Analytical Mean value Evaluating Technique(AMET).

3. MONTE CALRO SIMULATION

We examined a standard deviation of the evaluated value led from the AMET by the Monte Carlo simulation. Prompt gamma-rays broadened 470–485 keV yielded from

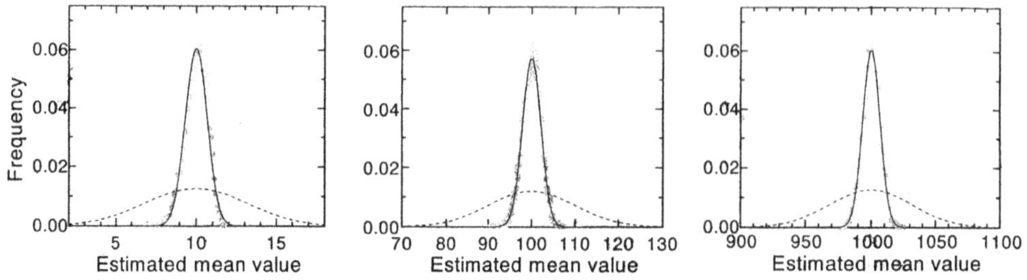

Figure 2. The frequency distribution of Monte Calro Simulation.

$^{10}B(n,\alpha\gamma)^{7}Li$ reactions are generated by Monte Calro method, then we get a random energy spectrum related to the ideal spectrum function. Here, when the Relativism is not considered, the ideal spectrum function is expressed as follows.

$$\frac{dN}{dE}=\frac{1}{2}\sqrt{1-\left(\frac{c}{V_{\mathrm{Li}}}\frac{E-E_0}{E}\right)^2} \tag{6}$$

Where c is the velocity of light, V_{Li} is the velocity of a lithium nucleus and equal to 4.804×10^6[m/s], and E_0 is the mean energy of the spectrum and equal to 477.612 keV.

We applied the AMET to the generated spectrum, and calculated a standard deviation of the evaluated value by iterating simulation 10,000 times. Figure 2 shows the result of the simulation for that α is 8 and N is 10, 100 and 1,000, respectively. In this figure, dots stand for the result of the simulation, solid line stands for the Gaussian distribution curve with the standard deviation led from Eq. 5, dotted line stands for the Gaussian distribution curve with the standard deviation of $\sqrt{N}$. We see that the Gaussian distribution curve of the solid line is well agree with the simulation in Figure 2.

If observed value is 100, the estimated mean value using the AMET fluctuates near around 100 with about 2% relative standard deviation. In conventional method, there is nothing else observed value to estimate a mean value, so the estimated mean value is assumed to be same with observed value and is supposed to have 10% relative statistical error under the assumption that an estimated mean value follows under the Poisson's law or the Gaussian law. From this difference, the AMET has better accuracy than the conventional method. Figure 3 shows an example of the difference visually. In this figure, N stands for observed value and also stands for the estimated mean value by the conventional method, x stands for the estimated mean value by the AMET. In general, x is not exactly same as N. The dotted line stands for the probability of the estimated mean value x, and the solid lines stand for the statistical error.

4. CONCLUSION

In this paper, we considered a new analytical technique, named Analytical Mean value Evaluating Technique(AMET), to reduce the statistical error for counting of prompt gamma-rays yielded from $^{10}B(n,\alpha\gamma)^{7}Li$ reaction at PG-SPECT measurement. We succeeded in the reduction of a standard deviation of the estimated mean value to about 1/5 by using the AMET.

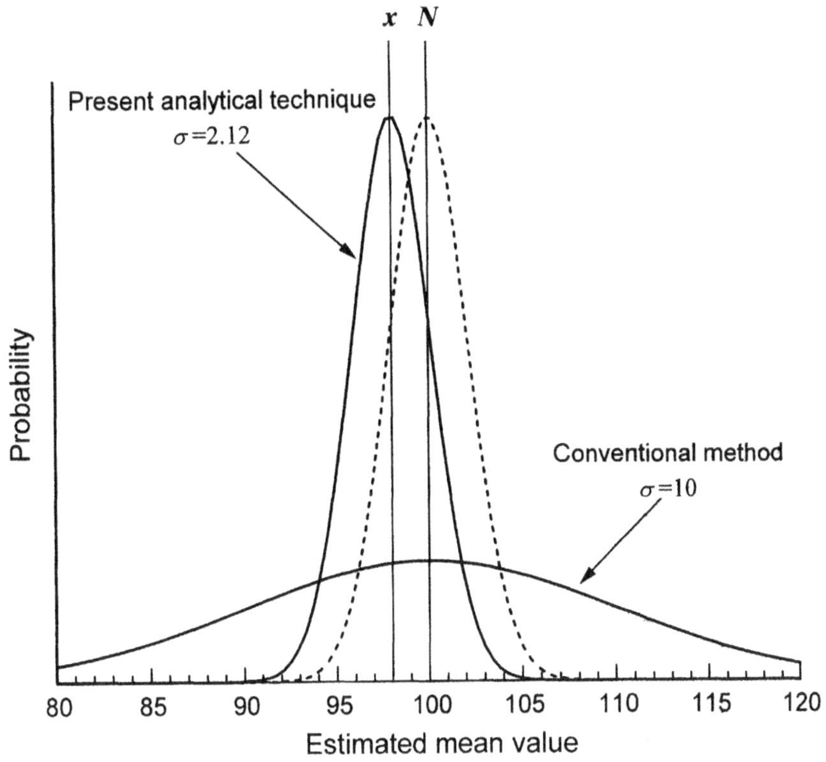

Figure 3. An example of the difference of estimated mean value.

The AMET is effective for all broadened energy spectrum that is caused in the generation process of photon and particles. In a practical use, we have to consider on background, the energy resolution of detector and etc., however, it seems that this analytical technique becomes one of promising techniques.

REFERENCES

1. T. Kobayashi, and K. Kanda, Microanalysis System of ppm-Order ^{10}B Concentrations in Tissue for Neutron Capture Therapy by Prompt Gamma-ray Sepctrometry, *Nucl. Instr. and Meth.*: 204, 525–531, 1983.
2. T. Kobayashi, Y. Sakurai, and M. Ishikawa, Real-time Absorbed Dose Estimation System for BNCT by PG-SPECT, *in* "Frontiers in Neutron Capture Therapy," M.F. Hawthorne, K. Shelly, R.W. Wiersema, eds., Kluwer Academic/Plenum Publishers, New York, 2001, pp. 641–645.
3. G.F. Knoll, "Radiation Detection And Measurement," John Wiley & Sons, Inc., New York, 1983, pp. 104–147.

MODELING THE IMPACT OF NEAR-THRESHOLD $^{7}Li(p,n)^{7}Be$ GENERATED, FAST NEUTRON BEAMS ON BORON DRUG DISTRIBUTION REQUIREMENTS FOR BNCT

T. R. LaHann, M. Griebenow, and J. Larsen

Department of Pharmaceutical Sciences
Idaho State University
Pocatello, Idaho

1. INTRODUCTION

One approach to developing a neutron source for BNCT is to use a monoenergetic proton beam operating near the threshold (1.881 MeV) for the $^{7}Li(p,n)^{7}Be$ reaction. The appeal of this approach lies in the assumption that the forward directed, compact, neutron-energy spectrum would be directly useful for BNCT, without moderation and the resulting beam degradation. Our study assumes that the "near threshold" accelerator concept can produce a neutron flux of E+9 n/cm^{2}s and evaluates the clinical merit of the achievable spectra. Our study also provides guidance to both boron-drug and neutron source developers by identifying the interactive performance of boron and neutron distributions believed to be necessary for clinical success.

2. APPROACH

BNCT was modeled as a single field treatment (12 cm diameter) to a 3-dimensional "head" constructed from magnetic resonance images (MRI) of an adult human male. Tissue regions corresponding to skin, bone, sinus and brain were defined from the MRI and a 3 cm, spherical tumor was modeled in brain with the deepest tumor volume 6 cm beneath the skin surface. Target volume in brain was defined as a sphere with radius 2 cm greater than that of the tumor.

Dose distributions were calculated for five idealized monoenergetic, monodirectional (MM) and two polyenergetic, polydirectional (PP) neutron beams (Table 1). Boron distribution was modeled as that achievable with boronophenylalanine-fructose (BPA-F)

Frontiers in Neutron Capture Therapy, edited by Hawthorne *et al.*
Kluwer Academic / Plenum Publishers, New York, 2001.

Table 1. Neutron source

	Neutron parameters	Gamma parameters
Flux	1E + 9 n/cm^{2}s	5E + 8 γ/cm^{2}s
Angular distribution	monoenergetic, monodirectional; polyenergetic, polydirectional	isotropic
Energy	monoenergetic 2, 20, 30, 50 or 70 keV; polyenergetic (BMRR or Kononov [1])	monoenergetic, 478 keV

or the ideal boron delivery drug. The 3-dimensional dose to tissue (Gy-eq) was calculated using rtt_MC, V1.06, the simulation program developed at the Idaho National Engineering and Environmental Laboratory. Rtt_MC is used for patient treatment planning both in the U.S. and Europe. For a given boron distribution and neutron source, the dose distributions from the $^1H(n,\gamma)^2H$ reaction, the $^{10}B(n,\alpha)^7Li$ reaction, the $^{14}N(n,p)^{14}C$ reaction, recoil protons produced by fast neutrons, and gamma photons contaminating the neutron beam[1] were each calculated.

The physical dose deposited in 0.125 cm^3 "voxels" was calculated for each component and after applying RBE or compound factors (CF), the biologically weighted (Gy-eq) doses were summed. RBE values for fast neutron, nitrogen-capture and γ radiation (Table 2) were extracted from the literature.[2,3,4,5,6] Compound factors (Table 2) and boron distributions of BPA-F were consistent with those determined in clinical BNCT studies.[3] The minimum radiation dose required for tumor control was defined as 35 Gy-eq to the deepest tumor volume, as suggested by the work of Laramore *et al.*[7] Based on the literature,[2,4,8,9,10,11] maximally tolerated radiation doses were defined as 15 Gy-eq to any 1 cm^3 of normal brain, 22 Gy-eq to any 1 cm^3 of skin and 20 Gy-eq to any 1 cm^3 of bone.

3. RESULTS

The kinematics of the near-threshold $^7Li(p,n)^7Be$ reaction are such that the lowest achievable mean energy of an unmoderated neutron beam is about 30 keV. Neutron energy and yield rise rapidly as proton energy rises; for a 1.885 MeV proton beam, mean and maximum neutron energies of about 30 keV and 52 keV were calculated;[12] for a 1.91 MeV beam, calculated values were about 40 keV and 120 keV.[13] Our study modeled monoenergetic neutron beams of 2, 20, 30, 50 and 70 keV. Although accelerators using the "near-threshold" approach cannot produce 2 or 20 keV beams without a moderator, these energies were evaluated for comparative purposes because 1) the monoenergetic 2 keV beam has been identified as the "ideal" beam for BNCT,[14] and 2) 20 keV is an intermediate energy that can be produced by other concepts. Brookhaven's epithermal beam (BMRR) and the PP neutron beam generated by a 1.885 MeV, $^7Li(p,n)^7Be$ reaction[12] were

Table 2. RBE and BPA-F compound factors

	Radiation type			
Tissue	Fast neutron	Nitrogen capture	Boron capture	Gamma
Skin	4.5	2.7	2.5	1
Bone	3.2	3.2	2.5	1
Brain	3.9	3.2	1.3	1
Target	3.9	3.2	1.3	1
Tumor	3.9	3.2	3.8	1

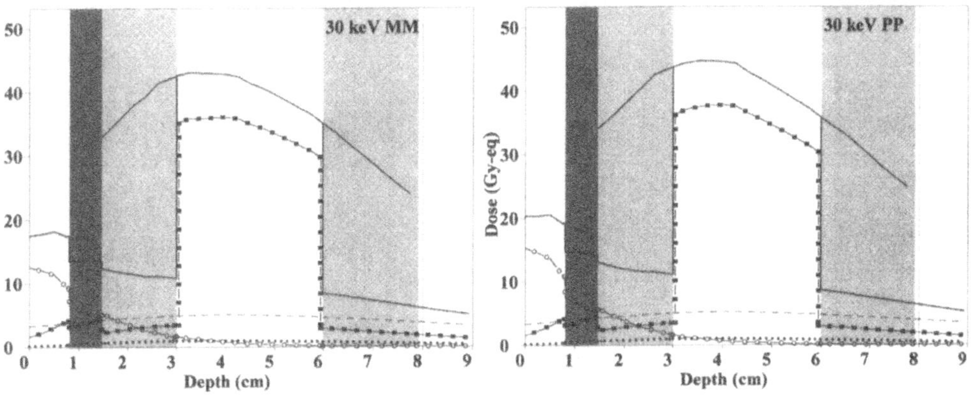

Figure 1. Beam line plot of Gy-eq dose for conditions delivering at least 35 Gy-eq to tumor volume. Comparison of a 30 keV monoenergetic, monodirectional beam (30 keV-MM) with the 30 keV, polyenergetic, polydirectional beam (30 keV-PP) of Kononov.[12] Shading patterns identify regions of skin, bone, target, tumor, target and normal brain. total dose: —— boron capture dose: -■- fast neutron dose: -O- gamma dose: - — - nitrogen capture dose: ••••• Upper solid lines in target region indicate dose to cancer cells which may have migrated into this region.

modeled to provide a baseline reference from which the clinical merits of other beams can be estimated.

Figure 1 shows the dose distributions calculated along beam line for a MM 30 keV neutron beam and the PP 30 keV beam of Kononov. In Figure 1 (and Figure 2), the upper solid lines in the target regions (2 cm on either side of the tumor) indicate the dose delivered to cancerous guerilla cells that likely occupy this region. We assume that these guerilla cells accumulate boron exactly like the main tumor mass and that the CF for BPA-F in these cells is the same as for the main tumor mass (i.e., 3.8). The "Kononov"

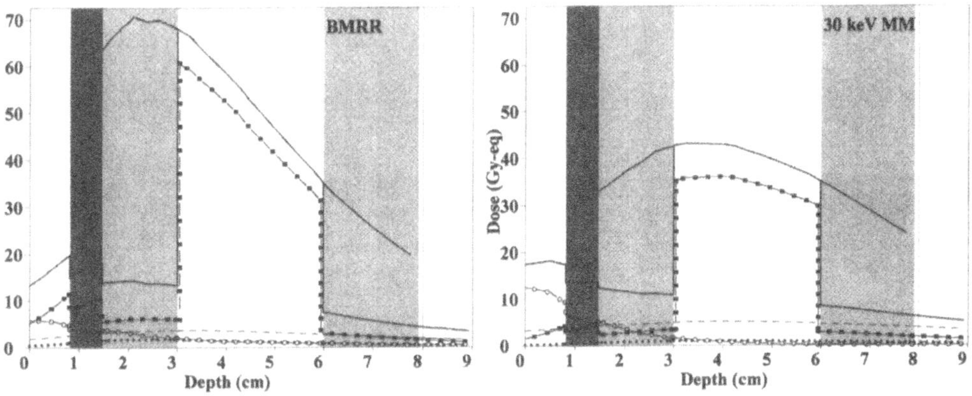

Figure 2. Beam line plot of Gy-eq dose for conditions delivering at least 35 Gy-eq to tumor volume. Comparison of a 30 keV monoenergetic, monodirectional beam (30 keV-MM) with the epithermal beam of the Brookhaven Medical Research Reactor (BMRR). Shading patterns identify regions of skin, bone, target, tumor, target and normal brain. total dose: —— boron capture dose: -■- fast neutron dose: -O- gamma dose: - — - nitrogen capture dose: ••••• Upper solid lines in target region indicate dose to cancer cells which may have migrated into this region.

Table 3. Peak Gy-eq dose to tissues under conditions delivering a minimum of 35 Gy-eq to deepest tumor volume[a]

Neutron energy spectrum	Maximum radiation dose (Gy-eq) Brain (limit: 15 Gy-eq)	Skin (limit: 22 Gy-eq)	Bone (limit: 20 Gy-eq)	Tumor
BMRR	13.8	19.6	17.6	67.9
2 keV monoenergetic	10.7	10.1	11.0	49.6
20 keV monoenergetic	10.4	15.3	12.0	44.8
30 keV monoenergetic	11.8	18.1	13.1	42.8
30 keV polyenergetic	13.1	20.6	14.5	44.1
50 keV monoenergetic	15.5	25.2	17.3	42.9
70 keV monoenergetic	19.6	33.9	22.0	42.3

[a]Blood [B] = 15 ppm; bone, brain and target [B] = blood [B]; tumor [B] = 3.5 × blood; skin [B] = 1.5 × blood.

beam is, in essence, the most compact neutron spectrum and the most forward directed, unmoderated beam that can be generated using the near-threshold concept. Both Figure 1 and Table 3 indicate that the simple case of the MM beam represents a "best-case" scenario for the more complex, PP beam. If a monoenergetic beam looks promising, then modeling of the more complex PP beam is warranted. However, if a monoenergetic beam is not promising, then the PP beam will also fail the specified criteria.

Figure 2 compares the dose distribution along beam line for the 30 keV MM beam and the BMRR epithermal beam. The superficial radiation dose due to recoil protons (fast neutrons) is much larger in the 30 keV beam, and skin is the dose-limiting tissue for this beam, while brain is the dose-limiting tissue for the BMRR beam. This is better appreciated by examining Table 3.

Neutron fluence for each beam is adjusted to deliver 35 Gy-eq to deepest tumor volume. For a boron distribution of 15 ppm in brain, and 52.5 ppm in tumor (mimicking Brookhaven clinical trials), neither the 50 nor the 70 keV beams can deliver 35 Gy-eq to deepest tumor without also causing clinically unacceptable skin damage.

Table 4 lists the minimum blood boron concentrations needed to deliver 35 Gy-eq to deepest tumor volume without exceeding tissue tolerance in skin, bone or brain. Table 4 illustrates that 1) the near-threshold concept becomes more viable as the absolute amount of boron delivered to tissues increases, and 2) for all unmoderated beams generated by the "near-threshold" approach, the skin is the dose-limiting tissue. One caveat

Table 4. Blood boron concentrations required to deliver a minimum of 35 Gy-eq to tumor volume without exceeding normal tissue tolerance[a]

Beam	Blood [B], ppm	Maximum Gy-eq dose delivered to 1 cm^3 Skin	Brain	Bone	Tumor
BMRR	12.5	20.8	15	18.6	67.7
2 keV monoenergetic	7.4	12.7	14.9	14.2	48.5
20 keV monoenergetic	8.6	21.9	14.5	15.9	44.5
30 keV monoenergetic	11.3	21.9	14.2	15.7	42.9
30 keV polyenergetic	13.7	21.9	13.9	14.6	44.2
50 keV monoenergetic	18.1	21.9	13.6	15.2	42.8
70 keV monoenergetic	25.7	21.9	13.2	14.9	41.6

[a]Bone, brain and target [B] = blood [B]; tumor [B] = 3.5 × blood; skin [B] = 1.5 × blood.

Table 5. Impact of the ideal boron drug: blood boron concentrations required to deliver a minimum of 35 Gy-eq to tumor volume without exceeding normal tissue tolerance

Beam	Tumor [B], ppm	[B] Concentration ratio (Beam: BMRR)	Dose limiting tissue
BMRR	6.1	1.0	brain
2 kev monoenergetic	4.5	0.7	brain
20 keV monoenergetic	7.0	1.1	skin
30 keV monoenergetic	9.2	1.5	skin
50 keV monoenergetic	14.8	2.4	skin
70 keV monoenergetic	21.3	3.5	skin

to Table 4 is that the relative distribution of boron in tissues affects the modeled outcomes. Modeling a 50 keV beam and BPA-F with a 1:3.5 blood:tumor ratio indicates that a blood boron level of at least 18.1 ppm is needed to deliver effective BNCT. Modeling the same scenario, but assuming a blood:tumor ratio of 1:3 yields a minimum blood boron level of 21.7 ppm (a 20% increase in boron).

Although BPA-F is useful for delivering boron to brain tumors, development of better drugs can be anticipated. Thus, we modeled BNCT using the ideal boron drug—a drug that accumulates only in cancer cells (normal tissue [B] = 0), and that localizes exclusively in the cell nucleus. We calculate that this ideal drug has a CF of 14.8. Our calculation is based on the following assumptions: 1) when boron is localized to a cell's nucleus, the probability of a boron capture damaging nuclear DNA is 2.87 times greater than when boron is homogeneously distributed throughout tissue,[15] 2) an ionizing particle coming from a capture reaction in the nucleus has a lower cell-kill efficacy (86%) than an ionizing particle hitting the cell nucleus from a distance[16] 3) boric acid distributes homogeneously through the cell. Our review of the literature indicated that the highest reported RBE for boric acid mediated BNCT is 6.01.[17] Thus, the CF for an ideal boron delivery drug is calculated as (RBE of boric acid)(nuclear localization factor) (nuclear efficiency factor) or (6.01) (2.87) (0.86) or 14.8. The CF for BPA-F in tumor cells has been reported as 3.8.[3] Thus, the ideal boron delivery drug will, at most, be a factor of 4 "better" than BPA-F. This indicates a need to counterbalance emphasis on subcellular "targeting" of new boron drugs with additional emphasis on developing new boron drugs that deliver large amounts of boron to tumor cells, irrespective of their subcellular distribution.

Table 5 summarizes the minimum tumor levels of boron needed when using the ideal boron drug. Compared to the existing BMRR beam, the 30 keV MM beam demands 1.5 times the boron to achieve the same minimum radiation dose to tumor. The 50 keV and 70 keV beams demand almost 2.5 times (50 keV) and 3.5 times (70 keV) the boron needed by the BMRR beam. The neutron beams generated by the near-threshold $^7Li(p,n)^7Be$ reaction clearly place greater demands on pharmaceutical systems. As the CF of drug delivery systems improve, the value of using an epithermal beam increases, while the value of a low energy fast neutron beam decreases.

4. CONCLUSIONS

Our results indicate that: 1) The rtt_MC simulation program is useful for estimating interactive neutron-beam and boron-drug design requirements. 2) Skin tolerance will limit clinical effectiveness of neutrons in the energy range produced by the unmoderated, near-threshold $^{7}Li(p,n)^{7}Be$ reaction. 3) When a neutron flux of E + 9 n/cm^{2}s and current BPA-F dosing regimens are assumed, the unmoderated neutron beam produced by a proton accelerator operating at 1.885 MeV will provide the same clinical outcome as the present BMRR neutron beam. More energetic proton beams will be clinically inferior. 4) If boron drugs significantly superior to BPA-F can be identified, accelerators operating above 1.885 MeV may duplicate BMRR clinical results. 5) when used with boron drugs superior to BPA-F, neutrons in the 2 to 20 keV range will be clinically superior to any unmoderated neutron beam generated by the $^{7}Li(p,n)^{7}Be$ reaction. 6) An ideal drug will have a compound-factor advantage only 4 times greater than BPA-F. Our results emphasize the need to develop neutron beams in the 2 to 20 keV energy range, but do not support continued development of unmoderated, near-threshold $^{7}Li(p,n)^{7}Be$ accelerator concepts.

ACKNOWLEDGMENTS

This study was supported by funds from the Idaho State University Research Committee and by the Hiram Cochran Fund of the Metropolitan Atlanta Community Foundation.

REFERENCES

1. D. Allen, and T. Beynon. A design for an accelerator-based epithermal neutron beam for BNCT. Phys. Med. Bio. 40:807–821, 1995.
2. J. Archambeau, R. Fairchild, and H. Brenneis. The response of the skin of swine to increasing absorbed doses of radiation from a thermal neutron beam, a degraded fission neutron beam and the $^{10}B(n,\alpha)^{7}Li$ Reaction. Radiation Research 45:145–165, 1971.
3. J. Coderre, E. Elowitz, M. Chadha, *et al.* BNCT for glioblastoma multiforme using p-BPA and epithermal neutrons: trial design and early clinical results. J. Neuro-Oncology 33:141–152, 1997.
4. H. Fukuda, J. Hiratsuka, C. Honda, *et al.* BNCT of malignant melanoma using ^{10}B-BPA with special reference to evaluation of radiation dose and damage to the normal skin. Radiation Research 138:435–442, 1994.
5. D. Gabel, K. Philipp, F. Wheeler, *et al.* The compound factor of the $^{10}B(n,\alpha)^{7}Li$ reaction from borocaptate sodium and the relative biological effectiveness of recoil protons for induction of brain damage in BNCT. Radiation Research 149:378–386, 1998.
6. P. Gavin, R. Huiskamp, F. Wheeler, *et al.* Large animal normal tissue tolerance using an epithermal neutron beam and borocaptate sodium. Strahlentherapie und Onkologie 169:48–56, 1993.
7. G. Laramore, F. Wheeler, D. Wessol, *et al.* A tumor control curve for malignant gliomas derived from fast neutron radiotherapy data: implications for treatment delivery and compound development. In: Advances in Neutron Capture Therapy Vol II, B. Larsson ed., Elsevier, Amsterdam, 1997, pp. 580–587.
8. J. Fike, C. Cann, K. Lurowski, *et al.* Radiation dose response of normal brain. Int. J. Radiation Oncol. Biol. Phys. 14:63–70, 1988.
9. R. Huiskamp, J. Hopewell, G. Morris, *et al.* Tolerance of the healthy nervous system after BNCT: remaining questions. In: Advances in Neutron Capture Therapy Vol II, B. Larsson ed., Elsevier, Amsterdam, 1997, pp. 687–691.
10. J. Hopewell, D. Barnes, M. Robbins, *et al.* The RBE of fractionated doses of fast neutrons for normal tissues in the pig. II. Late effects on cutaneous and subcutaneous tissues. Brit. J. Radiol 63:760–770, 1990.
11. R. Hattner, J. Hartmeyer, and W. Wara. Characterization of radiation-induced photopenic abnormalities on bone scans. Radiology 145:161–166, 1982.
12. V. Kononov, P. Androsenko, M. Bohovko, *et al.* $^{7}Li(p,n)^{7}Be$ reaction near the threshold: the prospective neutron source for BNCT. In: Proceeding of the First International Workshop on Accelerator-Based

Neutron Sources for BNCT, Vol II, R. Wiersema, ed., U.S. Department of Energy conf-940976, 1994 pp. 477–483.

13. R. Bartholomay. BNCT Benchmark Experiments for Near Lithium Threshold Target Performance. Master Thesis, University Microfilms International, London, 1998, pp. 1–174.
14. R. Moss, O. Aizawa, D. Deynon, *et al.* The requirements and development of neutron beams for neutron capture therapy of brain cancer. J. Neuro-Oncology 33:27–40, 1997.
15. C. Ceberg, A. Persson, A. Brun, *et al.* A stochastic model for subcellular dosimetry in BNCT. Phys. Med. Biol. 40:1819–1830, 1995.
16. R. Verrijk, R. Huiskamp, A. Begg, *et al.* A comprehensive PC-based computer model for microdosimetry of BNCT. Int. J. Radiat. Biol. 65:241–253, 1994.
17. H. Fukuda, T. Kobayashi, T. Matsuzawa, *et al.* RBE of a thermal neutron beam and the $^{10}B(n,\alpha)^{7}Li$ reaction on cultured B-16 melanoma cells. Int. J. Radiat. Biol. 51:167–175, 1987.

COMPUTATIONAL DOSIMETRY FOR BORON NEUTRON CAPTURE SYNOVECTOMY

D. W. Nigg[1], C. A. Wemple[1], F. J. Wheeler[1], D. E. Wessol[1], J. C. Yanch[2], and C. D. Scoville[3]

[1]Idaho National Engineering and Environmental Laboratory
Idaho Falls, Idaho
[2]Massachusetts Institute of Technology
Cambridge, Massachusetts
[3]Institute of Arthritis Research
Idaho Falls, Idaho

1. INTRODUCTION

Rheumatoid Arthritis (RA) is an autoimmune disease characterized by swollen, inflamed, and painful joints. It affects as many as 3% of U.S. adults and is more common in women than in men.[1,2] The precise cause is unknown. Recurrent inflammation of the synovial membrane (or synovium), the inner layer of the joint, is the source of pain and disability associated with RA. This inflammation leads to eventual enzymatic destruction of the cartilage in the joint, which in turn causes progressive loss of joint function and permanent disability. The source of the destructive enzymes is believed to be the synovium itself.

The current primary treatments for RA include administration of drugs, surgical removal of the inflamed synovium, and radiation synovectomy using internally-administered radiopharmaceuticals. Because of the difficulties with these standard treatments in some cases, it has been proposed that techniques originally developed for Boron Neutron Capture Therapy (BNCT) of certain malignant tumors be adapted for application to RA.[3] This approach, referred to as Boron Neutron Capture Synovectomy, or BNCS, is currently under investigation in a collaborative effort of the Massachusetts Institute of Technology, Harvard University Medical School, and the Idaho National Engineering and Environmental Laboratory (INEEL). The basic procedure involves injection of a non-radioactive compound tagged with boron-10 into the joint space, followed by local irradiation of the treatment volume using a well-collimated neutron beam having a suitable energy spectrum. Neutrons from the beam are moderated in the joint space and are absorbed by the boron, which leads to the release of two high-LET, high-RBE charged particles (Li and He) which have ranges on the order of the size of a typical

Frontiers in Neutron Capture Therapy, edited by Hawthorne *et al.*
Kluwer Academic / Plenum Publishers, New York, 2001.

cell. This, in turn, causes highly-selective tissue damage only in the tissue regions where both boron-10 and neutrons are present. The advantages of radiation synovectomy are thus attainable with far less concomitant healthy-tissue exposure, relative to the situation with internally administered radionuclides.

As in all radiotherapies, it is important to perform extensive computational dosimetry studies on a patient-specific basis in order to deliver an optimum treatment. The field size, number of fields, angles of incidence, and the spectrum of the therapy beam are all adjustable parameters that must be determined for optimal BNCS. An appropriate neutron beam for treatment of a knee joint will, for example, not necessarily be appropriate for treatment of the joints of the fingers, especially with respect to the energy spectrum of the neutron flux. The BNCT Radiotherapy Planning Environment (BNCT_rtpe) software,[4] originally developed by the INEEL for use in computational dosimetry and treatment planning for BNCT of brain and, more recently, lung tumors, is also well-suited for application to BNCS. This article describes an initial demonstration, on a non-clinical basis, of the application of BNCT_rtpe for this type of computation.

2. METHODS

A transaxial magnetic resonance (MR) image of the knee of a female RA patient, age 58, is shown in Figure 1. The full set of MR images to which this image belongs served as the basis for this study. Severe inflammation and edema in the region posterior to the patella is evident. In addition, a Baker's cyst is evident in the region posterior to

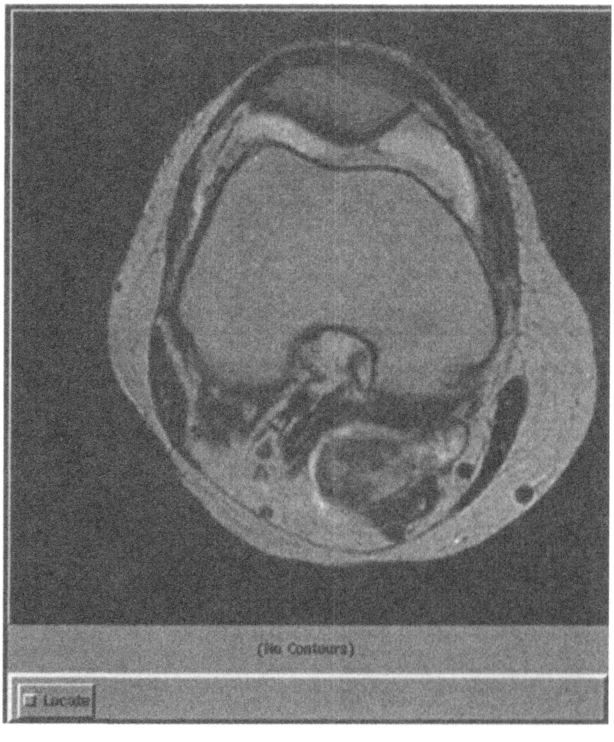

Figure 1. Transaxial magnetic resonance image of the knee joint of a rheumatoid arthritis patient.

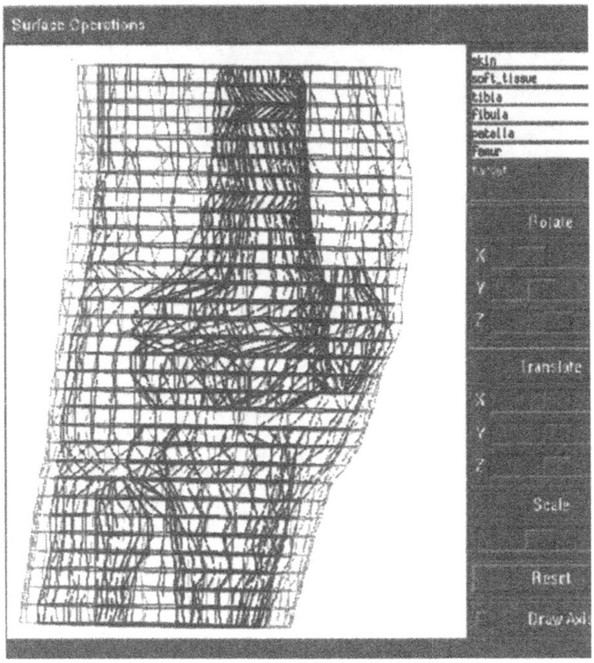

Figure 2. Non-uniform rational b-spline reconstruction of the knee of a rheumatoid arthritis patient.

the joint itself. These regions contain inflamed synovium and served as the hypothetical target volume for BNCS. The entire transverse image set for the knee region was input to BNCT_rtpe. The various regions of interest (bone, skin, target volume, etc.) were outlined on each transverse image using non-uniform rational b-spline curves in the usual manner.[4] The resulting three-dimensional spline geometric reconstruction of the knee is shown in Figure 2. This served as the geometric description for subsequent radiation transport computations to estimate the hypothetical BNCS radiation dose deposition profiles in the treatment volume under various conditions.

The neutron source spectrum requirements for BNCS are not necessarily the same as for BNCT. And, as mentioned previously, a different neutron spectrum will also be required for effective therapeutic coverage of a knee as opposed to, for example a finger joint. To bracket the range of neutron energies that may be useful for BNCS, calculations were performed for a series of monoenergetic neutron beams in this study. The assumed energies were 0.01, 0.1, 1.0, 10.0, 10^2, 10^3, 10^4, 10^5, and 10^6 eV. In each case the beam was assumed to be incident on the anterior surface of the knee near the center of the patella. All tissues in the target volume were assumed to be suffused with a uniform concentration of 50 parts per million boron-10. Neutron and gamma fluxes and radiation dose components were computed for each assumed source using the rtt_MC module of the BNCT_rtpe package.

3. RESULTS AND DISCUSSION

Figure 3 shows some typical total dose contours in four transaxial planes at various locations in the joint. These results are for the hypothetical 1 keV incident neutron beam.

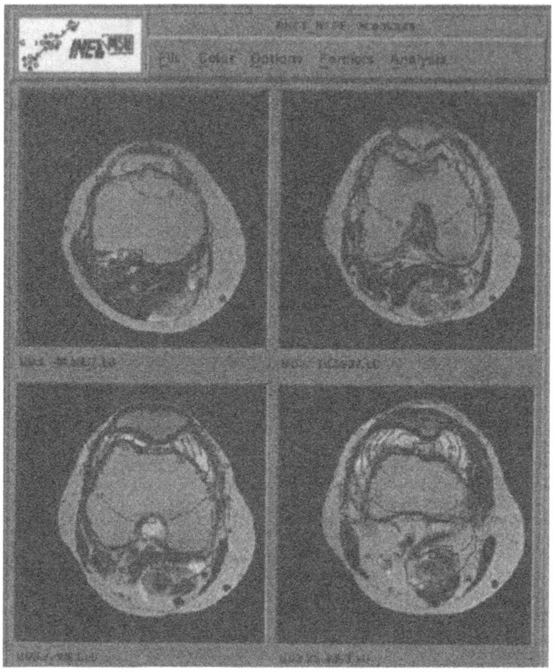

Figure 3. Isodose contours for monoenergetic 1 keV synovectomy beam. Contours assume he BNL healthy tissue RBE values, and 50 ppm ^{10}B in the target region only.

In this particular case this was the most appropriate energy in terms of providing good coverage of the anterior target volume. A second field, incident on the posterior surface of the joint would probably also be indicated in this case. In addition, in an actual clinical application the incident neutron beam would have a spectrum of energies, and additional constraints (e.g. dose to adjacent healthy tissue) would be factored into the selection of an optimal treatment plan for a given situation.

The results for total dose in Figure 3 include the boron-neutron capture component at 50 ppm, the background neutron dose (primarily from hydrogen-recoil), the nitrogen (n,p) dose, and the background gamma dose, primarily from neutron capture in hydrogen. Relative Biological Effectiveness (RBE) weightings that are typically used for brain tumor treatment plans have been applied to the various dose components that make up the total doses illustrated in Figure 3. This is, of course, for demonstration purposes only. In an actual clinical application, RBE factors appropriate for the specific tissues in the treatment volume would be required.

The results reported here demonstrate the applicability of the BNCT_rtpe computational dosimetry system for BNCS in a clinically-realistic situation. Some difficulty relative to the more-routine situation where BNCT_rtpe is used for treatment planning of BNCT or brain tumors was experienced in fitting the spline surfaces to the desired anatomical compartments. This was a result of the convoluted shape of these compartments and the very close spacing of the regions being modeled. A new geometric modeling capability that should eliminate this difficulty, based on direct pixel-by-pixel geometric reconstruction rather than spline surfaces is currently in clinical testing.[5]

REFERENCES

1. E.D. Harris Jr, "Rheumatoid Arthritis, The Clinical Spectrum", In: W.N. Kelly, E.D. Harris, S. Ruddy, C.B. Sledge, *Textbook of Rheumatology*, W.B. Sanders, Philadelphia, 1981, Page 930.
2. P.E. Lipsky, "Rheumatoid Arthritis", In: J.D. Wilson *et al.* (Eds.), *Harrison's Priciples of Internal Medicine*, 12th Edition, McGraw-Hill, New York, 1991, Pages 1437–1443.
3. J.C. Yanch, S. Shortkroff, R.E. Shefer, E. Binello, D. Gierga, A.G. Jones, G. Young, C. Vivieros, and B. Blackburn, Progress in the development of boron neutron capture synovectomy for the treatment of rheumatoid arthritis, In: M.F. Hawthorne, K. Shelly, R.W. Wiersema, eds., "Frontiers in Neutron Capture Therapy," Kluwer Academic/Plenum Publishers, New York, 2001, pp. 1389–1397.
4. D.W. Nigg, F.J. Wheeler, D.E. Wessol, J. Capala, and M. Chadha, "Computational Dosimetry and treatment Planning for Boron neutron Capture Therapy", Journal of Neuro-Oncology **33**:93–103, (1997).
5. M.W. Frandsen, D.E. Wessol, F.J. Wheeler, and D. Starkey, Rapid Geometry Interrogation for Uniform Volume Element-Based BNCT Monte Carlo Particle Transport Simulation, Program and Abstracts of the Eighth International Symposium on NCT for Cancer, La Jolla, CA, 1998, p. 97.

APPLICATION OF THERMOLUMINESCENT DOSIMETERS TO MIXED NEUTRON-GAMMA DOSIMETRY FOR BNCT

C. Kessler[1,2], F. Stecher-Rasmussen[3], J. Rassow[1], S. Garbe[1], and W. Sauerwein[1]

[1]Universitätsklinikum Essen
Klinik und Poliklinik für Strahlentherapie
Essen, Germany
[2]Comisión Nacional de Energía Atómica
Laboratorio Secundario de Calibración Dosimétrica
Av del Libertador 8250, (1429) Buenos Aires, Argentina
[3]Netherlands Energy Research Foundation ECN
Petten, The Netherlands

1. INTRODUCTION

Thermoluminescence dosimeters have become a current dosiemtric tool in photon, electron and neutron radiation beams for dose determination in "*in vivo*" dosimetry, as well as in phantoms simulating patient treatments.

The application of TL detectors in neutron beams, commonly refered to as mixed beams due to the unavoidably presence of gamma rays, requires a detail study of the detector response to these beams in order to determine the sensitivity to each component.

This work is specially oriented to determine the dosimetric response of four types of commercial TL detectors to thermal, epithermal and fission neutron-gamma beams in order to discriminate both dose components.

Using the fact that CaF_2: Tm (TLD 300) has a LET dependence of the glow peak structure, they have been used to discriminate between gamma and neutron dose components in fast neutron fields. The applicability of the two peak method in the Petten reactor neutron beams is investigated.

Detectors with a high sensitivity to thermal neutrons ^{6}LiF: Mg,Ti (TLD 600) and those with a low one, ^{7}LiF: Mg,Ti (TLD 700) and ^{7}LiF: Mg,Cu,P (MCP-7), are included in this investigation to analyze the possibility of decomposing the absorbed dose from

Frontiers in Neutron Capture Therapy, edited by Hawthorne *et al.*
Kluwer Academic / Plenum Publishers, New York, 2001.

the mixed fields into the gamma and neutron components. The response of BeO to these mixed fields is also studied.

Reference detectors irradiated only by ^{60}Co γ-rays allow the study of radiation damage after neutron irradiations.

2. MATERIALS AND METHODS

2.1. Thermoluminescence Detectors

The following commercial TL detectors were investigated in this work:

1. LiF:Mg,Ti (TLD 600 and TLD 700), rods ($1 \times 1 \times 6\,mm^3$) from Harshaw Chem.Co.
2. CaF_2:Tm (TLD 300), chips ($3.1 \times 3.1 \times 1\,mm^3$) from Harshaw Chem.Co.
3. LiF:Mg,Cu,P (MCP-7, ^{7}Li-enriched), pellets (4.5 mm diameter and 0.9 mm thick) from TLD Niewiadomski & Co, Poland.
4. BeO, pellets (5.0 mm diameter, 0.9 mm thick) from Brush Beryllium Co, Canada.

2.2. Annealing Procedure

The pre-irradiation annealing procedure for TLD 700, TLD 600 and TLD 300, was 1 hour at 400 °C, followed by 3 hours at 100 °C; for MCP detectors, 10 minutes at 240 °C followed by rapid cooling down to room temperature; and for BeO was 15 minutes at 600 °C.

A post-irradiation annealing at 100 °C for 10 minutes was applied for all the detectors to remove low temperature peaks.

2.3. Read-out System

The TLDs were measured with a commercial Harshaw 3500 TLD reader. A constant heating rate of $10\,°C s^{-1}$, a preheat of 80 °C and a maximum temperature of 400 °C were used for all the detectors, except for the MCP ones, which were heated up to 250 °C. The glow curve and the total area of it were obtained for evaluation.

2.4. TLD Calibration

Prior to neutron irradiation, the TLD batches were calibrated by gamma-ray irradiation in a PMMA phantom at the ^{60}Co source of the Strahlenklinik Universitätsklinikum-Essen. An absorbed dose of 1 Gy was delivered to the TLDs. Each batch was irradiated five times. The average of the five calibration factors determined in each irradiation and the corresponding standard deviation were calculated. The TLD, whose calibration factor had a standard deviation less than 2%, were selected for our study, except for BeO, which had a calibration factor with a standard deviation between 3% and 4%.

From each batch of TLDs, nine detectors were separated in order to be irradiated only at the ^{60}Co source, and to keep them as reference detectors with the purpose of determining changes in the detector sensitivity after neutron irradiation and also for neutron TL response normalization to ^{60}Co γ-rays response.

2.5. TLD Irradiations

The neutron irradiations were performed in Petten using the HB11 epithermal neutron beam facility of the High Flux Reactor (HFR) of the European Commission, and the thermal and fission beams from the ECN Argonaut Reactor.

For each neutron irradiation, 3 TL detectors of each type were placed in a polyethylene phantom, consisting on 10 plates of $120 \times 120 \times 10\,mm$,[3] with milled holes for each detector.

The TLDs were irradiated at several depths, for the three beams, seeking the variations of the resulting signal on the corresponding relative gamma-ray absorbed dose component.

One day after neutron irradiations, the reference detectors were irradiated in the Co-60 beam with three different absorbed doses (0.5, 1.0, 1.5 Gy), for a calibration curve and to normalize the results obtained in the mixed beams.

After evaluation, the TLDs were irradiated in the ^{60}Co beam, under the same conditions of calibration, to study the possible loose of sensitivity due to the neutrons of the mixed fields.

2.6. Neutron Beam Calibration

The neutron beams were calibrated with a magnesium ionization chamber flushed with argon gas [Mg(Ar) ionization chamber] for the determination of the gamma-ray dose and a tissue-equivalent ionization chamber flushed with tissue-equivalent gas [TE(TE) ionization chamber] for the determination of neutron dose.

For the determination of the neutron fluence, two sets of AuAl, Cu and MnNi activation foils were used for each beam facility.

Using the nomenclature recommended by ICRU for mixed beams, and including as a third component the thermal neutron fluence, the responses of the ionization chambers can be expressed as:

$$R_u = h_u \cdot D_\gamma + k_u \cdot D_n + k_u' \cdot \Phi \quad \text{for Mg(Ar) ionisation chamber}$$
$$R_t = h_t \cdot D_\gamma + k_t \cdot D_n + k_t' \cdot \Phi \quad \text{for TE(TE) ionisation chamber}$$

where R_u and R_t are the readings of the dosimeters in the mixed field relative to their response to the gamma rays used for calibration; h_u and h_t are the responses of the dosimeters to the gamma rays in the mixed field relative to their response to the gamma rays used for calibration; k_u and k_t are the responses of dosimeters to the epithermal neutrons in the mixed field relative to their response to the gamma rays used for calibration; D_γ and D_n are the gamma and epithermal neutron absorbed dose, respectively; k_u' and k_t' are the readings of the dosimeters due to the thermal neutrons in the mixed field per unit of neutron fluence relative to their responses to the gamma rays used for calibration, and Φ is the thermal neutron fluence.

Values for h_u, h_t, k_u, k_t, k_u' and k_t' were taken from Raaijmakers, *et al.*[1,2]

3. RESULTS

Our first attempt was to analize the possibility of discriminating both dose components using paired detectors of which one is mainly sensitive to gamma and the other to both neutron and gamma rays. TLD 700, considered to be sensitive only to gamma rays and TLD 600 for neutron dose determination after substracting the gamma contribution to the total thermoluminescence, were selected to be irradiated in the three beams described previously.

The glow curves of the TLD 700 obtained after neutron and ^{60}Co γ-rays irradiations show a large difference produced by the neutrons of the mixed fields, as can be seen in Figure 1.

Because of this, the response to thermal and epithermal neutrons must be taken into account for an accurate gamma dose determination. The response of the detector can be expressed with the same nomenclature used for the ionization chamber:

$$R_u = h_u \cdot D_\gamma + k_u \cdot D_n + k_u' \cdot \Phi$$

Following the argument explained by Raaijmakers, *et al.*[1,2] we assume $h_u = 1$ and $k_u = 0$. Table 1 shows the values for k_u' calculated for different irradiation depths in the thermal and epithermal neutron beam.

The response of the TLD 600 to the mixed fields was 3 orders of magnitude higher than the TLD 700 response, as can be seen in Figure 2.

As these detectors showed a remarkable decrease of their sensitivity after neutron irradiations, which could not be restored with annealing procedure, they became in non reliable detectors. Because of this, no further investigation was performed.

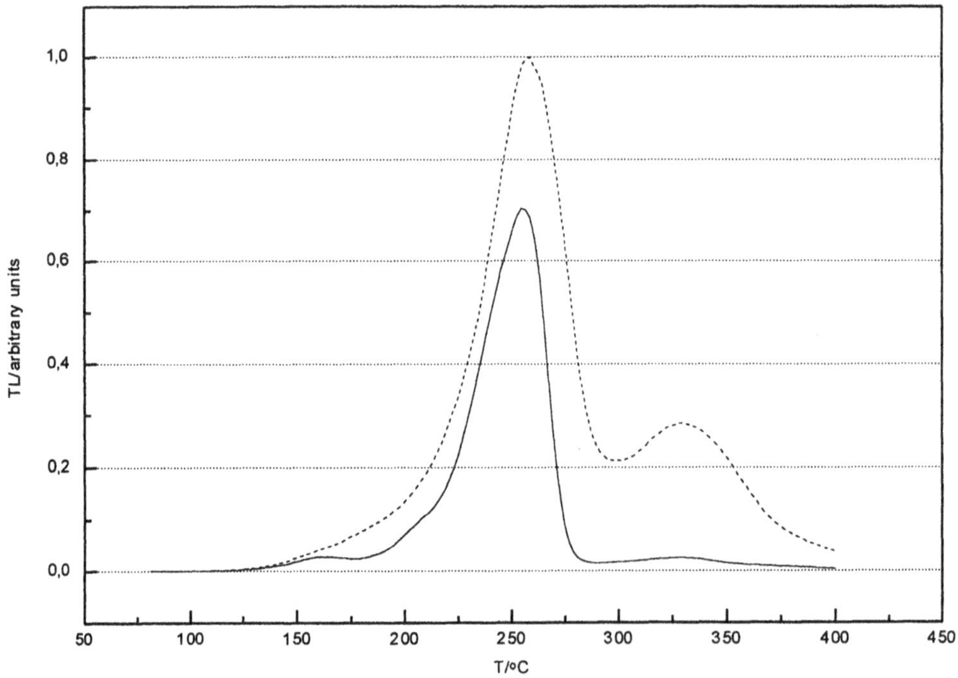

Figure 1. Glow curves for TLD 700 for an absorbed dose of 1 Gy. ——^{60}Co γ-rays, - - - -epithermal n.

Table 1. k'_u values for TLD 700

Neutron beam	Depth	R_u/Gy	D_γ/Gy	$\Phi/10^{11}\,cm^{-2}$	$k'_u/10^{-12}\,cm^2\,Gy$
Thermal	1 cm	3.99	0.83	8.60	3.7
Thermal	4 cm	1.10	0.50	1.90	3.2
Epithermal	1 cm	2.53	0.58	5.13	3.8
Epithermal	4 cm	3.68	1.28	6.65	3.6

Average value of $k'_u = 3.6 \pm 0.4$ Gy per neutron fluence of $10^{12}\,cm^{-2}$.

In order to study the application of the two peak method[3] for the TLD 300 in neutron beams, they were irradiated at several depths, placed in the polyethylene phantom. The peak ratio doesn't show appreciable variations, for all depths in the thermal and epithermal neutron beams, with no significant difference between the reactor and cobalt measurements as can be seen in Figure 3. Because of this, the method could not be used for any of the three neutron beams, in spite of the variations observed in the peak ratio corresponding to the fission neutron beam, which were not enough for evaluating the absorbed dose for neutron and gamma separately.

TLDs BeO and LiF:Mg,Cu,P were included in our last epithermal neutron beam irradiation.

The glow curves for these detectors are shown in Figures 4 and 5, respectively. It can be seen in Figure 5 that the glow curves corresponding to neutron irradiation differ from the ^{60}Co γ-rays one due to the neutron contribution.

Table 2 shows the absorbed dose values in the mixed field measured with the TE(TE) and Mg(Ar) ionization chambers in the second and third columns respectively; the R_u values for TLD 700 and MCP detectors, in the following two columns; the R_u

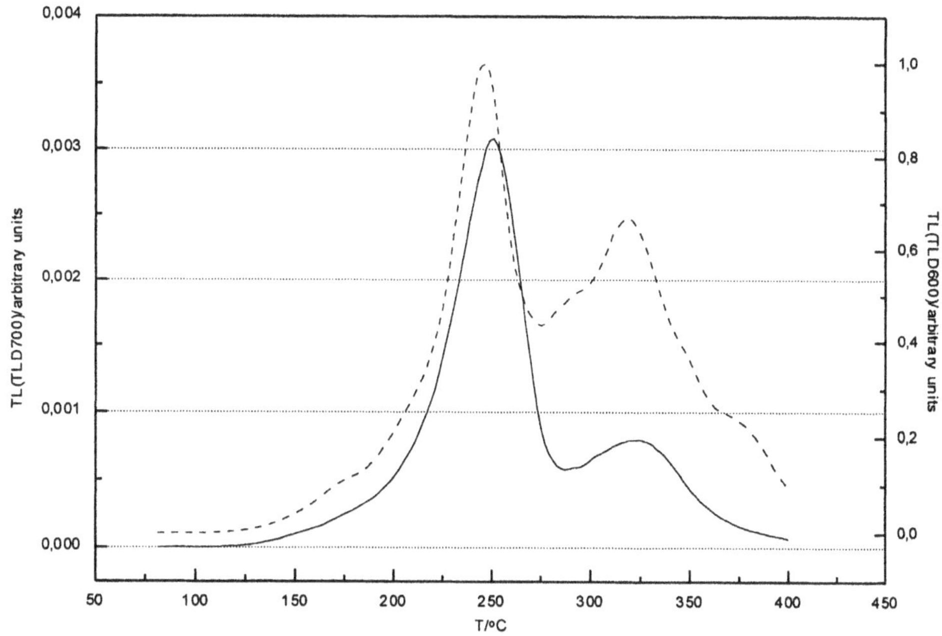

Figure 2. Glow curves after epithermal neutron irradiation for an absorbed dose of 1 Gy. ——TLD 700, - - - TLD 600.

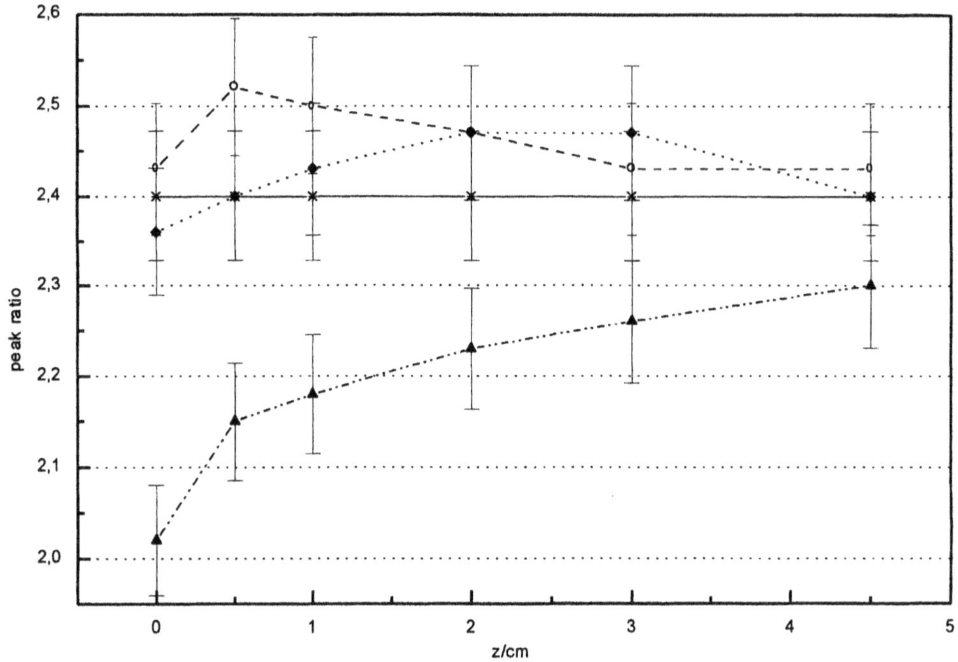

Figure 3. Peak ratio for TLD 300. – –O- . thermal n. · ·◆· · epithermal n. —*—^{60}Co γ-rays. – · ·▲– · · fission n.

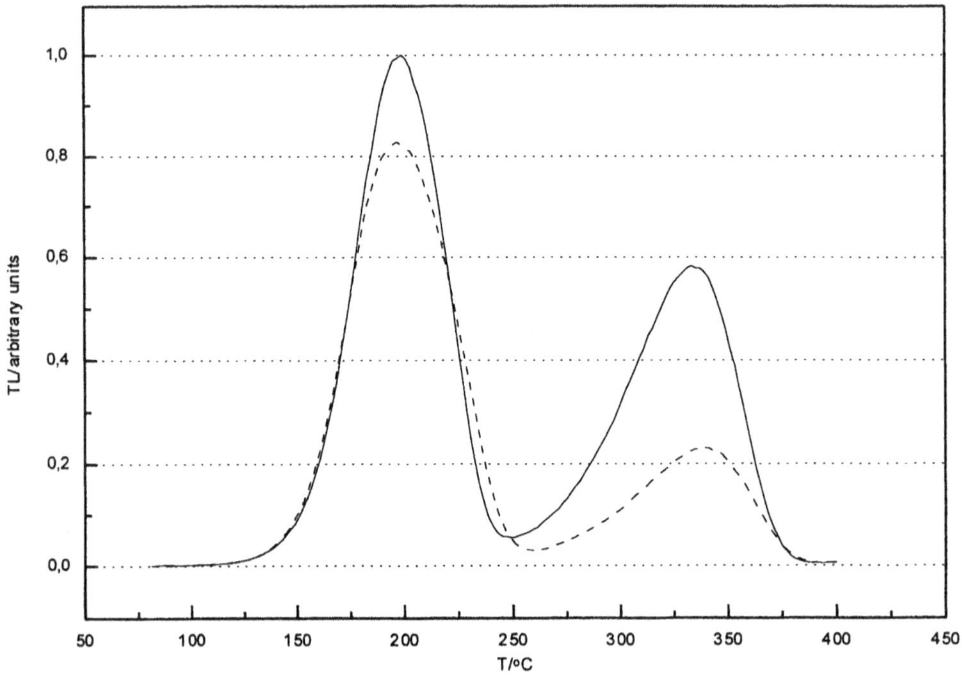

Figure 4. Glow curves for BeO for an absorbed dose of 1 Gy. ——^{60}Co γ-rays, – – – - epithermal n.

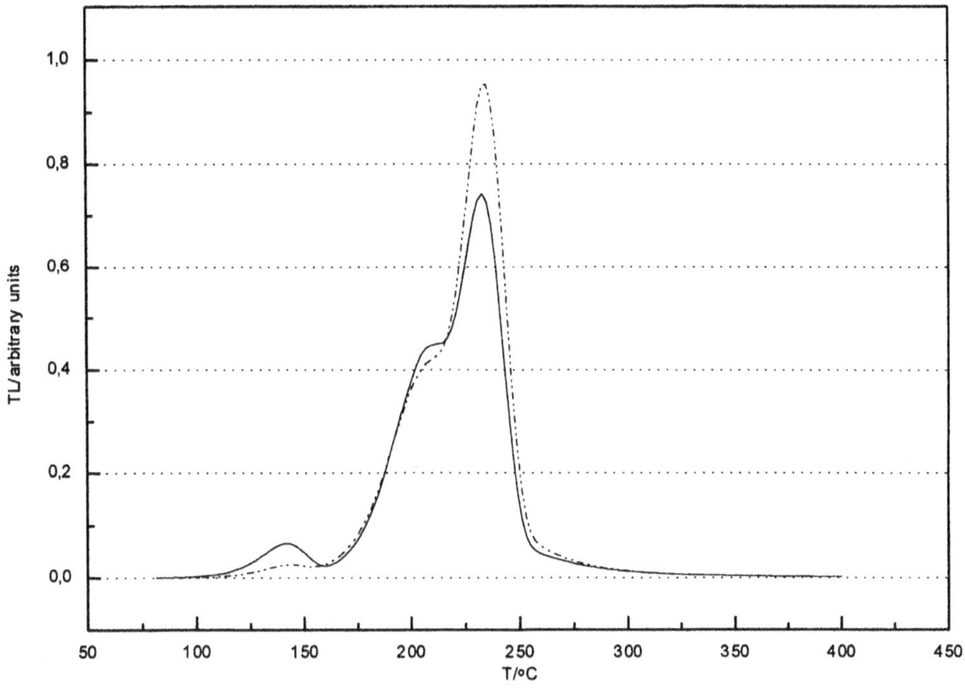

Figure 5. Glow curves for MCP for an absorbed dose of 1 Gy. ——^{60}Co γ-rays, – · · – · · - epithermal n.

values per unit of γ-ray absorbed dose in the following two columns and the R_u values per unit of gamma + neutron absorbed dose in last two columns.

It can be seen from column 5 that MCP detectors are less sensitive to neutrons than TLD 700 (column 4). In spite of this, for an accurate assessment of the gamma dose component it is necessary to determine their sensitivity to the neutron component, as it was done for TLD 700.

Figure 6 compares the R_u values for TLD 300 and BeO with the γ-ray absorbed dose values measured with Mg(Ar) ionisation chamber, for several irradiation depths at the three beams, while Table 3 shows the R_u values per unit of γ-ray absorbed dose. For γ-ray absorbed dose determinations by interpolation in a ^{60}Co γ-ray calibration curve using these TLD, a factor $h_u = 0.87 \pm 0.08$ for TLD300 and $h_u = 0.85 \pm 0.07$ for BeO must be applied.

Table 2. Results of TLD 700 and MCP detector irradiations in epithermal neutron beam

Depth	TE(TE) ionisation	Mg(Ar) chamber	R_u/Gy		R_u/D_γ		$R_u/D_{n+\gamma}$	
z/cm	$D_{n+\gamma}$/Gy	D_γ/Gy	700	MCP	700	MCP	700	MCP
1	0.99	0.58	2.53		4.36		2.56	
2	1.04	0.75	2.97		3.96		2.86	
3	0.97	0.76	2.71		3.57		2.79	
4	1.56	1.28	3.68		2.88		2.36	
1	1.03	0.60	2.45	1.33	4.08	2.22	2.37	1.29
2	1.07	0.77	2.82	1.55	3.66	2.01	2.64	1.45
3	1.01	0.77	2.59	1.44	3.36	1.87	2.56	1.43
4	0.88	0.71	2.09	1.21	2.94	1.70	2.38	1.38

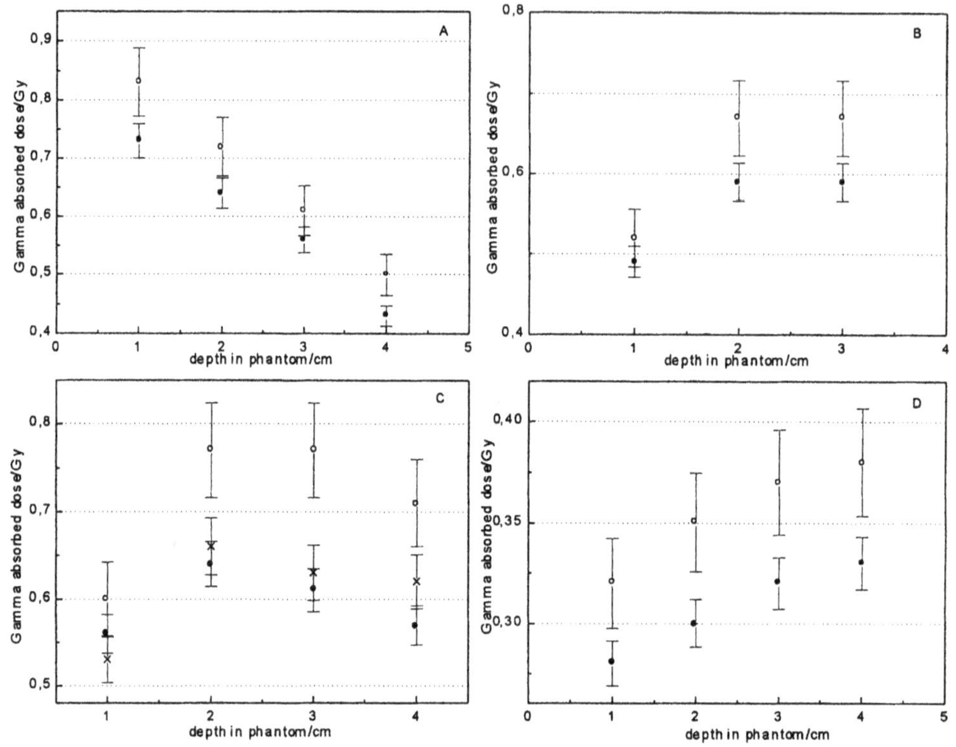

Figure 6. Comparison between gamma absorbed dose measured with Mg(Ar)on.chamber (◦) and R_u values for TLD 300 (•), and for BeO (×) A: thermal beam; B: 1st irradiation epithermal beam; C: 2nd irradiation epithermal beam; D: fission beam.

4. CONCLUSIONS

The application of TL detectors for dosimetry in mixed neutron-gamma fields requires knowledge of their response to neutron beams.

The limitation of the paired LiF thermoluminescence technique for the determination of gamma and neutron absorbed dose in mixed radiation fields is the important loss of sensitivity of the TLD 600 after neutron irradiation and the non negligible response of the TLD 700 to this component. The ^{6}Li impurity in these TLD can significantly affect their sensitivity in the mixed radiation fields and minute variations in the ^{6}Li content may cause large variations in the behaviour of the TL sensitivity to neutrons from batch to batch.

Table 3. Determination of h_u values for TLD 300 and BeO detectors

Beam	Depth	1 cm	2 cm	3 cm	4 cm	Average value
	thermal	0.88	0.89	0.91	0.86	
TLD 300	epithermal	0.94	0.89	0.88	0.84	0.87 (0.08)
		0.93	0.83	0.79	0.81	
	fission	0.88	0.86	0.86	0.87	
BeO	epithermal	0.87	0.86	0.81	0.87	0.85 (0.07)

TLD 700 as well as MCP detectors can be used for gamma absorbed dose measurements if their sensitivity to the neutron component is taken into account, and the thermal neutron fluence is determined.

For the mixed radiation fields investigated, no discrimination of neutron and gamma doses is possible with TLD 300, due to the low sensitivity of these detectors to low energy neutrons.

TLD 300 and BeO constitute an appropiate dosimetric system for the gamma dose determination in the mixed radiation beam by interpolation in a ^{60}Co γ-ray calibration curve, applying the corresponding h_u factors.

ACKNOWLEDGMENTS

To the staff of the Low Flux Reactor for their collaboration during the irradiations, providing, counting and analysing the activation foils and Dr Bos, for his suggestions and recommendations about TLD.

REFERENCES

1. C.P.J. Raaijmakers *et al.*, The neutron sensitivity of dosimeters applied to boron neutron capture therapy. Med. Phys. 23:1581–1589, 1996.
2. C.P.J. Raaijmakers *et al.*, Determination of dose components in phantoms irradiated with an epithermal neutron beam for neutron capture therapy, Med. Phys. 22:312–329, 1995.
3. J. Rassow, TLD-300 detectors for separate measurements of total and gamma absorbed dose distributions of single, multiple and moving-field neutron treatment—a new method of clinical dosimetry for fast neutron therapy, Technical Report Series, IAEA, Vienna, 1984, pp. 141–173.
4. Y.S. Horowitz, Thermoluminescence and thermoluminescent dosimetry. Vol I and II. CRC Press, Inc. Boca Raton, Florida.
5. G. Gambarini *et al.*, Thermoluminescent dosimeters in high fluxes of thermal neutrons, Advances in Neutron Capture Therapy. Vol 1, Medicine and Physics.
6. Y.S. Horowitz *et al.*, Limitation of the paired LiF TLD-600, 700 technique for the estimation of gamma ray dose in mixed n-γ radiation fields: the effect of thermal neutrons. Nucl. Instr. Meth. 160:317–320, 1979.

SEMICONDUCTOR DETECTORS FOR IN-PHANTOM THERMAL NEUTRON FLUX AND BORON DOSE MEASUREMENTS IN BNCT AND FAST NEUTRON THERAPY

Greg I. Kaplan[1], Anatoly B. Rosenfeld[2], Barry J. Allen[3], Jeffrey A. Coderre[4], Tooru Kobayashi[5], Richard L. Maughan[6], Chandrasekhar Kota[6], Mark Yudelev[6], Vladimir I. Khivrich[7], and Petro G. Litovchenko[7]

[1]Radiation Physics Group
University of Wollongong
Wollongong NSW 2522, Australia and
Westmead Institute for Cancer Research
University of Sydney at Westmead Hospital
NSW 2145, Australia
[2]Radiation Physics Group
University of Wollongong
Wollongong NSW 2522, Australia
[3]St. George Cancer Care Centre
Kogarah NSW 2217, Australia
[4]Medical Department
Brookhaven National Laboratory
Upton, New York 11973
[5]Research Reactor Institute
Kyoto University
Osaka 590-04, Japan
[6]Gershenson Radiation Oncology Centre
Harper Hospital
Detroit, Michigan
[7]Institute for Nuclear Research
Kiev, Ukraine

1. INTRODUCTION

A detailed knowledge of the thermal flux distribution in a phantom is important for modeling the boron dose in Boron Neutron Capture Therapy (BNCT) and augmented

Frontiers in Neutron Capture Therapy, edited by Hawthorne *et al.*
Kluwer Academic / Plenum Publishers, New York, 2001.

boron dose in Fast Neutron Therapy (FNT). Monte Carlo simulations of the thermal neutron flux with MCNP code need experimental verification, moreover such simulations in the case of FNT are difficult due to the lack of neutron cross-sections information for high energies of neutrons. In this circumstances a separate measurement of the thermal neutron flux is required. This measurement is complicated by a presence of low LET radiation in a mixed neutron-gamma field of a nuclear reactor or accelerator beam.

The neutron flux at a BNCT facility is measured by a variety of methods including TLD, a foil activation cadmium difference method and ionisation chambers with deposited fission converter material. The cadmium difference method requires two separate irradiations at a nuclear reactor and the post irradiation measurements of the induced gamma activity.[1] Cadmium cut off energy is also used as neutron energy separator in a Monte Carlo simulation package for the thermal neutron flux calculation. The energy dependence of cadmium neutron capture cross-section differs from the ^{10}B cross-section.

The TLD dosimetry[2] is not a direct method of the thermal neutron flux measurement but requires an independent calibration. Unshielded fission counters or ^{3}He-filled ion chambers are used at MIT research reactor to monitor the residual thermal neutron flux.[3] BF_3 proportional counter has been recently used at BMRR.[4] The reactor power was as low as only 50 W and no comparison with Monte-Carlo calculation of the thermal neutron flux was made.

In this paper we report on the application of a silicon detector with a ^{235}U converter for experimental measurement of the thermal neutron flux in epithermal and thermal BNCT and in FNT. Additional data and a more detailed analysis of thermal neutron flux in the epithermal neutron beam only are presented in.[5] The fission detector makes a direct and accurate measurement of thermal neutron flux. It requires a single irradiation with no post-irradiation measurement. It also allows on-line monitoring of thermal neutron flux.

2. MATERIALS AND METHODS

A standard spectroscopy setup consisting of a preamplifier, main amplifier, detector bias supply and a multi channel analyser was used for energy spectrum measurements. Epithermal BNCT experiments were performed in a $15 \times 15 \times 15\,cm^3$ perspex slab phantom. Position of a detector along the central axis of the phantom can be changed with a step of 1 mm by rearranging the slabs. The experimental measurements in thermal BNCT beam were performed in a cylindrical water filled head phantom 20 cm long and 18 cm in diameter. In FNT beam the measurement of thermal neutron flux were performed in $30 \times 30 \times 40\,cm^3$ water tank.

2.1. Radiation Facilities

Measurements of the thermal neutron flux with a fission detector were performed at the epithermal neutron facility at Brookhaven Medical Research Reactor (BMRR), BNL,[6] and at the quasi epithermal beam at the modernised reactor at Kyoto University, Osaka.[7] Patient irradiation at BNL is run at the full reactor power of 3 MW. A very high count rate caused the dead time of the detector to approach 100%. It made impossible to conduct spectroscopy measurements at full reactor power. A reactor power of 20 kW was used for silicon detectors. In Osaka the reactor power was set to 5 MW and was not variable. As in a clinical mode the cadmium filter was closed. The neutron flux was reduced by introducing 30 cm of heavy water filter. In Osaka and BNL the beam diameter was 10 cm.

For investigation of feasibility of boron enhancement of FNT, thermal neutron flux was measured at the superconducting cyclotron at Harper Hospital, Detroit.[8] Measurements with a fission detector were performed at a reduced beam current of 0.7 μA in a $15 \times 15\,cm^2$ field.

2.2. Radiation Detectors

The fission detector was produced by putting an aluminium foil deposited with 150 μm thick ^{235}U enriched uranium oxide in a contact on the surface of a $3 \times 3\,mm^2$ ion implanted silicon detector. Energy dependence of ^{235}U neutron cross-section is similar to the one of ^{10}B. It mostly captures thermal neutrons. After neutron capture, the uranium nucleus decays with 85% probability releasing two heavy fragments which share a total kinetic energy of 162 MeV between them.[9] One of the fragments enters the sensitive layer of the silicon detector and produces a pulse on the pulse height spectrum. Due to their high energy and high linear energy transfer (LET) the fragments are completely stopped within the sensitive layer and the pulse is well separated from the low LET radiation. In the absence of a neutron radiation field ^{235}U spontaneously decays releasing a 4.5 MeV alpha particle. The alpha particle spectrum, measured by the silicon detector, was used for energy calibration of the detector. Intrinsic alpha activity of the converter was used to calculate the effective number of ^{235}U nuclei, N_U, in the converter. Alpha activity was determined by calculating the number of alpha pulses per second, as measured by the silicon detector in 2π geometry of the uranium converter laying on the detector. N_U was adjusted to the difference in the range of 4.5 MeV alpha particle and a heavy ion like a typical uranium fragment ^{87}Br.[10] For the fission converter used in this study $N_U = 1.72 \times 10^{17}$ nuclei. The thermal neutron flux, Φ, is proportional to the experimentally measured flux of uranium neutron capture decay fragments, F, and inversely proportional to uranium neutron capture cross-section, σ, and probability of decay, p.

$$\Phi = \frac{F}{N_U \cdot \sigma \cdot p}$$

A more accurate formula was derived in.[5] For measurement of the boron dose enhancement the method of paired MOSFET detector[11] was used. One of the MOSFETs was covered by a boronated plastic while the other was not.

3. RESULTS AND DISCUSSION

In Figure 1(a) and (b) two pulse height spectra measured by the fission detector in epithermal BNCT beam of BMMR are presented. The extended wide peak of uranium fission fragments is clearly visible. For ensuring that in the absence of the fission converter the high energy pulses are not observed, the pulse height spectrum was measured by an identical silicon detector without the fission converter at the same depths in the phantom. No pulse with energy above 2 MeV was observed (Figure 1c). The absolute height of the low energy peak for the measurements without converter was higher then

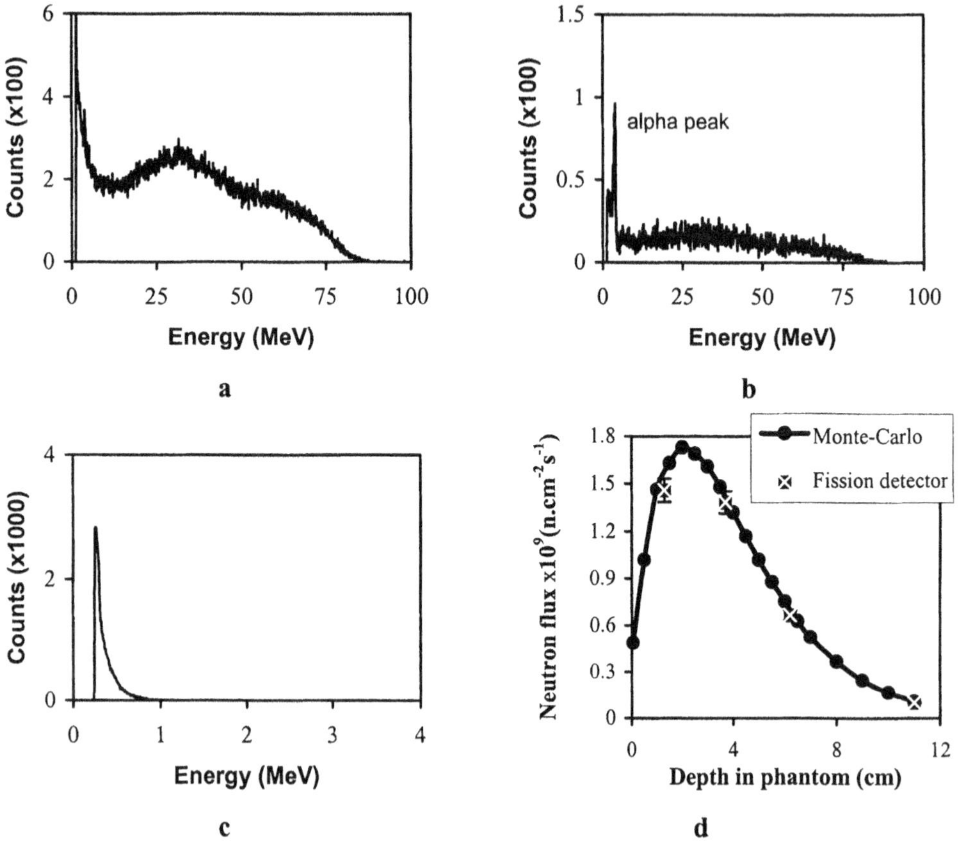

Figure 1. In-phantom measurements in BNCT beam at BMRR. Fission detector pulse height spectra at 3.7 cm (a) and 11 cm (b) depth in a perspex slab cube phantom. Pulse height spectrum measured by a silicon detector without the fission converter at 3.8 cm depth (c). Monte Carlo calculated and fission detector measured thermal neutron flux in the perspex slab cube phantom.

for the fission converter measurements due to the difference in setting of the low energy discriminator. The spectrum in Figure 1b was measured at the depth of 11 cm in the perspex phantom. As the thermal neutron flux was lower than at the depth of 3.7 cm (Figure 1a) the intensity of the broad fragment peak became lower and alpha particle peak of ^{235}U spontaneous decay became a prominent feature of the spectrum.

Thermal neutron flux measured by the fission detector was compared with the same calculated by Monte Carlo MCNP4a code.[12] Calculations were performed along the central axis of the $15 \times 15 \times 15\,cm^3$ perspex slab phantom at a reactor power of 3 MW. A numerical factor of 150 was used for conversion of experimental data measured at 20 kW reactor power to the full reactor power of 3 MW. The agreement between calculated and experimental data was good (Figure 1d). The peak in thermal neutron flux was observed in the depth 2 to 4 cm. There is also a clear thermal neutron flux build up region due to thermalisation of epithermal neutrons in the reactor beam. The shape of thermal neutron flux depth distribution corresponds to the shape of boron dose depth distribution (Figure 2) measured by paired MOSFET method.

The thermal neutron flux depth distribution was slightly different in the filtered BNCT beam of the Osaka reactor (Figure 3a). The build up region is narrower and the

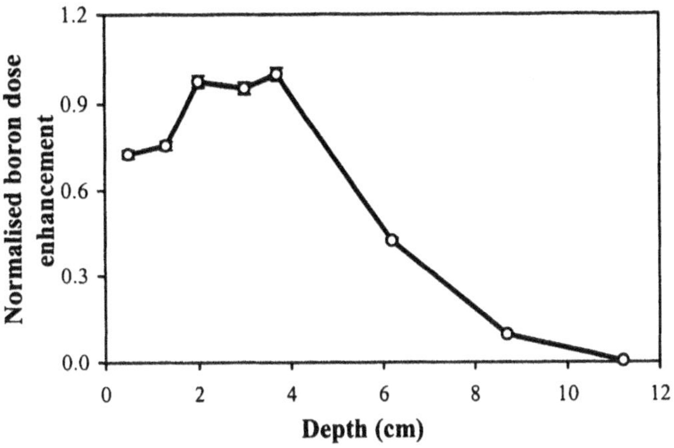

Figure 2. Relative in-phantom boron dose in BNCT beam of BMRR.

decrease in the flux with depth is steeper. The diameter of the beam was 10 cm but when we measured lateral change in the thermal neutron flux we found that the beam extends beyond 5 cm (Figure 3b) and the flux decreases 50% from the beam axis to the distance 4 cm off the axis.

The same fission detector and silicon detector probes were used for thermal neutron flux mapping in perspex and water phantoms exposed to a Be(48.5 MeV)d FNT beam. The radiation field was $10 \times 10\,cm^2$. In case of opened collimator the thermal neutron fluences per 1 monitoring unit (MU) were 1.98×10^8, 1.95×10^8, 1.74×10^8 at depths of 1.3, 6.2, and 8.7 cm in the perspex phantom. The experiments with 15×15 cm field and partially closed collimator with tungsten rods showed a relative increase of thermal neutron flux of about 30% at a depth 5 cm in water due to moderation of the input neutron spectra. From these data, it is clear that in case of FNT for one fractionation (150 MU), the absolute value of the thermal neutron fluence is 50–70 times less than maximum fluence in the case of BNCT at BMRR during the 20 minutes of treatment.

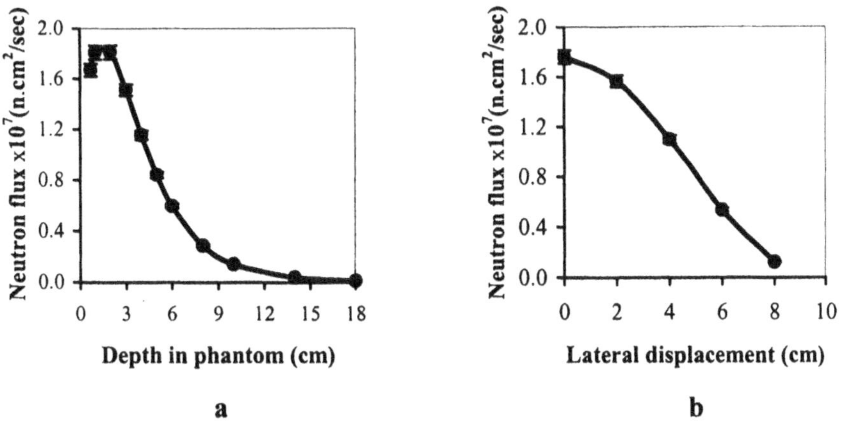

Figure 3. Thermal neutron flux measured by the fission detector in 30 cm D_2O filtered BNCT beam at Osaka reactor. Flux depth measurement along the phantom axis (a) and in the lateral direction at 2 cm depth. The beam diameter was 10 cm.

4. CONCLUSIONS

The fission detector is proven to be a valuable method for in-phantom thermal neutron flux measurements in BNCT and FNT beams. It is also possible to use it for on-line monitoring of the beam quality. To increase accuracy, reliability and to make measurements possible under full reactor power, the silicon detector should be less then $1 \times 1\,mm^2$ in size and the uranium converter should not be thicker than 2–3 μm.

ACKNOWLEDGMENT

We would like to thank Dr. H.B. Liu for the support with Monte Carlo calculations and Mr. Peter Bradley for his contribution to collecting of experimental data at Osaka reactor.

REFERENCES

1. H.B. Liu, D.D. Greenberg, J. Capala, and F.J. Wheeler, "An Improved Neutron Collimator for Brain Tumour Irradiation in Clinical BNCT," Medical Physics **23**, 2051–2060 (1996).
2. J.-P. Pignol, P. Cuendent, N. Brassart, G. Fares, F. Colomb, C.M.B. Diop, R. Sabattier, A. Hachem, and G. Prevot, "Combined use of FLUKA and MCNP-4A for the Monte-Carlo simulation of the dosimetry of ^{10}B capture enhancement of fast neutron irradiation," Medical Physics **25**(6), 885–891 (1998).
3. G. Solares, D. Katz, O. Harling, and R. Zamenhof, "On-Line Dosimetry for Boron Neutron Capture Therapy at the MIT Research Reactor," in *Advances in Neutron Capture Therapy*, edited by B. Larsson, J. Crawford, and R. Weinreich (Elsevier, Lausanne, 1997), Vol. 1, pp. 147–152.
4. D.E. Alburger, D. Raparia, and M.S. Zucker, "Phantoms with $^{10}BF_3$ detectors for boron neutron capture therapy applications," Medical Physics **25**(9), 1735–1738 (1998).
5. G.I. Kaplan, A.B. Rosenfeld, B.J. Allen, J.A. Coderre, and H.B. Liu, "Fission converter and metal-oxide-semiconductor field effect transistor study of thermal neutron flux distribution in an epithermal neutron therapy beam," Med. Phys. **26**(9), 1989–1994 (1999).
6. H.B. Liu, R.M. Brugger, D.D. Greenberg, D.C. Roger, J.P. Hu, and H.M. Hauptman, "Enhancement of the Epithermal Neutron Beam Used for BNCT at BMRR," Int. J. of Radiat. Oncol. Biol. Phys. **28**, 1149–1156 (1994).
7. T. Kobayashi, K. Kanda, Y. Fujita, Y. Sakurai, and K. Ono, "The upgrade of the heavy water facility of the Kyoto University Reactor for neutron capture therapy," in *Advances in Neutron Capture Therapy*, edited by B. Larsson, J. Crawford, and R. Weinreich (Elseiver, Lausanne, 1997), Vol. 1, pp. 321–325.
8. R.L. Maughan, and M. Yudelev, "Physical Characteristics of a Clinical d(48.5)+Be Neutron Therapy Beam Produced by a Superconducting Cyclotron," Medical Physics **22**, 1459–1465 (1995).
9. J.E. Turner, *Atoms, Radiation and Radiation Protection* (McGraw-Hill Inc., 1992).
10. O.F. Nemec, and Y.V. Hofman, *Nuclear Physics Handbook* (Naukova Dumka (in Russian), Kiev, 1975).
11. A.B. Rosenfeld, G.I. Kaplan, M.G. Carolan, B.J. Allen, R.L. Maughan, M. Yudelev, C. Kota, and J.A. Coderre, "Simultaneous Macro and Micro Dosimetry with MOSFETs," IEEE Trans. Nucl. Sci. **43**, 2693–2700 (1996).
12. J.F. Briesmeister, "MCNP-4A General Monte Carlo Code for Neutron and Photon Transport," Los Alamos National Lab, LA-7396-M, Rev. 2, 1986.

DETERMINATION OF AN RBE FOR AN EPITHERMAL NEUTRON BEAM FOR BNCT FROM MICRODOSIMETRIC MEASUREMENTS

Stuart Green[1], Mark Gainey[2], and Cecile Wojnecki[2]

[1]Department of Medical Physics and Clinical Engineering
Queen Elizabeth Hospital
University Hospital Birmingham NHS Trust
Edgbaston, Birmingham, B15 2TH, The United Kingdom
[2]School of Physics and Astronomy
The University of Birmingham
Edgbaston, Birmingham, B15 2TT, The United Kingdom

INTRODUCTION

As the field of BNCT matures, one key element must be the development of a mechanism which allows the clinical experience gained on the beams at the pioneering centres, to be transferred to newer centres. In this way, new centres will have the potential to begin therapy immediately at the optimum dose level, and dose escalation studies may not be necessary. A greater role for microdosimetry in the specification of neutron beam quality for fast neutron therapy has been suggested.[1] This paper explores a formalism through which measurements based on the proportional counter techniques can provide RBE data for epithermal neutron beams.

A number of authors have reported the use of the microdosimetric technique for BNCT dosimetry.[2,3] A microdosimetric measurement provides information on the distribution in linear energy transfer (LET) of the ionisation events which make up the radiation dose. It has been suggested[4] that this microdosimetric spectrum may be related to the RBE of the beam for a given biological end-point via a Biological Weighting Function.

METHODS

Microdosimetric Measurements

Measurements of microdosimetric spectra in tissue-equivalent material (A150 plastic) have been made at 4 depths in a rectangular section Perspex enclosed water

Frontiers in Neutron Capture Therapy, edited by Hawthorne *et al.*
Kluwer Academic / Plenum Publishers, New York, 2001.

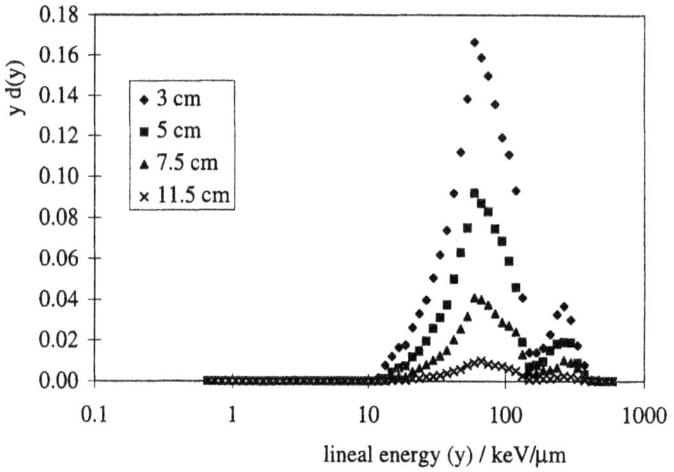

Figure 1. y·d(y) distributions at four depths in a rectangular water phantom.

phantom when exposed to the Birmingham epithermal neutron beam. The measurements are shown in Figure 1 as a function of depth, where the data which would normally appear below approximately 10 keV/μm (which is due mainly to photon interactions) has been subtracted. The relative proportion of this photon component increases with depth from 82% of the absorbed dose at 3 cm deep, to 95% at 11.5 cm deep. (These percentages are expressed in terms of physical dose deposited by the beam alone, without consideration of boron related dose components).

Figure 2 shows the same data as in Figure 1, except that it is now arbitrarily renormalised to have the same area beneath each curve. Hence the shapes of the curves may be directly compared. The shapes of these curves do not change significantly with depth in phantom (for depths from 3 to 11.5 cm).

Biological Weighting Function

The work of Loncol *et al.*[4] has shown that beam RBE for a specific biological end-point for a range of proton and neutron therapy beams, may be related to the measured microdosimetric spectra through a biological weighting function. This function is derived from a data-set of measured RBEs for a biological end-point concerned with crypt cell regeneration in mice, and is therefore only strictly valid for this end-point. This end-point may be an indicator of early radiation damage.

Once the biological weighting function is established across the LET interval covered by a new therapy beam for a particular biological end-point, it may be used in conjunction with measured microdosimetric spectra to predict an RBE for the new beam. In order to avoid confusion this predicted RBE is referred to as "R". The function derived by Loncol *et al.* is shown in Figure 3, and it is clear from comparison of Figures 2 and 3 that the Birmingham BNCT beam has an LET distribution in the region where the biological weighting function is changing rapidly.

In order to use the data in Figures 2 and 3 to make a prediction of "R" for the Birmingham beam, we must use equation 1.

$$\text{"R"} = \frac{\int r(y) \cdot d(y) dy}{\int d(y) dy} \qquad (1)$$

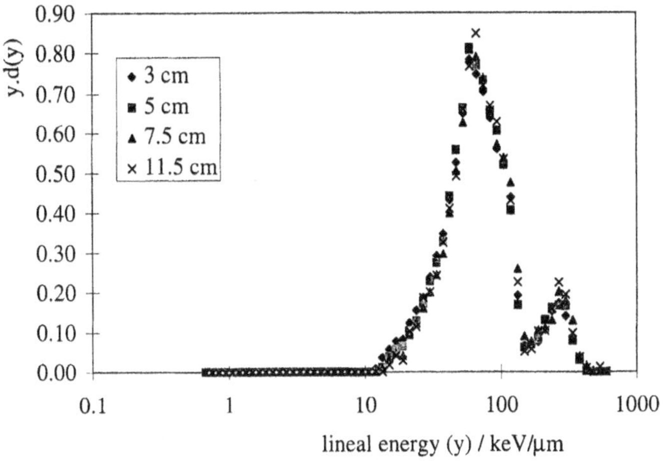

Figure 2. y·d(y) distributions from Figure 1, normalised to the same area.

where yd(y) is a microdosimetric spectrum taken from Figure 2 and r(y) is the biological weighting function in Figure 3.

Hence, all that is required is a convolution of the measured microdosimetric spectrum and the biological weighting function. Since all the measured microdosimetric spectra have been shown to be very similar, we may obtain a representative "R" value by selecting the measurement at 5 cm depth.

RESULTS

The result is a neutron RBE of 3.4 ± 0.1, which is shown in Table 1 together with other estimates for epithermal neutron beams for BNCT.

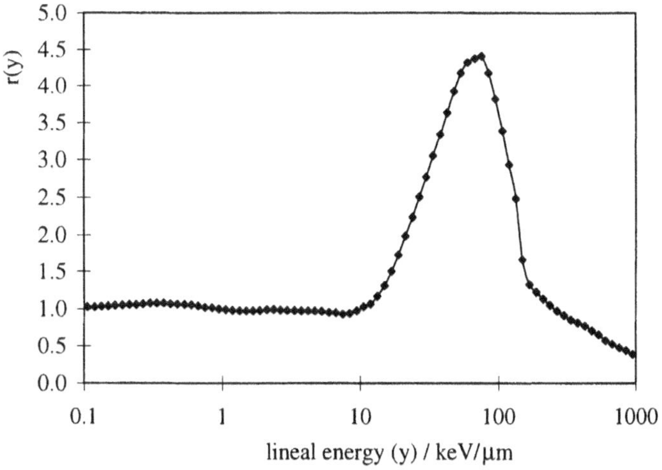

Figure 3. Biological weighting function.[4]

Table 1. Biological parameters for epithermal BNCT beams

Component	Published data	Nigg[7]	Birmingham (90° D_2O/Li)
Beam protons	3.3[5]	2.0–5.0	
$^{14}N(n,p)$	3.0[6]	2.5–4.0	
Calculated "R"			3.4 ± 0.1[#]

[#]This uncertainty is related to the ability of this kind of approach to predict RBE in fast neutron therapy beams.[1]

DISCUSSION

Caution should be exercised in interpreting the data in Table 1 for a number of reasons. The most significant of these is that the biological endpoints are not the same. The figure for $^{14}N(n,p)$ from reference 6 is derived for moist desquamation which is an early skin reaction. The figure from reference 5 is related to MRI detected damage to the white matter of dogs irradiated with an epithermal beam, and the Loncol *et al.* data[4] relates to an end-point concerned with crypt cell regeneration in mice which may be an indicator of early radiation damage.

In addition the following items should also be considered:

1. The Birmingham figure is derived from data which includes some component of $^{14}N(n,p)$ but this component is different from that which will occur in brain tissue because of the difference in nitrogen content of A-150 plastic (3.51%) and brain tissue (1.28%). This may not impact too strongly on the relative change in microdosimetric spectra with depth as at depths greater than around 3 cm the dose deposited is dominated by nitrogen capture events. At depths greater than the thermal peak, the thermal fluence is reduced with increasing depth. Hence the number of nitrogen capture reactions also reduces, but the characteristics of the energy deposition, as indicated by the shape of the microdosimetric spectrum, will not change.
2. The microdosimetric technique as used here makes no allowance for variations in dose-rate, or for the modifying influence of fractionation.
3. The calculated "R" is derived from consideration of energy deposition due to recoil protons, alphas and heavy nuclei in tissue equivalent plastic, while the quoted RBE values specify "beam protons".
4. The clinical RBE may be determined by other factors which are not accounted for in the approach outlined here.

If we assume there may be a predictable relationship between measured microdosimetric spectra and RBE, then we must conclude that if the measured microdosimetric spectra are reasonably constant with depth (as shown in Figure 2 for the depths from 3 to 11.5 cm), the RBE should be similarly constant over this range of depths. This analysis of course cannot address potential RBE changes at depths less than 3 cm.

It is also possible to make partial use of data such as those presented here. Rather than making a prediction of RBE from the microdosimetric data alone, it may be possible to use the microdosimetric data to inform the standard techniques. These techniques depend for their assessment of neutron beam biological parameters, on assumptions made concerning the photon RBE. This microdosimetric technique described here will not only provide separate data on neutron and photon dose components, but also on their respective biological characteristics.

ACKNOWLEDGMENTS

Thanks are due to the Dynamitron operating staff, to Dr Gerard Morris for his very useful comments on the original manuscript, and to other members of the Birmingham BNCT group.

REFERENCES

1. Wambersie A., "Contribution of microdosimetry to the specification of neutron beam quality for the choice of the 'Clinical RBE' in fast neutron therapy". *Radiation Protection Dosimetry* 52, No 1–4, 453–460, 1994.
2. Wuu S., Amols H.I., Kliauga P., Reinstein L.E., and Saraf S., "Microdosimetry for boron neutron capture therapy", *Radiation Research* 130, 355–359, 1992.
3. Kota C., and Maughan R.L., "A dosimetry system for boron neutron capture therapy based on the dual counter microdosimetric technique". *Bulletin Cancer/Radiotherapy* 83 (Suppl 1) 173s–175s, 1996.
4. Loncol T., Cosgrove V., Denis J.M., Gueulette J., Mazal A., Menzel H.G., Pihet P., and Sabatier R., "Radiobiological effectiveness of radiation beams with broad LET spectra: Microdosimetric analysis using biological weighting functions". *Radiation Protection Dosimetry* 52, Nos 1–4, 347–352, 1994.
5. Huiskamp R., Gavin P.R., Coderre J.A., Phillip K.H.I., and Wheeler F.J., "Brain tolerance in dogs to BNCT with borocaptate sodium (BSH) or boronphenylalanine (BPA)", *in*: Mishima Y. (ed) Neutron Capture Therapy for Cancer. Plenum Press, New York.
6. Gavin P.R., Wheeler F.J., Huiskamp R., Siefert A., Kraft S., and De Haan C., "Large animal studies of normal tissue tolerance using an epithermal neutron beam and borocaptate sodium", *in:* In Gabel D., and Moss R. (eds) Towards Clinical Trials of Glioma. Plenum Press, New York, 1992, pp. 197–209
7. Nigg D.W., "Methods for radiation dose distribution analysis and treatment planning in BNCT". *International Journal of Radiation Oncology, Biology and Physics* Vol 28, No 5, 1121–1134, 1994.

EFFECTS OF NEUTRON COLLIMATION, BEAM SIZE, AND SPECTRUM ON IN-PHANTOM PERFORMANCE OF REALISTIC EPITHERMAL NEUTRON BEAMS FOR BNCT

W. S. Kiger III, S. Sakamoto, and O. K. Harling

Nuclear Reactor Laboratory
Massachusetts Institute of Technology

1. INTRODUCTION

Sensitivity studies of the collimator and fast neutron filter/moderator designs of the MIT fission converter epithermal neutron beam have been conducted to examine the effects of beam collimation or directionality, beam size (i.e., neutron spatial distribution), and neutron spectrum on dosimetric performance in a water-filled ellipsoidal head phantom. The effects of these beam parameters have already been studied using ideal beams, i.e., monoenergetic, photon-free neutron beams with varying maximum neutron emission angle and beam size.[1] In those ideal beam studies, the effect of each parameter was examined separately by varying one parameter with the others fixed. However, in actual beam design, these three parameters, especially beam directionality and beam size, (and intensity as well), vary simultaneously, resulting in a much more complex optimization problem. The current study examines the effects of these beam characteristics using realistic epithermal neutron beams to evaluate their relative importance in treating brain tumors.

2. METHOD OF ANALYSIS

The beam used in this computational study, the MIT fission converter beam, has been well-optimized through an extensive series of design studies, with several designs producing high intensity (epithermal neutron flux in excess of 1.0×10^{10} n/cm^{2}s) and high quality (specific fast neutron and photon doses below 2.0×10^{-11} cGy cm^2/n) epithermal neutron beams.[2,3] With BPA-F, these beams also produce excellent in-phantom performance with high dose rates (400–600 RBE cGy/min), high advantage ratios[4] (~5), and

Frontiers in Neutron Capture Therapy, edited by Hawthorne *et al.*
Kluwer Academic / Plenum Publishers, New York, 2001.

deep effective penetration (advantage depths of ~9.5 cm) despite having directionality somewhat lower (current to flux ratio ~0.65) than that recommended by some researchers.[5]

The primary computational tool employed in this study is an MCNP[6] model of the fission converter beam indirectly coupled with the MITR-II reactor core.[7] The methodology for efficient computation used in this work has been previously described.[8]

Beam directionality and size were modified by using different collimator configurations, by using an absorbing lithiated polyethylene delimiter (containing 7% ^{6}Li by mass), and by changing the physical size of the beam aperture. Two collimator configurations were considered in this study. Horizontal cross-sectional views of these designs are shown in Figure 1a (funnel type) and 1b (truncated pyramid type). The funnel-shaped collimator consists of a four sided inwardly sloped component lined with a 15 cm thick lead reflector and a 20 cm long straight channel of with parallel sides surrounded by a Li-poly delimiter covering the whole beam cross section except for the beam aperture. The truncated pyramid-shaped collimator has a 15 cm thick lead reflector, where the last t cm (t = 0 or 10) long part of the lead reflector was replaced with Li-poly. For the physical beam aperture size, two different sizes, 20 cm × 20 cm and 10 cm × 10 cm, were considered. The variation of beam directionality was monitored by tallying the neutron angular distribution and calculating the neutron current to flux ratio (J/ϕ). The variation of effective beam size was also monitored by tallying the neutron spatial distribution at the exit aperture and by calculating the full width-half maximum of the vertical neutron flux profile.

Changing the neutron spectrum was effected by varying the filter/moderator length. In this analysis, the filter/moderator was composed of a 71 cm block of aluminum followed by a block of AlF_3 varying in length from 17 to 27 cm in 5 cm increments.

For all of the calculations, the target position was fixed at the beam's entrance to the medical irradiation room, 303 cm from the reactor core centerline, and the collimator length is fixed at 102 cm. For analysis of both beam directionality and spectrum, in-air figures of merit were tallied at the patient or target position and in-phantom parameters were calculated for unilateral irradiations of a water-filled ellipsoidal head phantom, represented by Eqn. 1, along its 14 cm x-axis. Assumptions for RBEs and ^{10}B uptake for calculation of dose profiles, noted in the caption of Table 1, are those used clinically at Harvard-MIT[9] and BNL for BNCT with BPA-F.[10]

$$\left(\frac{x}{7}\right)^2+\left(\frac{y}{10}\right)^2+\left(\frac{z}{8.5}\right)^2 \leq 1 \qquad (1)$$

3. RESULTS AND DISCUSSION

This section presents results of sensitivity studies of the collimator and filter/moderator designs.

3.1. Beam Directionality and Size

Results of the analyses of different collimator designs, delimiters, and aperture sizes are reported in Table 1. Interpretation of these results is not straightforward due, in some comparisons, to the unavoidable simultaneous variation of more than one parameter. Nevertheless, several general conclusions can be drawn. Comparison of intensity and

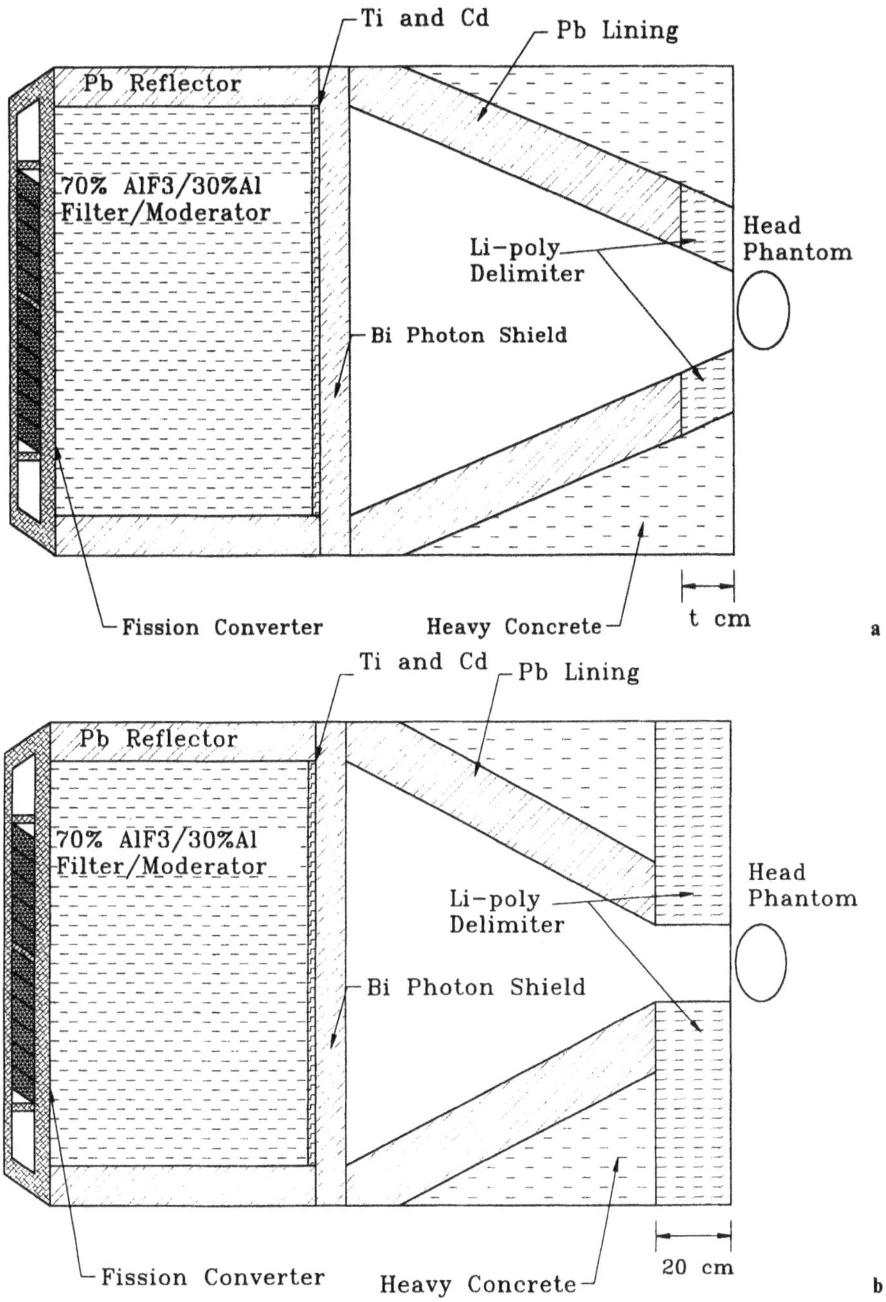

Figure 1. (a) Cross-sectional view of a fission converter beam using a four sided pyramidal collimator design. The collimator has a 15 cm thick lead lining, where the last *t* cm (t = 0 or 10) of the lead lining in front of the patient position was replaced with Li-poly. (b) Cross-sectional view of the fission converter beam with a funnel-shaped collimator design. This type of collimator consists of a four sided, inwardly sloped component lined with 15 cm thick lead followed by a 20 cm long portion with parallel sides surrounded by a Li-poly delimiter, which covers the whole beam cross section except for the beam aperture.

Table 1. Comparison of in-phantom figures of merit and in-air parameters produced by different collimator designs[a]

Physical beam description				In-air parameters					In-phantom figures of merit		
Design	Collimator shape	Length of Li-poly portion (cm)	Beam aperture size (cm)	Beam FWHM (cm)	ϕ_{epi} (n/cm^2 s)	J_{epi}/ϕ_{epi}	D_{fn}/ϕ_{epi} (cGy cm^2/n)	D_γ/ϕ_{epi} (cGy cm^2/n)	AD (cm)	ADDR (RBE cGy/min)	AR
1	Pyramid	0	20 × 20	24.2	1.30E + 10	0.67	1.33E − 11	1.02E − 11	9.5	435	4.9
					1.3%	1.7%	1.9%	2.0%	0.6%	1.3%	0.6%
2	Pyramid	0	10 × 10	14.6	1.28E + 10	0.68	1.29E − 11	1.06E − 11	9.0	362	5.0
					1.6%	2.0%	2.5%	2.4%	0.5%	0.9%	0.4%
3	Pyramid	10	20 × 20	20.7	9.33E + 09	0.77	1.34E − 11	1.46E − 11	9.3	356	4.9
					0.5%	0.7%	0.7%	1.0%	0.4%	0.7%	0.3%
4	Funnel	20	20 × 20	20.0	4.45E + 09	0.87	1.49E − 11	2.28E − 11	9.1	206	4.9
					0.6%	0.9%	0.8%	1.0%	0.4%	0.9%	0.4%
5	Funnel	20	10 × 10	9.1	2.19E + 09	0.91	1.62E − 11	3.26E − 11	8.5	79	4.9
					1.5%	2.1%	1.8%	2.2%	1.0%	1.8%	0.8%

[a]The effective beam size is reported as the full width half maximum (FWHM) of the vertical profile of the total neutron flux. For calculation of dose profiles, boron uptake was assumed at 40 ppm ^{10}B in tumor with a tumor-to-normal tissue uptake ratio of 3.5 : 1. The RBE's are 3.2 for neutrons, 1.0 for photons, 3.8 for boron in tumor and 1.35 for boron in normal tissue. The funnel-shaped collimator (Designs 4 and 5) consists of a four sided inwardly sloping portion with a 15 cm thick lead lining and a 20 cm long Li-poly delimiter with parallel sides. The pyramidal collimator is lined with 15 cm thick lead, where the last t cm (t = 0 for Designs 1 and 2 and t = 10 for Design 3) long parts of the lining in front of the patient position was replaced with Li-poly. These calculations used a fission converter composed of eleven burned MITR-II fuel elements (312 g ^{235}U per element) with D_2O cooling (producing 78.1 kW), which is followed by a fast neutron filter/moderator composed of 68 cm 70% AlF_3/30% Al − 2 cm Ti, a 0.04 cm Cd thermal neutron filter and an 8 cm Bi photon shield. The filter/moderator is surrounded by a 10 cm thick lead reflector. The statistical uncertainty of each quantity (one standard deviation) is listed immediately below the value as a percentage.

directionality (J/ϕ) for the different collimator designs illustrates the expected trade-off: reduction of the large angle components of the neutron flux necessary to achieve high directionality represents a large penalty in terms of diminished beam intensity. Design 1, which has similar collimation ($J/\phi \cong 0.67$) to Design 2 but an effective beam size ~10 cm larger, penetrates ~5 mm more deeply than the smaller beam. This result is consistent with the ideal beam studies of Ref. 1 and a study of neutron collimators at BNL,[11] where expansion of isodose contours was observed when the diameter of the irradiation aperture of the collimator was enlarged from 8 cm to 12 cm. The trend of decreasing penetration (AD) with increasing J/ϕ is contrary to the common perception that increased directionality provides better beam penetration.[5] This effect is more likely due to reduced beam size rather than directionality. Ref. 1 demonstrates that for ideal beams in the energy range of interest with sizes greater than or equal to the target size, penetration does not vary significantly for different neutron angular distributions with maximum emission angles of 60° or less. With 50–80% of the neutrons in these beams entering the exit aperture within a cone of 32°, these beams are very forward oriented and fall into this category. Computation of treatment plans for brain irradiations with low and high collimation beams of similar effective width (Designs 1 and 4) using the Harvard-MIT treatment planning software MacNCTPlan[12] has confirmed this behavior; tumor isodose contours normalized to the maximum normal tissue dose lie ~5 mm deeper with the low directionality beam.

For the size and shape of target studied here, collimation is of low importance for beam optimization in comparison to other in-air figures of merit such as specific fast neutron and photon dose, intensity, and beam size, due to the low significance of its impact on in-phantom performance. Furthremore, a high degree of collimation may be undesirable due to large intensity losses required to produce beams with very high directionality. However, neutronic performance in-phantom is not the only aspect of NCT irradiations affected by collimation. Collimation can significantly reduce the collateral dose incurred during NCT irradiations. A high degree of collimation can also ease the constraints of patient positioning. For instance, at the High Flux Reactor in Petten, with most of the neutrons in the beam falling within 2° of the beam axis, patients are positioned 30 cm from the exit port.[13]

3.2. Beam Spectrum

Table 2 lists in-air and in-phantom beam performance results produced by 3 different filter/moderators. The 3 different spectra produced all result in specific fast neutron dose less than the irreducible background dose from hydrogen capture gammas and the $^{14}N(n,p)$ reaction of $\sim 2.0 \times 10^{-10}$ cGy cm^2/n. Compared to the beam design with the most filtration (71 cm Al – 17 cm AlF_3), the design with the least filtration produces 2.1 times more fast neutron contamination and a beam 50% more intense. In spite of the increased contamination, the advantage ratio (AR) and advantage depth (AD) are as high or higher than the other beams. Penetration is slightly enhanced as a result of the harder spectrum.

4. CONCLUSIONS

Analysis of the collimator designs showed that, although the use of an absorbing beam delimiter increases directionality through removal of neutrons impinging on the target at large angles relative to the beam axis, it does not improve beam penetration; on the contrary, the advantage depth is decreased due to the reduced effective beam size.

Table 2. In-air and in-phantom beam performance for different fast neutron filter/moderator designs[a]

		In-air parameters				In-phantom figures of merit		
Design	Fast neutron filter/moderator	ϕ_{epi} n/cm²s	D_{fn}/ϕ_{epi} cGy cm²/n	D_{γ}/ϕ_{epi} cGy cm²/n	J_{epi}/ϕ_{epi}	AD cm	ADDR RBE cGy/min	AR
6	71 cm Al –	1.91E + 10	2.46E – 11	5.10E – 12	0.64	9.8	605	5.0
	17 cm AlF_3	0.4%	0.6%	0.8%	0.5%			
7	71 cm Al –	1.60E + 10	1.65E – 11	5.85E – 12	0.64	9.7	501	5.0
	22 cm AlF_3	0.4%	0.7%	0.8%	0.6%			
8	71 cm Al –	1.28E + 10	1.16E – 11	7.12E – 12	0.64	9.7	402	5.0
	27 cm AlF_3	0.5%	0.8%	1.1%	0.6%			

[a]These calculations used a D_2O cooled fission converter composed of eleven burned MITR-II fuel elements (fuel loading is 312 g ^{235}U per element) at a MITR-II reactor power of 5 MW. The fast neutron filter/moderator is surrounded by a 10 cm thick lead reflector. It is followed by a 0.04 cm Cd thermal neutron filter, an 8 cm Bi photon shield and a rectangular collimator with a 15 cm thick lead lining and an exit size of 20 cm × 20 cm which defines the beam at the target. The statistical uncertainty of each quantity (one standard deviation) is listed immediately below the value as a percentage.

Furthermore, the elimination of large angle components of the flux can severely reduce beam intensity. The effects of beam size and directionality on in-phantom performance are dependent on the size and shape of the target so that care must be exercised in applying these results, derived for a water-filled ellipsoidal head phantom and a model of a human head, to the analysis of other targets which have different shapes and sizes. However, it appears that better collimation does not invariably produce better beam performance.

Changing the neutron spectrum, effected by varying the filter/moderator length, showed that excellent beam quality parameters (advantage depth and advantage ratio) are retained if fast neutron contamination is maintained at a low level relative to the inherent background dose. However, the results also showed that strictly minimizing the specific fast neutron dose may not produce an optimum beam design; hardening the spectrum slightly can enhance beam penetration and, as a by-product, significantly increase intensity without adversely affecting the average therapeutic ratio.

ACKNOWLEDGMENTS

This research was supported by the U.S. Department of Energy under contract DE-FG02-97ER62193. One of the authors, S. Sakamoto, gratefully acknowledges the support of the Science and Technology Agency of Japan.

REFERENCES

1. J.C. Yanch, and O.K. Harling, Dosimetric Effects of Beam Size and Collimation of Epithermal Neutrons for Boron Neutron Capture Therapy, "Rad. Res." 135:131–145, 1993.
2. W.S. Kiger, III, Neutronic Design of a Fission Converter-Based Epithermal Beam for Neutron Capture Therapy, Nucl. Eng. Thesis, Massachusetts Institute of Technology, 1996.
3. S. Sakamoto, Sensitivity Studies of the Neutronic Design of a Fission Converter-Based Epithermal Beam for Boron Neutron Capture Therapy, Nucl. Eng. Thesis, Massachusetts Institute of Technology, 1997.

4. R.G. Zamenhof, B.W. Murray, G.L. Brownell, G.R. Wellum, and E.I. Tolpin, Boron Neutron Capture Therapy for the Treatment of Cerebral Gliomas. I. Theoretical Evaluation of the Efficacy of Various Neutron Beams, "Med. Phys.," 2:47–60, 1975.
5. R.L. Moss, O. Aizawa, D. Beynon, R. Brugger, G. Constantine, O. Harling, H.B. Liu, and P. Watkins, The Requirements and Development of Neutron Beams for Neutron Capture Therapy of Brain Cancer, "Journal of Neuro-oncology," 33:27–40, 1997.
6. J.F. Briesmeister, ed, MCNP—A General Monte Carlo N-Particle Transport Code (Version 4A), Los Alamos National Laboratory, LA-12625 (1993).
7. E.L. Redmond II, J.C. Yanch, and O.K. Harling, Monte Carlo Simulation of the Massachusetts Institute of Technology Research Reactor, "Nuclear Technology," 106: April 1994.
8. W.S. Kiger, III, Sakamoto, and O.K. Harling, Neutronic Design of a Fission Converter-Based Epithermal Neutron Beam for Neutron Capture Therapy, "Nuclear Science and Engineering," 131:1–22, 1999.
9. P.M. Busse, R.G. Zamenhof, O.K. Harling, I. Kaplan, J. Kaplan, C.F. Chuang, J.T. Goorley, W.S. Kiger, K.J. Riley, L. Tang, G.R. Solares, and M.R. Palmer, The Harvard-MIT BNCT program: overview of the clinical trials and translational research, *in* "Frontiers in Neutron Capture Therapy," M.F. Hawthorne, K. Shelly, R.W. Wiersema, eds., Kluwer Academic/Plenum Publishers, New York, 2001, pp. 37–60.
10. J.A. Coderre, E.H. Elowitz, M. Chadha, R. Bergland, J. Capala, D.D. Joel, H.B. Liu, D.N. Slatkin, and A.D. Chanana, Boron Neutron Capture Therapy for Glioblastoma using p-Boronophenylalanine and Epithermal Neutrons: Trial Design and Early Clinical Results, "Journal of Neuro-Oncology," 33: 141–152, 1997.
11. H.B. Liu, D.D. Greenberg, and J. Capala, An Improved Neutron Collimator for Brain Tumor Irradiations in Clinical Boron Neutron Capture Therapy, Med. Phys., 23:2051–2060, 1996.
12. R.G. Zamenhof, G.R. Solares, W.S. Kiger, III, E.L. Redmond II, P.M. Busse, and C.-S. Yam, MacNCTPlan: an improved Macintosh-based treatment planning program for boron neutron capture therapy in "Advances in Neutron Capture Therapy: Volume 1, Medicine and Physics," edited by B. Larsson *et al.*, Elsevier Science B.V., New York, 1997.
13. Personal communication from C. Vroegindeweij and W.F.A.R. Verbakel.

DOSE DISTRIBUTION IN ^{157}Gd AND ^{10}B CONTAINING PHANTOMS SUBSEQUENT TO IRRADIATION WITH THERMAL NEUTRONS

V. F. Khokhlov[1], A. I. Shaks[1], V. L. Gozenbuk[1], I. N. Sheino[1], V. L. Mazanov[1], K. N. Zaitsev[2], V. K. Sakharov[2], and R. Lawaczeck[3]

[1]Russian State Scientific Center «Institute of Biophysics»
Moscow, Russia
[2]Moscow State Institute of Engineering Physics
Moscow, Russia
[3]Institute for Diagnostic Research
Berlin, Germany

INTRODUCTION

Studies on the distribution of irradiation doses in phantoms with gadolinium have been carried out at the research nuclear reactor IRT MEPhI in Moscow in accordance with the Russian project «Neutron Capture Therapy» for malignant tumors.

The thermal neutron flux of 10^9 n/cm^{2}s K-4 reactor canal and phantoms from polyethylene in the form of parallelepiped and sphere were used. Tumors were modeled by inserted containers filled with Gadolinium/Boron containing solutions of various concentrations.

The distributions of thermal and fast neutrons, gamma radiation, and conversion electrons as function of the phantom depth were determined.

Gamma-radiation was monitored by detectors on the basis of manganese-activated aluminophosphate glass with diameter of 8 mm and thickness of 1 mm.[1] The thermal neutron fluxes were measured using detectors based on ^{6}Li-enriched lithium-containing glass.[1] The fast neutron fluxes were measured by track detector on the basis of ^{237}Np.[2]

Distributions comparable to the experimental doses were calculated using the MCNP-4a code.[3]

PHANTOMS INVESTIGATION

The exact quantitative estimation of a dose of thermal neutrons is possible provided the overall dimensions and mass of irradiated biological object are known. In

Frontiers in Neutron Capture Therapy, edited by Hawthorne *et al.*
Kluwer Academic / Plenum Publishers, New York, 2001.

greater extent it concerns an estimation of doses on models of a tumors. The dose in a tumor depends on its size, depth of its localization in the body of an experimental animal, the concentration of the neutron-capture element in the tumor, etc. Accordingly, the studies of the dose fields for neutrons and γ-radiation in models of laboratory animals bodies, and of fragments of a human body (phantoms), both, with and without physical modeling of a malignant tumor were carried out for various depths of its localization in the phantom.

For detailed studies of radiation distributions in a tumor with gadolinium, the tumor was simulated by four polyethylene packages filled with the Gd-DTPA solution. The packages dimensions were: area $20 \times 25\,mm^2$, thickness about 2 mm, i.e. total thickness of the "tumor model" was about 8 mm. The models with two values of Gd-DTPA (Magnevist®) concentrations were used: 70 μl/ml and 170 μl/ml. The tumor model was placed into a rat phantom. Figure 1 shows the phantom sizes, the tumor model arrangement and its orientation with respect to the neutron beam.

The tumor model was located at the depth of 2 mm from the front surface of the phantom, symmetrically with respect to the beam axis.

Phantoms were placed into the irradiation box, being aligned (centered) with respect to the canal axis, at a distance of 8 cm from an external surface of the reactor shield.

Figure 2 shows the depth distribution of the γ-dose rate (primary beam and generated by neutron capture reactions) for various concentrations of Gd-DTPA in the tumor model. All irradiations were performed at the nominal reactor power of 2.5 MW. The thermal neutrons beam flux was about $1.0 \times 10^9\,n/(cm^2 s)$, and the dose rate of the accompanying γ-radiation was 0.14 cGy/s.

The calculations were carried out using the computer code MCNP 4a,[3] for the geometrical conditions corresponding to the experiment.

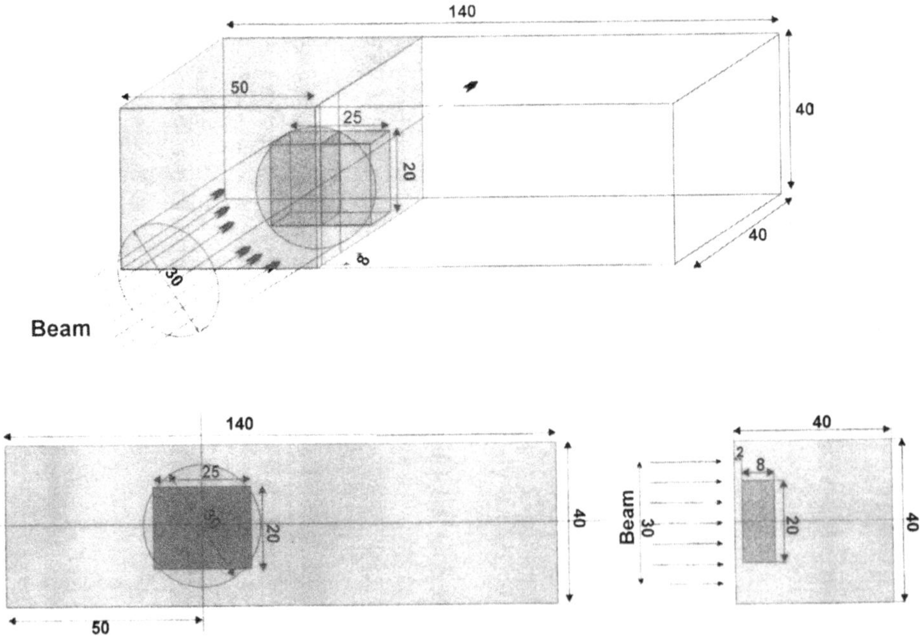

Figure 1. Tumor model arrangement in the rat phantom.

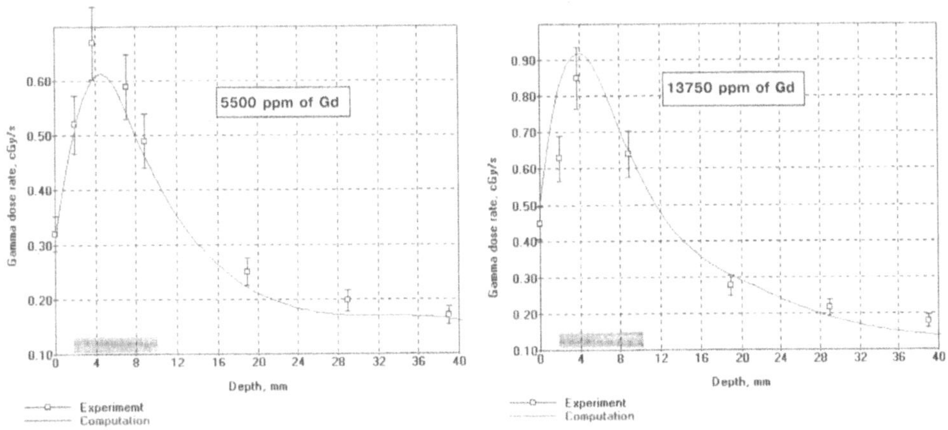

Figure 2. Distribution of the total gamma dose rate along central axis of beam in the phantom for various concentrations of Gd-DTPA in the tumor.

DISTRIBUTION OF A DOSE COMPONENTS

Thermal neutron capture by the nucleus of ^{157}Gd is considered as the ^{157}Gd(n,γ)^{158}Gd reaction. In the reaction a certain fraction of the excitation energy is released through the emission of internal conversion and Auger electrons.[4]

By definition, specific kerma (kerma per flux unit) of neutrons with energy E for nuclide A in a tissue is:

$$k_A(E) = 1.6 \cdot 10^{-10} n_A \sum_i \sigma_{Ai}(E) \cdot E_{Ai} \quad \text{Gy cm}^2/\text{neutron}$$

where

$1.6\ 10^{-10}$—the factor of translation from 1 MeV to 1 Gy·g;
n_A —number of nuclides A in 1 g of tissue;
σ_{Ai} —neutron microscopic cross-section of i-th reaction for the nuclide A_i at energy E, cm^2;
E_{Ai} —average energy, transmitted to secondary charged particles, MeV.

Table 1 contains the specific kerma values for estimation of the secondary charged particles dose in the tumor, containing ^{157}Gd and ^{10}B in concentrations of 70 μl/ml and 35 μg/g, respectively.

It is seen from the data of Table 1 that the presence of ^{157}Gd in a tissue with Magnevist® (Gd-DTPA) in concentration of 5,500 ppm (70 μl/ml) increases the dose from thermal neutrons by a factor of more than 30 due to the electron emission.

The estimations show, that only in the tumor model containing 70 μL/ml Gd-DTPA the dose rate from the conversion and Auger electrons is comparable to the γ-dose rate (primary beam and generated by the ^{1}H(n,γ)^{2}H neutron capture reaction), and in the first layers exceeds it (see Figure 3).

It is important to note, that the dose from electrons in a tumor with Gd everywhere surpasses a dose from γ-radiation, and in the obverse layers of a tumor the dose of electrons almost a factor of 2 higher than that for γ-radiation. This result is fundamentally important for perspectives of the use of Gd in preparations for NCT.

Table 1. Calculation of the specific kerma for thermal neutrons for nuclides ^{157}Gd and ^{10}B

Characteristics	^{157}Gd	^{10}B
Total microscopic cross-section, barn	254,000	3,837
Microscopic absorption cross-section, barn	240,000	3,595
Nuclear concentration in the tumor, 10^{24} nucl./cm^3	**$3 \cdot 10^{-6}$**	**$2.1 \cdot 10^{-6}$**
Total energy of the α and Li—particles yield, MeV	—	2.35
Total energy of the conversion electrons yield, MeV	0.0454	—
Total energy of the Auger electrons yield, MeV	0.0089	—
Total energy of the charged particles yield, MeV	0.0543	2.35
Specific kerma of the α particles, Gy·cm^2/neutron	—	**$2.84 \cdot 10^{-12}$**
Specific kerma of the conversion electrons, Gy·cm^2/neutron	**$5.23 \cdot 10^{-12}$**	—
Specific kerma of the Auger electrons, Gy·cm^2/neutron	**$1.03 \cdot 10^{-12}$**	—
Specific kerma of the nuclide, Gy·cm^2/neutron	**$6.26 \cdot 10^{-12}$**	**$2.84 \cdot 10^{-12}$**
Specific kerma for soft biologic tissues, Gy·cm^2/neutron [5]	**$2 \cdot 10^{-13}$**	**$2 \cdot 10^{-13}$**
Total specific kerma (nuclide + biologic tissues), Gy·cm^2/neutron	**$6.46 \cdot 10^{-12}$**	**$3.04 \cdot 10^{-12}$**

Earlier it was supposed that the dose in a tumor with Gd irradiated by thermal neutrons is determined mainly by the neutron-capture γ-radiation.[6,7]

Obviously, it is explained by difficulties of the electron dose determination on the background of the neutron capture γ-radiation in the tumor model with Gd, if one takes into account, that the average energy of these electrons is only about 45 keV.

At the following stages of investigations the attempt will be realized to carry out conversion electrons dose measurements by thin TTLD-580 detectors[8] (thin-layer thermoluminescence detector on the basis of MgB_4O_7 with diameter of 15 mm and thickness of about 80 μm). For this purpose a special container simulating a tumor in the phantom will be made to which thin detectors may be fastened.

The calculation of the dose distribution from charged particles produced in the reaction $^{10}B(n,\alpha)^7Li$ in a tumor does not represent appreciable difficulties, since it practically coincides with the distribution of thermal neutrons in a tumor, and is zero outside the localization of ^{10}B since all energy of charged particles is released at the point of absorption of the thermal neutron by ^{10}B. In Figure 4 such a distribution is presented.

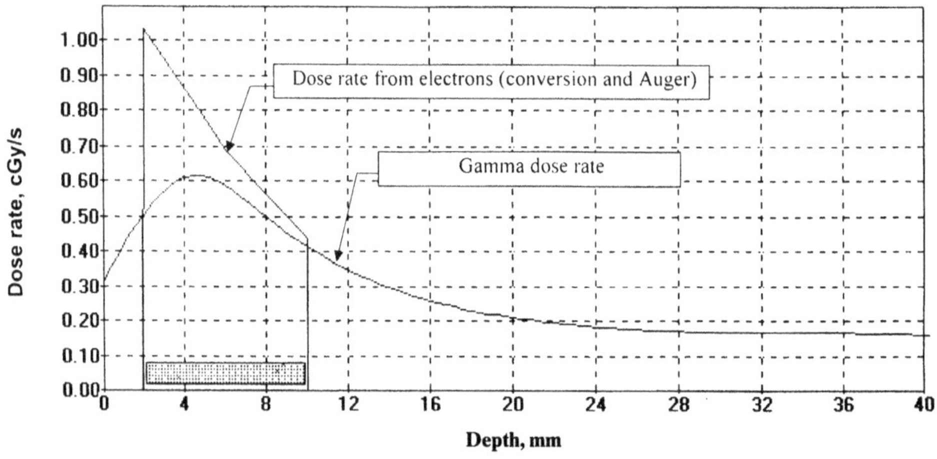

Figure 3. Distribution of the absorbed dose rate from gamma-rays and electrons along the beam axis in the phantom at 5,500 ppm ^{157}Gd in the tumor model.

Table 2. Average value of a dose rate and its components calculated for the model of a tumor in the rat phantom on the K-4 neutron beam

Doses	Without Gd and B			^{157}Gd (70 μl/ml Gd-DTPA)			^{10}B (35 mg/g)	
	γ-radiation from primary beam and generated by $^{1}H(n,\gamma)^{2}H$	Protons by thermal neutrons capture $^{10}N(n,p)^{14}C$	Delivered by epithermal and fast neutrons	γ-radiation generated by reaction: $^{157}Gd(n,\gamma)^{158}Gd$	Electrons	The sum of a dose	Charged particles α and ^{7}Li	The sum of a dose
Absorbed dose rate, cGy/s	0.17	0.03	0.2	0.52	0.75	1.67	0.42	0.82
RBE[9]	1	2	2	1	1.2		2.5	
Dose-equivalent rate, cGy-Eq/s	0.17	0.06	0.4	0.52	0.9	2.05	1.05	1.68

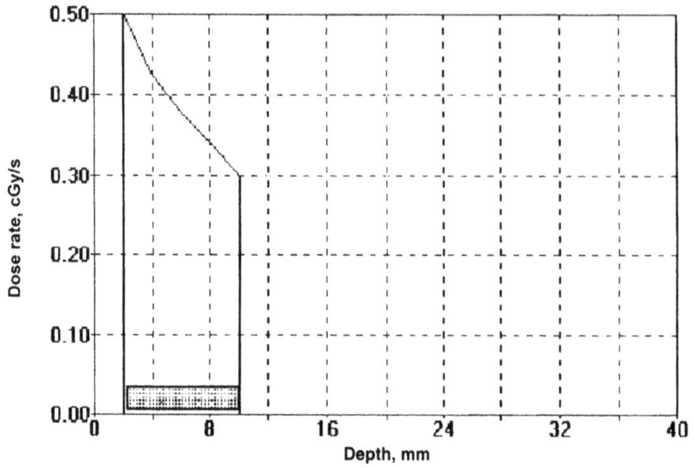

Figure 4. Distribution of the dose rate produced by charged particles from $^{10}B(n,\alpha)^7Li$ -reaction in the rat phantom at 35 ppm of ^{10}B in the tumor model.

It should be noted that the dose from the charged particles falls down in a tumor with ^{10}B slower, than the dose from electrons arising at the thermal neutron capture by Gd due to lower absorption probability of thermal neutrons by ^{10}B and, accordingly, with slower recession of the thermal neutrons flux.

For understanding the mechanism of the radiation dose formation in a tumor with ^{157}Gd and ^{10}B, the separate components of physical and effective doses of radiations in a tumor model in the rat phantom are presented in Table 2.

The average value of an effective dose of radiation in a tumor with Gd irradiated by the thermal neutron flux of 10^9 neutron/cm^2·s produces 2.05 cGy-Eq/s, of which 44% is formed by conversion and Auger-electrons. In the presence ^{10}B in a tumor the complete effective dose rate created by the thermal neutron capture will produce 1.68 cGy-Eq/s, of which 62.5% are formed by charged particles. In the absence of gadolinium and boron the main contribution into the dose is due to fast and epithermal neutrons. Obviously, the presence of gadolinium or boron in a tumor does not influence significantly this latter contribution.

CONCLUSION

The results of the experiments and calculations are in good agreement. It was shown that already at the gadolinium concentration of 5,500 ppm, the dose of conversion electrons in the tumor model is comparable with the dose from secondary gamma radiation and in front layers exceeds the latter.

REFERENCES

1. I.A. Bochvar, T.I. Gimadova, I.B. Keyrim-Markus, *et al.* "The Method of IKS Dosimetry," Atomizdat, Moscow, 1977, pp. 41–49, 58.
2. S.N. Kraitor. "Dosimetry for Radiation Accidents," I.B. Keyrim-Markus (ed), Atomizdat, Moscow, 1979, pp. 167–246.

3. Briesmeister J.F. MCNP-4A General Monte-Carlo N-Particle Transport Code. Version 4A, Los Alamos National Laboratory report, LA-12625, 1995.
4. J. Stepanek. Radiation spectrum of ^{158}Gd and radial dose distribution. Advances in Neutron Capture Therapy. Volume II, Chemistry and Biology. Proceeding of the Seventh International Symposium on Neutron Capture Therapy for Cancer. Switzerland 4–7 September 1996. Elsevier Science B.V. 1997. pp. 425–429.
5. ICRU report No. 33. Washington, 1985.
6. T. Matsumoto, K. Kato, Y. Sakuma. Phantom experiment of depth-dose distributions for gadolinium neutron capture therapy. Proccedings of the Fifth International Symposium on Advanced Nuclear Energy Research, 1993, pp. 892–898.
7. Y. Akine, N. Tokita *et al.* Gadolinium neutron capture therapy: preclinical studius. Proccedings of the Fifth International Symposium on Advanced Nuclear Energy Research, 1993, pp. 878–885.
8. D.P. Osanov, V.P. Krjuchkov, A.I. Shaks. Determination of beta radiation doses received by personel involved in the mitigation of the chernobyl accident. The Chernobyl Papers. V.1. Research Enerpr. N.-Y. 1990, pp. 313–346.
9. J. Coderre, D. Glass *et al.* Neutron capture therapy for melanoma: Plenum Press , New York, Clinical aspects of neutron capture therapy, 1989, pp. 219–233.

^{252}CF MIXED-FIELD DOSIMETRY FOR A VARIETY OF SOURCE GEOMETRIES AND PHANTOM MATERIALS FOR CLINICAL BRACHYTHERAPY

Comparison of Measurements and Calculations

Mark J. Rivard[1], Mark Yudelev[2], Frank Van den Heuvel[2],
Jacek G. Wierzbicki[3], Rodger C. Martin[4], and Robert R. McMahon[4]

[1]Department of Radiation Oncology
Tufts University School of Medicine
New England Medical Center #246
750 Washington Street
Boston, Massachusetts 02111
[2]Department of Radiation Oncology
Wayne State University
3990 John R. Road
Detroit, Michigan 48201
[3]Cancer Treatment Center
Saint Mary's Medical Center
830 South Jefferson Road
Saginaw, Michigan 48602
[4]Radiochemical Engineering Development Center
Chemical Technology Division
Oak Ridge National Laboratory
Oak Ridge, Tennessee 37831

1. INTRODUCTION

The purpose of this study was to compare experimental and calculated mixed-field dosimetry results for ^{252}Cf Applicator Tube (AT) neutron emitting brachytherapy sources. This subject has not been critically examined in many years,[1–6] and is timely as Oak Ridge National Laboratory (ORNL) is now able to fabricate AT sources, and eventually high dose rate (HDR) clinical sources, which would increase availability of high LET

Frontiers in Neutron Capture Therapy, edited by Hawthorne *et al.*
Kluwer Academic / Plenum Publishers, New York, 2001.

radiation sources. ^{252}Cf is unique as it has fast neutrons of relatively low energy and an appreciable photon dose component. Ion chambers and a GM counter were used to measure the total and photon dose, respectively, close to ^{252}Cf AT type sources. A brachytherapy dosimetry protocol was formulated similar to the external beam neutron formalism of ICRU 45. Comparisons of experimental dosimetry were made with results of Colvett *et al.*[4] Neutron kerma in a variety of materials and for various source geometries was calculated using Monte Carlo (MCNP)[7] methods and compared with other neutron sources.[8]

2. MATERIALS AND METHODS

2.1. Experimental Studies

Two types of TE ion chambers and a miniature GM counter were used as the dosimetry equipment for the experimental studies. Dosimetry measurements of ^{252}Cf AT sources made by Savannah River Laboratory (SRL) and ORNL were performed using a brachytherapy dosimetry protocol based on the ICRU 45 protocol for external neutron beam dosimetry.

2.1.1. TE Ion Chambers. Two types of TE ion chambers were used; the first type was a Far West Technology Inc. (FWT, model IC-17) with a collecting volume of approximately 1 cm^3. The second chamber (Exradin, model T1) had a 0.05 cm^3 collecting volume. Both chambers were composed of A-150 TE plastic with methane-based TE gas (3.2% N_2, 63.8% CH_4, and 33.0% CO_2) flowed through each chamber. A single Matheson type flow meter was used with a setting of 9.1 cm^3/minute for controlling and monitoring the TE gas flow through each chamber to displace air yet not cause over-pressurization. Each chamber was placed in a 54 liter (28.6 × 30.8 × 61.1 cm^3) thin-walled plastic water phantom. A plastic jig was used to centrally position each ion chamber among a circumferential array of AT sources with the chamber stem and source transverse axes parallel. A + 500 voltage bias was obtained from a power supply (Canberra, model #3102). An electrometer (Keithley, model 35617EBS) was used to measure the integrated charge over time. One minute readings were taken repeatedly for 15 to 45 minutes to establish reproducibility. The Nx values (R/nC) for two IC-17 and T1 chamber were: 2.933 ± 0.029 and 45.53 ± 0.46, respectively.

2.1.2. Miniature GM Counter. A miniature GM counter (TGM Inc., model N115, ZP 1301 equivalent) was used to discriminate photon dose from the total dose as determined with the TE chambers. Due to its construction and ^{252}Cf neutron energies, the GM counter was less sensitive to neutrons than photons.[9] GM counter photon sensitivity was determined through calibration with a ^{137}Cs source. Counts were integrated using a scaler, and a ^{6}LiF cap minimized the effect of thermal neutrons. This approach, as compared to utilizing paired chambers such as A-150/TE and Mg/Ar chambers was chosen as the GM counter was less sensitive to low energy neutrons than a Mg chamber. The true count rate was calculated from the measured count rate and knowledge of the dead-time using the non-paralyzable model.[10] A calibration factor of $5.10 \pm 0.11 \times 10^{-7}$ cGy/ct, was determined with $\Gamma(^{137}Cs) = 3.25$ R-cm^2/mCi-h, μ_{en}/ρ muscle to air ratio of 1.10, dead-time of 30.6 μs, and photon W/e value of 33.97 J/C.

2.2. ^{252}Cf Brachytherapy Dosimetry Protocol

A modern dosimetry protocol similar to ICRU 45 was formulated (Eq. 1) with parameters explicitly determined for ^{252}Cf mixed-field dosimetry.

$$D_T = M_T(\Pi k_M)_T d_T G N_X A_{WALL}(f_t)_c \frac{1}{k_T}\frac{1}{1+\delta} \tag{1}$$

$$D_T = 0.967 M_T(\Pi k_M)_T N_X \tag{2}$$

2.3. Monte Carlo Calculational Dosimetry

Neutron kerma was calculated with a distributed computing environment[11] using Monte Carlo methods (MCNP)[7] for eleven materials. These materials were: water, A-150 plastic, PMMA, brain, muscle, fat, pancreas, lung, bone, skin, and blood. The elemental composition and mass densities of the materials were taken from ICRU 44 and the CRC Handbook. Each material subtended a 15 cm diameter spherical phantom in which a centrally placed isotropic neutron point source was positioned. Neutron kerma for each material was calculated for radii of 0.1 to 5.0 cm. Transport of at least 1 million particles was necessary to achieve relative errors (1 σ) of 0.1%. The neutron source was modeled as a Maxwellian spectrum with T = 1.42 MeV.

3. RESULTS AND DISCUSSION

3.1. Experimental Studies

Many parameters recommended by ICRU 45 are independent of the radiation source. An A_{WALL} value of 0.983 for the FWT IC-17 chambers and 0.992 for the Exradin T1 chamber is recommended.[12] The exposure-to-absorbed dose to reference tissue correction factor $(f_t)_C$ was 0.966. Also according to ICRU 45, the value of the μ_{en}/ρ ratio for ICRU muscle and A-150 plastic is 1.001. A value of $(r_{m,g})_n/[(L/\rho)_m/(L/\rho)_g]_c = 1.00 \pm 0.02$ is recommended by ICRU 45 for neutrons. The product of total correction factors $(\Pi\ k_M)_T$ was reduced to 1.0291, with C_{TP} (1.025) and Cel (1.004). The energy necessary to produce an ion pair (W_C) in methane-based gas when irradiated by ^{60}Co is 29.3 eV. For recoil protons from ^{252}Cf neutrons, the energy per ion pair (W_N) was calculated to be 31.65 eV based on convolving the ^{252}Cf neutron energy in ICRU muscle with methane-based TE gas data from Goodman and Coyne[13] and the W_N/W_C ratio was 1.080 for ^{252}Cf in methane-based TE gas. The ratio of the kerma for ICRU muscle to A-150 plastic was calculated to be 0.958 which did not significantly change for source distances of 0.5 to 5 cm. Using the above parameters in Equation 9, a value of 0.969 ± 0.02 for k_T was obtained, which was within 1% of a calculated k_T value of 0.96 for neutrons with energy between zero and 5 MeV determined by Waterman *et al.*[14] The ratio of photon dose to total dose ranged from 25% to 40%, with higher values occurring at larger distances. For distances of 1.0, 2.0, 3.0, and 5.0 cm, the ratio of photon to total dose was approximately 32, 33, 36, and 44% with values of 0.010, 0.011, 0.012, and 0.014 for δ. The value of h_t was unity. Using a fixed value of 0.011 for δ, the above parameters may be incorporated into Eq. 1 with no significant radial dependence. For low energy neutrons and small ion

Table 1. ^{252}Cf total transverse axis dose

Distance (cm)	IC-17 #1 (cGy/h-μg)	IC-17 #2 (cGy/h-μg)	T1 (cGy/h-μg)	Colvett *et al.* (cGy/h-μg)	Krishnaswamy (cGy/h-μg)
1.0			2.582	2.895	2.809
1.5			1.264	1.357	1.324
2.0	0.659	0.654	0.663	0.769	0.765
2.5			0.439	0.489	0.493
3.0	0.296	0.291	0.290	0.337	0.331
3.5			0.218	0.244	0.238
4.0			0.164	0.186	0.179
4.5			0.131	0.146	0.142
5.0	0.102	0.099	0.103	0.118	0.112

chambers, d_T is unity.[15] Transverse axis total dose was calculated for each TE chamber (Table 1) with results of Colvett *et al.* and Krishnaswamy.[3] Dosimetry results obtained by Colvett *et al.* and Krishnaswamy were for an SRL *needle* source which was constructed in a different manner than that of the SRL or ORNL AT sources.

The T1 chamber had the smallest collecting volume of the chambers available, and was used to measure off-axis dosimetry using the same dosimetry formalism as for the transverse axis.

For both the on- and off-axis measurements, the total dose measured with the T1 chamber was (11.3 ± 2.0)% less than that determined by Colvett *et al.*[4] However, this discrepancy was readily explained when details of each experimental setup were examined. Anderson[5] derived a 7% dose over-estimation obtained by Colvett *et al.*, accounted for in part by exclusion of $K_{MUSCLE}/K_{A\text{-}150}$. An additional discrepancy of 2% is expected when accounting for the difference in the calculated ratio of W/e for neutrons and photons (1.057) of Anderson and Colvett *et al.* versus the value used herein (1.080). Finally, an approximate 2% under-estimation in source strength was derived from the difference in half-life used by Colvett *et al.* and that currently accepted. The average of these ratios was 0.977 which may be considered good agreement due to the aforementioned differences.

At an energy of approximately 4 MeV, an average k_U value of (0.6 ± 0.2)% was adopted for five different GM counter types studied.[16] Results by Jones[16] determined the k_U value (0.05 ± 0.05)% of a single GM counter type (ZP1320) for energies less than 1 MeV. From the studies of Lewis and Hunt[17] and Jones[16] and knowledge of the ^{252}Cf neutron energy spectrum, the N115 GM k_U value may be inferred as (0.2 ± 0.2)%. When accounting for measurement reproducibility and systematic uncertainties, a k_U value of zero may be used. Results are shown in Table 3.

Table 2. ^{252}Cf total off-axis dose

Position (cm) along × away	T1 (cGy/h-μg)	Colvett *et al.* (cGy/h-μg)	Krishnaswamy (cGy/h-μg)
2.0 × 2.0	0.336	0.386	0.393
4.0 × 2.0	0.128	0.143	0.140
2.0 × 4.0	0.131	0.147	0.143
5.0 × 3.0	0.075	0.0839	0.079
3.0 × 5.0	0.077	0.0858	0.082
3.0 × 3.0	0.142	0.163	0.159
4.0 × 4.0	0.080	0.0902	0.086

Table 3. ^{252}Cf AT photon transverse axis dose

Distance (cm)	N115 (cGy/h-μg)	Colvett *et al.* (cGy/h-μg)	Krishnaswamy (cGy/h-μg)
1.0		0.845	0.925
2.0	0.2248	0.242	0.263
3.0	0.1093	0.118	0.120
4.0	0.0658	0.072	0.072
5.0	0.0436	0.050	0.049
6.0	0.0314	0.037	
7.0	0.0238		

Table 4. Measured ^{252}Cf AT neutron transverse axis dose

Distance (cm)	T1 – N115 (cGy/h-μg)	IC-17 – N115 (cGy/h-μg)
2.0	0.438	0.432
3.0	0.181	0.184
4.0	0.098	
5.0	0.059	0.0572

The ^{252}Cf neutron dose as presented in Table 4 was obtained by subtracting the photon dose measurements from the N115 GM counter from Table 3 from the total dose.

3.2. Calculative Results

Table 5 presents the calculated neutron kerma for 11 materials of interest at a clinically relevant radial distance of 1.0 cm. ^{252}Cf neutron kerma is normalized to muscle at 1.0 cm for comparison with dosimetry properties of other neutron sources normalized to muscle. Due to the large statistics, relative errors of 0.1% resulted from the calculations. However, larger systematic errors may have been incorporated due to the choice of the ^{252}Cf neutron energy spectrum or from the ENDF-VI neutron cross-section libraries.

From ICRU 44, a 10.6 and 10.3% mass hydrogen content was used for pancreas and lung, respectively. As cited in Awschalom *et al.*,[8] the pancreas and lung compositions

Table 5. ^{252}Cf neutron kerma at 1 cm, and normalized kerma for other sources

	This study		Awschalom *et al.*		
Material	^{252}Cf kerma (cGy/h-μg)	^{252}Cf kerma muscle'	d(16) + Be muscle'	d + T muscle'	p(41) + Be muscle'
Muscle	1.907 ± 0.0011	1.000 ± 0.0006	1.00	1.00	1.00
Water	2.064 ± 0.0012	1.083 ± 0.0007	1.08	1.06	1.05
Brain	2.009 ± 0.0012	1.054 ± 0.0006	1.05	1.04	1.04
Skin	1.910 ± 0.0011	1.002 ± 0.0006	0.99	0.99	1.00
Blood	1.928 ± 0.0012	1.011 ± 0.0007	0.98	0.98	0.98
Fat	2.183 ± 0.0013	1.145 ± 0.0007	1.13	1.13	1.16
Pancreas	1.997 ± 0.0012	1.047 ± 0.0006	0.96	0.97	0.96
Lung	1.943 ± 0.0012	1.019 ± 0.0006	0.97	0.98	0.98
Bone	0.809 ± 0.0005	0.424 ± 0.0003	0.44	0.52	0.52
A-150	2.005 ± 0.0012	1.053 ± 0.0006	1.02	1.04	1.07
PMMA	1.654 ± 0.0010	0.868 ± 0.0005	0.83	0.88	0.90

were 9.7 and 9.9% mass hydrogen, respectively, as taken from ICRP 23. These differences in hydrogen content had the correct sign and magnitude as would be expected from the differences in kerma between the neutron sources. The pancreas and lung were the only materials with different ICRU 44 and ICRP 23 hydrogen mass content greater than 0.2%. The 5% kerma enhancement between brain and muscle due to a 5% increase in hydrogen content may not be considered detrimental for future applications such as cerebral ^{252}Cf brachytherapy.

There also were differences between the Monte Carlo results obtained herein and those obtained by Krishnaswamy.[3] A Maxwellian spectrum was used in this work to model the ^{252}Cf neutrons while Krishnaswamy employed a Watt fission spectrum. Comparing the moderated neutron kerma in muscle using the Watt fission spectra of Krishnaswamy with the kerma determined when using a Maxwellian spectrum, a difference in neutron kerma to muscle was expected. At radii of 0.5, 1.0, 2.0, and 5.0 cm, the correction factors used to compare kerma calculated using the two different neutron energy spectra were 1.049, 1.050, 1.055, and 1.076, respectively. Like Colvett *et al.*, Krishnaswamy employed a different specific source strength than the currently accepted value which resulted in a 1.1% correction factor. As Krishnaswamy modeled an extended source, instead of a point source, geometry factors were necessary to compare the two studies. A simple correction (1.029) was applied to compare the hydrogen mass percent. Finally, Krishnaswamy used excessively large voxels near the ^{252}Cf source where volume averaging within the voxel would markedly decrease the results.[5] The effect of these corrections was 8.3, 2.1, 0.5, and 0.1% for radial distances of 0.5, 1.0, 2.0, and 5.0 cm, respectively.[5] The ratio of MCNP results to those of Krishnaswamy after implementing the corrections were 1.012, 0.969, 0.970, and 1.017 at radii of 0.5, 1, 2, and 5 cm, respectively, or on average 0.992 ± 0.027 which was within the Monte Carlo uncertainties.[3]

3.3. Comparison of Calculated and Experimental Neutron Dose Rates

Using a 15 mm geometry factor to compare point source and extended source measurements one may convert the MCNP calculated neutron dose rates to that for an extended source such as the AT. This is presented in Table 6 where the ratio of geometry factors for a point source and 15 mm long extended source were 0.9567, 0.9799, 0.9885, and 0.9926 for distances of 2.0, 3.0, 4.0, and 5.0 cm, respectively. For comparison, measured neutron rates of AT sources are presented in Table 6, as are ratios of experimental neutron dose rates to those calculated using MCNP.

From inspection of Table 6, the neutron dose rate as derived using the combination of FWT chambers and GM counter were in approximately the same level of agreement with MCNP results as the neutron dose rate derived using the combination of

Table 6. ^{252}Cf AT transverse axis neutron dose rate

Distance (cm)	T1 – N115 (cGy/h-μg)	IC-17 – N115 (cGy/h-μg)	MCNP (cGy/h-μg)	T1 – N115 ratio to MCNP	IC-17 – N115 ratio to MCNP
2.0	0.438 ± 0.021	0.432 ± 0.011	0.44030 ± 0.00035	0.995 ± 0.050	0.982 ± 0.025
3.0	0.181 ± 0.009	0.184 ± 0.005	0.18764 ± 0.00017	0.965 ± 0.048	0.981 ± 0.025
4.0	0.098 ± 0.005		0.097809 ± 0.00011	1.002 ± 0.050	
5.0	0.059 ± 0.003	0.057 ± 0.001	0.056656 ± 0.000068	1.041 ± 0.052	1.005 ± 0.025

Exradin chamber and GM counter. The average ratio for the FWT chambers and GM counter combination was 0.989 while the average ratio for the Exradin chamber and GM counter combination was 1.001. Measurements with both chamber combinations agreed with the calculated neutron dose rates at all distances within the uncertainties.

4. CONCLUSION

A dosimetry protocol similar to ICRU 45 was applied to ^{252}Cf brachytherapy sources. Measurements using a GM counter and two types of TE chambers were used to determine the mixed-field dosimetry of ^{252}Cf AT sources. Comparisons were made between results of Colvett *et al.*, and Krishnaswamy with factors derived to permit reasonable comparisons. Due to improved measurements of nuclear data used for determination of factors such as $K_{MUSCLE}/K_{A\text{-}150}$, $N(E_N)$, and $(W/e)_N/(W/e)_C$, and application of a modern dosimetry formalism, the dosimetry results obtained herein were determined to be more reliable than those cited. Kerma of ^{252}Cf neutrons, as determined using Monte Carlo methods, was analyzed for a variety of clinically relevant tissues and dosimetry media. Comparisons of measurements and calculated neutron dose results revealed no significant differences between Monte Carlo derived neutron dosimetry and that determined using a GM counter and TE chambers.

REFERENCES

1. C.C. Schlea, and D.H. Stoddard, Californium isotopes proposed for intracavitary and interstitial radiation therapy with neutrons, *Nature*, 206:1058–1059, 1965.
2. V. Krishnaswamy, Calculation of the dose distribution about ^{252}Cf needles in tissue, *Radiol*, 98:155–160, 1971.
3. V. Krishnaswamy, Calculated depth dose tables for ^{252}Cf sources in tissue, *Phys. Med. Biol.*, 17:56–63, 1972.
4. R.D. Colvett, H.H. Rossi, and V. Krishnaswamy, Dose distribution around a ^{252}Cf needle, *Phys. Med. Biol.*, 17:356–364, 1972.
5. L.L. Anderson, Status of dosimetry for ^{252}Cf medical neutron sources, *Phys. Med. Biol.*, 18:779–799, 1973.
6. L.L. Anderson, ^{252}Cf physics and dosimetry, *Nuc. Sci. App.*, 2:273–281, 1986.
7. J.F. Briesmeister, Monte Carlo N-Particle transport code system, *in:* "MCNP4B User's Manual," LANL, Los Alamos, 1997.
8. M. Awschalom, I. Rosenberg, and A. Mravca, Kerma for various substances averaged over the energy spectra of fast neutron therapy beams: a study in uncertainties, *Med. Phys.*, 10:395–409, 1983.
9. V.E. Lewis, and J.B. Hunt, Fast neutron sensitivities of Geiger-Mueller counter gamma dosemeters, *Phys. Med. Biol.*, 23:888–893, 1978.
10. G.F. Knoll, "General properties of radiation detectors," in *Radiation detection and measurement* 2nd edition (John Wiley & Sons, New York, 1989), pp. 103–130.
11. A. Geist, A. Beguelin, J. Dongarra, W. Jiang, R. Manchek, and V. Sunderam. "PVM: Parallel Virtual Machine- A User's Guide and Tutorial for Networked Parallel Computing," The MIT Press, Cambridge, MA, 1994.
12. R. Gastorf, L. Humphries, and M. Rozenfeld, Cylindrical chamber dimensions and the corresponding values of A_{WALL} and $N_{GAS}/(N_X A_{ion})$, *Med. Phys.*, 13:751–754, 1986.
13. L.J. Goodman, and J.J. Coyne, Wn and neutron kerma for methane-based tissue-equivalent gas, *Radiat. Res.*, 83:491, 1980.
14. F.M. Waterman, F.T. Kuchnir, L.S. Skaggs, R.T. Kouzes, and W.H. Moore, Energy dependence of the neutron sensitivity of $C\text{-}CO_2$, Mg-Ar, and TE-TE ionisation chambers, *Phys. Med. Biol.*, 24:721, 1979.

15. J. Zoetelief, C.A. Engels, J.J. Broerse, and B.J. Mijnheer, Effect of finite size of ion chambers used for neutron dosimetry, *Phys. Med. Biol.*, 25:1121, 1980.
16. D.T.L. Jones, The neutron sensitivity of a GM counter between 0.5 and 8 MeV, *in:* "Radiation Protection, 4th Symposium on Neutron Dosimetry," edited by G. Burger and H.G. Erbert ed., EUR 7448, Munich, 1981, pp. 409–419.
17. V.E. Lewis and J.B. Hunt, Fast neutron sensitivities of Geiger-Mueller counter gamma dosemeters, *Phys. Med. Biol.*, 23:888–893, 1978.

^{157}Gd PHOTON DOSE ENHANCEMENT FROM A NEUTRON EMITTING ^{252}Cf POINT SOURCE

Mark J. Rivard, Mark Yudelev, and Jacek G. Wierzbicki

Department of Radiation Oncology
Wayne State University
3990 John R. Road
Detroit, Michigan 48201

1. INTRODUCTION

Though the efficacy of the ^{157}Gd(n,γ)^{158*}Gd neutron capture reaction (GdNCR) has yet to be demonstrated in humans, there is merit to studying the potential benefits of this modality. In this study, the photon dosimetry of GdNCR enhanced ^{252}Cf fast neutron brachytherapy was examined.

2. MATERIALS AND METHODS

Calculations were performed using Monte Carlo methods (MCNP)[1] A distributed computing environment[2,3] to permit brisk calculation of neutron, photon, and electron (n-p-e) transport in the determination of GdNCR photon (64157.55c) dose enhancement. The solid-state S(α,β) neutron scattering library (lwtr.01t) was used to model low energy neutron transport on hydrogen within ^{157}Gd-loaded media.[2] Parameters such as the ^{157}Gd loading, phantom composition, and size were varied. As a means to calculate absorbed photon dose from the GdNCR directly, a single-event photon dosimetry kernel was produced for the ^{158*}Gd photons in water and ICRU 44 brain. The merit of this approach is that convolution of the GdNCR distribution with the photon dosimetry kernel, or future electron dosimetry kernels, may provide accurate dose distributions in regions distant from boundaries or tissue inhomogeneities. In addition to calculation of absorbed dose from the ^{158*}Gd photons, the neutron and photon energy spectra were also evaluated for each ^{157}Gd loading and radial position. The neutron energy spectrum was divided into energy bins from 1 meV to 15.85 MeV with 10 energy bins per decade of energy while the photon energy spectrum was examined from zero to 9 MeV with 0.1 MeV increments. The entire volume was divided into radial bins using spheres of

Frontiers in Neutron Capture Therapy, edited by Hawthorne *et al.*
Kluwer Academic / Plenum Publishers, New York, 2001.

varying radii to permit spatial resolution. ^{157}Gd loadings ranged from 0 to 5,000 μg ^{157}Gd per gram (ppm) of material. An unencapsulated ^{252}Cf point source was positioned at the center of each sphere to mimic a ^{252}Cf HDR source. As recommended by Anderson,[4] the neutron energy spectrum was modeled as an isotropic Maxwellian distribution with an energy fitting parameter of 1.42 MeV.

3. RESULTS

3.1. Calculated GdNCR Photon Dose Rate

The GdNCR photon dose rate enhancement was calculated in brain material with ^{157}Gd loadings of 100 and 1,000 ppm. Results are presented in addition to the background induced photon dose in Figure 1. The relative error (10^6 particles) was approximately 10.7 and 2.5% at 1 and 5 cm for both loadings. An additional calculation was performed for the 1,000 ppm loading where the computer simulation was run for 8 continuous weeks (7×10^7 particles) to obtain the data set labeled: *1,000 ppm Gd-157, BIG.* At distances of 1 and 5 cm, the relative errors were 1.3 and 0.3%, respectively. It was impractical to perform direct GdNCR photon dose enhancement calculations using the aforementioned computer environment.

3.2. GdNCR Photon Dose Kernel

Single-event dosimetry kernels were produced for the ^{158*}Gd photons in 15 cm diameter spheres of water and brain material. The calculated data are presented in Figure 2, and then fit to Equation 1, r > 0.999, with parameters listed in Table 1. Use of an offset,

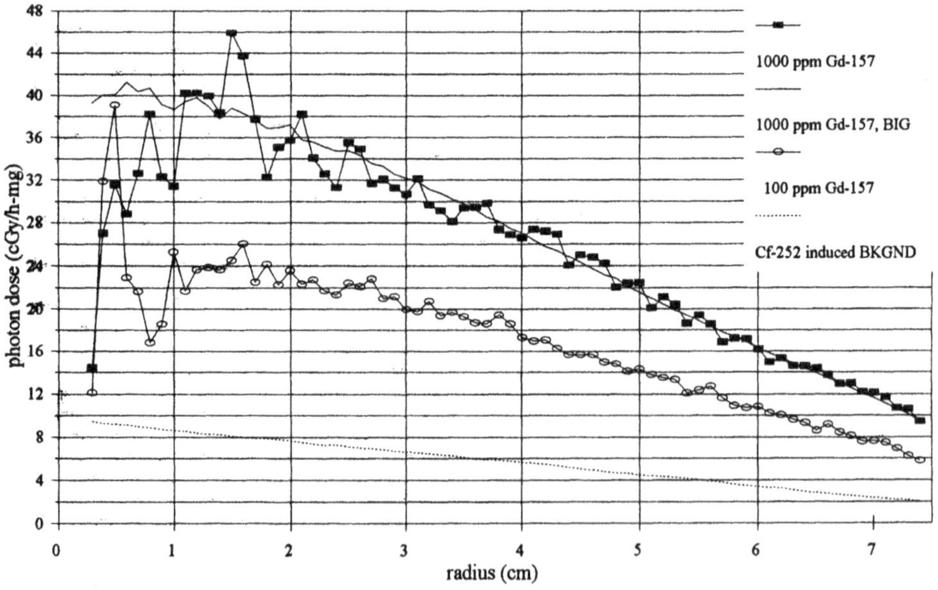

Figure 1. GdNCR photon dose enhancement in brain for 100 and 1,000 ppm ^{157}Gd.

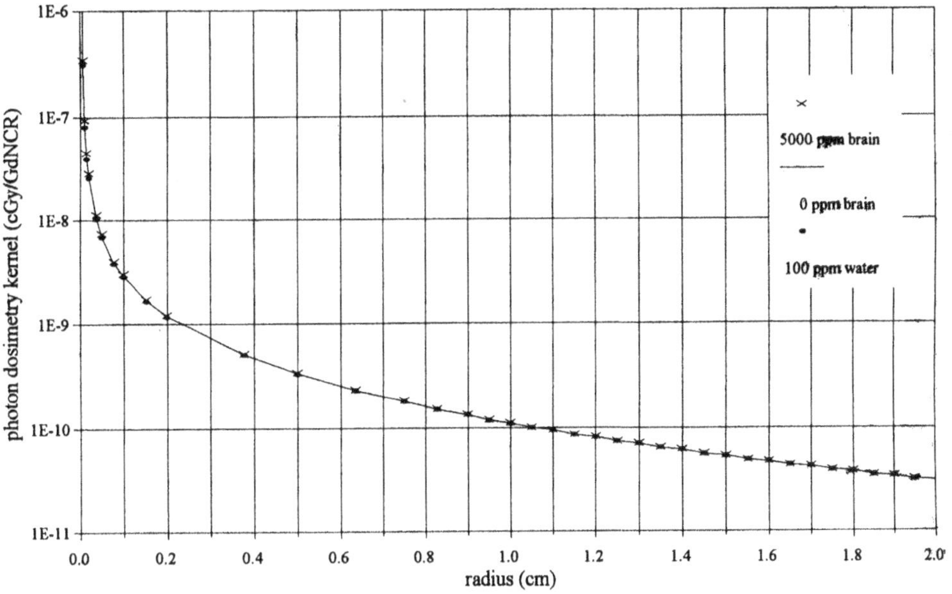

Figure 2. GdNCR photon dose kernel in water and brain.

δ, produced finite dosimetry results at the origin while still providing an accurate curve fit to the calculated data. A conversion efficiency of 2.207 was used for M,[5] and the attenuation length, μ_1, was similar to the $0.0282\,cm^{-1}$ value cited in the Materials and Methods section to justify the chord length estimate.

$$D(r)_{Gd-158^{*}photons} = \frac{M\alpha e^{-\mu_1 r}(1-\beta e^{-\mu_2 r})}{(r-\delta)^2}\left[\frac{MeV}{g-\gamma}\right] \qquad (1)$$

where

D(r)	=	absorbed dose in water or brain from GdNCR photons	[cGy/NCR]
M	=	conversion efficiency, number of photons per ^{157}GdNCR	[(γ/NCR]
α	=	the radial distance from source center to point of interest	[cGy/γ]
β	=	scattering intensity	[dimensionless]
δ	=	radial offset	[cm]
μ_1	=	attenuation length	[cm^{-1}]
μ_2	=	scattering length	[cm^{-1}]
r	=	the radial distance from source center to point of interest	[cm]

Table 1. GdNCR photon dosimetry kernel parameters for ^{252}Cf brachytherapy

Material / ^{157}Gd loading	α (cGy/γ)	β	δ (cm)	μ_1 (cm^{-1})	μ_2 (cm^{-1})
Water / 100 ppm	2.686×10^{-11}	0.967	0.0001	0.0235	2.15
Brain / 0 ppm	2.715×10^{-11}	0.967	0.0001	0.0225	2.15
Brain / 5,000 ppm	2.715×10^{-11}	0.967	0.0001	0.0225	2.15

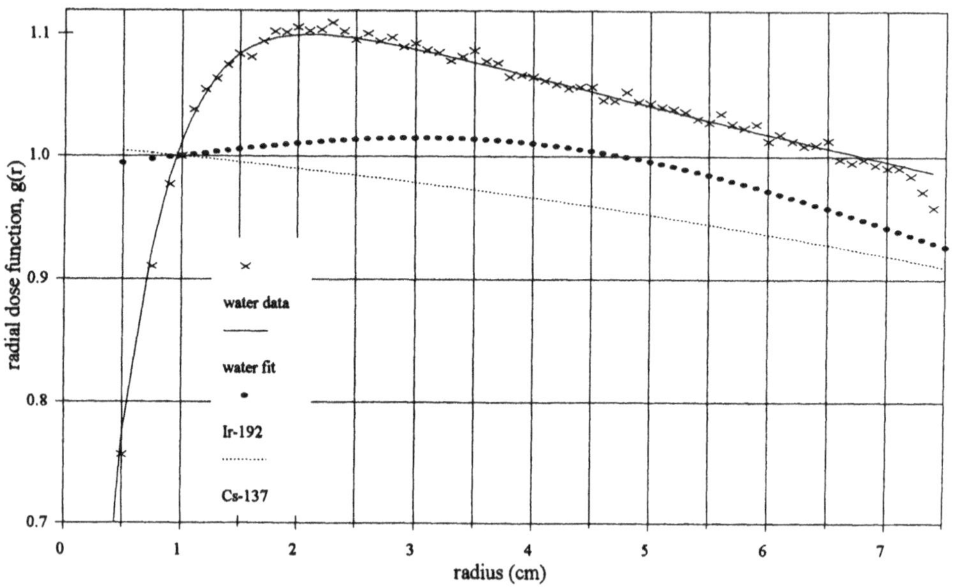

Figure 3. Radial dose functions for GdNCR photons and brachytherapy sources.

In Table 1, it appears that the μ_l ratio (1.044) between water and brain was proportional to the ratio of mass densities, 1.00 and 1.04 g/cm^3, respectively, while the α ratios (0.989) were approximately proportional to the hydrogen mass density of water (0.112 g/cm^3) and brain (0.111 g/cm^3). Using an identical formalism to the radial dose function, g(r), as defined in TG-43, a r^{-2} factor was removed from the GdNCR photon dose kernel due to the decrease expected because of diminishing solid angle with increasing distance.[6] In Figure 3, the radial dose functions of common brachytherapy sources in water are presented alongside that for the GdNCR photon dose kernel.[6,7] There were no significant differences between the radial dose functions in water and brain, and between the two brain materials with different loadings. The radial dose function values are significantly larger and fall-off significantly slower for increasing depths in comparison to that from the common brachytherapy sources. This was expected as GdNCR photons are of higher energy than common brachytherapy sources, and the radial dose function exhibits a buildup due to increasing scatter and attenuation with increasing distance from the point source.

3.3. Neutron Energy Spectra in ^{157}Gd-loaded Brain Material

The moderated ^{252}Cf neutron energy spectrum is presented Figure 4 at 0.5, 1, 2, and 5 cm for brain with a ^{157}Gd loading of 1,000 ppm. It is evident from this figure that only thermalized neutrons, those less than 1 eV, were significantly affected by the presence of ^{157}Gd, and that the fast neutrons were unaffected. Figure 5 shows the spatial distribution of ^{157}Gd neutron capture reactions for ^{157}Gd loadings of 100 and 1,000 ppm in brain material. GdNCR density was non-linear for this factor of ten increase in ^{157}Gd loading. On average, the 1,000 ppm GdNCR density was only a factor of 1.5 ± 0.1 greater than the 100 ppm GdNCR density over all radii.

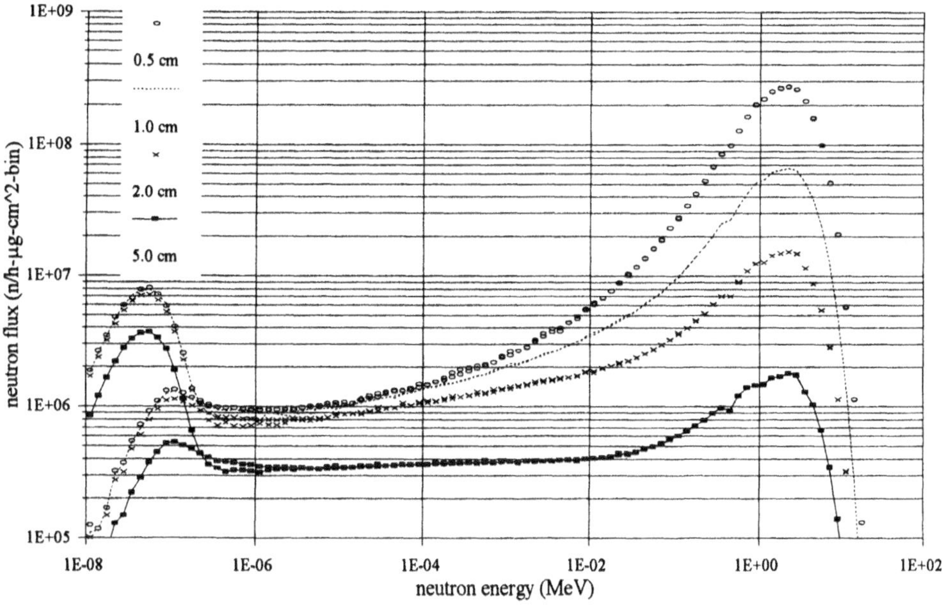

Figure 4. Moderated ^{252}Cf neutron energy spectrum in brain, with / without 1,000 ppm ^{157}Gd.

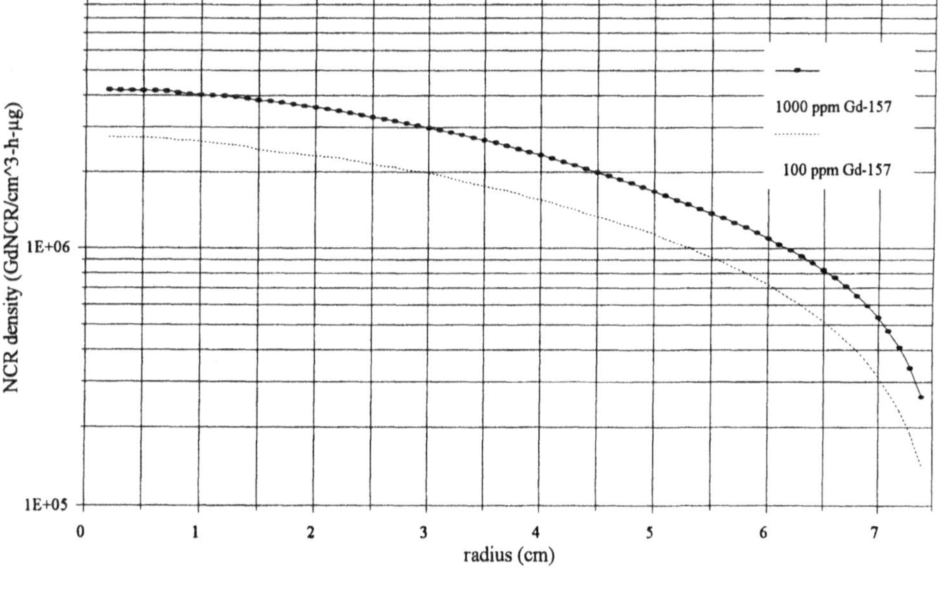

Figure 5. Spatial distribution of GdNCR in brain for ^{157}Gd loadings of 100 and 1,000 ppm.

Table 2. Photon dose enhancement for 160 ppm ^{157}Gd

Radii (cm)	Wierzbicki *et al.* (cGy/h-mg)	this study, calculation (cGy/h-mg)
1	71 ± 5%	26 ± 8%
2	84 ± 5%	24 ± 5%
3	68 ± 5%	21 ± 4%
4	54 ± 5%	18 ± 4%
5	52 ± 5%	15 ± 3%
6	39 ± 5%	11 ± 3%
7	29 ± 5%	8 ± 4%
8	27 ± 5%	—
9	19 ± 5%	—
10	18 ± 5%	—

4. DISCUSSION

4.1. Comparison of GdNCR Photon Dose Enhancement Results with Others

Wierzbicki *et al.*[8] calculated the photon dose enhancement of 160 ppm ^{157}Gd based on experimental measurements of the thermal neutron flux around a 6 mg ^{252}Cf source using gold foils in a water phantom. Calculations were based on results of Brugger and Shih.[9] A comparison of these results for 160 ppm ^{157}Gd are made in Table 2. Results for ^{157}Gd loading of 160 ppm for this study were determined through linear interpolation between 100 and 1,000 ppm ^{157}Gd dose enhancement results. Differences between results of Wierzbicki *et al.* and those obtained herein were attributed to differences in geometrical setup as Wierzbicki *et al.*, did not include gadolinium in the actual setup and derivation of the GdNCR photon dose enhancement was based largely on calculation of thermal neutron flux measurements. Consequently, Wierzbicki *et al.* overestimated the GdNCR photon dose enhancement due to exclusion of thermal neutron burnup and ^{157}Gd self-shielding, as well as attributing all the energy from the GdNCR to photon production.

4.2. GdNCR Photon Dose Estimation and Results of Stepanek

It was determined that the over 75% (0.754) of the energy from a 1.5 MeV photon would be deposited outside of a sphere of water with a 7.5 cm radius. If a photon μ_{en} value of 0.0280 cm^{-1} at 1.5 MeV for muscle is used with a 10 cm radius sphere, one determines that 68.8% of the energy is deposited outside the sphere. Stepanek determined a similar result (68.0%) after performing Monte Carlo calculations using the GEANT transport code.[5] Therefore, the GdNCR photon average energy may be well approximated as 1.5 MeV.

5. CONCLUSION

Calculations of the ^{157}Gd NCR photon dose enhancement for ^{252}Cf brachytherapy were performed, and determined to be infeasible due to time constraints inherent to the computer system available to the author. However, results from a limited number of calculations revealed the majority of the energy released by ^{158*}Gd photons was not deposited in a 15 cm diameter sphere of either water or brain material. Analysis of a GdNCR photon dosimetry kernel revealed only slight differences in water brain materials. This was expected due to the similar average atomic numbers of water and brain material. Spatial resolution of the GdNCR with 100 and 1,000 ppm ^{157}Gd in brain was similar in shape to the ^{1}H(n,γ)^{2}H NCR spatial distribution. There was significant non-linearity of the GdNCR with increasing ^{157}Gd loadings due to thermal neutron burnup.

REFERENCES

1. J.F. Briesmeister, Monte Carlo N-Particle transport code system, *in*: "MCNP4B User's Manual," LANL, Los Alamos, 1997.
2. M.J. Rivard, J.G. Wierzbicki, and F. Van den Heuvel, Calculations of the ^{252}Cf neutron spectrum in water for various positions and loadings of ^{10}B and ^{157}Gd, *in*: "American Nuclear Society Radiation Protection and Shielding Division: Technologies for the New Century," D.T. Ingersoll ed., ANS Inc., La Grange Park, IL, 1998, pp. 211–218.
3. A. Geist, A. Beguelin, J. Dongarra, W. Jiang, R. Manchek, and V. Sunderam. "PVM: Parallel Virtual Machine- A User's Guide and Tutorial for Networked Parallel Computing," The MIT Press, Cambridge, MA, 1994.
4. L.L. Anderson, Status of dosimetry for ^{252}Cf medical neutron sources, *Phys. Med. Biol.*, 18:779–799, 1973.
5. J. Stepanek, Radiation spectrum of ^{158}Gd and radial dose distribution, *in*: "Advances in Neutron Capture Therapy", B. Larsson, J. Crawford, and R. Wienreich eds., Elsevier Science, Netherlands, 1997, pp. 425–429.
6. R. Nath, L.L. Anderson, G. Luxton, K.A. Weaver, J.F. Williamson, and A.S. Meigooni, Dosimetry of interstitial brachytherapy sources: Recommendations of the AAPM Radiation Therapy Committee Task Group No. 43, *Med. Phys.*, 22:209–234, 1995.
7. L.L. Meisberger, R. Keller, and R.J. Shalek, The effective attenuation in water of the gamma rays of ^{198}Au, ^{192}Ir, ^{137}Cs, ^{226}Ra, and ^{60}Co, *Radiol.*, 90:953–957, 1968.
8. J.G. Wierzbicki, Y. Maruyama, and A.T. Porter, Measurement of augmentation of ^{252}Cf implant by ^{10}B and ^{157}Gd neutron capture", *Med. Phys.*, 21:787–790, 1994.
9. R.M. Brugger and J.A. Shih, Evaluation of gadolinium-157 as a neutron capture therapy agent" *Strahlentherapie Onkol.*, 165:153, 1989.

THEORETICAL CELL SURVIVAL PROBABILITY FOR PRECLINICAL EVALUATION OF BNCT TREATMENT CONDITIONS

Sung-Joon Ye and Karl O. Ott

Medical Physics, Rush-Presbyterian-St. Luke's Medical Center
1653 W. Congress Parkway, Chicago, Illinois 60612

1. INTRODUCTION

Conventionally, cell survival parameters can be experimentally measured in cell culture system. Using these parameters, cell survival curves *vs.* absorbed dose are plotted. However, this classical approach cannot explicitly describe cell survival variation with individual treatment conditions, such as beam spectra and field sizes. The emphasis in this paper is on the development of a computational protocol to *directly* calculate the *theoretical cell survival probability regarding BNCT treatment conditions.* Macroscopic dosimetry is in principle not sufficient to accurately evaluate BNCT therapeutic efficacy, which depends on ^{10}B-distributions at the cellular and subcellular levels.[1] Therefore, in this paper a treatment condition is characterized by the choice of boron carrier drugs (boronophenylalanine, BPA, or boron sulfhydryl hydride, BSH), two different neutron beam spectra, and single or bilateral exposures. The *radiation transport from the neutron source to a patient's tumor* is calculated using Monte Carlo methods. With an assumed tolerance dose (10.5 Gy-Eq), a cell survival probability is determined for a given treatment condition. The cell survival probability applies to a simple model of tumor control probability (TCP) by radiotherapy,[2] based on the principle that every tumor clonogen must be killed to achieve tumor control.[3]

2. MATHEMATICAL MODELS OF CELL SURVIVAL AND TUMOR CONTROL PROBABILITY

2.1. Cell Survival Probability

With high-LET particles from $^{10}B(n,\alpha)^{7}Li$ reactions, cell killing is a stochastic process that is governed by the Poisson probability distribution, *i.e.*, the probability for

Frontiers in Neutron Capture Therapy, edited by Hawthorne *et al.*
Kluwer Academic / Plenum Publishers, New York, 2001.

the occurrence of a particular number of events with the average number of events.[4] For simplicity, we call every particle track that would kill a cell a lethal hit. With a being the average number of lethal hits per cell, the cell survival probability (P_s) which corresponds to no lethal hit in the Poisson probability distribution is reduced to

$$P_s = e^{-a}. \tag{1}$$

The average number of lethal hits per cell (a) varies with treatment conditions that are characterized by neutron beam attributes, treatment planning, and the pharmacokinetics of the boron carrier drug. In cells irradiated by a certain treatment condition, a is proportional to the neutron fluence, the (n,α) reaction density per fluence, and the cell-killing yield of the two released particles. Hence, a can be expressed as follows:

$$a = y_k \times r_{(n,\alpha)} \times f_M \times V_{cell}, \tag{2}$$

where y_k = cell-killing yield due to one $^{10}B(n,\alpha)^7Li$ reaction in a cell, $r_{(n,\alpha)}$ = (n,α) reaction density per incident neutron fluence [(n,α)/cm/neutron], f_M = tolerable neutron fluence [neutrons/cm^2], V_{cell} = total cellular volume [cm^3]. In Eq. 2, V_{cell} is known from the geometrical cell models in the literature.[5] The y_k for individual molecules of a boron carrier drug may have its own distribution between the theoretical minimum (value of 0) and maximum (value of 2), but in radiotherapy only the average value of y_k is important. Most boron molecules accumulate in the cytoplasm and the extracellular region of tissues, with fewer than 10% of boron molecules found in the cell nuclei.[5] Therefore, the probable range of y_k in this paper is assumed to be 0.1–0.6. The tolerable neutron fluence is defined as the incident neutron fluence at the beam aperture that can deliver the normal brain tolerance dose (herein, 10.5 Gy-Eq as a peak dose at a certain depth). It depends on the treatment conditions. The number of lethal hits, a is directly proportional to $r_{(n,\alpha)}$. Since the $^{10}B(n,\alpha)^7Li$ reaction has a large cross-section (3,840 barns) only in the thermal range, the spatial variation of $r_{(n,\alpha)}$ is represented by the thermal neutron flux distribution along the beam center line. The thermal neutron flux forms by the moderation of incident epithermal neutrons in the first few cm, and then exponentially attenuates due to scatterings and $^{10}B(n,\alpha)^7Li$ reactions. Therefore, tumor cell survival probabilities (tumor lethal hits) also have a particular distribution across tumors, depending on the location in the brain.

2.2. Tumor Control Probability

A model that accurately predicts the ability of radiation therapy to control tumors from measurable parameters of the disease and known parameters of treatment would be a tumor control probability (TCP).[3] The cell survival probability will apply to the tumor control probability as follows:

$$TCP = e^{-MCR}, \tag{3}$$

where MCR = mean clonogen residue after irradiation. But MCR can be expressed:

$$MCR = n_0 \times V_{tumor} \times P_s, \tag{4}$$

where n_0 is the initial number density of clonogens before treatment, V_{tumor} is the pretreatment tumor volume, and P_s is the average cell survival.[2] The initial clonogen density could be determined by fitting these model to clinical data. Such fits suggest that n_0 is a small fraction of the metabolically active tumor cell density. The pretreatment tumor clonogen number is proportional to pretreatment tumor volume.[3]

3. METHODS AND MATERIALS

The computation consists of three kinds of Monte Carlo radiation transport calculations: 1) modeling a TRIGA Mark I reactor and its filter/moderator structure to yield the neutron beam library, 2) dosimetry to determine the tolerable neutron fluence, and 3) neutron transport for the $^{10}B(n,\alpha)^{7}Li$ reaction density in tumors. The Monte Carlo code, Monte Carlo for *N*-Particle Transport (MCNP), is an excellent choice for all these calculations.[6] The number of histories is chosen so that the statistical errors of the neutron beam are below 10% at any energy or angular bin. But, in the calculations of dose, cell survival and tumor control, these statistical errors may be negligible, compared to the large uncertainties of y_k, n_0, and dose modification factors.

4. RESULTS

4.1. Cell Survival Probabilities

In this paper, single exposure was applied to shallow surface tumor treatments, while bilateral exposure was applied to deep center tumor treatments. Lower tumor cell survival values can be achieved by the improved beam (Ω_{imp}). The degree of its reduction, when using Ω_{imp} instead of Ω_{org}, is approximately a factor of 10 ($y_k = 0.6$) – 1.5 ($y_k = 0.1$). The cell survival with BPA is lower than with BSH by a factor of about 1.1–6, primarily depending on y_k. This is because the ^{10}B concentration with BPA is 50% higher than with BSH, even though the tolerable neutron fluence with BSH is approximately 25%–40% higher than with BPA. However, the direct comparison of tumor cell survival probabilities of BPA and BSH at the same y_k is not plausible because values of y_k for both boron drugs would be somewhat different. This is because BSH is likely to have a better microdistribution for the two released particles to hit a cell nucleus than BPA.

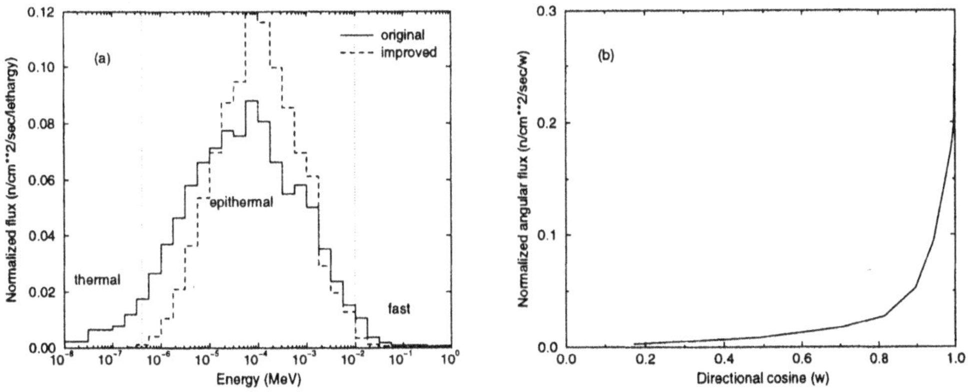

Figure 1. Normalized energy spectra (a) and angular distribution (b) for the original neutron beam (Ω_{org}) and the improved neutron beam (Ω_{imp}).

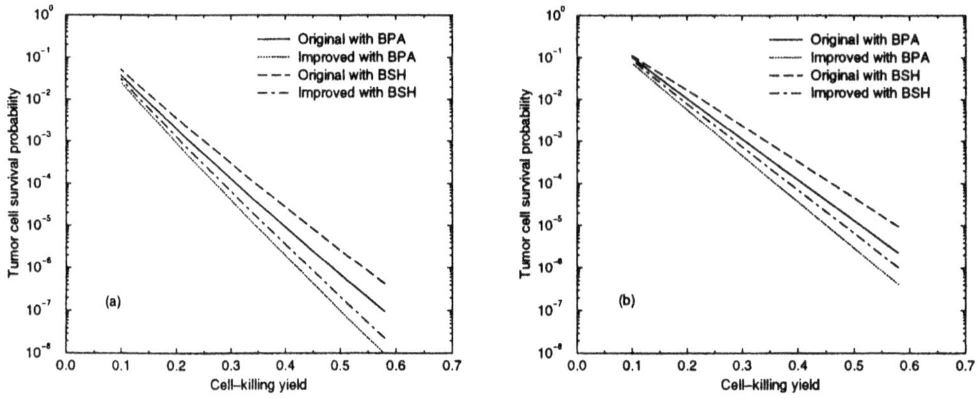

Figure 2. Tumor cell survival for four treatment conditions *vs.* cell-killing yield in shallow surface tumor treatments by single exposure (a) and in center tumor treatments by bilateral exposure (b).

Considering this situation, the predicted lower survival probability of BPA at the same y_k might not be a real case, but at most probable ranges of y_k (0.2–0.45) for both boron drugs, cell survival probabilities would range 10^{-3}–10^{-5} for the shallow surface tumor and 10^{-2}–10^{-4} for the deep center tumor. Because of this large overlap which primarily results form the large uncertainty of y_k, it is not clear from this study which of the two boron drugs would be superior.

4.2. Tumor Control Probabilities

Even though there are large variations of the initial clonogen density with the type of tumor,[2] Laramore's empirical data ($n_0 = 3.2 \times 10^5/cm^3$) for malignant brain tumors were well matched with the TCP relationship.[7] This value and P_s calculated in section 4.1 was applied to obtain the mean clonogen residue after irradiation (*MCR*). Figure 3 (a)

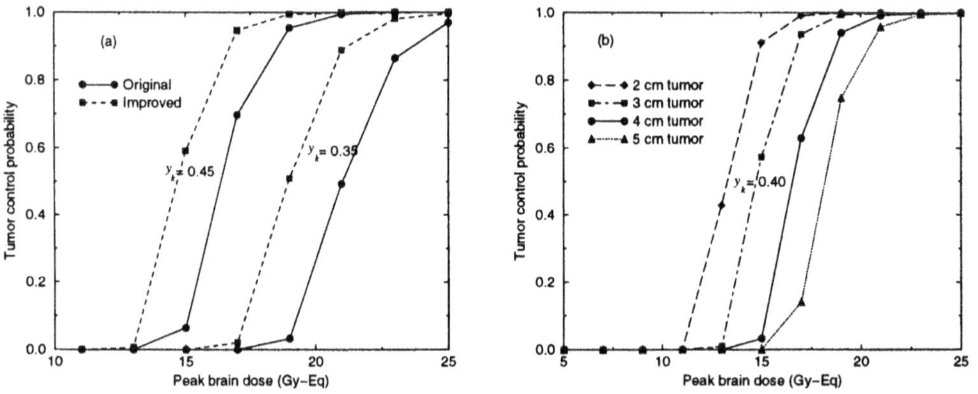

Figure 3. Tumor control probabilities for two values of $y_k = 0.35$ and 0.45 (a) and for four different diameters of the tumor (b) *vs.* peak brain dose. Here, all calculated values are for the shallow surface tumors exposed by Ω_{imp} and BPA.

shows a comparison of TCP results of the shallow surface tumor model for two different values of y_k. The uncertainty of y_k imposes the large variation of TCP. However, a boron carrier drug with a higher y_k definitely decreases the peak brain dose for a given value of TCP. Also, the improved beam (Ω_{imp}) contributes to decreasing the peak brain dose. As y_k varies from 0.45 to 0.35, the TCD_{50} values of Ω_{org} and Ω_{imp} are 16.3–21.1 Gy-Eq and 14.6–18.9 Gy-Eq, respectively. This means that using Ω_{imp} instead of Ω_{org}, the peak brain dose could be reduced by about 10%–12%. Figure 3 (b) shows the difficulty of tumor control as the tumor volume is increasing.

5. DISCUSSION AND CONCLUSIONS

The calculated cell survival and tumor control probabilities can only be interpreted in a relative sense because in practice, it is impossible to obtain the TCP parameters for individual patients. This is simply to estimate the potential improvement of therapeutic efficacy by various combinations of BNCT treatment conditions. Very pure epithermal beams, in which thermal, low epithermal, and fast components, were almost eliminated could considerably decrease the potential of tumor recurrence after treatments. The enhancement of neutron beam qualities such as pure epithermal and forward-directed beams could decrease a potential of tumor recurrence by factors of 2–10, depending the other treatment conditions. The malignant region of a deep-seated tumor could not be effectively treated by single exposure because a high potential of tumor cell regrowth would be expected. Therefore, treatments of a deep-seated tumor whose boundary is beyond about 5–6 cm depth from the skin would rather rely on multi-directional exposure such as bilateral exposure. In addition, a possible improvement of ^{10}B-carrier drugs for a higher y_k would rather contribute to decreasing tumor recurrence rate. However, the uncertainties of the cell-killing yield and the compound factor (CBE) are so large that the tumor cell survival probabilities for BPA and BSH overlap. It was assumed that only contributors to cell-kills were the two released high-LET particles from $^{10}B(n,\alpha)^7Li$ reactions. This should be a good assumption in tumor cells, but perhaps not in normal brain cells because the (n,α) dose in the brain is only a certain fraction of the total dose. Nevertheless, the theoretical cell survival model developed in this paper will serve a protocol to evaluate the preclinical BNCT efficacy, as well as to provide the basis for design goals of neutron beams and boron carrier drugs.

REFERENCES

1. D. Gabel, V.P. Bond, J. Kalef-Ezra, and R.G. Fairchild, The necessity of stochastic radiobiology in BNCT or what is the RBE of the $^{10}B(n,\alpha)^7Li$ reaction, *in*: "Process in Neutron Capture Therapy for Cancer," B. J. Allen *et al.*, ed., Plenum, New York, 1992, pp. 345–348.
2. E.R. Blazek, Mechanistic prediction of tumor control from pretreatment tumor volume, *in vitro* radiosensitivity, and cell kinetics parameters, *in*: "Volume and Kinetics in tumor Control & Normal Tissue Complications," B. Paliwal *et al.*, ed., Springer, New York, 1998, pp. 462–473.
3. D.J. Brenner, Dose, volume, and tumor-control predictions in radiotherapy, *Int. J. Radiat. Oncol. Biol. Phys.*, 26:171–179, 1993.
4. D.E. Charlton, and B.J. Allen, Monte Carlo calculation of ion passages through brain endothelial nuclei during boron neutron capture therapy, *Int. J. radiat. Biol.*, 64:739–747, 1993.

5. D. Gabel, S. Foster, R.G. Fairchild, The Monte Carlo simulation of the biological effect of the $^{10}B(n,\alpha)^{7}Li$ reaction in cells and tissues and its implication for boron neutron capture therapy, *Radiat. Res.*, 111:14–25, 1987.
6. J.F. Briesmeister, "*MCNP A General Monte Carlo N-Particle Transport Code Version 4A*," Los Alamos National Laboratory, LA-12625-M, 1993.
7. G.E. Laramore, *et al.*, A tumor control curve for malignant gliomas derived from fast neutron radiotherapy data: implication for treatment delivery and compound selection, *in*: "Advances in Neutron Capture Therapy, Vol II, Chemistry and Biology", B. Larsson, ed., Plenum, New York, 1997, pp. 580–587.

A MONTE CARLO CALCULATION OF THE CHANGE IN DOSIMETRIC PROPERTIES ON THE USAGE OF HEAVY WATER FOR BORON NEUTRON CAPTURE THERAPY

Wanno Lee[1], Eun-Hee Kim[2], Gyuseoung Cho[1], and Seung-Hoon Lee[3]

[1]Department of Nuclear Engineering
Korea Advanced Institute of Science and Technology
Teajon 305-701, Republic of Korea
[2]Laboratory of Cyclotron Application and Nuclear Medicine
[3]Department of Neurosurgery
Korea Cancer Center Hospital
Seoul 139-706, Republic of Korea

1. INTRODUCTION

Clinical results and theoretical calculations by many Boron Neutron Capture Therapy (BNCT) study groups have indicated that the usage of epithermal neutrons in BNCT allows treating the deep-seated tumors in brain. Also, Hatanaka[1] in Japan has reported that the therapeutic effect of the BNCT using the epithermal neutrons is improved when the normal water in brain tissue is replaced by the heavy water. Heavy water is considered to reduce the attenuation of thermal neutrons and thus facilitate penetration of thermal neutrons into deep-seated tumor. In this study, the change in dosimetric properties due to the replacement of the brain tissue water with heavy water is assessed.

2. METHODS

A general Monte Carlo N-Particle transport code MCNP4B[2] is used to simulate the neutron and the induced gamma-ray transport. A neutron beam of epithermal energy is modeled to expose onto the planar surface of a cylindrical phantom of 16 cm in diameter and 16 cm in height as shown in Figure 1. The phantom is composed of brain-equivalent materials based on 50%/50% weight average of gray and white matter.

Frontiers in Neutron Capture Therapy, edited by Hawthorne *et al.*
Kluwer Academic / Plenum Publishers, New York, 2001.

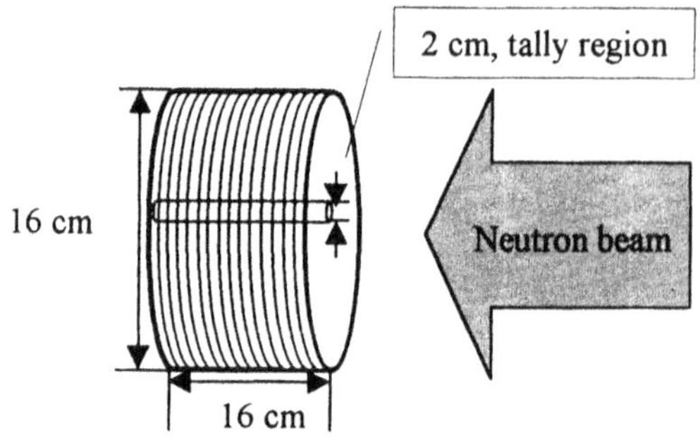

Figure 1. Schematic geometry of the cylindrical phantom for MCNP simulation.

Simulation is performed for the brain phantom having tissue water replaced with heavy water by various proportions ranging from 0% to 100%.

Dose is estimated by calculating the neutron and the induced gamma-ray flux at a site of interest and multiplying it by fluence-to-dose conversion factor.[3] Dose is calculated as a function of depth from the phantom surface along the central axis of the phantom using a disk target of 2 cm in diameter and 1 cm in thickness. Dose to the target includes four reaction components: $^{1}H(n, n')^{1}H$, $^{14}N(n, p)^{14}C$, $^{1}H(n, \gamma)^{2}H$ and $^{10}B(n, \alpha)^{7}Li$ reactions. In estimating the dose for the $^{10}B(n, \alpha)^{7}Li$ reaction, the uniform boron concentration of 30 ppm and 3 ppm is assumed in tumor and in the healthy tissue, respectively.

The merit of replacing the brain tissue water with heavy water in terms of the therapeutic benefit can be assessed by two quantities: the advantage depth (AD) and the advantage ratio (AR).[4] The advantage depth, defined as the depth in tissue at which the total dose to tumor equals the maximum dose to the healthy tissue, is a measure of the lesion depth where the neutron beam is still useful. The advantage ratio, on the other hand, represents the ratio of the integral tumor dose to the integral healthy tissue dose as measured from the tissue surface to the advantage depth.

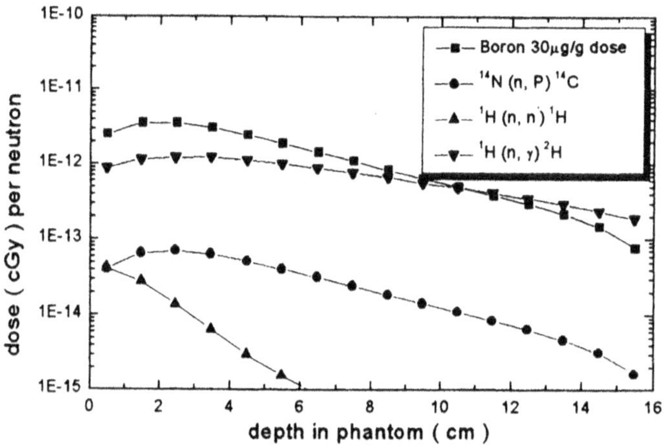

Figure 2. Dose components due to a single neutron irradiation as a function of depth from the phantom surface when 100% of the brain tissue water is light water.

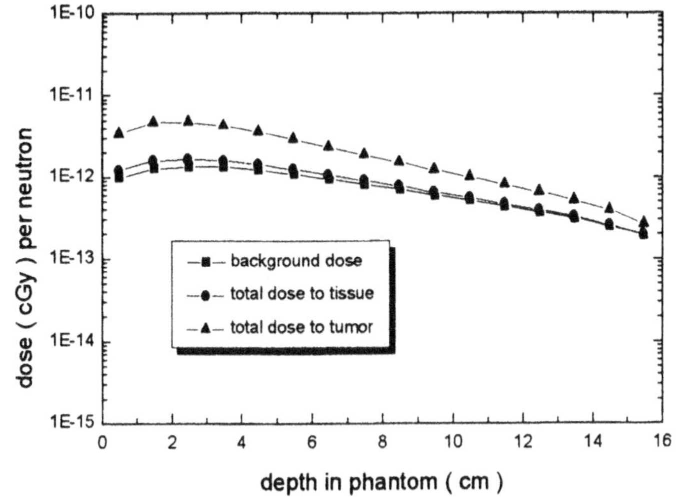

Figure 3. The total dose due to a single neutron irradiation as a function of depth from the phantom surface when 100% of the brain tissue water is light water.

3. RESULTS

The individual dose component is shown in Figure 2 as a function of depth from the phantom surface when the tissue water is not replaced by heavy water at all. Dose component of $^{10}B(n, \alpha)^{7}Li$ reaction is given for 30 μg/g of ^{10}B concentration. Shown in Figure 3 are the total healthy tissue dose and the total tumor dose including the background dose. Dose due to $^{1}H(n, n')^{1}H$, $^{14}N(n, p)^{14}C$, and $^{1}H(n, \gamma)^{2}H$ reactions makes up the background dose. The difference in dose between the normal tissue and the tumor is due to $^{10}B(n, \gamma)^{7}Li$ reaction component.

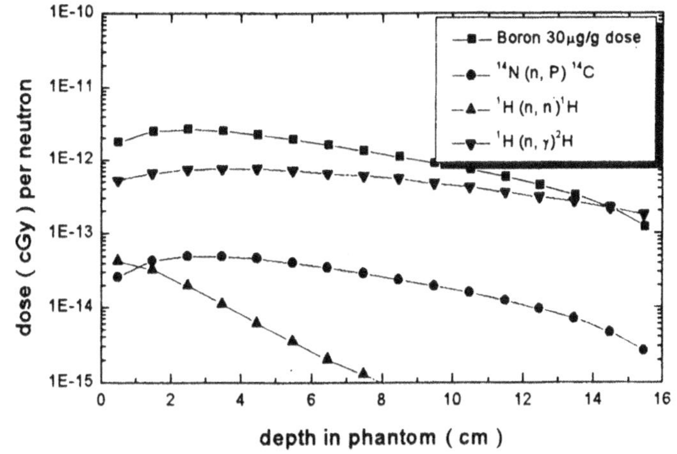

Figure 4. Dose components due to a single neutron irradiation as a function of depth from the phantom surface when 50% of the brain tissue water is replaced with heavy water.

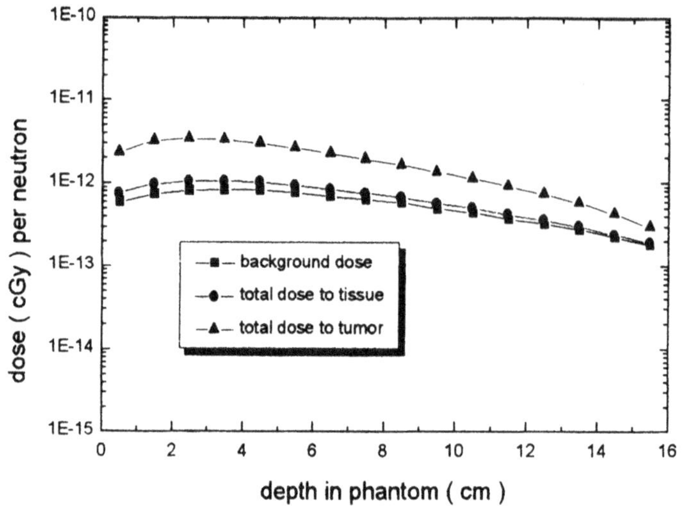

Figure 5. The total dose due to a single neutron irradiation as a function of depth from the phantom surface when 50% of the brain tissue water is replaced with heavy water.

When half of the brain tissue water is replaced with heavy water, the individual dose component and the combined dose due to a single neutron irradiation are illustrated in Figures 4 and 5.

The minimum advantage depth AD_{min} is calculated based on 10:1 concentration ratio of ^{10}B in tumor (30 µg/g), and healthy tissue (3 µg/g). The maximum advantage depth AD_{max} is calculated on the assumption of an infinite tumor-to-healthy tissue ^{10}B concentration ratio, that is 30 ug/g ^{10}B in tumor and no ^{10}B in the healthy tissue. As shown in Figure 6, both AD_{max} and AD_{min} increase as the greater portion of the brain tissue water is replaced with heavy water. The advantage depth increases by up to 71% (the case of 100% water replacement) as compared to that for normal brain tissue. Advantage ratio also increases by up to 41% as shown in Figure 7.

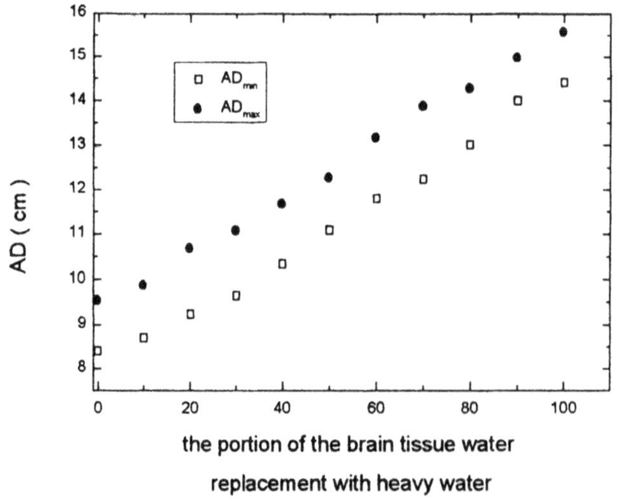

Figure 6. Advantage depth versus the percent replacement of the brain tissue water with heavy water.

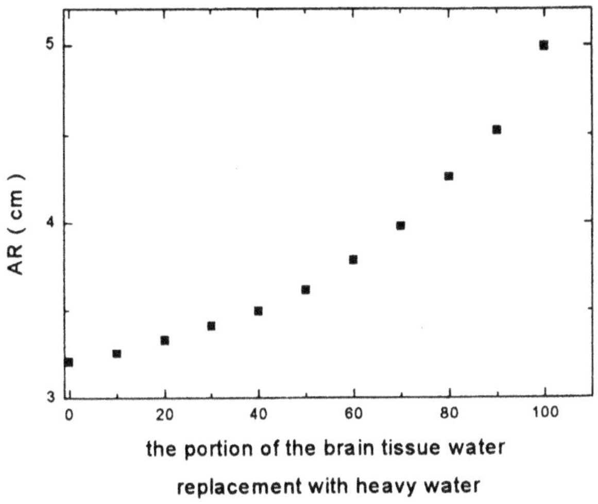

Figure 7. Advantage ratio versus the percent replacement of the brain tissue water with heavy water.

4. CONCLUSIONS

This study adds the theoretical explanation to the clinical results of Dr. Hatanaka showing that the replacement of brain tissue water with heavy water in BNCT is beneficial for treating deep-seated tumor. The greater advantage depth and the greater advantage ratio have been estimated when a greater portion of brain tissue water is replaced with heavy water.

REFERENCES

1. H. Hatanaka, Clinical results of boron neutron capture therapy, *in:* "Neutron Beam Design, Development, and Performance for Neutron Capture Therapy," O.K. Harling, *et al.*, ed., Plenum Press, New Work, 1990, pp. 15–21.
2. J.F. Briesmeister, "MCNP—a General Monte Carlo N-Particle Transport Code Version 4B", Los Alamos National Laboratory, New Mexico, 1977.
3. J.C. Yanch, X-L. Zhou, and G.L. Brownell, A Monte Carlo investigation of the dosimetric properties of monoenergetic neutron beams for neutron capture therapy, *Rad. Res.*, 126:1–20, 1991.
4. S.D. Clement, J.R Chot, R.G. Zamenhof, J.C. Yanch, and O.K. Harling, "Monte Carlo methods of neutron beam design for neutron therapy at the MIT research reactor", *in:* "Neutron Beam Design, Development, and Performance for Neutron Capture Therapy," O.K. Harling, *et al.*, ed., Plenum Press, New Work, 1990, pp. 51–69.

PRECLINICAL RADIOBIOLOGY

LARGE ANIMAL MODEL FOR HEALTHY TISSUE TOLERANCE STUDY IN BNCT

J. Benczik[1,2], T. Seppälä[2,3], J. Hopewell[4], M. Snellman[1], I. Auterinen[5], R. Joensuu[6], U. Abo Ramadan[3], J. Tanttu[7], J. Vähätalo[2,8], M. Kulvik[2], H. Joensuu[9], M. Tenhunen[9], S. Savolainen[3,6], M. Kallio[10], and M. Färkkilä[10]

[1]Faculty of Veterinary Medicine, Diagnostic Imaging
University of Helsinki, Finland
[2]Clinical Research Institute
Helsinki University Central Hospital
Helsinki, Finland
[3]Department of Physics
University of Helsinki, Finland
[4]Research Institute (University of Oxford)
The Churchill Hospital
Oxford, UK
[5]VTT Chemical Technology
Helsinki-Espoo, Finland
[6]Department of Radiology
Helsinki University Central Hospital
Helsinki, Finland
[7]Picker Nordstar Inc.
Helsinki, Finland
[8]Laboratory of Radiochemistry
University of Helsinki, Finland
[9]Department of Oncology
Helsinki University Central Hospital
Helsinki, Finland
[10]Department of Neurology
Helsinki University Central Hospital
Helsinki, Finland

1. INTRODUCTION

In radiotherapy where normal brain tissue is in the treatment field there is a risk of developing late radiation side effects. Consequently, evaluation of normal tissue

Frontiers in Neutron Capture Therapy, edited by Hawthorne *et al.*
Kluwer Academic / Plenum Publishers, New York, 2001.

tolerance is critical for the application of safe and effective treatment. The biological effectiveness of the calculated physical doses must be estimated before boron neutron capture therapy (BNCT) is applied clinically. Relative biological effectiveness (RBE) is a function of the energy of particles. As a consequence of the difference in neutron spectrums in different BNCT facilities the RBE of each beam may be different. The aim of the healthy tissue tolerance study at the Finnish Research Reactor (FiR 1) was to obtain a complete dose effect curve for beam characterisation as part of the safety studies. The effects of high linear energy transfer (LET) radiation from the $^{10}B(n,\alpha)^{7}Li$ reaction were compared with the effects of the beam-only irradiation. A large animal model (the dog brain) was used for the estimation of the probability of late radiation reactions as a function of dose. The endpoint of the study was visible magnetic resonance image (MRI) changes in the brain.

2. MATERIALS AND METHODS

2.1. Animals

Nineteen adult Beagle dogs of both sexes, bred for research purposes were used in the study. The appropriate ethics committees approved the experiment. The dogs were divided into three groups according to the volume and the area of their head perpendicular to the beam, specified after measurements of the size (width, length and height) of the cranium on the transversal and sagittal MRIs. Dogs were selected to the different dose groups from all the three subgroups of head size such that both sexes were represented in each dose group.

2.2. Dose Planning

The individual dose plans were made with the 3D-treatment planning software[1] BNCT Rtpe (version 2.2) using low-field MRIs. In the model the brain, the frontal sinus and the eyes were segmented separately. For intercomparison of the human and the dog head model the composition of the different tissues (brain, eyes and the rest as muscle) were defined after the ICRU Report[2] for the neutron and gamma fluence calculation. The neutron and gamma fluences were also calculated with the values shown by Moran[3] that gave 1% difference at the thermal neutron maximum compared to the calculations based on ICRU. The fluence-to-dose conversion was calculated according to brain tissue composition given by Brooks.[4]

The estimated weighted doses, expressed in weighted units (Gy(W)) for the three beam-only dose groups using the 90% isodose curve were 13.1, 14.1 and 16 Gy(W). Four Beagles were irradiated after boronphenylalanine-fructose (BPA-F) infusion to the planned dose of 14.1 Gy(W).

The total weighted dose was calculated according to the following equation:[5]

$$D_w = W_{fn}*D_{fn} + W_N*D_N + W_\gamma*D_\gamma + CF_{brain}*k_{brain}*c_{blood}*D_{B,ppm}$$

D_w:total weighted dose; W_{fn}:weighting factor for the fast neutron dose = 3.2; D_{fn}:physical fast neutron dose; W_N:weighting factor for the nitrogen dose = 3.2; D_N:physical dose from the nitrogen capture reactions; W_γ:weighting factor for the gamma dose = 1.0; D_γ:physical gamma dose; CF_{brain}:compound factor for brain tissue = 1.3; k_{brain}:^{10}B

Table 1. Dose calculation for the irradiation with an 11-cm beam aperture after a 60-min long BPA-F (700 mg BPA/kg bw) infusion[a]

At the 100% isodose	Total	Boron	Gamma	N-14	Fast
Physical dose rate, Gy/h	10.2	5.4	3.9	0.7	0.2
Weighting factors		1.3	1.0	3.2	3.2
Weighted dose rate, Gy (W)/h	13.9	7.0	3.9	2.2	0.7

[a]Average ^{10}B concentration in blood during irradiation: 13 ppm; total weighted dose at 90% isodose: 14.1 Gy (W); irradiation time: 65.3–71 min.

concentration ratio of the brain and blood = 1 : 1; c_{blood}:^{10}B concentration in blood; $D_{B,ppm}$:physical ^{10}B dose if there is 1 ppm ^{10}B in the tissue.

2.3. Irradiations

Groups of five dogs were irradiated with 14.1 or 16 Gy (W), four dogs with 13.1 Gy (W) and one dog with 10 Gy (W) to the 90% isodose with the epithermal beam at FiR 1. In addition four Beagles received a 60 minute long infusion of 700-mg BPA/kg body weight. The infusion was applied with an infusion pump to maintain a constant rate. The blood boron level was measured by inductively coupled plasma-mass spectrometry[6] (ICP-MS) every 15 minutes after the end of infusion. The plan was to irradiate at a blood ^{10}B content in the 11–14 ppm range. The measured ^{10}B level varied between 29–40 (38;29;32.4;40) ppm at the end of the infusion. It dropped in 60 to 92 (92;75;60;91) minutes to 13.3–14.9 (13.3;14.9;13.4;14) ppm. The blood boron levels were 9.97–11.5 (10.6;11.5;9.97;10.8) ppm after completion of irradiation. The dogs were positioned in lateral recumbence with their head extended on an individually prepared polystyrene vacuum pillow in front of the beam. The beam centerline and the head was modelled to help appropriate positioning. For induction and maintenance of anaesthesia propofol and isoflurane/O_2 were used. During irradiation the dogs were monitored with a pulse oximeter.

2.4. MRI Follow Up

MRI examinations were carried out prior to and every 4–6 weeks starting from 4 months after irradiation with an 0.1 T low-field imager.* The image quality was enhanced using a specially designed receiver coil for signal detection.[7] After a sagittal sequence for localisation, a T2-weighted sequence, and a T1-weighted before and after iv Gd-DTPA contrast (0.2-ml/kg bw Magnevist^R 469 mg/ml) are made. Additional T2-weighted sequence and a T1-weighted with and without contrast are made with a high field (1.0T) imager† using an extremity coil before and at 6 and 12 months after irradiation.

3. RESULTS AND DISCUSSION

3.1. Dose Planning

The dose rate varied according to the size of the head and the frontal sinus. The average dose rate during beam only irradiation was 7 Gy (W)/h with the minimum

---

* MeritTM Picker.
† SignaTM GE.

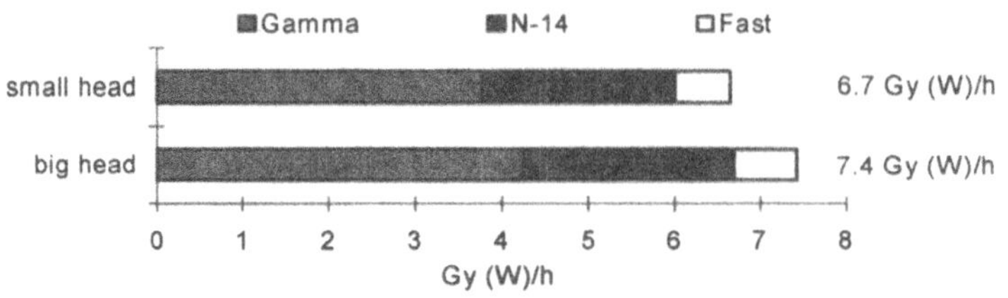

Figure 1. The influence of the head size on the dose rate during beam only irradiation.

(smallest head size) of 6.7 Gy (W)/h and the maximum (largest head size) of 7.4 Gy (W)/h. In the case of equal head size the dose rate was lower for the one with bigger frontal sinus.

At the fluence-to-dose conversion the nitrogen dose component is directly proportional to the atomic density of N-14 in tissue. The ICRU value[2] for the nitrogen content in brain tissue is 20% higher than that defined by Brooks,[4] resulting in approximately 1 Gy(W) difference at the 14.1 Gy(W) dose level. This could be critical because of the steepness of the dose response curve.

In the dose plan 90% of the brain volume was included by the 70% isodose curve. The dose, 14.1 Gy(W) for the dogs which received BPA-F was chosen to show the effect if 50% of the received dose is coming from the weighted boron dose compared to the corresponding 14.1 Gy(W) beam-only group.

3.2. Irradiations

For the large animal irradiations a beam with 11-cm diameter circular aperture was selected because this gave similar ratios between the different dose components and dose rate to the patient treatment situation. The average irradiation time was 134 min in the 14.1 Gy(W) beam-only group. The variation of the irradiation time arising from individual head and frontal sinus sizes was 12 min in this dose group.

3.3. MRI Follow Up

Images of low-field-strength MR units can provide satisfactory anatomic details of the canine brain. At the moment the follow up is too short to evaluate the MRI changes in the brain.

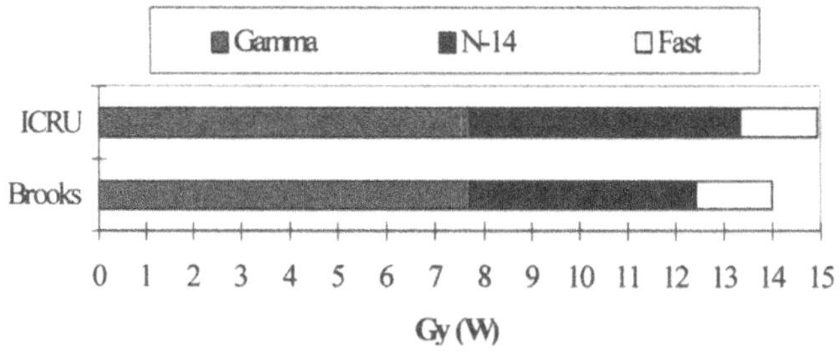

Figure 2. The influence of variation in the nitrogen (N-14) dose on the weighted total dose for the beam only irradiation. Gamma is the gamma and fast is the fast neutron dose. ICRU = 2.2 w-% nitrogen[2]; Brooks = 1.84 w-% nitrogen[4]. (Length of irradiation: 134 min; W_N and $W_n = 3.2$; gamma $W_\gamma = 1$ at 14 Gy (W) dose level).

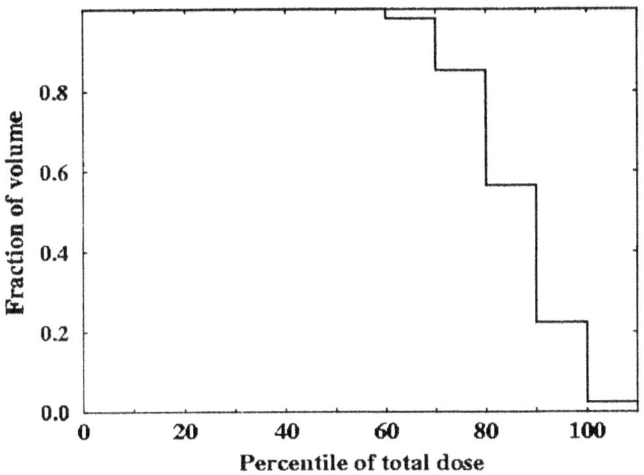

Figure 3. Weighted dose/volume histogram for a representative beam only irradiation. "x" axis is the % of the total weighted dose, "y" axis is the brain volume in %.

4. CONCLUSIONS

The values for the weighting factors of dose components, the CF for BPA in the brain, the blood/brain ^{10}B concentration ratio, and the atomic nitrogen content of brain tissue used in our dose calculations were the same as in the large animal radiobiological studies carried out elsewhere for the sake of comparison. Due to the low fast neutron content of this epithermal neutron beam, beam-only irradiation was possible to brain tolerance without severe skin reactions. The dogs have showed no early acute radiation damage with exception of corneal oedema (due to impaired lacrimal production) and mucositis in the mouth. Follow up is continuing, to evaluate late radiation damage to the brain.

ACKNOWLEDGMENTS

Research Institute of the University of Oxford, Netherlands Energy Research Foundation, Brookhaven National Laboratory, and Washington State University provided details about their radiobiological studies for comparison to our model. The work was supported by the Technical Development Centre of Finland, Research Foundation of Helsinki University Central Hospital (HUCH), Research grant of the Departments of Radiology and Neurology of HUCH, Instrumentarium Science Foundation, VTT Chemical Technology, University of Helsinki, The Finnish Academy of Sciences, and Finnish Centre for International Exchange Programmes.

REFERENCES

1. D.E. Wessol, *et al.*, http://id.inel.gov/cart/rtpe-manual/.
2. ICRU, Photon, electron, proton, and neutron interaction data for body tissues, ICRU Report 46, 1992.

3. J.M. Moran, D. Nigg, F.J. Wheeler, *et al.*, Macroscopic geometric heterogeneity effects in radiation dose distribution analysis for boron neutron capture therapy, *Med Phys.*, 19(3):723–732, 1992.
4. R.A. Brooks, G. Di Chiro, and M.R. Keller, Explanation of cerebral white-gray contrast in computer tomography, *J Comput Assist Tomo.*, 4(4):489–491, 1980.
5. W. Sauerwein, J. Rassow, and B. Mijnheer, Considerations about specification and reporting of dose in BNCT, Advances in Neutron Capture Therapy. Vol 2 pp. 531–534.
6. M. Kulvik, J. Laakso, J. Vähätalo, and R. Zilliacus, On-line Boron-10 determination from blood samples by ICP-MS, *in* "Frontiers in Neutron Capture Therapy," M.F. Hawthorne, K. Shelly, R.W. Wiersema, eds., Kluwer Academic/Plenum Publishers, New York, 2001, pp. 849–851.
7. M. Snellman, J. Benczik, R. Joensuu, *et al.*, Low-field magnetic resonance imaging of Beagle brain with a dedicated receiver coil, *Vet Radiol Ultrasoun.*, 39(5), 1998.

A LARGE ANIMAL MODEL STUDY INVESTIGATING POTENTIAL BNCT RETREATMENT USING BSH

Patrick R. Gavin[1], Hege Kippenes[1], and Judit Benczik[2]

[1]College of Veterinary Medicine
Veterinary Clinical Sciences
Washington State University
Pullman, WA 99164-6610
[2]Clinical Research Institute
Central Hospital, Helsinki University
Helsinki, Finland

INTRODUCTION

Boron neutron capture therapy (BNCT) treatments utilizing epithermal neutron beams and available boron compounds result in a mixed irradiation field of low and high linear energy components (LET). Previous studies attempting to measure the repair of the low LET component, comparing a multifraction study to single fraction study failed to identify sufficient repair in normal tissue tolerance studies of the dog brain utilizing borocaptate sodium (BSH) and epithermal neutron irradiation.[1] Many current patients in clinical trials relapse approximately 6 months after their initial BNCT treatment. Many have considered the possibility of retreatment due to the theoretical dose localization in BNCT. A six-month interval would not be sufficient to allow retreatment of a full course of conventional irradiation. Therefore, a study was undertaken to give purpose-bred laboratory dogs an initial BNCT to near tolerance levels followed in six-month by a second identical dose.

MATERIAL AND METHODS

Purpose-bred beagle-type laboratory dogs were used. The dogs were judged normal following physical and neurologic examinations, complete blood count, serum chemistry profile, and magnetic resonance imaging of the brain. Seven dogs were used. The dogs were separated into two groups. Four dogs were given a peak target dose of 11.5 Gy-

Frontiers in Neutron Capture Therapy, edited by Hawthorne *et al.*
Kluwer Academic / Plenum Publishers, New York, 2001.

equivalent. This is the peak dose received by a small fraction of the dorsal cerebral cortex. The dose descriptions of the various beam components and compound factors are the same as previously reported in this laboratory.[2,3] The Boron compound used was BSH enriched with ^{10}Boron to 95%. The target peak blood-boron concentration was 50 µg/g. This was obtained following administration of the compound at a dosage of 50 mg boron/kg bw intravenously over one hour, diluted in 11 ml saline/kg bw. The irradiation was accomplished using the Brookhaven Medical Research Reactor (BMRR), Brookhaven National Laboratory. A single dorsal field was utilized with an 8-cm circular field centered over the calvarium.

Three additional dogs were given a target dose of 12.5 Gy-equivalent. Six months after the initial irradiation, the examinations, as listed before, were repeated. The animals were returned to BMRR and received a second, basically identical, dose of BNCT utilizing the parameters mentioned above. The animals were returned to Washington State University for clinical and neurologic follow-up.

At the time of neurologic abnormality or at predetermined post-irradiation intervals (6 months and 12 months) any severe neurologic impairment or large change in the magnetic resonance image resulted in humane euthanasia of the dog followed by complete gross and histopathologic examination of the irradiated tissues.

RESULTS

11.5 Gy-Equivalent Group

All dogs remained neurologically normal and had no abnormalities in the blood parameters. All dogs were judged normal following magnetic resonance imaging at 6 months post-irradiation. Following the second irradiation, 2 dogs developed neurologic disease 3 months later. Magnetic resonance imaging, subsequent gross and histopathologic examination revealed severe edema and necrosis primarily of the paraventricular white matter. Five months following the second irradiation, the other 2 dogs developed neurologic disease that was confirmed with magnetic resonance imaging and pathologic examination to be cerebral edema and necrosis. Necrosis was judged to be due to hemorrhagic events and had an appearance similar to that previously reported with this model.

12.5 Gy-Equivalent Group

In the 6 months after the first dose, all dogs remained neurologically normal. None of the animals developed any abnormalities in their blood parameters. Two of the dogs remained normal on magnetic resonance imaging. One dog developed a lesion in the peripheral cortex of the right midbrain that exhibited ring enhancement following gadolinium contrast administration (Figure 1). The lesion was judged to be compatible with a cerebral infarct. Within 2 months of the second irradiation, all 3 dogs developed neurologic disease and edema, and cerebral necrosis was confirmed with magnetic resonance imaging and pathologic examination (Figure 2).

DISCUSSION

This study was based on prior work that has been published.[1,3,4] BSH was chosen as the boron agent due to the increased confidence in the normal tissue tolerance values

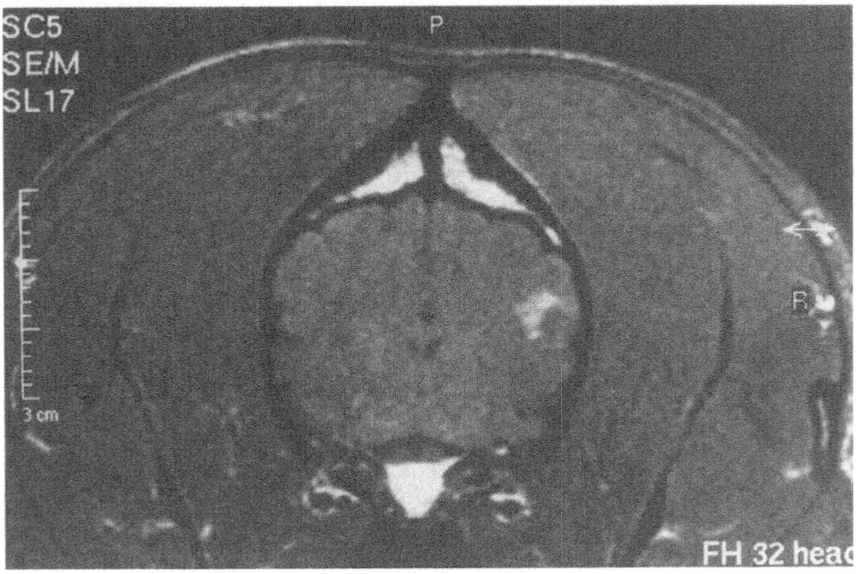

Figure 1. 12.5 Gy-equivalent dog, 6 months after T1-weighted axial image following Gd-DTPA administration. Ring-enhanced lesion in R mid-cerebrum.

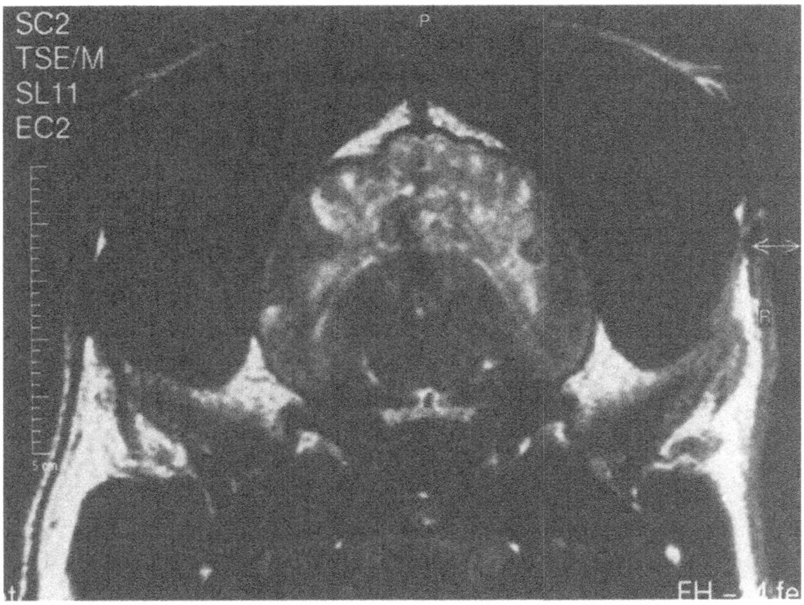

Figure 2. 12.5 Gy-equivalent dog, 2 months after 2nd irradiation. T2-weighted axial image revealed cerebral edema of the white matter tracts.

utilizing that compound. Blood boron concentration of 50 μg/g was chosen to reach a clinically reasonable level and maximize the high LET component of the mixed field irradiation. The dosages chosen were in keeping with peak normal brain values used in our treatment of canine brain tumors.[4] Only 5–10% of the brain receives this peak dose. With both the normal tissue tolerance studies and the spontaneous canine brain tumor treatments, 12.5 Gy-equivalent is approximately the dose required to cause magnetic resonance visible lesions in 50% of the animals. This dose has not been associated with severe neurologic impairment.[3] Therefore, some of the 12.5 Gy-equivalent dogs should have had magnetic resonance visible lesions following the first irradiation. This was confirmed in our study with 1 in 3 dogs being affected. 11.5 Gy-equivalent has generally been insufficient to cause magnetic resonance visible lesions or obvious neurologic impairment. In this study, none of the 4 animals revealed any abnormalities after 6 months following the first irradiation.

Magnetic resonance visible lesions only, or neurologic impairment with magnetic resonance visible lesions have previously been seen as early as 4 months post-irradiation. In general, the most common post-irradiation time for the development of these abnormalities is 5–6 months. Following the second irradiation, all dogs developed severe neurologic disease. Neurologic disease consisted of seizures and ataxia. The signs could not be reversed with the administration of corticosteroids. Subsequent magnetic resonance imaging revealed severe cerebral edema of the white matter and focal contrast enhancing lesions compatible with cerebral infarction. Pathologic examination revealed hemorrhagic infarction, necrosis, and cerebral edema. The changes were identical to that previously seen in our previous normal tissue tolerance studies. There was an accelerated onset of these lesions following the second irradiation dose. In the lower dose group (11.5 Gy-equivalent) 2 dogs were affected after only 3 months and 2 dogs at 5 months post-irradiation. In the higher dose group (12.5 Gy-equivalent), the 3 dogs were affected at 7–8 weeks post-irradiation. This study does not address patients that have undergone a partial course of conventional irradiation, focal irradiation delivered by radioisotope implants, or radiosurgery. The study also does not address what would be the minimal safe dose of a second course of BNCT treatment after a significant portion of the brain has been treated to tolerance, or to the tolerance of the tissue following BNCT to a focal irradiation, via radioisotope implants or radiosurgery. The study was only designed to investigate the potential use of BNCT retreatment to clinically applicable tolerance levels. Due to the accelerated development of the radiation necrosis, at this time, it was concluded that BNCT retreatment of patients with brain tumors should not be performed. It was concluded that BNCT retreatment of patients with brain tumors utilizing BSH as boron carrier should not be performed with similar doses to the primary treatment.

ACKNOWLEDGMENTS

The assistance of Ms. Peggy Micca and Dr. Jeff Coderre at Brookhaven National Laboratory is warmly acknowledged. This work was supported by a grant from the Department of Energy, DE-FG03-98ER62567/A000.

REFERENCES

1. P.R. Gavin, F.J. Wheeler, and C.D. Swartz, Fractionation effects of BNCT on normal tissue tolerance of the canine brain, *in* "Advances in Neutron Capture Therapy, volume II, Chemistry and Biology," B. Larsson, J. Crawford, and R. Weinreich, eds., Elsevier, Amsterdam, 1997, pp. 670–675.

2. P.R. Gavin, S.L. Kraft, C.E. DeHaan, C.D. Swartz, and M.L. Griebenow, Large animal normal tissue tolerance with boron neutron capture, *Int. J. Radiation Oncology Biol. Phys.* 28:1099–1106, 1994.
3. P.R. Gavin, S.L. Kraft, R. Huiskamp, and J.A. Coderre, A review: CNS effects and normal tissue tolerance in dogs, *J Neuro-Oncol*, 33:71–80, 1997.
4. P.R. Gavin, S.L. Kraft, C.D. Swartz, and F.J. Wheeler, Boron neutron capture therapy of spontaneous intracranial tumors in dogs using BSH and epithermal neutrons. Cancer Neutron Capture Therapy, Y. Mishima ed., Plenum Press, New York, 1996, pp. 763–768.

DOSIMETRY CHAIN FOR THE DOGS IRRADIATED IN THE EPITHERMAL NEUTRON BEAM AT THE FINNISH BNCT FACILITY

Sauli Savolainen[1,3], Iiro Auterinen[2], Carita Aschan[1,3], Pekka Hiismäki[2], Mika Kortesniemi[2,3,4], Antti Kosunen[5], Petri Kotiluoto[2,3], Juha Lampinen[1,3], Tiina Seppälä[2,3,4], Tom Serén[2], Vesa Tanner[2], and Matti Toivonen[5]

[1]Department of Radiology
Helsinki University Central Hospital
[2]VTT Chemical Technology
[3]Department of Physics
University of Helsinki
[4]Clinical Research Institute
Helsinki University Central Hospital
[5]STUK-Radiation and Nuclear Safety Authority

1. INTRODUCTION

In any radiotherapy the physical dose must have a metrologically traceable link to the national and international dosimetry standards. In addition, the uncertainty of the dose delivered to the patient has to be low enough for estimating the effects of the treatment beforehand and analyze then afterwards. Accurately characterized spectrum, fluence and dose distributions in a phantom form the basis for a reliable dose delivery to a patient and is essential also for biological studies at the BNCT beam. There is an obvious need for standardization in medical application of BNCT.[1-5]

The dog irradiations for the healthy tissue tolerance study at the Finnish BNCT facility FiR 1 was completed at the end of March 1998. Included were 4 dogs with BPA infusion and 15 without.[6] The dose to the target volume, the brain, were calculated individually in the dose planning. Doses to the target and to the body were monitored individually using *in vivo* dosimeters. The dogs were anesthetized and the *in vivo* dosimeters fixed. Dogs were positioned into the beam as indicated in the dose plan using cardboard profiles and laser pointers. Dose delivery was controlled using the beam monitoring system. For the dogs with BPA, boron concentration in blood was measured from blood

Frontiers in Neutron Capture Therapy, edited by Hawthorne *et al.*
Kluwer Academic / Plenum Publishers, New York, 2001.

samples (serum) that were taken before and after irradiation. After irradiation the *in vivo* dosimeters were removed and the induced activity of the dogs was monitored.

In this paper the dosimetry chain for the dogs irradiated at the Finnish BNCT facility is described. The dosimetry procedure is divided into three sections: pre-irradiation methods (measurements and calculations), dosimetric monitoring during the irradiation (*in vivo* dosimetry), and the post-irradiation measurements.

2. BEAM MONITORING

The epithermal neutron beam[7] from the reactor was monitored with three pulse-mode ^{235}U-fission chambers for epithermal neutron fluence rate and one current mode ionization chamber for gamma dose rate. The detectors were monitored on line with a PC-program based on virtual instrumentation to give the primary reference dose. Additional dose monitoring was accomplished by observing stand-alone (NIM) counters.[8] The relation between the absorbed dose to the tissue in the reference point in the reference geometry and the beam monitor readings was obtained with comprehensive phantom measurements.[9] The relation between the desired absorbed dose (D) and the monitor units (MU) can be calculated as follows:[10]

$$MU = \frac{D}{\dot{D}_{Ref}^{calc}} * \dot{MU}_{Ref}$$

Subscript *Ref* refers to reference beam intensity and superscript *calc* refers to the calculated values. The relation is based on the measured beam monitor (epithermal channel) count rate vs. measured (Au foil) Au reaction rate. Using this method the neutron fluence in the reference geometry can be administered with a precision better than 1%. To verify the stability of the beam monitors an (IC) chamber neutron dose measurement in the reference phantom was performed daily after the irradiations. The combined uncertainty was estimated to be 6.3% (1 SD%).

3. SPECTRUM CHARCTERIZATION

The description of the neutron beam in dose planning system was verified by measuring the neutron spectrum using activation detectors with different response characteristics, most of them from the ECN Petten set.[11] Complementary measurements have been performed by INEEL.[12] The measured reaction rates and calculated neutron spectra with uncertainties were used as input to the least-squares adjustment code LSL-M2.[13] A cross section library, mainly based on IRDF-90 (IAEA), was compiled and condensed to the BUGLE group structure. The uncertainties are of the order 20%, 5% and 30% for the thermal, epithermal and fast flux in the air, respectively.[7,10] The calculated and measured reaction rates inside the cylindrical phantoms were in so good agreement that the calculated beam model was only scaled to reference monitor units without any spectral adjustment. Furthermore other measurements (ionization chambers, TLD, SiLi) in the phantoms supported the use of this beam model.[10]

Table 1. An example of the calculated doses for a dog with estimated boron concentration of 13 ppm (BPA as a boron carrier)

	Dose rate, Gy/h					Total dose	
	Total	Boron	Gamma	N-14	Fast	Peak physical [Gy]	Weighted (90% isodose) Gy (W)
ICRU (N-2.20%,H-10.70%)	10.4	5.4	3.9	0.8	0.2	11.7	—
Brooks (N-1.84%,H-10.57%)	10.2	5.4	3.9	0.7	0.2	**11.5**	
Weighting factors		1.3	1.0	3.2	3.2		
Brooks weighted dose rate [Gy (W)]	13.9	7.0	3.9	2.2	0.7		**14.1**

4. DOSE PLANNING

The three-dimensional software BNCT_Rtpe/rtt_MC was used for dose planning.[14] The MR image of each dog was used to construct a three-dimensional model of a dog's head.[6] The three-dimensional neutron-photon transport was performed using Monte Carlo simulation with cross sections based on the ENDF/B (IV/V) library.[15] The calculated doses were verified with comprehensive phantom measurements[9] and with two other radiation transport calculation codes:[16] DORT (Two-Dimensional Discrete Ordinates Transport Code)[17] and MCNP (Monte Carlo N-Particle Transport Code System).[18]

For each dog a dose plan as shown as an example in Table I was produced prior to irradiation. In this case the weighted dose of 14.1 Gy (W) [see Appendix 1] is achieved in the time of 68 minutes with the reference beam intensity which corresponds to 141726407 monitor counts in the reference channel, which is the primary irradiation counter.

5. DOSIMETRY

5.1. Phantom Measurements

The accuracy of dose calculations (dose planning) was verified by phantom measurements beforehand. Activation foils (Au and Mn for thermal and epithermal neutron fluence), twin ionization chamber technique (tissue equivalent chamber for neutrons and Mg(Ar) for gamma), thermoluminescent dosimeters (LiF:Mg,Ti for (relative) neutron fluence and ^{7}LiF:Mg,Cu,P for gamma) and a SiLi-detector for thermal neutrons were used in an extensive series of dosimetric phantom measurements.[1,9,19] Standard sources (^{60}Co and ^{137}Cs) of the Secondary Standard Laboratory of STUK have been used for calibration and testing of the ionization chambers and TL dosimeters. The air kerma rates of the standard sources are traceable to the International Standard at Bureau International des Poids et Mesures in France and National Physical Laboratory in UK. The uncertainties of the measured rates (with IC) of absorbed dose to brain tissue in a water phantom were approx. 6% for gamma, and approx. 20% for neutrons at the depth of 25 mm in the phantom when using beam aperture 14 cm.

Based on the tests of reference phantom material it has been concluded that water is a suitable brain tissue substitute for dose calibration and routine quality control of an epithermal neutron beam in NCT, although for epithermal neutron dosimetry more accurate brain tissue substitutes can be produced.[20]

5.2. *In Vivo* Dosimetry

TL detector MCP-7s (^{7}LiF:Mg,Cu,P) of TLD Niewiadomski & Co. (Krakow, Poland) was used for gamma detection.[19] The individual dose monitoring was performed at several measurement points as reported by Aschan *et al.*[21] The corrections for the thermal neutron sensitivity of the dosimeters were made based on the fluence measured with Mn-Al (1%, Ø 12 mm × 0.2 mm) foils obtained from ECN Petten.[11] The foils were measured with a Ge detector and the saturation activities were calculated with the SAMPO program.[13] In general, for all the dogs, the measured and calculated absorbed gamma doses and thermal neutron fluences correlated well. The average gamma dose of the body was less than 1 Gy, and the uncertainty of measurements was approx. 10% (1 S.D.).

5.2.1. Boron Analysis. During the irradiation it is difficult to directly measure the ^{10}B concentration. For this reason, it is of great importance to be able to estimate the ^{10}B concentration in the tissue while it is being irradiated. The ^{10}B concentration in the blood after 4-dihydroxyborylphenylalanine fructose complex (BPA-F) infusion were modelled by a bi-exponential fit.[22] ^{10}B concentration in blood samples (serum) were measured by an inductively coupled plasma-mass spectrometry, ICP-MS.[23] In serum the detection limit of boron in sample solution is 5 ppb. The differences between the model based estimations of ^{10}B concentration in the blood and the post-irradiation measured values were lower than 0.7 ppm inducing the maximum error of 3% into the calculated doses.

5.3. Post-Irradiation Activity Measurements

Post-irradiation activities in dogs were measured with a portable HPGe-detector at 33 cm distance (whole-body) and directly on head. The measurements were performed on an inactive table immediately after removing the dogs from the irradiation room. The counting times were about 1 min. The principal induced activities were ^{24}Na, ^{38}Cl and ^{42}K. The statistical accuracy is approx. 3% for ^{24}Na, 7% for ^{38}Cl and 10% for ^{42}K. The calibration accuracy is about 30%. Based on the phantom measurements and calculations with the MIRD schema[24] it was estimated that the dose of the dogs due to the activated ^{24}Na, ^{38}Cl and ^{42}K nuclei were 0.1, 0.05 and 0.5 mGy, respectively. Overall the average dose of the dogs due to the radioactive isotopes of the body was less than 1 mGy. Dose due to non-detectable β-emitters, e.g. ^{32}P and ^{36}Cl, was neglected in these considerations. The contribution of these isotopes may be regarded as small, ^{32}P is localized mainly in bone and the range of β-radiation is less than 1 cm. In case of ^{36}Cl, the half-life is about 300,000 years (compared to 37 minutes of ^{38}Cl).

6. DISCUSSION AND CONCLUSIONS

Quality assurance for clinical dosimetry and standardization in medical application of BNCT are urgently needed. The ICRU recommendation of reporting of BNCT irradiation should be applied in order to evaluate the factors interfering with comparison of radiation doses to patients in different facilities.[2–5] The European project "A code of practice for the dosimetry of BNCT in Europe" is one step towards the standards in BNCT dosimetry.[25] Accurately characterized spectrum, fluence and dose distributions in phantoms form the basis for the dosimetry of BNCT.

The utilized dosimetry procedure at FiR 1 can be considered to guarantee that the delivered physical doses in treatments at the Finnish BNCT facility have metrologically traceable links to the international dosimetry standards. The uncertainties in the physical dose components in the phantom irradiations can be estimated. In the treatment of patients the most inaccurate dose component is due to the boron as the boron distribution in tissues during the irradiation cannot be monitored *in vivo*, and the (micro)distribution of the used boron compounds are only estimates based on boron concentration in the blood. In addition, the weighting factor of the boron dose is inaccurate as it is a function of the boron compound and the tissue characteristics. Due to the short range of the α-induced dose, the boron distribution should be known in cellular, and preferable, in subcellular level to define the accurate weighting factors in different tissues for the boron dose.

APPENDIX 1

The weighted dose D_W in BNCT is a sum of physical dose components (D_i) multiplied by the weighting factors (w_i) of each dose component.[26] The weighting factors are related to the relative biological effectiveness (RBE) of each radiation quality (LET characteristics) and, in case of boron, also to the pharmacokinetics (microdistribution) of the boron compound. The weighted dose can be written:

$$D_W = w_\gamma D_\gamma + w_B D_B + w_N D_N + w_{fast\_n} D_{fast\_n}.$$

The unit for physical dose (i.e. absorbed dose) is Gray [1 Gy = 1 J/kg]. As the weighting factors are dimensionless both the physical and the weighted dose are measured in same units i.e. in Grays. To illustrate the difference in the two concepts the letter W in parenthesis is added to the symbol Gy writing one space between the symbol and the additional specification for the weighted dose D_W.

ACKNOWLEDGMENTS

The work was supported by the Technology Development Centre of Finland, Academy of Finland, University of Helsinki, and the State Subsidy for University Hospitals, Helsinki University Central Hospital, research grant of the Departments of Radiology and Neurology.

REFERENCES

1. A. Kosunen, M. Kortesniemi, H. Ylä-Mella *et al.*, Twin ionization chambers for dose determination in phantom in an epithermal neutron beam, *Radiat. Prot. Dosim.* 82:65–69, 1999.
2. J. Rassow, F. Stecher-Rasmussen, K. Ravensberg *et al.*, Quality assurance for clinical dosimetry of the European trial on BNCT in Petten, *in:* "Frontiers in Neutron Capture Therapy," M.F. Hawthorne, K. Shelly, R.W. Wiersema, eds., Kluwer Academic/Plenum Publishers, New York, 2001, pp. 161–171.
3. W. Sauerwein, K. Hideghéty, M. de Vries *et al.*, The necessity and possibilities for standardization in medical application of BNCT, Program and Abstracts of the Eighth International Symposium on NCT for Cancer, La Jolla, CA, 1998, p. 102.
4. R. Gahbauer, N. Gupta, T. Blue, W. Sauerwein, and A. Wambersie, Reporting of BNCT irradiation application of the ICRU recommendation to the specific situation in BNCT, *in:* "Frontiers in Neutron Capture Therapy," M.F. Hawthorne, K. Shelly, R.W. Wiersema, eds., Kluwer Academic/Plenum Publishers, New York, 2001, pp. 565–569.

5. C. Vroegindeweij, J. Capala, R. Huiskamp, A.Z. Diaz, W. Sauerwein, and A.D. Chanana, Factors interfering with comparison of radiation doses to GBM patient at Petten and Brookhaven, Program and Abstracts of the Eighth International Symposium on NCT for Cancer, La Jolla, CA, 1998, p. 143.
6. J. Benczik, T. Seppälä, J. Hopewell *et al.*, Large animal model for healthy tissue tolerance study in BNCT, *in:* "Frontiers in Neutron Capture Therapy," M.F. Hawthorne, K. Shelly, R.W. Wiersema, eds., Kluwer Academic/Plenum Publishers, New York, 2001, pp. 1233–1238.
7. I. Auterinen, C. Aschan, P. Hiismäki *et al.*, Metamorphosis of a 35 years old TRIGA reactor into a modern BNCT facility, *in:* "Frontiers in Neutron Capture Therapy," M.F. Hawthorne, K. Shelly, R.W. Wiersema, eds., Kluwer Academic/Plenum Publishers, New York, 2001, pp. 267–275.
8. V. Tanner, I. Auterinen, J. Helin, and S. Savolainen, On line neutron beam monitoring of the Finnish BNCT Facility, *Nucl. Instrum. Meth. in Phys. Res.* A422:101–105, 1999.
9. M. Kortesniemi, A. Kosunen, T. Serén *et al.*, Phantom dose distributions in the beam of Finnish BNCT facility, *in:* "Frontiers in Neutron Capture Therapy," M.F. Hawthorne, K. Shelly, R.W. Wiersema, eds., Kluwer Academic/Plenum Publishers, New York, 2001, pp. 659–664.
10. T. Seppälä, T. Serén, and I. Auterinen, Source characterization for the rtt_MC treatment planning program at FiR 1, *in:* "Frontiers in Neutron Capture Therapy," M.F. Hawthorne, K. Shelly, R.W. Wiersema, eds., Kluwer Academic/Plenum Publishers, New York, 2001, pp. 219–224.
11. P. Watkins, Y. Harker, C. Amaro *et al.*, Nuclear Characterisation of the HFR Petten BNCT Facility, *in*: "Advances in Neutron Capture Therapy", A.H. Soloway *et al.*, eds., Plenum Press, New York 1993, pp. 59–65.
12. D.W. Nigg, Y.D. Harker, J.K. Hartwell *et al.*, Collaborative Spectral Characterization of the Finnish Epithermal-Neutron Beam Facility for BNCT, *in:* "INEEL BNCT Research Program Annual Report 1996, INEL/EXT-97-00319" J.R. Venhuizen, ed., Idaho Falls 1997, pp. 15–32.
13. W. Stallmann, LSL-M2: A Computer Program for Least-Squares Logarithmic Adjustment of Neutron Spectra, NUREG/CR-4349, Oak Ridge 1986.
14. D.E. Wessol, R.S. Babcock, F.J. Wheeler *et al.*, BNCT_Rtpe: BNCT Radiation Treatment Planning Environment User's Manual, Version 2.2, http://id.inel.gov/cart/rtpe-manual, February 21, 1997.
15. M.K. Drake, Data Formats and Procedures for the ENDF Neutron Cross Section Library, BNL 50274 (CT-601), 1970.
16. J. Lampinen, T. Seppälä, I. Auterinen, and S. Savolainen, Three dose calculation codes applied to neutron transport in BNCT, *in:* "Frontiers in Neutron Capture Therapy," M.F. Hawthorne, K. Shelly, R.W. Wiersema, eds., Kluwer Academic/Plenum Publishers, New York, 2001, pp. 213–217.
17. W.A. Rhoades, and R.L. Childs, The DORT Two-Dimensional Discrete Ordinates Transport Code. *Nucl. Sci & Engr.* 99:88–89, 1988.
18. J.F. Briesmeister, ed., MCNP-A General Monte Carlo N-Particle Transport Code. Los Alamos National Laboratory Report, LA-12625-M 1997.
19. C. Aschan, M. Toivonen, S. Savolainen, T. Seppälä, and I. Auterinen, Epithermal neutron beam dosimetry with TL dosemeters for boron neutron capture therapy *Radiat. Prot. Dosim.* 81:47–56, 1999.
20. T. Seppälä, J. Vähätalo, I. Auterinen *et al.*, Modelling of brain tissue substitutes for phantom materials in neutron capture therapy (NCT) dosimetry *Radiat. Phys. Chem.* 55:239–246, 1999.
21. C. Aschan, T. Serén, T. Seppälä, *et al.*, *In vivo* dosimetry of the dog irradiations at the Finnish BNCT facility, *in:* "Frontiers in Neutron Capture Therapy," M.F. Hawthorne, K. Shelly, R.W. Wiersema, eds., Kluwer Academic/Plenum Publishers, New York, 2001, pp. 1251–1256.
22. P. Ryynänen, S. Savolainen, J. Benczik, M. Kulvik, J. Vähätalo, and M. Snellman, Compartmental and non-compartmental methods in studying the kinetics of boron-10 after boronophenylalanine fructose complex (BPA-F)-infusion in dogs, *in:* "Frontiers in Neutron Capture Therapy," M.F. Hawthorne, K. Shelly, R.W. Wiersema, eds., Kluwer Academic/Plenum Publishers, New York, 2001, pp. 959–963.
23. M. Kulvik, J. Laakso, J. Vähätalo, and R. Zilliacus, On-line boron-10 determination from blood samples by ICP-MS, *in:* "Frontiers in Neutron Capture Therapy," M.F. Hawthorne, K. Shelly, R.W. Wiersema, eds., Kluwer Academic/Plenum Publishers, New York, 2001, pp. 849–851.
24. W.S. Snyder, M.R. Ford, C.G. Warner, and S.B. Watson, "S", Absorbed dose per unit cumulative activity for selected radionuclides and organs. MIRD Pamphlet No. 11, New York: The Society of Nuclear Medicine 1975.
25. F. Stecher-Rasmussen, I. Auterinen, A. Beddoe *et al.*, A code of practice for dosimetry of BNCT in Europe, *in:* "Advances in Neutron Capture Therapy", B. Larsson, J. Grawford, R. Weinreich, eds., Elsevier Science B.V., Amsterdam, 1998, pp. 237–240.
26. D.L. Bleuel, R.J. Donahue, B.A. Ludewigt, and J. Vujic, Designing accelerator-based epithermal neutron beams for boron neutron capture therapy, *Med. Phys.* 25:1725–1734, 1998.

IN VIVO DOSIMETRY OF THE DOG IRRADIATIONS AT THE FINNISH BNCT FACILITY

Carita Aschan[1,2], Tom Serén[3], Tiina Seppälä[1,3,4], Judit Benczik[4,5], Matti Toivonen[6], Iiro Auterinen[3], and Sauli Savolainen[1,2]

[1]Department of Physics
University of Helsinki
[2]Department of Radiology
Helsinki University Central Hospital
[3]VTT Chemical Technology
Technical Research Centre of Finland
[4]Clinical Research Institute
Helsinki University Central Hospital
[5]Faculty of Veterinary Medicine
University of Helsinki
[6]STUK · Radiation and Nuclear Safety Authority

1. INTRODUCTION

A healthy tissue tolerance study[1,2] with Beagle dogs was carried out at the Finnish BNCT facility FiR 1 during February–March 1998. Three dose groups of dogs without boron carrier, each comprising five Beagles, and one dose group with BPA-F infusion were irradiated. Individual dose plans[1] were done for each dog. The absorbed doses, which were actually delivered to healthy tissues, were monitored *in vivo* by placing the dosimeters on the shaved skin of the dogs or in the accessible cavities. In this study, absorbed gamma doses were monitored with thermoluminescent (TL) dosimeters, and neutron fluences were determined with activation foils. The obtained results were compared to the calculated values.

Frontiers in Neutron Capture Therapy, edited by Hawthorne *et al.*
Kluwer Academic / Plenum Publishers, New York, 2001.

2. MATERIAL AND METHODS

2.1. Dose Planning

Three groups of dogs, each comprising five Beagles, were irradiated without a boron carrier to the weighted doses of 13.1 (group I), 14.1 (group II) and 16.0 Gy (W)[2] (group III). The doses were estimated using the 90% isodose curve. In addition, four Beagles (group B) were irradiated after BPA-F infusion to the weighted dose of 14.1 Gy (W). Because the weighted boron dose was about 50% of the total weighted dose, the BPA-F infused dogs were irradiated in average half of the time that the dogs without boron infusion. Individual dose plans[1] of the dogs' head were done with the three-dimensional treatment planning software BNCT_Rtpe/rtt_MC (version 2.2)[3] based on the Monte Carlo method. The gamma doses and thermal neutron fluences for each measurement position in each dog were extracted from the individual dose plans using the line edit feature. Because the dose plan covered only the head, calculated values for the measurement points in the other parts of the body are not available.

2.2. *In Vivo* Dosimetry

In the irradiations, individual dose monitoring with TL dosimeters and activation foils was performed at several measurement points. Au and Mn activation foils were positioned at the top of the head on the central axis of the beam (#1, Figure 1a and b), and between the eyes on the midline (#2, Figure 1a and b). Mn foils and TL detectors were fixed on the outer surface of the endotracheal tube at the border of the soft and hard palate (#4, Figure 1b) and at the level of the canine teeth (#5, Figure 1b). Also, for the determination of the body dose arising from gamma radiation, TL detectors were inserted 3–5 cm cranial to the anus in the rectum (#6, Figure 1c), in front of manubrium (#7, Figure 1c) and on the back between the processus spinosus of thoracic 1–2 (Th1–Th2) vertebrae (#3, Figure 1a). As a bolus, a skin layer of approx. 2–4 mm has been folded (fixed with suture) over the TL detectors situated in front of the manubrium and between the shoulders.

The activation detectors used for neutron fluence determination were diluted (1%, Ø 12 mm × 0.2 mm) Mn-Al and Au-Al foils obtained from ECN Petten.[4] The foils were measured with a carefully calibrated Ge detector system, and the saturation activities were calculated with the SAMPO program.[5]

A TL material ^{7}LiF:Mg,Cu,P was selected for *in vivo* gamma detection because it has shown to be less sensitive for thermal neutrons, and therefore better, than the conventional ^{7}LiF:Mg,Ti.[6,7] In this study, a TL detector MCP-7s (^{7}LiF:Mg,Cu,P) from TLD Niewiadomski & Co. (Krakow, Poland) was used. The corrections for the thermal neutron sensitivity of the MCP-7s TL detectors[7] were made based on the neutron fluences measured with activation foils at the same place (measurement points #4, #5 in Figure 1b).

3. RESULTS AND DISCUSSION

The measured and calculated thermal neutron fluences and absorbed gamma doses at the measurement points are presented in the Table 1 as an average for the each dose group of dogs (groups I, II, III, B). The measured results, reflecting the gamma dose of the total body (measurement points #3, #6, #7), do not differ clearly between the

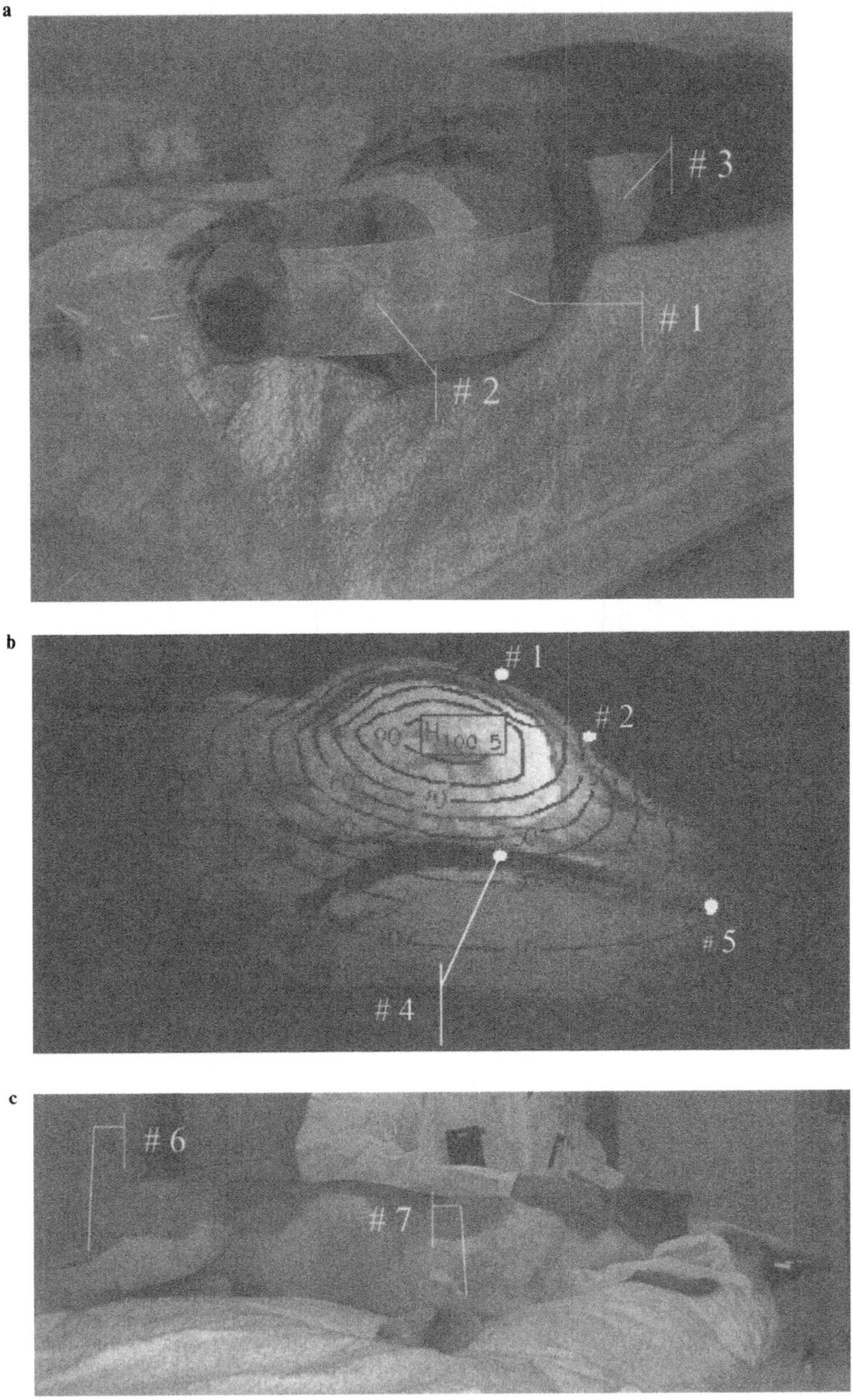

Figure 1. Measurement points of the dog irradiations.

Table 1. Measured and calculated (a) thermal neutron fluences [n cm^{-2}] and (b) absorbed gamma doses [Gy] as an average for the dose groups I, II, III, and B at the various measurement points. Figure 1 is the key for the positioning of the measurement points

(a)

Group	#1 Measured	#1 Calculated	#2 Measured	#2 Calculated	#4 Measured	#4 Calculated	#5 Measured	#5 Calculated
I	3.4E + 12	4.0E + 12	2.6E + 12	3.7E + 12	1.8E + 12	5.6E + 12	1.8E + 12	1.4E + 12
II	3.8E + 12	4.9E + 12	2.6E + 12	4.8E + 12	2.1E + 12	6.5E + 12	1.7E + 12	1.6E + 12
III	4.3E + 12	5.5E + 12	2.6E + 12	5.1E + 12	2.2E + 12	7.1E + 12	1.6E + 12	1.6E + 12
B	1.9E + 12	2.6E + 12	1.3E + 12	2.4E + 12	8.9E + 11	3.0E + 12	9.9E + 11	1.0E + 12

(b)

Group	#3 Measured	#4 Measured	#4 Calculated	#5 Measured	#5 Calculated	#6 Measured	#7 Measured
I	0.43	1.6	5.3	1.8	1.6	0.13	0.41
II	0.40	2.7	6.2	1.9	1.7	0.12	0.38
III	0.49	2.5	6.8	2.1	1.8	0.13	0.52
B	0.21	1.2	3.1	0.5	1.1	0.06	0.23

different dose groups (groups I, II, III). When comparing the body doses of the dogs belonging to the group B (14.1 Gy (W)) with those belonging to the group II (14.1 Gy (W)), a difference of approx. 50% occurs. The lower body doses in the group B are due to the shorter irradiation time. Therefore, no effect in the gamma dose of the total body is found to arise from the boron neutron capture reaction.

It appears from Table 1, that for all the dogs, the measured and calculated thermal neutron fluences and absorbed gamma doses correlate well except at the caudal measurement point in the mouth (#4, Figure 1b). According to the phantom measurements, rtt_MC overestimates the doses and fluences at the surface of phantoms.[8] The large discrepancy at the above mentioned caudal measurement point is assumed to arise from the used head model, in which normal brain, frontal sinus, eyes and rest of the head have been segmented, but not the mouth neither the endotracheal tube were included. The validation of the individual dose planning with *in vivo* dose measurements in dogs' heads in this work proved the importance of *in vivo* dose measurements in patients.

The uncertainty of the neutron fluence determination is a combination of the reaction rate and effective cross section uncertainties, and is roughly estimated to be 5% (1 S.D.). The uncertainties present in the absorbed gamma dose measurements can be combined from two types of uncertainties: (i) the uncertainty of the actual TL process[6] and (ii) the uncertainty of the thermal neutron sensitivity of the MCP-7s TL detectors.[7] The total uncertainty of gamma dose measurements are approx. 10% (1 S.D.) in those measurement points, in which thermal neutron fluence was negligible and, therefore, no correction for thermal neutron sensitivity was performed (#3, #6, #7), and approx. 20–30% (1 S.D.) to the measurement points #4 and #5 (Figure 1b). According to the phantom measurements and comparable calculations, the uncertainties of the dose calculations inside the model are 5% (1 S.D.) surrounding the dose maximum, and at the surface rtt_MC overestimates the dose 10–25%.[8,9]

4. CONCLUSIONS

For all the dogs, the measured and calculated doses correlate well except at the caudal measurement point in the mouth of the dogs. Therefore, the dog head model used in the dose planning should be developed further for a proper analysis of the measurement situation. This study proved the importance of the individual dose planning, but also that of the *in vivo* dose measurements at multiple locations (intraorally in addition to the skin) during the irradiation.

REFERENCES

1. J. Benczik, T. Seppälä, J. Hopewell *et al.*, Large animal model for healthy tissue tolerance study in BNCT, *in* "Frontiers in Neutron Capture Therapy," M.F. Hawthorne, K. Shelly, R.W. Wiersema, eds., Kluwer Academic/Plenum Publishers, New York, 2001, pp. 1233–1238.
2. S. Savolainen, I. Auterinen, C. Aschan *et al.*, Dosimetry chain of the dogs irradiated in the epithermal neutron beam at the Finnish BNCT facility, *in:* "Frontiers in Neutron Capture Therapy," M.F. Hawthorne, K. Shelly, R.W. Wiersema, eds., Kluwer Academic/Plenum Publishers, New York, 2001, pp. 1245–1250.
3. D.E. Wessol *et al.*, BNCT_Rtpe: BNCT Radiation treatment planning environment user's manual, Version 2.2. 1997; http://id.inel.gov/cart/rtpe-manual/.
4. P. Watkins *et al.*, Nuclear characterisation of the HFR Petten BNCT facility, *in*: Advances in Neutron Capture Therapy, A.H. Soloway *et al.* ed., Plenum Press, New York, 1993, pp. 59–65.

5. P. Aarnio, M. Nikkinen, and J. Routti, SAMPO 90—High Resolution Interactive Gamma Spectrum Analysis Including Automation with Macros, Report TKK-F-A682, Helsinki University of Technology, Espoo, April 1991.
6. C. Aschan, M. Toivonen, S. Savolainen, T. Seppälä, and I. Auterinen, Epithermal neutron beam dosimetry with TL dosemeters for boron neutron capture therapy, *Radiat. Prot. Dosim.*, (in press).
7. C. Aschan, M. Toivonen, S. Savolainen, and F. Stecher-Rasmussen, Experimental correction for thermal neutron sensitivity of gamma-ray TL dosemeters irradiated at BNCT beams, *Radiat. Prot. Dosim.*, (submitted).
8. T. Seppälä, T. Serén, and I. Auterinen, Source characterization for the rtt••MC treatment planning program at FiR 1, *in:* "Frontiers in Neutron Capture Therapy," M.F. Hawthorne, K. Shelly, R.W. Wiersema, eds., Kluwer Academic/Plenum Publishers, New York, 2001, pp. 219–224.
9. J.S. Lampinen, T. Seppälä, I. Auterinen, and S. Savolainen, Three dose calculation codes applied to neutron transport in BNCT, *in:* "Frontiers in Neutron Capture Therapy," M.F. Hawthorne, K. Shelly, R.W. Wiersema, eds., Kluwer Academic/Plenum Publishers, New York, 2001, pp. 213–217.

SURVIVAL AFTER BNCT IN COMBINATION WITH SURGERY FOR DOGS WITH SPONTANEOUS BRAIN TUMORS

Rodney S. Bagley, Patrick R. Gavin, Michael P. Moore, Gena M. Silver, Michael L. Harrington, and Hege Kippenes

College of Veterinary Medicine
Washington State University
Pullman, WA 99164-7060

Abstract

Dogs develop spontaneous brain tumors as in humans and serve as a large animal model for evaluation of BNCT for treatment of brain tumor. In this study, 13 dogs with spontaneous brain tumors were treated surgery (SX) followed by BNCT and the median survival of was determined. The median survival was also determined for a group of dogs with spontaneous brain tumors previously treated with BNCT alone (n = 20). All dogs had histologic confirmation of tumor type either at the time of surgery or at necropsy. Of the dogs treated surgically, BNCT followed 3 to 6 weeks after surgery. Dogs of each group had various tumor types, however, the majority were meningiomas (13/20 BNCT; 9/13 SX + BNCT). Of the dogs in the surgery group, all had incomplete resections of the tumor primarily due to tumor location or infiltration of surrounding brain. In both groups, BSH or p-BPA were used for the boron agent. Peak brain dose was 10 to 12.5 Gy- equivalent.

All dogs in the BNCT only group and 11/13 dogs in the SX + BNCT group died or were euthanized due to progression of intracranial signs attributed to progressive tumor. Two dogs in the SX + BNCT group are alive at the time of this writing (540 and 660 days after therapy).

Median survival in days was determined for both groups. Dogs in the historical group receiving BNCT treatment alone had median survival of 90 days (range 1 to 1,860 days). Dogs receiving surgical biopsy/debulking had a median survival of 300 days (range 17 to 690 days). When dogs with meningiomas were analyzed independently within each group, median survivals for the BNCT group was 75 days (range 1 to 180 days) and those in the SX + BNCT group was 300 days (range 14 to 690 days). These results may suggest

Frontiers in Neutron Capture Therapy, edited by Hawthorne *et al.*
Kluwer Academic / Plenum Publishers, New York, 2001.

that BNCT used following incomplete surgical brain tumor removal may improve survival over those previously reported.

Brain tumors occur in dogs with a frequency of 14.5 per 100,000 at risk.[1] Of 95 dogs with brain tumors evaluated at our hospital, the majority were older (median age 9 years; range 4 to 13 years) and had meningiomas (47%).[2] Meningioma is the most common primary brain tumor in dogs and cats.[3–5] The diagnosis of an intracranial mass is readily made using an advanced imaging modalities such as computed tomography (CT) or magnetic resonance (MR) imaging. Features of primary brain tumor have been reviewed.[6–8]

Treatments for brain tumors in dogs depend upon tumor type, tumor location, natural history of the tumor, associated morbidity/mortality of the treatment modality, and cost. Surgical removal is ideal for superficially located, encapsulated, relatively small, benign tumors of which meningiomas would be the most common example. Unfortunately, even when these tumors are histologically benign, they are often not well encapsulated and hence difficult to differentiate from surrounding non-tumorous brain. Brain damage in areas adjacent to the tumor can make gross delineation of tumor borders difficult if not impossible.

Limited studies have been published regarding survival in dogs with spontaneous brain tumors. Studies of survival after brain tumor treatment in dogs include small patient numbers and include a diverse patient population. Median survival in dogs with brain tumors treated with surgery alone vary but tend to cluster around 140 to 150 days.[9,10] As an adjunctive therapy to surgery, or as a primary therapy for tumors that are difficult to access surgically, conventional radiation therapy has been shown effective for some brain tumor in animals.[9,11,12] Treatment protocols and survival analysis for radiation therapy of brain tumors has been reviewed previously.[11] From these limited studies, median survival in dogs treated with conventional radiation therapy survival is also between 150 and 300 days.

Boron Neutron Capture Therapy (BNCT) has been performed on dogs with spontaneous brain tumors for over 10 years at our institution. From 1988, animals with radiologic or surgically confirmed neoplasms were entered in the BNCT protocol with the client's consent. From 1988 through 1992, dogs were treated with BNCT as the primary modality of therapy. From 1992 to the time of this writing BNCT has been used as therapy following surgical debulking. In an effort to better understand the role of surgery and BNCT therapy, median survival times were retrospectively calculated for two groups of dogs with spontaneous brain tumors. No significant numbers of control groups consisting of dogs with brain tumors not treated or dogs with brain tumors only surgically treated were available from our hospital population for statistical comparisons. The survival for the groups reported, however, provide some pilot information regarding surgery combined with BNCT for dogs with brain tumor.

EXPERIMENTAL DESIGN

Dogs with spontaneous brain tumors who received either BNCT alone or BNCT following surgical debulking were retrospectively grouped for analysis. Dogs were entered into either treatment group based upon client consent and therefore, were not randomized. Dogs were entered into the BNCT only group prior to the availability of surgical treatment. After 1992, dogs entered into the surgery and BNCT group had to 1) have a surgically confirmed brain tumor, and 2) live more than 21 days after surgery in order to have the BNCT therapy performed and to 3) die or be euthanized as a direct result of progression of their brain tumor. Dogs that did no live greater than 21 days after surgery were not included in survival times.

Thirty-three dogs with spontaneous brain tumors fulfilled these criteria and were treated with either BNCT (n = 20) or surgery (SX) followed by BNCT (n = 13). The effect on median survival of each treatment group was determined. Clinical signs varied at presentation but represented clinical abnormalities of intracranial dysfunction. Clinical signs primarily included seizures, behavior abnormalities, and, paresis. Tumors were presumptively diagnosed based upon imaging characteristics of computed tomography (CT) or magnetic resonance (MR) imaging. All dogs had histologic confirmation of tumor type either at the time of surgery or at necropsy. Of the dogs treated surgically, BNCT followed 3 to 6 weeks after surgery. Of the dogs in the surgery group, all had incomplete resections of the tumor primarily due to tumor location or infiltration of surrounding brain. In both groups, BSH or p-BPA were used for the boron agent. Dogs in the BNCT only group were treated with a radiation dose calculated to be equivalent to 10 or 12.5 Gy.[13] These dogs were all placed into the BNCT only group. Dogs in the BNCT + surgery group were treated with a radiation dose calculated to be equivalent to 12.5 Gy. Due to the size of the dogs' heads, all tumors were within 7 cm of the surface. The radiation beams were configured to provide the optimized target dose in relation to the peak normal brain dose.

RESULTS

Dogs of each group had various tumor types, however, the majority were meningiomas (13/20 BNCT; 9/13 SX + BNCT). All dogs in the BNCT only group and 11/13 dogs in the SX + BNCT group died or were euthanized due to progression of intracranial signs attributed to progressive tumor. Two dogs in the SX + BNCT group are alive at the time of this writing (540 and 660 days after therapy).

Data was analyzed and median survival in days was determined for both groups. Dogs receiving BNCT treatment alone had a median survival of 90 days (range 1 to 1860 days). Dogs receiving surgical biopsy/debulking followed by BNCT therapy had a median survival of 300 days (range 17 to 690 days). When dogs with meningiomas were analyzed independently, median survivals for the BNCT group were 75 days (range 1 to 180 days) and for the SX + BNCT group were 300 days (range 14 to 690 days). Direct comparisons of survival between groups was not performed as animals were not randomized into either treatment group, and therefore, contain different patient groups.

DISCUSSION

In this study, only the peak brain dose was calculated from measured blood boron levels. Radiation dose to the tumor was unknown. Estimates of tumor boron concentrations utilized by other investigators ignores the reality of heterogeneity of boron distribution. Importantly, dogs were treated to tolerance as calculated from their blood boron concentration.[14] Tumor doses, while unknown, could not be increased in these dogs given the boron compounds and the administration schedule utilized.

As this is a retrospective study of two separate groups of dogs, numerous variables and biases could have been introduced due primarily to the separate time eras of treatment. One obvious bias is that dogs with brain tumors that had significant morbidity or mortality associated with surgery were excluded from further therapy (follow-up BNCT), and therefore could have introduced a selection bias into the SX + BNCT group for those animals with the potential for longer survival. Additionally, treatments were not randomized and no control group of untreated dogs with spontaneous brain tumors were available in our hospital population.

Information from these dogs, however, suggests that the combination of surgery and BNCT therapy may provide for better outcomes in dogs with brain tumors than have historically been published in the veterinary literature. Some of this benefits may come from surgical stabilization of increasing intracranial pressure or other intracranial physiologic derangements due to the expanding brain mass. Alterations in tumor vascular flow dynamics may also be possible. The growth characteristics of the predominant tumor type may have also played a role. Additionally, it is possible that the benefits in survival were attributable to improvements in general patient management between the two eras. Regardless, these preliminary results suggest that BNCT used as therapy following incomplete surgical brain tumor removal may improve outcome in dogs. Median survival times with the combination of BNCT and surgery were approximately double those historically obtained for dogs with similar tumor and clinical characteristics treated with either surgery or surgery and conventional radiation therapy.

ACKNOWLEDGMENT

The authors gratefully acknowledge the assistance of Brookhaven National Laboratory, especially Dr. Jeff Coderre and Ms. Peggy Micca. The authors also thank Ms. Cyndi Johnson for her contributions. This work was supported in part by a grant from the Department of Energy, DE# FG03-98ER62567/AOO.

REFERENCES

1. Vandevelde M.: Brain tumors in domestic animals: An overview. Abstracts for Conference on Brain Tumors in Man and Animals. National Institute of Environmental Health Sciences, Research Triangle Park, NC, Sept 1984.
2. Bagley R.S., and Gavin P.R.: Unpublished observations, 1998.
3. Bagley R.S., Kornegay J.N., Page R.L., *et al.*: Central Nervous System. In: Slatter D., ed, Textbook of Small Animal Surgery, 2nd ed, Philadelphia: WB Saunders, 1993, pp. 2137–2166.
4. Patnaik A.K., Kay W.J., and Hurvitz A.I.: Intracranial meningioma: A comparative pathologic study of 28 dogs. Vet Path 23:369–373, 1986.
5. Braund K.G., and Ribas J.L.: Central nervous system meningiomas. Comp Contin Ed 8:241–248, 1986.
6. Tucker R.L., and Gavin P.G.: Brain imaging. Vet Clinic North Am 26:735, 1996.
7. Turrel J.M., Fike J.R., LeCouteur R.A., *et al.*: Computed tomographic characteristics of primary brain tumors in 50 dogs. J Am Vet Med Assoc 188:851–856, 1986.
8. Kraft S.L., Gavin P.R., DeHaan C., *et al.*: Retrospective review of canine intracranial tumors evaluated by magnetic resonance imaging. J Vet Int Med 11:218–225, 1997.
9. Heidner G.L., Kornegay J.N., Page R.L., *et al.*: Analysis of survival in a retrospective study of 86 dogs with brain tumor. J Vet Int Med 5:219–226, 1991.
10. Niebauer G.W., Dayrell-Hart B.L., and Speciale J.: Evaluation of craniotomy in dogs and cats. J Am Vet Med Assoc 198:89–95, 1991.
11. Gavin P.R., Fike J.R., and Hoopes P.J.: Central nervous system tumors. Seminars Vet Med Surg 10:180–189, 1995.

12. Brearley M.J., Jeffery N.D., Phillips S.M., and Dennis R.: Hypofractionated radiation therapy of brain masses in dogs: a retrospective analysis of survival of 83 cases (1991–1996). J Vet Int Med 13:408–412, 1999.
13. Gavin P.R., Wheeler F.J., Huiskamp R., DeHaan C.E., and Philipp K.H.I.: Large animal normal tissue tolerance to boron neutron capture: Comparision of empirical results with a microdosimetric model. Zamenhof R., Solares G., Harling O., eds, Topics in Dosimetry and Treatment Planning for Neurton Capture Therapy, Advanced Medical Publishing, Madison, WI, 1994, 205–216.
14. Gavin P.R., Kraft S.L., DeHaan C.E., Swartz C.D., and Griebenow M.L.: Large animal normal tissue tolerance with boron neutron capture. In J Rad Oncol Biol Phys 28:1099–1106, 1994.

QUANTITATIVE MRI IN ASSESSING IRRADIATION EFFECT IN DOG BRAIN

U. Abo Ramadan[1,2], J. Benczik[1,3], H. Joensuu[2], R. Joensuu[2], M. Kallio[2], L. Kankaanranta[2], S. Savolainen[1,2], M. Snellman[1], J. Tanttu[4], and M. Tenhunen[2]

[1]University of Helsinki, Finland
[2]Helsinki University Central Hospital, Finland
[3]Research Institute of Helsinki University Central Hospital, Finland
[4]Picker Nordstar, Inc., Finland

1. INTRODUCTION

The effect of radiation to the central nervous system has been studied frequently.[1–5] Early delayed injury occurs a few weeks to 3 months after therapy. Late radiation injury, which is observed a few months to 10 or more years after radiotherapy, is often irreversible, progressive, and is seen as focal injury or diffuse white matter injury. The degree of lesion and its characteristics depend upon the radiation dose, the time within which this dose is delivered, and the type of the target cell. The radiation response of normal brain has been investigated in a variety of animal models. Available data suggest that non-human primates and dogs are the best models of the human brain, with similar dose requirements to induce damage, and similar radiographic and histopathologic changes.[6]

Magnetic resonance (MR) imaging has provided new insights into the damage detection induced by radiation to the brain. One of the applications is the T_1 and T_2 relaxation time measurements or the change in signal intensity of T_1- and T_2-weighted images, respectively, within specific regions of interest in the organs (ROI).[2–4] Previous follow-up studies of normal brain tissue after radiation injury revealed that T_1 and T_2 relaxation times increased within the radiation targets.[2]

Whole brain irradiation of dogs with photons was carried out to establish adequate controls for normal tissue tolerance studies with boron neutron capture therapy (BNCT).[7,8] The aim of the study was to examine the time course of changes of T_2 relaxation times after different single doses of radiation to dogs' brain.

Frontiers in Neutron Capture Therapy, edited by Hawthorne *et al.*
Kluwer Academic / Plenum Publishers, New York, 2001.

2. MATERIALS AND METHODS

2.1. Animal Model

Fifteen adult Beagle dogs were irradiated with a 6 MV linear accelerator† and another five dogs served as controls.[9] Computerized tomographic images were used for treatment planning. The dogs were divided into three different dose groups with five dogs in each group. Doses of 10, 12 and 14 Gy to the 100% isodose were applied to the whole brain in a single fraction. The study protocol was approved by the appropriate ethics committees.

2.2. MR Imaging

All imaging was performed using a clinical 0.1 T MR system‡. A Beagle specific head coil was constructed for this purpose.[10] The sedated dogs were positioned in sternal recumbency during imaging.

After sagittal localization the spin-spin relaxation times (T_2) were measured using the spin echo sequence with TR of 1,500 and TE varying from 40 to 200 ms (five values). The matrix size was 180 × 256 and the field of view 180 × 256 with slice thickness of 7 mm. T_2 relaxation time measurements by using spin echo technique have limited applicability because of the effect of molecular diffusion. However, the T_2 relaxation times of the brain tissues are not long and therefore, the influence of diffusion on T_2 relaxation time measurements may be considered minor.

Standard ROI technique was employed for the calculations. For the measurements, the following signal intensity values were obtained from both the left and right hemispheres: white matter, and cortical gray matter ROIs at the level of hypophysis, and muscle tissue ROIs of the head. MR images were performed before and at 12, 18, 24, 30, 36, 42, 48, 56 and 68 weeks after irradiation to examine the time course of changes in T_2 relaxation times.

2.3. Data Analysis

With the use of a spin echo sequence, the tissue signal intensity may be described by the following equation providing the repetition time TR > 3 T_1 (spin-lattice relaxation time):

$$S \propto N_e^{-\frac{TE}{T2}} \tag{1}$$

where N is the spin density and TE is the echo time. T_2 were calculated by fitting the measured intensities to the exponential relaxation curves using the least-squares method.

---

†Varian Clinac 600c.

‡Merit, Picker Norolstar, Inc., Helsinki, Finland.

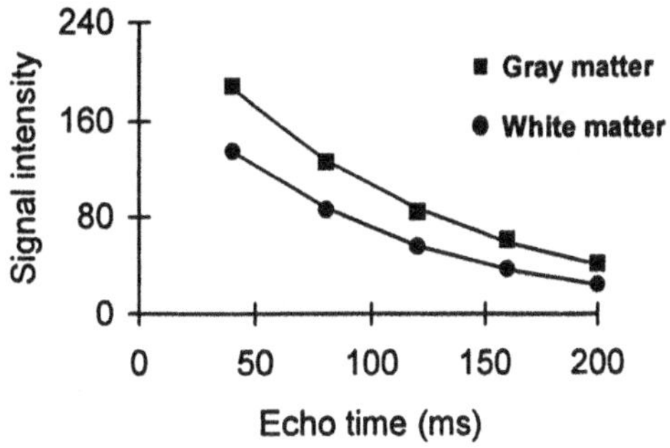

Figure 1. The signal intensity as a function of echo time in white- and gray matter of one control dog.

2.4. Statistical Analysis

Repeated measures analysis of variance method was used to compare T_2 values of white matter, gray matter and muscle before and at different times after irradiation.

3. RESULTS

Signal intensity of the spin echo sequence versus echo time in white and gray matter of one of the control dogs is illustrated in Figure 1. The T_2 of white matter, gray matter and muscle tissue of control dogs were 98 ± 6, 99 ± 6 and 34 ± 2 ms, respectively. The variation of T_2 ratios (T_2 after irradiation/T_2 before irradiation) in white matter is shown in Figure 2. T_2 of white matter varied significantly ($p < 0.05$) as a function of time in the dose groups of 12 and 14 Gy, but not in the 10 Gy group ($p > 0.05$). There is a significant interaction between doses and time ($p < 0.05$) which show different behavior for

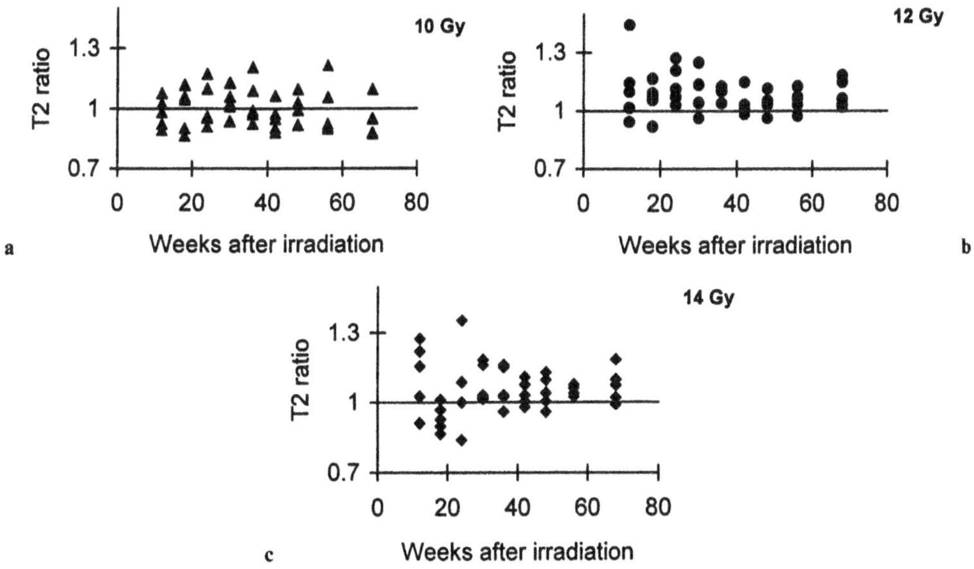

Figure 2a, b, and c. T_2 ratios as a function of time in brain white matter for doses of 10, 12, 14 Gy.

different doses as a function of time. There were no significant changes of T_2 in gray matter or muscle tissue in all the groups as a function of time ($p > 0.05$). All animals' neurological, blood and behavioral tests were normal during the 20 months follow up that is still continues.

4. DISCUSSION AND CONCLUSIONS

In biological tissues, the major determinants of the relaxation times are the tissue water content, and the degree and nature of interaction between water protons and protons bound in tissue macromolecules. Increase in tissue water content causes increase in mobile hydrogen density and prolongation of T_1 and T_2 relaxation times. Miot-Noirault *et al.*[2] have irradiated pig brain with a single dose of 25 Gy and the T_2 kinetic was analyzed as having three phases where the second and third phases were correlated with the detection of ventricular compression and necrosis, respectively. They suggested to detect a significant change in T_2 values close to the dose of 20 Gy. However, Fike *et al.*[5] demonstrated the existence of a tissue change that provided a sensitive measure of radiation damage over a dose range of 11.5–14.3 Gy with computerized tomography. In our work T_2 values of white and gray matter in brain and muscle tissue of all dogs before irradiation were calculated to establish normal base line values, and were in agreement with those found in the literature.[11] In our study the follow up to 20 months after irradiation shows significant change in T_2 of white matter for doses of 12 Gy or higher. T_2 relaxation times were demonstrated to be quantitative parameters which can be used for detecting lesions in the brain. The unclear pattern of alteration in T_2 may be related to the fact that the irradiation doses used (12–14 Gy) are in the threshold for MRI detectable tissue reactions due to irradiation.

REFERENCES

1. P.E. Valk, and W.P. Dillon, Radiation injury of the brain, *AJR.*, 12:45–62, 1991.
2. E. Miot-Noirault, S. Akokaka, D. Hoffschir, *et al.*, Potential of T2 relaxation time measurement for early detection of radiation injury to the brain: experimental study in Pigs, *AJNR.*, 17:907–912, 1996.
3. E.H. Lo, R.L. Delapaz, K.A. Frankel, *et al.*, MRI and PET of delayed heavy ion radiation injury in the rabit brain, *Int. J. Radiat. Oncol. Biol. Phys.*, 20:689–696, 1991.
4. K.M. Brennan, M.S. Roos, T.F. Budinger, *et al.*, A study of radiation necrosis and edema in the canine brain using positron emission tomography and magnetic resonance imaging, *Radiat. Res.*, 134:43–53, 1993.
5. J.R. Fike, E.C. Christopher, K. Turowski, *et al.*, Radiation dose response of normal brain, *Int. J. Radiat. Oncol. Biol. Phys.*, 14:63–70, 1988.
6. J.R. Fike, and G.T. Gobbel, Central nervous system radiation injury in large animal models, "Radiation Injury to the Nervous System", edited by P.H. Gutin, S.A. Leibel, and G.E. Sheline, Raven Press, New York, 1991, pp. 113–135.
7. R.F. Barth, A.H. Soloway, R.G. Fairchild, and R.M. Brugger, Boron Neutron Capture Therapy for Cancer, Realities and Prospects, *Cancer.*, 15:2995–3007, 1992.
8. J. Benczik, T. Seppälä, J. Hopewell, *et al.*, Large animal model for healthy tissue tolerance study in BNCT, *in* "Frontiers in Neutron Capture Therapy," M.F. Hawthorne, K. Shelly, R.W. Wiersema, eds., Kluwer Academic/Plenum Publishers, New York, 2001, pp. 1233–1238.
9. J. Benczik, M. Tenhunen, J. Hopewell, *et al.*, Irradiation of dog brain with single doses of X-rays, *in* "Frontiers in Neutron Capture Therapy," M.F. Hawthorne, K. Shelly, R.W. Wiersema, eds., Kluwer Academic/Plenum Publishers, New York, 2001, pp. 1269–1272.

10. M. Snellman, J. Benczik, R. Joensuu, U. Abo Ramàdan, J. Tanttu, and S. Savolainen, Low-field magnetic resonance imaging of the beagle brain with a dedicated receiver coil, *Vet. Radiol. & Ultras.*, 39(5), 1998.
11. Bottomley, T.H. Foster, R.E. Argersinger, and L.M. Pfeifer, A review of normal tissue hydrogen NMR relaxation times and relaxation mechanisms from 1–100 MHz: Dependence on tissue type, NMR frequency, temperature, species, excision, and age, *Med. Phys.*, 11:425–448, 1984.

IRRADIATION OF DOG BRAIN WITH SINGLE DOSES OF X-RAYS

J. Benczik[1,2], M. Tenhunen[3], J. Hopewell[4], M. Snellman[1], M. Kallio[5], L. Kankaanranta[3], and H. Joensuu[3]

[1]Faculty of Veterinary Medicine
Diagnostic Imaging
University of Helsinki, Finland
[2]Clinical Research Institute
Helsinki University Central Hospital
Helsinki, Finland
[3]Department of Oncology
Helsinki University Central Hospital
Helsinki, Finland
[4]Research Institute (University of Oxford)
The Churchill, Hospital, Oxford, UK
[5]Department of Neurology
Helsinki University Central Hospital
Helsinki, Finland

INTRODUCTION

It is not possible to selectively irradiate a single hemisphere of the dog brain with epithermal neutrons. Therefore, a whole brain irradiation study with photons was carried out to establish adequate controls for the normal tissue tolerance studies with boron neutron capture irradiation at the Finnish Research Reactor.[1] The primary objective of the study was to obtain a radiation-induced dose effect curve for visible changes on computerised tomography after photon irradiation and to calculate the ED_{50} to this change. Computerised tomography (CT), with contrast enhancement, was used to detect the changes in the brain, induced by radiation, after 20 months follow up. A secondary objective of the present study was to estimate possible differences in response between half brain and whole brain irradiation. The results of whole brain irradiation in dogs in the present study were compared to those obtained by Fike using half brain irradiation.[2,3] For immobilisation of the dogs during high-energy photon irradiation an anaesthetic agent that does not suppress the arterial oxygen saturation (SpO_2) is needed. In this study we used general

Frontiers in Neutron Capture Therapy, edited by Hawthorne *et al.*
Kluwer Academic / Plenum Publishers, New York, 2001.

anaesthesia accomplished with intravenous (iv.) propofol as the induction agent and isoflurane as the maintenance agent to ensure the oxygen saturation of haemoglobin.

MATERIALS AND METHODS

Twenty adult Beagle dogs, of both sexes and bred for research purposes, were used in the study. The appropriate ethics committees approved the experiment. Groups of five dogs were irradiated with single doses of 10, 12 or 14 Gy to the 100% isodose, using a Varian Clinac 600 C linear accelerator with a nominal dose rate of 2 Gy/minute. Five unirradiated dogs acted as controls for the follow up studies. CT images were used for radiation treatment planning. Only one treatment plan was used because there was no significant difference due to variation in head size. The dogs were positioned in sternal recumbence with a polyurethane pillow, both at the CT and the treatment facility. The doses delivered to the whole brain (10, 12, 14 Gy to the 100% isodose) in a single fraction, were comparable to the doses (10, 11.5, 13.5 Gy to the 95% isodose) used by Fike.[2,3] The treatment plan was designed to include the whole brain (the chiasma opticum and hypophysis, plus a small part of the spinal cord and the oral cavity). Other healthy tissues such as the eyes and most of the oral cavity were excluded. The dose uniformity was within ±5% in the brain region. The dose (100%) was prescribed at the isocenter.

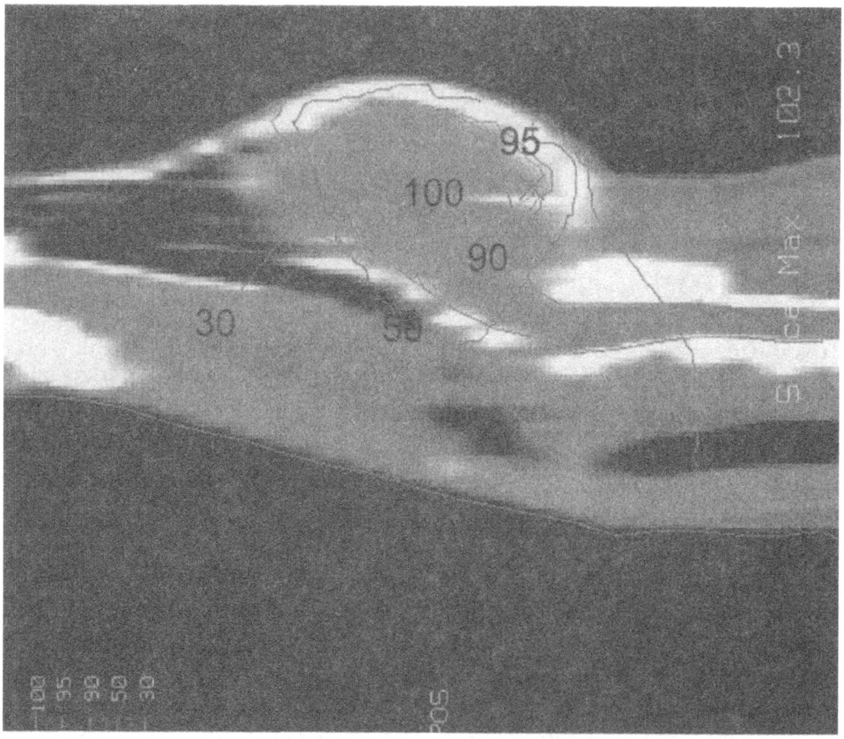

Figure 1. Sagittal view, showing the field of irradiation. The isodose curves are from the center: 100, 95, 90, 50 and 30%. The brain with the cranial part of the spinal cord is outlined.

For induction of general anaesthesia, propofol (Diprivan^R 10 mg/ml) was used as a 7 to 14 ml (7 mg/kg bw for 60 seconds) bolus iv. injection. After positioning (approximately 10 minutes) anaesthesia was maintained with isoflurane (3–1.8 volume %) in oxygen. The oxygen flow rate was 2 L/min, and the ventilation was spontaneous. SpO_2 and the pulse rate were recorded continuously with a pulse oximeter during the entire length of anaesthesia. The transmitting probe was placed on the tongue of each dog. The SpO_2 reading was considered to be accurate if the pulse oximeter's pulse rate was the same with the patient's palpable pulse rate. The colour and capillary refill time of mucous membranes, respiratory rate, auscultation of heart and lungs was followed until inhalation anaesthesia started.

For the follow up imaging a Philips SR 4,000 CT scanner was available. After a sagittal view for localisation, a non-contrast and contrast (iv. iopamidol, Iopamiro^R 370 mg I/ml, at a dose of 600 mg iodine/kg bw) scans of the whole brain were performed with 125 kV tube voltage. Transversal scans with a thickness of 5-mm contiguous slices were obtained beginning at the occipital protuberance and proceeding in a rostral direction to the cribriform plate. The matrix size was of 512 × 512 and the field of view 25-cm. There was a waiting time of 5 minutes after the injection of the contrast to let the non-ionic iodinated contrast agent to evenly distribute in the tissues. This allowed the detection of focal areas of retention of contrast media they were associated with the disruption of the blood-brain barrier or necrosis due to irradiation.[4] Assessments were made at 20 months after irradiation.

RESULTS

Propofol (7 mg/kg bw, iv.) provided adequate induction of anaesthesia for easy intubation. Anaesthesia was rapid and smooth without excitatory effects. The maintenance of iv. propofol anaesthesia until the start of inhalation anaesthesia was well tolerated by the dogs. Oxygen saturation was consistently greater than 75% during induction and maintenance of anaesthesia. Oxygen saturation recorded by the pulse oximeter was 100% in less than five minutes after transition to isoflurane anaesthesia. Maximal SpO_2 is usually observed in dogs during inhalant anaesthesia.[5]

There were no visible changes in CT of dogs in the two lowest dose groups. Two out of five dogs that received 14 Gy had well defined lesions visible in the 20 months postirradiation CT scans. One of these animals had a localised contrast enhanced region in the area of corpus striatum of the left hemisphere after iv. iopamidol contrast, surrounded by a low density ring. The other dog had a white matter lesion, dorso-lateral to the frontal part of the right centrum semiovale. This showed radiopacity on the non-contrast scan. After contrast application the radiodensity of the lesion increased. The neurological, blood and behavioural tests of all animals were normal during the 20 months follow up period. The degree of depigmentation of the hair in the irradiated area was greatest in the higher dose.

DISCUSSION AND CONCLUSIONS

For photon irradiation therapy the oxygenation of the tissues is critical. The oxygen saturation of haemoglobin may be slightly under the normal physiologic range (94.6 ± 2.0%) for a short time period during induction of anaesthesia with iv. propofol. The SpO_2

returns to the physiologic level during positioning of the animals. The fast return of maximal oxygen saturation is partly due to the rapid metabolism of the drug. Propofol, due to its favourable pharmacokinetic and dynamic profiles, has very mild and short cardiovascular effects (an arrhythmogen and negative inotropic effect). Propofol is a good hypnotic agent for induction of anaesthesia prior to maintenance with isoflurane for radiobiological studies.

The positive contrast enhanced changes probably results from a disruption of the blood-brain barrier, and a decreased density from oedema. At the 20 months follow up contrast enhancement, using iodinated contrast agent, was detected in two dogs', which received 14 Gy to the 100% isodose in the whole brain irradiation study. Fike reported similar changes in 2 dogs, which received 13.5 Gy to the 95% isodose to hemibrain[3] at 12 months follow up. Neurological examination or CT detected no differences in the two lower dose groups in the present study. However more subtle changes may be detectable with more sensitive techniques such as Flair, T2, or even gadolinium enhancement on magnetic resonance images. An additional group of five Beagles has been irradiated with 16 Gy to the 100% isodose in order to obtain a complete dose effect curve and to calculate an ED_{50} value for the photon irradiation study of whole brain. At present the follow up of the 16 Gy dose group is too short (3 months) to evaluate changes in the brain with non-invasive imaging methods.

ACKNOWLEDGMENTS

The work was supported by the Technical Development Centre of Finland, Research Foundation of Helsinki University Central Hospital (HUCH), Research grant of the Departments of Oncology and Neurology of HUCH, University of Helsinki, The Finnish Academy of Sciences and Finnish Centre for International Exchange Programmes.

REFERENCES

1. J. Benczik, T. Seppälä, J. Hopewell, M. Snellman, *et al.*, Large animal model for healthy tissue tolerance study in BNCT, *in* "Frontiers in Neutron Capture Therapy," M.F. Hawthorne, K. Shelly, R.W. Wiersema, eds., Kluwer Academic/Plenum Publishers, New York, 2001, pp. 1233–1238.
2. J.R. Fike, C.E. Cann, R.L. Davis, *et al.*, Computed tomography analysis of the canine brain: effects of hemibrain X irradiation, *Radiat Res.*, 99:294–310, 1984.
3. J.R. Fike, C.E. Cann, K. Turowski, *et al.*, Radiation dose response of normal brain, *Int J Radiat Oncol.*, 14(1):63–70, 1988.
4. M.Q. Bailey, Diagnostic imaging of intracranial lesions, Seminars in Veterinary Medicine and Surgery (Small Animal), Vol. 5, No. 4, pp. 232–236, 1990.
5. A. Bufalari, C.E. Short, C. Giannoni, *et al.*, Evaluation of selected cardiopulmonary and cerebral responses during medetomidine, propofol, and halothane anesthesia for laparoscopy in dogs, *Am J Vet Res.*, 58:1448, 1997.

A RAT MODEL OF ORAL MUCOSAL RESPONSE TO BORON NEUTRON CAPTURE IRRADIATION

G. M. Morris[1], J. A. Coderre[2], D. R. Smith[3], and J. W. Hopewell[1]

[1]Research Institute (University of Oxford)
Churchill Hospital, Oxford U.K.
[2]Medical Department
Brookhaven National Laboratory
Upton, New York
[3]Chemistry Department
Baker Laboratory
Cornell University
Ithaca, New York

INTRODUCTION

Oral mucositis is a frequent side effect during conventional radiotherapy for advanced head and neck tumours. While acute radiation damage to the oral mucosa is not a problem for the majority of glioma patients involved in the current boron neutron capture therapy (BNCT) clinical trial at BNL, a subset of patients with tumours located towards the front of the brain are presenting with transient acute effects in the oral mucosa. This suggests that the oral mucosa could become a potential dose limiting tissue in future treatment protocols involving higher doses and larger irradiation field sizes. The objective of the present study was to evaluate dose-effect relationships and estimate relative biological effectiveness (RBE) and compound biological effectiveness (CBE) factors for this critical normal tissue. The model selected for study was rat ventral tongue mucosa. Ventral tongue offers the most readily accessible area of representative multilayered oral epithelium. It can be locally irradiated (and readily scored), greatly reducing the radiation burden to the head.

MATERIALS AND METHODS

Experimental procedures were carried out on 12-week-old male Fischer 344 rats (260–290 g). The boron delivery agent used was p-boronophenylalanine (BPA)

Frontiers in Neutron Capture Therapy, edited by Hawthorne *et al.*
Kluwer Academic / Plenum Publishers, New York, 2001.

containing 4.9% boron-10 (^{10}B). It was dissolved in a fructose solution, as described previously (Coderre *et al.*, 1994), and administered by intraperitoneal (i.p.) injection. Both tissue sampling and irradiations were carried out at 3 h after administration of 700 mg/kg of BPA.

Irradiations with 100 kVp x-rays were carried out using a Philips RT-100 radiotherapy unit at a focus to tongue distance of 20 cm and 1.7 mm Al filtration. The circular irradiation field was restricted to a 20 mm diameter by lead shielding. The x-ray dose rate was 218 cGy/min.

BNC irradiation was carried out on the thermal beam of the Brookhaven Medical Research Reactor (BMRR). The physical dose rates (cGy/MWmin) were 2.84 (per $\mu g^{10}B/g$) for the $^{10}B(n,\alpha)^{7}Li$ reaction; 9.90 for the fast neutron interaction with hydrogen $^{1}H(n,n)p$; 7.00 for the nitrogen neutron capture reaction $^{14}N(n,p)^{14}C$; and 4.80 for the gamma ray component.

Boron levels in blood and oral mucosal tissue were measured using prompt gamma or inductively coupled atomic emission spectrometry. The microlocalisation of ^{10}B in mucosal tissue was evaluated using ion microscopy.

A 0.5 ml blood sample was taken immediately prior to thermal neutron irradiation, and the boron content determined. On average, 8 rats were irradiated per dose group. The various dose groups are detailed in Table 1. The irradiation field (10 mm diameter) was located on the ventral surface of the tongue. Ulceration was the biological endpoint used to establish dose effect curves. The doses required to produce a 50% incidence of ulceration were calculated from these curves using probit analysis. These ED_{50} values were used to estimate biological effectiveness. For all data presented, errors indicate the standard error of the mean (±SE).

Table 1. Thermal beam exposure times, blood ^{10}B content at the time of irradiation and physical absorbed doses for the various treatment groups[a]

Irradiation	Thermal beam exposure time (MW min)	Blood ^{10}B content (μg/g)	Total dose (Gy)
X-rays	—	—	11.6
X-rays	—	—	13.1
X-rays	—	—	13.8
X-rays	—	—	14.5
X-rays	—	—	16.0
Beam-only	14.5	—	3.2
Beam-only	16.1	—	3.5
Beam-only	17.9	—	3.9
Beam-only	19.7	—	4.3
Beam-only	21.5	—	4.7
Beam-only	30.2	—	6.6
Beam + BPA	2.8	19.4 ± 0.4	2.2
Beam + BPA	3.3	23.1 ± 0.7	2.9
Beam + BPA	3.8	20.8 ± 0.8	3.1
Beam + BPA	4.3	19.7 ± 0.6	3.3
Beam + BPA	4.8	20.4 ± 0.5	3.8
Beam + BPA	5.2	21.0 ± 0.5	4.2
Beam + BPA	6.6	17.3 ± 0.6	4.7
Beam + BPA	7.6	20.6 ± 1.6	6.1

[a]Errors indicate ± SE.

Table 2. Evaluation of ^{10}B microdistribution in ventral tongue mucosa using ion microscopy

Boron delivery agent	Blood (μg/g)	Epithelium (μg/g)	Lamina propria (μg/g)	Muscle (μg/g)	Blood vessel wall (μg/g)
BPA	20.7 ± 0.7	21.5 ± 0.5	31.1 ± 1.2	15.6 ± 1.7	19.2 ± 1.6

RESULTS AND DISCUSSION

Blood ^{10}B levels ranged from (17–23 μg/g) throughout the study (Table 1). At ~3 h after the i.p. injection of BPA the following concentrations of ^{10}B (μg/g) were measured for oral tissues: ventral tongue 22.7 ± 0.6; dorsal tongue 22.9 ± 0.5; soft palate mucosa 22.3 ± 0.7; buccal mucosa 24.1 ± 0.6. The microdistribution of ^{10}B was evaluated using ion microscopy (Table 2). Levels of ^{10}B were comparable in the lumens and walls of blood vessels and also in the epithelium. The concentration of ^{10}B in the lamina propria was significantly higher ($p > 0.1$, Student's t test). The ^{10}B partition ratio (tissue : blood) were 1 : 1 and 1.5 : 1 for epithelium and lamina propria, respectively.

The time of onset of tongue ulceration was 6 days, irrespective of the irradiation modality. The percentage of tongue fields developing ulceration was dose dependent, and the peak incidence of ulceration was between 6 and 9 days after irradiation. The dose-related incidences of ulceration are shown in Figure 1. ED_{50} values calculated from these dose-effect curves are listed in Table 3. The relative biological effectiveness (RBE) of the thermal neutron beam was calculated using the appropriate ED_{50} values for the beam-only and X-irradiation groups. In defining the biological effects of the $^{10}B(n,\alpha)^7Li$ neutron capture reaction relative to photons, the term compound biological effectiveness factor was used as an alternative to RBE. CBE factor = [X-ray ED_{50} – (thermal beam component of ED_{50} xRBE)]/$^{10}B(n,\alpha)^7Li$ component of ED_{50}.

The thermal beam had a relatively high RBE of 3.2 due to the fast neutron component of the beam and the induction of protons in the mucosal tissue. The CBE factor for BPA was 4.9.

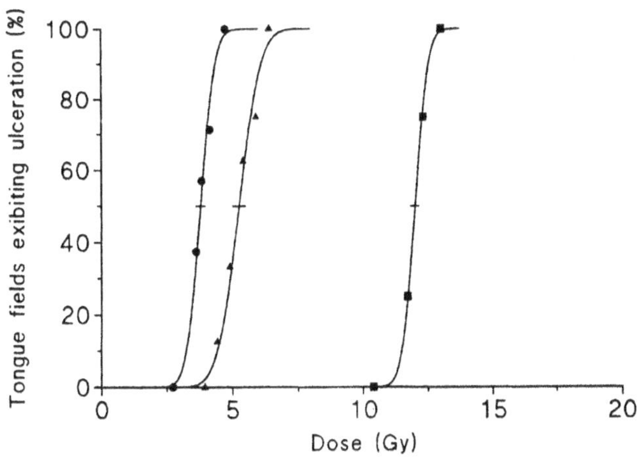

Figure 1. Dose related changes in the incidence of tongue ulceration. ■ = x-rays; ▲ = beam-only; ● = beam + BPA.

Table 3. ED_{50}, RBE and CBE factor values for the irradiation modalities using tongue ulceration as the end point

Irradiation modality	ED_{50}	RBE/CBE factor
X-rays	13.4 ± 0.2	—
Thermal beam	4.2 ± 0.1	3.2 ± 0.1
Beam + BPA	3.0 ± 0.1	4.9 ± 0.2

Others have reported similar findings to those of the present study, with regard to the timing and incidence of tongue ulceration. In the case of the ventral tongue of the mouse, single exposures of 29 kV x-rays at doses in excess of 15 Gy, resulted in the development of ulcers by days 9 and 10, with peak incidence on days 10 and 13 (Moses and Kummermehr, 1986; Dorr and Kummermehr, 1990). Mucositis was observed in mouse lips at 9 and 10 days, with peak incidence at 10–13 days, after single dose irradiation with 250 kV rays or ^{60}Co (Parkins *et al.*, 1983; Ang *et al.*, 1984).

The present study is the first to obtain a CBE factor estimate for oral mucosa. A limited number of reports have documented RBE and CBE factors for skin, using moist desquamation as the endpoint. For thermal beams, RBE values in the range 2.7–3.9 have been reported (Yamamoto, 1961; Archambeau, 1989; Morris *et al.*, 1994). Employing BPA as the boron delivery agent, CBE factors of ~2.5 have been estimated for hamster and human skin (Hiratsuka *et al.*, 1991; Fukuda *et al.*, 1994). This is a somewhat lower estimate than that reported for rat skin (3.7) using BPA (Morris *et al.*, 1994a). Data from a parallel ion microscopy study using borocaptate sodium (BSH) indicate that considerably less ^{10}B accumulates in the mucosal epithelium, relative to BPA. Indeed epithelium : blood ^{10}B partition ratios are 1 : 1 for BPA and 1 : 0.2 for BSH. This essentially accounts for the fact that the estimated CBE factor for BSH (~0.5) is considerably lower than that for BPA.

CONCLUSIONS

It can be concluded that the rat ventral tongue mucosa is a representative model for the oral mucosa. Comparison of previously published data from clinical studies (Parkins *et al.*, 1983; van der Schuren *et al.*, 1983) where patients were given a small number of fractions given within a week, suggest that the response of the oral mucosa to photon irradiation is comparable in the rat and man. The CBE factor calculated in the present study on oral mucosa, using BPA as the boron delivery agent, is therefore likely to be clinically applicable. This is, of course, assuming that the biodistribution profiles of BPA are similar in these two species. From a clinical perspective, the present findings indicate that oral mucosal tolerance to BPA mediated BNCT is relatively low. This will need to be taken into consideration in future clinical treatment protocols involving higher doses or larger field sizes.

ACKNOWLEDGMENTS

This work was supported by the Cancer Research Campaign. Additional support was provided by the Office of Health and Environmental Research, US Department of Energy under contract number DE-EACO276CH00016.

REFERENCES

Ang K.K., Xu F.X., Vanuytsel L., and van den Schueren E. Repopulation kinetics in irradiated mouse lip mucosa: the relative importance of treatment protraction and time distribution of irradiation. Radiat. Res. 1985; 101:162–169.

Archambeau J.O. A model to evaluate dose recovery from different radiation. In: Fairchild R.G., Bond V.P., and Woodhead A.D. (eds). Clinical Aspects of Neutron Capture Therapy. New York: Plenum Press, 1989, 9–20.

Coderre J.A., Button T.M., Micca P.L., Fisher C.D., Nawrocky M.M., and Liu H.B. Neutron capture therapy of the 9L rat gliosarcoma using the p-boronophenylalanine-fructose complex. Int. J. Radiat. Oncol. Biol. Phys. 1994; 3:643–652.

Dorr W., and Kummermehr J. Accelerated repopulation of mouse tongue epithelium during fractionated irradiations or following single doses. Radiother. Oncol. 1990; 17:249–259.

Fukuda H., Hiratsuka J., Honda C., Kobayeshi T., Yoshino K., Karashima H., *et al.* Boron neutron capture therapy of malignant melanoma using ^{10}B-paraboronophenylalanine with special reference to evaluation of radiation dose and damage to the skin. Radiat. Res. 1994; 138:435–442.

Hiratsuka J., Fukuda H., Kobayashi T., Karashima H., Yoshino K., Imajo Y., and Mishima Y. The relative biological effectiveness of ^{10}B neutron capture therapy for early reactions in the hamster. Radiat. Res. 1991; 128:186–191.

Morris G.M., Coderre J.A., Hopewell J.W., Micca P.L., and Rezvani M. Response of rat skin to boron neutron capture therapy with p-boronophenylalanine or borocaptate sodium. Radiother. Oncol. 1994; 32:144–153.

Moses R., and Kummermehr J. Radiation response of the mouse tongue epithelium. Br. J. Cancer. 1986; 53 Suppl. VII:12–15.

Parkins C., Fowler J.F., and Shen Y. A murine model of lip epidermal/mucosal reactions to X-irradiation. Radiother. Oncol. 1983; 1:159–165.

Yamamoto Y.L. The biological effectiveness of thermal neutrons and of the heavy particles from the $^{10}B(n,\alpha)^{7}Li$ reaction of the rabbit's ear and its utilisation for neutron capture therapy. Yokshoma Medical Bulletin. 1961; 12:4–22.

van der Schueren E., van den Bogaert W., and Ang K.K. Radiotherapy with multiple fractions per day. In: The Biological Basis of Radiotherapy. Edited by G.G. Steel, G. Adams, and M. Peckam (Elsevier Science: Amsterdam), pp. 195–210.

SPLIT-DOSE BNCT IRRADIATIONS OF RAT ORAL MUCOSA AND RAT BRAIN TUMOR USING BPA

Effects on Tolerance and Survival

Jeffrey A. Coderre[1], Gerard M. Morris[2], Ruimei Ma[1], Peggy L. Micca[1], Diana T. Lombardo[1], Kellyann M. Youngs[1], and John Kalef-Ezra[3]

[1]Medical Department
Brookhaven National Laboratory
Upton, New York 11973
[2]Research Institute
University of Oxford
Churchill Hospital
Oxford OX3 7LJ, UK
[3]Medical Physics Laboratory
Medical School
University of Ioannina
Ioannina, Greece

1. INTRODUCTION

Some BNCT patients irradiated with anterolateral beams have experienced mild, transient side effects in the oral mucosa, the ear mucosa and the parotid salivary gland. As the BPA-based BNCT clinical dose escalation trial continues, the possibility arises that oral mucosa or soft tissues in the head and neck, may become the critical dose-limiting tissue in some cases. The rapid turnover of mucosal epithelium suggests that a sparing effect could be achieved during fractionated BNCT through a repopulation mechanism. The effect of fractionated BNCT on the tumor must also be considered. Retargeting of boron to a different subset of tumor cells during the second fraction may make the two fraction treatment more effective than the single treatment. Tumor regrowth during the interfraction interval could make the two-fraction therapy less effective than the single treatment. Rat ventral tongue mucosa was chosen as a model system for oral

Frontiers in Neutron Capture Therapy, edited by Hawthorne *et al.*
Kluwer Academic / Plenum Publishers, New York, 2001.

mucosa radiobiology. Using the thermal neutron beam of the Brookhaven Medical Research Reactor, the single-fraction dose-response of the oral mucosa and the intracranial rat 9L gliosarcoma was compared to the same total dose given in two fractions separated by 3, 5, 7, or 9 days.

2. MATERIALS AND METHODS

Male Fischer 344 rats, approximately 12 weeks of age and weighing 260–290 grams, were used for both the oral mucosa and the brain tumor studies. Rats were anesthetized by i.p. injection of ketamine (54 mg/kg) and xylazine (9 mg/kg). *p*-Boronophenylalanine (BPA) containing 4.9% ^{10}B (Boron Biologicals, Inc., Raleigh, NC, USA) was dissolved in a fructose solution and administered by intraperitoneal (i.p.) injection, as described previously.[1] The collimator used for both the oral mucosa and the brain tumor irradiations consisted of Li_2CO_3 dispersed in polyethylene, 93% enriched in ^6Li, 7.5% total Li by weight, 10.2 cm thick, with a conical aperture 12 cm in diameter at the reactor side tapering to 2 cm diameter at the rat irradiation point.[2]

2.1. Oral Mucosa Model

Rat ventral tongue mucosa was chosen as a model system for oral mucosa radiobiology. Rats were anesthetized, the tongue was gently extruded and the distal 1 cm was fixed between two pieces of tape. A blood sample (≈0.5 ml) was taken from the retro-orbital sinus of each rat immediately prior to thermal neutron irradiation, and the boron content determined using prompt gamma spectrometry. On average, 8 rats were irradiated per dose group. Irradiation commenced between 3–4 h after BPA administration. Each rat was positioned on its back parallel to the beam, with the head angled to one side such that the ventral surface of the taped tongue could be positioned over the collimator aperture. The tongue reaction was observed daily, under light O_2/CO_2 anaesthesia. The incidence of blister formation was used as the primary endpoint and was evident by 6–7 days post-irradiation. A scoring system that included the response of the tongue top, tongue bottom and the degree of salivation/mucous formation was also employed. Each of the three sites was scored on a scale from 0 (no reaction) to 3 (severe reaction) making a score of 9 the maximum possible for a given day.

2.2. Brain Tumor Model

Procedures for implantation of the intracranial 9L gliosarcoma have been previously described.[1,3] Irradiation commenced approximately 3 hours after BPA-F injection. Therapy irradiations were done 14 days post-implantation when tumors ranged in weight from 40 to 50 mg. For the split-dose irradiations, the first fraction was delivered on day 14 post-implantation. Sixty eight rats were used for these irradiations: 9 controls, 8 for single dose, 14 for 3 day split, 11 for 5 day split, 13 for 7 day split and 13 for 9 day split.

3. RESULTS AND DISCUSSION

3.1. BPA-F Biodistribution

Ten Fischer 344 rats were given a single i.p. injection (4 ml) of BPA-fructose (50 mg BPA/ml) that delivered approximately 700 mg BPA/kg. Blood boron concentration at 3 hours after injection was 20.8 ± 1.2 μg ^{10}B/g. At ~3 h after the i.p. injection of BPA the following concentrations (μg ^{10}B/g) were measured for oral tissues: ventral tongue, 22.7 ± 1.8; dorsal tongue, 22.9 ± 1.7; soft palate mucosa, 22.3 ± 2.0; buccal mucosa, 24.1 ± 2.1. Measured blood ^{10}B levels from 33 groups of rats averaged 19.6 ± 1.6 (mean ± SEM). For estimation of the dose to the oral mucosa, the boron concentration in mucosa was assumed to be 23 μg ^{10}B/g.

For BNCT of intracerebral tumors, rats received a single i.p. injection (0.012 ml/gram body weight) of BPA-fructose (100 mg BPA/ml) that delivered 1200 mg BPA/kg. A previous report of boron concentrations under these conditions showed 89.6 ± 7.6, 27.7 ± 2.8, and 17.5 ± 1.5 μg ^{10}B/g in tumor, blood and brain, respectively (Coderre, 1994). The average blood boron concentration at the time of BNCT for all of the rats in the brain tumor experiments (both irradiations in split-dose groups) was 24.2 ± 2.8 (mean ± SD, n = 113). A tumor boron concentration of 89.6 μg/g was assumed for the estimation of tumor dose.[1]

3.2. Oral Mucosa Irradiations

The thermal neutron fluence at the ventral tongue surface was 5.46×10^9 n/cm^2/s, with the reactor running at 1 MW power. The boron dose was calculated from the following equation: boron dose (Gy) = (8.66 × 10E-8)(F)(flux) where F = weight fraction of boron in the tissue (g ^{10}B/g tissue) and flux = n/cm^2. The physical dose rates at the tongue ventral surface (Gy/MW-min) were 2.84 (per μg ^{10}B/g) for the $^{10}B(n,\alpha)^7Li$ reaction; 9.9 for the fast neutron interaction with hydrogen $^1H(n,n)p$; 7.0 for the nitrogen neutron capture reaction ($^{14}N(n,p)^{14}C$); and 4.8 for the total gamma component. For 23 μg ^{10}B/g in the tongue, the total dose rate was 87 cGy/MW-min (~75% from the boron capture reaction, and ~94.5% from heavy charged particles.

For all split-dose irradiations, the first fraction was 2.25 Gy, and the second fraction was varied to deliver the indicated total dose. Acute reactions were evident in the tongue by 6 days after the second fraction. They were characterized by erythema, edema, an area of mucosa with a white appearance, loose skin, and at higher doses ulceration. Ulcers tended to be located toward the base of the tongue, and less frequently the tip. They did not extend over the entire irradiation field. The degree of incidence and severity of ulceration was dose-related. Using ulceration on the tongue ventral surface as the endpoint, the ED_{50} was 3.8 Gy when the irradiation was given in a single fraction and 4.8 Gy when the irradiation is given in 2 fractions with a 9 day interval between the fractions. There was considerable scatter in the data from the split dose groups with 3, 5, and 7 day splits, making it impossible to estimate ED_{50} doses for these conditions. These experiments are being repeated. The use of an ulceration endpoint does not provide any indication of the severity of the reaction. Animals with a very mild response are scored the same as animals with severe ulceration. The use of a response index produced a better indication of the dose response. The reaction of the oral mucosa was scored using the top of the tongue, the bottom of the tongue and the degree of salivation/mucous

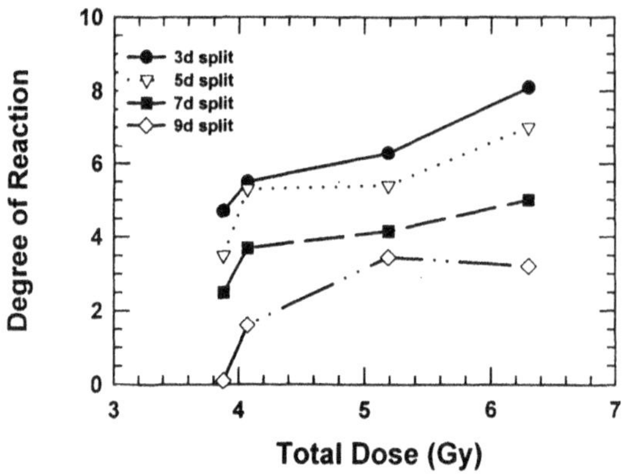

Figure 1. Dose response following BNCT irradiation of the rat ventral tongue using thermal neutrons plus BPA.

formation. Figure 1 shows that there is a progressive increase in the degree of reaction as a function of dose, and that increasing the time interval between the two doses produces a sparing effect.

3.3. Rat Brain Tumor Irradiations

The irradiation geometry and the platform to hold the rat have been reported.[2] At 1 MW reactor power, the entrance thermal neutron flux was 7.17 × $10^9 cm^{-2}s^{-1}$. Assuming that the tumor extends from 3 mm to 7 mm beneath the skin surface, the geometrical mean of the thermal neutron fluence rate in the tumor volume was 5.1 × $10^9 cm^{-2}s^{-1}$ and the minimum fluence rate was 4.4 × $10^9 cm^{-2}s^{-1}$. At 1 MW reactor power, the mean dose rates to the tumor volume were: boron capture, 2.65 cGy/min/μg ^{10}B; fast neutrons, 9.0 cGy/min; total gamma, 4.6 cGy/min; nitrogen capture, 6.1 cGy/min. Assuming 89.6 μg ^{10}B/g tumor,[1] the mean total dose rate was 2.57 Gy/MW-min, ~92% due to boron capture reactions, and ~98% due to heavy charged particles. All rats received 4 MW-min of reactor exposure: the single fraction group was irradiated for 4 min at 1 MW reactor power; the split dose groups were irradiated for 4 minutes at 500 kW reactor power for each dose. This 4 MW-min exposure corresponds to 10.28 Gy mean dose to the tumor. If the biological effectiveness factors derived in this brain tumor model (3.8 for the boron capture dose, 3.2 for fast neutrons, 3.2 for the nitrogen capture dose) are used to estimate the photon-equivalent (Gy-Eq) doses,[1,4] the tumor dose of 10.28 Gy corresponds to a dose of 38.2 Gy-Eq. Survival following a single fraction of BPA-based BNCT consisting of 4 MW-min of reactor exposure produced 50% long-term survival (Figure 2). A survival level in the 50% range was used in order to assess both potential positive and negative effects of split-dose irradiations on the overall survival. There was no trend either towards improved survival or towards poorer survival when the same total dose was divided into 2 equal fractions separated by 3, 5, 7 or 9 days (Figure 2).

These preliminary studies indicate that split-dose irradiation is a viable option for reducing BNCT side effects in mucosa of the head and neck.

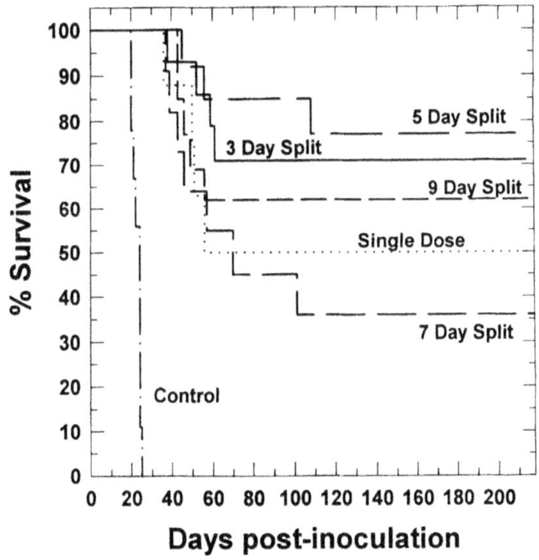

Figure 2. Percent survival of rats with intracerebral 9L gliosarcomas following BPA-based BNCT. The same total reactor exposure of 4 MW-min was delivered either in a single fraction or in two equal fractions divided by 3, 5, 7, or 9 days.

ACKNOWLEDGMENTS

This work was supported, in part, by the U.S. Department of Energy under Contract # DE-AC02-98CH10886.

REFERENCES

1. J.A. Coderre, T.M. Button, P.L. Micca, C.D. Fisher, M.M. Nawrocky, and H.B. Liu, Neutron capture therapy of the 9L rat gliosarcoma using the *p*-boronophenylalanine-fructose complex, *Int. J. Rad. Oncol. Biol. Phys.*, 30:643–652, 1994.
2. H.B. Liu, D.D. Joel, D.N. Slatkin, and J.A. Coderre, Improved apparatus for neutron capture therapy of rat brain tumors, *Int. J. Radiat. Oncol. Biol. Phys.*, 28:1167–1173, 1994.
3. D.D. Joel, R.G. Fairchild, J.A. Laissue, S.K. Saraf, J.A. Kalef-Ezra, and D.N. Slatkin, Boron neutron capture therapy of intracerebral rat gliosarcomas. *Proc. Natl. Acad. Sci., USA* 87: 9808–9812, 1990.
4. J.A. Coderre, M.S. Makar, P.L. Micca, M.M. Nawrocky, D.D. Joel, D.N. Slatkin, and H.I. Amols, Derivations of relative biological effectiveness for the high-LET radiations produced during boron neutron capture irradiations of the 9L rat gliosarcoma *in vitro* and *in vivo*, *Int. J. Rad. Oncol. Biol. Phys.*, 27:1121–1129, 1993.

EFFICACY OF BORON NEUTRON-CAPTURE THERAPY USING A CARBORANE-CONTAINING TETRAPHENYLPORPHYRIN IN MICE BEARING EMT-6 TUMORS

Michiko Miura[1], Peggy L. Micca[1], Diana T. Lombardo[1], Kelly M. Youngs[1], Gerard M. Morris[2], John A. Kalef-Ezra[3], Ruimei Ma[1], and Jeffrey A. Coderre[1]

[1]Medical Department
Brookhaven National Laboratory
Upton, New York 11973
[2]Research Institute
Churchill Hospital
Oxford, OX37LJ, United Kingdom
[3]Medical School
University of Ioannina
Ioannina, Greece

1. INTRODUCTION

Although numerous boron-containing compounds have been synthesized for boron neutron-capture therapy (BNCT), only a small fraction have been tested *in vivo*. Of this fraction, only a handful have shown biological properties favorable enough to warrant a therapeutic study. BNCT studies of tumors in rodent models have only been reported using BPA,[1,2] BSH,[3] and BSSB.[4] Herein, we describe a study using porphyrin-mediated BNCT to palliate tumors in mice.

As a general rule, biodistribution studies from animal studies should show an average of >30 μg B/g in tumor with a 5:1 tumor:normal tissue boron concentration ratio for BNCT to be effective. Requirements may vary depending on the intracellular localization properties. Porphyrins are of interest because of their use in photodynamic therapy (PDT), another bimodal cancer therapy where they are used as photosensitizers. However, 100 to 1000 times more compound is typically needed for BNCT compared to PDT, so that toxicity, particularly thrombocytopenia, becomes a major issue. Thrombocytopenia, as well as abnormal hepatic and renal enzyme levels have been observed with VCDP,[5] BOPP,[6] and BTPP[7] when therapeutic doses are administered.

Frontiers in Neutron Capture Therapy, edited by Hawthorne *et al.*
Kluwer Academic / Plenum Publishers, New York, 2001.

NiTCP was the first porphyrin to deliver 40 μg B/g or greater to tumor tissue and not cause overt toxicity, when given to mice or rats at total doses up to ≈250 mg/kg body weight.[8] Because NiTCP is lipophilic, it was emulsified in Cremophor EL (CRM), propylene glycol (PrG) and saline for *in vivo* experiments. It was recently shown that emulsifying porphyrins with CRM and PrG does not necessarily render them non-toxic.[9] It was also shown that another porphyrin structure NiTCP-H, and similarly CuTCP-H, showed even higher boron uptake than NiTCP in the EMT-6 tumor.[9] Boron concentrations of 100 μg B/g were typically found in EMT-6 tumor tissue following administration of ≈200 μg CuTCPH/g body weight.

We now report the results of BNCT using CuTCPH. The ^{10}B-enriched porphyrin was administered to mice bearing EMT-6 tumors on the leg, which were then irradiated with thermal neutrons at the Brookhaven Medical Research Reactor (BMRR). Controls for the BNCT group consisted of similar tumor-bearing mice that 1) received no treatment; 2) received thermal neutron irradiation only at different doses; or 3) received 100 kVp x-irradiation at different doses.

2. MATERIALS AND METHODS

2.1. Compounds

The synthesis of CuTCP-H was similar to that described for NiTCP-H[9]. The only difference was that copper (II) acetate was used in the metallation step instead of nickel (II) acetate. Metallation progress was monitored using optical spectroscopy and thin layer chromatography. The desired porphyrin was purified by flash chromatography using silica and 1:1 hexane:dichloromethane as eluent. A 9% CRM and 18% PrG solution of CuTCP-H (3.7 mg/mL) was prepared as described.[9]

2.2. Animals

Female BALB/c mice weighing 18–22 g were implanted subcutaneously on the leg with EMT-6 tumor fragments. After ≈10 days, mice were sorted into 11 groups, each with a similar distribution of tumor volumes. Five groups received ^{10}B-enriched CuTCP-H at a total dose of 186 μg/gbw via serial i.p. injections (6 over 32 hr). Two days later, 4 groups were irradiated with thermal neutrons at the BMRR at 15, 20, 25, and 30 MW-min (≈39, 53, 65, 79 Gy). The 5th group was euthanized for determination of boron concentration in various tissues.

Three groups of mice that were given no porphyrin were irradiated with thermal neutrons only at the BMRR at: 30, 40, and 50 MW-min (≈7.0, 9.3, 11.6 Gy).

Four groups that were given no porphyrin were irradiated with 100 kVp xrays (Philips RT-100 therapy unit) at doses of 21, 28, 35, and 42 Gy.

Tumors were measured twice a week and mice were euthanized when tumor volumes reached 500 mm^3 or when severe radiation damage was observed.

2.3. Reactor Irradiation Procedure

Mice were anesthetized with sodium pentobarbitol (≈60 μg/g body weight by i.p. injection). Each mouse was then supported on a masking tape sling with the tumor-bearing leg extended across the 2-cm aperture of the collimator. The 25 × 25 cm thermal port at the BMRR is fitted with a 10-cm thick lithiated polyethylene collimator, which has an aperture of 12-cm diameter at the port face, narrowing to 2.0 cm. Irradiations were carried out at 3 MW reactor power.

2.4. X-ray Irradiation Procedure

Similar to the reactor irradiation procedure, mice were anesthetized and positioned behind a 1-cm thick lead shield with the tumor-bearing leg extended across a 2-cm aperture in the lead shield. Mice were irradiated at 2.10 Gy/min with a Philips RT-100 therapy unit at 100 kV/8mA using 1.7 mm aluminum filtration and a focus-to-skin distance of 20 cm.

2.5. Dosimetry

An MDH Radcal xray monitor and ion chamber, and aluminum foils were used for dosimetry of the Philips RT-100 xray unit. The mean tumor dose rate determined for xrays was 2.10 Gy/min (HVL—2.7 mm Al). Ion chambers, thermoluminescent dosimeters and gold-foils for activation were used for dosimetry of the BMRR.[10] The BMRR dose components for a 2-mm-thick leg tumor are estimated as (cGy/MW-min): gamma photons 4.9; fast neutrons 9.6, nitrogen capture 8.8 and boron capture 240 (using 85 μg ^{10}B/g).

3. RESULTS AND DISCUSSION

Tissues of mice that were euthanized 2 days after the last injection, the same day the BNCT mice were irradiated, showed average boron concentrations of 85 ± 13 μg B/g in tumor, 13 ± 3 μg B/g in skin, and 4 ± 2 μg B/g in blood.

3.1. BNCT Groups

The survival curves for the x-irradiated groups are shown in Figure 1a, those for the neutron-only irradiated groups are shown in Figure 1b, and the CuTCPH-mediated BNCT groups are shown in Figure 1c. The biological effects exhibited by mice are summarized in Table 1. The greatest tumor control was demonstrated by CuTCPH-mediated BNCT at 65 Gy (25 MW-min) where 5 of 7 mice treated showed no signs of tumor regrowth after more than 200 days after treatment. The remaining 2 mice were euthanized within 90 days postirradiation due to tumor overgrowth. No mice in this group showed any signs of radiation syndrome. A 20% increase in neutron-irradiation exposure (30 MW-min, estimated dose of 79 Gy) showed a dramatic whole-body dose response wherein 4 of 7 mice died within 1 week after irradiation, most likely from gastrointestinal radiation syndrome. When mice appeared ill ≈6 days postirradiation, they were given nutritional supplements (FreAmine and dextrose) by daily i.p. injections for about 1 week. A 20% decrease in the neutron-irradiation exposure (20 MW-min, estimated dose of 53 Gy) showed a substantial increase in the number of mice euthanized for tumor overgrowth, allowing survival of only 2 in 7 mice.

The survivors of the 65-Gy BNCT group showed no signs of skin damage other than temporary alopecia. There were no signs of moist desquamation nor dermal necrosis in any of the BNCT-treated mice at follow-up times of ≈200 days.

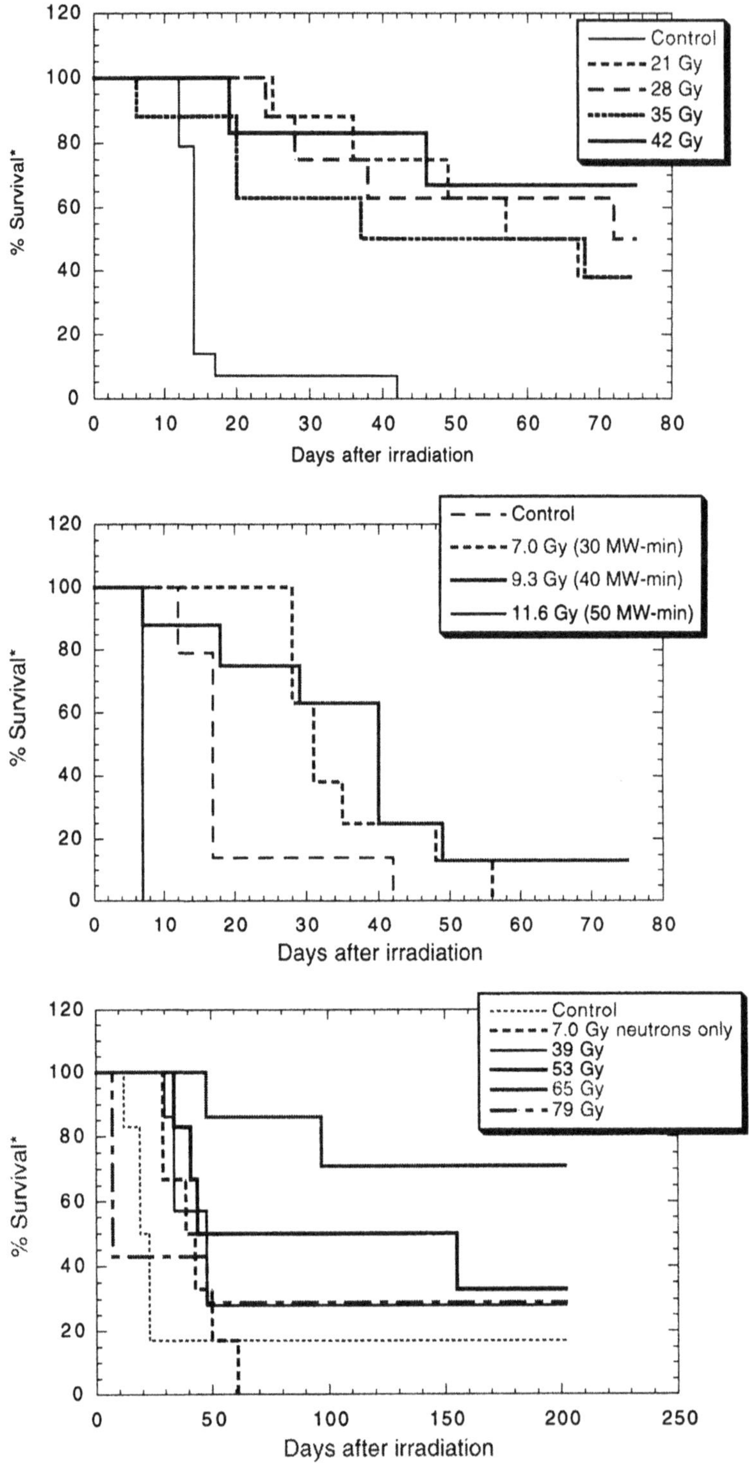

Figure 1. Survial curves for: a) (top) x-irradiated mice; b) (middle) neutrons-only irradiated mice; and c) (bottom) BNCT-treated mice.

Table 1. Comparison of survival and radiation effects from xrays, neutron irradiation and from CuTCPH-mediated BNCT

Porphyrin given[1]	Estimated absorbed dose (Gy)	BMRR MW-min	Type of radiation	Controlled tumors	Number of mice[2]	GI syndrome deaths	Severe radiation damage
yes	39	15	BNC	2	7	0	0
yes	53	20	BNC	2	7	0	0
yes	65	25	BNC	5	7	0	0
yes	79	30	BNC	2	3	4	0
no	7.0	30	neutrons	0	14	0	0
no	9.3	40	neutrons	1	7	1	1
no	11.6	50	neutrons	—	—	8	—
no	21	—	xrays	2	8	0	0
no	28	—	xrays	3	8	0	1
no	35	—	xrays	3	7	0	3
no	42	—	xrays	4	6	0	1

[1]186 μg CuTCPH/g body weight given via 6 i.p. injections over 32 hours.
[2]Number of mice does not include those that died from GI radiation syndrome.

3.2. Neutron-Irradiated Groups

The lowest neutrons-only dose group given 7.0 Gy and the highest BNCT-treated group of 79 Gy correspond to the same total reactor exposure of 30 MW-min. By 56 days postirradiation, all mice in the 7.0 Gy group were euthanized due to tumor overgrowth. At the same neutron-irradiation dose, but with the addition of ^{10}B-CuTCPH, the majority of animals died from gastrointestinal radiation syndrome stemming from the boron neutron-capture reaction in normal tissues. This is most likely due to the greater concentrations of boron in and around the gastrointestinal region. With the exception of normal brain tissue, porphyrins tend to accumulate in normal tissues, particularly those with reticuloendothelial cells, e.g. spleen and liver.

3.3. X-Irradiated Groups

The highest dose group (42 Gy), showed survival comparable to the 65-Gy BNCT group with 4 in 6 mice surviving; however, the experiment has only progressed 80 days compared to >200 days for the BNCT groups. Moreover, radiation damage to the leg was observed in both the 42-Gy and the 35-Gy xray groups and no such damage was observed in any of the BNCT groups at any time.

Extrapolation of our data shows that 50% tumor control using CuTCPH-mediated BNCT corresponds to approximately 59 Gy.

Estimation of a compound-biological effect (CBE)[11] factor takes into consideration the relative biological effect (RBE) of the high linear-energy transfer (LET) radiation from the neutron-capture reaction and the microlocalization of the compound in the cell, which is unique to the compound and its delivery vehicle. Using tumor control as an endpoint, and disregarding side effects to skin, the xray group at 42 Gy produced a tumor response similar to the 65-Gy CuTCPH-mediated BNCT group. The CuTCPH CBE factor in tumor appears to be less than 1. However, the CBE factors for skin and muscle also appear to be considerably less than 1, in addition to the fact that these tissues contain low boron concentrations (<20 μg B/g). Xray doses of 28 Gy and above, produced some incidence of skin and muscle damage, whereas CuTCPH-mediated BNCT, at estimated doses of up to 79 Gy, did not. This difference in radiation sensitivity and the high tumor : normal tissue boron concentration ratios make it possible for CuTCPH-mediated BNCT to produce a significant improvement in therapeutic gain.

In conclusion, CuTCPH-mediated BNCT of EMT-6 leg tumors in mice using 186 μg CuTCPH/g body weight at 25 MW-min (65-Gy estimated absorbed dose) effected 86% tumor control (6/7 mice) 80 days postirradiation with no damage to skin. 100 kVp xrays, estimated physical dose of 41 Gy, effected 67% tumor control also at 80 days. Although only one mouse showed signs of radiation damage to the leg in the 41-Gy xray group, 3 mice showed signs of radiation damage in the 35-Gy group. At 200 days postirradiation, the BNCT group showed 71% controlled tumors and no animals showed serious damage to leg.

ACKNOWLEDGMENTS

The authors wish to thank Michael Makar for maintaining the EMT-6 tumors and to the U.S. Department of Energy (Contract No. DE-AC02-98CH10886) for their support.

REFERENCES

1. J.A. Coderre, J.A. Kalef-Ezra, R.G. Fairchild, P.L. Micca, L.E. Reinstein, and J.D. Glass, Boron neutron capture therapy of a murine melanoma. *Cancer Res.*, 48:6313–6316, 1988.
2. J.A. Coderre, T.M. Button, P.L. Micca, C. Fisher, M.M. Nawrocky, and H.B. Liu, Neutron capture therapy of the 9L gliosarcoma using the *p*-boronophenylalanine-fructose complex. *Int. J. Radiat. Oncol. Biol. Phys.*, 30:643–652, 1994.
3. W. Yang, R.F. Barth, J.H. Rotaru, M.L. Moeschberger, D.D. Joel, M.M. Nawrocky, J.H. Goodman, and A.H. Soloway, Boron neutron capture of brain tumors: Enhanced survival following intracarotid injection of sodium borocaptate with or without blood-brain barrier disruption. *Int. J. Radiat. Oncol. Biol. Phys.*, 37:663–672, 1997.
4. D.D. Joel, R.G. Fairchild, J.A. Laissue, S.K. Saraf, J.A. Kalef-Ezra, and D.N. Slatkin, Boron neutron capture therapy of intracerebral rat gliosarcomas. *Proc. Natl. Acad. Sci. USA*, 87:9808–9812, 1990.
5. M. Miura, P.L. Micca, J.C. Heinrichs, D. Gabel, R.G. Fairchild, and D.N. Slatkin, Biodistribution and toxicity of 2,4-divinyl-nido-*o*-carboranyl-deuteroporphyrin IX in mice. *Biochem. Pharmacol.* 43:467–476, 1992.
6. J.S. Hill, S.B. Kahl, S.S. Stylli, Y. Nakamura, M.-S. Koo, and A.H. Kaye, Selective tumor kill of cerebral glioma by photodynamic therapy using a boronated porphyrin photosensitizer. *Proc. Natl. Acad. Sci. USA*, 92:12126–12130, 1995.
7. S.B. Kahl, D.D. Joel, M.M. Nawrocky, P.L. Micca, K.P. Tran, G.C. Finkel, and D.N. Slatkin, Uptake of a *nido*-carboranylporphyrin by human glioma xenografts in athymic nude mice and by syngeneic ovarian carcinomas in immunocompetent mice. *Proc. Natl. Acad. Sci. USA*, 87:7265–7269, 1990.
8. M. Miura, P.L. Micca, C.D. Fisher, J.C. Heinrichs, J.A. Donaldson, G.C. Finkel, and D.N. Slatkin, Synthesis of a nickel tetracarboranylphenylporphyrin for boron neutron-capture therapy: Biodistribution and toxicity in tumor-bearing mice. *Int. J. Cancer*, 68:114–119, 1996.
9. M. Miura, P.L. Micca, C.D. Fisher, C.R. Gordon, J.C. Heinrichs, and D.N. Slatkin, Evaluation of carborane-containing porphyrins as a tumour targeting agents for boron neutron capture therapy. *Br. J. Radiol.*, 71:773–781, 1998.

10. S.K. Saraf, J. Kalef-Ezra, R.G. Fairchild, B.H. Laster, S. Fiarman, and E. Ramsey, Epithermal beam development at the BMRR: Dosimetric evaluation, in "Neutron Beam Design, Development, and Performance for Neutron Capture Therapy", O.K. Harling, ed., Plenum Press, New York, 1990, pp. 307–316.
11. G.M. Morris, J.A. Coderre J.W. Hopewell, P.L. Micca, and M. Rezvani, Response of rat skin to boron neutron capture therapy with *p*-boronophenylalanine or borocaptate sodium. *Radiother. Oncol.*, 32:144–153, 1994.

APOPTOSIS OF TUMOR CELLS IN RESPONSE TO BORON NEUTRON CAPTURE THERAPY

Tetsuya Yamamoto[1], Akira Matsumura[1], Yasushi Shibata[1], Fumiyo Yoshida[1], Kei Nakai[1], Tadao Nose[1], Youji Horiguchi[2], Masao Aizawa[2], Susumu Nakajima[3], and Isao Sakata[4]

[1]Department of Neurosurgery
Institute of Clinical Medicine
University of Tsukuba
Tsukuba City, Ibaraki
[2]Japan Atomic Energy Institute
[3]Tokai, Ibaraki, Asahikawa Medical College
[4]Asahikawa, Hokkaido, Toyo Hakka Co., Ltd.
Satosho, Okayama, Japan

Abstract

We examined the incidence of radiation-induced apoptosis in 9L gliosarcoma cells in response to gamma-ray irradiation (single doses of 1, 10, or 20 Gy), neutron irradiation, and BNCT using BSH or boronated metalloporphyrin (STA-BX909). In control (unirradiated) cells, the incidence of apoptosis was below 1%. In the gamma-ray group, no significant increase of apoptotic cells was observed at 1 Gy. In the neutron and the 20 Gy gamma-ray groups, a significant increase (5.5%) of apoptosis was noted. There were few detectable apoptotic cells in the BNCT groups. Differences were seen not only in the cell-killing efficacy but also in the mode of cell death among *in vitro* gamma irradiation, neutron irradiation, and BNCT. Apoptosis is a possible biological marker of heterogeneous response of tumor cells following BNCT.

INTRODUCTION

Previous reports from our institute have demonstrated that both boronated metalloporphyrins and BSH preferentially distributed within tumor tissue in the brain and enhanced the cell-killing effect *via* the boron neutron capture reaction.[1,2,3] Recently, considerable attention has been focused on apoptosis in tumor cells, which increases

Frontiers in Neutron Capture Therapy, edited by Hawthorne *et al.*
Kluwer Academic / Plenum Publishers, New York, 2001.

following exposure to ionizing irradiation and to a variety of chemotherapeutic drugs.[4,5] Although apoptosis has been well recognized in tumors treated by ionizing irradiation, apoptosis induced by boron neutron capture therapy (BNCT) has not been evaluated. In the present report, we describe the radiobiological response of tumor cells that were treated by gamma-ray, neutron, and BNCT from the standpoint of the mode of cell death.

MATERIALS AND METHODS

Rat 9L gliosarcoma cells, which have a doubling time of 18 hr,[6] were incubated with RPMI 1640 medium (GIBCO BRL Co., Ltd.), 7.5% fetal bovine serum, and erythromycin. In the BNCT groups, sodium borocaptate (BSH) or 13,17-dipropionic acid—8- dodecaboranyl [$^{10}B_{12}$] thioethyl carbamoyl amidinoiminomethyl- 3- ethenyl-porphyrinato- 2, 7, 12, 18- tetramethyl manganese (III) (STA-BX909, MW; 894.80, Toyo Hakka Co., Ltd.) was added to the conditioned medium. After 24 hr of incubation, flasks were filled with the conditioned medium without boron compounds. For comparison with the BSH- or STA-BX909-BNCT groups, the cells that were irradiated with neutron without boron (neutron group) and those incubated without any additional treatment (control group) were used.

Gamma-ray irradiation was carried out with the Gamma Cell 40 system, which has ^{137}Cs unit (125 cGy min^{-1}). A single dose of 1, 10, or 20 Gy was given to the cultured cells in each gamma-ray subgroup.

Neutron irradiation was performed at the JRR-2 Tokai in the Japan Atomic Energy Institute (JAERI), which has a thermal neutron flux of $1.1 \times 10^9 cm^{-2} sec^{-1}$. The surviving fraction of 9L cells following neutron irradiation was approximately 0.58, 0.017, and 0.0001 for neutron, BSH- and STA-BX909–BNCT groups, respectively. The cytocidal effects for neutron, BSH- and STA-BX909–BNCT groups were equivalent to gamma-ray of approximately 3, 10, and 20 Gy, respectively.

Apoptosis was assessed immunohistochemically using an ApopTag kit (Oncor, Inc.). This assay detects nucleosome-sized DNA fragments by tagging its 3′-OH ends with digoxigenin-deoxyribonucleotide triphosphate (dUTP) and terminal deoxynucleotidyl transferase (TdT). The DNA fragments that were tagged with dUTP were then incubated with an antidigoxygenin antibody conjugated with peroxide. The reaction was visualized by using diaminobenzodine (DAB). At least 400 cells were counted for each data point, and cells having ApopTag positive, pyknotic, or fragmented nuclei with sharply delineated cell borders were quantified under microscopy at ×400 magnification (Figure 1).

RESULTS

The difference in the incidence of apoptotic cells in each group is shown in Figure 2. In control (unirradiated) cells, the incidence of apoptosis during 12 to 168 hr incubation was below 1%. In the gamma-ray group, there was no significant increase in the incidence of apoptotic cells at 1 Gy compared to the unirradiated control group. At 10 Gy, the incidence of apoptotic cells was from 1.5% to 2.5%. The maximal increase (5.5%) was noted at 20 Gy at 168 hr following irradiation.

In the neutron group, the maximal increase (5.5%) was noted at 168 hr. There were few detectable apoptotic cells in both BSH- and STA-BX909-BNCT groups.

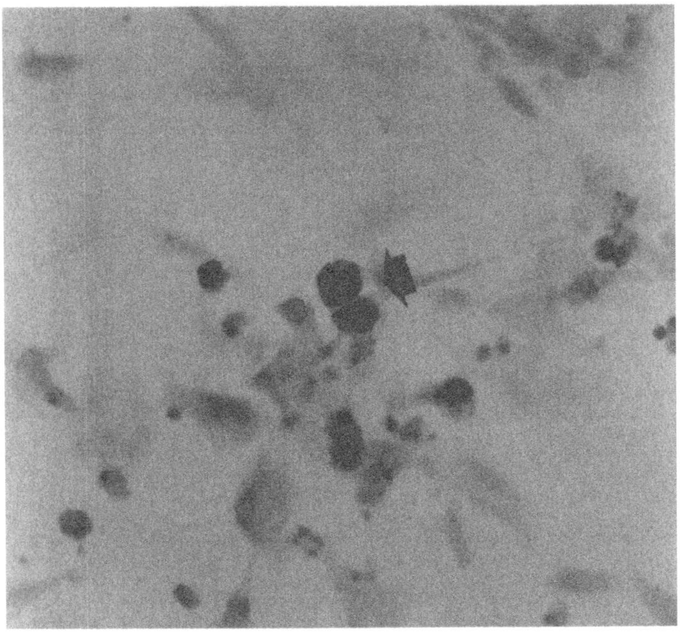

Figure 1. Microphotograph of 9L glioblastoma cells stained at 72 hr following neutron exposure, showing a cell with ApopTag positive nuclei (arrow).

DISCUSSION

Cell death can be divided into two classes by morphologic and biochemical characteristics–necrosis and apoptosis. While necrosis is a passive phenomenon due to external nonphysiological disturbances, apoptosis is thought to be an active process of gene-mediated cell suicide that requires *de novo* gene expression and protein synthesis.[6] Apoptosis is recognized as a fundamental mechanism of cell deletion in immune system maturation and normal animal development.[7,8] However, apoptosis is also a primary mode of cell death in the pathologic response to gamma-ray irradiation and to a variety of cytotoxic agents.[4,5]

The dose-dependent response and independence of LET in early interphase apoptosis were reported, and these suggested that initial DNA damage dominated the induction of apoptosis compared to DNA repair.[9] In the present study, the incidence of apoptotic 9L cells showed a dose-dependent response following gamma-ray irradiation. However, the incidence of apoptosis (5.5%) at 168 hr following neutron irradiation is similar to that of a higher equivalent dose of 20 Gy gamma-ray at this time period. Flow cytometric analysis revealed that a G2/M block occurred in 9L cells following neutron irradiation, and the release from the G2/M block with a smaller DNA peak of apoptosis was observed after 72 hr (delayed apoptosis).[3] A much lower level of delayed apoptosis was induced in the BNCT groups compared with the other groups. These observations bring out some assumptions on the mode of cell-killing: 1) Not only irradiation dose but also LET and/or difference in radiation quality affect the incidence of apoptosis; 2) Reproductive cell death is prominent when the boron compounds are added to thermal neutron irradiation; 3) Cell damage causes apoptosis of 9L cells following the neutron exposure without boron; and 4) In 9L cells treated by BNCT, cell damage may not be

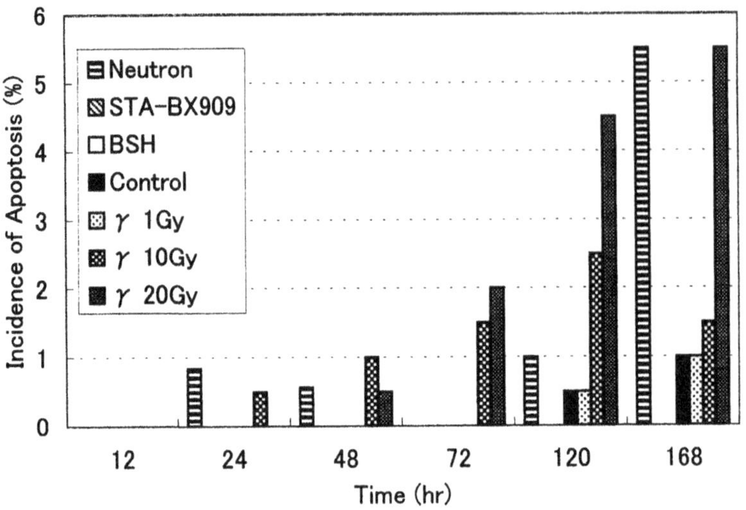

Figure 2. The incidence of apoptotic cells in unirradiated control and irradiated 9L gliosarcoma cells.

brought over from the first cell cycle, or severe, lethal cell damage may take place unless undergoing apoptosis. This suggests that the lower level of apoptosis at 12 to 168 hr following BNCT is mainly based on $^{10}B(n, \alpha)$ reaction and consequent dense double-strand breaks. The initial DNA damage may be beyond repair, and the cells may be unable to enter into the next cell cycle and/or apoptosis which require *de novo* gene transcription.

In conclusion, differences were seen not only in the cell-killing efficacy but also in the mode of cell death among *in vitro* gamma irradiation, neutron irradiation, and BNCT. Apoptosis is a possible biological marker of heterogeneous response of tumor cells following BNCT.

ACKNOWLEDGMENTS

This study was supported in part by University of Tsukuba Research Projects, by a Research Grant-in-Aid for Scientific Research from the Ministry of Education, Science and Culture of Japan (07557091), by a charitable trust of the Matsuoka Memorial Fund for Medical Aid to the Citizens in Ibaraki Prefecture, and by the Fund in Trust for the Association for Nuclear Technology in Medicine, Japan.

REFERENCES

1. T. Yamamoto, A. Matsumura, Y. Shibata, T. Yoshizawa, H. Fujimori, K. Nakagawa, K. Nakai, Y. Yoshii, T. Nose, I. Sakata, and S. Nakajima, MR imaging using boronated Mn-metalloporphyrin (STA-BX900) in BNCT, *in:* Advances in Neutron Capture Therapy—Vol. 2 Chemistry and Biology, Larsson, Crawford and Weinreich ed., Elsevier, 1997, pp. 267–270.
2. A. Matsumura, Y. Shibata, T. Yamamoto, T. Yamada, H. Fujimori, K. Nakai, I. Sakata, N. Miwa, S. Nakajima, and T. Nose, Pharmacokinetic study of boronated metalloporphyrin (STA-BX900) and BSH in the rat glioma model *in:* Advances in Neutron Capture Therapy—Vol. 2 Chemistry and Biology, Larsson, Crawford and Weinreich ed., Elsevier, 1997, pp. 158–162.

3. A. Matsumura, Y. Shibata, T. Yamamoto, F. Yoshida, K. Nakai, S. Kiriya, N. Shimojo, T. Nose, I. Sakata, S. Nakajima, and N. Miwa, Radiobiological Effects of Boronated porphyrin (STA-BX909) as a compound for neutron capture therapy. An *In vitro* study, *in:* Recent Progress of Photodynamic Therapy. Proceedings of the 7th Annual Conference of JCIPA, Nishisaka ed., Ionics, Tokyo, 1997, pp. 125–130.
4. M. Tada, R. Matsumoto, R.D. Iggo, R. Onimaru, H. Shirato, Y. Sawamura, and Y. Shinohe, Selective sensitivity to radiation of cerebral glioblastomas harboring p53 mutations, *Cancer Res*, 58:1793–1797, 1998.
5. B.A. Rupnow, A.D. Murtha, R.M. Alarcon, A.J. Giaccia, and S.J. Knox, Direct evidence that apoptosis enhances tumor responses to fractionated radiotherapy, *Cancer Res*, 58:1779–1784, 1998.
6. M. Weizsaecker, D.F. Deen, M.L. Rosenblum, T. Hoshino, P.H. Gutin, and M. Barker. The 9L rat brain tumor description and application of an animal model, *J Neurol*, 224: 183–192, 1981.
7. R.A. Schwartzman, and J.A. Cidlowski, Apoptosis: The biochemistry and molecular biology of programmed cell death, *Endocrine Rev*, 14:133–151, 1993.
8. H.M. Ellis, and H.R. Horvitz, Genetic control of programmed cell death in nematode C. elegans. *Cell*, 44:817–829, 1986.
9. A. Vral, M. Cornelissen, H. Thierens, H. Louagie, Philipp, K. Strijckmans, and L. DeRidder. Apoptosis induce by fast neutrons versus 60Co gamma-rays in human peripheral blood lymphocytes. Int J Radiat Biol 73:289–295, 1998.

APOPTOSIS INDUCED BY BORON NEUTRON CAPTURE THERAPY FOR BRAIN TUMORS

Michael A. Domet[1], Birte Otersen[1], Rolf F. Barth[1], Weilian Yang[1], Joan Rotaru[1], Zunyan Dai[1], James R. Lang[2], and Marta M. Nawrocky[3]

Departments of Pathology[1] and Otolaryngology[2]
The Ohio State University
Columbus, Ohio 43210
[3]Medical Department
Brookhaven National Laboratory
Upton, New York 11973

1. INTRODUCTION

Apoptosis, or programmed cell death,[1] is a process that can be initiated by a variety of external stimuli, including drugs[2] and ionizing radiation,[3] to activate a complex sequence of intracellular events that lead to death.[4] Once initiated, both the nucleus and cytoplasm condense, the nucleus fragments and apoptotic bodies are formed.[1] Since this process can be completed within as few as 4 hours,[4] a small time window exists for determining whether a specific stimulus can induce apoptosis. *In vivo*, a few apoptotic bodies may underepresent a much larger fraction of apoptotic cells due to the rapidity with which they are removed.[4] Thus, if apoptotic cells are detected, it is likely that an even larger, undetected population of cells already may have undergone programmed cell death. Up until recently, it was thought that ionizing radiation only produced tumor cell death by the induction of double strand chromosome breaks, which were either lethal or reproductively sterilized the cell.[5] There now is a growing body of evidence that ionizing radiation also can activate factors within the cell that initiate genetically programmed cell death or apoptosis.[3,6,7]

The F98 glioma model has been used by us for a variety of BNCT studies over the past 10 years, and as few as 10–100 tumor cells, implanted intracerebrally into syngeneic Fischer rats, are invariably fatal. Up until recently,[8] the F98 glioma has been refractory to chemotherapy, radiotherapy, or BNCT. However, using blood brain barrier disruption in combination with intracarotid injection of BPA and BSH, and as reported elsewhere,[9] as well as in these proceedings, it now has been possible to cure 25% of animals. There

Frontiers in Neutron Capture Therapy, edited by Hawthorne *et al.*
Kluwer Academic / Plenum Publishers, New York, 2001.

has been considerable controversy as to whether ^{10}B must be delivered to each individual tumor cell to produce a lethal $^{10}B(n,\alpha)^{7}$ Li capture reaction or whether some other mechanism(s) also might produce tumor cell death. The purpose of the present study was to determine if a second lethal event, apoptosis, also could produce death of F98 glioma cells in brain tumor bearing rats following BNCT using boronophenylalanine (BPA) as the capture agent.

2. MATERIALS AND METHODS

2.1. *In Vitro* Induction of Apoptosis and Detection by TUNEL Assay

In order to determine if apoptosis could be induced in F98 glioma cells, an *in vitro* study was carried out using the antibiotic staurosporine, which is a potent inhibitor of protein kinase C and an inducer of programmed cell death. Aliquots of F98 cells were seeded into 6 well macroplates and one day later the media was changed and media either with or without 50 nM of staurosporine was added to test and control wells. Cells were harvested either immediately (t = 0) or at 3, 6, 9 and 24 hours later. Two positive DNase controls also were carried out at time 0 and the other at 24 h. Apoptotic cells were detected by means of a terminal deoxynucleotidyl transferase mediated dUTP-fluorescein nick-end immunolabeling (TUNEL) assay[10] using a commercially available kit (Boehringer Mannheim Corp., Indianapolis, IN). Apoptotic cells were enumerated using a Coulter EPICS Elite flow cytometer (Coulter Corporation, Miami, FL) equipped with a 488 nm, 15 mW air-cooled argon laser. Calibration of the optical laser's alignment was performed using Coulter's DNA-Check EPICS alignment fluorosphere beads with coefficient of variations <2%. Immuno-4 software was used to compare staurosporine-treated cells to control cells at time 0. This program overlays these two one parameter histograms and gives the percent positive staining cells. Cell cycle analysis was performed using Multicycle and Modfit software to estimate DNA content and ploidy status.

In addition, *in vivo* cell cycle analysis also was carried out by means of an immunohistochemical method to identify F98 glioma cells that had taken up bromodeoxyuridine (BUdR).[11] Briefly, F98 glioma bearing rats were injected i.v. with a solution containing BUdR (4.7 mg/ml) and fluorodeoxyuridine (FUdR 0.47 mg/ml) at 2 h intervals, for a total time period of 16 h. Animals were euthanized at 2 h intervals, 30 min after each injection of BudR, their brains were removed, fixed in 70% ethanol, embedded in paraffin and then sectioned at 4 μ. BUdR immunostaining was performed using a commercially available kit (Zymed Laboratories Inc., South San Francisco, CA).

2.2. Determination of Soluble Fas Expression

Fas (CD95/APO1) is a cell surface receptor and a member of the nerve growth factor (NGF)/tumor necrosis factor (TNF) receptor superfamily of genes that mediates apoptosis in sensitive cells. Susceptibility to apoptosis in part is dependent upon Fas/APO1, and therefore it was of interest to determine if F98 glioma cells expressed Fas.[12] Western blot analysis was used to detect soluble Fas expression using F98 cells, which had been grown *in vitro*.

2.3. *In Vivo* Detection of Apoptotic Cells Following BNCT

Rats were stereotactically implanted with 100,000 F98 glioma cells into their right caudate nucleus using a procedure described by us in detail elsewhere.[8] Twelve days after tumor implantation, the rats were transported to the Brookhaven National Laboratory, Upton, New York for irradiation, which was carried out on day 14, as previously described.[8] Rats were anesthetized and were randomly assigned to three groups as follows: (1) was injected via the internal carotid artery (i.c.) with 500 mg/kg body weight ^{10}B-enriched BPA and 2.5 hours later they were irradiated at the Brookhaven Medical Research Reactor (BMRR); (2) did not receive BPA but was irradiated in order to control for apoptosis resulting from non-BNCT related radiation; (3) was untreated in order to determine the level of background apoptosis in non-irradiated tumor bearing rats. Animals were euthanized 3, 6, 9, 24 and 72 h after irradiation with a lethal dose of halothane and their brains were removed and fixed in 4% paraformaldehyde in PBS (pH 7.4). Apoptosis was assessed in tissue sections[13] by using terminal deoxynucleotidyl transferase mediated dUTP nick-end immunolabeling (TUNEL) assay (*in situ* cell death detection kit from Boehringer Mannheim) to directly detect double strand breaks in DNA. Tissue sections were de-parafinized by heating at 60 °C followed by washes in xylene and a graded series of ethanol solutions down to H_2O, and then were incubated with proteinase K (20 µg/ml in 10 mM TRIS.HCl, pH 7.6) for 15 min following which the slides were washed in PBS, and 50 µl of the TUNEL reaction mixture was overlaid on each tissue section, and incubated at 60 °C in a humified chamber. Following washing, they were exposed to an antifade reagent prior to cover slipping. Tissue sections also were stained with hematoxylin and eosin to allow for morphologic assessment of apoptosis by light microscopy. Apoptotic cells were detected by microscopic examination of coronal brain sections with a video camera interfaced with an image analysis system. A semi-quantitative system was used to enumerate apoptotic cells by examining 5 non-necrotic areas, which were randomly selected from each coronal section and 200 nuclei from each field were characterized as normal or apoptotic. The morphologic criteria for apoptosis were pyknosis, nuclear fragmentation, crescent shaped nuclei and membrane bound bodies containing pyknotic nuclei that were smaller in size than surrounding normal cells.

3. RESULTS

3.1. *In Vitro* Induction of Apoptosis and Cell Cycle Analysis

F98 glioma cells treated with 50 nM staurosporine showed TUNEL positive apoptotic cells at each time interval. The percent positive cells were 50.7, 76.5, 63.7 and 74.2% at 3, 6, 9, and 24 h, respectively (Figure 1).

Cell cycle analysis data revealed that the percent of untreated control cells in the G_1 phase of the cell cycle were 57.4, 67.7, 45.9 and 49.6% respectively, at 0, 6, 9 and 24 h. In S phase, there were 35, 34.2, 44.4 and 29.3% respectively, and in G_2 there were 7.7, 15.2, 9.7 and 21%, respectively. The percent of staurosporine treated cells at 6, 9, and 12 hours in G_1 were 30.5, 32.5 and 37.3%; in S there were 2.4, 5.5 and 7.9% and in G_2 were 67.1, 62.0 and 54.8% at 6, 9 and 24 h, respectively. As determined by BUdR uptake, the *in vivo* duration of the cell cycle (T_c) was 18.9 h, S phase (T_s) was 6.9 h, the labeling index at 12 h (LI_{12}) was 8%, and the growth fraction (GF) was 22%. As would have been

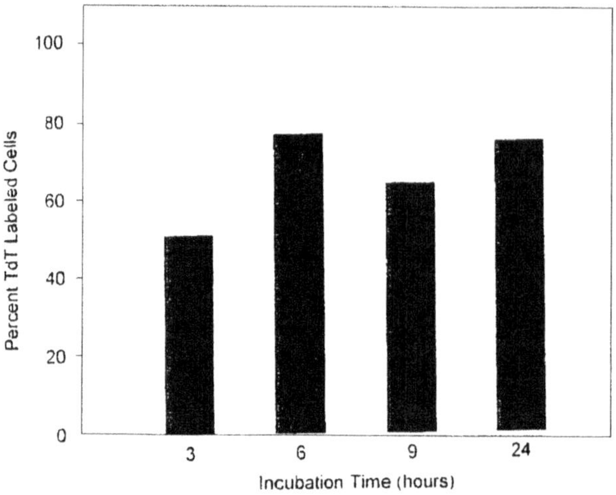

Figure 1. Staurosporine induced apoptosis of F98 glioma cells.

predicted from the survival data obtained from animals challenged with logarithmicaly increasing numbers of tumor cells, these data indicate that the F98 glioma is a rapidly proliferating tumor.

3.2. Fas Expression

F98 wild type, as well as cells that had been transfected with the gene encoding for the B7.1 co-stimulatory molecule (F98 B7.1 transfectants), human monocyte chemoattractant protein (F98/hMCP), and the cytomegalovirus promoter (F98/PC-CMV) transfectants were all found to have a band at 35 kD indicating that they all produced soluble Fas/APO-1.

3.3. *In Vivo* Induction of Apoptosis by BNCT

As determined by direct current plasma-atomic emission spectroscopy,[8] the estimated boron concentrations in tumor, normal brain and blood were 42.7 ± 4.5, 7.9 ± 3.2 and $5.8 \pm 1.5\,\mu g/g$, respectively. To estimate the biological equivalent doses delivered to the tumor, brain and blood the following RBE values were used: $^{14}N(n,p)^{14}C$ reaction, 3.2; $^{10}B(n,\alpha)^7Li$, 3.8 for BPA in blood, and 1.3 for brain; 1.0 for γ-photons, and 2.0 for fast neutrons. At a depth of 4–5 mm in the brain, the dose rates were: 0.27 Gy/min for fast neutrons and 0.11 Gy/min for γ-photons. The dose rates for the $^{14}N(n,p)^{14}C$ capture reaction was 0.076 Gy/min for blood and 0.052 Gy/min for brain, and for the $^{10}B(n,\alpha)^7Li$ reaction it was 0.034 Gy/min in blood, brain and tumor. The RBE equivalent doses delivered to these sites were 58.8 Gy for tumor, 17.6 Gy for brain and 15.1 Gy for blood. The physical doses delivered to these tissues were 17.7 Gy for tumor, 6.9 Gy for brain, and 6.2 Gy for blood. Rats were euthanized at 3, 6, 9, 24 and 72 h following irradiation and at each time interval the average number of apoptotic cells captured per field (~10–90/ per cross section) were 17, 82, 42, 37 and 18 for BNCT treated animals, respectively (Figure 2). There were 16 apoptotic cells per high power field in neutron irradiated controls and 9 in unirradiated controls.

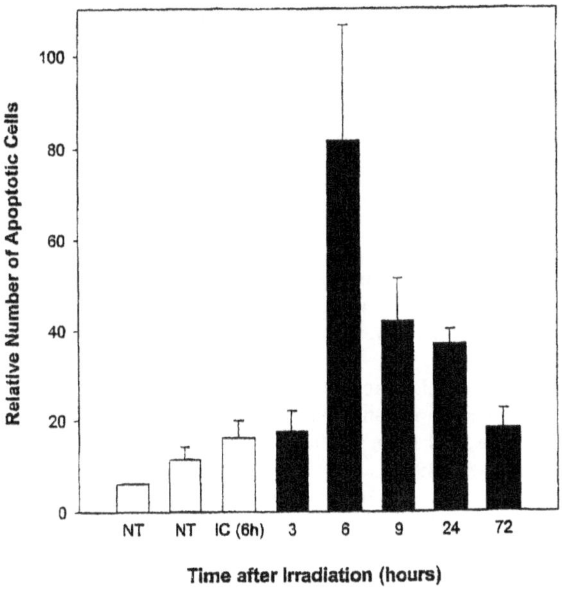

Figure 2. BNCT induced apoptosis in F98 glioma bearing rats.

4. DISCUSSION AND CONCLUSIONS

Apoptosis was first described by Kerr *et al.*, in 1972,[1] but it only has been in the present decade that its biological significance has been fully appreciated. Although it has been recognized that ionizing radiation can produce apoptosis,[3,4,6] the present study is the first to demonstrate that it can occur following BNCT. First, we established that apoptosis could be induced chemically in F98 glioma cells by exposing them to staurosporine, a protein kinase inhibitor, and a potent inducer of apoptosis. *In vitro*, staurosporine induced apoptosis in the G_2 phase of the cell cycle, which was markedly increased in treated cells. *In vivo* cell cycle analysis revealed that the F98 glioma rapidly progressed through the cell cycle, but did not provide data on the duration of the G_2 phase. Western blot analysis showed that the F98 glioma produced soluble Fas, but attempts to demonstrate cell surface membrane expression were unsuccessful due to limitations of the antibody that was used.

Our *in vivo* studies unequivocally demonstrated that BNCT produced an increased number of apoptotic cells compared to that found in irradiated controls. This might provide a partial answer to the question originally posed, as to whether it was necessary to deliver a critical number of ^{10}B atoms to each and every tumor cell to cure a brain tumor bearing animal. BNCT induced apoptosis may be an important amplifying mechanism by which "bystander" tumor cells, which contain $<10^9$ ^{10}B per cell might be killed.

ACKNOWLEDGMENTS

This work has been supported in part by Department of Energy grant DE-FG02-95ER62059. Michael Domet was the recipient of a summer fellowship from the Samuel J. Roessler Memorial Scholarship Fund and Zunyan Dai was the recipient of a

Presidential Fellowship from The Ohio State University. Dr. Birte Otersen received a fellowship from the Deutsche Akademische Austauschdienst. The expert secretarial assistance of Mrs. Linda Carlos is gratefully acknowledged in the preparation of this manuscript.

REFERENCES

1. J.F.R. Kerr, A.H. Wyllie, and A.R. Currie, Apoptosis: A basic biological phenomenon with wide-ranging implications in tissue kinetics. Br J Cancer 26:239–257, 1972.
2. J.L.-S. Au, N. Panchal, D. Li, and Y. Gan, Apoptosis: A new pharmacodynamic endpoint. Pharm Res 14:1659–1671, 1997.
3. K.R. Blank, M.S. Rudoltz, G.D. Kao, R.J. Muschel, and W.G. McKenna, The molecular regulation of apoptosis and implications for radiation oncology. Int J Radiat Biol 71:455–466, 1997.
4. F.T. Bosman, B.C. Visser, and J. van Oeveren, Apoptosis: Pathophysiology of programmed cell death. Path Res Pract 192:676–683, 1996.
5. E.J. Hall: "Radiobiology for the Radiologist." 4th ed., Lippincott Co., Phila., PA., 1994.
6. W.C. Dewey, C.C. Ling, and R.E. Meyn, Radiation-induced apoptosis: Relevance to radiotherapy. Int J Radiat Oncol Biol Phys 33:781–796, 1995.
7. G.T. Williams, and C.A. Smith, Molecular regulation of apoptosis: Genetic controls on cell death. Cell 74:777–779, 1993.
8. R.F. Barth, W. Yang, J.C. Rotaru, M.L. Moeschberger, D.D. Joel, M.M. Nawrocky, J.H. Goodman, and A.H. Soloway, Boron neutron capture therapy of brain tumors: Enhanced survival following intracarotid injection of either sodium borocaptate or borophenylalanine with or without blood-brain barrier disruption. Cancer Res 57:1129–1136, 1997.
9. R.F. Barth, A.H. Soloway, J.H. Goodman, R.E. Gahbauer, N. Gupta, T.E. Blue, W. Yang, and W. Tjarks, Boron neutron capture therapy of brain tumors: an emerging therapeutic modality. Neurosurgery, March 1999, (In Press).
10. R. Gold, M. Schmied, G. Giegerich, H. Breitschopf, H.P. Hartung, K.V. Toyka, and H. Lassmann, Differentiation between cellular apoptosis and necrosis by the combined use of *in situ* tailing and nick translation techniques. Lab Invest 71:219–225, 1994.
11. R.S. Nowakowski, S.B. Lewin, and M.W. Miller, Bromodeoxyuridine immunohistochemical determination of the lengths of the cell cycle and the DNA-synthetic phase for an anatomically defined population. J Neurocytol 18:311–318, 1989.
12. K. Sakata, A. Sakata, L. Kong, H. Dang, and N. Talal, Short analytical review: Role of Fas/FasL interaction in physiology and pathology: the good and the bad. Clin Immunol Immunopathol 87:1–7, 1998.
13. F. Labat-Moleur, C. Guillermet, P. Lorimier, C. Robert, S. Lantuejoul, E. Brambilla, and A. Nogoescu, TUNEL apoptotic cell detection in tissue sections: critical evaluation and improvement. J Histochem Cytochem 46:327–334, 1998.

CELL CYCLE CHANGE OF 9L GLIOMA CELL AFTER *IN VITRO* BORON NEUTRON CAPTURE THERAPY

Yasushi Shibata[1], Akira Matsumura[1], Fumiyo Yoshida[2], Tetsuya Yamamoto[1], Kei Nakai[1], Tadao Nose[1], Isao Sakata[3], and Susumu Nakajima[4]

[1]Department of Neurosurgery and
[2]College of Medical Technology
University of Tsukuba
Tsukuba, Ibaraki, 305, Japan
[3]Toyo Hakka Co. Ltd.
Okayama, 719-03, Japan
[4]Department of Operative Surgery
Asahikawa Medical College
Asahikawa, Hokkaido, 078, Japan

1. INTRODUCTION

Proliferating tumor cell progress through the phases of each cell cycle. Cell cycle change such as G2/M arrest is known to occur in tumor cells after γ ray radiation.[1,2] However, cell cycle change of tumor cell after neutron capture therapy has not been reported. We irradiated thermal neutron to 9L gliosarcoma cell *in vitro* with or without boron compounds and examined cell cycle changes of tumor cells with flow cytometry.

2. MATERIALS AND METHOD

2.1. Cell Preparation

Established cell line of 9L gliosarcoma cell was cultured in RPMI medium or MEM solution(Gibco, New York, USA) at 37 °C in an atmosphere of 5% CO_2 in air. The cell was incubated in boron containing medium for 3 hours. We used 2 boron compounds, sodium borocaptate (BSH) and boron porphyrin compound STA-BX909.[3] BSH is $Na_2B_{12}H_{11}SH$ (Boron biologicals, Inc. NC, USA). STA-BX909 is $C_{37}H_{49}N_8O_6{}^{10}B_{12}Mn$

Frontiers in Neutron Capture Therapy, edited by Hawthorne *et al.*
Kluwer Academic / Plenum Publishers, New York, 2001.

H_2O (Toyo Hakka Co. Ltd. Okayama, Japan). STA-BX909 contains 12 boron atoms and the molecular weight of STA-BX909 is 894.8 gm/mol. Boron accounts for 13% of the weight of the compound. The boron concentration in the medium was 30 μgB/ml. Control medium which contained no boron compound was also prepared. Each cell suspension was conditioned at the concentration of 3×10^5 cells/ml.

2.2. Neutron Irradiation

Thermal neutron was irradiated at Japan Research Reactor 2 (JRR2) of Japan Atomic Energy Research Institute (JAERI). Neutron fluence was arranged with the distance between cell sample and the radiaton port. Each flask containing cells in the medium was placed at 150 cm interior of the irradiation port (high dose) or 160 cm exterior of the irradiation port (low dose). All cell was irradiated with thermal neutron for 3 hours. Neutron fluence and incidental gamma ray dose were measured with gold wire and thermal luminescence dosimeter (TLD). High dose group recieved 20 × 10E12 (n/cm2)of thermal neutron and low dose group recieved 0.44 × 10E12. The cells were not rinsed before neutron irradiation. The boron compounds were present in the medium while neutron irradiation, because it was considered to make the result clear.

2.3. Cell Cycle Analysis

Each cell suspension were centrifuged and pellets were fixed with ice-cooled 70% ethanol at 6, 12, 24, 48, 72, 120, 168 hours after the irradiation. Fixed cells were stained with Chromomycin A3 at the concentration of 20μg/ml for 1 hours at 4 °C. Cell solution were centrifuged and each pellets were suspended in Phosphate buffered Saline (PBS). Cell suspension were filtered through the mesh of 2 mm (Becton Dickinson, CA, USA) to remove debris or cluster of cells. The flow cytometric data acquisition was conducted using FACSort (Becton Dickinson, CA, USA). The activated wave length of Chromomycin A3 is 400–460 nm, peak at 420 nm. The fluorescent wave length of Chromomycin A3 is 490–610 nm, peak at 560 nm. DNA content stained with Chromomycin A3 was measured with FL2 range. Each data were analyzed with Lysis 2 soft ware. The cells in G0/G1 and G2/M phase were identified from the histogram of FL2 and the percentage of G0/G1 or G2/M were calculated on the histogram of FL2.

3. RESULTS

The difference of cell cycle changes between 2 boron neutron capture (BNC) group (BSH and STA-BX909) was not recognized. In high dose irradiation groups, the percentage of G0/G1 was decreased with time (Figure 1). This tendency was dominant at BNC group. In low dose irradiation groups, the percentage of G0/G1 transiently decreased at 12 hours after irradiation. In neutron group, this transient decrease of G0/G1 returned to normal at 24 hours after the irradiation. In BNC group this recovery from the transient decrease of G0/G1 fraction was delayed to 120 hours after irradiation. In high dose neutron and BNC groups, the G2/M fraction continuously increased and this tendency was dominant for BNC group than neutron group (Figure 2). The DNA histogram of high dose neutron group showed euploidy, however that of high dose BNC group showed much aneuploidy other than tetraploidy (data not shown).

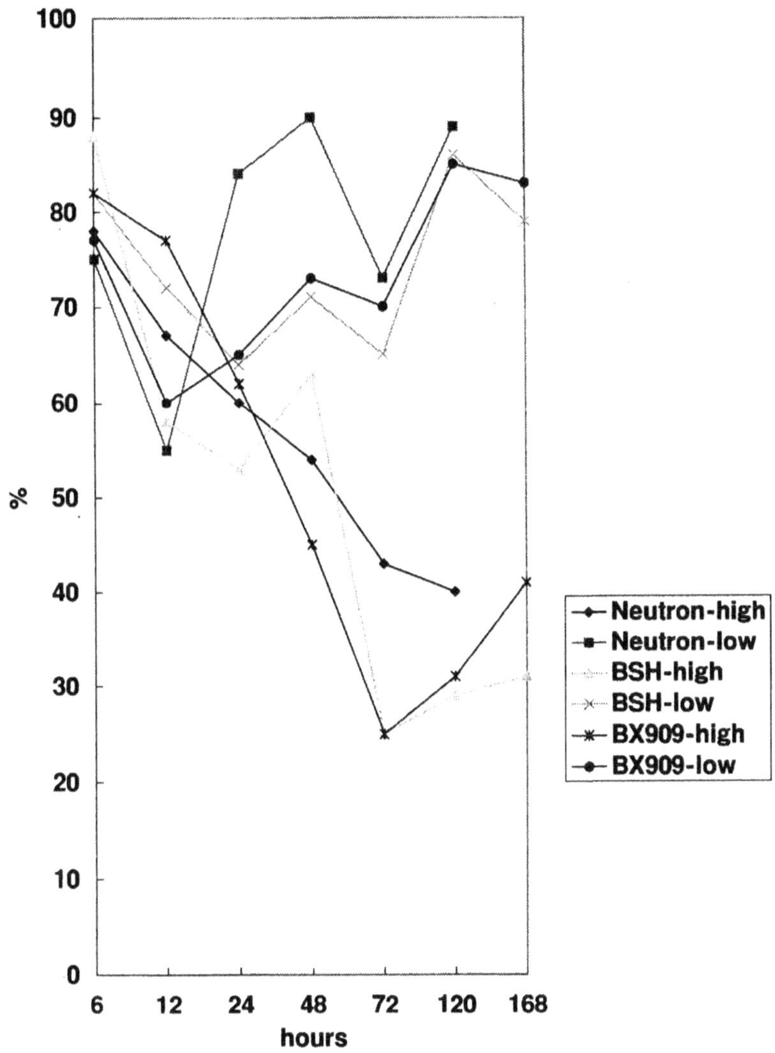

Figure 1. Shows the serial changes of the percentage of cells in G0/G1 phase. High radiation dose group shows continuous decrease of the percentage of cells in G0/G1 phase. Low radiation dose group shows transient decrease of these. Horizontal axis is the time after the radiation.

4. DISCUSSION

Proliferating tumor cell progress through the phases of each cell cycle. Cell cycle change such as G2/M arrest or block are known to occur in tumor cells after the gamma ray radiation.[1,2] The mechanism of G2/M arrest after irradiation is not fully understood, however it was thought for the repair of DNA damage and the check point gene is responsible. The length of G2/M arrest *in vitro* after irradiation depends on the radiation dose.[4,5]

In our results, transient G2/M arrest was seen in low dose group and progressive G2/M arrest was seen in high dose and BNC group. This results means the reversible

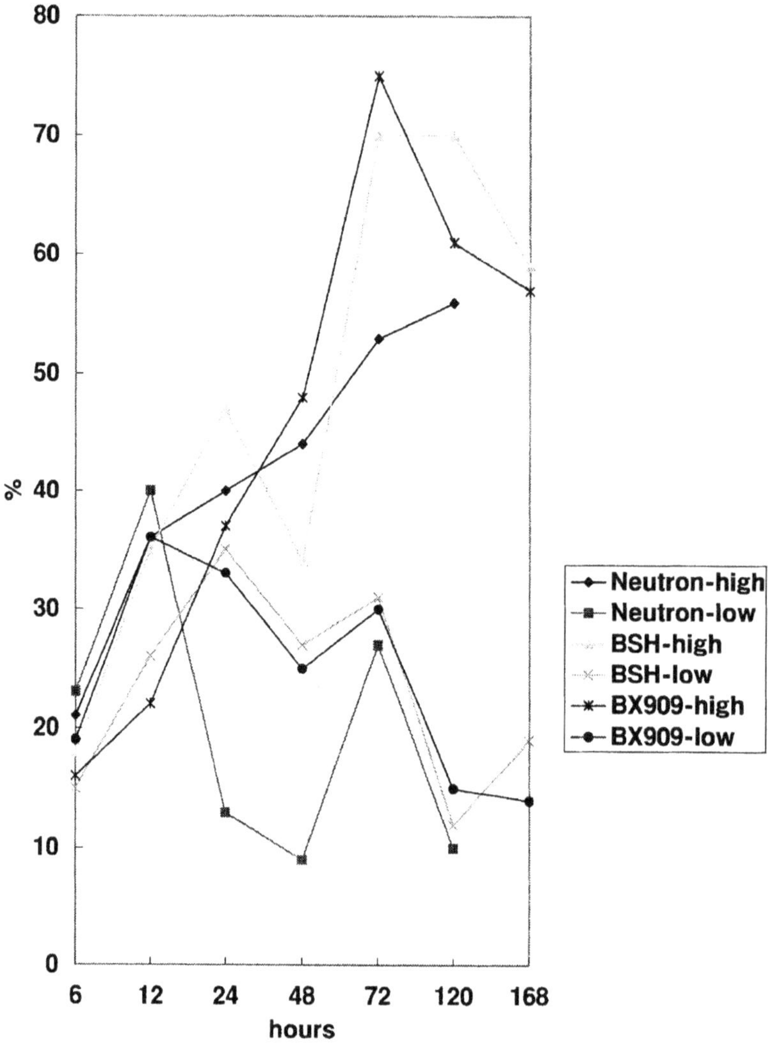

Figure 2. Showed the serial changes of the percentage of cells in G2/M phase. High radiation dose group shows continuous increase of the percentage of cells in G2/M phase. Low radiation dose group shows transient increase of these.

damage to tumor cell occurred in low dose group, and irreversible tumor cell damage occurred with sufficient radiation dose and boron compounds. Using such flowcytometry study, the characteristic radiobiological effect of BNC may be more precisely investigated.

5. CONCLUSION

In low neutron radiation, transient G2/M arrest of tumor cell was observed. In high dose neutron capture therapy progressive irreversible G2/M arrest was observed.

REFERENCES

1. J.-H. Hong, R.A. Gatti, Y.K. Huo, C.-S. Chiang, and W.H. McBride, G2/M-phase arrest and release in ataxia telangiectasia and normal cells after exposure to ionizing radiation. Radiat. Res., 140:17–23, 1994.
2. R. Villa, N. Zaffaroni, A. Bearzatto, A. Costa, A. Sichirollo, and R. Silvestrini, Effect of ionizing radiation on cell-cycle progression and cyclin B1 expression in human melanoma cells. Int. J. Cancer, 66:104–109, 1996.
3. A. Matsumura, Y. Shibata, T. Yamamoto, F. Yoshida, K. Nakai, I. Sakata, S. Nakajima, M. Okumura, and T. Nose, *In vivo* and *in vitro* pharmacokinetics of boronated porphyrins for neutron capture therapy. *In* "Frontiers in Neutron Capture Therapy," M.F. Hawthorne, K. Shelly, R.W. Wiersema, eds., Kluwer Academic/Plenum Publishers, New York, 2001, pp. 1015–1020.
4. D.B. Leeper, M.H. Schneiderman, and W.C. Dewey, Radiation-induced division delay in synchronized Chinese hamster ovary cells in monolayer culture. Radiat. Res., 50:401–417, 1972.
5. W.C. Dewey, and D.P. Highfield, G2 block in Chinese hamster cells induced by X irradiation, hyperthermia, cycloheximide, or actinomycin-D. Radiat. Res., 65:511–528, 1976.

RADIOBIOLOGICAL EFFECTS OF BORONATED PORPHYRIN (STA-BX909) COMPARED TO BSH IN CULTURED GLIOMA CELL LINE

Akira Matsumura[1], Yasushi Shibata[1], Tetsuya Yamamoto[1], Fumiyo Yoshida[1], Yoshinori Hayakawa[2], Tomonori Isobe[1], Kei Nakai[1], Isao Sakata[4], Susumu Nakajima[4], Naoto Miwa[5], Koji Ono[6], and Tadao Nose[1]

[1]Department of Neurosurgery
[2]Proton Medical Center
University of Tsukuba, Ibaraki, Japan
[3]Toyo Hakka Co. Ltd.
Okayama, Japan
[4]Central Division of Surgery
Asahikawa Medical College
Hokkaido, Japan
[5]Schering Japan Research Institute
Osaka, Japan
[6]Kyoto University Research Reactor Institute
Osaka, Japan

1. INTRODUCTION

Porphyrin is known as a tumor localizing agent that is clinically used in photodynamic therapy for cancers.[1] Boronated porphyrins have also been developed[2-7] and are now being investigated for possible clinical use.[2]

Our recent work has shown promising boron delivery to the tumor using boronated porphyrin[3,4,6,7] without phototoxicity. Other favorable facts indicate that porphyrins may be incorporated into the cells in all of the cell cycle,[9] and porphyrins are preferentially taken up in tumors but not in vasogenic edemas[9] or cytotoxic edema tissue.[10] Our manganese metalloporphyrin was also visualized on MRI as T1 contrast agent.[9]

Based on such benefits, we have further developed a tumor-localizing manganese metalloporphyrin, STA-BX909. In this study, we evaluate the radiobiological effects after neutron irradiation *in vitro*.

Frontiers in Neutron Capture Therapy, edited by Hawthorne *et al.*
Kluwer Academic / Plenum Publishers, New York, 2001.

2. MATERIALS AND METHODS

2.1. Cell Culture

9L gliosarcoma cell line was used for this study. The cells were incubated with RPMI 1640 supplemented with 7.5% FBS and erythromycin in culture flasks (25 cm^2), and the experiment was performed when they became semiconfluent.

2.2. Boron Compounds

In a boron neutron capture therapy (BNCT) group, tumor cells were incubated for 24 hours in boron compound-added condition medium of either borocaptate sodium (BSH) or boronated porphyrin (STA-BX909) with a concentration of 30 μg ^{10}B/ml. During the transportation to Kyoto University Research Reactor Institute (KURRI), the culture flasks were filled up with a condition medium (without boron compounds) to protect the cells from detachment. As a control, in the neutron group, tumor cells were irradiated with thermal neutrons without boron compounds.

2.3. Neutron Irradiation

The co-cultured flasks with or without boron compounds were transferred to the heavy water facility at KURRI. Thermal neutron fluence was arranged according to the irradiation time (0, 15, 30, 45, or 60 min). Neutron fluence and incidental gamma dose were measured with gold foil and TLD. The time interval after dilution of the condition medium until irradiation (the transportation time) was approximately 5 hours.

2.4. Surviving Assay

After neutron irradiation, the cells were dispersed with 0.25% trypsin in 1 mM EDTA and then counted. Cells were seeded onto 6-cm-diameter tissue culture discs. These discs were cultured for 2 weeks in a CO_2 incubator (5% CO_2, 37 °C). The colonies were counted after staining with 0.25% methylene blue in 90% ethanol. The surviving fractions were calculated by dividing the colony number of each disc by those of unirradiated control discs (n = 3 for each dose). The unirradiated control discs were treated in the same manner as the other discs with respect to drug application, transportation, and other procedures. The surviving fractions were plotted and fitted to an exponential curve.

3. RESULTS

The colony-forming assay showed that the cytocidal effect was most prominent in STA-BX909 followed by BSH and least prominent in the neutron-only (Figure 1). Comparing the survival rate of 10% (D_{10}) according to neutron fluence, BSH required 3.89×10^{12} n/cm^2 for D_{10} and 1.8×10^{12} n/cm^2 in STA-BX909. The enhancement ratio of STA-BX909 over BSH at D_{10} was 1.88. The given neutron fluence and gamma dose in each group were, respectively, 1.33×10^{12} n/cm^2 and 34.7 cGy for the 15-min group, 2.87×10^{12} n/cm^2 and 76.3 cGy for the 30-min group, 3.91×10^{12} n/cm^2 and 104 cGy for the 45-min group, and 5.61×10^{12} n/cm^2 and 133 cGy for the 60-min group (Figures 2, 3).

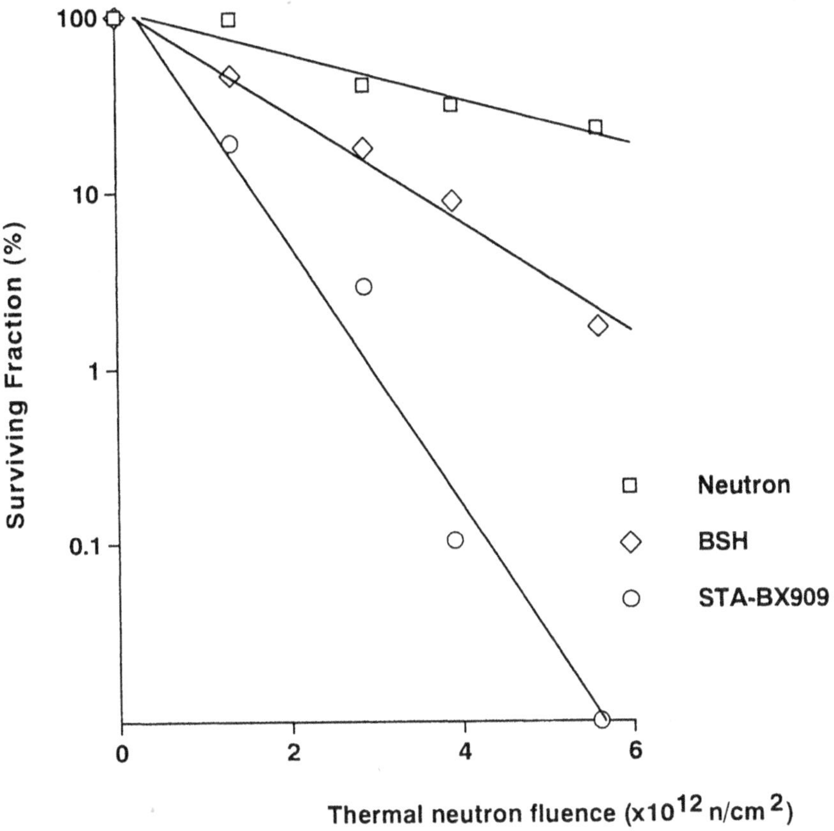

Figure 1. Surviving fraction of 9L cells after thermal neutron irradiation.

4. DISCUSSION

Porphyrins have been investigated for photodynamic agents[1] as well as for their capacity as boron carriers with some promising results. BOPP[2,11] was recently approved for phase I study as a photodynamic agent in Australia (Kahl, personal communication). Other nickel or copper tetracarboranyl-phenylporphryin has also been developed and succeeded in obtaining a therapeutic level of boron concentration.[5] In our study, a prominent *in vitro* radiobiological effect was demonstrated by using boronated manganese metalloporphyrin STA-BX909. Our compounds contain manganese in the center of the porphyrin, which eliminates phototoxicity due to its very short triplet lifetime.[12] This non-phototoxic property is a great advantage for clinical use.

From the three STA-BX series compounds, STA-BX909 was chosen for *in vitro* irradiation study due to its high tumor uptake *in vivo*. The gradients in surviving fractions between BSH and STA-BX909 were assumed to be due to the differences in concentration and distribution of boron within or around the tumor cells. The *in vitro* uptake study (boron uptake in cultured cells) has shown that boron concentrations of the cells increased as a function of time in STA-BX909 while it remained low in BSH[4] (also described in another chapter in these proceedings).

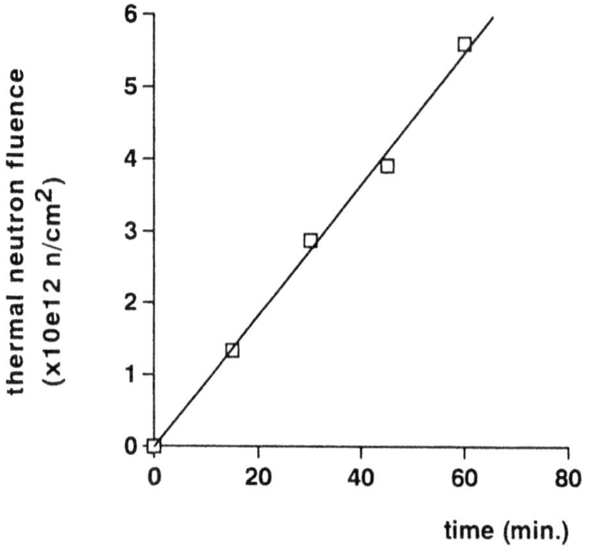

Figure 2. Correlation between neutron fluence and exposure time.

STA-BX909 showed accumulation into the cells, which resulted in more-prominent cell damage than BSH when cultured with the same boron concentration in the medium.[4] The irradiation study was performed using the condition medium concentration of about $3\,\mu g\,^{10}B/ml$, which is a 10-times dilution of the boron concentration (due to the cell protection during transportation as described above). Due to technical difficulties, the time lag between irradiation and culture medium washing could not be shortened in the present study. There was approximately 5 hours between medium dilution and irradiation. This may have caused boron leakage out of the cells. Nevertheless, there was a marked difference in the cell damage effect. This indicates that STA-BX909 is firmly

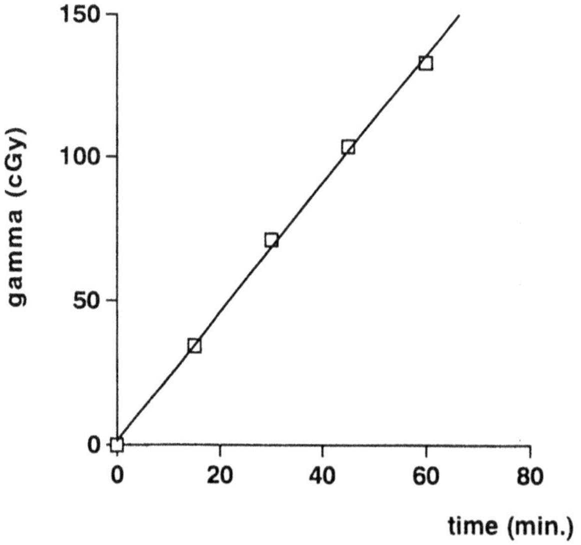

Figure 3. Correlation between gamma dose and exposure time.

incorporated in the cells. Fairchild,[11] in comparing BSH and boronated porphyrins, has reported the same tendency. The washing procedure decreased the cytocidal effect more in BSH than in boronated porphyrins, which is in line with our observation. We also proved the boron uptake by light microscopic examination in observing the brownish coloration of STA-BX909 within the cytoplasm (data not shown). However, whether or not STA-BX909 was incorporated in the nucleus could not be confirmed.

In future studies, the time interval between irradiation and the washing procedure should be shortened in order to eliminate the wash-out effect.

In conclusion, boronated porphyrin (STA-BX909) showed cytocidal effectiveness *in vitro*. However, since the *in vitro* and *in vivo* conditions differed, further study should be performed to prove the efficacy of STA-BX909 as a potential boron delivery agent for BNCT. *In vivo* study should be performed in the future to determine the tumor control effect.

ACKNOWLEDGMENTS

Part of this research was performed at the heavy water reactor facility at KURRI, and the authors are grateful to Dr. S. Masunaga, Dr. Y. Sakurai and Dr. T. Kobayashi of KURRI for their valuable support. This study was supported in part by the University of Tsukuba Research Projects and by a Research Grant-in-Aid for Scientific Research from the Japanese Ministry of Education, Science and Culture (07557091).

REFERENCES

1. T.J. Dougherty, Photodynamic therapy, *Adv Exp Med Biol*, 193:313–328, 1985.
2. S.B. Kahl, M.S, Koo, B.H. Laster, and R.G. Fairchild, Boronated porphyrins in NCT: Results with a new potent tumor localizer, *Strahlenther Onkol*, 165:134–137, 1989.
3. A. Matsumura, Y. Shibata, T. Yamamoto, H. Fujimori, K. Nakai, I. Sakata, N. Miwa, S. Nakajima, and T. Nose, Pharmacokinetic Study of Boronated Metalloporphyrin (STA-BX900) and BSH in the 9L Rat Glioma Model, *in* "Advances in Neutron Capture Therapy", B. Larsson, J. Crawford, and R., Weinrich ed. Vol. II, Elsevier, Amsterdam, pp. 158–162, 1997.
4. A. Matsumura, Y. Shibata, T. Yamamoto, F. Yoshida, K. Nakai, M. Kiriya, N. Shimojo, Y. Hayakawa, T. Nose, S. Nakajima, I. Sakata, N. Miwa, and Y. Itoh, Radiobiological effects of boronated porphyrin (STA-BX909) as a compound for neutron capture therapy. An *in vitro* study, *in* "Recent Progress of Photodynamic Therapy. Proceedings of the 7th Annual Conference of JCIPA". T. Nishisaka ed., Ionics, Tokyo, pp. 125–130, 1997.
5. M. Miura, P.L. Micca, C.D. Fisher, C.R. Gordon, J.C. Heinrichs, and D.N. Slatkin, Evaluation of carborane-containing porphyrins as tumour targeting agents for boron neutron capture therapy. *Br J Rad*, 71:773–781, 1998.
6. Y. Shibata, A. Matsumura, T. Yamamoto, K. Nakagawa, Y. Yoshii, T. Nose, I. Sakata, S. Nakajima, Y. Hayakawa, and K. Ono, Histologic changes in the rat 9L glioma after BNCT with a BSH-porphyrin compound, *in* "Advances in Neutron Capture Therapy", B. Larsson, J. Crawford, and R. Weinrich ed. Vol. II, Elsevier, Amsterdam, pp. 501–506, 1997.
7. T. Yamamoto, A. Matsumura, Y. Shibata, T. Yoshizawa, H. Fujimori, K. Nakagawa, K. Nakai, Y. Yoshii, T. Nose, I. Sakata, and S. Nakajima, MR imaging using boronated Mn-metalloporphyrin (STA-BX900) for BNCT, *in* "Advances in Neutron Capture Therapy", B. Larsson, J. Crawford, and R. Weinrich ed. Vol. II, Elsevier, Amsterdam, pp. 267–270, 1997.
8. Y. Shibata, A. Matsumura, F. Yoshida, T. Yamamoto, K. Nakai, T. Nose, I. Sakata, and S. Nakajima, Cell cycle dependency of porphyrin uptake in glioma cell line: Flow cytometry analysis. *Cancer Lett*, 129:77–85, 1998.

9. T. Yamamoto, A. Matsumura, Y. Shibata, H. Fujimori, K. Nakai, F. Yoshida, T. Nose, I. Sakata, S. Nakajima, and N. Miwa, Manganese-metalloporphyrin (ATN-10) as a tumor-localizing agent: magnetic resonance imaging and inductively coupled plasma atomic emission spectroscopy study with experimental brain tumors. *Neurosurgery*, 42:1332–1338, 1998.
10. H. Fujimori, A. Matsumura, T. Yamamoto, Y. Shibata, T. Yoshizawa, K. Nakagawa, Y. Yoshii, T. Nose, I. Sakata, and S. Nakajima, Tumor Specific Contrast Enhancement Study of Mn-Metalloporphyrin (ATN-10). Comparison of Rat Brain Tumor Model, Cytotoxic and Vasogenic Models, *Acta Neurochir Suppl* 70:167–169, 1997.
11. R.G. Fairchild, S.B. Kahl, B.H. Laster, J. Kalef-Ezra, and E.A. Popenoe, *In vitro* Determination of Uptake, Retention, Distribution, Biological Efficacy, and Toxicity of Boronated Compounds for Neutron Capture Therapy: A Comparison of Porphyrins with Sulfhydryl Boron Hydrides, *Cancer Res*, 50:4860–4865, 1990.
12. T. Takemura, N. Ohta, S. Nakajima, and I. Sakata, Critical importance of the triplet lifetime of photosensitizer in photodynamic therapy of the tumor. *Photochem Photobiol*, 50:339–344, 1989.

MUTAGENIC EFFECTS AT HPRT LOCUS IN CHINESE HAMSTER OVARY (CHO) CELLS INDUCED BY EPI-THERMAL AND THERMAL NEUTRONS

Yuko Kinashi, Yoshinori Sakurai, Shinichiro Masunaga,
Minoru Suzuki, Masao Takagaki, and Koji Ono

Research reactor Institute, Kyoto University
Kumatori-cho, Sennan-gun, Osaka 590-0494, Japan

INTRODUCTION

We previously reported the mutagenic effects of thermal neutrons in Chinese hamster ovary (CHO) cells, presenting evidence for increased the mutagenicity of thermal neutrons in the presence of boron.[1] The remodeling construction of the KUR heavy water thermal neutron irradiation facility was completed in 1996. Neutron irradiation with several energy spectra ranging from nearly pure thermal neutrons to epi-thermal neutrons can be provided by update facility.[2,3] The purpose of the present investigation is to examine the cell killing effects and the mutagenicity of an epi-thermal neutron beam from the KUR reactor.

MATERIALS AND METHODS

Cell Culture

CHO K-1(wild type) cells were cultured at 37 °C in a humidified 5% CO_2 atmosphere in MEM medium supplemented with 10% heat-inactivated calf serum. The cells were pre-incubated with ^{10}B-boric acid (^{10}B; 10 μg/ml) for 1 hour at 37 °C before neutron irradiation.

Radiation Sources

The Heavy Water Column of the Kyoto University Research Reactor (KUR) was used.

Frontiers in Neutron Capture Therapy, edited by Hawthorne *et al.*
Kluwer Academic / Plenum Publishers, New York, 2001.

Table 1. Neutron fluence (n/cm^2 min) and dose rates for each neutron irradiation mode

Irradiation mode	Thermal neutrons (n/cm^2 min) dose rates (cGy/min.)	Epi-thermal neutrons[a] (n/cm^2 min) dose rates (cGy/min.)	Fast neutrons[a] (n/cm^2 min) dose rates (cGy/min.)	γ-ray dose rates (cGy/min.)
Thermal mode	1.2×10^{11}	0.14×10^{10}	4.0×10^{8}	
	1.6	0.01	0.22	2.0
Mixed mode	3.6×10^{11}	6.61×10^{11}	2.8×10^{9}	
	4.7	5.3	0.15	6.2
Epi-thermal	none	6.15×10^{11}	2.8×10^{9}	
mode	0	4.9	0.15	1.2

[a]Measurements was carried out using the "irradiation rail device".
It is assumed that epithermal neutrons have a pure 1/E spectrum.

We choose three conditions, 1) thermal neutron mode, 2) mixed mode with thermal and epi-thermal neutron, 3) epi-thermal neutron mode. The thermal neutron were measured by gold foil activation analysis. The measurement of epi-thermal neutron and fast neutron fluences and dose rates were carried out using the "irradiation rail device" and estimated by the measured values of the neutron fluxes at the bismuth surface reported previously.[2,3] The gamma ray dose was measured with thermoluminescence dosimeters (Table 1).

Cell Survival Assay

Survival curves were obtained by standard colony formation assay. The survival fraction was calculated and the survival curves were obtained for each cell line.

Hypoxanthine Phosphoribosil Transferase (HPRT) Locus Mutation Assays

After thermal neutron exposure, to determine mutation frequencies, each treated cell culture was incubated with non-selective medium for 7 to 9 days to allow phenotype expression. Then, 2×10^5 cells were added to each of 20, P-100mm Petri dishes containing 5μg/ml of 6-thioguanine (6TG) and incubated for 10 days after which time the mutant colonies were counted. The mutation frequency is expressed as the number of resistant colonies divided by the total number of viable cells as determined by the cloning efficiency at the time of selection.

RESULTS

Induced mutant fraction normalized to the survival level is shown in Figure 1, and Table 2. In the Table 2, we compared the mutant frequencies induced by neutrons to those by gamma-rays. In the thermal neutron mode, clear differences were not observed among the frequencies of thermal neutron-induced mutations with or without boron. In the mixed mode with thermal and epi-thermal neutrons, mutation frequencies by thermal and epi-thermal neutron exposure, comparing at the 10% survival level, were 1.7 times

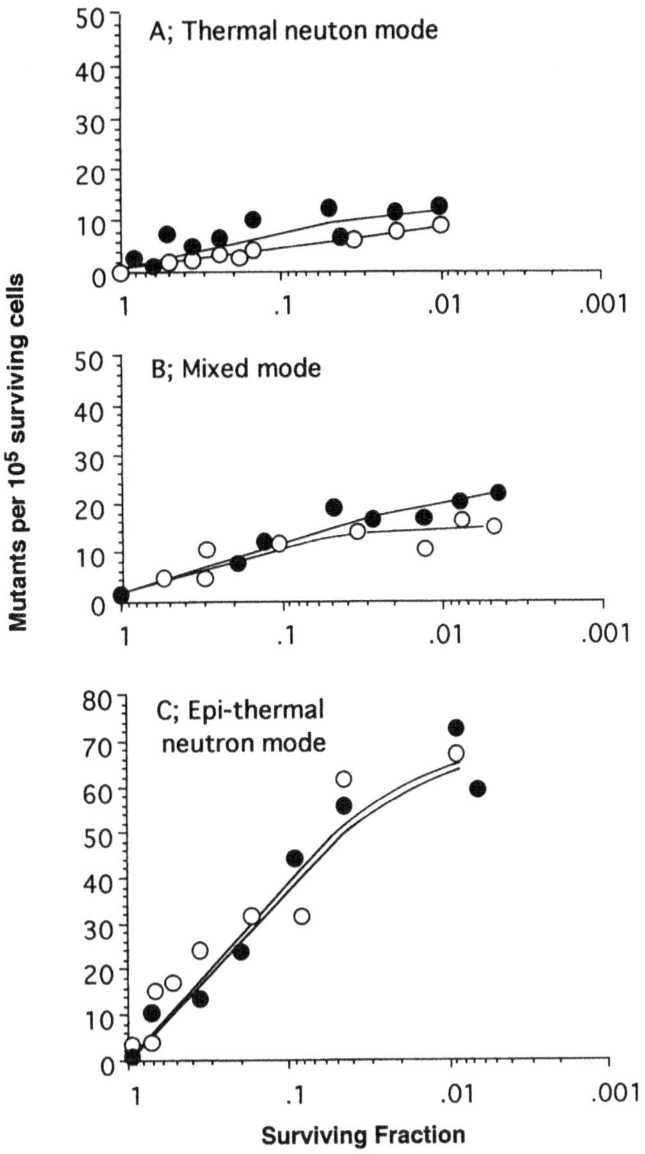

Figure 1. Induced mutant fraction normalized to the survival level at the HPRT locus in CHO cells by thermal and epithermal neutrons in three different modes by KUR reactor. Boron concentration was 0 ppm (○), 10 ppm (●). Lines were fitted by eyes.

higher than the thermal neutron mode in the absence of boron. The slightly higher mutagenicity was observed in the presence than in the absence of boron. In the epi-thermal neutron mode, mutation frequencies by epi-thermal neutron exposure showed 3.8 and 6.3 times higher mutagenicity than thermal neutron mode and mixed mode in the absence of boron, respectively. No clear differences were observed among the frequencies of epithermal neutron-induced mutations with or without boron.

Table 2. RBE value of HPRT mutation induction obtained from mutant fraction curves of CHO cells irradiated by thermal and epi-thermal neutron beam

Radiatios	Boron concentration	Mutants per 10^5 cells at SF was 0.1[a]	RBE[b]
^{60}Co γ-rays	—	6.0	1.0
Neutron			
Thermal neutron	$^{10}B = 0\,ppm$	6.0	1.0
mode	$^{10}B = 10\,ppm$	10.0	1.7
Mixed mode	$^{10}B = 0\,ppm$	10.0	1.7
	$^{10}B = 10\,ppm$	12.0	2.0
Epi-thermal neutron	$^{10}B = 0\,ppm$	38.0	6.3
mode	$^{10}B = 10\,ppm$	38.0	6.3

[a]Mean of mutant frequency from separated three experiments.
[b]RBE (relative biological effectiveness) was obtained by ratio of mean value of number of mutants compared to that of γ-rays at the 0.1 survival level.

DISCUSSION

Our result shows that 1) the mutagenicity of epi-thermal neutron was much higher than that of the other modes and 2) boron did not increased the mutant frequency significantly.

1) The differences observed in the mutation frequencies may be due to the different mixtures of dose components. The dose rate for each of the different radiation components in the three beams were shown in Table 1. In the epi-thermal mode, the fraction of epi-thermal neutrons was 78%, in the total dose. In the mixed mode, the fraction of epi-thermal neutrons was 32%, in the total dose. There was a correlation between the mutagenicity and the dose contribution of epi-thermal neutron component in the neutron beams. The high mutagenicity of epi-thermal neutrons compared to the cell killing efficiency might be explained by the production of large numbers of double strand breaks. Various investigators have measured the double strand breaks frequency of fast neutrons and reported the double strand breaks amount by fast neutrons were 1.1-fold larger than that of gamma rays by sucrose gradient method[4] and 1.0 or 1.6-fold larger by neutral elution method,[5,6] respectively. With fission neutrons, the yield of strand breaks per unit dose were reported to be the same for α-particles as for gamma-rays.[7] We propose that epi-thermal neutrons produced double strand breaks at high frequency and neutron-induced double strand breaks have a higher probability of being lethal because they are more likely to be misrepaired.
2) In the presence of boron, the energy of ^4He and ^7Li ions produced by the $^{10}B(n,\gamma)^7Li$ reaction is increased to 1.47 MeV(α-particle) and 0.84 MeV(^7Li-particle). The ranges of LET for α-particle and ^7Li -particle are 50–231 keV/μm and 65–266 keV/μm, respectively. In general, the "overkill " effect of high-LET irradiation is well known[8] and this effect might explain our result that boron does not increase the mutant frequency significantly. In BNCT, the cells which take up boron get the large total absorbed doses and the cell-killing effect and mutagenicity of neutrons are enhanced by the concentration of boron.

However, at the same survival level, our results suggested that boron did not increased the mutagenicity of neutrons. Therefore, the effect of boron as in BNCT is minimal in terms of mutation induction.

ACKNOWLEDGMENTS

This study was supported by Grant-in-Aid for Scientific Research of the Ministry of Education, Science and Culture of Japan.

REFERENCES

1. Y. Kinashi, S. Masunaga, M. Takagaki, and K. Ono, Mutagenic effects at HPRT locus induced in Chinese hamster ovary cells by thermal neutrons with or without boron compound. Mutation Res., 377:211–215, 1997.
2. T. Kobayashi, K. Kanda, Y. Fujita, Y. Sakurai, and K. Ono, The upgrade of the heavy water facility of the Kyoto University Reactor for neutron capture therapy. In: Advances in Neutron Capture Therapy. Volume I, Medicine and Physics. B. Larsson, J. Crawford, and R. Weinreich, ed., Elsevier, Amsterdam, pp. 321–325, 1997.
3. Y. Sakurai, T. Kobayashi, K. Kannda, and K. Ono, The irradiation characteristics of the upgrade heavy water facility of the Kyoto University Reactor. In: Advances in Neutron Capture Therapy. Volume I, Medicine and Physics. B. Larsson, J. Crawford, and R. Weinreich, ed., Elsevier, Amsterdam, pp. 316–320, 1997.
4. I. Furuno, T. Yada, H. Matsudaira, and T. Maruyama, Induction and repair of DNA strand breaks in cultured mammalian cells following fast neutron irradiation. Int. J. Radiat. Biol., 36:639–648, 1979.
5. G.P. Van Der Schans, M.C. Paterson, and W.G. Cross, DNA strand breaks and rejoining in cultured human fibroblasts exposed to fast neutrons or gamma rays. Int. J. Radiat. Biol., 44:75–85, 1983.
6. B.P. Kysela, J.E. Arrand, and B.P. Michael, Relative contributions of levels of initial damage and repair of double-strand breaks to the ionizing radiation-sensitive phenotype of the Chinese hamster cell mutant, XR-V15B. Part II. Neutrons. Int. J. Radiat. Biol., 64:531–538, 1993.
7. G.D.D. Jones, J.R. Milligan, J.F. Ward, P.M. Calablo-Jones, and J. Aguilera, Yield of strand breaks as a function of scavenger concentration and LET for SV 40 irradiated with ^{4}He ions. Radiation. Res., 136:190–196, 1993.
8. E.J. Hall, Radiobiology for the Radiologist Third Edition, Lippincott, Philadelphia, pp. 170–172, 1988.

EVALUATION OF NEUROTOXICITY AND RADIATION EFFECTS OF BNCT IN RATS FOLLOWING BLOOD-BRAIN BARRIER DISRUPTION AND INTRACAROTID INJECTION OF BPA OR BSH

Weilian Yang[1], Rolf F. Barth[1], Joan H. Rotaru[1],
Jacqueline C. Bresnahan[2], Maria Hadjiconstantinous[3], Darrel D. Joel[4],
and Marta M. Nawrocky[4]

[1]Department of Pathology
[2]Department of Cell Biology and Neurobiology
[3]Department of Pharmacology
The Ohio State University
Columbus, OH 43210
[4]Medical Department
Brookhaven National Laboratory
Upton, New York 11973

1. INTRODUCTION

The blood-brain barrier (BBB), which is formed by tight junctions of endothelial cells lining cerebral blood vessels and capillaries, effectively limits the delivery of most water soluble chemotherapeutic agents to brain tumors and surrounding brain tissue.[1] Using the F98 rat glioma model, we recently have reported that osmotic blood-brain barrier disruption (BBB-D), produced by the intracarotid (i.c.) infusion of a hyperosmotic solution of mannitol, followed by i.c. administration of BPA or BSH significantly enhanced tumor boron delivery and the efficacy of BNCT in F98 glioma bearing rats.[2-4] BPA and BSH have not been shown to have any significant short term neurotoxic and long term adverse radiation effects at doses that have been used clinically for BNCT.[5,6] Since osmotic BBB-D transiently opens tight endothelial cell junctions of capillaries in both normal brain and tumor of the infused cerebral hemisphere,[7] there is an increase in normal brain boron levels. However, by 2.5 hours these fall to values similar to those obtained following i.v. injection.[2-4] The purpose of present study was to evaluate the short term neurotoxic and the long term neuropathologic and functional effects of BBB-D and

Frontiers in Neutron Capture Therapy, edited by Hawthorne *et al.*
Kluwer Academic / Plenum Publishers, New York, 2001.

i.c. infusion of BPA or BSH in non-tumor bearing rats following BNCT studies similar to those carried out in tumor bearing animals.[2-4]

2. MATERIALS AND METHODS

2.1. Compound Administration, BBB-D and Evaluation of Neurotoxicity

Ten to 12 weeks old male Fischer rats weighing 200–220 g were anesthetized using a mixture of 5% isoflurane and 0_2, and maintained with 1% isoflurane, 70% nitrous oxide, and 29% 0_2 and then placed supine with all limbs restrained. The procedure used for cannulation of the external carotid artery has been described in detail elsewhere.[3] BBB-D was carried out by the i.c. infusion of a 25% solution of mannitol (1.373 mOsmol/ml) via the internal carotid artery at a rate of 0.12 ml/sec for 30 seconds.[4] Either BPA (25–45 mg ^{10}B/kg) or BSH (30–50 mg ^{10}B/kg) was infused for 5 min. immediately after BBB-D. Each agent was passed through a 0.2 μ filter and pre-warmed to 37 °C prior to infusion. To evaluate the possible neurotoxic effects of BPA or BSH following i.c. administration and BBB-D, rats were sacrificed at 2 and 30 days after surgery and brains were removed for neuropathological examination.

2.2. Neutron Iradiation and Neuropathologic Evaluation

Three groups of 5 animals each had BBB-D, followed by i.c. injection of saline, BPA (25 mg ^{10}B/kg) or BSH (30 mg ^{10}B/kg). They were irradiated 2.5 hours later at the Brookhaven Medical Research Reactor (BMRR) at 2 megawatts (MW) power for 4 minutes to give a dose of 8 MW-minutes. A blood sample for boron determination was taken from each animal prior to irradiation. All animals were weighed and their neurological status was evaluated at regular intervals for one year. A complete ophthalmic examination was performed on animals 10 months following BNCT. All rats were euthanized 12 months after irradiation and neuropathological changes were evaluated. Prior to euthanization, the animals were anesthetized, as described in the preceding section, and then perfused through the left ventricle of the heart with Cajal's formol-bromide fixative (2% w/v ammonium bromide in 14% formalin) using a pump system.[8] The brains subsequently were removed and kept for at least 24 hours in the same fixative, following which they were cut coronally at the level of optic chiasm, and 2 mm anterior and posterior to it. Six micrometer sections were cut and stained with hematoxylin and eosin and selected sections were stained with luxol fast blue and PAS. Neuropathologic changes were scored microscopically, as described in the results section, as follows: 0, none; 1, minimal; 2, mild; 3, moderate; and 4, severe.

2.3. Evaluation of Motor and Cognitive Function

Motor function of irradiated rats was assessed at 12 months using a quantitative locomotor grading scale.[9] Cognitive function, as measured by spatial learning and memory, was evaluated at the same time using the Morris water maze for open field testing of rats.[10] Animals always were tested in the same order, using the same maze and spatial enviroment, and at approximately the same time each day. Assessment of behavioral performance was made with the assistance of an automated computer program

(Videomex-V Water Maze Program, Version 2.11, Columbus Instruments, Columbus, OH). The screening of animals in the pre-test included blocks of four trials daily for 5 consecutive days. The initial interval was 60 seconds and on the last day, a spatial probe test was conducted following four training trials using a hidden platform, which was removed from the pool, and each animal was allowed to swim for 60 sec. starting from the north quadrant.

3. RESULTS AND DISSCUSSION

3.1. Short Term Neuropathological Effects

Short term neuropathological changes seen 2 days following i.c. administration of BPA or BSH with BBB-D are summarized in Table 1.

Only minor neuropathological changes were noted in the barrrier disrupted cerebral hemispheres of a few of the rats that received i.c. BPA or BSH at a dose of 25–30 mg ^{10}B/kg. However, scattered foci neuronal degeneration and hemorrhogic foci were seen in the barrier disrupted hemispheres of approximately half the rats that received the higher dose of BPA or BSH (45–50 mg^{10}B/kg).

3.2. Late Radiation Effects

Late effects were assessed 12 months after irradiation and the neuroanatomic sites of primary interest were the fimbriae, choroid plexus and leptomeninges. No pathological changes were seen in the leptomeninges and there was no evidence of neuronal loss or gliosis in the fimbriae. Small foci of mineralization were observed in the basal ganglia and thalamus. Neuropathological changes seen in the choroid plexus are summarized in Table 2. These included dilation of blood vessels and atrophy, flattening and vacuolation of choroidal epithelial cells, which ranged from minimal (1+) to moderate (2–3+) in 2 of 5 rats that received BPA. The mild neuropathological changes suggest that BBB-D followed by i.c. BPA or BSH and BNCT were well tolerated by the brain.

Table 1. Neuropathological changes in the brains of rats after i.c. injection of BPA or BSH following BBB-D[a]

Group	^{10}B Dose (mg/kg b.w.)	Inflammation[b]	Neuronal degeneration	Hemorrhage
BPA	25	0 (4/4)	0 + (2/4) 2 + (2/4)	0 (4/4)
BPA	45	0 (4/4)	0 (2/4) 2–3 + (2/4)	0 (2/4) 2 + (1/4) 4 + (1/4)
BSH	30	0 (4/4)	0 (3/4) 2 + (1/4)	0 (4/4)
BSH	50	0 (4/4)	1 + (2/4) 4 + (2/4)	0 (4/4)
Control	None	0	0	0

[a]Changes were scored as follows: 0 no apparent change; 1+ minimal; 2+ mild, 3+ moderate, and 4+ severe.
[b]The number of animals showing changes relative to the total number in the group is indicated in parentheses.

Table 2. Pathological changes in the choroid plexus of rats following BNCT or irradiation

Group	^{10}B dose (mg/kg b.w.)	Physical dose (Gy)[a]	Dilation of blood vessels[b]	Epithelial cell atrophy	Flattening epithelium	Vacuolation of epithelial cells
BPA	25	8.10	1 + (4/5)	1 + (3/5)	1 + (3/5)	0 (3/5)
			2–3 + (1/5)	2–3 + (2/5)	2–3 + (2/5)	1 + (2/5)
BSH	30	5.72	1 + (4/5)	1 + (5/5)	1 + (5/5)	0 (3/5)
			3 + (1/5)			1 + (2/5)
Control[c]	None	4.42	0 (4/5)	0 (5/5)	0 (3/5)	0 (4/5)
			1–2 + (1/5)		1 + (2/5)	1 + (1/5)

[a]All doses were calculated ~ at a depth of 4–5 mm in the brain.
[b]Changes were scored as follows: 0, no apparent change, 1+ minimal, 2+ mild, 3+ moderate, and 4+ severe. The number of animals showing the change relative to the total number in the group is indicated in parentheses.
[c]Four of five animals from this group showed changes in white and deep gray matter (basal ganglia and thalamus). These included gliosis in the basal ganglia 4 + (1/5); focal necrosis in the thalamus 2 + (1/5); old focal infarcts in the basal ganglia 1 + (1/5) and white matter 1 + (1/5).

3.3. Motor and Cognitive Function

None of the rats developed any clinically significant neurological deficits. Some animals had mild locomotor deficits, as evidenced by slightly decreased trunk stability and ability to maintain foot position throughout the stance phase. Locomotor rating scores ranged from 18–21 in animals one year after irradiation. Only minor differences were seen in scores of irradiated controls and of those rats that received either BPA or BSH (18.5, 18.9 and 19.3, respectively). Cognitive function, as measured by performance the Morris water maze, revealed that by the end of the five day test both control and BNCT-treated groups improved their ability to spatially navigate with a reduction in both escape latencies and swim distance scores (Figure 1).

4. CONCLUSIONS

In the present study, we have established that BBB-D combined with i.c. administration of either BSH or BPA and followed by BNCT produced no significant short term neuropathologic effects, and only minimal to mild changes were seen at 12 months in the chroid plexus, a critical radiation sensitive region of the brain. Locomotor and cognitive function test scores were similar to those determined in age-matched control rats. It can

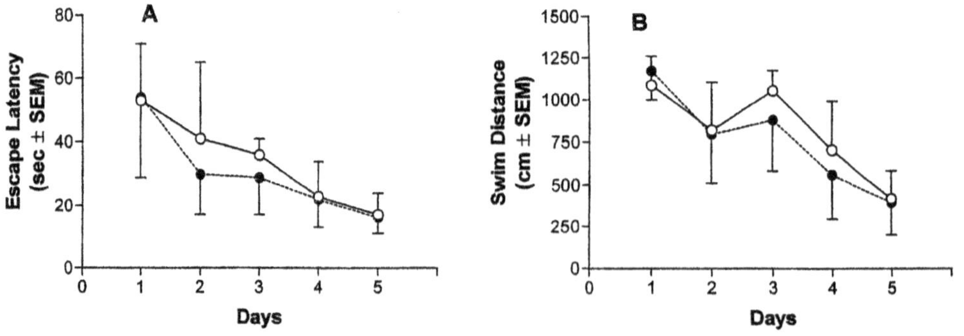

Figure 1. Escape Latency scores (A) and Swim Distance scores (B) for control (-O-) and BNCT-treated (-●-) rats.

be concluded that BBB-D combined with i.c. administration of either BPA or BSH and neutron irradiation was well tolerated in rats. Since optimizing the delivery of these two compounds has such a significant effect on enhancing survival in brain tumor bearing rats[2–4] and does not produce significant adverse effects, carefully performed clinical studies should be considered.

ACKNOWLEDGMENTS

We thank Ms. Dianne Adams for technical assistance, Ms. Amy Tovar, Neurosciences Spinal Injury Laboratory for assistance in carrying out the locomotor function test and Mrs. Linda Carlos for secretarial assistance in the preparation of this manuscript. The work described in this report has been supported by the United States Department of Energy Grants DE-AC02-76CH000016 and DE-FG02-95ER62059.

REFERENCES

1. N.H. Greig: Drug delivery to the brain by blood-brain barrier circumvention and drug modification. In: Implications of the Blood-Brain Barrier and its Manipulation (A.E. Neuwelt, ed). Plenum Press, New York, pp. 311–357, 1994.
2. R.F. Barth, W. Yang, J.H. Rotaru, M.L. Moeschberger, D.D. Joel, M.M. Nawrocky, J.H. Goodman, and A.H. Soloway: Boron neutron capture therapy of brain tumors: enhanced survival following intracarotid injection of either sodium borocaptate or boronophenylalanine with or without blood-brain barrier disruption. Cancer Res. 57:1129–1136, 1997.
3. W. Yang, R., F. Barth, J.H. Rotaru, M.L. Moeschberger, D.D. Joel, M.M. Nawrocky, J.H. Goodman, and A.H. Soloway: Boron neutron capture therapy of brain tumors: enhanced survival following intracarotid injection of sodium borocaptate with or without blood-brain barrier disruption. Int. J. Radiat. Oncol. Biol. Phys. 37:663–672, 1997.
4. W. Yang, R.F. Barth, J.H. Rotaru, D.D. Joel, M.M. Nawrocky, and J.H. Goodman: Enhanced survival of glioma bearing rats following boron neutron capture therapy with osmotic blood-brain barrier disruption and intracarotid injection of boronophenylalanine. J. Neurooncol 32:59–70, 1997.
5. Y. Nakagawa, and H. Hatanaka: Boron neutron capture therapy: Clinical brain tumor studies. J. Neurooncol 33:105–115, 1997.
6. J.A. Coderre, E.H. Elowitz, M. Chadha, R. Bergland, J. Capala, D.D. Joel, H.B. Liu, D.N. Slatkin, and A.D. Chanana: Boron neutron capture therapy for gliobastoma multiforme using *p*-boronophenylalanine and epithermal neutrons: Trial design and early clinical results. J. Neurooncol 33:141–152, 1997.
7. E.A. Neuwelt, M. Pagel, P. Barnett, M. Glassberg, and E.P. Frenkel: Pharmacology and toxicity of intracarotid adriamycin administration following osmotic blood-brain barrier modification. Cancer Res. 42:4466–4470, 1981.
8. K.Z. Matalka, R.F. Barth, M.Q. Bailey, D.A. Wilkie, A. Koestner, and J.W. Hopewell: Radiation effects of boron neutron capture therapy on brain, skin and eye of rats. Int. J. Radiation Oncology Biol. Phy. 28:1089–1097, 1994.
9. D.M. Basso, M.S. Beattie, and J.C. Bresnahan: A sensitive and reliable locomotor rating scale for open field testing in rats. J. Neurotrauma 12:1–21, 1995.
10. R. Morris: Developments of a water-maze procedure for studying spatial learning in the rat. J. Neuroscience Methods 11:61–67, 1984.

CELL SURVIVAL FOLLOWING *IN VITRO* IRRADIATION AT DEPTH IN A LUCITE PHANTOM AS A MEASURE OF EPITHERMAL BEAM RBE

Jeffrey A. Coderre, Jacek Capala, Ruimei Ma, and Michael Makar

Medical Department
Brookhaven National Laboratory
Upton, New York 11973

1. INTRODUCTION

The radiation field produced in tissue during BNCT consists of a mixture of components with differing linear energy transfer (LET) characteristics. In addition to the high-LET products of the $^{10}B(n,\alpha)^{7}Li$ reaction, the interaction of the neutron beam with the nuclei of elements in tissue will deliver an unavoidable, non-specific background dose, from a mixture of high- and low-LET radiation components, to both tumor and normal tissue. Thermal neutron capture by hydrogen releases a gamma ray through the $^{1}H(n,\gamma)^{2}H$ reaction. The capture of thermal neutrons by nitrogen in tissue, the $^{14}N(n,p)^{14}C$ reaction, releases a high-LET proton with an energy of 590 keV. Contaminating fast neutrons (those with kinetic energies >10 keV) in the epithermal neutron beam produce high-LET recoil protons through collisions with hydrogen nuclei ($^{1}H(n,n')p$ reaction) in tissue. Because the energies of the nitrogen capture proton and the fast neutron recoil proton tend to be in the same range, the biological effects of these high-LET components of the dose are most conveniently measured as a combined "proton dose". This high-LET "proton dose" must be multiplied by an experimentally determined factor for relative biological effectiveness (RBE) in order to express the total BNCT dose in photon-equivalent units.

Experimental determination of the RBE factors for the BNCT dose components has, for the most part, been carried out in thermal neutron beam experiments, either *in vitro*, or in small animals.[1] The exception is the dog work by Gavin[2] using the epithermal neutron beams at the High Flux Reactor, Petten, The Netherlands, and at the Brookhaven Medical Research Reactor. A direct measure of epithermal beam RBE in small animals

Frontiers in Neutron Capture Therapy, edited by Hawthorne *et al.*
Kluwer Academic / Plenum Publishers, New York, 2001.

is difficult due to the high whole body exposure. Build-up material would be required to thermalize the incident neutron beam.

In this report we present preliminary data from a model system consisting of cells in culture placed at increasing depths in a lucite phantom in an effort to provide a direct measurement of the RBE of the epithermal neutron beam at the BMRR.

2. METHODS AND MATERIALS

The GS-9L rat gliosarcoma cell line originated from an *N*-nitrosomethylurea-induced primary glioma.[3] Cells are maintained in Dulbecco's Modified Eagle Medium supplemented with 5% fetal bovine serum (inactivated). Colony forming assay methods have been described.[4] The 14 × 14 × 14 cm lucite cube used for epithermal beam dosimetry was used with removable 1 cm-diameter vertical rods such that vials containing cells in growth medium could be positioned along the beam axis center line at various depths from the surface. No boron was present during the irradiations. The clinical BNCT treatment planning software[5] was used to simulate the lucite cube and provide all dose components in the 1 cm^3 volume containing the bottom of the cell vial, and the cells.

3. RESULTS AND DISCUSSION

3.1. Dosimetry

At 3 MW reactor power, the total physical dose rates at depths of 1.0, 2.0, 3.5, and 7.0 cm in the lucite cube phantom were 9.5, 10.2, 9.1, and 4.7 cGy/min, respectively. The fraction of the total dose attributable to gamma radiation was 68% at 1.0 cm, 73% at 2.0 cm, 79% at 3.5 cm and 86% at a depth of 7.0 cm (Figure 1). Fast neutrons comprised 22%, 14%, 10% and 6% of the total physical dose at these same depths. The nitrogen capture component of the dose comprised 11%, 13%, 12% and 8% of the total dose at these depths. Cs-137 gamma irradiation was carried out at a dose rate (8.3 cGy/min) that was similar to the total dose rate used in the reactor experiments.

3.2. RBE Calculations from Cell Survival Data

Figure 2 shows the survival of 9L gliosarcoma cells as a function of reactor exposure time, expressed in MW-min (eg., irradiation for 1 minute at 3 MW reactor power yields 3 MW-min). The survival curve for the 9L gliosarcoma cells exposed to Cs-137 gamma rays is not shown, but Cs-137 doses of 650 cGy, 825 cGy and 1335 cGy reduced survival to 20%, 10% and 1%, respectively. Comparison of isoeffective doses (10% and 1% cell survival) of epithermal neutron beam and Cs-137 gammas allowed direct estimation of the beam RBE (see Table 1). The "beam RBE" is defined as the combined effect of all dose components, both high- and low-LET, compared to the reference Cs-137 dose response curve. Beam RBE is a useful measure for comparison of the effects of different beams, where the relative proportion of dose components may vary.

A measure of the RBE for the high-LET components of the beam (the combined effect of the protons from nitrogen capture and fast neutron recoil) can be obtained in the absence of boron by comparing the neutron beam dose with the Cs-137 gamma dose sufficient to reduce the cell survival to the same level. The result can be expressed as in

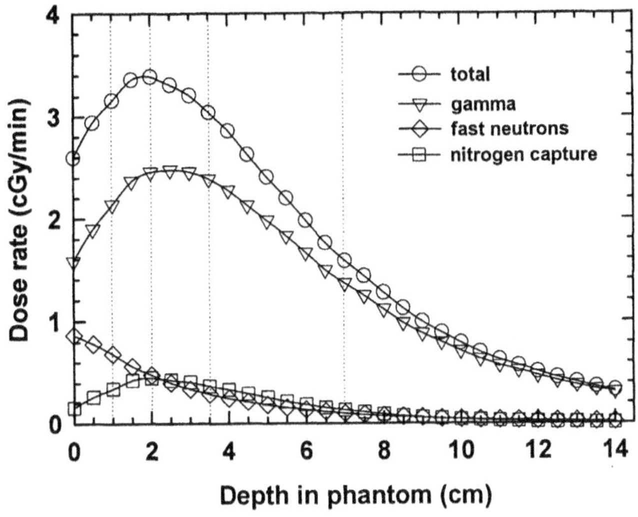

Figure 1. Dose components of the BMRR epithermal neutron beam as a function of depth in the lucite cube phantom. The vertical dotted lines indicate the cell irradiation depths of 1.0, 2.0, 3.5 and 7.0 cm. The dose rates shown are for 1 MW reactor power.

[Eq 1.], where the total physical dose on each side of the equation produces the same biological effect, and assuming that the RBE of the gamma component is 1.

$$\text{[``proton'' dose, in Gy][``proton'' RBE]} + \text{[gamma dose, in Gy]} = \text{Cs gamma dose, in Gy} \quad (1)$$

The RBE of the combined effect of the high-LET beam components is determined by rearranging as in Eq. 2. The results are summarized in Table 1.

$$\text{``proton'' RBE} = \frac{\text{Cs gamma dose, in Gy} - \text{gamma dose, in Gy}}{\text{proton dose, in Gy}} \quad (2)$$

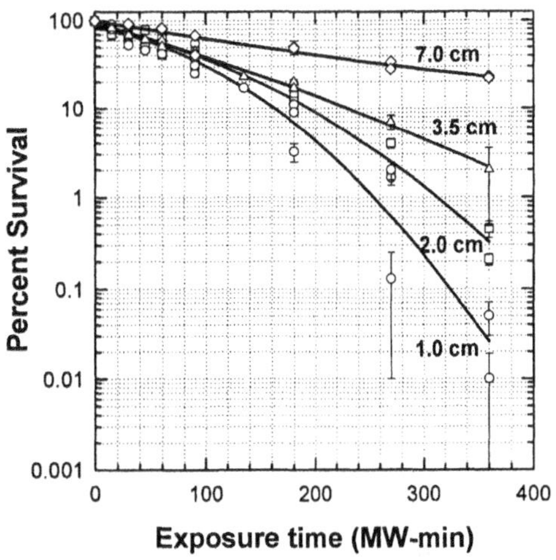

Figure 2. Survival fraction of 9L gliosarcoma cells as a function of increasing exposure time in the BMRR epithermal neutron beam. Cell vials were irradiated at depths of 1.0, 2.0, 3.5 and 7.0 cm in the lucite cube phantom.

Table 1. Calculation of the RBE of the BMRR epithermal neutron beam, and the RBE of the high-LET components of the beam, from cell survival data

	10% Survival endpoint			1% Survival endpoint		
Irradiation location	Dose (cGy)	Beam RBE	High-LET RBE	Dose (cGy)	Beam RBE	High-LET RBE
Epithermal beam 1.0 cm depth	490	1.7	3.1	774	1.7	3.2
Epithermal beam 2.0 cm depth	712	1.2	1.6	1138	1.2	1.7
Epithermal beam 3.5 cm depth	724	1.1	1.6	1259	1.1	1.3
Epithermal beam 7.0 cm depth*	600	1.1	1.5	n.d.**	n.d.	n.d.
Cs-137 gamma	825	1.0		1335	1.0	

*At 7.0 cm depth, the calculations were done using 20% survival as the endpoint. **n.d., not determined.

Beam RBE values derived at depths of 2.0, 3.5, and 7.0 cm were similar: RBE = 1.1 ± 0.05 (Table 1). The beam RBE derived at a depth of 1.0 cm, however, was significantly higher: RBE = 1.7. A similar effect was observed in the calculated RBEs for the high-LET proton component of the dose: the RBE value at 1.0 cm was ≈3.1 and was significantly greater than the value obtained at all deeper positions (RBE = 1.5 ± 0.15). Clearly, the higher proportion of fast neutrons at 1.0 cm contributes to the greater beam RBE at this depth, but the data suggest that the RBE of the fast neutrons may be higher at the surface and decrease as a function of depth. Further studies are planned which are intended to suppress the nitrogen capture dose and obtain direct measurements of the fast neutron RBE.

3.3. Conclusions

These preliminary studies have shown that the technique of using cell survival at depth in a phantom in the epithermal beam can detect differences in beam RBE as a function of depth. A method for direct measurement of the RBE of epithermal neutron beams could be of possible use in a number of applications such as: comparison of the beam RBE from different reactors or accelerator sources; investigation of the influence of dose per fraction on the beam RBE; investigation of whether beam RBE changes as a function of depth; and characterization of the high-LET beam components. Further studies are under way to refine this technique and to define its limitations.

ACKNOWLEDGMENTS

Supported by the U.S. Department of Energy under contract no. DE-AC02-98CH10886.

REFERENCES

1. J.A. Coderre, and G.M. Morris, Review article: The radiation biology of boron neutron capture therapy. *Radiat. Res.*, 151, 1–18, 1999.
2. P.R. Gavin, S.L. Kraft, R. Huiskamp, and J.A. Coderre, A review: CNS effects and normal tissue tolerance in dogs, *J. Neuro-Oncol.*, 33, 71–80, 1997.

3. H.H. Schmidek, S.L. Nielsen, A.L. Schiller, and J. Messer, Morphological studies of rat brain tumors induced by N-nitrosomethylurea, *J. Neurosurg.*, 34, 335–340, 1971.
4. J.A. Coderre, M.S. Makar, P.L. Micca, M.M. Nawrocky, D.D. Joel, D.N. Slatkin, and H.I. Amols, Derivations of relative biological effectiveness for the high-LET radiations produced during boron neutron capture irradiations of the 9L rat gliosarcoma *in vitro* and *in vivo*, *Int. J. Rad. Oncol. Biol. Phys.*, 27, 1121–1129, 1993.
5. D.W. Nigg, F.J. Wheeler, D.E. Wessol, J. Capala, and M. Chadha, Computational dosimetry and treatment planning for boron neutron capture therapy. *J. Neuro-Oncol.*, 33, 93–103, 1997.

APPLICABILITY OF COMBINATION WITH TIRAPAZAMINE IN BORON NEUTRON CAPTURE THERAPY

Concerning Effectiveness on Quiescent Tumor Cell Populations within Solid Tumors

Shin-ichiro Masunaga[1], Koji Ono[1], Yoshinori Sakurai[2], Hitoshi Hori[3], Masao Takagaki[1], Toru Kobayashi[2], Minoru Suzuki[1], Yuko Kinashi[1], and Mitsuhiko Akaboshi[2]

[1]Radiation Oncology Research Laboratory
[2]Division of Radiation Life Science
Research Reactor Institute, Kyoto University
Noda, Kumatori-cho, Sennan-gun, Osaka 590-0494, Japan
[3]Department of Biological Science and Technology
Faculty of Engineering, University of Tokushima
2-1, Minamijosanjima-cho, Tokushima 770, Japan

INTRODUCTION

Boron neutron capture therapy (BNCT) represents a promising modality for selective irradiation of tumor tissue, which selectively and maximally damages malignant cells while sparing normal tissue. This treatment modality selectively concentrates a boron compound in malignant cells, and then "activates" this compound with slow thermal neutrons resulting in a lethal event within the cancer cell.

On the other hand, it is well known that many tumor cells are not proliferating (quiescent) *in situ*, but are still clonogenic.[1] Therefore, to improve treatment against cancer, it is also necessary to determine the response of quiescent (Q) tumor cells to the treatment.

Here, we made an analysis of the killing effect of thermal neutron capture reaction on Q cells in solid tumors in terms of the micronucleus (MN) frequency, in comparison with that on the total tumor cell population, and we attempted to evaluate the usefulness of tirapazamine (TPZ), a novel bioreductive agent with selective

Frontiers in Neutron Capture Therapy, edited by Hawthorne *et al.*
Kluwer Academic / Plenum Publishers, New York, 2001.

cytotoxicity against hypoxic tumor cells, combined with ^{10}B-compound in BNCT for murine solid tumors.

MATERIALS AND METHODS

Labeling with 5-bromo-2′-deoxyuridine (BrdU)

SCC VII tumor cells were inoculated subcutaneously into both hind legs of female mice. Fourteen days later, the tumor reached 1 cm in diameter. Nine days after inoculation, BrdU was administered continuously for 5 days using subcutaneously implanted mini-osmotic pumps to label all proliferating cells in the tumors. Fourteen days after inoculation, the tumor was treated and its diameter was 1 cm. The labeling index after continuous administration of BrdU was 55.3 ± 4.5% (mean ± SD). In other words, the growth fraction of the tumors were considered to be 55.3 ± 4.5%. Therefore, tumor cells not incorporating BrdU after continuous administration were regarded as Q cells for all practical purposes.

Treatment

Twenty minutes after intraperitoneal injection of sodium borocaptate-^{10}B (BSH) of 125 mg/kg (71.4 mg/kg of ^{10}B) or 3 hours after oral administration of *dl-p*-boronophenylalanine-^{10}B (BPA) of 1,500 mg/kg (72.1 mg/kg of ^{10}B), the tumors were irradiated with thermal neutrons. For combination with TPZ, TPZ of 50 mg/kg was intraperitoneally injected 30 min before neutron irradiation. Each treatment group included both mice with and without BrdU administration.

Thermal neutrons were generated by Kyoto University Reactor. We used thermal neutron beams with a cadmium ratio of more than 5,000. Therefore, the amount of contaminated fast neutrons was considered to be negligible. The thermal neutron fluence was measured with a phantom using gold foils. Gamma-ray contamination from the reactor was measured with thermoluminescence dosimeter powder.

On the other hand, some tumors that did not receive radiation were used for the determination of intratumor ^{10}B concentration. This ^{10}B concentration was measured by prompt gamma-ray spectrometry.

Immunofluorescence Staining and Micronucleus Assay

Immediately after irradiation, tumors from mice given BrdU were excised and trypsinized. Tumor cell suspensions were incubated in tissue culture dishes containing complete medium and 1.0 μg/ml of cytochalasin-B to inhibit cytokinesis while preserving nuclear division. 48 hours later, the cells were trypsinized and single cell suspensions were fixed with 70% ethanol. After centrifugation, the cell pellet was resuspended with cold modified Carnoy's fluid (ethanol : acetic acid = 3 : 1), and placed on glass microscope slides and dried at room temperature. The slides were treated with 2 M hydrochloric acid for 30 min at room temperature to dissociate the histones and partially denature the DNA. The slides were then immersed in borax-borate buffer (pH 8.5) to neutralize the acid. BrdU-labeled cells were detected by indirect immunofluorescence using monoclonal anti-BrdU antibody and fluorescein isothiocyanate (FITC)-conjugated anti-mouse IgG.

To observe the double-staining of tumor cells with FITC and propidium iodide (PI), cells on the slides were treated with PI (1–5 $\mu g \cdot ml^{-1}$ in PBS) and monitored under a fluorescence microscope. In this manner, we were able to distinguish between cells incorporating BrdU, in which at least part of the nucleus or MN showed green fluorescence, and cells not incorporating BrdU in which the nucleus and MN showed only red fluorescence. It was then possible to selectively obtain the MN frequency of BrdU-unlabeled cells by counting the micronuclei in binuclear cells that showed only red fluorescence. The MN frequency was defined as the ratio of the number of micronuclei in binuclear cells to the total number of binuclear cells observed.

Micronuclei were also scored from tumors not pretreated with BrdU. MN frequency determined from these tumors was defined as the MN frequency of total tumor cells. More than 300 binuclear cells were counted to determine the MN frequency.

RESULTS

^{10}B concentrations in tumors in BSH and BPA administration groups were 10.2 ± 1.3 $\mu g \cdot g^{-1}$ and 10.4 ± 2.3 $\mu g \cdot g^{-1}$, respectively, and this difference was not significant. The average neutron fluence and gamma-ray dose rate were $(7.54 \pm 1.89) \times 10^9 n \cdot cm^{-2} \cdot s^{-1}$ and $(1.72 \pm 0.56) \times 10^{-3} Gy \cdot s^{-1}$ at 5 MW, respectively. The calculated radiation doses to BSH and BPA administered tumors attributable to the $^{10}B(n,\alpha)^7Li$ capture reaction were 1.6 ± 0.4 Gy, 3.2 ± 0.8 Gy and 4.8 ± 1.2 Gy, and 1.6 ± 0.4 Gy, 3.3 ± 0.8 Gy and 4.9 ± 1.2 Gy, respectively. Table 1 shows the MN frequencies without neutron radiation for total tumor cells and for Q cells. With or without drugs, MN frequency of Q cells was higher than that of total cells when no neutrons were delivered.

Figures 1 and 2 show the changes in the normalized MN frequency of each cell population treated with either BSH (Figure 1), or BPA (Figure 2), in combination with TPZ administration. When a drug was administered before tumor excision, even if no radiation was applied, the MN frequency was higher than when no drug was administered because of the genotoxicity of the drug (Table 1). Therefore, for baseline correction, we used the normalized MN frequency to exclude the effects of genotoxicity of the drug on the MN frequency. The normalized MN frequency was the MN frequency in the irradiated tumors minus the MN frequency in the nonirradiated tumors.

Table 1. Micronucleus frequency without neutron irradiation

	Total tumor cells	Quiescent cells
No drug	0.025 ± 0.003[a]	0.053 ± 0.004
TPZ[b]	0.040 ± 0.019	0.069 ± 0.021
BSH[c]	0.034 ± 0.004	0.071 ± 0.010
BPA[d]	0.048 ± 0.009	0.068 ± 0.010
BSH + TPZ	0.110 ± 0.042	0.121 ± 0.045
BPA + TPZ	0.098 ± 0.035	0.111 ± 0.039

[a]Mean ± 1 SD.
[b]Tirapazamine.
[c]Sodium borocaptate-^{10}B.
[d]*dl-p*-Boronophenylalanine-^{10}B.

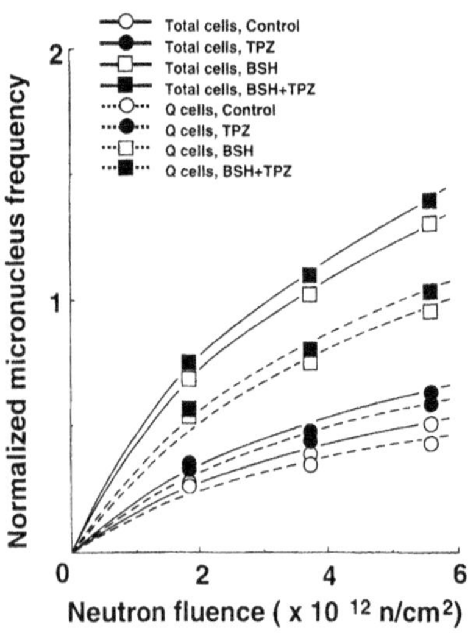

Figure 1. Dose-response curves for normalized micronucleus (MN) frequency (MN frequency-C, where C is the MN frequency in cells from tumors in nonirradiated animals) of tumors treated with BSH (sodium borocaptate-^{10}B) and/or TPZ (tirapazamine). To avoid confusion, only mean values are shown. In BSH-treated tumors and in those treated with BSH and TPZ, there were significant differences between total tumor cells and Q cells ($P < 0.05$). In each cell population, BSH treatment and combined treatment with BSH and TPZ produced significant increases in normalized MN frequencies ($P < 0.05$). In Q cells, TPZ treatment produced a significant increase in normalized MN frequency ($P < 0.05$), but with the addition to BSH treatment, the increase by addition of TPZ was not significant.

With or without ^{10}B-compounds, the normalized MN frequency of Q cells was lower than that of the total tumor cells. Neutron fluence modifying factors under each irradiation condition, which compared the neutron fluence to obtain various normalized MN frequencies in Q cells with that in the total tumor cells, were calculated using the values of the data in Figures 1 and 2 (Table 2). The use of ^{10}B-compounds increased the difference in normalized MN frequency between total and Q cells, especially when BPA was used. TPZ injection reduced the difference in MN frequency between total and Q cells induced by ^{10}B-compound administration, especially in the case of administration of BPA. TPZ treatment alone also reduced the difference in MN frequency between total and Q cells observed when no drugs were administered, mainly by increasing the MN frequency of Q cells.

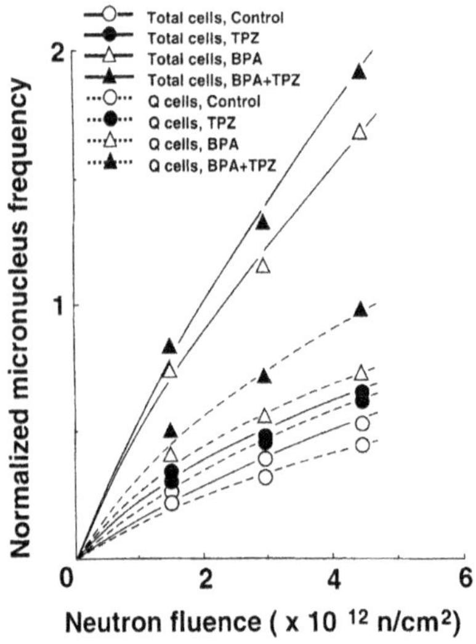

Figure 2. Dose-response curves for normalized micronucleus (MN) frequency (MN frequency-C, where C is the MN frequency in cells from tumors in nonirradiated animals) of tumors treated with BPA (*dl-p*-Boronophenylalanine-^{10}B) and/or TPZ (tirapazamine). To avoid confusion, only mean values are shown. In BPA-treated tumors and in those treated with BPA and TPZ, there were significant differences between total tumor cells and Q cells ($P < 0.05$). In each cell population, BPA treatment and combined treatment with BPA and TPZ produced significant increases in normalized MN frequencies ($P < 0.05$). With the exception of the total tumor cells from tumors treated with both BPA and TPZ, TPZ treatment produced significant increases in normalized MN frequency in each cell population regardless of combination with BPA treatment ($P < 0.05$).

Table 2. Neutron fluence modifying factors[a] for quiescent cells relative to the total tumor cells

Normalized MN freq.[b]	No drug	^{10}B-compound	TPZ[c]	^{10}B-compound +TPZ
^{10}B-compound = BSH[d]				
1.0	—	1.7 (1.5–1.9)[e]	—	1.7 (1.5–1.9)
0.8	—	1.7 (1.5–1.9)	—	1.6 (1.5–1.7)
0.4	1.2 (1.1–1.3)	1.6 (1.4–1.8)	1.1 (1.0–1.2)	1.5 (1.4–1.6)
0.2	1.2 (1.1–1.3)	1.6 (1.4–1.8)	1.1 (1.0–1.2)	1.5 (1.4–1.6)
^{10}B-compound = BPA[f]				
0.8	—	—	—	2.1 (1.7–2.5)
0.4	1.3 (1.2–1.4)	2.2 (2.0–2.4)	1.2 (1.1–1.3)	1.8 (1.6–2.0)
0.2	1.3 (1.2–1.4)	2.0 (1.8–2.2)	1.2 (1.1–1.3)	1.6 (1.4–1.8)

[a]Neutron fluence required to obtain each normalized micronucleus frequency in quiescent cells in relation to that required to obtain each normalized micronucleus frequency in total cells.
[b]Normalized micronucleus frequency, micronucleus frequency minus the micronucleus frequency in unirradiated tumors.
[c]Tirapazamine.
[d]Sodium borocaptate-^{10}B.
[e]95% confidence limit.
[f]*dl-p*-Boronophenylalanine-^{10}B.

To evaluate the effects of TPZ on the MN frequency in total and Q cells, the enhancement ratio (ER) by TPZ was calculated at various normalized MN frequencies using the data shown in Figs. 1 and 2 (Table 3). TPZ enhanced the sensitivity of tumor cells, including Q cells, even when no ^{10}B-compound was administered. In combination with ^{10}B-compound, all ER values decreased. However, TPZ showed a greater effect on Q cells when BPA was administered, but TPZ showed no apparent effect on each cell population when BSH was administered.

Table 3. The effects of TPZ[a] on neutron fluence ratio[b]

	Total tumor cells		Quiescent cells	
Normalized MN freq.[c]	No drug	^{10}B-compound	No drug	^{10}B-compound
^{10}B-compound = BSH[d]				
1.0	—	1.1 (1.0–1.2)[e]	—	1.2 (1.0–1.4)
0.8	—	1.1 (1.0–1.2)	—	1.1 (1.0–1.2)
0.4	1.4 (1.2–1.6)	1.1 (1.0–1.2)	1.6 (1.4–1.8)	1.1 (1.0–1.2)
0.2	1.2 (1.1–1.3)	1.2 (1.0–1.4)	1.4 (1.2–1.6)	1.2 (1.0–1.4)
^{10}B-compound = BPA[f]				
0.8	—	1.2 (1.0–1.4)	—	—
0.4	1.4 (1.2–1.6)	1.1 (1.0–1.2)	1.6 (1.4–1.8)	1.4 (1.3–1.5)
0.2	1.4 (1.2–1.6)	1.1 (1.0–1.2)	1.6 (1.4–1.8)	1.3 (1.2–1.4)

[a]Tirapazamine.
[b]Ratio of neutron fluence required to obtain each normalized micronucleus frequency without TPZ in relation to the neutron fluence required to obtain each normalized micronucleus frequency with TPZ.
[c]Normalized micronucleus frequency, micronucleus frequency minus the micronucleus frequency in unirradiated tumors.
[d]Sodium borocaptate-^{10}B.
[e]95% confidence limit.
[f]*dl-p*-Boronophenylalanine-^{10}B.

DISCUSSION

Previously Ono *et al.*[2] showed that more than 90% of SCC VII tumor cells divided at least once following neutron capture irradiation and that the probability of completing the first post-treatment mitosis was almost equivalent in SCC VII cells regardless of neutron capture compound administration. In addition, our preliminary experiment showed that, in total cell populations of SCC VII tumors, the correlations between normalized MN frequency and the surviving fraction for each of 6 conditions could be expressed as one regression line whether irradiated with or without drugs (Figure 3). Consequently, the sensitivity of tumor cells to neutron capture irradiation was thought to be reflected reasonably well by their normalized MN frequency instead of by their loss of clonogenicity.

Recent studies[3] have suggested that tumor hypoxia could be manipulated by the use of bioreductive agents which preferentially kill hypoxic cells. TPZ (SR 4233, WIN 59075, 3-amino-1,2,4-benzotriazine-1,4-dioxide) showed highly selective toxicity toward hypoxic cells in both *in vitro* and *in vivo* murine tumor systems. Severe hypoxia was not required and the level of hypoxia found in many human tumors was sufficient for the toxicity. We previously found that within SCC VII tumors Q cells include a larger hypoxic fraction than the total cells (Q cells: 53 ± 2% vs total cells: 11 ± 2%).[4] Therefore, TPZ showed a greater cytotoxicity on Q cells than the total cells (Table 3). Hence, TPZ treatment alone could reduce the difference in MN frequency between total and Q cells observed with neutrons irradiation without ^{10}B-compound (Table 2).

Meanwhile, ^{10}B-compounds increased the sensitivity of both Q and total cell populations, especially that of total cells (Figures 1 and 2); i.e., the difference in sensitivity between total and Q cells was widened by ^{10}B-compound (Table 2). It is thought that this event was induced mainly because Q cells could not take up ^{10}B as efficiently as P cells.[5]

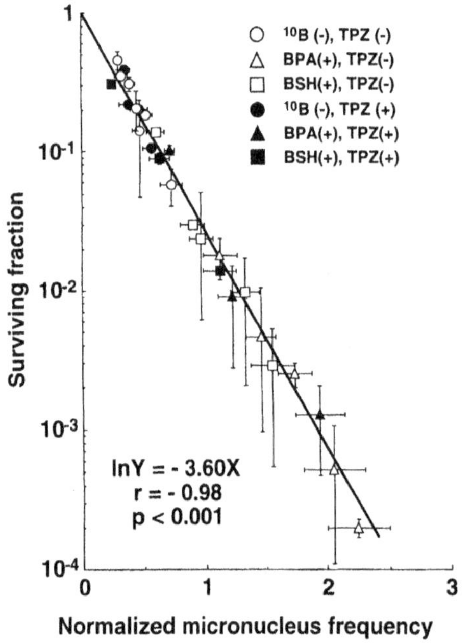

Figure 3. Correlation between normalized micronucleus (MN) frequency and surviving fraction for each of 6 conditions in total cell populations of SCC VII tumors. Six regression lines for each condition were statistically identical ($P < 0.05$). The regression line calculated from all these data was $\ln Y = -3.60 X$ (X = normalized MN frequency, Y = surviving fraction), and significant positive correlation of $r = -0.98$ was observed ($P < 0.001$). Thus, there is a close relationship between MN frequency and surviving fraction, and the sensitivity of SCC VII tumor cells to neutron capture irradiation can be expressed in terms of the MN frequency instead of surviving fraction. Bars represent SDs. BSH: sodium borocaptate-^{10}B, BPA: *dl-p*-Boronophenylalanine-^{10}B, TPZ: tirapazamine, and ^{10}B(−): no ^{10}B-compound.

Once ^{10}B is distributed in tumor cells, ^{10}B itself could enhance their sensitivity much more markedly than TPZ because of the heavily charged high linear energy transfer particles produced by neutron capture reaction (Figures 1 and 2). Therefore, TPZ could show larger ERs for Q cells than for total cells, especially when BPA, which showed poorer distribution in Q cells than BSH,[5] was administered (Table 3). Consequently, TPZ could reduce the difference in MN frequency between total and Q cells induced by ^{10}B-compound administration, especially in the case of BPA administration (Table 2).

With regard to TPZ effect, the curves of normalized MN frequency according to neutron fluence with and without TPZ were almost parallel (Figures 1 and 2). Therefore, the effect of TPZ was thought to be additive.

It has been reported that Q cells have lower radiosensitivity to photons than proliferating (P) cells in solid tumors *in vivo*.[4] Thus, more Q cells can survive after radiotherapy than P cells. Therefore, the control of Q cells, some of which will still be clonogenic,[1] is thought to greatly influence the outcome of anticancer radiotherapy. However, ^{10}B-compound administration in BNCT expanded the difference in sensitivity between total and Q cells compared with no drug administration (Table 2). From the viewpoint of tumor curability, including Q tumor cell control, the use of TPZ in BNCT appears to be encouraging even if TPZ effect is additive. Above all, TPZ combination with BPA may be effective and advantageous.

REFERENCES

1. R.C. Jackson, The problem of the quiescent cancer cells, *Enzyme Regul.*, 29:27–46, 1989.
2. M. Ono, S. Masunaga, Y. Kinashi, M. Takagaki, M. Akaboshi, T. Kobayashi, and K. Akuta, Radiobiological evidence suggesting heterogeneous microdistribution of boron compounds in tumors: its relation to quiescent cell population and tumor cure in neutron capture therapy, *Int. J. Raditat. Oncol. Biol. Phys.*, 34:1081–1086, 1996.
3. J.M. Brown, and B.G. Siim, Hypoxia-specific cytotoxins in cancer therapy. *Semin. Radiat. Oncol.*, 6:22–36, 1996.
4. S. Masunaga, K. Ono, and M. Abe, A method for selective measurement of the radiosensitivity of quiescent cells in solid tumors—combination of immunofluorescence staining to BrdU and micronucleus assay, *Radiat. Res.*, 125:243–247, 1991.
5. S. Masunaga, K. Ono, M. Suzuki, M. Takagaki, Y. Sakurai, T. Kobayashi, K. Akuta, M. Akaboshi, Y. Kinashi, and M. Abe, Modification of the response of a quiescent cell population within a solid tumour to boron neutron capture irradiation: studies with nicotinamide and hyperthermia. *Br. J. Radiol.*, 70:391–398, 1997.

BEHAVIORS OF CELLS IN DIFFERENT PROLIFERATIVE STATES WITHIN SOLID TUMORS IN BORON AND GADOLINIUM NEUTRON CAPTURE THERAPY WHEN IRRADIATED WITH TWO DIFFERENT CADMIUM RATIO NEUTRONS

Shin-ichiro Masunaga[1], Koji Ono[1], Yoshinori Sakurai[2], Minoru Suzuki[1], Masao Takagaki[1], Toru Kobayashi[2], Yuko Kinashi[1], and Mitsuhiko Akaboshi[2]

[1]Radiation Oncology Research Laboratory
[2]Division of Radiation Life Science
Research Reactor Institute, Kyoto University
Noda, Kumatori-cho, Sennan-gun, Osaka 590-0494, Japan

INTRODUCTION

Many intratumor cells are quiescent, but are still clonogenic.[1] These cells are thought to be more resistant to radiation and chemotherapy than proliferating (P) cells partly due to hypoxia and nutrient depletion and partly due to poor vascular supply. Therefore, many clonogenic quiescent (Q) tumor cells can survive after these therapy. Consequently, also in neutron capture therapy, whose effectiveness depends on tumor distribution of neutron capture compound and irradiated neutron distribution, controlling Q tumor cells is critical for therapeutic gain.

We analyzed the killing effects of neutron capture reaction using neutrons with two different energy spectra on Q cells within murine solid SCC VII squamous carcinoma, in terms of the MN frequency, in comparison with those in the total (P + Q) tumor cell populations. The reaction was performed with two ^{10}B-enriched compounds, sodium borocaptate-^{10}B (BSH) and *dl-p*-boronophenylalanine-^{10}B (BPA). Non-enriched Gd in the form of Omniscan (gadodiamide hydrate), which has been clinically used as a non-ionized contrast agent for magnetic resonance imaging (MRI), was also used.

Frontiers in Neutron Capture Therapy, edited by Hawthorne *et al.*
Kluwer Academic / Plenum Publishers, New York, 2001.

MATERIALS AND METHODS

Tumors, Mice, and Labeling with BrdU

SCC VII tumor cells (1.0×10^5) from C3H/He mice were inoculated subcutaneously into the hind legs of syngeneic female mice. Nine days after inoculation, 5-bromo-2′-deoxyuridine (BrdU) dissolved in physiological saline (250 mg/kg) was administered continuously for 5 days using subcutaneously implanted mini-osmotic pumps, to label all P cells in the tumors. Fourteen days after inoculation, the tumor was treated, and was 1 cm in diameter at the start of treatment. The labeling index after 5 days of continuous administration of BrdU was $55.3 \pm 4.5\%$ (mean ± SD), and reached a plateau level at this stage. In this study, tumor cells not incorporating BrdU after continuous administration were regarded as Q cells for all practical purposes, and the growth fraction of the tumors were considered to be $55.3 \pm 4.5\%$.

Treatment

Twenty min after the intraperitoneal injection of BSH (125 mg/kg), 3 h after oral administration of BPA (1,500 g/kg), or 30 min after the intraperitoneal injection of Omniscan (40 ml/kg, 12.92 g/kg), the tumors were irradiated with a neutron beam generated by biomedical radiation field at Kyoto University Reactor (KUR). We used neutron beams with two different cadmium (Cd) ratios which have become available through remodeling of the heavy water facility at KUR from August, 1995 through May, 1996. Cd ratios of employed neutron beams were 148 and 1.0. The neutron fluence was measured with a phantom by radioactivation of gold foil both at front and back sides of tumors. γ-Ray doses, including secondary γ-rays, were measured with thermoluminescence dosimeter powder at the back side of tumors. For estimation of neutron energy spectra, eight kinds of activation foil and fourteen kinds of nuclear reaction were used. Neutron absorbed dose was calculated using the flux-to-dose conversion factor.[2] Weight percentage of the tumors was assumed to be H (10.7%), C (12.1%), N (2%), O (71.4%) and others (3.8%). Since the tumors were small and located just beneath the surface, intratumor neutron fluence was assumed to be linearly decreased from the front to back side of the tumors. Thus, we used the averaged neutron fluence from the measured values at the front and back sides of tumors. Averaged neutron fluence and Kerma dose rate, and measured γ-ray dose rate for each irradiation mode are shown in Table 1. Meanwhile, for γ-ray radiation as a control, we used a cobalt-60 γ-ray irradiator at a dose rate of 5.97 Gy/min. Each treatment group included mice treated or not treated with BrdU.

Some of the tumors that were not irradiated were used to determine the ^{10}B concentration in the tumors. The ^{10}B concentration in the tumors was measured by prompt γ-ray spectrometry.

Immunofluorescence Staining and the Micronucleus Assay

After the above treatments, tumors from mice given BrdU were excised, minced, and trypsinized at 37 °C for 15 min. Tumor cell suspensions were incubated in tissue culture dishes containing complete media and 1.0 μg/ml of cytochalasin-B, to inhibit

Table 1. Neutron fluence (n/cm^{2}s) and Kerma dose rate (cGy/h) for each neutron irradiation mode

Cadmium ratio	Thermal neutrons (–0.6 eV)	Epithermal neutrons (0.6 eV – 10 keV)	Fast neutrons (10 keV–)	γ-Rays dose rate (cGy/h)
148	2.0×10^9	2.8×10^7	6.6×10^6	
	96	1.03	28.4	120
1.0	—	8.0×10^8	4.7×10^7	
	—	23	160	70

cytokinesis while allowing nuclear division. The cytokinesis-blocked cells were easily recognized as binuclear cells, and their percentage reached a maximum 48 hours after the beginning of culture. The cells were then trypsinized and single cell suspensions were fixed overnight at 4 °C with 70% ethanol. After centrifugation, the cell pellet was resuspended with cold modified Carnoy's fluid (ethanol : acetic acid = 3 : 1), and placed on a glass microslide, and dried at room temperature.

After denaturation treatment with 2 M hydrochloric acid for 45 min at room temperature and the neutralization with borax-borate buffer (pH 8.5), BrdU-labeled cells were detected by indirect immunofluorescence staining using monoclonal anti-BrdU antibody and fluorescein isothiocyanate (FITC)-conjugated anti-mouse IgG.

To observe the double-staining of tumor cells with FITC and propidium iodide (PI), cells on the slides were treated with PI (1–5 μg/ml in phosphate-buffered saline) under a fluorescence microscope. When the intensity of the red fluorescence produced by PI became similar to the intensity of the green fluorescence in nuclei prestained with FITC (P cells), the treatment was stopped by rinsing the slides with water. The MN frequency in BrdU-unlabeled cells (Q cells) could be examined by counting the micronuclei in those binuclear cells that showed only red fluorescence. The MN frequency was defined as the ratio of the number of micronuclei in the binuclear cells to the total number of binuclear cells observed. The ratio obtained in tumors not pretreated with BrdU indicated the MN frequency of all phases of the total tumor (P + Q) cell populations.

RESULTS

^{10}B concentration in tumors in BPA and BSH administration groups were 10.7 ± 2.8 μg/g (1.1 ± 0.3 mM) and 11.2 ± 1.2 μg/g (1.1 ± 0.1 mM), respectively, and this difference was not significant. ^{157}Gd concentration in tumors for Omniscan administration was 468.5 ± 52.8 μg/g (3.0 ± 0.3 mM). Table 2 shows the MN frequencies without radiation for total (P + Q) tumor cells and for Q cells.

Figures 1 and 2 show the normalized MN frequencies for each irradiation condition as a function of the absorbed radiation dose in total tumor cells (Figure 1) and in Q cells (Figure 2), respectively. When a neutron capture compound was administered before tumor excision, even if no radiation was given, the MN frequency was higher than when no drug was administered, because of the genotoxicity of the drug (Table 2). Therefore, for baseline correction, we used the normalized MN frequency to exclude the effects of the genotoxicity of the neutron capture compound on the MN frequency. The normalized MN frequency is the MN frequency in the irradiated tumors minus the MN frequency in the nonirradiated tumors.

Table 2. Micronucleus frequency at 0 Gy

Cell fraction No drug	BSH[a]	BPA[b]	Omniscan[c]
Total tumor cells 0.025 ± 0.003	0.034 ± 0.004	0.048 ± 0.009	0.031 ± 0.004
Quiescent tumor cells 0.053 ± 0.004	0.071 ± 0.010	0.068 ± 0.010	0.060 ± 0.008

[a]sodium borocaptate-^{10}B.
[b]*dl-p*-boronophenylalanine-^{10}B.
[c]gadodiamide hydrate.

Without neutron capture compounds, the normalized MN frequency in Q cells was lower than that in the total cell population, especially in the case of γ-ray irradiation. The values of dose-modifying factor (DMF) of Q cells in tumors not treated with any drugs for γ-rays were greater than 2.00, and significantly higher than 1.00 ($p < 0.05$). In contrast, with neutrons of each Cd ratio, the values were closer to 1.00 than those for γ-rays (Table 3). To examine the relative biological effectiveness (RBE) of irradiation with neutrons, DMFs, which compare the radiation doses necessary to obtain various normalized MN frequencies in tumors irradiated with γ-rays with those in tumors irradiated with neutrons without neutron capture compounds, were calculated using the mean values of the data shown in Figures 1 and 2 (Table 4). These values of DMFs were equivalent to those of RBE. All were significantly greater than 1.00 ($p < 0.05$), and the values for Q cells were significantly larger than those for total tumor cells ($p < 0.05$). Moreover, in each cell population, although not significantly, the values for low Cd ratio neutrons were larger than those for high Cd ratio neutrons.

With neutron capture compounds, the normalized MN frequency for each cell population was increased. This increase in the normalized MN frequency was marked when high Cd ratio neutrons were used. In general, the values of enhancement ratios (ERs) for total cells and high Cd ratio neutrons tended to be larger than those for Q cells and low

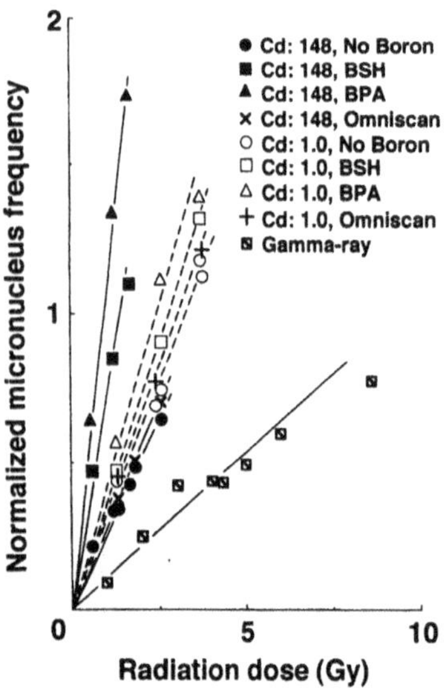

Figure 1. Dose-response curves of normalized MN frequency as a function of physically absorbed dose for total tumor cell populations. Tumors were irradiated using neutrons with a Cd ratio of 148 (●, ▲, ■, ×) and 1.0 (○, △, □, +). As a control irradiation, other tumors were irradiated with gamma-rays alone (◪). ●, ○ without neutron capture drug; ▲, △ with BPA administration; ■, □ with BSH administration; ×, + with omniscan administration. Only mean values are shown, to avoid confusion.

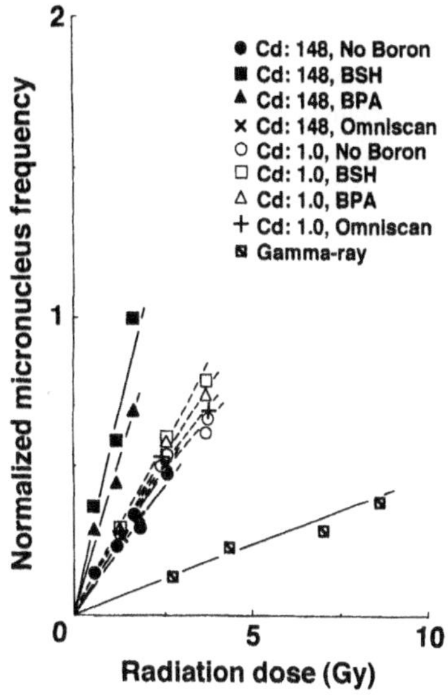

Figure 2. Dose-response curves of normalized MN frequency as a function of physically absorbed dose for quiescent tumor cell populations. Tumors were irradiated using neutrons with a Cd ratio of 148 (●, ▲, ■, ×) and 1.0 (○, △, □, +). As a control irradiation, other tumors were irradiated with gamma-rays alone (⊠). ●, ○ without neutron capture drug; ▲, △ with BPA administration; ■, □ with BSH administration; ×, + with omniscan administration. Only mean values are shown, to avoid confusion.

Table 3. Dose modifying factors[a] for quiescent cells relative to the total tumor cell populations

	Neutrons		
Normalized MN freq.[b]	Cd[c] : 148	Cd : 1.0	γ-Rays
0.5	1.5	1.6	—
0.35	1.5	1.5	2.4
0.25	1.5	1.6	2.4

[a]Radiation dose required to obtain each normalized micronucleus frequency in quiescent cells / radiation dose required to obtain each normalized micronucleus frequency in total tumor cells.
[b]Normalized micronucleus frequency, micronucleus frequency minus the micronucleus frequency in unirradiated tumors.
[c]Cadmium ratio.

Table 4. Relative biological effectiveness[a] in total tumor and quiescent cell populations

		Neutrons	
Cell fraction	Normalized MN freq.[b]	Cd[c] : 148	Cd : 1.0
Total tumor cells	0.75	—	2.9
	0.5	2.4	2.9
	0.35	2.4	2.8
	0.25	2.3	2.9
Quiescent tumor cells	0.35	3.7	4.5
	0.25	3.6	4.4

[a]Radiation dose required to obtain each normalized micronucleus frequency with γ-rays / radiation dose required to obtain each normalized micronucleus frequency with neutrons.
[b]Normalized micronucleus frequency, micronucleus frequency minus the micronucleus frequency in unirradiated tumors.
[c]Cadmium ratio.

Table 5. Enhancement ratios[a]

Cell fraction Cadmium ratio	Normalized MN freq.[b]	BSH[c]	Drugs BPA[d]	Omniscan[e]
Total tumor cells				
Cd ratio: 148				
	0.5	3.7	5.4	1.2
	0.35	3.6	5.1	1.2
	0.25	3.7	5.5	1.2
Cd ratio: 1.0				
	1.0	1.2	1.3	1.1
	0.75	1.2	1.4	1.1
	0.5	1.2	1.3	1.1
	0.35	1.2	1.4	1.1
	0.25	1.2	1.3	1.1
Quiescent tumor cells				
Cd ratio: 148				
	0.5	3.2	2.4	1.2
	0.35	3.3	2.4	1.2
	0.25	3.4	2.4	1.2
Cd ratio: 1.0				
	0.5	1.2	1.1	1.1
	0.35	1.2	1.1	1.1
	0.25	1.2	1.2	1.1

[a]Radiation dose required to obtain each normalized micronucleus frequency without drug / radiation dose required to obtain each normalized micronucleus frequency with a drug.
[b]Normalized micronucleus frequency, micronucleus frequency minus the micronucleus frequency in unirradiated tumors.
[c]Sodium borocaptate-^{10}B.
[d]*dl-p*-boronophenylalanine-^{10}B.
[e]gadodiamide hydrate.

Cd ratio neutrons, respectively (Table 5). With neutrons of each Cd ratio, in total tumor cells, the ER values for BPA administration were larger than those for any other drug. In contrast, in Q cells, the values for BSH were the largest. This tendency was clearly observed in higher Cd ratio neutrons. Our previous *in vitro* experiment using single cell suspensions of the SCC VII tumor cell line and the two kinds of Cd ratio (148 and 1.0) neutrons showed that Omniscan with 3.0 mM ^{157}Gd could increase the sensitivity of cultured tumor cells to the same extent as boric acid enriched with 1.0 mM ^{10}B (Figure 3). Similarly to this experiment, the tumors treated with ^{10}B-compounds or Omniscan also contained 1.0 mM ^{10}B or 3.0 mM ^{157}Gd. Nevertheless, Omniscan only slightly increased MN frequency both in total and Q cells, compared with the sole irradiation without drugs.

DISCUSSION

High linear energy transfer neutrons, irrespective of neutron Cd ratio, should be applied to reduce the difference in sensitivity between total and Q cells.

The values of RBE for Q cells were markedly larger than those for total cells. This was mainly because the difference in γ-ray sensitivity between total and Q cells was much wider than that in sensitivity to neutrons. In both cell populations, the RBE values for

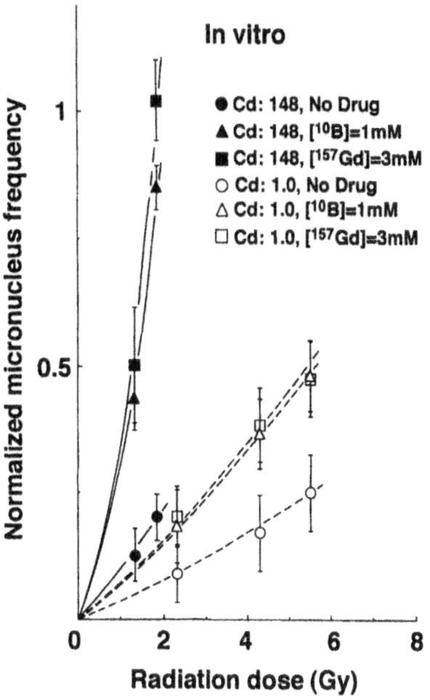

Figure 3. Dose-response curves of normalized MN frequency as a function of physically absorbed dose for cultured SCC VII tumor cells. Cultured cells were irradiated using neutrons with a Cd ratio of 148 (●, ▲, ■) and 1.0 (○, △, □). ●, ○ without neutron capture drug; ▲, △ with boric acid-^{10}B of ^{10}B-concentration of 1 mM; ■, □ with omniscan of ^{157}Gd-concentration of 3 mM. Bars represent SDs.

low Cd ratio neutrons tended to be larger than those for high Cd ratio neutrons. This might have been partially because the low Cd ratio neutrons applied here included a higher proportion of fast neutrons (Table 1), which include a larger proportion of 24 keV neutrons which are considered to induce chromosome abberations at high efficiency.[3,4]

When BPA was used, ^{10}B could be distributed to total tumor cells as a whole as well as when BSH was used. However, less ^{10}B could be localized into Q cells than when BSH was administered. This indicates that the distribution of ^{10}B, from BPA, in tumor cells is more dependent on the ^{10}B uptake ability of the tumor cells than that from BSH. When BSH was used, the ER values for total cells were almost the same as those for Q cells. That is, the distribution of ^{10}B from BSH is more dependent on the diffusion of the drug than that from BPA. Omniscan was too water-soluble to be accumulated in tumor tissue for a sufficient period during neutron irradiation. The water-soluble Gd-containing contrast media for MRI was not suitable as a neutron capture compound in neutron capture therapy.

The use of BSH combined with high Cd ratio neutron irradiation is effective in the control of radioresistant Q cells and that BPA with high Cd ratio neutrons is effective for controlling the total cells. From the viewpoint of tumor curability, combination of BSH and BPA may be useful in neutron capture therapy using high Cd ratio neutrons.[5]

REFERENCES

1. R.C. Jackson, The problem of the quiescent cancer cells, *Enzyme Regul.*, 29:27–46, 1989.
2. R.S. Caswell, J.J. Coyne, and M.L. Randolph, Kerma factors for neutron energies below 30 MeV, *Radiat. Res.*, 83:217–254, 1980.
3. S.Z. Aghamohammadi, D.T. Goodhead, and J.R. Savage, Production of chromosome aberrations, micronuclei, and sister-chromatid exchanges by 24-keV epithermal neutrons in human G0 lymphocytes, *Mutat. Res.*, 211:225–230, 1989.

4. A.J. Hill, G.R. Morgan, and S.M. Newman, Cell survival measurements in an argon, alminium and sulphur filtered neutron beam: a comparison with 24 keV neutrons and relevance to boron neutron capture therapy, *Br. J. Radiol.*, 67:1008–1016, 1994.
5. K. Ono, S. Masunaga, Y. Kinashi, M. Takagaki, and T. Kobayashi, Combined effect of BPA and BSH in boron neutron capture therapy for murine tumors, in "Kyoto University Research Reactor Institute Progress Report 1995," Res. React. Inst. Kyoto Univ., ed., Kumatori, Osaka, 1996, pp. 183.

DEPTH-DOSE EVALUATION FOR LIVER CANCER TREATMENT BY BNCT AND GDNCT USING AN EPITHERMAL NEUTRON BEAM

Tetsuo Matsumoto

Atomic Energy Research Laboratory
Musashi Institute of Technology
Ozenji 971, Asao-ku, Kawasaki-shi
215 Japan

INTRODUCTION

Primary and metastatic liver cancers are inevitably fatal. Survival rate of 5 years is only 3% in Japan. For neutron capture therapy (NCT) to enjoy the same recognition as conventional radiotherapy, it should be applied in order to treat not only brain tumors but also various cancers. An epi-thermal neutron beam could be now expected to apply for a deep and widespread tumor. This paper covers comparative evaluation of depth-dose distributions using boron (BNCT) and gadolinium (GdNCT) for possible treatment of liver cancers.

METHODS

The ICRU computational model of ADAM & EVA[1] was used as a liver phantom with simulating tumor ($3 \times 3 \times 3\,cm^3$) at the depth of 6 cm, in a central region of liver as shown in Figure 1a, b, and c. The epithermal neutron beam proposed at the Musashi reactor[2] was used to model the primary neutron beam for the depth-dose calculation (Figure 2). The boron and gadolinium concentrations in the tumor were assumed to be 30 ppm and 1,000 ppm, respectively. The calculations were performed using the three-dimensional continuous-energy Monte Carlo code MCNP4A.[3] The physical dose, without considering the values of RBE for each radiation, was calculated for the comparison of the dose distribution between BNCT and GdNCT.

Frontiers in Neutron Capture Therapy, edited by Hawthorne *et al.*
Kluwer Academic / Plenum Publishers, New York, 2001.

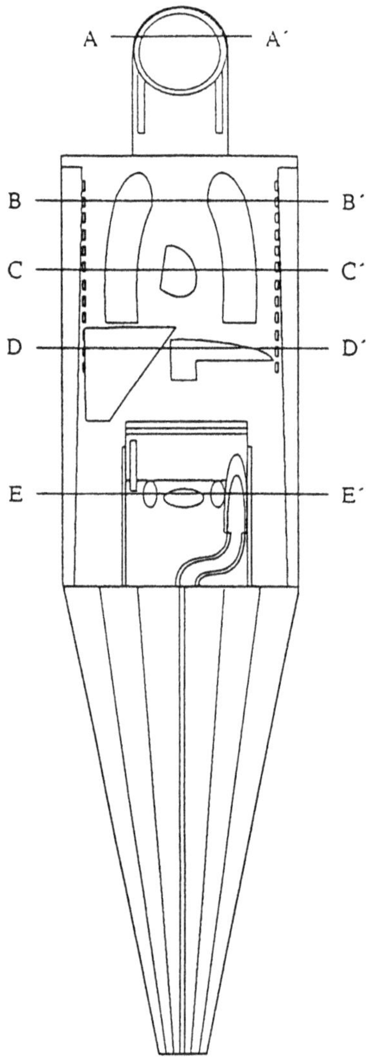

Figure 1a. ICRU computational model (EVA phantom).

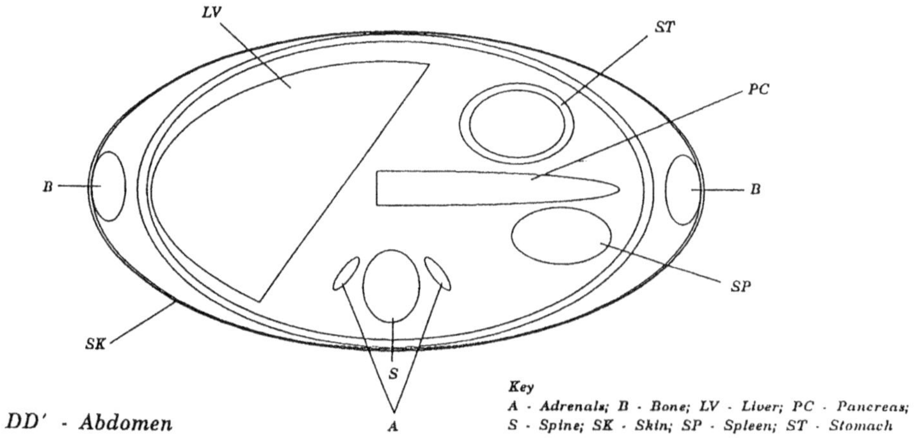

Figure 1b. D-D′ cross-sectional view of the EVA phantom.

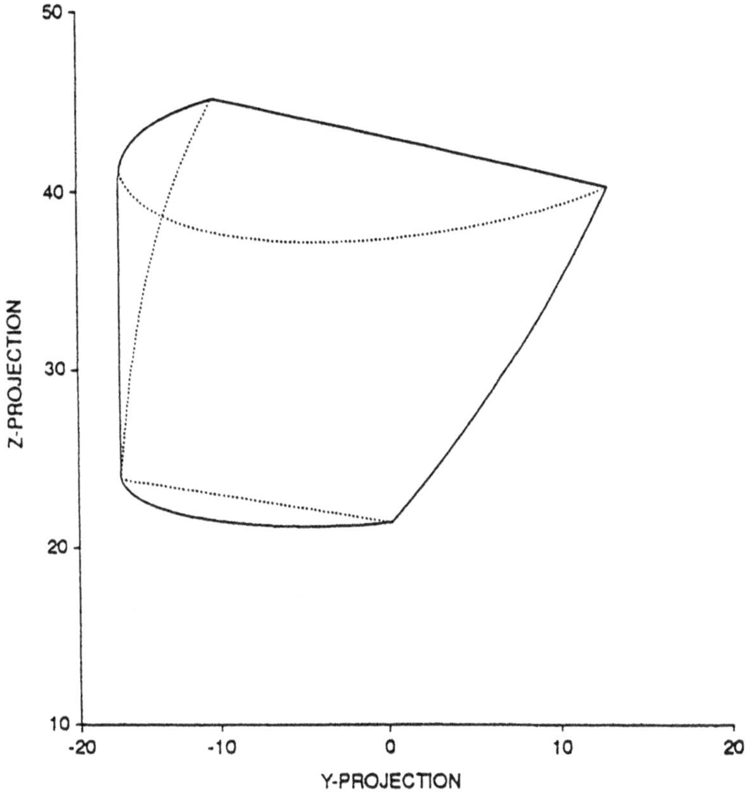

Figure 1c. ADAM's liver of ICRU model.

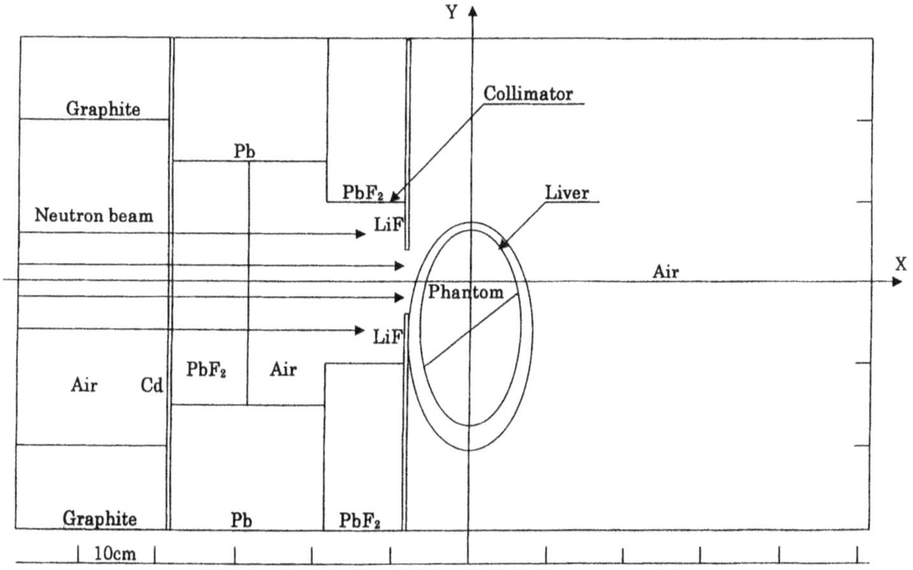

Figure 2. Phantom calculation in the irradiation field at the proposed Musashi reactor.

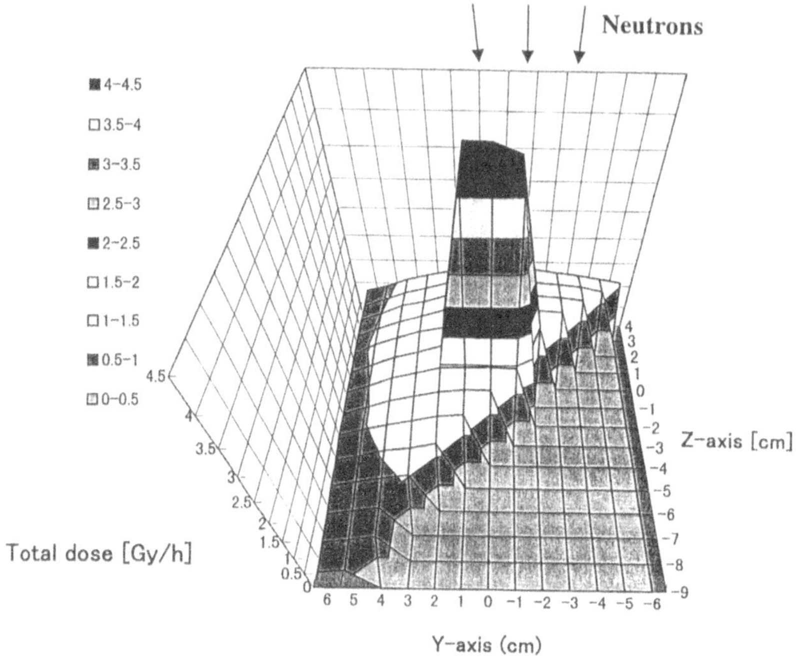

a

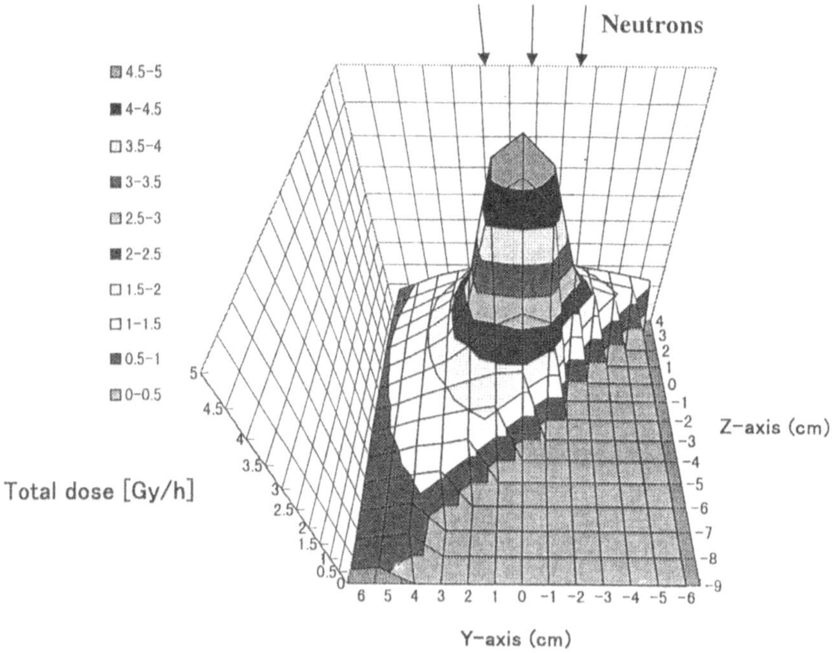

b

Figure 3. Total dose distributions in the liver by BNCT (a) and GdNCT (b) where 30 ppm boron and 1,000 ppm Gd are accumulated in the tumor, respectively.

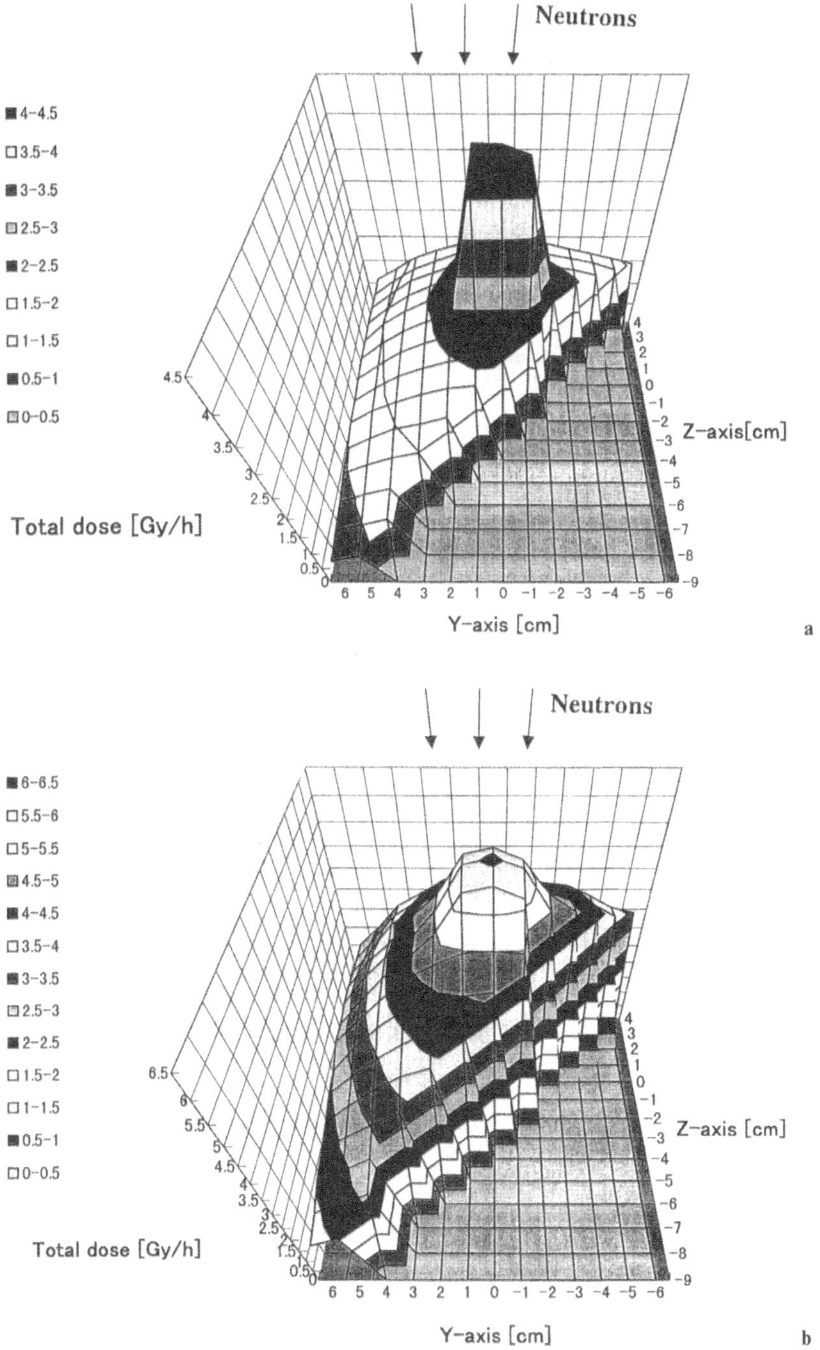

Figure 4. Total dose distributions in the liver by BNCT (a) and GdNCT (b) where the normal tissues contain 1/4 the tumor concentration, ie, 7.5 ppm boron and 250 ppm Gd, respectively.

RESULTS AND DISCUSSIONS

Figures 3a and 3b show the calculated three-dimensional depth-dose distributions of the total dose rates for BNCT and GdNCT. The dose observed for both NCTs is bumped over the tumor region but the dose in GdNCT is not tumor-specific as compared with that from BNCT because of gamma-rays from Gd neutron capture reaction. The average physical dose rates in the tumor were 4 Gy/h and 5 Gy/h, respectively, for 30 ppm boron and 1,000 ppm Gs when irradiated with an epithermal neutron flux of $5 \times 10^8 ncm^{-2}s^{-1}$. The tumor-to-normal tissue dose ratios were 2.7 and 2.5, respectively, for boron and for Gd assuming no boron or Gd in normal tissues. A lethal dose of 50 Gy to the liver cancer can be accomplished under conditions where the tolerance dose of 25 Gy to the normal tissues has not yet been reached. This result is very encouraging and seems to indicate that both BNCT and GdNCT could be applied for liver cancer treatment under above conditions. However, if the normal tissues contain 1/4 tumor concentration (7.5 ppm boron or 250 ppm Gd) as shown in Figure 4a and 4b, the dose ratios decreased to 1.8 and 1.2, respectively, for the BNCT and for the GdNCT. When considering a large RBE value for $^{10}B(n,\alpha)$ reactions, the BNCT would be still possible but the Gd NCT would be impossible for deep liver treatment.

CONCLUSION

The dose evaluation suggests that the BNCT and GdNCT could be applied for the treatment of liver cancer under the conditions studied in this paper. However, if normal tissues contain 1/4 the tumor concentration of boron or Gd, the BNCT would still possible when considering a large RBE value for $^{10}B(n,\alpha)$ reaction but the GdNCT would be impossible for deep liver treatment.

REFERENCES

1. ICRU report 48: Phantoms and computational models in therapy (1992).
2. Tetsuo Matsumoto: Design optimization of thermal and epithermal neutron beams and depth-dose evaluation at the proposed Musashi reactor, Advances in Neutron Capture Therapy, Vol. 1, p. 424 (1997).
3. J.F. Briesmeister editions: MCNP4 Monte Carlo neutron and photon transport code system, CCC200A/B (1991).

A COMPACT CELL CULTURE DISC FOR NEUTRON CAPTURE RADIOGRAPHY

G. Kühne, N. E. A. Crompton, C. Higgs, H. Reist, E. Lehmann, J. Stepanek, T. Suda, H. Blattmann, and B. Larsson

Paul Scherrer Institute
CH-5232 Villigen PSI, Switzerland

NEUTRON IRRADIATION FACILITY NCR AT PSI

At the Paul Scherrer Institute (PSI) in Switzerland a neutron irradiation facility for Neutron Capture Radiography (NCR) has been established (see Table 1). It is situated at the end of a cold neutron guide at the spallation neutron source SINQ.[1] We have also developed a new cell culture system based on a compact cell culture disc (CCCD), which can been integrated into the Neutron Capture Radiography facility. The CCCD is useful for individual testing of radiotherapy patient response and also for basic studies of cellular response to cytotoxic therapies (see the CASE-MATE contribution by Crompton *et al.*, in these proceedings).

To achieve homogeneous irradiation the samples are moved in front of the neutron guide opening in the vertical (y) direction and in the horizontal (x) direction. The scanning speed in the x–direction is controlled by the actual online measured neutron flux.

The maximum number of 30 cm × 30 cm samples which can be irradiated (one after the other) in the same run is 10.

The irradiation itself is semi–automatic. In a first step the operator has to define 1) the start- and the endpoint for the scanning path using a video camera and 2) the desired neutron flux for each sample. After recording all relevant data in a PC, the irradiation can be started and the scan runs automatically—controlled by the PC.

The start of operation is planned for the summer 1999.

COMPACT CELL CULTURE DISK (CCCD)

We have developed a new cell culture system based on a Compact Cell Culture Disc (CCCD) intended for pretherapeutic assays applied to individual patients. It is being manufactured by available Si-microtechnology, permits cultivation, microcloning,

Frontiers in Neutron Capture Therapy, edited by Hawthorne *et al.*
Kluwer Academic / Plenum Publishers, New York, 2001.

Table 1. Physical properties of the NCR facility

Neutron flux density	$2.5 \cdot 10^8 cm^{-2} \cdot s^{-1}$
Neutron energy	2 meV
Neutron cross section	$\sim 3 \cdot \sigma_{therm}$
Neutron guide opening	3 cm × 12 cm
Maximal sample size	30 cm × 30 cm

instrumental or visual observation, neutron capture microradiography for boron, and eventual storage (*in vitro* or after fixation) of the biological specimens. It contains a multitude of miniature culture chambers set into a planar compact disc suitable for microscopic study of small biological specimens. A tumour biopsy containing typically some 100,000 to 1,000,000 cells could then be conveniently analysed as several thousand individually isolated tumour and stroma cell clones in primary cell culture.

Single cells derived from biopsy or surgery specimens could be implanted individually, each one in its own separate "micropit". The latter consists of a shallow (typically

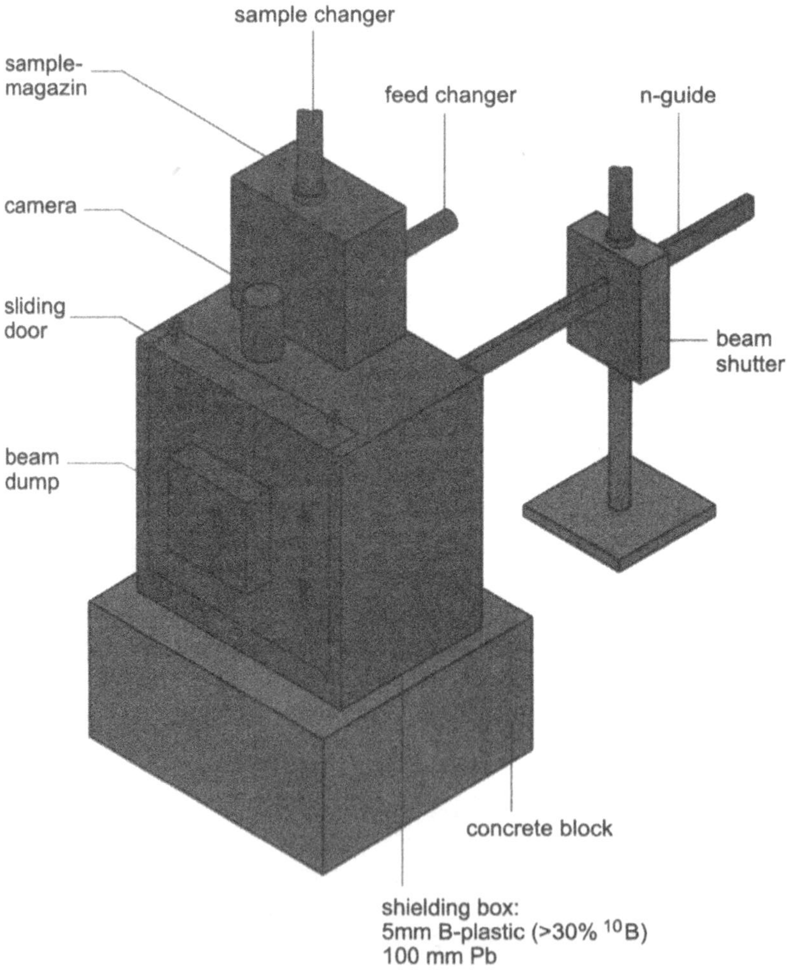

Figure 1. Schematic of the irradiation chamber.

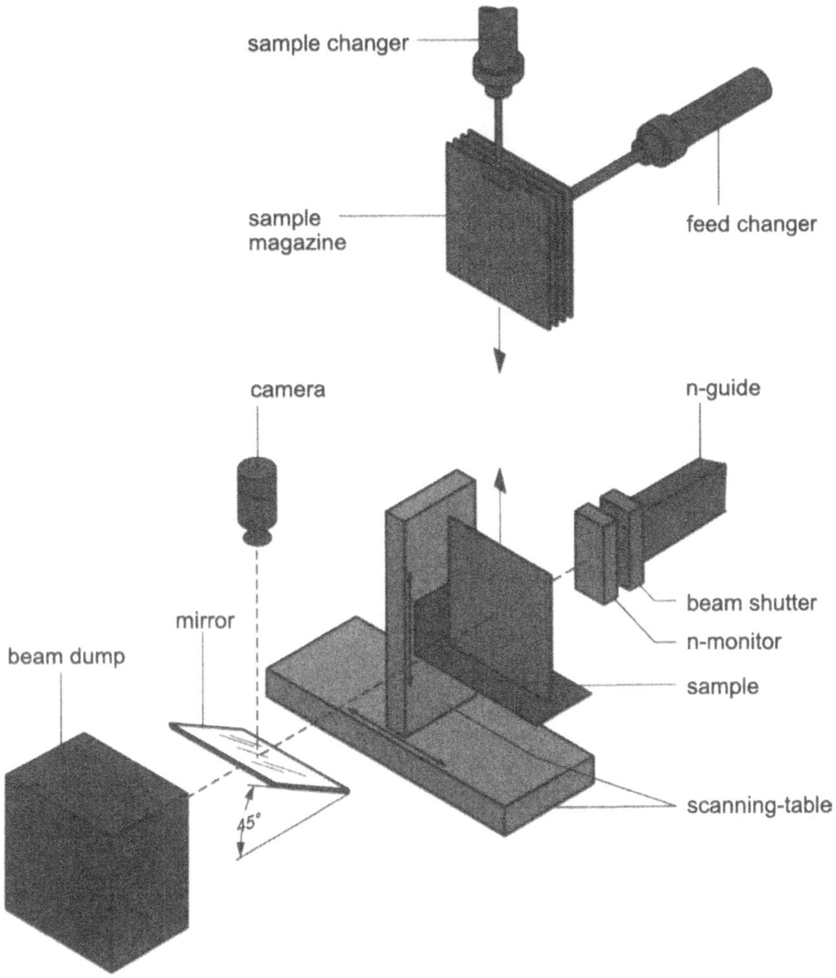

Figure 2. Schematic of the components inside the irradiation chamber.

10–100 μm deep), cylindrical depression defined by a hole (typically 500–5,000 μm in diameter) in a biochemically inert foil, firmly attached to a bottom plate made of silicon, quartz, glass, or plastic. After implantation ("cell seeding") the micropits can be covered by a thin plastic membrane forming a semipermeable cover foil. In this way it is possible to achive optimized and uniform supplies of respiratory gases, nutrients, antibiotics, reagents etc. to the microchamber.

Each such microchamber is a closed sterile compartment inaccessible to virus or bacteria. The internal microchamber environment is determined by the permeability characteristics of the cover foil. Boron compounds under study could either be injected before biopsy or surgery (for pretherapeutic studies of the behaviour of boronated drugs in the patient) or into the micropits (for comparative studies *in vitro*). The microchambers offer advantageous conditions for culture of daughter cell populations (up to 1,000–10,000 cells per microchamber) from single cells (cell cloning). The large number of cell clones produced on one single CCCD could be studied individually or in rapid succession. The uniform and reproducible culture conditions thus provided are of

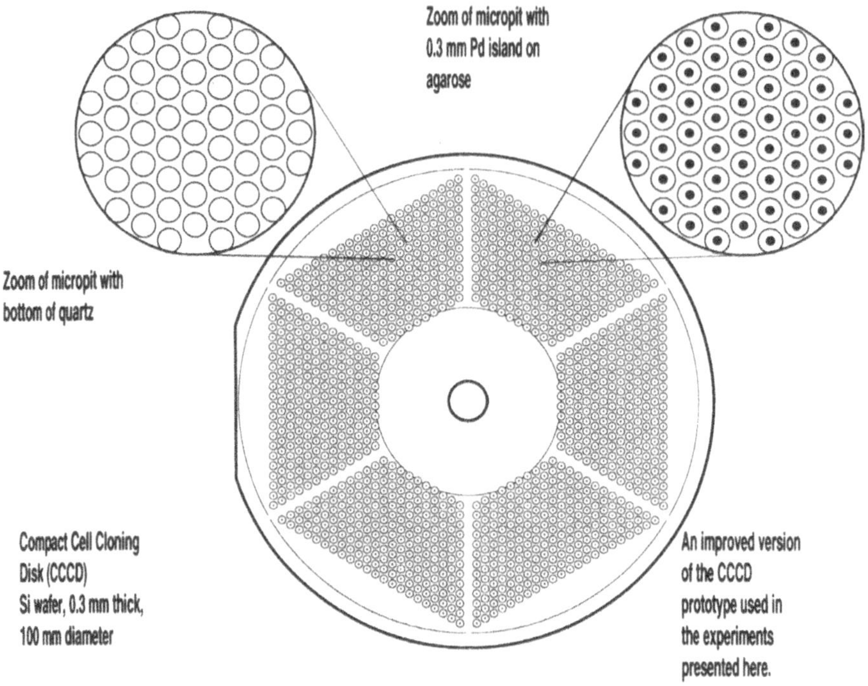

Figure 3. Plan view of the 100 mm silicon disk.

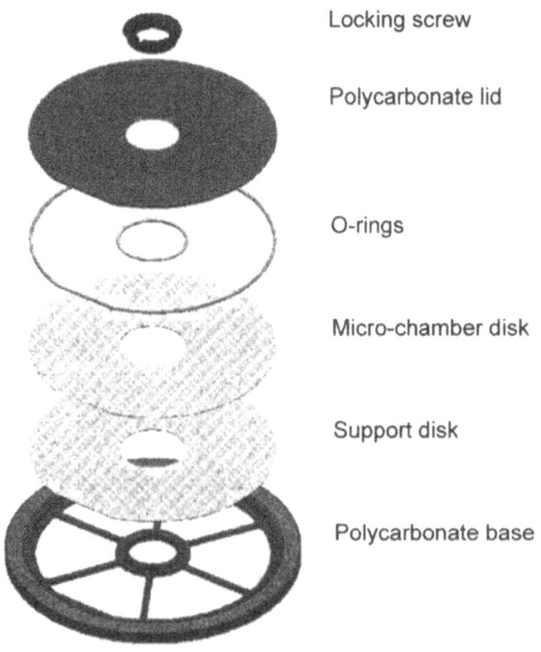

Figure 4. The polycarbonate "sandwick holder". The embedment of a 10 cm CCCD in a 12 cm CD structure provides a common source of low-molecular weight medium components for the microchambers, mechanical protection and a convenient structure for optomechanical data storage.

significant advantage, especially for the study of subcellular boron topography by neutron capture radiography techniques. To this end, single dispersed cells from the tumour can be guided into individual CCCD micropits by computer-assisted cell seeding. (Alternatively, one can use randomized seeding from a diluted tumour cell suspension, a technique that allows implantation and cloning of 1 or 0 cells per micropit.) The particle detector, typically polycarbonate, is separated from the biological specimen by a thin quartz or plastic window, that also permits microscopic observation of the cells in the transmission mode.

REFERENCES

1. S. Teichmann, J.F. Crawford, and B. Larsson: "A neutron capture radiography facility at PSI", Proceedings of the Seventh International Symposium in Neutron Capture Therapy for Cancer, Zurich, Switzerland, 4–7 September 1996, Volume I, Medicine and Physics.

FUTURE DIRECTIONS AND EMERGING APPLICATIONS

CURRENT CLINICAL PARADIGMS IN MELANOMA BNCT

Leading to Cell Modulation Era Teamed with Passive and Active ^{10}B-Biodelivery Systems

Yutaka Mishima[1], Hirofumi Kondoh[1], Kazuo Yoshino[2], and Junichi Hiratsuka[3]

[1]Mishima Institute for Dermatological Research
1-4-32 Sowa-cho, Nada-ku, Kobe 657-0063 Japan
[2]Department of Chemistry
Shinshu University, Nagano
[3]Department of Radiation Oncology
Kawasaki Medical School, Kurashiki

1. INTRODUCTION

At the last ISNCT meeting in Zurich (September, 1996), we presented our new findings on melanogenic gene transfer, converting fibroblasts to actively melanin synthesizing cells, and proposed applying this gene transfer to BNCT, called gBNCT, as a way of attacking a wider range of cancers.[1]

Over the past two years, we have extended our research and developed the concept of cell modulation BNCT teamed with both passive and active ^{10}B-biodelivery systems as a way of increasing efficacy and expanding the spectrum of target cancers in BNCT.

In this paper, we present first an overview of what we now believe to be the critical requirements for clinical success through our melanoma BNCT, which has also laid the groundwork for breakthroughs in the emerging field of cell modulation BNCT. Included is discussion on the importance of both chemistry and detailed, specific oncological expertise. We then describe our procedure for treating representative types of melanoma and give an updated analysis of melanoma patients treated by my team. Finally, we discuss a paradigm shift which we feel is necessary for the further development of cancer BNCT.

Frontiers in Neutron Capture Therapy, edited by Hawthorne *et al.*
Kluwer Academic / Plenum Publishers, New York, 2001.

2. PREREQUISITES FOR CLINICAL SUCCESS IN BNCT

We can pinpoint three basic criteria for clinical success in the treatment of human melanomas and these points will apparently hold true for future developments. Namely, we need to first develop new ideas, then proceed with multi-dimensional research, and finally pursue solid scientific and clinical achievements to demonstrate validity.

2.1. Critical Role of Chemistry for Boron Delivery into Subcellular Compartments in Cancer Cells

Figure 1 summarizes our concept of how ^{10}B-p-boronophenylalanine (^{10}B-BPA) is attracted into subcellular compartments. Briefly, from left to right, ribosomes first synthesize T-2 tyrosinase and DCTase, DOPAchrome tautomerase, which are greatly enhanced in melanoma. These enzymes are then transferred to the Golgi-associated endoplasmic reticulum of lysosomes (GERL), and are maturated by glycosylation to form T-1 tyrosinase. Then, GERL secrete coated vesicles (CV), intracellular transporting vesicles, which travel into premelanosomes where melanin polymer synthesis proceeds, becoming melanosomes.[2] These melanogenic compartments are schematically shown in Figure 2.

Stepwise, our major findings of why BPA accumulates within these subcellular compartments of malignant melanoma cells are summarized as follows:

A) In addition to tyrosinase, the molecules of two kinds of melanin monomers, namely 5,6-dihydroxyindole-2-carboxylic acid (DHICA) and dihydroxyindole (DHI), are highest within CV.[3]
B) CV are indeed boron-localizing organelles which accumulate ^{10}B-BPA initially in highest amounts among subcellular compartments in melanoma cells.
C) This led my team, including Dr. Yoshino, to look for possible chemical complex formation[4] of ^{10}B-BPA with monomers and dimers, rich in CV.

These melanogenesis related organelles can be isolated by density-gradient ultracentrifugation method.[5] Figure 3 is an electron micrograph of isolated CV used for pharmacokinetic studies. Following intra-peritoneal injection of ^{10}B-BPA into

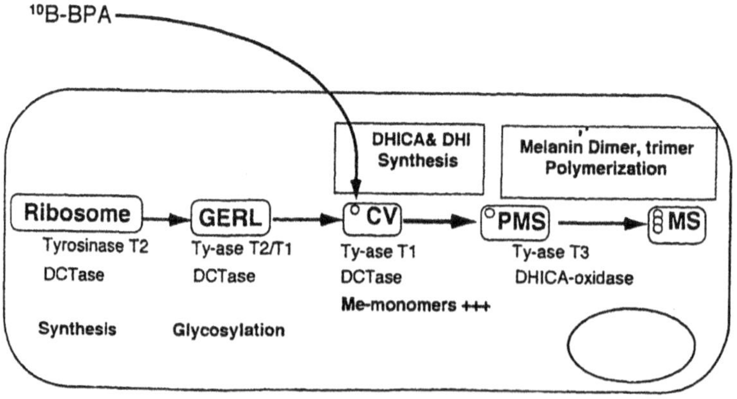

Figure 1. ^{10}B-BPA-attracting subcellular compartments, coated vesicle (CV), premelanosomes (PMS) and melanosomes (MS) which are rich in melanin monomers, dimers, trimers and polymer within target melanoma cells.

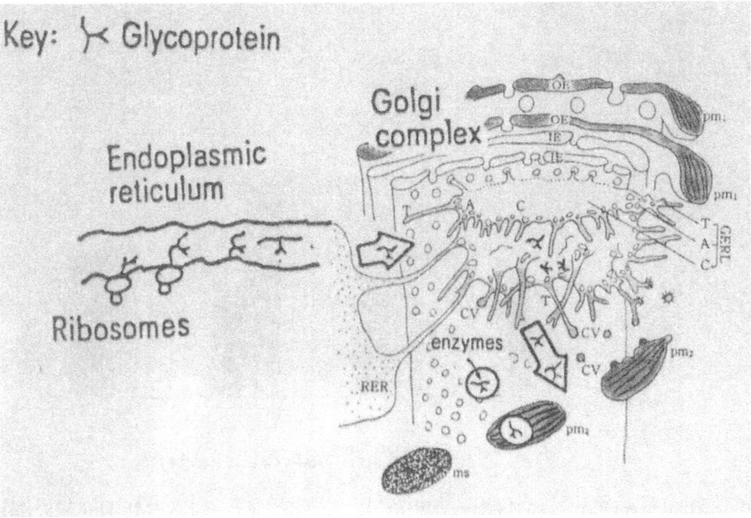

Figure 2. Three-dimensional scheme showing melanogenic compartments within melanoma cells. The process of melanogenesis showing glycosylation of tyrosinase in the Golgi-associated endoplasmic reticulum of lysosomes (GERL) and transfer of glycocylated tyrosinase and melanin monomers by coated vesicles (CV) into premelanosomes (pm). BPA is first highly incorporated into CV.

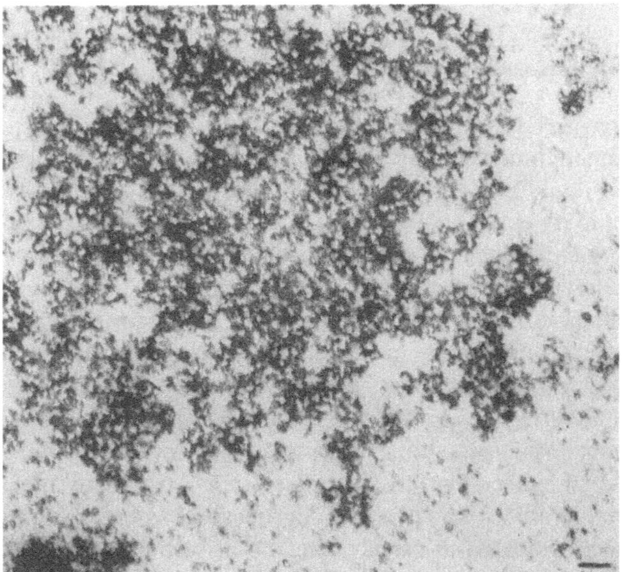

Figure 3. Electron microscopy of isolated CV used for pharmacokinetic studies. Bar: 0.1 μm.

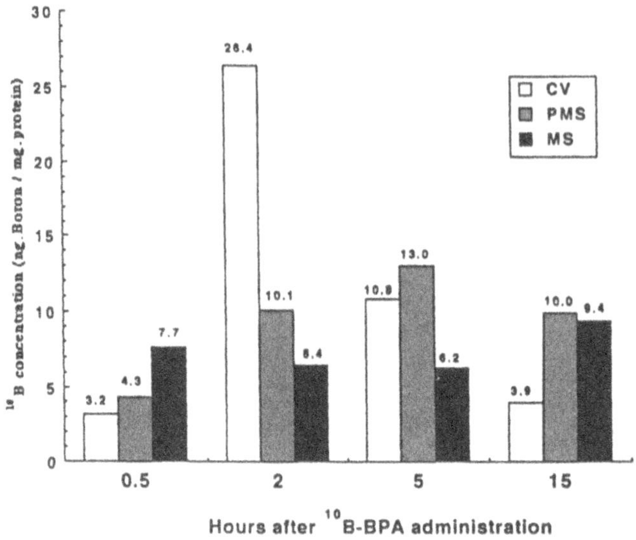

Figure 4. Pharmacokinetics of early highest ^{10}B-uptake in CV, followed by PMS and MS within melanoma proliferating subcutaneously in hamsters after ^{10}B-BPA *i.p.* injection.

melanoma-bearing hamsters, we have found the early highest boron uptake in CV followed by it's movement into premelanosomes and then melanosomes (Figure 4).

Pharmacokinetics studies of boron in human melanoma, which highly synthesizes melanin monomers and polymers, have shown that uptake of melanoma was 3 to 4 times greater than skin and blood.[6]

2.2. Role of Specialized Oncology

Without extensive background and experience in treating specific target cancers, including a profound understanding of the human organs in which these cancers develop, successful BNCT could be elusive. Here, we briefly discuss melanoma as an example.

First, melanoma is not a single entity, but includes 5 different diseases (Figure 5), as found by Clark and myself, with different ontogeny, pathology, prolifiration patterns, therapeutic response and prognosis.[7]

These differences stand out when comparing nodular melanoma with it's reversed pyramid vertical invasion (Figure 6a) with lentigo maligna melanoma, with its marked lateral spread. Microscopic pathology reveals vividly the distinction of these two types. Superficially proliferating lentigo maligna (Figure 6b) is less invasive even after developing into melanoma. When irradiating target melanomas with thermal or epithermal neutrons, we should remember that all four major types, excluding malignant blue nevus, originate and proliferate within epidermis together with later invasion into deep dermis. There is no intact epidermis in primary melanoma lesions, thus epidermal erosion by BNCT is often unavoidable.

There are also many other critical roles played by specialized oncology, including prediction of vertical and horizontal margin of invasion which helps us to determine corresponding optimal beam collimation, curable radiation doses vs. skin tolerance, and ^{10}B-BPA administration protocol.

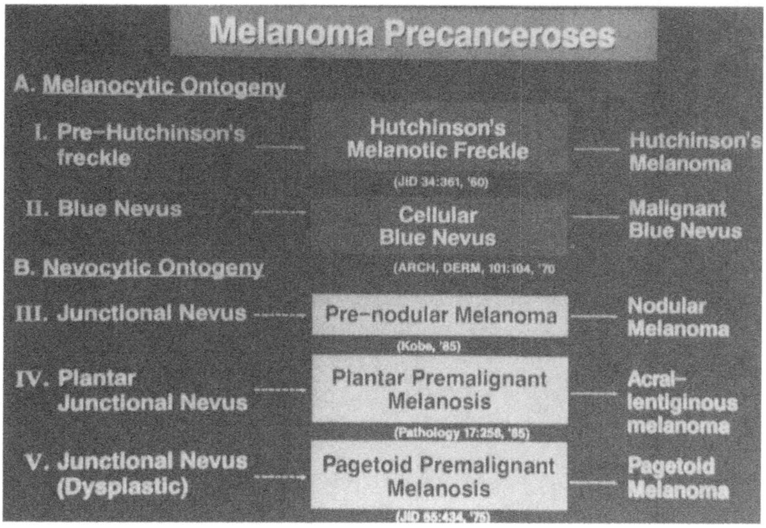

Figure 5. Malignant melanoma is not a single disease, but consists of five different types with distinctive ontogeny, pathology and prognosis. Melanoma BNCT protocol thus should be designed accordingly. Hutchinson's melanoma is synonymous with lentigo maligna melanoma.

Figure 7 summarizes our treatment protocol, the details of which can be found elsewhere.[6,8] Above the horizontal line are prerequisite research areas in forming protocol. Below the line are actual steps for melanoma cure by single irradiation.

It should be emphasized that the exact determination of ^{10}B-concentration, subsequent appropriate ^{10}B-BPA administration, and estimation of absorbed dose at deepest portion of melanoma after optimal patient positioning with beam collimation, each require precise diagnosis concerning type and progression stage of target melanoma.

3. CURRENT PROCEDURE FOR MELANOMA

3.1. Lentigo Maligna Melanoma

First, for lentigo maligna melanoma BNCT, since this occurs mostly on the face with minimal invasion, irradiation dose below skin tolerance with precise collimation is of primary importance. Thus, administration of large amounts of ^{10}B is often necessary in order to obtain sufficient absorbed radiation energy with a rather short irradiation time. We have previously reported on the successful cure of lentigo maligna melanoma occurring on the face of a 64 year-old woman engaged in agriculture with frequent sun-exposure. Details on optimal positioning, collimation, ^{10}B pharmacokinetics, neutron radiation doses and curing process of the lesion can be found elsewhere.[2]

3.2. Nodular Melanoma

Next, we discuss salient points of our NCT to cure nodular melanoma, usually having the worst prognosis with conventional therapy. Since nodular melanoma proliferates vertically and is so highly invasive, absorbed radiation energy should be determined

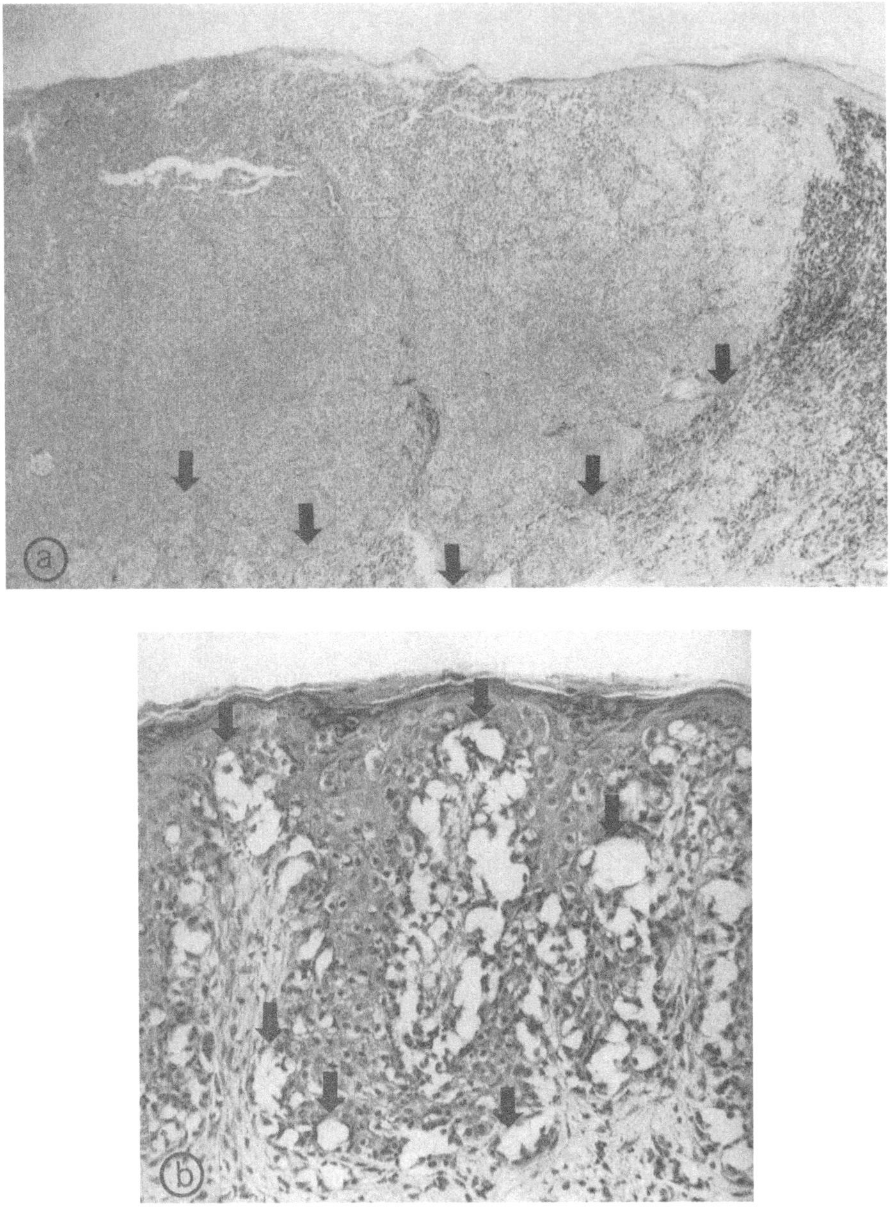

Figure 6. Pathology of most highly invasive nodular melanoma proliferating in a reversed pyramid vertical pattern (arrows) (a), and of less invasive lentigo maligna proliferating laterally (arrows) (b).

at the deepest portion rather than surface, particularly for huge advanced cases for which bolus as well as MRI are applied.

Figure 8a is the clinical appearance of nodular melanoma on the heel of an 85 year-old housewife before treatment. Figure 8b was taken 2 years and 3 months after BNCT, showing permanent regression. Bluish spots observed at that time were phagocitized melanin tattooing. Ten years have now passed without recurrence. Prior to this treatment, in order to precisely determine the vertical and horizontal distribution of absorbed dose

I. **Precise Diagnosis and *Selection of Melanoma Patient***

II. ***^{10}B-Pharmacokinetics* in Mm, Skin, Blood:**
1. Prompt-γ in-situ assay
2. Chemical assay
3. ICP-AES, ICP-MS

III. **Radiation *Dosimetry (n, γ)***
1. Life-size phantom - vertical and horizontal
2. Human melanoma - *15 min estimation at Patient Positioning*

IV. **Estimate *Absorbed Dose : Mm surface & bottom (using bolus)***

V. **Administrate *^{10}B-BPA: Dose & Methods***

VI. ***Collimate Beam* with Optimal Patient Positioning & Irradiate *Thermal-N***

VII. **Post-NCT *Follow Up***

Figure 7. Our treatment protocol for melanoma BNCT.

in the tumor and surrounding heel, we prepared and irradiated a life-size phantom of her heel with the melanoma nodule. Adopting these findings, we made final individualized collimation for her lesion, with neutron and gamma detectors (Figure 8c), and determined the optimal dose of ^{10}B-BPA to be 170 mg/kg BW administrated as fructose complex and thermal neutrons to be 1.2×10^{13}.

We later improved our treatment protocol for advanced huge nodular melanoma by taking into account: 1) the vertical margin below MRI observable lesions and 2) findings observed using a soft-tissue equivalent gelatin bolus.

The extremely fast vertical proliferation on the right sole of a 67 year-old housewife, underscored the necessity of not only eradicating the melanoma visible above the skin surface, but also that invading deeply below. Melanoma thickness above the surface of skin was 15 mm with vertical invasion of 5 mm at M4 revealed by MRI examination (Figure 9). A further 5 mm vertical safety margin at M5, shown as white reverse pyramid, was added as a therapeutic target based on our clinical and pathological evidence discussed earlier. Fluence at M4 was estimated in individualized life-size phantom studies to be 70% of M1 and at M5 to be 80% of M4. Actual fluence and calculated RBE-Gy on surface of melanoma, M1, and surrounding normal skin, S2, were obtained after irradiation for 118 minutes.

Secondly, we also used a soft-tissue equivalent gelatin bolus to obtain actual fluence within melanoma at a depth equal to skin surface. Details can be found elsewhere.[6] This greatly helps in determining both irradiation doses and obtaining better BNCT results.

4. UPDATED PATIENT STATUS

Dr. Hiratsuka and I recently re-examined the present status of patients whose observation period is in range of 2.5 to 7.6 years with median 3.6 years (Figure 10, left). Among 20 lesions, complete regression (CR) is 73%. For non-nodular types of melanoma, such as lentigo maligna (Hutchinson's), acral lentiginous, and superficial spreading (pagetoid) type, CR is 100% for 9 lesions.

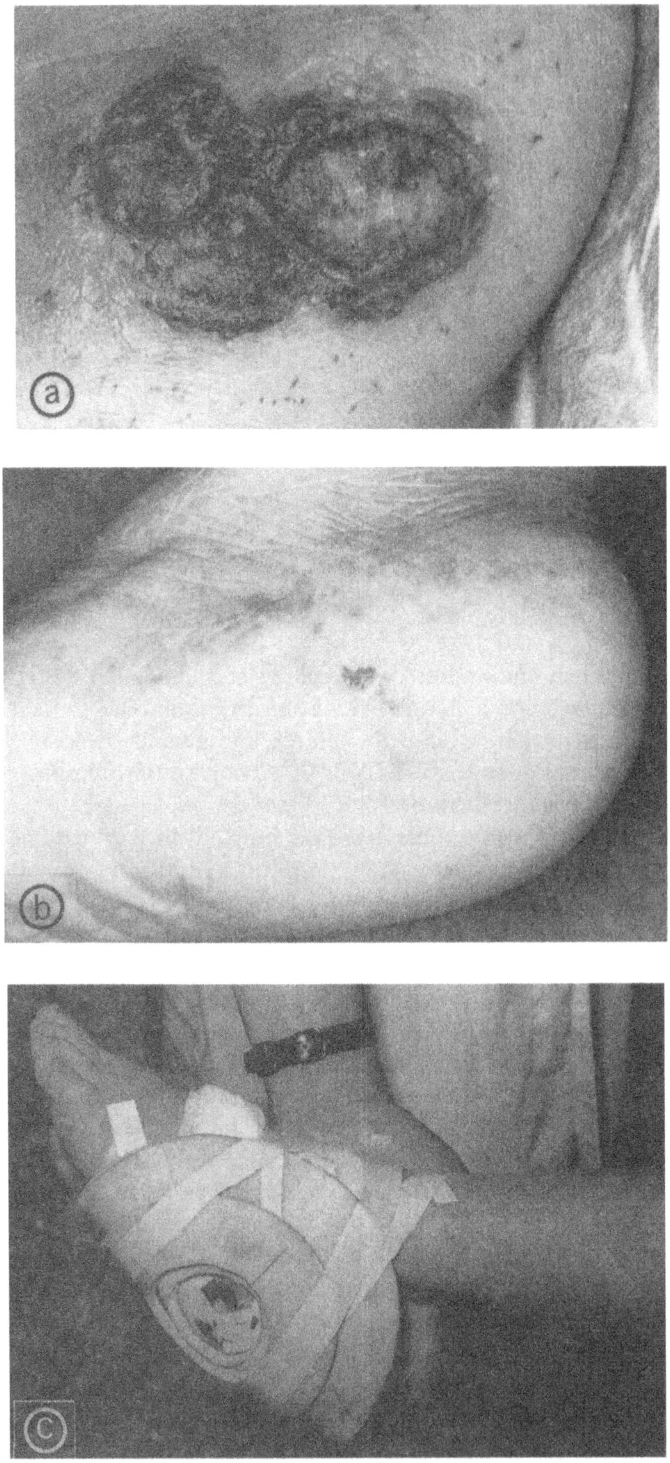

Figure 8. Clinical appearance of nodular melanoma occuring on the heel of an 85 year-old housewife before BNCT (a). Complete regression was observed 2.5 months after single BNCT. Note that the remaining discolored spot is the result of phagocitized melanin tatooing (b) and no recurrence has been observed for more than 10 years up to present. Example of individual collimation with neutron and γ-detectors designed for this patient (c).

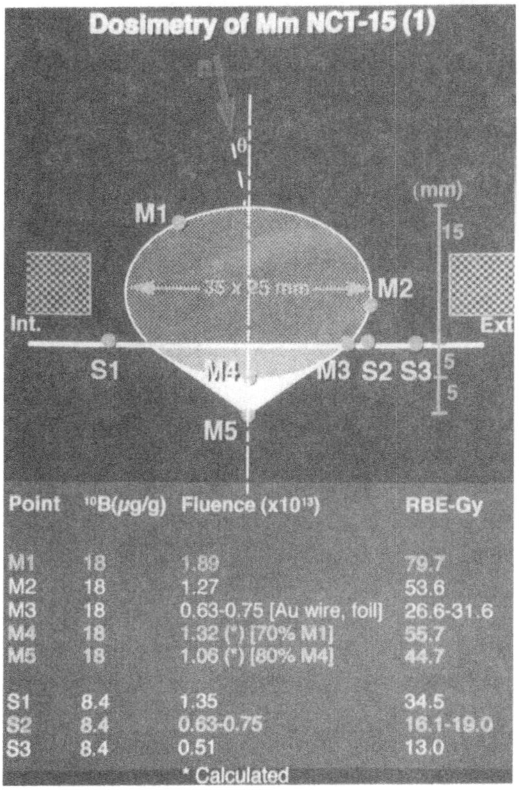

Figure 9. Results of dosimetry study on nodular melanoma deeply invading the right sole of a 67 year-old housewife.

We summarize local skin reaction for the 20 lesions we treated with our BNCT (Figure 10, right). Despite selective deposition of radiation energy at the cellular level, most still developed moist erosion since melanoma cells originate and proliferate within the epidermis along with dermal invasion as discussed earlier. Such erosion can be cured through dermatological treatment alone. Only 2 cases required skin grafts.

	Cases	CR
Total lesions	20	12 (60%)
Primary / Meta		
Primary	15	11 (73%)
Meta	5	1 (20%)
Type		
Nodular M	6	2 (33%)
non-Nodular M	9	9 (100%)

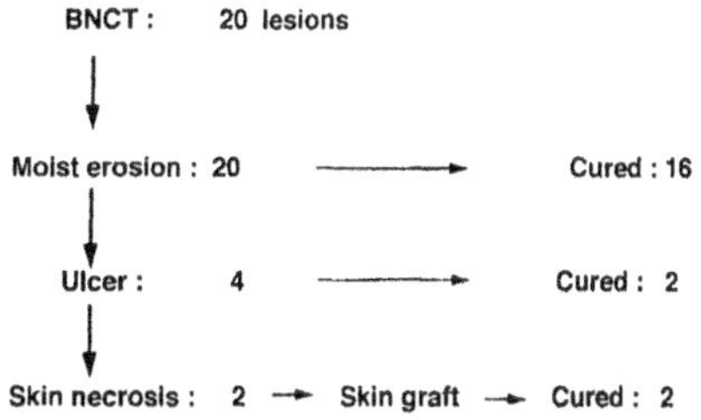

Figure 10. Evaluation of updated patient status revealed percentage of complete remission (CR) in all treated lesions (left) and local skin reaction (right).

5. CANCER CELL MODULATION BNCT

At the Seventh International Symposium on Neutron Capture Therapy for Cancer (Zurich, 1996), I proposed a strategy of gene transfer-applied BNCT.[1] Here, we discuss further developments and findings in this area of gBNCT leading to our most recent concept of cell modulation BNCT as a third generation BNCT.

First generation BNCT utilizes passive ^{10}B accumulation into glioblastoma through the widened blood brain barrier,[9] while second generation BNCT depends on active ^{10}B-biodelivery into target cancers.[6] In the later, we developed ^{10}B-BPA possessing the chemical property to seek target melanomas through their pre-existing metabolic activity, accentuated melanogenesis. Specifically, we synthesized ^{10}B-compound analogous to melanogenic substrate, DOPA.

As the third generation, instead of relying on pre-existing cancer cell properties when searching for increased therapeutic efficacy with current or new ^{10}B compounds, we have now developed the idea of modulating cancer cell properties to enhance their ^{10}B-compound-attracting ability. In addition, when studying the application of BNCT toward new cancers other than glioblastoma and melanoma, such cell modulation also presents additional possibilities. We need to stress that cell modulation can also be teamed with current passive and active ^{10}B-biodelivery to increase further overall efficiency after disclosing the critical molecules responsible in the process of active ^{10}B-biodelivery. An important first step along our journey toward cell modulation BNCT using DOPA analogue,[6] ^{10}B-BPA, was gaining a better understanding of the mechanisms of melanogenesis and its relation to the uptake of boron compounds. This provided a basis for further steps toward successful BNCT for melanoma and our concept of gBNCT leading to cell modulation BNCT which can be applied for a wider spectrum of cancers.

5.1. Melanin Polymer-Forming Pathways

Our current concept holds that DOPAquinone is formed by tyrosinase on tyrosine and DOPA, followed by the non-enzymic synthesis of DOPAchrome and then two melanin monomers, DHI and DHICA.[5] In contrast to the non-enzymic DHI-pathway, the DHICA-pathway is specifically controlled by two other enzymes, DOPAchrome tautomerase, controlled by the TRP-2 gene and then DHICA-oxidase, controlled by the TRP-1 gene. Thereafter, eumelanin-polymer is formed within melanosomes (Figures 1 and 11).

5.2. Cellular Compartment, CVs, Available for ^{10}B-Localization

These three enzymes and their products, melanin monomers, were found by us to be most highly concentrated in CV (Figure 3), which first pick-up the highest amounts of ^{10}B within melanoma cells after administration of ^{10}B-BPA into melanoma-bearing subjects (Figure 4). CVs then transfer their tyrosinase, monomers, and BPA into premelanosomes, where polymerization of melanin monomers proceeds resulting in the formation of melanosomes. Electron microscopy revealed that CVs fuse to the surface of premelanosomes.[10] Also, both organelles are strongly DOPA-uptaking and result in the formation of DOPA-melanin.

These findings hinted to us that the high affinity of BPA to melanoma is primarily due to the chemical complex formation of BPA with monomers. Thus, in cooperation with Dr. Yoshino's group, chemical complex formation of BPA with monomers were

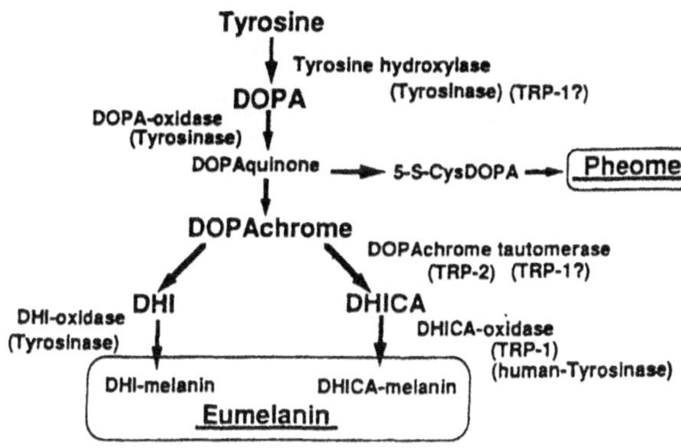

Figure 11. Melanin biosynthetic pathway.

investigated in chemical systems, and then in melanoma-bearing animal systems using ^{11}B-NMR analyses.[4]

5.3. Melanin Monomers and Dimers as Critical Molecules for Accumulation of ^{10}B into Melanoma Cells: Chemical Complex Formation with ^{10}B-BPA

We soon found that melanin monomers, DHICA and DHI, form chemical complexes with BPA in *in vitro* chemical systems and to some extent also *in vivo* using melanoma bearing hamster systems. We reported these results at the 7th ISNCT in Zurich.[1,4]

After obtaining this evidence, we came upon the idea that BPA can further form chemical complexes with melanin dimers and trimers during the process of melanin polymer formation. These dimers and trimers are kindly synthesized by Professor Shosuke Ito and Dr. Kazumasa Wakamatsu. Addition of BPA into DHI dimers first results in complex formation of BPA-DHI dimers at 30 min. After 5 and 24 hours, we detected subsequent complex formation between melanin dimers and boric acid, which is secondarily produced from BPA-dimer complexing (Figure 12a). Details can also be found elsewhere.[11] Figure 12b is the presumed structure of a boric acid-dimer complex. We note specifically that one boric acid forms a complex with two DHI-dimers. Figure 12c is the current presumed sequential chain reactions of BPA with monomers, dimers and further trimers. Borate released from the BPA-monomer complex, may further form complexes with dimers and trimers. During these chain reactions, some amount of boron may be incorporated or trapped into melanin polymer since boron is uptaken into monomer-dimer rich CV which subsequently fuse with melanosomes (Figure 1).

Since these sequential chain reactions occur within CV and melanosomes, membrane-enclosed microcosmos acting as boron localizing subcellular compartments, the ^{10}B involved in these chain reactions should easily be incorporated into melanin polymer within melanosomes. Professor Hawthorne also vividly emphasized the important role of subcellular compartments available for boron localization.[12]

5.4. Gene Transfer-Induced Cancer Cell Modulation BNCT

Since we have found that the critical molecules for accumulating ^{10}B-BPA into target melanoma cells are melanin monomers, DHICA and DHI, and subsequent melanin

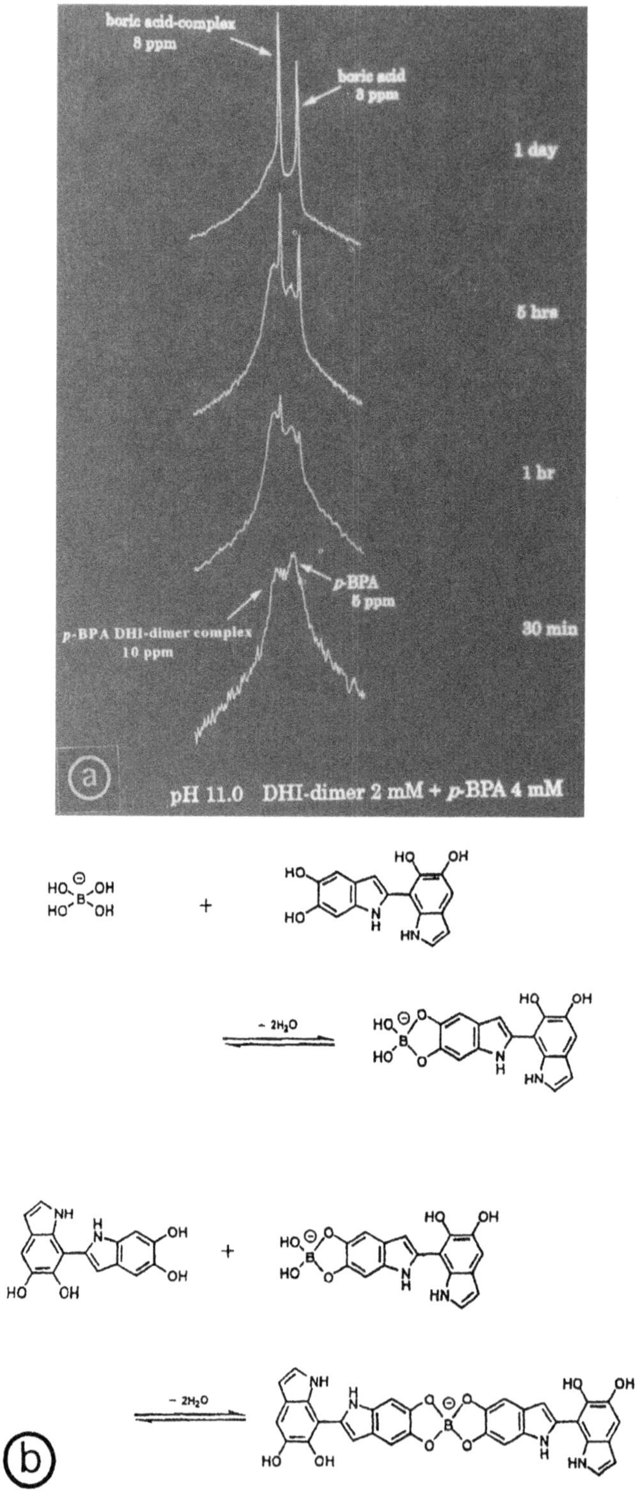

Figure 12. Addition of BPA into DHI dimers first results in the formation of the BPA-DHI dimer complex at 30 min. After 5 and 24 hours, second complexes are subsequently formed between melanin dimers and boric acid, which are produced from BPA-dimer complexing (a). Presumed chemical structure of a boric acid-dimer complex. Note that one boric acid can form a complex with two DHI-dimers (b). The current presumed sequential chain reaction of BPA with monomers, dimers and further trimers. Borate released from the BPA-monomer complex, may further form complexes with dimers and trimers (c).

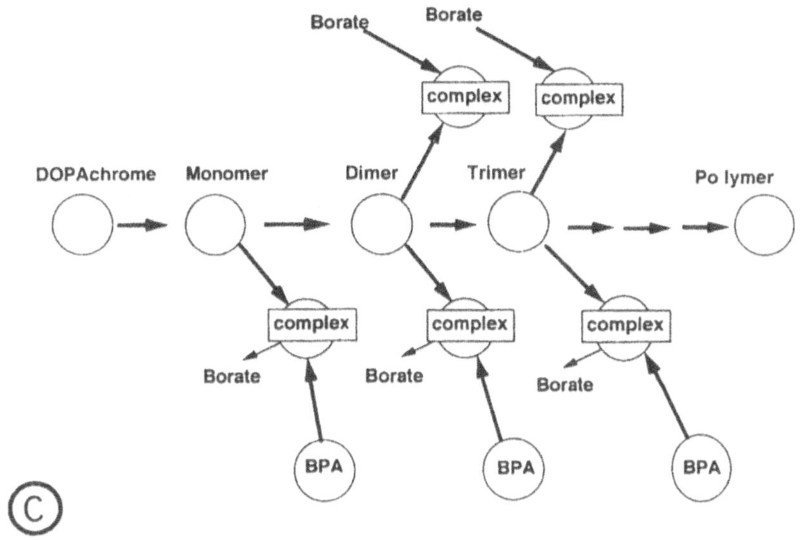

Figure 12. *Continued.*

dimers and trimers, we have investigated cell modulation of poorly melanin synthesizing amelanotic melanoma cells to highly melanin synthesizing melanoma cells by melanogenic gene-transfer as a prototype. This cell modulation successfully enhanced the uptake of ^{10}B-BPA into amelanotic melanoma cells and increased therapeutic efficacy on tumors consisting of these cells using BNCT.

Schematic representation of our prototype cancer cell modulation BNCT using melanogenic gene-transfer is shown in Figure 13.

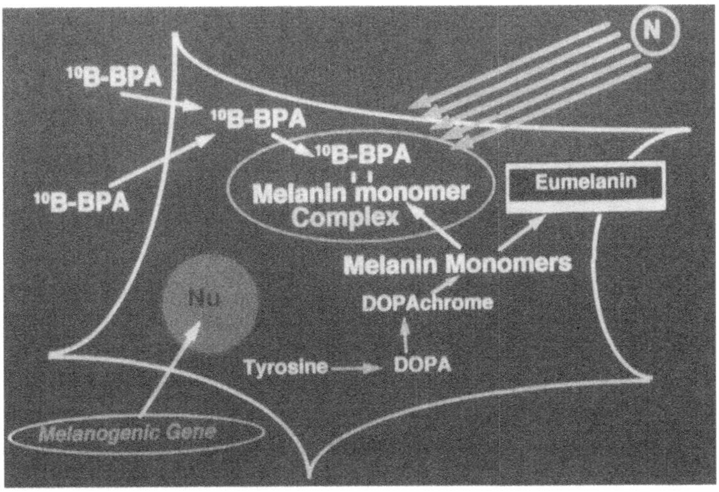

Figure 13. Our prototype cancer cell modulation BNCT using melanogenic gene-transfer. In this case, the critical molecules for accumulating ^{10}B-BPA into target cells are melanin monomers, DHICA and DHI, and subsequent melanin dimers and trimers. These molecules can be induced by melanogenic gene transfer for optimal specific BNCT.

Step 1. Transfer melanogenic genes into target cells in culture by physical method such as electroporation, or biochemical method such as using liposome, or lipofectin.
Step 2. Transfered melanogenic genes are transcripted and synthesize melanogenic enzymes.
Step 3. These enzymes induce the melanin monomer/polymer forming process.
Step 4. Biosynthesized melanin monomers form chemical complexes with BPA and accumulate ^{10}B within target cells.
Step 5. Thermal neutron irradiation selectively kills cells which have accumulated ^{10}B.

5.5. Prototype Experiments of Cell Modulation BNCT

5.5.1. Induction of Marked ^{10}B Uptake into Amelanotic Melanoma Cells by Melanogenic Gene-transfer: Culture Cell System Experiments. We have established amelanotic melanoma cell line A1059 which is a spontaneous amelanotic derivative of B16 mouse melanoma cells possesing expression of melanogenic genes except tyrosinase. A1059 cells were transfected with human tyrosinase expression plasmid BCMGS-NHT2[13] by electroporation. Several highly melanogenic stable clones were selected by using G418 and established (Figure 14a). These clones were evaluated for their eumelanin content, tyrosinase as well as DOPAchrome tautomerase activities, doubling time and expression of tyrosinase, TRP-1, TRP-2. These cells acquired new cellular properties, higher expression of human tyrosinase than that of avarage B16F10 cells. In addition, secondary upregulation of DOPAchrome tautomerase (TRP-2) has also been found in these cells.

The effect of this cell modulation on the uptake of ^{10}B-BPA has been evaluated quantatively as well as comparatively. As shown in Figure 14b, TA1059 cells, which are tyrosinase gene-transfected A1059 cells, exhibited twice the boron-uptake as parental A1059 cells and equivalent to melanotic B16F10 cells.

5.5.2. Induction of Marked ^{10}B Uptake and Killing Effect in Amelanotic Melanoma Cells by Melanogenic Gene-transfer: Tumor-Bearing Mouse Experiments. C57/BL male mice, 6 weeks old, were implanted subcutenously in buttocks with 1×10^7 cells of A1059 and of TA1059-1 each. When the tumors grew to about 5 mm in diameter, these mice were injected intraperitoneally with 400 mg ^{10}B-BPA·HCl/kg body weight. After 2 hrs, assay of boron uptake in TA1059-1 tumors has revealed approximately a fourfold increase as compared to A1059 tumors. Based on these results, we proceeded to undertake *in vivo* therapeutic experiments.[14] Each melanoma-bearing region of the buttocks was irradiated at KUR with thermal neutrons at 5 MW for 60 min. Therapeutic results were analyzed by measuring 3-dimensional tumor size twice a week. An example of tumor response after BNCT is shown in Figure 14c. In this experiment, fluence was in the range of 0.8 to 1.0×10^{13} neutrons/cm^2 measured by gold foil analysis. In contrast to A1059, TA1059-1 melanomas exhibited substantially higher regression up to day 53 following BNCT irradiation.Tumors disappeared macroscopically in 2 of 3 mice in this experiment.

5.5.3. Sequential Changes in Cellular Properties Modulated by a Melanogenic Gene. We would like to underscore that secondary sequential changes can occur following the transfer of a single melanogenic gene. Such secondary upregulation can further provide

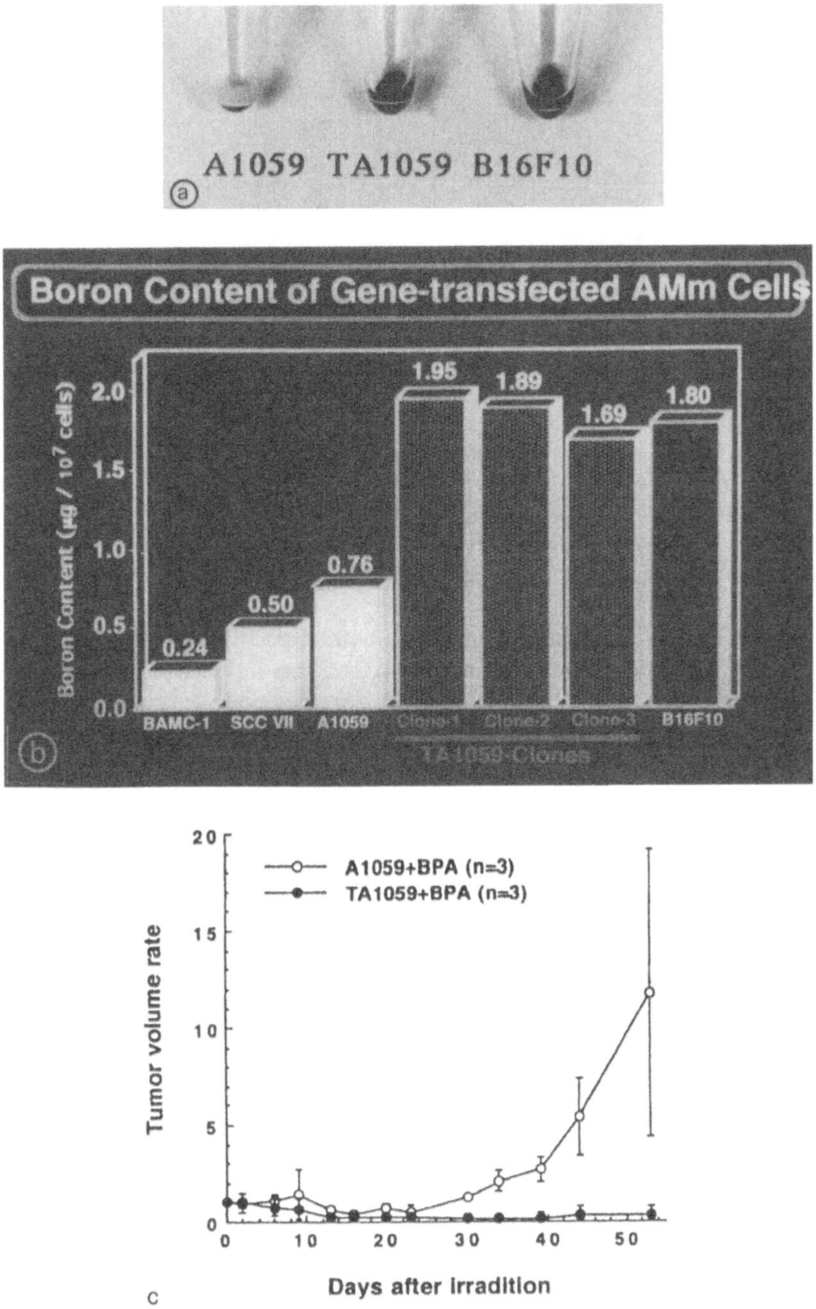

Figure 14. Tyrosinase gene transfection induces active melanogenesis in amelanotic melanoma cells (a). A1059: parental amelanotic melanoma, TA1059: human tyrosinase gene-transfected cells, B16F10: melanotic melanoma TA1059 exhibits over twice the boron-uptake than that of parental A1059 cells and equivalent to B16F10 cells (b). Note that non-pigment cell cancers (BAMC-1 and SCC-VII) exhibit lower uptake of boron than A1059. BAMC-1: mouse fibrosarcoma, SCC-VII: mouse squamous cell carcinoma. Enhanced killing effect of BNCT on gene-transfered amelanotic melanoma (c).

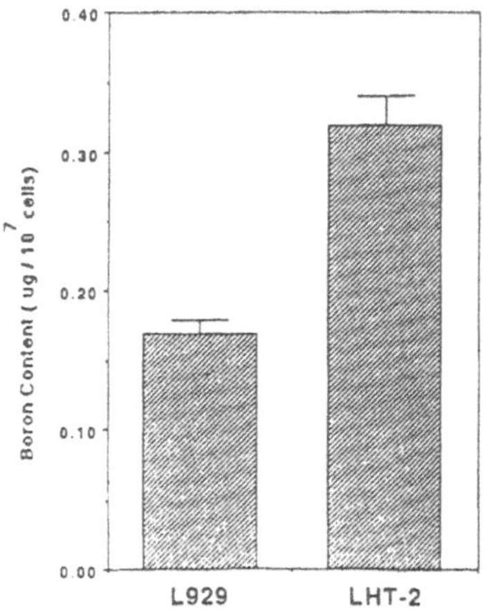

Figure 15. Induced biosynthesis of melanin-monomers and -polymer by tyrosinase gene-transfer increases boron uptake within non-pigment cell fibroblasts. These cells have exhibited high uptake of BPA, more than double that of non-transfected cells. L929: parental mouse fibroblast cells, LHT-2: human tyrosinase-transfected mouse fibroblast cells.

increased effect for cell modulation BNCT. We have found that tyrosinase gene-transfer into tyrosinase expression deficient A1059 cells has induced interestingly the increased DOPAchrome tautomerase activity, appoximately 1.4 times, in the established cell line, TA1059.

Furthermore TRP-2 gene transfer into human poorly melanotic melanoma cells (Ihara[2]) with low DOPAchrome tautomerase activity has induced higher tyrosinase activity, approximately 1.6 times, and BPA uptake, approximately 1.3 times, as compared to mock-transfected cells.[15] Similarly, the transfection of TRP-1 into TRP-1 deficient amelanotic melanoma also induced elavation of tyrosinase activity.[16]

5.5.4. Induction of Melanin Biosynthesis within Fibroblasts by Tyrosinase Gene-Transfer. We have further induced melanin monomer and polymer formation in fibroblasts as a step toward fibrosarcoma BNCT. Although fibroblasts do not form premelanosomes, electron microscopy of melanin-producing fibroblasts has revealed that the induction of melanin polymer formation is occuring in lysosomes.[2,17] These newly melanin polymer forming organelles have strong hydrolytic enzymes as well as phagocytic ability. As shown in Figure 15, these cells have exhibited high up-take of BPA, more than double that of non-transfected cells. Since lysosomes can be formed in most cancer cells, there is a good possibility of wide application of cell modulation BNCT using BPA.

6. CONCLUSION

We have presented here principal examples for success of gene-induced cell modulation BNCT (Figure 16, left). In essence:

1. Affinity of ^{10}B-BPA to melanoma is largely dependent on biosynthesis of critical molecules within target cells, in our case, melanin monomers and dimers.
2. Synthesis of such critical molecules in target cancer cells has been induced or enhanced by melanogenic genes.
3. These transfected target cells have exhibited induced or enhanced BPA-uptake resulting in higher killing efficiency of these cancer by BNCT.

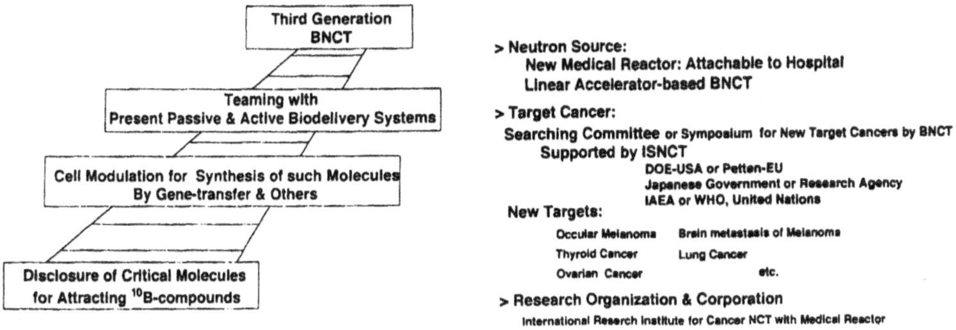

Figure 16. Disclosure of critical molecules for attracting ^{10}B-compounds into melanoma led to the new concept of third generation BNCT (left). The prospects for cancer BNCT (right).

We should add that cell modulation can also be induced by drugs or hormones, which would be categorized as drug-induced cell modulation BNCT. We would like to close by emphasizing that cell modulation, as exemplified here, should be teamed with currently used passive and active ^{10}B-biodelivery systems leading to third generation BNCT.

Our view on the prospects of cancer BNCT research are listed (Figure 16, right). Among three aspects, we would like to underscore for your consideration the formation of a searching committee or symposium for new target cancers for BNCT by at least the next 9th ISNCT meeting to be organized by Prof. Kanda. DOE, Petten, the Japanese government, IAEA, WHO, or the United Nations, could also be supporters.

ACKNOWLEDGMENTS

We wish to thank the various collaborators who have contributed over the years to the success we have had in treating human melanoma. Of special mention are H. Fukuda, H. Kakihana, T. Nakagawa, C. Honda, N. Wadabayashi, M. Shiono, M. Ichihashi, H. Obara, J. Shirakawa, H. Karashima, T. Kobayashi, T. Nozaki, O. Aizawa, T. Sato, K. Kanda, and T. Tsuboi. I also thank my close assistants K. Ohzu and B. Jones.

REFERENCES

1. Y. Mishima, Melanoma and nonmelanoma neutron capture therapy using gene therapy: overview, *in*: "Advances in Neutron Capture Therapy," B. Larsson, J. Crawford, and R. Weinreich, eds. Elsevier, Amsterdam, 1997, pp. 10–25.
2. Y. Mishima, Molecular and biological control of melanogenesis through tyrosinase genes, and intrinsic and extrinsic regulatory factors, *Pigment Cell Res.* 7:376–387, 1994.
3. S. Hatta, Y. Mishima, M. Ichihashi, and S. Ito, Melanin monomers within coated vesicles and premelanosomes in melanin synthesizing cells, *J. Invest Dermatol.* 91:181–184, 1988.
4. K. Yoshino, Y. Mishima, M. Kimura, J. Hiratsuka, Y. Mori, S. Ito, and H. Kakihana, Capture of *p*-bornophenylalanine in malignant melanoma cells by complex formation with melanin monomers, DOPA, DHI and DHICA, *in*: "Advances in Neutron Capture Therapy," B. Larsson, J. Crawford, and R. Weinreich, eds. Elsevier, Amsterdam, 1997, pp. 234–238.
5. Y. Mishima, A post melanosomal era: Control of melanogenesis and melanoma growth, *Pigment Cell Res. Suppl.* 2:3–16, 1992.

6. Y. Mishima, Selective thermal neutron capture therapy of cancer cells using their specific metabolic activities–melanoma as prototype, *in*: "Cancer Neutron Capture Therapy," Y. Mishima, ed. Plenum, New York, 1996, pp. 1–26.
7. K. Yamamura, and Y. Mishima, Antigen dynamics in melanocytic and nevocytic melanoma oncogenesis: Antiganglioside and anti-ras p21 antibodies as markers of tumor progression, *J. Invest. Dermatol.* 94:174–182, 1990.
8. Y. Mishima, Melanoma Neutron Capture Therapy, Proceedings of second Japan-Australia International Workshop on BNCT for malignant melanoma, *Pigment Cell Res.* 2:223–285, 1989.
9. H. Hatanaka, and Y. Nakagawa, Clinical results of long-surviving brain tumor patients who underwent boron neutron capture therapy, *Int. J. Radiation Oncology Biol. Phys.* 28:1061–1066, 1994.
10. Y. Mishima, and G. Imokawa, Role of Glycosylation in Inititial Melanogenesis: Post-Inhibition Dynamics, *in*: "Biological, Molecular and Clinical Aspects of Pigmentation," J. Bagnara *et al.*, eds. Tokyo, 1985, pp. 17–30.
11. K. Yoshino, M. Kamijo, M. Kimura, Y. Mishima, H. Kondoh, M. Kadota, S. Ito, K. Wakamatsu, and H. Kakihana, A father mechanism study of boron accumulation to and release from melanoma with p-boronophenylalanine, *in:* "Frontiers in Neutron Capture Therapy," M.F. Hawthorne, K. Shelly, R.W. Wiersema, eds., Kluwer Academic/Plenum Publishers, New York, 2001, pp. 985–989.
12. M.F. Hawthorne, The role of chemistry in the development of boron neutron capture therapy of cancer, *Angew. Chem. Int. Ed. Engl.* 32:950–984, 1993.
13. O. Ando, S. Hanada, Y. Suemoto, J. Atobe, M. Kurimoto, and Y. Mishima, Analyses of mixed melanogenesis in tyrosinase cDNA-transfected human amelanotic melanoma cells, *J. Invest. Dermatol.* 101:864–870, 1993.
14. T. Tsuboi, H. Kondoh, J. Hiratsuka, and Y. Mishima, Enhanced melanogenesis induced by tyrosinase gene-transfer increase boron-uptake and killing effect of boron neutron capture therapy for amelanotic melanoma, *Pigment Cell Res.*, in press, 1998.
15. H. Kondoh, K. Yamaguchi, J. Hiratsuka, and Y. Mishima, Enhancement of BPA-uptake in melanotic melanoma by DOPAchrome tautomerase gene-transfer, Program and Abstracts of the Eighth International Symposium on NCT for Cancer, La Jolla, CA, 1998, p. 97.
16. H. Zhao, D.J. Eling, E.E. Medrano, and R. Boissy, Retroviral infection with human Tyrosinase-related protein-1(TRP-1) cDNA upregulates tyrosinase activity and melanin synthesis in a TRP-1-deficient melanoma cell line, *J. Invest. Dermatol.* 106:744–752, 1996.
17. H. Kondoh, A. Wilczek, S. Narimizu, and Y. Mishima, Mouse fibroblast expressing human tyrosinase with DHICA-oxidase activity produces predominantly pheomelanin deposit in lysosome, *Zoological Sci.* 13:825–831, 1996.

BNCT BEYOND GLIOMA TREATMENT

Where are the Problems?

Detlef Gabel

Department of Chemistry
University of Bremen, Germany

1. INTRODUCTION

Boron neutron capture therapy has been used in Japan for a number of decades for the treatment of glioblastoma and since around a decade for the treatment of melanoma. In both cases, therapy is confined to the surface of the exposed tissue, as thermal neutrons are used. In the glioma treatment, intraoperative BNCT is therefore carried out.

Recently, epithermal neutron beams have become available for therapy and further epithermal neutron sources are under construction and consideration.

These beams allow the treatment of deeper-seated tumors, and no longer require intraoperative treatment. At present, they are used for the treatment of glioblastoma, in Phase I clinical trials. Here, BPA and BSH are being used as boron carriers.

In this paper, the steps required for the application of BNCT to other tumor entities are considered, both for the drugs presently already in clinical trials and for drugs under development.

2. REQUIREMENTS FOR SUCCESSFUL THERAPY

2.1. Basic Considerations

For successful therapy with BNCT, it is necessary that boron is accumulated or retained in the target tissues more selectively than in the surrounding healthy tissues, and in concentrations sufficient to overcome the neutron capture reactions in hydrogen and nitrogen. It has been estimated that a minimum of 10 μg boron per gram tissue (10 ppm) are required for this.[1] The analysis of the radiobiology of boron compounds[2,3] indicates that BNCT can be applied more successfully by delivering more compound, once the

Frontiers in Neutron Capture Therapy, edited by Hawthorne *et al.*
Kluwer Academic / Plenum Publishers, New York, 2001.

boron distribution is in favor of the target tissue. The radiobiological effectiveness of boron in tumor tissue and in the dose-limiting healthy tissue can be vastly different,[2,4] and will have to be considered both in the pre-treatment evaluation of a compound and the dose prescription in the treatment phase.

Successful therapy demands that the dose ratio between tumor tissue and surrounding healthy tissue is greater than unity, and that the radiation dose to the healthy tissue approaches the tolerance dose.

The first demand is largely determined by the compound chosen. The way of administration, time point of administration in relation to therapy, and the amount administered are important factors. Additionally, the treatment planning plays a role in maximizing the dose ratio. The second demand is determined by the dose prescribed and the treatment plan. Only this aspect is similar to conventional radiotherapy.

In this paper, the requirements for new compounds in BNCT are analyzed, and the use of existing compounds for new indications is considered. This is done with the background of the approach to clinical trials as practiced by the European Organization for Research and Treatment of Cancer (EORTC).[5]

2.2. Requirements for New Compounds

A number of new compounds have been suggested for BNCT. These include porphyrins, sugars and a great number of other drugs. One porphyrin, VCDP,[6,7] has been investigated concerning its toxicity and biodistribution in mice. It is taken here as an example to illustrate the procedure to be taken for a new compound in BNCT.

VCDP shows marked toxicity in mice. The LD_{50} value of the drug is around 30 µg VCDP per gram body weight as a single dose, the LD_{10} is around 20 µg/g. Repeated administration at this dose level is tolerated. Thrombocytopenia, granulocytosis and altered liver enzyme levels were measured in the blood. With repeated administration of VCDP (three injections of 20 µg VCDP each per day over four days), tumor concentrations of 30 ppm boron are reached, with blood concentrations of around 10 ppm. Boron uptake in the tumor is highly heterogeneous, as observed for a number of other porphyrins. Therapy could be carried out successfully even without optimization of the irradiation procedure and maximization of the radiation dose.[7]

For any compound to be useful in patient trials, its availability must be adequate. For VCDP, at an injected dose of a total of 250 µg porphyrin per gram body weight, around 20 g porphyrin would be required per patient. The preparation of these large amounts of the porphyrin would require methods different from those originally proposed and represent a major industrial effort.

2.3. Toxicity

Following the establishment of an LD_{10} in mice, EORTC regulations[5] foresee that the drug can be administered to patients without the necessity for further toxicity investigations in larger animals. The starting dose SD of such a toxicity Phase I study is 10% of LD_{10} in mice. For VCDP, a total of around 200 g VCDP would be required if toxicity levels are similar to those observed in mice.

In these patients, information about the pharmacokinetics of the compound can also be gained. It has been found that data cannot necessarily be extrapolated from mice to patients.[8]

2.4. Tissue Distribution

In judging the suitability of a new compound for BNCT, it is important to gain information about the tissue distribution in patients both in healthy tissue exposed to the beam, and in the tumor which is to eventually be treated. This aspect can differ from the normally prevailing schemes for a Phase I study, where patients eligible for participation must have a tumor which is no longer amenable to treatment with established forms of treatment. Usually, drugs are tested for which there is a possible (though in the early stage of the Phase I trial small) benefit, as the drug is chosen because it might influence the tumor growth.

In the case of BNCT, the drug is usually not active on its own, but is intended only to accumulate in the tumor to be treated. Therefore, there is no potential benefit from the drug alone. In addition, the patient must have a tumor of the type intended for treatment, and which can at least partially be removed surgically. If such patients cannot be operated upon, they cannot be considered for this step.

It is in this aspect important to realize that the tissue distribution can be very heterogeneous. Analysis of boron by bulk methods (such as ICP-OES or prompt-gamma) will have to be complemented by methods with higher resolution,[9,10] in order to estimate the potential radiobiological effect of boron in healthy and tumor tissue. Biopsy material might be too small to indicate whether the drug concentrations are highly variable of fairly homogeneous throughout a larger tumor.

2.5. Tolerance of the Healthy Tissue

The dose in BNCT is a composite of several dose components, namely gamma radiation both from the reactor and the $^{1}H(n,\gamma)^{2}H$ reactions, recoil protons from fast neutrons and the $^{14}N(n,p)^{14}C$ reaction, and the products of the $^{10}B(n,\alpha)^{7}Li$ reaction.[4]

Whereas good empirical data exist to correlate doses from gamma and recoil protons with biological effects, this is not true for the $^{10}B(n,\alpha)^{7}Li$ reaction. The biological effectiveness can vary by a factor of ten,[2] and between compounds it has been found to vary by a factor of four.[4,11]

For some tissues, data are or will be available concerning the tolerance to BNCT of the tissue in question, when BPA or BSH are taken as boron drugs. These tissues are mostly skin and bone, and healthy brain tissue. Little data exist for these compounds concerning lipoid tissue and muscle, as well as other tissues which might be exposed. For BPA or BSH, some kind of radiation dose escalation study must be performed on the tissue to be exposed. This study could be modelled after similar studies for brain tumor treatment.

For new drugs, no data on the tolerance of healthy tissue are available. For these drugs, it therefore appears highly desirable, if not mandatory, to perform tolerance studies in suitable test systems, before a radiation dose escalation study in patients is considered.

3. PHASE I CLINICAL TREATMENT STUDY

A Phase I clinical treatment study in BNCT differs from a Phase I study with a new chemotherapeutic agent insofar as an eligible patient must have a specific tumor, which has not been treated previously by radiotherapy. Furthermore, it is not the drug dose itself but the treatment dose which is escalated until tolerance is reached. This can, but does not necessarily have to, include an escalation of the drug. Indeed, due to the fact

Table 1. Preparatory steps for clinical trials in BNCT

Step	Requirement	VCDP	BSH, BPA
Availability	>3 g boron per patient per dose	not readily available	commercial product
Toxicity in small animals	EORTC: mice	known	known
Toxicity in large animals	EORTC: not required	not known	known
Toxicity in patients	Phase I dose escalation study	not yet performed	known
Uptake in tumor	>10 ppm boron	in human tissue not yet performed	known for brain tumors in patients known for some other tumors in animals
Concentration in and microdistribution in tumor and surrounding healthy tissue, including blood	Dose ratio tumor : surrounding tissue >1	achievable	yes
Tolerance of healthy tissue to treatment	known to within 30% from animal studies	not known for any tissue	known for skin, bone, nervous tissue

that BNCT with any given drug is better with high absolute concentrations of boron in tissue, a drug escalation study should be performed prior to a radiation dose escalation study. The necessary steps are summarized in Table 1.

Due to the complexity of the total effective dose in BNCT, Phase I clinical trials must be approached with caution. Side effects from the drug alone might be seen rather soon. Radiation damage might occur with delay, and without early signs of tissue reaction. Depending on the type of tissue at risk, rescue therapies might or might not be available. For new indications, the use of the two known compounds would enable a quick start of a Phase I clinical treatment trial.

New compounds, while in principle more promising, pose a considerable problem in all aspects of the clinical trial. The example of VCDP shows the problems of acute toxicity (which is considerable), of late phototoxicity of the skin (which has not been analyzed), the administration of the compound (which needs to be optimized) and the availability of the compound (which is difficult to prepare in large quantities). Whereas these problems are not insurmountable for any new compound, ethical and practical questions must be answered satisfactorily before a clinical trial protocol can be prepared.

ACKNOWLEDGMENTS

The work of the European Collaboration for Neutron Capture Therapy is supported by the European Commission under the Biomed II program. Work by the author on chemistry is supported by the Deutsche Forschungsgemeinschaft and the Fonds der Chemischen Industrie.

REFERENCES

1. Fairchild R.G., and Bond V.P. Current status of ^{10}B-neutron capture therapy: Enhancement of tumor dose via beam filtration and dose rate, and the effects of these parameters on minimum boron content: A theoretical evaluation. *Int J Radiat Oncol Biol Phys*, 11:831–840, 1985.

2. Gabel D., Foster S., and Fairchild R.G. The Monte-Carlo simulation of the biological effect of the $^{10}B(n,\alpha)^{7}Li$ reaction in cells and tissue and its implication for boron neutron capture therapy. *Radiat Res*, 111:14–25, 1987.
3. Morris G.M., Coderre J.A., Hopewell J.W., Micca P.L., and Fisher C. Boron neutron capture irradiation of the rat spinal cord: Effects of variable doses of borocaptate sodium. *Radiother Oncol* 39:253–259, 1996.
4. Gabel D., Philipp K.I.H., Wheeler F.J., and Huiskamp R. The compound factor of the $^{10}B(n,\alpha)^{7}Li$ reaction from borocaptate sodium and the relative biological effectiveness of recoil protons for induction of brain damage in boron neutron capture therapy. *Radiat Res*, 149:378–386, 1998.
5. European Organization for Research and Treatment of Cancer: A practical guide to EORTC studies. Brussels, 1996.
6. Miura M., Micca P.L., Heinrichs J.C., Gabel D., Fairchild R.G., and Slatkin D.N. Biodistribution and toxicity of 2,4-divinyl-nido-o-carboranyldeuteroporphyrin IX in mice. *Biochem Pharmacol*, 43:467–476, 1992.
7. Miura M., Micca P., Gabel D., Fairchild R.G., and Slatkin D. Biodistribution, toxicity and efficacy of a boronated porphyrin for boron neutron capture therapy. *in*: "Progress in Neutron Capture Therapy for Cancer" B.J. Allen, D.E. Moore, B.V Harrington, eds. Plenum Press, New York 1992, pp. 455–458.
8. Mehta S.C., and Lu D.R. Interspecies pharmacokinetic scaling of BSH in mice, rats, rabbits, and humans. *Biopharmaceutics & Drug Disposition*, 16:735–744, 1995.
9. Haritz D., Gabel D., and Huiskamp R. Clinical phase-I-study of $Na_2B_{12}H_{11}SH$ (BSH) in patients with malignant glioma as precondition for boron neutron capture therapy (BNCT). *Int J Radiat Oncol Biol Phys*, 28:1175–1181, 1994.
10. Otersen B., Haritz D., Grochulla F., Bergmann M., Sierralta W., and Gabel D. Binding and distribution of $Na_2B_{12}H_{11}SH$ on cellular and subcellular level in tumor tissue of glioma patients in boron neutron capture therapy. *J Neurooncol* 33:131–139, 1997.
11. Morris G.M., Coderre J.A., Hopewell J.W., Micca P.L., and Rezvani M. Response of rat skin to boron neutron capture therapy with p-boronophenylalanine or borocaptate sodium. *Radiother Oncol* 32:144–153, 1994.

PROGRESS IN THE DEVELOPMENT OF BORON NEUTRON CAPTURE SYNOVECTOMY FOR THE TREATMENT OF RHEUMATOID ARTHRITIS

J. C. Yanch[1], S. Shortkroff[2], R. E. Shefer[3], E. Binello[1], D. Gierga[1], A. G. Jones[4], G. Young[2], C. Vivieros[2], and B. Blackburn[1]

[1]Department of Nuclear Engineering and
Whitaker College of Health Sciences and Technology, MIT
[2]Department of Orthopedic Surgery
Brigham and Women's Hospital, Boston, MA
[3]Newton Scientific Incorporated
Cambridge, MA
[4]Department of Radiology
Brigham and Women's Hospital and Harvard Medical School
Boston, MA

1. INTRODUCTION

The potential of the $^{10}B(n,\alpha)^{7}Li$ nuclear reaction to be effective in the treatment of rheumatoid arthritis (RA) is under investigation at MIT's Laboratory for Accelerator Beam Applications (LABA) in collaboration with researchers at Newton Scientific, Incorporated and Brigham & Women's Hospital, Boston.[1] The goal of this approach is the ablation of the synovial membrane of rheumatic articulating joints and hence the proposed treatment is termed Boron Neutron Capture Synovectomy, or BNCS. Development of this potential modality has proceeded in a number of areas including *in vitro*, *ex vivo* and *in vivo* boron uptake studies,[2-4] neutron beam design,[5] construction, and installation, characterization of neutron beam assemblies, and preliminary evaluation of the efficacy of BNCS in an animal model of arthritis. These developments will be described here and results of initial studies of BNCS to treat arthritis in a rabbit model will be presented.

Frontiers in Neutron Capture Therapy, edited by Hawthorne *et al.*
Kluwer Academic / Plenum Publishers, New York, 2001.

2. RHEUMATOID ARTHRITIS AND BNCS

Rheumatoid arthritis is an autoimmune disease of unknown origin suffered by 1–3% of the general population.[6,7] Although it is a systemic disease, it is common for symptoms to arise only in articulating joints. Symptoms, including pain and even disability, are a result of inflammation of the synovium, the membrane lining the inner joint capsule. To relieve the symptoms, the patient is provided with anti-inflammatory drugs. This approach is usually effective in reducing the inflammation and the pain. In roughly 10% of patients, however, at least one joint remains unresponsive.[8] Thus, alternative treatment options are needed for 230,000–460,000 patients in the US alone.

Two alternative options exist, only one of which is available to patients in the US. Surgical removal of the synovium (surgical synovectomy) is an effective means of providing temporary symptomatic relief in 80% of patients.[9] Because only the symptoms are being treated and the cause of the disease is not addressed by this approach, symptoms typically reappear after 2–5 years. Another therapeutic approach involves the injection of beta particle emitters into the joint space; this is referred to as radiation synovectomy. The beta particle energy deposited in the synovium kills the tissue and the symptoms are thereby alleviated. As is the case with surgical synovectomy, new synovium quickly grows and roughly 80% of patients remain asymptomatic for 2–5 years.[10] This procedure is less invasive and more cost effective than surgery with similar results. Radiation synovectomy, while practiced in Europe, Canada, Australia and elsewhere, is not carried out in the US due to concerns of healthy tissue irradiation if the beta particle emitter leaks from the joint space.

Boron Neutron Capture Synovectomy involves the intra-articular injection of a boronated compound directly into the joint space, as illustrated in Figure 1. After waiting a designated interval for the compound to be taken up by the synovium, the joint is irradiated with a low energy, epithermal neutron beam. BNCS is, in effect, a radiation synovectomy but would overcome the problem of radioactive material leaking from the joint since the boron-labeled pharmaceutical is stable except during neutron irradiation of the joint.

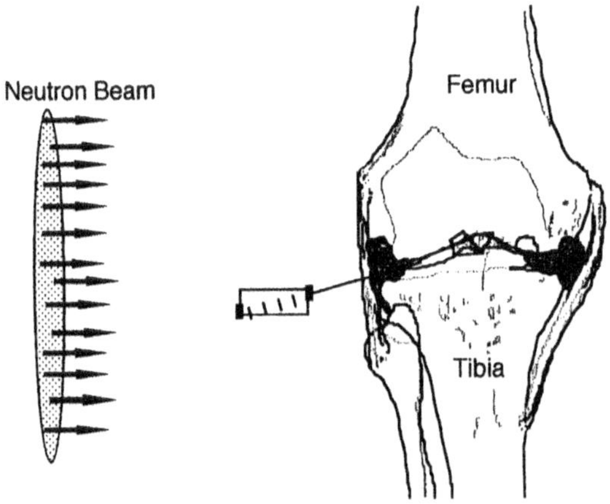

Figure 1. Schematic illustration of BNCS.

BNCS differs from BNCT of brain tumors in many ways. First, in BNCS the target tissue lies between 0.2 and 2.0 cm below the skin surface. For BNCT, neutron beams are designed to be able to provide dosimetric advantage to brain tumors at depths of up to 7 or 8 cm. Second, as a result of direct intra-articular injection of the boron-labeled compound, extremely high boron uptake levels are obtained in the synovium. For BNCT, systemic administration leads to lower boron uptake levels in the target tissue. Third, most articulating joints are located far from radiosensitive organs. Even cartilage, which is adjacent to the synovial fluid, is one of the most radio-resistant structures in the body. On the other hand, BNCT procedures for brain tumors run the risk of damaging many radiosensitive regions in the brain.

3. BORON UPTAKE STUDIES

3.1. *Ex Vivo* Boron Evaluation

As a first step in evaluating the potential usefulness of a variety of boron-labelled compounds for BNCS, *ex vivo* studies were carried out in which samples of human arthritic synovium were obtained from the operating room and incubated, under various conditions, in the presence of medium containing one of a number of boronated compounds.[2–4] [Synovial tissue remains viable in culture for at least one week.] Uptake was determined[11] using the prompt gamma neutron activation analysis (PGNAA) facility at the MIT Research Reactor.

Boron metal particulate and a number of water soluble compounds have been examined. Future work will focus on the evaluation of targeted compounds. The most promising compound to date, however, is $K_2B_{12}H_{12}$, the salt form of a polyhedral closo boronate ion. Table 1 shows uptake in human synovium[3] following incubation in medium containing various concentrations of $K_2B_{12}H_{12}$. The tissue was rinsed in saline and evaluated for boron uptake via PGNAA. Boron concentration in the tissue was found to be roughly 40–60% of the concentration in the incubation medium, regardless of incubation concentration. [Note that both unenriched and ^{10}B enriched $K_2B_{12}H_{12}$ have been used, however all data presented here are scaled to 100% enrichment in ^{10}B.]

In subsequent experiments the potential for washout of the compound from the tissue was investigated.[4] Results are shown in Table 2. After incubation for one hour in 5,000 ppm boron in $K_2B_{12}H_{12}$, the tissue was rinsed in saline and either immediately stored for PGNAA or re-incubated in boron-free medium for various times. Washout of approximately 75% of the compound from the tissue takes place within the first 30 minutes. Thereafter, further incubation results in no further loss of compound from the tissue.

Table 1. Boron uptake in human rheumatoid synovium following incubation in medium containing various concentrations of boron in $K_2B_{12}H_{12}$ for one hour

Incubation concentration (ppm)	Uptake in synovium (ppm)
500	160–205
2,500	1,125–2,565
5,000	2,030–2,725

Table 2. Boron uptake in human rheumatoid synovium following incubation with medium containing 5,000 ppm boron in $K_2B_{12}H_{12}$ for one hour, then rinsed and re-incubated for varying times in boron-free medium

Reincubation time (minutes)	Uptake (ppm)
0	2,370 (N = 3)
30	315–770 (N = 3)
60	360–715 (N = 3)
120	285–530 (N = 3)

3.2. *In Vivo* Boron Evaluation

The *ex vivo* results, while very encouraging, may not provide an accurate prediction of the synovial uptake *in vivo* where an abundance of blood vessels near the synovial lining provides another pathway for compound egress from the joint. Accordingly, a series of *in vivo* uptake studies was carried out in an animal model of arthritis. The antigen-induced arthritis (AIA) model in the New Zealand rabbit bears a close resemblance to rheumatoid arthritis in humans and has been used extensively in the evaluation of agents for beta-particle radiation synovectomy.[12,13] Based on an antigen-antibody reaction, it has the advantage over other arthritis models that not all joints are simultaneously affected. The procedure consists of two sensitizations over a six week period, followed by the antigen injection, which is made directly into the desired joint, or joints, as needed.

Once synovitis had begun (3–5 days after the induction with the antigen in the knee joint), a 0.25 ml injection of the boron compound was made into the knee of the anesthetized animal.[14] The knee was then subjected to two ranges-of-motion in an attempt to distribute the compound throughout as much of the synovial cavity as possible. Investigations were carried out using either 5,000 or 150,000 ppm boron in ^{10}B-enriched $K_2B_{12}H_{12}$. At various times following injection the animals were killed, the synovium and other tissues were dissected, rinsed in saline, and evaluated for boron uptake via PGNAA at the MITR. Results are shown in Table 3.

The data in Table 3 indicate that, for the anticipated duration of neutron irradiation (see below), synovial boron levels are extremely high, especially compared with boron concentrations achieved in BNCT. Uptake also appears to be roughly linearly related to injected concentration over the range studied. Note that a minimum of three rabbits were used for each time point. The range of values at a given time point is large due to incomplete distribution of the compound around the joint, and to difficulty in separating synovium from other (non-boronated) tissues during dissection.

Table 3. *In vivo* synovial uptake of boron following intra-articular injection of $K_2B_{12}H_{12}$ at either 5,000 or 150,000 ppm boron

Time (min)	5,000 ppm Boron uptake (ppm)	150,000 ppm Boron uptake (ppm)	150,000 ppm Boron uptake (ppm)
5	—	14,000–24,000	8,000–22,000
15	265–950	14,000–38,000	20,000–25,000
30	35–190	4,000–31,000	20,000–24,000
60	30–50	—	4,000–9,000
240	<10	—	—

Uptake data as a function of time indicate that the compound does not have a long residence time in the joint; the time course for egress appears to be longer for the higher concentrations. The evaluation of ^{10}B in other tissues and fluids in the rabbits suggests that most of the compound leaves the joint via the blood supply and is eventually excreted by the kidneys. Given the very high uptake concentrations, however, and the epithermal neutron fluxes available, the residence time of the boronated compound in the synovium is more than sufficient to easily allow a therapeutic dose to be safely delivered to the diseased tissue.

4. NEUTRON BEAM DESIGN

A detailed ideal beam study to determine the most useful range of neutron energies for BNCS led to the conclusion that neutrons in the thermal to 0.5 or 1.0 keV range provided optimum synovium to skin and synovium to bone surface dose ratios.[3] This energy range is considerably lower than the ideal BNCT neutron beam energies and is a result of the shallow depth of the synovium relative to the depth of a deep-seated tumor, the target of BNCT beam design.

The design goal of a very low energy neutron beam led to the early adoption of a D_2O moderator for rapid neutron energy loss and a graphite reflector to provide both neutron reflection and additional moderation.[15] Further improvements to this configuration[5] resulted in construction and installation of a moderator/reflector assembly on a dedicated accelerator beamline at MIT LABA.

The moderator/reflector assembly comprises a 23 cm long, 9 cm diameter D_2O moderator inside an 18 cm thick cylindrical graphite reflector. The graphite was resin-treated so that it would hold the D_2O. The accelerator target and cooling configuration can be placed at any position inside the moderator to harden or soften the resultant epithermal beam. Figure 2 shows a 2D plot of the MCNP model of the entire assembly including moderator, reflector, target, piping, channels for coolant flow and light water coolant. The 2.8 cm beryllium target will be cooled using the method of submerged jet impingement. At a flow rate of 6 gal/min, an estimated 1 kW of power can be safely removed from this target configuration. If necessary, the flow rate can be increased to remove higher heat loads.[16]

Also shown in Figure 2 is a cross-section through a rabbit knee joint phantom and "side" and "back" reflectors placed on three sides of the knee. The graphite reflectors reflect neutrons back into the knee and result in improvements in both therapy time and therapeutic ratios.[5,17] This MCNP model was used to evaluate the dosimetry of various epithermal neutron beams in the rabbit knee phantom. Table 4 provides therapy time, maximum skin dose and synovium-bone dose ratios in the rabbit phantom based on a neutron beam created by 1.5 MeV deuterons on a beryllium target. The target was placed 8 cm from the exit port of the moderator/reflector assembly. A concentration of

Table 4. Treatment parameters for rabbit knee joint irradiations based on a D_2O moderated Be(d,n) neutron beam, $E_d = 1.5$ MeV

Therapy time (min mA)	Maximum skin dose RBE-cGy	Min. bone/synovium dose ratio
0.96	69	208

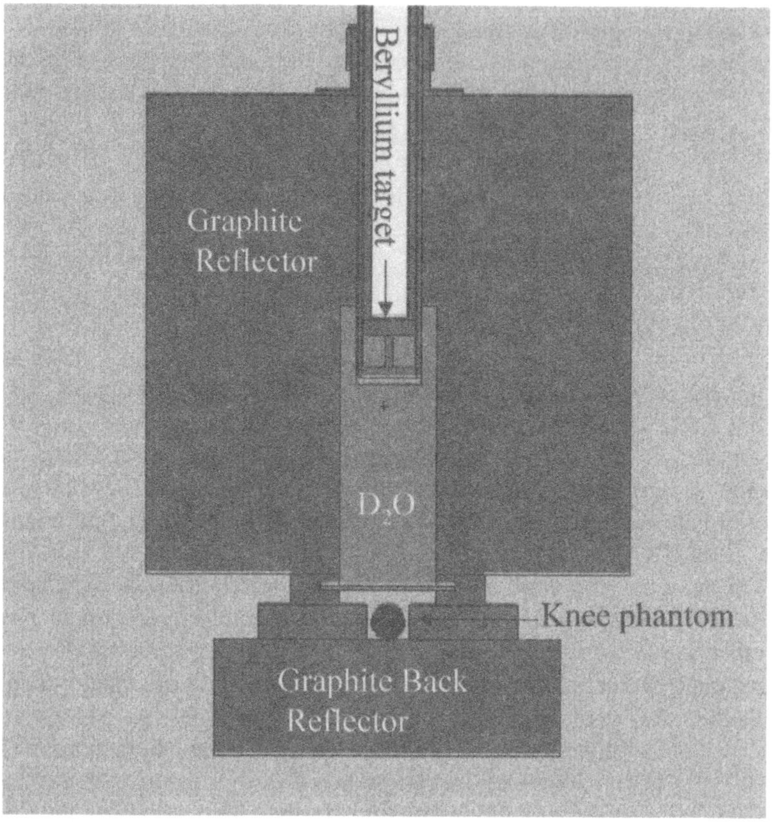

Figure 2. Two-dimensional plot through the MCNP model of the BNCS moderator/reflector assembly including target, cooling configuration, phantom and graphite side and back reflectors around the phantom.

19,000 ppm ^{10}B was assumed in the synovium, based on the *in vivo* uptake studies described in section 3.2.

Therapy times listed in Table 4 are based on a 1 mA particle current. They were obtained by using published values of neutron yield from the Be(d,n) reaction at 1.5 MeV[18,19] and determining the time require to deliver a minimum dose of 10,000 RBEcGy to all parts of the synovium. A synovial dose of approximately 10,000 cGy has been found to be clinically effective in studies of radiation synovectomy using beta-particle emitters.[20] For BNCS, however, since the $^{10}B(n,\alpha)^{7}Li$ reaction products and neutrons of all energies are known to be more biologically effective in creating damage than high energy beta particles, Relative Biologial Effectiveness factors (RBE) should be applied to individual dose components in order to arrive at a more realistic estimate of synovial dose as a function of neutron flux. However, no data related to the relevant endpoints for radiation synovectomy are available for the radiations found in BNCT. Therefore, RBE values currently in use for clinical trials of BNCT have been used in this study (4.0, 3.8 and 1.0 for the ^{10}B reaction products, neutrons and photons, respectively). In the future, once synovial necrosis can be correlated with physical dose in animal experiments, the use of RBE values will likely not be necessary.

Figure 3 shows the complete neutron irradiation assembly fully installed on a dedicated beamline on the 4.1 MeV tandem electrostatic accelerator at MIT LABA. The

Figure 3. Photograph taken inside the radiation shielding vault at MIT LABA. The BNCS moderator/reflector is installed on beamline 1 and is seen in the foreground.

accelerator, designed by Newton Scientific Incorporated, accelerates beams of protons, deuterons, or alpha particles. A wide-bore switching magnet allows installation of five different experimental endstations, The switching magnet and endstations are located in a shielded radiation vault which is next to the accelerator room. The accelerator beam tube extends into the vault through a hole in the four-foot thick concrete wall. Neutron beam characterization was carried out in a water-filled phantom and compared with MCNP dosimetry calculations using a model of the same phantom.[17] This step allowed the scaling of synovial dose in the rabbit knee phantom to compensate for uncertainty in the neutron yield for the Be(d,n) reaction as reported in the literature.[18,19]

5. EVALUATION OF BNCS EFFICACY IN AN ANIMAL MODEL

The availability of both a suitable boronated compound showing excellent synovial uptake characteristics and intense epithermal neutron beams using the LABA accelerator allowed initiation of an experimental evaluation of the efficacy of BNCS in an animal model.

Antigen-induced arthritis was induced in a total of thirty New Zealand rabbits. A 0.25 ml saline solution containing 150,000 ppm ^{10}B in $K_2B_{12}H_{12}$ was injected into the joint space and the knees again subjected to two ranges-of-motion. Boron uptake in the synovium was assumed to be constant at 19,000 ppm for the time period between 5 min and 25 min post injection. To overcome the uncertainty in knowing the precise time course of the compound in the joint throughout this time period, all irradiation times were fixed at 25 minutes and accelerator current adjusted so that a wide range of doses could be delivered during this period.

The 1.5 MeV Be(d,n) neutron beam, moderated by 8 cm D_2O was used. Animals were irradiated to doses ranging between 1,200 RBE-cGy and 100,000 RBE-cGy with three animals per dose point. One control animal received compound but no neutron irradiation, another received no compound but its knee was irradiated to the same neutron dose as the 100,000 RBE-cGy animal. All animals were killed at either 4 or 72 hours and their knees dissected. At each dose point, two animal knees were kept whole and decalcified for future whole-knee examination of synovial damage, and the third knee was dissected, synovium removed, sectioned and processed for histology. The tissue sections were stained with hematoxylin and eosin and examined for signs of tissue of necrosis, cellular debris and inflammatory cell infiltrates.

No healthy tissue effects, including erythema, were evident at 4 or 72 hours post irradiation. To date, only synovial tissue sections are available for examination. A histological evaluation of synovium treated to the highest doses show extensive radiation damage including distinct acellularity, pyknotic nuclei and karyorrhetic fragments; this damage is not seen in the controls. Some of these tissue changes are also seen at the lower treatment doses however the damage is not as extensive. Further histological evaluation is underway.

Although the results of the animal studies described here are preliminary and more detailed information will be available when the results of whole-knee histology are available, the data presented indicate that BNCS is successful in creating synovial necrosis in arthritic joints, in effect, in carrying out synovectomy.

ACKNOWLEDGMENTS

The support of the INEEL University Research Consortium, the US Department of Energy (Award number DE-FG02-89ER60874) and the National Institutes of Health (Award Number R44AR43680) is gratefully acknowledged.

REFERENCES

1. J.C. Yanch, S. Shortkroff, R.E. Shefer, S. Johnson, E. Binello, D. Gierga, A.G. Jones, G. Young, C. Vivieros, A. Davison, and C. Sledge, "Boron Neutron Capture Synovectomy: Treatment of the $^{10}B(n,\alpha)^{7}Li$ Nuclear Reaction" submitted for publication to Medical Physics, October 1998.
2. L.S. Johnson, J.C. Yanch, S. Shortkroff, and C. Sledge, "Temporal and Spatial Distribution of Boron Uptake in Excised Human Synovium," *Cancer Neutron Capture Therapy*, Y. Mishima, Ed. (Plenum Press, NY, 1996) pp. 183–188.
3. E. Binello, S. Shortkoff, A. Jones, C. Viveiros, A. Ly, C.B. Sledge, A. Davison, R.E. Shefer, and J.C. Yanch, "Research in Boron Neutron Capture Synovectomy," in *International Conference: Neutrons in Research and Industry*, George Vourvopoulos, Editor, Proc.SPIE 2867, pp. 68–71. 1997.
4. E. Binello, J.C. Yanch, S. Shortkroff, C. Vivieros, G. Young, A.G. Jones, C.B. Sledge, and A. Davison, "*In Vitro* Analysis of ^{10}B Uptake for Boron Neutron Capture Synovectomy," *Advances in Neutron Capture Therapy*, Vol. II (Elsevier, Amsterdam, 1997), pp. 609–613.
5. D.P. Gierga, J.C. Yanch, and R.E. Shefer," Design of Multidirectional Neutron Beams for Boron Neutron Capture Synovectomy," *Transactions of the American Nuclear Society*, TANSAO Vol. **77**, pp. 51–52, 1997.

6. E.D. Harris, Jr. *Rheumatoid arthritis. The clinical spectrum.* W.N. Kelley, E.D. Harris, S. Ruddy, and C.B. Sledge, in *Textbook of Rheumatology* (W.B. Saunders, Philadelphia 930, 1981).
7. P.E. Lipsky, Rheumatoid arthritis. In *Harrison's Principles of Internal Medicine*, 12th Ed., edited by J.D. Wilson *et al.* (McGraw-Hill, New York, 1991) pp. 1437–1443.
8. A. Fleming, R.T. Benn, M. Corbett, and P.H. Wood, "Early rheumatoid disease: patterns of joint involvement," Ann Rheum Dis 35:361–364, 1976.
9. C.A. Larin, J. Desmarchais, L. Daziano, *et al.*, "Long-term results of synovectomy of the knee in rheumatoid patient," J Bone Joint Surg 56(A):521–531, 1974.
10. Symposium on radioactive colloids in the treatment of arthritis. Ann Rheum Dis 32(Supplemental 6), 1973.
11. E. Binello, S. Shortkroff, and J.C. Yanch, "The PGNAA of Human Arthritic Synovium for Boron Neutron Capture Synovectomy," *Transactions of the American Nuclear Society*, TANSAO Vol. **77**, pp. 548–549, 1997.
12. T.D.V. Cooke, "Antigen-induced arthritis, polyarthritis, and tenosynovitis" CRC Handbook of Animal Models for the Rheumatic Diseases," R.A. Greenwald, and H.S. Diamond, Eds (CRC Press, Inc., Boca Raton, FL, 1988) pp. 53–81.
13. J.D. Zuckerman, C.B. Sledge, S. Shortkroff, and P. Venkatesan, "Treatment of antigen-induced arthritis in rabbits with dysprosium-165-ferric hydroxide macroaggregates,"*J Orthop Res* 7:50–60, 1989.
14. E. Binello, S. Shortkroff, R.E. Shefer, A.G. Jones, A. Davison, T. Thornhill, and J.C. Yanch, "Analysis of In Vivo Uptake for Boron Neutron Capture Synovectomy" *in* "Frontiers in Neutron Capture Therapy," M.F. Hawthorne, K. Shelly, R.W. Wiersema, eds., Kluwer Academic/Plenum Publishers, New York, 2001, pp. 1063–1066.
15. J.C. Yanch, R.E. Shefer, and E. Binello, "Design of Low-energy Neutron Beams for Boron Neutron Capture Synovectomy," in *International Conference: Neutrons in Research and Industry*, George Vourvopoulos, Editor, Proc. SPIE 2867, pp. 31–40, 1997.
16. B.W. Blackburn, J.C. Yanch, and R.E. Klinkowstein, "Development of a High-Power Water Cooled Beryllium Target for Use in Accelerator-based Boron Neutron Capture Therapy (ABNCT)," *Medical Physics*, 25(10), pp. 1967–1974, 1998.
17. D.P. Gierga, J.C. Yanch, R.E. Shefer, and B.W. Blackburn, "Construction of an accelerator based neutron beamline for treatment of arthritic joints" *in* "Frontiers in Neutron Capture Therapy," M.F. Hawthorne, K. Shelly, R.W. Wiersema, eds., Kluwer Academic/Plenum Publishers, New York, 2001, pp. 461–467.
18. S. Whittlestone, "Neutron distributions from the deuteron bombardment of a thick beryllium target" J Phys D, Appl. Phys. 10:1715–1723, 1977.
19. J. Guzek, U.A.S. Topper, W.R. McMurray, and J.I.W. Watterson, "Characteristic of the $^{9}Be(d,n)^{10}B$ reaction as a source of neutrons employing commercially available radio frequency quadrupole (RFQ) linacs" in *International Conference: Neutrons in Research and Industry*, George Vourvopoulos, Editor, Proc. SPIE 2867, pp. 509–512, 1997.
20. E. Deutsch, J.W. Brodack, and K.R. Deutsch, "Radiation Synovectomy Revisited," Eur. J. Nucl. Med. 20:1113–1127, 1993.

WHICH IS THE BEST NUCLIDE FOR NCT?

J. F. Crawford and D. Suwannakachorn

P.S.I. and I.M.R
CH-5232 Villigen, Switzerland

1. INTRODUCTION

Many agents have been proposed for neutron capture therapy (NCT), beginning with the first suggestion in 1936 by G. L. Locher.[1] Presumably because the significance of isotopes was not yet appreciated, he described only natural elements as active constituents, ranging from lithium and boron to thorium. Because the discovery of nuclear fission was still some years in the future, he did not discuss uranium.

Since then a great deal has been learned about nuclear physics, and it has been argued that other nuclides might be more suitable than ^{10}B for NCT, notably ^{157}Gd because of its much larger cross-section, or ^{235}U, because of its very high energy release.

In conventional ^{10}B NCT (BNCT), on average 2.35 MeV is released locally following the neutron capture. Depending on the exact location of the capture event within the cell, one such capture has a significant chance to kill the cell. On the other hand, because of the implications of Poisson statistics, an average of several dozen captures per cell are required for a reasonable chance of killing all tumour cells, and thus sterilising the tumour.

2. THE CHOICE OF NUCLIDE

We distinguish here between two cases:

- Charged particle emitters. This case can be sub-divided into:
  - Protons from nitrogen, α particles from ^{10}B and triton-α pairs from ^{6}Li.
  - Electrons, i.e. Auger and Coster-Kronig processes; although they are normally thought of as low LET particles, at the low energies and correspondingly short ranges characteristic of these processes their LET is as high as hundreds of keV per μ.
  - Fission; i.e. heavy ions.
- γ emitters.

Frontiers in Neutron Capture Therapy, edited by Hawthorne *et al.*
Kluwer Academic / Plenum Publishers, New York, 2001.

Table 1. Properties, including the FoM for some nuclides. As discussed in the text, ^{157}Gd is interesting both because of its (n,γ) reaction (row 3 in the table) and because it emits Auger and Coster-Kronig electrons (row 4). Gadolinium is also in routine use in NMR imaging

Nuclide	Capture cross-section (barns)	Energy release (MeV)	Atomic mass (grams)	FoM (gray × cm^2)
^{10}B	3,837	2.35	10.01	0.77×10^{-7}
^{14}N	1.81	0.63	14.00	0.70×10^{-11}
^{157}Gd	254*k*	7.94	156.9	1.09×10^{-6}
^{157}Gd	254*k*	0.004	156.9	0.55×10^{-9}
^{235}U	580	167	235.04	0.35×10^{-7}
^{251}Cf	4,300	ca 200	251.08	2.9×10^{-7}

At the energies considered here, a charged particle has a well-defined range in tissue which is small compared with the dimensions of a tumour; γ s of course do not have ranges in the sense that a proton or an α particle does.

As is well known, the relevant physical properties of nuclides are capture cross-section, energy release following capture and atomic mass. We can conveniently combine these into a figure of merit (FoM) that for charged particle emission under certain conditions can be regarded as a kerma factor: $N \times \sigma \times E/A$ where N is Avogadro's number, σ the capture cross-section, E the energy released by the capture and A the atomic mass. For ^{10}B this becomes $0.602 \times 10^{24} \times 3{,}837 \times 10^{-24}\,\text{cm}^2 \times 2.35\,\text{MeV}/10.01\,\text{g} = 542\,\text{MeV cm}^2/\text{g}$ $= 0.87 \times 10^{-7}$ gray × cm^2 . This is in fact an over-estimate, because thermalised neutrons have a Maxwellian energy distribution which is not symmetric, and because capture cross-sections vary inversely with the neutron velocity. It follows from the properties of Γ functions that a correction factor of $\sqrt{\pi}/2 = 0.886$ is required. This reduces the FoM for ^{10}B to 0.77×10^{-7} gray × cm^2.

In fact even this value is not quite correct, since slow neutrons are captured more quickly, the neutron energy spectrum is hardened, and the capture rate is reduced.[4] The effect—which is of the order of only a few percent in tissue loaded with therapeutic quantities of ^{10}B (say 30 ppmw)—is neglected here.

Flux is also depressed by the presence of an absorber such as ^{10}B, by an amount which depends on the volume of the tissue in question. Specifically, 30 ppmw of ^{10}B in a two-cm cube has been calculated to reduce the flux by 3%; in a four-cm cube the reduction is 10%.[3]

Neglecting the above corrections, the FoM is applied as follows. Given a ^{10}B load of 30 ppmw, a fluence of 10^{13} neutrons per cm^2 will result in a dose of $0.77 \times 10^{-7} \times 30 \times 10^{-6} \times 10^{13} = 23.1$ gray. Table 1 shows the FoM for a number of nuclides relevant to NCT, either as candidates for therapy or as backgrounds.

3. AUGER AND COSTER-KRONIG EMITTERS

Auger and Coster-Kronig electrons are ejected from atomic orbits following a nuclear reaction in the atom. The amount of energy involved is of the order of keV—much smaller than in typical nuclear processes. Despite the very large capture cross-section of 254 kilobarns in ^{157}Gd, the most promising candidate for this approach, the figure of merit is therefore much smaller than for other candidates (see Table 1). It is true that since the ^{157}Gd can be located in the cell DNA, the dose can be extremely effective, to the extent that one capture in a cell will most probably break enough DNA strands to kill it.

On the other hand, even if a single ^{157}Gd capture event in the DNA can be relied upon to kill the cell, the problem persists of ensuring that enough target cells are killed. Thus, if we consider a tumour containing $N \sim 10^{10}$ cells (10 gram), it follows from the properties of Poisson statistics that we need on average $log_e(N) \sim 23$ captures per cell in order to achieve a reasonable probability (37% in this example) of killing all the cells in the tumour.* Now, given the capture cross-section of ~254 kilobarns and a therapeutic fluence of 10^{13} neutrons per cm^2, it follows that we need 0.91×10^7 ^{157}Gd atoms per cell. This is much less than the 10^9 or so required for ^{10}B, and from this point of view is certainly attractive. However, 0.91×10^7 ^{157}Gd atoms weigh 2.37×10^{-15} grams, and taking the cell to contain 12 picograms of DNA, it follows that 2×10^{-4} or 200 ppmw of the DNA must be ^{157}Gd.

4. FISSION

Fission capture on ^{235}U releases 198 MeV. About 167 MeV appears in the form of the kinetic energy of two recoiling fragments; the remaining energy is taken up by neutron, γ and neutrino emission. The most probable fission product mass numbers are 95 and 139, with energies of 99 and 68 MeV and ranges in water of about 27 and 19 micron respectively. For comparison, the range of a 4 MeV α particle is about 24 micron.[2,5]

Taking a biological cell to have a mass of 1 nanogram and corresponding linear dimensions of 10 micron, it follows that almost all of the energy from ^{235}U fission is deposited within three cell diameters of the capture event, and only about six neighbouring cells will be hit by a fragment. If we further assume that the nucleus—the target of any radiotherapy—makes up one quarter of the mass of the cell, a simple estimate based on solid angles shows that only about one cell nucleus will be hit per fission event.

Experience with heavy ions is that a cell whose nucleus is hit by a medium-weight ion with an order of magnitude more energy than a fission fragment will be killed typically one third of the time.[6] Since LET increases with the inverse of the energy of the ion, it can be assumed that the cell kill by the fission fragments will be at least that high; however for the purposes of this paper we assume the same value of one third. It then follows that, despite the very high energy release, and the high mean LET of 3,800 keV/micron, less than one cell is likely to be killed by each fission event. This is comparable with the cell kill from BNCT.

Following a similar line of argument to that made above in connection with electron emission, we consider a tumour containing $N \sim 10^{10}$ cells (10 gram). We again need on average $\log_e(N) \sim 23$ hits per cell for a 37% chance of killing all the cells in the tumour. Now, given the capture cross-section of ~580 barns, and a therapeutic fluence of 10^{13} neutrons per cm^2, it follows that we need about 4×10^9 ^{235}U atoms per cell—even more than in the case of ^{10}B. This corresponds to 0.16% (1,600 ppmw), and would seem to be difficult or impossible in practice.

---

*This probability increases very quickly with the number of hits per cell; an increase to only 25 hits on average leads to 87% chance of killing all tumour cells.

Putting the argument in another way, a capture on ^{10}B has a significant chance of killing the cell in which it takes place, for an expenditure of 2.35 MeV; on the other hand, capture on ^{235}U liberates two orders of magnitude more energy but kills about the same number of cells, for a much less efficient use of energy.* Again, an average of 23 fissions per cell implies $23 \times 167 \times 10^6/10^{-9} = 3.84 \times 10^{18}$ eV/g = 615 gray—a very high dose.

^{251}Cf has a much larger fission cross-section of 4,300 barns. Assuming that its fission fragments behave in the same way as those from ^{235}U, and that we have the same therapeutic fluence of 10^{13} neutrons per cm^2, it follows that we need about 5.3×10^8 ^{251}Cf atoms per cell. Although this corresponds to 0.022% (220 ppmw), it is less than the 10^9 ^{10}B atoms per cell assumed to be necessary for BNCT, and is at first glance quite attractive. On the other hand the difficulties of preparing therapeutic quantities of ^{251}Cf are almost certainly prohibitive.

A further problem is presented by guerilla cells. In the extreme case, the nucleus of a single tumour cell which is surrounded by healthy cells that are not loaded with fissionable atoms, can be hit by a fragment only from a capture within that cell. The probability of this from simple geometrical arguments is only about 0.2 per fragment, i.e. 0.4. But from Poisson statistics there is a certain chance that there is no capture within a cell; if we assume that the expected number of fission captures per cell is 23 as before, the expected number of nuclear hits is $23 \times 0.4 = 9.2$, and the fraction of guerilla cells that contain no hit is $e^{-9.2} = 10^{-4}$. While guerilla cells can be expected to cluster to some degree so that there will be some amount of cross-fire, this number is distinctly ominous.

5. γ EMITTERS

Because γ s are not short-range particles, the FoM must be interpreted differently. It can be regarded as a kerma factor only within a target volume that is large compared with the mean free path of the γ, i.e. some tens of cm at the MeV energies concerned. Since this is larger than most tumours, the γ s will deposit about as much dose outside the tumour as within it. Therefore, even if the γ-emitter is confined to the tumour, the dose from it will not be.

In addition to its emission of Auger and Coster-Kronig electrons, ^{157}Gd has been studied as an NCT γ-emitter,[7] perhaps in combination with ^{10}B.[8] ^{151}Eu has been studied as a combined γ and electron emitter by others.[9] who concluded that it compared unfavourably with ^{10}B as an NCT agent.

We note in passing that ^{135}Xe has a capture cross-section ten times larger than ^{157}Gd, the largest in Table 1. But an inert gas with a half-life of only 9.2 hours cannot be regarded as anything more than a curiosity.

6. CHARGED PARTICLE EMITTERS

The nitrogen and ^{10}B capture reactions are too well known to need further consideration here. For completeness, we discuss the $^6Li\ (n,t)\ ^4He$ reaction. While this is

* This estimate ignores the fact that the fragments will not be fully ionised, at least at first. Fragments that are less than fully ionised will have longer ranges, improving the cell kill to some extent.

at first attractive, with a cross-section of 940 barns and an energy release of 4.8 MeV, the chemical properties of lithium render it impractical for NCT.

ACKNOWLEDGMENTS

It is a pleasure to acknowledge informative and stimulating discussions with Profs. Börje Larsson and Jürgen Kiefer, and Drs. Nigel Crompton, Hans-Walter Reist and Regin Weinreich.

REFERENCES

1. G.L. Locher, Am. J. Roent. and Rad. Therapy, 36, 1–13 (1936).
2. L.C. Northcliffe, and R.F. Schilling, Nuclear Data Tables, *A7*, pp. 233–463 (1970).
3. F. Pöller, Dissertation, University of Essen, F.R.G., February 1993.
4. K.H. Beckurts, and K. Wirtz, "Neutron Physics", Springer Verlag, Berlin, 1964.
5. S. Glasstone, and A. Sesonske, "Nuclear Reactor Engineering", third edition, Van Nostrand Reinhold Co., 1981.
6. Professor Jürgen Kiefer, Giessen, F.R.G., personal communication.
7. M.J. Rivard *et al.*, "Advances in Neutron Capture Therapy", B. Larsson, J. Crawford and R. Weinreich (Editors), Volume II, pp. 430–435, and references therein.
8. Y. Zhao *et al.*, *et al.*, ibid, Volume II, pp. 451–455.
9. J.G. Wierzbicki *et al.*, ibid, Volume II, pp. 456–459.

NCT WITH SEVERAL AGENTS

R. M. Brugger

Nuclear Engineer Program
University of Missouri
Columbia, Missouri

1. INTRODUCTION

Several protocols are being followed presently to demonstrate the effectiveness of BNCT for treating brain tumors. In these, a compound containing B is injected into the patient and the B selectively collects in the tumor prior to neutron irradiation. For deep brain tumors, when the dose to healthy brain is taken to a maximum limit of tolerance = 12.5 Gy-Eq, the dose to the tumor at the midline of the head may be marginal for sterilizing all cancerous cells. Sometimes, injecting more of the same B compound to increase the dose to the tumor is not practical. The present study considers injecting a second, but different, compound with a different agent which targets the tumor.[1] Here, agents means different elements and not several compounds containing B.

Gadolinium is a possible NCT agent[2] and its uptake can be imaged by MRI. The uptake can be quantified by CAT.[3] Several Gd compounds which are used for image enhancement of MRI have been shown to target brain tumors and are approved, or are being approved,[4,5] for human use. In this study, calculations of the dose to a tumor in a head phantom have been made to evaluate whether the addition of such Gd compounds to BNCT would enhance the dose to a tumor at its deepest point while holding the dose to healthy brain to the tolerable limit.

2. METHOD

The Monte Carlo neutron transport code (MCNP) has been used to calculate the dose delivered to a phantom head model. The phantom is an ellipsoid of revolution with a length of 19 cm and a radius of revolution of 7.5 cm. A cylindrical tumor with the axis pointed along a radius of revolution and toward the neutron beam is placed with its outer edge 4.5 cm inside the surface of the phantom and its inner edge at the midline of the phantom. The tumor is 4.0 cm in diameter and 3 cm long giving a volume of 37.7 cm^3.

Frontiers in Neutron Capture Therapy, edited by Hawthorne *et al.*
Kluwer Academic / Plenum Publishers, New York, 2001.

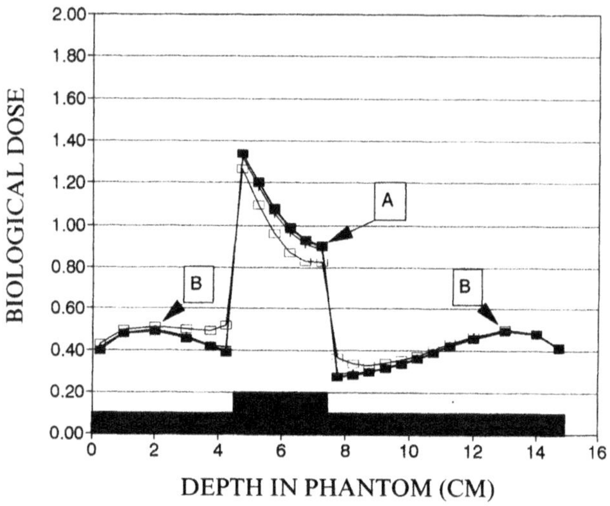

Figure 1. Biological doses in head phantom and tumor. 10 ppm of B-10 in brain and 30 ppm in tumor. Open squares = 118 ppm of natural Gd m tumor, solid squares = 1,000 ppm. RBE of gamma rays = 1. Point A is dose at back of tumor and point B is maximum dose in brain.

The material of the phantom is brain tissue and of the tumor is tumor tissue; no account is taken of skull or skin. All B is ^{10}B and all Gd is natural Gd.

The phantom is irradiated with a flux of neutrons with an energy spectrum and directionality similar to the present collimated NCT neutron beam at BMRR.[6] The direction of the irradiating beam is along the axis of the tumor cylinder. The phantom is irradiated parallel and opposed for equal times. The purpose of the calculations is to obtain the ratio of the dose in cell A to that in cell B (see Figures 1 and 2), that is, the dose in the cell in the tumor next to the midline to the dose in the cell in brain with the largest dose.

To evaluate the effect of adding another compound, biological doses must be estimated and for this, compound factors, or effectiveness factors, or RBEs, must be selected. In this study, the compound factors being used for B are ones similar to ones that the BNL group have determined for BPA[6] (see foot note in Table 1). Several different RBEs for gamma rays were tried. One Gd-compound[5] has shown an X-ray sensitization of 2 and Gd may have an Auger contribution to the gamma dose, so RBEs of 2 and 3 were tried for the gamma dose. Since fractionation reduces the effective RBE of gamma dose to healthy tissue, RBEs of 0.5 and 0.25 were used to provide a matrix showing what might be the effects of fractionation. If the maximum limit of the biological dose to healthy brain (point B) is taken to be 12.5 Gy-Eq, then a ratio of A/B of 2.4 is needed to reach 30 Gy-Eq to the deepest part of the tumor (point A).

3. RESULTS

Figure 1 shows the calculated biological dose distribution through the head phantom when both B and Gd compounds are present. Table I lists the dose ratios, A/B, that have been calculated for a number of cases. All have 10 ppm of B in healthy brain and 30 ppm in the tumor, but different RBEs for Gd. One notes that for B with no Gd, the ratio is only 1.82 with an RBE = 1 for gamma rays and only 2.01 if a radiosensitizer

Table 1. Ratio of biological doses for a B-compound like BPA[1] plus a Gd-compound

Gd-nat. in tumor	Gamma ray RBE for tumor	Dose ratio, A/B
0	1	1.82
0	2	2.07
25	1	1.83
25	2	2.08
25	3	2.32
118	1	1.79
118	2	2.10
118	3	2.42
1,000	1	1.06
1,000	2	1.70
1,000	3	2.34
Gd brachy sheet	1	1.76
Gd brachy sheet	2	2.79

[1]10 ppm of B in brain, 30 ppm in tumor. Compound factors; fast neutrons = 3.2, thermal neutrons in brain = 1.3, thermal neutrons in tumor = 3.8, gamma rays as shown.

Table 2. Ratio of biological doses for Gd-compounds only

Gd-nat. in tumor	Gamma ray RBE for brain	Gamma ray RBE for Tumor	Dose ratio, A/B
0	1	1	0.61
0	1	2	0.94
118	1	1	0.76
118	1	2	1.27
118	1	3	1.77
1,000	1	1	1.28
1,000	1	2	2.26
1,000	1	3	3.29
0	0.5	1	0.69
0	0.5	2	1.09
118	0.5	1	0.91
118	0.5	2	1.52
118	0.5	3	2.13
1,000	0.5	1	1.54
1,000	0.5	2	2.72
1,000	0.5	3	3.96
0	0.25	1	0.74
0	0.25	2	1.16
118	0.25	1	1.02
118	0.25	2	1.70
118	0.25	3	2.37
1,000	0.25	1	1.71
1,000	0.25	2	3.02
1,000	0.25	3	4.40

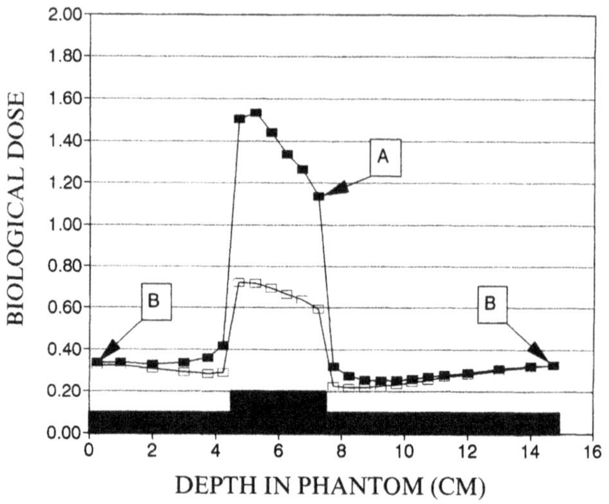

Figure 2. Biological doses in head phantom and tumor. No B-10 in brain or tumor. Open squares = 118 ppm of natural Gd in tumor, solid squares = 1,000 ppm in tumor. RBEs for gamma rays = 3. Point A is dose at back of tumor and point B is maximum dose in brain.

containing no Gd and having a radiosensitization of 2 collects in the tumor. From Figure 1 and Table 1, one notes that, when for gamma rays the RBE = 1, the biological dose decreases as more Gd compound collects in the tumor.

The results for the calculations for Gd compounds only is shown in Figure 2 and Table 2. When the RBEs for the compound in the tumor are 1, the doses delivered to the tumor are relatively small. When radiation sensitization and the possible effects of Auger electrons are included, as in Figure 2, the doses to the tumor can be large. With only Gd compounds, fractionation of the neutron irradiation can be used and the dose to healthy brain can be reduced. This is represented in Table 2 by RBEs for gamma rays in brain of 0.5 and 0.25. Fractionation increases the ration A/B to well above 2.4 for some cases.

4. CONCLUSIONS

Gadolinium is a second agent that might be used with BPA to enhance the dose to the tumor. However, the calculated doses are not as encouraging as expected. Gadolinium robs neutrons and reduces the dose from B. For Gd-compounds to help B-compounds, RBEs greater than 1 are needed for the radiation from the Gd gamma rays. Radiation enhancement and/or an additional dose from Auger electrons is needed. Porphyrins, i.e., GdBOPP[4] and Texaphyrin,[5] might provide such enhancement. On the other hand, Gd compounds alone, with radiation enhancement and Auger enhancement, look promising.

REFERENCES

1. A.H. Soloway, W. Tjarks, B.A. Barnum, F.-G. Rong, R.F. Barth, I.M. Codogni, and J.G. Wilson, "The chemistry of neutron capture therapy", Chem. Rev., 98:1515–1562, 1998.

2. R.M. Brugger, S. Gibb, J.-L. Shih, S. Pickup, X. Luo, J. Oro, and B. Laster, Estimated doses by gadolinium neutron capture therapy, "Topics in Dosimetry & Treatment Planning for Neutron Capture Therapy", R.G. Zamenhof, G.R. Solares, and O.K. Harling, ed., Advanced Medical Publications, Madison, Wisconsin, 1994.
3. M.A. Meyer, R.M. Brugger, J.C. Lattimer, K. Kutikkad, A.R. Schoone, and G. Ehrhardt, Non-invasive quantification of Gd uptake in tumors, *in* "Frontiers in Neutron Capture Therapy," M.F. Hawthorne, K. Shelly, R.W. Wiersema, eds., Kluwer Academic/Plenum Publishers, New York, 2001, pp. 923–931.
4. B. Laster, S. Kahl, J. Kirz, L. Warkentein, S. Wirrick, and G. Shani, Receptor-mediated deliver of In and Gd into the tumor cell nucleus by BOPP", Program and Abstracts of the Eighth International Symposium on NCT for Cancer, La Jolla, CA, 1998, p. 31.
5. S.W. Young, F. Qing, A. Harriman, J.L. Sessler, W.C. Dow, T.D. Mody, G.W. Hemmi, Y. Hao, and R.A. Miller, Gadolinium(III) texaphyrin: A tumor selective radiation sensitizer that is detectable by MRI, Proc. Natl. Acad. Sci. USA, 936:610–615, 1996, Medical Science.
6. A.Z. Diaz, A.D, Chanana, J. Capala, M. Chadha, J.A. Coderre, E.H. Elowitz, J. Iwai, D.D. Joel, H.B. Liu, R. Ma, N. Pendzick, N.S. Press, M.S. Shady, D.N. Slatkin, G.W. Tyson, and L. Wielpolski, Boron neutron capture therapy for glioblastoma multiforme: results from the initial Phase I/II dose escalation studies, *in* "Frontiers in Neutron Capture Therapy," M.F. Hawthorne, K. Shelly, R.W. Wiersema, eds., Kluwer Academic/Plenum Publishers, New York, 2001, pp. 61–72.

ENERGY DEPOSITED BY ELECTRONS IN DNA FOLLOWING n+Gd INTERACTION

S. Mark[1], I. Orion[2], B. Laster[1,2], and G. Shani[1]

[1]Department of Nuclear Engineering
Ben-Gurion University
POB 653, Beer Sheva 84105, Israel
[2]Medical Department
Brookhaven National Laboratory
11973 Upton, N.Y.

Neutron capture therapy treatment of brain tumors is generally done by introducing boron atoms into the tumor. The patient is then irradiated with thermal neutrons causing the emission of alpha particles and Li ions that produce high LET radiation damage. A newer approach is to use Auger electrons to impart similar damage. The superiority of Auger electrons over α-particles for cell lethality was demonstrated by Laster *et al.*,[1] when Gd atoms were bound to the DNA of V79 Chinese Hamster cells. The dose values for 10% survival in Megawat-Minutes were as follows: 9.1 for control cells, 4.8 for gadolinium ambient, 3.3 for BOPP and 1.8 for GdBOPP. The range of Auger electrons in the cell is of the order of tens of nanometers, but because they originate immediately at the critical target (the DNA), and because their range is so short, they are very efficient in damaging the DNA. In the present work, the energy deposition from Auger electrons in DNA was calculated using the EGS4 Monte Carlo code. The DNA was modeled using six nucleosomes symmetrically arranged in an appropriately-dimensioned circular array. Each nucleosome was modeled as two DNA helices wrapped 1.5 times around each central histone cylinder. The model included chemical composition and density of each region. The DNA model is shown in Figure 1. Since the exact location of the Gd atoms in the DNA is not known, the calculations were done assuming a random electron source location over the model described. The model used for electron energy deposition calculation, was a set of six nucleosomes as shown in Figure 2. A, B, C, and are source locations used for the various calculations. Energy deposition was calculated for several different electron sources and ranged between 0.5 and 30 keV. Some of the results are shown here. When a 5 keV electron source is adjacent to the DNA double helix strands, about 720 eV of each electron's energy is absorbed in the DNA strands closest to the source (that is the total energy absorbed in many nucleosomes), and about 260 eV in the

Frontiers in Neutron Capture Therapy, edited by Hawthorne *et al.*
Kluwer Academic / Plenum Publishers, New York, 2001.

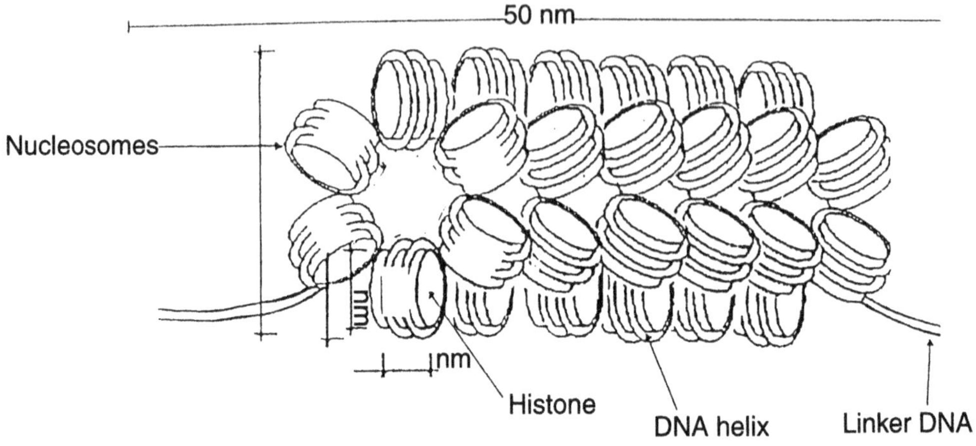

Figure 1. DNA model.

strands behind the histone cylinders. When the electron energy is 10 keV, the energy deposition in the helices closer to the source is 360 eV and 260 eV behind the histones. The electrons (or secondary electrons) retain enough energy to travel further and produce additional damage. About 10 Auger electrons in that energy range are emitted following the $^{157}Gd(n,\gamma)$ reaction; therefore, the amount of energy absorbed in DNA is considerable. It has been reported[2] that ~500 eV or more are needed for producing double strand

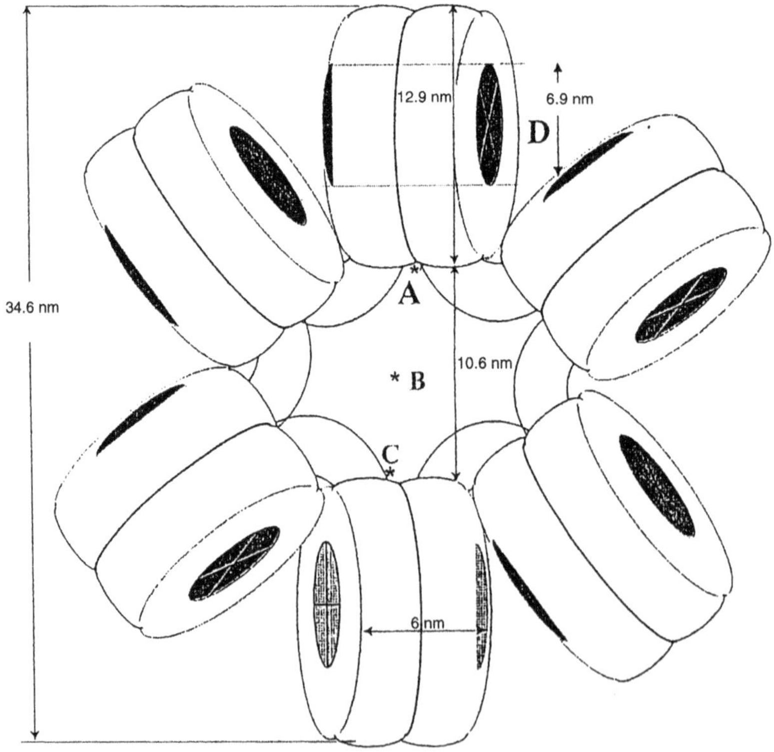

Figure 2. Model used for the calculations.

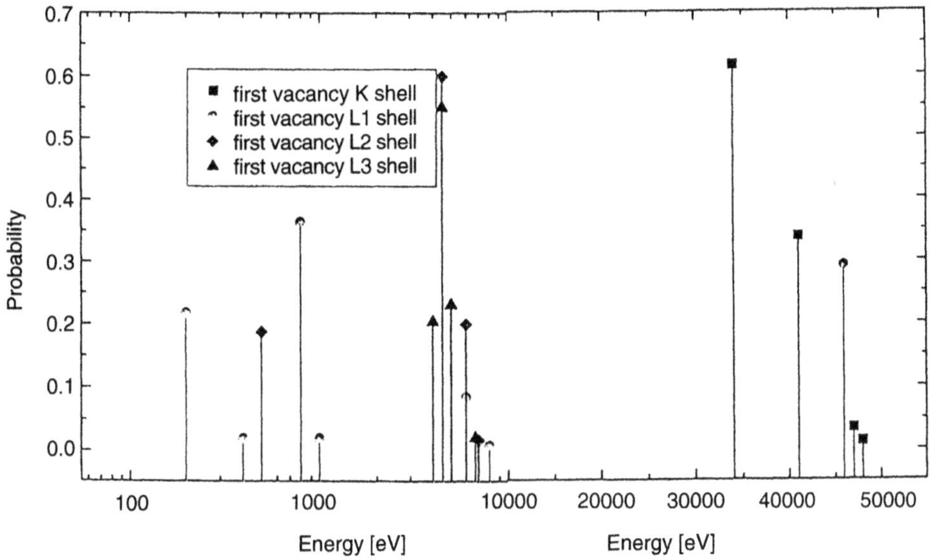

Figure 3. Calculated probability of auger electron emission from ^{158}Gd.

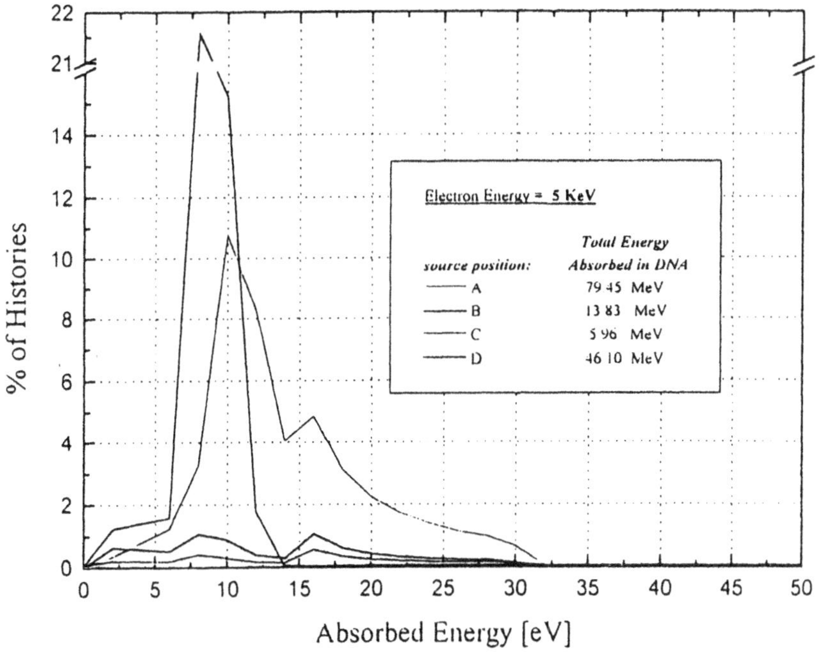

Figure 4. Energy deposited in DNA per 10^7 histories.

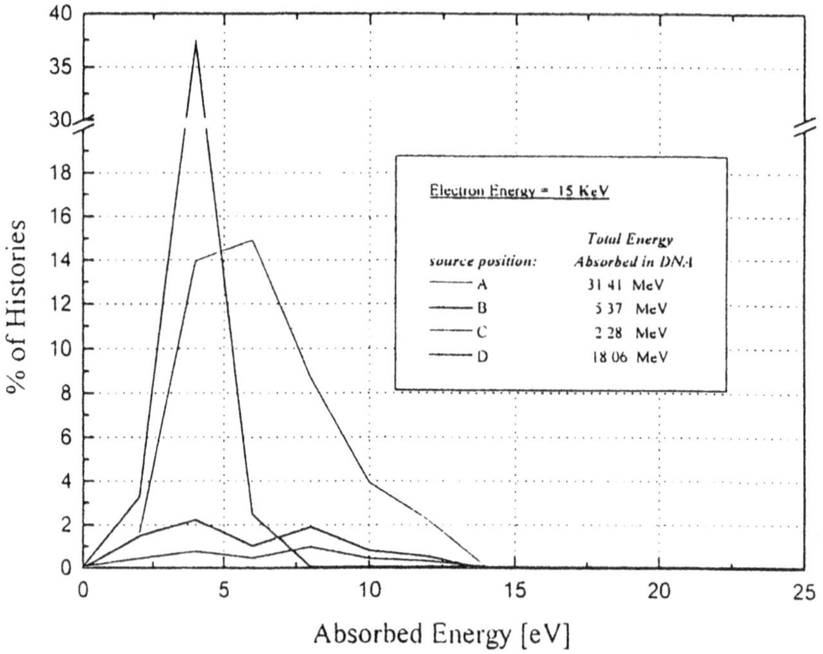

Figure 5. Energy deposited in DNA per 10^7 histories.

breakage. The numbers above indicate that emitted Auger electrons have a high probability of cell killing. Results are available for other electron energies and various source locations in the DNA.

Figure 3 shows calculated probability of Auger electron emission, following creation of a vacancy in the K shell or the L shell of a Gd atom. As can be seen, a vacancy in the K shell contributes a few electrons in the 30–50 keV range. Vacancy in L_I causes the emission of electrons at 46 keV as well as at 1 keV and lower. Other vacancies in the L shell cause emission of electrons in the 4–8 keV range. Probabilities due to vacancies in the M and N shells are not shown, that would add more electrons in the keV range and lower. The electron source can be located at any position in the DNA. We chose four positions as shown in the Figure 2. Position A is on the surface of the double helix (seen as a tubing wrapped around the histone, forming together a nucleosome). The nucleosoms were assumed to be arranged in hexagonal arrangement. About 6–8 such hexagons are grouped together between linker DNA, as shown in Figure 1. Position B is located

Table 1. Electron energy absorbed in cell

Electron energy (keV)	Energy absorbed in 2.5 nm of DNA	Energy absorbed in 3.5 nm of histones	Energy absorbed in 10 nm of water
2	21.3 eV	44.0 eV	40.8 eV
3	13.7 eV	34.0 eV	24.9 eV
5	9.9 eV	20.0 eV	18.0 eV
8	6.0 eV	12.0 eV	10.7 eV
15	4.2 eV	8.2 eV	6.0 eV
30	2.3 eV	4.5 eV	4.1 eV

at the center of the hexagon. Position C is symmetrical to position A, at the bottom of the hexagon. Position D is at the center of the flat side of the histone (the cylinder base). The energy absorbed in the DNA of the nucleosome adjacent to positions A (top side of the hexagon) was calculated. Figures 4–5 show absorbed energy spectra in DNA as mentioned above, for 5 and 15 keV source electrons. As can be seen in the table in each figure, the highest energy absorbed in the DNA is due to the source in position A. The graph describing the energy absorbed due to source at this position has two peaks. This is due to electrons passing through the histones, lose some of their energy, and deposit more energy in the top DNA. The energy absorbed due to a source at position D is the second highest, again because the source is close to the DNA. It is seen in the figures that although the number of electrons absorbed due to the source at this location is large, the energy deposited in the DNA is smaller. A possible explanation is that the path length of the electrons in the DNA is short. Sources at positions B and C deposit much less energy in that DNA section. To study energy absorption in the various components, energy absorbed in layers of DNA, histone and water is shown in Table 1. As seen, the lower energy electrons deposit more energy over a short range and therefore are the most damaging to the cell.

REFERENCES

1. B. Laster *et al.*, The Biological Effect of Auger Electrons Compared to a particles and Li Ions. Acta Oncologica Vol. 35 No 7 pp. 917–923, 1996.
2. D. Charlton, DNA Damage by Auger Emitters, Eds. Baverstock and Charlton. Taylor and Francis pp. 89–100 (1988).

NEUTRON CAPTURE THERAPY (BNCT) AND PHOTODYNAMIC THERAPY (PDT) CAN BOTH BE ACCOMPLISHED WITH A SINGLE COMPOUND

Boronated Protoporphyrin (BOPP)

N. E. A. Crompton[1], H. Walt[2], S. B. Kahl[3], S. D. Klein[2], Y. Shi[1], H. Blattmann[1], U. Haller[2], and B. Larsson[1]

[1]Tumour Therapy Evaluation, Radiation Medicine
Paul Scherrer Institute, Villigen, Switzerland
[2]Research Division of Gynecology, Dept. of OB/GYN
University Hospital Zurich, Switzerland
[3]Dept. Pharmaceutical Chemistry
University of California
San Francisco, California

OBJECTIVES

Using the Leukocyte Apoptosis Assay[1,2] and a standard colony formation assay, we examined induction of cytotoxicity induced by BOPP [2,4-bis-(α,β-dihydroxylethyl) deutero-porphyrin IX] in various human cell lines. The potential of this compound for use in both photodynamic therapy (PDT: red light activating the protophorphyrin group) and neutron capture therapy (NCT: neutrons activating the boron atoms—40 per molecule) was examined. The relative biological effect of the NCT reaction is very high. These experiments were also performed in order to examine whether NCT can induce programmed cell death, apoptosis, which is a physiological cellular response.

MATERIAL AND METHODS

MCF-7, a human breast cancer cell line, TK6 a human B-lymphoblastoid cell line, and CD4 and CD8 peripheral blood T-lymphocytes, were pre-treated for 24 h with

Frontiers in Neutron Capture Therapy, edited by Hawthorne *et al.*
Kluwer Academic / Plenum Publishers, New York, 2001.

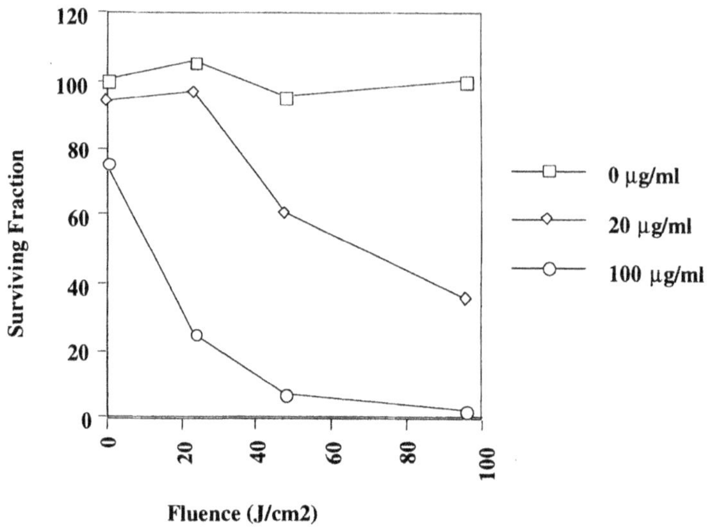

Figure 1. Photodynamic therapy at increasing dose of BOPP and increasing fluence of red light is increasingly cytotoxic. Colony formation assay with MCF-7 human breast cancer cells.

BOPP and their sensitivities to subsequent treatment with either neutrons or red light. The neutrons were obtained from one of the cold neutron guides at SINQ at the Paul Scherrer Institute. The 628 nm red light was obtained from a dye laser system. After treatment BOPP was washed from the cells. The cells were examined for survival using either a standard colony formation assay or an apoptosis assay based on flow cytometry.[1,2]

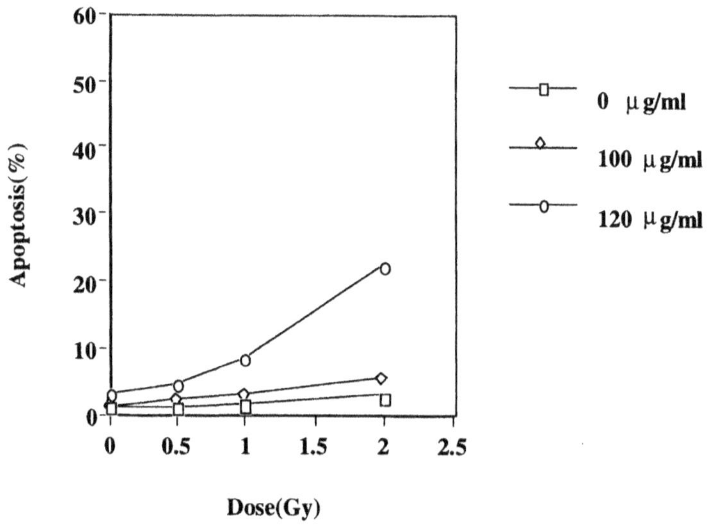

Figure 2. Apoptotic killing (0, 0.5, 1 and 2 Gy neutrons) of cells loaded with increasing concentrations of non-^{10}B-enriched BOPP. Studies with TK6 human B-lymphoblastoid cells.

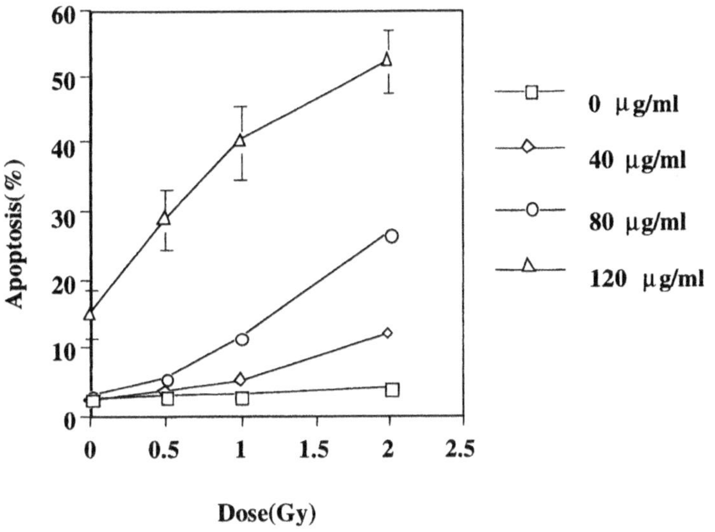

Figure 3. Apoptotic killing (0, 0.5, 1 and 2 Gy neutrons) of cells loaded with increasing concentrations of ^{10}B-enriched BOPP. Studies with TK6 human B-lymphoblastoid cells.

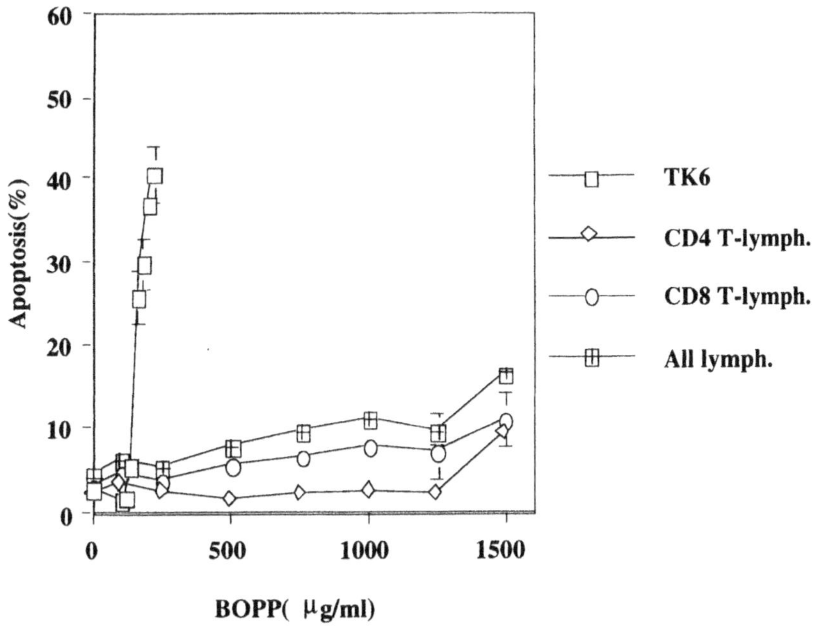

Figure 4. Apoptotic killing of ^{10}B-enriched BOPP in the absence of neutrons. Dividing cells are far more sensitive to the action of BOPP toxicity than non-dividing cells.

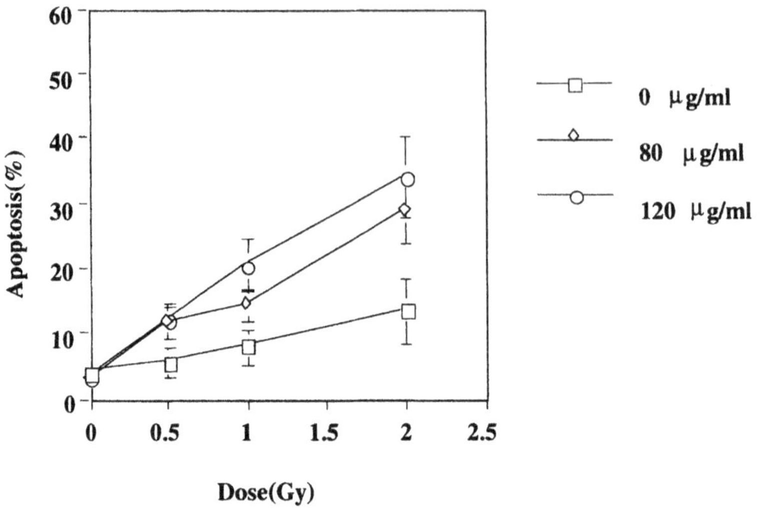

Figure 5. Apoptotic killing (0, 0.5, 1 and 2 Gy neutrons) of CD4 T-lymphocytes loaded with increasing concentrations of ^{10}B-enriched BOPP.

RESULTS

BOPP alone displayed marked selective cytotoxicity towards cycling TK6 lymphoblastoid cells but very little towards non-cycling lymphocytes even at high doses. Both PDT and NCT proved to be effective cytotoxic agents with enhanced synergistic cell kill at non-cytotoxic neutron/photon doses and at concentrations of BOPP which were also non-cytotoxic. At toxic concentrations, proliferative cells were far more sensitive to the action of BOPP than quiescent cells. Finally, both proliferative lymphoblastoid cells and

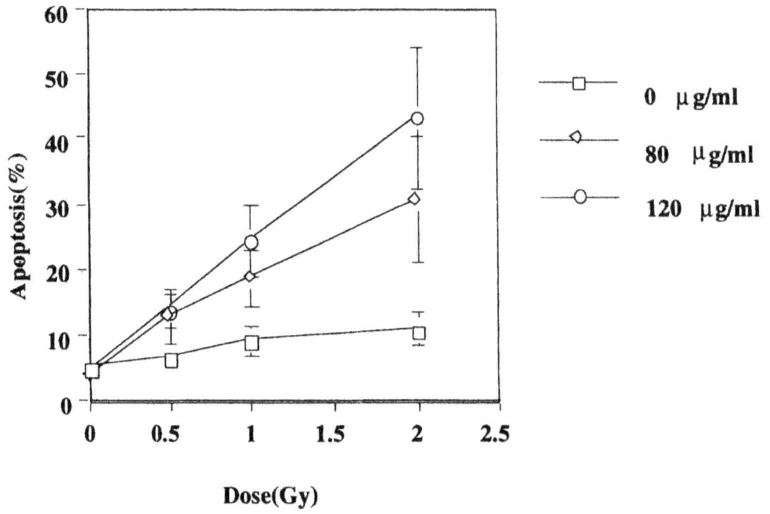

Figure 6. Apoptotic killing (0, 0.5, 1 and 2 Gy neutrons) of CD8 T-lymphocytes loaded with increasing concentrations of ^{10}B-enriched BOPP.

quiescent lymphoblasts were able to mount a physiological response to NCT-induced cell damage.

CONCLUSIONS

Our data indicate that BOPP can be used for both NCT and PDT. This dual treatment potentiality has various applications for conformal therapy, and mixed superficial and invasive tumour (e.g. melanoma) therapy. Physiological cell responses, apoptosis, can be induced even after NCT which has a high RBE.

REFERENCES

1. Crompton, N.E.A., and Ozsahin, M. A versatile and rapid radiosensitivity assay of peripheral blood leukocytes based on DNA and surface marker assessment of cytotoxicity. Radiat. Res., 147, 55–60 (1997).
2. Crompton, N.E.A., Miralbell, R., Rutz, H.-P., Ersoy, F., Wellmann, D., Bieri, S., Coucke, P.A., Emery, G.C., Shi, Y., Blattmann, H., Sanal, O., and Ozsahin, M. Altered apoptotic profiles in irradiated patients with increased toxicity. Int. J. Radiat. Oncol. Biol. Phys., 45, 707–714 (1999).

BNCT FOR THE PREVENTION OF RESTENOSIS FOLLOWING ANGIOPLASTY

J. C. Yanch and D. P. Gierga

Department of Nuclear Engineering and
Whitaker College of Health Sciences and Technology
Massachusetts Institute of Technology, Cambridge, MA

1. INTRODUCTION

The potential of Boron Neutron Capture Therapy for the prevention of restenosis following balloon angioplasty is under investigation at MIT's Laboratory for Accelerator Beam Applications (LABA). Angioplasty is a treatment for atherosclerosis, a progressive vascular disease characterized by 'plaque' within the intima of large and medium sized arteries.[1] Plaque is a result of (i) smooth muscle cell proliferation, (ii) extensive connective tissue including collagen and elastic fibers, and (iii) accumulation of lipid in the form of cholesterol. Plaque build-up results in occlusion of the artery and reduced oxygen delivery to the heart muscle. If untreated, this will be followed by reduced ejection fraction, muscle atrophy and eventual myocardial infarction.

One treatment for artherosclerosis involves the insertion of a balloon dilation catheter into the stenotic region of the occluded artery. The balloon is then inflated to dilate the catheter. This technique, called angioplasty, is safe and effective in improving blood flow. Unfortunately, even with the use of a stent in an attempt to maintain the diameter of the vessel, 30–60% of all balloon angioplasty patients exhibit restenosis within six months; the exact mechanism of this is not known although the primary mechanism is thought to be abnormal proliferation of the smooth muscle cells in the artery.

Radiation delivered to the vessel wall during or immediately after angioplasty has been shown to reduce or inhibit the restenosis. A recent double-blind, placebo-controlled, randomized clinical trial demonstrated that use of catheter-based intracoronary radiotherapy (using ^{192}Ir) resulted in a substantially reduced rate of restenosis.[2] Other radiation sources have been evaluated including radioactive fluid-filled balloons, external photon beams, beta-emitting sources placed in the artery via catheter, and radioactive stents; the optimum method for delivering the radiation is still under investigation.[3] In this paper the potential for delivering a therapeutic dose to a stenotic lesion via Boron

Frontiers in Neutron Capture Therapy, edited by Hawthorne *et al.*
Kluwer Academic / Plenum Publishers, New York, 2001.

Neutron Capture therapy while maintaining acceptably low doses to adjacent healthy tissue (skin, heart and lungs) is investigated.

2. METHOD

Lesion and healthy tissue doses were evaluated via Monte Carlo simulation. The MCNP code was used to model the dosimetric effects of three accelerator-based neutron beams in an anthropomorphic phantom[4] (see Figure 1). The phantom, representing an adult male, is based on a combination of MIRD,[4] reference man,[6] and Visible Man geometries;[7] the heart is based on the detailed MCAT phantom generated by B. Tsui *et al.*[8] A 0.31 cm diameter lesion[8] on the left anterior descending coronary artery was modeled on the surface of the heart closest to the skin of the chest.

The neutron beams evaluated are based on two different charged-particle reactions: ^{9}Be(p,n) at Ep = 4.0 MeV, and ^{9}Be(d,n) at Ed = 2.6 MeV. In each case, neutrons emerging from the target interact with a moderator/reflector configuration comprising a 23 cm long, 9 cm diameter D_2O moderator surrounded by an 18 cm thick graphite reflector.[10] The target was placed inside the moderator, either 23 cm or 15 cm from the patient position. The moderator/reflector configuration was centered over the lesion and directly abutted the skin. Neutron and photon fluences were determined in the lesion, the heart, the lungs and in the 0.2 cm thick skin in the 9 cm diameter region immediately next to the moderator. Fluences were modified by neutron, photon and ^{10}B fluence-to-kerma conversion factors and resultant doses were modified by RBE values of 4.0, 3.8 and 1.0 for ^{10}B reaction products, neutrons, and photons, respectively. A boronated compound showing 75% clearance from the blood pool was assumed (i.e. 4:1 lesion to blood pool ratio). Since the lung and, to some extent, the heart, are well perfused organs, a lesion to heart or lung boron uptake ratio of 4:1 was assumed.

The skin is the dose limiting tissue for this therapeutic application. Therefore results were analyzed as follows. First, the time required to deliver a skin dose of 800 RBE-cGy was determined. This is the dose threshold at which a mild but temporary skin erythema

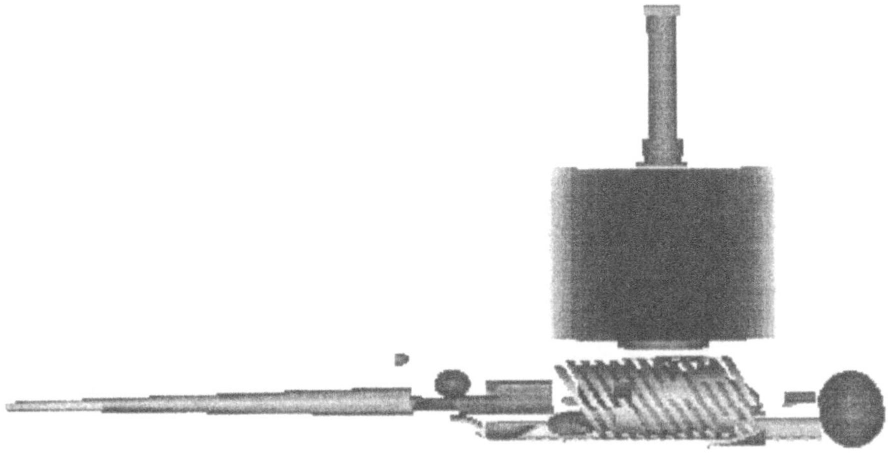

Figure 1. The phantom and irradiation set-up modeled in MCNP are illustrated. Skin and soft tissue have been made transparent.

Table 1. Lesion and healthy tissue doses, maximum therapy time and required boron uptake in lesion based on 2 mA particle beam current

	Lesion dose RBE-cGy	Skin dose RBE-cGy	Heart dose RBE-cGy	Lung dose RBE-cGy	Max. therapy time (min)	^{10}B conc. in lesion (ppm)
4.0 MeV Be(p,n) 23 cm D_2O	2,000	800	1,171	83	91	13
4.0 MeV Be(p,n) 15 cm D_2O	2,000	800	635	85	25.5	28
2.6 MeV Be(d,n) 23 cm D_2O	2,000	800	936	80	35.5	28

occurs.[11] Next, the concentration of boron needed to result in the delivery of 2,000 RBE-cGy to the lesion was calculated for the therapy time determined above. Finally, the heart and lung doses were calculated based on therapy time, boron uptake and assumed uptake ratios.

3. RESULTS

Results using three different accelerator-based epithermal beams are shown in Table 1. Note that *maximum* therapy times are listed. This corresponds to the delivery of 800 RBE-cGy to the skin and could certainly be lowered with a corresponding increase in boron concentration in the lesion. All three epithermal neutron beams result in healthy tissue doses that are below threshold for acute or late radiation effects. Heart and lung doses are especially low given the single fraction tolerance doses for these tissues of 1,500 cGy and 1,200 cGy respectively.

4. DISCUSSION AND CONCLUSION

Preliminary calculations demonstrate that delivery of radiation dose to stenotic lesions is dosimetrically possible with BNCT. However, a number of issues must be considered. The first is the feasibility of developing appropriate boron delivery agents. In the newly-emerging field of atherosclerotic lesion imaging a number of compounds labeled with photon emitters have been developed for the *in vivo* imaging of atherosclerotic lesions;[12,13] with additional effort it may be possible to achieve lesion uptake levels of the 30 ppm needed here.

The second issue stems from the uncertainty in knowing precisely which cells need to be damaged in order to prevent restenosis. This is not a problem for beta or gamma emitters but, given the very short path lengths of the ^{10}B reaction products, BNCT could be rendered completely ineffective if the targeting pharmaceutical is taken up by the wrong cells.

Third, it is known that irradiation of the vessel epithelium can actually create new stenotic lesions. It will therefore be essential that the boronated compound have a high specificity for the target lesion.

Finally, a number of different approaches to delivering radiation to the vessel wall during or following balloon angioplasty are currently under investigation.[3] These include radioactive fluid-filled balloons, external photon beams, insertion of beta or photon emitters into the vessel via catheter, and radioactive stents. It is unlikely that an approach based on BNCT will be the optimum choice for vessel irradiation in terms of cost, healthy tissue exposure, availability and clinical efficacy.

REFERENCES

1. Kent, K. "Restenosis After Percutaneous Transluminal Coronary Angioplasty," *American Journal of Cardiology* 1988; 61:67G–70G.
2. Teirstein, P.S. *et al.*, "Catheter-Based Radiotherapy to Inhibit Restenosis after Coronary Stenting," New England Journal of Medicine, 1997; 336(24):1697–1703.
3. Amols, H. Intravascular Brachytherapy Physics: Review of Radiation Sources and Techniques, in *Vascular Brachytherapy*, Second Edition, R. Waksman, Editor, Future Publishing Company, Armonk, New York, 1999.
4. Lambeth, M. "Development of a Computerized Anthropomorphic Phantom for Determination of Organ Doses for Diagnostic Radiology," Massachusetts Institute of Technology B.S. Thesis, 1997.
5. International Commission on Radiation Units and Measurements. *Phantoms and Computational Models in Therapy, Diagnosis, and Protection*. ICRU Report 48, Bethesda, MD: 1992.
6. International Commission on Radiological Protection. *Report of the Task Group on Reference Man*. W.S. Snyder, chairman; ICRP Report 23. New York: Pergamon Press, 1975.
7. Lorenson, W. "Marching Through the Visible Man." In: *Proceedings of Visualization '95*. IEEE Press: October 1995.
8. Tsui, B.M.W., Zha, X.D., Gregoriou, G.K., Lalushl, D.S., Frey, E.C., Johnston, R.E., and McCartney, W.H. "Quantitative Cardiac SPECT Reconstruction with Reduced Image Degradation due to Patient Anatomy." *IEEE Transactions on Nuclear Science 41*, no. 6, pt. 2, 1995:2838–2844.
9. Di Mario, C. *et al.* "Quantitative Assessment With Intracoronary Ultrasound of the Mechanisms of Restenosis After Percutaneous Transluminal Coronary Angioplasty and Directional Coronary Atherectomy," *American Journal of Cardiology* 1995; 75:772–777.
10. Gierga, D.P, Yanch, J.C., Shefer, R.E., and Blackburn, B.W. "Construction of an Accelerator-based Neutron Beamline for Treatment of Arthritic Joints," These Proceedings.
11. Nias, A.H.W. *An Introduction to Radiobiology* John Wiley & Sons, Chichester, 1990.
12. Roberts, A.B. *et al.*, "Selective accumulation of low density lipoproteins in damaged arterial wall," *J. Lipid Research* 1983; 24:1160–1167.
13. Narula J. *et al.*, "Localization of Experimental Atherosclerotic Lesions with Negatively-charged Polymer-modified Chimeric Antibody Specific for Proliferating Neointimal Smooth Muscle Cells," *Circulation* 1993; 88(4):I-250.

OPERATIVE MODALITIES AND EFFECTS OF BNCT ON LIVER METASTASES OF COLON ADENOCARCINOMA

A Microscopical and Ultrastructural Study in the Rat*

T. Pinelli[1], S. Altieri[1], F. Fossati[1], A. Zonta[2], C. Ferrari[2], U. Prati[2], L. Roveda[2], S. Ngnitejeu Tata[2], S. Barni[3], P. Chiari[3], R. Nano[3], and D. M. Ferguson[4]

[1]INFN, Pavia, Italy
Dept. of Nuclear and Theoretical Physics, University of Pavia, Italy
[2]Dept. of Surgery, Div. of General Surgery, University of Pavia, Italy
[3]Dept. of Animal Biology, University of Pavia, Italy, Centre of Study for Histochemistry, C.N.R., Pavia, Italy
[4]Mayo Clinic, Rochester, Minnesota

1. INTRODUCTION

During the biennial ISNCTC Symposia held in Bremen (1988)[1] and Kobe (1994)[2] we introduced the idea of a novel BNCT application. Its basic concept is the thermal neutron treatment of an explanted organ previously isolated and maintained in an extra-corporeal condition before reimplantation in the same donor patient (organ autotransplant). In those years some of us were refining the liver autotransplant technique later applied to several clinical cases.[2] Therefore our efforts were directed to the liver metastases therapy as the first application of the method;[2,16,17] in the future such a therapy could be extended to every organ suitable for autotransplant surgery.

The colon adenocarcinoma is one of the most frequent tumours afflicting the human population. There is an estimated number of nearly 600,000 cases per year in the world and 152,000 of those cases are ascertained in USA.[3] In one third of such cases the diagnosis is late and the survival rate 5 years after diagnosis is not higher

*The research is supported by I.N.F.N. since 1988. Support is also given by C.N.R. (A.C.R.O project) and I.R.C.C.S. Policlinic S. Matteo, Pavia, Italy.

Frontiers in Neutron Capture Therapy, edited by Hawthorne *et al.*
Kluwer Academic / Plenum Publishers, New York, 2001.

than 40–50%.[4,5] Liver metastases are responsible of this poor clinical outcome in two thirds of the cases. They are often multiple and diffuse, as a consequence surgically unresectable. Particularly disappointing is the consideration that in most cases the primary tumour in the colon can be excised easily while the liver is the only site of metastatic diffusion.

The irradiation of the entire isolated liver offers the following major positive aspects:

- An accurate knowledge of the tumour distribution is not necessary. The method guarantees the treatment of the whole organ with the same efficacy, even in the site of a microscopic infiltration. The probability of recurrent tumours is reduced to the minimum.
- The entire body of the patient (except liver) is preserved from any radiation damage.
- The liver, being washed before the irradiation, does not suffer any damage deriving from neutron interaction with the boron lasting in the blood.
- The radioactive dose absorbed during the treatment can be made substantially uniform in the entire liver volume. In fact, due to the isotropic nature of the neutron field inside the thermal column, it's not difficult to realize a satisfactory flat distribution of the neutron flux by an opportune arrangement of scattering and absorbing materials around the irradiation position (section 2).

Nevertheless these important advantages are associated to the crucial problem of preserving healthy tissues. After the therapy, the liver should support all its functions.

The feasibility of our project, beyond the surgery problems, has been defined by the following assumptions:

1) The total radioactive dose released in tumour must be larger than in normal tissues by a factor 2.5 at least.
2) The total dose given to the healthy tissues must be lower than the tolerance level (15 Gy-Eq).
3) The lowest local dose absorbed by the tumour must exceed the level of 40 Gy-Eq.
4) The treatment length must be largely compatible with the survival time of the patient in anhepatic phase.

The tolerance level of the hepatic parenchyma treated by BNCT has not been investigated until now. Values from 25 to 30 Gy are reported[6,7,8] for the case of photon single dose given to the whole liver. During the current clinical trials of BNCT for brain tumours, the upper limit of radiation dose to normal brain varied from 8.9 to 14.8 Gy-Eq.[9] We assumed as tolerance dose 15 Gy-Eq. This value represents a conservative choice being the RBE of 14.3 MeV neutron for liver tissues[10] more than 30% larger in comparison with that of thermal neutron beam for brain tissue.[11] The lower dose to be given to the tumour represents a median value when compared with those used to treat brain tumours in the current clinic trials.[12]

Our preliminary calculations and measurements[2,16] evidenced, as necessary conditions to proceed according to these assumptions, the following protocol prescriptions:

A) The thermal neutron flux in the whole volume of the liver undergoing the neutron irradiation must be uniform with a minimum value of $10^9 cm^{-2} sec^{-1}$.

B) The boron distribution in the liver must realize the condition $T = \frac{C_T}{C_H} \geq 4$. C_T and C_H represent ^{10}B concentration in tumour and normal tissues respectively.

T, the tumour over healthy tissue ^{10}B concentration ratio, is a time function whose behaviour is of the maximum importance in order to satisfy this second prescription in the most convenient manner.

C) The ^{10}B concentration in tumour must be larger than 20 ppm. This lower threshold allows the tumour tissues to absorb more than 40 Gy-Eq in a short time fitting an adequate and safe treatment.

2. METHODS AND RESULTS

The original structure of our thermal column was modified in order to realize a channel with a well-moderated neutron field. A strong reduction of gamma background has been obtained by bismuth screens in front of this channel (upper left side of Figure 1). We simulated the irradiation of a typical size liver shielded by a heavy water wall containing 35 ppm of ^6Li as represented in the upper right side of the same figure.

Figure 1 describes the main features of the neutron field in the volume of the liver model used both for experimental measures and simulation by the MCNP-4B code. The model results have been preliminary standardized and validated by absolute flux measurements inside a pig liver situated at the irradiation position.

The neutron space distribution along the longitudinal axis (x) of the reactor thermal column is shown in the figure for 4 vertical levels (Z = 0.5, 1.5, 2.5, 4.5 cm). Correspondent cross distributions are not reported being substantially constant inside the entire volume of our liver model (a hepatic-equivalent spherical bowl having a 6 cm height and 30 cm base diameter).

The obtained volume distribution evidences a satisfactory uniformity. The peak and lowest values are 8.2 and 4.5 $10^9 cm^{-2} sec^{-1}$. Obviously the dose distribution has the same behaviour. In addition at half-time of the one-shot treatment we have the possibility to rotate the organ by a 180° angle around z axis, so producing an improved uniformity; the difference between the peak and lowest values is reduced to 20% peak dose.

Results reported in Figure 1 evidence that our irradiation facility can largely satisfy the protocol prescription A. In particular, the neutron flux value allows to conclude the irradiation in about 12 min. Such a time is negligible when compared to the survival of the patient in anhepatic state (longer than 8 hours).[2]

Due to the crucial prescription B, we dedicated noticeable efforts to the study of boron uptake by hepatic tissues. Using a rat model we performed a large statistical analysis on a population of 100 rats (of these 83 resulted useful for our analysis). Details of methods concerning tumour induction, boron administration, boron concentration measurements in tumour and healthy hepatic tissues were described elsewhere[1,2,16,17] as well as the other operative instruments and procedures we utilized during this particular work. We recall the following protocol[2] adopted to study the liver and tumour boron uptake:

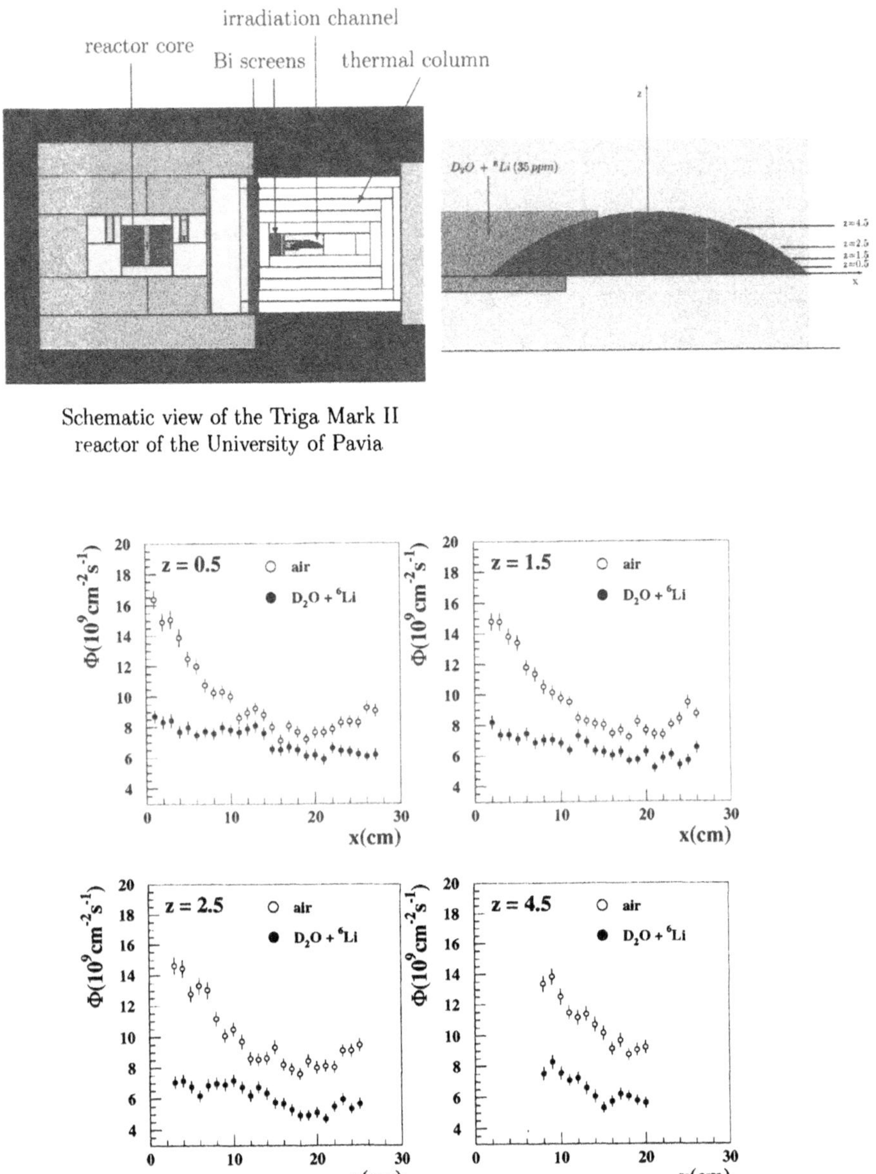

Figure 1. Neutron flux distribution in the longitudinal symmetry plane of a liver model (upper right). Data as obtained by a MCNP-4B code have been checked by experimental measurements. Open circles show the neutron flux behaviour when the D_2O–${}^6Li$ solution is replaced by air. On the upper left side of the figure the modified thermal column is shown inside a sketch of the reactor structure.

a) Tumour induction in the rat by inoculating 10^6 cells of colon carcinoma (section 3) by an intrasplenic injection.
b) 10 days later a dose of 300 mg ${}^{10}B - BPA$/Kg body weight (section 3) is intravenously injected.
c) After a time t from the injection the animal is sacrificed and its liver is extracted, then washed.

d) Samples of both tumour and normal tissues are cut in thin samples and submitted to histologic analysis and computerized imaging to evaluate the tumour percentage in each sample.
e) The boron concentration is evaluated with a 0.5 ppm resolution by a spectrometric analysis of alpha particles released by the $^{10}B(n,\alpha)^{7}Li$ inside the reactor thermal column.

Figure 2a,b,c detail the definitive data we obtained on boron uptake by liver tissues.

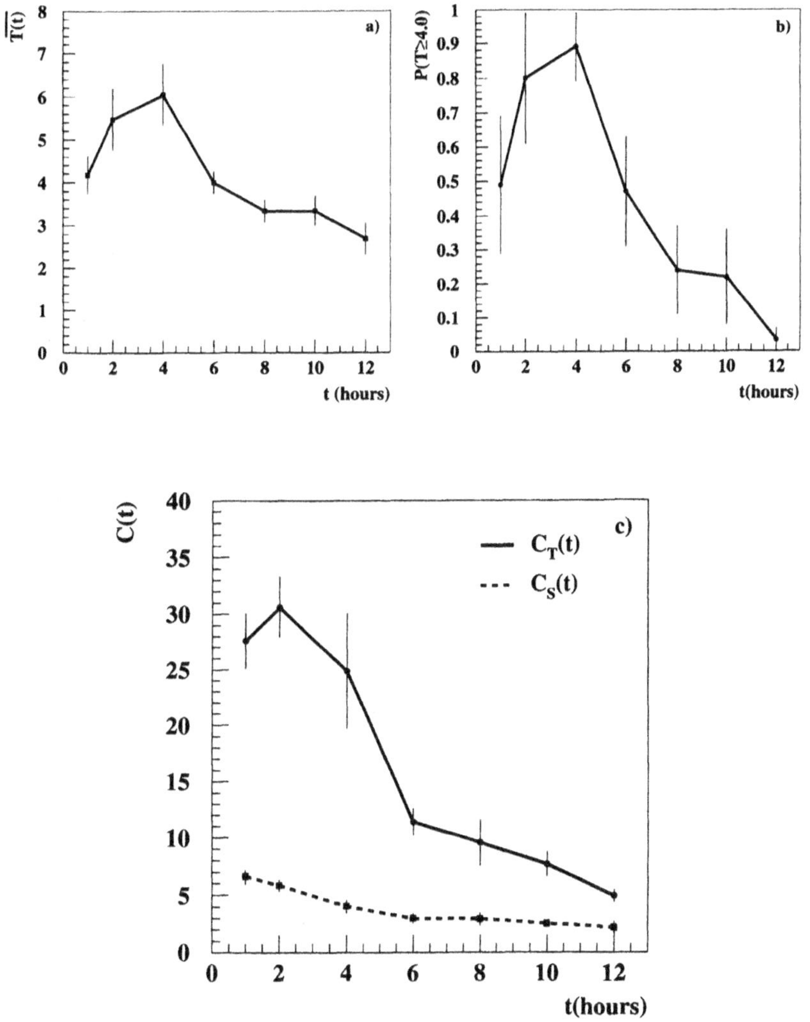

Figure 2. a) Time distribution of $\overline{T}$, the mean value of boron concentration ratio of tumour over healthy tissues $\left(T = \frac{C_T}{C_H}\right)$. The variable t is the time elapsed between the BPA infusion and the sacrifice of experimental animals. b) Time distribution of the probability to get values T ≥ 4 at the time t after BPA infusion (300 mg/Kg dose). c) Time distributions of boron concentration in tumour (red) and healthy tissues (blue). T, t and the BPA dose are the same as in Figures 2a and b.

Figure 2a shows the time distribution of $\overline{T}$(t), the mean value of boron concentration ratio in tumour over healthy tissues. We see that in the time interval from 1 to 6 hours the prescription B is largely fulfilled.

In Figure 2b we can consider in probabilistic terms the possibility to get satisfactory values of T (T ≥ 4) at a certain value of t. When the variable t is included in the interval 2–4 hours we meet high probabilities, going from 80 to 90%, to achieve values of T larger than 4. More precisely, inside this interval, the mean value $\overline{T}$ ranges from 4 to 6 (Figure 2a).

In Figure 2c we observe the diagrams describing the time behaviour of ^{10}B concentration in both cases of tumour and healthy tissues (C_T,C_H). The remarkable values assumed by the neoplastic tissues in the interval $1 \leq t \leq 4$ hours make the results fully compatible with the prescription C of the treatment protocol.

By assuming a neutron fluence of $5 \cdot 10^{12}\,cm^{-2}$ we can calculate the tumour and healthy tissue doses using the following equations:

$$D_H = 7.3 + 0.59 \cdot C_H$$

$$D_T = 7.3 + 1.71 \cdot C_T$$

where 7.3 Gy-Eq is the background dose, 1.71 Gy-Eq/ppm is the dose produced by the $^{10}B(n,\alpha)^7Li$ process when the boron concentration in tumour is 1 ppm, 0.59 Gy-Eq/ppm is referred to the concentration of 1 ppm in the normal tissue. The biological effectiveness of the undue radiations and the products of $^{10}B(n,\alpha)^7Li$ reaction are given in ref. 13 for brain normal and tumour tissues.

RBE values of particles released in Boron Neutron Capture process are not available for liver parenchyma, nevertheless a significant check of the project feasibility can be realized by using reasonable approximations. We find in literature[10,14] the RBE of 14 MeV neutron for rat hepatic parenchyma and intestine; in both cases the value is about 2 while for rat spinal cord RBE of the thermal beam (single dose) is 1.4. In addition the RBE value of neutron beam and the CBE value of BNC irradiation for brain white matter are roughly equal.[15] We may conclude that our dose calculations seem to be useful to verify the project feasibility.

With the sake of evaluating the RBE correct value for the normal liver tissues, we are experimentally studying the dose-effect relation by the usual *ex vivo-in vitro* method.

From Figures 2b and 2c we realize that the treatment is made in conditions of feasibility with 75–85% probability when t value is inside the interval from 2 to 4 hours.

The conclusive data elaboration of the complete experiment is in good agreement with the preliminary results we gave progressively in various occasions.[2,16,17]

The situation determined by our study on the treatment feasibility is graphically resumed in Figure 3. Here five straight lines are drawn. Three of these lines represent the assumptions 1, 2 and 3 of section 1.

The remaining two are the limits of that portion of the plane where T assumes the favorable values from 4 to 6. Then the dashed area represents the polygon whose internal points are associated to a practicable application of our method. Let's now consider the realistic example relative to the irradiation starting at t = 2 hours. The comparison between Figures 2a and 2b indicates a probability around 80% to get a $\overline{T}$ value larger than 5. The same probability can be attributed to the typical values of boron concentration $C_H = 5.8$, $C_T = 30.1$ ppm (Figures 2a and 2c). In the plane of Figure 3 the assumed example is represented by the point P, conveniently positioned inside the dashed polygon.

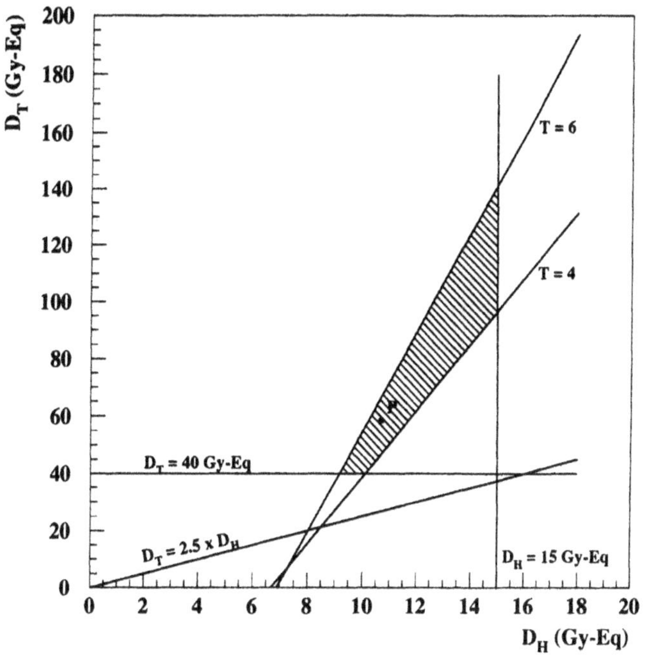

Figure 3. Graphical representation of the prescriptions defining the feasibility of our BNCT application. The dashed polygon is the locus of points where the feasibility prescriptions are satisfied. Point P is associated to our irradiation protocol (neutron fluence $5 \cdot 10^{12} cm^{-2}$, irradiation length 12 minutes, BPA dose 300 mg/Kg, irradiation beginning 2 hours after BPA infusion).

This is a synthetic indication that we are in condition to satisfy the feasibility protocol with a probability larger than 80%. Though satisfactory, this situation could be improved eventually by a convenient increase of the BPA dose up to 400–450 mg per Kg.

3. EXPERIMENTAL BNCT

After the positive results obtained in determining the feasibility parameters of the treatment we passed to analyse, by "*in vitro*" and "*in vivo*" methods, the effects induced on liver tissues by the BNCT procedure.

We considered as essential this third phase of our project for the following reasons:

i) To get a first realistic check on the correspondence between the damages induced by the treatment and those expected according to our project protocol. Of particular importance is the comparison between the radiation damage produced by BNCT in tumour and normal hepatic tissues.
ii) To get, by direct biological observation, a first answer for the problem regarding the boron uptake into the deepest tissue of those metastases having a relevant size. This problem could be originated by the poor blood circulation through such tissues.
iii) To realize a first approaching of our project to the clinical application.
   - ***In vitro* method** consisted in the analysis of the viability of boron enriched colon carcinoma DHDK cell line, after neutron irradiation.

The viability was determined using a technique that is a measure of the membrane permeability associated with cell death. The viability probe used is the anti-cytokeratin (CK) MoAb.[18]

Flasks containing 10^7 cells of colon carcinoma DHDK 12 TRb cell line in 20 ml of medium were treated for two hours with ^{10}BPA-fructose complex containing 60 mg D,L, p-boronphenylalaninehydrocloride (100 mg BPA combined with 2.0 ml fructose in 0.3 M solution. The pH is adjusted to 7.4–7.5 with NaOH 2N solution). The BPA was supplied by BBI, Raleigh, NC, USA with a ^{10}B enrichment more than 95%.

Harvested cells were washed three times in medium and then split in two aliquots. One was irradiated with the neutron fluence of $7 \cdot 10^{12}\,cm^{-2}$, while the other one was used as control. The equivalent dose corresponding to the given fluence is 58.1 Gy-Eq.

Soon after the irradiation, 10^6 cells were treated for 15 min with 200 µl of 1:10 diluted MoAb anti-CK MNF116 (DAKO). After washing, the cells were incubated with the secondary antibody IgG FITC (SIGMA) conjugated for 15 min, then fixed in 10 ml of 70% ethanol. One hour before analysis the cells, washed twice in PBS, were counter-stained with 1 ml of 5 µg/ml Propidium Iodide (PI). A two-parameter analysis of the DNA content (PI-red fluorescence) versus CK positive dead cells (FITC-green fluorescence) was performed on a FACStar (Becton Dickinson) equipped with a 5 W argon ion laser. The same procedure was used for the control cells. Moreover, aliquots of $2 \cdot 10^6$ cells were cultured again and analyzed after 24 and 48 hours with the previous described method.

The two-parameter analysis shows that the percentage of CK+ dead cells evaluated on the irradiated and non irradiated cell lines did not evidence any difference. Such a result proves that the neutron irradiation cannot be considered responsible of observable membrane damages. More interesting information (Figure 4) is obtained from the DNA content histograms deduced from the same analysis. Immediately after irradiation (t = 0) no difference can be noticed between the DNA histograms concerning irradiated and control cells respectively. After t = 24 hours we observe in both cases cells with DNA stainability lower than that of G_0G_1 cells; such findings could represent debris and/or apoptotic cells. This situation in the non irradiated cells might be caused by cell suffering due to the procedure they have been submitted. The third histograms in figure (t = 48 hours) show a different result. A permanent signal of suffering is still evident in the irradiated population while control cells have almost completely restored the initial condition (t = 0).

The radioactive dose we administered results adequate to induce severe and probably irreversible damages in neoplastic cells during their thermal neutron treatment.

The above results are to be intended as preliminary ones. More experiments are planned for extending our observations to longer time after treatment and using more sensitive reagents for apoptotic cells.

- ***In vivo* method** to analyze the effects of neutron irradiation in tumour and normal tissue consisted in the application of our entire procedure in a rat model. The objectives of the experiment were:
- Developing a technique to optimize the induction of hepatic metastases.
- Setting up a microsurgical technique of liver transplantation.
- Completing a whole procedure of boron compound infusion.
- Irradiated liver grafting into a syngeneic/allogeneic rat after the hepatectomy.
- Evaluating the cellular damage induced by the administered radioactive dose.

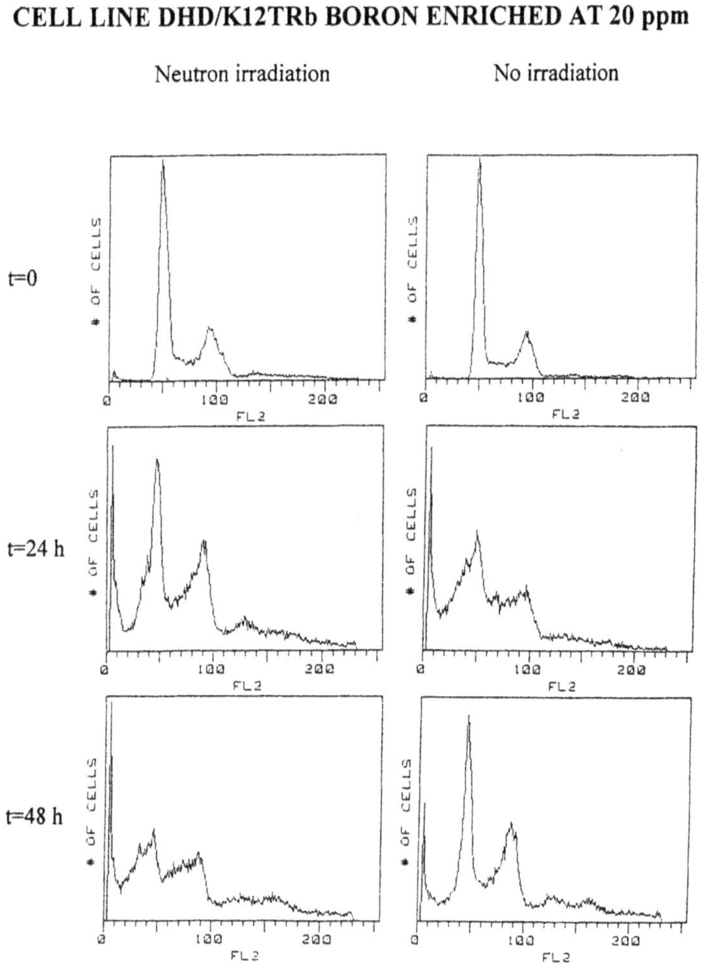

Figure 4. DNA content histograms showing the condition of irradiated and non irradiated colon carcinoma cells soon after irradiation (t = 0) and at the successive times t = 24 and t = 48 hours.

A colon carcinoma was induced in the experimental BD-IX strain rat by inoculation in the spleen of $2 \cdot 10^7$ cells from the DHDK cell line according to the method described by Caignard and coll.[19] During the splenic inoculation the left branch of the portal vein was clamped. In this way each animal gave samples of both tumour and healthy tissues.

Multiple and sometimes confluent liver metastases were induced in the right lobe of syngeneic rats during the following days. Ten days after the inoculation a D,L p-^{10}BPA dose of 300 mg/Kg was infused through the dorsal penis vein over a time of about 5 minutes. The used BPA-fructose complex was the same as in "*in vitro*" trials.

After 2 hours from boron infusion, the animal was sacrificed and its liver extracted, perfused with Beltzer solution, then irradiated inside the thermal neutron facility of our reactor. As in the case of the "*in vitro*" analysis the liver received a neutron fluence of $7 \cdot 10^{12} cm^{-2}$. Assuming boron concentration $C_T = 30.1$ *and* $C_H = 5.8 ppm$ in tumour and normal tissues (see section 3), the corresponding doses D_T *and* D_H are 82.5 and 15.0 Gy-Eq respectively.

Table 1. Experimental rats as stratified according to the "*in vivo*" treatment

Treatment	Group A (healthy rats)	Group B (rats with liver metastases)
Total	12	27
Liver transplant only (control)	9	1
BPA + liver transplant (control)	1	7
BPA + neutron irradiation + liver transplant	2	19

Hepatectomy was previously performed on a syngeneic or allogeneic recipient rat maintained in general anesthesia. After having been irradiated, the isolated liver was then transplanted orthotopically in this last animal using a modification of the Kamada method.[20] After a preset time the animal was sacrificed and liver rapidly removed. Samples of both tumour and healthy tissues were cut according to a common geometry and submitted to either histological or ultrastructural analysis.

The "*in vivo*" analysis schedule included 39 experiments variously designed in order to control each important step of the whole procedure.

All donor animals were BD-IX strain rats. The experimental animals were included into two groups. Group A consisted of 12 healthy rats, while group B contained 27 rats with liver metastases. Each group was then stratified according to the treatment as reported in Table 1.

Morphologic observations were performed both in light and electron microscopy. They were all made within 6 days from the liver transplant so to avoid the consequences of immunological intolerance in allogeneic grafts. Such observations were extended to a statistically significant number of both healthy and tumoral samples, before and after the neutron irradiation.

The results of the entire analysis, after comparison with the control population, can be resumed with the following overall conclusions:

- Cancerous samples from non irradiated animals evidence with a noticeable continuity good intrusive properties of the inoculated tumour. Such an indication is suggested by a significant amount of colon carcinoma cells observed both in free diffusive state and in aggregate nodules (Figure 5a and b). Moreover, the neoplastic cells have been seen either as a proliferating undifferentiated phenotype or as a pseudo-differentiated phenotype (Figure 5c and d).

Figure 6 gives a typical view of samples obtained from the liver submitted to the complete procedure (Metastases induction, BPA infusion, hepatectomy on the donor, neutron irradiation, transplant in the recipient animal). In particular this figure refers to a recipient animal sacrificed 4 days after the liver transplant.

Electron microscopy shows, with a significant frequency, severe damages both in neoplastic sinusoidal cells and nodules while the parenchyma cells (hepatocytes) are only occasionally found in a suffering state whose seriousness is difficult to establish.

4. DISCUSSION

The modified structure of the neutron facility together with the definitive results of our boron uptake study have determined an irradiation protocol largely practicable for our BNCT novel application:

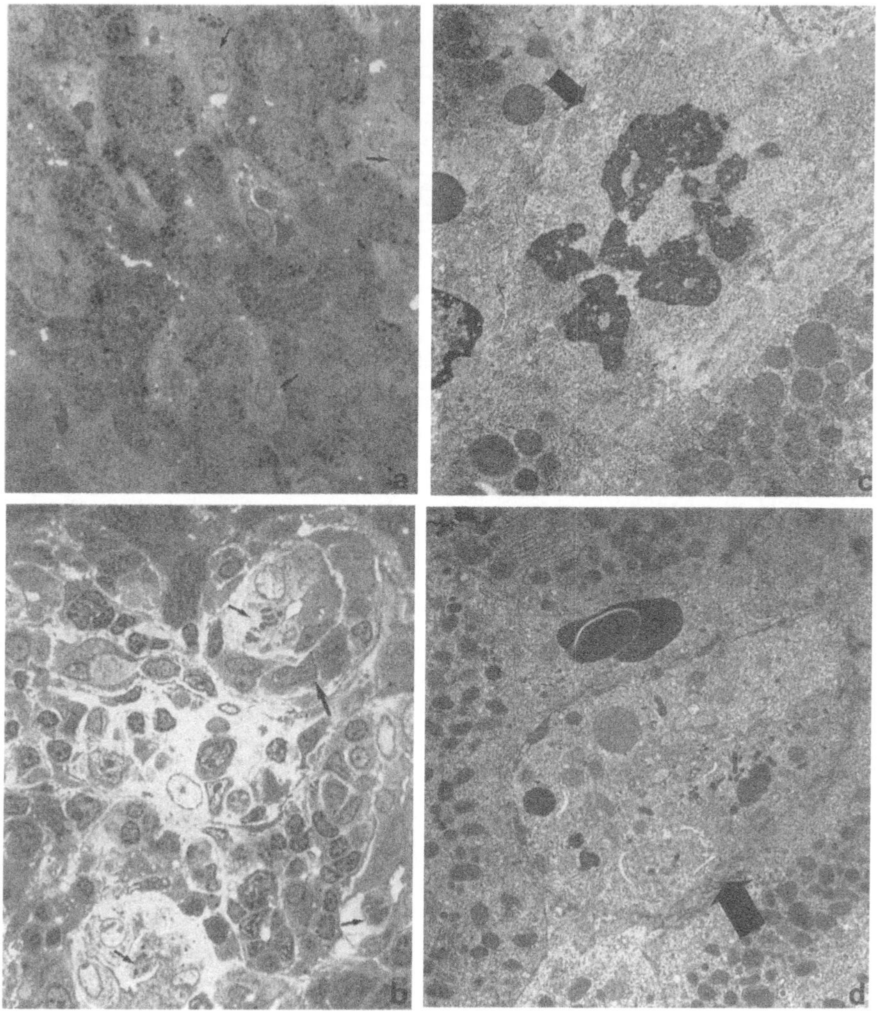

Figure 5. Light (a,b) and electron microscopy (c,d) of liver samples in pre-irradiation condition (after tumour inoculation and boron administration). Tumour cells lasting within the sinusoidal space after the liver washing are indicated by arrows in a,c and d micrographs. Still evidenced by arrows we can observe various tumour nodules in micrograph b.

- BPA dose: D,L p-^{10}BPA dose of 300 mg/Kg body weight
- Neutron irradiation starting time from BPA infusion: 2 hours
- Neutron fluence over the liver surface: $5 \cdot 10^{12} cm^{-2}$
- Neutron irradiation length: 12 min

This protocol allows, with a probability near 80%, to give doses of 10.7 and 59.0 Gy-Eq to healthy and tumour tissues respectively. The value of tumor over healthy tissue ratio $\frac{D_T}{D_H}$ results to be 5.5, more than the double we require for a safe treatment (2.5 at least).

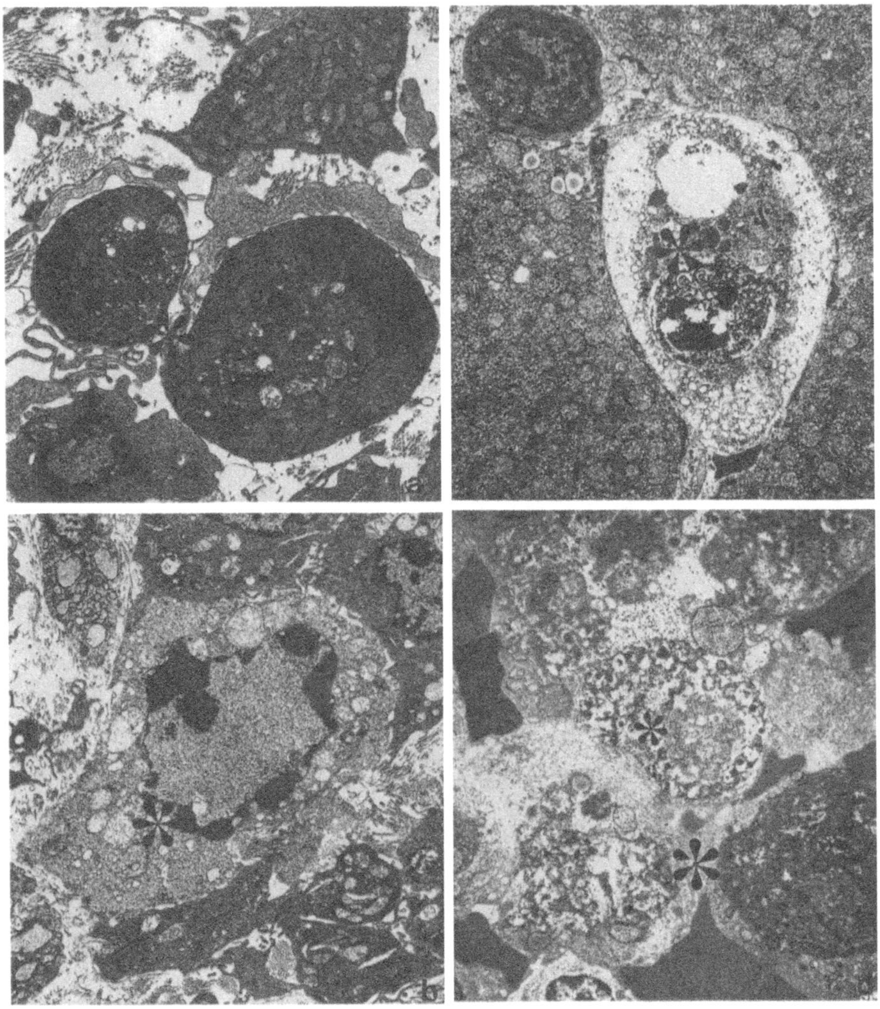

Figure 6. Electron microscopy of metastatic liver samples obtained 4 days after treatment and reimplantation. The four micrographs clearly prove the treatment efficiency for both sinusoidal and nodular tumour cells. Severe damages are evidenced by two kinds of cell death: apoptosis (a,b) and necrosis (c,d). Micrographs a, c refer to sinusoidal cells affected by apoptosis (a) and necrosis (c). Micrographs b, d give the important proof of the treatment efficiency for the deepest tissues inside the tumour nodules: apoptosis (b) and necrosis (d) are well evident in samples containing the most internal tissues of tumour nodules. Such severe injuries have never been observed in hundreds micrographs concerning normal hepatic parenchyma.

The "*in vitro*" analysis of radiation effects on colon carcinoma tumour cells, points out a different result between irradiated cells and the control population. After 48 hours from irradiation, while control cells show a complete recovery of the initial conditions, a permanent state of suffering is still evident in the irradiated population. This could be indication of debris and/or apoptotic cells due to severe damages induced by the neutron irradiation.

Interesting findings of our preclinical "*in vivo*" trials derive from microscopic and ultrastructural analyses. In particular 4–6 days from the irradiation the neoplastic cells, either singly dispersed or structured in nodules, appeared, with a discrete frequency, strongly damaged and in many cases in apoptotic state. The normal hepatocytes, on the other side, were never observed in condition clearly indicative of irreversible damages.

ACKNOWLEDGMENT

The authors wish to thank Mr. P. Bruschi for his precious and indispensable technical support along the entire development of the project. The Triga Mark II reactor staff is also acknowledged for the kind collaboration.

REFERENCES

1. S. Altieri, A. Braghieri, F. Fossati, P. Pedroni, T. Pinelli, A. Perotti, M. Specchiarello, G. Perlini, H. Rief, D. Chiaraviglio, F. De Grazia, and A. Zonta, "Evaluation of a selective boron absorption in liver tumours", Proceedings of the Third International Symposium on Neutron Capture Therapy for cancer", Bremen, May 31st–June 3rd 1988. *Strahlentherapie und Onkologie*, 165:170–172, 1989.
2. T. Pinelli, S. Altieri, F. Fossati, A. Zonta, D. Cossard, U. Prati, L. Roveda, G. Ricevuti, and R. Nano, "Development of a method to use Boron Neutron Capture Therapy for diffused tumours of liver (Taormina Project)", Proceedings of the Sixth International Symposium on Neutron Capture Therapy for Cancer, Kobe (Japan), Oct. 31st–Nov. 4th 1994, Y. Mishima Ed. Plenum Press, New York, 1996, pp. 783–794.
3. A.M. Cohen, S.Y. Winauer, M.A. Friedman *et al.*, "Cancer of the colon, rectum and anus" *in* "Epidemiology", D. Schottenfeld Ed. McGraw-Hill, New York, 1995, pp. 11–24.
4. W. Lineweavern, "Staging colon cancer", *Contemp. Surg.*, 25:19, 1984.
5. M. Huguier, F. Depoux, S. Houry *et al.*, "Adenocarcinoma of the rectum treated by abdominoperineal excision: multivariate analysis of prognostic factors", *Int. J. Colorect. Dis.*, 5:144–147, 1990.
6. Hassan K. Awwad, "Radiation Oncology—Radiobiological and Physiological Perspective: The Boundary-Zone between Clinical Radiotherapy and Fundamental Radiobiology and Physiology", Kluwer Academic Publishers, 1990, p. 421.
7. C. Martinenghi, "Basi fisiche e biologiche della radioterapia e della radioprotezione", Edizioni Libreria Cortini, Milano, 1984.
8. M. Cromheecke, A.J. Grond, B.C. Szabo, and H.J. Hoekstra, "Short and long-term histopathological changes in the canine liver following single high-dose intraoperative radiation therapy (IORT)", *Int. J. Radiat. Biol.*, 75:1437–1448, 1999.
9. A.D. Chanana, J. Capala, M. Chadha, J.A. Coderre A.Z. Diaz, E.H. Elowitz, J. Iwai, D.D. Joel, H.B. Liu, R. Ma, N. Pendzick, M.S. Peress, M.S. Shady, D.N. Slatkin, G.W. Tyson, and L. Wielopolski, "Boron Neutron Capture Therapy for Glioblastoma Multiforme: interim results from the Phase I/II dose-escalation studies", *Neurosurgery*, 44:1182–1192, 1999.
10. R.L. Jirtle, P.M. DeLuca, W.M. Hinshaw, M.S. Gould, and M.N. Gould, "Survival of parenchymal hepatocytes irradiated with 14.3 MeV neutrons", *Int. J. Radiation Oncology Biol. Phys.*, 10:895–899, 1984.
11. G.M. Morris, J.A. Coderre, J.W. Hopewell, P.L. Micca, M.M. Nawrocky, H.B. Liu, and A. Bywaters, "Response of the central nervous system to Boron Neutron Capture Irradiation: evaluation using rat spinal cord model", *Radiother. Oncol.*, 32:249–255, 1994.
12. J. Capala, J.A. Coderre, A.Z. Diaz, R. Ma, and A.D. Chanana, "Physical and nominal effective radiation doses to the tumour and different parts of the normal brain during BNCT for Glioblastoma Multiforme", Eighth Intern. Symp. On Neutron Capture Therapy for Cancer, Program and abstracts, V-6 1998, p. 168.
13. J.A. Coderre, R. Bergland, M. Chadha, A.D. Chanana, E. Elowitz, D.D. Joel, H.B. Liu, D.N. Slatkin, and L. Wielopolski, "Boron neutron capture therapy of glioblastoma multiforme using the p-boronophenylalanine-fructose complex and epithermal neutrons", Proceedings of the 6th International Symposium on Neutron Capture Therapy for Cancer, Kobe (Japan), Oct. 31st–Nov. 4th 1994, Y. Mishima Ed, Plenum Press, New York, 1996, pp. 553–561.
14. F. Zywietz, H. Jung, A. Hess, and H.D., "Response of mouse intestine to 14 Mev neutrons" *Int. J. Radiat. Biol. Relat. Stud. Phys. Chem. Med.*, 35:63–72, 1979.

15. J.A. Coderre, and G.M. Morris, " The radiation biology of Boron Neutron Capture Therapy", *Radiat. Res.*, 151:1–18, 1999.
16. T. Pinelli, S. Altieri, F. Fossati, A. Zonta, U. Prati, L. Roveda, and R. Nano, "Development of a method to extend, by boron neutron capture process, the therapeutic possibilities of the liver autograft", Proc. Int. Conf.: Neutrons in Research and Industry, G. Vourvopoulos Ed. Crete 1996, SPIE Editor 2867, 1997, pp. 92–96.
17. T. Pinelli, S. Altieri, F. Fossati, A. Zonta, D. Cossard, U. Prati, L. Roveda, G. Ricevuti, and R. Nano," ^{10}B neutron capture therapy to treat the human liver affected by tumour metastases (Taormina project)", Proceedings of the 1st National Joint Congress 'Radiations: from theory to multidisciplinary applications', Pisa Nov. 24–26 1994, Felici ed., Pisa, 1996, pp. 23–30.
18. M.C. O'Brien, and W.E. Bolton, "Comparison of cell viability probes compatible with fixation and permeabilization for combined surface and intracellular staining in flow cytometry", *Cytometry*, 19:243–255, 1995.
19. A. Caignard, M.S. Martin, M.F. Michel, and F. Martin, "Interaction between 2 cellular subpopulation of a rat colonic carcinoma when inoculated to the syngeneic host", *Int. J. Cancer*, 36:273–279, 1985.
20. R. Steffen, D.M. Ferguson, and R.A.F. Krom," A new method for orthotopic liver transplantation with arterial cuff anastomosis to the recipient common hepatic artery", *Transplantation*, 47:166–168, 1989.

PARTICIPANTS

Symposium Participants by Country of Origin

USA	148	Australia	3
Japan	45	France	3
Germany	20	Slovenia	3
Finland	18	Israel	2
Sweden	11	Taiwan	2
United Kingdom	9	Korea	2
Italy	8	Belgium	1
Czech Republic	7	Canada	1
Switzerland	7	China	1
Argentina	6	Portugal	1
The Netherlands	6	Thailand	1
Russia	5	Turkey	1

Total Attendance: 311 Total Countries: 24

Usama Abo Ramadan
Department of Physics
Helsinki University
P.O. Box 9
00014 University of Helsinki
Helsinki, Finland
Tel: +358-9-1918370
Fax: +358-9-1918378
Email: aboramad@penger.helsinki.fi

Dianne Adams
Department of Pathology
Ohio State University
168 Hamilton Hall
1645 Neil Avenue
Columbus, OH 43210 USA
Tel: +1-614-292-0693
Fax: +1-614-292-7072
Email: adams.21@osu.edu

Nabil Adnani
Medical Physics Unit
Montreal General Hospital
1650 Cedar Ave.
Montreal H3G 1A4, Canada
Tel: +1-514-934-8052
Fax: +1-514-934-8229
Email: adnana@mrs.mni.mcgill.ca

Stefano Agosteo
Dipartimento di Ingegneria Nucleare
Politecnico di Milano
via Ponzio 34/3
20133 Milano, Italy
Tel: +39-02-2399-6318
Fax: +39-02-2399-6309
Email: Stefano.Agosteo@polimi.it

Fazlul Alam
U.S. Borax
26877 Tourney Rd.
Valencia, CA 91355 USA
Tel: +1-805-287-6088
Fax: +1-805-287-6014
Email: alam007@aol.com

Brent J. Albertson
Nuclear Engineering Program
The Ohio State University
206 W. 18th Ave.
Columbus, OH 43210 USA
Tel: +1-614-292-1074
Fax: +1-614-292-3163
Email: brent@bullmoose.eng.ohio-state.edu

Corey Albright
Montana State University
357 EPS Bldg.
Bozeman MT 59717 USA
Tel: +1-406-994-7808
Fax: +1-406-994-4376
Email: albright@cs.montana.edu

Dennis A. Allen
School of Physics & Astronomy
University of Birmingham
Edgbaston
Birmingham B15 2TT U.K.
Tel: +44-121-4144690
Fax: +44-121-4144725
Email: dennis@aps6.ph.bham.ac.uk

Barry Allen
St. George Cancer Care Center
St. George Hospital
Gray Street
Kogarah 2217 NSW
Australia
Tel: +61-2-3503855
Fax: +61-2-3502456
Email: b.allen@unsw.edu.au

Kuniaki Amemiya
Dept. of Quantum Eng. & Systems Science
University of Tokyo
7-3-1 Hongo, Bunkyo-ku
Tokyo 113-8656 Japan
Tel: +81-3-3812-2111 ex.6974
Fax: +81-3-3818-3455
Email: kumi@susie.q.t.u-tokyo.ac.jp

Shigehiro An
Association for Nuclear Technology in Medicine
5 Mori Bldg.
1-17-1 Minato-ku, Toranomon
Tokyo, Japan
Tel: +81-03-3504-3961
Fax: +81-03-3504-1390

Carita Aschan
Department of Physics
University of Helsinki
P.O. Box 9
University of Helsinki
FIN-00014 HU, Finland
Tel: +358-9-191-8370
Fax: +358-9-191-8378
Email: carita.aschan@helsinki.fi

Iiro Auterinen
VTT Chemical Technology
P.O. Box 1404
FIN-02044 Finland
Tel: +358-9-456-6353
Fax: +358-9-456-6390
Email: Iiro.Auterinan@vtt.fi

Susan A. Autry-Conwell
Department of Neurosurgery
University of California Davis
4860 Y Street
Suite 3740
Sacramento, CA USA
Tel: +1-916-734-8120
Fax: +1-916-452-2580
Email: saautry@ucdavis.edu

Tariq Aziz
Dept. of Pathology
SUNY Stony Brook
Stony Brook, NY 11794 USA
Tel: +1-631-444-7631
Fax: +1-631-444-3419

Rodney S. Bagley
Dept. of Veterinary Clinical Sciences
Washington State University
College of Veterinary Medicine
Pullman, WA 99164-6610 USA
Tel: +1-509-335-6060
Fax: +1-509-335-0880
Email: bagley@vetmed.wsu.edu

Beverly Barnum
Fluorochem, Inc.
680 S. Ayon Ave.
Azusa, CA USA
Tel: +1-626-334-6714
Fax: +1-626-334-6714
Email: barnum22@hotmail.com

Rolf F. Barth
Department of Pathology
The Ohio State University
165 Hamilton Hall
1645 Neil Avenue
Columbus, OH 43210 USA
Tel: +1-614-292-2177
Fax: +1-614-292-7072
Email: barth.1@osu.edu

Roger W. Bartholomay
Department of Physics
Idaho State University
Campus Box 8106
Pocatello, ID 83209 USA
Email: rogerb@physics.isu.edu

John P. Barton
NRE Inc.
1422 Vue Du Bay Court
San Diego, CA 92109 USA
Tel: +1-858-488-8810
Fax: +1-858-488-8812

William F. Bauer
INEEL
P.O. Box 1625
Idaho Falls, ID 83402-2208 USA
Tel: +1-208-526-1180
Fax: +1-208-526-8541
Email: wlb@inel.gov

Claudia Bauer
Department of Chemistry
University of Bremen
Leobener Strasse
D-28359 Bremen
Germany
Tel: +49-421-2182119
Fax: +49-421-2182871

Judit Benczik
Faculty of Veterinary Medicine
University of Helsinki
Diagnostic Imaging
P.O. Box 57
FIN-00014 HU Helsinki, Finland
Tel: +358-9-70849620
Fax: +358-9-70849670
Email: judit.benczik@helsinki.fi

Peter Bendel
Dept. of Chemical Services—MR Center
The Weizmann Institute of Science
Rehovot 76100 Israel
Tel: +972-89343186
Fax: +972-89472218
Email: cvbendel@wis.weizmann.ac.il

John A. Bernard
Nuclear Reactor Laboratory
Massachusetts Institute of Technology
Mail Stop NW12-208
138 Albany Street
Cambridge, MA 02139 USA
Tel: +1-617-253-4202
Fax: +1-617-253-7300
Email: jbernard@mit.edu

Thomas D. Beynon
School of Physics & Space Research
University of Birmingham
Edgbaston
Birmingham B15 2TT U.K.
Tel: +44-121-4144694
Fax: +44-121-4144725
Email: t.d.beynon@bham.ac.uk

Emanuela Binello
Dept. of Nuclear Engineering
Massachusetts Institute of Technology
138 Albany Street
Room NW13-222
Cambridge, MA 02139 USA
Tel: +1-617-253-1749
Fax: +1-617-253-0760
Email: ebinello@mit.edu

Brandon Blackburn
Dept. of Nuclear Engineering
Massachusetts Institute of Technology
E25-330
45 Carleton St.
Cambridge, MA 02139 USA
Tel: +1-617-258-6999
Fax: +1-617-253-0760
Email: aggies@mit.edu

Eleanor Blakely
Lawrence Berkeley National Laboratory
One Cyclotron Road
Mailstop 70A-1118
Berkeley, CA 94720 USA
Tel: +1-510-486-6595
Fax: +1-510-486-4475
Email: eablakely@lbl.gov

Herman Blaumann
Centro Atomico Bariloche
Comision Nacional Energia Atomica
S.C. Bariloche 8400
Rio Negro, Argentina
Tel: +54-944-45247
Fax: +54-944-45225
Email: blaumann@cab.cnea.edu.ar

Darren Bleuel
Lawrence Berkeley National Laboratory
1 Cyclotron Road, M/S 71-259
Berkeley, CA 94720 USA
Tel: +1-510-486-7516
Fax: +1-510-486-7304
Email: dlbleuel@lbl.gov

Thomas E. Blue
Robinson Laboratory
The Ohio State University
206 W. 18th Ave.
Columbus, OH 43210 USA
Tel: +1-614-457-7269
Fax: +1-614-292-3163
Email: blue.1@osu.edu

Linden Blue
Neutron Therapies LLC
5340 Eastgate Mall
San Diego, CA 92121 USA
Tel: +1-858-678-8683
Fax: +1-858-678-8615

Jorgen Boethius
Karolinska Hospital
Box 130
S-171 76 Stockholm
Sweden
Tel: +468-51772095
Fax: +468-307091
Email: jorgenb@neuro.ks.se

Peter Bradley
Department of Engineering Physics
University of Wollongong
Northfields Avenue
Wollongong NSW 2522
Australia
Tel: +61-2-42214897
Fax: +61-2-42215944
Email: peter_bradley@uow.edu.au

Vladimir I. Bregadze
Nesmeyanov Inst. of Organoelem. Compunds
Russian Academy of Sciences
Vavilov Street 28, B-334
Moscow 117813, Russia
Tel: +7-095-135-7405
Fax: +7-095-135-5085
Email: bre@ineos.ac.ru

Sarah L. Brooke
School of Physics and Astonomy
University of Birmingham
Edgbaston
Birmingham B15 2TT
U.K.
Tel: +44-121-414-4689
Fax: +44-121-414-4725
Email: sarah@aps4.ph.bham.ac.uk

Robert M. Brugger
1012 Bourn Ave.
Columbia, MO 65203-1455 USA
Tel: +1-573-445-6580
Fax: +1-573-445-6580
Email: nucrb@showme.missouri.edu

Jiri Burian
NRI Dept. 803
Nuclear Research Institute
CZ-25068 Rez near Prague
Czech Republic
Tel: +420-2-6617-2465
Fax: +420-2-20940191
Email: Burian@nri.cz

Jay Burmeister
Gershenson Radiation Oncology Center
Harper Hospital / Wayne State University
Karmanos Cancer Institute
3990 John R.
Detroit, MI 48201 USA
Tel: +1-313-745-9693
Fax: +1-313-745-2314
Email: burmeist@kci.wayne.edu

Paul Busse
Joint Center for Radiation Therapy
Harvard Medical School
330 Brookline Ave.
Boston, MA 02215 USA
Tel: +1-617-632-8502
Fax: +1-617-632-0339
Email: busse@jcrt.harvard.edu

Thomas E. Byrne
Biology Department
Roane State Community College
276 Patton Lane
Harriman, TN 37748-5011 USA
Tel: +1-423-882-4533
Fax: +1-423-882-4686
Email: byrne_t@a1.rscc.tn.us

Osvaldo Calzetta Larrieu
Comision Nacional Energia Atomica (CNEA)
Centro Atómico Bariloche
8400 S.C. de Bariloche
Rio Negro, Argentina
Tel: +54-944-45240
Fax: +54-944-45225
Email: calzetta@cab.cnea.edu.ar

Jacek Capala
Medical Department
Brookhaven National Laboratory
Bldg. 490, 30 Bell Ave.
P.O. Box 5000
Upton, NY 11973-5000 USA
Tel: +1-631-344-2693
Fax: +1-631-344-1349
Email: capala@bnl.gov

Joseph R. Castro
Lawerence Berkeley Laboratory
University of California San Francisco
55-121
Berkeley, CA 94720 USA
Tel: +1-510-486-6325
Fax: +1-510-486-4170
Email: jrcastro@lbl.gov

Nicola Cerullo
DITEC—Università di Genova
Via All'Opera Pia, 15/A
I-16145 Genoa, Italy
Tel: +39-010-3532869/61
Fax: +39-010-311870
Email: cerullo@ditec.unige.it

Subhash Chandra
Department of Chemistry
Cornell University
Baker Laboratory, Cornell University
Ithaca, NY 14853-1301 USA
Tel: +1-607-255-3884
Fax: +1-607-255-4137
Email: sc40@cornell.edu

Fong-in Chou
Nuclear Science Tech. Development Center
National Tsing Hua University
Hsinchu 30043
Taiwan
Tel: +886-3-5-721962
Fax: +886-3-5-725974
Email: fichou@nstdc.nthu.edu.tw

William Chu
Lawrence Berkeley National Laboratory
One Cyclotron Road
Mailstop 71-259
Berkeley, CA 94720 USA
Tel: +1-510-486-7735
Fax: +1-510-486-5788
Email: wtchu@lbl.gov

Cynthia Chuang
Nuclear Reactor Laboratory
Massachusetts Institute of Technology
138 Albany St., NW13-252
Cambridge, MA 02139 USA
Tel: +1-617-667-0179
Fax: +1-617-975-5233
Email: cfchuang@mit.edu

Jeffrey Coderre
Medical Department
Brookhaven National Laboratory
Bldg. 490, 30 Bell Ave.
P.O. Box 5000
Upton, NY 11973-5000 USA
Tel: +1-631-344-2298
Fax: +1-631-344-1349
Email: coderre@bnl.gov

Pascal Coholis
Ion Beam Applications s.a.
Chemin du Cyclotron 3
B-1348 Louvain-la-Neuve
Belgium
Tel: +32-10-475878
Fax: +32-10-475810
Email: coholis@iba.be

Pietro Colangelo
Istituto Nazionale di Fisica Nucleare
Sezione di Bari
Via Amendola N.173
70126 Bari, Italy
Tel: +39-80-5443210
Fax: +39-80-5442470
Email: colangelo@bari.infn.it

Florence Colomb
Commissariat à L'Energie Atomique/SACLAY
CEA DRN/DRE/SIREN
Bât 635
Gif-sur-Yvette Cédex 91191 France
Tel: +33-1-69-08-34-35
Fax: +22-1-69-08-90-08
Email: colomb@orgue.saclay.cea.fr

John Crawford
Paul Scherrer Institute
CH-5232 Villigen PSI
Switzerland
Tel: +41-56-310-2080
Fax: +41-56-310-4412
Email: john.crawford@.psi.ch

Nigel Crompton
Paul Scherrer Institute
OFLD/O02
CH-5232 Villigen PSI
Switzerland
Tel: +41-56-3102871
Fax: +41-56-3103515
Email: crompton@psi.ch

Vladimir Dbaly
Neurosurgical Department
Homolka Hospital
Roentgenova 2
151 19 Prague 5
Czech Republic

Gelsomina De Stasio
University of Wisconsin—SRC
3731 Schneider Drive
Stoughton, WI USA
Tel: +1-608-877-2000
Fax: +1-608-877-2001
Email: pupa@src.wisc.edu

Aidnag Z. Diaz
Medical Department
Brookhaven National Laboratory
Bldg. 490, 30 Bell Ave.
P.O. Box 5000
Upton, NY 11973-5000 USA
Tel: +1-631-344-2773
Fax: +1-631-344-1349
Email: azdiaz@bnl.gov

George Dilworth
BioNeutrics, Inc.
10317 Technology Drive
Suite 1
Knoxville, TN 37932 USA
Tel: +1-423-675-5627
Fax: +1-423-675-3677
Email: georged@bioneutrics.com

Udo Doerfler
Department of Chemistry
University of Bremen
Leobener Strasse
D-28359 Bremen, Germany
Tel: +49-421-2182119
Fax: +49-421-2182871
Email: doerfler@chemie.uni-bremen.de

Richard J. Donahue
Lawrence Berkeley National Laboratory
1 Cyclotron Road
Berkeley, CA 94720 USA
Tel: +1-510-486-5597
Fax: +1-510-486-7304
Email: RJDonahue@LBL.GOV

Deborah de F. dos Santos
Institute of Organic Chemistry
University of Zurich
Winterthurerstrasse 190
CH-8057 Switzerland
Tel: +41-1-635-4247
Fax: +41-1-635-6812
Email: dfsantos@oci.unizh.ch

Katarina Edwards
Dept. of Physical Chemistry
Uppsala University
Box 532
S-75121 Uppsala, Sweden
Tel: +46-18-4713668
Fax: +46-18-508542
Email: edwards@fki.uu.se

Eric Elowitz
Dept. of Neurosurgery
Beth Israel Medical Center
170 East End Avenue
New York, NY 10128 USA
Tel: +1-212-870-9650
Fax: +1-212-870-9396
Email: eelowitz@bethisraelny.org

Jeffrey F. Evans
Robinson Laboratory
The Ohio State University
206 W. 18th Ave.
Columbus, OH 43210 USA
Tel: +1-614-457-7269
Fax: +1-614-292-3163
Email: evans.266@osu.edu

Heinz Fankhauser
Cabinet Neurochirurgical
Croix-Rouges 2
CH-1007 Lausanne
Switzerland
Tel: +4121-341-9393
Fax: +4121-341-9392
Email: heinz.fankhauser@nch.unil.ch

Markus Färkkilä
Dept. of Neurology
Hesinki University Central Hospital
Haartmaninkatu 4
FIN-00290 HYKS
Helsinki, Finland
Tel: +358-9-4712505
Fax: +358-9-4714009
Email: markus.farkkila@huch.fi

Debra Feakes
Department of Chemistry
Southwest Texas State University
601 University Drive
San Marcos, TX 78666 USA
Tel: +1-512-245-7609
Fax: +1-512-245-2374
Email: DF10@swt.edu

Peter Fessenden
Stanford University
27415 Ashland Dr.
Pioneer, CA 95666 USA
Tel: +1-209-295-7797
Email: fess@goldrush.com

Hiroshi Fukuda
Dept. of Nuclear Medicine and Radiology
Tohoku University
Institute of Development, Aging, and Cancer
4-1 Seiryo-machi, Aoba-ku
Sendai 980-8575 Japan
Tel: +81-22-717-8556
Fax: +81-22-717-8560
Email: hiro@idac.tohoku.ac.jp

Yoshinobu Fukumori
Faculty of Pharmaceutical Science
Kobe Gaukin University
Arise 518, Ikawadani-cho
Nishi-ku, Kobe 651-2180
Japan
Tel: +81-78-974-1551 (X2471)
Fax: +81-78-974-5689
Email: fukumori@pharm.kobegakuin.ac.jp

Detlef Gabel
Department of Chemistry
University of Bremen
P.O. Box 330440
D-28334 Bremen, Germany
Tel: +49-421-2182200
Fax: +49-421-2182871
Email: gabel@chemie.uni-bremen.de

Reinhard Gahbauer
Division of Radiation Oncology
Arthur James Cancer Hospital & Research Inst.
The Ohio State University
300 W. 10th Avenue
Columbus, OH 43210 USA
Tel: +1-614-293-8415
Fax: +1-614-293-4042
Email: gahbauer.1@osu.edu

Grazia Gambarini
Dipartimento di Fisica
Università di Milano
Via Celoria n.16
20133 Milano, Italy
Tel: +39-02-2392-243
Fax: +39-02-2392-630
Email: grazia.gambarini@mi.infn.it

Stephan Garbe
Radiologische Universitätsuling
Medizinische Einrichtungen Bonn
Sigmund-Freud-Str. 25
D-53105 Bonn, Germany
Tel: +49-228-287-6301
Fax: +49-228-287-6093
Email: garbe@uni-bonn.de

Patrick Gavin
Dept. of Veterinary Clinical Sciences
Washington State University
McCoy Hall
Pullman, WA 99164-6610 USA
Tel: +1-510-335-0764
Fax: +1-510-335-0880
Email: prg@vetmed.wsu.edu

Hadi Ghaneolhosseini
Dept. of Organic Chemistry
Uppsala University
Institute of Chemistry
Box 531
S-751 21 Uppsala, Sweden
Tel: +46-18 4713798
Fax: +46-18 508542
Email: Hadi@kemi.uu.se

Senait Ghirmai
Dept. of Organic Chemistry
Uppsala University
Institute of Chemistry
Box 531
S-751 21 Uppsala, Sweden
Tel: +46-18 4713798
Fax: +46-18 508542
Email: Senait@kemi.uu.se

David Gierga
Dept. of Nuclear Engineering
Massachusetts Institute of Technology
E25-330
45 Carleton St.
Cambridge, MA 02139 USA
Tel: +1-617-258-0752
Fax: +1-617-253-0760
Email: gierga@mit.edu

Benjamin Gilbert
Ecole Polytchnique Federal de Lausanne
CH-1015Lausanne
Switzerland
Tel: +41-21-693-4536
Fax: +41-21-693-4666
Email: bgilbert@dpmail.epfl.ch

Lucien Gilles
CEA-Saclay / DMT
Commissariat à L'Energie Atomique
Bât 635
Gif-sur-Yvette Cédex 91191 France
Tel: +33-1-6908-9516
Fax: +33-1-6908-6642
Email: l.gilles@lemel.fr

John Glass
Medical Department
Brookhaven National Laboratory
Bldg. 490, 30 Bell Ave.
P.O. Box 5000
Upton, NY 11973-5000 USA
Tel: +1-631-344-3612
Fax: +1-631-344-5311
Email: glass@bnl.gov

Steven Goetsch
San Diego Medical Physics
837 Son Rodolfo
Suite 124-113
Solana Beach, CA USA
Tel: +1-858-452-5020
Fax: +1-858-452-5672
Email: sjgphd@connectnet.com

Gerald Goldstein
10038 Maple Leaf Dr.
Gaithersberg, MD 20886 USA
Tel: +1-301-977-3891
Email: jerrypat@bellatlantic.net

Jamie G. Gomez
Health International Services
7040 W. Palmetto Pk. Rd.
Suite 2-549
Boca Raton, FL 33433 USA
Tel: +1-561-443-3705
Fax: +1-561-443-3706
Email: jimgom@pol.net

Isabel Caetano Gonçalves
Instituto Tecnológico e Nuclear / Reactor
Estrada Nacional 10
2686 Sacavém Codex
Portugal
Tel: +351-1-9550021
Fax: +351-1-9941039
Email: isabelcg@itn1.itn.pt

Joseph H. Goodman
Ohio State University
1011 North Dean Hall
410 West Tenth Avenue
Columbus, OH 43065 USA
Tel: +1-614-293-8587
Fax: +1-614-293-4281

Timothy Goorley
Nuclear Reactor Laboratory
Massachusetts Institute of Technology
NW-12 Receiving
138 Albany Street
Cambridge, MA 02139 USA
Tel: +1-617-253-5720
Fax: +1-617-253-7300
Email: jgoorley@mit.edu

John Grecula
A. James Cancer Hospital
Ohio State University
300 W. 10th Avenue
Columbus, OH 43210 USA
Tel: +1-614-293-3244
Fax: +1-614-293-4044
Email: grecula.2@osu.edu

Stuart Green
Dept. of Medical Physics & Clinical Engineering
Queen Elizabeth Hospital, Edgbaston
Birmingham
B15 2TH U.K.
Tel: +44-121-627-2090
Fax: +44-121-472-0393
Email: stuart@aps6.ph.bham.ac.uk

Ehud Greenspan
Dept. of Nuclear Engineering
University of California
Berkeley, CA 94720 USA
Tel: +1-510-643-9983
Fax: +1-510-643-9685
Email: gehud@nuc.berkeley.edu

Merle Griebenow
436 N 55 W
Idaho Falls, ID 83402 USA
Tel: +1-208-523-2043
Fax: +1-208-523-7906
Email: mlg@srv.net

Nilendu Gupta
061 James Cancer Hospital
Ohio State University
300 W. 10th Avenue
Columbus, OH 43210 USA
Tel: +1-614-293-4204
Fax: +1-614-293-4044
Email: gupta.6@osu.edu

David Halpern
3270 Hartridge Drive
Alpharetta, GA 30022 USA
Tel: +1-770-642-0306
Fax: +1-770-642-6020

Yale Harker
INEEL
P.O. Box 1625
Idaho Falls, ID 83415-7113 USA
Tel: +1-208-526-0707
Fax: +1-208-526-5208
Email: yoh@inel.gov

Otto Harling
Nuclear Reactor Laboratory
Masachusetts Institute of Technology
Rm. NW13-200
138 Albany St.
Cambridge, MA 02139 USA
Tel: +1-617-253-4201
Fax: +1-617-253-7300
Email: oharling@mit.edu

Frank Harmon
Idaho State University
785 S. 8th Ave,. P.O. Box 8106
Idaho State University
Pocatello, ID 83209 USA
Tel: +1-208-236-3150
Fax: +1-208-236-4649
Email: harmon@physics.isu.edu

Jack Hartwell
INEEL
P.O. Box 1625
Idaho Falls, ID 83415 USA
Tel: +1-208-526-9366
Fax: +1-208-526-9267
Email: jkh2@inel.gov

Toshinari Hasegawa
Veterinary Surgery
Yamaguchi University
1677-1 Yoshida
Yamaguchi 753-8515
Japan
Tel: +81-839-33-5898
Fax: +81-839-33-5930

M. Frederick Hawthorne
Department of Chemistry and Biochemistry
University of California Los Angeles
405 Hilgard Ave.
Los Angeles, CA 90095-1569 USA
Tel: +1-310-206-3182
Fax: +1-310-825-5490
Email: mfh@chem.ucla.edu

Volker Heinzel
Inst. für Reaktorsicherheit
University of Karlsruhe
P.O. Box 3640
D-76012 Karlsruhe
Germany
Tel: +49-7247-82-3761
Fax: +49-7247-82-3718
Email: heinzel@hdi.kfk.de

Cecile Henssen
Dept. of Organic Chemistry
Uppsala University
Institute of Chemistry
Box 531
S-751 21 Uppsala, Sweden
Tel: +46-18 4713798
Fax: +46-18 508542
Email: Ceclieh@kemi.uu.se

Katalin Hideghèty
Klinik und Poliklinik für Strahlentherapie
Universitätsklinikum Essen
Hufelandstrasse 55
D-45122 Essen
Germany
Tel: +49-201-723-2059
Fax: +49-201-723-5908
Email: kati.hide@uni-essen.de

Pekka Hiismäki
VTT The Technical Research Center of Finland
P.O. Box 1404
FIN-02044 VTT
Espoo, Finland
Tel: +358-9-4566320
Fax: +358-9-4566390
Email: pekka.hiismaki@vtt.fi

Junichi Hiratsuka
Dept. of Radiation Oncology
Kawasaki Medical School
577 Matsushima
Kurashiki City
Okayama 701-01, Japan
Tel: +81-86-462-1111
Fax: +81-86-462-1199
Email: hiratsuka@med.kawasaki-u.ac.jp

René Huiskamp
Radiobiology & Radioecology
Netherlands Energy Research Foundation
P.O. Box 1
1755 ZG Petten
The Netherlands
Tel: +31-224-564069
Fax: +31-224-563491
Email: huiskamp@ecn.nl

Tatsuo Ido
Cyclotron and Radioisotope Center
Tohoku University
Aoba, Aramaki Aoba-ku
Sendai 980-8578 Japan
Tel: +81-22-217-7797
Fax: +81-22-217-3485
Email: tido@cyric.tohuku.ac.jp

Daniel Ingersoll
Oak Ridge National Laboratory
P.O. Box 2008
Oak Ridge, TN 37831 USA
Tel: +1-423-574-6102
Fax: +1-423-574-9619
Email: dti@ornl.gov

Masayori Ishikawa
Graduate School of Energy Science
Kyoto University
Kumatori-cho
Sennan-gun
Osaka 590-0494 Japan
Tel: +81-724-51-2375
Fax: +81-724-51-2603
Email: masayori@post3.rri.kyoto-u.ac.jp

Nicholas D. James
University of Birmingham
Edgbaston
Birmingham B152TT U.K.
Tel: +44-121-414-4097
Fax: +44-121-414-3700
Email: n.d.james@bham.ac.edu

Cynthia Johnson
Dept. of Veterinary Clinical Sciences
Washington State University
McCoy Hall
Pullman, WA 99164-6610 USA
Tel: +1-509-335-0721
Fax: +1-509-335-0880
Email: clj@vetmed.wsu.edu

Markus Johnsson
Dept. of Physical Chemistry
Uppsala University
Box 532
S-75121 Uppsala, Sweden
Tel: +46-18-4713661
Fax: +46-18-508542
Email: markus.johnsson@fki.uu.se

George W. Kabalka
Department of Radiology
University of Tennessee
University of Tennessee Medical Center
Knoxville, TN 37920 USA
Tel: +1-423-544-9670
Fax: +1-423-974-3260
Email: kabalka@utk.edu

Teruyoshi Kageji
Dept. of Neurosurgery
University of Tokushima
Kuramoto-cho 2-50
Tokushima 770 Japan
Tel: +0886-33-7149
Fax: +0886-32-9464
Email: kageji@clin.med.tokushima-u.ac.jp

Stephen B. Kahl
Dept. of Pharmaceutical Chemistry
University of California San Francisco
513 Parnassus Ave.
San Francisco, CA 94143-0446 USA
Tel: +1-415-476-4684
Fax: +1-425-476-0688
Email: sbkahl@itsa.ucsf.edu

Merja Kallio
HUCH Research Instittute
Tukholmank 8F
FIN-00290 Helsinki
Finland
Tel: +358-9-4712139
Fax: +358-9-4715550
Email: merja.kallio@huch.fi

Shuji Kamano
Division of Neurosurgery
Inagi Municipal Hospital
Inagi City, Tokyo 2060801 Japan
Tel: +81-3-3467-8203
Fax: +81-3-3467-8203
Email: skamano@mb.kcom.ne.jp

Keiji Kanda
Research Reactor Institute
Kyoto University
Kumatori-cho
Sennan-gun
Osaka 590-0494 Japan
Tel: +81-724-51-2145
Fax: +81-724-51-2604
Email: kanda@kuca.rri.kyoto-u.ac.jp

Robert R. Kane
Department of Chemistry
Baylor University
P.O. Box 97348
Waco, TX 96798-7348 USA
Tel: +1-254-710-4556
Fax: +1-254-710-2403
Email: bob_kane@baylor.edu

Leena Kankaanranta
Dept. of Oncology/Basic Research
Helsinki University Central Hospital
FIN-00029 HYKS
Helsinki, Finland
Tel: +358–9-4711
Fax: +358-9-471-4201
Email: leena.kankaanranta@helsinki.fi

Irving Kaplan
Joint Center for Radiation Therapy
Harvard Medical School
330 Brookline Avenue
Boston, MA 02215 USA
Tel: +1-617-632-8502
Fax: +1-617-632-0339
Email: ikaplan@jcrt.harvard.edu

Jody Kaplan
Radiation Oncology
Harvard-MIT BNCT Program
198 Pilgrim Rd.
Boston, MA USA
Tel: +1-617-632-8502
Fax: +1-617-632-0339
Email: jakaplan@mediaone.net

Cecilia Kessler
Centro Regional de Referencia para Dosimetría
Comisión Nacional de Energía Atómica
Av. del Libertador 8250
1429 Buenos Aires, Argentina
Tel: +54-11-4379-8232
Fax: +54-11-4379-8228
Email: kessler@cae.cnea.gov.ar

Viatchesla F. Khokhlov
Institute of Biophysics
Russian State Scientific Center
Zhivopisnaja 46
123182 Moscow, Russia
Tel: +7-095-193-0186
Fax: +7-095-193-1102
Email: bre@ineos.ac.ru

W. Stead Kiger
Nuclear Reactor Laboratory
Massachusetts Institute of Technology
138 Albany Street
Bldg. NW13-252
Cambridge, MA 02139 USA
Tel: +1-617-253-5720
Fax: +1-617-253-7300
Email: wskiger@mit.edu

Yuko Kinashi
Research Reactor Institute
Kyoto University
Radiation Oncology Research Laboratory
Kumatori-cho, Sennan-gun
Osaka 590-0451 Japan
Tel: +81-0724-51-2407
Fax: +81-0724-51-2627
Email: kinashi@rri.kyoto-u.ac.jp

Peter Kirchner
Medical Sciences Division
U.S. Department of Energy
Office of Energy Research, SC-73
19901 Germantown Road
Germantown, MD 20874 USA
Tel: +1-301-903-9106
Fax: +1-301-903-0567
Email: peter.kirchner@uiowa.edu

Mitsunori Kirihata
Dept. of Applied Biochemistry
Osaka Prefecture University
1-1 Gakuencho, Sakai
Osaka 599-8523 Japan
Tel: +0722-54-9470
Fax: +0722-54-9470
Email: kirihata@biochem.osakafu-u.ac.jp

Toshiaki Kishi
Department of Research Reactor
JAERI, Tokai Research Establishment
2-4 Shirakata-shirane
Tokai-mura, Naka-gun
Ibaraki 319-1195 Japan
Tel: +81-29-282-5593
Fax: +81-29-282-6763
Email: kisi@jrr2ws1.tokai.jaeri.gp.jp

Yoshinori Kitaoka
Research Reactor Institute
Kyoto University
Noda, Kumatori-cho, Sennan-gun
Osaka 590-0494 Japan
Tel: +81-724-51-2440
Fax: +81-724-51-2622
Email: kitaoka@hl.rri.kyoto-u.ac.jp

Miyuki Kitazawa
Dept. of Chemistry, Faculty of Science
Shinshu University
Asahi 3-1-1
Matsumoto 390-8621 Japan
Tel: +81-263-37-2558
Fax: +81-263-37-2559
Email: c97sa302@ripms.shinshu-u.ac.jp

Robert Klinkowstein
Newton Scientific, Inc.
245 Bent Street
Cambridge, MA 02141 USA
Tel: +1-617-354-9469
Fax: +1-617-354-9479
Email: rek@world.std.com

Mitsue Kobayashi
Research Reactor Institute
Kyoto University
Kumatori-cho, Sennan-gun
Osaka 590-04 Japan
Tel: +81-724-53-3520
Fax: +81-724-53-3520
Email: hqp01574@niftyserve.or.jp

Tooru Kobayashi
Research Reactor Institute
Kyoto University
Noda, Kumatori-cho, Sennan-gun
Osaka 590-0494 Japan
Tel: +81-724-51-2430
Fax: +81-724-51-2620
Email: kobato@rri.kyoto-u.ac.jp

Hirofumi Kondoh
Mishima Institute for Dermatological Research
1-4-32 Sowa-cho, Nada-ku
Kobe 657-0063 Japan
Tel: +81-78-842-6342
Fax: +81-78-842-6340
Email: gch07434@niftyserve.or.jp

Mika Kortesniemi
Clinical Research Institute
Helsinki University Central Hospital
P.O. Box 105
FIN-00029 HYKS
Helsinki, Finland
Tel: +358-9-456-7171
Fax: +358-9-456-6390
Email: mika.kortesniemi@huch.fi

Antti Kosunen
STUK Radiation and Nuclear Safety Authority
P.O. Box 14
FIN-00881 Helsinki, Finland
Tel: +358-9-75988446
Fax: +358-9-75988450
Email: antti.kosunen@stuk.fi

Chandra Kota
Gershenson Radiation Oncology Center
Harper Hospital / Wayne State University
Karmanos Cancer Institute
3990 John R.
Detroit, MI 48201 USA
Tel: +1-313-745-2522
Fax: +1-313-745-2314
Email: kotac@kci.wayne.edu

Petri Kotiluoto
VTT Chemical Technology
Otakaari 3 A, Espoo
P.O. Box 1404
FIN-02044 VTT Finland
Tel: +358-9-456-6357
Fax: +358-9-456-6390
Email: petri.kotiluoto@vtt.fi

Andrès Juan Kreiner
Dept. of Physics
Argentine Atomic Energy Commission
Av. del Libertador 8250
Buenos Aires 1429 Argentina
Tel: +54-1-754-7151
Fax: +54-1-754-7121
Email: kreiner@tandar.cnea.edu.ar

Otomar Kriz
Katchem Ltd.
E. Krasnohorske 6
110 00 Praha, Czech Republic
Tel: +420-2-66173118
Fax: +420-2-6858116
Email: okriz@uachr.iic.cas.cz

Guido Kuehne-Cavelti
Paul Scherrer Institute
CH-5232 Villigen PSI
Switzerland

V. N. Kulakov
Institute of Biophysics
Russian State Research Center
Zhivopisnaya St. 46
123182 Moscow, Russia
Fax: +7-095-135-5085
Email: bre@ineos.ac.ru

Martti Kulvik
Clinical Research Institute
Hesinki University Central Hospital
P.O. Box 105
FIN-00029 HYKS
Helsinki, Finland
Tel: +358-9-4712138
Fax: +358-9-4715550
Email: martti.kulvik@huch.fi

Hiroaki Kumada
Department of Research Reactor
JAERI, Tokai Research Establishment
2-4 Shirakata-shirane
Tokai-mura, Naka-gun
Ibaraki 319-1195 Japan
Tel: +81-29-282-5593
Fax: +81-29-282-6763
Email: kumada@jrr3fep2.tokai.jaeri.gp.jp

Thomas LaHann
Department of Pharmaceutical Sciences
Idaho State University
970 S. 5th St.
Campus Box 8334
Pocatello, ID 83209-8334 USA
Tel: +1-208-236-4456
Fax: +1-208-236-4482
Email: lahann@otc.isu.edu

George Laramore
Dept. of Radiation Oncology
University of Washington
Box 356043
Seattle, WA 98195-6043 USA
Tel: +1-206-548-4110
Fax: +1-206-548-6218
Email: george@radonc.washington.edu

Edward Larsen
Dept. of Nuclear Eng. & Rad. Science
University of Michigan
Ann Arbor, MI 48109 USA
Tel: +1-734-936-0124
Email: edlarsen@engin.umich.edu

Jason Larson
College of Pharmacy
Idaho State University
970 S. 5th St.
Campus Box 8335
Pocatello, ID 83209-8334 USA
Tel: +1-208-236-2682
Fax: +1-208-236-4482
Email: lahann@otc.isu.edu

Brenda Laster
Medical Department
Brookhaven National Laboratory
490 Bell Street
Upton, NY 11973-5000 USA
Tel: +1-631-344-4836
Fax: +1-631-344-5311
Email: laster@bnl.gov

Chad Lee
Los Alamos National Laboratory
X-CI, MS F663
Los Alamos, NM 87545 USA
Tel: +1-505-665-2849
Fax: +1-505-665-8329
Email: chadlee@lanl.gov

Wanno Lee
Department of Nuclear Engineering
Korea Advanced Institute of Science
Teajon,
305-701 Republic of Korea
Tel: +82-42-869-3821
Fax: +82-42-869-3810
Email: petor@cais.kaist.ac.kr

Byung Chul Lee
Research Reactor Analysis Dept.
Korea Atomic Energy Research Inst.
HANARO Center
P.O. Box 105, Yusong
Taejon 305-600 Korea
Tel: +82-42-868-8489
Fax: +82-42-868-8341
Email: bjjun@nanum.kaeri.re.kr

Peter Lemmen
Inst. f. Org. Chemie u. Biochemie
Techn. Univ. Muenchen
Lichtenbergstr. 4
D-85747 Garching, Germany
Tel: +49-89-289-13319
Fax: +49-89-289-13329
Email: plemmen@nucleus.org.chemie.tu-muenchen.de

Arlene Lennox
Neutron Therapy Facility
Fermilab
P.O. Box 500
Mail Stop 301
Batavia, IL 60510 USA
Tel: +1-630-840-3865
Fax: +1-630-840-8766
Email: alennox@fnal.gov

Ka-Ngo Leung
Lawrence Berkeley National Laboratory
University of California Berkeley
1 Cyclotron Road
Berkeley, CA 94720 USA
Tel: +1-510-486-7918
Fax: +1-510-486-5105
Email: KNLeung@lbl.gov

Fangbiao Li
Dept. of Chemistry & Biochemistry
University of California Los Angeles
405 Hilgard Ave.
Los Angeles, CA 90095-1569 USA
Tel: +1-310-206-9444
Fax: +1-310-825-5490
Email: lif@chem.ucla.edu

Sara J. Liberman
Comision Nacional de Energia Atomica
Avda del Libertador 8250
1429 Buenos Aires, Argentina
Tel: +541-754-1349
Fax: +541-704-1193
Email: liberman@cnea.edu.ar

Victor M. Litjaev
Inst. of Physics, Power, & Engineering
249020 Bondarenko Sq. 1
Obninsk, Kaluga Region
Russia
Tel: +08439-9-83-56
Fax: +095-230-23-26
Email: korob@ippe.rssi.ru (or soloviev@ippe.rssi.ru)

Hungyuan B. Liu
McClellan Nuclear Radiation Center
University of California Davis
P.O. Box 2192
North Highlands, CA 95660 USA
Tel: +1-916-643-6251
Fax: +1-916-643-0176
Email: liu.ben@email.mcclellan.af.mil

Yen-wan Liu
Dept. of Engineering and System Science
National Tsing Hua University
Hsinchu 30043
Tiwan
Tel: +886-3-574-2669
Fax: +886-3-572-0724
Email: ywhliu@faculty.nthu.edu.tw

Bernhard Ludewigt
Lawrence Berkeley National Laboratory
One Cyclotron Road
Mailstop 71-259
Berkeley, CA 94720 USA
Tel: +1-510-486-7733
Fax: +1-510-486-5788
Email: Bernhard_Ludewigt@lbl.gov

Silke Lutz
Department of Chemistry
University of Bremen
Leobener Strasse
D-28359 Bremen
Germany
Tel: +49-421-218-2381
Fax: +49-421-218-2871
Email: lutz@chemie.uni-bremen.de

Ruimei Ma
Medical Department
Brookhaven National Laboratory
Bldg. 490, 30 Bell Ave.
P.O. Box 5000
Upton, NY 11973-5000 USA
Tel: +1-631-344-5180
Fax: +1-631-344-5311
Email: ma2@bnl.gov

Shin-ichiro Masunaga
Research Reactor Institute
Kyoto University
Radiation Oncology Res. Lab.
Noda, Kumatori-cho, Sennan-gun
Osaka 590-0494 Japan
Tel: +81-724-51-2406
Fax: +81-724-51-2627
Email: smasuna@rri.kyoto-u.ac.jp

Francis J. Mahoney
National Cancer Institute
6130 Executive Blvd., EPN 800
Rockville, MD USA
Tel: +1-301-496-9360
Fax: +1-301-480-5785
Email: mahoneyf@dtpepn.nci.nih.gov

Tetsuo Matsumoto
Atomic Energy Research Laboratory
Musashi Institute of Technology
971 Ozenji, Asao-ku,
Kawasaki-shi, 215 Japan
Tel: +81-44-966-6131
Fax: +81-44-955-6071
Email: mtetsuo@atom.musashi-tech.ac.jp

Caitlin Mansfield
School of Physics and Astronomy
University of Birmingham
Edgbaston
Birmingham B15 2TT U.K.
Tel: +44-121-414-4689
Fax: +44-121-414-4725
Email: caitlin@aps4.ph.bham.ac.uk

Akira Matsumura
Dept. of Neurosurgery
University of Tsukuba
Institute of Clinical Medicine
Tenno-dai 1-1-1
Tsukuba, Ibaraki 305-8575 Japan
Tel: +81-298-53-3168
Fax: +81-298-53-3214
Email: matsumur@igaku.md.tsukuba.ac.jp

Milan Marek
NRI Dept. 803
Nuclear Research Institute
CZ-25068 Rez near Prague
Czech Republic
Tel: +420-2-6617-2368
Fax: +420-2-20940191
Email: marek@nri.cz

Reiko Matsuoka
Assoc. for Nuclear Technology in Medicine
5 Mori Bldg.
1-17-1 Toranomon, Minato-ku
Tokyo 105-0001 Japan
Tel: +81-3-3504-3961
Fax: +81-3-3504-1390
Email: JDS00335@nifty.ne.jp

Rodger C. Martin
Oak Ridge National Laboratory
P.O. Box 2008
MS-6385
Oak Ridge, TN 37831-6385 USA
Tel: +1-423-576-2280
Fax: +1-423-241-2338
Email: m2m@ornl.gov

Marko Maucec
Reactor Physics Division
Jozef Stefan Institute
Jamova 39
SLO-1000 Ljubljana, Slovenia
Tel: +386-61-1885-206
Fax: +386-61-1612-335
Email: marko.maucec@ijs.si

Michael Meyer
Dept. of Neurology
University of Missouri
M741 University Hospital
One Hospital Drive
Columbia, MO 65212 USA
Tel: +1-573-882-3133
Fax: +1-573-884-4249
Email: mmeyer@brain.missouri.edu

Peggy Micca
Medical Department
Brookhaven National Laboratory
Bldg. 490, 30 Bell Ave.
P.O. Box 5000
Upton, NY 11973-5000 USA
Tel: +1-631-344-3566
Fax: +1-631-344-5311
Email: pmicca@bnl.gov

Laurence Miller
Dept. of Nuclear Engineering
University of Tennessee
Knoxville, TN 37996-2300 USA
Tel: +1-423-974-5048
Fax: +1-423-974-0668
Email: lfmiller@utk.edu

Yutaka Mishima
Mishima Institute for Dermatological Research
1-4-32 Sowa-cho, Nada-ku
Kobe 657-0063 Japan
Tel: +81-78-842-6342
Fax: +81-78-842-6340
Email: rxm02257@niftyserve.or.jp

Hannah Mitchell
Siemens Medical Systems, Inc.
4040 Nelson Ave.
Concord, CA 94520 USA
Tel: +1-925-246-8383
Fax: +1-925-246-8370
Email: hannah@smsocs.com

Masahito Miyamoto
Faculty of Pharmaceutical Science
Kobe Gaukin University
Arise 518, Ikawadani-cho
Nishi-ku, Kobe 651-2180
Japan
Tel: +81-78-974-1551 (X2470)
Fax: +81-78-974-5689
Email: miyamoto@pharm.kobegakuin.ac.jp

Gerard Michael Morris
Research Institute
University of Oxford
Churchill Hospital
Headington, Oxford
OX37LJ U.K.
Tel: +44-1865-225825
Fax: +44-1865-225847
Email: gmorris@radius.JR2.ox.ac.uk

Raymond Moss
HFR Division
Joint Research Center—IAM
Westerduinweg 3
NL-1755 ZG Petten
The Netherlands
Tel: +31-224-565126
Fax: +31-224-561449
Email: moss@jrc.nl

Laura Murphy
Dept. of Nuclear Engineering
Massachusetts Institute of Technology
E25-330
45 Carleton St.
Cambridge, MA 02139 USA
Tel: +1-617-253-6412
Fax: +1-617-253-0760
Email: lgmurphy@mit.edu

Fiona Murphy
Neutron Therapies LLC
5340 Eastgate Mall
San Diego, CA 92121 USA
Tel: +1-858-678-8683
Fax: +1-858-678-8615

Charlotta Naeslund
Dept. of Organic Chemistry
Uppsala University
Institute of Chemistry
Box 531
S-751 21 Uppsala, Sweden
Tel: +46-18 4713798
Fax: +46-18 508542
Email: LottaN@kemi.uu.se

Yoshinobu Nakagawa
Dept. of Neurosurgery
National Kagawa Children's Hospital
Zentsuji-cho 2603
Zentsuji City
Kagawa 765 Japan
Tel: +81-877-62-0885
Fax: +81-877-62-5384
Email: nakagawa@kagawasyo.hosp.go.jp

Kei Nakai
Dept. of Neurosurgery
Tsukuba University
Institute of Clinical Medicine
Tenno-dai 1-1-1
Tsukuba City, 305-8575 Japan
Tel: +81-298-53-3220
Fax: +81-298-53-3214
Email: LV5K-NKI@asahi-net.or.jp

Munekazu Nakaichi
Veterinary Surgery
Yamaguchi University
1677-1 Yoshida
Yamaguchi 753-8515
Japan
Tel: +81-839-33-5898
Fax: +81-839-33-5930
Email: nakaichi@agr.yamaguchi-u.ac.jp

Takehiro Nakano
Dept. of Applied Biochemistry
Osaka Prefecture University
1-1 Gakuencho Sakai
Osaka 599-8231 Japan
Tel: +0722-54-9469
Fax: +0722-54-9470
Email: takehiro@biochem.osakafu-u.ac.jp

Marta Nawrocky
Medical Department
Brookhaven National Laboratory
Bldg. 490, 30 Bell Ave.
P.O. Box 5000
Upton, NY 11973-5000 USA
Tel: +1-631-344-3592
Fax: +1-631-344-5311
Email: marta@bnl.gov

Michael Neumann
Department of Chemistry
University of Bremen
Leobener Strasse
D-28359 Bremen, Germany
Tel: +49-421-218-2381
Fax: +49-421-218-2871
Email: neumann@chemie.uni-bremen.de

Trent Nichols
UT Medical Center
University of Tennessee
1932 Alcoa Highway
Knoxville, TN 37920 USA
Tel: +1-423-544-6500
Fax: +1-423-544-6509
Email: tnichol2@utk.edu

David Nigg
INEEL
P.O. Box 1625, M/S 3855
EROB Building, 2525 North Fremont
Idaho Falls, ID 83415-3855 USA
Tel: +1-208-526-7627
Fax: +1-208-526-0528
Email: dwn@inel.gov

Frantisek Novy
Dept. of Oncology
Charles University
Unemocnice 2
128 08 Prague 2
Czech Republic
Tel: +420-2-294304
Fax: +420-2-298490

Richard B. Oberer
Neely Nuclear Research Center
Georgia Institute of Technology
900 Atlantic Drive
GIT, Atlanta, GA 30332-0425 USA
Email: chris.wang@me.gatech.edu

Koji Ono
Research Reactor Institute
Kyoto University
Radiation Oncology Res. Lab.
Kumatori-cho
Sennan-gun, Osaka 590-0494 Japan
Tel: +81-724-51-2475
Fax: +81-724-51-2627
Email: onokoji@rri.kyoto-u.ac.jp

Michael Orr
Department of Nuclear & Mechanical Engineering
The Ohio State University
2039 Robinson Lab
206 W. 18th Avenue
Columbus, OH 43210-1107 USA
Tel: +1-614-292-1074
Fax: +1-614-292-3163
Email: orr.91@osu.edu

Jens Osterloh
Department of Chemistry
University of Bremen
Leobener Strasse
D-28359 Bremen
Germany
Tel: +49-421-2182119
Fax: +49-421-2182871
Email: osterloh@chemie.uni-bremen.de

Karl O. Ott
School of Nuclear Engineering
Purdue University
West Lafayette, IN 47907-1290 USA
Tel: +1-765-494-5739
Fax: +1-765-494-9570

Matthew Palmer
Departmernt of Radiology
Beth Israel Deaconess Medical Center
1 Deaconess Road
Boston, MA 02215 USA
Tel: +1-617-667-0176
Fax: +1-617-975-5233
Email: mpalmer@caregroup.harvard.edu

Hemant Patel
Research Institute
University of Oxford
Churchill Hospital
Headington, Oxford
OX37LJ U.K.
Tel: +44-1865-225-827
Fax: +44-1865-225-847
Email: hpatel@radius.JR2.ox.ac.uk

Toralf Peymann
Dept. of Chemistry & Biochemistry
University of California Los Angeles
405 Hilgard Ave.
Los Angeles, CA 90095-1569 USA
Tel: +1-310-206-9444
Fax: +1-310-825-5490
Email: tope@chem.ucla.edu

Katharina Philipp
GKSS Research Center
Dept. WNS, Geb. 57
Max Planck Str.
D-21502 Geesthacht, Germany
Tel: +49-4152-2806
Fax: +49-4152-2840
Email: philipp@gkss.de

Theodore L. Phillips
Dept. of Radiation Oncology
University of California San Francisco
P.O. Box 0226
San Francisco, CA USA
Tel: +1-415-476-5356
Fax: +1-415-476-9069
Email: phillips@radonc17.ucsf.edu

Jean-Philippe Pignol
Radiothérapie-Oncologie
Centre Hospitalier de Mulhouse
87 Avenue d'Altkirch
68051 Mulhouse, France
Tel: +33-389-647520
Fax: +33-389-647532
Email: jp.pignol@infonie.fr

Tazio Pinelli
Dept. of Theoretical and Nuclear Physics
University of Pavia
Via Bassi 6
27100 Pavia, Italy
Tel: +01139-382-507437
Fax: +01139-382-526938
Email: pinelli@pv.infn.it

Mario Alberto Pisarev
Dept. Radiobiologia
Comision Nacional de Energia Atomica
Avda del Libertador 8250
1429 Buenos Aires, Argentina
Tel: +541-754-7184
Fax: +541-754-7121
Email: pisarev@cnea.edu.ar

Gena Poe
Georgia Institute of Technology
3114 Woodlands Drive
Smyrna, GA
USA
Tel: +1-770-803-0874
Email: gt5888a@prism.gatech.edu

Lowell Potiker
Neutron Therapies Inc.
3366 N. Torrey Pines Ct. Suite 210
La Jolla, CA
USA
Tel: +1-858-657-9400
Fax: +1-858-657-9031

Karel Prokes
Dept. of Oncology
Charles University
Unemocnice 2
128 08 Prague 2
Czech Republic
Tel: +420-2-294304
Fax: +420-2-298490
Email: kprok@lfl.cuni.cz

Joseph Rant
Jozef Stefan Institute
Jamova 39
1000 Ljubljana, Slovenia
Tel: +386-61-1885357
Fax: +386-61-1612335
Email: joze.rant@ijs.sl

Juergen Rassow
Institute for Medical Radiation Physics
University Hospital Essen
Hufelandstrasse 55
D-45122 Essen, Germany
Tel: +49-201-7234170
Fax: +49-201-7235965

Junaid Razvi
General Atomics
10240 Flanders Court
San Diego, CA 92121 USA
Tel: +1-858-457-8850
Fax: +1-858-457-8786
Email: razvi@gat.com

Louis L. Reginato
Lawrence Berkeley National Laboratory
One Cyclotron Road
Berkeley, CA 94720 USA
Tel: +1-510-486-5352
Fax: +1-510-486-5392
Email: llriginaato@lbl.gov

Iouri S. Riaboukhine
Blokhin Oncological Science Center
d.35, k.4, kv.3243
ul. akad. Millionschikova
115446 Moscow, Russia
Tel: +7-095-112-61-20
Fax: +7-095-324-16-70
Email: ab7706@mail.sitek.ru

Wade Richards
McClellan Air Force Base
5235 Price Ave.
Sacramento, CA
USA
Tel: +1-916-643-1024
Fax: +1-916-643-0176

Mark Rivard
Dept. of Radiation Oncology
Tufts University School of Medicine
New England Medical Center
750 Washington Street
Boston, MA 02111 USA
Tel: +1-617-636-1680
Fax: +1-617-636-7621
Email: mrivard@lifespan.org

David C. Rorer
Medical Department
Brookhaven National Laboratory
Bldg. 490, 30 Bell Ave.
P.O. Box 5000
Upton, NY 11973-5000 USA
Tel: +1-631-344-4056
Fax: +1-631-344-4915
Email: rorer@bnl.gov

Anatoly Rosenfeld
Department of Engineering Physics
University of Wollongong
Wollongong NSW 2522
Australia
Tel: +61-2-42219574
Fax: +61-2-42215944
Email: a.rozenfeld@uow.edu.au

Giancarlo Rosi
ENEA
Via Anguillarese 301-00060 S.M. Di Galeria
Rome, Italy
Tel: +6-304-83418
Fax: +6-304-86436

Joan Rotaru
Department of Pathology
Ohio State University
168 Hamilton Hall
1645 Neil Avenue
Columbus, OH 43210 USA
Tel: +1-614-292-0693
Fax: +1-614-292-7072
Email: rotaru.1@osu.edu

Silke Ruf
Department of Chemistry
University of Bremen
Leobener Strasse
D-28359 Bremen
Germany
Tel: +49-421-2182119
Fax: +49-421-2182871
Email: ruf@chemie.uni-bremen.de

Roger Rydin
Dept. of Mech., Aero. & Nuclear Engineering
University of Virginia
626 Cabell Ave.
Charlottesville, VA 22903 USA
Tel: +1-804-982-5440
Fax: +1-804-982-5473
Email: rar@virginia.edu

Fumihko Saito
Graduate School of Energy Science
Kyoto University
Kumatori-cho, Sennan-gun
Osaka 590-0494 Japan
Tel: +81-0724-51-2375
Fax: +81-0724-51-2603
Email: fsaito@post3.rri.kyoto-u.ac.jp

Kenji Sakae
Kyoto Prefectural University of Medicine
Kajii-cho 465, Hirokoji-agaru
Kawaramachidori, Kamikyo-ku
Kyoto-shi, Kyoto, Japan
Tel: +81-75-251-5543
Fax: +81-75-251-5544
Email: syougoingtr@msn.com

Yoshinori Sakurai
Research Reactor Institute
Kyoto University
Noda, Kumatori-cho, Sennan-gun
Osaka 590-0494 Japan
Tel: +81-0724-51-2468
Fax: +81-0724-51-2620
Email: ysakurai@kyoto-u.ac.jp

Wolfgang Sauerwein
Klinik und Poliklinik für Strahlentherapie
Universitätsklinikum Essen
Hufelandstrasse 55
D-45122 Essen, Germany
Tel: +49-201-7232052
Fax: +49-201-7235908
Email: w.sauerwein@uni-essen.de

Sauli Savolainen
Dept. of Radiology
Hesinki University Central Hospital
P.O. Box 380
FIN-00029 HYKS
Helsinki, Finland
Tel: +358-9-4712595
Fax: +358-9-4714404
Email: sauli.savolainen@huch.fi

Manuela Schaffrath
Department of Chemistry
University of Bremen
Leobener Strasse
D-28359 Bremen
Germany
Tel: +49-421-2182119
Fax: +49-421-2182871

Raymond F. Schinazi
VA Medical Center
Emory University
1670 Clairmont Road
Med. Rsch. 151
Decatur, GA 30033 USA
Tel: +1-404-728-7711
Fax: +1-404-728-7726
Email: rschina@emory.edu

Hermann Schweickert
Forschungzentrum Karlsruhe
P.O. Box 3640
Karlsruhe D-76021 Germany
Tel: +49-07247-82-2433
Fax: +49-07247-82-3156
Email: schweick@hzy.fzk.de

Amanda Schwint
Dept. Radiobiologia
Comision Nacional de Energia Atomica
Avda del Libertador 8250
1429 Buenos Aires, Argentina
Tel: +541-754-7149
Fax: +541-754-7121
Email: schwint@cnea.edu.ar

Tiina Seppälä
Clinical Research Institute
Helsinki University Central Hospital
VTT Chemical Technology
P.O. Box 1404
Espoo FIN-02044 VTT, Finland
Tel: +358-9-4566351
Fax: +358-9-4566390
Email: tiina.seppala@vtt.fi

Jeffrey Shafer
Midwest Inst. for Neutron Therapy
Fermilab
P.O. Box 5004
Batavia, IL 60510-0500 USA
Tel: +1-630-840-3865
Fax: +1-630-840-8766

Gad Shani
Dept. of Nuclear Engineering
Ben-Gurion University
P.O. Box 653
Beer-Sheva, Israel
Tel: +972-7-6461296
Fax: +972-7-6472955
Email: gshani@bgumail.bgu.ac.il

Ruth Shefer
Newton Scientific, Inc.
245 Bent Street
Cambridge, MA 02141 USA
Tel: +1-617-354-9469
Fax: +1-617-354-9479
Email: res@world.std.com

Ken Shelly
Dept. of Chemistry & Biochemistry
University of California Los Angeles
405 Hilgard Ave.
Los Angeles, CA 90095-1569 USA
Tel: +1-310-794-5228
Fax: +1-310-825-5490
Email: shelly@chem.ucla.edu

Yasushi Shibata
Department of Neurosurgery
University of Tsukuba
1-1-1 Tennodai
Tsukuba City
Ibaraki 305-0006 Japan
Tel: +81-298-533220
Fax: +81-298-533214
Email: y-shibata@amy.hi-ho.ne.jp

Axel Siefert
Dept. of Radiotherapy
LM University Munich
Maschioninistr. 48
81377 Munich, Germany
Tel: +49-89-7095-3770
Fax: +49-89-7095-3845
Email: siefert@radonc.med.uni-muenchen.de

Mats Silvander
Dept. of Physical Chemistry
Uppsala University
Box 532
S-75121 Uppsala, Sweden
Tel: +46-18-4713661
Fax: +46-18-508542
Email: mats.silvander@fki.uu.se

Mark A. Sitek
Dept. of Nuclear Engineering
Massachusetts Institute of Technology
E25-330
45 Carleton St.
Cambridge, MA 02139 USA
Tel: +1-617-253-1930
Fax: +1-617-253-0760
Email: sitek@mit.edu

Stefan Sjöberg
Dept. of Organic Chemistry
Uppsala University
Institute of Chemistry
P.O. Box 531
S-751 21 Uppsala, Sweden
Tel: +46-18 4713796
Fax: +46-18 508542
Email: ssj@kemi.uu.se

Kurt Sköld
Studsvik Neutron Research Laboratory
Uppsala University
S-61182 Nyköping, Sweden
Tel: +46-155-221849
Fax: +46-155-263001
Email: kurt.skold@studsvik.uu.se

Janez Skrk
Office of the Director
Institute of Oncology
Zaloska 2
SLO-1000 lLjubljana
Slovenia
Tel: +386-61-323063
Fax: +386-61-1314180
Email: jskrk@onko-i.si

Duane R. Smith
Department of Chemistry
Cornell University
Baker Labboratory, Cornell University
Ithaca, NY 14853-1301 USA
Tel: +1-607-255-3884
Fax: +1-607-255-4137
Email: ds50@cornell.edu

Robin Smith
Ryscor Science, Inc.
3209 Gresham Lake Rd.
Suite 147
Raleigh, NC 27615 USA
Tel: +1-919-875-1982
Fax: +1-919-954-6197
Email: robin@ryscor.com

Mark Smuckler
Dept. of Chemistry & Biochemistry
University of California Los Angeles
405 Hilgard Ave.
Los Angeles, CA 90095-1569 USA
Tel: +1-310-206-4068
Fax: +1-310-825-5490
Email: mbs@chem.ucla.edu

Albert Soloway
College of Pharmacy
The Ohio State University
500 West 12th Ave. Rm. 336
Columbus, OH 43210 USA
Tel: +1-614-292-2509
Fax: +1-614-292-2435
Email: soloway.1@osu.edu

Vollkan Sozer
Yildiz University
2 Kisim L 54/2 Atakoy
Istanbul, Turkey
Tel: +90-212-559-7750
Email: vsozer1@akback.com.tr

James Spencer
RxD Inc.
192 Sand Hill Circle
Menlo Park, CA USA
Tel: +1-650-854-2682
Email: jus@slac.stanford.edu

Bernard Spielvogel
Boron Biologicals, Inc.
3209 Gresham Lake Rd.
Suite 147
Raleigh, NC 27615 USA
Tel: +1-919-790-7990
Fax: +1-919-954-6197
Email: info@boronbiologicals.com

Rajiv R. Srivastava
Dept. of Radiology, Chemistry
University of Tennessee
Knoxville, TN 37996-1600 USA
Tel: +1-423-946-4086
Email: srivastava@novell.chem.utk.edu

Finn Stecher-Rasmussen
Netherlands Energy Research Foundation
P.O. Box 1
NL-1755 ZG Petten
Netherlands
Tel: +31-224-56-4537
Fax: +31-224-56-3491
Email: rasmussen@ecn.nl

Stefanie Steen-Ecke
Department of Chemistry
University of Bremen
Leobener Strasse
D-28359 Bremen
Germany
Tel: +49-421-2182119
Fax: +49-421-2182871
Email: steen@chemie.uni-bremen.de

Keith Stelzer
Dept. of Radiation Oncology
University of Washington
Medical Center
P.O. Box 356043
Seattle, WA 98195-6043 USA
Tel: +1-206-548-4115
Fax: +1-206-548-6218
Email: keith@radonc.washington.edu

Tom Strike
National Cancer Institute
6130 Executive Blvd.
EPN-800
Rockville, MD USA
Tel: +1-301-496-9360
Fax: +1-301-480-5785
Email: striket@dtpepn.nci.nih.gov

Marika Suominen
Department of Chemistry
University of Helsinki
P.O. Box 55
FIN-00014 UH
Helsinki, Finland
Tel: +358-9-191-40431
Fax: +358-9-191-40366
Email: marika.suominen@helsinki.fi

Minoru Suzuki
Research Reactor Institute
Kyoto University
Kumatori-cho, Sennan-gun
Osaka 590-04 Japan
Tel: +81-724-51-2475
Fax: +81-724-51-2527
Email: msuzuki@post3.rri.kyoto-u.ac.jp

Donald Swenson
Linac Systems
2167 N. Highway 77
Waxahachie, TX 75165 USA
Tel: +1-972-937-3283
Fax: +1-972-938-8817
Email: DASwenson@aol.com

Leena Tähtinen
Department of Chemistry
University of Helsinki
P.O. Box 55
FIN-00014 Helsinki
Finland
Tel: +358-9-191 40430
Fax: +358-9-191 40366
Email: leena.tahtinen@helsinki.fi

Yukiko Takahashi
Research Reactor Institute
Kyoto University
Kumatori-cho
Sennan-gun
Osaka 590-0494 Japan
Email: yuki@rri.kyoto-u.ac.jp

Kenichi Tanaka
Research Reactor Institute
Kyoto University
Dept. of Nuclear Technology
Noda, Kumatori-cho, Sennan-gun
Osaka 590-0494 Japan
Tel: +81-724-51-2468
Fax: +81-724-51-2620
Email: ktanaka@post3.rri.kyoto-u.ac.jp

Scott Taylor
Lawrence Berkeley National Laboratory
1 Cyclotron Road
MS 55-121
Berkeley, CA 94720 USA
Tel: +1-510-486-4103
Fax: +1-510-486-4867
Email: setaylr@lbl.gov

Jason Thomas
Dept. of Chemistry & Biochemistry
University of California Los Angeles
405 Hilgard Ave.
Los Angeles, CA 90095-1569 USA
Tel: +1-310-206-9444
Fax: +1-310-825-5490
Email: jasont@chem.ucla.edu

Brian E. Thurgood
General Atomics
10240 Flanders Court
San Diego, CA 92121 USA
Tel: +1-858-457-8796
Fax: +1-858-457-8786
Email: brian.thurgood@gat.com

Renato Tinti
Dipartimento Energia
ENEA
Divisione SIEC
Via Martiri di Monte Sole,4
40129 Bologna, Italy
Tel: +39-051-6098470
Fax: +39-051-6098594
Email: tinti@risc990.bologna.enea.it

Werner Tjarks
College of Pharmacy
Ohio State University
500 W. 12th Ave.
Columbus, OH 43210 USA
Tel: +1-614-292-7624
Fax: +1-614-292-2435
Email: tjarks@dendrite.pharmacy.ohio-state.edu

Matti Toivonen
Radiation and Nuclear Authority
STUK
P.O. Box 14
FIN-00881 Helsinki
Finland
Tel: +358-9-7598 8603
Fax: +358-9-7598 8589
Email: matti.toivonen@stuk.fi

Hiroyuki Tokumitsu
Faculty of Pharmaceutical Science
Kobe Gaukin University
Arise 518, Ikawadani-cho
Nishi-ku, Kobe 651-2180
Japan
Tel: +81-78-974-1551 (X2470)
Fax: +81-78-974-5689
Email: tokumitu@pharm.kobegakuin.ac.jp

Frantisek Tovarys
Neurosurgical Department
Hospital Na Homolce
Roentgenova 2
151 19 Prague 5
Czech Republic
Tel: +02-52923091
Fax: +02-52923091
Email: Frantisek.Tovarys@Homolka.cz

Larry Tummel
BioNeutrics, Inc.
10317 Technology Drive
Suite 1
Knoxville, TN 37932 USA
Tel: +1-423-675-5627
Fax: +1-423-675-3677
Email: larryt@bioneutrics.com

Ron J. Twilegar
Neutron Technology
1205 N. 11th St.
Boise, ID 83702 USA
Tel: +1-208-345-3460
Fax: +1-208-336-1195
Email: twilorific@aol.com

Jyrki Vähätalo
Laboratory of Radiochemistry
University of Helsinki
P.O. Box 55
A.I. Virtasen Aukio 1
FIN-00014 HU Finland
Tel: +358-9-191-40154
Fax: +358-9-191-40121
Email: jyrki.vahatalo@helsinki.fi

James Venhuizen
INEEL
P.O. Box 1625
Idaho Falls, ID 83415 USA
Tel: +1-208-526-9421
Fax: +1-208-526-0528
Email: jrv@inel.gov

Wilko Verbakel
Nuclear Research & Consultancy Group (NRG)
Netherlands Energy Research Foundation
P.O. Box 25
NL-1755 ZG Petten
The Netherlands
Tel: +31-224-564376
Fax: +31-224-564457
Email: verbakel@nrg-nl.com

Jerome Verbeke
Departmentof Nuclear Engineering
University of California Berkeley
Lawrence Berkeley National Laboratory
Berkeley, CA 94720 USA
Tel: +1-510-495-2792
Fax: +1-510-486-5105
Email: JMVerbeke@lbl.gov

Maria Vicente
Department of Chemistry
University of California Davis
Davis, CA USA
Tel: +1-530-752-7160
Fax: +1-530-752-8995
Email: vicente@chem.ucdavis.edu

Corine Vroegindeweij
High Flux Reactor Unit
Joint Research Center—IAM
P.O. Box 2
NL-1755 ZG Petten
The Netherlands
Tel: +31-224-565035
Fax: +31-224-561449
Email: corine@jrc.nl

Corey Waller
Department of Chemistry
Southwest Texas State University
601 University Drive
San Marcos, TX 78666 USA
Tel: +1-512-245-8283
Fax: +1-512-245-2374

C-K Chris Wang
School of Mechanical Engineering
Georgia Institute of Technology
Nuclear Engineering & Health Physics Program
Atlanta, GA 30332-0225 USA
Tel: +1-404-894-3727
Fax: +1-404-894-3733
Email: Chris.Wang@me.gatech.edu

Somsak Wanwilairat
Department of Radiology
Chiang Mai University
Faculty of Medicine
Chiang Mai 50200 Thailand
Tel: +66-53-945491
Fax: +66-53-217144
Email: somsakw@istrd.cmu.ac.th

Peter Watkins
Joint Research Center—IAM
P.O. Box 2
NL-1755 ZG Petten
The Netherlands
Tel: +31-224-565120
Fax: +31-224-561251
Email: watkins@jrc.nl

Zhou Wei
Department of Radiation Oncology
China-Japan Friendship Hospital
He Ping Li
Beijing 100029 China
Tel: +86-1-64221122
Fax: +86-1-64217749

Regin Weinreich
Paul Scherrer Institute
CH-5232 Villigen PSI
Switzerland
Tel: +41-56-310-4070
Fax: +41-56-310-4435
Email: regin.weinreich@imr.psi.ch

Lothar Weissfloch
Medical Department
Brookhaven National Laboratory
Bldg. 490, 30 Bell Ave.
P.O. Box 5000
Upton, NY 11973-5000 USA
Tel: +1-631-344-5556
Fax: +1-631-344-1349
Email: weissfloch@bnl.gov

Charles Wemple
INEEL
P.O. Box 1625
Idaho Falls, ID 83415 USA
Tel: +1-208-526-7667
Fax: +1-208-526-0528
Email: cew@inel.gov

Daniel Wessol
Computer Science Department
INEEL/Montana State University
357 EPS Bldg.
Bozeman, MT 59717 USA
Tel: +1-208-526-9421
Email: dew@inel.gov

Floyd Wheeler
INEEL
P.O. Box 1625
Idaho Falls, ID 83415-3855 USA
Tel: +1-208-526-7641
Fax: +1-208-526-0528
Email: fjw@inel.gov

Susan White
Dept. of Nuclear Engineering
Massachusetts Institute of Technology
E25-330
45 Carleton St.
Cambridge, MA 02139 USA
Tel: +1-617-258-8825
Fax: +1-617-253-0760
Email: smwhite@mit.edu

William Whittemore
Trigatech
P.O. Box 269025
La Jolla, CA 92126 USA
Tel: +1-858-456-2108
Fax: +1-858-457-8789

Lucian Wielopolski
Department of Applied Science
Brookhaven National Laboratory
Bldg. 490-D, 30 Bell Ave.
P.O. Box 5000
Upton, NY 11973-5000 USA
Tel: +1-631-344-3656
Fax: +1-631-344-7244
Email: lwielo@bnl.gov

Erik Wiener
Beckman Institute
University of Illinois Urbana-Champaign
405 North Matthews
Urbana, IL 61801 USA
Tel: +1-217-244-7193
Fax: +1-217-244-1330
Email: e-wiener@uiuc.edu

Chris Wiersema
Neutron Therapies LLC
5340 Eastgate Mall
San Diego, CA 92121 USA
Tel: +1-858-678-8683
Fax: +1-858-678-8615
Email: chrisnti@pacbell.net

Richard Wiersema
Neutron Therapies LLC
5340 Eastgate Mall
San Diego, CA 92121 USA
Tel: +1-858-678-8683
Fax: +1-858-678-8615
Email: wiersema@pacbell.net

Albert L. Wiley, Jr.
Radiation Therapy Department
Watson Clinic LLP
1600 Lakeland Hills Blvd.
P.O. Box 95000
Lakeland, FLA 33805 USA
Tel: +1-941-680-7100
1-941-680-7952
Fax: +941-680-7952
Email: wcphysics@juno.com

Karl-Johan Winberg
Dept. of Organic Chemistry
Uppsala University
Institute of Chemistry
Box 531
S-751 21 Uppsala, Sweden
Tel: +46-18 4713798
Fax: +46-18 508542
Email: Karlj@kemi.uu.se

Andrea Wittig
Department of Radiation Oncology
University of Essen
Hufelandstrasse 55
45122 Essen, Germany
Tel: +49-201-723-2052
Fax: +49-201-723-5908
Email: wittig@ph-cip.uni-koeln.de

Cecile Wojnecki
School of Physics and Astonomy
University of Birmingham
Edgbaston
Birmingham B15 2TT
U.K.
Tel: +44-121-414-4714
Fax: +44-121-414-4725
Email: cecile@aps7.ph.bham.ac.uk

Genevieve Wu
Dept. of Biomedical Physics
University of California Los Angeles
Box 951721
Los Angeles, CA 90095 USA
Tel: +1-310-825-7811
Email: nwu@ucla.edu

Tetsuya Yamamoto
Medical Sciences, Doctoral Program
Tsukuba University
Tenno-dai 1-1-1
Tsukuba City, 305-0006, Japan
Tel: +81-298-53-3220
Fax: +81-298-53-3214
Email: md985371@igaku.tsukuba.ac.jp

Yoshinori Yamamoto
Department of Chemistry
Tohoku University
Graduate School of Science
Sendai 980-8578 Japan
Tel: +81-22-217-6581
Fax: +81-22-217-6784
Email: yoshi@yamamoto1.chem.tohoku.ac.jp

Hironobu Yanagië
Department of Surgery
University of Tokyo
Institute of Medical Science
4-6-1 Shiroganedai. Minato-ku
Tokyo 108-0071 Japan
Tel: +81-3-5449-5352
Fax: +81-3-5449-5439
Email: yanagie@kiwi.co.jp

Jacquelyn C. Yanch
Dept. of Nuclear Engineering
Massachusetts Institute of Technology
E25-330
45 Carleton St.
Cambridge, MA 02139 USA
Tel: +1-617-258-6999
Fax: +1-617-253-0760
Email: jcy@imager.mit.edu

Tanaka Yasuhito
Dept. of Chemistry, Faculty of Science
Shinshu University
Asahi 3-1-1
Matsumoto 390-8621 Japan
Tel: +81-263-37-2558
Fax: +81-263-37-2559
Email: sa98310@ripus1.shinshu-u.ac.jp

Sung-Joon Ye
Department of Medical Physics
Rush-Presbyterian-St. Luke's Medical Center
Rush University
1653 Congress Parkway
Chicago, IL 60612 USA
Tel: +1-312-942-5751
Fax: +1-312-942-2339
Email: yesj@therad.rpslmc.edu

Kazuo Yoshino
Department of Chemistry
Shinshu University
Faculty of Science
Asahi 3-1-1
Matsumoto 390-8621 Japan
Tel: +81-263-37-2473
Fax: +81-263-37-2559
Email: kyoshino@ripms.shinshu-u.ac.jp

Robert Zamenhof
Department of Radiology
Harvard Institutes of Medicine
HIM-146
4 Blackfan Circle
Boston, MA 02215 USA
Tel: +1-617-667-0175
Fax: +1-617-975-5233
Email: zamenhof@mit.edu

Jin-cong Zhuo
College of Pharmacy
The Ohio State University
437 Parks Hall
500 West 12th Ave.
Columbus, OH 43210 USA
Tel: +1-614-688-3149
Fax: +1-614-292-2435
Email: zhou.1@osu.edu

Aris Zonta
Dept. of General Surgery
University of Pavia
Piazzale Golgi 2
27100 Pavia, Italy
Tel: +01139-382-526214
Fax: +01139-382-502999
Email: palumbo@unipv.it

AUTHOR INDEX, Volume 2

KEYWORD INDEX, Volume 2

Batch number: 10838292

Printed by Printforce, the Netherlands